Encyclopedia of Physical Sciences and Engineering Information Sources

Other Subject-Arranged Guides to Information Sources from Gale

Encyclopedia of Business Information Sources. Contains some 20,000 entries arranged under more than 1,000 business-related subjects. Entries provide details on both live and print sources of information as well as commercially available online data bases. Also available: interedition supplement.

Encyclopedia of Geographic Information Sources: U.S. Volume. Provides 11,000 live and print sources of business information on more than 380 U.S. cities, states, and regions. Emphasis is on sources of economic and financial data, such as analytical reports, demographic studies, trade directories, and economic planning documents.

Encyclopedia of Geographic Information Sources: International Volume. Covers 75 nations, six multinational regions, and 81 major industrial cities, with 12,000 entries providing the primary live and print information sources of business and economic data worldwide.

Encyclopedia of Health Information Sources. Contains over 15,000 print, electronic, and live information sources in the field of health and medicine. Citations to published works are limited to recent publications and classics in the field.

Encyclopedia of Legal Information Sources. Cites over 12,000 sources of live and print sources on 500 legal topics. Publisher names and addresses, publication dates, and frequency (as applicable) are included.

Encyclopedia of Public Affairs Information Sources. Covers thousands of live, print, and electronic information sources in nearly 300 subject sections. This volume is a guide to sources of up-to-date facts, figures, and opinions on social, political, and economic subjects of current concern.

Encyclopedia of Senior Citizens Information Sources. Provides access to a wide range of live, print, and electronic imformation sources on more than 300 different topics of immediate concern to the nation's senior citizens.

Other Related Titles from Gale

Engineering Research Centres. This international directory lists over 8,000 research and technology laboratories, industrial research centers, and educational establishments with research and development activity. (Distributed in the United States and Canada by Gale.)

Scientific and Technical Organizations and Agencies Directory. Over 15,000 entries furnish details on national and international organizations and agencies active in the physical and applied sciences, including engineering.

Encyclopedia of Physical Sciences and Engineering Information Sources

A Bibliographic Guide to Approximately 16,000 Citations for Publications, Organizations, and Other Sources of Information on 425 Subjects Relating to the Physical Sciences and Engineering

Includes: Abstract Services and Indexes; Annual Reviews and Yearbooks; Associations and Professional Societies; Bibliographies; Directories and Biographical Sources; Encyclopedias and Dictionaries; General Works; Handbooks and Manuals; Online Data Bases; Periodicals; Research Centers and Institutes; Specifications and Standards; Statistics Sources; and Other Sources of Information on Each Topic

FIRST EDITION

**Steven Wasserman,
Martin A. Smith,
and Susan Mottu,**
Editors

Gale Research Inc. • Book Tower • Detroit, Michigan 48226

Editors: Steven Wasserman, Martin A. Smith, Susan Mottu
Editorial Consultant: Paul Wasserman
Editorial Production Assistants: Carla Rose, Barbara Smith
Word Processing Consultant: Richard Rose

Gale Research Inc. Staff

Senior Editor: Donna Wood
Project Coordinator: Janice A. DeMaggio

Production Manager: Mary Beth Trimper
External Production Assistants: Linda A. Davis, Anthony J. Scolaro

Art Director: Arthur Chartow
Graphic Designer: Cynthia D. Baldwin

Production Supervisor: Laura Bryant
Internal Production Associate: Louise Gagné

Library of Congress Cataloging-in-Publication Data

Encyclopedia of physical sciences and engineering information sources:
a bibliographic guide to approximately 16,000 citations for
publications, organizations, and other sources of information on
425 subjects relating to the physical sciences and engineering /
Steven Wasserman, Martin A. Smith, and Susan Mottu, editors. —
1st ed.
　　p.　cm.
　ISBN 0-8103-2498-9
　1. Science—Bibliography.　2. Engineering—Bibliography.
3. Science—Information services.　4. Engineering—Information services.
I. Wasserman, Steven R.　II. Smith, Martin A.　III. Mottu, Susan.
Z7401.E56　1989
[Q158.5]　　　　　　　　　　　　　　　　　　　　　　88-38935
016.5—dc 19　　　　　　　　　　　　　　　　　　　　　　CIP

Copyright © 1989 by Gale Research Inc.

Printed in the United States of America

Contents

Introduction vii
Outline of Contents xi
Physical Sciences and Engineering Information Sources 1

Introduction

Physical scientists and engineers need an easy-to-use tool that leads to the books, periodicals, data bases, and organizations that can answer their technical questions. Those outside the physical science and engineering professions may also require a tool that directs them to more basic scientific books and to the organizations of, and related to, these particular areas.

To meet the specific and unique needs of both groups, the *Encyclopedia of Physical Sciences and Engineering Information Sources (EPSEIS)* serves as a starting point for any research project involving the physical sciences or engineering. It provides comprehensive coverage of sources of information in these areas for the non-specialist as well as the specialist working in another field.

Method of Compilation

Extensive research was required in order to gather the sources of information on the hundreds of subjects included in *EPSEIS*. A wide range of sources was consulted, including online data bases, bibliographies, indexes of science literature, and publishers' promotional brochures. In each instance, the details have been carefully verified and corroborated.

The editors have placed emphasis on the most recent written works in a field, but have included important older works where it was thought relevant. All printed works were available at the time the guide was compiled. Where possible telephone numbers for the publishers have been included, as well as toll-free "800" numbers, if available. Prices quoted are intended only as a guide to the relative cost of an item or service.

Subject Headings Narrowly Defined

So that information can be located easily, subject headings are narrowly defined and specific. This edition of *EPSEIS* includes approximately 425 topics and is further supplemented by numerous cross-references. For each subject, a bibliography of live, print, and electronic sources of information is provided.

Subjects selected for inclusion in this directory were compiled from several widely-used science and technology dictionaries, encyclopedias, and indexing services commonly found in college, research, and public libraries. All major established subjects such as chemistry, geology, physics, and civil engineering are covered, as well as many of the new areas of scientific inquiry and applied research, exemplified by such topics as artificial intelligence, chaos theory, superconductivity, and the like.

"Outline of Contents" Speeds Access to Topics

For the most efficient use of this publication, do *not* go directly to the text to look up the subject on which information is sought. Instead, find the appropriate term in the extensive and detailed Outline of Contents, where it is possible to scan the list of headings quickly to determine the exact form of the subject term that has been used and to be guided by cross-references to related subjects. If the term being sought has not been used as a subject heading, the Outline of Contents will refer to the term actually used.

Bibliography Covers 14 Kinds of Sources

Under each heading, citations are found for a wide variety of sources, both print and nonprint, on a particular topic. A complete list of the kinds of sources cited, represented as subheadings in *EPSEIS,* is outlined below, in the sequence in which they appear (although not all types of sources are listed for each subject).

Abstract Services and Indexes
Annual Reviews and Yearbooks
Associations and Professional Societies
Bibliographies
Directories and Biographical Sources
Encyclopedias and Dictionaries
General Works
Handbooks and Manuals
Online Data Bases
Other Sources
Periodicals
Research Centers and Institutes
Specifications and Standards
Statistics Sources

With regard to the online data bases included, many of them are available through one or more of the following major information vendors:

BRS Information Technologies
1200 Route 7
Latham, NY 12110
(800) 345-4277 or (518) 783-1161

DIALOG Information Services
3460 Hillview Avenue
Palo Alto, CA 94304
(800) 334-2564 or (415) 858-2700

Pergamon ORBIT InfoLine, Inc.
8000 Westpark Dr.
McLean, VA 22101
(800) 456-7248 or (703) 442-0900

Arrangement and Form of Entries

Entries are arranged alphabetically by ① subject, and further subdivided by ② type of source, and ③ publication title or organization name. Complete citations are provided for each information source. Citations for publications include title, author or editor, publisher's name and address, publication date or frequency, telephone number, and price. For data bases, citations include data base name, producer, producer's address, and telephone number. Citations for organizations include name, address, and telephone number. For example:

---

① **ELECTRON OPTICS**

See also: ELECTRON MICROSCOPY, ELECTRON SPECTROS-
COPY, OPTICS, OPTOELECTRONICS

② ABSTRACT SERVICES AND INDEXES

③ APPLIED SCIENCE AND TECHNOLOGY INDEX. H.W. Wilson and Company, 950 University Avenue, Bronx, NY 10452. (800) 367-6670 or (212) 588-8400. Monthly. Inquire as to cost and availability.

CURRENT CONTENTS: ENGINEERING, TECHNOLOGY AND APPLIED SCIENCES. Institute for Scientific Information, 3501 Market Street, Philadelphia, PA 19104. (800) 523-1850 or (215) 386-0100. Weekly. $275.00 per year.

ASSOCIATIONS AND PROFESSIONAL SOCIETIES

ELECTRON MICROSCOPY SOCIETY OF AMERICA. c/o Linda L. Horton, 1497 Chain Bridge Road, Suite 104, McLean, VA 22101. (703) 827-0498.

Acknowledgments

Cooperation, assistance, and encouragement were received from a number of colleagues and associates in the preparation of *EPSEIS*. In particular, G. Marvin Tatum contributed greatly in the early stages of discussion and planning which led to the final work.

Suggestions Are Welcome

Considerable care has been taken to keep errors and inconsistencies to a minimum, but they no doubt will occur. It would be appreciated if users would send to the editor any information, suggestions, comments, or corrections that might improve future editions. Please address remarks to:

>Editor
>*Encyclopedia of Physical Sciences
>   and Engineering Information Sources*
>Gale Research Inc.
>Book Tower
>Detroit, MI 48226

Outline of Contents

Abrasives - see Machining
Absorbers 1
Abstract Algebra - see Algebra
Accelerators - see Particle Accelerators
Accident - see Safety Engineering
Acetylene Welding - see Welding
Acid Rain..................................... 2
Acids and Bases 3
Acoustics 4
Acrylic Resins - see Plastics
Ada - see Programming Languages
Adhesives 6
Adiabatic Processes - see Thermodynamics
Aerial Photography 7
Aerodynamics 9
Aeronautical Engineering 11
Aeronautics 13
Aerosols 16
Aerospace Engineering 17
Aggregates 19
Agricultural Engineering 21
AI - see Artificial Intelligence
Air - see Meteorology
Air Conditioning 22
Air Currents - see Meteorology
Air Masses - see Meteorology
Air Pollution 23
Air Traffic Control 25
Airborne Radar - see Radar
Aircraft 26
Airfoils - see Aeronautical Engineering
Airplanes - see Aircraft
Airports 29
Airships 30
Alarm Systems 32
Alcohol 32
Algebra 34
Algebraic Number Theory - see Mathematics
ALGOL - see Programming Languages
Alkalis - see Acids and Bases
Alloys 36
Alpha Particle - see Particle Physics
Alternating Current - see Electricity
Aluminum 38
AM Radio - see Radio
Amateur Radio 40
Ammonia 41
Amplifiers 42
Analog Computers - see Computers

Analytic Geometry 44
Analytical Chemistry......................... 45
Antennas 47
Anthracite Coal - see Coal
Applied Mathematics - see Mathematics
Aquifer - see Groundwater
Arc Welding - see Welding
Arch - see Structural Engineering
Architectural Engineering 49
Argon Laser - see Lasers
Arithmetic - see Mathematics
Aromatization - see Chemical Engineering
Artesian Aquifer - see Groundwater
Artificial Intelligence 51
Artificial Satellites 53
Asbestos..................................... 54
Asphalt 55
Asteroids - see Solar System
Astrogeology 56
Astrometry and Astronomical Photometry -
  see Astronomy
Astronautical Engineering - see Aerospace
  Engineering
Astronautics - see Aeronautics
Astronomical Observatories - see Observatories
Astronomy.................................... 58
Astrophysics................................. 60
Atmosphere - see Meteorology
Atom - see Physics
Atomic Energy - see Nuclear Energy
Atomic Fission - see Fission
Atomic Physics - see Physics
Atomic Power - see Energy
Atomic Spectroscopy - see Spectroscopy
Atomic Theory - see Physics
Audio Amplifier - see Amplifiers
Audio Engineering............................ 62
Automatic Control Systems - see Automation
Automation 64
Automobile Transmissions - see Automotive
  Engineering
Automobiles - see Automotive Engineering
Automotive Engineering 67
Avionics 68
Ballistic Missiles - see Guided Missiles
Balloons - see Aeronautics
Banach Algebra - see Algebra
Band Spectra - see Spectroscopy
Band Theory of Solids - see Solid State Physics

Outline of Contents

Barometer - see Meteorology
Baryon - see Particle Physics
Bases - see Acids and Bases
BASIC - see Programming Languages
Batholith - see Volcanology
Bathymetry - see Oceanography
Batteries 71
Beaches - see Coasts
Beacons - see Navigation
Beam Columns - see Structural Engineering
Beam Foil Spectroscopy - see Spectroscopy
Beams - see Structural Engineering
Bearing Capacity - see Structural Engineering
Bearings and Ball Bearings 73
Bedrock 74
Benzene 76
Benzene Ring - see Organic Chemistry
Beryllium 78
Bessemer Process - see Steel and Steel Making
Beta Rays - see Particle Physics
Betatron - see Particle Accelerators
Bidirectional Transistor - see Transistors
Big Bang Theory 79
Binary Stars 81
Black Holes 83
Blast Furnace - see Steel and Steel Making
Blasting - see Explosives
Blimp - see Airships
Bohr Theory - see Physical Chemistry
Boilers 85
Booster Rockets - see Rockets
Boring - see Petroleum Engineering
Boron 87
Boron Steel - see Ferroalloys
Bosons - see Particle Physics
Bouger Gravity Anomaly - see Geophysics
Boundary Layer Flow - see Fluid Mechanics
Boundary Wave - see Fluid Mechanics
Brakes 88
Brass and Bronze 89
Brazing - see Welding
Breeder Reactors 91
Brick - see Building Materials
Bridges 93
British Thermal Unit - see Air Conditioning
Brittleness - see Materials Science
Broadcast Band - see Radio
Bromine 94
Bronze - see Brass and Bronze
Building Materials 96
Buildings - see Construction Engineering
Buoys 98
Bushings - see Bearings and Ball Bearings
Cable - see Wire and Cable
Cable Television - see Television
CAD (Computer-Aided Design) 99
Cadmium 101
CAE (Computer-Aided Engineering) - see CAD (Computer-Aided Design)
Caissons - see Civil Engineering
Calcium 103
Calculus 104
CAM (Computer-Aided Manufacturing) 106
Cameras 108

Canals 109
Cantilevers - see Structural Engineering
Capacitors 111
Carbon 112
Carbon-14 Dating - see Radiocarbon Dating
Carbonates 114
Carburetors - see Automative Engineering
Cartography 115
Cassegrain Telescopes - see Telescopes
Cast Iron 117
Catalysis 119
Catalytic Converter - see Air Pollution Control
Cathode Ray Tube (CRT) - see Computer Communications
Caves and Caving 120
Celestial Mechanics 121
Celestial Navigation - see Navigation
Cement 123
Ceramics 124
Chain Reaction - see Nuclear Energy
Channels 126
Chaos 127
Charm - see Particle Physics
Chemical Bonding 128
Chemical Elements - see Elements (Chemical)
Chemical Engineering 130
Chemical Equilibrium - see Physical Chemistry
Chemical Formulas - see Chemistry
Chemical Microscopy - see Microscopy
Chemical Separation Techniques - see Analytical Chemistry
Chemistry 132
Chips - see Integrated Circuits
Chlorine 134
Chromatography 136
Chromium 137
Chromium Steel - see Steel and Steel Making
Cinderblocks - see Building Materials
Cinematography - see Photography
Ciphers - see Computer Security
Circuit Breakers 139
Circulation - see Meteorology
Cirrus Clouds - see Clouds
Civil Engineering 141
Clay 143
Climatology 144
Cloth - see Textiles
Cloud Physics - see Clouds
Cloud Seeding - see Weather Modification
Clouds 146
Coal 148
Coal Gasification and Liquification 149
Coal Mining 151
Coastal Engineering 153
Coasts 154
Cobalt 155
COBOL - see Programming Languages
Codes and Ciphers - see Computer Security
Cofferdam - see Dams
Coke and Coking - see Steel and Steel Making
Colloids 157
Color 16 160
Color Film - see Photographic Film
Color Television - see Television

Colorimetry - see Color
Columns - see Structural Engineering
Combinatorial Theory - see Mathematics
Combustion - see Chemistry, Explosives
Comets 162
Communication Satellites - see Artificial Satellites
Compasses 163
Composites 165
Computer Architecture - see Computers
Computer Communications 166
Computer Graphics 169
Computer Hackers - see Computer Security
Computer Memory and Storage 171
Computer Operating Systems............... 174
Computer Programming 176
Computer Programs - see Software
Computer Security 178
Computer Vision 180
Computers 183
Concrete 185
Condensers - see Boilers
Conduction - see Heat Transfer
Connection Machine - see Parallel Computers
Constellations - see Stars
Construction Engineering 186
Continental Drift 188
Continental Margins 189
Control Systems - see Automation
Convection - see Heat Transfer
Cooling Towers 191
Copper 192
Core Drilling - see Drilling
Corona - see Sun
Corrosion 194
Cosmic Rays 196
Cosmochemistry.............................. 197
Cosmogony - see Cosmology
Cosmology 199
Coulometric Analysis - see Analytical Chemistry
Covalent Bond - see Chemical Bonding
Cracking - see Chemical Engineering
Craters - see Solar System
Critical Mass - see Nuclear Energy
Crossover Networks - see Computer Communications
Cruise Missile - see Guided Missiles
Crust - see Geology
Cryogenic Engineering - see Cryogenics
Cryogenics 202
Cryptography and Cryptology - see Computer Security
Crystal Optics - see Crystallography
Crystallography 203
Cumulus Clouds - see Clouds
Currents - see Ocean Currents; Meteorology
Cybernetics 205
Cyclones 207
Cyclotron - see Particle Accelerators
Dams 209
Dark Matter 210
Data Base Management - see Data Bases
Data Bases 212
Data Compression - see Data Bases
Data Display - see Terminals
Data Entry Terminal - see Data Processing
Data Processing 214
Data Retrieval - see Data Bases
Decimal System - see Mathematics
Decision Theory - see Probability
Decoding - see Computer Security
Deep Sea Platform - see Ocean Engineering
Deep Sea Trenches - see Oceanography
Depressions - see Meteorology
Desertification 216
Desiccation - see Industrial Engineering
Design Engineering - see CAD (Computer-Aided Design)
Detonations - see Explosives
Deuterium - see Hydrogen
Diamonds (Industrial) - see Machining
Dielectrics 218
Diesel Engines 220
Differential Calculus - see Calculus
Differential Equations 221
Diffraction Grating - see Spectrography
Digital Circuits - see Electronic Circuits and Components
Digital Computers - see Computers
Dikes - see Dams
Diodes 222
Direct Current - see Electricity
Dirigible - see Airships
Display Terminals - see Computer Communications
Distillation - see Chemical Engineering
Distributed Computing - see Networks (Computer), Local Area Networks
Dolby System - see Audio Engineering
Domes - see Structural Engineering
Doping - see Semiconductors
Doppler Radar - see Radar
Dosimetry - see Radiation
Drafting - see CAD (Computer-Aided Design)
Drag - see Aerodynamics
Drawbridges - see Bridges
Dredges and Dredging 225
Drilling - see Petroleum Engineering
Drought 226
Drying - see Industrial Engineering
Dust Storms 227
Dwarf Galaxy - see Galaxies
Dynamics - see Mechanics
Dynamos - see Generators
Earth - see Solar System
Earth Sciences - see Geology, Meteorology, and Oceanography
Earthquake Engineering 229
Earthquake Prediction - see Earthquakes
Earthquakes 231
Echo - see Radar, Sonar
Eclipses 232
EDP (Electronic Data Processing) - see Data Processing
Effluent - see Environmental Engineering
El Nino - see Oceanography
Elasticity - see Mechanics
Electric Current - see Electricity
Electric Motors 234
Electric Power Engineering 235

Electric Arc Welding - see Welding
Electrical Codes 237
Electrical Engineering 238
Electrical Impedance - see Electricity
Electrical Insulation - see Electricity
Electrical Resistance - see Electricity
Electrically Powered Vehicles - see Automotive Engineering
Electricity 240
Electroacoustics - see Acoustics
Electrochemistry 242
Electrodes - see Electrochemistry
Electrodynamics - see Electricity
Electrolysis - see Electrochemistry
Electrolytes - see Electrochemistry
Electromagnetism 244
Electron Lens - see Electron Optics
Electron Microscopy 246
Electron Optics 247
Electron Spectroscopy 248
Electronic Circuits and Components 250
Electronic Mail - see Computer Communications
Electronic Security Systems - see Alarm Systems
Electronics 252
Electronics Engineering 254
Electroplating - see Electrochemistry
Electrostatics - see Electricity
Elementary Particles - see Particle Physics
Elements (Chemical) 257
Elevators 258
Elliptical Galaxies - see Galaxies
Emission Electron Microscopes - see Electron Microscopy
Emission Spectroscopy - see Spectroscopy
Emulsions - see Colloids
Energy 259
Engineering Design - see CAD (Computer-Aided Design)
Engineering Geology - see Geotechnical Engineering
Engines 262
Entropy - see Thermodynamics
Environmental Engineering 263
Epoxy Resins - see Plastics
Epsilon Meson - see Particle Physics
Equations - see Algebra, Mathematics
Ergonomics 265
Erosion - see Geology
ERTS (Earth Resources Technology Satellites) - see Artificial Satellites
Ethyl Alcohol - see Alcohol
Euclidean Geometry - see Geometry
Expansion Joints - Structural Engineering
Experimental Design - see CAD (Computer-Aided Design)
Expert Systems - see Artificial Intelligence
Explosives 267
Extraction Processes - see Metallurgical Engineering
Extrusion - see Metallurgical Engineering
Fabrics - see Textiles
Factor Analysis - see Mathematics
Failure Analysis 269
Fallout - see Radiation
Fanjet - see Jet Propulsion

Fatigue - see Failure Analysis
Faults - see Physical Geology
Feedback Control Systems - see Automation
Feedforward Control Systems - see Automation
Ferroalloys 271
Ferroelectrics - see Electricity, Electromagnetism
Fiber Optics 273
Fibers-Textiles - see Textiles
Field-Effect Transistors - see Semiconductors
Fire Control - see Fire Protection
Fire Protection 274
Fission 276
Fission Reactors - see Nuclear Reactors
Flame Emission Spectroscopy - see Spectroscopy
Flame Lasers - see Lasers
Flash Welding - see Welding
Flight - see Aeronautics
Flight Dynamics - see Aerodynamics
Flood Control 278
Floppy Disks - see Computer Memory and Storage
Flow - see Fluid Mechanics
Flow Welding - see Welding
Fluid Dynamics 279
Fluid Mechanics 281
Fluid Statics - see Fluid Mechanics
Fluidics 283
Fluorescent Lighting - see Illumination
Fluorine 285
Fluorocarbons - see Halides
FM - see Radio
Foam - see Plastics
Focal Length - see Optics
Fog 288
Folds (Geology) - see Physical Geology
FORTRAN - see Programming Languages
Foundries 289
Fourier Analysis - see Mathematics
Fractionating Columns - see Chemical Engineering
Fracture Mechanics 291
Frequency Modulation - see Radio
Friction 292
Front (Meteorology) - see Meteorology
Fuel Cells - see Electrochemistry
Fuels 294
Fundamental Particles - see Particle Physics
Fuses (Electrical) - see Circuit Breakers
Fusion 296
Fusion Reactor - see Nuclear Reactors
Fusion Welding - see Welding
Galaxies 299
Game Theory - see Mathematics
Gamma-Ray Astronomy 301
Gamma-Ray Laser - see Lasers
Gamma-Ray Spectroscopy - see Spectroscopy
Gamma-Rays 303
Garbage Disposal - see Solid Waste Disposal
Gas Chromatography - see Chromatography
Gas Injection Wells - see Petroleum Engineering
Gas Laser - see Lasers
Gas Masers - see Lasers
Gas Pipelines - see Pipeline Technology
Gas Turbines 304
Gasification - see Coal Gasification and Liquification

Gasohol - see Fuels
Gasoline - see Fuels
Gears and Gearing - see Machinery
Gems - see Minerology
Generators 305
Geochemistry 307
Geochronology 308
Geodesy 310
Geology 311
Geomagnetism 313
Geometry 314
Geomorphology 316
Geophysical Engineering - see Geotechnical
   Engineering
Geophysics 317
Geostationary Satellites - see Artificial Satellites
Geotechnical Engineering 319
Geothermal Energy 321
Giant Stars - see Stars
Glacial Geology 322
Glass 323
Glues - see Adhesives
Gold and Gold Mining 325
Graph Theory - see Mathematics
Grating - see Spectroscopy
Gravimetric Analysis - see Analytical Chemistry
Gravitation 327
Gravitational Red Shift - see Astrophysics
Gravity - see Gravitation
Grease - see Lubrication
Greenhouse Effect 328
Grinding and Polishing - see Machining
Ground Effect Machines - see Aeronautics
Groundwater 330
Groundwater Pollution 332
Guidance Systems 333
Guided Missiles 334
Gyrocompasses - see Compasses
Gyroscopes 336
Hackers - see Computer Security
Hadrons - see Particle Physics
Hail 339
Half-life - see Radiation
Halides 340
Ham Radio - see Amateur Radio
Harbors - see Coastal Engineering
Hard Disks - see Computer Memory and Storage
Heat Exchangers 342
Heat Pipes 343
Heat Pumps 345
Heat Resistant Materials - see Materials Science
Heat Transfer 346
Heating and Ventilation - see Air Conditioning
Heavy Metal Alloys - see Alloys
Heavy Water - see Hydrogen
Helicopters 347
Helium 349
Herschel-Cassegrain Telescopes - see Telescopes
Heterocyclic Chemistry 351
High Carbon Steel - see Steel and Steel Making
High Energy Physics - see Particle Physics
High Explosives - see Explosives
High Fidelity - see Audio Engineering
High Temperature Materials - see Materials Science

High Voltage 353
Highway Engineering 354
Holography 356
Homing - see Navigation
Hovercraft - see Aeronautics
Hulls - see Naval Architecture
Human Engineering - see Ergonomics
Hurricanes 358
Hydraulic Engineering 359
Hydraulics 361
Hydrocarbons - see Petroleum Chemistry
Hydrodynamics 363
Hydroelectric Power 365
Hydrofoil Craft - see Naval Architecture
Hydrogen 367
Hydrogenation 369
Hydrogeology 369
Hydromechanics - see Hydrodynamics
Hydrostatics - see Hydrodynamics
Hyperbolic Navigation System - see Navigation
Hypersonic Flight - see Aerodynamics
Hypersonics - see Acoustics
Ice 373
Icebergs - see Ice
Igneous Rocks 374
Illumination 376
Image Analysis - see CAD (Computer-Aided
   Design)
Image Processing - see CAD (Computer-Aided
   Design)
Impact Testing 377
Industrial Engineering 379
Industrial Robots - see Robotics
Inertial Guidance Systems - see Guidance Systems
Inertial Navigation - see Navigation
Inertial Welding - see Welding
Infrared Astronomy 381
Infrared Photography 382
Infrared Spectroscopy - see Spectroscopy
Infrared Stars - see Stars
Inorganic Chemistry 384
Instrument Flight - see Aeronautics
Instrumentation 386
Integral Calculus - see Calculus
Integrated Circuits 387
Internal Combustion Engine - see Engines
Interstellar Matter 389
Iodine 391
Ion Accelerator - see Particle Accelerators
Ion Laser - see Lasers
Ionization 393
Ionosphere - see Meteorology, Geophysics
Ions - see Ionization
Iron - see Cast Iron; Steel and Steel Making
Isomers - see Organic Chemistry
Isotopes - see Ions
J Particle - see Particle Physics
Jahn-Teller Effect 397
Jet Propulsion 398
Jet Stream 400
Jewel Bearings - see Bearings and Ball Bearings
Joints 401
Josephson Effect - see Cryogenics
Jupiter 402

Karst	405
Kinematics	406
Kinetics	407
LANDSAT - see Artificial Satellites	
Laser Disks - see Optical Disks	
Lasers	411
Lenses - see Optics	
Leptons - see Particle Physics	
Lighting - see Illumination	
Lightning	413
Linear Accelerators - see Particle Accelerators	
Linear Algebra - see Algebra	
Linear Programming - see Mathematics	
Local Area Networks	414
LORAN - see Navigation	
Low-Temperature Physics - see Cryogenics	
Lubrication	416
Lunar Geology - see Astrogeology; Planetary Sciences	
Machine Design	419
Machinery	421
Machining	423
Magnesium	424
Magnetic Resonance - see Nuclear Magnetic Resonance	
Magnetism - see Electromagnetism	
Magnetohydrodynamic Generator - see Magnetohydrodynamics	
Magnetohydrodynamics	426
Mainframe Computers - see Computers	
Manned Space Flight	427
Mapping - see Cartography	
Marine Engineering	429
Marine Navigation - see Navigation	
Markov Processes	430
Mars	432
Maser - see Lasers	
Masking - see Semiconductors	
Mass Spectrometry	433
Materials Handling	434
Materials Science	435
Mathematical Physics	437
Mathematics (Applied)	438
Mechanical Engineering	440
Mechanics	442
Mercury	444
Meson - see Particle Physics	
Metallurgical Engineering	445
Metallurgy	446
Metals and Metalworking	448
Metamorphic Rocks	450
Meteorites - see Meteors	
Meteorological Satellites - see Artificial Satellites	
Meteorology	452
Meteors	454
Metrology	455
MHD - see Magnetohydrodynamics	
Microcircuitry - see Microelectronics	
Microcomputers - see Computers	
Microelectronics	457
Microprocessors	459
Microscopy	461
Microwave Spectroscopy - see Spectroscopy	
Microwaves	463
Milky Way	464
Mineral Exploration	466
Mineralogy	468
Minicomputers - see Computers	
Mining Engineering	469
Mirror Optics - see Optics	
Missiles - see Guided Missiles	
MODEMS - see Computer Communications	
Molecular Physics	471
Monopole - see Electromagnetism	
Monorail Technology - see Railroad Engineering	
Monte Carlo Method - see Statistical Methods	
Moon - see Satellites (Natural)	
Mossbauer Spectroscopy - see Spectroscopy	
Motors	473
Multivariate Analysis - see Mathematics	
Natural Gas	477
Naval Architecture	478
Navigation	480
NDT - see Nondestructive Testing	
Neptune	481
Networks (Computer)	482
Neutrinos - see Particle Physics	
Neutron Optics - see Optics	
Neutron Spectroscopy - see Spectroscopy	
Neutron Stars - see Stars	
NMR - see Nuclear Magnetic Resonance	
Noise Control	484
Nondestructive Testing	485
Nuclear Chemistry	486
Nuclear Energy	488
Nuclear Engineering	490
Nuclear Fuels - see Nuclear Energy	
Nuclear Magnetic Resonance	492
Nuclear Physics - see Nuclear Energy	
Nuclear Power Plants - see Nuclear Energy	
Nuclear Reactors	493
Nuclear Spectroscopy - see Spectroscopy	
Nucleonics - see Nuclear Physics	
Numerical Analysis - see Mathematics	
Observatories	497
Ocean-Atmosphere Boundary - see Meteorology	
Ocean Currents - see Oceanography	
Ocean Engineering	498
Ocean Trenches - see Oceanography	
Oceanography	500
Offshore Engineering - see Ocean Engineering	
Optical Communications	502
Optical Disks	504
Optical Fibers - see Fiber Optics	
Optics	506
Optoelectronics	508
Organic Chemistry	509
Organometallic Compounds - see Organic Chemistry	
Oscilloscope - see Instrumentation	
Ozone	511
Paper Chemistry	515
Parallel Computers	516
Particle Accelerators	518
Particle Physics	519
Pattern Recognition - see Computer Vision	
PASCAL - see Programming Languages	
Petroleum Engineering	521

Petroleum Exploration - see Mineral Exploration
Petroleum Geology 522
Petrology - see Geology; Igneous Rocks;
    Metamorphic Rocks; Sedimentary Rocks
Phonographs - see Audio Engineering
Photochemistry 524
Photoelectricity 526
Photogrammetry. 527
Photographic Film 529
Photography 530
Photometry - see Astronomy
Physical Chemistry 532
Physical Geology 534
Physics 536
Piezoelectricity 539
Pipeline Technology 540
Planets - see Solar System
Planetary Sciences 541
Plasma Physics 543
Plastics 544
Plate Tectonics 546
Pluto 548
Plutonium 549
Polymers 550
Portland Cement - see Cement
Powder Metallurgy 552
Power Systems - see Electric Power Engineering
Precipitation - see Hail, Rain, Snow
Pressure Vessels 554
Prestressed Concrete - see Concrete
Probability 555
Production Engineering - see Industrial Engineering
Programming - see Computer Programming,
    Computer Languages
Programming Languages 557
Programs - see Software
Projective Geometry - see Geometry
Propellants 560
Propjet - see Jet Propulsion
Propulsion Systems 561
Prospecting - see Mineral Exploration
Pulsars 562
Pumping Machinery 564
Pumps - see Pipeline Technology; Pumping
    Machinery
Pyrometallurgy - see Metallurgy
Quality Control Engineering 567
Quantum Chemistry 569
Quantum Electrodynamics - see Quantum
    Mechanics
Quantum Mechanics 571
Quantum Theory - see Quantum Mechanics
Quarks - see Particle Physics
Quarrying 573
Quasars 574
Quaternions - see Particle Physics
Quenching - see Metallurgy
Radar 577
Radiation 579
Radiation Shielding 580
Radio 581
Radio Astronomy 583
Radio Telescopes 585
Radioactivity - see Radiation

Radiocarbon Dating 586
Radiochemistry - see Nuclear Chemistry
Railroad Engineering 587
Rain 588
Raman Spectroscopy - see Spectroscopy
Ramjets - see Jet Propulsion
Reactors - see Nuclear Reactors
Recursive Programming - see Computer
    Programming
Red Dwarfs - see Stars
Red Giants - see Stars
Red Shift - see Astrophysics
Refrigeration - see Air Conditioning
Relativity - see Celestial Mechanics, Physics
Relays 590
Reliability Engineering - see Industrial Engineering,
    Quality Control Engineering
Remote Control Systems - see Control Systems
Remote Sensing 591
Reservoirs - see Dams
Resins - see Plastics
Resistors 593
Rheology - see Fluid Mechanics
Rings (Planetary) - see Planetary Sciences
Road Building - see Highway Engineering
Robotics 595
Rock Mechanics 597
Rocket Astronomy - see Astronomy
Rocket Engines - see Rockets
Rockets 598
Rotary Engines - see Automotive Engineering
Safety Engineering 601
Salt Domes - see Petroleum Geology
Sanitary Engineering - see Environmental
    Engineering
Satellites (Natural) 602
Saturn 604
Scientific Instrumentation - see Instrumentation,
    Metrology
Scintillation - see Radiation
Scramjets - see Jet Propulsion
Scrubbers - see Air Pollution
Sea Ice - see Ice
Sea Water - see Oceanography
Sedimentary Rocks 606
Seismology 607
Semiconductors 609
Servomechanisms - see Robotics
Sewage Treatment 612
Ship Design - see Naval Architecture
Shipbuilding 614
Side Looking Radar - see Radar
Signal Processing 615
Silicon 617
Silicon Chips - see Intigrated Circuits
Sintering - see Powder Metallurgy
Sludge 618
Smelting - see Metallurgy
Snow 620
Software 622
Software Engineering 624
Soil Chemistry 627
Soil Mechanics 628
Soil Science 630

Solar Energy 632
Solar Physics - see Sun
Solar System 633
Solar Wind 635
Solid State Chemistry - see Physical Chemistry;
  Solid State Physics
Solid State Physics 637
Solid Waste Disposal 639
Sonar 641
Sound Recording - see Audio Engineering
Space Shuttle 643
Space Station 644
Space Structures - see Space Station
Spacecraft 645
Spacecraft Propulsion System - see Spacecraft
Spectroscopy 647
Spectrum Analysis - see Spectroscopy
Speleology 649
Spin Glasses - see Solid State Physics
Standards 650
Stars 651
Statics - see Mechanics
Statistical Mechanics - see Mechanics
Statistical Thermodynamics - see Mechanics
Statistics 654
Steam 655
Steam Engines 657
Steam Turbines 658
Steel and Steel Makings 659
Stereochemistry - see Physical Chemistry
Stochastic Processes - see Markov Processes;
  Statistics
Strength of Materials - see Materials Science
Stress and Strain 661
Strip Mining - see Mining Engineering
Structural Engineering 663
Structural Geology - see Physical Geology
Structural Materials - see Structural Engineering
Structural Petrology - see Petrology
Structures 665
Sun 667
Superconducting Super Collider - see Particle
  Accelerators
Superconductivity 669
Supernova - see Stars
Surface Chemistry 671
Surveying 673
Switching Theory - see Electrical Engineering
Synthetic Fuels 675
Systems Analysis 677
Systems Engineering 679
Tectonics - see Plate Tectonics
Tektites 683
Telecommunications 684
Telegraphy - see Telecommunications
Telephones - see Telecommunications
Telescopes 686
Television 687
Temperature Measurement - see Thermometers
Terrain Sensing - see Remote Sensing
Terrestrial Magnetism - see Geophysics
Textiles 689
Theoretical Physics - see Physics
Thermochemistry - see Physical Chemistry

Thermocouples 690
Thermodynamics 691
Thermoelectricity - see Electricity
Thermometers - see Instrumentation
Thermonuclear Fusion - see Fusion
Thermoplastics 693
Thin Films - see Microelectronics
Thunderstorms 695
Tidal Waves - see Tsunamis
Tin 696
Titanium 698
Topographic Mapping 700
Tornados 702
Trace Analysis - see Analytical Chemistry
Traffic Engineering - see Highway Engineering
Transformers - see Electric Power Engineering
Transistors - see Semiconductors
Transmission Lines - see Electric Power Engineering
Transmissions - see Automotive Engineering
Tribology - see Lubrication
Trigonometry - see Mathematics
Tsunamis 703
Turbines 705
Turboprop - see Jet Propulsion
Ultra Large Scale Integration - see Very Large Scale
  Integration
Ultralight Aircraft - see Aircraft
Ultrasonics - see Acoustics
Ultraviolet Astronomy - see Astronomy
Ultraviolet Spectroscopy - see Spectroscopy
Uncertainty Principle - see Quantum Mechanics
Underwater Photography 707
UNIX - see Computer Operating Systems
Upper Atmosphere - see Meteorology
Uranus 708
Vacuum Metallurgy - see Metallurgy
Ventilation - see Air Conditioning
Venus 711
Vertical Takeoff and Landing (VTOL) - see Aircraft
Very Large Array - see Radio Telescopes
Very Large Scale Integration (VLSI) 712
Vibration 714
Video Disks - see Video Technology
Video Tape - see Video Technology
Video Technology 716
Volcanology 718
Wankel Engine - see Engines
Water Analysis - see Environmental Engineering
Water Pollution 721
Water Resources 723
Water Treatment 724
Wave Optics - see Optics
Waves - see Oceanography, Tsunamis
Weather - see Meteorology
Weather Forecasting - see Meteorology
Weather Modification 726
Welding 728
Wells - see Hydrology, Petroleum Engineering
Wind Power - see Energy
Wind Tunnels - see Aeronautical Engineering
Windmills - see Energy
Wire and Cable 729
Wiring, Electrical 731
X-Ray Astronomy - see Astronomy

X-Rays 733
Zinc 735
Zirconium - see Elements (Chemical)

Encyclopedia of Physical Sciences and Engineering Information Sources

A

ABRASIVES

See: MACHINING

ABSORBERS

See also: CHEMICAL ENGINEERING, DRYING

ABSTRACT SERVICES AND INDEXES

APPLIED SCIENCE AND TECHNOLOGY INDEX. H.W. Wilson and Company, 950 University Avenue, Bronx, NY 10452. (800) 367-6670 or (212) 588-8400. Monthly. Inquire as to cost and availability.

CHEMICAL ABSTRACTS. American Chemical Society, Chemical Abstracts Service, Box 3012, Columbus, OH 43210. (614) 421-3600. 1907 to present. Weekly. $9500.00 per year.

CURRENT CONTENTS: ENGINEERING, TECHNOLOGY AND APPLIED SCIENCES. Institute for Scientific Information, 3501 Market Street, Philadelphia, PA 19104. (800) 523-1850 or (215) 386-0100. Weekly. $275.00 per year.

ENGINEERING INDEX MONTHLY AND AUTHOR INDEX. Engineering Information Inc., 345 East 47th Street, New York, NY 10017. (212) 705-7600. Monthly. $1560.00 per year.

SCIENCE CITATION INDEX. Institute for Scientific Information, 3501 Market Street, Philadelphia, PA 19104. (800) 523-1850 or (215) 386-0100. Six times per year. $6200.00 per year.

ASSOCIATIONS AND PROFESSIONAL SOCIETIES

AMERICAN CHEMICAL SOCIETY. 1155 16th Street, N.W., Washington, DC 20036. (202) 872-4600.

AMERICAN INSTITUTE OF CHEMICAL ENGINEERS. 345 East 47th Street, New York, NY 10017. (212) 705-7338.

ASSOCIATION OF OFFICIAL ANALYTICAL CHEMISTS. 1111 North 19th Street, Suite 210, Arlington, VA 22209. (703) 522-3032.

DIRECTORIES AND BIOGRAPHICAL SOURCES

WHO'S WHO IN ENGINEERING. Gordon Davis, editor. Hemisphere Publishing Corporation, 1010 Vermont Avenue, N.W., Washington, DC 20005. (800) 526-0275. 6th edition. 1985. $200.00.

GENERAL WORKS

ABSORPTION, SURFACE AREA AND POROSITY. S.J. Gregg and K.S. Sing. Academic Press, Inc., 6277 Sea Harbor Drive, Orlando, FL 32821. (800) 321-5068. 1982. $60.50.

HANDBOOKS AND MANUALS

PROCESS ENGINEER'S ABSORPTION POCKET HANDBOOK. R.N. Maddox. Gulf Publishing Company, P.O. Box 2608, Houston, TX 77001. (713) 520-4444. 1985. $15.00.

ONLINE DATA BASES

CA SEARCH. Chemical Abstracts Service, P.O. Box 3012, Columbus, OH 43120. (800) 848-6538 or (614) 421-3600. Comprehensive guide to chemical literature, 1972 to present. Inquire as to online cost and availability.

COMPENDEX. Engineering Information, Inc., 345 East 47th Street, New York, NY 10017. (800) 221-1044 or (212) 705-7615. Engineering and technical literature, 1975 to present. Inquire as to online cost and availability.

NTIS. National Technical Information Service, 5285 Port Royal Road, Springfield, VA 22161. (703) 487-4630. Broad coverage of government sponsored research reports, 1964 to present. Inquire as to online cost and availability.

SCISEARCH. Institute for Scientific Information, 3501 Market Street, Philadelphia, PA 19104. (800) 523-1850 or (215) 386-0100. Broad multidisciplinary title and author index to the international literature of science and technology, 1974 to present. Inquire as to online cost and availability.

WILSONLINE. H.W. Wilson and Company, 950 University Avenue, Bronx, NY 10452. (800) 367-6770 or (212) 588-8400. Makes available online versions of the H.W. Wilson indexes including Applied Science and Technology Index, Business Periodicals Index and Readers' Guide to Periodical Literature. Approximately 1980 to present. Inquire as to online cost and availability.

PERIODICALS

CHEMICAL ENGINEERING PROGRESSS. American Institute of Chemical Engineers, 345 East 47th Street, New York, NY 10017. (212) 705-7338. 1947 to present. Monthly. $35.00 per year.

ABSTRACT ALGEBRA

See: ALGEBRA

ACCELERATORS

See: PARTICLE ACCELERATORS

ACCIDENTS

See: SAFETY ENGINEERING

ACETYLENE WELDING

See: WELDING

ACID RAIN

ABSTRACT SERVICES AND INDEXES

ACID RAIN ABSTRACTS. EIC/Intelligence, Incorporated, 48 West 38th Street, New York, NY 10018. (800) 223-6275 or (212) 944-8500. Bimonthly. $345.00 per year.

ENVIRONMENTAL PERIODICALS BIBLIOGRAPHY. Environmental Studies Institute, 2074 Alameda Padre Serra, Santa Barbara, CA 93103. (805) 965-5010.

POLLUTION ABSTRACTS. Cambridge Scientific Abstracts, 5161 River Road, Bethesda, MD 20816. (800) 638-8076 or (301) 951-1400.

BIBLIOGRAPHIES

ACID PRECIPITATION: A COMPILATION OF WORLDWIDE LITERATURE - A BIBLIOGRAPHY. U.S. Department of Energy Technical Information Center. 1983. $30.00 Available from: National Technical Information Service, 5285 Port Royal Road, Springfield, VA 22161. (703) 487-4630.

ACID RAIN IN CANADA: A SELECTED BIBLIOGRAPHY (CPL BIBLIOGRAPHY NO. 124). John J. Miletich. CPL Bibliographies, 1313 East 60th Street, Merriam Center, Chicago, IL 60637. (312) 947-2007.

DIRECTORIES AND BIOGRAPHICAL SOURCES

INTERNATIONAL DIRECTORY OF ACID DEPOSITION RESEARCHERS, NORTH AMERICAN AND EUROPEAN EDITION ENVIRONMENTAL RESEARCH LABORATORY, OFFICE OF RESEARCH AND DEVELOPMENT. Environmental Protection Agency, Corvallis, OR 97330. 1983. $16.00. Order from National Technical Information Service, 5285 Port Royal Road, Springfield, VA 22161.

GENERAL WORKS

ACID RAIN AND TRANSPORTED AIR POLLUTANTS: IMPLICATIONS FOR PUBLIC POLICY. U.S. Office of Technology Assessment. Unipub, P.O. Box 1222, Ann Arbor, MI 48106. (800) 521-8110 or (313) 761-4700.

A KILLING RAIN: THE GLOBAL THREAT OF ACID RAIN PRECIPITATION. Thomas A. Pawlick. Sierra Club Books, 2034 Fillmore Street, San Francisco, CA 94115. (415) 931-7950.

TROUBLED SKIES, TROUBLED WATERS: THE STORY OF ACID RAIN. Jon R. Luoma. Viking-Penguin, Incorporated, 40 West 23rd Street, New York, NY 10010. (212) 807-7300.

HANDBOOKS AND MANUALS

ACID RAIN INFORMATION BOOK. Second Edition. David V. Bubenick. Noyes Data Corporation, Mill Road at Grand Avenue, Park Ridge, NJ 07656. (201) 391-8484.

ONLINE DATA BASES

ENVIROLINE. EIC/Intelligence, Incorporated, 48 West 38th Street, New York, NY 10018. (800) 223-6275 or (212) 944-8500. Indexes and abstracts both English and Non-English language publications relating to the environment, 1971 to date. Inquire as to cost and availability.

POLLUTION ABSTRACTS (ONLINE). Cambridge Scientific Abstracts, 5161 River Road, Bethesda, MD 20816. (800) 638-8076 or (301) 951-1400. Provides indexing and abstracting of international environmental literature, 1970 to date. Inquire as to cost and availability.

OTHER SOURCES

ACID PRECIPITATION: A CURRENT AWARENESS BULLETIN. U.S. Department of Energy, Office of Scientific and Technical Information. Available from: National Technical Information Service, Springfield, Va 22161. Semimonthly. $40.00 per year.

PERIODICALS

ACID PRECIPITATION DIGEST. (Acid Rain Information Clearinghouse). Center for Environmental Information, Incorporated, 33 South Washington Street, Rochester, NY 14608. (716) 546-3796. Monthly. $30.00 per year.

AMBIO: A JOURNAL OF THE HUMAN ENVIRONMENT RESEARCH AND MANAGEMENT. Royal Swedish Academy of Sciences, Box 50005, S-104 05 Stockholm, Sweden. Bimonthly. $85.00 per year.

ATMOSPHERIC ENVIRONMENT. Pergamon Press Incorporated, Journals Division, Maxwell House, Fairview Park, Elmsford, NY 10523. (914) 592-7700. Monthly. $540.00 per year.

WATER, AIR AND SOIL POLLUTION. D. Reidel Publishing Company, 190 Old Derby Street, Hingham, MA 02043. (617) 749-5262. Twenty times per year. $520.00 per year.

RESEARCH CENTERS AND INSTITUTES

NEW JERSEY DEPARTMENT OF ENVIRONMENTAL PROTECTION. Environmental Quality Division, Labor and Industry Building, John Fitch Plaza, CN 027, Trenton, NJ 08625. (609) 292-6704.

OHIO STATE UNIVERSITY. Laboratory for Environmental Studies, Ohio Agricultural R&D Center, Madison, OH 44691. (216) 263-3720.

STATE UNIVERSITY OF NEW YORK AT PLATTSBURGH. Center for Earth and Environmental Science, Plattsburgh, NY 12901. (518) 564-2028.

UNIVERSITY OF RHODE ISLAND. Center for Atmospheric Chemistry Studies, Narragansett, RI 02882. (401) 792-6256.

ACIDS AND BASES

See also: CHEMISTRY, ORGANIC CHEMISTRY

ABSTRACT SERVICES AND INDEXES

APPLIED SCIENCE AND TECHNOLOGY INDEX. H.W. Wilson Company, 950 University Avenue, Bronx, NY 10452. (800) 367-6670 or (212) 588-8400. Inquire as to cost and availability.

CHEMICAL ABSTRACTS. Chemical Abstracts Service, 2540 Olentangy Road, P.O. Box 3012, Columbus, OH 43210. (800) 848-6538 or (614) 421-3600. Weekly. $9200.00 per year.

PHYSICS ABSTRACTS. Institute of Electrical Engineers, London, United Kingdom. Available from: Institute of Electrical and Electronic Engineers (IEEE), 345 East 47th Street, New York, NY 10017. (212) 705-7900.

SCIENCE CITATION INDEX. Institute for Scientific Information, 3501 Market Street, Philadelphia, PA 19104. (800) 523-1850 or (215) 386-0100.

ANNUAL REVIEWS AND YEARBOOKS

ANNUAL REVIEW OF PHYSICAL CHEMISTRY. Annual Reviews Incorporated, 4139 El Camino Way, Palo Alto, CA 94306. (415) 493-4400. Annual.

ASSOCIATIONS AND PROFESSIONAL SOCIETIES

AMERICAN CHEMICAL SOCIETY. 1155 16th Street, NW, Washington, DC 20036. (202) 872-4600.

ASSOCIATION OF CONSULTING CHEMISTS AND CHEMICAL ENGINEERS. 50 East 41st Street, Suite 92, New York, NY 10017. (212) 684-6255.

BIBLIOGRAPHIES

SCIENCE BOOKS AND FILMS. American Association for the Advancement of Science, 1333 H Street, NW, Washington, DC 20005.

SCIENTIFIC AND TECHNICAL BOOKS IN PRINT; AN INDEX TO LITERATURE IN SCIENCE AND TECHNOLOGY. R.R. Bowker Company, 205 East 42nd Street, New York, NY 10017. (800) 521-8110 or (212) 916-1600.

DIRECTORIES AND BIOGRAPHICAL SOURCES

AMERICAN INSTITUTE OF CHEMISTS. American Institute of Chemists, 7315 Wisconsin Avenue, Bethesda, MD 20814. (301) 652-2447. 1986. $35.00.

BIOGRAPHICAL DICTIONARY OF SCIENTISTS: CHEMISTS. David Abbott, editor. P. Bedrick Books, 125 East 23rd Street, New York, NY 10010. (212) 777-1187. 1984. $18.95.

CHEMICAL WEEK - BUYERS GUIDE ISSUE. McGraw-Hill Book Company, 1221 Avenue of the Americas, New York, NY 10020. (800) 628-0004. Annual, October. $50.00.

CONSULTING SERVICES: CHEMISTS AND CHEMICAL ENGINEERS. Association of Consulting Chemists and Chemical Engineers, 50 East 41st Street, New York, NY 10017. (212) 684-6255. Annual. 1986. $45.00.

RESEARCH CENTERS DIRECTORY. Gale Research Company, Book Tower, Detroit, MI 48226. (800) 521-0707. Eleventh edition. 1987. $355.00.

SCIENTIFIC AND TECHNICAL ORGANIZATIONS AND AGENCIES DIRECTORY. Gale Research Company, Book Tower, Detroit, MI 48226. (800) 521-0707. 1985. $150.00.

WORLD GUIDE TO SCIENTIFIC ASSOCIATIONS AND LEARNED SOCIETIES. K.G. Saur Incorporated, 175 Fifth Avenue, New York, NY 10010. (800) 521-0707 or (212) 982-1302. Fourth edition. 1984. $112.00.

WHO'S WHO IN FRONTIER SCIENCE AND TECHNOLOGY. Marquis Who's Who, Incorporated, 200 East Ohio Street, Chicago, IL 60611. (800) 428-3898 or (312) 787-2008.

WHO'S WHO IN TECHNOLOGY TODAY. Reston Publishing Company, Incorporated, c/o Prentice-Hall, Incorporated, Englewood Cliffs, NJ 07632. (800) 262-6868. Biennial. Five volumes. $425.00. Covers the fields of electronics, computer science, physics, optics, chemistry, biotechnology, mechanics, energy, and earth science.

ENCYCLOPEDIAS AND DICTIONARIES

CONCISE ENCYCLOPEDIA OF CHEMICAL TECHNOLOGY. Kirk Othmer. John Wiley and Sons, Incorporated, 605 Third Avenue, New York, NY 10158. (800) 526-5368 or (212) 850-6000. Third edition. 1985. $129.95.

CONDENSED CHEMICAL DICTIONARY. Gessner Hawley. Van Nostrand Reinhold, 115 Fifth Avenue, New York, NY 10003. Tenth edition. 1981. $49.95.

ENCYCLOPEDIA OF PHYSICAL SCIENCE AND TECHNOLOGY. Academic Press, Incorporated, Orlando, FL 32887. (800) 321-5068 or (305) 345-2734. Fifteen volumes, 1986.

GLOSSARY OF CHEMICAL TERMS. Clifford A. Hampel and Gessner G. Hawley. Van Nostrand Reinhold Company, 115 Fifth Avenue, New York, NY 10003. (800) 543-2681 or (212) 254-3232. Second edition. 1982. $21.95.

MCGRAW-HILL ENCYCLOPEDIA OF SCIENCE AND TECHNOLOGY. McGraw-Hill Book, Incorporated, 1221 Avenue of the Americas, New York, NY 10020. (212) 997-3675.

VAN NOSTRAND REINHOLD ENCYCLOPEDIA OF CHEMISTRY. Douglas M. Considine and Glenn D. Considine. Van Nostrand Reinhold Publishing Company, Incorporated, 115 Fifth Avenue, New York, NY 10003. (800) 543-2681 or (212) 254-3232. 1984. $97.95.

GENERAL WORKS

ANALYTICAL CHEMISTRY. Gary D. Christian. John Wiley and Sons, Incorporated, 605 Third Avenue, New York, NY 10158. (800) 526-5368 or (212) 850-6000. Fourth edition. 1986. $37.95.

CATALYSIS BY ACIDS AND BASES. B. Imelik. Elsevier Science Publishing Company, Incorporated, 52 Vanderbilt Avenue, New York, NY 10017. (212) 370-5520. 1985. $94.50.

CHEMISTRY OF THE ELEMENTS. N.N. Greenwood and A. Earnshaw. Pergamon Publishing, Incorporated, Maxwell House, Fairview Park, Elmsford, NY 10523. (914) 592-7700. 1984. $143.00.

HARD AND SOFT ACIDS AND BASES. R.G. Pearson, editor. Van Nostrand Reinhold Company, Incorporated, 135 West 50th Street, New York, NY 10020. (212) 265-8700. 1973. $60.00.

HAZARDOUS AND TOXIC CHEMICAL: SAFE HANDLING AND DISPOSAL. Howard Fawcett. John Wiley and Sons, Incorporated, 605 Third Avenue, New York, NY 10158. (800) 526-5368 or (212) 850-6000. 1984. $37.00.

INVESTIGATION OF RATES AND MECHANISMS OF REACTIONS, PART 1. Claude F. Bernasconi, editor. John Wiley and Sons, Incorporated, 605 Third Avenue, New York,

NY 10158. (800) 526-5368 or (212) 850-6000. Fourth edition. 1986. $175.00.

SAFE STORAGE OF LABORATORY CHEMICALS. David A. Piptone. John Wiley and Sons, Incorporated, 605 Third Avenue, New York, NY 10158. (800) 526-5368 or (212) 850-6000. 1984. $60.00.

HANDBOOKS AND MANUALS

THE CHEMIST'S COMPANION: A HANDBOOK OF PRACTICAL DATA, TECHNIQUES, AND REFERENCES. Arnold J. Gordon and Richard A. Ford. (800) 526-5368. 1973. $49.95.

CRC HANDBOOK OF CHEMISTRY AND PHYSICS. CRC Press, Incorporated, 2000 Corporate Boulevard, NW, Boca Raton, FL 33431. Sixty-seventh edition. 1986. $69.95.

HANDBOOK OF APPLIED CHEMISTRY: FACTS FOR ENGINEERS, SCIENTISTS, TECHNICIANS, AND TECHNICAL MANAGERS. Vollrath Hopp and Ingp Hennig. McGraw-Hill Book Company, 1221 Avenue of the Americas, New York, NY 10020. (800) 628-0004. 1983. $54.00.

LANGE'S HANDBOOK OF CHEMISTRY. John A. Dean, editor. McGraw-Hill Book Company, 1221 Avenue of the Americas, New York, NY 10020. (800) 628-0004. 1985. $59.50.

CA SEARCH. Chemical Abstracts Service, P.O. Box 3012, Columbus, OH 43210. Guide to chemical literatue, 1972 to present. Inquire as to cost and availability.

DISSERTATION ABSTRACTS ONLINE. University Microfilms International, 300 North Zeeb Road, Ann Arbor, MI 48106. (800) 521-0600 or (313) 761-4700. Scope includes virtually all doctoral dissertations accepted at accredited American institutions from 1861 to present in 252 subject areas. Inquire as to online cost and availability.

INSPEC. INSPEC Marketing Department, Institute of Electrical and Electronics Engineers, Incorporated, IEEE Service Department, 445 Hoes Lane, Piscataway, NJ 08854. (201) 981-0060. Inquire as to online cost and availability.

NTIS. National Technical Information Service, 5285 Port Royal Road, Springfield, VA 22161. (703) 487-4630. Broad coverate of government sponsored research reports, 1964 to present. Inquire as to online cost and availability.

SCISEARCH. Institute for Scientific Information, 3501 Market Street, Philadelphia, PA 19104. (800) 523-1850 or (215) 386-0100. Broad multidisciplinary title and author index to the international literature of science and technology, 1974 to present. Inquire as to online cost and availability.

OTHER SOURCES

CHEMICAL NOMENCLATURE USEAGE. Ronald Lees and Arthur F. Smith. John Wiley and Sons, Incorporated, 605 Third Avenue, New York, NY 10158. (800) 526-5368 or (212) 850-6000. 1983. $52.95.

GUIDE TO BASIC INFORMATION SOURCES IN CHEMISTRY. Arthur Antony. John Wiley and Sons, Incorporated, 605 Third Avenue, New York, NY 10158. (800) 526-5368 or (212) 850-6000. 1979. $26.95.

HOW TO FIND CHEMICAL INFORMATION: A GUIDE FOR PRACTICING CHEMISTS, TEACHERS, AND STUDENTS. John Wiley and Sons, Incorporated, 605 Third Avenue, New York, NY 10158. (800) 526-5368 or (212) 850-6000. 1986. $35.00.

PERIODICALS

ANALYTICAL CHEMISTRY. American Chemical Society, 1155 Sixteenth Street, NW, Washington, DC 20036. (800) 424-6747 or (202) 872-4700. Monthly. $33.00 per year.

BIOCHEMISTRY. American Chemical Society, 1155 Sixteenth Street, NW, Washington, DC 20036. (800) 424-6747 or (202) 872-4700. Biweekly. $303.00 per year.

CHEMICAL REVIEWS. American Chemical Society, 1155 Sixteenth Street, NW, Washington, DC 20036. (800) 424-6747 or (202) 872-4700. Bimonthly. $83.00 per year.

CHEMICAL WEEK. McGraw-Hill Publishing Company, 1221 Avenue of the Americas, New York, NY 10020. (212) 512-2000. Weekly. $30.00 per year.

CHEMTECH. American Chemical Society, 1155 Sixteenth Street, NW, Washington, DC 20036. (800) 424-6747 or (202) 872-4700. Monthly. $40.00 per year to individuals.

INORGANIC CHEMISTRY. American Chemical Society, 1155 Sixteenth Street, NW, Washington, DC 20036. (800) 424-6747 or (202) 872-4700. Monthly. $400.00 per year.

JOURNAL OF CHEMICAL EDUCATION. American Chemical Society, 1155 Sixteenth Street, NW, Washington, DC 20036. (800) 424-6747 or (202) 872-4700. Monthly. $40.00 per year.

JOURNAL OF ORGANIC CHEMISTRY. American Chemical Society, 1155 Sixteenth Street, NW, Washington, DC 20036. (800) 424-6747 or (202) 872-4700. Biweekly. $265.00 per year.

JOURNAL OF PHYSICAL CHEMISTRY. American Chemical Society, 1155 Sixteenth Street, NW, Washington, DC 20036. (800) 424-6747 or (202) 872-4700. Biweekly. $369.00 per year.

JOURNAL OF THE AMERICAN CHEMICAL SOCIETY. American Chemical Society, 1155 Sixteenth Street, NW, Washington, DC 20036. (800) 424-6747 or (202) 872-4700. Biweekly. $330.00 per year.

TETRAHEDRON. Pergamon Journals, Incorporated, Maxwell House, Fairview Park, Elmsford, NY 10523. (914) 592-7700. Biweekly. $1400.00 per year.

RESEARCH CENTERS AND INSTITUTES

HARVARD UNIVERSITY. Chemical Laboratories, Oxford Street, Cambridge, MA 02138. (617) 495-4283.

RENSSELAER POLYTECHNICAL INSTITUTE. Chemistry Laboratories, Cogswell Laboratory, Troy, NY 13181. (518) 266-8462.

SOUTHERN ILLINOIS UNIVERSITY AT CARBONDALE. Research Program in Chemistry and Biochemistry, Carbondale, IL 62901. (618) 453-5721.

UNIVERSITY OF WISCONSIN, MADISON. Theoretical Chemistry Institute, 1101 University Avenue, Madison, WI 53706. (608) 262-1511.

ACOUSTICS

See also: AUDIO ENGINEERING, PHONOGRAPHS, SIGNAL PROCESSING, SOUND RECORDING

ABSTRACT SERVICES AND INDEXES

ACOUSTICS ABSTRACTS. Multiscience Publishing Company, Limited, 42-45 New Broad Street, London, EC2M 1QY, England. 1967 to present. Monthly. $295.00 per year.

APPLIED SCIENCE AND TECHNOLOGY INDEX. H.W. Wilson and Company, 950 University Avenue, Bronx, NY 10452. (800) 367-6670 or (212) 588-8400. Monthly. Inquire as to cost and availability.

CURRENT CONTENTS: ENGINEERING, TECHNOLOGY AND APPLIED SCIENCES. Institute for Scientific Information, 3501 Market Street, Philadelphia, PA 19104. (800) 523-1850 or (215) 386-0100. Weekly. $275.00 per year.

ELECTRONICS AND COMMUNICATIONS ABSTRACTS JOURNAL. Cambridge Scientific Abstracts. 5161 River Road, Bethesda, MD 20816. (301) 951-1400. Bimonthly. Inquire as to cost and availability.

ENGINEERING INDEX MONTHLY AND AUTHOR INDEX. Engineering Information Inc., 345 East 47th Street, New York, NY 10017. (212) 705-7600. Monthly. $1560.00 per year.

GENERAL SCIENCE INDEX. H.W. Wilson and Company, 950 University Avenue, Bronx, NY 10452. (800) 367-6670 or (212) 588-8400. 1978 to present. Monthly. Inquire as to cost and availability.

INDEX TO SCIENTIFIC AND TECHNICAL PROCEEDINGS. Institute for Scientific Information, 3501 Market Street, Philadelphia, PA 19104. (800) 523-1850 or (215) 386-0100. 1978 to present. Monthly. $775.00 per year.

INDEX TO SCIENTIFIC REVIEWS. Institute for Scientific Information, 3501 Market Street, Philadelphia, PA 19104. (800) 523-1850 or (215) 386-0100. 1974 to present. Semi-annual. $550.00 per year.

PHYSICS ABSTRACTS. Institution of Electrical Engineers. Available from: IEEE Service Center, 445 Hoes Lane, Piscataway, NJ 08854. 1898 to present. Bimonthly. $1700.00 per year.

PHYSICS BRIEFS. Physik Verlag GmbH, Postfach 1260/1280, D-6940 Weinheim, West Germany. (212) 661-9404. 1920 to present. Twenty-six times per year. $1250.00 per year.

SCIENCE CITATION INDEX. Institute for Scientific Information, 3501 Market Street, Philadelphia, PA 19104. (800) 523-1850 or (215) 386-0100. Six times per year. $6200.00 per year.

ANNUAL REVIEWS AND YEARBOOKS

PHYSICAL ACOUSTICS. Academic Press, Inc., 6277 Sea Harbor Drive, Orlando, FL 32821. (800) 321-5068. Irregular. Price varies, inquire.

ASSOCIATIONS AND PROFESSIONAL SOCIETIES

AMERICAN INSTITUTE OF PHYSICS. 335 East 45th Street, New York, NY 10017. (212) 661-9494.

AUDIO ENGINEERING SOCIETY. 60 East 42nd Street, New York, NY 10165. (212) 661-2355.

INSTITUTE OF ELECTRICAL AND ELECTRONICS ENGINEERS. 345 East 47th Street, New York, NY 10017. (212) 705-7900.

DIRECTORIES AND BIOGRAPHICAL SOURCES

AMERICAN MEN AND WOMEN OF SCIENCE. R.R. Bowker, Inc., Order Department, 245 West 17th Street, New York, NY 10011. (800) 521-8110. Eight volumes. 1986. $595.00 for set.

IEEE MEMBERSHIP DIRECTORY. Institute of Electrical and Electronics Engineers. IEEE Service Center, 445 Hoes Lane, Piscataway, NJ 08854. Annual. $7.00.

INTERNATIONAL RESEARCH CENTERS DIRECTORY 1988-89. Darren L. Smith, editor. Gale Research Company, Book Tower, Detroit, MI 48226. (800) 521-0707. 4th edition. 1987. $360.00.

1987 DIRECTORY OF ENGINEERING SOCIETIES AND RELATED ORGANIZATIONS. Gordon Davis, editor. Hemisphere Publishing Corporation, 1010 Vermont Avenue, NW, Washington, DC 20005. (800) 526-0275. 12th edition. 1987. $100.00.

RESEARCH CENTERS DIRECTORY 1988. Gale Research Company, Book Tower, Detroit, MI 48226. (800) 521-0707. 12th edition. 1987. $365.00 for set.

SCIENTIFIC AND TECHNICAL ORGANIZATIONS AND AGENCIES DIRECTORY. Margaret Labash Young, editor. Gale Research Company, Book Tower, Detroit, MI 48226. (800) 521-0707. 2nd edition. 1987. $185.00.

WHO'S WHO IN ENGINEERING. Gordon Davis, editor. Hemisphere Publishing Corporation, 1010 Vermont Avenue, NW, Washington, DC 20005. (800) 526-0275. 6th edition. 1985. $200.00.

ENCYCLOPEDIAS AND DICTIONARIES

CONCISE SCIENCE DICTIONARY. Oxford University Press, 200 Madison Avenue, New York, NY 10016. (800) 458-5833. 1987. $9.95 in paper.

DICTIONARY OF THE PHYSICAL SCIENCES: TERMS, FORMULAS, DATA. Cesare Emiliani. Oxford University Press, 200 Madison Avenue, New York, NY 10016. (800) 458-5833. 1987. $19.95 in paper.

ENCYCLOPEDIA OF PHYSICS. A.M. Prokhorov and S. Chomet. Hemisphere Publishing Corporation, 79 Madison Avenue, New York, NY 10016-7892. (800) 821-8312. 1988. $155.00.

MCGRAW-HILL ENCYCLOPEDIA OF SCIENCE AND TECHNOLOGY. McGraw-Hill Book Company, 1221 Avenue of the Americas, New York, NY 10020. (212) 512-2000. 6th edition. 1987. $1600.00.

THESAURUS OF SCIENTIFIC, TECHNICAL, AND ENGINEERING TERMS. Hemisphere Publishing Corporation, 1010 Vermont Avenue, NW, Washington, DC 20005. (800) 526-0275. 1988. $125.00.

GENERAL WORKS

ACOUSTIC DESIGN. Duncan Templeton and David Saunders. Van Nostrand Reinhold Company, Inc., 135 West 50th Street, New York, NY 10020. (800) 543-2681. 1987. $36.95.

FUNDAMENTALS OF ACOUSTICS. L.E. Kinsler and others. John Wiley and Sons, Inc., 605 Third Avenue, New York, NY 10158. (800) 526-5368. Third edition. 1982. $45.00.

PRINCIPLES OF UNDERWATER SOUND. R.J. Urick. McGraw-Hill Book Company, 1221 Avenue of the Americas, New York, NY 10020. (212) 512-2000. Third edition. 1983. $48.00.

SOUND AND SOURCES OF SOUND. A.P. Dowling and J.E. Efowcs-Williams. John Wiley and Sons, Inc., 605 Third Avenue, New York, NY 10158. (800) 526-5368. 1983. $30.00 in paper.

ONLINE DATA BASES

COMPENDEX. Engineering Information, Inc., 345 East 47th Street, New York, NY 10017. (800) 221-1044 or (212) 705-7615. Engineering and technical literature, 1975 to present. Inquire as to online cost and availability.

DISSERTATION ABSTRACTS ONLINE. University Microfilms International, 300 North Zeeb Road, Ann Arbor, MI 48106. (800) 521-0600 or (313) 761-4700. Scope includes virtually all doctoral dissertations accepted at accredited American institutions from

1861 to present in over 250 subject areas. Inquire as to online cost and availability.

INSPEC. INSPEC Marketing Department, Institution of Electrical Engineers. Available from IEEE Service Center, 445 Hoes Lane, Piscataway, NJ 08854. (201) 981-0060. Online version of Physics Abstracts. Inquire as to online cost and availability.

NTIS. National Technical Information Service, 5285 Port Royal Road, Springfield, VA 22161. (703) 487-4630. Broad coverage of government sponsored research reports, 1964 to present. Inquire as to online cost and availability.

SCISEARCH. Institute for Scientific Information, 3501 Market Street, Philadelphia, PA 19104. (800) 523-1850 or (215) 386-0100. Broad multidisciplinary title and author index to the international literature of science and technology, 1974 to present. Inquire as to online cost and availability.

WILSONLINE. H.W. Wilson and Company, 950 University Avenue, Bronx, NY 10452. (800) 367-6770 or (212) 588-8400. Makes available online versions of the H.W. Wilson indexes including Applied Science and Technology Index, Business Periodicals Index and Readers' Guide to Periodical Literature. Approximately 1980 to present. Inquire as to online cost and availability.

OTHER SOURCES

WHAT EVERY ENGINEER SHOULD KNOW ABOUT ENGINEERING SOURCES. Marcel Dekker Inc., 270 Madison Avenue, New York, NY 10016. (800) 228-1160. 1984. $24.95.

PERIODICALS

ACOUSTICAL SOCIETY OF AMERICA JOURNAL. American Institute of Physics, 335 East 45th Street, New York, NY 10017. (212) 661-9494. Monthly. $300.00 per year.

AUDIO ENGINEERING SOCIETY JOURNAL. Audio Engineering Society, 60 East 42nd Street, New York, NY 10165. (212) 661-2355. Ten times per year. $70.00 per year.

APPLIED ACOUSTICS. Elsevier Science Publishing Company, Inc., 52 Vanderbilt Avenue, New York, NY 10017. (212) 370-5520. Six times per year. $166.00 per year.

DB: THE SOUND ENGINEERING MAGAZINE. Sagamore Publishing Company, Inc., 1120 Old Country Road, Plainview, NY 11803. (516) 433-6530. Bimonthly. $15.00 per year.

IEEE ACOUSTICS, SPEECH AND SIGNAL PROCESSING MAGAZINE. Institute of Electrical and Electronics Engineers. IEEE Service Center, 445 Hoes Lane, Piscataway, NJ 08854. Quarterly. $40.00 per year.

IEEE TRANSACTIONS ON ULTRASONICS, FERROELECTRICS AND FREQUENCY CONTROL. Institute of Electrical and Electronics Engineers. IEEE Service Center, 445 Hoes Lane, Piscataway, NJ 08854. 1954 to present. Bimonthly. $100.00 per year.

JOURNAL OF LOW FREQUENCY NOISE AND VIBRATION. Multiscience Publishing Company, Limited, 42-45 New Broad Street, London, EC2M 1QY, England. 1982 to present. Quarterly. $100.00 per year.

SOUND AND VIBRATION. Acoustical Publications, Inc., Box 40416, Bay Village, OH 44140. 1967 to present. Monthly. $10.00 per year.

ULTRASONICS. Butterworth Publishing, 80 Montvale Avenue, Stoneham, MA 02180. (800) 325-4177. 1963 to present. Bimonthly. $155.00 per year.

RESEARCH CENTERS AND INSTITUTES

ACOUSTICS AND VIBRATION LABORATORY. Massachusetts Institute of Technology, Building 3, Room 366, Cambridge, MA 02139. (617) 253-2214.

ACOUSTICS DIVISION, NAVAL RESEARCH LABORATORY. U.S. Department of the Navy, 4555 Overlook Avenue, S.W., Washington, DC 20375. (202) 767-3482.

APPLIED ULTRASONICS AND ELECTROMAGNETIC SIGNAL PROCESSING LABORATORY. Purdue University, School of Electrical Engineering, West Lafayette, IN 47906. (317) 494-3563.

CENTER FOR SOUND AND VIBRATION. North Carolina State University, Campus Box 7910, Raleigh, NC 27695. (919) 737-3024.

ACRYLIC RESINS

See: PLASTICS

ADA

See: PROGRAMMING LANGUAGES

ADHESIVES

See also: CHEMICAL ENGINEERING, CHEMISTRY, PLASTICS

ABSTRACT SERVICES AND INDEXES

APPLIED SCIENCE AND TECHNOLOGY INDEX. H.W. Wilson and Company, 950 University Avenue, Bronx, NY 10452. (800) 367-6670 or (212) 588-8400. Monthly. Inquire as to cost and availability.

CHEMICAL ABSTRACTS. American Chemical Society, Chemical Abstracts Service, Box 3012, Columbus, OH 43210. (614) 421-3600. 1907 to present. Weekly. $9500.00 per year.

CURRENT CONTENTS: ENGINEERING, TECHNOLOGY AND APPLIED SCIENCES. Institute for Scientific Information, 3501 Market Street, Philadelphia, PA 19104. (800) 523-1850 or (215) 386-0100. Weekly. $275.00 per year.

ENGINEERING INDEX MONTHLY AND AUTHOR INDEX. Engineering Information Inc., 345 East 47th Street, New York, NY 10017. (212) 705-7600. Monthly. $1560.00 per year.

INSTITUTE OF PAPER CHEMISTRY ABSTRACT BULLETIN. Institute of Paper Chemistry, 1043 East South River Street, Appleton, WI 54912. (414) 734-9251. 1930 to present. Monthly. Inquire as to price and availability.

SCIENCE CITATION INDEX. Institute for Scientific Information, 3501 Market Street, Philadelphia, PA 19104. (800) 523-1850 or (215) 386-0100. Six times per year. $6200.00 per year.

ASSOCIATIONS AND PROFESSIONAL SOCIETIES

ADHESIVE AND SEALANT COUNCIL. 1600 Wilson Boulevard, Suite 910, Arlington, VA 22209. (703) 841-1112.

ADHESIVES MANUFACTURING ASSOCIATION OF AMERICA. 111 East Wacker Drive, Chicago, IL 60601. (312) 644-6610.

AMERICAN CHEMICAL SOCIETY. 1155 16th Street, N.W., Washington, DC 20036. (800) 424-6747.

GUMMED INDUSTRIES ASSOCIATION. 380 North Broadway, Jericho, NY 11753. (516) 822-8948.

DIRECTORIES AND BIOGRAPHICAL SOURCES

ADHESIVES. International Plastics Selector, Inc., 9889 Willow Creek Road, San Diego, CA 92126. (619) 578-3910. Annual. $100.00.

ADHESIVES RED BOOK. 6285 Barfield Road, Atlanta, GA 30328. (404) 256-9800. Annual. $35.00.

ADHESIVES TECHNOLOGY. Noyes Data Corporation, Mill Road at Grand Avenue, Park Ridge, NJ 07656. (201) 391-8484. 1983. $50.00.

RESEARCH CENTERS DIRECTORY 1988. Gale Research Company, Book Tower, Detroit, MI 48226. (800) 521-0707. 12th edition. 1987. $365.00 for set.

WHO'S WHO IN ENGINEERING. Gordon Davis, editor. Hemisphere Publishing Corporation, 1010 Vermont Avenue, NW, Washington, DC 20005. (800) 526-0275. 6th edition. 1985. $200.00.

GENERAL WORKS

ADHESIVE AND SEALANT COMPOUND FORMULATIONS. E.W. Flick. Noyes Data Corporation, Mill Road at Grand Avenue, Park Ridge, NJ 07656. (201) 391-8484. Second Edition. 1984. $48.00.

ADHESIVES IN ENGINEERING DESIGN. W.A. Lees. Springer-Verlag New York, Inc., 175 Fifth Avenue, New York, NY 10010. (800) 526-7254. 1985. $28.00.

INDUSTRIAL APPLICATIONS OF ADHESIVE BONDING. M.M. Sadek, editor. Elsevier Science Publishing Company, Inc., 52 Vanderbilt Avenue, New York, NY 10017. (212) 370-5520. 1987. $61.25.

STRUCTURAL ADHESIVES: CHEMISTRY AND TECHNOLOGY. S.R. Hartshorn, editor. Plenum Publishing Corporation, 233 Spring Street, New York, NY 10013. (800) 221-9369. 1986. $75.00.

HANDBOOKS AND MANUALS

ADHESIVES TECHNOLOGY HANDBOOK. Arthur H. Landrock. Noyes Data Corporation, Mill Road at Grand Avenue, Park Ridge, NJ 07656. (201) 391-8484. 1986. $64.00.

CONSTRUCTION SEALANTS AND ADHESIVES. Julian R. Panek and John P. Cook. John Wiley and Sons, Inc., 605 Third Avenue, New York, NY 10158. (800) 526-5368. Second edition. 1984. $39.95.

HANDBOOK OF ADHESIVE BONDING. Charles V. Cagle. McGraw-Hill Book Company, 1221 Avenue of the Americas, New York, NY 10020. (212) 512-2000. 1982. $65.00.

HANDBOOK OF ADHESIVE RAW MATERIALS. Ernest W. Flick. Noyes Data Corporation, Mill Road at Grand Avenue, Park Ridge, NJ 07656. (201) 391-8484. 1982. $45.00.

HANDBOOK OF PRESSURE-SENSITIVE ADHESIVE TECHNOLOGY. Van Nostrand Reinhold Company, Inc., 135 West 50th Street, New York, NY 10020. (800) 543-2681. 1982. $37.50.

ONLINE DATA BASES

CA SEARCH. Chemical Abstracts Service, P.O. Box 3012, Columbus, OH 43120. (800) 848-6538 or (614) 421-3600. Comprehensive guide to chemical literature, 1972 to present. Inquire as to online cost and availability.

COMPENDEX. Engineering Information, Inc., 345 East 47th Street, New York, NY 10017. (800) 221-1044 or (212) 705-7615. Engineering and technical literature, 1975 to present. Inquire as to online cost and availability.

DISSERTATION ABSTRACTS ONLINE. University Microfilms International, 300 North Zeeb Road, Ann Arbor, MI 48106. (800) 521-0600 or (313) 761-4700. Scope includes virtually all doctoral dissertations accepted at accredited American institutions from 1861 to present in over 250 subject areas. Inquire as to online cost and availability.

NTIS. National Technical Information Service, 5285 Port Royal Road, Springfield, VA 22161. (703) 487-4630. Broad coverage of government sponsored research reports, 1964 to present. Inquire as to online cost and availability.

SCISEARCH. Institute for Scientific Information, 3501 Market Street, Philadelphia, PA 19104. (800) 523-1850 or (215) 386-0100. Broad multidisciplinary title and author index to the international literature of science and technology, 1974 to present. Inquire as to online cost and availability.

WILSONLINE. H.W. Wilson and Company, 950 University Avenue, Bronx, NY 10452. (800) 367-6770 or (212) 588-8400. Makes available online versions of the H.W. Wilson indexes including Applied Science and Technology Index, Business Periodicals Index and Readers' Guide to Periodical Literature. Approximately 1980 to present. Inquire as to online cost and availability.

PERIODICALS

ADHESIVES AGE. Communication Channels, Inc., 6255 Barfield Road, Atlanta, GA 30328. (404) 256-9800. 1958 to present. Monthly. $26.00 per year.

INTERNATIONAL JOURNAL OF ADHESION AND ADHESIVES. Butterworth's Publishing, 80 Montvale Avenue, Stoneham, MA 02180. (800) 325-4177. 1980 to present. Quarterly. $155.00 per year.

JOURNAL OF ADHESION. Gordon and Breach Science Publishers, Inc., 50 West 23rd Street, New York, NY 10010. (212) 206-8900. 1969 to present. Eight times per year. $260.00 per year.

RESEARCH CENTERS AND INSTITUTES

CENTER FOR ADHESIVES, SEALANTS AND COATINGS. Case Western Reserve University, Department of Chemistry, Cleveland, OH 44106. (216) 368-5030.

LABORATORY FOR SURFACE SCIENCE AND TECHNOLOGY. University of Maine, Barrows Hall, Orono, ME 04469. (207) 581-2254.

ADIABATIC PROCESSES

See: THERMODYNAMICS

AERIAL PHOTOGRAPHY

See also: CARTOGRAPHY, PHOTOGRAMMETRY, REMOTE SENSING, SURVEYING

ABSTRACT SERVICES AND INDEXES

APPLIED SCIENCE AND TECHNOLOGY INDEX. H.W. Wilson and Company, 950 University Avenue, Bronx, NY 10452. (800) 367-6670 or (212) 588-8400. Monthly. Inquire as to cost and availability.

BIBLIOGRAPHY AND INDEX OF GEOLOGY. American Geological Institute, 4220 King Street, Alexandria, VA 22302. (703) 379-2480. 1969 to present. Monthly. $1100.00 per year.

CHEMICAL ABSTRACTS. American Chemical Society, Chemical Abstracts Service, Box 3012, Columbus, OH 43210. (614) 421-3600. 1907 to present. Weekly. $9500.00 per year.

CURRENT CONTENTS: ENGINEERING, TECHNOLOGY AND APPLIED SCIENCES. Institute for Scientific Information, 3501 Market Street, Philadelphia, PA 19104. (800) 523-1850 or (215) 386-0100. Weekly. $275.00 per year.

ENGINEERING INDEX MONTHLY AND AUTHOR INDEX. Engineering Information Inc., 345 East 47th Street, New York, NY 10017. (212) 705-7600. Monthly. $1560.00 per year.

METEOROLOGICAL AND GEOASTROPHYSICAL ABSTRACTS. American Meteorological Society, 45 Beacon Street, Boston, MA 02108. (617) 227-2425. 1950 to present. Monthly. $450.00 per year.

PHYSICS ABSTRACTS. Institution of Electrical Engineers. Available from: IEEE Service Center, 445 Hoes Lane, Piscataway, NJ 08854. 1898 to present. Bimonthly. $1700.00 per year.

REMOTE SENSING OF NATURAL RESOURCES: A QUARTERLY LITERATURE REVIEW. University of New Mexico, Technology Application Center, Albuquerque, NM 87131. (505) 277-3622. 1974 to present. Quarterly. $150.00. Available to qualified agencies only.

SCIENCE CITATION INDEX. Institute for Scientific Information, 3501 Market Street, Philadelphia, PA 19104. (800) 523-1850 or (215) 386-0100. Six times per year. $6200.00 per year.

ASSOCIATIONS AND PROFESSIONAL SOCIETIES

AMERICAN SOCIETY FOR PHOTOGRAMMETRY AND REMOTE SENSING. 210 Little Falls Street, Falls Church, VA 22046-4398. (703) 534-6617.

OPTICAL SOCIETY OF AMERICA. 1816 Jefferson Place, N.W., Washington, DC 20036. (202) 223-8130.

SOCIETY OF PHOTOGRAPHIC SCIENTISTS AND ENGINEERS. 7003 Kilworth Lane, Springfield, VA 22151. (703) 642-9090.

SPIE - THE INTERNATIONAL SOCIETY FOR OPTICAL ENGINEERING. P.O. Box 10, 1022 19th Street, Bellingham, WA 98227. (206) 676-3290.

DIRECTORIES AND BIOGRAPHICAL SOURCES

INTERNATIONAL RESEARCH CENTERS DIRECTORY 1988-89. Darren L. Smith, editor. Gale Research Company, Book Tower, Detroit, MI 48226. (800) 521-0707. 4th edition. 1987. $360.00.

1987 DIRECTORY OF ENGINEERING SOCIETIES AND RELATED ORGANIZATIONS. Gordon Davis, editor. Hemisphere Publishing Corporation, 1010 Vermont Avenue, NW, Washington, DC 20005. (800) 526-0275. 12th edition. 1987. $100.00.

RESEARCH CENTERS DIRECTORY 1988. Gale Research Company, Book Tower, Detroit, MI 48226. (800) 521-0707. 12th edition. 1987. $365.00 for set.

SCIENTIFIC AND TECHNICAL ORGANIZATIONS AND AGENCIES DIRECTORY. Margaret Labash Young, editor. Gale Research Company, Book Tower, Detroit, MI 48226. (800) 521-0707. 2nd edition. 1987. $185.00.

WHO'S WHO IN ENGINEERING. Gordon Davis, editor. Hemisphere Publishing Corporation, 1010 Vermont Avenue, NW, Washington, DC 20005. (800) 526-0275. 6th edition. 1985. $200.00.

ENCYCLOPEDIAS AND DICTIONARIES

THESAURUS OF PHOTOGRAPHIC SCIENCE AND ENGINEERING. Society of Photographic Scientists and Engineers. Books on Demand, 300 North Zeeb Road, Ann Arbor, MI 48106. (313) 761-4700. $34.50 in paper.

THESAURUS OF SCIENTIFIC, TECHNICAL, AND ENGINEERING TERMS. Hemisphere Publishing Corporation, 1010 Vermont Avenue, NW, Washington, DC 20005. (800) 526-0275. 1988. $125.00.

GENERAL WORKS

AERIAL PHOTOGRAPHY AND IMAGE INTERPRETATION FOR RESOURCE MANAGEMENT. David P. Paine. John Wiley and Sons, Inc., 605 Third Avenue, New York, NY 10158. (800) 526-5368. 1981. $45.95.

CLOSE-RANGE PHOTOGRAMMETRY AND SURVEYING: STATE OF THE ART. American Society for Photogrammetry and Remote Sensing. 210 Little Falls Street, Falls Church, VA 22046-4398. (703) 534-6617. 1985. $65.00 in paper.

ELEMENTS OF PHOTOGRAMMETRY. P.R. Wolf. McGraw-Hill Book Company, 1221 Avenue of the Americas, New York, NY 10020. (212) 512-2000. 2nd edition. 1983. $49.95.

INTERPRETATION OF AERIAL PHOTOGRAPHS. Thomas E. Avery and others. Burgess Publishing Company, Ohms Lane, Minneapolis, MN 55435. (612) 831-1344. Fourth edition. 1985. Inquire.

MAPPING FROM AERIAL PHOTOGRAPHS. C.D. Burside. Halsted Press, a division of John Wiley and Sons, Inc., 605 Third Avenue, New York, NY 10158. (800) 526-5368. 1985. $49.95.

PHOTOGRAMMETRY. Francis H. Moffitt and Edward M. Mikhail. Harper and Row Publishers, Inc., 10 East 53rd Street, New York, NY 10022. (212) 207-7655. 3rd edition. 1980. $41.95.

PRINCIPLES OF REMOTE SENSING. Paul Curran. Halstead Press, division of John Wiley and Sons, Inc., 605 Third Avenue, New York, NY 10158. (800) 526-5368. 1986. $35.95.

REMOTE SENSING. Floyd F. Sabins. W.H. Freeman and Company, 41 Madison Avenue, New York, NY 10010. (212) 532-7660. 2nd edition. 1986. $39.95.

REMOTE SENSING METHODS AND APPLICATIONS. R. Hord. John Wiley and Sons, Inc., 605 Third Avenue, New York, NY 10158. (800) 526-5368. 1986. $39.95.

HANDBOOKS AND MANUALS

MANUAL OF AERIAL PHOTOGRAPHY. R. Graham and W. Reed. Focal Press, 80 Montvale Avenue, Stoneham, MA 02180. (617) 438-8464. 1986. $79.95.

MANUAL OF PHOTOGRAMMETRY. Chester C. Slama, editor. American Society for Photogrammetry and Remote Sensing. 210 Little Falls Street, Falls Church, VA 22046-4398. (703) 534-6617. 4th edition. 1980. $59.00.

MANUAL OF REMOTE SENSING. Robert N. Colwell, editor. American Society for Photogrammetry and Remote Sensing. 210 Little Falls Street, Falls Church, VA 22046-4398. (703) 534-6617. 2nd edition. 1983. $106.00 for set.

ONLINE DATA BASES

CA SEARCH. Chemical Abstracts Service, P.O. Box 3012, Columbus, OH 43120. (800) 848-6538 or (614) 421-3600. Comprehensive guide to chemical literature, 1972 to present. Inquire as to online cost and availability.

COMPENDEX. Engineering Information, Inc., 345 East 47th Street, New York, NY 10017. (800) 221-1044 or (212) 705-7615. Engineering and technical literature, 1975 to present. Inquire as to online cost and availability.

GEOREF. Online version of the BIBLIOGRAPHY AND INDEX OF GEOLOGY. American Geological Institute, 4220 King Street, Alexandria, VA 22302. (703) 379-2480. 1969 to present. Inquire as to online cost and availability.

NTIS. National Technical Information Service, 5285 Port Royal Road, Springfield, VA 22161. (703) 487-4630. Broad coverage of government sponsored research reports, 1964 to present. Inquire as to online cost and availability.

SCISEARCH. Institute for Scientific Information, 3501 Market Street, Philadelphia, PA 19104. (800) 523-1850 or (215) 386-0100. Broad multidisciplinary title and author index to the international literature of science and technology, 1974 to present. Inquire as to online cost and availability.

WILSONLINE. H.W. Wilson and Company, 950 University Avenue, Bronx, NY 10452. (800) 367-6770 or (212) 588-8400. Makes available online versions of the H.W. Wilson indexes including Applied Science and Technology Index, Business Periodicals Index and Readers' Guide to Periodical Literature. Approximately 1980 to present. Inquire as to online cost and availability.

PERIODICALS

IEEE TRANSACTIONS ON GEOSCIENCE AND REMOTE SENSING. IEEE Geoscience and Remote Sensing Society. Institute of Electrical and Electronics Engineers, 345 East 47th Street, New York, NY 10017. (212) 705-7900. Order from: IEEE Service Center, 445 Hoes Lane, Piscataway, NJ 08854. 1963 to present. Bimonthly. $110.00 per year.

JOURNAL OF IMAGING SCIENCE. Society of Photographic Scientists and Engineers. 7003 Kilworth Lane, Springfield, VA 22151. (703) 642-9090. 1956 to present. Bimonthly. $70.00 per year.

JOURNAL OF IMAGING TECHNOLOGY. Society of Photographic Scientists and Engineers. 7003 Kilworth Lane, Springfield, VA 22151. (703) 642-9090. 1975 to present. Bimonthly. $70.00 per year.

PHOTOGRAMMETRIA. Elsevier Science Publishing Company, Inc., 52 Vanderbilt Avenue, New York, NY 10017. (212) 370-5520. 1949 to present. Quarterly. $65.00 per year.

PHOTOGRAMMETRIC ENGINEERING AND REMOTE SENSING. American Society for Photogrammetry and Remote Sensing. 210 Little Falls Street, Falls Church, VA 22046-4398. (703) 534-6617. Order from: Allen Press, Inc., 1041 New Hampshire Street, Box 368, Lawrence, KS 66044. 1934 to present. Monthly. $80.00 per year.

PHOTOGRAMMETRIC RECORD. Photogrammetry Society, Department of Photogrammetry and Surveying, University College London, Gower Street, London WC1E 6BT, England. Semiannual. $37.50 per year.

SCIENTIFIC AND APPLIED PHOTOGRAPHY AND CINEMATOGRAPHY. Gordon and Breach Science Publishers, Inc., 50 West 23rd Street, New York, NY 10010. (212) 206-8900. 12 times per year. $496.00 per year.

REMOTE SENSING OF ENVIRONMENT. Elsevier Science Publishing Company, Inc., 52 Vanderbilt Avenue, New York, NY 10017. (212) 370-5520. 1968 to present. Six times per year. $210.00 per year.

REMOTE SENSING REVIEWS. Harwood Academic Publishers, 50 West 23rd Street, New York, NY 10010. (212) 206-8900. Quarterly. $160.00 per year.

RESEARCH CENTERS AND INSTITUTES

CENTER FOR REMOTE SENSING AND CARTOGRAPHY. 420 Chipta Way, Salt Lake City, UT 84112. (801) 581-8218.

GEOPHOTOGRAPHY AND REMOTE SENSING CENTER. University of Idaho, Geology Department, Moscow, ID 83843. (208) 885-7977.

NATIONAL RESEARCH COUNCIL OF CANADA, DIVISION OF PHYSICS. Ottawa, ON, Canada K1A OR6. (613) 993-1053.

PHOTOGRAMMETRY AND REMOTE SENSING SECTION. Tennessee Valley Authority, Office of Natural Resources and Economic Development, Haney Building, Chattanooga, TN 37401. (615) 755-2148.

AERODYNAMICS

See also: AERONAUTICAL ENGINEERING
AERONAUTICS, AEROSPACE ENGINEERING

ABSTRACT SERVICES AND INDEXES

ALLOYS INDEX. American Society for Metals, 9639 Kinsman Road, Metals Park, OH 44073. (216) 338-5151. 1974 to present. Monthly. $225.00 per year.

APPLIED MECHANICS REVIEWS. American Society of Mechanical Engineers, 345 East 47th Street, New York, NY 10017. (212) 705-7703. 1948 to present. Monthly. $360.00 per year.

APPLIED SCIENCE AND TECHNOLOGY INDEX. H.W. Wilson Company, 950 University Avenue, Bronx, NY 10452. (800) 367-6670 or (212) 588-8400. 1958 to present. Monthly. Inquire as to cost and availability.

CHEMICAL ABSTRACTS. Chemical Abstracts Service, Box 3012, Columbus, OH 43210. (614) 421-3600. 1907 to present. Weekly. $9200.00 per year.

CURRENT CONTENTS: ENGINEERING, TECHNOLOGY AND APPLIED SCIENCES. Institute for Scientific Information, 3501 Market Street, Philadelphia, PA 19104. (800) 523-1850 or (215) 386-0100. 1970 to present. Weekly. $272.00 per year.

ENGINEERING INDEX MONTHLY AND AUTHOR INDEX. Engineering Information, Incorporated, 345 East 47th Street, New York, NY 10017. (800) 221-1044 or (212) 705-7600. Monthly, with annual cumulation. $1560.00 per year.

INTERNATIONAL AEROSPACE ABSTRACTS. AIAA/TIS, 1633 Broadway, New York, NY 10019. (212) 581-4300. Semimonthly. $700.00 per year.

METALS ABSTRACTS. American Society for Metals, 9639 Kinsman Road, Metals Park, OH 44073. (216) 338-5151. Monthly. $890.00.

METALS ABSTRACTS INDEX. American Society for Metals, 9639 Kinsman Road, Metals Park, OH 44073. (216) 338-5151. Monthly. (Sold only to subscribers of Metals Abstracts).

SCIENCE ABSTRACTS. Section A: Physics; Section B: Electrical and Electronics Abstracts; Section C: Computer and Control Abstracts. Institute of Electrical Engineers, London, United Kingdom. Available from: Institute of Electrical and Electronic Engineers (IEEE), 445 Hoes Lane, Piscataway, NJ 08854. Inquire as to cost and availability.

SCIENCE CITATION INDEX. Institute for Scientific Information, 3501 Market Street, Philadelphia, PA 19104. (800) 523-1850 or (215) 386-0100.

STAR. (Scientific and Technical Aerospace Reports). U.S. National Aeronautics and Space Administration, Scientific and

AERODYNAMICS

Technical Information Facility, Box 8757, Baltimore-Washington International Airport, MD 21240. (202) 755-2210. Semimonthly, with semiannual and annual indexes. $85.00 per year.

WORLD ALUMINUM ABSTRACTS. Aluminum Association, Incorporated, 818 Connecticut Avenue, NW, Washington, DC 20006. (202) 862-5156. 1968 to present. Monthly. $240.00.

ASSOCIATIONS AND PROFESSIONAL SOCIETIES

AMERICAN HELICOPTER SOCIETY, Incorporated 217 North Washington Street, Alexandria, VA 22314. (703) 684-6777.

AMERICAN INSTITUTE OF AERONAUTICS AND ASTRONAUTICS. 1633 Broadway, New York, NY 10019. (212) 581-4300.

FLIGHT SAFETY FOUNDATION, Incorporated 5510 Columbia Pike, Arlington, VA 22204-3194. (703) 820-2777.

SOCIETY OF AUTOMOTIVE ENGINEERS. SAE, Incorporated, 400 Commonwealth Dr., Warrendale, PA 15096. (412) 776-4841.

BIBLIOGRAPHIES

AERONAUTICAL ENGINEERING: A SPECIAL BIBLIOGRAPHY WITH INDEXES. U.S. National Aeronautics and Space Administration, Washington, DC 20546. (202) 755-2320. Available from National Technical Information Service, Springfield, Va 22161. 1970 to present. Monthly. Inquire as to cost and availability.

NATIONAL AIR AND SPACE MUSEUM AND SMITHSONIAN INSTITUTION. Aerospace Periodical Index, 1973-1982. 1983. G.K. Hall and Company, 70 Lincoln Street, Boston, MA 02111. (800) 343-2806. $100.00.

SCIENTIFIC AND TECHNICAL BOOKS AND SERIALS IN PRINT, 1988: AN INDEX TO LITERATURE IN SCIENCE AND TECHNOLOGY. R.R. Bowker Co., 205 E. 42nd Street, New York, NY 10017. (800) 521-8110 or (212) 916-1600. $175.00.

DIRECTORIES AND BIOGRAPHICAL SOURCES

AAS DIRECTORY. American Astronautical Society, 6212-B Old Keene Mill Court, Springfield, VA 22152. (703) 866-0020. Annual. $35.00 per year.

RESEARCH CENTERS DIRECTORY. Gale Research Company, Detroit, MI 48226. (800) 521-0707. Eleventh edition. 1987.

WHO'S WHO IN ENGINEERING. Engineers Joint Council, 345 East 47th Street, New York, NY 10017. (212) 705-7010. 1985. $200.00.

WORLD AVIATION DIRECTORY. Murdoch Magazines, 1156 15th Street, NW, Washington, DC 20005. (202) 822-4600. Semiannual. $75.00 per year.

ENCYCLOPEDIAS AND DICTIONARIES

DICTIONARY OF AEROSPACE ENGINEERING. M.G. Kotik. Elsevier Science Publishing Company, Incorporated, 52 Vanderbilt Avenue, New York, NY 10017. (212) 370-5520. 1986. $170.00.

ENCYCLOPEDIA OF PHYSICAL SCIENCE AND TECHNOLOGY. Academic Press, Incorporated, Orlando, FL 32887. (800) 321-5068 or (305) 345-2734. Fifteen volumes, 1986.

JANE'S AEROSPACE DICTIONARY. Jane's Publishing Incorporated, 135 West 50th Street, New York, NY 10020. (212) 586-7745. Second edition. 1986. $39.95.

MCGRAW-HILL DICTIONARY OF ENGINEERING. Sybil P. Parker, editor. McGraw-Hill Book Company, 1221 Avenue of the Americans, New York, NY 10020. (212) 512-2000. 1984. $39.95.

MCGRAW-HILL ENCYCLOPEDIA OF SCIENCE AND TECHNOLOGY. McGraw-Hill Book, 1221 Avenue of the Americas, New York, NY 10020. (212) 512-2000.

GENERAL WORKS

AIRCRAFT BASIC SCIENCE. R.D. Bent and J.L. McKinley. McGraw-Hill Book Company, 1221 Avenue of the Americas, New York, NY 10020. (212) 512-2000. 1980. $32.95.

ENGINEERING FLUID MECHANICS. John J. Bertin. Prentice-Hall, Incorporated, Englewood Cliffs, NJ 07632. (201) 592-2000. Second edition. 1987. $34.95.

FOUNDATIONS OF AERODYNAMICS: BASES OF AERODYNAMIC DESIGN. A.M. Kuethe and C. Chow. John Wiley and Sons, Incorporated, 605 Third Avenue, New York, NY 10158. (800) 526-5368 or (212) 850-6000. Fourth edition. 1986. $43.95.

ILLUSTRATED GUIDE TO AERODYNAMICS. Hubert Smith. Tab Books, Incorporated, Monterey Lane, Blue Ridge Summit, PA 17214. (717) 794-2191. 1986. $14.95.

AN INFORMAL INTRODUCTION TO THEORETICAL FLUID MECHANICS. James Lighthill. Oxford University Press, 200 Madison Avenue, New York, NY 10016. (212) 679-7300. 1986. $35.00.

INTRODUCTION AEROSPACE STRUCTURAL ANALYSIS. David H. Allen and Walter E. Haisler. John Wiley and Sons, Incorporated, 605 Third Avenue, New York, NY 10158. (800) 526-5368 or (212) 850-6000. 1985. $42.95.

INTRODUCTION TO AIRCRAFT PERFORMANCE, SELECTION AND DESIGN. F.J. Hale. John Wiley and Sons, Incorporated, 605 Third Avenue, New York, NY 10158. (800) 526-5368 or (212) 850-6000. 1984. $39.95.

INTRODUCTION TO THEORETICAL AND COMPUTATIONAL AERODYNAMICS. Jack Moran. John Wiley and Sons, Incorporated, 605 Third Avenue, New York, NY 10158. (800) 526-5368 or (212) 850-6000. 1984. $40.95.

THEORY OF WING SECTIONS: INCLUDING A SUMMARY OF AIRFOIL DATA. Dover Publications, Incorporated, 31 East Second Street, Mineola, NY 11501. (516) 294-7000. 1949. $10.95.

HANDBOOKS AND MANUALS

STANDARD HANDBOOK OF ENGINEERING CALCULATIONS. Tyler G. Hicks, editor. McGraw-Hill Book Company, 1221 Avenue of the Americas, New York, NY 10020. (212) 512-2000. Second edition. 1984. $59.50.

THE WILEY ENGINEER'S DESK REFERENCE. Sanford I. Heisler. John Wiley and Sons, Incorporated, 605 Third Avenue, New York, NY 10158. (800) 526-5368 or (212) 850-6418. $36.00.

ONLINE DATA BASES

COMPENDEX. Engineering Information, Incorporated, 345 East 47th Street, New York, NY 10017. (800) 221-1044 or (212) 705-7615. Engineering and technical literature, 1975 to present. Inquire as to cost and availability.

NASA. National Aeronautics and Space Administration, Scientific and Technical Information Branch, 300 7th Street, SW, Washington, DC 20546. Citations and abstracts of aerospace literature, 1962 to present. Inquire as to cost and availability.

NTIS. National Technical Information Service, 5285 Port Royal Road, Springfield, VA 22161. (703) 487-4630. Broad coverage of government sponsored research reports, 1964 to present. Inquire as to cost and availability.

PERIODICALS

AERONAUTICAL JOURNAL. Aeronautical Society, 4 Hamilton Place, London W1V 0BQ, England. Ten times per year. $175.00 per year.

AEROSPACE AMERICA. American Institute of Aeronautics and Astronautics, 1633 Broadway, New York, New York 10019. (212) 581-4300. Monthly. $51.00 per year.

AEROSPACE ENGINEERING MAGAZINE (SOCIETY OF AUTOMOTIVE ENGINEERS). SAE, Incorporated, 400 Commonwealth Drive, Warrendale, PA 15096. (412) 776-4841. Monthly. $30.00 per year.

AIAA JOURNAL. AIAA/TIS, 1633 Broadway, New York, NY 10019. (212) 581-4300. Monthly. $205.00 per year.

AIRCRAFT ENGINEERING. Bunhill Publications Limited, 127 Stanstead Road, Forest Hill, London, SE23 JE1, England. 1929 to present. Monthly. $51.00 per year.

AMERICAN HELICOPTER SOCIETY JOURNAL. American Helicopter Society, Incorporated, 217 North Washington Street, Alexandria, VA 22314. (703) 684-6777. Quarterly. $25.00 per year.

AVIATION WEEK AND SPACE TECHNOLOGY. McGraw-Hill Book Company, Incorporated, 1221 Avenue of the Americas, New York, New York 10020. (212) 512-2000. Weekly. $55.00 per year.

CANADIAN AERONAUTICS AND SPACE JOURNAL. Canadian Aeronautics and Space Institute, 222 Somerset Street, West, Suite 601, Ottawa, ON K2P 0J1, Canada. (613) 234-0191. Quarterly. $35.00 per year.

FLYING. CBS Magazines, 1515 Broadway, New York, NY 10036. (212) 503-4200. Monthly. $14.00 per year.

IEEE TRANSACTIONS ON AEROSPACE. Institute of Electrical and Electronics Engineers, 345 East 47th Street, New York, NY 10017. Bimonthly. $108.00 per year.

INTERNATIONAL JOURNAL OF TURBO AND JET ENGINES. Martinus Nijhoff Publishers. Available from: Kluwer Academic Publishers Group, Distribution Centre, Postbus 322, 3300 AH Dordrecht, The Netherlands. Quarterly. $125.00 per year.

JOURNAL OF AIRCRAFT. American Institute of Aeronautics and Astronautics, AIAA/TIS, 1633 Broadway, New York, NY 10019. (212) 581-4300. Monthly. $185.00 per year.

JOURNAL OF GUIDANCE AND CONTROL. American Institute of Aeronautics and Astronautics, AIAA/TIS, 1633 Broadway, New York, NY 10019. (212) 581-4300. Bimonthly. $95.00 per year.

JOURNAL OF PROPULSION AND POWER. American Institute of Aeronautics and Astronautics, AIAA/TIS, 1633 Broadway, New York, NY 10019. (212) 581-4300. Bimonthly. $170.00 per year.

ROTOR AND WING INTERNATIONAL. PJS Publications, Incorporated, New Plaza, Box 1790, Peoria, IL 61656. (309) 682-6626. Monthly. $24.00 per year.

VERTICA: THE INTERNATIONAL JOURNAL OF ROTORCRAFT AND POWERLIFT AIRCRAFT. Pergamon Press, Incorporated, Journals Division, Maxwell House, Fairview Park, Elmsford, NY 10523. (914) 592-7700. Quarterly. $135.00 per year.

VERTIFLITE. American Helicopter Society, Incorporated, 217 North Washington Street, Alexandria, VA 22314. (703) 684-6777. Bimonthly. $25.00 per year.

RESEARCH CENTERS AND INSTITUTES

FLIGHT RESEARCH LABORATORY. University of Kansas, Raymond Nichols Hall, Lawrence, KS 66045. (913) 864-3043.

OHIO STATE UNIVERSITY. Aeronautical and Astronautical Research Laboratory, 2300 West Case Road, Columbus, OH 43220. (614) 422-1241.

STANFORD UNIVERSITY. Aero Structures Laboratory, Department of Aeronautics and Astronomy, Stanford, CA 94305. (415) 497-3317.

UNIVERSITY OF ARIZONA. Computational Fluid Mechanics Laboratory, Building #16, Room 312, Tucson, AZ 85721. (602) 621-4423.

UNIVERSITY OF TEXAS AT AUSTIN. Aeronautical Research Center, Aerospace Engineering and Engineering Mechanics, WRW 217, Austin, TX 78712. (512) 471-5962.

UNIVERSITY OF WASHINGTON. Aeronautical Laboratory, FS-10, Seattle, WA 98105. (206) 543-0439.

AERONAUTICAL ENGINEERING

See also: AERODYNAMICS, AERONAUTICS, AEROSPACE ENGINEERING

ABSTRACT SERVICES AND INDEXES

ALLOYS INDEX. American Society for Metals, 9639 Kinsman Road, Metals Park, OH 44073. (216) 338-5151. 1974 to present. Monthly. $225.00 per year.

APPLIED MECHANICS REVIEWS. American Society of Mechanical Engineers, 345 East 47th Street, New York, NY 10017. (212) 705-7703. 1948 to present. Monthly. $360.00 per year.

APPLIED SCIENCE AND TECHNOLOGY INDEX. H.W. Wilson Company, 950 University Avenue, Bronx, NY 10452. (800) 367-6670 or (212) 588-8400. 1958 to present. Monthly. Inquire as to cost and availability.

CHEMICAL ABSTRACTS. Chemical Abstracts Service, Box 3012, Columbus, OH 43210. (614) 421-3600. 1907 to present. Weekly. $9200.00 per year.

CURRENT CONTENTS: ENGINEERING, TECHNOLOGY AND APPLIED SCIENCES. Institute for Scientific Information, 3501 Market Street, Philadelphia, PA 19104. (800) 523-1850 or (215) 386-0100. 1970 to present. Weekly. $272.00 per year.

ENGINEERING INDEX MONTHLY AND AUTHOR INDEX. Engineering Information, Incorporated, 345 East 47th Street, New York, NY 10017. (800) 221-1044 or (212) 705-7600. Monthly, with annual cumulation. $1560.00 per year.

INTERNATIONAL AEROSPACE ABSTRACTS. AIAA/TIS, 1633 Broadway, New York, NY 10019. (212) 581-4300. Semimonthly. $700.00 per year.

METALS ABSTRACTS. American Society for Metals, 9639 Kinsman Road, Metals Park, OH 44073. (216) 338-5151. Monthly. $890.00.

METALS ABSTRACTS INDEX. American Society for Metals, 9639 Kinsman Road, Metals Park, OH 44073. (216) 338-5151. Monthly. (Sold only to subscribers of Metals Abstracts).

SCIENCE ABSTRACTS. Section A: Physics; Section B: Electrical and Electronics Abstracts; Section C: Computer and Control Abstracts. Institute of Electrical Engineers, London, United Kingdom. Available from: Institute of Electrical and Electronic Engineers (IEEE), 445 Hoes Lane, Piscataway, NJ 08854. Inquire as to cost and availability.

SCIENCE CITATION INDEX. Institute for Scientific Information, 3501 Market Street, Philadelphia, PA 19104. (800) 523-1850 or (215) 386-0100.

STAR. (Scientific and Technical Aerospace Reports). U.S. National Aeronautics and Space Administration, Scientific and Technical Information Facility, Box 8757, Baltimore-Washington International Airport, MD 21240. (202) 755-2210. Semimonthly, with semiannual and annual indexes. $85.00 per year.

WORLD ALUMINUM ABSTRACTS. Aluminum Association, Incorporated, 818 Connecticut Avenue, NW, Washington, DC 20006. (202) 862-5156. 1968 to present. Monthly $240.00.

ASSOCIATIONS AND PROFESSIONAL SOCIETIES

AIRCRAFT ELECTRONICS ASSOCIATION. Box 1981, Independence, MO 64055. (816) 373-6565.

AMERICAN HELICOPTER SOCIETY, Incorporated 217 North Washington Street, Alexandria, VA 22314. (703) 684-6777.

AMERICAN INSTITUTE OF AERONAUTICS AND ASTRONAUTICS. 1633 Broadway, New York, NY 10019. (212) 581-4300.

FLIGHT SAFETY FOUNDATION, Incorporated 5510 Columbia Pike, Arlington, VA 22204-3194. (703) 820-2777.

SOCIETY OF AUTOMOTIVE ENGINEERS. SAE, Incorporated, 400 Commonwealth Drive, Warrendale, PA 15096. (412) 776-4841.

BIBLIOGRAPHIES

AERONAUTICAL ENGINEERING: A SPECIAL BIBLIOGRAPHY WITH INDEXES. U.S. National Aeronautics and Space Administration, Washington, DC 20546. (202) 755-2320. Available from National Technical Information Service, Springfield, VA 22161. 1970 to present. Monthly. Inquire as to cost and availability.

AIRCRAFT, ENGINES, AND AIRMEN: A SELECTIVE REVIEW OF THE PERIODICAL LITERATURE, 1930-1969. A. Hanniball. Scarecrow Press, Incorporated, 52 Liberty Street, Methuchen, NJ 08840. (201) 548-8600. 1972. $39.50.

NATIONAL AIR AND SPACE MUSEUM AND SMITHSONIAN INSTITUTION. Aerospace Periodical Index, 1973-1982. 1983. G.K. Hall and Company, 70 Lincoln Street, Boston, MA 02111. (800) 343-2806. $100.00.

SCIENTIFIC AND TECHNICAL BOOKS AND SERIALS IN PRINT, 1988; AN INDEX TO LITERATURE IN SCIENCE AND TECHNOLOGY. R.R. Bowker Company, 205 East 42nd Street, New York, NY 10017. (800) 521-8110 or (212) 916-1600. $175.00.

DIRECTORIES AND BIOGRAPHICAL SOURCES

AAS DIRECTORY. American Astronautical Society, 6212-B Old Keene Mill Court, Springfield, VA 22152. (703) 866-0020. Annual. $35.00 per year.

RESEARCH CENTERS DIRECTORY. Gale Research Company, Detroit, MI 48226. (800) 521-0707. Eleventh edition. 1987.

WHO'S WHO IN ENGINEERING. Engineers Joint Council, 345 East 47th Street, New York, NY 10017. (212) 705-7010. $200.00.

WORLD AVIATION DIRECTORY. Murdoch Magazines, 1156 15th STreet, NW, Washington, DC 20005. (202) 822-4600. Semiannual. $75.00 per year.

ENCYCLOPEDIAS AND DICTIONARIES

DICTIONARY OF AEROSPACE ENGINEERING. M.G. Kotik. Elsevier Science Publishing Company, Incorporated, 52 Vanderbilt Avenue, New York, NY 10017. (212) 370-5520. 1986. $170.00.

ENCYCLOPEDIA OF PHYSICAL SCIENCE AND TECHNOLOGY. Academic Press, Incorporated, Orlando, Florida 32887. (800) 321-5068 or (305) 345-2734. Fifteen volumes, 1986.

JANE'S AEROSPACE DICTIONARY. Jane's Publishing Incorporated, 135 West 50th Street, New York, NY 10020. (212) 586-7745. Second edition. 1986. $39.95.

MCGRAW-HILL DICTIONARY OF ENGINEERING. Sybil P. Parker, editor. McGraw-HIll Book Company, 1221 Avenue of the Americas, New York, NY 10020. (212) 512-2000. 1984. $39.95.

MCGRAW-HILL ENCYCLOPEDIA OF SCIENCE AND TECHNOLOGY. McGraw-Hill Book, 1221 Avenue of the Americas, New York, NY 10020. (212) 512-2000.

GENERAL WORKS

AIRCRAFT BASIC SCIENCE. R.D. Bent and J.L. Mckinley. McGraw-Hill Book Company, 1221 Avenue of the Americas, New York, NY 10020. (212) 512-2000. 1980. $32.95.

DIGITAL AVIONICS SYSTEMS. Cary Spitzer. Prentice-Hall Press, Incorporated, Englewood Cliffs, NJ 07632. (800) 562-0245. 1987. $39.95.

FOUNDATIONS OF AERODYNAMICS: BASES OF AERODYNAMIC DESIGN. A.M. Kuethe and C. Chow. John Wiley and Sons, Incorporated, 605 Third Avenue, New York, NY 10158. (800) 526-5368 or (212) 850-6000. Fourth edition. 1986. $43.95.

ILLUSTRATED GUIDE TO AERODYNAMICS. Hubert Smith. Tab Books, Incorporated, Monterey Lane, Blue Ridge Summit, PA 17214. (717) 794-2191. 1986. $14.95.

INTRODUCTION AEROSPACE STRUCTURAL ANALYSIS. David H. Allen and Walter E. Haisler. John Wiley and Sons, Incorporated, 605 Third Avenue, New York, NY 10158. (800) 526-5368 or (212) 850-6000. 1985. $42.95.

INTRODUCTION TO AIRCRAFT PERFORMANCE. SELECTION AND DESIGN. F.J. Hale. John Wiley and Sons, Incorporated, 605 Third Avenue, New York, NY 10158. (800) 526-5368 or (212) 850-6000. 1984. $39.95.

INTRODUCTION TO THEORETICAL AND COMPUTATIONAL AERODYNAMICS. Jack Moran. John Wiley and Sons, Incorporated, 605 Third Avenue, New York, NY 10158. (800) 526-5368 or (212) 850-6000. 1984. $40.95.

THEORY OF WING SECTIONS; INCLUDING A SUMMARY OF AIRFOIL DATA. Dover Publications, Incorporated, 31 East Second Street, Mineola, NY 11501. (516) 294-7000. 1949. $10.95.

HANDBOOKS AND MANUALS

JANE'S ALL THE WORLD'S AIRCRAFT 1987-88. John W.R. Taylor. Jane's Publishing, Incorporated, 115 Fifth Avenue, New York, NY 10003. (214) 254-9097. 1987.

JANE'S AVIONICS 1987-88. Stephen R. Broadbent. Jane's Publishing, Incorporated, 115 Fifth Avenue, New York, NY 10003. (214) 254-9097. 1987.

STANDARD HANDBOOK OF ENGINEERING CALCULATIONS. Tyler G. Hicks, editor. McGraw-Hill Book Company, 1221 Avenue of the Americas, New York, NY 10020. (212) 512-2000. Second edition. 1984. $59.50.

THE WILEY ENGINEER'S DESK REFERENCE. Sanford I. Heisler. John Wiley and Sons, Incorporated, 605 Third Avenue, New York, NY 10158. (800) 526-5368 or (212) 850-6418. 1984. $36.00.

ONLINE DATA BASES

COMPENDEX. Engineering Information, Incorporated, 345 East 47th Street, New York, NY 10017. (800) 221-1044 or (212) 705-7615. Engineering and technical literature, 1975 to present. Inquire as to cost and availability.

NASA. National Aeronautics and Space Administration, Scientific and Technical Information Branch, 300 7th Street, SW, Washington, DC 20546. Citations and abstracts of aerospace literature, 1962 to present. Inquire as to cost and availability.

NTIS. National Technical Information Service, 5285 Port Royal Road, Springfield, VA 22161. (703) 487-4630. Broad coverage of government sponsored research reports, 1964 to present. Inquire as to cost and availability.

PERIODICALS

AERONAUTICAL JOURNAL. Aeronautical Society, 4 Hamilton Place, London W1V 0BQ, England. Ten times per year. $175.00 per year.

AEROSPACE AMERICA. American Institute of Aeronautics and Astronautics, 1633 Broadway, New York, NY 10019. (212) 581-4300. Monthly. $51.00 per year.

AEROSPACE ENGINEERING MAGAZINE (SOCIETY OF AUTOMOTIVE ENGINEERS). SAE, Incorporated, 400 Commonwealth Drive, Warrendale, PA 15096. (412) 776-4841. Monthly. $30.00 per year.

AIAA JOURNAL. AIAA/TIS, 1633 Broadway, New York, NY 10019. (212) 581-4300. Monthly. $205.00 per year.

AIRCRAFT ENGINEERING. Bunhill Publications Limited, 127 Stanstead Road, Forest Hill, London, SE23 JE1, England. 1929 to present. Monthly. $51.00 per year.

AMERICAN HELICOPTER SOCIETY JOURNAL. American Helicopter Society, Incorporated, 217 North Washington Street, Alexandria, Va 22314. (703) 684-6777. Quarterly. $25.00 per year.

AVIATION MECHANICS BULLETIN. Flight Safety Foundation, Incorporated, 5510 Columbia Pike, Arlington, VA 22204-3194. (703) 820-2777. Bimonthly. $6.50 per year.

AVIATION WEEK AND SPACE TECHNOLOGY. McGraw-Hill Book Company, Incorporated, 1221 Avenue of the Americas, New York, NY 10020. (212) 512-2000. Weekly. $55.00 per year.

AVIONICS. Atlantic Communications, Incorporated, Box 5100, Westport, CT 06881. (203) 227-2280. Monthly. $36.00 per year.

AVIONICS NEWS MAGAZINE. Aircraft Electronics Association, Box 1981, Independence, MO 64055. (816) 373-6565. Monthly. Free.

CANADIAN AERONAUTICS AND SPACE JOURNAL. Canadian Aeronautics and Space Institute, 222 Somerset Street, West, Suite 601, Ottawa, ON K2P 0J1, Canada. (613) 234-0191. Quarterly. $35.00 per year.

FLYING. CBS Magazines, 1515 Broadway, New York, NY 10036. (212) 503-4200. Monthly. $14.00 per year.

IEEE TRANSACTIONS ON AEROSPACE. Institute of Electrical and Electronics Engineers, 345 East 47th Street, New York, NY 10017. Bimonthly. $108.00 per year.

INTERNATIONAL JOURNAL OF TURBO AND JET ENGINES. Martinus Nijhoff Publishers. Available from: Kluwer Academic Publishers Group, Distribution Centre, Postbus 322, 3300 AH Dordrecht, The Netherlands. Quarterly. $125.00 per year.

JOURNAL OF AIRCRAFT. American Institute of Aeronautics and Astronautics, AIAA/TIS, 1633 Broadway, New York, NY 10019. (212) 581-4300. Monthly. $185.00 per year.

JOURNAL OF GUIDANCE AND CONTROL. American Institute of Aeronautics and Astronautics, AIAA/TIS, 1633 Broadway, New York, NY 10019. (212) 581-4300. Bimonthly. $95.00 per year.

JOURNAL OF PROPULSION AND POWER. American Institute of Aeronautics and Astronautics, AIAA/TIS, 1633 Broadway, New York, NY 10019. (212) 581-4300. Bimonthly. $170.00 per year.

ROTOR AND WING INTERNATIONAL. PJS Publications, Incorporated, New Plaza, Box 1790, Peoria, IL 61656. (309) 682-6626. Monthly. $24.00 per year.

VERTICA: THE INTERNATIONAL JOURNAL OF ROTORCRAFT AND POWERLIFT AIRCRAFT. Pergamon Press, Incorporated, Journals Division, Maxwell House, Fairview Park, elmsford, NY 10523. (914) 592-7700. Quarterly. $135.00 per year.

VERTIFLITE. American Helicopter Society, Incorporated, 217 North Washington Street, Alexandria, VA 22314. (703) 684-6777. Bimonthly. $25.00 per year.

RESEARCH CENTERS AND INSTITUTES

FLIGHT RESEARCH LABORATORY. University of Kansas, Raymond Nichols Hall, Lawrence, KS 66045. (913) 864-3043.

OHIO STATE UNIVERSITY. Aeronautical and Astronautical Research Laboratory, 2300 West Case Road, Columbus, OH 43220. (614) 422-1241.

STANFORD UNIVERSITY. Aero Structures Laboratory, Department of Aeronautics and Astronomy, Stanford, Ca 94305. (415) 497-3317.

UNIVERSITY OF TEXAS AT AUSTIN. Aeronautical Research Center, Aerospace Engineering and Engineering Mechanics, WRW 217, Austin, TX 78712. (512) 471-5962.

UNIVERSITY OF WASHINGTON. Aeronautical Laboratory, FS-10, Seattle, Wa 98105. (206) 543-0439.

AERONAUTICS

See also: AERONAUTICAL ENGINEERING, AEROSPACE ENGINEERING

ABSTRACT SERVICES AND INDEXES

ALLOYS INDEX. American Society for Metals, 9639 Kinsman Road, Metals Park, OH 44073. (216) 338-5151. 1974 to present. Monthly. $225.00 per year.

APPLIED MECHANICS REVIEWS. American Society of Mechanical Engineers, 345 East 47th Street, New York, NY 10017. (212) 705-7703. 1948 to present. Monthly. $360.00 per year.

APPLIED SCIENCE AND TECHNOLOGY INDEX. H.W. Wilson Company, 950 University Avenue, Bronx, NY 10452. (800) 367-6670 or (212) 588-8400. 1958 to present. Monthly. Inquire as to cost and availability.

CHEMICAL ABSTRACTS. Chemical Abstracts Service, Box 3012, Columbus, OH 43210. (614) 421-3600. 1907 to present. Weekly. $9200.00 per year.

CURRENT CONTENTS: ENGINEERING, TECHNOLOGY AND APPLIED SCIENCES. Institute for Scientific Information, 3501 Market Street, Philadelphia, PA 19104. (800) 523-1850 or (215) 386-0100. 1970 to present. Weekly. $272.00 per year.

ENGINEERING INDEX MONTHLY AND AUTHOR INDEX. Engineering Information, Incorporated, 345 East 47th Street, New York, NY 10017. (800) 221-1044 or (212) 705-7600. Monthly, with annual cumulation. $1560.00 per year.

INTERNATIONAL AEROSPACE ABSTRACTS. AIAA/TIS, 1633 Broadway, New York, NY 10019. (212) 581-4300. Semimonthly. $700.00 per year.

METALS ABSTRACTS. American Society for Metals, 9639 Kinsman Road, Metals Park, OH 44073. (216) 338-5151. Monthly. $890.00.

METALS ABSTRACTS INDEX. American Society for Metals, 9639 Kinsman Road, Metals Park, OH 44073. (216) 338-5151. Monthly. (Sold only to subscribers of Metals Abstracts).

SCIENCE ABSTRACTS. Section A: Physics; Section B: Electrical and Electronics Abstracts; Section C: Computer and Control Abstracts. Institute of Electrical Engineers, London, United Kingdom. Available from: Institute of Electrical and Electronic Engineers (IEEE), 445 Hoes Lane, Piscataway, NJ 08854. Inquire as to cost and availability.

SCIENCE CITATION INDEX. Institute for Scientific Information, 3501 Market Street, Philadelphia, PA 19104. (800) 523-1850 or (215) 386-0100.

STAR. (Scientific and Technical Aerospace Reports). U.S. National Aeronautics and Space Administration, Scientific and Technical Information Facility, Box 8757, Baltimore-Washington International Airport, MD 21240. (202) 755-2210. Semimonthly, with semiannual and annual indexes. $85.00 per year.

WORLD ALUMINUM ABSTRACTS. Aluminum Association, Incorporated, 818 Connecticut Avenue, NW, Washington, DC 20006. (202) 862-5156. 1968 to present. Monthly $240.00.

ASSOCIATIONS AND PROFESSIONAL SOCIETIES

AIRCRAFT ELECTRONICS ASSOCIATION. Box 1981, Independence, MO 64055. (816) 373-6565.

AMERICAN HELICOPTER SOCIETY, Incorporated 217 North Washington Street, Alexandria, VA 22314. (703) 684-6777.

AMERICAN INSTITUTE OF AERONAUTICS AND ASTRONAUTICS. 1633 Broadway, New York, NY 10019. (212) 581-4300.

FLIGHT SAFETY FOUNDATION, Incorporated 5510 Columbia Pike, Arlington, VA 22204-3194. (703) 820-2777.

SOCIETY OF AUTOMOTIVE ENGINEERS. SAE, Incorporated, 400 Commonwealth Drive, Warrendale, PA 15096. (412) 776-4841.

ANNUAL REVIEWS AND YEARBOOKS

JANE'S AVIATION REVIEW. Michael Taylor, editor. Jane's Publishing, Incorporated, 135 West 50th Street, New York, NY 10020. (212) 586-7745. Fourth edition. 1985. $14.95.

BIBLIOGRAPHIES

AERONAUTICAL ENGINEERING: A SPECIAL BIBLIOGRAPHY WITH INDEXES. U.S. National Aeronautics and Space Administration, Washington, DC 20546. (202) 755-2320. Available from National Technical Information Service, Springfield, VA 22161. 1970 to present. Monthly. Inquire as to cost and availability.

AIRCRAFT, ENGINES, AND AIRMEN: A SELECTIVE REVIEW OF THE PERIODICAL LITERATURE, 1930-1969. A. Hanniball. Scarecrow Press, Incorporated, 52 Liberty Street, Methuchen, NJ 08840. (201) 548-8600. 1972. $39.50.

NATIONAL AIR AND SPACE MUSEUM AND SMITHSONIAN INSTITUTION. Aerospace Periodical Index, 1973-1982. 1983. G.K. Hall and Company, 70 Lincoln Street, Boston, MA 02111. (800) 343-2806. $100.00.

SCIENTIFIC AND TECHNICAL BOOKS AND SERIALS IN PRINT, 1988; AN INDEX TO LITERATURE IN SCIENCE AND TECHNOLOGY. R.R. Bowker Company, 205 East 42nd Street, New York, NY 10017. (800) 521-8110 or (212) 916-1600. $175.00.

DIRECTORIES AND BIOGRAPHICAL SOURCES

AAS DIRECTORY. American Astronautical Society, 6212-B Old Keene Mill Court, Springfield, VA 22152. (703) 866-0020. Annual. $35.00 per year.

RESEARCH CENTERS DIRECTORY. Gale Research Company, Detroit, MI 48226. (800) 521-0707. Eleventh edition. 1987.

WHO'S WHO IN ENGINEERING. Engineers Joint Council, 345 East 47th Street, New York, NY 10017. (212) 705-7010. $200.00.

WORLD AVIATION DIRECTORY. Murdoch Magazines, 1156 15th STreet, NW, Washington, DC 20005. (202) 822-4600. Semiannual. $75.00 per year.

ENCYCLOPEDIAS AND DICTIONARIES

DICTIONARY OF AEROSPACE ENGINEERING. M.G. Kotik. Elsevier Science Publishing Company, Incorporated, 52 Vanderbilt Avenue, New York, NY 10017. (212) 370-5520. 1986. $170.00.

ENCYCLOPEDIA OF PHYSICAL SCIENCE AND TECHNOLOGY. Academic Press, Incorporated, Orlando, Florida 32887. (800) 321-5068 or (305) 345-2734. Fifteen volumes, 1986.

JANE'S AEROSPACE DICTIONARY. Jane's Publishing Incorporated, 135 West 50th Street, New York, NY 10020. (212) 586-7745. Second edition. 1986. $39.95.

MCGRAW-HILL DICTIONARY OF ENGINEERING. Sybil P. Parker, editor. McGraw-HIll Book Company, 1221 Avenue of the Americas, New York, NY 10020. (212) 512-2000. 1984. $39.95.

MCGRAW-HILL ENCYCLOPEDIA OF SCIENCE AND TECHNOLOGY. McGraw-Hill Book, 1221 Avenue of the Americas, New York, NY 10020. (212) 512-2000.

GENERAL WORKS

AIRCRAFT BASIC SCIENCE. R.D. Bent and J.L. Mckinley. McGraw-Hill Book Company, 1221 Avenue of the Americas, New York, NY 10020. (212) 512-2000. 1980. $32.95.

AIR CRASHES: WHAT GOES WRONG, WHY AND WHAT CAN BE DONE ABOUT IT. Richard L. Collins. Macmillan Publishing Company, Incorporated, 866 Third Avenue, New York, NY 10022. (212) 702-2000. 1986. $17.95.

AIRPORT OPERATIONS. Norman Ashford and others. John Wiley and Sons, Incorporated, 605 Third Avenue, New York, NY 10158. (800) 526-5368 or (212) 850-6000. 1984. $49.95.

FOUNDATIONS OF AERODYNAMICS: BASES OF AERODYNAMIC DESIGN. A.M. Kuethe and C. Chow. John Wiley and Sons, Incorporated, 605 Third Avenue, New York, NY 10158. (800) 526-5368 or (212) 850-6000. Fourth edition. 1986. $43.95.

ILLUSTRATED GUIDE TO AERODYNAMICS. Hubert Smith. Tab Books, Incorporated, Monterey Lane, Blue Ridge Summit, PA 17214. (717) 794-2191. 1986. $14.95.

INTRODUCTION AEROSPACE STRUCTURAL ANALYSIS. David H. Allen and Walter E. Haisler. John Wiley and Sons, Incorporated, 605 Third Avenue, New York, NY 10158. (800) 526-5368 or (212) 850-6000. 1985. $42.95.

INTRODUCTION TO AIRCRAFT PERFORMANCE. Selection and Design. F.J. Hale. John Wiley and Sons, Incorporated, 605 Third Avenue, New York, NY 10158. (800) 526-5368 or (212) 850-6000. 1984. $39.95.

INTRODUCTION TO THEORETICAL AND COMPUTATIONAL AERODYNAMICS. Jack Moran. John Wiley and Sons, Incorporated, 605 Third Avenue, New York, NY 10158. (800) 526-5368 or (212) 850-6000. 1984. $40.95.

HANDBOOKS AND MANUALS

STANDARD HANDBOOK OF ENGINEERING CALCULATIONS. Tyler G. Hicks, editor. McGraw-Hill Book Company, 1221 Avenue of the Americas, New York, NY 10020. (212) 512-2000. Second edition. 1984. $59.50.

THE WILEY ENGINEER'S DESK REFERENCE. Sanford I. Heisler. John Wiley and Sons, Incorporated, 605 Third Avenue, New York, NY 10158. (800) 526-5368 or (212) 850-6418. 1984. $36.00.

ONLINE DATA BASES

COMPENDEX. Engineering Information, Incorporated, 345 East 47th Street, New York, NY 10017. (800) 221-1044 or (212) 705-7615. Engineering and technical literature, 1975 to present. Inquire as to cost and availability.

NASA. National Aeronautics and Space Administration, Scientific and Technical Information Branch, 300 7th Street, SW, Washington, DC 20546. Citations and abstracts of aerospace literature, 1962 to present. Inquire as to cost and availability.

NTIS. National Technical Information Service, 5285 Port Royal Road, Springfield, VA 22161. (703) 487-4630. Broad coverage of government sponsored research reports, 1964 to present. Inquire as to cost and availability.

PERIODICALS

AERONAUTICAL JOURNAL. Aeronautical Society, 4 Hamilton Place, London W1V 0BQ, England. Ten times per year. $175.00 per year.

AEROSPACE AMERICA. American Institute of Aeronautics and Astronautics, 1633 Broadway, New York, NY 10019. (212) 581-4300. Monthly. $51.00 per year.

AEROSPACE ENGINEERING MAGAZINE (SOCIETY OF AUTOMOTIVE ENGINEERS). SAE, Incorporated, 400 Commonwealth Drive, Warrendale, PA 15096. (412) 776-4841. Monthly. $30.00 per year.

AIAA JOURNAL. AIAA/TIS, 1633 Broadway, New York, NY 10019. (212) 581-4300. Monthly. $205.00 per year.

AIRCRAFT ENGINEERING. Bunhill Publications Limited, 127 Stanstead Road, Forest Hill, London, SE23 JE1, England. 1929 to present. Monthly. $51.00 per year.

AMERICAN HELICOPTER SOCIETY JOURNAL. American Helicopter Society, Incorporated, 217 North Washington Street, Alexandria, Va 22314. (703) 684-6777. Quarterly. $25.00 per year.

AVIATION MECHANICS BULLETIN. Flight Safety Foundation, Incorporated, 5510 Columbia Pike, Arlington, VA 22204-3194. (703) 820-2777. Bimonthly. $6.50 per year.

AVIATION WEEK AND SPACE TECHNOLOGY. McGraw-Hill Book Company, Incorporated, 1221 Avenue of the Americas, New York, NY 10020. (212) 512-2000. Weekly. $55.00 per year.

AVIONICS. Atlantic Communications, Incorporated, Box 5100, Westport, CT 06881. (203) 227-2280. Monthly. $36.00 per year.

AVIONICS NEWS MAGAZINE. Aircraft Electronics Association, Box 1981, Independence, MO 64055. (816) 373-6565. Monthly. Free.

CANADIAN AERONAUTICS AND SPACE JOURNAL. Canadian Aeronautics and Space Institute, 222 Somerset Street, West, Suite 601, Ottawa, ON K2P 0J1, Canada. (613) 234-0191. Quarterly. $35.00 per year.

FLYING. CBS Magazines, 1515 Broadway, New York, NY 10036. (212) 503-4200. Monthly. $14.00 per year.

IEEE TRANSACTIONS ON AEROSPACE. Institute of Electrical and Electronics Engineers, 345 East 47th Street, New York, NY 10017. Bimonthly. $108.00 per year.

INTERNATIONAL JOURNAL OF TURBO AND JET ENGINES. Martinus Nijhoff Publishers. Available from: Kluwer Academic Publishers Group, Distribution Centre, Postbus 322, 3300 AH Dordrecht, The Netherlands. Quarterly. $125.00 per year.

JOURNAL OF AIRCRAFT. American Institute of Aeronautics and Astronautics, AIAA/TIS, 1633 Broadway, New York, NY 10019. (212) 581-4300. Monthly. $185.00 per year.

JOURNAL OF GUIDANCE AND CONTROL. American Institute of Aeronautics and Astronautics, AIAA/TIS, 1633 Broadway, New York, NY 10019. (212) 581-4300. Bimonthly. $95.00 per year.

JOURNAL OF PROPULSION AND POWER. American Institute of Aeronautics and Astronautics, AIAA/TIS, 1633 Broadway, New York, NY 10019. (212) 581-4300. Bimonthly. $170.00 per year.

ROTOR AND WING INTERNATIONAL. PJS Publications, Incorporated, New Plaza, Box 1790, Peoria, IL 61656. (309) 682-6626. Monthly. $24.00 per year.

VERTICA: THE INTERNATIONAL JOURNAL OF ROTORCRAFT AND POWERLIFT AIRCRAFT. Pergamon Press, Incorporated, Journals Division, Maxwell House, Fairview Park, elmsford, NY 10523. (914) 592-7700. Quarterly. $135.00 per year.

VERTIFLITE. American Helicopter Society, Incorporated, 217 North Washington Street, Alexandria, VA 22314. (703) 684-6777. Bimonthly. $25.00 per year.

RESEARCH CENTERS AND INSTITUTES

FLIGHT RESEARCH LABORATORY. University of Kansas, Raymond Nichols Hall, Lawrence, KS 66045. (913) 864-3043.

JOINT INSTITUTE FOR ADVANCEMENT OF FLIGHT SCIENCE. Langley Research Center, Mail Stop 269, Hampton, VA 23665. (804) 865-3124.

OHIO STATE UNIVERSITY. Aeronautical and Astronautical Research Laboratory, 2300 West Case Road, Columbus, OH

43220. (614) 422-1241.

STANFORD UNIVERSITY. Aero Structures Laboratory, Department of Aeronautics and Astronomy, Stanford, Ca 94305. (415) 497-3317.

UNIVERSITY OF TEXAS AT AUSTIN. Aeronautical Research Center, Aerospace Engineering and Engineering Mechanics, WRW 217, Austin, TX 78712. (512) 471-5962.

UNIVERSITY OF WASHINGTON. Aeronautical Laboratory, FS-10, Seattle, Wa 98105. (206) 543-0439.

AEROSOLS

See also: AIR POLLUTION, COLLOIDS

ABSTRACT SERVICES AND INDEXES

APPLIED SCIENCE AND TECHNOLOGY INDEX. H.W. Wilson Company, 950 University Avenue, Bronx, NY 10452. (800) 367-6670 or (212) 588-8400. Inquire as to cost and availability.

CHEMICAL ABSTRACTS. Chemical Abstracts Service, 2540 Olentangy Road, Post Office Box 3012, Columbus, OH 43210. (800) 848-6538 or (614) 421-3600. Weekly. $9200.00 per year.

GENERAL SCIENCE INDEX. H.W. Wilson Company, 950 University Avenue, Bronx, NY 10452. (800) 367-6770 or (212) 588-8400. Inquire as to cost and availability.

PHYSICS ABSTRACTS. Institute of Electrical Engineers, London, United Kingdom. Available from: Institute of Electrical and Electronic Engineers (IEEE), 345 East 47th Street, New York, NY 10017. (212) 705-7900.

SCIENCE CITATION INDEX. Institute for Scientific Information, 3501 Market Street, Philadelphia, PA 19104. (800) 523-1850 or (215) 386-0100.

ASSOCIATIONS AND PROFESSIONAL SOCIETIES

AMERICAN ASSOCIATION FOR AEROSOL RESEARCH. c/o Dr. S.K. Friedlander, Chemical Engineering Department, 5531 Boetler Hall, University of California at Los Angeles, Los Angeles, CA 90024. (213) 825-2206.

AMERICAN INSTITUTE OF CHEMICAL ENGINEERS. 345 East 47th Street, New York, NY 10017. (212) 705-7338.

AMERICAN CHEMICAL SOCIETY. 1155 Sixteenth Street, NW, Washington, DC 20036. (202) 872-4600.

ASSOCIATION OF CONSULTING CHEMISTS AND CHEMICAL ENGINEERS. 50 41st Street, Suite 92, New York, NY 10017. (212) 684-6255.

DIRECTORIES AND BIOGRAPHICAL SOURCES

AMERICAN INSTITUTE OF CHEMISTS. American Institute of Chemists, 7315 Wisconsin Avenue, Bethesda, MD 20814. (301) 652-2447. 1986. $35.00.

AMERICAN MEN AND WOMEN OF SCIENCE. Physical and Biological Sciences. Fifteenth edition. R.R. Bowker Company, 205 East 42nd Street, New York, NY 10017. (800) 521-8110 or (212) 916-1600.

BIOGRAPHICAL DICTIONARY OF SCIENTISTS: CHEMISTS. David Abbott, editor. P. Bedrick Books, 125 East 23rd Street, New York, NY 10010. (212) 777-1187. 1984. $18.95.

CONSULTING SERVICES: CHEMISTS AND CHEMICAL ENGINEERS. Association of Consulting Chemists and Chemical Engineers, 50 East 41st Street, New York, NY 10017. (212) 684-6255. Annual. 1986. $45.00.

GOVERNMENT RESEARCH DIRECTORY. Gale Research Company, Book Tower, Detroit, MI 48226. (800) 521-0707. Fourth edition. 1987. $350.00.

INTERNATIONAL RESEARCH CENTERS DIRECTORY 1986-1987. Gale Research Company, Book Tower, Detroit, MI 48226. (800) 521-0707. Third edition. 1986. $330.00.

RESEARCH CENTERS DIRECTORY. Gale Research Company, Book Tower, Detroit, MI 48226. (800) 521-0707. Eleventh edition. 1987. $355.00.

SCIENTIFIC AND TECHNICAL ORGANIZATIONS AND AGENCIES DIRECTORY. Gale Research Company, Book Tower, Detroit, MI 48226. (800) 521-0707. 1985. $150.00.

WHO'S WHO IN FRONTIER SCIENCE AND TECHNOLOGY. Marquis Who's Who, Incorporated, 200 East Ohio Street, Chicago, Illinois 60611. (800) 428-3898 or (312) 787-2008.

WHO'S WHO IN TECHNOLOGY TODAY. Reston Publishing Company, Incorporated, c/o Prentice-Hall, Incorporated, Englewood Cliffs, NJ 07632. (800) 262-6868. Biennial. Five volumes. $425.00. Covers the fields of electronics, computer science, physics, optics, chemistry, biotechnology, mechanics, energy, and earth science.

ENCYCLOPEDIAS AND DICTIONARIES

CONCISE ENCYCLOPEDIA OF CHEMICAL TECHNOLOGY. Kirk Othmer. John Wiley and Sons, Incorporated, 605 Third Avenue, New York, NY 10158. (800) 526-5368 or (212) 850-6000. Third edition. 1985. $129.95.

CONDENSED CHEMICAL DICTIONARY. Gessner Hawley. Van Nostrand Reinhold, 115 Fifth Avenue, New York, NY 10003. Tenth edition. 1981. $49.95.

ENCYCLOPEDIA OF PHYSICAL SCIENCE AND TECHNOLOGY. Academic Press, Incorporated, Orlando, FL 32887. (800) 321-5068 or (305) 345-2734. Fifteen volumes, 1986.

GLOSSARY OF CHEMICAL TERMS. Clifford A. Hampel and Gessner G. Hawley. Van Nostrand Reinhold Company, 115 Fifth Avenue, New York, NY 10003. (800) 543-2681 or (212) 254-3232. Second edition. 1982. $21.95.

MCGRAW-HILL ENCYCLOPEDIA OF SCIENCE AND TECHNOLOGY. McGraw-Hill Book, Incorporated, 1221 Avenue of the Americas, New York, NY 10020. (212) 997-3675.

VAN NOSTRAND REINHOLD ENCYCLOPEDIA OF CHEMISTRY. Douglas M. Considine and Glenn D. Considine. Van Nostrand Reinhold Publishing Company, Incorporated, 115 Fifth Avenue, New York, NY 10003. (800) 543-2681 or (212) 254-3232. 1984. $97.95.

GENERAL WORKS

AEROSOL SCIENCE. Charles N. Davies, editor. Academic Press, Incorporated, Orlando, FL 32887. (800) 321-5068 or (305) 345-2734. 1967. $81.00.

AEROSOL TECHNOLOGY IN HAZARD EVALUATION. T.T. Mercer. Academic Press, Incorporated, Orlando, FL 32887. (800) 321-5068 or (305) 345-2734. 1973. $39.50.

AEROSOL TECHNOLOGY: PROPERTIES, BEHAVIOR, AND MEASUREMENT OF AIRBORNE PARTICLES. William C. Hinds. John Wiley and Sons, Incorporated, 605 Third Avenue, New York, NY 10158. (800) 526-5368 or (212) 850-6000. 1982. $49.00.

AEROSOLS: AN INDUSTRIAL AND ENVIRONMENTAL SCIENCE. George M. Hidy. Academic Press, Incorporated, Orlando, FL 32887. (800) 321-5068 or (305) 345-2734. 1984.

$89.00.

AEROSOLS: SCIENCE, TECHNOLOGY, AND INDUSTRIAL APPLICATIONS OF AIRBORNE PARTICLES. B. Lui and others. Elsevier Science Publishing Company, Incorporated, 52 Vanderbilt Avenue, New York, NY 10017. (212) 370-5520. 1984. $98.00.

ATMOSPHERIC CHEMISTRY: FUNDAMENTALS AND EXPERIMENTAL TECHNIQUES. B.J. Finlayson-Pitts and J.N. Pitts. John Wiley and Sons, Incorporated, 605 Third Avenue, New York, NY 10158. (800) 526-5368 or (212) 850-6000. 1986. $59.95.

HANDBOOKS AND MANUALS

AEROSOL HANDBOOK. Montfort A. Johnson. Dorland Publishing Company, Box 264, Mendham, NJ 07945. (201) 543-2694.

CRC HANDBOOK OF CHEMISTRY AND PHYSICS. CRC Press, Incorporated, 2000 Corporate Boulevard, NW, Boca Raton, FL 33431. Sixty-seventh edition. 1986. $69.95.

HANDBOOK OF APPLIED CHEMISTRY: FACTS FOR ENGINEERS, SCIENTISTS, TECHNICIANS, AND TECHNICAL MANAGERS. Vollrath Hopp and Ingp Hennig. McGraw-Hill Book Company, 1221 Avenue of the Americas, New York, NY 10020. (800) 628-0004. 1983. $54.00.

ONLINE DATA BASES

CA SEARCH. Chemical Abstracts Service, Post Office Box 3012, Columbus, OH 43210. Guide to chemical literature, 1972 to present. Inquire as to cost and availability.

INSPEC. INSPEC Marketing Department, Institute of Electrical and Electronics Engineers, Incorporated, IEEE Service Department, 445 Hoes Lane, Piscataway, NJ 08854. (201) 981-0060. Inquire as to online cost and availability.

NTIS. National Technical Information Service, 5285 Port Royal Road, Springfield, Va 22161. (703) 487-4630. Broad coverage of government sponsored research reports, 1964 to present. Inquire as to online cost and availability.

SCISEARCH. Institute for Scientific Information, 3501 Market Street, Philadelphia, PA 19104. (800) 523-1850 or (215) 386-0100. Broad mulitidisciplinary title and author index to the international literature of science and technology, 1974 to present. Inquire as to online cost and availability.

PERIODICALS

AEROSOL AGE MAGAZINE. Industry Publications, Incorporated, 389 Passaic Avenue, Fairfield, NJ 07006. (201) 227-5151. Monthly. $14.00.

AEROSOL SCIENCE AND TECHNOLOGY. Elsevier Science Publishing Company, Incorporated, 52 Vanderbilt Avenue, New York, NY 10017. (212) 370-5520. Quarterly. $137.00 per year.

ANALYTICAL CHEMISTRY. American Chemical Society, 1155 Sixteenth Street, NW, Washington, DC 20036. (800) 424-6747 or (202) 872-4700. Monthly. $33.00 per year.

CHEMICAL REVIEWS. American Chemical Society, 1155 Sixteenth Street, NW, Washington, DC 20036. (800) 424-6747 or (202) 872-4700. Bimonthly. $83.00 per year.

CHEMICAL WEEK. McGraw-Hill Publishing Company, 1221 Avenue of the Americas, New York, NY 10020. (212) 512-2000. Weekly. $30.00 per year.

CHEMTECH. American Chemical Society, 1155 Sixteenth Street, NW, Washington, DC 20036. (800) 424-6747 or (202) 872-4700. Monthly. $40.00 per year to individuals.

INORGANIC CHEMISTRY. American Chemical Society, 1155 Sixteenth Street, NW, Washington, DC 20036. (800) 424-6747 or (202) 872-4700. Monthly. $400.00 per year.

JOURNAL OF ORGANIC CHEMISTRY. American Chemical Society, 1155 Sixteenth Street, NW, Washington, DC 20036. (800) 424-6747 or (202) 872-4700. Biweekly. $265.00 per year.

JOURNAL OF PHYSICAL CHEMISTRY. American Chemical Society, 1155 Sixteenth Street, NW, Washington, DC 20036. (800) 424-6747 or (202) 872-4700. Biweekly. $369.00 per year.

JOURNAL OF THE AMERICAN CHEMICAL SOCIETY. American Chemical Society, 1155 Sixteenth Street, NW, Washington, DC 20036. (800) 424-6747 or (202) 872-4700. Biweekly. $330.00 per year.

RESEARCH CENTERS AND INSTITUTES

ATMOSPHERIC SCIENCES RESEARCH CENTER. State University of New York at Albany, Room 324, Earth Science Building, 1400 Washington Avenue, Albany, NY 12222. (518) 442-3819.

FLUID DYNAMICS AND DIFFUSION LABORATORY. Colorado State University, Foothills Campus, Fort Collins, CO 80523. (303) 491-8574.

UNIVERSITY OF ARIZONA. Engineering Experiment Station, 206 Civil Engineering Building, Tucson, AZ 85721. (602) 621-6601.

AEROSPACE ENGINEERING

See also: AERONAUTICAL ENGINEERING

ABSTRACT SERVICES AND INDEXES

ALLOYS INDEX. American Society for Metals, 9639 Kinsman Road, Metals Park, OH 44073. (216) 338-5151. 1974 to present. Monthly. $225.00 per year.

APPLIED MECHANICS REVIEWS. American Society of Mechanical Engineers, 345 East 47th Street, New York, NY 10017. (212) 705-7703. 1948 to present. Monthly. $360.00 per year.

APPLIED SCIENCE AND TECHNOLOGY INDEX. H.W. Wilson Company, 950 University Avenue, Bronx, NY 10452. (800) 367-6670 or (212) 588-8400. 1958 to present. Monthly. Inquire as to cost and availability.

CHEMICAL ABSTRACTS. Chemical Abstracts Service, Box 3012, Columbus, OH 43210. (614) 421-3600. 1907 to present. Weekly. $9200.00 per year.

CURRENT CONTENTS: ENGINEERING, TECHNOLOGY AND APPLIED SCIENCES. Institute for Scientific Information, 3501 Market Street, Philadelphia, PA 19104. (800) 523-1850 or (215) 386-0100. 1970 to present. Weekly. $272.00 per year.

ENGINEERING INDEX MONTHLY AND AUTHOR INDEX. Engineering Information, Incorporated, 345 East 47th Street, New York, NY 10017. (800) 221-1044 or (212) 705-7600. Monthly, with annual cumulation. $1560.00 per year.

INTERNATIONAL AEROSPACE ABSTRACTS. AIAA/TIS, 1633 Broadway, New York, NY 10019. (212) 581-4300. Semimonthly. $700.00 per year.

METALS ABSTRACTS. American Society for Metals, 9639 Kinsman Road, Metals Park, OH 44073. (216) 338-5151. Monthly. $890.00.

METALS ABSTRACTS INDEX. American Society for Metals, 9639 Kinsman Road, Metals Park, OH 44073. (216) 338-5151.

Monthly. (Sold only to subscribers of Metals Abstracts).

SCIENCE ABSTRACTS. Section A: Physics; Section B: Electrical and Electronics Abstracts; Section C: Computer and Control Abstracts. Institute of Electrical Engineers, London, United Kingdom. Available from: Institute of Electrical and Electronic Engineers (IEEE), 445 Hoes Lane, Piscataway, NJ 08854. Inquire as to cost and availability.

SCIENCE CITATION INDEX. Institute for Scientific Information, 3501 Market Street, Philadelphia, PA 19104. (800) 523-1850 or (215) 386-0100.

STAR. (Scientific and Technical Aerospace Reports). U.S. National Aeronautics and Space Administration, Scientific and Technical Information Facility, Box 8757, Baltimore-Washington International Airport, MD 21240. (202) 755-2210. Semimonthly, with semiannual and annual indexes. $85.00 per year.

WORLD ALUMINUM ABSTRACTS. Aluminum Association, Incorporated, 818 Connecticut Avenue, NW, Washington, DC 20006. (202) 862-5156. 1968 to present. Monthly $240.00.

ASSOCIATIONS AND PROFESSIONAL SOCIETIES

AIRCRAFT ELECTRONICS ASSOCIATION. Box 1981, Independence, MO 64055. (816) 373-6565.

AMERICAN INSTITUTE OF AERONAUTICS AND ASTRONAUTICS. 1633 Broadway, New York, NY 10019. (212) 581-4300.

SOCIETY OF AUTOMOTIVE ENGINEERS. SAE, Incorporated, 400 Commonwealth Drive, Warrendale, PA 15096. (412) 776-4841.

BIBLIOGRAPHIES

AERONAUTICAL ENGINEERING: A SPECIAL BIBLIOGRAPHY WITH INDEXES. U.S. National Aeronautics and Space Administration, Washington, DC 20546. (202) 755-2320. Available from National Technical Information Service, Springfield, VA 22161. 1970 to present. Monthly. Inquire as to cost and availability.

NATIONAL AIR AND SPACE MUSEUM AND SMITHSONIAN INSTITUTION. Aerospace Periodical Index, 1973-1982. 1983. G.K. Hall and Company, 70 Lincoln Street, Boston, MA 02111. (800) 343-2806. $100.00.

SCIENTIFIC AND TECHNICAL BOOKS AND SERIALS IN PRINT, 1988; AN INDEX TO LITERATURE IN SCIENCE AND TECHNOLOGY. R.R. Bowker Company, 205 East 42nd Street, New York, NY 10017. (800) 521-8110 or (212) 916-1600. $175.00.

DIRECTORIES AND BIOGRAPHICAL SOURCES

AAS DIRECTORY. American Astronautical Society, 6212-B Old Keene Mill Court, Springfield, VA 22152. (703) 866-0020. Annual. $35.00 per year.

RESEARCH CENTERS DIRECTORY. Gale Research Company, Detroit, MI 48226. (800) 521-0707. Eleventh edition. 1987.

WHO'S WHO IN ENGINEERING. Engineers Joint Council, 345 East 47th Street, New York, NY 10017. (212) 705-7010. $200.00.

ENCYCLOPEDIAS AND DICTIONARIES

DICTIONARY OF AEROSPACE ENGINEERING. M.G. Kotik. Elsevier Science Publishing Company, Incorporated, 52 Vanderbilt Avenue, New York, NY 10017. (212) 370-5520. 1986. $170.00.

ENCYCLOPEDIA OF PHYSICAL SCIENCE AND TECHNOLOGY. Academic Press, Incorporated, Orlando, Florida 32887. (800) 321-5068 or (305) 345-2734. Fifteen volumes, 1986.

JANE'S AEROSPACE DICTIONARY. Jane's Publishing Incorporated, 135 West 50th Street, New York, NY 10020. (212) 586-7745. Second edition. 1986. $39.95.

MCGRAW-HILL DICTIONARY OF ENGINEERING. Sybil P. Parker, editor. McGraw-HIll Book Company, 1221 Avenue of the Americas, New York, NY 10020. (212) 512-2000. 1984. $39.95.

MCGRAW-HILL ENCYCLOPEDIA OF SCIENCE AND TECHNOLOGY. McGraw-Hill Book, 1221 Avenue of the Americas, New York, NY 10020. (212) 512-2000.

GENERAL WORKS

FOUNDATIONS OF AERODYNAMICS: BASES OF AERODYNAMIC DESIGN. A.M. Kuethe and C. Chow. John Wiley and Sons, Incorporated, 605 Third Avenue, New York, NY 10158. (800) 526-5368 or (212) 850-6000. Fourth edition. 1986. $43.95.

ILLUSTRATED GUIDE TO AERODYNAMICS. Hubert Smith. Tab Books, Incorporated, Monterey Lane, Blue Ridge Summit, PA 17214. (717) 794-2191. 1986. $14.95.

INTRODUCTION AEROSPACE STRUCTURAL ANALYSIS. David H. Allen and Walter E. Haisler. John Wiley and Sons, Incorporated, 605 Third Avenue, New York, NY 10158. (800) 526-5368 or (212) 850-6000. 1985. $42.95.

INTRODUCTION TO AIRCRAFT PERFORMANCE. Selection and Design. F.J. Hale. John Wiley and Sons, Incorporated, 605 Third Avenue, New York, NY 10158. (800) 526-5368 or (212) 850-6000. 1984. $39.95.

INTRODUCTION TO THEORETICAL AND COMPUTATIONAL AERODYNAMICS. Jack Moran. John Wiley and Sons, Incorporated, 605 Third Avenue, New York, NY 10158. (800) 526-5368 or (212) 850-6000. 1984. $40.95.

THEORY OF WING SECTIONS; INCLUDING A SUMMARY OF AIRFOIL DATA. Dover Publications, Incorporated, 31 East Second Street, Mineola, NY 11501. (516) 294-7000. 1949. $10.95.

HANDBOOKS AND MANUALS

STANDARD HANDBOOK OF ENGINEERING CALCULATIONS. Tyler G. Hicks, editor. McGraw-Hill Book Company, 1221 Avenue of the Americas, New York, NY 10020. (212) 512-2000. Second edition. 1984. $59.50.

THE WILEY ENGINEER'S DESK REFERENCE. Sanford I. Heisler. John Wiley and Sons, Incorporated, 605 Third Avenue, New York, NY 10158. (800) 526-5368 or (212) 850-6418. 1984. $36.00.

ONLINE DATA BASES

COMPENDEX. Engineering Information, Incorporated, 345 East 47th Street, New York, NY 10017. (800) 221-1044 or (212) 705-7615. Engineering and technical literature, 1975 to present. Inquire as to cost and availability.

NASA. National Aeronautics and Space Administration, Scientific and Technical Information Branch, 300 7th Street, SW, Washington, DC 20546. Citations and abstracts of aerospace literature, 1962 to present. Inquire as to cost and availability.

NTIS. National Technical Information Service, 5285 Port Royal Road, Springfield, VA 22161. (703) 487-4630. Broad coverage of government sponsored research reports, 1964 to present. Inquire as to cost and availability.

PERIODICALS

AERONAUTICAL JOURNAL. Aeronautical Society, 4 Hamilton Place, London W1V 0BQ, England. Ten times per year. $175.00 per year.

AEROSPACE. Royal Aeronautical Society, 4 Hamilton Place, London W1V OBQ, England. Ten times per year. $35.00 per year.

AEROSPACE AMERICA. American Institute of Aeronautics and Astronautics, 1633 Broadway, New York, NY 10019. (212) 581-4300. Monthly. $51.00 per year.

AEROSPACE ENGINEERING MAGAZINE (SOCIETY OF AUTOMOTIVE ENGINEERS). SAE, Incorporated, 400 Commonwealth Drive, Warrendale, PA 15096. (412) 776-4841. Monthly. $30.00 per year.

AIAA JOURNAL. AIAA/TIS, 1633 Broadway, New York, NY 10019. (212) 581-4300. Monthly. $205.00 per year.

AVIATION WEEK AND SPACE TECHNOLOGY. McGraw-Hill Book Company, Incorporated, 1221 Avenue of the Americas, New York, NY 10020. (212) 512-2000. Weekly. $55.00 per year.

AVIONICS. Atlantic Communications, Incorporated, Box 5100, Westport, CT 06881. (203) 227-2280. Monthly. $36.00 per year.

AVIONICS NEWS MAGAZINE. Aircraft Electronics Association, Box 1981, Independence, MO 64055. (816) 373-6565. Monthly. Free.

BRITISH PLANETARY SOCIETY JOURNAL. British Planetary Society, 27-29 South Lambeth Road, London, SW8 1SZ, England. Monthly. $120.00 per year.

CANADIAN AERONAUTICS AND SPACE JOURNAL. Canadian Aeronautics and Space Institute, 222 Somerset Street, West, Suite 601, Ottawa, ON K2P 0J1, Canada. (613) 234-0191. Quarterly. $35.00 per year.

IEEE TRANSACTIONS ON AEROSPACE. Institute of Electrical and Electronics Engineers, 345 East 47th Street, New York, NY 10017. Bimonthly. $108.00 per year.

INTERNATIONAL JOURNAL OF TURBO AND JET ENGINES. Martinus Nijhoff Publishers. Available from: Kluwer Academic Publishers Group, Distribution Centre, Postbus 322, 3300 AH Dordrecht, The Netherlands. Quarterly. $125.00 per year.

JOURNAL OF GUIDANCE AND CONTROL. American Institute of Aeronautics and Astronautics, AIAA/TIS, 1633 Broadway, New York, NY 10019. (212) 581-4300. Bimonthly. $95.00 per year.

JOURNAL OF PROPULSION AND POWER. American Institute of Aeronautics and Astronautics, AIAA/TIS, 1633 Broadway, New York, NY 10019. (212) 581-4300. Bimonthly. $170.00 per year.

JOURNAL OF SPACECRAFT. American Institute of Aeronautics and Astronautics, AIAA/TIS, 1633 Broadway, New York, NY 10019. (212) 581-4300. Bimonthly. $95.00 per year.

PROGRESS IN AEROSPACE SCIENCES. Pergamon Press, Incorporated, Journals Division, Maxwell House, Fairview Park, Elmsford, NY 10523. (914) 592-7700. Quarterly. $120.00 per year.

SPACEFLIGHT. British Planetary Society, 27-29 South Lambeth Road, London, SW8 1SZ, England. Ten times per year. $50.00 per year.

RESEARCH CENTERS AND INSTITUTES

NEW MEXICO STATE UNIVERSITY. Physical Science Laboratory, P.O. Box 3548, Las Cruces, NM 88003. (505) 522-9100.

OHIO STATE UNIVERSITY. Aeronautical and Astronautical Research Laboratory, 2300 West Case Road, Columbus, OH 43220. (614) 422-1241.

STANFORD UNIVERSITY. Aero Structures Laboratory, Department of Aeronautics and Astronomy, Stanford, Ca 94305. (415) 497-3317.

UNIVERSITY OF COLORADO - BOULDER. Engineering Research Center, Boulder CO 80309. (303) 492-7427.

AGGREGATES

See also: BUILDING MATERIALS, CEMENT, CONCRETE, PRESTRESSED CONCRETE

ABSTRACT SERVICES AND INDEXES

APPLIED SCIENCE AND TECHNOLOGY INDEX. H.W. Wilson Company, 950 University Avenue, Bronx, NY 10452. (800) 367-6670 or (212) 588-8400. Inquire as to cost and availability.

CONCRETE ABSTRACTS. American Concrete Institute, P.O. Box 19150, Redford Station, Detroit, MI 48219. (313) 532-2600. Bimonthly. $122.00 per year.

ENGINEERING INDEX MONTHLY. Engineering Information, Incorporated, 345 East 47th Street, New York, NY 10017. (800) 221-1044 or (212) 705-7600. Monthly with annual cumulation. $1560.00 per year.

ANNUAL REVIEWS AND YEARBOOKS

CONCRETE INDUSTRIES YEARBOOK. Pit and Quarry Publications, Incorporated, 105 West Adams Street, Chicago, IL 60603. $35.00.

MINERALS YEARBOOK. U.S. Bureau of Mines. Annual. $18.00. Send orders to: Government Printing Office, Washington, DC 20402.

ASSOCIATIONS AND PROFESSIONAL SOCIETIES

AMERICAN BUILDING STONE INSTITUTE. 420 Lexington Avenue, New York, NY 10170. (212) 490-2530.

AMERICAN CONCRETE INSTITUTE. P.O. Box 19150, Redford Station, Detroit, MI 48219. (313) 532-2600.

EXPANDED CLAY AND SLATE INSTITUTE. 6218 Montrose Road, Rockville, MD 20852. (301) 231-9497.

LIGHTWEIGHT AGGREGATE PRODUCERS ASSOCIATION. P.O. Box 1111, B & B Building, 546 Hamilton Street, Allentown, PA 18105. (215) 435-9687.

NATIONAL SAND AND GRAVEL ASSOCIATION. 900 Spring Street, Silver Spring, MD 20910. (301) 587-1400.

BIBLIOGRAPHIES

AGGREGATES - BUILDING MATERIALS: MONOGRAPHS. Mary Vance. Vance Bibliographies, P.O. Box 229, 112 North Charter Street, Monticello, IL 61856. (217) 762-3831. 1985. $3.75 in paper.

DIRECTORIES AND BIOGRAPHICAL SOURCES

AMERICAN CEMENT DIRECTORY. Bradley Pulverizer Company, 123 South Third Street, Allentown, PA 18105. (215)

434-5191. Annual. $46.00.

AMERICAN CONCRETE INSTITUTE MEMBERSHIP DIRECTORY. American Concrete Institute, P.O. Box 19150, Redford Station, Detroit, MI 48219. (313) 532-2600. Biennial. $44.95.

SAND AND GRAVEL DIRECTORY. American Business Directories, Incorporated, Division of American Business Lists, Incorporated, 5707 South 86th Circle, Omaha, NE 68127. (402) 331-7169. Annual. $200.00.

WORLD CEMENT DIRECTORY. Cembureau. European Cement Association, 2 rue Saint-Charles, F-75740 Paris Cedex 15, France. Irregular. 1986. $200.00, prepayment required. List of manufactures of cement and cement products on more than 140 countries.

ENCYCLOPEDIAS AND DICTIONARIES

CEMENT AND CONCRETE TERMINOLOGY. American Concrete Institute, P.O. Box 19150, Redford Station, Detroit, MI 48219. (313) 532-2600. 1985.

MCGRAW-HILL DICTIONARY OF ENGINEERING. Sybil P. Parker, editor. McGraw-Hill Book Company, 1221 Avenue of the Americas, New York, NY 10020. (212) 512-2000. 1984. $39.95.

MCGRAW-HILL ENCYCLOPEDIA OF SCIENCE AND TECHNOLOGY. McGraw-Hill Book, 1221 Avenue of the Americas, New York, NY 10020. (212) 512-2000.

GENERAL WORKS

AGGREGATES FOR CONCRETE. American Concrete Institute, P.O. Box 19150, Redford Station, Detroit, MI 48219. (313) 532-2600. 1978. $13.75.

CEMENT CHEMISTRY AND PHYSICS FOR CIVIL ENGINEERS. W.H. Dipl-lng and W.H.C. Czernin. Heyden & Sons, Incorporated, 247 South 41st Street, Philadelphia, PA 19104. (800) 345-8112 or (215) 382-6673. 1980. $24.00.

CEMENT ENGINEERS HANDBOOK. B. Kohlhaas and others. Heyden & Sons, Incorporated, 247 South 42st Street, Philadelphia, PA 19104. (800) 345-8112 or (215) 382-6673. 1982. $90.00.

CONCRETE PRIMER. F.R. McMillan and Lewis H. Tuthill. American Concrete Institute, P.O. Box 19150, Redford Station, Detroit, MI 48219. (313) 532-2600. Third edition. 1973. $9.50.

CONCRETE SCIENCE: A TREATISE ON CURRENT RESEARCH. V.S. Ramachandran, R.F. Feldman, and J.J. Beaudoin. John Wiley & Sons, 605 Third Avenue, New York, NY 10158. (800) 526-5368 or (212) 850-6000. 1981. $74.95.

CONCRETE: STRUCTURE, PROPERTIES AND MATERIALS. P. Kumar Mehta. Prentice-Hall, Incorporated, Englewood Cliffs, NJ 07632. (800) 262-6868 or (201) 592-2000. 1986. $43.95.

ACI MANUAL OF CONCRETE PRACTICE. American Concrete Institute, P.O. Box 19150, Redford Station, Detroit, MI 48219. (313) 532-2600. Five volumes. Annual.

CONCRETE CONSTRUCTION HANDBOOK. Joseph J. Waddell, editor. McGraw-Hill Book Company, 1221 Avenue of the Americas, New York, NY 10020. (212) 512-2000. Second edition. 1974. $70.50.

CONCRETE MANUAL. John Wiley & Sons, 605 Third Avenue, New York, NY 10158. (800 526-5368 or (212) 850-6000. Eighth edition. 1983. $37.50.

HANDBOOK OF CONCRETE ENGINEERING. Mark Fintel, editor. Van Nostrand Reinhold Company, 115 Fifth Avenue, New York, NY 10003. (212) 254-3232. Second edition. 1985.

$89.50.

ONLINE DATA BASES

COMPENDEX. Engineering Information, Incorporated, 345 East 47th Street, New York, NY 10017. (800) 221-1044 or (212) 705-7615. Engineering and technical literature, 1975 to present. Inquire as to cost and availability.

NTIS. National Technical Information Service, 5285 Port Royal Road, Sprinfield, VA 22161. (703) 487-4630. Broad coverate of government sponsored research reports, 1964 to present. Inquire as to online cost and availability.

WILSONLINE. H.W. Wilson Company, 950 University Avenue, Bronx, NY 10452. (800) 367-6770 or (212) 588-8400. Makes available online versions of the printed H.W. Wilson indexes including Applied Science, and Technology Index, Business Periodicals Index, and Readers' Guide to Periodical Literature. Period covered is generally 1983 to present. Inquire as to cost and availability.

OTHER SOURCES

IEEE CEMENT INDUSTRY TECHNICAL CONFERENCE RECORD. IEEE, Incorporated, 345 East 47th Street, New York, NY 10017. (202) 705-7900. Annual.

PERIODICALS

AMERICAN CONCRETE INSTITUTE JOURNAL. American Concrete Institutes, P.O. Box 19150, Redford Station, Detroit, MI 48219. (313) 532-2600. Monthly. $69.00 per year.

CEMENT AND CONCRETE RESEARCH. Pergamon Press, Incorporated, Maxwell House, Fairview Park, Elmsford, NY 10523. (914) 592-7700. Bimonthly. $190.00 per year.

CONCRETE. 120 West Second Street, Duluth, MN 55802. (218) 723-9253. Monthly. $14.00 per year.

CONCRETE CONSTRUCTION. Concrete Construction Publications, Incorporated, 426 South Westgate, Addison, IL 60101. (312) 543-0870. Monthly. $12.00 per year.

CONCRETE INTERNATIONAL: DESIGN AND CONSTRUCTION. American Concrete Institute, P.O. Box 19150, Redford Station, Detroit, MI 48219. (313) 532-2600. Monthly. $69.00.

JOURNAL OF RCI. Prestressed Concrete Institute. 201 North Wells Street, Chicago, IL 60606. (312) 346-4071. Bimonthly.

JOURNAL OF THE AMERICAN CONCRETE INSTITUTE. American Concrete Institute, P.O. Box 19150, Redford Station, Detroit, MI 48219. (313) 532-2600. Monthly. $69.00 per year.

PIT AND QUARRY. 120 West Second Street, Duluth, MN 55802. (218) 723-9253. Monthly. $12.00 per year.

RESEARCH CENTERS AND INSTITUTES

AMERICAN SOCIETY FOR TESTING AND MATERIALS. 1916 Race Street, Philadelphia, PA 19103. (215) 299-5400.
CONCRETE MATERIALS RESEARCH COUNCIL. P.O. Box 19150, Detroit, MI 48219. (313) 532-2600.

CONCRETE RESEARCH LABORATORY. University of Illinois, 1211 Newmark Civil Engineering Laboratory, 208 North Romine Street, Urbana, IL 61801. (217) 333-3394.

CONCRETE RESEARCH LABORATORY. University of Michigan, G.G. Brown Laboratory, Ann Arbor, MI 48109. (313) 763-8077.

PORTLAND CEMENT ASSOCIATION. Construction Technology Laboratory, 5420 Old Orchard Road, Skokie, IL 60077. (312) 965-7500.

SPECIFICATIONS AND STANDARDS

CEMENT STANDARDS: EVOLUTION AND TRENDS - STP 663. American Society for Testing and Materials, 1916 Race Street, Philadelphia, PA 19103. (215) 299-5400. 1979. $20.00.

CEMENT STANDARDS OF THE WORLD. Edited by Cembureau (European Cement Association). International Publications Service Incorporated, P.O. Box 230, Accord, MA 02018. (617) 749-3628. Second edition. 1980. $75.00.

AGRICULTURAL ENGINEERING

ABSTRACT SERVICES AND INDEXES

AGRICULTURAL ENGINEERING ABSTRACTS. Bernan-Unipub, 4611-F Assembly Drive, Lanham, MD 20706-4391. 1976 to present. Monthly. $215.00 per year.

APPLIED SCIENCE AND TECHNOLOGY INDEX. H.W. Wilson Company, 950 University Avenue, Bronx, NY 10452. (800) 367-6670 or (212) 588-8400. Inquire as to cost and availability.

ENGINEERING INDEX MONTHLY. Engineering Information, Incorporated, 345 East 47th Street, New York, NY 10017. (800) 221-1044 or (212) 705-7600. Monthly, with annual cumulation. $1425.00 per year.

ASSOCIATIONS AND PROFESSIONAL SOCIETIES

AMERICAN SOCIETY OF AGRICULTURAL ENGINEERS. 2950 Niles Road, St. Joseph, MI 49085. (616) 429-0300.

DIRECTORIES AND BIOGRAPHICAL SOURCES

AGRICULTURAL ENGINEERING - TECHNOLOGY ISSUE. American Society of Agricultural Engineers, 2950 Niles Road, St. Joseph, MI 49085. (616) 429-0300. Annual. $20.00.

AGRICULTURAL RESEARCH CENTERS. Gale Research Company, Detroit, MI 48277-0748. (800) 521-0707. 9th edition. 1988. $395.00 for set.

INTERNATIONAL DIRECTORY OF AGRICULTURAL ENGINEERING INSTITUTIONS. Food and Agricultural Organization of the United Nations. Available from: Bernan-Unipub, 4611-F Assembly Drive, Lanham, MD 20706-4391. 1983. $35.00.

PROFESSIONAL WORKERS IN STATE AGRICULTURAL EXPERIMENT STATIONS AND OTHER COOPERATIVE STATE INSTITUTIONS. Science and Education Administration - Cooperative State Research Service, Department of Agriculture, JSM Building, Room 25, Washington, DC 20250. Annual. $7.00 per year. Send orders to Government Printing Office, Washington, DC 20402.

RESEARCH CENTERS DIRECTORY. Gale Research Company, Detroit, MI 48227-0748. (800) 223-GALE. Thirteenth edition. 1989. $365.00 for set.

WHO'S WHO IN ENGINEERING. Engineers Joint Council, 345 East 47th Street, New York, NY 10017. (212) 705-7010. 1985. $200.00.

SCIENTIFIC AND TECHNICAL ORGANIZATIONS AND AGENCIES DIRECTORY. Gale Research Company, Book Tower, Detroit, MI 48226. (800) 521-0707. 1985.

WHO'S WHO IN TECHNOLOGY TODAY. Reston Publishing Company, Incorporated, c/o Prentice-Hall, Incorporated, Englewood Cliffs, NJ 07632. (800) 262-6868. Biennial. Five volumes. $425.00. Covers the fields of electronics, computer science, physics, optics, chemistry, biotechnology, mechanics, energy, and earth science.

ENCYCLOPEDIAS AND DICTIONARIES

ENCYCLOPEDIA OF PHYSICAL SCIENCE AND TECHNOLOGY. Academic Press, Incorporated, Orlando, FL 32887. (800) 321-5068 or (305) 345-2734. Fifteen Volumes. 1986.

MCGRAW-HILL DICTIONARY OF ENGINEERING. Sybil P. Parker, editor. McGraw-Hill Book Company, 1221 Avenue of the Americas, New York, NY 10020. (212) 512-2000. 1984. $39.95.

GENERAL WORKS

ENGINEERING APPLICATIONS IN AGRICULTURE. W. Bowers, and others. Stipes Publishing Company, 1-12 Chester Street, Champaign, IL 1820. (217) 356-8391. 1986. $15.00.

INTRODUCTION TO AGRICULTURAL ENGINEERING. L.O. Roth and others. AVI Publishing Company, Incorporated, 250 Post Road East, Westport, CT 06881. (203) 226-0738.

HANDBOOKS AND MANUALS

AGRICULTURAL ENGINEER'S HANDBOOK. C.B. Richey and others. McGraw-Hill Book Company, 1221 Avenue of the Americas, New York, NY 10020. (212) 512-2000. 1961. $62.50.

STANDARD HANDBOOK OF ENGINEERING CALCULATIONS. Tyler G. Hicks, editor. McGraw-Hill Book Company, 1221 Avenue of the Americas, New York, NY 10020. (212) 512-2000. Second edition. 1984. $59.50.

THE WILEY ENGINEER'S DESK REFERENCE. Sanford I. Heisler. John Wiley and Sons, Incorporated, 605 Third Avenue, New York, NY 10158. (800) 526-5368 or (212) 850-6418. 1984. $36.00.

ONLINE DATA BASES

AGRICOLA. U.S. National Agricultural Library, Beltsville, MD 20705. Agricultural literature, 1970 to present. Inquire as to online cost and availability.

COMPENDEX. Engineering Information, Incorporated, 345 East 47th Street, New York, NY 10017. (800) 221-1044 or (212) 705-7615. Engineering and technical literature, 1975 to present. Inquire as to cost and availability.

NTIS. National Technical Information Service, 5284 Port Royal Road, Springfield, VA 22161. (703) 487-4630. Borad coverage of government sponsored research reports, 1964 to present. Inquire as to cost and availability.

PERIODICALS

AGRICULTURAL ENGINEERING. 2950 Niles Road, St. Joseph, MI 49085. (616) 429-0300. 1920 to present. Monthly. $25.00 per year.

AGRICULTURAL MACHINERY JOURNAL. Farmers Publishing Group, Surrey House, 1 Throwley Way, Sutton, Surrey SM1 400 England. 1946 to present. Monthly. $57.20 per year.

APPLIED ENGINEERING IN AGRICULTURE. 2950 Niles Road, St. Joseph, MI 49085. (616) 429-0300. Semiannual. $17.50 per year.

AGRICULTURAL ENGINEERING

CANADIAN AGRICULTURAL ENGINEERING. Canadian Society of Agricultural Engineering, 151 Slater Street, Suite 907, Ottawa, ON K1P 5H4, Canada. Semiannual. $30.00 per year.

FARM JOURNAL. Farm Journal, Incorporated, 230 West Washington Square, Philadelphia, PA 19105. (215) 574-1200. 1877 to present. Monthly. $8.00 per year.

IMPLEMENT AND TRACTOR. Intertec Publishing Company, 9221 Quivira Road, Overland Park, KS 66212. (913) 888-4664. 1986 to present. Monthly. $10.00.

POWER FARMING. Farmers Publishing Group, Surrey House, One Throwley Way, Sutton, Surrey SM1 4QQ England. 1941 to present. Monthly. $45.00 per year.

TRANSACTIONS OF THE AMERICAN SOCIETY OF AGRICULTURAL ENGINEERS. 2950 Niles Road, St. Joseph, MI 49085. (616) 429-0300. Bimonthly. $95.00 per year.

RESEARCH CENTERS AND INSTITUTES

AGRICULTURAL ENGINEERING DEPARTMENT. Michigan State University, East Lansing, MI 48824. (517) 353-7168.

ENGINEERING EXPERIMENT STATION. Purdue University, West Lafayette, IN 47907. (317) 494-5340.

ENGINEERING RESEARCH CENTER. New Mexico State University, P.O. Box 3449, Las Cruces, NM 88003. (505) 646-3421.

ENGINEERING RESEARCH CENTERS. University of Nebraska-Lincoln, W181 Nebraska Hall, Lincoln, NE 68588. (402) 472-3181.

AI

See: ARTIFICIAL INTELLIGENCE

AIR

See: METEOROLOGY

AIR CONDITIONING

See also: COOLING TOWERS; HEATING AND VENTILATION

ABSTRACT SERVICES AND INDEXES

APPLIED SCIENCE AND TECHNOLOGY INDEX. H.W. Wilson Company, 950 University Avenue, Bronx, NY 10452. (800) 367-6670 or (212) 588-8400. Inquire as to cost and availability.

CHEMICAL ABSTRACTS. Chemical Abstracts Service, Box 3012, Columbus, OH 43210. (614) 421-3600. 1907 to present. Weekly. $9200.00 per year.

ENGINEERING INDEX MONTHLY. Engineering Information, Incorporated, 345 East 47th Street, New York, NY 10017. (800) 221-1044 or (212) 705-7600. Monthly, with annual cumulation. $1560.00 per year.

ASSOCIATIONS AND PROFESSIONAL SOCIETIES

AIR CONDITIONING AND REFRIGERATION INSTITUTE. 1501 Wilson Boulevard, Arlington, VA 22209. (703) 524-8800.

AIR CONDITIONING CONTRACTORS OF AMERICA. 1228 17th Street, NW, Washington, DC 20036. (202) 296-7610.

AMERICAN SOCIETY OF HEATING, REFRIGERATING, AND AIR CONDITIONING ENGINEERS, Incorporated 1791 Tullie Circle, NE, Atlanta, GA 30329. (404) 636-8400.

DIRECTORIES AND BIOGRAPHICAL SOURCES

AIR CONDITIONING DEALERS AND CONTRACTORS DIRECTORY. American Business Directories, Incorporated, Division of American Business Lists, Incorporated, 5707 South 86th Circle, Omaha, NE 68172. (402) 331-7169. Annual. $295.00.

AIR CONDITIONING EQUIPMENT DIRECTORY. American Business Directories, Incorporated, Division of American Business Lists, Incorporated, 5707 South 86th Circle, Omaha, NE 68127. (402) 331-7169. 1986. $85.00.

AIR CONDITIONING, HEATING AND REFRIGERATION NEWS - DIRECTORY ISSUE. Business News Publishing Company, 755 West Big Beaver Road, Troy, MI 48084. (313) 362-3700. Annual. $15.00.

RESEARCH CENTERS DIRECTORY. Gale Research Company, Detroit, MI 48226. (800) 521-0707. Eleventh edition. 1987.

WHO'S WHO IN ENGINEERING. Engineers Joint Council, 345 East 47th Street, New York, NY 10017. (212) 705-7010. 1985. $200.00.

WHO'S WHO IN FRONTIER SCIENCE AND TECHNOLOGY. Marquis Who's Who, Incorporated, 200 East Ohio Street, Chicago, IL 60611. (800) 428-3898 or (312) 787-2008.

SCIENTIFIC AND TECHNICAL ORGANIZATIONS AND AGENCIES DIRECTORY. Gale Research Company, Book Tower, Detroit, MI 48226. (800) 521-0707. 1985.

WHO'S WHO IN TECHNOLOGY TODAY. Reston Publishing Company, Incorporated, c/o Prentice-Hall, Incorporated, Englewood Cliffs, NJ 07632. (800) 262-6868. Biennial. Five volumes. $425.00. Covers the fields of electronics, computer science, physics, optics, chemistry, biotechnology, mechanics, energy, and earth science.

ENCYCLOPEDIAS AND DICTIONARIES

DICTIONARY OF REFRIGERATION AND AIR CONDITIONING. K.M. Booth. Elsevier Science Publishing Company, Incorporated, 52 Vanderbilt Avenue, New York, NY 10017. (212) 370-5520. 1971. $27.75.

GENERAL WORKS

AIR CONDITIONING ENGINEERING. W.P. Jones. Crane, Russak and Company, Incorporated, Three East 44th Street, New York, NY 10017. (212) 867-1490. Third edition. 1985. $49.50.

COMMERCIAL, INDUSTRIAL, AND INSTITUTIONAL REFRIGERATION: DESIGN, INSTALLATION AND TROUBLESHOOTING. Prentice-Hall, Incorporated, Englewood Cliffs, NJ 07632. (800) 562-0245. 1987. $29.95.

CONTROL SYSTEMS FOR HEATING, VENTILATING AND AIR CONDITIONING. Roger W. Haines. Van Nostrand Reinhold Company, 115 Fifth Avenue, New York, NY 20003. (212) 254-3232. Third edition. 1983. $28.50.

HEATING, VENTILATING AND AIR CONDITIONING: ANALYSIS AND DESIGN. F.C. McQuiston and J.D. Parker. John Wiley and Sons, Incorporated, 605 Third Avenue, New York, NY 10158. (800) 526-5368 or (212) 850-6000. Second edition. 1982. $43.00.

HEATING, VENTILATING AND AIR CONDITIONING: DESIGN FOR BUILDING CONSTRUCTION. John E. Traister. Prentice-Hall, Incorporated, Englewood Cliffs, NJ 07632. (800) 562-0245. 1987. $29.95.

REFRIGERATION AND AIR CONDITIONING. Air Conditioning and Refrigeration Institute. Prentice-Hall, Incorporated, Englewood Cliffs, NJ 07632. (800) 562-0245. Second edition. 1987. $29.95.

REFRIGERATION AND AIR CONDITIONING. A.R. Trott. McGraw-Hill Book Company, 1221 Avenue of the Americas, New York, NY 10020. (212) 512-2000. 1981. $39.50.

ONLINE DATA BASES

COMPENDEX. Engineering Information, Incorporated, 345 East 47th Street, New York, NY 10017. (800) 221-1044 or (212) 705-7615. Engineering and technical literature, 1975 to present. Inquire as to cost and availability.

NTIS. National Technical Information Service, 5285 Port Royal Road, Springfield, Va 22161. (703) 487-4630. Broad coverage of government sponsored research reports, 1964 to present. Inquire as to cost and availability.

WILSONLINE. H.W. Wilson Company, 950 University Avenue, Bronx, NY 10452. (800) 367-6770 or (212) 588-8400. Makes available online versions of the printed H.W. Wilson indexes including Applied Science and Technology Index, Business Periodicals Index, and Readers' Guide to Periodical Literature. Period covered is generally 1983 to present. Inquire as to cost and availability.

PERIODICALS

ASHRAE JOURNAL: HEATING, REFRIGERATING, AIR CONDITIONING, VENTILATING. American Society of Heating, Refrigerating, and Air Conditioning Engineers, Incorporated, 1791 Tullie Circle, NE, Atlanta, GA 30329. (404) 636-8400. Monthly. $35.00 per year.

ASHRAE TRANSACTIONS. American Society of Heating, Refrigerating, and Air Conditioning Engineers, Incorporated, 1791 Tullie Circle, NE, Atlanta, Ga 30329. (404) 636-8400. Annual. $148.00 per year.

HEATING, PIPING, AIR CONDITIONING. Penton-IPC, Penton Plaza, 1111 Chester Avenue, Cleveland, OH 44114. (216) 696-7000. Monthly. $35.00 per year.

AIR CONDITIONING, HEATING AND REFRIGERATION NEWS. Business News Publishing Company, Box 2600, Troy, MI 48007. (313) 362-3700. Weekly. $44.00 per year.

HEATING, AIR CONDITIONING, AND PLUMBING PRODUCTS. Gordon Publications Incorporated, Box 1952, Dover, NJ 07801. (201) 361-9060. Bimonthly. $6.00 per year.

RESEARCH CENTERS AND INSTITUTES

KANSAS STATE UNIVERSITY. Mechanical Engineering Research Laboratory, Duland Hall, Manhattan, KS 66506. (913) 532-5610.

LEHIGH UNIVERSITY. Institute of Thermo-Fluid Engineering and Science, Whitaker Laboratory, Bethlehem, PA 18015. (215) 861-4091.

REFRIGERATION RESEARCH FOUNDATION. 7315 Wisconsin Avenue, Bethesda, MD 20814. (301) 652-5674.

TEXAS A & M UNIVERSITY. Energy Systems Laboratory, College Station, TX 77843. (409) 845-6402.

AIR CURRENTS

See: METEOROLOGY

AIR MASSES

See: METEOROLOGY

AIR POLLUTION

See also: ACID RAIN

ABSTRACT SERVICES AND INDEXES

ACID RAIN ABSTRACTS. EIC/Intelligence, Incorporated, 48 West 38th Street, New York, NY 10018. (800) 223-6275 or (212) 944-8500. Bimonthly. 1985 to present. $345.00 per year.

AIR POLLUTION ABSTRACTS. Air Pollution Technical Information Center. Available from U.S. Government Printing Office, Washington, DC 20402. Monthly. Inquire as to cost and availability.

AIR POLLUTION TITLES. Pennsylvania State University, Center for Environmental Studies, 226 Fenske Laboratory, University Park, PA 16802. (814) 865-1415. 1965 to present. Bimonthly. $95.00 per year.

AIR QUALITY CONTROL DIGEST. University Digest Services, Box 343, Troy, MI 48099. (313) 651-2528. 1969 to present. Bimonthly. $87.00 per year.

APPLIED SCIENCE AND TECHNOLOGY INDEX. H.W. Wilson Company, 950 University Avenue, Bronx, NY 10452. (800) 367-6670 or (212) 588-8400. Inquire as to cost and availability.

CHEMICAL ABSTRACTS. Chemical Abstracts Service, 2540 Olentangy Road, P.O. Box 3012, Columbus, OH 43210. (800) 848-6538 or (614) 421-3600. Weekly. $9200.00 per year.

ENVIRONMENTAL PERIODICALS BIBLIOGRAPHY. Environmental Studies Institute, 2074 Alameda Padre Serra, Santa Barbara, CA 93103. (805) 965-5010. 1972 to present. Bimonthly. Inquire as to cost and availability.

METEOROLOGICAL AND GEOASTROPHYSICAL ABSTRACTS. American Meteorological Society, 45 Beacon Street, Boston, MA 02108. (617) 227-2425.

POLLUTION ABSTRACTS. Cambridge Scientific Abstracts, 5161 River Road, Bethesda, MD 20816. (800) 638-8076 or (301) 951-1400. 1970 to present. Bimonthly. $464.00 per year.

ANNUAL REVIEWS AND YEARBOOKS

AIR POLLUTION. Arthur C. Stern, editor. Academic Press, Incorporated, 6277 Sea Harbor Drive, Orlando, FL 32821. (800) 321-5068. Price varies.

ASSOCIATIONS AND PROFESSIONAL SOCIETIES

AIR POLLUTION CONTROL ASSOCIATION. Box 2861, Pittsburgh, PA 15230. (412) 232-3444.

NATIONAL CLEAN AIR COALITION. 530 Seventh Street, SE, Washington, DC. (202) 543-8200.

BIBLIOGRAPHIES

ENVIRONMENTAL POLLUTION. National Technical Information Service, 5285 Port Royal Road, Springfield,

VA 22161. (703) 487-4929. Produced weekly from the NTIS databases. $109.00 per year.

SCIENTIFIC AND TECHNICAL BOOKS IN PRINT; AN INDEX TO LITERATURE IN SCIENCE AND TECHNOLOGY. R.R. Bowker Co., 205 E. 42nd Street, New York, NY 10017. (800) 521-8110 or (212) 916-1600.

DIRECTORIES AND BIOGRAPHICAL SOURCES

GOVERNMENT RESEARCH DIRECTORY. Gale Research Company, Book Tower, Detroit, MI 48226. (800) 521-0707. Third edition. 1985.

JOURNAL OF AIR POLLUTION CONTROL ASSOCIATION. Directory of Governmental Air Pollution Control Agencies Issue. Air Pollution Control Association, Box 2861, Pittsburgh, PA 15230. (412) 232-3444. Annual. $12.50.

RESEARCH CENTERS DIRECTORY. Gale Research Company, Detroit, MI 48226. (800) 521-0707. Eleventh edition. 1987.

SCIENTIFIC AND TECHNICAL ORGANIZATIONS AND AGENCIES DIRECTORY. Gale Research Company, Book Tower, Detroit, MI 48226. (800) 521-0707. 1985.

WHO'S WHO IN TECHNOLOGY TODAY. Reston Publishing Company, Incorporated, c/o Prentice-Hall, Incorporated, Englewood Cliffs, NJ 07632. (800) 262-6868. Biennial. Five volumes. $425.00. Covers the fields of electronics, computer science, physics, optics, chemistry, biotechnology, mechanics, energy, and earth science.

WORLD GUIDE TO SCIENTIFIC ASSOCIATIONS AND LEARNED SOCIETIES. K.G. Saur, Incorporated, 175 Fifth Avenue, New York, NY 10010. (800) 521-0707 or (212) 982-1302. Fourth edition. 1984. $112.00.

ENCYCLOPEDIAS AND DICTIONARIES

ENCYCLOPEDIA OF PHYSICAL SCIENCE AND TECHNOLOGY. Academic Press, Incorporated, Orlando, FL 32887. (800) 321-5068 or (305) 345-2734. Fifteen volumes, 1986.

MCGRAW-HILL ENCYCLOPEDIA OF SCIENCE AND TECHNOLOGY. McGraw-Hill Book, Incorporated, 1221 Avenue of the Americas, New York, NY 10020. (212) 997-3675.

GENERAL WORKS

AEROSOLS AND ATMOSPHERIC CHEMISTRY. G.H. Hidy, editor. Academic Press, Incorporated, 6277 Sea Harbor Drive, Orlando, FL 32821. (800) 321-5068. 1972. $56.00.

FUNDAMENTALS OF AIR POLLUTION. Arthur C. Stern. Academic Press, Incorporated, 6277 Sea Harbor Drive, Orlando, FL 32821. (800) 321-5068. Second edition. 1984. $39.50.
INDOOR AIR QUALITY. Philip J. Walsh, editor. CRC Press, Incorporated, 2000 Corporate Boulevard, NW, Boca Raton, FL 33431. 1983. $68.00.

HANDBOOKS AND MANUALS

CRC HANDBOOK OF CHEMISTRY AND PHYSICS. CRC Press, Incorporated, 2000 Corporate Boulevard, NW, Boca Raton, FL 33431. Sixty-seventh edition. 1986. $69.95.

HANDBOOK OF AIR POLLUTION ANALYSIS. Roy M. Harrison and Roger Perry, editors. Methuen, Incorporated, 29 West 35th Street, New York, NY 10001. (212) 244-3336. 1986. $79.95.

ONLINE DATA BASES

CA SEARCH. Chemical Abstracts Service, P.O. Box 3012, Columbus, OH 43210. Guide to chemical literature, 1972 to present. Inquire as to cost and availability.

ENVIROLINE. EIC/Intelligence, Incorporated, 48 West 38th Street, New York, NY 10018. (800) 223-6275 or (212) 944-8500. Indexes and abstracts both English and non-English language publications relating to the environment, 1971 to date. Inquire as to cost and availability.

NTIS. National Technical Information Service, 5285 Port Royal Road, Springfield, VA 22161. (703) 487-4630. Broad coverage of government sponsored research reports, 1964 to present. Inquire as to cost and availability.

POLLUTION ABSTRACTS (ONLINE). Cambridge Scientific Abstracts, 5161 River Road, Bethesda, MD 20816. (800) 638-8076 or (301) 951-1400. Provides indexing and abstracting of international environmental literature, 1970 to date. Inquire as to cost and availability.

SCISEARCH. Institute for Scientific Information, 3501 Market Street, Philadelphia, PA 19104. (800) 523-1850 or (215) 386-0100. Broad multidisciplinary title and author index to the international literature of science and technology, 1974 to present. Inquire as to cost and availability.

OTHER SOURCES

AIR POLLUTION CONTROL. Bureau of National Affairs, Incorporated, 1231 25th Street, NW, Washington, DC 20037. (202) 452-4200. Biweekly. $356.00 per year.

PERIODICALS

AEROSOL SCIENCE AND TECHNOLOGY. Elsevier Science Publishing Company, Incorporated, 52 Vanderbilt Avenue, New York, NY 10017. (212) 370-5520. Quarterly. $137.00 per year.

EPA JOURNAL. U.S. Environmental Protection Agency, 401 M Street, SW, Washington, DC 20460. (202) 755-0736. Ten times per year. $12.00 per year.

FILTRATION NEWS. Filtration News, 31505 Grand River, Door 1, Farmington, MI 48024. (313) 474-0002. Bimonthly. Free.

JOURNAL OF AIR POLLUTION CONTROL ASSOCIATION. Air Pollution Control Association, Box 2861, Pittsburgh, PA 15230. (412) 232-3444. Monthly. $68.00 per year.

POLLUTION ENGINEERING. Pudvan Publishing Company, 1935 Shermer Road, Northbrook, IL 60092. (312) 498-9840. Monthly.

RESEARCH CENTERS AND INSTITUTES

ARGONNE NATIONAL LABORATORY. 9700 South Case Avenue, Argonne, IL 60439. (312) 972-2000.

CALIFORNIA AIR RESOURCES BOARD, RESEARCH DIVISION. P.O. Box 2815, Sacramento, CA 95812. (916) 445-0753.

ENVIRONMENTAL PROTECTION AGENCY. Public Information Center, 401 M Street, Washington, DC 20460.

UNIVERSITY OF MICHIGAN. Air Pollution Modeling and Monitoring Laboratory, 2213 Space Research Building, Ann Arbor, MI 48109.

AIR TRAFFIC CONTROL

See also: AERONAUTICAL ENGINEERING, AERONAUTICS, AEROSPACE ENGINEERING, AIRPORTS

ABSTRACT SERVICES AND INDEXES

APPLIED MECHANICS REVIEWS. American Society or Mechanical Engineers, 345 East 47th Street, New York, NY 10017. (212) 705-7703. 1948 to present. Monthly. $360.00 per year.

APPLIED SCIENCE AND TECHNOLOGY INDEX. H.W. Wilson Company, 950 University Avenue, Bronx, NY 10452. (800) 367-6670 or (212) 588-8400. 1958 to present. Monthly. Inquire as to cost and availability

CURRENT LITERATURE IN TRAFFIC AND TRANSPORTATION. Northwestern University Transportation Library, Deering 303, Evanston, IL 60201. (312) 492-7287. Monthly. $15.00 per year.

ENGINEERING INDEX MONTHLY AND AUTHOR INDEX. Engineering Information, Incorporated, 345 East 47th Street, New York, NY 10017. (800) 221-1044 or (212) 705-7600. Monthly, with annual cumulation. $1560.00 per year.

INTERNATIONAL AEROSPACE ABSTRACTS. AIAA/TIS, 1633 Broadway, New York, NY 10019. (212) 581-4300. Semimonthly. $700.00 per year.

STAR. (Scientific and Technical Aerospace Reports). U.S. National Aeronautics and Space Administration, Scientific and Technical Information Facility, Box 8757, Baltimore-Washington International Airport, MD 21240. (202) 755-2210. Semimonthly, with semiannual and annual indexes. $85.00 per year.

WEEKLY ABSTRACT NEWSLETTER: Transportation. National Technical Information Service, 5285 Port Royal Road, Springfield, VA 22161. (703) 487-4600. Weekly. $80.00 per year.

ASSOCIATIONS AND PROFESSIONAL SOCIETIES

AIR TRANSPORT ASSOCIATION OF AMERICA. 1709 New York Avenue, NW, Washington, DC 20006. (202) 626-4000.

AIRCRAFT OWNERS AND PILOTS ASSOCIATIONS, 421 Aviation Way, Fredrick, MD 21701. (301) 695-2000.

AIRPORT OPERATORS COUNCIL INTERNATIONAL. 1220 19th Street, NW, Suite 800, Washington, DC 20036. (202) 293-8500.

AIRPORT SECURITY COUNCIL, 570 Elmont Road, Elmont, NY 11003. (516) 328-2990.

AMERICAN ASSOCIATION OF AIRPORT EXECUTIVES. 4224 King Street, Alexandria, VA 22302. (703) 824-0500.

AMERICAN HELICOPTER SOCIETY, Incorporated 217 North Washington Street, Alexandria, VA 22314. (703) 684-6777.

AMERICAN INSTITUTE OF AERONAUTICS AND ASTRONAUTICS. 1633 Broadway, New York, NY 10019. (212) 581-4300.

FLINT SAFETY FOUNDATION, Incorporated 5510 Columbia Pike, Arlington, VA 22204-3194. (703) 820-2777.

SOCIETY OF AUTOMOTIVE ENGINEERS. SAE, Incorporated, 400 Commonwealth Drive, Warrendale, PA 15096. (412) 776-4841.

BIBLIOGRAPHIES

AERONAUTICAL ENGINEERING: A SPECIAL BIBLIOGRAPHY WITH INDEXES. U.S. National Aeronautics and Space Administration, Washington, DC 20546. (202) 755-2320. Available from National Technical Information Service, Springfield, VA 22161. 1970 to present. Monthly. Inquire as to cost and availability.

SCIENTIFIC AND TECHNICAL BOOKS AND SERIALS IN PRINT, 1988; AN INDEX TO LITERATURE IN SCIENCE AND TECHNOLOGY. R.R. Bowker Company, 205 East 42nd Street, New York, NY 10017. (800) 521-8110 or (212) 916-1600. $175.00.

DIRECTORIES AND BIOGRAPHICAL SOURCES

AAS DIRECTORY. American Astronautical Society, 6212-B Old Keene Mill Court, Springfield, VA 22152. (703) 866-0020. Annual. $35.00 per year.

AIRPORT FACILITY DIRECTORY. National Ocean Survey, National Oceanic and Atmospheric Administration, U.S. Department of Commerce, Riverdale, MD 20737. (301) 436-6990. Seven regional volumes published every eight weeks. Complete set, $75.00 per year. Single region, $18.00 per year. Contains non-military airport information for pilots.

AOPA'S AIRPORTS USA. Aircraft Owners and Pilots Association, 421 Aviation Way, Fredrick, MD 21701. (301) 695-2000. Annual. $24.95.

AVIATION DIRECTORY. Aviation Directory, Incorporated, Downtown Airport Terminal, St. Paul, MN 55107. (612) 224-7715. Annual. $19.95.

RESEARCH CENTERS DIRECTORY. Gale Research Company, Detroit, MI 48226. (800) 521-0707. Eleventh edition. 1987.

WHO'S WHO IN ENGINEERING. Engineers Joint Council, 345 East 47th Street, New York, NY 10017. (212) 705-7010. 1985. $200.00.

WORLD AVIATION DIRECTORY. Murdoch Magazines, 1156 15th Street, NW, Washington, DC 20005. (202) 822-4600. Semiannual. $75.00 per year.

ENCYCLOPEDIAS AND DICTIONARIES

DICTIONARY OF AEROSPACE ENGINEERING. M.G. Kotik. Elsevier Science Publishing Company, Incorporated, 52 Vanderbilt Avenue, New York, NY 10017. (212) 370-5520. 1986. $170.00.

JANE'S AEROSPACE DICTIONARY. Jane's Publishing Incorporated, 135 West 50th Street, New York, NY 10020. (212) 586-7745. Second edition. 1986. $39.95.

GENERAL WORKS

AIR TRAFFIC CONTROLLER. James E. Turner. Arco Publishing, Incorporated, 215 Park Avenue, South, New York, NY 10003. (212) 777-6300. 1986. $14.95 in paper.

AIRCRAFT BASIC SCIENCE. R.D. Bent and J.L. McKinley. McGraw-Hill Book Company, 1221 Avenue of the Americas, New York, NY 10020. (212) 512-2000. 1980. $32.95.

AIRPORT ENGINEERING. Norman Ashford and Paul H. Wright. John Wiley and Sons, Incorporated, 605 Third Avenue, New York, NY 10158. (800) 526-5368 or (212) 850-6000. Second edition. 1984. $38.95.

AIRPORT TERMINAL DESIGN. Walter Hart. John Wiley and Sons, Incorporated, 605 Third Avenue, New York, NY 10158. (800) 526-5368 or (212) 850-6000. 1985. $49.95.

INTERNATIONAL AIR TRAFFIC CONTROL: MANAGEMENT OF THE WORLD'S AIR SPACE. Arnold F. Obe. Pergamon Press, Incorporated, Maxwell House, Fairview Park, Elmsford, NY 10523. (914) 592-7700. Second edition. 1985. $25.00.

PLANNING AND DESIGN OF AIRPORTS. R. Horonjeff and F.X. McKelvey. McGraw-Hill Book Company, 1221 Avenue of the Americas, New York, NY 10020. (212) 512-2000. Third edition. 1983. $48.95.

ROCK THE TOWER. Uriel Flax. Vantage Press, Incorporated, 516 West 34th Street, New York, NY 10001. (212) 736-1767. 1985. $10.95.

ONLINE DATA BASES

COMPENDEX. Engineering Information, Incorporated, 345 East 47th Street, New York, NY 10017. (800) 221-1044 or (212) 705-7615. Engineering and technical literature, 1975 to present. Inquire as to cost and availability.

NASA. National Aeronautics and Space Administration, Scientific and Technical Information Branch, 300 7th Street, SW, Washington, DC 20546. Citations and abstracts of aerospace literature, 1962 to present. Inquire as to cost and availability.

NTIS. National Technical Information Service, 5285 Port Royal Road, Springfield, VA 22161. (703) 487-4630. Broad coverage of government sponsored research reports, 1964 to present. Inquire as to cost and availability.

TRIS (TRANSPORTATION RESEARCH INFORMATION SERVICE). U.S. Department of Transportation, Washington, DC 20590. (202) 426-0975. Citations and abstracts of transportation literature, 1968 to present. Inquire as to online cost and availability.

PERIODICALS

AEROSPACE AMERICA. American Institute of Aeronautics and Astronautics, 1633 Broadway, New York, NY 10019. (212) 581-4300. Monthly. $51.00 per year.

AEROSPACE ENGINEERING MAGAZINE (SOCIETY OF AUTOMOTIVE ENGINEERS). SAE, Incorporated, 400 Commonwealth Drive, Warrendale, PA 15096. (412) 776-4841. Monthly. $30.00 per year.

AIAA JOURNAL. AIAA/TIS, 1633 Broadway, New York, NY 10019. (212) 581-4300. Monthly. $205.00 per year.

AIR LINE PILOT. Air Line Pilots Association, 535 Herndon Parkway, Herndon, VA 22070. (703) 689-4176. Monthly. $15.00 per year.

AIRPORT JOURNAL. Airport Journal, Incorporated, P.O. Box 273, Clarendon Hills, IL 60514. (312) 986-8132. Monthly. $15.00 per year.

AIRPORT SAFETY BULLETIN. Flight Safety Foundation, Incorporated, 5510 Columbia Pike, Arlington, VA 22204-3194. (703) 820-2777. Bimonthly. $30.00 per year.

AMERICAN AIRPORT MANAGEMENT. Camrus Publishers, Incorporated, Woodward Building, 733 15th Street, NW, Suite 1036, Washington, DC 20005. (202) 462-1997. Monthly. Inquire as to cost and availability.

AMERICAN HELICOPTER SOCIETY JOURNAL. American Helicopter Society, Incorporated, 217 North Washington Street, Alexandria, VA 22314. (703) 684-6777. Quarterly. $25.00 per year.

AVIATION MECHANICS BULLETIN. Flight Safety Foundation, Incorporated, 5510 Columbia Pike, Arlington, VA 22204-3194. (703) 820-2777. Bimonthly. $6.50 per year.

AVIATION WEEK AND SPACE TECHNOLOGY. McGraw-Hill Book Company, Incorporated, 1221 Avenue of the Americas, New York, NY 10020. (212) 512-2000. Weekly. $55.00 per year.

CANADIAN AERONAUTICS AND SPACE JOURNAL. Canadian Aeronautics and Space Institute, 222 Somerset Street, West, Suite 601, Ottawa, ON K2P OJ1, Canada. (613) 234-0191. Quarterly. $35.00 per year.

FLYING. CBS Magazines, 1515 Broadway, New York, NY 10036. (212) 503-4200. Monthly. $14.00 per year.

IEEE TRANSACTIONS ON AEROSPACE. Institute of Electrical and Electronics Engineers, 345 East 47th Street, New York, NY 10017. Bimonthly. $108.00 per year.

JOURNAL OF AIRCRAFT. American Institute of Aeronautics and Astronautics, AIAA/TIS, 1633 Broadway, New York, NY 10019. (212) 581-4300. Monthly. $185.00 per year.

VERTIFLITE. American Helicopter Society, Incorporated, 217 North Washington Street, Alexandria, VA 22314. (703) 684-6777. Bimonthly. $25.00 per year.

RESEARCH CENTERS AND INSTITUTES

ALASKA DEPARTMENT OF TRANSPORTATION AND PUBLIC FACILITIES. Statewide Research Section, 2301 Peger Roas, Fairbanks, AK 99701. (907) 479-2241.

ATLANTA UNIVERSITY. Dolphus E. Milligan Science Research Institute, 440 Westview Drive, SW, Atlanta, GA 30310. (404) 523-5148.

AUBURN UNIVERSITY. Aviation Management Program, Aerospace Engineering, Auburn University, AL 36849. (205) 826-4874.

TEXAS A & M UNIVERSITY. Texas Transportation Institute, Economics and Planning Division, College Station, Tx 77843. (409) 845-5814.

AIRBORNE RADAR

See: RADAR

AIRCRAFT

See also: AERODYNAMICS, AERONAUTICAL ENGINEERING, AERONAUTICS

ABSTRACT SERVICES AND INDEXES

ALLOYS INDEX. American Society for Metals, 9639 Kinsman Road, Metals Park, OH 44073. (216) 338-5151. 1974 to present. Monthly. $225.00 per year.

APPLIED MECHANICS REVIEWS. American Society of Mechanical Engineers, 345 East 47th Street, New York, NY 10017. (212) 705-7703. 1948 to present. Monthly. $360.00 per year.

APPLIED SCIENCE AND TECHNOLOGY INDEX. H.W. Wilson Company, 950 University Avenue, Bronx, NY 10452. (800) 367-6670 or (212) 588-8400. 1985 to present. Monthly. Inquire as to cost and availability.

CURRENT CONTENTS: ENGINEERING, TECHNOLOGY AND APPLIED SCIENCES. Institute for Scientific Information, 3501 Market Street, Philadelphia, PA 19104. (800) 523-1850 or (215) 386-0100. 1970 to present. Weekly. $272.00 per year.

ENGINEERING INDEX MONTHLY AND AUTHOR INDEX. Engineering Information, Incorporated 345 East 47th Street, New York, NY 10017. (800) 221-1044 or (212) 705-7600. Monthly, with annual cumulation. $1560.00 per year.

INTERNATIONAL AEROSPACE ABSTRACTS. AIAA/TIS, 1633 Broadway, New York, NY 10019. (212) 581-4300. Semimonthly. $700.00 per year.

METALS ABSTRACTS INDEX. American Society for Metals, 9639 Kinsman Road, Metals Park, OH 44073. (216) 338-5151. Monthly. (Sold only to subscribers of Metals Abstracts).

SCIENCE ABSTRACTS. Section A: Physics; Section B: Electrical and Electronics Abstracts; Section C: Computer and Control Abstracts. Institute of Electrical Engineers, London, United Kingdom. Available from: Institute of Electrical and Electronic Engineers (IEEE), 445 Hoes Lane, Piscataway, NJ 08854. Inquire as to cost and availability.

SCIENCE CITATION INDEX. Institute for Scientific Information, 3501 Market Street, Philadelphia, PA 19104. (800) 523-1850 or (215) 386-0100.

STAR. (Scientific and Technical Aerospace Reports). U.S. National Aeronautics and Space Administration, Scientific and Technical Information Facility, Box 8757, Baltimore-Washington International Airport, MD 21240. (202) 755-2210. Semimonthly, with semiannual and annual indexes. $85.00 per year.

WORLD ALUMINUM ABSTRACTS. Aluminum Association, Incorporated, 818 Connecticut Avenue, NW, Washington, DC 20006. (202) 862-5156. 1968 to present. Monthly. $240.00.

ASSOCIATIONS AND PROFESSIONAL SOCIETIES

AIRCRAFT ELECTRONICS ASSOCIATION. Box 1981, Independence, MO 64055. (816) 373-6565.

AMERICAN HELICOPTER SOCIETY, Incorporated 217 North Washington Street, Alexandria, VA 22314. (703) 684-6777.

AMERICAN INSTITUTE OF AERONAUTICS AND ASTRONAUTICS. 1633 Broadway, New York, NY 10019. (212) 581-4300.

FLIGHT SAFETY FOUNDATION, Incorporated 5510 Columbia Pike, Arlington, VA 22204-3194. (703) 820-2777.

SOCIETY OF AUTOMOTIVE ENGINEERS. SAE, Incorporated, 400 Commonwealth Drive, Warrendale, PA 15096. (412) 776-4841.

BIBLIOGRAPHIES

AERONAUTICAL ENGINEERING: A SPECIAL BIBLIOGRAPHY WITH INDEXES. U.S. National Aeronautics and Space Administration, Washington, DC 20546. (202) 755-2320. Available from National Technical Information Service, Springfield, VA 22161. 1970 to present. Monthly. Inquire as to cost and availability.

AIRCRAFT, ENGINES, AND AIRMEN: A SELECTIVE REVIEW OF THE PERIODICAL LITERATURE, 1930-1969. A. Hanniball. Scarecrow Press, Incorporated, 52 Liberty Street, Methuchen, NJ 08840. (201) 548-8600. 1972. $39.50.

NATIONAL AIR AND SPACE MUSEUM AND SMITHSONIAN INSTITUTION. Aerospace Periodical Index, 1973-1982. 1983. G.K. Hall and Company, 70 Lincoln Street, Boston, MA 02111. (800) 343-2806. $100.00.

SCIENTIFIC AND TECHNICAL BOOKS AND SERIALS IN PRINT, 1988; AN INDEX TO LITERATURE IN SCIENCE AND TECHNOLOGY. R.R. Bowker Co., 205 E. 42nd Street, New York, NY 10017. (800) 521-8110 or (212) 916-1600. $175.00.

DIRECTORIES AND BIOGRAPHICAL SOURCES

AAS DIRECTORY. American Astronautical Society, 6212-B Old Keene Mill Court, Springfield, VA 22152. (703) 866-0020. Annual. $35.00 per year.

RESEARCH CENTERS DIRECTORY. Gale Research Company, Detroit, MI 48226. (800) 521-0707. Eleventh edition. 1987.

WHO'S WHO IN ENGINEERING. Engineers Joint Council, 345 East 47th Street, New York, NY 10017. (212) 705-7010. 1985. $200.00.

WORLD AVIATION DIRECTORY. Murdoch Magazines, 1156 15th Street, NW, Washington, DC 20005. (202) 822-4600. Semiannual. $75.00 per year.

ENCYCLOPEDIAS AND DICTIONARIES

DICTIONARY OF AEROSPACE ENGINEERING. M.G. Kotik. Elsevier Science Publishing Company, Incorporated, 52 Vanderbilt Avenue, New York, NY 10017. (212) 370-5520. 1986. $170.00.

ENCYCLOPEDIA OF PHYSICAL SCIENCE AND TECHNOLOGY. Academic Press, Incorporated, Orlando, FL 32887. (800) 321-5068 or (305) 345-2734. Fifteen volumes, 1986.

JANE'S AEROSPACE DICTIONARY. Jane's Publishing Incorporated, 135 West 50th Street, New York, NY 10020. (212) 586-7745. Second edition. 1986. $39.95.

MCGRAW-HILL DICTIONARY OF ENGINEERING. Sybil P. Parker, editor. McGraw-Hill Book Company, 1221 Avenue of the Americas, New York, NY 10020. (212) 512-2000. 1984. $39.95.

MCGRAW-HILL ENCYCLOPEDIA OF SCIENCE AND TECHNOLOGY. McGraw-Hill Book Company, 1221 Avenue of the Americas, New York, NY 10020. (212) 512-2000.

GENERAL WORKS

AIRCRAFT BASIC SCIENCE. R.D. Bent and J.L. McKinley. McGraw-Hill Book Company, 1221 Avenue of the Americas, New York, NY 10020. (212) 512-2000. 1980. $32.95.

AIRCRAFT STRUCTURES. D.J. Peery and J.J. Azar. McGraw-Hill Book Company, 1221 Avenue of the Americas, New York, NY 10020. (212) 512-2000. 1982. $44.00.

FOUNDATIONS OF AERODYNAMICS: BASES OF AERODYNAMIC DESIGN. A.M. Kuethe and C. Chow. John Wiley and Sons, Incorporated, 605 Third Avenue, New York, NY 10158. (800) 526-5368 or (212) 850-6000. Fourth edition. 1986. $43.95.

ILLUSTRATED GUIDE TO AERODYNAMICS. Hubert Smith. Tab Books, Incorporated, Monterey Lane, Blue Ridge Summit, PA 17214. (717) 794-2191. 1986. $14.95.

INTRODUCTION AEROSPACE STRUCTURAL ANALYSIS. David H. Allen and Walter E. Haisler. John Wiley and Sons, Incorporated, 605 Third Avenue, New York, NY 10158. (800) 526-5368 or (212) 850-6000. 1985. $42.95.

INTRODUCTION TO AIRCRAFT PERFORMANCE, SELECTION AND DESIGN. F.J. Hale. John Wiley and Sons, Incorporated, 605 Third Avenue, New York, NY 10158. (800) 526-5368 or (212) 850-6000. 1984. $39.95.

INTRODUCTION TO THEORETICAL AND COMPUTATIONAL AERODYNAMICS. Jack Moran. John Wiley and Sons, Incorporated, 605 Third Avenue, New York, NY 10158. (800) 526-5368 or (212) 850-6000. 1984. $40.95.

THEORY OF WING SECTIONS: INCLUDING A SUMMARY OF AIRFOIL DATA. Dover Publications, Incorporated, 31 East

Second Street, Mineola, NY 11501. (516) 294-7000. 1949. $10.95.

HANDBOOKS AND MANUALS

JANE'S ALL THE WORLD'S AIRCRAFT 1986-87. John W. Taylor, editor. Jane's Publishing Incorporated, 135 West 50th Street, New York, NY 10010. (212) 586-7745. 1986. $141.50.

STANDARD HANDBOOK OF ENGINEERING CALCULATIONS. Tyler G. Hicks, editor. McGraw-Hill Book Company, 1221 Avenue of the Americas, New York, NY 10020. (212) 512-2000. Second edition. 1984. $59.50.

THE WILEY ENGINEER'S DESK REFERENCE. Sanford I. Heisler. John Wilet and Sons, Incorporated, 605 Third Avenue, New York, NY 10158. (800) 526-5368 or (212) 850-6418. $36.00.

ONLINE DATA BASES

COMPENDEX. Engineering Information, Incorporated, 345 East 47th Street, New York, NY 10017. (800) 221-1044 or (212) 705-7615. Engineering and technical literature, 1975 to present. Inquire as to cost and availability.

NASA. National Aeronautics and Space Administration, Scientific and Technical Information Branch, 300 7th Street, SW, Washington, DC 20546. Citations and abstracts of aerospace literature, 1962 to present. Inquire as to cost and availability.

NTIS. National Technical Information Service, 5285 Port Royal Road, Springfield, VA 22161. (703) 487-4630. Broad coverage of government sponsored research reports, 1964 to present. Inquire as to cost and availability.

PERIODICALS

AERONAUTICAL JOURNAL. Aeronautical Society, 4 Hamilton Place, London W1V 0BQ, England. Ten times per year. $175.00 per year.

AEROSPACE AMERICA. American Institute of Aeronautics and Astronautics, 1633 Broadway, New York, New York 10019. (212) 581-4300. Monthly. $51.00 per year.

AEROSPACE ENGINEERING MAGAZINE (SOCIETY OF AUTOMOTIVE ENGINEERS). SAE, Incorporated, 400 Commonwealth Drive, Warrendale, PA 15096. (412) 776-4841. Monthly. $30.00 per year.

AIAA JOURNAL. AIAA/TIS, 1633 Broadway, New York, NY 10019. (212) 581-4300. Monthly. $205.00 per year.

AIRCRAFT ENGINEERING. Bunhill Publications Limited, 127 Stanstead Road, Forest Hill, London, SE23 JE1, England. 1929 to present. Monthly. $51.00 per year.

AMERICAN HELICOPTER SOCIETY JOURNAL. American Helicopter Society, Incorporated, 217 North Washington Street, Alexandria, VA 22314. (703) 684-6777. Quarterly. $25.00 per year.

AVIATION MECHANICS BULLETIN. Flight Safety Foundation, Incorporated, 5510 Columbia Pike, Arlington, VA 22204-3194. (703) 820-2777. Bimonthly. $6.50 per year.

AVIATION WEEK AND SPACE TECHNOLOGY. McGraw-Hill Book Company, Incorporated, 1221 Avenue of the Americas, New York, New York 10020. (212) 512-2000. Weekly. $55.00 per year.

AVIONICS. Atlantic Communications, Incorporated, Box 5100, Westport, CT 06881. (203) 227-2280. Monthly. $36.00 per year.

AVIONICS NEWS MAGAZINE. Aircraft Electronics Association, Box 1981, Independence, MO 64055. (816) 373-6565. Monthly. Free.

CANADIAN AERONAUTICS AND SPACE JOURNAL. Canadian Aeronautics and Space Institute, 222 Somerset Street, West, Suite 601, Ottawa, ON K2P 0J1, Canada. (613) 234-0191. Quarterly. $35.00 per year.

FLYING. CBS Magazines, 1515 Broadway, New York, NY 10036. (212) 503-4200. Monthly. $14.00 per year.

IEEE TRANSACTIONS ON AEROSPACE. Institute of Electrical and Electronics Engineers, 345 East 47th Street, New York, NY 10017. Bimonthly. $108.00 per year.

INTERNATIONAL JOURNAL OF TURBO AND JET ENGINES. Martinus Nijhoff Publishers. Available from: Kluwer Academic Publishers Group, Distribution Centre, Postbus 322, 3300 AH Dordrecht, The Netherlands. Quarterly. $125.00 per year.

JOURNAL OF AIRCRAFT. American Institute of Aeronautics and Astronautics, AIAA/TIS, 1633 Broadway, New York, NY 10019. (212) 581-4300. Monthly. $185.00 per year.

JOURNAL OF GUIDANCE AND CONTROL. American Institute of Aeronautics and Astronautics, AIAA/TIS, 1633 Broadway, New York, NY 10019. (212) 581-4300. Bimonthly. $95.00 per year.

JOURNAL OF PROPULSION AND POWER. American Institute of Aeronautics and Astronautics, AIAA/TIS, 1633 Broadway, New York, NY 10019. (212) 581-4300. Bimonthly. $170.00 per year.

ROTOR AND WING INTERNATIONAL. PJS Publications, Incorporated, New Plaza, Box 1790, Peoria, IL 61656. (309) 682-6626. Monthly. $24.00 per year.

VERTICA: THE INTERNATIONAL JOURNAL OF ROTORCRAFT AND POWERLIFT AIRCRAFT. Pergamon Press, Incorporated, Journals Division, Maxwell House, Fairview Park, Elmsford, NY 10523. (914) 592-7700. Quarterly. $135.00 per year.

VERTIFLITE. American Helicopter Society, Incorporated, 217 North Washington Street, Alexandria, VA 22314. (703) 684-6777. Bimonthly. $25.00 per year.

RESEARCH CENTERS AND INSTITUTES

FLIGHT RESEARCH LABORATORY. University of Kansas, Raymond Nichols Hall, Lawrence, KS 66045. (913) 864-3043.

GEORGIA INSTITUTE OF TECHNOLOGY. Center for Rotary Wing Aircraft Technology, Atlanta, GA 30332. (404) 894-3005.

OHIO STATE UNIVERSITY. Aeronautical and Astronautical Research Laboratory, 2300 West Case Road, Columbus, OH 43220. (614) 422-1241.

PRINCETON UNIVERSITY. Flight Mechanics Laboratory, James Forrestal Campus, Princeton, NJ 08540. (609) 452-5149.

STANFORD UNIVERSITY. Aero Structures Laboratory, Department of Aeronautics and Astronomy, Stanford, CA 94305. (415) 497-3317.

UNIVERSITY OF TEXAS AT AUSTIN. Aeronautical Research Center, Aerospace Engineering and Engineering Mechanics, WRW 217, Austin, TX 78712. (512) 471-5962.

UNIVERSITY OF WASHINGTON. Aeronautical Laboratory, FS-10, Seattle, WA 98105. (206) 543-0439.

AIRFOILS

See: AERONAUTICAL ENGINEERING

AIRPLANES

See: AIRCRAFT

AIRPORTS

See also: AERODYNAMICS,
AERONAUTICAL ENGINEERING,
AERONAUTICS, AEROSPACE ENGINEERING

ABSTRACT SERVICES AND INDEXES

APPLIED MECHANICS REVIEWS. American Society of Mechanical Engineers, 345 East 47th Street, New York, NY 10017. (212) 705-7703. 1948 to present. Monthly. $360.00 per year.

APPLIED SCIENCE AND TECHNOLOGY INDEX. H.W. Wilson Company, 950 University Avenue, Bronx, NY 10452. (800) 367-6670 or (212) 588-8400. 1958 to present. Monthly. Inquire as to cost and availability.

CURRENT LITERATURE IN TRAFFIC AND TRANSPORTATION. Northwestern University Transportation Library, Deering 303, Evanston, IL 60201. (312) 492-7287. Monthly. $15.00 per year.

ENGINEERING INDEX MONTHLY AND AUTHOR INDEX. Engineering Information, Incorporated, 345 East 47th Street, New York, NY 10017. (800) 221-1044 or (212) 705-7600. Monthly, with annual cumulations. $1560.00 per year.

INTERNATIONAL AEROSPACE ABSTRACTS. AIAA/TIS, 1633 Broadway, New York, NY 10019. (212) 581-4300. Semimonthly. $700.00 per year.

STAR. (Scientific and Technical Aerospace Reports). U.S. National Aeronautics and Space Administration, Scientific and Technical Information Facility, Box 8757, Baltimore-Washington International Airport, MD 21240. (202) 755-2210. Semimonthly with semiannual and annual indexes. $85.00 per year.

WEEKLY ABSTRACT NEWSLETTER: TRANSPORTATION. National Technical Information Service, 5285 Port Royal Road, Springfield, VA 22161. (703) 487-4600. Weekly. $80.00 per year.

ASSOCIATIONS AND PROFESSIONAL SOCIETIES

AIR TRANSPORT ASSOCIATION OF AMERICA. 1709 New York Avenue, NW, Washington, DC 20006. (202) 626-4000.

AIRCRAFT OWNERS AND PILOTS ASSOCIATION. 421 Aviation Way, Fredrick, MD 21701. (301) 695-2000.

AIRPORT OPERATORS COUNCIL INTERNATIONAL, 1220 19th Street, NW, Suite 800, Washington, DC 20036. (202) 293-8500.

AIRPORT OPERATORS COUNCIL INTERNATIONAL, 1220 19th Street, NW, Suite 800, Washington, DC 20036. (202) 293-8500.

AIRPORT SECURITY COUNCIL, 570 Elmont Road, Elmont, NY 11003. (516) 328-2990.

AMERICAN ASSOCIATION OF AIRPORT EXECUTIVES. 4224 King Street, Alexandria, VA 22302. (703) 824-0500.

AMERICAN HELICOPTER SOCIETY, Incorporated 217 North Washington Street, Alexandria, Va 22314. (703) 684-6777.

AMERICAN INSTITUTE OF AERONAUTICS AND ASTRONAUTICS. 1633 Broadway, New York, NY 20019. (212) 581-4300.

FLIGHT SAFETY FOUNDATION, Incorporated 5510 Columbia Pike, Arlington, VA 22204-3194. (703) 820-2777.

SOCIETY OF AUTOMOTIVE ENGINEERS. SAE, Incorporated, 400 Commonwealth Drive, Warrendale, PA 15096. (412) 776-4841.

BIBLIOGRAPHIES

AERONAUTICAL ENGINEERING: A SPECIAL BIBLIOGRAPHY WITH INDEXES. U.S. National Aeronautics and Space Administration, Washington, DC 20546. (202) 755-2320. Available from National Technical Information Service, Springfield, Va 22161. 1970 to present. Monthly. Inquire as to cost and availability.

SCIENTIFIC AND TECHNICAL BOOKS AND SERIALS IN PRINT, 1988; AN INDEX TO LITERATURE IN SCIENCE AND TECHNOLOGY. R.R. Bowker Company, 205 East 42nd Street, New York, NY 10017. (800) 521-8110 or (212) 916-1600. $175.00.

DIRECTORIES AND BIOGRAPHICAL SOURCES

AAS DIRECTORY. American Astronautical Society, 6212-B Old Keene Mill Court, Springfield, VA 22152. (703) 866-0020. Annual. $35.00 per year.

AIRPORT FACILITY DIRECTORY. National Ocean Survey, National Oceanic and Atmospheric Administration, U.S. Department of Commerce, Riverdale, MD 20737. (301) 436-6990. Seven regional volumes published every eight weeks. Complete set, $75.00 per year. Single region, $18.00 per year. Contains non-military airport information for pilots.

AOPA'S AIRPORTS USA. Aircraft Owners and Pilots Association, 421 Aviation Way, Fredrick, MD 21701. (301) 695-2000. Annual. $24.95.

AVIATION DIRECTORY. Aviation Directory, Incorporated, Downtown Airport Terminal, St. Paul, MN 55107. (612) 224-7715. Annual. $19.95.

RESEARCH CENTERS DIRECTORY. Gale Research Company, Detroit, MI 48226. (800) 521-0707. Eleventh edition. 1987.

WHO'S WHO IN ENGINEERING. Engineers Joint Council, 345 East 47th Street, New York, NY 10017. (212) 705-7010. 1985. $200.00.

WORLD AVIATION DIRECTORY. Murdoch Magazines, 1156 15th Street, NW, Washington, DC 20005. (202) 822-4600. Semiannual. $75.00 per year.

ENCYCLOPEDIAS AND DICTIONARIES

DICTIONARY OF AEROSPACE ENGINEERING. M.G. Kotik. Elsevier Science Publishing Company, Incorporated, 52 Vanderbilt Avenue, New York, NY 10017. (212) 370-5520. 1986. $170.00.

JANE'S AEROSPACE DICTIONARY. Jane's Publishing Incorporated, 135 West 50th Street, New York, NY 10020. (212) 586-7745. Second edition. $1986. $39.95.

GENERAL WORKS

AIRCRAFT BASIC SCIENCE. R.D. Bent and J.L. McKinley. McGrawn-Hill Book Company, 1221 Avenue of the Americas, New York, NY 10020. (212) 512-2000. 1980. $32.95.

AIRPORT ENGINEERING. Norman Ashford and Paul H. Wright. John Wiley and Sons, Incorporated, 605 Third Avenue, New York, NY 10158. (800) 526-5368 or (212) 850-6000. Second edition. 1984. $38.95.

AIRPORT TERMINAL DESIGN. Walter Hart, John Wiley and Sons, Incorporated, 605 Third Avenue, New York, NY 10158. (800) 526-5368 or (212) 850-6000. 1985. $49.95.

FOUNDATIONS OF AERODYNAMICS: BASES OF AERODYNAMIC DESIGN. A.M. Kuethe and C. Chow. John Wiley and Sons, Incorporated, 605 Third Avenue, New York, NY 10158. (800) 526-5368 or (212) 850-6000. Fourth edition. 1986. $43.95.

ILLUSTRATED GUIDE TO AERODYNAMICS. Hubert Smith. Tab Books, Incorporated, Monterey Lane, Blue Ridge Summit, PA 17214. (717) 794-2191. 1986. $14.95.

INTRODUCTION TO AIRCRAFT PERFORMANCE, SELECTION AND DESIGN. F.J. Hale. John Wiley and Sons, Incorporated, 605 Third Avenue, New York, NY 10158. (800) 526-5368 or (212) 850-6000. 1984. $39.95.

INTRODUCTION TO THEORETICAL AND COMPUTATIONAL AERODYNAMICS. Jack Moran. John Wiley and Sons, Incorporated, 605 Third Avenue, New York, NY 10158. (800) 526-5368 or (212) 850-6000. 1984. $40.95.

PLANNING AND DESIGN OF AIRPORTS. R. Horonjeff and F.X. McKelvey. McGraw-Hill Book Company, 1221 Avenue of the Americas, New York, NY 10020. (212) 512-2000. Third edition. 1983. $48.95.

ONLINE DATA BASES

COMPENDEX. Engineering Information, Incorporated, 345 East 47th Street, New York, NY 10017. (800) 221-1044 or (212) 705-7615. Engineering and technical literature, 1975 to present. Inquire as to cost and availability.

NASA. National Aeronautics and Space Administration, Scientific and Technical Information Branch, 300 7th Street, SW, Washington, DC 20546. Citations and abstracts of aerospace literature, 1962 to present. Inquire as to cost and availability.

NTIS. National Technical Information Service, 5285 Port Royal Road, Springfield, VA 22161. (703) 487-4630. Broad coverage of government sponsored research reports, 1964 to present. Inquire as to cost and availability.

PERIODICALS

AERONATUICAL JOURNAL. Aeronautical Society, 4 Hamilton Place, London W1V OBQ, England. Ten times per year. $175.00 per year.

AEROSPACE AMERICA. American Institute of Aeronautics and Astronautics, 1633 Broadway, New York, NY 10019. (212) 581-4300. Monthly. $51.00 per year.

AEROSPACE ENGINEERING MAGAZINE (SOCIETY OF AUTOMOTIVE ENGINEERS). SAE, Incorporated, 400 Commonwealth Drive, Warrendale, PA 15096. (412) 776-4841. Monthly. $30.00 per year.

AIAA JOURNAL. AIAA/TIS, 1633 Broadway, New York, NY 20019. (212) 581-4300. Monthly. $205.00 per year.

AIRCRAFT ENGINEERING. Bunhill Publications Limited, 127 Stanstead Road, Forest Hill, London, SE23 JE1, England. 1929 to present. Monthly. $51.00 per year.

AIRPORT JOURNAL. Airport Journal, Incorporated, P.O. Box 273, Clarendon Hills, IL 60514. (312) 986-8132. Monthly. $15.00 per year.

AMERICAN AIRPORT MANAGEMENT. Camrus Publishers, Incorporated, Woodward Building, 733 15th Street, NW, Suite 1036, Washington, DC 20005. (202) 462-1997. Monthly. Inquire as to cost and availability.

AMERICAN HELICOPTER SOCIETY JOURNAL. American Helicopter Society, Incorporated, 217 North Washington Street, Alexandria, VA 22314. (703) 684-6777. Quarterly. $25.00 per year.

AVIATION MECHANICS BULLETIN. Flight Safety Foundation, Incorporated, 5510 Columbia Pike, Arlington, VA 22204-3194. (703) 820-2777. Bimonthly. $6.50 per year.

AVIATION WEEK AND SPACE TECHNOLOGY. McGraw-Hill Book Company, Incorporated, 1221 Avenue of the Americas, New York, NY 10020. (212) 512-2000. Weekly. $55.00 per year.

CANADIAN AERONAUTICS AND SPACE JOURNAL. Canadian Aeronautics and Space Institute, 222 Somerset Street, West, Suite 601, Ottawa, ON K2P OJ1, Canada. (613) 234-0191. Quarterly. $35.00 per year.

FLYING. CBS Magazines, 1515 Broadway, New York, NY 10036. (212) 503-4200. Monthly. $14.00 per year.

IEEE TRANSACTIONS ON AEROSPACE. Institute of Electrical and Electronics Engineers, 345 East 47th Street, New York, NY 10017. Bimonthly. $108.00 per year.

JOURNAL OF AIRCRAFT. American Institute of Aeronautics and Astronautics, AIAA/TIS, 1633 Broadway, New York, NY 10019. (212) 581-4300. Monthly. $185.00 per year.

JOURNAL OF GUIDANCE AND CONTROL. American Institute of Aeronautics and Astronautics, AIAA/TIS, 1633 Broadway, New York, NY 10019. (212) 581-4300. Bimonthly. $95.00 per year.

ROTOR AND WING INTERNATIONAL. PJS Publications, Incorporated, New Plaza, Box 1790, Peoria, IL 61656. (309) 682-6626. Monthly. $24.00 per year.

VERTICA: THE INTERNATIONAL JOURNAL OF ROTORCRAFT AND POWERLIFT AIRCRAFT. Pergamon Press, Incorporated, Journals Division, Maxwell House, Fairview Park, Elmsford, NY 10523. (014) 592-7700. Quarterly. $135.00 per year.

VERTIFLITE. American Helicopter Society, Incorporated, 217 North Washington Street, Alexandria, Va 22314. (703) 684-6777. Bimonthly. $25.00 per year.

RESEARCH CENTERS AND INSTITUTES

ALASKA DEPARTMENT OF TRANSPORTATION AND PUBLIC FACILITIES, STATEWIDE RESEARCH SECTION. 2301 Peger Roas, Fairbanks, AK 99701. (907) 479-2241.

ATLANTA UNIVERSITY. Dolphus E. Milligan Science Research Institute, 440 Westview Drive, SW, Atlanta, GA 30310. (404) 523-5148.

AUBURN UNIVERSITY. Aviation Management Program, Aerospace Engineering, Auburn University, AL 36849. (205) 826-4874.

TEXAS A & M UNIVERSITY. Texas Transportation Institute, Economics and Planning Division, College Station, TX 77843. (409) 845-5814.

AIRSHIPS

See also: AERODYNAMICS,
AERONAUTICAL ENGINEERING, AERONAUTICS,
AIRCRAFT

ABSTRACT SERVICES AND INDEXES

APPLIED SCIENCE AND TECHNOLOGY INDEX. H.W. Wilson Company, 950 University Avenue, Bronx, NY 10452. (800) 367-6670 or (212) 588-8400. 1958 to present. Monthly. Inquire as to cost and availability.

ENGINEERING INDEX MONTHLY AND AUTHOR INDEX. Engineering Information, Incorporated, 345 East 47th Street, New York, NY 10017. (800) 221-1044 or (212) 705-7600. Monthly, with annual cumulation. $1560.00 per year.

INTERNATIONAL AEROSPACE ABSTRACTS. AIAA/TIS, 1633 Broadway, New York, NY 10019. (212) 581-4300. Semimonthly. $700.00 per year.

STAR. (Scientific and Technical Aerospace Reports). U.S. National Aeronautics and Space Administration, Scientific and Technical Information Facility, Box 8757, Baltimore-Washington International Airport, MD 21240. (202) 755-2210. Semimonthly, with semiannual and annual index. $85.00 per year.

ASSOCIATIONS AND PROFESSIONAL SOCIETIES

AIRSHIP ASSOCIATION. P.O. Box 43184, Washington, DC 20010. (301) 986-9202.

AMERICAN INSTITUTE OF AERONAUTICS AND ASTRONAUTICS. 1633 Broadway, New York, NY 10019. (212) 581-4300.

ASSOCIATION OF BALLOON AND AIRSHIP CONSTRUCTORS. P.O. Box 7, Rosemead, CA 91770. (818) 918-0298.

BIBLIOGRAPHIES

AERONAUTICAL ENGINEERING: A SPECIAL BIBLIOGRAPHY WITH INDEXES. U.S. National Aeronautics and Space Administration, Washington, DC 20546. (202) 755-2320. Available from National Technical Information Service, Springfield, VA 22161. 1970 to present. Monthly. Inquire as to cost and availability.

AIRCRAFT, ENGINES, AND AIRMEN: A SELECTIVE REVIEW OF THE PERIODICAL LITERATURE, 1930-1969. A. Hanniball. Scarecrow Press, Incorporated, 52 Liberty Street, Methuchen, NJ 08840. (201) 548-8600. 1972. $39.50.

NATIONAL AIR AND SPACE MUSEUM AND SMITHSONIAN INSTITUTION. Aerospace Periodical Index, 1973-1982. 1983. G.K. Hall and Company, 70 Lincoln Street, Boston, MA 02111. (800) 343-2806. $100.00.

SCIENTIFIC AND TECHNICAL BOOKS AND SERIALS IN PRINT, 1988; AN INDEX TO LITERATURE IN SCIENCE AND TECHNOLOGY. R.R. Bowker Company, 205 East 42nd Street, New York, NY 10017. (800) 521-8110 or (212) 916-1600. $175.00.

DIRECTORIES AND BIOGRAPHICAL SOURCES

AAS DIRECTORY. American Astronautical Society, 6212-B Old Keene Mill Court, Springfield, VA 22152. (703) 866-0020. Annual. $35.00 per year.

RESEARCH CENTERS DIRECTORY. Gale Research Company, Detroit, MI 48226. (800) 521-0707. Eleventh edition. 1987.

WHO'S WHO IN ENGINEERING. Engineers Joint Council, 345 East 47th Street, New York, NY 10017. (212) 705-7010. 1985. $200.00.

WORLD AVIATION DIRECTORY. Murdoch Magazines, 1156 15th Street, NW, Washington, DC 20005. (202) 822-4600. Semiannual. $75.00 per year.

ENCYCLOPEDIAS AND DICTIONARIES

DICTIONARY OF AEROSPACE ENGINEERING. M.G. Kotik. Elsevier Science Publishing Company, Incorporated, 52 Vanderbilt Avenue, New York, NY 10017. (212) 370-5520. 1986. $170.00.

ENCYCLOPEDIA OF PHYSICAL SCIENCE AND TECHNOLOGY. Academic Press, Incorporated, Orlando, FL 32887. (800) 321-5068 or (305) 345-2734. Fifteen volumes, 1986.

JANE'S AEROSPACE DICTIONARY. Jane's Publishing Incorporated, 135 West 50th Street, New York, NY 10020. (212) 586-7745. Second edition. 1986. $39.95.

MCGRAW-HILL DICTIONARY OF ENGINEERING. Sybil P. Parker, editor. Mcgraw-Hill Book Company, 1221 Avenue of the Americas, New York, NY 10020. (212) 512-2000. 1984. $39.95.

MCGRAW-HILL ENCYCLOPEDIA OF SCIENCE AND TECHNOLOGY. McGraw-Hill Book, 1221 Avenue of the Americas, New York, NY 10020. (212) 512-2000.

GENERAL WORKS

THE COMPLETE BOOK OF AIRSHIPS: DIRIGIBLES, BLIMPS AND HOT AIR BALLOONS. Don Dwiggins. Tab Brooks, Incorporated, Monterey Lane, Blue Ridge Summit, PA 17214. (717) 794-2191. 1982. $16.95.

THE DIRIGIBLE AND THE FUTURE. Henry J. Ambers. Edelweiss Press, 124 Front Street, Massapequa Park, NY 11762. (516) 799-1150. 1981. $5.00.

THE GOLDEN AGE OF THE GREAT PASSENGER AIRSHIPS: GRAF ZEPPLIN AND HINDENBERG. Harold G. Dick and Douglas H. Robinson. Smithsonian Institution Press, 955 L'Enfant Plaza, Suite 2100, Washington, DC 20560. (202) 287-3388. 1985. $24.95.

INTRODUCTION AEROSPACE STRUCTURAL ANALYSIS. David H. Allen and Walter E. Haisler. John Wiley and Sons, Incorporated, 605 Third Avenue, New York, NY 10158. (800) 526-5368 or (212) 850-6000. 1985. $42.95.

ONLINE DATA BASES

COMPENDEX. Engineering Information, Incorporated, 345 East 47th Street, New York, NY 10017. (800) 221-1044 or (212) 705-7615. Engineering and technical literature, 1975 to present. Inquires as to cost and availability.

NASA. National Aeronautics and Space Administration, Scientific and Technical Information Branch, 300 7th Street, SW, Washington, DC 20546. Citations and abstracts of aerospace literature, 1962 to present. Inquire as to cost and availability.

NTIS. National Technical Information Service, 5285 Port Royal Road, Springfield, Va 22161. (703) 487-4630. Broad coverage of government sponsored research reports, 1964 to present. Inquire as to cost and availability.

PERIODICALS

AEROSPACE AMERICA. American Institute of Aeronautics and Astronautics, 1633 Broadway, New York, NY 10019. (212) 581-4300. Monthly. $51.00 per year.

AEROSPACE ENGINEERING MAGAZINE (SOCIETY OF AUTOMOTIVE ENGINEERS). SAE, Incorporated, 400 Commonwealth Drive, Warrendale, PA 15096. (412) 776-4841. Monthly. $30.00 per year.

AIAA JOURNAL. AIAA/TIS, 1633 Broadway, New York, NY 10019. (212) 581-4300. Monthly. $205.00 per year.

AIRCRAFT ENGINEERING. Bunhill Publications Limited, 127 Stanstead Road, Forest Hill, London, SE23 JE1, England. 1929 to present. Monthly. $51.00 per year.

JOURNAL OF AIRCRAFT. American Institute of Aeronautics and Astronautics, AIAA/TIS, 1633 Broadway, New York, NY 10019. (212) 581-4300. Monthly. $185.00 per year.

ALARM SYSTEMS

See also: ELECTRONIC SECURITY SYSTEMS

ABSTRACT SERVICES AND INDEXES

APPLIED SCIENCE AND TECHNOLOGY INDEX. H.W. Wilson Company, 950 University Avenue, Bronx, NY 10452. (800) 367-6670 or (212) 588-8400. Inquire as to cost and availability.

ASSOCIATIONS AND PROFESSIONAL SOCIETIES

AMERICAN SOCIETY FOR INDUSTRIAL SECURITY. 1655 Fort Myer Drive, Arlington, VA 22209. (703) 522-5800.

AUTOMATIC FIRE ALARM ASSOCIATION. 14 McIntire Street, Lowell, MA 01851. (617) 459-4461.

NATIONAL BURGLAR AND FIRE ALARM ASSOCIATION. 1120 19th Street, NW, Washington, DC 20036. (202) 296-9595.

SECURITY EQUIPMENT INDUSTRY ASSOCIATION. 2665 30th Street, Santa Monica, CA 90405. (213) 450-4141.

DIRECTORIES AND BIOGRAPHICAL SOURCES

AUTOMOTIVE, BURGLARY PROTECTION AND MECHANICAL EQUIPMENT DIRECTORY. Underwriters Laboratories, Incorporated, 333 Pfingsten Road, Northbrook, IL 60062. (312) 272-8800. Annual. $2.50. Lists manufacturers qualified to use UL label.

FIRE ALARM SYSTEMS DEALERS DIRECTORY. American Business Directories, Incorporated, 5639 South 86th Circle, Omaha, NE 68127. (402) 331-7293. Annual. $84.00.

FIRE AND SECURITY DIRECTORY. Penton/IPC, Incorporated, 1111 Chester Avenue, Cleveland, OH 44114. (216) 696-7000. Annual. $4.00.

SECURITY SYSTEMS ADMINISTRATION: GOLD BOOK DIRECTORY OF PRODUCTS AND SERVICES ISSUE. PTN Publishing Corporation, 101 Crossways Park West, Woodbury, NY 11797. (516) 496-8000. Annual. $2.00.

SECURITY WORLD: PRODUCT DIRECTORY ISSUE. Cahners Publishing Company, 1350 East Touhy Avenue, Des Plaines, IL 60018. (312) 635-8800. Annual. $25.00.

GENERAL WORKS

ALARM SYSTEMS AND THEFT PREVENTION. T. Weber. Butterworth's, 80 Montvale Avenue, Stoneham, MA 02180. (617) 438-8464. 1985. $24.95.

SECURITY ELECTRONICS. John Cunningham. Sams, 4300 West 62nd Street, Indianapolis, IN 46268. (800) 428-7267 or (317) 298-5564. 1983. $13.95.

HANDBOOKS AND MANUALS

CONTROL SYSTEMS ENGINEERING. W.J. Palm. John Wiley and Sons, Incorporated, 605 Third Avenue, New York, NY 10158. (800) 526-5368 or (212) 850-6000. 1986. $44.95.

ELECTRONIC SECURITY SYSTEMS. Butterworth's, 80 Montvale Avenue, Stoneham, MA 02180. (617) 438-8464. 1983. $29.95.

ONLINE DATA BASES

NTIS. National Technical Information Service, 5285 Port Royal Road, Springfield, VA 22161. (703) 487-4630. Broad coverage of government sponsored research reports, 1964 to present. Inquire as to cost and availability.

WILSONLINE. H.W. Wilson Company, 950 University Avenue, Bronx, NY 10452. (800) 367-6770 or (212) 588-8400. Makes available onlines versions of the printed H.W. Wilson indexes including Applied Science and Technology Index, Business Periodicals Index, and Readers Guide to Periodical Literature. Period covered is generally 1983 to present. Inquire as to cost and availability.

PERIODICALS

DATA PROCESSING AND COMMUNICATIONS SECURITY. Territorial Imperative, Incorporated, P.O. Box 5323, Madison, WI 53705. (608) 231-3817. Bimonthly. $24.00 per year.

SECURITY MANAGEMENT. American Society for Industrial Security, 1655 Fort Myer Drive, Arlington, VA 22209. (703) 522-5800. Monthly. $27.00 per year.

SECURITY SYSTEMS ADMINISTRATION. PTN Publishing Corporation, 101 Crossways Park West, Woodbury, NY 11797. (516) 496-8000. Monthly. $10.00 per year.

SECURITY WORLD. Cahners Publishing Company, 1350 East Touhy Avenue, Des Plaines, IL 60018. (312) 635-8800. Monthly. $35.00 per year.

ALCOHOL

See also: CHEMICAL ENGINEERING, CHEMISTRY, ORGANIC CHEMISTRY

ABSTRACT SERVICES AND INDEXES

CHEMICAL ABSTRACTS. American Chemical Society, Chemical Abstracts Service, Box 3012, Columbus, OH 43210. (614) 421-3600. 1907 to present. Weekly. $9500.00 per year.

CONFERENCE PAPERS INDEX. Cambridge Scientific Abstracts, 5161 River Road, Bethesda, MD 20816. 1972 to present. Monthly. Inquire as to cost and availability.

CURRENT CONTENTS: PHYSICAL, CHEMICAL AND EARTH SCIENCES. Institute for Scientific Information, 3501 Market Street, Philadelphia, PA 19104. (800) 523-1850 or (215) 386-0100. Weekly. $275.00 per year.

GENERAL SCIENCE INDEX. H.W. Wilson and Company, 950 University Avenue, Bronx, NY 10452. (800) 367-6670 or (212) 588-8400. 1978 to present. Monthly. Inquire as to cost and availability.

INDEX TO SCIENTIFIC AND TECHNICAL PROCEEDINGS. Institute for Scientific Information, 3501 Market Street, Philadelphia, PA 19104. (800) 523-1850 or (215) 386-0100. 1978 to present. Monthly. $775.00 per year.

INDEX TO SCIENTIFIC REVIEWS. Institute for Scientific Information, 3501 Market Street, Philadelphia, PA 19104. (800) 523-1850 or (215) 386-0100. 1974 to present. Semi-annual. $550.00 per year.

SCIENCE CITATION INDEX. Institute for Scientific Information, 3501 Market Street, Philadelphia, PA 19104. (800) 523-1850 or (215) 386-0100. Six times per year. $6200.00 per year.

ANNUAL REVIEWS AND YEARBOOKS

ADVANCES IN PHYSICAL ORGANIC CHEMISTRY. Academic Press, Inc., 6277 Sea Harbor Drive, Orlando, FL 32821. (800) 321-5068. 1963 to present. Irregular. Price varies, inquire.

STUDIES IN ORGANIC CHEMISTRY. Marcel Dekker Inc., 270 Madison Avenue, New York, NY 10016. (800) 228-1160. 1973 to present. Irregular. Price varies, inquire.

ASSOCIATIONS AND PROFESSIONAL SOCIETIES

AMERICAN CARBON SOCIETY. The Stackpole Corporation, St. Marys, PA 15857. (814) 781-8410.

AMERICAN CHEMICAL SOCIETY. 1155 16th Street, N.W., Washington, DC 20036. (202) 872-4600.

AMERICAN INSTITUTE OF CHEMICAL ENGINEERS. 345 East 47th Street, New York, NY 10017. (212) 705-7338.

AMERICAN OIL CHEMISTS' SOCIETY. 508 South Sixth Street, Champaign, IL 61820. (217) 359-2344.

ASSOCIATION OF OFFICIAL ANALYTICAL CHEMISTS. 1111 North 19th Street, Suite 210, Arlington, VA 22209. (703) 522-3032.

DIRECTORIES AND BIOGRAPHICAL SOURCES

AMERICAN MEN AND WOMEN OF SCIENCE. R.R. Bowker, Inc., Order Department, 245 West 17th Street, New York, NY 10011. (800) 521-8110. Eight volumes. 1986. $595.00 for set.

INTERNATIONAL RESEARCH CENTERS DIRECTORY 1988-89. Darren L. Smith, editor. Gale Research Company, Book Tower, Detroit, MI 48226. (800) 521-0707. 4th edition. 1987. $360.00.

1987 DIRECTORY OF ENGINEERING SOCIETIES AND RELATED ORGANIZATIONS. Gordon Davis, editor. Hemisphere Publishing Corporation, 1010 Vermont Avenue, NW, Washington, DC 20005. (800) 526-0275. 12th edition. 1987. $100.00.

RESEARCH CENTERS DIRECTORY 1988. Gale Research Company, Book Tower, Detroit, MI 48226. (800) 521-0707. 12th edition. 1987. $365.00 for set.

SCIENTIFIC AND TECHNICAL ORGANIZATIONS AND AGENCIES DIRECTORY. Margaret Labash Young, editor. Gale Research Company, Book Tower, Detroit, MI 48226. (800) 521-0707. 2nd edition. 1987. $185.00.

WHO'S WHO IN ENGINEERING. Gordon Davis, editor. Hemisphere Publishing Corporation, 1010 Vermont Avenue, NW, Washington, DC 20005. (800) 526-0275. 6th edition. 1985. $200.00.

ENCYCLOPEDIAS AND DICTIONARIES

CONCISE SCIENCE DICTIONARY. Oxford University Press, 200 Madison Avenue, New York, NY 10016. (800) 458-5833. 1987. $9.95 in paper.

DICTIONARY OF ORGANIC COMPOUNDS. John Buckingham, editor. Methuen, Inc., 29 West 35th Street, New York, NY 10001. (212) 244-3336. Fifth edition. 1988. $675.00.

ENCYCLOPEDIA OF CHEMISTRY. Douglas M. Considine, editor. Van Nostrand Reinhold Company, Inc., 135 West 50th Street, New York, NY 10020. (800) 543-2681. 4th edition. 1984. $98.95.

MCGRAW-HILL ENCYCLOPEDIA OF SCIENCE AND TECHNOLOGY. McGraw-Hill Book Company, 1221 Avenue of the Americas, New York, NY 10020. (212) 512-2000. 6th edition. 1987. $1600.00.

THESAURUS OF SCIENTIFIC, TECHNICAL, AND ENGINEERING TERMS. Hemisphere Publishing Corporation, 1010 Vermont Avenue, NW, Washington, DC 20005. (800) 526-0275. 1988. $125.00.

GENERAL WORKS

ALCOHOLS AS MOTOR FUELS. Society of Automotive Engineers (SAE), 400 Commonwealth Drive, Warrendale, PA 15096. (412) 776-4841. 1980. $47.00.

BASIC ORGANIC CHEMISTRY. F.L. Wiseman. McGraw-Hill Book Company, 1221 Avenue of the Americas, New York, NY 10020. (212) 512-2000. 1988. $37.95.

MONOHYDRIC ALCOHOLS: MANUFACTURE, APPLICATIONS, AND CHEMISTRY. Zakhari Samit and others. American Chemical Society, 1155 16th Street, N.W., Washington, DC 20036. (800) 424-6747. 1981. $33.95.

ORGANIC SYNTHESIS REACTIONS AND MECHANISMS. B. Christoph and others. Springer-Verlag New York, Inc., 175 Fifth Avenue, New York, NY 10010. (800) 526-7254. 1986. $60.00.

REDUCTIONS IN ORGANIC CHEMISTRY. M. Hudlicky. John Wiley and Sons, Inc., 605 Third Avenue, New York, NY 10158. (800) 526-5368. 1984. $45.00.

RODD'S CHEMISTRY OF CARBON COMPOUNDS. E.H. Rodd. Elsevier Science Publishing Company, Inc., 52 Vanderbilt Avenue, New York, NY 10017. (212) 370-5520. Three volumes in 20 parts. 1964-1979. Price varies for each part, inquire.

HANDBOOKS AND MANUALS

CRC HANDBOOK OF CHEMISTRY AND PHYSICS. Robert C. Weast, editor. CRC Press, 2000 Corporate Boulevard, Boca Raton, FL 33431. (800) 272-7737. 68th edition. 1987. $69.95.

CRC HANDBOOK OF DATA ON ORGANIC COMPOUNDS. R.C. Weast and M.J. Astle, editors. CRC Press, 2000 Corporate Boulevard, Boca Raton, FL 33431. (800) 272-7737. 1985. Two volumes. $270.00 for set.

A GUIDEBOOK TO MECHANISMS IN ORGANIC CHEMISTRY. P. Sykes. John Wiley and Sons, Inc., 605 Third Avenue, New York, NY 10158. (800) 526-5368. 6th edition. 1986. $21.95.

HANDBOOK OF APPLIED CHEMISTRY. Vollrath Hopp and Ingo Hennig. Hemisphere Publishing Corporation, 79 Madison Avenue, New York, NY 10016-7892. Order from: Taylor and Francis/Hemisphere Distribution Center, 242 Cherry Street, Philadelphia, PA 19106-1906. (800) 821-8312. 1983. $49.95.

HANDBOOK OF ORGANIC CHEMISTRY. John A. Dean. McGraw-Hill Book Company, 1221 Avenue of the Americas, New York, NY 10020. (212) 512-2000. 1987. $64.50.

LANGE'S HANDBOOK OF CHEMISTRY. John A. Dean, editor. McGraw-Hill Book Company, 1221 Avenue of the Americas, New York, NY 10020. (212) 512-2000. 1985. $59.50.

ONLINE DATA BASES

CA SEARCH. Chemical Abstracts Service, P.O. Box 3012, Columbus, OH 43120. (800) 848-6538 or (614) 421-3600. Comprehensive guide to chemical literature, 1972 to present. Inquire as to online cost and availability.

COMPENDEX. Engineering Information, Inc., 345 East 47th Street, New York, NY 10017. (800) 221-1044 or (212) 705-7615. Engineering and technical literature, 1975 to present. Inquire as to online cost and availability.

DISSERTATION ABSTRACTS ONLINE. University Microfilms International, 300 North Zeeb Road, Ann Arbor, MI 48106. (800) 521-0600 or (313) 761-4700. Scope includes virtually all doctoral dissertations accepted at accredited American institutions from 1861 to present in over 250 subject areas. Inquire as to online cost and availability.

NTIS. National Technical Information Service, 5285 Port Royal Road, Springfield, VA 22161. (703) 487-4630. Broad coverage of government sponsored research reports, 1964 to present. Inquire as to online cost and availability.

SCISEARCH. Institute for Scientific Information, 3501 Market Street, Philadelphia, PA 19104. (800) 523-1850 or (215) 386-0100. Broad multidisciplinary title and author index to the international literature of science and technology, 1974 to present. Inquire as to online cost and availability.

PERIODICALS

AMERICAN CHEMICAL SOCIETY JOURNAL. American Chemical Society, 1155 16th Street, N.W., Washington, DC 20036. (800) 424-6747. 1879. Semi-monthly. $56.00 per year.

AMERICAN OIL CHEMISTS' SOCIETY. JOURNAL. American Oil Chemists' Society, 508 South Sixth Street, Champaign, IL 61820. (217) 359-2344. 1917 to present. Monthly. $60.00 per year.

CARBON. Pergamon Press, Inc., Maxwell House, Fairview Park, Elmsford, NY 10523. (914) 592-7700. 1963 to present. Bimonthly. $235.00 per year.

HYDROCARBON PROCESSING. Gulf Publishing Company, P.O. Box 2608, Houston, TX 77001. (713) 520-4444. Monthly. $10.00 per year.

JOURNAL OF HETEROCYCLIC CHEMISTRY. Hetero-Corporation, Box 16000 MH, Tampa, FL 33687. 1964 to present. 6 times per year. $160.00 per year.

JOURNAL OF ORGANIC CHEMISTRY. American Chemical Society, 1155 16th Street, N.W., Washington, DC 20036. (800) 424-6747. 1936 to present. Semi-monthly. $218.00 per year.

JOURNAL OF POLYMER SCIENCE. POLYMER CHEMISTRY EDITION. John Wiley and Sons, Inc., 605 Third Avenue, New York, NY 10158. (800) 526-5368. 1962 to present. Monthly. $895.00 per year, includes all editions.

TETRAHEDRON. Pergamon Press, Inc., Maxwell House, Fairview Park, Elmsford, NY 10523. (914) 592-7700. 1957 to present. 24 times per year. $1400.00 per year.

TETRAHEDRON LETTERS. Pergamon Press, Inc., Maxwell House, Fairview Park, Elmsford, NY 10523. (914) 592-7700. 1959 to present. Weekly. $1500.00 per year.

RESEARCH CENTERS AND INSTITUTES

CHEMICAL LABORATORIES. Harvard University, Oxford Street, Cambridge, MA 02138. (617) 495-4283.

CHEMICAL RESEARCH LABORATORY. Brown University, Providence, RI 02912. (401) 863-2256.

UNIVERSITY/INDUSTRY CHEMICAL RESEARCH CENTER. Mississippi State University, Department of Chemistry, P.O. Drawer CH, Mississippi State, MS 39762. (601) 325-3584.

ALGEBRA

See also: CALCULUS, GEOMETRY, MATHEMATICS

ABSTRACT SERVICES AND INDEXES

APPLIED MECHANICS REVIEW. American Society of Mechanical Engineers, 345 East 47th Street, New York, NY 10017. (212) 705-7703. 1948 to present. Monthly. $360.00 per year.

APPLIED SCIENCE AND TECHNOLOGY INDEX. H.W. Wilson and Company, 950 University Avenue, Bronx, NY 10452. (800) 367-6670 or (212) 588-8400. Monthly. Inquire as to cost and availability.

CHEMICAL ABSTRACTS. American Chemical Society, Chemical Abstracts Service, Box 3012, Columbus, OH 43210. (614) 421-3600. 1907 to present. Weekly. $9500.00 per year.

COMPUMATH CITATION INDEX. Institute for Scientific Information, 3501 Market Street, Philadelphia, PA 19104. (800) 523-1850 or (215) 386-0100. Three times per year. $875.00 per year.

CURRENT MATHEMATICAL PUBLICATIONS. American Mathematical Society, P.O. Box 6248, Providence, RI 02940. (800) 556-7774 or (401) 272-9500. 1969 to present. Seventeen times per year. $230.00 per year.

ENGINEERING INDEX MONTHLY AND AUTHOR INDEX. Engineering Information Inc., 345 East 47th Street, New York, NY 10017. (212) 705-7600. Monthly. $1560.00 per year.

GENERAL SCIENCE INDEX. H.W. Wilson and Company, 950 University Avenue, Bronx, NY 10452. (800) 367-6670 or (212) 588-8400. 1978 to present. Monthly. Inquire as to cost and availability.

INDEX TO SCIENTIFIC REVIEWS. Institute for Scientific Information, 3501 Market Street, Philadelphia, PA 19104. (800) 523-1850 or (215) 386-0100. 1974 to present. Semi-annual. $550.00 per year.

MATHEMATICAL REVIEWS: A REVIEWING JOURNAL COVERING THE WORLD LITERATURE OF MATHEMATICAL RESEARCH. American Mathematical Society, P.O. Box 6248, Providence, RI 02940. (800) 7774 or (401) 272-9500. 1940 to present. Monthly. $2800.00 per year.

PHYSICS ABSTRACTS. Institution of Electrical Engineers. Available from: IEEE Service Center, 445 Hoes Lane, Piscataway, NJ 08854. 1898 to present. Bimonthly. $1700.00 per year.

SCIENCE CITATION INDEX. Institute for Scientific Information, 3501 Market Street, Philadelphia, PA 19104. (800) 523-1850 or (215) 386-0100. Six times per year. $6200. 00 per year.

ANNUAL REVIEWS AND YEARBOOKS

ADVANCES IN APPLIED MATHEMATICS. Academic Press, Inc., 6277 Sea Harbor Drive, Orlando, FL 32821. (800) 321-5068. Irregular. Price varies, inquire.

ASSOCIATIONS AND PROFESSIONAL SOCIETIES

AMERICAN MATHEMATICAL SOCIETY. P.O. Box 6248, Providence, RI 02940. (401) 272-9500.

MATHEMATICAL ASSOCIATION OF AMERICA. 1529 18th Street, N.W., Washington, DC 20036. (202) 387-5200.

SOCIETY FOR INDUSTRIAL AND APPLIED MATHEMATICS. 1400 Architects Building, 117 South 17th Street, Philadelphia, PA 19103. (215) 564-2929.

DIRECTORIES AND BIOGRAPHICAL SOURCES

INTERNATIONAL RESEARCH CENTERS DIRECTORY 1988-89. Darren L. Smith, editor. Gale Research Company, Book Tower, Detroit, MI 48226. (800) 521-0707. 4th edition. 1987. $360.00.

RESEARCH CENTERS DIRECTORY 1988. Gale Research Company, Book Tower, Detroit, MI 48226. (800) 521-0707. 12th edition. 1987. $365.00 for set.

SCIENTIFIC AND TECHNICAL ORGANIZATIONS AND AGENCIES DIRECTORY. Margaret Labash Young, editor. Gale

Research Company, Book Tower, Detroit, MI 48226. (800) 521-0707. 2nd edition. 1987. $185.00.

ENCYCLOPEDIAS AND DICTIONARIES

ENCYCLOPEDIC DICTIONARY OF MATHEMATICS. Kiyosi Ito, editor. MIT Press, 55 Howard Street, Cambridge, MA 02142. (617) 253-2884. 2nd edition. 1987. Four volumes. $350.00 for set.

GENERAL WORKS

FUNDAMENTALS OF MATHEMATICS. H. Behnke, F. Bachman, K. Fladt, W. Suss, and H. Kunle, editors. MIT Press, 55 Howard Street, Cambridge, MA 02142. (617) 253-2884. Volume 1, Foundations of Mathematics: Real Number System and Algebra; Volume 2, Geometry; Volume 3, Analysis. Inquire.

INTERMEDIATE ALGEBRA: A STRAIGHTFORWARD APPROACH. M.M. Zuckerman. John Wiley and Sons, Inc., 605 Third Avenue, New York, NY 10158. (800) 526-5368. Third edition. 1986. $31.95.

INTERMEDIATE REAL ANALYSIS. Emanuel Fischer. Springer-Verlag New York, Inc., 175 Fifth Avenue, New York, NY 10010. (800) 526-7254. 1983. $30.00.

LINEAR ALGEBRA. Charles W. Curtis. Springer-Verlag New York, Inc., 175 Fifth Avenue, New York, NY 10010. (800) 526-7254. 1984. $25.95.

NUMERICAL METHODS IN ENGINEERING AND APPLIED SCIENCE. Bruce Irons. John Wiley and Sons, Inc., 605 Third Avenue, New York, NY 10158. (800) 526-5368. 1987. $32.95.

A SURVEY OF MODERN ALGEBRA. Garrett Birkhoff and Saunders MacLane. Macmillan Publishing Company, Inc., 866 Third Avenue, New York, NY 10022. (800) 257-5755. Fourth edition. 1977. $45.95.

HANDBOOKS AND MANUALS

CRC HANDBOOK OF MATHEMATICAL SCIENCES. William H. Beyer, editor. CRC Press, 2000 Corporate Boulevard, Boca Raton, FL 33431. (800) 272-7737. 6th edition. 1987. $64.95.

HANDBOOK OF MATHEMATICAL FUNCTIONS WITH FORMULAS, GRAPHS, AND MATHEMATICAL TABLES. John Wiley and Sons, Inc., 605 Third Avenue, New York, NY 10158. (800) 526-5368. 1964. $47.50.

ONLINE DATA BASES

CA SEARCH. Chemical Abstracts Service, P.O. Box 3012, Columbus, OH 43120. (800) 848-6538 or (614) 421-3600. Comprehensive guide to chemical literature, 1972 to present. Inquire as to online cost and availability.

COMPENDEX. Engineering Information, Inc., 345 East 47th Street, New York, NY 10017. (800) 221-1044 or (212) 705-7615. Engineering and technical literature, 1975 to present. Inquire as to online cost and availability.

DISSERTATION ABSTRACTS ONLINE. University Microfilms International, 300 North Zeeb Road, Ann Arbor, MI 48106. (800) 521-0600 or (313) 761-4700. Scope includes virtually all doctoral dissertations accepted at accredited American institutions from 1861 to present in over 250 subject areas. Inquire as to online cost and availability.

INSPEC. INSPEC Marketing Department, Institution of Electrical Engineers. Available from IEEE Service Center, 445 Hoes Lane, Piscataway, NJ 08854. (201) 981-0060. Online version of Physics Abstracts. Inquire as to online cost and availability.

MATHFILE. American Mathematical Society, P.O. Box 6248, Providence, RI 02940. (800) 556-7774 or (401) 272-9500. An online version of Mathematical Reviews. 1973 to present. Inquire as to online cost and availability.

NTIS. National Technical Information Service, 5285 Port Royal Road, Springfield, VA 22161. (703) 487-4630. Broad coverage of government sponsored research reports, 1964 to present. Inquire as to online cost and availability.

SCISEARCH. Institute for Scientific Information, 3501 Market Street, Philadelphia, PA 19104. (800) 523-1850 or (215) 386-0100. Broad multidisciplinary title and author index to the international literature of science and technology, 1974 to present. Inquire as to online cost and availability.

WILSONLINE. H.W. Wilson and Company, 950 University Avenue, Bronx, NY 10452. (800) 367-6770 or (212) 588-8400. Makes available online versions of the H.W. Wilson indexes including Applied Science and Technology Index, Business Periodicals Index and Readers' Guide to Periodical Literature. Approximately 1980 to present. Inquire as to online cost and availability.

PERIODICALS

AMERICAN JOURNAL OF MATHEMATICS. Johns Hopkins University Press, Journals Publishing Division, 701 West 40th Street, Suite 275, Baltimore, MD 21211. (301) 338-7864. Bimonthly. 1878 to present. $115.00 per year.

AMERICAN MATHEMATICAL MONTHLY. Mathematical Association of America, 1529 18th Street, N.W., Washington, DC 20036. (202) 387-5200. 1894 to present. Ten times per year. $70.00 per year.

APPLIED MATHEMATICS AND COMPUTATION. Elsevier Science Publishing Company, Inc., 52 Vanderbilt Avenue, New York, NY 10017. (212) 370-5520. 1975 to present. Monthly. $355.00 per year.

COLLEGE MATHEMATICS JOURNAL. Mathematical Association of America, 1529 18th Street, N.W., Washington, DC 20036. Five times per year. $35.00 per year.

COMMUNICATIONS IN ALGEBRA. Marcel Dekker Inc., 270 Madison Avenue, New York, NY 10016. (800) 228-1160. 1974 to present. 10 times per year. $475.00 per year.

JOURNAL OF ALGEBRA. Academic Press, Inc., 6277 Sea Harbor Drive, Orlando, FL 32821. (800) 321-5068. 1964 to present. Fourteen times per year. $700.00 per year.

JOURNAL OF PURE AND APPLIED ALGEBRA. Elsevier Science Publishing Company, Inc., 52 Vanderbilt Avenue, New York, NY 10017. (212) 370-5520. 1971 to present. Fifteen times per year. $330.00 per year.

LINEAR ALGEBRA AND ITS APPLICATIONS. Elsevier Science Publishing Company, Inc., 52 Vanderbilt Avenue, New York, NY 10017. (212) 370-5520. 1968 to present. 36 times per year. $800.00 per year.

SIAM JOURNAL ON ALGEBRAIC AND DISCRETE METHODS. Society for Industrial and Applied Mathematics, 1405 Architects Building, 117 South 17th Street. Philadelphia, PA 19103. (215) 564-2929. 1980 to present. Quarterly. $60.00 per year.

SIAM JOURNAL OF APPLIED MATHEMATICS. Society for Industrial and Applied Mathematics, 1405 Architects Building, 117 South 17th Street. Philadelphia, PA 19103. (215) 564-2929. 1953 to present. Bimonthly. $130.00 per year.

SIAM JOURNAL OF MATHEMATICAL ANALYSIS. Society for Industrial and Applied Mathematics, 1405 Architects Building, 117 South 17th Street. Philadelphia, PA 19103. (215) 564-2929. 1964 to present. Bimonthly. $120.00 per year.

SIAM REVIEW. Society for Industrial and Applied Mathematics, 1405 Architects Building, 117 South 17th Street. Philadelphia, PA 19103. (215) 564-2929. 1959 to present. Quarterly. $82.00 per

ALGEBRA

year.

RESEARCH CENTERS AND INSTITUTES

CENTER FOR APPLIED MATHEMATICS. University of Georgia, Tucker Hall, Athens, GA 30602. (404) 542-3491.

CENTER FOR MATHEMATICAL SCIENCE RESEARCH. Rutgers University, New Brunswick, NJ 08903. (201) 932-3117.

CENTER FOR PURE AND APPLIED MATHEMATICS. University of Californai, Berkeley, 977 Evans Hall, Berkeley, CA 94720. (415) 642-0116.

INSTITUTE FOR MATHEMATICS AND ITS APPLICATIONS. University of Minnesota, 514 Vincent, 206 Church Street, S.E., Minneapolis, MN 55455. (612) 624-6066.

INSTITUTE OF APPLIED MATHEMATICS. University of Missouri - Rolla, Rolla, MO 65401. (314) 341-4151.

ALGEBRAIC NUMBER THEORY

See: MATHEMATICS

ALGOL

See: PROGRAMMING LANGUAGES

ALKALIS

See: ACIDS AND BASES

ALLOYS

See also: BRASS AND BRONZE, FERROALLOYS, MATERIALS SCIENCE, METALLURGY

ABSTRACT SERVICES AND INDEXES

ALLOYS INDEX. American Society for Metals, 9639 Kinsman Road, Metals Park, OH 44073. (216) 338-5151. $130.00 per year.

APPLIED MECHANICS REVIEWS. American Society of Mechanical Engineers, 345 East 47th Street, New York, NY 10017. (212) 705-7722. Monthly. $380.00 per year. Critical reviews of the world literature in applied mechanics and related engineering science.

APPLIED SCIENCE AND TECHNOLOGY INDEX. H.W. Wilson Company, 950 University Avenue, Bronx, NY 10452. (800) 367-6670 or (212) 588-8400. Inquire as to cost and availability.

CHEMICAL ABSTRACTS. Chemical Abstracts Service, 2540 Olentangy Road, Post Office Box 3012, Columbus, OH 43210. (800) 848-6538 or (614) 421-3600. Weekly. $9200.00 per year.

CURRENT CONTENTS: ENGINEERING, TECHNOLOGY. Institute for Scientific Information, 3501 Market Street, Philadelphia, PA 19104. (800) 523-1850 or (215) 386-0100. $272.00 per year.

ENGINEERING INDEX MONTHLY. Engineering Information, Incorporated, 345 East 47th Street, New York, NY 10017. (800) 221-1044 or (212) 705-7600. Monthly, with annual cumulation. $1560.00 per year.

Ency. of Physical Sciences & Engineering Info. Sources

INTERNATIONAL COPPER INFORMATION BULLETIN. Copper Development Association, Orchard House, Mutton Lane, Potters Bar, Herts, EN6 3AP, England. 1976 to present. Quarterly. $35.00 per year.

METALS ABSTRACTS. American Society for Metals, 9639 Kinsman Road, Metals Park, OH 44073. (216) 338-5151. Monthly. $890.00.

METALS ABSTRACTS INDEX. American Society for Metals, 9639 Kinsman Road, Metals Park, OH 44073. (216) 338-5151. Monthly. (Sold only to subscribers of Metals Abstracts.)

WORLD ALUMINUM ABSTRACTS. Aluminum Association, 818 Connecticut Avenue, NW, Washington, DC 20006. (202) 862-5100. 1968 to present. Monthly. $165.00 per year.

ANNUAL REVIEWS AND YEARBOOKS

ANNUAL REVIEW OF MATERIALS SCIENCE. Annual Reviews, Incorporated, 4139 El Camino Way, Palo Alto, CA 94306. (415) 493-4400. Annual.

ASSOCIATIONS AND PROFESSIONAL SOCIETIES

ALUMINUM ASSOCIATION. 818 Connecticut Avenue, NW, Washington, DC 20006. (202) 862-5100.

AMERICAN SOCIETY FOR METALS. Metals Park, OH 44073. (216) 338-5151.

METALLURGICAL SOCIETY OF THE AIME (AMERICAN INSTITUTE OF MINING, METALLURGICAL AND PETROLEUM ENGINEERS). 420 Commonwealth Drive, Warrendale, PA 15086. (412) 776-9080.

BIBLIOGRAPHIES

SCIENTIFIC AND TECHNICAL BOOKS AND SERIALS IN PRINT 1988; AN INDEX TO LITERATURE IN SCIENCE AND TECHNOLOGY. R.R. Bowker Company, 205 East 42nd Street, New York, NY 10017. (800) 521-8110 or (212) 916-1600. $175.00.

SUPERALLOYS: SOURCE BOOKE. M.J. Donachie, Jr., editor. American Society for Metals, Metals Park, OH 44073. (216) 5151. 1983. $60.00.

DIRECTORIES AND BIOGRAPHICAL SOURCES

DUN'S INDUSTRIAL GUIDE - THE METAL WORKING DIRECTORY. Dun and Bradstreet, Incorporated, Three Century Drive, Parsippany, NJ 07054. (201) 455-0900. Annual. $550.00.

INTERNATIONAL RESEARCH CENTERS DIRECTORY 1988-1989. Gale Research Company, Book Tower, Detroit, MI 48226. (800) 521-0707. Fourth edition. 1988. $360.00.

METAL PRODUCTS DIRECTORY. American Business Directories, Incorporated, Division of American Business Lists, Incorporated, 5707 South 86th Circle, Omaha, NE 68127. (402) 331-7169. 1986. $80.00.

RESEARCH CENTERS DIRECTORY. Gale Research Company, Book Tower, Detroit, MI 48226. (800) 521-0707. Thirteenth edition. 1988. $365.00.

RESEARCH SERVICES DIRECTORY. Robert J. Huffman and Mary M. Watkins, editors. Gale Research Company, Book Tower, Detroit, MI 48226. (800) 521-0707. Third edition. 1987. $290.00.

SCIENTIFIC AND TECHNICAL ORGANIZATIONS AND AGENCIES DIRECTORY. Gale Research Company, Book Tower, Detroit, MI 48226. (800) 521-0707. 1985.

WHO'S WHO IN ENGINEERING. Engineers Joint Council, 345 East 47th Street, New York, NY 10017. (212) 705-7010. 1985. $200.00.

WHO'S WHO IN FRONTIER SCIENCE AND TECHNOLOGY. Marquis Who's Who, Incorporated, 200 East Ohio Street, Chicago, IL 60611. (800) 428-3898 or (312) 787-2008.

WHO'S WHO IN TECHNOLOGY TODAY. Reston Publishing Company, Incorporated, c/o Prentice-Hall, Incorporated, Englewood Cliffs, NJ 07632. (800) 262-6868. Biennial. Five volumes. $425.00. Covers the fields of electronics, computer science, physics, optics, chemistry, biotechnology, mechanics, energy, and earth science.

ENCYCLOPEDIAS AND DICTIONARIES

ENCYCLOPEDIA OF MATERIALS SCIENCE AND ENGINEERING. Michael B. Bever, editor. MIT Press, 28 Carlton Street, Cambridge, MA 02142. (617) 253-5646. Eight volumes. 1986. $1950.00.

ENCYCLOPEDIA OF PHYSICAL SCIENCE AND TECHNOLOGY. Academic Press, Incorporated, Orlando, FL 32887. (800) 321-5068 or (305) 345-2734. Fifteen volumes, 1986.

MCGRAW-HILL DICTIONARY OF ENGINEERING. Sybil P. Parker, editor. McGraw-Hill Book Company, 1221 Avenue of the Americas, New York, NY 10020. (212) 512-2000. 1984. $39.95.

MCGRAW-HILL ENCYCLOPEDIA OF SCIENCE AND TECHNOLOGY. McGraw-Hill Book Company, 1221 Avenue of the Americas, New York, NY 10020. (212) 512-2000.

GENERAL WORKS

ALUMINUM: PROPERTIES AND PHYSICAL METALLURGY. John E. Hatch, editor. American Society for Metals, Metals Park, OH 44073. (216) 338-5151. 1984. $75.00.

ESSENTIAL METALLURGY FOR ENGINEERS. W. Alexander, G. Davies, and K. Reynolds. Van Nostrand Reinhold, 115 Fifth Avenue, New York, NY 10003. (800) 543-2681. 1985. $17.95.

PHYSICAL METALLURGY. P. Haasen. Cambridge University Press, 32 East 57th Street, New York, NY 10022. (212) 688-8885. 1986. $24.95 in paper.

METALLURGY BASICS. D.V. Brown. Van Nostrand Reinhold, 115 Fifth Avenue, New York, NY 10003. (800) 543-2681. 1985. $17.95.

STRUCTURE AND PROPERTIES OF ENGINEERING ALLOYS. W.F. Smith. McGraw-Hill Book Company, 1221 Avenue of the Americas, New York, NY 10020. (212) 512-2000. 1980. $46.95.

SUPERALLOYS. C.T. Sims and W.C. Hagel. John Wiley and Sons, Incorporated, 605 Third Avenue, New York, NY 10158. (212) 850-6000. 1972. $74.95.

THEORY OF STRUCTURAL TRANSFORMATIONS IN SOLIDS. A.G. Khachaturyan. John Wiley and Sons, Incorporated, 605 Third Avenue, New York, NY 10158. (800) 526-5368 or (212) 850-6000. 1983. $69.95.

HANDBOOKS AND MANUALS

CRC HANDBOOK OF CHEMISTRY AND PHYSICS. CRC Press, Incorporated, 2000 Corporate Boulevard, Boca Raton, Florida 33341. (305) 994-0555. Sixth-seventh edition. 1986. $69.95.

SMITHELL'S METALS REFERENCE BOOK. Eric A. Brandes, editor. Butterworth Publishers, 80 Montvale Avenue, Stoneham, MA 02180. (800) 325-4177. Sixth edition. 1983. $210.00.

ONLINE DATA BASES

COMPENDEX. Engineering Information, Incorporated, 345 East 47th Street, New York, NY 10017. (800) 221-1044 or (212) 705-7615. Engineering and technical literature, 1975 to present. Inquire as to cost and availability.

INSPEC. INSPEC Marketing Department, Institute of Electrical and Electronics Engineers, Incorporated, IEEE Service Department, 445 Hoes Lane, Piscataway, NJ 08854. (201) 981-0060. Inquire as to on-line cost and availability.

METADEX. Metals Information, American Society for Metals, Metals Park, OH 44073. (216) 338-5151. (Metals Abstracts/Alloys Index). A worldwide literature on the science and practice of metallurgy, 1966 to present. Inquire as to online cost and availability.

NASA. National Aeronautics and Space Administration, Scientific and Technical Information Branch, 300 7th Street, SW, Washington, DC 20546. Citations and abstracts of aerospace literature, 1962 to present. Inquire as to cost and availability.

NON-FERROUS METALS ABSTRACTS. British Non-Ferrous Metals Technology Centre, Grove Laboratories, Denchworth Road, Wantage, Oxfordshire, England OX12 9 BJ. Citations and abstracts on non-ferrous metallurgy and technology, 1961 to present. Inquire as to online cost and availability.

NTIS. National Technical Information Service, 5285 Port Royal Road, Springfield, VA 22161. (703) 487-4630. Broad coverage of government sponsored research reports, 1964 to present. Inquire as to cost and availability.

WILSONLINE. H.W. Wilson Company, 950 University Avenue, Bronx, NY 10452. (800) 367-6770 or (212) 588-8400. Makes available online versions of the printed H.W. Wilson Indexes including Applied Science and Technology Index, Business Periodicals Index, and Reader's Guide to Periodical Literature. Period covered is generally 1983 to present. Inquire as to cost and availability.

OTHER SOURCES

INTERDOC: DIRECTORY OF PUBLISHED PROCEEDINGS, SERIES. SEMT-Science/Engineering/Medicine/Technology. Interdoc Corporation, 173 Halstead Avenue, Box 326, Harrison, NY 10528. (014) 835-3506. Ten times per year. $325.00 per year.

MATERIALS SCIENCE AND METALLURGY. National Technical Information Service (NTIS), 5285 Port Royal Road, Springfield, VA 22161. (703) 487-4630. Translations and abstracts of foreign language technical media. Irregular. $40.00 per year.

PERIODICALS

ALLOY DIGEST. Engineering Publications Incorporated, Box 823, Upper Montclair, NJ 07043. (201) 746-7930. Monthly. $50.00 per year.

BULLETIN OF ALLOY PHASE DIAGRAMS. American Society for Metals, Metals Park, OH 44073. (216) 338-5151. Bimonthly. $90.00 per year.

JOURNAL OF METALS. Metallurgical Society of the AIME (American Institute of Mining, Metallurgical and Petroleum Engineers), 420 Commonwealth Drive, Warrendale, PA 15086. (412) 776-9080. Monthly. $40.00 per year.

LIGHT METAL AGE. Fellom Publishing Company, 693 Mission Street, San Francisco, CA 94105. (415) 781-1431. Bimonthly. $20.00 per year.

METALLUGICAL TRANSACTIONS. Metallurgical Society of the AIME (American Institute of Mining, Metallurgical and Petroleum Engineers), 420 Commonwealth Drive, Warrendale,

PA 15086. (412) 776-9080. Monthly. $95.00 per year.

METALS WEEK. McGraw-Hill Book Company, 1221 Avenue of the Americas, New York, NY 10020. (212) 997-2823. Weekly. $527.00 per year.

RESEARCH CENTERS AND INSTITUTES

COLORADO SCHOOL OF MINES. Steel Research Center, Golden, CO 80401. (303) 273-3774.

GENERAL ELECTRIC COMPANY. Research and Development Center, Post Office Box 8, Schenectady, NY 12301. (518) 385-8415.

LAWRENCE BERKELEY LABORATORY. Center for Advanced Materials, 1 Cyclotron Road, Berkeley, CA 94720. (415) 486-4755.

PHYSICAL METALLURGY RESEARCH LABORATORIES. Centre for Mineral and Energy Technology, 555 Booth Street, Ottawa, ON Canada K1A 0G1. (613) 995-4807.

TEXAS A & M UNIVERSITY. Mechanics and Materials Center, ERC Building, College Station, TX 77843. (409) 845-7512.

UNIVERSITY OF CONNECTICUT. Institute of Materials Science, Storrs, CT 06268. (203) 486-4623.

UNIVERSITY OF FLORIDA. Department of Materials Science and Engineering, Gainesville, FL 32601. (904) 392-1454.

UNIVERSITY OF MINNESOTA. Corrosion Research Center, 112 Mines and Metallurgy Building, Minneapolis, MN 55455. (612) 373-4864.

UNIVERSITY OF WISCONSIN AT MADISON. Cast Metals Laboratory, 1509 University Avenue, Madison, WI 53706. (608) 262-2562.

ALPHA PARTICLES

See: PARTICLE PHYSICS

ALTNERNATING CURRENT

See: ELECTRICITY

ALUMINUM

See also: ALLOYS, METALLURGY

ABSTRACT SERVICES AND INDEXES

ALLOYS INDEX. American Society for Metals, 9639 Kinsman Road, Metals Park, OH 44073. (216) 338-5151. $130.00 per year.

APPLIED MECHANICS REVIEWS. American Society of Mechanical Engineers, 345 East 47th Street, New York, NY 10017. (212) 705-7722. Monthly. $380.00 per year. Critical reviews of the world literature in applied mechanics and related engineering science.

APPLIED SCIENCE AND TECHNOLOGY INDEX. H.W. Wilson Company, 950 University Avenue, Bronx, NY 10452. (800) 367-6670 or (212) 588-8400. Inquire as to cost and availability.

CURRENT CONTENTS: ENGINEERING, TECHNOLOGY. Institute for Scientific Information, 3501 Market Street, Philadelphia, PA 19104. (800) 523-1850 or (215) 386-0100. $272.00 per year.

ENGINEERING INDEX MONTHLY. Engineering Information, Incorporated, 345 East 47th Street, New York, NY 10017. (800) 221-1044 or (212) 705-7600. Monthly, with annual cumulation. $1560.00 per year.

METALS ABSTRACTS. American Society for Metals, 9639 Kinsman Road, Metals Park, OH 44073. (216) 338-5151. Monthly. $890.00.

METALS ABSTRACTS INDEX. American Society for Metals, 9639 Kinsman Road, Metals Park, OH 44073. (216) 338-5151. Monthly. (Sold only to subscribers of Metals Abstracts.)

WORLD ALUMINUM ABSTRACTS. Aluminum Association, 818 Connecticut Avenue, NW, Washington, DC 20006. (202) 862-5100. 1968 to present. Monthly. $165.00 per year.

ANNUAL REVIEWS AND YEARBOOKS

ANNUAL REVIEW OF MATERIALS SCIENCE. Annual Reviews, Incorporated, 4139 El Camino Way, Palo Alto, CA 94306. (415) 493-4400. Annual.

ASSOCIATIONS AND PROFESSIONAL SOCIETIES

ALUMINUM ASSOCIATION. 818 Connecticut Avenue, NW, Washington, DC 20006. (202) 862-5100.

ALUMINUM EXTRUDERS COUNCIL. 4300-L Lincoln Avenue, Rolling Meadows, IL 60008. (312) 359-8160.

ALUMINUM RECYCLING ASSOCIATION. 900 17th Street, NW, Suite 504, Washington, DC 20006. (202) 785-0550.

AMERICAN SOCIETY FOR METALS. Metals Park, OH 44073. (216) 338-5151.

DIRECTORIES AND BIOGRAPHICAL SOURCES

DUN'S INDUSTRIAL GUIDE - THE METAL WORKING DIRECTORY. Dun and Bradstreet, Incorporated, Three Century Drive, Parsippany, NJ 07054. (201) 455-0900. Annual. $550.00.

RESEARCH CENTERS DIRECTORY. Gale Research Company, Book Tower, Detroit, MI 48226. (800) 521-0707. Thirteenth edition. 1988. $365.00.

SCIENTIFIC AND TECHNICAL ORGANIZATIONS AND AGENCIES DIRECTORY. Gale Research Company, Book Tower, Detroit, MI 48226. (800) 521-0707. 1985.

WHO'S WHO IN TECHNOLOGY TODAY. Reston Publishing Company, Incorporated, c/o Prentice-Hall, Incorporated, Englewood Cliffs, NJ 07632. (800) 262-6868. Biennial. Five volumes. $425.00. Covers the fields of electronics, computer science, physics, optics, chemistry, biotechnology, mechanics, energy, and earth science.

ENCYCLOPEDIAS AND DICTIONARIES

ENCYCLOPEDIA OF PHYSICAL SCIENCE AND TECHNOLOGY. Academic Press, Incorporated, Orlando, FL 32887. (800) 321-5068 or (305) 345-2734. Fifteen volumes, 1986.

MCGRAW-HILL DICTIONARY OF ENGINEERING. Sybil P. Parker, editor. McGraw-Hill Book Company, 1221 Avenue of the Americas, New York, NY 10020. (212) 512-2000. 1984. $39.95.

MCGRAW-HILL ENCYCLOPEDIA OF SCIENCE AND TECHNOLOGY. McGraw-Hill Book Company, 1221 Avenue of the Americas, New York, NY 10020. (212) 512-2000.

GENERAL WORKS

ALUMINUM: PROPERTIES AND PHYSICAL METALLURGY. John E. Hatch, editor. American Society for Metals, Metals Park, OH 44073. (216) 338-5151. 1984. $75.00.

ESSENTIAL METALLURGY FOR ENGINEERS. W. Alexander, G. Davies, and K. Reynolds. Van Nostrand Reinhold, 115 Fifth Avenue, New York, NY 10003. (800) 543-2681. 1985. $17.95.

METALLURGY BASICS. D.V. Brown. Van Nostrand Reinhold, 115 Fifth Avenue, New York, NY 10003. (800) 543-2681. 1985. $17.95.

SUPERALLOYS. C.T. Sims and W.C. Hagel. John Wiley and Sons, Incorporated, 605 Third Avenue, New York, NY 10158. (212) 850-6000. 1972. $74.95.

HANDBOOKS AND MANUALS

CRC HANDBOOK OF CHEMISTRY AND PHYSICS. CRC Press, Incorporated, 2000 Corporate Boulevard, Boca Raton, Florida 33341. (305) 994-0555. Sixth-seventh edition. 1986. $69.95.

SMITHELL'S METALS REFERENCE BOOK. Eric A. Brandes, editor. Butterworth Publishers, 80 Montvale Avenue, Stoneham, MA 02180. (800) 325-4177. Sixth edition. 1983. $210.00.

ONLINE DATA BASES

COMPENDEX. Engineering Information, Incorporated, 345 East 47th Street, New York, NY 10017. (800) 221-1044 or (212) 705-7615. Engineering and technical literature, 1975 to present. Inquire as to cost and availability.

INSPEC. INSPEC Marketing Department, Institute of Electrical and Electronics Engineers, Incorporated, IEEE Service Department, 445 Hoes Lane, Piscataway, NJ 08854. (201) 981-0060. Inquire as to on-line cost and availability.

METADEX. Metals Information, American Society for Metals, Metals Park, OH 44073. (216) 338-5151. (Metals Abstracts/Alloys Index). A worldwide literature on the science and practice of metallurgy, 1966 to present. Inquire as to online cost and availability.

NASA. National Aeronautics and Space Administration, Scientific and Technical Information Branch, 300 7th Street, SW, Washington, DC 20546. Citations and abstracts of aerospace literature, 1962 to present. Inquire as to cost and availability.

NON-FERROUS METALS ABSTRACTS. British Non-Ferrous Metals Technology Centre, Grove Laboratories, Denchworth Road, Wantage, Oxfordshire, England OX12 9 BJ. Citations and abstracts on non-ferrous metallurgy and technology, 1961 to present. Inquire as to online cost and availability.

NTIS. National Technical Information Service, 5285 Port Royal Road, Springfield, VA 22161. (703) 487-4630. Broad coverage of government sponsored research reports, 1964 to present. Inquire as to cost and availability.

OTHER SOURCES

ALUMINUM DEVELOPMENTS DIGEST. Aluminum Association, Incorporated, 818 Connecticut Avenue, NW, Washington, DC 20006. (202) 862-5100. Three times per year.

MATERIALS SCIENCE AND METALLURGY. National Technical Information Service (NTIS), 5285 Port Royal Road, Springfield, VA 22161. (703) 487-4630. Translations and abstracts of foreign language technical media. Irregular. $40.00 per year.

PERIODICALS

ALLOY DIGEST. Engineering Publications Incorporated, Box 823, Upper Montclair, NJ 07043. (201) 746-7930. Monthly. $50.00 per year.

E&MJ: ENGINEERING AND MINING JOURNAL. McGraw-Hill Book Company, 1221 Avenue of the Americas, New York, NY 10020. (212) 512-2000. Monthly. $20.00 per year.

JOURNAL OF METALS. Metallurgical Society of the AIME (American Institute of Mining, Metallugical and Petroleum Engineers), 420 Commonwealth Drive, Warrendale, PA 15086. (412) 776-9080. Monthly. $40.00 per year.

LIGHT METAL AGE. Fellom Publishing Company, 693 Mission Street, San Francisco, CA 94105. (415) 781-1431. Bimonthly. $20.00 per year.

RESEARCH CENTERS AND INSTITUTES

GENERAL ELECTRIC COMPANY. Research and Development Center, Post Office Box 8, Schenectady, NY 12301. (518) 385-8415.

LAWRENCE BERKELEY LABORATORY. Center for Advanced Materials, 1 Cyclotron Road, Berkeley, CA 94720. (415) 486-4755.

PHYSICAL METALLURGY RESEARCH LABORATORIES. Centre for Mineral and Energy Technology, 555 Booth Street, Ottawa, ON Canada K1A 0G1. (613) 995-4807.

TEXAS A & M UNIVERSITY. Mechanics and Materials Center, ERC Building, College Station, TX 77843. (409) 845-7512.

UNIVERSITY OF CONNECTICUT. Institute of Materials Science, Storrs, CT 06268. (203) 486-4623.

UNIVERSITY OF FLORIDA. Department of Materials Science and Engineering, Gainesville, FL 32601. (904) 392-1454.

UNIVERSITY OF WISCONSIN AT MADISON. Cast Metals Laboratory, 1509 University Avenue, Madison, WI 53706. (608) 262-2562.

SPECIFICATIONS AND STANDARDS

ALUMINUM STANDARDS AND DATA. Aluminum Association, Incorporated, 818 Connecticut Avenue, NW, Washington, DC 20006. (202) 862-5100. Biennial. $12.00.

STATISTICS SOURCES

ALUMINUM INGOT. U.S. Bureau of the Census, U.S. Department of Commerce, Washington, DC 20233. (301) 763-7800. Monthly. $20.50 per year.

ALUMINUM SITUATION. Aluminum Association, Incorporated, 818 Connecticut Avenue, NW, Washington, DC 20006. (202) 862-5100. Monthly.

ALUMINUM STATISTICAL REVIEW. Aluminum Association, Incorporated, 818 Connecticut Avenue, NW, Washington, DC 20006. (202) 862-5100. Annual. $25.00.

AMERICAN BUREAU OF METAL STATISTICS YEAR BOOK. American Bureau of Metal Statistics, Post Office Box 1405, 400 Plaza Drive, Secaucus, NJ 07094. Annual.

AM RADIO

See: RADIO

AMATEUR RADIO

See also: ANTENNAS, ELECTRICAL ENGINEERING, ELECTRONIC CIRCUITS AND COMPONENTS, ELECTRONICS, ELECTRONICS ENGINEERING, RADAR, RADIO, TELEVISION

ABSTRACT SERVICES AND INDEXES

APPLIED SCIENCE AND TECHNOLOGY INDEX. H.W. Wilson and Company, 950 University Avenue, Bronx, NY 10452. (800) 367-6670 or (212) 588-8400. Monthly. Inquire as to cost and availability.

ELECTRICAL AND ELECTRONICS ABSTRACTS. Institution of Electrical Engineers. Available from: Institute of Electrical and Electronics Engineers. IEEE Service Center, 445 Hoes Lane, Piscataway, NJ 08854. Monthly. $1250.00 per year.

ELECTRONICS AND COMMUNICATIONS ABSTRACTS. Cambridge Scientific Abstracts, 5161 River Road, Bethesda, MD 20816. (301) 951-1400. Bimonthly. Inquire as to cost and availability.

ENGINEERING INDEX MONTHLY AND AUTHOR INDEX. Engineering Information Inc., 345 East 47th Street, New York, NY 10017. (212) 705-7600. Monthly. $1560.00 per year.

IEEE PUBLICATIONS BULLETIN. Institute of Electrical and Electronics Engineers. Institute of Electrical and Electronics Engineers. IEEE Service Center, 445 Hoes Lane, Piscataway, NJ 08854. Quarterly. Free.

PHYSICS ABSTRACTS. Institution of Electrical Engineers. Available from: IEEE Service Center, 445 Hoes Lane, Piscataway, NJ 08854. 1898 to present. Bimonthly. $1700.00 per year.

PHYSICS BRIEFS. Physik Verlag GmbH, Postfach 1260/1280, D-6940 Weinheim, West Germany. (212) 661-9404. 1920 to present. Twenty-six times per year. $1250.00 per year.

ASSOCIATIONS AND PROFESSIONAL SOCIETIES

AMERICAN ELECTRONICS ASSOCIATION. P.O. Box 10045, 2670 Hanover Street, Palo Alto, CA 94303. (415) 857-9300.

ELECTRONICS INDUSTRIES ASSOCIATION. 2001 Eye Street, N.W., Washington, DC 20006. (202) 457-4900.

INSTITUTE OF ELECTRICAL AND ELECTRONICS ENGINEERS. 345 East 47th Street, New York, NY 10017. (212) 705-7900.

INSTITUTE OF RADIO ENGINEERS. Institute of Electrical and Electronics Engineers, 345 East 47th Street, New York, NY 10017. (212) 705-7900.

NATIONAL ASSOCIATION OF RADIO AND TELE-COMMUNICATIONS ENGINEERS. P.O. Box 15029, Salem, OR 97309. (503) 581-7653.

RADIO AMATEUR SATELLITE CORPORATION. P.O. Box 27, Washington, DC 20044. (301) 589-6062.

DIRECTORIES AND BIOGRAPHICAL SOURCES

BROADCAST ENGINEERING BUYERS GUIDE/SPEC BOOK ISSUE. Intertec Publishing Corporation, Box 12901, Overland Park, KS 66212. (913) 888-4664. Annual. $20.00 per year.

IEEE MEMBERSHIP DIRECTORY. Institute of Electrical and Electronics Engineers. IEEE Service Center, 445 Hoes Lane, Piscataway, NJ 08854. Annual. $7.00.

RADIO AMATEUR CALLBOOK. Radio Amateur Callbook, Inc., 925 Sherwood Drive, Lake Bluff, IL 60044. (312) 234-6600. Annual. North American Edition, $22.00. International Edition, $21.00, plus shipping and handling.

RESEARCH CENTERS DIRECTORY 1988. Gale Research Company, Book Tower, Detroit, MI 48226. (800) 521-0707. 12th edition. 1987. $365.00 for set.

WHO'S WHO IN ELECTRONICS. Harris Publishing Company, 2057-2 Aurora Road, Twinsburg, OH 44087. (216) 425-9143. Annual. $90.00.

ENCYCLOPEDIAS AND DICTIONARIES

DICTIONARY OF AUDIO, RADIO AND VIDEO. R.S. Roberts. Butterworth's Publishing, 80 Montvale Avenue, Stoneham, MA 02180. (800) 325-4177. 1981. $45.00.

IEEE STANDARD DICTIONARY OF ELECTRICAL AND ELECTRONICS TERMS. Frank Jay, editor. John Wiley and Sons, Inc., 605 Third Avenue, New York, NY 10158. (800) 526-5368. 3rd edition. 1984. $49.95.

GENERAL WORKS

AMATEUR RADIO EQUIPMENT FUNDAMENTALS. Albert D. Helfrick. Prentice-Hall Publishing, Inc., Englewood Cliffs, NJ 07632. (800) 562-0245. 1982. $31.95.

AMATEUR RADIO, SUPER HOBBY: WHAT IS IT, WHO WE ARE, HOW TO JOIN. Vince Luciani. McGraw-Hill Book Company, 1221 Avenue of the Americas, New York, NY 10020. (212) 512-2000. 1984. $9.95 in paper.

ELECTROMAGNETIC CONCEPTS AND PRINCIPLES. Stanley V. Marshall and Gabriel G. Skitek. Prentice-Hall Publishing, Inc., Englewood Cliffs, NJ 07632. (800) 562-0245. 2nd edition. 1987. $42.95.

SOLID STATE RADIO ENGINEERING. Herbert L. Krauss and Charles W. Bostian. John Wiley and Sons, Inc., 605 Third Avenue, New York, NY 10158. (800) 526-5368. 1980. $45.00.

UNDERSTANDING RADIO ELECTRONICS. M. Kaufmann and others. McGraw-Hill Book Company, 1221 Avenue of the Americas, New York, NY 10020. (212) 512-2000. Fourth edition. 1972. $33.50.

HANDBOOKS AND MANUALS

ELECTRONIC ENGINEERS HANDBOOK. Donald G. Fink, editor. McGraw-Hill Book Company, 1221 Avenue of the Americas, New York, NY 10020. (212) 512-2000. 2nd edition. 1982. $89.00.

HANDBOOK FOR RADIO ENGINEERING. John F. Ross. Butterworth's Publishing, 80 Montvale Avenue, Stoneham, MA 02180. (800) 325-4177. 1980. $125.00.

HANDBOOK OF MODERN ELECTRONICS AND ELECTRICAL ENGINEERING. Charles Belove, editor. John Wiley and Sons, Inc., 605 Third Avenue, New York, NY 10158. (800) 526-5368. 1986. $88.95.

RADIO AMATEUR'S HANDBOOK. A. Frederick Collins and Robert Herzberg. Harper and Row Publishers, Inc., 10 East 53rd Street, New York, NY 10022. (800) 242-7737. Fifteenth revised edition. 1983. $15.00.

ONLINE DATA BASES

COMPENDEX. Engineering Information, Inc., 345 East 47th Street, New York, NY 10017. (800) 221-1044 or (212) 705-7615. Engineering and technical literature, 1975 to present. Inquire as to online cost and availability.

INSPEC. INSPEC Marketing Department, Institution of Electrical Engineers. Available from IEEE Service Center, 445 Hoes Lane, Piscataway, NJ 08854. (201) 981-0060. Online version of Physics Abstracts. Inquire as to online cost and availability.

NTIS. National Technical Information Service, 5285 Port Royal Road, Springfield, VA 22161. (703) 487-4630. Broad coverage of government sponsored research reports, 1964 to present. Inquire as to online cost and availability.

WILSONLINE. H.W. Wilson and Company, 950 University Avenue, Bronx, NY 10452. (800) 367-6770 or (212) 588-8400. Makes available online versions of the H.W. Wilson indexes including Applied Science and Technology Index, Business Periodicals Index and Readers' Guide to Periodical Literature. Approximately 1980 to present. Inquire as to online cost and availability.

PERIODICALS

BROADCASTER ENGINEERING. Intertec Publishing Corporation, Box 12901, Overland Park, KS 66212. (913) 888-4664. 1959 to present. Monthly. $25.00 per year.

COMMUNICATIONS: FOR THE PROFESSIONAL IN LAND MOBILE RADIO. Cardiff Publishing Company, 6530 South Yosemite, Enlewood, CO 80111. (303) 694-1522. 1963 to present. Monthly. $20.00 per year.

CQ; THE RADIO AMATEUR'S JOURNAL. CQ Publishing, Inc., 76 North Broadway, Hicksville, NY 11801. (516) 681-2922. 1945 to present. Monthly. $16.00 per year.

ELECTRONIC DESIGN. Hayden Publishing Company, 10 Mulholland Drive, Hasbrouck Heights, NJ 07604. (201) 288-7520. 1952 to present. Biweekly. $40.00 per year.

ELECTRONICS. McGraw-Hill Book Company, 1221 Avenue of the Americas, New York, NY 10020. (212) 512-2000. 1930 to present. Weekly. $32.00 per year.

ELECTRONICS AND WIRELESS WORLD. I.P.C. Electrical-Electronic Press, Ltd., Quadrant House, The Quadrant, Sutton, Surrey, SM2 5AS England. 1911 to present. Monthly. $105.00 per year.

IEEE TRANSACTIONS ON BROADCASTING. Institute of Electrical and Electronics Engineers. IEEE Service Center, 445 Hoes Lane, Piscataway, NJ 08854. 1955 to present. Quarterly. $37.00 per year.

AMMONIA

See also: CHEMICAL ENGINEERING, CHEMISTRY

ABSTRACT SERVICES AND INDEXES

CHEMICAL ABSTRACTS. American Chemical Society, Chemical Abstracts Service, Box 3012, Columbus, OH 43210. (614) 421-3600. 1907 to present. Weekly. $9500.00 per year.

CURRENT CONTENTS: PHYSICAL, CHEMICAL AND EARTH SCIENCES. Institute for Scientific Information, 3501 Market Street, Philadelphia, PA 19104. (800) 523-1850 or (215) 386-0100. Weekly. $275.00 per year.

GENERAL SCIENCE INDEX. H.W. Wilson and Company, 950 University Avenue, Bronx, NY 10452. (800) 367-6670 or (212) 588-8400. 1978 to present. Monthly. Inquire as to cost and availability.

SCIENCE CITATION INDEX. Institute for Scientific Information, 3501 Market Street, Philadelphia, PA 19104. (800) 523-1850 or (215) 386-0100. Six times per year. $6200.00 per year.

ASSOCIATIONS AND PROFESSIONAL SOCIETIES

AMERICAN CHEMICAL SOCIETY. 1155 16th Street, N.W., Washington, DC 20036. (202) 872-4600.

AMERICAN INSTITUTE OF CHEMICAL ENGINEERS. 345 East 47th Street, New York, NY 10017. (212) 705-7338.

ASSOCIATION OF OFFICIAL ANALYTICAL CHEMISTS. 1111 North 19th Street, Suite 210, Arlington, VA 22209. (703) 522-3032.

FERTILIZER INSTITUTE. 1015 18th Street, N.W., Washington, DC 20036. (202) 861-4900.

INTERNATIONAL INSTITUTE OF AMMONIA REFRIGERATION. 111 East Wacker Drive, Chicago, IL 60601. (312) 644-6610.

DIRECTORIES AND BIOGRAPHICAL SOURCES

AMERICAN MEN AND WOMEN OF SCIENCE. R.R. Bowker, Inc., Order Department, 245 West 17th Street, New York, NY 10011. (800) 521-8110. Eight volumes. 1986. $595.00 for set.

INTERNATIONAL RESEARCH CENTERS DIRECTORY 1988-89. Darren L. Smith, editor. Gale Research Company, Book Tower, Detroit, MI 48226. (800) 521-0707. 4th edition. 1987. $360.00.

1987 DIRECTORY OF ENGINEERING SOCIETIES AND RELATED ORGANIZATIONS. Gordon Davis, editor. Hemisphere Publishing Corporation, 1010 Vermont Avenue, NW, Washington, DC 20005. (800) 526-0275. 12th edition. 1987. $100.00.

RESEARCH CENTERS DIRECTORY 1988. Gale Research Company, Book Tower, Detroit, MI 48226. (800) 521-0707. 12th edition. 1987. $365.00 for set.

SCIENTIFIC AND TECHNICAL ORGANIZATIONS AND AGENCIES DIRECTORY. Margaret Labash Young, editor. Gale Research Company, Book Tower, Detroit, MI 48226. (800) 521-0707. 2nd edition. 1987. $185.00.

WHO'S WHO IN ENGINEERING. Gordon Davis, editor. Hemisphere Publishing Corporation, 1010 Vermont Avenue, NW, Washington, DC 20005. (800) 526-0275. 6th edition. 1985. $200.00.

ENCYCLOPEDIAS AND DICTIONARIES

ENCYCLOPEDIA OF CHEMISTRY. Douglas M. Considine, editor. Van Nostrand Reinhold Company, Inc., 135 West 50th Street, New York, NY 10020. (800) 543-2681. 4th edition. 1984. $98.95.

GENERAL WORKS

AMMONIA AND SYNTHESIS GAS: RECENT AND ENERGY-SAVING PROCESSES. F.J. Brykowski. Noyes Data Corporation, Mill Road at Grand Avenue, Park Ridge, NJ 07656. (201) 391-8484. 1982. $48.00.

AMMONIA PLANT SAFETY. American Institute of Chemical Engineers, 345 East 47th Street, New York, NY 10017. (212) 705-7338. 1985. $36.00.

TECHNOLOGY AND MANUFACTURE OF AMMONIA. Samuel Strelzoff. John Wiley and Sons, Inc., 605 Third Avenue, New York, NY 10158. (800) 526-5368. 1981. $91.00.

HANDBOOKS AND MANUALS

CRC HANDBOOK OF CHEMISTRY AND PHYSICS. Robert C. Weast, editor. CRC Press, 2000 Corporate Boulevard, Boca Raton, FL 33431. (800) 272-7737. 68th edition. 1987. $69.95.

HANDBOOK OF APPLIED CHEMISTRY. Vollrath Hopp and Ingo Hennig. Hemisphere Publishing Corporation, 79 Madison Avenue, New York, NY 10016-7892. Order from: Taylor and Francis/Hemisphere Distribution Center, 242 Cherry Street, Philadelphia, PA 19106-1906. (800) 821-8312. 1983. $49.95.

AMMONIA

LANGE'S HANDBOOK OF CHEMISTRY. John A. Dean, editor. McGraw-Hill Book Company, 1221 Avenue of the Americas, New York, NY 10020. (212) 512-2000. 1985. $59.50.

ONLINE DATA BASES

CA SEARCH. Chemical Abstracts Service, P.O. Box 3012, Columbus, OH 43120. (800) 848-6538 or (614) 421-3600. Comprehensive guide to chemical literature, 1972 to present. Inquire as to online cost and availability.

COMPENDEX. Engineering Information, Inc., 345 East 47th Street, New York, NY 10017. (800) 221-1044 or (212) 705-7615. Engineering and technical literature, 1975 to present. Inquire as to online cost and availability.

NTIS. National Technical Information Service, 5285 Port Royal Road, Springfield, VA 22161. (703) 487-4630. Broad coverage of government sponsored research reports, 1964 to present. Inquire as to online cost and availability.

SCISEARCH. Institute for Scientific Information, 3501 Market Street, Philadelphia, PA 19104. (800) 523-1850 or (215) 386-0100. Broad multidisciplinary title and author index to the international literature of science and technology, 1974 to present. Inquire as to online cost and availability.

PERIODICALS

COMMENTS ON INORGANIC CHEMISTRY. Gordon and Breach Science Publishers, Inc., 50 West 23rd Street, New York, NY 10010. (212) 206-8900. Twelve times per year. $166.00 per year.

INORGANIC CHEMISTRY. American Chemical Society, 1155 16th Street, N.W., Washington, DC 20036. (800) 424-6747. 1962 to present. Biweekly. $300.00 per year.

RESEARCH CENTERS AND INSTITUTES

CHEMICAL LABORATORIES. Harvard University, Oxford Street, Cambridge, MA 02138. (617) 495-4283.

CHEMICAL RESEARCH LABORATORY. Brown University, Providence, RI 02912. (401) 863-2256.

AMPLIFIERS

See also: AUDIO ENGINEERING, ELECTRICAL ENGINEERING, ELECTRONIC CIRCUITS AND COMPONENTS, ELECTRONICS, ELECTRONICS ENGINEERING, MICROELECTRONICS, RADIO

ABSTRACT SERVICES AND INDEXES

APPLIED SCIENCE AND TECHNOLOGY INDEX. H.W. Wilson and Company, 950 University Avenue, Bronx, NY 10452. (800) 367-6670 or (212) 588-8400. Monthly. Inquire as to cost and availability.

CHEMICAL ABSTRACTS. American Chemical Society, Chemical Abstracts Service, Box 3012, Columbus, OH 43210. (614) 421-3600. 1907 to present. Weekly. $9500.00 per year.

CURRENT CONTENTS: ENGINEERING, TECHNOLOGY AND APPLIED SCIENCES. Institute for Scientific Information, 3501 Market Street, Philadelphia, PA 19104. (800) 523-1850 or (215) 386-0100. Weekly. $275.00 per year.

ELECTRICAL AND ELECTRONICS ABSTRACTS. Institution of Electrical Engineers. Available from: Institute of Electrical and Electronics Engineers. IEEE Service Center, 445 Hoes Lane, Piscataway, NJ 08854. Monthly. $1250.00 per year.

ELECTRONICS AND COMMUNICATIONS ABSTRACTS. Cambridge Scientific Abstracts, 5161 River Road, Bethesda, MD 20816. (301) 951-1400. Bimonthly. Inquire as to cost and availability.

ENGINEERING INDEX MONTHLY AND AUTHOR INDEX. Engineering Information Inc., 345 East 47th Street, New York, NY 10017. (212) 705-7600. Monthly. $1560.00 per year.

IEEE PUBLICATIONS BULLETIN. Institute of Electrical and Electronics Engineers. Institute of Electrical and Electronics Engineers. IEEE Service Center, 445 Hoes Lane, Piscataway, NJ 08854. Quarterly. Free.

PHYSICS ABSTRACTS. Institution of Electrical Engineers. Available from: IEEE Service Center, 445 Hoes Lane, Piscataway, NJ 08854. 1898 to present. Bimonthly. $1700.00 per year.

PHYSICS BRIEFS. Physik Verlag GmbH, Postfach 1260/1280, D-6940 Weinheim, West Germany. (212) 661-9404. 1920 to present. Twenty-six times per year. $1250.00 per year.

SCIENCE CITATION INDEX. Institute for Scientific Information, 3501 Market Street, Philadelphia, PA 19104. (800) 523-1850 or (215) 386-0100. Six times per year. $6200.00 per year.

SOLID STATE ABSTRACTS: AN ABSTRACT JOURNAL INVOLVING THE PHYSICS, METALLURGY, CRYSTALLOGRAPHY, CHEMISTRY, AND DEVICE TECHNOLOGY OF SOLIDS. Cambridge Scientific Abstracts, 5161 River Road, Bethesda, MD 20816. (301) 951-1400. 1957 to present. Bimonthly. $550.00 per year.

ANNUAL REVIEWS AND YEARBOOKS

ADVANCES IN ELECTRONICS AND ELECTRON PHYSICS. Academic Press, Inc., 6277 Sea Harbor Drive, Orlando, FL 32821. (800) 321-5068. Irregular. Approximately $80.00 per volume.

ASSOCIATIONS AND PROFESSIONAL SOCIETIES

AMERICAN ELECTRONICS ASSOCIATION. P.O. Box 10045, 2670 Hanover Street, Palo Alto, CA 94303. (415) 857-9300.

AMERICAN INSTITUTE OF PHYSICS. 335 East 45th Street, New York, NY 10017. (212) 661-9494.

ELECTRONICS INDUSTRIES ASSOCIATION. 2001 Eye Street, N.W., Washington, DC 20006. (202) 457-4900.

INSTITUTE OF ELECTRICAL AND ELECTRONICS ENGINEERS. 345 East 47th Street, New York, NY 10017. (212) 705-7900.

INTERNATIONAL SOCIETY FOR HYBRID MICRO-ELECTRONICS. P.O. Box 2698, 1861 Wiehle Avenue, Suite 340, Reston, VA 22090. (703) 471-0066.

DIRECTORIES AND BIOGRAPHICAL SOURCES

AMERICAN MEN AND WOMEN OF SCIENCE. R.R. Bowker, Inc., Order Department, 245 West 17th Street, New York, NY 10011. (800) 521-8110. Eight volumes. 1986. $595.00 for set.

IEEE MEMBERSHIP DIRECTORY. Institute of Electrical and Electronics Engineers. IEEE Service Center, 445 Hoes Lane, Piscataway, NJ 08854. Annual. $7.00.

INTERNATIONAL RESEARCH CENTERS DIRECTORY 1988-89. Darren L. Smith, editor. Gale Research Company, Book Tower, Detroit, MI 48226. (800) 521-0707. 4th edition. 1987. $360.00.

1987 DIRECTORY OF ENGINEERING SOCIETIES AND RELATED ORGANIZATIONS. Gordon Davis, editor. Hemisphere Publishing Corporation, 1010 Vermont Avenue, NW, Washington, DC 20005. (800) 526-0275. 12th edition. 1987. $100.00.

RESEARCH CENTERS DIRECTORY 1988. Gale Research Company, Book Tower, Detroit, MI 48226. (800) 521-0707. 12th edition. 1987. $365.00 for set.

SCIENTIFIC AND TECHNICAL ORGANIZATIONS AND AGENCIES DIRECTORY. Margaret Labash Young, editor. Gale Research Company, Book Tower, Detroit, MI 48226. (800) 521-0707. 2nd edition. 1987. $185.00.

WHO'S WHO IN ELECTRONICS. Harris Publishing Company, 2057-2 Aurora Road, Twinsburg, OH 44087. (216) 425-9143. Annual. $90.00.

WHO'S WHO IN ENGINEERING. Gordon Davis, editor. Hemisphere Publishing Corporation, 1010 Vermont Avenue, NW, Washington, DC 20005. (800) 526-0275. 6th edition. 1985. $200.00.

ENCYCLOPEDIAS AND DICTIONARIES

IEEE STANDARD DICTIONARY OF ELECTRICAL AND ELECTRONICS TERMS. Frank Jay, editor. John Wiley and Sons, Inc., 605 Third Avenue, New York, NY 10158. (800) 526-5368. 3rd edition. 1984. $49.95.

GENERAL WORKS

ELECTRONIC INVENTIONS AND DISCOVERIES: ELECTRONICS FROM ITS EARLIEST BEGINNINGS TO PRESENT DAY. G.W.A. Dummer. Pergamon Press, Inc., Maxwell House, Fairview Park, Elmsford, NY 10523. (914) 592-7700. 1983. $49.50.

HIGH FREQUENCY AMPLIFIERS. R.S. Carson. John Wiley and Sons, Inc., 605 Third Avenue, New York, NY 10158. (800) 526-5368. Second edition. 1982. $35.95.

OPERATIONAL AMPLIFIERS: THE DEVICES AND THEIR APPLICATIONS. C.F. Wojslaw and E.A. Moustakas. John Wiley and Sons, Inc., 605 Third Avenue, New York, NY 10158. (800) 526-5368. 1986. $38.95.

HANDBOOKS AND MANUALS

ELECTRONIC ENGINEERS HANDBOOK. Donald G. Fink, editor. McGraw-Hill Book Company, 1221 Avenue of the Americas, New York, NY 10020. (212) 512-2000. 2nd edition. 1982. $89.00.

HANDBOOK OF MODERN ELECTRONICS AND ELECTRICAL ENGINEERING. Charles Belove, editor. John Wiley and Sons, Inc., 605 Third Avenue, New York, NY 10158. (800) 526-5368. 1986. $88.95.

ONLINE DATA BASES

CA SEARCH. Chemical Abstracts Service, P.O. Box 3012, Columbus, OH 43120. (800) 848-6538 or (614) 421-3600. Comprehensive guide to chemical literature, 1972 to present. Inquire as to online cost and availability.

COMPENDEX. Engineering Information, Inc., 345 East 47th Street, New York, NY 10017. (800) 221-1044 or (212) 705-7615. Engineering and technical literature, 1975 to present. Inquire as to online cost and availability.

DISSERTATION ABSTRACTS ONLINE. University Microfilms International, 300 North Zeeb Road, Ann Arbor, MI 48106. (800) 521-0600 or (313) 761-4700. Scope includes virtually all doctoral dissertations accepted at accredited American institutions from 1861 to present in over 250 subject areas. Inquire as to online cost and availability.

INSPEC. INSPEC Marketing Department, Institution of Electrical Engineers. Available from IEEE Service Center, 445 Hoes Lane, Piscataway, NJ 08854. (201) 981-0060. Online version of Physics Abstracts. Inquire as to online cost and availability.

NTIS. National Technical Information Service, 5285 Port Royal Road, Springfield, VA 22161. (703) 487-4630. Broad coverage of government sponsored research reports, 1964 to present. Inquire as to online cost and availability.

SCISEARCH. Institute for Scientific Information, 3501 Market Street, Philadelphia, PA 19104. (800) 523-1850 or (215) 386-0100. Broad multidisciplinary title and author index to the international literature of science and technology, 1974 to present. Inquire as to online cost and availability.

WILSONLINE. H.W. Wilson and Company, 950 University Avenue, Bronx, NY 10452. (800) 367-6770 or (212) 588-8400. Makes available online versions of the H.W. Wilson indexes including Applied Science and Technology Index, Business Periodicals Index and Readers' Guide to Periodical Literature. Approximately 1980 to present. Inquire as to online cost and availability.

OTHER SOURCES

A GUIDE TO THE LITERATURE OF ELECTRICAL AND ELECTRONICS ENGINEERING. Susan B. Ardis. Libraries Unlimited Inc., P.O. Box 263, Littleton, CO 80160. (303) 770-1220. 1987. $37.50.

PERIODICALS

ELECTRONIC DESIGN. Hayden Publishing Company, 10 Mulholland Drive, Hasbrouck Heights, NJ 07604. (201) 288-7520. 1952 to present. Biweekly. $40.00 per year.

ELECTRONICS. McGraw-Hill Book Company, 1221 Avenue of the Americas, New York, NY 10020. (212) 512-2000. 1930 to present. Weekly. $32.00 per year.

IEEE CIRCUITS AND DEVICES MAGAZINE. Institute of Electrical and Electronics Engineers. IEEE Service Center, 445 Hoes Lane, Piscataway, NJ 08854. Bimonthly. $70.00 per year.

IEEE JOURNAL OF SOLID STATE CIRCUITS. Institute of Electrical and Electronics Engineers. IEEE Service Center, 445 Hoes Lane, Piscataway, NJ 08854. 1966 to present. Bimonthly. $113.00 per year.

IEEE TRANSACTIONS ON ELECTRON DEVICES. Institute of Electrical and Electronics Engineers. IEEE Service Center, 445 Hoes Lane, Piscataway, NJ 08854. 1959 to present. Monthly. $175.00 per year.

INSTITUTE OF ELECTRICAL AND ELECTRONICS ENGINEERS PROCEEDINGS. Institute of Electrical and Electronics Engineers. IEEE Service Center, 445 Hoes Lane, Piscataway, NJ 08854. 1913 to present. Monthly. $140.00 per year.

SEMICONDUCTOR INTERNATIONAL. Cahners Publishing Company, Inc., Cahners Plaza, 1350 East Touhy Avenue, Des Plaines, IL 60018. (312) 635-8800. 1978 to present. Monthly. $55.00 per year.

SOLID STATE ELECTRONICS. Pergamon Press, Inc., Maxwell House, Fairview Park, Elmsford, NY 10523. (914) 592-7700. 1960 to present. Monthly. $330.00 per year.

RESEARCH CENTERS AND INSTITUTES

ELECTRICAL ENGINEERING RESEARCH LABORATORIES. Purdue University, Electrical Engineering Building, West Lafayette, IN 47907. (317) 494-3536.

ELECTRONICS RESEARCH CENTER. University of Texas at Austin, 132 Engineering Science Building, Austin, TX 78712. (512) 471-3954.

ELECTRONICS RESEARCH LABORATORY. University of California, Berkeley, 253 Cory Hall, Berkeley, CA 94720. (415) 642-2301.

LABORATORY FOR ELECTROMAGNETIC AND ELECTRONIC SYSTEMS. Massachusetts Institute of Technology, 77 Massachusetts Avenue, Cambridge, MA 02139. (617) 253-4631.

ANALOG COMPUTERS

See: COMPUTERS

ANALYTIC GEOMETRY

See also: ALGEBRA, CALCULUS, MATHEMATICS

ABSTRACT SERVICES AND INDEXES

COMPUMATH CITATION INDEX. Institute for Scientific Information, 3501 Market Street, Philadelphia, PA 19104. (800) 523-1850 or (215) 386-0100. Three times per year. $875.00 per year.

CURRENT MATHEMATICAL PUBLICATIONS. American Mathematical Society, P.O. Box 6248, Providence, RI 02940. (800) 556-7774 or (401) 272-9500. 1969 to present. Seventeen times per year. $230.00 per year.

ENGINEERING INDEX MONTHLY AND AUTHOR INDEX. Engineering Information Inc., 345 East 47th Street, New York, NY 10017. (212) 705-7600. Monthly. $1560.00 per year.

GENERAL SCIENCE INDEX. H.W. Wilson and Company, 950 University Avenue, Bronx, NY 10452. (800) 367-6670 or (212) 588-8400. 1978 to present. Monthly. Inquire as to cost and availability.

MATHEMATICAL REVIEWS: A REVIEWING JOURNAL COVERING THE WORLD LITERATURE OF MATHEMATICAL RESEARCH. American Mathematical Society, P.O. Box 6248, Providence, RI 02940. (800) 7774 or (401) 272-9500. 1940 to present. Monthly. $2800.00 per year.

ASSOCIATIONS AND PROFESSIONAL SOCIETIES

AMERICAN MATHEMATICAL SOCIETY. P.O. Box 6248, Providence, RI 02940. (401) 272-9500.

MATHEMATICAL ASSOCIATION OF AMERICA. 1529 18th Street, N.W., Washington, DC 20036. (202) 387-5200.

SOCIETY FOR INDUSTRIAL AND APPLIED MATHEMATICS. 1400 Architects Building, 117 South 17th Street, Philadelphia, PA 19103. (215) 564-2929.

DIRECTORIES AND BIOGRAPHICAL SOURCES

INTERNATIONAL RESEARCH CENTERS DIRECTORY 1988-89. Darren L. Smith, editor. Gale Research Company, Book Tower, Detroit, MI 48226. (800) 521-0707. 4th edition. 1987. $360.00.

RESEARCH CENTERS DIRECTORY 1988. Gale Research Company, Book Tower, Detroit, MI 48226. (800) 521-0707. 12th edition. 1987. $365.00 for set.

SCIENTIFIC AND TECHNICAL ORGANIZATIONS AND AGENCIES DIRECTORY. Margaret Labash Young, editor. Gale Research Company, Book Tower, Detroit, MI 48226. (800) 521-0707. 2nd edition. 1987. $185.00.

GENERAL WORKS

ANALYTIC GEOMETRY. G. Fuller. Addison-Wesley Publishing Company, Inc., 1 Jacob Way, Reading, MA 01867. (617) 944-3700. Sixth edition. 1986. $29.95.

CALCULUS AND ANALYTIC GEOMETRY. George B. Thomas. Addison-Wesley Publishing Company, Inc., 1 Jacob Way, Reading, MA 01867. (617) 944-3700. Sixth edition. 1984. $41.95.

CALCULUS WITH ANALYTIC GEOMETRY. H. Anton. John Wiley and Sons, Inc., 605 Third Avenue, New York, NY 10158. (800) 526-5368. Second edition. 1984. $42.50.

HANDBOOKS AND MANUALS

CRC HANDBOOK OF MATHEMATICAL SCIENCES. William H. Beyer, editor. CRC Press, 2000 Corporate Boulevard, Boca Raton, FL 33431. (800) 272-7737. 6th edition. 1987. $64.95.

ONLINE DATA BASES

COMPENDEX. Engineering Information, Inc., 345 East 47th Street, New York, NY 10017. (800) 221-1044 or (212) 705-7615. Engineering and technical literature, 1975 to present. Inquire as to online cost and availability.

DISSERTATION ABSTRACTS ONLINE. University Microfilms International, 300 North Zeeb Road, Ann Arbor, MI 48106. (800) 521-0600 or (313) 761-4700. Scope includes virtually all doctoral dissertations accepted at accredited American institutions from 1861 to present in over 250 subject areas. Inquire as to online cost and availability.

MATHFILE. American Mathematical Society, P.O. Box 6248, Providence, RI 02940. (800) 556-7774 or (401) 272-9500. An online version of Mathematical Reviews. 1973 to present. Inquire as to online cost and availability.

NTIS. National Technical Information Service, 5285 Port Royal Road, Springfield, VA 22161. (703) 487-4630. Broad coverage of government sponsored research reports, 1964 to present. Inquire as to online cost and availability.

WILSONLINE. H.W. Wilson and Company, 950 University Avenue, Bronx, NY 10452. (800) 367-6770 or (212) 588-8400. Makes available online versions of the H.W. Wilson indexes including Applied Science and Technology Index, Business Periodicals Index and Readers' Guide to Periodical Literature. Approximately 1980 to present. Inquire as to online cost and availability.

PERIODICALS

APPLIED MATHEMATICS AND COMPUTATION. Elsevier Science Publishing Company, Inc., 52 Vanderbilt Avenue, New York, NY 10017. (212) 370-5520. 1975 to present. Monthly. $355.00 per year.

COLLEGE MATHEMATICS JOURNAL. Mathematical Association of America, 1529 18th Street, N.W., Washington, DC 20036. Five times per year. $35.00 per year.

MATHEMATICAL INTELLIGENCER. Springer-Verlag New York, Inc., 175 Fifth Avenue, New York, NY 10010. (800) 526-7254. 1978 to present. Quarterly. $25.00 per year.

SIAM JOURNAL OF APPLIED MATHEMATICS. Society for Industrial and Applied Mathematics, 1405 Architects Building, 117 South 17th Street. Philadelphia, PA 19103. (215) 564-2929. 1953 to present. Bimonthly. $130.00 per year.

SIAM JOURNAL OF MATHEMATICAL ANALYSIS. Society for Industrial and Applied Mathematics, 1405 Architects Building, 117 South 17th Street. Philadelphia, PA 19103. (215) 564-2929. 1964 to present. Bimonthly. $120.00 per year.

SIAM REVIEW. Society for Industrial and Applied Mathematics, 1405 Architects Building, 117 South 17th Street. Philadelphia, PA 19103. (215) 564-2929. 1959 to present. Quarterly. $82.00 per year.

RESEARCH CENTERS AND INSTITUTES

CENTER FOR APPLIED MATHEMATICS. University of Georgia, Tucker Hall, Athens, GA 30602. (404) 542-3491.

CENTER FOR MATHEMATICAL SCIENCE RESEARCH. Rutgers University, New Brunswick, NJ 08903. (201) 932-3117.

CENTER FOR PURE AND APPLIED MATHEMATICS. University of California, Berkeley, 977 Evans Hall, Berkeley, CA 94720. (415) 642-0116.

INSTITUTE FOR MATHEMATICS AND ITS APPLICATIONS. University of Minnesota, 514 Vincent, 206 Church Street, S.E., Minneapolis, MN 55455. (612) 624-6066.

INSTITUTE OF APPLIED MATHEMATICS. University of Missouri - Rolla, Rolla, MO 65401. (314) 341-4151.

ANALYTICAL CHEMISTRY

See also: CHEMISTRY

ABSTRACT SERVICES AND INDEXES

APPLIED SCIENCE AND TECHNOLOGY INDEX. H.W. Wilson Company, 950 University Avenue, Bronx, NY 10452. (800) 367-6670 or (212) 588-8400. Inquire as to cost and availability.

CHEMICAL ABSTRACTS. Chemical Abstracts Service, 2540 Olentangy Road, P.O. Box 3012, Columbus, OH 43210. (800) 848-6538 or (614) 421-3600. Weekly. $9200.00 per year.

GENERAL SCIENCE INDEX. H.W. Wilson Company, 950 Unviersity Avenue, Bronx, NY 10452. (800) 367-6770 or (212) 588-8400. Inquire as to cost and availability.

PHYSICS ABSTRACTS. Institute of Electrical Engineers, London, United Kingdom. Available from: Institute of Electrical and Electronic Engineers (IEEE), 345 East 47th Street, New York, NY 10017. (212) 705-7900.

SCIENCE CITATION INDEX. Institute for Scientific Information, 3501 Market Street, Philadelphia, PA 19104. (800) 523-1850 or (215) 386-0100.

ASSOCIATIONS AND PROFESSIONAL SOCIETIES

AMERICAN CHEMICAL SOCIETY. 1155 16th Street, NW, Washington, DC 20036. (202) 872-4600.

ASSOCIATION OF CONSULTING CHEMISTS AND CHEMICAL ENGINEERS. 50 East 41st Street, Suite 92, New York, NY 10017. (212) 684-6255.

ASSOCIATION OF OFFICIAL ANALYTICAL CHEMISTS. 1111 North 19th Street, Suite 210, Arlington, VA 22209. (703) 522-3032.

BIBLIOGRAPHIES

SCIENCE BOOKS AND FILMS. American Association for the Advancement of Science, 1333 H Street, NW, Washington, DC 20005.

SCIENTIFIC AND TECHNICAL BOOKS IN PRINT; AN INDEX TO LITERATURE IN SCIENCE AND TECHNOLOGY. R.R. Bowker Company, 205 East 42nd Street, New York, NY 10017. (800) 521-8110 or (212) 916-1600.

DIRECTORIES AND BIOGRAPHICAL SOURCES

AMERICAN INSTITUTE OF CHEMISTS. American Institute of Chemists, 7315 Wisconsin Avenue, Bethesda, MD 20814. (301) 652-2447. 1986. $35.00.

AMERICAN MEN AND WOMEN OF SCIENCE. Physical and Biological Sciences. Fifteenth edition. R.R. Bowker Company, 205 East 42nd Street, New York, NY 10017. (800) 521-8110 or (212) 916-1600.

BIOGRAPHICAL DICTIONARY OF SCIENTISTS: CHEMISTS. David Abbott, editor. P. Bedrick Books, 125 East 23rd Street, New York, NY 10010. (212) 777-1187. 1984. $18.95.

CHEMICAL WEEK - BUYERS GUIDE ISSUE. McGraw-Hill Book Company, 1221 Avenue of the Americas, New York, NY 10020. (800) 628-0004. Annual, October. $50.00.

CONSULTING SERVICES: CHEMISTS AND CHEMICAL ENGINEERS. Association of Consulting Chemists and Chemical Engineers, 50 East 41st Street, New York, NY 10017. (212) 684-6255. Annual. 1986. $45.00.

GOVERNMENT RESEARCH DIRECTORY. Gale Research Company, Book Tower, Detroit, MI 48226. (800) 521-0707. Fourth edition. 1987. $350.00.

INTERNATIONAL RESEARCH CENTERS DIRECTORY 1986-1987. Gale Research Company, Book Tower, Detroit, MI 48226. (800) 521-0707. Third edition. 1986. $330.00.

RESEARCH CENTERS DIRECTORY. Gale Research Company, Book Tower, Detroit, MI 48226. (800) 521-0707. Eleventh edition. 1987. $355.00.

SCIENTIFIC AND TECHNICAL ORGANIZATIONS AND AGENCIES DIRECTORY. Gale Research Company, Book Tower, Detroit, MI 48226. (800) 521-0707. 1985. $150.00.

WORLD GUIDE TO SCIENTIFIC ASSOCIATIONS AND LEARNED SOCIETIES. K.G. Saur Incorporated, 175 Fifth Avenue, New York, NY 10010. (800) 521-0707 or (212) 982-1302. Fourth edition. 1984. $112.00.

WHO'S WHO IN FRONTIER SCIENCE AND TECHNOLOGY. Maruis Who's Who, Incorporated, 200 East Ohio Street, Chicago, IL 60611. (800) 428-3898 or (312) 787-2008.

WHO'S WHO IN TECHNOLOGY TODAY. Reston Publishing Company, Incorporated, c/o Prentice-Hall, Incorporated, Englewood Cliffs, NJ 07632. (800) 262-6868. Biennial. Five volumes. $425.00. Covers the fields of electronics, computer science, physics, optics, chemistry, biotechnology, mechanics, energy, and earth science.

ENCYCLOPEDIAS AND DICTIONARIES

CONCISE ENCYCLOPEDIA OF CHEMICAL TECHNOLOGY. Kirk-Othmer. John Wiley and Sons, Incorporated, 605 Third Avenue, New York, NY 10158. (800) 526-5368 or (212) 850-6000. Third edition. 1985. $129.95.

CONDENSED CHEMICAL DICTIONARY. Gessner Hawley. Van Nostrand Reinhold, 115 Fifth Avenue, New York, NY 10003. Tenth edition. 1981. $49.95.

ENCYCLOPEDIA OF PHYSICAL SCIENCE AND TECHNOLOGY. Academic Press, Incorporated, Orlando, FL 32887. (800) 321-5068 or (305) 345-2734. Fifteen volumes, 1986.

GLOSSARY OF CHEMICAL TERMS. Clifford A. Hampel and Gessner G. Hawley. Van Nostrand Reinhold Company, 115 Fifth Avenue, New York, NY 10003. (800) 543-2681 or (212) 254-3232. Second edition. 1982. $21.95.

MC-GRAW-HILL ENCYCLOPEDIA OF SCIENCE AND TECHNOLOGY. McGraw-Hill Book, Incorporated, 1221 Avenue of the Americas, New York, NY 10020. (212) 997-3675.

VAN NOSTRAND REINHOLD ENCYCLOPEDIA OF CHEMISTRY. Douglas M. Considine and Glenn D. Considine. Van Nostrand Reinhold Publishing Company, Incorporated, 115 Fifth Avenue, New York, NY 10003. (800) 543-2681 or (212) 254-3232. 1984. $97.95.

GENERAL WORKS

ANALYTICAL CHEMISTRY. Gary D. Christian. John Wiley and Sons, Incorporated, 605 Third Avenue, New York, NY 10158. (800) 526-5368 or (212) 850-6000. Fourth edition. 1986. $37.95.

BASIC ANALYTICAL CHEMISTRY. L. Pataki and E. Zapp. Pergamon Press, Maxwell House, Fairview Park, Elmsford, NY 10523. (914) 592-7700. 1981. $31.00 in paper.

CHEMICAL DERIVATIZATION IN ANALYTICAL CHEMISTRY. R.W. Frei, editor. Volume 1: Chromatography. 1981. $49.50. Volume 2: Separation and Continuous Flow Techniques. 1982. $45.00. Plenum Publishing Company, 233 Spring Street, New York, NY 10013. (800) 221-9369.

CHEMISTRY OF THE ELEMENTS. N.N. Greenwood and A. Earnshaw. Pergamon Publishing, Incorporated, Maxwell House, Fairview Park, Elmsford, NY 10523. (914) 592-7700. 1984. $143.00.

COMPUTERS IN ANALYTICAL CHEMISTRY. P.G. Baker. Pergamon Publishing, Incorporated, Maxwell House, Fairview Park, Elmsford, NY 10523. (914) 592-7700. 1983. $91.30.

ELECTROANALYTICAL CHEMISTRY: BASIC PRINCIPLES AND APPLICATIONS. J.A. Plembeck. John Wiley and Sons, Incorporated, 605 Third Avenue, New York, NY 10158. (800) 526-5368 or (212) 850-6000. 1982. $46.00.

HANDBOOKS AND MANUALS

THE CHEMIST'S COMPANION: A HANDBOOK OF PRACTICAL DATA, TECHNIQUES, AND REFERENCES. Arnold J. Gordon and Richard A. Ford. John Wiley and Sons, Incorporated, 605 Third Avenue, New York, NY 10158. (800) 526-5368. 1973. $49.95.

CRC HANDBOOK OF CHEMISTRY AND PHYSICS. CRC Press, Incorporated, 2000 Corporate Boulevard, NW, Boca Raton, FL 33431. (305) 994-0555. Sixty-seventh edition. 1986. $69.95.

HANDBOOK OF APPLIED CHEMISTRY: FACTS FOR ENGINEERS, SCIENTISTS, TECHNICIANS, AND TECHNICAL MANAGERS. Vollrath Hopp and Ingp Hennig. McGraw-Hill Book Company, 1221 Avenue of the Americas, New York, NY 10020. (800) 628-0004. 1983. $54.00.

LANGE'S HANDBOOK OF CHEMISTRY. John A. Dean, editor. McGraw-Hill Book Company, 1221 Avenue of the Americas, New York, NY 10020. (800) 628-0004. 1985. $59.50.

ONLINE DATA BASES

CA SEARCH. Chemical Abstracts Service, P.O. Box 3012, Columbus, OH 43210. Guide to chemical literature, 1972 to present. Inquire as to cost and availability.

DISSERTATION ABSTRACTS ONLINE. University Microfilms International, 300 North Zeeb Road, Ann Arbor, MI 48106. (800) 521-0600 or (313) 761-4700. Scope includes virtually all doctoral dissertations accepted at accredited American institutions from 1861 to present in 252 subject areas. Inquire as to online cost and availability.

INSPEC. INSPEC Marketing Department, Institute of Electrical and Electronics Engineers, Incorporated, IEEE Service Department, 445 Hoes Lane, Piscataway, NJ 08854. (201) 981-0060. Inquire as to online cost and availability.

NTIS. National Technical Information Service, 5285 Port Royal Road, Springfield, VA 22161. (703) 487-4630. Broad coverage of government sponsored research reports, 1964 to present. Inquire as to online cost and availability.

OTHER SOURCES

ACS ANALYTICAL CHEMISTRY LAB GUIDE. American Chemical Society, 1155 Sixteenth Street, NW, Washington, DC 20036. (800) 424-6747 or (202) 872-4700. Annual. $5.00.

GUIDE TO BASIC INFORMATION SOURCES IN CHEMISTRY. Arthur Antony. John Wiley and Sons, Incorporated, 605 Third Avenue, New York, NY 10158.

HOW TO FIND CHEMICAL INFORMATION: A GUIDE FOR PRACTICING CHEMISTS, TEACHERS, AND STUDENTS. John Wiley and Sons, Incorporated, 605 Third Avenue, New York, NY 10158. (800) 526-5368 or (212) 850-6000. 1986. $35.00.

PERIODICALS

AMERICAN LABORATORY. International Scientific Communications, Incorporated, 808 Kings Highway, Fairfield, CT 06430. (203) 576-0500. Monthly. $108.00 per year.

ANALYST. Royal Society of Chemistry, Burlington House, London, W1V OBN, England. Monthly. $285.00 per year.

ANALYTICA CHIMICA ACTA. Elsevier Science Publishing Company, Incorporated, 52 Vanderbilt Avenue, New York, NY 10017. (212) 370-5520. Fifteen times per year. $1050.00 per year.

ANALYTICAL CHEMISTRY. American Chemical Society, 1155 Sixteenth Street, NW, Washington, DC 20036. (800) 424-6747 or (202) 872-4700. Monthly. $33.00 per year.

ANALYTICAL LETTERS: CHEMICAL ANALYSIS/CLINICAL AND BIOMEDICAL ANALYSIS. Marcel Dekker Journals, 270 Madison Avenue, New York, NY 10016. (800) 228-1160. Twenty-four times per year. $535.00 per year.

ANALYTICAL PROCEEDINGS. Royal Society of Chemistry, Burlington House, London, W1V OBN, England. Monthly. $135.00 per year,

ASSOCIATION OF OFFICIAL ANALYTICAL CHEMISTS. Journal. Association of Official Analytical Chemists, 1111 North 19th Street, Suite 210, Arlington, VA 22209. (703) 522-3032. Bimonthly. $90.00 per year.

CRC CRITICAL REVIEWS IN ANALYTICAL CHEMISTRY. CRC Press, Incorporated, 2000 Corporate Boulevard, NW, Boca Raton, FL 33431. (305) 994-0555. Quarterly. $104.00 per year.

FRESENIUS ZEITSCHRIFT FUR ANALYTISCHE CHEMIE. Springer-Verlag New York, Incorporated, 175 Fifth Avenue, New York, NY 10010. (212) 460-1500. Twenty-five times per year. $683.00 per year.

INTERNATIONAL LABORATORY. International Scientific Communications, Incorporated, 808 Kings Highway, Fairfield, CT 06430. (203) 576-0500. Ten times per year. $135.00 per year.

JOURNAL OF CHROMATOGRAPHIC SCIENCE. Preston Publications, Incorporated, 7800 Merrimac Avenue, Niles, IL 60648. (312) 965-0566. Monthly. $97.00 per year.

JOURNAL OF CHROMATOGRAPHY. Elsevier Science Publishing Company, Incorporated, 52 Vanderbilt Avenue, New York, NY 10017. (212) 370-5520. Fifty-six times per year. $725.00 per year.

MIKROCHIMICA ACTA. Springer-Verlag New York, Incorporated, 175 Fifth Avenue, New York, NY 10010. (212)

460-1500. Eighteen times per year. $420.00 per year.

SEPARATION SCIENCE AND TECHNOLOGY. Marcel Dekker Journals, 270 Madison Avenue, New York, NY 10016. (800) 228-1160. Ten times per year. $400.00 per year.

TRENDS IN ANALYTICAL CHEMISTRY. Elsevier Science Publishing Company, Incorporated, 52 Vanderbilt Avenue, New York, NY 10017. (212) 370-5520. Ten times per year. $40.00 per year.

RESEARCH CENTERS AND INSTITUTES

DEPARTMENT OF AGRICULTURE. Analytical Chemistry Laboratory, Beltsville Agricultural Research Center - East, Building 306, Beltsville, MD 20705. (301) 344-2495.

NATIONAL BUREAU OF STANDARDS. Center for Analytical Chemistry, Gaithersburg, MD 20899. (301) 921-2851.

NORTHEASTERN UNIVERSITY. Barnett Institute of Chemical Analysis and Materials Science, 360 Huntington Avenue, Boston, MA 02115. (617) 437-2864.

TEXAS A & M UNIVERSITY. Center for Chemical Characterization and Analysis, College Station, TX 77843. (409) 845-2341.

UNIVERSITY OF WASHINGTON. Center for Process Analytical Chemistry, BG-10, Seattle, WA 98195. (206) 545-2326.

ANTENNAS

See also: ELECTRONIC CIRCUITS AND COMPONENTS, ELECTRONICS, ELECTRONICS ENGINEERING, MICROWAVES, RADIO, SIGNAL PROCESSING

ABSTRACT SERVICES AND INDEXES

APPLIED SCIENCE AND TECHNOLOGY INDEX. H.W. Wilson and Company, 950 University Avenue, Bronx, NY 10452. (800) 367-6670 or (212) 588-8400. Monthly. Inquire as to cost and availability.

CHEMICAL ABSTRACTS. American Chemical Society, Chemical Abstracts Service, Box 3012, Columbus, OH 43210. (614) 421-3600. 1907 to present. Weekly. $9500.00 per year.

CURRENT CONTENTS: ENGINEERING, TECHNOLOGY AND APPLIED SCIENCES. Institute for Scientific Information, 3501 Market Street, Philadelphia, PA 19104. (800) 523-1850 or (215) 386-0100. Weekly. $275.00 per year.

ELECTRICAL AND ELECTRONICS ABSTRACTS. Institution of Electrical Engineers. Available from: Institute of Electrical and Electronics Engineers. IEEE Service Center, 445 Hoes Lane, Piscataway, NJ 08854. Monthly. $1250.00 per year.

ELECTRONICS AND COMMUNICATIONS ABSTRACTS. Cambridge Scientific Abstracts, 5161 River Road, Bethesda, MD 20816. (301) 951-1400. Bimonthly. Inquire as to cost and availability.

ENGINEERING INDEX MONTHLY AND AUTHOR INDEX. Engineering Information Inc., 345 East 47th Street, New York, NY 10017. (212) 705-7600. Monthly. $1560.00 per year.

IEEE PUBLICATIONS BULLETIN. Institute of Electrical and Electronics Engineers. Institute of Electrical and Electronics Engineers. IEEE Service Center, 445 Hoes Lane, Piscataway, NJ 08854. Quarterly. Free.

PHYSICS ABSTRACTS. Institution of Electrical Engineers. Available from: IEEE Service Center, 445 Hoes Lane, Piscataway, NJ 08854. 1898 to present. Bimonthly. $1700.00 per year.

PHYSICS BRIEFS. Physik Verlag GmbH, Postfach 1260/1280, D-6940 Weinheim, West Germany. (212) 661-9404. 1920 to present. Twenty-six times per year. $1250.00 per year.

SCIENCE CITATION INDEX. Institute for Scientific Information, 3501 Market Street, Philadelphia, PA 19104. (800) 523-1850 or (215) 386-0100. Six times per year. $6200.00 per year.

ASSOCIATIONS AND PROFESSIONAL SOCIETIES

AMERICAN ELECTRONICS ASSOCIATION. P.O. Box 10045, 2670 Hanover Street, Palo Alto, CA 94303. (415) 857-9300.

AMERICAN INSTITUTE OF PHYSICS. 335 East 45th Street, New York, NY 10017. (212) 661-9494.

ELECTRONICS INDUSTRIES ASSOCIATION. 2001 Eye Street, N.W., Washington, DC 20006. (202) 457-4900.

INSTITUTE OF ELECTRICAL AND ELECTRONICS ENGINEERS. 345 East 47th Street, New York, NY 10017. (212) 705-7900.

DIRECTORIES AND BIOGRAPHICAL SOURCES

AMERICAN MEN AND WOMEN OF SCIENCE. R.R. Bowker, Inc., Order Department, 245 West 17th Street, New York, NY 10011. (800) 521-8110. Eight volumes. 1986. $595.00 for set.

IEEE MEMBERSHIP DIRECTORY. Institute of Electrical and Electronics Engineers. IEEE Service Center, 445 Hoes Lane, Piscataway, NJ 08854. Annual. $7.00.

INTERNATIONAL RESEARCH CENTERS DIRECTORY 1988-89. Darren L. Smith, editor. Gale Research Company, Book Tower, Detroit, MI 48226. (800) 521-0707. 4th edition. 1987. $360.00.

1987 DIRECTORY OF ENGINEERING SOCIETIES AND RELATED ORGANIZATIONS. Gordon Davis, editor. Hemisphere Publishing Corporation, 1010 Vermont Avenue, NW, Washington, DC 20005. (800) 526-0275. 12th edition. 1987. $100.00.

RESEARCH CENTERS DIRECTORY 1988. Gale Research Company, Book Tower, Detroit, MI 48226. (800) 521-0707. 12th edition. 1987. $365.00 for set.

SCIENTIFIC AND TECHNICAL ORGANIZATIONS AND AGENCIES DIRECTORY. Margaret Labash Young, editor. Gale Research Company, Book Tower, Detroit, MI 48226. (800) 521-0707. 2nd edition. 1987. $185.00.

WHO'S WHO IN ELECTRONICS. Harris Publishing Company, 2057-2 Aurora Road, Twinsburg, OH 44087. (216) 425-9143. Annual. $90.00.

WHO'S WHO IN ENGINEERING. Gordon Davis, editor. Hemisphere Publishing Corporation, 1010 Vermont Avenue, NW, Washington, DC 20005. (800) 526-0275. 6th edition. 1985. $200.00.

ENCYCLOPEDIAS AND DICTIONARIES

IEEE STANDARD DICTIONARY OF ELECTRICAL AND ELECTRONICS TERMS. Frank Jay, editor. John Wiley and Sons, Inc., 605 Third Avenue, New York, NY 10158. (800) 526-5368. 3rd edition. 1984. $49.95.

GENERAL WORKS

ANTENNAS AND RADIO WAVE PROPAGATION. R.E. Collin. McGraw-Hill Book Company, 1221 Avenue of the Americas, New York, NY 10020. (212) 512-2000. 1985. $49.95.

PRINCIPLES OF ANTENNA THEORY. K.F. Lee. John Wiley and Sons, Inc., 605 Third Avenue, New York, NY 10158. (800) 526-5368. 1984. $32.95.

UNDERSTANDING ANTENNAS FOR RADAR, COMMUNICATIONS AND AVIONICS. B. Rulf and G.A. Robertshaw. Van Nostrand Reinhold Company, Inc., 135 West 50th Street, New York, NY 10020. (800) 543-2681. 1987. $46.95.

HANDBOOKS AND MANUALS

ANTENNA HANDBOOK: THEORY, APPLICATIONS AND DESIGN. Y.T. Lo and S.W. Lee. Van Nostrand Reinhold Company, Inc., 135 West 50th Street, New York, NY 10020. (800) 543-2681. 1988. $129.95.

ELECTRONIC ENGINEERS HANDBOOK. Donald G. Fink, editor. McGraw-Hill Book Company, 1221 Avenue of the Americas, New York, NY 10020. (212) 512-2000. 2nd edition. 1982. $89.00.

HANDBOOK OF MODERN ELECTRONICS AND ELECTRICAL ENGINEERING. Charles Belove, editor. John Wiley and Sons, Inc., 605 Third Avenue, New York, NY 10158. (800) 526-5368. 1986. $88.95.

ONLINE DATA BASES

CA SEARCH. Chemical Abstracts Service, P.O. Box 3012, Columbus, OH 43120. (800) 848-6538 or (614) 421-3600. Comprehensive guide to chemical literature, 1972 to present. Inquire as to online cost and availability.

COMPENDEX. Engineering Information, Inc., 345 East 47th Street, New York, NY 10017. (800) 221-1044 or (212) 705-7615. Engineering and technical literature, 1975 to present. Inquire as to online cost and availability.

DISSERTATION ABSTRACTS ONLINE. University Microfilms International, 300 North Zeeb Road, Ann Arbor, MI 48106. (800) 521-0600 or (313) 761-4700. Scope includes virtually all doctoral dissertations accepted at accredited American institutions from 1861 to present in over 250 subject areas. Inquire as to online cost and availability.

INSPEC. INSPEC Marketing Department, Institution of Electrical Engineers. Available from IEEE Service Center, 445 Hoes Lane, Piscataway, NJ 08854. (201) 981-0060. Online version of Physics Abstracts. Inquire as to online cost and availability.

NTIS. National Technical Information Service, 5285 Port Royal Road, Springfield, VA 22161. (703) 487-4630. Broad coverage of government sponsored research reports, 1964 to present. Inquire as to online cost and availability.

SCISEARCH. Institute for Scientific Information, 3501 Market Street, Philadelphia, PA 19104. (800) 523-1850 or (215) 386-0100. Broad multidisciplinary title and author index to the international literature of science and technology, 1974 to present. Inquire as to online cost and availability.

WILSONLINE. H.W. Wilson and Company, 950 University Avenue, Bronx, NY 10452. (800) 367-6770 or (212) 588-8400. Makes available online versions of the H.W. Wilson indexes including Applied Science and Technology Index, Business Periodicals Index and Readers' Guide to Periodical Literature. Approximately 1980 to present. Inquire as to online cost and availability.

OTHER SOURCES

A GUIDE TO THE LITERATURE OF ELECTRICAL AND ELECTRONICS ENGINEERING. Susan B. Ardis. Libraries Unlimited Inc., P.O. Box 263, Littleton, CO 80160. (303) 770-1220. 1987. $37.50.

PERIODICALS

ELECTRONIC DESIGN. Hayden Publishing Company, 10 Mulholland Drive, Hasbrouck Heights, NJ 07604. (201) 288-7520. 1952 to present. Biweekly. $40.00 per year.

ELECTRONICS. McGraw-Hill Book Company, 1221 Avenue of the Americas, New York, NY 10020. (212) 512-2000. 1930 to present. Weekly. $32.00 per year.

IEE PROCEEDINGS PART H: MICROWAVES, ANTENNAS AND PROPAGATION. Institution of Electrical Engineers (London). Available from: Institute of Electrical and Electronics Engineers. IEEE Service Center, 445 Hoes Lane, Piscataway, NJ 08854. 1980 to present. Bimonthly. Inquire.

IEEE CIRCUITS AND DEVICES MAGAZINE. Institute of Electrical and Electronics Engineers. IEEE Service Center, 445 Hoes Lane, Piscataway, NJ 08854. Bimonthly. $70.00 per year.

IEEE TRANSACTIONS ON ANTENNAS AND PROPAGATION. Institute of Electrical and Electronics Engineers. IEEE Service Center, 445 Hoes Lane, Piscataway, NJ 08854. 1952 to present. Monthly. $140.00 per year.

IEEE TRANSACTIONS ON ELECTRON DEVICES. Institute of Electrical and Electronics Engineers. IEEE Service Center, 445 Hoes Lane, Piscataway, NJ 08854. 1959 to present. Monthly. $175.00 per year.

INSTITUTE OF ELECTRICAL AND ELECTRONICS ENGINEERS PROCEEDINGS. Institute of Electrical and Electronics Engineers. IEEE Service Center, 445 Hoes Lane, Piscataway, NJ 08854. 1913 to present. Monthly. $140.00 per year.

RESEARCH CENTERS AND INSTITUTES

ELECTRICAL ENGINEERING RESEARCH LABORATORIES. Purdue University, Electrical Engineering Building, West Lafayette, IN 47907. (317) 494-3536.

ELECTRONICS RESEARCH CENTER. University of Texas at Austin, 132 Engineering Science Building, Austin, TX 78712. (512) 471-3954.

ELECTRONICS RESEARCH LABORATORY. University of California, Berkeley, 253 Cory Hall, Berkeley, CA 94720. (415) 642-2301.

LABORATORY FOR ELECTROMAGNETIC AND ELECTRONIC SYSTEMS. Massachusetts Institute of Technology, 77 Massachusetts Avenue, Cambridge, MA 02139. (617) 253-4631.

ANTHRACITE COAL

See: COAL

APPLIED MATHEMATICS

See: MATHEMATICS

AQUIFER

See: GROUNDWATER

ARC WELDING

See: WELDING

ARCH

See: STRUCTURAL ENGINEERING

ARCHITECTURAL ENGINEERING

See also: BRIDGES, BUILDING MATERIALS, CIVIL ENGINEERING, CONCRETE, CONSTRUCTION ENGINEERING, STRUCTURAL ENGINEERING, STRUCTURES

ABSTRACT SERVICES AND INDEXES

ABSTRACT JOURNAL IN EARTHQUAKE ENGINEERING. University of California at Berkeley, Earthquake Engineering Research Center, 1301 South 46th Street, Richmond, CA 94804. (415) 231-9413. Semiannual. $70.00 per copy.

ABSTRACT NEWSLETTER: BUILDING INDUSTRY TECHNOLOGY. National Technical Information Service, 5285 Port Royal Road, Springfield, VA 22161. (703) 487-4929. Weekly. $75.00 per year.

APPLIED SCIENCE AND TECHNOLOGY INDEX. H.W. Wilson and Company, 950 University Avenue, Bronx, NY 10452. (800) 367-6670 or (212) 588-8400. Monthly. Inquire as to cost and availability.

CONCRETE ABSTRACTS. American Concrete Institute, P.O. Box 19150, Detroit, MI 48219. (513) 532-2600. 1972 to present. Bimonthly. $125.00 per year.

CURRENT CONTENTS: ENGINEERING, TECHNOLOGY AND APPLIED SCIENCES. Institute for Scientific Information, 3501 Market Street, Philadelphia, PA 19104. (800) 523-1850 or (215) 386-0100. Weekly. $275.00 per year.

ENGINEERING INDEX MONTHLY AND AUTHOR INDEX. Engineering Information Inc., 345 East 47th Street, New York, NY 10017. (212) 705-7600. Monthly. $1560.00 per year.

INDEX TO SCIENTIFIC AND TECHNICAL PROCEEDINGS. Institute for Scientific Information, 3501 Market Street, Philadelphia, PA 19104. (800) 523-1850 or (215) 386-0100. 1978 to present. Monthly. $775.00 per year.

INTERNATIONAL CIVIL ENGINEERING ABSTRACTS. CITIS Limited, 2 Rosemount Terrace, Blackrock, Dublin, Ireland. 1974 to present. Monthly. $350.00 per year.

INTERNATIONAL STRUCTURAL ENGINEERING ABSTRACTS. CITIS Limited, 2 Rosemount Terrace, Blackrock, Dublin, Ireland. 1986 to present. Quarterly. $95.00 per year.

PUBLICATIONS INFORMATION. American Society of Civil Engineers, 345 East 47th Street, New York, NY 10017. (212) 705-7420. Abstracts, subject and author indexes to the publications of the American Society of Civil Engineers. Bimonthly. $80.00 per year.

SCIENCE CITATION INDEX. Institute for Scientific Information, 3501 Market Street, Philadelphia, PA 19104. (800) 523-1850 or (215) 386-0100. Six times per year. $6200.00 per year.

ASSOCIATIONS AND PROFESSIONAL SOCIETIES

AMERICAN INSTITUTE OF ARCHITECTS. 1735 New York Avenue, N.W., Washington, DC 20006. (202) 626-7300.

AMERICAN INSTITUTE OF BUILDING DESIGN. 1412 19th Street, Sacramento, CA 95814. (916) 447-2422.

AMERICAN SOCIETY FOR TESTING AND MATERIALS. 1916 Race Street, Philadelphia, PA 19103. (215) 299-5400.

AMERICAN SOCIETY OF CIVIL ENGINEERS. 345 East 47th Street, New York, NY 10017. (212) 705-7420.

COUNCIL ON TALL BUILDINGS AND URBAN HABITAT. Fritz Engineering Laboratory, Building 13, Lehigh University, Bethlehem, PA 18015. (215) 861-3515.

NATIONAL SOCIETY OF ARCHITECTURAL ENGINEERS. P.O. Box 395, Lawrence, KS 66044. (913) 864-3434.

DIRECTORIES AND BIOGRAPHICAL SOURCES

AMERICAN SOCIETY OF CIVIL ENGINEERS DIRECTORY: OFFICIAL REGISTER. 345 East 47th Street, New York, NY 10017. (212) 705-7420. Annual. Free.

ENGINEERING NEWS-RECORD TOP 500 DESIGN FIRMS ISSUE. McGraw-Hill Book Company, 1221 Avenue of the Americas, New York, NY 10020. (212) 512-2000. Annual. $10.00.

ENGINEERS - CIVIL. American Business Directories, Inc., 5707 South 86th Circle, Omaha, NE 68127. Annual. $80.00.

INTERNATIONAL RESEARCH CENTERS DIRECTORY 1988-89. Darren L. Smith, editor. Gale Research Company, Book Tower, Detroit, MI 48226. (800) 521-0707. 4th edition. 1987. $360.00.

1987 DIRECTORY OF ENGINEERING SOCIETIES AND RELATED ORGANIZATIONS. Gordon Davis, editor. Hemisphere Publishing Corporation, 1010 Vermont Avenue, NW, Washington, DC 20005. (800) 526-0275. 12th edition. 1987. $100.00.

RESEARCH CENTERS DIRECTORY 1988. Gale Research Company, Book Tower, Detroit, MI 48226. (800) 521-0707. 12th edition. 1987. $365.00 for set.

SCIENTIFIC AND TECHNICAL ORGANIZATIONS AND AGENCIES DIRECTORY. Margaret Labash Young, editor. Gale Research Company, Book Tower, Detroit, MI 48226. (800) 521-0707. 2nd edition. 1987. $185.00.

WHO'S WHO IN ENGINEERING. Gordon Davis, editor. Hemisphere Publishing Corporation, 1010 Vermont Avenue, NW, Washington, DC 20005. (800) 526-0275. 6th edition. 1985. $200.00.

ENCYCLOPEDIAS AND DICTIONARIES

CONSTRUCTION GLOSSARY: AN ENCYCLOPEDIC REFERENCE AND MANUAL. J.S. Stein. John Wiley and Sons, Inc., 605 Third Avenue, New York, NY 10158. (800) 526-5368. 1986. $35.95.

DICTIONARY OF CIVIL ENGINEERING. John S. Scott. John Wiley and Sons, Inc., 605 Third Avenue, New York, NY 10158. (800) 526-5368. Third edition. 1981. $26.95.

GENERAL WORKS

APPLIED STRUCTURAL STEEL DESIGN. L. Spiegel and G.F. Limbrunner. Prentice-Hall Publishing, Inc., Englewood Cliffs, NJ 07632. (800) 562-0245. 1986. $34.95.

DESIGN OF REINFORCED CONCRETE. Samuel E. French. Prentice-Hall Publishing, Inc., Englewood Cliffs, NJ 07632. (800) 562-0245. 1987. $25.00.

INTRODUCTION TO EARTHQUAKE ENGINEERING. S. Okamoto. Columbia University Press, 562 West 113th Street, New York, NY 10025. (212) 316-7100. 1985. $75.00.

SCIENCE AND TECHNOLOGY OF BUILDING MATERIALS. Henry Cowan and Peter Smith. Van Nostrand Reinhold Company, Inc., 135 West 50th Street, New York, NY 10020. (800) 543-2681. 1988. $34.95.

STRUCTURAL CONCEPTS AND SYSTEMS FOR ARCHITECTS AND ENGINEERS. T.Y. Lin and Sydney Statesbury. Van Nostrand Reinhold Company, Inc., 135 West 50th Street, New York, NY 10020. (800) 543-2681. Revised edition. 1988. $42.95.

STRUCTURAL ENGINEERING FOR ARCHITECTS. K.R. Lauer. McGraw-Hill Book Company, 1221 Avenue of the Americas, New York, NY 10020. (212) 512-2000. 1981. $44.95.

WIND EFFECTS ON STRUCTURES: AN INTRODUCTION TO WIND ENGINEERING. Emil Simiu and Robert H. Scanlan. John Wiley and Sons, Inc., 605 Third Avenue, New York, NY 10158. (800) 526-5368. Second edition. 1985. $49.95.

HANDBOOKS AND MANUALS

ARCHITECTURAL AND ENGINEERING CALCULATIONS MANUAL. R.B. Butler. McGraw-Hill Book Company, 1221 Avenue of the Americas, New York, NY 10020. (212) 512-2000. 1984. $28.50.

BUILDING STRUCTURES HANDBOOK. R.N. White and C.G. Salmon. John Wiley and Sons, Inc., 605 Third Avenue, New York, NY 10158. (800) 526-5368. 1986. $79.95.

CIVIL ENGINEERING CALCULATIONS REFERENCE GUIDE. Tyler G. Hicks, editor. McGraw-Hill Book Company, 1221 Avenue of the Americas, New York, NY 10020. (212) 512-2000. 1987. $29.50.

CIVIL ENGINEERING PRACTICE. Paul N. Cheremisinoff and others, editors. Technomic Publishing Company, Inc., 851 Holland Avenue, Box 3535, Lancaster, PA 17604. (800) 233-9936. Five volumes. 1987-1988. $750.00 for set.

HANDBOOK OF CONCRETE ENGINEERING. Mark Fintel. Van Nostrand Reinhold Company, Inc., 135 West 50th Street, New York, NY 10020. (800) 543-2681. 1986. $89.95.

HANDBOOK OF MECHANICS, MATERIALS, AND STRUCTURES. A. Blake. John Wiley and Sons, Inc., 605 Third Avenue, New York, NY 10158. (800) 526-5368. 1985. $64.50.

STANDARD HANDBOOK FOR CIVIL ENGINEERS. F. S. Merritt, editor. McGraw-Hill Book Company, 1221 Avenue of the Americas, New York, NY 10020. (212) 512-2000. Third edition. 1983. $89.50.

STRUCTURAL ENGINEERING HANDBOOK. E.H. Gaylord and C.N. Gaylord, editors. McGraw-Hill Book Company, 1221 Avenue of the Americas, New York, NY 10020. (212) 512-2000. Second edition. 1979. $76.50.

ONLINE DATA BASES

COMPENDEX. Engineering Information, Inc., 345 East 47th Street, New York, NY 10017. (800) 221-1044 or (212) 705-7615. Engineering and technical literature, 1975 to present. Inquire as to online cost and availability.

DISSERTATION ABSTRACTS ONLINE. University Microfilms International, 300 North Zeeb Road, Ann Arbor, MI 48106. (800) 521-0600 or (313) 761-4700. Scope includes virtually all doctoral dissertations accepted at accredited American institutions from 1861 to present in over 250 subject areas. Inquire as to online cost and availability.

NTIS. National Technical Information Service, 5285 Port Royal Road, Springfield, VA 22161. (703) 487-4630. Broad coverage of government sponsored research reports, 1964 to present. Inquire as to online cost and availability.

SCISEARCH. Institute for Scientific Information, 3501 Market Street, Philadelphia, PA 19104. (800) 523-1850 or (215) 386-0100. Broad multidisciplinary title and author index to the international literature of science and technology, 1974 to present. Inquire as to online cost and availability.

TRIS. National Academy of Sciences. Transportation Research, 2101 Constitution Avenue, N.W., Washington, DC 20418. (202) 334-2000. Covers highway and transportation research. 1968 to present. Inquire as to cost and availability.

WILSONLINE. H.W. Wilson and Company, 950 University Avenue, Bronx, NY 10452. (800) 367-6770 or (212) 588-8400. Makes available online versions of the H.W. Wilson indexes including Applied Science and Technology Index, Business Periodicals Index and Readers' Guide to Periodical Literature. Approximately 1980 to present. Inquire as to online cost and availability.

PERIODICALS

AMERICAN CONCRETE INSTITUTE JOURNAL. P.O. Box 19150, Redford Station, Detroit, MI 48219. (513) 532-2600. 1929 to present. Bimonthly. $69.00 per year.

ARCHITECTURAL TECHNOLOGY. American Institute of Architects, 1735 New York Avenue, N.W., Washington, DC 20006. (202) 626-7300. 1983 to present. Six times per year. $24.00 per year.

BUILDING DESIGN AND CONSTRUCTION. Cahners Publishing Company, Inc., Cahners Plaza, Box 5080, 1350 East Touhy Avenue, Des Plaines, IL 60018. (312) 635-8800. 1958 to present. Monthly. $50.00 per year.

CIVIL ENGINEERING. American Society of Civil Engineers, 345 East 47th Street, New York, NY 10017. (212) 705-7420. 1930 to present. Monthly. $48.00 per year.

CONCRETE. Harcourt, Brace Jovanovich, Inc., 7500 Old Oak Boulevard, Cleveland, OH 44130. 1937 to present. Monthly. $25.00 per year.

JOURNAL OF STRUCTURAL ENGINEERING. American Society of Civil Engineers, 345 East 47th Street, New York, NY 10017. (212) 705-7420. 1956 to present. Monthly. $140.00 per year.

STRUCTURAL ENGINEERING PRACTICE: ANALYSIS, DESIGN, MANAGEMENT. Marcel Dekker Inc., 270 Madison Avenue, New York, NY 10016. (800) 228-1160. 1981 to present. Quarterly. $75.00 per year.

RESEARCH CENTERS AND INSTITUTES

COUNCIL ON TALL BUILDINGS AND URBAN HABITAT. Fritz Engineering Laboratory, Building 13, Lehigh University, Bethleham, PA 18015. (215) 861-3515.

PHIL M. FERGUSON STRUCTURAL ENGINEERING LABORATORY. University of Texas at Austin, Balcones Research Center, 10100 Burnet Road, Building 24, Austin, TX 78758. (512) 471-3062.

STRUCTURAL ENGINEERING LABORATORY. University of Michigan, 2340 G.G. Brown Building, Ann Arbor, MI 48109. (313) 763-3046.

STRUCTURAL ENGINEERING MATERIALS LABORATORY. University of California, Berkeley, Davis Hall, Berkeley, CA 94720. (415) 642-3434.

STRUCTURAL STABILITY RESEARCH COUNCIL. Fritz Engineering Laboratory No. 13, Lehigh University, Bethlehem, PA 18105. (215) 861-3519.

SPECIFICATIONS AND STANDARDS

SPECIFICATIONS FOR ARCHITECTURE, ENGINEERING, AND CONSTRUCTION. C. Ayers. McGraw-Hill Book Company, 1221 Avenue of the Americas, New York, NY 10020. (212) 512-2000. Second edition. 1984. $47.50.

STRUCTURAL DESIGN GUIDE TO THE ACI BUILDING CODE. Paul Rice. Van Nostrand Reinhold Company, Inc., 135 West 50th Street, New York, NY 10020. (800) 543-2681. Third edition. 1985. $47.95.

ARGON LASER

See: LASERS

ARITHMETIC

See: MATHEMATICS

AROMATIZATION

See: CHEMICAL ENGINEERING

ARTESIAN AQUIFER

See: GROUNDWATER

ARTIFICIAL INTELLIGENCE

See also: COMPUTER PROGRAMMING, EXPERT SYSTEMS, ROBOTICS

ABSTRACT SERVICES AND INDEXES

APPLIED SCIENCE AND TECHNOLOGY INDEX. H.W. Wilson Company, 950 University Avenue, Bronx, NY 10452. (800) 367-6670 or (212) 588-8400. Inquire as to cost and availability.

EIC/ARTIFICIAL INTELLIGENCE. EIC/Intelligence, Incorporated, 48 West 38th Street, New York, NY 10018. (800) 223-6275 or (212) 944-8500. Inquire as to cost and availability.

GENERAL SCIENCE INDEX. H.W. Wilson Company, 950 University Avenue, Bronx, NY 10452. (800) 367-6770 or (212) 588-8400.

MATHEMATICAL REVIEWS. American Mathematical Society, P.O. Box 6248, Providence, RI 02940. (800) 556-7774 or (401) 272-9500.

PHYSICS ABSTRACTS. Institute of Electrical Engineers, London, United Kingdom. Available from: Institute of Electrical and Electronic Engineers (IEEE), 345 East 47th Street, New York, NY 10017. (212) 705-7900.

SCIENCE CITATION INDEX. Institute for Scientific Information, 3501 Market Street, Philadelphia, PA 19104. (800) 523-1850 or (215) 386-0100. Inquire as to cost and availability.

ANNUAL REVIEWS AND YEARBOOKS

ADVANCES IN ARTIFICIAL INTELLIGENCE. JAI Press, Incorporated, 36 Sherwood Place, Greenwich, CT 06836. (203) 661-7602. Annual. Price varies.

ASSOCIATIONS AND PROFESSIONAL SOCIETIES

AMERICAN ASSOCIATION FOR ARTIFICIAL INTELLIGENCE. 445 Burgess Drive, Menlo Park, Ca 94025-3496. (415) 328-3123.

ASSOCIATION FOR COMPUTING MACHINERY. Special Interest Group on Artificial Intelligence, 11 West 42nd Street, New York, NY 10036.

BIBLIOGRAPHIES

ARTIFICIAL INTELLIGENCE BIBLIOGRAPHIC SUMMARIES OF THE SELECT LITERATURE. Henry M. Rylko, editor. Reort Store, 910 Massachusetts Street, Suite 503B, Lawrence, KS 66044. (913) 842-7348. 1985. Two volumes. 1985. $325.00.

SCIENTIFIC AND TECHNICAL BOOKS IN PRINT; AN INDEX TO LITERATURE IN SCIENCE AND TECHNOLOGY. R.R. Bowker Company, 205 East 42nd Street, New York, NY 10017. (800) 521-8110 or (212) 916-1600. 1987. Three Volumes. $159.95.

DIRECTORIES AND BIOGRAPHICAL SOURCES

DIRECTORY OF CONSULTANTS IN COMPUTER SYSTEMS. Research Publications, Twelve Lunar Drive, Woodbridge, CT 06525. (203) 397-2600. Annual. $75.00.

GOVERNMENT RESEARCH DIRECTORY. Gale Research Company, Book Tower, Detroit, MI 48226. (800) 521-0707. Third edition. 1985.

INTERNATIONAL RESEARCH CENTERS DIRECTORY 1986-1987. Gale Research Company, Book Tower, Detroit, MI 48226. (800) 521-0707. Third edition. 1986.

RESEARCH CENTERS DIRECTORY. Gale Research Company, Detroit, MI 48226. (800) 521-0707. Eleventh edition. 1987.

SCIENTIFIC AND TECHNICAL ORGANIZATIONS AND AGENCIES DIRECTORY. Gale Research Company, Book Tower, Detroit, MI 48226. (800) 521-0707. 1985.

WHO'S WHO IN ARTIFICIAL INTELLIGENCE. Alan Kernoff. P.O. Box 620098, Woodside, CA 94062. (415) 965-4561. Approximately $100.00.

WHO'S WHO IN TECHNOLOGY TODAY. Reston Publishing Company, Incorporated, c/o Prentice-Hall, Incorporated, Englewood Cliffs, NJ 07632. (800) 262-6868. Biennial. Five volumes. $425.00. Covers the fields of electronics, computer science, physics, optics, chemistry, biotechnology, mechanics, energy, and earth science.

ENCYCLOPEDIAS AND DICTIONARIES

ENCYCLOPEDIA OF ARTIFICIAL INTELLIGENCE. Stuart C. Shapiro, editor. John Wiley and Sons, Incorporated, 605 Third Avenue, New York, NY 10158. (800) 526-5368. Two volumes. 1987. $149.95 for set.

ENCYCLOPEDIA OF PHYSICAL SCIENCE AND TECHNOLOGY. Academic Press, Incorporated, Orlando, FL 32887. (800) 321-5068 or (305) 345-2734. Fifteen Volumes, 1986.

MCGRAW-HILL ENCYCLOPEDIA OF SCIENCE AND TECHNOLOGY. McGraw-Hill Book, Incorporated, 1221 Avenue of the Americas, New York, NY 10020. (212) 997-3675.

GENERAL WORKS

ACTORS: A MODEL OF CONCURRENT COMPUTATION IN DISTRIBUTED SYSTEMS. MIT Press, 28 Carleton Street, Cambridge, MA 02142. (617) 253-2884. 1986. $25.00.

APPLICATIONS OF ARTIFICIAL INTELLIGENCE TO ENGINEERING PROBLEMS. R.A. Adey and D. Sriram, editors. Springer-Verlag, New York, 175 Fifth Avenue, New York, NY 10010. (800) 526-7264. Two volumes. 1986. $196.00.

ARTIFICIAL INTELLIGENCE. Patrick Henry Winston. Addison-Wesley Publishing Company, Incorporated, 1 Jacob Way, Reading, MA 01867. (617) 944-3700. Second edition. 1984.

DYNAMIC MEMORY: A THEORY OF REMINDING AND LEARNING IN COMPUTERS AND PEOPLE. Roger C. Schank. Cambridge University Press, 32 East 57th Street, New York, NY 10022. (800) 431-1580 or (212) 688-8888. 1983. $12.95 in paper.

INTRODUCTION TO ARTIFICIAL INTELLIGENCE. Eugene Charniak and Drew McDermott. Addison-Wesley Publishing Company, Incorporated, 1 Jacob Way, Reading, MA 01867. (617) 944-3700. 1985. $36.95.

MACHINERY OF THE MIND: INSIDE THE NEW SCIENCE OF ARTIFICIAL INTELLIGENCE. George Johnson. Times Books, 201 East 50th Street, New York, NY 10022. (800) 242-7737 or (212) 751-2000. 1986. $19.95.

MAN-MADE MINDS: THE PROMISE OF ARTIFICIAL INTELLIGENCE. M. Waldrop. Walker and Company, 720 Fifth Avenue, New York, NY 10019. (212) 265-3632. 1987. $19.95.

RULE-BASED EXPERT SYSTEMS. Bruce G. Buchanan and Edward H. Shortliffe. Addison-Wesley, 1 Jacob Way, Reading, MA 10867. (617) 944-3700. 1985.

UNDERSTANDING COMPUTERS AND COGNITION. T. Winograd and F. Flores. Ablex Publishing Corporation, 355 Chestnut Street, Norwood, NJ 07648. (201) 767-8450. 1985. $24.95.

HANDBOOKS AND MANUALS

HANDBOOK OF ARTIFICIAL INTELLIGENCE. Avrom Barr and Edward A. Feigenbaum. William Kaufmann, Incorporated, 95 First Street, Los Altos, CA 94022. (415) 948-5810. Three volumes. $79.95 in paper.

ONLINE DATA BASES

DISSERTATION ABSTRACTS ONLINE. University Microfilms International, 300 North Zeeb Road, Ann Arbor, MI 48106. (800) 521-0600 or (313) 761-4700. Scope includes virtually all doctoral dissertations accepted at accredited American institutions from 1861 to present in 252 subject areas. Inquire as to cost and availability.

INSPEC. INSPEC Marketing Department, Institute of Electrical and Electronics Engineers, Incorporated, IEEE Service Department, 445 Hoes Lane, Piscataway, NJ 08854. (201) 981-0600. Inquire as to on-line cost and availability.

MATHFILE. American Mathematical Society, P.O. Box 6248, Providence, RI 02940. (800) 556-7774 or (401) 272-9500. Scope includes pure and applied mathematics and related areas of physics, statistics, engineering, computer science, and operations research literature since 1973. Inquire as to cost and availability.

NTIS. National Technical Information Service, 5285 Port Royal Road, Springfield, VA 22161. (703) 487-4630. Broad coverage of government sponsored research reports, 1964 to present. Inquire as to cost and availability.

SCISEARCH. Institute for Scientific Information, 3501 Market Street, Philadelphia, PA 19104. (800) 523-1850 or (215) 386-0100. Broad multidisciplinary title and author index to the international literature of science and technology, 1974 to present. Inquire as to cost and availability.

WILSONLINE. H.W. Wilson Company, 950 University Avenue, Bronx, NY 10452. (800) 367-6770 or (212) 588-8400. Makes available online versions of the printed H.W. Wilson indexes including Applied Science and Technology Index, Business Periodicals Index, and Readers' Guide to Periodical Literature. Period covered is generally 1983 to present. Inquire as to cost and availability.

OTHER SOURCES

ARTIFICIAL INTELLIGENCE ACTIVITY SOURCEBOOK. Artificial Intelligence Research Service, 3857 Birch Street, Suite 464, Newport Beach, CA 92660. (714) 642-8978. Contains in-depth analysis of 130 artificial intelligence tools. Approximately $100.00.

ARTIFICIAL INTELLIGENCE AND EXPERT SYSTEMS SOURCEBOOK. Daniel V. Hunt. Chapman and Hall, 29 West 35th Street, New York, NY 10001. (212) 244-3336. 1986. $34.50.

PERIODICALS

AI MAGAZINE. American Association for Artificial Intelligence, 445 Burgess Drive, Menlo Park, CA 94025-3496. (415) 328-3123. Five times per year. $50.00 per year.

APPLIED ARTIFICIAL INTELLIGENCE REPORTER. University of Miami Computer Systems Research Institute, Box 248235, Coral Gables, FL 33124. (284-5195. Monthly. $49.00 per year.

ARTIFICIAL INTELLIGENCE. Elsevier Science Publishing Company, Incorporated, 52 Vanderbilt Avenue, New York, NY 10017. (212) 370-5520. Nine times per year. $176.00.

INTERNATIONAL JOINT CONFERENCE ON ARTIFICIAL INTELLIGENCE. Advance Papers of the Conference. International Joint Conference on Artificial Intelligence, c/o Louis Robinsonanon, AAAI Office, 445 Burgess Drive, Menlo Park, Ca 94025. Semi-annual. $25.00.

INTERNATIONAL JOURNAL FOR ARTIFICIAL INTELLIGENCE IN ENGINEERING. CML Publications, Suite 6200, 400 West Cummings Park, Woburn, MA 01801. Quarterly. $130.00 per year.

ROBOTICA: INTERNATIONAL JOURNAL OF INFORMATION, EDUCATION AND RESEARCH IN ROBOTICS AND ARTIFICIAL INTELLIGENCE. Cambridge University Press, 32 East 57th Street, New York, NY 10022. (800) 872-7423. Quarterly. $50.00.

SIGART NEWSLETTER. Association for Computing Machinery, Special Interest Group on Artificial Intelligence, 11 West 42nd Street, New York, NY 10036. Quarterly. $15.00.

RESEARCH CENTERS AND INSTITUTES

MASSACHUSETTS INSTITUTE OF TECHNOLOGY. Artificial Intelligence Laboratory, 545 Technology Square, Cambridge, MA 02139. (617) 253-6218.

SRI INTERNATIONAL CENTER FOR INTELLIGENT COMPUTER SYSTEMS. 333 Ravenswood Avenue, Menlo Park, CA 94025. (415) 859-4771.

UNIVERSITY OF MARYLAND. Systems Research Center, College Park, MD 20742. (301) 454-6876.

UNIVERSITY OF MIAMI. Intelligent Computer Systems Research Institute, P.O. Box 248235, Coral Gables, FL 33124. (305) 284-5195.

UNIVERSITY OF PENNSYLVANIA. Center for Artificial Intelligence, Computer and Information Science Department, Moore School of Electrical Engineering/D2, Philadelphia, PA 19104. (215) 898-6222.

ARTIFICIAL SATELLITES

See also: RADIO, REMOTE SENSING, TELECOMMUNICATIONS, TELEVISION

ABSTRACT SERVICES AND INDEXES

APPLIED SCIENCE AND TECHNOLOGY INDEX. H.W. Wilson and Company, 950 University Avenue, Bronx, NY 10452. (800) 367-6670 or (212) 588-8400. Monthly. Inquire as to cost and availability.

ASTRONOMY AND ASTROPHYSICS ABSTRACTS. Springer-Verlag New York, Inc., 175 Fifth Avenue, New York, NY 10010. (800) 526-7254. 1969 to present. Approximately $70.00 per year.

ASTRONOMY AND ASTROPHYSICS MONTHLY INDEX. Olivetree Associates, P.O. Box 236, Sierre Madre, CA 91024. $220.00 per year. Complementary copies available on request.

CURRENT CONTENTS: ENGINEERING, TECHNOLOGY AND APPLIED SCIENCES. Institute for Scientific Information, 3501 Market Street, Philadelphia, PA 19104. (800) 523-1850 or (215) 386-0100. Weekly. $275.00 per year.

ENGINEERING INDEX MONTHLY AND AUTHOR INDEX. Engineering Information Inc., 345 East 47th Street, New York, NY 10017. (212) 705-7600. Monthly. $1560.00 per year.

GENERAL SCIENCE INDEX. H.W. Wilson and Company, 950 University Avenue, Bronx, NY 10452. (800) 367-6670 or (212) 588-8400. 1978 to present. Monthly. Inquire as to cost and availability.

INDEX TO SCIENTIFIC AND TECHNICAL PROCEEDINGS. Institute for Scientific Information, 3501 Market Street, Philadelphia, PA 19104. (800) 523-1850 or (215) 386-0100. 1978 to present. Monthly. $775.00 per year.

INDEX TO SCIENTIFIC REVIEWS. Institute for Scientific Information, 3501 Market Street, Philadelphia, PA 19104. (800) 523-1850 or (215) 386-0100. 1974 to present. Semi-annual. $550.00 per year.

METEOROLOGICAL AND GEOASTROPHYSICAL ABSTRACTS. American Meteorological Society, 45 Beacon Street, Boston, MA 02108. (617) 227-2425. 1950 to present. Monthly. $450.00 per year.

OCEAN ABSTRACTS. Cambridge Scientific Abstracts, 5161 River Road, Bethesda, MD 20816. (301) 951-1400. 1963 to present. Bimonthly. $450.00 per year.

PHYSICS ABSTRACTS. Institution of Electrical Engineers. Available from: IEEE Service Center, 445 Hoes Lane, Piscataway, NJ 08854. 1898 to present. Bimonthly. $1700.00 per year.

PHYSICS BRIEFS. Physik Verlag GmbH, Postfach 1260/1280, D-6940 Weinheim, West Germany. (212) 661-9404. 1920 to present. Twenty-six times per year. $1250.00 per year.

SCIENCE CITATION INDEX. Institute for Scientific Information, 3501 Market Street, Philadelphia, PA 19104. (800) 523-1850 or (215) 386-0100. Six times per year. $6200.00 per year.

ASSOCIATIONS AND PROFESSIONAL SOCIETIES

AMERICAN INSTITUTE OF PHYSICS. 335 East 45th Street, New York, NY 10017. (212) 661-9494.

INSTITUTE OF ELECTRICAL AND ELECTRONICS ENGINEERS. IEEE Service Center, 445 Hoes Lane, Piscataway, NJ 08854.

INTERNATIONAL TELECOMMUNICATIONS SATELLITE ORGANIZATION. 3400 International Drive, N.W., Washington, DC 20008. (202) 944-6800.

NATIONAL ENVIRONMENTAL SATELLITE, DATA, AND INFORMATION SERVICE. 3300 Whitehaven Street, N.W., Washington, DC 20235. (202) 634-7318.

RADIO AMATEUR SATELLITE CORPORATION. P.O. Box 27, Washington, DC 20044. (301) 589-6062.

DIRECTORIES AND BIOGRAPHICAL SOURCES

INTERNATIONAL RESEARCH CENTERS DIRECTORY 1988-89. Darren L. Smith, editor. Gale Research Company, Book Tower, Detroit, MI 48226. (800) 521-0707. 4th edition. 1987. $360.00.

INTERNATIONAL SATELLITE DIRECTORY. S.F.P. Designs, Inc., 369 Redwood Avenue, Corte Madera, CA 96925. (415) 927-0379.

1987 DIRECTORY OF ENGINEERING SOCIETIES AND RELATED ORGANIZATIONS. Gordon Davis, editor. Hemisphere Publishing Corporation, 1010 Vermont Avenue, NW, Washington, DC 20005. (800) 526-0275. 12th edition. 1987. $100.00.

RESEARCH CENTERS DIRECTORY 1988. Gale Research Company, Book Tower, Detroit, MI 48226. (800) 521-0707. 12th edition. 1987. $365.00 for set.

SATELLITE COMMUNICATIONS, SATELLITE INDUSTRY DIRECTORY ISSUE. Cardiff Publishing Company, 6530 South Yosemite Street, Englewood, CO 80111. (303) 694-1522.

SATELLITE DIRECTORY. Phillips Publishing, Inc., 7811 Montrose Road, Potomac, MD 20854. (301) 340-2100.

SCIENTIFIC AND TECHNICAL ORGANIZATIONS AND AGENCIES DIRECTORY. Margaret Labash Young, editor. Gale Research Company, Book Tower, Detroit, MI 48226. (800) 521-0707. 2nd edition. 1987. $185.00.

WHO'S WHO IN ENGINEERING. Gordon Davis, editor. Hemisphere Publishing Corporation, 1010 Vermont Avenue, NW, Washington, DC 20005. (800) 526-0275. 6th edition. 1985. $200.00.

GENERAL WORKS

INTRODUCTION TO SATELLITE COMMUNICATIONS. G.B. Bleazard. John Wiley and Sons, Inc., 605 Third Avenue, New York, NY 10158. (800) 526-5368. 1985. $39.95.

SATELLITE OCEANOGRAPHY: AN INTRODUCTION FOR OCEANOGRAPHERS AND REMOTE SENSING SCIENTIST. I. Robinson. John Wiley and Sons, Inc., 605 Third Avenue, New York, NY 10158. (800) 526-5368. 1985. $59.95.

THE VERSATILE SATELLITE. Richard W. Porter. Oxford University Press, 200 Madison Avenue, New York, NY 10016. (800) 458-5833. 1977. $16.95.

ONLINE DATA BASES

COMPENDEX. Engineering Information, Inc., 345 East 47th Street, New York, NY 10017. (800) 221-1044 or (212) 705-7615. Engineering and technical literature, 1975 to present. Inquire as to online cost and availability.

DISSERTATION ABSTRACTS ONLINE. University Microfilms International, 300 North Zeeb Road, Ann Arbor, MI 48106. (800) 521-0600 or (313) 761-4700. Scope includes virtually all doctoral dissertations accepted at accredited American institutions from 1861 to present in over 250 subject areas. Inquire as to online cost and availability.

INSPEC. INSPEC Marketing Department, Institution of Electrical Engineers. Available from IEEE Service Center, 445 Hoes Lane, Piscataway, NJ 08854. (201) 981-0060. Online version of Physics Abstracts. Inquire as to online cost and availability.

NTIS. National Technical Information Service, 5285 Port Royal Road, Springfield, VA 22161. (703) 487-4630. Broad coverage of

government sponsored research reports, 1964 to present. Inquire as to online cost and availability.

SCISEARCH. Institute for Scientific Information, 3501 Market Street, Philadelphia, PA 19104. (800) 523-1850 or (215) 386-0100. Broad multidisciplinary title and author index to the international literature of science and technology, 1974 to present. Inquire as to online cost and availability.

WILSONLINE. H.W. Wilson and Company, 950 University Avenue, Bronx, NY 10452. (800) 367-6770 or (212) 588-8400. Makes available online versions of the H.W. Wilson indexes including Applied Science and Technology Index, Business Periodicals Index and Readers' Guide to Periodical Literature. Approximately 1980 to present. Inquire as to online cost and availability.

PERIODICALS

IEEE COMMUNICATIONS MAGAZINE. Institute of Electrical and Electronics Engineers. IEEE Service Center, 445 Hoes Lane, Piscataway, NJ 08854. 1953 to present. Monthly. $10.00 per year.

IEEE TRANSACTIONS IN COMMUNICATIONS TECHNOLOGY. Institute of Electrical and Electronics Engineers. IEEE Service Center, 445 Hoes Lane, Piscataway, NJ 08854. 1953 to present. Monthly. $115.00 per year.

IEEE TRANSACTIONS ON GEOSCIENCE AND REMOTE SENSING. IEEE Geoscience and Remote Sensing Society. Institute of Electrical and Electronics Engineers, 345 East 47th Street, New York, NY 10017. (212) 705-7900. Order from: IEEE Service Center, 445 Hoes Lane, Piscataway, NJ 08854. 1963 to present. Bimonthly. $110.00 per year.

PHOTOGRAMMETRIC ENGINEERING AND REMOTE SENSING. American Society for Photogrammetry and Remote Sensing. 210 Little Falls Street, Falls Church, VA 22046-4398. (703) 534-6617. Order from: Allen Press, Inc., 1041 New Hampshire Street, Box 368, Lawrence, KS 66044. 1934 to present. Monthly. $80.00 per year.

REMOTE SENSING OF ENVIRONMENT. Elsevier Science Publishing Company, Inc., 52 Vanderbilt Avenue, New York, NY 10017. (212) 370-5520. 1968 to present. Six times per year. $210.00 per year.

REMOTE SENSING REVIEWS. Harwood Academic Publishers, 50 West 23rd Street, New York, NY 10010. (212) 206-8900. Quarterly. $160.00 per year.

SATELLITE COMMUNICATIONS. Cardiff Publishing Company, 6530 South Yosemite, Englewood, CO 80111. (303) 694-1522. 1977 to present. Monthly. $27.00 per year.

RESEARCH CENTERS AND INSTITUTES

CENTER FOR SPACE RESEARCH. Massachusetts Institute of Technology, 77 Massachusetts Avenue, Cambridge, MA 02139. (617) 253-7501.

ASBESTOS

ABSTRACT SERVICES AND INDEXES

APPLIED SCIENCE AND TECHNOLOGY INDEX. H.W. Wilson and Company, 950 University Avenue, Bronx, NY 10452. (800) 367-6670 or (212) 588-8400. Monthly. Inquire as to cost and availability.

CHEMICAL ABSTRACTS. American Chemical Society, Chemical Abstracts Service, Box 3012, Columbus, OH 43210. (614) 421-3600. 1907 to present. Weekly. $9500.00 per year.

CONFERENCE PAPERS INDEX. Cambridge Scientific Abstracts, 5161 River Road, Bethesda, MD 20816. (301) 951-1400. 1972 to present. Monthly. Inquire as to cost and availability.

ENGINEERING INDEX MONTHLY AND AUTHOR INDEX. Engineering Information Inc., 345 East 47th Street, New York, NY 10017. (212) 705-7600. Monthly. $1560.00 per year.

GENERAL SCIENCE INDEX. H.W. Wilson and Company, 950 University Avenue, Bronx, NY 10452. (800) 367-6670 or (212) 588-8400. 1978 to present. Monthly. Inquire as to cost and availability.

BIBLIOGRAPHY AND INDEX OF GEOLOGY. American Geological Institute, 4220 King Street, Alexandria, VA 22302. (703) 379-2480. 1969 to present. Monthly. $1100.00 per year.

POLLUTION ABSTRACTS. Cambridge Scientific Abstracts, 5161 River Road, Bethesda, MD 20816. (301) 951-1400. Six times per year. Inquire as to cost and availability.

SCIENCE CITATION INDEX. Institute for Scientific Information, 3501 Market Street, Philadelphia, PA 19104. (800) 523-1850 or (215) 386-0100. Six times per year. $6200.00 per year.

ASSOCIATIONS AND PROFESSIONAL SOCIETIES

ASBESTOS INFORMATION ASSOCIATION OF NORTH AMERICA. 1745 Jefferson Davis Highway, Suite 509, Arlington, VA 22202. (703) 979-1150.

DIRECTORIES AND BIOGRAPHICAL SOURCES

AMERICAN MEN AND WOMEN OF SCIENCE. R.R. Bowker, Inc., Order Department, 245 West 17th Street, New York, NY 10011. (800) 521-8110. Eight volumes. 1986. $595.00 for set.

INTERNATIONAL RESEARCH CENTERS DIRECTORY 1988-89. Darren L. Smith, editor. Gale Research Company, Book Tower, Detroit, MI 48226. (800) 521-0707. 4th edition. 1987. $360.00.

1987 DIRECTORY OF ENGINEERING SOCIETIES AND RELATED ORGANIZATIONS. Gordon Davis, editor. Hemisphere Publishing Corporation, 1010 Vermont Avenue, NW, Washington, DC 20005. (800) 526-0275. 12th edition. 1987. $100.00.

RESEARCH CENTERS DIRECTORY 1988. Gale Research Company, Book Tower, Detroit, MI 48226. (800) 521-0707. 12th edition. 1987. $365.00 for set.

SCIENTIFIC AND TECHNICAL ORGANIZATIONS AND AGENCIES DIRECTORY. Margaret Labash Young, editor. Gale Research Company, Book Tower, Detroit, MI 48226. (800) 521-0707. 2nd edition. 1987. $185.00.

WHO'S WHO IN ENGINEERING. Gordon Davis, editor. Hemisphere Publishing Corporation, 1010 Vermont Avenue, NW, Washington, DC 20005. (800) 526-0275. 6th edition. 1985. $200.00.

ENCYCLOPEDIAS AND DICTIONARIES

ENCYCLOPEDIA OF HEALTH INFORMATION SOURCES. Paul Wasserman, editor. Gale Research Company, Book Tower, Detroit, MI 48226. (800) 521-0707. 1987. $135.00.

ENCYCLOPEDIA OF OCCUPATIONAL HEALTH AND SAFETY. Luigi Parmeggiani, editor. Gale Research Company, Book Tower, Detroit, MI 48226. (800) 521-0707. Third edition. 1983. $195.00 for set.

GENERAL WORKS

AIR QUALITY. T. Godish. Lewis Publishers, Inc., 121 South Main Street, P.O. Box 519, Chelsea, MI 48118. (313) 475-8619. 1985. $41.95.

ASBESTOS: PROPERTIES, APPLICATIONS AND HAZARDS. L. Michaels and S.S. Chissick. John Wiley and Sons, Inc., 605 Third Avenue, New York, NY 10158. (800) 526-5368. 1983. Two volumes. $225.00 for set.

ONLINE DATA BASES

CA SEARCH. Chemical Abstracts Service, P.O. Box 3012, Columbus, OH 43120. (800) 848-6538 or (614) 421-3600. Comprehensive guide to chemical literature, 1972 to present. Inquire as to online cost and availability.

COMPENDEX. Engineering Information, Inc., 345 East 47th Street, New York, NY 10017. (800) 221-1044 or (212) 705-7615. Engineering and technical literature, 1975 to present. Inquire as to online cost and availability.

DISSERTATION ABSTRACTS ONLINE. University Microfilms International, 300 North Zeeb Road, Ann Arbor, MI 48106. (800) 521-0600 or (313) 761-4700. Scope includes virtually all doctoral dissertations accepted at accredited American institutions from 1861 to present in over 250 subject areas. Inquire as to online cost and availability.

ENVIROLINE. EIC Intelligence, Inc., 48 West 38th Street, New York, NY 10018. (212) 944-8500. Worldwide environmental literature, 1970 to present. Inquire as to online cost and availability.

GEOREF. Online version of the BIBLIOGRAPHY AND INDEX OF GEOLOGY. American Geological Institute, 4220 King Street, Alexandria, VA 22302. (703) 379-2480. 1969 to present. Inquire as to online cost and availability.

INSPEC. INSPEC Marketing Department, Institution of Electrical Engineers. Available from IEEE Service Center, 445 Hoes Lane, Piscataway, NJ 08854. (201) 981-0060. Online version of Physics Abstracts. Inquire as to online cost and availability.

NTIS. National Technical Information Service, 5285 Port Royal Road, Springfield, VA 22161. (703) 487-4630. Broad coverage of government sponsored research reports, 1964 to present. Inquire as to online cost and availability.

SCISEARCH. Institute for Scientific Information, 3501 Market Street, Philadelphia, PA 19104. (800) 523-1850 or (215) 386-0100. Broad multidisciplinary title and author index to the international literature of science and technology, 1974 to present. Inquire as to online cost and availability.

WILSONLINE. H.W. Wilson and Company, 950 University Avenue, Bronx, NY 10452. (800) 367-6770 or (212) 588-8400. Makes available online versions of the H.W. Wilson indexes including Applied Science and Technology Index, Business Periodicals Index and Readers' Guide to Periodical Literature. Approximately 1980 to present. Inquire as to online cost and availability.

OTHER SOURCES

MINERALS YEARBOOK. Bureau of Mines, U.S. Department of the Interior. Available from: U.S. Government Printing Office, Washington, DC 20402. (202) 783-3238. Annual. Three volumes. $45.00.

PERIODICALS

ASBESTOS. Box B. Lakeville, PA 18438. Monthly. $6.00 per year.

ASBESTOS WORKER. International Association of Asbestos Workers, Machinists Building, 1300 Connecticut Avenue, N.W., Washington, DC 20036. (202) 785-2388. 1916 to present. Quarterly. Free.

ENGINEERING AND MINING JOURNAL. McGraw-Hill Publishing Company, 1221 Avenue of the Americas, New York, NY 10020. (212) 512-2000. Monthly. $20.00 per year.

RESEARCH CENTERS AND INSTITUTES

ASBESTOS HAZARDS ABATEMENT ASSISTANCE. U.S. Environmental Protection Agency, Office of Pesticides and Toxic Substances, 401 M Street, S.W., Washington, DC 20640. (202) 382-3949.

ASPHALT

See also: HIGHWAY ENGINEERING

ABSTRACT SERVICES AND INDEXES

APPLIED SCIENCE AND TECHNOLOGY INDEX. H.W. Wilson Company, 950 University Avenue, Bronx, NY 10452. (800) 367-6670 or (212) 588-8400. Inquire as to cost and availability.

CHEMICAL ABSTRACTS. Chemical Abstracts Service, 2540 Olentangy Road, Post Office Box 3012, Columbus, OH 43210. (800) 848-6538 or (614) 421-3600. Weekly. $9200.00 per year.

ENGINEERING INDEX MONTHLY. Engineering Information, Incorporated 345 East 47th Street, New York, NY 10017. (800) 221-1044 or (212) 705-7600. Monthly, with annual cumulation. $1560.00 per year.

ASSOCIATIONS AND PROFESSIONAL SOCIETIES

ASPHALT EMULSION MANUFACTURES ASSOCIATION. 1133 15th Street, NW, Washington, DC 20005. (202) 429-9440.

ASPHALT INSTITUTE. Asphalt Institute Building, College Park, MD 20742. (301) 277-4258.

ASPHALT RECYCLING AND RECLAIMING ASSOCIATION. Three Church Circle, Suite 250, Annapolis, MD 21401. (301) 267-0023.

ASSOCIATION OF ASPHALT PAVING TECHNOLOGISTS. Civil and Mineral Engineering Building, Room 134, University of Minnesota, Minneapolis, MN 55455. (612) 644-2996.

BIBLIOGRAPHIES

LIBRARY PATHFINDER: ASPHALT. Addison-Wesley Publishing Company, Reading, MA 01867. (617) 944-3700. $1.50.

LIST OF ASPHALT INSTITUTE PUBLICATIONS. Asphalt Institute, Asphalt Institute Building, College Park, MD 20742. (301) 277-4258. Annual. Free.

DIRECTORIES AND BIOGRAPHICAL SOURCES

ASPHALT/ASPHALT PRODUCTS DIRECTORY. American Business Directories, Incorporated, Division of American Business Lists, Incorporated, 5707 South 86th Circle, Omaha, NE 68127. (402) 331-7169. $125.00.

ASPHALT EMULSION MANUFACTURES ASSOCIATION - MEMBERSHIP DIRECTORY. Asphalt Emulsion Manufactures Association, 1133 15th Street, NW, Washington, DC 20005. (202) 429-9440. Annual. Free.

ASPHALT PAVING TECHNOLOGISTS DIRECTORY. Association of Asphalt Paving Technologists, Civil and Mineral Engineering Building, Room 134, University of Minnesota, Minneapolis, MN 55455. (612) 644-2996. Annual. $50.00.

ASPHALT RECYCLING AND RECLAIMING ASSOCIATION MEMBERSHIP DIRECTORY. Asphalt Recycling and Reclaiming Association, Three Church Circle, Suite 250, Annapolis, MD 21401. (301) 267-0023. Available to members only.

ASPHALT

RESEARCH CENTERS DIRECTORY. Gale Research Company, Book Tower, Detroit, MI 48226. (800) 521-0707. Thirteenth edition. 1988. $365.00.

WHO'S WHO IN ENGINEERING. Engineers Joint Council, 345 East 47th Street, New York, NY 10017. (212) 705-7010. 1985. $200.00.

ENCYCLOPEDIAS AND DICTIONARIES

ENCYCLOPEDIA OF PHYSICAL SCIENCE AND TECHNOLOGY. Academic Press, Incorporated, Orlando, FL 32887. (800) 321-5068 or (305) 345-2734. Fifteen volumes, 1986. Inquire as to cost and availability.

MCGRAW-HILL DICTIONARY OF ENGINEERING. Sybil P. Parker, editor. McGraw-Hill Book Company, 1221 Avenue of the Americas, New York, NY 10020. (212) 512-2000. 1984. $39.95.

MCGRAW-HILL ENCYCLOPEDIA OF SCIENCE AND TECHNOLOGY. McGraw-Hill Book, 1221 Avenue of the Americas, New York, NY 10020. (212) 512-2000. Inquire as to cost and availability.

GENERAL WORKS

ASPHALT PAVEMENT ENGINEERING. Hugh A. Wallace and J.R. Martin. McGraw-Hill Book Company, 1221 Avenue of the Americas, New York, NY 10020. (800) 628-0004. 1967. $46.50.

ASPHALT SCIENCE AND TECHNOLOGY. Edwin J. Barth. Gordon and Breach Science Publishers, Incorporated, 50 West 23rd Street, New York, NY 10010. (212) 206-8900. 1962. $160.75.

ASPHALT AND ROAD MATERIALS: MODERN TECHNOLOGY. J.E. Parson. Noyes Data Corporation, Mill Road at Grand Avenue, Park Ridge, NJ 07656. (201) 391-8484. 1977. $36.00.

HANDBOOKS AND MANUALS

ASPHALT COLD-MIX MANUAL. Asphalt Institute, Asphalt Institute Building, College Park, MD 20742. (301) 277-4258. 1977. $5.00 in paper.

ASPHALT PAVING MANUAL. Asphalt Institute, Asphalt Institute Building, College Park, MD 20742. (301) 277-4258. 1983. $6.00.

ASPHALT PLANT MANUAL. Asphalt Institute, Asphalt Institute Building, College Park, MD 20742. (302) 277-4258. 1983. $6.00.

ASPHALT POCKETBOOK OF USEFUL INFORMATION. Asphalt Institute Building, College Park, MD 20742. (301) 277-4258. 1982. $5.00.

ONLINE DATA BASES

CA SEARCH. Chemical Abstracts Service, Post Office Box 3012, Columbus, OH 43210. Guide to chemical Literature, 1972 to present. Inquire as to cost and availability.

COMPENDEX. Engineering Information, Incorporated, 345 East 47th Street, New York, NY 10017. (800) 221-1044 or (212) 705-7615. Engineering and technical literature, 1975 to present. Inquire as to cost and availability.

NTIS. National Technical Information Service, 5285 Port Royal Road, Springfield, VA 22161. (703) 487-4630. Broad coverage of government sponsored research reports, 1964 to present. Inquire as to cost and availability.

PERIODICALS

ASPHALT EMULSION MANUFACTURERS ASSOCIATION NEWSLETTER. Asphalt Emulsion Manufactures Association, 1133 15th Street, NW, Washington, DC 20005. (202) 429-9440. Bimonthly. $10.00.

ASPHALT PAVING TECHNOLOGY. Association of Asphalt Paving Technologists, Civil and Mineral Engineering Building, Room 134, University of Minnesota, Minneapolis, MN 55455. (612) 644-2996. Inquire as to cost and availability.

ASPHALT: THE MAGAZINE OF THE ASPHALT INSTITUTE. Asphalt Institute, Asphalt Institute Building, College Park, MD 20742. (301) 277-4258. Three times per year. Inquire as to cost and availability.

ASPHALTNEWS. Asphalt Institute, Asphalt Institute Building, College Park, MD 20742. (301) 277-4258. Quarterly. $5.00 per year.

RESEARCH CENTERS AND INSTITUTES

ASPHALT INSTITUTE. Asphalt Institute Building, College Park, MD 20742. (301) 277-4258.

PENNSYLVANIA STATE UNIVERSITY. Civil Engineering Materials Laboratory, 212 Sackett Building, University Park, PA 16802. (814) 865-4682.

TEXAS A & M UNIVERSITY. Texas Transportation Institute, Materials, Pavement and Construction Division. Engineering Research Center, College Station, TX 77843. (409) 845-8212.

ASTEROIDS

See: SOLAR SYSTEM

ASTROGEOLOGY

See also: ASTEROIDS, ASTRONOMY, ASTROPHYSICS, JUPITER, MARS, MERCURY, METEORITICS, NEPTUNE, PLANETARY SCIENCES, PLUTO, SOLAR SYSTEM, SATELLITES, SATURN, URANUS, VENUS

ABSTRACT SERVICES AND INDEXES

ASTRONOMY AND ASTROPHYSICS ABSTRACTS. Springer-Verlag New York, Inc., 175 Fifth Avenue, New York, NY 10010. (800) 526-7254. 1969 to present. Approximately $70.00 per year.

ASTRONOMY AND ASTROPHYSICS MONTHLY INDEX. Olivetree Associates, P.O. Box 236, Sierre Madre, CA 91024. $220.00 per year. Complementary copies available on request.

CHEMICAL ABSTRACTS. American Chemical Society, Chemical Abstracts Service, Box 3012, Columbus, OH 43210. (614) 421-3600. 1907 to present. Weekly. $10,000.00 per year.

CURRENT CONTENTS: PHYSICAL, CHEMICAL, AND EARTH SCIENCES. Institute for Scientific Information, 3501 Market Street, Philadelphia, PA 19104. (800) 523-1850 or (215) 386-0100. Weekly. $275.00 per year.

BIBLIOGRAPHY AND INDEX OF GEOLOGY. American Geological Institute, 4220 King Street, Alexandria, VA 22302. (703) 379-2480. 1969 to present. Monthly. $1100.00 per year.

METEOROLOGICAL AND GEOASTROPHYSICAL ABSTRACTS. American Meteorological Society, 45 Beacon Street, Boston, MA 02108. (617) 227-2425. 1950 to present. Monthly. $450.00 per year.

SCIENCE CITATION INDEX. Institute for Scientific Information, 3501 Market Street, Philadelphia, PA 19104. (800) 523-1850 or (215) 386-0100. Six times per year. $6200.00 per year.

STAR. SCIENTIFIC AND TECHNICAL AEROSPACE REPORTS. U.S. National Aeronautics and Space Administration, Scientific and Technical Information Facility, Box 8757, Baltimore-Washington International Airport, MD 21240. (202) 755-2210. Semimonthly, with semiannual and annual indexes. $85.00 per year.

ANNUAL REVIEWS AND YEARBOOKS

ANNUAL REVIEW OF EARTH AND PLANETARY SCIENCES. Annual Reviews, Inc., 4139 El Camino Way, Palo Alto, CA 94306. (415) 493-4400. Annual. Inquire as to cost and availability.

ASSOCIATIONS AND PROFESSIONAL SOCIETIES

AMERICAN ASTRONOMICAL SOCIETY. 1816 Jefferson Place, N.W., Washington, DC 20036. (202) 659-0134.

AMERICAN GEOPHYSICAL UNION. 2000 Florida Avenue, N.W., Washington, DC 20009. (202) 462-6903.

ASTRONOMICAL SOCIETY OF THE PACIFIC. 1290 24th Avenue, San Francisco, CA 94122. (415) 661-8660.

GEOLOGICAL SOCIETY OF AMERICA. 3300 Penrose Place, Boulder, CO 80301. (303) 447-2020.

PLANETARY SOCIETY. 65 North Catalina Avenue, Pasadena, CA 91106. (818) 793-5100.

DIRECTORIES AND BIOGRAPHICAL SOURCES

AMERICAN ASTRONOMICAL SOCIETY MEMBERSHIP DIRECTORY. 1816 Jefferson Place, N.W., Washington, DC 20036. (202) 659-0134. Annual. Inquire.

AMERICAN MEN AND WOMEN OF SCIENCE. R.R. Bowker, Inc., Order Department, 245 West 17th Street, New York, NY 10011. (800) 521-8110. Eight volumes. 1986. $595.00 for set.

INTERNATIONAL RESEARCH CENTERS DIRECTORY 1988-89. Darren L. Smith, editor. Gale Research Company, Book Tower, Detroit, MI 48226. (800) 521-0707. 4th edition. 1987. $360.00.

RESEARCH CENTERS DIRECTORY 1988. Gale Research Company, Book Tower, Detroit, MI 48226. (800) 521-0707. 12th edition. 1987. $365.00 for set.

SCIENTIFIC AND TECHNICAL ORGANIZATIONS AND AGENCIES DIRECTORY. Margaret Labash Young, editor. Gale Research Company, Book Tower, Detroit, MI 48226. (800) 521-0707. 2nd edition. 1987. $185.00.

GENERAL WORKS

THE BIRTH OF THE EARTH. David E. Fisher. Columbia University Press, 562 West 113th Street, New York, NY 10025. (212) 316-7100. 1987. $24.95.

CHEMISTRY AND PHYSICS OF THE TERRESTRIAL PLANETS. Surendra K. Saxena. Springer-Verlag New York, Inc., 175 Fifth Avenue, New York, NY 10010. (800) 526-7254. 1986. $59.00.

GEOLOGY OF THE TERRESTRIAL PLANETS. M. Carr, editor. National Aeronautics and Space Administration Special Publication 469. National Technical Information Service, 5285 Port Royal Road, Springfield, VA 22161. (703) 487-4929. 1984. $16.00.

MERCURY: THE ELUSIVE PLANET. Robert Strom. Cambridge University Press, 32 East 57th Street, New York, NY 10022. (800) 872-7423. 1986. $30.00.

METEORITES: THEIR RECORD OF EARLY SOLAR-SYSTEM HISTORY. J. Wasson. W.H. Freeman and Company, 41 Madison Avenue, New York, NY 10010. (212) 532-7660. 1985. $34.95.

THE PLANETARY SYSTEM. Tobias Owen and David Morrison. Addison-Wesley Publishing Company, Inc., 1 Jacob Way, Reading, MA 01867. (617) 944-3700. 1987. $39.95.

THE STORY OF THE EARTH. P. Cattermole and P. Moore. Cambridge University Press, 32 East 57th Street, New York, NY 10022. (800) 872-7423. 1985. $29.95.

URANUS AND NEPTUNE. J. Bergstrahl, editor. National Aeronautics and Space Administration Conference Paper 2330. National Technical Information Service, 5285 Port Royal Road, Springfield, VA 22161. (703) 487-4929. 1984. $25.00.

VENUS: AN ARRANT TWIN. E. Burgess. Columbia University Press, 562 West 113th Street, New York, NY 10025. (212) 316-7100. 1985. $29.95.

ONLINE DATA BASES

CA SEARCH. Chemical Abstracts Service, P.O. Box 3012, Columbus, OH 43120. (800) 848-6538 or (614) 421-3600. Comprehensive guide to chemical literature, 1972 to present. Inquire as to online cost and availability.

DISSERTATION ABSTRACTS ONLINE. University Microfilms International, 300 North Zeeb Road, Ann Arbor, MI 48106. (800) 521-0600 or (313) 761-4700. Scope includes virtually all doctoral dissertations accepted at accredited American institutions from 1861 to present in over 250 subject areas. Inquire as to online cost and availability.

GEOREF. Online version of the BIBLIOGRAPHY AND INDEX OF GEOLOGY. American Geological Institute, 4220 King Street, Alexandria, VA 22302. (703) 379-2480. 1969 to present. Inquire as to online cost and availability.

INSPEC. INSPEC Marketing Department, Institution of Electrical Engineers. Available from IEEE Service Center, 445 Hoes Lane, Piscataway, NJ 08854. (201) 981-0060. Online version of Physics Abstracts. Inquire as to online cost and availability.

NTIS. National Technical Information Service, 5285 Port Royal Road, Springfield, VA 22161. (703) 487-4630. Broad coverage of government sponsored research reports, 1964 to present. Inquire as to online cost and availability.

SCISEARCH. Institute for Scientific Information, 3501 Market Street, Philadelphia, PA 19104. (800) 523-1850 or (215) 386-0100. Broad multidisciplinary title and author index to the international literature of science and technology, 1974 to present. Inquire as to online cost and availability.

OTHER SOURCES

ATLAS OF THE SOLAR SYSTEM. P. Moore and G. Hunt. Rand McNally and Company, P.O. Box 7600, Chicago, IL 60680. (800) 323-4070. 1983. $40.00.

CAMBRIDGE PHOTOGRAPHIC ATLAS OF THE PLANETS. Cambridge University Press, 32 East 57th Street, New York, NY 10022. (800) 872-7423. 1986.

PERIODICALS

ASTRONOMICAL JOURNAL. American Astronomical Society. Available from: American Institute of Physics, 335 East 45th Street, New York, NY 10017. (212) 661-9494. Monthly. $125.00 per year.

ASTRONOMICAL SOCIETY OF THE PACIFIC PUBLICATIONS. Astronomical Society of the Pacific, 1290 24th Avenue, San Francisco, CA 94122. (415) 661-8660. Monthly. $40.00 per year.

ASTRONOMY. AstroMedia Corporation, 625 E Street, Box 92788, Milwaukee, WI 53202. (414) 276-2689. 1973 to present. Monthly. $21.00 per year.

ASTRONOMY AND ASTROPHYSICS. Springer-Verlag New York, Inc., 175 Fifth Avenue, New York, NY 10010. (800) 526-7254. Monthly. $680.00 per year.

EARTH, MOON AND PLANETS. D. Reidel Publishing Company, 190 Old Derby Street, Hingham, MA 02043. 1969 to present. Nine times per year. $275.00 per year.

GEOCHEMICA ET COSMOCHIMICA ACTA. Pergamon Press, Inc., Maxwell House, Fairview Park, Elmsford, NY 10523. (914) 592-7700. 1950 to present. Monthly. $340.00 per year.

LUNAR AND PLANETARY INFORMATION BULLETIN. Lunar and Planetary Institute, 3303 NASA Road One, Houston, TX 77058-4399. (713) 486-2135. 1970 to present. Three times per year. Free.

MERCURY. Astronomical Society of the Pacific, 1290 24th Avenue, San Francisco, CA 94122. (415) 661-8660. 1972 to present. Bimonthly. $21.00 per year.

METEORITICS. Center for Meteorite Studies, Arizona State University, Tempe, AZ 85287. (602) 965-3576. 1955 to present. Quarterly. $40.00 per year.

PLANETARY AND SPACE SCIENCE. Pergamon Press, Inc., Maxwell House, Fairview Park, Elmsford, NY 10523. (914) 592-7700. 1959 to present. Monthly. $430.00 per year.

PLANETARY REPORT. Planetary Society, 65 North Catalina, Pasadena, CA 91106-2301. 1980 to present. Bimonthly. $20.00 per year.

SKY AND TELESCOPE. Sky Publishing Corporation, 49 Bay State Road, Cambridge, MA 02238. (617) 864-7360. Monthly. $18.00 per year.

RESEARCH CENTERS AND INSTITUTES

CENTER FOR EARTH AND PLANETARY STUDIES. National Air and Space Museum, Smithsonian Institution, Washington, DC 20560. 357-1424.

LABORATORY FOR PLANETARY GEOLOGY. Arizona State University, Department of Geology, Tempe, AZ 85281. (602) 965-7029.

LABORATORY FOR PLANETARY STUDIES. Cornell University, 302 Space Sciences Building, Ithaca, NY 14853. (607) 256-4971.

LUNAR AND PLANETARY INSTITUTE. 3303 NASA Road One, Houston, TX 77058. (713) 486-2139.

ASTROMETRY AND ASTRONOMICAL PHOTOMETRY

See: ASTRONOMY

ASTRONAUTICAL ENGINEERING

See: AEROSPACE ENGINEERING

ASTRONAUTICS

See: AERONAUTICS

ASTRONOMICAL OBSERVATORIES

See: OBSERVATORIES

ASTRONOMY

See also: ASTROPHYSICS

ABSTRACT SERVICES AND INDEXES

ASTRONOMY AND ASTROPHYSICS ABSTRACTS. Springer-Verlag New York, Incorporated, 175 Fifth Avenue, New York, NY 10010. (212) 460-1500. $70.00 per year.

ASTRONOMY AND ASTROPHYSICS MONTHLY INDEX. Olivetree Associates, P.O. Box 236, Sierre Madre, Ca 91024. $212.00 per year. Complimentary copies available on request.

GENERAL SCIENCE INDEX. H.W. Wilson Company, 950 University Avenue, Bronx, NY 10452. (800) 367-6770 or (212) 588-8400. Inquire as to cost and availability.

PHYSICS ABSTRACTS. Institute of Electrical Engineers, London, United Kingdom. Available from: Institute of Electrical and Electronic Engineers (IEEE), 345 East 47th Street, New York, NY 10017. (212) 705-7900.

SCIENCE CITATION INDEX. Institute for Scientific Information, 3501 Market Street, Philadelphia, PA 19104. (800) 523-1850 or (215) 386-0100.

STAR. (SCIENTIFIC AND TECHNICAL AEROSPACE REPORTS). U.S. National Aeronautics and Space Administration, Scientific and Technical Information Facility, Box 8757, Baltimore-Washington International Airport, MD 21240. (202) 755-2210. Semimonthly, with semiannual and annual indexes. $85.00 per year.

ANNUAL REVIEWS AND YEARBOOKS

THE ASTRONOMICAL ALMANAC. Superintendent of Documents, U.S. Government Printing Office, Washington, DC 20402. (202) 783-3238. Yearly.

ANNUAL REVIEW OF ASTRONOMY AND ASTROPHYSICS. Annual Reviews, Incorporated, 4139 El Camino Way, Palo Alto, CA 94306. (415) 493-4400.

ANNUAL REVIEW OF EARTH AND PLANETARY SCIENCES. Annual Reviews, Incorporated, 4139 El Camino Way, Palo Alto, CA 94306. (415) 493-4400.

ASSOCIATIONS AND PROFESSIONAL SOCIETIES

AMERICAN ASTRONOMICAL SOCIETY. 2000 Florida Avenue, NW, Suite 300, Washington, DC 20009. (202) 659-0134.

AMERICAN ASSOCIATION OF VARIABLE STAR OBSERVERS. 187 Concord Avenue, Cambridge, Ma 02138. (617) 354-0484.

ASTRONOMICAL LEAGUE. P.O. Box 12821, Tucson, Az 85732. (602) 790-8471.

ASTRONOMICAL SOCIETY OF THE PACIFIC. 1290 24th Avenue, San Francisco, CA 94122. (415) 661-8660.

BIBLIOGRAPHIES

A BIBLIOGRAPHY OF ASTRONOMY, 1970-1979. R.A. Seal and S.S. Martin. Libraries Unlimited, Incorporated, Littleton, CO 80160. 1982. $37.50.

SCIENCE BOOKS AND FILMS. American Association for the Advancement of Science, 1333 H Street, NW, Washington, DC 20005.

SCIENTIFIC AND TECHNICAL BOOKS IN PRINT; AN INDEX TO LITERATURE IN SCIENCE AND TECHNOLOGY. R.R. Bowker Company, 205 East 42nd Street, New York, NY 10017. (800) 521-8110 or (212) 916-1600.

DIRECTORIES AND BIOGRAPHICAL SOURCES

AMERICAN MEN AND WOMEN OF SCIENCE. Physical and Biological Sciences. Fifteenth edition. R.R. Bowker Company, 205 East 42nd Street, New York, NY 10017. (800) 521-8110 or (212) 916-1600.

THE BIOGRAPHICAL DICTIONARY OF SCIENTISTS: ASTRONOMERS. D. Abbott, editor. Peter Bedrick Books, 125 East 23rd Street, New York, NY 10010. 1984.

DIRECTORY OF PHYSICS AND ASTRONOMY STAFF MEMBERS. American Institute of Physics, 335 East 45th Street, New York, NY 10017. Annual.

RESEARCH CENTERS DIRECTORY. Gale Research Company, Detroit, MI 48226. Eleventh edition, 1987. (800) 521-0707. $340.00.

WHO'S WHO IN FRONTIER SCIENCE AND TECHNOLOGY. Marquis Who's Who, Incorporated, 200 East Ohio Street, Chicago, IL 60611. (800) 428-3898 or (312) 787-2008.

ENCYCLOPEDIAS AND DICTIONARIES

ENCYCLOPEDIA OF PHYSICAL SCIENCE AND TECHNOLOGY. Academic Press, Incorporated, Orlando, FL 32887. (800) 321-5068 or (305) 345-2734.

MCGRAW-HILL ENCYCLOPEDIA OF SCIENCE AND TECHNOLOGY. McGraw-Hill Book, Incorporated, 1221 Avenue of the Americas, New York, NY 10020. (212) 997-3675. Fifth edition, 15 volumes. $1100.00.

GENERAL WORKS

THE CAMBRIDGE ASTRONOMY GUIDE: A PRACTICAL INTRODUCTION TO ASTRONOMY. Cambridge University Press, 32 East 57th Street, New York, NY 10022. (212) 688-8888. 1985. $24.95.

THE COSMOS FROM SPACE. David H. Clark. Crown Publishers, Incorporated, 34 Engelhard Avenue, Avenel, NJ 07001. 1987. $14.95

ONLINE DATA BASES

CA SEARCH. Chemical Abstracts Service, P.O. Box 3012, Columbus, Oh 43210. Guide to chemical literature, 1972 to present. Inquire as to cost and availability.

DISSERTATION ABSTRACTS ONLINE. University Microfilms International, 300 North Zeeb Road, Ann Arbor, MI 48106. (800) 521-0600 or (313) 761-4700. Scope includes virtually all doctoral dissertations accepted at accredited American institutions from 1861 to present in 252 subject areas. Inquire as to cost and availability.

INSPEC. INSPEC Marketing Department, Institute of Electrical and Electronics Engineers, Incorporated, IEEE Service Department, 445 Hoes Lane, Piscataway, NJ 08854. (201) 981-0060. Inquire as to on-line cost and availability.

MATHFILE. American Mathematical Society, P.O. Box 6248, Providence, RI 02940. (800) 556-7774 or (401) 272-9500. Scope includes pure and applied mathematics and related areas of physics, statistics, engineering, computer science, and operations research literature since 1973. Inquire as to cost and availability.

NASA. National Aeronautics and Space Administration, Scientific and Technical Information Branch, 300 7th Street, SW, Washington, DC 20546. Citations and abstracts of aerospace literature, 1962 to present. Inquire as to cost and availability.

NTIS. National Technical Information Service, 5285 Port Royal Road, Springfield, Va 22161. (703) 487-4630. Broad coverage of government sponsored research reports, 1964 to present. Inquire as to cost and availability.

SCISEARCH. Institute for Scientific Information, 3501 Market Street, Philadelphia, PA 19104. (800) 523-1850 or (215) 386-0100. Broad multidisciplinary title and author index to the international literature of science and technology, 1974 to present. Inquire as to cost and availability.

WILSONLINE. H.W. Wilson Company, 950 University Avenue, Bronx, NY 10452. (800) 367-6770 or (212) 588-8400. Makes available online versions of the printed H.W. Wilson indexes including Applied Science and Technology Index, Business Periodicals Index, and Readers' Guide to Periodical Literature. Period covered is generally 1983 to present. Inquire as to cost and availability.

OTHER SOURCES

ATLAS OF DEEP-SKY SPLENDORS. H. Vehrenberg. Cambridge University Press, 32 East 57th Street, New York, NY 10022. (800) 431-1580 or (212) 688-8888. Fourth edition, 1984. $44.50.

THE BRIGHT STAR CATALOGUE AND SUPPLEMENT. D. Hoffleit and Carlos Jaschek. Yale University Observatory, P.O. Box 6666, New Haven, CT 06511. (203) 436-3460. Fourth edition, 1982. $35.00.

SKY CATALOGUE 2000.0. A. Hirshfeld and R. Sinnott. Cambridge University Press, 32 East 57th Street, New York, NY 10022. (800) 431-1580 or (212) 688-8888.

PERIODICALS

ASTRONOMICAL JOURNAL. American Astronomical Society. Available from: American Institute of Physics, 335 East 45th Street, New York, NY 10017. (212) 661-9404. $125.00 per year.

ASTRONOMICAL SOCIETY OF THE PACIFIC. Publications. Astronomical Society of the Pacific, 1290 24th Avenue, San Francisco, CA 94122. (415) 661-8660. Monthly. $38.00.

ASTRONOMY. Astro Media Corporation, 625 East Paul Avenue, Milwaukee, WI 53202. Monthly. 418.00 per year.

ASTRONOMY AND ASTROPHYSICS. Springer-Verlag New York, Incorporated, 175 Fifth Avenue, New York, NY 10010. (800) 526-7254 or (212) 460-1500. $680.00 per year.

ASTROPHYSICAL JOURNAL. American Astronomical Society, University of Chicago Press, 5801 Ellis Avenue, Chicago, IL 60637. Biweekly. $305.00 per year.

ASTROPHYSICS AND SPACE SCIENCE. D. Reidel Publishing Company, 190 Old Derby Street, Hingham, MA 02043. Monthly. $101.00 per year.

CELESTIAL MECHANICS; AN INTERNATIONAL JOURNAL OF SPACE DYNAMICS. D. Reidel Publishing Company, 190 Old Derby Street, Hingham, MA 02043. Monthly. $310.00 per year.

EARTH, MOON AND PLANETS; AN INTERNATIONAL JOURNAL OF COMPARATIVE PLANETOLOGY. D. Reidel Publishing Company, 190 Old Derby Street, Hingham, MA 02043. Nine times per year. $275.00 per year.

ICARUS: INTERNATIONAL JOURNAL OF THE SOLAR SYSTEM STUDIES. Academic Press, Incorporated, Orlando, FL 32887. (305) 345-4100. Monthly. $484.00 per year.

MERCURY. Astronomical Society of the Pacific, 1290 24th Avenue, San Francisco, CA 94122. (15) 661-8660. Bimonthly. $21.00 per year.

MONTHLY NOTICES OF THE ROYAL ASTRONOMICAL SOCIETY. Blackwell Science Publications, Incorporated, 667 Lytton Avenue, Palo Alto, CA 94301. (415) 324-1688. Monthly. $850.00 per year.

PLANETARY AND SPACE SCIENCE. Pergamon Press, Incorporated, Maxwell House, Fairview Park, Elmsford, NY 10523. (914) 592-7700. Monthly. $430.00 per year.

SKY AND TELESCOPE. Sky Publishing Corporation, 49 Bay State Road, Cambridge, MA 02238. (617) 864-7360. Monthly. $18.00 per year.

SOLAR PHYSICS. D. Reidel Publishing Company, 190 Old Derby Street, Hingham, MA 02043. Monthly. $620.00 per year.

SOVIET ASTRONOMY (TRANSLATION OF ASTRO-NOMICHESKII ZHURNALL). American Institute of Physics, 335 East 45th Street, New York, NY 10017. (212) 661-9404. Bimonthly. $425.00 per year.

SPACE SCIENCE REVIEWS. D. Reidel Publishing Company, 190 Old Derby Street, Hingham, MA 02043. Monthly. $305.00 per year.

VISTAS IN ASTRONOMY. Pergamon Press, Incorporated, Maxwell House, Fairview Park, Elmsford, NY 10523. (914) 592-7700. Quarterly. $145.00 per year.

RESEARCH CENTERS AND INSTITUTES

NATIONAL ASTRONOMY AND IONOSPHERE CENTER. Cornell University, Space Sciences Building, Ithaca, NY 14853. (607) 256-3734.

HARVARD-SMITHSONIAN CENTER FOR ASTROPHYSICS. 60 Garden Street, Cambridge, MA 02138. (617) 495-7461.

NATIONAL OPTICAL ASTRONOMY OBSERVATIORIES. 1002 North Warren Avenue, Tucson, AZ 85719. (602) 325-9230.

NATIONAL RADIO ASTRONOMY OBSERVATORY. Edgemont Road, Charlottesville, VA 22903. (804) 296-0211.

SPACE TELESCOPE SCIENCE INSITUTE. 3700 San Martin Drive, Baltimore, MD 21218. (301) 338-4700.

ASTROPHYSICS

See also: ASTRONOMY, BIG BANG, BLACKHOLES, COSMOLOGY, GRAVITATION

ABSTRACT SERVICES AND INDEXES

ASTRONOMY AND ASTROPHYSICS ABSTRACTS. Springer-Verlag New York, Incorporated, 175 Fifth Avenue, New York, NY 10010. (212) 460-1500. $70.00 per year.

ASTRONOMY AND ASTROPHYSICS MONTHLY INDEX. Olivetree Associates, P.O. Box 236, Sierre Madre, Ca 91024. $212.00 per year. Complimentary copies available on request.

GENERAL SCIENCE INDEX. H.W. Wilson Company, 950 University Avenue, Bronx, NY 10452. (800) 367-6770 or (212) 588-8400. Inquire as to cost and availability.

PHYSICS ABSTRACTS. Institute of Electrical Engineers, London, United Kingdom. Available from: Institute of Electrical and Electronic Engineers (IEEE), 345 East 47th Street, New York, NY 10017. (212) 705-7900.

SCIENCE CITATION INDEX. Institute for Scientific Information, 3501 Market Street, Philadelphia, PA 19104. (800) 523-1850 or (215) 386-0100.

STAR. (SCIENTIFIC AND TECHNICAL AEROSPACE REPORTS). U.S. National Aeronautics and Space Administration, Scientific and Technical Information Facility, Box 8757, Baltimore-Washington International Airport, MD 21240. (202) 755-2210. Semimonthly, with semiannual and annual indexes. $85.00 per year.

ANNUAL REVIEWS AND YEARBOOKS

THE ASTRONOMICAL ALMANAC. Superintendent of Documents, U.S. Government Printing Office, Washington, DC 20402. (202) 783-3238. Yearly.

ANNUAL REVIEW OF ASTRONOMY AND ASTROPHYSICS. Annual Reviews, Incorporated, 4139 El Camino Way, Palo Alto, CA 94306. (415) 493-4400.

ANNUAL REVIEW OF EARTH AND PLANETARY SCIENCES. Annual Reviews, Incorporated, 4139 El Camino Way, Palo Alto, CA 94306. (415) 493-4400.

ASSOCIATIONS AND PROFESSIONAL SOCIETIES

AMERICAN ASTRONOMICAL SOCIETY. 2000 Florida Avenue, NW, Suite 300, Washington, DC 20009. (202) 659-0134.

AMERICAN ASSOCIATION OF VARIABLE STAR OBSERVERS. 187 Concord Avenue, Cambridge, Ma 02138. (617) 354-0484.

AMERICAN INSTITUTE OF PHYSICS. 335 East 45th Street, New York, NY 10017. (800) 247-7497.

ASTRONOMICAL LEAGUE. P.O. Box 12821, Tucson, Az 85732. (602) 790-8471.

ASTRONOMICAL SOCIETY OF THE PACIFIC. 1290 24th Avenue, San Francisco, CA 94122. (415) 661-8660.

BIBLIOGRAPHIES

A BIBLIOGRAPHY OF ASTRONOMY, 1970-1979. R.A. Seal and S.S. Martin. Libraries Unlimited, Incorporated, Littleton, CO 80160. 1982. $37.50.

SCIENCE BOOKS AND FILMS. American Association for the Advancement of Science, 1333 H Street, NW, Washington, DC 20005.

SCIENTIFIC AND TECHNICAL BOOKS IN PRINT; AN INDEX TO LITERATURE IN SCIENCE AND TECHNOLOGY. R.R. Bowker Company, 205 East 42nd Street, New York, NY 10017. (800) 521-8110 or (212) 916-1600.

DIRECTORIES AND BIOGRAPHICAL SOURCES

AMERICAN MEN AND WOMEN OF SCIENCE. Physical and Biological Sciences. Fifteenth edition. R.R. Bowker Company, 205 East 42nd Street, New York, NY 10017. (800) 521-8110 or (212) 916-1600.

THE BIOGRAPHICAL DICTIONARY OF SCIENTISTS: ASTRONOMERS. D. Abbott, editor. Peter Bedrick Books, 125 East 23rd Street, New York, NY 10010. 1984.

DIRECTORY OF PHYSICS AND ASTRONOMY STAFF MEMBERS. American Institute of Physics, 335 East 45th Street, New York, NY 10017. Annual.

RESEARCH CENTERS DIRECTORY. Gale Research Company, Detroit, MI 48226. Eleventh edition, 1987. (800) 521-0707. $340.00.

WHO'S WHO IN FRONTIER SCIENCE AND TECHNOLOGY. Marquis Who's Who, Incorporated, 200 East Ohio Street, Chicago, IL 60611. (800) 428-3898 or (312) 787-2008.

ENCYCLOPEDIAS AND DICTIONARIES

ENCYCLOPEDIA OF PHYSICAL SCIENCE AND TECHNOLOGY. Academic Press, Incorporated, Orlando, FL 32887. (800) 321-5068 or (305) 345-2734.

MCGRAW-HILL ENCYCLOPEDIA OF SCIENCE AND TECHNOLOGY. McGraw-Hill Book, Incorporated, 1221 Avenue of the Americas, New York, NY 10020. (212) 997-3675. Fifth edition, 15 volumes. $1100.00.

GENERAL WORKS

ASTROPHYSICAL CONCEPTS. Martin Harwit. Concepts, P.O. Box 6750, Ithaca, NY 14851. (607) 272-3346. 1984. $25.00.

ASTROPHYSICS TODAY. A.G. Cameron, editor. American Institute of Physics, 335 East 45th Street, New York, NY 10017. (800) 247-7497. 1984. $25.00.

ONE HUNDRED BILLION SUNS: THE BIRTH, LIFE AND DEATH OF STARS. Rudolf Klippenhahn. Basic Books, Incorporated, 10 East 53rd Street, New York, NY 10022. (212) 207-7292. 1983. $25.00.

PHYSICS OF STARS. S.A. Kaplan. John Wiley and Sons, Incorporated, 605 Third Avenue, New York, NY 10158. (800) 526-5368 or (212) 850-6000. 1983. $51.95.

HANDBOOKS AND MANUALS

ASTROPHYSICAL TECHNIQUES. C.R. Kitchin. International Publishing Service, inc., P.O. box 230, Accord, MA 02018. (617) 749-3628.

HANDBOOK OF SPACE ASTRONOMY AND ASTROPHYSICS. Martin V. Zombeck. Cambridge University Press, 32 East 57th Street, New York, NY 10022. (800) 872-7423. 1983. (29.95.

ONLINE DATA BASES

CA SEARCH. Chemical Abstracts Service, P.O. Box 3012, Columbus, Oh 43210. Guide to chemical literature, 1972 to present. Inquire as to cost and availability.

DISSERTATION ABSTRACTS ONLINE. University Microfilms International, 300 North Zeeb Road, Ann Arbor, MI 48106. (800) 521-0600 or (313) 761-4700. Scope includes virtually all doctoral dissertations accepted at accredited American institutions from 1861 to present in 252 subject areas. Inquire as to cost and availability.

INSPEC. INSPEC Marketing Department, Institute of Electrical and Electronics Engineers, Incorporated, IEEE Service Department, 445 Hoes Lane, Piscataway, NJ 08854. (201) 981-0060. Inquire as to on-line cost and availability.

MATHFILE. American Mathematical Society, P.O. Box 6248, Providence, RI 02940. (800) 556-7774 or (401) 272-9500. Scope includes pure and applied mathematics and related areas of physics, statistics, engineering, computer science, and operations research literature since 1973. Inquire as to cost and availability.

NASA. National Aeronautics and Space Administration, Scientific and Technical Information Branch, 300 7th Street, SW, Washington, DC 20546. Citations and abstracts of aerospace literature, 1962 to present. Inquire as to cost and availability.

NTIS. National Technical Information Service, 5285 Port Royal Road, Springfield, Va 22161. (703) 487-4630. Broad coverage of government sponsored research reports, 1964 to present. Inquire as to cost and availability.

SCISEARCH. Institute for Scientific Information, 3501 Market Street, Philadelphia, PA 19104. (800) 523-1850 or (215) 386-0100. Broad multidisciplinary title and author index to the international literature of science and technology, 1974 to present. Inquire as to cost and availability.

WILSONLINE. H.W. Wilson Company, 950 University Avenue, Bronx, NY 10452. (800) 367-6770 or (212) 588-8400. Makes available online versions of the printed H.W. Wilson indexes including Applied Science and Technology Index, Business Periodicals Index, and Readers' Guide to Periodical Literature. Period covered is generally 1983 to present. Inquire as to cost and availability.

PERIODICALS

ASTRONOMICAL JOURNAL. American Astronomical Society. Available from: American Institute of Physics, 335 East 45th Street, New York, NY 10017. (212) 661-9404. $125.00 per year.

ASTRONOMICAL SOCIETY OF THE PACIFIC. Publications. Astronomical Society of the Pacific, 1290 24th Avenue, San Francisco, CA 94122. (415) 661-8660. Monthly. $38.00.

ASTRONOMY. Astro Media Corporation, 625 East Paul Avenue, Milwaukee, WI 53202. Monthly. 418.00 per year.

ASTRONOMY AND ASTROPHYSICS. Springer-Verlag New York, Incorporated, 175 Fifth Avenue, New York, NY 10010. (800) 526-7254 or (212) 460-1500. $680.00 per year.

ASTROPHYSICAL JOURNAL. American Astronomical Society, University of Chicago Press, 5801 Ellis Avenue, Chicago, IL 60637. Biweekly. $305.00 per year.

ASTROPHYSICS AND SPACE SCIENCE. D. Reidel Publishing Company, 190 Old Derby Street, Hingham, MA 02043. Monthly. $101.00 per year.

MONTHLY NOTICES OF THE ROYAL ASTRONOMICAL SOCIETY. Blackwell Science Publications, Incorporated, 667 Lytton Avenue, Palo Alto, CA 94301. (415) 324-1688. Monthly. $850.00 per year.

SKY AND TELESCOPE. Sky Publishing Corporation, 49 Bay State Road, Cambridge, MA 02238. (617) 864-7360. Monthly. $18.00 per year.

RESEARCH CENTERS AND INSTITUTES

NATIONAL ASTRONOMY AND IONOSPHERE CENTER. Cornell University, Space Sciences Building, Ithaca, NY 14853. (607) 256-3734.

HARVARD-SMITHSONIAN CENTER FOR ASTROPHYSICS. 60 Garden Street, Cambridge, MA 02138. (617) 495-7461.

NATIONAL OPTICAL ASTRONOMY OBSERVATORIES. 1002 North Warren Avenue, Tucson, AZ 85719. (602) 325-9230.

NATIONAL RADIO ASTRONOMY OBSERVATORY. Edgemont Road, Charlottesville, VA 22903. (804) 296-0211.

SPACE TELESCOPE SCIENCE INSITUTE. 3700 San Martin Drive, Baltimore, MD 21218. (301) 338-4700.

ATMOSPHERE

See: METEOROLOGY

ATOM

See: PHYSICS

ATOMIC ENERGY

See: NUCLEAR ENERGY

ATOMIC FISSION

See: FISSION

ATOMIC PHYSICS

See: PHYSICS

ATOMIC POWER

See: ENERGY

ATOMIC SPECTROSCOPY

See: SPECTROSCOPY

ATOMIC THEORY

See: PHYSICS

AUDIO AMPLIFIER

See: AMPLIFIERS

AUDIO ENGINEERING

See also: ACOUSTICS, ELECTRICAL ENGINEERING, ELECTRICITY, ELECTRONIC CIRCUITS AND COMPONENTS, ELECTRONIC ENGINEERING

ABSTRACT SERVICES AND INDEXES

ACOUSTICS ABSTRACTS. Multiscience Publishing Company, Limited, 42-45 New Broad Street, London, EC2M 1QY, England. Monthly. 1967 to present. $295.00 per year.

APPLIED SCIENCE AND TECHNOLOGY INDEX. H.W. Wilson Company, 950 University Avenue, Bronx, NY 10452. (800) 367-6670 or (212) 588-8400. Inquire as to cost and availability.

ELECTRONICS AND COMMUNICATIONS ABSTRACTS JOURNAL. Cambridge Scientific Abstracts, 5161 River Road, Bethesda, MD 20816. (301) 951-1400. Bimonthly. Inquire as to cost and availability.

ENGINEERING INDEX MONTHLY. Engineering Information, Incorporated, 345 East 47th Street, New York, NY 10017. (800) 221-1044 or (212) 705-7600. Monthly, with annual cumulation. $1560.00 per year.

PHYSICS ABSTRACTS. Institute of Electrical Engineers, London, United Kingdom. Available from: Institute of Electrical and Electronic Engineers (IEEE), 345 East 47th Street, New York, NY 10017. (212) 705-7900.

SCIENCE CITATION INDEX. Institute for Scientific Information, 3501 Market Street, Philadelphia, PA 19104. (800) 523-1850 or (215) 286-0100.

ANNUAL REVIEWS AND YEARBOOKS

ADVANCES IN ELECTRONICS AND ELECTRON PHYSICS. Academic Press, Incorporated, 6277 Sea Harbor Drive, Orlando, FL 32821. (800) 321-5068. Irregular. Approximately $80.00 per volume.

PHYSICAL ACOUSTICS. Academic Press, Incorporated, 6277 Sea Harbor Drive, Orlando, FL 32821. (800) 321-5068. Irregular. Inquire as to cost and availability.

ASSOCIATIONS AND PROFESSIONAL SOCIETIES

AMERICAN ELECTRONICS ASSOCIATION. Post Office Box 10045, 2670 Hanover Street, Palo Alto, CA 94303. (415) 857-9300.

AUDIO ENGINEERING SOCIETY. 60 East 42nd Street, New York, NY 10165. (212) 661-2355.

ELECTRONIC INDUSTRIES ASSOCIATION. 2001 Eye Street, NW, Washington, DC 20006. (202) 457-4900.

IEEE (INSTITUTE OF ELECTRICAL AND ELECTRONICS ENGINEERS). 345 East 47th Street, New York, NY 10017. (212) 705-7900.

BIBLIOGRAPHIES

HANDBOOKS AND TABLES IN SCIENCE AND TECHNOLOGY. Russell H. Powell, editor. Oryx Press, 2214 North Central Avenue, Phoenix, AZ 85004-1483. (602) 254-6156. Second edition. 1983. $55.00.

SCIENTIFIC AND TECHNICAL BOOKS IN PRINT; AN INDEX TO LITERATURE IN SCIENCE AND TECHNOLOGY. R.R. Bowker Company, 205 East 42nd Street, New York, NY 10017. (800) 521-8110 or (212) 916-1600.

DIRECTORIES AND BIOGRAPHICAL SOURCES

IEEE MEMBERSHIP DIRECTORY. Institute of Electrical and Electronics Engineers, IEEE Service Center, 445 Hoes Lane, Piscataway, NJ 08854. (212) 705-7900. Annual. $7.00.

RESEARCH CENTERS DIRECTORY. Gale Research Company, Book Tower, Detroit, MI 48226. (800) 521-0707. Thirteenth edition. 1988. $365.00.

WHO'S WHO IN ELECTRONICS. Harris Publishing Company, 2057-2 Aurora Road, Twinsburg, OH 44087. (216) 425-9000. Annual. $89.00.

WHO'S WHO IN ENGINEERING. Engineers Joint Council, 345 East 47th Street, New York, NY 10017. (212) 705-7010. 1985. $200.00.

WHO'S WHO IN FRONTIER SCIENCE AND TECHNOLOGY. Marquis Who's Who, Incorporated, 200 East Ohio Street, Chicago, IL 60611. (800) 428-3898 or (312) 787-2008.

SCIENTIFIC AND TECHNICAL ORGANIZATIONS AND AGENCIES DIRECTORY. Gale Research Company, Book Tower, Detroit, MI 48226. (800) 521-0707. 1985.

WHO'S WHO IN TECHNOLOGY TODAY. Reston Publishing Company, Incorporated, c/o Prentice-Hall, Incorporated,

Englewood Cliffs, NJ 07632. (800) 262-6868. Biennial. Five volumes. $425.00. Covers the fields of electronics, computer science, physics, optics, chemistry, biotechnology, mechanics, energy, and earth science.

ENCYCLOPEDIAS AND DICTIONARIES

ENCYCLOPEDIA OF PHYSICAL SCIENCE AND TECHNOLOGY. Academic Press, Incorporated, Orlando, FL 32887. (800) 321-5068 or (305) 345-2734. Fifteen volumes, 1986.

IEEE STANDARD DICTIONARY OF ELECTRICAL AND ELECTRONICS TERMS. Frank Jay, editor. John Wiley and Sons, Incorporated, 605 Third Avenue, New York, NY 10158. (800) 526-5368 or (212) 850-6000. Third edition. 1984. $49.95.

MCGRAW-HILL ENCYCLOPEDIA OF SCIENCE AND TECHNOLOGY. McGraw-Hill Book, 1221 Avenue of the Americas, New York, NY 10020. (212) 512-2000.

GENERAL WORKS

CIRCUITS AND SOFTWARE FOR ELECTRONICS ENGINEERS. Howard Bierman. McGraw-Hill Book Company, 1221 Avenue of the Americas, New York, NY 10020. (800) 628-0004. 1984. $39.50.

CIRCUITS, DEVICES AND SYSTEMS: A FIRST COURSE IN ELECTRICAL ENGINEERING. R.J. Smith. John Wiley and Sons, Incorporated, 605 Third Avenue, New York, NY 10158. (800) 526-5368 or (212) 850-6000. Fourth edition. 1984. $41.45.

CIRCUITS, SIGNALS AND SYSTEMS. William M. Siebert. McGraw-Hill Book Company, 1221 Avenue of the Americas, New York, NY 10020. (800) 628-0004. 1986. $37.95.

ELECTRICAL AND ELECTRONIC INSTRUMENTATION. Hai Hung Chiang. John Wiley and Sons, Incorporated, 605 Third Avenue, New York, NY 10158. (800) 526-5368 or (212) 850-6000. 1984. $64.95.

ELECTROMAGNETIC WAVE THEORY. J.A. Kong. John Wiley and Sons, Incorporated, 605 Third Avenue, New York, NY 10158. (800) 526-5368 or (212) 850-6000. 1985. $34.95.

ELECTRONIC INVENTIONS AND DISCOVERIES: ELECTRONICS FROM ITS EARLIEST BEGINNINGS TO THE PRESENT DAY. G.W.A. Dummer. Pergamon Press, Incorporated, Maxwell House, Fairview Park, Elmsford, NY 10523. (914) 592-7700. 1983. $48.50.

INTRODUCTION TO RANDOM PROCESSES: WITH APPLICATION TO SIGNALS AND SYSTEMS. William A. Gardner. Macmillan Publishing Company, Incorporated, 866 Third Avenue, New York, NY 10022. (800) 257-5755 or (212) 935-2000. 1986. $34.95.

SEMICONDUCTOR CIRCUIT APPROXIMATIONS: AN INTRODUCTION TO TRANSISTORS AND INTEGRATED CIRCUITS. Albert Paul Malvino. McGraw-Hill Book Company, 1221 Avenue of the Americas, New York, NY 10020. (800) 628-0004. Fourth edition. 1985. $34.95.

TELECOMMUNICATION ENGINEERING: ANALOG AND DIGITAL NETWORK DESIGN. R.L. Freeman. John Wiley and Sons, Incorporated, 605 Third Avenue, New York, NY 10158. (800) 526-5368 or (212) 850-6000. 1980. $48.95.

HANDBOOKS AND MANUALS

CONTEMPORARY ELECTRONICS CIRCUITS DESKBOOK. Harry Helms. McGraw-Hill Book Company, 1221 Avenue of the Americas, New York, NY 10020. (800) 628-0004. 1986. $29.95.

CRC HANDBOOK OF TABLES FOR APPLIED ENGINEERING SCIENCE. R.E. Bolz and G.L. Tuve, editors. CRC Press, Incorporated, 2000 Corporate Boulevard, NW, Boca Raton, FL 33431. Second edition. 1973. $69.00.

ELECTRONIC ENGINEERS HANDBOOK. Donald G. Fink, editor. McGraw-Hill Book Company, 1221 Avenue of the Americas, New York, NY 10020. (800) 628-0004. Second edition. 1982. $89.00.

HANDBOOK OF ELECTRONICS INDUSTRY COST ESTIMATING DATA. Theodore Taylor. John Wiley and Sons, Incorporated, 605 Third Avenue, New York, NY 10158. (800) 526-5368 or (212) 850-6000. 1985. $54.95.

HANDBOOK OF ELECTRONICS MANUFACTURING ENGINEERING. B.S. Matisoff. Van Nostrand Reinhold Company, Incorporated, 115 Fifth Avenue, New York, NY 10003. (800) 543-2681. Second edition. 1986. $52.95.

HANDBOOK OF MODERN ELECTRONICS AND ELECTRICAL ENGINEERING. Charles Belove. John Wiley and Sons, Incorporated, 605 Third Avenue, New York, NY 10158. (800) 526-5368 or (212) 850-6000. 1986. $85.00.

REFERENCE MANUAL FOR TELECOMMUNICATIONS. R.L. Freeman. John Wiley and Sons, Incorporated, 605 Third Avenue, New York, NY 10158. (800) 526-5368 or (212) 850-6000. 1985. $79.95.

STANDARD HANDBOOK FOR ELECTRICAL ENGINEERS. Donald G. Fink and H. Wayne Beaty. McGraw-Hill Book Company, 1221 Avenue of the Americas, New York, NY 10020. (212) 512-2000. Eleventh edition. 1978. $85.00.

VLSI HANDBOOK. Norman G. Einspruch, editor. Academic Press, Incorporated, 6277 Sea Harbor Drive, Orlando, FL 32821. (800) 321-5068. 1985. $125.00.

THE WILEY ENGINEER'S DESK REFERENCE. Snaford I. Heisler. John Wilet and Sons, Incorporated, 605 Third Avenue, New York, NY 10158. (800) 526-5368 or (212) 850-6418. 1984. $36.00.

ONLINE DATA BASES

COMPENDEX. Engineering Information, Incorporated, 345 East 47th Street, New York, NY 10017. (800) 221-1044 or (212) 705-7615. Engineering and technical literature, 1975 to present. Inquire as to cost and availability.

DISSERTATION ABSTRACTS ONLINE. University Microfilms International, 300 North Zeeb Road, Ann Arbor, MI 48106. (800) 521-0600 or (313) 761-4700. Scope includes virtually all doctoral dissertations accepted at accredited American institutions from 1861 to present in 252 subject areas. Inquire as to cost and availability.

INSPEC. INSPEC Marketing Department, Institute of Electrical and Electronics Engineers, Incorporated, IEEE Service Department, 445 Hoes Lane, Piscataway, NJ 08854. (201) 981-0060. Inquire as to on-line cost and availability.

NTIS. National Technical Information Service, 5285 Port Royal Road, Springfield, VA 22161. (703) 487-4630. Broad coverage of government sponsored research reports, 1964 to present. Inquire as to cost and availability.

WILSONLINE. H.W. Wilson Company, 950 University Avenue, Bronx, NY 10452. (800) 367-6770 or (212) 588-8400. Makes available online versions of the printed H.W. Wilson Indexes including Applied Science and Technology Index, Business Periodicals Index, and Readers' Guide to Periodical Literature. Period covered is generally 1983 to present. Inquire as to cost and availability.

PERIODICALS

ACOUSTICAL SOCIETY OF AMERICA. Journal. American Institute of Physics, 335 East 45th Street, New York, NY 10017. (212) 661-9404. Monthly. $300.00 per year.

AUDIO ENGINEERING SOCIETY. Journal. Audio Engineering Society, 60 East 42nd Street, New York, NY 10165. (212) 661-2355. Ten times per year. $70.00 per year.

APPLIED ACOUSTICS. Elsevier Applied Science Publishers Limited, Crown House, Linton Road, Barking, Essex, IG11 8JU, England. Six times per year. $166.00 per year.

DB, THE SOUND ENGINEERING MAGAZINE. Sagamore Publishing Company, Incorporated, 1120 Old Country Road, Plainview, NY 11803. (516) 433-6530. Bimonthly. $15.00 per year.

IEEE ACOUSTICS, SPEECH AND SIGNAL PROCESSING MAGAZINE. Institute of Electrical and Electronics Engineers, IEEE Service Center, 445 Hoes Lane, Piscataway, NJ 08854. (212) 705-7900. Quarterly. $40.00 per year.

IEEE TRANSACTIONS ON ULTRASONICS, FERROELECTRICS AND FREQUENCY CONTROL. Institute of Electrical and Electronics Engineers, IEEE Service Center, 445 Hoes Lane, Piscataway, NJ 08854. (212) 705-7900. 1954 to present. Bimonthly. $100.00 per year.

JOURNAL OF LOW FREQUENCY NOISE AND VIBRATION. Multiscience Publishing Company, Limited, 42-45 New Broad Street, London, EC2M 1QY, England. 1982 to present. Quarterly. $100.00 per year.

SOUND AND VIBRATION. Acoustical Publications, Incorporated, Box 40416, Bay Village, OH 44140. 1967 to present. Monthly. $10.00 per year.

ULTRASONICS. Butterworth Scientific, Limited, Post Office Box 63, Westbury House, Bury Street, Guildford, Surrey, GU2 5BH, England. 1963 to present. Bimonthly. $155.00 per year.

RESEARCH CENTERS AND INSTITUTES

DEPARTMENT OF THE NAVY. Naval Research Laboratory, Acoustics Division, 4555 Overlook Avenue, SW, Washington, DC 20375. (202) 767-3482.

MASSACHUSETTS INSTITUTE OF TECHNOLOGY. Acoustics and Vibration Laboratory, 77 Massachusetts Avenue, Building 3, Room 366, Cambridge, MA 02139. (617) 253-2214.

NORTH CAROLINA STATE UNIVERSITY. Center for Sound and Vibration, Campus Box 7910, Raleigh, NC 27695. (919) 737-3024.

PURDUE UNIVERSITY. Applied Ultrasonics and Electromagnetic Signal Processing Laboratory, School of Electrical Engineering, West Lafayette, IN 47906. (317) 494-3563.

TEXAS A & M UNIVERSITY. Aeroacoustics Laboratory, Aerospace Engineering Department, College Station, TX 77843. (704) 845-1649.

UNIVERSITY OF MICHIGAN. Communications and Signal Processing Laboratory, 2355 Bonisteel Boulevard, North Campus, Ann Arbor, MI 48109. (313) 764-5210.

AUTOMATIC CONTROL SYSTEMS

See: AUTOMATION

AUTOMATION

See also: COMPUTER MEMORY, COMPUTER OPERATING SYSTEMS, COMPUTER VISION, COMPUTERS, CONTROL SYSTEMS, INDUSTRIAL ENGINEERING, PARALLEL COMPUTERS, PROGRAMMING LANGUAGES, SOFTWARE, SOFTWARE ENGINEERING

ABSTRACT SERVICES AND INDEXES

APPLIED SCIENCE AND TECHNOLOGY INDEX. H.W. Wilson Company, 950 University Avenue, Bronx, NY 10452. (800) 367-6670 or (212) 588-8400. Inquire as to cost and availability.

COMPUTER ABSTRACTS. Technical Information Company, Limited, Post Office Box 59, Saint Helier, Jersey British Channel Inlands, England. Monthly. $310.00 per year.

COMPUTER AND CONTROL ABSTRACTS. Institute of Electrical Engineers, London, United Kingdom. Available from: IEEE Service Center, 445 Hoes Lane, Piscataway, NJ 08854. (201) 981-0060. Semimonthly. $775.00 per year.

COMPUTER AND INFORMATION SYSTEMS: AN ABSTRACT JOURNAL PERTAINING TO THE THEORY, DESIGN, FABRICATION AND APPLICATION OF COMPUTER AND INFORMATION SYSTEMS. Cambridge Scientific Abstracts, Incorporated, 5161 River Road, Bethesda, MD 20816. (301) 951-1400. Semimonthly. $590.00 per year.

COMPUTER CONTENTS: THE BIWEEKLY COMPILATION OF TABLES OF CONTENTS FROM COMPUTER, ELECTRONIC AND TELECOMMUNICATIONS MAGAZINES, JOURNALS AND TRANSACTIONS. Find/SVP, 500 Fifth Avenue, New York, NY 10110. (800) 346-3787 or (212) 354-2424. Biweekly. $115.00 per year.

COMPUTER PROGRAM ABSTRACTS. U.S. National Aeronautics and Space Administration. Available from U.S. Government Printing Office, Washington, DC 20402. Quarterly. $6.50 per year.

COMPUTING REVIEWS. Association for Computing Machinery, 11 West 42nd Street, New York, NY 10036. (212) 869-7440. Monthly. $60.00 per year.

ENGINEERING INDEX MONTHLY. Engineering Information, Incorporated, 345 East 47th Street, New York, NY 10017. (800) 221-1044 or (212) 705-7600. Monthly, with annual cumulation. $1560.00 per year.

MATHEMATICAL REVIEWS. American Mathematical Society, Post Office Box 6248, Providence, RI 02940. (800) 556-7774 or (401) 272-9500.

SCIENCE CITATION INDEX. Institute for Scientific Information, 3501 Market Street, Philadelphia, PA 19104. (800) 523-1850 or (215) 386-0100.

ANNUAL REVIEWS AND YEARBOOKS

ADVANCES IN COMPUTERS. Academic Press, Incorporated, 6277 Sea Harbor Drive, Orlando, FL 32821. (800) 321-5068. Yearly. Approximately $50.00 per volume.

ASSOCIATIONS AND PROFESSIONAL SOCIETIES

AMERICAN FEDERATION OF INFORMATION PROCESSING SOCIETIES. 1899 Preston White Drive, Reston, VA 22091. (703) 620-8900.

AMERICAN SOCIETY FOR CYBERNETICS. Department of Decision Sciences, George Mason University, Fairfax, VA 22030. (703) 323-2738.

ASSOCIATION OF COMPUTING MACHINERY (ACM). 11 West 42nd Street, New York, NY 10036. (212) 896-7440.

ASSOCIATION OF COMPUTER PROGRAMMERS AND ANALYSTS. 2108-C Gallows Road, Vienna, VA 22180. (703) 790-0490.

COMPUTER AND AUTOMATION SYSTEMS ASSOCIATION OF SME (SOCIETY OF MANUFACTURING ENGINEERS). One SME Drive, Box 930, Dearborn, Mi 48121. (313) 271-1500.

IEEE COMPUTER SOCIETY. 1730 Massachusetts Avenue, NW, Washington, DC 20036. (202) 371-0101.

INSTITUTE OF ELECTRICAL AND ELECTRONIC ENGINEERS (IEEE). 345 East 47th Street, New York, NY 10017. (212) 705-7900.

LIBRARY AND INFORMATION TECHNOLOGY ASSOCIATION (DIVISION OF AMERICAN LIBRARY ASSOCIATION). 50 East Huron Street, Chicago, IL 60611. (312) 944-6780.

MACHINE VISION ASSOCIATION. Post Office Box 930, One SME Drive, Dearborn, MI 48121. (313) 271-1500.

OFFICE AUTOMATION SOCIETY INTERNATIONAL. 15269 Mimosa Trial, Dumfries, VA 22026. (703) 690-3880.

SOCIETY FOR COMPUTER SIMULATION. Post Office Box 17900. San Diego, CA 92117. (619) 277-3888.

BIBLIOGRAPHIES

COMPUTER LITERATURE INDEX. Applied Computer Research, Incorporated, Post Office Box 9280, Phoenix, AZ 85068. (602) 995-5929. Quarterly. $125.00 per year.

SCIENCE BOOKS AND FILMS. American Association for the Advancement of Science, 1333 H Street, NW, Washington, DC 20005.

SCIENTIFIC AND TECHNICAL BOOKS AND SERIALS IN PRINT 1988; AN INDEX TO LITERATURE IN SCIENCE AND TECHNOLOGY. R.R. Bowker Company, 205 East 42nd Street, New York, NY 10017. (800) 521-8110 or (212) 916-1600. $175.00.

DIRECTORIES AND BIOGRAPHICAL SOURCES

AMERICAN MEN AND WOMEN OF SCIENCE. Physical and Biological Sciences. Fifteenth edition. R.R. Bowker Company, 205 East 42nd Street, New York, NY 10017. (800) 521-8110 or (212) 916-1600.

AMERICAN SOCIETY FOR INFORMATION SCIENCE, HANDBOOK AND DIRECTORY. American Society for Information Science, 1424 16th Street, NW, Suite 404, Washington, DC 20036. (202) 462-1000. $50.00.

COMPUTER REVIEW. GML Information Services, 594 Marrett Road, Lexington, MA 02173. (617) 861-0515. Two issues per year. $195.00 per year. Directory of computer main frame and minicomputer manufacturers.

COMPUTERS AND COMPUTING INFORMATION RESOURCES DIRECTORY. Gale Research Company, Book Tower, Detroit, MI 48226. (800) 521-0707. 1986. $160.00.

DIRECTORY OF CONSULTANTS IN COMPUTER SYSTEMS. Research Publications, Incorporated, 12 Lunar Drive, Woodbridge, CT 06525. (203) 397-2600. Annual. $85.00 per year.

RESEARCH CENTERS DIRECTORY. Gale Research Company, Book Tower, Detroit, MI 48226. (800) 521-0707. Thirteenth edition. 1988. $365.00.

ROBOTICS TECHNICAL DIRECTORY. Instrument Society of America, 67 Alexander Drive, Research Triangle Park, NC 27709. (919) 549-8411. Annual. $49.95.

ROBOTICS WORLD DIRECTORY. Communication Channels, Incorporated, 6255 Barfield Road, Atlanta, GA 30328. (404) 256-9800. Annual. $35.00.

WHO'S WHO IN FRONTIER SCIENCE AND TECHNOLOGY. Marquis Who's Who, Incorporated, 200 East Ohio Street, Chicago, IL 60611. (800) 428-3898 or (312) 787-2008.

WHO'S WHO IN TECHNOLOGY TODAY. Reston Publishing Company, Incorporated, c/o Prentice-Hall, Incorporated, Englewood Cliffs, NJ 07632. (800) 262-6868. Biennial. Five volumes. $425.00. Covers the fields of electronics, computer science, physics, optics, chemistry, biotechnology, mechanics, energy, and earth science.

ENCYCLOPEDIAS AND DICTIONARIES

COMPUTER AND TELECOMMUNICATIONS ACRONYMS. Julie E. Towell and Helen E. Sheppard, editors. Gale Research Company, Book Tower, Detroit, MI 48226. (800) 521-0707. 1986. $60.00.

COMPUTER DICTIONARY. Charles J. Sippl. Howard W. Sams and Company, Incorporated, 4300 West 62nd Street, Indianapolis, IN 46268. (800) 428-7267 or (317) 298-5564. Fourth edition. 1985. $17.95.

DICTIONARY OF COMPUTING. Oxford University Press, 200 Madison Avenue, New York, NY 10016. (212) 679-7300. Second edition. 1986. $29.95.

ENCYCLOPEDIA OF COMPUTER SCIENCE AND ENGINEERING. Anthony Ralston, editor. Van Nostrand Reinhold Book Company, 115 Fifth Avenue, New York, NY 10003. (800) 543-2681. Second edition. 1982. $89.95.

ENCYCLOPEDIA OF COMPUTER SCIENCE AND TECHNOLOGY. Jack Belzer, Albert G. Holzman, and Allan Kent. Marcel Dekker, Incorporated, 270 Madison Avenue, New York, NY 10016. (212) 696-9000. Sixteen volumes. $115.00 per volume.

MCGRAW-HILL DICTIONARY OF COMPUTERS. McGraw-Hill Book Company, 1221 Avenue of the Americas, New York, NY 10020. (212) 512-2000. 1985. $17.50.

GENERAL WORKS

THE COMPUTER PIONEERS: THE MAKING OF THE MODERN COMPUTER. David Ritchie. Simon and Schuster, Incorporated, 1230 Avenue of the Americas, New York, NY 10020. (800) 223-2336 or (212) 245-6400. 1986. $17.95.

COMPUTERS AND DATA PROCESSING. H.L. Capron and Brian K. Williams. Benjamin/Cummings Publishing Company, Incorporated, 2727 Sand Hill Road, Menlo Park, CA 94025. (415) 854-6020. Second edition. 1984. $28.95.

FIFTH GENERATION COMPUTERS. Richard K. Miller, editor. The Fairmont Press, Incorporated, 4025 Pleasantdale Road, Atlanta, GA 30340. (404) 447-5314.

INFORMATION SYSTEMS: THEORY AND PRACTICE. J. Burch and G. Grudnitski. John Wiley and Sons, Incorporated, 605 Third Avenue, New York, NY 10158. Fourth edition. 1986. $30.95.

INTRODUCTION TO COMPUTER ARCHITECTURE AND ORGANIZATION. H. Lorin. John Wiley and Sons, Incorporated, 605 Third Avenue, New York, NY 10158. 1982. $30.95.

AUTOMATION

HANDBOOKS AND MANUALS

FUNDAMENTALS HANDBOOK OF ELECTRICAL AND COMPUTER ENGINEERING. Sheldon S.L. Chang, editor. John Wiley and Sons, Incorporated, 605 Third Avenue, New York, NY 10158. (800) 526-5368 or (212) 850-6000. Three volumes. 1982. $180.00 set price.

HANDBOOK OF COMPUTERS AND COMPUTING. Arthur H. Seidman, editor. Van Nostrand Reinhold Book Company, 115 Fifth Avenue, New York, NY 10003. (800) 543-2681. 1984. $79.95.

HANDBOOK OF SOFTWARE ENGINEERING. Charles R. Vick, editor. Van Nostrand Reinhold Book Company, 115 Fifth Avenue, New York, NY 10003. (800) 543-2681. 1984. $66.95.

MCGRAW-HILL COMPUTER HANDBOOK. Harry Helms, editor. McGraw-Hill Book Company, 1221 Avenue of the Americas, New York, NY 10020. (212) 512-2000. 1983. $84.50.

ONLINE DATA BASES

COMPENDEX. Engineering Information, Incorporated, 345 East 47th Street, New York, NY 10017. (800) 221-1044 or (212) 705-7615. Engineering and technical literature, 1975 to present. Inquire as to cost and availability.

DISSERTATION ABSTRACTS ONLINE. University Microfilms International, 300 North Zeeb Road, Ann Arbor, MI 48106. (800) 521-0600 or (313) 761-4700. Scope includes virtually all doctoral dissertations accepted at accredited American institutions from 1861 to present in 252 subject areas. Inquire as to cost and availability.

INSPEC. INSPEC Marketing Department, Institute of Electrical and Electronics Engineers, Incorporated, IEEE Service Department, 445 Hoes Lane, Piscataway, NJ 08854. (201) 981-0060. Inquire as to on-line cost and availability.

MATHFILE. American Mathematical Society, Post Office Box 6248, Providence, RI 02940. (800) 556-7774 or (401) 272-9500. Scope includes pure and applied mathematics and related areas of physics, statistics, engineering, computer science, and operations research literature since 1973. Inquire as to cost and availability.

NTIS. National Technical Information Service, 5285 Port Royal Road, Springfield, VA 22161. (703) 487-4630. Broad coverage of government sponsored research reports, 1964 to present. Inquire as to cost and availability.

SCISEARCH. Institute for Scientific Information, 3501 Market Street, Philadelphia, PA 19104. (800) 523-1850 or (215) 386-0100. Broad multidisciplinary title and author index to the international literature of science and technology, 1974 to present. Inquire as to cost and availability.

OTHER SOURCES

COMPUTER PUBLISHERS AND PUBLICATIONS 1988-89: AN INTERNATIONAL DIRECTORY AND YEARBOOK. Edited by Efrem Sigel and Frederica Evan. Communications Trends, Incorporated. Distributed by Gale Research Company, Book Tower, Detroit, MI 48226. (800) 521-0707. Third edition. 1988. $140.00. Provides information on publishers of computer books and periodicals, recommended titles for libraries and bookstores, industry trends and statistics, and computer manufacturers as publishers.

PERIODICALS

ACM TRANSACTIONS ON COMPUTER SYSTEMS. Association for Computing Machinery, 11 West 42nd Street, New York, NY 10036. (212) 869-7440. Quarterly. $70.00 per year.

BYTE. Byte Publications, Incorporated, 70 Main Street, Petersborough, NJ 03458. (603) 924-9281. Monthly. $21.00 per year.

COMMUNICATIONS OF THE ACM. Association for Computing Machinery, 11 West 42nd Street, New York, NY 10036. (212) 869-7440. Monthly. $78.00 per year.

COMPUTER. Institute of Electrical and Electronic Engineers (IEEE), IEEE Service Center, 445 Hoes Lane, Piscataway, NJ 08854. (201) 981-0060. Monthly. $90.00 per year.

COMPUTER DESIGN. PennWell Directories, Incorporated, Post Office Box 1260, Tulsa, OK 74101. (918) 835-3161. Monthly. $60.00 per year.

COMPUTER MAGAZINE. IEEE Computer Society, 1109 Spring Street, Suite 300, Silver Spring, MD 20910. (301) 589-8142. Monthly. $34.00 per year.

COMPUTERS AND ELECTRONICS. Ziff-Davis Publishing Company, 3460 Wilshire Boulevard, Los Angeles, CA 90010. Monthly. $17.00 per year.

COMPUTERWORLD. C.W. Communications, 375 Cochituate Road, Famington, MA 01701. (617) 879-0700. Weekly. $44.00 per year.

CREATIVE COMPUTING. Creative Computing, Post Office Box 5214, Boulder, CO 80321. Monthly. $25.00 per year.

DATAMATION. Technical Publishing Company, 875 Third Avenue, New York, NY 10022. (212) 605-9400. Semi-monthly. $50.00 per year.

DR. DOBB'S JOURNAL OF SOFTWARE TOOLS. M & T Publishing, Incorporated, 2464 Embarcadero Way, Palo Alto, CA 94303. (415) 424-0600. Monthly. $25.00 per year.

IEEE TRANSACTIONS ON COMPUTERS. Institute of Electrical and Electronic Engineers (IEEE), IEEE Service Center, 445 Hoes Lane, Piscataway, NJ 08854. (201) 981-0060. Monthly. $130.00 per year.

INTERNATIONAL JOURNAL OF COMPUTER AND INFORMATION SCIENCE. Plenum Publishing Corporation, 233 Spring Street, New York, NY 10013. (800) 221-9369. Bimonthly. $195.00 per year.

JOURNAL OF THE ASSOCIATION FOR COMPUTING MACHINERY (ACM). Association for Computing Machinery, 11 West 42nd Street, New York, NY 10036. (212) 869-7440. Quarterly. $60.00 per year.

SIGARCH COMPUTER ARCHITECTURE NEWS. Association for Computing Machinery, 11 West 42nd Street, New York, NY 10036. (212) 869-7440. Bimonthly. $15.00 per year.

RESEARCH CENTERS AND INSTITUTES

CASE WESTERN UNIVERSITY. Center for Automation and Intelligent Systems Research, 2040 Adelbert Road, Cleveland, OH 44106. (216) 368-4040.

OAKLAND UNIVERSITY. Center for Robotics and Advanced Automation, Rochester, MI 48063. (313) 370-2233.

UNIVERSITY OF MARYLAND. Center for Automation Research, College Park, MD 20742. (301) 454-4526.

UNIVERSITY OF MISSOURI - ROLLA. Institute for Flexible Manufacturing and Industrial Automation, Mechanical Engineering Department, Rolla, MO 65401. (314) 341-4614.

AUTOMOBILE TRANSMISSIONS

See: AUTOMOTIVE ENGINEERING

AUTOMOBILES

See: AUTOMOTIVE ENGINEERING

AUTOMOTIVE ENGINEERING

See also: BRAKES, DIESEL ENGINES

ABSTRACT SERVICES AND INDEXES

APPLIED SCIENCE AND TECHNOLOGY INDEX. H.W. Wilson Company, 950 University Avenue, Bronx, NY 10452. (800) 367-6670 or (212) 588-8400. Inquire as to cost and availability.

CHEMICAL ABSTRACTS. Chemical Abstracts Service, Box 3012, Columbus, OH 43210. (614) 421-3600. 1907 to present. Weekly. $9200.00 per year.

ENGINEERING INDEX MONTHLY. Engineering Information, Incorporated, 345 East 47th Street, New York, NY 10017. (800) 221-1044 or (212) 705-7600. Monthly, with annual cumulation. $1560.00 per year.

PHYSICS ABSTRACTS. Institute of Electrical Engineers, London, United Kingdom. Available from: Institute of Electrical and Electronic Engineers (IEEE), 345 East 47th Street, New York, NY 10017. (212) 705-7900.

SAE TECHNICAL LITERATURE ABSTRACTS. Society of Automotive Engineers, 400 Commonwealth Drive, Warrendale, PA 15096. (412) 776-4841. Quarterly. $55.00 per year.

ASSOCIATIONS AND PROFESSIONAL SOCIETIES

SOCIETY OF AUTOMOTIVE ENGINEERS. 400 Commonwealth Drive, Warrendale, PA 15096. (412) 776-4841.

BIBLIOGRAPHIES

SCIENTIFIC AND TECHNICAL BOOKS IN PRINT; AN INDEX TO LITERATURE IN SCIENCE AND TECHNOLOGY. R.R. Bowker Company, 205 East 42nd Street, New York, NY 10017. (800) 521-8110 or (212) 916-1600.

DIRECTORIES AND BIOGRAPHICAL SOURCES

AUTOMOTIVE CONSULTANTS DIRECTORY. Society of Automotive Engineers, 400 Commonwealth Drive, Warrendale, PA 15096. (412) 776-4841. Annual. $16.00.

AUTOMOTIVE ENGINEERING - DIRECTORY AND CATALOG FILE. Society of Automotive Engineers, 400 Commonwealth Drive, Warrendale, PA 15096. (412) 776-4841. Annual. Available only to advertisers and exhibitors in SAE journals and expositions.

RESEARCH CENTERS DIRECTORY. Gale Research Company, Detroit, MI 48226. (800) 521-0707. Eleventh edition. 1987.

WHO'S WHO IN ENGINEERING. Engineers Joint Council, 345 East 47th Street, New York, NY 10017. (212) 705-7010. 1985. $200.00.

WHO'S WHO IN FRONTIER SCIENCE AND TECHNOLOGY. Marquis Who's Who Incorporated, 200 East Ohio Street, Chicago, IL 60611. (800) 428-3898 or (312) 787-2008.

SCIENTIFIC AND TECHNICAL ORGANIZATIONS AND AGENCIES DIRECTORY. Gale Research Company, Book Tower, Detroit, MI 48226. (800) 521-0707. 1985.

WHO'S WHO IN TECHNOLOGY TODAY. Reston Publishing Company, Incorporated, c/o Prentice-Hall, Incorporated, Englewood Cliffs, NJ 07632. (800) 262-6868. Biennial. Five volumes. $425.00. Covers the fields of electronics, computer science, physics, optics, chemistry, biotechnology, mechanics, energy, and earth science.

ENCYCLOPEDIAS AND DICTIONARIES

ENCYCLOPEDIA OF PHYSICAL SCIENCE AND TECHNOLOGY. Academic Press, Incorporated, Orlando, FL 32887. (800) 321-5068 or (305) 345-2734. Fifteen volumes, 1986.

MCGRAW-HILL DICTIONARY OF ENGINEERING. Sybil P. Parker, editor. McGraw-Hill Book Company, 1221 Avenue of the Americas, New York, NY 10020. (212) 512-2000. 1984. $39.95.

MCGRAW-HILL ENCYCLOPEDIA OF SCIENCE AND TECHNOLOGY. McGraw-Hill Book, 1221 Avenue of the Americas, New York, NY 10020. (212) 512-2000.

SAE MOTOR VEHICLE, SAFETY AND ENVIRONMENTAL TERMINOLOGY. Society of Automotive Engineers, 400 Commonwealth Drive, Warrendale, PA 15096. (412) 776-4841. $10.00 in paper.

GENERAL WORKS

AUTOMOTIVE AERODYNAMICS. Society of Automotive Engineers, 400 Commonwealth Drive, Warrendale, PA 15096. (412) 776-4841. 1978. $38.00.

AUTOMOTIVE COMPUTERS AND CONTROL SYSTEMS. Tom Weathers and Claud Hunter. Prentice-Hall Publishing Company, Englewood Cliffs, NJ 07632. (201) 592-2352. 1984. $24.95.

AUTOMOTIVE ENGINE DESIGN. William H. Crouse and D.L. Anglin. McGraw-Hill Book Company, 1221 Avenue of the Americas, New York, NY 10020. (800) 628-0004. 1970. $21.60.

HANDBOOKS AND MANUALS

SAE HANDBOOK. SAE, Incorporated, 400 Commonwealth Drive, Warrendale, PA 15096. (412) 776-4841. $70.00 members; $140.00 non-members.

STANDARD HANDBOOK OF ENGINEERING CALCULATIONS. Tyler G. Hicks, editor. McGraw-Hill Book Company, 1221 Avenue of the Americas, New York, NY 10020. (212) 512-2000. Second edition. 1984. $59.50.

THE WILEY ENGINEER'S DESK REFERENCE. Sanford I. Heisler. John Wiley and Sons, Incorporated, 605 Third Avenue, New York, NY 10158. (800) 526-5368 or (212) 850-6418. 1984. $36.00.

ONLINE DATA BASES

COMPENDEX. Engineering Information, Incorporated, 345 East 47th Street, New York, NY 10017. (800) 221-1044 or (212) 705-7615. Engineering and technical literature, 1975 to present. Inquire as to cost and availability.

NTIS. National Technical Information Service, 5285 Port Royal Road, Springfield, VA 22161. (703) 487-4630. Broad coverage of government sponsored research reports, 1964 to present. Inquire as to cost and availability.

SAE DATABASES. SAE, Incorporated, 400 Commonwealth Drive, Warrendale, PA 15096. (412) 776-4841. Citations to literature on all types of self-propelled vehicles, 1965 to present. Inquire as to on-line cost and availability.

WILSONLINE. H.W. Wilson Company, 950 University Avenue, Bronx, NY 10452. (800) 367-6770 or (212) 588-8400. Makes available online versions of the printed H.W. Wilson indexes including Applied Science and Technology Index, Business Periodicals Index, and Readers' Guide to Periodical Literature. Period covered is generally 1983 to present. Inquire as to cost and availability.

OTHER SOURCES

FEDERAL MOTOR VEHICLE SAFETY STANDARDS AND REGULATIONS. U.S. National Highway Traffic Safety Administration, Department of Transportation, 400 7th Street, SW, Washington, DC 20590. (202) 426-2768. Base volume plus monthly updates. $120.00 per year.

SAE CUMULATIVE INDEX 1965-1983. Society of Automotive Engineers, 400 Commonwealth Drive, Warrendale, PA 15096. (412) 776-4841. 1984. $75.00.

PERIODICALS

AUTOMOTIVE ENGINEER. Institution of Mechanical Engineers, Mechanical Engineering Publications, Limited, Box 24, Northgate Avenue, Bury Street, Edmunds, Suffolk IP32 6BW, England. Six times per year. $75.00.

AUTOMOTIVE ENGINEERING MAGAZINE. Society of Automotive Engineers, 400 Commonwealth Drive, Warrendale, PA 15096. (412) 776-4841. Monthly. $42.00 per year.

AUTOMOTIVE NEWS. Crain Communications Incorporated, 1400 Woodbridge Avenue, Detroit, MI 48207. (313) 446-6000. Weekly. $45.00 per year.

JOURNAL OF TERRAMECHANICS. Pergamon Journals, Incorporated, Maxwell House, Fairview Park, Elmsford, NY 10523. (603) 646-4100. Deals with off-road vehicles. Quarterly. $100.00 per year.

RESEARCH CENTERS AND INSTITUTES

COORDINATING RESEARCH COUNCIL, Incorporated 219 Perimeter Center Parkway, Atlanta, GA 30346. (404) 396-3400.

UNIVERSITY OF MICHIGAN. Automotive Engineering Laboratory, Department of Mechanical Engineering and Applied Mechanics, Ann Arbor, MI. (313) 764-4256.

WAYNE STATE UNIVERSITY. Center for Automotive Research, 5050 Anthony Wayne Drive, Detroit, MI 48202. (313) 577-3887.

AVIONICS

See also: AERONAUTICAL ENGINEERING, AEROSPACE ENGINEERING

ABSTRACT SERVICES AND INDEXES

APPLIED SCIENCE AND TECHNOLOGY INDEX. H.W. Wilson Company, 950 University Avenue, Bronx, NY 10452. (800) 367-6670 or (212) 588-8400. 1958 to present. Monthly. Inquire as to cost and availability.

CURRENT CONTENTS: ENGINEERING, TECHNOLOGY AND APPLIED SCIENCES. Institute for Scientific Information, 3501 Market Street, Philadelphia, PA 19104. (800) 523-1850 or (215) 386-0100. 1970 to present. Weekly. $272.00 per year.

ENGINEERING INDEX MONTHLY AND AUTHOR INDEX. Engineering Information, Incorporated, 345 East 47th Street, New York, NY 10017. (800) 221-1044 or (212) 705-7600. Monthly, with annual cumulation. $1560.00 per year.

INTERNATIONAL AEROSPACE ABSTRACTS. AIAA/TIS, 1633 Broadway, New York, NY 10019. (212) 581-4300. Semimonthly. $700.00 per year.

SCIENCE ABSTRACTS. Section A: Physics; Section B: Electrical and Electronics Abstracts; Section C: Computer and Control Abstracts. Institute of Electrical Engineers, London, United Kingdom. Available from: Institute of Electrical and Electronic Engineers (IEEE), 445 Hoes Lane, Piscataway, NJ 08854. Inquire as to cost and availability.

SCIENCE CITATION INDEX. Institute for Scientific Information, 3501 Market Street, Philadelphia, PA 19104. (800) 523-1850 or (215) 386-0100.

STAR. (Scientific and Technical Aerospace Reports). U.S. National Aeronautics and Space Administration, Scientific and Technical Information Facility, Box 8757, Baltimore-Washington International Airport, MD 21240. (202) 755-2210. Semimonthly, with semiannual and annual indexes. $85.00 per year.

ANNUAL REVIEWS AND YEARBOOKS

JANE'S AIRPORT EQUIPMENT 1987. David Rider, editor. Jane's Publishing Incorporated, 135 West 50th Street, New York, NY 10020. (212) 586-7745. Fifth edition. 1987. $135.00.

JANE'S AVIONICS 1987-88. Stephen R. Broadbent, editor. Jane's Publishing Incorporated, 135 West 50th Street, New York, NY 10020. (212) 586-7745. Sixth edition. 1987. $125.00.

JANE'S MILITARY COMMUNICATIONS 1987. R.J. Raggett, editor. Jane's Publishing Incorporated, 135 West 50th Street, New York, NY 10020. (212) 586-7745. Eighth edition. 1987. $135.00.

ASSOCIATIONS AND PROFESSIONAL SOCIETIES

AIRCRAFT ELECTRONICS ASSOCIATION. Box 1981, Independence, MO 64055. (816) 373-6565.

AMERICAN HELICOPTER SOCIETY, Incorporated 217 North Washington Street, Alexandria, VA 22314. (703) 684-6777.

AMERICAN INSTITUTE OF AERONAUTICS AND ASTRONAUTICS. 1633 Broadway, New York, NY 10019. (212) 581-4300.

FLIGHT SAFETY FOUNDATION, Incorporated 5510 Columbia Pike, Arlington, VA 22204-3194. (703) 820-2777.

INSTITUTE OF ELECTRICAL AND ELECTRONICS ENGINEERS (IEEE). 345 East 47th Street, New York, NY 10017.

SOCIETY OF AUTOMOTIVE ENGINEERS. SAE, Incorporated, 400 Commonwealth Drive, Warrendale, PA 15096. (412) 776-4841.

BIBLIOGRAPHIES

AERONAUTICAL ENGINEERING: A SPECIAL BIBLIOGRAPHY WITH INDEXES. U.S. National Aeronautics and Space Administration, Washington, DC 20546. (202) 755-2320. Available from National Technical Information Service, Springfield, VA 22161. 1970 to present. Monthly. Inquire as to cost and availability.

NATIONAL AIR AND SPACE MUSEUM AND SMITHSONIAN INSTITUTION. Aerospace Periodical Index, 1973-1982. 1983. G.K. Hall and Company, 70 Lincoln Street, Boston, MA 02111. (800) 343-2806. $100.00.

SCIENTIFIC AND TECHNICAL BOOKS AND SERIALS IN PRINT, 1988; AN INDEX TO LITERATURE IN SCIENCE AND TECHNOLOGY. R.R. Bowker Company, 205 East 42nd Street, New York, NY 10017. (800) 521-8110 or (212) 916-1600. $175.00.

DIRECTORIES AND BIOGRAPHICAL SOURCES

AAS DIRECTORY. American Astronautical Society, 6212-B Old Keene Mill Court, Springfield, VA 22152. (703) 866-0020. Annual. $35.00 per year.

RESEARCH CENTERS DIRECTORY. Gale Research Company, Detroit, MI 48226. (800) 521-0707. Eleventh edition. 1987.

WHO'S WHO IN ENGINEERING. Engineers Joint Council, 345 East 47th Street, New York, NY 10017. (212) 705-7010. $200.00.

WORLD AVIATION DIRECTORY. Murdoch Magazines, 1156 15th STreet, NW, Washington, DC 20005. (202) 822-4600. Semiannual. $75.00 per year.

ENCYCLOPEDIAS AND DICTIONARIES

DICTIONARY OF AEROSPACE ENGINEERING. M.G. Kotik. Elsevier Science Publishing Company, Incorporated, 52 Vanderbilt Avenue, New York, NY 10017. (212) 370-5520. 1986. $170.00.

ENCYCLOPEDIA OF PHYSICAL SCIENCE AND TECHNOLOGY. Academic Press, Incorporated, Orlando, Florida 32887. (800) 321-5068 or (305) 345-2734. Fifteen volumes, 1986.

JANE'S AEROSPACE DICTIONARY. Jane's Publishing Incorporated, 135 West 50th Street, New York, NY 10020. (212) 586-7745. Second edition. 1986. $39.95.

GENERAL WORKS

AIRCRAFT BASIC SCIENCE. R.D. Bent and J.L. Mckinley. McGraw-Hill Book Company, 1221 Avenue of the Americas, New York, NY 10020. (212) 512-2000. 1980. $32.95.

APPLIED AVIONICS. John L. Dearness and Robert B. Angus. Bowen's Publishing Division, P.O. Box 270, Bedford, MA 01730-0270. (617) 275-1660. Third edition. 1982. $40.00.

DIGITAL AVIONICS SYSTEMS. Cary Spitzer. Prentice-Hall Press, Incorporated, Englewood Cliffs, NJ 07632. (800) 562-0245. 1987. $39.95.

FOUNDATIONS OF AERODYNAMICS: BASES OF AERODYNAMIC DESIGN. A.M. Kuethe and C. Chow. John Wiley and Sons, Incorporated, 605 Third Avenue, New York, NY 10158. (800) 526-5368 or (212) 850-6000. Fourth edition. 1986. $43.95.

MODERN AVIATION ELECTRONICS. Albert D. Helfrick. Prentice-Hall Publishing, Incorporated, Englewood Cliffs, NJ 07632. (800) 562-0245. 1984. $34.95.

HANDBOOKS AND MANUALS

JANE'S ALL THE WORLD'S AIRCRAFT 1987-88. John W.R. Taylor. Jane's Publishing, Incorporated, 115 Fifth Avenue, New York, NY 10003. (214) 254-9097. 1987.

STANDARD HANDBOOK OF ENGINEERING CALCULATIONS. Tyler G. Hicks, editor. McGraw-Hill Book Company, 1221 Avenue of the Americas, New York, NY 10020. (212) 512-2000. Second edition. 1984. $59.50.

THE WILEY ENGINEER'S DESK REFERENCE. Sanford I. Heisler. John Wiley and Sons, Incorporated, 605 Third Avenue, New York, NY 10158. (800) 526-5368 or (212) 850-6418. 1984. $36.00.

ONLINE DATA BASES

COMPENDEX. Engineering Information, Incorporated, 345 East 47th Street, New York, NY 10017. (800) 221-1044 or (212) 705-7615. Engineering and technical literature, 1975 to present. Inquire as to cost and availability.

NASA. National Aeronautics and Space Administration, Scientific and Technical Information Branch, 300 7th Street, SW, Washington, DC 20546. Citations and abstracts of aerospace literature, 1962 to present. Inquire as to cost and availability.

NTIS. National Technical Information Service, 5285 Port Royal Road, Springfield, VA 22161. (703) 487-4630. Broad coverage of government sponsored research reports, 1964 to present. Inquire as to cost and availability.

PERIODICALS

AEROSPACE AMERICA. American Institute of Aeronautics and Astronautics, 1633 Broadway, New York, NY 10019. (212) 581-4300. Monthly. $51.00 per year.

AEROSPACE ENGINEERING MAGAZINE (SOCIETY OF AUTOMOTIVE ENGINEERS). SAE, Incorporated, 400 Commonwealth Drive, Warrendale, PA 15096. (412) 776-4841. Monthly. $30.00 per year.

AIAA JOURNAL. AIAA/TIS, 1633 Broadway, New York, NY 10019. (212) 581-4300. Monthly. $205.00 per year.

AMERICAN HELICOPTER SOCIETY JOURNAL. American Helicopter Society, Incorporated, 217 North Washington Street, Alexandria, Va 22314. (703) 684-6777. Quarterly. $25.00 per year.

AVIATION MECHANICS BULLETIN. Flight Safety Foundation, Incorporated, 5510 Columbia Pike, Arlington, VA 22204-3194. (703) 820-2777. Bimonthly. $6.50 per year.

AVIATION WEEK AND SPACE TECHNOLOGY. McGraw-Hill Book Company, Incorporated, 1221 Avenue of the Americans, New York, NY 10020. (212) 512-2000. Weekly. $55.00 per year.

AVIONICS. Atlantic Communications, Incorporated, Box 5100, Westport, CT 06881. (203) 227-2280. Monthly. $36.00 per year.

AVIONICS NEWS MAGAZINE. Aircraft Electronics Association, Box 1981, Independence, MO 64055. (816) 373-6565. Monthly. Free.

CANADIAN AERONAUTICS AND SPACE JOURNAL. Canadian Aeronautics and Space Institute, 222 Somerset Street, West, Suite 601, Ottawa, ON K2P 0J1, Canada. (613) 234-0191. Quarterly. $35.00 per year.

FLYING. CBS Magazines, 1515 Broadway, New York, NY 10036. (212) 503-4200. Monthly. $14.00 per year.

IEEE TRANSACTIONS ON AEROSPACE. Institute of Electrical and Electronics Engineers, 345 East 47th Street, New York, NY 10017. Bimonthly. $108.00 per year.

JOURNAL OF AIRCRAFT. American Institute of Aeronautics and Astronautics, AIAA/TIS, 1633 Broadway, New York, NY 10019. (212) 581-4300. Monthly. $185.00 per year.

JOURNAL OF GUIDANCE AND CONTROL. American Institute of Aeronautics and Astronautics, AIAA/TIS, 1633 Broadway, New York, NY 10019. (212) 581-4300. Bimonthly. $95.00 per year.

ROTOR AND WING INTERNATIONAL. PJS Publications, Incorporated, New Plaza, Box 1790, Peoria, IL 61656. (309) 682-6626. Monthly. $24.00 per year.

VERTIFLITE. American Helicopter Society, Incorporated, 217 North Washington Street, Alexandria, VA 22314. (703) 684-6777. Bimonthly. $25.00 per year.

B

BALLISTIC MISSILES
See: GUIDED MISSILES

BALLOONS
See: AERONAUTICS

BANACH ALGEBRA
See: ALGEBRA

BAND SPECTRA
See: SPECTROSCOPY

BAND THEORY OF SOLIDS
See: SOLID STATE PHYSICS

BAROMETER
See: METEOROLOGY

BARYON
See: PARTICLE PHYSICS

BASES
See: ACIDS AND BASES

BASIC
See: PROGRAMMING LANGUAGES

BATHOLITH
See: VOLCANOLOGY

BATHYMETRY
See: OCEANOGRAPHY

BATTERIES
See also: ELECTROCHEMISTRY

ABSTRACT SERVICES AND INDEXES

APPLIED SCIENCE AND TECHNOLOGY INDEX. H.W. Wilson Company, 950 University Avenue, Bronx, NY 10452. (800) 367-6670 or (212) 588-8400. Inquire as to cost and availability.

CHEMICAL ABSTRACTS. Chemical Abstracts Service, 2540 Olentangy Road, P.O. Box 3012, Columbus, OH 43210. (800) 848-6538 or (614) 421-3600. Weekly. $9200.00 per year.

PHYSICS ABSTRACTS. Institute of Electrical Engineers, London, United Kingdom. Available from: Institute of Electrical and Electronic Engineers (IEEE), 345 East 47th Street, NEw York, NY 10017. (212) 705-7900.

SCIENCE CITATION INDEX. Institute for Scientific Information, 3501 Market Street, Philadelphia, PA 19104. (800) 523-1850 or (215) 386-0100.

ANNUAL REVIEWS AND YEARBOOKS

ADVANCES IN ELECTROCHEMISTRY AND ELECTROCHEMICAL ENGINEERING. John Wiley and Sons, Incorporated, 605 Third Avenue, New York, NY 10158. (800) 526-5368 or (212) 850-6000. Irregular. Approximately $60.00.

ASSOCIATIONS AND PROFESSIONAL SOCIETIES

AMERICAN CHEMICAL SOCIETY. 1155 16th Street, NW, Washington, DC 20036. (202) 872-4600.

ASSOCIATION OF CONSULTING CHEMISTS AND CHEMICAL ENGINEERS. 50 East 41st Street, Suite 92, New York, NY 10017. (212) 684-6255.

BATTERY COUNCIL INTERNATIONAL. 111 East Wacker Drive, Chicago, IL 60601. (312) 644-6610.

ELECTROCHEMICAL SOCIETY. Ten South Main Street, Pennington, NJ 08534. (609) 737-1902.

BATTERIES

BIBLIOGRAPHIES

SCIENCE BOOKS AND FILMS. American Association for the Advancement of Science, 1333 H Street, NW, Washington, DC 20005.

SCIENTIFIC AND TECHNICAL BOOKS AND SERIALS IN PRINT 1988; AN INDEX TO LITERATURE IN SCIENCE AND TECHNOLOGY. R.R. Bowker Company, 205 East 42nd Street, New York, NY 10017. (800) 521-8110 or (212) 916-1600. $175.00.

DIRECTORIES AND BIOGRAPHICAL SOURCES

ELECTROCHEMICAL SOCIETY MEMBERSHIP DIRECTORY. Electrochemical Society, Ten South Main Street, Pennington, NJ 08534. (609) 737-1902. Annual. Available to members only.

RESEARCH CENTERS DIRECTORY. Gale Research Company, Book Tower, Detroit, MI 48226. (800) 521-0707. Eleventh edition. 1987.

SCIENTIFIC AND TECHNICAL ORGANIZATIONS AND AGENCIES DIRECTORY. Gale Research Company, Book Tower, Detroit, MI 48226. (800) 521-0707. 1985.

SLIG BUYERS' GUIDE. Independent Battery Manufacturers Association, 100 Larchwood Drive, Largo, FL 33540. Biennial. $8.00.

WHO'S WHO IN TECHNOLOGY TODAY. Reston Publishing Company, Incorporated, c/o Prentice-Hall, Incorporated, Englewood Cliffs, NJ 07632. (800) 262-6868. Biennial. Five volumes. $425.00. Covers the fields of electronics, computer science, physics, optics, chemistry, biotechnology, mechanics, energy, and earth science.

ENCYCLOPEDIAS AND DICTIONARIES

ENCYCLOPEDIA OF PHYSICAL SCIENCE AND TECHNOLOGY. Academic Press, Incorporated, Orlando, FL 32887. (800) 321-5068 or (305) 345-2734. Fifteen volumes, 1986.

IEEE STANDARD DICTIONARY OF ELECTRICAL AND ELECTRONIC TERMS. John Wiley and Sons, Incorporated, 605 Third Avenue, New York, NY 10158. (800) 526-5368 or (212) 850-6000. 1984. $49.95.

MCGRAW-HILL ENCYCLOPEDIA OF SCIENCE AND TECHNOLOGY. McGraw-Hill Book, Incorporated, 1221 Avenue of the Americas, New York, NY 10020. (212) 997-3675.

GENERAL WORKS

BATTERIES AND ENERGY SYSTEMS. Charles L. Mantell, editor. McGraw-Hill Book Company, 1221 Avenue of the Americas, New York, NY 10020. (800) 628-0004. 1982. $39.95.

ELECTROCHEMICAL, ELECTRICAL AND MAGNETIC STORAGE OF ENERGY. W.V. Hassenzahl, editor. Van Nostrand Reinhold Company, Incorporated, 115 Fifth Avenue, New York, NY 10003. (800) 543-2681. 1982. $52.95.

ELECTROCHEMISTRY. Journal of Chemical Education, 238 Kent Road, Springfield, PA 19064. 1983. $6.90.

ELECTRODE PROCESSES AND ELECTROCHEMICAL ENGINEERING. Fumio Hine. Plenum Publishing Corporation, 233 Spring Street, New York, NY 10013. (800) 221-9369. 1985. $55.00.

HANDBOOKS AND MANUALS

HANDBOOK OF BATTERIES AND FUEL CELLS. David Linden. McGraw-Hill Book Company, 1221 Avenue of the Americas, New York, NY 10020. (800) 628-0004. 1984. $82.50.

ONLINE DATA BASES

CA SEARCH. Chemical Abstracts Service, P.O. Box 3012, Columbus, OH 43210. (614) 421-3600. Guide to chemical literature, 1972 to present. Inquire as to cost and availability.

INSPEC. INSPEC Marketing Department, Institute of Electrical and Electronics Engineers, Incorporated, IEEE Service Department, 445 Hoes Lane, Piscataway, NJ 08854. (201) 981-0060. Inquire as to on-line cost and availability.

NTIS. National Technical Information Service, 5285 Port Royal Road, Springfield, VA 22161. (703) 487-4630. Broad coverage of government sponsored research reports, 1964 to present. Inquire as to cost and availability.

SCISEARCH. Institute for Scientific Information, 3501 Market Street, Philadelphia, PA 19104. (800) 523-1850 or (215) 386-0100. Broad multidisciplinary title and author index to the international literature of science and technology, 1974 to present. Inquire as to cost and availability.

PERIODICALS

ADVANCED BATTERY TECHNOLOGY. Advanced Battery Technology, P.O. Box 98, Dana Point, CA 92629. (714) 496-2574. Monthly. $72.00 per year.

ELECTRIC VEHICLE/BATTERY TECHNOLOGY. Business Communications, Incorporated, 9 Viaduct Road, Box 2070C, Stamford, CT 06906. (203) 325-2208. Monthly. $275.00 per year.

ELECTROCHEMICAL REACTIONS. Chemical Abstracts Services, P.O. Box 3012, Columbus, OH 43210. (614) 421-3600. Bimonthly. $110.00 per year.

ELECTROCHEMICAL SOCIETY JOURNAL. Electrochemical Society, Ten South Main Street, Pennington, NJ 08534. (609) 737-1902.

ELECTROCHIMICA ACTA. Pergamon Journals, Incorporated, Maxwell House, Fairview Park, Elmsford, NY 10523. (914) 592-7700. Monthly. $430.00 per year.

ELECTROLYSIS WORLD. American Electrolysis Association, 710 Tennent Road, Englishtown, NJ 07826. (201) 536-6477. Bimonthly.

ELECTRONIC CHEMICALS AND MATERIALS. Chemical Abstracts Service, P.O. Box 3012, Columbus, OH 43210. (614) 421-3600. Bimonthly. $120.00 per year.

JOURNAL OF THE ELECTROCHEMICAL SOCIETY. Electrochemical Society Incorporated, 10 South Main Street, Pennington, NJ 08534. (609) 737-1902. Monthly. $125.00 per year.

RESEARCH CENTERS AND INSTITUTES

ELECTROCHEMICAL ANALYSIS DIAGNOSTIC LABORATORY. Argonne National Laboratory, 9700 South Cass Avenue, Argonne, IL 60439. (312) 972-7764.

LAWRENCE BERKELEY LABORATORY. Applied Science Division, 1 Cyclotron Road, Berkeley, CA 94720. (415) 486-5001.

BEACHES

See: COASTS

BEACONS

See: NAVIGATION

BEAM COLUMNS

See: STRUCTURAL ENGINEERING

BEAM FOIL SPECTROSCOPY

See: SPECTROSCOPY

BEAMS

See: STRUCTURAL ENGINEERING

BEARING CAPACITY

See: STRUCTURAL ENGINEERING

BEARINGS AND BALL BEARINGS

ABSTRACT SERVICES AND INDEXES

ALLOYS INDEX. American Society for Metals, 9639 Kinsman Road, Metals Park, OH 44073. (216) 338-5151. $130.00 per year.

APPLIED MECHANICS REVIEWS. American Society of Mechanical Engineers, 345 East 47th Street, New York, NY 10017. (212) 705-7722. Monthly. $380.00 per year. Critical reviews of the world literatue in applied mechanics and related engineering science.

APPLIED SCIENCE AND TECHNOLOGY INDEX. H.W. Wilson Company, 950 University Avenue, Bronx, NY 10452. (800) 367-6670 or (212) 588-8400. Inquire as to cost and availability.

CHEMICAL ABSTRACTS. Chemical Abstracts Service, Box 3012, Columbus, OH 43210. (614) 421-3600. 1907 to present. Weekly. $9200.00 per year.

ENGINEERING INDEX MONTHLY. Engineering Information, Incorporated, 345 East 47th Street, New York, NY 10017. (800) 221-1044 or (212) 705-7600. Monthly, with annual cumulation. $1560.00 per year.

ISMEC BULLETIN (INFORMATION SERVICE IN MECHANICAL ENGINEERING). Cambridge Scientific Abstracts, 5161 River Road, Bethesda, MD 20816. (800) 638-8076 or (301) 951-1400. Monthly. $320.00 per year.

METALS ABSTRACTS. American Society for Metals, 9639 Kinsman Road, Metals Park, OH 44073. (216) 338-5151. Monthly. $890.00.

METALS ABSTRACTS INDEX. American Society for Metals, 9639 Kinsman Road, Metals Park, OH 44073. (216) 338-5151. Monthly. (Sold only to subscribers of Metals Abstracts).

PHYSICS ABSTRACTS. Institute of Electrical Engineers, London, United Kingdom. Available from: Institute of Electrical and Electronic Engineers (IEEE), 345 East 47th Street, New York, NY 10017. (212) 705-7900.

ASSOCIATIONS AND PROFESSIONAL SOCIETIES

ANTIFRICTION BEARING MANUFACTURERS ASSOCIATION. Crystal Gateway, Suite 704, 1235 Jefferson Davis Highway, Arlington, VA 22202. (703) 979-1261.

BEARINGS SPECIALISTS ASSOCIATION. 221 North LaSalle Street, Chicago, IL 60601. (312) 346-1600.

CAST BRONZE BEARING INSTITUTE. 221 North LaSalle Street, Suite 2026, Chicago, IL 60601. (312) 346-1600.

ROLLER BEARING ENGINEERS COMMITTEE. 1235 Jefferson Davis Highway, Suite 704, Arlington, VA 22202. (703) 979-1261.

BIBLIOGRAPHIES

BIBLIOGRAPHICAL GUIDE TO TECHNOLOGY. G.K. Hall, 70 Lincoln Street, Boston, MA 02111. (617) 423-3990. Lists technology materials cataloged by the New York Public Library. $175.00 per year.

DIRECTORIES AND BIOGRAPHICAL SOURCES

RESEARCH CENTERS DIRECTORY. Gale Research Company, Book Tower, Detroit, MI 48226. (800) 521-0707. Eleventh edition. 1987.

WHO'S WHO IN ENGINEERING. Engineers Joint Council, 345 East 47th Street, New York, NY 10017. (212) 705-7010. 1985. $200.00.

WORLD LIST OF BALL AND ROLLER BEARING MANUFACTURERS. Antifriction Bearing Manufacturers Association. Crystal Gateway, Suite 704, 1235 Jefferson Davis Highway, Arlington, VA 22202. (703) 979-1261. 1984. $100.00.

ENCYCLOPEDIAS AND DICTIONARIES

ENCYCLOPEDIA OF PHYSICAL SCIENCE AND TECHNOLOGY. Academic Press, Incorporated, Orlando, FL 32887. (800) 321-5068 or (305) 345-2734. Fifteen volumes, 1986.

MCGRAW-HILL DICTIONARY OF ENGINEERING. Sybil P. Parker, editor. McGraw-Hill Book Company, 1221 Avenue of the Americas, New York, NY 10020. (212) 512-2000. 1984. $39.95.

MCGRAW-HILL ENCYCLOPEDIA OF SCIENCE AND TECHNOLOGY. McGraw-Hill Book Company, 1221 Avenue of the Americas, New York, NY 10020. (212) 512-2000.

GENERAL WORKS

BALL AND ROLLER BEARINGS: THEORY, DESIGN, AND APPLICATION. Eschmann, Hasbargen and Weigand. John Wiley and Sons, Incorporated, 605 Third Avenue, New York, NY 10158. (800) 526-5368 or (212) 850-6000. Second edition. 1985. $54.95.

BALL BEARING LUBRICATION: THE ELASTO-HYDRODYNAMICS OF ELLIPTICAL CONTACTS. B.J. Hamrock and D. Dowson. John Wiley and Sons, Incorporated, 605 Third Avenue, New York, NY 10158. (800) 526-5368 or (212) 850-6000. 1981. $67.95.

BEARING DESIGN: HISTORICAL ASPECTS, PRESENT TECHNOLOGY, AND FUTURE PROBLEMS. W.J. Anderson, editor. American Society of Mechanical Engineers, 345 East 47th Street, New York, NY 10017. (212) 705-7722. 1980. $30.00.

BEARING SYSTEMS: PRINCIPLES AND PRACTICES. F.T. Barwell. Oxford University Press, 200 Madison Avenue, New York, NY 10016. (212) 679-7300. 1979. $78.00.

HANDBOOKS AND MANUALS

IBI GUIDE (INTERNATIONAL BEARING EXCHANGE). Interchange Incorporated, P.O. Box 16012, Saint Louis Park, MN 55416. (612) 929-6669. Biennial. $120.00.

POWER TRANSMISSION AND BEARING HANDBOOK. Industrial Publishing Company, 1111 Chester Avenue, Cleveland, OH 44114. (216) 696=7000. Biennial. $15.00.

STANDARDS FOR BALL AND ROLLER STEEL BALLS AND MAINTENANCE MANUAL. Anti-Friction Bearing Manufacturers Association, Century Building, Suite 1015, 2341 Jefferson Davis Highway, Arlington, VA 22202.

ONLINE DATA BASES

COMPENDEX. Engineering Information, Incorporated, 345 East 47th Street, New York, NY 10017. (800) 221-1044 or (212) 705-7615. Engineering and technical literature, 1975 to present. Inquire as to cost and availability.

INSPEC. INSPEC Marketing Department, Institute of Electrical and Electronics Engineers, Incorporated, IEEE Service Department, 445 Hoes Lane, Piscataway, NJ 08854. (201) 981-0060. Inquire as to on-line cost and availability.

ISMEC. Cambridge Scientific Abstracts, 5161 River Road, Bethesda, MD 20816. (800) 638-8076 or (301) 951-1400. (Information Service in Mechanical Engineering). Literature of mechanical and production engineering, 1973 to present. Inquire as to cost and availability.

METADEX. Metals Information, American Society for Metals, Metals Park, OH 44073. (216) 338-5151. Worldwide literature on the science and practice of metallurgy, 1966 to present. Inquire as to online cost and availability.

NTIS. National Technical Information Service, 5285 Port Royal Road, Springfield, VA 22161. (703) 487-4630. Broad coverage of government sponsored research reports, 1964 to present. Inquire as to cost and availability.

WILSONLINE. H.W. Wilson Company, 950 University Avenue, Bronx, NY 10452. (800) 367-6770 or (212) 588-8400. Makes available online versions of the printed H.W. Wilson indexes including Applied Science and Technology Index, Business Periodicals Index, and Readers' Guide to Periodical Literature. Period covered is generally 1983 to present. Inquire as to cost and availability.

PERIODICALS

JOURNAL OF APPLIED MECHANICS. American Society of Mechanical Engineers, 345 East 47th Street, New York, NY 10017. (212) 705-7722. Quarterly. $100.00 per year.

JOURNAL OF TRIBOLOGY. (Journal of Lubrication Technology). American Society of Mechanical Engineers, 345 East 47th Street, New York, NY 10017. (212) 705-7722. Quarterly. $80.00 per year.

RESEARCH CENTERS AND INSTITUTES

FRANKLIN INSTITUTE. Benjamin Franklin Parkway at 20th Street, Philadelphia, PA 19103. (215) 448-1000.

BEDROCK

See also: CRUST, EARTH SCIENCES, GEOCHEMISTRY, GEOLOGY, GEOPHYSICS, PHYSICAL GEOLOGY

ABSTRACT SERVICES AND INDEXES

BIBLIOGRAPHY AND INDEX OF GEOLOGY. American Geological Institute, 4220 King Street, Alexandria, VA 22302. (703) 379-2480.

CHEMICAL ABSTRACTS. Chemical Abstracts Service, 2540 Olentangy Road, P.O. Box 3012, Columbus, OH 43210. (800) 848-6538 or (614) 421-3600. Weekly. $6200.00 per year.

DEEP-SEA RESEARCH WITH OCEANOGRAPHIC LITERATURE REVIEW. Pergamon Press, Incorporated, Maxwell House, Fairview Park, Elmsford, NY 10523. (914) 592-7700. Twenty-four times per year. $600.00 per year.

MINERALOGICAL ABSTRACTS. Mineralogical Society and the Mineralogical Society of America, 41 Queen's Gate, London, SW7 5HR, England. Quarterly. $190.00 per year.

OCEANIC ABSTRACTS. Cambridge Scientific Abstracts, 5161 River Road, Bethesda, MD 20816. (301) 951-1400. Bimonthly. $652.00 per year.

SCIENCE CITATION INDEX. Institute for Scientific Information, 3501 Market Street, Philadelphia, PA 19104. (800) 523-1850 or (215) 386-0100. Inquire as to price and availability.

GENERAL SCIENCE INDEX. H.W. Wilson Company, 950 University Avenue, Bronx, NY 10452. (800) 367-6770 or (212) 588-8400. Inquire as to cost and availability.

ANNUAL REVIEWS AND YEARBOOKS

ADVANCES IN GEOPHYSICS. Academic Press, Incorporated, 6277 Sea Harbor Drive, Orlando, FL 32821. (800) 321-5068. Irregular. $62.00 per volume.

ANNUAL REVIEW AND EARTH AND PLANETARY SCIENCES. Annual Reviews, Incorporated, 4139 El Camino Way, Palo Alto, CA 94306. (415) 493-4400.

MINERALS YEARBOOK. Bureau of Mines, U.S. Department of the Interior. Available from U.S. Government Printing Office, Washington, DC 20402. (202) 783-3238. Annual. Three volumes. $45.00.

ASSOCIATIONS AND PROFESSIONAL SOCIETIES

AMERICAN ASSOCIATION OF PETROLEUM GEOLOGISTS. P.O. Box 979, Tulsa, OK 74101. (918) 584-2555.

AMERICAN GEOLOGICAL INSTITUTE. 4220 King Street, Alexandria, VA 22302. (703) 379-2480.

AMERICAN GEOPHYSICAL UNION. 2000 Florida Avenue, NW, Washington, DC 20009. (202) 462-6903.

AMERICAN INSTITUTE OF PROFESSIONAL GEOLOGISTS. 7828 Vance Drive, Suite 103, Arvada, CO 80003. (303) 431-0831.

ASSOCIATION OF ENGINEERING GEOLOGISTS. P.O. Box 1068, Brentwood, TN 37027. (615) 377-3578.

GEOLOGICAL SOCIETY OF AMERICA. P.O. BOX 9140, 3300 Penrose Place, Boulder, CO 80301. (303) 447-2020.

BIBLIOGRAPHIES

SCIENCE BOOKS AND FILMS. American Association for the Advancement of Science, 1333 H Street, NW, Washington, DC

20005.

SCIENTIFIC AND TECHNICAL BOOKS IN PRINT; AN INDEX TO LITERATURE IN SCIENCE AND TECHNOLOGY. R.R. Bowker Company, 205 East 42nd Street, New York, NY 10017. (800) 521-8110 or (212) 916-1600.

DIRECTORIES AND BIOGRAPHICAL SOURCES

AMERICAN MEN AND WOMEN OF SCIENCE: PHYSICAL AND BIOLOGICAL SCIENCES. Sixteenth edition. R.R. Bowker Company, 205 East 42nd Street, New York, NY 10017. (800) 521-8810 or (212) 916-1600. $595.00.

AMERICAN INSTITUTE OF PROFESSIONAL GEOLOGISTS. Membership Directory. American Institute of Professional Geologists, 7828 Vance Drive, Suite 103, Arvada, CO 80003. (303) 431-0831. Annual. $15.00.

ASSOCIATION OF ENGINEERING GEOLOGISTS DIRECTORY. Association of Engineering Geologists, Dr. G. Lee Christensen, Civil Engineering Department, Villanova University, Villanova, PA 19085. (215) 645-4960. Annual. $15.00.

GEOLOGICAL SOCIETY OF AMERICA. Membership Directory. Geological Society of America, 3300 Penrose Place, Boulder, CO 80301. (303) 447-2020. Annual. Available to members only.

RESEARCH CENTERS DIRECTORY. Gale Research Company, Book Tower, Detroit, MI 48226. Eleventh edition, 1987. (800) 521-0707.

WHO'S WHO IN FRONTIER SCIENCE AND TECHNOLOGY. Marquis Who's Who, Incorporated, 200 East Ohio Street, Chicago, IL 60611. (800) 428-3898 or (312) 787-2008.

GENERAL WORKS

ATLAS OF IGNEOUS ROCKS AND THEIR TEXTURES. W.S. Mackenzie, C.H. Donaldson, and C. Guilford. John Wiley and Sons, Incorporated, 605 Third Avenue, New York, NY 10158. (800) 526-5368 or (212) 850-6000. 1982. $29.95.

INTRODUCTION TO ROCK FORMING MINERALS. W.A. Deer, J. Zussman and R.A. Howie. John Wiley and Sons, Incorporated, 605 Third Avenue, New York, NY 10158. (800) 526-5368 or (212) 850-6000. 1966. $29.95.

PHYSICAL GEOLOGY. R.F. Flint and B.J. Skinner. John Wiley and Sons, Incorporated, 605 Third Avenue, New York, NY 10158. (800) 526-5368 or (212) 850-6000. Second edition. 1977. $40.95.

PLATE TECTONICS AND CRUSTAL EVOLUTION. K.C. Condie. Pergamon Press, Incorporated, Maxwell House, Fairview Park, Elmsford, NY 10523. (914) 592-7700. 1982. $39.95 in paper.

STRUCTURAL GEOLOGY OF ROCKS AND REGIONS. George H. Davis. John Wiley and Sons, Incorporated, 605 Third Avenue, New York, NY 10158. (800) 526-5368 or (212) 850-6000. 1984. $39.95.

HANDBOOKS AND MANUALS

GEOLOGY IN THE FIELD. R.R. Compton. John Wiley and Sons, Incorporated, 605 Third Avenue, New York, NY 10158. (800) 526-5368 or (212) 850-6418. 1985. $23.95.

ONLINE DATA BASES

GEOREF. American Geological Institute, 4220 King Street, Alexandria, VA 22302. (800) 336-4764 or (703) 379-2480. Geology and geosciences literature, 1961 to present. Inquire as to online cost and availability.

GEOARCHIVE. Geosystems, P.O. Box 1024, Westminster, London, England, SW1 P 2JL. Citations to literature on geoscience, 1669 to present. Inquire as to online cost and availability.

NTIS. National Technical Information Service, 5285 Port Royal Road, Springfield, VA 22161. (703) 487-4630. Broad coverage of government sponsored research reports, 1964 to present. Inquire as to online cost and availability.

SCISEARCH. Institute for Scientific Information, 3501 Market Street, Philadelphia, PA 19104. (800) 523-1850 or (215) 386-0100. Broad interdisciplinary index to the literature of science and technology, 1965 to present. Inquire as to online cost and availability.

PERIODICALS

AARG BULLETIN. American Association of Petroleum Geologists, P.O. Box 979, Tulsa, OK 74101. (918) 584-2555.

AMERICAN JOURNAL OF SCIENCE. Kline Geology Laboratory, Yale University, New Haven, CT 06520. Ten times per year. $80.00 per year.

ECONOMIC GEOLOGY. Society of Economic Geologists, P.O. Box 571, Golden, CO 80402. (303) 279-1899. Eight times per year. $25.00.

GEOLOGICAL MAGAZINE. Cambridge University Press, 32 East 57th Street, New York, NY 10022. (800) 872-7423. Bimonthly. $165.00 per year.

GEOLOGICAL SOCIETY JOURNAL. Geological Society of London. Blackwell Scientific Publications, Incorporated, 667 Lytton Avenue, Palo Alto, CA 94301. (415) 324-1688. Six times per year. $290.00.

GEOLOGICAL SOCIETY OF AMERICA BULLETIN. P.O. Box 9140, 3300 Penrose Place, Boulder, CO 80301. (303) 447-2020. Monthly. $80.00.

GEOLOGY. Geological Society of America. P.O. Box 9140, 3300 Penrose Place, Boulder, CO 80301. (303) 447-2020. Monthly. $55.00 per year.

GEOPHYSICS. Society of Exploration Geophysicists, P.O. Box 702740, Tulsa, OK 74170. (918) 493-3516. Monthly. $45.00 per year.

GEOTIMES. American Geological Institute, 4220 King Street, Alexandria, VA 22032. Monthly. $18.00 per year.

JOURNAL OF GEOLOGY. University of Chicago Press, 5801 South Ellis Street, Chicago, IL 60637. (312) 962-6700. Bimonthly. $30.00 per year.

JOURNAL OF GEOPHYSICAL RESEARCH. American Geophysical Union, 2000 Florida Avenue, NW, Washington, DC 20009. (202) 462-6903. Weekly. $680.00 per year to individuals.

JOURNAL OF GEOPHYSICS. Springer-Verlag, Incorporated, 175 Fifth Avenue, New York, NY 10010. (800) 526-7254 or (212) 460-1500. Bimonthly. $175.00 per year.

JOURNAL OF STRUCTURAL GEOLOGY. Pergamon Journals, Incorporated, Maxwell House, Fairview Park, Elmsford, NY 10523. (914) 592-7700. Eight times per year. $160.00 per year.

LITHOS; AN INTERNATIONAL JOURNAL OF MINERALOGY, PETROLOGY, AND GEOCHEMISTRY. Elsevier Science Publishers, P.O. Box 330, Irving-on-Hudson, NY 10533. Quarterly. $73.00 per year.

MOUNTAIN GEOLOGIST. Rocky Mountain Association of Geologists, 4201 West 51st Avenue, Denver, CO 80212-2902. Quarterly. $15.00 per year.

PROFESSIONAL GEOLOGIST. American Institute of Professional Geologists, 7828 Vance Drive, Suite 103, Arvada, CO 80003. Monthly.

REVIEWS OF GEOPHYSICS. American Geophysical Union, 2000 Florida Avenue, NW, Washington, DC 20009. (202) 462-6903. Quarterly. $240.00 per year.

ROCK MECHANICS AND ROCK ENGINEERING. Springer-Verlag New York, Incorporated, 175 Fifth Avenue, New York, NY 10010. (800) 526-7254 or (212) 460-1500. Quarterly. $61.00 per year.

TECTONICS. American Geophysical Union, 2000 Florida Avenue, NW, Washington, DC 20009. (202) 462-6903. Bimonthly. $30.00 per year to individuals.

TECTONOPHYSICS; AN INTERNATIONAL JOURNAL OF GEOTECTONICS AND THE GEOLOGY AND PHYSICS OF THE INTERIOR OF THE EARTH. Elsevier Science Publishers, P.O. Box 330, Irving-on-Hudson, NY 10533. Forty-four times per year. $870.00 per year.

VOLCANOLOGY AND SEISMOLOGY. Gordon and Breach Science Publishers, 50 West 23rd Street, New York, NY 10010. (212) 206-8900. Monthly. $498.00 per year.

RESEARCH CENTERS AND INSTITUTES

U.S. GEOLOGICAL SURVEY. National Center, 12201 Sunrise Valley Drive, Reston, VA 22092. The major geological research agency of the federal government conducting research in most areas of pure and applied research in the geosciences.

BENZENE

See also: CHEMISTRY, CHEMICAL ENGINEERING, HETEROCYCLIC CHEMISTRY, ORGANIC CHEMISTRY, PLASTICS, PETROLEUM CHEMISTRY

ABSTRACT SERVICES AND INDEXES

CHEMICAL ABSTRACTS. American Chemical Society, Chemical Abstracts Service, Box 3012, Columbus, OH 43210. (614) 421-3600. 1907 to present. Weekly. $9500.00 per year.

CONFERENCE PAPERS INDEX. Cambridge Scientific Abstracts, 5161 River Road, Bethesda, MD 20816. 1972 to present. Monthly. Inquire as to cost and availability.

CURRENT CONTENTS: PHYSICAL, CHEMICAL AND EARTH SCIENCES. Institute for Scientific Information, 3501 Market Street, Philadelphia, PA 19104. (800) 523-1850 or (215) 386-0100. Weekly. $275.00 per year.

GENERAL SCIENCE INDEX. H.W. Wilson and Company, 950 University Avenue, Bronx, NY 10452. (800) 367-6670 or (212) 588-8400. 1978 to present. Monthly. Inquire as to cost and availability.

INDEX TO SCIENTIFIC AND TECHNICAL PROCEEDINGS. Institute for Scientific Information, 3501 Market Street, Philadelphia, PA 19104. (800) 523-1850 or (215) 386-0100. 1978 to present. Monthly. $775.00 per year.

INDEX TO SCIENTIFIC REVIEWS. Institute for Scientific Information, 3501 Market Street, Philadelphia, PA 19104. (800) 523-1850 or (215) 386-0100. 1974 to present. Semi-annual. $550.00 per year.

SCIENCE CITATION INDEX. Institute for Scientific Information, 3501 Market Street, Philadelphia, PA 19104. (800) 523-1850 or (215) 386-0100. Six times per year. $6200.00 per year.

ANNUAL REVIEWS AND YEARBOOKS

ADVANCES IN PHYSICAL ORGANIC CHEMISTRY. Academic Press, Inc., 6277 Sea Harbor Drive, Orlando, FL 32821. (800) 321-5068. 1963 to present. Irregular. Price varies, inquire.

STUDIES IN ORGANIC CHEMISTRY. Marcel Dekker Inc., 270 Madison Avenue, New York, NY 10016. (800) 228-1160. 1973 to present. Irregular. Price varies, inquire.

ASSOCIATIONS AND PROFESSIONAL SOCIETIES

AMERICAN CARBON SOCIETY. The Stackpole Corporation, St. Marys, PA 15857. (814) 781-8410.

AMERICAN CHEMICAL SOCIETY. 1155 16th Street, N.W., Washington, DC 20036. (202) 872-4600.

AMERICAN INSTITUTE OF CHEMICAL ENGINEERS. 345 East 47th Street, New York, NY 10017. (212) 705-7338.

AMERICAN OIL CHEMISTS' SOCIETY. 508 South Sixth Street, Champaign, IL 61820. (217) 359-2344.

ASSOCIATION OF OFFICIAL ANALYTICAL CHEMISTS. 1111 North 19th Street, Suite 210, Arlington, VA 22209. (703) 522-3032.

DIRECTORIES AND BIOGRAPHICAL SOURCES

AMERICAN MEN AND WOMEN OF SCIENCE. R.R. Bowker, Inc., Order Department, 245 West 17th Street, New York, NY 10011. (800) 521-8110. Eight volumes. 1986. $595.00 for set.

INTERNATIONAL RESEARCH CENTERS DIRECTORY 1988-89. Darren L. Smith, editor. Gale Research Company, Book Tower, Detroit, MI 48226. (800) 521-0707. 4th edition. 1987. $360.00.

1987 DIRECTORY OF ENGINEERING SOCIETIES AND RELATED ORGANIZATIONS. Gordon Davis, editor. Hemisphere Publishing Corporation, 1010 Vermont Avenue, NW, Washington, DC 20005. (800) 526-0275. 12th edition. 1987. $100.00.

RESEARCH CENTERS DIRECTORY 1988. Gale Research Company, Book Tower, Detroit, MI 48226. (800) 521-0707. 12th edition. 1987. $365.00 for set.

SCIENTIFIC AND TECHNICAL ORGANIZATIONS AND AGENCIES DIRECTORY. Margaret Labash Young, editor. Gale Research Company, Book Tower, Detroit, MI 48226. (800) 521-0707. 2nd edition. 1987. $185.00.

WHO'S WHO IN ENGINEERING. Gordon Davis, editor. Hemisphere Publishing Corporation, 1010 Vermont Avenue, NW, Washington, DC 20005. (800) 526-0275. 6th edition. 1985. $200.00.

ENCYCLOPEDIAS AND DICTIONARIES

DICTIONARY OF ORGANIC COMPOUNDS. John Buckingham, editor. Methuen, Inc., 29 West 35th Street, New York, NY 10001. (212) 244-3336. Fifth edition. 1988. $675.00.

DICTIONARY OF ORGANOMETALLIC COMPOUNDS. J.E. MacIntyre, editor. Methuen, Inc., 29 West 35th Street, New York, NY 10001. (212) 244-3336. 1988. $325.00.

ENCYCLOPEDIA OF CHEMISTRY. Douglas M. Considine, editor. Van Nostrand Reinhold Company, Inc., 135 West 50th Street, New York, NY 10020. (800) 543-2681. 4th edition. 1984. $98.95.

THESAURUS OF SCIENTIFIC, TECHNICAL, AND ENGINEERING TERMS. Hemisphere Publishing Corporation, 1010 Vermont Avenue, NW, Washington, DC 20005. (800) 526-0275. 1988. $125.00.

GENERAL WORKS

BASIC ORGANIC CHEMISTRY. F.L. Wiseman. McGraw-Hill Book Company, 1221 Avenue of the Americas, New York, NY 10020. (212) 512-2000. 1988. $37.95.

BENZENE: BASIC AND HAZARDOUS PROPERTIES. M. Cherimisinoff. Marcel Dekker Inc., 270 Madison Avenue, New York, NY 10016. (800) 228-1160. 1979. $49.75.

BENZENE: SCIENTIFIC UPDATE. Myron A. Mehlman. Alan R. Liss, Inc., 150 Fifth Avenue, New York, NY 10011. (212) 741-2515. 1985. $24.00.

BENZOPYRENES. M.R. Osborne and N.T. Crosby. Cambridge University Press, 32 East 57th Street, New York, NY 10022. (800) 872-7423. 1987. $79.50.

CONTEMPORARY HETEROCYCLIC CHEMISTRY: SYNTHESIS, REACTIONS AND APPLICATIONS. G.R. Newkome and W.W. Paudler. John Wiley and Sons, Inc., 605 Third Avenue, New York, NY 10158. (800) 526-5368. 1982. $39.50.

FUNDAMENTALS OF ORGANIC CHEMISTRY. T.W.G. Solomon. John Wiley and Sons, Inc., 605 Third Avenue, New York, NY 10158. (800) 526-5368. 2nd edition. 1986. $39.95.

ORGANIC SYNTHESIS REACTIONS AND MECHANISMS. B. Christoph and others. Springer-Verlag New York, Inc., 175 Fifth Avenue, New York, NY 10010. (800) 526-7254. 1986. $60.00.

REDUCTIONS IN ORGANIC CHEMISTRY. M. Hudlicky. John Wiley and Sons, Inc., 605 Third Avenue, New York, NY 10158. (800) 526-5368. 1984. $45.00.

HANDBOOKS AND MANUALS

CRC HANDBOOK OF CHEMISTRY AND PHYSICS. Robert C. Weast, editor. CRC Press, 2000 Corporate Boulevard, Boca Raton, FL 33431. (800) 272-7737. 69th edition. 1988. $94.00.

CRC HANDBOOK OF DATA ON ORGANIC COMPOUNDS. R.C. Weast and M.J. Astle, editors. CRC Press, 2000 Corporate Boulevard, Boca Raton, FL 33431. (800) 272-7737. 1985. Two volumes. $270.00 for set.

A GUIDEBOOK TO MECHANISMS IN ORGANIC CHEMISTRY. P. Sykes. John Wiley and Sons, Inc., 605 Third Avenue, New York, NY 10158. (800) 526-5368. 6th edition. 1986. $21.95.

HANDBOOK OF APPLIED CHEMISTRY. Vollrath Hopp and Ingo Hennig. Hemisphere Publishing Corporation, 79 Madison Avenue, New York, NY 10016-7892. Order from: Taylor and Francis/Hemisphere Distribution Center, 242 Cherry Street, Philadelphia, PA 19106-1906. (800) 821-8312. 1983. $49.95.

HANDBOOK OF ORGANIC CHEMISTRY. John A. Dean. McGraw-Hill Book Company, 1221 Avenue of the Americas, New York, NY 10020. (212) 512-2000. 1987. $64.50.

HAZARDOUS CHEMICALS DESK REFERENCE. N.I. Sax and R.J. Lewis, editors. Van Nostrand Reinhold Company, Inc., 135 West 50th Street, New York, NY 10020. (800) 543-2681. 1987. $69.95.

LANGE'S HANDBOOK OF CHEMISTRY. John A. Dean, editor. McGraw-Hill Book Company, 1221 Avenue of the Americas, New York, NY 10020. (212) 512-2000. 1985. $59.50.

ONLINE DATA BASES

CA SEARCH. Chemical Abstracts Service, P.O. Box 3012, Columbus, OH 43120. (800) 848-6538 or (614) 421-3600. Comprehensive guide to chemical literature, 1972 to present. Inquire as to online cost and availability.

COMPENDEX. Engineering Information, Inc., 345 East 47th Street, New York, NY 10017. (800) 221-1044 or (212) 705-7615. Engineering and technical literature, 1975 to present. Inquire as to online cost and availability.

DISSERTATION ABSTRACTS ONLINE. University Microfilms International, 300 North Zeeb Road, Ann Arbor, MI 48106. (800) 521-0600 or (313) 761-4700. Scope includes virtually all doctoral dissertations accepted at accredited American institutions from 1861 to present in over 250 subject areas. Inquire as to online cost and availability.

NTIS. National Technical Information Service, 5285 Port Royal Road, Springfield, VA 22161. (703) 487-4630. Broad coverage of government sponsored research reports, 1964 to present. Inquire as to online cost and availability.

SCISEARCH. Institute for Scientific Information, 3501 Market Street, Philadelphia, PA 19104. (800) 523-1850 or (215) 386-0100. Broad multidisciplinary title and author index to the international literature of science and technology, 1974 to present. Inquire as to online cost and availability.

PERIODICALS

AMERICAN OIL CHEMISTS' SOCIETY. JOURNAL. American Oil Chemists' Society, 508 South Sixth Street, Champaign, IL 61820. (217) 359-2344. 1917 to present. Monthly. $60.00 per year.

CARBON. Pergamon Press, Inc., Maxwell House, Fairview Park, Elmsford, NY 10523. (914) 592-7700. 1963 to present. Bimonthly. $235.00 per year.

HYDROCARBON PROCESSING. Gulf Publishing Company, P.O. Box 2608, Houston, TX 77001. (713) 520-4444. Monthly. $10.00 per year.

JOURNAL OF HETEROCYCLIC CHEMISTRY. Hetero-Corporation, Box 16000 MH, Tampa, FL 33687. 1964 to present. 6 times per year. $160.00 per year.

JOURNAL OF ORGANIC CHEMISTRY. American Chemical Society, 1155 16th Street, N.W., Washington, DC 20036. (800) 424-6747. 1936 to present. Semi-monthly. $218.00 per year.

JOURNAL OF POLYMER SCIENCE. POLYMER CHEMISTRY EDITION. John Wiley and Sons, Inc., 605 Third Avenue, New York, NY 10158. (800) 526-5368. 1962 to present. Monthly. $895.00 per year, includes all editions.

ORGANOMETALLICS. American Chemical Society, 1155 16th Street, N.W., Washington, DC 20036. (800) 424-6747. 1982 to present. Monthly. $195.00 per year.

TETRAHEDRON. Pergamon Press, Inc., Maxwell House, Fairview Park, Elmsford, NY 10523. (914) 592-7700. 1957 to present. 24 times per year. $1400.00 per year.

TETRAHEDRON LETTERS. Pergamon Press, Inc., Maxwell House, Fairview Park, Elmsford, NY 10523. (914) 592-7700. 1959 to present. Weekly. $1500.00 per year.

RESEARCH CENTERS AND INSTITUTES

CHEMICAL LABORATORIES. Harvard University, Oxford Street, Cambridge, MA 02138. (617) 495-4283.

CHEMICAL RESEARCH LABORATORY. Brown University, Providence, RI 02912. (401) 863-2256.

BENZENE RING

See: ORGANIC CHEMISTRY

BERYLLIUM

See also: ALLOYS, COPPER, METALLURGICAL ENGINEERING, METALLURGY, METALS AND METALWORKING

ABSTRACT SERVICES AND INDEXES

ALLOYS INDEX. American Society for Metals, Metals Park, OH 44073. (216) 338-5151. 1974 to present. Monthly. $225.00.

APPLIED MECHANICS REVIEW. American Society of Mechanical Engineers, 345 East 47th Street, New York, NY 10017. (212) 705-7703. 1948 to present. Monthly. $360.00 per year.

APPLIED SCIENCE AND TECHNOLOGY INDEX. H.W. Wilson and Company, 950 University Avenue, Bronx, NY 10452. (800) 367-6670 or (212) 588-8400. Monthly. Inquire as to cost and availability.

CHEMICAL ABSTRACTS. American Chemical Society, Chemical Abstracts Service, Box 3012, Columbus, OH 43210. (614) 421-3600. 1907 to present. Weekly. $9500.00 per year.

CORROSION ABSTRACTS. National Association of Corrosion Engineers, Box 218340, Houston, TX 77218. (713) 492-0535. 1962 to present. Bimonthly. $200.00 per year.

CURRENT CONTENTS: ENGINEERING, TECHNOLOGY AND APPLIED SCIENCES. Institute for Scientific Information, 3501 Market Street, Philadelphia, PA 19104. (800) 523-1850 or (215) 386-0100. Weekly. $275.00 per year.

ENGINEERING INDEX MONTHLY AND AUTHOR INDEX. Engineering Information Inc., 345 East 47th Street, New York, NY 10017. (212) 705-7600. Monthly. $1560.00 per year.

ISMEC BULLETIN (Information Service in Mechanical Engineering). Cambridge Scientific Abstracts, 5161 River Road, Bethesda, MD 20816. (301) 951-1400. 1973 to present. Monthly. $450.00 per year.

METALS ABSTRACTS AND METALS ABSTRACTS INDEX. American Society for Metals, Metals Park, OH 44073. (216) 338-5151. 1968 to present. Monthly. Abstracts are $1100.00 per year and Index is $460.00 per year.

PHYSICS ABSTRACTS. Institution of Electrical Engineers. Available from: IEEE Service Center, 445 Hoes Lane, Piscataway, NJ 08854. 1898 to present. Bimonthly. $1700.00 per year.

SCIENCE CITATION INDEX. Institute for Scientific Information, 3501 Market Street, Philadelphia, PA 19104. (800) 523-1850 or (215) 386-0100. Six times per year. $6200.00 per year.

WORLD ALUMINUM ABSTRACTS. Aluminum Association, 818 Connecticut Avenue, NW, Washington, DC 20006. (202) 862-5156. 1968 to present. Monthly. $240.00 per year.

ASSOCIATIONS AND PROFESSIONAL SOCIETIES

AMERICAN INSTITUTE OF MINING, METALLURGICAL AND PETROLEUM ENGINEERS (AIME). 420 Commonwealth Drive, Warrendale, PA 15086. (412) 776-9086.

AMERICAN POWDER METALLURGY INSTITUTE. 105 College Road, East, Princeton, NJ 08540. (609) 452-7700.

AMERICAN SOCIETY FOR METALS. Metals Park, OH 44073. (216) 338-5151.

AMERICAN SOCIETY FOR TESTING AND MATERIALS. 1916 Race Street, Philadelphia, PA 19103. (215) 299-5400.

AMERICAN SOCIETY OF MECHANICAL ENGINEERS. 345 47th Street, New York, NY 10017. (212) 705-7722.

THE METALLURGICAL SOCIETY. 420 Commonwealth Drive, Warrendale, PA 15086. (412) 776-9000.

NATIONAL ASSOCIATION OF CORROSION ENGINEERS. Box 218340, Houston, TX 77218. (713) 492-0535.

DIRECTORIES AND BIOGRAPHICAL SOURCES

DUN'S INDUSTRIAL GUIDE: THE METALWORKING DIRECTORY. Dun and Bradstreet Corporation, 49 Old Bloomfield Road, Mountain Lakes, NJ 07046. (201) 953-0300. Annual. $610.00.

INDUSTRIAL EQUIPMENT AND SUPPLIES DIRECTORY. American Business Directories, Inc., 5707 South 86th Circle, Omaha, NE 68127. (402) 331-7169.

METALLURGICAL SOCIETY OF AIME - MEMBERSHIP LIST. American Institute of Mining, Metallurgical and Petroleum Engineers (AIME). 345 East 47th Street, New York, NY 10017. (212) 705-7695. 1984.

1987 DIRECTORY OF ENGINEERING SOCIETIES AND RELATED ORGANIZATIONS. Gordon Davis, editor. Hemisphere Publishing Corporation, 79 Madison Avenue, New York, NY 10016-7892. (800) 821-8312. 12th edition. 1987. $100.00.

RESEARCH CENTERS DIRECTORY 1988. Gale Research Company, Book Tower, Detroit, MI 48226. (800) 521-0707. 12th edition. 1987. $365.00 for set.

SCIENTIFIC AND TECHNICAL ORGANIZATIONS AND AGENCIES DIRECTORY. Margaret Labash Young, editor. Gale Research Company, Book Tower, Detroit, MI 48226. (800) 521-0707. 2nd edition. 1987. $185.00.

WHO'S WHO IN ENGINEERING. Gordon Davis, editor. Hemisphere Publishing Corporation, 79 Madison Avenue, New York, NY 10016-7892. (800) 821-8312. 6th edition. 1985. $200.00.

GENERAL WORKS

BERYLLIUM. George E. Darwin and J.H. Buddery. Books on Demand, 300 North Zeeb Road, Ann Arbor, MI 48106. (313) 761-4700. $100.50.

BERYLLIUM SCIENCE AND TECHNOLOGY. D. Webster and others, editors. Plenum Publishing Corporation, 233 Spring Street, New York, NY 10013. (800) 221-9369. Two volumes. 1979. Volume 1 $49.50, volume 2 $69.50.

ESSENTIAL METALLURGY FOR ENGINEERS. William Alexander. Van Nostrand Reinhold Company, Inc., 135 West 50th Street, New York, NY 10020. (800) 543-2681. 1985. $17.95.

METALLURGY BASICS. Donald V. Brown. Van Nostrand Reinhold Company, Inc., 135 West 50th Street, New York, NY 10020. (800) 543-2681. 1983. $19.95 in paper.

METALS ENGINEERING: A TECHNICAL GUIDE. L.E. Samuals. American Society for Metals, Metals Park, OH 44073. (216) 338-5151. 1988. $68.00.

HANDBOOKS AND MANUALS

HANDBOOK OF METAL FORMING. Kurt Lange, editor. John Wiley and Sons, Inc., 605 Third Avenue, New York, NY 10158. (800) 526-5368. 1985. $89.50.

HANDBOOK OF METAL FORMING PROCESSES. B. Avitzur. John Wiley and Sons, Inc., 605 Third Avenue, New York, NY 10158. (800) 526-5368. 1983. $105.00.

METALS HANDBOOK. American Society for Metals, Metals Park, OH 44073. (216) 338-5151. 9th edition. 14 volumes. 1988. $1310.00 for set.

WOLDMAN'S ENGINEERING ALLOYS. Robert C. Gibbons, editor. American Society for Metals, Metals Park, OH 44073. (216) 338-5151. 6th edition. 1979. $112.00.

ONLINE DATA BASES

CA SEARCH. Chemical Abstracts Service, P.O. Box 3012, Columbus, OH 43120. (800) 848-6538 or (614) 421-3600. Comprehensive guide to chemical literature, 1972 to present. Inquire as to online cost and availability.

COMPENDEX. Engineering Information, Inc., 345 East 47th Street, New York, NY 10017. (800) 221-1044 or (212) 705-7615. Engineering and technical literature, 1975 to present. Inquire as to online cost and availability.

INSPEC. INSPEC Marketing Department, Institution of Electrical Engineers. Available from IEEE Service Center, 445 Hoes Lane, Piscataway, NJ 08854. (201) 981-0060. Online version of Physics Abstracts. Inquire as to online cost and availability.

ISMEC. Cambridge Scientific Abstracts, 5161 River Road, Besthda, MD 20816. (800) 638-8076 or (301) 951-1400. Literature of mechanical and production engineering, 1973 to present. Inquire as to online cost and availability.

NTIS. National Technical Information Service, 5285 Port Royal Road, Springfield, VA 22161. (703) 487-4630. Broad coverage of government sponsored research reports, 1964 to present. Inquire as to online cost and availability.

SCISEARCH. Institute for Scientific Information, 3501 Market Street, Philadelphia, PA 19104. (800) 523-1850 or (215) 386-0100. Broad multidisciplinary title and author index to the international literature of science and technology, 1974 to present. Inquire as to online cost and availability.

WILSONLINE. H.W. Wilson and Company, 950 University Avenue, Bronx, NY 10452. (800) 367-6770 or (212) 588-8400. Makes available online versions of the H.W. Wilson indexes including Applied Science and Technology Index, Business Periodicals Index and Readers' Guide to Periodical Literature. Approximately 1980 to present. Inquire as to online cost and availability.

PERIODICALS

ALLOY DIGEST. Engineering Alloys Digest, Inc., Box 823, Upper Montclair, NJ 07043 . (201) 746-7930. 1952 to present. Monthly. $50.00 per year.

JOURNAL OF APPLIED MECHANICS. American Society of Mechanical Engineers, 345 East 47th Street, New York, NY 10017. (212) 705-7703. 1935 to present. Quarterly. $100.00 per year.

JOURNAL OF ENGINEERING FOR INDUSTRY. American Society of Mechanical Engineers, 345 East 47th Street, New York, NY 10017. (212) 705-7703. 1970 to present. Quarterly. $100.00 per year.

JOURNAL OF METALS. American Institute of Mining, Metallurgical, and Petroleum Engineers, Inc., Metallurgical Society, 420 Commonwealth Drive, Warrendale, PA 15086. (412) 776-9086. 1949 to present. Monthly. $40.00 per year.

LIGHT METAL AGE. Fellom Publishing Company, 693 Mission Street, San Francisco, CA 94105. (415) 781-1431. 1942 to present. Bimonthly. $20.00.

METAL PROGRESS. American Society for Metals, Metals Park, OH 44073. (216) 338-5151. 1930 to present. Monthly. $40.00.

METALLURGICAL TRANSACTIONS. Metallurgical Society of AIME, 420 Commonwealth Drive, Warrendale, PA 15086. (412) 776-9080. 1970 to present. Monthly. $95.00 per year.

METALS WEEK. McGraw-Hill Book Company, 1221 Avenue of the Americas, New York, NY 10020. (212) 997-2823. 1930 to present. Weekly. $597.00 per year.

METALWORKING DIGEST. Philos Publications, Inc., 1 East Chase Street, Baltimore, MD 21202. (301) 361-9060. 1969 to present. Nine times per year. $9.00 per year.

METIFAX MAGAZINE. Huebner Publications Inc., 6521 Davis Industrial Parkway, Solon, OH 44139. (216) 248-1125. 1956 to present. Monthly. $40.00 per year.

RESEARCH CENTERS AND INSTITUTES

CANADIAN INSTITUTE OF METALWORKING. 1276 Sandhill Drive, P.O. Box 7317, Ancaster, ON, Canada L9G 3N6.

COOPERATIVE PROGRAM IN METALLURGY. Pennsylvania State University, 208A Steidle Building, University Park, PA 16802. (814) 865-5446.

DEPARTMENT OF MATERIALS SCIENCES AND ENGINEERING. University of Florida, Gainesville, FL 32601. (904) 392-1454.

INSTITUTE OF MATERIALS SCIENCE. University of Connecticut, Storrs, CT 06268. (203) 486-4623.

BESSEMER PROCESS

See: STEEL AND STEEL MAKING

BETA RAYS

See: PARTICLE PHYSICS

BETATRON

See: PARTICLE ACCELERATORS

BIDIRECTIONAL TRANSISTOR

See: TRANSISTORS

BIG BANG THEORY

See also: ASTRONOMY, ASTROPHYSICS, COSMOCHEMISTRY, COSMOLOGY

ABSTRACT SERVICES AND INDEXES

ASTRONOMY AND ASTROPHYSICS ABSTRACTS. Springer-Verlag New York, Incorporated, 175 Fifth Avenue, New York, NY 10010. (212) 460-1500. $70.00 per year.

ASTRONOMY AND ASTROPHYSICS MONTHLY INDEX. Olivetree Associations, P.O. Box 236, Sierre Madre, CA 91024. $212.00 per year. Complimentary copies available on request.

GENERAL SCIENCE INDEX. H.W. Wilson Company, 950 University Avenue, Bronx, NY 10452. (800) 367-6770 or (212) 588-8400. Inquire as to cost and availability.

PHYSICS ABSTRACTS. Institute of Electrical Engineers, London, United Kingdom. Available from: Institute of Electrical and Electronic Engineers (IEEE), 345 East 47th Street,

New York, NY 10017. (212) 705-7900.

SCIENCE CITATION INDEX. Institute for Scientific Information, 3501 Market Street, Philadelphia, PA 19104. (800) 523-1850 or (215) 386-0100.

STAR. (Scientific and Technical Aerospace Reports). U.S. National Aeronautics and Space Administration, Scientific and Technical Information Facility, Box 8757, Baltimore-Washington International Airport, MD 21240. (202) 755-2210. Semimonthly, with semiannual and annual indexes. $85.00 per year.

ANNUAL REVIEWS AND YEARBOOKS

ANNUAL REVIEW OF ASTRONOMY AND ASTROPHYSICS. Annual Reviews, Incorporated, 4139 El Camino Way, Palo Alto, CA 94306. (415) 493-4400.

ASSOCIATIONS AND PROFESSIONAL SOCIETIES

AMERICAN ASTRONOMICAL SOCIETY. 2000 Florida Avenue, NW, Suite 300, Washington, DC 20009. (202) 659-0134.

AMERICAN ASSOCIATION OF VARIABLE STAR OBSERVERS. 187 Concord Avenue, Cambridge, MA 02138. (617) 354-0484.

ASTRONOMICAL LEAGUE. P.O. Box 12821, Tucson, AZ 85732. (602) 790-8471.

ASTRONOMICAL SOCIETY OF THE PACIFIC. 1290 24th Avenue, San Francisco, Ca 94122. (415) 661-8660.

BIBLIOGRAPHIES

A BIBLIOGRAPHY OF ASTRONOMY, 1970-1979. R.A. Seal and S.S. Martin. Libraries Unlimited, Incorporated, Littleton, CO 80160. 1982. $37.50.

SCIENCE BOOKS AND FILMS. American Association for the Advancement of Science, 1333 H Street, NW, Washington, DC 20005.

SCIENTIFIC AND TECHNICAL BOOKS IN PRINT; AN INDEX TO LITERATURE IN SCIENCE AND TECHNOLOGY. R.R. Bowker Company, 205 East 42nd Street, New York, NY 10017. (800) 521-8110 or (212) 916-1600.

DIRECTORIES AND BIOGRAPHICAL SOURCES

AMERICAN MEN AND WOMEN OF SCIENCE. Physical and Biological Sciences. Fifteenth edition. R.R. Bowker Company, 205 East 42nd Street, New York, NY 10017. (800) 521-8110 or (212) 916-1600.

THE BIOGRAPHICAL DICTIONARY OF SCIENTISTS: ASTRONOMERS. D. Abbott, editor. Peter Bedrick Books, 125 East 23rd Street, New York, NY 10010. 1984.

DIRECTORY OF PHYSICS AND ASTRONOMY STAFF MEMBERS. American Institute of Physics, 335 East 45th Street, New York, NY 10017. Annual.

RESEARCH CENTERS DIRECTORY. Gale Research Company, Book Tower, Detroit, MI 48226. Eleventh edition, 1987. (800) 521-0707. $340.00.

WHO'S WHO IN FRONTIER SCIENCE AND TECHNOLOGY. Marquis Who's Who, Incorporated, 200 East Ohio Street, Chicago, IL 60611. (800) 428-3898 or (312) 787-2008.

ENCYCLOPEDIAS AND DICTIONARIES

ENCYCLOPEDIA OF PHYSICAL SCIENCE AND TECHNOLOGY. Academic Press, Incorporated, Orlando, FL 32887. (800) 321-5068 or (305) 345-2734.

ILLUSTRATED ENCYCLOPEDIA OF THE UNIVERSE: UNDERSTANDING AND EXPLORING THE COSMOS. Richard S. Lewis. Crown Publishers, Incorporated, 1 Park Avenue, New York, NY 10016. (800) 526-4264. 1986. $24.95.

MCGRAW-HILL ENCYCLOPEDIA OF SCIENCE AND TECHNOLOGY. McGraw-Hill Book Company, 1221 Avenue of the Americas, New York, NY 10020. (212) 997-3675. Fifth edition, 15 volumes. $1100.00.

GENERAL WORKS

THE BIG BANG: THE CREATION AND EVOLUTION OF THE UNIVERSE. Joseph Silk. W.H. Freeman and Company, 41 Madison Avenue, New York, NY 10010. (212) 532-7660. 1980. $13.95 in paper.

CONSTRUCTING THE UNIVERSE. David Layzer. W.H. Freeman and Company, 41 Madison Avenue, New York, NY 10010. (212) 532-6770. 1984. $29.95.

COSMOLOGICAL CONSTANTS: PAPERS IN MODERN COSMOLOGY. Jeremy Bernstein and Gerald Feinberg. Columbia University Press, 562 West 113th Street, New York, NY 10025. (212) 316-7100. 1986. $38.00.

COSMOLOGY, PHYSICS AND PHILOSOPHY. Bernard Gal-Or. Springer-Verlag New York, Incorporated, 175 Fifth Avenue, New York, NY 10010. (800) 526-7254. 1981. $34.00.

FIRST THREE MINUTES: A MODERN VIEW OF THE ORIGIN OF THE UNIVERSE. Steven Weinberg. Basic Books, Incorporated, 10 Eat 53rd Street, New York, NY 10022. (800) 242-7737. 1976. $14.95.

THE FORMATION AND EVOLUTION OF GALAXIES AND LARGE STRUCTURES IN THE UNIVERSE. Jean Audouze and Thanh Van Tran. Kluwer Academic Publishers, 190 Old Derby Street, Hingham, MA 02043. (617) 749-5262. 1984. $58.00.

LARGE SCALE STRUCTURE OF THE UNIVERSE. P.J. Peebles. Princeton University Press, 41 William Street, Princeton, NJ 08540. (609) 452-4122. 1980. $15.95 in paper.

THE LEFT HAND OF CREATION: THE ORIGIN AND EVOLUTION OF THE EXPANDING UNIVERSE. John D. Barrow and Joseph Silk. Basic Books, Incorporated, 10 East 53rd Street, New York, NY 10022. (800) 242-7737. 1986. $7.05 in paper.

MODERN COSMOLOGY. Dennis Sciama. Cambridge University Press, 32 East 57th Street, New York, NY 10022. (800) 872-7423. 1982. $15.95 in paper.

MOMENT OF CREATION: BIG BANK PHYSICS FROM BEFORE THE FIRST MILLISECOND TO THE PRESENT UNIVERSE. James S. Trefil. Charles Scribner's and Sons, 115 Fifth Avenue, New York, NY 10003. (800) 257-5755. 1983. $15.95 in paper.

RELATIVISTIC COSMOLOGY: AN INTRODUCTION. J. Heidmann. Springer-Verlag New York, Incorporated, 175 Fifth Avenue, New York, NY 10010. (800) 526-7254. 1980. $28.00 in paper.

SPACE, TIME AND GRAVITY: THE THEORY OF THE BIG BANG AND BLACK HOLES. Robert M. Wald. University of Chicago Press, 5801 Ellis Avenue, Chicago, IL 60637. (312) 568-1550. 1977. $5.95 in paper.

THE VERY EARLY UNIVERSE. G.W. Gibbons and others, editors. Cambridge University Press, 32 East 57th Street, New York, NY 10022. (800) 872-7423. 1985. $24.95.

ONLINE DATA BASES

CA SEARCH. Chemical Abstracts Service, P.O. Box 3012, Columbus, OH 43210. Guide to chemical literature, 1972 to present. Inquire as to cost and availability.

DISSERTATION ABSTRACTS ONLINE. University Microfilms International, 300 North Zeeb Road, Ann Arbor, MI 48106. (800) 521-0600 or (313) 761-4700. Scope includes virtually all doctoral dissertations accepted at accredited American institutions from 1861 to present in 252 subject areas. Inquire as to cost and availability.

INSPEC. INSPEC Marketing Department, Institute of Electrical and Electronics Engineers, Incorporated, IEEE Service Department, 445 Hoes Lane, Piscataway, NJ 08854. (201) 981-0060. Inquire as to on-line cost and availability.

MATHFILE. American Mathematical Society, P.O. Box 6248, Providence, RI 02940. (800) 556-7774 or (401) 272-9500. Scope includes pure and applied mathematics and related areas of physics, statistics, engineering, computer science, and operations research literature since 1973. Inquire as to cost and availability.

NASA. National Aeronautics and Space Administration, Scientific and Technical Information Branch, 300 7th Street, SW, Washington, DC 20546. Citations and abstracts of aerospace literature, 1962 to present. Inquire as to cost and availability.

SCISEARCH. Institute for Scientific Information, 3501 Market Street, Philadelphia, PA 19104. (800) 523-1850 or (215) 386-0100. Broad multidisciplinary title and author index to the international literature of science and technology, 1974 to present. Inquire as to cost and availability.

WILSONLINE. H.W. Wilson Company, 950 University Avenue, Bronx, NY 10452. (800) 367-6770 or (212) 588-8400. Makes available online versions of the printed H.W. Wilson indexes including Applied Science and Technology Index, Business Periodicals Index, and Readers' Guide to Periodical Literature. Period covered is generally 1983 to present. Inquire as to cost and availability.

OTHER SOURCES

ATLAS OF DEEP-SKY SPLENDORS. H. Vehrenberg. Cambridge University Press, 32 East 57th Street, New York, NY 10022. (800) 431-1580 or (212) 688-8888. Fourth edition, 1984. $44.50.

SKY CATALOGUE 2000.0. A. Hirshfeld and R. Sinnott. Cambridge University Press, 32 East 57th Street, New York, NY 10022. (800) 431-1580 or (212) 688-8888.

PERIODICALS

ASTRONOMICAL JOURNAL. American Astronomical Society. Available from: American Institute of Physics, 335 East 45th Street, New York, NY 10017. (212) 661-9404. $125.00 per year.

ASTRONOMICAL SOCIETY OF THE PACIFIC. Publications. Astronomical Society of the Pacific, 1290 24th Avenue, San Francisco, Ca 94122. (415) 661-8660. Monthly. $38.00.

ASTRONOMY. Astro Media Corporation, 625 East Paul Avenue, Milwaukee, WI 53202. Monthly. $18.00 per year.

ASTRONOMY AND ASTROPHYSICS. Springer-Verlag New York, Incorporated, 175 Fifth Avenue, New York, NY 10010. (800) 526-7254 or (212) 460-1500. $680.00 per year.

ASTROPHYSICAL JOURNAL. American Astronomical Society, University of Chicago Press, 5801 Ellis Avenue, Chicago, IL 60637. Biweekly. $305.00 per year.

ASTROPHYSICS AND SPACE SCIENCE. D.Reidel Publishing Company, 190 Old Derby Street, Hingham, MA 02043. Monthly. $101.00 per year.

CELESTIAL MECHANICS; AN INTERNATIONAL JOURNAL OF SPACE DYNAMICS. D. Reidel Publishing Company, 190 Old Derby Street, Hingham, MA 02043. Monthly. $310.00 per year.

FUNDAMENTALS OF COSMIC PHYSICS. Gordon and Breach Science Publishers, Limited, P.O. Box 197, London, WC2E 9FX. Monthly. $335.00 per year.

MERCURY. Astronomical Society of the Pacific, 1290 24th Avenue, San Francisco, Ca 94122. (415) 661-8660. Bimonthly. $21.00 per year.

MONTHLY NOTICES OF THE ROYAL ASTRONOMICAL SOCIETY. Blackwell Science Publications, Incorporated, 667 Lytton Avenue, Palo Alto, CA 94301. (415) 324-1688. Monthly. $850.00 per year.

SKY AND TELESCOPE. Sky Publishing Corporation, 49 Bay State Road, Cambridge, Ma 02238. (617) 864-7360. Monthly. $18.00 per year.

SOVIET ASTRONOMY (TRANSLATION OF ASTRO-NOMICHESKII ZHURNAL). American Institute of Physics, 335 East 45th Street, New York, NY 10017. (212) 661-9404. Bimonthly. $425.00 per year.

SPACE SCIENCE REVIEWS. D. Reidel Publishing Company, 190 Old Derby Street, Hingham, MA 02043. Monthly. $305.00 per year.

VISTAS IN ASTRONOMY. Pergamon Press, Incorporated, Maxwell House, Fairview Park, Elmsford, NY 10523. (914) 592-7700. Quarterly. $145.00 per year.

RESEARCH CENTERS AND INSTITUTES

HARVARD-SMITHSONIAN CENTER FOR ASTROPHYSICS. 60 Garden Street, Cambridge, MA 02138. (617) 495-7461.

MOUNT WILSON AND LAS CAMPANAS OBSERVATORIES. 813 Santa Barbara Street, Pasadena, Ca 91101. (818) 577-1122.

NATIONAL OPTICAL ASTRONOMY OBSERVATORIES. 1002 North Warren Avenue, Tucson, AZ 85719. (602) 325-9230.

STANFORD UNIVERSITY. Center for Space Science and Astrophysics, 325 Durand Building, Stanford, Ca 94305. (415) 497-3582.

UNIVERSITY OF CHICAGO. Yerkes Observatory, Williams Bay, WI. (414) 245-5555.

YALE UNIVERSITY OBSERVATORY. P.O. Box 6666, New Haven, CT 06511. (203) 436-3460.

BINARY STARS

See also: ASTRONOMY, ASTROPHYSICS, GALAXIES, STARS

ABSTRACT SERVICES AND INDEXES

ASTRONOMY AND ASTROPHYSICS ABSTRACTS. Springer-Verlag New York, Incorporated, 175 Fifth Avenue, New York, NY 10010. (212) 460-1500. $70.00 per year.

ASTRONOMY AND ASTROPHYSICS MONTHLY INDEX. Olivetree Associates, Post Office Box 236, Sierra Madre, CA 91024. $212.00 per year. Complimentary copies available on request.

GENERAL SCIENCE INDEX. H.W. Wilson Company, 950 University Avenue, Bronx, NY 10452. (800) 367-6770 or (212)

588-8400. Inquire as to cost and availability.

PHYSICS ABSTRACTS. Institute of Electrical Engineers, London, United Kingdom. Available from: Institute of Electrical and Electronic Engineers (IEEE), 345 East 47th Street, New York, NY 10017. (212) 705-7900. 1898 to present. Monthly. $1670.00 per year.

SCIENCE CITATION INDEX. Institute for Scientific Information, 3501 Market Street, Philadelphia, PA 19104. (800) 523-1850 or (215) 386-0100. Inquire as to cost and availability.

STAR. (Scientific and Technical Aerospace Reports. United States National Aeronautics and Space Administration, Scientific and Technical Information Facility, Box 8757, Baltimore-Washington International Airport, MD 21240. (202) 755-2210. Semimonthly, with semiannual and annual indexes. $85.00 per year.

ANNUAL REVIEWS AND YEARBOOKS

THE ASTRONOMICAL ALMANAC. Superintendent of Documents, United States Government Printing Office, Washington, DC 20402. (202) 783-3238. Yearly.

ANNUAL REVIEW OF ASTRONOMY AND ASTROPHYSICS. Annual Reviews, Incorporated, 4139 El Camino Way, Palo Alto, CA 94306. (415) 493-4400.

ASSOCIATIONS AND PROFESSIONAL SOCIETIES

AMERICAN ASTRONOMICAL SOCIETY. 2000 Florida Avenue, NW, Suite 300, Washington, DC 20009. (202) 659-0134.

AMERICAN ASSOCIATION OF VARIABLE STAR OBSERVERS. 187 Concord Avenue, Cambridge, MA 02138. (617) 354-0484.

ASTRONOMICAL LEAGUE. Post Office Box 12821, Tucson, AZ 85732. (602) 790-8471.

ASTRONOMICAL SOCIETY OF THE PACIFIC. 1290 24th Avenue, San Francisco, CA 94122. (415) 661-8660.

BIBLIOGRAPHIES

BIBLIOGRAPHY OF ASTRONOMY, 1970-1979. R.A. Seal and S.S. Martin. Libraries Unlimited, Incorporated, Littleton, CO 80160. 1982. $37.50.

SCIENCE BOOKS AND FILMS. American Association for the Advancement of Science, 1333 H Street, NW, Washington, DC 20005. 1965 to present. Five times per year. $20.00 per year.

SCIENTIFIC AND TECHNICAL BOOKS AND SERIALS IN PRINT 1988; AN INDEX TO LITERATURE IN SCIENCE AND TECHNOLOGY. R.R. Bowker Company, 205 East 42nd Street, New York, NY 10017. (800) 521-8110 or (212) 916-1600. $175.00.

DIRECTORIES AND BIOGRAPHICAL SOURCES

AMERICAN MEN AND WOMEN OF SCIENCE. Physical and Biological Sciences. Sixteenth edition. R.R. Bowker Company, 205 East 42nd Street, New York, NY 10017. (800) 521-8110 or (212) 916-1600. 1987. $595.00.

THE BIOGRAPHICAL DICTIONARY OF SCIENTISTS: ASTRONOMERS. D. Abbott, editor. Peter Bedrick Books, 125 East 23rd Street, New York, NY 10010. 1984.

DIRECTORY OF PHYSICS AND ASTRONOMY STAFF MEMBERS. American Institute of Physics, 335 East 45th Street, New York, NY 10017. Annual.

RESEARCH CENTERS DIRECTORY. Gale Research Company, Book Tower, Detroit, MI 48226. Twelfth edition, 1988. (800) 521-0707. $240.00.

WHO'S WHO IN FRONTIER SCIENCE AND TECHNOLOGY. Marquis Who's Who, Incorporated, 200 East Ohio Street, Chicago, IL 60611. (800) 428-3898 or (312) 787-2008.

ENCYCLOPEDIAS AND DICTIONARIES

ENCYCLOPEDIA OF PHYSICAL SCIENCE AND TECHNOLOGY. Academic Press, Incorporated, Orlando, FL 32887. (800) 321-5068 or (305) 345-2734. Inquire as to cost and availability.

MCGRAW-HILL ENCYCLOPEDIA OF SCIENCE AND TECHNOLOGY. McGraw-Hill Book, Incorporated, 1221 Avenue of the Americas, New York, NY 10020. (212) 997-3675. Fifth edition, 15 volumes. $1100.00.

GENERAL WORKS

THE CAMBRIDGE ASTRONOMY GUIDE: A PRACTICAL INTRODUCTION TO ASTRONOMY. Cambridge University Press, 32 East 57th Street, New York, NY 10022. (212) 688-8888. 1985. $24.95.

INTERACTING BINARY STARS. J. Sahade and F. Wood. Pergamon Press, Maxwell House, Fairview Park, Elmsford, NY 10523. (914) 592-7700. 1978. $35.00.

OBSERVING VISUAL DOUBLE STARS. Paul Couteau. MIT Press, 28 Carlton Street, Cambridge, MA 02142. (617) 253-5646. 1981. $25.00.

ONE HUNDRED BILLION SUNS: THE BIRTH, LIFE AND DEATH OF THE STARS. Rudolf Kippenhahn. Basic Books, Incorporated, 10 East 53rd Street, New York, NY 10022. (800) 242-7737.

PHYSICS OF THE STARS. S.A. Kaplan. John Wiley and Sons, Incorporated, 605 Third Avenue, New York, NY 10158. (800) 526-5368 or (212) 850-6000. 1982. $34.95.

PROTOSTARS AND PLANETS: STUDIES OF STAR FORMATION AND THE ORIGIN OF THE SOLAR SYSTEM. Tom Gehrels, editor. University of Arizona Press, 1615 East Speedway, Tucson, AZ 85719. (602) 621-1441. 1978. $160.00.

PROTOSTARS AND PLANETS II. David C. Black and Mildred S. Matthews, editors. University of Arizona Press, 1615 East Speedway, Tucson, AZ 85719. (602) 621-1441. 1985. $45.00.

THE QUEST FOR SS433. David H. Clark. Penguin Books, Incorporated, 40 West 23rd Street, New York, NY 10010. (800) 631-3577. 1986. $6.95 in paper.

ONLINE DATA BASES

CA SEARCH. Chemical Abstracts Service, Post Office Box 3012, Columbus, OH 43210. Guide to chemical literature, 1972 to present. Inquire as to cost and availability.

DISSERTATION ABSTRACTS ONLINE. University Microfilms International, 300 North Zeeb Road, Ann Arbor, MI 48106. (800) 521-0600 or (313) 761-4700. Scope includes virtually all doctoral dissertations accepted at accredited American institutions from 1861 to present in 252 subject areas. Inquire as to cost and availability.

INSPEC. INSPEC Marketing Department, Institute of Electrical and Electronics Engineers, Incorporated, IEEE Service Department, 445 Hoes Lane, Piscataway, NJ 08854. (201) 981-0060. Inquire as to on-line cost and availability.

NASA. National Aeronautics and Space Administration, Scientific and Technical Information Branch, 300 7th Street, SW, Washington, DC 20546. Citations and abstracts of aerospace literature, 1962 to present. Inquire as to cost and availability.

NTIS. National Technical Information Service, 5285 Port Royal Road, Springfield, VA 22161. (703) 487-4630. Broad coverage of government sponsored research reports, 1964 to present. Inquire as to cost and availability.

SCISEARCH. Institute for Scientific Information, 3501 Market Street, Philadelphia, PA 19104. (800) 523-1850 or (215) 386-0100. Broad multidisciplinary title and author index to the international literature of science and technology, 1974 to present. Inquire as to cost and availability.

OTHER SOURCES

ATLAS OF DEEP-SKY SPLENDORS. H. Vehrenberg. Cambridge University Press, 32 East 57th Street, New York, NY 10022. (800) 431-1580 or (212) 688-8888. Fourth edition, 1984. $47.50.

THE BRIGHT STAR CATALOGUE AND SUPPLEMENT. D. Hoffleit and Carlos Jaschek. Yale University Observatory, Post Office Box 6666, New Haven, CT 06511. (203) 436-3460. Fourth edition, 1982. $35.00.

THE CAMBRIDGE ATLAS OF ASTRONOMY. Jean Audouze and Guy Isreal, editors. Cambridge University Press, 32 East 57th Street, New York, NY 10022. (212) 688-8888. 1985. $75.00.

SKY CATALOGUE 2000.0. A. Hirshfeld and R. Sinnott. Cambridge University Press, 32 East 57th Street, New York, NY 10022. (800) 431-1580 or (212) 688-8888.

STAR MAPS FOR BEGINNERS. I.M. Levitt and Roy K. Marshall. Simon and Schuster, Incorporated, 1230 Avenue of the Americas, New York, NY 10020. (800) 223-2336. 1983. $7.95.

STARS, GALAXIES, COSMOS: A CATALOG OF ASTRONOMICAL ANOMALIES. W.R. Corliss, compiler. Sourcebook Project, Post Office Box 107, Glen Arm, MD 21057. (301) 668-6047. 1987. $17.95.

PERIODICALS

ASTRONOMICAL JOURNAL. American Astronomical Society. Available from: American Institute of Physics, 335 East 45th Street, New York, NY 10017. (212) 661-9404. $125.00 per year.

ASTRONOMICAL SOCIETY OF THE PACIFIC PUBLICATIONS. Astronomical Society of the Pacific, 1290 24th Avenue, San Francisco, CA 94122. (415) 661-8660. Monthly. $38.00.

ASTRONOMY. Astro Media Corporation, 625 East Paul Avenue, Milwaukee, WI 53202. Monthly. $18.00 per year.

ASTRONOMY AND ASTROPHYSICS. Springer-Verlag New York, Incorporated, 175 Fifth Avenue, New York, NY 10010. (800) 526-7254 or (212) 460-1500. $680.00 per year.

ASTROPHYSICAL JOURNAL. American Astronomical Society, University of Chicago Press, 5801 Ellis Avenue, Chicago, IL 60637. Biweekly. $305.00 per year.

ASTROPHYSICS AND SPACE SCIENCE. D. Reidel Publishing Company, 190 Old Derby Street, Hingham, MA 02043. Monthly. $101.00 per year.

CELESTIAL MECHANICS: AN INTERNATIONAL JOURNAL OF SPACE DYNAMICS. D. Reidel Publishing Company, 190 Old Derby Street, Hingham, MA 02043. Monthly. $310.00 per year.

MERCURY. Astronomical Society of the Pacific, 1290 245h Avenue, San Francisco, CA 94122. (415) 661-8660. Bimonthly. $21.00 per year.

MONTHLY NOTICES OF THE ROYAL ASTRONOMICAL SOCIETY. Blackwell Science Publications, Incorporated, 667 Lytton Avenue, Palo Alto, CA 94301. (415) 324-1688. Monthly. $134.00 per year.

SKY AND TELESCOPE. Sky Publishing Corporation, 49 Bay State Road, Cambridge, MA 02238. (617) 864-7360. Monthly. $18.00 per year.

SOLAR PHYSICS. D. Reidel Publishing Company, 190 Old Derby Street, Hingham, MA 02043. Monthly. $620.00 per year.

SOVIET ASTRONOMY (TRANSLATION OF ASTRO-NOMICHESKII ZHURNAL). American Institute of Physics, 335 East 45th Street, New York, NY 10017. (212) 661-9404. Bimonthly. $425.00 per year.

SPACE SCIENCE REVIEWS. D. Reidel Publishing Company, 190 Old Derby Street, Hingham, MA 02043. Monthly. $305.00 per year.

VISTAS IN ASTRONOMY. Pergamon Press, Incorporated, Maxwell House, Fairview Park, Elmsford, NY 10523. (914) 592-7700. Quarterly. $145.00 per year.

RESEARCH CENTERS AND INSTITUTES

CALIFORNIA INSTITUTE OF TECHNOLOGY. Palomar Observatory, 105-24, Pasadena, CA 91125. (818) 356-4033.

HARVARD-SMITHSONIAN CENTER FOR ASTROPHYSICS. 60 Garden Street, Cambridge, MA 02138. (617) 495-7461.

LOWELL OBSERVATORY. Mars Hill Road, 1400 West, Flagstaff, AZ 86001. (602) 774-3358.

NATIONAL OPTICAL ASTRONOMY OBSERVATORIES. 1002 North Warren Avenue, Tucson, AZ 85719. (602) 325-9230.

UNIVERSITY OF TOLEDO. Ritter Astrophysical Research Center, 2801 West Bancroft Street, Toledo, OH 43606. (419) 537-2276.

VANDERBILT UNIVERSITY. Arthur J. Dyer Observatory, Box 1803, Station B, Nashville, TN 37235.

BLACK HOLES

See also: ASTRONOMY, ASTROPHYSICS
COSMOLOGY

ABSTRACT SERVICES AND INDEXES

ASTRONOMY AND ASTROPHYSICS ABSTRACTS. Springer-Verlag New York, Incorporated, 175 Fifth Avenue, New York, NY 10010. (212) 460-1500. $70.00 per year.

ASTRONOMY AND ASTROPHYSICS MONTHLY INDEX. Olivetree Associations, P.O. Box 236, Sierre Madre, CA 91024. $212.00 per year. Complimentary copies available on request.

GENERAL SCIENCE INDEX. H.W. Wilson Company, 950 University Avenue, Bronx, NY 10452. (800) 367-6770 or (212) 588-8400. Inquire as to cost and availability.

PHYSICS ABSTRACTS. Institute of Electrical Engineers, London, United Kingdom. Available from: Institute of Electrical and Electronic Engineers (IEEE), 345 East 47th Street, New York, NY 10017. (212) 705-7900.

SCIENCE CITATION INDEX. Institute for Scientific Information, 3501 Market Street, Philadelphia, PA 19104. (800) 523-1850 or (215) 386-0100.

STAR. (Scientific and Technical Aerospace Reports). U.S. National Aeronautics and Space Administration, Scientific and Technical Information Facility, Box 8757, Baltimore-Washington International Airport, MD 21240. (202) 755-2210. Semimonthly, with semiannual and annual indexes. $85.00 per year.

ANNUAL REVIEWS AND YEARBOOKS

THE ASTRONOMICAL ALMANAC. Superintendent of Documents, U.S. Government Printing Office, Washington, DC 20402. (202) 783-3238. Yearly.

ANNUAL REVIEW OF ASTRONOMY AND ASTROPHYSICS. Annual Reviews, Incorporated, 4139 El Camino Way, Palo Alto, CA 94306. (415) 493-4400.

ASSOCIATIONS AND PROFESSIONAL SOCIETIES

AMERICAN ASTRONOMICAL SOCIETY. 2000 Florida Avenue, NW, Suite 300, Washington, DC 20009. (202) 659-0134.

AMERICAN ASSOCIATION OF VARIABLE STAR OBSERVERS. 187 Concord Avenue, Cambridge, MA 02138. (617) 354-0484.

ASTRONOMICAL LEAGUE. P.O. Box 12821, Tucson, AZ 85732. (602) 790-8471.

ASTRONOMICAL SOCIETY OF THE PACIFIC. 1290 24th Avenue, San Francisco, Ca 94122. (415) 661-8660.

BIBLIOGRAPHIES

A BIBLIOGRAPHY OF ASTRONOMY, 1970-1979. R.A. Seal and S.S. Martin. Libraries Unlimited, Incorporated, Littleton, CO 80160. 1982. $37.50.

BLACK HOLES: AN ANNOTATED BIBLIOGRAPHY, 1975-1983. Steven I. Danko. The Scarecrow Press, Incorporated, Metuchen, NJ.

SCIENCE BOOKS AND FILMS. American Association for the Advancement of Science, 1333 H Street, NW, Washington, DC 20005.

SCIENTIFIC AND TECHNICAL BOOKS IN PRINT; AN INDEX TO LITERATURE IN SCIENCE AND TECHNOLOGY. R.R. Bowker Company, 205 East 42nd Street, New York, NY 10017. (800) 521-8110 or (212) 916-1600.

DIRECTORIES AND BIOGRAPHICAL SOURCES

AMERICAN MEN AND WOMEN OF SCIENCE. Physical and Biological Sciences. Fifteenth edition. R.R. Bowker Company, 205 East 42nd Street, New York, NY 10017. (800) 521-8110 or (212) 916-1600.

THE BIOGRAPHICAL DICTIONARY OF SCIENTISTS: ASTRONOMERS. D. Abbott, editor. Peter Bedrick Books, 125 East 23rd Street, New York, NY 10010. 1984.

DIRECTORY OF PHYSICS AND ASTRONOMY STAFF MEMBERS. American Institute of Physics, 335 East 45th Street, New York, NY 10017. Annual.

RESEARCH CENTERS DIRECTORY. Gale Research Company, Book Tower, Detroit, MI 48226. Eleventh edition, 1987. (800) 521-0707. $340.00.

WHO'S WHO IN FRONTIER SCIENCE AND TECHNOLOGY. Marquis Who's Who, Incorporated, 200 East Ohio Street, Chicago, IL 60611. (800) 428-3898 or (312) 787-2008.

ENCYCLOPEDIAS AND DICTIONARIES

ENCYCLOPEDIA OF PHYSICAL SCIENCE AND TECHNOLOGY. Academic Press, Incorporated, Orlando, FL 32887. (800) 321-5068 or (305) 345-2734.

MCGRAW-HILL ENCYCLOPEDIA OF SCIENCE AND TECHNOLOGY. McGraw-Hill Book Company, 1221 Avenue of the Americas, New York, NY 10020. (212) 997-3675. Fifth edition, 15 volumes. $1100.00.

GENERAL WORKS

BLACK HOLES: THE MEMBRANE PARADIGM. Kip S. Thorne, Richard H. Price and Douglas A. Macdonald, editors. Yale University Press, 302 Temple Street, New Haven, CT 06520. (203) 436-7584. $14.95 in paper.

BLACK HOLES, WHITE DWARFS, AND NEUTRON STARS: THE PHYSICS OF COMPACT OBJECTS. Stuart Shapiro and Saul A. Teukolsky. John Wiley and Sons, Incorporated, 605 Third Avenue, New York, NY 10158. (800) 526-5368. 1983. $26.95 in paper.

THE EDGE OF INFINITY: NAKED SINGULARITIES AND THE DESTRUCTION OF SPACETIME. Paul Davies. Simon and Schuster, New York, NY. (800) 223-2336.

GRAVITY, BLACK HOLES AND THE UNIVERSE. Iain Nicolson. Halsted Press, New York, NY. (800) 526-5368.

THE LIFE AND DEATH OF STARS. Donald Cooke. Crown Publishers, New York, NY. $29.95.

MATHEMATICAL THEORY OF BLACK HOLES. Subrahmanyan Chandrasekhar. Oxford University Press, 200 Madison Avenue, New York, NY 10016. (800) 458-5833. 1982. $89.00.

SPACEWARPS. John Gribbin. Delacorte Press. New York, NY. (800) 645-6156.

THE ULTIMATE FATE OF THE UNIVERSE. Jamal Islam. Cambridge University Press, New York, NY. (800) 431-1580.

ONLINE DATA BASES

CA SEARCH. Chemical Abstracts Service, P.O. Box 3012, Columbus, OH 43210. Guide to chemical literature, 1972 to present. Inquire as to cost and availability.

DISSERTATION ABSTRACTS ONLINE. University Microfilms International, 300 North Zeeb Road, Ann Arbor, MI 48106. (800) 521-0600 or (313) 761-4700. Scope includes virtually all doctoral dissertations accepted at accredited American institutions from 1861 to present in 252 subject areas. Inquire as to cost and availability.

INSPEC. INSPEC Marketing Department, Institute of Electrical and Electronics Engineers, Incorporated, IEEE Service Department, 445 Hoes Lane, Piscataway, NJ 08854. (201) 981-0060. Inquire as to on-line cost and availability.

MATHFILE. American Mathematical Society, P.O. Box 6248, Providence, RI 02940. (800) 556-7774 or (401) 272-9500. Scope includes pure and applied mathematics and related areas of physics, statistics, engineering, computer science, and operations research literature since 1973. Inquire as to cost and availability.

NASA. National Aeronautics and Space Administration, Scientific and Technical Information Branch, 300 7th Street, SW, Washington, DC 20546. Citations and abstracts of aerospace literature, 1962 to present. Inquire as to cost and availability.

NTIS. National Technical Information Service, 5285 Port Royal Road, Springfield, Va 22161. (703) 487-4630. Broad coverage of government sponsored research reports, 1964 to present. Inquire as to cost and availability.

SCISEARCH. Institute for Scientific Information, 3501 Market Street, Philadelphia, PA 19104. (800) 523-1850 or (215) 386-0100. Broad multidisciplinary title and author index to the international literature of science and technology, 1974 to present. Inquire as to cost and availability.

WILSONLINE. H.W. Wilson Company, 950 University Avenue, Bronx, NY 10452. (800) 367-6770 or (212) 588-8400. Makes available online versions of the printed H.W. Wilson indexes including Applied Science and Technology Index, Business Periodicals Index, and Readers' Guide to Periodical Literature. Period covered is generally 1983 to present. Inquire as to cost and availability.

OTHER SOURCES

ATLAS OF DEEP-SKY SPLENDORS. H. Vehrenberg. Cambridge University Press, 32 East 57th Street, New York, NY 10022. (800) 431-1580 or (212) 688-8888. Fourth edition, 1984. $44.50.

THE BRIGHT STAR CATALOGUE AND SUPPLEMENT. D. Hoffleit and Carlos Jaschek. Yale University Observatory, P.O. Box 6666, New Haven, CT 06511. (203) 436-3460. Fourth edition, 1982. $35.00.

SKY CATALOGUE 2000.0. A. Hirshfeld and R. Sinnott. Cambridge University Press, 32 East 57th Street, New York, NY 10022. (800) 431-1580 or (212) 688-8888.

PERIODICALS

ASTRONOMICAL JOURNAL. American Astronomical Society. Available from: American Institute of Physics, 335 East 45th Street, New York, NY 10017. (212) 661-9404. $125.00 per year.

ASTRONOMICAL SOCIETY OF THE PACIFIC. Publications. Astronomical Society of the Pacific, 1290 24th Avenue, San Francisco, Ca 94122. (415) 661-8660. Monthly. $38.00.

ASTRONOMY. Astro Media Corporation, 625 East Paul Avenue, Milwaukee, WI 53202. Monthly. $18.00 per year.

ASTRONOMY AND ASTROPHYSICS. Springer-Verlag New York, Incorporated, 175 Fifth Avenue, New York, NY 10010. (800) 526-7254 or (212) 460-1500. $680.00 per year.

ASTROPHYSICAL JOURNAL. American Astronomical Society, University of Chicago Press, 5801 Ellis Avenue, Chicago, IL 60637. Biweekly. $305.00 per year.

ASTROPHYSICS AND SPACE SCIENCE. D.Reidel Publishing Company, 190 Old Derby Street, Hingham, MA 02043. Monthly. $101.00 per year.

CELESTIAL MECHANICS; AN INTERNATIONAL JOURNAL OF SPACE DYNAMICS. D. Reidel Publishing Company, 190 Old Derby Street, Hingham, MA 02043. Monthly. $310.00 per year.

MERCURY. Astronomical Society of the Pacific, 1290 24th Avenue, San Francisco, Ca 94122. (415) 661-8660. Bimonthly. $21.00 per year.

MONTHLY NOTICES OF THE ROYAL ASTRONOMICAL SOCIETY. Blackwell Science Publications, Incorporated, 667 Lytton Avenue, Palo Alto, CA 94301. (415) 324-1688. Monthly. $850.00 per year.

PLANETARY AND SPACE SCIENCE. Pergamon Press, Incorporated, Maxwell House, Fairview Park, Elmsford, NY 10523. (914) 592-7700. Monthly. $430.00 per year.

SKY AND TELESCOPE. Sky Publishing Corporation, 49 Bay State Road, Cambridge, Ma 02238. (617) 864-7360. Monthly. $18.00 per year.

SOVIET ASTRONOMY (TRANSLATION OF ASTRO-NOMICHESKII ZHURNAL). American Institute of Physics, 335 East 45th Street, New York, NY 10017. (212) 661-9404. Bimonthly. $425.00 per year.

SPACE SCIENCE REVIEWS. D. Reidel Publishing Company, 190 Old Derby Street, Hingham, MA 02043. Monthly. $305.00 per year.

VISTAS IN ASTRONOMY. Pergamon Press, Incorporated, Maxwell House, Fairview Park, Elmsford, NY 10523. (914) 592-7700. Quarterly. $145.00 per year.

RESEARCH CENTERS AND INSTITUTES

NATIONAL ASTRONOMY AND IONOSPHERE CENTER. Cornell University, Space Sciences Building, Ithaca, NY 14853. (607) 256-3734.

HARVARD-SMITHSONIAN CENTER FOR ASTROPHYSICS. 60 Garden Street, Cambridge, MA 02138. (617) 495-7461.

NATIONAL OPTICAL ASTRONOMY OBSERVATORIES. 1002 North Warren Avenue, Tucson, AZ 85719. (602) 325-9230.

NATIONAL RADIO ASTRONOMY OBSERVATORY. Edgemont Road, Charlottesville, VA 22903. (804) 296-0211.

SPACE TELESCOPE SCIENCE INSTITUTE. 3700 San Martin Drive, Baltimore, MD 21218. (301) 338-4700.

BLAST FURNACE

See: STEEL AND STEEL MAKING

BLASTING

See: EXPLOSIVES

BLIMP

See: AIRSHIPS

BOHR THEORY

See: PHYSICAL CHEMISTRY

BOILERS

See also: ELECTRIC POWER ENGINEERING, HEAT EXCHANGERS, HEAT TRANSFER, MECHANICAL ENGINEERING, NUCLEAR ENGINEERING, PRESSURE VESSELS, STEAM, THERMODYNAMICS

ABSTRACT SERVICES AND INDEXES

APPLIED MECHANICS REVIEW. American Society of Mechanical Engineers, 345 East 47th Street, New York, NY 10017. (212) 705-7703. 1948 to present. Monthly. $360.00 per year.

APPLIED SCIENCE AND TECHNOLOGY INDEX. H.W. Wilson and Company, 950 University Avenue, Bronx, NY 10452. (800) 367-6670 or (212) 588-8400. Monthly. Inquire as to cost and availability.

CHEMICAL ABSTRACTS. American Chemical Society, Chemical Abstracts Service, Box 3012, Columbus, OH 43210. (614) 421-3600. 1907 to present. Weekly. $9500.00 per year.

CURRENT CONTENTS: ENGINEERING, TECHNOLOGY AND APPLIED SCIENCES. Institute for Scientific Information, 3501 Market Street, Philadelphia, PA 19104. (800) 523-1850 or (215) 386-0100. Weekly. $275.00 per year.

ENGINEERING INDEX MONTHLY AND AUTHOR INDEX. Engineering Information Inc., 345 East 47th Street, New York, NY 10017. (212) 705-7600. Monthly. $1560.00 per year.

ISMEC BULLETIN (Information Service in Mechanical Engineering). Cambridge Scientific Abstracts, 5161 River Road, Bethesda, MD 20816. (301) 951-1400. 1973 to present. Monthly. $450.00 per year.

ASSOCIATIONS AND PROFESSIONAL SOCIETIES

AMERICAN INSTITUTE OF CHEMICAL ENGINEERS. 345 East 47th Street, New York, NY 10017. (212) 705-7703.

AMERICAN INSTITUTE OF PLANT ENGINEERS. 3975 Erie Avenue, Cincinnati, OH 45208. (513) 561-6000.

AMERICAN SOCIETY OF HEATING, REFRIGERATING AND AIR CONDITIONING ENGINEERS, INC., 1791 Tullie Circle, Atlanta, GA 30329. (404) 636-8400.

AMERICAN SOCIETY OF MECHANICAL ENGINEERS. 345 East 47th Street, New York, NY 10017. (212) 705-7703.

NATIONAL ASSOCIATION OF POWER ENGINEERS. 2350 East Devon Avenue, Suite 115, Des Plaines, IL 60018. (312) 298-0600.

POWER ENGINEERING SOCIETY. Institute of Electrical and Electronics Engineers, 345 East 47th Street, New York, NY 10017. (212) 705-7900.

DIRECTORIES AND BIOGRAPHICAL SOURCES

1987 DIRECTORY OF ENGINEERING SOCIETIES AND RELATED ORGANIZATIONS. Gordon Davis, editor. Hemisphere Publishing Corporation, 1010 Vermont Avenue, NW, Washington, DC 20005. (800) 526-0275. 12th edition. 1987. $100.00.

RESEARCH CENTERS DIRECTORY 1988. Gale Research Company, Book Tower, Detroit, MI 48226. (800) 521-0707. 12th edition. 1987. $365.00 for set.

WHO'S WHO IN ENGINEERING. Gordon Davis, editor. Hemisphere Publishing Corporation, 1010 Vermont Avenue, NW, Washington, DC 20005. (800) 526-0275. 6th edition. 1985. $200.00.

GENERAL WORKS

FUNDAMENTALS OF HEAT EXCHANGER AND PRESSURE VESSEL TECHNOLOGY. J.P. Gupta. Hemisphere Publishing Corporation, 79 Madison Avenue, New York, NY 10016-7892. (800) 821-8312. 1986. $49.95.

HEAT TRANSFER EQUIPMENT DESIGN. R.K. Shah and others. Hemisphere Publishing Corporation, 79 Madison Avenue, New York, NY 10016-7892. 1988. $135.00.

INTRODUCTION TO HEAT TRANSFER. F. Incropera and D. Dewitt. John Wiley and Sons, Inc., 605 Third Avenue, New York, NY 10158. (800) 526-5368. 1985. $39.95.

MANAGING STEAM: AN ENGINEERING GUIDE TO INDUSTRIAL, COMMERCIAL AND UTILITY SYSTEMS. Jason Makansi, editor. Hemisphere Publishing Corporation, 79 Madison Avenue, New York, NY 10016-7892. Order from: Taylor and Francis/Hemisphere Distribution Center, 242 Cherry Street, Philadelphia, PA 19106-1906. (800) 821-8312. 1986. $35.00.

POWER PLANT SYSTEM DESIGN. K.W. Li and A.P. Priddy. John Wiley and Sons, Inc., 605 Third Avenue, New York, NY 10158. (800) 526-5368. 1985. $44.50.

PRESSURE VESSEL SYSTEMS; A USER'S GUIDE TO SAFE OPERATIONS AND MAINTENANCE. Anthony L. Kohan. McGraw-Hill Book Company, 1221 Avenue of the Americas, New York, NY 10020. (212) 512-2000. 1987. $42.50.

STANDARD METHODS FOR HYDRAULIC DESIGN FOR POWER BOILERS. V.A. Lokshin and others. Hemisphere Publishing Corporation, 79 Madison Avenue, New York, NY 10016-7892. (800) 821-8312. 1987. $59.95.

THEORY AND DESIGN OF PRESSURE VESSELS. John F. Harvey. Van Nostrand Reinhold Company, Inc., 135 West 50th Street, New York, NY 10020. (800) 543-2681. 1985. $51.95.

HANDBOOKS AND MANUALS

HANDBOOK OF HEAT TRANSFER APPLICATIONS. W.M. Rohsenow and others. McGraw-Hill Book Company, 1221 Avenue of the Americas, New York, NY 10020. (212) 512-2000. Second edition. 1986. $79.50.

HEAT EXCHANGER SOURCEBOOK. J.W. Palen, editor. Hemisphere Publishing Corporation, 79 Madison Avenue, New York, NY 10016-7892. (800) 821-8312. 1986. $59.95.

PRESSURE DESIGN HANDBOOK. Henry Bedner. McGraw-Hill Book Company, 1221 Avenue of the Americas, New York, NY 10020. (212) 512-2000. Second edition. 1985. $49.95.

STEAM TABLES: THERMODYNAMIC PROPERTIES OF WATER INCLUDING VAPOR, LIQUID, AND SOLID PHASES. J.H. Keenan and others. John Wiley and Sons, Inc., 605 Third Avenue, New York, NY 10158. (800) 526-5368. 1969. $44.95.

ONLINE DATA BASES

CA SEARCH. Chemical Abstracts Service, P.O. Box 3012, Columbus, OH 43120. (800) 848-6538 or (614) 421-3600. Comprehensive guide to chemical literature, 1972 to present. Inquire as to online cost and availability.

COMPENDEX. Engineering Information, Inc., 345 East 47th Street, New York, NY 10017. (800) 221-1044 or (212) 705-7615. Engineering and technical literature, 1975 to present. Inquire as to online cost and availability.

NTIS. National Technical Information Service, 5285 Port Royal Road, Springfield, VA 22161. (703) 487-4630. Broad coverage of government sponsored research reports, 1964 to present. Inquire as to online cost and availability.

WILSONLINE. H.W. Wilson and Company, 950 University Avenue, Bronx, NY 10452. (800) 367-6770 or (212) 588-8400. Makes available online versions of the H.W. Wilson indexes including Applied Science and Technology Index, Business Periodicals Index and Readers' Guide to Periodical Literature. Approximately 1980 to present. Inquire as to online cost and availability.

PERIODICALS

ASHRAE JOURNAL. American Society of Heating, Refrigerating and Air Conditioning Engineers, Inc., 1791 Tullie Circle, Atlanta, GA 30329. (404) 636-8400. 1914 to present. Monthly. $35.00 per year.

HEAT TRANSFER ENGINEERING. Hemisphere Publishing Corporation, 79 Madison Avenue, New York, NY 10016-7892. (800) 821-8312. 1979 to present. Quarterly. $97.50.

HEATING/PIPING/AIR CONDITIONING. Penton-IPC, Reinhold Publishing Division, 600 Summer Street, Box 1361, Stamford, CT 06904. (203) 348-7531. 1929 to present. Monthly. $35.00 per year.

JOURNAL OF HEAT TRANSFER. American Society of Mechanical Engineers, 345 East 47th Street, New York, NY 10017. (212) 705-7703. 1970 to present. Quarterly. $100.00 per year.

JOURNAL OF PRESSURE VESSEL TECHNOLOGY. American Society of Mechanical Engineers, 345 East 47th Street, New York, NY 10017. (212) 705-7703. 1974 to present. Quarterly. $80.00 per year.

PLANT ENGINEERING. Technical Publishing, Box 1030, 1301 South Grove Avenue, Barrington, IL 60010. (312) 381-1840. 1947 to present. Semi-monthly. $50.00 per year.

POWER. McGraw-Hill Book Company, 1221 Avenue of the Americas, New York, NY 10020. (212) 512-2000. 1882 to present. Monthly. $12.00 per year.

STEAM POWER. Kirk Enterprises, Limited, Midlands Steam Centre, 106a Derby Road, Loughborough LE11 OAG, England. 1949 to present. Quarterly. $20.00 to present.

RESEARCH CENTERS AND INSTITUTES

HEAT TRANSFER LABORATORY. Massachusetts Institute of Technology, 77 Massachusetts Avenue, Cambridge, MA 02139. (716) 253-2248.

HEAT TRANSFER RESEARCH. University of Wisconsin, Milwaukee, P.O. Box 784, Milwaukee, WI 53201. (414) 963-5001.

HEAT TRANSFER RESEARCH FACILITY. Columbia University, 632 West 125th Street, New York, NY 10027. (212) 280-4163.

INSTITUTE OF THERMO-FLUID ENGINEERING AND SCIENCE. Lehigh University, Whitaker Laboratory, Bethlehem, PA 18015. (215) 861-4091.

SPECIFICATIONS AND STANDARDS

PRESSURE VESSELS: THE ASME CODE SIMPLIFIED. Robert Chuse. McGraw-Hill Book Company, 1221 Avenue of the Americas, New York, NY 10020. (212) 512-2000. Sixth edition. 1984. $35.00.

BOOSTER ROCKETS

See: ROCKETS

BORING

See: PETROLEUM ENGINEERING

BORON

See also: ALLOYS, CHEMISTRY, INORGANIC CHEMISTRY, ORGANIC CHEMISTRY, SILICON, STEEL AND STEEL MAKING

ABSTRACT SERVICES AND INDEXES

APPLIED MECHANICS REVIEW. American Society of Mechanical Engineers, 345 East 47th Street, New York, NY 10017. (212) 705-7703. 1948 to present. Monthly. $360.00 per year.

APPLIED SCIENCE AND TECHNOLOGY INDEX. H.W. Wilson and Company, 950 University Avenue, Bronx, NY 10452. (800) 367-6670 or (212) 588-8400. Monthly. Inquire as to cost and availability.

CHEMICAL ABSTRACTS. American Chemical Society, Chemical Abstracts Service, Box 3012, Columbus, OH 43210. (614) 421-3600. 1907 to present. Weekly. $9500.00 per year.

CONFERENCE PAPERS INDEX. Cambridge Scientific Abstracts, 5161 River Road, Bethesda, MD 20816. 1972 to present. Monthly. Inquire as to cost and availability.

CURRENT CONTENTS: ENGINEERING, TECHNOLOGY AND APPLIED SCIENCES. Institute for Scientific Information, 3501 Market Street, Philadelphia, PA 19104. (800) 523-1850 or (215) 386-0100. Weekly. $275.00 per year.

ENGINEERING INDEX MONTHLY AND AUTHOR INDEX. Engineering Information Inc., 345 East 47th Street, New York, NY 10017. (212) 705-7600. Monthly. $1560.00 per year.

SCIENCE CITATION INDEX. Institute for Scientific Information, 3501 Market Street, Philadelphia, PA 19104. (800) 523-1850 or (215) 386-0100. Six times per year. $6200.00 per year.

ASSOCIATIONS AND PROFESSIONAL SOCIETIES

AMERICAN CHEMICAL SOCIETY. 1155 16th Street, N.W., Washington, DC 20036. (800) 424-6747.

AMERICAN INSTITUTE OF MINING, METALLURGICAL AND PETROLEUM ENGINEERS. 345 East 47th Street, New York, NY 10017. (212) 705-7695.

AMERICAN SOCIETY FOR METALS. Metals Park, OH 44073. (216) 338-5151.

AMERICAN SOCIETY FOR TESTING AND MATERIALS. 1916 Race Street, Philadelphia, PA 19103. (215) 299-5400.

ASSOCIATION OF IRON AND STEEL ENGINEERS. Three Gateway Center, Suite 2350, Pittsburgh, PA 15222. (412) 281-6323.

THE METALLURICAL SOCIETY. 400 Commonwealth Drive, Warrendale, PA 15096. (412) 776-4841.

DIRECTORIES AND BIOGRAPHICAL SOURCES

AMERICAN MEN AND WOMEN OF SCIENCE. R.R. Bowker, Inc., Order Department, 245 West 17th Street, New York, NY 10011. (800) 521-8110. Eight volumes. 1986. $595.00 for set.

INTERNATIONAL RESEARCH CENTERS DIRECTORY 1988-89. Darren L. Smith, editor. Gale Research Company, Book Tower, Detroit, MI 48226. (800) 521-0707. 4th edition. 1987. $360.00.

1987 DIRECTORY OF ENGINEERING SOCIETIES AND RELATED ORGANIZATIONS. Gordon Davis, editor. Hemisphere Publishing Corporation, 1010 Vermont Avenue, NW, Washington, DC 20005. (800) 526-0275. 12th edition. 1987. $100.00.

RESEARCH CENTERS DIRECTORY 1988. Gale Research Company, Book Tower, Detroit, MI 48226. (800) 521-0707. 12th edition. 1987. $365.00 for set.

SCIENTIFIC AND TECHNICAL ORGANIZATIONS AND AGENCIES DIRECTORY. Margaret Labash Young, editor. Gale Research Company, Book Tower, Detroit, MI 48226. (800) 521-0707. 2nd edition. 1987. $185.00.

WHO'S WHO IN ENGINEERING. Gordon Davis, editor. Hemisphere Publishing Corporation, 1010 Vermont Avenue, NW, Washington, DC 20005. (800) 526-0275. 6th edition. 1985. $200.00.

GENERAL WORKS

BORON-RICH SOLIDS. D. Emin and others, editors. American Institute of Physics, 335 East 45th Street, New York, NY 10017. (212) 661-9494. AIP Conference Proceeding No. 140. 1986. $57.75.

BORONIZING. Alfred G. Von Matushka. John Wiley and Sons, Inc., 605 Third Avenue, New York, NY 10158. (800) 526-5368. 1981. $48.95.

BORON

METAL INTERACTIONS WITH BORON CLUSTERS. Russell Grimes. Plenum Publishing Corporation, 233 Spring Street, New York, NY 10013. (800) 221-9369. 1982. $49.50.

STRUCTURAL CHEMISTY OF BORON AND SILICON. Springer-Verlag New York, Inc., 175 Fifth Avenue, New York, NY 10010. (800) 526-7254. 1986. $56.00.

HANDBOOKS AND MANUALS

CRC HANDBOOK OF CHEMISTRY AND PHYSICS. Robert C. Weast, editor. CRC Press, 2000 Corporate Boulevard, Boca Raton, FL 33431. (800) 272-7737. 68th edition. 1987. $69.95.

METALS HANDBOOK. American Society for Metals, Metals Park, OH 44073. (216) 338-5151. Fourteen volumes. Ninth edition. 1987. $1310.00 for set.

ONLINE DATA BASES

CA SEARCH. Chemical Abstracts Service, P.O. Box 3012, Columbus, OH 43120. (800) 848-6538 or (614) 421-3600. Comprehensive guide to chemical literature, 1972 to present. Inquire as to online cost and availability.

COMPENDEX. Engineering Information, Inc., 345 East 47th Street, New York, NY 10017. (800) 221-1044 or (212) 705-7615. Engineering and technical literature, 1975 to present. Inquire as to online cost and availability.

DISSERTATION ABSTRACTS ONLINE. University Microfilms International, 300 North Zeeb Road, Ann Arbor, MI 48106. (800) 521-0600 or (313) 761-4700. Scope includes virtually all doctoral dissertations accepted at accredited American institutions from 1861 to present in over 250 subject areas. Inquire as to online cost and availability.

NTIS. National Technical Information Service, 5285 Port Royal Road, Springfield, VA 22161. (703) 487-4630. Broad coverage of government sponsored research reports, 1964 to present. Inquire as to online cost and availability.

SCISEARCH. Institute for Scientific Information, 3501 Market Street, Philadelphia, PA 19104. (800) 523-1850 or (215) 386-0100. Broad multidisciplinary title and author index to the international literature of science and technology, 1974 to present. Inquire as to online cost and availability.

WILSONLINE. H.W. Wilson and Company, 950 University Avenue, Bronx, NY 10452. (800) 367-6770 or (212) 588-8400. Makes available online versions of the H.W. Wilson indexes including Applied Science and Technology Index, Business Periodicals Index and Readers' Guide to Periodical Literature. Approximately 1980 to present. Inquire as to online cost and availability.

PERIODICALS

AMERICAN CHEMICAL SOCIETY JOURNAL. American Chemical Society, 1155 16th Street, N.W., Washington, DC 20036. (800) 424-6747. 1879 to present. Semi-monthly. $275.00 per year.

INDUSTRIAL AND ENGINEERING CHEMISTRY FUNDAMENTALS. American Chemical Society, 1155 16th Street, N.W., Washington, DC 20036. (800) 424-6747. 1962 to present. Quarterly. $60.00 per year.

INORGANIC CHEMISTRY. American Chemical Society, 1155 16th Street, N.W., Washington, DC 20036. (800) 424-6747. 1962 to present. Biweekly. $300.00 per year.

RESEARCH CENTERS AND INSTITUTES

UNIVERSITY/INDUSTRY CHEMICAL RESEARCH CENTER. Mississippi State University, Department of Chemistry, P.O. Drawer CH, Mississippi State, MS 39762. (601) 325-3584.

BORON STEEL

See: FERROALLOYS

BOSONS

See: PARTICLE PHYSICS

BOUGER GRAVITY ANOMALY

See: GEOPHYSICS

BOUNDARY LAYER FLOW

See: FLUID MECHANICS

BOUNDARY WAVE

See: FLUID MECHANICS

BRAKES

ABSTRACT SERVICES AND INDEXES

APPLIED SCIENCE AND TECHNOLOGY INDEX. H.W. Wilson Company, 950 University Avenue, Bronx, NY 10452. (800) 367-6670 or (212) 588-8400. Inquire as to cost and availability.

ENGINEERING IDEX MONTHLY. Engineering Information, Incorporated, 345 East 47th Street, New York, NY 10017. (800) 221-1044 or (212) 705-7600. Monthly, with annual cumulation. $1560.00 per year.

SAE ABSTRACTS INDEX OF TECHNICAL PAPERS. SAE, Incorporated, 400 Commonwealth Drive, Warrendale, PA 15096. (412) 776-4841.

ASSOCIATIONS AND PROFESSIONAL SOCIETIES

AIR BRAKE ASSOCIATION. P.O. Box 1, Wilmerding, PA 15148. (412) 825-1465.

FRICTION MATERIALS STANDARDS INSTITUTE. Route 4, E-210, Paramus, NJ 07652. (201) 845-0440.

SOCIETY OF AUTOMOTIVE ENGINEERS. 400 Commonwealth Drive, Warrendale, PA 15096. (412) 776-4841.

DIRECTORIES AND BIOGRAPHICAL SOURCES

THE DIRECTORY OF DIRECTORIES. Cecilia Ann Marlow and Robert C. Thomas, editors. Gale Research Company, Book Tower, Detroit, MI 48226. (800) 521-0707. Fourth edition. 1987.

RESEARCH CENTERS DIRECTORY. Gale Research Company, Book Tower, Detroit, MI 48226. (800) 521-0707. Eleventh edition. 1987.

WHO'S WHO IN ENGINEERING. Engineers Joint Council, 345 East 47th Street, New York, NY 10017. (212) 705-7010. 1985. $200.00.

GENERAL WORKS

BRAKING: RECENT DEVELOPMENTS. SAE, Incorporated, 4000 Commonwealth Drive, Warrendale, PA 15096. (412) 776-4841. 1984. $18.00.

HANDBOOKS AND MANUALS

SAE HANDBOOK. SAE, Incorporated, 400 Commonwealth Drive, Warrendale, PA 15096. (412) 776-4841. $70.00 members; $140.00 non-members.

ONLINE DATA BASES

COMPENDEX. Engineering Information, Incorporated, 345 East 47th Street, New York, NY 10017. (800) 221-1044 or (212) 705-7615. Engineering and technical literature, 1975 to present. Inquire as to cost and availability.

NTIS. National Technical Information Service, 5285 Port Royal Road, Springfield, VA 22161. (703) 487-4630. Broad coverage of government sponsored research reports, 1964 to present. Inquire as to cost and availability.

SAE DATABASES. SAE, Incorporated, 400 Commonwealth Drive, Warrendale, PA 15096. (412) 776-4841. Citations to literature on all types of self-propelled vehicles, 1965 to present. Inquire as to on-line cost and availability.

PERIODICALS

AMERICAN RAILWAY ENGINEERING ASSOCIATION BULLETIN. 2000 L Street, NW, Washington, DC 20036. (202) 835-9334. Quarterly. $48.00 per year.

AUTOMOTIVE ENGINEERING. SAE, Incorporated, 400 Commonwealth Drive, Warrendale, PA 15096. (412) 776-4841. Monthly. $42.00 per year.

FRICTION MATERIALS STANDARDS INSTITUTE, BULLETIN. Friction Materials Standards Institute, Route 4, E-210, Paramus, NJ 07652. (201) 845-0440.

RESEARCH CENTERS AND INSTITUTES

COORDINATING RESEARCH COUNCIL, Incorporated 219 Perimeter Center Parkway, Atlanta, Ga 30346. (404) 396-3400.

UNIVERSITY OF MICHIGAN. Automotive Engineering Laboratory, Department of Mechanical Engineering and Applied Mechanics, Ann Arbor, MI. (313) 764-4256.

WAYNE STATE UNIVERSITY. Center for Automotive Research, 5050 Anthony Wayne Drive, Detroit, MI 48202. (313) 577-3887.

BRASS AND BRONZE

See also: ALLOYS, FERROALLOYS, HEAVY METAL ALLOYS, MATERIALS SCIENCE, METALLURGY

ABSTRACT SERVICES AND INDEXES

ALLOYS INDEX. American Society for Metals, 9639 Kinsman Road, Metals Park, OH 44073. (216) 338-5151. $130.00 per year.

APPLIED MECHANICS REVIEWS. American Society of Mechanical Engineers, 345 East 47th Street, New York, NY 10017. (212) 705-7722. Monthly. $380.00 per year. Critical reviews of the world literature in applied mechanics and related engineering science.

APPLIED SCIENCE AND TECHNOLOGY INDEX. H.W. Wilson Company, 950 University Avenue, Bronx, NY 10452. (800) 367-6670 or (212) 588-8400. Inquire as to cost and availability.

CHEMICAL ABSTRACTS. Chemical Abstracts Service, 2540 Olentangy Road, Post Office Box 3012, Columbus, OH 43210. (800) 848-6538 or (614) 421-3600. Weekly. $9200.00 per year.

CURRENT CONTENTS: ENGINEERING, TECHNOLOGY. Institute for Scientific Information, 3501 Market Street, Philadelphia, PA 19104. (800) 523-1850 or (215) 386-0100. $272.00 per year.

ENGINEERING INDEX MONTHLY. Engineering Information, Incorporated, 345 East 47th Street, New York, NY 10017. (800) 221-1044 or (212) 705-7600. Monthly. with annual cumulation. $1560.00 per year.

INTERNATIONAL COPPER INFORMATION BULLETIN. Copper Development Association, Orchard House, Mutton Lane, Potters Bar, Herts, EN6 3AP, England. 1976 to present. Quarterly. $35.00 per year.

METALS ABSTRACTS. American Society for Metals, 9639 Kinsman Road, Metals Park, OH 44073. (216) 338-5151. Monthly. $890.00.

METALS ABSTRACTS INDEX. American Society for Metals, 9639 Kinsman Road, Metals park, OH 44073. (216) 338-5151. Monthly. (Sold only to subscribers of Metals Abstracts.)

ZINC ABSTRACTS. Zinc Development Association, 34 Berkeley Square, London W1X 6AJ, England. 1943 to present. Quarterly. $100.00 per year.

ANNUAL REVIEWS AND YEARBOOKS

ANNUAL REVIEW OF MATERIALS SCIENCE. Annual Reviews, Incorporated, 4139 El Camino Way, Palo Alto, CA 94306. (415) 493-4400.

ASSOCIATIONS AND PROFESSIONAL SOCIETIES

AMERICAN SOCIETY FOR METALS. Metals Park, OH 44073. (216) 338-5151.

BRASS AND BRONZE INGOT INSTITUTE. 33 North LaSalle Street, Room 3500, Chicago, IL 60602. (312) 236-2715.

COPPER DEVELOPMENT ASSOCIATION. Box 1840, Greenwich Office Park 2, Greenwich, CT 06836. (203) 625-8210.

METALLURGICAL SOCIETY OF THE AIME (AMERICAN INSTITUTE OF MINING, METALLURGICAL AND PETROLEUM ENGINEERS). 420 Commonwealth Drive, Warrendale, PA 15086. (412) 776-9080.

DIRECTORIES AND BIOGRAPHICAL SOURCES

DUN'S INDUSTRIAL GUIDE - THE METAL WORKING DIRECTORY. Dun and Bradstreet, Incorporated, Three Century Drive, Parsippany, NJ 07054. (201) 455-0900. Annual. $550.00.

INTERNATIONAL RESEARCH CENTERS DIRECTORY 1986-1987. Gale Research Company, Book Tower, Detroit, MI 48226. (800) 521-0707. Third edition. 1986.

RESEARCH CENTERS DIRECTORY. Gale Research Company, Detroit, MI 48226. (800) 521-0707. Eleventh edition. 1987.

SCIENTIFIC AND TECHNICAL ORGANIZATIONS AND AGENCIES DIRECTORY. Gale Research Company, Book Tower, Detroit, MI 48226. (800) 521-0707. 1985.

WHO'S WHO IN ENGINEERING. Engineers Joint Council, 345 East 47th Street, New York, NY 10017. (212) 705-7010. 1985. $200.00.

WHO'S WHO IN FRONTIER SCIENCE AND TECHNOLOGY. Marquis Who's Who, Incorporated, 200 East Ohio Street, Chicago, IL 60611. (800) 428-3898 or (312) 787-2008.

WHO'S WHO IN TECHNOLOGY TODAY. Reston Publishing Company, Incorporated, c/o Prentice-Hall, Incorporated, Englewood Cliffs, NJ 07632. (800) 262-6868. Biennial. Five volumes. $425.00. Covers the fields of electronics, computer science, physics, optics, chemistry, biotechnology, mechanics, energy, and earth science.

ENCYCLOPEDIAS AND DICTIONARIES

ENCYCLOPEDIA OF PHYSICAL SCIENCE AND TECHNOLOGY. Academic Press, Incorporated, Orlando FL 32887. (800) 321-5068 or (305) 345-2734. Fifteen volumes, 1986.

MCGRAW-HILL DICTIONARY OF ENGINEERING. Sybil P. Parker, editor. McGraw-Hill Book Company, 1221 Avenue of the Americas, New York, NY 10020. (212) 512-2000. 1984. $39.95.

GENERAL WORKS

ESSENTIAL METALLURGY FOR ENGINEERS. W. Alexander, G. Davies, and K. Reynolds. Van Nostrand Reinhold, 115 Fifth Avenue, New York, NY 10003. (800) 543-2681. 1985. $17.95.

PHYSICAL METALLURGY. P. Hassen. Cambridge University Press, 32 East 57th Street, New York, NY 10022. (212) 688-8885. 1986. $24.95 in paper.

METALLURGY BASICS. D.V. Brown. Van Nostrand Reinhold, 115 Fifth Avenue, New York, NY 10003. (800) 543-2681. 1985. $17.95.

STRUCTURE AND PROPERTIES OF ENGINEERING ALLOYS. W.F. Smith. McGraw-Hill Book Company, Incorporated, 1221 Avenue of the Americas, New York, NY 10020. (212) 512-2000. 1980. $46.95.

SUPERALLOYS. C.T. Sims and W.C. Hagel. John Wiley and Sons, Incorporated, 605 Third Avenue, New York, NY 10158. (212) 850-6000. 1972. $74.95.

THEORY OF STRUCTURAL TRANSFORMATIONS IN SOLIDS. A.G. Khachaturyan. John Wiley and Sons, Incorporated, 605 Third Avenue, New York, NY 10158. (800) 526-5368 or (212) 850-6000. 1983. $69.95.

HANDBOOKS AND MANUALS

SMITHELL'S METALS REFERENCE BOOK. Eric A. Brandes, editor. Butterworth Publishers, 80 Montvale Avenue, Stoneham, MA 02180. (800) 325-4177. Sixth edition. 1983. $210.00.

ONLINE DATA BASES

COMPENDEX. Engineering Information, Incorporated, 345 East 47th Street, New York, NY 10017. (800) 221-1044 or (212) 705-7615. Engineering and technical literature, 1975 to present. Inquire as to cost and availability.

INSPEC. INSPEC Marketing Department, Institute of Electrical and Electronics Engineers, Incorporated, IEEE Service Department, 445 Hoes Lane, Piscataway, NJ 08854. (201) 981-0060. Inquire as to on-line cost and availability.

METADEX. Metals Information, American Society for Metals, Metals Park, OH 44073. (216) 338-5151. (Metals Abstracts/Alloys Index). A worldwide literature on the science and practice of metallurgy, 1966 to present. Inquire as to online cost and availability.

NASA. National Aeronautics and Space Administration, Scientific and Technical Information Branch, 300 7th Street, SW, Washington, DC 20546. Citations and abstracts of aerospace literature, 1962 to present. Inquire as to cost and availability.

NON-FERROUS METALS ABSTRACTS. British Non-Ferrous Metals Technology Centre, Grove Laboratories, Denchworth Road, Wantage, Oxfordshire, England OX12 9 BJ. Citations and abstracts on non-ferrous metallurgy and technology, 1961 to present. Inquire as to online cost and availability.

NTIS. National Technical Information Service, 5285 Port Royal Road, Springfield, VA 22161. (703) 487-4630. Broad coverage of government sponsored research reports, 1964 to present. Inquire as to cost and availability.

WILSONLINE. H.W. Wilson Company, 950 University Avenue, Bronx, NY 10452. (800) 367-6770 or (212) 588-8400. Makes available online versions of the printed H.W. Wilson Indexes including Applied Science and Technology Index, Business Periodicals Index, and Readers' Guide to Periodical Literature. Period covered is generally 1983 to present. Inquire as to cost and availability.

OTHER SOURCES

MATERIALS SCIENCE AND METALLURGY. National Technical Information Service (NTIS), 5285 Port Royal Road, Springfield, Va 22161. (703) 487-4630. Translations and abstracts of foreign language technical media. Irregular. $40.00 per year.

PERIODICALS

ALLOY DIGEST. Engineering Publications Incorporated, Box 823, Upper Montclair, NJ 07043. (201) 746-7930. Monthly. $50.00 per year.

CASTING ENGINEERING AND FOUNDRY WORLD. Continental Communications, Incorporated, 1115 Main Street, Bridgeport, CT 06604. (203) 377-5566.

JOURNAL OF METALS. Metallurgical Society of the AIME (American Institute of Mining, Metallurgical and Petroleum Engineers), 420 Commonwealth Drive, Warrendale, PA 15086. (412) 776-9080. Monthly. $40.00 per year.

METALLURGICAL TRANSACTIONS. Metallurgical Society of the AIME (American Institute of Mining, Metallurgical and Petroleum Engineers), 420 Commonwealth Drive, Warrendale, PA 15086. (412) 776-9080. Monthly. $95.00 per year.

METALS WEEK. McGraw-Hill Book Company, Incorporated, 1221 Avenue of the Americas, New York, NY 10020. (212) 997-2823. Weekly. $527.00 per year. Tin International. Tin Publications, Limited, 222 Strand, London, WC2R 1BA, England. Monthly. $85.00 per year.

RESEARCH CENTERS AND INSTITUTES

COPPER DEVELOPMENT ASSOCIATION. Box 1840, Greenwich Office Park 2, Greenwich, CT 06836. (203) 625-8210.

PHYSICAL METALLURGY RESEARCH LABORATORIES. Canada Centre for Mineral and Energy Technology, 555 Booth Street, Ottawa, ON Canada K1A 0G1. (613) 995-4807.

TEXAS A & M UNIVERSITY. Mechanics and Materials Center, ERC Building, College Station, TX 77843. (409) 845-7512.

UNIVERSITY OF CONNECTICUT. Institute of Materials Science, Storrs, CT 06268. (203) 486-4623.

UNIVERSITY OF FLORIDA. Department of Materials Science and Engineering, Gainesville, FL 32601. (904) 392-1454.

UNIVERSITY OF WISCONSIN AT MADISON. Cast Metals Laboratory, 1509 University Avenue, Madison, WI 53706. (608) 262-2562.

STATISTICAL SOURCES

METALS STATISTICS. Fairchild Publications, 7 East 12th Street, New York, NY 10003. (212) 741-4426. Annual. $50.00.

BRAZING

See: WELDING

BREEDER REACTORS

See also: NUCLEAR ENERGY, NUCLEAR ENGINEERING, NUCLEAR PHYSICS, NUCLEAR REACTORS, PLUTONIUM

ABSTRACT SERVICES AND INDEXES

APPLIED MECHANICS REVIEW. American Society of Mechanical Engineers, 345 East 47th Street, New York, NY 10017. (212) 705-7703. 1948 to present. Monthly. $360.00 per year.

APPLIED SCIENCE AND TECHNOLOGY INDEX. H.W. Wilson and Company, 950 University Avenue, Bronx, NY 10452. (800) 367-6670 or (212) 588-8400. Monthly. Inquire as to cost and availability.

CHEMICAL ABSTRACTS. American Chemical Society, Chemical Abstracts Service, Box 3012, Columbus, OH 43210. (614) 421-3600. 1907 to present. Weekly. $9500.00 per year.

CURRENT CONTENTS: ENGINEERING, TECHNOLOGY AND APPLIED SCIENCES. Institute for Scientific Information, 3501 Market Street, Philadelphia, PA 19104. (800) 523-1850 or (215) 386-0100. Weekly. $275.00 per year.

ENGINEERING INDEX MONTHLY AND AUTHOR INDEX. Engineering Information Inc., 345 East 47th Street, New York, NY 10017. (212) 705-7600. Monthly. $1560.00 per year.

ISMEC BULLETIN (Information Service in Mechanical Engineering). Cambridge Scientific Abstracts, 5161 River Road, Bethesda, MD 20816. (301) 951-1400. 1973 to present. Monthly. $450.00 per year.

PHYSICS ABSTRACTS. Institution of Electrical Engineers. Available from: IEEE Service Center, 445 Hoes Lane, Piscataway, NJ 08854. 1898 to present. Bimonthly. $1700.00 per year.

SCIENCE CITATION INDEX. Institute for Scientific Information, 3501 Market Street, Philadelphia, PA 19104. (800) 523-1850 or (215) 386-0100. Six times per year. $6200.00 per year.

ANNUAL REVIEWS AND YEARBOOKS

ADVANCES IN NUCLEAR SCIENCE AND TECHNOLOGY. J.Lewins, editor. Plenum Publishing Corporation, 233 Spring Street, New York, NY 10013. (800) 221-9369. 1977 to presnet. Irregular. Inquire as to cost and availability.

ASSOCIATIONS AND PROFESSIONAL SOCIETIES

AMERICAN NUCLEAR SOCIETY. 555 North Kensington Avenue, La Grange, IL 60525. (312) 352-6611.

FUSION ENERGY FOUNDATION. P.O. Box 17149, Washington, DC 20041. (703) 689-2490.

INSTITUTE OF NUCLEAR MATERIALS MANAGEMENT. 60 Revere Drive, Northbrook, IL 60062. (312) 480-9080.

BIBLIOGRAPHIES

SCIENTIFIC AND TECHNICAL BOOKS AND SERIALS IN PRINT 1988; AN INDEX TO LITERATURE IN SCIENCE AND TECHNOLOGY. R.R. Bowker Company, 205 East 42nd Street, New York, NY 10017. (800) 521-8110. $175.00.

DIRECTORIES AND BIOGRAPHICAL SOURCES

ENERGY INFORMATION CENTERS DIRECTORY. Public Affairs and Information Program, Atomic Industrial Forum, 7101 Wisconsin Avenue, Bethesda, MD 20814. (301) 654-9260. 1985. Free.

INTERNATIONAL DIRECTORY OF NUCLEAR UTILITIES. Lotte, Limited, Box 237, Contract Station 27, Lakewood, CO 80215. (303) 232-3026. Annual. $160.00.

INTERNATIONAL RESEARCH CENTERS DIRECTORY 1988-89. Darren L. Smith, editor. Gale Research Company, Book Tower, Detroit, MI 48226. (800) 521-0707. 4th edition. 1987. $360.00.

1987 DIRECTORY OF ENGINEERING SOCIETIES AND RELATED ORGANIZATIONS. Gordon Davis, editor. Hemisphere Publishing Corporation, 1010 Vermont Avenue, NW, Washington, DC 20005. (800) 526-0275. 12th edition. 1987. $100.00.

NUCLEAR REACTORS BUILT, BEING BUILT, OR PLANNED IN THE UNITED STATES. Office of Scientific and Technical Information, Department of Energy, Box 62, Oak Ridge, TN 37831. (615) 576-5637. Annual. $11.00. Send orders to: National Technical Information Service, Springfield, VA 22161.

RESEARCH CENTERS DIRECTORY 1988. Gale Research Company, Book Tower, Detroit, MI 48226. (800) 521-0707. 12th edition. 1987. $365.00 for set.

SCIENTIFIC AND TECHNICAL ORGANIZATIONS AND AGENCIES DIRECTORY. Margaret Labash Young, editor. Gale Research Company, Book Tower, Detroit, MI 48226. (800) 521-0707. 2nd edition. 1987. $185.00.

WHO'S WHO IN ENGINEERING. Gordon Davis, editor. Hemisphere Publishing Corporation, 1010 Vermont Avenue, NW, Washington, DC 20005. (800) 526-0275. 6th edition. 1985. $200.00.

ENCYCLOPEDIAS AND DICTIONARIES

THESAURUS OF SCIENTIFIC, TECHNICAL, AND ENGINEERING TERMS. Hemisphere Publishing Corporation, 1010 Vermont Avenue, NW, Washington, DC 20005. (800) 526-0275. 1988. $125.00.

GENERAL WORKS

FAST BREEDER REACTORS. A. Waltar and J. Reynolds. Pergamon Press, Inc., Maxwell House, Fairview Park, Elmsford, NY 10523. (914) 592-7700. 1981. $27.50.

INTRODUCTION TO NUCLEAR POWER. John G. Collier. Hemisphere Publishing Corporation, 1010 Vermont Avenue, NW, Washington, DC 20005. (800) 526-0275. 1987. $49.95.

NUCLEAR ENERGY TECHNOLOGY. Ronald Allen Knief. Hemisphere Publishing Corporation, 1010 Vermont Avenue, NW, Washington, DC 20005. (800) 526-0275. 1983. $48.00.

NUCLEAR FISSION REACTORS. I.R. Cameron. Plenum Publishing Corporation, 233 Spring Street, New York, NY 10013. (800) 221-9369. 1982. $49.50.

NUCLEAR FISSION REACTORS: POTENTIAL ROLE AND RISK OF CONVERTERS AND BREEDERS. G. Kessler. Springer-Verlag New York, Inc., 175 Fifth Avenue, New York, NY 10010. (800) 526-7254. 1983. $42.95.

NUCLEAR PHYSICS FOR ENGINEERS AND SCIENTISTS: LOW ENERGY THEORY WITH APPLICATIONS INCLUDING REACTORS AND THEIR ENVIRONMENTAL IMPACT. S.E. Hunt. John Wiley and Sons, Inc., 605 Third Avenue, New York, NY 10158. (800) 526-5368. 1987. $129.95.

NUCLEAR REACTOR ENGINEERING. Samuel Glasstone. Van Nostrand Reinhold Company, Inc., 135 West 50th Street, New York, NY 10020. (800) 543-2681. 1980. $49.95.

NUCLEAR MATERIALS AND APPLICATIONS. Benjamin M. Ma. Van Nostrand Reinhold Company, Inc., 135 West 50th Street, New York, NY 10020. (800) 543-2681. 1982. $45.95.

HANDBOOKS AND MANUALS

CRC HANDBOOK OF NUCLEAR REACTORS CALCULATIONS. CRC Press, 2000 Corporate Boulevard, Boca Raton, FL 33431. (800) 272-7737. Three volumes. 1986. $750.00 for set.

A GUIDE TO NUCLEAR POWER TECHNOLOGY: A RESOURCE FOR DECISION MAKING. F.J. Rahn and others. John Wiley and Sons, Inc., 605 Third Avenue, New York, NY 10158. (800) 526-5368. 1984. $85.95.

NUCLEAR ENGINEERING DATA BASES, STANDARDS AND NUMERICAL ANALYSIS. Jack Jedruch. Van Nostrand Reinhold Company, Inc., 135 West 50th Street, New York, NY 10020. (800) 543-2681. 1985. $59.95.

ONLINE DATA BASES

CA SEARCH. Chemical Abstracts Service, P.O. Box 3012, Columbus, OH 43120. (800) 848-6538 or (614) 421-3600. Comprehensive guide to chemical literature, 1972 to present. Inquire as to online cost and availability.

COMPENDEX. Engineering Information, Inc., 345 East 47th Street, New York, NY 10017. (800) 221-1044 or (212) 705-7615. Engineering and technical literature, 1975 to present. Inquire as to online cost and availability.

DOE ENERGY DATA BASE. U.S. Department of Energy, Office of Scientific and Technical Information, P.O. Box 62, Oak Ridge, TN 37831. (615) 576-6837. A database that covers all aspects of energy including the science and technology of energy. 1948 to present. Available through the DIALOG search service or DOE/RECON. Inquire as to online cost and availability.

ENERGYLINE. EIC/Intelligence, Inc., 48 West 38th Street, New York, NY 10018. (212) 944-8500. A database of resources on the scientific, engineering, political, and socioeconomic aspects of energy resources. 1976 to present. Inquire as to online cost and availability.

INSPEC. INSPEC Marketing Department, Institution of Electrical Engineers. Available from IEEE Service Center, 445 Hoes Lane, Piscataway, NJ 08854. (201) 981-0060. Online version of Physics Abstracts. Inquire as to online cost and availability.

NTIS. National Technical Information Service, 5285 Port Royal Road, Springfield, VA 22161. (703) 487-4630. Broad coverage of government sponsored research reports, 1964 to present. Inquire as to online cost and availability.

SCISEARCH. Institute for Scientific Information, 3501 Market Street, Philadelphia, PA 19104. (800) 523-1850 or (215) 386-0100. Broad multidisciplinary title and author index to the international literature of science and technology, 1974 to present. Inquire as to online cost and availability.

WILSONLINE. H.W. Wilson and Company, 950 University Avenue, Bronx, NY 10452. (800) 367-6770 or (212) 588-8400. Makes available online versions of the H.W. Wilson indexes including Applied Science and Technology Index, Business Periodicals Index and Readers' Guide to Periodical Literature. Approximately 1980 to present. Inquire as to online cost and availability.

PERIODICALS

AMERICAN NUCLEAR SOCIETY TRANSACTIONS. American Nuclear Society, 555 North Kensington Avenue, La Grange, IL 60525. (312) 352-6611. 1958 to present. Semiannual. $255.00 per year.

ANNALS OF NUCLEAR ENERGY. Pergamon Press, Inc., Maxwell House, Fairview Park, Elmsford, NY 10523. (914) 592-7700. 1954 to present. Monthly. $280.00 per year.

BULLETIN OF THE ATOMIC SCIENTISTS. Educational Foundation for Nuclear Science, 5801 South Kenwood Avenue, Chicago, IL 60637. (312) 363-5225. 1945 to present. Ten times per year. $22.50 per year.

FUSION TECHNOLOGY. American Nuclear Society, 555 North Kensington Avenue, La Grange, IL 60525. (312) 352-6611. 1981 to present. Bimonthly. $250.00 per year.

JOURNAL OF FUSION ENERGY. Plenum Publishing Corporation, 233 Spring Street, New York, NY 10013. (800) 221-9369. 1981 to present. Bimonthly. $105.00 per year.

NUCLEAR ENGINEER. Institution of Nuclear Engineers, 1 Penerley Road, Nondon SE6 2LQ, England. 1959 to present. Bimonthly. $110.00 per year.

NUCLEAR ENGINEERING AND DESIGN. Elsevier Science Publishing Company, Inc., 52 Vanderbilt Avenue, New York, NY 10017. (212) 370-5520. 1965 to present. $160.00 per year.

NUCLEAR ENGINEERING INTERNATIONAL. Electrical-Electronic Press, Quadrant House, The Quadrant, Sutton, Surrey, SM2 5AS, England. 1956 to present. Monthly. $210.00 per year.

NUCLEAR SCIENCE AND ENGINEERING. American Nuclear Society, 555 North Kensington Avenue, La Grange, IL 60525. (312) 352-6611. 1956 to present. Monthly. $220.00 per year.

NUCLEAR TECHNOLOGY. American Nuclear Society, 555 North Kensington Avenue, La Grange, IL 60525. (312) 352-6611. 1965 to present. Monthly. $345.00 per year.

RESEARCH CENTERS AND INSTITUTES

ASSISTANT SECRETARY FOR NUCLEAR ENERGY. U.S. Department of Energy, 1000 Independence Avenue, SW, Washington, DC 20585. (202) 252-6450.

DEPARTMENT OF NUCLEAR ENGINEERING AND ENGINEERING PHYSICS. University of Wisconsin - Madison, 1500 Johnson Drive, Madison, WI 53706. (608) 263-1648.

INSTITUTE OF NUCLEAR SCIENCE AND ENGINEERING. Oregon State University, Radiation Center, 35th and Jefferson Streets, Corvallis, OR 97331. (503) 754-2341.

LABORATORY OF BASIC AND APPLIED NUCLEAR RESEARCH. University of Cincinnati, Department of Chemistry, Cincinnati, OH 45221. (513) 475-3652.

WHITESHELL NUCLEAR RESEARCH ESTABLISHMENT. Research Company, Atomic Energy of Canada Limited, Pinawa,

MB, Canada ROE 1LO. (204) 753-2311.

SPECIFICATIONS AND STANDARDS

INFORMATION CENTER ON NUCLEAR STANDARDS. American Nuclear Society, 555 North Kensington Avenue, La Grange, IL 60525. (312) 352-6611. Standards for all aspects of the design, construction, operation, and maintenance of nuclear power plants.

BRICK

See: BUILDING MATERIALS

BRIDGES

See also: CIVIL ENGINEERING, STRUCTURAL ENGINEERING

ABSTRACT SERVICES AND INDEXES

APPLIED MECHANICS REVIEW. American Society of Mechanical Engineers, 345 East 47th Street, New York, NY 10017. (212) 705-7703. Monthly. $360.00 per year.

APPLIED SCIENCE AND TECHNOLOGY INDEX. H.W. Wilson Company, 950 University Avenue, Bronx, NY 10452. (800) 367-6670 or (212) 588-8400. Inquire as to cost and availability.

ENGINEERING INDEX MONTHLY. Engineering Information, Incorporated, 345 East 47th Street, New York, NY 10017. (800) 221-1044 or (212) 705-7600. Monthly, with annual cumulation. $1560.00 per year.

INTERNATIONAL STRUCTURAL ENGINEERING ABSTRACTS. CITIS Limited, 2 Rosemount Terrace, Blackrock, Dublin, Ireland. Quarterly. $95.00 per year.

SCIENCE CITATION INDEX. Institute for Scientific Information, 3501 Market Street, Philadelphia, PA 19104. (800) 523-1850 or (215) 386-0100.

ASSOCIATIONS AND PROFESSIONAL SOCIETIES

AMERICAN RAILWAY BRIDGE AND BUILDING ASSOCIATION. 18154 Harwood Avenue, Homewood, IL 60430. (312) 799-4650.

AMERICAN SOCIETY OF CIVIL ENGINEERS. 345 East 47th Street, New York, NY 10017. (212) 705-7496.

ASSOCIATION FOR BRIDGE CONSTRUCTION AND DESIGN. Post Office Box 11054, Pittsburgh, PA 15237. (412) 261-0101.

INTERNATIONAL ASSOCIATION FOR BRIDGE AND STRUCTURAL ENGINEERING. ETH-Honggerberg, Zurich CH-8093, Switzerland.

DIRECTORIES AND BIOGRAPHICAL SOURCES

AMERICAN SOCIETY OF CIVIL ENGINEERS - OFFICIAL REGISTER. American Society of Civil Engineers, 345 East 47th Street, New York, NY 10017. (212) 705-7496. Annual. Free.

ENGINEERS - CIVIL. American Business Directories, Incorporated, 5707 South 86th Circle, Omaha, NE 68127. (402) 331-7169. Annual. $80.00.

RESEARCH CENTERS DIRECTORY. Gale Research Company, Book Tower, Detroit, MI 48226. (800) 521-0707. Twelfth edition. $365.00.

WHO'S WHO IN ENGINEERING. Engineers Joint Council, 345 East 47th Street, New York, NY 10017. (212) 705-7010. 1985. $200.00.

WHO'S WHO IN FRONTIER SCIENCE AND TECHNOLOGY. Marquis Who's Who, Incorporated, 200 East Ohio Street, Chicago, IL 60611. (800) 428-3898 or (312) 787-2008.

WHO'S WHO IN TECHNOLOGY TODAY. Reston Publishing Company, Incorporated, c/o Prentice-Hall, Incorporated, Englewood Cliffs, NJ 07632. (800) 262-6868. Biennial. Five volumes. $425.00. Covers the fields of electronics, computer science, physics, optics, chemistry, biotechnology, mechanics, energy, and earth science.

GENERAL WORKS

ANALYSIS AND DESIGN OF BRIDGES. Cetin Vilmas, editor. Kluwer Academic Publishers, 190 Old Derby Street, Hingham, MA 02043. (617) 749-5262. 1984. $57.00.

ARCHES AND SHORT SPAN BRIDGES. S. Leliavsky. Methuen, Incorporated, 29 West 35th Street, New York, NY 10001. (212) 244-3336. 1982. $45.00.

BRIDGES. Fritz Leonhardt. MIT Press, 28 Carleton Street, Cambridge, MA 02142. (617) 253-2884. 1984. $55.00.

CANTILEVER CONSTRUCTION OF PRESTRESSED CONCRETE BRIDGES. Jacques Mathivat. John Wiley and Sons, Incorporated, 605 Third Avenue, New York, NY 10158. (800) 526-5368 or (212) 850-6000. 1984. $102.00.

DESIGN OF MODERN STEEL BRIDGES. Sukhen Chatterjee. Sheridan House, Incorporated, 145 Palisade Street, Dobbs Ferry, NY 10522. (914) 693-2410. 1987. $55.00.

HANDBOOKS AND MANUALS

BRIDGE ANALYSIS SIMPLIFIED. Baidar Bakht and Leslie G. Jaeger. McGraw-Hill Book Company, 1221 Avenue of the Americas, New York, NY 10020. (800) 628-0004. 1985. $49.95.

CRC HANDBOOK OF TABLES FOR APPLIED ENGINEERING SCIENCE. R.E. Bolz and G.L. Tuve, editors. CRC Press, Incorporated, 2000 Corporate Boulevard, NW, Boca Raton, FL 33431. Second edition. 1973. $69.00.

STANDARD HANDBOOK OF ENGINEERING CALCULATIONS. Tyler G. Hicks, editor. McGraw-Hill Book Company, 1221 Avenue of the Americas, New York, NY 10020. (212) 512-2000. Second edition. 1984. $59.50.

THE WILEY ENGINEER'S DESK REFERENCE. Sanford I. Heisler. John Wiley and Sons, Incorporated, 605 Third Avenue, New York, NY 10158. (800) 526-5368 or (212) 850-6418. 1984. $36.00.

ONLINE DATA BASES

COMPENDEX. Engineering Information, Incorporated, 345 East 47th Street, New York, NY 10017. (800) 221-1044 or (212) 705-7615. Engineering and technical literature, 1975 to present. Inquire as to cost and availability.

DISSERTATION ABSTRACTS ONLINE. University Microfilms International, 300 North Zeeb Road, Ann Arbor, MI 48106. (800) 521-0600 or (313) 761-4700. Scope includes virtually all doctoral dissertations accepted at accredited American institutions from 1861 to present in 252 subject areas. Inquire as to cost and availability.

INSPEC. INSPEC Marketing Department, Institute of Electrical and Electronics Engineers, Incorporated, IEEE Service Department, 445 Hoes Lane, Piscataway, NJ 08854. (201) 981-0060. Inquire as to on-line cost and availability.

NTIS. National Technical Information Service, 5285 Port Royal Road, Springfield, VA 22161. (703) 487-4630. Broad coverage of government sponsored research reports, 1964 to present. Inquire as to cost and availability.

WILSONLINE. H.W. Wilson Company, 950 University Avenue, Bronx, NY 10452. (800) 367-6770 or (212) 588-8400. Makes available online versions of the printed H.W. Wilson Indexes including Applied Science and Technology Index, Business Periodicals Index, and Readers' Guide to Periodical Literature. Period covered is generally 1983 to present. Inquire as to cost and availability.

PERIODICALS

JOURNAL OF STRUCTURAL ENGINEERING. American Society of Civil Engineers, 345 East 47th Street, New York, NY 10017. (212) 705-7275. Monthly. $140.00 per year.

JOURNAL OF STRUCTURAL MECHANICS. Marcel Dekker Journals, 270 Madison Avenue, New York, NY 10016. Quarterly. $75.00 per year.

STRUCTURAL ENGINEER. Institution of Structural Engineers, 11 Upper Belgrave Street, London, SW1X 8BH, England. Monthly. $95.00 per year.

RESEARCH CENTERS AND INSTITUTES

IOWA STATE UNIVERSITY. Structural Research Laboratory, Ames, IA 50011.

NEW MEXICO STATE UNIVERSITY. Center for Transportation Research, College of Engineering, Box 3CE, Las Cruces, NM 88003-0083. (505) 646-3135.

NORTH CAROLINA STATE UNIVERSITY. Center for Transportation Engineering Studies, Department of Civil Engineering, Box 7908, Raleigh, NC 27695-7908. (919) 737-2331.

UNIVERSITY OF MARYLAND. Computer Aided Design Laboratory, Engineering Research Center, College Park, MD 20742. (301) 454-7941.

BRITISH THERMAL UNIT

See: AIR CONDITIONING

BRITTLENESS

See: MATERIALS SCIENCE

BROADCAST BAND

See: RADIO

BROMINE

See also: CHLORINE, FLUORINE, HALIDES, IODINE

ABSTRACT SERVICES AND INDEXES

APPLIED SCIENCE AND TECHNOLOGY INDEX. H.W. Wilson Company, 950 University Avenue, Bronx, NY 10452. (800) 367-6670 or (212) 588-8400. Inquire as to cost and availability.

CHEMICAL ABSTRACTS. Chemical Abstracts Service, 2540 Olentangy Road, Post Office Box 3012, Columbus, OH 43210. (800) 848-6538 or (614) 421-3600. Weekly. $9200.00 per year.

GENERAL SCIENCE INDEX. H.W. Wilson Company, 950 University Avenue, Bronx, NY 10452. (800) 367-6770 or (212) 588-8400. Inquire as to cost and availability.

PHYSICS ABSTRACTS. Institute of Electrical Engineers, London, United Kingdom. Available from: Institute of Electrical and Electronic Engineers (IEEE), 345 East 47th Street, New York, NY 10017. (212) 705-7900.

SCIENCE CITATION INDEX. Institute for Scientific Information, 3501 Market Street, Philadelphia, PA 19104. (800) 523-1850 or (215) 386-0100.

ASSOCIATIONS AND PROFESSIONAL SOCIETIES

AMERICAN CHEMICAL SOCIETY. 1155 16th Street, NW, Washington, DC 20036. (202) 872-4600.

ASSOCIATION OF CONSULTING CHEMISTS AND CHEMICAL ENGINEERS. 50 East 41st Street, Suite 92, New York, NY 10017. (212) 684-6255.

BIBLIOGRAPHIES

SCIENTIFIC AND TECHNICAL BOOKS AND SERIALS IN PRINT 1988: AN INDEX TO LITERATURE IN SCIENCE AND TECHNOLOGY. R.R. Bowker Company, 205 East 42nd Street, New York, NY 10017. (800) 521-8110 or (212) 916-1600. $175.00.

DIRECTORIES AND BIOGRAPHICAL SOURCES

AMERICAN INSTITUTE OF CHEMISTS. American Institute of Chemists, 7315 Wisconsin Avenue, Bethesda, MD 20814. (301) 652-2447. 1986. $35.00.

AMERICAN MEN AND WOMEN OF SCIENCE. Physical and Biological Sciences. Fifteenth edition. R.R. Bowker Company, 205 East 42nd Street, New York, NY 10017. (800) 521-8110 or (212) 916-1600.

BIOGRAPHICAL DICTIONARY OF SCIENTISTS: CHEMISTS. David Abbott, editor. P. Bedrick Books, 125 East 23rd Street, New York, NY 10010. (212) 777-1187. 1984. $18.95.

CHEMICAL WEEK - BUYERS GUIDE ISSUE. McGraw-Hill Book Company, 1221 Avenue of the Americas, New York, NY 10020. (800) 628-0004. Annual, October. $50.00.

CONSULTING SERVICES: CHEMISTS AND CHEMICAL ENGINEERS. Association of Consulting Chemists and Chemical Engineers, 50 East 41st Street, New York, NY 10017. (212) 684-6255. Annual. 1986. $45.00.

GOVERNMENT RESEARCH DIRECTORY. Gale Research Company, Book Tower, Detroit, MI 48226. (800) 521-0707. Fourth edition. 1987. $350.00.

INTERNATIONAL RESEARCH CENTERS DIRECTORY 1988-1989. Gale Research Company, Book Tower, Detroit, MI 48226. (800) 521-0707. Fourth edition. 1987. $360.00.

RESEARCH CENTERS DIRECTORY. Gale Research Company, Book Tower, Detroit, MI 48226. (800) 521-0707. Twelfth edition. 1987. $365.00.

SCIENTIFIC AND TECHNICAL ORGANIZATIONS AND AGENCIES DIRECTORY. Gale Research Company, Book Tower, Detroit, MI 48226. (800) 521-0707. Second edition. 1987. $185.00.

WHO'S WHO IN FRONTIER SCIENCE AND TECHNOLOGY. Marquis Who's Who, Incorporated, 200 East Ohio Street,

Chicago, IL 60611. (800) 428-3898 or (312) 787-2008.

WHO'S WHO IN TECHNOLOGY TODAY. Reston Publishing Company, Incorporated, c/o Prentice-Hall, Incorporated, Englewood Cliffs, NJ 07632. (800) 262-6868. Biennial. Five volumes. $425.00. Covers the fields of electronics, computer science, physics, optics, chemistry, biotechnology, mechanics, energy, and earth science.

ENCYCLOPEDIAS AND DICTIONARIES

CONCISE ENCYCLOPEDIA OF CHEMICAL TECHNOLOGY. Kirk Othmer. John Wiley and Sons, Incorporated, 605 Third Avenue, New York, NY 10158. (800) 526-5368 or (212) 850-6000. Third edition. 1985. $129.95.

CONDENSED CHEMICAL DICTIONARY. Gessner Hawley. Van Nostrand Reinhold, 115 Fifth Avenue, New York, NY 10003. Tenth edition. 1981. $49.95.

ENCYCLOPEDIA OF PHYSICAL SCIENCE AND TECHNOLOGY. Academic Press, Incorporated, Orlando, FL 32887. (800) 321-5068 or (305) 345-2734. Fifteen volumes, 1986. Inquire as to cost and availability.

GLOSSARY OF CHEMICAL TERMS. Clifford A. Hampel and Gessner G. Hawley. Van Nostrand Reinhold Company, 115 Fifth Avenue, New York, NY 10003. (800) 543-2681 or (212) 254-3232. Second edition. 1982. $21.95.

MCGRAW-HILL ENCYCLOPEDIA OF SCIENCE AND TECHNOLOGY. McGraw-Hill Book, Incorporated, 1221 Avenue of the Americas, New York, NY 10020. (212) 997-3675. Inquire as to cost and availability.

VAN NOSTRAND REINHOLD ENCYCLOPEDIA OF CHEMISTRY. Douglas M. Considine and Glenn D. Considine. Van Nostrand Reinhold Publishing Company, Incorporated, 115 Fifth Avenue, New York, NY 10003. (800) 543-2681 or (212) 254-3232. 1984. $97.95.

GENERAL WORKS

CHEMISTRY OF HALIDES, PSEUDOHALIDES AND AZIDES: SUPPLEMENT D. CHEMISTRY OF FUNCTIONAL GROUPS SERIES, PART 1 AND 2. Saul Patai. John Wiley and Sons, Incorporated, 605 Third Avenue, New York, NY 10158. (800) 526-5368 or (212) 850-6000. 1983. $745.00 set.

CHEMISTRY OF THE ELEMENTS. N.N. Greenwood and A. Earnshaw. Pergamon Publishing, Incorporated, Maxwell House, Fairview Park, Elmsford, NY 10523. (914) 592-7700. 1984. $143.00.

HALOGEN CHEMISTRY. V. Gutman, editor. Academic Press, Incorporated, 6277 Sea Harbor Drive, Orlando, FL 32821. (800) 321-5068. Three volumes. 1967. $90.00 each.

HAZARDOUS AND TOXIC CHEMICAL: SAFE HANDLING AND DISPOSAL. Howard Fawcett. John Wiley and Sons, Incorporated, 605 Third Avenue, New York, NY 10158. (800) 526-5368 or (212) 850-6000. 1984. $37.00.

SAFE STORAGE OF LABORATORY CHEMICALS. David A. Pipitone. John Wiley and Sons, Incorporated, 605 Third Avenue, New York, NY 10158. (800) 526-5368 or (212) 850-6000. 1984. $60.00.

TOXICOLOGY OF HALOGENATED HYDROCARBONS: HEALTH AND ECOLOGICAL EFFECTS. M.A. Khan and R.H. Stanton, editors. Pergamon Press, Incorporated, Maxwell House, Fairview Park, Elmsford, NY 10523. (914) 592-7700. 1981. $72.50.

HANDBOOKS AND MANUALS

THE CHEMIST'S COMPANION: A HANDBOOK OF PRACTICAL DATA, TECHNIQUES, AND REFERENCES. Arnold J. Gordon and Richard A. Ford. John Wiley and Sons, Incorporated, 605 Third Avenue, New York, NY 10158. (800) 526-5368. 1973. $49.95.

CRC HANDBOOK OF CHEMISTRY AND PHYSICS. CRC Press, Incorporated, 2000 Corporate Boulevard, NW, Boca Raton, FL 33431. Sixty-seventh edition. 1986. $69.95.

HANDBOOK OF APPLIED CHEMISTRY: FACTS FOR ENGINEERS, SCIENTISTS, TECHNICIANS, AND TECHNICAL MANAGERS. Vollrath Hopp and Ingp Hennig. McGraw-Hill Book Company, 1221 Avenue of the Americas, New York, NY 10020. (800) 628-0004. 1983. $54.00.

HANDBOOK OF COMPUTATIONAL CHEMISTRY: A PRACTICAL GUIDE TO CHEMICAL STRUCTURE AND ENERGY CALCULATIONS. Tim Clark. John Wiley and Sons, Incorporated, 605 Third Avenue, New York, NY 10158. (800) 526-5368 or (212) 850-6000. 1985. $35.00.

LANGE'S HANDBOOK OF CHEMISTRY. John A. Dean, editor. Mcgraw-Hill Book Company, 1221 Avenue of the Americas, New York, NY 10020. (800) 628-0004. 1985. $59.50.

ONLINE DATA BASES

CA SEARCH. Chemical Abstracts Service, Post Office Box 3012, Columbus, OH 43210. Guide to chemical literature, 1972 to present. Inquire as to cost and availability.

DISSERTATION ABSTRACTS ONLINE. University Microfilms International, 300 North Zeeb Road, Ann Arbor, MI 48106. (800) 521-0600 or (313) 761-4700. Scope includes virtually all doctoral dissertations accepted at accredited American institutions from 1861 to present in 252 subject areas. Inquire as to online cost and availability.

INSPEC. INSPEC Marketing Department, Institute of Electrical and Electronics Engineers, Incorporated, IEEE Service Department, 445 Hoes Lane, Piscataway, NJ 08854. (201) 981-0060. Inquire as to online cost and availability.

NTIS. National Technical Information Service, 5285 Port Royal Road, Springfield, VA 22161. (703) 487-4630. Broad coverage of government sponsored research reports, 1964 to present. Inquire as to online cost and availability.

SCISEARCH. Institute for Scientific Information, 3501 Market Street, Philadelphia, PA 19104. (800) 523-1850 or (215) 386-0100. Broad multidisciplinary title and author index to the international literature of science and technology. 1974 to present. Inquire as to online cost and availability.

OTHER SOURCES

ANNUAL ENERGY REVIEW. United States Department of Energy, Energy Information Administration, Washington, DC 20585. Annual.

CHEMICAL NOMENCLATURE USEAGE. Ronald Less and Arthur F. Smith. John Wiley and Sons, Incorporated, 605 Third Avenue, New York, NY 10158. (800) 526-5368 or (212) 850-6000. 1983. $52.95.

GUIDE TO BASIC INFORMATION SOURCES IN CHEMISTRY. Arthur Antony. John Wiley and Sons, Incorporated, 605 Third Avenue, New York, NY 10158. (800) 526-5368 or (212) 850-6000. 1979. $26.95.

HOW TO FIND CHEMICAL INFORMATION: A GUIDE FOR PRACTICING CHEMISTS, TEACHERS, AND STUDENTS. John Wiley and Sons, Incorporated, 605 Third Avenue, New York, NY 10158. (800) 526-5368 or (212) 850-6000. 1986. $35.00.

MINERALS YEARBOOK. United States Department of the Interior, Bureau of Mines, C Street between Eighteenth and Ninteenth Streets, NW, Washington, DC 20240. Annual.

PERIODICALS

ANALYTICAL CHEMISTRY. American Chemical Society, 1155 Sixteenth Street, NW, Washington, DC 20036. (800) 424-6747 or (202) 872-4700. Monthly. $33.00 per year.

ANDEWANDTE CHEMIE (GESPELLSCHAFT DEUTSCHER CHEMIKER, GW). V C H Verlagsgesellschaft mbH, Pappelallee 3, Postfach 1260, 6940 Weinheim, West Germany, Monthly. $300.00 per year.

CHEMICAL REVIEWS. American Chemical Society, 1155 Sixteenth Street, NW, Washington, DC 20036. (800) 424-6747 or (202) 872-4700. Bimonthly. $83.00 per year.

CHEMICAL WEEK. McGraw-Hill Publishing Company, 1221 Avenue of the Americas, New York, NY 10020. (212) 512-2000. Weekly. $30.00 per year.

CHEMTECH. American Chemical Society, 1155 Sixteenth Street, NW, Washington, DC 20036. (800) 424-6747 or (202) 872-4700. Monthly. $40.00 per year to individuals.

INORGANIC CHEMISTRY. American Chemical Society, 1155 Sixteenth Street, NW, Washington, DC 20036. (800) 424-6747 or (202) 872-4700. Monthly. $400.00 per year.

JOURNAL OF THE AMERICAN CHEMICAL SOCIETY. American Chemical Society, 1155 Sixteenth Street, NW, Washington, DC 20036. (800) 424-6747 or (202) 872-4700. Biweekly. $330.00 per year.

POLYHEDRON. Pergamon Journals, Incorporated, Maxwell House, Fairview Park, Elmsford, NY 10523. (914) 592-7700. Monthly. $595.00 per year.

RESEARCH CENTERS AND INSTITUTES

HARVARD UNIVERSITY. Chemical Laboratories, Oxford Street, Cambridge, MA 02138. (617) 495-5283.

RENSSELAER POLYTECHNICAL INSTITUTE. Chemistry Laboratories, Cogswell Laboratory, Troy, NY 13181. (518) 266-8462.

SOUTHERN ILLINOIS UNIVERSITY AT CARBONDALE. Research Program in Chemistry and Bicohemistry, Carbondale, IL 62901. (618) 453-5721.

UNIVERSITY OF WISCONSIN, MADISON. Theoretical Chemistry Institute, 1101 University Avenue, Madison, WI 53706. (608) 262-1511.

BRONZE

See: BRASS AND BRONZE

BUILDING MATERIALS

See also: CEMENT, CONCRETE, PRESTRESSED CONCRETE

ABSTRACT SERVICES AND INDEXES

APPLIED SCIENCE AND TECHNOLOGY INDEX. H.W. Wilson Company, 950 University Avenue, Bronx, NY 10452. (800) 367-6670 OR (212) 588-8400. Inquire as to cost and availability.

CONCRETE ABSTRACTS. American Concrete Institute, P.O. Box 19150, Redford Station, Detroit, MI 48219. (313) 532-2600. Bimonthly. $122.00 per year.

ENGINEERING INDEX MONTHLY. Engineering Information, Incorporated, 345 East 47th Street, New York, NY 10017. (800) 221-1044 or (212) 705-7600. Monthly, with annual cumulation. $1560.00 per year.

ANNUAL REVIEWS AND YEARBOOKS

CONCRETE INDUSTRIES YEARBOOK. Pit and Quarry Publications, Incorporated, 105 West Adams Street, Chicago, IL 60603. $35.00.

MINERALS YEARBOOK. U.S. Bureau of Mines. Annual. $18.00. Send orders to: Government Printing Office, Washington, DC 20402.

ASSOCIATIONS AND PROFESSIONAL SOCIETIES

AMERICAN CONCRETE INSTITUTE. P.O. BOX 19150, Redford Station, Detroit, MI 48219. (313) 532-2600.

BRICK INSTITUTE OF AMERICA. 11490 Commerce Park Drive, Suite 300, Reston, VA 22091. (703) 620-0010.

NATIONAL BUILDING MATERIAL DISTRIBUTORS ASSOCIATION. 1710 Lake Avenue, Suite 170, Glenview, IL 60025. (312) 724-6900.

PRESTRESSED CONCRETE INSTITUTE. 201 North Wells Street, Chicago, IL 60606. (312) 346-4071.

DIRECTORIES AND BIOGRAPHICAL SOURCES

AMERICAN CEMENT DIRECTORY. Bradley Pulverizer Company, 123 South Third Street, Allentown, PA 18105. (215) 434-5191. Annual. $46.00.

AMERICAN CONCRETE INSITUTE MEMBERSHIP DIRECTORY. American Concrete Institute, P.O. Box 19150, Redford Station, Detroit, MI 48219. (313) 532-2600. Biennial. $44.95.

BUILDING MATERIALS - WHOLESALERS DIRECTORY. American Business Directories, Incorporated, Division of American Business Lists, Incorporated, 5707 South 86th Circle, Omaha, NE 68127. (402) 331-7169. 1986.

ENCYCLOPEDIAS AND DICTIONARIES

CEMENT AND CONCRETE TERMINOLOGY. American Concrete Institute, P.O. Box 19150, Redford Station, Detroit, MI 48219. (313) 532-2600. 1985.

CONSTRUCTION GLOSSARY: AN ENCYCLOPEDIC REFERENCE AND MANUAL. J.S. Stein. John Wiley and Sons, Incorporated, 605 Third Avenue, New York, NY 10158. (800) 526-5368 or (212) 850-6000. 1980. $85.95.

DICTIONARY OF ARCHITECTURE AND CONSTRUCTION. Cyril M. Harris. McGraw-Hill Book Company, 1221 Avenue of the Americas, New York, NY 10020. (800) 628-0004. 1975. $49.50.

MCGRAW-HILL DICTIONARY OF ENGINEERING. Sybil P. Parker, editor. McGraw-Hill Book Company, 1221 Avenue of the Americas, New York, NY 10020. (212) 512-2000. 1984. $39.95.

MCGRAW-HILL ENCYCLOPEDIA OF SCIENCE AND TECHNOLOGY. McGraw-Hill Book Company, 1221 Avenue of the Americas, New York, NY 10020. (212) 512-2000.

GENERAL WORKS

BASIC CONSTRUCTION MATERIALS. Charles Herubin and Theodore Marotta. Reston Publishing Company, Incorporated, 200 Old Tappan Road, Old Tappan, NJ 07675. (201) 592-2352. Second edition. 1981. $27.95.

CONCRETE: STRUCTURE, PROPERTIES AND MATERIALS. P. Kumar Mehta. Prentice-Hall, Incorporated, Englewood Cliffs, NJ 07632. (800) 262-6868 or (201) 592-2000. 1986. $43.95.

CONSTRUCTION MATERIALS EVALUATION AND SELECTION: A SYSTEMATIC APPROACH. Harold J. Rosen and Phillip M. Bennett. John Wiley and Sons, Incorporated, 605 Third Avenue, New York, NY 10158. (800) 526-5368 or (212) 850-6000. 1979. $34.95.

CONSTRUCTION MATERIALS FOR ARCHITECTURE. Harold J. Rosen. John Wiley and Sons, Incorporated, 605 Third Avenue, New York, NY 10158. (800) 526-5368 or (212) 850-6000. 1985. $42.95.

MATERIALS OF CONSTRUCTION. Frank Dagostino. Reston Publishing Company, Incorporated, 200 Old Tappan Road, Old Tappan, NJ 07675. (201) 592-2352. 1981. $28.95.

HANDBOOKS AND MANUALS

ACI MANUAL OF CONCRETE PRACTICE. American Concrete Institute, P.O. Box 19150, Redford Station, Detroit, MI 48219. (313) 532-2600. Five volumes. Annual.

BUILDING MATERIALS EVALUATION HANDBOOK. Forrest Wilson. Van Nostrand Reinhold Company, 115 Fifth Avenue, New York, NY 10003. (800) 543-2681. 1984. $38.95.

BUILDING STRUCTURAL DESIGN HANDBOOK. Richard White and Charles G. Salmon. John Wiley and Sons, Incorporated, 605 Third Avenue, New York, NY 10158. (800) 526-5368 or (212) 850-6000. 1986. $74.95.

COMPLETE CONCRETE, MASONRY AND BRICK HANDBOOK. J.T. Adams. Van Nostrand Reinhold Company, 115 Fifth Avenue, New York, NY 10003. (800) 543-2681. 1983. $23.95 per year.

CONCRETE CONSTRUCTION HANDBOOK. Joseph J. Waddell, editor. McGraw-Hill Book Company, 1221 Avenue of the Americas, New York, NY 10020. (212) 512-2000. Second edition. 1974. $70.50.

CONSTRUCTION MATERIALS READY-REFERENCE MANUAL. J.J. WADDELL. McGraw-Hill Book Company, 1221 Avenue of the Americas, New York, NY 10020. (212) 512-2000. 1984. $29.50.

ONLINE DATA BASES

COMPENDEX. Engineering Information, Incorporated, 345 East 47th Street, New York, NY 10017. (800) 221-1044 or (212) 705-7615. Engineering and technical literature, 1975 to present. Inquire as to cost and availability.

NTIS. National Technical Information Service, 5285 Port Royal Road, Springfield, VA 22161. (703) 487-4630. Broad coverage of government sponsored research reports, 1964 to present. Inquire as to online cost and availability.

U.S. ARMY CORPS OF ENGINEERS. Waterways Experiment Station, Concrete Technology Information Analysis Center, P.O. Box 631, Vicksburg, MS 39180. (601) 634-3264. Priority is given to U.S. Department of Defense personnel, but services are also available to others in the scientific community as time permits.

WILSONLINE. H.W. Wilson Company, 950 University Avenue, Bronx, NY 10452. (800) 367-6770 or (212) 588-8400. Makes available online versions of the printed H.W Wilson indexes including Applied Science and Technology Index, Business Periodicals Index, and Readers Guide to Periodical Literature. Period covered is generally 1983 to present. Inquire as to cost and availability.

PERIODICALS

CONCRETE CONSTRUCTION. Concrete Construction Publications, Incorporated, 426 South Westgate, Addison, IL 60101. (312) 543-0870. Monthly. $12.00 per year.

CONCRETE INTERNATIONAL: DESIGN AND CONSTRUCTION. American Concrete Institute, P.O. Box 19150, Redford Station, Detroit, MI 48219. (313) 532-2600. Monthly. $69.00.

JOURNAL OF COMPOSITE MATERIALS. Technomic Publishing Company Incorporated, 851 New Holland Avenue, Box 3535, Lancaster, PA 17604. (717) 291-5609. Bimonthly. $185.00 per year.

JOURNAL OF PCI. Prestressed Concrete Institute, 201 North Wells Street, Chicago, IL 60606. (312) 346-4071. Bimonthly.

JOURNAL OF THE AMERICAN CONCRETE INSTITUTE. American Concrete Institute, P.O. Box 19150, Redford Station, Detroit, MI 48219. (313) 532-2600. Monthly. $69.00 per year.

MASONRY. Mason Contractors Association of America, 17 West 601 14th Street, Oakbrook Terrace, IL 60181. (312) 620-6767. Bimonthly. $6.50 per year.

RESEARCH CENTERS AND INSTITUTES

AMERICAN INSTITUTE OF TIMBER CONSTRUCTION. 333 West Hampden Avenue, Suite 712, Englewood, CO 80110. (303) 761-3312.

AMERICAN IRON AND STEEL INSTITUTE. 1000 16th Street, NW, Washington, DC 20036. (202) 452-7100.

AMERICAN SOCIETY FOR TESTING AND MATERIALS. 1916 Race Street, Philadelphia, PA 19103. (215) 299-5400.

NAHB National Research Center (National Association of Home Builders). 400 Prince Georges Center Boulevard, Upper Marlboro, MD 20772. (301) 249-4000.

SPECIFICATIONS AND STANDARDS

BLENDED CEMENTS. STP 897. G. Frohnsdorff, editor. American Society for Testing and Materials (ASTM), 1916 Race Street, Philadelphia, PA 19103. (215) 299-5400. 1986. $26.00.

CEMENT STANDARDS: EVOLUTION AND TRENDS - STP 663. American Society for Testing and Materials, 1916 Race Street, Philadelphia, PA 19103. (215) 299-5400. 1979. $20.00.

CEMENT STANDARDS OF THE WORLD. Edited by Cembureau (European Cement Association). International Publications Service Incorporated, P.O. Box 230, Accord, MA 02018. (617) 749-3628. Second edition. 1980. $75.00.

MASONRY: RESEARCH, APPLICATION, AND PROBLEMS. STP 871. American Society for Testing and Materials (ASTM), 1916 Race Street, Philadelphia, PA 19103. (215) 299-5400. 1985. $36.00.

TEMPERATURE EFFECTS ON CONCRETE. STP 858. T.R. Naik, editor. American Society for Testing and Materials (ASTM), 1916 Race Street, Philadelphia, PA 19103. (215) 299-5400. 1985. $280.

BUILDINGS

See: CONSTRUCTION ENGINEERING

BUOYS

See also: NAVIGATION, OCEAN ENGINEERING, RADAR

ABSTRACT SERVICES AND INDEXES

APPLIED SCIENCE AND TECHNOLOGY INDEX. H.W. Wilson and Company, 950 University Avenue, Bronx, NY 10452. (800) 367-6670 or (212) 588-8400. Monthly. Inquire as to cost and availability.

ENGINEERING INDEX MONTHLY AND AUTHOR INDEX. Engineering Information Inc., 345 East 47th Street, New York, NY 10017. (212) 705-7600. Monthly. $1560.00 per year.

GENERAL SCIENCE INDEX. H.W. Wilson and Company, 950 University Avenue, Bronx, NY 10452. (800) 367-6670 or (212) 588-8400. 1978 to present. Monthly. Inquire as to cost and availability.

ASSOCIATIONS AND PROFESSIONAL SOCIETIES

INSTITUTE OF NAVIGATION. 815 15th Street, N.W., Suite 832, Washington, DC 20005. (202) 783-4121.

INTERNATIONAL OMEGA ASSOCIATION. P.O. Box 2324, 1720 South Eads Street, Arlington, VA 22202. (301) 593-4144.

PERMANENT INTERNATIONAL ASSOCIATION OF NAVIGATION CONGRESSES, UNITED STATES SECTION. c/o U.S. Army Corps of Engineers, Water Resources Support Center, Casey Building, Fort Belvoir, VA 22060. (202) 355-2096.

WILD GOOSE ASSOCIATION. P.O. Box 556, Bedford, MA 01730.

DIRECTORIES AND BIOGRAPHICAL SOURCES

INTERNATIONAL RESEARCH CENTERS DIRECTORY 1988-89. Darren L. Smith, editor. Gale Research Company, Book Tower, Detroit, MI 48226. (800) 521-0707. 4th edition. 1987. $360.00.

1987 DIRECTORY OF ENGINEERING SOCIETIES AND RELATED ORGANIZATIONS. Gordon Davis, editor. Hemisphere Publishing Corporation, 1010 Vermont Avenue, NW, Washington, DC 20005. (800) 526-0275. 12th edition. 1987. $100.00.

RESEARCH CENTERS DIRECTORY 1988. Gale Research Company, Book Tower, Detroit, MI 48226. (800) 521-0707. 12th edition. 1987. $365.00 for set.

SCIENTIFIC AND TECHNICAL ORGANIZATIONS AND AGENCIES DIRECTORY. Margaret Labash Young, editor. Gale Research Company, Book Tower, Detroit, MI 48226. (800) 521-0707. 2nd edition. 1987. $185.00.

WHO'S WHO IN ENGINEERING. Gordon Davis, editor. Hemisphere Publishing Corporation, 79 Madison Avenue, New York, NY 10016-7892. (800) 821-8312. Sixth edition. 1985. $200.00.

GENERAL WORKS

BUOY ENGINEERING. H.O. Berteaux. Books on Demand, 300 North Zeeb Road, Ann Arbor, MI 48106. (313) 761-4700. $83.00 in paper.

DUTTON'S NAVIGATION AND PILOTING. Elbert S. Maloney. Naval Institute Press, U.S. Naval Institute, Annapolis, MD 21402. (301) 268-6110. Fourth edition. 1985. $32.95.

MARINE ELECTRONIC NAVIGATION. S.F. Appleyard. Methuen, Inc., 29 West 35th Street, New York, NY 10001. (212) 244-3336. 1980. $34.95.

SEAMANSHIP: FUNDAMENTALS FOR THE DECK OFFICER. David O. Dodge and S.E. Kyriss. Naval Institute Press, U.S. Naval Institute, Annapolis, MD 21402. (301) 268-6110. 1981. $16.95.

ONLINE DATA BASES

COMPENDEX. Engineering Information, Inc., 345 East 47th Street, New York, NY 10017. (800) 221-1044 or (212) 705-7615. Engineering and technical literature, 1975 to present. Inquire as to online cost and availability.

NTIS. National Technical Information Service, 5285 Port Royal Road, Springfield, VA 22161. (703) 487-4630. Broad coverage of government sponsored research reports, 1964 to present. Inquire as to online cost and availability.

WILSONLINE. H.W. Wilson and Company, 950 University Avenue, Bronx, NY 10452. (800) 367-6770 or (212) 588-8400. Makes available online versions of the H.W. Wilson indexes including Applied Science and Technology Index, Business Periodicals Index and Readers' Guide to Periodical Literature. Approximately 1980 to present. Inquire as to online cost and availability.

PERIODICALS

JOURNAL OF NAVIGATION. Royal Institute of Navigation. Cambridge University Press, 32 East 57th Street, New York, NY 10022. (800) 872-7423. 1947 to present. Three per year. $100.00 per year.

JOURNAL OF SHIP RESEARCH. Society of Naval Architects and Marine Engineers, One World Trade Center, Suite 1369, New York, NY 10048. (212) 432-0310. 1957 to present. Quarterly. $40.00 per year.

RESEARCH CENTERS AND INSTITUTES

CHARLES STARK DRAPER LABORATORY, INC. 555 Technology Square, Cambridge, MA 02139. (617) 258-1000.

OCEAN ENGINEERING LABORATORY. University of Washington, Mail Stop FU-10, Mechanical Engineering Building, Seattle, WA 98195. (206) 543-7446.

BUSHINGS

See: BEARINGS AND BALL BEARINGS

C

CABLE

See: WIRE AND CABLE

CABLE TELEVISION

See: TELEVISION

CAD (COMPUTER-AIDED DESIGN)

See also: AUTOMATION, CAM, COMPUTERS, ROBOTICS

ABSTRACT SERVICES AND INDEXES

APPLIED SCIENCE AND TECHNOLOGY INDEX. H. W. Wilson Company, 950 University Avenue, Bronx, NY 10452. (800) 367-6670 or (212) 588-8400. Inquire as to cost and availability.

COMPUTER ABSTRACTS. Technical Information Company, Limited, P.O. Box 59, Saint Helier, Jersey British Channel Inlands, England. Monthly. $310.00 per year.

COMPUTER AND CONTROL ABSTRACTS. Institute of Electrical Engineers, London, United Kingdom. Available from: IEEE Service Center, 445 Hoes Lane, Piscataway, NJ 08854. (201) 981-0060. Semimonthly. $775.00 per year.

COMPUTER AND INFORMATION SYSTEMS: AN ABSTRACT JOURNAL PERTAINING TO THE THEORY, DESIGN, FABRICATION AND APPLICATION OF COMPUTER AND INFORMATION SYSTEMS. Cambridge Scientific Abstracts, Incorporated, 5161 River Road, Bethesda, MD 20816. (301) 951-1400. Semimonthly. $590.00 per year.

COMPUTING REVIEWS. Association for Computing Machinery, 11 West 42nd Street, New York, NY 10036. (212) 869-7440. Monthly. $60.00 per year.

CAD/CAM ABSTRACTS. EIC/Intelligence, Incorporated, 48 West 38th Street, New York, NY 10018. (800) 223-6275 or (212) 944-8500. Monthly. $345.00 per year.

ENGINEERING INDEX MONTHLY. Engineering Information, Incorporated 345 East 47th Street, New York, NY 10017. (800) 221-1044 or (212) 705-7600. Monthly, with annual cumulation. $1560.00 per year.

SCIENCE CITATION INDEX. Institute for Scientific Information, 3501 Market St., Philadelphia, PA 19104. (800) 523-1850 or (215) 386-0100. Bimonthly. $6200.00 per year.

ANNUAL REVIEWS AND YEARBOOKS

ADVANCES IN COMPUTER-AIDED ENGINEERING DESIGN. JAI Press, Incorporated, 36 Sherwood Place, Greenwich, CT 06836. (203) 661-7602. Annual. $52.50 per year.

ASSOCIATIONS AND PROFESSIONAL SOCIETIES

AMERICAN AUTOMATIC CONTROL COUNCIL. 1051 Camino Velasquez, Green Valley, AZ 85614. (602) 625-0401.

ASSOCIATION OF COMPUTING MACHINERY (ACM). 11 West 42nd Street, New York, NY 10036. (212) 896-7440.

COMPUTER AND AUTOMATION SYSTEMS ASSOCIATION OF SME. One SME Drive, Box 930, Dearborn, MI 48121. (313) 271-1500.

INSTITUTE OF ELECTRICAL AND ELECTRONIC ENGINEERS (IEEE). 345 East 47th Street, New York, NY 10017. (212) 705-7900.

NUMERICAL CONTROL SOCIETY. 111 East Wasker Drive, Suite 600, Chicago, IL 60601. (312) 644-6610.

SOCIETY FOR COMPUTER APPLICATIONS IN ENGINEERING, PLANNING AND ARCHITECTURE. 358 Hungerford Drive, Rockville, MD 20850. (301) 762-6070.

BIBLIOGRAPHIES

COMPUTER LITERATURE INDEX. Applied Computer Research, Incorporated, P.O. Box 9280, Phoenix, AZ 85068. (602) 995-5929. Quarterly. $125.00 per year.

SCIENTIFIC AND TECHNICAL BOOKS IN PRINT; AN INDEX TO LITERATURE IN SCIENCE AND TECHNOLOGY. R.R. Bowker Company, 205 E. 42nd Street, New York, NY 10017. (800) 521-8110 or (212) 916-1600.

DIRECTORIES AND BIOGRAPHICAL SOURCES

CAD/CAM SOFTWARE DIRECTORY. Technical Database Corporation, 1300 South Frazier, Conroe, TX 77305. (409) 539-9688. Annual. $35.00.

COMPUTER-AIDED DESIGN (CAD) DIRECTORY. Technical Database Corporation, 1300 South Frazier, Conroe, TX 77305. (409) 539-9688. Annual. $40.00.

COMPUTER PERIPHERALS REVIEW. GML Information Services, 594 Marrett Road, Lexington, MA 02173. (617) 861-0515. Two issues per year. $215.00 per year. Directory of computer peripheral equipment manufacturers.

COMPUTERS AND COMPUTING INFORMATION RESOURCES DIRECTORY. Gale Research Company, Book Tower, Detroit, MI 48226. (800) 521-0707. 1986. $160.00.

CAD (COMPUTER-AIDED DESIGN)

DIRECTORY OF CONSULTANTS IN COMPUTER SYSTEMS. Research Publications, Incorporated, 12 Lunar Drive, Woodbridge, CT 06525. (203) 397-2600. Annual. $85.00 per year.

RESEARCH CENTERS DIRECTORY. Gale Research Company, Detroit, MI 48226. (800) 521-0707. Eleventh edition. 1987.

ROBOTICS-CAD/CAM DIRECTORY. Technical Data Publishing Corporation, 53 Lake Shore Drive, Rockaway, NJ 07866. (201) 625-9647. Annual. $3.00.

WHO'S WHO IN FRONTIER SCIENCE AND TECHNOLOGY. Marquis Who's Who, Incorporated, 200 East Ohio St., Chicago, IL 60611. (800) 428-3898 or (312) 787-2008.

WHO'S WHO IN TECHNOLOGY TODAY. Reston Publishing Company, Incorporated, c/o Prentice-Hall, Incorporated, Englewood Cliffs, NJ 07632. (800) 262-6868. Biennial. Five volumes. $425.00. Covers the fields of electronics, computer science, physics, optics, chemistry, biotechnology, mechanics, energy, and earth science.

ENCYCLOPEDIAS AND DICTIONARIES

COMPUTER DICTIONARY. Charles J. Sippl. Howard W. Sams and Company, Incorporated, 4300 West 62nd Street, Indianapolis, IN 46268. (800) 428-7267 or (317) 298-5564. Fourth edition. 1985. $17.95.

DICTIONARY OF COMPUTING. Oxford University Press, 200 Madison Avenue, New York, NY 10016. (212) 679-7300. Second edition. 1986. $29.95.

ENCYCLOPEDIA OF COMPUTER SCIENCE AND ENGINEERING. Anthony Ralston, editor. Van Nostrand Reinhold Book Company, 115 Fifth Avenue, New York, NY 10003. (800) 543-2681. Second edition. 1982. $89.95.

ENCYCLOPEDIA OF COMPUTER SCIENCE AND TECHNOLOGY. Jack Belzer, Albert G. Holzman, and Allan Kent. Marcel Dekker, Incorporated, 270 Madison Avenue, New York, NY 10016. (212) 696-9000. Sixteen volumes. $115.00 per volume.

MCGRAW-HILL DICTIONARY OF COMPUTERS. McGraw-Hill Book Company, 1221 Avenue of the Americas, New York, NY 10020. (212) 512-2000. 1985. $17.50.

GENERAL WORKS

CAD-CAM: COMPUTER-AIDED DESIGN AND MANUFACTURING. Mikell P. Groover and Emory W. Zimmers. Prentice-Hall, Incorporated, Englewood Cliffs, NJ 07632. (800) 562-0245 or (201) 599-2000. 1984. $36.95.

CAD-CAM. SOCIETY OF MANUFACTURING ENGINEERS. One SME Drive, Box 930, Dearborn, MI 48121. (313) 271-1500. Second edition. 1985. $49.00.

CAD-CAM: INTEGRATION AND INNOVATION. Society of Manufacturing Engineers. One SME Drive, Box 930, Dearborn, MI 48121. (313) 271-1500. 1985. $49.00.

CAD-CAM WITH PERSONAL COMPUTERS. Patrick R. Carbury. TAB Books, Incorporated, Monterey Lue, Blue Ridge Summit, PA 17214. (717) 794-2191. 1985. $21.95.

CAD: PRINCIPLES AND APPLICATIONS. Paul Barr, et al. Prentice-Hall, Incorporated, Englewood Cliffs, NJ 07632. (800) 562-0245 or (201) 599-2000. 1985. $21.95.

COMPUTER-AIDED DESIGN AND COMPUTER-AIDED MANUFACTURING: THE CAD-CAM REVOLUTION. John K. Krouse, editor. (What Every Engineer Should Know Series, Volume 10). Marcel Dekker, Incorporated, 270 Madison Avenue, New York, NY 10016. (800) 228-1160 or (212) 696-9000. 1982. $24.75.

COMPUTER-AIDED DESIGN AND MANUFACTURE. C. B. Besant. Halsted Press, Division of John Wiley and Sons, Incorporated, 605 Third Avenue, New York, NY 10158. (800) 526-5368 or (212) 850-6000. Second edition. 1982. $26.95 in paper.

HANDBOOKS AND MANUALS

CAD/CAM HANDBOOK. Eric Teicholz. McGraw-Hill Book Company, 1221 Avenue of the Americas, New York, NY 10020. (800) 628-0004. 1985. $54.00.

HANDBOOK OF COMPUTERS AND COMPUTING. Arthur H. Seidman, editor. Van Nostrand Reinhold Book Company, 115 Fifth Avenue, New York, NY 10003. (800) 543-2681. 1984. $79.95.

MCGRAW-HILL COMPUTER HANDBOOK. Harry Helms, editor. McGraw-Hill Book Company, 1221 Avenue of the Americas, New York, NY 10020. (212) 512-2000. 1983. $84.50.

ONLINE DATA BASES

COMPENDEX. Engineering Information, Incorporated, 345 East 47th Street, New York, NY 10017. (800) 221-1044 or (212) 705-7615. Engineering and technical literature, 1975 to present. Inquire as to cost and availability.

DISSERTATION ABSTRACTS ONLINE. University Microfilms International, 300 North Zeeb Road, Ann Arbor, MI 48106. (800) 521-0600 or (313) 761-4700. Scope includes virtually all doctoral dissertations accepted at accredited American institutions from 1861 to present in 252 subject areas. Inquire as to cost and availability.

INSPEC. INSPEC Marketing Department, Institute of Electrical and Electronics Engineers, Incorporated, IEEE Service Department, 445 Hoes Lane, Piscataway, NJ 08854. (201) 981-0060. Inquire as to on-line cost and availability.

NTIS. National Technical Information Service, 5285 Port Royal Road, Springfield, VA 22161. (703) 487-4630. Broad coverage of government sponsored research reports, 1964 to present. Inquire as to cost and availability.

SCISEARCH. Institute for Scientific Information, 3501 Market Street, Philadelphia, PA 19104. (800) 523-1850 or (215) 386-0100. Broad multidisciplinary title and author index to the international literature of science and technology, 1974 to present. Inquire as to cost and availability.

OTHER SOURCES

ALL ABOUT CAD/CAM SYSTEMS. Datapro Research Corporation, 1805 Underwood Boulevard, Delran, NJ 08075. (800) 257-9406 or (609) 764-0100. Annual. $25.00.

ALL ABOUT GRAPHIC DISPLAY DEVICES. Datapro Research Corporation, 1805 Underwood Boulevard, Delran, NJ 08075. (800) 257-9406 or (609) 764-0100. Annual. $15.00.

COMPUTER DISPLAY REVIEW. GML Information Services, 594 Marrett Road, Lexington, MA 02173. (617) 861-0515. $650.00 per year. Loose-leaf volumes with periodic supplements. Gives specifications on computer dislplay systems.

PERIODICALS

ACM TRANSACTIONS ON COMPUTER SYSTEMS. Association for Computing Machinery, 11 West 42nd Street, New York, NY 10036. (212) 869-7440. Quarterly. $70.00 per year.

ACM TRANSACTIONS ON GRAPHICS. Association for Computing Machinery, 11 West 42nd Street, New York, NY 10036. (212) 869-7440. Quarterly. $65.00 per year.

CAD/CAM AND ROBOTICS. Kerrwill Publications, Limited, 501 Oakdale Downsview, Ontario M3N 1W7, Canada. Quarterly. $27.00 per year.

CADFILE. Northwest Computer Society, Box 4193, Seattle, WA 98104. Monthly. $39.00 per year.

CAE: COMPUTER-AIDED ENGINEERING; DATA BASE APPLICATIONS IN DESIGN AND MANUFACTURING. Penton/IPC, Penton Plaza, 1111 Chester Avenue, Cleveland, OH 44114. (216) 696-7000. Monthly. $30.00 per year.

CIM TECHNOLOGY. Society of Manufacturing Engineers. One SME Drive, Box 930, Dearborn, MI 48121. (313) 271-1500. Quarterly. $60.00 per year.

CIME: COMPUTERS IN MECHANICAL ENGINEERING. American Society of Mechanical Engineers, 345 East 47th Street, New York, NY 10017. (212) 705-7722. Bimonthly. $35.00 per year.

COMPUTER-AIDED DESIGN, ENGINEERING, AND DRAFTING. Auerbach Publishers, Incorporated, 6560 North Park Drive, Pennsauken, NJ 08109. (609) 662-2070. Quarterly. $125.00 per year.

COMPUTERS FOR DESIGN AND CONSTRUCTION. Meta Data Publishing Corporation, 310 East 44th Street, Suite 1124, New York, NY 10017. (212) 867-2080. Bimonthly. $48.00 per year.

DESIGN COMPUTING. John Wiley and Sons, Incorporated, 605 Third Avenue, New York, NY 10158. (800) 526-5368 or (212) 850-6000. Quarterly. $100.00 per year.

IEEE COMPUTER GRAPHICS AND APPLICATIONS. Institute of Electrical and Electronic Engineers. IEEE Service Center, 445 Hoes Lane, Piscataway, NJ 08854. (201) 981-0060. Monthly. $155.00 per year.

IEEE JOURNAL OF ROBOTICS AND AUTOMATION. Institute of Electrical and Electronic Engineers. IEEE Service Center, 445 Hoes Lane, Piscataway, NJ 08854. (201) 981-0060. Quarterly. $65.00 per year.

S. KLEIN NEWSLETTER ON COMPUTER GRAPHICS (HARVARD NEWSLETTER ON COMPUTER GRAPHICS). Technology and Business Communications, Incorporated, 730 Boston Post Road, Sudbury, MA 01776. (617) 443-4671. Semimonthly. $178.00 per year.

RESEARCH CENTERS AND INSTITUTES

ALBERT H. CASE CENTER FOR COMPUTER-AIDED DESIGN. Michigan State University, College of Engineering, 236 Engineering Building, East Lansing, MI 48824. (517) 355-6453.

CENTER FOR CAD/CAM. Purdue University, 799 West Michigan Street, Indianapolis, IN 46202. (317) 264-8627.

CENTER FOR COMPUTER-AIDED ENGINEERING (CCAE). University of Virginia, School of Engineering and Applied Science, Charlottesville, VA 22901. (804) 924-6217.

COMPUTER-AIDED DESIGN LABORATORY (CAD LAB). University of Maryland, Engineering Research Center, College of Engineering, College Park, MD 20742. (301) 454-7941.

SRC-CMU RESEARCH CENTER FOR COMPUTER-AIDED DESIGN. Carnegie-Mellon University, Pittsburgh, PA 15213. (412) 578-8889.

CADMIUM

See also: ALLOYS, ALUMINUM, BRASS AND BRONZE, COPPER, MATERIALS SCIENCE, METALLURGY

ABSTRACT SERVICES AND INDEXES

ALLOYS INDEX. American Society for Metals, 9639 Kinsman Road, Metals Park, OH 44073. (216) 338-5151. $130.00 per year.

APPLIED MECHANICS REVIEWS. American Society of Mechanical Engineers, 345 East 47th Street, New York, NY 10017. (212) 705-7722. Monthly. $380.00 per year. Critical reviews of the world literature in applied mechanics and related engineering science.

APPLIED SCIENCE AND TECHNOLOGY INDEX. H.W. Wilson Company, 950 University Avenue, Bronx, NY 10452. (800) 367-6670 or (212) 588-8400. Inquire as to cost and availability.

CADMIUM ABSTRACTS. Cadmium Association, 34 Berkeley Square, London, W1X 6AJ, England. 1977 to present. Quarterly $71.50 per year.

CHEMICAL ABSTRACTS. Chemical Abstracts Service, 2540 Olentangy Road, Post Office Box 3012, Columbus, OH 43210. (800) 848-6538 or (614) 421-3600. Weekly. $9200.00 per year.

CORROSION ABSTRACTS: ABSTRACTS OF THE WORLD'S LITERATURE ON CORROSION AND CORROSION MITIGATION. Association of Corrosion Engineers, Box 218340, Houston, TX 77218. (713) 492-0535. 1962 to present. Bimonthly. $250.00 per year.

CURRENT CONTENTS: ENGINEERING, TECHNOLOGY. Institute for Scientific Information, 3501 Market Street, Philadelphia, PA 19104. (800) 523-1850 or (215) 386-0100. $272.00 per year.

ENGINEERING INDEX MONTHLY. Engineering Information, Incorporated 345 East 47th Street, New York, NY 10017. (800) 221-1044 or (212) 705-7600. Monthly, with annual cumulation. $1560.00 per year.

INTERNATIONAL AEROSPACE ABSTRACTS. American Institute of Aeronautics and Astronaustics, Technical Information Service, 555 West 57th Street, New York, NY 10019. (212) 247-6500. 1961 to present. Semi-monthly. $950.00 per year with indexes.

METALS ABSTRACTS. American Society for Metals, 9639 Kinsman Road, Metals Park, OH 44073. (216) 338-5151. Monthly. $890.00.

METALS ABSTRACTS INDEX. American Society for Metals, 9639 Kinsman Road, Metals Park, OH 44073. (216) 338-5151. Monthly. (Sold only to subscribers of Metals Abstracts).

WORLD ALUMINUM ABSTRACTS. Aluminum Association, 818 Connecticut Avenue, NW, Washington, DC 20006. (202) 862-5100. 1968 to present. Monthly. $165.00 per year.

ASSOCIATIONS AND PROFESSIONAL SOCIETIES

AMERICAN SOCIETY FOR METALS. Metals Park, Oh 44073. (216) 338-5151.

CADMIUM COUNCIL. 292 Madison Avenue, New York, NY 10017. (212) 578-4750.

METALLURGICAL SOCIETY OF THE AIME (AMERICAN INSTITUTE OF MINING, METALLURGICAL AND PETROLEUM ENGINEERS). 420 Commonwealth Drive, Warrendale, PA 15086. (412) 776-9080.

NATIONAL ASSOCIATION OF CORROSION ENGINEERS. Box 218340, Houston, TX 77218. (713) 492-0535.

DIRECTORIES AND BIOGRAPHICAL SOURCES

DUN'S INDUSTRIAL GUIDE - THE METAL WORKING DIRECTORY. Dun and Bradstreet, Incorporated, Three Century Drive, Parsippany, NJ 07054. (201) 455-0900. Annual. $550.00.

INTERNATIONAL RESEARCH CENTERS DIRECTORY 1986-1987. Gale Research Company, Book Tower, Detroit, MI 48226. (800) 521-0707. Third edition. 1986.

MATERIALS PERFORMANCE - NACE CORROSION ENGINEERING BUYER'S GUIDE ISSUE. National Association of Corrosion Engineers, Box 218340, Houston, TX 77218. (713) 492-0535. Annual, July. $15.00.

METAL PRODUCTS DIRECTORY. American Business Directories, Incorporated, Division of American Business Lists, Incorporated, 5707 South 86th Circle, Omaha, NE 68127. (402) 331-7169. 1986. $80.00.

RESEARCH CENTERS DIRECTORY. Gale Research Company, Book Tower, Detroit, MI 48226. (800) 521-0707. Eleventh edition. 1987.

SCIENTIFIC AND TECHNICAL ORGANIZATIONS AND AGENCIES DIRECTORY. Gale Research Company, Book Tower, Detroit, MI 48226. (800) 521-0707. 1985.

WHO'S WHO IN ENGINEERING. Engineers Joint Council, 345 East 47th Street, New York, NY 10017. (212) 705-7010. 1985. $200.00.

WHO'S WHO IN FRONTIER SCIENCE AND TECHNOLOGY. Marquis Who's Who, Incorporated, 200 East Ohio Street, Chicago, IL 60611. (800) 428-3898 or (312) 787-2008.

WHO'S WHO IN TECHNOLOGY TODAY. Reston Publishing Company, Incorporated, c/o Prentice-Hall, Incorporated, Englewood Cliffs, NJ 07632. (800) 262-6868. Biennial. Five volumes. $425.00. Covers the fields of electronics, computer science, physics, optics, chemistry, biotechnology, mechanics, energy, and earth science.

ENCYCLOPEDIAS AND DICTIONARIES

ENCYCLOPEDIA OF PHYSICAL SCIENCE AND TECHNOLOGY. Academic Press, Incorporated, Orlando, FL 32887. (800) 321-5068 or (305) 345-2734. Fifteen volumes, 1986.

MCGRAW-HILL DICTIONARY OF ENGINEERING. Sybil P. Parker, editor. McGraw-Hill Book Company, 1221 Avenue of the Americas, New York, NY 10020. (212) 512-2000. 1984. $39.95.

MCGRAW-HILL ENCYCLOPEDIA OF SCIENCE AND TECHNOLOGY. McGraw-Hill Book Company, 1221 Avenue of the Americas, New York, NY 10020. (212) 512-2000.

GENERAL WORKS

BASIC CORROSION AND OXIDATION. John M. West. John Wiley and Sons, Incorporated, 605 Third Avenue, New York, NY 10158. (800) 526-5368 or (212) 850-6000. 1986. $44.95.

CADMIUM IN THE ENVIRONMENT. Mislin and Ravera. Birkhauser Boston, Incorporated, 380 Green Street, Cambridge, MA 02139. (617) 876-2333. 1986. $34.00.

CHEMISTRY OF THE ELEMENTS. N.N. Greenwood and A. Earnshaw. Pergamon Publishing, Incorporated, Maxwell House, Fairview Park, Elmsford, NY 10523. (914) 592-7700. 1984. $143.00.

CORROSION AND CORROSION CONTROL. H.H. Uhlig and R. Winston Revie. John Wiley and Sons, Incorporated, 605 Third Avenue, New York, NY 10158. (800) 526-5368 or (212) 850-6000. Third edition. 1985. $42.50.

ESSENTIAL METALLURGY FOR ENGINEERS. W. Alexander, G. Davies, and K. Reynolds. Van Nostrand Reinhold, 115 Fifth Avenue, New York, NY 10003. (800) 543-2681. 1985. $17.95.

PHYSICAL METALLURGY. P. Haasen. Cambridge University Press, 32 East 57th Street, New York, NY 10022. (212) 688-8885. 1986. $24.95 in paper.

HANDBOOKS AND MANUALS

CORROSION RESIDENT MATERIALS HANDBOOK. D.J. De Renzo, editor. Noyes Data Corporation, Mill Road at Grand Avenue, Park Ridge, NJ 07656. (201) 391-8484. 1986. $125.00.

SMITHELL'S METALS REFERENCE BOOK. Eric A. Brandes, editor. Butterworth Publishers, 80 Montvale Avenue, Stoneham, MA 02180. (800) 325-4177. Sixth edition. 1983. $210.00.

ONLINE DATA BASES

COMPENDEX. Engineering Information, Incorporated, 345 East 47th Street, New York, NY 10017. (800) 221-1044 or (212) 705-7615. Engineering and technical literature, 1975 to present. Inquire as to cost and availability.

CORROSION DATA BASE. Marcel Dekker, Incorporated, 270 Madison Avenue, New York, NY 10016. (212) 696-9000. Available online through System Development Corporation (SDC). Inquire as to online cost and availability.

INSPEC. INSPEC Marketing Department, Institute of Electrical and Electronics Engineers, Incorporated, IEEE Service Department, 445 Hoes Lane, Piscataway, NJ 08854. (201) 981-0060. Inquire as to on-line cost and availability.

METADEX. Metals Information, American Society for Metals, Metals Park, OH 44073. (216) 338-5151. (Metals Abstracts/Alloys Index). A worldwide literature on the science and practice of metallurgy, 1966 to present. Inquire as to online cost and availability.

NASA. National Aeronautics and Space Administration, Scientific and Technical Information Branch, 300 7th Street, SW, Washington, DC 20546. Citations and abstracts of aerospace literature, 1962 to present. Inquire as to cost and availability.

NON-FERROUS METALS ABSTRACTS. British Non-Ferrous Metals Technology Centre, Grove Laboratories, Denchworth Road, Wantage, Oxfordshire, England OX12 9 BJ. Citations and abstracts on non-online cost and availability.

NTIS. National Technical Information Service, 5285 Port Royal Road, Springfield, VA 22161. (703) 487-4630. Broad coverage of government sponsored research reports, 1964 to present. Inquire as to cost and availability.

WILSONLINE. H.W. Wilson Company, 950 University Avenue, Bronx, NY 10452. (800) 367-6770 or (212) 588-8400. Makes available online versions of the printed H.W. Wilson Indexes including Applied Science and Technology Index, Business Periodicals Index, and Readers' Guide to Periodical Literature. Period covered is generally 1983 to present. Inquire as to cost and availability.

OTHER SOURCES

MATERIALS SCIENCE AND METALLURGY. National Technical Information Service (NTIS), 5285 Port Royal Road, Springfield, Va 22161. (703) 487-4630. Translations and abstracts of foreign language technical media. Irregular. $40.00 per year.

PERIODICALS

ALLOYS DIGEST. Engineering Publications Incorporated, Box 823, Upper Montclair, NJ 07043. (201) 746-7930. Monthly. $50.00 per year.

CORROSION. National Association of Corrosion Engineers, Box 218340, Houston, TX 77218. (713) 492-0535. 1945 to present. Monthly. $75.00 per year.

CORROSION PREVENTION AND CONTROL. Scientific Surveys, Limited, Box 21, Beaconsfield, Bucks, HP9 1NS, England. 1954 to present. Bimonthly. $65.00 per year.

CORROSION SCIENCE. Pergamon Press, Incorporated, Journals Division. Maxwell House. Fairview Park, Elmsford, NY 10523. (914) 592-7700. 1961 to present. Monthly. $300.00 per year.

JOURNAL OF METALS. Metallurgical Society of the AIME (American Institute of Mining, Metallurgical and Petroleum Engineers), 420 Commonwealth Drive, Warrendale, PA 15086. (412) 776-9080. Monthly. $40.00 per year.

LIGHT METAL AGE. Fellom Publishing Company, 693 Mission Street, San Francisco, CA 94105. (415) 781-1431. Bimonthly. $20.00 per year.

METALLURGICAL TRANSACTIONS. Metallurgical Society of the AIME (American Institute of Mining, Metallurgical and Petroleum Engineers), 420 Commonwealth Drive, Warrendale, PA 15086. (412) 776-9080. Monthly. $95.00 per year.

METALS WEEK. McGraw-Hill Book Company, Incorporated, 1221 Avenue of the Americas, New York, NY 10020. (212) 997-2823. Weekly. $527.00 per year.

RESEARCH CENTERS AND INSTITUTES

INTERNATIONAL LEAD-ZINC RESEARCH ORGANIZATION. 292 Madison Avenue, New York, NY 10017. (212) 532-2373.

LAWRENCE BERKELEY LABORATORY. Center for Advanced Materials, 1 Cyclotron Road, Berkeley, CA 94720. (415) 486-4755.

PHYSICAL METALLURGY RESEARCH LABORATORIES. Canada Centre for Mineral and Energy Technology, 555 Booth Street, Ottawa, ON Canada K1A 0G1. (613) 995-4807.

TEXAS A AND M UNIVERSITY. Mechanics and Materials Center, ERC Building, College Station, TX 77843. (409) 845-7512.

UNIVERSITY OF CONNECTICUT. Institute of Materials Science, Storrs, CT 06268. (203) 486-4623.

UNIVERSITY OF FLORIDA. Department of Materials Science and Engineering, Gainesville, FL 32601. (904) 392-1454.

UNIVERSITY OF MINNESOTA. Corrosion Research Center, 112 Mines and Metallurgy Building, Minneapolis, MN 55455. (612) 373-4864.

UNIVERSITY OF WASHINGTON. Institute of Environmental Studies, Engineering Annex, FM-12, Seattle, WA 98195. (206) 543-1812.

CAE (COMPUTER-AIDED ENGINEERING)

See: CAD (COMPUTER AIDED DESIGN)

CAISSONS

See: CIVIL ENGINEERING

CALCIUM

See also: ANALYTICAL CHEMISTRY, CARBONATES, CHEMISTRY, ELEMENTS, GEOCHEMISTRY, INORGANIC CHEMISTRY, METALLURGY

ABSTRACT SERVICES AND INDEXES

APPLIED SCIENCE AND TECHNOLOGY INDEX. H.W. Wilson and Company, 950 University Avenue, Bronx, NY 10452. (800) 367-6670 or (212) 588-8400. Monthly. Inquire as to cost and availability.

CHEMICAL ABSTRACTS. American Chemical Society, Chemical Abstracts Service, Box 3012, Columbus, OH 43210. (614) 421-3600. 1907 to present. Weekly. $9500.00 per year.

CONFERENCE PAPERS INDEX. Cambridge Scientific Abstracts, 5161 River Road, Bethesda, MD 20816. 1972 to present. Monthly. Inquire as to cost and availability.

CURRENT CONTENTS: PHYSICAL, CHEMICAL AND EARTH SCIENCES. Institute for Scientific Information, 3501 Market Street, Philadelphia, PA 19104. (800) 523-1850 or (215) 386-0100. Weekly. $275.00 per year.

GENERAL SCIENCE INDEX. H.W. Wilson and Company, 950 University Avenue, Bronx, NY 10452. (800) 367-6670 or (212) 588-8400. 1978 to present. Monthly. Inquire as to cost and availability.

INDEX TO SCIENTIFIC AND TECHNICAL PROCEEDINGS. Institute for Scientific Information, 3501 Market Street, Philadelphia, PA 19104. (800) 523-1850 or (215) 386-0100. 1978 to present. Monthly. $775.00 per year.

INDEX TO SCIENTIFIC REVIEWS. Institute for Scientific Information, 3501 Market Street, Philadelphia, PA 19104. (800) 523-1850 or (215) 386-0100. 1974 to present. Semi-annual. $550.00 per year.

SCIENCE CITATION INDEX. Institute for Scientific Information, 3501 Market Street, Philadelphia, PA 19104. (800) 523-1850 or (215) 386-0100. Six times per year. $6200.00 per year.

ASSOCIATIONS AND PROFESSIONAL SOCIETIES

AMERICAN CHEMICAL SOCIETY. 1155 16th Street, N.W., Washington, DC 20036. (202) 872-4600.

AMERICAN INSTITUTE OF CHEMICAL ENGINEERS. 345 East 47th Street, New York, NY 10017. (212) 705-7338.

THE METALLURICAL SOCIETY. 420 Commonwealth Drive, Warrendale, PA 15096. (412) 776-9000.

AMERICAN SOCIETY FOR METALS. Metals Park, OH 44073. (216) 338-5151.

ASSOCIATION OF OFFICIAL ANALYTICAL CHEMISTS. 1111 North 19th Street, Suite 210, Arlington, VA 22209. (703) 522-3032.

IRON AND STEEL SOCIETY OF AIME. 410 Commonwealth Drive, Warrendale, PA 15086. (412) 776-1535.

DIRECTORIES AND BIOGRAPHICAL SOURCES

AMERICAN MEN AND WOMEN OF SCIENCE. R.R. Bowker, Inc., Order Department, 245 West 17th Street, New York, NY 10011. (800) 521-8110. Eight volumes. 1986. $595.00 for set.

INTERNATIONAL RESEARCH CENTERS DIRECTORY 1988-89. Darren L. Smith, editor. Gale Research Company, Book Tower, Detroit, MI 48226. (800) 521-0707. 4th edition. 1987. $360.00.

1987 DIRECTORY OF ENGINEERING SOCIETIES AND RELATED ORGANIZATIONS. Gordon Davis, editor. Hemisphere

Publishing Corporation, 1010 Vermont Avenue, NW, Washington, DC 20005. (800) 526-0275. 12th edition. 1987. $100.00.

RESEARCH CENTERS DIRECTORY 1988. Gale Research Company, Book Tower, Detroit, MI 48226. (800) 521-0707. 12th edition. 1987. $365.00 for set.

SCIENTIFIC AND TECHNICAL ORGANIZATIONS AND AGENCIES DIRECTORY. Margaret Labash Young, editor. Gale Research Company, Book Tower, Detroit, MI 48226. (800) 521-0707. 2nd edition. 1987. $185.00.

WHO'S WHO IN ENGINEERING. Gordon Davis, editor. Hemisphere Publishing Corporation, 1010 Vermont Avenue, NW, Washington, DC 20005. (800) 526-0275. 6th edition. 1985. $200.00.

ENCYCLOPEDIAS AND DICTIONARIES

ENCYCLOPEDIA OF CHEMISTRY. Douglas M. Considine, editor. Van Nostrand Reinhold Company, Inc., 135 West 50th Street, New York, NY 10020. (800) 543-2681. 4th edition. 1984. $98.95.

MCGRAW-HILL ENCYCLOPEDIA OF SCIENCE AND TECHNOLOGY. McGraw-Hill Book Company, 1221 Avenue of the Americas, New York, NY 10020. (212) 512-2000. 6th edition. 1987. $1600.00.

THESAURUS OF SCIENTIFIC, TECHNICAL, AND ENGINEERING TERMS. Hemisphere Publishing Corporation, 1010 Vermont, NW, Washington, DC 20005. (800) 526-0275. 1988. $125.00.

GENERAL WORKS

CHEMISTRY OF THE ELEMENTS. N.N. Greenwood and A. Earnshaw. Pergamon Press, Inc., Maxwell House, Fairview Park, Elmsford, NY 10523. (914) 592-7700. 1984. $145.00.

INORGANIC CHEMISTRY. William W. Porterfield. Addison-Wesley Publishing Company, Inc., 1 Jacob Way, Reading, MA 01867. (617) 944-3700. 1983. $36.95.

HANDBOOKS AND MANUALS

CRC HANDBOOK OF CHEMISTRY AND PHYSICS. Robert C. Weast, editor. CRC Press, 2000 Corporate Boulevard, Boca Raton, FL 33431. (800) 272-7737. 68th edition. 1987. $69.95.

HANDBOOK OF APPLIED CHEMISTRY. Vollrath Hopp and Ingo Hennig. Hemisphere Publishing Corporation, 79 Madison Avenue, New York, NY 10016-7892. Order from: Taylor and Francis/Hemisphere Distribution Center, 242 Cherry Street, Philadelphia, PA 19106-1906. (800) 821-8312. 1983. $49.95.

LANGE'S HANDBOOK OF CHEMISTRY. John A. Dean, editor. McGraw-Hill Book Company, 1221 Avenue of the Americas, New York, NY 10020. (212) 512-2000. 1985. $59.50.

ONLINE DATA BASES

CA SEARCH. Chemical Abstracts Service, P.O. Box 3012, Columbus, OH 43120. (800) 848-6538 or (614) 421-3600. Comprehensive guide to chemical literature, 1972 to present. Inquire as to online cost and availability.

COMPENDEX Engineering Information, Inc., 345 East 47th Street, New York, NY 10017. (800) 221-1044 or (212) 705-7615. Engineering and technical literature, 1975 to present. Inquire as to online cost and availability.

DISSERTATION ABSTRACTS ONLINE. University Microfilms International, 300 North Zeeb Road, Ann Arbor, MI 48106. (800) 521-0600 or (313) 761-4700. Scope includes virtually all doctoral dissertations accepted at accredited American institutions from 1861 to present in over 250 subject areas. Inquire as to online cost and availability.

NTIS. National Technical Information Service, 5285 Port Royal Road, Springfield, VA 22161. (703) 487-4630. Broad coverage of government sponsored research reports, 1964 to present. Inquire as to online cost and availability.

SCISEARCH. Institute for Scientific Information, 3501 Market Street, Philadelphia, PA 19104. (800) 523-1850 or (215) 386-0100. Broad multidisciplinary title and author index to the international literature of science and technology, 1974 to present. Inquire as to online cost and availability.

PERIODICALS

AMERICAN CHEMICAL SOCIETY JOURNAL. American Chemical Society, 1155 16th Street, N.W., Washington, DC 20036. (800) 424-6747. 1879 to present. Biweekly. $275.00 per year.

INORGANIC CHEMISTRY. American Chemical Society, 1155 16th Street, N.W., Washington, DC 20036. (800) 424-6747. 1962 to present. Biweekly. $300.00 per year.

INORGANICA CHIMICA ACTA. Elsevier Science Publishing Company, Inc., 52 Vanderbilt Avenue, New York, NY 10017. (212) 370-5520. 1967 to present. Biweekly. $1500.00 per year.

JOURNAL OF THE CHEMICAL SOCIETY. CHEMICAL COMMUNICATIONS. Royal Society of Chemistry, Burlington House, London, W1V 0BN, England. 1965 to present. Semimonthly. $385.00 per year.

POLYHEDRON. Pergamon Press, Inc., Maxwell House, Fairview Park, Elmsford, NY 10523. (914) 592-7700. 1982 to present. Monthly. $595.00.

ROYAL SOCIETY OF CHEMISTRY. JOURNAL. DALTON TRANSACTIONS. Royal Society of Chemistry, Burlington House, London, W1V 0BN, England. 1972 to present. Monthly. $585.00 per year.

RESEARCH CENTERS AND INSTITUTES

CHEMICAL LABORATORIES. Harvard University, Oxford Street, Cambridge, MA 02138. (617) 495-4283.

CHEMICAL RESEARCH LABORATORY. Brown University, Providence, RI 02912. (401) 863-2256.

UNIVERSITY/INDUSTRY CHEMICAL RESEARCH CENTER. Mississippi State University, Department of Chemistry, P.O. Drawer CH, Mississippi State, MS 39762. (601) 325-3584.

CALCULUS

See also: ANALYTICAL GEOMETRY, MATHEMATICS

ABSTRACT SERVICES AND INDEXES

ABSTRACTS OF PAPERS PRESENTED TO THE AMERICAN MATHEMATICAL SOCIETY. American Mathematical Society, Post Office Box 6248, Providence, RI 02940. (800) 556-7774 or (401) 272-9500. Semimonthly. $42.00 per year.

APPLIED SCIENCE AND TECHNOLOGY INDEX. H.W. Wilson Company, 950 University Avenue, Bronx, NY 10452. (800) 367-6670 or (212) 588-8400. Inquire as to cost and availability.

COMPUMATH CITATION INDEX. Institute for Scientific Information, 3501 Market Street, Philadelphia, PA 19104. (800) 523-1850 or (215) 386-0100. Three times per year. $875.00 per year.

CURRENT MATHEMATICAL PUBLICATIONS. American Mathematical Society, Post Office Box 6248, Providence, RI 02940. (800) 556-7774 or (401) 272-9500. Seventeen times per year. $220.00 per year.

MATHEMATICAL REVIEWS. American Mathematical Society, Post Office Box 6248, Providence, RI 02940. (800) 556-7774 or (401) 272-9500. $2750.00 per year.

PHYSICS ABSTRACTS. Institute of Electrical Engineers, London, United Kingdom. Available from: Institute of Electrical and Electronic Engineers (IEEE), 345 East 47th Street, New York, NY 10017. (212) 705-7900. Inquire as to cost and availability.

SCIENCE CITATION INDEX. Institute for Scientific Information, 3501 Market Street, Philadelphia, PA 19104. (800) 523-1850 or (215) 386-0100. Inquire as to cost and availability.

ANNUAL REVIEWS AND YEARBOOKS

ADVANCE IN APPLIED MATHEMATICS. Academic Press, Incorporated, 6277 Sea Harbor Drive, Orlando, FL 32821. (800) 321-5068.

ADVANCES IN MATHEMATICS. Academic Press, Incorporated, 6277 Sea Harbor Drive, Orlando, FL 32821. (800) 321-5068. Inquire as to cost and availability.

ASSOCIATIONS AND PROFESSIONAL SOCIETIES

AMERICAN MATHEMATICAL SOCIETY. Post Office Box 6248, Providence, RI 02940. (401) 272-9500.

MATHEMATICAL ASSOCIATION OF AMERICA. 1529 18th Street, NW, Washington, DC 20036. (202) 387-5200.

SOCIETY FOR INDUSTRIAL AND APPLIED MATHEMATICS. 1400 Architects Building, 117 South 17th Street, Philadelphia, PA 19103. (215) 564-2929.

BIBLIOGRAPHIES

SCIENCE BOOKS AND FILMS. American Association for the Advancement of Science, 1333 H Street, NW, Washington, DC 20005. Quarterly. $20.00 per year.

SCIENTIFIC AND TECHNICAL BOOKS AND SERIALS IN PRINT 1988: AN INDEX TO LITERATURE IN SCIENCE AND TECHNOLOGY. R.R. Bowker Company, 205 East 42nd Street, New York, NY 10017. (800) 521-8110 or (212) 916-1600. $175.00.

DIRECTORIES AND BIOGRAPHICAL SOURCES

AMERICAN MEN AND WOMEN OF SCIENCE. Physical and Biological Sciences. Fifteenth edition. R.R. Bowker Company, 205 East 42nd Street, New York, NY 10017. (800) 521-0110 or (212) 916-1600. $565.00.

BIOGRAPHICAL DICTIONARY OF SCIENTISTS. T.I. Williams. Halsted Press, 605 Third Avenue, New York, NY 10158. (800) 526-5368 or (212) 850-6418. Third edition. 1982. $29.95.

WHO'S WHO IN FRONTIER SCIENCE AND TECHNOLOGY. Marquis Who's Who, Incorporated, 200 East Ohio Street, Chicago, IL 60611. (800) 428-3898 or (312) 787-2008.

WHO'S WHO IN TECHNOLOGY TODAY. Reston Publishing Company, Incorporated, c/o Prentice-Hall, Incorporated, Englewood Cliffs, NJ 07632. (800) 262-6868. Biennial. Five volumes. $425.00. Covers the fields of electronics, computer science, physics, optics, chemistry, biotechnology, mechanics, energy, and earth science.

DIRECTORIES

INTERNATIONAL RESEARCH CENTERS DIRECTORY 1988-1989. Gale Research Company, Book Tower, Detroit, MI 48226. (800) 521-0707. Fourth edition. 1987. $360.00.

RESEARCH CENTERS DIRECTORY. Gale Research Company, Book Tower, Detroit, MI 48226. (800) 521-0707. Twelfth edition. 1987. $365.00 per set.

SCIENTIFIC AND TECHNICAL ORGANIZATIONS AND AGENCIES DIRECTORY. Gale Research Company, Book Tower, Detroit, MI 48226. (800) 521-0707. Second edition. 1987. $185.00.

WORLD GUIDE TO SCIENTIFIC ASSOCIATIONS AND LEARNED SOCIETIES. K.G. Saur Incorporated, 175 Fifth Avenue, New York, NY 10010. (800) 521-0707 or (212) 982-1302. Fourth edition. 1984. $112.00.

ENCYCLOPEDIAS AND DICTIONARIES

ENCYCLOPEDIA OF PHYSICAL SCIENCE AND TECHNOLOGY. Academic Press, Incorporated, Orlando, FL 32887. (800) 321-5068 or (305) 345-2734. Fifteen volumes, 1986.

ENCYCLOPEDIC DICTIONARY OF MATHEMATICS. Kiyosi Ito, editor. MIT Press, 55 Hayward Street, Cambridge, MA 02142. (617) 253-2884. Second edition. 1987. Four volumes. $350.00.

MCGRAW-HILL ENCYCLOPEDIA OF SCIENCE AND TECHNOLOGY. McGraw-Hill Book Company, Incorporated, 1221 Avenue of the Americas, New York, NY 10020. (212) 997-3675.

GENERAL WORKS

ANALYTICAL GEOMETRY AND THE CALCULUS. A.W. Goodman. Macmillan Publishing Company, Incorporated, 866 Third Avenue, New York, NY 10022. (800) 257-5755. $41.95.

ASPECTS OF CALCULUS. Gabriel Klambauer. Springer-Verlag, Incorporated, 175 Fifth Avenue, New York, NY 10010. (800) 526-7254. 1986. $38.00.

BRIEF CALCULUS AND ITS APPLICATIONS. Larry J. Goldstein and other. Prentice-Hall, Incorporated, Englewood Cliffs, NJ 07632. (201) 592-2000. 1987. $32.95.

POWER OF CALCULUS. Kenneth L. Whipkey and Mary N. Whipkey. John Wiley and Sons, Incorporated, 605 Third Avenue, New York, NY 10158. (800) 526-5368 or (212) 850-6000. Fourth Edition. 1986. $31.00.

HANDBOOKS AND MANUALS

CRC HANDBOOK OF MATHEMATICAL SCIENCES. William H. Beyer, editor. CRC Press, Incorporated, 2000 Corporate Boulevard, Boca Raton, FL 33341. (305) 994-0555. Sixth Edition. 1987. $91.95.

CRC STANDARD MATHEMATICAL TABLES. W.H. Beyer, editor. CRC Press, Incorporated, 2000 Corporate Boulevard, Boca Raton, FL 33341. (305) 994-0555. Twenty-eighth edition. 1987. $39.95.

ONLINE DATA BASES

DISSERTATION ABSTRACTS ONLINE. University Microfilms International, 300 North Zeeb Road, Ann Arbor, MI 48106. (800) 521-0600 or (313) 761-4700. Scope includes virtually all doctoral dissertations accepted at accredited American institutions from 1861 to present in 252 subject areas. Inquire as to cost and availability.

CALCULUS

INSPEC. INSPEC Marketing Department, Institute of Electrical and Electronics Engineers, Incorporated, IEEE Service Department, 445 Hoes Lane, Piscataway, NJ 08854. (201) 981-0060. Inquire as to on-line cost and availability.

MATHFILE. American Mathematical Society, Post Office Box 6248, Providence, RI 20940. (800) 556-7774 or (401) 272-9500. Scope includes pure and applied mathematics and related areas of physics, statistics, engineering, computer science, and operations research literature since 1973. Inquire as to cost and availability.

SCISEARCH. Institute for Scientific Information, 3501 Market Street, Philadelphia, PA 19104. (800) 523-1850 or (215) 386-0100. Broad multidisciplinary title and author index to the international literature of science and technology, 1974 to present. Inquire as to cost and availability.

OTHER SOURCES

FROM ONE TO ZERO: UNIVERSAL HISTORY OF NUMBERS. Georges Ifrah. Viking-Penquin, Incorporated, 40 West 23rd Street, New York, NY 10010. (800) 631-3577 or (212) 807-7300. 1986. $35.00.

MATHEMATICS AND THE SEARCH FOR KNOWLEDGE. Morris Kline. Oxford University Press, Incorporated, 200 Madison Avenue, New York, NY 10016. (212) 564-6680. 1986. $19.95.

PERIODICALS

AMERICAN JOURNAL OF MATHEMATICS. Johns Hopkins University Press, Journals Publishing Division, 701 West 40th Street, Suite 275, Baltimore, MD 21211. (301) 338-7864. Bimonthly. $40.00.

AMERICAN MATHEMATICAL MONTHLY. Mathematical Association of America, 1529 18th Street, NW, Washington, DC 20036. (202) 387-5200. Monthly. $55.00.

COLLEGE MATHEMATICS JOURNAL. Mathematical Association of America, 1529 18th Street, NW, Washington, DC 20036. Five issues per year. $40.00.

JOURNAL OF COMPUTATIONAL AND APPLIED MATHEMATICS. Elsevier Science Publishers B.V., Box 211, 1000 AE Amsterdam, The Netherlands. Nine times per year. $76.00 per year.

MATHEMATICS MAGAZINE. Mathematical Association of America, 1529 18th Street, NW, Washington, DC 20036. (202) 387-5200. Five issues per year. $35.00.

SIAM JOURNAL OF APPLIED MATHEMATICS. Society for Industrial and Applied Mathematics, 117 South 17th Street, Suite 1400, Philadelphia, PA 19103. (215) 564-2929. Bimonthly. $34.00 per year to individuals.

RESEARCH CENTERS AND INSTITUTES

UNIVERSITY OF CALIFORNIA, BERKELEY. Center for Pure and Applied Mathematics, 977 Evans Hall, Berkeley, CA 94720. (415) 642-3865.

UNIVERSITY OF FLORIDA. Center for Applied Mathematics, Gainesville, FL 32611. (904) 392-0281.

CAM (COMPUTER-AIDED MANUFACTURING)

See also: AUTOMATION, CAD, COMPUTERS, ROBOTICS

ABSTRACT SERVICES AND INDEXES

APPLIED SCIENCE AND TECHNOLOGY INDEX. H. W. Wilson Company, 950 University Avenue, Bronx, NY 10452. (800) 367-6670 or (212) 588-8400. Inquire as to cost and availability.

COMPUTER ABSTRACTS. Technical Information Company, Limited, P.O. Box 59, Saint Helier, Jersey British Channel Inlands, England. Monthly. $310.00 per year.

COMPUTER AND CONTROL ABSTRACTS. Institute of Electrical Engineers, London, United Kingdom. Available from: IEEE Service Center, 445 Hoes Lane, Piscataway, NJ 08854. (201) 981-0060. Semimonthly. $775.00 per year.

COMPUTER AND INFORMATION SYSTEMS: AN ABSTRACT JOURNAL PERTAINING TO THE THEORY, DESIGN, FABRICATION AND APPLICATION OF COMPUTER AND INFORMATION SYSTEMS. Cambridge Scientific Abstracts, Incorporated, 5161 River Road, Bethesda, MD 20816. (301) 951-1400. Semimonthly. $590.00 per year.

COMPUTING REVIEWS. Association for Computing Machinery, 11 West 42nd Street, New York, NY 10036. (212) 869-7440. Monthly. $60.00 per year.

CAD/CAM ABSTRACTS. EIC/Intelligence, Incorporated, 48 West 38th Street, New York, NY 10018. (800) 223-6275 or (212) 944-8500. Monthly. $345.00 per year.

ENGINEERING INDEX MONTHLY. Engineering Information, Incorporated 345 East 47th Street, New York, NY 10017. (800) 221-1044 or (212) 705-7600. Monthly, with annual cumulation. $1560.00 per year.

SCIENCE CITATION INDEX. Institute for Scientific Information, 3501 Market St., Philadelphia, PA 19104. (800) 523-1850 or (215) 386-0100. Bimonthly. $6200.00 per year.

ASSOCIATIONS AND PROFESSIONAL SOCIETIES

AMERICAN AUTOMATIC CONTROL COUNCIL. 1051 Camino Velasquez, Green Valley, AZ 85614. (602) 625-0401.

ASSOCIATION OF COMPUTING MACHINERY (ACM). 11 West 42nd Street, New York, NY 10036. (212) 896-7440.

COMPUTER AND AUTOMATION SYSTEMS ASSOCIATION OF SME. One SME Drive, Box 930, Dearborn, MI 48121. (313) 271-1500.

INSTITUTE OF ELECTRICAL AND ELECTRONIC ENGINEERS (IEEE). 345 East 47th Street, New York, NY 10017. (212) 705-7900.

NUMERICAL CONTROL SOCIETY. 111 East Wasker Drive, Suite 600, Chicago, IL 60601. (312) 644-6610.

SOCIETY FOR COMPUTER APPLICATIONS IN ENGINEERING, PLANNING AND ARCHITECTURE. 358 Hungerford Drive, Rockville, MD 20850. (301) 762-6070.

SOCIETY OF MANUFACTURING ENGINEERS. One SME Drive, P.O. Box 930, Dearborn, MI 48121. (313) 271-1500.

BIBLIOGRAPHIES

COMPUTER LITERATURE INDEX. Applied Computer Research, Incorporated, P.O. Box 9280, Phoenix, AZ 85068. (602) 995-5929.

Quarterly. $125.00 per year.

SCIENTIFIC AND TECHNICAL BOOKS IN PRINT; AN INDEX TO LITERATURE IN SCIENCE AND TECHNOLOGY. R.R. Bowker Company, 205 E. 42nd Street, New York, NY 10017. (800) 521-8110 or (212) 916-1600.

DIRECTORIES AND BIOGRAPHICAL SOURCES

CAD/CAM SOFTWARE DIRECTORY. Technical Database Corporation, 1300 South Frazier, Conroe, TX 77305. (409) 539-9688. Annual. $35.00.

COMPUTER-AIDED DESIGN (CAD) DIRECTORY. Technical Database Corporation, 1300 South Frazier, Conroe, TX 77305. (409) 539-9688. Annual. $40.00.

COMPUTER PERIPHERALS REVIEW. GML Information Services, 594 Marrett Road, Lexington, MA 02173. (617) 861-0515. Two issues per year. $215.00 per year. Directory of computer peripheral equipment manufacturers.

COMPUTERS AND COMPUTING INFORMATION RESOURCES DIRECTORY. Gale Research Company, Book Tower, Detroit, MI 48226. (800) 521-0707. 1986. $160.00.

DIRECTORY OF CONSULTANTS IN COMPUTER SYSTEMS. Research Publications, Incorporated, 12 Lunar Drive, Woodbridge, CT 06525. (203) 397-2600. Annual. $85.00 per year.

RESEARCH CENTERS DIRECTORY. Gale Research Company, Detroit, MI 48226. (800) 521-0707. Eleventh edition. 1987.

ROBOTICS-CAD/CAM DIRECTORY. Technical Data Publishing Corporation, 53 Lake Shore Drive, Rockaway, NJ 07866. (201) 625-9647. Annual. $3.00.

WHO'S WHO IN FRONTIER SCIENCE AND TECHNOLOGY. Marquis Who's Who, Incorporated, 200 East Ohio St., Chicago, IL 60611. (800) 428-3898 or (312) 787-2008.

WHO'S WHO IN TECHNOLOGY TODAY. Reston Publishing Company, Incorporated, c/o Prentice-Hall, Incorporated, Englewood Cliffs, NJ 07632. (800) 262-6868. Biennial. Five volumes. $425.00. Covers the fields of electronics, computer science, physics, optics, chemistry, biotechnology, mechanics, energy, and earth science.

ENCYCLOPEDIAS AND DICTIONARIES

COMPUTER DICTIONARY. Charles J. Sippl. Howard W. Sams and Company, Incorporated, 4300 West 62nd Street, Indianapolis, IN 46268. (800) 428-7267 or (317) 298-5564. Fourth edition. 1985. $17.95.

DICTIONARY OF COMPUTING. Oxford University Press, 200 Madison Avenue, New York, NY 10016. (212) 679-7300. Second edition. 1986. $29.95.

ENCYCLOPEDIA OF COMPUTER SCIENCE AND ENGINEERING. Anthony Ralston, editor. Van Nostrand Reinhold Book Company, 115 Fifth Avenue, New York, NY 10003. (800) 543-2681. Second edition. 1982. $89.95.

ENCYCLOPEDIA OF COMPUTER SCIENCE AND TECHNOLOGY. Jack Belzer, Albert G. Holzman, and Allan Kent. Marcel Dekker, Incorporated, 270 Madison Avenue, New York, NY 10016. (212) 696-9000. Sixteen volumes. $115.00 per volume.

MCGRAW-HILL DICTIONARY OF COMPUTERS. McGraw-Hill Book Company, 1221 Avenue of the Americas, New York, NY 10020. (212) 512-2000. 1985. $17.50.

GENERAL WORKS

CAD-CAM: COMPUTER-AIDED DESIGN AND MANUFACTURING. Mikell P. Groover and Emory W. Zimmers. Prentice-Hall, Incorporated, Englewood Cliffs, NJ 07632. (800) 562-0245 or (201) 599-2000. 1984. $36.95.

CAD-CAM. SOCIETY OF MANUFACTURING ENGINEERS. One SME Drive, Box 930, Dearborn, MI 48121. (313) 271-1500. Second edition. 1985. $49.00.

CAD-CAM: INTEGRATION AND INNOVATION. Society of Manufacturing Engineers. One SME Drive, Box 930, Dearborn, MI 48121. (313) 271-1500. 1985. $49.00.

CAD-CAM WITH PERSONAL COMPUTERS. Patrick R. Carbury. TAB Books, Incorporated, Monterey Lue, Blue Ridge Summit, PA 17214. (717) 794-2191. 1985. $21.95.

CAD: PRINCIPLES AND APPLICATIONS. Paul Barr, et al. Prentice-Hall, Incorporated, Englewood Cliffs, NJ 07632. (800) 562-0245 or (201) 599-2000. 1985. $21.95.

COMPUTER-AIDED DESIGN AND COMPUTER-AIDED MANUFACTURING: THE CAD-CAM REVOLUTION. John K. Krouse, editor. (What Every Engineer Should Know Series, Volume 10). Marcel Dekker, Incorporated, 270 Madison Avenue, New York, NY 10016. (800) 228-1160 or (212) 696-9000. 1982. $24.75.

COMPUTER-AIDED DESIGN AND MANUFACTURE. C. B. Besant. Halsted Press, Division of John Wiley and Sons, Incorporated, 605 Third Avenue, New York, NY 10158. (800) 526-5368 or (212) 850-6000. Second edition. 1982. $26.95 in paper.

HANDBOOKS AND MANUALS

CAD/CAM HANDBOOK. Eric Teicholz. McGraw-Hill Book Company, 1221 Avenue of the Americas, New York, NY 10020. (800) 628-0004. 1985. $54.00.

HANDBOOK OF COMPUTERS AND COMPUTING. Arthur H. Seidman, editor. Van Nostrand Reinhold Book Company, 115 Fifth Avenue, New York, NY 10003. (800) 543-2681. 1984. $79.95.

MCGRAW-HILL COMPUTER HANDBOOK. Harry Helms, editor. McGraw-Hill Book Company, 1221 Avenue of the Americas, New York, NY 10020. (212) 512-2000. 1983. $84.50.

ONLINE DATA BASES

COMPENDEX. Engineering Information, Incorporated, 345 East 47th Street, New York, NY 10017. (800) 221-1044 or (212) 705-7615. Engineering and technical literature, 1975 to present. Inquire as to cost and availability.

DISSERTATION ABSTRACTS ONLINE. University Microfilms International, 300 North Zeeb Road, Ann Arbor, MI 48106. (800) 521-0600 or (313) 761-4700. Scope includes virtually all doctoral dissertations accepted at accredited American institutions from 1861 to present in 252 subject areas. Inquire as to cost and availability.

INSPEC. INSPEC Marketing Department, Institute of Electrical and Electronics Engineers, Incorporated, IEEE Service Department, 445 Hoes Lane, Piscataway, NJ 08854. (201) 981-0060. Inquire as to on-line cost and availability.

NTIS. National Technical Information Service, 5285 Port Royal Road, Springfield, VA 22161. (703) 487-4630. Broad coverage of government sponsored research reports, 1964 to present. Inquire as to cost and availability.

SCISEARCH. Institute for Scientific Information, 3501 Market Street, Philadelphia, PA 19104. (800) 523-1850 or (215) 386-0100. Broad multidisciplinary title and author index to the

international literature of science and technology, 1974 to present. Inquire as to cost and availability.

OTHER SOURCES

ALL ABOUT CAD/CAM SYSTEMS. Datapro Research Corporation, 1805 Underwood Boulevard, Delran, NJ 08075. (800) 257-9406 or (609) 764-0100. Annual. $25.00.

PERIODICALS

CAD/CAM AND ROBOTICS. Kerrwill Publications, Limited, 501 Oakdale Downsview, Ontario M3N 1W7, Canada. Quarterly. $27.00 per year.

CAE: COMPUTER-AIDED ENGINEERING; DATA BASE APPLICATIONS IN DESIGN AND MANUFACTURING. Penton/IPC, Penton Plaza, 1111 Chester Avenue, Cleveland, OH 44114. (216) 696-7000. Monthly. $30.00 per year.

CIME: COMPUTERS IN MECHANICAL ENGINEERING. American Society of Mechanical Engineers, 345 East 47th Street, New York, NY 10017. (212) 705-7722. Bimonthly. $35.00 per year.

COMPUTER-INTEGRATED MANUFACTURING REVIEW. Auerbach Publishers, Incorporated, 6560 North Park Drive, Pennasuken, NJ 08109. (609) 662-2060. Quarterly. $76.00 per year.

COMPUTERS AND INDUSTRIAL ENGINEERING. Pergamon Journals, Maxwell House, Fairview Park, Elmsford, NY 10523. (914) 592-7700. Quarterly. $210.00 per year.

COMPUTERS FOR DESIGN AND CONSTRUCTION. Meta Data Publishing Corporation, 310 East 44th Street, Suite 1124, New York, NY 10017. (212) 867-2080. Bimonthly. $48.00 per year.

DESIGN COMPUTING. John Wiley and Sons, Incorporated, 605 Third Avenue, New York, NY 10158. (800) 526-5368 or (212) 850-6000. Quarterly. $100.00 per year.

IEEE COMPUTER GRAPHICS AND APPLICATIONS. Institute of Electrical and Electronic Engineers. IEEE Service Center, 445 Hoes Lane, Piscataway, NJ 08854. (201) 981-0060. Monthly. $155.00 per year.

IEEE JOURNAL OF ROBOTICS AND AUTOMATION. Institute of Electrical and Electronic Engineers. IEEE Service Center, 445 Hoes Lane, Piscataway, NJ 08854. (201) 981-0060. Quarterly. $65.00 per year.

ROBOTICS ENGINEERING. Robotics Age, Incorporated, 174 Concord Street, Peterborough, NH 03458. (603) 924-7136. Monthly. $24.00 per year.

ROBOTICS TODAY. Society of Manufacturing Engineers, One SME Drive, P.O. Box 930, Dearborn, MI 48121. (313) 271-1500. Bimonthly. $60.00 per year.

ROBOTICS WORLD. Communication Channels, Incorporated, 6255 Barfield Road, Atlanta GA 30328-4305. (404) 256-9800. Monthly. $27.00 per year.

RESEARCH CENTERS AND INSTITUTES

CENTER FOR CAD/CAM. Purdue University, 799 West Michigan Street, Indianapolis, IN 46202. (317) 264-8627.

CENTER FOR RESEARCH IN INTEGRATED MANUFACTURING. University of Michigan, 251 Chrysler Center, Ann Arbor, MI 48109. (313) 764-8496.

CENTER FOR ROBOTICS AND ADVANCED AUTOMATION. Oakland University, Rochester, MI 48063. (313) 37002233.

CORNELL MANUFACTURING ENGINEERING AND PRODUCTIVITY PROGRAM. Cornell University, 319 Upson Hall, Ithaca, NY 14853. (607) 255-4856.

MANUFACTURING PRODUCTIVITY CENTER. Illinois Institute of Technology, 10 West 35th Street, Chicago, IL 60616. (312) 567-4800.

CAMERAS

See also: AERIAL PHOTOGRAPHY, COLOR, OPTICS, PHOTOGRAMMETRY, PHOTOGRAPHIC FILM, PHOTOGRAPHY

ABSTRACT SERVICES AND INDEXES

APPLIED SCIENCE AND TECHNOLOGY INDEX. H.W. Wilson and Company, 950 University Avenue, Bronx, NY 10452. (800) 367-6670 or (212) 588-8400. Monthly. Inquire as to cost and availability.

CHEMICAL ABSTRACTS. American Chemical Society, Chemical Abstracts Service, Box 3012, Columbus, OH 43210. (614) 421-3600. 1907 to present. Weekly. $9500.00 per year.

CURRENT CONTENTS: ENGINEERING, TECHNOLOGY AND APPLIED SCIENCES. Institute for Scientific Information, 3501 Market Street, Philadelphia, PA 19104. (800) 523-1850 or (215) 386-0100. Weekly. $275.00 per year.

ENGINEERING INDEX MONTHLY AND AUTHOR INDEX. Engineering Information Inc., 345 East 47th Street, New York, NY 10017. (212) 705-7600. Monthly. $1560.00 per year.

PHOTOGRAPHIC ABSTRACTS. Royal Photographic Society of Great Britain, Scientific and Technical Group, 62 Chelmsford Road, Shenfield, Brentwood, Essex, England. 1921 to present. Six times per year. $140.00 per year.

PHYSICS ABSTRACTS. Institution of Electrical Engineers. Available from: IEEE Service Center, 445 Hoes Lane, Piscataway, NJ 08854. 1898 to present. Bimonthly. $1700.00 per year.

SCIENCE CITATION INDEX. Institute for Scientific Information, 3501 Market Street, Philadelphia, PA 19104. (800) 523-1850 or (215) 386-0100. Six times per year. $6200.00 per year.

ASSOCIATIONS AND PROFESSIONAL SOCIETIES

AMERICAN SOCIETY FOR PHOTOGRAMMETRY AND REMOTE SENSING. 210 Little Falls Street, Falls Church, VA 22046-4398. (703) 534-6617.

OPTICAL SOCIETY OF AMERICA. 1816 Jefferson Place, N.W., Washington, DC 20036. (202) 223-8130.

SOCIETY OF PHOTOGRAPHIC SCIENTISTS AND ENGINEERS. 7003 Kilworth Lane, Springfield, VA 22151. (703) 642-9090.

SPIE - THE INTERNATIONAL SOCIETY FOR OPTICAL ENGINEERING. P.O. Box 10, 1022 19th Street, Bellingham, WA 98227. (206) 676-3290.

DIRECTORIES AND BIOGRAPHICAL SOURCES

INTERNATIONAL RESEARCH CENTERS DIRECTORY 1988-89. Darren L. Smith, editor. Gale Research Company, Book Tower, Detroit, MI 48226. (800) 521-0707. 4th edition. 1987. $360.00.

1987 DIRECTORY OF ENGINEERING SOCIETIES AND RELATED ORGANIZATIONS. Gordon Davis, editor. Hemisphere Publishing Corporation, 1010 Vermont Avenue, NW, Washington, DC 20005. (800) 526-0275. 12th edition. 1987. $100.00.

RESEARCH CENTERS DIRECTORY 1988. Gale Research Company, Book Tower, Detroit, MI 48226. (800) 521-0707. 12th edition. 1987. $365.00 for set.

SCIENTIFIC AND TECHNICAL ORGANIZATIONS AND AGENCIES DIRECTORY. Margaret Labash Young, editor. Gale Research Company, Book Tower, Detroit, MI 48226. (800) 521-0707. 2nd edition. 1987. $185.00.

WHO'S WHO IN ENGINEERING. Gordon Davis, editor. Hemisphere Publishing Corporation, 1010 Vermont Avenue, NW, Washington, DC 20005. (800) 526-0275. 6th edition. 1985. $200.00.

ENCYCLOPEDIAS AND DICTIONARIES

THESAURUS OF PHOTOGRAPHIC SCIENCE AND ENGINEERING. Society of Photographic Scientists and Engineers. Books on Demand, 300 North Zeeb Road, Ann Arbor, MI 48106. (313) 761-4700. $34.50 in paper.

THESAURUS OF SCIENTIFIC, TECHNICAL, AND ENGINEERING TERMS. Hemisphere Publishing Corporation, 1010 Vermont Avenue, NW, Washington, DC 20005. (800) 526-0275. 1988. $125.00.

GENERAL WORKS

THE CAMERA BOOK. Michael J. Langford, editor. Watson-Guptill Publications, Inc., 1 Astor Place, 1515 Broadway, New York, NY 10036. (800) 526-3641. 1985. $16.95 in paper.

LIGHT AND COLOR. R.D. Overheim and D.L. Wagner. John Wiley and Sons, Inc., 605 Third Avenue, New York, NY 10158. (800) 526-5368. 1982. $28.50.

PHOTOGRAPHIC SCIENCE. Earl. N. Mitchell. John Wiley and Sons, Inc., 605 Third Avenue, New York, NY 10158. (800) 526-5368. 1984. $37.50.

PHOTOGRAPHY FOR THE SCIENTIST. Richard A. Morton. Academic Press, Inc., 6277 Sea Harbor Drive, Orlando, FL 32821. (800) 321-5068. 2nd edition. 1984. $102.50.

HANDBOOKS AND MANUALS

HANDBOOK OF PHOTOGRAPHIC SCIENCE AND ENGINEERING. Society of Photographic Scientists and Engineers. 7003 Kilworth Lane, Springfield, VA 22151. (703) 642-9090. Inquire.

ONLINE DATA BASES

CA SEARCH. Chemical Abstracts Service, P.O. Box 3012, Columbus, OH 43120. (800) 848-6538 or (614) 421-3600. Comprehensive guide to chemical literature, 1972 to present. Inquire as to online cost and availability.

COMPENDEX. Engineering Information, Inc., 345 East 47th Street, New York, NY 10017. (800) 221-1044 or (212) 705-7615. Engineering and technical literature, 1975 to present. Inquire as to online cost and availability.

INSPEC. INSPEC Marketing Department, Institution of Electrical Engineers. Available from IEEE Service Center, 445 Hoes Lane, Piscataway, NJ 08854. (201) 981-0060. Online version of Physics Abstracts. Inquire as to online cost and availability.

NTIS. National Technical Information Service, 5285 Port Royal Road, Springfield, VA 22161. (703) 487-4630. Broad coverage of government sponsored research reports, 1964 to present. Inquire as to online cost and availability.

SCISEARCH. Institute for Scientific Information, 3501 Market Street, Philadelphia, PA 19104. (800) 523-1850 or (215) 386-0100. Broad multidisciplinary title and author index to the international literature of science and technology, 1974 to present. Inquire as to online cost and availability.

WILSONLINE. H.W. Wilson and Company, 950 University Avenue, Bronx, NY 10452. (800) 367-6770 or (212) 588-8400. Makes available online versions of the H.W. Wilson indexes including Applied Science and Technology Index, Business Periodicals Index and Readers' Guide to Periodical Literature. Approximately 1980 to present. Inquire as to online cost and availability.

PERIODICALS

BRITISH JOURNAL OF PHOTOGRAPHY. Henry Greenwood and Company, Limited, 28 Great James Street, London WC1N 3HL, England. 1854 to present. Weekly. $70.00 per year.

FUNCTIONAL PHOTOGRAPHY. PTN Publishing Corporation, 210 Crossways Park Drive, Woodbury, NY 11797. (516) 496-8000. 1967 to present. Bimonthly. $7.50 per year.

INDUSTRIAL PHOTOGRAPHY. United Business Publications, Inc., 475 Park Avenue South, New York, NY 10016. (212) 725-2300. 1952 to present. Monthly. $15.00 per year.

JOURNAL OF IMAGING SCIENCE. Society of Photographic Scientists and Engineers. 7003 Kilworth Lane, Springfield, VA 22151. (703) 642-9090. 1956 to present. Bimonthly. $70.00 per year.

JOURNAL OF IMAGING TECHNOLOGY. Society of Photographic Scientists and Engineers. 7003 Kilworth Lane, Springfield, VA 22151. (703) 642-9090. 1975 to present. Bimonthly. $70.00 per year.

JOURNAL OF PHOTOGRAPHIC SCIENCE. Royal Photographic Society of Great Britain, 7 Ladbroke Walk, London W11, England. 1953 to present. Bimonthly. $40.00 per year.

PHOTOGRAMMETRIC ENGINEERING AND REMOTE SENSING. American Society for Photogrammetry and Remote Sensing. 210 Little Falls Street, Falls Church, VA 22046-4398. (703) 534-6617. Order from: Allen Press, Inc., 1041 New Hampshire Street, Box 368, Lawrence, KS 66044. 1934 to present. Monthly. $80.00 per year.

SCIENTIFIC AND APPLIED PHOTOGRAPHY AND CINEMATOGRAPHY. Gordon and Breach Science Publishers, Inc., 50 West 23rd Street, New York, NY 10010. (212) 206-8900. 12 times per year. $496.00 per year.

CANALS

See also: CHANNELS, CIVIL ENGINEERING, DREDGES AND DREDGING, GEOTECHNICAL ENGINEERING, HYDRAULIC ENGINEERING, OCEAN ENGINEERING

ABSTRACT SERVICES AND INDEXES

APPLIED SCIENCE AND TECHNOLOGY INDEX. H.W. Wilson and Company, 950 University Avenue, Bronx, NY 10452. (800) 367-6670 or (212) 588-8400. Monthly. Inquire as to cost and availability.

ENGINEERING INDEX MONTHLY AND AUTHOR INDEX. Engineering Information Inc., 345 East 47th Street, New York, NY 10017. (212) 705-7600. Monthly. $1560.00 per year.

INTERNATIONAL CIVIL ENGINEERING ABSTRACTS. CITIS Limited, 2 Rosemount Terrace, Blackrock, Dublin, Ireland. 1974 to present. Monthly. $350.00 per year.

PUBLICATIONS INFORMATION. American Society of Civil Engineers, 345 East 47th Street, New York, NY 10017. (212) 705-7420. Abstracts, subject and author indexes to the publications of the American Society of Civil Engineers. Bimonthly. $80.00

per year.

SCIENCE CITATION INDEX. Institute for Scientific Information, 3501 Market Street, Philadelphia, PA 19104. (800) 523-1850 or (215) 386-0100. Six times per year. $6200.00 per year.

ASSOCIATIONS AND PROFESSIONAL SOCIETIES

AMERICAN GEOLOGICAL INSTITUTE. 4220 King Street, Alexandria, VA 22302. (703) 379-2480.

AMERICAN SOCIETY OF CIVIL ENGINEERS. 345 East 47th Street, New York, NY 10017. (212) 705-7420.

DIRECTORIES AND BIOGRAPHICAL SOURCES

1987 DIRECTORY OF ENGINEERING SOCIETIES AND RELATED ORGANIZATIONS. Gordon Davis, editor. Hemisphere Publishing Corporation, 1010 Vermont Avenue, NW, Washington, DC 20005. (800) 526-0275. 12th edition. 1987. $100.00.

RESEARCH CENTERS DIRECTORY 1988. Gale Research Company, Book Tower, Detroit, MI 48226. (800) 521-0707. 12th edition. 1987. $365.00 for set.

SCIENTIFIC AND TECHNICAL ORGANIZATIONS AND AGENCIES DIRECTORY. Margaret Labash Young, editor. Gale Research Company, Book Tower, Detroit, MI 48226. (800) 521-0707. 2nd edition. 1987. $185.00.

WHO'S WHO IN ENGINEERING. Gordon Davis, editor. Hemisphere Publishing Corporation, 1010 Vermont Avenue, NW, Washington, DC 20005. (800) 526-0275. 6th edition. 1985. $200.00.

ENCYCLOPEDIAS AND DICTIONARIES

DICTIONARY OF CIVIL ENGINEERING. John S. Scott. John Wiley and Sons, Inc., 605 Third Avenue, New York, NY 10158. (800) 526-5368. Third edition. 1981. $26.95.

GENERAL WORKS

CANAL AND RIVER LEVEES. P. Peter. Elsevier Science Publishing Company, Inc., 52 Vanderbilt Avenue, New York, NY 10017. (212) 370-5520. 1982. $104.25.

CHANNELS AND CHANNEL CONTROL STRUCTURES. K.V. Smith, editor. Springer-Verlag New York, Inc., 175 Fifth Avenue, New York, NY 10010. (800) 526-7254. 1985. $37.70.

FLUID FLOW: PUMPS, PIPES, AND CHANNELS. N.P. Chermisinoff. Butterworth's Publishing, 80 Montvale Avenue, Stoneham, MA 02180. (800) 325-4177. 1982. $59.95.

OPEN-CHANNEL HYDRAULICS. R.H. French. McGraw-Hill Book Company, 1221 Avenue of the Americas, New York, NY 10020. (212) 512-2000. 1985. $49.95.

HANDBOOKS AND MANUALS

CIVIL ENGINEERING CALCULATIONS REFERENCE GUIDE. Tyler G. Hicks, editor. McGraw-Hill Book Company, 1221 Avenue of the Americas, New York, NY 10020. (212) 512-2000. 1987. $29.50.

CIVIL ENGINEERING PRACTICE. Paul N. Cheremisinoff and others, editors. Technomic Publishing Company, Inc., 851 Holland Avenue, Box 3535, Lancaster, PA 17604. (800) 233-9936. Five volumes. 1987-1988. $750.00 for set.

STANDARD HANDBOOK FOR CIVIL ENGINEERS. F. S. Merritt, editor. McGraw-Hill Book Company, 1221 Avenue of the Americas, New York, NY 10020. (212) 512-2000. Third edition. 1983. $89.50.

SURVEYING READY-REFERENCE MANUAL. Guy O. Stenstrom. McGraw-Hill Book Company, 1221 Avenue of the Americas, New York, NY 10020. (212) 512-2000. 1987. $26.50.

ONLINE DATA BASES

COMPENDEX. Engineering Information, Inc., 345 East 47th Street, New York, NY 10017. (800) 221-1044 or (212) 705-7615. Engineering and technical literature, 1975 to present. Inquire as to online cost and availability.

GEOREF. Online version of the BIBLIOGRAPHY AND INDEX OF GEOLOGY. American Geological Institute, 4220 King Street, Alexandria, VA 22302. (703) 379-2480. 1969 to present. Inquire as to online cost and availability.

NTIS. National Technical Information Service, 5285 Port Royal Road, Springfield, VA 22161. (703) 487-4630. Broad coverage of government sponsored research reports, 1964 to present. Inquire as to online cost and availability.

SCISEARCH. Institute for Scientific Information, 3501 Market Street, Philadelphia, PA 19104. (800) 523-1850 or (215) 386-0100. Broad multidisciplinary title and author index to the international literature of science and technology, 1974 to present. Inquire as to online cost and availability.

WILSONLINE. H.W. Wilson and Company, 950 University Avenue, Bronx, NY 10452. (800) 367-6770 or (212) 588-8400. Makes available online versions of the H.W. Wilson indexes including Applied Science and Technology Index, Business Periodicals Index and Readers' Guide to Periodical Literature. Approximately 1980 to present. Inquire as to online cost and availability.

PERIODICALS

CIVIL ENGINEERING. American Society of Civil Engineers, 345 East 47th Street, New York, NY 10017. (212) 705-7420. 1930 to present. Monthly. $48.00 per year.

JOURNAL OF CONSTRUCTION ENGINEERING AND MANAGEMENT. American Society of Civil Engineers, 345 East 47th Street, New York, NY 10017. (212) 705-7420. 1956 to present. Quarterly. $48.00 per year.

JOURNAL OF SURVEYING ENGINEERING. American Society of Civil Engineers, 345 East 47th Street, New York, NY 10017. (212) 705-7420. 1956 to present. Three times per year. $35.00 per year.

JOURNAL OF GEOTECHNICAL ENGINEERING. American Society of Civil Engineers, 345 East 47th Street, New York, NY 10017. (212) 705-7420. 1956 to present. Monthly. $96.00 per year.

JOURNAL OF HYDRAULIC ENGINEERING. American Society of Civil Engineers, 345 East 47th Street, New York, NY 10017. (212) 705-7420. 1956 to present. Monthly. $112.00 per year.

JOURNAL OF IRRIGATION AND DRAINAGE. American Society of Civil Engineers, 345 East 47th Street, New York, NY 10017. (212) 705-7420. 1956 to present. Quarterly. $45.00 per year.

JOURNAL OF WATER RESOURCES PLANNING AND MANAGEMENT. American Society of Civil Engineers, 345 East 47th Street, New York, NY 10017. (212) 705-7420. 1956 to present. Quarterly. $56.00 per year.

JOURNAL OF WATERWAY, PORT, COASTAL AND OCEAN ENGINEERING. American Society of Civil Engineers, 345 East 47th Street, New York, NY 10017. (212) 705-7420. 1956 to present. Bimonthly. $72.00 per year.

RESEARCH CENTERS AND INSTITUTES

PORTS AND WATERWAYS INSTITUTE. Louisiana State University, 60 University Lakeshore Drive, Baton Rouge, LA

70803. (504) 388-2772.

CANTILEVERS

See: STRUCTURAL ENGINEERING

CAPACITORS

See also: ELECTRICAL ENGINEERING, ELECTRONIC CIRCUITS AND COMPONENTS, ELECTRONICS, ELECTRONICS ENGINEERING

ABSTRACT SERVICES AND INDEXES

APPLIED SCIENCE AND TECHNOLOGY INDEX. H.W. Wilson and Company, 950 University Avenue, Bronx, NY 10452. (800) 367-6670 or (212) 588-8400. Monthly. Inquire as to cost and availability.

CHEMICAL ABSTRACTS. American Chemical Society, Chemical Abstracts Service, Box 3012, Columbus, OH 43210. (614) 421-3600. 1907 to present. Weekly. $9500.00 per year.

CURRENT CONTENTS: ENGINEERING, TECHNOLOGY AND APPLIED SCIENCES. Institute for Scientific Information, 3501 Market Street, Philadelphia, PA 19104. (800) 523-1850 or (215) 386-0100. Weekly. $275.00 per year.

ELECTRICAL AND ELECTRONICS ABSTRACTS. Institution of Electrical Engineers. Available from: Institute of Electrical and Electronics Engineers. IEEE Service Center, 445 Hoes Lane, Piscataway, NJ 08854. Monthly. $1250.00 per year.

ELECTRONICS AND COMMUNICATIONS ABSTRACTS. Cambridge Scientific Abstracts, 5161 River Road, Bethesda, MD 20816. (301) 951-1400. Bimonthly. Inquire as to cost and availability.

ENGINEERING INDEX MONTHLY AND AUTHOR INDEX. Engineering Information Inc., 345 East 47th Street, New York, NY 10017. (212) 705-7600. Monthly. $1560.00 per year.

IEEE PUBLICATIONS BULLETIN. Institute of Electrical and Electronics Engineers. Institute of Electrical and Electronics Engineers. IEEE Service Center, 445 Hoes Lane, Piscataway, NJ 08854. Quarterly. Free.

PHYSICS ABSTRACTS. Institution of Electrical Engineers. Available from: IEEE Service Center, 445 Hoes Lane, Piscataway, NJ 08854. 1898 to present. Bimonthly. $1700.00 per year.

PHYSICS BRIEFS. Physik Verlag GmbH, Postfach 1260/1280, D-6940 Weinheim, West Germany. (212) 661-9404. 1920 to present. Twenty-six times per year. $1250.00 per year.

SCIENCE CITATION INDEX. Institute for Scientific Information, 3501 Market Street, Philadelphia, PA 19104. (800) 523-1850 or (215) 386-0100. Six times per year. $6200.00 per year.

SOLID STATE ABSTRACTS: AN ABSTRACT JOURNAL INVOLVING THE PHYSICS, METALLURGY, CRYSTALLOGRAPHY, CHEMISTRY, AND DEVICE TECHNOLOGY OF SOLIDS. Cambridge Scientific Abstracts, 5161 River Road, Bethesda, MD 20816. (301) 951-1400. 1957 to present. Bimonthly. $550.00 per year.

ANNUAL REVIEWS AND YEARBOOKS

ADVANCES IN ELECTRONICS AND ELECTRON PHYSICS. Academic Press, Inc., 6277 Sea Harbor Drive, Orlando, FL 32821. (800) 321-5068. Irregular. Approximately $80.00 per volume.

ASSOCIATIONS AND PROFESSIONAL SOCIETIES

AMERICAN ELECTRONICS ASSOCIATION. P.O. Box 10045, 2670 Hanover Street, Palo Alto, CA 94303. (415) 857-9300.

AMERICAN INSTITUTE OF PHYSICS. 335 East 45th Street, New York, NY 10017. (212) 661-9494.

ELECTRONICS INDUSTRIES ASSOCIATION. 2001 Eye Street, N.W., Washington, DC 20006. (202) 457-4900.

INSTITUTE OF ELECTRICAL AND ELECTRONICS ENGINEERS. 345 East 47th Street, New York, NY 10017. (212) 705-7900.

INTERNATIONAL SOCIETY FOR HYBRID MICROELECTRONICS. P.O. Box 2698, 1861 Wiehle Avenue, Suite 340, Reston, VA 22090. (703) 471-0066.

DIRECTORIES AND BIOGRAPHICAL SOURCES

AMERICAN MEN AND WOMEN OF SCIENCE. R.R. Bowker, Inc., Order Department, 245 West 17th Street, New York, NY 10011. (800) 521-8110. Eight volumes. 1986. $595.00 for set.

IEEE MEMBERSHIP DIRECTORY. Institute of Electrical and Electronics Engineers. IEEE Service Center, 445 Hoes Lane, Piscataway, NJ 08854. Annual. $7.00.

INTERNATIONAL RESEARCH CENTERS DIRECTORY 1988-89. Darren L. Smith, editor. Gale Research Company, Book Tower, Detroit, MI 48226. (800) 521-0707. 4th edition. 1987. $360.00.

1987 DIRECTORY OF ENGINEERING SOCIETIES AND RELATED ORGANIZATIONS. Gordon Davis, editor. Hemisphere Publishing Corporation, 1010 Vermont Avenue, NW, Washington, DC 20005. (800) 526-0275. 12th edition. 1987. $100.00.

RESEARCH CENTERS DIRECTORY 1988. Gale Research Company, Book Tower, Detroit, MI 48226. (800) 521-0707. 12th edition. 1987. $365.00 for set.

SCIENTIFIC AND TECHNICAL ORGANIZATIONS AND AGENCIES DIRECTORY. Margaret Labash Young, editor. Gale Research Company, Book Tower, Detroit, MI 48226. (800) 521-0707. 2nd edition. 1987. $185.00.

WHO'S WHO IN ELECTRONICS. Harris Publishing Company, 2057-2 Aurora Road, Twinsburg, OH 44087. (216) 425-9143. Annual. $90.00.

WHO'S WHO IN ENGINEERING. Gordon Davis, editor. Hemisphere Publishing Corporation, 1010 Vermont Avenue, NW, Washington, DC 20005. (800) 526-0275. 6th edition. 1985. $200.00.

ENCYCLOPEDIAS AND DICTIONARIES

IEEE STANDARD DICTIONARY OF ELECTRICAL AND ELECTRONICS TERMS. Frank Jay, editor. John Wiley and Sons, Inc., 605 Third Avenue, New York, NY 10158. (800) 526-5368. 3rd edition. 1984. $49.95.

GENERAL WORKS

CAPACITORS. D.S. Campbell. Gordon and Breach Science Publishers, Inc., 50 West 23rd Street, New York, NY 10010. (212) 206-8900. 1987. $39.95.

CIRCUITS, DEVICES AND SYSTEMS: A FIRST COURSE IN ELECTRICAL ENGINEERING. R.J. Smith. John Wiley and Sons, Inc., 605 Third Avenue, New York, NY 10158. (800) 526-5368. Fourth edition. 1984. $42.95.

ELECTRONIC INVENTIONS AND DISCOVERIES: ELECTRONICS FROM ITS EARLIEST BEGINNINGS TO PRESENT DAY. G.W.A. Dummer. Pergamon Press, Inc., Maxwell House, Fairview Park, Elmsford, NY 10523. (914) 592-7700. 1983. $49.50.

HANDBOOKS AND MANUALS

ELECTRONIC ENGINEERS HANDBOOK. Donald G. Fink, editor. McGraw-Hill Book Company, 1221 Avenue of the Americas, New York, NY 10020. (212) 512-2000. 2nd edition. 1982. $89.00.

HANDBOOK OF MODERN ELECTRONICS AND ELECTRICAL ENGINEERING. Charles Belove, editor. John Wiley and Sons, Inc., 605 Third Avenue, New York, NY 10158. (800) 526-5368. 1986. $88.95.

ONLINE DATA BASES

CA SEARCH. Chemical Abstracts Service, P.O. Box 3012, Columbus, OH 43120. (800) 848-6538 or (614) 421-3600. Comprehensive guide to chemical literature, 1972 to present. Inquire as to online cost and availability.

COMPENDEX. Engineering Information, Inc., 345 East 47th Street, New York, NY 10017. (800) 221-1044 or (212) 705-7615. Engineering and technical literature, 1975 to present. Inquire as to online cost and availability.

DISSERTATION ABSTRACTS ONLINE. University Microfilms International, 300 North Zeeb Road, Ann Arbor, MI 48106. (800) 521-0600 or (313) 761-4700. Scope includes virtually all doctoral dissertations accepted at accredited American institutions from 1861 to present in over 250 subject areas. Inquire as to online cost and availability.

INSPEC. INSPEC Marketing Department, Institution of Electrical Engineers. Available from IEEE Service Center, 445 Hoes Lane, Piscataway, NJ 08854. (201) 981-0060. Online version of Physics Abstracts. Inquire as to online cost and availability.

NTIS. National Technical Information Service, 5285 Port Royal Road, Springfield, VA 22161. (703) 487-4630. Broad coverage of government sponsored research reports, 1964 to present. Inquire as to online cost and availability.

SCISEARCH. Institute for Scientific Information, 3501 Market Street, Philadelphia, PA 19104. (800) 523-1850 or (215) 386-0100. Broad multidisciplinary title and author index to the international literature of science and technology, 1974 to present. Inquire as to online cost and availability.

WILSONLINE. H.W. Wilson and Company, 950 University Avenue, Bronx, NY 10452. (800) 367-6770 or (212) 588-8400. Makes available online versions of the H.W. Wilson indexes including Applied Science and Technology Index, Business Periodicals Index and Readers' Guide to Periodical Literature. Approximately 1980 to present. Inquire as to online cost and availability.

OTHER SOURCES

A GUIDE TO THE LITERATURE OF ELECTRICAL AND ELECTRONICS ENGINEERING. Susan B. Ardis. Libraries Unlimited Inc., P.O. Box 263, Littleton, CO 80160. (303) 770-1220. 1987. $37.50.

PERIODICALS

ELECTRONIC DESIGN. Hayden Publishing Company, 10 Mulholland Drive, Hasbrouck Heights, NJ 07604. (201) 288-7520. 1952 to present. Biweekly. $40.00 per year.

ELECTRONICS. McGraw-Hill Book Company, 1221 Avenue of the Americas, New York, NY 10020. (212) 512-2000. 1930 to present. Weekly. $32.00 per year.

IEEE CIRCUITS AND DEVICES MAGAZINE. Institute of Electrical and Electronics Engineers. IEEE Service Center, 445 Hoes Lane, Piscataway, NJ 08854. Bimonthly. $70.00 per year.

IEEE JOURNAL OF SOLID STATE CIRCUITS. Institute of Electrical and Electronics Engineers. IEEE Service Center, 445 Hoes Lane, Piscataway, NJ 08854. 1966 to present. Bimonthly. $113.00 per year.

IEEE TRANSACTIONS ON ELECTRON DEVICES. Institute of Electrical and Electronics Engineers. IEEE Service Center, 445 Hoes Lane, Piscataway, NJ 08854. 1959 to present. Monthly. $175.00 per year.

INSTITUTE OF ELECTRICAL AND ELECTRONICS ENGINEERS PROCEEDINGS. Institute of Electrical and Electronics Engineers. IEEE Service Center, 445 Hoes Lane, Piscataway, NJ 08854. 1913 to present. Monthly. $140.00 per year.

SEMICONDUCTOR INTERNATIONAL. Cahners Publishing Company, Inc., Cahners Plaza, 1350 East Touhy Avenue, Des Plaines, IL 60018. (312) 635-8800. 1978 to present. Monthly. $55.00 per year.

SOLID STATE ELECTRONICS. Pergamon Press, Inc., Maxwell House, Fairview Park, Elmsford, NY 10523. (914) 592-7700. 1960 to present. Monthly. $330.00 per year.

RESEARCH CENTERS AND INSTITUTES

ELECTRICAL ENGINEERING RESEARCH LABORATORIES. Purdue University, Electrical Engineering Building, West Lafayette, IN 47907. (317) 494-3536.

ELECTRONICS RESEARCH CENTER. University of Texas at Austin, 132 Engineering Science Building, Austin, TX 78712. (512) 471-3954.

ELECTRONICS RESEARCH LABORATORY. University of California, Berkeley, 253 Cory Hall, Berkeley, CA 94720. (415) 642-2301.

LABORATORY FOR ELECTROMAGNETIC AND ELECTRONIC SYSTEMS. Massachusetts Institute of Technology, 77 Massachusetts Avenue, Cambridge, MA 02139. (617) 253-4631.

CARBON

See also: CHEMISTRY, ORGANIC CHEMISTRY, PETROLEUM CHEMISTRY, PETROLEUM ENGINEERING, PLASTICS, POLYMERS

ABSTRACT SERVICES AND INDEXES

CHEMICAL ABSTRACTS. American Chemical Society, Chemical Abstracts Service, Box 3012, Columbus, OH 43210. (614) 421-3600. 1907 to present. Weekly. $9500.00 per year.

CONFERENCE PAPERS INDEX. Cambridge Scientific Abstracts, 5161 River Road, Bethesda, MD 20816. 1972 to present. Monthly. Inquire as to cost and availability.

CURRENT CONTENTS: PHYSICAL, CHEMICAL AND EARTH SCIENCES. Institute for Scientific Information, 3501 Market Street, Philadelphia, PA 19104. (800) 523-1850 or (215) 386-0100. Weekly. $275.00 per year.

GENERAL SCIENCE INDEX. H.W. Wilson and Company, 950 University Avenue, Bronx, NY 10452. (800) 367-6670 or (212) 588-8400. 1978 to present. Monthly. Inquire as to cost and availability.

INDEX TO SCIENTIFIC AND TECHNICAL PROCEEDINGS. Institute for Scientific Information, 3501 Market Street, Philadelphia, PA 19104. (800) 523-1850 or (215) 386-0100. 1978 to present. Monthly. $775.00 per year.

INDEX TO SCIENTIFIC REVIEWS. Institute for Scientific Information, 3501 Market Street, Philadelphia, PA 19104. (800) 523-1850 or (215) 386-0100. 1974 to present. Semi-annual. $550.00

per year.

SCIENCE CITATION INDEX. Institute for Scientific Information, 3501 Market Street, Philadelphia, PA 19104. (800) 523-1850 or (215) 386-0100. Six times per year. $6200.00 per year.

ANNUAL REVIEWS AND YEARBOOKS

CHEMISTRY AND PHYSICS OF CARBON. Philip J. Walker and Peter A. Thrower, editors. Marcel Dekker Inc., 270 Madison Avenue, New York, NY 10016. (800) 228-1160. 1966 to 1984. Irregular. Price varies. Inquire.

STUDIES IN ORGANIC CHEMISTRY. Marcel Dekker Inc., 270 Madison Avenue, New York, NY 10016. (800) 228-1160. 1973 to present. Irregular. Price varies, inquire.

ASSOCIATIONS AND PROFESSIONAL SOCIETIES

AMERICAN CARBON SOCIETY. The Stackpole Corporation, St. Marys, PA 15857. (814) 781-8410.

AMERICAN CHEMICAL SOCIETY. 1155 16th Street, N.W., Washington, DC 20036. (202) 872-4600.

AMERICAN INSTITUTE OF CHEMICAL ENGINEERS. 345 East 47th Street, New York, NY 10017. (212) 705-7338.

AMERICAN OIL CHEMISTS' SOCIETY. 508 South Sixth Street, Champaign, IL 61820. (217) 359-2344.

ASSOCIATION OF OFFICIAL ANALYTICAL CHEMISTS. 1111 North 19th Street, Suite 210, Arlington, VA 22209. (703) 522-3032.

DIRECTORIES AND BIOGRAPHICAL SOURCES

AMERICAN MEN AND WOMEN OF SCIENCE. R.R. Bowker, Inc., Order Department, 245 West 17th Street, New York, NY 10011. (800) 521-8110. Eight volumes. 1986. $595.00 for set.

INTERNATIONAL RESEARCH CENTERS DIRECTORY 1988-89. Darren L. Smith, editor. Gale Research Company, Book Tower, Detroit, MI 48226. (800) 521-0707. 4th edition. 1987. $360.00.

1987 DIRECTORY OF ENGINEERING SOCIETIES AND RELATED ORGANIZATIONS. Gordon Davis, editor. Hemisphere Publishing Corporation, 1010 Vermont Avenue, NW, Washington, DC 20005. (800) 526-0275. 12th edition. 1987. $100.00.

RESEARCH CENTERS DIRECTORY 1988. Gale Research Company, Book Tower, Detroit, MI 48226. (800) 521-0707. 12th edition. 1987. $365.00 for set.

SCIENTIFIC AND TECHNICAL ORGANIZATIONS AND AGENCIES DIRECTORY. Margaret Labash Young, editor. Gale Research Company, Book Tower, Detroit, MI 48226. (800) 521-0707. 2nd edition. 1987. $185.00.

WHO'S WHO IN ENGINEERING. Gordon Davis, editor. Hemisphere Publishing Corporation, 1010 Vermont Avenue, NW, Washington, DC 20005. (800) 526-0275. 6th edition. 1985. $200.00.

ENCYCLOPEDIAS AND DICTIONARIES

DICTIONARY OF ORGANIC COMPOUNDS. John Buckingham, editor. Methuen, Inc., 29 West 35th Street, New York, NY 10001. (212) 244-3336. Fifth edition. 1988. $675.00.

ENCYCLOPEDIA OF CHEMISTRY. Douglas M. Considine, editor. Van Nostrand Reinhold Company, Inc., 135 West 50th Street, New York, NY 10020. (800) 543-2681. 4th edition. 1984. $98.95.

GENERAL WORKS

BASIC ORGANIC CHEMISTRY. F.L. Wiseman. McGraw-Hill Book Company, 1221 Avenue of the Americas, New York, NY 10020. (212) 512-2000. 1988. $37.95.

FUNDAMENTALS OF ORGANIC CHEMISTRY. T.W.G. Solomon. John Wiley and Sons, Inc., 605 Third Avenue, New York, NY 10158. (800) 526-5368. 2nd edition. 1986. $39.95.

PETROLEUM-DERIVED CARBONS. John D. Bacha and others. American Chemical Society, 1155 16th Street, N.W., Washington, DC 20036. (800) 424-6747. 1986. $75.00.

HANDBOOKS AND MANUALS

CRC HANDBOOK OF CHEMISTRY AND PHYSICS. Robert C. Weast, editor. CRC Press, 2000 Corporate Boulevard, Boca Raton, FL 33431. (800) 272-7737. 68th edition. 1987. $69.95.

CRC HANDBOOK OF DATA ON ORGANIC COMPOUNDS. R.C. Weast and M.J. Astle, editors. CRC Press, 2000 Corporate Boulevard, Boca Raton, FL 33431. (800) 272-7737. 1985. Two volumes. $270.00 for set.

A GUIDEBOOK TO MECHANISMS IN ORGANIC CHEMISTRY. P. Sykes. John Wiley and Sons, Inc., 605 Third Avenue, New York, NY 10158. (800) 526-5368. 6th edition. 1986. $21.95.

HANDBOOK OF APPLIED CHEMISTRY. Vollrath Hopp and Ingo Hennig. Hemisphere Publishing Corporation, 79 Madison Avenue, New York, NY 10016-7892. Order from: Taylor and Francis/Hemisphere Distribution Center, 242 Cherry Street, Philadelphia, PA 19106-1906. (800) 821-8312. 1983. $49.95.

HANDBOOK OF ORGANIC CHEMISTRY. John A. Dean. McGraw-Hill Book Company, 1221 Avenue of the Americas, New York, NY 10020. (212) 512-2000. 1987. $64.50.

LANGE'S HANDBOOK OF CHEMISTRY. John A. Dean, editor. McGraw-Hill Book Company, 1221 Avenue of the Americas, New York, NY 10020. (212) 512-2000. 1985. $59.50.

ONLINE DATA BASES

CA SEARCH. Chemical Abstracts Service, P.O. Box 3012, Columbus, OH 43120. (800) 848-6538 or (614) 421-3600. Comprehensive guide to chemical literature, 1972 to present. Inquire as to online cost and availability.

COMPENDEX. Engineering Information, Inc., 345 East 47th Street, New York, NY 10017. (800) 221-1044 or (212) 705-7615. Engineering and technical literature, 1975 to present. Inquire as to online cost and availability.

DISSERTATION ABSTRACTS ONLINE. University Microfilms International, 300 North Zeeb Road, Ann Arbor, MI 48106. (800) 521-0600 or (313) 761-4700. Scope includes virtually all doctoral dissertations accepted at accredited American institutions from 1861 to present in over 250 subject areas. Inquire as to online cost and availability.

NTIS. National Technical Information Service, 5285 Port Royal Road, Springfield, VA 22161. (703) 487-4630. Broad coverage of government sponsored research reports, 1964 to present. Inquire as to online cost and availability.

SCISEARCH. Institute for Scientific Information, 3501 Market Street, Philadelphia, PA 19104. (800) 523-1850 or (215) 386-0100. Broad multidisciplinary title and author index to the international literature of science and technology, 1974 to present. Inquire as to online cost and availability.

PERIODICALS

AMERICAN OIL CHEMISTS' SOCIETY. JOURNAL. American Oil Chemists' Society, 508 South Sixth Street, Champaign, IL

61820. (217) 359-2344. 1917 to present. Monthly. $60.00 per year.

CARBON. Pergamon Press, Inc., Maxwell House, Fairview Park, Elmsford, NY 10523. (914) 592-7700. 1963 to present. Bimonthly. $235.00 per year.

HYDROCARBON PROCESSING. Gulf Publishing Company, P.O. Box 2608, Houston, TX 77001. (713) 520-4444. Monthly. $10.00 per year.

JOURNAL OF HETEROCYCLIC CHEMISTRY. Hetero-Corporation, Box 16000 MH, Tampa, FL 33687. 1964 to present. 6 times per year. $160.00 per year.

JOURNAL OF ORGANIC CHEMISTRY. American Chemical Society, 1155 16th Street, N.W., Washington, DC 20036. (800) 424-6747. 1936 to present. Semi-monthly. $218.00 per year.

JOURNAL OF POLYMER SCIENCE. POLYMER CHEMISTRY EDITION. John Wiley and Sons, Inc., 605 Third Avenue, New York, NY 10158. (800) 526-5368. 1962 to present. Monthly. $895.00 per year, includes all editions.

ORGANOMETALLICS. American Chemical Society, 1155 16th Street, N.W., Washington, DC 20036. (800) 424-6747. 1982 to present. Monthly. $195.00 per year.

TETRAHEDRON. Pergamon Press, Inc., Maxwell House, Fairview Park, Elmsford, NY 10523. (914) 592-7700. 1957 to present. 24 times per year. $1400.00 per year.

TETRAHEDRON LETTERS. Pergamon Press, Inc., Maxwell House, Fairview Park, Elmsford, NY 10523. (914) 592-7700. 1959 to present. Weekly. $1500.00 per year.

RESEARCH CENTERS AND INSTITUTES

CHEMICAL LABORATORIES. Harvard University, Oxford Street, Cambridge, MA 02138. (617) 495-4283.

CHEMICAL RESEARCH LABORATORY. Brown University, Providence, RI 02912. (401) 863-2256.

UNIVERSITY/INDUSTRY CHEMICAL RESEARCH CENTER. Mississippi State University, Department of Chemistry, P.O. Drawer CH, Mississippi State, MS 39762. (601) 325-3584.

SPECIFICATIONS AND STANDARDS

ASTM STANDARDS ON MANUFACTURED CARBON AND GRAPHITE PRODUCTS. American Society for Testing and Materials, 1916 Race Street, Philadelphia, PA 19103. (215) 299-5400. 1981. $3.50.

CARBON-14 DATING

See: RADIOCARBON DATING

CARBONATES

See also: CAVES AND CAVING, GEOLOGY, GEOMORPHOLOGY, KARST, PETROLEUM GEOLOGY, SEDIMENTARY ROCKS

ABSTRACT SERVICES AND INDEXES

BIBLIOGRAPHY AND INDEX OF GEOLOGY. American Geological Institute, 4220 King Street, Alexandria, VA 22302. (703) 379-2480. 1969 to present. Monthly. $1100.00 per year.

CHEMICAL ABSTRACTS. American Chemical Society, Chemical Abstracts Service, Box 3012, Columbus, OH 43210. (614) 421-3600. 1907 to present. Weekly. $9500.00 per year.

GENERAL SCIENCE INDEX. H.W. Wilson and Company, 950 University Avenue, Bronx, NY 10452. (800) 367-6670 or (212) 588-8400. 1978 to present. Monthly. Inquire as to cost and availability.

SCIENCE CITATION INDEX. Institute for Scientific Information, 3501 Market Street, Philadelphia, PA 19104. (800) 523-1850 or (215) 386-0100. Six times per year. $6200.00 per year.

ASSOCIATIONS AND PROFESSIONAL SOCIETIES

AMERICAN GEOLOGICAL INSTITUTE. 4220 King Street, Alexandria, VA 22302. (703) 379-2480.

GEOLOGICAL SOCIETY OF AMERICA. 3300 Penrose Place, Boulder, CO 80301. (303) 447-2020.

NATIONAL SPELEOLOGICAL SOCIETY. Cave Avenue, Huntsville, AL 35810. (205) 852-1300.

SOCIETY OF ECONOMIC PALEONTOLOGISTS AND MINERALOGISTS. Box 4756, Tulsa, OK 74159. (918) 743-9765.

DIRECTORIES AND BIOGRAPHICAL SOURCES

AMERICAN MEN AND WOMEN OF SCIENCE. R.R. Bowker, Inc., Order Department, 245 West 17th Street, New York, NY 10011. (800) 521-8110. Eight volumes. 1986. $595.00 for set.

INTERNATIONAL RESEARCH CENTERS DIRECTORY 1988-89. Darren L. Smith, editor. Gale Research Company, Book Tower, Detroit, MI 48226. (800) 521-0707. 4th edition. 1987. $360.00.

NATIONAL SPELEOLOGICAL SOCIETY MEMBERSHIP LIST. Cave Avenue, Huntsville, AL 35810. (205) 852-1300. Annual. Inquire.

RESEARCH CENTERS DIRECTORY 1988. Gale Research Company, Book Tower, Detroit, MI 48226. (800) 521-0707. 12th edition. 1987. $365.00 for set.

SCIENTIFIC AND TECHNICAL ORGANIZATIONS AND AGENCIES DIRECTORY. Margaret Labash Young, editor. Gale Research Company, Book Tower, Detroit, MI 48226. (800) 521-0707. 2nd edition. 1987. $185.00.

SOCIETY OF ECONOMIC PALEONTOLOGISTS AND MINERALOGISTS MEMBERSHIP LIST. Box 4756, Tulsa, OK 74159. (918) 743-9765. Biennial. Inquire.

GENERAL WORKS

CARBONATE SEDIMENTS AND THEIR DIAGENESIS. R.G. Bathurst. Elsevier Science Publishing Company, Inc., 52 Vanderbilt Avenue, New York, NY 10017. (212) 370-5520. Second edition. 1975. $29.75 in paper.

INTRODUCTION TO CARBONATE SEDIMENTS AND ROCKS. T.P. Scoffin. Methuen, Inc., 29 West 35th Street, New York, NY 10001. (212) 244-3336. 1987. $17.95 in paper.

KARST GEOMORPHOLOGY. J.N. Jennings. Blackwell Scientific Publications, Inc., 52 Beacon Street, Boston, MA 02108. (800) 325-4177. 1985. $14.95 in paper.

SEDIMENTARY ENVIRONMENTS AND FACIES. H.G. Reading, editors. Blackwell Scientific Publications, Inc., 52 Beacon Street, Boston, MA 02108. (800) 325-4177. 2nd edition. 1986. $40.00 in paper.

HANDBOOKS AND MANUALS

ENCYCLOPEDIA OF FIELD GEOLOGY. Charles W. Finkl, editor. Van Nostrand Reinhold Company, Inc., 135 West 50th

Street, New York, NY 10020. (800) 543-2681. 1988. $89.95.

MANUAL OF CARBONATE SEDIMENTOLOGY: A LEXIGRAPHICAL APPROACH. T.J.A.Reijers and K.J. Hsu, editors. Academic Press, Inc., 6277 Sea Harbor Drive, Orlando, FL 32821. (800) 321-5068. 1986. $49.95.

ONLINE DATA BASES

CA SEARCH. Chemical Abstracts Service, P.O. Box 3012, Columbus, OH 43120. (800) 848-6538 or (614) 421-3600. Comprehensive guide to chemical literature, 1972 to present. Inquire as to online cost and availability.

GEOREF. Online version of the BIBLIOGRAPHY AND INDEX OF GEOLOGY. American Geological Institute, 4220 King Street, Alexandria, VA 22302. (703) 379-2480. 1969 to present. Inquire as to online cost and availability.

NTIS. National Technical Information Service, 5285 Port Royal Road, Springfield, VA 22161. (703) 487-4630. Broad coverage of government sponsored research reports, 1964 to present. Inquire as to online cost and availability.

SCISEARCH. Institute for Scientific Information, 3501 Market Street, Philadelphia, PA 19104. (800) 523-1850 or (215) 386-0100. Broad multidisciplinary title and author index to the international literature of science and technology, 1974 to present. Inquire as to online cost and availability.

WILSONLINE. H.W. Wilson and Company, 950 University Avenue, Bronx, NY 10452. (800) 367-6770 or (212) 588-8400. Makes available online versions of the H.W. Wilson indexes including Applied Science and Technology Index, Business Periodicals Index and Readers' Guide to Periodical Literature. Approximately 1980 to present. Inquire as to online cost and availability.

PERIODICALS

EARTH SURFACE PROCESSES AND LANDFORMS. British Geomorphological Research Group. Distributed by: John Wiley and Sons, Inc., 605 Third Avenue, New York, NY 10158. (800) 526-5368. 1976 to present. Bimonthly. $280.00 per year.

GEOLOGICAL SOCIETY OF AMERICA. BULLETIN. Geological Society of America. 3300 Penrose Place, Boulder, CO 80301. (303) 447-2020. 1888 to present. Monthly. $80.00 per year.

JOURNAL OF GEOLOGY. University of Chicago Press, 5801 Ellis Avenue, Chicago, IL 60637. (800) 621-2736. 1893 to present. Bimonthly. $30.00 per year.

JOURNAL OF SEDIMENTARY PETROLOGY. Society of Economic Paleontologists and Mineralogists, Box 4756, Tulsa, OK 74159. (918) 743-9765. 1931 to present. Bimonthly. $41.00 per year to individuals.

SEDIMENTARY GEOLOGY. Elsevier Science Publishing Company, Inc., 52 Vanderbilt Avenue, New York, NY 10017. (212) 370-5520. 1967 to present. Twenty times per year. $420.00 per year.

SEDIMENTOLOGY. Blackwell Scientific Publications, Inc., 52 Beacon Street, Boston, MA 02108. (800) 325-4177. 1952 to present. 6 times per year. $184.00 per year.

RESEARCH CENTERS AND INSTITUTES

BUREAU OF ECONOMIC GEOLOGY. University of Texas at Austin, Box X, University Station, Austin, TX 78713. (512) 471-7721.

CENTER FOR SEDIMENTOLOGY. Texas A&M University, College of Geosciences, College Station, TX 77843. (409) 845-2460.

NEW MEXICO MINES AND MINERAL RESOURCES. Institute of Mining and Technology, Campus Station, Socorro, NM 87801. (505) 5420.

CARBURETORS

See: AUTOMOTIVE ENGINEERING

CARTOGRAPHY

See also: AERIAL PHOTOGRAPHY, ARTIFICIAL SATELLITES, GEODESY, PHOTOGRAMMETRY, REMOTE SENSING, SURVEYING

ABSTRACT SERVICES AND INDEXES

APPLIED SCIENCE AND TECHNOLOGY INDEX. H.W. Wilson and Company, 950 University Avenue, Bronx, NY 10452. (800) 367-6670 or (212) 588-8400. Monthly. Inquire as to cost and availability.

BIBLIOGRAPHY AND INDEX OF GEOLOGY. American Geological Institute, 4220 King Street, Alexandria, VA 22302. (703) 379-2480. 1969 to present. Monthly. $1100.00 per year.

CURRENT CONTENTS: ENGINEERING, TECHNOLOGY AND APPLIED SCIENCES. Institute for Scientific Information, 3501 Market Street, Philadelphia, PA 19104. (800) 523-1850 or (215) 386-0100. Weekly. $275.00 per year.

ENGINEERING INDEX MONTHLY AND AUTHOR INDEX. Engineering Information Inc., 345 East 47th Street, New York, NY 10017. (212) 705-7600. Monthly. $1560.00 per year.

METEOROLOGICAL AND GEOASTROPHYSICAL ABSTRACTS. American Meteorological Society, 45 Beacon Street, Boston, MA 02108. (617) 227-2425. 1950 to present. Monthly. $450.00 per year.

REMOTE SENSING OF NATURAL RESOURCES: A QUARTERLY LITERATURE REVIEW. University of New Mexico, Technology Application Center, Albuquerque, NM 87131. (505) 277-3622. 1974 to present. Quarterly. $150.00. Available to qualified agencies only.

SCIENCE CITATION INDEX. Institute for Scientific Information, 3501 Market Street, Philadelphia, PA 19104. (800) 523-1850 or (215) 386-0100. Six times per year. $6200.00 per year.

ASSOCIATIONS AND PROFESSIONAL SOCIETIES

AMERICAN ASSOCIATION FOR GEODETIC SURVEYING. c/o American Congress on Surveying and Mapping, 210 Little Falls Street, Falls Church, VA 22046. (703) 241-2446.

AMERICAN CARTOGRAPHIC ASSOCIATION. c/o American Congress on Surveying and Mapping, 210 Little Falls Street, Falls Church, VA 22046. (703) 241-2446.

AMERICAN CONGRESS ON SURVEYING AND MAPPING. 210 Little Falls Street, Falls Church, VA 22046. (703) 241-2446.

AMERICAN SOCIETY FOR PHOTOGRAMMETRY AND REMOTE SENSING. 210 Little Falls Street, Falls Church, VA 22046-4398. (703) 534-6617.

NATIONAL ASSOCIATION OF PROFESSIONAL SURVEYORS. c/o American Congress on Surveying and Mapping, 210 Little Falls Street, Falls Church, VA 22046. (703) 241-2446.

NORTH AMERICAN CARTOGRAPHIC ASSOCIATION. 6010 Executive Boulevard, Suite 100, Rockville, MD 20852. (301) 443-8075.

OPTICAL SOCIETY OF AMERICA. 1816 Jefferson Place, N.W., Washington, DC 20036. (202) 223-8130.

SOCIETY OF PHOTOGRAPHIC SCIENTISTS AND ENGINEERS. 7003 Kilworth Lane, Springfield, VA 22151. (703) 642-9090.

SPIE - THE INTERNATIONAL SOCIETY FOR OPTICAL ENGINEERING. P.O. Box 10, 1022 19th Street, Bellingham, WA 98227. (206) 676-3290.

BIBLIOGRAPHIES

BIBLIOGRAPHY OF CARTOGRAPHY. Library of Congress Geography and Map Division. G.K. Hall and Company, 70 Lincoln Street, Boston, MA 02111. (800) 343-2806. Five volumes and supplements. 1973-1979. Inquire for set price.

DIRECTORIES AND BIOGRAPHICAL SOURCES

INTERNATIONAL RESEARCH CENTERS DIRECTORY 1988-89. Darren L. Smith, editor. Gale Research Company, Book Tower, Detroit, MI 48226. (800) 521-0707. 4th edition. 1987. $360.00.

1987 DIRECTORY OF ENGINEERING SOCIETIES AND RELATED ORGANIZATIONS. Gordon Davis, editor. Hemisphere Publishing Corporation, 1010 Vermont Avenue, NW, Washington, DC 20005. (800) 526-0275. 12th edition. 1987. $100.00.

RESEARCH CENTERS DIRECTORY 1988. Gale Research Company, Book Tower, Detroit, MI 48226. (800) 521-0707. 12th edition. 1987. $365.00 for set.

SCIENTIFIC AND TECHNICAL ORGANIZATIONS AND AGENCIES DIRECTORY. Margaret Labash Young, editor. Gale Research Company, Book Tower, Detroit, MI 48226. (800) 521-0707. 2nd edition. 1987. $185.00.

WHO'S WHO IN ENGINEERING. Gordon Davis, editor. Hemisphere Publishing Corporation, 1010 Vermont Avenue, NW, Washington, DC 20005. (800) 526-0275. 6th edition. 1985. $200.00.

GENERAL WORKS

CLOSE-RANGE PHOTOGRAMMETRY AND SURVEYING: STATE OF THE ART. American Society for Photogrammetry and Remote Sensing. 210 Little Falls Street, Falls Church, VA 22046-4398. (703) 534-6617. 1985. $65.00 in paper.

ELEMENTS OF CARTOGRAPHY. Arthur H. Robinson and others. John Wiley and Sons, Inc., 605 Third Avenue, New York, NY 10158. (800) 526-5368. 5th edition. 1984. $35.50.

ELEMENTS OF PHOTOGRAMMETRY. P.R. Wolf. McGraw-Hill Book Company, 1221 Avenue of the Americas, New York, NY 10020. (212) 512-2000. 2nd edition. 1983. $49.95.

INTRODUCTORY CARTOGRAPHY. John Campbell. Prentice-Hall Publishing, Inc., Englewood Cliffs, NJ 07632. (800) 562-0245. 1984. $37.95.

PHOTOGRAMMETRY. Francis H. Moffitt and Edward M. Mikhail. Harper and Row Publishers, Inc., 10 East 53rd Street, New York, NY 10022. (212) 207-7655. 3rd edition. 1980. $41.95.

PRINCIPLES OF REMOTE SENSING. Paul Curran. Halstead Press, division of John Wiley and Sons, Inc., 605 Third Avenue, New York, NY 10158. (800) 526-5368. 1986. $35.95.

REMOTE SENSING. Floyd F. Sabins. W.H. Freeman and Company, 41 Madison Avenue, New York, NY 10010. (212) 532-7660. 2nd edition. 1986. $39.95.

REMOTE SENSING METHODS AND APPPLICATIONS. R. Hord. John Wiley and Sons, Inc., 605 Third Avenue, New York, NY 10158. (800) 526-5368. 1986. $39.95.

SURVEY OF THE PROFESSION: PHOTOGRAMMETRY, SURVEYING, MAPPING, REMOTE SENSING. American Society for Photogrammetry and Remote Sensing. 210 Little Falls Street, Falls Church, VA 22046-4398. (703) 534-6617. 1982. $35.00 in paper.

HANDBOOKS AND MANUALS

MANUAL OF PHOTOGRAMMETRY. Chester C. Slama, editor. American Society for Photogrammetry and Remote Sensing. 210 Little Falls Street, Falls Church, VA 22046-4398. (703) 534-6617. 4th edition. 1980. $59.00.

MANUAL OF REMOTE SENSING. Robert N. Colwell, editor. American Society for Photogrammetry and Remote Sensing. 210 Little Falls Street, Falls Church, VA 22046-4398. (703) 534-6617. 2nd edition. 1983. $106.00 for set.

ONLINE DATA BASES

CA SEARCH. Chemical Abstracts Service, P.O. Box 3012, Columbus, OH 43120. (800) 848-6538 or (614) 421-3600. Comprehensive guide to chemical literature, 1972 to present. Inquire as to online cost and availability.

COMPENDEX. Engineering Information, Inc., 345 East 47th Street, New York, NY 10017. (800) 221-1044 or (212) 705-7615. Engineering and technical literature, 1975 to present. Inquire as to online cost and availability.

GEOREF. Online version of the BIBLIOGRAPHY AND INDEX OF GEOLOGY. American Geological Institute, 4220 King Street, Alexandria, VA 22302. (703) 379-2480. 1969 to present. Inquire as to online cost and availability.

NTIS. National Technical Information Service, 5285 Port Royal Road, Springfield, VA 22161. (703) 487-4630. Broad coverage of government sponsored research reports, 1964 to present. Inquire as to online cost and availability.

SCISEARCH. Institute for Scientific Information, 3501 Market Street, Philadelphia, PA 19104. (800) 523-1850 or (215) 386-0100. Broad multidisciplinary title and author index to the international literature of science and technology, 1974 to present. Inquire as to online cost and availability.

WILSONLINE. H.W. Wilson and Company, 950 University Avenue, Bronx, NY 10452. (800) 367-6770 or (212) 588-8400. Makes available online versions of the H.W. Wilson indexes including Applied Science and Technology Index, Business Periodicals Index and Readers' Guide to Periodical Literature. Approximately 1980 to present. Inquire as to online cost and availability.

PERIODICALS

AMERICAN CARTOGRAPHER. American Congress on Surveying and Mapping, 210 Little Falls Street, Falls Church, VA 22046. (703) 241-2446. 1974 to present. Quarterly. $60.00 per year.

AMERICAN CONGRESS OF SURVEYING AND MAPPING. BULLETIN. American Congress on Surveying and Mapping, 210 Little Falls Street, Falls Church, VA 22046. (703) 241-2446. 1950 to present. Bimonthly. $50.00 per year.

AMERICAN CONGRESS OF SURVEYING AND MAPPING. PROCEEDINGS. American Congress on Surveying and Mapping, 210 Little Falls Street, Falls Church, VA 22046. (703) 241-2446. 1942 to present. Semi-annual. $25.00 per year.

ASSOCIATION OF AMERICAN GEOGRAPHERS. ANNALS. Association of American Geographers, 1710 16th Street, N.W., Washington, DC 20009. (202) 234-1450. 1911 to present. Quarterly. $45.00 per year.

IEEE TRANSACTIONS ON GEOSCIENCE AND REMOTE SENSING. IEEE Geoscience and Remote Sensing Society. Institute of Electrical and Electronics Engineers, 345 East 47th Street, New York, NY 10017. (212) 705-7900. Order from: IEEE Service Center, 445 Hoes Lane, Piscataway, NJ 08854. 1963 to present. Bimonthly. $110.00 per year.

JOURNAL OF IMAGING SCIENCE. Society of Photographic Scientists and Engineers. 7003 Kilworth Lane, Springfield, VA 22151. (703) 642-9090. 1956 to present. Bimonthly. $70.00 per year.

JOURNAL OF IMAGING TECHNOLOGY. Society of Photographic Scientists and Engineers. 7003 Kilworth Lane, Springfield, VA 22151. (703) 642-9090. 1975 to present. Bimonthly. $70.00 per year.

MARINE GEODESY: AN INTERNATIONAL JOURNAL OF OCEAN SURVEYS, MAPPING, AND SENSING. Crane Russak and Company, Inc., 3 East 44th Street, New York, NY 10017. (212) 867-1490. 1977 to present. Quarterly. $86.00 per year.

PHOTOGRAMMETRIA. Elsevier Science Publishing Company, Inc., 52 Vanderbilt Avenue, New York, NY 10017. (212) 370-5520. 1949 to present. Quarterly. $65.00 per year.

PHOTOGRAMMETRIC ENGINEERING AND REMOTE SENSING. American Society for Photogrammetry and Remote Sensing. 210 Little Falls Street, Falls Church, VA 22046-4398. (703) 534-6617. Order from: Allen Press, Inc., 1041 New Hampshire Street, Box 368, Lawrence, KS 66044. 1934 to present. Monthly. $80.00 per year.

PHOTOGRAMMETRIC RECORD. Photogrammetry Society, Department of Photogrammetry and Surveying, University College London, Gower Street, London WC1E 6BT, England. Semiannual. $37.50 per year.

SCIENTIFIC AND APPLIED PHOTOGRAPHY AND CINEMATOGRAPHY. Gordon and Breach Science Publishers, Inc., 50 West 23rd Street, New York, NY 10010. (212) 206-8900. 12 times per year. $496.00 per year.

REMOTE SENSING OF ENVIRONMENT. Elsevier Science Publishing Company, Inc., 52 Vanderbilt Avenue, New York, NY 10017. (212) 370-5520. 1968 to present. Six times per year. $210.00 per year.

REMOTE SENSING REVIEWS. Harwood Academic Publishers, 50 West 23rd Street, New York, NY 10010. (212) 206-8900. Quarterly. $160.00 per year.

RESEARCH CENTERS AND INSTITUTES

CARTOGRAPHIC CENTER. Ohio University, Porter Hall, Athens, OH 45701. (614) 593-1150.

CENTER FOR REMOTE SENSING AND CARTOGRAPHY. 420 Chipta Way, Salt Lake City, UT 84112. (801) 581-8218.

GEOPHOTOGRAPHY AND REMOTE SENSING CENTER. University of Idaho, Geology Department, Moscow, ID 83843. (208) 885-7977.

NATIONAL RESEARCH COUNCIL OF CANADA, DIVISION OF PHYSICS. Ottawa, ON, Canada K1A OR6. (613) 993-1053.

PHOTOGRAMMETRY AND REMOTE SENSING SECTION. Tennessee Valley Authority, Office of Natural Resources and Economic Development, Haney Building, Chattanooga, TN 37401. (615) 755-2148.

CASSEGRAIN TELESCOPES

See: TELESCOPES

CAST IRON

See also: METALLURGY, STEEL AND STEEL MAKING

ABSTRACT SERVICES AND INDEXES

ALLOYS INDEX. American Society for Metals, 9639 Kinsman Road, Metals Park, OH 44073. (216) 338-5151. $130.00 per year.

APPLIED MECHANICS REVIEWS. American Society of Mechanical Engineers, 345 East 47th Street, New York, NY 10017. (212) 705-7722. Monthly. $380.00 per year. Critical reviews of the world literature in applied mechanics and related engineering science.

APPLIED SCIENCE AND TECHNOLOGY INDEX. H.W. Wilson Company, 950 University Avenue, Bronx, NY 10452. (800) 367-6670 or (212) 588-8400. Inquire as to cost and availability.

CHEMICAL ABSTRACTS. Chemical Abstracts Service, 2540 Olentangy Road, Post Office Box 3012, Columbus, OH 43210. (800) 848-6538 or (614) 421-3600. Weekly. $9200.00 per year.

CURRENT CONTENTS: ENGINEERING, TECHNOLOGY. Institute for Scientific Information, 3501 Market Street, Philadelphia, PA 19104. (800) 523-1850 or (215) 386-0100. $272.00 per year.

ENGINEERING INDEX MONTHLY. Engineering Information, Incorporated, 345 East 47th Street, New York, NY 10017. (800) 221-1044 or (212) 705-7600. Monthly, with annual cumulation. $1425.00 per year.

METALS ABSTRACTS. American Society for Metals, 9639Kinsman Road Metals Park, OH 44073. (216) 338-5151. Monthly. $890.00.

METALS ABSTRACT INDEX. American Society for Metals, 9639 Kinsman Road, Metals Park, OH 44073. (216) 338-5151. Monthly. (Sold only to subscribers of Metals Abstracts).

ANNUAL REVIEWS AND YEARBOOKS

ANNUAL REVIEW OF MATERIALS SCIENCE. Annual Reviews, Incorporated 4139 El Camino Way, Palo Alto, CA 94306. (415) 493-4400. Inquire.

ASSOCIATION AND PROFESSIONAL SOCIETIES

AMERICAN FOUNDRYMAN'S SOCIETY, INCORPORATED. Golf and Wolf Roads, Des Plaines, IL 60016. (312) 824-0181.

AMERICAN IRON AND STEEL INSTITUTE. 1000 Sixteenth Street, NW, Washington, DC 20036. (202) 452-7100.

AMERICAN SOCIETY FOR METALS. Metals Park, OH 44073. (216) 338-5151.

IRON CASTINGS SOCIETY. 455 State Street, Des Plaines, IL 60016. (312) 299-9160.

METALLURGICAL SOCIETY OF THE AIME (AMERICAN INSTITUTE OF MINING, METALLURGICAL AND PETROLEUM ENGINEERS). 420 Commonwealth Drive, Warrendale, PA 15086. (412) 776-9080.

BIBLIOGRAPHIES

SCIENTIFIC AND TECHNICAL BOOKS IN PRINT: AN INDEX TO LITERATURE IN SCIENCE AND TECHNOLOGY. R.R. Bowker Company, 205 East 42nd Street, New York, NY 10017. (800) 521-8110 or (212) 916-1600.

CAST IRON

DIRECTORIES AND BIOGRAPHICAL SOURCES

DUN'S INDUSTRIAL GUIDE - THE METAL WORKING DIRECTORY. Dun and Bradstreet, Incorporated, Three Century Drive, Parsippany, NJ 07054. (201) 455-0900. Annual. $550.00.

INTERNATIONAL RESEARCH CENTERS DIRECTORY 1986-1987. Gale Research Company, Book Tower, Detroit, MI 48226. (800) 521-0707. Third edition. 1986.

RESEARCH CENTERS DIRECTORY. Gale Research Company, Book Tower, Detroit, MI 48226. (800) 521-0707. Eleventh edition. 1987.

SCIENTIFIC AND TECHNICAL ORGANIZATIONS AND AGENCIES DIRECTORY. Gale Research Company, Book Tower, Detroit, MI 48226. (800) 521-0707. 1985.

WHO'S WHO IN ENGINEERING. Engineers Joint Council, 345 East 47th Street, New York, NY 10017. (212) 705-7010. 1985. $200.00.

WHO'S WHO IN FRONTIER SCIENCE AND TECHNOLOGY. Marquis Who's Who, Incorporated, 200 East Ohio Street, Chicago, IL 60611. (800) 428-3898 or (312) 787-2008.

WHO'S WHO IN TECHNOLOGY TODAY. Reston Publishing Company, Incorporated, c/o Prentice-Hall, Incorporated, Englewood Cliffs, NJ 07632. (800) 262-6868. Biennial. Five volumes. $425.00. Covers the fields of electronics, computer science, physics, optics, chemistry, biotechnology, mechanics, energy, and earth science.

ENCYCLOPEDIAS AND DICTIONARIES

ENCYCLOPEDIA OF PHYSICAL SCIENCE AND TECHNOLOGY. Academic Press, Incorporated, Orlando, FL 32887. (800) 321-5068 or (305) 345-2734. Fifteen volumes, 1986.

MCGRAW-HILL DICTIONARY OF ENGINEERING. Sybil P. Parker, editor. McGraw-Hill Book Company, 1221 Avenue of the Americas, New York, NY 10020. (212) 512-2000. 1984. $39.95.

MCGRAW-HILL ENCYCLOPEDIA OF SCIENCE AND TECHNOLOGY. McGraw-Hill Book Company, 1221 Avenue of the Americas, New York, NY 10020. (212) 512-2000.

GENERAL WORKS

CAST IRON: PHYSICAL AND ENGINEERING PROPERTIES. H.T. Angus. Butterworths Publishing, 80 Montvale Avenue, Stoneham, MA 02180. (617) 438-8464. Second edition. 1978. $160.00.

ESSENTIAL METALLURGY FOR ENGINEERS. W. Alexander, G. Davies, and K. Reynolds. Van Nostrand Reinhold, 115 Fifth Avenue, New York, NY 10003. (800) 543-2681. 1985. $17.95.

METALLURGY BASICS. D.V. Brown. Van Nostrand Reinhold, 115 Fifth Avenue, New York, NY 10003. (800) 543-2681. 1985. $17.95.

PHYSICAL METALLURGY. P. Haasen. Cambridge University Press, 32 East 57th Street, New York, NY 10022. (212) 688-8885. 1986. $24.95 in paper.

PHYSICAL METALLURGY OF CAST IRON: MATERIALS RESEARCH SOCIETY SYMPOSIA PROCEEDINGS, VOLUME 34. H. Fredriksson and M. Hillert, editors. Elsevier Science Publishing Company, Incorporated, 52 Vanderbilt Avenue, New York, NY 10017. (212) 370-5520. 1985. $90.00.

HANDBOOKS AND MANUALS

SMITHELL'S METALS REFERENCE BOOK. Eric A. Brandes, editor. Butterworth Publishers, 80 Montvale Avenue, Stoneham, MA 02180. (800) 325-4177. Sixth edition. 1983. $210.00.

ONLINE DATA BASES

COMPENDEX. Engineering Information, Incorporated, 345 East 47th Street, New York, NY 10017. (800) 221-1044 or (212) 705-7615. Engineering and technical literature, 1975 to present. Inquire as to cost and availability.

INSPEC. INSPEC Marketing Department, Institute of Electrical and Electronics Engineers, Incorporated, IEEE Service Department, 445 Hoes Lane, Piscataway, NJ 08854. (201) 981-0060. Inquire as to on-line cost and availability.

METADEX. Metals Information, American Society for Metals, Metals Park, Oh 44073. (216) 338-5151. (Metals Abstracts/Alloys Index). A worldwide literature on the science and practice of metallurgy, 1966 to present. Inquire as to online cost and availability.

NASA. National Aeronautics and Space Administration, Scientific and Technical Information Branch, 300 7th Street, SW, Washington, DC 20546. Citations and abstracts of aerospace literature, 1962 to present. Inquire as to cost and availability.

NTIS. National Technical Information Service, 5285 Port Royal Road, Springfield, VA 22161. (703) 487-4630. Broad coverage of government sponsored research reports, 1964 to present. Inquire as to cost and availability.

WILSONLINE. H.W. Wilson Company, 950 University Avenue, Bronx, NY 10452. (800) 367-6770 or (212) 588-8400. Makes available online versions of the printed H.W. Wilson Indexes included Applied Science and Technology Index, Business Periodicals Index, and Readers' Guide to Periodical Literature. Period covered is generally 1983 to present. Inquire as to cost and available.

OTHER SOURCES

INTERDOC: DIRECTORY OF PUBLISHED PROCEEDINGS, SERIES. SEMT-Science/Engineering/Medicine/Technology. Interdoc Corporation, 173 Halstead Avenue, Box 326, Harrison, NY 10528. (914) 835-3506. Ten times per year. $325.00 per year.

MATERIALS SCIENCE AND METALLURGY. National Technical Information Service (NTIS), 5285 Port Royal Road, Springfield, VA 22161. (703) 487-4630. Translations and abstracts of foreign language technical media. Irregular. $40.00 per year.

PERIODICALS

CASTING DIGEST. American Society for Metals, 9639 Kinsman Road, Metals Park, OH 44073. (216) 338-5151. Monthly. $90.00 per year.

CASTING WORLD. Continental Communications, Incorporated, 1115 Main Street, Bridgeport, CT 06604. (203) 377-5566. Quarterly. $10.00 per year.

FOUNDRY MANAGEMENT AND TECHNOLOGY. Penton-IPC, 1100 Superior Avenue, Cleveland, OH 44114. (216) 696-7000. Monthly. $35.00 per year.

IRONCASTER. Iron Castings Society, 455 State Street, Des Plaines, IL 60016. (312) 299-9160. Bimonthly. $14.50 per year.

JOURNAL OF METALS. Metallurgical Society of the AIME (American Institute of Mining, Metallurgical and Petroleum Engineers), 420 Commonwealth Drive, Warrendale, PA 15086. (412) 776-9080. Monthly. $40.00 per year.

METALLURGICAL TRANSACTIONS. Metallurgical Society of the AIME (American Institute of Mining, Metallurgical and Petroleum Engineers), 420 Commonwealth Drive, Warrendale,

PA 15086. (412) 776-9080. Monthly. $95.00 per year.

METALS WEEK. McGraw-Hill Book Company, Incorporated, 1221 Avenue of the Americas, New York, NY 10020. (212) 997-2823. Weekly. $527.00 per year.

MODERN CASTING. American Foundrymen's Society, Incorporated, Golf and Wolf Roads, Des Plaines, IL 60016. (312) 824-0181. Monthly. $25.00 per year.

RESEARCH CENTERS AND INSTITUTES

AMERICAN IRON AND STEEL INSTITUTE. 1000 Sixteenth Street, NW, Washington, DC 20036. (202) 452-7100.

CARNEGIE-MELLON UNIVERSITY. Center for Iron and Steel Making Research, MEMS Department, Pittsburgh, PA 15312. (412) 578-2677.

COLORADO SCHOOL OF MINES. Steel Research Center, Golden, CO 80401. (303) 273-3774.

IRON CASTING RESEARCH INSTITUTE, INCORPORATED. 870 West Third Avenue, Columbus, OH 43212. (614) 299-3336.

UNIVERSITY OF CONNECTICUT. Institute of Materials Science, Storrs, CT 06268. (203) 486-4623.

UNIVERSITY OF WISCONSIN AT MADISON. Cast Metals Laboratory, 1509 University Avenue, Madison, WI 53706. (608) 262-2562.

CATALYSIS

ABSTRACT SERVICES AND INDEXES

APPLIED SCIENCE AND TECHNOLOGY INDEX. H.W. Wilson Company, 950 University Avenue, Bronx, NY 10452. (800) 367-6670 or (212) 588-8400. Inquire as to cost and availability.

CHEMICAL ABSTRACTS. Chemical Abstracts Service, 2540 Olentangy Road, P.O. Box 3012, Columbus, OH 43210. (800) 848-6538 or (614) 421-3600. Weekly. $9200.00 per year.

CHEMICAL INDUSTRY NOTES. Chemical Abstracts Service, P.O. Box 3012, Columbus, OH 43210. (800) 848-6538 or (614) 421-3600. Weekly. $2000.00 per year; $1175.00 with index volume.

ENGINEERING INDEX MONTHLY. Engineering Information, Incorporated, 345 East 47th Street, New York, NY 10017. (800) 221-1044 or (212) 705-7600. Monthly, with annual cumulation. $1560.00 per year.

GENERAL SCIENCE INDEX. H.W. Wilson Company, 950 University Avenue, Bronx, NY 10452. (800) 367-6770 or (212) 588-8400.

PROCESS ENGINEERING INDEX. Technical Indexes Limited, Willougby Road, Bracknell, Berks, England RG12 4DW. Semiannual.

ANNUAL REVIEWS AND YEARBOOKS

ADVANCES IN CATALYSIS AND RELATED SUBJECTS. Academic Press, Incorporated, Orlando, FL 32887. (800) 321-5068. (302) 345-2000. Annual. Inquire as to price.

ADVANCES IN CHEMICAL ENGINEERING. Academic Press, Incorporated, Orlando, FL 32887. (800) 321-5068. (302) 345-2000. Annual. Inquire as to price.

ASSOCIATIONS AND PROFESSIONAL SOCIETIES

AMERICAN CHEMICAL SOCIETY. 1155 Sixteenth Street, NW, Washington, DC 20036. (202) 872-4600.

AMERICAN INSTITUTE OF CHEMICAL ENGINEERS. 345 East 47th Street, New York, NY 10017. (212) 705-7338.

CATALYSIS SOCIETY OF NORTH AMERICA. c/o Dr. Williams J. Linn, E.I. Dupont Experiment Station, Wilmington, DE 19898. (302) 772-4655.

DIRECTORIES AND BIOGRAPHICAL SOURCES

CONSULTING SERVICES DIRECTORY. Association of Consulting Chemists and Chemical Engineers, 50 East 41st Street, New York, NY 10017. (212) 686-6255. Annual. $45.00.

RESEARCH CENTERS DIRECTORY. Gale Research Company, Detroit, MI 48226. (800) 521-0707. Twelfth edition. 1988.

WHO'S WHO IN FRONTIER SCIENCE AND TECHNOLOGY. Marquis Who's Who, Incorporated, 200 East Ohio Street, Chicago, IL 60611. (800) 428-3898 or (312) 787-2008.

WHO'S WHO IN TECHNOLOGY TODAY. Reston Publishing Company, Incorporated, c/o Prentice-Hall, Incorporated, Englewood Cliffs, NJ 07632. (800) 262-6868. Biennial. Five volumes. $425.00. Covers the fields of electronics, computer science, physics, optics, chemistry, biotechnology, mechanics, energy, and earth science.

ENCYCLOPEDIAS AND DICTIONARIES

DICTIONARY OF CHEMICAL TERMINOLOGY. D. Kryt. Elsevier Publishing Company, Incorporated, 52 Vanderbilt Avenue, New York, NY 10017. (212) 370-5520. 1980. $83.00.

ENCYCLOPEDIA OF PHYSICAL SCIENCE AND TECHNOLOGY. Academic Press, Incorporated, Orlando, FL 32887. (800) 321-5068 or (305) 345-2734. Fifteen volumes. 1986.

MCGRAW-HILL ENCYCLOPEDIA OF SCIENCE AND TECHNOLOGY. McGraw-Hill Book, Incorporated, 1221 Avenue of the Americas, New York, NY 10020. (212) 997-3675.

GENERAL WORKS

APPLIED INDUSTRIAL CATALYSIS. Bruce E. Leach, editor. Academic Press, Incorporated, 6277 Sea Harbor Drive, Orlando, FL 32821. (800) 321-5068. Two Volumes. 1983. $115.00 set.

CATALYSIS AND CHEMICAL PROCESSES. Ronald Pearce and William R. Patterson. Halsted Press, Division of John Wiley and Sons, 605 Third Avenue, New York, NY 10158. (800) 526-5368 or (212) 850-6000. 1981. $79.95.

CATALYSIS: SCIENCE AND TECHNOLOGY. J.R. Anderson and M. Boudart, editors. Spring-Verlag New York, 175 Fifth Avenue, New York, NY 10010. (212) 460-1500. Seven volumes. Price varies.

ONLINE DATA BASES

CA SEARCH. Chemical Abstracts Service, P.O. Box 3012, Columbus, OH 43210. Guide to chemical literature, 1972 to present. Inquire as to cost and availability.

COMPENDEX. Engineering Information, Incorporated, 345 East 47th Street, New York, NY 10017. (800) 221-1044 or (212) 705-7615. Engineering and technical literature, 1975 to present. Inquire as to cost and availability.

NTIS. National Technical Information Service, 5285 Port Royal Road, Springfield, VA 22161. (703) 487-4630. Broad coverage of government sponsored research reports, 1964 to present.

Inquire as to cost and availability.

PERIODICALS

AICHE JOURNAL. American Institute of Chemical Engineers, 345 East 47th Street, New York, NY 10017. (212) 705-7321. Bimonthly. $60.00 per year.

CATALYSIS TODAY. Elsevier Science Publishing Company, 52 Vanderbilt Avenue, New York, NY 10017. Five times per year. $200.00 per year.

CATALYSIS REVIEWS - SCIENCE AND ENGINEERING. Marcel Dekker, Incorporated, 270 Madison Avenue, New York, NY 10016. (800) 228-1160 or (212) 696-9000. Quarterly. $200.00 per year.

CATALYSIS SOCIETY NEWSLETTER. Catalysis Society of North America, Dupont Experiment Station, Wilmington, DE 19898. (302) 772-2622. Inquire as to cost and availability.

CHEMICAL ENGINEERING. McGraw-Hill Book Company, 1221 Avenue of the Americas, New York, NY 10020. (212) 512-2000. Biweekly. $24.50 per year.

RESEARCH CENTERS AND INSTITUTES

NATIONAL SCIENCE FOUNDATION. Kinetics, Catalysis, and Reaction Engineering Program, Chemical and Process Engineering Division, 1800 G Street, NW, Washington, DC 20550. (202) 357-9624.

NORTHWESTERN UNIVERSITY. Catalysis Center, 2145 Sheraton Road, Evanston, IL 60201. (312) 491-4354.

UNIVERSITY OF DELAWARE. Center for Catalytic Science and Technology, Departments of Chemical Engineering and Chemistry, Newark, DE 19716. (302) 451-8056.

UNIVERSITY OF FLORIDA. Center for Catalysis, Gainesville, Fl 32601. (904) 392-6043.

CATALYTIC CONVERTER

See: AIR POLLUTION CONTROL

CATHODE RAY TUBE (CRT)

See: COMPUTER COMMUNICATIONS

CAVES AND CAVING

See also: GEOLOGY, GEOMORPHOLOGY, KARST, SEDIMENTARY ROCKS, SPELEOLOGY

ABSTRACT SERVICES AND INDEXES

BIBLIOGRAPHY AND INDEX OF GEOLOGY. American Geological Institute, 4220 King Street, Alexandria, VA 22302. (703) 379-2480. 1969 to present. Monthly. $1100.00 per year.

CHEMICAL ABSTRACTS. American Chemical Society, Chemical Abstracts Service, Box 3012, Columbus, OH 43210. (614) 421-3600. 1907 to present. Weekly. $9500.00 per year.

GENERAL SCIENCE INDEX. H.W. Wilson and Company, 950 University Avenue, Bronx, NY 10452. (800) 367-6670 or (212) 588-8400. 1978 to present. Monthly. Inquire as to cost and availability.

SCIENCE CITATION INDEX. Institute for Scientific Information, 3501 Market Street, Philadelphia, PA 19104. (800) 523-1850 or (215) 386-0100. Six times per year. $6200.00 per year.

ANNUAL REVIEWS AND YEARBOOKS

CAVE RESEARCH FOUNDATION ANNUAL REPORT. 1019 Maplewood Drive, #211, Cedar Falls, IA 50613. Available from: Cave Books MO, 756 Harvard Avenue, St. Louis, MO 63130. (314) 862-7646. Annual. Inquire as to price and availability.

ASSOCIATIONS AND PROFESSIONAL SOCIETIES

AMERICAN GEOLOGICAL INSTITUTE. 4220 King Street, Alexandria, VA 22302. (703) 379-2480.

ASSOCIATION FOR MEXICAN CAVE STUDIES. P.O. Box 7037, Austin, TX 78712. (512) 847-2709.

CAVE RESEARCH FOUNDATION. 1019 Maplewood Drive, #211, Cedar Falls, IA 50613.

GEOLOGICAL SOCIETY OF AMERICA. 3300 Penrose Place, Boulder, CO 80301. (303) 447-2020.

NATIONAL SPELEOLOGICAL SOCIETY. Cave Avenue, Huntsville, AL 35810. (205) 852-1300.

DIRECTORIES AND BIOGRAPHICAL SOURCES

AMERICAN MEN AND WOMEN OF SCIENCE. R.R. Bowker, Inc., Order Department, 245 West 17th Street, New York, NY 10011. (800) 521-8110. Eight volumes. 1986. $595.00 for set.

INTERNATIONAL RESEARCH CENTERS DIRECTORY 1988-89. Darren L. Smith, editor. Gale Research Company, Book Tower, Detroit, MI 48226. (800) 521-0707. 4th edition. 1987. $360.00.

NATIONAL SPELEOLOGICAL SOCIETY MEMBERSHIP LIST. Cave Avenue, Huntsville, AL 35810. (205) 852-1300. Annual. Inquire.

RESEARCH CENTERS DIRECTORY 1988. Gale Research Company, Book Tower, Detroit, MI 48226. (800) 521-0707. 12th edition. 1987. $365.00 for set.

SCIENTIFIC AND TECHNICAL ORGANIZATIONS AND AGENCIES DIRECTORY. Margaret Labash Young, editor. Gale Research Company, Book Tower, Detroit, MI 48226. (800) 521-0707. 2nd edition. 1987. $185.00.

GENERAL WORKS

CAVERS, CAVES AND CAVING. Bruce Sloane. Rutgers University Press, 30 College Avenue, New Brunswick, NJ 08903. (201) 932-7764. 1977. $25.00.

EXPLORING CAVES: A GUIDE TO THE UNDERGROUND WILDERNESS. David McClurg. Stackpole Books, Inc., P.O. Box 1831, Harrisburg, PA 17105. (800) 732-3669. 1980. $11.95.

INTRODUCTION TO CARBONATE SEDIMENTS AND ROCKS. T.P. Scoffin. Methuen, Inc., 29 West 35th Street, New York, NY 10001. (212) 244-3336. 1987. $17.95 in paper.

KARST GEOMORPHOLOGY. J.N. Jennings. Blackwell Scientific Publications, Inc., 52 Beacon Street, Boston, MA 02108. (800) 325-4177. 1985. $14.95 in paper.

KARST LANDFORMS. M.M. Sweeting. Columbia University Press, 562 West 113th Street, New York, NY 10025. (212) 316-7100. 1973. $60.00.

SCIENCE OF SPELEOLOGY. T.D. Ford, editor. Academic Press, Inc., 6277 Sea Harbor Drive, Orlando, FL 32821. (800) 321-5068. 1976. $60.50.

SEDIMENTARY ENVIRONMENTS AND FACIES. H.G. Reading, editors. Blackwell Scientific Publications, Inc., 52 Beacon Street, Boston, MA 02108. (800) 325-4177. 2nd edition. 1986. $40.00 in paper.

SPELEOLOGY: THE STUDY OF CAVES. George W. Moore and G.N. Sullivan. Cave Books, 756 Harvard Avenue, St. Louis, MO 63130. (314) 862-7646. 1981. $5.95.

HANDBOOKS AND MANUALS

CAVE RESEARCH FOUNDATION PERSONNEL MANUAL. Diana O. Daunt-Mergens, editor. Available from: Cave Books, 756 Harvard Avenue, St. Louis, MO 63130. (314) 862-7646. 1981. $5.00.

CAVES AND CAVING: A HANDBOOK AND GUIDE TO AMERICAN CAVES. Don Jacobson and Lee Stral. Harbor House Publications, 221 Water Street, Boyne City, MI 49712. (616) 582-2814. 1986. $10.95.

ENCYCLOPEDIA OF FIELD GEOLOGY. Charles W. Finkl, editor. Van Nostrand Reinhold Company, Inc., 135 West 50th Street, New York, NY 10020. (800) 543-2681. 1988. $89.95.

ONLINE DATA BASES

CA SEARCH. Chemical Abstracts Service, P.O. Box 3012, Columbus, OH 43120. (800) 848-6538 or (614) 421-3600. Comprehensive guide to chemical literature, 1972 to present. Inquire as to online cost and availability.

GEOREF. Online version of the BIBLIOGRAPHY AND INDEX OF GEOLOGY. American Geological Institute, 4220 King Street, Alexandria, VA 22302. (703) 379-2480. 1969 to present. Inquire as to online cost and availability.

NTIS. National Technical Information Service, 5285 Port Royal Road, Springfield, VA 22161. (703) 487-4630. Broad coverage of government sponsored research reports, 1964 to present. Inquire as to online cost and availability.

SCISEARCH. Institute for Scientific Information, 3501 Market Street, Philadelphia, PA 19104. (800) 523-1850 or (215) 386-0100. Broad multidisciplinary title and author index to the international literature of science and technology, 1974 to present. Inquire as to online cost and availability.

WILSONLINE. H.W. Wilson and Company, 950 University Avenue, Bronx, NY 10452. (800) 367-6770 or (212) 588-8400. Makes available online versions of the H.W. Wilson indexes including Applied Science and Technology Index, Business Periodicals Index and Readers' Guide to Periodical Literature. Approximately 1980 to present. Inquire as to online cost and availability.

PERIODICALS

CAVES AND CAVING. British Cave Research Association, 30 Main Road, Westonzoyland, Bridgewater, Somerset TA7 OEB, England. 1973 to present. Quarterly. $10.00 per year.

EARTH SURFACE PROCESSES AND LANDFORMS. British Geomorphological Research Group. Distributed by: John Wiley and Sons, Inc., 605 Third Avenue, New York, NY 10158. (800) 526-5368. 1976 to present. Bimonthly. $280.00 per year.

GEOLOGICAL SOCIETY OF AMERICA BULLETIN. Geological Society of America. 3300 Penrose Place, Boulder, CO 80301. (303) 447-2020. 1888 to present. Monthly. $80.00 per year.

JOURNAL OF GEOLOGY. University of Chicago Press, 5801 Ellis Avenue, Chicago, IL 60637. (800) 621-2736. 1893 to present. Bimonthly. $30.00 per year.

NATIONAL SPELEOLOGICAL SOCIETY BULLETIN. National Speleological Society, Cave Avenue, Huntsville, AL 35810. (205) 852-1300. 1941 to present. Semi-annual. $15.00 per year.

NATIONAL SPELEOLOGICAL SOCIETY NEWS. National Speleological Society, Cave Avenue, Huntsville, AL 35810. (205) 852-1300. 1943 to present. Monthly. $15.00 per year.

PROGRESS IN PHYSICAL GEOGRAPHY. Cambridge University Press, 32 East 57th Street, New York, NY 10022. (800) 872-7423. 1976 to present. $50.00 per year.

SEDIMENTARY GEOLOGY. Elsevier Science Publishing Company, Inc., 52 Vanderbilt Avenue, New York, NY 10017. (212) 370-5520. 1967 to present. Twenty times per year. $420.00 per year.

SEDIMENTOLOGY. Blackwell Scientific Publications, Inc., 52 Beacon Street, Boston, MA 02108. (800) 325-4177. 1952 to present. 6 times per year. $184.00 per year.

RESEARCH CENTERS AND INSTITUTES

ASSOCIATION FOR MEXICAN CAVE STUDIES. P.O. Box 7037, Austin, TX 78712. (512) 847-2709.

CAVE RESEARCH FOUNDATION. 1019 Maplewood Drive, #211, Cedar Falls, IA 50613.

CELESTIAL MECHANICS

See also: ASTRONOMY, ASTROPHYSICS

ABSTRACT SERVICES AND INDEXES

ASTRONOMY AND ASTROPHYSICS ABSTRACTS. Springer-Verlag New York, Incorporated, 175 Fifth Avenue, New York, NY 10010. (212) 460-1500. $70.00 per year.

ASTRONOMY AND ASTROPHYSICS MONTHLY INDEX. Olivetree Associates, Post Office Box 236, Sierra Madre, CA 91024. $212.00 per year. Complimentary copies available on request.

PHYSICS ABSTRACTS. Institute of Electrical Engineers, London, United Kingdom. Available from: Institute of Electrical and Electronic Engineers (IEEE), 345 East 47th Street, New York, NY 10017. (212) 705-7900. 1898 to present. Monthly. $1670.00 per year.

SCIENCE CITATION INDEX. Institute for Scientific Information, 3501 Market Street, Philadelphia, PA 19104. (800) 523-1850 or (215) 386-0100. Inquire as to cost and availability.

STAR. (Scientific and Technical Aerospace Reports. United States National Aeronautics and Space Administration, Scientific and Technical Information Facility, Box 8757, Baltimore-Washington International Airport, MD 21240. (202) 755-2210. Semimonthly, with semiannual and annual indexes. $85.00 per year.

ANNUAL REVIEWS AND YEARBOOKS

THE ASTRONOMICAL ALMANAC. Superintendent of Documents, United States Government Printing Office, Washington, DC 20402. (202) 783-3238. Yearly.

ANNUAL REVIEW OF ASTRONOMY AND ASTROPHYSICS. Annual Reviews, Incorporated, 4139 El Camino Way, Palo Alto, CA 94306. (415) 493-4400.

ASSOCIATIONS AND PROFESSIONAL SOCIETIES

AMERICAN ASTRONOMICAL SOCIETY. 2000 Florida Avenue, NW, Suite 300, Washington, DC 20009. (202) 659-0134.

AMERICAN ASSOCIATION OF VARIABLE STAR OBSERVERS. 187 Concord Avenue, Cambridge, MA 02138. (617) 354-0484.

ASTRONOMICAL LEAGUE. Post Office Box 12821, Tucson, AZ 85732. (602) 790-8471.

ASTRONOMICAL SOCIETY OF THE PACIFIC. 1290 24th Avenue, San Francisco, CA 94122. (415) 661-8660.

BIBLIOGRAPHIES

A BIBLIOGRAPHY OF ASTRONOMY, 1970-1979. R.A. Seal and S.S. Martin. Libraries Unlimited, Incorporated, Littleton, CO 80160. 1982. $37.50.

SCIENTIFIC AND TECHNICAL BOOKS AND SERIALS IN PRINT 1988; AN INDEX TO LITERATURE IN SCIENCE AND TECHNOLOGY. R.R. Bowker Company, 205 East 42nd Street, New York, NY 10017. (800) 521-8110 or (212) 916-1600. $175.00.

DIRECTORIES AND BIOGRAPHICAL SOURCES

AMERICAN MEN AND WOMEN OF SCIENCE. Physical and Biological Sciences. Sixteenth edition. R.R. Bowker Company, 205 East 42nd Street, New York, NY 10017. (800) 521-8110 or (212) 916-1600. 1987. $595.00.

THE BIOGRAPHICAL DICTIONARY OF SCIENTISTS: ASTRONOMERS. D. Abbott, editor. Peter Bedrick Books, 125 East 23rd Street, New York, NY 10010. 1984.

DIRECTORY OF PHYSICS AND ASTRONOMY STAFF MEMBERS. American Institute of Physics, 335 East 45th Street, New York, NY 10017. Annual.

RESEARCH CENTERS DIRECTORY. Gale Research Company, Book Tower, Detroit, MI 48226. Twelfth edition, 1988. (800) 521-0707. $240.00.

WHO'S WHO IN FRONTIER SCIENCE AND TECHNOLOGY. Marquis Who's Who, Incorporated, 200 East Ohio Street, Chicago, IL 60611. (800) 428-3898 or (312) 787-2008.

GENERAL WORKS

THE CAMBRIDGE ASTRONOMY GUIDE: A PRACTICAL INTRODUCTION TO ASTRONOMY. Cambridge University Press, 32 East 57th Street, New York, NY 10022. (212) 688-8888. 1985. $24.95.

INTRODUCTION TO CELESTIAL MECHANICS. Forest R. Moulton. Dover Publishers, Incorporated, 31 East Second Street, Mineola, NY 11501. (516) 294-7000. 1984. $8.95 in paper.

SPACE DYNAMICS AND CELESTIAL MECHANICS. K.B. Bhatnagar, editor. Kluwer Academic Publishers, 190 Old Derby Street, Hingham, MA 02043. (617) 749-5262. 1986. $99.00.

ONLINE DATA BASES

DISSERTATION ABSTRACTS ONLINE. University Microfilms International, 300 North Zeeb Road, Ann Arbor, MI 48106. (800) 521-0600 or (313) 761-4700. Scope includes virtually all doctoral dissertations accepted at accredited American institutions from 1861 to present in 252 subject areas. Inquire as to cost and availability.

INSPEC. INSPEC Marketing Department, Institute of Electrical and Electronics Engineers, Incorporated, IEEE Service Department, 445 Hoes Lane, Piscataway, NJ 08854. (201) 981-0060. Inquire as to on-line cost and availability.

MATHFILE. American Mathematical Society, Post Office Box 6248, Providence, RI 02940. (800) 556-7774 or (401) 272-9500. Scope includes pure and applied mathematics and related areas of physics, statistics, engineering, computer science, and operations research literature since 1973. Inquire as to cost and availability.

NASA. National Aeronautics and Space Administration, Scientific and Technical Information Branch, 300 7th Street, SW, Washington, DC 20546. Citations and abstracts of aerospace literature, 1962 to present. Inquire as to cost and availability.

NTIS. National Technical Information Service, 5285 Port Royal Road, Springfield, VA 22161. (703) 487-4630. Broad coverage of government sponsored research reports, 1964 to present. Inquire as to cost and availability.

SCISEARCH. Institute for Scientific Information, 3501 Market Street, Philadelphia, PA 19104. (800) 523-1850 or (215) 386-0100. Broad multidisciplinary title and author index to the international literature of science and technology, 1974 to present. Inquire as to cost and availability.

OTHER SOURCES

SKY CATALOGUE 2000.0. A. Hirshfeld and R. Sinnott. Cambridge University Press, 32 East 57th Street, New York, NY 10022. (800) 431-1580 or (212) 688-8888.

PERIODICALS

ASTRONOMICAL JOURNAL. American Astronomical Society. Available from: American Institute of Physics, 335 East 45th Street, New York, NY 10017. (212) 661-9404. $125.00 per year.

ASTRONOMICAL SOCIETY OF THE PACIFIC PUBLICATIONS. Astronomical Society of the Pacific, 1290 24th Avenue, San Francisco, CA 94122. (415) 661-8660. Monthly. $38.00.

ASTRONOMY. Astro Media Corporation, 625 East Paul Avenue, Milwaukee, WI 53202. Monthly. $18.00 per year.

ASTRONOMY AND ASTROPHYSICS. Springer-Verlag New York, Incorporated, 175 Fifth Avenue, New York, NY 10010. (800) 526-7254 or (212) 460-1500. $680.00 per year.

ASTROPHYSICAL JOURNAL. American Astronomical Society, University of Chicago Press, 5801 Ellis Avenue, Chicago, IL 60637. Biweekly. $305.00 per year.

ASTROPHYSICS AND SPACE SCIENCE. D. Reidel Publishing Company, 190 Old Derby Street, Hingham, MA 02043. Monthly. $101.00 per year.

CELESTIAL MECHANICS: AN INTERNATIONAL JOURNAL OF SPACE DYNAMICS. D. Reidel Publishing Company, 190 Old Derby Street, Hingham, MA 02043. Monthly. $310.00 per year.

EARTH, MOON AND PLANETS: AN INTERNATIONAL JOURNAL OF COMPARATIVE PLANETOLOGY. D. Reidel Publishing Company, 190 Old Derby Street, Hingham, MA 02043. Nine times per year. $275.00 per year.

ICARUS: INTERNATIONAL JOURNAL OF THE SOLAR SYSTEM STUDIES. Academic Press, Incorporated, Orlando, FL 32887. (305) 345-4100. Monthly. $484.00 per year.

MERCURY. Astronomical Society of the Pacific, 1290 245h Avenue, San Francisco, CA 94122. (415) 661-8660. Bimonthly. $21.00 per year.

MONTHLY NOTICES OF THE ROYAL ASTRONOMICAL SOCIETY. Blackwell Science Publications, Incorporated, 667 Lytton Avenue, Palo Alto, CA 94301. (415) 324-1688. Monthly. $134.00 per year.

PLANETARY AND SPACE SCIENCE. Pergamon Press, Incorporated, Maxwell House, Fairview Park, Elmsford, NY 10523. (914) 592-7700. Monthly. $430.00 per year.

SKY AND TELESCOPE. Sky Publishing Corporation, 49 Bay State Road, Cambridge, MA 02238. (617) 864-7360. Monthly. $18.00 per year.

SPACE SCIENCE REVIEWS. D. Reidel Publishing Company, 190 Old Derby Street, Hingham, MA 02043. Monthly. $305.00 per year.

VISTAS IN ASTRONOMY. Pergamon Press, Incorporated, Maxwell House, Fairview Park, Elmsford, NY 10523. (914) 592-7700. Quarterly. $145.00 per year.

RESEARCH CENTERS AND INSTITUTES

NATIONAL ASTRONOMY AND IONOSPHERE CENTER. Cornell University, Space Sciences Building, Ithaca, NY 14853. (607) 256-3734.

HARVARD-SMITHSONIAN CENTER FOR ASTROPHYSICS. 60 Garden Street, Cambridge, Ma 02138. (617) 495-7461.

NATIONAL OPTICAL ASTRONOMY OBSERVATORIES. 1002 North Warren Avenue, Tucson, AZ 85719. (602) 325-9230.

CELESTIAL NAVIGATION

See: NAVIGATION

CEMENT

See also: BUILDING MATERIALS, CONCRETE, PRESTRESSED CONCRETE

ABSTRACT SERVICES AND INDEXES

APPLIED SCIENCE AND TECHNOLOGY INDEX. H.W. Wilson Company, 950 University Avenue, Bronx, NY 10452. (800) 367-6670 or (212) 588-8400. Inquire as to cost and availability.

CONCRETE ABSTRACTS. American Concrete Institute, P.O. Box 19150, Redford Station, Detroit, MI 48219. (313) 532-2600. Bimonthly. $122.00 per year.

ENGINEERING INDEX MONTHLY. Engineering Information, Incorporated, 345 East 47th Street, New York, NY 10017. (800) 221-1044 or (212) 705-7600. Monthly, with annual cumulation. $1560.00 per year.

ANNUAL REVIEWS AND YEARBOOKS

CONCRETE INDUSTRIES YEARBOOK. Pit and Quarry Publications, Incorporated, 105 West Adams Street, Chicago, IL 60603. $35.00.

MINERALS YEARBOOK. U.S. Bureau of Mines. Annual. $18.00. Send orders to: Government Printing Office, Washington, DC 20402.

ASSOCIATIONS AND PROFESSIONAL SOCIETIES

AMERICAN CONCRETE INSTITUTE. P.O. Box 19150, Redford Station, Detroit, MI 48219. (313) 532-2600.

PRESTRESSED CONCRETE INSTITUTE. 201 North Wells Street, Chicago, IL 60606. (312) 346-4071.

BIBLIOGRAPHIES

SCIENTIFIC AND TECHNICAL BOOKS IN PRINT; AN INDEX TO LITERATURE IN SCIENCE AND TECHNOLOGY. R.R. Bowker Company, 205 East 42nd Street, New York, NY 10017. (800) 521-8110 or (212) 916-1600.

DIRECTORIES AND BIOGRAPHICAL SOURCES

AMERICAN CEMENT DIRECTORY. Bradley Pulverizer Company, 123 South Third Street, Allentown, PA 18105. (215) 434-5191. Annual. $46.00.

AMERICAN CONCRETE INSTITUTE MEMBERSHIP DIRECTORY. American Concrete Institute, P.O. Box 19150, Redford Station, Detroit, MI 48219. (313) 532-2600. Biennial. $44.95.

WHO'S WHO IN ENGINEERING. Engineers Joint Council, 345 East 47th Street, New York, NY 10017. (212) 705-7010. 1985. $200.00.

WHO'S WHO IN FRONTIER SCIENCE AND TECHNOLOGY. Marquis Who's Who, Incorporated, 200 East Ohio Street, Chicago, IL 60611. (800) 428-3898 or (312) 787-2008.

WORLD CEMENT DIRECTORY. Cembureau. European Cement Association, 2 rue Saint-Charles, F-75740 Paris Cedex 15, France. Irregular. 1986. $200.00, prepayment required. List of manufactures of cement and cement products on more than 140 countries.

ENCYCLOPEDIAS AND DICTIONARIES

CEMENT AND CONCRETE TERMINOLOGY. American Concrete Institute, P.O. Box 19150, Redford Station, Detroit, MI 48219. (313) 532-2600. 1985.

ENCYCLOPEDIA OF PHYSICAL SCIENCE AND TECHNOLOGY. Academic Press, Incorporated, Orlando, FL 32887. (800) 321-5068 or (305) 345-2734. Fifteen volumes, 1986.

MCGRAW-HILL DICTIONARY OF ENGINEERING. Sybil P. Parker, editor. McGraw-Hill Book Company, 1221 Avenue of the Americas, New York, NY 10020. (212) 512-2000. 1984. $39.95.

MCGRAW-HILL ENCYCLOPEDIA OF SCIENCE AND TECHNOLOGY. McGraw-Hill Book Company, 1221 Avenue of the Americas, New York, NY 10020. (212) 512-2000.

GENERAL WORKS

CEMENT CHEMISTRY AND PHYSICS FOR CIVIL ENGINEERS. W.H. Dipl-Ing and W.H.C. Czernin. Heyden & Sons, Incorporated, 247 South 41st Street, Philadelphia, PA 19104. (800) 345-8112 or (215) 382-6673. 1980. $24.00.

CEMENT ENGINEERS HANDBOOK. B. Kohlhaas and others. Heyden & Sons, Incorporated, 247 South 41st Street, Philadelphia, PA 19104. (800) 345-8112 or (215) 382-6673. 1982. $90.00.

CONCRETE PRIMER. F.R. McMillan and Lewis H. Tuthill. American Concrete Institute, P.O. Box 19150, Redford Station, Detroit, MI 48219. (313) 532-2600. Third edition. 1973. $9.50.

CONCRETE SCIENCE: A TREATISE ON CURRENT RESEARCH. V.S. Ramachandran, R.F. Feldman, and J.J. Beaudoin. John Wiley & Sons, 605 Third Avenue, New York, NY 10158. (800) 526-5368 or (212) 850-6000. 1981. $74.95.

CONCRETE: STRUCTURE, PROPERTIES AND MATERIALS. P. Kumar Mehta. Prentice-Hall, Incorporated, Englewood Cliffs, NJ 07632. (800) 262-6868 or (201) 592-2000. 1986. $43.95.

HANDBOOKS AND MANUALS

ACI MANUAL OF CONCRETE PRACTICE. American Concrete Institute, P.O. Box 19150, Redford Station, Detroit, MI 48219.

(313) 532-2600. Five volumes. Annual.

CONCRETE CONSTRUCTION HANDBOOK. Joseph J. Waddell, editor. McGraw-Hill Book Company, 1221 Avenue of the Americas, New York, NY 10020. (212) 512-2000. Second edition. 1974. $70.50.

CONCRETE MANUAL. John Wiley & Sons, 605 Third Avenue, New York, NY 10158. (800) 526-5368 or (212) 850-6000. Eighth edition. 1983. $37.50.

HANDBOOK OF CONCRETE ENGINEERING. Mark Fintel, editor. Van Nostrand Reinhold Company, 115 Fifth Avenue, New York, NY 10003. (212) 254-3232. Second edition. 1985. $89.50.

STRUCTURAL DESIGN GUIDE TO THE ACI BUILDING CODE. Paul F. Rice and Edward S. Hoffman. Van Nostrand Reinhold Company, 115 Fifth Avenue, New York, NY 10003. (212) 254-3232. Third edition. 1985. $44.50.

ONLINE DATA BASES

COMPENDEX. Engineering Information, Incorporated, 345 East Street, New York, NY 10017. (800) 221-1044 or (212) 705-7615. Engineering and technical literature, 1975 to present. Inquire as to cost and availability.

NTIS. National Technical Information Service, 5285 Port Royal Road, Springfield, VA 22161. (703) 487-4630. Broad coverage of government sponsored research reports, 1964 to present. Inquire as to online cost and availability.

U.S. ARMY CORPS OF ENGINEERS, WATERWAYS EXPERIMENT STATION, CONCRETE TECHNOLOGY INFORMATION ANALYSIS CENTER. P.O. Box 631, Vicksburg, MS 39180. (601) 634-3264. Priority is given to U.S. Department of Defense personnel, but services are also available to others in the scientific community as time permits.

WILSONLINE. H.W. Wilson Company, 950 University Avenue, Bronx, NY 10452. (800) 367-6770 or (212) 588-8400. Makes available online versions of the printed H.W. Wilson indexes including Applied Science and Technology Index, Business Periodicals Index, and Readers Guide to Periodical Literature. Period covered is generally 1983 to present. Inquire as to cost and availability.

OTHER SOURCES

IEEE CEMENT INDUSTRY TECHNICAL CONFERENCE RECORD. IEEE, Incorporated, 345 East 47th Street, New York, NY 10017. (202) 705-7900. Annual.

PERIODICALS

AMERICAN CONCRETE INSTITUTE JOURNAL. American Concrete Institute, P.O. Box 19150, Redford Station, Detroit, MI 48219. (313) 532-2600. Monthly. $69.00 per year.

CEMENT AND CONCRETE RESEARCH. Pergamon Press, Incorporated, Maxwell House, Fairview Park, Elmsford, NY 10523. (914) 592-7700. Bimonthly. $190.00 per year.

CONCRETE. 120 West Second Street, Duluth, MN 55802. (218) 723-9253. Monthly. $14.00 per year.

CONCRETE CONSTRUCTION. Concrete Construction Publications, Incorporated, 426 South Westgage, Addison, IL 60101. (312) 543-0870. Monthly. $12.00 per year.

CONCRETE INTERNATIONAL: DESIGN AND CONSTRUCTION. American Concrete Institute, P.O. Box 19150, Redford Station, Detroit, MI 48219. (313) 532-2600. Monthly. $69.00.

JOURNAL OF PCI. Prestressed Concrete Institute, 201 North Wells Street, Chicago, IL 60606. (312) 346-4071. Bimonthly.

JOURNAL OF THE AMERICAN CONCRETE INSTITUTE. American Concrete Institute, P.O. Box 19150, Redford Station, Detroit, MI 48219. (313) 532-2600. Monthly. 469.00 per year.

RESEARCH CENTERS AND INSTITUTES

AMERICAN SOCIETY FOR TESTING AND MATERIALS. 1916 Race Street, Philadelphia, PA 19103. (215) 299-5400.

CONCRETE MATERIALS RESEARCH COUNCIL. P.O. Box 19150, Detroit, MI 48219. (313) 532-2600.

CONCRETE RESEARCH LABORATORY. University of Illinois, 1211 Newmark Civil Engineering Laboratory, 208 North Romine Street, Urbana, IL 61801. (217) 333-3394.

CONCRETE RESEARCH LABORATORY. University of Michigan, G.G. Brown Laboratory, Ann Arbor, MI 48109. (313) 763-8077.

PORTLAND CEMENT ASSOCIATION. Construction Technology Laboratory, 5420 Old Orchard Road, Skokie, IL 60077. (312) 965-7500.

SPECIFICATIONS AND STANDARDS

BLENDED CEMENTS. STP 897. G. Frohnsdorff, editor. American Society for Testing and Materials (ASTM), 1916 Race Street, Philadelphia, PA 19103. (215) 299-5400. 1986. $26.00.

CEMENT STANDARDS: EVOLUTION AND TRENDS - STP 663. American Society for Testing and Materials, 1916 Race Street, Philadelphia, PA 19103. (215) 299-5400. 1979. $20.00.

MASONRY: RESEARCH, APPLICATION, AND PROBLEMS. STP 871. American Society for Testing and Materials (ASTM), 1916 Race Street, Philadelphia, PA 19103. (215) 299-5400. 1985. $36.00.

CERAMICS

See also: CLAY, MATERIALS SCIENCE

ABSTRACT SERVICES AND INDEXES

APPLIED SCIENCE AND TECHNOLOGY INDEX. H.W. Wilson Company, 950 University Avenue, Bronx, NY 10452. (800) 367-6670 or (212) 588-8400. Inquire as to cost and availability.

CERAMIC ABSTRACTS. American Ceramic Society, 65 Ceramic Drive, Columbus, OH 43214. (614) 268-8645. 1922 to present. Bimonthly. $185.00 per year.

CHEMICAL ABSTRACTS. Chemical Abstracts Service, 2540 Olentangy Road, P.O. Box 3012, Columbus, OH 43210. (800) 848-6538 or (614) 421-3600. Weekly. $9200.00 per year.

ENGINEERING INDEX MONTHLY. Engineering Information, Incorporated, 345 East 47th Street, New York, NY 10017. (800) 221-1044 or (212) 705-7600. Monthly, with annual cumulation. $1560.00 per year.

PHYSICS ABSTRACTS. Institute of Electrical Engineers, London, United Kingdom. Available from: Institute of Electrical and Electronic Engineers (IEEE), 345 East 47th Street, New York, NY 10017. (212) 705-7900.

ASSOCIATIONS AND PROFESSIONAL SOCIETIES

AMERICAN CERAMIC SOCIETY. 65 Ceramic Drive, Columbus, OH 43214. (614) 268-8645.

NATIONAL INSTITUTE OF CERAMIC ENGINEERS. 65 Ceramic Drive, Columbus, OH 43214. (614) 268-8645.

BIBLIOGRAPHIES

A BIBLIOGRAPHY OF CERAMICS AND GLASS. L.L. Mench and B.A. McEldowney. American Ceramic Society, 65 Ceramic Drive, Columbus, OH 43214. (614) 268-8645.

DIRECTORIES AND BIOGRAPHICAL SOURCES

AMERICAN MEN AND WOMEN OF SCIENCE. Physical and Biological Sciences. Fifteenth edition. R.R. Bowker Company, 205 East 42nd Street, New York, NY 10017. (800) 521-8110 or (212) 916-1600.

CERAMIC COMPANY DIRECTORY. American Ceramic Society, 65 Ceramic Drive, Columbus, OH 43214. (614) 268-8645. Annual. $10.00.

GOVERNMENT RESEARCH DIRECTORY. Gale Research Company, Book Tower, Detroit, MI 48226. (800) 521-0707. Third edition. 1985.

INTERNATIONAL RESEARCH CENTERS DIRECTORY 1986-1987. Gale Research Company, Book Tower, Detroit, MI 48226. (800) 521-0707. Third edition. 1986.

MATERIALS RESEARCH CENTERS: A WORLD DIRECTORY OF ORGANIZATIONS AND PROGRAMS IN MATERIAL SCIENCE. Longman Group Limited. Distributed by Gale Research Company, Book Tower, Detroit, MI 48226. (800) 521-0707. 1983. $160.00.

RESEARCH CENTERS DIRECTORY. Gale Research Company, Book Tower, Detroit, MI 48226. (800) 521-0707. Eleventh edition. 1987.

SCIENTIFIC AND TECHNICAL ORGANIZATIONS AND AGENCIES DIRECTORY. Gale Research Company, Book Tower, Detroit, MI 48226. (800) 521-0707. 1985.

WHO'S WHO IN TECHNOLOGY TODAY. Reston Publishing Company, Incorporated, c/o Prentice-Hall, Incorporated, Englewood Cliffs, NJ 07632. (800) 262-6868. Biennial. Five volumes. $425.00. Covers the fields of electronics, computer science, physics, optics, chemistry, biotechnology, mechanics, energy, and earth science.

ENCYCLOPEDIAS AND DICTIONARIES

CERAMIC GLOSSARY. Walter W. Perkins, editor. American Ceramic Society, 65 Ceramic Drive, Columbus, OH 43214. (614) 268-8645. 1984. $12.00.

ENCYCLOPEDIA OF PHYSICAL SCIENCE AND TECHNOLOGY. Academic Press, Incorporated, Orlando, FL 32887. (800) 321-5068 or (305) 345-2734. Fifteen volumes, 1986.

MCGRAW-HILL ENCYCLOPEDIA OF SCIENCE AND TECHNOLOGY. McGraw-Hill Book, Incorporated, 1221 Avenue of the Americas, New York, NY 10020. (212) 997-3675.

GENERAL WORKS

CERAMIC MATERIALS: AN INTRODUCTION TO THEIR PROPERTIES. R. Pam Puch. Elsevier Science Publishing Company, Incorporated, 52 Vanderbilt Avenue, New York, NY 10017. (212) 370-5520. 1976. $57.50.

CERAMIC RAW MATERIALS. W.E. Worrall. Pergamon Press, Maxwell House, Fairview Park, Elmsford, NY 10523. (914) 592-7700. Second edition. 1982. $10.00 paper.

INTRODUCTION TO CERAMICS. W.D. Kingery, H.K. Bowen, D.R. Uhlmann. John Wiley & Sons, Incorporated, 605 Third Avenue, New York, NY 10158. (800) 526-5368 or (212) 850-6000. 1976. $72.95.

HANDBOOKS AND MANUALS

CERAMIC SCIENCE FOR MATERIALS TECHNOLOGISTS. I.J. McColm. Metheun, Incorporated, 29 West 35th Street, New York, NY 10001. (212) 244-3336. 1983. $85.95.

ONLINE DATA BASES

CA SEARCH. Chemical Abstracts Service, P.O. Box 3012, Columbus, OH 43210. Guide to chemical literature, 1972 to present. Inquire as to cost and availability.

INSPEC. INSPEC Marketing Department, Institute of Electrical and Electronics Engineers, Incorporated, IEEE Service Department, 445 Hoes Lane, Piscataway, NJ 08854. (201) 981-0060. Inquire as to on-line cost and availability.

NTIS. National Technical Information Service, 5285 Port Royal Road, Springfield, VA 22161. (703) 487-4630. Broad coverage of government sponsored research reports, 1964 to present. Inquire as to cost and availability.

PERIODICALS

AMERICAN CERAMIC SOCIETY BULLETIN. American Ceramic Society, 65 Ceramic Drive, Columbus, OH 43214. (614) 268-8645. Monthly. $12.50 per year.

AMERICAN CERAMIC SOCIETY JOURNAL WITH COMMUNICATIONS. American Ceramic Society, 65 Ceramic Drive, Columbus, OH 43214. (614) 268-8645. Monthly. $185.00 (includes Bulletin and Abstracts).

BRICK AND CLAY. Cahners Publishing Company, Incorporated, Cahners Plaza, 1350 East Touhy Avenue, Des Plaines, IL 60018. (312) 685-8800. Monthly. $30.00 per year.

CANADIAN CERAMICS QUARTERLY (CANADIAN CERAMICS SOCIETY). Taylor Enterprises, Limited, Suite 110, 2175 Sheppard Avenue, East, Willowdale, Ontario, M2J 1W8, Canada. (416) 491-3556. Quarterly. $12.00.

CERAMIC ENGINEERING AND SCIENCE PROCEEDINGS. American Ceramic Society, 65 Ceramic Drive, Columbus, OH 43214. (614) 268-8645. Bimonthly. $60.00 per year.

CERAMIC INDUSTRY. Cahners Publishing Company, Incorporated, Cahners Plaza, 1350 East Touhy Avenue, Des Plaines, IL 60018. (312) 685-8800. Monthly. $35.00 per year.

RESEARCH CENTERS AND INSTITUTES

CENTER FOR CERAMICS RESEARCH. Rutgers University, College of Engineering. Brett and Bowser Roads, P.O. Box 909, Piscataway, NJ 08854. (201) 932-2724.

CERAMIC RESEARCH CENTER. Ohio State University, 2041 North College Road, Columbus, OH 43210. (614) 422-2060.

EDWARD ORTON, JUNIOR, CERAMIC FOUNDATION. 6991 Old 3C Highway, P.O. Box 460, Westerville, OH 43081. (614) 895-2663.

CHAIN REACTION

See: NUCLEAR ENERGY

CHANNELS

See also: CANALS, CIVIL ENGINEERING, DREDGES AND DREDGING, GEOTECHNICAL ENGINEERING, HYDRAULIC ENGINEERING, OCEAN ENGINEERING

ABSTRACT SERVICES AND INDEXES

APPLIED SCIENCE AND TECHNOLOGY INDEX. H.W. Wilson and Company, 950 University Avenue, Bronx, NY 10452. (800) 367-6670 or (212) 588-8400. Monthly. Inquire as to cost and availability.

ENGINEERING INDEX MONTHLY AND AUTHOR INDEX. Engineering Information Inc., 345 East 47th Street, New York, NY 10017. (212) 705-7600. Monthly. $1560.00 per year.

INTERNATIONAL CIVIL ENGINEERING ABSTRACTS. CITIS Limited, 2 Rosemount Terrace, Blackrock, Dublin, Ireland. 1974 to present. Monthly. $350.00 per year.

PUBLICATIONS INFORMATION. American Society of Civil Engineers, 345 East 47th Street, New York, NY 10017. (212) 705-7420. Abstracts, subject and author indexes to the publications of the American Society of Civil Engineers. Bimonthly. $80.00 per year.

SCIENCE CITATION INDEX. Institute for Scientific Information, 3501 Market Street, Philadelphia, PA 19104. (800) 523-1850 or (215) 386-0100. Six times per year. $6200.00 per year.

ASSOCIATIONS AND PROFESSIONAL SOCIETIES

AMERICAN GEOLOGICAL INSTITUTE. 4220 King Street, Alexandria, VA 22302. (703) 379-2480.

AMERICAN SOCIETY OF CIVIL ENGINEERS. 345 East 47th Street, New York, NY 10017. (212) 705-7420.

DIRECTORIES AND BIOGRAPHICAL SOURCES

1987 DIRECTORY OF ENGINEERING SOCIETIES AND RELATED ORGANIZATIONS. Gordon Davis, editor. Hemisphere Publishing Corporation, 1010 Vermont Avenue, NW, Washington, DC 20005. (800) 526-0275. 12th edition. 1987. $100.00.

RESEARCH CENTERS DIRECTORY 1988. Gale Research Company, Book Tower, Detroit, MI 48226. (800) 521-0707. 12th edition. 1987. $365.00 for set.

SCIENTIFIC AND TECHNICAL ORGANIZATIONS AND AGENCIES DIRECTORY. Margaret Labash Young, editor. Gale Research Company, Book Tower, Detroit, MI 48226. (800) 521-0707. 2nd edition. 1987. $185.00.

WHO'S WHO IN ENGINEERING. Gordon Davis, editor. Hemisphere Publishing Corporation, 1010 Vermont Avenue, NW, Washington, DC 20005. (800) 526-0275. 6th edition. 1985. $200.00.

ENCYCLOPEDIAS AND DICTIONARIES

DICTIONARY OF CIVIL ENGINEERING. John S. Scott. John Wiley and Sons, Inc., 605 Third Avenue, New York, NY 10158. (800) 526-5368. Third edition. 1981. $26.95.

GENERAL WORKS

CANAL AND RIVER LEVEES. P. Peter. Elsevier Science Publishing Company, Inc., 52 Vanderbilt Avenue, New York, NY 10017. (212) 370-5520. 1982. $104.25.

CHANNELS AND CHANNEL CONTROL STRUCTURES. K.V. Smith, editor. Springer-Verlag New York, Inc., 175 Fifth Avenue, New York, NY 10010. (800) 526-7254. 1985. $37.70.

FLUID FLOW: PUMPS, PIPES, AND CHANNELS. N.P. Cheremisinoff. Butterworth's Publishing, 80 Montvale Avenue, Stoneham, MA 02180. (800) 325-4177. 1982. $59.95.

OPEN-CHANNEL HYDRAULICS. R.H. French. McGraw-Hill Book Company, 1221 Avenue of the Americas, New York, NY 10020. (212) 512-2000. 1985. $49.95.

HANDBOOKS AND MANUALS

CIVIL ENGINEERING CALCULATIONS REFERENCE GUIDE. Tyler G. Hicks, editor. McGraw-Hill Book Company, 1221 Avenue of the Americas, New York, NY 10020. (212) 512-2000. 1987. $29.50.

CIVIL ENGINEERING PRACTICE. Paul N. Cheremisinoff and others, editors. Technomic Publishing Company, Inc., 851 Holland Avenue, Box 3535, Lancaster, PA 17604. (800) 233-9936. Five volumes. 1987-1988. $750.00 for set.

STANDARD HANDBOOK FOR CIVIL ENGINEERS. F. S. Merritt, editor. McGraw-Hill Book Company, 1221 Avenue of the Americas, New York, NY 10020. (212) 512-2000. Third edition. 1983. $89.50.

SURVEYING READY-REFERENCE MANUAL. Guy O. Stenstrom. McGraw-Hill Book Company, 1221 Avenue of the Americas, New York, NY 10020. (212) 512-2000. 1987. $26.50.

ONLINE DATA BASES

COMPENDEX. Engineering Information, Inc., 345 East 47th Street, New York, NY 10017. (800) 221-1044 or (212) 705-7615. Engineering and technical literature, 1975 to present. Inquire as to online cost and availability.

GEOREF. Online version of the BIBLIOGRAPHY AND INDEX OF GEOLOGY. American Geological Institute, 4220 King Street, Alexandria, VA 22302. (703) 379-2480. 1969 to present. Inquire as to online cost and availability.

NTIS. National Technical Information Service, 5285 Port Royal Road, Springfield, VA 22161. (703) 487-4630. Broad coverage of government sponsored research reports, 1964 to present. Inquire as to online cost and availability.

SCISEARCH. Institute for Scientific Information, 3501 Market Street, Philadelphia, PA 19104. (800) 523-1850 or (215) 386-0100. Broad multidisciplinary title and author index to the international literature of science and technology, 1974 to present. Inquire as to online cost and availability.

WILSONLINE. H.W. Wilson and Company, 950 University Avenue, Bronx, NY 10452. (800) 367-6770 or (212) 588-8400. Makes available online versions of the H.W. Wilson indexes including Applied Science and Technology Index, Business Periodicals Index and Readers' Guide to Periodical Literature. Approximately 1980 to present. Inquire as to online cost and availability.

PERIODICALS

CIVIL ENGINEERING. American Society of Civil Engineers, 345 East 47th Street, New York, NY 10017. (212) 705-7420. 1930 to present. Monthly. $48.00 per year.

JOURNAL OF CONSTRUCTION ENGINEERING AND MANAGEMENT. American Society of Civil Engineers, 345 East 47th Street, New York, NY 10017. (212) 705-7420. 1956 to present. Quarterly. $48.00 per year.

JOURNAL OF GEOTECHNICAL ENGINEERING. American Society of Civil Engineers, 345 East 47th Street, New York, NY 10017. (212) 705-7420. 1956 to present. Monthly. $96.00 per year.

JOURNAL OF HYDRAULIC ENGINEERING. American Society of Civil Engineers, 345 East 47th Street, New York, NY 10017. (212) 705-7420. 1956 to present. Monthly. $112.00 per year.

JOURNAL OF IRRIGATION AND DRAINAGE. American Society of Civil Engineers, 345 East 47th Street, New York, NY 10017. (212) 705-7420. 1956 to present. Quarterly. $45.00 per year.

JOURNAL OF SURVEYING ENGINEERING. American Society of Civil Engineers, 345 East 47th Street, New York, NY 10017. (212) 705-7420. 1956 to present. Three times per year. $35.00 per year.

JOURNAL OF WATER RESOURCES PLANNING AND MANAGEMENT. American Society of Civil Engineers, 345 East 47th Street, New York, NY 10017. (212) 705-7420. 1956 to present. Quarterly. $56.00 per year.

JOURNAL OF WATERWAY, PORT, COASTAL AND OCEAN ENGINEERING. American Society of Civil Engineers, 345 East 47th Street, New York, NY 10017. (212) 705-7420. 1956 to present. Bimonthly. $72.00 per year.

RESEARCH CENTERS AND INSTITUTES

PORTS AND WATERWAYS INSTITUTE. Louisiana State University, 60 University Lakeshore Drive, Baton Rouge, LA 70803. (504) 388-2772.

CHAOS

See also: MATHEMATICS

ABSTRACT SERVICES AND INDEXES

CHEMICAL ABSTRACTS. Chemical Abstracts Service, 2540 Olentangy Road, Post Office Box 3012, Columbus, OH 43210. (800) 848-6538 or (614) 421-3600. Weekly. $9200.00 per year.

CURRENT CONTENTS: PHYSICAL, CHEMICAL AND EARTH SCIENCES. Institute for Scientific Information, 3501 Market Street, Philadelphia, PA 19104. (800) 523-1850 or (215) 386-0100. Weekly. $272.00 per year.

CURRENT PAPERS IN PHYSICS. Available from: IEEE Service Center, 445 Hoes Lane, Piscataway, NJ 08854. Biweekly. $215.00 per year.

GENERAL PHYSICS ADVANCE ABSTRACTS. American Institute of Physics, 335 East 45th Street, New York, NY 10017. (212) 661-9404. 1985 to present. Semimonthly. $150.00 per year.

GENERAL SCIENCE INDEX. H.W. Wilson Company, 950 University Avenue, Bronx, NY 10452. (800) 367-6770 or (212) 588-8400.

MATHEMATICAL REVIEWS. American Mathematical Society, Post Office Box 6248, Providence, RI 02940. (800) 556-7774 or (401) 272-9500.

PHYSICAL REVIEW ABSTRACTS. American Institute of Physics, 335 East 45th Street, New York, NY 10017. (212) 661-9404. 1970 to present. Semimonthly. $140.00 per year.

PHYSICS ABSTRACTS. Institute of Electrical Engineers, London, United Kingdom. Available from: Institute of Electrical and Electronic Engineers (IEEE), 345 East 47th Street, New York, NY 10017. (212) 705-7900.

PHYSICS BRIEFS. Physics Verlag GmbH, Postfach 1260/1280, D-6940, Weinheim, West Germany. 1920 to present. Twenty-six times per year. $1200.00 per year.

SCIENCE CITATION INDEX. Institute for Scientific Information, 3501 Market Street, Philadelphia, PA 19104. (800) 523-1850 or (215) 386-0100.

SOLID STATE ABSTRACTS: AN ABSTRACT JOURNAL INVOLVING THE PHYSICS, METALLURGY, CRYSTALLOGRAPHY, CHEMISTRY, AND DEVICE TECHNOLOGY OF SOLIDS. Cambridge Scientific Abstracts, 5161 River Road, Bethesda, MD 20816. (301) 951-1400. 1957 to present. Bimonthly. $550.00 per year.

ASSOCIATIONS AND PROFESSIONAL SOCIETIES

AMERICAN INSTITUTE OF PHYSICS. 335 East 45th Street, New York, NY 10017. (212) 661-9404.

AMERICAN MATHEMATICAL SOCIETY. Post Office Box 6248, Providence, RI 02940. (401) 272-9500.

ENCYCLOPEDIAS AND DICTIONARIES

ENCYCLOPEDIA OF PHYSICS. Robert M. Besancon, editor. Van Nostrand Reinhold, Incorporated, 115 Fifth Avenue, New York, NY 10003. (800) 543-2681. Third edition. 1985. $99.95.

GENERAL WORKS

CHAOTIC DYNAMICS AND FRACTALS. Michael F. Barnsley and Stephen G. Demko. Academic Press, Incorporated, 6277 Sea Harbor Drive, Orlando, FL 32821. (800) 321-5068. 1986. $29.95 in paper.

CHAOS. Arun V. Holden, editor. Princeton University Press, Princeton, Nj 08540. (609) 452-4900. 1986. $19.95 in paper.

CHAOS: MAKING A NEW SCIENCE. James Gleick. Viking-Penguin, Incorporated, 40 West 23rd Street, New York, NY 10010. (800) 631-3577. 1987. $19.95.

DETERMINISTIC CHAOS: AN INTRODUCTION. H.G. Schuster. VCH Publishers, Incorporated, 3030 NW 12th Avenue, Deerfield Beach, FL 33442-1705. (305) 428-5566. 1985. $43.75.

DYNAMICS: THE GEOMETRY OF BEHAVIOR. Ralph Abraham and Chris Shaw. Aerial Press, Post Office Box 1360, Santa Cruz, CA 95061. Three volumes. 1982-85. Part 1, $32.00; Part 2, $26.00; Part 3, $26.00.

THE EUDAEMONIC PIE: OR WHY WOULD ANYONE PLAY ROULETTE WITHOUT A COMPUTER IN HIS SHOE? Thomas A. Bass. Houghton Mifflin Company, 2 Park Street, Boston, MA 02108. (800) 225-3362. 1985. $15.95.

ORDER WITHIN CHAOS: TOWARDS A DETERMINISTIC APPROACH TO TURBULENCE. Pierre Berge and others. John Wiley and Sons, Incorporated, 605 Third Avenue, New York, NY 10158. (800) 526-5368 or (212) 850-6000. 1986. $54.00.

ONLINE DATA BASES

CA SEARCH. Chemical Abstracts Service, Post Office Box 3012, Columbus, OH 43210. Guide to chemical literature, 1972 to present. Inquire as to cost and availability.

DISSERTATION ABSTRACTS ONLINE. University Microfilms International, 300 North Zeeb Road, Ann Arbor, MI 48106. (800) 521-0600 or (313) 761-4700. Scope includes virtually all doctoral dissertations accepted at accredited American institutions from 1861 to present in 252 subject areas. Inquire as to cost and availability.

INSPEC. INSPEC Marketing Department, Institute of Electrical and Electronics Engineers, Incorporated, IEEE Service Department, 445 Hoes Lane, Piscataway, NJ 08854. (201) 981-0060. Inquire as to on-line cost and availability.

MATHFILE. American Mathematical Society, Post Office Box 6248, Providence, RI 02940. (800) 556-7774 or (401) 272-9500.

Scope includes pure and applied mathematics and related areas of physics, statistics, engineering, computer science, and operations research literature since 1973. Inquire as to cost and availability.

NTIS. National Technical Information Service, 5285 Port Royal Road, Springfield, VA 22161. (703) 487-4630. Broad coverage of government sponsored research reports 1964 to present. Inquire as to cost and availability.

SCISEARCH. Institute for Scientific Information, 3501 Market Street, Philadelphia, PA 19104. (800) 523-1850 or (215) 386-0100. Broad multidisciplinary title and author index to the present. Inquire as to cost and availability.

PERIODICALS

CHAOS. Res Bureaux, Box 1598, Kingston, ON K7L 5CB, Canada. (613) 542-7277. Eight times per year. Canada $16.00 per year.

CHARM

See: PARTICLE PHYSICS

CHEMICAL BONDING

See also: CHEMISTRY, PHYSICAL CHEMISTRY

ABSTRACT SERVICES AND INDEXES

APPLIED SCIENCE AND TECHNOLOGY INDEX. H.W. Wilson Company, 950 University Avenue, Bronx, NY 10452. (800) 367-6670 or (212) 588-8400. Inquire as to cost and availability.

CHEMICAL ABSTRACTS. Chemical Abstracts Service, 1540 Olentangy Road, Post Office Box 3012, Columbus, OH 43210. (800) 848-6538 or (614) 421-3600. Weekly. $9200.00 per year.

GENERAL SCIENCE INDEX. H.W. Wilson Company, 950 University Avenue, Bronx, NY 10452. (800) 367-6770 or (212) 588-8400. Inquire as to cost and availability.

PHYSICS ABSTRACTS. Institute of Electrical Engineers, London United Kingdom. Available from: Institute of Electrical and Electronic Engineers (IEEE), 345 East 47th Street, New York, NY 10017. (212) 705-7900.

SCIENCE CITATION INDEX. Institute for Scientific Information, 3501 Market Street, Philadelphia, PA 19104. (800) 523-1850 or (215) 386-0100.

ANNUAL REVIEWS AND YEARBOOKS

ANNUAL REVIEW OF PHYSICAL CHEMISTRY. Annual Reviews Incorporated, 4139 El Camino Way, Palo Alto, CA 94306. (415) 493-4400. Annual.

ASSOCISTIONS AND PROFESSIONAL SOCIETIES

AMERICAN CHEMICAL SOCIETY. 1155 16th Street, NW, Washington, DC 20036. (202) 872-4600.

ASSOCIATION OF CONSULTING CHEMISTS AND CHEMICAL ENGINEERS. 50 East 41st Street, Suite 92, New York, NY 10017. (212) 684-6255.

BIBLIOGRAPHIES

SCIENCE BOOKS AND FILMS. American Association for the Advancement of Science, 1333 H Street, NW, Washington, DC 20005.

SCIENTIFIC AND TECHNICAL BOOKS AND SERIALS IN PRINT 1988: AN INDEX TO LITERATURE IN SCIENCE AND TECHNOLOGY. R.R. Bowker Company, 205 East 42nd Street, New York, NY 10017. (800) 521-8110 or (212) 916-1600. $175.00.

DIRECTORIES AND BIOGRAPHICAL SOURCES

AMERICAN INSTITUTE OF CHEMISTS. American Institute of Chemists, 7315 Wisconsin Avenue, Bethesda, MD 20814. (301) 652-2447. 1986. $35.00.

AMERICAN MEN AND WOMEN OF SCIENCE. Physical and Biological Sciences. R.R. Bowker Company, 205 East 42nd Street, New York, NY 10017. (800) 521-8110 or (212) 916-1600. Fifteenth edition. $565.00.

BIOGRAPHICAL DICTIONARY OF SCIENTISTS: CHEMISTS. David Abbott, editor. P. Bedrick Books, 125 East 23rd Street, New York, NY 10010. (212) 777-1187. 1984. $18.95.

CHEMICAL WEEK - BUYERS GUIDE ISSUE. McGraw-Hill Book Company, 1221 Avenue of the Americas, New York, NY 10020. (800) 628-0004. Annual, October. $50.00.

CONSULTING SERVICES: CHEMISTS AND CHEMICAL ENGINEERS. Association of Consulting Chemists and Chemical Engineers, 50 East 41st Street, New York, NY 10017. (212) 684-6255. Annual. 1986. $45.00.

GOVERNMENT RESEARCH DIRECTORY. Gale Research Company, Book Tower, Detroit, MI 48226. (800) 521-0707. Fourth edition. 1987. $350.00.

INTERNATIONAL RESEARCH CENTERS DIRECTORY 1986-1987. Gale Research Company, Book Tower, Detroit, MI 48226. (800) 521-0707. Third edition. 1986. $330.00.

RESEARCH CENTERS DIRECTORY. Gale Research Company, Book Tower, Detroit, MI 48226. (800) 521-0707. Eleventh edition. 1987. $355.00.

RESEARCH SERVICES DIRECTORY. Robert J. Huffman and Mary M. Watkins, editors. Gale Research Company, Book Tower, Detroit, MI 48226. (800) 521-0707. Third edition. $290.00.

SCIENTIFIC AND TECHNICAL ORGANIZATIONS AND AGENCIES DIRECTORY. Gale Research Company, Book Tower, Detroit, MI 48226. (800) 521-0707. 1985. $150.00.

WHO'S WHO IN FRONTIER SCIENCE AND TECHNOLOGY. Marquis Who's Who, Incorporated, 200 East Ohio Street, Chicago, IL 60611. (800) 428-3898 or (312) 787-2008.

WHO'S WHO IN TECHNOLOGY TODAY. Reston Publishing Company, Incorporated. c/o Prentice-Hall, Incorporated, Englewood Cliffs, NJ 07632. (800) 262-6868. Biennial. Five volumes. $425.00. Covers the fields of electronics, computer science, physics, optics, chemistry, biotechnology, mechanics, energy, and earth science.

WORLD GUIDE TO SCIENTIFIC ASSOCIATIONS AND LEARNED SOCIETIES. K.G. Saur Incorporated, 175 Fifth Avenue, New York, NY 10010. (800) 521-0707 or (212) 982-1302. Fourth edition. 1984. $112.00.

ENCYCLOPEDIAS AND DICTIONARIES

CONCISE ENCYCLOPEDIA OF CHEMICAL TECHNOLOGY. Kirk Othmer. John Wiley and Sons, Incorporated, 605 Third Avenue, New York, NY 10158. (800) 526-5368 or (212) 850-6000.

Third edition. 1985. $129.95.

CONDENSED CHEMICAL DICTIONARY. Gessner Hawley. Van Nostrand Reinhold, 115 Fifth Avenue, New York, NY 10003. Tenth edition. 1981. $49.95.

ENCYCLOPEDIA OF PHYSICAL SCIENCE AND TECHNOLOGY. Academic Press, Incorporated, Orlando, FL 32887. (800) 321-5068 or (305) 345-2734. Fifteen volumes, 1986.

GLOSSARY OF CHEMICAL TERMS. Clifford A. Hampel and Gessner G. Hawley. Van Nostrand Reinhold Company, 115 Fifth Avenue, New York, NY 10003. (800) 543-2681 or (212) 254-3232. Second edition. 1982. $21.95.

HAWLEY'S CONDENSED CHEMICAL DICTIONARY. N. Irving Sax and Richard J. Lewis. Sr., editors. Van Nostrand Reinhold, Incorporated, 115 Fifth Avenue, New York, NY 10003. (800) 543-2681. Eleventh edition. 1987. $52.95.

MCGRAW-HILL ENCYCLOPEDIA OF SCIENCE AND TECHNOLOGY. McGraw-Hill Book, Incorporated, 1221 Avenue of the Americas, New York, NY 10020. (212) 997-3675.

VAN NOSTRAND REINHOLD ENCYCLOPEDIA OF CHEMISTRY. Douglas M. Considine and Glenn D. Considine. Van Nostrand Reinhold Publishing Company, Incorporated, 115 Fifth Avenue, New York, NY 10003. (800) 543-2681 or (212) 254-3232. 1984. $97.95.

GENERAL WORKS

THE CHEMICAL BOND. John N. Murrell and others. John Wiley and Sons, Incorporated, 605 Third Avenue, New York, NY 10158. (800) 526-5368 or (212) 850-6000. 1985. $31.95.

CHEMICAL BONDING MODELS, VOLUME 1. J.F. Liebman and A. Greenberg, editors. VCH Publishing, Incorporated, 303 NW 12th Avenue, Deerfield Beach, FL 33442-1705. (305) 428-5566. 1986. $77.50.

CHEMISTRY. Charles E. Mortimer. Wadsworth Publishing Company, 10 Davis Drive. Belmont, CA 94002. (415) 595-2350. Sixth edition. 1986. $49.95.

CHEMISTRY OF THE ELEMENTS. N.N. Greenwood and A. Earnshaw. Pergamon Publishing, Incorporated, Maxwell House, Fairview Park, Elmsford, NY 10523. (914) 592-7700. 1984. $143.00.

INORGANIC CHEMISTRY: PRINCIPLES OF STRUCTURE AND REACTIVITY. James E. Huheey. Harper and Row Publishers Incorporated, 10 East 53rd Street, New York, NY 10022. (212) 207-7655. Third edition. $36.50.

INTRODUCTION TO PHYSICAL CHEMISTRY. M.F.C. Ladde and W.H. Lee. Cambridge University Press, 32 East 57th Street, New York, NY 10022. (212) 688-8885. 1986. $19.95 in paper.

INTRODUCTION TO SYMMETRY IN BONDING AND SPECTRA. Bodie Douglas and Charles A. Hollingsworth. Academic Press, Incorporated, 6277 Sea Harbor Drive, Orlando, FL 32821. (800) 321-5068. 1985. $40.00.

INVESTIGATION OF RATES AND MECHANISMS OF REACTIONS, PART 1. Claude F. Bernasconi, editor. John Wiley and Sons, Incorporated, 605 Third Avenue, New York, NY 10158. (800) 526-5368 or (212) 850-6000. Fourth edition. 1986. $175.00.

QUANTUM CHEMISTRY OF ATOMS AND MOLECULES. Philip S. C. Matthews. Cambridge University Press, 32 East 57th Street, New York, NY 10022. (212) 688-8885. 1986. $44.50.

SYMMETRY AND STRUCTURE. S.A. Kettle. John Wiley and Sons, Incorporated, 605 Third Avenue, New York, NY 10158. (800) 526-5368 or (212) 850-6000. 1985. $37.95.

HANDBOOKS AND MANUALS

THE CHEMIST'S COMPANION: A HANDBOOK OF PRACTICAL DATA, TECHNIQUES, AND REFERENCES. Arnold J. Gordon and Richard A. Ford. John Wiley and Sons, Incorporated, 605 Third Avenue, New York, NY 10158. (800) 526-5368. 1973. $49.95.

CRC HANDBOOK OF CHEMISTRY AND PHYSICS. CRC Press, Incorporated, 2000 Corporate Boulevard, NW, Boca Raton, FL 33431. Sixty-seventh edition. 1986. $69.95.

HANDBOOK OF APPLIED CHEMISTRY: FACTS FOR ENGINEERS, SCIENTISTS, TECHNICIANS, AND TECHNICAL MANAGERS. Vollrath Hopp and Ingp Hennig. McGraw-Hill Book Company, 1221 Avenue of the Americas, New York, NY 10020. (800) 628-0004. 1983. $54.00.

HANDBOOK OF COMPUTATIONAL CHEMISTRY: A PRACTICAL GUIDE TO CHEMICAL STRUCTURE AND ENERGY CALCULATIONS. Tim Clark. John Wiley and Sons, Incorporated, 605 Third Avenue, New York, NY 10158. (800) 526-5368 or (212) 850-6000. 1985. $35.00.

LANGE'S HANDBOOK OF CHEMISTRY. John A. Dean, editor. McGraw-Hill Book Company, 1221 Avenue of the Americas, New York, NY 10020. (800) 628-0004. 1985. $59.50

TABLES OF PHYSICAL AND CHEMICAL CONSTANTS; AND SOME MATHEMATICAL FUNCTIONS. G.W.C. Kaye and T.H. Laby, editors. Longman, Incorporated, 95 Church Street, White Plains, NY 10601. (014) 993-5000. 1986. $39.95.

ONLINE DATA BASES

CA SEARCH. Chemical Abstracts Service, Post Office Box 3012, Columbus, OH 43210. Guide to chemical literature, 1972 to present. Inquire as to cost and availability.

DISSERTATION ABSTRACTS ONLINE. University Microfilms International, 300 North Zeeb Road, Ann Arbor, MI 48106. (800) 521-0600 or (313) 761-4700. Scope includes virtually all doctoral dissertations accepted at accredited American institutions from 1861 to present in 252 subject areas. Inquire as to cost and availability.

INSPEC. INSPEC Marketing Department, Institute of Electrical and Electronics Engineers, Incorporated, IEEE Service Department, 445 Hoes Lane, Piscataway, NJ 08854. (201) 981-0060. Inquire as to on-line cost and availability.

NTIS. National Technical Information Service, 5285 Port Royal Road, Springfield, VA 22161. (703) 487-4630. Broad coverage of government sponsored research reports 1964 to present. Inquire as to cost and availability.

SCISEARCH. Institute for Scientific Information, 3501 Market Street, Philadelphia, PA 19104. (800) 523-1850 or (215) 386-0100. Broad multidisciplinary title and author index to the present. Inquire as to cost and availability.

OTHER SOURCES

GUIDE TO BASIC INFORMATION SOURCES IN CHEMISTRY. Arthur Antony. John Wiley and Sons, Incorporated, 605 Third Avenue, New York, NY 10158. (800) 526-5368 or (212) 850-6000. 1979. $26.95.

HOW TO FIND CHEMICAL INFORMATION: A GUIDE FOR PRACTICING CHEMISTS, TEACHERS, AND STUDENTS. John Wiley and Sons, Incorporated, 605 Third Avenue, New York, NY 10158. (800) 526-5368 or (212) 850-6000. 1986. $35.00.

PERIODICALS

ANALYTICAL CHEMISTRY. American Chemical Society, 1155 Sixteenth Street, NW, Washington, DC 20036. (800) 424-6747 or

(202) 872-4700. Monthly. $33.00 per year.

ANDEWANDTE CHEMIE (GESELLSCHAFT DEUTSCHER CHEMIKER, GW). V C H Verlagsgessellschaft mbH, Pappelallee 3, Postfach 1260, 6940 Weinheim, West Germany. Monthly. $300.00 per year.

BIOCHEMISTRY. American Chemical Society, 1155 Sixteenth Street, NW, Washington, DC 20036. (800) 424-6747 or (202) 872-4700. Biweekly. $303.00 per year.

CHEMICAL PHYSICS. Elsevier Science Publishers B.V., Box 211, 1000 AE Amsterdam, The Netherlands. Thirty times per year. $1200.00 per year.

CHEMICAL PHYSICS LETTERS. Elsevier Science Publishers B.V., Box 211, 1000 AE Amsterdam, The Netherlands. Sixty-six times per year. $1500.00 per year.

CHEMICAL REVIEWS. American Chemical Society, 1155 Sixteenth Street, NW, Washington, DC 20036. (800) 424-6747 or (202) 872-4700. Bimonthly. $83.00 per year.

CHEMICAL WEEK. McGraw-Hill Publishing Company, 1221 Avenue of the Americas, New York, NY 10020. (212) 512-2000. Weekly. $30.00 per year.

CHEMTECH. American Chemical Society, 1155 Sixteenth Street, NW, Washington, DC 20036. (800) 424-6747 or (202) 872-4700. Monthly. $40.00 per year to individuals.

INORGANIC CHEMISTRY. American Chemical Society, 1155 Sixteenth Street, NW, Washington, DC 20036. (800) 424-6747 or (202) 872-4700. Monthly. $400.00 per year.

INTERNATIONAL REVIEWS IN PHYSICAL CHEMISTRY. Taylor and Francis, Limited, 242 Cherry Street, Philadelphia, PA 19106-1906. Three times per year. $143.00 per year.

JOURNAL OF ORGANIC CHEMISTRY. American Chemical Society, 1155 Sixteenth Street, NW, Washington, DC 20036. (800) 424-6747 or (202) 872-4700. Biweekly. $265.00 per year.

JOURNAL OF PHYSICAL CHEMISTRY. American Chemical Society, 1155 Sixteenth Street, NW, Washington, DC 20036. (800) 424-6747 or (202) 872-4700. Biweekly. $369.00 per year.

JOURNAL OF THE AMERICAN CHEMICAL SOCIETY. American Chemical Society, 1155 Sixteenth Street, NW, Washington, DC 20036. (800) 424-6747 or (202) 872-4700. Biweekly. $330.00 per year.

RESEARCH CENTERS AND INSTITUTES

BROWN UNIVERSITY. Chemical Research Laboratory, Providence, RI 02912. (401) 863-2256.

CORNELL UNIVERSITY. Materials Science Center, Clark Hall of Science, Ithaca, NY 14853. (607) 255-4272.

HARVARD UNIVERSITY. Chemical Laboratories, Oxford Street, Cambridge, MA 02138. (617) 495-4283.

UNIVERSITY OF PENNSYLVANIA. Laboratory for Research on the Structure of Matter, 3231 Walnut Street, Philadelphia, PA 19104. (215) 898-8571.

CHEMICAL ELEMENTS

See: ELEMENTS (CHEMICAL)

CHEMICAL ENGINEERING

ABSTRACT SERVICES AND INDEXES

APPLIED SCIENCE AND TECHNOLOGY INDEX. H.W. Wilson Company, 950 University Avenue, Bronx, NY 10452. (800) 367-6670 or (212) 588-8400. Inquire as to cost and availability.

CHEMICAL ABSTRACTS. Chemical Abstracts Service, 2540 Olentangy Road, P.O. Box 3012, Columbus, OH 43210. (800) 848-6538 or (614) 421-3600. Weekly. $9200.00 per year.

CHEMICAL INDUSTRY NOTES. Chemical Abstracts Service, P.O. Box 3012, Columbus, OH 43210. (800) 848-6538 or (614) 421-3600. Weekly. $1000.00 per year; $1175.00 with index volume.

ENGINEERING INDEX MONTHLY. Engineering Information, Incorporated, 345 East 47th Street, New York, NY 10017. (800) 221-1044 or (212) 705-7600. Monthly, with annual cumulation. $1560.00 per year.

PHYSICS ABSTRACTS. Institute of Electrical Engineers, London, United Kingdom. Available from: Institute of Electrical and Electronic Engineers (IEEE), 345 East 47th Street, New York, NY 10017. (212) 705-7900.

PROCESS ENGINEERING INDEX. Technical Indexes Limited, Willougby Road, Bracknell, Berks, England RG12 4DW. Semiannual.

ANNUAL REVIEWS AND YEARBOOKS

ADVANCES IN CHEMICAL ENGINEERING. Academic Press, Incorporated, Orlando, FL 32887. (800) 321-5068. (302) 345-2000. Annual. Inquire as to price.

ASSOCIATIONS AND PROFESSIONAL SOCIETIES

AMERICAN CHEMICAL SOCIETY. 1155 Sixteenth Street, NW, Washington, DC 20036. (202) 872-4600.

AMERICAN INSTITUTE OF CHEMICAL ENGINEERS. 345 East 47th Street, New York, NY 10017. (212) 705-7338.

ASSOCIATION OF CONSULTING CHEMISTS AND CHEMICAL ENGINEERS. 50 East 41st Street, Suite 92, New York, NY 10017. (212) 684-6255.

DIRECTORIES AND BIOGRAPHICAL SOURCES

AMERICAN INSTITUTE OF CHEMICAL ENGINEERS. Directory. American Institute of Chemical Engineers, 345 East 47th Street, New York, NY 10017. (212) 705-7338. Annual.

CHEMICAL ENGINEERING FACULTIES. American Institute of Chemical Engineers, 345 East 47th Street, New York, NY 10017. (212) 705-7338. Annual. $40.00 per year.

CONSULTING SERVICE DIRECTORY. Association of Consulting Chemists Engineers, 50 East 41st Street, New York, NY 10017. (212) 686-6255. Annual. $45.00.

RESEARCH CENTERS DIRECTORY. Gale Research Company, Detroit, MI 48226. (800) 521-0707. Eleventh edition. 1987.

WHO'S WHO IN FRONTIER SCIENCE AND TECHNOLOGY. Marquis Who's Who, Incorporated, 200 East Ohio Street, Chicago, IL 60611. (800) 428-3898 or (312) 787-2008.

WHO'S WHO IN TECHNOLOGY TODAY. Reston Publishing Company, Incorporated, c/o Prentice-Hall, Incorporated, Englewood Cliffs, NJ 07632. (800) 262-6868. Biennial. Five volumes. $425.00. Covers the fields of electronics, computer science, physics, optics, chemistry, biotechnology, mechanics, energy, and earth science.

ENCYCLOPEDIAS AND DICTIONARIES

DICTIONARY OF CHEMICAL TERMINOLOGY. D. Kryt. Elsevier Publishing Company, Incorporated, 52 Vanderbilt Avenue, New York, NY 10017. (212) 370-5520. 1980. $83.00.

ENCYCLOPEDIA OF PHYSICAL SCIENCE AND TECHNOLOGY. Academic Press, Incorporated, Orlando, FL 32887. (800) 321-5068 or (305) 345-2734. Fifteen volumes, 1986.

MCGRAW-HILL ENCYCLOPEDIA OF SCIENCE AND TECHNOLOGY. McGraw-Hill Book, Incorporated, 1221 Avenue of the Americas, New York, NY 10020. (212) 997-3675.

GENERAL WORKS

APPLIED CHEMICAL PROCESS DESIGN. Frank Aerstin and Gary Street. Plenum Publishing Company, 233 Spring Street, New York, NY 10013. (800) 221-9369. 1978. $32.50.

ELEMENTARY PRINCIPLES OF CHEMICAL PROCESSES. R.M. Felder and R.W. Rousseau. John Wiley and Son, Incorporated, 605 Third Avenue, New York, NY 10158. (800) 526-5368 or (212) 850-6000. Second edition. 1987. $43.95.

GUIDE TO CHEMICAL ENGINEERING PROCESS DESIGN AND ECONOMICS. G.D. Ulrich. John Wiley and Son, Incorporated, 605 Third Avenue, New York, NY 10158. (800) 526-5368 or (212) 850-6000. 1984. $43.50.

HANDBOOKS AND MANUALS

CHEMICAL ENGINEERS HANDBOOK. Robert H. Perry and Don W. Green, editors. McGraw-Hill Book Company, 1221 Avenue of the Americas, New York, NY 10020. (212) 512-2000. Sicth edition. 1984. $95.00.

GAS TABLES: THERMODYNAMIC PROPERTIES OF AIR PRODUCTS OF COMBUSTION AND COMPONENT GASES, COMPRESSIBLE FLOW FUNCTIONS. John Wiley and Son, Incorporated, 605 Third Avenue, New York, NY 10158. (800) 526-5368 or (212) 850-6000. Second Edition. 1986. $82.00.

HANDBOOK OF CHEMICAL ENGINEERING CALCULATIONS. N.P. Chopey and T.G. Hicks. McGraw-Hill Book Company, 1221 Avenue of the Americans, New York, NY 10020. (212) 512-2000. 1984. $49.50.

HANDBOOK OF THERMODYNAMIC TABLES AND CHARTS. K. Raznjevic. McGraw-Hill Book Company, 1221 Avenue of the Americas, New York, NY 10020. (212) 512-2000. 1976. $67.50.

STEAM TABLES: THERMODYNAMIC PROPERTIES OF WATER, INCLUDING VAPOR, LIQUID, AND SOLID PHASES. J.H. Keenan, F.G. Keyes, P.G. Hill, and J.G. Moore. John Wiley and Sons, Incorporated, 605 Third Avenue, New York, NY 10158. (800) 526-5368 or (212) 850-6000. 1986. Two volume set. $76.50.

ONLINE DATA BASES

CA SEARCH. Chemical Abstracts Service, P.O. Box 3012, Columbus, OH 43210. Guide to chemical literature, 1972 to present. Inquire as to cost and availability.

COMPENDEX. Engineering Information, Incorporated, 345 East 47th Street, New York, NY 10017. (800) 221-1044 or (212) 705-7615. Engineering and technical literature, 1975 to present. Inquire as to cost and availability.

NTIS. National Technical Information Service, 5285 Port Royal Road, Springfield, VA 22161. (703) 487-4630. Broad coverage of government sponsored research reports, 1964 to present. Inquire as to cost and availability.

PERIODICALS

AICHE JOURNAL. American Institute of Chemical Engineers, 345 East 47th Street, New York, NY 10017. (212) 705-7321. Bimonthly. $60.00 per year.

APPLIED CATALYSIS. Elsevier Science Publishing Company, Incorporated, 52 Vanderbilt Avenue, New York, NY 10017. (212) 370-5520. Monthly. $200.00 per year.

CHEMICAL AND ENGINEERING NEWS. American Chemical Society, 1155 Sixteenth Street, NW, Washington, DC 20036. (202) 872-4600. Weekly. $37.00 per year.

CHEMICAL ENGINEERING. McGraw-Hill Book Company, 1221 Avenue of the Americas, New York, NY 10020. (212) 512-2000. Biweekly. $27.50 per year.

CHEMICAL ENGINEERING COMMUNICATIONS. Gordon and Breach Science Publishers, Box 786, Cooper Station, New York, NY 10276. (212) 206-8900. Quarterly. $278.00 per year.

CHEMICAL ENGINEERING PROGRESS. American Institute of Chemical Engineers, 345 East 47th Street, New York, NY 10017. (212) 705-7321. Monthly. $35.00 per year.

CHEMICAL ENGINEERING SCIENCE. Pergamon Journals, Maxwell House, Fairview Park, Elmsford, NY 10523. (914) 592-7700. Monthly. $435.00 per year.

CHEMICAL PROCESSING. Putman Publishing Company, 302 East Erie Street, Chicago, IL 60611. (312) 644-2020. Fifteen times per year. $24.00 per year.

CHEMICAL WEEK. McGraw-Hill Publishing Company, 1221 Avenue of the Americas, New York, NY 10020. (212) 512-2000. Weekly. 430.00 per year.

CHEMIE-INGENIEUR-TECHNIK. Verlag Chemie International, 303 Northwest 12th Street, Deerfield Beach, FL 33441. (305) 428-5566. Monthly. $178.00 per year.

INDUSTRIAL AND ENGINEERING FUNDAMENTALS. American Chemical Society, 1155 16th Street, NW, Washington, DC 20036. (202) 872-4600. Quarterly. $71.00 per year.

INDUSTRIAL AND ENGINEERING PROCESS, DESIGN AND DEVELOPMENT. American Chemical Society, 1155 16th Street, NW, Washington, DC 20036. (202) 872-4600. Quarterly. $99.00 per year.

INDUSTRIAL AND ENGINEERING PRODUCT RESEARCH AND DEVELOPMENT. American Chemical Society, 1155 16th Street, NW, Washington, DC 20036. (202) 872-4600. Quarterly. $76.00 per year.

INTERNATIONAL CHEMICAL ENGINEERING. American Institute of Chemical Engineers, 345 East 47th Street, New York, NY 10017. (212) 705-7321. Quarterly.

POLYMER PROCESS ENGINEERING. Marcel Dekker, Incorporated, 270 Madison Avenue, New York, NY 10016. (212) 696-9000. Quarterly. 484.00 per year.

THEORETICAL FOUNDATIONS OF CHEMICAL ENGINEERING. Plenum Publishing Company, 233 Spring Street, New York, NY 10013. (800) 221-9369. Bimonthly. $625.00 per year.

RESEARCH CENTERS AND INSTITUTES

KANSAS STATE UNIVERSITY. Institute for Systems Design and Optimization, Department of Chemical Engineering, Manhattan, KS 66506. (913) 532-5584.

PENNSYLVANIA STATE UNIVERSITY. Engineering Research Program, 101 Hammond Building, University Park, PA 16802. (814) 865-4542.

PURDUE UNIVERSITY. Engineering Experiment Station, West Lafayette, IN 47907. (317) 494-5340.

CHEMICAL EQUILIBRIUM

See: PHYSICAL CHEMISTRY

CHEMICAL FORMULAS

See: CHEMISTRY

CHEMICAL MICROSCOPY

See: MICROSCOPY

CHEMICAL SEPARATION TECHNIQUES

See: ANALYTICAL CHEMISTRY

CHEMISTRY

See also: INORGANIC CHEMISTRY, ORGANIC CHEMISTRY, PHYSICAL CHEMISTRY

ABSTRACT SERVICES AND INDEXES

APPLIED SCIENCE AND TECHNOLOGY INDEX. H.W. Wilson Company, 950 University Avenue, Bronx, NY 10452. (800) 367-6670 or (212) 588-8400. Inquire as to cost and availability.

CHEMICAL ABSTRACTS. Chemical Abstracts Service, 1540 Olentangy Road, Post Office Box 3012, Columbus, OH 43210. (800) 848-6538 or (614) 421-3600. Weekly. $9200.00 per year.

GENERAL SCIENCE INDEX. H.W. Wilson Company, 950 University Avenue, Bronx, NY 10452. (800) 367-6770 or (212) 588-8400. Inquire as to cost and availability.

PHYSICS ABSTRACTS. Institute of Electrical Engineers, London United Kingdom. Available from: Institute of Electrical and Electronic Engineers (IEEE), 345 East 47th Street, New York, NY 10017. (212) 705-7900.

SCIENCE CITATION INDEX. Institute for Scientific Information, 3501 Market Street, Philadelphia, PA 19104. (800) 523-1850 or (215) 386-0100.

ANNUAL REVIEWS AND YEARBOOKS

ANNUAL REVIEW OF PHYSICAL CHEMISTRY. Annual Reviews Incorporated, 4139 El Camino Way, Palo Alto, CA 94306. (415) 493-4400. Annual.

ASSOCIATIONS AND PROFESSIONAL SOCIETIES

AMERICAN CHEMICAL SOCIETY. 1155 16th Street, NW, Washington, DC 20036. (202) 872-4600.

ASSOCIATION OF CONSULTING CHEMISTS AND CHEMICAL ENGINEERS. 50 East 41st Street, Suite 92, New York, NY 10017. (212) 684-6255.

BIBLIOGRAPHIES

SCIENCE BOOKS AND FILMS. American Association for the Advancement of Science, 1333 H Street, NW, Washington, DC 20005.

SCIENTIFIC AND TECHNICAL BOOKS AND SERIALS IN PRINT 1988: AN INDEX TO LITERATURE IN SCIENCE AND TECHNOLOGY. R.R. Bowker Company, 205 East 42nd Street, New York, NY 10017. (800) 521-8110 or (212) 916-1600. $175.00.

DIRECTORIES AND BIOGRAPHICAL SOURCES

AMERICAN INSTITUTE OF CHEMISTS. American Institute of Chemists, 7315 Wisconsin Avenue, Bethesda, MD 20814. (301) 652-2447. 1986. $35.00.

AMERICAN MEN AND WOMEN OF SCIENCE. Physical and Biological Sciences. R.R. Bowker Company, 205 East 42nd Street, New York, NY 10017. (800) 521-8110 or (212) 916-1600. Fifteenth edition. $565.00.

BIOGRAPHICAL DICTIONARY OF SCIENTISTS: CHEMISTS. David Abbott, editor. P. Bedrick Books, 125 East 23rd Street, New York, NY 10010. (212) 777-1187. 1984. $18.95.

CHEMICAL WEEK - BUYERS GUIDE ISSUE. McGraw-Hill Book Company, 1221 Avenue of the Americas, New York, NY 10020. (800) 628-0004. Annual, October. $50.00.

CONSULTING SERVICES: CHEMISTS AND CHEMICAL ENGINEERS. Association of Consulting Chemists and Chemical Engineers, 50 East 41st Street, New York, NY 10017. (212) 684-6255. Annual. 1986. $45.00.

GOVERNMENT RESEARCH DIRECTORY. Gale Research Company, Book Tower, Detroit, MI 48226. (800) 521-0707. Fourth edition. 1987. $350.00.

INTERNATIONAL RESEARCH CENTERS DIRECTORY 1986-1987. Gale Research Company, Book Tower, Detroit, MI 48226. (800) 521-0707. Third edition. 1986. $330.00.

RESEARCH CENTERS DIRECTORY. Gale Research Company, Book Tower, Detroit, MI 48226. (800) 521-0707. Eleventh edition. 1987. $355.00.

RESEARCH SERVICES DIRECTORY. Robert J. Huffman and Mary M. Watkins, editors. Gale Research Company, Book Tower, Detroit, MI 48226. (800) 521-0707. Third edition. $290.00.

SCIENTIFIC AND TECHNICAL ORGANIZATIONS AND AGENCIES DIRECTORY. Gale Research Company, Book Tower, Detroit, MI 48226. (800) 521-0707. 1985. $150.00.

WHO'S WHO IN FRONTIER SCIENCE AND TECHNOLOGY. Marquis Who's Who, Incorporated, 200 East Ohio Street, Chicago, IL 60611. (800) 428-3898 or (312) 787-2008.

WHO'S WHO IN TECHNOLOGY TODAY. Reston Publishing Company, Incorporated. c/o Prentice-Hall, Incorporated, Englewood Cliffs, NJ 07632. (800) 262-6868. Biennial. Five volumes. $425.00. Covers the fields of electronics, computer science, physics, optics, chemistry, biotechnology, mechanics, energy, and earth science.

WORLD GUIDE TO SCIENTIFIC ASSOCIATIONS AND LEARNED SOCIETIES. K.G. Saur Incorporated, 175 Fifth Avenue, New York, NY 10010. (800) 521-0707 or (212) 982-1302. Fourth edition. 1984. $112.00.

ENCYCLOPEDIAS AND DICTIONARIES

CONCISE ENCYCLOPEDIA OF CHEMICAL TECHNOLOGY. Kirk Othmer. John Wiley and Sons, Incorporated, 605 Third Avenue, New York, NY 10158. (800) 526-5368 or (212) 850-6000.

Third edition. 1985. $129.95.

CONDENSED CHEMICAL DICTIONARY. Gessner Hawley. Van Nostrand Reinhold, 115 Fifth Avenue, New York, NY 10003. Tenth edition. 1981. $49.95.

ENCYCLOPEDIA OF PHYSICAL SCIENCE AND TECHNOLOGY. Academic Press, Incorporated, Orlando, FL 32887. (800) 321-5068 or (305) 345-2734. Fifteen volumes, 1986.

GLOSSARY OF CHEMICAL TERMS. Clifford A. Hampel and Gessner G. Hawley. Van Nostrand Reinhold Company, 115 Fifth Avenue, New York, NY 10003. (800) 543-2681 or (212) 254-3232. Second edition. 1982. $21.95.

HAWLEY'S CONDENSED CHEMICAL DICTIONARY. N. Irving Sax and Richard J. Lewis. Sr., editors. Van Nostrand Reinhold, Incorporated, 115 Fifth Avenue, New York, NY 10003. (800) 543-2681. Eleventh edition. 1987. $52.95.

MCGRAW-HILL ENCYCLOPEDIA OF SCIENCE AND TECHNOLOGY. McGraw-Hill Book, Incorporated, 1221 Avenue of the Americas, New York, NY 10020. (212) 997-3675.

VAN NOSTRAND REINHOLD ENCYCLOPEDIA OF CHEMISTRY. Douglas M. Considine and Glenn D. Considine. Van Nostrand Reinhold Publishing Company, Incorporated, 115 Fifth Avenue, New York, NY 10003. (800) 543-2681 or (212) 254-3232. 1984. $97.95.

GENERAL WORKS

ANALYTICAL CHEMISTRY. Gary D. Christian. John Wiley and Sons, Incorporated, 605 Third Avenue, New York, NY 10158. (800) 526-5368 or (212) 850-6000. Fourth edition. 1986. $37.95.

CHEMISTRY OF THE ELEMENTS. N.N. Greenwood and A. Earnshaw. Pergamon Publishing, Incorporated, Maxwell House, Fairview Park, Elmsford, NY 10523. (914) 592-7700. 1984. $143.00.

COMPUTERS IN ANALYTICAL CHEMISTRY. P.G. Baker. Pergamon Publishing, Incorporated, Maxwell House, Fairview Park, Elmsford, NY 10523. (914) 592-7700. 1983. $91.30.

HAZARDOUS AND TOXIC CHEMICAL: SAFE HANDLING AND DISPOSAL. Howard Fawcett. John Wiley and Sons, Incorporated, 605 Third Avenue, New York, NY 10158. (800) 526-5368 or (212) 850-6000. 1984. $37.00.

INVESTIGATION OF RATES AND MECHANISMS OF REACTIONS, PART 1. Claude F. Bernasconi, editor. John Wiley and Sons, Incorporated, 605 Third Avenue, New York, NY 10158. (800) 526-5368 or (212) 850-6000. Fourth edition. 1986. $175.00.

SAFE STORAGE OF LABORATORY CHEMICALS. David A. Piptone. John Wiley and Sons, Incorporated, 605 Third Avenue, New York, NY 10158. (800) 526-5368 or (212) 850-6000. 1984. $60.00.

HANDBOOKS AND MANUALS

THE CHEMIST'S COMPANION: A HANDBOOK OF PRACTICAL DATA, TECHNIQUES, AND REFERENCES. Arnold J. Gordon and Richard A. Ford. John Wiley and Sons, Incorporated, 605 Third Avenue, New York, NY 10158. (800) 526-5368. 1973. $49.95.

CRC HANDBOOK OF CHEMISTRY AND PHYSICS. CRC Press, Incorporated, 2000 Corporate Boulevard, NW, Boca Raton, FL 33431. Sixty-seventh edition. 1986. $69.95.

HANDBOOK OF APPLIED CHEMISTRY: FACTS FOR ENGINEERS, SCIENTISTS, TECHNICIANS, AND TECHNICAL MANAGERS. Vollrath Hopp and Ingp Hennig. McGraw-Hill Book Company, 1221 Avenue of the Americas, New York, NY 10020. (800) 628-0004. 1983. $54.00.

HANDBOOK OF COMPUTATIONAL CHEMISTRY: A PRACTICAL GUIDE TO CHEMICAL STRUCTURE AND ENERGY CALCULATIONS. Tim Clark. John Wiley and Sons, Incorporated, 605 Third Avenue, New York, NY 10158. (800) 526-5368 or (212) 850-6000. 1985. $35.00.

HANDBOOK OF HETEROCYCLIC CHEMISTRY. F.R.S. Katritzky. Pergamon Publishing, Incorporated, Maxwell House, Fairview Park, Elmsford, NY 10523. (914) 592-7700. 1984. $88.00.

LANGE'S HANDBOOK OF CHEMISTRY. John A. Dean, editor. McGraw-Hill Book Company, 1221 Avenue of the Americas, New York, NY 10020. (800) 628-0004. 1985. $59.50.

TABLES OF PHYSICAL AND CHEMICAL CONSTANTS; AND SOME MATHEMATICAL FUNCTIONS. G.W.C. Kaye and T.H. Laby, editors. Longman, Incorporated, 95 Church Street, White Plains, NY 10601. (014) 993-5000. 1986. $39.95.

ONLINE DATA BASES

CA SEARCH. Chemical Abstracts Service, Post Office Box 3012, Columbus, OH 43210. Guide to chemical literature, 1972 to present. Inquire as to cost and availability.

DISSERTATION ABSTRACTS ONLINE. University Microfilms International, 300 North Zeeb Road, Ann Arbor, MI 48106. (800) 521-0600 or (313) 761-4700. Scope includes virtually all doctoral dissertations accepted at accredited American institutions from 1861 to present in 252 subject areas. Inquire as to cost and availability.

INSPEC. INSPEC Marketing Department, Institute of Electrical and Electronics Engineers, Incorporated, IEEE Service Department, 445 Hoes Lane, Piscataway, NJ 08854. (201) 981-0060. Inquire as to on-line cost and availability.

NTIS. National Technical Information Service, 5285 Port Royal Road, Springfield, VA 22161. (703) 487-4630. Broad coverage of government sponsored research reports 1964 to present. Inquire as to cost and availability.

SCISEARCH. Institute for Scientific Information, 3501 Market Street, Philadelphia, PA 19104. (800) 523-1850 or (215) 386-0100. Broad multidisciplinary title and author index to the present. Inquire as to cost and availability.

OTHER SOURCES

ANNUAL ENERGY REVIEW. United States Department of Energy, Energy Information Administration, Washington, DC 20585. Annual.

CHEMICAL NOMENCLATURE USEAGE. Ronald Lees and Arthur F. Smith. John Wiley and Sons, Incorporated, 605 Third Avenue, New York, NY 10158. (800) 526-5368 or (212) 850-6000. 1979. $26.95.

GUIDE TO BASIC INFORMATION SOURCES IN CHEMISTRY. Arthur Antony. John Wiley and Sons, Incorporated, 605 Third Avenue, New York, NY 10158. (800) 526-5368 or (212) 850-6000. 1979. $26.95.

HOW TO FIND CHEMICAL INFORMATION: A GUIDE FOR PRACTICING CHEMISTS, TEACHERS, AND STUDENTS. John Wiley and Sons, Incorporated, 605 Third Avenue, New York, NY 10158. (800) 526-5368 or (212) 850-6000. 1986. $35.00.

MINERALS YEARBOOK. United States Department of the Interior, Bureau of Mines, C Street between Eighteenth and Ninteenth Streets, NW, Washington, DC 20240. Annual.

CHEMISTRY

PERIODICALS

ANALYTICAL CHEMISTRY. American Chemical Society, 1155 Sixteenth Street, NW, Washington, DC 20036. (800) 424-6747 or (202) 872-4700. Monthly. $33.00 per year.

ANDEWANDTE CHEMIE (GESELLSCHAFT DEUTSCHER CHEMIKER, GW). V C H Verlagsgessellschaft mbH, Pappelallee 3, Postfach 1260, 6940 Weinheim, West Germany. Monthly. $300.00 per year.

BIOCHEMISTRY. American Chemical Society, 1155 Sixteenth Street, NW, Washington, DC 20036. (800) 424-6747 or (202) 872-4700. Biweekly. $303.00 per year.

CHEMICAL REVIEWS. American Chemical Society, 1155 Sixteenth Street, NW, Washington, DC 20036. (800) 424-6747 or (202) 872-4700. Bimonthly. $83.00 per year.

CHEMICAL WEEK. McGraw-Hill Publishing Company, 1221 Avenue of the Americas, New York, NY 10020. (212) 512-2000. Weekly. $30.00 per year.

CHEMTECH. American Chemical Society, 1155 Sixteenth Street, NW, Washington, DC 20036. (800) 424-6747 or (202) 872-4700. Monthly. $40.00 per year to individuals.

INORGANIC CHEMISTRY. American Chemical Society, 1155 Sixteenth Street, NW, Washington, DC 20036. (800) 424-6747 or (202) 872-4700. Monthly. $400.00 per year.

JOURNAL OF ORGANIC CHEMISTRY. American Chemical Society, 1155 Sixteenth Street, NW, Washington, DC 20036. (800) 424-6747 or (202) 872-4700. Biweekly. $265.00 per year.

JOURNAL OF PHYSICAL CHEMISTRY. American Chemical Society, 1155 Sixteenth Street, NW, Washington, DC 20036. (800) 424-6747 or (202) 872-4700. Biweekly. $369.00 per year.

JOURNAL OF THE AMERICAN CHEMICAL SOCIETY. American Chemical Society, 1155 Sixteenth Street, NW, Washington, DC 20036. (800) 424-6747 or (202) 872-4700. Biweekly. $330.00 per year.

TETRAHEDRON. Pergamon Journals, Incorporated, Maxwell House, Fairview Park, Elmsford, NY 10523. (914) 592-7700.

CHIPS

See: INTEGRATED CIRCUITS

CHLORINE

See also: BROMINE, FLUORIDE, IODINE, HALIDES

ABSTRACT SERVICES AND INDEXES

APPLIED SCIENCE AND TECHNOLOGY INDEX. H.W. Wilson Company, 950 University Avenue, Bronx, NY 10452. (800) 367-6670 or (212) 588-8400. Inquire as to cost and availability.

CHEMICAL ABSTRACTS. Chemical Abstracts Service, 1540 Olentangy Road, Post Office Box 3012, Columbus, OH 43210. (800) 848-6538 or (614) 421-3600. Weekly. $9200.00 per year.

GENERAL SCIENCE INDEX. H.W. Wilson Company, 950 University Avenue, Bronx, NY 10452. (800) 367-6770 or (212) 588-8400. Inquire as to cost and availability.

PHYSICS ABSTRACTS. Institute of Electrical Engineers, London United Kingdam. Available from: Institute of Electrical and Electronic Engineers (IEEE), 345 East 47th Street, New York, NY 10017. (212) 705-7900.

SCIENCE CITATION INDEX. Institute for Scientific Information, 3501 Market Street, Philadelphia, PA 19104. (800) 523-1850 or (215) 386-0100.

ASSOCISTIONS AND PROFESSIONAL SOCIETIES

AMERICAN CHEMICAL SOCIETY. 1155 16th Street, NW, Washington, DC 20036. (202) 872-4600.

ASSOCIATION OF CONSULTING CHEMISTS AND CHEMICAL ENGINEERS. 50 East 41st Street, Suite 92, New York, NY 10017. (212) 684-6255.

CHLORINE INSTITUTE. 70 West 40th Street, New York, NY 10018. (212) 819-1677.

BIBLIOGRAPHIES

CHLORINE: AN ANNOTATED BIBLIOGRAPHY. Ralph G. Smith. Chlorine Institute, 70 West 40th Street, New York, NY 10018. (212) 819-1677. 1971. $12.00.

CHLORINE: AN ANNOTATED BIBLIOGRAPHY SUPPLEMENT. Ralph G. Smith. Chlorine Institute, 70 West 40th Street, New York, NY 10018. (212) 819-1677. 1983. $12.00 in paper.

DIRECTORIES AND BIOGRAPHICAL SOURCES

AMERICAN INSTITUTE OF CHEMISTS. American Institute of Chemists, 7315 Wisconsin Avenue, Bethesda, MD 20814. (301) 652-2447. 1986. $35.00.

AMERICAN MEN AND WOMEN OF SCIENCE. Physical and Biological Sciences. R.R. Bowker Company, 205 East 42nd Street, New York, NY 10017. (800) 521-8110 or (212) 916-1600. Fifteenth edition. $565.00.

BIOGRAPHICAL DICTIONARY OF SCIENTISTS: CHEMISTS. David Abbott, editor. P. Bedrick Books, 125 East 23rd Street, New York, NY 10010. (212) 777-1187. 1984. $18.95.

CHEMICAL WEEK - BUYERS GUIDE ISSUE. McGraw-Hill Book Company, 1221 Avenue of the Americas, New York, NY 10020. (800) 628-0004. Annual, October. $50.00.

CONSULTING SERVICES: CHEMISTS AND CHEMICAL ENGINEERS. Association of Consulting Chemists and Chemical Engineers, 50 East 41st Street, New York, NY 10017. (212) 684-6255. Annual. 1986. $45.00.

GOVERNMENT RESEARCH DIRECTORY. Gale Research Company, Book Tower, Detroit, MI 48226. (800) 521-0707. Fourth edition. 1987. $350.00.

INTERNATIONAL RESEARCH CENTERS DIRECTORY 1986-1987. Gale Research Company, Book Tower, Detroit, MI 48226. (800) 521-0707. Third edition. 1986. $330.00.

RESEARCH CENTERS DIRECTORY. Gale Research Company, Book Tower, Detroit, MI 48226. (800) 521-0707. Eleventh edition. 1987. $355.00.

SCIENTIFIC AND TECHNICAL ORGANIZATIONS AND AGENCIES DIRECTORY. Gale Research Company, Book Tower, Detroit, MI 48226. (800) 521-0707. 1985. $150.00.

WHO'S WHO IN FRONTIER SCIENCE AND TECHNOLOGY. Marquis Who's Who, Incorporated, 200 East Ohio Street, Chicago, IL 60611. (800) 428-3898 or (312) 787-2008.

WHO'S WHO IN TECHNOLOGY TODAY. Reston Publishing Company, Incorporated. c/o Prentice-Hall, Incorporated, Englewood Cliffs, NJ 07632. (800) 262-6868. Biennial. Five volumes. $425.00. Covers the fields of electronics, computer science, physics, optics, chemistry, biotechnology, mechanics, energy, and earth science.

WORLD GUIDE TO SCIENTIFIC ASSOCIATIONS AND LEARNED SOCIETIES. K.G. Saur Incorporated, 175 Fifth Avenue, New York, NY 10010. (800) 521-0707 or (212) 982-1302. Fourth edition. 1984. $112.00.

ENCYCLOPEDIAS AND DICTIONARIES

CONCISE ENCYCLOPEDIA OF CHEMICAL TECHNOLOGY. Kirk-Othmer. John Wiley and Sons, Incorporated, 605 Third Avenue, New York, NY 10158. (800) 526-5368 or (212) 850-6000. Third edition. 1985. $129.95.

CONDENSED CHEMICAL DICTIONARY. Gessner Hawley. Van Nostrand Reinhold, 115 Fifth Avenue, New York, NY 10003. Tenth edition. 1981. $49.95.

ENCYCLOPEDIA OF PHYSICAL SCIENCE AND TECHNOLOGY. Academic Press, Incorporated, Orlando, FL 32887. (800) 321-5068 or (305) 345-2734. Fifteen volumes, 1986.

GLOSSARY OF CHEMICAL TERMS. Clifford A. Hampel and Gessner G. Hawley. Van Nostrand Reinhold Company, 115 Fifth Avenue, New York, NY 10003. (800) 543-2681 or (212) 254-3232. Second edition. 1982. $21.95.

MCGRAW-HILL ENCYCLOPEDIA OF SCIENCE AND TECHNOLOGY. McGraw-Hill Book, Incorporated, 1221 Avenue of the Americas, New York, NY 10020. (212) 997-3675.

VAN NOSTRAND REINHOLD ENCYCLOPEDIA OF CHEMISTRY. Douglas M. Considine and Glenn D. Considine. Van Nostrand Reinhold Publishing Company, Incorporated, 115 Fifth Avenue, New York, NY 10003. (800) 543-2681 or (212) 254-3232. 1984. $97.95.

GENERAL WORKS

CHEMISTRY OF HALIDES, PSEUDOHALIDES AND AZIDES: SUPPLEMENT D. CHEMISTRY OF FUNCTIONAL GROUPS SERIES, PART 1 AND 2. Saul Patai. John Wiley and Sons, Incorporated, 605 Third Avenue, New York, NY 10158. (800) 526-5368 or (212) 850-6000. 1983. $745.00 set.

CHEMISTRY OF THE ELEMENTS. N.N. Greenwood and A. Earnshaw. Pergamon Publishing, Incorporated, Maxwell House, Fairview Park, Elmsford, NY 10523. (914) 592-7700. 1984. $143.00.

CHLORINE: ITS MANUFACTURE, PROPERTIES AND USES. J.S. Sconse, editor. Robert E. Krieger Publishing Company, Incorporated, Box 9542, Melbourne, FL 32902-9542. (305) 724-9542. 1972. $59.50.

HALOGEN CHEMISTRY. V. Gutman, editor. Academic Press, Incorporated, 6277 Sea Harbor Drive, Orlando, FL 32821. (800) 321-5068. Three volumes. 1967. $90.00 each.

PROPERTIES OF CHLORINE IN SI UNITS. Chlorine Institute, 70 West 40th Street, New York, NY 10018. (212) 819-1677. 1981. $23.00.

SULFUR DIOXIDE, CHLORINE, FLUORINE AND CHLORINE OXIDES. A.S. Young. Pergamon Publishing, Incorporated, Maxwell House, Fairview Park, Elmsford, NY 10523. (014) 592-7700. 1983. $100.00.

TOXICOLOGY OF HALOGENATED HYDROCARBONS: HEALTH AND ECOLOGICAL EFFECTS. M.A. Khan and R.H. Stanton, editors. Pergamon Press, Incorporated, Maxwell House, Fairview Park, Elmsford, NY 10523. (014) 592-7700. 1981. $72.50.

HANDBOOKS AND MANUALS

THE CHEMIST'S COMPANION: A HANDBOOK OF PRACTICAL DATA, TECHNIQUES, AND REFERENCES. Arnold J. Gordon and Richard A. Ford. John Wiley and Sons, Incorporated, 605 Third Avenue, New York, NY 10158. (800) 526-5368. 1973. $49.95.

CRC HANDBOOK OF CHEMISTRY AND PHYSICS. CRC Press, Incorporated, 2000 Corporate Boulevard, NW, Boca Raton, FL 33431. Sixty-seventh edition. 1986. $69.95.

HANDBOOK OF APPLIED CHEMISTRY: FACTS FOR ENGINEERS, SCIENTISTS, TECHNICIANS, AND TECHNICAL MANAGERS. Vollrath Hopp and Ingp Hennig. McGraw-Hill Book Company, 1221 Avenue of the Americas, New York, NY 10020. (800) 628-0004. 1983. $54.00.

HANDBOOK OF COMPUTATIONAL CHEMISTRY: A PRACTICAL GUIDE TO CHEMICAL STRUCTURE AND ENERGY CALCULATIONS. Tim Clark. John Wiley and Sons, Incorporated, 605 Third Avenue, New York, NY 10158. (800) 526-5368 or (212) 850-6000. 1985. $35.00.

LANGE'S HANDBOOK OF CHEMISTRY. John A. Dean, editor. McGraw-Hill Book Company, 1221 Avenue of the Americas, New York, NY 10020. (800) 628-0004. 1985. $59.50

ONLINE DATA BASES

CA SEARCH. Chemical Abstracts Service, Post Office Box 3012, Columbus, OH 43210. Guide to chemical literature, 1972 to present. Inquire as to cost and availability.

DISSERTATION ABSTRACTS ONLINE. University Microfilms International, 300 North Zeeb Road, Ann Arbor, MI 48106. (800) 521-0600 or (313) 761-4700. Scope includes virtually all doctoral dissertations accepted at accredited American institutions from 1861 to present in 252 subject areas. Inquire as to cost and availability.

INSPEC. INSPEC Marketing Department, Institute of Electrical and Electronics Engineers, Incorporated, IEEE Service Department, 445 Hoes Lane, Piscataway, NJ 08854. (201) 981-0060. Inquire as to on-line cost and availability.

NTIS. National Technical Information Service, 5285 Port Royal Road, Springfield, VA 22161. (703) 487-4630. Broad coverage of government sponsored research reports 1964 to present. Inquire as to cost and availability.

SCISEARCH. Institute for Scientific Information, 3501 Market Street, Philadelphia, PA 19104. (800) 523-1850 or (215) 386-0100. Broad multidisciplinary title and author index to the present. Inquire as to cost and availability.

OTHER SOURCES

ANNUAL ENERGY REVIEW. United States Department of Energy, Energy Information Administration, Washington, DC 20585. Annual.

CHEMICAL NOMENCLATURE USEAGE. Ronald Lees and Arthur F. Smith. John Wiley and Sons, Incorporated, 605 Third Avenue, New York, NY 10158. (800) 526-5368 or (212) 850-6000. 1979. $26.95.

GUIDE TO BASIC INFORMATION SOURCES IN CHEMISTRY. Arthur Antony. John Wiley and Sons, Incorporated, 605 Third Avenue, New York, NY 10158. (800) 526-5368 or (212) 850-6000. 1979. $26.95.

HOW TO FIND CHEMICAL INFORMATION: A GUIDE FOR PRACTICING CHEMISTS, TEACHERS, AND STUDENTS. John Wiley and Sons, Incorporated, 605 Third Avenue, New York, NY 10158. (800) 526-5368 or (212) 850-6000. 1986. $35.00.

MINERALS YEARBOOK. United States Department of the Interior, Bureau of Mines, C Street between Eighteenth and Ninteenth Streets, NW, Washington, DC 20240. Annual.

CHLORINE

PERIODICALS

ANALYTICAL CHEMISTRY. American Chemical Society, 1155 Sixteenth Street, NW, Washington, DC 20036. (800) 424-6747 or (202) 872-4700. Monthly. $33.00 per year.

ANDEWANDTE CHEMIE (GESELLSCHAFT DEUTSCHER CHEMIKER, GW). V C H Verlagsgesellschaft mbH, Pappelallee 3, Postfach 1260, 6940 Weinheim, West Germany. Monthly. $300.00 per year.

CHEMICAL REVIEWS. American Chemical Society, 1155 Sixteenth Street, NW, Washington, DC 20036. (800) 424-6747 or (202) 872-4700. Bimonthly. $83.00 per year.

CHEMICAL WEEK. McGraw-Hill Publishing Company, 1221 Avenue of the Americas, New York, NY 10020. (212) 512-2000. Weekly. $30.00 per year.

CHEMTECH. American Chemical Society, 1155 Sixteenth Street, NW, Washington, DC 20036. (800) 424-6747 or (202) 872-4700. Monthly. $40.00 per year to individuals.

INORGANIC CHEMISTRY. American Chemical Society, 1155 Sixteenth Street, NW, Washington, DC 20036. (800) 424-6747 or (202) 872-4700. Monthly. $400.00 per year.

JOURNAL OF THE AMERICAN CHEMICAL SOCIETY. American Chemical Society, 1155 Sixteenth Street, NW, Washington, DC 20036. (800) 424-6747 or (202) 872-4700. Biweekly. $330.00 per year.

POLYHEDRON. Pergamon Journals, Incorporated, Maxwell House, Fairview Park, Elmsford, NY 10523. (014) 592-7700. Monthly. $595.00 per year.

RESEARCH CENTERS AND INSTITUTES

HARVARD UNIVERSITY. Chemical Laboratories, Oxford Street, Cambridge, MA 02138. (617) 495-4283.

RENSSELAER POLYTECHNICAL INSTITUTE. Chemistry Laboratories, Cogswell Laboratory, Troy, NY 13181. (518) 266-8462.

SOUTHERN ILLINOIS UNIVERSITY AT CARBONDALE. Research Program in Chemistry and Biochemistry, Carbondale, IL 62901. (618) 453-5721.

UNIVERSITY OF WISCONSIN, MADISON. Theoretical Chemistry Institute, 1101 University Avenue, Madison, WI 53706. (608) 262-1511.

CHROMATOGRAPHY

See also: ANALYTICAL CHEMISTRY, CHEMICAL ENGINEERING, CHEMISTRY, ORGANIC CHEMISTRY, SPECTROSCOPY

ABSTRACT SERVICES AND INDEXES

APPLIED SCIENCE AND TECHNOLOGY INDEX. H.W. Wilson and Company, 950 University Avenue, Bronx, NY 10452. (800) 367-6670 or (212) 588-8400. Monthly. Inquire as to cost and availability.

CHEMICAL ABSTRACTS. American Chemical Society, Chemical Abstracts Service, Box 3012, Columbus, OH 43210. (614) 421-3600. 1907 to present. Weekly. $9500.00 per year.

CONFERENCE PAPERS INDEX. Cambridge Scientific Abstracts, 5161 River Road, Bethesda, MD 20816. 1972 to present. Monthly. Inquire as to cost and availability.

CURRENT CONTENTS: PHYSICAL, CHEMICAL, AND EARTH SCIENCES. Institute for Scientific Information, 3501 Market Street, Philadelphia, PA 19104. (800) 523-1850 or (215) 386-0100. Weekly. $275.00 per year.

GENERAL SCIENCE INDEX. H.W. Wilson and Company, 950 University Avenue, Bronx, NY 10452. (800) 367-6670 or (212) 588-8400. 1978 to present. Monthly. Inquire as to cost and availability.

INDEX TO SCIENTIFIC AND TECHNICAL PROCEEDINGS. Institute for Scientific Information, 3501 Market Street, Philadelphia, PA 19104. (800) 523-1850 or (215) 386-0100. 1978 to present. Monthly. $775.00 per year.

INDEX TO SCIENTIFIC REVIEWS. Institute for Scientific Information, 3501 Market Street, Philadelphia, PA 19104. (800) 523-1850 or (215) 386-0100. 1974 to present. Semi-annual. $550.00 per year.

SCIENCE CITATION INDEX. Institute for Scientific Information, 3501 Market Street, Philadelphia, PA 19104. (800) 523-1850 or (215) 386-0100. Six times per year. $6200.00 per year.

ANNUAL REVIEWS AND YEARBOOKS

ADVANCES IN CHROMATOGRAPHY. Marcel Dekker Inc., 270 Madison Avenue, New York, NY 10016. (800) 228-1160. Irregular. Price varies.

ASSOCIATIONS AND PROFESSIONAL SOCIETIES

AMERICAN CHEMICAL SOCIETY. 1155 16th Street, N.W., Washington, DC 20036. (800) 424-6747.

ASSOCIATION OF CONSULTING CHEMISTS AND CHEMICAL ENGINEERS. 50 East 41st Street, Suite 92, New York, NY 10017. (212) 684-6255.

ASSOCIATION OF OFFICIAL ANALYTICAL CHEMISTS. 1111 North 19th Street, Suite 210, Arlington, VA 22209. (703) 522-3032.

DIRECTORIES AND BIOGRAPHICAL SOURCES

AMERICAN INSTITUTE OF CHEMISTS DIRECTORY. American Institute of Chemists, 7315 Wisconsin Avenue, Bethesda, MD 20814. (301) 652-2447. 1986. $35.00.

AMERICAN MEN AND WOMEN OF SCIENCE. R.R. Bowker, Inc., Order Department, 245 West 17th Street, New York, NY 10011. (800) 521-8110. Eight volumes. 1986. $595.00 for set.

BIOGRAPHICAL DICTIONARY OF SCIENTISTS: CHEMISTS. David Abbott, editor. P. Bedrick Books, 125 East 23rd Street, New York, NY 10010. (212) 777-1187. 1984. $18.95.

CONSULTING SERVICES: CHEMISTS AND CHEMICAL ENGINEERS. Association of Consulting Chemists and Chemical Engineers, 50 East 41st Street, Suite 92, New York, NY 10017. (212) 684-6255. Annual. 1986. $45.00.

INTERNATIONAL RESEARCH CENTERS DIRECTORY 1988-89. Darren L. Smith, editor. Gale Research Company, Book Tower, Detroit, MI 48226. (800) 521-0707. 4th edition. 1987. $360.00.

RESEARCH CENTERS DIRECTORY 1988. Gale Research Company, Book Tower, Detroit, MI 48226. (800) 521-0707. 12th edition. 1987. $365.00 for set.

SCIENTIFIC AND TECHNICAL ORGANIZATIONS AND AGENCIES DIRECTORY. Margaret Labash Young, editor. Gale Research Company, Book Tower, Detroit, MI 48226. (800) 521-0707. 2nd edition. 1987. $185.00.

ENCYCLOPEDIAS AND DICTIONARIES

CONCISE ENCYCLOPEDIA OF CHEMICAL TECHNOLOGY. Kirk Othmer. John Wiley and Sons, Inc., 605 Third Avenue,

New York, NY 10158. (800) 526-5368. 1985. $129.95.

DICTIONARY OF CHROMATOGRAPHY. Ronald C. Denney. John Wiley and Sons, Inc., 605 Third Avenue, New York, NY 10158. (800) 526-5368. Second edition. 1982. $51.00.

VAN NOSTRAND REINHOLD ENCYCLOPEDIA OF CHEMISTRY. Douglas M. Considine and Glenn D. Considine. Van Nostrand Reinhold Company, Inc., 135 West 50th Street, New York, NY 10020. (800) 543-2681. 1984. $98.95.

GENERAL WORKS

CONTEMPORARY PRACTICE OF CHROMATOGRAPHY. C.F. Poole and S.A. Schuette. Elsevier Science Publishing Company, Inc., 52 Vanderbilt Avenue, New York, NY 10017. (212) 370-5520. 1984. $59.00.

HANDBOOKS AND MANUALS

CRC HANDBOOK OF CHROMATOGRAPHY: INORGANICS. M. Qureshi, editor. CRC Press, 2000 Corporate Boulevard, Boca Raton, FL 33431. (800) 272-7737. 1986. $155.00.

ION CHROMATOGRAPHY APPLICATIONS. Robert E. Smith. CRC Press, 2000 Corporate Boulevard, Boca Raton, FL 33431. (800) 272-7737. 1988. $97.50.

ONLINE DATA BASES

CA SEARCH. Chemical Abstracts Service, P.O. Box 3012, Columbus, OH 43120. (800) 848-6538 or (614) 421-3600. Comprehensive guide to chemical literature, 1972 to present. Inquire as to online cost and availability.

COMPENDEX. Engineering Information, Inc., 345 East 47th Street, New York, NY 10017. (800) 221-1044 or (212) 705-7615. Engineering and technical literature, 1975 to present. Inquire as to online cost and availability.

DISSERTATION ABSTRACTS ONLINE. University Microfilms International, 300 North Zeeb Road, Ann Arbor, MI 48106. (800) 521-0600 or (313) 761-4700. Scope includes virtually all doctoral dissertations accepted at accredited American institutions from 1861 to present in over 250 subject areas. Inquire as to online cost and availability.

NTIS. National Technical Information Service, 5285 Port Royal Road, Springfield, VA 22161. (703) 487-4630. Broad coverage of government sponsored research reports, 1964 to present. Inquire as to online cost and availability.

SCISEARCH. Institute for Scientific Information, 3501 Market Street, Philadelphia, PA 19104. (800) 523-1850 or (215) 386-0100. Broad multidisciplinary title and author index to the international literature of science and technology, 1974 to present. Inquire as to online cost and availability.

WILSONLINE. H.W. Wilson and Company, 950 University Avenue, Bronx, NY 10452. (800) 367-6770 or (212) 588-8400. Makes available online versions of the H.W. Wilson indexes including Applied Science and Technology Index, Business Periodicals Index and Readers' Guide to Periodical Literature. Approximately 1980 to present. Inquire as to online cost and availability.

PERIODICALS

AMERICAN LABORATORY. International Scientific Communications, Inc., 808 Kings Highway, Fairfield, CT 06430. (203) 576-0500. Monthly. $108.00 per year.

ANALYTICAL CHEMISTRY. American Chemical Society, 1155 16th Street, N.W., Washington, DC 20036. (800) 424-6747. Monthly. $35.00 per year.

ANALYTICAL LETTERS: CHEMICAL ANALYSIS/CLINICAL AND BIOMEDICAL ANALYSIS. Marcel Dekker Inc., 270 Madison Avenue, New York, NY 10016. (800) 228-1160. 24 times per year. $535.00 per year.

ASSOCIATION OF OFFICIAL ANALYTICAL CHEMISTS. JOURNAL. Association of Official Analytical Chemists, 1111 North 19th Street, Suite 210, Arlington, VA 22209. (703) 522-3032. Bimonthly. $100.00 per year.

CRC CRITICAL REVIEWS IN ANALYTICAL CHEMISTRY. CRC Press, 2000 Corporate Boulevard, Boca Raton, FL 33431. (800) 272-7737. Quarterly. $105.00 per year.

JOURNAL OF CHROMATOGRAPHIC SCIENCE. Preston Publications, Inc., 7800 Merrimac Avenue, Niles, IL 60648. (312) 965-0566. Monthly. $98.00 per year.

JOURNAL OF CHROMATOGRAPHY. Elsevier Science Publishing Company, Inc., 52 Vanderbilt Avenue, New York, NY 10017. (212) 370-5520. Fifty-six times per year. $725.00 per year.

JOURNAL OF LIQUID CHROMATOGRAPHY. Marcel Dekker Inc., 270 Madison Avenue, New York, NY 10016. (800) 228-1160. 1978 to present. 16 times per year. $450.00 per year.

SEPARATION SCIENCE AND TECHNOLOGY. Marcel Dekker Inc., 270 Madison Avenue, New York, NY 10016. (800) 228-1160. Ten times per year. $400.00 per year.

RESEARCH CENTERS AND INSTITUTES

BARNETT INSTITUTE OF CHEMICAL ANALYSIS AND MATERIALS SCIENCE. Mortheastern University, 360 Huntington Avenue, Boston, MA 02115. (617) 437-2864.

CHROMATOGRAPHY LABORATORY. University of Wisconsin - Madison, Department of Chemical Engineering, 1415 Johnson Drive, Madison, WI 53706. (608) 262-1092.

CHROMATOGRAPHY-MASS SPECTROSCOPY FACILITY. University of Missouri - Columbia, Room 4, Agricultural Building, Columbia, MO 65211. (314) 882-2608.

SPECIFICATIONS AND STANDARDS

ASTM STANDARDS ON CHROMATOGRAPHY. American Society for Testing and Materials, 1916 Race Street, Philadelphia, PA 19103. (215) 299-5400. 1981. $40.00.

CHROMIUM

See also: ALLOYS, FERROALLOYS, IRON, MATERIALS SCIENCE, METALLURGY, STEEL AND STEEL MAKING

ABSTRACT SERVICES AND INDEXES

ALLOYS INDEX. American Society for Metals, 9639 Kinsman Road, Metals Park, OH 44073. (216) 338-5151. $130.00 per year.

APPLIED MECHANICS REVIEWS. American Society of Mechanical Engineers, 345 East 47th Street, New York, NY 10017. (212) 705-7722. Monthly. $380.00 per year. Critical reviews of the world literature in applied mechanics and related engineering science.

APPLIED SCIENCE AND TECHNOLOGY INDEX. H.W. Wilson Company, 950 University Avenue, Bronx, NY 10452. (800) 367-6670 or (212) 588-8400. Inquire as to cost and availability.

CHEMICAL ABSTRACTS. Chemical Abstracts Service, 2540 Olentangy Road, Post Office Box 3012, Columbus, OH 43210. (800) 848-6538 or (614) 421-3600. Weekly. $9200.00 per year.

CHROMIUM

CURRENT CONTENTS: ENGINEERING, TECHNOLOGY. Institute for Scientific Information, 3501 Market Street, Philadelphia, PA 19104. (800) 523-1850 or (215) 386-0100. $272.00 per year.

ENGINEERING INDEX MONTHLY. Engineering Information, Incorporated, 345 East 47th Street, New York, NY 10017. (800) 221-1044 or (212) 705-7600. Monthly, with annual cumulation. $1425.00 per year.

METALS ABSTRACTS. American Society for Metals, 9639 Kinsman Road Metals Park, OH 44073. (216) 338-5151. Monthly. $890.00.

METALS ABSTRACT INDEX. American Society for Metals, 9639 Kinsman Road, Metals Park, OH 44073. (216) 338-5151. Monthly. (Sold only to subscribers of Metals Abstracts).

ANNUAL REVIEWS AND YEARBOOKS

ANNUAL REVIEW OF MATERIALS SCIENCE. Annual Reviews, Incorporated 4139 El Camino Way, Palo Alto, CA 94306. (415) 493-4400. Inquire.

ASSOCIATION AND PROFESSIONAL SOCIETIES

AMERICAN FOUNDRYMAN'S SOCIETY, INCORPORATED. Golf and Wolf Roads, Des Plaines, IL 60016. (312) 824-0181.

AMERICAN IRON AND STEEL INSTITUTE. 1000 Sixteenth Street, NW, Washington, DC 20036. (202) 452-7100.

AMERICAN SOCIETY FOR METALS. Metals Park, OH 44073. (216) 338-5151.

METALLURGICAL SOCIETY OF THE AIME (AMERICAN INSTITUTE OF MINING, METALLURGICAL AND PETROLEUM ENGINEERS). 420 Commonwealth Drive, Warrendale, PA 15086. (412) 776-9080.

BIBLIOGRAPHIES

SCIENTIFIC AND TECHNICAL BOOKS IN PRINT: AN INDEX TO LITERATURE IN SCIENCE AND TECHNOLOGY. R.R. Bowker Company, 205 East 42nd Street, New York, NY 10017. (800) 521-8110 or (212) 916-1600.

DIRECTORIES AND BIOGRAPHICAL SOURCES

DUN'S INDUSTRIAL GUIDE - THE METAL WORKING DIRECTORY. Dun and Bradstreet, Incorporated, Three Century Drive, Parsippany, NJ 07054. (201) 455-0900. Annual. $550.00.

INTERNATIONAL RESEARCH CENTERS DIRECTORY 1986-1987. Gale Research Company, Book Tower, Detroit, MI 48226. (800) 521-0707. Third edition. 1986.

METAL PRODUCTS DIRECTORY. American Business Directories, Incorporated, Division of American Business Lists, Incorporated, 5707 South 86th Circle, Omaha, NE 68127. (402) 331-7169. 1986. $80.00.

RESEARCH CENTERS DIRECTORY. Gale Research Company, Book Tower, Detroit, MI 48226. (800) 521-0707. Eleventh edition. 1987.

SCIENTIFIC AND TECHNICAL ORGANIZATIONS AND AGENCIES DIRECTORY. Gale Research Company, Book Tower, Detroit, MI 48226. (800) 521-0707. 1985.

WHO'S WHO IN ENGINEERING. Engineers Joint Council, 345 East 47th Street, New York, NY 10017. (212) 705-7010. 1985. $200.00.

WHO'S WHO IN FRONTIER SCIENCE AND TECHNOLOGY. Marquis Who's Who, Incorporated, 200 East Ohio Street, Chicago, IL 60611. (800) 428-3898 or (312) 787-2008.

WHO'S WHO IN TECHNOLOGY TODAY. Reston Publishing Company, Incorporated, c/o Prentice-Hall, Incorporated, Englewood Cliffs, NJ 07632. (800) 262-6868. Biennial. Five volumes. $425.00. Covers the fields of electronics, computer science, physics, optics, chemistry, biotechnology, mechanics, energy, and earth science.

ENCYCLOPEDIAS AND DICTIONARIES

ENCYCLOPEDIA OF MATERIALS SCIENCE AND ENGINEERING. Michael B. Bever, editor. MIT Press, 28 Carlton Street, Cambridge, MA 02142. (617) 253-5646. Eight volumes. 1986. $1950.00.

ENCYCLOPEDIA OF PHYSICAL SCIENCE AND TECHNOLOGY. Academic Press, Incorporated, Orlando, FL 32887. (800) 321-5068 or (305) 345-2734. Fifteen volumes, 1986.

MCGRAW-HILL ENCYCLOPEDIA OF SCIENCE AND TECHNOLOGY. McGraw-Hill Book Company, 1221 Avenue of the Americas, New York, NY 10020. (212) 512-2000.

GENERAL WORKS

CHEMISTRY OF THE ELEMENTS. N.N. Greenwood and A. Earnshaw. Pergamon Publishing, Incorporated, Maxwell House, Fairview Park, Elmsford, NY 10523. (914) 592-7700. 1984. $143.00.

ESSENTIAL METALLURGY FOR ENGINEERS. W. Alexander, G. Davies, and K. Reynolds. Van Nostrand Reinhold, 115 Fifth Avenue, New York, NY 10003. (800) 543-2681. 1985. $17.95.

FERROALLOYS AND OTHER ADDITIVES TO LIQUID IRON AN STEEL. American Society for Testing and Materials, 1916 Race Street, Philadelphia, PA 19103. (215) 299-5400. 1981. $24.75.

METALLURGY BASICS. D.V. Brown. Van Nostrand Reinhold, 115 Fifth Avenue, New York, NY 10003. (800) 543-2681. 1985. $17.95.

NICKEL AND CHROMIUM PLATING. J.K. Dennis and T.E. Such. Butterworths, Incorporated, 80 Montvale Avenue, Stoneham, MA 02180. (617) 438-8464. 1986. $120.00.

PHYSICAL METALLURGY. P. Haasen. Cambridge University Press, 32 East 57th Street, New York, NY 10022. (212) 688-8885. 1986. $24.95 in paper.

PROPERTIES OF SELECTED FERROUS ALLOYING ELEMENTS. Y.S. Touloukian and C.Y. Ho. McGraw-Hill Book Company, Incorporated, 1221 Avenue of the Americas, New York, NY 10020. (212) 512-2000. 1981. $56.95.

STAINLESS STEEL. R.A. Lula. American Society for Testing and Materials, 1916 Race Street, Philadelphia, PA 19103. 9215) 299-5400. 1985. $47.00.

HANDBOOKS AND MANUALS

SMITHELL'S METALS REFERENCE BOOK. Eric A. Brandes, editor. Butterworth Publishers, 80 Montvale Avenue, Stoneham, MA 02180. (800) 325-4177. Sixth edition. 1983. $210.00.

ONLINE DATA BASES

COMPENDEX. Engineering Information, Incorporated, 345 East 47th Street, New York, NY 10017. (800) 221-1044 or (212) 705-7615. Engineering and technical literature, 1975 to present. Inquire as to cost and availability.

INSPEC. INSPEC Marketing Department, Institute of Electrical and Electronics Engineers, Incorporated, IEEE Service

Department, 445 Hoes Lane, Piscataway, NJ 08854. (201) 981-0060. Inquire as to on-line cost and availability.

METADEX. Metals Information, American Society for Metals, Metals Park, Oh 44073. (216) 338-5151. (Metals Abstracts/Alloys Index). A worldwide literature on the science and practice of metallurgy, 1966 to present. Inquire as to online cost and availability.

NASA. National Aeronautics and Space Administration, Scientific and Technical Information Branch, 300 7th Street, SW, Washington, DC 20546. Citations and abstracts of aerospace literature, 1962 to present. Inquire as to cost and availability.

NTIS. National Technical Information Service, 5285 Port Royal Road, Springfield, VA 22161. (703) 487-4630. Broad coverage of government sponsored research reports, 1964 to present. Inquire as to cost and availability.

WILSONLINE. H.W. Wilson Company, 950 University Avenue, Bronx, NY 10452. (800) 367-6770 or (212) 588-8400. Makes available online versions of the printed H.W. Wilson Indexes included Applied Science and Technology Index, Business Periodicals Index, and Readers' Guide to Periodical Literature. Period covered is generally 1983 to present. Inquire as to cost and available.

OTHER SOURCES

INTERDOC: DIRECTORY OF PUBLISHED PROCEEDINGS, SERIES. SEMT-Science/Engineering/Medicine/Technology. Interdoc Corporation, 173 Halstead Avenue, Box 326, Harrison, NY 10528. (914) 835-3506. Ten times per year. $325.00 per year.

MATERIALS SCIENCE AND METALLURGY. National Technical Information Service (NTIS), 5285 Port Royal Road, Springfield, VA 22161. (703) 487-4630. Translations and abstracts of foreign language technical media. Irregular. $40.00 per year.

PERIODICALS

ALLOY DIGEST. Engineering Publications Incorporated, Box 823, Upper Montclair, NJ 07043. (201) 746-7930. Monthly. $50.00 per year.

BULLETIN OF ALLOY PHASE DIAGRAMS. American Society for Metals, Metals Park, OH 44073. (216) 338-5151. Bimonthly. $90.00 per year.

IRON AND STEEL TECHNOLOGY INSIGHTS. Merton Allen Associates, 2307 Dean Street, Schenectady, NY 12309. (518) 393-1933. Bimonthly. $49.00 per year.

JOURNAL OF METALS. Metalurgical Society of the AIME (American Institute of Mining, Metalurgical and Petroleum Engineers), 420 Commonwealth Drive, Warrendale, PA 15086. (412) 776-9080. Monthly. $40.00 per year.

METALLURGICAL TRANSACTIONS. Metallurgical Society of the AIME (American Institute of Mining, Metalurgical and Petroleum Engineers), 420 Commonwealth Drive, Warrendale, PA 15086. (412) 776-9080. Monthly. $95.00 per year.

METALS WEEK. McGraw-Hill Book Company, Incorporated, 1221 Avenue of the Americas, New York, NY 10020. (212) 997-2823. Weekly. $527.00 per year.

MODERN METALS. Modern Metals Publishing Company, 211 East Chicago Avenue, Chicago, IL 60611. (312) 337-0638. Monthly. $23.00 per year.

MOLYBDENUM MOSAIC. AMAX, 1600 Huron Parkway, Ann Arbor, MI 48105. (313) 761-2300. Quarterly. Inquire as to cost and availability.

STAINLESS STEELS DIGEST. American Society for Metals, 9639 Kinsman Road, Metals Park, OH 44073. (216) 338-5151. Monthly. $90.00 per year.

RESEARCH CENTERS AND INSTITUTES

AMERICAN IRON AND STEEL INSTITUTE. 1000 Sixteenth Street, NW, Washington, DC 20036. (202) 452-7100.

CARNEGIE-MELLON UNIVERSITY. Center for Iron and Steel Making Research, MEMS Department, Pittsburgh, PA 15312. (412) 578-2677.

COLORADO SCHOOL OF MINES. Steel Research Center, Golden, CO 80401. (303) 273-3774.

PHYSICAL METALLURGY RESEARCH LABORATORIES. Canada Centre for Mineral and Energy Technology, 555 Booth Street, Ottawa, ON Canada K1A OG1. (613) 995-4807.

TEXAS A AND M UNIVERSITY. Mechanics and Materials Center, ERC Building, College Station, TX 77843. (409) 845-7512.

UNIVERSITY OF CONNECTICUT. Institute of Meterials Science, Storrs, CT 06268. (203) 486-4623.

UNIVERSITY OF FLORIDA. Department of Materials Science and Engineering, Gainesville, FL 32601. (904) 392-1454.

UNIVERSITY OF MINNESOTA. Corrosion Research Center, 112 Mines and Metallurgy Building, Minneapolis, MN 55455. (612) 373-4864.

UNIVERSITY OF WISCONSIN AT MADISON. Cast Metals Laboratory, 1509 University Avenue, Madison, WI 53706. (608) 262-2562.

CHROMIUM STEEL

See: STEEL AND STEEL MAKING

CINDERBLOCKS

See: BUILDING MATERIALS

CINEMATOGRAPHY

See: PHOTOGRAPHY

CIPHERS

See: COMPUTER SECURITY

CIRCUIT BREAKERS

See also: ELECTRICAL ENGINEERING, ELECTRICITY, ELECTROMAGNETISM, ELECTRONIC CIRCUITS AND COMPONENTS, ELECTRONICS, ELECTRONICS ENGINEERING, RELAYS

ABSTRACT SERVICES AND INDEXES

APPLIED SCIENCE AND TECHNOLOGY INDEX. H.W. Wilson and Company, 950 University Avenue, Bronx, NY 10452. (800)

367-6670 or (212) 588-8400. Monthly. Inquire as to cost and availability.

CURRENT CONTENTS: ENGINEERING, TECHNOLOGY AND APPLIED SCIENCES. Institute for Scientific Information, 3501 Market Street, Philadelphia, PA 19104. (800) 523-1850 or (215) 386-0100. Weekly. $275.00 per year.

ELECTRIC POWER INDUSTRY ABSTRACTS. Edison Electric Institute, c/o Utility Data Institute, 2011 I Street, N.W., Suite 700, Washington, DC 20006. 1975 to present. Bimonthly. Inquire as to cost and availability.

ELECTRICAL AND ELECTRONICS ABSTRACTS. Institution of Electrical Engineers. Available from: Institute of Electrical and Electronics Engineers. IEEE Service Center, 445 Hoes Lane, Piscataway, NJ 08854. Monthly. $1250.00 per year.

ENGINEERING INDEX MONTHLY AND AUTHOR INDEX. Engineering Information Inc., 345 East 47th Street, New York, NY 10017. (212) 705-7600. Monthly. $1560.00 per year.

IEEE PUBLICATIONS BULLETIN. Institute of Electrical and Electronics Engineers. Institute of Electrical and Electronics Engineers. IEEE Service Center, 445 Hoes Lane, Piscataway, NJ 08854. Quarterly. Free.

ASSOCIATIONS AND PROFESSIONAL SOCIETIES

AMERICAN ELECTRONICS ASSOCIATION. P.O. Box 10045, 2670 Hanover Street, Palo Alto, CA 94303. (415) 857-9300.

AMERICAN INSTITUTE OF PHYSICS. 335 East 45th Street, New York, NY 10017. (212) 661-9494.

EDISON ELECTRIC INSTITUTE. 1111 19th Street, N.W., Washington, DC 20036. (202) 828-7400.

INSTITUTE OF ELECTRICAL AND ELECTRONICS ENGINEERS. 345 East 47th Street, New York, NY 10017. (212) 705-7900.

DIRECTORIES AND BIOGRAPHICAL SOURCES

1987 DIRECTORY OF ENGINEERING SOCIETIES AND RELATED ORGANIZATIONS. Gordon Davis, editor. Hemisphere Publishing Corporation, 79 Madison Avenue, New York, NY 10016-7892. (800) 821-8312. 12th edition. 1987. $100.00.

RESEARCH CENTERS DIRECTORY 1988. Gale Research Company, Book Tower, Detroit, MI 48226. (800) 521-0707. 12th edition. 1987. $365.00 for set.

SCIENTIFIC AND TECHNICAL ORGANIZATIONS AND AGENCIES DIRECTORY. Margaret Labash Young, editor. Gale Research Company, Book Tower, Detroit, MI 48226. (800) 521-0707. 2nd edition. 1987. $185.00.

WHO'S WHO IN ELECTRONICS. Harris Publishing Company, 2057-2 Aurora Road, Twinsburg, OH 44087. (216) 425-9143. Annual. $90.00.

WHO'S WHO IN ENGINEERING. Gordon Davis, editor. Hemisphere Publishing Corporation, 79 Madison Avenue, New York, NY 10016-7892. (800) 821-8312. 6th edition. 1985. $200.00.

GENERAL WORKS

BASIC ELECTRIC CIRCUITS. Donald P. Leach. John Wiley and Sons, Inc., 605 Third Avenue, New York, NY 10158. (800) 526-5368. 1984. $34.95.

ELECTRICAL ENGINEERING. W.H. Roadstrum and Dan H. Wolaver. Harper and Row Publishers, Inc., 10 East 53rd Street, New York, NY 10022. (800) 242-7737. 1986. $37.50.

ELECTRICITY AND MAGNETISM. M.H. Nayfeh and M.K. Brussel. John Wiley and Sons, Inc., 605 Third Avenue, New York, NY 10158. (800) 526-5368. 1985. $32.95.

FUNDAMENTALS OF ELECTRIC CIRCUITS. David A. Bell. Prentice-Hall Publishing, Inc., Englewood Cliffs, NJ 07632. (800) 562-0245. 4th edition. 1988. $30.25.

INTRODUCTION TO ELECTRICITY AND ELECTRONICS FOR THE COMPUTER AGE. R. Rosen. John Wiley and Sons, Inc., 605 Third Avenue, New York, NY 10158. (800) 526-5368. 1987. $35.95.

PROTECTIVE RELAYS: THEIR THEORY AND PRACTICE. A.R. Warrington. Methuen, Inc., 29 West 35th Street, New York, NY 10001. (212) 244-3336. Two volumes. Volume one, second edition, 1968, $44.95. Volume two, third edition, 1978, $44.95.

HANDBOOKS AND MANUALS

HANDBOOK OF MODERN ELECTRONICS AND ELECTRICAL ENGINEERING. Charles Belove, editor. John Wiley and Sons, Inc., 605 Third Avenue, New York, NY 10158. (800) 526-5368. 1986. $88.95.

ONLINE DATA BASES

COMPENDEX. Engineering Information, Inc., 345 East 47th Street, New York, NY 10017. (800) 221-1044 or (212) 705-7615. Engineering and technical literature, 1975 to present. Inquire as to online cost and availability.

INSPEC. INSPEC Marketing Department, Institution of Electrical Engineers. Available from IEEE Service Center, 445 Hoes Lane, Piscataway, NJ 08854. (201) 981-0060. Online version of Physics Abstracts. Inquire as to online cost and availability.

NTIS. National Technical Information Service, 5285 Port Royal Road, Springfield, VA 22161. (703) 487-4630. Broad coverage of government sponsored research reports, 1964 to present. Inquire as to online cost and availability.

SCISEARCH. Institute for Scientific Information, 3501 Market Street, Philadelphia, PA 19104. (800) 523-1850 or (215) 386-0100. Broad multidisciplinary title and author index to the international literature of science and technology, 1974 to present. Inquire as to online cost and availability.

WILSONLINE. H.W. Wilson and Company, 950 University Avenue, Bronx, NY 10452. (800) 367-6770 or (212) 588-8400. Makes available online versions of the H.W. Wilson indexes including Applied Science and Technology Index, Business Periodicals Index and Readers' Guide to Periodical Literature. Approximately 1980 to present. Inquire as to online cost and availability.

OTHER SOURCES

A GUIDE TO THE LITERATURE OF ELECTRICAL AND ELECTRONICS ENGINEERING. Susan B. Ardis. Libraries Unlimited Inc., P.O. Box 263, Littleton, CO 80160. (303) 770-1220. 1987. $37.50.

PERIODICALS

ELECTRIC LIGHT AND POWER. Technical Publishing Company, 875 Third Avenue, New York, NY 10022. (212) 605-9400. 1922 to present. Monthly. $38.00 per year.

ELECTRIC MACHINES AND POWER SYSTEMS. Hemisphere Publishing Corporation, 1010 Vermont Avenue, NW, Washington, DC 20005. (800) 526-0275. 1976 to present. Bimonthly. $134.50 per year.

ELECTRICAL WORLD. McGraw-Hill Book Company, 1221 Avenue of the Americas, New York, NY 10020. (212) 512-2000. 1874 to present. Monthly. $11.00 per year.

ELECTRONIC DESIGN. Hayden Publishing Company, 10 Mulholland Drive, Hasbrouck Heights, NJ 07604. (201) 288-7520. 1952 to present. Biweekly. $40.00 per year.

ELECTRONICS. McGraw-Hill Book Company, 1221 Avenue of the Americas, New York, NY 10020. (212) 512-2000. 1930 to present. Weekly. $32.00 per year.

IEEE POWER ENGINEERING REVIEW. Institute of Electrical and Electronics Engineers. IEEE Service Center, 445 Hoes Lane, Piscataway, NJ 08854. 1981 to present. Monthly. $75.00 per year.

IEEE TRANSACTIONS ON POWER DELIVERY. Institute of Electrical and Electronics Engineers. IEEE Service Center, 445 Hoes Lane, Piscataway, NJ 08854. 1986 to present. Quarterly. $100.00 per year.

INSTITUTE OF ELECTRICAL AND ELECTRONICS ENGINEERS PROCEEDINGS. Institute of Electrical and Electronics Engineers. IEEE Service Center, 445 Hoes Lane, Piscataway, NJ 08854. 1913 to present. Monthly. $140.00 per year.

RESEARCH CENTERS AND INSTITUTES

EDISON ELECTRIC INSTITUTE. 1111 19th Street, N.W., Washington, DC 20036. (202) 778-6778.

ELECTRIC POWER INSTITUTE. Texas A&M University, Department of Electrical Engineering, College Station, TX 77843.

ELECTRICAL ENGINEERING RESEARCH LABORATORIES. Purdue University, Electrical Engineering Building, West Lafayette, IN 47907. (317) 494-3536.

LABORATORY FOR ELECTROMAGNETIC AND ELECTRONIC SYSTEMS. 77 Massachusetts Avenue, Cambridge, MA 02139. (617) 253-4631.

CIRCULATION

See: METEOROLOGY

CIRRUS CLOUDS

See: CLOUDS

CIVIL ENGINEERING

See also: CONSTRUCTION ENGINEERING, GEOTECHNICAL ENGINEERING, HIGHWAY ENGINEERING, OCEAN ENGINEERING, STRUCTURAL ENGINEERING, SURVEYING

ABSTRACT SERVICES AND INDEXES

APPLIED SCIENCE AND TECHNOLOGY INDEX. H.W. Wilson and Company, 950 University Avenue, Bronx, NY 10452. (800) 367-6670 or (212) 588-8400. Monthly. Inquire as to cost and availability.

CURRENT CONTENTS: ENGINEERING, TECHNOLOGY AND APPLIED SCIENCES. Institute for Scientific Information, 3501 Market Street, Philadelphia, PA 19104. (800) 523-1850 or (215) 386-0100. Weekly. $275.00 per year.

ENGINEERING INDEX MONTHLY AND AUTHOR INDEX. Engineering Information Inc., 345 East 47th Street, New York, NY 10017. (212) 705-7600. Monthly. $1560.00 per year.

INDEX TO SCIENTIFIC AND TECHNICAL PROCEEDINGS. Institute for Scientific Information, 3501 Market Street, Philadelphia, PA 19104. (800) 523-1850 or (215) 386-0100. 1978 to present. Monthly. $775.00 per year.

INTERNATIONAL CIVIL ENGINEERING ABSTRACTS. CITIS Limited, 2 Rosemount Terrace, Blackrock, Dublin, Ireland. 1974 to present. Monthly. $350.00 per year.

INTERNATIONAL STRUCTURAL ENGINEERING ABSTRACTS. CITIS Limited, 2 Rosemount Terrace, Blackrock, Dublin, Ireland. 1986 to present. Quarterly. $95.00 per year.

PUBLICATIONS INFORMATION. American Society of Civil Engineers, 345 East 47th Street, New York, NY 10017. (212) 705-7420. Abstracts, subject and author indexes to the publications of the American Society of Civil Engineers. Bimonthly. $80.00 per year.

PHYSICS ABSTRACTS. Institution of Electrical Engineers. Available from: IEEE Service Center, 445 Hoes Lane, Piscataway, NJ 08854. 1898 to present. Bimonthly. $1700.00 per year.

PHYSICS BRIEFS. Physik Verlag GmbH, Postfach 1260/1280, D-6940 Weinheim, West Germany. (212) 661-9404. 1920 to present. Twenty-six times per year. $1250.00 per year.

SCIENCE CITATION INDEX. Institute for Scientific Information, 3501 Market Street, Philadelphia, PA 19104. (800) 523-1850 or (215) 386-0100. Six times per year. $6200.00 per year.

ASSOCIATIONS AND PROFESSIONAL SOCIETIES

AMERICAN SOCIETY FOR TESTING AND MATERIALS. 1916 Race Street, Philadelphia, PA 19103. (215) 299-5400.

AMERICAN SOCIETY OF CIVIL ENGINEERS. 345 East 47th Street, New York, NY 10017. (212) 705-7420.

BIBLIOGRAPHIES

NEW TECHNICAL BOOKS: A SELECTIVE LIST WITH DESCRIPTIVE ANNOTATIONS. New York Public Library, Science and Technology Research Center, Fifth Avenue and 42nd Street, New York, NY 10018. (212) 930-0800. 1915 to present. Monthly. $15.00 per year.

SCIENCE BOOKS AND FILMS. American Association for the Advancement of Science, 1333 H Street, NW, Washington, DC 20005. (202) 326-6454. Five times per year. $20.00 per year.

SCIENTIFIC AND TECHNICAL BOOKS AND SERIALS IN PRINT 1988; AN INDEX TO LITERATURE IN SCIENCE AND TECHNOLOGY. R.R. Bowker Company, 205 East 42nd Street, New York, NY 10017. (800) 521-8110. $175.00.

DIRECTORIES AND BIOGRAPHICAL SOURCES

AMERICAN MEN AND WOMEN OF SCIENCE. R.R. Bowker, Inc., Order Department, 245 West 17th Street, New York, NY 10011. (800) 521-8110. Eight volumes. 1986. $595.00 for set.

INTERNATIONAL RESEARCH CENTERS DIRECTORY 1988-89. Darren L. Smith, editor. Gale Research Company, Book Tower, Detroit, MI 48226. (800) 521-0707. 4th edition. 1987. $360.00.

1987 DIRECTORY OF ENGINEERING SOCIETIES AND RELATED ORGANIZATIONS. Gordon Davis, editor. Hemisphere Publishing Corporation, 1010 Vermont Avenue, NW, Washington, DC 20005. (800) 526-0275. 12th edition. 1987. $100.00.

RESEARCH CENTERS DIRECTORY 1988. Gale Research Company, Book Tower, Detroit, MI 48226. (800) 521-0707. 12th edition. 1987. $365.00 for set.

SCIENTIFIC AND TECHNICAL ORGANIZATIONS AND AGENCIES DIRECTORY. Margaret Labash Young, editor. Gale Research Company, Book Tower, Detroit, MI 48226.

WHO'S WHO IN ENGINEERING. Gordon Davis, editor. Hemisphere Publishing Corporation, 1010 Vermont Avenue, NW, Washington, DC 20005. (800) 526-0275. 6th edition. 1985. $200.00.

ENCYCLOPEDIAS AND DICTIONARIES

CONSTRUCTION GLOSSARY: AN ENCYCLOPEDIC REFERENCE AND MANUAL. J.S. Stein. John Wiley and Sons, Inc., 605 Third Avenue, New York, NY 10158. (800) 526-5368. 1986. $35.95.

DICTIONARY OF CIVIL ENGINEERING. John S. Scott. John Wiley and Sons, Inc., 605 Third Avenue, New York, NY 10158. (800) 526-5368. Third edition. 1981. $26.95.

MCGRAW-HILL ENCYCLOPEDIA OF SCIENCE AND TECHNOLOGY. McGraw-Hill Book Company, 1221 Avenue of the Americas, New York, NY 10020. (212) 512-2000. 6th edition. 1987. $1600.00.

THESAURUS OF SCIENTIFIC, TECHNICAL, AND ENGINEERING TERMS. Hemisphere Publishing Corporation, 1010 Vermont Avenue, NW, Washington, DC 20005. (800) 526-0275. 1988. $125.00.

GENERAL WORKS

APPLIED STRUCTURAL STEEL DESIGN. L. Spiegel and G.F. Limbrunner. Prentice-Hall Publishing, Inc., Englewood Cliffs, NJ 07632. (800) 562-0245. 1986. $34.95.

DESIGN OF REINFORCED CONCRETE. Samuel E. French. Prentice-Hall Publishing, Inc., Englewood Cliffs, NJ 07632. (800) 562-0245. 1987. $25.00.

ELEMENTARY THEORY OF STRUCTURES. Yu H. Yuan. Prentice-Hall Publishing, Inc., Englewood Cliffs, NJ 07632. (800) 562-0245. Third edition. 1988. $42.95.

FOUNDATION DESIGN AND CONSTRUCTION. M.J. Tomlinson. John Wiley and Sons, Inc., 605 Third Avenue, New York, NY 10158. (800) 526-5368. Fifth edition. 1986. $45.95.

HIGHWAY ENGINEERING. Paul H. Paquette and Randor J. Paquette. John Wiley and Sons, Inc., 605 Third Avenue, New York, NY 10158. (800) 526-5368. Fifth edition. 1986. $45.00.

HIGHWAYS: HIGHWAY ENGINEERING. C.A. O'Flaherty. Edward Arnold Publishers, Limited, 300 North Charles Street, Baltimore, MD 21201. (301) 539-1529. Volume 2. Third edition. 1987. $29.95 in paper.

MATERIALS FOR CIVIL AND HIGHWAY ENGINEERS. K.N. Derucher and G.P. Korgiatis. Prentice-Hall Publishing, Inc., Englewood Cliffs, NJ 07632. (800) 562-0245. Second edition. 1988. $42.50.

MICROCOMPUTER-AIDED ENGINEERING: STRUCTURAL DYNAMICS. Mario Paz. Van Nostrand Reinhold Company, Inc., 135 West 50th Street, New York, NY 10020. (800) 543-2681. 1986. $49.95.

PROBABILISTIC FRACTURE MECHANICS AND RELIABILITY. James W. Provan, editor. Martinus Nijhoff. Distributed by Kluwer Academic Publishers, 190 Old Derby Street, Hingham, MA 02043. (617) 749-5262. 1987. $135.50.

STRUCTURAL ANALYSIS. R.C. Coates and others. Van Nostrand Reinhold Company, Inc., 135 West 50th Street, New York, NY 10020. (800) 543-2681. Third edition. 1987. $47.95 in paper.

STRUCTURAL ENGINEERING ANALYSIS ON PERSONAL COMPUTERS. John F. Fleming. McGraw-Hill Book Company, 1221 Avenue of the Americas, New York, NY 10020. (212) 512-2000. 1986. $19.95.

STRUCTURAL STABILITY. Wai-Fah Chen and E.M. Lui. Elsevier Science Publishing Company, Inc., 52 Vanderbilt Avenue, New York, NY 10017. (212) 370-5520. 1987. $49.50.

HANDBOOKS AND MANUALS

BUILDING STRUCTURES HANDBOOK. R.N. White and C.G. Salmon. John Wiley and Sons, Inc., 605 Third Avenue, New York, NY 10158. (800) 526-5368. 1986. $79.95.

CIVIL ENGINEERING CALCULATIONS REFERENCE GUIDE. Tyler G. Hicks, editor. McGraw-Hill Book Company, 1221 Avenue of the Americas, New York, NY 10020. (212) 512-2000. 1987. $29.50.

CIVIL ENGINEERING PRACTICE. Paul N. Cheremisinoff and others, editors. Technomic Publishing Company, Inc., 851 Holland Avenue, Box 3535, Lancaster, PA 17604. (800) 233-9936. Five volumes. 1987-1988. $750.00 for set.

HANDBOOK OF CONCRETE ENGINEERING. Mark Fintel. Van Nostrand Reinhold Company, Inc., 135 West 50th Street, New York, NY 10020. (800) 543-2681. 1986. $89.95.

HANDBOOK OF MECHANICS, MATERIALS, AND STRUCTURES. A. Blake. John Wiley and Sons, Inc., 605 Third Avenue, New York, NY 10158. (800) 526-5368. 1985. $64.50.

STANDARD HANDBOOK FOR CIVIL ENGINEERS. F. S. Merritt, editor. McGraw-Hill Book Company, 1221 Avenue of the Americas, New York, NY 10020. (212) 512-2000. Third edition. 1983. $89.50.

STRUCTURAL ENGINEERING HANDBOOK. E.H. Gaylord and C.N. Gaylord, editors. McGraw-Hill Book Company, 1221 Avenue of the Americas, New York, NY 10020. (212) 512-2000. Second edition. 1979. $76.50.

SURVEYING READY-REFERENCE MANUAL. Guy O. Stenstrom. McGraw-Hill Book Company, 1221 Avenue of the Americas, New York, NY 10020. (212) 512-2000. 1987. $26.50.

ONLINE DATA BASES

COMPENDEX. Engineering Information, Inc., 345 East 47th Street, New York, NY 10017. (800) 221-1044 or (212) 705-7615. Engineering and technical literature, 1975 to present. Inquire as to online cost and availability.

DISSERTATION ABSTRACTS ONLINE. University Microfilms International, 300 North Zeeb Road, Ann Arbor, MI 48106. (800) 521-0600 or (313) 761-4700. Scope includes virtually all doctoral dissertations accepted at accredited American institutions from 1861 to present in over 250 subject areas. Inquire as to online cost and availability.

GEOREF. Online version of the BIBLIOGRAPHY AND INDEX OF GEOLOGY. American Geological Institute, 4220 King Street, Alexandria, VA 22302. (703) 379-2480. 1969 to present. Inquire as to online cost and availability.

INSPEC. INSPEC Marketing Department, Institution of Electrical Engineers. Available from IEEE Service Center, 445 Hoes Lane, Piscataway, NJ 08854. (201) 981-0060. Online version of Physics Abstracts. Inquire as to online cost and availability.

NTIS. National Technical Information Service, 5285 Port Royal Road, Springfield, VA 22161. (703) 487-4630. Broad coverage of government sponsored research reports, 1964 to present. Inquire as to online cost and availability.

SCISEARCH. Institute for Scientific Information, 3501 Market Street, Philadelphia, PA 19104. (800) 523-1850 or (215) 386-0100. Broad multidisciplinary title and author index to the international literature of science and technology, 1974 to present. Inquire as to online cost and availability.

WILSONLINE. H.W. Wilson and Company, 950 University Avenue, Bronx, NY 10452. (800) 367-6770 or (212) 588-8400. Makes available online versions of the H.W. Wilson indexes including Applied Science and Technology Index, Business Periodicals Index and Readers' Guide to Periodical Literature. Approximately 1980 to present. Inquire as to online cost and availability.

OTHER SOURCES

WHAT EVERY ENGINEER SHOULD KNOW ABOUT ENGINEERING SOURCES. Marcel Dekker Inc., 270 Madison Avenue, New York, NY 10016. (800) 228-1160. 1984. $24.95.

PERIODICALS

AMERICAN CONCRETE INSTITUTE JOURNAL. P.O. Box 19150, Redford Station, Detroit, MI 48219. (513) 532-2600. 1929 to present. Bimonthly. $69.00 per year.

CIVIL ENGINEERING. American Society of Civil Engineers, 345 East 47th Street, New York, NY 10017. (212) 705-7420. 1930 to present. Monthly. $48.00 per year.

CONCRETE. Harcourt, Brace Jovanovich, Inc., 7500 Old Oak Boulevard, Cleveland, OH 44130. 1937 to present. Monthly. $25.00 per year.

HIGHWAY AND HEAVY CONSTRUCTION. Technical Publishing Company, 875 Third Avenue, New York, NY 10022. (212) 605-9400. 1892 to present. Monthly. $35.00 per year.

JOURNAL OF CONSTRUCTION ENGINEERING AND MANAGEMENT. American Society of Civil Engineers, 345 East 47th Street, New York, NY 10017. (212) 705-7420. 1956 to present. Quarterly. $48.00 per year.

JOURNAL OF ENGINEERING MECHANICS. American Society of Civil Engineers, 345 East 47th Street, New York, NY 10017. (212) 705-7420. 1956 to present. Monthly. $120.00 per year.

JOURNAL OF STRUCTURAL ENGINEERING. American Society of Civil Engineers, 345 East 47th Street, New York, NY 10017. (212) 705-7420. 1956 to present. Monthly. $140.00 per year.

JOURNAL OF STRUCTURAL MECHANICS. Marcel Dekker Inc., 270 Madison Avenue, New York, NY 10016. (800) 228-1160. 1972 to present. Quarterly. $75.00 per year.

JOURNAL OF SURVEYING ENGINEERING. American Society of Civil Engineers, 345 East 47th Street, New York, NY 10017. (212) 705-7420. 1956 to present. Three times per year. $35.00 per year.

STRUCTURAL ENGINEER. PART A AND B. Institution of Structural Engineers, 11 Upper Belgrave Street, London SW1X 8BH, England. 1908 to present. Monthly. $140.00 per year.

STRUCTURAL ENGINEERING PRACTICE: ANALYSIS, DESIGN, MANAGEMENT. Marcel Dekker Inc., 270 Madison Avenue, New York, NY 10016. (800) 228-1160. 1981 to present. Quarterly. $75.00 per year.

RESEARCH CENTERS AND INSTITUTES

STRUCTURAL STABILITY RESEARCH COUNCIL. Fritz Engineering Laboratory No. 13, Lehigh University, Bethlehem, PA 18105. (215) 861-3519.

SPECIFICATIONS AND STANDARDS

STRUCTURAL DESIGN GUIDE TO THE ACI BUILDING CODE. Paul Rice. Van Nostrand Reinhold Company, Inc., 135 West 50th Street, New York, NY 10020. (800) 543-2681. Third edition. 1985. $47.95.

CLAY

See also: BRICKS, CERAMICS, GLASS

ABSTRACT SERVICES AND INDEXES

APPLIED SCIENCE AND TECHNOLOGY INDEX. H.W. Wilson Company, 950 University Avenue, Bronx, NY 10452. (800) 367-6670 or (212) 588-8400. Inquire as to cost and availability.

BIBLIOGRAPHY AND INDEX OF GEOLOGY. American Geological Institute, 4220 King Street, Alexandria, Va 22302. (703) 379-2480.

CERAMIC ABSTRACTS. American Ceramic Society, Incorporated, 65 Ceramic Drive, Columbus, OH 43214. (614) 268-8645. Semimonthly. $185.00 per year.

CHEMICAL ABSTRACTS. Chemical Abstracts Service, 2540 Olentangy Road, P.O. Box 3012, Columbus, OH 43210. (800) 848-6538 or (614) 421-3600. Weekly. $9200.00 per year.

ASSOCIATIONS AND PROFESSIONAL SOCIETIES

CLAY MINERALS SOCIETY, 22149 North Pet Lane, Prairie View, IL 60069. (312) 321-1515.

SOCIETY OF ECONOMIC GEOLOGISTS. P.O. Box 571, Golden, CO 80402. (303) 279-1899.

DIRECTORIES AND BIOGRAPHICAL SOURCES

AMERICAN MEN AND WOMEN OF SCIENCE. Physical and Biological Sciences. Fifteenth edition. R.R. Bowker Company, 205 East 42nd Street, New York, NY 10017. (800) 521-8110 or (212) 916-1600.

BIOGRAPHICAL DICTIONARY OF SCIENTISTS. T.I. Williams. Halsted Press, 605 Third Avenue, New York, NY 10158. (800) 526-5368 or (212) 850-6418. Third edition. 1982. $29.95.

GOVERNMENT RESEARCH DIRECTORY. Gale Research Company, Book Tower, Detroit, MI 48226. (800) 521-0707. Third edition. 1985.

INTERNATIONAL RESEARCH CENTERS DIRECTORY 1986-1987. Gale Research Company, Book Tower, Detroit, MI 48226. (800) 521-0707. Third edition. 1986.

RESEARCH CENTERS DIRECTORY. Gale Research Company, Detroit, MI 48226. (800) 521-0707. Eleventh edition. 1987.

SCIENTIFIC AND TECHNICAL ORGANIZATIONS AND AGENCIES DIRECTORY. Gale Research Company, Book Tower, Detroit, MI 48226. (800) 521-0707. 1985.

WHO'S WHO IN FRONTIER SCIENCE AND TECHNOLOGY. Marquis Who's Who, Incorporated, 200 East Ohio Street, Chicago, IL 60611. (800) 428-3898 or (312) 787-2008.

WHO'S WHO IN TECHNOLOGY TODAY. Reston Publishing Company, Incorporated, c/o Prentice-Hall, Incorporated, Englewood Cliffs, NJ 07632. (800) 262-6868. Biennial. Five volumes. $425.00. Covers the fields of electronics, computer science, physics, optics, chemistry, biotechnology, mechanics, energy, and earth science.

ENCYCLOPEDIAS AND DICTIONARIES

ENCYCLOPEDIA OF GLASS, CERAMICS, CLAY AND CEMENT. Martin Grayson, editor. John Wiley and Sons, Incorporated, 605 Third Avenue, New York, NY 10158. (800) 526-5368 or (212) 850-6000. 1985. $89.95.

ENCYCLOPEDIA OF PHYSICAL SCIENCE AND TECHNOLOGY. Academic Press, Incorporated, Orlando, FL

CLAY

32887. (800) 321-5068 or (305) 345-2734. Fifteen volumes, 1986.

MCGRAW-HILL ENCYCLOPEDIA OF SCIENCE AND TECHNOLOGY. McGraw-Hill Book, Incorporated, 1221 Avenue of the Americas, New York, NY 10020. (212) 997-3675.

GENERAL WORKS

CHEMISTRY OF CLAYS AND CLAY MINERALS. A.C. Newman. John Wiley and Sons, Incorporated, 605 Third Avenue, New York, NY 10158. (800) 526-5368 or (212) 850-6000. 1986. $110.00.

CLAY MINERALS: A PHYSIO-CHEMICAL EXPLANATION OF THEIR OCCURRENCE. B. Velde. Elsevier Science Publishing Company, Incorporated, 52 Vanderbilt Avenue, New York, NY 10017. (212) 370-5520. 1985. $59.95.

CLAY MINERALS: THEIR STRUCTURE, BEHAVIOR AND USE. Leslie Powden, editor. Scholium International, Incorporated, 265 Great Neck Road, Great Neck, NY 11021. (516) 466-5181. 1984. $70.00.

CRYSTAL STRUCTURES OF CLAY MINERALS AND THEIR X-RAY IDENTIFICATION. G.W. Brindley and G. Brown, editors. Brookfield Publishing Company, Old Post Road, Brookfield, VT 05036. (802) 276-3162. 1982. $80.00.

INTRODUCTION TO CLAY COLLOID CHEMISTRY. H. Van Olphen. John Wiley and Sons, Incorporated, 605 Third Avenue, New York, NY 10158. (800) 526-5368 or (212) 850-6000. 1977. $50.00.

HANDBOOKS AND MANUALS

DATA HANDBOOK FOR CLAY MINERALS AND OTHER NON-METALLIC MINERALS. H. Van Olphen and J.J. Fripiat, editors. Pergamon Press, MAxwell House, Fairview Park, Elmsford, NY 10523. (914) 592-7700. 1979. $90.00.

HANDBOOK OF DETERMINATIVE METHODS IN CLAY MINERALOGY. M.J. Wilson, editor. Metheum Company, Incorporated, 29 West 35th Street, New York, NY 10001. (212) 244-3336. 1986. $69.95.

ONLINE DATA BASES

CA SEARCH. Chemical Abstracts Service, P.O. Box 3012, Columbus, OH 43210. Guide to chemical literature, 1972 to present. Inquire as to cost and availability.

GEOREF. American Geological Institute, 4220 King Street, Alexandria, VA 22302. (800) 336-4764 or (703) 379-2480. Geology and geosciences literature, 1961 to present. Inquire as to online cost and availability.

GEOARCHIVE. Geosystems, P.O. Box 1024, Westminister, London, England, SW1 P 2JL. Citations to literature on geoscience, 1969 to present. Inquire as to online cost and availability.

OTHER SOURCES

CLAY CONSTRUCTION PRODUCTS. U.S. Bureau of the Census. U.S. Department of Commerce, Washington, DC 20233. Monthly. $20.00 per year.

PERIODICALS

BRICK AND CLAY RECORD. Cahner's Publishing Company, Box 5080, Des Plaines, IL 60018. (312) 685-8800. Monthly. $30.00 per year.

CLAY MINERALS SOCIETY NEWSLETTER. Clay Minerals Society, P.O. Box 2295, Bloomington, IN 47402. Bimonthly.

CLAYS AND CLAY MINERALS. Clay Minerals Society, P.O. Box 2295, Bloomington, IN 47402. Bimonthly. $96.00 per year.

ECONOMIC GEOLOGY. Society of Economic Geologists, P.O. Box 571, Golden, CO 80402. (303) 279-1899. Eight times per year. $25.00.

RESEARCH CENTERS AND INSTITUTES

GEOTECHNICAL/BITUMINOUS RESEARCH LABORATORIES. Iowa State University, Ames, IA 50011. (515) 294-7689.

STATISTICAL SOURCES

CLAYS. Bureau of Mines, U.S. Department of the Interior, 4800 Forbes Avenue, Pittsburgh, PA 15213. Annual. Inquire as to cost and availability.

MINERALS YEARBOOK. Bureau of Mines, U.S. Department of the Interior. Available from U.S. Government Printing Office, Washington, DC 20402. (202) 783-3238. Annual. Three volumes. $45.00.

CLIMATOLOGY

See also: DESERTIFICATION, GREENHOUSE EFFECT, METEOROLOGY

ABSTRACT SERVICES AND INDEXES

APPLIED SCIENCE AND TECHNOLOGY INDEX. H.W. Wilson Company, 950 University Avenue, Bronx, NY 10452. (800) 367-6670 or (212) 588-8400. Inquire as to cost and availability.

BIBLIOGRAPHY AND INDEX OF GEOLOGY. American Geological Institute, 4220 King Street, Alexandria, VA 22302. (703) 379-2480. 1969 to present. Monthly. $1100.00 per year.

CHEMICAL ABSTRACTS. Chemical Abstracts Service, 2540 Olentangy Road, Post Office Box 3012, Columbus, OH 43210. (800) 848-6538 or (614) 421-3600. Weekly. $9200.00 per year.

CURRENT CONTENTS: PHYSICAL, CHEMICAL AND EARTH SCIENCES. Institute for Scientific Information, 3501 Market Street, Philadelphia, PA 19104. (800) 523-1850 or (215) 386-0100. $272.00 per year.

GENERAL SCIENCE INDEX. H.W. Wilson Company, 950 University Avenue, Bronx, NY 10452. (800) 367-6770 or (212) 588-8400.

METEOROLOGICAL AND GEOASTROPHYSICAL ABSTRACTS. American Meteorological Society, 45 Beacon Street, Boston, MA 02108. (617) 227-2425. 1950 to present. Monthly. $450.00 per year.

OCEANIC ABSTRACTS. Cambridge Scientific Abstracts, Incorporated, 5161 River Road, Bethesda, MD 20816. (301) 951-1400. 1964 to present. Bimonthly. $660.00 per year.

OCEANOGRAPHIC LITERATURE REVIEW. Pergamon Press Incorporated, Maxwell House, Fairview Park, Elmsford, NY 10523. (014) 592-7700. 1979 to present. Inquire as to cost and availability.

PHYSICS ABSTRACTS. Institute of Electrical Engineers, London, United Kingdom. Available from: Institute of Electrical and Electronic Engineers (IEEE), 345 East 47th Street, New York, NY 10017. (212) 705-7900.

SCIENCE CITATION INDEX. Institute for Scientific Information, 3501 Market Street, Philadelphia, PA 19104. (800) 523-1850 or (215) 386-0100. Inquire as to cost and availability.

ANNUAL REVIEWS AND YEARBOOKS

OCEAN YEARBOOK. Elizabeth M. Borgese and Norton Ginsberg, editors. University of Chicago Press, 5801 Ellis Avenue, Chicago, IL 60637. (312) 961-7906. 1979 to present. $49.00 per volume.

ASSOCIATIONS AND PROFESSIONAL SOCIETIES

AMERICAN ASSOCIATION STATE CLIMATOLOGISTS. c/o Professor John Griffiths, Meteorology Department, O and M Building, Texas A and M University, College Station, TX 77843. (409) 845-7320.

AMERICAN METEOROLOGICAL SOCIETY. 45 Beacon Street, Boston, MA 02108. (617) 227-2425.

INTERNATIONAL ASSOCIATION OF METEOROLOGY AND ATMOSPHERIC PHYSICS. UCAR, Post Office Box 3000, Boulder, CO 80307.

NATIONAL ENVIRONMENTAL SATELLITE DATA, AND INFORMATION SERVICE. 3300 Whitehaven Street, NW, Washington, DC 20235. (202) 634-7318.

NATIONAL WEATHER ASSOCIATION. 4400 Stamp Road, Room 404, Temple Hills, MD 20748. (301) 899-3784.

UNIVERSITY CORPORATION FOR ATMOSPHERIC RESEARCH. Box 3000, 1850 Table Mesa Drive, Boulder, CO 80307. (303) 497-1000.

WEATHER MODIFICATION ASSOCIATION. Post Office Box 8116, Fresno, CA 93747. (209) 291-8466.

BIBLIOGRAPHIES

SCIENCE BOOKS AND FILMS. American Association for the Advancement of Science, 1333 H Street, NW, Washington, DC 20005. Five times per year. $20.00 per year.

SCIENTIFIC AND TECHNICAL BOOKS AND SERIALS IN PRINT 1988: AN INDEX TO LITERATURE IN SCIENCE AND TECHNOLOGY. R.R. Bowker Company, 205 East 42nd Street, New York, NY 10017. (800) 521-8110 or (212) 916-1600. $175.00.

DIRECTORIES AND BIOGRAPHICAL SOURCES

AMERICAN MEN AND WOMEN OF SCIENCE. Physical and Biological Sciences. Sixteenth edition. R.R. Bowker Company, 205 East 42nd Street, New York, NY 10017. (800) 521-8110 or (212) 916-1600. 1986. $595.00.

BIOGRAPHICAL DICTIONARY OF SCIENTISTS. T.I. Williams. Third edition. Halsted Press, 605 Third Avenue, New York, NY 10158. (800) 526-5368 or (212) 850-6418. 1982. $29.95.

GOVERNMENT RESEARCH DIRECTORY. Gale Research Company, Book Tower, Detroit, MI 48226. (800) 521-0707. Fourth edition. 1987. $350.00.

METEOROLOGICAL SERVICES OF THE WORLD. World Meteorological Organization. Available from: American Meteorological Society, 45 Beacon Street, Boston, MA 02108. (617) 227-2425. Annual. $35.00.

NATIONAL WEATHER SERVICE OFFICES AND STATIONS. National Oceanic and Atmospheric Administration, Department of Commerce, Silver Spring, MD 20910. (301) 427-7698. Annual. Free.

RESEARCH CENTERS DIRECTORY. Gale Research Company, Book Tower, Detroit, MI 48226. (800) 521-0707. Twelfth edition. 1987. $365.00 for set.

SCIENTIFIC AND TECHNICAL ORGANIZATIONS AND AGENCIES DIRECTORY. Gale Research Company, Book Tower, Detroit, MI 48226. (800) 521-0707. Second edition. 1987. $185.00.

WHO'S WHO IN FRONTIER SCIENCE AND TECHNOLOGY. Marquis Who's Who, Incorporated, 200 East Ohio Street, Chicago, IL 60611. (800) 428-3898 or (312) 787-2008.

WHO'S WHO IN TECHNOLOGY TODAY. Reston Publishing Company, Incorporated, c/o Prentice-Hall, Incorporated, Englewood Cliffs, NJ 07632. (800) 262-6868. Biennial. Five volumes. $425.00. Covers the fields of electronics, computer science, physics, optics, chemistry, biotechnology, mechanics, energy, and earth science.

WORLD GUIDE TO SCIENTIFIC ASSOCIATIONS AND LEARNED SOCIETIES. K.G. Saur Incorporated, 175 Fifth Avenue, New York, NY 10010. (800) 521-0707 or (212) 982-1302. Fourth edition. 1984. $112.00.

ENCYCLOPEDIAS AND DICTIONARIES

ENCYCLOPEDIA OF CLIMATOLOGY. John E. Oliver and Rhodes W. Fairbridge, editors. Van Nostrand Reinhold, Incorporated, 115 Fifth Avenue, New York, NY 10003. (800) 543-2681. 1987. $89.95.

ENCYCLOPEDIA OF PHYSICAL SCIENCE AND TECHNOLOGY. Academic Press, Incorporated, Orlando, FL 32887. (800) 321-5068 or (305) 345-2734. Fifteen volumes, 1986.

MCGRAW-HILL ENCYCLOPEDIA OF SCIENCE AND TECHNOLOGY. McGraw-Hill Book, Incorporated, 1221 Avenue of the Americas, New York, NY 10020. (212) 997-3675. Fifth edition, 15 volumes. $1100.00.

GENERAL WORKS

FUTURE WEATHER AND THE GREENHOUSE EFFECT. John Gribbin. Delacorte Press, 1 Dag Hammarskjold Plaza, New York, NY 10017. (212) 605-3000. 1982. $15.95.

GENERAL CLIMATOLOGY. Howard J. Critchfield. Prentice-Hall publishers, Incorporated, Englewood Cliffs, NJ 07632. (800) 562-0245. Fourth edition. 1983. Inquire.

THE GLOBAL CLIMATE. John T. Houghton, editor. Cambridge University Press, 32 East 57th Street, New York, NY 10022. (800) 872-7423. 1984. $55.00.

THE GREENHOUSE EFFECT: CLIMATIC CHANGE AND ECOSYSTEMS. Bert Bolin and Bo R. Doos. John Wiley and Sons, Incorporated, 605 Third Avenue, New York, NY 10158. (800) 526-5368 or (212) 850-6000. 1986. Inquire.

ICE SHEETS AND CLIMATES. J. Oerlemans and C.J. Van Der Veen. Kluwer Academic Publishers, 101 Philip Drive, Norwell, MA 02061. (617) 871-6600. 1984. $35.50.

INTRODUCTION TO CLIMATE. Glenn T. Trewartha and L.H. Horn. McGraw-Hill Book Company, Incorporated, 1221 Avenue of the Americas, New York, NY 10020. (212) 997-3675. Fifth edition. 1980. $39.95.

INTRODUCTION TO THE THEORY OF CLIMATE. A.S. Monin. Kluwer Academic Publishers, 101 Philip Drive, Norwell, MA 02061. (617) 871-6600. 1985. $65.00.

OUR THREATENED CLIMATE. Wilfrid Bach. Kluwer Academic Publishers, 101 Philip Drive, Norwell, MA 02061. (617) 871-6600. 1983. $29.00.

ONLINE DATA BASES

CA SEARCH. Chemical Abstracts Service, Post Office Box 3012, Columbus, OH 43210. Guide to chemical literature, 1972 to present. Inquire as to cost and availability.

DISSERTATION ABSTRACTS ONLINE. University Microfilms International, 300 North Zeeb Road, Ann Arbor, MI 48106. (800) 521-0600 or (313) 761-4700. Scope includes virtually all doctoral dissertations accepted at accredited American institutions from 1861 to present in 252 subject areas. Inquire as to cost and availability.

INSPEC. INSPEC Marketing Department, Institute of Electrical and Electronics Engineers, Incorporated, IEEE Service Department, 445 Hoes Lane, Piscataway, NJ 08854. (201) 981-0060. Inquire as to on-line cost and availability.

METEOROLOGICAL AND GEOASTROPHYSICAL ABSTRACTS. American Meteorological Society, 45 Beacon Street, Boston, MA 02108. (617) 227-2425. 1950 to present. Monthly. $450.00 per year.

NTIS. National Technical Information Service, 5285 Port Royal Road, Springfield, VA 22161. (703) 487-4630. Broad coverage of government sponsored research reports, 1964 to present. Inquire as to cost and availability.

OCEANIC ABSTRACTS. Cambridge Scientific Abstracts, Incorporated, 5161 River Road, Bethesda, MD 20816. (301) 951-1400. 1964 to present. Inquire as to online cost and availability.

SCISEARCH. Institute for Scientific Information, 3501 Market Street, Philadelphia, PA 19104. (800) 523-1850 or (215) 386-0100. Broad multidisciplinary title and author index to the international literature of science and technology, 1974 to present. Inquire as to cost and availability.

WILSONLINE. H.W. Wilson Company, 950 University Avenue, Bronx, NY 10452. (800) 367-6770 or (212) 588-8400. Makes available online versions of the printed H.W. Wilson Indexes including Applied Science and Technology Index, Business Periodicals Index, and Readers' Guide to Periodical Literature. Period covered is generally 1983 to present. Inquire as to cost and availability.

PERIODICALS

AGRICULTURAL AND FOREST METEOROLOGY. Elsevier Science Publishing Company, Incorporated, 52 Vanderbilt Avenue, New York, NY 10017. (212) 370-5520. 1964 to present. Monthly. $260.00 per year.

AMERICAN METEOROLOGICAL SOCIETY BULLETIN. American Meteorological Society, 45 Beacon Street, Boston, MA 02108. (617) 227-2425.

BOUNDARY-LAYER METEOROLOGY: AN INTERNATIONAL JOURNAL OF PHYSICAL AND BIOLOGICAL PROCESSES IN THE ATMOSPHERIC BOUNDARY LAYER. D. Reidel Publishing Company, 190 Old Derby Street, Hingham, MA 02043. (617) 871-6600. 1970 to present. Sixteen times per year. $425.00 per year.

CLIMATIC CHANGE: AN INTERDISCIPLINARY, INTERNATIONAL JOURNAL DEVOTED TO THE DESCRIPTION, CAUSES AND IMPLICATIONS OF CLIMATIC CHANGE. D. Reidel Publishing Company, 190 Old Derby Street, Hingham, MA 02043. (617) 871-6600. 1977 to present. Six times per year. $125.00 per year.

DYNAMICS OF ATMOSPHERES AND OCEANS. Elsevier Science Publishing Company, Incorporated, 52 Vanderbilt Avenue, New York, NY 10017. (212) 370-5520. 1977 to present. Quarterly. $90.00 per year.

JOURNAL OF ATMOSPHERIC AND OCEANIC TECHNOLOGY. American Meteorological Society, 45 Beacon Street, Boston, MA 02108. (617) 227-2425. 1984 to present. Quarterly. $80.00 per year.

JOURNAL OF CLIMATE AND APPLIED METEOROLOGY. American Meteorological Society, 45 Beacon Street, Boston, MA 02108. (617) 227-2425. 1962 to present. Monthly. $120.00 per year.

JOURNAL OF PHYSICAL OCEANOGRAPHY. American Meteorological Society, 45 Beacon Street, Boston, MA 02108. (617) 227-2425. 1971 to present. Monthly. $120.00 per year.

JOURNAL OF THE ATMOSPHERIC SCIENCES. American Meteorological Society, 45 Beacon Street, Boston, MA 02108. (617) 227-2425. 1944 to present. Semimonthly. $220.00 per year.

NATIONAL WEATHER DIGEST. National Weather Association, 4400 Stamp Road, Room 404, Temple Hills, MD 20748. (301) 899-3784. 1976 to present. Quarterly. $20.00 per year.

WEATHER. Royal Meteorological Society, James Glaisher House, Grenville Place, Bracknell Berkshire, RG12 1BX, England. 1946 to present. $30.00 per year.

WEATHERWISE. Heldref Publications, 4000 Albemarle Street, NW, Washington, DC 20016. (202) 362-6445. 1948 to present. Bimonthly. $20.00 per year.

RESEARCH CENTERS AND INSTITUTES

CENTER FOR CLIMATE RESEARCH. Columbia University, Lamont-Doherty Geological Observatory, Palisades, NY 10964. (914) 359-2900.

CENTER FOR CLIMATIC RESEARCH. University of Delaware, Department of Geography, Newark, DE 19716. (302) 451-8998.

FLORIDA STATE CLIMATE CENTER. 305A Love Building, Florida State University, Tallahassee, FL 32306. (904) 644-3417.

MOUNT WASHINGTON OBSERVATORY. 1 Washington Street, Gorham, NH 03581. (603) 466-3388.

NATIONAL CENTER FOR ATMOSPHERIC RESEARCH. Box 3000, Boulder, CO 80307. (303) 497-1000.

CLOTH

See: TEXTILES

CLOUD PHYSICS

See: CLOUDS

CLOUD SEEDING

See: WEATHER MODIFICATION

CLOUDS

See also: AEROSOLS, FOG, METEOROLOGY, RAIN, WEATHER MODIFICATION

ABSTRACT SERVICES AND INDEXES

CHEMICAL ABSTRACTS. Chemical Abstracts Service, 2540 Olentangy Road, Post Office Box 3012, Columbus, OH 43210. (800) 848-6538 or (614) 421-3600. Weekly. $9200.00 per year.

METEOROLOGICAL AND GEOASTROPHYSICAL ABSTRACTS. American Meteorological Society, 45 Beacon Street, Boston, MA 02108. (617) 227-2425. 1950 to present. Monthly. $450.00 per year.

OCEANIC ABSTRACTS. Cambridge Scientific Abstracts, Incorporated, 5161 River Road, Bethesda, MD 20816. (301) 951-1400. 1964 to present. Inquire as to online cost and availability.

ASSOCIATIONS AND PROFESSIONAL SOCIETIES

AMERICAN METEOROLOGICAL SOCIETY. 45 Beacon Street, Boston, MA 02108. (617) 227-2425.

INTERNATIONAL ASSOCIATION OF METEOROLOGY AND ATMOSPHERIC PHYSICS. UCAR, Post Office Box 3000, Boulder, CO 80307.

NATIONAL WEATHER SERVICE OFFICES AND STATIONS. National Oceanic and Atmospheric Administration, Department of Commerce, Silver Spring, MD 20910. (301) 427-7698. Annual. Free.

UNIVERSITY CORPORATION FOR ATMOSPHERIC RESEARCH. Box 3000, 1850 Table Mesa Drive, Boulder, CO 80307. (303) 497-1000.

WEATHER MODIFICATION ASSOCIATION. Post Office Box 8116, Fresno, Ca 93747. (209) 291-8466.

DIRECTORIES AND BIOGRAPHICAL SOURCES

GOVERNMENT RESEARCH DIRECTORY. Gale Research Company, Book Tower, Detroit, Mi 48226. (800) 521-0707. Fourth edition. 1987. $350.00.

METEOROLOGICAL SERVICES OF THE WORLD. World Meteorological Organization. Available from: American Meteorological Society, 45 Beacon Street, Boston, MA 02108. (617) 227-2425. Annual. $35.00.

NATIONAL WEATHER SERVICE OFFICES AND STATIONS. National Oceanic and Atmospheric Administration, Department of Commerce, Silver Spring, MD 20910. (301) 427-7698. Annual. Free.

RESEARCH CENTERS DIRECTORY. Gale Research Company, Book Tower, Detroit, MI 48226. (800) 521-0707. Twelfth edition. 1987. $365.00 for set.

SCIENTIFIC AND TECHNICAL ORGANIZATIONS AND AGENCIES DIRECTORY. Gale Research Company, Book Tower, Detroit, MI 48226. (800) 521-0707. Second edition. 1987. $185.00.

GENERAL WORKS

THE ATMOSPHERE: AN INTRODUCTION TO METEOROLOGY. Frederick K. Lutgens and Edward J. Tarbuck. Prentice-Hall Publishing, Incorporated, Englewood Cliffs, NJ 07632. (800) 526-0245. Third edition. 1986. $34.95.

ATMOSPHERIC CHEMISTRY: FUNDAMENTALS AND EXPERIMENTAL TECHNIQUES. B.J. Finlayson-Pitts and J.N. Pitts. John Wiley and Sons, Incorporated, 605 Third Avenue, New York, NY 10158. (800) 526-5368 or (212) 850-6000. 1986. $59.95.

CLOUD TYPES FOR OBSERVERS. Kraus International Publications, 1 Water Street, White Plains, NY 10601. (914) 761-9600. Second edition. 1983. $19.95 in paper.

CLOUDS: THEIR FORMATION, OPTICAL PROPERTIES, AND EFFECTS. Peter Hobbs and A. Deepak, editors. Academic Press, Incorporated, Orlando, FL 32887. (800) 321-5068. 1981. $55.00.

PHYSICAL METEOROLOGY. Henry G. Houghton. MIT Press, 28 Carlton Street, Cambridge, MA 02142. (617) 253-5646. 1985. $37.50.

ONLINE DATA BASES

CA SEARCH. Chemical Abstracts Service, Post Office Box 3012, Columbus, OH 43210. Guide to chemical literature, 1972 to present. Inquire as to cost and availability.

INSPEC. INSPEC Marketing Department, Institute of Electrical and Electronics Engineers, Incorporated, IEEE Service Department, 445 Hoes Lane, Piscataway, NJ 08854. (201) 981-0060. Inquire as to on-line cost and availability.

METEOROLOGICAL AND GEOASTROPHYSICAL ABSTRACTS. American Meteorological Society, 45 Beacon Street, Boston, MA 02108. (617) 227-2425. 1950 to present. Monthly. $450.00 per year.

NTIS. National Technical Information Service, 5285 Port Royal Road, Springfield, VA 22161. (703) 487-4630. Broad coverage of government sponsored research reports, 1964 to present. Inquire as to cost and availability.

OCEANIC ABSTRACTS. Cambridge Scientific Abstracts, Incorporated, 5161 River Road, Bethesda, MD 20816. (301) 951-1400. 1964 to present. Inquire as to online cost and availability.

SCISEARCH. Institute for Scientific Information, 3501 Market Street, Philadelphia, PA 19104. (800) 523-1850 or (215) 386-0100. Broad multidisciplinary title and author index to the international literature of science and technology, 1974 to present. Inquire as to cost and availability.

PERIODICALS

AMERICAN METEOROLOGICAL SOCIETY BULLETIN. American Meteorological Society, 45 Beacon Street, Boston, MA 02108. (617) 227-2425.

DYNAMICS OF ATMOSPHERES AND OCEANS. Elsevier Science Publishing Company, Incorporated, 52 Vanderbilt Avenue, New York, NY 10017. (212) 370-5520. 1977 to present. Quarterly. $90.00 per year.

JOURNAL OF CLIMATE AND APPLIED METEOROLOGY. American Meteorological Society, 45 Beacon Street, Boston, MA 02108. (617) 227-2425. 1962 to present. Monthly. $120.00 per year.

JOURNAL OF PHYSICAL OCEANOGRAPHY. American Meteorological Society, 45 Beacon Street, Boston, MA 02108. (617) 227-2425. 1971 to present. Monthly. $120.00 per year.

JOURNAL OF THE ATMOSPHERIC SCIENCES. American Meteorological Society, 45 Beacon Street, Boston, MA 02108. (617) 227-2425. 1944 to present. Semimonthly. $220.00 per year.

MONTHLY WEATHER REVIEW. American Meteorological Society, 45 Beacon Street, Boston, MA 02108. (617) 227-2425. 1872 to present. Monthly. $120.00 per year.

NATIONAL WEATHER DIGEST. National Weather Association, 4400 Stamp Road, Room 404, Temple Hills, MD 20748. (301) 899-3784. 1976 to present. Quarterly. $20.00 per year.

WEATHER. Royal Meteorological Society, James Glaisher House, Grenville Place, Bracknell Berkshire, RG12 1BX, England. 1946 to present. $30.00 per year.

WEATHERWISE. Heldref Publications, 4000 Albemarle Street, NW, Washington, DC 20016. (202) 362-6445. 1948 to present. Bimonthly. $20.00 per year.

RESEARCH CENTERS AND INSTITUTES

CLOUD SIMULATION AND AEROSOL LABORATORY. Colorado State University, Fort Collins, CO 80523. (030) 491-8667.

COOPERATIVE INSTITUTE FOR MESOSCALE METEOROLOGICAL STUDIES. University of Oklahoma, 401 East Boyd, Norman, OK 73019. (405) 325-3041.

GRADUATE CENTER FOR CLOUD PHYSICS RESEARCH. University of Missouri - Rolla, Rolla, MO 65401. (314) 341-4332.

NATIONAL CENTER FOR ATMOSPHERIC RESEARCH. Box 3000, Boulder, CO 80307. (303) 497-1000.

UNIVERSITY OF ARIZONA. Institute of Atmospheric Physics, Tucson, AZ 85721. (602) 626-6831.

COAL

See also: COAL GASIFICATION AND LIQUIFACTION, COAL MINING

ABSTRACT SERVICES AND INDEXES

APPLIED SCIENCE AND TECHNOLOGY INDEX. H.W. Wilson Company, 950 University Avenue, Bronx, NY 10452. (800) 367-6670 or (212) 588-8400. Inquire as to cost and availability.

CHEMICAL ABSTRACTS. Chemical Abstracts Service, 2540 Olentangy Road, P.O. Box 3012, Columbus, OH 43210. (800) 848-6538 or (614) 421-3600. Weekly. $9200.00 per year.

COAL ABSTRACTS. I.E.A. Coal Research, 14/15 Lower Grosvenor Place, London, SW1W OEX, England. From 1977 to present. Monthly. $145.00 per year.

ENERGY INDEX. EIC Intelligence, Incorporated, 48 West 38th Street, New York, NY 10018. (212) 944-8500. Annual. $245.00 per year.

ENERGY INFORMATION ABSTRACTS. EIC Intelligence, Incorporated, 48 West 38th Street, New York, NY 10018. (212) 944-8500. Monthly. $845.00 per year.

ENGINEERING INDEX MONTHLY. Engineering Information, Incorporated, 345 East 47th Street, New York, NY 10017. (800) 221-1044 or (212) 705-7600. Monthly, with annual cumulation. $1560.00 per year.

GENERAL SCIENCE INDEX. H.W. Wilson Company, 950 University Avenue, Bronx, NY 10452. (800) 367-6770 or (212) 588-8400.

ASSOCIATIONS AND PROFESSIONAL SOCIETIES

NATIONAL COAL ASSOCIATION. Coal Building, 1130 17th Street, NW, Washington, DC 20036. (202) 463-2625.

NATIONAL ENERGY RESOURCES ORGANIZATION. 1713 Birch Road, McLean, VA 22101. (703) 734-2720.

NATIONAL INDEPENDENT COAL OPERATORS ASSOCIATION. P.O. Box 354, Richlands, VA 24641. (703) 963-9011.

SOCIETY OF MINING ENGINEERS OF THE AMERICAN INSTITUTE OF MINING METALLURGICAL AND PETROLEUM ENGINEERS. 8307 Shaffer Parkway, Littleton, CO 80127. (303) 973-9550.

BIBLIOGRAPHIES

COAL INFORMATION SOURCES AND DATA BASES. Carolyn C. Bloch. Noyes Data Corporation, Mill Road at Grand Avenue, Park Ridge, Nj 07656. (201) 391-8484. 1981. $24.00.

IN-SITU COAL GASIFICATION: A BIBLIOGRAPHY. Catherine M. Grissom, editor. U.S. Department of Energy. Available from: National Technical Information Service, 5285 Port Royal Road, Springfield, Va 22161. (703) 487-4630. 1984. $25.00 per year.

DIRECTORIES AND BIOGRAPHICAL SOURCES

COAL MINE DIRECTORY. McGraw-Hill Book Company, 1221 Avenue of the Americas, New York, NY 10020. (212) 512-2000. Annual. $45.00.

GOVERNMENT RESEARCH DIRECTORY. Gale Research Company, Book Tower, Detroit, MI 48226. (800) 521-0707. Third edition. 1985.

KEYSTONE COAL INDUSTRY MANUAL. McGraw-Hill Book Company, 1221 Avenue of the Americas, New York, NY 10020. (212) 512-2000. Annual. $125.00.

WHO'S WHO IN TECHNOLOGY TODAY. Reston Publishing Company, Incorporated, c/o Prentice-Hall, Incorporated, Englewood Cliffs, NJ 07632. (800) 262-6868. Biennial. Five volumes. $425.00. Covers the fields of electronics, computer science, physics, optics, chemistry, biotechnology, mechanics, energy, and earth science.

ENCYCLOPEDIAS AND DICTIONARIES

ENCYCLOPEDIA OF PHYSICAL SCIENCE AND TECHNOLOGY. Academic Press, Incorporated, Orlando, FL 32887. (800) 321-5068 or (305) 345-2734. Fifteen volumes, 1986.

MCGRAW-HILL ENCYCLOPEDIA OF SCIENCE AND TECHNOLOGY. McGraw-Hill Book, Incorporated, 1221 Avenue of the Americas, New York, NY 10020. (212) 997-3675.

GENERAL WORKS

COAL GEOLOGY AND COAL TECHNOLOGY: EXPLORATION, MINING AND PREPARATION. Colin R. Ward, editor. Blackwell Scientific Publications, Incorporated, 52 Beacon Street, Boston, MA 02108. (800) 325-4177 or (617) 720-0761. 1985. $40.00 in paper.

COAL MINING TECHNOLOGY: THEORY AND PRACTICE. Robert Stefanko. Society of Mining Engineers of the American Institute of Mining, Metallurgical and Petroleum Engineers, 8307 Shaffer Parkway, Littleton, CO 80127. (303) 973-9550. 1983. $45.00.

COAL STRUCTURE. Martin L. Gorbaty and K. Ouchi, editors. American Chemical Society Advances in Chemistry Series, volume 192. American Chemical Society, 1155 16th Street, NW, Washington, DC 20036. (202) 872-4600. 1981. $39.95.

COAL: TOPOLOGY, CHEMISTRY, PHYSICS AND CONSTITUTION. Van Krevelen. Elsevier Scinece Publishing Company, Incorporated, 52 Vanderbilt Avenue, New York, NY 10017. (212) 370-5520. 1981. $106.00.

CONSTRUCTION ENGINEERING IN UNDERGROUND COAL MINES. Scott G. Britton. Society of Mining Engineers of the American Institute of Mining, Metallurgical and Petroleum Engineers, 8307 Shaffer Parkway, Littleton, CO 80127. (303) 973-9550. 1983. $48.00.

ECONOMICS OF THE MINERALS INDUSTRIES. William A. Vogely, editor. Society of Mining Engineers of the American Institute of Mining, Metallurgical and Petroleum Engineers, 8307 Shaffer Parkway, Littleton, CO 80127. (303) 973-9550. Fourth edition. 1985. $50.00.

ELEMENTS OF PRACTICAL COAL MINING. Douglas F. Crickner and David A. Zegeer, editors. Society of Mining Engineers of the American Institute of Mining, Metallurgical and Petroleum Engineers, 8307 Shaffer Parkway, Littleton, CO 80127. (303) 973-9550. Second edition. 1981. $45.00.

SURFACE MINING. Society of Mining Engineers of the American Institute of Mining, Metallurgical and Petroleum Engineers, 8307 Shaffer Parkway, Littleton, CO 80127. (303) 973-9550. 1968. $30.00.

HANDBOOKS AND MANUALS

CANADIAN MINES HANDBOOK. Northern Miner Press, Limited, 7 Labatt Street, Toronto, Ontario, M5A 3P2, Canada. Annual. $24.00.

COAL HANDBOOK. Robert A. Meyers. Marcel Dekker, Incorporated, 270 Madison Avenue, New York, NY 10016. (212) 696-9000. 1981. $79.75.

ONLINE DATA BASES

CA SEARCH. Chemical Abstracts Service, P.O. Box 3012, Columbus, OH 43210. Guide to chemical literature, 1972 to present. Inquire as to cost and availability.

COAL DATA BANKS. U.S. Department of Energy, 12th and Pennsylvania Avenue, NW, Washington, DC 20461. Inquire as to online cost and availability.

COMPENDEX. Engineering Information, Incorporated, 345 East 47th Street, New York, NY 10017. (800) 221-1044 or (212) 705-7615. Engineering and technical literature, 1975 to present. Inquire as to cost and availability.

MINTEC. Energy, Mines and Resources, CANMET, Technology Information Division, 555 Booth Street, Ottawa, Ontario, Canada K1A OG1. (613) 995-4029. Citations and abstracts of literature on mining technology, 1968 to present. Inquire as to online cost and availability.

NATIONAL COAL LIQUEFACTION TECHNOLOGY DATA BASE. U.S. Department of Energy, Pittsburgh Energy Technology Center, P.O. Box 10940, Pittsburgh, PA 15236. (412) 675-6000. Inquire as to online cost and availability.

NTIS. National Technical Information Service, 5285 Port Royal Road, Springfield, VA 22161. (703) 487-4630. Broad coverage of government sponsored research reports, 1964 to present. Inquire as to cost and availability.

PERIODICALS

AMERICAN MINING CONGRESS JOURNAL. American Mining Congress, 1920 N Street, NW, Suite 300, Washington, DC 20036. (202) 861-2800. Semimonthly. $50.00 per year.

CANADIAN INSTITUTE OF MINING AND METALLURGY. Bulletin. 400-1130 Sherbrooke West, Montreal, Quebec, H3A 2M8, Canada. Monthly. $90.00 per year.

COAL AGE. McGraw-Hill Book Company, 1221 Avenue of the Americas, New York, NY 10020. (212) 512-2000. Monthly. $20.00 per year.

COAL MINING AND PROCESSING. Maclean-Hunter Publishing Corporation, 300 West Adams Street, Chicago, IL 60606. (312) 726-2802. Monthly. $35.00 per year.

COAL OUTLOOK. Pasha Publications, 1401 Wilson Boulevard, Arlington, VA 22209. (703) 528-1244. Weekly. $327.00 per year.

COAL WEEK. McGraw-Hill Publications, 1120 Vermont Avenue, NW, Washington, DC 20005. (202) 463-1600. Weekly. $597.00 per year.

ENGINEERING AND MINING JOURNAL. McGraw-Hill Book Company, 1221 Avenue of the Americas, New York, NY 10020. (800) 628-0004. Monthly. $20.00.

MINES MAGAZINE. Colorado School of Mines, Golden, CO 80401. (303) 273-3607. Bimonthly. $12.00 per year.

MINING ENGINEERING. Society of Mining Engineers of the American Institute of Mining, Metallurgical and Petroleum Engineers, 8307 Shaffer Parkway, Littleton, CO 80127. (303) 973-9550. Monthly. $40.00.

NORTHERN MINER. Northern Miner Press, Limited, 7 Labatt Street, Toronto, Ontario, M5A 3P2, Canada. Weekly. $40.00.

PIT AND QUARRY. Harcourt, Brace and Jovanovich, Incorporated, 7500 Old Oak Boulevard, Cleveland, OH 44130. Monthly. $15.00.

RESEARCH CENTERS AND INSTITUTES

COAL EXTRACTION AND UTILIZATION RESEARCH CENTER. Southern Illinois University at Carbondale, Carbondale, IL 62901. (618) 536-5521.

COAL RESEARCH CENTER. Purdue University, Potter Engineering Center, West Lafayette, In 47907. (317) 494-7037.

COAL RESEARCH STATION. Pennyslvania State University, 513 Deike Building, University Park, PA 16802. (814) 865-6544.

ENERGY RESEARCH CENTER. Lehigh University, Bethlehem, PA 18015. (215) 861-4090.

U.S. DEPARTMENT OF ENERGY. Office of Surface Coal Gasification, Washington, DC 20545. (301) 353-3498.

U.S. DEPARTMENT OF ENERGY. Underground Coal Gasification Program, Washington, DC 20545. (301) 353-2724.

STATISTICAL SOURCES

COAL: BITUMINOUS AND LIGNITE. Bureau of Mines, U.S. Department of the Interior, 4800 Forbes Avenue, Pittsburgh, PA 15213. Annual. Inquire as to cost and availability.

COAL DATA. National Coal Association, Coal Building, 1130 17th Street, NW, Washington, Dc 20036. (202) 463-2625. Annual. $75.00.

MINERALS YEARBOOK. Bureau of Mines, U.S. Department of the Interior. Available from U.S. Government Printing Office, Washington, DC 20402. (202) 783-3238. Annual. Three volumes. $45.00.

COAL GASIFICATION AND LIQUIFICATION

See also: COAL, COAL MINING

ABSTRACT SERVICES AND INDEXES

APPLIED SCIENCE AND TECHNOLOGY INDEX. H.W. Wilson Company, 950 University Avenue, Bronx, NY 10452. (800) 367-6670 or (212) 588-8400. Inquire as to cost and availability.

CHEMICAL ABSTRACTS. Chemical Abstracts Service, 2540 Olentangy Road, P.O. Box 3012, Columbus, OH 43210. (800) 848-6538 or (614) 421-3600. Weekly. $9200.00 per year.

ENERGY INDEX. EIC Intelligence, Incorporated, 48 West 38th Street, New York, NY 10018. (212) 944-8500. Annual. $245.00

per year.

ENERGY INFORMATION ABSTRACTS. EIC Intelligence, Incorporated, 48 West 38th Street, New York, NY 10018. (212) 944-8500. Monthly. $845.00 per year.

ENGINEERING INDEX MONTHLY. Engineering Information, Incorporated, 345 East 47th Street, New York, NY 10017. (800) 221-1044 or (212) 705-7600. Monthly, with annual cumulation. $1560.00 per year.

GENERAL SCIENCE INDEX. H.W. Wilson Company, 950 University Avenue, Bronx, NY 10452. (800) 367-6770 or (212) 588-8400.

ASSOCIATIONS AND PROFESSIONAL SOCIETIES

NATIONAL COAL ASSOCIATION. Coal Building, 1130 17th Street, NW, Washington, DC 20036. (202) 463-2625.

NATIONAL ENERGY RESOURCES ORGANIZATION. 1713 Birch Road, McLean, VA 22101. (703) 734-2720.

NATIONAL INDEPENDENT COAL OPERATORS ASSOCIATION. P.O. Box 354, Richlands, VA 24641. (703) 963-9011.

BIBLIOGRAPHIES

COAL INFORMATION SOURCES AND DATA BASES. Carolyn C. Bloch. Noyes Data Corporation, Mill Road at Grand Avenue, Park Ridge, Nj 07656. (201) 391-8484. 1981. $24.00.

IN-SITU COAL GASIFICATION: A BIBLIOGRAPHY. Catherine M. Grissom, editor. U.S. Department of Energy. Available from: National Technical Information Service, 5285 Port Royal Road, Springfield, Va 22161. (703) 487-4630. 1984. $25.00 per year.

DIRECTORIES AND BIOGRAPHICAL SOURCES

COAL MINE DIRECTORY. McGraw-Hill Book Company, 1221 Avenue of the Americas, New York, NY 10020. (212) 512-2000. Annual. $45.00.

GOVERNMENT RESEARCH DIRECTORY. Gale Research Company, Book Tower, Detroit, MI 48226. (800) 521-0707. Third edition. 1985.

KEYSTONE COAL INDUSTRY MANUAL. McGraw-Hill Book Company, 1221 Avenue of the Americas, New York, NY 10020. (212) 512-2000. Annual. $125.00.

WHO'S WHO IN TECHNOLOGY TODAY. Reston Publishing Company, Incorporated, c/o Prentice-Hall, Incorporated, Englewood Cliffs, NJ 07632. (800) 262-6868. Biennial. Five volumes. $425.00. Covers the fields of electronics, computer science, physics, optics, chemistry, biotechnology, mechanics, energy, and earth science.

ENCYCLOPEDIAS AND DICTIONARIES

ENCYCLOPEDIA OF PHYSICAL SCIENCE AND TECHNOLOGY. Academic Press, Incorporated, Orlando, FL 32887. (800) 321-5068 or (305) 345-2734. Fifteen volumes, 1986.

MCGRAW-HILL ENCYCLOPEDIA OF SCIENCE AND TECHNOLOGY. McGraw-Hill Book, Incorporated, 1221 Avenue of the Americas, New York, NY 10020. (212) 997-3675.

GENERAL WORKS

CARBON AND COAL GASIFICATION: SCIENCE AND TECHNOLOGY. J.L. Figueiredo and J.A. Moulijn, editors. Kluwer Academic Publishers, 190 Old Derby Street, Hingham, MA 02043. (617) 749-5262. 1986. $106.00.

COAL COMBUSTION AND GASIFICATION. Douglas I. Smoot and Philip J. Smith. Plenum Publishing Corporation, 233 Spring Street, New York, NY 10013. (800) 221-9369. 1985. $59.50.

COAL GASIFICATION PROCESSES. Perry Nowacki, editor. Noyes Data Corporation, Mill Road at Grand Avenue, Park Ridge, NJ 07656. (201) 391-8484. 1982. $48.00.

COAL GEOLOGY AND COAL TECHNOLOGY: EXPLORATION, MINING AND PREPARATION. Colin R. Ward, editor. Blackwell Scientific Publications, Incorporated, 52 Beacon Street, Boston, MA 02108. (800) 325-4177 or (617) 720-0761. 1985. $40.00 in paper.

COAL LIQUEFACTION FUNDAMENTALS. D.D. Whitehurst, editor. American Chemical Society, 1155 16th Street, NW, Washington, DC 20036. (202) 872-4600. 1980. $39.95.

COAL STRUCTURE. Martin L. Gorbaty and K. Ouchi, editors. American Chemical Society Advances in Chemistry Series, Volume 192. American Chemical Society, 1155 16th Street, NW, Washington, DC 20036. (202) 872-4600. 1981. $39.95.

COAL: TOPOLOGY, CHEMISTRY, PHYSICS AND CONSTITUTION. Van Krevelen. Elsevier Scinece Publishing Company, Incorporated, 52 Vanderbilt Avenue, New York, NY 10017. (212) 370-5520. 1981. $106.00.

THE FISCHER-TROPSCH SYNTHESIS. Robert B. Anderson. Academic Press, Incorporated, Orlando, FL 32887. (800) 321-5068 or (305) 345-2734. 1984. $60.50.

FREE RADICALS IN COALS AND SYNTHETIC FUELS. L. Petrakais and D.W. Grandy. Elsevier Science Publishing Company, Incorporated, 52 Vanderbilt Avenue, New York, NY 10017. (212) 370-5520. 1983. $66.00.

HANDBOOKS AND MANUALS

COAL HANDBOOK. Robert A. Meyers. Marcel Dekker, Incorporated, 270 Madison Avenue, New York, NY 10016. (212) 696-9000. 1981. $79.75.

ONLINE DATA BASES

CA SEARCH. Chemical Abstracts Service, P.O. Box 3012, Columbus, OH 43210. Guide to chemical literature, 1972 to present. Inquire as to cost and availability.

COAL DATA BANKS. U.S. Department of Energy, 12th and Pennsylvania Avenue, NW, Washington, DC 20461. Inquire as to online cost and availability.

COMPENDEX. Engineering Information, Incorporated, 345 East 47th Street, New York, NY 10017. (800) 221-1044 or (212) 705-7615. Engineering and technical literature, 1975 to present. Inquire as to cost and availability.

MINTEC. Energy, Mines and Resources, CANMET, Technology Information Division, 555 Booth Street, Ottawa, Ontario, Canada K1A OG1. (613) 995-4029. Citations and abstracts of literature on mining technology, 1968 to present. Inquire as to online cost and availability.

NATIONAL COAL LIQUIFACTION TECHNOLOGY DATA BASE. U.S. Department of Energy, Pittsburgh Energy Technology Center, P.O. Box 10940, Pittsburgh, PA 15236. (412) 675-6000. Inquire as to online cost and availability.

NTIS. National Technical Information Service, 5285 Port Royal Road, Springfield, VA 22161. (703) 487-4630. Broad coverage of government sponsored research reports, 1964 to present. Inquire as to cost and availability.

PERIODICALS

AMERICAN MINING CONGRESS JOURNAL. American Mining Congress, 1920 N Street, NW, Suite 300, Washington, DC 20036. (202) 861-2800. Semimonthly. $50.00 per year.

COAL AGE. McGraw-Hill Book Company, 1221 Avenue of the Americas, New York, NY 10020. (212) 512-2000. Monthly. $20.00 per year.

COAL MINING AND PROCESSING. Maclean-Hunter Publishing Corporation, 300 West Adams Street, Chicago, IL 60606. (312) 726-2802. Monthly. $35.00 per year.

COAL OUTLOOK. Pasha Publications, 1401 Wilson Boulevard, Arlington, VA 22209. (703) 528-1244. Weekly. $327.00 per year.

COAL WEEK. McGraw-Hill Publications, 1120 Vermont Avenue, NW, Washington, DC 20005. (202) 463-1600. Weekly. $597.00 per year.

FUEL PROCESSING TECHNOLOGY. Elsevier Science Publishers B.V., Box 211, 1000 AE, Amsterdam, The Netherlands. Nine times per year.

FUEL SCIENCE AND TECHNOLOGY. Marcel Dekker Journals, 270 Madison Avenue, New York, NY 10016. (212) 696-9000. Six times per year. $175.00 per year.

RESEARCH CENTERS AND INSTITUTES

COAL EXTRACTION AND UTILIZATION RESEARCH CENTER. Southern Illinois University at Carbondale, Carbondale, IL 62901. (618) 536-5521.

COAL RESEARCH CENTER. Purdue University, Potter Engineering Center, West Lafayette, In 47907. (317) 494-7037.

COAL RESEARCH STATION. Pennyslvania State University, 513 Deike Building, University Park, PA 16802. (814) 865-6544.

ENERGY RESEARCH CENTER. Lehigh University, Bethlehem, PA 18015. (215) 861-4090.

U.S. DEPARTMENT OF ENERGY. Office of Surface Coal Gasification, Washington, DC 20545. (301) 353-3498.

U.S. DEPARTMENT OF ENERGY. Underground Coal Gasification Program, Washington, DC 20545. (301) 353-2724.

STATISTICAL SOURCES

COAL: BITUMINOUS AND LIGNITE. Bureau of Mines, U.S. Department of the Interior, 4800 Forbes Avenue, Pittsburgh, PA 15213. Annual. Inquire as to cost and availability.

COAL DATA. National Coal Association, Coal Building, 1130 17th Street, NW, Washington, Dc 20036. (202) 463-2625. Annual. $75.00.

MINERALS YEARBOOK. Bureau of Mines, U.S. Department of the Interior. Available from U.S. Government Printing Office, Washington, DC 20402. (202) 783-3238. Annual. Three volumes. $45.00.

COAL MINING

See also: COAL, COAL GASIFICATION AND LIQUIFACTION

ABSTRACT SERVICES AND INDEXES

APPLIED SCIENCE AND TECHNOLOGY INDEX. H.W. Wilson Company, 950 University Avenue, Bronx, NY 10452. (800) 367-6670 or (212) 588-8400. Inquire as to cost and availability.

CHEMICAL ABSTRACTS. Chemical Abstracts Service, 2540 Olentangy Road, P.O. Box 3012, Columbus, OH 43210. (800) 848-6538 or (614) 421-3600. Weekly. $9200.00 per year.

ENERGY INDEX. EIC Intelligence, Incorporated, 48 West 38th Street, New York, NY 10018. (212) 944-8500. Annual. $245.00 per year.

ENERGY INFORMATION ABSTRACTS. EIC Intelligence, Incorporated, 48 West 38th Street, New York, NY 10018. (212) 944-8500. Monthly. $845.00 per year.

ENGINEERING INDEX MONTHLY. Engineering Information, Incorporated, 345 East 47th Street, New York, NY 10017. (800) 221-1044 or (212) 705-7600. Monthly, with annual cumulation. $1560.00 per year.

GENERAL SCIENCE INDEX. H.W. Wilson Company, 950 University Avenue, Bronx, NY 10452. (800) 367-6770 or (212) 588-8400.

ASSOCIATIONS AND PROFESSIONAL SOCIETIES

NATIONAL COAL ASSOCIATION. Coal Building, 1130 17th Street, NW, Washington, DC 20036. (202) 463-2625.

NATIONAL ENERGY RESOURCES ORGANIZATION. 1713 Birch Road, McLean, VA 22101. (703) 734-2720.

NATIONAL INDEPENDENT COAL OPERATORS ASSOCIATION. P.O. Box 354, Richlands, VA 24641. (703) 963-9011.

SOCIETY OF MINING ENGINEERS OF THE AMERICAN INSTITUTE OF MINING METALLURGICAL AND PETROLEUM ENGINEERS. 8307 Shaffer Parkway, Littleton, CO 80127. (303) 973-9550.

BIBLIOGRAPHIES

COAL INFORMATION SOURCES AND DATA BASES. Carolyn C. Bloch. Noyes Data Corporation, Mill Road at Grand Avenue, Park Ridge, Nj 07656. (201) 391-8484. 1981. $24.00.

IN-SITU COAL GASIFICATION: A BIBLIOGRAPHY. Catherine M. Grissom, editor. U.S. Department of Energy. Available from: National Technical Information Service, 5285 Port Royal Road, Springfield, Va 22161. (703) 487-4630. 1984. $25.00 per year.

DIRECTORIES AND BIOGRAPHICAL SOURCES

COAL MINE DIRECTORY. McGraw-Hill Book Company, 1221 Avenue of the Americas, New York, NY 10020. (212) 512-2000. Annual. $45.00.

GOVERNMENT RESEARCH DIRECTORY. Gale Research Company, Book Tower, Detroit, MI 48226. (800) 521-0707. Third edition. 1985.

KEYSTONE COAL INDUSTRY MANUAL. McGraw-Hill Book Company, 1221 Avenue of the Americas, New York, NY 10020. (212) 512-2000. Annual. $125.00.

WHO'S WHO IN TECHNOLOGY TODAY. Reston Publishing Company, Incorporated, c/o Prentice-Hall, Incorporated, Englewood Cliffs, NJ 07632. (800) 262-6868. Biennial. Five volumes. $425.00. Covers the fields of electronics, computer science, physics, optics, chemistry, biotechnology, mechanics, energy, and earth science.

ENCYCLOPEDIAS AND DICTIONARIES

ENCYCLOPEDIA OF PHYSICAL SCIENCE AND TECHNOLOGY. Academic Press, Incorporated, Orlando, FL 32887. (800) 321-5068 or (305) 345-2734. Fifteen volumes, 1986.

MCGRAW-HILL ENCYCLOPEDIA OF SCIENCE AND TECHNOLOGY. McGraw-Hill Book, Incorporated, 1221 Avenue of the Americas, New York, NY 10020. (212) 997-3675.

GENERAL WORKS

COAL GEOLOGY AND COAL TECHNOLOGY: EXPLORATION, MINING AND PREPARATION. Colin R. Ward, editor. Blackwell Scientific Publications, Incorporated, 52 Beacon Street, Boston, MA 02108. (800) 325-4177 or (617) 720-0761. 1985. $40.00 in paper.

COAL MINING TECHNOLOGY: THEORY AND PRACTICE. Robert Stefanko. Society of Mining Engineers of the American Institute of Mining, Metallurgical and Petroleum Engineers, 8307 Shaffer Parkway, Littleton, CO 80127. (303) 973-9550. 1983. $45.00.

COAL STRUCTURE. Martin L. Gorbaty and K. Ouchi, editors. American Chemical Society Advances in Chemistry Series, volume 192. American Chemical Society, 1155 16th Street, NW, Washington, DC 20036. (202) 872-4600. 1981. $39.95.

COAL: TOPOLOGY, CHEMISTRY, PHYSICS AND CONSTITUTION. Van Krevelen. Elsevier Scinece Publishing Company, Incorporated, 52 Vanderbilt Avenue, New York, NY 10017. (212) 370-5520. 1981. $106.00.

CONSTRUCTION ENGINEERING IN UNDERGROUND COAL MINES. Scott G. Britton. Society of Mining Engineers of the American Institute of Mining, Metallurgical and Petroleum Engineers, 8307 Shaffer Parkway, Littleton, CO 80127. (303) 973-9550. 1983. $48.00.

ECONOMICS OF THE MINERALS INDUSTRIES. William A. Vogely, editor. Society of Mining Engineers of the American Institute of Mining, Metallurgical and Petroleum Engineers, 8307 Shaffer Parkway, Littleton, CO 80127. (303) 973-9550. Fourth edition. 1985. $50.00.

ELEMENTS OF PRACTICAL COAL MINING. Douglas F. Crickner and David A. Zegeer, editors. Society of Mining Engineers of the American Institute of Mining, Metallurgical and Petroleum Engineers, 8307 Shaffer Parkway, Littleton, CO 80127. (303) 973-9550. Second edition. 1981. $45.00.

SURFACE MINING. Society of Mining Engineers of the American Institute of Mining, Metallurgical and Petroleum Engineers, 8307 Shaffer Parkway, Littleton, CO 80127. (303) 973-9550. 1968. $30.00.

HANDBOOKS AND MANUALS

CANADIAN MINES HANDBOOK. Northern Miner Press, Limited, 7 Labatt Street, Toronto, Ontario, M5A 3P2, Canada. Annual. $24.00.

COAL HANDBOOK. Robert A. Meyers. Marcel Dekker, Incorporated, 270 Madison Avenue, New York, NY 10016. (212) 696-9000. 1981. $79.75.

ONLINE DATA BASES

CA SEARCH. Chemical Abstracts Service, P.O. Box 3012, Columbus, OH 43210. Guide to chemical literature, 1972 to present. Inquire as to cost and availability.

COAL DATA BANKS. U.S. Department of Energy, 12th and Pennsylvania Avenue, NW, Washington, DC 20461. Inquire as to online cost and availability.

COMPENDEX. Engineering Information, Incorporated, 345 East 47th Street, New York, NY 10017. (800) 221-1044 or (212) 705-7615. Engineering and technical literature, 1975 to present. Inquire as to cost and availability.

MINTEC. Energy, Mines and Resources, CANMET, Technology Information Division, 555 Booth Street, Ottawa, Ontario, Canada K1A OG1. (613) 995-4029. Citations and abstracts of literature on mining technology, 1968 to present. Inquire as to online cost and availability.

NATIONAL COAL LIQUIFACTION TECHNOLOGY DATA BASE. U.S. Department of Energy, Pittsburgh Energy Technology Center, P.O. Box 10940, Pittsburgh, PA 15236. (412) 675-6000. Inquire as to online cost and availability.

NTIS. National Technical Information Service, 5285 Port Royal Road, Springfield, VA 22161. (703) 487-4630. Broad coverage of government sponsored research reports, 1964 to present. Inquire as to cost and availability.

PERIODICALS

AMERICAN MINING CONGRESS JOURNAL. American Mining Congress, 1920 N Street, NW, Suite 300, Washington, DC 20036. (202) 861-2800. Semimonthly. $50.00 per year.

CANADIAN INSTITUTE OF MINING AND METALLURGY. Bulletin. 400-1130 Sherbrooke West, Montreal, Quebec, H3A 2M8, Canada. Monthly. $90.00 per year.

COAL AGE. McGraw-Hill Book Company, 1221 Avenue of the Americas, New York, NY 10020. (212) 512-2000. Monthly. $20.00 per year.

COAL MINING AND PROCESSING. Maclean-Hunter Publishing Corporation, 300 West Adams Street, Chicago, IL 60606. (312) 726-2802. Monthly. $35.00 per year.

COAL OUTLOOK. Pasha Publications, 1401 Wilson Boulevard, Arlington, VA 22209. (703) 528-1244. Weekly. $327.00 per year.

COAL WEEK. McGraw-Hill Publications, 1120 Vermont Avenue, NW, Washington, DC 20005. (202) 463-1600. Weekly. $597.00 per year.

ENGINEERING AND MINING JOURNAL. McGraw-Hill Book Company, 1221 Avenue of the Americas, New York, NY 10020. (800) 628-0004. Monthly. $20.00.

MINES MAGAZINE. Colorado School of Mines, Golden, CO 80401. (303) 273-3607. Bimonthly. $12.00 per year.

MINING ENGINEERING. Society of Mining Engineers of the American Institute of Mining, Metallurgical and Petroleum Engineers, 8307 Shaffer Parkway, Littleton, CO 80127. (303) 973-9550. Monthly. $40.00.

NORTHERN MINER. Northern Miner Press, Limited, 7 Labatt Street, Toronto, Ontario, M5A 3P2, Canada. Weekly. $40.00.

PIT AND QUARRY. Harcourt, Brace and Jovanovich, Incorporated, 7500 Old Oak Boulevard, Cleveland, OH 44130. Monthly. $15.00.

RESEARCH CENTERS AND INSTITUTES

COAL EXTRACTION AND UTILIZATION RESEARCH CENTER. Southern Illinois University at Carbondale, Carbondale, IL 62901. (618) 536-5521.

COAL RESEARCH CENTER. Purdue University, Potter Engineering Center, West Lafayette, In 47907. (317) 494-7037.

COAL RESEARCH STATION. Pennyslvania State University, 513 Deike Building, University Park, PA 16802. (814) 865-6544.

ENERGY RESEARCH CENTER. Lehigh University, Bethlehem, PA 18015. (215) 861-4090.

U.S. DEPARTMENT OF ENERGY. Office of Surface Coal Gasification, Washington, DC 20545. (301) 353-3498.

U.S. DEPARTMENT OF ENERGY. Underground Coal Gasification Program, Washington, DC 20545. (301) 353-2724.

STATISTICAL SOURCES

COAL: BITUMINOUS AND LIGNITE. Bureau of Mines, U.S. Department of the Interior, 4800 Forbes Avenue, Pittsburgh, PA 15213. Annual. Inquire as to cost and availability.

COAL DATA. National Coal Association, Coal Building, 1130 17th Street, NW, Washington, Dc 20036. (202) 463-2625. Annual. $75.00.

MINERALS YEARBOOK. Bureau of Mines, U.S. Department of the Interior. Available from U.S. Government Printing Office, Washington, DC 20402. (202) 783-3238. Annual. Three volumes. $45.00.

COASTAL ENGINEERING

See also: CIVIL ENGINEERING, COASTS, CONTINENTAL MARGINS, OCEAN ENGINEERING

ABSTRACT SERVICES AND INDEXES

APPLIED SCIENCE AND TECHNOLOGY INDEX. H.W. Wilson and Company, 950 University Avenue, Bronx, NY 10452. (800) 367-6670 or (212) 588-8400. Monthly. Inquire as to cost and availability.

BIBLIOGRAPHY AND INDEX OF GEOLOGY. American Geological Institute, 4220 King Street, Alexandria, VA 22302. (703) 379-2480. 1969 to present. Monthly. $1100.00 per year.

CHEMICAL ABSTRACTS. American Chemical Society, Chemical Abstracts Service, Box 3012, Columbus, OH 43210. (614) 421-3600. 1907 to present. Weekly. $9500.00 per year.

CURRENT CONTENTS: ENGINEERING, TECHNOLOGY AND APPLIED SCIENCES. Institute for Scientific Information, 3501 Market Street, Philadelphia, PA 19104. (800) 523-1850 or (215) 386-0100. Weekly. $275.00 per year.

ENGINEERING INDEX MONTHLY AND AUTHOR INDEX. Engineering Information Inc., 345 East 47th Street, New York, NY 10017. (212) 705-7600. Monthly. $1560.00 per year.

OCEAN ABSTRACTS. Cambridge Scientific Abstracts, 5161 River Road, Bethesda, MD 20816. (301) 951-1400. 1963 to present. Bimonthly. $450.00 per year.

SCIENCE CITATION INDEX. Institute for Scientific Information, 3501 Market Street, Philadelphia, PA 19104. (800) 523-1850 or (215) 386-0100. Six times per year. $6200. 00 per year.

ANNUAL REVIEWS AND YEARBOOKS

COASTAL ENGINEERING. American Society of Civil Engineers, 345 East 47th Street, New York, NY 10017. (212) 705-7420. 1974 to 1980. Multiple volume sets for each year. Price varies, inquire.

ASSOCIATIONS AND PROFESSIONAL SOCIETIES

AMERICAN SOCIETY OF CIVIL ENGINEERS. 345 East 47th Street, New York, NY 10017. (212) 705-7420.

COASTAL ENGINEERING RESEARCH COUNCIL. 207 East Bay Street, Suite 311, Charleston, SC 29401. (803) 723-4864.

THE COASTAL SOCIETY. 5410 Grosvenor Lane, Suite 110, Bethesda, MD 20814. (301) 897-8616.

GEOLOGICAL SOCIETY OF AMERICA. 3300 Penrose Place, Boulder, CO 80301. (303) 447-2020.

DIRECTORIES AND BIOGRAPHICAL SOURCES

1987 DIRECTORY OF ENGINEERING SOCIETIES AND RELATED ORGANIZATIONS. Gordon Davis, editor. Hemisphere Publishing Corporation, 1010 Vermont Avenue, NW, Washington, DC 20005. (800) 526-0275. 12th edition. 1987. $100.00.

RESEARCH CENTERS DIRECTORY 1988. Gale Research Company, Book Tower, Detroit, MI 48226. (800) 521-0707. 12th edition. 1987. $365.00 for set.

SCIENTIFIC AND TECHNICAL ORGANIZATIONS AND AGENCIES DIRECTORY. Margaret Labash Young, editor. Gale Research Company, Book Tower, Detroit, MI 48226. (800) 521-0707. 2nd edition. 1987. $185.00.

WHO'S WHO IN ENGINEERING. Gordon Davis, editor. Hemisphere Publishing Corporation, 1010 Vermont Avenue, NW, Washington, DC 20005. (800) 526-0275. 6th edition. 1985. $200.00.

GENERAL WORKS

BASIC COASTAL ENGINEERING. R.M. Sorensen. John Wiley and Sons, Inc., 605 Third Avenue, New York, NY 10158. (800) 526-5368. 1978. $39.95.

ENVIRONMENTAL OCEANOGRAPHY: AN INTRODUCTION TO THE BEHAVIOUR OF COASTAL WATERS. T. Beer. Pergamon Press, Inc., Maxwell House, Fairview Park, Elmsford, NY 10523. (914) 592-7700. 1983. $13.00 in paper.

PHYSICAL OCEANOGRAPHY OF COASTAL WATERS. K.F. Bowden. John Wiley and Sons, Inc., 605 Third Avenue, New York, NY 10158. (800) 526-5368. 1984. $74.95.

ONLINE DATA BASES

CA SEARCH. Chemical Abstracts Service, P.O. Box 3012, Columbus, OH 43120. (800) 848-6538 or (614) 421-3600. Comprehensive guide to chemical literature, 1972 to present. Inquire as to online cost and availability.

COMPENDEX. Engineering Information, Inc., 345 East 47th Street, New York, NY 10017. (800) 221-1044 or (212) 705-7615. Engineering and technical literature, 1975 to present. Inquire as to online cost and availability.

DISSERTATION ABSTRACTS ONLINE. University Microfilms International, 300 North Zeeb Road, Ann Arbor, MI 48106. (800) 521-0600 or (313) 761-4700. Scope includes virtually all doctoral dissertations accepted at accredited American institutions from 1861 to present in over 250 subject areas. Inquire as to online cost and availability.

GEOREF. Online version of the BIBLIOGRAPHY AND INDEX OF GEOLOGY. American Geological Institute, 4220 King Street, Alexandria, VA 22302. (703) 379-2480. 1969 to present. Inquire as to online cost and availability.

INSPEC. INSPEC Marketing Department, Institution of Electrical Engineers. Available from IEEE Service Center, 445 Hoes Lane, Piscataway, NJ 08854. (201) 981-0060. Online version of Physics Abstracts. Inquire as to online cost and availability.

NTIS. National Technical Information Service, 5285 Port Royal Road, Springfield, VA 22161. (703) 487-4630. Broad coverage of government sponsored research reports, 1964 to present. Inquire

as to online cost and availability.

SCISEARCH. Institute for Scientific Information, 3501 Market Street, Philadelphia, PA 19104. (800) 523-1850 or (215) 386-0100. Broad multidisciplinary title and author index to the international literature of science and technology, 1974 to present. Inquire as to online cost and availability.

WILSONLINE. H.W. Wilson and Company, 950 University Avenue, Bronx, NY 10452. (800) 367-6770 or (212) 588-8400. Makes available online versions of the H.W. Wilson indexes including Applied Science and Technology Index, Business Periodicals Index and Readers' Guide to Periodical Literature. Approximately 1980 to present. Inquire as to online cost and availability.

PERIODICALS

COASTAL ENGINEERING: AN INTERNATIONAL JOURNAL FOR COASTAL, HARBOUR, AND OFFSHORE ENGINEERS. Elsevier Science Publishing Company, Inc., 52 Vanderbilt Avenue, New York, NY 10017. (212) 370-5520. 1977 to present. Quarterly. $95.00 per year.

COASTAL RESEARCH. Florida State University, Geology Department, Tallahassee, FL 32306. (904) 644-5860. 1962 to present. Three times per year. $5.00 per year.

COASTAL ZONE MANAGEMENT JOURNAL. Crane Russak and Company, Inc., 3 East 44th Street, New York, NY 10017. (212) 867-1490. 1973 to present. Quarterly. $72.00 per year.

COASTWATCH. University of North Carolina, Sea Grant College Program, Box 8605, North Carolina State University, Raleigh, NC 27695-8605. (919) 737-2454. 1970 to present. Monthly. Free.

ESTUARINE, COASTAL AND SHELF SCIENCE. Academic Press, Inc., 6277 Sea Harbor Drive, Orlando, FL 32821. (800) 321-5068. 1973 to present. Monthly. $315.00 per year.

JOURNAL OF WATERWAY, PORT, COASTAL AND OCEAN ENGINEERING. American Society of Civil Engineers, 345 East 47th Street, New York, NY 10017. (212) 705-7420. 1956 to present. Bimonthly. $72.00 per year.

OCEAN ENGINEERING. Pergamon Press, Inc., Maxwell House, Fairview Park, Elmsford, NY 10523. (914) 592-7700. 1968 to present. Bimonthly. $200.00 per year.

OCEAN SCIENCE AND ENGINEERING. Marcel Dekker Inc., 270 Madison Avenue, New York, NY 10016. (800) 228-1160. 1974 to present. Quarterly. $165.00 per year.

U.S. COASTAL ENGINEERING RESEARCH CENTER. QUARTERLY CIRCULAR INFORMATION BULLETIN. U.S. Coastal Engineering Research Center, Kingman Building, Fort Belvoir, VA 22060. (703) 325-7429. Quarterly. Inquire.

RESEARCH CENTERS AND INSTITUTES

COASTAL AND OCEANOGRAPHIC ENGINEERING LABORATORY. University of Florida, 336 Weil Hall, Gainesville, FL 32607. (904) 392-1436.

COASTAL ENGINEERING RESEARCH COUNCIL. 207 East Bay Street, Suite 311, Charleston, SC 29401. (803) 723-4864.

COASTAL LABORATORY. Queen's University at Kingston, Ellis Hall, Kingston, ON, Canada K7L 3N6. (613) 545-6265.

COASTS

See also: CIVIL ENGINEERING, COASTAL ENGINEERING, CONTINENTAL MARGINS, OCEAN ENGINEERING, OCEANOGRAPHY

ABSTRACT SERVICES AND INDEXES

APPLIED SCIENCE AND TECHNOLOGY INDEX. H.W. Wilson and Company, 950 University Avenue, Bronx, NY 10452. (800) 367-6670 or (212) 588-8400. Monthly. Inquire as to cost and availability.

BIBLIOGRAPHY AND INDEX OF GEOLOGY. American Geological Institute, 4220 King Street, Alexandria, VA 22302. (703) 379-2480. 1969 to present. Monthly. $1100.00 per year.

CHEMICAL ABSTRACTS. American Chemical Society, Chemical Abstracts Service, Box 3012, Columbus, OH 43210. (614) 421-3600. 1907 to present. Weekly. $9500.00 per year.

CONFERENCE PAPERS INDEX. Cambridge Scientific Abstracts, 5161 River Road, Bethesda, MD 20816. 1972 to present. Monthly. Inquire as to cost and availability.

CURRENT CONTENTS: ENGINEERING, TECHNOLOGY AND APPLIED SCIENCES. Institute for Scientific Information, 3501 Market Street, Philadelphia, PA 19104. (800) 523-1850 or (215) 386-0100. Weekly. $275.00 per year.

ENGINEERING INDEX MONTHLY AND AUTHOR INDEX. Engineering Information Inc., 345 East 47th Street, New York, NY 10017. (212) 705-7600. Monthly. $1560.00 per year.

GENERAL SCIENCE INDEX. H.W. Wilson and Company, 950 University Avenue, Bronx, NY 10452. (800) 367-6670 or (212) 588-8400. 1978 to present. Monthly. Inquire as to cost and availability.

METEOROLOGICAL AND GEOASTROPHYSICAL ABSTRACTS. American Meteorological Society, 45 Beacon Street, Boston, MA 02108. (617) 227-2425. 1950 to present. Monthly. $450.00 per year.

OCEAN ABSTRACTS. Cambridge Scientific Abstracts, 5161 River Road, Bethesda, MD 20816. (301) 951-1400. 1963 to present. Bimonthly. $450.00 per year.

SCIENCE CITATION INDEX. Institute for Scientific Information, 3501 Market Street, Philadelphia, PA 19104. (800) 523-1850 or (215) 386-0100. Six times per year. $6200.00 per year.

ASSOCIATIONS AND PROFESSIONAL SOCIETIES

AMERICAN SOCIETY OF CIVIL ENGINEERS. 345 East 47th Street, New York, NY 10017. (212) 705-7420.

COASTAL ENGINEERING RESEARCH COUNCIL. 207 East Bay Street, Suite 311, Charleston, SC 29401. (803) 723-4864.

THE COASTAL SOCIETY. 5410 Grosvenor Lane, Suite 110, Bethesda, MD 20814. (301) 897-8616.

GEOLOGICAL SOCIETY OF AMERICA. 3300 Penrose Place, Boulder, CO 80301. (303) 447-2020.

DIRECTORIES AND BIOGRAPHICAL SOURCES

1987 DIRECTORY OF ENGINEERING SOCIETIES AND RELATED ORGANIZATIONS. Gordon Davis, editor. Hemisphere Publishing Corporation, 1010 Vermont Avenue, NW, Washington, DC 20005. (800) 526-0275. 12th edition. 1987. $100.00.

RESEARCH CENTERS DIRECTORY 1988. Gale Research Company, Book Tower, Detroit, MI 48226. (800) 521-0707. 12th edition. 1987. $365.00 for set.

SCIENTIFIC AND TECHNICAL ORGANIZATIONS AND AGENCIES DIRECTORY. Margaret Labash Young, editor. Gale Research Company, Book Tower, Detroit, MI 48226. (800) 521-0707. 2nd edition. 1987. $185.00.

WHO'S WHO IN ENGINEERING. Gordon Davis, editor. Hemisphere Publishing Corporation, 1010 Vermont Avenue, NW, Washington, DC 20005. (800) 526-0275. 6th edition. 1985. $200.00.

ENCYCLOPEDIAS AND DICTIONARIES

THE ENCYCLOPEDIA OF BEACHES AND COASTAL ENVIRONMENTS. M.L. Schwartz, editor. McGraw-Hill Book Company, 1221 Avenue of the Americas, New York, NY 10020. (212) 512-2000. 1983. $110.95.

GENERAL WORKS

COASTS. Eric Bird. Blackwell Scientific Publications, Inc., 52 Beacon Street, Boston, MA 02108. (800) 325-4177. Third edition. 1984. $35.95.

ENVIRONMENTAL OCEANOGRAPHY: AN INTRODUCTION TO THE BEHAVIOUR OF COASTAL WATERS. T. Beer. Pergamon Press, Inc., Maxwell House, Fairview Park, Elmsford, NY 10523. (914) 592-7700. 1983. $13.00 in paper.

PHYSICAL OCEANOGRAPHY OF COASTAL WATERS. K.F. Bowden. John Wiley and Sons, Inc., 605 Third Avenue, New York, NY 10158. (800) 526-5368. 1984. $74.95.

THE WORLD'S COASTLINE. E.C. Bird and M.L. Schwartz, editors. Van Nostrand Reinhold Company, Inc., 135 West 50th Street, New York, NY 10020. (800) 543-2681. 1985. $97.50.

ONLINE DATA BASES

CA SEARCH. Chemical Abstracts Service, P.O. Box 3012, Columbus, OH 43120. (800) 848-6538 or (614) 421-3600. Comprehensive guide to chemical literature, 1972 to present. Inquire as to online cost and availability.

COMPENDEX. Engineering Information, Inc., 345 East 47th Street, New York, NY 10017. (800) 221-1044 or (212) 705-7615. Engineering and technical literature, 1975 to present. Inquire as to online cost and availability.

DISSERTATION ABSTRACTS ONLINE. University Microfilms International, 300 North Zeeb Road, Ann Arbor, MI 48106. (800) 521-0600 or (313) 761-4700. Scope includes virtually all doctoral dissertations accepted at accredited American institutions from 1861 to present in over 250 subject areas. Inquire as to online cost and availability.

GEOREF. Online version of the BIBLIOGRAPHY AND INDEX OF GEOLOGY. American Geological Institute, 4220 King Street, Alexandria, VA 22302. (703) 379-2480. 1969 to present. Inquire as to online cost and availability.

INSPEC. INSPEC Marketing Department, Institution of Electrical Engineers. Available from IEEE Service Center, 445 Hoes Lane, Piscataway, NJ 08854. (201) 981-0060. Online version of Physics Abstracts. Inquire as to online cost and availability.

NTIS. National Technical Information Service, 5285 Port Royal Road, Springfield, VA 22161. (703) 487-4630. Broad coverage of government sponsored research reports, 1964 to present. Inquire as to online cost and availability.

SCISEARCH. Institute for Scientific Information, 3501 Market Street, Philadelphia, PA 19104. (800) 523-1850 or (215) 386-0100. Broad multidisciplinary title and author index to the international literature of science and technology, 1974 to present. Inquire as to online cost and availability.

WILSONLINE. H.W. Wilson and Company, 950 University Avenue, Bronx, NY 10452. (800) 367-6770 or (212) 588-8400. Makes available online versions of the H.W. Wilson indexes including Applied Science and Technology Index, Business Periodicals Index and Readers' Guide to Periodical Literature. Approximately 1980 to present. Inquire as to online cost and availability.

PERIODICALS

COASTAL ENGINEERING: AN INTERNATIONAL JOURNAL FOR COASTAL, HARBOUR, AND OFFSHORE ENGINEERS. Elsevier Science Publishing Company, Inc., 52 Vanderbilt Avenue, New York, NY 10017. (212) 370-5520. 1977 to present. Quarterly. $95.00 per year.

COASTAL RESEARCH. Florida State University, Geology Department, Tallahassee, FL 32306. (904) 644-5860. 1962 to present. Three times per year. $5.00 per year.

COASTAL ZONE MANAGEMENT JOURNAL. Crane Russak and Company, Inc., 3 East 44th Street, New York, NY 10017. (212) 867-1490. 1973 to present. Quarterly. $72.00 per year.

COASTWATCH. University of North Carolina, Sea Grant College Program, Box 8605, North Carolina State University, Raleigh, NC 27695-8605. (919) 737-2454. 1970 to present. Monthly. Free.

CONTINENTAL SHELF RESEARCH. Pergamon Press, Inc., Maxwell House, Fairview Park, Elmsford, NY 10523. (914) 592-7700. 1982 to present. Bimonthly. $1200.00 per year. Includes Oceanographic Literature Review.

ESTUARINE, COASTAL AND SHELF SCIENCE. Academic Press, Inc., 6277 Sea Harbor Drive, Orlando, FL 32821. (800) 321-5068. 1973 to present. Monthly. $315.00 per year.

JOURNAL OF WATERWAY, PORT, COASTAL AND OCEAN ENGINEERING. American Society of Civil Engineers, 345 East 47th Street, New York, NY 10017. (212) 705-7420. 1956 to present. Bimonthly. $72.00 per year.

OCEAN ENGINEERING. Pergamon Press, Inc., Maxwell House, Fairview Park, Elmsford, NY 10523. (914) 592-7700. 1968 to present. Bimonthly. $200.00 per year.

OCEAN SCIENCE AND ENGINEERING. Marcel Dekker Inc., 270 Madison Avenue, New York, NY 10016. (800) 228-1160. 1974 to present. Quarterly. $165.00 per year.

U.S. COASTAL ENGINEERING RESEARCH CENTER. QUARTERLY CIRCULAR INFORMATION BULLETIN. U.S. Coastal Engineering Research Center, Kingman Building, Fort Belvoir, VA 22060. (703) 325-7429. Quarterly. Inquire.

RESEARCH CENTERS AND INSTITUTES

COASTAL AND OCEANOGRAPHIC ENGINEERING LABORATORY. University of Florida, 336 Weil Hall, Gainesville, FL 32607. (904) 392-1436.

COASTAL ENGINEERING RESEARCH COUNCIL. 207 East Bay Street, Suite 311, Charleston, SC 29401. (803) 723-4864.

COASTAL LABORATORY. Queen's University at Kingston, Ellis Hall, Kingston, ON, Canada K7L 3N6. (613) 545-6265.

COBALT

See also: ALLOYS, ELEMENTS, FERROALLOYS, INORGANIC CHEMISTRY, METALLURGICAL ENGINEERING, METALS AND METALWORKING

ABSTRACT SERVICES AND INDEXES

ALLOYS INDEX. American Society for Metals, Metals Park, OH 44073. (216) 338-5151. 1974 to present. Monthly. $225.00.

APPLIED SCIENCE AND TECHNOLOGY INDEX. H.W. Wilson and Company, 950 University Avenue, Bronx, NY 10452. (800) 367-6670 or (212) 588-8400. Monthly. Inquire as to cost and availability.

CHEMICAL ABSTRACTS. American Chemical Society, Chemical Abstracts Service, Box 3012, Columbus, OH 43210. (614) 421-3600. 1907 to present. Weekly. $9500.00 per year.

CURRENT CONTENTS: ENGINEERING, TECHNOLOGY AND APPLIED SCIENCES. Institute for Scientific Information, 3501 Market Street, Philadelphia, PA 19104. (800) 523-1850 or (215) 386-0100. Weekly. $275.00 per year.

ENGINEERING INDEX MONTHLY AND AUTHOR INDEX. Engineering Information Inc., 345 East 47th Street, New York, NY 10017. (212) 705-7600. Monthly. $1560.00 per year.

ISMEC BULLETIN (Information Service in Mechanical Engineering). Cambridge Scientific Abstracts, 5161 River Road, Bethesda, MD 20816. (301) 951-1400. 1973 to present. Monthly. $450.00 per year.

LEAD ABSTRACTS. Lead Development Association, 34 Berkeley Square, London, W1X 6AJ, England. 1958 to present. Quarterly. $70.00 per year.

METALS ABSTRACTS AND METALS ABSTRACTS INDEX. American Society for Metals, Metals Park, OH 44073. (216) 338-5151. 1968 to present. Monthly. Abstracts are $1100.00 per year and Index is $460.00 per year.

PHYSICS ABSTRACTS. Institution of Electrical Engineers. Available from: IEEE Service Center, 445 Hoes Lane, Piscataway, NJ 08854. 1898 to present. Bimonthly. $1700.00 per year.

SCIENCE CITATION INDEX. Institute for Scientific Information, 3501 Market Street, Philadelphia, PA 19104. (800) 523-1850 or (215) 386-0100. Six times per year. $6200.00 per year.

WORLD ALUMINUM ABSTRACTS. Aluminum Association, 818 Connecticut Avenue, NW, Washington, DC 20006. (202) 862-5156. 1968 to present. Monthly. $240.00 per year.

ASSOCIATIONS AND PROFESSIONAL SOCIETIES

AMERICAN INSTITUTE OF MINING, METALLURGICAL AND PETROLEUM ENGINEERS (AIME). 420 Commonwealth Drive, Warrendale, PA 15086. (412) 776-9086.

AMERICAN POWDER METALLURGY INSTITUTE. 105 College Road, East, Princeton, NJ 08540. (609) 452-7700.

AMERICAN SOCIETY FOR METALS. Metals Park, OH 44073. (216) 338-5151.

AMERICAN SOCIETY FOR TESTING AND MATERIALS. 1916 Race Street, Philadelphia, PA 19103. (215) 299-5400.

AMERICAN SOCIETY OF MECHANICAL ENGINEERS. 345 47th Street, New York, NY 10017. (212) 705-7722.

ASSOCIATION OF IRON AND STEEL ENGINEERS. Three Gateway Center, Suite 2350, Pittsburgh, PA 15222. (412) 281-6323.

THE METALLURGICAL SOCIETY. 420 Commonwealth Drive, Warrendale, PA 15086. (412) 776-9000.

NATIONAL ASSOCIATION OF CORROSION ENGINEERS. Box 218340, Houston, TX 77218. (713) 492-0535.

DIRECTORIES AND BIOGRAPHICAL SOURCES

DUN'S INDUSTRIAL GUIDE: THE METALWORKING DIRECTORY. Dun and Bradstreet Corporation, 49 Old Bloomfield Road, Mountain Lakes, NJ 07046. (201) 953-0300. Annual. $610.00.

INDUSTRIAL EQUIPMENT AND SUPPLIES DIRECTORY. American Business Directories, Inc., 5707 South 86th Circle, Omaha, NE 68127. (402) 331-7169.

METALLURGICAL SOCIETY OF AIME - MEMBERSHIP LIST. American Institute of Mining, Metallurgical and Petroleum Engineers (AIME). 345 East 47th Street, New York, NY 10017. (212) 705-7695. 1984.

1987 DIRECTORY OF ENGINEERING SOCIETIES AND RELATED ORGANIZATIONS. Gordon Davis, editor. Hemisphere Publishing Corporation, 1010 Vermont Avenue, NW, Washington, DC 20005. (800) 526-0275. 12th edition. 1987. $100.00.

RESEARCH CENTERS DIRECTORY 1988. Gale Research Company, Book Tower, Detroit, MI 48226. (800) 521-0707. 12th edition. 1987. $365.00 for set.

SCIENTIFIC AND TECHNICAL ORGANIZATIONS AND AGENCIES DIRECTORY. Margaret Labash Young, editor. Gale Research Company, Book Tower, Detroit, MI 48226. (800) 521-0707. 2nd edition. 1987. $185.00.

WHO'S WHO IN ENGINEERING. Gordon Davis, editor. Hemisphere Publishing Corporation, 1010 Vermont Avenue, NW, Washington, DC 20005. (800) 526-0275. 6th edition. 1985. $200.00.

ENCYCLOPEDIAS AND DICTIONARIES

THESAURUS OF SCIENTIFIC, TECHNICAL, AND ENGINEERING TERMS. Hemisphere Publishing Corporation, 1010 Vermont Avenue, NW, Washington, DC 20005. (800) 526-0275. 1988. $125.00.

GENERAL WORKS

COBALT AND ITS ALLOYS. W. Betteridge. John Wiley and Sons, Inc., 605 Third Avenue, New York, NY 10158. (800) 526-5368. 1982. $57.95.

ESSENTIAL METALLURGY FOR ENGINEERS. William Alexander. Van Nostrand Reinhold Company, Inc., 135 West 50th Street, New York, NY 10020. (800) 543-2681. 1985. $17.95.

METALLURGY BASICS. Donald V. Brown. Van Nostrand Reinhold Company, Inc., 135 West 50th Street, New York, NY 10020. (800) 543-2681. 1983. $19.95 in paper.

METALS ENGINEERING: A TECHNICAL GUIDE. L.E. Samuals. American Society for Metals, Metals Park, OH 44073. (216) 338-5151. 1988. $68.00.

HANDBOOKS AND MANUALS

HANDBOOK OF METAL FORMING. Kurt Lange, editor. John Wiley and Sons, Inc., 605 Third Avenue, New York, NY 10158. (800) 526-5368. 1985. $89.50.

HANDBOOK OF METAL FORMING PROCESSES. B. Avitzur. John Wiley and Sons, Inc., 605 Third Avenue, New York, NY 10158. (800) 526-5368. 1983. $105.00.

METALS HANDBOOK. American Society for Metals, Metals Park, OH 44073. (216) 338-5151. 9th edition. 14 volumes. 1988. $1310.00 for set.

WOLDMAN'S ENGINEERING ALLOYS. Robert C. Gibbons, editor. American Society for Metals, Metals Park, OH 44073. (216) 338-5151. 6th edition. 1979. $112.00.

ONLINE DATA BASES

CA SEARCH. Chemical Abstracts Service, P.O. Box 3012, Columbus, OH 43120. (800) 848-6538 or (614) 421-3600. Comprehensive guide to chemical literature, 1972 to present.

Inquire as to online cost and availability.

COMPENDEX. Engineering Information, Inc., 345 East 47th Street, New York, NY 10017. (800) 221-1044 or (212) 705-7615. Engineering and technical literature, 1975 to present. Inquire as to online cost and availability.

INSPEC. INSPEC Marketing Department, Institution of Electrical Engineers. Available from IEEE Service Center, 445 Hoes Lane, Piscataway, NJ 08854. (201) 981-0060. Online version of Physics Abstracts. Inquire as to online cost and availability.

ISMEC. Cambridge Scientific Abstracts, 5161 River Road, Besthda, MD 20816. (800) 638-8076 or (301) 951-1400. Literature of mechanical and production engineering, 1973 to present. Inquire as to online cost and availability.

NTIS. National Technical Information Service, 5285 Port Royal Road, Springfield, VA 22161. (703) 487-4630. Broad coverage of government sponsored research reports, 1964 to present. Inquire as to online cost and availability.

SCISEARCH. Institute for Scientific Information, 3501 Market Street, Philadelphia, PA 19104. (800) 523-1850 or (215) 386-0100. Broad multidisciplinary title and author index to the international literature of science and technology, 1974 to present. Inquire as to online cost and availability.

WILSONLINE. H.W. Wilson and Company, 950 University Avenue, Bronx, NY 10452. (800) 367-6770 or (212) 588-8400. Makes available online versions of the H.W. Wilson indexes including Applied Science and Technology Index, Business Periodicals Index and Readers' Guide to Periodical Literature. Approximately 1980 to present. Inquire as to online cost and availability.

PERIODICALS

ALLOY DIGEST. Engineering Alloys Digest, Inc., Box 823, Upper Montclair, NJ 07043 . (201) 746-7930. 1952 to present. Monthly. $50.00 per year.

IRON AND STEEL ENGINEER. Association of Iron and Steel Engineers, Suite 2350, Three Gateway Center, Pittsburgh, PA 15222. (412) 281-6323. 1924 to present. Monthly. $34.00 per year.

JOURNAL OF METALS. American Institute of Mining, Metallurgical, and Petroleum Engineers, Inc., Metallurgical Society, 420 Commonwealth Drive, Warrendale, PA 15086. (412) 776-9086. 1949 to present. Monthly. $40.00 per year.

METAL PROGRESS. American Society for Metals, Metals Park, OH 44073. (216) 338-5151. 1930 to present. Monthly. $40.00.

METALLURGICAL TRANSACTIONS. Metallurgical Society of AIME, 420 Commonwealth Drive, Warrendale, PA 15086. (412) 776-9080. 1970 to present. Monthly. $95.00 per year.

METALS WEEK. McGraw-Hill Book Company, 1221 Avenue of the Americas, New York, NY 10020. (212) 997-2823. 1930 to present. Weekly. $597.00 per year.

METALWORKING DIGEST. Philos Publications, Inc., 1 East Chase Street, Baltimore, MD 21202. (301) 361-9060. 1969 to present. Nine times per year. $9.00 per year.

METIFAX MAGAZINE. Huebner Publications Inc., 6521 Davis Industrial Parkway, Solon, OH 44139. (216) 248-1125. 1956 to present. Monthly. $40.00 per year.

RESEARCH CENTERS AND INSTITUTES

COOPERATIVE PROGRAM IN METALLURGY. Pennsylvania State University, 208A Steidle Building, University Park, PA 16802. (814) 865-5446.

DEPARTMENT OF MATERIALS SCIENCES AND ENGINEERING. University of Florida, Gainesville, FL 32601.
(904) 392-1454.

INSTITUTE OF MATERIALS SCIENCE. University of Connecticut, Storrs, CT 06268. (203) 486-4623.

COBOL

See: PROGRAMMING LANGUAGES

CODES AND CIPHERS

See: COMPUTER SECURITY

COFFERDAM

See: DAMS

COKE AND COKING

See: STEEL AND STEEL MAKING

COLLOIDS

See also: CATALYSIS, CHEMICAL BONDING, CHEMISTRY, ELECTROCHEMISTRY, PHOTO-CHEMISTRY, PHYSICAL CHEMISTRY, POLYMERS, SURFACE CHEMISTRY

ABSTRACT SERVICES AND INDEXES

APPLIED SCIENCE AND TECHNOLOGY INDEX. H.W. Wilson Company, 950 University Avenue, Bronx, NY 10452. (800) 367-6670 or (212) 588-8400. Inquire as to cost and availability.

CHEMICAL ABSTRACTS. Chemical Abstracts Service, 1540 Olentangy Road, Post Office Box 3012, Columbus, OH 43210. (800) 848-6538 or (614) 421-3600. Weekly. $9200.00 per year.

GENERAL SCIENCE INDEX. H.W. Wilson Company, 950 University Avenue, Bronx, NY 10452. (800) 367-6770 or (212) 588-8400. Inquire as to cost and availability.

PHYSICS ABSTRACTS. Institute of Electrical Engineers, London United Kingdom. Available from: Institute of Electrical and Electronic Engineers (IEEE), 345 East 47th Street, New York, NY 10017. (212) 705-7900.

SCIENCE CITATION INDEX. Institute for Scientific Information, 3501 Market Street, Philadelphia, PA 19104. (800) 523-1850 or (215) 386-0100.

ANNUAL REVIEWS AND YEARBOOKS

ANNUAL REVIEW OF PHYSICAL CHEMISTRY. Annual Reviews Incorporated, 4139 El Camino Way, Palo Alto, CA 94306. (415) 493-4400. Annual.

ASSOCIATIONS AND PROFESSIONAL SOCIETIES

AMERICAN CHEMICAL SOCIETY. 1155 16th Street, NW, Washington, DC 20036. (202) 872-4600.

ASSOCIATION OF CONSULTING CHEMISTS AND CHEMICAL ENGINEERS. 50 East 41st Street, Suite 92, New York, NY 10017. (212) 684-6255.

DIVISION OF PHYSICAL CHEMISTRY (A DIVISION OF THE AMERICAN CHEMICAL SOCIETY). c/o Dr. James Kinsey, Chemistry Department, Room 6-215, Massachusetts Institute of Technology, Cambridge, MA 02139.

BIBLIOGRAPHIES

SCIENCE BOOKS AND FILMS. American Association for the Advancement of Science, 1333 H Street, NW, Washington, DC 20005.

SCIENTIFIC AND TECHNICAL BOOKS AND SERIALS IN PRINT 1988: AN INDEX TO LITERATURE IN SCIENCE AND TECHNOLOGY. R.R. Bowker Company, 205 East 42nd Street, New York, NY 10017. (800) 521-8110 or (212) 916-1600. $175.00.

DIRECTORIES AND BIOGRAPHICAL SOURCES

AMERICAN INSTITUTE OF CHEMISTS. American Institute of Chemists, 7315 Wisconsin Avenue, Bethesda, MD 20814. (301) 652-2447. 1986. $35.00.

AMERICAN MEN AND WOMEN OF SCIENCE. Physical and Biological Sciences. R.R. Bowker Company, 205 East 42nd Street, New York, NY 10017. (800) 521-8110 or (212) 916-1600. Fifteenth edition. $565.00.

BIOGRAPHICAL DICTIONARY OF SCIENTISTS: CHEMISTS. David Abbott, editor. P. Bedrick Books, 125 East 23rd Street, New York, NY 10010. (212) 777-1187. 1984. $18.95.

CHEMICAL WEEK - BUYERS GUIDE ISSUE. McGraw-Hill Book Company, 1221 Avenue of the Americas, New York, NY 10020. (800) 628-0004. Annual, October. $50.00.

CONSULTING SERVICES: CHEMISTS AND CHEMICAL ENGINEERS. Association of Consulting Chemists and Chemical Engineers, 50 East 41st Street, New York, NY 10017. (212) 684-6255. Annual. 1986. $45.00.

GOVERNMENT RESEARCH DIRECTORY. Gale Research Company, Book Tower, Detroit, MI 48226. (800) 521-0707. Fourth edition. 1987. $350.00.

INTERNATIONAL RESEARCH CENTERS DIRECTORY 1986-1987. Gale Research Company, Book Tower, Detroit, MI 48226. (800) 521-0707. Third edition. 1986. $330.00.

RESEARCH CENTERS DIRECTORY. Gale Research Company, Book Tower, Detroit, MI 48226. (800) 521-0707. Eleventh edition. 1987. $355.00.

RESEARCH SERVICES DIRECTORY. Robert J. Huffman and Mary M. Watkins, editors. Gale Research Company, Book Tower, Detroit, MI 48226. (800) 521-0707. Third edition. $290.00.

SCIENTIFIC AND TECHNICAL ORGANIZATIONS AND AGENCIES DIRECTORY. Gale Research Company, Book Tower, Detroit, MI 48226. (800) 521-0707. 1985. $150.00.

WHO'S WHO IN FRONTIER SCIENCE AND TECHNOLOGY. Marquis Who's Who, Incorporated, 200 East Ohio Street, Chicago, IL 60611. (800) 428-3898 or (312) 787-2008.

WHO'S WHO IN TECHNOLOGY TODAY. Reston Publishing Company, Incorporated. c/o Prentice-Hall, Incorporated, Englewood Cliffs, NJ 07632. (800) 262-6868. Biennial. Five volumes. $425.00. Covers the fields of electronics, computer science, physics, optics, chemistry, biotechnology, mechanics, energy, and earth science.

WORLD GUIDE TO SCIENTIFIC ASSOCIATIONS AND LEARNED SOCIETIES. K.G. Saur Incorporated, 175 Fifth Avenue, New York, NY 10010. (800) 521-0707 or (212) 982-1302. Fourth edition. 1984. $112.00.

ENCYCLOPEDIAS AND DICTIONARIES

CONCISE ENCYCLOPEDIA OF CHEMICAL TECHNOLOGY. Kirk-Othmer. John Wiley and Sons, Incorporated, 605 Third Avenue, New York, NY 10158. (800) 526-5368 or (212) 850-6000. Third edition. 1985. $129.95.

CONDENSED CHEMICAL DICTIONARY. Gessner Hawley. Van Nostrand Reinhold, 115 Fifth Avenue, New York, NY 10003. Tenth edition. 1981. $49.95.

ENCYCLOPEDIA OF PHYSICAL SCIENCE AND TECHNOLOGY. Academic Press, Incorporated, Orlando, FL 32887. (800) 321-5068 or (305) 345-2734. Fifteen volumes, 1986.

GLOSSARY OF CHEMICAL TERMS. Clifford A. Hampel and Gessner G. Hawley. Van Nostrand Reinhold Company, 115 Fifth Avenue, New York, NY 10003. (800) 543-2681 or (212) 254-3232. Second edition. 1982. $21.95.

HAWLEY'S CONDENSED CHEMICAL DICTIONARY. N. Irving Sax and Richard J. Lewis. Sr., editors. Van Nostrand Reinhold, Incorporated, 115 Fifth Avenue, New York, NY 10003. (800) 543-2681. Eleventh edition. 1987. $52.95.

MCGRAW-HILL ENCYCLOPEDIA OF SCIENCE AND TECHNOLOGY. McGraw-Hill Book, Incorporated, 1221 Avenue of the Americas, New York, NY 10020. (212) 997-3675.

VAN NOSTRAND REINHOLD ENCYCLOPEDIA OF CHEMISTRY. Douglas M. Considine and Glenn D. Considine. Van Nostrand Reinhold Publishing Company, Incorporated, 115 Fifth Avenue, New York, NY 10003. (800) 543-2681 or (212) 254-3232. 1984. $97.95.

GENERAL WORKS

CHEMISTRY OF THE ELEMENTS. N.N. Greenwood and A. Earnshaw. Pergamon Publishing, Incorporated, Maxwell House, Fairview Park, Elmsford, NY 10523. (914) 592-7700. 1984. $143.00.

COLLOID PHENOMENA: ADVANCED TOPICS. C.S. Hirtzel and Raj Rajagopalan. Noyes Data Corporation, Mill Road at Grand Avenue, Park Ridge, NJ 07656. 1985. $36.00.

INORGANIC CHEMISTRY: PRINCIPLES OF STRUCTURE AND REACTIVITY. James E. Huheey. Harper and Row Publishers Incorporated, 10 East 53rd Street, New York, NY 10022. (212) 207-7655. Third edition. $36.50.

INTRODUCTION TO COLLOID AND SURFACE CHEMISTRY. D.J. Shaw. Butterworths Publishing, 80 Montvale Avenue, Stoneham, MA 02180. (617) 720-0761.

INTRODUCTION TO PHYSICAL CHEMISTRY. M.F.C. Ladde and W.H. Lee. Cambridge University Press, 32 East 57th Street, New York, NY 10022. (212) 688-8885. 1986. $19.95 in paper.

MODERN TRENDS OF COLLOID SCIENCE IN CHEMISTRY AND BIOLOGY. Hans-Friedrich Eicke, editor. Birkhauser Boston, Incorporated, 380 Green Street, Cambridge, MA 02139. (617) 876-2333. 1985. $34.95.

PHYSICAL CHEMISTRY. Robert A. Alberty. John Wiley and Sons, Incorporated, 605 Third Avenue, New York, NY 10158. (800) 526-5368 or (212) 850-6000. 1987. Inquire as to cost and availability.

SCIENCE AND TECHNOLOGY OF POLYMER COLLOIDS. R. Buscall and others, editors. Elsevier Science Publishing Company, Incorporated, 52 Vanderbilt Avenue, New York, NY 10017. (212) 370-5520. 1985. $63.00.

HANDBOOKS AND MANUALS

THE CHEMIST'S COMPANION: A HANDBOOK OF PRACTICAL DATA, TECHNIQUES, AND REFERENCES. Arnold J. Gordon and Richard A. Ford. John Wiley and Sons, Incorporated, 605 Third Avenue, New York, NY 10158. (800) 526-5368. 1973. $49.95.

CRC HANDBOOK OF CHEMISTRY AND PHYSICS. CRC Press, Incorporated, 2000 Corporate Boulevard, NW, Boca Raton, FL 33431. Sixty-seventh edition. 1986. $69.95.

HANDBOOK OF APPLIED CHEMISTRY: FACTS FOR ENGINEERS, SCIENTISTS, TECHNICIANS, AND TECHNICAL MANAGERS. Vollrath Hopp and Ingp Hennig. McGraw-Hill Book Company, 1221 Avenue of the Americas, New York, NY 10020. (800) 628-0004. 1983. $54.00.

HANDBOOK OF COMPUTATIONAL CHEMISTRY: A PRACTICAL GUIDE TO CHEMICAL STRUCTURE AND ENERGY CALCULATIONS. Tim Clark. John Wiley and Sons, Incorporated, 605 Third Avenue, New York, NY 10158. (800) 526-5368 or (212) 850-6000. 1985. $35.00.

LANGE'S HANDBOOK OF CHEMISTRY. John A. Dean, editor. McGraw-Hill Book Company, 1221 Avenue of the Americas, New York, NY 10020. (800) 628-0004. 1985. $59.50

TABLES OF PHYSICAL AND CHEMICAL CONSTANTS; AND SOME MATHEMATICAL FUNCTIONS. G.W.C. Kaye and T.H. Laby, editors. Longman, Incorporated, 95 Church Street, White Plains, NY 10601. (014) 993-5000. 1986. $39.95.

ONLINE DATA BASES

CA SEARCH. Chemical Abstracts Service, Post Office Box 3012, Columbus, OH 43210. Guide to chemical literature, 1972 to present. Inquire as to cost and availability.

DISSERTATION ABSTRACTS ONLINE. University Microfilms International, 300 North Zeeb Road, Ann Arbor, MI 48106. (800) 521-0600 or (313) 761-4700. Scope includes virtually all doctoral dissertations accepted at accredited American institutions from 1861 to present in 252 subject areas. Inquire as to cost and availability.

INSPEC. INSPEC Marketing Department, Institute of Electrical and Electronics Engineers, Incorporated, IEEE Service Department, 445 Hoes Lane, Piscataway, NJ 08854. (201) 981-0060. Inquire as to on-line cost and availability.

NTIS. National Technical Information Service, 5285 Port Royal Road, Springfield, VA 22161. (703) 487-4630. Broad coverage of government sponsored research reports 1964 to present. Inquire as to cost and availability.

SCISEARCH. Institute for Scientific Information, 3501 Market Street, Philadelphia, PA 19104. (800) 523-1850 or (215) 386-0100. Broad multidisciplinary title and author index to the present. Inquire as to cost and availability.

OTHER SOURCES

GUIDE TO BASIC INFORMATION SOURCES IN CHEMISTRY. Arthur Antony. John Wiley and Sons, Incorporated, 605 Third Avenue, New York, NY 10158. (800) 526-5368 or (212) 850-6000. 1979. $26.95.

HOW TO FIND CHEMICAL INFORMATION: A GUIDE FOR PRACTICING CHEMISTS, TEACHERS, AND STUDENTS. John Wiley and Sons, Incorporated, 605 Third Avenue, New York, NY 10158. (800) 526-5368 or (212) 850-6000. 1986. $35.00.

PERIODICALS

ADVANCES IN COLLOID AND INTERFACE SCIENCE. Elsevier Science Publishers B.V., Box 211, 1000 AE Amsterdam, The Netherlands. Eight times per year. $190.00 per year.

ANDEWANDTE CHEMIE (GESELLSCHAFT DEUTSCHER CHEMIKER, GW). V C H Verlagsgessellschaft mbH, Pappelallee 3, Postfach 1260, 6940 Weinheim, West Germany. Monthly. $300.00 per year.

CHEMICAL PHYSICS. Elsevier Science Publishers B.V., Box 211, 1000 AE Amsterdam, The Netherlands. Thirty times per year. $1200.00 per year.

CHEMICAL PHYSICS LETTERS. Elsevier Science Publishers B.V., Box 211, 1000 AE Amsterdam, The Netherlands. Sixty-six times per year. $1500.00 per year.

COLLOID AND POLYMER SCIENCE. Dr. Dietrich Steinkopff Verlag, Saalbaustrr 12, Postfach, Darmstadt 11, West Germany. Monthly. $355.00 per year.

COLLOIDS AND SURFACE. Elsevier Science Publishers B.V., Box 211, 1000 AE Amsterdam, The Netherlands. Twenty-four times per year. $540.00 per year.

INORGANIC CHEMISTRY. American Chemical Society, 1155 Sixteenth Street, NW, Washington, DC 20036. (800) 424-6747 or (202) 872-4700. Monthly. $400.00 per year.

INTERNATIONAL REVIEWS IN PHYSICAL CHEMISTRY. Taylor and Francis, Limited, 242 Cherry Street, Philadelphia, PA 19106-1906. Three times per year. $143.00 per year.

JOURNAL OF CATALYSIS. Academic Press, Incorporated, Journals Division, 1250 Sixth Avenue, San Diego, CA 92101. (619) 230-1840. Monthly. $654.00 per year.

JOURNAL OF COLLOID AND INTERFACE SCIENCE. Academic Press, Incorporated, Journals Division, 1250 Sixth Avenue, San Diego, CA 92101. (619) 230-1840. Monthly. $667.00 per year.

JOURNAL OF PHYSICAL CHEMISTRY. American Chemical Society, 1155 Sixteenth Street, NW, Washington, DC 20036. (800) 424-6747 or (202) 872-4700. Biweekly. $369.00 per year.

JOURNAL OF SOLID STATE CHEMISTRY. Academic Press, Incorporated, Journals Division, 1250 Sixth Avenue, San Diego, CA 92101. (619) 230-1840. Fifteen times per year. $530.00 per year.

JOURNAL OF THE AMERICAN CHEMICAL SOCIETY. American Chemical Society, 1155 Sixteenth Street, NW, Washington, DC 20036. (800) 424-6747 or (202) 872-4700. Biweekly. $330.00 per year.

RESEARCH CENTERS AND INSTITUTES

BROWN UNIVERSITY. Chemical Research Laboratory, Providence, RI 02912. (401) 863-2256.

CLARKSON UNIVERSITY. Institute of Colloid and Surface Science, Potsdam, NY 13676. (315) 268-2353.

COATINGS INDUSTRY EDUCATION FUNDS. 1315 Walnut Street, Philadelphia, PA 19107. (215) 545-1507.

LEHIGH UNIVERSITY. Center for Surface and Coatings Research, Bethlehem, PA 18015. (215) 861-3570.

UNIVERSITY OF MICHIGAN. Colloid Stability Laboratorym, 3095 Dow Building, Ann Arbor, MI 48109. (313) 764-4313.

UNIVERSITY OF TEXAS AT AUSTIN. Materials Research Group, Department of Chemistry, Austin, TX 78712. (512) 471-3704.

COLOR

See also: CAMERAS, OPTICS, PHOTOCHEMISTRY, PHOTOGRAPHIC FILM, PHOTOGRAPHY, SPECTROSCOPY

ABSTRACT SERVICES AND INDEXES

APPLIED SCIENCE AND TECHNOLOGY INDEX. H.W. Wilson and Company, 950 University Avenue, Bronx, NY 10452. (800) 367-6670 or (212) 588-8400. Monthly. Inquire as to cost and availability.

CHEMICAL ABSTRACTS. American Chemical Society, Chemical Abstracts Service, Box 3012, Columbus, OH 43210. (614) 421-3600. 1907 to present. Weekly. $9500.00 per year.

ENGINEERING INDEX MONTHLY AND AUTHOR INDEX. Engineering Information Inc., 345 East 47th Street, New York, NY 10017. (212) 705-7600. Monthly. $1560.00 per year.

GENERAL SCIENCE INDEX. H.W. Wilson and Company, 950 University Avenue, Bronx, NY 10452. (800) 367-6670 or (212) 588-8400. 1978 to present. Monthly. Inquire as to cost and availability.

PHOTOGRAPHIC ABSTRACTS. Royal Photographic Society of Great Britain, Scientific and Technical Group, 62 Chelmsford Road, Shenfield, Brentwood, Essex, England. 1921 to present. Six times per year. $140.00 per year.

PHYSICS ABSTRACTS. Institution of Electrical Engineers. Available from: IEEE Service Center, 445 Hoes Lane, Piscataway, NJ 08854. 1898 to present. Bimonthly. $1700.00 per year.

SCIENCE CITATION INDEX. Institute for Scientific Information, 3501 Market Street, Philadelphia, PA 19104. (800) 523-1850 or (215) 386-0100. Six times per year. $6200.00 per year.

ASSOCIATIONS AND PROFESSIONAL SOCIETIES

AMERICAN INSTITUTE OF PHYSICS. 335 East 45th Street, New York, NY 10017. (212) 661-9494.

OPTICAL SOCIETY OF AMERICA. 1816 Jefferson Place, N.W., Washington, DC 20036. (202) 223-8130.

SOCIETY OF PHOTOGRAPHIC SCIENTISTS AND ENGINEERS. 7003 Kilworth Lane, Springfield, VA 22151. (703) 642-9090.

SPIE - THE INTERNATIONAL SOCIETY FOR OPTICAL ENGINEERING. P.O. Box 10, 1022 19th Street, Bellingham, WA 98227. (206) 676-3290.

BIBLIOGRAPHIES

NEW TECHNICAL BOOKS: A SELECTIVE LIST WITH DESCRIPTIVE ANNOTATIONS. New York Public Library, Science and Technology Research Center, Fifth Avenue and 42nd Street, New York, NY 10018. (212) 930-0800. 1915 to present. Monthly. $15.00 per year.

SCIENTIFIC AND TECHNICAL BOOKS AND SERIALS IN PRINT 1988; AN INDEX TO LITERATURE IN SCIENCE AND TECHNOLOGY. R.R. Bowker Company, 205 East 42nd Street, New York, NY 10017. (800) 521-8110. $175.00.

DIRECTORIES AND BIOGRAPHICAL SOURCES

INTERNATIONAL RESEARCH CENTERS DIRECTORY 1988-89. Darren L. Smith, editor. Gale Research Company, Book Tower, Detroit, MI 48226. (800) 521-0707. 4th edition. 1987. $360.00.

1987 DIRECTORY OF ENGINEERING SOCIETIES AND RELATED ORGANIZATIONS. Gordon Davis, editor. Hemisphere Publishing Corporation, 1010 Vermont Avenue, NW, Washington, DC 20005. (800) 526-0275. 12th edition. 1987. $100.00.

RESEARCH CENTERS DIRECTORY 1988. Gale Research Company, Book Tower, Detroit, MI 48226. (800) 521-0707. 12th edition. 1987. $365.00 for set.

SCIENTIFIC AND TECHNICAL ORGANIZATIONS AND AGENCIES DIRECTORY. Margaret Labash Young, editor. Gale Research Company, Book Tower, Detroit, MI 48226. (800) 521-0707. 2nd edition. 1987. $185.00.

WHO'S WHO IN ENGINEERING. Gordon Davis, editor. Hemisphere Publishing Corporation, 1010 Vermont Avenue, NW, Washington, DC 20005. (800) 526-0275. 6th edition. 1985. $200.00.

ENCYCLOPEDIAS AND DICTIONARIES

DICTIONARY OF THE PHYSICAL SCIENCES: TERMS, FORMULAS, DATA. Cesare Emiliani. Oxford University Press, 200 Madison Avenue, New York, NY 10016. (800) 458-5833. 1987. $19.95 in paper.

MCGRAW-HILL ENCYCLOPEDIA OF SCIENCE AND TECHNOLOGY. McGraw-Hill Book Company, 1221 Avenue of the Americas, New York, NY 10020. (212) 512-2000. 6th edition. 1987. $1600.00.

THESAURUS OF PHOTOGRAPHIC SCIENCE AND ENGINEERING. Society of Photographic Scientists and Engineers. Books on Demand, 300 North Zeeb Road, Ann Arbor, MI 48106. (313) 761-4700. $34.50 in paper.

THESAURUS OF SCIENTIFIC, TECHNICAL, AND ENGINEERING TERMS. Hemisphere Publishing Corporation, 1010 Vermont Avenue, NW, Washington, DC 20005. (800) 526-0275. 1988. $125.00.

GENERAL WORKS

COLOR SCIENCE: CONCEPTS AND METHODS, QUANTITATIVE DATA AND FORMULAE. Gunter Wyszecki and W.S. Stiles. John Wiley and Sons, Inc., 605 Third Avenue, New York, NY 10158. (800) 526-5368. Second edition. 1982. $87.95.

LIGHT AND COLOR. R.D. Overheim and D.L. Wagner. John Wiley and Sons, Inc., 605 Third Avenue, New York, NY 10158. (800) 526-5368. 1982. $28.50.

PHOTOGRAPHIC SCIENCE. Earl. N. Mitchell. John Wiley and Sons, Inc., 605 Third Avenue, New York, NY 10158. (800) 526-5368. 1984. $37.50.

PHOTOGRAPHY FOR THE SCIENTIST. Richard A. Morton. Academic Press, Inc., 6277 Sea Harbor Drive, Orlando, FL 32821. (800) 321-5068. 2nd edition. 1984. $102.50.

THE PHYSICS AND CHEMISTRY OF COLOR: THE FIFTEEN CAUSES OF COLOR. Kurt Nassau. John Wiley and Sons, Inc., 605 Third Avenue, New York, NY 10158. (800) 526-5368. 1983. $48.50.

HANDBOOKS AND MANUALS

HANDBOOK OF PHOTOGRAPHIC SCIENCE AND ENGINEERING. Society of Photographic Scientists and Engineers. 7003 Kilworth Lane, Springfield, VA 22151. (703) 642-9090. Inquire.

KODAK PROFESSIONAL PHOTOGUIDE. Carolyn Grimes, editor. Eastman Kodak Company, 343 State Street, Rochester, NY 14650. (716) 724-4000. 1986. $19.95 in paper.

ONLINE DATA BASES

CA SEARCH. Chemical Abstracts Service, P.O. Box 3012, Columbus, OH 43120. (800) 848-6538 or (614) 421-3600. Comprehensive guide to chemical literature, 1972 to present. Inquire as to online cost and availability.

COMPENDEX. Engineering Information, Inc., 345 East 47th Street, New York, NY 10017. (800) 221-1044 or (212) 705-7615. Engineering and technical literature, 1975 to present. Inquire as to online cost and availability.

DISSERTATION ABSTRACTS ONLINE. University Microfilms International, 300 North Zeeb Road, Ann Arbor, MI 48106. (800) 521-0600 or (313) 761-4700. Scope includes virtually all doctoral dissertations accepted at accredited American institutions from 1861 to present in over 250 subject areas. Inquire as to online cost and availability.

INSPEC. INSPEC Marketing Department, Institution of Electrical Engineers. Available from IEEE Service Center, 445 Hoes Lane, Piscataway, NJ 08854. (201) 981-0060. Online version of Physics Abstracts. Inquire as to online cost and availability.

NTIS. National Technical Information Service, 5285 Port Royal Road, Springfield, VA 22161. (703) 487-4630. Broad coverage of government sponsored research reports, 1964 to present. Inquire as to online cost and availability.

SCISEARCH. Institute for Scientific Information, 3501 Market Street, Philadelphia, PA 19104. (800) 523-1850 or (215) 386-0100. Broad multidisciplinary title and author index to the international literature of science and technology, 1974 to present. Inquire as to online cost and availability.

WILSONLINE. H.W. Wilson and Company, 950 University Avenue, Bronx, NY 10452. (800) 367-6770 or (212) 588-8400. Makes available online versions of the H.W. Wilson indexes including Applied Science and Technology Index, Business Periodicals Index and Readers' Guide to Periodical Literature. Approximately 1980 to present. Inquire as to online cost and availability.

PERIODICALS

APPLIED OPTICS. Optical Society of America. 1816 Jefferson Place, N.W., Washington, DC 20036. (202) 223-8130. Order from: American Institute of Physics, 335 East 45th Street, New York, NY 10017. (212) 661-9404. 1962 to present. Semi-monthly. $330.00 per year.

APPLIED SPECTROSCOPY REVIEWS. Marcel Dekker Inc., 270 Madison Avenue, New York, NY 10016. (800) 228-1160. 1964 tp present. Quarterly. $185.00 per year.

BRITISH JOURNAL OF PHOTOGRAPHY. Henry Greenwood and Company, Limited, 28 Great James Street, London WC1N 3HL, England. 1854 to present. Weekly. $70.00 per year.

FIBER AND INTEGRATED OPTICS. Crane Russak and Company, Inc., 3 East 44th Street, New York, NY 10017. (212) 867-1490. 1977 to present. Quarterly. $86.00 per year.

FUNCTIONAL PHOTOGRAPHY. PTN Publishing Corporation, 210 Crossways Park Drive, Woodbury, NY 11797. (516) 496-8000. 1967 to present. Bimonthly. $7.50 per year.

INDUSTRIAL PHOTOGRAPHY. United Business Publications, Inc., 475 Park Avenue South, New York, NY 10016. (212) 725-2300. 1952 to present. Monthly. $15.00 per year.

JOURNAL OF IMAGING SCIENCE. Society of Photographic Scientists and Engineers. 7003 Kilworth Lane, Springfield, VA 22151. (703) 642-9090. 1956 to present. Bimonthly. $70.00 per year.

JOURNAL OF IMAGING TECHNOLOGY. Society of Photographic Scientists and Engineers. 7003 Kilworth Lane, Springfield, VA 22151. (703) 642-9090. 1975 to present. Bimonthly. $70.00 per year.

JOURNAL OF LIGHTWAVE TECHNOLOGY. Institute of Electrical and Electronics Engineers. IEEE Service Center, 445 Hoes Lane, Piscataway, NJ 08854. 1983 to present. Monthly. $145.00 per year.

JOURNAL OF PHOTOGRAPHIC SCIENCE. Royal Photographic Society of Great Britain, 7 Ladbroke Walk, London W11, England. 1953 to present. Bimonthly. $40.00 per year.

LASER FOCUS. PennWell Publishing Company, 119 Russell Street, Littleton, MA 01460. (617) 486-9501. 1965 to present. Monthly. $55.00 per year.

OPTICAL ENGINEERING. Society of Photo-Optical Instrumentation Engineers (SPIE), P.O. Box 10, 1022 19th Street, Bellingham, WA 98227. (206) 676-3290. 1962 to present. Monthly. $95.00 per year.

OPTICAL SOCIETY OF AMERICA, JOURNAL, PARTS A AND B. Optical Society of America. 1816 Jefferson Place, N.W., Washington, DC 20036. (202) 223-8130. Order from: American Institute of Physics, 335 East 45th Street, New York, NY 10017. (212) 661-9404. 1917 to present. Monthly. $180.00 each part per year.

OPTICS COMMUNICATIONS. Elsevier Science Publishing Company, Inc., 52 Vanderbilt Avenue, New York, NY 10017. (212) 370-5520. 1969 to present. 24 times per year. $425.00 per year.

OPTICS LETTERS. Optical Society of America. Order from: American Institute of Physics, 335 East 45th Street, New York, NY 10017. (212) 661-9404. 1977 to present. Monthly. $150.00 per year.

SOCIETY OF PHOTO-OPTICAL INSTRUMENTATION ENGINEERS (SPIE) PROCEEDINGS. Society of Photo-Optical Instrumentation Engineers (SPIE), P.O. Box 10, 1022 19th Street, Bellingham, WA 98227. (206) 676-3290. 1963 to present. Approximately 50 numbers per year. Approximately $40.00 per number.

RESEARCH CENTERS AND INSTITUTES

CENTER FOR APPLIED OPTICS. University of Alabama in Huntsville, Research Institute Building, Huntsville, AL 35899. (205) 895-6102.

CENTER FOR APPLIED OPTICS. University of Texas at Dallas, P.O. Box 830688, Richardson, TX 75083. (214) 690-2868.

INSTITUTE OF OPTICS. University of Rochester, Rochester, NY 14627. (716) 275-2314.

OPTICAL PHYSICS LABORATORY. University of Miami, Coral Gables, FL 33124. (305) 284-2324.

COLOR FILM

See: PHOTOGRAPHIC FILM

COLOR TELEVISION

See: TELEVISION

COLORIMETRY

See: COLOR

COLUMNS

See: STRUCTURAL ENGINEERING

COMBINATORIAL THEORY

See: MATHEMATICS

COMBUSTION

See: CHEMISTRY, EXPLOSIVES

COMETS

See also: ASTRONOMY

ABSTRACT SERVICES AND INDEXES

ASTRONOMY AND ASTROPHYSICS ABSTRACTS. Springer-Verlag New York, Incorporated, 175 Fifth Avenue, New York, NY 10010. (212) 460-1500. $70.00 per year.

ASTRONOMY AND ASTROPHYSICS MONTHLY INDEX. Olivetree Associates, P.O. Box 236, Sierre Madre, Ca 91024. $212.00 per year. Complimentary copies available on request.

GENERAL SCIENCE INDEX. H.W. Wilson Company, 950 University Avenue, Bronx, NY 10452. (800) 367-6770 or (212) 588-8400. Inquire as to cost and availability.

PHYSICS ABSTRACTS. Institute of Electrical Engineers, London, United Kingdom. Available from: Institute of Electrical and Electronic Engineers (IEEE), 345 East 47th Street, New York, NY 10017. (212) 705-7900.

SCIENCE CITATION INDEX. Institute for Scientific Information, 3501 Market Street, Philadelphia, PA 19104. (800) 523-1850 or (215) 386-0100.

STAR. (Scientific and Technical Aerospace Reports). U.S. National Aeronautics and Space Administration, Scientific and Technical Information Facility, Box 8757, Baltimore-Washington International Airport, Md 21240. (202) 755-2210. Semimonthly, with semiannual and annual indexes. $85.00 per year.

ANNUAL REVIEWS AND YEARBOOKS

THE ASTRONOMICAL ALMANAC. Superintendent of Documents, U.S. Government Printing Office, Washington, DC 20402. (202) 783-3238. Yearly.

ANNUAL REVIEW OF ASTRONOMY AND ASTROPHYSICS. Annual Reviews, Incorporated, 4139 El Camino Way, Palo Alto, CA 94306. (415) 493-4400.

ANNUAL REVIEW OF EARTH AND PLANETARY SCIENCES. Annual Reviews, Incorporated, 4139 East Camino Way, Palo Alto, CA 94306. (415) 493-4400.

ASSOCIATIONS AND PROFESSIONAL SOCIETIES

AMERICAN ASTRONOMICAL SOCIETY. 2000 Florida Avenue, NW, Suite 300, Washington, DC 20009. (202) 659-0134.

AMERICAN ASSOCIATION OF VARIABLE STAR OBSERVERS. 187 Concord Avenue, Cambridge, MA 02138. (617) 354-0484.

ASTRONOMICAL LEAGUE. P.O. Box 12821, Tucson, AZ 85732. (602) 790-8471.

ASTRONOMICAL SOCIETY OF THE PACIFIC. 1290 24th Avenue, San Francisco, Ca 94122. (415) 661-8660.

BIBLIOGRAPHIES

A BIBLIOGRAPHY OF ASTRONOMY, 1970-1979. R.A. Seal and S.S. Martin. Libraries Unlimited, Incorporated, Littleton, CO 80160. 1982. $37.50.

HALLEY'S COMET: A BIBLIOGRAPHY. Ruth Freitag. Library of Congress, Washington, DC. References to Comet Halley from 1495 to 1983 with annotations. 1984. $26.00.

HALLEY'S COMET, 1755-1984: A BIBLIOGRAPHY. B. Morton. Greenwood Press. 1985. $35.00.

DIRECTORIES AND BIOGRAPHICAL SOURCES

AMERICAN MEN AND WOMEN OF SCIENCE. Physical and Biological Sciences. Fifteenth edition. R.R. Bowker Company, 205 East 42nd Street, New York, NY 10017. (800) 521-8110 or (212) 916-1600.

THE BIOGRAPHICAL DICTIONARY OF SCIENTISTS: ASTRONOMERS. D. Abbott, editor. Peter Bedrick Books, 125 East 23rd Street, New York, NY 10010. 1984.

DIRECTORY OF PHYSICS AND ASTRONOMY STAFF MEMBERS. American Institite of Physics, 335 East 45th Street, New York, NY 10017. Annual.

RESEARCH CENTERS DIRECTORY. Gale Research Company, Detroit, MI 48226. Eleventh edition, 1987. (800) 521-0707. $340.00.

WHO'S WHO IN FRONTIER SCIENCE AND TECHNOLOGY. Marquis Who's Who, Incorporated, 200 East Ohio Street, Chicago, IL 60611. (800) 428-3898 or (312) 787-2008.

ENCYCLOPEDIAS AND DICTIONARIES

ENCYCLOPEDIA OF PHYSICAL SCIENCE AND TECHNOLOGY. Academic Press, Incorporated, Orlando, FL 32887. (800) 321-5068 or (305) 345-2734.

MCGRAW-HILL ENCYCLOPEDIA OF SCIENCE AND TECHNOLOGY. McGraw-Hill Book, Incorporated, 1221 Avenue of the Americas, New York, NY 10020. (212) 997-3675. Fifth edition, 15 volumes. $1100.00.

GENERAL WORKS

COMET HALLEY: ONCE IN A LIFETIME. M. Littman and D. Yeomans. American Chemical Society, 1155 16th Street, NW, Washington, DC 20036.

COMETS: A DESCRIPTIVE CATALOG. G. Kronk. Enslow, New York. Has detailed information on over 650 comets found between 372 B.C. and 1982 A.D.

CATALOG OF COMETARY ORBITS. Brian Marsden. Enslow, New York. Tabulates information about the orbits of 1,109 comets from 240 B.C. to May 1982.

OBSERVE COMETS. S. Edberg and D. Levy. Astronomical League, write for information to: Michele Sherlin, 1001 South Cornelia Street, Sioux City, IA 51106. A 56 page guide to comet hunting and observing.

COMETS, METEORS, AND ASTEROIDS. Stan Gibilisco. TAB Books Incorporated, Monterey Lane, Blue Ridge Summit, PA 17214. (717) 794-2191. $12.95 in paperback.

ONLINE DATA BASES

CA SEARCH. Chemical Abstracts Service, P.O. Box 3012, Columbus, OH 43210. Guide to chemical literature, 1972 to present. Inquire as to cost and availability.

DISSERTATION ABSTRACTS ONLINE. University Microfilms International, 300 North Zeeb Road, Ann Arbor, MI 48106. (800) 521-0600 or (313) 761-4700. Scope includes virtually all doctoral dissertations accepted at accredited American institutions from 1861 to present in 252 subject areas. Inquire as to cost and availability.

INSPEC. INSPEC Marketing Department, Institute of Electrical and Electronics Engineers, Incorporated, IEEE Service Department, 445 Hoes Lane, Piscataway, NJ 08854. (201) 981-0060. Inquire as to on-line cost and availability.

NTIS. National Technical Information Service, 5285 Port Royal Road, Springfield, Va 22161. (703) 487-4630. Broad coverage of government sponsored research reports, 1964 to present. Inquire as to cost and availability.

SCISEARCH. Institute for Scientific Information, 3501 Market Street, Philadelphia, PA 19104. (800) 523-1850 or (215) 386-0100. Broad multidisciplinary title and author index to the international literature of science and technology, 1974 to present. Inquire as to cost and availability.

OTHER SOURCES

FIRE AND ICE: A HISTORY OF COMETS IN ART. R. Olson. Walker Press.

ASTROPHOTOGRAPHY: A STEP BY STEP APPROACH. Robert Little. Macmillan Publishing Company, 866 Third Avenue, New York, NY 10022. (800) 257-5755. $19.95.

CAPTURE A COMET: AN AMATEUR PHOTOGRAPHER'S GUIDE. M. Roberts. Aurorae Publishing. 1985. $12.95.

PERIODICALS

ASTRONOMICAL JOURNAL. American Astronomical Society. Available from: American Institute of Physics, 335 East 45th Street, New York, NY 10017. (212) 661-9404. $125.00 per year.

ASTRONOMICAL SOCIETY OF THE PACIFIC. Publications. Astronomical Society of the Pacific, 1290 24th Avenue, San Francisco, Ca 94122. (415) 661-8660. Monthly. $38.00.

ASTRONOMY. Astro Media Corporation, 625 East Paul Avenue, Milwaukee, WI 53202. Monthly. $18.00 per year.

ASTRONOMY AND ASTROPHYSICS. Springer-Verlag New York, Incorporated, 175 Fifth Avenue, New York, NY 10010. (800) 526-7254 or (212) 460-1500. $680.00 per year.

ASTROPHYSICAL JOURNAL. American Astronomical Society, University of Chicago Press, 5801 Ellis Avenue, Chicago, IL 60637. Biweekly. $305.00 per year.

ASTROPHYSICS AND SPACE SCIENCE. D. Reidel Publishing Company, 190 Old Derby Street, Hingham, MA 02043. Monthly. $101.00 per year.

CELESTIAL MECHANICS; AN INTERNATIONAL JOURNAL OF SPACE DYNAMICS. D. Reidel Publishing Company, 190 Old Derby Street, Hingham, MA 02043. Monthly. $310.00 per year.

EARTH, MOON AND PLANETS; AN INTERNATIONAL JOURNAL OF COMPARATIVE PLANETOLOGY. D. Reidel Publishing Company, 190 Old Derby Street, Hingham, MA 02043. Nine times per year. $275.00 per year.

ICARUS: INTERNATIONAL JOURNAL OF THE SOLAR SYSTEM STUDIES. Academic Press, Incorporated, Orlando, FL 32887. (305) 345-4100. Monthly. $484.00 per year.

MERCURY. Astronomical Society of the Pacific, 1290 24th Avenue, San Francisco, CA 94122. (415) 661-8660. Bimonthly. $21.00 per year.

MONTHLY NOTICES OF THE ROYAL ASTRONOMICAL SOCIETY. Blackwell Science Publications, Incorporated, 667 Lytton Avenue, Palo Alto, CA 94301. (415) 324-1688. Monthly. $134.00 per year.

PLANETARY AN SPACE SCIENCE. Pergamon Press, Incorporated, Maxwell House, Fairview Park, Elmsford, NY 10523. (914) 592-7700. Monthly. $430.00 per year.

ROYAL ASTRONOMICAL SOCIETY. Monthly Notices. Blackwell Scientific Publications, Incorporated, 52 Beacon Street, Boston, MA 02108. (617) 720-0761. Monthly. $850.00 per year.

SKY AND TELESCOPE. Sky Publishing Corporation, 49 Bay State Road, Cambridge, Ma 02238. (617) 864-7360. Monthly. $18.00 per year.

SOVIET ASTRONOMY (TRANSLATION OF ASTRO-NOMICHESKII ZHURNAL). American Institute of Physics, 335 East 45th Street, New York, NY 10017. (212) 661-9404. Bimonthly. $425.00 per year.

SPACE SCIENCE REVIEWS. D. Reidel Publishing Company, 190 Old Derby Street, Hingham, MA 02043. Monthly. $305.00 per year.

VISTAS IN ASTRONOMY. Pergamon Press, Incorporated, Maxwell House, Fairview Park, Elmsford, NY 10523. (914) 592-7700. Quarterly. $145.00 per year.

RESEARCH CENTERS AND INSTITUTES

INSTITUTE FOR ASTRONOMY. University of Hawaii, 2680 Woodlawn Drive, Honolulu, HI 96822. (808) 948-8566.

JOINT OBSERVATORY FOR COMETARY RESEARCH. New Mexico Institute of Mining and Technology, Physics Department, Socorro, NM 87801. (505) 835-5431.

LOWELL OBSERVATORY. Box 1269, Flagstaff, AZ 86002. (602) 774-3358.

LUNAR AND PLANETARY LABORATORY. University of Arizona, Tucson, Az 85721. (602) 621-6962.

COMMUNICATION SATELLITES

See: ARTIFICIAL SATELLITES

COMPASSES

See also: AVIONICS, BUOYS, GUIDANCE SYSTEMS, GYROSCOPES, NAVIGATION, RADAR

ABSTRACT SERVICES AND INDEXES

APPLIED SCIENCE AND TECHNOLOGY INDEX. H.W. Wilson and Company, 950 University Avenue, Bronx, NY 10452. (800) 367-6670 or (212) 588-8400. Monthly. Inquire as to cost and availability.

ENGINEERING INDEX MONTHLY AND AUTHOR INDEX. Engineering Information Inc., 345 East 47th Street, New York, NY 10017. (212) 705-7600. Monthly. $1560.00 per year.

INTERNATIONAL AEROSPACE ABSTRACTS. American Institute of Aeronautics and Astronautics, Technical Information

Service, 370 L'Enfant Promenade, S.W., Washington, DC 20024. (202) 646-7400. 1961 to present. Semi-monthly. $700.00 per year.

ASSOCIATIONS AND PROFESSIONAL SOCIETIES

INSTITUTE OF NAVIGATION. 815 15th Street, N.W., Suite 832, Washington, DC 20005. (202) 783-4121.

INTERNATIONAL OMEGA ASSOCIATION. P.O. Box 2324, 1720 South Eads Street, Arlington, VA 22202. (301) 593-4144.

PERMANENT INTERNATIONAL ASSOCIATION OF NAVIGATION CONGRESSES, UNITED STATES SECTION. c/o U.S. Army Corps of Engineers, Water Resources Support Center, Casey Building, Fort Belvoir, VA 22060. (202) 355-2096.

WILD GOOSE ASSOCIATION. P.O. Box 556, Bedford, MA 01730.

DIRECTORIES AND BIOGRAPHICAL SOURCES

1987 DIRECTORY OF ENGINEERING SOCIETIES AND RELATED ORGANIZATIONS. Gordon Davis, editor. Hemisphere Publishing Corporation, 1010 Vermont Avenue, NW, Washington, DC 20005. (800) 526-0275. 12th edition. 1987. $100.00.

RESEARCH CENTERS DIRECTORY 1988. Gale Research Company, Book Tower, Detroit, MI 48226. (800) 521-0707. 12th edition. 1987. $365.00 for set.

WHO'S WHO IN ENGINEERING. Gordon Davis, editor. Hemisphere Publishing Corporation, 79 Madison Avenue, New York, NY 10016-7892. (800) 821-8312. Sixth edition. 1985. $200.00.

GENERAL WORKS

AUTOMATIC CONTROL OF AIRCRAFT AND MISSILES. John H. Blakelock. John Wiley and Sons, Inc., 605 Third Avenue, New York, NY 10158. (800) 526-5368. 1965. $64.50.

AUTOMATIC FLIGHT CONTROL. E.H. Pallett. Sheridan Publishers, Inc., 145 Palisade Street, Dobbs Ferry, NY 10522. (914) 693-2410. Second edition. 1983. $32.50.

COCKPIT COMPUTERS AND NAVIGATION AVIONICS. P. Garrison. McGraw-Hill Book Company, 1221 Avenue of the Americas, New York, NY 10020. (212) 512-2000. 1982. $29.95.

DUTTON'S NAVIGATION AND PILOTING. Elbert S. Maloney. Naval Institute Press, U.S. Naval Institute, Annapolis, MD 21402. (301) 268-6110. Fourth edition. 1985. $32.95.

GYROSCOPIC THEORY. George Greehill. Chelsea Publishing Company, 15 East 26th Street, New York, NY 10010. (212) 889-8095. $22.50.

MARINE ELECTRONIC NAVIGATION. S.F. Appleyard. Methuen, Inc., 29 West 35th Street, New York, NY 10001. (212) 244-3336. 1980. $34.95.

NOTES ON COMPASS WORK. J.F. Kemp and P. Young. Sheridan Publishers, Inc., 145 Palisade Street, Dobbs Ferry, NY 10522. (914) 693-2410. 1972. $10.95.

SEAMANSHIP: FUNDAMENTALS FOR THE DECK OFFICER. David O. Dodge and S.E. Kyriss. Naval Institute Press, U.S. Naval Institute, Annapolis, MD 21402. (301) 268-6110. 1981. $16.95.

ONLINE DATA BASES

COMPENDEX. Engineering Information, Inc., 345 East 47th Street, New York, NY 10017. (800) 221-1044 or (212) 705-7615. Engineering and technical literature, 1975 to present. Inquire as to online cost and availability.

NTIS. National Technical Information Service, 5285 Port Royal Road, Springfield, VA 22161. (703) 487-4630. Broad coverage of government sponsored research reports, 1964 to present. Inquire as to online cost and availability.

WILSONLINE. H.W. Wilson and Company, 950 University Avenue, Bronx, NY 10452. (800) 367-6770 or (212) 588-8400. Makes available online versions of the H.W. Wilson indexes including Applied Science and Technology Index, Business Periodicals Index and Readers' Guide to Periodical Literature. Approximately 1980 to present. Inquire as to online cost and availability.

PERIODICALS

AEROSPACE AMERICA. American Institute of Aeronautics and Astronautics, Technical Information Service, 370 L'Enfant Promenade, S.W., Washington, DC 20024. (202) 646-7400. 1932 to present. Monthly. $56.00 per year.

AEROSPACE ENGINEERING MAGAZINE. Society of Automotive Engineers, 400 Commonwealth Drive, Warrendale, PA 15096. (412) 776-4841. 1981 to present. Monthly. $24.00 per year.

AIAA JOURNAL. American Institute of Aeronautics and Astronautics, Technical Information Service, 370 L'Enfant Promenade, S.W., Washington, DC 20024. (202) 646-7400. 1963 to present. Monthly. $205.00 per year.

AVIATION WEEK AND SPACE TECHNOLOGY. McGraw-Hill Book Company, 1221 Avenue of the Americas, New York, NY 10020. (212) 512-2000. 1916 to present. Weekly. $56.00 per year.

JOURNAL OF NAVIGATION. Royal Institute of Navigation. Cambridge University Press, 32 East 57th Street, New York, NY 10022. (800) 872-7423. 1947 to present. Three per year. $100.00 per year.

JOURNAL OF SHIP RESEARCH. Society of Naval Architects and Marine Engineers, One World Trade Center, Suite 1369, New York, NY 10048. (212) 432-0310. 1957 to present. Quarterly. $40.00 per year.

JOURNAL OF SPACECRAFT AND ROCKETS. American Institute of Aeronautics and Astronautics, Technical Information Service, 370 L'Enfant Promenade, S.W., Washington, DC 20024. (202) 646-7400. 1964 to present. Bimonthly. $95.00 per year.

RESEARCH CENTERS AND INSTITUTES

CHARLES STARK DRAPER LABORATORY, INC. 555 Technology Square, Cambridge, MA 02139. (617) 258-1000.

GODDARD SPACE FLIGHT CENTER. National Aeronautics and Space Administration, Greenbelt Road, Greenbelt, MD 20771. (301) 344-5121.

INSTRUMENTATION AND CONTROL LABORATORY. Princeton University, Department of MAE, English Quadrangle, Princeton, NJ 08544.

MARSHALL SPACE FLIGHT CENTER. National Aeronautics and Space Administration, Huntsville, AL 35812. (205) 453-2121.

STANFORD ELECTRONICS LABORATORIES. Stanford University, Stanford, CA 94305. (415) 723-1804.

COMPOSITES

See also: MATERIALS SCIENCE

ABSTRACT SERVICES AND INDEXES

APPLIED MECHANICS REVIEW. American Society of Mechanical Engineers, 345 East 47th Street, New York, NY 10017. (212) 705-7703. From 1948 to present, Monthly. $360.00 per year.

APPLIED SCIENCE AND TECHNOLOGY INDEX. H.W. Wilson Company, 950 University Avenue, Bronx, NY 10452. (800) 367-6670 or (212) 588-8400. Inquire as to cost and availability.

CHEMICAL ABSTRACTS. Chemical Abstracts Service, 2540 Olentangy Road, P.O. Box 3012, Columbus, OH 43210. (800) 848-6538 or (614) 421-3600. Weekly. $9200.00 per year.

ENGINEERING INDEX MONTHLY. Engineering Information, Incorporated, 345 East 47th Street, New York, NY 10017. (800) 221-1044 or (212) 705-7600. Monthly, with annual cumulation. $1560.00 per year.

GENERAL SCIENCE INDEX. H.W. Wilson Company, 950 University Avenue, Bronx, NY 10452. (800) 367-6770 or (212) 588-8400.

INTERNATIONAL AEROSPACE ABSTRACTS. American Institution of Aeronautics and Astronautics, Technical Information Service, 555 West 57th Street, New York, NY 10019. (212) 247-6500. 1961 to present. Monthly. $700.00 per year.

METALS ABSTRACTS. American Society for Metals, 9639 Kinsman Road, Metals Park, OH 44073. (216) 338-5151. Monthly. $890.00.

METALS ABSTRACTS INDEX. American Society for Metals, 9639 Kinsman Road, Metals Park, OH 44073. (216) 338-5151. Monthly. (Sold only to subscribers of Metals Abstracts)

STAR. (Scientific and Technical Aerospace Reports). U.S. National Aeronautics and Space Administration, Scientific and Technical Information Facility, Box 8757, Baltimore-Washington International Airport, MD 21240. (202) 755-2210. Semimonthly, with semiannual and annual indexes. $85.00 per year.

ASSOCIATIONS AND PROFESSIONAL SOCIETIES

ASTM (AMERICAN SOCIETY FOR TESTING AND MATERIALS). 1916 Race Street, Philadelphia, PA 19103. (215) 299-5400.

AMERICAN SOCIETY FOR METALS. Metals Park, OH 44073. (216) 338-5151.

SOCIETY FOR THE ADVANCEMENT OF MATERIAL AND PROCESS ENGINEERING (SAMPE). P.O. Box 2459, Covina, Ca 91722. (818) 331-0616.

BIBLIOGRAPHIES

SCIENCE BOOKS AND FILMS. American Association for the Advancement of Science, 1333 H Street, NW, Washington, DC 20005.

SCIENTIFIC AND TECHNICAL BOOKS IN PRINT; AN INDEX TO LITERATURE IN SCIENCE AND TECHNOLOGY. R.R. Bowker Company, 205 East 42nd Street, New York, NY 10017. (800) 521-8110 or (212) 916-1600.

DIRECTORIES AND BIOGRAPHICAL SOURCES

MATERIALS RESEARCH CENTERS: A WORLD DIRECTORY OF ORGANIZATIONS AND PROGRAMMES IN MATERIALS SCIENCE. Longman Group Limited, Westgate House, The High, Harlow, Essex CM20 1NE United Kingdom. Distributed by Gale Research Company, Book Tower, Detroit, MI 48226. (800) 521-0707. 1983. $160.00.

MECHANICS OF COMPOSITE MATERIALS DIRECTORY. Air Force Wright Aeronautical Laboratories, Wright-Patterson Air Force Base, OH 45433. 1986. $18.95.

RESEARCH CENTERS DIRECTORY. Gale Research Company, Book Tower, Detroit, MI 48226. (800) 521-0707. Eleventh edition. 1987.

WHO'S WHO IN FRONTIER SCIENCE AND TECHNOLOGY. Marquis Who's Who, Incorporated, 200 East Ohio Street, Chicago, IL 60611. (800) 428-3898 or (312) 787-2008.

WHO'S WHO IN TECHNOLOGY TODAY. Reston Publishing Company, Incorporated, c/o Prentice-Hall, Incorporated, Englewood Cliffs, NJ 07632. (800) 262-6868. Biennial. Five volumes. $425.00. Covers the fields of electronics, computer science, physics, optics, chemistry, biotechnology, mechanics, energy, and earth science.

ENCYCLOPEDIAS AND DICTIONARIES

COMPOSITE MATERIALS GLOSSARY. C.M. Bower, editor. T-C Publications, P.O. Box 842, El Sequndo, CA 90245. (213) 938-6923. 1985. $18.00.

ENCYCLOPEDIA OF COMPOSITE MATERIALS AND COMPONENTS. Martin Grayson editor. John Wiley and Sons, 605 Third Avenue, New York, NY 10158. (800) 526-5368 or (212) 850-6418. $125.00. 1983.

ENCYCLOPEDIA OF PHYSICAL SCIENCE AND TECHNOLOGY. Academic Press, Incorporated, Orlando, FL 32887. (800) 321-5068 or (305) 345-2734. Fifteen volumes, 1986.

MCGRAW-HILL ENCYCLOPEDIA OF SCIENCE AND TECHNOLOGY. McGraw-Hill Book, Incorporated, 1221 Avenue of the Americas, New York, NY 10020. (212) 997-3675.

GENERAL WORKS

THE BEHAVIOR OF STRUCTURES COMPOSED OF COMPOSITE MATERIALS. Jack R. Vinson and R.L. Sierakowski. Martinus Nijhoff, Dordrecht, The Netherlands. Distributed by Kluwer Academic Publishers, 190 Old Derby Street, Hingham, Ma 02043. (617) 749-5262. 1986. $59.50.

COMPOSITE MATERIALS: MECHANICS, MECHANICAL PROPERTIES AND FABRICATION. K. Kawata and T. Akasaka, editors. Elsevier Science Publishing Company, Incorporated, 52 Vanderbilt Avenue, New York, NY 10017. (212) 370-5520. 1982. $68.50.

AN INTRODUCTION TO COMPOSITE MATERIALS. Derek Hull. Cambridge University Press, 32 East 57th Street, New York, NY 10022. (800) 872-7423. 1981. $19.95 in paper.

HANDBOOKS AND MANUALS

COMPOSITE MATERIALS HANDBOOK. Mel M. Schwartz. McGraw-Hill Book Company, 1221 Avenue of the Americas, New York, NY 10020. (212) 512-2000. 1984.

HANDBOOK OF COMPOSITES. George Lubin, editor. Van Nostrand Reinhold Company, Incorporated, 135 West 50th Street, New York, NY 10020. (212) 265-8700. 1982. $74.50.

FABRICATION OF COMPOSITE MATERIALS, SOURCE BOOK. American Society for Metals, Metals Park, OH 44073. (216) 338-5151. 1985. $62.00.

COMPOSITES

ONLINE DATA BASES

CA SEARCH. Chemical Abstracts Service, P.O. Box 3012, Columbus, OH 43210. Guide to chemical literature, 1972 to present. Inquire as to cost and availability.

COMPENDEX. Engineering Information, Incorporated, 345 East 47th Street, New York, NY 10017. (800) 221-1044 or (212) 705-7615. Engineering and technical literature, 1975 to present. Inquire as to cost and availability.

NASA. National Aeronautics and Space Administration, Scientific and Technical Information Branch, 300 7th Street, SW, Washington, DC 20546. Citations and abstracts of aerospace literature, 1962 to present. Inquire as to cost and availability.

NTIS. National Technical Information Service, 5285 Port Royal Road, Springfield, Va 22161. (703) 487-4630. Broad coverage of government sponsored research reports, 1964 to present. Inquire as to cost and availability.

PERIODICALS

AVIATION WEEK AND SPACE TECHNOLOGY. McGraw-Hill, Incorporated, 1221 Avenue of the Americas, New York, NY 10020. (212) 512-2000. Weekly. $55.00 per year.

COMPOSITES. Butterworth Scientific Limited, P.O. Box 63, Westbury House, Bury Street, Guildford, Surrey, GU2 5BH, England. quarterly. $166.00 per year.

COMPOSITES SCIENCE AND TECHNOLOGY. Elsevier Applied Science Publishers, Limited, Crown House, Linton Road, Barking, Essex IG11 8JU, England, Eight times per year. $230.00 per year.

JOURNAL OF APPLIED POLYMER SCIENCE. John Wiley and Sons, Incorporated, 605 Third Avenue, New York, NY 10158. (800) 526-5368 or (212) 850-6418. Monthly. $375.00 per year.

JOURNAL OF COMPOSITE MATERIALS. Technomic Publishing Company, Incorporated, 851 New Holland Avenue, Lancaster, PA 17604. (717) 291-5609. Bimonthly. $155.00 per year.

JOURNAL OF COMPOSITES TECHNOLOGY AND RESEARCH. ASTM (American Society for Testing and Materials), 1916 Race Street, Philadelphia, PA 19103. (215) 299-5400. Quarterly. $38.00 per year.

JOURNAL OF ELASTOMERS AND PLASTICS. Technomic Publishing Company, Incorporated, 851 New Holland Avenue, Lancaster, PA 17604. (717) 291-5609. Quarterly. $90.00 per year.

MATERIALS ENGINEERING. Penton-IPC, Penton Plaza, 1111 Chester Avenue, Cleveland, OH 44114. (216) 579-6333. Monthly. 430.00.

RESEARCH CENTERS AND INSTITUTES

LAWRENCE BERKELEY LABORATORY. Center for Advanced Materials, 1 Cyclotron Road, Berkeley, CA 94720. (415) 486-4755.

MICHIGAN STATE UNIVERSITY. Composite Materials and Structures Center, College of Engineering, East Lansing, MI 48824. (517) 353-5466.

UNIVERSITY OF FLORIDA. Center for Studies of Advanced Structural Composites, Engineering Sciences Department, 231 Aerospace Building, Gainesville, FL 32611. (904) 392-0961.

COMPUTER ARCHITECTURE

See: COMPUTERS

COMPUTER COMMUNICATIONS

See also: AUTOMATION, COMPUTER MEMORY, COMPUTER OPERATING SYSTEMS, COMPUTER PROGRAMMING, COMPUTER VISION, COMPUTERS, PARALLEL COMPUTERS, PROGRAMMING LANGUAGES, SOFTWARE

ABSTRACT SERVICES AND INDEXES

APPLIED SCIENCE AND TECHNOLOGY INDEX. H.W. Wilson Company, 950 University Avenue, Bronx, NY 10452. (800) 367-6670 or (212) 588-8400. Inquire as to cost and availability.

COMPUMATH CITATION INDEX. Institute for Scientific Information, 3501 Market Street, Philadelphia, PA 19104. (800) 523-1850 or (215) 386-0100. Three times per year. $875.00 per year.

COMPUTER ABSTRACTS. Technical Information Company, Limited, Post Office Box 59, Saint Helier, Jersey British Channel Inlands, England. Monthly. $310.00 per year.

COMPUTER AND CONTROL ABSTRACTS. Institute of Electrical Engineers, London, United Kingdom. Available from: IEEE Service Center, 445 Hoes Lane, Piscataway, NJ 08854. (201) 981-0060. Semimonthly. $775.00 per year.

COMPUTER AND INFORMATION SYSTEMS: AN ABSTRACT JOURNAL PERTAINING TO THE THEORY, DESIGN, FABRICATION AND APPLICATION OF COMPUTER AND INFORMATION SYSTEMS. Cambridge Scientific Abstracts, Incorporated, 5161 River Road, Bethesda, MD 20816. (301) 951-1400. Semimonthly. $590.00 per year.

COMPUTER CONTENTS: THE BIWEEKLY COMPILATION OF TABLES OF CONTENTS FROM COMPUTER, ELECTRONIC AND TELECOMMUNICATIONS MAGAZINES, JOURNALS AND TRANSACTIONS. Find/SVP, 500 Fifth Avenue, New York, NY 10110. (800) 346-3787 or (212) 354-2424. Biweekly. $115.00 per year.

COMPUTER PROGRAM ABSTRACTS. U.S. National Aeronautics and Space Administration. Available from U.S. Government Printing Office, Washington, DC 20402. Quarterly. $6.50 per year.

COMPUTING REVIEWS. Association for Computing Machinery, 11 West 42nd Street, New York, NY 10036. (212) 869-7440. Monthly. $60.00 per year.

ENGINEERING INDEX MONTHLY. Engineering Information, Incorporated, 345 East 47th Street, New York, NY 10017. (800) 221-1044 or (212) 705-7600. Monthly, with annual cumulation. $1560.00 per year.

MATHEMATICAL REVIEWS. American Mathematical Society, Post Office Box 6248, Providence, RI 02940. (800) 556-7774 or (401) 272-9500.

SCIENCE CITATION INDEX. Institute for Scientific Information, 3501 Market Street, Philadelphia, PA 19104. (800) 523-1850 or (215) 386-0100.

ANNUAL REVIEWS AND YEARBOOKS

ADVANCES IN COMPUTERS. Academic Press, Incorporated, 6277 Sea Harbor Drive, Orlando, FL 32821. (800) 321-5068. Yearly. Approximately $50.00 per volume.

ASSOCIATIONS AND PROFESSIONAL SOCIETIES

AMERICAN FEDERATION OF INFORMATION PROCESSING SOCIETIES. 1899 Preston White Drive, Reston, VA 22091. (703) 620-8900.

AMERICAN SOCIETY FOR CYBERNETICS. Department of Decision Sciences, George Mason University, Fairfax, VA 22030. (703) 323-2738.

ASSOCIATION OF COMPUTING MACHINERY (ACM). 11 West 42nd Street, New York, NY 10036. (212) 896-7440.

ASSOCIATION OF COMPUTER PROGRAMMERS AND ANALYSTS. 2108-C Gallows Road, Vienna, VA 22180. (703) 790-0490.

COMPUTER AND AUTOMATION SYSTEMS ASSOCIATION OF SME (SOCIETY OF MANUFACTURING ENGINEERS). One SME Drive, Box 930, Dearborn, Mi 48121. (313) 271-1500.

IEEE COMPUTER SOCIETY. 1730 Massachusetts Avenue, NW, Washington, DC 20036. (202) 371-0101.

INSTITUTE OF ELECTRICAL AND ELECTRONIC ENGINEERS (IEEE). 345 East 47th Street, New York, NY 10017. (212) 705-7900.

INSTITUTE FOR PERSONAL COMPUTING. Post Office Box 558250, Miami, FL 33155. (305) 274-7440.

MACHINE VISION ASSOCIATION. Post Office Box 930, One SME Drive, Dearborn, MI 48121. (313) 271-1500.

SOCIETY FOR COMPUTER SIMULATION. Post Office Box 17900. San Diego, CA 92117. (619) 277-3888.

SOCIETY FOR INFORMATION DISPLAY. 8055 Manchester Avenue, Suite 615, Playa Del Rey, CA 90293. (213) 305-1502.

BIBLIOGRAPHIES

COMPUTER LITERATURE INDEX. Applied Computer Research, Incorporated, Post Office Box 9280, Phoenix, AZ 85068. (602) 995-5929. Quarterly. $125.00 per year.

SCIENCE BOOKS AND FILMS. American Association for the Advancement of Science, 1333 H Street, NW, Washington, DC 20005.

SCIENTIFIC AND TECHNICAL BOOKS AND SERIALS IN PRINT 1988; AN INDEX TO LITERATURE IN SCIENCE AND TECHNOLOGY. R.R. Bowker Company, 205 East 42nd Street, New York, NY 10017. (800) 521-8110 or (212) 916-1600. $175.00.

DIRECTORIES AND BIOGRAPHICAL SOURCES

AMERICAN MEN AND WOMEN OF SCIENCE. Physical and Biological Sciences. Fifteenth edition. R.R. Bowker Company, 205 East 42nd Street, New York, NY 10017. (800) 521-8110 or (212) 916-1600.

AMERICAN SOCIETY FOR INFORMATION SCIENCE, HANDBOOK AND DIRECTORY. American Society for Information Science, 1424 16th Street, NW, Suite 404, Washington, DC 20036. (202) 462-1000. $50.00.

COMPUTER REVIEW. GML Information Services, 594 Marrett Road, Lexington, MA 02173. (617) 861-0515. Two issues per year. $195.00 per year. Directory of computer main frame and minicomputer manufacturers.

COMPUTERS AND COMPUTING INFORMATION RESOURCES DIRECTORY. Gale Research Company, Book Tower, Detroit, MI 48226. (800) 521-0707. 1986. $160.00.

COMPUTER PERIPHERALS REVIEW. GML Information Services, 594 Marrett Road, Lexington, MA 02173. (617) 861-0515. Two issues per year. $215.00 per year. Directory of computer peripheral equipment manufacturers.

DATAPRO REPORTS ON DATA COMMUNICATIONS. Datapro Research Corporation, 1805 Underwood Boulevard, Delran, NJ 08075. Three base volumes, with monthly updates. $780.00.

DATAPRO 70. Datapro Research Corporation, 1805 Underwood Boulevard, Delran, NJ 08075. Twelve monthly updates. $803.00 per year. List of about 1200 companies which offer data processing equipment and services.

DIRECTORY OF CONSULTANTS IN COMPUTER SYSTEMS. Research Publications, Incorporated, 12 Lunar Drive, Woodbridge, CT 06525. (203) 397-2600. Annual. $85.00 per year.

RESEARCH CENTERS DIRECTORY. Gale Research Company, Book Tower, Detroit, MI 48226. (800) 521-0707. Eleventh edition. 1987.

WHO'S WHO IN FRONTIER SCIENCE AND TECHNOLOGY. Marquis Who's Who, Incorporated, 200 East Ohio Street, Chicago, IL 60611. (800) 428-3898 or (312) 787-2008.

WHO'S WHO IN TECHNOLOGY TODAY. Reston Publishing Company, Incorporated, c/o Prentice-Hall, Incorporated, Englewood Cliffs, NJ 07632. (800) 262-6868. Biennial. Five volumes. $425.00. Covers the fields of electronics, computer science, physics, optics, chemistry, biotechnology, mechanics, energy, and earth science.

ENCYCLOPEDIAS AND DICTIONARIES

COMPUTER AND TELECOMMUNICATIONS ACRONYMS. Julie E. Towell and Helen E. Sheppard, editors. Gale Research Company, Book Tower, Detroit, MI 48226. (800) 521-0707. 1986. $60.00.

COMPUTER DICTIONARY. Charles J. Sippl. Howard W. Sams and Company, Incorporated, 4300 West 62nd Street, Indianapolis, IN 46268. (800) 428-7267 or (317) 298-5564. Fourth edition. 1985. $17.95.

DICTIONARY OF COMPUTING. Oxford University Press, 200 Madison Avenue, New York, NY 10016. (212) 679-7300. Second edition. 1986. $29.95.

ENCYCLOPEDIA OF COMPUTER SCIENCE AND ENGINEERING. Anthony Ralston, editor. Van Nostrand Reinhold Book Company, 115 Fifth Avenue, New York, NY 10003. (800) 543-2681. Second edition. 1982. $89.95.

ENCYCLOPEDIA OF COMPUTER SCIENCE AND TECHNOLOGY. Jack Belzer, Albert G. Holzman, and Allan Kent. Marcel Dekker, Incorporated, 270 Madison Avenue, New York, NY 10016. (212) 696-9000. Sixteen volumes. $115.00 per volume.

MCGRAW-HILL DICTIONARY OF COMPUTERS. McGraw-Hill Book Company, 1221 Avenue of the Americas, New York, NY 10020. (212) 512-2000. 1985. $17.50.

PRENTICE-HALL ENCYCLOPEDIA OF INFORMATION TECHNOLOGY. Robert A. Edmunds. Prentice-Hall, Incorporated, Englewood Cliffs, NJ 07632. (201) 592-2000. 1987. $49.95.

GENERAL WORKS

COMPUTERS AND DATA PROCESSING. H.L. Capron and Brian K. Williams. Benjamin/Cummings Publishing Company, Incorporated, 2727 Sand Hill Road, Menlo Park, CA 94025. (415) 854-6020. Second edition. 1984. $28.95.

DATA COMMUNICATIONS AND TELEPROCESSING SYSTEMS. Trevor Housley. Prentice-Hall, Incorporated, Englewood Cliffs, NJ 07632. (201) 592-2000. Second edition. 1987. $38.95.

DATA COMMUNICATIONS NETWORKING DEVICES: CHARACTERISTICS, OPERATIONS, APPLICATIONS. Gilbert Held. John Wiley and Sons, Incorporated, 605 Third Avenue, New York, NY 10158. (800) 526-5368 or (212) 850-

6000. 1986. $29.95.

DIGITAL COMMUNICATIONS SYSTEMS. J.R. Peebles and Z. Peyton. Prentice-Hall, Incorporated, Englewood Cliffs, NJ 07632. (201) 592-2000. 1987. $39.95.

ELECTRONIC SPEECH RECOGNITION: TECHNIQUES, TECHNOLOGY, AND APPLICATIONS. Geoff Bristow, editor. McGraw-Hill Book Company, 1221 Avenue of the Americas, New York, NY 10020. (212) 512-2000. 1986. $48.50.

INFORMATION SYSTEMS: THEORY AND PRACTICE. J. Burch and G. Grudnitski. John Wiley and Sons, Incorporated, 605 Third Avenue, New York, NY 10158. Fourth edition. 1986. $30.95.

INTRODUCTION TO DATA COMMUNICATIONS AND COMPUTER NETWORKS. Fred Halsall. Addison-Wesley Publishing Company, Incorporated, 1 Jacob Way, Reading, MA 01867. (617) 944-3700. 1986. $31.95.

HANDBOOKS AND MANUALS

FUNDAMENTALS HANDBOOK OF ELECTRICAL AND COMPUTER ENGINEERING. Sheldon S.L. Chang, editor. John Wiley and Sons, Incorporated, 605 Third Avenue, New York, NY 10158. (800) 526-5368 or (212) 850-6000. Three volumes. 1982. $180.00 set price.

HANDBOOK OF COMPUTERS AND COMPUTING. Arthur H. Seidman, editor. Van Nostrand Reinhold Book Company, 115 Fifth Avenue, New York, NY 10003. (800) 543-2681. 1984. $79.95.

HANDBOOK OF DATA COMMUNICATION AND COMPUTER NETWORKS. Dimitris N. Chorafas. Petrocelli Books, 251 Wall Street, Princeton, NJ 08540. (609) 924-5851. 1985. $59.95.

HANDBOOK OF SOFTWARE ENGINEERING. Charles R. Vick, editor. Van Nostrand Reinhold Book Company, 115 Fifth Avenue, New York, NY 10003. (800) 543-2681. 1984. $66.95.

MCGRAW-HILL COMPUTER HANDBOOK. Harry Helms, editor. McGraw-Hill Book Company, 1221 Avenue of the Americas, New York, NY 10020. (212) 512-2000. 1983. $84.50.

ONLINE DATA BASES

COMPENDEX. Engineering Information, Incorporated, 345 East 47th Street, New York, NY 10017. (800) 221-1044 or (212) 705-7615. Engineering and technical literature, 1975 to present. Inquire as to cost and availability.

DISSERTATION ABSTRACTS ONLINE. University Microfilms International, 300 North Zeeb Road, Ann Arbor, MI 48106. (800) 521-0600 or (313) 761-4700. Scope includes virtually all doctoral dissertations accepted at accredited American institutions from 1861 to present in 252 subject areas. Inquire as to cost and availability.

INSPEC. INSPEC Marketing Department, Institute of Electrical and Electronics Engineers, Incorporated, IEEE Service Department, 445 Hoes Lane, Piscataway, NJ 08854. (201) 981-0060. Inquire as to on-line cost and availability.

MATHFILE. American Mathematical Society, Post Office Box 6248, Providence, RI 02940. (800) 556-7774 or (401) 272-9500. Scope includes pure and applied mathematics and related areas of physics, statistics, engineering, computer science, and operations research literature since 1973. Inquire as to cost and availability.

NTIS. National Technical Information Service, 5285 Port Royal Road, Springfield, VA 22161. (703) 487-4630. Broad coverage of government sponsored research reports, 1964 to present. Inquire as to cost and availability.

SCISEARCH. Institute for Scientific Information, 3501 Market Street, Philadelphia, PA 19104. (800) 523-1850 or (215) 386-0100. Broad multidisciplinary title and author index to the international literature of science and technology, 1974 to present. Inquire as to cost and availability.

OTHER SOURCES

COMPUTER PUBLISHERS AND PUBLICATIONS 1985-86: AN INTERNATIONAL DIRECTORY AND YEARBOOK. Edited by Efrem Sigel and Frederica Evan. Communications Trends, Incorporated. Distributed by Gale Research Company, Book Tower, Detroit, MI 48226. (800) 521-0707. Second edition. 1985. $95.00. Supplement available for 1986, $35.00. Provides information on publishers of computer books and periodicals, recommended titles for libraries and bookstores, industry trends and statistics, and computer manufacturers as publishers.

PERIODICALS

ACM TRANSACTIONS ON COMPUTER SYSTEMS. Association for Computing Machinery, 11 West 42nd Street, New York, NY 10036. (212) 869-7440. Quarterly. $70.00 per year.

COMMUNICATIONS OF THE ACM. Association for Computing Machinery, 11 West 42nd Street, New York, NY 10036. (212) 869-7440. Monthly. $78.00 per year.

COMPUTER COMMUNICATIONS: THE INTERNATIONAL JOURNAL FOR THE COMPUTER AND TELECOMMUNICATIONS INDUSTRY. Butterworth Scientific, Limited, Post Office Box 63, Westbury House, Bury Street, Guildford, Surrey, GU2 5BH, England. Bimonthly. $169.00 per year.

COMPUTER NETWORKS AND ISDN SYSTEMS. Elsevier Science Publishers B.V., Box 211, 1000 AE Amsterdam, The Netherlands. Ten times per year. $72.00 per year.

DATA COMMUNICATIONS. McGraw-Hill Information Systems Company, 1221 Avenue of the Americas, New York, NY 10020. (212) 512-2000. Monthly. $30.00 per year.

DATAMATION. Technical Publishing Company, 875 Third Avenue, New York, NY 10022. (212) 605-9400. Semi-monthly. $50.00 per year.

INTERNATIONAL JOURNAL OF COMPUTER AND INFORMATION SCIENCE. Plenum Publishing Corporation, 233 Spring Street, NEw York, NY 10013. (800) 221-9369. Bimonthly. $195.00 per year.

JOURNAL OF TELECOMMUNICATION NETWORKS. Computer Science Press, Incorporated, 1803 Research Boulevard, Suite 500, Rockville, MD 20850-3155. (301) 251-9050. Quarterly. $100.00 per year.

JOURNAL OF THE ASSOCIATION FOR COMPUTING MACHINERY (ACM). Association for Computing Machinery, 11 West 42nd Street, New York, NY 10036. (212) 869-7440. Quarterly. $60.00 per year.

MICRO COMMUNICATIONS. Miller Freeman Publications, Incorporated, 500 Howard Street, San Francisco, CA 94105. (415) 397-1881. Bimonthly. $28.00 per year.

RESEARCH CENTERS AND INSTITUTES

GEORGIA INSTITUTE OF TECHNOLOGY. School of Information and Computer Science, 258 Fourth Street, NW, Atlanta, GA 30332. (404) 894-3180.

SRI INTERNATIONAL CENTER FOR INTELLIGENT COMPUTER SYSTEMS. 333 Ravenswood Avenue, Menlo Park, CA 94025. (415) 859-4771.

UNIVERSITY OF MARYLAND. Center for Automation Research, College Park, MD 20742. (301) 454-4526.

UNIVERSITY OF WATERLOO. Computer Communication Network Group, CPH 2369A, Waterloo, ON, Canada N2L 3G1. (519) 885-1211.

COMPUTER GRAPHICS

See also: AUTOMATION, COMPUTER MEMORY, COMPUTER OPERATING SYSTEMS, COMPUTER PROGRAMMING, COMPUTER VISION, PARALLEL COMPUTERS, PROGRAMMING LANGUAGES, SOFTWARE

ABSTRACT SERVICES AND INDEXES

APPLIED SCIENCE AND TECHNOLOGY INDEX. H.W. Wilson Company, 950 University Avenue, Bronx, NY 10452. (800) 367-6670 or (212) 588-8400. Inquire as to cost and availability.

COMPUTER ABSTRACTS. Technical Information Company, Limited, Post Office Box 59, Saint Helier, Jersey British Channel Inlands, England. Monthly. $310.00 per year.

COMPUTER AND CONTROL ABSTRACTS. Institute of Electrical Engineers, London, United Kingdom. Available from: IEEE Service Center, 445 Hoes Lane, Piscataway, NJ 08854. (201) 981-0060. Semimonthly. $775.00 per year.

COMPUTER AND INFORMATION SYSTEMS: AN ABSTRACT JOURNAL PERTAINING TO THE THEORY, DESIGN, FABRICATION AND APPLICATION OF COMPUTER AND INFORMATION SYSTEMS. Cambridge Scientific Abstracts, Incorporated, 5161 River Road, Bethesda, MD 20816. (301) 951-1400. Semimonthly. $590.00 per year.

COMPUTER CONTENTS: THE BIWEEKLY COMPILATION OF TABLES OF CONTENTS FROM COMPUTER, ELECTRONIC AND TELECOMMUNICATIONS MAGAZINES, JOURNALS AND TRANSACTIONS. Find/SVP, 500 Fifth Avenue, New York, NY 10110. (800) 346-3787 or (212) 354-2424. Biweekly. $115.00 per year.

COMPUTER LITERATURE INDEX. Applied Computer Research, Incorporated, Post Office Box 9280, Phoenix, AZ 85068. (602) 995-5929. Quarterly. $125.00 per year.

COMPUTER PROGRAM ABSTRACTS. U.S. National Aeronautics and Space Administration. Available from U.S. Government Printing Office, Washington, DC 20402. Quarterly. $6.50 per year.

COMPUTING REVIEWS. Association for Computing Machinery, 11 West 42nd Street, New York, NY 10036. (212) 869-7440. Monthly. $60.00 per year.

ENGINEERING INDEX MONTHLY. Engineering Information, Incorporated, 345 East 47th Street, New York, NY 10017. (800) 221-1044 or (212) 705-7600. Monthly, with annual cumulation. $1560.00 per year.

MATHEMATICAL REVIEWS. American Mathematical Society, Post Office Box 6248, Providence, RI 02940. (800) 556-7774 or (401) 272-9500.

SCIENCE CITATION INDEX. Institute for Scientific Information, 3501 Market Street, Philadelphia, PA 19104. (800) 523-1850 or (215) 386-0100.

ANNUAL REVIEWS AND YEARBOOKS

ADVANCES IN COMPUTERS. Academic Press, Incorporated, 6277 Sea Harbor Drive, Orlando, FL 32821. (800) 321-5068. Yearly. Approximately $50.00 per volume.

ASSOCIATIONS AND PROFESSIONAL SOCIETIES

AMERICAN FEDERATION OF INFORMATION PROCESSING SOCIETIES. 1899 Preston White Drive, Reston, VA 22091. (703) 620-8900.

ASSOCIATION FOR COMPUTING MACHINERY (ACM). 11 West 42nd Street, New York, NY 10036. (212) 896-7440.

ASSOCIATION OF COMPUTER PROGRAMMERS AND ANALYSTS. 2108-C Gallows Road, Vienna, VA 22180. (703) 790-0490.

COMPUTER AND AUTOMATION SYSTEMS ASSOCIATION OF SME (SOCIETY OF MANUFACTURING ENGINEERS). One SME Drive, Box 930, Dearborn, MI 48121. (313) 271-1500.

GRAPHIC COMMUNICATIONS ASSOCIATION. 1730 North Lynn Street, Suite 604, Arlington, VA 22209. (703) 841-8160.

IEEE COMPUTER SOCIETY. 1730 Massachusetts Avenue, NW, Washington, DC 20036. (202) 371-0101.

INSTITUTE OF ELECTRICAL AND ELECTRONIC ENGINEERS (IEEE). 345 East 47th Street, New York, NY 10017. (212) 705-7900.

INSTITUTE FOR PERSONAL COMPUTING. Post Office Box 558250, Miami, FL 33155. (305) 274-7440.

MACHINE VISION ASSOCIATION. Post Office Box 930, One SME Drive, Dearborn, MI 48121. (313) 271-1500.

NATIONAL COMPUTER GRAPHICS ASSOCIATION. 2722 Merrilee Drive, Suite 200, Fairfax, VA 22031. (703) 698-9600.

SOCIETY FOR COMPUTER SIMULATION. Post Office Box 17900. San Diego, CA 92117. (619) 277-3888.

SOCIETY FOR INFORMATION DISPLAY. 8055 Manchester Avenue, Suite 615, Playa Del Rey, CA 90293. (213) 305-1502.

SPECIAL INTEREST GROUP ON COMPUTER GRAPHICS (SIGGRAPH). c/o Association for Computing Machinery (ACM). 11 West 42nd Street, New York, NY 10036. (212) 896-7440.

WORLD COMPUTER GRAPHICS ASSOCIATION. 2033 M Street, Suite 399, Washington, DC 20036. (202) 775-9556.

BIBLIOGRAPHIES

COMPUTER BOOKS AND SERIALS IN PRINT, 1986-1987. R.R. Bowker Company, 205 East 42nd Street, New York, NY 10017. (800) 521-8110 or (212) 916-1600. $69.95.

SCIENCE BOOKS AND FILMS. American Association for the Advancement of Science, 1333 H Street, NW, Washington, DC 20005.

SCIENTIFIC AND TECHNICAL BOOKS AND SERIALS IN PRINT 1988; AN INDEX TO LITERATURE IN SCIENCE AND TECHNOLOGY. R.R. Bowker Company, 205 East 42nd Street, New York, NY 10017. (800) 521-8110 or (212) 916-1600. $175.00.

DIRECTORIES AND BIOGRAPHICAL SOURCES

AMERICAN MEN AND WOMEN OF SCIENCE. Physical and Biological Sciences. Fifteenth edition. R.R. Bowker Company, 205 East 42nd Street, New York, NY 10017. (800) 521-8110 or (212) 916-1600.

AMERICAN SOCIETY FOR INFORMATION SCIENCE, HANDBOOK AND DIRECTORY. American Society for Information Science, 1424 16th Street, NW, Suite 404, Washington, DC 20036. (202) 462-1000. $50.00.

COMPUTERS AND COMPUTING INFORMATION RESOURCES DIRECTORY. Gale Research Company, Book Tower, Detroit, MI 48226. (800) 521-0707. 1986. $160.00.

COMPUTER PERIPHERALS REVIEW. GML Information Services, 594 Marrett Road, Lexington, MA 02173. (617) 861-0515. Two issues per year. $215.00 per year. Directory of computer peripheral equipment manufacturers.

COMPUTER REVIEW. GML Information Services, 594 Marrett Road, Lexington, MA 02173. (617) 861-0515. Two issues per year. $195.00 per year. Directory of computer main frame and minicomputer manufacturers.

DATAPRO 70. Datapro Research Corporation, 1805 Underwood Boulevard, Delran, NJ 08075. Twelve monthly updates. $803.00 per year. List of about 1200 companies which offer data processing equipment and services.

DIRECTORY OF CONSULTANTS IN COMPUTER SYSTEMS. Research Publications, Incorporated, 12 Lunar Drive, Woodbridge, CT 06525. (203) 397-2600. Annual. $85.00 per year.

RESEARCH CENTERS DIRECTORY. Gale Research Company, Book Tower, Detroit, MI 48226. (800) 521-0707. Eleventh edition. 1987.

WHO'S WHO IN FRONTIER SCIENCE AND TECHNOLOGY. Marquis Who's Who, Incorporated, 200 East Ohio Street, Chicago, IL 60611. (800) 428-3898 or (312) 787-2008.

WHO'S WHO IN TECHNOLOGY TODAY. Reston Publishing Company, Incorporated, c/o Prentice-Hall, Incorporated, Englewood Cliffs, NJ 07632. (800) 262-6868. Biennial. Five volumes. $425.00. Covers the fields of electronics, computer science, physics, optics, chemistry, biotechnology, mechanics, energy, and earth science.

ENCYCLOPEDIAS AND DICTIONARIES

COMPUTER AND TELECOMMUNICATIONS ACRONYMS. Julie E. Towell and Helen E. Sheppard, editors. Gale Research Company, Book Tower, Detroit, MI 48226. (800) 521-0707. 1986. $60.00.

COMPUTER DICTIONARY. Charles J. Sippl. Howard W. Sams and Company, Incorporated, 4300 West 62nd Street, Indianapolis, IN 46268. (800) 428-7267 or (317) 298-5564. Fourth edition. 1985. $17.95.

DICTIONARY OF COMPUTING. Oxford University Press, 200 Madison Avenue, New York, NY 10016. (212) 679-7300. Second edition. 1986. $29.95.

ENCYCLOPEDIA OF COMPUTER SCIENCE AND ENGINEERING. Anthony Ralston, editor. Van Nostrand Reinhold Book Company, 115 Fifth Avenue, New York, NY 10003. (800) 543-2681. Second edition. 1982. $89.95.

ENCYCLOPEDIA OF COMPUTER SCIENCE AND TECHNOLOGY. Jack Belzer, Albert G. Holzman, and Allan Kent. Marcel Dekker, Incorporated, 270 Madison Avenue, New York, NY 10016. (212) 696-9000. Sixteen volumes. $115.00 per volume.

MCGRAW-HILL DICTIONARY OF COMPUTERS. McGraw-Hill Book Company, 1221 Avenue of the Americas, New York, NY 10020. (212) 512-2000. 1985. $17.50.

GENERAL WORKS

COMPUTER GRAPHICS. David Darling. Dillion Press, Incorporated, 242 Portland Avenue, South, Minneapolis, MN 55415. (612) 333-2691. 1987. $11.95 in paper.

FUNDAMENTALS OF ENGINEERING DRAWING WITH AN INTRODUCTION TO INTERACTIVE COMPUTER GRAPHICS FOR DESIGN AND PRODUCTION. Warren J. Luzadder. Prentice-Hall Press, Incorporated, Englewood Cliffs, NJ 07632. (201) 592-2000. Nineth edition. 1986. $37.95.

FRACTAL GEOMETRY OF NATURE. Benoit B. Mandelbrot. W.H. Freeman and Company, Incorporated, 41 Madison Avenue, New York, NY 10010. (212) 532-7660. 1982. $36.95.

MICROCOMPUTER DISPLAYS, GRAPHICS, AND ANIMATION. Bruce A. Artwick. Prentice-Hall Press, Incorporated, Englewood Cliffs, NJ 07632. (201) 592-2000. 1985. $21.95 in paper.

PROGRAMMING PRINCIPLES IN COMPUTER GRAPHICS. L. Ammeraal. John Wiley and Sons, Incorporated, 605 Third Avenue, New York, NY 10158. (800) 526-5368 or (212) 850-6000. 1986. $19.95 in paper.

HANDBOOKS AND MANUALS

FUNDAMENTALS HANDBOOK OF ELECTRICAL AND COMPUTER ENGINEERING. Sheldon S.L. Chang, editor. John Wiley and Sons, Incorporated, 605 Third Avenue, New York, NY 10158. (800) 526-5368 or (212) 850-6000. Three volumes. 1982. $180.00 set price.

HANDBOOK OF COMPUTERS AND COMPUTING. Arthur H. Seidman, editor. Van Nostrand Reinhold Book Company, 115 Fifth Avenue, New York, NY 10003. (800) 543-2681. 1984. $79.95.

HANDBOOK OF SOFTWARE ENGINEERING. Charles R. Vick, editor. Van Nostrand Reinhold Book Company, 115 Fifth Avenue, New York, NY 10003. (800) 543-2681. 1984. $66.95.

MCGRAW-HILL COMPUTER HANDBOOK. Harry Helms, editor. McGraw-Hill Book Company, 1221 Avenue of the Americas, New York, NY 10020. (212) 512-2000. 1983. $84.50.

ONLINE DATA BASES

COMPENDEX. Engineering Information, Incorporated, 345 East 47th Street, New York, NY 10017. (800) 221-1044 or (212) 705-7615. Engineering and technical literature, 1975 to present. Inquire as to cost and availability.

DISSERTATION ABSTRACTS ONLINE. University Microfilms International, 300 North Zeeb Road, Ann Arbor, MI 48106. (800) 521-0600 or (313) 761-4700. Scope includes virtually all doctoral dissertations accepted at accredited American institutions from 1861 to present in 252 subject areas. Inquire as to cost and availability.

INSPEC. INSPEC Marketing Department, Institute of Electrical and Electronics Engineers, Incorporated, IEEE Service Department, 445 Hoes Lane, Piscataway, NJ 08854. (201) 981-0060. Inquire as to on-line cost and availability.

MATHFILE. American Mathematical Society, Post Office Box 6248, Providence, RI 02940. (800) 556-7774 or (401) 272-9500. Scope includes pure and applied mathematics and related areas of physics, statistics, engineering, computer science, and operations research literature since 1973. Inquire as to cost and availability.

NTIS. National Technical Information Service, 5285 Port Royal Road, Springfield, VA 22161. (703) 487-4630. Broad coverage of government sponsored research reports, 1964 to present. Inquire as to cost and availability.

SCISEARCH. Institute for Scientific Information, 3501 Market Street, Philadelphia, PA 19104. (800) 523-1850 or (215) 386-0100. Broad multidisciplinary title and author index to the international literature of science and technology, 1974 to present. Inquire as to cost and availability.

OTHER SOURCES

COMPUTER PUBLISHERS AND PUBLICATIONS 1985-86: AN INTERNATIONAL DIRECTORY AND YEARBOOK. Edited by Efrem Sigel and Frederica Evan. Communications Trends, Incorporated. Distributed by Gale Research Company, Book Tower, Detroit, MI 48226. (800) 521-0707. Second edition. 1985. $95.00. Supplement available for 1986, $35.00. Provides information on publishers of computer books and periodicals, recommended titles for libraries and bookstores, industry trends and statistics, and computer manufacturers as publishers.

PERIODICALS

ACM TRANSACTIONS ON COMPUTER SYSTEMS. Association for Computing Machinery, 11 West 42nd Street, New York, NY 10036. (212) 869-7440. Quarterly. $70.00 per year.

COMPUTER-AIDED DESIGN. Butterworth Scientific Limited, Post Office Box 63, Westbury House, Bury Street, Guildford, Surrey, GU2 5BH, England. Ten times per year. $285.00 per year.

COMPUTER GRAPHICS. Association for Computing Machinery, 11 West 42nd Street, New York, NY 10036. (212) 869-7440. Quarterly. $25.00 per year.

COMPUTER GRAPHICS FORUM. Elsevier Science Publishers B.V., Box 211, 1000 AE Amsterdam, The Netherlands. Four times per year. $85.00 per year.

COMPUTER GRAPHICS TODAY. Media Horizons, Incorporated, 475 Park Avenue, South, New York, NY 10016. (212) 725-2300. Monthly. $15.00 per year.

COMPUTER GRAPHICS WORLD. Pennwell Publishing Company, Advanced Technology Group, 1714 Stockton Street, San Francisco, CA 94133. (415) 398-7151. Monthly. $30.00 per year.

COMPUTERS AND GRAPHICS. Pergamon Press, Incorporated, Journals Division, Maxwell House, Fairview Park, Elmsford, NY 10523. (914) 592-7700. Quarterly. $200.00 per year.

CREATIVE COMPUTING. Creative Computing, Post Office Box 5214, Boulder, CO 80321. Monthly. $25.00 per year.

DISPLAYS: TECHNOLOGY AND APPLICATIONS. Butterworth Scientific Limited, Post Office Box 63, Westbury House, Bury Street, Guildford, Surrey, GU2 5BH, England. Quarterly. $133.00 per year.

IEEE COMPUTER GRAPHICS AND APPLICATIONS. Institute of Electrical and Electronic Engineers (IEEE), IEEE Service Center, 445 Hoes Lane, Piscataway, NJ 08854. (201) 981-0060. Monthly. $72.00 per year.

IEEE TRANSACTIONS ON PATTERN ANALYSIS AND MACHINE INTELLIGENCE. Institute of Electrical and Electronic Engineers (IEEE), IEEE Service Center, 445 Hoes Lane, Piscataway, NJ 08854. (201) 981-0060. Bimonthly. $125.00 per year.

IMAGE AND VISION COMPUTING. Butterworth Scientific Limited, Post Office Box 63, Westbury House, Bury Street, Guildford, Surrey, GU2 5BH, England, Quarterly. $175.00 per year.

PATTERN RECOGNITION LETTERS. Elsevier Science Publishers B.V., Box 211, 1000 AE Amsterdam, The Netherlands. Bimonthly. $90.00 per year.

RECOGNITION TECHNOLOGIES TODAY. Recognition Technologies Users Association, Box 2016, Battenkill Building, Manchester, VT 05255. (802) 362-4151. Bimonthly. Inquire as to cost and availability.

S. KLEIN NEWSLETTER ON COMPUTER GRAPHICS. Technology and Business Communications, Incorporated, 730 Boston Post Road, Box 915, Sudbury, MA 01776. (617) 443-4671. Semimonthly. $178.00 per year.

RESEARCH CENTERS AND INSTITUTES

NEW YORK INSTITUTE OF TECHNOLOGY. Computer Graphics Laboratory, Old Westbury, NY 11568. (516) 686-7644.

OHIO STATE UNIVERSITY. Computer Graphics Research Group, Cranston Center, 1501 Neil Avenue, Columbus, OH 43210. (614) 422-3416.

PURDUE UNIVERSITY. Computer Aided Design and Graphics Laboratory, 134 Potter Engineering Center, West Lafayette, IN 47907. (317) 494-5944.

RENSSELAER POLYTECHNIC INSTITUTE. Center for Interactive Computer Graphics, CC 121, Troy, NY 12180-3590. (518) 266-6751.

UNIVERSITY OF MINNESOTA. Computer Graphics and Computer Aided Engineering Laboratory, 111 Church Street, SE, Minneapolis, MN 55455. (612) 373-2977.

COMPUTER HACKERS

See: COMPUTER SECURITY

COMPUTER MEMORY AND STORAGE

See also: AUTOMATION, COMPUTER OPERATING SYSTEMS, COMPUTER PROGRAMMING, COMPUTER VISION, PARALLEL COMPUTERS, PROGRAMMING LANGUAGES, SOFTWARE

ABSTRACT SERVICES AND INDEXES

APPLIED SCIENCE AND TECHNOLOGY INDEX. H.W. Wilson Company, 950 University Avenue, Bronx, NY 10452. (800) 367-6670 or (212) 588-8400. Inquire as to cost and availability.

COMPUTER ABSTRACTS. Technical Information Company, Limited, Post Office Box 59, Saint Helier, Jersey British Channel Inlands, England. Monthly. $310.00 per year.

COMPUTER AND CONTROL ABSTRACTS. Institute of Electrical Engineers, London, United Kingdom. Available from: IEEE Service Center, 445 Hoes Lane, Piscataway, NJ 08854. (201) 981-0060. Semimonthly. $775.00 per year.

COMPUTER AND INFORMATION SYSTEMS: AN ABSTRACT JOURNAL PERTAINING TO THE THEORY, DESIGN, FABRICATION AND APPLICATION OF COMPUTER AND INFORMATION SYSTEMS. Cambridge Scientific Abstracts, Incorporated, 5161 River Road, Bethesda, MD 20816. (301) 951-1400. Semimonthly. $590.00 per year.

COMPUTER CONTENTS: THE BIWEEKLY COMPILATION OF TABLES OF CONTENTS FROM COMPUTER, ELECTRONIC AND TELECOMMUNICATIONS MAGAZINES, JOURNALS AND TRANSACTIONS. Find/SVP, 500 Fifth Avenue, New York, NY 10110. (800) 346-3787 or (212) 354-2424. Biweekly. $115.00 per year.

COMPUTER LITERATURE INDEX. Applied Computer Research, Incorporated, Post Office Box 9280, Phoenix, AZ 85068. (602) 995-5929. Quarterly. $125.00 per year.

COMPUTER PROGRAM ABSTRACTS. U.S. National Aeronautics and Space Administration. Available from U.S. Government Printing Office, Washington, DC 20402. Quarterly. $6.50 per year.

COMPUTING REVIEWS. Association for Computing Machinery, 11 West 42nd Street, New York, NY 10036. (212) 869-7440. Monthly. $60.00 per year.

ENGINEERING INDEX MONTHLY. Engineering Information, Incorporated, 345 East 47th Street, New York, NY 10017. (800) 221-1044 or (212) 705-7600. Monthly, with annual cumulation. $1560.00 per year.

SCIENCE CITATION INDEX. Institute for Scientific Information, 3501 Market Street, Philadelphia, PA 19104. (800) 523-1850 or (215) 386-0100.

ANNUAL REVIEWS AND YEARBOOKS

ADVANCES IN COMPUTERS. Academic Press, Incorporated, 6277 Sea Harbor Drive, Orlando, FL 32821. (800) 321-5068. Yearly. Approximately $50.00 per volume.

ASSOCIATIONS AND PROFESSIONAL SOCIETIES

AMERICAN FEDERATION OF INFORMATION PROCESSING SOCIETIES. 1899 Preston White Drive, Reston, VA 22091. (703) 620-8900.

ASSOCIATION OF COMPUTING MACHINERY (ACM). 11 West 42nd Street, New York, NY 10036. (212) 896-7440.

ASSOCIATION OF COMPUTER PROGRAMMERS AND ANALYSTS. 2108-C Gallows Road, Vienna, VA 22180. (703) 790-0490.

ASSOCIATION OF MINICOMPUTER USERS. Two Frederick Street, Framingham, MA 01701. (617) 879-5955.

COMPUTER AND AUTOMATION SYSTEMS ASSOCIATION OF SME (SOCIETY OF MANUFACTURING ENGINEERS). One SME Drive, Box 930, Dearborn, Mi 48121. (313) 271-1500.

IEEE COMPUTER SOCIETY. 1730 Massachusetts Avenue, NW, Washington, DC 20036. (202) 371-0101.

INSTITUTE OF ELECTRICAL AND ELECTRONIC ENGINEERS (IEEE). 345 East 47th Street, New York, NY 10017. (212) 705-7900.

INSTITUTE FOR PERSONAL COMPUTING. Post Office Box 558250, Miami, FL 33155. (305) 274-7440.

BIBLIOGRAPHIES

COMPUTER BOOKS AND SERIALS IN PRINT, 1986-1987. 1986. R.R. Bowker Company, 205 East 42nd Street, New York, NY 10017. (800) 521-8110 or (212) 916-1600. $175.00.

SCIENTIFIC AND TECHNICAL BOOKS AND SERIALS IN PRINT 1988; AN INDEX TO LITERATURE IN SCIENCE AND TECHNOLOGY. R.R. Bowker Company, 205 East 42nd Street, New York, NY 10017. (800) 521-8110 or (212) 916-1600. $175.00.

DIRECTORIES AND BIOGRAPHICAL SOURCES

AMERICAN SOCIETY FOR INFORMATION SCIENCE, HANDBOOK AND DIRECTORY. American Society for Information Science, 1424 16th Street, NW, Suite 404, Washington, DC 20036. (202) 462-1000. $50.00.

COMPUTERS AND COMPUTING INFORMATION RESOURCES DIRECTORY. Gale Research Company, Book Tower, Detroit, MI 48226. (800) 521-0707. 1986. $160.00.

COMPUTER PERIPHERALS REVIEW. GML Information Services, 594 Marrett Road, Lexington, MA 02173. (617) 861-0515. Two issues per year. $215.00 per year. Directory of computer peripheral equipment manufacturers.

COMPUTER REVIEW. GML Information Services, 594 Marrett Road, Lexington, MA 02173. (617) 861-0515. Two issues per year. $195.00 per year. Directory of computer main frame and minicomputer manufacturers.

DATAPRO 70. Datapro Research Corporation, 1805 Underwood Boulevard, Delran, NJ 08075. Twelve monthly updates. $803.00 per year. List of about 1200 companies which offer data processing equipment and services.

DIRECTORY OF CONSULTANTS IN COMPUTER SYSTEMS. Research Publications, Incorporated, 12 Lunar Drive, Woodbridge, CT 06525. (203) 397-2600. Annual. $85.00 per year.

RESEARCH CENTERS DIRECTORY. Gale Research Company, Book Tower, Detroit, MI 48226. (800) 521-0707. Eleventh edition. 1987.

WHO'S WHO IN FRONTIER SCIENCE AND TECHNOLOGY. Marquis Who's Who, Incorporated, 200 East Ohio Street, Chicago, IL 60611. (800) 428-3898 or (312) 787-2008.

WHO'S WHO IN TECHNOLOGY TODAY. Reston Publishing Company, Incorporated, c/o Prentice-Hall, Incorporated, Englewood Cliffs, NJ 07632. (800) 262-6868. Biennial. Five volumes. $425.00. Covers the fields of electronics, computer science, physics, optics, chemistry, biotechnology, mechanics, energy, and earth science.

ENCYCLOPEDIAS AND DICTIONARIES

COMPUTER DICTIONARY. Charles J. Sippl. Howard W. Sams and Company, Incorporated, 4300 West 62nd Street, Indianapolis, IN 46268. (800) 428-7267 or (317) 298-5564. Fourth edition. 1985. $17.95.

DICTIONARY OF COMPUTING. Oxford University Press, 200 Madison Avenue, New York, NY 10016. (212) 679-7300. Second edition. 1986. $29.95.

ENCYCLOPEDIA OF COMPUTER SCIENCE AND ENGINEERING. Anthony Ralston, editor. Van Nostrand Reinhold Book Company, 115 Fifth Avenue, New York, NY 10003. (800) 543-2681. Second edition. 1982. $89.95.

ENCYCLOPEDIA OF COMPUTER SCIENCE AND TECHNOLOGY. Jack Belzer, Albert G. Holzman, and Allan Kent. Marcel Dekker, Incorporated, 270 Madison Avenue, New York, NY 10016. (212) 696-9000. Sixteen volumes. $115.00 per volume.

MCGRAW-HILL DICTIONARY OF COMPUTERS. McGraw-Hill Book Company, 1221 Avenue of the Americas, New York, NY 10020. (212) 512-2000. 1985. $17.50.

GENERAL WORKS

COMPUTERS AND DATA PROCESSING. H.L. Capron and Brian K. Williams. Benjamin/Cummings Publishing Company, Incorporated, 2727 Sand Hill Road, Menlo Park, CA 94025. (415) 854-6020. Second edition. 1984. $28.95.

DIGITAL COMPUTER'S MEMORY TECHNOLOGY. Majumda Dutta and J. Das. John Wiley and Sons, Incorporated, 605 Third Avenue, New York, NY 10158. (800) 526-5368 or (212) 850-6000. Second edition. 1984. $31.95.

FIFTH GENERATION COMPUTERS. Richard K. Miller, editor. The Fairmont Press, Incorporated, 4025 Pleasantdale Road, Atlanta, GA 30340. (404) 447-5314.

INFORMATION SYSTEMS: THEORY AND PRACTICE. J. Burch and G. Grudnitski. John Wiley and Sons, Incorporated, 605 Third Avenue, New York, NY 10158. Fourth edition. 1986. $30.95.

INTRODUCTION TO COMPUTER ARCHITECTURE AND ORGANIZATION. H. Lorin. John Wiley and Sons, Incorporated, 605 Third Avenue, New York, NY 10158. 1982. $30.95.

OPTICAL DISKS FOR DATA AND DOCUMENTS STORAGE. William Saffady. Meckler Publishers, 11 Ferry Lane West, Westport, CT 06880. (203) 226-6967. 1986. $29.95.

HANDBOOKS AND MANUALS

FUNDAMENTALS HANDBOOK OF ELECTRICAL AND COMPUTER ENGINEERING. Sheldon S.L. Chang, editor. John Wiley and Sons, Incorporated, 605 Third Avenue, New York, NY 10158. (800) 526-5368 or (212) 850-6000. Three volumes. 1982. $180.00 set price.

HANDBOOK OF COMPUTERS AND COMPUTING. Arthur H. Seidman, editor. Van Nostrand Reinhold Book Company, 115 Fifth Avenue, New York, NY 10003. (800) 543-2681. 1984. $79.95.

HANDBOOK OF SOFTWARE ENGINEERING. Charles R. Vick, editor. Van Nostrand Reinhold Book Company, 115 Fifth Avenue, New York, NY 10003. (800) 543-2681. 1984. $66.95.

MCGRAW-HILL COMPUTER HANDBOOK. Harry Helms, editor. McGraw-Hill Book Company, 1221 Avenue of the Americas, New York, NY 10020. (212) 512-2000. 1983. $84.50.

MEMORY COMPONENTS HANDBOOK. Intel Staff. Intel Corporation, 3065 Bowers Avenue, Santa Clara, Ca 95051. (408) 496-8225. 1986. $18.00 in paper.

ONLINE DATA BASES

COMPENDEX. Engineering Information, Incorporated, 345 East 47th Street, New York, NY 10017. (800) 221-1044 or (212) 705-7615. Engineering and technical literature, 1975 to present. Inquire as to cost and availability.

DISSERTATION ABSTRACTS ONLINE. University Microfilms International, 300 North Zeeb Road, Ann Arbor, MI 48106. (800) 521-0600 or (313) 761-4700. Scope includes virtually all doctoral dissertations accepted at accredited American institutions from 1861 to present in 252 subject areas. Inquire as to cost and availability.

INSPEC. INSPEC Marketing Department, Institute of Electrical and Electronics Engineers, Incorporated, IEEE Service Department, 445 Hoes Lane, Piscataway, NJ 08854. (201) 981-0060. Inquire as to on-line cost and availability.

MATHFILE. American Mathematical Society, Post Office Box 6248, Providence, RI 02940. (800) 556-7774 or (401) 272-9500. Scope includes pure and applied mathematics and related areas of physics, statistics, engineering, computer science, and operations research literature since 1973. Inquire as to cost and availability.

NTIS. National Technical Information Service, 5285 Port Royal Road, Springfield, VA 22161. (703) 487-4630. Broad coverage of government sponsored research reports, 1964 to present. Inquire as to cost and availability.

SCISEARCH. Institute for Scientific Information, 3501 Market Street, Philadelphia, PA 19104. (800) 523-1850 or (215) 386-0100. Broad multidisciplinary title and author index to the international literature of science and technology, 1974 to present. Inquire as to cost and availability.

PERIODICALS

ACM TRANSACTIONS ON COMPUTER SYSTEMS. Association for Computing Machinery, 11 West 42nd Street, New York, NY 10036. (212) 869-7440. Quarterly. $70.00 per year.

BYTE. Byte Publications, Incorporated, 70 Main Street, Petersborough, NH 03458. (603) 924-9281. Monthly. $21.00 per year.

COMMUNICATIONS OF THE ACM. Association for Computing Machinery, 11 West 42nd Street, New York, NY 10036. (212) 869-7440. Monthly. $78.00 per year.

COMPUTER. Institute of Electrical and Electronic Engineers (IEEE), IEEE Service Center, 445 Hoes Lane, Piscataway, NJ 08854. (201) 981-0060. Monthly. $90.00 per year.

COMPUTER DESIGN. PennWell Directories, Incorporated, Post Office Box 1260, Tulsa, OK 74101. (018) 835-3161. Monthly. $60.00 per year.

COMPUTER MAGAZINE. IEEE Computer Society, 1109 Spring Street, Suite 300, Silver Spring, MD 20910. (301) 589-8142. Monthly. $34.00 per year.

COMPUTERS AND ELECTRONICS. Ziff-Davis Publishing Company, 3460 Wilshire Boulevard, Los Angeles, Ca 90010. Monthly. $17.00 per year.

COMPUTERWORLD. C.W. Communications, 375 Cochituate Road, Farmington, MA 01701. (617) 879-0700. $44.00 per year.

CREATIVE COMPUTING. Creative Computing, Post Office Box 5214, Boulder, CO 80321. Monthly. $25.00 per year.

DATAMATION. Technical Publishing Company, 875 Third Avenue, New York, NY 10022. (212) 605-9400. Semi-monthly. $50.00 per year.

IEEE TRANSACTIONS ON COMPUTERS. Institute of Electrical and Electronic Engineers (IEEE), IEEE Service Center, 445 Hoes Lane, Piscataway, NJ 08854. (201) 981-0060. Monthly. $130.00 per year.

INTERNATIONAL JOURNAL OF COMPUTER AND INFORMATION SCIENCE. Plenum Publishing Corporation, 233 Spring Street, NEw York, NY 10013. (800) 221-9369. Bimonthly. $195.00 per year.

JOURNAL OF THE ASSOCIATION FOR COMPUTING MACHINERY (ACM). Association for Computing Machinery, 11 West 42nd Street, New York, NY 10036. (212) 869-7440. Quarterly. $60.00 per year.

SIGARCH COMPUTER ARCHITECTURE NEWS. Association for Computing Machinery, 11 West 42nd Street, New York, NY 10036. (212) 869-7440. Bimonthly. $15.00 per year.

RESEARCH CENTERS AND INSTITUTES

MICROELECTRONICS CENTER OF NORTH CAROLINA. Post Office Box 12889, Research Triangle Park, NC 27709. (010) 248-1800.

STANDARD UNIVERSITY. Center for Integrated Systems, Stanford, CA 94305. (415) 725-3620.

SYRACUSE UNIVERSITY. New York State Center for Advanced Technology in Computer Application and Software, 120 Hinds Hall, Syracuse, NY 13244-1190. (315) 423-1064.

UNIVERSITY OF ILLINOIS. Center for Supercomputing Research and Development, 305 Talbot Laboratory, 104 South Right Street, Urbana, IL 61801. (217) 333-6223.

COMPUTER OPERATING SYSTEMS

See also: ARTIFICIAL INTELLIGENCE, COMPUTER PROGRAMMING, COMPUTERS, MICROPROCESSORS, PARALLEL COMPUTERS, SOFTWARE, SOFTWARE ENGINEERING, SYSTEMS ANALYSIS, SYSTEMS ENGINEERING

ABSTRACT SERVICES AND INDEXES

APPLIED SCIENCE AND TECHNOLOGY INDEX. H.W. Wilson and Company, 950 University Avenue, Bronx, NY 10452. (800) 367-6670 or (212) 588-8400. Monthly. Inquire as to cost and availability.

COMPUTER AND CONTROL ABSTRACTS. Institute of Electrical Engineers. Available from: Institute of Electrical and Electronics Engineers. IEEE Service Center, 445 Hoes Lane, Piscataway, NJ 08854. Semimonthly. $775.00 per year.

COMPUTER AND INFORMATION SYSTEMS: AN ABSTRACT JOURNAL PERTAINING TO THE THEORY, DESIGN, FABRICATION AND APPLICATION OF COMPUTER AND INFORMATION SYSTEMS. Cambridge Scientific Abstracts, 5161 River Road, Bethesda, MD 20816. 1972 to present. Semi-monthly. Inquire as to cost and availability.

COMPUTER CONTENTS: THE BIWEEKLY COMPILATION OF TABLES OF CONTENTS FROM COMPUTER, ELECTRONIC AND TELECOMMUNICATIONS MAGAZINES, JOURNALS AND TRANACTIONS. Find/SVP, 500 Fifth Avenue, New York, NY 101110. (800) 346-3787 or (212) 354-2424. Biweekly. $115.00 per year.

COMPUTER LITERATURE INDEX. Applied Computer Research, Inc., P.O. Box 9280, Phoenix, AZ 85068. (602) 995-5929. Quarterly. $125.00 per year.

COMPUTER PROGRAMS ABSTRACTS. U.S. National Aeronautics and Space Administration. Available from: U.S. Government Printing Office, Washington, DC 20402. Quarterly. $10.00 per year.

COMPUTING REVIEWS. Association of Computing Machinery, 11 West 42nd Street, New York, NY 10036. (212) 869-7440. Monthly. $60.00 per year.

CURRENT CONTENTS: ENGINEERING, TECHNOLOGY AND APPLIED SCIENCES. Institute for Scientific Information, 3501 Market Street, Philadelphia, PA 19104. (800) 523-1850 or (215) 386-0100. Weekly. $275.00 per year.

ENGINEERING INDEX MONTHLY AND AUTHOR INDEX. Engineering Information Inc., 345 East 47th Street, New York, NY 10017. (212) 705-7600. Monthly. $1560.00 per year.

INDEX TO SCIENTIFIC AND TECHNICAL PROCEEDINGS. Institute for Scientific Information, 3501 Market Street, Philadelphia, PA 19104. (800) 523-1850 or (215) 386-0100. 1978 to present. Monthly. $775.00 per year.

INDEX TO SCIENTIFIC REVIEWS. Institute for Scientific Information, 3501 Market Street, Philadelphia, PA 19104. (800) 523-1850 or (215) 386-0100. 1974 to present. Semi-annual. $550.00 per year.

PCR-2: PERSONAL COMPUTER REVIEW - SQUARED. Toolbox Publications, Inc., P.O. Box 5451, 2514 Birch Creek Lane, Orchard Lake, MI 48033. 1987 to present. Bimonthly. $60.00 per year.

PHYSICS ABSTRACTS. Institution of Electrical Engineers. Available from: IEEE Service Center, 445 Hoes Lane, Piscataway, NJ 08854. 1898 to present. Bimonthly. $1700.00 per year.

SCIENCE CITATION INDEX. Institute for Scientific Information, 3501 Market Street, Philadelphia, PA 19104. (800) 523-1850 or (215) 386-0100. Six times per year. $6200.00 per year.

ANNUAL REVIEWS AND YEARBOOKS

ADVANCES IN COMPUTERS. Academic Press, Inc., 6277 Sea Harbor Drive, Orlando, FL 32821. (800) 321-5068. Yearly. Approximately $50.00 per volume.

ASSOCIATIONS AND PROFESSIONAL SOCIETIES

AMERICAN FEDERATION OF INFORMATION PROCESSING SOCIETIES. 1899 Preston White Drive, Reston, VA 22091. (703) 620-8900.

ASSOCIATION OF COMPUTER PROGRAMMERS AND ANALYSTS. 2108-C Gallows Road, Vienna, VA 22180. (703) 790-0490.

ASSOCIATION OF COMPUTING MACHINERY (ACM). 11 West 42nd Street, New York, NY 10036. (212) 869-7440.

IEEE COMPUTER SOCIETY. 1730 Massachusetts Avenue, N.W., Washington, DC 20036. (202) 371-0101.

INSTITUTE OF ELECTRICAL AND ELECTRONICS ENGINEERS. IEEE Service Center, 445 Hoes Lane, Piscataway, NJ 08854.

SOCIETY FOR COMPUTER SIMULATION. P.O. Box 17900, San Diego, CA 92117. (619) 277-3888.

DIRECTORIES AND BIOGRAPHICAL SOURCES

AMERICAN MEN AND WOMEN OF SCIENCE. R.R. Bowker, Inc., Order Department, 245 West 17th Street, New York, NY 10011. (800) 521-8110. Eight volumes. 1986. $595.00 for set.

AMERICAN SOCIETY FOR INFORMATION SCIENCE HANDBOOK AND DIRECTORY. American Society for Information Science, 1424 16th Street, N.W., Suite 404, Washington, DC 20036. (202) 462-1000. $50.00.

COMPUTERS AND COMPUTING INFORMATION RESOURCES DIRECTORY. Martin Connors, editor. Gale Research Company, Book Tower, Detroit, MI 48226. (800) 521-0707. 1987. $165.00. Supplement available at $85.00.

INTERNATIONAL RESEARCH CENTERS DIRECTORY 1988-89. Darren L. Smith, editor. Gale Research Company, Book Tower, Detroit, MI 48226. (800) 521-0707. 4th edition. 1987. $360.00.

1987 DIRECTORY OF ENGINEERING SOCIETIES AND RELATED ORGANIZATIONS. Gordon Davis, editor. Hemisphere Publishing Corporation, 1010 Vermont Avenue, NW, Washington, DC 20005. (800) 526-0275. 12th edition. 1987. $100.00.

RESEARCH CENTERS DIRECTORY 1988. Gale Research Company, Book Tower, Detroit, MI 48226. (800) 521-0707. 12th edition. 1987. $365.00 for set.

SCIENTIFIC AND TECHNICAL ORGANIZATIONS AND AGENCIES DIRECTORY. Margaret Labash Young, editor. Gale Research Company, Book Tower, Detroit, MI 48226. (800) 521-0707. 2nd edition. 1987. $185.00.

WHO'S WHO IN ENGINEERING. Gordon Davis, editor. Hemisphere Publishing Corporation, 1010 Vermont Avenue, NW, Washington, DC 20005. (800) 526-0275. 6th edition. 1985. $200.00.

ENCYCLOPEDIAS AND DICTIONARIES

COMPUTER AND TELECOMMUNICATIONS ACRONYMS. Julie E. Towell and Helen E. Sheppard, editors. Gale Research Company, Book Tower, Detroit, MI 48226. (800) 521-0707. 1986. $60.00.

DICTIONARY OF COMPUTING. Oxford University Press, 200 Madison Avenue, New York, NY 10016. (800) 458-5833. Second edition. 1986. $29.95.

ENCYCLOPEDIA OF INFORMATION SYSTEMS AND SERVICES 1988. Amy Lucas and Annette Novallo, editors. Gale Research Company, Book Tower, Detroit, MI 48226. (800) 521-0707. 8th edition. 1987. $400.00 for set.

PRENTICE-HALL ENCYCLOPEDIA OF INFORMATION TECHNOLOGY. Robert A. Edmunds. Prentice-Hall Publishing, Inc., Englewood Cliffs, NJ 07632. (800) 562-0245. 1987. $49.95.

SOFTWARE ENCYCLOPEDIA. R.R. Bowker Company, 205 East 42nd Street, New York, NY 10017. (800) 521-8110. Tow volumes. 1987. $125.00 for set.

GENERAL WORKS

COMPUTER ORGANIZATION: HARDWARE/SOFTWARE. G.W. Gorsline. Prentice-Hall Publishing, Inc., Englewood Cliffs, NJ 07632. (800) 562-0245. Second edition. 1986. $40.95.

CONCURRENT PROGRAMMING FOR SOFTWARE ENGINEERS. Dick Whiddett. Halsted Press, available from: John Wiley and Sons, Inc., 605 Third Avenue, New York, NY 10158. (800) 526-5368. 1987. $39.95.

DESIGN OF DISTRIBUTED OPERATING SYSTEMS. P.J. Fortier. McGraw-Hill Book Company, 1221 Avenue of the Americas, New York, NY 10020. (212) 512-2000. 1986. $42.95.

MICROPROCESSORS AND MICROCOMPUTERS: HARDWARE AND SOFTWARE. R.J. Tocci and L.P. Laskowski. Prentice-Hall Publishing, Inc., Englewood Cliffs, NJ 07632. (800) 562-0245. Third edition. 1987. $37.95.

OPERATING SYSTEMS CONCEPTS AND DESIGN. M. Milenkovic. McGraw-Hill Book Company, 1221 Avenue of the Americas, New York, NY 10020. (212) 512-2000. 1987. $40.95.

OPERATING SYSTEMS: STRUCTURES AND MECHANISMS. Philippe Janson. Academic Press, Inc., 6277 Sea Harbor Drive, Orlando, FL 32821. (800) 321-5068. 1985. $32.50.

THE UNIX OPERATING SYSTEM. K. Christian. John Wiley and Sons, Inc., 605 Third Avenue, New York, NY 10158. (800) 526-5368. 1983. $21.95 in paper.

HANDBOOKS AND MANUALS

HANDBOOK OF SOFTWARE ENGINEERING. Charles R. Vick, editor. Van Nostrand Reinhold Company, Inc., 135 West 50th Street, New York, NY 10020. (800) 543-2681. 1984. $66.95.

ONLINE DATA BASES

COMPENDEX. Engineering Information, Inc., 345 East 47th Street, New York, NY 10017. (800) 221-1044 or (212) 705-7615. Engineering and technical literature, 1975 to present. Inquire as to online cost and availability.

DISSERTATION ABSTRACTS ONLINE. University Microfilms International, 300 North Zeeb Road, Ann Arbor, MI 48106. (800) 521-0600 or (313) 761-4700. Scope includes virtually all doctoral dissertations accepted at accredited American institutions from 1861 to present in over 250 subject areas. Inquire as to online cost and availability.

INSPEC. INSPEC Marketing Department, Institution of Electrical Engineers. Available from IEEE Service Center, 445 Hoes Lane, Piscataway, NJ 08854. (201) 981-0060. Online version of Physics Abstracts. Inquire as to online cost and availability.

NTIS. National Technical Information Service, 5285 Port Royal Road, Springfield, VA 22161. (703) 487-4630. Broad coverage of government sponsored research reports, 1964 to present. Inquire as to online cost and availability.

SCISEARCH. Institute for Scientific Information, 3501 Market Street, Philadelphia, PA 19104. (800) 523-1850 or (215) 386-0100. Broad multidisciplinary title and author index to the international literature of science and technology, 1974 to present. Inquire as to online cost and availability.

WILSONLINE. H.W. Wilson and Company, 950 University Avenue, Bronx, NY 10452. (800) 367-6770 or (212) 588-8400. Makes available online versions of the H.W. Wilson indexes including Applied Science and Technology Index, Business Periodicals Index and Readers' Guide to Periodical Literature. Approximately 1980 to present. Inquire as to online cost and availability.

PERIODICALS

ACM TRANSACTIONS ON PROGRAMMING LANGUAGES AND SYSTEMS. Association of Computing Machinery, 11 West 42nd Street, New York, NY 10036. (212) 869-7440. 1979 to present. Quarterly. $55.00 per year.

ADVANCES IN ENGINEERING SOFTWARE. CML Publications, 400 West Cummings Park, Suite 6200, Woburn, MA 01801. (617) 933-7374. 1979 to present. Quarterly. $130.00 per year.

BYTE. Byte Publications, Inc., 70 Main Street, Petersborough, NH 03458. (603) 924-9281. Monthly. $21.00 per year.

COMMUNICATIONS OF THE ACM. Association of Computing Machinery, 11 West 42nd Street, New York, NY 10036. (212) 869-7440. Monthly. $80.00 per year.

COMPUTER. Institute of Electrical and Electronics Engineers. IEEE Service Center, 445 Hoes Lane, Piscataway, NJ 08854. 1966 to present. Monthly. $90.00 per year.

IEEE TRANSACTIONS ON SOFTWARE ENGINEERING. Institute of Electrical and Electronics Engineers. IEEE Service Center, 445 Hoes Lane, Piscataway, NJ 08854. 1975 to present. Monthly. $160.00 per year.

JOURNAL OF SYSTEMS AND SOFTWARE. Elsevier Science Publishing Company, Inc., 52 Vanderbilt Avenue, New York, NY 10017. (212) 370-5520. 1979 to present. Quarterly. $95.00 per year.

MICROPROCESSORS AND MICROSYSTEMS. Butterworth's Publishing, 80 Montvale Avenue, Stoneham, MA 02180. (800) 325-4177. 1976 to present. Ten times per year. $160.00 per year.

MINI-MICRO SYSTEMS. Cahners Publishing Company, Inc., 275 Washington Street, Newton, MA 02158. (617) 964-3030. 1968 to present. Monthly. $65.00.

SIGSOFT SOFTWARE ENGINEERING NOTICES. Association of Computing Machinery Special Interest Group on Software Engineering. 11 West 42nd Street, New York, NY 10036. (212) 869-7440. Quarterly. $12.00 per year.

SOFTWARE DEVELOPER'S MONTHLY. SourceView Press, 835 Castro Street, Martinez, CA 94553. (415) 228-6220. 1985 to present. Monthly. $144.00 per year.

SOFTWARE ENGINEERING JOURNAL. Institute of Electrical Engineers, Savoy Place, London, WC2R OBL, England. 1981 to present. Bimonthly. $85.00 per year.

SOFTWARE PRACTICE AND EXPERIENCE. John Wiley and Sons, Inc., 605 Third Avenue, New York, NY 10158. (800) 526-5368. 1971 to present. Monthly. $260.00 per year.

UNIX REVIEW (SAN FRANCISCO). Miller Freeman Publications, Inc., 500 Howard Street, San Francisco, CA 94105. (415) 397-1881. 1983 to present. Monthly. $35.00 per year.

RESEARCH CENTERS AND INSTITUTES

INSTITUTE FOR INFORMATION SCIENCE AND TECHNOLOGY. George Washington University, 801 22nd Street, N.W., Washington, DC 20052. (202) 676-4921.

RESEARCH INSTITUTE FOR COMPUTING AND INFORMATION SYSTEMS. University of Houston at Clear Lake, 2700 Bay Area Boulevard, Houston, TX 77058. (713) 488-9392.

SOFTWARE ENGINEERING INSTITUTE. Carnegie-Mellon University, Pittsburgh, PA 15213. (412) 268-6700.

SOFTWARE ENGINEERING RESEARCH INSTITUTE. Georgia Institute of Technology, 258 Fourth Street, N.W., Atlanta, GA 30332. (404) 894-3180.

COMPUTER PROGRAMMING

See also: ARTIFICIAL INTELLIGENCE, COMPUTERS, COMPUTER OPERATING SYSTEMS, PARALLEL COMPUTERS, PROGRAMMING LANGUAGES, SOFTWARE, SOFTWARE ENGINEERING, SYSTEMS ANALYSIS, SYSTEMS ENGINEERING

ABSTRACT SERVICES AND INDEXES

APPLIED SCIENCE AND TECHNOLOGY INDEX. H.W. Wilson and Company, 950 University Avenue, Bronx, NY 10452. (800) 367-6670 or (212) 588-8400. Monthly. Inquire as to cost and availability.

COMPUTER AND CONTROL ABSTRACTS. Institute of Electrical Engineers. Available from: Institute of Electrical and Electronics Engineers. IEEE Service Center, 445 Hoes Lane, Piscataway, NJ 08854. Semimonthly. $775.00 per year.

COMPUTER AND INFORMATION SYSTEMS: AN ABSTRACT JOURNAL PERTAINING TO THE THEORY, DESIGN, FABRICATION AND APPLICATION OF COMPUTER AND INFORMATION SYSTEMS. Cambridge Scientific Abstracts, 5161 River Road, Bethesda, MD 20816. 1972 to present. Semi-monthly. Inquire as to cost and availability.

COMPUTER CONTENTS: THE BIWEEKLY COMPILATION OF TABLES OF CONTENTS FROM COMPUTER, ELECTRONIC AND TELECOMMUNICATIONS MAGAZINES, JOURNALS AND TRANACTIONS. Find/SVP, 500 Fifth Avenue, New York, NY 101110. (800) 346-3787 or (212) 354-2424. Biweekly. $115.00 per year.

COMPUTER LITERATURE INDEX. Applied Computer Research, Inc., P.O. Box 9280, Phoenix, AZ 85068. (602) 995-5929. Quarterly. $125.00 per year.

COMPUTER PROGRAMS ABSTRACTS. U.S. National Aeronautics and Space Administration. Available from: U.S. Government Printing Office, Washington, DC 20402. Quarterly. $10.00 per year.

COMPUTING REVIEWS. Association of Computing Machinery, 11 West 42nd Street, New York, NY 10036. (212) 869-7440. Monthly. $60.00 per year.

CONFERENCE PAPERS INDEX. Cambridge Scientific Abstracts, 5161 River Road, Bethesda, MD 20816. 1972 to present. Monthly. Inquire as to cost and availability.

CURRENT CONTENTS: ENGINEERING, TECHNOLOGY AND APPLIED SCIENCES. Institute for Scientific Information, 3501 Market Street, Philadelphia, PA 19104. (800) 523-1850 or (215) 386-0100. Weekly. $275.00 per year.

CURRENT CONTENTS: PHYSICAL, CHEMICAL AND EARTH SCIENCES. Institute for Scientific Information, 3501 Market Street, Philadelphia, PA 19104. (800) 523-1850 or (215) 386-0100. Weekly. $275.00 per year.

CURRENT MATHEMATICAL PUBLICATIONS. American Mathematical Society, P.O. Box 6248, Providence, RI 02940. (800) 556-7774 or (401) 272-9500. 1969 to present. Seventeen times per year. $230.00 per year.

ENGINEERING INDEX MONTHLY AND AUTHOR INDEX. Engineering Information Inc., 345 East 47th Street, New York, NY 10017. (212) 705-7600. Monthly. $1560.00 per year.

GENERAL SCIENCE INDEX. H.W. Wilson and Company, 950 University Avenue, Bronx, NY 10452. (800) 367-6670 or (212) 588-8400. 1978 to present. Monthly. Inquire as to cost and availability.

INDEX TO SCIENTIFIC AND TECHNICAL PROCEEDINGS. Institute for Scientific Information, 3501 Market Street, Philadelphia, PA 19104. (800) 523-1850 or (215) 386-0100. 1978 to present. Monthly. $775.00 per year.

INDEX TO SCIENTIFIC REVIEWS. Institute for Scientific Information, 3501 Market Street, Philadelphia, PA 19104. (800) 523-1850 or (215) 386-0100. 1974 to present. Semi-annual. $550.00 per year.

MATHEMATICAL REVIEWS: A REVIEWING JOURNAL COVERING THE WORLD LITERATURE OF MATHEMATICAL RESEARCH. American Mathematical Society, P.O. Box 6248, Providence, RI 02940. (800) 7774 or (401) 272-9500. 1940 to present. Monthly. $2800.00 per year.

PCR-2: PERSONAL COMPUTER REVIEW - SQUARED. Toolbox Publications, Inc., P.O. Box 5451, 2514 Birch Creek Lane, Orchard Lake, MI 48033. 1987 to present. Bimonthly. $60.00 per year.

PHYSICS ABSTRACTS. Institution of Electrical Engineers. Available from: IEEE Service Center, 445 Hoes Lane, Piscataway, NJ 08854. 1898 to present. Bimonthly. $1700.00 per year.

PHYSICS BRIEFS. Physik Verlag GmbH, Postfach 1260/1280, D-6940 Weinheim, West Germany. (212) 661-9404. 1920 to present. Twenty-six times per year. $1250.00 per year.

SCIENCE CITATION INDEX. Institute for Scientific Information, 3501 Market Street, Philadelphia, PA 19104. (800) 523-1850 or (215) 386-0100. Six times per year. $6200. 00 per year.

ANNUAL REVIEWS AND YEARBOOKS

ADVANCES IN COMPUTERS. Academic Press, Inc., 6277 Sea Harbor Drive, Orlando, FL 32821. (800) 321-5068. Yearly. Approximately $50.00 per volume.

COMPUTER PUBLISHERS AND PUBLICATIONS 1988-89: AN INTERNATIONAL DIRECTORY AND YEARBOOK. Efrem Sigel and Frederica Evan, editors. Gale Research Company, Book Tower, Detroit, MI 48226. (800) 521-0707. Third edition. $140.00.

ASSOCIATIONS AND PROFESSIONAL SOCIETIES

AMERICAN FEDERATION OF INFORMATION PROCESSING SOCIETIES. 1899 Preston White Drive, Reston, VA 22091. (703) 620-8900.

ASSOCIATION OF COMPUTER PROGRAMMERS AND ANALYSTS. 2108-C Gallows Road, Vienna, VA 22180. (703) 790-0490.

ASSOCIATION OF COMPUTING MACHINERY (ACM). 11 West 42nd Street, New York, NY 10036. (212) 869-7440.

IEEE COMPUTER SOCIETY. 1730 Massachusetts Avenue, N.W., Washington, DC 20036. (202) 371-0101.

INSTITUTE OF ELECTRICAL AND ELECTRONICS ENGINEERS. IEEE Service Center, 445 Hoes Lane, Piscataway, NJ 08854.

MACHINE VISION ASSOCIATION. P.O. Box 930, One SME Drive, Dearborn, MI 48121. (313) 271-1500.

SOCIETY FOR COMPUTER SIMULATION. P.O. Box 17900, San Diego, CA 92117. (619) 277-3888.

SOCIETY FOR INFORMATION DISPLAY. 8055 Manchester Avenue, Suite 615, Playa Del Rey, CA 90293. (213) 305-1502.

BIBLIOGRAPHIES

NEW TECHNICAL BOOKS: A SELECTIVE LIST WITH DESCRIPTIVE ANNOTATIONS. New York Public Library, Science and Technology Research Center, Fifth Avenue and 42nd Street, New York, NY 10018. (212) 930-0800. 1915 to present. Monthly. $15.00 per year.

SCIENCE BOOKS AND FILMS. American Association for the Advancement of Science, 1333 H Street, NW, Washington, DC 20005. (202) 326-6454. Five times per year. $20.00 per year.

SCIENTIFIC AND TECHNICAL BOOKS AND SERIALS IN PRINT 1988; AN INDEX TO LITERATURE IN SCIENCE AND TECHNOLOGY. R.R. Bowker Company, 205 East 42nd Street, New York, NY 10017. (800) 521-8110. $175.00.

DIRECTORIES AND BIOGRAPHICAL SOURCES

AMERICAN MEN AND WOMEN OF SCIENCE. R.R. Bowker, Inc., Order Department, 245 West 17th Street, New York, NY 10011. (800) 521-8110. Eight volumes. 1986. $595.00 for set.

AMERICAN SOCIETY FOR INFORMATION SCIENCE HANDBOOK AND DIRECTORY. American Society for Information Science, 1424 16th Street, N.W., Suite 404, Washington, DC 20036. (202) 462-1000. $50.00.

COMPUTERS AND COMPUTING INFORMATION RESOURCES DIRECTORY. Martin Connors, editor. Gale Research Company, Book Tower, Detroit, MI 48226. (800) 521-0707. 1987. $165.00. Supplement available at $85.00.

INTERNATIONAL RESEARCH CENTERS DIRECTORY 1988-89. Darren L. Smith, editor. Gale Research Company, Book Tower, Detroit, MI 48226. (800) 521-0707. 4th edition. 1987. $360.00.

1987 DIRECTORY OF ENGINEERING SOCIETIES AND RELATED ORGANIZATIONS. Gordon Davis, editor. Hemisphere Publishing Corporation, 1010 Vermont Avenue, NW, Washington, DC 20005. (800) 526-0275. 12th edition. 1987. $100.00.

RESEARCH CENTERS DIRECTORY 1988. Gale Research Company, Book Tower, Detroit, MI 48226. (800) 521-0707. 12th edition. 1987. $365.00 for set.

SCIENTIFIC AND TECHNICAL ORGANIZATIONS AND AGENCIES DIRECTORY. Margaret Labash Young, editor. Gale Research Company, Book Tower, Detroit, MI 48226. (800) 521-0707. 2nd edition. 1987. $185.00.

WHO'S WHO IN ENGINEERING. Gordon Davis, editor. Hemisphere Publishing Corporation, 1010 Vermont Avenue, NW, Washington, DC 20005. (800) 526-0275. 6th edition. 1985. $200.00.

ENCYCLOPEDIAS AND DICTIONARIES

COMPUTER AND TELECOMMUNICATIONS ACRONYMS. Julie E. Towell and Helen E. Sheppard, editors. Gale Research Company, Book Tower, Detroit, MI 48226. (800) 521-0707. 1986. $60.00.

DICTIONARY OF COMPUTING. Oxford University Press, 200 Madison Avenue, New York, NY 10016. (800) 458-5833. Second edition. 1986. $29.95.

ENCYCLOPEDIA OF INFORMATION SYSTEMS AND SERVICES 1988. Amy Lucas and Annette Novallo, editors. Gale Research Company, Book Tower, Detroit, MI 48226. (800) 521-0707. 8th edition. 1987. $400.00 for set.

PRENTICE-HALL ENCYCLOPEDIA OF INFORMATION TECHNOLOGY. Robert A. Edmunds. Prentice-Hall Publishing, Inc., Englewood Cliffs, NJ 07632. (800) 562-0245. 1987. $49.95.

SOFTWARE ENCYCLOPEDIA. R.R. Bowker Company, 205 East 42nd Street, New York, NY 10017. (800) 521-8110. Tow volumes. 1987. $125.00 for set.

GENERAL WORKS

CONCURRENT PROGRAMMING FOR SOFTWARE ENGINEERS. Dick Whiddett. Halsted Press, available from: John Wiley and Sons, Inc., 605 Third Avenue, New York, NY 10158. (800) 526-5368. 1987. $39.95.

GENERAL PURPOSE PROGRAMMING LANGUAGES. John V. Cugini. Petrocelli Books, available from: Van Nostrand Reinhold Company, Inc., 135 West 50th Street, New York, NY 10020. (800) 543-2681. 1986. $24.95.

HIGH LEVEL LANGUAGE AND SOFTWARE APPLICATIONS REFERENCE. W.J. Birnes. McGraw-Hill Book Company, 1221 Avenue of the Americas, New York, NY 10020. (212) 512-2000. 1988. $29.95.

MICROPROCESSORS AND MICROCOMPUTERS: HARDWARE AND SOFTWARE. R.J. Tocci and L.P. Laskowski. Prentice-Hall Publishing, Inc., Englewood Cliffs, NJ 07632. (800) 562-0245. Third edition. 1987. $37.95.

PRINCIPLES OF PROGRAMMING LANGUAGES: DESIGN, EVALUATION AND IMPLEMENTATION. Bruce MacLennan. Holt, Rinehart and Winston, Inc., 383 Madison Avenue, New York, NY 10017. (212) 872-2000. Second edition. 1986. $38.75.

PROGRAMMING LANGUAGES. A.B. Tucker. McGraw-Hill Book Company, 1221 Avenue of the Americas, New York, NY 10020. (212) 512-2000. Second edition. 1986. $43.95.

PROGRAMMING LANGUAGES: A GRAND TOUR. Ellis Horowitz, editor. Computer Science Press, 11 Taft Court, Rockville, MD 20850. (800) 242-7737. Third Edition. 1987. $39.95.

SOFTWARE ENGINEERING IN C. Philip E. Margolis and Peter A. Darnell. Springer-Verlag New York, Inc., 175 Fifth Avenue, New York, NY 10010. (800) 526-7254. 1988. $49.50.

HANDBOOKS AND MANUALS

HANDBOOK OF SOFTWARE ENGINEERING. Charles R. Vick, editor. Van Nostrand Reinhold Company, Inc., 135 West 50th Street, New York, NY 10020. (800) 543-2681. 1984. $66.95.

ONLINE DATA BASES

COMPENDEX. Engineering Information, Inc., 345 East 47th Street, New York, NY 10017. (800) 221-1044 or (212) 705-7615. Engineering and technical literature, 1975 to present. Inquire as to online cost and availability.

DISSERTATION ABSTRACTS ONLINE. University Microfilms International, 300 North Zeeb Road, Ann Arbor, MI 48106. (800) 521-0600 or (313) 761-4700. Scope includes virtually all doctoral dissertations accepted at accredited American institutions from 1861 to present in over 250 subject areas. Inquire as to online cost and availability.

INSPEC. INSPEC Marketing Department, Institution of Electrical Engineers. Available from IEEE Service Center, 445 Hoes Lane, Piscataway, NJ 08854. (201) 981-0060. Online version of Physics Abstracts. Inquire as to online cost and availability.

MATHFILE. American Mathematical Society, P.O. Box 6248, Providence, RI 02940. (800) 556-7774 or (401) 272-9500. An online version of Mathematical Reviews. 1973 to present. Inquire as to online cost and availability.

NTIS. National Technical Information Service, 5285 Port Royal Road, Springfield, VA 22161. (703) 487-4630. Broad coverage of

government sponsored research reports, 1964 to present. Inquire as to online cost and availability.

SCISEARCH. Institute for Scientific Information, 3501 Market Street, Philadelphia, PA 19104. (800) 523-1850 or (215) 386-0100. Broad multidisciplinary title and author index to the international literature of science and technology, 1974 to present. Inquire as to online cost and availability.

WILSONLINE. H.W. Wilson and Company, 950 University Avenue, Bronx, NY 10452. (800) 367-6770 or (212) 588-8400. Makes available online versions of the H.W. Wilson indexes including Applied Science and Technology Index, Business Periodicals Index and Readers' Guide to Periodical Literature. Approximately 1980 to present. Inquire as to online cost and availability.

PERIODICALS

ACM TRANSACTIONS ON MATHEMATICAL SOFTWARE. Association of Computing Machinery, 11 West 42nd Street, New York, NY 10036. (212) 869-7440. 1975 to present. Quarterly. $55.00 per year.

ACM TRANSACTIONS ON PROGRAMMING LANGUAGES AND SYSTEMS. Association of Computing Machinery, 11 West 42nd Street, New York, NY 10036. (212) 869-7440. 1979 to present. Quarterly. $55.00 per year.

ADVANCES IN ENGINEERING SOFTWARE. CML Publications, 400 West Cummings Park, Suite 6200, Woburn, MA 01801. (617) 933-7374. 1979 to present. Quarterly. $130.00 per year.

BYTE. Byte Publications, Inc., 70 Main Street, Petersborough, NH 03458. (603) 924-9281. Monthly. $21.00 per year.

C JOURNAL. InfoPro Systems, 3108 Route 10, Denville, NJ 07834. (201) 989-0570. 1985 to present. Quarterly. $28.00 per year.

COMMUNICATIONS OF THE ACM. Association of Computing Machinery, 11 West 42nd Street, New York, NY 10036. (212) 869-7440. Monthly. $80.00 per year.

COMPUTER LANGUAGES. Pergamon Press, Inc., Maxwell House, Fairview Park, Elmsford, NY 10523. (914) 592-7700. 1976 to present. Quarterly. $195.00 per year.

DR. DOBB'S JOURNAL OF SOFTWARE TOOLS. M & T Publishing, Inc., 2464 Embarcadero Way, Palo Alto, CA 94303. (415) 424-0600. 1976 to present. Monthly. $25.00 per year.

IEEE SOFTWARE. Institution of Electrical and Electronics Engineers. IEEE Service Center, 445 Hoes Lane, Piscataway, NJ 08854. 1984 to present. Quarterly. $15.00 per issue.

IEEE TRANSACTIONS ON SOFTWARE ENGINEERING. Institute of Electrical and Electronics Engineers. IEEE Service Center, 445 Hoes Lane, Piscataway, NJ 08854. 1975 to present. Monthly. $160.00 per year.

INTERFACE. International Computer Programs, Inc., 9000 Keystone Crossing, Indianapolis, IN 46240. (317) 844-7461. 1975 to present. Quarterly. $10.00 per year.

JOURNAL OF LOGIC PROGRAMMING. Elsevier Science Publishing Company, Inc., 52 Vanderbilt Avenue, New York, NY 10017. (212) 370-5520. 1984 to present. Quarterly. $95.00 per year.

JOURNAL OF PASCAL, ADA, AND MODULA-2. John Wiley and Sons, Inc., 605 Third Avenue, New York, NY 10158. (800) 526-5368. 1982 to present. Bimonthly. $20.00 per year.

JOURNAL OF SYSTEMS AND SOFTWARE. Elsevier Science Publishing Company, Inc., 52 Vanderbilt Avenue, New York, NY 10017. (212) 370-5520. 1979 to present. Quarterly. $95.00 per year.

PASCAL AND MODULA-2. Pascal Users Group, Box 538, Chesterland, OH 44026. (216) 729-3227. Quarterly. $25.00 per year.

SIGPLAN NOTICES. Association of Computing Machinery Special Interest Group on Programming Languages, 11 West 42nd Street, New York, NY 10036. (212) 869-7440. 1965 to present. Monthly. $25.00 per year.

SIGSOFT SOFTWARE ENGINEERING NOTICES. Association of Computing Machinery Special Interest Group on Software Engineering. 11 West 42nd Street, New York, NY 10036. (212) 869-7440. Quarterly. $12.00 per year.

SOFTWARE DEVELOPER'S MONTHLY. SourceView Press, 835 Castro Street, Martinez, CA 94553. (415) 228-6220. 1985 to present. Monthly. $144.00 per year.

SOFTWARE ENGINEERING JOURNAL. Institute of Electrical Engineers, Savoy Place, London, WC2R OBL, England. 1981 to present. Bimonthly. $85.00 per year.

SOFTWARE PRACTICE AND EXPERIENCE. John Wiley and Sons, Inc., 605 Third Avenue, New York, NY 10158. (800) 526-5368. 1971 to present. Monthly. $260.00 per year.

UNIX REVIEW (SAN FRANCISCO). Miller Freeman Publications, Inc., 500 Howard Street, San Francisco, CA 94105. (415) 397-1881. 1983 to present. Monthly. $35.00 per year.

RESEARCH CENTERS AND INSTITUTES

INSTITUTE FOR INFORMATION SCIENCE AND TECHNOLOGY. George Washington University, 801 22nd Street, N.W., Washington, DC 20052. (202) 676-4921.

RESEARCH INSTITUTE FOR COMPUTING AND INFORMATION SYSTEMS. University of Houston at Clear Lake, 2700 Bay Area Boulevard, Houston, TX 77058. (713) 488-9392.

SOFTWARE ENGINEERING INSTITUTE. Carnegie-Mellon University, Pittsburgh, PA 15213. (412) 268-6700.

SOFTWARE ENGINEERING RESEARCH INSTITUTE. Georgia Institute of Technology, 258 Fourth Street, N.W., Atlanta, GA 30332. (404) 894-3180.

COMPUTER PROGRAMS

See: SOFTWARE

COMPUTER SECURITY

See also: AUTOMATION, COMPUTER MEMORY, COMPUTER OPERATING SYSTEMS, COMPUTER PROGRAMMING, COMPUTERS

ABSTRACT SERVICES AND INDEXES

APPLIED SCIENCE AND TECHNOLOGY INDEX. H.W. Wilson Company, 950 University Avenue, Bronx, NY 10452. (800) 367-6670 or (212) 588-8400. Inquire as to cost and availability.

COMPUTER ABSTRACTS. Technical Information Company, Limited, Post Office Box 59, Saint Helier, Jersey British Channel Inlands, England. Monthly. $310.00 per year.

COMPUTER AND CONTROL ABSTRACTS. Institute of Electrical Engineers, London, United Kingdom. Available from: IEEE Service Center, 445 Hoes Lane, Piscataway, NJ 08854. (201) 981-0060. Semimonthly. $775.00 per year.

COMPUTER AND INFORMATION SYSTEMS: AN ABSTRACT JOURNAL PERTAINING TO THE THEORY, DESIGN, FABRICATION AND APPLICATION OF COMPUTER AND INFORMATION SYSTEMS. Cambridge Scientific Abstracts, Incorporated, 5161 River Road, Bethesda, MD 20816. (301) 951-1400. Semimonthly. $590.00 per year.

COMPUTER CONTENTS: THE BIWEEKLY COMPILATION OF TABLES OF CONTENTS FROM COMPUTER, ELECTRONIC AND TELECOMMUNICATIONS MAGAZINES, JOURNALS AND TRANSACTIONS. Find/SVP, 500 Fifth Avenue, New York, NY 10110. (800) 346-3787 or (212) 354-2424. Biweekly. $115.00 per year.

COMPUTER LITERATURE INDEX. Applied Computer Research, Incorporated, Post Office Box 9280, Phoenix, AZ 85068. (602) 995-5929. Quarterly. $125.00 per year.

COMPUTER PROGRAM ABSTRACTS. U.S. National Aeronautics and Space Administration. Available from U.S. Government Printing Office, Washington, DC 20402. Quarterly. $6.50 per year.

COMPUTING REVIEWS. Association for Computing Machinery, 11 West 42nd Street, New York, NY 10036. (212) 869-7440. Monthly. $60.00 per year.

ENGINEERING INDEX MONTHLY. Engineering Information, Incorporated, 345 East 47th Street, New York, NY 10017. (800) 221-1044 or (212) 705-7600. Monthly, with annual cumulation. $1560.00 per year.

ASSOCIATIONS AND PROFESSIONAL SOCIETIES

AMERICAN FEDERATION OF INFORMATION PROCESSING SOCIETIES. 1899 Preston White Drive, Reston, VA 22091. (703) 620-8900.

ASSOCIATION FOR COMPUTING MACHINERY (ACM). 11 West 42nd Street, New York, NY 10036. (212) 896-7440.

ASSOCIATION OF MINICOMPUTER USERS. Two Frederick Street, Framingham, MA 01701. (617) 879-5955.

COMPUTER AND AUTOMATION SYSTEMS ASSOCIATION OF SME (SOCIETY OF MANUFACTURING ENGINEERS). One SME Drive, Box 930, Dearborn, MI 48121. (313) 271-1500.

COMPUTER SECURITY INSTITUTE. 43 Boston Post Road, Northboro, MA 01532. (617) 845-5050.

IEEE COMPUTER SOCIETY. 1730 Massachusetts Avenue, NW, Washington, DC 20036. (202) 371-0101.

INSTITUTE OF ELECTRICAL AND ELECTRONIC ENGINEERS (IEEE). 345 East 47th Street, New York, NY 10017. (212) 705-7900.

INSTITUTE FOR PERSONAL COMPUTING. Post Office Box 558250, Miami, FL 33155. (305) 274-7440.

SPECIAL INTEREST GROUP ON SECURITY, AUDIT, AND CONTROL. c/o Association for Computing Machinery (ACM). 11 West 42nd Street, New York, NY 10036. (212) 896-7440.

BIBLIOGRAPHIES

COMPUTER BOOKS AND SERIALS IN PRINT, 1986-1987. 1986. R.R. Bowker Company, 205 East 42nd Street, New York, NY 10017. (800) 521-8110 or (212) 916-1600. $175.00.

SCIENTIFIC AND TECHNICAL BOOKS AND SERIALS IN PRINT 1988; AN INDEX TO LITERATURE IN SCIENCE AND TECHNOLOGY. R.R. Bowker Company, 205 East 42nd Street, New York, NY 10017. (800) 521-8110 or (212) 916-1600. $175.00.

DIRECTORIES AND BIOGRAPHICAL SOURCES

COMPUTERS AND COMPUTING INFORMATION RESOURCES DIRECTORY. Gale Research Company, Book Tower, Detroit, MI 48226. (800) 521-0707. 1986. $160.00.

COMPUTER SECURITY, AUDITING AND CONTROLS, COMSAC. Management Advisory Publications, 57 Greylock, Box 151, Wellesley Hills, MA 02181. (617) 235-2895. Semiannual. $55.00 per year.

COMPUTER SECURITY BUYERS GUIDE. Computer Security Institute, 43 Boston Post Road, Northboro, MA 01532. (617) 845-5050. Annual.

DIRECTORY OF CONSULTANTS IN COMPUTER SYSTEMS. Research Publications, Incorporated, 12 Lunar Drive, Woodbridge, CT 06525. (203) 397-2600. Annual. $85.00 per year.

RESEARCH CENTERS DIRECTORY. Gale Research Company, Book Tower, Detroit, MI 48226. (800) 521-0707. Eleventh edition. 1987.

ENCYCLOPEDIAS AND DICTIONARIES

COMPUTER AND TELECOMMUNICATIONS ACRONYMS. Julie E. Towell and Helen E. Sheppard, editors. Gale Research Company, Detroit, MI 48226. (800) 521-0707. 1986. $60.00.

COMPUTER DICTIONARY. Charles J. Sippl. Howard W. Sams and Company, Incorporated, 4300 West 62nd Street, Indianapolis, IN 46268. (800) 428-7267 or (317) 298-5564. Fourth edition. 1985. $17.95.

DICTIONARY OF COMPUTING. Oxford University Press, 200 Madison Avenue, New York, NY 10016. (212) 679-7300. Second edition. 1986. $29.95.

ENCYCLOPEDIA OF COMPUTER SCIENCE AND ENGINEERING. Anthony Ralston, editor. Van Nostrand Reinhold Book Company, 115 Fifth Avenue, New York, NY 10003. (800) 543-2681. Second edition. 1982. $89.95.

ENCYCLOPEDIA OF COMPUTER SCIENCE AND TECHNOLOGY. Jack Belzer, Albert G. Holzman, and Allan Kent. Marcel Dekker, Incorporated, 270 Madison Avenue, New York, NY 10016. (212) 696-9000. Sixteen volumes. $115.00 per volume.

MCGRAW-HILL DICTIONARY OF COMPUTERS. McGraw-Hill Book Company, 1221 Avenue of the Americas, New York, NY 10020. (212) 512-2000. 1985. $17.50.

GENERAL WORKS

COMPUTER SECURITY: A GLOBAL CHALLENGE. J.H. Finch and E.G. Dougall, editors. Elsevier Science Publishing Company, Incorporated, 52 Vanderbilt Avenue, New York, NY 10017. (212) 370-5520. 1985. $50.00.

COMPUTER SECURITY TECHNOLOGY. James A. Cooper. Lexington Books, 125 Spring Street, Lexington, MA 02173. (617) 862-6650. 1984. $28.00.

OUT OF THE INNER CIRCLE: A HACKER'S GUIDE TO COMPUTER SECURITY. Bill Landreth. Microsoft Press, 10700 Northup Way, Bellevue, Wa 98004. (206) 828-8080. 1985. $9.95 in paper.

HANDBOOKS AND MANUALS

COMPUTER SECURITY HANDBOOK. Richard H. Baker. Tab Books, Incorporated, Monterey Lane, Blue Ridge Summit, PA 17214. (717) 794-2191. 1985. $24.00.

COMPUTER SECURITY HANDBOOK: STRATEGIES AND TECHNIQUES FOR PREVENTING DATA LOSS OR THEFT. Rolf T. Moulton. Prentice-Hall Press, Incorporated, Englewood Cliffs, NJ 07632. (201) 592-2000. 1986. $29.95.

FUNDAMENTALS HANDBOOK OF ELECTRICAL AND COMPUTER ENGINEERING. Sheldon S.L. Chang, editor. John Wiley and Sons, Incorporated, 605 Third Avenue, New York, NY 10158. (800) 526-5368 or (212) 850-6000. Three volumes. 1982. $180.00 set price.

HANDBOOK OF COMPUTERS AND COMPUTING. Arthur H. Seidman, editor. Van Nostrand Reinhold Book Company, 115 Fifth Avenue, New York, NY 10003. (800) 543-2681. 1984. $79.95.

MCGRAW-HILL COMPUTER HANDBOOK. Harry Helms, editor. McGraw-Hill Book Company, 1221 Avenue of the Americas, New York, NY 10020. (212) 512-2000. 1983. $84.50.

ONLINE DATA BASES

COMPENDEX. Engineering Information, Incorporated, 345 East 47th Street, New York, NY 10017. (800) 221-1044 or (212) 705-7615. Engineering and technical literature, 1975 to present. Inquire as to cost and availability.

INSPEC. INSPEC Marketing Department, Institute of Electrical and Electronics Engineers, Incorporated, IEEE Service Department, 445 Hoes Lane, Piscataway, NJ 08854. (201) 981-0060. Inquire as to on-line cost and availability.

NTIS. National Technical Information Service, 5285 Port Royal Road, Springfield, VA 22161. (703) 487-4630. Broad coverage of government sponsored research reports, 1964 to present. Inquire as to cost and availability.

SCISEARCH. Institute for Scientific Information, 3501 Market Street, Philadelphia, PA 19104. (800) 523-1850 or (215) 386-0100. Broad multidisciplinary title and author index to the international literature of science and technology, 1974 to present. Inquire as to cost and availability.

OTHER SOURCES

COMPUTER PUBLISHERS AND PUBLICATIONS 1985-86: AN INTERNATIONAL DIRECTORY AND YEARBOOK. Edited by Efrem Sigel and Frederica Evan. Communications Trends, Incorporated. Distributed by Gale Research Company, Book Tower, Detroit, MI 48226. (800) 521-0707. Second edition. 1985. $95.00. Supplement available for 1986, $35.00 Provides information on publishers of computer books and periodicals, recommended titles for libraries and bookstores, industry trends and statistics, and computer manufacturers as publishers.

PERIODICALS

ACM TRANSACTIONS ON COMPUTER SYSTEMS. Association for Computing Machinery, 11 West 42nd Street, New York, NY 10036. (212) 869-7440. Quarterly. $70.00 per year.

BYTE. Byte Publications, Incorporated, 70 Main Street, Petersborough, NH 03458. (603) 924-9281. Monthly. $21.00 per year.

COMMUNICATIONS OF THE ACM. Association for Computing Machinery, 11 West 42nd Street, New York, NY 10036. (212) 869-7440. Monthly. $78.00 per year.

COMPUTER. Institute of Electrical and Electronic Engineers (IEEE), IEEE Service Center, 445 Hoes Lane, Piscataway, NJ 08854. (201) 981-0060. Monthly. $90.00 per year.

COMPUTER DESIGN. PennWell Directories, Incorporated, Post Office Box 1260, Tulsa, OK 74101. (018) 835-3161. Monthly. $60.00 per year.

COMPUTER MAGAZINE. IEEE Computer Society, 1109 Spring Street, Suite 300, Silver Spring, MD 20910. (301) 589-8142. Monthly. $34.00 per year.

COMPUTER SECURITY DIGEST. Computer Protection Systems, Incorporated, 150 North Main Street, Plymouth, MI 48170. (313) 459-8787. Monthly. $90.00 per year.

COMPUTER SECURITY JOURNAL. Computer Security Institute, 43 Boston Post Road, Northboro, MA 01532. (617) 845-5050. Semi-annual. $65.00 per year.

COMPUTER SECURITY PRODUCTS REPORT. Assets Protection Publishing, Box 5323, Madison, WI 53704. (608) 274-7751. Quarterly. $30.00 per year.

COMPUTERS AND SECURITY. Elsevier/North Holland, Incorporated, 52 Vanderbilt Avenue, New York, NY 10017. Quarterly. $68.00 per year.

COMPUTERWORLD. C.W. Communications, 375 Cochituate Road, Farmington, MA 01701. (617) 879-0700. Weekly. $44.00 per year.

DATAMATION. Technical Publishing Company, 875 Third Avenue, New York, NY 10022. (212) 605-9400. Semi-monthly. $50.00 per year.

COMPUTER VISION

See also: AUTOMATION, COMPUTER MEMORY, COMPUTER OPERATING SYSTEMS, COMPUTER PROGRAMMING, PARALLEL COMPUTERS, PROGRAMMING LANGUAGES, ROBOTICS, SOFTWARE

ABSTRACT SERVICES AND INDEXES

APPLIED SCIENCE AND TECHNOLOGY INDEX. H.W. Wilson Company, 950 University Avenue, Bronx, NY 10452. (800) 367-6670 or (212) 588-8400. Inquire as to cost and availability.

COMPUTER ABSTRACTS. Technical Information Company, Limited, Post Office Box 59, Saint Helier, Jersey British Channel Inlands, England. Monthly. $310.00 per year.

COMPUTER AND CONTROL ABSTRACTS. Institute of Electrical Engineers, London, United Kingdom. Available from: IEEE Service Center, 445 Hoes Lane, Piscataway, NJ 08854. (201) 981-0060. Semimonthly. $775.00 per year.

COMPUTER AND INFORMATION SYSTEMS: AN ABSTRACT JOURNAL PERTAINING TO THE THEORY, DESIGN, FABRICATION AND APPLICATION OF COMPUTER AND INFORMATION SYSTEMS. Cambridge Scientific Abstracts, Incorporated, 5161 River Road, Bethesda, MD 20816. (301) 951-1400. Semimonthly. $590.00 per year.

COMPUTER CONTENTS: THE BIWEEKLY COMPILATION OF TABLES OF CONTENTS FROM COMPUTER, ELECTRONIC AND TELECOMMUNICATIONS MAGAZINES, JOURNALS AND TRANSACTIONS. Find/SVP, 500 Fifth Avenue, New York, NY 10110. (800) 346-3787 or (212) 354-2424. Biweekly. $115.00 per year.

COMPUTER LITERATURE INDEX. Applied Computer Research, Incorporated, Post Office Box 9280, Phoenix, AZ 85068. (602) 995-5929. Quarterly. $125.00 per year.

COMPUTER PROGRAM ABSTRACTS. U.S. National Aeronautics and Space Administration. Available from U.S. Government Printing Office, Washington, DC 20402. Quarterly. $6.50 per year.

COMPUTING REVIEWS. Association for Computing Machinery, 11 West 42nd Street, New York, NY 10036. (212) 869-7440. Monthly. $60.00 per year.

ENGINEERING INDEX MONTHLY. Engineering Information, Incorporated, 345 East 47th Street, New York, NY 10017. (800) 221-1044 or (212) 705-7600. Monthly, with annual cumulation. $1560.00 per year.

SCIENCE CITATION INDEX. Institute for Scientific Information, 3501 Market Street, Philadelphia, PA 19104. (800) 523-1850 or (215) 386-0100.

ANNUAL REVIEWS AND YEARBOOKS

ADVANCES IN COMPUTERS. Academic Press, Incorporated, 6277 Sea Harbor Drive, Orlando, FL 32821. (800) 321-5068. Yearly. Approximately $50.00 per volume.

ASSOCIATIONS AND PROFESSIONAL SOCIETIES

AMERICAN FEDERATION OF INFORMATION PROCESSING SOCIETIES. 1899 Preston White Drive, Reston, VA 22091. (703) 620-8900.

ASSOCIATION FOR COMPUTING MACHINERY (ACM). 11 West 42nd Street, New York, NY 10036. (212) 896-7440.

ASSOCIATION OF COMPUTER PROGRAMMERS AND ANALYSTS. 2108-C Gallows Road, Vienna, VA 22180. (703) 790-0490.

COMPUTER AND AUTOMATION SYSTEMS ASSOCIATION OF SME (SOCIETY OF MANUFACTURING ENGINEERS). One SME Drive, Box 930, Dearborn, MI 48121. (313) 271-1500.

GRAPHIC COMMUNICATIONS ASSOCIATION. 1730 North Lynn Street, Suite 604, Arlington, VA 22209. (703) 841-8160.

IEEE COMPUTER SOCIETY. 1730 Massachusetts Avenue, NW, Washington, DC 20036. (202) 371-0101.

INSTITUTE OF ELECTRICAL AND ELECTRONIC ENGINEERS (IEEE). 345 East 47th Street, New York, NY 10017. (212) 705-7900.

INSTITUTE FOR PERSONAL COMPUTING. Post Office Box 558250, Miami, FL 33155. (305) 274-7440.

MACHINE VISION ASSOCIATION. Post Office Box 930, One SME Drive, Dearborn, MI 48121. (313) 271-1500.

NATIONAL COMPUTER GRAPHICS ASSOCIATION. 2722 Merrilee Drive, Suite 200, Fairfax, VA 22031. (703) 698-9600.

SOCIETY FOR COMPUTER SIMULATION. Post Office Box 17900. San Diego, CA 92117. (619) 277-3888.

SOCIETY FOR INFORMATION DISPLAY. 8055 Manchester Avenue, Suite 615, Playa Del Rey, CA 90293. (213) 305-1502.

SPECIAL INTEREST GROUP ON COMPUTER GRAPHICS (SIGGRAPH). c/o Association for Computing Machinery (ACM), 11 West 42nd Street, New York, NY 10036. (212) 896-7440.

WORLD COMPUTER GRAPHICS ASSOCIATION. 2033 M Street, Suite 399, Washington, DC 20036. (202) 775-9556.

BIBLIOGRAPHIES

COMPUTER BOOKS AND SERIALS IN PRINT, 1986-1987. 1986. R.R. Bowker Company, 205 East 42nd Street, New York, NY 10017. (800) 521-8110 or (212) 916-1600. $69.95.

SCIENCE BOOKS AND FILMS. American Association for the Advancement of Science, 1333 H Street, NW, Washington, DC 20005.

SCIENTIFIC AND TECHNICAL BOOKS AND SERIALS IN PRINT 1988; AN INDEX TO LITERATURE IN SCIENCE AND TECHNOLOGY. R.R. Bowker Company, 205 East 42nd Street, New York, NY 10017. (800) 521-8110 or (212) 916-1600. $175.00.

DIRECTORIES AND BIOGRAPHICAL SOURCES

COMPUTER PERIPHERALS REVIEW. GML Information Services, 594 Marrett Road, Lexington, MA 02173. (617) 861-0515. Two issues per year. $215.00 per year. Directory of computer peripheral equipment manufacturers.

COMPUTERS AND COMPUTING INFORMATION RESOURCES DIRECTORY. Gale Research Company, Book Tower, Detroit, MI 48226. (800) 521-0707. 1986. $160.00.

DATAPRO 70. Datapro Research Corporation, 1805 Underwood Boulevard, Delran, NJ 08075. Twelve monthly updates. $803.00 per year. List of about 1200 companies which offer data processing equipment and services.

DIRECTORY OF CONSULTANTS IN COMPUTER SYSTEMS. Research Publications, Incorporated, 12 Lunar Drive, Woodbridge, CT 06525. (203) 397-2600. Annual. $85.00 per year.

RESEARCH CENTERS DIRECTORY. Gale Research Company, Book Tower, Detroit, MI 48226. (800) 521-0707. Eleventh edition. 1987.

WHO'S WHO IN FRONTIER SCIENCE AND TECHNOLOGY. Marquis Who's Who, Incorporated, 200 East Ohio Street, Chicago, IL 60611. (800) 428-3898 or (312) 787-2008.

WHO'S WHO IN TECHNOLOGY TODAY. Reston Publishing Company, Incorporated, c/o Prentice-Hall, Incorporated, Englewood Cliffs, NJ 07632. (800) 262-6868. Biennial. Five volumes. $425.00. Covers the fields of electronics, computer science, physics, optics, chemistry, biotechnology, mechanics, energy, and earth science.

ENCYCLOPEDIAS AND DICTIONARIES

COMPUTER AND TELECOMMUNICATIONS ACRONYMS. Julie E. Towell and Helen E. Sheppard, editors. Gale Research Company, Book Tower, Detroit, MI 48226. (800) 521-0707. 1986. $60.00.

COMPUTER DICTIONARY. Charles J. Sippl. Howard W. Sams and Company, Incorporated, 4300 West 62nd Street, Indianapolis, IN 46268. (800) 428-7267 or (317) 298-5564. Fourth edition. 1985. $17.95.

DICTIONARY OF COMPUTING. Oxford University Press, 200 Madison Avenue, New York, NY 10016. (212) 679-7300. Second edition. 1986. $29.95.

ENCYCLOPEDIA OF COMPUTER SCIENCE AND ENGINEERING. Anthony Ralston, editor. Van Nostrand Reinhold Book Company, 115 Fifth Avenue, New York, NY 10003. (800) 543-2681. Second edition. 1982. $89.95.

ENCYCLOPEDIA OF COMPUTER SCIENCE AND TECHNOLOGY. Jack Belzer, Albert G. Holzman, and Allan Kent. Marcel Dekker, Incorporated, 270 Madison Avenue, New York, NY 10016. (212) 696-9000. Sixteen volumes. $115.00 per volume.

MCGRAW-HILL DICTIONARY OF COMPUTERS. McGraw-Hill Book Company, 1221 Avenue of the Americas, New York, NY 10020. (212) 512-2000. 1985. $17.50.

GENERAL WORKS

ARTIFICIAL INTELLIGENCE. Patrick Henry Winston. Addison-Wesley Publishing Company, Incorporated, 1 Jacob Way, Reading, MA 01867. (617) 944-3700. Second edition. 1984. $35.95.

COMPUTER VISION. Dana H. Ballard and Christopher M. Brown. Prentice-Hall, Incorporated, Englewood Cliffs, NJ 07632. (800) 562-0245. 1982. $46.95.

COMPUTER VISION. M. Brady, editor. Elsevier Science Publishing Company, Incorporated, 52 Vanderbilt Avenue, New York, NY 10017. (212) 370-5520. 1984. $30.00 in paper.

COMPUTER VISION: AN OVERVIEW. William B. Gevarter. Business Technology Books, Post Office Box 574, Orinda, CA 94563. (415) 839-3370. 1984. $34.50 in paper.

INDUSTRIAL ROBOTS: COMPUTER INTERFACING AND CONTROL. Wesley E. Snyder. Prentice-Hall, Incorporated, Englewood Cliffs, NJ 07632. (800) 562-0245. 1985. $32.95.

PATTERN RECOGNITION: HUMAN AND MECHANICAL. S. Watanabe. John Wiley and Sons, Incorporated, 605 Third Avenue, New York, NY 10158. (800) 526-5368 or (212) 850-6000. 1985. $44.95.

HANDBOOKS AND MANUALS

FUNDAMENTALS HANDBOOK OF ELECTRICAL AND COMPUTER ENGINEERING. Sheldon S.L. Chang, editor. John Wiley and Sons, Incorporated, 605 Third Avenue, New York, NY 10158. (800) 526-5368 or (212) 850-6000. Three volumes. 1982. $180.00 set price.

HANDBOOK OF COMPUTERS AND COMPUTING. Arthur H. Seidman, editor. Van Nostrand Reinhold Book Company, 115 Fifth Avenue, New York, NY 10003. (800) 543-2681. 1984. $79.95.

HANDBOOK OF SOFTWARE ENGINEERING. Charles R. Vick, editor. Van Nostrand Reinhold Book Company, 115 Fifth Avenue, New York, NY 10003. (800) 543-2681. 1984. $66.95.

MCGRAW-HILL COMPUTER HANDBOOK. Harry Helms, editor. McGraw-Hill Book Company, 1221 Avenue of the Americas, New York, NY 10020. (212) 512-2000. 1983. $84.50.

ONLINE DATA BASES

COMPENDEX. Engineering Information, Incorporated, 345 East 47th Street, New York, NY 10017. (800) 221-1044 or (212) 705-7615. Engineering and technical literature, 1975 to present. Inquire as to cost and availability.

DISSERTATION ABSTRACTS ONLINE. University Microfilms International, 300 North Zeeb Road, Ann Arbor, MI 48106. (800) 521-0600 or (313) 761-4700. Scope includes virtually all doctoral dissertations accepted at accredited American institutions from 1861 to present in 252 subject areas. Inquire as to cost and availability.

INSPEC. INSPEC Marketing Department, Institute of Electrical and Electronics Engineers, Incorporated, IEEE Service Department, 445 Hoes Lane, Piscataway, NJ 08854. (201) 981-0060. Inquire as to on-line cost and availability.

MATHFILE. American Mathematical Society, Post Office Box 6248, Providence, RI 02940. (800) 556-7774 or (401) 272-9500. Scope includes pure and applied mathematics and related areas of physics, statistics, engineering, computer science, and operations research literature since 1973. Inquire as to cost and availability.

NTIS. National Technical Information Service, 5285 Port Royal Road, Springfield, VA 22161. (703) 487-4630. Broad coverage of government sponsored research reports, 1964 to present. Inquire as to cost and availability.

SCISEARCH. Institute for Scientific Information, 3501 Market Street, Philadelphia, PA 19104. (800) 523-1850 or (215) 386-0100. Broad multidisciplinary title and author index to the international literature of science and technology, 1974 to present. Inquire as to cost and availability.

PERIODICALS

ACM TRANSACTIONS ON COMPUTER SYSTEMS. Association for Computing Machinery, 11 West 42nd Street, New York, NY 10036. (212) 869-7440. Quarterly. $70.00 per year.

COMPUTER-AIDED DESIGN. Butterworth Scientific Limited, Post Office Box 63, Westbury House, Bury Street, Guildford, Surrey, GU2 5BH, England. Ten times per year. $285.00 per year.

COMPUTER GRAPHICS. Association for Computing Machinery, 11 West 42nd Street, New York, NY 10036. (212) 869-7440. Quarterly. $25.00 per year.

COMPUTER GRAPHICS FORUM. Elsevier Science Publishers B.V., Box 211, 1000 AE Amsterdam, The Netherlands. Four times per year. $85.00 per year.

COMPUTER VISION, GRAPHICS, AND IMAGE PROCESSING. Academic Press, Incorporated, Journals Division, 1250 Sixth Avenue, San Diego, Ca 92101. (619) 230-1840. Monthly. $328.00 per year.

COMPUTERS AND GRAPHICS. Pergamon Press, Incorporated, Journals Division, Maxwell House, Fairview Park, Elmsford, NY 10523. (914) 592-7700. Quarterly. $200.00 per year.

IEEE COMPUTER GRAPHICS AND APPLICATIONS. Institute of Electrical and Electronic Engineers (IEEE), IEEE Service Center, 445 Hoes Lane, Piscataway, NJ 08854. (201) 981-0060. Monthly. $72.00 per year.

IEEE TRANSACTIONS ON PATTERN ANALYSIS AND MACHINE INTELLIGENCE. Institute of Electrical and Electronic Engineers (IEEE), IEEE Service Center, 445 Hoes Lane, Piscataway, NJ 08854. (201) 981-0060. Bimonthly. $125.00 per year.

IMAGE AND VISION COMPUTING. Butterworth Scientific Limited, Post Office Box 63, Westbury House, Bury Street, Guildford, Surrey, GU2 5BH, England, Quarterly. $175.00 per year.

PATTERN RECOGNITION LETTERS. Elsevier Science Publishers B.V., Box 211, 1000 AE Amsterdam, The Netherlands. Bimonthly. $90.00 per year.

RECOGNITION TECHNOLOGIES TODAY. Recognition Technologies Users Association, Box 2016, Battenkill Building, Manchester, VT 05255. (802) 362-4151. Bimonthly. Inquire as to cost and availability.

S. KLEIN NEWSLETTER ON COMPUTER GRAPHICS. Technology and Business Communications, Incorporated, 730 Boston Post Road, Box 915, Sudbury, MA 01776. (617) 443-4671. Semimonthly. $178.00 per year.

VISION COMPUTER (COMPUTER GRAPHICS SOCIETY). Springer-Verlag, 175 Fifth Avenue, New York, NY 10010. (212) 460-1500. Six times per year. $119.00 per year

RESEARCH CENTERS AND INSTITUTES

CARNEGIE-MELLON UNIVERSITY. Center for Excellence in Optical Data Processing, Department of Electrical and Computer Engineering, Pittsburgh, PA 15213. (412) 268-2464.

MASSACHUSETTS INSTITUTE OF TECHNOLOGY. Artificial Intelligence Laboratory, 545 Technology Square, Cambridge, MA 02139. (617) 253-6218.

OAKLAND UNIVERSITY. Center for Robotics and Advanced Automation, Rochester, MI 48309-4401. (313) 370-2233.

UNIVERSITY OF ARIZONA. Digital Image Analysis Laboratory, Department of Electrical and Computer Engineering, Tucson, Az 85721. (602) 621-4554.

UNIVERSITY OF CALIFORNIA, LOS ANGELES. Machine Perception Laboratory, UCLA Computer Science Department, 405 Hilgard Avenue, Los Angeles, CA 90024. (213) 825-6121.

UNIVERSITY OF TEXAS AT AUSTIN. Artificial Intelligence Laboratory, Taylor Hall 2.124, Austin, TX 78712-1188. (512) 471-9567.

COMPUTERS

See also: AUTOMATION, COMPUTER MEMORY, COMPUTER OPERATING SYSTEMS, COMPUTER PROGRAMMING, COMPUTER VISION, MICROCOMPUTERS, PARALLEL COMPUTERS, PROGRAMMING LANGUAGES, SOFTWARE ENGINEERING

ABSTRACT SERVICES AND INDEXES

APPLIED SCIENCE AND TECHNOLOGY INDEX. H.W. Wilson Company, 950 University Avenue, Bronx, NY 10452. (800) 367-6670 or (212) 588-8400. Inquire as to cost and availability.

COMPUTER ABSTRACTS. Technical Information Company, Limited, P.O. Box 59, Saint Helier, Jersey British Channel Inlands, England. Monthly. $310.00 per year.

COMPUTER AND CONTROL ABSTRACTS. Institute of Electrical Engineers, London, United Kingdom. Available from: IEEE Service Center, 445 Hoes Lane, Piscataway, NJ 08854. (201) 981-0060. Semimonthly. $775.00 per year.

COMPUTER AND INFORMATION SYSTEMS: AN ABSTRACT JOURNAL PERTAINING TO THE THEORY, DESIGN, FABRICATION AND APPLICATION OF COMPUTER AND INFORMATION SYSTEMS. Cambridge Scientific Abstracts, Incorporated, 5161 River Road, Bethesda, MD 20816. (301) 951-1400. Semimonthly. $590.00 per year.

COMPUTER CONTENTS: THE BIWEEKLY COMPILATION OF TABLES OF CONTENTS FROM COMPUTER, ELECTRONIC AND TELECOMMUNICATIONS MAGAZINES, JOURNALS AND TRANSACTIONS. Find/SVP, 500 Fifth Avenue, New York, NY 10110. (800) 346-3787 or (212) 354-2424. Biweekly. $115.00 per year.

COMPUTER PROGRAM ABSTRACTS. U.S. National Aeronautics and Space Administration. Available from U.S. Government Printing Office, Washington, DC 20402. Quarterly. $6.50 per year.

COMPUTING REVIEWS. Association for Computing Machinery, 11 West 42nd Street, New York, NY 10036. (212) 869-7440. Monthly. $60.00 per year.

ENGINEERING INDEX MONTHLY. Engineering Information, Incorporated, 345 East 47th Street, New York, NY 10017. (800) 221-1044 or (212) 705-7600. Monthly, with annual cumulation. $1560.00 per year.

MATHEMATICAL REVIEWS. American Mathematical Society, P. O. Box 6248, Providence, RI 02940. (800) 556-7774 or (401) 272-9500.

SCIENCE CITATION INDEX. Institute for Scientific Information, 3501 Market St., Philadelphia, PA 19104. (800) 523-1850 or (215) 386-0100.

ANNUAL REVIEWS AND YEARBOOKS

ADVANCES IN COMPUTERS. Academic Press, Incorporated, 6277 Sea Harbor Drive, Orlando, FL 32821. (800) 321-5068. Yearly. Approximately $50.00 per volume.

ASSOCIATIONS AND PROFESSIONAL SOCIETIES

AMERICAN FEDERATION OF INFORMATION PROCESSING SOCIETIES. 1899 Preston White Drive, Reston, VA 22091. (703) 620-8900.

ASSOCIATION OF COMPUTING MACHINERY (ACM). 11 West 42nd Street, New York, NY 10036. (212) 896-7440.

ASSOCIATION OF COMPUTER PROGRAMMERS AND ANALYSTS. 2108-C Gallows Road, Vienna, VA 22180. (703) 790-0490.

BOSTON COMPUTER SOCIETY. One Center Plaza, Boston, MA 02108.

COMPUTER AND AUTOMATION SYSTEMS ASSOCIATION OF SME. One SME Drive, Box 930, Dearborn, MI 48121. (313) 271-1500.

IEEE COMPUTER SOCIETY. 1730 Massachusetts Avenue, NW, Washington, DC 20036. (202) 371-0101.

INSTITUTE OF ELECTRICAL AND ELECTRONIC ENGINEERS (IEEE). 345 East 47th Street, New York, NY 10017. (212) 705-7900.

BIBLIOGRAPHIES

COMPUTER LITERATURE INDEX. Applied Computer Research, Incorporated, P.O. Box 9280, Phoenix, AZ 85068. (602) 995-5929. Quarterly. $125.00 per year.

SCIENCE BOOKS AND FILMS. American Association for the Advancement of Science, 1333 H Street, NW, Washington, DC 20005.

SCIENTIFIC AND TECHNICAL BOOKS IN PRINT; AN INDEX TO LITERATURE IN SCIENCE AND TECHNOLOGY. R. R. Bowker Company, 205 E. 42nd Street, New York, NY 10017. (800) 521-8110 or (212) 916-1600.

DIRECTORIES AND BIOGRAPHICAL SOURCES

AMERICAN MEN AND WOMEN OF SCIENCE. Physical and Biological Sciences. Fifteenth edition. R. R. Bowker Company, 205 E. 42nd Street, New York, NY 10017. (800) 521-8110 or (212) 916-1600.

AMERICAN SOCIETY FOR INFORMATION SCIENCE, HANDBOOK AND DIRECTORY. American Society for Information Science, 1424 16th Street, NW, Suite 404, Washington, DC 20036. (202) 462-1000. $50.00.

COMPUTER REVIEW. GML Information Services, 594 Marrett Road, Lexington, MA 02173. (617) 861-0515. Two issues per year. $195.00 per year. Directory of computer main frame and minicomputer manufacturers.

COMPUTER PERIPHERALS REVIEW. GML Information Services, 594 Marrett Road, Lexington, MA 02173. (617) 861-0515. Two issues per year. $215.00 per year. Directory of computer peripheral equipment manufacturers.

COMPUTERS AND COMPUTING INFORMATION RESOURCES DIRECTORY. Gale Research Company, Book Tower, Detroit, MI 48226. (800) 521-0707. 1986. $160.00.

DATAPRO 70. Datapro Research Corporation, 1805 Underwood Boulevard, Delran, NJ 08075. Twelve monthly updates. $803.00 per year. List of about 1200 companies which offer data processing equipment and services.

DIRECTORY OF CONSULTANTS IN COMPUTER SYSTEMS. Research Publications, Incorporated, 12 Lunar Drive, Woodbridge, CT 06525. (203) 397-2600. Annual. $85.00 per year.

RESEARCH CENTERS DIRECTORY. Gale Research Company, Detroit, MI 48226. (800) 521-0707. Eleventh edition. 1987.

WHO'S WHO IN FRONTIER SCIENCE AND TECHNOLOGY. Marquis Who's Who, Incorporated, 200 East Ohio St., Chicago, IL 60611. (800) 428-3898 or (312) 787-2008.

WHO'S WHO IN TECHNOLOGY TODAY. Reston Publishing Company, Incorporated, c/o Prentice-Hall, Incorporated, Englewood Cliffs, NJ 07632. (800) 262-6868. Biennial. Five volumes. $425.00. Covers the fields of electronics, computer science, physics, optics, chemistry, biotechnology, mechanics, energy, and earth science.

ENCYCLOPEDIAS AND DICTIONARIES

COMPUTER DICTIONARY. Charles J. Sippl. Howard W. Sams and Company, Incorporated, 4300 West 62nd Street, Indianapolis, IN 46268. (800) 428-7267 or (317) 298-5564. Fourth edition. 1985. $17.95.

DICTIONARY OF COMPUTING. Oxford University Press, 200 Madison Avenue, New York, NY 10016. (212) 679-7300. Second edition. 1986. $29.95.

ENCYCLOPEDIA OF COMPUTER SCIENCE AND ENGINEERING. Anthony Ralston, editor. Van Nostrand Reinhold Book Company, 115 Fifth Avenue, New York, NY 10003. (800) 543-2681. Second edition. 1982. $89.95.

ENCYCLOPEDIA OF COMPUTER SCIENCE AND TECHNOLOGY. Jack Belzer, Albert G. Holzman, and Allan Kent. Marcel Dekker, Incorporated, 270 Madison Avenue, New York, NY 10016. (212) 696-9000. Sixteen volumes. $115.00 per volume.

MCGRAW-HILL DICTIONARY OF COMPUTERS. McGraw-Hill Book Company, 1221 Avenue of the Americas, New York, NY 10020. (212) 512-2000. 1985. $17.50.

GENERAL WORKS

THE COMPUTER PIONEERS: THE MAKING OF THE MODERN COMPUTER. David Ritchie. Simon and Schuster, Incorporated, 1230 Avenue of the Americas, New York, NY 10020. (800) 223-2336 or (212) 245-6400. 1986. $17.95.

COMPUTERS AND DATA PROCESSING. H. L. Capron and Brian K. Williams. Benjamin/Cummings Publishing Company, Incorporated, 2727 Sand Hill Road, Menlo Park, CA 94025. (415) 854-6020. Second edition. 1984. $28.95.

INFORMATION SYSTEMS: THEORY AND PRACTICE. J. Burch and G. Grudnitski. John Wiley and Sons, Incorporated, 605 Third Avenue, New York, NY 10158. Fourth edition. 1986. $30.95.

INTRODUCTION TO COMPUTER ARCHITECTURE AND ORGANIZATION. H. Lorin. John Wiley and Sons, Incorporated, 605 Third Avenue, New York, NY 10158. 1982. $30.95.

HANDBOOKS AND MANUALS

FUNDAMENTALS HANDBOOK OF ELECTRICAL AND COMPUTER ENGINEERING. Sheldon S.L. Chang, editor. John Wiley and Sons, Incorporated, 605 Third Avenue, New York, NY 10158. (800) 526-5368 or (212) 850-6000. Three volumes. 1982. $180.00 set price.

HANDBOOK OF COMPUTERS AND COMPUTING. Arthur H. Seidman, editor. Van Nostrand Reinhold Book Company, 115 Fifth Avenue, New York, NY 10003. (800) 543-2681. 1984. $79.95.

HANDBOOK OF SOFTWARE ENGINEERING. Charles R. Vick, editor. Van Nostrand Reinhold Book Company, 115 Fifth Avenue, New York, NY 10003. (800) 543-2681. 1984. $66.95.

MCGRAW-HILL COMPUTER HANDBOOK. Harry Helms, editor. McGraw-Hill Book Company, 1221 Avenue of the Americas, New York, NY 10020. (212) 512-2000. 1983. $84.50.

ONLINE DATA BASES

COMPENDEX. Engineering Information, Incorporated, 345 East 47th Street, New York, NY 10017. (800) 221-1044 or (212) 705-7615. Engineering and technical literature, 1975 to present. Inquire as to cost and availability.

DISSERTATION ABSTRACTS ONLINE. University Microfilms International, 300 North Zeeb Road, Ann Arbor, MI 48106. (800) 521-0600 or (313) 761-4700. Scope includes virtually all doctoral dissertations accepted at accredited American institutions from 1861 to present in 252 subject areas. Inquire as to cost and availability.

INSPEC. INSPEC Marketing Department, Institute of Electrical and Electronics Engineers, Incorporated, IEEE Service Department, 445 Hoes Lane, Piscataway, NJ 08854. (201) 981-0060. Inquire as to on-line cost and availability.

MATHFILE. American Mathematical Society, Post Office Box 6248, Providence, RI 02940. (800) 556-7774 or (401) 272-9500. Scope includes pure and applied mathematics and related areas of physics, statistics, engineering, computer science, and operations research literature since 1973. Inquire as to cost and availability.

NTIS. National Technical Information Service, 5285 Port Royal Road, Springfield, VA 22161. (703) 487-4630. Broad coverage of government sponsored research reports, 1964 to present. Inquire as to cost and availability.

SCISEARCH. Institute for Scientific Information, 3501 Market Street, Philadelphia, PA 19104. (800) 523-1850 or (215) 386-0100. Broad multidisciplinary title and author index to the international literature of science and technology, 1974 to present. Inquire as to cost and availability.

OTHER SOURCES

COMPUTER PUBLISHERS AND PUBLICATIONS 1985-86: AN INTERNATIONAL DIRECTORY AND YEARBOOK. Edited by Efrem Sigel and Frederica Evan. Communications Trends, Incorporated. Distributed by Gale Research Company, Book Tower, Detroit, MI 48226. (800) 521-0707. Second edition. 1985. $95.00. Supplement available for 1986, $35.00. Provides information on publishers of computer books and periodicals, recommended titles for libraries and bookstores, industry trends and statistics, and computer manufacturers as publishers.

PERIODICALS

ACM TRANSACTIONS ON COMPUTER SYSTEMS. Association for Computing Machinery, 11 West 42nd Street, New York, NY 10036. (212) 869-7440. Quarterly. $70.00 per year.

BYTE. Byte Publications, Incorporated, 70 Main Street, Petersborough, NH 03458. (603) 924-9281. Monthly. $21.00 per year.

COMMUNICATIONS OF THE ACM. Association for Computing Machinery, 11 West 42nd Street, New York, NY 10036. (212) 869-7440. Monthly. $78.00 per year.

COMPUTER. Institute of Electrical and Electronic Engineers (IEEE), IEEE Service Center, 445 Hoes Lane, Piscataway, NJ 08854. (201) 981-0060. Monthly. $90.00 per year.

COMPUTER DESIGN. PennWell Directories, Incorporated, Post Office Box 1260, Tulsa, OK 74101. (918) 835-3161. Monthly. $60.00 per year.

COMPUTER MAGAZINE. IEEE Computer Society, 1109 Spring Street, Suite 300, Silver Spring, MD 20910. (301) 589-8142. Monthly. $34.00 per year.

COMPUTERS AND ELECTRONICS. Ziff-Davis Publishing Company, 3460 Wilshire Boulevard, Los Angeles, Ca 90010. Monthly. $17.00 per year.

COMPUTERWORLD. C.W. Communications, 375 Cochituate Road, Farmington, MA 01701. (617) 879-0700. Weekly. $44.00 per year.

CREATIVE COMPUTING. Creative Computing, Post Office Box 5214, Boulder, CO 80321. Monthly. $25.00 per year.

DATAMATION. Technical Publishing Company, 875 Third Avenue, New York, NY 10022. (212) 605-9400. Semi-monthly. $50.00 per year.

DR. DOBB'S JOURNAL OF SOFTWARE TOOLS. M & T Publishing, Incorporated, 2464 Embarcadero Way, Palo Alto, CA 94303. (415) 424-0600. Monthly. $25.00 per year.

IEEE TRANSACTIONS ON COMPUTERS. Institute of Electrical and Electronic Engineers (IEEE), IEEE Service Center, 445 Hoes Lane, Piscataway, NJ 08854. (201) 981-0060. Monthly. $130.00 per year.

INTERNATIONAL JOURNAL OF COMPUTER AND INFORMATION SCIENCE. Plenum Publishing Corporation, 233 Spring Street, New York, NY 10013. (800) 221-9369. Bimonthly. $195.00 per year.

JOURNAL OF THE ASSOCIATION FOR COMPUTING MACHINERY (ACM). Association for Computing Machinery, 11 West 42nd Street, New York, NY 10036. (212) 869-7440. Quarterly. $60.00 pr year.

SIGARCH COMPUTER ARCHITECTURE NEWS. Association for Computing Machinery, 11 West 42nd Street, New York, NY 10036. (212) 869-7440. Bimonthly. $15.00 per year.

CONCRETE

See also: BUILDING MATERIALS, CEMENT, PRESTRESSED CONCRETE

ABSTRACT SERVICES AND INDEXES

APPLIED SCIENCE AND TECHNOLOGY INDEX. H.W. Wilson Company, 950 University Avenue, Bronx, NY 10452. (800) 367-6670 or (212) 588-8400. Inquire as to cost and availability.

CONCRETE ABSTRACTS. American Concrete Institute, Post Office Box 19150, Redford Station, Detroit, MI 48219. (313) 532-2600. Bimonthly. $122.00 per year.

ENGINEERING INDEX MONTHLY. Engineering Information, Incorporated, 345 East 47th Street, New York, NY 10017. (800) 221-1044 or (212) 705-7600. Monthly, with annual cumulation. $1560.00 per year.

ANNUAL REVIEWS AND YEARBOOKS

CONCRETE INDUSTRIES YEARBOOK. Pit and Quarry Publications, Incorporated, 105 West Adams Street, Chicago, IL 60603. $35.00.

ASSOCIATIONS AND PROFESSIONAL SOCIETIES

AMERICAN CONCRETE INSTITUTE. Post Office Box 19150, Redford Station, Detroit, MI 48219. (313) 532-2600.

PRESTRESSED CONCRETE INSTITUTE. 201 North Wells Street, Chicago, IL 60606. (312) 346-4071.

BIBLIOGRAPHIES

SCIENTIFIC AND TECHNICAL BOOKS IN PRINT; AN INDEX TO LITERATURE IN SCIENCE AND TECHNOLOGY. R.R. Bowker Company, 205 East 42nd Street, New York, NY 10017. (800) 521-8110 or (212) 916-1600.

DIRECTORIES AND BIOGRAPHICAL SOURCES

AMERICAN CONCRETE INSTITUTE MEMBERSHIP DIRECTORY. American Concrete Institute, P.O. Box 19150, Redford Station, Detroit, MI 48219. (313) 532-2600. Biennial. $44.95.

WHO'S WHO IN ENGINEERING. Engineers Joint Council, 345 East 47th Street, New York, NY 10017. (212) 705-7010. 1985. $200.00.

WHO'S WHO IN FRONTIER SCIENCE AND TECHNOLOGY. Marquis Who's Who, Incorporated, 200 East Ohio Street, Chicago, IL 60611. (800) 428-3898 or (312) 787-2008.

WORLD CEMENT DIRECTORY. Cembureau. European Cement Association, 2 rue Saint-Charles, F-75740 Paris Cedex 15, France. Irregular. 1986. $200.00, prepayment required. List of manufactures of cement and cement products on more than 140 countries.

ENCYCLOPEDIAS AND DICTIONARIES

CEMENT AND CONCRETE TERMINOLOGY. American Concrete Institute, P.O. Box 19150, Redford Station, Detroit, MI 48219. (313) 532-2600. 1985.

ENCYCLOPEDIA OF PHYSICAL SCIENCE AND TECHNOLOGY. Academic Press, Incorporated, Orlando, FL 32887. (800) 321-5068 or (305) 345-2734. Fifteen volumes, 1986.

MCGRAW-HILL DICTIONARY OF ENGINEERING. Sybil P. Parker, editor. McGraw-Hill Book Company, 1221 Avenue of the Americas, New York, NY 10020. (212) 512-2000. 1984. $39.95.

MCGRAW-HILL ENCYCLOPEDIA OF SCIENCE AND TECHNOLOGY. McGraw-Hill Book Company, 1221 Avenue of the Americas, New York, NY 10020. (212) 512-2000.

GENERAL WORKS

CONCRETE PRIMER. F.R. McMillan and Lewis H. Tuthill. American Concrete Institute, P.O. Box 19150, Redford Station, Detroit, MI 48219. (313) 532-2600. Third edition. 1973. $9.50.

CONCRETE SCIENCE: A TREATISE ON CURRENT RESEARCH. V.S. Ramachandran, R.F. Feldman, and J.J. Beaudoin. John Wiley & Sons, 605 Third Avenue, New York, NY 10158. (800) 526-5368 or (212) 850-6000. 1981. $74.95.

CONCRETE: STRUCTURE, PROPERTIES AND MATERIALS. P. Kumar Mehta. Prentice-Hall, Incorporated, Englewood Cliffs, NJ 07632. (800) 262-6868 or (201) 592-2000. 1986. $43.95.

DESIGN OF CONCRETE STRUCTURES. George Winter and others. McGraw-Hill Book Company, 1221 Avenue of the Americas, New York, NY 10020. (212) 512-2000. Tenth edition. 1986. $42.95.

HANDBOOKS AND MANUALS

ACI MANUAL OF CONCRETE PRACTICE. American Concrete Institute, P.O. Box 19150, Redford Station, Detroit, MI 48219. (313) 532-2600. Five volumes. Annual.

CONCRETE CONSTRUCTION HANDBOOK. Joseph J. Waddell, editor. McGraw-Hill Book Company, 1221 Avenue of the Americas, New York, NY 10020. (212) 512-2000. Second edition. 1974. $70.50.

CONCRETE MANUAL. John Wiley & Sons, 605 Third Avenue, New York, NY 10158. (800) 526-5368 or (212) 850-6000. Eighth edition. 1983. $37.50.

HANDBOOK OF CONCRETE ENGINEERING. Mark Fintel, editor. Van Nostrand Reinhold Company, 115 Fifth Avenue, New York, NY 10003. (212) 254-3232. Second edition. 1985. $89.50.

STRUCTURAL DESIGN GUIDE TO THE ACI BUILDING CODE. Paul F. Rice and Edward S. Hoffman. Van Nostrand Reinhold Company, 115 Fifth Avenue, New York, NY 10003. (212) 254-3232. Third edition. 1985. $44.50.

ONLINE DATA BASES

COMPENDEX. Engineering Information, Incorporated, 345 East Street, New York, NY 10017. (800) 221-1044 or (212) 705-7615. Engineering and technical literature, 1975 to present. Inquire as to cost and availability.

NTIS. National Technical Information Service, 5285 Port Royal Road, Springfield, VA 22161. (703) 487-4630. Broad coverage of government sponsored research reports, 1964 to present. Inquire as to online cost and availability.

U.S. ARMY CORPS OF ENGINEERS, WATERWAYS EXPERIMENT STATION, CONCRETE TECHNOLOGY INFORMATION ANALYSIS CENTER. P.O. Box 631, Vicksburg, MS 39180. (601) 634-3264. Priority is given to U.S. Department of Defense personnel, but services are also available to others in the scientific community as time permits.

WILSONLINE. H.W. Wilson Company, 950 University Avenue, Bronx, NY 10452. (800) 367-6770 or (212) 588-8400. Makes available online versions of the printed H.W. Wilson indexes including Applied Science and Technology Index, Business Periodicals Index, and Readers Guide to Periodical Literature. Period covered is generally 1983 to present. Inquire as to cost and availability.

PERIODICALS

CEMENT AND CONCRETE RESEARCH. Pergamon Press, Incorporated, Maxwell House, Fairview Park, Elmsford, NY 10523. (914) 592-7700. Bimonthly. $190.00 per year.

CONCRETE. 120 West Second Street, Duluth, MN 55802. (218) 723-9253. Monthly. $14.00 per year.

CONCRETE CONSTRUCTION. Concrete Construction Publications, Incorporated, 426 South Westgage, Addison, IL 60101. (312) 543-0870. Monthly. $12.00 per year.

CONCRETE INTERNATIONAL: DESIGN AND CONSTRUCTION. American Concrete Institute, P.O. Box 19150, Redford Station, Detroit, MI 48219. (313) 532-2600. Monthly. $69.00.

JOURNAL OF PCI. Prestressed Concrete Institute, 201 North Wells Street, Chicago, IL 60606. (312) 346-4071. Bimonthly.

JOURNAL OF THE AMERICAN CONCRETE INSTITUTE. American Concrete Institute, P.O. Box 19150, Redford Station, Detroit, MI 48219. (313) 532-2600. Monthly. 469.00 per year.

RESEARCH CENTERS AND INSTITUTES

AMERICAN SOCIETY FOR TESTING AND MATERIALS. 1916 Race Street, Philadelphia, PA 19103. (215) 299-5400.

CONCRETE MATERIALS RESEARCH COUNCIL. P.O. Box 19150, Detroit, MI 48219. (313) 532-2600.

CONCRETE RESEARCH LABORATORY. University of Illinois, 1211 Newmark Civil Engineering Laboratory, 208 North Romine Street, Urbana, IL 61801. (217) 333-3394.

CONCRETE RESEARCH LABORATORY. University of Michigan, G.G. Brown Laboratory, Ann Arbor, MI 48109. (313) 763-8077.

PORTLAND CEMENT ASSOCIATION. Construction Technology Laboratory, 5420 Old Orchard Road, Skokie, IL 60077. (312) 965-7500.

SPECIFICATIONS AND STANDARDS

BLENDED CEMENTS. STP 897. G. Frohnsdorff, editor. American Society for Testing and Materials (ASTM), 1916 Race Street, Philadelphia, PA 19103. (215) 299-5400. 1986. $26.00.

MASONRY: RESEARCH, APPLICATION, AND PROBLEMS. STP 871. American Society for Testing and Materials (ASTM), 1916 Race Street, Philadelphia, PA 19103. (215) 299-5400. 1985. $36.00.

TEMPERATURE EFFECTS ON CONCRETE. STP 858. T.R. Naik, editor. American Society for Testing and Materials (ASTM), 1916 Race Street, Philadelphia, PA 19103. (215) 299-5400. 1985. $28.00.

CONDENSERS

See: BOILERS

CONDUCTION

See: HEAT TRANSFER

CONNECTION MACHINE

See: PARALLEL COMPUTERS

CONSTELLATIONS

See: STARS

CONSTRUCTION ENGINEERING

See also: ARCHITECTURAL ENGINEERING, BUILDING MATERIALS, CIVIL ENGINEERING, CONCRETE, HIGHWAY ENGINEERING, STRUCTURAL ENGINEERING

ABSTRACT SERVICES AND INDEXES

APPLIED SCIENCE AND TECHNOLOGY INDEX. H.W. Wilson and Company, 950 University Avenue, Bronx, NY 10452. (800) 367-6670 or (212) 588-8400. Monthly. Inquire as to cost and availability.

CURRENT CONTENTS: ENGINEERING, TECHNOLOGY AND APPLIED SCIENCES. Institute for Scientific Information, 3501 Market Street, Philadelphia, PA 19104. (800) 523-1850 or (215) 386-0100. Weekly. $275.00 per year.

ENGINEERING INDEX MONTHLY AND AUTHOR INDEX. Engineering Information Inc., 345 East 47th Street, New York, NY 10017. (212) 705-7600. Monthly. $1560.00 per year.

SCIENCE CITATION INDEX. Institute for Scientific Information, 3501 Market Street, Philadelphia, PA 19104. (800) 523-1850 or (215) 386-0100. Six times per year. $6200.00 per year.

ASSOCIATIONS AND PROFESSIONAL SOCIETIES

AMERICAN CONCRETE INSTITUTE. P.O. Box 19150, Detroit, MI 48219. (313) 532-2600.

AMERICAN SOCIETY FOR TESTING AND MATERIALS. 1916 Race Street, Philadelphia, PA 19103. (215) 299-5400.

AMERICAN SOCIETY OF CIVIL ENGINEERS. 345 East 47th Street, New York, NY 10017. (212) 705-7420.

NATIONAL INSTITUTE OF BUILDING SCIENCES. 1015 15th Street, N.W., Suite 700, Washington, DC 20005. (202) 347-5710.

DIRECTORIES AND BIOGRAPHICAL SOURCES

INTERNATIONAL RESEARCH CENTERS DIRECTORY 1988-89. Darren L. Smith, editor. Gale Research Company, Book Tower, Detroit, MI 48226. (800) 521-0707. 4th edition. 1987. $360.00.

1987 DIRECTORY OF ENGINEERING SOCIETIES AND RELATED ORGANIZATIONS. Gordon Davis, editor. Hemisphere Publishing Corporation, 1010 Vermont Avenue, NW, Washington, DC 20005. (800) 526-0275. 12th edition. 1987. $100.00.

PUBLIC WORKS MANUAL. Public Works Journal Corporation, 200 South Broad Street, Ridgewood, NJ 07451. (201) 445-5800. Annual. $20.00.

RESEARCH CENTERS DIRECTORY 1988. Gale Research Company, Book Tower, Detroit, MI 48226. (800) 521-0707. 12th edition. 1987. $365.00 for set.

SCIENTIFIC AND TECHNICAL ORGANIZATIONS AND AGENCIES DIRECTORY. Margaret Labash Young, editor. Gale Research Company, Book Tower, Detroit, MI 48226. (800) 521-0707. 2nd edition. 1987. $185.00.

WHO'S WHO IN ENGINEERING. Gordon Davis, editor. Hemisphere Publishing Corporation, 1010 Vermont Avenue, NW, Washington, DC 20005. (800) 526-0275. 6th edition. 1985. $200.00.

ENCYCLOPEDIAS AND DICTIONARIES

CONSTRUCTION GLOSSARY: AN ENCYCLOPEDIC REFERENCE AND MANUAL. J.S. Stein. John Wiley and Sons, Inc., 605 Third Avenue, New York, NY 10158. (800) 526-5368. 1986. $35.95.

DICTIONARY OF BUILDING. J.S. Scott. John Wiley and Sons, Inc., 605 Third Avenue, New York, NY 10158. (800) 526-5368. Third edition. 1984. $26.95.

MCGRAW-HILL ENCYCLOPEDIA OF SCIENCE AND TECHNOLOGY. McGraw-Hill Book Company, 1221 Avenue of the Americas, New York, NY 10020. (212) 512-2000. 6th edition. 1987. $1600.00.

THESAURUS OF SCIENTIFIC, TECHNICAL, AND ENGINEERING TERMS. Hemisphere Publishing Corporation, 1010 Vermont Avenue, NW, Washington, DC 20005. (800) 526-0275. 1988. $125.00.

GENERAL WORKS

APPLIED STRUCTURAL STEEL DESIGN. L. Spiegel and G.F. Limbrunner. Prentice-Hall Publishing, Inc., Englewood Cliffs, NJ 07632. (800) 562-0245. 1986. $34.95.

BUILDING CONSTRUCTION, DRAFTING AND DESIGN. John Molnar. Van Nostrand Reinhold Company, Inc., 135 West 50th Street, New York, NY 10020. (800) 543-2681. 1986. $32.95.

DESIGN OF REINFORCED CONCRETE. Samuel E. French. Prentice-Hall Publishing, Inc., Englewood Cliffs, NJ 07632. (800) 562-0245. 1987. $25.00.

ELEMENTARY THEORY OF STRUCTURES. Yu H. Yuan. Prentice-Hall Publishing, Inc., Englewood Cliffs, NJ 07632. (800) 562-0245. Third edition. 1988. $42.95.

FOUNDATION DESIGN AND CONSTRUCTION. M.J. Tomlinson. John Wiley and Sons, Inc., 605 Third Avenue, New York, NY 10158. (800) 526-5368. Fifth edition. 1986. $45.95.

HIGHWAY ENGINEERING. Paul H. Paquette and Randor J. Paquette. John Wiley and Sons, Inc., 605 Third Avenue, New York, NY 10158. (800) 526-5368. Fifth edition. 1986. $45.00.

HIGHWAYS: HIGHWAY ENGINEERING. C.A. O'Flaherty. Edward Arnold Publishers, Limited, 300 North Charles Street, Baltimore, MD 21201. (301) 539-1529. Volume 2. Third edition. 1987. $29.95 in paper.

MATERIALS FOR CIVIL AND HIGHWAY ENGINEERS. K.N. Derucher and G.P. Korgiatis. Prentice-Hall Publishing, Inc., Englewood Cliffs, NJ 07632. (800) 562-0245. Second edition. 1988. $42.50.

MICROCOMPUTER-AIDED ENGINEERING: STRUCTURAL DYNAMICS. Mario Paz. Van Nostrand Reinhold Company, Inc., 135 West 50th Street, New York, NY 10020. (800) 543-2681. 1986. $49.95.

STRUCTURAL STABILITY. Wai-Fah Chen and E.M. Lui. Elsevier Science Publishing Company, Inc., 52 Vanderbilt Avenue, New York, NY 10017. (212) 370-5520. 1987. $49.50.

HANDBOOKS AND MANUALS

BUILDING DESIGN AND CONSTRUCTION HANDBOOK. F.S. Merritt, editor. McGraw-Hill Book Company, 1221 Avenue of the Americas, New York, NY 10020. (212) 512-2000. 1982. $89.50.

BUILDING STRUCTURES HANDBOOK. R.N. White and C.G. Salmon. John Wiley and Sons, Inc., 605 Third Avenue, New York, NY 10158. (800) 526-5368. 1986. $79.95.

CIVIL ENGINEERING CALCULATIONS REFERENCE GUIDE. Tyler G. Hicks, editor. McGraw-Hill Book Company, 1221 Avenue of the Americas, New York, NY 10020. (212) 512-2000. 1987. $29.50.

CIVIL ENGINEERING PRACTICE. Paul N. Cheremisinoff and others, editors. Technomic Publishing Company, Inc., 851 Holland Avenue, Box 3535, Lancaster, PA 17604. (800) 233-9936. Five volumes. 1987-1988. $750.00 for set.

HANDBOOK OF CONCRETE ENGINEERING. Mark Fintel. Van Nostrand Reinhold Company, Inc., 135 West 50th Street, New York, NY 10020. (800) 543-2681. 1986. $89.95.

HANDBOOK OF MECHANICS, MATERIALS, AND STRUCTURES. A. Blake. John Wiley and Sons, Inc., 605 Third Avenue, New York, NY 10158. (800) 526-5368. 1985. $64.50.

STANDARD HANDBOOK FOR CIVIL ENGINEERS. F. S. Merritt, editor. McGraw-Hill Book Company, 1221 Avenue of the Americas, New York, NY 10020. (212) 512-2000. Third edition. 1983. $89.50.

STRUCTURAL ENGINEERING HANDBOOK. E.H. Gaylord and C.N. Gaylord, editors. McGraw-Hill Book Company, 1221 Avenue of the Americas, New York, NY 10020. (212) 512-2000. Second edition. 1979. $76.50.

SURVEYING READY-REFERENCE MANUAL. Guy O. Stenstrom. McGraw-Hill Book Company, 1221 Avenue of the Americas,

CONSTRUCTION ENGINEERING

New York, NY 10020. (212) 512-2000. 1987. $26.50.

ONLINE DATA BASES

COMPENDEX. Engineering Information, Inc., 345 East 47th Street, New York, NY 10017. (800) 221-1044 or (212) 705-7615. Engineering and technical literature, 1975 to present. Inquire as to online cost and availability.

DISSERTATION ABSTRACTS ONLINE. University Microfilms International, 300 North Zeeb Road, Ann Arbor, MI 48106. (800) 521-0600 or (313) 761-4700. Scope includes virtually all doctoral dissertations accepted at accredited American institutions from 1861 to present in over 250 subject areas. Inquire as to online cost and availability.

GEOREF. Online version of the BIBLIOGRAPHY AND INDEX OF GEOLOGY. American Geological Institute, 4220 King Street, Alexandria, VA 22302. (703) 379-2480. 1969 to present. Inquire as to online cost and availability.

INSPEC. INSPEC Marketing Department, Institution of Electrical Engineers. Available from IEEE Service Center, 445 Hoes Lane, Piscataway, NJ 08854. (201) 981-0060. Online version of Physics Abstracts. Inquire as to online cost and availability.

NTIS. National Technical Information Service, 5285 Port Royal Road, Springfield, VA 22161. (703) 487-4630. Broad coverage of government sponsored research reports, 1964 to present. Inquire as to online cost and availability.

SCISEARCH. Institute for Scientific Information, 3501 Market Street, Philadelphia, PA 19104. (800) 523-1850 or (215) 386-0100. Broad multidisciplinary title and author index to the international literature of science and technology, 1974 to present. Inquire as to online cost and availability.

WILSONLINE. H.W. Wilson and Company, 950 University Avenue, Bronx, NY 10452. (800) 367-6770 or (212) 588-8400. Makes available online versions of the H.W. Wilson indexes including Applied Science and Technology Index, Business Periodicals Index and Readers' Guide to Periodical Literature. Approximately 1980 to present. Inquire as to online cost and availability.

PERIODICALS

AMERICAN CONCRETE INSTITUTE JOURNAL. P.O. Box 19150, Redford Station, Detroit, MI 48219. (513) 532-2600. 1929 to present. Bimonthly. $69.00 per year.

CIVIL ENGINEERING. American Society of Civil Engineers, 345 East 47th Street, New York, NY 10017. (212) 705-7420. 1930 to present. Monthly. $48.00 per year.

CONCRETE. Harcourt, Brace Jovanovich, Inc., 7500 Old Oak Boulevard, Cleveland, OH 44130. 1937 to present. Monthly. $25.00 per year.

ENGINEERING NEWS-RECORD. McGraw-Hill Book Company, 1221 Avenue of the Americas, New York, NY 10020. (212) 512-2000. 1874 to present. Weekly. $35.00 per year.

HIGHWAY AND HEAVY CONSTRUCTION. Technical Publishing Company, 875 Third Avenue, New York, NY 10022. (212) 605-9400. 1892 to present. Monthly. $35.00 per year.

JOURNAL OF CONSTRUCTION ENGINEERING AND MANAGEMENT. American Society of Civil Engineers, 345 East 47th Street, New York, NY 10017. (212) 705-7420. 1956 to present. Quarterly. $48.00 per year.

JOURNAL OF STRUCTURAL ENGINEERING. American Society of Civil Engineers, 345 East 47th Street, New York, NY 10017. (212) 705-7420. 1956 to present. Monthly. $140.00 per year.

JOURNAL OF SURVEYING ENGINEERING. American Society of Civil Engineers, 345 East 47th Street, New York, NY 10017. (212) 705-7420. 1956 to present. Three times per year. $35.00 per year.

STRUCTURAL ENGINEERING PRACTICE: ANALYSIS, DESIGN, MANAGEMENT. Marcel Dekker Inc., 270 Madison Avenue, New York, NY 10016. (800) 228-1160. 1981 to present. Quarterly. $75.00 per year.

RESEARCH CENTERS AND INSTITUTES

BUILDING RESEARCH BOARD. 2101 Constitution Avenue, N.W., Washington, DC 20418. (202) 334-3376.

STRUCTURAL STABILITY RESEARCH COUNCIL. Fritz Engineering Laboratory No. 13, Lehigh University, Bethlehem, PA 18105. (215) 861-3519.

SPECIFICATIONS AND STANDARDS

STRUCTURAL DESIGN GUIDE TO THE ACI BUILDING CODE. Paul Rice. Van Nostrand Reinhold Company, Inc., 135 West 50th Street, New York, NY 10020. (800) 543-2681. Third edition. 1985. $47.95.

CONTINENTAL DRIFT

See also: CONTINENTAL MARGINS, GEOLOGY, PHYSICAL GEOLOGY, PLATE TECTONICS

ABSTRACT SERVICES AND INDEXES

BIBLIOGRAPHY AND INDEX OF GEOLOGY. American Geological Institute, 4220 King Street, Alexandria, VA 22302. (703) 379-2480. 1969 to present. Monthly. $1100.00 per year.

CONFERENCE PAPERS INDEX. Cambridge Scientific Abstracts, 5161 River Road, Bethesda, MD 20816. 1972 to present. Monthly. Inquire as to cost and availability.

CURRENT CONTENTS: PHYSICAL, CHEMICAL, AND EARTH SCIENCES. Institute for Scientific Information, 3501 Market Street, Philadelphia, PA 19104. (800) 523-1850 or (215) 386-0100. Weekly. $275.00 per year.

GENERAL SCIENCE INDEX. H.W. Wilson and Company, 950 University Avenue, Bronx, NY 10452. (800) 367-6670 or (212) 588-8400. 1978 to present. Monthly. Inquire as to cost and availability.

OCEAN ABSTRACTS. Cambridge Scientific Abstracts, 5161 River Road, Bethesda, MD 20816. (301) 951-1400. 1963 to present. Bimonthly. $450.00 per year.

SCIENCE CITATION INDEX. Institute for Scientific Information, 3501 Market Street, Philadelphia, PA 19104. (800) 523-1850 or (215) 386-0100. Six times per year. $6200.00 per year.

ASSOCIATIONS AND PROFESSIONAL SOCIETIES

AMERICAN GEOLOGICAL INSTITUTE. 4220 King Street, Alexandria, VA 22302. (703) 379-2480.

AMERICAN GEOPHYSICAL UNION. 2000 Florida Avenue, N.W., Washington, DC 20009. (202) 462-6903.

GEOLOGICAL SOCIETY OF AMERICA. 3300 Penrose Place, Boulder, CO 80301. (303) 447-2020.

DIRECTORIES AND BIOGRAPHICAL SOURCES

AMERICAN MEN AND WOMEN OF SCIENCE. R.R. Bowker, Inc., Order Department, 245 West 17th Street, New York, NY 10011. (800) 521-8110. Eight volumes. 1986. $595.00 for set.

INTERNATIONAL RESEARCH CENTERS DIRECTORY 1988-89. Darren L. Smith, editor. Gale Research Company, Book Tower, Detroit, MI 48226. (800) 521-0707. 4th edition. 1987. $360.00.

1987 DIRECTORY OF ENGINEERING SOCIETIES AND RELATED ORGANIZATIONS. Gordon Davis, editor. Hemisphere Publishing Corporation, 1010 Vermont Avenue, NW, Washington, DC 20005. (800) 526-0275. 12th edition. 1987. $100.00.

RESEARCH CENTERS DIRECTORY 1988. Gale Research Company, Book Tower, Detroit, MI 48226. (800) 521-0707. 12th edition. 1987. $365.00 for set.

SCIENTIFIC AND TECHNICAL ORGANIZATIONS AND AGENCIES DIRECTORY. Margaret Labash Young, editor. Gale Research Company, Book Tower, Detroit, MI 48226. (800) 521-0707. 2nd edition. 1987. $185.00.

GENERAL WORKS

CONTINENTAL DRIFT. James H. Shea, editor. Van Nostrand Reinhold Company, Inc., 135 West 50th Street, New York, NY 10020. (800) 543-2681. 1985. $49.50.

MECHANISMS OF CONTINENTAL DRIFT AND PLATE TECTONICS. P.A. Davies and S.K. Runcorn, editors. Academic Press, Inc., 6277 Sea Harbor Drive, Orlando, FL 32821. (800) 321-5068. 1981. $75.50.

NEW VIEWS ON AN OLD PLANET: CONTINENTAL DRIFT AND THE HISTORY OF THE EARTH. T.H. Van Andel. Cambridge University Press, 32 East 57th Street, New York, NY 10022. (800) 872-7423. 1985. $19.95.

ORIGIN OF CONTINENTS AND OCEANS. Alfred Wegener. Dover Publications, Inc., 31 East Second Street, Mineola, NY 11501. (516) 294-7000. 1966. $5.95.

WANDERING CONTINENTS AND SPREADING SEA FLOORS ON AN EXPANDING EARTH. Lester C. King. John Wiley and Sons, Inc., 605 Third Avenue, New York, NY 10158. (800) 526-5368. 1984. $47.95.

ONLINE DATA BASES

DISSERTATION ABSTRACTS ONLINE. University Microfilms International, 300 North Zeeb Road, Ann Arbor, MI 48106. (800) 521-0600 or (313) 761-4700. Scope includes virtually all doctoral dissertations accepted at accredited American institutions from 1861 to present in over 250 subject areas. Inquire as to online cost and availability.

GEOREF. Online version of the BIBLIOGRAPHY AND INDEX OF GEOLOGY. American Geological Institute, 4220 King Street, Alexandria, VA 22302. (703) 379-2480. 1969 to present. Inquire as to online cost and availability.

NTIS. National Technical Information Service, 5285 Port Royal Road, Springfield, VA 22161. (703) 487-4630. Broad coverage of government sponsored research reports, 1964 to present. Inquire as to online cost and availability.

SCISEARCH. Institute for Scientific Information, 3501 Market Street, Philadelphia, PA 19104. (800) 523-1850 or (215) 386-0100. Broad multidisciplinary title and author index to the international literature of science and technology, 1974 to present. Inquire as to online cost and availability.

WILSONLINE. H.W. Wilson and Company, 950 University Avenue, Bronx, NY 10452. (800) 367-6770 or (212) 588-8400. Makes available online versions of the H.W. Wilson indexes including Applied Science and Technology Index, Business Periodicals Index and Readers' Guide to Periodical Literature. Approximately 1980 to present. Inquire as to online cost and availability.

PERIODICALS

CONTINENTAL SHELF RESEARCH. Pergamon Press, Inc., Maxwell House, Fairview Park, Elmsford, NY 10523. (914) 592-7700. 1982 to present. Bimonthly. $1200.00 per year. Includes Oceanographic Literature Review.

GEO-MARINE LETTERS. Springer-Verlag New York, Inc., 175 Fifth Avenue, New York, NY 10010. (800) 526-7254. 1981 to present. Bimonthly. $100.00 per year.

MARINE GEOLOGY; INTERNATIONAL JOURNAL OF MARINE GEOLOGY, GEOCHEMISTRY AND GEOPHYSICS. Elsevier Science Publishing Company, Inc., 52 Vanderbilt Avenue, New York, NY 10017. (212) 370-5520. 1964 to present. Twenty-four times per year. $510.00 per year.

TECTONOPHYSICS; INTERNATIONAL JOURNAL OF GEOTECTONICS AND THE GEOLOGY AND PHYSICS OF THE INTERIOR OF THE EARTH. Elsevier Science Publishing Company, Inc., 52 Vanderbilt Avenue, New York, NY 10017. (212) 370-5520. Forty-four times per year. $900.00 per year.

RESEARCH CENTERS AND INSTITUTES

GEOLOGICAL RESEARCH DIVISION. University of California, San Diego, A-020, Scripps Institution of Oceanography, La Jolla, CA 92093. (619) 534-1830.

INSTITUTE FOR THE STUDY OF THE CONTINENTS. Cornell University, 3120 Snee Hall, Ithaca, NY 14853-1504. (607) 255-3474.

LAMONT-DOHERTY GEOLOGICAL OBSERVATORY. Columbia University, Palisades, NY 10964. (914) 359-2900.

CONTINENTAL MARGINS

See also: CIVIL ENGINEERING, COASTAL ENGINEERING, COASTS, GEOLOGY, OCEAN ENGINEERING, PHYSICAL GEOLOGY

ABSTRACT SERVICES AND INDEXES

APPLIED SCIENCE AND TECHNOLOGY INDEX. H.W. Wilson and Company, 950 University Avenue, Bronx, NY 10452. (800) 367-6670 or (212) 588-8400. Monthly. Inquire as to cost and availability.

BIBLIOGRAPHY AND INDEX OF GEOLOGY. American Geological Institute, 4220 King Street, Alexandria, VA 22302. (703) 379-2480. 1969 to present. Monthly. $1100.00 per year.

CHEMICAL ABSTRACTS. American Chemical Society, Chemical Abstracts Service, Box 3012, Columbus, OH 43210. (614) 421-3600. 1907 to present. Weekly. $9500.00 per year.

CONFERENCE PAPERS INDEX. Cambridge Scientific Abstracts, 5161 River Road, Bethesda, MD 20816. 1972 to present. Monthly. Inquire as to cost and availability.

CURRENT CONTENTS: ENGINEERING, TECHNOLOGY AND APPLIED SCIENCES. Institute for Scientific Information, 3501 Market Street, Philadelphia, PA 19104. (800) 523-1850 or (215) 386-0100. Weekly. $275.00 per year.

ENGINEERING INDEX MONTHLY AND AUTHOR INDEX. Engineering Information Inc., 345 East 47th Street, New York, NY 10017. (212) 705-7600. Monthly. $1560.00 per year.

GENERAL SCIENCE INDEX. H.W. Wilson and Company, 950 University Avenue, Bronx, NY 10452. (800) 367-6670 or (212) 588-8400. 1978 to present. Monthly. Inquire as to cost and availability.

METEOROLOGICAL AND GEOASTROPHYSICAL ABSTRACTS. American Meteorological Society, 45 Beacon Street, Boston, MA

02108. (617) 227-2425. 1950 to present. Monthly. $450.00 per year.

OCEAN ABSTRACTS. Cambridge Scientific Abstracts, 5161 River Road, Bethesda, MD 20816. (301) 951-1400. 1963 to present. Bimonthly. $450.00 per year.

SCIENCE CITATION INDEX. Institute for Scientific Information, 3501 Market Street, Philadelphia, PA 19104. (800) 523-1850 or (215) 386-0100. Six times per year. $6200.00 per year.

ASSOCIATIONS AND PROFESSIONAL SOCIETIES

AMERICAN SOCIETY OF CIVIL ENGINEERS. 345 East 47th Street, New York, NY 10017. (212) 705-7420.

COASTAL ENGINEERING RESEARCH COUNCIL. 207 East Bay Street, Suite 311, Charleston, SC 29401. (803) 723-4864.

THE COASTAL SOCIETY. 5410 Grosvenor Lane, Suite 110, Bethesda, MD 20814. (301) 897-8616.

GEOLOGICAL SOCIETY OF AMERICA. 3300 Penrose Place, Boulder, CO 80301. (303) 447-2020.

DIRECTORIES AND BIOGRAPHICAL SOURCES

1987 DIRECTORY OF ENGINEERING SOCIETIES AND RELATED ORGANIZATIONS. Gordon Davis, editor. Hemisphere Publishing Corporation, 1010 Vermont Avenue, NW, Washington, DC 20005. (800) 526-0275. 12th edition. 1987. $100.00.

RESEARCH CENTERS DIRECTORY 1988. Gale Research Company, Book Tower, Detroit, MI 48226. (800) 521-0707. 12th edition. 1987. $365.00 for set.

SCIENTIFIC AND TECHNICAL ORGANIZATIONS AND AGENCIES DIRECTORY. Margaret Labash Young, editor. Gale Research Company, Book Tower, Detroit, MI 48226. (800) 521-0707. 2nd edition. 1987. $185.00.

WHO'S WHO IN ENGINEERING. Gordon Davis, editor. Hemisphere Publishing Corporation, 1010 Vermont Avenue, NW, Washington, DC 20005. (800) 526-0275. 6th edition. 1985. $200.00.

GENERAL WORKS

CONTINENTAL DRIFT. James H. Shea, editor. Van Nostrand Reinhold Company, Inc., 135 West 50th Street, New York, NY 10020. (800) 543-2681. 1985. $49.50.

THE GEOLOGY OF THE CONTINENTAL MARGINS. C.A. Burk and C.L. Drake, editors. Springer-Verlag New York, Inc., 175 Fifth Avenue, New York, NY 10010. (800) 526-7254. 1974. $63.00.

THE OCEAN BASINS AND MARGINS. A.E.M. Nairn, editor. Plenum Publishing Corporation, 233 Spring Street, New York, NY 10013. (800) 221-9369. 1973-1985. Eight volumes. Price varies, inquire.

PHYSICAL OCEANOGRAPHY OF COASTAL WATERS. K.F. Bowden. John Wiley and Sons, Inc., 605 Third Avenue, New York, NY 10158. (800) 526-5368. 1984. $74.95.

THE WORLD'S COASTLINE. E.C. Bird and M.L. Schwartz, editors. Van Nostrand Reinhold Company, Inc., 135 West 50th Street, New York, NY 10020. (800) 543-2681. 1985. $99.95.

ONLINE DATA BASES

CA SEARCH. Chemical Abstracts Service, P.O. Box 3012, Columbus, OH 43120. (800) 848-6538 or (614) 421-3600. Comprehensive guide to chemical literature, 1972 to present. Inquire as to online cost and availability.

COMPENDEX. Engineering Information, Inc., 345 East 47th Street, New York, NY 10017. (800) 221-1044 or (212) 705-7615. Engineering and technical literature, 1975 to present. Inquire as to online cost and availability.

DISSERTATION ABSTRACTS ONLINE. University Microfilms International, 300 North Zeeb Road, Ann Arbor, MI 48106. (800) 521-0600 or (313) 761-4700. Scope includes virtually all doctoral dissertations accepted at accredited American institutions from 1861 to present in over 250 subject areas. Inquire as to online cost and availability.

GEOREF. Online version of the BIBLIOGRAPHY AND INDEX OF GEOLOGY. American Geological Institute, 4220 King Street, Alexandria, VA 22302. (703) 379-2480. 1969 to present. Inquire as to online cost and availability.

INSPEC. INSPEC Marketing Department, Institution of Electrical Engineers. Available from IEEE Service Center, 445 Hoes Lane, Piscataway, NJ 08854. (201) 981-0060. Online version of Physics Abstracts. Inquire as to online cost and availability.

NTIS. National Technical Information Service, 5285 Port Royal Road, Springfield, VA 22161. (703) 487-4630. Broad coverage of government sponsored research reports, 1964 to present. Inquire as to online cost and availability.

SCISEARCH. Institute for Scientific Information, 3501 Market Street, Philadelphia, PA 19104. (800) 523-1850 or (215) 386-0100. Broad multidisciplinary title and author index to the international literature of science and technology, 1974 to present. Inquire as to online cost and availability.

WILSONLINE. H.W. Wilson and Company, 950 University Avenue, Bronx, NY 10452. (800) 367-6770 or (212) 588-8400. Makes available online versions of the H.W. Wilson indexes including Applied Science and Technology Index, Business Periodicals Index and Readers' Guide to Periodical Literature. Approximately 1980 to present. Inquire as to online cost and availability.

PERIODICALS

COASTAL ENGINEERING: AN INTERNATIONAL JOURNAL FOR COASTAL, HARBOUR, AND OFFSHORE ENGINEERS. Elsevier Science Publishing Company, Inc., 52 Vanderbilt Avenue, New York, NY 10017. (212) 370-5520. 1977 to present. Quarterly. $95.00 per year.

COASTAL RESEARCH. Florida State University, Geology Department, Tallahassee, FL 32306. (904) 644-5860. 1962 to present. Three times per year. $5.00 per year.

COASTWATCH. University of North Carolina, Sea Grant College Program, Box 8605, North Carolina State University, Raleigh, NC 27695-8605. (919) 737-2454. 1970 to present. Monthly. Free.

CONTINENTAL SHELF RESEARCH. Pergamon Press, Inc., Maxwell House, Fairview Park, Elmsford, NY 10523. (914) 592-7700. 1982 to present. Bimonthly. $1200.00 per year. Includes Oceanographic Literature Review.

ESTUARINE, COASTAL AND SHELF SCIENCE. Academic Press, Inc., 6277 Sea Harbor Drive, Orlando, FL 32821. (800) 321-5068. 1973 to present. Monthly. $315.00 per year.

JOURNAL OF WATERWAY, PORT, COASTAL AND OCEAN ENGINEERING. American Society of Civil Engineers, 345 East 47th Street, New York, NY 10017. (212) 705-7420. 1956 to present. Bimonthly. $72.00 per year.

OCEAN ENGINEERING. Pergamon Press, Inc., Maxwell House, Fairview Park, Elmsford, NY 10523. (914) 592-7700. 1968 to present. Bimonthly. $200.00 per year.

OCEAN SCIENCE AND ENGINEERING. Marcel Dekker Inc., 270 Madison Avenue, New York, NY 10016. (800) 228-1160. 1974 to present. Quarterly. $165.00 per year.

RESEARCH CENTERS AND INSTITUTES

EARTH RESOURCES INSTITUTE. Texas A&M University, College Station, TX 77843. (409) 845-3651.

GEOLOGICAL RESEARCH DIVISION. University of California, San Diego, A-020, Scripps Institution of Oceanography, La Jolla, CA 92093. (619) 534-1830.

INSTITUTE FOR THE STUDY OF THE CONTINENTS. Cornell University, 3120 Snee Hall, Ithaca, NY 14853-1504. (607) 255-3474.

CONTROL SYSTEMS

See: AUTOMATION

CONVECTION

See: HEAT TRANSFER

COOLING TOWERS

See also: AIR CONDITIONING, HEAT TRANSFER, MECHANICAL ENGINEERING, NUCLEAR ENERGY, PRESSURE VESSELS, THERMODYNAMICS

ABSTRACT SERVICES AND INDEXES

APPLIED MECHANICS REVIEW. American Society of Mechanical Engineers, 345 East 47th Street, New York, NY 10017. (212) 705-7703. 1948 to present. Monthly. $360.00 per year.

APPLIED SCIENCE AND TECHNOLOGY INDEX. H.W. Wilson and Company, 950 University Avenue, Bronx, NY 10452. (800) 367-6670 or (212) 588-8400. Monthly. Inquire as to cost and availability.

CHEMICAL ABSTRACTS. American Chemical Society, Chemical Abstracts Service, Box 3012, Columbus, OH 43210. (614) 421-3600. 1907 to present. Weekly. $9500.00 per year.

CURRENT CONTENTS: ENGINEERING, TECHNOLOGY AND APPLIED SCIENCES. Institute for Scientific Information, 3501 Market Street, Philadelphia, PA 19104. (800) 523-1850 or (215) 386-0100. Weekly. $275.00 per year.

ENGINEERING INDEX MONTHLY AND AUTHOR INDEX. Engineering Information Inc., 345 East 47th Street, New York, NY 10017. (212) 705-7600. Monthly. $1560.00 per year.

ISMEC BULLETIN (Information Service in Mechanical Engineering). Cambridge Scientific Abstracts, 5161 River Road, Bethesda, MD 20816. (301) 951-1400. 1973 to present. Monthly. $450.00 per year.

ASSOCIATIONS AND PROFESSIONAL SOCIETIES

AMERICAN INSTITUTE OF CHEMICAL ENGINEERS. 345 East 47th Street, New York, NY 10017. (212) 705-7703.

AMERICAN SOCIETY OF CIVIL ENGINEERS. 345 East 47th Street, New York, NY 10017. (212) 705-7420.

AMERICAN SOCIETY OF MECHANICAL ENGINEERS. 345 East 47th Street, New York, NY 10017. (212) 705-7703.

DIRECTORIES AND BIOGRAPHICAL SOURCES

1987 DIRECTORY OF ENGINEERING SOCIETIES AND RELATED ORGANIZATIONS. Gordon Davis, editor. Hemisphere Publishing Corporation, 1010 Vermont Avenue, NW, Washington, DC 20005. (800) 526-0275. 12th edition. 1987. $100.00.

RESEARCH CENTERS DIRECTORY 1988. Gale Research Company, Book Tower, Detroit, MI 48226. (800) 521-0707. 12th edition. 1987. $365.00 for set.

WHO'S WHO IN ENGINEERING. Gordon Davis, editor. Hemisphere Publishing Corporation, 1010 Vermont Avenue, NW, Washington, DC 20005. (800) 526-0275. 6th edition. 1985. $200.00.

GENERAL WORKS

COOLING TOWER PERFORMANCE. Donald R. Baker. Chemical Publishing Company, Inc., 80 8th Avenue, New York, NY 10011. (212) 255-1950. 1984. $40.00.

COOLING TOWERS: SELECTION, DESIGN, AND PRACTICE. N.P. Cheremisinoff and P.N. Cheremisinoff. Butterworth's Publishing, 80 Montvale Avenue, Stoneham, MA 02180. (800) 325-4177. 1981. $59.95.

FUNDAMENTALS OF HEAT EXCHANGER AND PRESSURE VESSEL TECHNOLOGY. J.P. Gupta. Hemisphere Publishing Corporation, 79 Madison Avenue, New York, NY 10016-7892. (800) 821-8312. 1986. $49.95.

HEAT TRANSFER EQUIPMENT DESIGN. R.K. Shah and others. Hemisphere Publishing Corporation, 79 Madison Avenue, New York, NY 10016-7892. 1988. $135.00.

HANDBOOKS AND MANUALS

HANDBOOK OF HEAT TRANSFER APPLICATIONS. W.M. Rohsenow and others. McGraw-Hill Book Company, 1221 Avenue of the Americas, New York, NY 10020. (212) 512-2000. Second edition. 1986. $79.50.

HEAT EXCHANGER SOURCEBOOK. J.W. Palen, editor. Hemisphere Publishing Corporation, 79 Madison Avenue, New York, NY 10016-7892. (800) 821-8312. 1986. $59.95.

ONLINE DATA BASES

CA SEARCH. Chemical Abstracts Service, P.O. Box 3012, Columbus, OH 43120. (800) 848-6538 or (614) 421-3600. Comprehensive guide to chemical literature, 1972 to present. Inquire as to online cost and availability.

COMPENDEX. Engineering Information, Inc., 345 East 47th Street, New York, NY 10017. (800) 221-1044 or (212) 705-7615. Engineering and technical literature, 1975 to present. Inquire as to online cost and availability.

NTIS. National Technical Information Service, 5285 Port Royal Road, Springfield, VA 22161. (703) 487-4630. Broad coverage of government sponsored research reports, 1964 to present. Inquire as to online cost and availability.

WILSONLINE. H.W. Wilson and Company, 950 University Avenue, Bronx, NY 10452. (800) 367-6770 or (212) 588-8400. Makes available online versions of the H.W. Wilson indexes including Applied Science and Technology Index, Business Periodicals Index and Readers' Guide to Periodical Literature. Approximately 1980 to present. Inquire as to online cost and availability.

PERIODICALS

HEAT TRANSFER ENGINEERING. Hemisphere Publishing Corporation, 79 Madison Avenue, New York, NY 10016-7892.

(800) 821-8312. 1979 to present. Quarterly. $97.50.

INTERNATIONAL COMMUNICATIONS IN HEAT AND MASS TRANSFER. Pergamon Press, Inc., Maxwell House, Fairview Park, Elmsford, NY 10523. (914) 592-7700. 1974 to present. Bimonthly. $160.00 per year.

INTERNATIONAL JOURNAL OF HEAT AND MASS TRANSFER. Pergamon Press, Inc., Maxwell House, Fairview Park, Elmsford, NY 10523. (914) 592-7700. 1960 to present. Monthly. $500.00 per year.

JOURNAL OF HEAT TRANSFER. American Society of Mechanical Engineers, 345 East 47th Street, New York, NY 10017. (212) 705-7703. 1970 to present. Quarterly. $100.00 per year.

RESEARCH CENTERS AND INSTITUTES

HEAT TRANSFER LABORATORY. Massachusetts Institute of Technology, 77 Massachusetts Avenue, Cambridge, MA 02139. (716) 253-2248.

HEAT TRANSFER RESEARCH. University of Wisconsin, Milwaukee, P.O. Box 784, Milwaukee, WI 53201. (414) 963-5001.

HEAT TRANSFER RESEARCH FACILITY. Columbia University, 632 West 125th Street, New York, NY 10027. (212) 280-4163.

COPPER

See also: BRASS AND BRONZE, MATERIALS SCIENCE, METALLURGY

ABSTRACT SERVICES AND INDEXES

ALLOYS INDEX. American Society for Metals, 9639 Kinsman Road, Metals Park, OH 44073. (216) 338-5151. $130.00 per year.

APPLIED SCIENCE AND TECHNOLOGY INDEX. H.W. Wilson Company, 950 University Avenue, Bronx, NY 10452. (800) 367-6670 or (212) 588-8400. Inquire as to cost and availability.

CHEMICAL ABSTRACTS. Chemical Abstracts Service, 2540 Olentangy Road, Post Office Box 3012, Columbus, OH 43210. (800) 848-6538 or (614) 421-3600. Weekly. $9200.00 per year.

CURRENT CONTENTS: ENGINEERING, TECHNOLOGY. Institute for Scientific Information, 3501 Market Street, Philadelphia, PA 19104. (800) 523-1850 or (215) 386-0100. $272.00 per year.

ENGINEERING INDEX MONTHLY. Engineering Information, Incorporated, 345 East 47th Street, NEw York, NY 10017. (800) 221-1044 or (212) 705-7600. Monthly, with annual cumulation. $1560.00 per year.

INTERNATIONAL COPPER INFORMATION BULLETIN. Copper Development Association, Orchard House, Mutton Lane, Potters Bar, Herts. EN6 3AP, England. 1976 to present. Quarterly. $35.00 per year.

METALS ABSTRACTS. American Society for Metals, 9639 Kinsman Road, Metals Park, OH 44073. (216) 338-5151. Monthly. $890.00.

METALS ABSTRACTS INDEX. American Society for Metals, 9639 Kinsman Road, Metals Park, OH 44073. (216) 338-5151. Monthly. (Sold only to subscribers of Metals Abstracts).

ANNUAL REVIEWS AND YEARBOOKS

ANNUAL REVIEW OF MATERIALS SCIENCE. Annual Reviews, Incorporated, 4139 El Camino Way, Palo Alto, CA 94306. (415) 493-4400.

ASSOCIATIONS AND PROFESSIONAL SOCIETIES

AMERICAN SOCIETY FOR METALS. Metals Park, OH 44073. (216) 338-5151.

COPPER DEVELOPMENT ASSOCIATION. Box 1840, Greenwich Office Park 2, Greenwich, CT 06836. (203) 625-8210.

INTERNATIONAL COPPER RESEARCH ASSOCIATION. 708 Third Avenue, New York, NY 10017. (212) 697-9355.

METALLURGICAL SOCIETY OF THE AIME (AMERICAN INSTITUTE OF MINING, METALLURGICAL AND PETROLEUM ENGINEERS). 420 Commonwealth Drive, Warrendale, PA 15086. (412) 776-9080.

BIBLIOGRAPHIES

SCIENTIFIC AND TECHNICAL BOOKS IN PRINT: AN INDEX TO LITERATURE IN SCIENCE AND TECHNOLOGY. R.R. Bowker Company, 205 East 42nd Street, New York, NY 10017. (800) 521-8110 or (212) 916-1600.

DIRECTORIES AND BIOGRAPHICAL SOURCES

COPPER WORLD SURVEY. Metal Bulletin Books, Limited, Park House, Park Terrace, Worcester Park, Surrey KT4 7HY, England. Available in the United States from: Metals Bulletin, Incorporated, 708 Third Avenue, New York, NY 10017. $48.60 (1980 edition).

DUN'S INDUSTRIAL GUIDE - THE METAL WORKING DIRECTORY. Dun and Bradstreet, Incorporated, Three Century Drive, Parsippany, NJ 07054. (201) 455-0900. Annual. $550.00.

INTERNATIONAL RESEARCH CENTERS DIRECTORY 1986-1987. Gale Research Company, Book Tower, Detroit, MI 48226. (800) 521-0707. Third edition. 1986.

RESEARCH CENTERS DIRECTORY. Gale Research Company, Book Tower, Detroit, MI 48226. (800) 521-0707. Eleventh edition. 1987.

SCIENTIFIC AND TECHNICAL ORGANIZATIONS AND AGENCIES DIRECTORY. Gale Research Company, Book Tower, Detroit, MI 48226. (800) 521-0707. 1985.

WHO'S WHO IN ENGINEERING. Engineers Joint Council, 345 East 47th Street, New York, NY 10017. (212) 705-7010. 1985. $200.00.

WHO'S WHO IN FRONTIER SCIENCE AND TECHNOLOGY. Marquis Who's Who, Incorporated, 200 East Ohio Street, Chicago, IL 60611. (800) 428-3898 or (312) 787-2008.

WHO'S WHO IN TECHNOLOGY TODAY. Reston Publishing Company, Incorporated, c/o Prentice-Hall, Incorporated, Englewood Cliffs, NJ 07632. (800) 262-6868. Biennial. Five volumes. $425.00. Covers the fields of electronics, computer science, physics, optics, chemistry, biotechnology, mechanics, energy, and earth science.

ENCYCLOPEDIAS AND DICTIONARIES

ENCYCLOPEDIA OF PHYSICAL SCIENCE AND TECHNOLOGY. Academic Press, Incorporated, Orlando, FL 32887. (800) 321-5068 or (305) 345-2734. Fifteen volumes, 1986.

MCGRAW-HILL DICTIONARY OF ENGINEERING. Sybil P. Parker, editor. McGraw-Hill Book Company, 1221 Avenue of the Americas, New York, NY 10020. (212) 512-2000. 1984. $39.95.

MCGRAW-HILL ENCYCLOPEDIA OF SCIENCE AND TECHNOLOGY. McGraw-Hill Book Company, 1221 Avenue of the Americas, New York, NY 10020. (212) 512-2000.

THESAURUS OF TERMS ON COPPER TECHNOLOGY. Copper Development Association, Orchard House, Mutton Lane, Potters Bar, Herts, EN6 3AP, England. $40.00.

GENERAL WORKS

COPPER AND ITS ALLOYS. E.G. West. John Wiley and Sons, Incorporated, 605 Third Avenue, New York, NY 10158. (800) 526-5368 or (212) 850-6000. 1982. $74.95.

COPPER IN IRON AND STEEL. I. Lemay and L.M. Schetky. John Wiley and Sons, Incorporated, 605 Third Avenue, New York, NY 10158. (800) 526-5368 or (212) 850-6000. 1983. $59.50.

ESSENTIAL METALLURGY FOR ENGINEERS. W. Alexander, G. Davies, and K. Reynolds. Van Nostrand Reinhold, 115 Fifth Avenue, New York, NY 10003. (800) 543-2681. 1985. $17.95.

EXTRACTIVE METALLURGY OF COPPER. Pergamon Press, Incorporated, Maxwell House, Fairview Park, Elmsford, NY 10523. (914) 592-7700. Second edition. 1980. $56.00.

GEOLOGY AND METALLOGENY OF COPPER DEPOSTIS. G. Friedrich and others, editors. Springer-Verlag New York, 175 Fifth Avenue, New York, NY 10010. (800) 526-7254. 1986. $90.00.

LEACHING AND RECOVERING COPPER FROM AS-MINED MATERIALS. W.J. Schlitt, editor. Society of Mining Engineers (American Institute of Mining, Metallurgical and Petroleum Engineers), 420 Commonwealth Drive, Warrendale, PA 15086. (412) 776-9080. 1980. $20.00.

METALLURGY BASICS. D.V. Brown. Van Nostrand Reinhold, 115 Fifth Avenue, New York, NY 10003. (800) 543-2681. 1985. $17.95.

PHYSICAL METALLURGY. P. Haasen. Cambridge University Press, 32 East 57th Street, New York, NY 10022. (212) 688-8885. 1986. $24.95 in paper.

STRUCTURE AND PROPERTIES OF ENGINEERING ALLOYS. W.F. Smith. McGraw-Hill Book Company, Incorporated, 1221 Avenue of the Americas, New York, NY 10020. (212) 512-2000. 1980. $46.95.

HANDBOOKS AND MANUALS

CRC HANDBOOK OF CHEMISTRY AND PHYSICS. CRC Press, Incorporated, 2000 Corporate Boulevard, Boca Raton, FL 33341. (305) 994-0555. Sixth-seventh edition. 1986. $69.95.

SMITHELL'S METALS REFERENCE BOOK. Eric A. Brandes, editor. Butterworth Publishers, 80 Montvale Avenue, Stoneham, MA 02180. (800) 325-4177. Sixth edition. 1983. $210.00.

ONLINE DATA BASES

COMPENDEX. Engineering Information, Incorporated, 345 East 47th Street, New York, NY 10017. (800) 221-1044 or (212) 705-7615. Engineering and technical literature, 1975 to present. Inquire as to cost and availability.

INSPEC. INSPEC Marketing Department, Institute of Electrical and Electronics Engineers, Incorporated, IEEE Service Department, 445 Hoes Lane, Piscataway, NJ 08854. (201) 981-0060. Inquire as to on-line cost and availability.

METADEX. Metals Information, American Society for Metals, Metals Park, OH 44073. (216) 338-5151. (Metals Abstracts/Alloys Index). A worldwide literature on the science and practice and metallurgy, 1966 to present. Inquire as to online cost and availability.

NASA. National Aeronautics and Space Administration. Scientific and Technical Information Branch, 300 7th Street, SW, Washington, DC 20546. Citations and abstracts of aerospace literature, 1962 to present. Inquire as to cost and availability.

NON-FERROUS METALS ABSTRACTS. British Non-Ferrous Metals Technology Centre, Grove Laboratories, Denchworth Road, Wantage, Oxfordshire, England OX12 9 BJ. Citations and abstracts on non-ferrous metallurgy and technology. 1961 to present. Inquire as to online cost and availability.

NTIS. National Technical Information Service, 5285 Port Royal Road, Springfield, VA 22161. (703) 487-4630. Broad coverage of government sponsored research reports, 1964 to present. Inquire as to cost and availability.

WILSONLINE. H.W. Wilson Company, 950 University Avenue, Bronx, NY 10452. (800) 367-6770 or (212) 588-8400. Makes available online versions of the printed H.W. Wilson Indexes including Applied Science and Technology Index, Business Periodicals Index, and Readers' Guide to Periodical Literature. Period covered is generally 1983 to present. Inquire as to cost and availability.

OTHER SOURCES

INTERDOC: DIRECTORY OF PUBLISHED PROCEEDINGS, SERIES. SEMT - Science/Engineering/Medicine/Technology. Interdoc Corporation, 173 Halstead Avenue, Box 326, Harrison, NY 10528. (014) 835-3506. Ten times per year. $325.00 per year.

MATERIALS SCIENCE AND METALLURGY. National Technical Information Service (NTIS), 5285 Port Royal Road, Springfield, VA 22161. (703) 487-4630. Translations and abstracts of foreign language technical media. Irregular. $40.00 per year.

PERIODICALS

ALLOY DIGEST. Engineering Publications Incorporated, Box 823, Upper Montclair, NJ 07043. (201) 746-7930. Monthly. $50.00 per year.

BULLETIN OF ALLOY PHASE DIAGRAMS. American Society for Metals, Metals Park, OH 44073. (216) 338-5151. Bimonthly. $90.00 per year.

JOURNAL OF METALS. Metallurgical Society of the AIME (American Institute of Mining, Metallurgical and Petroleum Engineers), 420 Commonwealth Drive, Warrendale, PA 15086. (412) 776-9080. Monthly. $40.00 per year.

LIGHT METAL AGE. Fellom Publishing Company, 693 Mission Street, San Francisco, CA 94105. (415) 781-1431. Bimonthly. $20.00 per year.

METALLURGICAL TRANSACTIONS. Metallurgical Society of the AIME (American Institute of Mining, Metallurgical and Petroleum Engineers), 420 Commonwealth Drive, Warrendale, PA 15086. (412) 776-9080. Monthly. $95.00 per year.

METALS WEEK. McGraw-Hill Book Company, Incorporated, 1221 Avenue of the Americas, New York, NY 10020. (212) 997-2823. Weekly. $527.00 per year.

WIRE JOURNAL INTERNATIONAL. Wire Journal Incorporated, 1570 Boston Post Road, Guilford, CT 06437. (203) 453-2777. Monthly. $40.00.

RESEARCH CENTERS AND INSTITUTES

GENERAL ELECTRIC COMPANY. Research and Development Center, Post Office Box 8, Schenectady, NY 12301. (518) 385-8415.

INTERNATIONAL COPPER RESEARCH ASSOCIATION. 708 Third Avenue, New York, NY 10017. (212) 697-9355.

NEW MEXICO INSTITUTE OF MINING AND TECHNOLOGY. Mining and Mineral Resources Research Institute, Campus Station, Socorro, NM 87801. (505) 835-5210.

PHYSICAL METALLURGY RESEARCH LABORATORIES. Canada Centre for Mineral and Energy Technology, 555 Booth Street, Ottawa, ON Canada K1A 0G1. (613) 995-4807.

UNIVERSITY OF CONNECTICUT. Institute of Materials Science, Storrs, CT 06268. (203) 486-4623.

UNIVERSITY OF FLORIDA. Center for Applied Thermodynamics and Corrosion, Department of Materials Science and Engineering, Gainesville, FL 32601. (904) 392-1454.

CORE DRILLING

See: DRILLING

CORONA

See: SUN

CORROSION

See also: ALLOYS, ALUMINUM, CADMIUM, ELECTROCHEMISTRY, FERROALLOYS, MATERIALS SCIENCE

ABSTRACT SERVICES AND INDEXES

ALLOYS INDEX. American Society for Metals, 9639 Kinsman Road, Metals Park, OH 44073. (216) 338-5151. $130.00 per year.

APPLIED MECHANICS REVIEWS. American Society of Mechanical Engineers, 345 East 47th Street, New York, NY 10017. (212) 705-7722. Monthly. $380.00 per year. Critical reviews of the world literature in applied mechanics and related engineering science.

APPLIED SCIENCE AND TECHNOLOGY INDEX. H.W. Wilson Company, 950 University Avenue, Bronx, NY 10452. (800) 367-6670 or (212) 588-8400. Inquire as to cost and availability.

CHEMICAL ABSTRACTS. Chemical Abstracts Service, 2540 Olentangy Road, Post Office Box 3012, Columbus, OH 43210. (800) 848-6538 or (614) 421-3600. Weekly. $9200.00 per year.

CORROSION ABSTRACTS: ABSTRACTS OF THE WORLD'S LITERATURE ON CORROSION AND CORROSION MITIGATION. Association of Corrosion Engineers, Box 218340, Houston, TX 77218. (713) 492-0535. 1962 to present. Bimonthly. $250.00 per year.

CORROSION CONTROL ABSTRACTS (ENGLISH TRANSLATION OF REFERATIVNYI ZHURNAL. Korroziya i Zashchita ot Korrozii). Scientific Information Consultants, Limited, 661 Finchley Road, London, NW2 2HN, England. 1966 to present. Monthly. $405.00 per year.

CORROSION REVIEWS. Freund Publishing House, Limited, 61 Nachmani Street, Tel Aviv, Israel. Text in English. 1972 to present. Quarterly. $90.00 per year.

CURRENT CONTENTS: ENGINEERING, TECHNOLOGY. Institute for Scientific Information, 3501 Market Street, Philadelphia, PA 19104. (800) 523-1850 or (215) 386-0100. $272.00 per year.

ENGINEERING INDEX MONTHLY. Engineering Information, Incorporated, 345 East 47th Street, NEw York, NY 10017. (800) 221-1044 or (212) 705-7600. Monthly, with annual cumulation. $1560.00 per year.

INTERNATIONAL AEROSPACE ABSTRACTS. American Institute of Aeronautics and Astronautics, Technical Information Service, 555 West 57th Street, New York, NY 10019. (212) 247-6500. 1961 to present. Semi-monthly. $950.00 per year with indexes.

METALS ABSTRACTS. American Society for Metals, 9639 Kinsman Road, Metals Park, OH 44073. (216) 338-5151. Monthly. $890.00.

METALS ABSTRACTS INDEX. American Society for Metals, 9639 Kinsman Road, Metals Park, OH 44073. (216) 338-5151. Monthly. (Sold only to subscribers of Metals Abstracts).

PETROLEUM ABSTRACTS. University of Tulsa, Information Services Division, 600 South College Street, Tulsa, OK 74104. (918) 592-6000. 1961 to present. Weekly. Inquire as to cost and availability.

WORLD ALUMINUM ABSTRACTS. Aluminum Association, 818 Connecticut Avenue, NW, Washington, DC 20006. (202) 862-5100. 1968 to present. Monthly. $165.00 per year.

ASSOCIATIONS AND PROFESSIONAL SOCIETIES

ALUMINUM ASSOCIATION. 818 Connecticut Avenue, NW, Washington, DC 20006. (202) 862-5100.

AMERICAN SOCIETY FOR METALS. Metals Park, OH 44073. (216) 338-5151.

METALLURGICAL SOCIETY OF THE AIME (AMERICAN INSTITUTE OF MINING, METALLURGICAL AND PETROLEUM ENGINEERS). 420 Commonwealth Drive, Warrendale, PA 15086. (412) 776-9080.

NATIONAL ASSOCIATION OF CORROSION ENGINEERS. Box 218340, Houston, TX 77218. (713) 492-0535.

DIRECTORIES AND BIOGRAPHICAL SOURCES

DUN'S INDUSTRIAL GUIDE - THE METAL WORKING DIRECTORY. Dun and Bradstreet, Incorporated, Three Century Drive, Parsippany, NJ 07054. (201) 455-0900. Annual. $550.00.

INTERNATIONAL RESEARCH CENTERS DIRECTORY 1986-1987. Gale Research Company, Book Tower, Detroit, MI 48226. (800) 521-0707. Third edition. 1986.

MATERIALS PERFORMANCE - NACE CORROSION ENGINEERING BUYER'S GUIDE ISSUE. National Association of Corrosion Engineers, Box 218340, Houston, TX 77218. (713) 492-0535. Annual, July. $15.00.

METAL PRODUCTS DIRECTORY. American Business Directories, Incorporated, Division of American Business Lists, Incorporated, 5707 South 86th Circle, Omaha, NE 68127. (402) 331-7169. 1986. $80.00.

RESEARCH CENTERS DIRECTORY. Gale Research Company, Book Tower, Detroit, MI 48226. (800) 521-0707. Eleventh edition. 1987.

SCIENTIFIC AND TECHNICAL ORGANIZATIONS AND AGENCIES DIRECTORY. Gale Research Company, Book Tower, Detroit, MI 48226. (800) 521-0707. 1985.

WHO'S WHO IN ENGINEERING. Engineers Joint Council, 345 East 47th Street, New York, NY 10017. (212) 705-7010. 1985. $200.00.

WHO'S WHO IN FRONTIER SCIENCE AND TECHNOLOGY. Marquis Who's Who, Incorporated, 200 East Ohio Street,

Chicago, IL 60611. (800) 428-3898 or (312) 787-2008.

WHO'S WHO IN TECHNOLOGY TODAY. Reston Publishing Company, Incorporated, c/o Prentice-Hall, Incorporated, Englewood Cliffs, NJ 07632. (800) 262-6868. Biennial. Five volumes. $425.00. Covers the fields of electronics, computer science, physics, optics, chemistry, biotechnology, mechanics, energy, and earth science.

ENCYCLOPEDIAS AND DICTIONARIES

ENCYCLOPEDIA OF PHYSICAL SCIENCE AND TECHNOLOGY. Academic Press, Incorporated, Orlando, FL 32887. (800) 321-5068 or (305) 345-2734. Fifteen volumes, 1986.

MCGRAW-HILL DICTIONARY OF ENGINEERING. Sybil P. Parker, editor. McGraw-Hill Book Company, 1221 Avenue of the Americas, New York, NY 10020. (212) 512-2000. 1984. $39.95.

MCGRAW-HILL ENCYCLOPEDIA OF SCIENCE AND TECHNOLOGY. McGraw-Hill Book Company, 1221 Avenue of the Americas, New York, NY 10020. (212) 512-2000.

GENERAL WORKS

ALUMINUM: PROPERTIES AND PHYSICAL METALLURGY. John E. Hatch, editor. American Society for Metals, Metals Park, OH 44073. (216) 338-5151. Third edition. 1984. $75.00.

ATMOSPHERIC CORROSION OF METALS. Dean and Rhea, editors. American Society for Testing and Materials, 1916 Race Street, Philadelphia, PA 19103. (215) 299-5400. STP Volume 767. 1982. $42.50.

BASIC CORROSION AND OXIDATION. John M. West. John Wiley and Sons, Incorporated, 605 Third Avenue, New York, NY 10158. (800) 526-5368 or (212) 850-6000. 1986. $44.95.

CORROSION AND CORROSION CONTROL. H.H. Uhlig and R. Winston Revie. John Wiley and Sons, Incorporated, 605 Third Avenue, New York, NY 10158. (800) 526-5368 or (212) 850-6000. Third edition. 1985. $42.50.

CORROSION BASICS: AN INTRODUCTION. National Association of Corrosion Engineers, Box 218340, Houston, TX 77218. (713) 492-0535. 1984. $40.00.

CORROSION ENGINEERING. M.G. Fontana. McGraw-Hill Book Company, 1221 Avenue of the Americas, New York, NY 10020. (212) 512-2000. Third edition. $46.95.

CORROSION FATIQUE: MECHANICS, METALLURGY, ELECTROCHEMISTRY, AND ENGINEERING. T.W. Crooker and B.N. Leis, editors. American Society for Testing and Materials, 1916 Race Street, Philadelphia, PA 19103. (215) 299-5400. STP Volume 801. 1983. $62.00.

CORROSION INHIBITORS. National Association of Corrosion Engineers, Box 218340. Houston, TX 77218. (713) 492-0535. 1983. $35.00.

ESSENTIAL METALLURGY FOR ENGINEERS. W. Alexander, G. Davies, and K. Reynolds. Van Nostrand Reinhold, 115 Fifth Avenue, New York, NY 10003. (800) 543-2681. 1985. $17.95.

PHYSICAL METALLURGY. P. Haasen. Cambridge University Press, 32 East 57th Street, New York, NY 10022. (212) 688-8885. 1986. $24.95 in paper.

METALLURGY BASICS. D.V. Brown. Van Nostrand Reinhold, 115 Fifth Avenue, New York, NY 10003. (800) 543-2681. 1985. $17.95.

HANDBOOKS AND MANUALS

CHEMICAL ENGINEERING GUIDE TO CORROSION IN THE PROCESS INDUSTRIES. Chemical Engineering Magazine. McGraw-Hill Book Company, 1221 Avenue of the Americas, New York, NY 10020. (212) 512-2000. 1985. $42.50.

CORROSION: SOURCE BOOK. Seymour K. Coburn, editor. American Society for Public Administration, 1120 G Street, NW, Suite 500, Washington, DC 20005. (212) 393-7878. 1984. $49.00.

CORROSION RESISTENCE TABLES. P. Schweitzer. Marcel Dekker, Incorporated, 270 Madison Avenue, New York, NY 10016. (212) 696-9000. Second edition. 1986. $145.00.

CORROSION RESISTANT MATERIALS HANDBOOK. D.J. De Renzo, editor. Noyes Data Corporation, Mill Road at Grand Avenue, Park Ridge, NJ 07656. (201) 391-8484. 1986. $125.00.

FORMS OF CORROSION RECOGNITION AND PREVENTION. (NACE Handbook 1). National Association of Corrosion Engineers, Box 218340, Houston, TX 77218. (713) 492-0535. 1982. $30.00.

SMITHELL'S METALS REFERENCE BOOK. Eric A. Brandes, editor. Butterworth Publishers, 80 Montvale Avenue, Stoneham, MA 02180. (800) 325-4177. Sixth edition. 1983. $210.00.

ONLINE DATA BASES

COMPENDEX. Engineering Information, Incorporated, 345 East 47th Street, New York, NY 10017. (800) 221-1044 or (212) 705-7615. Engineering and technical literature, 1975 to present. Inquire as to cost and availability.

CORROSION DATA BASE. Marcel Dekker, Incorporated, 270 Madison Avenue, New York, NY 10016. (212) 696-9000. Available online through System Development Corporation (SDC). Inquire as to online cost and availability.

INSPEC. INSPEC Marketing Department, Institute of Electrical and Electronics Engineers, Incorporated, IEEE Service Department, 445 Hoes Lane, Piscataway, NJ 08854. (201) 981-0060. Inquire as to on-line cost and availability.

METADEX. Metals Information, American Society for Metals, Metals Park, OH 44073. (216) 338-5151. (Metals Abstracts/Alloys Index). A worldwide literature on the science and practice and metallurgy, 1966 to present. Inquire as to online cost and availability.

NASA. National Aeronautics and Space Administration. Scientific and Technical Information Branch, 300 7th Street, SW, Washington, DC 20546. Citations and abstracts of aerospace literature, 1962 to present. Inquire as to cost and availability.

NON-FERROUS METALS ABSTRACTS. British Non-Ferrous Metals Technology Centre, Grove Laboratories, Denchworth Road, Wantage, Oxfordshire, England OX12 9 BJ. Citations and abstracts on non-ferrous metallurgy and technology. 1961 to present. Inquire as to online cost and availability.

NTIS. National Technical Information Service, 5285 Port Royal Road, Springfield, VA 22161. (703) 487-4630. Broad coverage of government sponsored research reports, 1964 to present. Inquire as to cost and availability.

WILSONLINE. H.W. Wilson Company, 950 University Avenue, Bronx, NY 10452. (800) 367-6770 or (212) 588-8400. Makes available online versions of the printed H.W. Wilson Indexes including Applied Science and Technology Index, Business Periodicals Index, and Readers' Guide to Periodical Literature. Period covered is generally 1983 to present. Inquire as to cost and availability.

CORROSION

OTHER SOURCES

INTERDOC: DIRECTORY OF PUBLISHED PROCEEDINGS, SERIES. SEMT - Science/Engineering/Medicine/Technology. Interdoc Corporation, 173 Halstead Avenue, Box 326, Harrison, NY 10528. (014) 835-3506. Ten times per year. $325.00 per year.

MATERIALS SCIENCE AND METALLURGY. National Technical Information Service (NTIS), 5285 Port Royal Road, Springfield, VA 22161. (703) 487-4630. Translations and abstracts of foreign language technical media. Irregular. $40.00 per year.

PERIODICALS

ALLOY DIGEST. Engineering Publications Incorporated, Box 823, Upper Montclair, NJ 07043. (201) 746-7930. Monthly. $50.00 per year.

CORROSION. National Association of Corrosion Engineers, Box 218340, Houston, TX 77218. (713) 492-0535. 1945 to present. Monthly. $75.00 per year.

CORROSION PREVENTION AND CONTROL. Scientific Surveys, Limited, Box 21, Beaconsfield, Bucks, HP9 1NS, England. 1954 to present. Monthly. $75.00 per year.

CORROSION PREVENTION/INHIBITION DIGEST. American Society for Metals, 9639 Kinsman Road, Metals Park, OH 44073. (216) 338-5151. 1976 to present. Monthly. $90.00 per year.

CORROSION SCIENCE. Pergamon Press, Incorporated, Journals Division, Maxwell House, Fairview Park, Elmsford, NY 10523. (914) 592-7700. 1961 to present. Monthly. $300.00 per year.

JOURNAL OF METALS. Metallurgical Society of the AIME (American Institute of Mining, Metallurgical and Petroleum Engineers), 420 Commonwealth Drive, Warrendale, PA 15086. (412) 776-9080. Monthly. $40.00 per year.

LIGHT METAL AGE. Fellom Publishing Company, 693 Mission Street, San Francisco, CA 94105. (415) 781-1431. Bimonthly. $20.00 per year.

METALLURGICAL TRANSACTIONS. Metallurgical Society of the AIME (American Institute of Mining, Metallurgical and Petroleum Engineers), 420 Commonwealth Drive, Warrendale, PA 15086. (412) 776-9080. Monthly. $95.00 per year.

METALS WEEK. McGraw-Hill Book Company, Incorporated, 1221 Avenue of the Americas, New York, NY 10020. (212) 997-2823. Weekly. $527.00 per year.

RESEARCH CENTERS AND INSTITUTES

COLORADO SCHOOL OF MINES. Steel Research Center, Golden, CO 80401. (303) 273-3774.

FAIRLEIGH DICKINSON UNIVERSITY. Corrosion and Surface Study Laboratory, 30 Van Orden Place, Hackensack, NJ 07601. (201) 692-2285.

OHIO STATE UNIVERSITY. Fontana Corrosion Center, 116 West 19th Avenue, Columbus, OH 43210. (614) 422-6255.

TEXAS A AND M UNIVERSITY. Corrosion Engineering Laboratory, Mechanical Engineering Department, College Station, TX 77843. (409) 845-2944.

UNIVERSITY OF FLORIDA. Center for Applied Thermodynamics and Corrosion, Department of Materials Science and Engineering, Gainesville, FL 32601. (904) 392-1454.

UNIVERSITY OF MINNESOTA. Corrosion Research Center, 112 Mines and Metallurgy Building, Minneapolis, MN 55455. (612) 373-4864.

COSMIC RAYS

See also: ASTRONOMY, ASTROPHYSICS, PARTICLE PHYSICS, PHYSICS, PLANETARY SCIENCE

ABSTRACT SERVICES AND INDEXES

ASTRONOMY AND ASTROPHYSICS ABSTRACTS. Springer-Verlag New York, Inc., 175 Fifth Avenue, New York, NY 10010. (800) 526-7254. 1969 to present. Approximately $70.00 per year.

ASTRONOMY AND ASTROPHYSICS MONTHLY INDEX. Olivetree Associates, P.O. Box 236, Sierre Madre, CA 91024. $220.00 per year. Complementary copies available on request.

CHEMICAL ABSTRACTS. American Chemical Society, Chemical Abstracts Service, Box 3012, Columbus, OH 43210. (614) 421-3600. 1907 to present. Weekly. $10,000.00 per year.

CONFERENCE PAPERS INDEX. Cambridge Scientific Abstracts, 5161 River Road, Bethesda, MD 20816. 1972 to present. Monthly. Inquire as to cost and availability.

INDEX TO SCIENTIFIC AND TECHNICAL PROCEEDINGS. Institute for Scientific Information, 3501 Market Street, Philadelphia, PA 19104. (800) 523-1850 or (215) 386-0100. 1978 to present. Monthly. $775.00 per year.

INDEX TO SCIENTIFIC REVIEWS. Institute for Scientific Information, 3501 Market Street, Philadelphia, PA 19104. (800) 523-1850 or (215) 386-0100. 1974 to present. Semi-annual. $550.00 per year.

METEOROLOGICAL AND GEOASTROPHYSICAL ABSTRACTS. American Meteorological Society, 45 Beacon Street, Boston, MA 02108. (617) 227-2425. 1950 to present. Monthly. $450.00 per year.

PHYSICS ABSTRACTS. Institution of Electrical Engineers. Available from: IEEE Service Center, 445 Hoes Lane, Piscataway, NJ 08854. 1898 to present. Bimonthly. $1700.00 per year.

PHYSICS BRIEFS. Physik Verlag GmbH, Postfach 1260/1280, D-6940 Weinheim, West Germany. (212) 661-9404. 1920 to present. Twenty-six times per year. $1250.00 per year.

SCIENCE CITATION INDEX. Institute for Scientific Information, 3501 Market Street, Philadelphia, PA 19104. (800) 523-1850 or (215) 386-0100. Six times per year. $6200.00 per year.

STAR. (SCIENTIFIC AND TECHNICAL AEROSPACE REPORTS). U.S. National Aeronautics and Space Administration, Scientific and Technical Information Facility, Box 8757, Baltimore-Washington International Airport, MD 21240. (202) 755-2210. Semimonthly, with semiannual and annual indexes. $85.00 per year.

ANNUAL REVIEWS AND YEARBOOKS

ANNUAL REVIEW OF ASTRONOMY AND ASTROPHYSICS. Annual Reviews, Inc., 4139 El Camino Way, Palo Alto, CA 94306. (415) 493-4400. Annual. Inquire as to cost and availability.

ASSOCIATIONS AND PROFESSIONAL SOCIETIES

AMERICAN ASTRONOMICAL SOCIETY. 1816 Jefferson Place, N.W., Washington, DC 20036. (202) 659-0134.

AMERICAN GEOPHYSICAL UNION. 2000 Florida Avenue, N.W., Washington, DC 20009. (202) 462-6903.

AMERICAN INSTITUTE OF PHYSICS. 335 East 45th Street, New York, NY 10017. (212) 661-9494.

ASTRONOMICAL SOCIETY OF THE PACIFIC. 1290 24th Avenue, San Francisco, CA 94122. (415) 661-8660.

DIRECTORIES AND BIOGRAPHICAL SOURCES

AMERICAN ASTRONOMICAL SOCIETY MEMBERSHIP DIRECTORY. 1816 Jefferson Place, N.W., Washington, DC 20036. (202) 659-0134. Annual. Inquire.

AMERICAN MEN AND WOMEN OF SCIENCE. R.R. Bowker, Inc., Order Department, 245 West 17th Street, New York, NY 10011. (800) 521-8110. Eight volumes. 1986. $595.00 for set.

THE BIOGRAPHICAL DICTIONARY OF SCIENTISTS: ASTRONOMERS. D. Abbott, editor. Peter Bedrick Books, 125 East 23rd Street, New York, NY 10010. 1984.

DIRECTORY OF PHYSICS AND ASTRONOMY STAFF MEMBERS. American Institute of Physics, 335 East 45th Street, New York, NY 10017. (212) 661-9494. Annual.

INTERNATIONAL RESEARCH CENTERS DIRECTORY 1988-89. Darren L. Smith, editor. Gale Research Company, Book Tower, Detroit, MI 48226. (800) 521-0707. 4th edition. 1987. $360.00.

RESEARCH CENTERS DIRECTORY 1988. Gale Research Company, Book Tower, Detroit, MI 48226. (800) 521-0707. 12th edition. 1987. $365.00 for set.

SCIENTIFIC AND TECHNICAL ORGANIZATIONS AND AGENCIES DIRECTORY. Margaret Labash Young, editor. Gale Research Company, Book Tower, Detroit, MI 48226. (800) 521-0707. 2nd edition. 1987. $185.00.

GENERAL WORKS

COMPOSITION OF COSMIC RADIATION. K.M. Apparao. Gordon and Breach Science Publishers, Inc., 50 West 23rd Street, New York, NY 10010. (212) 206-8900. 1975. $37.25.

COSMIC RAYS. L.I. Dorman, editor. Elsevier Science Publishing Company, Inc., 52 Vanderbilt Avenue, New York, NY 10017. (212) 370-5520. 1974. $92.50.

THE GALAXY AND THE SOLAR SYSTEM. Roman Smoluchowski and others. University of Arizona Press, 1615 East Speedway, Tucson, AZ 85719. (602) 621-1441. 1986. $29.95.

ONLINE DATA BASES

CA SEARCH. Chemical Abstracts Service, P.O. Box 3012, Columbus, OH 43120. (800) 848-6538 or (614) 421-3600. Comprehensive guide to chemical literature, 1972 to present. Inquire as to online cost and availability.

DISSERTATION ABSTRACTS ONLINE. University Microfilms International, 300 North Zeeb Road, Ann Arbor, MI 48106. (800) 521-0600 or (313) 761-4700. Scope includes virtually all doctoral dissertations accepted at accredited American institutions from 1861 to present in over 250 subject areas. Inquire as to online cost and availability.

INSPEC. INSPEC Marketing Department, Institution of Electrical Engineers. Available from IEEE Service Center, 445 Hoes Lane, Piscataway, NJ 08854. (201) 981-0060. Online version of Physics Abstracts. Inquire as to online cost and availability.

NTIS. National Technical Information Service, 5285 Port Royal Road, Springfield, VA 22161. (703) 487-4630. Broad coverage of government sponsored research reports, 1964 to present. Inquire as to online cost and availability.

SCISEARCH. Institute for Scientific Information, 3501 Market Street, Philadelphia, PA 19104. (800) 523-1850 or (215) 386-0100. Broad multidisciplinary title and author index to the international literature of science and technology, 1974 to present. Inquire as to online cost and availability.

PERIODICALS

ASTRONOMICAL JOURNAL. American Astronomical Society. Available from: American Institute of Physics, 335 East 45th Street, New York, NY 10017. (212) 661-9494. Monthly. $125.00 per year.

ASTRONOMICAL SOCIETY OF THE PACIFIC PUBLICATIONS. Astronomical Society of the Pacific, 1290 24th Avenue, San Francisco, CA 94122. (415) 661-8660. Monthly. $40.00 per year.

ASTRONOMY. AstroMedia Corporation, 625 E Street, Box 92788, Milwaukee, WI 53202. (414) 276-2689. 1973 to present. Monthly. $21.00 per year.

ASTRONOMY AND ASTROPHYSICS. Springer-Verlag New York, Inc., 175 Fifth Avenue, New York, NY 10010. (800) 526-7254. Monthly. $680.00 per year.

ASTROPHYSICAL JOURNAL. University of Chicago Press, 5801 Ellis Avenue, Chicago, IL 60637. (800) 621-2736. 1895 to present. Semi-monthly. $310.00 per year.

ASTROPHYSICS AND SPACE SCIENCE. D. Reidel Publishing Company, 190 Old Derby Street, Hingham, MA 02043. 1968 to present. 22 times per year. $1286.00 per year.

EARTH, MOON AND PLANETS. D. Reidel Publishing Company, 190 Old Derby Street, Hingham, MA 02043. 1969 to present. Nine times per year. $275.00 per year.

FUNDAMENTALS OF COSMIC PHYSICS. Gordon and Breach Science Publishers, Inc., 50 West 23rd Street, New York, NY 10010. (212) 206-8900. Six volumes per year, four numbers per volume. Inquire.

GEOCHEMICA ET COSMOCHIMICA ACTA. Pergamon Press, Inc., Maxwell House, Fairview Park, Elmsford, NY 10523. (914) 592-7700. 1950 to present. Monthly. $340.00 per year.

PLANETARY AND SPACE SCIENCE. Pergamon Press, Inc., Maxwell House, Fairview Park, Elmsford, NY 10523. (914) 592-7700. 1959 to present. Monthly. $430.00 per year.

RESEARCH CENTERS AND INSTITUTES

ASTRONOMY PROGRAM. University of Maryland, College Park, MD 20742. (301) 454-3005.

COSMIC RAY LABORATORY. University of Arizona, Department of Physics, Tucson, AZ 85721. (602) 621-6820.

COSMIC RAY RESEARCH PROJECT. University of Utah, 310 James Fletcher Building, Salt Lake City, UT 84112. (801) 581-6930.

SPACE SCIENCES LABORATORY. University of California, Berkeley, Berkeley, CA 94720. (415) 642-1364.

COSMOCHEMISTRY

See also: ASTRONOMY, ASTROPHYSICS, BIG BANG, COSMOLOGY

ABSTRACT SERVICES AND INDEXES

ASTRONOMY AND ASTROPHYSICS ABSTRACTS. Springer-Verlag New York, Incorporated, 175 Fifth Avenue, New York, NY 10010. (212) 460-1500. $70.00 per year.

ASTRONOMY AND ASTROPHYSICS MONTHLY INDEX. Olivetree Associates, P.O. Box 236, Sierre Madre, Ca 91024. $212.00 per year. Complimentary copies available on request.

GENERAL SCIENCE INDEX. H.W. Wilson Company, 950 University Avenue, Bronx, NY 10452. (800) 367-6770 or (212)

588-8400. Inquire as to cost and availability.

PHYSICS ABSTRACTS. Institute of Electrical Engineers, London, United Kingdom. Available from: Institute of Electrical and Electronic Engineers (IEEE), 345 East 47th Street, New York, NY 10017. (212) 705-7900.

SCIENCE CITATION INDEX. Institute for Scientific Information, 3501 Market Street, Philadelphia, PA 19104. (800) 523-1850 or (215) 386-0100.

STAR. (Scientific and Technical Aerospace Reports). U.S. National Aeronautics and Space Administration, Scientific and Technical Information Facility, Box 8757, Baltimore-Washington International Airport, Md 21240. (202) 755-2210. Semimonthly, with semiannual and annual indexes. $85.00 per year.

ANNUAL REVIEWS AND YEARBOOKS

ANNUAL REVIEW OF ASTRONOMY AND ASTROPHYSICS. Annual Reviews, Incorporated, 4139 El Camino Way, Palo Alto, CA 94306. (415) 493-4400.

ASSOCIATIONS AND PROFESSIONAL SOCIETIES

AMERICAN ASTRONOMICAL SOCIETY. 2000 Florida Avenue, NW, Suite 300, Washington, DC 20009. (202) 659-0134.

AMERICAN ASSOCIATION OF VARIABLE STAR OBSERVERS. 187 Concord Avenue, Cambridge, MA 02138. (617) 354-0484.

ASTRONOMICAL LEAGUE. P.O. Box 12821, Tucson, AZ 85732. (602) 790-8471.

ASTRONOMICAL SOCIETY OF THE PACIFIC. 1290 24th Avenue, San Francisco, Ca 94122. (415) 661-8660.

BIBLIOGRAPHIES

A BIBLIOGRAPHY OF ASTRONOMY, 1970-1979. R.A. Seal and S.S. Martin. Libraries Unlimited, Incorporated, Littleton, CO 80160. 1982. $37.50.

SCIENCE BOOKS AND FILMS. American Association for the Advancement of Science, 1333 H Street, NW, Washington, DC 20005. Five times per year. $20.00 per year.

SCIENTIFIC AND TECHNICAL BOOKS AND SERIALS IN PRINT 1988; AN INDEX TO LITERATURE IN SCIENCE AND TECHNOLOGY. R.R. Bowker Company, 205 East 42nd Street, New York, NY 10017. (800) 521-8110 or (212) 916-1600. $175.00.

DIRECTORIES AND BIOGRAPHICAL SOURCES

AMERICAN MEN AND WOMEN OF SCIENCE. Physical and Biological Sciences. Fifteenth edition. R.R. Bowker Company, 205 East 42nd Street, New York, NY 10017. (800) 521-8110 or (212) 916-1600.

THE BIOGRAPHICAL DICTIONARY OF SCIENTISTS: ASTRONOMERS. D. Abbott, editor. Peter Bedrick Books, 125 East 23rd Street, New York, NY 10010. 1984.

DIRECTORY OF PHYSICS AND ASTRONOMY STAFF MEMBERS. American Institite of Physics, 335 East 45th Street, New York, NY 10017. Annual.

RESEARCH CENTERS DIRECTORY. Gale Research Company, Detroit, MI 48226. Eleventh edition, 1987. (800) 521-0707. $340.00.

WHO'S WHO IN FRONTIER SCIENCE AND TECHNOLOGY. Marquis Who's Who, Incorporated, 200 East Ohio Street, Chicago, IL 60611. (800) 428-3898 or (312) 787-2008.

ENCYCLOPEDIAS AND DICTIONARIES

ENCYCLOPEDIA OF PHYSICAL SCIENCE AND TECHNOLOGY. Academic Press, Incorporated, Orlando, FL 32887. (800) 321-5068 or (305) 345-2734.

ILLUSTRATED ENCYCLOPEDIA OF THE UNIVERSE: UNDERSTANDING AND EXPLORING THE COSMOS. Richard S. Lewis. Crown Publishers, Incorporated, 1 Park Avenue, New York, NY 10016. (800) 526-4264. 1986. $24.95.

MCGRAW-HILL ENCYCLOPEDIA OF SCIENCE AND TECHNOLOGY. McGraw-Hill Book, Incorporated, 1221 Avenue of the Americas, New York, NY 10020. (212) 997-3675. Fifth edition, 15 volumes. $1100.00.

GENERAL WORKS

CONSTRUCTING THE UNIVERSE. David Layzer. W.H. Freeman and Company, 41 Madison Avenue, New York, NY 10010. (212) 532-7660. 1984. $29.95.

COSMO AND GEOCHEMISTRY. K.N. Raymond, editor. Springer-Verlag New York, Incorporated, 175 Fifth Avenue, New York, NY 10010. (800) 526-7254. 1981. $35.00.

COSMOCHEMISTRY. F. Boschke, editor. Springer-Verlag New York, Incorporated, 175 Fifth Avenue, New York, NY 10010. (800) 526-7254. 1974. $31.00.

COSMOCHEMISTRY AND THE ORIGIN OF LIFE. Cyril Ponnamperuma, editor. Kluwer Academic Publishers, 190 Old Derby Street, Hingham, MA 02043. (617) 749-5262. 1983. $63.00.

COSMOLOGICAL CONSTANTS: PAPERS IN MODERN COSMOLOGY. Jeremy Bernstein and Gerald Feinberg. Columbia University Press, 562 West 113th Street, New York, NY 10025. (212) 316-7100. 1986. $38.00.

FIRST THREE MINUTES: A MODERN VIEW OF THE ORIGIN OF THE UNIVERSE. Steven Weinberg. Basic Books, Incorporated, 10 East 53rd Street, New York, NY 10022. (800) 242-7737. 1976. $14.95.

THE FORMATION AND EVOLUTION OF GALAXIES AND LARGE STRUCTURES IN THE UNIVERSE. Jean Audouze and Thanh Van Tran. Kluwer Academic Publishers, 190 Old Derby Street, Hingham, MA 02043. (617) 749-5262. 1984. $58.00.

LARGE SCALE STRUCTURE OF THE UNIVERSE. P.J. Poebles. Princeton University Press, 41 William Street, Princeton, Nj 08540. (609) 452-4122. 1980. $15.95 in paper.

THE LEFT HAND OF CREATION: THE ORIGIN AND EVOLUTION OF THE EXPANDING UNIVERSE. John D. Barrow and Joseph Silk. Basic Books, Incorporated, 10 East 53rd Street, New York, NY 10022. (800) 242-7737. 1986. $7.95 in paper.

MODERN COSMOLOGY. Dennis Sciama. Cambridge University, 32 East 57th Street, New York, NY 10022. (800) 872-7423. 1982. $15.95 in paper.

THE VERY EARLY UNIVERSE. G.W. Gibbons and others, editors. Cambridge University Press, 32 East 57th Street, New York, NY 10022. (800) 872-7423. 1985. $24.95.

ONLINE DATA BASES

CA SEARCH. Chemical Abstracts Service, P.O. Box 3012, Columbus, OH 43210. Guide to chemical literature, 1972 to present. Inquire as to cost and availability.

DISSERTATION ABSTRACTS ONLINE. University Microfilms International, 300 North Zeeb Road, Ann Arbor, MI 48106. (800) 521-0600 or (313) 761-4700. Scope includes virtually all doctoral dissertations accepted at accredited American institutions from 1861 to present in 252 subject areas. Inquire as to cost and availability.

INSPEC. INSPEC Marketing Department, Institute of Electrical and Electronics Engineers, Incorporated, IEEE Service Department, 445 Hoes Lane, Piscataway, NJ 08854. (201) 981-0060. Inquire as to on-line cost and availability.

NASA. National Aeronautics and Space Administration, Scientific and Technical Information Branch, 300 7th Street, SW, Washington, Dc 20546. Citations and abstracts of aerospace literature, 1962 to present. Inquire as to cost and availability.

NTIS. National Technical Information Service, 5285 Port Royal Road, Springfield, Va 22161. (703) 487-4630. Broad coverage of Government sponsored research reports, 1964 to present. Inquire as to cost and availability.

SCISEARCH. Institute for Scientific Information, 3501 Market Street, Philadelphia, PA 19104. (800) 523-1850 or (215) 386-0100. Broad multidisciplinary title and author index to the international literature of science and technology, 1974 to present. Inquire as to cost and availability.

OTHER SOURCES

ATLAS OF DEEP-SKY SPLENDORS. H. Vehrenberg. Cambridge University Press, 32 East 57th Street, New York, NY 10022. (800) 431-1580 or (212) 688-8888. Fourth edition, 1984. $44.50.

SKY CATALOGUE 2000.0. A. Hirshfeld and R. Sinnott. Cambridge University Press, 32 East 57th Street, New York, NY 10022. (800) 431-1580 or (212) 688-8888.

PERIODICALS

ASTRONOMICAL JOURNAL. American Astronomical Society. Available from: American Institute of Physics, 335 East 45th Street, New York, NY 10017. (212) 661-9404. $125.00 per year.

ASTRONOMICAL SOCIETY OF THE PACIFIC. Publications. Astronomical Society of the Pacific, 1290 24th Avenue, San Francisco, Ca 94122. (415) 661-8660. Monthly. $38.00.

ASTRONOMY. Astro Media Corporation, 625 East Paul Avenue, Milwaukee, WI 53202. Monthly. $18.00 per year.

ASTRONOMY AND ASTROPHYSICS. Springer-Verlag New York, Incorporated, 175 Fifth Avenue, New York, NY 10010. (800) 526-7254 or (212) 460-1500. $680.00 per year.

ASTROPHYSICAL JOURNAL. American Astronomical Society, University of Chicago Press, 5801 Ellis Avenue, Chicago, IL 60637. Biweekly. $305.00 per year.

ASTROPHYSICS AND SPACE SCIENCE. D. Reidel Publishing Company, 190 Old Derby Street, Hingham, MA 02043. Monthly. $101.00 per year.

FUNDAMENTALS OF COSMIC PHYSICS. Gordon and Breach Science Publishers, Limited, P.O. Box 197, London, WC2E 9PX. Monthly. $335.00 per year.

GEOCHIMICA ET COSMOCHIMICA ACTA. Pergamon Press, Incorporated, Journals Division, Maxwell House, Fairview Park, Elmsford, NY 10523. (914) 592-7700. Monthly. $340.00 per year.

MERCURY. Astronomical Society of the Pacific, 1290 24th Avenue, San Francisco, CA 94122. (415) 661-8660. Bimonthly. $21.00 per year.

MONTHLY NOTICES OF THE ROYAL ASTRONOMICAL SOCIETY. Blackwell Science Publications, Incorporated, 667 Lytton Avenue, Palo Alto, CA 94301. (415) 324-1688. Monthly. $134.00 per year.

SKY AND TELESCOPE. Sky Publishing Corporation, 49 Bay State Road, Cambridge, Ma 02238. (617) 864-7360. Monthly. $18.00 per year.

SOVIET ASTRONOMY (TRANSLATION OF ASTRONOMICHESKII ZHURNAL). American Institute of Physics, 335 East 45th Street, New York, NY 10017. (212) 661-9404. Bimonthly. $425.00 per year.

SPACE SCIENCE REVIEWS. D. Reidel Publishing Company, 190 Old Derby Street, Hingham, MA 02043. Monthly. $305.00 per year.

VISTAS IN ASTRONOMY. Pergamon Press, Incorporated, Maxwell House, Fairview Park, Elmsford, NY 10523. (914) 592-7700. Quarterly. $145.00 per year.

RESEARCH CENTERS AND INSTITUTES

HARVARD-SMITHSONIAN CENTER FOR ASTROPHYSICS. 60 Garden Street, Cambridge, MA 02138. (617) 495-7461.

MOUNT WILSON AND LAS CAMPANAS OBSERVATORIES. 813 Santa Barbara Street, Pasadena, Ca 91101. (818) 577-1122.

NATIONAL OPTICAL ASTRONOMY OBSERVATORIES. 1002 North Warren Avenue, Tucson, Az 85719. (602) 325-9230.

STANFORD UNIVERSITY. Center for Space Science and Astrophysics, 325 Durand Building, Stanford, CA 94305. (415) 497-3582.

UNIVERSITY OF CHICAGO. Yerkes Observatory, William Bay, WI. (414) 245-5555.

YALE UNIVERSITY OBSERVATORY. P.O. Box 6666, New Haven, CT 06511. (203) 436-3460.

COSMOGONY

See: COSMOLOGY

COSMOLOGY

See also: ASTRONOMY, ASTROPHYSICS, BIG BANG, COSMOCHEMISTRY

ABSTRACT SERVICES AND INDEXES

ASTRONOMY AND ASTROPHYSICS ABSTRACTS. Springer-Verlag New York, Incorporated, 175 Fifth Avenue, New York, NY 10010. (212) 460-1500. $70.00 per year.

ASTRONOMY AND ASTROPHYSICS MONTHLY INDEX. Olivetree Associates, Post Office Box 236, Sierra Madre, CA 91024. $212.00 per year. Complimentary copies available on request.

GENERAL SCIENCE INDEX. H.W. Wilson Company, 950 University Avenue, Bronx, NY 10452. (800) 367-6770 or (212) 588-8400. Inquire as to cost and availability.

PHYSICS ABSTRACTS. Institute of Electrical Engineers, London, United Kingdom. Available from: Institute of Electrical and Electronic Engineers (IEEE), 345 East 47th Street, New York, NY 10017. (212) 705-7900. 1898 to present. Monthly. $1670.00 per year.

SCIENCE CITATION INDEX. Institute for Scientific Information, 3501 Market Street, Philadelphia, PA 19104. (800) 523-1850 or (215) 386-0100. Inquire as to cost and availability.

STAR. (Scientific and Technical Aerospace Reports. United States National Aeronautics and Space Administration, Scientific and Technical Information Facility, Box 8757, Baltimore-Washington International Airport, MD 21240. (202) 755-2210. Semimonthly, with semiannual and annual indexes. $85.00 per year.

ANNUAL REVIEWS AND YEARBOOKS

ANNUAL REVIEW OF ASTRONOMY AND ASTROPHYSICS. Annual Reviews, Incorporated, 4139 El Camino Way, Palo Alto, CA 94306. (415) 493-4400.

ASSOCIATIONS AND PROFESSIONAL SOCIETIES

AMERICAN ASTRONOMICAL SOCIETY. 2000 Florida Avenue, NW, Suite 300, Washington, DC 20009. (202) 659-0134.

AMERICAN ASSOCIATION OF VARIABLE STAR OBSERVERS. 187 Concord Avenue, Cambridge, MA 02138. (617) 354-0484.

ASTRONOMICAL LEAGUE. Post Office Box 12821, Tucson, AZ 85732. (602) 790-8471.

ASTRONOMICAL SOCIETY OF THE PACIFIC. 1290 24th Avenue, San Francisco, CA 94122. (415) 661-8660.

BIBLIOGRAPHIES

A BIBLIOGRAPHY OF ASTRONOMY, 1970-1979. R.A. Seal and S.S. Martin. Libraries Unlimited, Incorporated, Littleton, CO 80160. 1982. $37.50.

SCIENCE BOOKS AND FILMS. American Association for the Advancement of Science, 1333 H Street, NW, Washington, DC 20005. Five times per year. $20.00 per year.

SCIENTIFIC AND TECHNICAL BOOKS AND SERIALS IN PRINT 1988; AN INDEX TO LITERATURE IN SCIENCE AND TECHNOLOGY. R.R. Bowker Company, 205 East 42nd Street, New York, NY 10017. (800) 521-8110 or (212) 916-1600. $175.00.

DIRECTORIES AND BIOGRAPHICAL SOURCES

AMERICAN MEN AND WOMEN OF SCIENCE. Physical and Biological Sciences. Sixteenth edition. R.R. Bowker Company, 205 East 42nd Street, New York, NY 10017. (800) 521-8110 or (212) 916-1600. 1987. $595.00.

THE BIOGRAPHICAL DICTIONARY OF SCIENTISTS: ASTRONOMERS. D. Abbott, editor. Peter Bedrick Books, 125 East 23rd Street, New York, NY 10010. 1984.

DIRECTORY OF PHYSICS AND ASTRONOMY STAFF MEMBERS. American Institute of Physics, 335 East 45th Street, New York, NY 10017. Annual.

RESEARCH CENTERS DIRECTORY. Gale Research Company, Book Tower, Detroit, MI 48226. Twelfth edition, 1988. (800) 521-0707. $240.00.

WHO'S WHO IN FRONTIER SCIENCE AND TECHNOLOGY. Marquis Who's Who, Incorporated, 200 East Ohio Street, Chicago, IL 60611. (800) 428-3898 or (312) 787-2008.

ENCYCLOPEDIAS AND DICTIONARIES

ENCYCLOPEDIA OF PHYSICAL SCIENCE AND TECHNOLOGY. Academic Press, Incorporated, Orlando, FL 32887. (800) 321-5068 or (305) 345-2734. Inquire as to cost and availability.

ILLUSTRATED ENCYCLOPEDIA OF THE UNIVERSE: UNDERSTANDING AND EXPLORING THE COSMOS. Richard S. Lewis. Crown Publishers Incorporated, 1 Park Avenue, New York, NY 10016. (800) 526-4264. 1986. $24.95.

MCGRAW-HILL ENCYCLOPEDIA OF SCIENCE AND TECHNOLOGY. McGraw-Hill Book, Incorporated, 1221 Avenue of the Americas, New York, NY 10020. (212) 997-3675. Fifth edition, 15 volumes. $1100.00.

GENERAL WORKS

BEYOND EINSTEIN: THE COSMIC QUEST FOR THE THEORY OF THE UNIVERSE. Michio Kahu and Jennifer Trainer. Bantam Books, Incorporated, 666 Fifth Avenue, New York, NY 10019. (212) 765-6500. 1987. $9.95 in paper.

CONSTRUCTING THE UNIVERSE. David Layzer. W.H. Freeman and Company, 41 Madison Avenue, New York, NY 10010. (212) 532-7660. 1984. $29.95.

COSMOLOGICAL CONSTANTS: PAPERS IN MODERN COSMOLOGY. Jersey Bernstein and Gerald Feinberg. Columbia University Press, 562 West 113th Street, New York, NY 10025. (212) 316-7100. 1986. $38.00.

COSMOLOGY, PHYSICS AND PHILOSOPHY. Bernard Gal-Or. Springer-Verlag New York, Incorporated, 175 Fifth Avenue, New York, NY 10010. (800) 526-7254. 1981. $34.00.

FIRST THREE MINUTES: A MODERN VIEW OF THE ORIGIN OF THE UNIVERSE. Steven Weinberg. Basic Books, Incorporated, 10 East 53rd Street, New York, NY 10022. (800) 242-7737. 1976. $14.95.

THE FORMATION AND EVOLUTION OF GALAXIES AND LARGE STRUCTURES IN THE UNIVERSE. Jean Audouze and Thanh Van Tran. Kluwer Academic Publishers, 190 Old Derby Street, Hingham, MA 02043. (617) 749-5262. 1984. $58.00.

LARGE SCALE STRUCTURE OF THE UNIVERSE. P.J. Peebles. Princeton University Press, 41 William Street, Princeton, NJ 08540. (609) 452-4122. 1980. $15.95 in paper.

THE LEFT HAND OF CREATION: THE ORIGIN AND EVOLUTION OF THE EXPANDING UNIVERSE. John D. Barrow and Joseph Silk. Basic Books, Incorporated, 10 East 53rd Street, New York, NY 10022. (800) 242-7737. 1986. $7.95 in paper.

MODERN COSMOLOGY. Dennis Sciama. Cambridge University Press, 32 East 57th Street, New York, NY 10022. (800) 872-7423. 1982. $15.95 in paper.

QUARK TO QUASAR. Peter H. Cadogan. Cambridge University Press, 32 East 57th Street, New York, NY 10022. (800) 872-7423. 1985. $24.95.

RELATIVISTIC COSMOLOGY: AN INTRODUCTION. J. Heidmann. Springer-Verlag New York, Incorporated, 175 Fifth Avenue, New York, NY 10010. (800) 526-7254. 1980. $28.00 in paper.

THE VERY EARLY UNIVERSE. G.W. Gibbons and others, editors. Cambridge University Press, 32 East 57th Street, New York, NY 10022. (800) 872-7423. 1985. $24.95.

ONLINE DATA BASES

DISSERTATION ABSTRACTS ONLINE. University Microfilms International, 300 North Zeeb Road, Ann Arbor, MI 48106. (800) 521-0600 or (313) 761-4700. Scope includes virtually all doctoral dissertations accepted at accredited American institutions from 1861 to present in 252 subject areas. Inquire as to cost and availability.

INSPEC. INSPEC Marketing Department, Institute of Electrical and Electronics Engineers, Incorporated, IEEE Service Department, 445 Hoes Lane, Piscataway, NJ 08854. (201) 981-0060. Inquire as to on-line cost and availability.

NASA. National Aeronautics and Space Administration, Scientific and Technical Information Branch, 300 7th Street, SW, Washington, DC 20546. Citations and abstracts of aerospace literature, 1962 to present. Inquire as to cost and availability.

NTIS. National Technical Information Service, 5285 Port Royal Road, Springfield, VA 22161. (703) 487-4630. Broad coverage of government sponsored research reports, 1964 to present. Inquire as to cost and availability.

SCISEARCH. Institute for Scientific Information, 3501 Market Street, Philadelphia, PA 19104. (800) 523-1850 or (215) 386-0100. Broad multidisciplinary title and author index to the international literature of science and technology, 1974 to present. Inquire as to cost and availability.

OTHER SOURCES

ATLAS OF DEEP-SKY SPLENDORS. H. Vehrenberg. Cambridge University Press, 32 East 57th Street, New York, NY 10022. (800) 431-1580 or (212) 688-8888. Fourth edition, 1984. $44.50.

SKY CATALOGUE 2000.0. A. Hirshfeld and R. Sinnott. Cambridge University Press, 32 East 57th Street, New York, NY 10022. (800) 431-1580 or (212) 688-8888.

PERIODICALS

ASTRONOMICAL JOURNAL. American Astronomical Society. Available from: American Institute of Physics, 335 East 45th Street, New York, NY 10017. (212) 661-9404. $125.00 per year.

ASTRONOMICAL SOCIETY OF THE PACIFIC PUBLICATIONS. Astronomical Society of the Pacific, 1290 24th Avenue, San Francisco, CA 94122. (415) 661-8660. Monthly. $38.00.

ASTRONOMY. Astro Media Corporation, 625 East Paul Avenue, Milwaukee, WI 53202. Monthly. $18.00 per year.

ASTRONOMY AND ASTROPHYSICS. Springer-Verlag New York, Incorporated, 175 Fifth Avenue, New York, NY 10010. (800) 526-7254 or (212) 460-1500. $680.00 per year.

ASTROPHYSICAL JOURNAL. American Astronomical Society, University of Chicago Press, 5801 Ellis Avenue, Chicago, IL 60637. Biweekly. $305.00 per year.

ASTROPHYSICS AND SPACE SCIENCE. D. Reidel Publishing Company, 190 Old Derby Street, Hingham, MA 02043. Monthly. $101.00 per year.

CELESTIAL MECHANICS: AN INTERNATIONAL JOURNAL OF SPACE DYNAMICS. D. Reidel Publishing Company, 190 Old Derby Street, Hingham, MA 02043. Monthly. $310.00 per year.

FUNDAMENTALS OF COSMIC PHYSICS. Gordon and Breach Science Publishers, Limited, Post Office Box 197, London, WC2E 9PX. Monthly. $335.00 per year.

MERCURY. Astronomical Society of the Pacific, 1290 245h Avenue, San Francisco, CA 94122. (415) 661-8660. Bimonthly. $21.00 per year.

MONTHLY NOTICES OF THE ROYAL ASTRONOMICAL SOCIETY. Blackwell Science Publications, Incorporated, 667 Lytton Avenue, Palo Alto, CA 94301. (415) 324-1688. Monthly. $134.00 per year.

SKY AND TELESCOPE. Sky Publishing Corporation, 49 Bay State Road, Cambridge, MA 02238. (617) 864-7360. Monthly. $18.00 per year.

SOVIET ASTRONOMY (TRANSLATION OF ASTRO-NOMICHESKII ZHURNAL). American Institute of Physics, 335 East 45th Street, New York, NY 10017. (212) 661-9404. Bimonthly. $425.00 per year.

SPACE SCIENCE REVIEWS. D. Reidel Publishing Company, 190 Old Derby Street, Hingham, MA 02043. Monthly. $305.00 per year.

VISTAS IN ASTRONOMY. Pergamon Press, Incorporated, Maxwell House, Fairview Park, Elmsford, NY 10523. (914) 592-7700. Quarterly. $145.00 per year.

RESEARCH CENTERS AND INSTITUTES

HARVARD-SMITHSONIAN CENTER FOR ASTROPHYSICS. 60 Garden Street, Cambridge, Ma 02138. (617) 495-7461.

MOUNT WILSON AND LAS CAMPANAS OBSERVATORIES. 813 Santa Barbara Street, Pasadena, CA 91101. (818) 577-1122.

NATIONAL OPTICAL ASTRONOMY OBSERVATORIES. 1002 North Warren Avenue, Tucson, AZ 85719. (602) 325-9230.

STANFORD UNIVERSITY. Center for Space Science and Astrophysics, 325 Durand Building, Stanford, CA 94305. (415) 497-3582.

UNIVERSITY OF CHICAGO. Yerkes Observatory, Williams Bay, WI. (414) 245-5555.

YALE UNIVERSITY OBSERVATORY. Post Office Box 6666, New Haven, CT 06511. (203) 436-3460.

COULOMETRIC ANALYSIS

See: ANALYTICAL CHEMISTRY

COVALENT BOND

See: CHEMICAL BONDING

CRACKING

See: CHEMICAL ENGINEERING

CRATERS

See: SOLAR SYSTEM

CRITICAL MASS

See: NUCLEAR ENERGY

CROSSOVER NETWORKS

See: COMPUTER COMMUNICATIONS

CRUISE MISSILE

See: GUIDED MISSILES

CRUST

See: GEOLOGY

CRYOGENIC ENGINEERING

See: CRYOGENICS

CRYOGENICS

See also: CHEMICAL ENGINEERING, FLUID MECHANICS, HEAT TRANSFER, PHYSICAL CHEMISTRY, PHYSICS, THERMODYNAMICS

ABSTRACT SERVICES AND INDEXES

APPLIED MECHANICS REVIEW. American Society of Mechanical Engineers, 345 East 47th Street, New York, NY 10017. (212) 705-7703. 1948 to present. Monthly. $360.00 per year.

APPLIED SCIENCE AND TECHNOLOGY INDEX. H.W. Wilson and Company, 950 University Avenue, Bronx, NY 10452. (800) 367-6670 or (212) 588-8400. Monthly. Inquire as to cost and availability.

CHEMICAL ABSTRACTS. American Chemical Society, Chemical Abstracts Service, Box 3012, Columbus, OH 43210. (614) 421-3600. 1907 to present. Weekly. $9500.00 per year.

CONFERENCE PAPERS INDEX. Cambridge Scientific Abstracts, 5161 River Road, Bethesda, MD 20816. 1972 to present. Monthly. Inquire as to cost and availability.

CURRENT CONTENTS: ENGINEERING, TECHNOLOGY AND APPLIED SCIENCES. Institute for Scientific Information, 3501 Market Street, Philadelphia, PA 19104. (800) 523-1850 or (215) 386-0100. Weekly. $275.00 per year.

CURRENT CONTENTS: PHYSICAL, CHEMICAL AND EARTH SCIENCES. Institute for Scientific Information, 3501 Market Street, Philadelphia, PA 19104. (800) 523-1850 or (215) 386-0100. Weekly. $275.00 per year.

ENGINEERING INDEX MONTHLY AND AUTHOR INDEX. Engineering Information Inc., 345 East 47th Street, New York, NY 10017. (212) 705-7600. Monthly. $1560.00 per year.

ISMEC BULLETIN (Information Service in Mechanical Engineering). Cambridge Scientific Abstracts, 5161 River Road, Bethesda, MD 20816. (301) 951-1400. 1973 to present. Monthly. $450.00 per year.

INDEX TO SCIENTIFIC AND TECHNICAL PROCEEDINGS. Institute for Scientific Information, 3501 Market Street, Philadelphia, PA 19104. (800) 523-1850 or (215) 386-0100. 1978 to present. Monthly. $775.00 per year.

INDEX TO SCIENTIFIC REVIEWS. Institute for Scientific Information, 3501 Market Street, Philadelphia, PA 19104. (800) 523-1850 or (215) 386-0100. 1974 to present. Semi-annual. $550.00 per year.

PHYSICS ABSTRACTS. Institution of Electrical Engineers. Available from: IEEE Service Center, 445 Hoes Lane, Piscataway, NJ 08854. 1898 to present. Bimonthly. $1700.00 per year.

PHYSICS BRIEFS. Physik Verlag GmbH, Postfach 1260/1280, D-6940 Weinheim, West Germany. (212) 661-9404. 1920 to present. Twenty-six times per year. $1250.00 per year.

SCIENCE CITATION INDEX. Institute for Scientific Information, 3501 Market Street, Philadelphia, PA 19104. (800) 523-1850 or (215) 386-0100. Six times per year. $6200.00 per year.

ANNUAL REVIEWS AND YEARBOOKS

ADVANCES IN CRYOGENIC ENGINEERING. R.P. Reed and others, editors. Plenum Publishing Corporation, 233 Spring Street, New York, NY 10013. (800) 221-9369. 1960-1984. Irregular. Price varies, inquire.

ASSOCIATIONS AND PROFESSIONAL SOCIETIES

AMERICAN CHEMICAL SOCIETY. 1155 16th Street, N.W., Washington, DC 20036. (800) 424-6747.

AMERICAN INSTITUTE OF CHEMICAL ENGINEERS. 345 East 47th Street, New York, NY 10017. (212) 705-7338.

AMERICAN INSTITUTE OF PHYSICS. 335 East 45th Street, New York, NY 10017. (212) 661-9494.

AMERICAN SOCIETY OF MECHANICAL ENGINEERS. 345 East 47th Street, New York, NY 10017. (212) 705-7703.

CRYOGENIC ENGINEERING CONFERENCE. 73 Vassar Street, Room 41-204, Cambridge, MA 02139. (617) 253-2296.

CRYOGENIC SOCIETY OF AMERICA. c/o Huget Advertising, 1033 South Boulevard, Oak Park, IL 60302. (312) 383-7053.

DIRECTORIES AND BIOGRAPHICAL SOURCES

AMERICAN MEN AND WOMEN OF SCIENCE. R.R. Bowker, Inc., Order Department, 245 West 17th Street, New York, NY 10011. (800) 521-8110. Eight volumes. 1986. $595.00 for set.

INTERNATIONAL RESEARCH CENTERS DIRECTORY 1988-89. Darren L. Smith, editor. Gale Research Company, Book Tower, Detroit, MI 48226. (800) 521-0707. 4th edition. 1987. $360.00.

1987 DIRECTORY OF ENGINEERING SOCIETIES AND RELATED ORGANIZATIONS. Gordon Davis, editor. Hemisphere Publishing Corporation, 1010 Vermont Avenue, NW, Washington, DC 20005. (800) 526-0275. 12th edition. 1987. $100.00.

RESEARCH CENTERS DIRECTORY 1988. Gale Research Company, Book Tower, Detroit, MI 48226. (800) 521-0707. 12th edition. 1987. $365.00 for set.

WHO'S WHO IN ENGINEERING. Gordon Davis, editor. Hemisphere Publishing Corporation, 1010 Vermont Avenue, NW, Washington, DC 20005. (800) 526-0275. 6th edition. 1985. $200.00.

GENERAL WORKS

APPLIED THERMODYNAMICS FOR ENGINEERING TECHNOLOGIES. T.D. Eastop and A. McConkey. John Wiley and Sons, Inc., 605 Third Avenue, New York, NY 10158. (800) 526-5368. 1986. $29.95.

A COURSE IN THERMODYNAMICS. J. Kestin. Hemisphere Publishing Corporation, 79 Madison Avenue, New York, NY 10016-7892. (800) 821-8312. Two volumes. 1979. $45.00 each.

CRYOGENIC SYSTEMS. Randall F. Barron. Oxford University Press, 200 Madison Avenue, New York, NY 10016. (800) 458-5833. 1985. $59.00.

FUNDAMENTALS OF ENGINEERING THERMODYNAMICS. V. Hubka and others. McGraw-Hill Book Company, 1221 Avenue of the Americas, New York, NY 10020. (212) 512-2000. 1986. $39.95.

TECHNIQUES IN LOW TEMPERATURE PHYSICS. R.C. Richardson. Benjamin-Cummings Publishing Company, 2727 Sand Hill Road, Menlo Park, CA 94025. (415) 854-6020. 1987. $59.95.

ONLINE DATA BASES

CA SEARCH. Chemical Abstracts Service, P.O. Box 3012, Columbus, OH 43120. (800) 848-6538 or (614) 421-3600. Comprehensive guide to chemical literature, 1972 to present. Inquire as to online cost and availability.

COMPENDEX. Engineering Information, Inc., 345 East 47th Street, New York, NY 10017. (800) 221-1044 or (212) 705-7615. Engineering and technical literature, 1975 to present. Inquire as to online cost and availability.

DISSERTATION ABSTRACTS ONLINE. University Microfilms International, 300 North Zeeb Road, Ann Arbor, MI 48106. (800) 521-0600 or (313) 761-4700. Scope includes virtually all doctoral dissertations accepted at accredited American institutions from 1861 to present in over 250 subject areas. Inquire as to online cost and availability.

INSPEC. INSPEC Marketing Department, Institution of Electrical Engineers. Available from IEEE Service Center, 445 Hoes Lane, Piscataway, NJ 08854. (201) 981-0060. Online version of Physics Abstracts. Inquire as to online cost and availability.

NTIS. National Technical Information Service, 5285 Port Royal Road, Springfield, VA 22161. (703) 487-4630. Broad coverage of government sponsored research reports, 1964 to present. Inquire as to online cost and availability.

SCISEARCH. Institute for Scientific Information, 3501 Market Street, Philadelphia, PA 19104. (800) 523-1850 or (215) 386-0100. Broad multidisciplinary title and author index to the international literature of science and technology, 1974 to present. Inquire as to online cost and availability.

WILSONLINE. H.W. Wilson and Company, 950 University Avenue, Bronx, NY 10452. (800) 367-6770 or (212) 588-8400. Makes available online versions of the H.W. Wilson indexes including Applied Science and Technology Index, Business Periodicals Index and Readers' Guide to Periodical Literature. Approximately 1980 to present. Inquire as to online cost and availability.

PERIODICALS

CRYOGENIC INFORMATION REPORT. Technical Economic Associates, Box 1972, Estes Park, CO 80517. (303) 586-5636. 1963 to present. Ten times per year. $85.00 per year.

CRYOGENICS: THE INTERNATIONAL JOURNAL OF LOW TEMPERATURE ENGINEERING AND RESEARCH. Butterworth Publishing, 80 Montvale Avenue, Stoneham, MA 02180. (800) 325-4177. 1960 to present. Monthly. $335.00 per year.

EXPERIMENTAL THERMAL AND FLUID SCIENCE. Elsevier Science Publishing Company, Inc., 52 Vanderbilt Avenue, New York, NY 10017. (212) 370-5520. 1988 to present. Quarterly. $125.00 per year.

HEAT TRANSFER AND FLUID FLOW DIGEST. Hemisphere Publishing Corporation, 79 Madison Avenue, New York, NY 10016-7892. (800) 821-8312. 1968 to present. Monthly. $137.50 per year.

HEAT TRANSFER ENGINEERING. Hemisphere Publishing Corporation, 79 Madison Avenue, New York, NY 10016-7892. (800) 821-8312. 1979 to present. Quarterly. $97.50.

INTERNATIONAL COMMUNICATIONS IN HEAT AND MASS TRANSFER. Pergamon Press, Inc., Maxwell House, Fairview Park, Elmsford, NY 10523. (914) 592-7700. 1974 to present. Bimonthly. $160.00 per year.

INTERNATIONAL JOURNAL OF HEAT AND FLUID FLOW. Butterworth Publishing, 80 Montvale Avenue, Stoneham, MA 02180. (800) 325-4177. 1979 to present. Quarterly. $140.00 per year.

INTERNATIONAL JOURNAL OF HEAT AND MASS TRANSFER. Pergamon Press, Inc., Maxwell House, Fairview Park, Elmsford, NY 10523. (914) 592-7700. 1960 to present. Monthly. $500.00 per year.

INTERNATIONAL JOURNAL OF THERMOPHYSICS. Plenum Publishing Corporation, 233 Spring Street, New York, NY 10013. (800) 221-9369. 1980 to present. Six times per year. $150.00 per year.

JOURNAL OF HEAT TRANSFER. American Society of Mechanical Engineers, 345 East 47th Street, New York, NY 10017. (212) 705-7703. 1970 to present. Quarterly. $100.00 per year.

JOURNAL OF LOW TEMPERATURE PHYSICS. Plenum Publishing Corporation, 233 Spring Street, New York, NY 10013. (800) 221-9369. 1969 to present. 24 times per year. $750.00 per year.

JOURNAL OF SUPERCONDUCTIVITY. Plenum Publishing Corporation, 233 Spring Street, New York, NY 10013. (800) 221-9369. 1988 to present. Quarterly. $125.00 per year.

NUMERICAL HEAT TRANSFER: AN INTERNATIONAL JOURNAL OF COMPUTATION AND METHODLOGY. Hemisphere Publishing Corporation, 79 Madison Avenue, New York, NY 10016-7892. (800) 821-8312. 1978 to present. Monthly. $370.00 per year.

RESEARCH CENTERS AND INSTITUTES

CENTER FOR THERMODYNAMICS. Brigham Young University, 226 ESC, Provo, UT 84602. (801) 378-3668.

HEAT TRANSFER LABORATORY. Massachusetts Institute of Technology, 77 Massachusetts Avenue, Cambridge, MA 02139. (716) 253-2248.

THERMODYNAMICS RESEARCH CENTER. Texas A&M University, Texas Engineering Experiment Station, College Station, TX 77843-3111. (409) 845-4940.

CRYPTOGRAPHY AND CRYPTOLOGY

See: COMPUTER SECURITY

CRYSTAL OPTICS

See: CRYSTALLOGRAPHY

CRYSTALLOGRAPHY

See also: CHEMISTRY, GEOCHEMISTRY, GEOLOGY, MINERALOGY, SOLID STATE CHEMISTRY, SOLID STATE PHYSICS

ABSTRACT SERVICES AND INDEXES

BIBLIOGRAPHY AND INDEX OF GEOLOGY. American Geological Institute, 4220 King Street, Alexandria, VA 22302. (703) 379-2480. 1969 to present. Monthly. $1100.00 per year.

CHEMICAL ABSTRACTS. American Chemical Society, Chemical Abstracts Service, Box 3012, Columbus, OH 43210. (614) 421-3600. 1907 to present. Weekly. $9500.00 per year.

CONFERENCE PAPERS INDEX. Cambridge Scientific Abstracts, 5161 River Road, Bethesda, MD 20816. 1972 to present. Monthly. Inquire as to cost and availability.

CURRENT CONTENTS: PHYSICAL, CHEMICAL AND EARTH SCIENCES. Institute for Scientific Information, 3501 Market Street, Philadelphia, PA 19104. (800) 523-1850 or (215) 386-0100. Weekly. $275.00 per year.

ENGINEERING INDEX MONTHLY AND AUTHOR INDEX. Engineering Information Inc., 345 East 47th Street, New York, NY 10017. (212) 705-7600. Monthly. $1560.00 per year.

GENERAL SCIENCE INDEX. H.W. Wilson and Company, 950 University Avenue, Bronx, NY 10452. (800) 367-6670 or (212) 588-8400. 1978 to present. Monthly. Inquire as to cost and availability.

INDEX TO SCIENTIFIC AND TECHNICAL PROCEEDINGS. Institute for Scientific Information, 3501 Market Street, Philadelphia, PA 19104. (800) 523-1850 or (215) 386-0100. 1978 to present. Monthly. $775.00 per year.

INDEX TO SCIENTIFIC REVIEWS. Institute for Scientific Information, 3501 Market Street, Philadelphia, PA 19104. (800) 523-1850 or (215) 386-0100. 1974 to present. Semi-annual. $550.00 per year.

MINERALOGICAL ABSTRACTS: A QUARTERLY JOURNAL OF ABSTRACTS IN ENGLISH, COVERING THE WORLD LITERATURE AND RELATED SUBJECTS. Mineralogical Society, 41 Queen's Gate, London SW7 5HR, England. 1959 to present. Quarterly. $190.00 per year.

PHYSICS ABSTRACTS. Institution of Electrical Engineers. Available from: IEEE Service Center, 445 Hoes Lane, Piscataway, NJ 08854. 1898 to present. Bimonthly. $1700.00 per year.

PHYSICS BRIEFS. Physik Verlag GmbH, Postfach 1260/1280, D-6940 Weinheim, West Germany. (212) 661-9404. 1920 to present. Twenty-six times per year. $1250.00 per year.

SCIENCE CITATION INDEX. Institute for Scientific Information, 3501 Market Street, Philadelphia, PA 19104. (800) 523-1850 or (215) 386-0100. Six times per year. $6200.00 per year.

SOLID STATE ABSTRACTS: AN ABSTRACT JOURNAL INVOLVING THE PHYSICS, METALLURGY, CRYSTALLOGRAPHY, CHEMISTRY, AND DEVICE TECHNOLOGY OF SOLIDS. Cambridge Scientific Abstracts, 5161 River Road, Bethesda, MD 20816. (301) 951-1400. 1957 to present. Bimonthly. $550.00 per year.

ASSOCIATIONS AND PROFESSIONAL SOCIETIES

AMERICAN FEDERATION OF MINERALOGICAL SOCIETIES. 1203 East Hillsboro, Tampa, FL 33604. (813) 238-0427.

AMERICAN GEOLOGICAL INSTITUTE. 4220 King Street, Alexandria, VA 22302. (703) 379-2480.

GEOLOGICAL SOCIETY OF AMERICA. 3300 Penrose Place, Boulder, CO 80301. (303) 447-2020.

MINERALOGICAL SOCIETY OF AMERICA. 2000 Florida Avenue, N.W., Washington, DC 20009. (202) 462-6913.

DIRECTORIES AND BIOGRAPHICAL SOURCES

AMERICAN MEN AND WOMEN OF SCIENCE. R.R. Bowker, Inc., Order Department, 245 West 17th Street, New York, NY 10011. (800) 521-8110. Eight volumes. 1986. $595.00 for set.

GEOLOGICAL SOCIETY OF AMERICA MEMBERSHIP DIRECTORY. 3300 Penrose Place, Boulder, CO 80301. (303) 447-2020. Annual. Available to membership only.

INTERNATIONAL RESEARCH CENTERS DIRECTORY 1988-89. Darren L. Smith, editor. Gale Research Company, Book Tower, Detroit, MI 48226. (800) 521-0707. 4th edition. 1987. $360.00.

RESEARCH CENTERS DIRECTORY 1988. Gale Research Company, Book Tower, Detroit, MI 48226. (800) 521-0707. 12th edition. 1987. $365.00 for set.

SCIENTIFIC AND TECHNICAL ORGANIZATIONS AND AGENCIES DIRECTORY. Margaret Labash Young, editor. Gale Research Company, Book Tower, Detroit, MI 48226. (800) 521-0707. 2nd edition. 1987. $185.00.

ENCYCLOPEDIAS AND DICTIONARIES

DICTIONARY OF GEOLOGICAL TERMS. American Geological Institute. Doubleday and Company, Inc., 245 Park Avenue, New York, NY 10017. (800) 645-6156. Third edition. 1984. $19.95.

GENERAL WORKS

CRYSTAL STRUCTURE ANALYSIS: A PRIMER. G.P. Glusker and K.N. Trueblood. Oxford University Press, 200 Madison Avenue, New York, NY 10016. (800) 458-5833. Second edition. 1985. $18.95 in paper.

CRYSTALLOGRAPHY: AN INTRODUCTION FOR EARTH SCIENCE AND OTHER SOLID STATE STUDENTS. E.J. Whittake. Pergamon Press, Inc., Maxwell House, Fairview Park, Elmsford, NY 10523. (914) 592-7700. 1981. $19.95 in paper.

CRYSTALLOGRAPHY FOR SOLID STATE PHYSICS. A.R. Verma and others. John Wiley and Sons, Inc., 605 Third Avenue, New York, NY 10158. (800) 526-5368. 1982. $26.95.

GEOMETRIC CRYSTALLOGRAPHY: AN AXIOMATIC INTRODUCTION TO CRYSTALLOGRAPHY. Peter Engel. D. Reidel Publishing Company, 190 Old Derby Street, Hingham, MA 02043. 1986. $59.00.

HANDBOOKS AND MANUALS

CRC HANDBOOK OF CHEMISTRY AND PHYSICS. Robert C. Weast, editor. CRC Press, 2000 Corporate Boulevard, Boca Raton, FL 33431. (800) 272-7737. 68th edition. 1987. $69.95.

MANUAL OF MINERALOGY. Cornelis Klein and Cornelis S. Hurlbut, Jr. John Wiley and Sons, Inc., 605 Third Avenue, New York, NY 10158. (800) 526-5368. 20th edition. 1985. $40.95.

ONLINE DATA BASES

CA SEARCH. Chemical Abstracts Service, P.O. Box 3012, Columbus, OH 43120. (800) 848-6538 or (614) 421-3600. Comprehensive guide to chemical literature, 1972 to present. Inquire as to online cost and availability.

COMPENDEX. Engineering Information, Inc., 345 East 47th Street, New York, NY 10017. (800) 221-1044 or (212) 705-7615. Engineering and technical literature, 1975 to present. Inquire as to online cost and availability.

DISSERTATION ABSTRACTS ONLINE. University Microfilms International, 300 North Zeeb Road, Ann Arbor, MI 48106. (800) 521-0600 or (313) 761-4700. Scope includes virtually all doctoral dissertations accepted at accredited American institutions from 1861 to present in over 250 subject areas. Inquire as to online cost and availability.

GEOREF. Online version of the BIBLIOGRAPHY AND INDEX OF GEOLOGY. American Geological Institute, 4220 King Street, Alexandria, VA 22302. (703) 379-2480. 1969 to present. Inquire as to online cost and availability.

INSPEC. INSPEC Marketing Department, Institution of Electrical Engineers. Available from IEEE Service Center, 445 Hoes Lane,

Piscataway, NJ 08854. (201) 981-0060. Online version of Physics Abstracts. Inquire as to online cost and availability.

NTIS. National Technical Information Service, 5285 Port Royal Road, Springfield, VA 22161. (703) 487-4630. Broad coverage of government sponsored research reports, 1964 to present. Inquire as to online cost and availability.

SCISEARCH. Institute for Scientific Information, 3501 Market Street, Philadelphia, PA 19104. (800) 523-1850 or (215) 386-0100. Broad multidisciplinary title and author index to the international literature of science and technology, 1974 to present. Inquire as to online cost and availability.

WILSONLINE. H.W. Wilson and Company, 950 University Avenue, Bronx, NY 10452. (800) 367-6770 or (212) 588-8400. Makes available online versions of the H.W. Wilson indexes including Applied Science and Technology Index, Business Periodicals Index and Readers' Guide to Periodical Literature. Approximately 1980 to present. Inquire as to online cost and availability.

PERIODICALS

AMERICAN MINERALOGIST. Mineralogical Society of America, 2000 Florida Avenue, N.W., Washington, DC 20009. (202) 462-6913. Bimonthly. $115.00 per year.

CRYSTAL LATTICE DEFECTS AND AMORPHOUS MATERIALS. Gordon and Breach Science Publishers, Inc., 50 West 23rd Street, New York, NY 10010. (212) 206-8900. 1969. Quarterly. $280.00 per year.

GEOLOGICAL SOCIETY OF AMERICA BULLETIN. Geological Society of America, 3300 Penrose Place, Boulder, CO 80301. (303) 447-2020. 1888. Monthly. $80.00.

JOURNAL OF GEOLOGY. University of Chicago Press, 5801 South Ellis Avenue, Chicago, IL 60637. (312) 962-7600. 1893 to present. Bimonthly. $30.00 per year.

MINERALIUM DEPOSITA. Springer-Verlag New York, Inc., 175 Fifth Avenue, New York, NY 10010. (800) 526-7254. Quarterly. $120.00 per year.

ROCKS AND MINERALS. Heldraf Publications, 4000 Albermerle Street, N.W., Washington, DC 20016. (202) 362-6445. Bimonthly. $25.00 per year.

RESEARCH CENTERS AND INSTITUTES

CARNEGIE INSTITUTION OF WASHINGTON GEOPHYSICAL LABORATORY. 2810 Upton Street, N.W., Washington, DC 20008. (202) 966-0334.

INSTITUTE OF MATERIALS SCIENCE. University of Connecticut, Storrs, CT 06268. (203) 486-4623.

LIQUID CRYSTAL INSTITUTE. Kent State University, Kent, OH 44242. (216) 672-2654.

MATERIALS RESEARCH LABORATORY. University of Illinois, 23 Stadium, IL 61820. (217) 333-3190.

CUMULUS CLOUDS

See: CLOUDS

CURRENTS

See: OCEAN CURRENTS; METEOROLOGY

CYBERNETICS

See also: AUTOMATION, COMPUTER MEMORY, COMPUTER OPERATING SYSTEMS, COMPUTER PROGRAMMING, COMPUTER VISION, PARALLEL COMPUTERS, PROGRAMMING LANGUAGES, SOFTWARE

ABSTRACT SERVICES AND INDEXES

APPLIED SCIENCE AND TECHNOLOGY INDEX. H.W. Wilson Company, 950 University Avenue, Bronx, NY 10452. (800) 367-6670 or (212) 588-8400. Inquire as to cost and availability.

COMPUTER ABSTRACTS. Technical Information Company, Limited, Post Office Box 59, Saint Helier, Jersey British Channel Inlands, England. Monthly. $310.00 per year.

COMPUTER AND CONTROL ABSTRACTS. Institute of Electrical Engineers, London, United Kingdom. Available from: IEEE Service Center, 445 Hoes Lane, Piscataway, NJ 08854. (201) 981-0060. Semimonthly. $775.00 per year.

COMPUTER AND INFORMATION SYSTEMS: AN ABSTRACT JOURNAL PERTAINING TO THE THEORY, DESIGN, FABRICATION AND APPLICATION OF COMPUTER AND INFORMATION SYSTEMS. Cambridge Scientific Abstracts, Incorporated, 5161 River Road, Bethesda, MD 20816. (301) 951-1400. Semimonthly. $590.00 per year.

COMPUTER CONTENTS: THE BIWEEKLY COMPILATION OF TABLES OF CONTENTS FROM COMPUTER, ELECTRONIC AND TELECOMMUNICATIONS MAGAZINES, JOURNALS AND TRANSACTIONS. Find/SVP, 500 Fifth Avenue, New York, NY 10110. (800) 346-3787 or (212) 354-2424. Biweekly. $115.00 per year.

COMPUTER LITERATURE INDEX. Applied Computer Research, Incorporated, Post Office Box 9280, Phoenix, AZ 85068. (602) 995-5929. Quarterly. $125.00 per year.

COMPUTER PROGRAM ABSTRACTS. U.S. National Aeronautics and Space Administration. Available from U.S. Government Printing Office, Washington, DC 20402. Quarterly. $6.50 per year.

COMPUTING REVIEWS. Association for Computing Machinery, 11 West 42nd Street, New York, NY 10036. (212) 869-7440. Monthly. $60.00 per year.

ENGINEERING INDEX MONTHLY. Engineering Information, Incorporated, 345 East 47th Street, New York, NY 10017. (800) 221-1044 or (212) 705-7600. Monthly, with annual cumulation. $1560.00 per year.

MATHEMATICAL REVIEWS. American Mathematical Society, Post Office Box 6248, Providence, RI 02940. (800) 556-7774 or (401) 272-9500.

SCIENCE CITATION INDEX. Institute for Scientific Information, 3501 Market Street, Philadelphia, PA 19104. (800) 523-1850 or (215) 386-0100.

ANNUAL REVIEWS AND YEARBOOKS

ADVANCES IN COMPUTERS. Academic Press, Incorporated, 6277 Sea Harbor Drive, Orlando, FL 32821. (800) 321-5068. Yearly. Approximately $50.00 per volume.

ASSOCIATIONS AND PROFESSIONAL SOCIETIES

AMERICAN FEDERATION OF INFORMATION PROCESSING SOCIETIES. 1899 Preston White Drive, Reston, VA 22091. (703) 620-8900.

AMERICAN SOCIETY FOR CYBERNETICS. Department of Decision Sciences, George Mason University, Fairfax, VA 22030.

(703) 323-2738.

ASSOCIATION FOR COMPUTING MACHINERY (ACM). 11 West 42nd Street, New York, NY 10036. (212) 896-7440.

ASSOCIATION OF COMPUTER PROGRAMMERS AND ANALYSTS. 2108-C Gallows Road, Vienna, VA 22180. (703) 790-0490.

COMPUTER AND AUTOMATION SYSTEMS ASSOCIATION OF SME (SOCIETY OF MANUFACTURING ENGINEERS). One SME Drive, Box 930, Dearborn, MI 48121. (313) 271-1500.

IEEE COMPUTER SOCIETY. 1730 Massachusetts Avenue, NW, Washington, DC 20036. (202) 371-0101.

INSTITUTE OF ELECTRICAL AND ELECTRONIC ENGINEERS (IEEE). 345 East 47th Street, New York, NY 10017. (212) 705-7900.

SOCIETY FOR COMPUTER SIMULATION. Post Office Box 17900. San Diego, CA 92117. (619) 277-3888.

SYSTEMS, MAN AND CYBERNETICS COUNCIL. Institute of Electrical and Electronic Engineers (IEEE), 345 East 47th Street, New York, NY 10017. (212) 705-7900.

BIBLIOGRAPHIES

COMPUTER BOOKS AND SERIALS IN PRINT, 1986-1987. R.R. Bowker Company, 205 East 42nd Street, New York, NY 10017. (800) 521-8110 or (212) 916-1600. $69.95.

SCIENTIFIC AND TECHNICAL BOOKS AND SERIALS IN PRINT 1988; AN INDEX TO LITERATURE IN SCIENCE AND TECHNOLOGY. R.R. Bowker Company, 205 East 42nd Street, New York, NY 10017. (800) 521-8110 or (212) 916-1600. $175.00.

DIRECTORIES AND BIOGRAPHICAL SOURCES

AMERICAN MEN AND WOMEN OF SCIENCE. Physical and Biological Sciences. Fifteenth edition. R.R. Bowker Company, 205 East 42nd Street, New York, NY 10017. (800) 521-8110 or (212) 916-1600.

AMERICAN SOCIETY FOR INFORMATION SCIENCE, HANDBOOK AND DIRECTORY. American Society for Information Science, 1424 16th Street, NW, Suite 404, Washington, DC 20036. (202) 462-1000. $50.00.

COMPUTERS AND COMPUTING INFORMATION RESOURCES DIRECTORY. Gale Research Company, Book Tower, Detroit, MI 48226. (800) 521-0707. 1986. $160.00.

DIRECTORY OF CONSULTANTS IN COMPUTER SYSTEMS. Research Publications, Incorporated, 12 Lunar Drive, Woodbridge, CT 06525. (203) 397-2600. Annual. $85.00 per year.

RESEARCH CENTERS DIRECTORY. Gale Research Company, Book Tower, Detroit, MI 48226. (800) 521-0707. Eleventh edition. 1987.

WHO'S WHO IN FRONTIER SCIENCE AND TECHNOLOGY. Marquis Who's Who, Incorporated, 200 East Ohio Street, Chicago, IL 60611. (800) 428-3898 or (312) 787-2008.

WHO'S WHO IN TECHNOLOGY TODAY. Reston Publishing Company, Incorporated, c/o Prentice-Hall, Incorporated, Englewood Cliffs, NJ 07632. (800) 262-6868. Biennial. Five volumes. $425.00. Covers the fields of electronics, computer science, physics, optics, chemistry, biotechnology, mechanics, energy, and earth science.

ENCYCLOPEDIAS AND DICTIONARIES

COMPUTER AND TELECOMMUNICATIONS ACRONYMS. Julie E. Towell and Helen E. Sheppard, editors. Gale Research Company, Book Tower, Detroit, MI 48226. (800) 521-0707. 1986. $60.00.

COMPUTER DICTIONARY. Charles J. Sippl. Howard W. Sams and Company, Incorporated, 4300 West 62nd Street, Indianapolis, IN 46268. (800) 428-7267 or (317) 298-5564. Fourth edition. 1985. $17.95.

DICTIONARY OF COMPUTING. Oxford University Press, 200 Madison Avenue, New York, NY 10016. (212) 679-7300. Second edition. 1986. $29.95.

ENCYCLOPEDIA OF COMPUTER SCIENCE AND ENGINEERING. Anthony Ralston, editor. Van Nostrand Reinhold Book Company, 115 Fifth Avenue, New York, NY 10003. (800) 543-2681. Second edition. 1982. $89.95.

ENCYCLOPEDIA OF COMPUTER SCIENCE AND TECHNOLOGY. Jack Belzer, Albert G. Holzman, and Allan Kent. Marcel Dekker, Incorporated, 270 Madison Avenue, New York, NY 10016. (212) 696-9000. Sixteen volumes. $115.00 per volume.

MCGRAW-HILL DICTIONARY OF COMPUTERS. McGraw-Hill Book Company, 1221 Avenue of the Americas, New York, NY 10020. (212) 512-2000. 1985. $17.50.

GENERAL WORKS

THE COMPUTER PIONEERS: THE MAKING OF THE MODERN COMPUTER. David Ritchie. Simon and Schuster, Incorporated, 1230 Avenue of the Americas, New York, NY 10020. (800) 223-2336 or (212) 245-6400. 1986. $17.95.

COMPUTERS AND THE CYBERNETIC SOCIETY. Michael Arbib. Academic Press, Incorporated, 6277 Sea Harbor Drive, Orlando, FL 32821. (800) 321-5068. Second edition. 1984. $15.00.

CYBERNETICS: CONTROL AND COMMUNICATION IN THE ANIMAL AND THE MACHINE. Norbert Wiener. MIT Press, 28 Carlton Street, Cambridge, MA 02142. (617) 253-5646. Second edition. 1966. $7.95 in paper.

DESIGNING FOR HUMAN-COMPUTER COMMUNICATION. Max S. Sime, editor. Academic Press, Incorporated, 6277 Sea Harbor Drive, Orlando, FL 32821. (800) 321-5068. 1983. $50.00.

FIFTH GENERATION COMPUTERS. Richard K. Miller, editor. The Fairmont Press, Incorporated, 4025 Pleasantdale Road, Atlanta, GA 30340. (404) 447-5314.

INFORMATION SYSTEMS: THEORY AND PRACTICE. J. Burch and G. Grudnitski. John Wiley and Sons, Incorporated, 605 Third Avenue, New York, NY 10158. Fourth edition. 1986. $30.95.

INTRODUCTION TO COMPUTER ARCHITECTURE AND ORGANIZATION. H. Lorin. John Wiley and Sons, Incorporated, 605 Third Avenue, New York, NY 10158. 1982. $30.95.

HANDBOOKS AND MANUALS

FUNDAMENTALS HANDBOOK OF ELECTRICAL AND COMPUTER ENGINEERING. Sheldon S.L. Chang, editor. John Wiley and Sons, Incorporated, 605 Third Avenue, New York, NY 10158. (800) 526-5368 or (212) 850-6000. Three volumes. 1982. $180.00 set price.

HANDBOOK OF COMPUTERS AND COMPUTING. Arthur H. Seidman, editor. Van Nostrand Reinhold Book Company, 115 Fifth Avenue, New York, NY 10003. (800) 543-2681. 1984. $79.95.

HANDBOOK OF SOFTWARE ENGINEERING. Charles R. Vick, editor. Van Nostrand Reinhold Book Company, 115 Fifth Avenue, New York, NY 10003. (800) 543-2681. 1984. $66.95.

MCGRAW-HILL COMPUTER HANDBOOK. Harry Helms, editor. McGraw-Hill Book Company, 1221 Avenue of the Americas, New York, NY 10020. (212) 512-2000. 1983. $84.50.

ONLINE DATA BASES

COMPENDEX. Engineering Information, Incorporated, 345 East 47th Street, New York, NY 10017. (800) 221-1044 or (212) 705-7615. Engineering and technical literature, 1975 to present. Inquire as to cost and availability.

DISSERTATION ABSTRACTS ONLINE. University Microfilms International, 300 North Zeeb Road, Ann Arbor, MI 48106. (800) 521-0600 or (313) 761-4700. Scope includes virtually all doctoral dissertations accepted at accredited American institutions from 1861 to present in 252 subject areas. Inquire as to cost and availability.

INSPEC. INSPEC Marketing Department, Institute of Electrical and Electronics Engineers, Incorporated, IEEE Service Department, 445 Hoes Lane, Piscataway, NJ 08854. (201) 981-0060. Inquire as to on-line cost and availability.

MATHFILE. American Mathematical Society, Post Office Box 6248, Providence, RI 02940. (800) 556-7774 or (401) 272-9500. Scope includes pure and applied mathematics and related areas of physics, statistics, engineering, computer science, and operations research literature since 1973. Inquire as to cost and availability.

NTIS. National Technical Information Service, 5285 Port Royal Road, Springfield, VA 22161. (703) 487-4630. Broad coverage of government sponsored research reports, 1964 to present. Inquire as to cost and availability.

SCISEARCH. Institute for Scientific Information, 3501 Market Street, Philadelphia, PA 19104. (800) 523-1850 or (215) 386-0100. Broad multidisciplinary title and author index to the international literature of science and technology, 1974 to present. Inquire as to cost and availability.

PERIODICALS

ACM TRANSACTIONS ON COMPUTER SYSTEMS. Association for Computing Machinery, 11 West 42nd Street, New York, NY 10036. (212) 869-7440. Quarterly. $70.00 per year.

COMMUNICATIONS OF THE ACM. Association for Computing Machinery, 11 West 42nd Street, New York, NY 10036. (212) 869-7440. Monthly. $78.00 per year.

COMPUTER. Institute of Electrical and Electronic Engineers (IEEE), IEEE Service Center, 445 Hoes Lane, Piscataway, NJ 08854. (201) 981-0060. Monthly. $90.00 per year.

COMPUTER MAGAZINE. IEEE Computer Society, 1109 Spring Street, Suite 300, Silver Spring, MD 20910. (301) 589-8142. Monthly. $34.00 per year.

CYBERNETIC. American Society for Cybernetics, Department of Decision Sciences, George Mason University, Fairfax, VA 22030. (703) 323-2738. Quarterly. $30.00 per year.

IEEE TRANSACTIONS ON COMPUTERS. Institute of Electrical and Electronic Engineers (IEEE), IEEE Service Center, 445 Hoes Lane, Piscataway, NJ 08854. (201) 981-0060. Monthly. $130.00 per year.

IEEE TRANSACTIONS ON SYSTEMS, MAN AND CYBERNETICS. Institute of Electrical and Electronic Engineers (IEEE), IEEE Service Center, 445 Hoes Lane, Piscataway, NJ 08854. (201) 981-0060. Bimonthly. $113.00 per year.

INTERNATIONAL JOURNAL OF COMPUTER AND INFORMATION SCIENCE. Plenum Publishing Corporation, 233 Spring Street, New York, NY 10013. (800) 221-9369. Bimonthly. $195.00 per year.

INTERNATIONAL JOURNAL OF MAN-MACHINE STUDIES. Academic Press, Incorporated, 111 Fifth Avenue, New York, NY 10003. Monthly. $337.00 per year.

JOURNAL OF THE ASSOCIATION FOR COMPUTING MACHINERY (ACM). Association for Computing Machinery, 11 West 42nd Street, New York, NY 10036. (212) 869-7440. Quarterly. $60.00 per year.

NUMERICAL ENGINEERING. Numerical Engineering Society. Rochester House, 66 Little Ealing Lane, London W5 4XX, England. Bimonthly. $35.00 per year.

RESEARCH CENTERS AND INSTITUTES

PURDUE UNIVERSITY. Computer Aided Design and Graphics Laboratory, 134 Potter Engineering Center, West Lafayette, IN 47907. (317) 494-5944.

STANFORD UNIVERSITY. Stanford Electronics Laboratories. Stanford, CA 94305. (415) 497-1013.

UNIVERSITY OF TEXAS AT AUSTIN. Center for Cybernetic Studies, CBA 5.202, Austin, TX 78712. (512) 471-1821.

CYCLONES

See also: METEOROLOGY

ABSTRACT SERVICES AND INDEXES

GENERAL SCIENCE INDEX. H.W. Wilson Company, 950 University Avenue, Bronx, NY 10452. (800) 367-6770 or (212) 588-8400. Inquire as to cost and availability.

METEOROLOGICAL AND GEOASTROPHYSICAL ABSTRACTS. American Meteorological Society, 45 Beacon Street, Boston, MA 02108. (617) 227-2425. 1950 to present. Monthly. $450.00 per year.

SCIENCE CITATION INDEX. Institute for Scientific Information, 3501 Market Street, Philadelphia, PA 19104. (800) 523-1850 or (215) 386-0100. Inquire as to cost and availability.

ASSOCIATIONS AND PROFESSIONAL SOCIETIES

AMERICAN ASSOCIATION STATE CLIMATOLOGISTS. c/o Professor John Griffiths, Meteorology Department, O and M Building, Texas A and M University, College Station, TX 77843. (409) 845-7320.

AMERICAN METEOROLOGICAL SOCIETY. 45 Beacon Street, Boston, MA 02108. (617) 227-2425.

INTERNATIONAL ASSOCIATION OF METEOROLOGY AND ATMOSPHERIC PHYSICS. UCAR, Post Office Box 3000, Boulder, CO 80307.

NATIONAL ENVIRONMENTAL SATELLITE DATA, AND INFORMATION SERVICE. 3300 Whitehaven Street, NW, Washington, DC 20235. (202) 634-7318.

NATIONAL WEATHER ASSOCIATION. 4400 Stamp Road, Room 404, Temple Hills, MD 20748. (301) 899-3784.

UNIVERSITY CORPORATION FOR ATMOSPHERIC RESEARCH. Box 3000, 1850 Table Mesa Drive, Boulder, CO 80307. (303) 497-1000.

WEATHER MODIFICATION ASSOCIATION. Post Office Box 8116, Fresno, CA 93747. (209) 291-8466.

DIRECTORIES AND BIOGRAPHICAL SOURCES

AMERICAN MEN AND WOMEN OF SCIENCE. Physical and Biological Sciences. Sixteenth edition. R.R. Bowker Company, 205 East 42nd Street, New York, NY 10017. (800) 521-8110 or (212) 916-1600. 1986. $595.00.

GOVERNMENT RESEARCH DIRECTORY. Gale Research Company, Book Tower, Detroit, MI 48226. (800) 521-0707. Fourth edition. 1987. $350.00.

INTERDOC: DIRECTORY OF PUBLISHED PROCEEDINGS, SERIES. SEMT Science/Engineering/Medicine/Technology. Interdoc Corporation, 173 Halstead Avenue, Box 326, Harrison, NY 10528. (914) 835-3506. Ten times per year. $325.00 per year.

METEOROLOGICAL SERVICES OF THE WORLD. World Meteorological Organization. Available from: American Meteorological Society, 45 Beacon Street, Boston, MA 02108. (617) 227-2425. Annual. $35.00.

NATIONAL WEATHER SERVICE OFFICES AND STATIONS. National Oceanic and Atmospheric Administration, Department of Commerce, Silver Spring, MD 20910. (301) 427-7698. Annual. Free.

RESEARCH CENTERS DIRECTORY. Gale Research Company, Book Tower, Detroit, MI 48226. (800) 521-0707. Twelfth edition. 1987. $365.00 for set.

ENCYCLOPEDIAS AND DICTIONARIES

ENCYCLOPEDIA OF CLIMATOLOGY. John E. Oliver and Rhodes W. Fairbridge, editors. Van Nostrand Reinhold, Incorporated, 115 Fifth Avenue, New York, NY 10003. (800) 543-2681. 1987. $89.95.

ENCYCLOPEDIA OF PHYSICAL SCIENCE AND TECHNOLOGY. Academic Press, Incorporated, Orlando, FL 32887. (800) 321-5068 or (305) 345-2734. Inquire as to cost and availability.

GENERAL WORKS

OPERATIONAL TECHNIQUES FOR FORECASTING TROPICAL CYCLONE INTENSITY AND MOVEMENT. Unipub, 205 East 42nd Street, New York, NY 10017. (212) 916-1659. 1979. $40.00.

ONLINE DATA BASES

DISSERTATION ABSTRACTS ONLINE. University Microfilms International, 300 North Zeeb Road, Ann Arbor, MI 48106. (800) 521-0600 or (313) 761-4700. Scope includes virtually all doctoral dissertations accepted at accredited American institutions from 1861 to present in 252 subject areas. Inquire as to cost and availability.

METEOROLOGICAL AND GEOASTROPHYSICAL ABSTRACTS. American Meteorological Society, 45 Beacon Street, Boston, MA 02108. (617) 227-2425. 1950 to present. Monthly. $450.00 per year.

NTIS. National Technical Information Service, 5285 Port Royal Road, Springfield, VA 22161. (703) 487-4630. Broad coverage of government sponsored research reports, 1964 to present. Inquire as to cost and availability.

SCISEARCH. Institute for Scientific Information, 3501 Market Street, Philadelphia, PA 19104. (800) 523-1850 or (215) 386-0100. Broad multidisciplinary title and author index to the international literature of science and technology, 1974 to present. Inquire as to cost and availability.

PERIODICALS

AGRICULTURAL AND FOREST METEOROLOGY. Elsevier Science Publishing Company, Incorporated, 52 Vanderbilt Avenue, New York, NY 10017. (212) 370-5520. 1964 to present. Monthly. $260.00 per year.

AMERICAN METEOROLOGICAL SOCIETY BULLETIN. American Meteorological Society, 45 Beacon Street, Boston, MA 02108. (617) 227-2425.

CLIMATIC CHANGE: AN INTERDISCIPLINARY, INTERNATIONAL JOURNAL DEVOTED TO THE DESCRIPTION, CAUSES AND IMPLICATIONS OF CLIMATIC CHANGE. D. Reidel Publishing Company, 190 Old Derby Street, Hingham, MA 02043. (617) 871-6600. 1977 to present. Six times per year. $125.00 per year.

JOURNAL OF THE ATMOSPHERIC SCIENCES. American Meteorological Society, 45 Beacon Street, Boston, MA 02108. (617) 227-2425. 1944 to present. Semimonthly. $220.00 per year.

MONTHLY WEATHER REVIEW. American Meteorological Society, 45 Beacon Street, Boston, MA 02108. (617) 227-2425. 1872 to present. Monthly. $120.00 per year.

NATIONAL WEATHER DIGEST. National Weather Association, 4400 Stamp Road, Room 404, Temple Hills, MD 20748. (301) 899-3784. 1976 to present. Quarterly. $20.00 per year.

WEATHER. Royal Meteorological Society, James Glaisher House, Grenville Place, Bracknell Berkshire, RG12 1BX, England. 1946 to present. $30.00 per year.

WEATHERWISE. Heldref Publications, 4000 Albemarle Street, NW, Washington, DC 20016. (202) 362-6445. 1948 to present. Bimonthly. $20.00 per year.

RESEARCH CENTERS AND INSTITUTES

NATIONAL CENTER FOR ATMOSPHERIC RESEARCH. Box 3000, Boulder, CO 80307. (303) 497-1000.

NATIONAL SERVICE STORMS LABORATORY. 1313 Halley Circle, Norman, OK 73069. (405) 360-3620.

UNIVERSITY OF MIAMI. Remote Sensing Laboratory, Post Office Box 248003, Coral Gables, FL 33124. (305) 284-3881.

CYCLOTRON

See: PARTICLE ACCELERATORS

D

DAMS

See also: CHANNELS, CIVIL ENGINEERING, FLOOD CONTROL, GEOTECHNICAL ENGINEERING, HYDRAULIC ENGINEERING, HYDROELECTRIC POWER

ABSTRACT SERVICES AND INDEXES

APPLIED SCIENCE AND TECHNOLOGY INDEX. H.W. Wilson and Company, 950 University Avenue, Bronx, NY 10452. (800) 367-6670 or (212) 588-8400. Monthly. Inquire as to cost and availability.

CURRENT CONTENTS: ENGINEERING, TECHNOLOGY AND APPLIED SCIENCES. Institute for Scientific Information, 3501 Market Street, Philadelphia, PA 19104. (800) 523-1850 or (215) 386-0100. Weekly. $275.00 per year.

ENGINEERING INDEX MONTHLY AND AUTHOR INDEX. Engineering Information Inc., 345 East 47th Street, New York, NY 10017. (212) 705-7600. Monthly. $1560.00 per year.

INDEX TO SCIENTIFIC AND TECHNICAL PROCEEDINGS. Institute for Scientific Information, 3501 Market Street, Philadelphia, PA 19104. (800) 523-1850 or (215) 386-0100. 1978 to present. Monthly. $775.00 per year.

INTERNATIONAL CIVIL ENGINEERING ABSTRACTS. CITIS Limited, 2 Rosemount Terrace, Blackrock, Dublin, Ireland. 1974 to present. Monthly. $350.00 per year.

INTERNATIONAL STRUCTURAL ENGINEERING ABSTRACTS. CITIS Limited, 2 Rosemount Terrace, Blackrock, Dublin, Ireland. 1986 to present. Quarterly. $95.00 per year.

PUBLICATIONS INFORMATION. American Society of Civil Engineers, 345 East 47th Street, New York, NY 10017. (212) 705-7420. Abstracts, subject and author indexes to the publications of the American Society of Civil Engineers. Bimonthly. $80.00 per year.

ASSOCIATIONS AND PROFESSIONAL SOCIETIES

AMERICAN SOCIETY OF CIVIL ENGINEERS. 345 East 47th Street, New York, NY 10017. (212) 705-7420.

UNITED STATES COMMITTEE ON LARGE DAMS OF THE INTERNATIONAL COMMISSION ON LARGE DAMS. C/O C.T. Main Corporation, Prudential Center, Boston, MA 02199. (617) 262-3200.

DIRECTORIES AND BIOGRAPHICAL SOURCES

1987 DIRECTORY OF ENGINEERING SOCIETIES AND RELATED ORGANIZATIONS. Gordon Davis, editor. Hemisphere Publishing Corporation, 1010 Vermont Avenue, NW, Washington, DC 20005. (800) 526-0275. 12th edition. 1987. $100.00.

RESEARCH CENTERS DIRECTORY 1988. Gale Research Company, Book Tower, Detroit, MI 48226. (800) 521-0707. 12th edition. 1987. $365.00 for set.

SCIENTIFIC AND TECHNICAL ORGANIZATIONS AND AGENCIES DIRECTORY. Margaret Labash Young, editor. Gale Research Company, Book Tower, Detroit, MI 48226. (800) 521-0707. 2nd edition. 1987. $185.00.

WHO'S WHO IN ENGINEERING. Gordon Davis, editor. Hemisphere Publishing Corporation, 1010 Vermont Avenue, NW, Washington, DC 20005. (800) 526-0275. 6th edition. 1985. $200.00.

ENCYCLOPEDIAS AND DICTIONARIES

DICTIONARY OF CIVIL ENGINEERING. John S. Scott. John Wiley and Sons, Inc., 605 Third Avenue, New York, NY 10158. (800) 526-5368. Third edition. 1981. $26.95.

GENERAL WORKS

DAMS AND EARTHQUAKES. Institution of Civil Engineers Staff, editors. American Society of Civil Engineers, 345 East 47th Street, New York, NY 10017. (212) 705-7420. 1981. $62.50.

EARTH AND EARTH-ROCK DAMS: ENGINEERING PROBLEMS OF DESIGN AND CONSTRUCTION. James L. Sherard and others. John Wiley and Sons, Inc., 605 Third Avenue, New York, NY 10158. (800) 526-5368. 1963. $76.95.

EARTHQUAKE ENGINEERING FOR LARGE DAMS. Radu Priscu and others. John Wiley and Sons, Inc., 605 Third Avenue, New York, NY 10158. (800) 526-5368. 1985. $44.95.

EVALUATION OF DAM SAFETY. American Society of Civil Engineers, 345 East 47th Street, New York, NY 10017. (212) 705-7420. 1977. $16.00 in paper.

MATERIALS FOR CIVIL AND HIGHWAY ENGINEERS. K.N. Derucher and G.P. Korgiatis. Prentice-Hall Publishing, Inc., Englewood Cliffs, NJ 07632. (800) 562-0245. Second edition. 1988. $42.50.

HANDBOOKS AND MANUALS

CIVIL ENGINEERING CALCULATIONS REFERENCE GUIDE. Tyler G. Hicks, editor. McGraw-Hill Book Company, 1221 Avenue of the Americas, New York, NY 10020. (212) 512-2000. 1987. $29.50.

CIVIL ENGINEERING PRACTICE. Paul N. Cheremisinoff and others, editors. Technomic Publishing Company, Inc., 851 Holland Avenue, Box 3535, Lancaster, PA 17604. (800) 233-9936. Five volumes. 1987-1988. $750.00 for set.

HANDBOOK OF CONCRETE ENGINEERING. Mark Fintel. Van Nostrand Reinhold Company, Inc., 135 West 50th Street, New York, NY 10020. (800) 543-2681. 1986. $89.95.

HANDBOOK OF DAM ENGINEERING. Alfred R. Golze, editor. Van Nostrand Reinhold Company, Inc., 135 West 50th Street, New York, NY 10020. (800) 543-2681. 1977. $69.95.

HANDBOOK OF MECHANICS, MATERIALS, AND STRUCTURES. A. Blake. John Wiley and Sons, Inc., 605 Third Avenue, New York, NY 10158. (800) 526-5368. 1985. $64.50.

STANDARD HANDBOOK FOR CIVIL ENGINEERS. F. S. Merritt, editor. McGraw-Hill Book Company, 1221 Avenue of the Americas, New York, NY 10020. (212) 512-2000. Third edition. 1983. $89.50.

SURVEYING READY-REFERENCE MANUAL. Guy O. Stenstrom. McGraw-Hill Book Company, 1221 Avenue of the Americas, New York, NY 10020. (212) 512-2000. 1987. $26.50.

ONLINE DATA BASES

COMPENDEX. Engineering Information, Inc., 345 East 47th Street, New York, NY 10017. (800) 221-1044 or (212) 705-7615. Engineering and technical literature, 1975 to present. Inquire as to online cost and availability.

GEOREF. Online version of the BIBLIOGRAPHY AND INDEX OF GEOLOGY. American Geological Institute, 4220 King Street, Alexandria, VA 22302. (703) 379-2480. 1969 to present. Inquire as to online cost and availability.

NTIS. National Technical Information Service, 5285 Port Royal Road, Springfield, VA 22161. (703) 487-4630. Broad coverage of government sponsored research reports, 1964 to present. Inquire as to online cost and availability.

WILSONLINE. H.W. Wilson and Company, 950 University Avenue, Bronx, NY 10452. (800) 367-6770 or (212) 588-8400. Makes available online versions of the H.W. Wilson indexes including Applied Science and Technology Index, Business Periodicals Index and Readers' Guide to Periodical Literature. Approximately 1980 to present. Inquire as to online cost and availability.

PERIODICALS

CIVIL ENGINEERING. American Society of Civil Engineers, 345 East 47th Street, New York, NY 10017. (212) 705-7420. 1930 to present. Monthly. $48.00 per year.

CONCRETE. Harcourt, Brace Jovanovich, Inc., 7500 Old Oak Boulevard, Cleveland, OH 44130. 1937 to present. Monthly. $25.00 per year.

HIGHWAY AND HEAVY CONSTRUCTION. Technical Publishing Company, 875 Third Avenue, New York, NY 10022. (212) 605-9400. 1892 to present. Monthly. $35.00 per year.

JOURNAL OF CONSTRUCTION ENGINEERING AND MANAGEMENT. American Society of Civil Engineers, 345 East 47th Street, New York, NY 10017. (212) 705-7420. 1956 to present. Quarterly. $48.00 per year.

JOURNAL OF ENGINEERING MECHANICS. American Society of Civil Engineers, 345 East 47th Street, New York, NY 10017. (212) 705-7420. 1956 to present. Monthly. $120.00 per year.

JOURNAL OF GEOTECHNICAL ENGINEERING. American Society of Civil Engineers, 345 East 47th Street, New York, NY 10017. (212) 705-7420. 1956 to present. Monthly. $96.00 per year.

JOURNAL OF HYDRAULIC ENGINEERING. American Society of Civil Engineers, 345 East 47th Street, New York, NY 10017. (212) 705-7420. 1956 to present. Monthly. $112.00 per year.

JOURNAL OF IRRIGATION AND DRAINAGE. American Society of Civil Engineers, 345 East 47th Street, New York, NY 10017. (212) 705-7420. 1956 to present. Quarterly. $45.00 per year.

JOURNAL OF STRUCTURAL ENGINEERING. American Society of Civil Engineers, 345 East 47th Street, New York, NY 10017. (212) 705-7420. 1956 to present. Monthly. $140.00 per year.

JOURNAL OF STRUCTURAL MECHANICS. Marcel Dekker Inc., 270 Madison Avenue, New York, NY 10016. (800) 228-1160. 1972 to present. Quarterly. $75.00 per year.

JOURNAL OF SURVEYING ENGINEERING. American Society of Civil Engineers, 345 East 47th Street, New York, NY 10017. (212) 705-7420. 1956 to present. Three times per year. $35.00 per year.

JOURNAL OF WATER RESOURCES PLANNING AND MANAGEMENT. American Society of Civil Engineers, 345 East 47th Street, New York, NY 10017. (212) 705-7420. 1956 to present. Quarterly. $56.00 per year.

STRUCTURAL ENGINEERING PRACTICE: ANALYSIS, DESIGN, MANAGEMENT. Marcel Dekker Inc., 270 Madison Avenue, New York, NY 10016. (800) 228-1160. 1981 to present. Quarterly. $75.00 per year.

DARK MATTER

See also: ASTRONOMY, ASTROPHYSICS, COSMOLOGY, INTERSTELLAR MATTER

ABSTRACT SERVICES AND INDEXES

ASTRONOMY AND ASTROPHYSICS ABSTRACTS. Springer-Verlag New York, Incorporated, 175 Fifth Avenue, New York, NY 10010. (212) 460-1500. $70.00 per year.

ASTRONOMY AND ASTROPHYSICS MONTHLY INDEX. Olivetree Associates, Post Office Box 236, Sierra Madre, CA 91024. $212.00 per year. Complimentary copies available on request.

PHYSICS ABSTRACTS. Institute of Electrical Engineers, London, United Kingdom. Available from: Institute of Electrical and Electronic Engineers (IEEE), 345 East 47th Street, New York, NY 10017. (212) 705-7900. 1898 to present. Monthly. $1670.00 per year.

SCIENCE CITATION INDEX. Institute for Scientific Information, 3501 Market Street, Philadelphia, PA 19104. (800) 523-1850 or (215) 386-0100. Inquire as to cost and availability.

STAR. (Scientific and Technical Aerospace Reports. United States National Aeronautics and Space Administration, Scientific and Technical Information Facility, Box 8757, Baltimore-Washington International Airport, MD 21240. (202) 755-2210. Semimonthly, with semiannual and annual indexes. $85.00 per year.

ANNUAL REVIEWS AND YEARBOOKS

ANNUAL REVIEW OF ASTRONOMY AND ASTROPHYSICS. Annual Reviews, Incorporated, 4139 El Camino Way, Palo Alto, CA 94306. (415) 493-4400.

ASSOCIATIONS AND PROFESSIONAL SOCIETIES

AMERICAN ASTRONOMICAL SOCIETY. 2000 Florida Avenue, NW, Suite 300, Washington, DC 20009. (202) 659-0134.

AMERICAN ASSOCIATION OF VARIABLE STAR OBSERVERS. 187 Concord Avenue, Cambridge, MA 02138. (617) 354-0484.

ASTRONOMICAL LEAGUE. Post Office Box 12821, Tucson, AZ 85732. (602) 790-8471.

ASTRONOMICAL SOCIETY OF THE PACIFIC. 1290 24th Avenue, San Francisco, CA 94122. (415) 661-8660.

DIRECTORIES AND BIOGRAPHICAL SOURCES

AMERICAN MEN AND WOMEN OF SCIENCE. Physical and Biological Sciences. Sixteenth edition. R.R. Bowker Company, 205 East 42nd Street, New York, NY 10017. (800) 521-8110 or (212) 916-1600. 1986. $595.00.

THE BIOGRAPHICAL DICTIONARY OF SCIENTISTS: ASTRONOMERS. D. Abbott, editor. Peter Bedrick Books, 125 East 23rd Street, New York, NY 10010. 1984.

DIRECTORY OF PHYSICS AND ASTRONOMY STAFF MEMBERS. American Institute of Physics, 335 East 45th Street, New York, NY 10017. Annual.

RESEARCH CENTERS DIRECTORY. Gale Research Company, Book Tower, Detroit, MI 48226. (800) 521-0707. Twelfth edition. 1987. $365.00 for set.

WHO'S WHO IN FRONTIER SCIENCE AND TECHNOLOGY. Marquis Who's Who, Incorporated, 200 East Ohio Street, Chicago, IL 60611. (800) 428-3898 or (312) 787-2008.

GENERAL WORKS

THE CAMBRIDGE ASTRONOMY GUIDE: A PRACTICAL INTRODUCTION TO ASTRONOMY. Cambridge University Press, 32 East 57th Street, New York, NY 10022. (212) 688-8888. 1985. $24.95.

DARK MATTER IN THE UNIVERSE. J. Kormendy and G.R. Knapp, editors. D. Reidel, 101 Philip Drive, Assinippi Park, Norwell, MA 02061. (617) 871-6600. 1987. $104.00.

THE INTERACTION OF SUPERNOVA REMNANTS WITH THE INTERSTELLAR MEDIUM. R.S. Roger and T.L. Landecker, editors. Cambridge University Press, 32 East 57th Street, New York, NY 10022. (212) 688-8888. 1988. $59.95.

INTERSTELLAR CHEMISTRY. Walter Duley and David A. Williams. Academic Press, Incorporated, 6277 Sea Harbor Drive, Orlando, FL 32821. (800) 321-5068. 1984. $49.50.

NEW IDEAS IN ASTRONOMY. Barry F. Madore, editor. Cambridge University Press, 32 East Street, New York, NY 10022. (212) 688-8888. $49.95.

STARS, NEBULAE, AND THE INTERSTELLAR MEDIUM. C.R. Kitchen. Boston: Adam Hilger. 1987. $33.00.

ONLINE DATA BASES

CA SEARCH. Chemical Abstracts Service, Post Office Box 3012, Columbus, OH 43210. Guide to chemical literature, 1972 to present. Inquire as to cost and availability.

DISSERTATION ABSTRACTS ONLINE. University Microfilms International, 300 North Zeeb Road, Ann Arbor, MI 48106. (800) 521-0600 or (313) 761-4700. Scope includes virtually all doctoral dissertations accepted at accredited American institutions from 1861 to present in 252 subject areas. Inquire as to cost and availability.

INSPEC. INSPEC Marketing Department, Institute of Electrical and Electronics Engineers, Incorporated, IEEE Service Department, 445 Hoes Lane, Piscataway, NJ 08854. (201) 981-0060. Inquire as to on-line cost and availability.

NASA. National Aeronautics and Space Administration, Scientific and Technical Information Branch, 300 7th Street, SW, Washington, DC 20546. Citations and abstracts of aerospace literature, 1962 to present. Inquire as to cost and availability.

NTIS. National Technical Information Service, 5285 Port Royal Road, Springfield, VA 22161. (703) 487-4630. Broad coverage of government sponsored research reports, 1964 to present. Inquire as to cost and availability.

SCISEARCH. Institute for Scientific Information, 3501 Market Street, Philadelphia, PA 19104. (800) 523-1850 or (215) 386-0100. Broad multidisciplinary title and author index to the international literature of science and technology, 1974 to present. Inquire as to cost and availability.

OTHER SOURCES

ATLAS OF DEEP-SKY SPLENDORS. H. Vehrenberg. Cambridge University Press, 32 East 57th Street, New York, NY 10022. (800) 431-1580 or (212) 688-8888. Fourth edition, 1984. $47.50.

SKY CATALOGUE 2000.0. A. Hirshfeld and R. Sinnott. Cambridge University Press, 32 East 57th Street, New York, NY 10022. (800) 431-1580 or (212) 688-8888.

PERIODICALS

ASTRONOMICAL JOURNAL. American Astronomical Society. Available from: American Institute of Physics, 335 East 45th Street, New York, NY 10017. (212) 661-9404. $125.00 per year.

ASTRONOMICAL SOCIETY OF THE PACIFIC PUBLICATIONS. Astronomical Society of the Pacific, 1290 24th Avenue, San Francisco, CA 94122. (415) 661-8660. Monthly. $38.00.

ASTRONOMY. Astro Media Corporation, 625 East Paul Avenue, Milwaukee, WI 53202. Monthly. $18.00 per year.

ASTRONOMY AND ASTROPHYSICS. Springer-Verlag New York, Incorporated, 175 Fifth Avenue, New York, NY 10010. (800) 526-7254 or (212) 460-1500. $680.00 per year.

ASTROPHYSICAL JOURNAL. American Astronomical Society, University of Chicago Press, 5801 Ellis Avenue, Chicago, IL 60637. Biweekly. $305.00 per year.

ASTROPHYSICS AND SPACE SCIENCE. D. Reidel Publishing Company, 190 Old Derby Street, Hingham, MA 02043. Monthly. $101.00 per year.

CELESTIAL MECHANICS: AN INTERNATIONAL JOURNAL OF SPACE DYNAMICS. D. Reidel Publishing Company, 190 Old Derby Street, Hingham, MA 02043. Monthly. $310.00 per year.

ICARUS: INTERNATIONAL JOURNAL OF THE SOLAR SYSTEM STUDIES. Academic Press, Incorporated, Orlando, FL 32887. (305) 345-4100. Monthly. $484.00 per year.

MERCURY. Astronomical Society of the Pacific, 1290 245h Avenue, San Francisco, CA 94122. (415) 661-8660. Bimonthly. $21.00 per year.

MONTHLY NOTICES OF THE ROYAL ASTRONOMICAL SOCIETY. Blackwell Science Publications, Incorporated, 667 Lytton Avenue, Palo Alto, CA 94301. (415) 324-1688. Monthly. $134.00 per year.

PLANETARY AND SPACE SCIENCE. Pergamon Press, Incorporated, Maxwell House, Fairview Park, Elmsford, NY 10523. (014) 592-7700. Monthly. $430.00 per year.

SKY AND TELESCOPE. Sky Publishing Corporation, 49 Bay State Road, Cambridge, MA 02238. (617) 864-7360. Monthly. $18.00 per year.

SOVIET ASTRONOMY (TRANSLATION OF ASTRO-NOMICHESKII ZHURNAL). American Institute of Physics, 335 East 45th Street, New York, NY 10017. (212) 661-9404. Bimonthly. $425.00 per year.

SPACE SCIENCE REVIEWS. D. Reidel Publishing Company, 190 Old Derby Street, Hingham, MA 02043. Monthly. $305.00 per year.

VISTAS IN ASTRONOMY. Pergamon Press, Incorporated, Maxwell House, Fairview Park, Elmsford, NY 10523. (914) 592-7700. Quarterly. $145.00 per year.

RESEARCH CENTERS AND INSTITUTES

CALIFORNIA INSTITUTE OF TECHNOLOGY. Palomar Observatory, 105-24, Pasadena, CA 91125. (818) 356-4033.

NATIONAL ASTRONOMY AND IONOSPHERE CENTER. Cornell University, Space Sciences Building, Ithaca, NY 14853

HARVARD-SMITHSONIAN CENTER FOR ASTROPHYSICS. 60 Garden Street, Cambridge, MA 02138. (617) 495-7461.

NATIONAL OPTICAL ASTRONOMY OBSERVATORIES. 1002 North Warren Avenue, Tucson, AZ 85719. (602) 325-9230.

NATIONAL RADIO ASTRONOMY OBSERVATORY. Edgemont Road, Charlottesville, VA 22903. (804) 296-0211.

SPACE TELESCOPE SCIENCE INSTITUTE. 3700 San Martin Drive, Baltimore, MD 21218. (301) 338-4700.

UNIVERSITY OF MARYLAND. Astronomy Program, College Park, MD 20742. (301) 454-3005.

DATA BASE MANAGEMENT

See: DATA BASES

DATA BASES

See also: COMPUTER COMMUNICATIONS, COMPUTER PROGRAMMING, COMPUTER SECURITY, COMPUTERS, DATA PROCESSING, PROGRAMMING LANGUAGES, SOFTWARE, SOFTWARE ENGINEERING, SYSTEMS ANALYSIS, SYSTEMS ENGINEERING

ABSTRACT SERVICES AND INDEXES

APPLIED SCIENCE AND TECHNOLOGY INDEX. H.W. Wilson and Company, 950 University Avenue, Bronx, NY 10452. (800) 367-6670 or (212) 588-8400. Monthly. Inquire as to cost and availability.

COMPUTER AND CONTROL ABSTRACTS. Institute of Electrical Engineers. Available from: Institute of Electrical and Electronics Engineers. IEEE Service Center, 445 Hoes Lane, Piscataway, NJ 08854. Semimonthly. $775.00 per year.

COMPUTER AND INFORMATION SYSTEMS: AN ABSTRACT JOURNAL PERTAINING TO THE THEORY, DESIGN, FABRICATION AND APPLICATION OF COMPUTER AND INFORMATION SYSTEMS. Cambridge Scientific Abstracts, 5161 River Road, Bethesda, MD 20816. 1972 to present. Semi-monthly. Inquire as to cost and availability.

COMPUTER CONTENTS: THE BIWEEKLY COMPILATION OF TABLES OF CONTENTS FROM COMPUTER, ELECTRONIC AND TELECOMMUNICATIONS MAGAZINES, JOURNALS AND TRANACTIONS. Find/SVP, 500 Fifth Avenue, New York, NY 101110. (800) 346-3787 or (212) 354-2424. Biweekly. $115.00 per year.

COMPUTER LITERATURE INDEX. Applied Computer Research, Inc., P.O. Box 9280, Phoenix, AZ 85068. (602) 995-5929. Quarterly. $125.00 per year.

COMPUTER PROGRAMS ABSTRACTS. U.S. National Aeronautics and Space Administration. Available from: U.S. Government Printing Office, Washington, DC 20402. Quarterly. $10.00 per year.

COMPUTING REVIEWS. Association of Computing Machinery, 11 West 42nd Street, New York, NY 10036. (212) 869-7440. Monthly. $60.00 per year.

ENGINEERING INDEX MONTHLY AND AUTHOR INDEX. Engineering Information Inc., 345 East 47th Street, New York, NY 10017. (212) 705-7600. Monthly. $1560.00 per year.

PCR-2: PERSONAL COMPUTER REVIEW - SQUARED. Toolbox Publications, Inc., P.O. Box 5451, 2514 Birch Creek Lane, Orchard Lake, MI 48033. 1987 to present. Bimonthly. $60.00 per year.

SCIENCE CITATION INDEX. Institute for Scientific Information, 3501 Market Street, Philadelphia, PA 19104. (800) 523-1850 or (215) 386-0100. Six times per year. $6200.00 per year.

ANNUAL REVIEWS AND YEARBOOKS

ADVANCES IN COMPUTERS. Academic Press, Inc., 6277 Sea Harbor Drive, Orlando, FL 32821. (800) 321-5068. Yearly. Approximately $50.00 per volume.

COMPUTER PUBLISHERS AND PUBLICATIONS 1988-89: AN INTERNATIONAL DIRECTORY AND YEARBOOK. Efrem Sigel and Frederica Evan, editors. Gale Research Company, Book Tower, Detroit, MI 48226. (800) 521-0707. Third edition. $140.00.

ASSOCIATIONS AND PROFESSIONAL SOCIETIES

AMERICAN FEDERATION OF INFORMATION PROCESSING SOCIETIES. 1899 Preston White Drive, Reston, VA 22091. (703) 620-8900.

ASSOCIATION OF COMPUTER PROGRAMMERS AND ANALYSTS. 2108-C Gallows Road, Vienna, VA 22180. (703) 790-0490.

ASSOCIATION OF COMPUTING MACHINERY (ACM). 11 West 42nd Street, New York, NY 10036. (212) 869-7440.

IEEE COMPUTER SOCIETY. 1730 Massachusetts Avenue, N.W., Washington, DC 20036. (202) 371-0101.

INSTITUTE OF ELECTRICAL AND ELECTRONICS ENGINEERS. IEEE Service Center, 445 Hoes Lane, Piscataway, NJ 08854.

MACHINE VISION ASSOCIATION. P.O. Box 930, One SME Drive, Dearborn, MI 48121. (313) 271-1500.

SOCIETY FOR COMPUTER SIMULATION. P.O. Box 17900, San Diego, CA 92117. (619) 277-3888.

SOCIETY FOR INFORMATION DISPLAY. 8055 Manchester Avenue, Suite 615, Playa Del Rey, CA 90293. (213) 305-1502.

DIRECTORIES AND BIOGRAPHICAL SOURCES

AMERICAN SOCIETY FOR INFORMATION SCIENCE HANDBOOK AND DIRECTORY. American Society for Information Science, 1424 16th Street, N.W., Suite 404, Washington, DC 20036. (202) 462-1000. $50.00.

COMPUTERS AND COMPUTING INFORMATION RESOURCES DIRECTORY. Martin Connors, editor. Gale Research Company, Book Tower, Detroit, MI 48226. (800) 521-0707. 1987. $165.00. Supplement available at $85.00.

INTERNATIONAL RESEARCH CENTERS DIRECTORY 1988-89. Darren L. Smith, editor. Gale Research Company, Book Tower, Detroit, MI 48226. (800) 521-0707. 4th edition. 1987. $360.00.

1987 DIRECTORY OF ENGINEERING SOCIETIES AND RELATED ORGANIZATIONS. Gordon Davis, editor. Hemisphere Publishing Corporation, 1010 Vermont Avenue, NW, Washington, DC 20005. (800) 526-0275. 12th edition. 1987. $100.00.

RESEARCH CENTERS DIRECTORY 1988. Gale Research Company, Book Tower, Detroit, MI 48226. (800) 521-0707. 12th edition. 1987. $365.00 for set.

SCIENTIFIC AND TECHNICAL ORGANIZATIONS AND AGENCIES DIRECTORY. Margaret Labash Young, editor. Gale Research Company, Book Tower, Detroit, MI 48226. (800) 521-0707. 2nd edition. 1987. $185.00.

WHO'S WHO IN ENGINEERING. Gordon Davis, editor. Hemisphere Publishing Corporation, 1010 Vermont Avenue, NW, Washington, DC 20005. (800) 526-0275. 6th edition. 1985. $200.00.

ENCYCLOPEDIAS AND DICTIONARIES

COMPUTER AND TELECOMMUNICATIONS ACRONYMS. Julie E. Towell and Helen E. Sheppard, editors. Gale Research Company, Book Tower, Detroit, MI 48226. (800) 521-0707. 1986. $60.00.

DICTIONARY OF COMPUTING. Oxford University Press, 200 Madison Avenue, New York, NY 10016. (800) 458-5833. Second edition. 1986. $29.95.

ENCYCLOPEDIA OF INFORMATION SYSTEMS AND SERVICES 1988. Amy Lucas and Annette Novallo, editors. Gale Research Company, Book Tower, Detroit, MI 48226. (800) 521-0707. 8th edition. 1987. $400.00 for set.

PRENTICE-HALL ENCYCLOPEDIA OF INFORMATION TECHNOLOGY. Robert A. Edmunds. Prentice-Hall Publishing, Inc., Englewood Cliffs, NJ 07632. (800) 562-0245. 1987. $49.95.

SOFTWARE ENCYCLOPEDIA. R.R. Bowker Company, 205 East 42nd Street, New York, NY 10017. (800) 521-8110. Tow volumes. 1987. $125.00 for set.

GENERAL WORKS

DATA BASE MANAGEMENT SYSTEMS. P.B. Seybold and H. Uhler. McGraw-Hill Book Company, 1221 Avenue of the Americas, New York, NY 10020. (212) 512-2000. 1985. $19.95 in paper.

DATA COMPRESSION TECHNIQUES AND APPLICATIONS. T.J. Lynch. Van Nostrand Reinhold Company, Inc., 135 West 50th Street, New York, NY 10020. (800) 543-2681. 1985. $45.95.

DATA COMPRESSION: TECHNIQUES AND APPLICATIONS; HARDWARE AND SOFTWARE CONSIDERATIONS. G. Held. John Wiley and Sons, Inc., 605 Third Avenue, New York, NY 10158. (800) 526-5368. 1983. $34.95.

EFFECTIVE DATABASE MANAGEMENT. A. Gaydasch. Prentice-Hall Publishing, Inc., Englewood Cliffs, NJ 07632. (800) 562-0245. 1987. $37.95.

HIGH LEVEL LANGUAGE AND SOFTWARE APPLICATIONS REFERENCE. W.J. Birnes. McGraw-Hill Book Company, 1221 Avenue of the Americas, New York, NY 10020. (212) 512-2000. 1988. $29.95.

INTRODUCING RELATIONAL DATABASES. A. Mayne and M. Wood. John Wiley and Sons, Inc., 605 Third Avenue, New York, NY 10158. (800) 526-5368. 1983. $21.95 in paper.

RELATIONAL DATABASE DESIGN WITH MICROCOMPUTER APPLICATIONS. Glenn L. Jackson. Prentice-Hall Publishing, Inc., Englewood Cliffs, NJ 07632. (800) 562-0245. 1987. $20.00.

HANDBOOKS AND MANUALS

HANDBOOK OF SOFTWARE ENGINEERING. Charles R. Vick, editor. Van Nostrand Reinhold Company, Inc., 135 West 50th Street, New York, NY 10020. (800) 543-2681. 1984. $66.95.

ONLINE DATA BASES

COMPENDEX. Engineering Information, Inc., 345 East 47th Street, New York, NY 10017. (800) 221-1044 or (212) 705-7615. Engineering and technical literature, 1975 to present. Inquire as to online cost and availability.

DISSERTATION ABSTRACTS ONLINE. University Microfilms International, 300 North Zeeb Road, Ann Arbor, MI 48106. (800) 521-0600 or (313) 761-4700. Scope includes virtually all doctoral dissertations accepted at accredited American institutions from 1861 to present in over 250 subject areas. Inquire as to online cost and availability.

NTIS. National Technical Information Service, 5285 Port Royal Road, Springfield, VA 22161. (703) 487-4630. Broad coverage of government sponsored research reports, 1964 to present. Inquire as to online cost and availability.

SCISEARCH. Institute for Scientific Information, 3501 Market Street, Philadelphia, PA 19104. (800) 523-1850 or (215) 386-0100. Broad multidisciplinary title and author index to the international literature of science and technology, 1974 to present. Inquire as to online cost and availability.

WILSONLINE. H.W. Wilson and Company, 950 University Avenue, Bronx, NY 10452. (800) 367-6770 or (212) 588-8400. Makes available online versions of the H.W. Wilson indexes including Applied Science and Technology Index, Business Periodicals Index and Readers' Guide to Periodical Literature. Approximately 1980 to present. Inquire as to online cost and availability.

PERIODICALS

ADVANCES IN ENGINEERING SOFTWARE. CML Publications, 400 West Cummings Park, Suite 6200, Woburn, MA 01801. (617) 933-7374. 1979 to present. Quarterly. $130.00 per year.

BYTE. Byte Publications, Inc., 70 Main Street, Petersborough, NH 03458. (603) 924-9281. Monthly. $21.00 per year.

COMMUNICATIONS OF THE ACM. Association of Computing Machinery, 11 West 42nd Street, New York, NY 10036. (212) 869-7440. Monthly. $80.00 per year.

COMPUTER LANGUAGES. Pergamon Press, Inc., Maxwell House, Fairview Park, Elmsford, NY 10523. (914) 592-7700. 1976 to present. Quarterly. $195.00 per year.

DR. DOBB'S JOURNAL OF SOFTWARE TOOLS. M & T Publishing, Inc., 2464 Embarcadero Way, Palo Alto, CA 94303. (415) 424-0600. 1976 to present. Monthly. $25.00 per year.

IEEE SOFTWARE. Institution of Electrical and Electronics Engineers. IEEE Service Center, 445 Hoes Lane, Piscataway, NJ 08854. 1984 to present. Quarterly. $15.00 per issue.

IEEE TRANSACTIONS ON SOFTWARE ENGINEERING. Institute of Electrical and Electronics Engineers. IEEE Service Center, 445 Hoes Lane, Piscataway, NJ 08854. 1975 to present. Monthly. $160.00 per year.

JOURNAL OF SYSTEMS AND SOFTWARE. Elsevier Science Publishing Company, Inc., 52 Vanderbilt Avenue, New York, NY 10017. (212) 370-5520. 1979 to present. Quarterly. $95.00 per year.

SIGSOFT SOFTWARE ENGINEERING NOTICES. Association of Computing Machinery Special Interest Group on Software Engineering. 11 West 42nd Street, New York, NY 10036. (212) 869-7440. Quarterly. $12.00 per year.

SOFTWARE DEVELOPER'S MONTHLY. SourceView Press, 835 Castro Street, Martinez, CA 94553. (415) 228-6220. 1985 to present. Monthly. $144.00 per year.

SOFTWARE ENGINEERING JOURNAL. Institute of Electrical Engineers, Savoy Place, London, WC2R OBL, England. 1981 to present. Bimonthly. $85.00 per year.

SOFTWARE PRACTICE AND EXPERIENCE. John Wiley and Sons, Inc., 605 Third Avenue, New York, NY 10158. (800) 526-5368. 1971 to present. Monthly. $260.00 per year.

UNIX REVIEW (SAN FRANCISCO). Miller Freeman Publications, Inc., 500 Howard Street, San Francisco, CA 94105. (415) 397-1881. 1983 to present. Monthly. $35.00 per year.

RESEARCH CENTERS AND INSTITUTES

DATABASE SYSTEMS RESEARCH AND DEVELOPMENT CENTER. University of Florida, 512 Weil Hall, Gainesville, FL 32611. (904) 392-2371.

DATABASE SYSTEMS RESEARCH CENTER. University of Maryland, College of Business and Management, Tydings Hall, College Park, MD 20742. (301) 454-6258.

INSTITUTE FOR COMPUTER RESEARCH. University of Waterloo, 200 University Avenue, Waterloo, ON, Canada N2L 3G1. (519) 885-1211.

DATA COMPRESSION

See: DATA BASES

DATA DISPLAY

See: TERMINALS

DATA ENTRY TERMINAL

See: DATA PROCESSING

DATA PROCESSING

See also: COMPUTER COMMUNICATIONS, COMPUTER OPERATING SYSTEMS, COMPUTERS, DATA BASES, PROGRAMMING LANGUAGES, SOFTWARE, SOFTWARE ENGINEERING, SYSTEMS ANALYSIS, SYSTEMS ENGINEERING

ABSTRACT SERVICES AND INDEXES

APPLIED SCIENCE AND TECHNOLOGY INDEX. H.W. Wilson and Company, 950 University Avenue, Bronx, NY 10452. (800) 367-6670 or (212) 588-8400. Monthly. Inquire as to cost and availability.

COMPUTER AND CONTROL ABSTRACTS. Institute of Electrical Engineers. Available from: Institute of Electrical and Electronics Engineers. IEEE Service Center, 445 Hoes Lane, Piscataway, NJ 08854. Semimonthly. $775.00 per year.

COMPUTER AND INFORMATION SYSTEMS: AN ABSTRACT JOURNAL PERTAINING TO THE THEORY, DESIGN, FABRICATION AND APPLICATION OF COMPUTER AND INFORMATION SYSTEMS. Cambridge Scientific Abstracts, 5161 River Road, Bethesda, MD 20816. 1972 to present. Semi-monthly. Inquire as to cost and availability.

COMPUTER CONTENTS: THE BIWEEKLY COMPILATION OF TABLES OF CONTENTS FROM COMPUTER, ELECTRONIC AND TELECOMMUNICATIONS MAGAZINES, JOURNALS AND TRANACTIONS. Find/SVP, 500 Fifth Avenue, New York, NY 101110. (800) 346-3787 or (212) 354-2424. Biweekly. $115.00 per year.

COMPUTER LITERATURE INDEX. Applied Computer Research, Inc., P.O. Box 9280, Phoenix, AZ 85068. (602) 995-5929. Quarterly. $125.00 per year.

COMPUTER PROGRAMS ABSTRACTS. U.S. National Aeronautics and Space Administration. Available from: U.S. Government Printing Office, Washington, DC 20402. Quarterly. $10.00 per year.

COMPUTING REVIEWS. Association of Computing Machinery, 11 West 42nd Street, New York, NY 10036. (212) 869-7440. Monthly. $60.00 per year.

CONFERENCE PAPERS INDEX. Cambridge Scientific Abstracts, 5161 River Road, Bethesda, MD 20816. 1972 to present. Monthly. Inquire as to cost and availability.

CURRENT CONTENTS: ENGINEERING, TECHNOLOGY AND APPLIED SCIENCES. Institute for Scientific Information, 3501 Market Street, Philadelphia, PA 19104. (800) 523-1850 or (215) 386-0100. Weekly. $275.00 per year.

ENGINEERING INDEX MONTHLY AND AUTHOR INDEX. Engineering Information Inc., 345 East 47th Street, New York, NY 10017. (212) 705-7600. Monthly. $1560.00 per year.

PCR-2: PERSONAL COMPUTER REVIEW - SQUARED. Toolbox Publications, Inc., P.O. Box 5451, 2514 Birch Creek Lane, Orchard Lake, MI 48033. 1987 to present. Bimonthly. $60.00 per year.

SCIENCE CITATION INDEX. Institute for Scientific Information, 3501 Market Street, Philadelphia, PA 19104. (800) 523-1850 or (215) 386-0100. Six times per year. $6200. 00 per year.

ANNUAL REVIEWS AND YEARBOOKS

ADVANCES IN COMPUTERS. Academic Press, Inc., 6277 Sea Harbor Drive, Orlando, FL 32821. (800) 321-5068. Yearly. Approximately $50.00 per volume.

COMPUTER PUBLISHERS AND PUBLICATIONS 1988-89: AN INTERNATIONAL DIRECTORY AND YEARBOOK. Efrem Sigel and Frederica Evan, editors. Gale Research Company, Book Tower, Detroit, MI 48226. (800) 521-0707. Third edition. $140.00.

ASSOCIATIONS AND PROFESSIONAL SOCIETIES

AMERICAN FEDERATION OF INFORMATION PROCESSING SOCIETIES. 1899 Preston White Drive, Reston, VA 22091. (703) 620-8900.

ASSOCIATION OF COMPUTER PROGRAMMERS AND ANALYSTS. 2108-C Gallows Road, Vienna, VA 22180. (703) 790-0490.

ASSOCIATION OF COMPUTING MACHINERY (ACM). 11 West 42nd Street, New York, NY 10036. (212) 869-7440.

IEEE COMPUTER SOCIETY. 1730 Massachusetts Avenue, N.W., Washington, DC 20036. (202) 371-0101.

INSTITUTE OF ELECTRICAL AND ELECTRONICS ENGINEERS. IEEE Service Center, 445 Hoes Lane, Piscataway, NJ 08854.

MACHINE VISION ASSOCIATION. P.O. Box 930, One SME Drive, Dearborn, MI 48121. (313) 271-1500.

SOCIETY FOR COMPUTER SIMULATION. P.O. Box 17900, San Diego, CA 92117. (619) 277-3888.

SOCIETY FOR INFORMATION DISPLAY. 8055 Manchester Avenue, Suite 615, Playa Del Rey, CA 90293. (213) 305-1502.

DIRECTORIES AND BIOGRAPHICAL SOURCES

AMERICAN MEN AND WOMEN OF SCIENCE. R.R. Bowker, Inc., Order Department, 245 West 17th Street, New York, NY

10011. (800) 521-8110. Eight volumes. 1986. $595.00 for set.

AMERICAN SOCIETY FOR INFORMATION SCIENCE HANDBOOK AND DIRECTORY. American Society for Information Science, 1424 16th Street, N.W., Suite 404, Washington, DC 20036. (202) 462-1000. $50.00.

COMPUTERS AND COMPUTING INFORMATION RESOURCES DIRECTORY. Martin Connors, editor. Gale Research Company, Book Tower, Detroit, MI 48226. (800) 521-0707. 1987. $165.00. Supplement available at $85.00.

INTERNATIONAL RESEARCH CENTERS DIRECTORY 1988-89. Darren L. Smith, editor. Gale Research Company, Book Tower, Detroit, MI 48226. (800) 521-0707. 4th edition. 1987. $360.00.

1987 DIRECTORY OF ENGINEERING SOCIETIES AND RELATED ORGANIZATIONS. Gordon Davis, editor. Hemisphere Publishing Corporation, 1010 Vermont Avenue, NW, Washington, DC 20005. (800) 526-0275. 12th edition. 1987. $100.00.

RESEARCH CENTERS DIRECTORY 1988. Gale Research Company, Book Tower, Detroit, MI 48226. (800) 521-0707. 12th edition. 1987. $365.00 for set.

SCIENTIFIC AND TECHNICAL ORGANIZATIONS AND AGENCIES DIRECTORY. Margaret Labash Young, editor. Gale Research Company, Book Tower, Detroit, MI 48226. (800) 521-0707. 2nd edition. 1987. $185.00.

WHO'S WHO IN ENGINEERING. Gordon Davis, editor. Hemisphere Publishing Corporation, 1010 Vermont Avenue, NW, Washington, DC 20005. (800) 526-0275. 6th edition. 1985. $200.00.

ENCYCLOPEDIAS AND DICTIONARIES

COMPUTER AND TELECOMMUNICATIONS ACRONYMS. Julie E. Towell and Helen E. Sheppard, editors. Gale Research Company, Book Tower, Detroit, MI 48226. (800) 521-0707. 1986. $60.00.

DICTIONARY OF COMPUTING. Oxford University Press, 200 Madison Avenue, New York, NY 10016. (800) 458-5833. Second edition. 1986. $29.95.

ENCYCLOPEDIA OF INFORMATION SYSTEMS AND SERVICES 1988. Amy Lucas and Annette Novallo, editors. Gale Research Company, Book Tower, Detroit, MI 48226. (800) 521-0707. 8th edition. 1987. $400.00 for set.

PRENTICE-HALL ENCYCLOPEDIA OF INFORMATION TECHNOLOGY. Robert A. Edmunds. Prentice-Hall Publishing, Inc., Englewood Cliffs, NJ 07632. (800) 562-0245. 1987. $49.95.

SOFTWARE ENCYCLOPEDIA. R.R. Bowker Company, 205 East 42nd Street, New York, NY 10017. (800) 521-8110. Tow volumes. 1987. $125.00 for set.

GENERAL WORKS

GENERAL PURPOSE PROGRAMMING LANGUAGES. John V. Cugini. Petrocelli Books, available from: Van Nostrand Reinhold Company, Inc., 135 West 50th Street, New York, NY 10020. (800) 543-2681. 1986. $24.95.

HIGH LEVEL LANGUAGE AND SOFTWARE APPLICATIONS REFERENCE. W.J. Birnes. McGraw-Hill Book Company, 1221 Avenue of the Americas, New York, NY 10020. (212) 512-2000. 1988. $29.95.

INTRODUCTION TO DATA PROCESSING: MAINFRAMES, MINIS, AND MICROCOMPUTERS. M.L. Harris. John Wiley and Sons, Inc., 605 Third Avenue, New York, NY 10158. (800) 526-5368. Third edition. 1986. $12.95 in paper.

OPERATING SYSTEMS PRINCIPLES. Stanley A. Kurzban and others. Van Nostrand Reinhold Company, Inc., 135 West 50th Street, New York, NY 10020. (800) 543-2681. Second edition. 1984. $38.95.

PRINCIPLES OF PROGRAMMING LANGUAGES: DESIGN, EVALUATION AND IMPLEMENTATION. Bruce MacLennan. Holt, Rinehart and Winston, Inc., 383 Madison Avenue, New York, NY 10017. (212) 872-2000. Second edition. 1986. $38.75.

PROGRAMMING LANGUAGES. A.B. Tucker. McGraw-Hill Book Company, 1221 Avenue of the Americas, New York, NY 10020. (212) 512-2000. Second edition. 1986. $43.95.

PROGRAMMING LANGUAGES: A GRAND TOUR. Ellis Horowitz, editor. Computer Science Press, 11 Taft Court, Rockville, MD 20850. (800) 242-7737. Third Edition. 1987. $39.95.

HANDBOOKS AND MANUALS

PROFESSIONAL MICROCOMPUTER HANDBOOK. Ivan Flores. Van Nostrand Reinhold Company, Inc., 135 West 50th Street, New York, NY 10020. (800) 543-2681. 1985. $51.95.

ONLINE DATA BASES

COMPENDEX. Engineering Information, Inc., 345 East 47th Street, New York, NY 10017. (800) 221-1044 or (212) 705-7615. Engineering and technical literature, 1975 to present. Inquire as to online cost and availability.

DISSERTATION ABSTRACTS ONLINE. University Microfilms International, 300 North Zeeb Road, Ann Arbor, MI 48106. (800) 521-0600 or (313) 761-4700. Scope includes virtually all doctoral dissertations accepted at accredited American institutions from 1861 to present in over 250 subject areas. Inquire as to online cost and availability.

INSPEC. INSPEC Marketing Department, Institution of Electrical Engineers. Available from IEEE Service Center, 445 Hoes Lane, Piscataway, NJ 08854. (201) 981-0060. Online version of Physics Abstracts. Inquire as to online cost and availability.

NTIS. National Technical Information Service, 5285 Port Royal Road, Springfield, VA 22161. (703) 487-4630. Broad coverage of government sponsored research reports, 1964 to present. Inquire as to online cost and availability.

SCISEARCH. Institute for Scientific Information, 3501 Market Street, Philadelphia, PA 19104. (800) 523-1850 or (215) 386-0100. Broad multidisciplinary title and author index to the international literature of science and technology, 1974 to present. Inquire as to online cost and availability.

WILSONLINE. H.W. Wilson and Company, 950 University Avenue, Bronx, NY 10452. (800) 367-6770 or (212) 588-8400. Makes available online versions of the H.W. Wilson indexes including Applied Science and Technology Index, Business Periodicals Index and Readers' Guide to Periodical Literature. Approximately 1980 to present. Inquire as to online cost and availability.

PERIODICALS

ACM TRANSACTIONS ON PROGRAMMING LANGUAGES AND SYSTEMS. Association of Computing Machinery, 11 West 42nd Street, New York, NY 10036. (212) 869-7440. 1979 to present. Quarterly. $55.00 per year.

ADVANCES IN ENGINEERING SOFTWARE. CML Publications, 400 West Cummings Park, Suite 6200, Woburn, MA 01801. (617) 933-7374. 1979 to present. Quarterly. $130.00 per year.

BYTE. Byte Publications, Inc., 70 Main Street, Petersborough, NH 03458. (603) 924-9281. Monthly. $21.00 per year.

COMMUNICATIONS OF THE ACM. Association of Computing Machinery, 11 West 42nd Street, New York, NY 10036. (212) 869-7440. Monthly. $80.00 per year.

COMPUTER LANGUAGES. Pergamon Press, Inc., Maxwell House, Fairview Park, Elmsford, NY 10523. (914) 592-7700. 1976 to present. Quarterly. $195.00 per year.

COMPUTERWORLD. C.W. Communications, 375 Cochituate Road, Box 880, Farmingham, MA 01701. (617) 879-0700. 1967 to present. Weekly. $44.00 per year.

DATAMATION. Technical Publishing Company, 875 Third Avenue, New York, NY 10022. (212) 605-9400. 1957 to present. Semi-monthly. $50.00 per year.

IEEE SOFTWARE. Institution of Electrical and Electronics Engineers. IEEE Service Center, 445 Hoes Lane, Piscataway, NJ 08854. 1984 to present. Quarterly. $15.00 per issue.

IEEE TRANSACTIONS ON SOFTWARE ENGINEERING. Institute of Electrical and Electronics Engineers. IEEE Service Center, 445 Hoes Lane, Piscataway, NJ 08854. 1975 to present. Monthly. $160.00 per year.

INTERFACE. International Computer Programs, Inc., 9000 Keystone Crossing, Indianapolis, IN 46240. (317) 844-7461. 1975 to present. Quarterly. $10.00 per year.

JOURNAL OF LOGIC PROGRAMMING. Elsevier Science Publishing Company, Inc., 52 Vanderbilt Avenue, New York, NY 10017. (212) 370-5520. 1984 to present. Quarterly. $95.00 per year.

JOURNAL OF PASCAL, ADA, AND MODULA-2. John Wiley and Sons, Inc., 605 Third Avenue, New York, NY 10158. (800) 526-5368. 1982 to present. Bimonthly. $20.00 per year.

JOURNAL OF SYSTEMS AND SOFTWARE. Elsevier Science Publishing Company, Inc., 52 Vanderbilt Avenue, New York, NY 10017. (212) 370-5520. 1979 to present. Quarterly. $95.00 per year.

PC: THE INDEPENDENT GUIDE TO IBM PERSONAL COMPUTERS. Ziff-Davis Publishing, Computer Publications Division, One Park Avenue, New York, NY 10016. (212) 503-3500. Subscribe to: Box 2443, Boulder, CO 80321. 1982 to present. Twenty-two times per year. $35.00 per year.

SOFTWARE DEVELOPER'S MONTHLY. SourceView Press, 835 Castro Street, Martinez, CA 94553. (415) 228-6220. 1985 to present. Monthly. $144.00 per year.

SOFTWARE PRACTICE AND EXPERIENCE. John Wiley and Sons, Inc., 605 Third Avenue, New York, NY 10158. (800) 526-5368. 1971 to present. Monthly. $260.00 per year.

UNIX REVIEW (SAN FRANCISCO). Miller Freeman Publications, Inc., 500 Howard Street, San Francisco, CA 94105. (415) 397-1881. 1983 to present. Monthly. $35.00 per year.

RESEARCH CENTERS AND INSTITUTES

INSTITUTE FOR INFORMATION SCIENCE AND TECHNOLOGY. George Washington University, 801 22nd Street, N.W., Washington, DC 20052. (202) 676-4921.

RESEARCH INSTITUTE FOR COMPUTING AND INFORMATION SYSTEMS. University of Houston at Clear Lake, 2700 Bay Area Boulevard, Houston, TX 77058. (713) 488-9392.

SOFTWARE ENGINEERING INSTITUTE. Carnegie-Mellon University, Pittsburgh, PA 15213. (412) 268-6700.

SOFTWARE ENGINEERING RESEARCH INSTITUTE. Georgia Institute of Technology, 258 Fourth Street, N.W., Atlanta, GA 30332. (404) 894-3180.

DATA RETRIEVAL

See: DATA BASES

DECIMAL SYSTEM

See: MATHEMATICS

DECISION THEORY

See: PROBABILITY

DECODING

See: COMPUTER SECURITY

DEEP SEA PLATFORM

See: OCEAN ENGINEERING

DEEP SEA TRENCHES

See: OCEANOGRAPHY

DEPRESSIONS

See: METEOROLOGY

DESERTIFICATION

See also: CLIMATOLOGY, METEOROLOGY

ABSTRACT SERVICES AND INDEXES

APPLIED SCIENCE AND TECHNOLOGY INDEX. H.W. Wilson Company, 950 University Avenue, Bronx, NY 10452. (800) 367-6670 or (212) 588-8400. Inquire as to cost and availability.

CHEMICAL ABSTRACTS. Chemical Abstracts Service, 2540 Olentangy Road, Post Office Box 3012, Columbus, OH 43210. (800) 848-6538 or (614) 421-3600. Weekly. $9200.00 per year.

CURRENT CONTENTS: PHYSICAL AND CHEMICAL SCIENCES. Institute for Scientific Information, 3501 Market Street, Philadelphia, PA 19104. (800) 523-1850 or (215) 386-0100. $272.00 per year.

GENERAL SCIENCE INDEX. H.W. Wilson Company, 950 University Avenue, Bronx, NY 10452. (800) 367-6770 or (212) 588-8400. Inquire as to cost and availability.

METEOROLOGICAL AND GEOASTROPHYSICAL ABSTRACTS. American Meteorological Society, 45 Beacon Street, Boston, MA 02108. (617) 227-2425. 1950 to present. Monthly. $450.00 per year.

SCIENCE CITATION INDEX. Institute for Scientific Information, 3501 Market Street, Philadelphia, PA 19104. (800) 523-1850 or (215) 386-0100. Inquire as to cost and availability.

ASSOCIATIONS AND PROFESSIONAL SOCIETIES

AMERICAN ASSOCIATION STATE CLIMATOLOGISTS. c/o Professor John Griffiths. Meteorology Department, O and M Building, Texas A and M University, College Station, TX 77843. (409) 845-7320.

AMERICAN METEOROLOGICAL SOCIETY. 45 Beacon Street, Boston, MA 02108. (617) 227-2425.

INTERNATIONAL ASSOCIATION OF METEOROLOGY AND ATMOSPHERIC PHYSICS. UCAR, Post Office Box 3000, Boulder, CO 80307.

NATIONAL ENVIRONMENTAL SATELLITE DATA, AND INFORMATION SERVICE. 3300 Whitehaven Street, NW, Washington, DC 20235. (202) 634-7318.

NATIONAL WEATHER SERVICE OFFICES AND STATIONS. National Oceanic and Atmospheric Administration, Department of Commerce, Silver Spring, MD 20910. (301) 427-7698. Annual. Free.

UNIVERSITY CORPORATION FOR ATMOSPHERIC RESEARCH. Box 3000, 1850 Table Mesa Drive, Boulder, CO 80307. (303) 497-1000.

WEATHER MODIFICATION ASSOCIATION. Post Office Box 8116, Fresno, CA 93747. (209) 291-8466.

DIRECTORIES AND BIOGRAPHICAL SOURCES

AMERICAN MEN AND WOMEN OF SCIENCE. Physical and Biological Sciences. Sixteenth edition. R.R. Bowker Company, 205 East 42nd Street, New York, NY 10017. (800) 521-8110 or (212) 916-1600. 1986. $595.00.

GOVERNMENT RESEARCH DIRECTORY. Gale Research Company, Book Tower, Detroit, MI 48226. (800) 521-0707. Fourth edition. 1987. $350.00.

INTERDOC: DIRECTORY OF PUBLISHED PROCEEDINGS, SERIES. SEMT Science/Engineering/Medicine/Technology. Interdoc Corporation, 173 Halstead Avenue, Box 326, Harrison, NY 10528. (014) 835-3506. Ten times per year. $325.00 per year.

METEOROLOGICAL SERVICES OF THE WORLD. World Meteorological Services Organization. Available from: American Meteorological Society, 45 Beacon Street, Boston, MA 02108. (617) 227-2425. Annual. $35.00.

NATIONAL WEATHER SERVICE OFFICES AND STATIONS. National Oceanic and Atmospheric Administration, Department of Commerce, Silver Spring, MD 20910. (301) 427-7698. Annual. Free.

RESEARCH CENTERS DIRECTORY. Gale Research Company, Book Tower, Detroit, MI 48226. (800) 521-0707. Twelfth edition. 1987. $365.00 for set.

SCIENTIFIC AND TECHNICAL ORGANIZATIONS AND AGENCIES DIRECTORY. Gale Research Company, Book Tower, Detroit, MI 48226. (800) 521-0707. Second edition. 1987. $185.00.

WHO'S WHO IN FRONTIER SCIENCE AND TECHNOLOGY. Marquis Who's Who, Incorporated, 200 East Ohio Street, Chicago, IL 60611. (800) 428-3898 or (312) 787-2008.

WORLD GUIDE TO SCIENTIFIC ASSOCIATIONS AND LEARNED SOCIETIES. K.G. Saur Incorporated, 175 Fifth Avenue, New York, NY 10010. (800) 521-0707 or (212) 982-1302. Fourth edition. 1984. $112.00.

ENCYCLOPEDIAS AND DICTIONARIES

ENCYCLOPEDIA OF CLIMATOLOGY. John E. Oliver and Rhodes W. Fairbridge, editors. Van Nostrand Reinhold, Incorporated, 115 Fifth Avenue, New York, NY 10003. (800) 543-2681. 1987. $89.95.

GENERAL WORKS

CASE STUDIES ON DESERTIFICATION. J.A. Mabbutt and C. Floret, editors. Prepared by UNESCO-UNEP-UNDP. (Natural Resources Research Service: Number 18) Unipup, 205 East 42nd Street, New York, NY 10017. (212) 916-1659. 1981. $44.75.

DESERTIFICATION AND DEVELOPMENT: DRYLAND ECOLOGY IN SOCIAL PERSPECTIVE. B. Spooner and H.S. Mann. Academic Press, Incorporated, Orlando, FL 32887. (305) 345-4100. 1983. $65.50.

DESERTIFICATION: ASSOCIATED CASE STUDIES PREPARED FOR THE UNITED NATIONS CONFERENCE ON DESERTIFICATION. (Environmental Sciences and Applications Services: Volume 12). Pergamon Press, Incorporated, Maxwell House, Fairview Park, Elmsford, NY 10523. (914) 592-7700. 1980. $105.00.

DESERTIFICATION IN EUROPE. R. Fantechi and N.S. Margaris, editors. Kluwer Academic Publications, 190 Old Derby Street, Hingham, MA 02043. (617) 749-5262. 1986. $49.95.

PHYSICS OF DESERTIFICATION. El-Baz, and M.H. Hassan, editors. Kluwer Academic Publications, 190 Old Derby Street, Hingham, MA 02043. (617) 749-5262. 1986. $79.00.

ONLINE DATA BASES

CA SEARCH. Chemical Abstracts Service, Post Office Box 3012, Columbus, OH 43210. Guide to chemical literature, 1972 to present. Inquire as to cost and availability.

DISSERTATION ABSTRACTS ONLINE. University Microfilms International, 300 North Zeeb Road, Ann Arbor, MI 48106. (800) 521-0600 or (313) 761-4700. Scope includes virtually all doctoral dissertations accepted at accredited American institutions from 1861 to present in 252 subject areas. Inquire as to cost and availability.

METEOROLOGICAL AND GEOASTROPHYSICAL ABSTRACTS. American Meteorological Society, 45 Beacon Street, Boston, MA 02108. (617) 227-2425. 1950 to present. Monthly. $450.00 per year.

NTIS. National Technical Information Service, 5285 Port Royal Road, Springfield, VA 22161. (703) 487-4630. Broad coverage of government sponsored research reports, 1964 to present. Inquire as to cost and availability.

OCEANIC ABSTRACTS. Cambridge Scientific Abstracts, Incorporated, 5161 River Road, Bethesda, MD 20816. (301) 951-1400. 1964 to present. Inquire as to online cost and availability.

SCISEARCH. Institute for Scientific Information, 3501 Market Street, Philadelphia, PA 19104. (800) 523-1850 or (215) 386-0100. Broad multidisciplinary title and author index to the international literature of science and technology, 1974 to present. Inquire as to cost and availability.

PERIODICALS

AGRICULTURAL AND FOREST METEOROLOGY. Elsevier Science Publishing Company, Incorporated, 52 Vanderbilt Avenue, New York, NY 10017. (212) 370-5520. 1964 to present. Monthly. $260.00 per year.

AMERICAN METEOROLOGICAL SOCIETY BULLETIN. American Meteorological Society, 45 Beacon Street, Boston, MA

DESERTIFICATION

02108. (617) 227-2425.

CLIMATIC CHANGE: AN INTERDISCIPLINARY, INTERNATIONAL JOURNAL DEVOTED TO THE DESCRIPTION, CAUSES AND IMPLICATIONS OF CLIMATIC CHANGE. D. Reidel Publishing Company, 190 Old Derby Street, Hingham, MA 02043. (617) 871-6600. 1977 to present. Six times per year. $125.00 per year.

JOURNAL OF CLIMATE AND APPLIED METEOROLOGY. American Meteorological Society, 45 Beacon Street, Boston, MA 02108. (617) 227-2425. 1962 to present. Monthly. $120.00 per year.

JOURNAL OF THE ATMOSPHERIC SCIENCES. American Meteorological Society, 45 Beacon Street, Boston, MA 02108. (617) 227-2425. 1944 to present. Semimonthly. $220.00 per year.

MONTHLY WEATHER REVIEW. American Meteorological Society, 45 Beacon Street, Boston, MA 02108. (617) 227-2425. 1872 to present. Monthly. $120.00 per year.

RESEARCH CENTERS AND INSTITUTES

CENTER FOR CLIMATE RESEARCH. Columbia University, Lamont-Doherty Geological Observatory, Palisades, NY 10964.

CENTER FOR CLIMATIC RESEARCH. University of Delaware, Department of Geography, Newark, DE 19716. (302) 451-8998.

FLORIDA STATE CLIMATE CENTER. 305A Love Building, Florida State University, Tallahassee, FL 32306. (904) 644-3417.

NATIONAL CENTER FOR ATMOSPHERIC RESEARCH. Box 3000, Boulder, CO 80307. (303) 497-1000.

WORLD RESOURCES INSTITUTE. 1735 New York Avenue, NW, Suite 400, Washington, DC 20006. (202) 638-6300.

DESICCATION

See: INDUSTRIAL ENGINEERING

DESIGN ENGINEERING

See: CAD (COMPUTER-AIDED DESIGN)

DETONATIONS

See: EXPLOSIVES

DEUTERIUM

See: HYDROGEN

DIAMONDS (INDUSTRIAL)

See: MACHINING

DIELECTRICS

See also: ELECTRICAL ENGINEERING, ELECTROMAGNETISM, ELECTRONICS, ELECTRONICS ENGINEERING, MATERIALS SCIENCE, PIEZOELECTRICS, SOLID STATE PHYSICS

ABSTRACT SERVICES AND INDEXES

APPLIED MECHANICS REVIEW. American Society of Mechanical Engineers, 345 East 47th Street, New York, NY 10017. (212) 705-7703. 1948 to present. Monthly. $360.00 per year.

APPLIED SCIENCE AND TECHNOLOGY INDEX. H.W. Wilson and Company, 950 University Avenue, Bronx, NY 10452. (800) 367-6670 or (212) 588-8400. Monthly. Inquire as to cost and availability.

CHEMICAL ABSTRACTS. American Chemical Society, Chemical Abstracts Service, Box 3012, Columbus, OH 43210. (614) 421-3600. 1907 to present. Weekly. $9500.00 per year.

ELECTRIC POWER INDUSTRY ABSTRACTS. Edison Electric Institute, c/o Utility Data Institute, 2011 I Street, N.W., Suite 700, Washington, DC 20006. 1975 to present. Bimonthly. Inquire as to cost and availability.

ELECTRICAL AND ELECTRONICS ABSTRACTS. Institution of Electrical Engineers. Available from: Institute of Electrical and Electronics Engineers. IEEE Service Center, 445 Hoes Lane, Piscataway, NJ 08854. Monthly. $1250.00 per year.

ENGINEERING INDEX MONTHLY AND AUTHOR INDEX. Engineering Information Inc., 345 East 47th Street, New York, NY 10017. (212) 705-7600. Monthly. $1560.00 per year.

IEEE PUBLICATIONS BULLETIN. Institute of Electrical and Electronics Engineers. Institute of Electrical and Electronics Engineers. IEEE Service Center, 445 Hoes Lane, Piscataway, NJ 08854. Quarterly. Free.

PHYSICS ABSTRACTS. Institution of Electrical Engineers. Available from: IEEE Service Center, 445 Hoes Lane, Piscataway, NJ 08854. 1898 to present. Bimonthly. $1700.00 per year.

PHYSICS BRIEFS. Physik Verlag GmbH, Postfach 1260/1280, D-6940 Weinheim, West Germany. (212) 661-9404. 1920 to present. Twenty-six times per year. $1250.00 per year.

SCIENCE CITATION INDEX. Institute for Scientific Information, 3501 Market Street, Philadelphia, PA 19104. (800) 523-1850 or (215) 386-0100. Six times per year. $6200.00 per year.

SOLID STATE ABSTRACTS: AN ABSTRACT JOURNAL INVOLVING THE PHYSICS, METALLURGY, CRYSTALLOGRAPHY, CHEMISTRY, AND DEVICE TECHNOLOGY OF SOLIDS. Cambridge Scientific Abstracts, 5161 River Road, Bethesda, MD 20816. (301) 951-1400. 1957 to present. Bimonthly. $550.00 per year.

ASSOCIATIONS AND PROFESSIONAL SOCIETIES

AMERICAN ELECTRONICS ASSOCIATION. P.O. Box 10045, 2670 Hanover Street, Palo Alto, CA 94303. (415) 857-9300.

AMERICAN INSTITUTE OF PHYSICS. 335 East 45th Street, New York, NY 10017. (212) 661-9494.

AMERICAN SOCIETY FOR TESTING AND MATERIALS. 1916 Race Street, Philadelphia, PA 19103. (215) 299-5400.

EDISON ELECTRIC INSTITUTE. 1111 19th Street, N.W., Washington, DC 20036. (202) 828-7400.

INSTITUTE OF ELECTRICAL AND ELECTRONICS ENGINEERS. 345 East 47th Street, New York, NY 10017. (212) 705-7900.

DIRECTORIES AND BIOGRAPHICAL SOURCES

INTERNATIONAL RESEARCH CENTERS DIRECTORY 1988-89. Darren L. Smith, editor. Gale Research Company, Book Tower, Detroit, MI 48226. (800) 521-0707. 4th edition. 1987. $360.00.

1987 DIRECTORY OF ENGINEERING SOCIETIES AND RELATED ORGANIZATIONS. Gordon Davis, editor. Hemisphere Publishing Corporation, 1010 Vermont Avenue, NW, Washington, DC 20005. (800) 526-0275. 12th edition. 1987. $100.00.

RESEARCH CENTERS DIRECTORY 1988. Gale Research Company, Book Tower, Detroit, MI 48226. (800) 521-0707. 12th edition. 1987. $365.00 for set.

SCIENTIFIC AND TECHNICAL ORGANIZATIONS AND AGENCIES DIRECTORY. Margaret Labash Young, editor. Gale Research Company, Book Tower, Detroit, MI 48226. (800) 521-0707. 2nd edition. 1987. $185.00.

WHO'S WHO IN ELECTRONICS. Harris Publishing Company, 2057-2 Aurora Road, Twinsburg, OH 44087. (216) 425-9143. Annual. $90.00.

WHO'S WHO IN ENGINEERING. Gordon Davis, editor. Hemisphere Publishing Corporation, 1010 Vermont Avenue, NW, Washington, DC 20005. (800) 526-0275. 6th edition. 1985. $200.00.

GENERAL WORKS

ELECTRICITY AND MAGNETISM. M.H. Nayfeh and M.K. Brussel. John Wiley and Sons, Inc., 605 Third Avenue, New York, NY 10158. (800) 526-5368. 1985. $32.95.

ELECTROMAGNETIC CONCEPTS AND PRINCIPLES. Stanley V. Marshall and Gabriel G. Skitek. Prentice-Hall Publishing, Inc., Englewood Cliffs, NJ 07632. (800) 562-0245. 2nd edition. 1987. $42.95.

ELEMENTS OF ENGINEERING ELECTROMAGNETICS. N.N. Rao. Prentice-Hall Publishing, Inc., Englewood Cliffs, NJ 07632. (800) 562-0245. 2nd edition. 1987. $45.95.

ENGINEERING DIELECTRICS: VOLUME 1: CORONA MEASUREMENT AND INTERPRETATION. R.Bartnikas and E.J. McMahon, editors. American Society for Testing and Materials, 1916 Race Street, Philadelphia, PA 19103. (215) 299-5400. STP 669. 1979. $49.00.

FUNDAMENTALS OF ELECTRIC CIRCUITS. David A. Bell. Prentice-Hall Publishing, Inc., Englewood Cliffs, NJ 07632. (800) 562-0245. 4th edition. 1988. $30.25.

PHYSICS OF SOLID DIELECTRICS. I. Bunget and M. Popescu. Elsevier Science Publishing Company, Inc., 52 Vanderbilt Avenue, New York, NY 10017. (212) 370-5520. 1984. $92.00.

HANDBOOKS AND MANUALS

HANDBOOK OF MODERN ELECTRONICS AND ELECTRICAL ENGINEERING. Charles Belove, editor. John Wiley and Sons, Inc., 605 Third Avenue, New York, NY 10158. (800) 526-5368. 1986. $88.95.

ONLINE DATA BASES

CA SEARCH. Chemical Abstracts Service, P.O. Box 3012, Columbus, OH 43120. (800) 848-6538 or (614) 421-3600. Comprehensive guide to chemical literature, 1972 to present. Inquire as to online cost and availability.

COMPENDEX. Engineering Information, Inc., 345 East 47th Street, New York, NY 10017. (800) 221-1044 or (212) 705-7615. Engineering and technical literature, 1975 to present. Inquire as to online cost and availability.

INSPEC. INSPEC Marketing Department, Institution of Electrical Engineers. Available from IEEE Service Center, 445 Hoes Lane, Piscataway, NJ 08854. (201) 981-0060. Online version of Physics Abstracts. Inquire as to online cost and availability.

NTIS. National Technical Information Service, 5285 Port Royal Road, Springfield, VA 22161. (703) 487-4630. Broad coverage of government sponsored research reports, 1964 to present. Inquire as to online cost and availability.

SCISEARCH. Institute for Scientific Information, 3501 Market Street, Philadelphia, PA 19104. (800) 523-1850 or (215) 386-0100. Broad multidisciplinary title and author index to the international literature of science and technology, 1974 to present. Inquire as to online cost and availability.

WILSONLINE. H.W. Wilson and Company, 950 University Avenue, Bronx, NY 10452. (800) 367-6770 or (212) 588-8400. Makes available online versions of the H.W. Wilson indexes including Applied Science and Technology Index, Business Periodicals Index and Readers' Guide to Periodical Literature. Approximately 1980 to present. Inquire as to online cost and availability.

OTHER SOURCES

A GUIDE TO THE LITERATURE OF ELECTRICAL AND ELECTRONICS ENGINEERING. Susan B. Ardis. Libraries Unlimited Inc., P.O. Box 263, Littleton, CO 80160. (303) 770-1220. 1987. $37.50.

PERIODICALS

ELECTRONIC DESIGN. Hayden Publishing Company, 10 Mulholland Drive, Hasbrouck Heights, NJ 07604. (201) 288-7520. 1952 to present. Biweekly. $40.00 per year.

ELECTRONICS. McGraw-Hill Book Company, 1221 Avenue of the Americas, New York, NY 10020. (212) 512-2000. 1930 to present. Weekly. $32.00 per year.

IEEE CIRCUITS AND DEVICES MAGAZINE. Institute of Electrical and Electronics Engineers. IEEE Service Center, 445 Hoes Lane, Piscataway, NJ 08854. Bimonthly. $70.00 per year.

IEEE JOURNAL OF SOLID STATE CIRCUITS. Institute of Electrical and Electronics Engineers. IEEE Service Center, 445 Hoes Lane, Piscataway, NJ 08854. 1966 to present. Bimonthly. $113.00 per year.

IEEE TRANSACTIONS ON ELECTRON DEVICES. Institute of Electrical and Electronics Engineers. IEEE Service Center, 445 Hoes Lane, Piscataway, NJ 08854. 1959 to present. Monthly. $175.00 per year.

INSTITUTE OF ELECTRICAL AND ELECTRONICS ENGINEERS PROCEEDINGS. Institute of Electrical and Electronics Engineers. IEEE Service Center, 445 Hoes Lane, Piscataway, NJ 08854. 1913 to present. Monthly. $140.00 per year.

SEMICONDUCTOR INTERNATIONAL. Cahners Publishing Company, Inc., Cahners Plaza, 1350 East Touhy Avenue, Des Plaines, IL 60018. (312) 635-8800. 1978 to present. Monthly. $55.00 per year.

SOLID STATE ELECTRONICS. Pergamon Press, Inc., Maxwell House, Fairview Park, Elmsford, NY 10523. (914) 592-7700. 1960 to present. Monthly. $330.00 per year.

RESEARCH CENTERS AND INSTITUTES

EDISON ELECTRIC INSTITUTE. 1111 19th Street, N.W., Washington, DC 20036. (202) 778-6778.

ELECTRIC POWER INSTITUTE. Texas A&M University, Department of Electrical Engineering, College Station, TX 77843.

ELECTRICAL ENGINEERING RESEARCH LABORATORIES. Purdue University, Electrical Engineering Building, West Lafayette, IN 47907. (317) 494-3536.

LABORATORY FOR ELECTROMAGNETIC AND ELECTRONIC SYSTEMS. 77 Massachusetts Avenue, Cambridge, MA 02139. (617) 253-4631.

DIESEL ENGINES

See also: AUTOMOTIVE ENGINEERING, ENGINES

ABSTRACT SERVICES AND INDEXES

APPLIED SCIENCE AND TECHNOLOGY INDEX. H.W. Wilson Company, 950 University Avenue, Bronx, NY 10452. (800) 367-6670 or (212) 588-8400. Inquire as to cost and availability.

ENGINEERING INDEX MONTHLY. Engineering Information, Incorporated, 345 East 47th Street, New York, NY 10017. (800) 221-1044 or (212) 705-7600. Monthly, with annual cumulation. $1560.00 per year.

SAE ABSTRACTS INDEX OF TECHNICAL PAPERS. SAE, Incorporated, 400 Commonwealth Drive, Warrendale, PA 15096. (412) 776-4841. Annual. $40.00.

ASSOCIATION AND PROFESSIONAL SOCIETIES

DIESEL ENGINE MANUFACTURES ASSOCIATION. A.P. Wherry and Associates, Incorporated, 712 Lakewood Center North, Cleveland, OH 44107. (216) 226-7700.

ENGINE MANUFACTURES ASSOCIATION. 111 East Wacker Drive, Chicago, IL 60601. (312) 644-6610.

SOCIETY OF AUTOMOTIVE ENGINEERS. 400 Commonwealth Drive, Warrendale, PA 15096. (412) 776-4841.

DIRECTORIES AND BIOGRAPHICAL SOURCES

RESEARCH CENTERS DIRECTORY. Gale Research Company, Detroit, MI 48226. (800) 521-0707. Eleventh edition. 1987.

WHO'S WHO IN ENGINEERING. Engineers Joint Council, 345 East 47th Street, New York, NY 10017. (212) 705-7010. 1985. $200.00.

WHO'S WHO IN FRONTIER SCIENCE AND TECHNOLOGY. Marquis Who's Who, Incorporated, 200 East Ohio Street, Chicago, IL 60611. (800) 428-3898 or (312) 787-2008.

SCIENTIFIC AND TECHNICAL ORGANIZATIONS AND AGENCIES DIRECTORY. Gale Research Company, Book Tower, Detroit, MI 48226. (800) 521-0707. 1985.

ENCYCLOPEDIAS AND DICTIONARIES

ENCYCLOPEDIA OF PHYSICAL SCIENCE AND TECHNOLOGY. Academic Press, Incorporated, Orlando, FL 32887. (800) 321-5068 or (305) 345-2734. Fifteen volumes, 1986.

MCGRAW-HILL DICTIONARY OF ENGINEERING. Sybil P. Parker, editor. 1984. $39.95.

MCGRAW-HILL ENCYCLOPEDIA OF SCIENCE AND TECHNOLOGY. McGraw-Hill Book Company, 1221 Avenue of the Americas, New York, NY 10020. (212) 512-2000.

GENERAL WORKS

DESIGN AND APPLICATION IN DIESEL ENGINEERING. S.D. Haddad and N. Watson, editors. John Wiley and Sons, Incorporated, 605 Third Avenue, New York, NY 10158. (800) 526-5368 or (212) 850-6000. $65.95.

DESIGN AND DEVELOPMENT OF DIESEL ENGINEES AND COMPONENTS. Society of Automotive Engineers, 400 Commonwealth Drive, Warrendale, PA 15096. (412) 776-4841. 1985. $22.00.

DIESEL ENGINE COMBUSTION AND EMISSIONS. Society of Automotive Engineers, 400 Commonwealth Drive, Warrendale, PA 15096. (412) 776-4841. 1984. $28.00.

DIESEL ENGINES FOR AUTOMOBILES. Tom Weathers and C. Hunter. Reston Publishing Company, 200 Old Tappan Road, Old Tappan, NJ 07675. (201) 592-2352. 1985. $26.95.

DIESEL MECHANICS. Erich J. Schulz. McGraw-Hill Book Company, 1221 Avenue of the Americas, New York, NY 10020. (212) 512-2000. Second edition. 1983. $33.45.

HANDBOOKS AND MANUALS

DIESEL ENGINE MANUAL. Perry O. Black and William E. Schahill. Theodore Audel, distributed by G.K. Hall and Company, 70 Lincoln Street, Boston, MA 02111. (800) 343-2806. Fourth edition. 1983. $12.95.

DIESEL ENGINEERING HANDBOOK. Business Journals, 22 South Smith Street, Post Office Box 5550, East Norwalk, CT 06856. $28.50.

SAE HANDBOOK. SAE, Incorporated, 400 Commonwealth Drive, Warrendale, PA 15096. (412) 776-4841. $70.00 members; $140.00 non-members.

STANDARD HANDBOOK OF ENGINEERING CALCULATIONS. Tyler G. Hicks, editor. McGraw-Hill Book Company, 1221 Avenue of the Americas, New York, NY 10020. (212) 512-2000. Second edition. 1984. $59.50.

THE WILEY ENGINEER'S DESK REFERENCE. Sanford I. Heisler. John Wiley and Sons, Incorporated, 605 Third Avenue, New York, NY 10158. (800) 526-5368 or (212) 850-6418. 1984. $36.00.

ONLINE DATA BASES

COMPENDEX. Engineering Information, Incorporated, 345 East 47th Street, New York, NY 10017. (800) 221-1044 or (212) 705-7615. Engineering and technical literature, 1975 to present. Inquire as to cost and availability.

NTIS. National Technical Information Service, 5285 Port Royal Road, Springfield, VA 22161. (703) 487-4630. Broad coverage of government sponsored research reports, 1964 to present, Inquire as to cost and availability.

SAE DATABASES. Society of Automotive Engineers, 400 Commonwealth Drive, Warrendale, PA 15096. (412) 776-4841. Citations to literature on all types of self-propelled vehicles, 1965 to present. Inquire as to on-line cost and availability.

WILSONLINE. H.W. Wilson Company, 950 University Avenue, Bronx, NY 10452. (800) 367-6770 or (212) 588-8400. Makes available online versions of the printed H.W. Wilson Indexes including Applied Science and Technology Index, Business Periodicals Index, and Readers' Guide to Periodical Literature. Period covered is generally 1983 to present. Inquire as to cost and availability.

PERIODICALS

AUTOMOTIVE ENGINEERING. Society of Automotive Engineers, 400 Commonwealth Drive, Warrendale, PA 15096. (412) 776-4841. Monthly. $42.00 per year.

DIESEL AND GAS TURBINE WORLDWIDE. Diesel Engines Incorporated, 13555 Bishop's Court, Brookfield, WI 53005. Monthly, 10 per year. $45.00 per year.

DIESEL EQUIPMENT SUPERINTENDENT. Business Journals, Incorporated, 22 South Smith Street, Box 5550, Norwalk, CT 06856. (203) 853-6015. Monthly. $30.00.

DIESEL PROGRESS NORTH AMERICAN. Diesel and Gas Turbine Publications, 13555 Bishop's Court, Brookfield, WI 53005-6286. (414) 784-9177. Monthly. $35.00.

RESEARCH CENTERS AND INSTITUTES

COORDINATING RESEARCH COUNCIL, INCORPORATED. 219 Perimeter Center Parkway, Atlanta, GA 30346. (404) 396-3400.

UNIVERSITY OF MICHIGAN. Automotive Engineering Laboratory, Department of Mechanical Engineering and Applied Mechanics, Ann Arbor, MI. (313) 764-4256.

WAYNE STATE UNIVERSITY. Center for Automotive Research, 5050 Anthony Wayne Drive, Detroit, MI 48202. (313) 577-3887.

DIFFERENTIAL CALCULUS

See: CALCULUS

DIFFERENTIAL EQUATIONS

See also: CALCULUS, MATHEMATICS

ABSTRACT SERVICES AND INDEXES

ABSTRACTS OF PAPERS PRESENTED TO THE AMERICAN MATHEMATICAL SOCIETY. American Mathematical Society, Post Office Box 6248, Providence, RI 02940. (800) 556-7774 or (401) 272-9500. Semimonthly. $42.00 per year.

APPLIED SCIENCE AND TECHNOLOGY INDEX. H.W. Wilson Company, 950 University Avenue, Bronx, NY 10452. (800) 367-6670 or (212) 588-8400. Inquire as to cost and availability.

CURRENT MATHEMATICAL PUBLICATIONS. American Mathematical Society, Post Office Box 6248, Providence, RI 02940. (800) 556-7774 or (401) 272-9500. Seventeen times per year. $220.00 per year.

MATHEMATICAL REVIEWS. American Mathematical Society, Post Office Box 6248, Providence, RI 02940. (800) 556-7774 or (401) 272-9500.

PHYSICS ABSTRACTS. Institute of Electrical Engineers, London, United Kingdom. Available from: Institute of Electrical and Electronic Engineers (IEEE), 345 East 47th Street, New York, NY 10017. (212) 705-7900.

SCIENCE CITATION INDEX. Institute for Scientific Information, 3501 Market Street, Philadelphia, PA 19104. (800) 523-1850 or (215) 386-0100.

ANNUAL REVIEWS AND YEARBOOKS

ADVANCES IN APPLIED MATHEMATICS. Academic Press, Incorporated, 6277 Sea Harbor Drive, Orlando, FL 32821. (800) 321-5068. Price varies.

ADVANCES IN MATHEMATICS. Academic Press, Incorporated, 6277 Sea Harbor Drive, Orlando, FL 32821. (800) 321-5068. Inquire as to cost and availability.

ASSOCIATIONS AND PROFESSIONAL SOCIETIES

AMERICAN MATHEMATICAL SOCIETY. Post Office Box 6248, Providence, RI 02940. (401) 272-9500.

MATHEMATICAL ASSOCIATION OF AMERICA. 1529 18th Street, NW, Washington, DC 20036. (202) 387-5200.

SOCIETY FOR INDUSTRIAL AND APPLIED MATHEMATICS. 1400 Architects Building, 117 South 17th Street, Philadelphia, PA 19103. (215) 564-2929.

BIBLIOGRAPHIES

SCIENTIFIC AND TECHNICAL BOOKS AND SERIALS IN PRINT 1988: AN INDEX TO LITERATURE IN SCIENCE AND TECHNOLOGY. R.R. Bowker Company, 205 East 42nd Street, New York, NY 10017. (800) 521-8110 or (212) 916-1600. $175.00.

DIRECTORIES AND BIOGRAPHICAL SOURCES

AMERICAN MEN AND WOMEN OF SCIENCE. Physical and Biological Sciences. Fifteenth edition. R.R. Bowker Company, 205 East 42nd Street, New York, NY 10017. (800) 521-8110 or (212) 916-1600.

BIOGRAPHICAL DICTIONARY OF SCIENTISTS. T.I. Williams. Halsted Press, 605 Third Avenue, New York, NY 10158. (800) 526-5368 or (212) 850-6418. Third edition. 1982. $29.95.

RESEARCH CENTERS DIRECTORY. Gale Research Company, Book Tower, Detroit, MI 48226. (800) 521-0707. Twelfth edition. 1987. $365.00 for set.

WHO'S WHO IN FRONTIER SCIENCE AND TECHNOLOGY. Marquis Who's Who, Incorporated, 200 East Ohio Street, Chicago, IL 60611. (800) 428-3898 or (312) 787-2008.

WHO'S WHO IN TECHNOLOGY TODAY. Reston Publishing Company, Incorporated, c/o Prentice-Hall, Incorporated, Englewood Cliffs, NJ 07632. (800) 262-6868. Biennial. Five volumes. $425.00. Covers the fields of electronics, computer science, physics, optics, chemistry, biotechnology, mechanics, energy, and earth science.

WORLD GUIDE TO SCIENTIFIC ASSOCIATIONS AND LEARNED SOCIETIES. K.G. Saur, Incorporated, 175 Fifth Avenue, New York, NY 10010. (800) 521-0707 or (212) 982-1302. Fourth edition. 1984. $112.00.

ENCYCLOPEDIAS AND DICTIONARIES

ENCYCLOPEDIA OF PHYSICAL SCIENCE AND TECHNOLOGY. Academic Press, Incorporated, Orlando FL 32887. (800) 321-5068 or (305) 345-2734. Fifteen volumes, 1986.

ENCYCLOPEDIC DICTIONARY OF MATHEMATICS. Kiyosi Ito, editor. MIT Press, 55 Hayward Street, Cambridge, MA 02142. (617) 253-2884. Second edition. 1987. Four volumes. $350.00.

MCGRAW-HILL ENCYCLOPEDIA OF SCIENCE AND TECHNOLOGY. McGraw-Hill Book Company, Incorporated, 1221 Avenue of the Americas, New York, NY 10020. (212) 997-3675.

DIFFERENTIAL EQUATIONS

GENERAL WORKS

ADVANCED CALCULUS AND ITS APPLICATIONS TO THE ENGINEERING AND PHYSICAL SCIENCES. John C. Amazigo and Lester A. Rubenfeld. John Wiley and Sons, Incorporated, 605 Third Avenue, New York, NY 10158. (800) 526-5368 or (212) 850-6000. 1980. $36.45.

CALCULUS AND ANALYTIC GEOMETRY. George B. Thomas, Jr. and Ross L. Finney. Addison Wesley Publishing Company, Incorporated, 1 Jacob Way, Reading, MA 01867. (617) 944-3700. Sixth edition. 1984. $39.95.

FIRST COURSE IN DIFFERENTIAL EQUATIONS WITH APPLICATIONS. Dennis G. Zill. PWS Publications, 20 Park Plaza, Boston, MA 02116. (617) 482-2344. Third edition. 1985. Inquire.

PARTIAL DIFFERENTIAL EQUATIONS OF APPLIED MATHEMATICS. Erich Zauderer. John Wiley and Sons, Incorporated, 605 Third Avenue, New York, NY 10158. (800) 526-5368 or (212) 850-6000. 1983. $57.50.

HANDBOOKS AND MANUALS

CRC HANDBOOK OF MATHEMATICAL SCIENCES. William H. Beyer, editor. CRC Press, Incorporated, 2000 Corporate Boulevard, Boca Raton, FL 33341. (305) 994-0555. Sixth edition. 1987. $91.95.

CRC STANDARD MATHEMATICAL TABLES. W.H. Beyer, editor. CRC Press, Incorporated, 2000 Corporate Boulevard, Boca Raton, FL 33341. (305) 994-0555. Twenty-eighth edition. 1987. $39.95.

ONLINE DATA BASES

DISSERTATION ABSTRACTS ONLINE. University Microfilms International, 300 North Zeeb Road, Ann Arbor, MI 48106. (800) 521-0600 or (313) 761-4700. Scope includes virtually all doctoral dissertations accepted at accredited American institutions from 1861 to present in 252 subject areas. Inquire as to cost and availability.

INSPEC. INSPEC Marketing Department, Institute of Electrical and Electronics Engineers, Incorporated, IEEE Service Department, 445 Hoes Lane, Piscataway, NJ 08854. (201) 981-0060. Inquire as to on-line cost and availability.

MATHFILE. American Mathematical Society, Post Office Box 6248, Providence, RI 02940. (800) 556-7774 or (401) 272-9500. Scope includes pure and applied mathematics and related areas of physics, statistics, engineering, computer science, and operations research literature since 1973. Inquire as to cost and availability.

SCISEARCH. Institute for Scientific Information, 3501 Market Street, Philadelphia, PA 19104. (800) 523-1850 or (215) 386-0100. Broad multidisciplinary title and author index to the international literature of science and technology, 1974 to present. Inquire as to cost and availability.

OTHER SOURCES

FROM ONE TO ZERO: UNIVERSAL HISTORY OF NUMBERS. Georges Ifrah. Viking-Penquin, Incorporated, 40 West 23rd Street, New York, NY 10010. (800) 631-3577 or (212) 807-7300. 1986. $35.00.

MATHEMATICS AND THE SEARCH FOR KNOWLEDGE. Morris Kline. Oxford University Press, Incorporated, 200 Madison Avenue, New York, NY 10016. (212) 564-6680. 1986. $19.95.

PERIODICALS

AMERICAN JOURNALS OF MATHEMATICS. Johns Hopkins University Press, Journals Publishing Division, 701 West 40th Street, Suite 275, Baltimore, MD 21211. (301) 338-7864. Bimonthly. $40.00.

AMERICAN MATHEMATICAL MONTHLY. Mathematical Association of America, 1529 18th Street, NW, Washington, DC 20036. (202) 387-5200. Monthly. $55.00.

COLLEGE MATHEMATICS JOURNAL. Mathematical Association of America, 1529 18th Street, NW, Washington, DC 20036. Five issues per year. $40.00.

DIFFERENTIAL EQUATIONS. (Translation of Differentsial'nye Uravneniya). Plenum Publishing Corporation, 233 Spring Street, New York, NY 10013. (212) 620-8000. Monthly. $650.00 per year.

JOURNAL OF DIFFERENTIAL EQUATIONS. Academic Press, Incorporated, 111 Fifth Avenue, New York, NY 10003. Monthly. $490.00 per year.

MATHEMATICS MAGAZINE. Mathematical Association of America, 1529 18th Street, NW, Washington, DC 20036. (202) 387-5200. Five issues per year. $35.00.

NUMERICAL METHODS FOR PARTIAL DIFFERENTIAL EQUATIONS. John Wiley and Sons, Incorporated, 605 Third Avenue, New York, NY 10158. (800) 526-5368 or (212) 850-6000. Quarterly. $90.00.

SIAM JOURNAL OF APPLIED MATHEMATICS. Society for Industrial and Applied Mathematics, 117 South 17th Street, Suite 1400, Philadelphia, PA 19103. (215) 564-2929. Bimonthly. $34.00 per year to individuals.

DIFFERENTIAL GRATING

See: SPECTROSCOPY

DIGITAL CIRCUITS

See: ELECTRONIC CIRCUITS AND COMPONENTS

DIGITAL COMPUTERS

See: COMPUTERS

DIKES

See: DAMS

DIODES

See also: ELECTRICAL ENGINEERING, ELECTRICITY, ELECTRONIC CIRCUITS AND COMPONENTS, ELECTRONICS, ELECTRONICS ENGINEERING, MICROELECTRONICS

ABSTRACT SERVICES AND INDEXES

APPLIED SCIENCE AND TECHNOLOGY INDEX. H.W. Wilson and Company, 950 University Avenue, Bronx, NY 10452. (800) 367-6670 or (212) 588-8400. Monthly. Inquire as to cost and availability.

CHEMICAL ABSTRACTS. American Chemical Society, Chemical Abstracts Service, Box 3012, Columbus, OH 43210. (614) 421-3600. 1907 to present. Weekly. $9500.00 per year.

CURRENT CONTENTS: ENGINEERING, TECHNOLOGY AND APPLIED SCIENCES. Institute for Scientific Information, 3501 Market Street, Philadelphia, PA 19104. (800) 523-1850 or (215) 386-0100. Weekly. $275.00 per year.

ELECTRICAL AND ELECTRONICS ABSTRACTS. Institution of Electrical Engineers. Available from: Institute of Electrical and Electronics Engineers. IEEE Service Center, 445 Hoes Lane, Piscataway, NJ 08854. Monthly. $1250.00 per year.

ELECTRONICS AND COMMUNICATIONS ABSTRACTS. Cambridge Scientific Abstracts, 5161 River Road, Bethesda, MD 20816. (301) 951-1400. Bimonthly. Inquire as to cost and availability.

ENGINEERING INDEX MONTHLY AND AUTHOR INDEX. Engineering Information Inc., 345 East 47th Street, New York, NY 10017. (212) 705-7600. Monthly. $1560.00 per year.

IEEE PUBLICATIONS BULLETIN. Institute of Electrical and Electronics Engineers. Institute of Electrical and Electronics Engineers. IEEE Service Center, 445 Hoes Lane, Piscataway, NJ 08854. Quarterly. Free.

PHYSICS ABSTRACTS. Institution of Electrical Engineers. Available from: IEEE Service Center, 445 Hoes Lane, Piscataway, NJ 08854. 1898 to present. Bimonthly. $1700.00 per year.

PHYSICS BRIEFS. Physik Verlag GmbH, Postfach 1260/1280, D-6940 Weinheim, West Germany. (212) 661-9404. 1920 to present. Twenty-six times per year. $1250.00 per year.

SCIENCE CITATION INDEX. Institute for Scientific Information, 3501 Market Street, Philadelphia, PA 19104. (800) 523-1850 or (215) 386-0100. Six times per year. $6200.00 per year.

SOLID STATE ABSTRACTS: AN ABSTRACT JOURNAL INVOLVING THE PHYSICS, METALLURGY, CRYSTALLOGRAPHY, CHEMISTRY, AND DEVICE TECHNOLOGY OF SOLIDS. Cambridge Scientific Abstracts, 5161 River Road, Bethesda, MD 20816. (301) 951-1400. 1957 to present. Bimonthly. $550.00 per year.

ANNUAL REVIEWS AND YEARBOOKS

ADVANCES IN ELECTRONICS AND ELECTRON PHYSICS. Academic Press, Inc., 6277 Sea Harbor Drive, Orlando, FL 32821. (800) 321-5068. Irregular. Approximately $80.00 per volume.

CRC CRITICAL REVIEWS IN SOLID STATE AND MATERIALS SCIENCES. CRC Press, 2000 Corporate Boulevard, Boca Raton, FL 33431. (800) 272-7737. Irregular. Inquire.

ASSOCIATIONS AND PROFESSIONAL SOCIETIES

AMERICAN ELECTRONICS ASSOCIATION. P.O. Box 10045, 2670 Hanover Street, Palo Alto, CA 94303. (415) 857-9300.

AMERICAN INSTITUTE OF PHYSICS. 335 East 45th Street, New York, NY 10017. (212) 661-9494.

EDISON ELECTRIC INSTITUTE. 1111 19th Street, N.W., Washington, DC 20036. (202) 828-7400.

ELECTRONICS INDUSTRIES ASSOCIATION. 2001 Eye Street, N.W., Washington, DC 20006. (202) 457-4900.

INSTITUTE OF ELECTRICAL AND ELECTRONICS ENGINEERS. 345 East 47th Street, New York, NY 10017. (212) 705-7900.

INTERNATIONAL SOCIETY FOR HYBRID MICROELECTRONICS. P.O. Box 2698, 1861 Wiehle Avenue, Suite 340, Reston, VA 22090. (703) 471-0066.

DIRECTORIES AND BIOGRAPHICAL SOURCES

IEEE MEMBERSHIP DIRECTORY. Institute of Electrical and Electronics Engineers. IEEE Service Center, 445 Hoes Lane, Piscataway, NJ 08854. Annual. $7.00.

1987 DIRECTORY OF ENGINEERING SOCIETIES AND RELATED ORGANIZATIONS. Gordon Davis, editor. Hemisphere Publishing Corporation, 79 Madison Avenue, New York, NY 10016-7892. (800) 821-8312. 12th edition. 1987. $100.00.

RESEARCH CENTERS DIRECTORY 1988. Gale Research Company, Book Tower, Detroit, MI 48226. (800) 521-0707. 12th edition. 1987. $365.00 for set.

SCIENTIFIC AND TECHNICAL ORGANIZATIONS AND AGENCIES DIRECTORY. Margaret Labash Young, editor. Gale Research Company, Book Tower, Detroit, MI 48226. (800) 521-0707. 2nd edition. 1987. $185.00.

WHO'S WHO IN ELECTRONICS. Harris Publishing Company, 2057-2 Aurora Road, Twinsburg, OH 44087. (216) 425-9143. Annual. $90.00.

ENCYCLOPEDIAS AND DICTIONARIES

IEEE STANDARD DICTIONARY OF ELECTRICAL AND ELECTRONICS TERMS. Frank Jay, editor. John Wiley and Sons, Inc., 605 Third Avenue, New York, NY 10158. (800) 526-5368. 3rd edition. 1984. $49.95.

GENERAL WORKS

ELECTRONIC INVENTIONS AND DISCOVERIES: ELECTRONICS FROM ITS EARLIEST BEGINNINGS TO PRESENT DAY. G.W.A. Dummer. Pergamon Press, Inc., Maxwell House, Fairview Park, Elmsford, NY 10523. (914) 592-7700. 1983. $49.50.

ELECTRONICS. H. Kybett. John Wiley and Sons, Inc., 605 Third Avenue, New York, NY 10158. (800) 526-5368. Second edition. 1986. $12.95 in paper.

EXPERIMENTER'S GUIDE TO SOLID-STATE DIODES. Robert J. Traister. Prentice-Hall Publishing, Inc., Englewood Cliffs, NJ 07632. (800) 562-0245. 1985. $12.95 in paper.

INTRODUCTION TO ELECTRICITY AND ELECTRONICS FOR THE COMPUTER AGE. R. Rosen. John Wiley and Sons, Inc., 605 Third Avenue, New York, NY 10158. (800) 526-5368. 1987. $35.95.

LIGHT-EMITTING DIODES. Arpad Bergh and Paul J. Dean. Oxford University Press, 200 Madison Avenue, New York, NY 10016. (800) 458-5833. 1976. $89.00.

POWER ELECTRONICS. B.M. Bird and K.G. King. John Wiley and Sons, Inc., 605 Third Avenue, New York, NY 10158. (800) 526-5368. 1983. $29.95 in paper.

SWITCHING IN SEMICONDUCTOR DIODES. Y.R. Nosov. Plenum Publishing Corporation, 233 Spring Street, New York, NY 10013. (800) 221-9369. 1969. $29.50.

HANDBOOKS AND MANUALS

ELECTRONIC ENGINEERS HANDBOOK. Donald G. Fink, editor. McGraw-Hill Book Company, 1221 Avenue of the Americas, New York, NY 10020. (212) 512-2000. 2nd edition. 1982. $89.00.

HANDBOOK OF MODERN ELECTRONICS AND ELECTRICAL ENGINEERING. Charles Belove, editor. John Wiley and Sons, Inc., 605 Third Avenue, New York, NY 10158. (800) 526-5368. 1986. $88.95.

ONLINE DATA BASES

CA SEARCH. Chemical Abstracts Service, P.O. Box 3012, Columbus, OH 43120. (800) 848-6538 or (614) 421-3600. Comprehensive guide to chemical literature, 1972 to present. Inquire as to online cost and availability.

COMPENDEX. Engineering Information, Inc., 345 East 47th Street, New York, NY 10017. (800) 221-1044 or (212) 705-7615. Engineering and technical literature, 1975 to present. Inquire as to online cost and availability.

INSPEC. INSPEC Marketing Department, Institution of Electrical Engineers. Available from IEEE Service Center, 445 Hoes Lane, Piscataway, NJ 08854. (201) 981-0060. Online version of Physics Abstracts. Inquire as to online cost and availability.

NTIS. National Technical Information Service, 5285 Port Royal Road, Springfield, VA 22161. (703) 487-4630. Broad coverage of government sponsored research reports, 1964 to present. Inquire as to online cost and availability.

SCISEARCH. Institute for Scientific Information, 3501 Market Street, Philadelphia, PA 19104. (800) 523-1850 or (215) 386-0100. Broad multidisciplinary title and author index to the international literature of science and technology, 1974 to present. Inquire as to online cost and availability.

WILSONLINE. H.W. Wilson and Company, 950 University Avenue, Bronx, NY 10452. (800) 367-6770 or (212) 588-8400. Makes available online versions of the H.W. Wilson indexes including Applied Science and Technology Index, Business Periodicals Index and Readers' Guide to Periodical Literature. Approximately 1980 to present. Inquire as to online cost and availability.

OTHER SOURCES

A GUIDE TO THE LITERATURE OF ELECTRICAL AND ELECTRONICS ENGINEERING. Susan B. Ardis. Libraries Unlimited Inc., P.O. Box 263, Littleton, CO 80160. (303) 770-1220. 1987. $37.50.

PERIODICALS

ELECTRONIC DESIGN. Hayden Publishing Company, 10 Mulholland Drive, Hasbrouck Heights, NJ 07604. (201) 288-7520. 1952 to present. Biweekly. $40.00 per year.

ELECTRONICS. McGraw-Hill Book Company, 1221 Avenue of the Americas, New York, NY 10020. (212) 512-2000. 1930 to present. Weekly. $32.00 per year.

IEEE CIRCUITS AND DEVICES MAGAZINE. Institute of Electrical and Electronics Engineers. IEEE Service Center, 445 Hoes Lane, Piscataway, NJ 08854. Bimonthly. $70.00 per year.

IEEE JOURNAL OF SOLID STATE CIRCUITS. Institute of Electrical and Electronics Engineers. IEEE Service Center, 445 Hoes Lane, Piscataway, NJ 08854. 1966 to present. Bimonthly. $113.00 per year.

IEEE TRANSACTIONS ON ELECTRON DEVICES. Institute of Electrical and Electronics Engineers. IEEE Service Center, 445 Hoes Lane, Piscataway, NJ 08854. 1959 to present. Monthly. $175.00 per year.

INSTITUTE OF ELECTRICAL AND ELECTRONICS ENGINEERS PROCEEDINGS. Institute of Electrical and Electronics Engineers. IEEE Service Center, 445 Hoes Lane, Piscataway, NJ 08854. 1913 to present. Monthly. $140.00 per year.

SEMICONDUCTOR INTERNATIONAL. Cahners Publishing Company, Inc., Cahners Plaza, 1350 East Touhy Avenue, Des Plaines, IL 60018. (312) 635-8800. 1978 to present. Monthly. $55.00 per year.

SOLID STATE ELECTRONICS. Pergamon Press, Inc., Maxwell House, Fairview Park, Elmsford, NY 10523. (914) 592-7700. 1960 to present. Monthly. $330.00 per year.

RESEARCH CENTERS AND INSTITUTES

ELECTRICAL ENGINEERING RESEARCH LABORATORIES. Purdue University, Electrical Engineering Building, West Lafayette, IN 47907. (317) 494-3536.

ELECTRONICS RESEARCH CENTER. University of Texas at Austin, 132 Engineering Science Building, Austin, TX 78712. (512) 471-3954.

ELECTRONICS RESEARCH LABORATORY. University of California, Berkeley, 253 Cory Hall, Berkeley, CA 94720. (415) 642-2301.

LABORATORY FOR ELECTROMAGNETIC AND ELECTRONIC SYSTEMS. Massachusetts Institute of Technology, 77 Massachusetts Avenue, Cambridge, MA 02139. (617) 253-4631.

DIRECT CURRENT

See: ELECTRICITY

DIRIGIBLE

See: AIRSHIPS

DISPLAY TERMINALS

See: COMPUTER COMMUNICATIONS

DISTILLATION

See: CHEMICAL ENGINEERING

DISTRIBUTED COMPUTING

See: NETWORKS (COMPUTER), LOCAL AREA NETWORKS

DOLBY SYSTEM

SeeL AUDIO ENGINEERING

DOMES

See: STRUCTURAL ENGINEERING

DOPING

See: SEMICONDUCTORS

DOPPLER RADAR

See: RADAR

DOSIMETRY

See: RADIATION

DRAFTING

See: CAD (COMPUTER-AIDED DESIGN)

DRAG

See: AERODYNAMICS

DRAWBRIDGES

See: BRIDGES

DREDGES AND DREDGING

See also: CHANNELS, CIVIL ENGINEERING, COASTAL ENGINEERING, COASTS, GEOTECHNICAL ENGINEERING, NAVIGATION, OCEAN ENGINEERING

ABSTRACT SERVICES AND INDEXES

APPLIED SCIENCE AND TECHNOLOGY INDEX. H.W. Wilson and Company, 950 University Avenue, Bronx, NY 10452. (800) 367-6670 or (212) 588-8400. Monthly. Inquire as to cost and availability.

BIBLIOGRAPHY AND INDEX OF GEOLOGY. American Geological Institute, 4220 King Street, Alexandria, VA 22302. (703) 379-2480. 1969 to present. Monthly. $1100.00 per year.

CURRENT CONTENTS: ENGINEERING, TECHNOLOGY AND APPLIED SCIENCES. Institute for Scientific Information, 3501 Market Street, Philadelphia, PA 19104. (800) 523-1850 or (215) 386-0100. Weekly. $275.00 per year.

ENGINEERING INDEX MONTHLY AND AUTHOR INDEX. Engineering Information Inc., 345 East 47th Street, New York, NY 10017. (212) 705-7600. Monthly. $1560.00 per year.

OCEAN ABSTRACTS. Cambridge Scientific Abstracts, 5161 River Road, Bethesda, MD 20816. (301) 951-1400. 1963 to present. Bimonthly. $450.00 per year.

SCIENCE CITATION INDEX. Institute for Scientific Information, 3501 Market Street, Philadelphia, PA 19104. (800) 523-1850 or (215) 386-0100. Six times per year. $6200.00 per year.

ANNUAL REVIEWS AND YEARBOOKS

COASTAL ENGINEERING. American Society of Civil Engineers, 345 East 47th Street, New York, NY 10017. (212) 705-7420. 1974 to 1980. Multiple volume sets for each year. Price varies, inquire.

ASSOCIATIONS AND PROFESSIONAL SOCIETIES

AMERICAN SOCIETY OF CIVIL ENGINEERS. 345 East 47th Street, New York, NY 10017. (212) 705-7420.

COASTAL ENGINEERING RESEARCH COUNCIL. 207 East Bay Street, Suite 311, Charleston, SC 29401. (803) 723-4864.

THE COASTAL SOCIETY. 5410 Grosvenor Lane, Suite 110, Bethesda, MD 20814. (301) 897-8616.

GEOLOGICAL SOCIETY OF AMERICA. 3300 Penrose Place, Boulder, CO 80301. (303) 447-2020.

NATIONAL ASSOCIATION OF DREDGING CONTRACTORS. 1625 Eye Street, N.W., Suite 321, Washington, DC 20006. (202) 223-4820.

DIRECTORIES AND BIOGRAPHICAL SOURCES

1987 DIRECTORY OF ENGINEERING SOCIETIES AND RELATED ORGANIZATIONS. Gordon Davis, editor. Hemisphere Publishing Corporation, 1010 Vermont Avenue, NW, Washington, DC 20005. (800) 526-0275. 12th edition. 1987. $100.00.

RESEARCH CENTERS DIRECTORY 1988. Gale Research Company, Book Tower, Detroit, MI 48226. (800) 521-0707. 12th edition. 1987. $365.00 for set.

SCIENTIFIC AND TECHNICAL ORGANIZATIONS AND AGENCIES DIRECTORY. Margaret Labash Young, editor. Gale Research Company, Book Tower, Detroit, MI 48226. (800) 521-0707. 2nd edition. 1987. $185.00.

WHO'S WHO IN ENGINEERING. Gordon Davis, editor. Hemisphere Publishing Corporation, 1010 Vermont Avenue, NW, Washington, DC 20005. (800) 526-0275. 6th edition. 1985. $200.00.

GENERAL WORKS

BASIC COASTAL ENGINEERING. R.M. Sorensen. John Wiley and Sons, Inc., 605 Third Avenue, New York, NY 10158. (800) 526-5368. 1978. $39.95.

DREDGING AND ITS ENVIRONMENTAL EFFECTS. American Society of Civil Engineers, 345 East 47th Street, New York, NY 10017. (212) 705-7420. 1976. $30.00.

FUNDAMENTALS OF HYDRAULIC DREDGING. Thomas M. Turner. Cornell Maritime Press, Inc., P.O. Box 456, Centreville, MD 21617. (301) 758-1075. 1984. $16.00.

HYDRAULIC DREDGING: THEORECTICAL AND APPLIED. John Huston. Cornell Maritime Press, Inc., P.O. Box 456, Centreville, MD 21617. (301) 758-1075. 1970. $17.50.

ONLINE DATA BASES

COMPENDEX. Engineering Information, Inc., 345 East 47th Street, New York, NY 10017. (800) 221-1044 or (212) 705-7615. Engineering and technical literature, 1975 to present. Inquire as to online cost and availability.

GEOREF. Online version of the BIBLIOGRAPHY AND INDEX OF GEOLOGY. American Geological Institute, 4220 King Street, Alexandria, VA 22302. (703) 379-2480. 1969 to present. Inquire as to online cost and availability.

NTIS. National Technical Information Service, 5285 Port Royal Road, Springfield, VA 22161. (703) 487-4630. Broad coverage of government sponsored research reports, 1964 to present. Inquire as to online cost and availability.

SCISEARCH. Institute for Scientific Information, 3501 Market Street, Philadelphia, PA 19104. (800) 523-1850 or (215) 386-0100. Broad multidisciplinary title and author index to the

international literature of science and technology, 1974 to present. Inquire as to online cost and availability.

WILSONLINE. H.W. Wilson and Company, 950 University Avenue, Bronx, NY 10452. (800) 367-6770 or (212) 588-8400. Makes available online versions of the H.W. Wilson indexes including Applied Science and Technology Index, Business Periodicals Index and Readers' Guide to Periodical Literature. Approximately 1980 to present. Inquire as to online cost and availability.

PERIODICALS

COASTAL ENGINEERING: AN INTERNATIONAL JOURNAL FOR COASTAL, HARBOUR, AND OFFSHORE ENGINEERS. Elsevier Science Publishing Company, Inc., 52 Vanderbilt Avenue, New York, NY 10017. (212) 370-5520. 1977 to present. Quarterly. $95.00 per year.

COASTAL RESEARCH. Florida State University, Geology Department, Tallahassee, FL 32306. (904) 644-5860. 1962 to present. Three times per year. $5.00 per year.

COASTAL ZONE MANAGEMENT JOURNAL. Crane Russak and Company, Inc., 3 East 44th Street, New York, NY 10017. (212) 867-1490. 1973 to present. Quarterly. $72.00 per year.

COASTWATCH. University of North Carolina, Sea Grant College Program, Box 8605, North Carolina State University, Raleigh, NC 27695-8605. (919) 737-2454. 1970 to present. Monthly. Free.

ESTUARINE, COASTAL AND SHELF SCIENCE. Academic Press, Inc., 6277 Sea Harbor Drive, Orlando, FL 32821. (800) 321-5068. 1973 to present. Monthly. $315.00 per year.

JOURNAL OF WATERWAY, PORT, COASTAL AND OCEAN ENGINEERING. American Society of Civil Engineers, 345 East 47th Street, New York, NY 10017. (212) 705-7420. 1956 to present. Bimonthly. $72.00 per year.

OCEAN ENGINEERING. Pergamon Press, Inc., Maxwell House, Fairview Park, Elmsford, NY 10523. (914) 592-7700. 1968 to present. Bimonthly. $200.00 per year.

OCEAN SCIENCE AND ENGINEERING. Marcel Dekker Inc., 270 Madison Avenue, New York, NY 10016. (800) 228-1160. 1974 to present. Quarterly. $165.00 per year.

U.S. COASTAL ENGINEERING RESEARCH CENTER. QUARTERLY CIRCULAR INFORMATION BULLETIN. U.S. Coastal Engineering Research Center, Kingman Building, Fort Belvoir, VA 22060. (703) 325-7429. Quarterly. Inquire.

RESEARCH CENTERS AND INSTITUTES

CENTER FOR DREDGING STUDIES. Texas A&M University, Texas Engineering Experiment Station, College Station, TX 77843. (409) 845-4517.

COASTAL AND OCEANOGRAPHIC ENGINEERING LABORATORY. University of Florida, 336 Weil Hall, Gainesville, FL 32607. (904) 392-1436.

COASTAL ENGINEERING RESEARCH COUNCIL. 207 East Bay Street, Suite 311, Charleston, SC 29401. (803) 723-4864.

COASTAL LABORATORY. Queen's University at Kingston, Ellis Hall, Kingston, ON, Canada K7L 3N6. (613) 545-6265.

J.K.K. LOOK LABORATORY OF OCEANOGRAPHIC ENGINEERING. University of Hawaii, 811 Olomehani Street, Honolulu, HI 96813. (808) 533-6412.

DRILLING

See: PETROLEUM ENGINEERING

DROUGHT

See also: CLIMATE, DESERTIFICATION, METEOROLOGY, WATER RESOURCES

ABSTRACT SERVICES AND INDEXES

GENERAL SCIENCE INDEX. H.W. Wilson Company, 950 University Avenue, Bronx, NY 10452. (800) 367-6770 or (212) 588-8400.

METEOROLOGICAL AND GEOASTROPHYSICAL ABSTRACTS. American Meteorological Society, 45 Beacon Street, Boston, MA 02108. (617) 227-2425. 1950 to present. Monthly. $450.00 per year.

OCEANIC ABSTRACTS. Cambridge Scientific Abstracts, Incorporated, 5161 River Road, Bethesda, MD 20816. (301) 951-1400. 1964 to present. Inquire as to online cost and availability.

SCIENCE CITATION INDEX. Institute for Scientific Information, 3501 Market Street, Philadelphia, PA 19104. (800) 523-1850 or (215) 386-0100. Inquire as to cost and availability.

ASSOCIATIONS AND PROFESSIONAL SOCIETIES

AMERICAN ASSOCIATION STATE CLIMATOLOGISTS. c/o Professor John Griffiths, Meteorology Department, O and M Building, Texas A and M University, College Station, TX 77843. (409) 845-7320.

AMERICAN METEOROLOGICAL SOCIETY. 45 Beacon Street, Boston, MA 02108. (617) 227-2425.

NATIONAL ENVIRONMENTAL SATELLITE DATA, AND INFORMATION SERVICE. 3300 Whitehaven Street, NW, Washington, DC 20235. (202) 634-7318.

NATIONAL WEATHER ASSOCIATION. 4400 Stamp Road, Room 404, Temple Hills, MD 20748. (301) 899-3784.

WEATHER MODIFICATION ASSOCIATION. Post Office Box 8116, Fresno, CA 93747. (209) 291-8466.

DIRECTORIES AND BIOGRAPHICAL SOURCES

GOVERNMENT RESEARCH DIRECTORY. Gale Research Company, Book Tower, Detroit, MI 48226. (800) 521-0707. Fourth edition. 1987. $350.00.

INTERDOC: DIRECTORY OF PUBLISHED PROCEEDINGS, SERIES. SEMT-Science/Engineering/Medicine/Technology. Interdoc Corporation, 173 Halstead Avenue, Box 326, Harrison, NY 10528. (014) 835-3506. Ten times per year. $325.00 per year.

INTERNATIONAL RESEARCH CENTERS DIRECTORY 1988-1989. Gale Research Company, Book Tower, Detroit, MI 48226. (800) 521-0707. Fourth edition. 1987. $360.00.

METEOROLOGICAL SERVICES OF THE WORLD. World Meteorological Services Organization. Available from: American Meteorological Society, 45 Beacon Street, Boston, MA 02108. (617) 227-2425. Annual. $35.00.

NATIONAL WEATHER SERVICE OFFICES AND STATIONS. National Oceanic and Atmospheric Administration, Department of Commerce, Silver Spring, MD 20910. (301) 427-7698. Annual. Free.

RESEARCH CENTERS DIRECTORY. Gale Research Company, Book Tower, Detroit, MI 48226. (800) 521-0707. Twelfth edition. 1987. $365.00 for set.

ENCYCLOPEDIAS AND DICTIONARIES

ENCYCLOPEDIA OF CLIMATOLOGY. John E. Oliver and Rhodes W. Fairbridge, editors. Van Nostrand Reinhold, Incorporated, 115 Fifth Avenue, New York, NY 10003. (800) 543-2681. 1987. $89.95.

GENERAL WORKS

DROUGHT AND MAN: THE NINETEEN SEVENTY-TWO CASE HISTORY. Garcia, R.V. and Escudero, J. Pergamon Press, Incorporated. Maxwell House, Fairview Park, Elmsford, NY 10523. (914) 592-7700. 1986. $99.00.

WATER SHORTAGE: LESSONS IN CONSERVATION FROM THE GREAT CALIFORNIA DROUGHT, 1976-77. Berk, A., et al. University Press of America, 4720 Boston Way, Lanham, MD 20706. (301) 459-3366. 1984. $25.00.

ONLINE DATA BASES

DISSERTATION ABSTRACTS ONLINE. University Microfilms International, 300 North Zeeb Road, Ann Arbor, MI 48106. (800) 521-0600 or (313) 761-4700. Scope includes virtually all doctoral dissertations accepted at accredited American institutions from 1861 to present in 252 subject areas. Inquire as to cost and availability.

METEOROLOGICAL AND GEOASTROPHYSICAL ABSTRACTS. American Meteorological Society, 45 Beacon Street, Boston, MA 02108. (617) 227-2425. 1950 to present. Monthly. $450.00 per year.

NTIS. National Technical Information Service, 5285 Port Royal Road, Springfield, VA 22161. (703) 487-4630. Broad coverage of government sponsored research reports, 1964 to present. Inquire as to cost and availability.

SCISEARCH. Institute for Scientific Information, 3501 Market Street, Philadelphia, PA 19104. (800) 523-1850 or (215) 386-0100. Broad multidisciplinary title and author index to the international literature of science and technology, 1974 to present. Inquire as to cost and availability.

WILSONLINE. H.W. Wilson Company, 950 University Avenue, Bronx, NY 10452. (800) 367-6770 or (212) 588-8400. Makes available online versions of the printed H.W. Wilson Indexes including Applied Science and Technology Index, Business Periodicals Index, and Readers' Guide to Periodical Literature. Period covered is generally 1983 to present. Inquire as to cost and availability.

SCIENTIFIC AND TECHNICAL ORGANIZATIONS AND AGENCIES DIRECTORY. Gale Research Company, Book Tower, Detroit, MI 48226. (800) 521-0707. Second edition. 1987. $185.00.

PERIODICALS

AGRICULTURAL AND FOREST METEOROLOGY. Elsevier Science Publishing Company, Incorporated, 52 Vanderbilt Avenue, New York, NY 10017. (212) 370-5520. 1964 to present. Monthly. $260.00 per year.

AMERICAN METEOROLOGICAL SOCIETY BULLETIN. American Meteorological Society, 45 Beacon Street, Boston, Ma 02108. (617) 227-2425. 1920 to present. Monthly. $60.00 per year.

CLIMATIC CHANGE: AN INTERDISCIPLINARY, INTERNATIONAL JOURNAL DEVOTED TO THE DESCRIPTION, CAUSES AND IMPLICATIONS OF CLIMATIC CHANGE. D. Reidel Publishing Company, 190 Old Derby Street, Hingham, MA 02043. (617) 871-6600. 1977 to present. Six times per year. $125.00 per year.

JOURNAL OF CLIMATE AND APPLIED METEOROLOGY. American Meteorological Society, 45 Beacon Street, Boston, MA 02108. (617) 227-2425. 1962 to present. Monthly. $120.00 per year.

MONTHLY WEATHER REVIEW. American Meteorological Society, 45 Beacon Street, Boston, MA 02108. (617) 227-2425. 1872 to present. Monthly. $120.00 per year.

NATIONAL WEATHER DIGEST. National Weather Association, 4400 Stamp Road, Room 404, Temple Hills, MD 20748. (301) 899-3784. 1976 to present. Quarterly. $20.00 per year.

RESEARCH CENTERS AND INSTITUTES

CENTER FOR CLIMATE RESEARCH. Columbia University, Lamont-Doherty Geological Observatory, Palisades, NY 10964. (914) 359-2900.

CENTER FOR CLIMATIC RESEARCH. University of Delaware, Department of Geography, Newark, DE 19716. (302) 451-8998.

NATIONAL CENTER FOR ATMOSPHERIC RESEARCH. Box 3000, Boulder, CO 80307. (303) 497-1000.

DRYING

See: INDUSTRIAL ENGINEERING

DUST STORMS

See also: AIR POLLUTION, CLIMATE, DESERTIFICATION, DROUGHT, METEOROLOGY

ABSTRACT SERVICES AND INDEXES

GENERAL SCIENCE INDEX. H.W. Wilson Company, 950 University Avenue, Bronx, NY 10452. (800) 367-6770 or (212) 588-8400.

METEOROLOGICAL AND GEOASTROPHYSICAL ABSTRACTS. American Meteorological Society, 45 Beacon Street, Boston, MA 02108. (617) 227-2425. 1950 to present. Monthly. $450.00 per year.

SCIENCE CITATION INDEX. Institute for Scientific Information, 3501 Market Street, Philadelphia, PA 19104. (800) 523-1850 or (215) 386-0100. Inquire as to cost and availability.

ASSOCIATIONS AND PROFESSIONAL SOCIETIES

AMERICAN ASSOCIATION STATE CLIMATOLOGISTS. c/o Professor John Griffiths, Meteorology Department, O and M Building, Texas A and M University, College Station, TX 77843. (409) 845-7320.

AMERICAN METEOROLOGICAL SOCIETY. 45 Beacon Street, Boston, MA 02108. (617) 227-2425.

INTERNATIONAL ASSOCIATION OF METEOROLOGY AND ATMOSPHERIC PHYSICS. UCAR, Post Office Box 3000, Boulder, CO 80307.

NATIONAL ENVIRONMENTAL SATELLITE DATA, AND INFORMATION SERVICE. 3300 Whitehaven Street, NW, Washington, DC 20235. (202) 634-7318.

NATIONAL WEATHER ASSOCIATION. 4400 Stamp Road, Room 404, Temple Hills, MD 20748. (301) 899-3784.

UNIVERSITY CORPORATION FOR ATMOSPHERIC RESEARCH. Box 3000, 1850 Table Mesa Drive, Boulder, CO 80307. (303) 497-1000.

WEATHER MODIFICATION ASSOCIATION. Post Office Box 8116, Fresno, CA 93747. (209) 291-8466.

DIRECTORIES AND BIOGRAPHICAL SOURCES

AMERICAN MEN AND WOMEN OF SCIENCE. Physical and Biological Sciences. Sixteenth edition. R.R. Bowker Company, 205 East 42nd Street, New York, NY 10017. (800) 521-8110 or (212) 916-1600. 1986. $595.00.

INTERDOC: DIRECTORY OF PUBLISHED PROCEEDINGS, SERIES. SEMT-Science/Engineering/Medicine/Technology. Interdoc Corporation, 173 Halstead Avenue, Box 326, Harrison, NY 10528. (014) 835-3506. Ten times per year. $325.00 per year.

METEOROLOGICAL SERVICES OF THE WORLD. World Meteorological Services Organization. Available from: American Meteorological Society, 45 Beacon Street, Boston, MA 02108. (617) 227-2425. Annual. $35.00.

NATIONAL WEATHER SERVICE OFFICES AND STATIONS. National Oceanic and Atmospheric Administration, Department of Commerce, Silver Spring, MD 20910. (301) 427-7698. Annual. Free.

ENCYCLOPEDIAS AND DICTIONARIES

ENCYCLOPEDIA OF CLIMATOLOGY. John E. Oliver and Rhodes W. Fairbridge, editors. Van Nostrand Reinhold, Incorporated, 115 Fifth Avenue, New York, NY 10003. (800) 543-2681. 1987. $89.95.

ENCYCLOPEDIA OF PHYSICAL SCIENCE AND TECHNOLOGY. Academic Press, Incorporated, Orlando, FL 32887. (800) 321-5068 or (305) 345-2734. Fifteen volumes, 1986.

GENERAL WORKS

THE DUST BOWL: AN AGRICULTURAL AND SOCIAL HISTORY. Douglas R. Hurt. Nelson-Hall, Incorporated, 111 North Canal Street, Chicago, IL 60606. (312) 930-9446. 1981. $22.95.

ONLINE DATA BASES

DISSERTATION ABSTRACTS ONLINE. University Microfilms International, 300 North Zeeb Road, Ann Arbor, MI 48106. (800) 521-0600 or (313) 761-4700. Scope includes virtually all doctoral dissertations accepted at accredited American institutions from 1861 to present in 252 subject areas. Inquire as to cost and availability.

METEOROLOGICAL AND GEOASTROPHYSICAL ABSTRACTS. American Meteorological Society, 45 Beacon Street, Boston, MA 02108. (617) 227-2425. 1950 to present. Monthly. $450.00 per year.

NTIS. National Technical Information Service, 5285 Port Royal Road, Springfield, VA 22161. (703) 487-4630. Broad coverage of government sponsored research reports, 1964 to present. Inquire as to cost and availability.

SCISEARCH. Institute for Scientific Information, 3501 Market Street, Philadelphia, PA 19104. (800) 523-1850 or (215) 386-0100. Broad multidisciplinary title and author index to the international literature of science and technology, 1974 to present. Inquire as to cost and availability.

PERIODICALS

AGRICULTURAL AND FOREST METEOROLOGY. Elsevier Science Publishing Company, Incorporated, 52 Vanderbilt Avenue, New York, NY 10017. (212) 370-5520. 1964 to present. Monthly. $260.00 per year.

AMERICAN METEOROLOGICAL SOCIETY BULLETIN. American Meteorological Society, 45 Beacon Street, Boston, Ma 02108. (617) 227-2425. 1920 to present. Monthly. $60.00 per year.

CLIMATIC CHANGE: AN INTERDISCIPLINARY, INTERNATIONAL JOURNAL DEVOTED TO THE DESCRIPTION, CAUSES AND IMPLICATIONS OF CLIMATIC CHANGE. D. Reidel Publishing Company, 190 Old Derby Street, Hingham, MA 02043. (617) 871-6600. 1977 to present. Six times per year. $125.00 per year.

JOURNAL OF THE ATMOSPHERIC SCIENCES. American Meteorological Society, 45 Beacon Street, Boston, MA 02108. (617) 227-2425. 1944 to present. Semimonthly. $220.00 per year.

MONTHLY WEATHER REVIEW. American Meteorological Society, 45 Beacon Street, Boston, MA 02108. (617) 227-2425. 1872 to present. Monthly. $120.00 per year.

NATIONAL WEATHER DIGEST. National Weather Association, 4400 Stamp Road, Room 404, Temple Hills, MD 20748. (301) 899-3784. 1976 to present. Quarterly. $20.00 per year.

WEATHER. Royal Meteorological Society, James Glaisher House, Grenville Place, Bracknell Berkshire, RG12 1BX, England. 1946 to present. $30.00 per year.

WEATHERWISE. Heldref Publications, 4000 Albemarle Street, NW, Washington, DC 20016. (212) 362-6445. 1948 to present. Bimonthly. $20.00 per year.

RESEARCH CENTERS AND INSTITUTES

NATIONAL CENTER FOR ATMOSPHERIC RESEARCH. Box 3000, Boulder, CO 80307. (303) 497-1000.

NATIONAL SEVERE STORMS LABORATORY. 1313 Halley Circle, Norman, OK 73069. (405) 360-3620.

DWARF GALAXY

See: GALAXIES

DYNAMICS

See: MECHANICS

DYNAMOS

See: GENERATORS

E

EARTH

See: SOLAR SYSTEM

EARTH SCIENCES

See: GEOLOGY, METEOROLOGY, OCEANOGRAPHY

EARTHQUAKE ENGINEERING

See also: CIVIL ENGINEERING, CONSTRUCTION ENGINEERING, EARTHQUAKES, GEOPHYSICS, GEOTECHNICAL ENGINEERING, ROCK MECHANICS, SEISMOLOGY, STRUCTURAL ENGINEERING

ABSTRACT SERVICES AND INDEXES

ABSTRACT JOURNAL IN EARTHQUAKE ENGINEERING. Earthquake Engineering Research Institute, 2620 Telegraph Avenue, Berkeley, CA 94704. (415) 848-0972. Semi-annual. $70.00 per year.

APPLIED SCIENCE AND TECHNOLOGY INDEX. H.W. Wilson and Company, 950 University Avenue, Bronx, NY 10452. (800) 367-6670 or (212) 588-8400. Monthly. Inquire as to cost and availability.

BIBLIOGRAPHY AND INDEX OF GEOLOGY. American Geological Institute, 4220 King Street, Alexandria, VA 22302. (703) 379-2480. 1969 to present. Monthly. $1100.00 per year.

CHEMICAL ABSTRACTS. American Chemical Society, Chemical Abstracts Service, Box 3012, Columbus, OH 43210. (614) 421-3600. 1907 to present. Weekly. $9500.00 per year.

CONFERENCE PAPERS INDEX. Cambridge Scientific Abstracts, 5161 River Road, Bethesda, MD 20816. 1972 to present. Monthly. Inquire as to cost and availability.

CURRENT CONTENTS: ENGINEERING, TECHNOLOGY, AND APPLIED SCIENCES. Institute for Scientific Information, 3501 Market Street, Philadelphia, PA 19104. (800) 523-1850 or (215) 386-0100. Weekly. $295.00 per year.

INDEX TO SCIENTIFIC AND TECHNICAL PROCEEDINGS. Institute for Scientific Information, 3501 Market Street, Philadelphia, PA 19104. (800) 523-1850 or (215) 386-0100. 1978 to present. Monthly. $775.00 per year.

INTERNATIONAL CIVIL ENGINEERING ABSTRACTS. CITIS Limited, 2 Rosemount Terrace, Blackrock, Dublin, Ireland. 1974 to present. Monthly. $350.00 per year.

INTERNATIONAL STRUCTURAL ENGINEERING ABSTRACTS. CITIS Limited, 2 Rosemount Terrace, Blackrock, Dublin, Ireland. 1986 to present. Quarterly. $95.00 per year.

OCEAN ABSTRACTS. Cambridge Scientific Abstracts, 5161 River Road, Bethesda, MD 20816. (301) 951-1400. 1963 to present. Bimonthly. $450.00 per year.

PUBLICATIONS INFORMATION. American Society of Civil Engineers, 345 East 47th Street, New York, NY 10017. (212) 705-7420. Abstracts, subject and author indexes to the publications of the American Society of Civil Engineers. Bimonthly. $80.00 per year.

SCIENCE CITATION INDEX. Institute for Scientific Information, 3501 Market Street, Philadelphia, PA 19104. (800) 523-1850 or (215) 386-0100. Six times per year. $6200.00 per year.

ASSOCIATIONS AND PROFESSIONAL SOCIETIES

AMERICAN SOCIETY OF CIVIL ENGINEERS. 345 East 47th Street, New York, NY 10017. (212) 705-7420.

EARTHQUAKE ENGINEERING RESEARCH INSTITUTE. 2620 Telegraph Avenue, Berkeley, CA 94704. (415) 848-0972.

GEOLOGICAL SOCIETY OF AMERICA. 3300 Penrose Place, Boulder, CO 80301. (303) 447-2020.

SEISMOLOGICAL SOCIETY OF AMERICA. 6431 Fairmont Avenue, No. 7, El Cerrito, CA 94530. (415) 525-5474.

DIRECTORIES AND BIOGRAPHICAL SOURCES

INTERNATIONAL RESEARCH CENTERS DIRECTORY 1988-89. Darren L. Smith, editor. Gale Research Company, Book Tower, Detroit, MI 48226. (800) 521-0707. 4th edition. 1987. $360.00.

1987 DIRECTORY OF ENGINEERING SOCIETIES AND RELATED ORGANIZATIONS. Gordon Davis, editor. Hemisphere Publishing Corporation, 1010 Vermont Avenue, NW, Washington, DC 20005. (800) 526-0275. 12th edition. 1987. $100.00.

RESEARCH CENTERS DIRECTORY 1988. Gale Research Company, Book Tower, Detroit, MI 48226. (800) 521-0707. 12th edition. 1987. $365.00 for set.

SCIENTIFIC AND TECHNICAL ORGANIZATIONS AND AGENCIES DIRECTORY. Margaret Labash Young, editor. Gale Research Company, Book Tower, Detroit, MI 48226. (800) 521-0707. 2nd edition. 1987. $185.00.

WHO'S WHO IN ENGINEERING. Gordon Davis, editor. Hemisphere Publishing Corporation, 79 Madison Avenue, New York, NY 10016-7892. (800) 821-8312. Sixth edition. 1985. $200.00.

GENERAL WORKS

DAMS AND EARTHQUAKES. Institution of Civil Engineering (London). American Society of Civil Engineers, 345 East 47th Street, New York, NY 10017. (212) 705-7420. 1981. $62.50.

DESIGN OF EARTHQUAKE RESISTANT BUILDINGS. M. Wakabayashi. McGraw-Hill Book Company, 1221 Avenue of the Americas, New York, NY 10020. (212) 512-2000. 1986. $49.50.

EARTHQUAKE ENGINEERING: DAMAGE ASSESSMENT AND STRUCTURAL DESIGN. S.F. Borg. John Wiley and Sons, Inc., 605 Third Avenue, New York, NY 10158. (800) 526-5368. 1984. $39.95.

EARTHQUAKE PREDICTION. K. Shimazaki and W.D. Stuart, editors. Birkhauser Boston, Inc., 380 Green Street, Cambridge, MA 02139. (617) 876-2333. 1985. $24.95.

EARTHQUAKE RESISTANT BUILDING DESIGN AND CONSTRUCTION. Norman E. Green. Van Nostrand Reinhold Company, Inc., 135 West 50th Street, New York, NY 10020. (800) 543-2681. Second edition. 1981. $24.95.

GEOTECHNICAL ENGINEERING PRACTICES. R.E. Hunt. McGraw-Hill Book Company, 1221 Avenue of the Americas, New York, NY 10020. (212) 512-2000. 1986. $65.00.

INTRODUCTION TO EARTHQUAKE ENGINEERING. Shunzo Okamoto. Columbia University Press, 562 West 113th Street, New York, NY 10025. (212) 316-7100. 1985. $74.50.

PRACTICAL APPROACHES TO EARTHQUAKE PREDICTION AND WARNING. C. Kisslinger and T. Rikitake, editors. Kluwer Academic Publishers, 190 Old Derby Street, Hingham, MA 02043. (617) 749-5262. 1986. $79.00.

TERRA NON FIRMA: UNDERSTANDING AND PREPARING FOR EARTHQUAKES. James M. Gere and H.C. Shah. W.H. Freeman and Company, 41 Madison Avenue, New York, NY 10010. (212) 532-7660. 1984. $12.95.

HANDBOOKS AND MANUALS

BUILDING STRUCTURES HANDBOOK. R.N. White and C.G. Salmon. John Wiley and Sons, Inc., 605 Third Avenue, New York, NY 10158. (800) 526-5368. 1986. $79.95.

CIVIL ENGINEERING CALCULATIONS REFERENCE GUIDE. Tyler G. Hicks, editor. McGraw-Hill Book Company, 1221 Avenue of the Americas, New York, NY 10020. (212) 512-2000. 1987. $29.50.

CIVIL ENGINEERING PRACTICE. Paul N. Cheremisinoff and others, editors. Technomic Publishing Company, Inc., 851 Holland Avenue, Box 3535, Lancaster, PA 17604. (800) 233-9936. Five volumes. 1987-1988. $750.00 for set.

HANDBOOK OF CONCRETE ENGINEERING. Mark Fintel. Van Nostrand Reinhold Company, Inc., 135 West 50th Street, New York, NY 10020. (800) 543-2681. 1986. $89.95.

HANDBOOK OF MECHANICS, MATERIALS, AND STRUCTURES. A. Blake. John Wiley and Sons, Inc., 605 Third Avenue, New York, NY 10158. (800) 526-5368. 1985. $64.50.

STANDARD HANDBOOK FOR CIVIL ENGINEERS. F. S. Merritt, editor. McGraw-Hill Book Company, 1221 Avenue of the Americas, New York, NY 10020. (212) 512-2000. Third edition. 1983. $89.50.

STRUCTURAL ENGINEERING HANDBOOK. E.H. Gaylord and C.N. Gaylord, editors. McGraw-Hill Book Company, 1221 Avenue of the Americas, New York, NY 10020. (212) 512-2000. Second edition. 1979. $76.50.

ONLINE DATA BASES

CA SEARCH. Chemical Abstracts Service, P.O. Box 3012, Columbus, OH 43120. (800) 848-6538 or (614) 421-3600. Comprehensive guide to chemical literature, 1972 to present. Inquire as to online cost and availability.

COMPENDEX. Engineering Information, Inc., 345 East 47th Street, New York, NY 10017. (800) 221-1044 or (212) 705-7615. Engineering and technical literature, 1975 to present. Inquire as to online cost and availability.

DISSERTATION ABSTRACTS ONLINE. University Microfilms International, 300 North Zeeb Road, Ann Arbor, MI 48106. (800) 521-0600 or (313) 761-4700. Scope includes virtually all doctoral dissertations accepted at accredited American institutions from 1861 to present in over 250 subject areas. Inquire as to online cost and availability.

GEOREF. Online version of the BIBLIOGRAPHY AND INDEX OF GEOLOGY. American Geological Institute, 4220 King Street, Alexandria, VA 22302. (703) 379-2480. 1969 to present. Inquire as to online cost and availability.

NTIS. National Technical Information Service, 5285 Port Royal Road, Springfield, VA 22161. (703) 487-4630. Broad coverage of government sponsored research reports, 1964 to present. Inquire as to online cost and availability.

SCISEARCH. Institute for Scientific Information, 3501 Market Street, Philadelphia, PA 19104. (800) 523-1850 or (215) 386-0100. Broad multidisciplinary title and author index to the international literature of science and technology, 1974 to present. Inquire as to online cost and availability.

PERIODICALS

DISASTERS: THE INTERNATIONAL JOURNAL OF DISASTER STUDIES AND PRACTICE. International Disaster Institute. Subscribe to: IDI, 85 Maryleborn Hign Street, London, W1M 3DE, England. 1977 to present. Quarterly. $60.00 per year.

EARTHQUAKE ENGINEERING AND STRUCTURAL DYNAMICS. John Wiley and Sons, Inc., 605 Third Avenue, New York, NY 10158. (800) 526-5368. 1972 to present. Bimonthly. $350.00 per year.

EARTHQUAKE INFORMATION BULLETIN. U.S. Geological Survey, 12201 Sunrise Valley Drive, Reston, VA 22092. Order from: Superintendent of Documents, U.S. Government Printing Office, 20402. Bimonthly. $15.00 per year.

EARTHQUAKE NOTES. Seismological Society of America, c/o Wilbur Rinehart, 1320 Meadow Avenue, Boulder, CO 80302. 1929 to present. Quarterly. $10.00 per year.

EARTHQUAKE PREDICTION RESEARCH. D. Reidel Publishing Company, 190 Old Derby Street, Hingham, MA 02043. 1982 to present. Quarterly. $85.00 per year.

ENGINEERING GEOLOGY. Elsevier Science Publishing Company, Inc., 52 Vanderbilt Avenue, New York, NY 10017. (212) 370-5520. 1965 to present. Eight times per year. $140.00 per year.

ENGINEERING STRUCTURES: THE JOURNAL OF EARTHQUAKE, WIND AND OCEAN ENGINEERING. Butterworth's Publishing, 80 Montvale Avenue, Stoneham, MA 02180. (800) 325-4177. 1979 to present. Quarterly. $180.00 per year.

SEISMOLOGICAL SOCIETY OF AMERICA BULLETIN. Seismological Society of America, 6431 Fairmont Avenue, No. 7, El Cerrito, CA 94530. (415) 525-5474. 1911 to present. Bimonthly. $90.00 per year.

RESEARCH CENTERS AND INSTITUTES

CENTER FOR CONCRETE AND GEOMATERIALS. Northwestern University, Technological Institute, Room 2474, 2145 Sheridan Road, Evanston, IL 60201. (312) 491-3858.

EARTHQUAKE ENGINEERING RESEARCH INSTITUTE. 2620 Telegraph Avenue, Berkeley, CA 94704. (415) 848-0972.

SOIL MECANICS LABORATORY. Rensselaer Polytechnic Institute, Civil Engineering Department, Troy, NY 12180-3590. (518) 266-6213.

EARTHQUAKE PREDICTION

See: EARTHQUAKES

EARTHQUAKES

See also: EARTHQUAKE ENGINEERING, GEOLOGY, GEOPHYSICS, GEOTECHNICAL ENGINEERING, OCEANOGRAPHY, PLANETARY SCIENCE, PLATE TECTONICS, SEISMOLOGY, VOLCANOLOGY

ABSTRACT SERVICES AND INDEXES

BIBLIOGRAPHY AND INDEX OF GEOLOGY. American Geological Institute, 4220 King Street, Alexandria, VA 22302. (703) 379-2480. 1969 to present. Monthly. $1100.00 per year.

CHEMICAL ABSTRACTS. American Chemical Society, Chemical Abstracts Service, Box 3012, Columbus, OH 43210. (614) 421-3600. 1907 to present. Weekly. $9500.00 per year.

CONFERENCE PAPERS INDEX. Cambridge Scientific Abstracts, 5161 River Road, Bethesda, MD 20816. 1972 to present. Monthly. Inquire as to cost and availability.

CURRENT CONTENTS: PHYSICAL, CHEMICAL AND EARTH SCIENCES. Institute for Scientific Information, 3501 Market Street, Philadelphia, PA 19104. (800) 523-1850 or (215) 386-0100. Weekly. $295.00 per year.

INDEX TO SCIENTIFIC AND TECHNICAL PROCEEDINGS. Institute for Scientific Information, 3501 Market Street, Philadelphia, PA 19104. (800) 523-1850 or (215) 386-0100. 1978 to present. Monthly. $775.00 per year.

INDEX TO SCIENTIFIC REVIEWS. Institute for Scientific Information, 3501 Market Street, Philadelphia, PA 19104. (800) 523-1850 or (215) 386-0100. 1974 to present. Semi-annual. $550.00 per year.

METEOROLOGICAL AND GEOASTROPHYSICAL ABSTRACTS. American Meteorological Society, 45 Beacon Street, Boston, MA 02108. (617) 227-2425. 1950 to present. Monthly. $450.00 per year.

OCEAN ABSTRACTS. Cambridge Scientific Abstracts, 5161 River Road, Bethesda, MD 20816. (301) 951-1400. 1963 to present. Bimonthly. $450.00 per year.

PHYSICS ABSTRACTS. Institution of Electrical Engineers. Available from: IEEE Service Center, 445 Hoes Lane, Piscataway, NJ 08854. 1898 to present. Bimonthly. $1700.00 per year.

PHYSICS BRIEFS. Physik Verlag GmbH, Postfach 1260/1280, D-6940 Weinheim, West Germany. (212) 661-9404. 1920 to present. Twenty-six times per year. $1250.00 per year.

SCIENCE CITATION INDEX. Institute for Scientific Information, 3501 Market Street, Philadelphia, PA 19104. (800) 523-1850 or (215) 386-0100. Six times per year. $6200.00 per year.

ASSOCIATIONS AND PROFESSIONAL SOCIETIES

AMERICAN ASTRONOMICAL SOCIETY. 1816 Jefferson Place, N.W., Washington, DC 20036. (202) 659-0134.

AMERICAN GEOPHYSICAL UNION. 2000 Florida Avenue, N.W., Washington, DC 20009. (202) 462-6903.

AMERICAN INSTITUTE OF PHYSICS. 335 East 45th Street, New York, NY 10017. (212) 661-9494.

EARTHQUAKE ENGINEERING RESEARCH INSTITUTE. 2620 Telegraph Avenue, Berkeley, CA 94704. (415) 848-0972.

GEOLOGICAL SOCIETY OF AMERICA. 3300 Penrose Place, Boulder, CO 80301. (303) 447-2020.

SEISMOLOGICAL SOCIETY OF AMERICA. 6431 Fairmont Avenue, No. 7, El Cerrito, CA 94530. (415) 525-5474.

DIRECTORIES AND BIOGRAPHICAL SOURCES

AMERICAN MEN AND WOMEN OF SCIENCE. R.R. Bowker, Inc., Order Department, 245 West 17th Street, New York, NY 10011. (800) 521-8110. Eight volumes. 1986. $595.00 for set.

INTERNATIONAL RESEARCH CENTERS DIRECTORY 1988-89. Darren L. Smith, editor. Gale Research Company, Book Tower, Detroit, MI 48226. (800) 521-0707. 4th edition. 1987. $360.00.

1987 DIRECTORY OF ENGINEERING SOCIETIES AND RELATED ORGANIZATIONS. Gordon Davis, editor. Hemisphere Publishing Corporation, 1010 Vermont Avenue, NW, Washington, DC 20005. (800) 526-0275. 12th edition. 1987. $100.00.

RESEARCH CENTERS DIRECTORY 1988. Gale Research Company, Book Tower, Detroit, MI 48226. (800) 521-0707. 12th edition. 1987. $365.00 for set.

SCIENTIFIC AND TECHNICAL ORGANIZATIONS AND AGENCIES DIRECTORY. Margaret Labash Young, editor. Gale Research Company, Book Tower, Detroit, MI 48226. (800) 521-0707. 2nd edition. 1987. $185.00.

SOCIETY OF EXPLORATION GEOPHYSICISTS ROSTER. P.O. Box 702740, Tulsa, OK 74170. (918) 493-3516. Annual. Inquire.

WHO'S WHO IN ENGINEERING. Gordon Davis, editor. Hemisphere Publishing Corporation, 79 Madison Avenue, New York, NY 10016-7892. (800) 821-8312. Sixth edition. 1985. $200.00.

ENCYCLOPEDIAS AND DICTIONARIES

MCGRAW-HILL ENCYCLOPEDIA OF THE GEOLOGICAL SCIENCES. McGraw-Hill Book Company, 1221 Avenue of the Americas, New York, NY 10020. (212) 512-2000. Second edition. 1988. $85.00.

GENERAL WORKS

DAMS AND EARTHQUAKES. Institution of Civil Engineering (London). American Society of Civil Engineers, 345 East 47th Street, New York, NY 10017. (212) 705-7420. 1981. $62.50.

EARTHQUAKE PREDICTION. K. Shimazaki and W.D. Stuart, editors. Birkhauser Boston, Inc., 380 Green Street, Cambridge, MA 02139. (617) 876-2333. 1985. $24.95.

EARTHQUAKES, TIDES, UNIDENTIFIED SOUNDS AND RELATED PHENOMENA. William R. Corliss. The Sourcebook Project, P.O. Box 107, Glen Arm, MD 21057. (301) 668-6047. 1983. $12.95.

THE GREAT WAVES. Douglas Myles. McGraw-Hill Book Company, 1221 Avenue of the Americas, New York, NY 10020. (212) 512-2000. 1985. $16.95.

INSIDE THE EARTH: EVIDENCE FROM EARTHQUAKES. Bruce A. Bolt. W.H. Freeman and Company, 41 Madison Avenue, New York, NY 10010. (212) 532-7660. 1982. $24.95.

AN INTRODUCTION TO THE THEORY OF SEISMOLOGY. K.E. Bullen and A.B. Bolt. Cambridge University Press, 32 East 57th Street, New York, NY 10022. (800) 872-7423. 1985. $69.50.

PRACTICAL APPROACHES TO EARTHQUAKE PREDICTION AND WARNING. C. Kisslinger and T. Rikitake, editors. Kluwer Academic Publishers, 190 Old Derby Street, Hingham, MA 02043. (617) 749-5262. 1986. $79.00.

SEISMIC WAVES AND SOURCES. A. Ben-Menahem and S. Singh. Springer-Verlag New York, Inc., 175 Fifth Avenue, New York, NY 10010. (800) 526-7254. 1981. $99.00.

TERRA NON FIRMA: UNDERSTANDING AND PREPARING FOR EARTHQUAKES. James M. Gere and H.C. Shah. W.H. Freeman and Company, 41 Madison Avenue, New York, NY 10010. (212) 532-7660. 1984. $12.95.

ONLINE DATA BASES

CA SEARCH. Chemical Abstracts Service, P.O. Box 3012, Columbus, OH 43120. (800) 848-6538 or (614) 421-3600. Comprehensive guide to chemical literature, 1972 to present. Inquire as to online cost and availability.

COMPENDEX. Engineering Information, Inc., 345 East 47th Street, New York, NY 10017. (800) 221-1044 or (212) 705-7615. Engineering and technical literature, 1975 to present. Inquire as to online cost and availability.

DISSERTATION ABSTRACTS ONLINE. University Microfilms International, 300 North Zeeb Road, Ann Arbor, MI 48106. (800) 521-0600 or (313) 761-4700. Scope includes virtually all doctoral dissertations accepted at accredited American institutions from 1861 to present in over 250 subject areas. Inquire as to online cost and availability.

GEOREF. Online version of the BIBLIOGRAPHY AND INDEX OF GEOLOGY. American Geological Institute, 4220 King Street, Alexandria, VA 22302. (703) 379-2480. 1969 to present. Inquire as to online cost and availability.

INSPEC. INSPEC Marketing Department, Institution of Electrical Engineers. Available from IEEE Service Center, 445 Hoes Lane, Piscataway, NJ 08854. (201) 981-0060. Online version of Physics Abstracts. Inquire as to online cost and availability.

NTIS. National Technical Information Service, 5285 Port Royal Road, Springfield, VA 22161. (703) 487-4630. Broad coverage of government sponsored research reports, 1964 to present. Inquire as to online cost and availability.

SCISEARCH. Institute for Scientific Information, 3501 Market Street, Philadelphia, PA 19104. (800) 523-1850 or (215) 386-0100. Broad multidisciplinary title and author index to the international literature of science and technology, 1974 to present. Inquire as to online cost and availability.

PERIODICALS

DISASTERS: THE INTERNATIONAL JOURNAL OF DISASTER STUDIES AND PRACTICE. International Disaster Institute. Subscribe to: IDI, 85 Maryleborn High Street, London, W1M 3DE, England. 1977 to present. Quarterly. $60.00 per year.

EARTHQUAKE INFORMATION BULLETIN. U.S. Geological Survey, 12201 Sunrise Valley Drive, Reston, VA 22092. Order from: Superintendent of Documents, U.S. Government Printing Office, 20402. Bimonthly. $15.00 per year.

EARTHQUAKE NOTES. Seismological Society of America, c/o Wilbur Rinehart, 1320 Meadow Avenue, Boulder, CO 80302. 1929 to present. Quarterly. $10.00 per year.

EARTHQUAKE PREDICTION RESEARCH. D. Reidel Publishing Company, 190 Old Derby Street, Hingham, MA 02043. 1982 to present. Quarterly. $85.00 per year.

GEOPHYSICAL AND ASTROPHYSICAL FLUID DYNAMICS. Gordon and Breach Science Publishers, Inc., 50 West 23rd Street, New York, NY 10010. (212) 206-8900. 1970 to present. 16 times per year. $260.00 per year.

GEOPHYSICAL RESEARCH LETTERS. American Geophysical Union, 2000 Florida Avenue, N.W., Washington, DC 20009. (202) 462-6903. 1974 to present. Monthly. $185.00 per year.

JGR: JOURNAL OF GEOPHYSICAL RESEARCH: SOLID EARTH AND PLANETS. American Geophysical Union, 2000 Florida Avenue, N.W., Washington, DC 20009. (202) 462-6903. Monthly. $760.00 per year.

JOURNAL OF VOLCANOLOGY AND GEOTHERMAL RESEARCH. Elsevier Science Publishing Company, Inc., 52 Vanderbilt Avenue, New York, NY 10017. (212) 370-5520. 1976 to present. 16 times per year. $360.00 per year.

SEISMOLOGICAL SOCIETY OF AMERICA BULLETIN. Seismological Society of America, 6431 Fairmont Avenue, No. 7, El Cerrito, CA 94530. (415) 525-5474. 1911 to present. Bimonthly. $90.00 per year.

TECTONOPHYSICS: AN INTERNATIONAL JOURNAL OF GEOTECTONICS AND THE GEOLOGY AND PHYSICS OF THE INTERIOR OF THE EARTH. Elsevier Science Publishing Company, Inc., 52 Vanderbilt Avenue, New York, NY 10017. (212) 370-5520. 1964 to present. $1200.00 per year.

VOLCANOLOGY AND SEISMOLOGY. Gordon and Breach Science Publishers, Inc., 50 West 23rd Street, New York, NY 10010. (212) 206-8900. Monthly. $500.00 per year.

RESEARCH CENTERS AND INSTITUTES

EARTHQUAKE ENGINEERING RESEARCH INSTITUTE. 2620 Telegraph Avenue, Berkeley, CA 94704. (415) 848-0972.

LAMONT-DOHERTY GEOLOGICAL OBSERVATORY. Columbia University, Palisades, NY 10964. (914) 359-2900.

SEISMOGRAPHIC STATIONS. University of California at Berkeley, Berkeley, CA 94720. (415) 642-3977.

SEISMOLOGICAL LABORATORY. California Institute of Technology, Pasadena, CA 91125. (818) 356-6912.

ECHO

See: RADAR, SONAR

ECLIPSES

ABSTRACT SERVICES AND INDEXES

ASTRONOMY AND ASTROPHYSICS ABSTRACTS. Springer-Verlag New York, Incorporated, 175 Fifth Avenue, New York, NY 10010. (212) 460-1500. $70.00 per year.

ASTRONOMY AND ASTROPHYSICS MONTHLY INDEX. Olivetree Associates, Post Office Box 236, Sierra Madre, CA 91024. $212.00 per year. Complimentary copies available on request.

GENERAL SCIENCE INDEX. H.W. Wilson Company, 950 University Avenue, Bronx, NY 10452. (800) 367-6770 or (212) 588-8400. Inquire as to cost and availability.

SCIENCE CITATION INDEX. Institute for Scientific Information, 3501 Market Street, Philadelphia, PA 19104. (800) 523-1850 or (215) 386-0100. Inquire as to cost and availability.

STAR. (Scientific and Technical Aerospace Reports. United States National Aeronautics and Space Administration, Scientific and Technical Information Facility, Box 8757, Baltimore-Washington International Airport, MD 21240. (202) 755-2210. Semimonthly, with semiannual and annual indexes. $85.00 per year.

ANNUAL REVIEWS AND YEARBOOKS

THE ASTRONOMICAL ALMANAC. Superintendent of Documents, United States Government Printing Office, Washington, DC 20402. (202) 783-3238. Yearly.

ANNUAL REVIEW OF EARTH AND PLANETARY SCIENCES. Annual Reviews, Incorporated, 4139 East Camino Way, Palo Alto, CA 94306. (415) 493-4400.

ASSOCIATIONS AND PROFESSIONAL SOCIETIES

AMERICAN ASTRONOMICAL SOCIETY. 2000 Florida Avenue, NW, Suite 300, Washington, DC 20009. (202) 659-0134.

ASTRONOMICAL LEAGUE. Post Office Box 12821, Tucson, AZ 85732. (602) 790-8471.

ASTRONOMICAL SOCIETY OF THE PACIFIC. 1290 24th Avenue, San Francisco, CA 94122. (415) 661-8660.

BIBLIOGRAPHIES

A BIBLIOGRAPHY OF ASTRONOMY, 1970-1979. R.A. Seal and S.S. Martin. Libraries Unlimited, Incorporated, Littleton, CO 80160. 1982.

ENCYCLOPEDIAS AND DICTIONARIES

ENCYCLOPEDIA OF PHYSICAL SCIENCE AND TECHNOLOGY. Academic Press, Incorporated, Orlando, FL 32887. (800) 321-5068 or (305) 345-2734. Inquire as to cost and availability.

MCGRAW-HILL ENCYCLOPEDIA OF SCIENCE AND TECHNOLOGY. McGraw-Hill Book, Incorporated, 1221 Avenue of the Americas, New York, NY 10020. (212) 997-3675. Fifth edition, 15 volumes. $1100.00.

ONLINE DATA BASES

DISSERTATION ABSTRACTS ONLINE. University Microfilms International, 300 North Zeeb Road, Ann Arbor, MI 48106. (800) 521-0600 or (313) 761-4700. Scope includes virtually all doctoral dissertations accepted at accredited American institutions from 1861 to present in 252 subject areas. Inquire as to cost and availability.

NASA. National Aeronautics and Space Administration, Scientific and Technical Information Branch, 300 7th Street, SW, Washington, DC 20546. Citations and abstracts of aerospace literature, 1962 to present. Inquire as to cost and availability.

NTIS. National Technical Information Service, 5285 Port Royal Road, Springfield, VA 22161. (703) 487-4630. Broad coverage of government sponsored research reports, 1964 to present. Inquire as to cost and availability.

SCISEARCH. Institute for Scientific Information, 3501 Market Street, Philadelphia, PA 19104. (800) 523-1850 or (215) 386-0100. Broad multidisciplinary title and author index to the international literature of science and technology, 1974 to present. Inquire as to cost and availability.

WILSONLINE. H.W. Wilson Company, 950 University Avenue, Bronx, NY 10452. (800) 367-6770 or (212) 588-8400. Makes available online versions of the printed H.W. Wilson Indexes including Applied Science and Technology Index, Business Periodicals Index, and Readers' Guide to Periodical Literature. Period covered is generally 1983 to present. Inquire as to cost and availability.

PERIODICALS

ASTRONOMICAL JOURNAL. American Astronomical Society. Available from: American Institute of Physics, 335 East 45th Street, New York, NY 10017. (212) 661-9404. $125.00 per year.

ASTRONOMICAL SOCIETY OF THE PACIFIC. Publications. Astronomical Society of the Pacific, 1290 24th Avenue, San Francisco, CA 94122. (415) 661-8660. Monthly. $38.00.

EARTH, MOON AND PLANETS: AN INTERNATIONAL JOURNAL OF COMPARATIVE PLANETOLOGY. D. Publishing Company, 190 Old Derby Street, Hingham, MA 02043. Nine times per year. $275.00 per year.

ICARUS: INTERNATIONAL JOURNAL OF THE SOLAR SYSTEM STUDIES. Academic Press, Incorporated, Orlando, FL 32887. (305) 345-4100. Monthly. $484.00 per year.

MERCURY. Astronomical Society of the Pacific, 1290 245h Avenue, San Francisco, CA 94122. (415) 661-8660. Bimonthly. $21.00 per year.

MONTHLY NOTICES OF THE ROYAL ASTRONOMICAL SOCIETY. Blackwell Science Publications, Incorporated, 667 Lytton Avenue, Palo Alto, CA 94301. (415) 324-1688. Monthly. $134.00 per year.

PLANETARY AND SPACE SCIENCE. Pergamon Press, Incorporated, Maxwell House, Fairview Park, Elmsford, NY 10523. (014) 592-7700. Monthly. $430.00 per year.

VISTAS IN ASTRONOMY. Pergamon Press, Incorporated, Maxwell House, Fairview Park, Elmsford, NY 10523. (914) 592-7700. Quarterly. $145.00 per year.

EDP (ELECTRONIC DATA PROCESSING)

See: DATA PROCESSING

EFFLUENT

See: ENVIRONMENTAL ENGINEERING

EL NINO

See: OCEANOGRAPHY

ELASTICITY

See: MECHANICS

ELECTRIC CURRENT

See: ELECTRICITY

ELECTRIC MOTORS

See also: ELECTRIC POWER ENGINEERING, ELECTRICAL CODES, ELECTRICAL ENGINEERING, ELECTRICITY, ELECTROMAGNETISM, GENERATORS

ABSTRACT SERVICES AND INDEXES

APPLIED MECHANICS REVIEW. American Society of Mechanical Engineers, 345 East 47th Street, New York, NY 10017. (212) 705-7703. 1948 to present. Monthly. $360.00 per year.

APPLIED SCIENCE AND TECHNOLOGY INDEX. H.W. Wilson and Company, 950 University Avenue, Bronx, NY 10452. (800) 367-6670 or (212) 588-8400. Monthly. Inquire as to cost and availability.

CHEMICAL ABSTRACTS. American Chemical Society, Chemical Abstracts Service, Box 3012, Columbus, OH 43210. (614) 421-3600. 1907 to present. Weekly. $9500.00 per year.

CURRENT CONTENTS: ENGINEERING, TECHNOLOGY AND APPLIED SCIENCES. Institute for Scientific Information, 3501 Market Street, Philadelphia, PA 19104. (800) 523-1850 or (215) 386-0100. Weekly. $275.00 per year.

ELECTRIC POWER INDUSTRY ABSTRACTS. Edison Electric Institute, c/o Utility Data Institute, 2011 I Street, N.W., Suite 700, Washington, DC 20006. 1975 to present. Bimonthly. Inquire as to cost and availability.

ELECTRICAL AND ELECTRONICS ABSTRACTS. Institution of Electrical Engineers. Available from: Institute of Electrical and Electronics Engineers. IEEE Service Center, 445 Hoes Lane, Piscataway, NJ 08854. Monthly. $1250.00 per year.

ENGINEERING INDEX MONTHLY AND AUTHOR INDEX. Engineering Information Inc., 345 East 47th Street, New York, NY 10017. (212) 705-7600. Monthly. $1560.00 per year.

IEEE PUBLICATIONS BULLETIN. Institute of Electrical and Electronics Engineers. Institute of Electrical and Electronics Engineers. IEEE Service Center, 445 Hoes Lane, Piscataway, NJ 08854. Quarterly. Free.

PHYSICS ABSTRACTS. Institution of Electrical Engineers. Available from: IEEE Service Center, 445 Hoes Lane, Piscataway, NJ 08854. 1898 to present. Bimonthly. $1700.00 per year.

PHYSICS BRIEFS. Physik Verlag GmbH, Postfach 1260/1280, D-6940 Weinheim, West Germany. (212) 661-9404. 1920 to present. Twenty-six times per year. $1250.00 per year.

SCIENCE CITATION INDEX. Institute for Scientific Information, 3501 Market Street, Philadelphia, PA 19104. (800) 523-1850 or (215) 386-0100. Six times per year. $6200.00 per year.

ASSOCIATIONS AND PROFESSIONAL SOCIETIES

AMERICAN ELECTRONICS ASSOCIATION. P.O. Box 10045, 2670 Hanover Street, Palo Alto, CA 94303. (415) 857-9300.

EDISON ELECTRIC INSTITUTE. 1111 19th Street, N.W., Washington, DC 20036. (202) 828-7400.

INSTITUTE OF ELECTRICAL AND ELECTRONICS ENGINEERS. 345 East 47th Street, New York, NY 10017. (212) 705-7900.

NATIONAL ELECTRICAL MANUFACTURERS ASSOCIATION. 2101 L Street, N.W., Washington, DC. (202) 457-8400.

DIRECTORIES AND BIOGRAPHICAL SOURCES

ELECTRICAL WORLD DIRECTORY OF ELECTRICAL UTILITIES. McGraw-Hill Book Company, 1221 Avenue of the Americas, New York, NY 10020. (212) 512-2000. Annual. $275.00.

1987 DIRECTORY OF ENGINEERING SOCIETIES AND RELATED ORGANIZATIONS. Gordon Davis, editor. Hemisphere Publishing Corporation, 1010 Vermont Avenue, NW, Washington, DC 20005. (800) 526-0275. 12th edition. 1987. $100.00.

RESEARCH CENTERS DIRECTORY 1988. Gale Research Company, Book Tower, Detroit, MI 48226. (800) 521-0707. 12th edition. 1987. $365.00 for set.

SCIENTIFIC AND TECHNICAL ORGANIZATIONS AND AGENCIES DIRECTORY. Margaret Labash Young, editor. Gale Research Company, Book Tower, Detroit, MI 48226. (800) 521-0707. 2nd edition. 1987. $185.00.

WHO'S WHO IN ELECTRONICS. Harris Publishing Company, 2057-2 Aurora Road, Twinsburg, OH 44087. (216) 425-9143. Annual. $90.00.

WHO'S WHO IN ENGINEERING. Gordon Davis, editor. Hemisphere Publishing Corporation, 1010 Vermont Avenue, NW, Washington, DC 20005. (800) 526-0275. 6th edition. 1985. $200.00.

ENCYCLOPEDIAS AND DICTIONARIES

MCGRAW-HILL ENCYCLOPEDIA OF SCIENCE AND TECHNOLOGY. McGraw-Hill Book Company, 1221 Avenue of the Americas, New York, NY 10020. (212) 512-2000. 6th edition. 1987. $1600.00.

THESAURUS OF SCIENTIFIC, TECHNICAL, AND ENGINEERING TERMS. Hemisphere Publishing Corporation, 1010 Vermont Avenue, NW, Washington, DC 20005. (800) 526-0275. 1988. $125.00.

GENERAL WORKS

ELECTRIC MACHINERY. Peter F. Ryff. Prentice-Hall Publishing, Inc., Englewood Cliffs, NJ 07632. (800) 562-0245. 1988. $34.00.

ELECTRICAL ENGINEERING. W.H. Roadstrum and Dan H. Wolaver. Harper and Row Publishers, Inc., 10 East 53rd Street, New York, NY 10022. (800) 242-7737. 1986. $37.50.

ELECTRICAL MACHINES: AN INTRODUCTION TO PRINCIPLES AND CHARACTERISTICS. J.D. Edwards. Macmillan Publishing Company, Inc., 866 Third Avenue, New York, NY 10022. (800) 257-5755. 2nd edition. 1987. $32.50.

ELECTRICITY AND MAGNETISM. M.H. Nayfeh and M.K. Brussel. John Wiley and Sons, Inc., 605 Third Avenue, New York, NY 10158. (800) 526-5368. 1985. $32.95.

ELECTROMAGNETIC CONCEPTS AND PRINCIPLES. Stanley V. Marshall and Gabriel G. Skitek. Prentice-Hall Publishing, Inc., Englewood Cliffs, NJ 07632. (800) 562-0245. 2nd edition. 1987. $42.95.

ELEMENTS OF ENGINEERING ELECTROMAGNETICS. N.N. Rao. Prentice-Hall Publishing, Inc., Englewood Cliffs, NJ 07632. (800) 562-0245. 2nd edition. 1987. $45.95.

FUNDAMENTALS OF ELECTRIC CIRCUITS. David A. Bell. Prentice-Hall Publishing, Inc., Englewood Cliffs, NJ 07632. (800) 562-0245. 4th edition. 1988. $30.25.

INTRODUCTION TO ELECTRICITY AND ELECTRONICS FOR THE COMPUTER AGE. R. Rosen. John Wiley and Sons, Inc., 605 Third Avenue, New York, NY 10158. (800) 526-5368. 1987. $35.95.

HANDBOOKS AND MANUALS

HANDBOOK OF ELECTRIC MACHINES. Syed A. Nasar. McGraw-Hill Book Company, 1221 Avenue of the Americas, New York, NY 10020. (212) 512-2000. 1987. $59.50.

HANDBOOK OF MODERN ELECTRONICS AND ELECTRICAL ENGINEERING. Charles Belove, editor. John Wiley and Sons,

Inc., 605 Third Avenue, New York, NY 10158. (800) 526-5368. 1986. $88.95.

ONLINE DATA BASES

CA SEARCH. Chemical Abstracts Service, P.O. Box 3012, Columbus, OH 43120. (800) 848-6538 or (614) 421-3600. Comprehensive guide to chemical literature, 1972 to present. Inquire as to online cost and availability.

COMPENDEX. Engineering Information, Inc., 345 East 47th Street, New York, NY 10017. (800) 221-1044 or (212) 705-7615. Engineering and technical literature, 1975 to present. Inquire as to online cost and availability.

INSPEC. INSPEC Marketing Department, Institution of Electrical Engineers. Available from IEEE Service Center, 445 Hoes Lane, Piscataway, NJ 08854. (201) 981-0060. Online version of Physics Abstracts. Inquire as to online cost and availability.

NTIS. National Technical Information Service, 5285 Port Royal Road, Springfield, VA 22161. (703) 487-4630. Broad coverage of government sponsored research reports, 1964 to present. Inquire as to online cost and availability.

SCISEARCH. Institute for Scientific Information, 3501 Market Street, Philadelphia, PA 19104. (800) 523-1850 or (215) 386-0100. Broad multidisciplinary title and author index to the international literature of science and technology, 1974 to present. Inquire as to online cost and availability.

WILSONLINE. H.W. Wilson and Company, 950 University Avenue, Bronx, NY 10452. (800) 367-6770 or (212) 588-8400. Makes available online versions of the H.W. Wilson indexes including Applied Science and Technology Index, Business Periodicals Index and Readers' Guide to Periodical Literature. Approximately 1980 to present. Inquire as to online cost and availability.

OTHER SOURCES

ELECTROMECHANICAL BENCH REFERENCE. Barks Publications, Inc., 400 North Michigan Avenue, Chicago, IL 60611-4198. (312) 321-9440. Annual. $5.00.

A GUIDE TO THE LITERATURE OF ELECTRICAL AND ELECTRONICS ENGINEERING. Susan B. Ardis. Libraries Unlimited Inc., P.O. Box 263, Littleton, CO 80160. (303) 770-1220. 1987. $37.50.

MOTORS AND GENERATORS. U.S. Bureau of the Census, U.S. Department of Commerce, Washington, DC 20233. (301) 763-7800. Current industrial reports. Annual. $1.25.

PERIODICALS

ELECTRIC LIGHT AND POWER. Technical Publishing Company, 875 Third Avenue, New York, NY 10022. (212) 605-9400. 1922 to present. Monthly. $38.00 per year.

ELECTRIC MACHINES AND POWER SYSTEMS. Hemisphere Publishing Corporation, 1010 Vermont Avenue, NW, Washington, DC 20005. (800) 526-0275. 1976 to present. Bimonthly. $134.50 per year.

ELECTRICAL WORLD. McGraw-Hill Book Company, 1221 Avenue of the Americas, New York, NY 10020. (212) 512-2000. 1874 to present. Monthly. $11.00 per year.

ELECTRONIC DESIGN. Hayden Publishing Company, 10 Mulholland Drive, Hasbrouck Heights, NJ 07604. (201) 288-7520. 1952 to present. Biweekly. $40.00 per year.

ELECTRONICS. McGraw-Hill Book Company, 1221 Avenue of the Americas, New York, NY 10020. (212) 512-2000. 1930 to present. Weekly. $32.00 per year.

IEEE POWER ENGINEERING REVIEW. Institute of Electrical and Electronics Engineers. IEEE Service Center, 445 Hoes Lane, Piscataway, NJ 08854. 1981 to present. Monthly. $75.00 per year.

IEEE TRANSACTIONS ON POWER DELIVERY. Institute of Electrical and Electronics Engineers. IEEE Service Center, 445 Hoes Lane, Piscataway, NJ 08854. 1986 to present. Quarterly. $100.00 per year.

INSTITUTE OF ELECTRICAL AND ELECTRONICS ENGINEERS PROCEEDINGS. Institute of Electrical and Electronics Engineers. IEEE Service Center, 445 Hoes Lane, Piscataway, NJ 08854. 1913 to present. Monthly. $140.00 per year.

POWER TRANSMISSION DESIGN. Penton-IPC, 1100 Superior Avenue, Cleveland, OH 44114. (216) 696-7000. 1959 to present. Monthly. $35.00 per year.

RESEARCH CENTERS AND INSTITUTES

EDISON ELECTRIC INSTITUTE. 1111 19th Street, N.W., Washington, DC 20036. (202) 778-6778.

ELECTRIC POWER INSTITUTE. Texas A&M University, Department of Electrical Engineering, College Station, TX 77843.

ELECTRICAL ENGINEERING RESEARCH LABORATORIES. Purdue University, Electrical Engineering Building, West Lafayette, IN 47907. (317) 494-3536.

LABORATORY FOR ELECTROMAGNETIC AND ELECTRONIC SYSTEMS. 77 Massachusetts Avenue, Cambridge, MA 02139. (617) 253-4631.

ELECTRIC POWER ENGINEERING

See also: ELECTRIC MOTORS, ELECTRICAL ENGINEERING, ELECTRICITY, ELECTROMAGNETISM, GENERATORS, HYDROELECTRIC POWER

ABSTRACT SERVICES AND INDEXES

APPLIED MECHANICS REVIEW. American Society of Mechanical Engineers, 345 East 47th Street, New York, NY 10017. (212) 705-7703. 1948 to present. Monthly. $360.00 per year.

APPLIED SCIENCE AND TECHNOLOGY INDEX. H.W. Wilson and Company, 950 University Avenue, Bronx, NY 10452. (800) 367-6670 or (212) 588-8400. Monthly. Inquire as to cost and availability.

CHEMICAL ABSTRACTS. American Chemical Society, Chemical Abstracts Service, Box 3012, Columbus, OH 43210. (614) 421-3600. 1907 to present. Weekly. $9500.00 per year.

CURRENT CONTENTS: ENGINEERING, TECHNOLOGY AND APPLIED SCIENCES. Institute for Scientific Information, 3501 Market Street, Philadelphia, PA 19104. (800) 523-1850 or (215) 386-0100. Weekly. $275.00 per year.

ELECTRIC POWER INDUSTRY ABSTRACTS. Edison Electric Institute, c/o Utility Data Institute, 2011 I Street, N.W., Suite 700, Washington, DC 20006. 1975 to present. Bimonthly. Inquire as to cost and availability.

ELECTRICAL AND ELECTRONICS ABSTRACTS. Institution of Electrical Engineers. Available from: Institute of Electrical and Electronics Engineers. IEEE Service Center, 445 Hoes Lane, Piscataway, NJ 08854. Monthly. $1250.00 per year.

ENGINEERING INDEX MONTHLY AND AUTHOR INDEX. Engineering Information Inc., 345 East 47th Street, New York, NY 10017. (212) 705-7600. Monthly. $1560.00 per year.

IEEE PUBLICATIONS BULLETIN. Institute of Electrical and Electronics Engineers. Institute of Electrical and Electronics

Engineers. IEEE Service Center, 445 Hoes Lane, Piscataway, NJ 08854. Quarterly. Free.

PHYSICS ABSTRACTS. Institution of Electrical Engineers. Available from: IEEE Service Center, 445 Hoes Lane, Piscataway, NJ 08854. 1898 to present. Bimonthly. $1700.00 per year.

PHYSICS BRIEFS. Physik Verlag GmbH, Postfach 1260/1280, D-6940 Weinheim, West Germany. (212) 661-9404. 1920 to present. Twenty-six times per year. $1250.00 per year.

SCIENCE CITATION INDEX. Institute for Scientific Information, 3501 Market Street, Philadelphia, PA 19104. (800) 523-1850 or (215) 386-0100. Six times per year. $6200.00 per year.

ASSOCIATIONS AND PROFESSIONAL SOCIETIES

AMERICAN ELECTRONICS ASSOCIATION. P.O. Box 10045, 2670 Hanover Street, Palo Alto, CA 94303. (415) 857-9300.

EDISON ELECTRIC INSTITUTE. 1111 19th Street, N.W., Washington, DC 20036. (202) 828-7400.

INSTITUTE OF ELECTRICAL AND ELECTRONICS ENGINEERS. 345 East 47th Street, New York, NY 10017. (212) 705-7900.

DIRECTORIES AND BIOGRAPHICAL SOURCES

ELECTRICAL WORLD DIRECTORY OF ELECTRICAL UTILITIES. McGraw-Hill Book Company, 1221 Avenue of the Americas, New York, NY 10020. (212) 512-2000. Annual. $275.00.

INTERNATIONAL RESEARCH CENTERS DIRECTORY 1988-89. Darren L. Smith, editor. Gale Research Company, Book Tower, Detroit, MI 48226. (800) 521-0707. 4th edition. 1987. $360.00.

1987 DIRECTORY OF ENGINEERING SOCIETIES AND RELATED ORGANIZATIONS. Gordon Davis, editor. Hemisphere Publishing Corporation, 1010 Vermont Avenue, NW, Washington, DC 20005. (800) 526-0275. 12th edition. 1987. $100.00.

RESEARCH CENTERS DIRECTORY 1988. Gale Research Company, Book Tower, Detroit, MI 48226. (800) 521-0707. 12th edition. 1987. $365.00 for set.

SCIENTIFIC AND TECHNICAL ORGANIZATIONS AND AGENCIES DIRECTORY. Margaret Labash Young, editor. Gale Research Company, Book Tower, Detroit, MI 48226. (800) 521-0707. 2nd edition. 1987. $185.00.

WHO'S WHO IN ELECTRONICS. Harris Publishing Company, 2057-2 Aurora Road, Twinsburg, OH 44087. (216) 425-9143. Annual. $90.00.

WHO'S WHO IN ENGINEERING. Gordon Davis, editor. Hemisphere Publishing Corporation, 1010 Vermont Avenue, NW, Washington, DC 20005. (800) 526-0275. 6th edition. 1985. $200.00.

ENCYCLOPEDIAS AND DICTIONARIES

MCGRAW-HILL ENCYCLOPEDIA OF SCIENCE AND TECHNOLOGY. McGraw-Hill Book Company, 1221 Avenue of the Americas, New York, NY 10020. (212) 512-2000. 6th edition. 1987. $1600.00.

THESAURUS OF SCIENTIFIC, TECHNICAL, AND ENGINEERING TERMS. Hemisphere Publishing Corporation, 1010 Vermont Avenue, NW, Washington, DC 20005. (800) 526-0275. 1988. $125.00.

GENERAL WORKS

ELECTRIC MACHINERY. Peter F. Ryff. Prentice-Hall Publishing, Inc., Englewood Cliffs, NJ 07632. (800) 562-0245. 1988. $34.00.

ELECTRIC POWER TRANSMISSION SYSTEM ENGINEERING: ANALYSIS AND DESIGN. Turan Gonen. John Wiley and Sons, Inc., 605 Third Avenue, New York, NY 10158. (800) 526-5368. 1987. $75.00.

ELECTRICAL ENGINEERING. W.H. Roadstrum and Dan H. Wolaver. Harper and Row Publishers, Inc., 10 East 53rd Street, New York, NY 10022. (800) 242-7737. 1986. $37.50.

ELECTRICITY AND MAGNETISM. M.H. Nayfeh and M.K. Brussel. John Wiley and Sons, Inc., 605 Third Avenue, New York, NY 10158. (800) 526-5368. 1985. $32.95.

ELECTROMAGNETIC CONCEPTS AND PRINCIPLES. Stanley V. Marshall and Gabriel G. Skitek. Prentice-Hall Publishing, Inc., Englewood Cliffs, NJ 07632. (800) 562-0245. 2nd edition. 1987. $42.95.

ELEMENTS OF ENGINEERING ELECTROMAGNETICS. N.N. Rao. Prentice-Hall Publishing, Inc., Englewood Cliffs, NJ 07632. (800) 562-0245. 2nd edition. 1987. $45.95.

FUNDAMENTALS OF ELECTRIC CIRCUITS. David A. Bell. Prentice-Hall Publishing, Inc., Englewood Cliffs, NJ 07632. (800) 562-0245. 4th edition. 1988. $30.25.

INTRODUCTION TO ELECTRICITY AND ELECTRONICS FOR THE COMPUTER AGE. R. Rosen. John Wiley and Sons, Inc., 605 Third Avenue, New York, NY 10158. (800) 526-5368. 1987. $35.95.

HANDBOOKS AND MANUALS

HANDBOOK OF MODERN ELECTRONICS AND ELECTRICAL ENGINEERING. Charles Belove, editor. John Wiley and Sons, Inc., 605 Third Avenue, New York, NY 10158. (800) 526-5368. 1986. $88.95.

ONLINE DATA BASES

CA SEARCH. Chemical Abstracts Service, P.O. Box 3012, Columbus, OH 43120. (800) 848-6538 or (614) 421-3600. Comprehensive guide to chemical literature, 1972 to present. Inquire as to online cost and availability.

COMPENDEX. Engineering Information, Inc., 345 East 47th Street, New York, NY 10017. (800) 221-1044 or (212) 705-7615. Engineering and technical literature, 1975 to present. Inquire as to online cost and availability.

INSPEC. INSPEC Marketing Department, Institution of Electrical Engineers. Available from IEEE Service Center, 445 Hoes Lane, Piscataway, NJ 08854. (201) 981-0060. Online version of Physics Abstracts. Inquire as to online cost and availability.

NTIS. National Technical Information Service, 5285 Port Royal Road, Springfield, VA 22161. (703) 487-4630. Broad coverage of government sponsored research reports, 1964 to present. Inquire as to online cost and availability.

SCISEARCH. Institute for Scientific Information, 3501 Market Street, Philadelphia, PA 19104. (800) 523-1850 or (215) 386-0100. Broad multidisciplinary title and author index to the international literature of science and technology, 1974 to present. Inquire as to online cost and availability.

WILSONLINE. H.W. Wilson and Company, 950 University Avenue, Bronx, NY 10452. (800) 367-6770 or (212) 588-8400. Makes available online versions of the H.W. Wilson indexes including Applied Science and Technology Index, Business Periodicals Index and Readers' Guide to Periodical Literature. Approximately 1980 to present. Inquire as to online cost and availability.

OTHER SOURCES

A GUIDE TO THE LITERATURE OF ELECTRICAL AND ELECTRONICS ENGINEERING. Susan B. Ardis. Libraries

Unlimited Inc., P.O. Box 263, Littleton, CO 80160. (303) 770-1220. 1987. $37.50.

WHAT EVERY ENGINEER SHOULD KNOW ABOUT ENGINEERING SOURCES. Marcel Dekker Inc., 270 Madison Avenue, New York, NY 10016. (800) 228-1160. 1984. $24.95.

PERIODICALS

ELECTRIC LIGHT AND POWER. Technical Publishing Company, 875 Third Avenue, New York, NY 10022. (212) 605-9400. 1922 to present. Monthly. $38.00 per year.

ELECTRIC MACHINES AND POWER SYSTEMS. Hemisphere Publishing Corporation, 1010 Vermont Avenue, NW, Washington, DC 20005. (800) 526-0275. 1976 to present. Bimonthly. $134.50 per year.

ELECTRICAL WORLD. McGraw-Hill Book Company, 1221 Avenue of the Americas, New York, NY 10020. (212) 512-2000. 1874 to present. Monthly. $11.00 per year.

ELECTRONIC DESIGN. Hayden Publishing Company, 10 Mulholland Drive, Hasbrouck Heights, NJ 07604. (201) 288-7520. 1952 to present. Biweekly. $40.00 per year.

ELECTRONICS. McGraw-Hill Book Company, 1221 Avenue of the Americas, New York, NY 10020. (212) 512-2000. 1930 to present. Weekly. $32.00 per year.

IEEE POWER ENGINEERING REVIEW. Institute of Electrical and Electronics Engineers. IEEE Service Center, 445 Hoes Lane, Piscataway, NJ 08854. 1981 to present. Monthly. $75.00 per year.

IEEE TRANSACTIONS ON POWER DELIVERY. Institute of Electrical and Electronics Engineers. IEEE Service Center, 445 Hoes Lane, Piscataway, NJ 08854. 1986 to present. Quarterly. $100.00 per year.

INSTITUTE OF ELECTRICAL AND ELECTRONICS ENGINEERS PROCEEDINGS. Institute of Electrical and Electronics Engineers. IEEE Service Center, 445 Hoes Lane, Piscataway, NJ 08854. 1913 to present. Monthly. $140.00 per year.

RESEARCH CENTERS AND INSTITUTES

EDISON ELECTRIC INSTITUTE. 1111 19th Street, N.W., Washington, DC 20036. (202) 778-6778.

ELECTRIC POWER INSTITUTE. Texas A&M University, Department of Electrical Engineering, College Station, TX 77843.

ELECTRICAL ENGINEERING RESEARCH LABORATORIES. Purdue University, Electrical Engineering Building, West Lafayette, IN 47907. (317) 494-3536.

LABORATORY FOR ELECTROMAGNETIC AND ELECTRONIC SYSTEMS. 77 Massachusetts Avenue, Cambridge, MA 02139. (617) 253-4631.

ELECTRIC ARC WELDING

See: WELDING

ELECTRICAL CODES

See also: CONSTRUCTION ENGINEERING, ELECTRICAL ENGINEERING, ELECTRICITY

ABSTRACT SERVICES AND INDEXES

APPLIED SCIENCE AND TECHNOLOGY INDEX. H.W. Wilson and Company, 950 University Avenue, Bronx, NY 10452. (800) 367-6670 or (212) 588-8400. Monthly. Inquire as to cost and availability.

ELECTRIC POWER INDUSTRY ABSTRACTS. Edison Electric Institute, c/o Utility Data Institute, 2011 I Street, N.W., Suite 700, Washington, DC 20006. 1975 to present. Bimonthly. Inquire as to cost and availability.

ELECTRICAL AND ELECTRONICS ABSTRACTS. Institution of Electrical Engineers. Available from: Institute of Electrical and Electronics Engineers. IEEE Service Center, 445 Hoes Lane, Piscataway, NJ 08854. Monthly. $1250.00 per year.

ENGINEERING INDEX MONTHLY AND AUTHOR INDEX. Engineering Information Inc., 345 East 47th Street, New York, NY 10017. (212) 705-7600. Monthly. $1560.00 per year.

IEEE PUBLICATIONS BULLETIN. Institute of Electrical and Electronics Engineers. Institute of Electrical and Electronics Engineers. IEEE Service Center, 445 Hoes Lane, Piscataway, NJ 08854. Quarterly. Free.

ASSOCIATIONS AND PROFESSIONAL SOCIETIES

AMERICAN ELECTRONICS ASSOCIATION. P.O. Box 10045, 2670 Hanover Street, Palo Alto, CA 94303. (415) 857-9300.

EDISON ELECTRIC INSTITUTE. 1111 19th Street, N.W., Washington, DC 20036. (202) 828-7400.

INSTITUTE OF ELECTRICAL AND ELECTRONICS ENGINEERS. 345 East 47th Street, New York, NY 10017. (212) 705-7900.

INTERNATIONAL ASSOCIATION OF ELECTRICAL INSPECTORS. 930 Busse Highway, Park Ridge, IL 60068. (312) 696-1455.

NATIONAL ELECTRICAL CONTRACTORS ASSOCIATION. 7315 Wisconsin Avenue, Bethesda, MD 20814. (301) 657-3110.

DIRECTORIES AND BIOGRAPHICAL SOURCES

ELECTRICAL WORLD DIRECTORY OF ELECTRICAL UTILITIES. McGraw-Hill Book Company, 1221 Avenue of the Americas, New York, NY 10020. (212) 512-2000. Annual. $275.00.

WHO'S WHO IN ELECTRONICS. Harris Publishing Company, 2057-2 Aurora Road, Twinsburg, OH 44087. (216) 425-9143. Annual. $90.00.

WHO'S WHO IN ENGINEERING. Gordon Davis, editor. Hemisphere Publishing Corporation, 1010 Vermont Avenue, NW, Washington, DC 20005. (800) 526-0275. 6th edition. 1985. $200.00.

GENERAL WORKS

FUNDAMENTALS OF ELECTRIC CIRCUITS. David A. Bell. Prentice-Hall Publishing, Inc., Englewood Cliffs, NJ 07632. (800) 562-0245. 4th edition. 1988. $30.25.

HANDBOOKS AND MANUALS

AMERICAN ELECTRICIAN'S HANDBOOK. T. Croft and W. Summers. McGraw-Hill Book Company, 1221 Avenue of the Americas, New York, NY 10020. (212) 512-2000. 11th edition.

ELECTRICAL CODES

1987. $64.50.

HANDBOOK OF MODERN ELECTRONICS AND ELECTRICAL ENGINEERING. Charles Belove, editor. John Wiley and Sons, Inc., 605 Third Avenue, New York, NY 10158. (800) 526-5368. 1986. $88.95.

MCGRAW-HILL'S NATIONAL ELECTRICAL CODE HANDBOOK. J.F. McPartland. McGraw-Hill Book Company, 1221 Avenue of the Americas, New York, NY 10020. (212) 512-2000. 19th edition. 1987. $42.50.

ONLINE DATA BASES

COMPENDEX. Engineering Information, Inc., 345 East 47th Street, New York, NY 10017. (800) 221-1044 or (212) 705-7615. Engineering and technical literature, 1975 to present. Inquire as to online cost and availability.

NTIS. National Technical Information Service, 5285 Port Royal Road, Springfield, VA 22161. (703) 487-4630. Broad coverage of government sponsored research reports, 1964 to present. Inquire as to online cost and availability.

WILSONLINE. H.W. Wilson and Company, 950 University Avenue, Bronx, NY 10452. (800) 367-6770 or (212) 588-8400. Makes available online versions of the H.W. Wilson indexes including Applied Science and Technology Index, Business Periodicals Index and Readers' Guide to Periodical Literature. Approximately 1980 to present. Inquire as to online cost and availability.

OTHER SOURCES

A GUIDE TO THE LITERATURE OF ELECTRICAL AND ELECTRONICS ENGINEERING. Susan B. Ardis. Libraries Unlimited Inc., P.O. Box 263, Littleton, CO 80160. (303) 770-1220. 1987. $37.50.

PERIODICALS

ELECTRIC LIGHT AND POWER. Technical Publishing Company, 875 Third Avenue, New York, NY 10022. (212) 605-9400. 1922 to present. Monthly. $38.00 per year.

ELECTRIC MACHINES AND POWER SYSTEMS. Hemisphere Publishing Corporation, 1010 Vermont Avenue, NW, Washington, DC 20005. (800) 526-0275. 1976 to present. Bimonthly. $134.50 per year.

ELECTRICAL CONSTRUCTION AND MAINTENANCE. McGraw-Hill Book Company, 1221 Avenue of the Americas, New York, NY 10020. (212) 512-2000. 1901 to present. Monthly. $18.00 per year.

ELECTRICAL CONTRACTOR. National Electrical Contractors Association, 7315 Wisconsin Avenue, Bethesda, MD 20814. (301) 657-3110. 1939 to present. Monthly. Inquire.

ELECTRICAL WORLD. McGraw-Hill Book Company, 1221 Avenue of the Americas, New York, NY 10020. (212) 512-2000. 1874 to present. Monthly. $11.00 per year.

IAEI NEWS. International Association of Electrical Inspectors. 930 Busse Highway, Park Ridge, IL 60068. (312) 696-1455. 1929 to present. Bimonthly. $24.00 per year.

IEEE POWER ENGINEERING REVIEW. Institute of Electrical and Electronics Engineers. IEEE Service Center, 445 Hoes Lane, Piscataway, NJ 08854. 1981 to present. Monthly. $75.00 per year.

IEEE TRANSACTIONS ON POWER DELIVERY. Institute of Electrical and Electronics Engineers. IEEE Service Center, 445 Hoes Lane, Piscataway, NJ 08854. 1986 to present. Quarterly. $100.00 per year.

RESEARCH CENTERS AND INSTITUTES

EDISON ELECTRIC INSTITUTE. 1111 19th Street, N.W., Washington, DC 20036. (202) 778-6778.

ELECTRIC POWER INSTITUTE. Texas A&M University, Department of Electrical Engineering, College Station, TX 77843.

ELECTRICAL ENGINEERING

See also: ELECTRICITY, ELECTRONIC CIRCUITS AND COMPONENTS, ELECTRONIC ENGINEERING

ABSTRACT SERVICES AND INDEXES

APPLIED SCIENCE AND TECHNOLOGY INDEX. H.W. Wilson Company, 950 University Avenue, Bronx, NY 10452. (800) 367-6670 or (212) 588-8400. Inquire as to cost and availability.

ELECTRONICS AND COMMUNICATIONS ABSTRACTS JOURNAL. Cambridge Scientific Abstracts, 5161 River Road, Bethesda, MD 20816. (301) 951-1400. Bimonthly. Inquire as to cost and availability.

ENGINEERING INDEX MONTHLY. Engineering Information, Incorporated, 345 East 47th Street, New York, NY 10017. (800) 221-1044 or (212) 705-7600. Monthly, with annual cumulation. $1425.00 per year.

PHYSICS ABSTRACTS. Institute of Electrical Engineers, London, United Kingdom. Available from: Institute of Electrical and Electronic Engineers (Ieee), 345 East 47th Street, New York, NY 10017. (212) 705-7900.

SCIENCE CITATION INDEX. Institute for Scientific Information, 3501 Market Street, Philadelphia, PA 19104. (800) 523-1850 or (215) 386-0100. Inquire as to cost and availability.

ANNUAL REVIEWS AND YEARBOOKS

ADVANCES IN ELECTRONICS AND ELECTRON PHYSICS. Academic Press, Incorporated, 6277 Sea Harbor Drive, Orlando, FL 32821. (800) 321-5068. Irregular. Approximately $80.00 per volume.

ASSOCIATIONS AND PROFESSIONAL SOCIETIES

AMERICAN ELECTRONICS ASSOCIATION. Post Office Box 10045, 2670 Hanover Street, Palo Alto, CA 94303. (415) 857-9300.

ASSOCIATION OF OLD CROWS (ELECTRONIC WARFARE). 2300 Ninth Street, South, Suite 300, Arlington, VA 22204. (703) 920-1600.

EDISON ELECTRIC INSTITUTE. 1111 19th Street, NW, Washington, DC 20036. (202) 828-7400.

ELECTRONIC INDUSTRIES ASSOCIATION. 2001 Eye Street, NW, Washington, DC 20006. (202) 457-4900.

IEEE (INSTITUTE OF ELECTRICAL AND ELECTRONICS ENGINEERS). 345 East 47th Street, New York, NY 10017. (212) 705-7900.

INTERNATIONAL SOCIETY FOR HYBRID MICROELECTRONICS. Post Office Box 2698, 1861 Wiehle Avenue, Suite 340, Reston, VA 22090. (703) 471-0066.

BIBLIOGRAPHIES

HANDBOOKS AND TABLES IN SCIENCE AND TECHNOLOGY. Russell H. Powell, editor. Oryx Press, 2214 North Central Avenue, Phoenix, AZ 85004-1483. (602) 254-6156.

Second edition. 1983. $55.00.

SCIENTIFIC AND TECHNICAL BOOKS IN PRINT: AN INDEX TO LITERATURE IN SCIENCE AND TECHNOLOGY. R.R. Bowker Company, 205 East 42nd Street, New York, NY 10017. (800) 521-8110 or (212) 916-1600.

DIRECTORIES AND BIOGRAPHICAL SOURCES

IEEE MEMBERSHIP DIRECTORY. Institute of Electrical and Electronics Engineers, IEEE Service Center, 445 Hoes Lane, Piscataway, NJ 08854. (212) 705-7900. Annual. $7.00.

Research Centers Directory. Gale Research Company, Book Tower, Detroit, MI 48226. (800) 521-0707. Twelfth edition. 1987. $365.00 for set.

SCIENTIFIC AND TECHNICAL ORGANIZATIONS AND AGENCIES DIRECTORY. Gale Research Company, Book Tower, Detroit, MI 48226. (800) 521-0707. 1985.

WHO'S WHO IN ELECTRONICS. Harris Publishing Company, 2057-2 Aurora Road, Twinsburg, OH 44087. (216) 425-9000. Annual. $89.00.

WHO'S WHO IN ENGINEERING. Engineers Joint Council, 345 East 47th Street, New York, NY 10017. (212) 705-7010. 1985. $200.00.

WHO'S WHO IN FRONTIER SCIENCE AND TECHNOLOGY. Marquis Who's Who, Incorporated, 200 East Ohio Street, Chicago, IL 60611. (800) 428-3898 or (312) 787-2008.

WHO'S WHO IN TECHNOLOGY TODAY. Reston Publishing Company, Incorporated, c/o Prentice-Hall, Incorporated, Englewood Cliffs, NJ 07632. (800) 262-6868. Biennial. Five volumes. $425.00. Covers the fields of electronics, computer science, physics, optics, chemistry, biotechnology, mechanics, energy, and earth science.

ENCYCLOPEDIAS AND DICTIONARIES

ENCYCLOPEDIA OF PHYSICAL SCIENCE AND TECHNOLOGY. Academic Press, Incorporated, Orlando, FL 32887. (800) 321-5068 or (305) 345-2734. Fifteen volumes, 1986.

IEEE STANDARD DICTIONARY OF ELECTRICAL AND ELECTRONICS TERMS. Frank Jay, editor. John Wiley and Sons, Incorporated, 605 Third Avenue, New York, NY 10158. (800) 526-5368 or (212) 850-6000. Third edition. 1984. $49.95.

MCGRAW-HILL ENCYCLOPEDIA OF SCIENCE AND TECHNOLOGY. McGraw-Hill Book, Incorporated, 1221 Avenue of the Americas, New York, NY 10020. (212) 997-3675. Fifth edition, 15 volumes. $1100.00.

GENERAL WORKS

CIRCUITS AND SOFTWARE FOR ELECTRONICS ENGINEERS. Howard Bierman. McGraw-Hill Book Company, 1221 Avenue of the Americas, New York, NY 10020. (800) 628-0004. 1984. $39.50.

CIRCUITS, DEVICES AND SYSTEMS: A FIRST COURSE IN ELECTRICAL ENGINEERING. R.J. Smith. John Wiley and Sons, Incorporated, 605 Third Avenue, New York, NY 10158. (800) 526-5368 or (212) 850-6000. Fourth edition. 1984. $41.45.

CIRCUITS, SIGNALS AND SYSTEMS. William M. Siebert. McGraw-Hill Book Company, 1221 Avenue of the Americas, New York, NY 10020. (800) 628-0004. 1986. $37.95.

ELECTRIC UTILITY SYSTEMS AND PRACTICES. Homer M. Rustebakke, editor. John Wiley and Sons, Incorporated, 605 Third Avenue, New York, NY 10158. (800) 526-5368 or (212) 850-6000. Fourth edition. 1983. $45.95.

ELECTRICAL AND ELECTRONIC INSTRUMENTATION. Hai Hung Chiang. John Wiley and Sons, Incorporated, 605 Third Avenue, New York, NY 10158. (800) 526-5368 or (212) 850-6000. 1984. $64.95.

ELECTRICAL MACHINES AND THEIR APPLICATIONS. J. Hindmarsh. Pergamon Press, Incorporated, Maxwell House, Fairview Park, Elmsford, NY 10523. (914) 592-7700. Fourth edition. 1984. $50.65.

HIGH VOLTAGE ENGINEERING FUNDAMENTALS. E. Kuffel and W.S. Zaengl. Pergamon Press, Incorporated, Maxwell House, Fairview Park, Elmsford, NY 10523. (914) 592-7700. 1984. $55.00.

SEMICONDUCTOR CIRCUIT APPROXIMATIONS: AN INTRODUCTION TO TRANSISTORS AND INTEGRATED CIRCUITS. Albert Paul Malvino. McGraw-Hill Book Company, 1221 Avenue of the Americas, New York, NY 10020. (800) 628-0004. Fourth edition. 1985. $34.95.

TELECOMMUNICATIONS ENGINEERING. J. Dunlop and D.G. Smith. Van Nostrand Reinhold Book Company, 115 Fifth Avenue, New York, NY 10003. (800 543-2681. 1985. $19.95 in paper.

HANDBOOKS AND MANUALS

CRC HANDBOOK OF TABLES FOR APPLIED ENGINEERING SCIENCE. R.E. Bolz and G.L. Tuve, editors. CRC Press, Incorporated, 2000 Corporate Boulevard, NW, Boca Raton, FL 33431. Second edition. 1973. $69.00.

ELECTRONIC ENGINEERS HANDBOOK. Donald G. Fink, editor. McGraw-Hill Book Company, 1221 Avenue of the Americas, New York, NY 10020. (800) 628-0004. Second edition. 1982. $89.00.

HANDBOOK OF ELECTRONICS MANUFACTURING ENGINEERING. B.S. Matisoff. Van Nostrand Reinhold Company, Incorporated, 115 Fifth Avenue, New York, NY 10003. (800) 543-2681. Second edition. 1986. $52.95.

HANDBOOK OF MODERN ELECTRONICS AND ELECTRICAL ENGINEERING. Charles Belove. John Wiley and Sons, Incorporated, 605 Third Avenue, New York, NY 10158. (800) 526-5368 or (212) 850-6000. 1986. $85.00.

REFERENCE MANUAL FOR TELECOMMUNICATIONS. R.L. Freeman. John Wiley and Sons, Incorporated, 605 Third Avenue, New York, NY 10158. (800) 526-5368 or (212) 850-6000. 1985. $79.95.

STANDARD HANDBOOK FOR ELECTRICAL ENGINEERS. Donald G. Fink and H. Wayne Beaty. McGraw-Hill Book Company, 1221 Avenue of the Americas, New York, NY 10020. (212) 512-2000. Eleventh edition. 1978. $85.00.

VLSI HANDBOOK. Norman G. Einspruch, editor. Academic Press, Incorporated, 6277 Sea Harbor Drive, Orlando, FL 32821. (800) 321-5068. 1985. $125.00.

THE WILEY ENGINEER'S DESK REFERENCE. Sanford I. Heisler. John Wiley and Sons, Incorporated, 605 Third Avenue, New York, NY 10158. (800) 526-5368 or (212) 850-6418. 1984. $36.00.

ONLINE DATA BASES

COMPENDEX. Engineering Information, Incorporated, 345 East 47th Street, New York, NY 10017. (800) 221-1044 or (212) 705-7615. Engineering and technical literature, 1975 to present. Inquire as to cost and availability.

DISSERTATION ABSTRACTS ONLINE. University Microfilms International, 300 North Zeeb Road, Ann Arbor, MI 48106. (800) 521-0600 or (313) 761-4700. Scope includes virtually all doctoral dissertations accepted at accredited American institutions from 1861 to present in 252 subject areas. Inquire as to cost and availability.

INSPEC. INSPEC Marketing Department, Institute of Electrical and Electronics Engineers, Incorporated, IEEE Service Department, 445 Hoes Lane, Piscataway, NJ 08854. (201) 981-0060. Inquire as to on-line cost and availability.

NTIS. National Technical Information Service, 5285 Port Royal Road, Springfield, VA 22161. (703) 487-4630. Broad coverage of government sponsored research reports, 1964 to present. Inquire as to cost and availability.

WILSONLINE. H.W. Wilson Company, 950 University Avenue, Bronx, NY 10452. (800) 367-6770 or (212) 588-8400. Makes available online versions of the printed H.W. Wilson Indexes including Applied Science and Technology Index, Business Periodicals Index, and Readers' Guide to Periodical Literature. Period covered is generally 1983 to present. Inquire as to cost and availability.

OTHER SOURCES

A GUIDE TO THE LITERATURE OF ELECTRICAL AND ELECTRONIC ENGINEERING. Susan B. Ardis. Libraries Unlimited, Incorporated, Post Office Box 263, Littleton, CO 80160. (303) 770-1220. 1986. $37.50.

PERIODICALS

CANADAIAN ELECTRONICS ENGINEERING. Maclean Hunter Research Bureau, 777 Bay Street, Toronto, ON M5W 1A7 Canada. (416) 596-5729.

CIRCUITS, SYSTEMS AND SIGNAL PROCESSING. Birkhauser Boston, Incorporated, 380 Green Street, Post Office Box 2007, Cambridge, MA 02139. (617) 876-2333. Quarterly. $175.00 per year.

ELECTRONIC DESIGN. Heyden Publishing Company, Incorporated, 10 Mulholland Drive, Hasbrouck Heights, NJ 07604. (201) 393-6000. Biweekly. $65.00.

ELECTRONICS WEEK. McGraw-Hill Book Company, 1221 Avenue of the Americas, New York, NY 10020. (800) 628-0004. Weekly. $18.00.

IEEE CIRCUITS AND DEVICES MAGAZINE. Institute of Electrical and Electronics Engineers, IEEE Service Center, 445 Hoes Lane, Piscataway, NJ 08854. (212) 705-7900. Bimonthly. $70.00 per year.

IEEE SPECTRUM. Institute of Electrical and Electronics Engineers, IEEE Service Center, 445 Hoes Lane, Piscataway, NJ 08854. (212) 705-7900. Monthly. $88.00 per year.

IEEE TRANSACTIONS ON CIRCUITS AND SYSTEMS. Institute of Electrical and Electronics Engineers, IEEE Service Center, 445 Hoes Lane, Piscataway, NJ 08854. (212) 705-7900. Monthly. $108.00 per year.

IEEE TRANSACTIONS ON POWER APPARATUS AND SYSTEMS. Institute of Electrical and Electronics Engineers, IEEE Service Center, 445 Hoes Lane, Piscataway, NJ 08854. (212) 705-7900. Monthly. $193.00 per year.

PROCEEDINGS OF THE IEEE. Institute of Electrical and Electronics Engineers, IEEE Service Center, 445 Hoes Lane, Piscataway, NJ 08854. (212) 705-7900. Monthly. $120.00 per year.

RESEARCH CENTERS AND INSTITUTES

EDISON ELECTRICAL INSTITUTE. 1111 19th Street, NW, Washington, DC 20036. (202) 828-7585.

ELECTRIC POWER RESEARCH INSTITUTE. 3412 Hillview Avenue, Palo Alto, CA 94304. (415) 855-2000.

MASSACHUSETTS INSTITUTE OF TECHNOLOGY. Laboratory for Electromagnetic and Electronic Systems, 77 Massachusetts Avenue, Cambridge, MA 02139. (617) 253-4631.

PURDUE UNIVERSITY. Electrical Engineering Research Laboratories, Electrical Engineering Building, West Lafayette, IN 47907. (317) 494-3536.

TEXAS A AND M UNIVERSITY. Electrical Power Institute. Department of Electrical Engineering, College Station, TX 77843. (409) 845-1423.

ELECTRICAL IMPEDANCE

See: ELECTRICITY

ELECTRICAL INSULATION

See: ELECTRICITY

ELECTRICAL RESISTANCE

See: ELECTRICITY

ELECTRICALLY POWERED VEHICLES

See: AUTOMOTIVE ENGINEERING

ELECTRICITY

See also: ELECTRIC POWER ENGINEERING, ELECTRICAL ENGINEERING, ELECTROCHEMISTRY, ELECTROMAGNETISM, ELECTRONICS, ELECTRONICS ENGINEERING

ABSTRACT SERVICES AND INDEXES

APPLIED MECHANICS REVIEW. American Society of Mechanical Engineers, 345 East 47th Street, New York, NY 10017. (212) 705-7703. 1948 to present. Monthly. $360.00 per year.

APPLIED SCIENCE AND TECHNOLOGY INDEX. H.W. Wilson and Company, 950 University Avenue, Bronx, NY 10452. (800) 367-6770 or (212) 588-8400. Monthly. Inquire as to cost and availability.

CHEMICAL ABSTRACTS. American Chemical Society, Chemical Abstracts Service, Box 3012, Columbus, OH 43210. (614) 421-3600. 1907 to present. Weekly. $9500.00 per year.

CONFERENCE PAPERS INDEX. Cambridge Scientific Abstracts, 5161 River Road, Bethesda, MD 20816. 1972 to present. Monthly. Inquire as to cost and availability.

CURRENT CONTENTS: ENGINEERING, TECHNOLOGY AND APPLIED SCIENCES. Institute for Scientific Information, 3501 Market Street, Philadelphia, PA 19104. (800) 523-1850 or (215) 386-0100. Weekly. $275.00 per year.

ELECTRIC POWER INDUSTRY ABSTRACTS. Edison Electric Institute, c/o Utility DatA Institute, 2011 I Street, N.W., Suite 700, Washington, DC 20006. 1975 to present. Bimonthly. Inquire as to cost and availability.

ELECTRICAL AND ELECTRONICS ABSTRACTS. Institution of Electrical Engineers. Available from: Institute of Electrical and Electronics Engineers. IEEE Service Center, 445 Hoes Lane, Piscataway, NJ 08854. Monthly. $1250.00 per year.

ENGINEERING INDEX MONTHLY AND AUTHOR INDEX. Engineering Information Inc., 345 East 47th Street, New York, NY 10017. (212) 705-7600. Monthly. $1560.00 per year.

GENERAL SCIENCE INDEX. H.W. Wilson and Company, 950 University Avenue, Bronx, NY 10452. (800) 367-6670 or (212) 588-8400. 1978 to present. Monthly. Inquire as to cost and availability.

INDEX TO SCIENTIFIC AND TECHNICAL PROCEEDINGS. Institute for Scientific Information, 3501 Market Street, Philadelphia, PA 19104. (800) 523-1850 or (215) 386-0100. 1978 to present. Monthly. $775.00 per year.

INDEX TO SCIENTIFIC REVIEWS. Institute for Scientific Information, 3501 Market Street, Philadelphia, PA 19104. (800) 523-1850 or (215) 386-0100. 1974 to present. Semi-annual. $550.00 per year.

IEEE PUBLICATIONS BULLETIN. Institute of Electrical and Electronics Engineers. Institute of Electrical and Electronics Engineers. IEEE Service Center, 445 Hoes Lane, Piscataway, NJ 08854. Quarterly. Free.

PHYSICS ABSTRACTS. Institution of Electrical Engineers. Available from: IEEE Service Center, 445 Hoes Lane, Piscataway, NJ 08854. 1898 to present. Bimonthly. $1700.00 per year.

PHYSICS BRIEFS. Physik Verlag GmbH, Postfach 1260/1280, D-6940 Weinheim, West Germany. (212) 661-9404. 1920 to present. Twenty-six times per year. $1250.00 per year.

SCIENCE CITATION INDEX. Institute for Scientific Information, 3501 Market Street, Philadelphia, PA 19104. (800) 523-1850 or (215) 386-0100. Six times per year. $6200.00 per year.

ASSOCIATIONS AND PROFESSIONAL SOCIETIES

AMERICAN ELECTRONICS ASSOCIATION. P.O. Box 10045, 2670 Hanover Street, Palo Alto, CA 94303. (415) 857-9300.

AMERICAN INSTITUTE OF PHYSICS. 335 East 45th Street, New York, NY 10017. (212) 661-9494.

EDISON ELECTRIC INSTITUTE. 1111 19th Street, N.W., Washington, DC 20036. (202) 828-7400.

INSTITUTE OF ELECTRICAL AND ELECTRONICS ENGINEERS. 345 East 47th Street, New York, NY 10017. (212) 705-7900.

BIBLIOGRAPHIES

NEW TECHNICAL BOOKS: A SELECTIVE LIST WITH DESCRIPTIVE ANNOTATIONS. New York Public Library, Science and Technology Research Center, Fifth Avenue and 42nd Street, New York, NY 10018. (212) 930-0800. 1915 to present. Monthly. $15.00 per year.

SCIENCE BOOKS AND FILMS. American Association for the Advancement of Science, 1333 H Street, NW, Washington, DC 20005. (202) 326-6454. Five times per year. $20.00 per year.

SCIENTIFIC AND TECHNICAL BOOKS AND SERIALS IN PRINT 1988; AN INDEX TO LITERATURE IN SCIENCE AND TECHNOLOGY. R.R. Bowker Company, 205 East 42nd Street, New York, NY 10017. (800) 521-8110. $175.00.

DIRECTORIES AND BIOGRAPHICAL SOURCES

AMERICAN MEN AND WOMEN OF SCIENCE. R.R. Bowker, Inc., Order Department, 245 West 17th Street, New York, NY 10011. (800) 521-8110. Eight volumes. 1986. $595.00 for set.

ELECTRICAL WORLD DIRECTORY OF ELECTRICAL UTILITIES. McGraw-Hill Book Company, 1221 Avenue of the Americas, New York, NY 10020. (212) 512-2000. Annual. $275.00.

INTERNATIONAL RESEARCH CENTERS DIRECTORY 1988-89. Darren L. Smith, editor. Gale Research Company, Book Tower, Detroit, MI 48226. (800) 521-0707. 4th edition. 1987. $360.00.

1987 DIRECTORY OF ENGINEERING SOCIETIES AND RELATED ORGANIZATIONS. Gordon Davis, editor. Hemisphere Publishing Corporation, 1010 Vermont Avenue, NW, Washington, DC 20005. (800) 526-0275. 12th edition. 1987. $100.00.

RESEARCH CENTERS DIRECTORY 1988. Gale Research Company, Book Tower, Detroit, MI 48226. (800) 521-0707. 12th edition. 1987. $365.00 for set.

SCIENTIFIC AND TECHNICAL ORGANIZATIONS AND AGENCIES DIRECTORY. Margaret Labash Young, editor. Gale Research Company, Book Tower, Detroit, MI 48226. (800) 521-0707. 2nd edition. 1987. $185.00.

WHO'S WHO IN ELECTRONICS. Harris Publishing Company, 2057-2 Aurora Road, Twinsburg, OH 44087. (216) 425-9143. Annual. $90.00.

WHO'S WHO IN ENGINEERING. Gordon Davis, editor. Hemisphere Publishing Corporation, 1010 Vermont Avenue, NW, Washington, DC 20005. (800) 526-0275. 6th edition. 1985. $200.00.

ENCYCLOPEDIAS AND DICTIONARIES

CONCISE SCIENCE DICTIONARY. Oxford University Press, 200 Madison Avenue, New York, NY 10016. (800) 458-5833. 1987. $9.95 in paper.

DICTIONARY OF THE PHYSICAL SCIENCES: TERMS, FORMULAS, DATA. Cesare Emiliani. Oxford University Press, 200 Madison Avenue, New York, NY 10016. (800) 458-5833. 1987. $19.95 in paper.

MCGRAW-HILL ENCYCLOPEDIA OF SCIENCE AND TECHNOLOGY. McGraw-Hill Book Company, 1221 Avenue of the Americas, New York, NY 10020. (212) 512-2000. 6th edition. 1987. $1600.00.

THESAURUS OF SCIENTIFIC, TECHNICAL, AND ENGINEERING TERMS. Hemisphere Publishing Corporation, 1010 Vermont Avenue, NW, Washington, DC 20005. (800) 526-0275. 1988. $125.00.

GENERAL WORKS

ELECTRIC MACHINERY. Peter F. Ryff. Prentice-Hall Publishing, Inc., Englewood Cliffs, NJ 07632. (800) 562-0245. 1988. $34.00.

ELECTRIC POWER TRANSMISSION SYSTEM ENGINEERING: ANALYSIS AND DESIGN. Turan Gonen. John Wiley and Sons, Inc., 605 Third Avenue, New York, NY 10158. (800) 526-5368. 1987. $75.00.

ELECTRICAL ENGINEERING. W.H. Roadstrum and Dan H. Wolaver. Harper and Row Publishers, Inc., 10 East 53rd Street, New York, NY 10022. (800) 242-7737. 1986. $37.50.

ELECTRICITY AND MAGNETISM. M.H. Nayfeh and M.K. Brussel. John Wiley and Sons, Inc., 605 Third Avenue, New York, NY 10158. (800) 526-5368. 1985. $32.95.

ELECTROMAGNETIC CONCEPTS AND PRINCIPLES. Stanley V. Marshall and Gabriel G. Skitek. Prentice-Hall Publishing, Inc., Englewood Cliffs, NJ 07632. (800) 562-0245. 2nd edition. 1987. $42.95.

ELEMENTS OF ENGINEERING ELECTROMAGNETICS. N.N. Rao. Prentice-Hall Publishing, Inc., Englewood Cliffs, NJ 07632. (800) 562-0245. 2nd edition. 1987. $45.95.

FUNDAMENTALS OF ELECTRIC CIRCUITS. David A. Bell. Prentice-Hall Publishing, Inc., Englewood Cliffs, NJ 07632. (800) 562-0245. 4th edition. 1988. $30.25.

INTRODUCTION TO ELECTRICITY AND ELECTRONICS FOR THE COMPUTER AGE. R. Rosen. John Wiley and Sons, Inc., 605 Third Avenue, New York, NY 10158. (800) 526-5368. 1987. $35.95.

HANDBOOKS AND MANUALS

HANDBOOK OF MODERN ELECTRONICS AND ELECTRICAL ENGINEERING. Charles Belove, editor. John Wiley and Sons, Inc., 605 Third Avenue, New York, NY 10158. (800) 526-5368. 1986. $88.95.

ONLINE DATA BASES

CA SEARCH. Chemical Abstracts Service, P.O. Box 3012, Columbus, OH 43120. (800) 848-6538 or (614) 421-3600. Comprehensive guide to chemical literature, 1972 to present. Inquire as to online cost and availability.

COMPENDEX. Engineering Information, Inc., 345 East 47th Street, New York, NY 10017. (800) 221-1044 or (212) 705-7615. Engineering and technical literature, 1975 to present. Inquire as to online cost and availability.

DISSERTATION ABSTRACTS ONLINE. University Microfilms International, 300 North Zeeb Road, Ann Arbor, MI 48106. (800) 521-0600 or (313) 761-4700. Scope includes virtually all doctoral dissertations accepted at accredited American institutions from 1861 to present in over 250 subject areas. Inquire as to online cost and availability.

INSPEC. INSPEC Marketing Department, Institution of Electrical Engineers. Available from IEEE Service Center, 445 Hoes Lane, Piscataway, NJ 08854. (201) 981-0060. Online version of Physics Abstracts. Inquire as to online cost and availability.

NTIS. National Technical Information Service, 5285 Port Royal Road, Springfield, VA 22161. (703) 487-4630. Broad coverage of government sponsored research reports, 1964 to present. Inquire as to online cost and availability.

SCISEARCH. Institute for Scientific Information, 3501 Market Street, Philadelphia, PA 19104. (800) 523-1850 or (215) 386-0100. Broad multidisciplinary title and author index to the international literature of science and technology, 1974 to present. Inquire as to online cost and availability.

WILSONLINE. H.W. Wilson and Company, 950 University Avenue, Bronx, NY 10452. (800) 367-6770 or (212) 588-8400. Makes available online versions of the H.W. Wilson indexes including Applied Science and Technology Index, Business Periodicals Index and Readers' Guide to Periodical Literature. Approximately 1980 to present. Inquire as to online cost and availability.

OTHER SOURCES

A GUIDE TO THE LITERATURE OF ELECTRICAL AND ELECTRONICS ENGINEERING. Susan B. Ardis. Libraries Unlimited Inc., P.O. Box 263, Littleton, CO 80160. (303) 770-1220. 1987. $37.50.

WHAT EVERY ENGINEER SHOULD KNOW ABOUT ENGINEERING SOURCES. Marcel Dekker Inc., 270 Madison Avenue, New York, NY 10016. (800) 228-1160. 1984. $24.95.

PERIODICALS

ELECTRIC LIGHT AND POWER. Technical Publishing Company, 875 Third Avenue, New York, NY 10022. (212) 605-9400. 1922 to present. Monthly. $38.00 per year.

ELECTRIC MACHINES AND POWER SYSTEMS. Hemisphere Publishing Corporation, 1010 Vermont Avenue, NW, Washington, DC 20005. (800) 526-0275. 1976 to present. Bimonthly. $134.50 per year.

ELECTRICAL WORLD. McGraw-Hill Book Company, 1221 Avenue of the Americas, New York, NY 10020. (212) 512-2000. 1874 to present. Monthly. $11.00 per year.

ELECTRONIC DESIGN. Hayden Publishing Company, 10 Mulholland Drive, Hasbrouck Heights, NJ 07604. (201) 288-7520. 1952 to present. Biweekly. $40.00 per year.

ELECTRONICS. McGraw-Hill Book Company, 1221 Avenue of the Americas, New York, NY 10020. (212) 512-2000. 1930 to present. Weekly. $32.00 per year.

IEEE POWER ENGINEERING REVIEW. Institute of Electrical and Electronics Engineers. IEEE Service Center, 445 Hoes Lane, Piscataway, NJ 08854. 1981 to present. Monthly. $75.00 per year.

IEEE TRANSACTIONS ON POWER DELIVERY. Institute of Electrical and Electronics Engineers. IEEE Service Center, 445 Hoes Lane, Piscataway, NJ 08854. 1986 to present. Quarterly. $100.00 per year.

INSTITUTE OF ELECTRICAL AND ELECTRONICS ENGINEERS PROCEEDINGS. Institute of Electrical and Electronics Engineers. IEEE Service Center, 445 Hoes Lane, Piscataway, NJ 08854. 1913 to present. Monthly. $140.00 per year.

RESEARCH CENTERS AND INSTITUTES

EDISON ELECTRIC INSTITUTE. 1111 19th Street, N.W., Washington, DC 20036. (202) 778-6778.

ELECTRIC POWER INSTITUTE. Texas A&M University, Department of Electrical Engineering, College Station, TX 77843.

ELECTRICAL ENGINEERING RESEARCH LABORATORIES. Purdue University, Electrical Engineering Building, West Lafayette, IN 47907. (317) 494-3536.

LABORATORY FOR ELECTROMAGNETIC AND ELECTRONIC SYSTEMS. 77 Massachusetts Avenue, Cambridge, MA 02139. (617) 253-4631.

ELECTROACOUSTICS

See: ACOUSTICS

ELECTROCHEMISTRY

ABSTRACT SERVICES AND INDEXES

APPLIED SCIENCE AND TECHNOLOGY INDEX. H. W. Wilson Company, 950 University Avenue, Bronx, NY 10452. (800) 367-6670 or (212) 588-8400. Inquire as to cost and availability.

CHEMICAL ABSTRACTS. Chemical Abstracts Service, 2540 Olentangy Road, P.O. Box 3012, Columbus, OH 43210. (800) 848-6538 or (614) 421-3600. Weekly. $9,200.00 per year.

PHYSICS ABSTRACTS. Institute of Electrical Engineers, London, United Kingdom. Available from: Institute of Electrical and Electronic Engineers (IEEE), 345 47th Street, New York, NY 10017. (212) 705-7900.

SCIENCE CITATION INDEX. Institute for Scientific Information, 3501 Market Street, Philadelphia, PA 19104. (800) 523-1850 or (215) 386-0100.

ANNUAL REVIEWS AND YEARBOOKS

ADVANCES IN ELECTROCHEMISTRY AND ELECTROCHEMICAL ENGINEERING. John Wiley and Sons, Inc., 605 Third Avenue, New York, NY 10158. (800) 526-5368 or (212) 850-6000. Irregular. Approximately $60.00.

ASSOCIATIONS AND PROFESSIONAL SOCIETIES

AMERICAN CHEMICAL SOCIETY. 1155 16th Street, NW, Washington, DC 20036. (202) 872-4600.

ASSOCIATION OF CONSULTING CHEMISTS AND CHEMICAL ENGINEERS. 50 East 41st Street, Suite 92, New York, NY 10017. (212) 684-6255.

ELECTROCHEMICAL SOCIETY. Ten South Main Street, Pennington, NJ 08534. (609) 737-1902.

BIBLIOGRAPHIES

SCIENCE BOOKS AND FILMS. American Association for the Advancement of Science, 1333 H Street, NW, Washington, DC 20005. Five times per year. $20.00 per year.

SCIENTIFIC AND TECHNICAL BOOKS AND SERIALS IN PRINT 1988; An Index to Literature in Science and Technology. R. R. Bowker Company, 205 East 42nd Street, New York, NY 10017. (800) 521-8110 or (212) 916-1600. $175.00

DIRECTORIES AND BIOGRAPHICAL SOURCES

AMERICAN MEN AND WOMEN OF SCIENCE. Physical and Biological Sciences. Fifteenth Edition. R. R. Bowker Company, 205 East 42nd Street, New York, NY 10017. (800) 521-8110 or (212) 916-1600.

BIOGRAPHICAL DICTIONARY OF SCIENTISTS: CHEMISTS. David Abbott, Editor. P. Bedrick Books, 125 East 23rd Street, New York, NY 10010. (212) 777-1187. 1984. $18.95.

ELECTROCHEMICAL SOCIETY MEMBERSHIP DIRECTORY. Electrochemical Society, Ten South Main Street, Pennington, NJ 08534. (609) 737-1902. Annual. Available to members only.

GOVERNMENT RESEARCH DIRECTORY. Gale Research Company, Book Tower, Detroit, MI 48226. (800) 521-0707. Third Edition. 1985.

RESEARCH CENTERS DIRECTORY. Gale Research Company, Book Tower, Detroit, MI 48226. Twelfth Edition. 1988. $240.00.

WHO'S WHO IN FRONTIER SCIENCE AND TECHNOLOGY. Marquis Who's Who, Inc., 200 East Ohio Street, Chicago, IL 60611. (800) 428-3898 or (312) 787-2008.

WHO'S WHO IN TECHNOLOGY TODAY. Reston Publishing Company, Inc., c/o Prentice-Hall, Inc., Englewood Cliffs, NJ 07632. (800) 262-6868. Biennial. Five volumes. $425.00. Covers the fields of elecctronics, computer science, physics, optics, chemistry, biotechnology, mechanics, energy, and earth science.

ENCYCLOPEDIAS AND DICTIONARIES

DICTIONARY OF ELECTROCHEMISTRY. D. B. Hibbert and A. M. James. John Wiley and Sons, Inc., 605 Third Avenue, New York, NY 10158. (800) 526-5368 or (212) 850-6000. Second Edition. 1985. $37.00.

ENCYCLOPEDIA OF PHYSICAL SCIENCE AND TECHNOLOGY. Academic Press, Inc., Orlando, FL 32887. (800) 321-5068 or (305) 345-2734. Fifteen volumes, 1986.

GLOSSARY OF CHEMICAL TERMS. Clifford A. Hampel and Gessner G. Hawley. Van Nostrand Reinhold Company, 115 Fifth Avenue, New York, NY 10003. (212) 254-3232. Second Edition. $21.95.

MCGRAW-HILL ENCYCLOPEDIA OF SCIENCE AND TECHNOLOGY. McGraw-Hill Book, Inc., 1221 Avenue of the Americas, New York, NY 10020. (212) 997-3675.

GENERAL WORKS

ELECTROCHEMICAL METHODS: FUNDAMENTALS AND APPLICATIONS. Allen J. Bard and Larry R. Faulkner. John Wiley and Sons, Inc., 605 Third Avenue, New York, NY 10158. (800) 526-5368 or (212) 850-6000. 1980. $49.95.

ELECTROCHEMISTRY. Journal of Chemical Education, 238 Kent Road, Springfield, PA 19064. 1983. $6.90.

ELECTROCHEMISTRY: THE INTERFACING SCIENCE. D. A. Rand and A. M. Bond. Elsevier Science Publishing Company, Inc., 52 Vanderbilt Avenue, New York, NY 10017. (212) 370-5520. 1984. $110.00.

ELECTRODE PROCESSES AND ELECTROCHEMICAL ENGINEERING. Fumio Hine. Plenum Publishing Corporation, 233 Spring Street, New York, NY 10013. (800) 221-9369. 1985. $55.00.

ONLINE DATA BASES

CA SEARCH. Chemical Abstracts Service, P.O. Box 3012, Columbus, OH 43210. (614) 421-3600. Guide to chemical literature, 1972 to present. Inquire as to cost and availability.

DISSERTATION ABSTRACTS ONLINE. University Microfilms International, 300 North Zeeb Road, Ann Arbor, MI 48106. (800) 521-0600 or (313) 761-4700. Scope includes virtually all doctoral dissertations accepted at accredited American institutions from 1861 to present in 252 subject areas. Inquire as to cost and availability.

INSPEC. INSPEC Marketing Department, Institute of Electrical and Electronics Engineers, Inc., IEEE Service Department, 445 Hoes Lane, Piscataway, NJ 08854. (201) 981-0060. Inquire as to on-line cost and availability.

NTIS. National Technical Information Service, 5285 Port Royal Road, Springfield, VA 22161. (703) 487-4630. Broad coverage of government sponsored research reports, 1964 to present. Inquire as to cost and availability.

SCISEARCH. Institute for Scientific Information, 3501 Market Street, Philadelphia, PA 19104. (800) 523-1850 or (215) 386-0100. Broad multidisciplinary title and author index to the international literature of science and technology, 1974 to present. Inquire as to cost and availability.

PERIODICALS

ELECTROCHEMICAL REACTIONS. Chemical Abstracts Service, P.O. Box 3012, Columbus, OH 43210. (614) 421-3600. Bimonthly. $110.00 per year.

ELECTROCHEMICAL SOCIETY JOURNAL. Electrochemical Society, Ten South Main Street, Pennington, NJ 08534. (609) 737-1902. Monthly. $195.00 per year.

ELECTROCHIMICA ACTA. Pergamon Journals, Inc., Maxwell House, Fairview Park, Elmsford, NY 10523. (914) 592-7700. Monthly. $430.00 per year.

ELECTROLYSIS WORLD. American Electrolysis Association, 710 Tennent Road, Englishtown, NJ 07826. (201) 536-6477. Bimonthly.

ELECTRONIC CHEMICALS AND MATERIALS. Chemical Abstracts Service, P.O. Box 3012, Columbus, OH 43210. (614) 421-3600. Bimonthly. $120.00 per year.

JOURNAL OF THE ELECTROCHEMICAL SOCIETY. Electrochemical Society, Inc., 10 South Main Street, Pennington, NJ 085434. (609) 737-1902. Monthly. $125.00 per year.

SOVIET ELECTROCHEMISTRY. Plenum Publishing Corporation, 233 Spring Street, New York, NY 10013. (800) 221-9369 or (212) 620-8000. Monthly. $795.00 per year. Translation.

RESEARCH CENTERS AND INSTITUTES

CASE WESTERN RESERVE UNIVERSITY. Case Center for Electrochemical Sciences, Cleveland, OH 44106. (216) 368-3626.

INTERNATIONAL LEAD-ZINC RESEARCH ORGANIZATION, 292 Madison Avenue, New York, NY 10017. (212) 532-2373.

PENNSYLVANIA STATE UNIVERSITY. Electrochemistry and Enthalpimetric Analysis Laboratory, 428B Davey Laboratory Building, University Park, PA 16802. (814) 865-2022.

UNIVERSITY OF TEXAS AT AUSTIN. Laboratory for Electrochemical Research, Austin, TX 78712. (512) 471-3761.

ELECTRODES

See: ELECTROCHEMISTRY

ELECTRODYNAMICS

See: ELECTRICITY

ELECTROLYSIS

See: ELECTROCHEMISTRY

ELECTROLYTES

See: ELECTROCHEMISTRY

ELECTROMAGNETISM

See also: ELECTRIC MOTORS, ELECTRICAL ENGINEERING, ELECTRICITY, ELECTRONICS, ELECTRONICS ENGINEERING, PHYSICS

ABSTRACT SERVICES AND INDEXES

APPLIED MECHANICS REVIEW. American Society of Mechanical Engineers, 345 East 47th Street, New York, NY 10017. (212) 705-7703. 1948 to present. Monthly. $360.00 per year.

APPLIED SCIENCE AND TECHNOLOGY INDEX. H.W. Wilson and Company, 950 University Avenue, Bronx, NY 10452. (800) 367-6670 or (212) 588-8400. Monthly. Inquire as to cost and availability.

CHEMICAL ABSTRACTS. American Chemical Society, Chemical Abstracts Service, Box 3012, Columbus, OH 43210. (614) 421-3600. 1907 to present. Weekly. $9500.00 per year.

CURRENT CONTENTS: ENGINEERING, TECHNOLOGY AND APPLIED SCIENCES. Institute for Scientific Information, 3501 Market Street, Philadelphia, PA 19104. (800) 523-1850 or (215) 386-0100. Weekly. $275.00 per year.

ELECTRIC POWER INDUSTRY ABSTRACTS. Edison Electric Institute, c/o Utility Data Institute, 2011 I Street, N.W., Suite 700, Washington, DC 20006. 1975 to present. Bimonthly. Inquire as to cost and availability.

ELECTRICAL AND ELECTRONICS ABSTRACTS. Institution of Electrical Engineers. Available from: Institute of Electrical and Electronics Engineers. IEEE Service Center, 445 Hoes Lane, Piscataway, NJ 08854. Monthly. $1250.00 per year.

ENGINEERING INDEX MONTHLY AND AUTHOR INDEX. Engineering Information Inc., 345 East 47th Street, New York, NY 10017. (212) 705-7600. Monthly. $1560.00 per year.

GENERAL SCIENCE INDEX. H.W. Wilson and Company, 950 University Avenue, Bronx, NY 10452. (800) 367-6670 or (212) 588-8400. 1978 to present. Monthly. Inquire as to cost and availability.

INDEX TO SCIENTIFIC AND TECHNICAL PROCEEDINGS. Institute for Scientific Information, 3501 Market Street, Philadelphia, PA 19104. (800) 523-1850 or (215) 386-0100. 1978 to present. Monthly. $775.00 per year.

INDEX TO SCIENTIFIC REVIEWS. Institute for Scientific Information, 3501 Market Street, Philadelphia, PA 19104. (800) 523-1850 or (215) 386-0100. 1974 to present. Semi-annual. $550.00 per year.

IEEE PUBLICATIONS BULLETIN. Institute of Electrical and Electronics Engineers. Institute of Electrical and Electronics Engineers. IEEE Service Center, 445 Hoes Lane, Piscataway, NJ 08854. Quarterly. Free.

PHYSICS ABSTRACTS. Institution of Electrical Engineers. Available from: IEEE Service Center, 445 Hoes Lane, Piscataway, NJ 08854. 1898 to present. Bimonthly. $1700.00 per year.

PHYSICS BRIEFS. Physik Verlag GmbH, Postfach 1260/1280, D-6940 Weinheim, West Germany. (212) 661-9404. 1920 to present. Twenty-six times per year. $1250.00 per year.

SCIENCE CITATION INDEX. Institute for Scientific Information, 3501 Market Street, Philadelphia, PA 19104. (800) 523-1850 or (215) 386-0100. Six times per year. $6200.00 per year.

ASSOCIATIONS AND PROFESSIONAL SOCIETIES

AMERICAN ELECTRONICS ASSOCIATION. P.O. Box 10045, 2670 Hanover Street, Palo Alto, CA 94303. (415) 857-9300.

AMERICAN INSTITUTE OF PHYSICS. 335 East 45th Street, New York, NY 10017. (212) 661-9494.

EDISON ELECTRIC INSTITUTE. 1111 19th Street, N.W., Washington, DC 20036. (202) 828-7400.

INSTITUTE OF ELECTRICAL AND ELECTRONICS ENGINEERS. 345 East 47th Street, New York, NY 10017. (212) 705-7900.

BIBLIOGRAPHIES

NEW TECHNICAL BOOKS: A SELECTIVE LIST WITH DESCRIPTIVE ANNOTATIONS. New York Public Library, Science and Technology Research Center, Fifth Avenue and 42nd Street, New York, NY 10018. (212) 930-0800. 1915 to present. Monthly. $15.00 per year.

SCIENCE BOOKS AND FILMS. American Association for the Advancement of Science, 1333 H Street, NW, Washington, DC 20005. (202) 326-6454. Five times per year. $20.00 per year.

SCIENTIFIC AND TECHNICAL BOOKS AND SERIALS IN PRINT 1988; AN INDEX TO LITERATURE IN SCIENCE AND TECHNOLOGY. R.R. Bowker Company, 205 East 42nd Street, New York, NY 10017. (800) 521-8110. $175.00.

DIRECTORIES AND BIOGRAPHICAL SOURCES

AMERICAN MEN AND WOMEN OF SCIENCE. R.R. Bowker, Inc., Order Department, 245 West 17th Street, New York, NY 10011. (800) 521-8110. Eight volumes. 1986. $595.00 for set.

ELECTRICAL WORLD DIRECTORY OF ELECTRICAL UTILITIES. McGraw-Hill Book Company, 1221 Avenue of the Americas, New York, NY 10020. (212) 512-2000. Annual. $275.00.

INTERNATIONAL RESEARCH CENTERS DIRECTORY 1988-89. Darren L. Smith, editor. Gale Research Company, Book Tower, Detroit, MI 48226. (800) 521-0707. 4th edition. 1987. $360.00.

1987 DIRECTORY OF ENGINEERING SOCIETIES AND RELATED ORGANIZATIONS. Gordon Davis, editor. Hemisphere Publishing Corporation, 1010 Vermont Avenue, NW, Washington, DC 20005. (800) 526-0275. 12th edition. 1987. $100.00.

RESEARCH CENTERS DIRECTORY 1988. Gale Research Company, Book Tower, Detroit, MI 48226. (800) 521-0707. 12th edition. 1987. $365.00 for set.

SCIENTIFIC AND TECHNICAL ORGANIZATIONS AND AGENCIES DIRECTORY. Margaret Labash Young, editor. Gale Research Company, Book Tower, Detroit, MI 48226. (800) 521-0707. 2nd edition. 1987. $185.00.

WHO'S WHO IN ELECTRONICS. Harris Publishing Company, 2057-2 Aurora Road, Twinsburg, OH 44087. (216) 425-9143. Annual. $90.00.

WHO'S WHO IN ENGINEERING. Gordon Davis, editor. Hemisphere Publishing Corporation, 1010 Vermont Avenue, NW, Washington, DC 20005. (800) 526-0275. 6th edition. 1985. $200.00.

ENCYCLOPEDIAS AND DICTIONARIES

CONCISE SCIENCE DICTIONARY. Oxford University Press, 200 Madison Avenue, New York, NY 10016. (800) 458-5833. 1987. $9.95 in paper.

DICTIONARY OF THE PHYSICAL SCIENCES: TERMS, FORMULAS, DATA. Cesare Emiliani. Oxford University Press, 200 Madison Avenue, New York, NY 10016. (800) 458-5833. 1987. $19.95 in paper.

MCGRAW-HILL ENCYCLOPEDIA OF SCIENCE AND TECHNOLOGY. McGraw-Hill Book Company, 1221 Avenue of the Americas, New York, NY 10020. (212) 512-2000. 6th edition. 1987. $1600.00.

THESAURUS OF SCIENTIFIC, TECHNICAL, AND ENGINEERING TERMS. Hemisphere Publishing Corporation, 1010 Vermont Avenue, NW, Washington, DC 20005. (800) 526-0275. 1988. $125.00.

GENERAL WORKS

ELECTRIC MACHINERY. Peter F. Ryff. Prentice-Hall Publishing, Inc., Englewood Cliffs, NJ 07632. (800) 562-0245. 1988. $34.00.

ELECTRICAL ENGINEERING. W.H. Roadstrum and Dan H. Wolaver. Harper and Row Publishers, Inc., 10 East 53rd Street, New York, NY 10022. (800) 242-7737. 1986. $37.50.

ELECTRICITY AND MAGNETISM. M.H. Nayfeh and M.K. Brussel. John Wiley and Sons, Inc., 605 Third Avenue, New York, NY 10158. (800) 526-5368. 1985. $32.95.

ELECTROMAGNETIC CONCEPTS AND PRINCIPLES. Stanley V. Marshall and Gabriel G. Skitek. Prentice-Hall Publishing, Inc., Englewood Cliffs, NJ 07632. (800) 562-0245. 2nd edition. 1987. $42.95.

ELEMENTS OF ENGINEERING ELECTROMAGNETICS. N.N. Rao. Prentice-Hall Publishing, Inc., Englewood Cliffs, NJ 07632. (800) 562-0245. 2nd edition. 1987. $45.95.

FUNDAMENTALS OF ELECTRIC CIRCUITS. David A. Bell. Prentice-Hall Publishing, Inc., Englewood Cliffs, NJ 07632. (800) 562-0245. 4th edition. 1988. $30.25.

INTRODUCTION TO ELECTRICITY AND ELECTRONICS FOR THE COMPUTER AGE. R. Rosen. John Wiley and Sons, Inc., 605 Third Avenue, New York, NY 10158. (800) 526-5368. 1987. $35.95.

HANDBOOKS AND MANUALS

HANDBOOK OF MODERN ELECTRONICS AND ELECTRICAL ENGINEERING. Charles Belove, editor. John Wiley and Sons, Inc., 605 Third Avenue, New York, NY 10158. (800) 526-5368. 1986. $88.95.

ONLINE DATA BASES

CA SEARCH. Chemical Abstracts Service, P.O. Box 3012, Columbus, OH 43120. (800) 848-6538 or (614) 421-3600. Comprehensive guide to chemical literature, 1972 to present. Inquire as to online cost and availability.

COMPENDEX. Engineering Information, Inc., 345 East 47th Street, New York, NY 10017. (800) 221-1044 or (212) 705-7615. Engineering and technical literature, 1975 to present. Inquire as to online cost and availability.

DISSERTATION ABSTRACTS ONLINE. University Microfilms International, 300 North Zeeb Road, Ann Arbor, MI 48106. (800) 521-0600 or (313) 761-4700. Scope includes virtually all doctoral dissertations accepted at accredited American institutions from 1861 to present in over 250 subject areas. Inquire as to online cost and availability.

INSPEC. INSPEC Marketing Department, Institution of Electrical Engineers. Available from IEEE Service Center, 445 Hoes Lane, Piscataway, NJ 08854. (201) 981-0060. Online version of Physics Abstracts. Inquire as to online cost and availability.

NTIS. National Technical Information Service, 5285 Port Royal Road, Springfield, VA 22161. (703) 487-4630. Broad coverage of government sponsored research reports, 1964 to present. Inquire as to online cost and availability.

SCISEARCH. Institute for Scientific Information, 3501 Market Street, Philadelphia, PA 19104. (800) 523-1850 or (215) 386-0100. Broad multidisciplinary title and author index to the international literature of science and technology, 1974 to present. Inquire as to online cost and availability.

WILSONLINE. H.W. Wilson and Company, 950 University Avenue, Bronx, NY 10452. (800) 367-6770 or (212) 588-8400. Makes available online versions of the H.W. Wilson indexes including Applied Science and Technology Index, Business Periodicals Index and Readers' Guide to Periodical Literature. Approximately 1980 to present. Inquire as to online cost and availability.

OTHER SOURCES

A GUIDE TO THE LITERATURE OF ELECTRICAL AND ELECTRONICS ENGINEERING. Susan B. Ardis. Libraries Unlimited Inc., P.O. Box 263, Littleton, CO 80160. (303) 770-1220. 1987. $37.50.

WHAT EVERY ENGINEER SHOULD KNOW ABOUT ENGINEERING SOURCES. Marcel Dekker Inc., 270 Madison Avenue, New York, NY 10016. (800) 228-1160. 1984. $24.95.

PERIODICALS

ELECTRIC LIGHT AND POWER. Technical Publishing Company, 875 Third Avenue, New York, NY 10022. (212) 605-9400. 1922 to present. Monthly. $38.00 per year.

ELECTROMAGNETISM

ELECTRIC MACHINES AND POWER SYSTEMS. Hemisphere Publishing Corporation, 1010 Vermont Avenue, NW, Washington, DC 20005. (800) 526-0275. 1976 to present. Bimonthly. $134.50 per year.

ELECTRICAL WORLD. McGraw-Hill Book Company, 1221 Avenue of the Americas, New York, NY 10020. (212) 512-2000. 1874 to present. Monthly. $11.00 per year.

ELECTRONIC DESIGN. Hayden Publishing Company, 10 Mulholland Drive, Hasbrouck Heights, NJ 07604. (201) 288-7520. 1952 to present. Biweekly. $40.00 per year.

ELECTRONICS. McGraw-Hill Book Company, 1221 Avenue of the Americas, New York, NY 10020. (212) 512-2000. 1930 to present. Weekly. $32.00 per year.

IEEE TRANSACTIONS ON ELECTROMAGNETIC COMPATIBILITY. Institute of Electrical and Electronics Engineers. IEEE Service Center, 445 Hoes Lane, Piscataway, NJ 08854. 1959 to present. quarterly. $72.00 per year.

IEEE TRANSACTIONS ON MAGNETICS. Institute of Electrical and Electronics Engineers. IEEE Service Center, 445 Hoes Lane, Piscataway, NJ 08854. 1965 to present. Monthly. $136.00.

INSTITUTE OF ELECTRICAL AND ELECTRONICS ENGINEERS PROCEEDINGS. Institute of Electrical and Electronics Engineers. IEEE Service Center, 445 Hoes Lane, Piscataway, NJ 08854. 1913 to present. Monthly. $140.00 per year.

RESEARCH CENTERS AND INSTITUTES

EDISON ELECTRIC INSTITUTE. 1111 19th Street, N.W., Washington, DC 20036. (202) 778-6778.

ELECTRIC POWER INSTITUTE. Texas A&M University, Department of Electrical Engineering, College Station, TX 77843.

ELECTRICAL ENGINEERING RESEARCH LABORATORIES. Purdue University, Electrical Engineering Building, West Lafayette, IN 47907. (317) 494-3536.

LABORATORY FOR ELECTROMAGNETIC AND ELECTRONIC SYSTEMS. 77 Massachusetts Avenue, Cambridge, MA 02139. (617) 253-4631.

ELECTRON LENS

See: ELECTRON OPTICS

ELECTRON MICROSCOPY

See also: MICROSCOPY, OPTICS

ABSTRACT SERVICES AND INDEXES

APPLIED SCIENCE AND TECHNOLOGY INDEX. H.W. Wilson and Company, 950 University Avenue, Bronx, NY 10452. (800) 367-6670 or (212) 588-8400. Monthly. Inquire as to cost and availability.

BIBLIOGRAPHY AND INDEX OF GEOLOGY. American Geological Institute, 4220 King Street, Alexandria, VA 22302. (703) 379-2480. 1969 to present. Monthly. $1100.00 per year.

CHEMICAL ABSTRACTS. American Chemical Society, Chemical Abstracts Service, Box 3012, Columbus, OH 43210. (614) 421-3600. 1907 to present. Weekly. $9500.00 per year.

CURRENT CONTENTS: ENGINEERING, TECHNOLOGY AND APPLIED SCIENCES. Institute for Scientific Information, 3501 Market Street, Philadelphia, PA 19104. (800) 523-1850 or (215) 386-0100. Weekly. $275.00 per year.

ENGINEERING INDEX MONTHLY AND AUTHOR INDEX. Engineering Information Inc., 345 East 47th Street, New York, NY 10017. (212) 705-7600. Monthly. $1560.00 per year.

GENERAL SCIENCE INDEX. H.W. Wilson and Company, 950 University Avenue, Bronx, NY 10452. (800) 367-6670 or (212) 588-8400. 1978 to present. Monthly. Inquire as to cost and availability.

PHYSICS ABSTRACTS. Institution of Electrical Engineers. Available from: IEEE Service Center, 445 Hoes Lane, Piscataway, NJ 08854. 1898 to present. Bimonthly. $1700.00 per year.

PHYSICS BRIEFS. Physik Verlag GmbH, Postfach 1260/1280, D-6940 Weinheim, West Germany. (212) 661-9404. 1920 to present. Twenty-six times per year. $1250.00 per year.

SCIENCE CITATION INDEX. Institute for Scientific Information, 3501 Market Street, Philadelphia, PA 19104. (800) 523-1850 or (215) 386-0100. Six times per year. $6200.00 per year.

ANNUAL REVIEWS AND YEARBOOKS

ADVANCES IN OPTICAL AND ELECTRON MICROSCOPY. V.E. Cosslett and R. Barer, editors. Academic Press, Inc., 6277 Sea Harbor Drive, Orlando, FL 32821. (800) 321-5068. 1966 to present. Irregular. Inquire.

ASSOCIATIONS AND PROFESSIONAL SOCIETIES

ELECTRON MICROSCOPY SOCIETY OF AMERICA. c/o Linda L. Horton, 1497 Chain Bridge Road, Suite 104, McLean, VA 22101. (703) 827-0498.

MICROBEAM ANALYSIS SOCIETY. c/o John Armstrong, Department of Geology, 170-25, California Institute of Technology, Pasadena, CA 91125. (818) 356-6253.

OPTICAL SOCIETY OF AMERICA. 1816 Jefferson Place, N.W., Washington, DC 20036. (202) 223-8130.

SPIE - THE INTERNATIONAL SOCIETY FOR OPTICAL ENGINEERING. P.O. Box 10, Bellington, WA 98227. (206) 676-3290.

DIRECTORIES AND BIOGRAPHICAL SOURCES

AMERICAN MEN AND WOMEN OF SCIENCE. R.R. Bowker, Inc., Order Department, 245 West 17th Street, New York, NY 10011. (800) 521-8110. Eight volumes. 1986. $595.00 for set.

INTERNATIONAL RESEARCH CENTERS DIRECTORY 1988-89. Darren L. Smith, editor. Gale Research Company, Book Tower, Detroit, MI 48226. (800) 521-0707. 4th edition. 1987. $360.00.

1987 DIRECTORY OF ENGINEERING SOCIETIES AND RELATED ORGANIZATIONS. Gordon Davis, editor. Hemisphere Publishing Corporation, 1010 Vermont Avenue, NW, Washington, DC 20005. (800) 526-0275. 12th edition. 1987. $100.00.

RESEARCH CENTERS DIRECTORY 1988. Gale Research Company, Book Tower, Detroit, MI 48226. (800) 521-0707. 12th edition. 1987. $365.00 for set.

SCIENTIFIC AND TECHNICAL ORGANIZATIONS AND AGENCIES DIRECTORY. Margaret Labash Young, editor. Gale Research Company, Book Tower, Detroit, MI 48226. (800) 521-0707. 2nd edition. 1987. $185.00.

SPIE - THE INTERNATIONAL SOCIETY FOR OPTICAL ENGINEERING ANNUAL MEMBERSHIP DIRECTORY. P.O. Box 10, Bellington, WA 98227. (206) 676-3290. Annual. Inquire.

WHO'S WHO IN ENGINEERING. Gordon Davis, editor. Hemisphere Publishing Corporation, 1010 Vermont Avenue, NW, Washington, DC 20005. (800) 526-0275. 6th edition. 1985. $200.00.

GENERAL WORKS

INTRODUCTION TO ELECTRON MICROSCOPY. Saul Wischnitzer. Pergamon Press, Inc., Maxwell House, Fairview Park, Elmsford, NY 10523. (914) 592-7700. 3rd edition. 1981. $43.00.

INTRODUCTION TO MICROSCOPY BY MEANS OF LIGHT, ELECTRONS, X-RAYS, OR ULTRASOUND. T.G. Rochow and E.G. Rochow. Plenum Publishing Corporation, 233 Spring Street, New York, NY 10013. (800) 221-9369. 1979. $32.50.

THEORY AND PRACTICE OF SCANNING OPTICAL MICROSCOPY. Tony C. Wilson and J.R. Sheppard. Academic Press, Inc., 6277 Sea Harbor Drive, Orlando, FL 32821. (800) 321-5068. 1984. $41.00.

ONLINE DATA BASES

CA SEARCH. Chemical Abstracts Service, P.O. Box 3012, Columbus, OH 43120. (800) 848-6538 or (614) 421-3600. Comprehensive guide to chemical literature, 1972 to present. Inquire as to online cost and availability.

COMPENDEX. Engineering Information, Inc., 345 East 47th Street, New York, NY 10017. (800) 221-1044 or (212) 705-7615. Engineering and technical literature, 1975 to present. Inquire as to online cost and availability.

DISSERTATION ABSTRACTS ONLINE. University Microfilms International, 300 North Zeeb Road, Ann Arbor, MI 48106. (800) 521-0600 or (313) 761-4700. Scope includes virtually all doctoral dissertations accepted at accredited American institutions from 1861 to present in over 250 subject areas. Inquire as to online cost and availability.

GEOREF. Online version of the BIBLIOGRAPHY AND INDEX OF GEOLOGY. American Geological Institute, 4220 King Street, Alexandria, VA 22302. (703) 379-2480. 1969 to present. Inquire as to online cost and availability.

INSPEC. INSPEC Marketing Department, Institution of Electrical Engineers. Available from IEEE Service Center, 445 Hoes Lane, Piscataway, NJ 08854. (201) 981-0060. Online version of Physics Abstracts. Inquire as to online cost and availability.

NTIS. National Technical Information Service, 5285 Port Royal Road, Springfield, VA 22161. (703) 487-4630. Broad coverage of government sponsored research reports, 1964 to present. Inquire as to online cost and availability.

SCISEARCH. Institute for Scientific Information, 3501 Market Street, Philadelphia, PA 19104. (800) 523-1850 or (215) 386-0100. Broad multidisciplinary title and author index to the international literature of science and technology, 1974 to present. Inquire as to online cost and availability.

WILSONLINE. H.W. Wilson and Company, 950 University Avenue, Bronx, NY 10452. (800) 367-6770 or (212) 588-8400. Makes available online versions of the H.W. Wilson indexes including Applied Science and Technology Index, Business Periodicals Index and Readers' Guide to Periodical Literature. Approximately 1980 to present. Inquire as to online cost and availability.

PERIODICALS

JOURNAL OF MICROSCOPY. Blackwell Scientific Publications, Inc., 52 Beacon Street, Boston, MA 02108. (800) 325-4177. 1878 to present. Monthly. $315.00 per year.

SCANNING ELECTRON MICROSCOPY. Scanning Microscopy Inc., Box 66507, AMF O'Hare, IL 60666. (312) 529-6677. 1968 to present. Quarterly. $109.00 per year.

RESEARCH CENTERS AND INSTITUTES

CENTER FOR ELECTRON MICROSCOPY. University of Illinois, 74 Bevier Hall, 905 South Goodwin Avenue, Urbana, IL 61801. (217) 333-2108.

ELECTRON MICROBEAM ANALYSIS LABORATORY. 413 Space Research Building, Ann Arbor, MI 48109. (313) 764-3360.

ELECTRON OPTICS

See also: ELECTRON MICROSCOPY, ELECTRON SPECTROSCOPY, OPTICS, OPTOELECTRONICS

ABSTRACT SERVICES AND INDEXES

APPLIED SCIENCE AND TECHNOLOGY INDEX. H.W. Wilson and Company, 950 University Avenue, Bronx, NY 10452. (800) 367-6670 or (212) 588-8400. Monthly. Inquire as to cost and availability.

CHEMICAL ABSTRACTS. American Chemical Society, Chemical Abstracts Service, Box 3012, Columbus, OH 43210. (614) 421-3600. 1907 to present. Weekly. $9500.00 per year.

CURRENT CONTENTS: ENGINEERING, TECHNOLOGY AND APPLIED SCIENCES. Institute for Scientific Information, 3501 Market Street, Philadelphia, PA 19104. (800) 523-1850 or (215) 386-0100. Weekly. $275.00 per year.

ENGINEERING INDEX MONTHLY AND AUTHOR INDEX. Engineering Information Inc., 345 East 47th Street, New York, NY 10017. (212) 705-7600. Monthly. $1560.00 per year.

GENERAL SCIENCE INDEX. H.W. Wilson and Company, 950 University Avenue, Bronx, NY 10452. (800) 367-6670 or (212) 588-8400. 1978 to present. Monthly. Inquire as to cost and availability.

PHYSICS ABSTRACTS. Institution of Electrical Engineers. Available from: IEEE Service Center, 445 Hoes Lane, Piscataway, NJ 08854. 1898 to present. Bimonthly. $1700.00 per year.

PHYSICS BRIEFS. Physik Verlag GmbH, Postfach 1260/1280, D-6940 Weinheim, West Germany. (212) 661-9404. 1920 to present. Twenty-six times per year. $1250.00 per year.

SCIENCE CITATION INDEX. Institute for Scientific Information, 3501 Market Street, Philadelphia, PA 19104. (800) 523-1850 or (215) 386-0100. Six times per year. $6200.00 per year.

ANNUAL REVIEWS AND YEARBOOKS

ADVANCES IN OPTICAL AND ELECTRON MICROSCOPY. V.E. Cosslett and R. Barer, editors. Academic Press, Inc., 6277 Sea Harbor Drive, Orlando, FL 32821. (800) 321-5068. 1966 to present. Irregular. Inquire.

ASSOCIATIONS AND PROFESSIONAL SOCIETIES

ELECTRON MICROSCOPY SOCIETY OF AMERICA. c/o Linda L. Horton, 1497 Chain Bridge Road, Suite 104, McLean, VA 22101. (703) 827-0498.

MICROBEAM ANALYSIS SOCIETY. c/o John Armstrong, Department of Geology, 170-25, California Institute of Technology, Pasadena, CA 91125. (818) 356-6253.

OPTICAL SOCIETY OF AMERICA. 1816 Jefferson Place, N.W., Washington, DC 20036. (202) 223-8130.

SPIE - THE INTERNATIONAL SOCIETY FOR OPTICAL ENGINEERING. P.O. Box 10, Bellington, WA 98227. (206) 676-3290.

DIRECTORIES AND BIOGRAPHICAL SOURCES

AMERICAN MEN AND WOMEN OF SCIENCE. R.R. Bowker, Inc., Order Department, 245 West 17th Street, New York, NY 10011. (800) 521-8110. Eight volumes. 1986. $595.00 for set.

INTERNATIONAL RESEARCH CENTERS DIRECTORY 1988-89. Darren L. Smith, editor. Gale Research Company, Book Tower, Detroit, MI 48226. (800) 521-0707. 4th edition. 1987. $360.00.

1987 DIRECTORY OF ENGINEERING SOCIETIES AND RELATED ORGANIZATIONS. Gordon Davis, editor. Hemisphere Publishing Corporation, 1010 Vermont Avenue, NW, Washington, DC 20005. (800) 526-0275. 12th edition. 1987. $100.00.

RESEARCH CENTERS DIRECTORY 1988. Gale Research Company, Book Tower, Detroit, MI 48226. (800) 521-0707. 12th edition. 1987. $365.00 for set.

SCIENTIFIC AND TECHNICAL ORGANIZATIONS AND AGENCIES DIRECTORY. Margaret Labash Young, editor. Gale Research Company, Book Tower, Detroit, MI 48226. (800) 521-0707. 2nd edition. 1987. $185.00.

SPIE - THE INTERNATIONAL SOCIETY FOR OPTICAL ENGINEERING ANNUAL MEMBERSHIP DIRECTORY. P.O. Box 10, Bellington, WA 98227. (206) 676-3290. Annual. Inquire.

WHO'S WHO IN ENGINEERING. Gordon Davis, editor. Hemisphere Publishing Corporation, 1010 Vermont Avenue, NW, Washington, DC 20005. (800) 526-0275. 6th edition. 1985. $200.00.

GENERAL WORKS

BASIC TECHNIQUES FOR TRANSMISSION ELECTRON MICROSCOPY. M.A. Hayat. Academic Press, Inc., 6277 Sea Harbor Drive, Orlando, FL 32821. (800) 321-5068. 1985. $58.50.

ELECTRO-OPTICS. Lewis J. Pinson. John Wiley and Sons, Inc., 605 Third Avenue, New York, NY 10158. (800) 526-5368. 1985. $35.95.

INTRODUCTION TO ELECTRON AND ION OPTICS. Paul Dahl. Academic Press, Inc., 6277 Sea Harbor Drive, Orlando, FL 32821. (800) 321-5068. 1973. $38.50.

INTRODUCTION TO OPTICAL ELECTRONICS. Kenneth Jones. Harper and Row Publishers, Inc., 10 East 53rd Street, New York, NY 10022. (800) 242-7737. 1986. $34.95.

ONLINE DATA BASES

CA SEARCH. Chemical Abstracts Service, P.O. Box 3012, Columbus, OH 43120. (800) 848-6538 or (614) 421-3600. Comprehensive guide to chemical literature, 1972 to present. Inquire as to online cost and availability.

COMPENDEX. Engineering Information, Inc., 345 East 47th Street, New York, NY 10017. (800) 221-1044 or (212) 705-7615. Engineering and technical literature, 1975 to present. Inquire as to online cost and availability.

DISSERTATION ABSTRACTS ONLINE. University Microfilms International, 300 North Zeeb Road, Ann Arbor, MI 48106. (800) 521-0600 or (313) 761-4700. Scope includes virtually all doctoral dissertations accepted at accredited American institutions from 1861 to present in over 250 subject areas. Inquire as to online cost and availability.

GEOREF. Online version of the BIBLIOGRAPHY AND INDEX OF GEOLOGY. American Geological Institute, 4220 King Street, Alexandria, VA 22302. (703) 379-2480. 1969 to present. Inquire as to online cost and availability.

INSPEC. INSPEC Marketing Department, Institution of Electrical Engineers. Available from IEEE Service Center, 445 Hoes Lane, Piscataway, NJ 08854. (201) 981-0060. Online version of Physics Abstracts. Inquire as to online cost and availability.

NTIS. National Technical Information Service, 5285 Port Royal Road, Springfield, VA 22161. (703) 487-4630. Broad coverage of government sponsored research reports, 1964 to present. Inquire as to online cost and availability.

SCISEARCH. Institute for Scientific Information, 3501 Market Street, Philadelphia, PA 19104. (800) 523-1850 or (215) 386-0100. Broad multidisciplinary title and author index to the international literature of science and technology, 1974 to present. Inquire as to online cost and availability.

WILSONLINE. H.W. Wilson and Company, 950 University Avenue, Bronx, NY 10452. (800) 367-6770 or (212) 588-8400. Makes available online versions of the H.W. Wilson indexes including Applied Science and Technology Index, Business Periodicals Index and Readers' Guide to Periodical Literature. Approximately 1980 to present. Inquire as to online cost and availability.

PERIODICALS

ELECTRO-OPTICS. PennWell Publishing Company, 119 Russell Street, Littleton, MA 01460. (617) 486-9501. Twenty times per year. $200.00 per year.

SCANNING ELECTRON MICROSCOPY. Scanning Microscopy Inc., Box 66507, AMF O'Hare, IL 60666. (312) 529-6677. 1968 to present. Quarterly. $109.00 per year.

RESEARCH CENTERS AND INSTITUTES

CENTER FOR ELECTROMAGNETICS RESEARCH. Northeastern University, 235 Forsyth Building, Boston, MA 02115. (617) 437-5110.

CENTER FOR TELECOMMUNICATIONS RESEARCH. Columbia University, Room 1220 S.W. Mudd Building, New York, NY 10027. (212) 280-2483.

ELECTRONICS RESEARCH LABORATORY. University of California at Berkeley, 253 Cory Hall, Berkeley, CA 94720. (415) 642-2301.

ELECTRON SPECTROSCOPY

See also: ANALYTICAL CHEMISTRY, MASS SPECTROMETRY, NUCLEAR MAGNETIC RESONANCE, SPECTROSCOPY

ABSTRACT SERVICES AND INDEXES

APPLIED SCIENCE AND TECHNOLOGY INDEX. H.W. Wilson and Company, 950 University Avenue, Bronx, NY 10452. (800) 367-6670 or (212) 588-8400. Monthly. Inquire as to cost and availability.

CHEMICAL ABSTRACTS. American Chemical Society, Chemical Abstracts Service, Box 3012, Columbus, OH 43210. (614) 421-3600. 1907 to present. Weekly. $9500.00 per year.

MASS SPECTROMETRY BULLETIN. Royal Society of Chemistry. Available from: Distribution Centre, Blackhorse Road, Letchworth, Herts SG6 1HN, England. 1966 to present. Monthly. $350.00 per year.

PHYSICS ABSTRACTS. Institution of Electrical Engineers. Available from: IEEE Service Center, 445 Hoes Lane, Piscataway, NJ 08854. 1898 to present. Bimonthly. $1700.00 per year.

SCIENCE CITATION INDEX. Institute for Scientific Information, 3501 Market Street, Philadelphia, PA 19104. (800) 523-1850 or (215) 386-0100. Six times per year. $6200.00 per year.

ANNUAL REVIEWS AND YEARBOOKS

APPLIED SPECTROSCOPY REVIEWS. Marcel Dekker Inc., 270 Madison Avenue, New York, NY 10016. (800) 228-1160. Price varies, inquire.

PROGRESS IN ANALYTICAL ATOMIC SPECTROSCOPY. Pergamon Press, Inc., Maxwell House, Fairview Park, Elmsford, NY 10523. (914) 592-7700. Irregular. Price varies, inquire.

ASSOCIATIONS AND PROFESSIONAL SOCIETIES

AMERICAN CHEMICAL SOCIETY. 1155 16th Street, N.W., Washington, DC 20036. (202) 872-4600.

AMERICAN SOCIETY FOR MASS SPECTROSCOPY. P.O. Box 1508, East Lansing, MI 48823. (517) 337-2548.

COBLENTZ SOCIETY. c/o Robert W. Hannah, Perkin-Elmer Corporation, Main Avenue, Norwalk, CT 06859. (203) 431-7797.

SOCIETY FOR APPLIED SPECTROSCOPY. P.O. Box 1438, Frederick, MD 21701. (301) 694-8122.

DIRECTORIES AND BIOGRAPHICAL SOURCES

INTERNATIONAL RESEARCH CENTERS DIRECTORY 1988-89. Darren L. Smith, editor. Gale Research Company, Book Tower, Detroit, MI 48226. (800) 521-0707. 4th edition. 1987. $360.00.

RESEARCH CENTERS DIRECTORY 1988. Gale Research Company, Book Tower, Detroit, MI 48226. (800) 521-0707. 12th edition. 1987. $365.00 for set.

SCIENTIFIC AND TECHNICAL ORGANIZATIONS AND AGENCIES DIRECTORY. Margaret Labash Young, editor. Gale Research Company, Book Tower, Detroit, MI 48226. (800) 521-0707. 2nd edition. 1987. $185.00.

SOCIETY FOR APPLIED SPECTROSCOPY MEMBERSHIP DIRECTORY. Society for Applied Spectroscopy, P.O. Box 1438, Frederick, MD 21701. (301) 694-8122. 1988. $75.00.

ENCYCLOPEDIAS AND DICTIONARIES

DICTIONARY OF SPECTROSCOPY. R.C. Denney. John Wiley and Sons, Inc., 605 Third Avenue, New York, NY 10158. (800) 526-5368. 1982. $42.00.

DICTIONARY OF THE PHYSICAL SCIENCES: TERMS, FORMULAS, DATA. Cesare Emiliani. Oxford University Press, 200 Madison Avenue, New York, NY 10016. (800) 458-5833. 1987. $19.95 in paper.

THESAURUS OF SCIENTIFIC, TECHNICAL, AND ENGINEERING TERMS. Hemisphere Publishing Corporation, 1010 Vermont Avenue, NW, Washington, DC 20005. (800) 526-0275. 1988. $125.00.

GENERAL WORKS

APPLIED ELECTRON SPECTROSCOPY FOR CHEMICAL ANALYSIS. H. Windawi and F.L. Ho, editors. John Wiley and Sons, Inc., 605 Third Avenue, New York, NY 10158. (800) 526-5368. 1982. $55.00.

ELECTRON SPECTROSCOPY: THEORY, TECHNIQUES AND APPLICATIONS. C.R. Brundle and A.D. Baker, editors. Academic Press, Inc., 6277 Sea Harbor Drive, Orlando, FL 32821. (800) 321-5068. 1981. $81.00.

HIGH RESOLUTION SPECTROSCOPY. J. Michael Hollas. Butterworth's Publishing, 80 Montvale Avenue, Stoneham, MA 02180. (800) 325-4177. 1982. $130.00.

INTRODUCTION TO APPLICATIONS OF SYMMETRY TO BONDING AND SPECTRA. B.E. Douglas and C.A. Hollingsworth. Academic Press, Inc., 6277 Sea Harbor Drive, Orlando, FL 32821. (800) 321-5068. 1985. $39.00.

MOLECULES AND RADIATION: AN INTRODUCTION TO MODERN MOLECULAR SPECTROSCOPY. Jeffrey I. Steinfeld. MIT Press, 28 Carleton Street, Cambridge, MA 02142. (617) 253-2884. 1985. $19.95.

PRINCIPLES OF SPECTROSCOPY. Raymond Chang. Robert E. Krieger Publishing Company, Inc., P.O. Box 9542, Melbourne, FL 32902-9542. (305) 724-9542. 1978. $21.50.

HANDBOOKS AND MANUALS

HANDBOOK OF SPECTROSCOPY. VOLUME 3. J.W. Robinson. CRC Press, 2000 Corporate Boulevard, Boca Raton, FL 33431. (800) 272-7737. 1981. $97.00.

LANGE'S HANDBOOK OF CHEMISTRY. John A. Dean, editor. McGraw-Hill Book Company, 1221 Avenue of the Americas, New York, NY 10020. (212) 512-2000. 13th edition. 1985. $59.50.

ONLINE DATA BASES

CA SEARCH. Chemical Abstracts Service, P.O. Box 3012, Columbus, OH 43120. (800) 848-6538 or (614) 421-3600. Comprehensive guide to chemical literature, 1972 to present. Inquire as to online cost and availability.

DISSERTATION ABSTRACTS ONLINE. University Microfilms International, 300 North Zeeb Road, Ann Arbor, MI 48106. (800) 521-0600 or (313) 761-4700. Scope includes virtually all doctoral dissertations accepted at accredited American institutions from 1861 to present in over 250 subject areas. Inquire as to online cost and availability.

INSPEC. INSPEC Marketing Department, Institution of Electrical Engineers. Available from IEEE Service Center, 445 Hoes Lane, Piscataway, NJ 08854. (201) 981-0060. Online version of Physics Abstracts. Inquire as to online cost and availability.

NTIS. National Technical Information Service, 5285 Port Royal Road, Springfield, VA 22161. (703) 487-4630. Broad coverage of government sponsored research reports, 1964 to present. Inquire as to online cost and availability.

SCISEARCH. Institute for Scientific Information, 3501 Market Street, Philadelphia, PA 19104. (800) 523-1850 or (215) 386-0100. Broad multidisciplinary title and author index to the international literature of science and technology, 1974 to present. Inquire as to online cost and availability.

WILSONLINE. H.W. Wilson and Company, 950 University Avenue, Bronx, NY 10452. (800) 367-6770 or (212) 588-8400. Makes available online versions of the H.W. Wilson indexes including Applied Science and Technology Index, Business Periodicals Index and Readers' Guide to Periodical Literature. Approximately 1980 to present. Inquire as to online cost and availability.

PERIODICALS

APPLIED SPECTROSCOPY. Society for Applied Spectroscopy, P.O. Box 1438, Frederick, MD 21701. (301) 694-8122. 1946 to present. Monthly. $145.00 per year.

APPLIED SPECTROSCOPY REVIEWS. Marcel Dekker Inc., 270 Madison Avenue, New York, NY 10016. (800) 228-1160. 1964 to present. Quarterly. $185.00 per year.

INTERNATIONAL JOURNAL OF MASS SPECTROMETRY AND ION PROCESSES. Elsevier Science Publishing Company, Inc., 52 Vanderbilt Avenue, New York, NY 10017. (212) 370-5520. 1968 to present. Twenty-one times per year. $380.00 per year.

JOURNAL OF ELECTRON SPECTROSCOPY AND RELATED PHENOMENA. Elsevier Science Publishing Company, Inc., 52 Vanderbilt Avenue, New York, NY 10017. (212) 370-5520. 1972 to present. Monthly. $255.00 per year.

JOURNAL OF MOLECULAR SPECTROSCOPY. Academic Press, Inc., 6277 Sea Harbor Drive, Orlando, FL 32821. (800) 321-5068. 1957 to present. Monthly. $630.00 per year.

JOURNAL OF QUANTITATIVE SPECTROSCOPY AND RADIATIVE TRANSFER. Pergamon Press, Inc., Maxwell House, Fairview Park, Elmsford, NY 10523. (914) 592-7700. 1961 to present. Monthly. $485.00 per year.

MASS SPECTROMETRY REVIEWS. John Wiley and Sons, Inc., 605 Third Avenue, New York, NY 10158. (800) 526-5368. 1982 to present. Quarterly. $135.00 per year.

SPECTROCHIMICA ACTA. A: MOLECULAR SPECTROSCOPY. Pergamon Press, Inc., Maxwell House, Fairview Park, Elmsford, NY 10523. (914) 592-7700. 1939 to present. Monthly. $400.00 per year.

SPECTROCHIMICA ACTA. B: ATOMIC SPECTROSCOPY. Pergamon Press, Inc., Maxwell House, Fairview Park, Elmsford, NY 10523. (914) 592-7700. 1939 to present. Monthly. $350.00 per year.

SPECTROSCOPY LETTERS. Marcel Dekker Inc., 270 Madison Avenue, New York, NY 10016. (800) 228-1160. 1968 to present. Ten times per year. $280.00 per year.

RESEARCH CENTERS AND INSTITUTES

ATOMIC, MOLECULAR AND CHEMICAL PHYSICS LABORATORY. University of Nevada - Reno, Reno, NV 89507. (702) 784-4920.

GEORGE R. HARRISON SPECTROSCOPY LABORATORY. Massachusetts Institute of Technology, 77 Massachusetts Avenue, Cambridge, MA 02139. (617) 253-7700.

INSTITUTE FOR PHYSICAL SCIENCE AND TECHNOLOGY. University of Maryland, College Park, MD 20742. (301) 454-2636.

INSTITUTE OF OPTICS. University of Rochester, Rochester, NY 14627. (716) 275-2314.

SOUTHERN CALIFORNIA REGIONAL CENTER FOR NUCLEAR MAGNETIC RESONANCE SPECTROSCOPY. California Institute of Technology, Mail Code 164-30, Pasadena, CA 91125. (818) 356-6241.

ELECTRONIC CIRCUITS AND COMPONENTS

See also: ELECTRICAL ENGINEERING, ELECTRICITY, ELECTRONIC ENGINEERING

ABSTRACT SERVICES AND INDEXES

APPLIED SCIENCE AND TECHNOLOGY INDEX. H.W. Wilson Company, 950 University Avenue, Bronx, NY 10452. (800) 367-6670 or (212) 588-8400. Inquire as to cost and availability.

ELECTRONICS AND COMMUNICATIONS ABSTRACTS JOURNAL. Cambridge Scientific Abstracts, 5161 River Road, Bethesda, MD 20816. (301) 951-1400. Bimonthly. Inquire as to cost and availability.

ENGINEERING INDEX MONTHLY. Engineering Information, Incorporated, 345 East 47th Street, New York, NY 10017. (800) 221-1044 or (212) 705-7600. Monthly, with annual cumulation. $1425.00 per year.

PHYSICS ABSTRACTS. Institute of Electrical Engineers, London, United Kingdom. Available from: Institute of Electrical and Electronic Engineers (Ieee), 345 East 47th Street, New York, NY 10017. (212) 705-7900.

SCIENCE CITATION INDEX. Institute for Scientific Information, 3501 Market Street, Philadelphia, PA 19104. (800) 523-1850 or (215) 386-0100. Inquire as to cost and availability.

ASSOCIATIONS AND PROFESSIONAL SOCIETIES

AMERICAN ELECTRONICS ASSOCIATION. Post Office Box 10045, 2670 Hanover Street, Palo Alto, CA 94303. (415) 857-9300.

ELECTRONIC INDUSTRIES ASSOCIATION. 2001 Eye Street, NW, Washington, DC 20006. (202) 457-4900.

IEEE (INSTITUTE OF ELECTRICAL AND ELECTRONICS ENGINEERS). 345 East 47th Street, New York, NY 10017. (212) 705-7900.

INTERNATIONAL SOCIETY FOR HYBRID MICROELECTRONICS. Post Office Box 2698, 1861 Wiehle Avenue, Suite 340, Reston, VA 22090. (703) 471-0066.

BIBLIOGRAPHIES

HANDBOOKS AND TABLES IN SCIENCE AND TECHNOLOGY. Russell H. Powell, editor. Oryx Press, 2214 North Central Avenue, Phoenix, AZ 85004-1483. (602) 254-6156. Second edition. 1983. $55.00.

SCIENTIFIC AND TECHNICAL BOOKS IN PRINT: AN INDEX TO LITERATURE IN SCIENCE AND TECHNOLOGY. R.R. Bowker Company, 205 East 42nd Street, New York, NY 10017. (800) 521-8110 or (212) 916-1600.

DIRECTORIES AND BIOGRAPHICAL SOURCES

AMERICAN ELECTRONICS ASSOCIATION DIRECTORY. American Electronics Association, Post Office Box 10045, 2670 Hanover Street, Palo Alto, CA 94303. (415) 857-9300. Annual.

EEM - ELECTRONIC ENGINEERS MASTER. Hearst Business Communications/UTP Division, 645 Stewart Avenue, Garden City, NY 11530. (516) 227-1300. Annual. $75.00 per copy.

IEEE MEMBERSHIP DIRECTORY. Institute of Electrical and Electronics Engineers, IEEE Service Center, 445 Hoes Lane, Piscataway, NJ 08854. (212) 705-7900. Annual. $7.00.

RESEARCH CENTERS DIRECTORY. Gale Research Company, Book Tower, Detroit, MI 48226. (800) 521-0707. Twelfth edition. 1987. $365.00 for set.

SCIENTIFIC AND TECHNICAL ORGANIZATIONS AND AGENCIES DIRECTORY. Gale Research Company, Book Tower, Detroit, MI 48226. (800) 521-0707. Second edition. 1987. $185.00.

WHO'S WHO IN ELECTRONICS. Harris Publishing Company, 2057-2 Aurora Road, Twinsburg, OH 44087. (216) 425-9000. Annual. $89.00.

WHO'S WHO IN ENGINEERING. Engineers Joint Council, 345 East 47th Street, New York, NY 10017. (212) 705-7010. 1985. $200.00.

WHO'S WHO IN FRONTIER SCIENCE AND TECHNOLOGY. Marquis Who's Who, Incorporated, 200 East Ohio Street, Chicago, IL 60611. (800) 428-3898 or (312) 787-2008.

WHO'S WHO IN TECHNOLOGY TODAY. Reston Publishing Company, Incorporated, c/o Prentice-Hall, Incorporated, Englewood Cliffs, NJ 07632. (800) 262-6868. Biennial. Five volumes. $425.00. Covers the fields of electronics, computer science, physics, optics, chemistry, biotechnology, mechanics, energy, and earth science.

ENCYCLOPEDIAS AND DICTIONARIES

ENCYCLOPEDIA OF PHYSICAL SCIENCE AND TECHNOLOGY. Academic Press, Incorporated, Orlando, FL 32887. (800) 321-5068 or (305) 345-2734. Fifteen volumes, 1986.

IEEE STANDARD DICTIONARY OF ELECTRICAL AND ELECTRONICS TERMS. Frank Jay, editor. John Wiley and Sons, Incorporated, 605 Third Avenue, New York, NY 10158. (800) 526-5368 or (212) 850-6000. Third edition. 1984. $49.95.

MCGRAW-HILL ENCYCLOPEDIA OF SCIENCE AND TECHNOLOGY. McGraw-Hill Book, Incorporated, 1221 Avenue of the Americas, New York, NY 10020. (212) 997-3675. Fifth edition, 15 volumes. $1100.00.

GENERAL WORKS

CIRCUITS AND SOFTWARE FOR ELECTRONICS ENGINEERS. Howard Bierman. McGraw-Hill Book Company, 1221 Avenue of the Americas, New York, NY 10020. (800) 628-0004. 1984. $39.50.

CIRCUITS, DEVICES AND SYSTEMS: A FIRST COURSE IN ELECTRICAL ENGINEERING. R.J. Smith. John Wiley and Sons, Incorporated, 605 Third Avenue, New York, NY 10158. (800) 526-5368 or (212) 850-6000. Fourth edition. 1984. $41.45.

CIRCUITS, SIGNALS AND SYSTEMS. William M. Siebert. McGraw-Hill Book Company, 1221 Avenue of the Americas, New York, NY 10020. (800) 628-0004. 1986. $37.95.

ELECTRICAL AND ELECTRONIC INSTRUMENTATION. Hai Hung Chiang. John Wiley and Sons, Incorporated, 605 Third Avenue, New York, NY 10158. (800) 526-5368 or (212) 850-6000. 1984. $64.95.

ELECTRONIC CIRCUIT ANALYSIS: BASIC PRINCIPLES. Roy A. Colclaser, Donald A. Neaman, Charles F. Hawkins. John Wiley and Sons, Incorporated, 605 Third Avenue, New York, NY 10158. (800) 526-5368 or (212) 850-6000. 1984. $42.50.

ELECTRONIC CIRCUITS AND APPLICATIONS. Stephen D. Senturia and B.D. Wedlock. John Wiley and Sons, Incorporated, 605 Third Avenue, New York, NY 10158. (800) 526-5368 or (212) 850-6000. 1975. $44.00.

ELECTRONIC DEVICES AND COMPONENTS. J. Seymour. John Wiley and Sons, Incorporated, 605 Third Avenue, New York, NY 10158. (800) 526-5368 or (212) 850-6000. 1981. $37.00 in paper.

ELECTRONIC INVENTIONS AND DISCOVERIES: ELECTRONICS FROM ITS EARLIEST BEGINNING TO THE PRESENT DAY. G.W.A. Dummer. Pergamon Press, Incorporated, Maxwell House, Fairview Park, Elmsford, NY 10523. (914) 592-7700. 1983. $48.50.

INTRODUCTION TO RANDOM PROCESSES: WITH APPLICATION TO SIGNALS AND SYSTEMS. William A. Gardner. Macmillan Publishing Company Incorporated, 866 Third Avenue, New York, NY 10022. (800) 257-5755 or (212) 935-2000. 1986. $34.95.

SEMICONDUCTOR CIRCUIT APPROXIMATIONS: AN INTRODUCTION TO TRANSISTORS AND INTEGRATED CIRCUITS. Albert Paul Malvino. McGraw-Hill Book Company, 1221 Avenue of the Americas, New York, NY 10020. (800) 628-0004. Fourth edition. 1985. $34.95.

HANDBOOKS AND MANUALS

CONTEMPORARY ELECTRONICS CIRCUITS DESKBOOK. Harry Helms. McGraw-Hill Book Company, 1221 Avenue of the Americas, New York, NY 10020. (800) 628-0004. 1986. $29.95.

CRC HANDBOOK OF TABLES FOR APPLIED ENGINEERING SCIENCE. R.E. Bolz and G.L. Tuve, editors. CRC Press, Incorporated, 2000 Corporate Boulevard, NW, Boca Raton, FL 33431. Second edition. 1973. $69.00.

ELECTRONIC ENGINEERS HANDBOOK. Donald G. Fink, editor. McGraw-Hill Book Company, 1221 Avenue of the Americas, New York, NY 10020. (800) 628-0004. Second edition. 1982. $89.00.

HANDBOOK OF ELECTRONICS INDUSTRY COST ESTIMATING DATA. Theodore Taylor. John Wiley and Sons, Incorporated, 605 Third Avenue, New York, NY 10158. (800) 526-5368 or (212) 850-6000. 1985. $54.95.

HANDBOOK OF ELECTRONICS MANUFACTURING ENGINEERING. B.S. Matisoff. Van Nostrand Reinhold Company, Incorporated, 115 Fifth Avenue, New York, NY 10003. (800) 543-2681. Second edition. 1986. $52.95.

HANDBOOK OF MODERN ELECTRONICS AND ELECTRICAL ENGINEERING. Charles Belove. John Wiley and Sons, Incorporated, 605 Third Avenue, New York, NY 10158. (800) 526-5368 or (212) 850-6000. 1986. $85.00.

REFERENCE MANUAL FOR TELECOMMUNICATIONS. R.L. Freeman. John Wiley and Sons, Incorporated, 605 Third Avenue, New York, NY 10158. (800) 526-5368 or (212) 850-6000. 1985. $79.95.

STANDARD HANDBOOK FOR ELECTRICAL ENGINEERS. Donald G. Fink and H. Wayne Beaty. McGraw-Hill Book Company, 1221 Avenue of the Americas, New York, NY 10020. (212) 512-2000. Eleventh edition. 1978. $85.00.

VLSI HANDBOOK. Norman G. Einspruch, editor. Academic Press, Incorporated, 6277 Sea Harbor Drive, Orlando, FL 32821. (800) 321-5068. 1985. $125.00.

THE WILEY ENGINEER'S DESK REFERENCE. Sanford I. Heisler. John Wiley and Sons, Incorporated, 605 Third Avenue, New York, NY 10158. (800) 526-5368 or (212) 850-6418. 1984. $36.00.

ONLINE DATA BASES

COMPENDEX. Engineering Information, Incorporated, 345 East 47th Street, New York, NY 10017. (800) 221-1044 or (212) 705-7615. Engineering and technical literature, 1975 to present. Inquire as to cost and availability.

DISSERTATION ABSTRACTS ONLINE. University Microfilms International, 300 North Zeeb Road, Ann Arbor, MI 48106. (800) 521-0600 or (313) 761-4700. Scope includes virtually all doctoral dissertations accepted at accredited American institutions from 1861 to present in 252 subject areas. Inquire as to cost and availability.

INSPEC. INSPEC Marketing Department, Institute of Electrical and Electronics Engineers, Incorporated, IEEE Service Department, 445 Hoes Lane, Piscataway, NJ 08854. (201) 981-0060. Inquire as to on-line cost and availability.

NTIS. National Technical Information Service, 5285 Port Royal Road, Springfield, VA 22161. (703) 487-4630. Broad coverage of government sponsored research reports, 1964 to present. Inquire as to cost and availability.

WILSONLINE. H.W. Wilson Company, 950 University Avenue, Bronx, NY 10452. (800) 367-6770 or (212) 588-8400. Makes available online versions of the printed H.W. Wilson Indexes including Applied Science and Technology Index, Business Periodicals Index, and Readers' Guide to Periodical Literature.

Period covered is generally 1983 to present. Inquire as to cost and availability.

OTHER SOURCES

A GUIDE TO THE LITERATURE OF ELECTRICAL AND ELECTRONIC ENGINEERING. Susan B. Ardis. Libraries Unlimited, Incorporated, Post Office Box 263, Littleton, CO 80160. (303) 770-1220. 1986. $37.50.

PERIODICALS

CANADAIAN ELECTRONICS ENGINEERING. Maclean Hunter Research Bureau, 777 Bay Street, Toronto, ON M5W 1A7 Canada. (416) 596-5729.

CIRCUITS, SYSTEMS AND SIGNAL PROCESSING. Birkhauser Boston, Incorporated, 380 Green Street, Post Office Box 2007, Cambridge, MA 02139. (617) 876-2333. Quarterly. $175.00 per year.

ELECTROCOMPONENT SCIENCE. Gordon Breach Science Publishers, Post Office Box 786, Cooper Station, New York, NY 10276. (212) 206-8900. Quarterly. $224.00 per year.

ELECTRONIC DESIGN. Heyden Publishing Company, Incorporated, 10 Mulholland Drive, Hasbrouck Heights, NJ 07604. (201) 393-6000. Biweekly. $65.00.

ELECTRONIC ENGINEERING TIMES. CMP Publications, Incorporated, 600 Community Drive, Manhasset, NY 11030. (516) 365-4600. Weekly. $65.00 per year.

ELECTRONICS WEEK. McGraw-Hill Book Company, 1221 Avenue of the Americas, New York, NY 10020. (800) 628-0004. Weekly. $18.00.

IEEE CIRCUITS AND DEVICES MAGAZINE. Institute of Electrical and Electronics Engineers, IEEE Service Center, 445 Hoes Lane, Piscataway, NJ 08854. (212) 705-7900. Bimonthly. $70.00 per year.

IEEE ELECTRON DEVICE LETTERS. Institute of Electrical and Electronics Engineers, IEEE Service Center, 445 Hoes Lane, Piscataway, NJ 08854. (212) 705-7900. Monthly. $70.00 per year.

IEEE JOURNAL OF SOLID STATE CIRCUITS. Institute of Electrical and Electronics Engineers, IEEE Service Center, 445 Hoes Lane, Piscataway, NJ 08854. (212) 705-7900. Bimonthly. $113.00 per year.

IEEE TRANSACTIONS ON CIRCUITS AND SYSTEMS. Institute of Electrical and Electronics Engineers, IEEE Service Center, 445 Hoes Lane, Piscataway, NJ 08854. (212) 705-7900. Monthly. $108.00 per year.

IEEE TRANSACTIONS ON ELECTRON DEVICES. Institute of Electrical and Electronics Engineers, IEEE Service Center, 445 Hoes Lane, Piscataway, NJ 08854. (212) 705-7900. Monthly. $159.00 per year.

RADIO ELECTRONICS. Gernsback Publications, Incorporated, 500-B BiCounty Boulevard, Farmingdale, NY 11735. (516) 293-3000. Monthly. $16.00 per year.

VLSI SYSTEMS DESIGN. CMP Publications, Incorporated, 600 Community Drive, Manhasset, NY 11030. (516) 365-4600. Monthly. $20.00 per year.

RESEARCH CENTERS AND INSTITUTES

MASSACHUSETTS INSTITUTE OF TECHNOLOGY. Laboratory for Electromagnetic and Electronic Systems, 77 Massachusetts Avenue, Cambridge, MA 02139. (617) 253-4631.

NORTH CAROLINA STATE UNIVERSITY. Solid State Electronics Laboratory, 432 Daniels Hall, Raleigh, NC 27695. (919) 737-2336.

OHIO STATE UNIVERSITY, ELECTROSCIENCE LABORATORY. 1320 Kinnear Road, Columbus, OH 43212. (614) 422-7981.

PENNSYLVANIA STATE UNIVERSITY. Solid State Device Laboratory, 210 Electrical Engineering, West Building, University Park, PA 16802. (814) 865-1666.

STANFORD UNIVERSITY. Stanford Electronics Laboratories, Stanford, CA 94305. (415) 497-1013.

ELECTRONIC MAIL

See: COMPUTER COMMUNICATIONS

ELECTRONIC SECURITY SYSTEMS

See: ALARM SYSTEMS

ELECTRONICS

See also: ELECTRICAL ENGINEERING, ELECTROMAGNETISM, ELECTRONIC CIRCUITS AND COMPONENTS, ELECTRONICS, ELECTRONICS ENGINEERING, MICROELECTRONICS, SOLID STATE PHYSICS

ABSTRACT SERVICES AND INDEXES

APPLIED SCIENCE AND TECHNOLOGY INDEX. H.W. Wilson and Company, 950 University Avenue, Bronx, NY 10452. (800) 367-6670 or (212) 588-8400. Monthly. Inquire as to cost and availability.

CHEMICAL ABSTRACTS. American Chemical Society, Chemical Abstracts Service, Box 3012, Columbus, OH 43210. (614) 421-3600. 1907 to present. Weekly. $9500.00 per year.

CONFERENCE PAPERS INDEX. Cambridge Scientific Abstracts, 5161 River Road, Bethesda, MD 20816. (301) 951-1400. 1972 to present. Monthly. Inquire as to cost and availability.

CURRENT CONTENTS: ENGINEERING, TECHNOLOGY AND APPLIED SCIENCES. Institute for Scientific Information, 3501 Market Street, Philadelphia, PA 19104. (800) 523-1850 or (215) 386-0100. Weekly. $275.00 per year.

ELECTRICAL AND ELECTRONICS ABSTRACTS. Institution of Electrical Engineers. Available from: Institute of Electrical and Electronics Engineers. IEEE Service Center, 445 Hoes Lane, Piscataway, NJ 08854. Monthly. $1250.00 per year.

ELECTRONICS AND COMMUNICATIONS ABSTRACTS. Cambridge Scientific Abstracts, 5161 River Road, Bethesda, MD 20816. (301) 951-1400. Bimonthly. Inquire as to cost and availability.

ENGINEERING INDEX MONTHLY AND AUTHOR INDEX. Engineering Information Inc., 345 East 47th Street, New York, NY 10017. (212) 705-7600. Monthly. $1560.00 per year.

GENERAL SCIENCE INDEX. H.W. Wilson and Company, 950 University Avenue, Bronx, NY 10452. (800) 367-6670 or (212) 588-8400. 1978 to present. Monthly. Inquire as to cost and availability.

INDEX TO SCIENTIFIC AND TECHNICAL PROCEEDINGS. Institute for Scientific Information, 3501 Market Street, Philadelphia, PA 19104. (800) 523-1850 or (215) 386-0100. 1978 to present. Monthly. $775.00 per year.

INDEX TO SCIENTIFIC REVIEWS. Institute for Scientific Information, 3501 Market Street, Philadelphia, PA 19104. (800) 523-1850 or (215) 386-0100. 1974 to present. Semi-annual. $550.00 per year.

IEEE PUBLICATIONS BULLETIN. Institute of Electrical and Electronics Engineers. Institute of Electrical and Electronics Engineers. IEEE Service Center, 445 Hoes Lane, Piscataway, NJ 08854. Quarterly. Free.

PHYSICS ABSTRACTS. Institution of Electrical Engineers. Available from: IEEE Service Center, 445 Hoes Lane, Piscataway, NJ 08854. 1898 to present. Bimonthly. $1700.00 per year.

PHYSICS BRIEFS. Physik Verlag GmbH, Postfach 1260/1280, D-6940 Weinheim, West Germany. (212) 661-9404. 1920 to present. Twenty-six times per year. $1250.00 per year.

SCIENCE CITATION INDEX. Institute for Scientific Information, 3501 Market Street, Philadelphia, PA 19104. (800) 523-1850 or (215) 386-0100. Six times per year. $6200.00 per year.

SOLID STATE ABSTRACTS: AN ABSTRACT JOURNAL INVOLVING THE PHYSICS, METALLURGY, CRYSTALLOGRAPHY, CHEMISTRY, AND DEVICE TECHNOLOGY OF SOLIDS. Cambridge Scientific Abstracts, 5161 River Road, Bethesda, MD 20816. (301) 951-1400. 1957 to present. Bimonthly. $550.00 per year.

ANNUAL REVIEWS AND YEARBOOKS

ADVANCES IN ELECTRONICS AND ELECTRON PHYSICS. Academic Press, Inc., 6277 Sea Harbor Drive, Orlando, FL 32821. (800) 321-5068. Irregular. Approximately $80.00 per volume.

CRC CRITICAL REVIEWS IN SOLID STATE AND MATERIALS SCIENCES. CRC Press, 2000 Corporate Boulevard, Boca Raton, FL 33431. (800) 272-7737. Irregular. Inquire.

ASSOCIATIONS AND PROFESSIONAL SOCIETIES

AMERICAN ELECTRONICS ASSOCIATION. P.O. Box 10045, 2670 Hanover Street, Palo Alto, CA 94303. (415) 857-9300.

AMERICAN INSTITUTE OF PHYSICS. 335 East 45th Street, New York, NY 10017. (212) 661-9494.

EDISON ELECTRIC INSTITUTE. 1111 19th Street, N.W., Washington, DC 20036. (202) 828-7400.

ELECTRONICS INDUSTRIES ASSOCIATION. 2001 Eye Street, N.W., Washington, DC 20006. (202) 457-4900.

INSTITUTE OF ELECTRICAL AND ELECTRONICS ENGINEERS. 345 East 47th Street, New York, NY 10017. (212) 705-7900.

INTERNATIONAL SOCIETY FOR HYBRID MICROELECTRONICS. P.O. Box 2698, 1861 Wiehle Avenue, Suite 340, Reston, VA 22090. (703) 471-0066.

BIBLIOGRAPHIES

NEW TECHNICAL BOOKS: A SELECTIVE LIST WITH DESCRIPTIVE ANNOTATIONS. New York Public Library, Science and Technology Research Center, Fifth Avenue and 42nd Street, New York, NY 10018. (212) 930-0800. 1915 to present. Monthly. $15.00 per year.

SCIENCE BOOKS AND FILMS. American Association for the Advancement of Science, 1333 H Street, NW, Washington, DC 20005. (202) 326-6454. Five times per year. $20.00 per year.

SCIENTIFIC AND TECHNICAL BOOKS AND SERIALS IN PRINT 1988; AN INDEX TO LITERATURE IN SCIENCE AND TECHNOLOGY. R.R. Bowker Company, 205 East 42nd Street, New York, NY 10017. (800) 521-8110. $175.00.

DIRECTORIES AND BIOGRAPHICAL SOURCES

AMERICAN MEN AND WOMEN OF SCIENCE. R.R. Bowker, Inc., Order Department, 245 West 17th Street, New York, NY 10011. (800) 521-8110. Eight volumes. 1986. $595.00 for set.

IEEE MEMBERSHIP DIRECTORY. Institute of Electrical and Electronics Engineers. IEEE Service Center, 445 Hoes Lane, Piscataway, NJ 08854. Annual. $7.00.

INTERNATIONAL RESEARCH CENTERS DIRECTORY 1988-89. Darren L. Smith, editor. Gale Research Company, Book Tower, Detroit, MI 48226. (800) 521-0707. 4th edition. 1987. $360.00.

1987 DIRECTORY OF ENGINEERING SOCIETIES AND RELATED ORGANIZATIONS. Gordon Davis, editor. Hemisphere Publishing Corporation, 1010 Vermont Avenue, NW, Washington, DC 20005. (800) 526-0275. 12th edition. 1987. $100.00.

RESEARCH CENTERS DIRECTORY 1988. Gale Research Company, Book Tower, Detroit, MI 48226. (800) 521-0707. 12th edition. 1987. $365.00 for set.

SCIENTIFIC AND TECHNICAL ORGANIZATIONS AND AGENCIES DIRECTORY. Margaret Labash Young, editor. Gale Research Company, Book Tower, Detroit, MI 48226. (800) 521-0707. 2nd edition. 1987. $185.00.

WHO'S WHO IN ELECTRONICS. Harris Publishing Company, 2057-2 Aurora Road, Twinsburg, OH 44087. (216) 425-9143. Annual. $90.00.

WHO'S WHO IN ENGINEERING. Gordon Davis, editor. Hemisphere Publishing Corporation, 1010 Vermont Avenue, NW, Washington, DC 20005. (800) 526-0275. 6th edition. 1985. $200.00.

ENCYCLOPEDIAS AND DICTIONARIES

IEEE STANDARD DICTIONARY OF ELECTRICAL AND ELECTRONICS TERMS. Frank Jay, editor. John Wiley and Sons, Inc., 605 Third Avenue, New York, NY 10158. (800) 526-5368. 3rd edition. 1984. $49.95.

MCGRAW-HILL ENCYCLOPEDIA OF SCIENCE AND TECHNOLOGY. McGraw-Hill Book Company, 1221 Avenue of the Americas, New York, NY 10020. (212) 512-2000. 6th edition. 1987. $1600.00.

THESAURUS OF SCIENTIFIC, TECHNICAL, AND ENGINEERING TERMS. Hemisphere Publishing Corporation, 1010 Vermont Avenue, NW, Washington, DC 20005. (800) 526-0275. 1988. $125.00.

GENERAL WORKS

ELECTROMAGNETIC CONCEPTS AND PRINCIPLES. Stanley V. Marshall and Gabriel G. Skitek. Prentice-Hall Publishing, Inc., Englewood Cliffs, NJ 07632. (800) 562-0245. 2nd edition. 1987. $42.95.

ELECTRONIC INVENTIONS AND DISCOVERIES: ELECTRONICS FROM ITS EARLIEST BEGINNINGS TO PRESENT DAY. G.W.A. Dummer. Pergamon Press, Inc., Maxwell House, Fairview Park, Elmsford, NY 10523. (914) 592-7700. 1983. $49.50.

ELEMENTS OF ENGINEERING ELECTROMAGNETICS. N.N. Rao. Prentice-Hall Publishing, Inc., Englewood Cliffs, NJ 07632. (800) 562-0245. 2nd edition. 1987. $45.95.

INTRODUCTION TO ELECTRICITY AND ELECTRONICS FOR THE COMPUTER AGE. R. Rosen. John Wiley and Sons, Inc., 605 Third Avenue, New York, NY 10158. (800) 526-5368. 1987. $35.95.

ELECTRONICS

MICROPROCESSOR TECHNOLOGY AND MICROCOMPUTERS. E.J. Pasahow. McGraw-Hill Book Company, 1221 Avenue of the Americas, New York, NY 10020. (212) 512-2000. 1987. $31.95.

HANDBOOKS AND MANUALS

ELECTRONIC ENGINEERS HANDBOOK. Donald G. Fink, editor. McGraw-Hill Book Company, 1221 Avenue of the Americas, New York, NY 10020. (212) 512-2000. 2nd edition. 1982. $89.00.

HANDBOOK OF MODERN ELECTRONICS AND ELECTRICAL ENGINEERING. Charles Belove, editor. John Wiley and Sons, Inc., 605 Third Avenue, New York, NY 10158. (800) 526-5368. 1986. $88.95.

ONLINE DATA BASES

CA SEARCH. Chemical Abstracts Service, P.O. Box 3012, Columbus, OH 43120. (800) 848-6538 or (614) 421-3600. Comprehensive guide to chemical literature, 1972 to present. Inquire as to online cost and availability.

COMPENDEX. Engineering Information, Inc., 345 East 47th Street, New York, NY 10017. (800) 221-1044 or (212) 705-7615. Engineering and technical literature, 1975 to present. Inquire as to online cost and availability.

DISSERTATION ABSTRACTS ONLINE. University Microfilms International, 300 North Zeeb Road, Ann Arbor, MI 48106. (800) 521-0600 or (313) 761-4700. Scope includes virtually all doctoral dissertations accepted at accredited American institutions from 1861 to present in over 250 subject areas. Inquire as to online cost and availability.

INSPEC. INSPEC Marketing Department, Institution of Electrical Engineers. Available from IEEE Service Center, 445 Hoes Lane, Piscataway, NJ 08854. (201) 981-0060. Online version of Physics Abstracts. Inquire as to online cost and availability.

NTIS. National Technical Information Service, 5285 Port Royal Road, Springfield, VA 22161. (703) 487-4630. Broad coverage of government sponsored research reports, 1964 to present. Inquire as to online cost and availability.

SCISEARCH. Institute for Scientific Information, 3501 Market Street, Philadelphia, PA 19104. (800) 523-1850 or (215) 386-0100. Broad multidisciplinary title and author index to the international literature of science and technology, 1974 to present. Inquire as to online cost and availability.

WILSONLINE. H.W. Wilson and Company, 950 University Avenue, Bronx, NY 10452. (800) 367-6770 or (212) 588-8400. Makes available online versions of the H.W. Wilson indexes including Applied Science and Technology Index, Business Periodicals Index and Readers' Guide to Periodical Literature. Approximately 1980 to present. Inquire as to online cost and availability.

OTHER SOURCES

A GUIDE TO THE LITERATURE OF ELECTRICAL AND ELECTRONICS ENGINEERING. Susan B. Ardis. Libraries Unlimited Inc., P.O. Box 263, Littleton, CO 80160. (303) 770-1220. 1987. $37.50.

WHAT EVERY ENGINEER SHOULD KNOW ABOUT ENGINEERING SOURCES. Marcel Dekker Inc., 270 Madison Avenue, New York, NY 10016. (800) 228-1160. 1984. $24.95.

PERIODICALS

ELECTRONIC DESIGN. Hayden Publishing Company, 10 Mulholland Drive, Hasbrouck Heights, NJ 07604. (201) 288-7520. 1952 to present. Biweekly. $40.00 per year.

Ency. of Physical Sciences & Engineering Info. Sources

ELECTRONICS. McGraw-Hill Book Company, 1221 Avenue of the Americas, New York, NY 10020. (212) 512-2000. 1930 to present. Weekly. $32.00 per year.

IEEE CIRCUITS AND DEVICES MAGAZINE. Institute of Electrical and Electronics Engineers. IEEE Service Center, 445 Hoes Lane, Piscataway, NJ 08854. Bimonthly. $70.00 per year.

IEEE JOURNAL OF SOLID STATE CIRCUITS. Institute of Electrical and Electronics Engineers. IEEE Service Center, 445 Hoes Lane, Piscataway, NJ 08854. 1966 to present. Bimonthly. $113.00 per year.

IEEE TRANSACTIONS ON ELECTRON DEVICES. Institute of Electrical and Electronics Engineers. IEEE Service Center, 445 Hoes Lane, Piscataway, NJ 08854. 1959 to present. Monthly. $175.00 per year.

INSTITUTE OF ELECTRICAL AND ELECTRONICS ENGINEERS PROCEEDINGS. Institute of Electrical and Electronics Engineers. IEEE Service Center, 445 Hoes Lane, Piscataway, NJ 08854. 1913 to present. Monthly. $140.00 per year.

SEMICONDUCTOR INTERNATIONAL. Cahners Publishing Company, Inc., Cahners Plaza, 1350 East Touhy Avenue, Des Plaines, IL 60018. (312) 635-8800. 1978 to present. Monthly. $55.00 per year.

SOLID STATE ELECTRONICS. Pergamon Press, Inc., Maxwell House, Fairview Park, Elmsford, NY 10523. (914) 592-7700. 1960 to present. Monthly. $330.00 per year.

RESEARCH CENTERS AND INSTITUTES

ELECTRICAL ENGINEERING RESEARCH LABORATORIES. Purdue University, Electrical Engineering Building, West Lafayette, IN 47907. (317) 494-3536.

ELECTRONICS RESEARCH CENTER. University of Texas at Austin, 132 Engineering Science Building, Austin, TX 78712. (512) 471-3954.

ELECTRONICS RESEARCH LABORATORY. University of California, Berkeley, 253 Cory Hall, Berkeley, CA 94720. (415) 642-2301.

LABORATORY FOR ELECTROMAGNETIC AND ELECTRONIC SYSTEMS. Massachusetts Institute of Technology, 77 Massachusetts Avenue, Cambridge, MA 02139. (617) 253-4631.

ELECTRONICS ENGINEERING

See also: ELECTRICAL ENGINEERING, ELECTRICITY, ELECTRONIC CIRCUITS AND COMPONENTS

ABSTRACT SERVICES AND INDEXES

APPLIED SCIENCE AND TECHNOLOGY INDEX. H.W. Wilson Company, 950 University Avenue, Bronx, NY 10452. (800) 367-6670 or (212) 588-8400. Inquire as to cost and availability.

ELECTRONICS AND COMMUNICATIONS ABSTRACTS JOURNAL. Cambridge Scientific Abstracts, 5161 River Road, Bethesda, MD 20816. (301) 951-1400. Bimonthly. Inquire as to cost and availability.

ENGINEERING INDEX MONTHLY. Engineering Information, Incorporated, 345 East 47th Street, New York, NY 10017. (800) 221-1044 or (212) 705-7600. Monthly, with annual cumulation. $1425.00 per year.

PHYSICS ABSTRACTS. Institute of Electrical Engineers, London, United Kingdom. Available from: Institute of Electrical and Electronic Engineers (Ieee), 345 East 47th Street, New York, NY 10017. (212) 705-7900.

SCIENCE CITATION INDEX. Institute for Scientific Information, 3501 Market Street, Philadelphia, PA 19104. (800) 523-1850 or (215) 386-0100. Inquire as to cost and availability.

SOLID STATE ABSTRACTS: AN ABSTRACT JOURNAL INVOLVING THE PHYSICS, METALLURGY, CRYSTALLOGRAPHY, CHEMISTRY, AND DEVICE TECHNOLOGY OF SOLIDS. Cambridge Scientific Abstracts, 5161 River Road, Bethesda, MD 20816. (301) 951-1400. 1957 to present. Bimonthly. $440.00 per year.

ANNUAL REVIEWS AND YEARBOOKS

ADVANCES IN ELECTRONICS AND ELECTRON PHYSICS. Academic Press, Incorporated, 6277 Sea Harbor Drive, Orlando, FL 32821. (800) 321-5068. Irregular. Approximately $80.00 per volume.

CRC CRITICAL REVIEWS IN SOLID STATE AND MATERIALS SCIENCES. CRC Press, Incorporated, 2000 Corporate Boulevard, NW, Boca Raton, FL 33431. Irregular. Inquire as to cost and availability.

ASSOCIATIONS AND PROFESSIONAL SOCIETIES

AMERICAN ELECTRONICS ASSOCIATION. Post Office Box 10045, 2670 Hanover Street, Palo Alto, CA 94303. (415) 857-9300.

ASSOCIATION OF OLD CROWS (ELECTRONIC WARFARE). 2300 Ninth Street, South, Suite 300, Arlington, VA 22204. (703) 920-1600.

ELECTRONIC INDUSTRIES ASSOCIATION. 2001 Eye Street, NW, Washington, DC 20006. (202) 457-4900.

IEEE (INSTITUTE OF ELECTRICAL AND ELECTRONICS ENGINEERS). 345 East 47th Street, New York, NY 10017. (212) 705-7900.

INTERNATIONAL SOCIETY FOR HYBRID MICROELECTRONICS. Post Office Box 2698, 1861 Wiehle Avenue, Suite 340, Reston, VA 22090. (703) 471-0066.

BIBLIOGRAPHIES

HANDBOOKS AND TABLES IN SCIENCE AND TECHNOLOGY. Russell H. Powell, editor. Oryx Press, 2214 North Central Avenue, Phoenix, AZ 85004-1483. (602) 254-6156. Second edition. 1983. $55.00.

SCIENTIFIC AND TECHNICAL BOOKS IN PRINT: AN INDEX TO LITERATURE IN SCIENCE AND TECHNOLOGY. R.R. Bowker Company, 205 East 42nd Street, New York, NY 10017. (800) 521-8110 or (212) 916-1600.

DIRECTORIES AND BIOGRAPHICAL SOURCES

IEEE MEMBERSHIP DIRECTORY. Institute of Electrical and Electronics Engineers, IEEE Service Center, 445 Hoes Lane, Piscataway, NJ 08854. (212) 705-7900. Annual. $7.00.

RESEARCH CENTERS DIRECTORY. Gale Research Company, Book Tower, Detroit, MI 48226. (800) 521-0707. Twelfth edition. 1987. $365.00 for set.

SCIENTIFIC AND TECHNICAL ORGANIZATIONS AND AGENCIES DIRECTORY. Gale Research Company, Book Tower, Detroit, MI 48226. (800) 521-0707. Second edition. 1987. $185.00.

WHO'S WHO IN ELECTRONICS. Harris Publishing Company, 2057-2 Aurora Road, Twinsburg, OH 44087. (216) 425-9000. Annual. $89.00.

WHO'S WHO IN ENGINEERING. Engineers Joint Council, 345 East 47th Street, New York, NY 10017. (212) 705-7010. 1985. $200.00.

WHO'S WHO IN FRONTIER SCIENCE AND TECHNOLOGY. Marquis Who's Who, Incorporated, 200 East Ohio Street, Chicago, IL 60611. (800) 428-3898 or (312) 787-2008.

WHO'S WHO IN TECHNOLOGY TODAY. Reston Publishing Company, Incorporated, c/o Prentice-Hall, Incorporated, Englewood Cliffs, NJ 07632. (800) 262-6868. Biennial. Five volumes. $425.00. Covers the fields of electronics, computer science, physics, optics, chemistry, biotechnology, mechanics, energy, and earth science.

ENCYCLOPEDIAS AND DICTIONARIES

ENCYCLOPEDIA OF PHYSICAL SCIENCE AND TECHNOLOGY. Academic Press, Incorporated, Orlando, FL 32887. (800) 321-5068 or (305) 345-2734. Fifteen volumes, 1986.

IEEE STANDARD DICTIONARY OF ELECTRICAL AND ELECTRONICS TERMS. Frank Jay, editor. John Wiley and Sons, Incorporated, 605 Third Avenue, New York, NY 10158. (800) 526-5368 or (212) 850-6000. Third edition. 1984. $49.95.

MCGRAW-HILL ENCYCLOPEDIA OF SCIENCE AND TECHNOLOGY. McGraw-Hill Book, Incorporated, 1221 Avenue of the Americas, New York, NY 10020. (212) 997-3675. Fifth edition, 15 volumes. $1100.00.

GENERAL WORKS

CIRCUITS AND SOFTWARE FOR ELECTRONICS ENGINEERS. Howard Bierman. McGraw-Hill Book Company, 1221 Avenue of the Americas, New York, NY 10020. (800) 628-0004. 1984. $39.50.

CIRCUITS, DEVICES AND SYSTEMS: A FIRST COURSE IN ELECTRICAL ENGINEERING. R.J. Smith. John Wiley and Sons, Incorporated, 605 Third Avenue, New York, NY 10158. (800) 526-5368 or (212) 850-6000. Fourth edition. 1984. $41.45.

CIRCUITS, SIGNALS AND SYSTEMS. William M. Siebert. McGraw-Hill Book Company, 1221 Avenue of the Americas, New York, NY 10020. (800) 628-0004. 1986. $37.95.

ELECTRICAL AND ELECTRONIC INSTRUMENTATION. Hai Hung Chiang. John Wiley and Sons, Incorporated, 605 Third Avenue, New York, NY 10158. (800) 526-5368 or (212) 850-6000. 1984. $64.95.

ELECTROMAGNETIC WAVE THEORY. J.A. Kong. John Wiley and Sons, Incorporated, 605 Third Avenue, New York, NY 10158. (800) 526-5368 or (212) 850-6000. 1985. $34.95.

ELECTRONIC INVENTIONS AND DISCOVERIES: ELECTRONICS FROM ITS EARLIEST BEGINNING TO THE PRESENT DAY. G.W.A. Dummer. Pergamon Press, Incorporated, Maxwell House, Fairview Park, Elmsford, NY 10523. (914) 592-7700. 1983. $48.50.

INTRODUCTION TO COMPUTER ENGINEERING: HARDWARE AND SOFTWARE. Taylor L. Booth. John Wiley and Sons, Incorporated, 605 Third Avenue, New York, NY 10158. (800) 526-5368 or (212) 850-6000. 1984. $40.95.

INTRODUCTION TO RANDOM PROCESSES: WITH APPLICATION TO SIGNALS AND SYSTEMS. William A. Gardner. Macmillan Publishing Company Incorporated, 866 Third Avenue, New York, NY 10022. (800) 257-5755 or (212) 935-2000. 1986. $34.95.

SEMICONDUCTOR CIRCUIT APPROXIMATIONS: AN INTRODUCTION TO TRANSISTORS AND INTEGRATED CIRCUITS. Albert Paul Malvino. McGraw-Hill Book Company, 1221 Avenue of the Americas, New York, NY 10020. (800) 628-0004. Fourth edition. 1985. $34.95.

TELECOMMUNICATION ENGINEERING: ANALOG AND DIGITAL NETWORK DESIGN. R.L. Freeman. John Wiley and Sons, Incorporated, 605 Third Avenue, New York, NY 10158. (800) 526-5368 or (212) 850-6000. 1980. $48.95.

HANDBOOKS AND MANUALS

CONTEMPORARY ELECTRONICS CIRCUITS DESKBOOK. Harry Helms. McGraw-Hill Book Company, 1221 Avenue of the Americas, New York, NY 10020. (800) 628-0004. 1986. $29.95.

CRC HANDBOOK OF TABLES FOR APPLIED ENGINEERING SCIENCE. R.E. Bolz and G.L. Tuve, editors. CRC Press, Incorporated, 2000 Corporate Boulevard, NW, Boca Raton, FL 33431. Second edition. 1973. $69.00.

ELECTRONIC ENGINEERS HANDBOOK. Donald G. Fink, editor. McGraw-Hill Book Company, 1221 Avenue of the Americas, New York, NY 10020. (800) 628-0004. Second edition. 1982. $89.00.

HANDBOOK OF ELECTRONICS INDUSTRY COST ESTIMATING DATA. Theodore Taylor. John Wiley and Sons, Incorporated, 605 Third Avenue, New York, NY 10158. (800) 526-5368 or (212) 850-6000. 1985. $54.95.

HANDBOOK OF ELECTRONICS MANUFACTURING ENGINEERING. B.S. Matisoff. Van Nostrand Reinhold Company, Incorporated, 115 Fifth Avenue, New York, NY 10003. (800) 543-2681. Second edition. 1986. $52.95.

HANDBOOK OF MODERN ELECTRONICS AND ELECTRICAL ENGINEERING. Charles Belove. John Wiley and Sons, Incorporated, 605 Third Avenue, New York, NY 10158. (800) 526-5368 or (212) 850-6000. 1986. $85.00.

REFERENCE MANUAL FOR TELECOMMUNICATIONS. R.L. Freeman. John Wiley and Sons, Incorporated, 605 Third Avenue, New York, NY 10158. (800) 526-5368 or (212) 850-6000. 1985. $79.95.

STANDARD HANDBOOK FOR ELECTRICAL ENGINEERS. Donald G. Fink and H. Wayne Beaty. McGraw-Hill Book Company, 1221 Avenue of the Americas, New York, NY 10020. (212) 512-2000. Eleventh edition. 1978. $85.00.

VLSI HANDBOOK. Norman G. Einspruch, editor. Academic Press, Incorporated, 6277 Sea Harbor Drive, Orlando, FL 32821. (800) 321-5068. 1985. $125.00.

THE WILEY ENGINEER'S DESK REFERENCE. Sanford I. Heisler. John Wiley and Sons, Incorporated, 605 Third Avenue, New York, NY 10158. (800) 526-5368 or (212) 850-6418. 1984. $36.00.

ONLINE DATA BASES

COMPENDEX. Engineering Information, Incorporated, 345 East 47th Street, New York, NY 10017. (800) 221-1044 or (212) 705-7615. Engineering and technical literature, 1975 to present. Inquire as to cost and availability.

DISSERTATION ABSTRACTS ONLINE. University Microfilms International, 300 North Zeeb Road, Ann Arbor, MI 48106. (800) 521-0600 or (313) 761-4700. Scope includes virtually all doctoral dissertations accepted at accredited American institutions from 1861 to present in 252 subject areas. Inquire as to cost and availability.

INSPEC. INSPEC Marketing Department, Institute of Electrical and Electronics Engineers, Incorporated, IEEE Service Department, 445 Hoes Lane, Piscataway, NJ 08854. (201) 981-0060. Inquire as to on-line cost and availability.

NTIS. National Technical Information Service, 5285 Port Royal Road, Springfield, VA 22161. (703) 487-4630. Broad coverage of government sponsored research reports, 1964 to present. Inquire as to cost and availability.

WILSONLINE. H.W. Wilson Company, 950 University Avenue, Bronx, NY 10452. (800) 367-6770 or (212) 588-8400. Makes available online versions of the printed H.W. Wilson Indexes including Applied Science and Technology Index, Business Periodicals Index, and Readers' Guide to Periodical Literature. Period covered is generally 1983 to present. Inquire as to cost and availability.

OTHER SOURCES

A GUIDE TO THE LITERATURE OF ELECTRICAL AND ELECTRONIC ENGINEERING. Susan B. Ardis. Libraries Unlimited, Incorporated, Post Office Box 263, Littleton, CO 80160. (303) 770-1220. 1986. $37.50.

PERIODICALS

CANADAIAN ELECTRONICS ENGINEERING. Maclean Hunter Research Bureau, 777 Bay Street, Toronto, ON M5W 1A7 Canada. (416) 596-5729.

CIRCUITS, SYSTEMS AND SIGNAL PROCESSING. Birkhauser Boston, Incorporated, 380 Green Street, Post Office Box 2007, Cambridge, MA 02139. (617) 876-2333. Quarterly. $175.00 per year.

ELECTRONIC DESIGN. Heyden Publishing Company, Incorporated, 10 Mulholland Drive, Hasbrouck Heights, NJ 07604. (201) 393-6000. Biweekly. $65.00.

ELECTRONICS WEEK. McGraw-Hill Book Company, 1221 Avenue of the Americas, New York, NY 10020. (800) 628-0004. Weekly. $18.00.

IEEE CIRCUITS AND DEVICES MAGAZINE. Institute of Electrical and Electronics Engineers, IEEE Service Center, 445 Hoes Lane, Piscataway, NJ 08854. (212) 705-7900. Bimonthly. $70.00 per year.

IEEE SPECTRUM. Institute of Electrical and Electronics Engineers, IEEE Service Center, 445 Hoes Lane, Piscataway, NJ 08854. (212) 705-7900. Monthly. $88.00 per year.

IEEE TRANSACTIONS ON CIRCUITS AND SYSTEMS. Institute of Electrical and Electronics Engineers, IEEE Service Center, 445 Hoes Lane, Piscataway, NJ 08854. (212) 705-7900. Monthly. $108.00 per year.

PROCEEDINGS OF THE IEEE. Institute of Electrical and Electronics Engineers, IEEE Service Center, 445 Hoes Lane, Piscataway, NJ 08854. (212) 705-7900. Monthly. $120.00 per year.

RESEARCH CENTERS AND INSTITUTES

MASSACHUSETTS INSTITUTE OF TECHNOLOGY. Laboratory for Electromagnetic and Electronic Systems, 77 Massachusetts Avenue, Cambridge, MA 02139. (617) 253-4631.

NORTH CAROLINA STATE UNIVERSITY. Solid State Electronics Laboratory, 432 Daniels Hall, Raleigh, NC 27695. (919) 737-2336.

OHIO STATE UNIVERSITY, ELECTROSCIENCE LABORATORY. 1320 Kinnear Road, Columbus, OH 43212. (614) 422-7981.

PENNSYLVANIA STATE UNIVERSITY. Solid State Device Laboratory, 210 Electrical Engineering, West Building, University Park, PA 16802. (814) 865-1666.

STANFORD UNIVERSITY. Stanford Electronics Laboratories, Stanford, CA 94305. (415) 497-1013.

ELECTROPLATING

See: ELECTROCHEMISTRY

ELECTROSTATICS

See: ELECTRICITY

ELEMENTARY PARTICLES

See: PARTICLE PHYSICS

ELEMENTS (CHEMICAL)

See also: ANALYTICAL CHEMISTRY, CHEMISTRY, COSMOCHEMISTRY, ELECTROCHEMISTRY, GEOCHEMISTRY, INORGANIC CHEMISTRY, ORGANIC CHEMISTRY, PHYSICAL CHEMISTRY

ABSTRACT SERVICES AND INDEXES

APPLIED SCIENCE AND TECHNOLOGY INDEX. H.W. Wilson and Company, 950 University Avenue, Bronx, NY 10452. (800) 367-6670 or (212) 588-8400. Monthly. Inquire as to cost and availability.

ASTRONOMY AND ASTROPHYSICS ABSTRACTS. Springer-Verlag New York, Inc., 175 Fifth Avenue, New York, NY 10010. (800) 526-7254. 1969 to present. Approximately $70.00 per year.

ASTRONOMY AND ASTROPHYSICS MONTHLY INDEX. Olivetree Associates, P.O. Box 236, Sierre Madre, CA 91024. $220.00 per year. Complementary copies available on request.

BIBLIOGRAPHY AND INDEX OF GEOLOGY. American Geological Institute, 4220 King Street, Alexandria, VA 22302. (703) 379-2480. 1969 to present. Monthly. $1100.00 per year.

CHEMICAL ABSTRACTS. American Chemical Society, Chemical Abstracts Service, Box 3012, Columbus, OH 43210. (614) 421-3600. 1907 to present. Weekly. $9500.00 per year.

CONFERENCE PAPERS INDEX. Cambridge Scientific Abstracts, 5161 River Road, Bethesda, MD 20816. 1972 to present. Monthly. Inquire as to cost and availability.

CURRENT CONTENTS: PHYSICAL, CHEMICAL AND EARTH SCIENCES. Institute for Scientific Information, 3501 Market Street, Philadelphia, PA 19104. (800) 523-1850 or (215) 386-0100. Weekly. $275.00 per year.

GENERAL SCIENCE INDEX. H.W. Wilson and Company, 950 University Avenue, Bronx, NY 10452. (800) 367-6670 or (212) 588-8400. 1978 to present. Monthly. Inquire as to cost and availability.

INDEX TO SCIENTIFIC AND TECHNICAL PROCEEDINGS. Institute for Scientific Information, 3501 Market Street, Philadelphia, PA 19104. (800) 523-1850 or (215) 386-0100. 1978 to present. Monthly. $775.00 per year.

INDEX TO SCIENTIFIC REVIEWS. Institute for Scientific Information, 3501 Market Street, Philadelphia, PA 19104. (800) 523-1850 or (215) 386-0100. 1974 to present. Semi-annual. $550.00 per year.

SCIENCE CITATION INDEX. Institute for Scientific Information, 3501 Market Street, Philadelphia, PA 19104. (800) 523-1850 or (215) 386-0100. Six times per year. $6200.00 per year.

ANNUAL REVIEWS AND YEARBOOKS

ADVANCES IN INORGANIC AND RADIOCHEMISTRY. Academic Press, Inc., 6277 Sea Harbor Drive, Orlando, FL 32821. (800) 321-5068. 1959 to present. Irregular. Price varies, inquire.

ASSOCIATIONS AND PROFESSIONAL SOCIETIES

AMERICAN CHEMICAL SOCIETY. 1155 16th Street, N.W., Washington, DC 20036. (202) 872-4600.

AMERICAN INSTITUTE OF CHEMICAL ENGINEERS. 345 East 47th Street, New York, NY 10017. (212) 705-7338.

AMERICAN INSTITUTE OF PHYSICS. 335 East 45th Street, New York, NY 10017. (212) 661-9494.

ASSOCIATION OF OFFICIAL ANALYTICAL CHEMISTS. 1111 North 19th Street, Suite 210, Arlington, VA 22209. (703) 522-3032.

BIBLIOGRAPHIES

NEW TECHNICAL BOOKS: A SELECTIVE LIST WITH DESCRIPTIVE ANNOTATIONS. New York Public Library, Science and Technology Research Center, Fifth Avenue and 42nd Street, New York, NY 10018. (212) 930-0800. 1915 to present. Monthly. $15.00 per year.

SCIENCE BOOKS AND FILMS. American Association for the Advancement of Science, 1333 H Street, NW, Washington, DC 20005. (202) 326-6454. Five times per year. $20.00 per year.

SCIENTIFIC AND TECHNICAL BOOKS AND SERIALS IN PRINT 1988; AN INDEX TO LITERATURE IN SCIENCE AND TECHNOLOGY. R.R. Bowker Company, 205 East 42nd Street, New York, NY 10017. (800) 521-8110. $175.00.

DIRECTORIES AND BIOGRAPHICAL SOURCES

AMERICAN MEN AND WOMEN OF SCIENCE. R.R. Bowker, Inc., Order Department, 245 West 17th Street, New York, NY 10011. (800) 521-8110. Eight volumes. 1986. $595.00 for set.

INTERNATIONAL RESEARCH CENTERS DIRECTORY 1988-89. Darren L. Smith, editor. Gale Research Company, Book Tower, Detroit, MI 48226. (800) 521-0707. 4th edition. 1987. $360.00.

1987 DIRECTORY OF ENGINEERING SOCIETIES AND RELATED ORGANIZATIONS. Gordon Davis, editor. Hemisphere Publishing Corporation, 79 Madison Avenue, New York, NY 10016-7892. (800) 821-8312. 12th edition. 1987. $100.00.

RESEARCH CENTERS DIRECTORY 1988. Gale Research Company, Book Tower, Detroit, MI 48226. (800) 521-0707. 12th edition. 1987. $365.00 for set.

SCIENTIFIC AND TECHNICAL ORGANIZATIONS AND AGENCIES DIRECTORY. Margaret Labash Young, editor. Gale Research Company, Book Tower, Detroit, MI 48226. (800) 521-0707. 2nd edition. 1987. $185.00.

WHO'S WHO IN ENGINEERING. Gordon Davis, editor. Hemisphere Publishing Corporation, 79 Madison Avenue, New York, NY 10016-7892. (800) 821-8312. 6th edition. 1985. $200.00.

ENCYCLOPEDIAS AND DICTIONARIES

ENCYCLOPEDIA OF CHEMISTRY. Douglas M. Considine, editor. Van Nostrand Reinhold Company, Inc., 135 West 50th Street, New York, NY 10020. (800) 543-2681. 4th edition. 1984. $98.95.

MCGRAW-HILL ENCYCLOPEDIA OF SCIENCE AND TECHNOLOGY. McGraw-Hill Book Company, 1221 Avenue of the Americas, New York, NY 10020. (212) 512-2000. 6th edition. 1987. $1600.00.

ELEMENTS (CHEMICAL)

THESAURUS OF SCIENTIFIC, TECHNICAL, AND ENGINEERING TERMS. Hemisphere Publishing Corporation, 1010 Vermont Avenue, NW, Washington, DC 20005. (800) 526-0275. 1988. $125.00.

GENERAL WORKS

ABUNDANCE OF CHEMICAL ELEMENTS. Viktor Cherdyntsev. Books on Demand, 300 North Zeeb Road, Ann Arbor, MI 48106. (313) 761-4700. $78.50.

BASIC ORGANIC CHEMISTRY. F.L. Wiseman. McGraw-Hill Book Company, 1221 Avenue of the Americas, New York, NY 10020. (212) 512-2000. 1988. $37.95.

CHEMISTRY OF THE ELEMENTS. N.N. Greenwood and A. Earnshaw. Pergamon Press, Inc., Maxwell House, Fairview Park, Elmsford, NY 10523. (914) 592-7700. 1984. $145.00.

CONCEPTS AND MODELS OF INORGANIC CHEMISTRY. B.E. Douglas and others. John Wiley and Sons, Inc., 605 Third Avenue, New York, NY 10158. (800) 526-5368. Second edition. 1982. $42.50.

INORGANIC CHEMISTRY. William W. Porterfield. Addison-Wesley Publishing Company, Inc., 1 Jacob Way, Reading, MA 01867. (617) 944-3700. 1983. $36.95.

INORGANIC CHEMISTRY: AN INTRODUCTION. T. Moeller. John Wiley and Sons, Inc., 605 Third Avenue, New York, NY 10158. (800) 526-5368. 1982. $44.95.

HANDBOOKS AND MANUALS

CRC HANDBOOK OF CHEMISTRY AND PHYSICS. Robert C. Weast, editor. CRC Press, 2000 Corporate Boulevard, Boca Raton, FL 33431. (800) 272-7737. 68th edition. 1987. $69.95.

HANDBOOK OF APPLIED CHEMISTRY. Vollrath Hopp and Ingo Hennig. Hemisphere Publishing Corporation, 79 Madison Avenue, New York, NY 10016-7892. Order from: Taylor and Francis/Hemisphere Distribution Center, 242 Cherry Street, Philadelphia, PA 19106-1906. (800) 821-8312. 1983. $49.95.

LANGE'S HANDBOOK OF CHEMISTRY. John A. Dean, editor. McGraw-Hill Book Company, 1221 Avenue of the Americas, New York, NY 10020. (212) 512-2000. 1985. $59.50.

ONLINE DATA BASES

CA SEARCH. Chemical Abstracts Service, P.O. Box 3012, Columbus, OH 43120. (800) 848-6538 or (614) 421-3600. Comprehensive guide to chemical literature, 1972 to present. Inquire as to online cost and availability.

COMPENDEX. Engineering Information, Inc., 345 East 47th Street, New York, NY 10017. (800) 221-1044 or (212) 705-7615. Engineering and technical literature, 1975 to present. Inquire as to online cost and availability.

DISSERTATION ABSTRACTS ONLINE. University Microfilms International, 300 North Zeeb Road, Ann Arbor, MI 48106. (800) 521-0600 or (313) 761-4700. Scope includes virtually all doctoral dissertations accepted at accredited American institutions from 1861 to present in over 250 subject areas. Inquire as to online cost and availability.

NTIS. National Technical Information Service, 5285 Port Royal Road, Springfield, VA 22161. (703) 487-4630. Broad coverage of government sponsored research reports, 1964 to present. Inquire as to online cost and availability.

SCISEARCH. Institute for Scientific Information, 3501 Market Street, Philadelphia, PA 19104. (800) 523-1850 or (215) 386-0100. Broad multidisciplinary title and author index to the international literature of science and technology, 1974 to present. Inquire as to online cost and availability.

PERIODICALS

AMERICAN CHEMICAL SOCIETY JOURNAL. American Chemical Society, 1155 16th Street, N.W., Washington, DC 20036. (800) 424-6747. 1879 to present. Biweekly. $275.00 per year.

INORGANIC CHEMISTRY. American Chemical Society, 1155 16th Street, N.W., Washington, DC 20036. (800) 424-6747. 1962 to present. Biweekly. $300.00 per year.

INORGANICA CHIMICA ACTA. Elsevier Science Publishing Company, Inc., 52 Vanderbilt Avenue, New York, NY 10017. (212) 370-5520. 1967 to present. Biweekly. $1500.00 per year.

JOURNAL OF THE CHEMICAL SOCIETY. CHEMICAL COMMUNICATIONS. Royal Society of Chemistry, Burlington House, London, W1V OBN, England. 1965 to present. Semimonthly. $385.00 per year.

POLYHEDRON. Pergamon Press, Inc., Maxwell House, Fairview Park, Elmsford, NY 10523. (914) 592-7700. 1982 to present. Monthly. $595.00.

ROYAL SOCIETY OF CHEMISTRY. JOURNAL. DALTON TRANSACTIONS. Royal Society of Chemistry, Burlington House, London, W1V OBN, England. 1972 to present. Monthly. $585.00 per year.

RESEARCH CENTERS AND INSTITUTES

CHEMICAL LABORATORIES. Harvard University, Oxford Street, Cambridge, MA 02138. (617) 495-4283.

CHEMICAL RESEARCH LABORATORY. Brown University, Providence, RI 02912. (401) 863-2256.

UNIVERSITY/INDUSTRY CHEMICAL RESEARCH CENTER. Mississippi State University, Department of Chemistry, P.O. Drawer CH, Mississippi State, MS 39762. (601) 325-3584.

ELEVATORS

ABSTRACT SERVICES AND INDEXES

APPLIED SCIENCE AND TECHNOLOGY INDEX. H.W. Wilson Company, 950 University Avenue, Bronx, NY 10452. (800) 367-6670 or (212) 588-8400. Inquire as to cost and availability.

ELEVATOR ABSTRACTS, INCLUDING ESCALATORS. G.C. Barney, editor. Horwood, distributed by John Wiley and Sons, Incorporated, 605 Third Avenue, New York, NY 10158. (800) 526-5368 or (212) 850-6000. Contains references from the mid-1960's to 1986. 1987. $69.95.

ENGINEERING INDEX MONTHLY. Engineering Information, Incorporated, 345 East 47th Street, New York, NY 10017. (800) 221-1044 or (212) 705-7600. Monthly, with annual cumulation. $1425.00 per year.

INTERNATIONAL AEROSPACE ABSTRACTS. AIAA/TIS, 1633 Broadway, New York, NY 10019. (212) 581-4300. Semimonthly. $700.00 per year.

STAR. (Scientific and Technical Aerospace Reports. United States National Aeronautics and Space Administration, Scientific and Technical Information Facility, Box 8757, Baltimore-Washington International Airport, MD 21240. (202) 755-2210. Semimonthly, with semiannual and annual indexes. $85.00 per year.

ANNUAL REVIEWS AND YEARBOOKS

NATIONAL ELEVATOR INDUSTRY YEARBOOK. National Elevator Industry, Incorporated, 600 Third Avenue, New York, NY 10016. (212) 986-1545. Annual.

ASSOCIATIONS AND PROFESSIONAL SOCIETIES

AMERICAN SOCIETY OF MECHANICAL ENGINEERS. 345 East 47th Street, New York, NY 10017. (212) 705-7722.

NATIONAL ASSOCIATION OF ELEVATOR CONTRACTORS. 2964 Peachtree Road, NW, Suite 665, Atlanta, Ga 30305. (404) 261-0166.

NATIONAL ELEVATOR INDUSTRY, INCORPORATED. 600 Third Avenue, New York, NY 10016. (212) 986-1545.

BIBLIOGRAPHIES

BIBLIOGRAPHICAL GUIDE TO TECHNOLOGY. G.K. Hall, 70 Lincoln Street, Boston, MA 02111. (617) 423-3990. Lists technology materials cataloged by the New York Public Library. $175.00 per year.

SCIENTIFIC AND TECHNICAL BOOKS IN PRINT: AN INDEX TO LITERATURE IN SCIENCE AND TECHNOLOGY. R.R. Bowker Company, 205 East 42nd Street, New York, NY 10017. (800) 521-8110 or (212) 916-1600.

DIRECTORIES AND BIOGRAPHICAL SOURCES

ELEVATOR WORLD DIRECTORY ISSUE. Elevator World, 345 Morgan Avenue, Mobile, AL 36606. (205) 479-4514. Annual. $25.00.

SCIENTIFIC AND TECHNICAL ORGANIZATIONS AND AGENCIES DIRECTORY. Gale Research Company, Book Tower, Detroit, MI 48226. (800) 521-0707. Second edition. 1987. $185.00.

WHO'S WHO IN ENGINEERING. Engineers Joint Council, 345 East 47th Street, New York, NY 10017. (212) 705-7010. 1985. $200.00.

GENERAL WORKS

ELEVATOR MECHANICAL DESIGN: PRINCIPLES AND CONCEPTS. Lubomir Janovsky. E. Hornwood. Distributed by: John Wiley and Sons, Incorporated, 605 Third Avenue, New York, NY 10158. (800) 526-5368 or (212) 850-6000. 1987. $49.95.

VERTICAL TRANSPORTATION: ELEVATORS AND ESCALATORS. George R. Strakosch. John Wiley and Sons, Incorporated, 605 Third Avenue, New York, NY 10158. (800) 526-5368 or (212) 850-6000. Second edition. 1983. $57.50.

HANDBOOKS AND MANUALS

SAFETY CODE FOR ELEVATORS AND ESCALATORS: HANDBOOK ON A17.1.1. E.A. Donoghue, editor. American Society of Mechanical Engineers, 345 East 47th Street, New York, NY 10017. (212) 705-7722. 1981. $50.00.

ONLINE DATA BASES

COMPENDEX. Engineering Information, Incorporated, 345 East 47th Street, New York, NY 10017. (800) 221-1044 or (212) 705-7615. Engineering and technical literature, 1975 to present. Inquire as to cost and availability.

NTIS. National Technical Information Service, 5285 Port Royal Road, Springfield, VA 22161. (703) 487-4630. Broad coverage of government sponsored research reports, 1964 to present. Inquire as to cost and availability.

PERIODICALS

ELEVATOR WORLD. Post Office Box 6506, Loop Branch, Mobile, AL 36606. (205) 479-4514. Monthly. $37.00 per year.

NATIONAL ELEVATOR INDUSTRY NEWSLETTER. National Elevator Industry, Incorporated, 600 Third Avenue, New York, NY 10016. (212) 986-1545. Biennial. Free.

ELLIPTICAL GALAXIES

See: GALAXIES

EMISSION ELECTRON MICROSCOPES

See: ELECTRON MICROSCOPY

EMISSION SPECTROSCOPY

See: SPECTROSCOPY

EMULSIONS

See: COLLOIDS

ENERGY

See also: COAL, ELECTRICAL POWER ENGINEERING, FUSION, GEOTHERMAL POWER, HYDROELECTRIC POWER, HYDROGEN, NATURAL GAS, NUCLEAR ENERGY, PETROLEUM, SOLAR ENERGY,

ABSTRACT SERVICES AND INDEXES

ABSTRACT NEWSLETTER: ENERGY. National Technical Information Service, 5285 Port Royal Road, Springfield, VA 22161. (703) 487-4929. Weekly. $95.00 per year.

APPLIED SCIENCE AND TECHNOLOGY INDEX. H.W. Wilson and Company, 950 University Avenue, Bronx, NY 10452. (800) 367-6670 or (212) 588-8400. Monthly. Inquire as to cost and availability.

CHEMICAL ABSTRACTS. American Chemical Society, Chemical Abstracts Service, Box 3012, Columbus, OH 43210. (614) 421-3600. 1907 to present. Weekly. $9500.00 per year.

CONFERENCE PAPERS INDEX. Cambridge Scientific Abstracts, 5161 River Road, Bethesda, MD 20816. (301) 951-1400. 1972 to present. Monthly. Inquire as to cost and availability.

CURRENT CONTENTS: ENGINEERING, TECHNOLOGY AND APPLIED SCIENCES. Institute for Scientific Information, 3501 Market Street, Philadelphia, PA 19104. (800) 523-1850 or (215) 386-0100. Weekly. $275.00 per year.

ELECTRICAL AND ELECTRONICS ABSTRACTS. Institution of Electrical Engineers. Available from: Institute of Electrical and Electronics Engineers. IEEE Service Center, 445 Hoes Lane, Piscataway, NJ 08854. Monthly. $1250.00 per year.

ENERGY RESEARCH ABSTRACTS. U.S. Department of Energy, Office of Scientific and Technical Information, Box 62, Oak Ridge, TN 37831. (615) 576-1155. Subscribe to: U.S. Superintendent of Documents, Washington, DC 20402. 1976 to present. Semimonthly. $146.00 per year.

ENGINEERING INDEX MONTHLY AND AUTHOR INDEX. Engineering Information Inc., 345 East 47th Street, New York, NY 10017. (212) 705-7600. Monthly. $1560.00 per year.

GENERAL SCIENCE INDEX. H.W. Wilson and Company, 950 University Avenue, Bronx, NY 10452. (800) 367-6670 or (212) 588-8400. 1978 to present. Monthly. Inquire as to cost and availability.

INDEX TO SCIENTIFIC AND TECHNICAL PROCEEDINGS. Institute for Scientific Information, 3501 Market Street, Philadelphia, PA 19104. (800) 523-1850 or (215) 386-0100. 1978 to present. Monthly. $775.00 per year.

INDEX TO SCIENTIFIC REVIEWS. Institute for Scientific Information, 3501 Market Street, Philadelphia, PA 19104. (800) 523-1850 or (215) 386-0100. 1974 to present. Semi-annual. $550.00 per year.

IEEE PUBLICATIONS BULLETIN. Institute of Electrical and Electronics Engineers. Institute of Electrical and Electronics Engineers. IEEE Service Center, 445 Hoes Lane, Piscataway, NJ 08854. Quarterly. Free.

PHYSICS ABSTRACTS. Institution of Electrical Engineers. Available from: IEEE Service Center, 445 Hoes Lane, Piscataway, NJ 08854. 1898 to present. Bimonthly. $1700.00 per year.

PHYSICS BRIEFS. Physik Verlag GmbH, Postfach 1260/1280, D-6940 Weinheim, West Germany. (212) 661-9404. 1920 to present. Twenty-six times per year. $1250.00 per year.

SCIENCE CITATION INDEX. Institute for Scientific Information, 3501 Market Street, Philadelphia, PA 19104. (800) 523-1850 or (215) 386-0100. Six times per year. $6200.00 per year.

ANNUAL REVIEWS AND YEARBOOKS

ADVANCES IN ENERGY SYSTEMS AND TECHNOLOGY. Academic Press, Inc., 6277 Sea Harbor Drive, Orlando, FL 32821. (800) 321-5068. 1979 to present. Irregular. Inquire.

ANNUAL REVIEW OF ENERGY. Annual Reviews, Inc., 4139 El Camino Way, Palo Alto, CA 94306. (415) 493-4400. Annual. Inquire.

ASSOCIATIONS AND PROFESSIONAL SOCIETIES

AMERICAN INSTITUTE OF PHYSICS. 335 East 45th Street, New York, NY 10017. (212) 661-9494.

AMERICAN SOLAR ENERGY SOCIETY. 2030 17th Street, Boulder, CO 80302. (303) 443-3130.

AMERICAN WIND ENERGY ASSOCIATION. 1516 King Street, Alexandria, VA 22314. (703) 684-5196.

ASSOCIATION OF ENERGY ENGINEERS. 4025 Pleasantdale Road, Suite 340, Atlanta, GA 30340. (404) 447-5083.

EDISON ELECTRIC INSTITUTE. 1111 19th Street, N.W., Washington, DC 20036. (202) 828-7400.

INSTITUTE OF ELECTRICAL AND ELECTRONICS ENGINEERS. 345 East 47th Street, New York, NY 10017. (212) 705-7900.

BIBLIOGRAPHIES

NEW TECHNICAL BOOKS: A SELECTIVE LIST WITH DESCRIPTIVE ANNOTATIONS. New York Public Library, Science and Technology Research Center, Fifth Avenue and 42nd Street, New York, NY 10018. (212) 930-0800. 1915 to present. Monthly. $15.00 per year.

SCIENCE BOOKS AND FILMS. American Association for the Advancement of Science, 1333 H Street, NW, Washington, DC 20005. (202) 326-6454. Five times per year. $20.00 per year.

SCIENTIFIC AND TECHNICAL BOOKS AND SERIALS IN PRINT 1988; AN INDEX TO LITERATURE IN SCIENCE AND TECHNOLOGY. R.R. Bowker Company, 205 East 42nd Street, New York, NY 10017. (800) 521-8110. $175.00.

DIRECTORIES AND BIOGRAPHICAL SOURCES

AMERICAN MEN AND WOMEN OF SCIENCE. R.R. Bowker, Inc., Order Department, 245 West 17th Street, New York, NY 10011. (800) 521-8110. Eight volumes. 1986. $595.00 for set.

ENERGY INFORMATION CENTERS DIRECTORY. Public Affairs and Information Program, Atomic Industrial Forum, 7101 Wisconsin Avenue, Bethesda, MD 20814. (301) 654-9260. 1985. Free.

IEEE MEMBERSHIP DIRECTORY. Institute of Electrical and Electronics Engineers. IEEE Service Center, 445 Hoes Lane, Piscataway, NJ 08854. Annual. $7.00.

INTERNATIONAL RESEARCH CENTERS DIRECTORY 1988-89. Darren L. Smith, editor. Gale Research Company, Book Tower, Detroit, MI 48226. (800) 521-0707. 4th edition. 1987. $360.00.

1987 DIRECTORY OF ENGINEERING SOCIETIES AND RELATED ORGANIZATIONS. Gordon Davis, editor. Hemisphere Publishing Corporation, 1010 Vermont Avenue, NW, Washington, DC 20005. (800) 526-0275. 12th edition. 1987. $100.00.

RESEARCH CENTERS DIRECTORY 1988. Gale Research Company, Book Tower, Detroit, MI 48226. (800) 521-0707. 12th edition. 1987. $365.00 for set.

SCIENTIFIC AND TECHNICAL ORGANIZATIONS AND AGENCIES DIRECTORY. Margaret Labash Young, editor. Gale Research Company, Book Tower, Detroit, MI 48226. (800) 521-0707. 2nd edition. 1987. $185.00.

WHO'S WHO IN ENGINEERING. Gordon Davis, editor. Hemisphere Publishing Corporation, 1010 Vermont Avenue, NW, Washington, DC 20005. (800) 526-0275. 6th edition. 1985. $200.00.

ENCYCLOPEDIAS AND DICTIONARIES

MCGRAW-HILL ENCYCLOPEDIA OF SCIENCE AND TECHNOLOGY. McGraw-Hill Book Company, 1221 Avenue of the Americas, New York, NY 10020. (212) 512-2000. 6th edition. 1987. $1600.00.

THESAURUS OF SCIENTIFIC, TECHNICAL, AND ENGINEERING TERMS. Hemisphere Publishing Corporation, 1010 Vermont Avenue, NW, Washington, DC 20005. (800) 526-0275. 1988. $125.00.

GENERAL WORKS

ENERGY ANALYSIS OF 108 INDUSTRIAL PROCESSES. H.L. Brown and others. Prentice-Hall Publishing, Inc., Englewood Cliffs, NJ 07632. (800) 562-0245. 1982. $45.00.

ENERGY FROM BIOMASS. G. Grassi and H. Zibetta, editors. Elsevier Science Publishing Company, Inc., 52 Vanderbilt Avenue, New York, NY 10017. (212) 370-5520. 1987. $88.25.

ENERGY FROM WAVES. D. Ross. Pergamon Press, Inc., Maxwell House, Fairview Park, Elmsford, NY 10523. (914) 592-7700. 2nd edition. 1981. $11.00 in paper.

ENERGY OPTIONS. David Merrick and Richard Marshall. John Wiley and Sons, Inc., 605 Third Avenue, New York, NY 10158. (800) 526-5368. 1984. $75.95.

ENERGY 2000: AN OVERVIEW OF THE WORLD'S ENERGY RESOURCES IN THE DECADES TO COME. Heinz Knoepfel. Gordon and Breach Science Publishers, Inc., 50 West 23rd Street, New York, NY 10010. (212) 206-8900. 1986. $42.00.

GEOTHERMAL SYSTEMS: PRINCIPLES AND CASE HISTORIES. L. Rybach and L.J.P. Muffler. John Wiley and Sons, Inc., 605 Third Avenue, New York, NY 10158. (800) 526-5368. 1981. $104.00.

A GUIDE TO NUCLEAR POWER TECHNOLOGY: A RESOURCE FOR DECISION MAKING. F.J. Rahn and others. John Wiley and Sons, Inc., 605 Third Avenue, New York, NY 10158. (800) 526-5368. 1984. $85.95.

PETROLEUM RESOURCES AND DEVELOPMENT. Kameel I. Kheir, editor. Columbia University Press, 562 West 113th Street, New York, NY 10025. (212) 316-7100. 1988. $30.00.

SUN POWER: AN INTRODUCTION TO THE APPLICATION OF SOLAR ENERGY. J.C. McVeigh. Pergamon Press, Inc., Maxwell House, Fairview Park, Elmsford, NY 10523. (914) 592-7700. 2nd edition. 1983. $48.40.

WIND POWER PLANTS: THEORY AND DESIGN. D. Le Gourieres. Pergamon Press, Inc., Maxwell House, Fairview Park, Elmsford, NY 10523. (914) 592-7700. 1982. $35.95 in paper.

HANDBOOKS AND MANUALS

HANDBOOK OF ENERGY SYSTEMS ENGINEERING. Leslie C. Wilbur, editor. John Wiley and Sons, Inc., 605 Third Avenue, New York, NY 10158. (800) 526-5368. 1985. $74.95.

HANDBOOK OF ENERGY TECHNOLOGY AND ECONOMICS. R.A. Meyers, editor. John Wiley and Sons, Inc., 605 Third Avenue, New York, NY 10158. (800) 526-5368. 1983. $79.95.

HANDBOOK OF GEOTHERMAL ENERGY. Lyman Edwards and others, editors. Gulf Publishing Company, P.O. Box 2608, Houston, TX 77001. (713) 520-4444. 1982. $85.00.

ONLINE DATA BASES

CA SEARCH. Chemical Abstracts Service, P.O. Box 3012, Columbus, OH 43120. (800) 848-6538 or (614) 421-3600. Comprehensive guide to chemical literature, 1972 to present. Inquire as to online cost and availability.

COMPENDEX. Engineering Information, Inc., 345 East 47th Street, New York, NY 10017. (800) 221-1044 or (212) 705-7615. Engineering and technical literature, 1975 to present. Inquire as to online cost and availability.

DISSERTATION ABSTRACTS ONLINE. University Microfilms International, 300 North Zeeb Road, Ann Arbor, MI 48106. (800) 521-0600 or (313) 761-4700. Scope includes virtually all doctoral dissertations accepted at accredited American institutions from 1861 to present in over 250 subject areas. Inquire as to online cost and availability.

DOE ENERGY DATA BASE. U.S. Department of Energy, Office of Scientific and Technical Information, P.O. Box 62, Oak Ridge, TN 37831. (615) 576-6837. A database that covers all aspects of energy including the science and technology of energy. 1948 to present. Available through the DIALOG search service or DOE/RECON. Inquire as to online cost and availability.

ENERGYLINE. EIC/Intelligence, Inc., 48 West 38th Street, New York, NY 10018. (212) 944-8500. A database of resources on the scientific, engineering, political, and socioeconomic aspects of energy resources. 1976 to present. Inquire as to online cost and availability.

INSPEC. INSPEC Marketing Department, Institution of Electrical Engineers. Available from IEEE Service Center, 445 Hoes Lane, Piscataway, NJ 08854. (201) 981-0060. Online version of Physics Abstracts. Inquire as to online cost and availability.

NTIS. National Technical Information Service, 5285 Port Royal Road, Springfield, VA 22161. (703) 487-4630. Broad coverage of government sponsored research reports, 1964 to present. Inquire as to online cost and availability.

SCISEARCH. Institute for Scientific Information, 3501 Market Street, Philadelphia, PA 19104. (800) 523-1850 or (215) 386-0100. Broad multidisciplinary title and author index to the international literature of science and technology, 1974 to present. Inquire as to online cost and availability.

WILSONLINE. H.W. Wilson and Company, 950 University Avenue, Bronx, NY 10452. (800) 367-6770 or (212) 588-8400. Makes available online versions of the H.W. Wilson indexes including Applied Science and Technology Index, Business Periodicals Index and Readers' Guide to Periodical Literature. Approximately 1980 to present. Inquire as to online cost and availability.

OTHER SOURCES

WHAT EVERY ENGINEER SHOULD KNOW ABOUT ENGINEERING SOURCES. Marcel Dekker Inc., 270 Madison Avenue, New York, NY 10016. (800) 228-1160. 1984. $24.95.

PERIODICALS

ALTERNATIVE SOURCES OF ENERGY. Alternative Sources of Energy, Inc., 107 South Central Avenue, Milaca, MN 56353. (612) 983-6892. 1971 to present. Ten times per year. $48.00 per year.

ENERGY AND FUELS. American Chemical Society, 1155 16th Street, N.W., Washington, DC 20036. (800) 424-6747. 1987 to present. Bimonthly. $269.00 per year.

ENERGY ENGINEERING. Association of Energy Engineers. Available from: Fairmont Press, Box 14227, Atlanta, GA 30324. (404) 447-5314. 1904 to present. Bimonthly. $44.00 per year.

ENERGY SOURCES. Crane Russak and Company, Inc., 3 East 44th Street, New York, NY 10017. (212) 867-1490. 1973. Quarterly. $73.00 per year.

INTERNATIONAL JOURNAL OF SOLAR ENERGY. Harwood Academic Publishers, 50 West 23rd Street, New York, NY 10010. (212) 206-8900. 1982 to present. 6 times per year. $132.00 per year.

JOURNAL OF ENERGY RESOURCES TECHNOLOGY. American Society of Mechanical Engineers, 345 East 47th Street, New York, NY 10017. (212) 705-7703. 1979 to present. $24.00 per year.

POWER ENGINEERING (NEW YORK). Technical Publishing Company, 875 Third Avenue, New York, NY 10022. (212) 605-9400. 1896 to present. Monthly. $24.00 per year.

SOLAR ENERGY. Pergamon Press, Inc., Maxwell House, Fairview Park, Elmsford, NY 10523. (914) 592-7700. 1957 to present. Monthly. $298.00 per year.

WIND POWER DIGEST. Wind Power Publishing, Box 700, Bascom, OH 44809. (419) 937-2299. 1975 to present. Quarterly. $12.00 per year.

RESEARCH CENTERS AND INSTITUTES

CENTER FOR ENERGY STUDIES. University of Texas at Austin, 1318 EME Building, 10100 Burnet Road, Austin, TX 78758. (512) 471-7792.

ENERGY CENTER. Stevens Institute of Technology, Department of Mechanical Energy, Castle Point Station, Hoboken, NJ 07030. (201) 420-5560.

ENERGY LABORATORY. Massachusetts Institute of Technology, 123 Amherst Street, Building E-40-455, Cambridge, MA 02139. (617) 253-3400.

OAK RIDGE NATIONAL LABORATORY. U.S. Department of Energy, P.O. Box X, Oak Ridge, TN 39831. (615) 576-2900.

ENGINEERING DESIGN

See: CAD (COMPUTER-AIDED DESIGN)

ENGINEERING GEOLOGY

See: GEOTECHNICAL ENGINEERING

ENGINES

See also: AUTOMOTIVE ENGINEERING, DIESEL ENGINES

ABSTRACT SERVICES AND INDEXES

APPLIED SCIENCE AND TECHNOLOGY INDEX. H. W. Wilson Company, 950 University Avenue, Bronx, NY 10452. (800) 367-6670 or (212) 588-8400. Inquire as to cost and availability.

ENGINEERING INDEX MONTHLY. Engineering Information, Inc., 345 East 47th Street, New York, NY 10017. (800) 221-1044 or (212) 705-7600. Monthly, with annual cumulation. $1,425.00 per year.

SAE ABSTRACTS INDEX OF TECHNICAL PAPERS. SAE, Inc., 400 Commonwealth Drive, Warrendale, PA 15096. (412) 776-4841. Annual. $40.00.

ASSOCIATIONS AND PROFESSIONAL SOCIETIES

DIESEL ENGINE MANUFACTURES ASSOCIATION. A. P. Wherry and Associates, Inc., 712 Lakewood Center North, Cleveland, OH 44107. (216) 226-7700.

ENGINE MANUFACTURES ASSOCIATION. 111 East Wacker Drive, Chicago, IL 60601. (312) 644-6610.

SOCIETY OF AUTOMOTIVE ENGINEERS, 400 Commonwealth Drive, Warrendale, PA 15096. (412) 776-4841.

DIRECTORIES AND BIOGRAPHICAL SOURCES

RESEARCH CENTERS DIRECTORY. Gale Research Company, Book Tower, Detroit, MI 48226. Eleventh Edition. 1987.

WHO'S WHO IN ENGINEERING. Engineering Joint Council, 345 East 47th Street, New York, NY 10017. (212) 705-7010. 1985. $200.00.

WHO'S WHO IN FRONTIER SCIENCE AND TECHNOLOGY. Marquis Who's Who, Inc., 200 East Ohio Street, Chicago, IL 60611. (800) 428-3898 or (312) 787-2008.

SCIENTIFIC AND TECHNICAL ORGANIZATIONS AND AGENCIES DIRECTORY. Gale Research Company, Book Tower, Detorit, MI 48226. (800) 521-0707. 1985.

ENCYCLOPEDIAS AND DICTIONARIES

ENCYCLOPEDIA OF PHYSICAL SCIENCE AND TECHNOLOGY. Academic Press, Inc., Orlando, FL 32887. (800) 321-5068 or (305) 345-2734. Fifteen volumes, 1986.

MCGRAW-HILL DICTIONARY OF ENGINEERING. Sybil P. Parker, Editor. McGraw-Hill Book Company, 1221 Avenue of the Americas, New York, NY 10020. (212) 512-2000. 1984. $39.95.

MCGRAW-HILL ENCYCLOPEDIA OF SCIENCE AND TECHNOLOGY. McGraw-Hill Book Company, 1221 Avenue of the Americas, New York, NY 10020. (212) 512-2000. Fifth Edition. 1982. $1,100.00 for set.

GENERAL WORKS

AUTOMOTIVE ENGINES. William C. Crouse and Donald L. Anglin. McGraw-Hill Book Company, 1221 Avenue of the Americas, New York, NY 10020. 1980. $24.95.

DESIGN AND APPLICATION IN DIESEL ENGINEERING. S. D. Haddad and N. Watson, Editors. John Wiley and Sons, Inc., 605 Third Avenue, New York, NY 10158. (800) 526-6368 or (212) 850-6000. 1984. $65.95.

DESIGN AND DEVELOPMENT OF DIESEL ENGINES AND COMPONENTS. Society of Automotive Engineers, 400 Commonwealth Drive, Warrendale, PA 15096. (412) 776-4841. 1985. $22.00.

DIESEL ENGINES FOR AUTOMOBILES. Tom Weathers and C. Hunter. Reston Publishing Company, 200 Old Tappan Road, Old Tappan, NJ 07675. (201) 592-2352. 1985. $26.95.

INTERNAL COMBUSTION ENGINES. Rowland S. Benson and N. D. Whitehouse. Pergamon Press, Inc., Maxwell House, Fairview Park, Elmsford, NY 10523. (914) 592-7700. Two volumes. 1984. $28.00 for set.

INTERNAL COMBUSTION ENGINES: APPLIED THERMOSCIENCES. Colin R. Ferguson. John Wiley and Sons, Inc., 605 Third Avenue, New York, NY 10158. (800) 526-5368 or (212) 850-6000. 1986. $42.95.

SMALL ENGINE MECHANICS. William C. Crouse and Donald L. Anglin. McGraw-Hill Book Company, 1221 Avenue of the Americas, New York, NY 10020. (212) 512-2000. Third Edition. 1986. $23.95.

HANDBOOKS AND MANUALS

DIESEL ENGINE MANUAL. Perry O. Black and William E. Schahill. Theodore Audel, distributed by G. K. Hall and Company, 70 Lincoln Street, Boston, MA 02111. (800) 343-2806. Fourth Edition. 1983. $12.95.

DIESEL ENGINEERING HANDBOOK. Business Journals, 22 South Smith Street, P.O. Box 5550, East Norwalk, CT 06856. $28.50.

SAE HANDBOOK. SAE, Inc., 400 Commonwealth Drive, Warrendale, PA 15096. (412) 776-4841. $70.00 members; $140.00 non-members.

STANDARD HANDBOOK OF ENGINEERING CALCULATIONS. Tyler G. Hicks, Editor. McGraw-Hill Book Company, 1221 Avenue of the Americas, New York, NY 10020. (212) 512-2000. Second Edition. 1984. $59.50.

THE WILEY ENGINEER'S DESK REFERENCE. Sanford I. Heisler. John Wiley and Sons, Inc., 605 Third Avenue, New York, NY 10158. (800) 526-5368 or (212) 850-6418. 1984. $36.00.

ONLINE DATA BASES

COMPENDEX. Engineering Information, Inc., 345 East 47th Street, New York, NY 10017. (800) 221-1044 or (212) 705-7615. Engineering and technical literature, 1975 to present. Inquire as to cost and availability.

NTIS. National Technical Information Service, 5285 Port Royal Road, Springfield, VA 22161. (703) 487-4630. Broad coverage of government sponsored research reports, 1964 to present. Inquire as to cost and availability.

SAE DATABASES. Society of Automotive Engineers, 400 Commonwealth Drive, Warrendale, PA 15096. (412) 776-4841. Citations to literature on all types of self-propelled vehicles,

1965 to present. Inquire as to on-line cost and availability.

WILSONLINE. H. W. Wilson Company, 950 University Avenue, Bronx, NY 10452. (800) 367-6770 or (212) 588-8400. Makes available online versions of the printed H. W. Wilson indexes including Applied Science and Technology Index, Business Periodicals Index, and Readers' Guide to Periodical Literature. Period covered is generally 1983 to present. Inquire as to cost and availability.

PERIODICALS

AUTOMOTIVE ENGINEERING. Society of Automotive Engineers, 400 Commonwealth Drive, Warrendale, PA 15096. (412) 776-4841. Monthly. $42.00 per year.

DIESEL AND GAS TURBINE WORLDWIDE. Diesel Engines, Inc., 13555 Bishop's Court, Brookfield, WI 53005. Monthly, 10 per year. $45.00 per year.

JOURNAL OF APPLIED MECHANICS. American Society of Mechanical Engineers, 345 East 47th Street, New York, NY 10017. (212) 705-7703. Quarterly. $100.00 per year.

JOURNAL OF ENGINEERING FOR GAS TURBINES AND POWER. American Society of Mechanical Engineers, 345 East 47th Street, New York, NY 10017. (212) 705-7703. Quarterly. $100.00 per year.

RESEARCH CENTERS AND INSTITUTES

UNIVERSITY OF FLORIDA. Internal Combustion Engines Laboratory, Department of Mechanical Engineering, Gainesville, FL 32611. (904) 392-7555.

UNIVERSITY OF MICHIGAN, AUTOMOTIVE ENGINEERING LABORATORY, Department of Mechanical Engineering and Applied Mechanics, Ann Arbor, MI. (313) 764-4256.

UNIVERSITY OF NEBRASKA. Center for Engine Technology, 255 WSEC, Lincoln, NE 68588-0525. (402) 472-2376.

WAYNE STATE UNIVERSITY, CENTER FOR AUTOMOTIVE RESEARCH, 5050 Anthony Wayne Drive, Detroit, MI 48202. (313) 577-3887.

ENTROPY

See: THERMODYNAMICS

ENVIRONMENTAL ENGINEERING

See also: GROUND WATER POLLUTION, SOLID WASTE DISPOSAL, SLUDGE, SEWAGE TREATMENT, WATER POLLUTION, WATER TREATMENT

ABSTRACT SERVICES AND INDEXES

CHEMICAL ABSTRACTS. Chemical Abstracts Service, 2540 Olentangy Road, Post Office Box 3012, Columbus, OH 43210. (800) 848-6538 or (614) 421-3600. Weekly. $9200.00 per year.

CURRENT CONTENTS: PHYSICAL AND CHEMICAL SCIENCES. Institute for Scientific Information, 3501 Market Street, Philadelphia, PA 19104. (800) 523-1850 or (215) 386-0100. $272.00 per year.

DELFT HYDROSCIENCE ABSTRACTS. Delft Hydraulics Laboratory, Rotterdamseweg 185, Postbus 177, NL-2600 MH Delft, The Netherlands. Monthly. $177.00 per year.

ENGINEERING INDEX MONTHLY AND AUTHOR INDEX. Engineering Information, Incorporatred, 345 East 47th Street, New York, NY 10017. (800) 221-1044 or (212) 705-7600. Monthly, with annual cumulation. $1560.00 per year.

HYDRO-ABSTRACTS. Environmental Hydrology Corporation, Box 14701, Minneapolis, MN 55414. (612) 379-0901. 1968 to present. Monthly. $120.00 per year.

POLLUTION ABSTRACTS. Cambridge Scientific Abstracts, 5161 River Road, Bethesda, Md 20816. (301) 951-1400. 1970 to present. Bimonthly. $465.00 per year.

SCIENCE CITATION INDEX. Institute for Scientific Information, 3501 Market Street, Philadelphia, PA 19104. (800) 523-1850 or (215) 386-0100. Inquire as to cost and availability.

SELECTED WATER RESOURCES ABSTRACTS. United States Geological Survey, Water Resources Scientific Information Center. Available from: National Technical Information Service, Springfield, VA 22161. (703) 860-7455. 1968 to present. Monthly. $115.00 per year.

WATER QUALITY CONTROL DIGEST. University Digest Services, Post Office Box 343, Troy, MI 48099. (313) 651-2528. 1969 to present. Bimonthly. $87.00 per year.

ASSOCIATIONS AND PROFESSIONAL SOCIETIES

AMERICAN PUBLIC WORKS ASSOCIATION. 1313 East 60th Street, Chicago, IL 60637. (312) 667-2200.

AMERICAN SOCIETY OF CIVIL ENGINEERS. Engineering Division, 345 East 47th Street, New York, NY 10017-2398. (212) 705-7520.

AMERICAN WATER RESOURCES ASSOCIATION. 5410 Grosvenor Lane, Suite, Bethesda, MD 20814. (301) 493-8600.

AMERICAN WATER WORKS ASSOCIATION. 6666 West Quincy Avenue, Denver, CO 80235. (303) 794-7711.

ASSOCIATION OF GROUND WATER SCIENTISTS AND ENGINEERS. 6375 Riverside Drive, Dublin, OH 43017. (614) 761-1711.

ASSOCIATION OF METROPOLITAN SEWERAGE AGENCIES. 1015 18th Street, NW, Suite 1002, Washington, DC 20036. (202) 659-9161.

GROUND WATER INSTITUTE. Post Office Box 981, Minneapolis, MN 55440. (612) 698-4395.

NATIONAL WATER RESOURCES ASSOCIATION. 955 L'Enfant Plaza, SW, Suite 1202N, Washington, DC 20024. (202) 488-0610.

NATIONAL WATER WELL ASSOCIATION. 6375 Riverside Drive, Dublin, OH 43017. (614) 761-1711.

WATER POLLUTION CONTROL FEDERATION. 2626 Pennsylvania Avenue, Washington, DC 20037. (202) 337-2500.

WATER QUALITY ASSOCIATION. 4151 Naperville Road, Lisle, IL 60532. (312) 369-1600.

BIBLIOGRAPHIES

STABILISATION, DISINFECTION, AND ORDOUR CONTROL IN SEWAGE SLUDGE TREATMENT: AN ANNOTATED BIBLIOGRAPHY COVERING THE PERIOD 1950-1983. E.S. Connor and A.M. Bruce. John Wiley and Sons, Incorporated, 605 Third Avenue, New York, NY 10158. (800) 526-5368 or (212) 850-6000. 1984. $63.95.

WATER POLLUTION: A GUIDE TO INFORMATION SOURCES. Allen W. Knight and Mary Ann Simmons, editors. Gale Research Company, Detroit, MI 48226. (800) 521-0707. 1980. $62.00.

ENVIRONMENTAL ENGINEERING

DIRECTORIES AND BIOGRAPHICAL SOURCES

AMERICAN WATER WORKS ASSOCIATION - OFFICERS AND COMMITTEE DIRECTORY. American Water Works Association, 6666 West Quincy Avenue, Denver, CO 80235. (303) 794-7711. Annual. $15.00.

JOURNAL OF THE AMERICAN WATER WORKS ASSOCIATION BUYERS GUIDE ISSUE. American Water Works Association, 6666 West Quincy Avenue, Denver, CO 80235. (303) 794-7711. Annual, with subscription.

PUBLIC WORKS MANUAL. Public Works Journal Corporation, 200 South Broad Street, Ridgewood, NJ 07451. (201) 445-5800. Annual. $20.00.

SCIENTIFIC AND TECHNICAL ORGANIZATIONS AND AGENCIES DIRECTORY. Gale Research Company, Detroit, MI 48226. (800) 521-0707. Second edition. Two volumes. 1987. $185.00 set.

WATER QUALITY ASSOCIATION DIRECTORY. Water Quality Association, 4151 Naperville Road, Lisle, IL 60532. (312) 369-1600. Annual. Available to members only.

GENERAL WORKS

APPLIED STREAM SANITATION. C.J. Velz. John Wiley and Sons, Incorporated, 605 Third Avenue, New York, NY 10158. (800) 526-5368 or (212) 850-6000. Second edition. 1984. $79.95.

CHEMICAL PROCESSES IN WASTE WATER TREATMENT. W.J. Eilbeck and G. Mattock. E. Horwood. Distributed by: John Wiley and Sons, Incorporated, 605 Third Avenue, New York, NY 10158. (800) 526-5368 or (212) 850-6000. 1987. $85.00.

ENVIRONMENTAL ENGINEERING AND SANITATION. J.A. Salvato. John Wiley and Sons, Incorporated, 605 Third Avenue, New York, NY 10158. (800) 526-5368 or (212) 850-6000. Third edition. 1982. $70.00.

GROUNDWATER CONTAMINATION. J.H. Guswa and others. Noyes Data Corporation, Mill Road at Grand Avenue, Park Ridge, NJ 07656. (201) 391-8484. 1985. $48.00.

GROUNDWATER CONTAMINATION. National Research Council. National Academy Press, 2101 Constitution Avenue, NW, Washington, DC 20418. (201) 334-3313. 1984. $17.95.

GROUND WATER POLLUTION CONTROL. Larry W. Canter and R.C. Knox. Lewis Publishing, Incorporated, 121 South Main Street, Chelsea, MI 48118. (313) 475-8619. 1985. $49.95.

GROUND WATER QUALITY. C.H. Ward and others. John Wiley and Sons, Incorporated, 605 Third Avenue, New York, NY 10158. (800) 526-5368 or (212) 850-6000. 1985. $45.00.

GROUND WATER QUALITY PROTECTION. L.W. Canter and others. Lewis Publishing, Incorporated, 121 South Main Street, Chelsea, MI 48118. (313) 475-8619. $49.95.

URBAN WATER INFRASTRUCTURE: PLANNING, MANAGEMENT, AND OPERATIONS. N.S. Grigg. John Wiley and Sons, Incorporated, 605 Third Avenue, New York, NY 10158. (800) 526-5368 or (212) 850-6000. 1986. $39.95.

WATER AND WASTE-WATER TECHNOLOGY. M.J. Hammer. John Wiley and Sons, Incorporated, 605 Third Avenue, New York, NY 10158. (800) 526-5368 or (212) 850-6000. Second edition. 1986. $35.95.

WATER SUPPLY AND POLLUTION CONTROL. Warren Viessman and Mark J. Hammer. Harper and Row Publishers, Incorporated, 10 East 53rd Street, New York, NY 10022. (800) 242-7737. Fourth edition. 1985. $45.00.

WATER TREATMENT PRINCIPLES AND DESIGN. J.M. Montgomery. John Wiley and Sons, Incorporated, 605 Third Avenue, New York, NY 10158. (800) 526-5368 or (212) 850-6000. 1985. $49.95.

HANDBOOKS AND MANUALS

HANDBOOK OF URBAN DRAINAGE AND SEWAGE DISPOSAL. Klaus Imhoff and others. John Wiley and Sons, Incorporated, 605 third Avenue, New York, NY 10158. (800) 526-5368 or (212) 850-6000. Fourth edition. 1987. $49.95.

SOLID WASTE HANDBOOK: A PRACTICAL GUIDE. W.D. Robinson. John Wiley and Sons, Incorporated, 605 Third Avenue, New York, NY 10158. (800) 526-5368 or (212) 850-6000. 1985. $75.95.

WATER TREATMENT HANDBOOK. Degremont Company. John Wiley and Sons, Incorporated, 605 Third Avenue, New York, NY 10158. (800) 526-5368 or (212) 850-6000. Fifth edition. 1979. $95.00.

ONLINE DATA BASES

COMPENDEX. Engineering Information, Incorporated, 345 East 47th Street, New York, NY 10017. (800) 221-1044 or (212) 705-7615. Engineering and technical literature, 1975 to present. Inquire as to cost and availability.

GEOREF. American Geological Institute, 4220 King Street, Alexandria, VA 22302. (703) 379-2480. 1967 to present. Inquire as to online cost and availability.

NTIS. National Technical Information Service, 5285 Port Royal Road, Springfield, VA 22161. (703) 487-4630. Broad coverage of government sponsored research reports, 1964 to present. Inquire as to cost and availability.

POLLUTION ABSTRACTS. Cambridge Scientific Abstracts, 5161 River Road, Bethesda, MD 20816. (301) 951-1400. 1970 to present. Available for online searching through DIALOG Information Services and BRS Information Technologies. Inquire as to online cost and availability.

WATER DATA BANK. United States Department of Agriculture, Agricultural Research Service, Hydrology Laboratory, Water Data Center, Room 139, Building 007, BARC-West, Beltsville, MD 20705. (301) 344-4411. Inquire as to online cost and availability.

WATERNET. American Water Works Association, 6666 West Quincy Avenue, Denver, CO 80235. (303) 794-7711. A data base providing abstracts and indexing of information published in the American Water Works Association Journal. 1971 to present. Inquire as to online cost and availability.

PERIODICALS

AMERICAN WATER WORKS ASSOCIATION. Journal. American Water Works Association, 6666 West Quincy Avenue, Denver, CO 80235. (303) 794-7711. Monthly. $50.00 per year.

ASSOCIATION OF METROPOLITAN SEWERAGE AGENCIES BULLETIN. Association of Metropolitan Sewerage Agencies, 1015 18th Street, NW, Suite 1002, Washington, DC 20036. (202) 659-9161. Monthly.

GROUND WATER. National Water Well Association. Water Well Journal Publishing Company, 500 West Wilson Bridge Road, Worthington, OH 43085. (614) 846-4967. Bimonthly. $53.00 per year.

JOURNAL OF ENVIRONMENTAL ENGINEERING. American Society of Civil Engineers, 345 East 47th Street, New York, NY 10017. (212) 705-7275. 1956 to present. Bimonthly. $80.00 per year.

JOURNAL OF ENVIRONMENTAL SCIENCES. Institute of Environmental Sciences, 940 East Northwest Highway, Mount

Prospect, IL 60056. (312) 255-1561. Bimonthly. $25.00 per year.

JOURNAL OF GROUND WATER. National Water Well Association. Water Well Journal Publishing Company, 500 West Wilson Bridge Road, Worthington, OH 43085. (614) 846-4967. Quarterly. $16.00 per year.

JOURNAL OF WATER RESOURCES PLANNING AND MANAGEMENT. American Society of Civil Engineers, 345 East 47th Street, New York, NY 10017-2398. (212) 705-7520. Quarterly. $56.00 per year.

WATER, AIR AND SOIL POLLUTION. D. Reidel Publishing Company, 190 Old Derby Street, Hingham, MA 02043. Twenty times per year. $550.00 per year.

WATER: ENGINEERING AND MANAGEMENT. Scranton Gillette Communications, Incorporated, 380 Northwest Highway, Des Plaines, IL 60016. (312) 298-6622. Monthly. $20.00 per year.

WATER POLLUTION CONTROL. The Bureau of National Affairs, Incorporated, 1231 25th Street, NW, Washington, DC 20037. (202) 452-4200. Biweekly. $360.00 per year.

WATER POLLUTION CONTROL FEDERATION. Journal. Water Pollution Control Federation, 2626 Pennsylvania Avenue, NW, Washington, DC 20037. (202) 337-2500. Monthly. $120.00 per year.

WATER RESEARCH. International Association on Water Pollution Research. Pergamon Press Incorporated, Journals Division, Maxwell House, Fairview Park, Elmsford, NY 10523. (914) 592-7700. Monthly. $550.00 per year.

WATER RESOURCES BULLETIN. American Water Resources Association, 5410 Grosvenor Lane, Suite 220, Bethesda, MD 20814. (301) 493-8600. Bimonthly. $65.00 per year.

WATER RESOURCES RESEARCH. American Geophysical Union, 2000 Florida Avenue, NW, Washington, DC 20009. (202) 462-6903. Monthly. $295.00 per year.

RESEARCH CENTERS AND INSTITUTES

CALIFORNIA INSTITUTE OF TECHNOLOGY. W.M. Keck Engineering Laboratory of Hydraulics and Water Resources, 1201 East California Boulevard, Pasadena, CA 91125. (818) 356-4404.

LENOX INSTITUTE FOR RESEARCH, INCORPORATED. 101 Yokum Avenue, Lenox, MA 01240. (413) 637-3025.

PENNSYLVANIA STATE UNIVERSITY. Environmental Engineering Laboratory, 212 Sackett Building, University Park, PA 16802. (814) 863-4385.

UNIVERSITY OF ILLINOIS. Advanced Environmental Control Technology Research Center, 3230 Newmark, C.E. Laboratory, 208 North Romine Street, Urbana, IL 61801. (217) 333-3822.

UNIVERSITY OF TEXAS. Institute of Environmental Health, Post Office Box 20186, Houston, TX 77225-0186. (713) 792-4425.

WASHINGTON STATE UNIVERSITY. Environmental Engineering Research Laboratory, 141 Sloan, Pullman, WA 99164. (509) 335-3175.

EPOXY RESINS

See: PLASTICS

EPSILON MESON

See: PARTICLE PHYSICS

EQUATIONS

See: ALGEBRA, MATHEMATICS

ERGONOMICS

See also: CAD (COMPUTER-AIDED DESIGN), INDUSTRIAL ENGINEERING, MECHANICAL ENGINEERING

ABSTRACT SERVICES AND INDEXES

APPLIED MECHANICS REVIEW. American Society of Mechanical Engineers, 345 East 47th Street, New York, NY 10017. (212) 705-7703. 1948 to present. Monthly. $360.00 per year.

APPLIED SCIENCE AND TECHNOLOGY INDEX. H.W. Wilson and Company, 950 University Avenue, Bronx, NY 10452. (800) 367-6670 or (212) 588-8400. Monthly. Inquire as to cost and availability.

CHEMICAL ABSTRACTS. American Chemical Society, Chemical Abstracts Service, Box 3012, Columbus, OH 43210. (614) 421-3600. 1907 to present. Weekly. $9500.00 per year.

CONFERENCE PAPERS INDEX. Cambridge Scientific Abstracts, 5161 River Road, Bethesda, MD 20816. 1972 to present. Monthly. Inquire as to cost and availability.

CURRENT CONTENTS: ENGINEERING, TECHNOLOGY AND APPLIED SCIENCES. Institute for Scientific Information, 3501 Market Street, Philadelphia, PA 19104. (800) 523-1850 or (215) 386-0100. Weekly. $275.00 per year.

ENGINEERING INDEX MONTHLY AND AUTHOR INDEX. Engineering Information Inc., 345 East 47th Street, New York, NY 10017. (212) 705-7600. Monthly. $1560.00 per year.

ERGONOMICS ABSTRACTS. Ergonomics Information Analysis Centre. Available from: Taylor and Francis Limited, Rankine Road, Basingstoke, Hants RG24 OPR, England. 1968 to present. Quarterly. $270.00 per year.

ISMEC BULLETIN (Information Service in Mechanical Engineering). Cambridge Scientific Abstracts, 5161 River Road, Bethesda, MD 20816. (301) 951-1400. 1973 to present. Monthly. $450.00 per year.

INDEX TO SCIENTIFIC AND TECHNICAL PROCEEDINGS. Institute for Scientific Information, 3501 Market Street, Philadelphia, PA 19104. (800) 523-1850 or (215) 386-0100. 1978 to present. Monthly. $775.00 per year.

INDEX TO SCIENTIFIC REVIEWS. Institute for Scientific Information, 3501 Market Street, Philadelphia, PA 19104. (800) 523-1850 or (215) 386-0100. 1974 to present. Semi-annual. $550.00 per year.

SCIENCE CITATION INDEX. Institute for Scientific Information, 3501 Market Street, Philadelphia, PA 19104. (800) 523-1850 or (215) 386-0100. Six times per year. $6200. 00 per year.

ASSOCIATIONS AND PROFESSIONAL SOCIETIES

AMERICAN SOCIETY OF MECHANICAL ENGINEERS. 345 East 47th Street, New York, NY 10017. (212) 705-7703.

HUMAN FACTORS SOCIETY. Box 1369, Santa Monica, CA 90406. (213) 394-1811.

ERGONOMICS

DIRECTORIES AND BIOGRAPHICAL SOURCES

AMERICAN MEN AND WOMEN OF SCIENCE. R.R. Bowker, Inc., Order Department, 245 West 17th Street, New York, NY 10011. (800) 521-8110. Eight volumes. 1986. $595.00 for set.

INTERNATIONAL RESEARCH CENTERS DIRECTORY 1988-89. Darren L. Smith, editor. Gale Research Company, Book Tower, Detroit, MI 48226. (800) 521-0707. 4th edition. 1987. $360.00.

1987 DIRECTORY OF ENGINEERING SOCIETIES AND RELATED ORGANIZATIONS. Gordon Davis, editor. Hemisphere Publishing Corporation, 1010 Vermont Avenue, NW, Washington, DC 20005. (800) 526-0275. 12th edition. 1987. $100.00.

RESEARCH CENTERS DIRECTORY 1988. Gale Research Company, Book Tower, Detroit, MI 48226. (800) 521-0707. 12th edition. 1987. $365.00 for set.

SCIENTIFIC AND TECHNICAL ORGANIZATIONS AND AGENCIES DIRECTORY. Margaret Labash Young, editor. Gale Research Company, Book Tower, Detroit, MI 48226. (800) 521-0707. 2nd edition. 1987. $185.00.

WHO'S WHO IN ENGINEERING. Gordon Davis, editor. Hemisphere Publishing Corporation, 79 Madison Avenue, New York, NY 10016-7892. (800) 821-8312. Sixth edition. 1985. $200.00.

ENCYCLOPEDIAS AND DICTIONARIES

THESAURUS OF SCIENTIFIC, TECHNICAL, AND ENGINEERING TERMS. Hemisphere Publishing Corporation, 1010 Vermont Avenue, NW, Washington, DC 20005. (800) 526-0275. 1988. $125.00.

GENERAL WORKS

HUMAN FACTORS IN ENGINEERING AND DESIGN. M. Sanders and E.J. McCormick. McGraw-Hill Book Company, 1221 Avenue of the Americas, New York, NY 10020. (212) 512-2000. 6th edition. 1987. $47.95.

METHODS OF APPLIED ERGONOMICS. E.N. Corbett and others. Taylor and Francis, Inc., 242 Cherry Street, Philadelphia, PA 19106-1906. (800) 821-8312. 1987. $45.00 in paper.

THE PRACTICE AND MANAGEMENT OF INDUSTRIAL ERGONOMICS. David C. Alexander. Prentice-Hall Publishing, Inc., Englewood Cliffs, NJ 07632. (800) 562-0245. 1986. $44.95.

HANDBOOKS AND MANUALS

APPLIED ERGONOMICS HANDBOOK. Ian A. Galer, editor. Butterworth's Publishing, 80 Montvale Avenue, Stoneham, MA 02180. (800) 325-4177. Second edition. 1987. $49.95.

ONLINE DATA BASES

CA SEARCH. Chemical Abstracts Service, P.O. Box 3012, Columbus, OH 43120. (800) 848-6538 or (614) 421-3600. Comprehensive guide to chemical literature, 1972 to present. Inquire as to online cost and availability.

COMPENDEX. Engineering Information, Inc., 345 East 47th Street, New York, NY 10017. (800) 221-1044 or (212) 705-7615. Engineering and technical literature, 1975 to present. Inquire as to online cost and availability.

DISSERTATION ABSTRACTS ONLINE. University Microfilms International, 300 North Zeeb Road, Ann Arbor, MI 48106. (800) 521-0600 or (313) 761-4700. Scope includes virtually all doctoral dissertations accepted at accredited American institutions from 1861 to present in over 250 subject areas. Inquire as to online cost and availability.

INSPEC. INSPEC Marketing Department, Institution of Electrical Engineers. Available from IEEE Service Center, 445 Hoes Lane, Piscataway, NJ 08854. (201) 981-0060. Online version of Physics Abstracts. Inquire as to online cost and availability.

NTIS. National Technical Information Service, 5285 Port Royal Road, Springfield, VA 22161. (703) 487-4630. Broad coverage of government sponsored research reports, 1964 to present. Inquire as to online cost and availability.

SCISEARCH. Institute for Scientific Information, 3501 Market Street, Philadelphia, PA 19104. (800) 523-1850 or (215) 386-0100. Broad multidisciplinary title and author index to the international literature of science and technology, 1974 to present. Inquire as to online cost and availability.

WILSONLINE. H.W. Wilson and Company, 950 University Avenue, Bronx, NY 10452. (800) 367-6770 or (212) 588-8400. Makes available online versions of the H.W. Wilson indexes including Applied Science and Technology Index, Business Periodicals Index and Readers' Guide to Periodical Literature. Approximately 1980 to present. Inquire as to online cost and availability.

OTHER SOURCES

WHAT EVERY ENGINEER SHOULD KNOW ABOUT ENGINEERING SOURCES. Marcel Dekker Inc., 270 Madison Avenue, New York, NY 10016. (800) 228-1160. 1984. $24.95.

PERIODICALS

APPLIED ERGONOMICS. Butterworth's Publishing, 80 Montvale Avenue, Stoneham, MA 02180. (800) 325-4177. 1969 to present. Quarterly. $95.00 per year.

ERGONOMICS: AN INTERNATIONAL JOURNAL ON THE SCIENTIFIC STUDY OF HUMAN FACTORS IN RELATION TO WORKING ENVIRONMENTS AND EQUIPMENT DESIGN. Taylor and Francis Limited, Rankine Road, Basingstoke, Hants RG24 OPR, England. 1957 to present. Monthly. $320.00 per year.

HUMAN FACTORS. Human Factors Society, Box 1369, Santa Monica, CA 90406. (213) 394-1811. 1977 to present. Bimonthly. $60.00 per year.

HUMAN FACTORS SOCIETY BULLETIN. Human Factors Society, Box 1369, Santa Monica, CA 90406. (213) 394-1811. 1958 to present. Monthly. $20.00 per year.

EROSION

See: GEOLOGY

ERTS (EARTH RESOURCES TECHNOLOGY SATELLITES)

See: ARTIFICIAL SATELLITES

ETHYL ALCOHOL

See: ALCOHOL

EUCLIDEAN GEOMETRY

See: GEOMETRY

EXPANSION JOINTS

See: STRUCTURAL ENGINEERING

EXPERIMENTAL DESIGN

See: CAD (COMPUTER-AIDED DESIGN)

EXPERT SYSTEMS

See: ARTIFICIAL INTELLIGENCE

EXPLOSIVES

See also: CHEMICAL ENGINEERING, FIRE PROTECTION, FUELS, PROPULSION SYSTEMS, PROPELLANTS, ROCKETS, SAFETY ENGINEERING

ABSTRACT SERVICES AND INDEXES

APPLIED SCIENCE AND TECHNOLOGY INDEX. H.W. Wilson and Company, 950 University Avenue, Bronx, NY 10452. (800) 367-6670 or (212) 588-8400. Monthly. Inquire as to cost and availability.

CHEMICAL ABSTRACTS. American Chemical Society, Chemical Abstracts Service, Box 3012, Columbus, OH 43210. (614) 421-3600. 1907 to present. Weekly. $9500.00 per year.

ENGINEERING INDEX MONTHLY AND AUTHOR INDEX. Engineering Information Inc., 345 East 47th Street, New York, NY 10017. (212) 705-7600. Monthly. $1560.00 per year.

INTERNATIONAL AEROSPACE ABSTRACTS. American Institute of Aeronautics and Astronautics, Technical Information Service, 370 L'Enfant Promenade, S.W., Washington, DC 20024. (202) 646-7400. 1961 to present. Semi-monthly. $700.00 per year.

SCIENCE CITATION INDEX. Institute for Scientific Information, 3501 Market Street, Philadelphia, PA 19104. (800) 523-1850 or (215) 386-0100. Six times per year. $6200.00 per year.

STAR (SCIENTIFIC AND TECHNICAL AEROSPACE REPORTS). National Aeronautics and Space Administration, Scientific and Technical Information Facility, Box 8757, Baltimore-Washington International Airport, MD 21240. (202) 453-8545. 1963 to present. Semi-monthly. $85.00 per year with indexes.

ANNUAL REVIEWS AND YEARBOOKS

EXPLOSIVES AND BLASTING TECHNICALS. Society of Explosives Engineers, P.O. Box 185, 6990 Summers Drive, Montville, OH 44064. (216) 474-8436. Annual. Inquire.

ASSOCIATIONS AND PROFESSIONAL SOCIETIES

AMERICAN ASTRONAUTICAL SOCIETY. 6212-B Old Keene Mill Court, Springfield, VA 22152. (703) 866-0020.

AMERICAN CHEMICAL SOCIETY. 1155 16th Street, N.W., Washington, DC 20036. (800) 424-6747.

AMERICAN INSTITUTE OF AERONAUTICS AND ASTRONAUTICS. 370 L'Enfant Promenade, S.W., Washington, DC 20024. (202) 646-7400.

AMERICAN INSTITUTE OF CHEMICAL ENGINEERS. 345 East 47th Street, New York, NY 10017. (212) 705-7338.

SOCIETY OF EXPLOSIVES ENGINEERS. P.O. Box 185, 6990 Summers Drive, Montville, OH 44064. (216) 474-8436.

DIRECTORIES AND BIOGRAPHICAL SOURCES

RESEARCH CENTERS DIRECTORY 1988. Gale Research Company, Book Tower, Detroit, MI 48226. (800) 521-0707. 12th edition. 1987. $365.00 for set.

SCIENTIFIC AND TECHNICAL ORGANIZATIONS AND AGENCIES DIRECTORY. Margaret Labash Young, editor. Gale Research Company, Book Tower, Detroit, MI 48226. (800) 521-0707. 2nd edition. 1987. $185.00.

SOCIETY OF EXPLOSIVES ENGINEERS DIRECTORY OF MEMBER SPECIALTIES. P.O. Box 185, 6990 Summers Drive, Montville, OH 44064. (216) 474-8436. Annual. Inquire.

WHO'S WHO IN ENGINEERING. Gordon Davis, editor. Hemisphere Publishing Corporation, 1010 Vermont Avenue, NW, Washington, DC 20005. (800) 526-0275. 6th edition. 1985. $200.00.

GENERAL WORKS

EXPLOSIVES AND PROPELLANTS FROM COMMONLY AVAILABLE MATERIALS. Gordon Press Publishers, P.O. Box 459, Bowling Green Station, New York, NY 10004. (212) 668-8819. 1986. $79.95.

PROPELLANTS: MANUFACTURE, HAZARDS AND TESTING. Carl Boyars and Karl Klager, editors. American Chemical Society, 1155 16th Street, N.W., Washington, DC 20036. (800) 424-6747. 1969. $34.95.

ONLINE DATA BASES

CA SEARCH. Chemical Abstracts Service, P.O. Box 3012, Columbus, OH 43120. (800) 848-6538 or (614) 421-3600. Comprehensive guide to chemical and related literature, 1972 to present. Inquire as to online cost and availability.

COMPENDEX. Engineering Information, Inc., 345 East 47th Street, New York, NY 10017. (800) 221-1044 or (212) 705-7615. Engineering and technical literature, 1975 to present. Inquire as to online cost and availability.

NTIS. National Technical Information Service, 5285 Port Royal Road, Springfield, VA 22161. (703) 487-4630. Broad coverage of government sponsored research reports, 1964 to present. Inquire as to online cost and availability.

SCISEARCH. Institute for Scientific Information, 3501 Market Street, Philadelphia, PA 19104. (800) 523-1850 or (215) 386-0100. Broad multidisciplinary title and author index to the international literature of science and technology, 1974 to present. Inquire as to online cost and availability.

WILSONLINE. H.W. Wilson and Company, 950 University Avenue, Bronx, NY 10452. (800) 367-6770 or (212) 588-8400. Makes available online versions of the H.W. Wilson indexes including Applied Science and Technology Index, Business Periodicals Index and Readers' Guide to Periodical Literature. Approximately 1980 to present. Inquire as to online cost and availability.

PERIODICALS

PROPELLANTS, EXPLOSIVES, PYROTECHNICS. International Pyrotechnics Society. Available from: VCH Publishers, Inc., 303 N.W. 12th Avenue, Deerfield Beach, FL 33441. 1976 to present. Bimonthly. $175.00 per year.

JOURNAL OF EXPLOSIVES ENGINEERING. Society of Explosives Engineers, P.O. Box 185, 6990 Summers Drive, Montville, OH 44064. (216) 474-8436. Bimonthly. $12.00 per year.

RESEARCH CENTERS AND INSTITUTES

CENTER FOR EXPLOSIVES TECHNOLOGY RESEARCH. New Mexico Institute of Mining and Technology, Campus Station, Socorro, NM 87801. (505) 835-5733.

ROCK MECHANICS AND EXPLOSIVES RESEARCH CENTER. University of Missouri - Rolla, Rolla, MO 65401. (314) 4364.

THERMAL SCIENCES AND PROPULSION CENTER. Purdue University, Lafayette, IN 47907. (317) 494-1500.

EXTRACTION PROCESSES

See: METALLURGICAL ENGINEERING

EXTRUSION

See: METALLURGICAL ENGINEERING

F

FABRICS

See: TEXTILES

FACTOR ANALYSIS

See: MATHEMATICS

FAILURE ANALYSIS

See also: AERONAUTICAL ENGINEERING, CIVIL ENGINEERING, FATIGUE, FRACTURE MECHANICS, MATERIALS SCIENCE, MECHANICAL ENGINEERING, NONDESTRUCTIVE TESTING, STRUCTURAL ENGINEERING

ABSTRACT SERVICES AND INDEXES

ABSTRACT JOURNAL IN EARTHQUAKE ENGINEERING. University of California at Berkeley, Earthquake Engineering Research Center, 1301 South 46th Street, Richmond, CA 94804. (415) 231-9413. Semiannual. $70.00 per copy.

APPLIED SCIENCE AND TECHNOLOGY INDEX. H.W. Wilson and Company, 950 University Avenue, Bronx, NY 10452. (800) 367-6670 or (212) 588-8400. Monthly. Inquire as to cost and availability.

CURRENT CONTENTS: ENGINEERING, TECHNOLOGY AND APPLIED SCIENCES. Institute for Scientific Information, 3501 Market Street, Philadelphia, PA 19104. (800) 523-1850 or (215) 386-0100. Weekly. $275.00 per year.

ENGINEERING INDEX MONTHLY AND AUTHOR INDEX. Engineering Information Inc., 345 East 47th Street, New York, NY 10017. (212) 705-7600. Monthly. $1560.00 per year.

INDEX TO SCIENTIFIC AND TECHNICAL PROCEEDINGS. Institute for Scientific Information, 3501 Market Street, Philadelphia, PA 19104. (800) 523-1850 or (215) 386-0100. 1978 to present. Monthly. $775.00 per year.

INTERNATIONAL CIVIL ENGINEERING ABSTRACTS. CITIS Limited, 2 Rosemount Terrace, Blackrock, Dublin, Ireland. 1974 to present. Monthly. $350.00 per year.

INTERNATIONAL STRUCTURAL ENGINEERING ABSTRACTS. CITIS Limited, 2 Rosemount Terrace, Blackrock, Dublin, Ireland. 1986 to present. Quarterly. $95.00 per year.

ISMEC BULLETIN (Information Service in Mechanical Engineering). Cambridge Scientific Abstracts, 5161 River Road, Bethesda, MD 20816. (301) 951-1400. 1973 to present. Monthly. $450.00 per year.

PUBLICATIONS INFORMATION. American Society of Civil Engineers, 345 East 47th Street, New York, NY 10017. (212) 705-7420. Abstracts, subject and author indexes to the publications of the American Society of Civil Engineers. Bimonthly. $80.00 per year.

SCIENCE CITATION INDEX. Institute for Scientific Information, 3501 Market Street, Philadelphia, PA 19104. (800) 523-1850 or (215) 386-0100. Six times per year. $6200.00 per year.

ASSOCIATIONS AND PROFESSIONAL SOCIETIES

AMERICAN SOCIETY FOR NONDESTRUCTIVE TESTING. 4153 Arlingate Plaza, Caller 28518, Columbus, OH 43228. (614) 274-6003.

AMERICAN SOCIETY FOR TESTING AND MATERIALS. 1916 Race Street, Philadelphia, PA 19103. (215) 299-5400.

AMERICAN SOCIETY OF CIVIL ENGINEERS. 345 East 47th Street, New York, NY 10017. (212) 705-7420.

AMERICAN SOCIETY OF MECHANICAL ENGINEERS. 345 East 47th Street, New York, NY 10017. (212) 705-7703.

DIRECTORIES AND BIOGRAPHICAL SOURCES

INTERNATIONAL RESEARCH CENTERS DIRECTORY 1988-89. Darren L. Smith, editor. Gale Research Company, Book Tower, Detroit, MI 48226. (800) 521-0707. 4th edition. 1987. $360.00.

1987 DIRECTORY OF ENGINEERING SOCIETIES AND RELATED ORGANIZATIONS. Gordon Davis, editor. Hemisphere Publishing Corporation, 1010 Vermont Avenue, NW, Washington, DC 20005. (800) 526-0275. 12th edition. 1987. $100.00.

RESEARCH CENTERS DIRECTORY 1988. Gale Research Company, Book Tower, Detroit, MI 48226. (800) 521-0707. 12th edition. 1987. $365.00 for set.

SCIENTIFIC AND TECHNICAL ORGANIZATIONS AND AGENCIES DIRECTORY. Margaret Labash Young, editor. Gale Research Company, Book Tower, Detroit, MI 48226. (800) 521-0707. 2nd edition. 1987. $185.00.

WHO'S WHO IN ENGINEERING. Gordon Davis, editor. Hemisphere Publishing Corporation, 1010 Vermont Avenue, NW, Washington, DC 20005. (800) 526-0275. 6th edition. 1985. $200.00.

GENERAL WORKS

ADVANCED MECHANICS OF MATERIALS. A.P. Boresi and O.M. Sidebottom. John Wiley and Sons, Inc., 605 Third Avenue, New York, NY 10158. (800) 526-5368. Fourth edition. 1985. $44.00.

ELEMENTARY ENGINEERING FRACTURE MECHANICS. David Broek. Martinus Nijhof. Distributed by Kluwer Academic

Publishers, 190 Old Derby Street, Hingham, MA 02043. (617) 749-5262. Fourth edition. 1986. $110.00.

ELEMENTARY THEORY OF STRUCTURES. Yu H. Yuan. Prentice-Hall Publishing, Inc., Englewood Cliffs, NJ 07632. (800) 562-0245. Third edition. 1988. $42.95.

FAILURE OF MATERIALS IN MECHANICAL DESIGN: ANALYSIS, PREDICTION, PREVENTION. J.A. Collins. John Wiley and Sons, Inc., 605 Third Avenue, New York, NY 10158. (800) 526-5368. 1981. $59.95.

FATIGUE AND FRACTURE IN STEEL BRIDGES: CASE STUDIES. John W. Fisher. John Wiley and Sons, Inc., 605 Third Avenue, New York, NY 10158. (800) 526-5368. 1984. $45.95.

FRACTURE AND FATIGUE CONTROL IN STRUCTURES: APPLICATIONS OF FRACTURE MECHANICS. John M. Barson and Stanley T. Rolfe. Prentice-Hall Publishing, Inc., Englewood Cliffs, NJ 07632. (800) 562-0245. 1987. $46.95.

PROBABILISTIC FRACTURE MECHANICS AND RELIABILITY. James W. Provan, editor. Martinus Nijhoff. Distributed by Kluwer Academic Publishers, 190 Old Derby Street, Hingham, MA 02043. (617) 749-5262. 1987. $135.50.

HANDBOOKS AND MANUALS

HANDBOOK OF MECHANICS, MATERIALS, AND STRUCTURES. A. Blake. John Wiley and Sons, Inc., 605 Third Avenue, New York, NY 10158. (800) 526-5368. 1985. $64.50.

MECHANICAL ENGINEER'S HANDBOOK. Myer Kutz, editor. John Wiley and Sons, Inc., 605 Third Avenue, New York, NY 10158. (800) 526-5368. 1986. $79.95.

ONLINE DATA BASES

COMPENDEX. Engineering Information, Inc., 345 East 47th Street, New York, NY 10017. (800) 221-1044 or (212) 705-7615. Engineering and technical literature, 1975 to present. Inquire as to online cost and availability.

DISSERTATION ABSTRACTS ONLINE. University Microfilms International, 300 North Zeeb Road, Ann Arbor, MI 48106. (800) 521-0600 or (313) 761-4700. Scope includes virtually all doctoral dissertations accepted at accredited American institutions from 1861 to present in over 250 subject areas. Inquire as to online cost and availability.

INSPEC. INSPEC Marketing Department, Institution of Electrical Engineers. Available from IEEE Service Center, 445 Hoes Lane, Piscataway, NJ 08854. (201) 981-0060. Online version of Physics Abstracts. Inquire as to online cost and availability.

NTIS. National Technical Information Service, 5285 Port Royal Road, Springfield, VA 22161. (703) 487-4630. Broad coverage of government sponsored research reports, 1964 to present. Inquire as to online cost and availability.

SCISEARCH. Institute for Scientific Information, 3501 Market Street, Philadelphia, PA 19104. (800) 523-1850 or (215) 386-0100. Broad multidisciplinary title and author index to the international literature of science and technology, 1974 to present. Inquire as to online cost and availability.

WILSONLINE. H.W. Wilson and Company, 950 University Avenue, Bronx, NY 10452. (800) 367-6770 or (212) 588-8400. Makes available online versions of the H.W. Wilson indexes including Applied Science and Technology Index, Business Periodicals Index and Readers' Guide to Periodical Literature. Approximately 1980 to present. Inquire as to online cost and availability.

PERIODICALS

JOURNAL OF ENGINEERING MECHANICS. American Society of Civil Engineers, 345 East 47th Street, New York, NY 10017. (212) 705-7420. 1956 to present. Monthly. $120.00 per year.

JOURNAL OF STRUCTURAL MECHANICS. Marcel Dekker Inc., 270 Madison Avenue, New York, NY 10016. (800) 228-1160. 1972 to present. Quarterly. $75.00 per year.

JOURNAL OF TESTING AND EVALUATION. American Society for Testing and Materials, 1916 Race Street, Philadelphia, PA 19103. (215) 299-5400. 1966 to present. Bimonthly. $40.00 per year.

MATERIALS ENGINEERING. Penton-IPC, 1100 Superior Avenue, Cleveland, OH 44114. (216)696-7000. 1929 to present. Monthly. $40.00 per year.

MATERIALS EVALUATION. American Society for Nondestructive Testing, 4153 Arlingate Plaza, Caller 28518, Columbus, OH 43228. 1942 to present. Monthly. $50.00 per year.

SAMPE JOURNAL. Society for the Advancement of Material and Process Engineering, 843 West Glentana, Box 2459, Covina, CA 91722. (818) 331-0610. 1965 to present. Bimonthly. $31.00 per year.

STRAIN. British Society for Strain Measurement, Suite 7, Exchange Building, Ouayside, Newcastle-upon-Tyne, NE1 3BJ, England. 1965 to present. Quarterly. $45.00 per year.

RESEARCH CENTERS AND INSTITUTES

CENTER ON QUALITY ENGINEERING AND FAILURE PREVENTION. Northwestern University, Technological Institute, Evanston, IL 60201. (312) 491-5527.

STRUCTURAL ENGINEERING MATERIALS LABORATORY. University of California, Berkeley, Davis Hall, Berkeley, CA 94720. (415) 642-3434.

STRUCTURAL STABILITY RESEARCH COUNCIL. Fritz Engineering Laboratory No. 13, Lehigh University, Bethlehem, PA 18105. (215) 861-3519.

WISCONSIN CENTER FOR STRUCTURAL AND MATERIALS TESTING. University of Wisconsin at Madison, 1415 Johnson Drive, Madison, WI 53706. (608) 262-3205.

SPECIFICATIONS AND STANDARDS

ELASTIC-PLASTIC FRACTURE MECHANICS TECHNOLOGY. J.C. Newman and F.J. Loss, editors. American Society for Testing and Materials, 1916 Race Street, Philadelphia, PA 19103. (215) 299-5400. STP No. 896. 1986. $30.00.

FALLOUT

See: RADIATION

FANJET

See: JET PROPULSION

FATIGUE

See: FAILURE ANALYSIS

FAULTS

See: PHYSICAL GEOLOGY

FEEDBACK CONTROL SYSTEMS

See: AUTOMATION

FEEDFORWARD CONTROL SYSTEMS

See: AUTOMATION

FERROALLOYS

See also: ALLOYS, CAST IRON, IRON, MATERIALS SCIENCE, STEEL

ABSTRACT SERVICES AND INDEXES

ALLOYS INDEX. American Society for Metals, 9639 Kinsman Road, Metals Park, OH 44073. (216) 338-5151. $130.00 per year.

APPLIED MECHANICS REVIEWS. American Society of Mechanical Engineers, 345 East 47th Street, New York, NY 10017. (212) 705-7722. Monthly. $380.00 per year. Critical reviews of the world literature in applied mechanics and related engineering science.

APPLIED SCIENCE AND TECHNOLOGY INDEX. H.W. Wilson Company, 950 University Avenue, Bronx, NY 10452. (800) 367-6770 or (212) 588-8400. Inquire as to cost and availability.

CHEMICAL ABSTRACTS. Chemical Abstracts Service, 2540 Olentangy Road, Post Office Box 3012, Columbus, OH 43210. (800) 848-6538 or (614) 421-3600. Weekly. $9200.00 per year.

CURRENT CONTENTS: ENGINEERING, TECHNOLOGY. Institute for Scientific Information, 3501 Market Street, Philadelphia, PA 19104. (800) 523-1850 or (215) 386-0100. $272.00 per year.

ENGINEERING INDEX MONTHLY. Engineering Information, Incorporated, 345 East 47th Street, New York, NY 10017. (800) 221-1044 or (212) 705-7600. Monthly, with annual cumulation. $1560.00 per year.

METALS ABSTRACTS. American Society for Metals, 9639 Kinsman Road, Metals Park, OH 44073. (216) 338-5151. Monthly. $890.00.

METALS ABSTRACTS INDEX. American Society for Metals, 9639 Kinsman Road, Metals Park, OH 44073. (216) 338-5151. Monthly. (Sold only to subscribers of Metals Abstracts.)

ANNUAL REVIEWS AND YEARBOOKS

ANNUAL REVIEW OF MATERIALS SCIENCE. Annual Reviews, Incorporated, 4139 El Camino Way, Palo Alto, CA 94306. (415) 493-4400.

ASSOCIATIONS AND PROFESSIONAL SOCIETIES

AMERICAN FOUNDRYMEN'S SOCIETY, INCORPORATED. Golf and Wolf Roads, Des Plaines, IL 60016. (312) 824-0181.

AMERICAN IRON AND STEEL INSTITUTE. 1000 Sixteenth Street, NW, Washington, DC 20036. (202) 452-7100.

AMERICAN SOCIETY FOR METALS. Metals Park, OH 44073. (216) 338-5151.

IRON CASTINGS SOCIETY. 455 State Street, Des Plaines, IL 60016. (312) 299-9160.

METALLURGICAL SOCIETY OF THE AIME (AMERICAN INSTITUTE OF MINING, METALLURGICAL AND PETROLEUM ENGINEERS). 420 Commonwealth Drive, Warrendale, PA 15086. (412) 776-9080.

BIBLIOGRAPHIES

SCIENTIFIC AND TECHNICAL BOOKS AND SERIALS IN PRINT 1988; AN INDEX TO LITERATURE IN SCIENCE AND TECHNOLOGY. R.R. Bowker Company, 205 East 42nd Street, New York, NY 10017. (800) 521-8110 or (212) 916-1600. $175.00.

DIRECTORIES AND BIOGRAPHICAL SOURCES

DUN'S INDUSTRIAL GUIDE - THE METAL WORKING DIRECTORY. Dun and Bradstreet, Incorporated, Three Century Drive, Parsippany, NJ 07054. (201) 455-0900. Annual. $550.00.

INTERNATIONAL RESEARCH CENTERS DIRECTORY 1986-1987. Gale Research Company, Book Tower, Detroit, MI 48226. (800) 521-0707. Third edition. 1986.

METAL PRODUCTS DIRECTORY. American Business Directories, Incorporated, Division of American Business Lists, Incorporated, 5707 South 86th Circle, Omaha, NE 68127. (402) 331-7169. 1986. $80.00.

RESEARCH CENTERS DIRECTORY. Gale Research Company, Book Tower, Detroit, MI 48226. (800) 521-0707. Eleventh edition. 1987.

SCIENTIFIC AND TECHNICAL ORGANIZATIONS AND AGENCIES DIRECTORY. Gale Research Company, Book Tower, Detroit, MI 48226. (800) 521-0707. 1985.

WHO'S WHO IN ENGINEERING. Engineers Joint Council, 345 East 47th Street, New York, NY 10017. (212) 705-7010. 1985. $200.00.

WHO'S WHO IN FRONTIER SCIENCE AND TECHNOLOGY. Marquis Who's Who, Incorporated, 200 East Ohio Street, Chicago, IL 60611. (800) 428-3898 or (312) 787-2008.

WHO'S WHO IN TECHNOLOGY TODAY. Reston Publishing Company, Incorporated, c/o Prentice-Hall, Incorporated, Englewood Cliffs, NJ 07632. (800) 262-6868. Biennial. Five volumes. $425.00. Covers the fields of electronics, computer science, physics, optics, chemistry, biotechnology, mechanics, energy, and earth science.

ENCYCLOPEDIAS AND DICTIONARIES

ENCYCLOPEDIA OF PHYSICAL SCIENCE AND TECHNOLOGY. Academic Press, Incorporated, Orlando, FL 32887. (800) 321-5068 or (305) 345-2734. Fifteen volumes, 1986.

MCGRAW-HILL DICTIONARY OF ENGINEERING. Sybil P. Parker, editor. McGraw-Hill Book Company, 1221 Avenue of the Americas, New York, NY 10020. (212) 512-2000. 1984. $39.95.

MCGRAW-HILL ENCYCLOPEDIA OF SCIENCE AND TECHNOLOGY. McGraw-Hill Book Company, Incorporated, 1221 Avenue of the Americas, New York, NY 10020. (212) 512-2000.

GENERAL WORKS

ESSENTIAL METALLURGY FOR ENGINEERS. W. Alexander, G. Davies, and K. Reynolds. Van Nostrand Reinhold, 115 Fifth

Avenue, New York, NY 10003. (800) 543-2681. 1985. $17.95.

FERROALLOYS AND OTHER ADDITIVES TO LIQUID IRON AND STEEL. American Society for Testing and Materials, 1916 Race Street, Philadelphia, PA 19103. (215) 299-5400. 1981. $24.75.

METTALLURGY BASICS. D.V. Brown. Van Nostrand Reinhold, 115 Fifth Avenue, New York, NY 10003. (800) 543-2681. 1985. $17.95.

PHYSICAL METALLURGY. P. Haasen. Cambridge University Press, 32 East 57th Street, New York, NY 10022. (212) 688-8885. 1986. $24.95 in paper.

PROPERTIES OF SELECTED FERROUS ALLOYING ELEMENTS. Y.S. Touloukian and C.Y. Ho. McGraw-Hill Book Company, Incorporated, 1221 Avenue of the Americas, New York, NY 10020. (212) 512-2000. 1981. $56.95.

HANDBOOKS AND MANUALS

CRC HANDBOOK OF CHEMISTRY AND PHYSICS. CRC Press, Incorporated, 2000 Corporate Boulevard, Boca Raton, Florida 33341. (305) 994-0555. Sixth-seventh edition. 1986. $69.95.

SMITHELL'S METALS REFERENCE BOOK. Eric A. Brandes, editor. Butterworth Publishers, 80 Montvale Avenue, Stoneham, MA 02180. (800) 325-4177. Sixth edition. 1983. $210.00.

ONLINE DATA BASES

COMPENDEX. Engineering Information, Incorporated, 345 East 47th Street, New York, NY 10017. (800) 221-1044 or (212) 705-7615. Engineering and technical literature, 1975 to present. Inquire as to cost and availability.

INSPEC. INSPEC Marketing Department, Institute of Electrical and Electronics Engineers, Incorporated, IEEE Service Department, 445 Hoes Lane, Piscataway, NJ 08854. (201) 981-0060. Inquire as to on-line cost and availability.

METADEX. Metals Information, American Society for Metals, Metals Park, OH 44073. (216) 338-5151. (Metals Abstracts/Alloys Index). A worldwide literature on the science and practice of metallurgy, 1966 to present. Inquire as to online cost and availability.

NASA. National Aeronautics and Space Administration, Scientific and Technical Information Branch, 300 7th Street, SW, Washington, DC 20546. Citations and abstracts of aerospace literature, 1962 to present. Inquire as to cost and availability.

NTIS. National Technical Information Service, 5285 Port Royal Road, Springfield, VA 22161. (703) 487-4630. Broad coverage of government sponsored research reports, 1964 to present. Inquire as to cost and availability.

WILSONLINE. H.W. Wilson Company, 950 University Avenue, Bronx, NY 10452. (800) 367-6770 or (212) 588-8400. Makes available online versions of the printed H.W. Wilson Indexes including Applied Science and Technology Index, Business Periodicals Index, and Reader's Guide to Periodical Literature. Period covered is generally 1983 to present. Inquire as to cost and availability.

OTHER SOURCES

INTERDOC: DIRECTORY OF PUBLISHED PROCEEDINGS, SERIES. SEMT-Science/Engineering/Medicine/Technology. Interdoc Corporation, 173 Halstead Avenue, Box 326, Harrison, NY 10528. (014) 835-3506. Ten times per year. $325.00 per year.

MATERIALS SCIENCE AND METALLURGY. National Technical Information Service (NTIS), 5285 Port Royal Road, Springfield, VA 22161. (703) 487-4630. Translations and abstracts of foreign language technical media. Irregular. $40.00 per year.

PERIODICALS

ALLOY DIGEST. Engineering Publications Incorporated, Box 823, Upper Montclair, NJ 07043. (201) 746-7930. Monthly. $50.00 per year.

BULLETIN OF ALLOY PHASE DIAGRAMS. American Society for Metals, Metals Park, OH 44073. (216) 338-5151. Bimonthly. $90.00 per year.

IRON AND STEEL TECHNOLOGY INSIGHTS. Merton Allen Associates, 2307 Dean Street, Schenectady, NY 12309. (518) 393-1933. Bimonthly. $49.00 per year.

JOURNAL OF METALS. Metallurgical Society of the AIME (American Institute of Mining, Metallurgical and Petroleum Engineers), 420 Commonwealth Drive, Warrendale, PA 15086. (412) 776-9080. Monthly. $40.00 per year.

METALLURGICAL TRANSACTIONS. Metallurgical Society of the AIME (American Institute of Mining, Metallurgical and Petroleum Engineers), 420 Commonwealth Drive, Warrendale, PA 15086. (412) 776-9080. Monthly. $95.00 per year.

METALS WEEK. McGraw-Hill Book Company, Incorporated, 1221 Avenue of the Americas, New York, NY 10020. (212) 997-2823. Weekly. $527.00 per year.

RESEARCH CENTERS AND INSTITUTES

AMERICAN IRON AND STEEL INSTITUTE. 1000 Sixteenth Street, NW, Washington, DC 20036. (202) 452-7100.

CARNEGIE-MELLON UNIVERSITY. Center for Iron and Steel Making Research, MEMS Department, Pittsburgh, PA 15312. (412) 578-2677.

COLORADO SCHOOL OF MINES. Steel Research Center, Golden, CO 80401. (303) 273-3774.

GENERAL ELECTRIC COMPANY. Research and Development Center, Post Office Box 8, Schenectady, NY 12301. (518) 385-8415.

LAWRENCE BERKELEY LABORATORY. Center for Advanced Materials, 1 Cyclotron Road, Berkeley, CA 94720. (415) 486-4755.

PHYSICAL METALLURGY RESEARCH LABORATORIES. Centre for Mineral and Energy Technology, 555 Booth Street, Ottawa, ON Canada K1A 0G1. (613) 995-4807.

TEXAS A & M UNIVERSITY. Mechanics and Materials Center, ERC Building, College Station, TX 77843. (409) 845-7512.

UNIVERSITY OF CONNECTICUT. Institute of Materials Science, Storrs, CT 06268. (203) 486-4623.

UNIVERSITY OF FLORIDA. Department of Materials Science and Engineering, Gainesville, FL 32601. (904) 392-1454.

UNIVERSITY OF MINNESOTA. Corrosion Research Center, 112 Mines and Metallurgy Building, Minneapolis, MN 55455. (612) 373-4864.

UNIVERSITY OF WISCONSIN AT MADISON. Cast Metals Laboratory, 1509 University Avenue, Madison, WI 53706. (608) 262-2562.

FERROELECTRIC

See: ELECTRICITY

FIBER OPTICS

See also: HOLOGRAPHY, OPTICAL COMMUNICATIONS, OPTICAL DISK TECHNOLOGY, OPTOELECTRONICS

ABSTRACT SERVICES AND INDEXES

APPLIED SCIENCE AND TECHNOLOGY INDEX. H.W. Wilson and Company, 950 University Avenue, Bronx, NY 10452. (800) 367-6670 or (212) 588-8400. Monthly. Inquire as to cost and availability.

CHEMICAL ABSTRACTS. American Chemical Society, Chemical Abstracts Service, Box 3012, Columbus, OH 43210. (614) 421-3600. 1907 to present. Weekly. $9500.00 per year.

CONFERENCE PAPERS INDEX. Cambridge Scientific Abstracts, 5161 River Road, Bethesda, MD 20816. 1972 to present. Monthly. Inquire as to cost and availability.

CURRENT CONTENTS: ENGINEERING, TECHNOLOGY AND APPLIED SCIENCES. Institute for Scientific Information, 3501 Market Street, Philadelphia, PA 19104. (800) 523-1850 or (215) 386-0100. Weekly. $275.00 per year.

ENGINEERING INDEX MONTHLY AND AUTHOR INDEX. Engineering Information Inc., 345 East 47th Street, New York, NY 10017. (212) 705-7600. Monthly. $1560.00 per year.

INDEX TO SCIENTIFIC AND TECHNICAL PROCEEDINGS. Institute for Scientific Information, 3501 Market Street, Philadelphia, PA 19104. (800) 523-1850 or (215) 386-0100. 1978 to present. Monthly. $775.00 per year.

PHYSICS ABSTRACTS. Institution of Electrical Engineers. Available from: IEEE Service Center, 445 Hoes Lane, Piscataway, NJ 08854. 1898 to present. Bimonthly. $1700.00 per year.

PHYSICS BRIEFS. Physik Verlag GmbH, Postfach 1260/1280, D-6940 Weinheim, West Germany. (212) 661-9404. 1920 to present. Twenty-six times per year. $1250.00 per year.

SCIENCE CITATION INDEX. Institute for Scientific Information, 3501 Market Street, Philadelphia, PA 19104. (800) 523-1850 or (215) 386-0100. Six times per year. $6200.00 per year.

ASSOCIATIONS AND PROFESSIONAL SOCIETIES

AMERICAN INSTITUTE OF PHYSICS. 335 East 45th Street, New York, NY 10017. (212) 661-9404.

INSTITUTION OF ELECTRICAL AND ELECTRONICS ENGINEERS (IEEE). 345 East 47th Street, New York, NY 10017. (212) 705-7900.

OPTICAL SOCIETY OF AMERICA. 1816 Jefferson Place, N.W., Washington, DC 20036. (202) 223-8130.

SOCIETY OF PHOTOGRAPHIC SCIENTISTS AND ENGINEERS. 7003 Kilworth Lane, Springfield, VA 22151. (703) 642-9090.

SOCIETY OF PHOTO-OPTICAL INSTRUMENTATION ENGINEERS - THE INTERNATIONAL SOCIETY OF OPTICAL ENGINEERING. P.O. Box 10, 1022 19th Street, Bellingham, WA 98227. (206) 676-3290.

DIRECTORIES AND BIOGRAPHICAL SOURCES

AMERICAN MEN AND WOMEN OF SCIENCE. R.R. Bowker, Inc., Order Department, 245 West 17th Street, New York, NY 10011. (800) 521-8110. Eight volumes. 1986. $595.00 for set.

INTERNATIONAL RESEARCH CENTERS DIRECTORY 1988-89. Darren L. Smith, editor. Gale Research Company, Book Tower, Detroit, MI 48226. (800) 521-0707. 4th edition. 1987. $360.00.

1987 DIRECTORY OF ENGINEERING SOCIETIES AND RELATED ORGANIZATIONS. Gordon Davis, editor. Hemisphere Publishing Corporation, 1010 Vermont Avenue, NW, Washington, DC 20005. (800) 526-0275. 12th edition. 1987. $100.00.

RESEARCH CENTERS DIRECTORY 1988. Gale Research Company, Book Tower, Detroit, MI 48226. (800) 521-0707. 12th edition. 1987. $365.00 for set.

SCIENTIFIC AND TECHNICAL ORGANIZATIONS AND AGENCIES DIRECTORY. Margaret Labash Young, editor. Gale Research Company, Book Tower, Detroit, MI 48226. (800) 521-0707. 2nd edition. 1987. $185.00.

WHO'S WHO IN ENGINEERING. Gordon Davis, Hemisphere Publishing Corporation, 79 Madison Avenue, New York, NY 10016-7892. (800) 821-8312. 6th edition. 1985. $200.00.

GENERAL WORKS

INTRODUCTION TO OPTICS. Frank L. Pedrotti and Leno S. Pedrotti. Prentice-Hall Publishing, Inc., Englewood Cliffs, NJ 07632. (800) 562-0245. 1987. $49.95.

LASERS: INVENTIONS TO APPLICATION. Jesse H. Ausubel and H.D. Langford, editors. National Academy Press, 2101 Constitution Avenue, Washington, DC 20418. (202) 334-3313. 1987. $14.95 in paper.

OPTICAL FIBRES. J. Geisler and others. Pergamon Press, Inc., Maxwell House, Fairview Park, Elmsford, NY 10523. (914) 592-7700. 1986. $165.00.

OPTICS. M.V. Klein and T.E. Furtak. John Wiley and Sons, Inc., 605 Third Avenue, New York, NY 10158. (800) 526-5368. 2nd edition. 1983. $44.95.

PRINCIPLES OF OPTICS. M. Born and E. Wolf. Pergamon Press, Inc., Maxwell House, Fairview Park, Elmsford, NY 10523. (914) 592-7700. 6th edition. 1980. $35.00 in paper.

HANDBOOKS AND MANUALS

HANDBOOK OF OPTICS. Optical Society of America. McGraw-Hill Book Company, 1221 Avenue of the Americas, New York, NY 10020. (212) 512-2000. 1978. $100.00.

ONLINE DATA BASES

CA SEARCH. Chemical Abstracts Service, P.O. Box 3012, Columbus, OH 43120. (800) 848-6538 or (614) 421-3600. Comprehensive guide to chemical literature, 1972 to present. Inquire as to online cost and availability.

COMPENDEX. Engineering Information, Inc., 345 East 47th Street, New York, NY 10017. (800) 221-1044 or (212) 705-7615. Engineering and technical literature, 1975 to present. Inquire as to online cost and availability.

DISSERTATION ABSTRACTS ONLINE. University Microfilms International, 300 North Zeeb Road, Ann Arbor, MI 48106. (800) 521-0600 or (313) 761-4700. Scope includes virtually all doctoral dissertations accepted at accredited American institutions from 1861 to present in over 250 subject areas. Inquire as to online cost and availability.

INSPEC. INSPEC Marketing Department, Institution of Electrical Engineers. Available from IEEE Service Center, 445 Hoes Lane, Piscataway, NJ 08854. (201) 981-0060. Online version of Physics Abstracts. Inquire as to online cost and availability.

NTIS. National Technical Information Service, 5285 Port Royal Road, Springfield, VA 22161. (703) 487-4630. Broad coverage of government sponsored research reports, 1964 to present. Inquire as to online cost and availability.

SCISEARCH. Institute for Scientific Information, 3501 Market Street, Philadelphia, PA 19104. (800) 523-1850 or (215) 386-0100. Broad multidisciplinary title and author index to the international literature of science and technology, 1974 to present. Inquire as to online cost and availability.

WILSONLINE. H.W. Wilson and Company, 950 University Avenue, Bronx, NY 10452. (800) 367-6770 or (212) 588-8400. Makes available online versions of the H.W. Wilson indexes including Applied Science and Technology Index, Business Periodicals Index and Readers' Guide to Periodical Literature. Approximately 1980 to present. Inquire as to online cost and availability.

PERIODICALS

APPLIED OPTICS. Optical Society of America. 1816 Jefferson Place, N.W., Washington, DC 20036. (202) 223-8130. Order from: American Institute of Physics, 335 East 45th Street, New York, NY 10017. (212) 661-9404. 1962 to present. Semi-monthly. $330.00 per year.

APPLIED SPECTROSCOPY REVIEWS. Marcel Dekker Inc., 270 Madison Avenue, New York, NY 10016. (800) 228-1160. 1964 tp present. Quarterly. $185.00 per year.

FIBER AND INTEGRATED OPTICS. Crane Russak and Company, Inc., 3 East 44th Street, New York, NY 10017. (212) 867-1490. 1977 to present. Quarterly. $86.00 per year.

JOURNAL OF LIGHTWAVE TECHNOLOGY. Institute of Electrical and Electronics Engineers. IEEE Service Center, 445 Hoes Lane, Piscataway, NJ 08854. 1983 to present. Monthly. $145.00 per year.

LASER FOCUS. PennWell Publishing Company, 119 Russell Street, Littleton, MA 01460. (617) 486-9501. 1965 to present. Monthly. $55.00 per year.

OPTICAL ENGINEERING. Society of Photo-Optical Instrumentation Engineers (SPIE), P.O. Box 10, 1022 19th Street, Bellingham, WA 98227. (206) 676-3290. 1962 to present. Monthly. $95.00 per year.

OPTICAL SOCIETY OF AMERICA, JOURNAL, PARTS A AND B. Optical Society of America. 1816 Jefferson Place, N.W., Washington, DC 20036. (202) 223-8130. Order from: American Institute of Physics, 335 East 45th Street, New York, NY 10017. (212) 661-9404. 1917 to present. Monthly. $180.00 each part per year.

OPTICS COMMUNICATIONS. Elsevier Science Publishing Company, Inc., 52 Vanderbilt Avenue, New York, NY 10017. (212) 370-5520. 1969 to present. 24 times per year. $425.00 per year.

OPTICS LETTERS. Optical Society of America. Order from: American Institute of Physics, 335 East 45th Street, New York, NY 10017. (212) 661-9404. 1977 to present. Monthly. $150.00 per year.

SOCIETY OF PHOTO-OPTICAL INSTRUMENTATION ENGINEERS (SPIE) PROCEEDINGS. Society of Photo-Optical Instrumentation Engineers (SPIE), P.O. Box 10, 1022 19th Street, Bellingham, WA 98227. (206) 676-3290. 1963 to present. Approximately 50 numbers per year. Approximately $40.00 per number.

RESEARCH CENTERS AND INSTITUTES

CENTER FOR APPLIED OPTICS. University of Alabama in Huntsville, Research Institute Building, Huntsville, AL 35899. (205) 895-6102.

CENTER FOR APPLIED OPTICS. University of Texas at Dallas, P.O. Box 830688, Richardson, TX 75083. (214) 690-2868.

INSTITUTE OF OPTICS. University of Rochester, Rochester, NY 14627. (716) 275-2314.

OPTICAL PHYSICS LABORATORY. University of Miami, Coral Gables, FL 33124. (305) 284-2324.

FIBERS—TEXTILES

See: TEXTILES

FIELD-EFFECT TRANSISTORS

See: SEMICONDUCTORS

FIRE CONTROL

See: FIRE PROTECTION

FIRE PROTECTION

See also: ALARM SYSTEMS, COMBUSTION, EXPLOSIVES, RADIOACTIVE AND TOXIC WASTES, SAFETY ENGINEERING

ABSTRACT SERVICES AND INDEXES

APPLIED SCIENCE AND TECHNOLOGY INDEX. H.W. Wilson and Company, 950 University Avenue, Bronx, NY 10452. (800) 367-6670 or (212) 588-8400. Monthly. Inquire as to cost and availability.

CHEMICAL ABSTRACTS. American Chemical Society, Chemical Abstracts Service, Box 3012, Columbus, OH 43210. (614) 421-3600. 1907 to present. Weekly. $9500.00 per year.

CURRENT CONTENTS: ENGINEERING, TECHNOLOGY AND APPLIED SCIENCES. Institute for Scientific Information, 3501 Market Street, Philadelphia, PA 19104. (800) 523-1850 or (215) 386-0100. Weekly. $275.00 per year.

ENGINEERING INDEX MONTHLY AND AUTHOR INDEX. Engineering Information Inc., 345 East 47th Street, New York, NY 10017. (212) 705-7600. Monthly. $1560.00 per year.

FIRE SCIENCE ABSTRACTS. Fire Research Station, Boreham Wood, Herts, WD6 2BL, England. 1947 to present. Quarterly. $36.00 per year.

SAFETY SCIENCE ABSTRACTS JOURNAL. Cambridge Scientific Abstracts, 5161 River Road, Bethesda, MD 20816. (301) 951-1400. 1973 to present. Quarterly. $365.00 per year.

ASSOCIATIONS AND PROFESSIONAL SOCIETIES

AMERICAN SOCIETY OF SAFETY ENGINEERS. 1800 East Oakton Street, Des Plaines, IL 60018-2187. (312) 692-4121.

NATIONAL FIRE PROTECTION ASSOCIATION. c/o Joseph Scaramozza, Batterymarch Park, Quiney, MA 02269. (617) 770-3000.

NATIONAL SAFETY COUNCIL. 444 North Michigan Avenue, Chicago, IL 60611. (312) 527-4800.

DIRECTORIES AND BIOGRAPHICAL SOURCES

1987 DIRECTORY OF ENGINEERING SOCIETIES AND RELATED ORGANIZATIONS. Gordon Davis, editor. Hemisphere Publishing Corporation, 1010 Vermont Avenue, NW, Washington, DC 20005. (800) 526-0275. 12th edition. 1987. $100.00.

RESEARCH CENTERS DIRECTORY 1988. Gale Research Company, Book Tower, Detroit, MI 48226. (800) 521-0707. 12th edition. 1987. $365.00 for set.

SCIENTIFIC AND TECHNICAL ORGANIZATIONS AND AGENCIES DIRECTORY. Margaret Labash Young, editor. Gale Research Company, Book Tower, Detroit, MI 48226. (800) 521-0707. 2nd edition. 1987. $185.00.

WHO'S WHO IN ENGINEERING. Gordon Davis, editor. Hemisphere Publishing Corporation, 79 Madison Avenue, New York, NY 10016-7892. (800) 821-8312. Sixth edition. 1985. $200.00.

GENERAL WORKS

FIRE INVESTIGATION. M.F. Dennett. Van Nostrand Reinhold Company, Inc., 135 West 50th Street, New York, NY 10020. (800) 543-2681. 1980. $12.95 in paper.

HANDBOOKS AND MANUALS

FIRE AND FLAMMABILITY HANDBOOK. Neil Schultz. Van Nostrand Reinhold Company, Inc., 135 West 50th Street, New York, NY 10020. (800) 543-2681. 1985. $52.95.

FIRE PROTECTION HANDBOOK. National Fire Protection Association, c/o Joseph Scaramozza, Batterymarch Park, Quiney, MA 02269. (617) 770-3000. 16th revised edition. 1986. $75.00.

ONLINE DATA BASES

CA SEARCH. Chemical Abstracts Service, P.O. Box 3012, Columbus, OH 43120. (800) 848-6538 or (614) 421-3600. Comprehensive guide to chemical literature, 1972 to present. Inquire as to online cost and availability.

COMPENDEX. Engineering Information, Inc., 345 East 47th Street, New York, NY 10017. (800) 221-1044 or (212) 705-7615. Engineering and technical literature, 1975 to present. Inquire as to online cost and availability.

DISSERTATION ABSTRACTS ONLINE. University Microfilms International, 300 North Zeeb Road, Ann Arbor, MI 48106. (800) 521-0600 or (313) 761-4700. Scope includes virtually all doctoral dissertations accepted at accredited American institutions from 1861 to present in over 250 subject areas. Inquire as to online cost and availability.

NFIRS. U.S. Fire Administration, Office of Fire Data and Analysis, Emmitsburg, MD 21727. (301) 447-6771. Data base of information on the characteristics, conditions, causes, and victims of fires. 1975 to present. Inquire as to availability and cost.

NTIS. National Technical Information Service, 5285 Port Royal Road, Springfield, VA 22161. (703) 487-4630. Broad coverage of government sponsored research reports, 1964 to present. Inquire as to online cost and availability.

SCISEARCH. Institute for Scientific Information, 3501 Market Street, Philadelphia, PA 19104. (800) 523-1850 or (215) 386-0100. Broad multidisciplinary title and author index to the international literature of science and technology, 1974 to present. Inquire as to online cost and availability.

WILSONLINE. H.W. Wilson and Company, 950 University Avenue, Bronx, NY 10452. (800) 367-6770 or (212) 588-8400. Makes available online versions of the H.W. Wilson indexes including Applied Science and Technology Index, Business Periodicals Index and Readers' Guide to Periodical Literature. Approximately 1980 to present. Inquire as to online cost and availability.

PERIODICALS

AMERICAN FIRE JOURNAL. Fire Publications, Inc., 9072 East Artesia Boulevard, Suite 7, Belliflower, CA 90706. (213) 866-1664. 1950 to present. Monthly. $15.00 per year.

ACCIDENT ANALYSIS AND PREVENTION. Pergamon Press, Inc., Maxwell House, Fairview Park, Elmsford, NY 10523. (914) 592-7700. 1969 to present. Bimonthly. $185.00.

CANADIAN FIREFIGHTER. The Canadian Firefighter Publishing Company, Limited, Box 37, Station M, Toronto, ON M6S 4T2, Canada. 1977 to present. Six times per year. $9.00 per year.

FIRE COMMAND. National Fire Protection Association, c/o Joseph Scaramozza, Batterymarch Park, Quiney, MA 02269. (617) 770-3000. 1933 to present. Monthly. $6.50 per year.

FIRE ENGINEERING: THE JOURNAL OF THE FIRE PROTECTION PROFESSION. Technical Publications Company, 875 Third Avenue, New York, NY 10022. (212) 605-9400. 1877 to present. Monthly. $16.95 per year.

FIRE SAFETY JOURNAL. Elsevier Science Publishing Company, Inc., 52 Vanderbilt Avenue, New York, NY 10017. (212) 370-5520. 1977 to present. Three times per year. $185.00 per year.

FIRE TECHNOLOGY: AN INTERNATIONAL JOURNAL OF FIRE PROTECTION RESEARCH AND ENGINEERING. National Fire Protection Association, c/o Joseph Scaramozza, Batterymarch Park, Quiney, MA 02269. (617) 770-3000. 1965 to present. Quarterly. $19.50 per year.

HAZARD PREVENTION. System Safety Society, Inc. Gallant Charger Publishing Company, 14252 Culva Drive, Suite A-261, Irvine, CA 92714. (714) 474-0330. 1965 to present. Bimonthly. $30.00 per year.

INDUSTRIAL SAFETY. (Industrial Safety Product News). Ames Publishing Company, 201 King of Prussia Road, Radnor, PA 19087. (215) 964-4000. 1967 to present. Monthly. $24.00 per year.

JOURNAL OF SAFETY RESEARCH. Pergamon Press, Inc., Maxwell House, Fairview Park, Elmsford, NY 10523. (914) 592-7700. 1982 to present. Quarterly. $75.00 per year.

NATIONAL SAFETY AND HEALTH NEWS. National Safety Council, 444 North Michigan Avenue, Chicago, IL 60611. (312) 527-4800. 1919 to present. Monthly. $26.00 per year.

PROFESSIONAL SAFETY. American Society of Safety Engineers, 1800 East Oakton Street, Des Plaines, IL 60018-2187. (312) 692-4121. 1956 to present. Monthly. $30.00 per year.

SOCIETY UPDATE. AMERICAN SOCIETY OF SAFETY ENGINEERS. 1800 East Oakton Street, Des Plaines, IL 60018-2187. (312) 692-4121. 1911 to present. Quarterly. Membership only, inquire.

RESEARCH CENTERS AND INSTITUTES

BUILDING RESEARCH LABORATORY. Ohio State University, 2001 Research Center, 1314 Kinnear Road, Columbus, OH 43212. (614) 292-6227.

FACTORY MUTUAL RESEARCH CORPORATION. 1151 Boston-Providence Turnpike, Norwood, MA 02062. (617) 762-4300.

INSTITUTE FOR RESEARCH IN CONSTRUCTION. National Research Council of Canada, Building M20, Montreal Road, Ottawa, ON, Canada K1A OR6. (613) 993-2607.

FISSION

See also: ENERGY, FUSION, NUCLEAR ENERGY, NUCLEAR ENGINEERING, NUCLEAR PHYSICS, NUCLEAR REACTORS

ABSTRACT SERVICES AND INDEXES

APPLIED MECHANICS REVIEW. American Society of Mechanical Engineers, 345 East 47th Street, New York, NY 10017. (212) 705-7703. 1948 to present. Monthly. $360.00 per year.

APPLIED SCIENCE AND TECHNOLOGY INDEX. H.W. Wilson and Company, 950 University Avenue, Bronx, NY 10452. (800) 367-6670 or (212) 588-8400. Monthly. Inquire as to cost and availability.

CHEMICAL ABSTRACTS. American Chemical Society, Chemical Abstracts Service, Box 3012, Columbus, OH 43210. (614) 421-3600. 1907 to present. Weekly. $9500.00 per year.

CURRENT CONTENTS: ENGINEERING, TECHNOLOGY AND APPLIED SCIENCES. Institute for Scientific Information, 3501 Market Street, Philadelphia, PA 19104. (800) 523-1850 or (215) 386-0100. Weekly. $275.00 per year.

ENGINEERING INDEX MONTHLY AND AUTHOR INDEX. Engineering Information Inc., 345 East 47th Street, New York, NY 10017. (212) 705-7600. Monthly. $1560.00 per year.

ISMEC BULLETIN (Information Service in Mechanical Engineering). Cambridge Scientific Abstracts, 5161 River Road, Bethesda, MD 20816. (301) 951-1400. 1973 to present. Monthly. $450.00 per year.

INDEX TO SCIENTIFIC AND TECHNICAL PROCEEDINGS. Institute for Scientific Information, 3501 Market Street, Philadelphia, PA 19104. (800) 523-1850 or (215) 386-0100. 1978 to present. Monthly. $775.00 per year.

PHYSICS ABSTRACTS. Institution of Electrical Engineers. Available from: IEEE Service Center, 445 Hoes Lane, Piscataway, NJ 08854. 1898 to present. Bimonthly. $1700.00 per year.

SCIENCE CITATION INDEX. Institute for Scientific Information, 3501 Market Street, Philadelphia, PA 19104. (800) 523-1850 or (215) 386-0100. Six times per year. $6200.00 per year.

ANNUAL REVIEWS AND YEARBOOKS

ADVANCES IN NUCLEAR SCIENCE AND TECHNOLOGY. J.Lewins, editor. Plenum Publishing Corporation, 233 Spring Street, New York, NY 10013. (800) 221-9369. 1977 to presnet. Irregular. Inquire as to cost and availability.

ASSOCIATIONS AND PROFESSIONAL SOCIETIES

AMERICAN NUCLEAR SOCIETY. 555 North Kensington Avenue, La Grange, IL 60525. (312) 352-6611.

FUSION ENERGY FOUNDATION. P.O. Box 17149, Washington, DC 20041. (703) 689-2490.

INSTITUTE OF NUCLEAR MATERIALS MANAGEMENT. 60 Revere Drive, Northbrook, IL 60062. (312) 480-9080.

DIRECTORIES AND BIOGRAPHICAL SOURCES

ENERGY INFORMATION CENTERS DIRECTORY. Public Affairs and Information Program, Atomic Industrial Forum, 7101 Wisconsin Avenue, Bethesda, MD 20814. (301) 654-9260. 1985. Free.

INTERNATIONAL DIRECTORY OF NUCLEAR UTILITIES. Lotte, Limited, Box 237, Contract Station 27, Lakewood, CO 80215. (303) 232-3026. Annual. $160.00.

INTERNATIONAL RESEARCH CENTERS DIRECTORY 1988-89. Darren L. Smith, editor. Gale Research Company, Book Tower, Detroit, MI 48226. (800) 521-0707. 4th edition. 1987. $360.00.

1987 DIRECTORY OF ENGINEERING SOCIETIES AND RELATED ORGANIZATIONS. Gordon Davis, editor. Hemisphere Publishing Corporation, 1010 Vermont Avenue, NW, Washington, DC 20005. (800) 526-0275. 12th edition. 1987. $100.00.

NUCLEAR REACTORS BUILT, BEING BUILT, OR PLANNED IN THE UNITED STATES. Office of Scientific and Technical Information, Department of Energy, Box 62, Oak Ridge, TN 37831. (615) 576-5637. Annual. $11.00. Send orders to: National Technical Information Service, Springfield, VA 22161.

RESEARCH CENTERS DIRECTORY 1988. Gale Research Company, Book Tower, Detroit, MI 48226. (800) 521-0707. 12th edition. 1987. $365.00 for set.

SCIENTIFIC AND TECHNICAL ORGANIZATIONS AND AGENCIES DIRECTORY. Margaret Labash Young, editor. Gale Research Company, Book Tower, Detroit, MI 48226. (800) 521-0707. 2nd edition. 1987. $185.00.

WHO'S WHO IN ENGINEERING. Gordon Davis, editor. Hemisphere Publishing Corporation, 1010 Vermont Avenue, NW, Washington, DC 20005. (800) 526-0275. 6th edition. 1985. $200.00.

ENCYCLOPEDIAS AND DICTIONARIES

THESAURUS OF SCIENTIFIC, TECHNICAL, AND ENGINEERING TERMS. Hemisphere Publishing Corporation, 1010 Vermont Avenue, NW, Washington, DC 20005. (800) 526-0275. 1988. $125.00.

GENERAL WORKS

INTRODUCTION TO NUCLEAR POWER. John G. Collier. Hemisphere Publishing Corporation, 1010 Vermont Avenue, NW, Washington, DC 20005. (800) 526-0275. 1987. $49.95.

NUCLEAR ENERGY TECHNOLOGY. Ronald Allen Knief. Hemisphere Publishing Corporation, 1010 Vermont Avenue, NW, Washington, DC 20005. (800) 526-0275. 1983. $48.00.

NUCLEAR FISSION REACTORS. I.R. Cameron. Plenum Publishing Corporation, 233 Spring Street, New York, NY 10013. (800) 221-9369. 1982. $49.50.

NUCLEAR PHYSICS FOR ENGINEERS AND SCIENTISTS: LOW ENERGY THEORY WITH APPLICATIONS INCLUDING REACTORS AND THEIR ENVIRONMENTAL IMPACT. S.E. Hunt. John Wiley and Sons, Inc., 605 Third Avenue, New York, NY 10158. (800) 526-5368. 1987. $129.95.

NUCLEAR REACTOR ENGINEERING. Samuel Glasstone. Van Nostrand Reinhold Company, Inc., 135 West 50th Street, New York, NY 10020. (800) 543-2681. 1980. $49.95.

NUCLEAR MATERIALS AND APPLICATIONS. Benjamin M. Ma. Van Nostrand Reinhold Company, Inc., 135 West 50th Street, New York, NY 10020. (800) 543-2681. 1982. $45.95.

HANDBOOKS AND MANUALS

A GUIDE TO NUCLEAR POWER TECHNOLOGY: A RESOURCE FOR DECISION MAKING. F.J. Rahn and others. John Wiley and Sons, Inc., 605 Third Avenue, New York, NY 10158. (800) 526-5368. 1984. $85.95.

NUCLEAR ENGINEERING DATA BASES, STANDARDS AND NUMERICAL ANALYSIS. Jack Jedruch. Van Nostrand Reinhold Company, Inc., 135 West 50th Street, New York, NY 10020. (800) 543-2681. 1985. $59.95.

ONLINE DATA BASES

CA SEARCH. Chemical Abstracts Service, P.O. Box 3012, Columbus, OH 43120. (800) 848-6538 or (614) 421-3600. Comprehensive guide to chemical literature, 1972 to present. Inquire as to online cost and availability.

COMPENDEX. Engineering Information, Inc., 345 East 47th Street, New York, NY 10017. (800) 221-1044 or (212) 705-7615. Engineering and technical literature, 1975 to present. Inquire as to online cost and availability.

DISSERTATION ABSTRACTS ONLINE. University Microfilms International, 300 North Zeeb Road, Ann Arbor, MI 48106. (800) 521-0600 or (313) 761-4700. Scope includes virtually all doctoral dissertations accepted at accredited American institutions from 1861 to present in over 250 subject areas. Inquire as to online cost and availability.

DOE ENERGY DATA BASE. U.S. Department of Energy, Office of Scientific and Technical Information, P.O. Box 62, Oak Ridge, TN 37831. (615) 576-6837. A database that covers all aspects of energy including the science and technology of energy. 1948 to present. Available through the DIALOG search service or DOE/RECON. Inquire as to online cost and availability.

ENERGYLINE. EIC/Intelligence, Inc., 48 West 38th Street, New York, NY 10018. (212) 944-8500. A database of resources on the scientific, engineering, political, and socioeconomic aspects of energy resources. 1976 to present. Inquire as to online cost and availability.

INSPEC. INSPEC Marketing Department, Institution of Electrical Engineers. Available from IEEE Service Center, 445 Hoes Lane, Piscataway, NJ 08854. (201) 981-0060. Online version of Physics Abstracts. Inquire as to online cost and availability.

NTIS. National Technical Information Service, 5285 Port Royal Road, Springfield, VA 22161. (703) 487-4630. Broad coverage of government sponsored research reports, 1964 to present. Inquire as to online cost and availability.

SCISEARCH. Institute for Scientific Information, 3501 Market Street, Philadelphia, PA 19104. (800) 523-1850 or (215) 386-0100. Broad multidisciplinary title and author index to the international literature of science and technology, 1974 to present. Inquire as to online cost and availability.

WILSONLINE. H.W. Wilson and Company, 950 University Avenue, Bronx, NY 10452. (800) 367-6770 or (212) 588-8400. Makes available online versions of the H.W. Wilson indexes including Applied Science and Technology Index, Business Periodicals Index and Readers' Guide to Periodical Literature. Approximately 1980 to present. Inquire as to online cost and availability.

PERIODICALS

AMERICAN NUCLEAR SOCIETY TRANSACTIONS. American Nuclear Society, 555 North Kensington Avenue, La Grange, IL 60525. (312) 352-6611. 1958 to present. Semiannual. $255.00 per year.

ANNALS OF NUCLEAR ENERGY. Pergamon Press, Inc., Maxwell House, Fairview Park, Elmsford, NY 10523. (914) 592-7700. 1954 to present. Monthly. $280.00 per year.

BULLETIN OF THE ATOMIC SCIENTISTS. Educational Foundation for Nuclear Science, 5801 South Kenwood Avenue, Chicago, IL 60637. (312) 363-5225. 1945 to present. Ten times per year. $22.50 per year.

FUSION TECHNOLOGY. American Nuclear Society, 555 North Kensington Avenue, La Grange, IL 60525. (312) 352-6611. 1981 to present. Bimonthly. $250.00 per year.

JOURNAL OF FUSION ENERGY. Plenum Publishing Corporation, 233 Spring Street, New York, NY 10013. (800) 221-9369. 1981 to present. Bimonthly. $105.00 per year.

NUCLEAR ENGINEER. Institution of Nuclear Engineers, 1 Penerley Road, Nondon SE6 2LQ, England. 1959 to present. Bimonthly. $110.00 per year.

NUCLEAR ENGINEERING AND DESIGN. Elsevier Science Publishing Company, Inc., 52 Vanderbilt Avenue, New York, NY 10017. (212) 370-5520. 1965 to present. $160.00 per year.

NUCLEAR ENGINEERING INTERNATIONAL. Electrical-Electronic Press, Quadrant House, The Quadrant, Sutton, Surrey, SM2 5AS, England. 1956 to present. Monthly. $210.00 per year.

NUCLEAR SCIENCE AND ENGINEERING. American Nuclear Society, 555 North Kensington Avenue, La Grange, IL 60525. (312) 352-6611. 1956 to present. Monthly. $220.00 per year.

NUCLEAR TECHNOLOGY. American Nuclear Society, 555 North Kensington Avenue, La Grange, IL 60525. (312) 352-6611. 1965 to present. Monthly. $345.00 per year.

RESEARCH CENTERS AND INSTITUTES

ASSISTANT SECRETARY FOR NUCLEAR ENERGY. U.S. Department of Energy, 1000 Independence Avenue, SW, Washington, DC 20585. (202) 252-6450.

DEPARTMENT OF NUCLEAR ENGINEERING AND ENGINEERING PHYSICS. University of Wisconsin - Madison, 1500 Johnson Drive, Madison, WI 53706. (608) 263-1648.

INSTITUTE OF NUCLEAR SCIENCE AND ENGINEERING. Oregon State University, Radiation Center, 35th and Jefferson Streets, Corvallis, OR 97331. (503) 754-2341.

LABORATORY OF BASIC AND APPLIED NUCLEAR RESEARCH. University of Cincinnati, Department of Chemistry, Cincinnati, OH 45221. (513) 475-3652.

OFFICE OF FUSION ENERGY. U.S. Department of Energy, Washington, DC 20545.

WHITESHELL NUCLEAR RESEARCH ESTABLISHMENT. Research Company, Atomic Energy of Canada Limited, Pinawa, MB, Canada ROE 1LO. (204) 753-2311.

SPECIFICATIONS AND STANDARDS

INFORMATION CENTER ON NUCLEAR STANDARDS. American Nuclear Society, 555 North Kensington Avenue, La Grange, IL 60525. (312) 352-6611. Standards for all aspects of the design, construction, operation, and maintenance of nuclear power plants.

FISSION REACTORS

See: NUCLEAR REACTORS

FLAME EMISSION SPECTROSCOPY

See: SPECTROSCOPY

FLAME LASERS

See: LASERS

FLASH WELDING

See: WELDING

FLIGHT

See: AERONAUTICS

FLIGHT DYNAMICS

See: AERODYNAMICS

FLOOD CONTROL

See also: CIVIL ENGINEERING, CHANNELS, DAMS, GEOTECHNICAL ENGINEERING, HYDRAULIC ENGINEERING, HYDROLOGY

ABSTRACT SERVICES AND INDEXES

APPLIED SCIENCE AND TECHNOLOGY INDEX. H.W. Wilson and Company, 950 University Avenue, Bronx, NY 10452. (800) 367-6670 or (212) 588-8400. Monthly. Inquire as to cost and availability.

CIVIL ENGINEERING HYDRAULICS ABSTRACTS. BHRA Fluid Engineering, Cranfield, Bedford, MK43 OAJ, England. Distributed by Learned Information Inc., 143 Old Marlton Pike, Medford, NJ 08055. 1968 to present. Monthly. $225.00 per year.

ENGINEERING INDEX MONTHLY AND AUTHOR INDEX. Engineering Information Inc., 345 East 47th Street, New York, NY 10017. (212) 705-7600. Monthly. $1560.00 per year.

INTERNATIONAL CIVIL ENGINEERING ABSTRACTS. CITIS Limited, 2 Rosemount Terrace, Blackrock, Dublin, Ireland. 1974 to present. Monthly. $350.00 per year.

INTERNATIONAL STRUCTURAL ENGINEERING ABSTRACTS. CITIS Limited, 2 Rosemount Terrace, Blackrock, Dublin, Ireland. 1986 to present. Quarterly. $95.00 per year.

PUBLICATIONS INFORMATION. American Society of Civil Engineers, 345 East 47th Street, New York, NY 10017. (212) 705-7420. Abstracts, subject and author indexes to the publications of the American Society of Civil Engineers. Bimonthly. $80.00 per year.

SELECTED WATER RESOURCES ABSTRACTS. U.S. Geological Survey, Water Resources Scientific Information Center. Available from: National Technical Information Service, 5285 Port Royal Road, Springfield, VA 22161. (703) 487-4929. 1968 to present. Monthly. $115.00 per year.

ASSOCIATIONS AND PROFESSIONAL SOCIETIES

AMERICAN INSTITUTE OF HYDROLOGY. P.O. Box 14251, St. Paul, MN 55114. (612) 379-1030.

AMERICAN SOCIETY OF CIVIL ENGINEERS. 345 East 47th Street, New York, NY 10017. (212) 705-7420.

BIBLIOGRAPHIES

FLOOD DAMAGE PREVENTION: MONOGRAPHS. Mary Vance. Vance Bibliographies, P.O. Box 229, 112 North Charter Street, Monticello, IL 61856. (217) 762-3831. 1985. $3.00.

DIRECTORIES AND BIOGRAPHICAL SOURCES

1987 DIRECTORY OF ENGINEERING SOCIETIES AND RELATED ORGANIZATIONS. Gordon Davis, editor. Hemisphere Publishing Corporation, 1010 Vermont Avenue, NW, Washington, DC 20005. (800) 526-0275. 12th edition. 1987. $100.00.

RESEARCH CENTERS DIRECTORY 1988. Gale Research Company, Book Tower, Detroit, MI 48226. (800) 521-0707. 12th edition. 1987. $365.00 for set.

SCIENTIFIC AND TECHNICAL ORGANIZATIONS AND AGENCIES DIRECTORY. Margaret Labash Young, editor. Gale Research Company, Book Tower, Detroit, MI 48226. (800) 521-0707. 2nd edition. 1987. $185.00.

WHO'S WHO IN ENGINEERING. Gordon Davis, editor. Hemisphere Publishing Corporation, 1010 Vermont Avenue, NW, Washington, DC 20005. (800) 526-0275. 6th edition. 1985. $200.00.

ENCYCLOPEDIAS AND DICTIONARIES

DICTIONARY OF CIVIL ENGINEERING. John S. Scott. John Wiley and Sons, Inc., 605 Third Avenue, New York, NY 10158. (800) 526-5368. Third edition. 1981. $26.95.

GENERAL WORKS

APPLIED HYDROGEOLOGY. Charles W. Fetter. Charles E. Merrill Publishing Company, 1300 Alum Creek Drive, Columbus, OH 43216. (614) 890-1111. 1980. $38.95.

CIVIL ENGINEERING HYDRAULICS. J.R. Francis and P. Minton. Edward Arnold Publishers, Limited, 300 North Charles Street, Baltimore, MD 21201. (301) 539-1529. Fifth edition. 1984. $24.95.

DAMS AND EARTHQUAKES. Institution of Civil Engineers Staff, editors. American Society of Civil Engineers, 345 East 47th Street, New York, NY 10017. (212) 705-7420. 1981. $62.50.

EARTH AND EARTH-ROCK DAMS: ENGINEERING PROBLEMS OF DESIGN AND CONSTRUCTION. James L. Sherard and others. John Wiley and Sons, Inc., 605 Third Avenue, New York, NY 10158. (800) 526-5368. 1963. $76.95.

EARTHQUAKE ENGINEERING FOR LARGE DAMS. Radu Priscu and others. John Wiley and Sons, Inc., 605 Third Avenue, New York, NY 10158. (800) 526-5368. 1985. $44.95.

EVALUATION OF DAM SAFETY. American Society of Civil Engineers, 345 East 47th Street, New York, NY 10017. (212) 705-7420. 1977. $16.00 in paper.

FLOODS AND RESERVOIR SAFETY: AN ENGINEERING GUIDE. Institution of Civil Engineers Staff, editors. American Society of Civil Engineers, 345 East 47th Street, New York, NY 10017. (212) 705-7420. 1978. $10.00.

FLOODS DUE TO HIGH WINDS AND TIDES. D.H. Peregrine, editor. Academic Press, Inc., 6277 Sea Harbor Drive, Orlando, FL 32821. (800) 321-5068. 1981. $28.50.

HYDROLOGY IN PRACTICE. Elizabeth M. Shaw. Van Nostrand Reinhold Company, Inc., 135 West 50th Street, New York, NY 10020. (800) 543-2681. 1983. $45.95.

INTRODUCTION TO HYDROMETEOROLOGY. J.P. Bruce and R.H Clark. Pergamon Press, Inc., Maxwell House, Fairview Park, Elmsford, NY 10523. (914) 592-7700. 1987. $50.00.

HANDBOOKS AND MANUALS

CIVIL ENGINEERING PRACTICE. Paul N. Cheremisinoff and others, editors. Technomic Publishing Company, Inc., 851 Holland Avenue, Box 3535, Lancaster, PA 17604. (800) 233-9936. Five volumes. 1987-1988. $750.00 for set.

HANDBOOK OF DAM ENGINEERING. Alfred R. Golze, editor. Van Nostrand Reinhold Company, Inc., 135 West 50th Street, New York, NY 10020. (800) 543-2681. 1977. $69.95.

STANDARD HANDBOOK FOR CIVIL ENGINEERS. F. S. Merritt, editor. McGraw-Hill Book Company, 1221 Avenue of the Americas, New York, NY 10020. (212) 512-2000. Third edition. 1983. $89.50.

ONLINE DATA BASES

COMPENDEX. Engineering Information, Inc., 345 East 47th Street, New York, NY 10017. (800) 221-1044 or (212) 705-7615. Engineering and technical literature, 1975 to present. Inquire as to online cost and availability.

GEOREF. Online version of the BIBLIOGRAPHY AND INDEX OF GEOLOGY. American Geological Institute, 4220 King Street, Alexandria, VA 22302. (703) 379-2480. 1969 to present. Inquire as to online cost and availability.

NTIS. National Technical Information Service, 5285 Port Royal Road, Springfield, VA 22161. (703) 487-4630. Broad coverage of government sponsored research reports, 1964 to present. Inquire as to online cost and availability.

WATER DATA BANK. U.S. Department of Agriculture, Agricultural Research Service. Hydrology Laboratory. Water Data Center, Room 139, Building 007, BARC-West, Beltsville, MD 20705. (301) 344-4411. Inquire as to online cost and availability.

WILSONLINE. H.W. Wilson and Company, 950 University Avenue, Bronx, NY 10452. (800) 367-6770 or (212) 588-8400. Makes available online versions of the H.W. Wilson indexes including Applied Science and Technology Index, Business Periodicals Index and Readers' Guide to Periodical Literature. Approximately 1980 to present. Inquire as to online cost and availability.

PERIODICALS

CIVIL ENGINEERING. American Society of Civil Engineers, 345 East 47th Street, New York, NY 10017. (212) 705-7420. 1930 to present. Monthly. $48.00 per year.

JOURNAL OF CONSTRUCTION ENGINEERING AND MANAGEMENT. American Society of Civil Engineers, 345 East 47th Street, New York, NY 10017. (212) 705-7420. 1956 to present. Quarterly. $48.00 per year.

JOURNAL OF GEOTECHNICAL ENGINEERING. American Society of Civil Engineers, 345 East 47th Street, New York, NY 10017. (212) 705-7420. 1956 to present. Monthly. $96.00 per year.

JOURNAL OF HYDROLOGY. Elsevier Science Publishing Company, Inc., 52 Vanderbilt Avenue, New York, NY 10017. (212) 370-5520. Thirty-two times per year. $675.00 per year.

JOURNAL OF HYDRAULIC ENGINEERING. American Society of Civil Engineers, 345 East 47th Street, New York, NY 10017. (212) 705-7420. 1956 to present. Monthly. $112.00 per year.

JOURNAL OF IRRIGATION AND DRAINAGE. American Society of Civil Engineers, 345 East 47th Street, New York, NY 10017. (212) 705-7420. 1956 to present. Quarterly. $45.00 per year.

JOURNAL OF SURVEYING ENGINEERING. American Society of Civil Engineers, 345 East 47th Street, New York, NY 10017. (212) 705-7420. 1956 to present. Three times per year. $35.00 per year.

JOURNAL OF WATER RESOURCES PLANNING AND MANAGEMENT. American Society of Civil Engineers, 345 East 47th Street, New York, NY 10017. (212) 705-7420. 1956 to present. Quarterly. $56.00 per year.

RESEARCH CENTERS AND INSTITUTES

W.M. KECK ENGINEERING LABORATORY OF HYDRAULICS AND WATER RESOURCES. California Institute of Technology, 1201 East California Boulevard, Pasadena, CA 91125. (818) 356-4404.

U.S. ARMY HYDROLOGIC ENGINEERING CENTER. 609 Center Street, Davis, CA 95616. (916) 440-3285.

FLOPPY DISKS

See: COMPUTER MEMORY AND STORAGE

FLOW

See: FLUID MECHANICS

FLOW WELDING

See: WELDING

FLUID DYNAMICS

See also: AERODYNAMICS, FLUID MECHANICS, FLUIDICS, HYDRAULIC ENGINEERING, HYDRAULICS, HYDRODYNAMICS

ABSTRACT SERVICES AND INDEXES

APPLIED MECHANICS REVIEWS. American Society of Mechanical Engineers, 345 East 47th Street, New York, NY 10017. (212) 705-7703. 1948 to present. Monthly. $360.00 per year.

APPLIED SCIENCE AND TECHNOLOGY INDEX. H.W. Wilson Company, 950 University Avenue, Bronx, NY 10452. (800) 367-6670 or (212) 588-8400. Inquire as to cost and availability.

CHEMICAL ABSTRACTS. Chemical Abstracts Service, 2540 Olentangy Road, Post Office Box 3012, Columbus, OH 43210. (800) 848-6538 or (614) 421-3600. Weekly. $9200.00 per year.

CIVIL ENGINEERING HYDRAULICS ABSTRACTS. BHRA Fluid Engineering, Cranfield, Bedford MK43 0AJ, England. Distributed by Learned Information Incorporated, 143 Old Marlton Pike, Medford, NJ 08055. 1968 to present. Monthly. $225.00 per year.

CURRENT CONTENTS: ENGINEERING, TECHNOLOGY AND APPLIED SCIENCES. Institute for Scientific Information, 3501 Market Street, Philadelphia, PA 19104. (800) 523-1850 or (215) 386-0100. 1970 to present. Weekly. $272.00 per year.

ENGINEERING INDEX MONTHLY AND AUTHOR INDEX. Engineering Information, Incorporated, 345 East 47th Street, New York, NY 10017. (800) 221-1044 or (212) 705-7600. Monthly, with annual cumulation. $1560.00 per year.

FLUID POWER ABSTRACTS. BHRA Fluid Engineering, Cranfield, Bedford MK43 0AJ, England. Distributed by Learned Information Incorporated, 143 Old Marlton Pike, Medford, NJ 08055. 1970 to present. Bimonthly. $196.00 per year.

INDUSTRIAL AERODYNAMICS ABSTRACTS. BHRA Fluid Engineering, Cranfield, Bedford MK43 0AJ, England. Disbtributed by Learned Information Incorporated, 143 Old Marlton Pike, Medford, NJ 08055. 1970 to present. Monthly. $196.00 per year.

INTERNATIONAL AEROSPACE ABSTRACTS. AIAA/TIS, 1633 Broadway, New York, NY 10019. (212) 581-4300. Semimonthly. $700.00 per year.

ISMEC BULLETIN. (Information Service in Mechanical Engineering). Cambridge Scientific Abstracts, 5161 River Road,

Bethesda, MD 20816. (301) 951-1400. 1973 to present. Monthly. $475.00 per year.

SCIENCE CITATION INDEX. Institute for Scientific Information, 3501 Market Street, Philadelphia, PA 19104. (800) 523-1850 or (215) 386-0100.

STAR. (Scientific and Technical Aerospace Reports. United States National Aeronautics and Space Administration, Scientific and Technical Information Facility, Box 8757, Baltimore-Washington International Airport, MD 21240. (202) 755-2210. Semimonthly, with semiannual and annual indexes. $85.00 per year.

ASSOCIATIONS AND PROFESSIONAL SOCIETIES

AMERICAN INSTITUTE OF AERONAUTICS AND ASTRONAUTICS. 1633 Broadway, New York, NY 10019. (212) 581-4300.

AMERICAN SOCIETY OF CIVIL ENGINEERS. 345 East 47th Street, New York, NY 10017-2398. (212) 705-7520.

AMERICAN SOCIETY OF MECHANICAL ENGINEERS. 345 East 47th Street, New York, NY 10017-2398. (212) 705-7722.

FLUID CONTROLS INSTITUTE. Post Office Box 9036, Morristown, NJ 07960. (201) 829-0990.

FLUID POWER SOCIETY. 3333 North Mayfair Road, Milwaukee, WI 53222. (414) 778-3377.

HYDRAULIC INSTITUTE. 712 Lakewood Center North, 14600 Detroit Avenue, Cleveland, OH 44107. (216) 226-7700.

INTERNATIONAL ASSOCIATION FOR HYDRAULIC RESEARCH. 185 Rotterdamseweg, Box 177, Delft, The Netherlands.

NATIONAL CONFERENCE ON FLUID POWER. IIT Research Institute, Ten West 35th Street, Chicago, IL 60616. (312) 567-4414.

NATIONAL FLUID POWER ASSOCIATION. 3333 North Mayfair Road, Suite 311, Milwaukee, WI 53222. (414) 778-3344.

SOCIETY OF AUTOMOTIVE ENGINEERS. SAE, Incorporated, 400 Commonwealth Drive, Warrendale, PA 15096. (412) 776-4841.

DIRECTORIES AND BIOGRAPHICAL SOURCES

AAS DIRECTORY. American Astronautical Society, 6212-B Old Keene Mill Court, Springfield, VA 22152. (703) 866-0020. Annual. $35.00 per year.

DIRECTORY OF HYDRAULIC RESEARCH INSTITUTES AND LABORATORIES. International Association for Hydraulic Research, 185 Rotterdamseweg, Box 177, Delft, The Netherlands. 1986. 60 Dutch florins.

FLUID POWER HANDBOOK AND DIRECTORY. Penton/IPC, Incorporated, 1111 Chester Avenue, Cleveland, OH 44114. (216) 606-7000. Biennial. $45.00.

FLUID POWER SOCIETY MEMBERSHIP DIRECTORY. Fluid Power Society. 3333 North Mayfair Road, Milwaukee, WI 53222. (414) 778-3377. Annual. $50.00.

NATIONAL FLUID POWER ASSOCIATION MEMBERSHIP DIRECTORY. 3333 North Mayfair Road, Suite 311, Milwaukee, WI 53222. (414) 778-3344. Annual. $150.00.

RESEARCH CENTERS DIRECTORY. Gale Research Company, Book Tower, Detroit, MI 48226. (800) 521-0707. Twelfth edition. 1987. $365.00 for set.

WHO'S WHO IN ENGINEERING. Engineers Joint Council, 345 East 47th Street, New York, NY 10017. (212) 705-7010. 1985. $200.00.

ENCYCLOPEDIAS AND DICTIONARIES

ENCYCLOPEDIA OF FLUID MECHANICS. Nicholas P. Cheremisinnoff, editor. Gulf Publishing Company, Book Division, Post Office Box 2608, Houston, TX 77001. (713) 520-4444. Six Volumes. 1987. $1000.00 for set.

ENCYCLOPEDIA OF PHYSICAL SCIENCE AND TECHNOLOGY. Academic Press, Incorporated, Orlando, FL 32887. (800) 321-5068 or (305) 345-2734. Inquire as to cost and availability.

MCGRAW-HILL DICTIONARY OF ENGINEERING. Sybil P. Parker, editor. McGraw-Hill Book Company, 1221 Avenue of the Americas, New York, NY 10020. (212) 512-2000. 1984. $39.95.

MCGRAW-HILL ENCYCLOPEDIA OF SCIENCE AND TECHNOLOGY. McGraw-Hill Book, Incorporated, 1221 Avenue of the Americas, New York, NY 10020. (212) 997-3675. Fifth edition, 15 volumes. $1100.00.

GENERAL WORKS

ENGINEERING FLUID MECHANICS. John J. Bertin. Prentice-Hall, Incorporated, Englewood Cliffs, NJ 07632. (201) 592-2000. Second edition. 1987. $34.95.

A FIRST COURSE IN FLUID DYNAMICS. A.R. Peterson. Cambridge University Press, 32 East 57th Street, New York, NY 10022. (212) 688-8885. 1984. $65.50.

FLOW VISUALISATION. Japan Society of Mechanical Engineers. Pergamon Press, Incorporated, Maxwell House, Fairview Park, Elmsford, NY 10523. (914) 592-7700. 1986. $65.00.

FOUNDATIONS OF AERODYNAMICS: BASES OF AERODYNAMIC DESIGN. A.M. Kuethe and C. Chow. John Wiley and Sons, Incorporated, 605 Third Avenue, New York, NY 10158. (800) 526-5368 or (212) 850-6000. Fourth edition. 1986. $43.95.

FUNDAMENTALS OF FLUID MECHANICS AND HYDRAULICS. J. Evett and C. Liu. McGraw-Hill Book Company, 1221 Avenue of the Americas, New York, NY 10020. (212) 512-2000. 1987. $49.95.

FUNDAMENTALS OF HYDRAULIC ENGINEERING SYSTEMS. Ned H. Hwang and Carlos E. Hita. Prentice-Hall, Incorporated, Englewood Cliffs, NJ 07632. (800) 562-0245. 1987. $41.95.

ILLUSTRATED GUIDE TO AERODYNAMICS. Hubert Smith. Tab Books, Incorporated, Monterey Lane, Blue Ridge Summit, PA 17214. (717) 794-2191. 1986. $14.95.

AN INFORMAL INTRODUCTION TO THEORETICAL FLUID MECHANICS. James Lighthill. Oxford University Press, 200 Madison Avenue, New York, NY 10016. (212) 679-7300. 1986. $35.00.

INTRODUCTION TO FLUID FLOW AND THE TRANSFER OF HEAT AND MASS. A.T. Olson and K.A. Shelstad. Prentice-Hall, Incorporated, Englewood Cliffs, NJ 07632. (201) 592-2000. 1987. $35.95.

INTRODUCTION TO FLUID MECHANICS. R.W. Fox and A.T. McDonald. John Wiley and Sons, Incorporated, 605 Third Avenue, New York, NY 10158. (800) 526-5368 or (212) 850-6000. Third edition. 1985. $45.00.

HANDBOOKS AND MANUALS

APPLIED FLUID DYNAMICS HANDBOOK. Robert D. Blevins. McGraw-Hill Book Company, Incorporated, 1221 Avenue of the Americas, New York, NY 10020. (212) 512-2000. 1984. $52.95.

HANDBOOK OF FLUIDS IN MOTION. Nicholas P. Cheremisinoff and Ramesh Gupta, editors. Butterworth's, 80 Montvale Avenue, Stoneham, MA 02180. (617) 438-8464. 1983. $84.95.

HANDBOOK OF HYDRAULIC RESISTANCE. I.E. Idelchik and others. Hemisphere Publishing Corporation, 79 Madison Avenue, New York, NY 10016. (800) 526-0275. Second edition. 1986. $89.95.

STANDARD HANDBOOK OF ENGINEERING CALCULATIONS. Tyler G. Hicks, editor. McGraw-Hill Book Company, Incorporated, 1221 Avenue of the Americas, New York, NY 10020. (212) 512-2000. Second edition. 1984. $59.50.

THE WILEY ENGINEER'S DESK REFERENCE. Sanford I. Heisler. John Wilet and Sons, Incorporated, 605 Third Avenue, New York, NY 10158. (800) 526-5368 or (212) 850-6418. 1984. $36.00.

ONLINE DATA BASES

COMPENDEX. Engineering Information, Incorporated, 345 East 47th Street, New York, NY 10017. (800) 221-1044 or (212) 705-7615. Engineering and technical literature, 1975 to present. Inquire as to cost and availability.

ISMEC (INFORMATION SERVICE IN MECHANICAL ENGINEERING). Cambridge Scientific Abstracts, 5161 River Road, Bethesda, MD 20816. (301) 951-1400. 1973 to present. Inquire as to online cost and availability.

NASA. National Aeronautics and Space Administration, Scientific and Technical Information Branch, 300 7th Street, SW, Washington, DC 20546. Citations and abstracts of aerospace literature, 1962 to present. Inquire as to cost and availability.

NTIS. National Technical Information Service, 5285 Port Royal Road, Springfield, VA 22161. (703) 487-4630. Broad coverage of government-sponsored research reports, 1964 to present. Inquire as to cost and availability.

PERIODICALS

AERONAUTICAL JOURNAL. Aeronautical Society, 4 Hamilton Place, London W1V 0BQ, England. Ten times per year. $175.00 per year.

EXPERIMENTS IN FLUIDS. Springer-Verlag New York, 175 Fifth Avenue, New York, NY 10010. (212) 460-1500. Bimonthly. $115.00 per year.

FLUID MECHANICS. Scripta Publishing Company, 7961 Eastern Avenue, Silver Spring, MD 20910. (301) 588-0484. Translation of Soviet research papers in the fields of applied and theoretical fluid mechanics. Bimonthly. $300.00 per year.

HYDRAULICS AND PNEUMATICS. Penton/IPC, Incorporated, 1100 Superior Avenue, Cleveland, OH 44114. (216) 696-7000. Monthly. $35.00 per year.

JOURNAL OF FLUID CONTROL (FLUIDICS QUARTERLY). Delbridge Publishing Company, Box 2989, Stanford, Ca 94305. (408) 446-3131. Quarterly. $125.00 per year.

JOURNAL OF FLUID MECHANICS. Cambridge University Press, 32 East 57th Street, New York, NY 10022. (800) 872-7423. Twelve times per year. $380.00 per year.

JOURNAL OF FLUID ENGINEERING. American Society of Mechanical Engineers, 345 East 47th Street, New York, NY 10017-2398. (212) 705-7722. Quarterly. $100.00 per year.

JOURNAL OF HYDRAULIC ENGINEERING. American Society of Civil Engineers, 345 East 47th Street, New York, NY 10017-2398. (212) 705-7520. Monthly. $115.00 per year.

RESEARCH CENTERS AND INSTITUTES

FLUID PROPERTIES RESEARCH, INCORPORATED. School of Chemical Engineering, Georgia Institute of Technology, Atlanta, GA 30332. (405) 894-3098.

LEHIGH UNIVERSITY. Institute of Thermo-Fluid Engineering and Science, Whitaker Laboratory, Bethlehem, PA 18015. (215) 861-4091.

PURDUE UNIVERSITY. Fluid Mechanics Laboratory, School of Mechanical Engineering, West Lafayette, IN 47907. (317) 494-5633.

STANFORD UNIVERSITY. Environmental Fluid Mechanics Laboratory, Department of Civil Engineering, Stanford, CA 94305. (415) 723-1825.

UNIVERSITY OF ARIZONA. Computational Fluid Mechanics Laboratory, Building #16, Room 312, Tucson, Az 85721. (602) 621-4423.

UNIVERSITY OF IOWA. Institute of Hydraulic Research, Iowa City, IA 52242. (319) 353-4679.

UNIVERSITY OF WASHINGTON. Aeronautical Laboratory, FS-10, Seattle, WA 98105. (206) 543-0439.

FLUID MECHANICS

See also: AERODYNAMICS, FLUID DYNAMICS, HYDRAULIC ENGINEERING, HYDRAULICS, HYDRODYNAMICS

ABSTRACT SERVICES AND INDEXES

APPLIED MECHANICS REVIEWS. American Society of Mechanical Engineers, 345 East 47th Street, New York, NY 10017. (212) 705-7703. 1948 to present. Monthly. $360.00 per year.

APPLIED SCIENCE AND TECHNOLOGY INDEX. H.W. Wilson Company, 950 University Avenue, Bronx, NY 10452. (800) 367-6670 or (212) 588-8400. Inquire as to cost and availability.

CHEMICAL ABSTRACTS. Chemical Abstracts Service, 2540 Olentangy Road, Post Office Box 3012, Columbus, OH 43210. (800) 848-6538 or (614) 421-3600. Weekly. $9200.00 per year.

CIVIL ENGINEERING HYDRAULICS ABSTRACTS. BHRA Fluid Engineering, Cranfield, Bedford MK43 0AJ, England. Distributed by Learned Information Incorporated, 143 Old Marlton Pike, Medford, NJ 08055. 1968 to present. Monthly. $225.00 per year.

CURRENT CONTENTS: ENGINEERING, TECHNOLOGY AND APPLIED SCIENCES. Institute for Scientific Information, 3501 Market Street, Philadelphia, PA 19104. (800) 523-1850 or (215) 386-0100. 1970 to present. Weekly. $272.00 per year.

ENGINEERING INDEX MONTHLY AND AUTHOR INDEX. Engineering Information, Incorporated, 345 East 47th Street, New York, NY 10017. (800) 221-1044 or (212) 705-7600. Monthly, with annual cumulation. $1560.00 per year.

FLUID POWER ABSTRACTS. BHRA Fluid Engineering, Cranfield, Bedford MK43 0AJ, England. Distributed by Learned Information Incorporated, 143 Old Marlton Pike, Medford, NJ 08055. 1970 to present. Bimonthly. $196.00 per year.

INDUSTRIAL AERODYNAMICS ABSTRACTS. BHRA Fluid Engineering, Cranfield, Bedford MK43 0AJ, England.

Disbtributed by Learned Information Incorporated, 143 Old Marlton Pike, Medford, NJ 08055. 1970 to present. Monthly. $196.00 per year.

INTERNATIONAL AEROSPACE ABSTRACTS. AIAA/TIS, 1633 Broadway, New York, NY 10019. (212) 581-4300. Semimonthly. $700.00 per year.

ISMEC BULLETIN. (Information Service in Mechanical Engineering). Cambridge Scientific Abstracts, 5161 River Road, Bethesda, MD 20816. (301) 951-1400. 1973 to present. Monthly. $475.00 per year.

STAR. (Scientific and Technical Aerospace Reports. United States National Aeronautics and Space Administration, Scientific and Technical Information Facility, Box 8757, Baltimore-Washington International Airport, MD 21240. (202) 755-2210. Semimonthly, with semiannual and annual indexes. $85.00 per year.

ASSOCIATIONS AND PROFESSIONAL SOCIETIES

AMERICAN INSTITUTE OF AERONAUTICS AND ASTRONAUTICS. 1633 Broadway, New York, NY 10019. (212) 581-4300.

AMERICAN SOCIETY OF CIVIL ENGINEERS. 345 East 47th Street, New York, NY 10017-2398. (212) 705-7520.

AMERICAN SOCIETY OF MECHANICAL ENGINEERS. 345 East 47th Street, New York, NY 10017-2398. (212) 705-7722.

FLUID CONTROLS INSTITUTE. Post Office Box 9036, Morristown, NJ 07960. (201) 829-0990.

FLUID POWER SOCIETY. 3333 North Mayfair Road, Milwaukee, WI 53222. (414) 778-3377.

HYDRAULIC INSTITUTE. 712 Lakewood Center North, 14600 Detroit Avenue, Cleveland, OH 44107. (216) 226-7700.

INTERNATIONAL ASSOCIATION FOR HYDRAULIC RESEARCH. 185 Rotterdamseweg, Box 177, Delft, The Netherlands.

NATIONAL CONFERENCE ON FLUID POWER. IIT Research Institute, Ten West 35th Street, Chicago, IL 60616. (312) 567-4414.

NATIONAL FLUID POWER ASSOCIATION. 3333 North Mayfair Road, Suite 311, Milwaukee, WI 53222. (414) 778-3344.

SOCIETY OF AUTOMOTIVE ENGINEERS. SAE, Incorporated, 400 Commonwealth Drive, Warrendale, PA 15096. (412) 776-4841.

DIRECTORIES AND BIOGRAPHICAL SOURCES

AAS DIRECTORY. American Astronautical Society, 6212-B Old Keene Mill Court, Springfield, VA 22152. (703) 866-0020. Annual. $35.00 per year.

DIRECTORY OF HYDRAULIC RESEARCH INSTITUTES AND LABORATORIES. International Association for Hydraulic Research, 185 Rotterdamseweg, Box 177, Delft, The Netherlands. 1986. 60 Dutch florins.

FLUID POWER HANDBOOK AND DIRECTORY. Penton/IPC, Incorporated, 1111 Chester Avenue, Cleveland, OH 44114. (216) 606-7000. Biennial. $45.00.

FLUID POWER SOCIETY MEMBERSHIP DIRECTORY. Fluid Power Society. 3333 North Mayfair Road, Milwaukee, WI 53222. (414) 778-3377. Annual. $50.00.

NATIONAL FLUID POWER ASSOCIATION MEMBERSHIP DIRECTORY. 3333 North Mayfair Road, Suite 311, Milwaukee, WI 53222. (414) 778-3344. Annual. $150.00.

RESEARCH CENTERS DIRECTORY. Gale Research Company, Book Tower, Detroit, MI 48226. (800) 521-0707. Twelfth edition. 1987. $365.00 for set.

WHO'S WHO IN ENGINEERING. Engineers Joint Council, 345 East 47th Street, New York, NY 10017. (212) 705-7010. 1985. $200.00.

ENCYCLOPEDIAS AND DICTIONARIES

ENCYCLOPEDIA OF FLUID MECHANICS. Nicholas P. Cheremisinnoff, editor. Gulf Publishing Company, Book Division, Post Office Box 2608, Houston, TX 77001. (713) 520-4444. Six Volumes. 1987. $1000.00 for set.

ENCYCLOPEDIA OF PHYSICAL SCIENCE AND TECHNOLOGY. Academic Press, Incorporated, Orlando, FL 32887. (800) 321-5068 or (305) 345-2734. Inquire as to cost and availability.

MCGRAW-HILL DICTIONARY OF ENGINEERING. Sybil P. Parker, editor. McGraw-Hill Book Company, 1221 Avenue of the Americas, New York, NY 10020. (212) 512-2000. 1984. $39.95.

MCGRAW-HILL ENCYCLOPEDIA OF SCIENCE AND TECHNOLOGY. McGraw-Hill Book, Incorporated, 1221 Avenue of the Americas, New York, NY 10020. (212) 997-3675. Fifth edition, 15 volumes. $1100.00.

GENERAL WORKS

ENGINEERING FLUID MECHANICS. John J. Bertin. Prentice-Hall, Incorporated, Englewood Cliffs, NJ 07632. (201) 592-2000. Second edition. 1987. $34.95.

FLOW VISUALISATION. Japan Society of Mechanical Engineers. Pergamon Press, Incorporated, Maxwell House, Fairview Park, Elmsford, NY 10523. (914) 592-7700. 1986. $65.00.

FOUNDATIONS OF AERODYNAMICS: BASES OF AERODYNAMIC DESIGN. A.M. Kuethe and C. Chow. John Wiley and Sons, Incorporated, 605 Third Avenue, New York, NY 10158. (800) 526-5368 or (212) 850-6000. Fourth edition. 1986. $43.95.

FUNDAMENTALS OF FLUID MECHANICS AND HYDRAULICS. J. Evett and C. Liu. McGraw-Hill Book Company, 1221 Avenue of the Americas, New York, NY 10020. (212) 512-2000. 1987. $49.95.

FUNDAMENTALS OF HYDRAULIC ENGINEERING SYSTEMS. Ned H. Hwang and Carlos E. Hita. Prentice-Hall, Incorporated, Englewood Cliffs, NJ 07632. (800) 562-0245. 1987. $41.95.

ILLUSTRATED GUIDE TO AERODYNAMICS. Hubert Smith. Tab Books, Incorporated, Monterey Lane, Blue Ridge Summit, PA 17214. (717) 794-2191. 1986. $14.95.

AN INFORMAL INTRODUCTION TO THEORETICAL FLUID MECHANICS. James Lighthill. Oxford University Press, 200 Madison Avenue, New York, NY 10016. (212) 679-7300. 1986. $35.00.

INTRODUCTION TO FLUID MECHANICS. R.W. Fox and A.T. McDonald. John Wiley and Sons, Incorporated, 605 Third Avenue, New York, NY 10158. (800) 526-5368 or (212) 850-6000. Third edition. 1985. $45.00.

HANDBOOKS AND MANUALS

APPLIED FLUID DYNAMICS HANDBOOK. Robert D. Blevins. McGraw-Hill Book Company, Incorporated, 1221 Avenue of the Americas, New York, NY 10020. (212) 512-2000. 1984. $52.95.

HANDBOOK OF HYDRAULIC RESISTENCE. I.E. Idelchik and others. Hemisphere Publishing Corporation, 79 Madison Avenue, New York, NY 10016. (800) 526-0275. Second edition. 1986. $89.95.

STANDARD HANDBOOK OF ENGINEERING CALCULATIONS. Tyler G. Hicks, editor. McGraw-Hill Book Company, Incorporated, 1221 Avenue of the Americas, New York, NY 10020. (212) 512-2000. Second edition. 1984. $59.50.

THE WILEY ENGINEER'S DESK REFERENCE. Sanford I. Heisler. John Wilet and Sons, Incorporated, 605 Third Avenue, New York, NY 10158. (800) 526-5368 or (212) 850-6418. 1984. $36.00.

ONLINE DATA BASES

COMPENDEX. Engineering Information, Incorporated, 345 East 47th Street, New York, NY 10017. (800) 221-1044 or (212) 705-7615. Engineering and technical literature, 1975 to present. Inquire as to cost and availability.

ISMEC (INFORMATION SERVICE IN MECHANICAL ENGINEERING). Cambridge Scientific Abstracts, 5161 River Road, Bethesda, MD 20816. (301) 951-1400. 1973 to present. Inquire as to online cost and availability.

NASA. National Aeronautics and Space Administration, Scientific and Technical Information Branch, 300 7th Street, SW, Washington, DC 20546. Citations and abstracts of aerospace literature, 1962 to present. Inquire as to cost and availability.

NTIS. National Technical Information Service, 5285 Port Royal Road, Springfield, VA 22161. (703) 487-4630. Broad coverage of government-sponsored research reports, 1964 to present. Inquire as to cost and availability.

PERIODICALS

AERONAUTICAL JOURNAL. Aeronautical Society, 4 Hamilton Place, London W1V 0BQ, England. Ten times per year. $175.00 per year.

EXPERIMENTS IN FLUIDS. Springer-Verlag New York, 175 Fifth Avenue, New York, NY 10010. (212) 460-1500. Bimonthly. $115.00 per year.

FLUID MECHANICS. Scripta Publishing Company, 7961 Eastern Avenue, Silver Spring, MD 20910. (301) 588-0484. Translation of Soviet research papers in the fields of applied and theoretical fluid mechanics. Bimonthly. $300.00 per year.

HYDRAULICS AND PNEUMATICS. Penton/IPC, Incorporated, 1100 Superior Avenue, Cleveland, OH 44114. (216) 696-7000. Monthly. $35.00 per year.

JOURNAL OF FLUID CONTROL (FLUIDICS QUARTERLY). Delbridge Publishing Company, Box 2989, Stanford, Ca 94305. (408) 446-3131. Quarterly. $125.00 per year.

JOURNAL OF FLUID MECHANICS. Cambridge University Press, 32 East 57th Street, New York, NY 10022. (800) 872-7423. Twelve times per year. $380.00 per year.

JOURNAL OF FLUID ENGINEERING. American Society of Mechanical Engineers, 345 East 47th Street, New York, NY 10017-2398. (212) 705-7722. Quarterly. $100.00 per year.

JOURNAL OF HYDRAULIC ENGINEERING. American Society of Civil Engineers, 345 East 47th Street, New York, NY 10017-2398. (212) 705-7520. Monthly. $115.00 per year.

RESEARCH CENTERS AND INSTITUTES

FLUID PROPERTIES RESEARCH, INCORPORATED. School of Chemical Engineering, Georgia Institute of Technology, Atlanta, GA 30332. (405) 894-3098.

LEHIGH UNIVERSITY. Institute of Thermo-Fluid Engineering and Science, Whitaker Laboratory, Bethlehem, PA 18015. (215) 861-4091.

PURDUE UNIVERSITY. Fluid Mechanics Laboratory, School of Mechanical Engineering, West Lafayette, IN 47907. (317) 494-5633.

STANFORD UNIVERSITY. Environmental Fluid Mechanics Laboratory, Department of Civil Engineering, Stanford, CA 94305. (415) 723-1825.

UNIVERSITY OF ARIZONA. Computational Fluid Mechanics Laboratory, Building #16, Room 312, Tucson, AZ 85721. (602) 621-4423.

UNIVERSITY OF IOWA. Institute of Hydraulic Research, Iowa City, IA 52242. (319) 353-4679.

UNIVERSITY OF WASHINGTON. Aeronautical Laboratory, FS-10, Seattle, WA 98105. (206) 543-0439.

FLUID STATICS

See: FLUID MECHANICS

FLUIDICS

See also: AERODYNAMICS, FLUID DYNAMICS, FLUID MECHANICS, HYDRAULIC ENGINEERING, HYDRAULICS, HYDRODYNAMICS

ABSTRACT SERVICES AND INDEXES

APPLIED MECHANICS REVIEWS. American Society of Mechanical Engineers, 345 East 47th Street, New York, NY 10017. (212) 705-7703. 1948 to present. Monthly. $360.00 per year.

APPLIED SCIENCE AND TECHNOLOGY INDEX. H.W. Wilson Company, 950 University Avenue, Bronx, NY 10452. (800) 367-6670 or (212) 588-8400. Inquire as to cost and availability.

CHEMICAL ABSTRACTS. Chemical Abstracts Service, 2540 Olentangy Road, Post Office Box 3012, Columbus, OH 43210. (800) 848-6538 or (614) 421-3600. Weekly. $9200.00 per year.

CIVIL ENGINEERING HYDRAULICS ABSTRACTS. BHRA Fluid Engineering, Cranfield, Bedford MK43 0AJ, England. Distributed by Learned Information Incorporated, 143 Old Marlton Pike, Medford, NJ 08055. 1968 to present. Monthly. $225.00 per year.

CURRENT CONTENTS: ENGINEERING, TECHNOLOGY AND APPLIED SCIENCES. Institute for Scientific Information, 3501 Market Street, Philadelphia, PA 19104. (800) 523-1850 or (215) 386-0100. 1970 to present. Weekly. $272.00 per year.

ENGINEERING INDEX MONTHLY AND AUTHOR INDEX. Engineering Information, Incorporated, 345 East 47th Street, New York, NY 10017. (800) 221-1044 or (212) 705-7600. Monthly, with annual cumulation. $1560.00 per year.

FLUID POWER ABSTRACTS. BHRA Fluid Engineering, Cranfield, Bedford MK43 0AJ, England. Distributed by Learned Information Incorporated, 143 Old Marlton Pike, Medford, NJ 08055. 1970 to present. Bimonthly. $196.00 per year.

INDUSTRIAL AERODYNAMICS ABSTRACTS. BHRA Fluid Engineering, Cranfield, Bedford MK43 0AJ, England. Disbtributed by Learned Information Incorporated, 143 Old Marlton Pike, Medford, NJ 08055. 1970 to present. Monthly. $196.00 per year.

INTERNATIONAL AEROSPACE ABSTRACTS. AIAA/TIS, 1633 Broadway, New York, NY 10019. (212) 581-4300. Semimonthly. $700.00 per year.

ISMEC BULLETIN. (Information Service in Mechanical Engineering). Cambridge Scientific Abstracts, 5161 River Road, Bethesda, MD 20816. (301) 951-1400. 1973 to present. Monthly. $475.00 per year.

SCIENCE CITATION INDEX. Institute for Scientific Information, 3501 Market Street, Philadelphia, PA 19104. (800) 523-1850 or (215) 386-0100.

STAR. (Scientific and Technical Aerospace Reports. United States National Aeronautics and Space Administration, Scientific and Technical Information Facility, Box 8757, Baltimore-Washington International Airport, MD 21240. (202) 755-2210. Semimonthly, with semiannual and annual indexes. $85.00 per year.

ASSOCIATIONS AND PROFESSIONAL SOCIETIES

AMERICAN INSTITUTE OF AERONAUTICS AND ASTRONAUTICS. 1633 Broadway, New York, NY 10019. (212) 581-4300.

AMERICAN SOCIETY OF CIVIL ENGINEERS. 345 East 47th Street, New York, NY 10017-2398. (212) 705-7520.

AMERICAN SOCIETY OF MECHANICAL ENGINEERS. 345 East 47th Street, New York, NY 10017-2398. (212) 705-7722.

FLUID CONTROLS INSTITUTE. Post Office Box 9036, Morristown, NJ 07960. (201) 829-0990.

FLUID POWER SOCIETY. 3333 North Mayfair Road, Milwaukee, WI 53222. (414) 778-3377.

HYDRAULIC INSTITUTE. 712 Lakewood Center North, 14600 Detroit Avenue, Cleveland, OH 44107. (216) 226-7700.

INTERNATIONAL ASSOCIATION FOR HYDRAULIC RESEARCH. 185 Rotterdamseweg, Box 177, Delft, The Netherlands.

NATIONAL CONFERENCE ON FLUID POWER. IIT Research Institute, Ten West 35th Street, Chicago, IL 60616. (312) 567-4414.

NATIONAL FLUID POWER ASSOCIATION. 3333 North Mayfair Road, Suite 311, Milwaukee, WI 53222. (414) 778-3344.

SOCIETY OF AUTOMOTIVE ENGINEERS. SAE, Incorporated, 400 Commonwealth Drive, Warrendale, PA 15096. (412) 776-4841.

DIRECTORIES AND BIOGRAPHICAL SOURCES

AAS DIRECTORY. American Astronautical Society, 6212-B Old Keene Mill Court, Springfield, VA 22152. (703) 866-0020. Annual. $35.00 per year.

DIRECTORY OF HYDRAULIC RESEARCH INSTITUTES AND LABORATORIES. International Association for Hydraulic Research, 185 Rotterdamseweg, Box 177, Delft, The Netherlands. 1986. 60 Dutch florins.

FLUID POWER HANDBOOK AND DIRECTORY. Penton/IPC, Incorporated, 1111 Chester Avenue, Cleveland, OH 44114. (216) 606-7000. Biennial. $45.00.

FLUID POWER SOCIETY MEMBERSHIP DIRECTORY. Fluid Power Society. 3333 North Mayfair Road, Milwaukee, WI 53222. (414) 778-3377. Annual. $50.00.

NATIONAL FLUID POWER ASSOCIATION MEMBERSHIP DIRECTORY. 3333 North Mayfair Road, Suite 311, Milwaukee, WI 53222. (414) 778-3344. Annual. $150.00.

RESEARCH CENTERS DIRECTORY. Gale Research Company, Book Tower, Detroit, MI 48226. (800) 521-0707. Twelfth edition. 1987. $365.00 for set.

WHO'S WHO IN ENGINEERING. Engineers Joint Council, 345 East 47th Street, New York, NY 10017. (212) 705-7010. 1985. $200.00.

ENCYCLOPEDIAS AND DICTIONARIES

ENCYCLOPEDIA OF FLUID MECHANICS. Nicholas P. Cheremisinnoff, editor. Gulf Publishing Company, Book Division, Post Office Box 2608, Houston, TX 77001. (713) 520-4444. Six Volumes. 1987. $1000.00 for set.

ENCYCLOPEDIA OF PHYSICAL SCIENCE AND TECHNOLOGY. Academic Press, Incorporated, Orlando, FL 32887. (800) 321-5068 or (305) 345-2734. Inquire as to cost and availability.

MCGRAW-HILL DICTIONARY OF ENGINEERING. Sybil P. Parker, editor. McGraw-Hill Book Company, 1221 Avenue of the Americas, New York, NY 10020. (212) 512-2000. 1984. $39.95.

MCGRAW-HILL ENCYCLOPEDIA OF SCIENCE AND TECHNOLOGY. McGraw-Hill Book, Incorporated, 1221 Avenue of the Americas, New York, NY 10020. (212) 997-3675. Fifth edition, 15 volumes. $1100.00.

GENERAL WORKS

BASIC FLUID POWER. Dudley Pease and John Pippenger. Prentice-Hall, Incorporated, Englewood Cliffs, NJ 07632. (201) 592-2000. Second edition. 1987. $33.95.

ENGINEERING FLUID MECHANICS. John J. Bertin. Prentice-Hall, Incorporated, Englewood Cliffs, NJ 07632. (201) 592-2000. Second edition. 1987. $34.95.

FLOW VISUALISATION. Japan Society of Mechanical Engineers. Pergamon Press, Incorporated, Maxwell House, Fairview Park, Elmsford, NY 10523. (914) 592-7700. 1986. $65.00.

FLUIDIC FLOW CONTROL. J.R. Tippetts. Heyden and Son, Incorporated, 247 South 41st Street, Philadelphia, PA 19104. (215) 382-6673. 1984. Inquire.

FOUNDATIONS OF AERODYNAMICS: BASES OF AERODYNAMIC DESIGN. A.M. Kuethe and C. Chow. John Wiley and Sons, Incorporated, 605 Third Avenue, New York, NY 10158. (800) 526-5368 or (212) 850-6000. Fourth edition. 1986. $43.95.

FUNDAMENTALS OF FLUID MECHANICS AND HYDRAULICS. J. Evett and C. Liu. McGraw-Hill Book Company, 1221 Avenue of the Americas, New York, NY 10020. (212) 512-2000. 1987. $49.95.

FUNDAMENTALS OF HYDRAULIC ENGINEERING SYSTEMS. Ned H. Hwang and Carlos E. Hita. Prentice-Hall, Incorporated, Englewood Cliffs, NJ 07632. (800) 562-0245. 1987. $41.95.

AN INFORMAL INTRODUCTION TO THEORETICAL FLUID MECHANICS. James Lighthill. Oxford University Press, 200 Madison Avenue, New York, NY 10016. (212) 679-7300. 1986. $35.00.

INTRODUCTION TO FLUID MECHANICS. R.W. Fox and A.T. McDonald. John Wiley and Sons, Incorporated, 605 Third Avenue, New York, NY 10158. (800) 526-5368 or (212) 850-6000. Third edition. 1985. $45.00.

HANDBOOKS AND MANUALS

APPLIED FLUID DYNAMICS HANDBOOK. Robert D. Blevins. McGraw-Hill Book Company, Incorporated, 1221 Avenue of the Americas, New York, NY 10020. (212) 512-2000. 1984. $52.95.

HANDBOOK OF HYDRAULIC RESISTENCE. I.E. Idelchik and others. Hemisphere Publishing Corporation, 79 Madison Avenue, New York, NY 10016. (800) 526-0275. Second edition. 1986. $89.95.

STANDARD HANDBOOK OF ENGINEERING CALCULATIONS. Tyler G. Hicks, editor. McGraw-Hill Book Company, Incorporated, 1221 Avenue of the Americas, New York, NY 10020. (212) 512-2000. Second edition. 1984. $59.50.

THE WILEY ENGINEER'S DESK REFERENCE. Sanford I. Heisler. John Wilet and Sons, Incorporated, 605 Third Avenue, New York, NY 10158. (800) 526-5368 or (212) 850-6418. 1984. $36.00.

ONLINE DATA BASES

COMPENDEX. Engineering Information, Incorporated, 345 East 47th Street, New York, NY 10017. (800) 221-1044 or (212) 705-7615. Engineering and technical literature, 1975 to present. Inquire as to cost and availability.

ISMEC (INFORMATION SERVICE IN MECHANICAL ENGINEERING). Cambridge Scientific Abstracts, 5161 River Road, Bethesda, MD 20816. (301) 951-1400. 1973 to present. Inquire as to online cost and availability.

NASA. National Aeronautics and Space Administration, Scientific and Technical Information Branch, 300 7th Street, SW, Washington, DC 20546. Citations and abstracts of aerospace literature, 1962 to present. Inquire as to cost and availability.

NTIS. National Technical Information Service, 5285 Port Royal Road, Springfield, VA 22161. (703) 487-4630. Broad coverage of government-sponsored research reports, 1964 to present. Inquire as to cost and availability.

PERIODICALS

EXPERIMENTS IN FLUIDS. Springer-Verlag New York, 175 Fifth Avenue, New York, NY 10010. (212) 460-1500. Bimonthly. $115.00 per year.

FLUID MECHANICS. Scripta Publishing Company, 7961 Eastern Avenue, Silver Spring, MD 20910. (301) 588-0484. Translation of Soviet research papers in the fields of applied and theoretical fluid mechanics. Bimonthly. $300.00 per year.

HYDRAULICS AND PNEUMATICS. Penton/IPC, Incorporated, 1100 Superior Avenue, Cleveland, OH 44114. (216) 696-7000. Monthly. $35.00 per year.

JOURNAL OF FLUID CONTROL (FLUIDICS QUARTERLY). Delbridge Publishing Company, Box 2989, Stanford, Ca 94305. (408) 446-3131. Quarterly. $125.00 per year.

JOURNAL OF FLUID MECHANICS. Cambridge University Press, 32 East 57th Street, New York, NY 10022. (800) 872-7423. Twelve times per year. $380.00 per year.

JOURNAL OF FLUID ENGINEERING. American Society of Mechanical Engineers, 345 East 47th Street, New York, NY 10017-2398. (212) 705-7722. Quarterly. $100.00 per year.

JOURNAL OF HYDRAULIC ENGINEERING. American Society of Civil Engineers, 345 East 47th Street, New York, NY 10017-2398. (212) 705-7520. Monthly. $115.00 per year.

RESEARCH CENTERS AND INSTITUTES

LEHIGH UNIVERSITY. Institute of Thermo-Fluid Engineering and Science, Whitaker Laboratory, Bethlehem, PA 18015. (215) 861-4091.

OHIO STATE UNIVERSITY. Fluid Power Laboratory, Mechanical Engineering Department, 206 West 18th Avenue, Columbus, OH 43210. (614) 422-2289.

OKLAHOMA STATE UNIVERSITY. Fluid Power Research Center, Stillwater, OK 74074. (405) 624-7375.

PURDUE UNIVERSITY. Fluid Mechanics Laboratory, School of Mechanical Engineering, West Lafayette, IN 47907. (317) 494-5633.

STANFORD UNIVERSITY. Environmental Fluid Mechanics Laboratory, Department of Civil Engineering, Stanford, CA 94305. (415) 723-1825.

UNIVERSITY OF ARIZONA. Computational Fluid Mechanics Laboratory, Building #16, Room 312, Tucson, Az 85721. (602) 621-4423.

FLUORINE

See also: BROMINE, CHLORINE, IODINE, HALIDES

ABSTRACT SERVICES AND INDEXES

APPLIED SCIENCE AND TECHNOLOGY INDEX. H. W. Wilson Company, 950 University Avenue, Bronx, NY 10452. (800) 367-6670 or (212) 588-8400. Inquire as to cost and availability.

CHEMICAL ABSTRACTS. Chemical Abstracts Service, 2540 Olentangy Road, P.O. Box 3012, Columbus, OH 43210. (800) 848-6538 or (614) 421-3600. Weekly. $9,200.00 per year.

GENERAL SCIENCE INDEX. H. W. Wilson Company, 950 University Avenue, Bronx, NY 10452. (800) 367-6770 or (212) 588-8400. Inquire as to cost and availability.

PHYSICS ABSTRACTS. Institute of Electrical Engineers, London, United Kingdom. Available from: Institute of Electrical and Electronic Engineers (IEEE), 345 East 47th Street, New York, NY 10017. (212) 705-7900.

SCIENCE CITATION INDEX. Institute for Scientific Information, 3501 Market Street, Philadelphia, PA 19104. (800) 523-1850 or (215) 386-0100.

ASSOCIATIONS AND PROFESSIONAL SOCIETIES

AMERICAN CHEMICAL SOCIETY. 1155 16th Street, NW, Washington, DC 20036. (202) 872-4600.

ASSOCIATION OF CONSULTING CHEMISTS AND CHEMICAL ENGINEERS. 50 East 41st Street, Suite 92, New York, NY 10017. (212) 684-6255.

INTERNATIONAL SOCIETY FOR FLUORIDE RESEARCH, P.O. Box 692, Warren, MI 48090.

BIBLIOGRAPHIES

SCIENTIFIC AND TECHNICAL BOOKS IN PRINT; An Index to Literature in Science and Technology. R. R. Bowker Company, 205 East 42nd Street, New York, NY 10017. (800) 521-8110 or (212) 916-1600.

DIRECTORIES AND BIOGRAPHICAL SOURCES

AMERICAN INSTITUTE OF CHEMISTS. American Institute of Chemists, 7315 Wisconsin Avenue, Bethesda, MD 20814. (301) 652-2447. 1986. $35.00.

AMERICAN MEN AND WOMEN OF SCIENCE. Physical and Biological Sciences. Fifteenth Edition. R. R. Bowker Company, 205 East 42nd Street, New York, NY 10017. (800) 521-8110 or (212) 916-1600.

BIOGRAPHICAL DICTIONARY OF SCIENTISTS: CHEMISTS. David Abbott, Editor. Bedrick Books, 125 East 23rd Street, New York, NY 10010. (212) 777-1187. 1984. $18.95.

CHEMICAL WEEK - BUYERS GUIDE ISSUE. McGraw-Hill Book Company, 1221 Avenue of the Americas, New York, NY 10020. (800) 628-0004. Annual, October. $50.00.

CONSULTING SERVICES: CHEMISTS AND CHEMICAL ENGINEERS. Association of Consulting Chemists and Chemical Engineers, 50 East 41st Street, New York, NY 10017. (212) 684-6255. Annual. 1986. $45.00.

GOVERNMENT RESEARCH DIRECTORY. Gale Research Company, Book Tower, Detorit, MI 48226. (800) 521-0707. Fourth Edition. 1987. $350.00.

INTERNATIONAL RESEARCH CENTERS DIRECTORY 1986-1987. Gale Research Company, Book Tower, Detorit, MI 48226. (800) 521-0707. Third Edition. 1986. $330.00.

RESEARCH CENTERS DIRECTORY. Gale Research Company, Book Tower, Detorit, MI 48226. (800) 521-0707. Eleventh Edition. 1987. $355.00.

SCIENTIFIC AND TECHNICAL ORGANIZATIONS AND AGENCIES DIRECTORY. Gale Research Company, Book Tower, Detorit, MI 48226. (800) 521-0707. 1985. $150.00.

WORLD GUIDE TO SCIENTIFIC ASSOCIATIONS AND LEARNED SOCIETIES. K. G. Saur Inc., 175 Fifth Avenue, New York, NY 10010. (800) 521-0707 or (212) 982-1302. Fourth Edition. 1984. $112.00

WHO'S WHO IN FRONTIER SCIENCE AND TECHNOLOGY. Marquis Who's Who, Inc., 200 East Ohio Street, Chicago, IL 60611. (800) 428-3898 or (312) 787-2008.

WHO'S WHO IN TECHNOLOGY TODAY. Reston Publishing Company, Inc., c/o Prentice-Hall, Inc., Englewood Cliffs, NJ 07632. (800) 262-6868. Biennial. Five volumes. $425.00. Covers the fields of elecctronics, computer science, physics, optics, chemistry, biotechnology, mechanics, energy, and earth science.

ENCYCLOPEDIAS AND DICTIONARIES

CONCISE ENCYCLOPEDIA OF CHEMICAL TECHNOLOGY. Kirk-Othmer. John Wiley and Sons, Inc., 605 Third Avenue, New York, NY 10158. (800) 526-5368 or (212) 850-6000. Third Edition. 1985. $129.95.

CONDENSED CHEMICAL DICTIONARY. Gessner Hawley. Van Nostrand Reinhold, 115 Fifth Avenue, New York, NY 10003. Tenth Edition. 1981. $49.95.

ENCYCLOPEDIA OF PHYSICAL SCIENCE AND TECHNOLOGY. Academic Press, Inc., Orlando, FL 32887. (800) 321-5068 or (305) 345-2734. Fifteen volumes, 1986. Inquire as to cost and availability.

GLOSSARY OF CHEMICAL TERMS. Clifford A. Hampel and Gessner G. Hawley. Van Nostrand Reinhold Company, 115 Fifth Avenue, New York, NY 10003. (800) 543-2681 or (212) 254-3232. Second Edition. 1982. $21.95.

MCGRAW-HILL ENCYCLOPEDIA OF SCIENCE AND TECHNOLOGY. McGraw-Hill Book Company, 1221 Avenue of the Americas, New York, NY 10020. (212) 997-3675. Inquire as to cost and availability.

VAN NOSTRAND REINHOLD ENCYCLOPEDIA OF CHEMISTRY. Douglas M. Considine and Glenn D. Considine. Van Nostrand Reinhold Publishing Company, Inc., 115 Fifth Avenue, New York, NY 10003. (800) 543-2681 or (212) 254-3232. 1984. $97.95.

GENERAL WORKS

CHEMISTRY OF FLORINE AND ITS COMPOUNDS. H. J. Emeleus. Academic Press, Inc., 6277 Sea Harbor Drive, Orlando, FL 32821. (800) 321-5068. 1969. $49.00.

CHEMISTRY OF HALIDES, PSEUDOHALIDES AND AZIDES: SUPPLEMENT D. CHEMISTRY OF FUNCTIONAL GROUPS SERIES, PART 1 AND 2. Saul Patai. John Wiley and Sons, Inc., 605 Third Avenue, New York, NY 10158. (800) 526-5368 or (212) 850-6000. 1983. $745.00 set.

CHEMISTRY OF THE ELEMENTS. N. N. Greenwood and A. Earnshaw. Pergamon Publishing, Inc., Maxwell House, Fairview Park, Elmsford, NY 10523. (914) 592-7700. 1984. $143.00.

HALOGEN CHEMISTRY. V. Gutman, Editor. Academic Press, Inc., 6277 Sea Harbor Drive, Orlando, FL 32821. (800) 321-5068. Three volumes. 1967. $90.00 each.

HAZARDOUS AND TOXIC CHEMICALS: SAFE HANDLING AND DISPOSAL. Howard Fawcett. John Wiley and Sons, Inc., 605 Third Avenue, New York, NY 10158. (800) 526-5368 or (212) 850-6000. 1984. $37.00.

SAFE STORAGE OF LABORATORY CHEMICALS. David A. Pipitone. John Wiley and Sons, Inc., 605 Third Avenue, New York, NY 10158. (800) 526-5368 or (212) 850-6000. 1984. $60.00.

SULFUR DIOXIDE, CHLORINE, FLUORINE AND CHLORINE OXIDES. A. S. Young. Pergamon Publishing, Inc., Maxwell House, Fairview Park, Elmsford, NY 10523. (914) 592-7700. 1983. $100.00.

TOXICOLOGY OF HALOGENATED HYDROCARBONS: HEALTH AND ECOLOGICAL EFFECTS. M. A. Khan and R. H. Stanton, Editors. Pergamon Press, Inc., Maxwell House, Fairview Park, Elmsford, NY 10523. (914) 592-7700. 1981. $72.50.

HANDBOOKS AND MANUALS

THE CHEMIST'S COMPANION: A HANDBOOK OF PRACTICAL DATA, TECHNIQUES, AND REFERENCES. Arnold J. Gordon and Richard A. Ford. John Wiley and Sons, Inc., 605 Third Avenue, New York, NY 10158. (800) 526-5368. 1973. $49.95.

CRC HANDBOOK OF CHEMISTRY AND PHYSICS. CRC Press, Inc., 2000 Corporate Boulevard, NW, Boca Raton, FL 33431. Sixty-seventh Edition. 1986. $69.95.

HANDBOOK OF APPLIED CHEMISTRY: FACTS FOR ENGINEERS, SCIENTISTS, TECHNICIANS, AND TECHNICAL MANAGERS. Vollrath Hopp and Ingp Hennig. McGraw-Hill Book Company, 1221 Avenue of the Americas, New York, NY 10020. (800) 628-0004. 1983. $54.00.

HANDBOOK OF COMPUTATIONAL CHEMISTRY: A PRACTICAL GUIDE TO CHEMICAL STRUCTURE AND ENERGY CALCULATIONS. Tim Clark. John Wiley and Sons, Inc., 605 Third Avenue, New York, NY 10158. (800) 526-5368 or (212) 850-6000. 1985. $35.00.

LANGE'S HANDBOOK OF CHEMISTRY. John A. Dean, Editor. McGraw-Hill Book Company, 1221 Avenue of the Americas, New York, NY 10020. (800) 628-0004. 1985. $59.50.

ONLINE DATA BASES

CA SEARCH. Chemical Abstracts Service, P.O. Box 3012, Columbus, OH 43210. Guide to chemical literature, 1972 to present. Inquire as to cost and availability.

DISSERTATION ABSTRACTS ONLINE. University Microfilms International, 300 North Zeeb Road, Ann Arbor, MI 48106. (800) 521-0600 or (313) 761-4700. Scope includes virtually all doctoral dissertations accepted at accredited American institutions from 1861 to present in 252 subject areas. Inquire as to online cost and availability.

INSPEC. INSPEC Marketing Department, Institute of Electrical and Electronics Engineers, Inc., IEEE Service Department, 445 Hoes Lane, Piscataway, NJ 08854. (201) 981-0060. Inquire as to online cost and availability.

NTIS. National Technical Information Service, 5285 Port Royal Road, Springfield, VA 22161. (703) 487-4630. Broad coverage of government sponsored research reports, 1964 to present. Inquire as to online cost and availability.

SCISEARCH. Institute for Scientific Information, 3501 Market Street, Philadelphia, PA 19104. (800) 523-1850 or (215) 386-0100. Broad multidisciplinary title and author index to the international literature of science and technology, 1974 to present. Inquire as to online cost and availability.

OTHER SOURCES

ANNUAL ENERGY REVIEW. U.S. Department of Energy, Energy Information Administration, Washington, DC 20585. Annual.

CHEMICAL NOMENCLATURE USEAGE. Ronald Lees and Arthur F. Smith. John Wiley and Sons, Inc., 605 Third Avenue, New York, NY 10158. (800) 526-5368 or (212) 850-6000. 1983. $52.95.

GUIDE TO BASIC INFORMATION SOURCES IN CHEMISTRY. Arthur Antony. John Wiley and Sons, Inc., 605 Third Avenue, New York, NY 10158. (800) 526-5368 or (212) 850-6000. 1979. $26.95.

HOW TO FIND CHEMICAL INFORMATION: A GUIDE FOR PRACTICING CHEMISTS, TEACHERS, AND STUDENTS. John Wiley and Sons, Inc., 605 Third Avenue, New York, NY 10158. (800) 526-5368 or (212) 850-6000. 1986. $35.00.

MINERALS YEARBOOK. U.S. Department of the Interior, Bureau of Mines, C Street between Eighteenth and Nineteenth Streets, NW, Washington, DC 20240.

PERIODICALS

ANALYTICAL CHEMISTRY. American Chemical Society, 1155 Sixteenth Street, NW, Washington, DC 20036. (800) 424-6747 or (202) 872-4700. Monthly. $33.00 per year.

ANDEWANDTE CHEMIE (Gesellschaft Deutscher Chemiker, GW). V C H Verlagsgesellschaft mbH, Pappelallee 3, Postfach 1260, 6940 Weinheim, West Germany. Monthly. $300.00 per year.

CHEMICAL REVIEWS. American Chemical Society, 1155 Sixteenth Street, NW, Washington, DC 20036. (800) 424-6747 or (202) 872-4700. Bimonthly. $83.00 per year.

CHEMICAL WEEK. McGraw-Hill Publishing Company, 1221 Avenue of the Americas, New York, NY 10020. (212) 512-2000. Weekly. $30.00 per year.

CHEMTECH. American Chemical Society, 1155 Sixteenth Street, NW, Washington, DC 20036. (800) 424-6747 or (202) 872-4700. Monthly. $40.00 per year to individuals.

FLOURIDE. International Society for Fluoride Research, P.O. Box 692, Warren, MI 48090. Quarterly. $30.00.

INORGANIC CHEMISTRY. American Chemical Society, 1155 Sixteenth Street, NW, Washington, DC 20036. (800) 424-6747 or (202) 872-4700. Monthly. $400.00 per year.

JOURNAL OF FLUORINE CHEMISTRY. Elesevier Sequoia, S.A., Box 851, CH-1001, Lausanne 1, Switzerland. Sixteen times per year. $460.00 per year.

JOURNAL OF THE AMERICAN CHEMICAL SOCIETY. American Chemical Society, 1155 Sixteenth Street, NW, Washington, DC 20036. (800) 424-6747 or (202) 872-4700. Biweekly. $330.00 per year.

POLYHEDRON. Pergamon Journals, Inc., Maxwell House, Fairview Park, Elmsford, NY 10523. (914) 592-7700. Monthly. $595.00 per year.

RESEARCH CENTERS AND INSTITUTES

HARVARD UNIVERSITY. Chemical Laboratories, Oxford Street, Cambridge, MA 02138. (617) 495-4283.

RENSSELAER POLYTECHNICAL INSTITUTE. Chemistry Laboratories, Cogswell Laboratory, Troy, NY 13181. (518) 266-8462.

SOUTHERN ILLINOIS UNIVERSITY AT CARBONDALE. Research Program in Chemistry and Biochemistry, Carbondale, IL 62901. (618) 453-5721.

UNIVERSITY OF WISCONSIN, MADISON. Theoretical Chemistry Institute, 1101 University Avenue, Madison, WI 53706. (608) 262-1511.

FLUORESCENT LIGHTING

See: ILLUMINATION

FLUOROCARBONS

See: HALIDES

FM

See: RADIO

FOAM

See: PLASTICS

FOCAL LENGTH

See: OPTICS

FOG

See also: AEROSOLS, CLOUDS, METEOROLOGY, WEATHER MODIFICATION

ABSTRACT SERVICES AND INDEXES

CHEMICAL ABSTRACTS. Chemical Abstracts Service, 2540 Olentangy Road, Post Office Box 3012, Columbus, OH 43210. (800) 848-6538 or (614) 421-3600. Weekly. $9200.00 per year.

METEOROLOGICAL AND GEOASTROPHYSICAL ABSTRACTS. American Meteorological Society, 45 Beacon Street, Boston, MA 02108. (617) 227-2425. 1950 to present. Monthly. $450.00 per year.

OCEANIC ABSTRACTS. Cambridge Scientific Abstracts, Incorporated, 5161 River Road, Bethesda, MD 20816. (301) 951-1400. 1964 to present. Inquire as to online cost and availability.

ASSOCIATIONS AND PROFESSIONAL SOCIETIES

AMERICAN METEOROLOGICAL SOCIETY. 45 Beacon Street, Boston, MA 02108. (617) 227-2425.

INTERNATIONAL ASSOCIATION OF METEOROLOGY AND ATMOSPHERIC PHYSICS. UCAR, Post Office Box 3000, Boulder, CO 80307.

NATIONAL WEATHER ASSOCIATION. 4400 Stamp Road, Room 404, Temple Hills, MD 20748. (301) 899-3784.

UNIVERSITY CORPORATION FOR ATMOSPHERIC RESEARCH. Box 3000, 1850 Table Mesa Drive, Boulder, CO 80307. (303) 497-1000.

WEATHER MODIFICATION ASSOCIATION. Post Office Box 8116, Fresno, CA 93747. (209) 291-8466.

DIRECTORIES AND BIOGRAPHICAL SOURCES

GOVERNMENT RESEARCH DIRECTORY. Gale Research Company, Book Tower, Detroit, MI 48226. (8000 521-0707. Fourth edition. 1987. $350.00.

METEOROLOGICAL SERVICES OF THE WORLD. World Meteorological Organization. Available from: American Meteorological Society, 45 Beacon Street, Boston, MA 02108. (617) 227-2425. Annual. $35.00.

NATIONAL WEATHER SERVICE OFFICES AND STATIONS. National Oceanic and Atmospheric Administration, Department of Commerce, Silver Spring, MD 20910. (301) 427-7698. Annual. Free.

RESEARCH CENTERS DIRECTORY. Gale Research Company, Book Tower, Detroit, MI 48226. (800) 521-0707. Twelfth edition. 1987. $365.00 for set.

SCIENTIFIC AND TECHNICAL ORGANIZATIONS AND AGENCIES DIRECTORY. Gale Research Company, Book Tower, Detroit, MI 48226. (800) 521-0707. Second edition. 1987. $185.00.

GENERAL WORKS

THE ATMOSPHERE: AN INTRODUCTION TO METEOROLOGY. Frederick K. Lutgens and Edward J. Tarbuck. Prentice-Hall Publishing, Incorporated, Englewood Cliffs, NJ 07632. (800) 562-0245. Third edition. 1986. $34.95.

ATMOSPHERIC CHEMISTRY: FUNDAMENTALS AND EXPERIMENTAL TECHNIQUES. B.J. Finlayson-Pitts and J.N. Pitts. John Wiley and Sons, Incorporated, 605 Third Avenue, New York, NY 10158. (800) 526-5368 or (212) 850-6000. 1986. $59.95.

CLOUDS: THEIR FORMATION, OPTICAL PROPERTIES, AND EFFECTS. Peter Hobbs and A. Deepak, editors. Academic Press, Incorporated, Orlando, FL 32887. (800) 321-5068. 1981. $55.00.

PHYSICAL METEOROLOGY. Henry G. Houghton. MIT Press, 28 Carlton Street, Cambridge, MA 02142. (617) 253-5646. 1985. $37.50.

THEORY OF FOG CONDENSATION. A.G. Amelin. Coronet Books, 311 Bainbridge Street, Philadelphia, PA 19147. (215) 925-2762. 1967. $50.00.

ONLINE DATA BASES

CA SEARCH. Chemical Abstracts Service, Post Office Box 3012, Columbus, OH 43210. Guide to chemical literature, 1972 to present. Inquire as to cost and availability.

INSPEC. INSPEC Marketing Department, Institute of Electrical and Electronics Engineers, Incorporated, IEEE Service Department, 445 Hoes Lane, Piscataway, NJ 08854. (201) 981-0060. Inquire as to on-line cost and availability.

METEOROLOGICAL AND GEOASTROPHYSICAL ABSTRACTS. American Meteorological Society, 45 Beacon Street, Boston, MA 02108. (617) 227-2425. 1950 to present. Monthly. $450.00 per year.

NTIS. National Technical Information Service, 5285 Port Royal Road, Springfield, VA 22161. (703) 487-4630. Broad coverage of government-sponsored research reports, 1964 to present. Inquire as to cost and availability.

OCEANIC ABSTRACTS. Cambridge Scientific Abstracts, Incorporated, 5161 River Road, Bethesda, MD 20816. (301) 951-1400. 1964 to present. Inquire as to online cost and availability.

SCISEARCH. Institute for Scientific Information, 3501 Market Street, Philadelphia, PA 19104. (800) 523-1850 or (215) 386-0100. Broad multidisciplinary title and author index to the international literature of science and technology, 1974 to present. Inquire as to cost and availability.

PERIODICALS

AMERICAN METEOROLOGICAL SOCIETY BULLETIN. American Meteorological Society, 45 Beacon Street, Boston, MA 02108. (617) 227-2425. 1920 to present. Monthly. $60.00 per year.

DYNAMICS OF ATMOSPHERES AND OCEANS. Elsevier Science Publishing Company, Incorporated, 52 Vanderbilt Avenue, New York, NY 10017. (212) 370-5520. 1977 to present. Quarterly. $90.00 per year.

JOURNAL OF CLIMATE AND APPLIED METEOROLOGY. American Meteorological Society, 45 Beacon Street, Boston, MA 02108. (617) 227-2425. 1962 to present. Monthly. $120.00 per year.

JOURNAL OF PHYSICAL OCEANOGRAPHY. American Meteorological Society, 45 Beacon Street, Boston, MA 02108. (617) 227-2425. 1971 to present. Monthly. $120.00 per year.

JOURNAL OF THE ATMOSPHERIC SCIENCES. American Meteorological Society, 45 Beacon Street, Boston, MA 02108. (617) 227-2425. 1944 to present. Semimonthly. $220.00 per year.

MONTHLY WEATHER REVIEW. American Meteorological Society, 45 Beacon Street, Boston, MA 02108. (617) 227-2425. 1872 to present. Monthly. $120.00 per year.

NATIONAL WEATHER DIGEST. National Weather Association, 4400 Stamp Raod, Room 404, Temple Hills, MD 20748. (301) 899-3784. 1976 to present. Quarterly. $20.00 per year.

WEATHER. Royal Meteorological Society, James Glaisher House, Grenville Place, Bracknell Berkshire, RG12 1BX, England. 1946 to present. $30.00 per year.

WEATHERWISE. Heldref Publications, 4000 Albemarle Street, NW, Washington, DC 20016. (202) 362-6445. 1948 to present. Bimonthly. $20.00 per year.

RESEARCH CENTERS AND INSTITUTES

CLOUD SIMULATION AND AEROSOL LABORATORY. Colorado State University, Fort Collins, CO 80523. (030) 491-8667.

COOPERATIVE INSTITUTE FOR MESOSCALE METEOROLOGICAL STUDIES. University of Oklahoma, 401 East Boyd, Norman, OK 73019. (405) 325-3041.

GRADUATE CENTER FOR CLOUD PHYSICS RESEARCH. University of Missouri - Rolla, Rolla, MO 65401. (314) 341-4332.

NATIONAL CENTER FOR ATMOSPHERIC RESEARCH. Box 3000, Boulder, CO 80307. (303) 497-1000.

UNIVERSITY OF ARIZONA. Institute of Atmospheric Physics, Tucson, AZ 85721. (602) 626-6831.

RESEARCH CENTERS AND INSTITUTES

LEHIGH UNIVERSITY. Institute of Thermo-Fluid Engineering and Science, Whitaker Laboratory, Bethlehem, PA 18015. (215) 861-4091.

OHIO STATE UNIVERSITY. Fluid Power Laboratory, Mechanical Engineering Department, 206 West 18th Avenue, Columbus, OH 43210. (614) 422-2289.

OKLAHOMA STATE UNIVERSITY. Fluid Power Research Center, Stillwater, OK 74074. (405) 624-7375.

PURDUE UNIVERSITY. Fluid Mechanics Laboratory, School of Mechanical Engineering, West Lafayette, IN 47907. (317) 494-5633.

STANFORD UNIVERSITY. Environmental Fluid Mechanics Laboratory, Department of Civil Engineering, Stanford, CA 94305. (415) 723-1825.

UNIVERSITY OF ARIZONA. Computational Fluid Mechanics Laboratory, Building #16, Room 312, Tucson, Az 85721. (602) 621-4423.

FOLDS (GEOLOGY)

See: PHYSICAL GEOLOGY

FORTRAN

See: PROGRAMMING LANGUAGES

FOUNDRIES

See also: ALLOYS, CAST IRON, FERROALLOYS, IRON, MACHINING, METALLURGICAL ENGINEERING, METALLURGY, METALS AND METALWORKING, STEEL AND STEEL MAKING

ABSTRACT SERVICES AND INDEXES

ALLOYS INDEX. American Society for Metals, Metals Park, OH 44073. (216) 338-5151. 1974 to present. Monthly. $225.00.

APPLIED SCIENCE AND TECHNOLOGY INDEX. H.W. Wilson and Company, 950 University Avenue, Bronx, NY 10452. (800) 367-6670 or (212) 588-8400. Monthly. Inquire as to cost and availability.

CHEMICAL ABSTRACTS. American Chemical Society, Chemical Abstracts Service, Box 3012, Columbus, OH 43210. (614) 421-3600. 1907 to present. Weekly. $9500.00 per year.

CORROSION ABSTRACTS. National Association of Corrosion Engineers, Box 218340, Houston, TX 77218. (713) 492-0535. 1962 to present. Bimonthly. $200.00 per year.

CURRENT CONTENTS: ENGINEERING, TECHNOLOGY AND APPLIED SCIENCES. Institute for Scientific Information, 3501 Market Street, Philadelphia, PA 19104. (800) 523-1850 or (215) 386-0100. Weekly. $275.00 per year.

ENGINEERING INDEX MONTHLY AND AUTHOR INDEX. Engineering Information Inc., 345 East 47th Street, New York, NY 10017. (212) 705-7600. Monthly. $1560.00 per year.

METALS ABSTRACTS AND METALS ABSTRACTS INDEX. American Society for Metals, Metals Park, OH 44073. (216) 338-5151. 1968 to present. Monthly. Abstracts are $1100.00 per year and Index is $460.00 per year.

SCIENCE CITATION INDEX. Institute for Scientific Information, 3501 Market Street, Philadelphia, PA 19104. (800) 523-1850 or (215) 386-0100. Six times per year. $6200. 00 per year.

ASSOCIATIONS AND PROFESSIONAL SOCIETIES

AMERICAN FOUNDRYMEN'S SOCIETY. Golf and Wolf Roads, Des Plaines, IL 60016. (312) 824-0181.

AMERICAN INSTITUTE OF MINING, METALLURGICAL AND PETROLEUM ENGINEERS (AIME). 420 Commonwealth Drive, Warrendale, PA 15086. (412) 776-9086.

AMERICAN IRON AND STEEL INSTITUTE. 1000 Sixteenth Street, N.W., Washington, DC 20036. (202) 452-7100.

AMERICAN POWDER METALLURGY INSTITUTE. 105 College Road, East, Princeton, NJ 08540. (609) 452-7700.

AMERICAN SOCIETY FOR METALS. Metals Park, OH 44073. (216) 338-5151.

AMERICAN SOCIETY FOR TESTING AND MATERIALS. 1916 Race Street, Philadelphia, PA 19103. (215) 299-5400.

ASSOCIATION OF IRON AND STEEL ENGINEERS. Three Gateway Center, Suite 2350, Pittsburgh, PA 15222. (412) 281-6323.

IRON CASTINGS SOCIETY. 455 State Street, Des Plaines, IL 60016. (312) 299-9160.

THE METALLURGICAL SOCIETY. 420 Commonwealth Drive, Warrendale, PA 15086. (412) 776-9000.

NATIONAL ASSOCIATION OF CORROSION ENGINEERS. Box 218340, Houston, TX 77218. (713) 492-0535.

FOUNDRIES

DIRECTORIES AND BIOGRAPHICAL SOURCES

DUN'S INDUSTRIAL GUIDE: THE METALWORKING DIRECTORY. Dun and Bradstreet Corporation, 49 Old Bloomfield Road, Mountain Lakes, NJ 07046. (201) 953-0300. Annual. $610.00.

INDUSTRIAL EQUIPMENT AND SUPPLIES DIRECTORY. American Business Directories, Inc., 5707 South 86th Circle, Omaha, NE 68127. (402) 331-7169.

MACHINERY BUYERS GUIDE. Findlay Publications Limited, Maitland House, Warrior Square, Southend-on-Sea, Essex SS5 5AR England. Annual. $45.00.

METALLURGICAL SOCIETY OF AIME - MEMBERSHIP LIST. American Institute of Mining, Metallurgical and Petroleum Engineers (AIME). 345 East 47th Street, New York, NY 10017. (212) 705-7695. 1984.

1987 DIRECTORY OF ENGINEERING SOCIETIES AND RELATED ORGANIZATIONS. Gordon Davis, editor. Hemisphere Publishing Corporation, 79 Madison Avenue, New York, NY 10016-7892. (800) 821-8312. 12th edition. 1987. $100.00.

RESEARCH CENTERS DIRECTORY 1988. Gale Research Company, Book Tower, Detroit, MI 48226. (800) 521-0707. 12th edition. 1987. $365.00 for set.

SCIENTIFIC AND TECHNICAL ORGANIZATIONS AND AGENCIES DIRECTORY. Margaret Labash Young, editor. Gale Research Company, Book Tower, Detroit, MI 48226. (800) 521-0707. 2nd edition. 1987. $185.00.

WHO'S WHO IN ENGINEERING. Gordon Davis, editor. Hemisphere Publishing Corporation, 1010 Vermont Avenue, NW, Washington, DC 20005. (800) 526-0275. 6th edition. 1985. $200.00.

GENERAL WORKS

ADVANCED CASTING TECHNOLOGY. J. Easwaran, editor. American Society for Metals, Metals Park, OH 44073. (216) 338-5151. 1987. $72.00.

FOUNDRY ENGINEERING. American Foundrymen's Society, Golf and Wolf Roads, Des Plaines, IL 60016. (312) 824-0181. Second edition. $32.95.

HANDBOOKS AND MANUALS

HANDBOOK OF METAL FORMING. Kurt Lange, editor. John Wiley and Sons, Inc., 605 Third Avenue, New York, NY 10158. (800) 526-5368. 1985. $89.50.

HANDBOOK OF METAL FORMING PROCESSES. B. Avitzur. John Wiley and Sons, Inc., 605 Third Avenue, New York, NY 10158. (800) 526-5368. 1983. $105.00.

METALS HANDBOOK. American Society for Metals, Metals Park, OH 44073. (216) 338-5151. 9th edition. 14 volumes. 1988. $1310.00 for set.

WOLDMAN'S ENGINEERING ALLOYS. Robert C. Gibbons, editor. American Society for Metals, Metals Park, OH 44073. (216) 338-5151. 6th edition. 1979. $112.00.

ONLINE DATA BASES

CA SEARCH. Chemical Abstracts Service, P.O. Box 3012, Columbus, OH 43120. (800) 848-6538 or (614) 421-3600. Comprehensive guide to chemical literature, 1972 to present. Inquire as to online cost and availability.

COMPENDEX. Engineering Information, Inc., 345 East 47th Street, New York, NY 10017. (800) 221-1044 or (212) 705-7615. Engineering and technical literature, 1975 to present. Inquire as to online cost and availability.

NTIS. National Technical Information Service, 5285 Port Royal Road, Springfield, VA 22161. (703) 487-4630. Broad coverage of government sponsored research reports, 1964 to present. Inquire as to online cost and availability.

SCISEARCH. Institute for Scientific Information, 3501 Market Street, Philadelphia, PA 19104. (800) 523-1850 or (215) 386-0100. Broad multidisciplinary title and author index to the international literature of science and technology, 1974 to present. Inquire as to online cost and availability.

WILSONLINE. H.W. Wilson and Company, 950 University Avenue, Bronx, NY 10452. (800) 367-6770 or (212) 588-8400. Makes available online versions of the H.W. Wilson indexes including Applied Science and Technology Index, Business Periodicals Index and Readers' Guide to Periodical Literature. Approximately 1980 to present. Inquire as to online cost and availability.

PERIODICALS

CASTING ENGINEER AND FOUNDRY WORLD. Continental Communications, Inc., 1115 Main Street, Bridgeport, CT 06604. (203) 377-5566. 1969 to present. Quarterly. $20.00 per year.

FOUNDRY MANAGEMENT AND TECHNOLOGY. Penton-IPC, 1100 Superior Avenue, Cleveland, OH 44114. (216) 696-7000. 1892 to present. Monthly. $35.00 per year.

INDUSTRY WEEK. Penton-IPC, 1100 Superior Avenue, Cleveland, OH 44114. (216) 696-7000. 1882 to present. 26 times per year. $50.00 per year.

IRON AND STEEL ENGINEER. Association of Iron and Steel Engineers, Suite 2350, Three Gateway Center, Pittsburgh, PA 15222. (412) 281-6323. 1924 to present. Monthly. $34.00 per year.

I. S. AND M. (IRON AND STEELMAKER). Iron and Steel Society, 410 Commonwealth Drive, Warrendale, PA 15096. (412) 776-1535. 1974 to present. Monthly. $35.00 per year.

IRONCASTER. Iron Castings Society, 455 State Street, Des Plaines, IL 60016. (312) 299-9160. 1947 to present. Bimonthly. $14.50 per year.

JOURNAL OF APPLIED METALWORKING. American Society for Metals, Metals Park, OH 44073. (216) 338-5151. 1979 to present. Semi-annual. $80.00 for two years.

JOURNAL OF ENGINEERING FOR INDUSTRY. American Society of Mechanical Engineers, 345 East 47th Street, New York, NY 10017. (212) 705-7703. 1970 to present. Quarterly. $100.00 per year.

JOURNAL OF METALS. American Institute of Mining, Metallurgical, and Petroleum Engineers, Inc., Metallurgical Society, 420 Commonwealth Drive, Warrendale, PA 15086. (412) 776-9086. 1949 to present. Monthly. $40.00 per year.

METAL PROGRESS. American Society for Metals, Metals Park, OH 44073. (216) 338-5151. 1930 to present. Monthly. $40.00.

METALLURGICAL TRANSACTIONS. Metallurgical Society of AIME, 420 Commonwealth Drive, Warrendale, PA 15086. (412) 776-9080. 1970 to present. Monthly. $95.00 per year.

METALS WEEK. McGraw-Hill Book Company, 1221 Avenue of the Americas, New York, NY 10020. (212) 997-2823. 1930 to present. Weekly. $597.00 per year.

METALWORKING DIGEST. Philos Publications, Inc., 1 East Chase Street, Baltimore, MD 21202. (301) 361-9060. 1969 to present. Nine times per year. $9.00 per year.

METIFAX MAGAZINE. Huebner Publications Inc., 6521 Davis Industrial Parkway, Solon, OH 44139. (216) 248-1125. 1956 to present. Monthly. $40.00 per year.

RESEARCH CENTERS AND INSTITUTES

ADVANCED STEEL PROCESSING AND PRODUCTS RESEARCH CENTER. Colorado School of Mines, Golden, CO 80401. (303) 273-3774.

AMERICAN IRON AND STEEL INSTITUTE. 1000 Sixteenth Street, N.W., Washington, DC 20036. (202) 452-7100.

CANADIAN INSTITUTE OF METALWORKING. 1276 Sandhill Drive, P.O. Box 7317, Ancaster, ON, Canada L9G 3N6.

CENTER FOR IRON AND STEEL MAKING RESEARCH. Carnegie-Mellon University, MEMS Department, Pittsburgh, PA 15213. (412) 268-2677.

COOPERATIVE PROGRAM IN METALLURGY. Pennsylvania State University, 208A Steidle Building, University Park, PA 16802. (814) 865-5446.

STATISTICS SOURCES

WORLD METAL STATISTICS. World Bureau of Metal Statistics, 41 Doughty Street, London, WC1N 2LF, England. 1948 to present. Monthly. $1250.00 per year.

FOURIER ANALYSIS

See: MATHEMATICS

FRACTIONATING COLUMNS

See: CHEMICAL ENGINEERING

FRACTURE MECHANICS

See also: AERONAUTICAL ENGINEERING, CIVIL ENGINEERING, FATIGUE, MATERIALS SCIENCE, MECHANICAL ENGINEERING, MECHANICS, NONDESTRUCTIVE TESTING, STRUCTURAL ENGINEERING

ABSTRACT SERVICES AND INDEXES

ABSTRACT JOURNAL IN EARTHQUAKE ENGINEERING. University of California at Berkeley, Earthquake Engineering Research Center, 1301 South 46th Street, Richmond, CA 94804. (415) 231-9413. Semiannual. $70.00 per copy.

APPLIED SCIENCE AND TECHNOLOGY INDEX. H.W. Wilson and Company, 950 University Avenue, Bronx, NY 10452. (800) 367-6670 or (212) 588-8400. Monthly. Inquire as to cost and availability.

CURRENT CONTENTS: ENGINEERING, TECHNOLOGY AND APPLIED SCIENCES. Institute for Scientific Information, 3501 Market Street, Philadelphia, PA 19104. (800) 523-1850 or (215) 386-0100. Weekly. $275.00 per year.

ENGINEERING INDEX MONTHLY AND AUTHOR INDEX. Engineering Information Inc., 345 East 47th Street, New York, NY 10017. (212) 705-7600. Monthly. $1560.00 per year.

INDEX TO SCIENTIFIC AND TECHNICAL PROCEEDINGS. Institute for Scientific Information, 3501 Market Street, Philadelphia, PA 19104. (800) 523-1850 or (215) 386-0100. 1978 to present. Monthly. $775.00 per year.

INTERNATIONAL CIVIL ENGINEERING ABSTRACTS. CITIS Limited, 2 Rosemount Terrace, Blackrock, Dublin, Ireland. 1974 to present. Monthly. $350.00 per year.

INTERNATIONAL STRUCTURAL ENGINEERING ABSTRACTS. CITIS Limited, 2 Rosemount Terrace, Blackrock, Dublin, Ireland. 1986 to present. Quarterly. $95.00 per year.

ISMEC BULLETIN (Information Service in Mechanical Engineering). Cambridge Scientific Abstracts, 5161 River Road, Bethesda, MD 20816. (301) 951-1400. 1973 to present. Monthly. $450.00 per year.

PUBLICATIONS INFORMATION. American Society of Civil Engineers, 345 East 47th Street, New York, NY 10017. (212) 705-7420. Abstracts, subject and author indexes to the publications of the American Society of Civil Engineers. Bimonthly. $80.00 per year.

SCIENCE CITATION INDEX. Institute for Scientific Information, 3501 Market Street, Philadelphia, PA 19104. (800) 523-1850 or (215) 386-0100. Six times per year. $6200.00 per year.

ASSOCIATIONS AND PROFESSIONAL SOCIETIES

AMERICAN SOCIETY FOR NONDESTRUCTIVE TESTING. 4153 Arlingate Plaza, Caller 28518, Columbus, OH 43228. (614) 274-6003.

AMERICAN SOCIETY FOR TESTING AND MATERIALS. 1916 Race Street, Philadelphia, PA 19103. (215) 299-5400.

AMERICAN SOCIETY OF CIVIL ENGINEERS. 345 East 47th Street, New York, NY 10017. (212) 705-7420.

AMERICAN SOCIETY OF MECHANICAL ENGINEERS. 345 East 47th Street, New York, NY 10017. (212) 705-7703.

DIRECTORIES AND BIOGRAPHICAL SOURCES

INTERNATIONAL RESEARCH CENTERS DIRECTORY 1988-89. Darren L. Smith, editor. Gale Research Company, Book Tower, Detroit, MI 48226. (800) 521-0707. 4th edition. 1987. $360.00.

1987 DIRECTORY OF ENGINEERING SOCIETIES AND RELATED ORGANIZATIONS. Gordon Davis, editor. Hemisphere Publishing Corporation, 1010 Vermont Avenue, NW, Washington, DC 20005. (800) 526-0275. 12th edition. 1987. $100.00.

RESEARCH CENTERS DIRECTORY 1988. Gale Research Company, Book Tower, Detroit, MI 48226. (800) 521-0707. 12th edition. 1987. $365.00 for set.

SCIENTIFIC AND TECHNICAL ORGANIZATIONS AND AGENCIES DIRECTORY. Margaret Labash Young, editor. Gale Research Company, Book Tower, Detroit, MI 48226. (800) 521-0707. 2nd edition. 1987. $185.00.

WHO'S WHO IN ENGINEERING. Gordon Davis, editor. Hemisphere Publishing Corporation, 1010 Vermont Avenue, NW, Washington, DC 20005. (800) 526-0275. 6th edition. 1985. $200.00.

GENERAL WORKS

ADVANCED MECHANICS OF MATERIALS. A.P. Boresi and O.M. Sidebottom. John Wiley and Sons, Inc., 605 Third Avenue, New York, NY 10158. (800) 526-5368. Fourth edition. 1985. $44.00.

ELEMENTARY ENGINEERING FRACTURE MECHANICS. David Broek. Martinus Nijhof. Distributed by Kluwer Academic Publishers, 190 Old Derby Street, Hingham, MA 02043. (617) 749-5262. Fourth edition. 1986. $110.00.

ELEMENTARY THEORY OF STRUCTURES. Yu H. Yuan. Prentice-Hall Publishing, Inc., Englewood Cliffs, NJ 07632. (800) 562-0245. Third edition. 1988. $42.95.

FAILURE OF MATERIALS IN MECHANICAL DESIGN: ANALYSIS, PREDICTION, PREVENTION. J.A. Collins. John Wiley and Sons, Inc., 605 Third Avenue, New York, NY 10158.

(800) 526-5368. 1981. $59.95.

FATIGUE AND FRACTURE IN STEEL BRIDGES: CASE STUDIES. John W. Fisher. John Wiley and Sons, Inc., 605 Third Avenue, New York, NY 10158. (800) 526-5368. 1984. $45.95.

FRACTURE AND FATIGUE CONTROL IN STRUCTURES: APPLICATIONS OF FRACTURE MECHANICS. John M. Barson and Stanley T. Rolfe. Prentice-Hall Publishing, Inc., Englewood Cliffs, NJ 07632. (800) 562-0245. 1987. $46.95.

PROBABILISTIC FRACTURE MECHANICS AND RELIABILITY. James W. Provan, editor. Martinus Nijhoff. Distributed by Kluwer Academic Publishers, 190 Old Derby Street, Hingham, MA 02043. (617) 749-5262. 1987. $135.50.

HANDBOOKS AND MANUALS

HANDBOOK OF MECHANICS, MATERIALS, AND STRUCTURES. A. Blake. John Wiley and Sons, Inc., 605 Third Avenue, New York, NY 10158. (800) 526-5368. 1985. $64.50.

MECHANICAL ENGINEER'S HANDBOOK. Myer Kutz, editor. John Wiley and Sons, Inc., 605 Third Avenue, New York, NY 10158. (800) 526-5368. 1986. $79.95.

ONLINE DATA BASES

COMPENDEX. Engineering Information, Inc., 345 East 47th Street, New York, NY 10017. (800) 221-1044 or (212) 705-7615. Engineering and technical literature, 1975 to present. Inquire as to online cost and availability.

DISSERTATION ABSTRACTS ONLINE. University Microfilms International, 300 North Zeeb Road, Ann Arbor, MI 48106. (800) 521-0600 or (313) 761-4700. Scope includes virtually all doctoral dissertations accepted at accredited American institutions from 1861 to present in over 250 subject areas. Inquire as to online cost and availability.

INSPEC. INSPEC Marketing Department, Institution of Electrical Engineers. Available from IEEE Service Center, 445 Hoes Lane, Piscataway, NJ 08854. (201) 981-0060. Online version of Physics Abstracts. Inquire as to online cost and availability.

NTIS. National Technical Information Service, 5285 Port Royal Road, Springfield, VA 22161. (703) 487-4630. Broad coverage of government sponsored research reports, 1964 to present. Inquire as to online cost and availability.

SCISEARCH. Institute for Scientific Information, 3501 Market Street, Philadelphia, PA 19104. (800) 523-1850 or (215) 386-0100. Broad multidisciplinary title and author index to the international literature of science and technology, 1974 to present. Inquire as to online cost and availability.

WILSONLINE. H.W. Wilson and Company, 950 University Avenue, Bronx, NY 10452. (800) 367-6770 or (212) 588-8400. Makes available online versions of the H.W. Wilson indexes including Applied Science and Technology Index, Business Periodicals Index and Readers' Guide to Periodical Literature. Approximately 1980 to present. Inquire as to online cost and availability.

PERIODICALS

JOURNAL OF ENGINEERING MECHANICS. American Society of Civil Engineers, 345 East 47th Street, New York, NY 10017. (212) 705-7420. 1956 to present. Monthly. $120.00 per year.

JOURNAL OF STRUCTURAL MECHANICS. Marcel Dekker Inc., 270 Madison Avenue, New York, NY 10016. (800) 228-1160. 1972 to present. Quarterly. $75.00 per year.

JOURNAL OF TESTING AND EVALUATION. American Society for Testing and Materials, 1916 Race Street, Philadelphia, PA 19103. (215) 299-5400. 1966 to present. Bimonthly. $40.00 per year.

MATERIALS ENGINEERING. Penton-IPC, 1100 Superior Avenue, Cleveland, OH 44114. (216)696-7000. 1929 to present. Monthly. $40.00 per year.

MATERIALS EVALUATION. American Society for Nondestructive Testing, 4153 Arlingate Plaza, Caller 28518, Columbus, OH 43228. 1942 to present. Monthly. $50.00 per year.

SAMPE JOURNAL. Society for the Advancement of Material and Process Engineering, 843 West Glentana, Box 2459, Covina, CA 91722. (818) 331-0610. 1965 to present. Bimonthly. $31.00 per year.

STRAIN. British Society for Strain Measurement, Suite 7, Exchange Building, Quayside, Newcastle-upon-Tyne, NE1 3BJ, England. 1965 to present. Quarterly. $45.00 per year.

RESEARCH CENTERS AND INSTITUTES

CENTER ON QUALITY ENGINEERING AND FAILURE PREVENTION. Northwestern University, Technological Institute, Evanston, IL 60201. (312) 491-5527.

STRUCTURAL ENGINEERING MATERIALS LABORATORY. University of California, Berkeley, Davis Hall, Berkeley, CA 94720. (415) 642-3434.

STRUCTURAL STABILITY RESEARCH COUNCIL. Fritz Engineering Laboratory No. 13, Lehigh University, Bethlehem, PA 18105. (215) 861-3519.

WISCONSIN CENTER FOR STRUCTURAL AND MATERIALS TESTING. University of Wisconsin at Madison, 1415 Johnson Drive, Madison, WI 53706. (608) 262-3205.

SPECIFICATIONS AND STANDARDS

ELASTIC-PLASTIC FRACTURE MECHANICS TECHNOLOGY. J.C. Newman and F.J. Loss, editors. American Society for Testing and Materials, 1916 Race Street, Philadelphia, PA 19103. (215) 299-5400. STP No. 896. 1986. $30.00.

FREQUENCY MODULATION

See: RADIO

FRICTION

See also: LUBRICATION, MECHANICAL ENGINEERING, MACHINERY, MECHANICS

ABSTRACT SERVICES AND INDEXES

APPLIED MECHANICS REVIEW. American Society of Mechanical Engineers, 345 East 47th Street, New York, NY 10017. (212) 705-7703. 1948 to present. Monthly. $360.00 per year.

APPLIED SCIENCE AND TECHNOLOGY INDEX. H.W. Wilson and Company, 950 University Avenue, Bronx, NY 10452. (800) 367-6670 or (212) 588-8400. Monthly. Inquire as to cost and availability.

CHEMICAL ABSTRACTS. American Chemical Society, Chemical Abstracts Service, Box 3012, Columbus, OH 43210. (614) 421-3600. 1907 to present. Weekly. $9500.00 per year.

CORROSION ABSTRACTS: ABSTRACTS OF THE WORLD'S LITERATURE ON CORROSION AND CORROSION MITIGATION. National Association of Corrosion Engineers, Box 218340, Houston, TX 77218. (713) 492-0535. 1962 to present.

Bimonthly. $250.00 per year.

CURRENT CONTENTS: ENGINEERING, TECHNOLOGY AND APPLIED SCIENCES. Institute for Scientific Information, 3501 Market Street, Philadelphia, PA 19104. (800) 523-1850 or (215) 386-0100. Weekly. $275.00 per year.

ENGINEERING INDEX MONTHLY AND AUTHOR INDEX. Engineering Information Inc., 345 East 47th Street, New York, NY 10017. (212) 705-7600. Monthly. $1560.00 per year.

ISMEC BULLETIN (Information Service in Mechanical Engineering). Cambridge Scientific Abstracts, 5161 River Road, Bethesda, MD 20816. (301) 951-1400. 1973 to present. Monthly. $450.00 per year.

SCIENCE CITATION INDEX. Institute for Scientific Information, 3501 Market Street, Philadelphia, PA 19104. (800) 523-1850 or (215) 386-0100. Six times per year. $6200.00 per year.

ASSOCIATIONS AND PROFESSIONAL SOCIETIES

AMERICAN SOCIETY FOR METALS. Metals Park, OH 44073. (216) 338-5151.

AMERICAN SOCIETY OF LUBRICATION ENGINEERS. 838 Busse Highway, Park Ridge, IL 60068. (312) 825-5536.

AMERICAN SOCIETY OF MECHANICAL ENGINEERS. 345 East 47th Street, New York, NY 10017. (212) 705-7703.

METALLURGICAL SOCIETY OF THE AIME (AMERICAN INSTITUTE OF MINING, METALLURGICAL AND PETROLEUM ENGINEERS). 420 Commonwealth Drive, Warrendale, PA 15086. (412) 776-9080.

NATIONAL ASSOCIATION OF CORROSION ENGINEERS. Box 218340, Houston, TX 77218. (713) 492-0535.

NATIONAL LUBRICATING GREASE INSTITUTE. 4635 Wyandotte Street, Kansas City, MO 64112. (816) 931-9480.

SOCIETY OF AUTOMOTIVE ENGINEERS (SAE). 400 Commonwealth Drive, Warrendale, PA 15096. (412) 776-4841.

DIRECTORIES AND BIOGRAPHICAL SOURCES

AMERICAN SOCIETY OF LUBRICATION ENGINEERS MEMBERSHIP ROSTER. American Society of Lubrication Engineers, 838 Busse Highway, Park Ridge, IL 60068. (312) 825-5536. Annual. $20.00.

1987 DIRECTORY OF ENGINEERING SOCIETIES AND RELATED ORGANIZATIONS. Gordon Davis, editor. Hemisphere Publishing Corporation, 1010 Vermont Avenue, NW, Washington, DC 20005. (800) 526-0275. 12th edition. 1987. $100.00.

RESEARCH CENTERS DIRECTORY 1988. Gale Research Company, Book Tower, Detroit, MI 48226. (800) 521-0707. 12th edition. 1987. $365.00 for set.

SCIENTIFIC AND TECHNICAL ORGANIZATIONS AND AGENCIES DIRECTORY. Margaret Labash Young, editor. Gale Research Company, Book Tower, Detroit, MI 48226. (800) 521-0707. 2nd edition. 1987. $185.00.

WHO'S WHO IN ENGINEERING. Gordon Davis, editor. Hemisphere Publishing Corporation, 1010 Vermont Avenue, NW, Washington, DC 20005. (800) 526-0275. 6th edition. 1985. $200.00.

ENCYCLOPEDIAS AND DICTIONARIES

THESAURUS OF SCIENTIFIC, TECHNICAL, AND ENGINEERING TERMS. Hemisphere Publishing Corporation, 1010 Vermont Avenue, NW, Washington, DC 20005. (800) 526-0275. 1988. $125.00.

GENERAL WORKS

BASIC LUBRICATION THEORY. A. Cameron and C.M. Ettles. Halsted Press, a division of John Wiley and Sons, Inc., 605 Third Avenue, New York, NY 10158. (800) 526-5368. Third edition. 1982. $37.95.

ENGINEERING MECHANICS. J.L. Meriam and L.G. Kraige. John Wiley and Sons, Inc., 605 Third Avenue, New York, NY 10158. (800) 526-5368. 2nd edition. 1986. $52.95.

LUBRICATION: A PRACTICAL GUIDE TO LUBRICANT SELECTION. A.R. Lansdown. Pergamon Press, Inc., Maxwell House, Fairview Park, Elmsford, NY 10523. (914) 592-7700. 1982. $19.95 in paper.

NEW DIRECTIONS IN LUBRICATION, MATERIALS, WEAR AND SURFACE INTERACTION: TRIBOLOGY IN THE 80'S. William R. Loomis, editor. Noyes Data Corporation, Mill Road at Grand Avenue, Park Ridge, NJ 07656. (201) 391-8484. 1985. $69.00.

THEORY AND PRACTICE OF LUBRICATION FOR ENGINEERS. D.D. Fuller. John Wiley and Sons, Inc., 605 Third Avenue, New York, NY 10158. (800) 526-5368. 2nd edition. 1984. $64.95.

HANDBOOKS AND MANUALS

FRICTION, WEAR, AND LUBRICATION: A COMPLETE HANDBOOK OF TRIBOLOGY. I.V. Kregelsky an V.V. Alisin, editors. Pergamon Press, Inc., Maxwell House, Fairview Park, Elmsford, NY 10523. (914) 592-7700. Three volumes. 1982. $130.00 for set.

HANDBOOK OF LUBRICATION. American Society of Lubrication Engineers, 838 Busse Highway, Park Ridge, IL 60068. (312) 825-5536. Three volumes. 1983. $315.00 for set.

ONLINE DATA BASES

APILIT. American Petroleum Institute, Central Abstracting and Indexing Service, 2101 L Street, N.W., Washington, DC 20037. (202) 682-8000. Worldwide petroleum literature, 1964 to present. Inquire as to online cost and availability.

CA SEARCH. Chemical Abstracts Service, P.O. Box 3012, Columbus, OH 43120. (800) 848-6538 or (614) 421-3600. Comprehensive guide to chemical literature, 1972 to present. Inquire as to online cost and availability.

COMPENDEX. Engineering Information, Inc., 345 East 47th Street, New York, NY 10017. (800) 221-1044 or (212) 705-7615. Engineering and technical literature, 1975 to present. Inquire as to online cost and availability.

NTIS. National Technical Information Service, 5285 Port Royal Road, Springfield, VA 22161. (703) 487-4630. Broad coverage of government sponsored research reports, 1964 to present. Inquire as to online cost and availability.

SCISEARCH. Institute for Scientific Information, 3501 Market Street, Philadelphia, PA 19104. (800) 523-1850 or (215) 386-0100. Broad multidisciplinary title and author index to the international literature of science and technology, 1974 to present. Inquire as to online cost and availability.

WILSONLINE. H.W. Wilson and Company, 950 University Avenue, Bronx, NY 10452. (800) 367-6770 or (212) 588-8400. Makes available online versions of the H.W. Wilson indexes including Applied Science and Technology Index, Business Periodicals Index and Readers' Guide to Periodical Literature. Approximately 1980 to present. Inquire as to online cost and availability.

OTHER SOURCES

WHAT EVERY ENGINEER SHOULD KNOW ABOUT ENGINEERING SOURCES. Marcel Dekker Inc., 270 Madison Avenue, New York, NY 10016. (800) 228-1160. 1984. $24.95.

PERIODICALS

JOURNAL OF APPLIED MECHANICS. American Society of Mechanical Engineers, 345 East 47th Street, New York, NY 10017. (212) 705-7703. 1935 to present. Quarterly. $100.00 per year.

JOURNAL OF ENGINEERING FOR INDUSTRY. American Society of Mechanical Engineers, 345 East 47th Street, New York, NY 10017. (212) 705-7703. 1970 to present. Quarterly. $100.00 per year.

JOURNAL OF FLUID CONTROL. Delbridge Publishing Company, Box 2989, Stanford, CA 94305. (408) 446-3131. 1967 to present. Quarterly. Inquire as to cost and availability.

JOURNAL OF MECHANISMS, TRANSMISSIONS AND AUTOMATION IN DESIGN. American Society of Mechanical Engineers, 345 East 47th Street, New York, NY 10017. (212) 705-7703. 1983 to present. Quarterly. $80.00 per year.

JOURNAL OF TRIBOLOGY. American Society of Mechanical Engineers, 345 East 47th Street, New York, NY 10017. (212) 705-7703. 1967 to present. Quarterly. $80.00 per year.

LUBRICATION ENGINEERING. American Society of Lubrication Engineers, 838 Busse Highway, Park Ridge, IL 60068. (312) 825-5536. Monthly. $39.00 per year.

MACHINE DESIGN. Penton-IPC, 1100 Superior Avenue, Cleveland, OH 44114. (216) 696-7000. 1929 to present. Twenty-eight times per year. $60.00 per year.

MECHANICAL ENGINEERING. American Society of Mechanical Engineers, 345 East 47th Street, New York, NY 10017. (212) 705-7722.

N.L.G.I. SPOKESMAN. National Lubrication Grease Institute. 4635 Wyandotte Street, Kansas City, MO 64112. (816) 931-9480. 1937 to present. Monthly. $10.00 per year.

RESEARCH CENTERS AND INSTITUTES

COORDINATING RESEARCH COUNCIL, INC. 219 Perimeter Center Parkway, Atlanta, GA 30346. (404) 396-3400.

FLUID POWER RESEARCH CENTER. Oklahoma State University, Stillwater, OK 74074. (405) 624-7375.

INSTITUTE FOR WEAR CONTROL AND TRIBOLOGY. Rensselaer Polytechnic Institute, Mechanical Engineering Department, Jonsson Engineering Center, Troy, NY 12181. (518) 266-6000.

FRONT (METEOROLOGY)

See: METEOROLOGY

FUEL CELLS

See: ELECTROCHEMISTRY

FUELS

See also: COAL, ENERGY, NATURAL GAS, NUCLEAR ENERGY, PETROLEUM, SYNTHETIC FUELS

ABSTRACT SERVICES AND INDEXES

ABSTRACT NEWSLETTER: ENERGY. National Technical Information Service, 5285 Port Royal Road, Springfield, VA 22161. (703) 487-4929. Weekly. $95.00 per year.

APPLIED SCIENCE AND TECHNOLOGY INDEX. H.W. Wilson and Company, 950 University Avenue, Bronx, NY 10452. (800) 367-6670 or (212) 588-8400. Monthly. Inquire as to cost and availability.

CHEMICAL ABSTRACTS. American Chemical Society, Chemical Abstracts Service, Box 3012, Columbus, OH 43210. (614) 421-3600. 1907 to present. Weekly. $9500.00 per year.

CONFERENCE PAPERS INDEX. Cambridge Scientific Abstracts, 5161 River Road, Bethesda, MD 20816. (301) 951-1400. 1972 to present. Monthly. Inquire as to cost and availability.

CURRENT CONTENTS: ENGINEERING, TECHNOLOGY AND APPLIED SCIENCES. Institute for Scientific Information, 3501 Market Street, Philadelphia, PA 19104. (800) 523-1850 or (215) 386-0100. Weekly. $275.00 per year.

ENERGY RESEARCH ABSTRACTS. U.S. Department of Energy, Office of Scientific and Technical Information, Box 62, Oak Ridge, TN 37831. (615) 576-1155. Subscribe to: U.S. Superintendent of Documents, Washington, DC 20402. 1976 to present. Semimonthly. $146.00 per year.

ENGINEERING INDEX MONTHLY AND AUTHOR INDEX. Engineering Information Inc., 345 East 47th Street, New York, NY 10017. (212) 705-7600. Monthly. $1560.00 per year.

GENERAL SCIENCE INDEX. H.W. Wilson and Company, 950 University Avenue, Bronx, NY 10452. (800) 367-6670 or (212) 588-8400. 1978 to present. Monthly. Inquire as to cost and availability.

INDEX TO SCIENTIFIC AND TECHNICAL PROCEEDINGS. Institute for Scientific Information, 3501 Market Street, Philadelphia, PA 19104. (800) 523-1850 or (215) 386-0100. 1978 to present. Monthly. $775.00 per year.

INDEX TO SCIENTIFIC REVIEWS. Institute for Scientific Information, 3501 Market Street, Philadelphia, PA 19104. (800) 523-1850 or (215) 386-0100. 1974 to present. Semi-annual. $550.00 per year.

PHYSICS ABSTRACTS. Institution of Electrical Engineers. Available from: IEEE Service Center, 445 Hoes Lane, Piscataway, NJ 08854. 1898 to present. Bimonthly. $1700.00 per year.

PHYSICS BRIEFS. Physik Verlag GmbH, Postfach 1260/1280, D-6940 Weinheim, West Germany. (212) 661-9404. 1920 to present. Twenty-six times per year. $1250.00 per year.

SCIENCE CITATION INDEX. Institute for Scientific Information, 3501 Market Street, Philadelphia, PA 19104. (800) 523-1850 or (215) 386-0100. Six times per year. $6200.00 per year.

ANNUAL REVIEWS AND YEARBOOKS

ADVANCES IN ENERGY SYSTEMS AND TECHNOLOGY. Academic Press, Inc., 6277 Sea Harbor Drive, Orlando, FL 32821. (800) 321-5068. 1979 to present. Irregular. Inquire.

ANNUAL REVIEW OF ENERGY. Annual Reviews, Inc., 4139 El Camino Way, Palo Alto, CA 94306. (415) 493-4400. Annual. Inquire.

ASSOCIATIONS AND PROFESSIONAL SOCIETIES

AMERICAN CHEMICAL SOCIETY. 1155 16th Street, N.W., Washington, DC 20036. (800) 424-6747.

AMERICAN INSTITUTE OF PHYSICS. 335 East 45th Street, New York, NY 10017. (212) 661-9494.

ASSOCIATION OF ENERGY ENGINEERS. 4025 Pleasantdale Road, Suite 340, Atlanta, GA 30340. (404) 447-5083.

EDISON ELECTRIC INSTITUTE. 1111 19th Street, N.W., Washington, DC 20036. (202) 828-7400.

INSTITUTE OF ELECTRICAL AND ELECTRONICS ENGINEERS. 345 East 47th Street, New York, NY 10017. (212) 705-7900.

BIBLIOGRAPHIES

NEW TECHNICAL BOOKS: A SELECTIVE LIST WITH DESCRIPTIVE ANNOTATIONS. New York Public Library, Science and Technology Research Center, Fifth Avenue and 42nd Street, New York, NY 10018. (212) 930-0800. 1915 to present. Monthly. $15.00 per year.

SCIENCE BOOKS AND FILMS. American Association for the Advancement of Science, 1333 H Street, NW, Washington, DC 20005. (202) 326-6454. Five times per year. $20.00 per year.

SCIENTIFIC AND TECHNICAL BOOKS AND SERIALS IN PRINT 1988; AN INDEX TO LITERATURE IN SCIENCE AND TECHNOLOGY. R.R. Bowker Company, 205 East 42nd Street, New York, NY 10017. (800) 521-8110. $175.00.

DIRECTORIES AND BIOGRAPHICAL SOURCES

AMERICAN MEN AND WOMEN OF SCIENCE. R.R. Bowker, Inc., Order Department, 245 West 17th Street, New York, NY 10011. (800) 521-8110. Eight volumes. 1986. $595.00 for set.

ENERGY INFORMATION CENTERS DIRECTORY. Public Affairs and Information Program, Atomic Industrial Forum, 7101 Wisconsin Avenue, Bethesda, MD 20814. (301) 654-9260. 1985. Free.

IEEE MEMBERSHIP DIRECTORY. Institute of Electrical and Electronics Engineers. IEEE Service Center, 445 Hoes Lane, Piscataway, NJ 08854. Annual. $7.00.

INTERNATIONAL RESEARCH CENTERS DIRECTORY 1988-89. Darren L. Smith, editor. Gale Research Company, Book Tower, Detroit, MI 48226. (800) 521-0707. 4th edition. 1987. $360.00.

1987 DIRECTORY OF ENGINEERING SOCIETIES AND RELATED ORGANIZATIONS. Gordon Davis, editor. Hemisphere Publishing Corporation, 79 Madison Avenue, New York, NY 10016-7892. (800) 821-8312. 12th edition. 1987. $100.00.

RESEARCH CENTERS DIRECTORY 1988. Gale Research Company, Book Tower, Detroit, MI 48226. (800) 521-0707. 12th edition. 1987. $365.00 for set.

SCIENTIFIC AND TECHNICAL ORGANIZATIONS AND AGENCIES DIRECTORY. Margaret Labash Young, editor. Gale Research Company, Book Tower, Detroit, MI 48226. (800) 521-0707. 2nd edition. 1987. $185.00.

WHO'S WHO IN ENGINEERING. Gordon Davis, editor. Hemisphere Publishing Corporation, 79 Madison Avenue, New York, NY 10016-7892. (800) 821-8312. 6th edition. 1985. $200.00.

GENERAL WORKS

ENERGY 2000: AN OVERVIEW OF THE WORLD'S ENERGY RESOURCES IN THE DECADES TO COME. Heinz Knoepfel. Gordon and Breach Science Publishers, Inc., 50 West 23rd Street, New York, NY 10010. (212) 206-8900. 1986. $42.00.

FUEL AND ENERGY. J.H. Harker and J. Backhurst. Academic Press, Inc., 6277 Sea Harbor Drive, Orlando, FL 32821. (800) 321-5068. 1981. $33.00 in paper.

FUELS AND FUEL TECHNOLOGY. W. Francis and M.C. Peters. Pergamon Press, Inc., Maxwell House, Fairview Park, Elmsford, NY 10523. (914) 592-7700. 1980. $42.00 in paper.

PETROLEUM RESOURCES AND DEVELOPMENT. Kameel I. Kheir, editor. Columbia University Press, 562 West 113th Street, New York, NY 10025. (212) 316-7100. 1988. $30.00.

THE SCIENCE AND TECHNOLOGY OF COAL AND COAL UTILIZATION. B.R. Cooper, editor. Plenum Publishing Corporation, 233 Spring Street, New York, NY 10013. (800) 221-9369. 1984. $85.00.

HANDBOOKS AND MANUALS

HANDBOOK OF ENERGY SYSTEMS ENGINEERING. Leslie C. Wilbur, editor. John Wiley and Sons, Inc., 605 Third Avenue, New York, NY 10158. (800) 526-5368. 1985. $74.95.

HANDBOOK OF ENERGY TECHNOLOGY AND ECONOMICS. R.A. Meyers, editor. John Wiley and Sons, Inc., 605 Third Avenue, New York, NY 10158. (800) 526-5368. 1983. $79.95.

ONLINE DATA BASES

CA SEARCH. Chemical Abstracts Service, P.O. Box 3012, Columbus, OH 43120. (800) 848-6538 or (614) 421-3600. Comprehensive guide to chemical literature, 1972 to present. Inquire as to online cost and availability.

COMPENDEX. Engineering Information, Inc., 345 East 47th Street, New York, NY 10017. (800) 221-1044 or (212) 705-7615. Engineering and technical literature, 1975 to present. Inquire as to online cost and availability.

DISSERTATION ABSTRACTS ONLINE. University Microfilms International, 300 North Zeeb Road, Ann Arbor, MI 48106. (800) 521-0600 or (313) 761-4700. Scope includes virtually all doctoral dissertations accepted at accredited American institutions from 1861 to present in over 250 subject areas. Inquire as to online cost and availability.

DOE ENERGY DATA BASE. U.S. Department of Energy, Office of Scientific and Technical Information, P.O. Box 62, Oak Ridge, TN 37831. (615) 576-6837. A database that covers all aspects of energy including the science and technology of energy. 1948 to present. Available through the DIALOG search service or DOE/RECON. Inquire as to online cost and availability.

ENERGYLINE. EIC/Intelligence, Inc., 48 West 38th Street, New York, NY 10018. (212) 944-8500. A database of resources on the scientific, engineering, political, and socioeconomic aspects of energy resources. 1976 to present. Inquire as to online cost and availability.

INSPEC. INSPEC Marketing Department, Institution of Electrical Engineers. Available from IEEE Service Center, 445 Hoes Lane, Piscataway, NJ 08854. (201) 981-0060. Online version of Physics Abstracts. Inquire as to online cost and availability.

NTIS. National Technical Information Service, 5285 Port Royal Road, Springfield, VA 22161. (703) 487-4630. Broad coverage of government sponsored research reports, 1964 to present. Inquire as to online cost and availability.

SCISEARCH. Institute for Scientific Information, 3501 Market Street, Philadelphia, PA 19104. (800) 523-1850 or (215) 386-0100. Broad multidisciplinary title and author index to the international literature of science and technology, 1974 to present. Inquire as to online cost and availability.

WILSONLINE. H.W. Wilson and Company, 950 University Avenue, Bronx, NY 10452. (800) 367-6770 or (212) 588-8400. Makes available online versions of the H.W. Wilson indexes

FUELS

including Applied Science and Technology Index, Business Periodicals Index and Readers' Guide to Periodical Literature. Approximately 1980 to present. Inquire as to online cost and availability.

OTHER SOURCES

WHAT EVERY ENGINEER SHOULD KNOW ABOUT ENGINEERING SOURCES. Marcel Dekker Inc., 270 Madison Avenue, New York, NY 10016. (800) 228-1160. 1984. $24.95.

PERIODICALS

ALTERNATIVE SOURCES OF ENERGY. Alternative Sources of Energy, Inc., 107 South Central Avenue, Milaca, MN 56353. (612) 983-6892. 1971 to present. Ten times per year. $48.00 per year.

ENERGY AND FUELS. American Chemical Society, 1155 16th Street, N.W., Washington, DC 20036. (800) 424-6747. 1987 to present. Bimonthly. $269.00 per year.

ENERGY ENGINEERING. Association of Energy Engineers. Available from: Fairmont Press, Box 14227, Atlanta, GA 30324. (404) 447-5314. 1904 to present. Bimonthly. $44.00 per year.

ENERGY SOURCES. Crane Russak and Company, Inc., 3 East 44th Street, New York, NY 10017. (212) 867-1490. 1973. Quarterly. $73.00 per year.

INTERNATIONAL JOURNAL OF SOLAR ENERGY. Harwood Academic Publishers, 50 West 23rd Street, New York, NY 10010. (212) 206-8900. 1982 to present. 6 times per year. $132.00 per year.

JOURNAL OF ENERGY RESOURCES TECHNOLOGY. American Society of Mechanical Engineers, 345 East 47th Street, New York, NY 10017. (212) 705-7703. 1979 to present. $24.00 per year.

POWER ENGINEERING (NEW YORK). Technical Publishing Company, 875 Third Avenue, New York, NY 10022. (212) 605-9400. 1896 to present. Monthly. $24.00 per year.

RESEARCH CENTERS AND INSTITUTES

CENTER FOR ENERGY STUDIES. University of Texas at Austin, 1318 EME Building, 10100 Burnet Road, Austin, TX 78758. (512) 471-7792.

ENERGY CENTER. Stevens Institute of Technology, Department of Mechanical Energy, Castle Point Station, Hoboken, NJ 07030. (201) 420-5560.

ENERGY LABORATORY. Massachusetts Institute of Technology, 123 Amherst Street, Building E-40-455, Cambridge, MA 02139. (617) 253-3400.

ENERGY RESEARCH/DEVELOPMENT CENTER. University of Kansas, 345 Nichols Hall, 2291 Irving Hill Drive, Lawrence, KS 66045. (913) 864-4079.

INSTITUTE OF GAS TECHNOLOGY. 3424 South State Street, Chicago, IL 60616. (312) 567-3650.

OAK RIDGE NATIONAL LABORATORY. U.S. Department of Energy, P.O. Box X, Oak Ridge, TN 39831. (615) 576-2900.

FUNDAMENTAL PARTICLES

See: PARTICLE PHYSICS

FUSES (ELECTRICAL)

See: CIRCUIT BREAKERS

FUSION

See also: ENERGY, FISSION, NUCLEAR ENERGY, NUCLEAR ENGINEERING, NUCLEAR PHYSICS, NUCLEAR REACTORS

ABSTRACT SERVICES AND INDEXES

APPLIED MECHANICS REVIEW. American Society of Mechanical Engineers, 345 East 47th Street, New York, NY 10017. (212) 705-7703. 1948 to present. Monthly. $360.00 per year.

APPLIED SCIENCE AND TECHNOLOGY INDEX. H.W. Wilson and Company, 950 University Avenue, Bronx, NY 10452. (800) 367-6670 or (212) 588-8400. Monthly. Inquire as to cost and availability.

CHEMICAL ABSTRACTS. American Chemical Society, Chemical Abstracts Service, Box 3012, Columbus, OH 43210. (614) 421-3600. 1907 to present. Weekly. $9500.00 per year.

CURRENT CONTENTS: ENGINEERING, TECHNOLOGY AND APPLIED SCIENCES. Institute for Scientific Information, 3501 Market Street, Philadelphia, PA 19104. (800) 523-1850 or (215) 386-0100. Weekly. $275.00 per year.

ENGINEERING INDEX MONTHLY AND AUTHOR INDEX. Engineering Information Inc., 345 East 47th Street, New York, NY 10017. (212) 705-7600. Monthly. $1560.00 per year.

ISMEC BULLETIN (Information Service in Mechanical Engineering). Cambridge Scientific Abstracts, 5161 River Road, Bethesda, MD 20816. (301) 951-1400. 1973 to present. Monthly. $450.00 per year.

INDEX TO SCIENTIFIC AND TECHNICAL PROCEEDINGS. Institute for Scientific Information, 3501 Market Street, Philadelphia, PA 19104. (800) 523-1850 or (215) 386-0100. 1978 to present. Monthly. $775.00 per year.

PHYSICS ABSTRACTS. Institution of Electrical Engineers. Available from: IEEE Service Center, 445 Hoes Lane, Piscataway, NJ 08854. 1898 to present. Bimonthly. $1700.00 per year.

SCIENCE CITATION INDEX. Institute for Scientific Information, 3501 Market Street, Philadelphia, PA 19104. (800) 523-1850 or (215) 386-0100. Six times per year. $6200. 00 per year.

ANNUAL REVIEWS AND YEARBOOKS

ADVANCES IN NUCLEAR SCIENCE AND TECHNOLOGY. J.Lewins, editor. Plenum Publishing Corporation, 233 Spring Street, New York, NY 10013. (800) 221-9369. 1977 to presnet. Irregular. Inquire as to cost and availability.

ASSOCIATIONS AND PROFESSIONAL SOCIETIES

AMERICAN NUCLEAR SOCIETY. 555 North Kensington Avenue, La Grange, IL 60525. (312) 352-6611.

FUSION ENERGY FOUNDATION. P.O. Box 17149, Washington, DC 20041. (703) 689-2490.

INSTITUTE OF NUCLEAR MATERIALS MANAGEMENT. 60 Revere Drive, Northbrook, IL 60062. (312) 480-9080.

DIRECTORIES AND BIOGRAPHICAL SOURCES

ENERGY INFORMATION CENTERS DIRECTORY. Public Affairs and Information Program, Atomic Industrial Forum, 7101 Wisconsin Avenue, Bethesda, MD 20814. (301) 654-9260. 1985. Free.

INTERNATIONAL DIRECTORY OF NUCLEAR UTILITIES. Lotte, Limited, Box 237, Contract Station 27, Lakewood, CO 80215. (303) 232-3026. Annual. $160.00.

INTERNATIONAL RESEARCH CENTERS DIRECTORY 1988-89. Darren L. Smith, editor. Gale Research Company, Book Tower, Detroit, MI 48226. (800) 521-0707. 4th edition. 1987. $360.00.

1987 DIRECTORY OF ENGINEERING SOCIETIES AND RELATED ORGANIZATIONS. Gordon Davis, editor. Hemisphere Publishing Corporation, 1010 Vermont Avenue, NW, Washington, DC 20005. (800) 526-0275. 12th edition. 1987. $100.00.

NUCLEAR REACTORS BUILT, BEING BUILT, OR PLANNED IN THE UNITED STATES. Office of Scientific and Technical Information, Department of Energy, Box 62, Oak Ridge, TN 37831. (615) 576-5637. Annual. $11.00. Send orders to: National Technical Information Service, Springfield, VA 22161.

RESEARCH CENTERS DIRECTORY 1988. Gale Research Company, Book Tower, Detroit, MI 48226. (800) 521-0707. 12th edition. 1987. $365.00 for set.

SCIENTIFIC AND TECHNICAL ORGANIZATIONS AND AGENCIES DIRECTORY. Margaret Labash Young, editor. Gale Research Company, Book Tower, Detroit, MI 48226. (800) 521-0707. 2nd edition. 1987. $185.00.

WHO'S WHO IN ENGINEERING. Gordon Davis, editor. Hemisphere Publishing Corporation, 1010 Vermont Avenue, NW, Washington, DC 20005. (800) 526-0275. 6th edition. 1985. $200.00.

ENCYCLOPEDIAS AND DICTIONARIES

THESAURUS OF SCIENTIFIC, TECHNICAL, AND ENGINEERING TERMS. Hemisphere Publishing Corporation, 1010 Vermont Avenue, NW, Washington, DC 20005. (800) 526-0275. 1988. $125.00.

GENERAL WORKS

FUSION ENERGY. Robert A. Gross. John Wiley and Sons, Inc., 605 Third Avenue, New York, NY 10158. (800) 526-5368. 1984. $41.95.

INTRODUCTION TO NUCLEAR POWER. John G. Collier. Hemisphere Publishing Corporation, 1010 Vermont Avenue, NW, Washington, DC 20005. (800) 526-0275. 1987. $49.95.

NUCLEAR ENERGY TECHNOLOGY. Ronald Allen Knief. Hemisphere Publishing Corporation, 1010 Vermont Avenue, NW, Washington, DC 20005. (800) 526-0275. 1983. $48.00.

NUCLEAR FISSION REACTORS. I.R. Cameron. Plenum Publishing Corporation, 233 Spring Street, New York, NY 10013. (800) 221-9369. 1982. $49.50.

NUCLEAR PHYSICS FOR ENGINEERS AND SCIENTISTS: LOW ENERGY THEORY WITH APPLICATIONS INCLUDING REACTORS AND THEIR ENVIRONMENTAL IMPACT. S.E. Hunt. John Wiley and Sons, Inc., 605 Third Avenue, New York, NY 10158. (800) 526-5368. 1987. $129.95.

NUCLEAR REACTOR ENGINEERING. Samuel Glasstone. Van Nostrand Reinhold Company, Inc., 135 West 50th Street, New York, NY 10020. (800) 543-2681. 1980. $49.95.

NUCLEAR MATERIALS AND APPLICATIONS. Benjamin M. Ma. Van Nostrand Reinhold Company, Inc., 135 West 50th Street, New York, NY 10020. (800) 543-2681. 1982. $45.95.

UNCONVENTIONAL APPROACHES TO FUSION. B. Brunelli and G.G. Leotta, editors. Plenum Publishing Corporation, 233 Spring Street, New York, NY 10013. (800) 221-9369. 1982. $85.00.

HANDBOOKS AND MANUALS

CRC HANDBOOK OF CHEMISTRY AND PHYSICS. Robert C. Weast, editor. CRC Press, 2000 Corporate Boulevard, Boca Raton, FL 33431. (800) 272-7737. 68th edition. 1987. $69.95.

A GUIDE TO NUCLEAR POWER TECHNOLOGY: A RESOURCE FOR DECISION MAKING. F.J. Rahn and others. John Wiley and Sons, Inc., 605 Third Avenue, New York, NY 10158. (800) 526-5368. 1984. $85.95.

NUCLEAR ENGINEERING DATA BASES, STANDARDS AND NUMERICAL ANALYSIS. Jack Jedruch. Van Nostrand Reinhold Company, Inc., 135 West 50th Street, New York, NY 10020. (800) 543-2681. 1985. $59.95.

ONLINE DATA BASES

CA SEARCH. Chemical Abstracts Service, P.O. Box 3012, Columbus, OH 43120. (800) 848-6538 or (614) 421-3600. Comprehensive guide to chemical literature, 1972 to present. Inquire as to online cost and availability.

COMPENDEX. Engineering Information, Inc., 345 East 47th Street, New York, NY 10017. (800) 221-1044 or (212) 705-7615. Engineering and technical literature, 1975 to present. Inquire as to online cost and availability.

DISSERTATION ABSTRACTS ONLINE. University Microfilms International, 300 North Zeeb Road, Ann Arbor, MI 48106. (800) 521-0600 or (313) 761-4700. Scope includes virtually all doctoral dissertations accepted at accredited American institutions from 1861 to present in over 250 subject areas. Inquire as to online cost and availability.

DOE ENERGY DATA BASE. U.S. Department of Energy, Office of Scientific and Technical Information, P.O. Box 62, Oak Ridge, TN 37831. (615) 576-6837. A database that covers all aspects of energy including the science and technology of energy. 1948 to present. Available through the DIALOG search service or DOE/RECON. Inquire as to online cost and availability.

ENERGYLINE. EIC/Intelligence, Inc., 48 West 38th Street, New York, NY 10018. (212) 944-8500. A database of resources on the scientific, engineering, political, and socioeconomic aspects of energy resources. 1976 to present. Inquire as to online cost and availability.

INSPEC. INSPEC Marketing Department, Institution of Electrical Engineers. Available from IEEE Service Center, 445 Hoes Lane, Piscataway, NJ 08854. (201) 981-0060. Online version of Physics Abstracts. Inquire as to online cost and availability.

NTIS. National Technical Information Service, 5285 Port Royal Road, Springfield, VA 22161. (703) 487-4630. Broad coverage of government sponsored research reports, 1964 to present. Inquire as to online cost and availability.

SCISEARCH. Institute for Scientific Information, 3501 Market Street, Philadelphia, PA 19104. (800) 523-1850 or (215) 386-0100. Broad multidisciplinary title and author index to the international literature of science and technology, 1974 to present. Inquire as to online cost and availability.

WILSONLINE. H.W. Wilson and Company, 950 University Avenue, Bronx, NY 10452. (800) 367-6770 or (212) 588-8400. Makes available online versions of the H.W. Wilson indexes including Applied Science and Technology Index, Business Periodicals Index and Readers' Guide to Periodical Literature. Approximately 1980 to present. Inquire as to online cost and availability.

PERIODICALS

AMERICAN NUCLEAR SOCIETY TRANSACTIONS. American Nuclear Society, 555 North Kensington Avenue, La Grange, IL 60525. (312) 352-6611. 1958 to present. Semiannual. $255.00 per year.

ANNALS OF NUCLEAR ENERGY. Pergamon Press, Inc., Maxwell House, Fairview Park, Elmsford, NY 10523. (914) 592-7700. 1954 to present. Monthly. $280.00 per year.

BULLETIN OF THE ATOMIC SCIENTISTS. Educational Foundation for Nuclear Science, 5801 South Kenwood Avenue, Chicago, IL 60637. (312) 363-5225. 1945 to present. Ten times per year. $22.50 per year.

FUSION TECHNOLOGY. American Nuclear Society, 555 North Kensington Avenue, La Grange, IL 60525. (312) 352-6611. 1981 to present. Bimonthly. $250.00 per year.

JOURNAL OF FUSION ENERGY. Plenum Publishing Corporation, 233 Spring Street, New York, NY 10013. (800) 221-9369. 1981 to present. Bimonthly. $105.00 per year.

NUCLEAR ENGINEER. Institution of Nuclear Engineers, 1 Penerley Road, Nondon SE6 2LQ, England. 1959 to present. Bimonthly. $110.00 per year.

NUCLEAR ENGINEERING AND DESIGN. Elsevier Science Publishing Company, Inc., 52 Vanderbilt Avenue, New York, NY 10017. (212) 370-5520. 1965 to present. $160.00 per year.

NUCLEAR ENGINEERING INTERNATIONAL. Electrical-Electronic Press, Quadrant House, The Quadrant, Sutton, Surrey, SM2 5AS, England. 1956 to present. Monthly. $210.00 per year.

NUCLEAR SCIENCE AND ENGINEERING. American Nuclear Society, 555 North Kensington Avenue, La Grange, IL 60525. (312) 352-6611. 1956 to present. Monthly. $220.00 per year.

NUCLEAR TECHNOLOGY. American Nuclear Society, 555 North Kensington Avenue, La Grange, IL 60525. (312) 352-6611. 1965 to present. Monthly. $345.00 per year.

RESEARCH CENTERS AND INSTITUTES

ASSISTANT SECRETARY FOR NUCLEAR ENERGY. U.S. Department of Energy, 1000 Independence Avenue, SW, Washington, DC 20585. (202) 252-6450.

DEPARTMENT OF NUCLEAR ENGINEERING AND ENGINEERING PHYSICS. University of Wisconsin - Madison, 1500 Johnson Drive, Madison, WI 53706. (608) 263-1648.

INSTITUTE OF NUCLEAR SCIENCE AND ENGINEERING. Oregon State University, Radiation Center, 35th and Jefferson Streets, Corvallis, OR 97331. (503) 754-2341.

LABORATORY OF BASIC AND APPLIED NUCLEAR RESEARCH. University of Cincinnati, Department of Chemistry, Cincinnati, OH 45221. (513) 475-3652.

OFFICE OF FUSION ENERGY. U.S. Department of Energy, Washington, DC 20545.

WHITESHELL NUCLEAR RESEARCH ESTABLISHMENT. Research Company, Atomic Energy of Canada Limited, Pinawa, MB, Canada ROE 1LO. (204) 753-2311.

SPECIFICATIONS AND STANDARDS

INFORMATION CENTER ON NUCLEAR STANDARDS. American Nuclear Society, 555 North Kensington Avenue, La Grange, IL 60525. (312) 352-6611. Standards for all aspects of the design, construction, operation, and maintenance of nuclear power plants.

FUSION REACTOR

See: NUCLEAR REACTORS

FUSION WELDING

See: WELDING

G

GALAXIES

See also: ASTRONOMY, ASTROPHYSICS, COSMOCHEMISTRY, COSMOLOGY

ABSTRACT SERVICES AND INDEXES

ASTRONOMY AND ASTROPHYSICS ABSTRACTS. Springer-Verlag New York, Inc., 175 Fifth Avenue, New York, NY 10010. (212) 460-1500. $70.00 per year.

ASTRONOMY AND ASTROPHYSICS MONTHLY INDEX. Olivetree Associates, P.O. Box 236, Sierre Madre, CA 91024. $212.00 per year. Complimentary copies available on request.

GENERAL SCIENCE INDEX. H.W. Wilson Co., 950 University Avenue, Bronx, NY 10452. (800) 367-6770 or (212) 588-8400. Inquire as to cost and availability.

PHYSICS ABSTRACTS. Institute of Electrical Engineers, London, United Kingdom. Available from: Institute of Electrical and Electronic Engineers (IEEE), 345 East 47th Street, New York, NY 10017. (212) 705-7900.

SCIENCE CITATION INDEX. Institute for Scientific Information 3501 Market Street, Philadelphia, PA 19104. (800) 523-1850 or (215) 386-0100.

STAR. (Scientific and Technical Aerospace Reports. U.S. National Aeronautics and Space Administration, Scientific and Technical Information Facility, Box 8757, Baltimore-Washington International Airport, MD 21240. (202) 755-2210. Semimonthly, with semiannual and annual indexes. $85.00 per year.

ANNUAL REVIEWS AND YEARBOOKS

ANNUAL REVIEW OF ASTRONOMY AND ASTROPHYSICS. Annual Reviews, Inc., 4139 El Camino Way, Palo Alto, CA 94306. (415) 493-4400.

ASSOCIATIONS AND PROFESSIONAL SOCIETIES

AMERICAN ASTRONOMICAL SOCIETY. 2000 Florida Avenue, NW, Suite 300, Washington, DC 20009. (202) 659-0134.

AMERICAN ASSOCIATION OF VARIABLE STAR ABSERVERS. 187 Concord Avenue, Cambridge, MA 02138. (617) 354-0484.

ASTRONOMICAL LEAGUE. P.O. Box 12821, Tucson, AZ 85732. (602) 790-8471.

ASTRONOMICAL SOCIETY OF THE PACIFIC. 1290 24th Avenue, San Francisco, CA 94122. (415) 661-8660.

BIBLIOGRAPHIES

A BIBLIOGRAPHY OF ASTRONOMY, 1970-1979. R.A. Seal and S.S. Martin. Libraries Unlimited, Inc., Littleton, Colorado 80160. 1982. $37.50.

SCIENCE BOOKS AND FILMS. American Association for the Advancement of Science, 1333 H Street, NW, Washington, DC 20005. Five times per year. $20.00 per year.

SCIENTIFIC AND TECHNICAL BOOKS AND SERIALS IN PRINT 1988; AN INDEX TO LITERATURE IN SCIENCE AND TECHNOLOGY. R.R. Bowker Company, 205 E. 42nd Street, New York, NY 10017. (800) 521-8110 or (212) 916-1600. $175.00.

DIRECTORIES AND BIOGRAPHICAL SOURCES

AMERICAN MEN AND WOMEN OF SCIENCE. Physical and Biological Sciences. Fifteenth edition. R.R. Bowker Company, 205 E. 42nd Street, New York, NY 10017 (800) 521-8110 or (212) 916-1600.

THE BIOGRAPHICAL DICTIONARY OF SCIENTISTS: ASTRONOMERS. D. Abbott, Editor. Peter Bedrick Books, 125 East 23rd Street, New York, NY 10010. 1984.

DIRECTORY OF PHYSICS AND ASTRONOMY STAFF MEMBERS. American Institute of Physics, 335 East 45th Street, New York, NY 10017. Annual.

RESEARCH CENTERS DIRECTORY. Gale Research Company, Detroit, MI 48226. Twelfth Edition, 1988. (800) 521-0707. $240.00.

WHO'S WHO IN FRONTIER SCIENCE AND TECHNOLOGY. Marquis Who's Who, Inc., 200 East Ohio Street, Chicago, IL 60611. (800) 428-3898 or (312) 787-2008.

ENCYCLOPEDIAS AND DICTIONARIES

ENCYCLOPEDIA OF PHYSICAL SCIENCE AND TECHNOLOGY. Academic Press, Inc., Orlando, FL 32887. (800) 321-5068 or (305) 345-2734.

ILLUSTRATED ENCYCLOPEDIA OF THE UNIVERSE: UNDERSTANDING AND EXPLORING THE COSMOS. Richard S. Lewis. Crown Publishers, Inc., 1 Park Avenue, New York, NY 10016. (800) 526-4264. 1986. $24.95.

MCGRAW-HILL ENCYCLOPEDIA OF SCIENCE AND TECHNOLOGY. McGraw-Hill Book, Inc., 1221 Avenue of the Americas, New York, NY 10020. (212) 997-3675. Fifth Edition, 15 volumes. $1100.00.

GENERAL WORKS

THE BIG BANG: THE CREATION AND EVOLUTION OF THE UNIVERSE. Joseph Silk. W.H. Freeman and Company, 41

Madison Avenue, New York, NY 10010. (212) 532-7660. 1980. $13.95 in paper.

CONSTRUCTING THE UNIVERSE. David Layzer. W.H. Freeman and Company, 41 Madison Avenue, New York, NY 10010. (212) 532-7660. 1984. $29.95.

COSMOLOGICAL CONSTANTS: PAPERS IN MODERN COSMOLOGY. Jeremy Bernstein and Gerald Feinberg. Columbia University Press, 562 West 113th Street, New York, NY 10025. (212) 316-7100. 1986. $38.00.

COSMOLOGY, PHYSICS AND PHILOSOPHY. Bernard Gal-Or. Springer-Verlag New York, Inc., 175 Fifth Avenue, New York, NY 10010. (800) 526-7254. 1981. $34.00.

FIRST THREE MINUTES: A MODERN VIEW OF THE ORIGIN OF THE UNIVERSE. Steven Weinberg. Basic Books, Inc., 10 East 53rd Street, New York, NY 10022. (800) 242-7737. 1976. $14.95.

THE FORMATION AND EVOLUTION OF GALAXIES AND LARGE STRUCTURES IN THE UNIVERSE. Jean Audouze and Thanh Van Tran. Kluwer Academic Publishers, 190 Old Derby Street, Hingham, MA 02043. (617) 749-5262. 1984. $58.00.

GALAXIES. Paul W. Hodge. Harvard University Press, 79 Garden Street, Cambridge, MA 02138. (617) 495-2600. 1986. $22.50.

LARGE SCALE STRUCTURE OF THE UNIVERSE. P.J. Peebles. Princeton University Press, 41 William Street, Princeton, NJ 08540. (609) 452-4122. 1980. $15.95 in paper.

THE LEFT HAND OF CREATION: THE ORIGIN AND EVOLUTION OF THE EXPANDING UNIVERSE. John D. Barrow and Joseph Silk. Basic Books, Inc., 10 East 53rd Street, New York, NY 10022. (800) 242-7737. 1986. $7.95 in paper.

MODERN COSMOLOGY. Dennis Sciama. Cambridge University Press, 32 East 57th Street, New York, NY 10022. (800) 872-7423. 1982. $15.95 in paper.

MOMENT OF CREATION: BIG BANG PHYSICS FROM BEFORE THE FIRST MILLISECOND TO THE PRESENT UNIVERSE. James S. Trefil. Charles Scribner's and Sons, 115 Fifth Avenue, New York, NY 10003. (800) 257-5755. 1983. $15.95 in paper.

RELATIVISTIC COSMOLOGY: AN INTRODUCTION. J. Heidmann. Springer-Verlag New York, Inc., 175 Fifth Avenue, New York, NY 10010. (800) 526-7254. 1980. $28.00 in paper.

SPACE, TIME AND GRAVITY: THE THEORY OF THE BIG BANG AND BLACK HOLES. Robert M. Wald. University of Chicago Press, 5801 Ellis Avenue, Chicago, IL 60637. (312) 568-1550. 1977. $5.95 in paper.

THE VERY EARLY UNIVERSE. G.W. Gibbons and others, editors. Cambridge University Press, 32 East 57th Street, New York, NY 10022. (800) 872-7423. 1985. $24.95.

ONLINE DATA BASES

CA SEARCH. Chemical Abstracts Service, P.O. Box 3012, Columbus, OH 43210. Guide to chemical literature, 1972 to present. Inquire as to cost and availability.

DISSERTATION ABSTRACTS ONLINE. University Microfilms International, 300 North Zeeb Road, Ann Arbor, MI 48106. (800) 521-0600 or (313) 761-4700. Scope includes virtually all doctoral dissertations accepted at accredited American institutions from 1861 to present in 252 subject areas. Inquire as to cost and availability.

INSPEC. INSPEC Marketing Department, Institute of Electrical and Electronics Engineers, Inc., IEEE Service Department, 445 Hoes Lane, Piscataway, NJ 08854. (201) 981-0060. Inquire as to on-line cost and availability.

NASA. National Aeronautics and Space Administration, Scientific and Technical Information Branch, 300 7th Street, SW, Washington, DC 20546. Citations and abstracts of aerospace literature, 1962 to present. Inquire as to cost and availability.

NTIS. National Technical Information Service, 5285 Port Royal Road, Springfield, VA 22161. (703) 487-4630. Broad coverage of government sponsored research reports, 1964 to present. Inquire as to cost and availability.

SCISEARCH. Institute for Scientific Information, 3501 Market Street, Philadelphia, PA 19104. (800) 523-1850 or (215) 386-0100. Broad multidisciplinary title and author index to the international literature of science and technology, 1974 to present. Inquire as to cost and availability.

ONLINE SOURCES

ATLAS OF DEEP-SKY SPLENDORS. H. Vehrenberg. Cambridge University Press, 32 East 57th Street, New York, NY 10022. (800) 431-1580 or (212) 688-8888. Fourth edition, 1984. $44.50.

SKY CATALOGUE 2000.0. A. Hirshfeld and R. Sinnott. Cambridge University Press, 32 East 57th Street, New York, NY 10022. (800) 431-1580 or (212) 688-8888.

PERIODICALS

ASTRONOMICAL JOURNAL. American Astronomical Society. Available from: American Institute of Physics, 335 East 45th Street, New York, NY 10017. (212) 661-9404. $125.00 per year.

ASTRONOMICAL SOCIETY OF THE PACIFIC. Publications. Astronomical Society of the Pacific, 1290 24th Avenue, San Francisco, CA 94122. (415) 661-8660. Monthly. $38.00.

ASTRONOMY. Astro Media Corporation, 625 East Paul Avenue, Milwaukee, WI 53202. Monthly. $18.00 per year.

ASTRONOMY AND ASTROPHYSICS. Springer-Verlag New York, Inc., 175 Fifth Avenue, New York, NY 10010. (800) 526-7254 or (212) 460-1500. $680.00 per year.

ASTROPHYSICAL JOURNAL. American Astronomical Society, University of Chicago Press, 5801 Ellis Avenue, Chicago, IL 60637. Biweekly. $305.00 per year.

ASTROPHYSICS AND SPACE SCIENCE. D. Reidel Publishing Company, 190 Old Derby Street, Hingham, MA 02043. Monthly. $101.00 per year.

CELESTIAL MECHANICS; AN INTERNATIONAL JOURNAL OF SPACE DYNAMICS. D. Reidel Publishing Company, 190 Old Derby Street, Hingham, MA 02043. Monthly. $310.00 per year.

FUNDAMENTALS OF COSMIC PHYSICS. Gordon and Breach Science Publishers, Limited, P.O. Box 197, London, WC2E 9PX. Monthly. $335.00 per year.

MERCURY. Astronomical Society of the Pacific, 1290 24th Avenue, San Francisco, CA 94122. (415) 661-8660. Bimonthly. $21.00 per year.

MONTHLY NOTICES OF THE ROYAL ASTRONOMICAL SOCIETY. Blackwell Science Publications, Inc., 667 Lytton Avenue, Palo Alto, CA 94301. (415) 324-1688. Monthly. $134.00 per year.

SKY AND TELESCOPE. Sky Publishing Corporation, 49 Bay State Road, Cambridge, MA 02238. (617) 864-7360. Monthly. $18.00 per year.

SOVIET ASTRONOMY (TRANSLATION OF ASTRO-NOMICHESKII ZHURNAL). American Institute of Physics, 335 East 45th Street, New York, NY 10017. (212) 661-9404. Bimonthly. $425.00 per year.

SPACE SCIENCE REVIEWS. D. Reidel Publishing Company, 190 Old Derby Street, Hingham, MA 02043. Monthly. $305.00 per year.

VISTAS IN ASTRONOMY. Pergamon Press, Inc., Maxwell House, Fairview Park, Elmsford, NY 10523. (914) 592-7700. Quarterly. $145.00 per year.

RESEARCH CENTERS AND INSTITUTES

HARVARD-SMITHOSONIAN CENTER FOR ASTROPHYSICS. 60 Garden Street, Cambridge, MA 02138. (617) 495-7461.

MOUNT WILSON AND LAS CAMPANAS OBSERVATORIES. 813 Santa Barbara Street, Pasadena, CA 91101. (818) 495-7461.

NATIONAL OPTICAL ASTRONOMY OBSERVATORIES. 1002 North Warren Avenue, Tucson, AZ 85719. (602) 325-3582.

STANFORD UNIDVERSITY. Center for Space Science and Astrophysics. 325 Durand Building, Stanford, Ca 94305. (415) 497-3582.

UNIVERSITY OF CHICAGO. Yerkes Observatory, Williams Bay, WI. (414) 245-5555.

GAME THEORY

See: MATHEMATICS

GAMMA-RAY ASTRONOMY

See also: ASTRONOMY, ASTROPHYSICS

ABSTRACT SERVICES AND INDEXES

ASTRONOMY AND ASTROPHYSICS ABSTRACTS. Springer-Verlag New York, Incorporated, 175 Fifth Avenue, New York, NY 10010. (212) 460-1500. $70.00 per year.

ASTRONOMY AND ASTROPHYSICS MONTHLY INDEX. Olivetree Associates, Post Office Box 236, Sierra Madre, CA 91024. $212.00 per year. Complimentary copies available on request.

GENERAL SCIENCE INDEX. H.W. Wilson Company, 950 University Avenue, Bronx, NY 10452. (800) 367-6770 or (212) 588-8400. Inquire as to cost and availability.

PHYSICS ABSTRACTS. Institute of Electrical Engineers, London, United Kingdom. Available from: Institute of Electrical and Electronic Engineers (IEEE), 345 East 47th Street, New York, NY 10017. (212) 705-7900. 1898 to present. Monthly. $1670.00 per year.

SCIENCE CITATION INDEX. Institute for Scientific Information, 3501 Market Street, Philadelphia, PA 19104. (800) 523-1850 or (215) 386-0100. Inquire as to cost and availability.

STAR. (Scientific and Technical Aerospace Reports. United States National Aeronautics and Space Administration, Scientific and Technical Information Facility, Box 8757, Baltimore-Washington International Airport, MD 21240. (202) 755-2210. Semimonthly, with semiannual and annual indexes. $85.00 per year.

ANNUAL REVIEWS AND YEARBOOKS

THE ASTRONOMICAL ALMANAC. Superintendent of Documents, United States Government Printing Office, Washington, DC 20402. (202) 783-3238. Yearly.

ANNUAL REVIEW OF ASTRONOMY AND ASTROPHYSICS. Annual Reviews, Incorporated, 4139 El Camino Way, Palo Alto, CA 94306. (415) 493-4400.

ANNUAL REVIEW OF EARTH AND PLANETARY SCIENCES. Annual Reviews, Incorporated, 4139 East Camino Way, Palo Alto, CA 94306. (415) 493-4400.

ASSOCIATIONS AND PROFESSIONAL SOCIETIES

AMERICAN ASTRONOMICAL SOCIETY. 2000 Florida Avenue, NW, Suite 300, Washington, DC 20009. (202) 659-0134.

AMERICAN ASSOCIATION OF VARIABLE STAR OBSERVERS. 187 Concord Avenue, Cambridge, MA 02138. (617) 354-0484.

ASTRONOMICAL LEAGUE. Post Office Box 12821, Tucson, AZ 85732. (602) 790-8471.

ASTRONOMICAL SOCIETY OF THE PACIFIC. 1290 24th Avenue, San Francisco, CA 94122. (415) 661-8660.

BIBLIOGRAPHIES

A BIBLIOGRAPHY OF ASTRONOMY, 1970-1979. R.A. Seal and S.S. Martin. Libraries Unlimited, Incorporated, Littleton, CO 80160. 1982. $37.50.

SCIENCE BOOKS AND FILMS. American Association for the Advancement of Science, 1333 H Street, NW, Washington, DC 20005. 1965 to present. Five times per year. $20.00 per year.

SCIENTIFIC AND TECHNICAL BOOKS AND SERIALS IN PRINT 1988; AN INDEX TO LITERATURE IN SCIENCE AND TECHNOLOGY. R.R. Bowker Company, 205 East 42nd Street, New York, NY 10017. (800) 521-8110 or (212) 916-1600. $175.00.

DIRECTORIES AND BIOGRAPHICAL SOURCES

AMERICAN MEN AND WOMEN OF SCIENCE. Physical and Biological Sciences. Sixteenth edition. R.R. Bowker Company, 205 East 42nd Street, New York, NY 10017. (800) 521-8110 or (212) 916-1600. 1987. $595.00.

THE BIOGRAPHICAL DICTIONARY OF SCIENTISTS: ASTRONOMERS. D. Abbott, editor. Peter Bedrick Books, 125 East 23rd Street, New York, NY 10010. 1984.

DIRECTORY OF PHYSICS AND ASTRONOMY STAFF MEMBERS. American Institute of Physics, 335 East 45th Street, New York, NY 10017. Annual.

RESEARCH CENTERS DIRECTORY. Gale Research Company, Book Tower, Detroit, MI 48226. Twelfth edition, 1988. (800) 521-0707. $240.00.

WHO'S WHO IN FRONTIER SCIENCE AND TECHNOLOGY. Marquis Who's Who, Incorporated, 200 East Ohio Street, Chicago, IL 60611. (800) 428-3898 or (312) 787-2008.

ENCYCLOPEDIAS AND DICTIONARIES

ENCYCLOPEDIA OF PHYSICAL SCIENCE AND TECHNOLOGY. Academic Press, Incorporated, Orlando, FL 32887. (800) 321-5068 or (305) 345-2734. Inquire as to cost and availability.

MCGRAW-HILL ENCYCLOPEDIA OF SCIENCE AND TECHNOLOGY. McGraw-Hill Book, Incorporated, 1221 Avenue of the Americas, New York, NY 10020. (212) 997-3675. Fifth edition, 15 volumes. $1100.00.

GAMMA-RAY ASTRONOMY

GENERAL WORKS

THE CAMBRIDGE ASTRONOMY GUIDE: A PRACTICAL INTRODUCTION TO ASTRONOMY. Cambridge University Press, 32 East 57th Street, New York, NY 10022. (212) 688-8888. 1985. $24.95.

GAMMA-RAY ASTRONOMY. Rodney R. Hillier. Oxford University Press, 200 Madison Avenue, New York, NY 10016. (212) 679-7300. 1984. $29.95.

GAMMA-RAY ASTRONOMY. Murthy P.V. Ramana and A. Wolfendale. Cambridge University Press, 32 East 57th Street, New York, NY 10022. (212) 688-8888. 1986. $49.50.

ONLINE DATA BASES

CA SEARCH. Chemical Abstracts Service, Post Office Box 3012, Columbus, OH 43210. Guide to chemical literature, 1972 to present. Inquire as to cost and availability.

DISSERTATION ABSTRACTS ONLINE. University Microfilms International, 300 North Zeeb Road, Ann Arbor, MI 48106. (800) 521-0600 or (313) 761-4700. Scope includes virtually all doctoral dissertations accepted at accredited American institutions from 1861 to present in 252 subject areas. Inquire as to cost and availability.

INSPEC. INSPEC Marketing Department, Institute of Electrical and Electronics Engineers, Incorporated, IEEE Service Department, 445 Hoes Lane, Piscataway, NJ 08854. (201) 981-0060. Inquire as to on-line cost and availability.

NASA. National Aeronautics and Space Administration, Scientific and Technical Information Branch, 300 7th Street, SW, Washington, DC 20546. Citations and abstracts of aerospace literature, 1962 to present. Inquire as to cost and availability.

NTIS. National Technical Information Service, 5285 Port Royal Road, Springfield, VA 22161. (703) 487-4630. Broad coverage of government-sponsored research reports, 1964 to present. Inquire as to cost and availability.

SCISEARCH. Institute for Scientific Information, 3501 Market Street, Philadelphia, PA 19104. (800) 523-1850 or (215) 386-0100. Broad multidisciplinary title and author index to the international literature of science and technology, 1974 to present. Inquire as to cost and availability.

PERIODICALS

ASTRONOMICAL JOURNAL. American Astronomical Society. Available from: American Institute of Physics, 335 East 45th Street, New York, NY 10017. (212) 661-9404. $125.00 per year.

ASTRONOMICAL SOCIETY OF THE PACIFIC. Publications. Astronomical Society of the Pacific, 1290 24th Avenue, San Francisco, CA 94122. (415) 661-8660. Monthly. $38.00.

ASTRONOMY. Astro Media Corporation, 625 East Paul Avenue, Milwaukee, WI 53202. Monthly. $18.00 per year.

ASTRONOMY AND ASTROPHYSICS. Springer-Verlag New York, Incorporated, 175 Fifth Avenue, New York, NY 10010. (800) 526-7254 or (212) 460-1500. $680.00 per year.

ASTROPHYSICAL JOURNAL. American Astronomical Society, University of Chicago Press, 5801 Ellis Avenue, Chicago, IL 60637. Biweekly. $305.00 per year.

ASTROPHYSICS AND SPACE SCIENCE. D. Reidel Publishing Company, 190 Old Derby Street, Hingham, MA 02043. Monthly. $101.00 per year.

CELESTIAL MECHANICS: AN INTERNATIONAL JOURNAL OF SPACE DYNAMICS. D. Reidel Publishing Company, 190 Old Derby Street, Hingham, MA 02043. Monthly. $310.00 per year.

EARTH, MOON AND PLANETS: AN INTERNATIONAL JOURNAL OF COMPARATIVE PLANETOLOGY. D. Reidel Publishing Company, 190 Old Derby Street, Hingham, MA 02043. Nine times per year. $275.00 per year.

ICARUS: INTERNATIONAL JOURNAL OF THE SOLAR SYSTEM STUDIES. Academic Press, Incorporated, Orlando, FL 32887. (305) 345-4100. Monthly. $484.00 per year.

MERCURY. Astronomical Society of the Pacific, 1290 245h Avenue, San Francisco, CA 94122. (415) 661-8660. Bimonthly. $21.00 per year.

MONTHLY NOTICES OF THE ROYAL ASTRONOMICAL SOCIETY. Blackwell Science Publications, Incorporated, 667 Lytton Avenue, Palo Alto, CA 94301. (415) 324-1688. Monthly. $134.00 per year.

PLANETARY AND SPACE SCIENCE. Pergamon Press, Incorporated, Maxwell House, Fairview Park, Elmsford, NY 10523. (914) 592-7700. Monthly. $430.00 per year.

SKY AND TELESCOPE. Sky Publishing Corporation, 49 Bay State Road, Cambridge, MA 02238. (617) 864-7360. Monthly. $18.00 per year.

SOLAR PHYSICS. D. Reidel Publishing Company, 190 Old Derby Street, Hingham, MA 02043. Monthly. $620.00 per year.

SOVIET ASTRONOMY (TRANSLATION OF ASTRONOMICHESKII ZHURNAL). American Institute of Physics, 335 East 45th Street, New York, NY 10017. (212) 661-9404. Bimonthly. $425.00 per year.

SPACE SCIENCE REVIEWS. D. Reidel Publishing Company, 190 Old Derby Street, Hingham, MA 02043. Monthly. $305.00 per year.

VISTAS IN ASTRONOMY. Pergamon Press, Incorporated, Maxwell House, Fairview Park, Elmsford, NY 10523. (914) 592-7700. Quarterly. $145.00 per year.

RESEARCH CENTERS AND INSTITUTES

NATIONAL SCIENTIFIC BALLOON FACILITY. Post Office Box 1175, F/M Road 3224, Palestine, TX 75801. (214) 729-0271.

SMITHSONIAN INSTITUTION. Fred L. Whipple Observatory, Post Office Box 97, Amado, AZ 85645. (602) 629-6741.

UNIVERSITY OF CALIFORNIA, SAN DIEGO. Center for Astrophysics and Space Sciences, C-011, La Jolla, CA 92093. ((619) 534-3933.

UNIVERSITY OF FLORIDA. Space Astronomy Laboratory, 1810 NW, 6th Street, Gainesville, FL 32609. (904) 392-5450.

GAMMA-RAY LASER

See: LASERS

GAMMA-RAY SPECTROSCOPY

See: SPECTROSCOPY

GAMMA-RAYS

See also: ASTRONOMY, ASTROPHYSICS, GAMMA-RAY ASTRONOMY, PHYSICS

ABSTRACT SERVICES AND INDEXES

ASTRONOMY AND ASTROPHYSICS ABSTRACTS. Springer-Verlag New York, Incorporated, 175 Fifth Avenue, New York, NY 10010. (212) 460-1500. $70.00 per year.

GENERAL SCIENCE INDEX. H.W. Wilson Company, 950 University Avenue, Bronx, NY 10452. (800) 367-6770 or (212) 588-8400. Inquire as to cost and availability.

PHYSICS ABSTRACTS. Institute of Electrical Engineers, London, United Kingdom. Available from: Institute of Electrical and Electronic Engineers (IEEE), 345 East 47th Street, New York, NY 10017. (212) 705-7900. 1898 to present. Monthly. $1670.00 per year.

SCIENCE CITATION INDEX. Institute for Scientific Information, 3501 Market Street, Philadelphia, PA 19104. (800) 523-1850 or (215) 386-0100. Inquire as to cost and availability.

STAR. (Scientific and Technical Aerospace Reports. United States National Aeronautics and Space Administration, Scientific and Technical Information Facility, Box 8757, Baltimore-Washington International Airport, MD 21240. (202) 755-2210. Semimonthly, with semiannual and annual indexes. $85.00 per year.

ASSOCIATIONS AND PROFESSIONAL SOCIETIES

AMERICAN ASTRONOMICAL SOCIETY. 2000 Florida Avenue, NW, Suite 300, Washington, DC 20009. (202) 659-0134.

AMERICAN NUCLEAR SOCIETY. 555 North Kensington Avenue, La Grange Park, IL 60525. (312) 352-6611.

AMERICAN SOCIETY OF NONDESTRUCTIVE TESTING. 4153 Arlingate Plaza, Caller #28518, Columbus, OH 43228. (614) 274-6003.

NATIONAL COUNCIL ON RADIATION PROTECTION AND MEASUREMENTS. 7910 Woodmont Avenue, Suite 1016. Bethesda, MD 10814. (301) 657-2652.

RADIATION RESEARCH SOCIETY. 925 Chestnut Street, Philadelphia, PA 19107. (215) 574-3153.

BIBLIOGRAPHIES

SCIENTIFIC AND TECHNICAL BOOKS AND SERIALS IN PRINT 1988; AN INDEX TO LITERATURE IN SCIENCE AND TECHNOLOGY. R.R. Bowker Company, 205 East 42nd Street, New York, NY 10017. (800) 521-8110 or (212) 916-1600. $175.00.

DIRECTORIES AND BIOGRAPHICAL SOURCES

AMERICAN MEN AND WOMEN OF SCIENCE. Physical and Biological Sciences. Sixteenth edition. R.R. Bowker Company, 205 East 42nd Street, New York, NY 10017. (800) 521-8110 or (212) 916-1600. 1987. $595.00.

DIRECTORY OF PHYSICS AND ASTRONOMY STAFF MEMBERS. American Institute of Physics, 335 East 45th Street, New York, NY 10017. Annual.

RESEARCH CENTERS DIRECTORY. Gale Research Company, Book Tower, Detroit, MI 48226. Twelfth edition, 1988. (800) 521-0707. $240.00.

WHO'S WHO IN FRONTIER SCIENCE AND TECHNOLOGY. Marquis Who's Who, Incorporated, 200 East Ohio Street, Chicago, IL 60611. (800) 428-3898 or (312) 787-2008.

ENCYCLOPEDIAS AND DICTIONARIES

ENCYCLOPEDIA OF PHYSICAL SCIENCE AND TECHNOLOGY. Academic Press, Incorporated, Orlando, FL 32887. (800) 321-5068 or (305) 345-2734. Inquire as to cost and availability.

MCGRAW-HILL ENCYCLOPEDIA OF SCIENCE AND TECHNOLOGY. McGraw-Hill Book, Incorporated, 1221 Avenue of the Americas, New York, NY 10020. (212) 997-3675. Fifth edition, 15 volumes. $1100.00.

GENERAL WORKS

GAMMA AND X-RAY SPECTROMETRY TECHNIQUES AND APPLICATIONS. K. Debertin and W.B. Mann. Pergamon Press, Incorporated, Maxwell House, Fairview Park, Elmsford, NY 10523. (014) 592-7700. 1983. $25.00.

GAMMA-RAY ASTRONOMY. Rodney R. Hillier. Oxford University Press, 200 Madison Avenue, New York, NY 10016. (212) 679-7300. 1984. $29.95.

GAMMA-RAY ASTRONOMY. Murthy P.V. Ramana and A. Wolfendale. Cambridge University Press, 32 East 57th Street, New York, NY 10022. (212) 688-8888. 1986. $49.50.

ONLINE DATA BASES

CA SEARCH. Chemical Abstracts Service, Post Office Box 3012, Columbus, OH 43210. Guide to chemical literature, 1972 to present. Inquire as to cost and availability.

DISSERTATION ABSTRACTS ONLINE. University Microfilms International, 300 North Zeeb Road, Ann Arbor, MI 48106. (800) 521-0600 or (313) 761-4700. Scope includes virtually all doctoral dissertations accepted at accredited American institutions from 1861 to present in 252 subject areas. Inquire as to cost and availability.

INSPEC. INSPEC Marketing Department, Institute of Electrical and Electronics Engineers, Incorporated, IEEE Service Department, 445 Hoes Lane, Piscataway, NJ 08854. (201) 981-0060. Inquire as to on-line cost and availability.

NASA. National Aeronautics and Space Administration, Scientific and Technical Information Branch, 300 7th Street, SW, Washington, DC 20546. Citations and abstracts of aerospace literature, 1962 to present. Inquire as to cost and availability.

NTIS. National Technical Information Service, 5285 Port Royal Road, Springfield, VA 22161. (703) 487-4630. Broad coverage of government-sponsored research reports, 1964 to present. Inquire as to cost and availability.

SCISEARCH. Institute for Scientific Information, 3501 Market Street, Philadelphia, PA 19104. (800) 523-1850 or (215) 386-0100. Broad multidisciplinary title and author index to the international literature of science and technology, 1974 to present. Inquire as to cost and availability.

PERIODICALS

ASTRONOMICAL JOURNAL. American Astronomical Society. Available from: American Institute of Physics, 335 East 45th Street, New York, NY 10017. (212) 661-9404. $125.00 per year.

ASTROPHYSICAL JOURNAL. American Astronomical Society, University of Chicago Press, 5801 Ellis Avenue, Chicago, IL 60637. Biweekly. $305.00 per year.

JOURNAL OF ENVIRONMENTAL RADIOACTIVITY. Elsevier Science Publishing Company, Incorporated, 52 Vanderbilt Avenue, New York, NY 10017. (212) 370-5520. Semimonthly.

$81.00 per year.

MONTHLY NOTICES OF THE ROYAL ASTRONOMICAL SOCIETY. Blackwell Science Publications, Incorporated, 667 Lytton Avenue, Palo Alto, CA 94301. (415) 324-1688. Monthly. $134.00 per year to individuals.

NUCLEAR SCIENCE AND ENGINEERING. American Nuclear Society, 555 North Kensington Avenue, La Grange Park, IL 60525. (312) 352-6611. Monthly. $340.00 per year.

RADIATION RESEARCH. Academic Press, Incorporated, 111 Fifth Avenue, New York, NY 10003. (212) 741-6800. Monthly. $300.00 per year.

SOLAR PHYSICS. D. Reidel Publishing Company, 190 Old Derby Street, Hingham, MA 02043. Monthly. $620.00 per year.

RESEARCH CENTERS AND INSTITUTES

NATIONAL SCIENTIFIC BALLOON FACILITY. Post Office Box 1175, F/M Road 3224, Palestine, TX 75801. (214) 729-0271.

SMITHSONIAN INSTITUTION. Fred L. Whipple Observatory, Post Office Box 97, Amado, AZ 85645. (602) 629-6741.

NORTH CAROLINA STATE UNIVERSITY. Center for Engineering Applications of Radioisotopes, Box 7909, Department of Nuclear Engineering, Raleigh, NC 27695-7909. (919) 737-3378.

UNIVERSITY OF CALIFORNIA, SAN DIEGO. Center for Astrophysics and Space Sciences, C-011, La Jolla, CA 92093. ((619) 534-3933.

UNIVERSITY OF MICHIGAN. Ford Nuclear Reactor - Phoenix Memorial Laboratory, 2301 Bonisteel Boulevard, Ann Arbor, MI 48109-2100. (313) 764-6223.

UNIVERSITY OF WASHINGTON. Nuclear Physics Laboratory, Seattle, WA 98195. (206) 543-4080.

GARBAGE DISPOSAL

See: SOLID WASTE DISPOSAL

GAS CHROMATOGRAPHY

See: CHROMATOGRAPHY

GAS INJECTION WELLS

See: PETROLEUM ENGINEERING

GAS LASER

See: LASERS

GAS MASERS

See: LASERS

GAS PIPELINES

See: PIPELINE TECHNOLOGY

GAS TURBINES

See also: AERONAUTICS, AUTOMOTIVE ENGINEERING, ENGINES, JET PROPULSION, STEAM ENGINES, STEAM TURBINES

ABSTRACT SERVICES AND INDEXES

APPLIED MECHANICS REVIEW. American Society of Mechanical Engineers, 345 East 47th Street, New York, NY 10017. (212) 705-7703. 1948 to present. Monthly. $360.00 per year.

APPLIED SCIENCE AND TECHNOLOGY INDEX. H.W. Wilson and Company, 950 University Avenue, Bronx, NY 10452. (800) 367-6670 or (212) 588-8400. Monthly. Inquire as to cost and availability.

ENGINEERING INDEX MONTHLY AND AUTHOR INDEX. Engineering Information Inc., 345 East 47th Street, New York, NY 10017. (212) 705-7600. Monthly. $1560.00 per year.

INTERNATIONAL AEROSPACE ABSTRACTS. American Institute of Aeronautics and Astronautics, Technical Information Service, 370 L'Enfant Promenade, S.W., Washington, DC 20024. (202) 646-7400. 1961 to present. Semi-monthly. $700.00 per year.

ISMEC BULLETIN (Information Service in Mechanical Engineering). Cambridge Scientific Abstracts, 5161 River Road, Bethesda, MD 20816. (301) 951-1400. 1973 to present. Monthly. $450.00 per year.

ASSOCIATIONS AND PROFESSIONAL SOCIETIES

AMERICAN INSTITUTE OF AERONAUTICS AND ASTRONAUTICS. 370 L'Enfant Promenade, S.W., Washington, DC 20024. (202) 646-7400.

AMERICAN SOCIETY OF MECHANICAL ENGINEERS. 345 East 47th Street, New York, NY 10017. (212) 705-7703.

INTERNATIONAL GAS TURBINE INSTITUTE. 4250 Perimeter Park, South, Atlanta, GA 30341. (404) 451-1905.

SOCIETY OF AUTOMOTIVE ENGINEERS (SAE). 400 Commonwealth Drive, Warrendale, PA 15096. (412) 776-4841.

NATIONAL ASSOCIATION OF POWER ENGINEERS. 2350 East Devon Avenue, Suite 115, Des Plaines, IL 60018. (312) 298-0600.

DIRECTORIES AND BIOGRAPHICAL SOURCES

ANNUAL WHO'S WHO IN GAS TURBINE TECHNOLOGY. International Gas Turbine Institute, 4250 Perimeter Park, South, Atlanta, GA 30341. (404) 451-1905. Annual. Inquire.

1987 DIRECTORY OF ENGINEERING SOCIETIES AND RELATED ORGANIZATIONS. Gordon Davis, editor. Hemisphere Publishing Corporation, 1010 Vermont Avenue, NW, Washington, DC 20005. (800) 526-0275. 12th edition. 1987. $100.00.

RESEARCH CENTERS DIRECTORY 1988. Gale Research Company, Book Tower, Detroit, MI 48226. (800) 521-0707. 12th edition. 1987. $365.00 for set.

SCIENTIFIC AND TECHNICAL ORGANIZATIONS AND AGENCIES DIRECTORY. Margaret Labash Young, editor. Gale Research Company, Book Tower, Detroit, MI 48226. (800) 521-0707. 2nd edition. 1987. $185.00.

WHO'S WHO IN ENGINEERING. Gordon Davis, editor. Hemisphere Publishing Corporation, 1010 Vermont Avenue, NW, Washington, DC 20005. (800) 526-0275. 6th edition. 1985. $200.00.

ENCYCLOPEDIAS AND DICTIONARIES

THESAURUS OF SCIENTIFIC, TECHNICAL, AND ENGINEERING TERMS. Hemisphere Publishing Corporation, 1010 Vermont Avenue, NW, Washington, DC 20005. (800) 526-0275. 1988. $125.00.

GENERAL WORKS

AERODYNAMICS OF TURBINES AND COMPRESSORS. W.R. Hawthorne, editor. Princeton University Press, 41 William Street, Princeton, NJ 08540. (609) 452-4122. 1964. $66.50.

THE DESIGN OF HIGH-EFFICIENCY TURBOMACHINERY AND GAS TURBINES. David G. Wilson. MIT Press, 28 Carleton Street, Cambridge, MA 02142. (617) 253-2884. 1983. $39.95.

TURBOMECHANICS: A GUIDE TO DESIGN, SELECTION AND THEORY. O.E. Balje, editor. John Wiley and Sons, Inc., 605 Third Avenue, New York, NY 10158. (800) 526-5368. 1981. $64.95.

TURBOMACHINERY. W. Logan. Marcel Dekker Inc., 270 Madison Avenue, New York, NY 10016. (800) 228-1160. 1981. $29.75.

HANDBOOKS AND MANUALS

SAWYER'S TURBOMECHINERY MAINTENANCE HANDBOOKS. John W. Sawyer and Kurt Hallberg, editors. Turbomechinery International Publications, P.O. Box 5550, Norwalk, CT 06856. (203) 853-6015. Three volumes. 1981. $180.00 for set.

ONLINE DATA BASES

COMPENDEX. Engineering Information, Inc., 345 East 47th Street, New York, NY 10017. (800) 221-1044 or (212) 705-7615. Engineering and technical literature, 1975 to present. Inquire as to online cost and availability.

NTIS. National Technical Information Service, 5285 Port Royal Road, Springfield, VA 22161. (703) 487-4630. Broad coverage of government sponsored research reports, 1964 to present. Inquire as to online cost and availability.

WILSONLINE. H.W. Wilson and Company, 950 University Avenue, Bronx, NY 10452. (800) 367-6770 or (212) 588-8400. Makes available online versions of the H.W. Wilson indexes including Applied Science and Technology Index, Business Periodicals Index and Readers' Guide to Periodical Literature. Approximately 1980 to present. Inquire as to online cost and availability.

PERIODICALS

INTERNATIONAL JOURNAL OF TURBO AND JET ENGINES. Kluwer Academic Publishers, 190 Old Derby Street, Hingham, MA 02043. (617) 749-5262. Quarterly. $125.00 per year.

JOURNAL OF ENGINEERING FOR GAS TURBINES AND POWER. American Society of Mechanical Engineers, 345 East 47th Street, New York, NY 10017. (212) 705-7703. 1970 to present. Quarterly. $100.00 per year.

JOURNAL OF PROPULSION AND POWER. American Institute of Aeronautics and Astronautics, 370 L'Enfant Promenade, S.W., Washington, DC 20024. (202) 646-7400. Bimonthly. $170.00 per year.

TURBOMACHINERY INTERNATIONAL. Turbomachinery International Publications, P.O. Box 5550, Norwalk, CT 06856. (203) 853-6015. 1959 To present. 9 times per year. $38.00 per year

RESEARCH CENTERS AND INSTITUTES

INTERNATIONAL GAS TURBINE INSTITUTE. 4250 Perimeter Park, South, Atlanta, GA 30341. (404) 451-1905.

TURBOMACHINERY LABORATORY. Texas A&M University, Mechanical Engineering Department, College Station, TX 77843. (409) 845-7417.

GASIFICATION

See: COAL GASIFICATION AND LIQUIFICATION

GASOHOL

See: FUELS

GASOLINE

See: FUELS

GEARS AND GEARING

See: MACHINERY

GEMS

See: MINERALOGY

GENERATORS

See also: ELECTRIC MOTORS, ELECTRIC POWER, ELECTRICAL ENGINEERING, ELECTRICITY, ELECTROMAGNETISM

ABSTRACT SERVICES AND INDEXES

APPLIED MECHANICS REVIEW. American Society of Mechanical Engineers, 345 East 47th Street, New York, NY 10017. (212) 705-7703. 1948 to present. Monthly. $360.00 per year.

APPLIED SCIENCE AND TECHNOLOGY INDEX. H.W. Wilson and Company, 950 University Avenue, Bronx, NY 10452. (800) 367-6670 or (212) 588-8400. Monthly. Inquire as to cost and availability.

CURRENT CONTENTS: ENGINEERING, TECHNOLOGY AND APPLIED SCIENCES. Institute for Scientific Information, 3501 Market Street, Philadelphia, PA 19104. (800) 523-1850 or (215) 386-0100. Weekly. $275.00 per year.

ELECTRIC POWER INDUSTRY ABSTRACTS. Edison Electric Institute, c/o Utility Date Institute, 2011 I Street, N.W., Suite 700, Washington, DC 20006. 1975 to present. Bimonthly. Inquire as to cost and availability.

ELECTRICAL AND ELECTRONICS ABSTRACTS. Institution of Electrical Engineers. Available from: Institute of Electrical and Electronics Engineers. IEEE Service Center, 445 Hoes Lane, Piscataway, NJ 08854. Monthly. $1250.00 per year.

ENGINEERING INDEX MONTHLY AND AUTHOR INDEX. Engineering Information Inc., 345 East 47th Street, New York, NY 10017. (212) 705-7600. Monthly. $1560.00 per year.

IEEE PUBLICATIONS BULLETIN. Institute of Electrical and Electronics Engineers. Institute of Electrical and Electronics Engineers. IEEE Service Center, 445 Hoes Lane, Piscataway, NJ 08854. Quarterly. Free.

SCIENCE CITATION INDEX. Institute for Scientific Information, 3501 Market Street, Philadelphia, PA 19104. (800) 523-1850 or (215) 386-0100. Six times per year. $6200.00 per year.

ASSOCIATIONS AND PROFESSIONAL SOCIETIES

AMERICAN ELECTRONICS ASSOCIATION. P.O. Box 10045, 2670 Hanover Street, Palo Alto, CA 94303. (415) 857-9300.

EDISON ELECTRIC INSTITUTE. 1111 19th Street, N.W., Washington, DC 20036. (202) 828-7400.

INSTITUTE OF ELECTRICAL AND ELECTRONICS ENGINEERS. 345 East 47th Street, New York, NY 10017. (212) 705-7900.

NATIONAL ELECTRICAL MANUFACTURERS ASSOCIATION. 2101 L Street, N.W., Washington, DC. (202) 457-8400.

DIRECTORIES AND BIOGRAPHICAL SOURCES

ELECTRICAL WORLD DIRECTORY OF ELECTRICAL UTILITIES. McGraw-Hill Book Company, 1221 Avenue of the Americas, New York, NY 10020. (212) 512-2000. Annual. $275.00.

1987 DIRECTORY OF ENGINEERING SOCIETIES AND RELATED ORGANIZATIONS. Gordon Davis, editor. Hemisphere Publishing Corporation, 1010 Vermont Avenue, NW, Washington, DC 20005. (800) 526-0275. 12th edition. 1987. $100.00.

RESEARCH CENTERS DIRECTORY 1988. Gale Research Company, Book Tower, Detroit, MI 48226. (800) 521-0707. 12th edition. 1987. $365.00 for set.

SCIENTIFIC AND TECHNICAL ORGANIZATIONS AND AGENCIES DIRECTORY. Margaret Labash Young, editor. Gale Research Company, Book Tower, Detroit, MI 48226. (800) 521-0707. 2nd edition. 1987. $185.00.

WHO'S WHO IN ELECTRONICS. Harris Publishing Company, 2057-2 Aurora Road, Twinsburg, OH 44087. (216) 425-9143. Annual. $90.00.

WHO'S WHO IN ENGINEERING. Gordon Davis, editor. Hemisphere Publishing Corporation, 1010 Vermont Avenue, NW, Washington, DC 20005. (800) 526-0275. 6th edition. 1985. $200.00.

GENERAL WORKS

ELECTRIC MACHINERY. Peter F. Ryff. Prentice-Hall Publishing, Inc., Englewood Cliffs, NJ 07632. (800) 562-0245. 1988. $34.00.

ELECTRICAL ENGINEERING. W.H. Roadstrum and Dan H. Wolaver. Harper and Row Publishers, Inc., 10 East 53rd Street, New York, NY 10022. (800) 242-7737. 1986. $37.50.

ELECTRICAL MACHINES: AN INTRODUCTION TO PRINCIPLES AND CHARACTERISTICS. J.D. Edwards. Macmillan Publishing Company, Inc., 866 Third Avenue, New York, NY 10022. (800) 257-5755. 2nd edition. 1987. $32.50.

ELECTRICITY AND MAGNETISM. M.H. Nayfeh and M.K. Brussel. John Wiley and Sons, Inc., 605 Third Avenue, New York, NY 10158. (800) 526-5368. 1985. $32.95.

ELECTROMAGNETIC CONCEPTS AND PRINCIPLES. Stanley V. Marshall and Gabriel G. Skitek. Prentice-Hall Publishing, Inc., Englewood Cliffs, NJ 07632. (800) 562-0245. 2nd edition. 1987. $42.95.

ELEMENTS OF ENGINEERING ELECTROMAGNETICS. N.N. Rao. Prentice-Hall Publishing, Inc., Englewood Cliffs, NJ 07632. (800) 562-0245. 2nd edition. 1987. $45.95.

FUNDAMENTALS OF ELECTRIC CIRCUITS. David A. Bell. Prentice-Hall Publishing, Inc., Englewood Cliffs, NJ 07632. (800) 562-0245. 4th edition. 1988. $30.25.

INTRODUCTION TO ELECTRICITY AND ELECTRONICS FOR THE COMPUTER AGE. R. Rosen. John Wiley and Sons, Inc., 605 Third Avenue, New York, NY 10158. (800) 526-5368. 1987. $35.95.

HANDBOOKS AND MANUALS

HANDBOOK OF ELECTRIC MACHINES. Syed A. Nasar. McGraw-Hill Book Company, 1221 Avenue of the Americas, New York, NY 10020. (212) 512-2000. 1987. $59.50.

HANDBOOK OF MODERN ELECTRONICS AND ELECTRICAL ENGINEERING. Charles Belove, editor. John Wiley and Sons, Inc., 605 Third Avenue, New York, NY 10158. (800) 526-5368. 1986. $88.95.

ONLINE DATA BASES

COMPENDEX. Engineering Information, Inc., 345 East 47th Street, New York, NY 10017. (800) 221-1044 or (212) 705-7615. Engineering and technical literature, 1975 to present. Inquire as to online cost and availability.

INSPEC. INSPEC Marketing Department, Institution of Electrical Engineers. Available from IEEE Service Center, 445 Hoes Lane, Piscataway, NJ 08854. (201) 981-0060. Online version of Physics Abstracts. Inquire as to online cost and availability.

NTIS. National Technical Information Service, 5285 Port Royal Road, Springfield, VA 22161. (703) 487-4630. Broad coverage of government sponsored research reports, 1964 to present. Inquire as to online cost and availability.

SCISEARCH. Institute for Scientific Information, 3501 Market Street, Philadelphia, PA 19104. (800) 523-1850 or (215) 386-0100. Broad multidisciplinary title and author index to the international literature of science and technology, 1974 to present. Inquire as to online cost and availability.

WILSONLINE. H.W. Wilson and Company, 950 University Avenue, Bronx, NY 10452. (800) 367-6770 or (212) 588-8400. Makes available online versions of the H.W. Wilson indexes including Applied Science and Technology Index, Business Periodicals Index and Readers' Guide to Periodical Literature. Approximately 1980 to present. Inquire as to online cost and availability.

OTHER SOURCES

ELECTROMECHANICAL BENCH REFERENCE. Barks Publications, Inc., 400 North Michigan Avenue, Chicago, IL 60611-4198. (312) 321-9440. Annual. $5.00.

A GUIDE TO THE LITERATURE OF ELECTRICAL AND ELECTRONICS ENGINEERING. Susan B. Ardis. Libraries Unlimited Inc., P.O. Box 263, Littleton, CO 80160. (303) 770-1220. 1987. $37.50.

MOTORS AND GENERATORS. U.S. Bureau of the Census, U.S. Department of Commerce, Washington, DC 20233. (301) 763-7800. Current industrial reports. Annual. $1.25.

PERIODICALS

ELECTRIC LIGHT AND POWER. Technical Publishing Company, 875 Third Avenue, New York, NY 10022. (212) 605-9400. 1922 to present. Monthly. $38.00 per year.

ELECTRIC MACHINES AND POWER SYSTEMS. Hemisphere Publishing Corporation, 1010 Vermont Avenue, NW, Washington, DC 20005. (800) 526-0275. 1976 to present. Bimonthly. $134.50 per year.

ELECTRICAL WORLD. McGraw-Hill Book Company, 1221 Avenue of the Americas, New York, NY 10020. (212) 512-2000. 1874 to present. Monthly. $11.00 per year.

ELECTRONIC DESIGN. Hayden Publishing Company, 10 Mulholland Drive, Hasbrouck Heights, NJ 07604. (201) 288-7520. 1952 to present. Biweekly. $40.00 per year.

ELECTRONICS. McGraw-Hill Book Company, 1221 Avenue of the Americas, New York, NY 10020. (212) 512-2000. 1930 to present. Weekly. $32.00 per year.

IEEE POWER ENGINEERING REVIEW. Institute of Electrical and Electronics Engineers. IEEE Service Center, 445 Hoes Lane, Piscataway, NJ 08854. 1981 to present. Monthly. $75.00 per year.

IEEE TRANSACTIONS ON POWER DELIVERY. Institute of Electrical and Electronics Engineers. IEEE Service Center, 445 Hoes Lane, Piscataway, NJ 08854. 1986 to present. Quarterly. $100.00 per year.

INSTITUTE OF ELECTRICAL AND ELECTRONICS ENGINEERS PROCEEDINGS. Institute of Electrical and Electronics Engineers. IEEE Service Center, 445 Hoes Lane, Piscataway, NJ 08854. 1913 to present. Monthly. $140.00 per year.

POWER TRANSMISSION DESIGN. Penton-IPC, 1100 Superior Avenue, Cleveland, OH 44114. (216) 696-7000. 1959 to present. Monthly. $35.00 per year.

RESEARCH CENTERS AND INSTITUTES

EDISON ELECTRIC INSTITUTE. 1111 19th Street, N.W., Washington, DC 20036. (202) 778-6778.

ELECTRIC POWER INSTITUTE. Texas A&M University, Department of Electrical Engineering, College Station, TX 77843.

ELECTRICAL ENGINEERING RESEARCH LABORATORIES. Purdue University, Electrical Engineering Building, West Lafayette, IN 47907. (317) 494-3536.

LABORATORY FOR ELECTROMAGNETIC AND ELECTRONIC SYSTEMS. 77 Massachusetts Avenue, Cambridge, MA 02139. (617) 253-4631.

GEOCHEMISTRY

See also: GEOLOGY, GEOPHYSICS, IGNEOUS ROCKS, METAMORPHIC ROCKS, PETROLEUM GEOLOGY, SEDIMENTARY ROCKS

ABSTRACT SERVICES AND INDEXES

BIBLIOGRAPHY AND INDEX OF GEOLOGY. American Geological Institute, 4220 King Street, Alexandria, VA 22302. (703) 379-2480. 1969 to present. Monthly. $1100.00 per year.

CHEMICAL ABSTRACTS. Chemical Abstracts Service, 2540 Olentangy Road, Post Office Box 3012, Columbus, OH 43210. (800) 848-6538 or (614) 421-3600. Weekly. $9200.00 per year.

DEEP-SEA RESEARCH WITH OCEANOGRAPHIC LITERATURE REVIEW. Pergamon Press, Incorporated, Maxwell House, Fairview Park, Elmsford, NY 10523. (914) 592-7700. 1953 to present. Twenty-four times per year. $600.00 per year.

GENERAL SCIENCE INDEX. H.W. Wilson Company, 950 University Avenue, Bronx, NY 10452. (800) 367-6770 or (212) 588-8400. Inquire as to cost and availability.

MINERALOGICAL ABSTRACTS. Mineralogical Society and the Mineralogical Society of America, 41 Queen's Gate, London, SW7 5HR, England. 1959 to present. Quarterly. $190.00 per year.

OCEANIC ABSTRACTS. Cambridge Scientific Abstracts, 5161 River Road, Bethesda, MD 20816. (301) 951-1400. 1964 to present. Bimonthly. $652.00 per year.

SCIENCE CITATION INDEX. Institute for Scientific Information, 3501 Market Street, Philadelphia, PA 19104. (800) 523-1850 or (215) 386-0100. Inquire as to cost and availability.

ANNUAL REVIEWS AND YEARBOOKS

ANNUAL REVIEW AND EARTH AND PLANETARY SCIENCES. Annual Reviews, Incorporated, 4139 El Camino Way, Palo Alto, CA 94306. (415) 493-4400.

MINERALS YEARBOOK. Bureau of Mines, United States Department of the Interior. Available from United States Government Printing Office, Washington, DC 20402. (202) 783-3238. Annual. Three volumes. $45.00.

ASSOCIATIONS AND PROFESSIONAL SOCIETIES

AMERICAN ASSOCIATION OF PETROLEUM GEOLOGISTS. Post Office Box 979, Tulsa, OK 74101. (918) 584-2555.

AMERICAN GEOLOGICAL INSTITUTE. 4220 King Street, Alexandria, VA 22302. (703) 379-2480.

AMERICAN GEOPHYSICAL UNION. 2000 Florida Avenue, NW, Washington, DC 20009. (202) 462-6903.

AMERICAN INSTITUTE OF PROFESSIONAL GEOLOGISTS. 7828 Vance Drive, Suite 103, Arvada, CO 80003. (303) 431-0831.

ASSOCIATION OF ENGINEERING GEOLOGISTS. Post Office Box 1068, Brentwood, TN 37027. (615) 377-3578.

GEOLOGICAL SOCIETY OF AMERICA. Box 9140, Boulder, CO 80301. (303) 447-2020.

SOCIETY OF ECONOMIC PALEONTOLOGISTS AND MINERALOGISTS. Box 4756, Tulsa, OK 74159. (918) 743-9765.

DIRECTORIES AND BIOGRAPHICAL SOURCES

AMERICAN MEN AND WOMEN OF SCIENCE. Physical and Biological Sciences. Sixteenth edition. R.R. Bowker Company, 205 East 42nd Street, New York, NY 10017. (800) 521-8110 or (212) 916-1600. 1986. $595.00.

AMERICAN INSTITUTE OF PROFESSIONAL GEOLOGISTS. Membership Directory. American Institute of Professional Geologists, 7828 Vance Drive, Suite 103, Arvada, CO 80003. (303) 431-0831. Annual. $15.00.

ASSOCIATION OF ENGINEERING GEOLOGISTS DIRECTORY. Association of Engineering Geologists, Dr. G. Lee Christensen, Civil Engineering Department, Villanova University, Villanova, PA 19085. (215) 645-4960. Annual. $15.00.

GEOLOGICAL SOCIETY OF AMERICA. Membership Directory. Geological Society of America, 3300 Penrose Place, Boulder, CO 80301. (303) 447-2020. Annual. Available to members only.

RESEARCH CENTERS DIRECTORY. Gale Research Company, Book Tower, Detroit, MI 48226. (800) 521-0707. Twelfth edition. 1987. $365.00 for set.

WHO'S WHO IN FRONTIER SCIENCE AND TECHNOLOGY. Marquis Who's Who, Incorporated, 200 East Ohio Street, Chicago, IL 60611. (800) 428-3898 or (312) 787-2008.

ENCYCLOPEDIAS AND DICTIONARIES

DICTIONARY OF GEOLOGICAL TERMS. American Geological Institute. Doubleday and Company, Incorporated, 245 Park Avenue, New York, NY 10017. (800) 645-6156 or (212) 953-4561. Third edition. 1984. $19.95.

GLOSSARY OF GEOLOGY. Robert L. Bates and Julia A. Jackson. American Geological Institute, 4220 King Street, Alexandria, VA 22032. (703) 379-2480. 1980. $60.00.

GENERAL WORKS

APPLIED GEOCHEMICAL ANALYSIS. C.O. Ingamells and Francis F. Pitard. John Wiley and Sons, Incorporated, 605 Third Avenue, New York, NY 10158. (800) 526-5368 or (212) 850-6000. 1986. $75.00.

GEOCHEMICAL THERMODYNAMICS. D.K. Nordstrom and J.L. Munoz. Blackwell Scientific Publications, Incorporated, 667 Lytton Avenue, Palo Alto, CA 94301. (415) 324-1688 or (415) 965-4081. 1986. $45.00.

INTRODUCTION TO ROCK FORMING MINERALS. W.A. Deer, J. Zussman and R.A. Howie. John Wiley and Sons, Incorporated, 605 Third Avenue, New York, NY 10158. (800) 526-5368 or (212) 850-6000. 1986. $29.95.

THE NATURAL ENVIRONMENT AND THE BIOGEOCHEMICAL CYCLES. O. Hutzinger. Springer-Verlag New York, Incorpoarted, 175 Fifth Avenue, New York, NY 10010. (212) 460-1500. 1985. $59.00.

HANDBOOKS AND MANUALS

GEOLOGY IN THE FIELD. R.R. Compton. John Wiley and Sons, Incorporated, 605 Third Avenue, New York, NY 10158. (800) 526-5368 or (212) 850-6418. 1985. $23.95.

ONLINE DATA BASES

CA SEARCH. Chemical Abstracts Service, Post Office Box 3012, Columbus, OH 43210. Guide to chemical literature, 1972 to present. Inquire as to cost and availability.

DISSERTATION ABSTRACTS ONLINE. University Microfilms International, 300 North Zeeb Road, Ann Arbor, MI 48106. (800) 521-0600 or (313) 761-4700. Scope includes virtually all doctoral dissertations accepted at accredited American institutions from 1861 to present in 252 subject areas. Inquire as to cost and availability.

GEOREF. American Geological Institute, 4220 King Street, Alexandria, VA 22302. (703) 379-2480. 1967 to present. Inquire as to online cost and availability.

GEOARCHIVE. Geosystems, Post Office Box 1024, Westminister, London, England, SW1 P 2JL. Citations to literature on geoscience, 1969 to present. Inquire as to online cost and availability.

NTIS. National Technical Information Service, 5285 Port Royal Road, Springfield, VA 22161. (703) 487-4630. Broad coverage of government-sponsored research reports, 1964 to present. Inquire as to cost and availability.

SCISEARCH. Institute for Scientific Information, 3501 Market Street, Philadelphia, PA 19104. (800) 523-1850 or (215) 386-0100. Broad multidisciplinary title and author index to the international literature of science and technology, 1974 to present. Inquire as to cost and availability.

PERIODICALS

AAPG BULLETIN. American Association of Petroleum Geologists, Post Office Box 979, Tulsa, Ok 74101. (918) 584-2555.

AMERICAN JOURNAL OF SCIENCE. Kline Geology Laboratory, Yale University, New Haven, CT 06520. Ten times per year. $80.00 per year.

ECONOMIC GEOLOGY. Society of Economic Geologists, Post Office Box 571, Golden, CO 80402. (303) 279-1899. 8 times per year. $25.00.

GEOCHIMICA ET COSMOCHIMICA ACTA. Pergamon Journals, Incorporated, Maxwell House, Fairview Park, Elmsford, NY 10523. (914) 592-7700. Monthly. $340.00 per year.

GEOLOGICAL MAGAZINE. Cambridge University Press, 32 East 57th Street, New York, NY 10022. (800) 872-7423. Bimonthly. $165.00 per year.

GEOLOGICAL SOCIETY JOURNAL. Geological Society of London, Blackwell Scientific Publications, Incorporated, 667 Lytton Avenue, Palo Alto, CA 94301. (415) 324-1688. 6 times per year. $290.00.

GEOLOGICAL SOCIETY OF AMERICA BULLETIN. Post Office Box 9140, 3300 Penrose Place, Boulder, CO 80301. (303) 447-2020. Monthly. $80.00.

GEOLOGY. Geological Society of America, Post Office Box 9140, 3300 Penrose Place, Boulder, CO 80301. (303) 447-2020. Monthly. $55.00 per year.

JOURNAL OF GEOLOGY. University of Chicago Press, 5801 South Ellis Street, Chicago, IL 60637. (312) 962-7600. Bimonthly. $30.00 per year.

LITHOS: AN INTERNATIONAL JOURNAL OF MINERALOGY, PETROLOGY, AND GEOCHEMISTRY. Elsevier Science Publishers, Post Office Box 330, Irving-on-Hudson, NY 10533. Quarterly. $73.00 per year.

RESEARCH CENTERS AND INSTITUTES

COLUMBIA UNIVERSITY. Lamont-Doherty Geological Observatory, Palisades, NY 10964. (914) 359-2900.

UNITED STATES GEOLOGICAL SURVEY. National Center, 12201 Sunrise Valley Drive, Reston, Va 22092. The major geological research agency of the federal government conducting research in most areas of pure and applied research in the geosciences.

GEOLOGICAL SOCIETY OF AMERICA. Post Office Box 9140, 3300 Penrose Place, Boulder, CO 80301. (303) 477-2020.

UNIVERSITY OF ARIZONA. Laboratory of Isotope Geochemistry, 136 Gould-Simpson Building, Tucson, Az 85721. (602) 621-6014.

UNIVERSITY OF CALIFORNIA, SAN DIEGO. Geological Research Division. A-020. Scripps Institution of Oceanography, San Diego, CA 92093. (619) 534-1830.

GEOCHRONOLOGY

See also: GEOLOGY, GEOPHYSICS, PALEONTOLOGY, SEDIMENTARY ROCKS

ABSTRACT SERVICES AND INDEXES

BIBLIOGRAPHY AND INDEX OF GEOLOGY. American Geological Institute, 4220 King Street, Alexandria, VA 22302. (703) 379-2480. 1969 to present. Monthly. $1100.00 per year.

CHEMICAL ABSTRACTS. Chemical Abstracts Service, 2540 Olentangy Road, Post Office Box 3012, Columbus, OH 43210. (800) 848-6538 or (614) 421-3600. Weekly. $9200.00 per year.

GENERAL SCIENCE INDEX. H.W. Wilson Company, 950 University Avenue, Bronx, NY 10452. (800) 367-6770 or (212) 588-8400. Inquire as to cost and availability.

MINERALOGICAL ABSTRACTS. Mineralogical Society and the Mineralogical Society of America, 41 Queen's Gate, London, SW7 5HR, England. 1959 to present. Quarterly. $190.00 per year.

OCEANIC ABSTRACTS. Cambridge Scientific Abstracts, 5161 River Road, Bethesda, MD 20816. (301) 951-1400. 1964 to present. Bimonthly. $652.00 per year.

SCIENCE CITATION INDEX. Institute for Scientific Information, 3501 Market Street, Philadelphia, PA 19104. (800) 523-1850 or (215) 386-0100. Inquire as to cost and availability.

ANNUAL REVIEWS AND YEARBOOKS

ANNUAL REVIEW AND EARTH AND PLANETARY SCIENCES. Annual Reviews, Incorporated, 4139 El Camino Way, Palo Alto, CA 94306. (415) 493-4400.

ASSOCIATIONS AND PROFESSIONAL SOCIETIES

AMERICAN ASSOCIATION OF PETROLEUM GEOLOGISTS. Post Office Box 979, Tulsa, OK 74101. (918) 584-2555.

AMERICAN GEOLOGICAL INSTITUTE. 4220 King Street, Alexandria, VA 22302. (703) 379-2480.

AMERICAN GEOPHYSICAL UNION. 2000 Florida Avenue, NW, Washington, DC 20009. (202) 462-6903.

AMERICAN INSTITUTE OF PROFESSIONAL GEOLOGISTS. 7828 Vance Drive, Suite 103, Arvada, CO 80003. (303) 431-0831.

ASSOCIATION OF ENGINEERING GEOLOGISTS. Post Office Box 1068, Brentwood, TN 37027. (615) 377-3578.

GEOLOGICAL SOCIETY OF AMERICA. Box 9140, Boulder, CO 80301. (303) 447-2020.

SOCIETY OF ECONOMIC PALEONTOLOGISTS AND MINERALOGISTS. Box 4756, Tulsa, OK 74159. (918) 743-9765.

DIRECTORIES AND BIOGRAPHICAL SOURCES

AMERICAN MEN AND WOMEN OF SCIENCE. Physical and Biological Sciences. Sixteenth edition. R.R. Bowker Company, 205 East 42nd Street, New York, NY 10017. (800) 521-8110 or (212) 916-1600. 1986. $595.00.

AMERICAN INSTITUTE OF PROFESSIONAL GEOLOGISTS. Membership Directory. American Institute of Professional Geologists, 7828 Vance Drive, Suite 103, Arvada, CO 80003. (303) 431-0831. Annual. $15.00.

ASSOCIATION OF ENGINEERING GEOLOGISTS DIRECTORY. Association of Engineering Geologists, Dr. G. Lee Christensen, Civil Engineering Department, Villanova University, Villanova, PA 19085. (215) 645-4960. Annual. $15.00.

GEOLOGICAL SOCIETY OF AMERICA. Membership Directory. Geological Society of America, 3300 Penrose Place, Boulder, CO 80301. (303) 447-2020. Annual. Available to members only.

RESEARCH CENTERS DIRECTORY. Gale Research Company, Book Tower, Detroit, MI 48226. (800) 521-0707. Twelfth edition. 1987. $365.00 for set.

WHO'S WHO IN FRONTIER SCIENCE AND TECHNOLOGY. Marquis Who's Who, Incorporated, 200 East Ohio Street, Chicago, IL 60611. (800) 428-3898 or (312) 787-2008.

ENCYCLOPEDIAS AND DICTIONARIES

DICTIONARY OF GEOLOGICAL TERMS. American Geological Institute. Doubleday and Company, Incorporated, 245 Park Avenue, New York, NY 10017. (800) 645-6156 or (212) 953-4561. Third edition. 1984. $19.95.

GLOSSARY OF GEOLOGY. Robert L. Bates and Julia A. Jackson. American Geological Institute, 4220 King Street, Alexandria, VA 22032. (703) 379-2480. 1980. $60.00.

HANDBOOKS AND MANUALS

GEOLOGY IN THE FIELD. R.R. Compton. John Wiley and Sons, Incorporated, 605 Third Avenue, New York, NY 10158. (800) 526-5368 or (212) 850-6418. 1985. $23.95.

ONLINE DATA BASES

CA SEARCH. Chemical Abstracts Service, Post Office Box 3012, Columbus, OH 43210. Guide to chemical literature, 1972 to present. Inquire as to cost and availability.

DISSERTATION ABSTRACTS ONLINE. University Microfilms International, 300 North Zeeb Road, Ann Arbor, MI 48106. (800) 521-0600 or (313) 761-4700. Scope includes virtually all doctoral dissertations accepted at accredited American institutions from 1861 to present in 252 subject areas. Inquire as to cost and availability.

GEOREF. American Geological Institute, 4220 King Street, Alexandria, VA 22302. (703) 379-2480. 1967 to present. Inquire as to online cost and availability.

GEOARCHIVE. Geosystems, Post Office Box 1024, Westminster, London, England, SW1 P 2JL. Citations to literature on geoscience, 1969 to present. Inquire as to online cost and availability.

NTIS. National Technical Information Service, 5285 Port Royal Road, Springfield, VA 22161. (703) 487-4630. Broad coverage of government-sponsored research reports, 1964 to present. Inquire as to cost and availability.

SCISEARCH. Institute for Scientific Information, 3501 Market Street, Philadelphia, PA 19104. (800) 523-1850 or (215) 386-0100. Broad multidisciplinary title and author index to the international literature of science and technology, 1974 to present. Inquire as to cost and availability.

PERIODICALS

AAPG BULLETIN. American Association of Petroleum Geologists, Post Office Box 979, Tulsa, OK 74101. (918) 584-2555.

AMERICAN JOURNAL OF SCIENCE. Kline Geology Laboratory, Yale University, New Haven, CT 06520. Ten times per year. $80.00 per year.

ECONOMIC GEOLOGY. Society of Economic Geologists, Post Office Box 571, Golden, CO 80402. (303) 279-1899. 8 times per year. $25.00.

GEOCHIMICA ET COSMOCHIMICA ACTA. Pergamon Journals, Incorporated, Maxwell House, Fairview Park, Elmsford, NY 10523. (914) 592-7700. Monthly. $340.00 per year.

GEOLOGICAL MAGAZINE. Cambridge University Press, 32 East 57th Street, New York, NY 10022. (800) 872-7423. Bimonthly. $165.00 per year.

GEOLOGICAL SOCIETY JOURNAL. Geological Society of London, Blackwell Scientific Publications, Incorporated, 667

Lytton Avenue, Palo Alto, CA 94301. (415) 324-1688. 6 times per year. $290.00.

GEOLOGICAL SOCIETY OF AMERICA BULLETIN. Post Office Box 9140, 3300 Penrose Place, Boulder, CO 80301. (303) 447-2020. Monthly. $80.00.

GEOLOGY. Geological Society of America, Post Office Box 9140, 3300 Penrose Place, Boulder, CO 80301. (303) 447-2020. Monthly. $55.00 per year.

JOURNAL OF GEOLOGY. University of Chicago Press, 5801 South Ellis Street, Chicago, IL 60637. (312) 962-7600. Bimonthly. $30.00 per year.

LITHOS: AN INTERNATIONAL JOURNAL OF MINERALOGY, PETROLOGY, AND GEOCHEMISTRY. Elsevier Science Publishers, Post Office Box 330, Irving-on-Hudson, NY 10533. Quarterly. $73.00 per year.

RESEARCH CENTERS AND INSTITUTES

COLUMBIA UNIVERSITY. Lamont-Doherty Geological Observatory, Palisades, NY 10964. (914) 359-2900.

UNITED STATES GEOLOGICAL SURVEY. National Center, 12201 Sunrise Valley Drive, Reston, Va 22092. The major geological research agency of the federal government conducting research in most areas of pure and applied research in the geosciences.

GEOLOGICAL SOCIETY OF AMERICA. Post Office Box 9140, 3300 Penrose Place, Boulder, CO 80301. (303) 477-2020.

GEODESY

See also: CARTOGRAPHY, REMOTE SENSING, SURVEYING

ABSTRACT SERVICES AND INDEXES

APPLIED SCIENCE AND TECHNOLOGY INDEX. H.W. Wilson Company, 950 University Avenue, Bronx, NY 10452. (800) 367-6670 or (212) 588-8400. Inquire as to cost and availability.

BIBLIOGRAPHY AND INDEX OF GEOLOGY. American Geological Institute, 4220 King Street, Alexandria, Va 22302. (703) 379-2480. 1969 to present. Monthly. $1100.00 per year.

ENGINEERING INDEX MONTHLY AND AUTHOR INDEX. Engineering Information, Incorporated, 345 East 47th Street, New York, NY 10017. (800) 221-1044 or (212) 705-7615. Engineering and technical literature. Monthly. $1560.00 per year.

GENERAL SCIENCE INDEX. H.W. Wilson Company, 950 University Avenue, Bronx, NY 10452. (800) 367-6770 or (212) 588-8400. Inquire as to cost and availability.

SCIENCE CITATION INDEX. Institute for Scientific Information, 3501 Market Street, Philadelphia, PA 19104. (800) 523-1850 or (215) 386-0100. Inquire as to cost and availability.

ASSOCIATIONS AND PROFESSIONAL SOCIETIES

AMERICAN CONGRESS OF SURVEYING AND MAPPING. 210 Little Falls Street, Falls Church, VA 22046. (703) 241-2446.

AMERICAN GEOPHYSICAL UNION. 2000 Florida Avenue, NW, Washington, DC 20009. (202) 462-6903.

ASSOCIATION OF AMERICAN GEOGRAPHERS. 1710 16th Street, NW, Washington, DC 20009. (202) 234-1450.

DIRECTORIES AND BIOGRAPHICAL SOURCES

AMERICAN MEN AND WOMEN OF SCIENCE. Physical and Biological Sciences. Sixteenth edition. R.R. Bowker Company, 205 East 42nd Street, New York, NY 10017. (800) 521-8110 or (212) 916-1600. 1986. $595.00.

ASSOCIATION OF ENGINEERING GEOLOGISTS DIRECTORY. Association of Engineering Geologists, Dr. G. Lee Christensen, Civil Engineering Department, Villanova University, Villanova, PA 19085. (215) 645-4960. Annual. $15.00.

GEOLOGICAL SOCIETY OF AMERICA. Membership Directory. Geological Society of America, 3300 Penrose Place, Boulder, CO 80301. (303) 447-2020. Annual. Available to members only.

INTERNATIONAL RESEARCH CENTERS DIRECTORY 1988-89. Darren L. Smith, editor. Gale Research Company, Book Tower, Detroit, MI 48226. (800) 521-0707. 4th edition. 1987. $360.00.

RESEARCH CENTERS DIRECTORY. Gale Research Company, Book Tower, Detroit, MI 48226. (800) 521-0707. Twelfth edition. 1987. $365.00 for set.

SCIENTIFIC AND TECHNICAL ORGANIZATIONS AND AGENCIES DIRECTORY. Margaret Labash Young, editor. Gale Research Company, Detroit, MI 48226. (800) 521-0707. Second edition. 1987. $185.00.

ENCYCLOPEDIAS AND DICTIONARIES

DICTIONARY OF GEOLOGICAL TERMS. American Geological Institute. Doubleday and Company, Incorporated, 245 Park Avenue, New York, NY 10017. (800) 645-6156 or (212) 953-4561. Third edition. 1984. $19.95.

GLOSSARY OF GEOLOGY. Robert L. Bates and Julia A. Jackson. American Geological Institute, 4220 King Street, Alexandria, VA 22032. (703) 379-2480. 1980. $60.00.

GENERAL WORKS

GEODESY. G. Bomford. Oxford University Press, Incorporated, 200 Madison Avenue, New York, NY 10016. (212) 564-6680; Orders to 16-00 Pollitt Drive, Fair Lawn, NJ 07410. (201) 796-8000. 1980. $119.00.

GEODESY: THE CONCEPTS. P. Vanicek and E.J. Krakiwsky. Elsevier Science Publishing Company, Incorporated, 52 Vanderbilt Avenue, New York, NY 10017. (212) 370-5520. 1986. $39.00.

GEODETIC REFRACTION. F.K. Brunner. Springer-Verlag New York, Incorporated, 175 Fifth Avenue, New York, NY 10010. (212) 460-1500. 1984. $17.00.

ONLINE DATA BASES

COMPENDEX. Engineering Information, Incorporated, 345 East 47th Street, New York, NY 10017. (800) 221-1044 or (212) 705-7615. Engineering and technical literature, 1975 to present. Inquire as to cost and availability.

DISSERTATION ABSTRACTS ONLINE. University Microfilms International, 300 North Zeeb Road, Ann Arbor, MI 48106. (800) 521-0600 or (313) 761-4700. Scope includes virtually all doctoral dissertations accepted at accredited American institutions from 1861 to present in 252 subject areas. Inquire as to cost and availability.

GEOREF. American Geological Institute, 4220 King Street, Alexandria, VA 22302. (703) 379-2480. 1967 to present. Inquire as to online cost and availability.

GEOARCHIVE. Geosystems, Post Office Box 1024, Westminister, London, England, SW1 P 2JL. Citations to literature on geoscience, 1969 to present. Inquire as to online cost and

availability.

NTIS. National Technical Information Service, 5285 Port Royal Road, Springfield, VA 22161. (703) 487-4630. Broad coverage of government-sponsored research reports, 1964 to present. Inquire as to cost and availability.

SCISEARCH. Institute for Scientific Information, 3501 Market Street, Philadelphia, PA 19104. (800) 523-1850 or (215) 386-0100. Broad multidisciplinary title and author index to the international literature of science and technology, 1974 to present. Inquire as to cost and availability.

PERIODICALS

MARINE GEODESY: AN INTERNATIONAL JOURNAL OF OCEAN SURVEYS, MAPPING AND SENSING. Crane, Russak and Company, 3 East 44th Street, New York, NY 10017. (212) 867-1490. 1977 to present. Quarterly. $90.00 per year.

SURVEYING AND MAPPING: DEVOTED TO THE ADVANCEMENT OF THE SCIENCES OF SURVEYING AND MAPPING. American Congress of Surveying and Mapping, 210 Little Falls Street, Falls Church, VA 22046. (703) 241-2446. 1941 to present. Quarterly. $60.00 per year.

RESEARCH CENTERS AND INSTITUTES

UNITED STATES GEOLOGICAL SURVEY. National Center, 12201 Sunrise Valley Drive, Reston, VA 22092. The major geological research agency of the federal government conducting research in most areas of pure and applied research in the geosciences.

UNIVERSITY OF HAWAII. Hawaii Institute of Geophysics, 2525 Correa Road, Honolulu, HI 96822. (808) 948-8760.

UNIVERSITY OF TORONTO. Survey Science. Erindale College, Mississauga Road, Mississauga, ON, Canada L5L 1C6. (416) 828-5298.

GEOLOGY

See also: CRUST, GEOCHEMISTRY, GEOPHYSICS, PHYSICAL GEOLOGY, SEISMOLOGY, VOLCANOLOGY

ABSTRACT SERVICES AND INDEXES

BIBLIOGRAPHY AND INDEX OF GEOLOGY. American Geological Institute, 4220 King Street, Alexandria, VA 22302. (703) 379-2480.

CHEMICAL ABSTRACTS. Chemical Abstracts Service, 2540 Olentangy Road, P.O. Box 3012, Columbus, OH 43210. (800) 848-6538 or (614) 421-3600. $6200.00 per year.

DEEP-SEA RESEARCH WITH OCEANOGRAPHIC LITERATURE REVIEW. Pergamon Press, Inc., Maxwell House, Fairview Park, Elmsford, NY 10523. (914) 592-7700. Twenty-four times per year. $600.00 per year.

MINERALOGICAL ABSTRACTS. Mineralogical Society and the Mineralogical Society of America, 41 Queen's Gate, London, SW7 5HR, England. Quarterly. $190.00 per year.

OCEANIC ABSTRACTS. Cambridge Scientific Abstracts, 5161 River Road, Bethesda, MD 20816. (301) 951-1400. Bimonthly. $652.00 per year.

SCIENCE CITATION INDEX. Institute for Scientific Information, 3501 Market Street, Philadelphia, PA 19104. (800) 523-1850 or (215) 386-0100. Inquire as to price and availability.

GENERAL SCIENCE INDEX. H.W. Wilson Co., 950 University Avenue, Bronx, NY 10452. (800) 367-6770 or (212) 588-8400. Inquire as to price and availability.

ANNUAL REVIEWS AND YEARBOOKS

ADVANCES IN GEOPHYSICS. Academic Press, Inc., 6277 Sea Harbor Drive, Orlando, FL 32821. (800) 321-5068. Irregular. $62.00 per volume.

ANNUAL REVIEW AND EARTH AND PLANETARY SCIENCES. Annual Reviews, Inc., 4139 El Camino Way, Palo Alto, CA 94306. (415) 493-4400.

MINERALS YEARBOOK. Bureau of Mines, U.S. Department of the Interior. Available from U.S. Government Printing Office, Washington, DC 20402. (202) 783-3238. Annual. Three volumes. $45.00.

ASSOCIATIONS AND PROFESSIONAL SOCIETIES

AMERICAN ASSOCIATION OF PETROLEUM GEOLOGISTS. P.O. Box 979, Tulsa, OK 74101. (918) 584-2555.

AMERICAN GEOLOGICAL INSTITUTE. 4220 King Street, Alexandria, VA 22302. (703) 379-2480.

AMERICAN GEOPHYSICAL UNION, 2000 Florida Avenue, N.W., Washington, DC 20009. (202) 462-6903.

AMERICAN INSTITUTE OF PROFESSIONAL GEOLOGISTS. 7828 Vance Drive, Suite 103, Arvada, CO 80003. (303) 431-0831.

ASSOCIATION OF ENGINEERING GEOLOGISTS. P.O Box 1068, Brentwood, TN 37027. (615) 377-3578.

BIBLIOGRAPHIES

SCIENCE BOOKS AND FILMS. American Association for the Advancement of Science, 1333 H Street, N.W., Washington, DC 20005.

SCIENTIFIC AND TECHNICAL BOOKS IN PRINT; An index to Literature in Science and Technology. R.R. Bowker Co., 205 E. 42nd Street, New York, NY 10017. (800) 521-8110 or (212) 916-1600.

DIRECTORIES AND BIOGRAPHICAL SOURCES

AMERICAN MEN AND WOMEN OF SCIENCE: Physical and Biological Sciences. Sixteenth edition. R.R. Bowker Company, 245 West 17th Street, New York, NY 10011. (800) 521-8810 or (212) 916-1600. $595.00.

AMERICAN INSTITUTE OF PROFESSIONAL GEOLOGISTS. Membership Directory. American Institute of Professional Geologists, 7828 Vance Drive, Suite 103, Arvada, CO 80003. (303) 431-0831. Annual, $15.00.

ASSOCIATION OF ENGINEERING GEOLOGISTS DIRECTORY. Association of Engineering Geologists, Dr. G. Lee Christensen, Civil Engineering Department, Villanova University, Villanova, PA 19085. (215) 645-4960. Annual. $15.00.

GEOLOGICAL SOCIETY OF AMERICA. Membership Directory. Geological Society of America, 3300 Penrose Place, Boulder, CO 80301. (303) 447-2020. Annual. Available to members only.

RESEARCH CENTERS DIRECTORY. Gale Research Company, Detroit, MI 48226. 11th edition, 1987. (800) 521-0707.

WHO'S WHO IN FRONTIER SCIENCE AND TECHNOLOGY. Marquis Who's Who, Inc., 200 East Ohio Street, Chicago, IL 60611. (800) 428-3898 or (312) 787-2008.

GEOLOGY

ENCYCLOPEDIAS AND DICTIONARIES

DICTIONARY OF GEOLOGICAL TERMS. American Geological Institute. Doubleday and Company, Inc., 245 Park Avenue, New York, NY 10017. (800) 645-6156 or (212) 953-4561. Third edition. 1984. $19.95.

DICTIONARY AND PETROLOGY. S.I. Tomkeieff. John Wiley and Sons, Inc., 605 Third Avenue, New York, NY 10158. (800) 526-5368 or (212) 850-6000. 1983. $140.00.

ENCYCLOPEDIA OF PHYSICAL SCIENCE AND TECHNOLOGY. Academic Pres, Inc., Orlando, FL 32887. (800) 321-5068.

GLOSSARY OF GEOLOGY. Robert L. Bates and Julia A. Jackson. American Geological Institute, 4220 King Street, Alexandria, VA 22032. (703) 379-2480. 1980. $60.00.

GENERAL WORKS

THE EARTH. Peter J. Smith. Macmillan Publishing Co., 866 Third Avenue, New York, NY 10022. (800) 257-5755 or (212) 702-2000. 1986. $40.00.

INTRODUCTION TO ROCK FORMING MINERALS. W.A. Deer, J. Zussman and R.A. Howie. John Wiley and Sons, Inc., 605 Third Avenue, New York, NY 10158. (800) 526-5368 or (212) 850-6000. 1966. $29.95.

INTRODUCTION TO SEDIMENTOLOGY. R.C. Selley. Academic Press, Inc., 6277 Sea Harbor Drive, Orlando, FL 32821. (800) 321-5068. Second edition. 1982. $47.50.

PHYSICAL GEOLOGY. R.F. Flint and B.J. Skinner. John Wiley and Sons, Inc., 605 Third Avenue, New York, NY 10158. (800) 526-5368 or (212) 850-6000. Second edition. 1977. $40.95.

HANDBOOKS AND MANUALS

GEOLOGY IN THE FIELD. R.R. Compton. John Wiley and Sons, Inc., 605 Third Avenue, New York, NY 10158. (800) 526-5368 or (212) 850-6418. 1985. $23.95.

ONLINE DATA BASES

GEOREF. American Geological Institute, 4220 King Street, Alexandria, VA 22302. (800) 336-4764 or (703) 379-2480. Geology and geosciences literature, 1961 to present. Inquire as to online cost and availability.

GEOARCHIVE. Geosystems, P.O. Box 1024, Westminster, London, England, SW1 P 2JL. Citations to literature on geoscience, 1969 to present. Inquire as to online cost and availability.

NTIS. National Technical Information Service, 5285 Port Royal Road, Springfield, VA 22161. (703) 487-4630. Broad coverage of government sponsored research reports, 1964 to present. Inquire as to online cost and availability.

SCISEARCH. Institute for Scientific Information, 3501 Market Street, Philadelphia, PA 19104. (800) 523-1850 or (215) 386-0100. Broad interdisciplinary index to the literature of science and technology, 1965 to present. Inquire as to online cost and availability.

PERIODICALS

AAPG BULLETIN. American Association of Petroleum Geologists, P.O. Box 979, Tulsa, OK 7401. (918) 584-2555.

AMERICAN JOURNAL OF SCIENCE. Kline Geology Laboratory, Yale University, New Haven, CT 06520. Ten times per year. $80.00 per year.

ECONOMIC GEOLOGY. Society of Economic Geologists, P.O. Box 571, Golden, CO 80402. (303) 279-1899. 8 times per year. $25.00.

GEOCHIMICA ET COSMOCHIMICA ACTA. Pergamon Journals, Inc., Maxwell House, Fairview Park, Elmsford, NY 10523. (914) 592-7700. Monthly. $340.00 per year.

GEOLOGICAL MAGAZINE. Cambridge University Press, 32 East 57th Street, New York, NY 10022. (800) 872-7423. Bimonthly. $165.00 per year.

GEOLOGICAL SOCIETY JOURNAL. Geological Society of London. Blackwell Scientific Publications, Inc., 667 Lytton Avenue, Palo Alto, CA 94301. (415) 324-1688. 6 times per year. $290.00.

GEOLOGICAL SOCIETY OF AMERICA BULLETIN. P.O. Box 9140, 3300 Penrose Place, Boulder, CO 80301. (303) 447-2020. Monthly. $80.00.

GEOLOGY. Geological Society of America. P.O. Box 9140, 3300 Penrose Place, Boulder, CO 80301. (303) 447-2020. Monthly. $55.00 per year.

GEOPHYSICS. Society of Exploration Geophysicists, P.O. Box 702740, Tulsa, OK 74170. (918) 493-3516. Monthly. $45.00 per year.

GEOTIMES. American Geological Institute, 4220 King Street, Alexandria, VA 22032. Monthly. $18.00 per year.

INTERNATIONAL ASSOCIATION FOR MATHEMATICAL GEOLOGY. Journal. Plenum Publishing Corporation, 233 Spring Street, New York, NY 10013. (800) 221-9369. Six times per year. $225.00 per year.

JOURNAL OF GEOLOGY. University of Chicago Press, 5801 South Ellis Street, Chicago, IL 60637. (312) 962-7600. Bimonthly. $30.00 per year.

JOURNAL OF GEOPHYSICAL RESEARCH. American Geophysical Union, 2000 Florida Avenue, N.W., Washington, DC 20009. (202) 462-6903. Weekly. $680.00 per year to individuals.

JOURNAL OF GEOPHYSICS. Springer-Verlag, Inc., 175 Fifth Avenue, New York, NY 10010. (800) 526-7254 or (212) 460-1500. Bimonthly. $175.00 per year.

JOURNAL OF SEDIMENTARY PETROLOGY. Society of Economic Paleontologists and Mineralogists, P.O. Box 4756, Tulsa, OK 74159. (918) 743-9765. Bimonthly.

JOURNAL OF STRUCTURAL GEOLOGY. Pergamon Journals, Inc., Maxwell House, Fairview Park, Elmsford, NY 10523. (914) 592-7700. Eight times per year. $160.00 per year.

LITHOS; an international journal of mineralogy, petrology, and geochemistry. Elsevier Science Publishers, P.O. Box 330, Irving-on-Hudson, NY 10533. Quarterly. $73.00 per year.

MOUNTAIN GEOLOGIST. Rocky Mountain Association of Geologists, 4201 West 51st Avenue, Denver, CO 80212-2902. Quarterly. $15.00 per year.

PROFESSIONAL GEOLGIST. American Institute of Professional Geologists, 7828 Vance Drive, Suite 103, Arvada, CO 80003. Monthly.

REVIEWS OF GEOPHYSICS. American Geophysical Union, 2000 Florida Avenue, N.W., Washington, DC 20009. (202) 462-6903. Quarterly. $240.00 per year.

ROCK MECHANICS AND ROCK ENGINEERING. Springer-Verlag New York, Inc., 175 Fifth Avenue, New York, NY 10010. (800) 526-7254 or (212) 460-1500. Quarterly. $61.00 per year.

SEDIMENTARY GEOLOGY. Elsevier Science Publishers, P.O. Box 330, Irving-on-Hudson, NY 10533. Twenty times per year. $421.00 per year.

SEDIMENTOLOGY. Blackwell Scientific Publications, Inc., 52 Beacon Street, Boston, MA 02108. (617) 720-0761. Six times per year. $185.00 per year.

TECTONICS. American Geophysical Union, 2000 Florida Avenue, N.W., Washington, DC 20009. (202) 462-6903. Bimonthly. $30.00 per year to individuals.

TECTONOPHYSICS; an international journal of geotectonics and the geology and physics of the interior of the Earth. Elsevier Science Publishers, P.O. Box 330, Irving-on-Hudson, NY 10533. Forty-four times per year. $870.00 per year.

VOLCANOLOGY AND SEISMOLOGY. Gordon and Breach Science Publishers, 50 West 23rd Street, New York, NY 10010. (212) 206-8900. Monthly. $498.00 per year.

RESEARCH CENTERS AND INSTITUTES

U.S. GEOLOGICAL SURVEY, Nationl Center, 12201 Sunrise Valley Drive, Reston, VA 22092. The major geological research agency of the federal government conducting research in most areas of pure and applied research in the geosciences.

GEOLOGICAL SOCIETY OF AMERICA. P.O. Box 9140, 3300 Penrose Place, Boulder, CO 80301. (303) 447-2020.

SOCIETY OF ECONOMIC GEOLOGISTS. P.O. Box 571, Golden, CO 80402.

SOCIETY OF ECONOMIC PALEONTOLOGISTS AND MINERALOGISTS. P.O. Box 4756, Tulsa, OK 74159. (918) 743-9765.

GEOMAGNETISM

See also: GEOLOGY, GEOPHYSICS

ABSTRACT SERVICES AND INDEXES

BIBLIOGRAPHY AND INDEX OF GEOLOGY. American Geological Institute, 4220 King Street, Alexandria, VA 22302. (703) 379-2480. 1969 to present. Monthly. $1100.00 per year.

CHEMICAL ABSTRACTS. Chemical Abstracts Service, 2540 Olentangy Road, Post Office Box 3012, Columbus, OH 43210. (800) 848-6538 or (614) 421-3600. Weekly. $9200.00 per year.

DEEP-SEA RESEARCH WITH OCEANOGRAPHIC LITERATURE REVIEW. Pergamon Press, Incorporated, Maxwell House, Fairview Park, Elmsford, NY 10523. (914) 592-7700. 1953 to present. Twenty-four times per year. $600.00 per year.

GENERAL SCIENCE INDEX. H.W. Wilson Company, 950 University Avenue, Bronx, NY 10452. (800) 367-6770 or (212) 588-8400. Inquire as to cost and availability.

MINERALOGICAL ABSTRACTS. Mineralogical Society and the Mineralogical Society of America, 41 Queen's Gate, London, SW7 5HR, England. 1959 to present. Quarterly. $190.00 per year.

OCEANIC ABSTRACTS. Cambridge Scientific Abstracts, 5161 River Road, Bethesda, MD 20816. (301) 951-1400. 1964 to present. Bimonthly. $652.00 per year.

PHYSICS ABSTRACTS. Institute of Electrical Engineers, London, United Kingdom. Available from: Institute of Electrical and Electronic Engineers (IEEE), 345 East 47th Street, New York, NY 10017. (212) 705-7900. 1898 to present. Monthly. $1670.00 per year.

SCIENCE CITATION INDEX. Institute for Scientific Information, 3501 Market Street, Philadelphia, PA 19104. (800) 523-1850 or (215) 386-0100. Inquire as to cost and availability.

ANNUAL REVIEWS AND YEARBOOKS

ANNUAL REVIEW AND EARTH AND PLANETARY SCIENCES. Annual Reviews, Incorporated, 4139 El Camino Way, Palo Alto, CA 94306. (415) 493-4400.

ASSOCIATIONS AND PROFESSIONAL SOCIETIES

AMERICAN ASSOCIATION OF PETROLEUM GEOLOGISTS. Post Office Box 979, Tulsa, OK 74101. (918) 584-2555.

AMERICAN GEOLOGICAL INSTITUTE. 4220 King Street, Alexandria, VA 22302. (703) 379-2480.

AMERICAN GEOPHYSICAL UNION. 2000 Florida Avenue, NW, Washington, DC 20009. (202) 462-6903.

AMERICAN INSTITUTE OF PROFESSIONAL GEOLOGISTS. 7828 Vance Drive, Suite 103, Arvada, CO 80003. (303) 431-0831.

ASSOCIATION OF ENGINEERING GEOLOGISTS. Post Office Box 1068, Brentwood, TN 37027. (615) 377-3578.

GEOLOGICAL SOCIETY OF AMERICA. Box 9140, Boulder, CO 80301. (303) 447-2020.

SOCIETY OF ECONOMIC PALEONTOLOGISTS AND MINERALOGISTS. Box 4756, Tulsa, OK 74159. (918) 743-9765.

DIRECTORIES AND BIOGRAPHICAL SOURCES

AMERICAN MEN AND WOMEN OF SCIENCE. Physical and Biological Sciences. Sixteenth edition. R.R. Bowker Company, 205 East 42nd Street, New York, NY 10017. (800) 521-8110 or (212) 916-1600. 1986. $595.00.

AMERICAN INSTITUTE OF PROFESSIONAL GEOLOGISTS. Membership Directory. American Institute of Professional Geologists, 7828 Vance Drive, Suite 103, Arvada, CO 80003. (303) 431-0831. Annual. $15.00.

ASSOCIATION OF ENGINEERING GEOLOGISTS DIRECTORY. Association of Engineering Geologists, Dr. G. Lee Christensen, Civil Engineering Department, Villanova University, Villanova, PA 19085. (215) 645-4960. Annual. $15.00.

GEOLOGICAL SOCIETY OF AMERICA. Membership Directory. Geological Society of America, 3300 Penrose Place, Boulder, CO 80301. (303) 447-2020. Annual. Available to members only.

RESEARCH CENTERS DIRECTORY. Gale Research Company, Book Tower, Detroit, MI 48226. (800) 521-0707. Twelfth edition. 1987. $365.00 for set.

WHO'S WHO IN FRONTIER SCIENCE AND TECHNOLOGY. Marquis Who's Who, Incorporated, 200 East Ohio Street, Chicago, IL 60611. (800) 428-3898 or (312) 787-2008.

ENCYCLOPEDIAS AND DICTIONARIES

DICTIONARY OF GEOLOGICAL TERMS. American Geological Institute. Doubleday and Company, Incorporated, 245 Park Avenue, New York, NY 10017. (800) 645-6156 or (212) 953-4561. Third edition. 1984. $19.95.

GLOSSARY OF GEOLOGY. Robert L. Bates and Julia A. Jackson. American Geological Institute, 4220 King Street, Alexandria, VA 22032. (703) 379-2480. 1980. $60.00.

GEOLOGY IN THE FIELD. R.R. Compton. John Wiley and Sons, Incorporated, 605 Third Avenue, New York, NY 10158. (800) 526-5368 or (212) 850-6418. 1985. $23.95.

GEOMAGNETISM

ONLINE DATA BASES

CA SEARCH. Chemical Abstracts Service, Post Office Box 3012, Columbus, OH 43210. Guide to chemical literature, 1972 to present. Inquire as to cost and availability.

DISSERTATION ABSTRACTS ONLINE. University Microfilms International, 300 North Zeeb Road, Ann Arbor, MI 48106. (800) 521-0600 or (313) 761-4700. Scope includes virtually all doctoral dissertations accepted at accredited American institutions from 1861 to present in 252 subject areas. Inquire as to cost and availability.

GEOREF. American Geological Institute, 4220 King Street, Alexandria, VA 22302. (703) 379-2480. 1967 to present. Inquire as to online cost and availability.

GEOARCHIVE. Geosystems, Post Office Box 1024, Westminister, London, England, SW1 P 2JL. Citations to literature on geoscience, 1969 to present. Inquire as to online cost and availability.

NTIS. National Technical Information Service, 5285 Port Royal Road, Springfield, VA 22161. (703) 487-4630. Broad coverage of government-sponsored research reports, 1964 to present. Inquire as to cost and availability.

SCISEARCH. Institute for Scientific Information, 3501 Market Street, Philadelphia, PA 19104. (800) 523-1850 or (215) 386-0100. Broad multidisciplinary title and author index to the international literature of science and technology, 1974 to present. Inquire as to cost and availability.

PERIODICALS

AAPG BULLETIN. American Association of Petroleum Geologists, Post Office Box 979, Tulsa, OK 74101. (918) 584-2555.

AMERICAN JOURNAL OF SCIENCE. Kline Geology Laboratory, Yale University, New Haven, CT 06520. Ten times per year. $80.00 per year.

ECONOMIC GEOLOGY. Society of Economic Geologists, Post Office Box 571, Golden, CO 80402. (303) 279-1899. 8 times per year. $25.00.

GEOCHIMICA ET COSMOCHIMICA ACTA. Pergamon Journals, Incorporated, Maxwell House, Fairview Park, Elmsford, NY 10523. (914) 592-7700. Monthly. $340.00 per year.

GEOLOGICAL MAGAZINE. Cambridge University Press, 32 East 57th Street, New York, NY 10022. (800) 872-7423. Bimonthly. $165.00 per year.

GEOLOGICAL SOCIETY JOURNAL. Geological Society of London, Blackwell Scientific Publications, Incorporated, 667 Lytton Avenue, Palo Alto, CA 94301. (415) 324-1688. 6 times per year. $290.00.

GEOLOGICAL SOCIETY OF AMERICA BULLETIN. Post Office Box 9140, 3300 Penrose Place, Boulder, CO 80301. (303) 447-2020. Monthly. $80.00.

GEOLOGY. Geological Society of America, Post Office Box 9140, 3300 Penrose Place, Boulder, CO 80301. (303) 447-2020. Monthly. $55.00 per year.

JOURNAL OF GEOLOGY. University of Chicago Press, 5801 South Ellis Street, Chicago, IL 60637. (312) 962-7600. Bimonthly. $30.00 per year.

LITHOS: AN INTERNATIONAL JOURNAL OF MINERALOGY, PETROLOGY, AND GEOCHEMISTRY. Elsevier Science Publishers, Post Office Box 330, Irving-on-Hudson, NY 10533. Quarterly. $73.00 per year.

RESEARCH CENTERS AND INSTITUTES

COLUMBIA UNIVERSITY. Lamont-Doherty Geological Observatory, Palisades, NY 10964. (914) 359-2900.

UNITED STATES GEOLOGICAL SURVEY. National Center, 12201 Sunrise Valley Drive, Reston, Va 22092. The major geological research agency of the federal government conducting research in most areas of pure and applied research in the geosciences.

GEOLOGICAL SOCIETY OF AMERICA. Post Office Box 9140, 3300 Penrose Place, Boulder, CO 80301. (303) 477-2020.

UNIVERSITY OF ARIZONA. Laboratory of Isotope Geochemistry, 136 Gould-Simpson Building, Tucson, Az 85721. (602) 621-6014.

UNIVERSITY OF CALIFORNIA, SAN DIEGO. Geological Research Division. A-020. Scripps Institution of Oceanography, San Diego, CA 92093. (619) 534-1830.

GEOMETRY

See also: MATHEMATICS

ABSTRACT SERVICES AND INDEXES

APPLIED SCIENCE AND TECHNOLOGY INDEX. H.W. Wilson Company, 950 University Avenue, Bronx, NY 10452. (800) 367-6670 or (212) 588-8400. Inquire as to cost and availability.

CHEMICAL ABSTRACTS. Chemical Abstracts Service, 2540 Olentangy Road, P.O. Box 3012, Columbus, OH 43210. (800) 848-6538 or (614) 421-3600. Weekly. $9200.00 per year.

GENERAL SCIENCE INDEX. H.W. Wilson Co. 950 University Avenue, Bronx, NY 10452. (800) 367-6770 or (212) 588-8400.

MATHEMATICAL REVIEWS. American Mathematical Society, P.O. Box 6248, Providence, RI 02940. (800) 556-7774 or (401) 272-9500.

PHYSICS ABSTRACTS. Institute of Electrical Engineers, London, United Kingdom. Available from: Institute of Electrical and Electronic Engineers (IEEE), 345 East 47th Street, New York, NY 10017. (212) 705-7900.

SCIENCE CITATION INDEX. Institute for Scientific Information, 3501 Market Street, Philadelphia, PA 19104. (800) 523-1850 or (215) 386-0100.

ASSOCIATIONS AND PROFESSIONAL SOCIETIES

AMERICAN MATHEMATICAL SOCIETY, P.O. Box 6248, Providence, RI 02940. (401) 272-9500.

MATHEMATICAL ASSOCIATION OF AMERICA, 1529 18th Street, N.W., Washington, DC 20036. (202) 387-5200.

SOCIETY FOR INDUSTRIAL AND APPLIED MATHEMATICS, 1400 Architects Building, 117 South 17th Street, Philadelphia, PA 19103. (215) 564-2929.

SOCIETY FOR INDUSTRIAL AND APPLIED MATHEMATICS, 1400 ARchitects Building, 117 South 17th Street, Philadelphia, PA 19103. (215) 564-2929.

BIBLIOGRAPHIES

SCIENCE BOOKS AND FILMS. American Association for the Advancement of Science, 1333 H Street, N.W., Washington, DC 20005.

SCIENTIFIC AND TECHNICAL BOOKS IN PRINT; An Index to Literature in Science and Technology. R.R. Bowker Co., 205 E. 42nd Street, New York, NY 10017. (800) 521-8110 or (212) 916-1600.

DIRECTORIES AND BIOGRAPHICAL SOURCES

AMERICAN MEN AND WOMEN OF SCIENCE. Physical and Biological Sciences. Fifteenth edition. R.R. Bowker Co., 205 E. 42nd Street, New York, NY 10017. (800) 521-8110 or (212) 916-1600.

BIOGRAPHICAL DICTIONARY OF SCIENTISTS. T.I. Williams. Halsted Press, 605 Third Avenue, New York, NY 10158. (800) 526-5368 or (212) 850-6418. Third edition. 1982. $29.95.

WHO'S WHO IN FRONTIER SCIENCE AND TECHNOLOGY. Marquis Who's Who, Inc., 200 East Ohio Street, Chicago, Illinois 60611. (800) 428-3898 or (312) 787-2008.

WHO'S WHO IN TECHNOLOGY TODAY. Reston Publishing Company, Inc., c/o Prentice-Hall, Inc., Englewood Cliffs, NJ 07632. (800) 262-6868. Biennial. Five volumes. $425.00. Covers the fields of electronics, computer science, physics, optics, chemistry, biotechnology, mechanics, energy, and earth science.

DIRECTORIES

INTERNATIONAL RESEARCH CENTERS DIRECTORY 1986-1987. Gale Research Company, Book Tower, Detroit, MI 48226. (800) 521-0707. Third edition. 1986.

RESEARCH CENTERS DIRECTORY. Gale Research Company, Book Tower, Detroit, Michigan 48226. (800) 521-0707. Eleventh edition. 1987.

SCIENTIFIC AND TECHNICAL ORGANIZATIONS AND AGENCIES DIRECTORY. Gale Research Company, Book Tower, Detroit, MI 48226. (800) 521-0707. 1985.

WORLD GUIDE TO SCIENTIFIC ASSOCIATIONS AND LEARNED SOCIETIES. K.G. Saur Inc., 175 Fifth Avenue, New York, NY 10010. (800) 521-0707 or (212) 982-1302. Fourth edition. 1984. $112.00.

ENCYCLOPEDIAS AND DICTIONAIRES

ENCYCLOPEDIA OF PHYSICAL SCIENCE AND TECHNOLOGY. Academic Press, Inc., Orlando, FL 32887. (800) 321-5068 or (305) 345-2734. Fifteen volumes, 1986.

ENCYCLOPEDIC DICTIONARY OF MATHEMATICS. Koyosi Ito, editor. MIT Press, 55 Hayward Street, Cambridge, MA 02142. (617) 253-2884. Second edition. 1987. Four volumes. $350.00.

MCGRAW-HILL ENCYCLOPEDIA OF SCIENCE AND TECHNOLOGY. McGraw-Hill Book, Inc., 1221 Avenue of the Americas, New York, NY 10020. (212) 997-3675.

GENERAL WORKS

APPLIED DIFFERENTIAL GEOMETRY. William L. Burke. Cambridge University Press, 32 East 57th Street, New York, NY 10022. (212) 316-8885. $55.50.

CALCULUS AND ANALYTIC GEOMETRY. Edwin J. Purcell and Jale Verberg. Prentice-Hall, Englewood Cliffs, NJ 07632. (201) 592-2000. Fifth edition. 1987. $45.95.

FRATICAL GEOMETRY OF NATURE. Benoit B. Mandelbrot. W.H. Freeman and Company, 41 Madison Avenue, New York, NY 10010. (212) 532-7660. 1982. $39.95.

FUNDAMENTALS OF MATHEMATICS. H. Behnke, F. Bachmann, K. Fladt, W. Suss, and H. Kunle, editors. MIT Press, 55 Hayward Street, Cambridge, MA 02142. (617) 253-2884. Volume 1, Foundations of Mathematics: Real Number System and Algebra; Volume 2, Geometry; Volume 3, Analysis. $65.00.

ELEMENTARY PLANE GEOMETRY. R.D. Gustafson and P.D. Frisk. John Wiley & Sons, 605 Third Avenue, New York, NY 10158. (800) 526-5368 or (212) 850-6000. Second edition. 1985. $29.50.

FUNDAMENTALS OF MATHEMATICS. H. Behnke, F. Bachmann, K. Fladt, W. Suss, and H. Kunle, editors. MIT Press, 55 Hayward Street, Cambridge, MA 02142. (617) 253-2884. Volume 1, Foundations of Mathematics: Real Number System and Algebra; Volume 2, Geometry; Volume 3, Analysis. $65.00.

GEOMETRY I AND II. Marcel Berger. Springer Verlag, 175 Fifth Avenue, New York, NY 10010. (800) 526-7254. 1986. Each volume $39.00 in paper.

INVITATION TO GEOMETRY. Z.A. Melzak. John Wiley & Sons, 605 Third Avenue, New York, NY 10158. (800) 526-5368 or (212) 850-6000. 1983. $36.95.

ONLINE DATA BASES

CA SEARCH. Chemical Abstracts Service, P.O. Box 3012, Columbus, OH 43210. Guide to chemical literature, 1972 to present. Inquire as to cost and availability.

DISSERTATION ABSTRACTS ONLINE. University Microfilms International, 300 North Zeeb Road, Ann Arbor, MI 48106. (800) 521-0600 or (313) 761-4700. Scope includes virtually all doctoral dissertations accepted at accredited American institutions from 1861 to present in 252 subject areas. Inquire as to cost and availability.

INSPEC. INSPEC Marketing Department, Institute of Electrical and Electronics Engineers, Inc., IEEE Service Department, 445 Hoes Lane, Piscataway, NJ 08854. (201) 981-0060. Inquire as to on-line cost and availability.

MATHFILE. American Mathematical Society, P.O. Box 6248, Providence, RI 02940. (800) 556-7774 or (401) 272-9500. Scope includes pure and applied mathematics and related areas of physics, statistics, engineering, computer science, and operations research literature since 1973. Inquire as to cost and availability.

NASA. National Aeronautics and Space Administration, Scientific and Technical Information Branch, 300 7th Street, S.W., Washington, DC 20546. Citations and abstracts of aerospace literature, 1962 to present. Inquire as to cost and availability.

NTIS. National Technical Information Service, 5285 Port Royal Road, Springfield, VA 22161. (703) 487-4630. Broad coverage of government sponsored research reports, 1964 to present. Inquire as to cost and availability.

SCISEARCH. Institute for Scientific Information, 3501 Market Street, Philadelphia, PA 19104. (800) 523-1850 or (215) 386-0100. Broad multidisciplinary title and author index to the international literature of science and technology, 1974 to present. Inquire as to cost and availability.

OTHER SOURCES

FROM ONE TO ZERO: UNIVERSAL HISTORY OF NUMBERS. Georges Ifrah. Viking-Penguin, Inc., 40 West 23rd Street, New York, NY 10010. (800) 631-3577 or (212) 807-7300. 1986. $35.00.

MATHEMATICS AND THE SEARCH FOR KNOWLEDGE. Morris Kline. Oxford University Press, Inc., 200 Madison Avenue, New York, NY 10016. (212) 564-6680. 1986. $19.95.

GEOMETRY

PERIODICALS

AMERICAN JOURNAL OF MATHEMATICS. Johns Hopkins University Press, Journals Publishing Division, 701 West 40th Street, Suite 275, Baltimore, MD 21211. (301) 338-7864. Bimonthly. $40.00.

AMERICAN MATHEMATICAL MONTHLY. Mathematical Association of America, 1529 18th Street, N.W., Washington, DC 20036. (202) 387-5200. Monthly. $55.00.

GEOMETRIAE DEDICATA. D. Reidel Publishing Company, 190 Old Derby Street, Hingham, MA 02043. Six times per year. $180.00 per year.

JOURNAL OF DIFFERNTIAL GEOMETRY. Lehigh University, Box F-44, Bethleham, PA 18015. Quarterly. $190.00 per year.

JOURNAL OF GEOMETRY. Birkauser Boston, Inc., 380 Green Street, Cambridge, MA 02139. (617) 876-2333. Quarterly. $75.00 per year.

MATHEMATICS MAGAZINE. Mathematical Association of America, 1529 18th Street, N.W., Washington, DC 20036. (202) 387-5200. Five issues per year. $28.00.

GEOMORPHOLOGY

See also: GEOLOGY, IGNEOUS ROCKS, METAMORPHIC ROCKS, PHYSICAL GEOLOGY, SEDIMENTARY ROCKS

ABSTRACT SERVICES AND INDEXES

BIBLIOGRAPHY AND INDEX OF GEOLOGY. American Geological Institute, 4220 King Street, Alexandria, VA 22302. (703) 379-2480. 1969 to present. Monthly. $1100.00 per year.

DEEP-SEA RESEARCH WITH OCEANOGRAPHIC LITERATURE REVIEW. Pergamon Press, Incorporated, Maxwell House, Fairview Park, Elmsford, NY 10523. (914) 592-7700. 1953 to present. Twenty-four times per year. $600.00 per year.

GENERAL SCIENCE INDEX. H.W. Wilson Company, 950 University Avenue, Bronx, NY 10452. (800) 367-6770 or (212) 588-8400. Inquire as to cost and availability.

MINERALOGICAL ABSTRACTS. Mineralogical Society and the Mineralogical Society of America, 41 Queen's Gate, London, SW7 5HR, England. 1959 to present. Quarterly. $190.00 per year.

OCEANIC ABSTRACTS. Cambridge Scientific Abstracts, 5161 River Road, Bethesda, MD 20816. (301) 951-1400. 1964 to present. Bimonthly. $652.00 per year.

SCIENCE CITATION INDEX. Institute for Scientific Information, 3501 Market Street, Philadelphia, PA 19104. (800) 523-1850 or (215) 386-0100. Inquire as to cost and availability.

ASSOCIATIONS AND PROFESSIONAL SOCIETIES

AMERICAN GEOLOGICAL INSTITUTE. 4220 King Street, Alexandria, VA 22302. (703) 379-2480.

AMERICAN GEOPHYSICAL UNION. 2000 Florida Avenue, NW, Washington, DC 20009. (202) 462-6903.

AMERICAN INSTITUTE OF PROFESSIONAL GEOLOGISTS. 7828 Vance Drive, Suite 103, Arvada, CO 80003. (303) 431-0831.

ASSOCIATION OF ENGINEERING GEOLOGISTS. Post Office Box 1068, Brentwood, TN 37027. (615) 377-3578.

GEOLOGICAL SOCIETY OF AMERICA. Box 9140, Boulder, CO 80301. (303) 447-2020.

DIRECTORIES AND BIOGRAPHICAL SOURCES

AMERICAN MEN AND WOMEN OF SCIENCE. Physical and Biological Sciences. Sixteenth edition. R.R. Bowker Company, 205 East 42nd Street, New York, NY 10017. (800) 521-8110 or (212) 916-1600. 1986. $595.00.

AMERICAN INSTITUTE OF PROFESSIONAL GEOLOGISTS. Membership Directory. American Institute of Professional Geologists, 7828 Vance Drive, Suite 103, Arvada, CO 80003. (303) 431-0831. Annual. $15.00.

ASSOCIATION OF ENGINEERING GEOLOGISTS DIRECTORY. Association of Engineering Geologists, Dr. G. Lee Christensen, Civil Engineering Department, Villanova University, Villanova, PA 19085. (215) 645-4960. Annual. $15.00.

GEOLOGICAL SOCIETY OF AMERICA. Membership Directory. Geological Society of America, 3300 Penrose Place, Boulder, CO 80301. (303) 447-2020. Annual. Available to members only.

RESEARCH CENTERS DIRECTORY. Gale Research Company, Book Tower, Detroit, MI 48226. (800) 521-0707. Twelfth edition. 1987. $365.00 for set.

ENCYCLOPEDIAS AND DICTIONARIES

DICTIONARY OF GEOLOGICAL TERMS. American Geological Institute. Doubleday and Company, Incorporated, 245 Park Avenue, New York, NY 10017. (800) 645-6156 or (212) 953-4561. Third edition. 1984. $19.95.

GLOSSARY OF GEOLOGY. Robert L. Bates and Julia A. Jackson. American Geological Institute, 4220 King Street, Alexandria, VA 22032. (703) 379-2480. 1980. $60.00.

GENERAL WORKS

CLIMACTIC GEOMORPHOLOGY. Julius Budel. Princeton University Press, William Street, Princeton, NJ 08540. (609) 452-4122. 1982. $58.00.

DEVELOPMENTS AND APPLICATIONS OF GEOMORPHOLOGY. J. Costa and P.J. Fleisher. Springer-Verlag New York, Incorporated, 175 Fifth Avenue, New York, NY 10010. (212) 460-1500. 1984. $44.00.

SOILS AND GEOMORPHOLOGY. Peter W. Birkeland. Oxford University Press, Incorporated, 200 Madison Avenue, New York, NY 10016. (212) 564-660; Orders to: 16-00 Pollitt Drive, Fair Lawn, NJ 07410. (201) 796-8000. 1984. $22.95.

HANDBOOKS AND MANUALS

GEOLOGY IN THE FIELD. R.R. Compton. John Wiley and Sons, Incorporated, 605 Third Avenue, New York, NY 10158. (800) 526-5368 or (212) 850-6418. 1985. $23.95.

GEOMORPHOLOGICAL FIELD MANUAL. V. Gardiner and R. Dackcombe. Allen & Unwin, Incorporated, 8 Winchester Place, Winchester, MA 01890. (800) 547-8889. 1982. $15.95.

A HANDBOOK OF ENGINEERING GEOMORPHOLOGY. P.G. Fookes and P.R. Vaughn, editors. Methuen, Incorporated, 29 West 35th Street, New York, NY 10001. (212) 244-3336. 1985. $110.00.

ONLINE DATA BASES

GEOREF. American Geological Institute, 4220 King Street, Alexandria, VA 22302. (703) 379-2480. 1967 to present. Inquire as to online cost and availability.

GEOARCHIVE. Geosystems, Post Office Box 1024, Westminister, London, England, SW1 P 2JL. Citations to literature on geoscience, 1969 to present. Inquire as to online cost and

availability.

NTIS. National Technical Information Service, 5285 Port Royal Road, Springfield, VA 22161. (703) 487-4630. Broad coverage of government-sponsored research reports, 1964 to present. Inquire as to cost and availability.

SCISEARCH. Institute for Scientific Information, 3501 Market Street, Philadelphia, PA 19104. (800) 523-1850 or (215) 386-0100. Broad multidisciplinary title and author index to the international literature of science and technology, 1974 to present. Inquire as to cost and availability.

PERIODICALS

AMERICAN JOURNAL OF SCIENCE. Kline Geology Laboratory, Yale University, New Haven, CT 06520. Ten times per year. $80.00 per year.

GEOLOGICAL MAGAZINE. Cambridge University Press, 32 East 57th Street, New York, NY 10022. (800) 872-7423. Bimonthly. $165.00 per year.

GEOLOGICAL SOCIETY JOURNAL. Geological Society of London, Blackwell Scientific Publications, Incorporated, 667 Lytton Avenue, Palo Alto, CA 94301. (415) 324-1688. 6 times per year. $290.00.

GEOLOGICAL SOCIETY OF AMERICA BULLETIN. Post Office Box 9140, 3300 Penrose Place, Boulder, CO 80301. (303) 447-2020. Monthly. $80.00.

GEOLOGY. Geological Society of America, Post Office Box 9140, 3300 Penrose Place, Boulder, CO 80301. (303) 447-2020. Monthly. $55.00 per year.

JOURNAL OF GEOLOGY. University of Chicago Press, 5801 South Ellis Street, Chicago, IL 60637. (312) 962-7600. Bimonthly. $30.00 per year.

SEDIMENTARY GEOLOGY. Elsevier Science Publishers, Post Office Box 330, Irving-on-Hudson, NY 10533. Twenty times per year. $421.00 per year.

SEDIMENTOLOGY. Blackwell Scientific Publications, Incorporated, 52 Beacon Street, Boston, MA 02108. (617) 720-0761. Six times per year. $185.00 per year.

RESEARCH CENTERS AND INSTITUTES

UNITED STATES GEOLOGICAL SURVEY. National Center, 12201 Sunrise Valley Drive, Reston, Va 22092. The major geological research agency of the federal government conducting research in most areas of pure and applied research in the geosciences.

GEOLOGICAL SOCIETY OF AMERICA. Post Office Box 9140, 3300 Penrose Place, Boulder, CO 80301. (303) 477-2020.

SOCIETY OF ECONOMIC GEOLOGISTS. Post Office Box 571, Golden, CO 80402. (303) 279-1899.

SOCIETY OF ECONOMIC PALEONTOLOGISTS AND MINERALOGISTS. Post Office Box 4756, Tulsa, OK 74159. (918) 743-9765.

GEOPHYSICAL ENGINEERING

See: GEOTECHNICAL ENGINEERING

GEOPHYSICS

See also: GEOLOGY, SEISMOLOGY

ABSTRACT SERVICES AND INDEXES

BIBLIOGRAPHY AND INDEX OF GEOLOGY. American Geological Institute, 4220 King Street, Alexandria, Va 22302. (703) 379-2480.

CHEMICAL ABSTRACTS. Chemical Abstracts Service, 2540 Olentangy Road, P.O. Box 3012, Columbus, OH 43210. (800) 848-6538 or (614) 421-3600. Weekly. $6200.00 per year.

SCIENCE CITATION INDEX. Institute for Scientific Information, 3501 Market Street, Philadelphia, PA 19104. (800) 523-1850 or (215) 386-0100. Inquire as to price and availability.

GENERAL SCIENCE INDEX. H.W. Wilson Co., 950 University Avenue, Bronx, NY 10452. (800) 367-6770 or (212) 588-8400. Inquire as to price and availability.

ANNUAL REVIEWS AND YEARBOOKS

ADVANCES IN GEOPHYSICS. Academic Press, Inc., 6277 Sea Harbor Drive, Orlando, FL 32821. (800) 321-5068. Irregular. $62.00 per volume.

ANNUAL REVIEW AND EARTH AND PLANETARY SCIENCES. Annual Reviews, Inc., 4139 El Camino Way, Palo Alto, CA 94306. (415) 493-4400.

ASSOCIATIONS AND PROFESSIONAL SOCIETIES

AMERICAN ASSOCIATION OF PETROLEUM GEOLOGISTS. P.O. Box 979, Tulsa, OK 74101. (918) 584-2555.

AMERICAN GEOLOGICAL INSTITUTE. 4220 King Street, Alexandria, VA 22302. (703) 379-2480.

AMERICAN GEOPHYSICAL UNION, 2000 Florida Avenue, N.W., Washington, DC 20009. (202) 462-6903.

AMERICAN INSTITUTE OF PROFESSIONAL GEOLOGISTS. 7828 Vance Drive, Suite 103, Arvada, CO 80003. (303) 431-0831.

ASSOCIATION OF ENGINEERING GEOLOGISTS. P.O. Box 1068, Brentwood, TN 37027. (615) 377-3578.

BIBLIOGRAPHIES

SCIENCE BOOKS AND FILMS. American Association for the Advancement of Science, 1333 H Street, N.W., Washington, DC 20005.

SCIENTIFIC AND TECHNICAL BOOKS IN PRINT; An index to Literature in Science and Technology. R.R. Bowker Co., E. 42nd Street, New York, NY 10017. (800) 521-8110 or (212) 916-1600.

DICTIONARIES AND BIOGRAPHICAL SOURCES

AMERICAN MEN AND WOMEN OF SCIENCE: PHYSICAL AND BIOLOGICAL SCIENCES. Sixteenth edition. R.R. Bowker Co., E. 42nd Street, New York, NY 10017. (800) 521-8110 or (212) 916-1600. $595.00.

AMERICAN INSTITUTE OF PROFESSIONAL GEOLOGISTS. Membership Directory. American Institute of Professional Geologists, 7828 Vance Drive, Suite 103, Arvada, CO 80003. (303) 431-0831. Annual, $15.00.

ASSOCIATION OF ENGINEERING GEOLOGISTS DIRECTORY. Association of Engineering Geologists, Dr. G. Lee Christensen, Civil Engineering Department, Villanova University, Villanova,

PA 19085. (215) 645-4960. Annual. $15.00.

GEOLOGICAL SOCIETY OF AMERICA. Membership Directory. Geological Society of America, 3300 Penrose Place, Boulder, CO 80301. (303) 447-2020. Annual. Available to members only.

RESEARCH CENTERS DIRECTORY. Gale Research Company, Book Tower, Detroit, MI 48226. 11th edition, 1987. (800) 521-0707.

WHO'S WHO IN FRONTIER SCIENCE AND TECHNOLOGY. Marquis Who's Who, Inc., 200 East Ohio Street, Chicago, IL 60611. (800) 428-3898 or (312) 787-2008.

ENCYCLOPEDIAS AND DICTIONARIES

DICTIONARY OF GEOLOGICAL TERMS. American Geological Institute. Doubleday and Company, Inc., 245 Park Avenue, New York, NY 10017. (800) 645-6156 or (212) 953-4561. Third edition. 1984. $19.95.

DICTIONARY OF PETROLOGY. S.I. Tomkeieff. John Wiley & Sons, 605 Third Avenue, New York, NY 10158. (800) 526-5368 or (212) 850-6000. 1983. $140.00.

ENCYCLOPEDIA OF PHYSICAL SCIENCE AND TECHNOLOGY. Academic Press, Inc., Orlando, FL 32887. (800) 321-5068.

GLOSSARY OF GEOLOGY. Robert L. Bates and Julia A. Jackson. American Geological Institute, 4220 King Street, Alexandria, Va 22032. (703) 379-2480. 1980. $60.00.

GENERAL WORKS

THE EARTH. Peter J. Smith. Macmillan Publishing Co., 866 Third Avenue, New York, NY 10022. (800) 257-5755 or (212) 702-2000. 1986. $40.00.

INTRODUCTION TO ROCK FORMING MINERALS. W.A. Deer, J. Zussman and R.A. Howie. John Wiley & Sons, 605 Third Avenue, New York, NY 10158. (800) 526-5368 or (212) 850-6000. 1966. $29.95.

INTRODUCTION TO SEDIMENTOLOGY. R.C. Selley. Academic Press, Inc., 6277 Sea Harbor Drive, Orlando, FL 32821. (800) 321-5068. Second edition. 1982. $47.50.

PHYSICAL GEOLOGY. R.F. Flint and B.J. Skinner. John Wiley & Sons, 605 Third Avenue, New York, NY 10158. (800) 526-5368 or (212) 850-6000. SEcond edition. 1977. $40.95.

HANDBOOKS AND MANUALS

GEOLOGY IN THE FIELD. R.R. Compton. John Wiley & Sons, 605 Third Avenue, New York, NY 10158. (800) 526-5368 or (212) 850-6000. 1985. $23.95.

ONLINE DATA BASES

GEOREF. American Geological Institute, 4220 King Street, Alexandria, Va 22302. (800) 336-4764 or (703) 379-2480. Geology and geosciences literature, 1961 to present. Inquire as to online cost and availability.

GEOARCHIVE. Geosystems, P.O. Box 1024, Westminster, London, England, SW1 P 2JL. Citations to literature on geoscience, 1969 to present. Inquire as to online cost and availability.

NTIS. National Technical Information Service, 5285 Port Royal Road, Springfield, VA 22161. (703) 487-4630. Broad coverage of government sponsored research reports, 1964 to present. Inquire as to online cost and availability.

SCISEARCH. Institute for Scientific Information, 3501 Market Street, Philadelphia, PA 19104. (800) 523-1850 or (215) 386-0100. Broad interdisciplinary index to the literature of science and technology, 1965 to present. Inquire as to online cost and availability.

PERIODICALS

AAPG BULLETIN. American Association of Petroleum Geologists, P.O. Box 979, Tulsa, OK 74101. (918) 584-2555.

AMERICAN JOURNAL OF SCIENCE. Kline Geology Laboratory, Yale University, New Haven, CT 06520. Ten times per year. $80.00 per year.

ECONOMIC GEOLOGY. Society of Economic Geologists, P.O. Box 571, Golden, CO 80402. (303) 279-1899. 8 times per year. $25.00.

GEOLOGICAL SOCIETY JOURNAL. Geological Society of London. Blackwell Scientific Publications, Inc., 667 Lytton Avenue, Palo Alto, CA 94301. (415) 324-1688. 6 times per year. $290.00.

GEOLOGICAL SOCIETY OF AMERICA BULLETIN. P.O. Box 9140, 3300 Penrose Place, Boulder, CO 80301. (303) 447-2020. Monthly. $80.00.

GEOLOGY. Geological Society of America. P.O. Box 9140, 3300 Penrose Place, Boulder, CO 80301. (303) 447-2020. Monthly. $55.00 per year.

GEOPHYSICS. Society of Exploration Geophysicists, P.O. Box 702740, Tulsa, OK 74170. (918) 493-3516. Monthly. $45.00 per year.

GEOTIMES. American Geological Institute, 4220 King Street, Alexandria, VA 22032. Monthly. $18.00 per year.

JOURNAL OF GEOLOGY. University of Chicago Press, 5801 South Ellis Street, Chicago, IL 60637. (312) 962-7600. Bimonthly. $30.00 per year.

JOURNAL OF GEOPHYSICAL RESEARCH. American Geophysical Union, 2000 Florida Avenue, N.W., Washington, DC 20009. (202) 462-6903. Weekly. $680.00 per year to individuals.

JOURNAL OF GEOPHYSICS. Springer Verlag, Inc., 175 Fifth Avenue, New York, NY 10010. (800) 526-7254 or (212) 460-1500. Bimonthly. $175.00 per year.

JOURNAL OF STRUCTURAL GEOLOGY. Pergamon Journals, Inc., Maxwell House, Fairview Park, Elmsford, NY 10523. (914) 592-7700. Eight times per year. $160.00 per year.

PROFESSIONAL GEOLOGIST. American Institute of Professional Geologists, 7828 Vance Drive, Suite 103, Arvada, CO 80003. Monthly.

REVIEWS OF GEOPHYSICS. American Geophysical Union, 2000 Florida Avenue, N.W., Washington, DC 20009. (202) 462-6903. Quarterly. $240.00 per year.

TECTONICS. American Geophysical Union, 2000 Florida Avenue, N.W., Washington, DC 20009. (202) 462-6903. Bimonthly. $30.00 per year to individuals.

RESEARCH CENTERS AND INSTITUTES

U.S. GEOLOGICAL SURVEY, National Center, 12201 Sunrise Valley Drive, Reston, VA 22092. The major geological research agency of the federal government conducting research in most areas of pure and applied research in the geosciences.

GEOLOGICAL SOCIETY OF AMERICA. P.O. Box 9140, 3300 Penrose Place, Boulder, CO 80301. (303) 447-2020.

SOCIETY OF ECONOMIC GEOLOGISTS. P.O. Box 571, Golden, CO 80402. (303) 279-1899.

SOCIETY OF ECONOMIC PALEONTOLOGISTS AND MINERALOGISTS. P.O. Box 4756, Tulsa, OK 74159. (918) 743-9765.

GEOSTATIONARY SATELLITES

See: ARTIFICIAL SATELLITES

GEOTECHNICAL ENGINEERING

See also: CIVIL ENGINEERING, PHYSICAL GEOLOGY, ROCK MECHANICS

ABSTRACT SERVICES AND INDEXES

BIBLIOGRAPHY AND INDEX OF GEOLOGY. American Geological Institute, 4220 King Street, Alexandria, VA 22302. (703) 379-2480.

CHEMICAL ABSTRACTS. Chemical Abstracts Service, 2540 Olentangy Road, P.O. Box 3012, Columbus, OH 43210. (800) 848-6538 or (614) 421-3600. Weekly. $6200.00 per year.

ENGINEERING INDEX MONTHLY. Engineering Information, Inc. 345 East 47th Street, New York, NY 10017. (800) 221-1044 or (212) 705-7600. Monthly, with annual cumulation. $1425.00 per year.

INTERNATIONAL JOURNAL OF ROCK MECHANICS AND MINING SCIENCES AND GEOMECHANICS ABSTRACTS. Pergamon Press, Inc., Maxwell House, Fairview Park, Elmsford, NY 10523. (914) 592-7700. From 1964 to present. Bimonthly. $385.00 per year.

MINERALOGICAL ABSTRACTS. Mineralogical Society and the Mineralogical Society of America, 41 Queen's Gate, London, SW7 5HR, England. Quarterly. $190.00 per year.

SCIENCE CITATION INDEX. Institute for Scientific Information, 3501 Market Street, Philadelphia, PA 19104. (800) 523-1850 or (215) 386-0100. Inquire as to price and availability.

ANNUAL REVIEWS AND YEARBOOKS

ADVANCES IN GEOPHYSICS. Academic Press, Inc., 6277 Sea Harbor Drive, Orlando, FL 32821. (800) 321-5068. Irregular. $62.00 per volume.

ANNUAL REVIEW AND EARTH AND PLANETARY SCIENCES. Annual Reviews, Inc., 4139 El Camino Way, Palo Alto, CA 94306. (415) 493-4400.

MINERALS YEARBOOK. Bureau of Mines, U.S. Department of the Interior. Available from U.S. Government Printing Office, Washington, DC 20402. (202) 783-3238. Annual. Three volumes. $45.00.

ASSOCIATIONS AND PROFESSIONAL SOCIETIES

AMERICAN ASSOCIATION OF PETROLEUM GEOLOGISTS. P.O. Box 979, Tulsa, OK 74101. (918) 584-2555.

AMERICAN GEOLOGICAL INSTITUTE. 4220 King Street, Alexandria, VA 22302. (703) 379-2480.

AMERICAN GEOPHYSICAL UNION, 2000 Florida Avenue, N.W., Washington, DC 20009. (202) 462-6903.

AMERICAN INSTITUTE OF PROFESSIONAL GEOLOGISTS. 7828 Vance Drive, Suite 103, Arvada, CO 80003. (303) 431-0831.

AMERICAN SOCIETY OF CIVIL ENGINEERS, 345 East 47th Street, New York, NY 10017-2398. (212) 705-7496.

ASSOCIATION OF ENGINEERING GEOLOGISTS. P.O. Box 1068, Brentwood, TN 37027. (615) 377-3578.

INTERNATIONAL SOCIETY FOR ROCK MECHANICS, c/o Laboratorio Nacional de Engenharia Civil, 101 Avenida do Brasil, P-1799 Lisbon Codex, Portugal.

BIBLIOGRAPHIES

SCIENCE BOOKS AND FILMS. American Association for the Advancement of Science, 1333 H Street, N.W., Washington, DC 20005.

SCIENTIFIC AND TECHNICAL BOOKS IN PRINT; An index to Literature in Science and TEchnology. R.R. Bowker Co., E. 42nd Street, New York, NY 10017. (800) 521-8110 or (212) 916-1600.

DIRECTORIES AND BIOGRAPHICAL SOURCES

AMERICAN MEN AND WOMEN OF SCIENCE: PHYSICAL AND BIOLOGICAL SCIENCES. Sixteenth edition. R.R. Bowker Co., E. 42nd Street, New York, NY 10017. (800) 521-8110 or (212) 916-1600. $595.00.

AMERICAN INSTITUTE OF PROFESSIONAL GEOLOGISTS. Membership Directory. American Institute of Professional Geologists, 7828 Vance Drive, Suite 103, Arvada, CO 80003. (303) 431-0831. Annual, $15.00.

ASSOCIATION OF ENGINEERING GEOLOGISTS DIRECTORY. Association of Engineering Geologists, Dr. G. Lee Christensen, Civil Engineering Department, Villanova University, Villanova, PA 19085. (215) 645-4960. Annual. $15.00.

GEOLOGICAL SOCIETY OF AMERICA. Membership Directory. Geological Society of America, 3300 Penrose Place, Boulder, CO 80301. (303) 447-2020. Annual. Available to members only.

RESEARCH CENTERS DIRECTORY. Gale Research Company, Detroit, MI 48226. 11th edition, 1987. (800) 521-0707.

WHO'S WHO IN FRONTIER SCIENCE AND TECHNOLOGY. Marquis Who's Who, Inc., 200 East Ohio Street, Chicago, IL 60611. (800) 428-3898 or (312) 787-2008.

ENCYCLOPEDIAS AND DICTIONAIRES

DICTIONARY OF GEOLOGICAL TERMS. American Geological Institute. Doubleday and Company, Inc., 245 Park Avenue, New York, NY 10017. (800) 645-6156 or (212) 953-4561. Third edition. 1984. $19.95.

ENCYCLOPEDIA OF PHYSICAL SCIENCE AND TECHNOLOGY. Academic Press, Inc., Orlando, FL 32887. (800) 321-5068.

GLOSSARY OF GEOLOGY. Robert L. Bates and Julia A. Jackson. American Geological Institute, 4220 King Street, Alexandria, VA 22032. (703) 379-2480. 1980. $60.00.

GENERAL WORKS

ANALYTICAL COMPUTATION AND METHODS IN ENGINEERING ROCK MECHANICS. E.T. Brown, editor. Allen and Unwin, Inc., 8 Winchester Place, Winchester, MA 01890. (800) 547-8889. 1987. $45.00.

BOUNDARY ELEMENT METHODS IN SOLID MECHANICS. Steven L. Crouch and A.M. Starfield. Allen and Unwin, Inc., 8

Winchester Place, Winchester, MA 01890. (800) 547-8889. 1983. $39.95.

GEOTECHNICAL ENGINEERING. John N. Cernica. Holt, Rinehart and Winston, Inc., 383 Madison Avenue, New York, NY 10017. (212) 872-2000. 1982. $40.95.

GEOTECHNICAL ENGINEERING PRACTICES. R.E. Hunt. McGraw-Hill Book Company, 1221 Avenue of the Americas, New York, NY 10020. (212) 935-2000. 1985. $59.95.

PHYSICAL GEOLOGY. R.F. Flint and B.J. Skinner. John Wiley & Sons, 605 Third Avenue, New York, NY 10158. (800) 526-5368 or (212) 850-6000. 1977. $40.95.

HANDBOOKS AND MANUALS

GEOLOGY IN THE FIELD. R.R. Compton. John Wiley & Sons, 605 Third Avenue, New York, NY 10158. (800) 526-5368 or (212) 850-6000. 1985. $23.95.

HANDBOOK OF GEOLOGY IN CIVIL ENGINEERING. Robert Legget and Paul F. Karrow. McGraw-Hill Book, Inc., 1221 Avenue of the Americas, New York, NY 10020. (212) 935-2000. 1982. $79.50.

ONLINE DATA BASES

COMPENDEX. Engineering Information, Inc., 345 East 47th Street, New York, NY 10017. (800) 221-1044 or (212) 705-7615. Engineering and technical literature, 1975 to present. Inquire as to cost and availability.

GEOREF. American Geological Institute, 4220 King Street, Alexandria, VA 22302. (800) 336-4764 or (703) 379-2480. Geology and geosciences literature, 1961 to present. Inquire as to online cost and availability.

GEOARCHIVE. Geosystems, P.O. Box 1024, Westminster, London, England, SW1 P 2JL. Citations to literature on geoscience, 1969 to present. Inquire as to online cost and availability.

NTIS. National Technical Information Service, 5285 Port Royal Road, Springfield, VA 22161. (703) 487-4630. Broad coverage of government sponsored research reports, 1964 to present. Inquire as to online cost and availability.

SCISEARCH. Institute for Scientific Information, 3501 Market Street, Philadelphia, PA 19104. (800) 523-1850 or (215) 386-0100. Broad interdisciplinary index to the literature of science and technology, 1965 to present. Inquire as to online cost and availability.

PERIODICALS

AMERICAN JOURNAL OF SCIENCE. Kline Geology Laboratory, Yale University, New Haven, CT 06520. TEn times per year. $80.00 per year.

GEOLOGICAL MAGAZINE. Cambridge University Press, 32 East 57th Street, New York, NY 10022. (800) 872-7423. Bimonthly. $165.00 per year.

GEOLOGICAL SOCIETY JOURNAL. Geological Society of London. Blackwell Scientific Publications, Inc., 667 Lytton Avenue, Palo Alto, CA 94301. (415) 324-1688. 6 times per year. $290.00.

GEOLOGICAL SOCIETY OF AMERICA BULLETIN. P.O. Box 9140, 3300 Penrose Place, Boulder, CO 80301. (303) 447-2020. Monthly. $80.00.

GEOLOGY. Geological Society of America. P.O. Box 9140, 3300 Penrose Place, Boulder, CO 80301. (303) 447-2020. Monthly. $55.00 per year.

GEOTECHNIQUE; INTERNATIONAL JOURNAL OF SOIL MECHANICS. Institution of Civil Engineers, Thomas Telford, Limited, 1-7 Great George Street, Westminster, London, SW1P 3AA. Quarterly. $140.00 per year.

INTERNATIONAL JOURNAL OF ROCK MECHANICS AND MINING SCIENCES AND GEOMECHANICS ABSTRACTS. Pergamon Press, Inc., Maxwell House, Fairview Park, Elmsford, NY 10523. (914) 592-7700. From 1964 to present. Bimonthly. $385.00 per year.

JOURNAL OF GEOLOGY. University of Chicago Press, 5801 South Ellis Street, Chicago, IL 60637. (312) 962-7600. Bimonthly. $30.00 per year.

JOURNAL OF GEOTECHNICAL ENGINEERING. American Society of Civil Engineers, 345 East 47th Street, New York, NY 10017-2398. (212) 705-7496. Monthly. $96.00 per year.

JOURNAL OF STRUCTURAL GEOLOGY. Pergamon Journals, Inc., Maxwell House, Fairview Park, Elmsford, NY 10523. (914) 592-7700. Eight times per year. $160.00 per year.

LITHOS; an international journal of mineralogy, petrology, and geochemistry. Elsevier Science Publishers, P.O. Box 330, Irving-on-Hudson, NY 10533. Quarterly. $73.00 per year.

MOUNTAIN GEOLOGIST. Rocky Mountain Association of GEologists, 4201 West 51st Avenue, Denver, CO 80212-2902. Quarterly. $15.00 per year.

PROFESSIONAL GEOLOGIST. American Institute of Professional Geologists, 7828 Vance Drive, Suite 103, Arvada, CO 80003. Monthly.

ROCK MECHANICS AND ROCK ENGINEERING. Springer-Verlag New York, Inc., 175 Fifth Avenue, New York, NY 10010. (800) 526-7254 or (212) 460-1500. Quarterly. $61.00 per year.

TECTONICS. American Geophysical Union, 2000 Florida Avenue, N.W., Washington, DC 20009. (202) 462-6903. Bimonthly. $30.00 per year to individuals.

TECTONOPHYSICS; an international journal of geotectonics and the geology and physics of the interior of the Earth. Elsevier Science Publishers, P.O. Box 330, Irving-on-Hudson, NY 10533. Forty-four times per year. $870.00 per year.

RESEARCH CENTERS AND INSTITUTES

U.S. GEOLOGICAL SURVEY, National Center, 12201 Sunrise Valley Drive, Reston, VA 22092. The major geological research agency of the federal government conducting research in most areas of pure and applied research in the geosciences.

GEOLOGICAL SOCIETY OF AMERICA. P.O. Box 9140, 3300 Penrose Place, Boulder, CO 80301. (303) 447-2020.

PENNSYLVANIA STATE UNIVERSITY. Rock Mechanics Laboratory, Room 117, Mineral Science Building, University Park, PA 16802. (814) 863-1620.

UNIVERSITY OF ARIZONA. Laboratory of Geomechanics, College of Mines, Tucson, AZ 85721. (602) 621-2501.

UNIVERSITY OF MISSOURI - ROLLA. Rock Mechanics and Explosives Research Center, Rolla, MO 65401. (314) 341-4365.

GEOTHERMAL ENERGY

See also: ENERGY, GEOLOGY, GEOPHYSICS, VOLCANOES

ABSTRACT SERVICES AND INDEXES

APPLIED SCIENCE AND TECHNOLOGY INDEX. H.W. Wilson Company, 950 University Avenue, Bronx, NY 10452. (800) 367-6670 or (212) 588-8400. Inquire as to cost and availability.

BIBLIOGRAPHY AND INDEX OF GEOLOGY. American Geological Institute, 4220 King Street, Alexandria, Va 22302. (703) 379-2480. 1969 to present. Monthly. $1100.00 per year.

CHEMICAL ABSTRACTS. Chemical Abstracts Service, 2540 Olentangy Road, Post Office Box 3012, Columbus, OH 43210. (800) 848-6538 or (614) 421-3600. Weekly. $9200.00 per year.

DEEP-SEA RESEARCH WITH OCEANOGRAPHIC LITERATURE REVIEW. Pergamon Press, Incorporated, Maxwell House, Fairview Park, Elmsford, NY 10523. (914) 592-7700. 1953 to present. Twenty-four times per year. $600.00 per year.

GENERAL SCIENCE INDEX. H.W. Wilson Company, 950 University Avenue, Bronx, NY 10452. (800) 367-6770 or (212) 588-8400. Inquire as to cost and availability.

MINERALOGICAL ABSTRACTS. Mineralogical Society and the Mineralogical Society of America, 41 Queen's Gate, London, SW7 5HR, England. 1959 to present. Quarterly. $190.00 per year.

OCEANIC ABSTRACTS. Cambridge Scientific Abstracts, Incorporated, 5161 River Road, Bethesda, MD 20816. (301) 951-1400. 1964 to present. Inquire as to online cost and availability.

SCIENCE CITATION INDEX. Institute for Scientific Information, 3501 Market Street, Philadelphia, PA 19104. (800) 523-1850 or (215) 386-0100. Inquire as to cost and availability.

ANNUAL REVIEWS AND YEARBOOKS

ANNUAL REVIEW AND EARTH AND PLANETARY SCIENCES. Annual Reviews, Incorporated, 4139 El Camino Way, Palo Alto, CA 94306. (415) 493-4400.

ASSOCIATIONS AND PROFESSIONAL SOCIETIES

AMERICAN ASSOCIATION OF PETROLEUM GEOLOGISTS. Post Office Box 979, Tulsa, OK 74101. (918) 584-2555.

AMERICAN GEOLOGICAL INSTITUTE. 4220 King Street, Alexandria, VA 22302. (703) 379-2480.

AMERICAN GEOPHYSICAL UNION. 2000 Florida Avenue, NW, Washington, DC 20009. (202) 462-6903.

AMERICAN INSTITUTE OF PROFESSIONAL GEOLOGISTS. 7828 Vance Drive, Suite 103, Arvada, CO 80003. (303) 431-0831.

ASSOCIATION OF ENGINEERING GEOLOGISTS. Post Office Box 1068, Brentwood, TN 37027. (615) 377-3578.

GEOLOGICAL SOCIETY OF AMERICA. Box 9140, Boulder, CO 80301. (303) 447-2020.

SOCIETY OF ECONOMIC PALEONTOLOGISTS AND MINERALOGISTS. Box 4756, Tulsa, OK 74159. (918) 743-9765.

DIRECTORIES AND BIOGRAPHICAL SOURCES

AMERICAN MEN AND WOMEN OF SCIENCE. Physical and Biological Sciences. Sixteenth edition. R.R. Bowker Company, 205 East 42nd Street, New York, NY 10017. (800) 521-8110 or (212) 916-1600. 1986. $595.00.

AMERICAN INSTITUTE OF PROFESSIONAL GEOLOGISTS. Membership Directory. American Institute of Professional Geologists, 7828 Vance Drive, Suite 103, Arvada, CO 80003. (303) 431-0831. Annual, $15.00.

ASSOCIATION OF ENGINEERING GEOLOGISTS DIRECTORY. Association of Engineering Geologists, Dr. G. Lee Christensen, Civil Engineering Department, Villanova University, Villanova, PA 19085. (215) 645-4960. Annual. $15.00.

GEOLOGICAL SOCIETY OF AMERICA. Membership Directory. Geological Society of America, 3300 Penrose Place, Boulder, CO 80301. (303) 447-2020. Annual. Available to members only.

RESEARCH CENTERS DIRECTORY. Gale Research Company, Book Tower, Detroit, MI 48226. (800) 521-0707. Twelfth edition. 1987. $365.00 for set.

WHO'S WHO IN FRONTIER SCIENCE AND TECHNOLOGY. Marquis Who's Who, Incorporated, 200 East Ohio Street, Chicago, IL 60611. (800) 428-3898 or (312) 787-2008.

ENCYCLOPEDIAS AND DICTIONARIES

DICTIONARY OF GEOLOGICAL TERMS. American Geological Institute. Doubleday and Company, Incorporated, 245 Park Avenue, New York, NY 10017. (800) 645-6156 or (212) 953-4561. Third edition. 1984. $19.95.

GLOSSARY OF GEOLOGY. Robert L. Bates and Julia A. Jackson. American Geological Institute, 4220 King Street, Alexandria, VA 22032. (703) 379-2480. 1980. $60.00.

GENERAL WORKS

GEOCHEMICAL THERMODYNAMICS. D.K. Nordstrom and J.L. Munoz. Blackwell Scientific Publications, Incorporated, 667 Lytton Avenue, Palo Alto, CA 94301. (415) 324-1688 or (415) 965-4081. 1986. $45.00.

GEOTHERMAL ENERGY: ITS PAST, PRESENT AND FUTURE CONTRIBUTIONS TO THE ENERGY NEEDS OF MAN. H.C. Armstead. Methuen, Incorporated, 29 West 35th Street, New York, NY 10001. (212) 244-3336. 1983. $59.95.

GEOTHERMAL RESEVOIR ENGINEERING. Malcolm A. Grant. Academic Press, Incorporated, Orlando, FL 32887. (305) 345-4100. 1983. $54.50.

INTERNATIONAL SYMPOSIUM ON GEOTHERMAL ENERGY. Claudia Stone, editor. Geothermal Resources Council, Post Office Box 1350, Davis, CA 95617. (916) 758-2360. 1985. Inquire as to cost.

HANDBOOKS AND MANUALS

GEOLOGY IN THE FIELD. R.R. Compton. John Wiley and Sons, Incorporated, 605 Third Avenue, New York, NY 10158. (800) 526-5368 or (212) 850-6418. 1985. $23.95.

ONLINE DATA BASES

CA SEARCH. Chemical Abstracts Service, Post Office Box 3012, Columbus, OH 43210. Guide to chemical literature, 1972 to present. Inquire as to cost and availability.

COMPENDEX. Engineering Information, Incorporated, 345 East 47th Street, New York, NY 10017. (800) 221-1044 or (212) 705-7615. Engineering and technical literature, 1975 to present. Inquire as to cost and availability.

DISSERTATION ABSTRACTS ONLINE. University Microfilms International, 300 North Zeeb Road, Ann Arbor, MI 48106. (800) 521-0600 or (313) 761-4700. Scope includes virtually all doctoral dissertations accepted at accredited American institutions from 1861 to present in 252 subject areas. Inquire as to cost and availability.

GEOREF. American Geological Institute, 4220 King Street, Alexandria, VA 22302. (703) 379-2480. 1967 to present. Inquire as to online cost and availability.

GEOARCHIVE. Geosystems, Post Office Box 1024, Westminister, London, England, SW1 P 2JL. Citations to literature on geoscience, 1969 to present. Inquire as to online cost and availability.

NTIS. National Technical Information Service, 5285 Port Royal Road, Springfield, VA 22161. (703) 487-4630. Broad coverage of government-sponsored research reports, 1964 to present. Inquire as to cost and availability.

SCISEARCH. Institute for Scientific Information, 3501 Market Street, Philadelphia, PA 19104. (800) 523-1850 or (215) 386-0100. Broad multidisciplinary title and author index to the international literature of science and technology, 1974 to present. Inquire as to cost and availability.

PERIODICALS

AMERICAN JOURNAL OF SCIENCE. Kline Geology Laboratory, Yale University, New Haven, CT 06520. Ten times per year. $80.00 per year.

ECONOMIC GEOLOGY. Society of Economic Geologists, Post Office Box 571, Golden, CO 80402. (303) 279-1899. 8 times per year. $25.00.

GEOCHIMICA ET COSMOCHIMICA ACTA. Pergamon Journals, Incorporated, Maxwell House, Fairview Park, Elmsford, NY 10523. (914) 592-7700. Monthly. $340.00 per year.

GEOLOGICAL MAGAZINE. Cambridge University Press, 32 East 57th Street, New York, NY 10022. (800) 872-7423. Bimonthly. $165.00 per year.

GEOLOGICAL SOCIETY JOURNAL. Geological Society of London. Blackwell Scientific Publications, Incorporated, 667 Lytton Avenue, Palo Alto, CA 94301. (415) 324-1688. 6 times per year. $290.00.

GEOLOGICAL SOCIETY OF AMERICA BULLETIN. Post Office Box 9140, 3300 Penrose Place, Boulder, CO 80301. (303) 447-2020. Monthly. $80.00.

GEOLOGY. Geological Society of America, Post Office Box 9140, 3300 Penrose Place, Boulder, CO 80301. (303) 447-2020. Monthly. $55.00 per year.

JOURNAL OF GEOLOGY. University of Chicago Press, 5801 South Ellis Street, Chicago, IL 60637. (312) 962-7600. Bimonthly. $30.00 per year.

RESEARCH CENTERS AND INSTITUTES

GEOTHERMAL RESOURCES COUNCIL. Post Office Box 1350, Davis, CA 95617-1350. (916) 758-2360.

INSTITUTE FOR THE STUDY OF EARTH AND MAN. Geothermal Laboratory, 217 Heroy Building, Southern Methodist University, Dallas, TX 75275. (214) 692-2745.

NEW MEXICO STATE UNIVERSITY. Engineering Research Center, Post Office Box 3449, Las Cruces, NM 88003. (505) 646-3421.

UNITED STATES GEOLOGICAL SURVEY. National Center, 12201 Sunrise Valley Drive, Reston, VA 22092. The major geological research agency of the federal government conducting research in most areas of pure and applied research in the geosciences.

GIANT STARS

See: STARS

GLACIAL GEOLOGY

See also: GEOLOGY, GEOMORPHOLOGY, IGNEOUS ROCKS, METAMORPHIC ROCKS, PHYSICAL GEOLOGY, SEDIMENTARY ROCKS

ABSTRACT SERVICES AND INDEXES

BIBLIOGRAPHY AND INDEX OF GEOLOGY. American Geological Institute, 4220 King Street, Alexandria, VA 22302. (703) 379-2480. 1969 to present. Monthly. $1100.00 per year.

GENERAL SCIENCE INDEX. H.W. Wilson Company, 950 University Avenue, Bronx, NY 10452. (800) 367-6770 or (212) 588-8400. Inquire as to cost and availability.

SCIENCE CITATION INDEX. Institute for Scientific Information, 3501 Market Street, Philadelphia, PA 19104. (800) 523-1850 or (215) 386-0100. Inquire as to cost and availability.

ASSOCIATIONS AND PROFESSIONAL SOCIETIES

AMERICAN GEOLOGICAL INSTITUTE. 4220 King Street, Alexandria, VA 22302. (703) 379-2480.

AMERICAN GEOPHYSICAL UNION. 2000 Florida Avenue, NW, Washington, DC 20009. (202) 462-6903.

AMERICAN INSTITUTE OF PROFESSIONAL GEOLOGISTS. 7828 Vance Drive, Suite 103, Arvada, CO 80003. (303) 431-0831.

ASSOCIATION OF ENGINEERING GEOLOGISTS. Post Office Box 1068, Brentwood, TN 37027. (615) 377-3578.

GEOLOGICAL SOCIETY OF AMERICA. Box 9140, Boulder, CO 80301. (303) 447-2020.

DIRECTORIES AND BIOGRAPHICAL SOURCES

AMERICAN MEN AND WOMEN OF SCIENCE. Physical and Biological Sciences. Sixteenth edition. R.R. Bowker Company, 205 East 42nd Street, New York, NY 10017. (800) 521-8110 or (212) 916-1600. 1986. $595.00.

AMERICAN INSTITUTE OF PROFESSIONAL GEOLOGISTS. Membership Directory. American Institute of Professional Geologists, 7828 Vance Drive, Suite 103, Arvada, CO 80003. (303) 431-0831. Annual. $15.00.

ASSOCIATION OF ENGINEERING GEOLOGISTS DIRECTORY. Association of Engineering Geologists, Dr. G. Lee Christensen, Civil Engineering Department, Villanova University, Villanova, PA 19085. (215) 645-4960. Annual. $15.00.

GEOLOGICAL SOCIETY OF AMERICA. Membership Directory. Geological Society of America, 3300 Penrose Place, Boulder, CO 80301. (303) 447-2020. Annual. Available to members only.

RESEARCH CENTERS DIRECTORY. Gale Research Company, Book Tower, Detroit, MI 48226. (800) 521-0707. Twelfth edition. 1987. $365.00 for set.

ENCYCLOPEDIAS AND DICTIONARIES

DICTIONARY OF GEOLOGICAL TERMS. American Geological Institute. Doubleday and Company, Incorporated, 245 Park Avenue, New York, NY 10017. (800) 645-6156 or (212) 953-4561. Third edition. 1984. $19.95.

GLOSSARY OF GEOLOGY. Robert L. Bates and Julia A. Jackson. American Geological Institute, 4220 King Street, Alexandria, VA 22032. (703) 379-2480. 1980. $60.00.

GENERAL WORKS

GLACIAL GEOLOGIC PROCESSES. David Drewry. London, Arnold. 1986. $60.00.

GLACIAL GEOLOGY: AN INTRODUCTION FOR ENGINEERS AND EARTH SCIENTISTS. N. Eyles, editor. Pergamon Press, Incorporated, Maxwell House, Fairview Park, Elmsford, NY 10523. (914) 592-7700. 1983. $66.00.

HANDBOOKS AND MANUALS

GEOLOGY IN THE FIELD. R.R. Compton. John Wiley and Sons, Incorporated, 605 Third Avenue, New York, NY 10158. (800) 526-5368 or (212) 850-6418. 1985. $23.95.

ONLINE DATA BASES

GEOREF. American Geological Institute, 4220 King Street, Alexandria, VA 22302. (703) 379-2480. 1967 to present. Inquire as to online cost and availability.

GEOARCHIVE. Geosystems, Post Office Box 1024, Westminster, London, England, SW1 P 2JL. Citations to literature on geoscience, 1969 to present. Inquire as to online cost and availability.

NTIS. National Technical Information Service, 5285 Port Royal Road, Springfield, VA 22161. (703) 487-4630. Broad coverage of government-sponsored research reports, 1964 to present. Inquire as to cost and availability.

SCISEARCH. Institute for Scientific Information, 3501 Market Street, Philadelphia, PA 19104. (800) 523-1850 or (215) 386-0100. Broad multidisciplinary title and author index to the international literature of science and technology, 1974 to present. Inquire as to cost and availability.

PERIODICALS

AMERICAN JOURNAL OF SCIENCE. Kline Geology Laboratory, Yale University, New Haven, CT 06520. Ten times per year. $80.00 per year.

GEOLOGICAL MAGAZINE. Cambridge University Press, 32 East 57th Street, New York, NY 10022. (800) 872-7423. Bimonthly. $165.00 per year.

GEOLOGICAL SOCIETY JOURNAL. Geological Society of London, Blackwell Scientific Publications, Incorporated, 667 Lytton Avenue, Palo Alto, CA 94301. (415) 324-1688. 6 times per year. $290.00.

GEOLOGICAL SOCIETY OF AMERICA BULLETIN. Post Office Box 9140, 3300 Penrose Place, Boulder, CO 80301. (303) 447-2020. Monthly. $80.00.

GEOLOGY. Geological Society of America, Post Office Box 9140, 3300 Penrose Place, Boulder, CO 80301. (303) 447-2020. Monthly. $55.00 per year.

JOURNAL OF GEOLOGY. University of Chicago Press, 5801 South Ellis Street, Chicago, IL 60637. (312) 962-7600. Bimonthly. $30.00 per year.

SEDIMENTARY GEOLOGY. Elsevier Science Publishers, Post Office Box 330, Irving-on-Hudson, NY 10533. Twenty times per year. $421.00 per year.

SEDIMENTOLOGY. Blackwell Scientific Publications, Incorporated, 52 Beacon Street, Boston, Ma 02108. (617) 720-0761. Six times per year. $185.00 per year.

RESEARCH CENTERS AND INSTITUTES

UNITED STATES GEOLOGICAL SURVEY. National Center, 12201 Sunrise Valley Drive, Reston, Va 22092. The major geological research agency of the federal government conducting research in most areas of pure and applied research in the geosciences.

GEOLOGICAL SOCIETY OF AMERICA. Post Office Box 9140, 3300 Penrose Place, Boulder, CO 80301. (303) 477-2020.

SOCIETY OF ECONOMIC GEOLOGISTS. Post Office Box 571, Golden, CO 80402. (303) 279-1899.

SOCIETY OF ECONOMIC PALEONTOLOGISTS AND MINERALOGISTS. Post Office Box 4756, Tulsa, OK 74159. (918) 743-9765.

UNIVERSITY OF IDAHO. Juneau Icefield Research Program. Gaciological and Arctic Sciences Institute, Moscow, ID 83843. (208) 885-6195.

GLASS

See also: CERAMICS, MATERIALS SCIENCE, OPTICS, SILICON

ABSTRACT SERVICES AND INDEXES

APPLIED SCIENCE AND TECHNOLOGY INDEX. H.W. Wilson and Company, 950 University Avenue, Bronx, NY 10452. (800) 367-6670 or (212) 588-8400. Monthly. Inquire as to cost and availability.

CERAMIC ABSTRACTS. American Ceramic Society, Inc., 65 Ceramic Drive, Columbus, OH 43214. (614) 268-8645. 1922 to present. Bimonthly. $185 per year.

CHEMICAL ABSTRACTS. American Chemical Society, Chemical Abstracts Service, Box 3012, Columbus, OH 43210. (614) 421-3600. 1907 to present. Weekly. $9500.00 per year.

CURRENT CONTENTS: ENGINEERING, TECHNOLOGY AND APPLIED SCIENCES. Institute for Scientific Information, 3501 Market Street, Philadelphia, PA 19104. (800) 523-1850 or (215) 386-0100. Weekly. $275.00 per year.

ENGINEERING INDEX MONTHLY AND AUTHOR INDEX. Engineering Information Inc., 345 East 47th Street, New York, NY 10017. (212) 705-7600. Monthly. $1560.00 per year.

PHYSICS ABSTRACTS. Institution of Electrical Engineers. Available from: IEEE Service Center, 445 Hoes Lane, Piscataway, NJ 08854. 1898 to present. Bimonthly. $1700.00 per year.

SCIENCE CITATION INDEX. Institute for Scientific Information, 3501 Market Street, Philadelphia, PA 19104. (800) 523-1850 or (215) 386-0100. Six times per year. $6200. 00 per year.

ASSOCIATIONS AND PROFESSIONAL SOCIETIES

AMERICAN CERAMIC SOCIETY, INC. 65 Ceramic Drive, Columbus, OH 43214. (614) 268-8645.

AMERICAN SOCIETY FOR NONDESTRUCTIVE TESTING. 4153 Arlingate Plaza, Caller #28518, Columbus, OH 43228. (614) 274-6003.

AMERICAN SOCIETY FOR TESTING AND MATERIALS. 1916 Race Street, Philadelphia, PA 19103. (215) 299-5400.

MATERIALS PROPERTIES COUNCIL. 345 East 47th Street, New York, NY 10017. (212) 705-7693.

MATERIALS RESEARCH SOCIETY. 9800 McKnight Road, Suite 327, Pittsburgh, PA 15237. (412) 367-3003.

NATIONAL INSTITUTE OF CERAMIC ENGINEERS. 65 Ceramic Drive, Columbus, OH 43214. (614) 268-8645.

NATIONAL MATERIALS ADVISORY BOARD. 2101 Constitution Avenue, NW, Washington, DC 20418. (202) 334-3505.

SOCIETY FOR THE ADVANCEMENT OF MATERIAL AND PROCESS ENGINEERING (SAMPE). P.O. Box 2459, 843 West Glentana Street, Covina, CA 91722. (818) 331-0616.

BIBLIOGRAPHIES

A BIBLIOGRAPHY OF CERAMICS AND GLASS. L.L. Mench and B.A. McEldowney. American Ceramic Society, 65 Ceramic Drive, Columbus, OH 43214. (614) 268-8645. Inquire.

DIRECTORIES AND BIOGRAPHICAL SOURCES

CERAMIC COMPANY DIRECTORY. American Ceramic Society, 65 Ceramic Drive, Columbus, OH 43214. (614) 268-8645. Annual. $10.00.

MATERIALS DIRECTORY. Cahners Publishing Company, 275 Washington Street, Newton, MA 02158. Annual. $15.00.

1987 DIRECTORY OF ENGINEERING SOCIETIES AND RELATED ORGANIZATIONS. Gordon Davis, editor. Hemisphere Publishing Corporation, 1010 Vermont Avenue, NW, Washington, DC 20005. (800) 526-0275. 12th edition. 1987. $100.00.

RESEARCH CENTERS DIRECTORY 1988. Gale Research Company, Book Tower, Detroit, MI 48226. (800) 521-0707. 12th edition. 1987. $365.00 for set.

SCIENTIFIC AND TECHNICAL ORGANIZATIONS AND AGENCIES DIRECTORY. Margaret Labash Young, editor. Gale Research Company, Book Tower, Detroit, MI 48226. (800) 521-0707. 2nd edition. 1987. $185.00.

WHO'S WHO IN ENGINEERING. Gordon Davis, editor. Hemisphere Publishing Corporation, 1010 Vermont Avenue, NW, Washington, DC 20005. (800) 526-0275. 6th edition. 1985. $200.00.

ENCYCLOPEDIAS AND DICTIONARIES

CERAMIC GLOSSARY. Walter W. Perkins, editor. American Ceramic Society, 65 Ceramic Drive, Columbus, OH 43214. (614) 268-8645. 1984. $12.00.

ENCYCLOPEDIA OF GLASS, CERAMICS, CLAY AND CEMENT. M. Grayson. John Wiley and Sons, Inc., 605 Third Avenue, New York, NY 10158. (800) 526-5368. 1985. $89.95.

THESAURUS OF SCIENTIFIC, TECHNICAL, AND ENGINEERING TERMS. Hemisphere Publishing Corporation, 1010 Vermont Avenue, NW, Washington, DC 20005. (800) 526-0275. 1988. $125.00.

GENERAL WORKS

CERAMIC RAW MATERIALS: AN INTRODUCTION TO THEIR PROPERTIES. R. Pampuch. Elsevier Science Publishing Company, Inc., 52 Vanderbilt Avenue, New York, NY 10017. (212) 370-5520. 1982. $10.00 in paper.

GLASS: SCIENCE AND TECHNOLOGY. D.R. Uhlmann and others, editors. Academic Press, Inc., 6277 Sea Harbor Drive, Orlando, FL 32821. (800) 321-5068. Volume 1, Glass-forming Systems; volume 2, Processing; volume 5, Elasticity and Strength in Glasses. 1980 to 1984. $78.00 per volume.

STRUCTURAL CHEMISTY OF BORON AND SILICON. Springer-Verlag New York, Inc., 175 Fifth Avenue, New York, NY 10010. (800) 526-7254. 1986. $56.00.

HANDBOOKS AND MANUALS

CERAMIC SCIENCE FOR MATERIALS TECHNOLOGISTS. I.J. McColm. Methuen, Inc., 29 West 35th Street, New York, NY 10001. (212) 244-3336. 1983. $85.95.

MATERIALS HANDBOOK. George S. Brady and Henry R. Clauser. McGraw-Hill Book Company, 1221 Avenue of the Americas, New York, NY 10020. (212) 512-2000. 12th edition. 1986. $59.50.

ONLINE DATA BASES

CA SEARCH. Chemical Abstracts Service, P.O. Box 3012, Columbus, OH 43120. (800) 848-6538 or (614) 421-3600. Comprehensive guide to chemical literature, 1972 to present. Inquire as to online cost and availability.

COMPENDEX. Engineering Information, Inc., 345 East 47th Street, New York, NY 10017. (800) 221-1044 or (212) 705-7615. Engineering and technical literature, 1975 to present. Inquire as to online cost and availability.

INSPEC. INSPEC Marketing Department, Institution of Electrical Engineers. Available from IEEE Service Center, 445 Hoes Lane, Piscataway, NJ 08854. (201) 981-0060. Online version of Physics Abstracts. Inquire as to online cost and availability.

NTIS. National Technical Information Service, 5285 Port Royal Road, Springfield, VA 22161. (703) 487-4630. Broad coverage of government sponsored research reports, 1964 to present. Inquire as to online cost and availability.

SCISEARCH. Institute for Scientific Information, 3501 Market Street, Philadelphia, PA 19104. (800) 523-1850 or (215) 386-0100. Broad multidisciplinary title and author index to the international literature of science and technology, 1974 to present. Inquire as to online cost and availability.

WILSONLINE. H.W. Wilson and Company, 950 University Avenue, Bronx, NY 10452. (800) 367-6770 or (212) 588-8400. Makes available online versions of the H.W. Wilson indexes including Applied Science and Technology Index, Business Periodicals Index and Readers' Guide to Periodical Literature. Approximately 1980 to present. Inquire as to online cost and availability.

PERIODICALS

AMERICAN CERAMIC SOCIETY BULLETIN. American Ceramic Society, 65 Ceramic Drive, Columbus, OH 43214. (614) 268-8645. Monthly. $12.50 per year.

CERAMIC ENGINEERING AND SCIENCE PROCEEDINGS. American Ceramic Society, 65 Ceramic Drive, Columbus, OH 43214. (614) 268-8645. Bimonthly. $60.00 per year.

JOURNAL OF MATERIALS RESEARCH. Materials Research Society. 9800 McKnight Road, Suite 327, Pittsburgh, PA 15237. (412) 367-3003.

MATERIALS ENGINEERING. Penton-IPC, 1100 Superior Avenue, Cleveland, OH 44114. (216)696-7000. 1929 to present. Monthly. $40.00.

MATERIALS EVALUATION. American Society for Nondestructive Testing, 4153 Arlingate Plaza, Caller #28518, Columbus, OH 43228. (614) 274-6003. 1942 to present. Monthly. $50.00 per year.

MATERIALS PERFORMANCE. National Association of Corrosion Engineers, P.O. Box 218340, Houston, TX 77218. (713) 492-0535. 1962 to present. Monthly. $50.00 per year.

SAMPE JOURNAL. Society for the Advancement of Material and Process Engineering (SAMPE), P.O. Box 2459, 843 West Glentana Street, Covina, CA 91722. (818) 331-0616. 1965 to present. Bimonthly. $31.00 per year.

RESEARCH CENTERS AND INSTITUTES

CENTER FOR MATERIALS SCIENCE AND ENGINEERING. Massachusetts Institute of Technology, Building 13, Room 2090, 77 Massachusetts Avenue, Cambridge, MA 02139. (617) 253-6801.

CENTER FOR RESEARCH IN MATERIALS SCIENCE AND ENGINEERING. University of Texas at Austin, ETC 9.104, Austin, TX 78712. (512) 471-1504.

CERAMIC RESEARCH CENTER. Ohio State University, 2041 North College Road, Columbus, OH 43210. (614) 422-2960.

MATERIALS RESEARCH LABORATORY. Ohio State University, 3063 Alpheus Smith Laboratory, 174 West 18th Avenue, Columbus, OH 43210. (614) 422-5190.

MATERIALS RESEARCH LABORATORY. University of Illinois, Urbana, IL 61801. (217) 333-1371.

MATERIALS SCIENCE CENTER. Cornell University, Clark Hall of Science, Ithaca, NY 14853. (607) 255-4272.

GLUES

See: ADHESIVES

GOLD AND GOLD MINING

See also: METALS AND METAL WORKING, MINING ENGINEERING

ABSTRACT SERVICES AND INDEXES

ALLOYS INDEX. American Society for Metals, 9639 Kinsman Road, Metals Park, OH 44073. (216) 338-5151. $230.00 per year.

APPLIED SCIENCE AND TECHNOLOGY INDEX. H.W. Wilson Company, 950 University Avenue, Bronx, NY 10452. (800) 367-6670 or (212) 588-8400. Inquire as to cost and availability.

CHEMICAL ABSTRACTS. Chemical Abstracts Service, 2540 Olentangy Road, Post Office Box 3012, Columbus, OH 43210. (800) 848-6538 or (614) 421-3600. Weekly. $9200.00 per year.

CURRENT CONTENTS: ENGINEERING, TECHNOLOGY AND APPLIED SCIENCES. Institute for Scientific Information, 3501 Market Street, Philadelphia, PA 19104. (800) 523-1850 or (215) 386-0100. $275.00 per year.

ENGINEERING INDEX MONTHLY. Engineering Information, Incorporated, 345 East 47th Street, New York, NY 10017. (800) 221-1044 or (212) 705-7600. Monthly, with annual cumulation. $1560.00 per year.

GOLD BULLETIN; RESEARCH ON GOLD AND ITS APPLICATION IN INDUSTRY. International Gold Corporation Limited, Box 61897, Marshalltown 2107, South Africa. 1968 to present. Quarterly. Free.

METALS ABSTRACTS. American Society for Metals, 9639 Kinsman Road, Metals Park, OH 44073. (216) 338-5151. 1968 to present. Monthly. $1100.00 per year.

METALS ABSTRACTS INDEX. American Society for Metals, 9639 Kinsman Road, Metals Park, OH 44073. (216) 338-5151. Monthly. (Sold only to subscribers of Metals Abstracts).

NEW SILVER TECHNOLOGY. Silver Institute, 1026 16th Street, NW, Suite 101, Washington, DC 20036. (202) 331-1485. 1974 to present. Quarterly. $45.00 per year.

WORLD ALUMINUM ABSTRACTS. Aluminum Association, 818 Connecticut Avenue, NW, Washington, DC 20006. (202) 862-5156. 1968 to present. Monthly. $150.00 per year.

ASSOCIATIONS AND PROFESSIONAL SOCIETIES

AMERICAN BUREAU OF METAL STATISTICS. Post Office Box 1405, 400 Plaza Drive, Secaucus, NJ 07094. (201) 863-6900.

AMERICAN SOCIETY FOR METALS. Metals Park, OH 44073. (216) 338-5151.

GOLD INSTITUTE. 1026 16th Street, NW, Suite 101, Washington, DC 20036. (202) 783-0500.

METALLURGICAL SOCIETY OF THE AIME (AMERICAN INSTITUTE OF MINING, METALLURGICAL AND PETROLEUM ENGINEERS). 420 Commonwealth Drive, Warrendale, PA 15086. (412) 776-9080.

DIRECTORIES AND BIOGRAPHICAL SOURCES

COPPER WORLD SURVEY. Metal Bulletin Books, Limited, Park House, Park Terrace, Worcester Park, Surrey KT4 7HY, England. Available in the United States from: Metals Bulletin, Incorporated, 708 Third Avenue, New York, NY 10017. $48.60 (1980 edition).

DUN'S INDUSTRIAL GUIDE - THE METAL WORKING DIRECTORY. Dun and Bradstreet, Incorporated, Three Century Drive, Parsippany, NJ 07054. (201) 455-0900. Annual. $550.00.

INTERNATIONAL RESEARCH CENTERS DIRECTORY 1986-1987. Gale Research Company, Book Tower, Detroit, MI 48226. (800) 521-0707. Third edition. 1986.

RESEARCH CENTERS DIRECTORY. Gale Research Company, Book Tower, Detroit, MI 48226. (800) 521-0707. Eleventh edition. 1987.

SCIENTIFIC AND TECHNICAL ORGANIZATIONS AND AGENCIES DIRECTORY. Gale Research Company, Book Tower, Detroit, MI 48226. (800) 521-0707. 1985.

WHO'S WHO IN ENGINEERING. Engineers Joint Council, 345 East 47th Street, New York, NY 10017. (212) 705-7010. 1985. $200.00.

WHO'S WHO IN FRONTIER SCIENCE AND TECHNOLOGY. Marquis Who's Who, Incorporated, 200 East Ohio Street, Chicago, IL 60611. (800) 428-3898 or (312) 787-2008.

WHO'S WHO IN TECHNOLOGY TODAY. Reston Publishing Company, Incorporated, c/o Prentice-Hall, Incorporated, Englewood Cliffs, NJ 07632. (800) 262-6868. Biennial. Five volumes. $425.00. Covers the fields of electronics, computer science, physics, optics, chemistry, biotechnology, mechanics, energy, and earth science.

ENCYCLOPEDIAS AND DICTIONARIES

ENCYCLOPEDIA OF PHYSICAL SCIENCE AND TECHNOLOGY. Academic Press, Incorporated, Orlando, FL 32887. (800) 321-5068 or (305) 345-2734. Fifteen volumes, 1986.

MCGRAW-HILL DICTIONARY OF ENGINEERING. Sybil P. Parker, editor. McGraw-Hill Book Company, 1221 Avenue of the Americas, New York, NY 10020. (212) 512-2000. 1984. $39.95.

MCGRAW-HILL ENCYCLOPEDIA OF SCIENCE AND TECHNOLOGY. McGraw-Hill Book Company, Incorporated, 1221 Avenue of the Americas, New York, NY 10020. (212) 512-2000.

GENERAL WORKS

ESSENTIAL METALLURGY FOR ENGINEERS. W. Alexander, G. Davies, and K. Reynolds. Van Nostrand Reinhold, 115 Fifth Avenue, New York, NY 10003. (800) 543-2681. 1985. $17.95.

GOLD: HISTORY AND GENESIS OF DEPOSITS. Robert W. Boyle. Van Nostrand Reinhold, VNR Order Processing, Post Office Box 668, Florence, KY 41042-9979. 1987. $49.95.

METALLURGY BASICS. D.V. Brown. Van Nostrand Reinhold, 115 Fifth Avenue, New York, NY 10003. (800) 543-2681. 1985. $17.95.

PHYSICAL METALLURGY. P. Haasen. Cambridge University Press, 32 East 57th Street, New York, NY 10022. (212) 688-8885. 1986. $24.95 in paper.

STRUCTURE AND PROPERTIES OF ENGINEERING ALLOYS. W.F. Smith. McGraw-Hill Book Company, Incorporated, 1221 Avenue of the Americas, New York, NY 10020. (212) 512-2000. 1980. $46.95.

WORLD GOLD DEPOSITS. J.J. Bache. Elsevier Science Publishing Company, Incorporated, 52 Vanderbilt Avenue, New York, NY 10017. (212) 370-5520. 1987. $45.00.

HANDBOOKS AND MANUALS

METALS HANDBOOK. American Society of Metals, Metals Park, OH 44073. (800) 268-9800. 14 volumes. Ninth edition. $1310.00 for set.

SMITHELL'S METALS REFERENCE BOOK. Eric A. Brandes, editor. Butterworth Publishers, 80 Montvale Avenue, Stoneham, MA 02180. (800) 325-4177. Sixth edition. 1983. $210.00.

ONLINE DATA BASES

COMPENDEX. Engineering Information, Incorporated, 345 East 47th Street, New York, NY 10017. (800) 221-1044 or (212) 705-7615. Engineering and technical literature, 1975 to present. Inquire as to cost and availability.

INSPEC. INSPEC Marketing Department, Institute of Electrical and Electronics Engineers, Incorporated, IEEE Service Department, 445 Hoes Lane, Piscataway, NJ 08854. (201) 981-0060. Inquire as to on-line cost and availability.

METADEX. Metals Information, American Society for Metals, Metals Park, OH 44073. (216) 338-5151. (Metals Abstracts/Alloys Index). A worldwide literature on the science and practice of metallurgy, 1966 to present. Inquire as to online cost and availability.

NASA. National Aeronautics and Space Administration, Scientific and Technical Information Branch, 300 7th Street, SW, Washington, DC 20546. Citations and abstracts of aerospace literature, 1962 to present. Inquire as to cost and availability.

NON-FERROUS METALS ABSTRACTS. British Non-Ferrous Metals Technology Centre, Grove Laboratories, Denchworth Road, Wantage, Oxfordshire, England OX12 (BJ. Citations and abstracts on non-ferrous metallurgy and technology. 1961 to present. Inquire as to online cost and availability.

NTIS. National Technical Information Service, 5285 Port Royal Road, Springfield, VA 22161. (703) 487-4630. Broad coverage of government sponsored research reports, 1964 to present. Inquire as to cost and availability.

WILSONLINE. H.W. Wilson Company, 950 University Avenue, Bronx, NY 10452. (800) 367-6770 or (212) 588-8400. Makes available online versions of the printed H.W. Wilson Indexes including Applied Science and Technology Index, Business Periodicals Index, and Readers' Guide to Periodical Literature. Period covered is generally 1983 to present. Inquire as to cost and availability.

OTHER SOURCES

INTERDOC: DIRECTORY OF PUBLISHED PROCEEDINGS, SERIES. SEMT-Science/Engineering/Medicine/Technology. Interdoc Corporation, 173 Halstead Avenue, Box 326, Harrison, NY 10528. (014) 835-3506. Ten times per year. $325.00 per year.

MATERIALS SCIENCE AND METALLURGY. National Technical Information Service (NTIS), 5285 Port Royal Road, Springfield, VA 22161. (703) 487-4630. Translations and abstracts of foreign language technical media. Irregular. $40.00 per year.

PERIODICALS

ALLOY DIGEST. Engineering Publications Incorporated, Box 823, Upper Montclair, NJ 07043. (201) 746-7930. Monthly. $50.00 per year.

ENGINEERING AND MINING JOURNAL. McGraw-Hill Publications Company, 1221 Avenue of the Americas, New York, NY 10020. 1866 to present. Monthly. $20.00.

JOURNAL OF METALS. Metallurgical Society of the AIME (American Institute of Mining, Metallurgical and Petroleum Engineers), 420 Commonwealth Drive, Warrendale, PA 15086. (412) 776-9080. Monthly. $40.00 per year.

METALLURGICAL TRANSACTIONS. Metallurgical Society of the AIME (American Institute of Mining, Metallurgical and Petroleum Engineers), 420 Commonwealth Drive, Warrendale, PA 15086. (412) 776-9080. Monthly. $95.00 per year.

METALS WEEK. McGraw-Hill Book Company, Incorporated, 1221 Avenue of the Americas, New York, NY 10020. (212) 997-2823. Weekly. $527.00 per year.

RESEARCH CENTERS AND INSTITUTES

INSTITUTE FOR THE STUDY OF MINERAL DEPOSITS. South Dakota School of Mines and Technology, 501 East St. Joseph Street, Rapid City, SD 57701-3995. (605) 394-6152.

NEW MEXICO INSTITUTE OF MINING AND TECHNOLOGY. Mining and Mineral Resources Research Institute, Campus Station, Socorro, NM 87801. (505) 835-5210.

PHYSICAL METALLURGY RESEARCH LABORATORIES. Canada Centre for Mineral and Energy Technology, 555 Booth Street, Ottawa, ON Canada K1A 0G1. (613) 995-4807.

GRAPH THEORY

See: MATHEMATICS

GRATING

See: SPECTROSCOPY

GRAVIMETRIC ANALYSIS

See: ANALYTICAL CHEMISTRY

GRAVITATION

ABSTRACT SERVICES AND INDEXES

APPLIED SCIENCE AND TECHNOLOGY INDEX. H.W. Wilson Company, 950 University Avenue, Bronx, NY 10452. (800) 367-6670 or (212) 588-8400. Inquire as to cost and availability.

GENERAL SCIENCE INDEX. H.W. Wilson Co., 950 University Avenue, Bronx, NY 10452. (800) 367-6770 or (212) 588-8400.

PHYSICS ABSTRACTS. Institute of Electricial Engineers, London, United Kingdom. Available from: Institute of Electrical and Electronic Engineers (IEEE), 345 East 47th Street, New York, NY 10017. (212) 705-7900.

PHYSICS BRIEFS. Physics Verlag GmbH, Postfach 1260/1280, D-6940, Weinheim, West Germany. Twenty-six times per year. $1200.00 per year.

SCIENCE CITATION INDEX. Institute for Scientific Information, 3501 Market Street, Philadelphia, PA 19104. (800) 523-1850 or (215) 386-0100.

ANNUAL REVIEWS AND YEARBOOKS

ADVANCES IN APPLIED MECHANICS. Academic Press, Inc., 6277 Sea Harbor Drive, Orlando, FL 32821. (800) 321-5068. Price varies.

ADVANCES IN ATOMIC AND MOLECULAR PHYSICS. Academic Press, Inc., 6277 Sea Harbor Drive, Orlando, FL 32821. (800) 321-5068. Price varies. Inquire.

ADVANCES IN ELECTRONICS AND ELECTRON PHYSICS. Academic Press, Inc., 6277 Sea Harbor Drive, Orlando, FL 32821. (800) 321-5068.

ADVANCES IN NUCLEAR PHYSICS. Plenum Publishing Corporation, 233 Spring Street, New York, NY 10013. (800) 221-9369. Inquire.

ADVANCES IN PHYSICS. Taylor and Francis, Inc., 242 Cherry Street, Philadelphia, PA 19106. (215) 238-0939. Price varies.

ANNUAL REVIEW OF NUCLEAR AND PARTICLE SCIENCE. Annual Reviews, Inc., 4139 El Camino Way, Palo Alto, CA 94306. (415) 493-4400.

PHYSICS REPORTS - REVIEW SECTION OF PHYSICS LETTERS. Elsevier Science Publishing Company, Inc., 52 Vanderbilt Avenue, New York, NY 10017. (212) 370-5520.

PHYSICS AND CHEMISTRY OF THE EARTH. Pergamon Press, Inc., Maxwell House, Fairview Park, Elmsford, NY 10523. (914) 592-7700.

TOPICS IN APPLIED PHYSICS. Springer-Verlag New York, Inc., 175 Fifth Avenue, New York, NY 10010. (800) 526-7254.

ASSOCIATIONS AND PROFESSIONAL SOCIETIES

AMERICAN INSTITUTE OF PHYSICS. 335 East 45th Street, New York, NY 10017. (212) 661-9404.

AMERICAN PHYSICAL SOCIETY. 335 East 45th Street, New York, NY 10017. (212) 682-7341.

BIBLIOGRAPHIES

SCIENTIFIC AND TECHNICAL BOOKS AND SERIALS IN PRINT 1988; An Index to Literature in Science and TEchnology. R.R. Bowker Co., E. 42nd Street, New York, NY 10017. (800) 521-8110 or (212) 916-1600. $175.00.

DIRECTORIES AND BIOGRAPHICAL SOURCES

AMERICAN ASSOCIATION OF PHYSICS TEACHERS - DIRECTORY OF MEMBERS. American Association Physics Teachers, 5110 Roanoke Place, College Park, MD 20740. Biennial. $2.00.

AMERICAN MEN AND WOMEN OF SCIENCE. Physical and Biological Sciences. Fifteenth edition. R.R. Bowker Co., E. 42nd Street, New York, NY 10017. (800) 521-8110 or (212) 916-1600.205 East 42nd Street, New York, NY 10017. (800) 521-8110 or (212) 916-1600.

AMERICAN PHYSICAL SOCIETY MEMBERSHIP DIRECTORY BULLETIN ISSUE. Physical Society. 335 East 45th Street, New York, New York 10017. (212) 682-7341. Biennial. $30.00.

BIOGRAPHICAL DICTIONARY OF SCIENTISTS. T.I. Williams. Halsted Press, 605 Third Avenue, New York, NY 10158. (800) 526-5368 or (212) 850-6418. Third edition. 1982. $29.95.

DIRECTORY OF PHYSICS AND ASTRONOMY STAFF MEMBERS. American Institute of Physics. 335 East 45th Street, New York, NY 10017. (212) 661-9404. Biennial. $30.00.

RESEARCH CENTERS DIRECTORY. Gale Research Company, Book Tower, Detroit, MI 48226. (800) 521-0707. Twelveth edition. 1987. $365.00.

SCIENTIFIC AND TECHNICAL ORGANIZATIONS AND AGENCIEDS DIRECTORY. Gale Research Company, Book Tower, Detroit, MI 48226. (800) 521-0707. Second edition. 1987. $185.00 for set.

WHO'S WHO IN FRONTEIR SCIENCE AND TECHNOLOGY. Marquis Who's Who, Inc., 200 East Ohio Street, Chicago, Illinois 60611. (800) 428-3898 or (312) 787-2008.

WHO'S WHO IN TECHNOLOGY TODAY. Reston Publishing Company, Inc., c/o Prentice-Hall, Inc., Englewood Cliffs, NJ 07632. (800) 262-6868. Biennial. Five volumes. $425.00. Covers the fields of elecctronics, computer science, physics, optics, chemistry, biotechnology, mechanics, energy, and earth science.

WORLD GUIDE TO SCIENTIFIC ASSOCIATIONS AND LEARNED SOCIETIES. K.G. Saur Inc., 175 Fifth Avenue, New York, NY 10010. (800) 521-0707 or (212) 982-1302. Fourth edition. 1984. $112.00.

ENCYCLOPEDIAS AND DICTIONAIRES

ENCYCLOPEDIA OF PHYSICAL SCIENCE AND TECHNOLOGY. Academic Press, Inc., Orlando, FL 32887. (800) 321-5068 or (305) 345-2734. Fifteen volumes, 1986.

ENCYCLOPEDIA OF PHYSICS. Robert M. Besancon, editor. Van Nostand Reinhold, Inc., 115 Fifth Avenue, New York, NY 10003. (800) 543-2681. Third edition. 1985. $99.95.

MCGRAW-HILL ENCYCLOPEDIA OF SCIENCE AND TECHNOLOGY. McGraw-Hill Book, Inc., 1221 Avenue of the Americas, New York, NY 10020. (212) 997-3675.

GENERAL WORKS

GENERAL RELATIVITY AND GRAVITATION. B. Bettotti, editor. Kluwer Academic Publishers, 190 Old Derby Street, Hingham, MA 02043. (617) 749-5262. 1984. $69.00.

GRAVITATION. George B. Airy and I. Bernard Cohen, editors. Anyer Company Publications, Inc., 382 Main Street, Salem, NH 03079. (603) 898-1200. 1981. $20.00.

GRAVITATION. Charles W. Misner. W.H. Freeman and Company, 41 Madison Avenue, New York, NY 10010. (212) 532-7660. 1973. $43.95.

GRAVITATION AND COSMOLOGY: PRINCIPLES AND APPLICATIONS OF THE GENERAL THEORY OF RELATIVITY. Steven Weinberg. John Wiley and Sons, Inc., 605 Third Avenue, New York, NY 10158. (800) 526-5368 or (212) 850-6000. 1972. $52.95.

INTRODUCTION TO GRAVITATION. M. Gasperini and V. DeSabbata. Taylor and Francis, Inc., 242 Cherry Street, Philadelphia, PA 19106-1906. (215) 238-0930. 1986. $42.00.

HANDBOOKS AND MANUALS

CRC HANDBOOK OF CHEMISTRY AND PHYSICS. CRC Press, Inc., 2000 Corporate Boulevard, N.W., Boca Raton, FL 33431. Sixty-seventh edition. 1986. $69.95.

HANDBOOKS AND TABLES IN SCIENCE AND TECHNOLOGY. Russell H. Powell, editor. Oryz Press, 2214 North Central Avenue, Phoenix, AZ 85004-1483. (602) 254-6156. Second edition. 1983. $55.00.

MATHEMATICAL METHODS FOR PHYSICISTS. George B. Arfken. Academic Press, Inc., 6277 Sea Harbor Drive, Orlando, FL 32821. (800) 321-5068. Third Edition. 1985. $44.50.

ONLINE DATA BASES

INSPEC. INSPECT Marketing Department, Institute of Electrical and Electronics Engineers, Inc., IEEE Service Department, 445 Hoes Lane, Piscataway, NJ 08854. (201) 981-0060. Inquire as to on-line cost and availability.

NASA. National Aeronautics and Space Administration, Scientific and Technical Information Branch, 300 7th Street, S.W., Washington, DC 20546. Citations and abstracts of aerospace literature, 1962 to present. Inquire as to cost and availability.

SCISEARCH. Institute for Scientific Information, 3501 Market Street, Philadelphia, PA 19104. (800) 523-1850 or (215) 386-0100. Broad multidisciplinary title and author index to the international literature of science and technology, 1974 to present. Inquire as to cost and availability.

WILSONLINE. H.W. Wilson Company, 950 University Avenue, Bronx, NY 10452. (800) 367-6770 or (212) 588-8400. Makes available online versions of the printed H.W. Wilson indexes including Applied Science and Technology Index, Business Periodicals Index, and Readers' Guide to Periodical Literature. Period covered is generally 1983 to present. Inquire as to cost and availability.

PERIODICALS

AMERICAN JOURNAL OF PHYSICS. American Association of Physics Teachers, 5100 Roanoke Place, College Park, MD 20740. Monthly. $104.00 per year.

CONTEMPORARY PHYSICS. Taylor and Francis, Inc., 242 Cherry Street, Philadelphia, PA 19106-1906. (215) 238-0939. Bimonthly. $184.00 per year.

GENERAL RELATIVITY AND GRAVITATION. Plenum Press, Inc., 233 Spring Street, New York, NY 10013. Monthly. $350.00 per year.

PHYSICAL REVIEW A (General Physics). American Institute of Physics, 335 East 45th Street, New York, NY 10017. (212) 661-9404. Monthly. $445.00 per year.

PHYSICAL REVIEW LETTERS. American Institute of Physics, 335 East 45th Street, New York, NY 10017. (212) 661-9404. Weekly. $430.00 per year.

REVIEWS OF MODERN PHYSICS. American Institute of Physics. 335 East 45th Street, New York, NY 10017. (212) 661-9404. Quarterly. $145.00 per year.

SOVIET PHYSICS (translation of Doklady Akademii Nauk SSSR). American Institute of Physics, 335 East 45th Street, New York, NY 10017. (212) 661-9404. Monthly. $4485.00 per year.

RESEARCH CENTERS AND INSTITUTES

JOINT INSTITUTE FOR LABORATORY ASTROPHYSICS. University of Colorado at Boulder, Boulder, CO 80309. (303) 492-7789.

MASSACHUSETTS INSTITUTE OF TECHNOLOGY. Center for Space Research, 77 Massachusetts Avenue, Cambridge, MA 02139. (617) 253-7501.

STANFORD UNIVERSITY. Center for Space Science and Astrophysics, 325 Durand Building, Stanford, CA 94305. (415) 497-3582.

UNIVERSITY OF TEXAS AT AUSTIN. Center for Relativity, RLM 9.224, Austin, TX 78712. (512) 471-5774.

GRAVITATIONAL RED SHIFT

See: ASTROPHYSICS

GRAVITY

See: GRAVITATION

GREASE

See: LUBRICATION

GREENHOUSE EFFECT

See also: CLIMATOLOGY, DESERTIFICATION, METEOROLOGY

ABSTRACT SERVICES AND INDEXES

CHEMICAL ABSTRACTS. Chemical Abstracts Service, 2540 Olentangy Road, Post Office Box 3012, Columbus, OH 43210. (800) 848-6538 or (614) 421-3600. Weekly. $9200.00 per year.

GENERAL SCIENCE INDEX. H.W. Wilson Company, 950 University Avenue, Bronx, NY 10452. (800) 367-6770 or (212) 588-8400. Inquire as to cost and availability.

METEOROLOGICAL SOCIETY. 45 Beacon Street, Boston, Ma 02108. (617) 227-2425. 1950 to present. Monthly. $450.00 per year.

OCEANIC ABSTRACTS. Cambridge Scientific Abstracts, Incorporated, 5161 River Road, Bethesda, MD 20816. (301) 951-1400. 1964 to present. Inquire as to online cost and availability.

OCEANOGRAPHIC LITERATURE REVIEW. Pergamon Press Incorporated, Maxwell House, Fairview Park, Elmsford, NY 10523. (914) 592-7700. 1979 to present. Inquire as to cost and availability.

SCIENCE CITATION INDEX. Institute for Scientific Information, 3501 Market Street, Philadelphia, PA 19104. (800) 523-1850 or (215) 386-0100. Inquire as to cost and availability.

ASSOCIATIONS AND PROFESSIONAL SOCIETIES

AMERICAN ASSOCIATION OF STATE CLIMATOLOGISTS. c/o Professor John Griffiths, Meteorology Department, O and M Building, Texas A and M University, College Station, TX 77843. (409) 845-7320.

AMERICAN METEOROLOGICAL SOCIETY. 45 Beacon Street, Boston, Ma 02108. (617) 227-2425.

INTERNATIONAL ASSOCIATION OF METEOROLOGY AND ATMOSPHERIC PHYSICS. UCAR, Post Office Box 3000, Boulder, CO 80307.

NATIONAL ENVIRONMENTAL SATELLITE DATA, AND INFORMATION SERVICE. 3300 Whitehaven Street, NW, Washington, DC 20235. (202) 634-7318.

NATIONAL WEATHER ASSOCIATION. 4400 Stamp Road, Room 404, Temple Hills, MD 20748. (301) 899-3784.

UNIVERSITY CORPORATION FOR ATMOSPHERIC RESEARCH. Box 3000, 1850 Table Mesa Drive, Boulder, CO 80307. (303) 497-1000.

WEATHER MODIFICATION ASSOCIATION. Post Office Box 8116, Fresno, CA 93747. (209) 291-8466.

DIRECTORIES AND BIOGRAPHICAL SOURCES

AMERICAN MEN AND WOMEN OF SCIENCE. Physical and Biological Sciences. Sixteenth edition. R.R. Bowker Company, 205 East 42nd Street, New York, NY 10017. (800) 521-8110 or (212) 916-1600. 1986. $595.00.

GOVERNMENT RESEARCH DIRECTORY. Gale Research Company, Book Tower, Detroit, MI 48226. (800) 521-0707. Fourth edition. 1987. $350.00.

INTERNATIONAL RESEARCH CENTERS DIRECTORY 1988-89. Gale Research Company, Book Tower, Detroit, MI 48226. (800) 521-0707. Fourth edition. 1987. $360.00.

METEOROLOGICAL SERVICES OF THE WORLD. World Meteorological Organization. Available from: American Meteorological Society, 45 Beacon Street, Boston, MA 02108. (617) 227-2425. Annual. $35.00.

NATIONAL WEATHER SERVICE OFFICES AND STATIONS. National Oceanic and Atmospheric Administration, Department of Commerce, Silver Spring, Md 20910. (301) 427-7698. Annual. Free.

WHO'S WHO IN FRONTIER SCIENCE AND TECHNOLOGY. Marquis Who's Who, Incorporated, 200 East Ohio Street, Chicago, IL 60611. (800) 428-3898 or (312) 787-2008.

ENCYCLOPEDIAS AND DICTIONARIES

ENCYCLOPEDIA OF CLIMATOLOGY. John E. Oliver and Rhodes W. Fairbridge, editors. Van Nostrand Reinhold, Incorporated, 115 Fifth Avenue, New York, NY 10003. (800) 543-2681. 1987. $89.95.

GENERAL WORKS

FUTURE WEATHER AND THE GREENHOUSE EFFECT. John Gribbin. Delacorte Press, 1 Dag Hammarskjold Plaza, New York, NY 10017. (212) 605-3000. 1982. $15.95.

THE GLOBAL CLIMATE. John T. Houghton, editor. Cambridge University Press, 32 East 57th Street, New York, NY 10022. (800) 872-7423. 1984. $55.00.

GREENHOUSE EFFECT AND SEA LEVEL RISE. Michael C. Barth and James G. Titus. Van Nostrand Reinhold, 135 West 50th L Street, New York, NY 10020. (212) 265-8700. 1984. $24.50.

THE GREENHOUSE EFFECT: CLIMATIC CHANGE AND ECOSYSTEMS. Bert Bolin and Bo R. Doos. John Wiley and Sons, incorporated, 605 Third Avenue, New York, NY 10158. (800) 526-5368 or (212) 850-6000. 1986. Inquire.

OUR THREATENED CLIMATE. Wilfrid Bach. Kluwer Academic Publishers, 101 Philip Drive, Norwell, MA 02061. (617) 871-6600. 1983. $29.00.

ONLINE DATA BASES

CA SEARCH. Chemical Abstracts Service, Post Office Box 3012, Columbus, OH 43210. Guide to chemical literature, 1972 to present. Inquire as to cost and availability.

DISSERTATION ABSTRACTS ONLINE. University Microfilms International, 300 North Zeeb Road, Ann Arbor, MI 48106. (800) 521-0600 or (313) 761-4700. Scope includes virtually all doctoral dissertations accepted at accredited American institutions from 1861 to present in 252 subject areas. Inquire as to cost and availability.

METEOROLOGICAL AND GEOASTROPHYSICAL ABSTRACTS. American Meteorological Society, 45 Beacon Street, Boston, MA 02108. (617) 227-2425. 1950 to present. Monthly. $450.00 per year.

NTIS. National Technical Information Service, 5285 Port Royal Road, Springfield, VA 22161. (703) 487-4630. Broad coverage of government-sponsored research reports, 1964 to present. Inquire as to cost and availability.

OCEANIC ABSTRACTS. Cambridge Scientific Abstracts, Incorporated, 5161 River Road, Bethesda, MD 20816. (301) 951-1400. 1964 to present. Inquire as to online cost and availability.

SCISEARCH. Institute for Scientific Information, 3501 Market Street, Philadelphia, PA 19104. (800) 523-1850 or (215) 386-0100. Broad multidisciplinary title and author index to the international literature of science and technology, 1974 to present. Inquire as to cost and availability.

PERIODICALS

AGRICULTURAL AND FOREST METEOROLOGY. Elsevier Science Publishing Company, Incorporated, 52 Vanderbilt Avenue, New York, NY 10017. (212) 370-5520. 1964 to present. Monthly. $260.00 per year.

AMERICAN METEOROLOGICAL SOCIETY BULLETIN. American Meteorological Society, 45 Beacon Street, Boston, MA 02108. (617) 227-2425. 1920 to present. Monthly. $60.00 per year.

CLIMATIC CHANGE: AN INTERDISCIPLINARY, INTERNATIONAL JOURNAL DEVOTED TO THE DESCRIPTION, CAUSES AND IMPLICATIONS OF CLIMATIC CHANGE. D. Reidel Publishing Company, 190 Old Derby Street, Hingham, MA 02043. (617) 871-6600. 1977 to present. Six times per year. $125.00 per year.

DYNAMICS OF ATMOSPHERES AND OCEANS. Elsevier Science Publishing Company, Incorporated, 52 Vanderbilt Avenue, New York, NY 10017. (212) 370-5520. 1977 to present. Quarterly. $90.00 per year.

JOURNAL OF ATMOSPHERIC AND OCEANIC TECHNOLOGY. American Meteorological Society, 45 Beacon Street, Boston, MA 02108. (617) 227-2425. 1984 to present. Quarterly. $80.00 per year.

JOURNAL OF PHYSICAL OCEANOGRAPHY. American Meteorological Society, 45 Beacon Street, Boston, MA 02108. (617) 227-2425. 1971 to present. Monthly. $120.00 per year.

JOURNAL OF THE ATMOSPHERIC SCIENCES. American Meteorological Society, 45 Beacon Street, Boston, MA 02108. (617) 227-2425. 1944 to present. Semimonthly. $220.00 per year.

MONTHLY WEATHER REVIEW. American Meteorological Society, 45 Beacon Street, Boston, MA 02108. (617) 227-2425. 1872 to present. Monthly. $120.00 per year.

NATIONAL WEATHER DIGEST. National Weather Association, 4400 Stamp Road, Room 404, Temple Hills, MD 20748. (301) 899-3784. 1976 to present. Quarterly. $20.00 per year.

WEATHER. Royal Meteorological Society, James Glaisher House, Grenville Place, Bracknell Berkshire, RG12 1BX, England. 1946 to present. $30.00 per year.

WEATHERWISE. Heldref Publications, 4000 Albemarle Street, NW, Washington, DC 20016. (202) 362-6445. 1948 to present. Bimonthly. $20.00 per year.

RESEARCH CENTERS AND INSTITUTES

CENTER FOR CLIMATE RESEARCH. Columbia University, Lamont-Doherty Geological Observatory, Palisades, NY 10964. (914) 359-2900.

CENTER FOR CLIMATIC RESEARCH. University of Delaware, Department of Geography, Newark, DE 19716. (302) 451-8998.

FLORIDA STATE CLIMATE CENTER. 305A Love Building, Florida State University, Tallahassee, FL 32306. (904) 644-3417.

NATIONAL CENTER FOR ATMOSPHERIC RESEARCH. Box 3000, Boulder, CO 80307. (303) 497-1000.

SAN DIEGO STATE UNIVERSITY. Systems Ecology Research Group, San Diego, CA 92182. (619) 265-6613.

UNIVERSITY OF NEW HAMPSHIRE. Cooperative Institute for the Remote Sensing of Biogeophysical Processes. Institute for the Study of Earth, Oceans and Space, Durham, NH 03824. (603) 862-1792.

GRINDING AND POLISHING

See: MACHINES

GROUND EFFECT MACHINES

See: AERONAUTICS

GROUNDWATER

See also: HYDROGEOLOGY, HYDROLOGY, HYDRODYNAMICS, WATER RESOURCES

ABSTRACT SERVICES AND INDEXES

BIBLIOGRAPHY AND INDEX OF GEOLOGY. American Geological Institute, 4220 King Street, Alexandria, VA 22302. (703) 379-2480. 1969 to present. Monthly. $1100.00 per year.

CHEMICAL ABSTRACTS. Chemical Abstracts Service, 2540 Olentangy Road, Post Office Box 3012, Columbus, OH 43210. (800) 848-6538 or (614) 421-3600. Weekly. $9200.00 per year.

CURRENT CONTENTS: PHYSICAL, CHEMICAL AND EARTH SCIENCES. Institute for Scientific Information, 3501 Market Street, Philadelphia, PA 19104. (800) 523-1850 or (215) 386-0100. 1970 to present. Weekly. $272.00 per year.

DELFT HYDROSCIENCE ABSTRACTS. Delft Hydraulics Laboratory, Rotterdamseweg 185, Postbus 177, NL-2600 MH Delft, The Netherlands. Monthly. $177.00 per year.

ENGINEERING INDEX MONTHLY AND AUTHOR INDEX. Engineering Information, Incorporated, 345 East 47th Street, New York, NY 10017. (800) 221-1044 or (212) 705-7600. Monthly, with annual cumulation. $1560.00 per year.

HYDRO-ABSTRACTS. Environmental Hydrology Corporation, Box 14701, Minneapolis, MN 55414. (612) 379-0901. 1968 to present. Monthly. $120.00 per year.

METEOROLOGICAL AND GEOASTROPHYSICAL ABSTRACTS. American Meteorological Society, 45 Beacon Street, Boston, MA 02108. (617) 227-2425. 1950 to present. Monthly. $450.00 per year.

SCIENCE CITATION INDEX. Institute for Scientific Information, 3501 Market Street, Philadelphia, PA 19104. (800) 523-1850 or (215) 386-0100. Inquire as to cost and availability.

SELECTED WATER RESOURCES ABSTRACTS. United States Geological Survey, Water Resources Scientific Information Center. Available from: National Technical Information Service, Springfield, VA 22161. (703) 860-7455. 1968 to present. Monthly. $115.00 per year.

WATER QUALITY CONTROL DIGEST. University Digest Services, Post Office Box 343, Troy, MI 48099. (313) 651-2528. 1969 to present. Bimonthly. $87.00 per year.

ANNUAL REVIEWS AND YEARBOOKS

ADVANCES IN HYDROSCIENCE. Ven Te Chow, editor. Academic Press, Incorporated, Orlando, FL 32887. (800) 321-5068 or (305) 345-2734. 1964-1981. Inquire, price varies.

ASSOCIATIONS AND PROFESSIONAL SOCIETIES

AMERICAN INSTITUTE OF HYDROLOGY. Post Office Box 14251, St. Paul, MN 55114. (612) 379-1030.

AMERICAN PUBLIC WORKS ASSOCIATION. 1313 East 60th Street, Chicago, IL 60637. (312) 667-2200.

AMERICAN SOCIETY OF CIVIL ENGINEERS. 345 East 47th Street, New York, NY 10017-2398. (212) 705-7520.

AMERICAN WATER RESOURCES ASSOCIATION. 5410 Grosvenor Lane, Suite, Bethesda, MD 20814. (301) 493-8600.

AMERICAN WATER WORKS ASSOCIATION. 666 West Quincy Avenue, Denver, CO 80235. (303) 794-7711.

ASSOCIATION OF GROUND WATER SCIENTISTS AND ENGINEERS. 6375 Riverside Drive, Dublin, OH 43017. (614) 761-1711.

GROUND WATER INSTITUTE. Post Office Box 981, Minneapolis, MN 55440. (612) 698-4395.

NATIONAL WATER RESOURCES ASSOCIATION. 955 L'Enfant Plaza, SW, Suite 1202N, Washington, DC 20024. (202) 488-0610.

NATIONAL WATER WELL ASSOCIATION. 6375 Riverside Drive, Dublin, OH 43017. (614) 761-1711.

WATER POLLUTION CONTROL FEDERATION. 2626 Pennsylvania Avenue, Washington, DC 20037. (337-2500.

WATER QUALITY ASSOCIATION. 4151 Naperville Road, Lisle, IL 60532. (312) 369-1600.

BIBLIOGRAPHIES

GERAGHTY AND MILLER'S GROUNDWATER BIBLIOGRAPHY. Frits Van Der Leeden. Water Information Center, Incorporated, 6800 Jericho Turnpike, Syosset, NY 11791. (516) 921-7690. Third edition. 1983. $22.00.

DIRECTORIES AND BIOGRAPHICAL SOURCES

AMERICAN WATER WORKS ASSOCIATION - OFFICERS AND COMMITTEE DIRECTORY. American Water Works Association, 6666 West Quincy Avenue, Denver, CO 80235. (303) 794-7711. Annual. $15.00.

JOURNAL OF THE AMERICAN WATER WORKS ASSOCIATION BUYERS GUIDE ISSUE. American Water Works Association, 6666 West Quincy Avenue, Denver, CO 80235. (303) 794-7711. Annual, with subscription.

PUBLIC WORKS MANUAL. Public Works Journal Corporation, 200 South Broad Street, Ridgewood, NJ 07451. (201) 445-5800. Annual. $20.00.

SCIENTIFIC AND TECHNICAL ORGANIZATIONS AND AGENCIES DIRECTORY. Gale Research Company, Book Tower, Detroit, MI 48226. (800) 521-0707. Second edition. Two volumes. 1987. $185.00 set.

WATER QUALITY ASSOCIATION DIRECTORY. Water Quality Association, 4151 Naperville Road, Lisle, IL 60532. (312) 369-1600. Annual. Available to members only.

GENERAL WORKS

BASIC HYDROLOGY. James J. Sharp and P.G. Sawden. Butterworth's Incorporated, 80 Montvale Avenue, Stoneham, Ma 02180. (617) 438-8464. 1984. $15.95 in paper.

GROUNDWATER. R. Bowen. Elsevier Science Publishing Company, Incorporated, 52 Vanderbilt Avenue, New York, NY 10017. (212) 370-5520. Second edition. 1986. $79.25.

GROUND WATER CHEMISTRY. Richard Rice. Lewis Publishing, Incorporated, 121 South Main Street, Chelsea, MI 48118. (313) 475-8619. 1986. $45.00.

GROUND WATER HYDROLOGY. David K. Todd. John Wiley and Sons, Incorporated, 605 Third Avenue, New York, NY 10158. (800) 526-5368 or (212) 850-6000. Second edition. 1980. $48.00.

GROUND WATER POLLUTION CONTROL. Larry W. Canter and R.C. Knox. Lewis Publishing, Incorporated, 121 South Main Street, Chelsea, MI 48118. (313) 475-8619. 1985. $49.95.

GROUND WATER QUALITY. C.H. Ward and others. John Wiley and Sons, Incorporated, 605 Third Avenue, New York, NY 10158. (800) 526-5368 or (212) 850-6000. 1985. $45.00.

GROUND WATER SYSTEMS PLANNING AND MANAGEMENT. Robert Willis. Prentice Hall, Incorporated, Englewood Cliffs, NJ 07632. (800) 562-0245. 1987. $49.00.

HYDROLOGY AND QUALITY OF WATER RESOURCES. Mark J. Hammer. John Wiley and Sons, Incorporated, 605 Third Avenue, New York, NY 10158. (800) 526-5368 or (212) 850-6000. Second edition. 1986. $42.95.

ONLINE DATA BASES

COMPENDEX. Engineering Information, Incorporated, 345 East 47th Street, New York, NY 10017. (800) 221-1044 or (212) 705-7615. Engineering and technical literature, 1975 to present. Inquire as to cost and availability.

GEOREF. American Geological Institute, 4220 King Street, Alexandria, VA 22302. (703) 379-2480. 1967 to present. Inquire as to online cost and availability.

NTIS. National Technical Information Service, 5285 Port Royal Road, Springfield, VA 22161. (703) 487-4630. Broad coverage of government-sponsored research reports, 1964 to present. Inquire as to cost and availability.

WATER DATA BANK. United States Department of Agriculture, Agricultural Research Service, Hydrology Laboratory, Water Data Center, Room 139, Building 007, BARC-West, Beltsville, MD 20705. (301) 344-4411. Inquire as to online cost and availability.

PERIODICALS

AMERICAN WATER WORKS ASSOCIATION. Journal. American Water Works Association, 6666 West Quincy Avenue, Denver, CO 80235. (303) 794-7711. Monthly. $50.00 per year.

GROUND WATER. National Water Well Association. Water Well Journal Publishing Company, 500 West Wilson Bridge Road, Worthington, OH 43085. (614) 846-4967. Bimonthly. $53.00 per year.

JOURNAL OF GROUND WATER. National Water Well Association. Water Well Journal Publishing Company, 500 West Wilson Bridge Road, Worthington, OH 43085. (614) 846-4967. Quarterly. $16.00 per year.

JOURNAL OF HYDROLOGY. Elsevier Science Publishers B.V., Post Office Box 211, 1000 AE Amsterdam, The Netherlands. Thirty-two times per year. $675.00 per year.

JOURNAL OF IRRIGATION AND DRAINAGE. American Society of Civil Engineers, 345 East 47th Street, New York, NY 10017-2398. (212) 705-7520. Quarterly. $44.00 per year.

JOURNAL OF WATER RESOURCES PLANNING AND MANAGEMENT. American Society of Civil Engineers, 345 East 47th Street, New York, NY 10017-2398. (212) 705-7520. Quarterly. $56.00 per year.

LIMNOLOGY AND OCEANOGRAPHY. American Society of Limnology and Oceanography, 1530 12th Avenue, Grafton, WI 53024. (414) 377-4871. Bimonthly. $60.00 per year.

WATER RESOURCES BULLETIN. American Water Resources Association, 5410 Grosvenor Lane, Suite 220, Bethesda, MD 20814. (301) 493-8600. Bimonthly. $65.00 per year.

WATER RESOURCES RESEARCH. American Geophysical Union, 2000 Florida Avenue, NW, Washington, DC 20009. (202) 462-6903. Monthly. $295.00 per year.

RESEARCH CENTERS AND INSTITUTES

CALIFORNIA INSTITUTE OF TECHNOLOGY. W.M. Keck Engineering Laboratory of Hydraulics and Water Resources, 1201 East California Boulevard, Pasadena, Ca 91125. (818) 356-4404.

ILLINOIS STATE GEOLOGICAL SURVEY. Natural Resources Building, 615 East Peabody Drive, Champaign, IL 61820. (217) 344-1481.

TEXAS A AND M UNIVERSITY. Center for Engineering Geosciences, College Station, TX 77843. (409) 845-3224.

UNITED STATES ARMY HYDROLOGIC ENGINEERING CENTER. 609 Center Street, Davis, CA 95616. (916) 440-3285.

UNITED STATES GEOLOGICAL SURVEY. Water Resources Division, 421 National Center, Reston, VA 22092. (703) 860-6031.

UNIVERSITY OF MINNESOTA. Minnesota Geological Survey, 2642 University Avenue, St. Paul, MN 55114. (612) 373-3372.

GROUNDWATER POLLUTION

See also: GROUNDWATER, HYDROGEOLOGY, HYDROLOGY, HYDRODYNAMICS, WATER POLLUTION, WATER RESOURCES

ABSTRACT SERVICES AND INDEXES

BIBLIOGRAPHY AND INDEX OF GEOLOGY. American Geological Institute, 4220 King Street, Alexandria, VA 22302. (703) 379-2480. 1969 to present. Monthly. $1100.00 per year.

CHEMICAL ABSTRACTS. Chemical Abstracts Service, 2540 Olentangy Road, Post Office Box 3012, Columbus, OH 43210. (800) 848-6538 or (614) 421-3600. Weekly. $9200.00 per year.

CURRENT CONTENTS: PHYSICAL, CHEMICAL AND EARTH SCIENCES. Institute for Scientific Information, 3501 Market Street, Philadelphia, PA 19104. (800) 523-1850 or (215) 386-0100. 1970 to present. Weekly. $272.00 per year.

DELFT HYDROSCIENCE ABSTRACTS. Delft Hydraulics Laboratory, Rotterdamseweg 185, Postbus 177, NL-2600 MH Delft, The Netherlands. Monthly. $177.00 per year.

ENGINEERING INDEX MONTHLY AND AUTHOR INDEX. Engineering Information, Incorporated, 345 East 47th Street, New York, NY 10017. (800) 221-1044 or (212) 705-7600. Monthly, with annual cumulation. $1560.00 per year.

HYDRO-ABSTRACTS. Environmental Hydrology Corporation, Box 14701, Minneapolis, MN 55414. (612) 379-0901. 1968 to present. Monthly. $120.00 per year.

POLLUTION ABSTRACTS. Cambridge Scientific Abstracts, 5161 River Road, Bethesda, MD 20816. (301) 951-1400. 1970 to present. Bimonthly. $465.00 per year.

SCIENCE CITATION INDEX. Institute for Scientific Information, 3501 Market Street, Philadelphia, PA 19104. (800) 523-1850 or (215) 386-0100. Inquire as to cost and availability.

SELECTED WATER RESOURCES ABSTRACTS. United States Geological Survey, Water Resources Scientific Information Center. Available from: National Technical Information Service, Springfield, VA 22161. (703) 860-7455. 1968 to present. Monthly. $115.00 per year.

WATER QUALITY CONTROL DIGEST. University Digest Services, Post Office Box 343, Troy, MI 48099. (313) 651-2528. 1969 to present. Bimonthly. $87.00 per year.

ANNUAL REVIEWS AND YEARBOOKS

ADVANCES IN HYDROSCIENCE. Ven Te Chow, editor. Academic Press, Incorporated, Orlando, FL 32887. (800) 321-5068 or (305) 345-2734. 1964-1981. Inquire, price varies.

ASSOCIATIONS AND PROFESSIONAL SOCIETIES

AMERICAN PUBLIC WORKS ASSOCIATION. 1313 East 60th Street, Chicago, IL 60637. (312) 667-2200.

AMERICAN SOCIETY OF CIVIL ENGINEERS. 345 East 47th Street, New York, NY 10017-2398. (212) 705-7520.

AMERICAN WATER RESOURCES ASSOCIATION. 5410 Grosvenor Lane, Suite, Bethesda, MD 20814. (301) 493-8600.

AMERICAN WATER WORKS ASSOCIATION. 666 West Quincy Avenue, Denver, CO 80235. (303) 794-7711.

ASSOCIATION OF GROUND WATER SCIENTISTS AND ENGINEERS. 6375 Riverside Drive, Dublin, OH 43017. (614) 761-1711.

GROUND WATER INSTITUTE. Post Office Box 981, Minneapolis, MN 55440. (612) 698-4395.

NATIONAL WATER RESOURCES ASSOCIATION. 955 L'Enfant Plaza, SW, Suite 1202N, Washington, DC 20024. (202) 488-0610.

NATIONAL WATER WELL ASSOCIATION. 6375 Riverside Drive, Dublin, OH 43017. (614) 761-1711.

WATER POLLUTION CONTROL FEDERATION. 2626 Pennsylvania Avenue, Washington, DC 20037. (337-2500.

WATER QUALITY ASSOCIATION. 4151 Naperville Road, Lisle, IL 60532. (312) 369-1600.

BIBLIOGRAPHIES

GERAGHTY AND MILLER'S GROUNDWATER BIBLIOGRAPHY. Frits Van Der Leeden. Water Information Center, Incorporated, 6800 Jericho Turnpike, Syosset, NY 11791. (516) 921-7690. Third edition. 1983. $22.00.

DIRECTORIES AND BIOGRAPHICAL SOURCES

AMERICAN WATER WORKS ASSOCIATION - OFFICERS AND COMMITTEE DIRECTORY. American Water Works Association, 6666 West Quincy Avenue, Denver, CO 80235. (303) 794-7711. Annual. $15.00.

JOURNAL OF THE AMERICAN WATER WORKS ASSOCIATION BUYERS GUIDE ISSUE. American Water Works Association, 6666 West Quincy Avenue, Denver, CO 80235. (303) 794-7711. Annual, with subscription.

PUBLIC WORKS MANUAL. Public Works Journal Corporation, 200 South Broad Street, Ridgewood, NJ 07451. (201) 445-5800. Annual. $20.00.

SCIENTIFIC AND TECHNICAL ORGANIZATIONS AND AGENCIES DIRECTORY. Gale Research Company, Book Tower, Detroit, MI 48226. (800) 521-0707. Second edition. Two volumes. 1987. $185.00 set.

WATER QUALITY ASSOCIATION DIRECTORY. Water Quality Association, 4151 Naperville Road, Lisle, IL 60532. (312) 369-1600. Annual. Available to members only.

GENERAL WORKS

BASIC HYDROLOGY. James J. Sharp and P.G. Sawden. Butterworth's Incorporated, 80 Montvale Avenue, Stoneham, Ma 02180. (617) 438-8464. 1984. $15.95 in paper.

GROUND WATER. R. Bowen. Elsevier Science Publishing Company, Incorporated, 52 Vanderbilt Avenue, New York, NY 10017. (212) 370-5520. Second edition. 1986. $79.25.

GROUND WATER CHEMISTRY. Richard Rice. Lewis Publishing, Incorporated, 121 South Main Street, Chelsea, MI 48118. (313) 475-8619. 1986. $45.00.

GROUND WATER CONTAMINATION. J.H. Guswa and others. Noyes Data Corporation, Mill Road at Grand Avenue, Park Ridge, NJ 07656. (201) 391-8484. 1985. $48.00.

GROUND WATER CONTAMINATION. National Research Council. National Academy Press, 2101 Constitution Avenue, NW, Washington, DC 20418. (202) 334-3313. 1984. $17.95.

GROUND WATER HYDROLOGY. David K. Todd. John Wiley and Sons, Incorporated, 605 Third Avenue, New York, NY 10158. (800) 526-5368 or (212) 850-6000. Second edition. 1980. $48.00.

GROUND WATER POLLUTION CONTROL. Larry W. Canter and R.C. Knox. Lewis Publishing, Incorporated, 121 South Main Street, Chelsea, MI 48118. (313) 475-8619. 1985. $49.95.

GROUND WATER QUALITY. C.H. Ward and others. John Wiley and Sons, Incorporated, 605 Third Avenue, New York, NY 10158. (800) 526-5368 or (212) 850-6000. 1985. $45.00.

GROUND WATER QUALITY PROTECTION. L.W. Canter and others. Lewis Publishing, Incorporated, 121 South Main Street, Chelsea, MI 48118. (313) 475-8619. 1986. $49.95.

GROUND WATER SYSTEMS PLANNING AND MANAGEMENT. Robert Willis. Prentice Hall, Incorporated, Englewood Cliffs, NJ 07632. (800) 562-0245. 1987. $49.00.

HYDROLOGY AND QUALITY OF WATER RESOURCES. Mark J. Hammer. John Wiley and Sons, Incorporated, 605 Third Avenue, New York, NY 10158. (800) 526-5368 or (212) 850-6000. Second edition. 1986. $42.95.

ONLINE DATA BASES

COMPENDEX. Engineering Information, Incorporated, 345 East 47th Street, New York, NY 10017. (800) 221-1044 or (212) 705-7615. Engineering and technical literature, 1975 to present. Inquire as to cost and availability.

GEOREF. American Geological Institute, 4220 King Street, Alexandria, VA 22302. (703) 379-2480. 1967 to present. Inquire as to online cost and availability.

NTIS. National Technical Information Service, 5285 Port Royal Road, Springfield, VA 22161. (703) 487-4630. Broad coverage of government-sponsored research reports, 1964 to present. Inquire as to cost and availability.

POLLUTION ABSTRACTS. Cambridge Scientific Abstracts, 5161 River Road, Bethesda, MD 20816. (301) 951-1400. 1970 to present. Available for online searching through DIALOG Information Services and BRS Information Technologies. Inquire as to online cost and availability.

WATER DATA BANK. United States Department of Agriculture, Agricultural Research Service, Hydrology Laboratory, Water Data Center, Room 139, Building 007, BARC-West, Beltsville, MD 20705. (301) 344-4411. Inquire as to online cost and availability.

WATERNET. American Water Works Association, 6666 West Quincy Avenue, Denver, CO 80235. (303) 794-7711. A data base providing abstracts and indexing of information published in the American Water Works Association Journal. 1971 to present. Inquire as to online cost and availability.

PERIODICALS

AMERICAN WATER WORKS ASSOCIATION. Journal. American Water Works Association, 6666 West Quincy Avenue, Denver, CO 80235. (303) 794-7711. Monthly. $50.00 per year.

GROUND WATER. National Water Well Association. Water Well Journal Publishing Company, 500 West Wilson Bridge Road, Worthington, OH 43085. (614) 846-4967. Bimonthly. $53.00 per year.

JOURNAL OF GROUND WATER. National Water Well Association. Water Well Journal Publishing Company, 500 West Wilson Bridge Road, Worthington, OH 43085. (614) 846-4967. Quarterly. $16.00 per year.

JOURNAL OF HYDROLOGY. Elsevier Science Publishers B.V., Post Office Box 211, 1000 AE Amsterdam, The Netherlands. Thirty-two times per year. $675.00 per year.

JOURNAL OF IRRIGATION AND DRAINAGE. American Society of Civil Engineers, 345 East 47th Street, New York, NY 10017-2398. (212) 705-7520. Quarterly. $44.00 per year.

JOURNAL OF WATER RESOURCES PLANNING AND MANAGEMENT. American Society of Civil Engineers, 345 East 47th Street, New York, NY 10017-2398. (212) 705-7520. Quarterly. $56.00 per year.

WATER POLLUTION CONTROL FEDERATION. Journal. Water Pollution Control Federation, 2626 Pennsylvania Avenue, NW, Washington, DC 20037. (202) 337-2500. Monthly. $120.00 per year.

WATER RESOURCES BULLETIN. American Water Resources Association, 5410 Grosvenor Lane, Suite 220, Bethesda, MD 20814. (301) 493-8600. Bimonthly. $65.00 per year.

WATER RESOURCES RESEARCH. American Geophysical Union, 2000 Florida Avenue, NW, Washington, DC 20009. (202) 462-6903. Monthly. $295.00 per year.

RESEARCH CENTERS AND INSTITUTES

CALIFORNIA INSTITUTE OF TECHNOLOGY. W.M. Keck Engineering Laboratory of Hydraulics and Water Resources, 1201 East California Boulevard, Pasadena, Ca 91125. (818) 356-4404.

ILLINOIS STATE GEOLOGICAL SURVEY. Natural Resources Building, 615 East Peabody Drive, Champaign, IL 61820. (217) 344-1481.

UNITED STATES GEOLOGICAL SURVEY. Water Resources Division, 421 National Center, Reston, VA 22092. (703) 860-6031.

UNIVERSITY OF ILLINOIS. Advanced Environmental Control Technology Research Center, 3230 Newmark C.E. Laboratory, 208 North Romine Street, Urbana, IL 61801. (217) 333-3822.

UNIVERSITY OF MINNESOTA. Minnesota Geological Survey, 2642 University Avenue, St. Paul, MN 55114. (612) 373-3372.

GUIDANCE SYSTEMS

See also: AEROSPACE ENGINEERING, GUIDED MISSILES, ORDNANCE, ROCKETS

ABSTRACT SERVICES AND INDEXES

APPLIED SCIENCE AND TECHNOLOGY INDEX. H.W. Wilson and Company, 950 University Avenue, Bronx, NY 10452. (800) 367-6670 or (212) 588-8400. Monthly. Inquire as to cost and availability.

ENGINEERING INDEX MONTHLY AND AUTHOR INDEX. Engineering Information Inc., 345 East 47th Street, New York, NY 10017. (212) 705-7600. Monthly. $1560.00 per year.

INTERNATIONAL AEROSPACE ABSTRACTS. American Institute of Aeronautics and Astronautics, Technical Information Service, 370 L'Enfant Promenade, S.W., Washington, DC 20024. (202) 646-7400. 1961 to present. Semi-monthly. $700.00 per year.

SCIENCE CITATION INDEX. Institute for Scientific Information, 3501 Market Street, Philadelphia, PA 19104. (800) 523-1850 or (215) 386-0100. Six times per year. $6200.00 per year.

STAR (SCIENTIFIC AND TECHNICAL AEROSPACE REPORTS). National Aeronautics and Space Administration, Scientific and Technical Information Facility, Box 8757, Baltimore-Washington International Airport, MD 21240. (202) 453-8545. 1963 to present. Semi-monthly. $85.00 per year with indexes.

ASSOCIATIONS AND PROFESSIONAL SOCIETIES

AMERICAN ASTRONAUTICAL SOCIETY. 6212-B Old Keene Mill Court, Springfield, VA 22152. (703) 866-0020.

AMERICAN INSTITUTE OF AERONAUTICS AND ASTRONAUTICS. 370 L'Enfant Promenade, S.W., Washington, DC 20024. (202) 646-7400.

AMERICAN SPACE FOUNDATION. 111 Massachusetts Avenue, N.W., Suite 200, Washington, DC 20002. (202) 289-2293.

NATIONAL SPACE INSTITUTE. West Wing, Suite 203, 600 Maryland Avenue, S.W., Washington, DC 20024. (202) 484-1111.

SOCIETY OF AUTOMOTIVE ENGINEERS. 400 Commonwealth Drive, Warrendale, PA 15096. (412) 776-4841.

SPACE STUDIES INSTITUTE. P.O. Box 82, 285 Rosedale, Road, Princeton, NJ 08540. (609) 921-0377.

DIRECTORIES AND BIOGRAPHICAL SOURCES

AMERICAN ASTRONAUTICAL SOCIETY DIRECTORY. American Astronautical Society, 6212-B Old Keene Mill Court, Springfield, VA 22152. (703) 866-0020. 1986. $30.00.

AMERICAN INSTITUTE OF AERONAUTICS AND ASTRONAUTICS ROSTER. American Institute of Aeronautics and Astronautics, 370 L'Enfant Promenade, S.W., Washington, DC 20024. (202) 646-7400. 1984. $75.00.

INTERNATIONAL RESEARCH CENTERS DIRECTORY 1988-89. Darren L. Smith, editor. Gale Research Company, Book Tower, Detroit, MI 48226. (800) 521-0707. 4th edition. 1987. $360.00.

RESEARCH CENTERS DIRECTORY 1988. Gale Research Company, Book Tower, Detroit, MI 48226. (800) 521-0707. 12th edition. 1987. $365.00 for set.

SCIENTIFIC AND TECHNICAL ORGANIZATIONS AND AGENCIES DIRECTORY. Margaret Labash Young, editor. Gale Research Company, Book Tower, Detroit, MI 48226. (800) 521-0707. 2nd edition. 1987. $185.00.

WHO'S WHO IN ENGINEERING. Gordon Davis, editor. Hemisphere Publishing Corporation, 1010 Vermont Avenue, NW, Washington, DC 20005. (800) 526-0275. 6th edition. 1985. $200.00.

GENERAL WORKS

AUTOMATIC CONTROL OF AIRCRAFT AND MISSILES. John H. Blakelock. John Wiley and Sons, Inc., 605 Third Avenue, New York, NY 10158. (800) 526-5368. 1965. $64.50.

AUTOMATIC FLIGHT CONTROL. E.H. Pallett. Sheridan Publishers, Inc., 145 Palisade Street, Dobbs Ferry, NY 10522. (914) 693-2410. Second edition. 1983. $32.50.

GUIDED WEAPONS. J.E. Lee. Pergamon Press, Inc., Maxwell House, Fairview Park, Elmsford, NY 10523. (914) 592-7700. 1983. $13.00 in paper.

RADAR HOMING GUIDANCE FOR TACTICAL MISSILES. D.A. James. Halsted Press, a division of John Wiley and Sons, Inc., 605 Third Avenue, New York, NY 10158. (800) 526-5368. 1986. $34.95.

ONLINE DATA BASES

CA SEARCH. Chemical Abstracts Service, P.O. Box 3012, Columbus, OH 43120. (800) 848-6538 or (614) 421-3600. Comprehensive guide to chemical and related literature, 1972 to present. Inquire as to online cost and availability.

COMPENDEX. Engineering Information, Inc., 345 East 47th Street, New York, NY 10017. (800) 221-1044 or (212) 705-7615. Engineering and technical literature, 1975 to present. Inquire as to online cost and availability.

INSPEC. INSPEC Marketing Department, Institution of Electrical Engineers. Available from IEEE Service Center, 445 Hoes Lane, Piscataway, NJ 08854. (201) 981-0060. Online version of Physics Abstracts. Inquire as to online cost and availability.

NTIS. National Technical Information Service, 5285 Port Royal Road, Springfield, VA 22161. (703) 487-4630. Broad coverage of government sponsored research reports, 1964 to present. Inquire as to online cost and availability.

SCISEARCH. Institute for Scientific Information, 3501 Market Street, Philadelphia, PA 19104. (800) 523-1850 or (215) 386-0100. Broad multidisciplinary title and author index to the international literature of science and technology, 1974 to present. Inquire as to online cost and availability.

WILSONLINE. H.W. Wilson and Company, 950 University Avenue, Bronx, NY 10452. (800) 367-6770 or (212) 588-8400. Makes available online versions of the H.W. Wilson indexes including Applied Science and Technology Index, Business Periodicals Index and Readers' Guide to Periodical Literature. Approximately 1980 to present. Inquire as to online cost and availability.

PERIODICALS

AEROSPACE AMERICA. American Institute of Aeronautics and Astronautics, Technical Information Service, 370 L'Enfant Promenade, S.W., Washington, DC 20024. (202) 646-7400. 1932 to present. Monthly. $56.00 per year.

AEROSPACE ENGINEERING MAGAZINE. Society of Automotive Engineers, 400 Commonwealth Drive, Warrendale, PA 15096. (412) 776-4841. 1981 to present. Monthly. $24.00 per year.

AIAA JOURNAL. American Institute of Aeronautics and Astronautics, Technical Information Service, 370 L'Enfant Promenade, S.W., Washington, DC 20024. (202) 646-7400. 1963 to present. Monthly. $205.00 per year.

AVIATION WEEK AND SPACE TECHNOLOGY. McGraw-Hill Book Company, 1221 Avenue of the Americas, New York, NY 10020. (212) 512-2000. 1916 to present. Weekly. $56.00 per year.

JOURNAL OF SPACECRAFT AND ROCKETS. American Institute of Aeronautics and Astronautics, Technical Information Service, 370 L'Enfant Promenade, S.W., Washington, DC 20024. (202) 646-7400. 1964 to present. Bimonthly. $95.00 per year.

RESEARCH CENTERS AND INSTITUTES

CHARLES STARK DRAPER LABORATORY, INC. 555 Technology Square, Cambridge, MA 02139. (617) 258-1000.

GODDARD SPACE FLIGHT CENTER. National Aeronautics and Space Administration, Greenbelt Road, Greenbelt, MD 20771. (301) 344-5121.

INSTRUMENTATION AND CONTROL LABORATORY. Princeton University, Department of MAE, English Quadrangle, Princeton, NJ 08544.

MARSHALL SPACE FLIGHT CENTER. National Aeronautics and Space Administration, Huntsville, AL 35812. (205) 453-2121.

STANFORD ELECTRONICS LABORATORIES. Stanford University, Stanford, CA 94305. (415) 723-1804.

GUIDED MISSILES

See also: AEROSPACE ENGINEERING, ORDNANCE, ROCKETS

ABSTRACT SERVICES AND INDEXES

APPLIED SCIENCE AND TECHNOLOGY INDEX. H.W. Wilson and Company, 950 University Avenue, Bronx, NY 10452. (800) 367-6670 or (212) 588-8400. Monthly. Inquire as to cost and availability.

ENGINEERING INDEX MONTHLY AND AUTHOR INDEX. Engineering Information Inc., 345 East 47th Street, New York, NY 10017. (212) 705-7600. Monthly. $1560.00 per year.

INTERNATIONAL AEROSPACE ABSTRACTS. American Institute of Aeronautics and Astronautics, Technical Information Service, 370 L'Enfant Promenade, S.W., Washington, DC 20024. (202) 646-7400. 1961 to present. Semi-monthly. $700.00 per year.

SCIENCE CITATION INDEX. Institute for Scientific Information, 3501 Market Street, Philadelphia, PA 19104. (800) 523-1850 or (215) 386-0100. Six times per year. $6200.00 per year.

STAR (SCIENTIFIC AND TECHNICAL AEROSPACE REPORTS). National Aeronautics and Space Administration, Scientific and Technical Information Facility, Box 8757, Baltimore-Washington International Airport, MD 21240. (202) 453-8545. 1963 to present. Semi-monthly. $85.00 per year with indexes.

ASSOCIATIONS AND PROFESSIONAL SOCIETIES

AMERICAN ASTRONAUTICAL SOCIETY. 6212-B Old Keene Mill Court, Springfield, VA 22152. (703) 866-0020.

AMERICAN INSTITUTE OF AERONAUTICS AND ASTRONAUTICS. 370 L'Enfant Promenade, S.W., Washington, DC 20024. (202) 646-7400.

AMERICAN SPACE FOUNDATION. 111 Massachusetts Avenue, N.W., Suite 200, Washington, DC 20002. (202) 289-2293.

NATIONAL SPACE INSTITUTE. West Wing, Suite 203, 600 Maryland Avenue, S.W., Washington, DC 20024. (202) 484-1111.

SOCIETY OF AUTOMOTIVE ENGINEERS. 400 Commonwealth Drive, Warrendale, PA 15096. (412) 776-4841.

SPACE STUDIES INSTITUTE. P.O. Box 82, 285 Rosedale, Road, Princeton, NJ 08540. (609) 921-0377.

DIRECTORIES AND BIOGRAPHICAL SOURCES

AMERICAN ASTRONAUTICAL SOCIETY DIRECTORY. American Astronautical Society, 6212-B Old Keene Mill Court, Springfield, VA 22152. (703) 866-0020. 1986. $30.00.

AMERICAN INSTITUTE OF AERONAUTICS AND ASTRONAUTICS ROSTER. American Institute of Aeronautics and Astronautics, 370 L'Enfant Promenade, S.W., Washington, DC 20024. (202) 646-7400. 1984. $75.00.

INTERNATIONAL RESEARCH CENTERS DIRECTORY 1988-89. Darren L. Smith, editor. Gale Research Company, Book Tower, Detroit, MI 48226. (800) 521-0707. 4th edition. 1987. $360.00.

RESEARCH CENTERS DIRECTORY 1988. Gale Research Company, Book Tower, Detroit, MI 48226. (800) 521-0707. 12th edition. 1987. $365.00 for set.

SCIENTIFIC AND TECHNICAL ORGANIZATIONS AND AGENCIES DIRECTORY. Margaret Labash Young, editor. Gale Research Company, Book Tower, Detroit, MI 48226. (800) 521-0707. 2nd edition. 1987. $185.00.

WHO'S WHO IN ENGINEERING. Gordon Davis, editor. Hemisphere Publishing Corporation, 1010 Vermont Avenue, NW, Washington, DC 20005. (800) 526-0275. 6th edition. 1985. $200.00.

GENERAL WORKS

CHEMICAL ROCKET PROPULSION AND COMBUSTION RESEARCH. S.S. Penner. Gordon and Breach Science Publishers, Inc., 50 West 23rd Street, New York, NY 10010. (212) 206-8900. 1962. $45.25.

GUIDED WEAPONS. J.E. Lee. Pergamon Press, Inc., Maxwell House, Fairview Park, Elmsford, NY 10523. (914) 592-7700. 1983. $13.00 in paper.

RADAR HOMING GUIDANCE FOR TACTICAL MISSILES. D.A. James. Halsted Press, a division of John Wiley and Sons, Inc., 605 Third Avenue, New York, NY 10158. (800) 526-5368. 1986. $34.95.

ROCKET PROPULSION ELEMENTS: AN INTRODUCTION TO THE ENGINEERING OF ROCKETS. George A. Sutton and Donald M. Ross. John Wiley and Sons, Inc., 605 Third Avenue, New York, NY 10158. (800) 526-5368. Fourth edition. 1976. $52.95.

ROCKET PROPULSION TECHNOLOGY. D.S. Carton and others, editors. Plenum Publishing Corporation, 233 Spring Street, New York, NY 10013. (800) 221-9369. 1961. $32.50.

ONLINE DATA BASES

CA SEARCH. Chemical Abstracts Service, P.O. Box 3012, Columbus, OH 43120. (800) 848-6538 or (614) 421-3600. Comprehensive guide to chemical and related literature, 1972 to present. Inquire as to online cost and availability.

COMPENDEX. Engineering Information, Inc., 345 East 47th Street, New York, NY 10017. (800) 221-1044 or (212) 705-7615. Engineering and technical literature, 1975 to present. Inquire as to online cost and availability.

INSPEC. INSPEC Marketing Department, Institution of Electrical Engineers. Available from IEEE Service Center, 445 Hoes Lane, Piscataway, NJ 08854. (201) 981-0060. Online version of Physics Abstracts. Inquire as to online cost and availability.

NTIS. National Technical Information Service, 5285 Port Royal Road, Springfield, VA 22161. (703) 487-4630. Broad coverage of government sponsored research reports, 1964 to present. Inquire as to online cost and availability.

SCISEARCH. Institute for Scientific Information, 3501 Market Street, Philadelphia, PA 19104. (800) 523-1850 or (215) 386-0100. Broad multidisciplinary title and author index to the international literature of science and technology, 1974 to present. Inquire as to online cost and availability.

WILSONLINE. H.W. Wilson and Company, 950 University Avenue, Bronx, NY 10452. (800) 367-6770 or (212) 588-8400. Makes available online versions of the H.W. Wilson indexes including Applied Science and Technology Index, Business Periodicals Index and Readers' Guide to Periodical Literature. Approximately 1980 to present. Inquire as to online cost and availability.

PERIODICALS

AEROSPACE AMERICA. American Institute of Aeronautics and Astronautics, Technical Information Service, 370 L'Enfant Promenade, S.W., Washington, DC 20024. (202) 646-7400. 1932 to present. Monthly. $56.00 per year.

AEROSPACE ENGINEERING MAGAZINE. Society of Automotive Engineers, 400 Commonwealth Drive, Warrendale, PA 15096. (412) 776-4841. 1981 to present. Monthly. $24.00 per year.

AIAA JOURNAL. American Institute of Aeronautics and Astronautics, Technical Information Service, 370 L'Enfant Promenade, S.W., Washington, DC 20024. (202) 646-7400. 1963 to present. Monthly. $205.00 per year.

AVIATION WEEK AND SPACE TECHNOLOGY. McGraw-Hill Book Company, 1221 Avenue of the Americas, New York, NY 10020. (212) 512-2000. 1916 to present. Weekly. $56.00 per year.

JOURNAL OF SPACECRAFT AND ROCKETS. American Institute of Aeronautics and Astronautics, Technical Information

Service, 370 L'Enfant Promenade, S.W., Washington, DC 20024. (202) 646-7400. 1964 to present. Bimonthly. $95.00 per year.

RESEARCH CENTERS AND INSTITUTES

CENTER FOR SPACE RESEARCH AND APPLICATIONS. University of Texas at Austin, WRW 402, Austin, TX 78712. (512) 471-1356.

GODDARD SPACE FLIGHT CENTER. National Aeronautics and Space Administration, Greenbelt Road, Greenbelt, MD 20771. (301) 344-5121.

GRADUATE AERONAUTICAL LABORATORIES. California Institute of Technology, 105-50, Pasadena, CA 91125. (818) 356-4551.

MARSHALL SPACE FLIGHT CENTER. National Aeronautics and Space Administration, Huntsville, AL 35812. (205) 453-2121.

SPACE SCIENCE AND ENGINEERING CENTER. University of Wisconsin, Madison, 1225 West Dayton Street, Madison, WI 53706. (608) 262-0544.

THERMAL SCIENCES AND PROPULSION CENTER. Purdue University, Lafayette, IN 47907. (317) 494-1500.

GYROCOMPASSES

See: COMPASSES

GYROSCOPES

See also: AEROSPACE ENGINEERING, COMPASSES, GUIDANCE SYSTEMS, GUIDED MISSILES, NAVIGATION, ORDNANCE, ROCKETS

ABSTRACT SERVICES AND INDEXES

APPLIED SCIENCE AND TECHNOLOGY INDEX. H.W. Wilson and Company, 950 University Avenue, Bronx, NY 10452. (800) 367-6670 or (212) 588-8400. Monthly. Inquire as to cost and availability.

ENGINEERING INDEX MONTHLY AND AUTHOR INDEX. Engineering Information Inc., 345 East 47th Street, New York, NY 10017. (212) 705-7600. Monthly. $1560.00 per year.

INTERNATIONAL AEROSPACE ABSTRACTS. American Institute of Aeronautics and Astronautics, Technical Information Service, 370 L'Enfant Promenade, S.W., Washington, DC 20024. (202) 646-7400. 1961 to present. Semi-monthly. $700.00 per year.

SCIENCE CITATION INDEX. Institute for Scientific Information, 3501 Market Street, Philadelphia, PA 19104. (800) 523-1850 or (215) 386-0100. Six times per year. $6200.00 per year.

STAR (SCIENTIFIC AND TECHNICAL AEROSPACE REPORTS). National Aeronautics and Space Administration, Scientific and Technical Information Facility, Box 8757, Baltimore-Washington International Airport, MD 21240. (202) 453-8545. 1963 to present. Semi-monthly. $85.00 per year with indexes.

ASSOCIATIONS AND PROFESSIONAL SOCIETIES

AMERICAN ASTRONAUTICAL SOCIETY. 6212-B Old Keene Mill Court, Springfield, VA 22152. (703) 866-0020.

AMERICAN INSTITUTE OF AERONAUTICS AND ASTRONAUTICS. 370 L'Enfant Promenade, S.W., Washington, DC 20024. (202) 646-7400.

AMERICAN SPACE FOUNDATION. 111 Massachusetts Avenue, N.W., Suite 200, Washington, DC 20002. (202) 289-2293.

NATIONAL SPACE INSTITUTE. West Wing, Suite 203, 600 Maryland Avenue, S.W., Washington, DC 20024. (202) 484-1111.

SOCIETY OF AUTOMOTIVE ENGINEERS. 400 Commonwealth Drive, Warrendale, PA 15096. (412) 776-4841.

SPACE STUDIES INSTITUTE. P.O. Box 82, 285 Rosedale, Road, Princeton, NJ 08540. (609) 921-0377.

DIRECTORIES AND BIOGRAPHICAL SOURCES

AMERICAN ASTRONAUTICAL SOCIETY DIRECTORY. American Astronautical Society, 6212-B Old Keene Mill Court, Springfield, VA 22152. (703) 866-0020. 1986. $30.00.

AMERICAN INSTITUTE OF AERONAUTICS AND ASTRONAUTICS ROSTER. American Institute of Aeronautics and Astronautics, 370 L'Enfant Promenade, S.W., Washington, DC 20024. (202) 646-7400. 1984. $75.00.

INTERNATIONAL RESEARCH CENTERS DIRECTORY 1988-89. Darren L. Smith, editor. Gale Research Company, Book Tower, Detroit, MI 48226. (800) 521-0707. 4th edition. 1987. $360.00.

RESEARCH CENTERS DIRECTORY 1988. Gale Research Company, Book Tower, Detroit, MI 48226. (800) 521-0707. 12th edition. 1987. $365.00 for set.

SCIENTIFIC AND TECHNICAL ORGANIZATIONS AND AGENCIES DIRECTORY. Margaret Labash Young, editor. Gale Research Company, Book Tower, Detroit, MI 48226. (800) 521-0707. 2nd edition. 1987. $185.00.

WHO'S WHO IN ENGINEERING. Gordon Davis, editor. Hemisphere Publishing Corporation, 1010 Vermont Avenue, NW, Washington, DC 20005. (800) 526-0275. 6th edition. 1985. $200.00.

GENERAL WORKS

AUTOMATIC CONTROL OF AIRCRAFT AND MISSILES. John H. Blakelock. John Wiley and Sons, Inc., 605 Third Avenue, New York, NY 10158. (800) 526-5368. 1965. $64.50.

AUTOMATIC FLIGHT CONTROL. E.H. Pallett. Sheridan Publishers, Inc., 145 Palisade Street, Dobbs Ferry, NY 10522. (914) 693-2410. Second edition. 1983. $32.50.

GUIDED WEAPONS. J.E. Lee. Pergamon Press, Inc., Maxwell House, Fairview Park, Elmsford, NY 10523. (914) 592-7700. 1983. $13.00 in paper.

GYROSCOPIC THEORY. George Greehill. Chelsea Publishing Company, 15 East 26th Street, New York, NY 10010. (212) 889-8095. $22.50.

SPINNING TOPS AND GYROSCOPIC MOTION. Harold Crabtree. Chelsea Publishing Company, 15 East 26th Street, New York, NY 10010. (212) 889-8095. 1977. $12.95.

ONLINE DATA BASES

CA SEARCH. Chemical Abstracts Service, P.O. Box 3012, Columbus, OH 43120. (800) 848-6538 or (614) 421-3600. Comprehensive guide to chemical and related literature, 1972 to present. Inquire as to online cost and availability.

COMPENDEX. Engineering Information, Inc., 345 East 47th Street, New York, NY 10017. (800) 221-1044 or (212) 705-7615. Engineering and technical literature, 1975 to present. Inquire as to online cost and availability.

INSPEC. INSPEC Marketing Department, Institution of Electrical Engineers. Available from IEEE Service Center, 445 Hoes Lane, Piscataway, NJ 08854. (201) 981-0060. Online version of Physics Abstracts. Inquire as to online cost and availability.

NTIS. National Technical Information Service, 5285 Port Royal Road, Springfield, VA 22161. (703) 487-4630. Broad coverage of government sponsored research reports, 1964 to present. Inquire as to online cost and availability.

SCISEARCH. Institute for Scientific Information, 3501 Market Street, Philadelphia, PA 19104. (800) 523-1850 or (215) 386-0100. Broad multidisciplinary title and author index to the international literature of science and technology, 1974 to present. Inquire as to online cost and availability.

WILSONLINE. H.W. Wilson and Company, 950 University Avenue, Bronx, NY 10452. (800) 367-6770 or (212) 588-8400. Makes available online versions of the H.W. Wilson indexes including Applied Science and Technology Index, Business Periodicals Index and Readers' Guide to Periodical Literature. Approximately 1980 to present. Inquire as to online cost and availability.

PERIODICALS

AEROSPACE AMERICA. American Institute of Aeronautics and Astronautics, Technical Information Service, 370 L'Enfant Promenade, S.W., Washington, DC 20024. (202) 646-7400. 1932 to present. Monthly. $56.00 per year.

AEROSPACE ENGINEERING MAGAZINE. Society of Automotive Engineers, 400 Commonwealth Drive, Warrendale, PA 15096. (412) 776-4841. 1981 to present. Monthly. $24.00 per year.

AIAA JOURNAL. American Institute of Aeronautics and Astronautics, Technical Information Service, 370 L'Enfant Promenade, S.W., Washington, DC 20024. (202) 646-7400. 1963 to present. Monthly. $205.00 per year.

AVIATION WEEK AND SPACE TECHNOLOGY. McGraw-Hill Book Company, 1221 Avenue of the Americas, New York, NY 10020. (212) 512-2000. 1916 to present. Weekly. $56.00 per year.

JOURNAL OF SPACECRAFT AND ROCKETS. American Institute of Aeronautics and Astronautics, Technical Information Service, 370 L'Enfant Promenade, S.W., Washington, DC 20024. (202) 646-7400. 1964 to present. Bimonthly. $95.00 per year.

RESEARCH CENTERS AND INSTITUTES

CHARLES STARK DRAPER LABORATORY, INC. 555 Technology Square, Cambridge, MA 02139. (617) 258-1000.

GODDARD SPACE FLIGHT CENTER. National Aeronautics and Space Administration, Greenbelt Road, Greenbelt, MD 20771. (301) 344-5121.

INSTRUMENTATION AND CONTROL LABORATORY. Princeton University, Department of MAE, English Quadrangle, Princeton, NJ 08544.

MARSHALL SPACE FLIGHT CENTER. National Aeronautics and Space Administration, Huntsville, AL 35812. (205) 453-2121.

STANFORD ELECTRONICS LABORATORIES. Stanford University, Stanford, CA 94305. (415) 723-1804.

H

HACKERS

See: COMPUTER SECURITY

HADRONS

See: PARTICLE PHYSICS

HAIL

See also: METEOROLOGY, THUNDERSTORMS

ABSTRACT SERVICES AND INDEXES

APPLIED SCIENCE AND TECHNOLOGY INDEX. H.W. Wilson Company, 950 University Avenue, Bronx, NY 10452. (800) 367-6670 or (212) 588-8400. Inquire as to cost and availability.

CHEMICAL ABSTRACTS. Chemical Abstracts Service, 2540 Olentangy Road, Post Office Box 3012, Columbus, OH 43210. (800) 848-6538 or (614) 421-3600. Weekly. $9200.00 per year.

CURRENT CONTENTS: PHYSICAL AND CHEMICAL SCIENCES. Institute for Scientific Information, 3501 Market Street, Philadelphia, PA 19104. (800) 523-1850 or (215) 386-0100. $272.00 per year.

GENERAL SCIENCE INDEX. H.W. Wilson Company, 950 University Avenue, Bronx, NY 10452. (800) 367-6770 or (212) 588-8400. Inquire as to cost and availability.

METEOROLOGICAL AND GEOASTROPHYSICAL ABSTRACTS. American Meteorological Society, 45 Beacon Street, Boston, MA 02108. (617) 227-2425. 1950 to present. Monthly. $450.00 per year.

SCIENCE CITATION INDEX. Institute for Scientific Information, 3501 Market Street, Philadelphia, PA 19104. (800) 523-1850 or (215) 386-0100. Inquire as to cost and availability.

ASSOCIATIONS AND PROFESSIONAL SOCIETIES

AMERICAN METEOROLOGICAL SOCIETY. 45 Beacon Street, Boston, MA 02108. (617) 227-2425.

INTERNATIONAL ASSOCIATION OF METEOROLOGY AND ATMOSPHERIC PHYSICS. UCAR, Post Office Box 3000, Boulder, CO 80307.

NATIONAL WEATHER ASSOCIATION. 4400 Stamp Road, Room 404, Temple Hills, MD 20748. (301) 899-3784.

UNIVERSITY CORPORATION FOR ATMOSPHERIC RESEARCH. Box 3000, 1850 Table Mesa Drive, Boulder, CO 80307. (303) 497-1000.

WEATHER MODIFICATION ASSOCIATION. Post Office Box 8116, Fresno, CA 93747. (209) 291-8466.

DIRECTORIES AND BIOGRAPHICAL SOURCES

AMERICAN MEN AND WOMEN OF SCIENCE. Physical and Biological Sciences. Sixteenth edition. R.R. Bowker Company, 205 East 42nd Street, New York, NY 10017. (800) 521-8110 or (212) 916-1600. 1986. $595.00.

INTERNATIONAL RESEARCH CENTERS DIRECTORY 1988-1989. Gale Research Company, Book Tower, Detroit, MI 48226. (800) 521-0707. Fourth edition. 1987. $360.00.

METEOROLOGICAL SERVICES OF THE WORLD. World Meteorological Organization. Available from: American Meteorological Society, 45 Beacon Street, Boston, MA 02108. (617) 227-2425. Annual. $35.00.

NATIONAL WEATHER SERVICE OFFICES AND STATIONS. National Oceanic and Atmospheric Administration, Department of Commerce, Silver Spring, MD 20910. (301) 427-7698. Annual. Free.

ENCYCLOPEDIAS AND DICTIONARIES

ENCYCLOPEDIA OF CLIMATOLOGY. John E. Oliver and Rhodes W. Fairbridge, editors. Van Nostrand Reinhold, Incorporated, 115 Fifth Avenue, New York, NY 10003. (800) 543-2681. 1987. $89.95.

GENERAL WORKS

HAILSTORMS AND HAILSTONE GROWTH. Narayan R. Gokhale. State University of New York Press, State University Plaza, Albany, NY 12246. (518) 472-5000. 1976. $49.50.

ONLINE DATA BASES

CA SEARCH. Chemical Abstracts Service, Post Office Box 3012, Columbus, OH 43210. Guide to chemical literature, 1972 to present. Inquire as to cost and availability.

DISSERTATION ABSTRACTS ONLINE. University Microfilms International, 300 North Zeeb Road, Ann Arbor, MI 48106. (800) 521-0600 or (313) 761-4700. Scope includes virtually all doctoral dissertations accepted at accredited American institutions from 1861 to present in 252 subject areas. Inquire as to cost and availability.

METEOROLOGICAL AND GEOASTROPHYSICAL ABSTRACTS. American Meteorological Society, 45 Beacon Street, Boston, MA 02108. (617) 227-2425. 1950 to present. Monthly. $450.00 per year.

SCISEARCH. Institute for Scientific Information, 3501 Market Street, Philadelphia, PA 19104. (800) 523-1850 or (215) 386-0100. Broad multidisciplinary title and author index to the international literature of science and technology, 1974 to present. Inquire as to cost and availability.

PERIODICALS

AGRICULTURAL AND FOREST METEOROLOGY. Elsevier Science Publishing Company, Incorporated, 52 Vanderbilt Avenue, New York, NY 10017. (212) 370-5520. 1964 to present. Monthly. $260.00 per year.

AMERICAN METEOROLOGICAL SOCIETY BULLETIN. American Meteorological Society, 45 Beacon Street, Boston, MA 02108. (617) 227-2425. 1920 to present. Monthly. $60.00 per year.

JOURNAL OF CLIMATE AND APPLIED METEOROLOGY. American Meteorological Society, 45 Beacon Street, Boston, MA 02108. (617) 227-2425. 1962 to present. Monthly. $120.00 per year.

JOURNAL OF THE ATMOSPHERIC SCIENCES. American Meteorological Society, 45 Beacon Street, Boston, MA 02108. (617) 227-2425. 1944 to present. Semimonthly. $220.00 per year.

MONTHLY WEATHER REVIEW. American Meteorological Society, 45 Beacon Street, Boston, MA 02108. (617) 227-2425. 1872 to present. Monthly. $120.00 per year.

NATIONAL WEATHER DIGEST. National Weather Association, 4400 Stamp Road, Room 404, Temple Hills, MD 20748. (301) 899-3784. 1976 to present. Quarterly. $20.00 per year.

WEATHER. Royal Meteorological Society, James Glaisher House, Grenville Place, Bracknell Berkshire, RG12 1BX, England. 1946 to present. $30.00 per year.

WEATHERWISE. Heldref Publications, 4000 Albemarle Street, NW, Washington, DC 20016. (202) 362-6445. 1948 to present. Bimonthly. $20.00 per year.

RESEARCH CENTERS AND INSTITUTES

NATIONAL CENTER FOR ATMOSPHERIC RESEARCH. Box 3000, Boulder, CO 80307. (303) 497-1000.

HALF-LIFE

See: RADIATION

HALIDES

See also: BROMINE, CHLORINE, FLUORINE, IODINE

ABSTRACT SERVICES AND INDEXES

APPLIED SCIENCE AND TECHNOLOGY INDEX. H.W. Wilson Company, 950 University Avenue, Bronx, NY 10452. (800) 367-6670 or (212) 588-8400. Inquire as to cost and availability.

CHEMICAL ABSTRACTS. Chemical Abstracts Service, 2540 Olentangy Road, P.O. Box 3012, Columbus, OH 43210. (800) 848-6538 or (614) 421-3600. Weekly. $9200.00 per year.

GENERAL SCIENCE INDEX. H.W. Wilson Co. 950 University Avenue, Bronx, NY 10452. (800) 367-6770 or (212) 588-8400. Inquire as to cost and availability.

PHYSICS ABSTRACTS. Institute of Electrical Engineers, London, United Kingdom. Available from: Institute of Electrical and Electronic Engineers (IEEE), 345 East 47th Street, New York, NY 10017. (212) 705-7900.

SCIENCE CITATION INDEX. Institute for Scientific Information, 3501 Market Street, Philadelphia, PA 19104. (800) 523-1850 or (215) 386-0100.

ASSOCIATIONS AND PROFESSIONAL SOCIETIES

AMERICAN CHEMICAL SOCIETY. 1155 16th Street, NW, Washington, DC 20036. (202) 872-4600.

ASSOCIATION OF CONSULTING CHEMISTS AND CHEMICAL ENGINEERS. 50 East 41st Street, Suite 92, New York, NY 10017. (212) 684-6255.

CHLORINE INSTITUTE, 70 West 40th Street, New York, NY 10018. (212) 819-1677.

INTERNATIONAL SOCIETY FOR FLUORIDE RESEARCH, P.O. Box 692, Warren, MI 48090.

BIBLIOGRAPHIES

SCIENTIFIC AND TECHNICAL BOOKS IN PRINT; An Index to Literature in Science and Technology. R.R. Bowker Co., E. 42nd Street, New York, NY 10017. (800) 521-8110 or (212) 916-1600.

DIRECTORIES AND BIOGRAPHICAL SOURCES

AMERICAN INSTITUTE OF CHEMISTS. American Institute of Chemists, 7315 Wisconsin Avenue, Bethesda, MD 20814. (301) 652-2447. 1986. $35.00.

AMERICAN MEN AND WOMEN OF SCIENCE. Physical and Biological Sciences. Fifteenth edition. R.R. Bowker Co., E. 42nd Street, New York, NY 10017. (800) 521-8110 or (212) 916-1600.

BIOGRAPHICAL DICTIONARY OF SCIENTISTS: CHEMISTS. David Abbott, editor. P. Bedrick Books, 125 East 23rd Street, New York, NY 10010. (212) 777-1187. 1984. $18.95.

CHEMICAL WEEK - BUYERS GUIDE ISSUE. McGraw-Hill Book Company, 1221 Avenue of the Americas, New York, NY 10020. (800) 628-0004. Annual, October. $50.00.

CONSULTING SERVICES: CHEMISTS AND CHEMICAL ENGINEERS. Association of Consulting Chemists and Chemical Engineers, 50 East 41st Street, New York, NY 10017. (212) 684-6255. Annual. 1986. $45.00.

GOVERNMENT RESEARCH DIRECTORY. Gale Research Company, Book Tower, Detroit, MI 48226. (800) 521-0707. Fourth edition. 1987. $350.00.

INTERNATIONAL RESEARCH CENTERS DIRECTORY 1986-1987. Gale Research Company, Book Tower, Detroit, MI 48226. (800) 521-0707. Third edition. 1986. $330.00.

RESEARCH CENTERS DIRECTORY. Gale Research Company, Book Tower, Detroit, MI 48226. (800) 521-0707. Elevent edition. 1987. $355.00.

SCIENTIFIC AND TECHNICAL ORGANIZATIONS AND AGENCIES DIRECTORY. Gale Research Company, Book Tower, Detroit, MI 48226. (800) 521-0707. 1985. $150.00.

WORLD GUIDE TO SCIENTIFIC ASSOCIATIONS AND LEARNED SOCIETIES. K.G. SAUR INC., 175 Fifth Avenue, New York, NY 10010. (800) 521-0707 or (212) 982-1302. Fourth edition. 1984. $112.00.

WHO'S WHO IN FRONTIER SCIENCE AND TECHNOLOGY. Marquis Who's Who, Inc., 200 East Ohio Street, Chicago, IL

60611. (800) 428-3898 or (312) 787-2008.

WHO'S WHO IN TECHNOLOGY TODAY. Reston Publishing Company, Inc., c/o Prentice-Hall, Inc., Englewood Cliffs, NJ 07632. (800) 262-6868. Biennial. Five volumes. $425.00. Covers the fields of electronics, computer science, physics, optics, chemistry, biotechnology, mechanics, energy, and earth science.

ENCYCLOPEDIAS AND DICTIONARIES

CONCISE ENCYCLOPEDIA OF CHEMICAL TECHNOLOGY. Kirk-Othmer. John Wiley & Sons, 605 Third Avenue, New York, NY 10158. (800) 526-5368 or (212) 850-6000. Third edition. 1985. $129.95.

CONDENSED CHEMICAL DICTIONARY. Gessner Hawley. Van Nostrand Reinhold, 115 Fifth Avenue, New York, NY 10003.
Tenth edition. 1981. $49.95.

ENCYCLOPEDIA OF PHYSICAL SCIENCE AND TECHNOLOGY. Academic Press, Inc., Orlando, FL 32887. (800) 321-5068 or (305) 345-2734. Fifteen volumes, 1986. Inquire as to cost and availability.

GLOSSARY OF CHEMICAL TERMS. Clifford A. Hampel and Gessner G. Hawley. Van Nostrand Reinhold Company, 115 Fifth Avenue, New York, NY 10003. (800) 543-2681 or (212) 254-3232. Second edition. 1982. $21.95.

MCGRAW-HILL ENCYCLOPEDIA OF SCIENCE AND TECHNOLOGY. McGraw-Hill Book, Inc., 1221 Avenue of the Americas, New York, NY 10020. (212) 997-3675. Inquire as to cost and availability.

VAN NOSTRAND REINHOLD ENCYCLOPEDIA OF CHEMISTRY. Douglas M. Considine and Glenn D. Considine. Van Nostrand Reinhold Publishing Company, Inc., 115 Fifth Avenue, New York, NY 10003. (800) 543-2681 or (212) 254-3232. 1984. $97.95.

GENERAL WORKS

CHEMISTRY OF FLOURINE AND ITS COMPOUNDS. H.J. Emeleus. Academic Press, Inc., 6277 Sea Harbor Drive, Orlando, FL 32821. (800) 321-5068. 1969. $49.00.

CHEMISTRY OF HALIDES, PSEUDOHALIDES AND AZIDES: SUPPLEMENT D. Chemistry of Functional Groups Series, Part 1 and 2. Saul Patai. John Wiley & Sons, 605 Third Avenue, New York, NY 10158. (800) 526-5368 or (212) 850-6000. 1983. $745.00 set.

CHEMISTRY OF THE ELEMENTS. N.N. Greenwood and A. Earnshaw. Pergamon Publishing, Inc., Maxwell House, Fairview Park, Elmsford, NY 10523. (914) 592-7700. 1984. $143.00.

HALOGEN CHEMISTRY. V. Gutman, editor. Academic Press, Inc., 6277 Sea Harbor Drive, Orlando, FL 32821. (800) 321-5068. Three volumes. 1967. $90.00 each.

HAZARDOUS AND TOXIC CHEMICAL: SAFE HANDLING AND DISPOSAL. Howard Fawcett. John Wiley & Sons, 605 Third Avenue, New York, NY 10158. (800) 526-5368 or (212) 850-6000. 1984. $37.00.

SAFETY STORAGE OF LABORATORY CHEMICALS. David A. Pipitone. John Wiley & Sons, 605 Third Avenue, New York, NY 10158. (800) 526-5368 or (212) 850-6000. 1984. $60.00.

SULFUR DIOXIDE, CHLORINE, FLUORINE AND CHLORINE OXIDES. A.S. Young. Pergamon Publishing, Inc., Maxwell House, Fairview Park, Elmsford, NY 10523. (914) 592-7700. 1983. $100.00.

TOXICOLOGY OF HALOGENATED HYDROCARBONS: HEALTH AND ECOLOGICAL EFFECTS. M.A. Khan and R.H. Stanton, editors. Pergamon Press, Inc., Maxwell House, Fairview Park, Elmsford, NY 10523. (914) 592-7700. 1981. $72.50.

HANDBOOKS AND MANUALS

THE CHEMIST'S COMPANION: A HANDBOOK OF PRACTICAL DATA, TECHNIQUES, AND REFERENCES. Arnold J. Gordon and Richard A. Ford. John Wiley & Sons, 605 Third Avenue, New York, NY 10158. (800) 526-5368. 1973. $49.95.

CRC HANDBOOK OF CHEMISTRY AND PHYSICS. CRC Press, Inc., 2000 Corporate Boulevard, N.W., Boca Raton, FL 33431. Sixty-seventh edition. 1986. $69.95.

HANDBOOK OF APPLIED CHEMISTRY: FACTS FOR ENGINEERS, SCIENTISTS, TECHNICIANS, AND TECHNICAL MANAGERS. Vollrath Hopp and Ingp Hennig. McGraw-Hill Book, Inc., 1221 Avenue of the Americas, New York, NY 10020. (800) 628-0004. 1983. $54.00.

HANDBOOK OF COMPUTATIONAL CHEMISTRY: A PRACTICAL GUIDE TO CHEMICAL STRUCTURE AND ENERGY CALCULATIONS. Tim Clark. John Wiley & Sons, 605 Third Avenue, New York, NY 10158. (800) 526-5368 or (212) 850-6000. 1985. $35.00.

LANGE'S HANDBOOK OF CHEMISTRY. John A. Dean, editor. McGraw-Hill Book, Inc., 1221 Avenue of the Americas, New York, NY 10020. (800) 628-0004. 1985. $59.50.

ONLINE DATA BASES

CA SEARCH. Chemical Abstracts Service, P.O. Box 3012, Columbus, OH 43210. Guide to chemical literature, 1972 to present. Inquire as to cost and availability.

DISSERTATION ABSTRACTS ONLINE. University Microfilms International, 300 North Zeeb Road, Ann Arbor, MI 48106. (800) 521-0600 or (313) 761-4700. Scope includes virtually all doctoral dissertations accepted at accredited American institutions from 1861 to present in 252 subject areas. Inquire as to online cost and availability.

INSPEC. INSPEC Marketing Department, Institute of Electrical and Electronics Engineers, Inc., IEEE Service Department, 445 Hoes Lane, Piscataway, NJ 08854. (201) 981-0060. Inquire as to online cost and availability.

NTIS. National Technical Information Service, 5285 Port Royal Road, Springfield, VA 22161. (703) 487-4630. Broad coverage of government sponsored research reports, 1964 to present. Inquire as to online cost and availability.

SCISEARCH. Institute for Scientific Information, 3501 Market Street, Philadelphia, PA 19104. (800) 523-1850 or (215) 386-0100. Broad multidisciplinary title and author index to the international literature of science and technology, 1974 to present. Inquire as to online cost and availability.

OTHER SOURCES

ANNUAL ENERGY REVIEW. U.S. Department of Energy, Energy Information Administration, Washington, DC 20585. Annual.

CHEMICAL NOMENCLATURE USEAGE. Ronald Lees and Arthur F. Smith. John Wiley & Sons, 605 Third Avenue, New York, NY 10158. (800) 526-5368 or (212) 850-6000. 1983. $52.95.

GUIDE TO BASIC INFORMATION SOURCES IN CHEMISTRY. Arthur Antony. John Wiley & Sons, 605 Third Avenue, New York, NY 10158. (800) 526-5368 or (212) 850-6000. 1979. $26.95.

HALIDES

GUIDE TO BASIC INFORMATION SOURCES IN CHEMISTRY. Arthur Antony. John Wiley & Sons, 605 Third Avenue, New York, NY 10158. (800) 526-5368 or (212) 850-6000. 1979. $26.95.

HOW TO FIND CHEMICAL INFORMATION: A GUIDE FOR PRACTICING CHEMISTS, TEACHERS, AND STUDENTS. John Wiley & Sons, 605 Third Avenue, New York, NY 10158. (800) 526-5368 or (212) 850-6000. 1986. $35.00.

MINERALS YEARBOOK. U.S. Department of the Interior, Bureau of Mines, C Street between Eighteenth and Nineteenth Streets, N.W., Washington, DC 20240. Annual.

PERIODICALS

ANALYTICAL CHEMISTRY. American Chemical Society, 1155 Sixteenth Street, N.W., Washington, DC 20036. (800) 424-6747 or (202) 872-4700. Monthly. $33.00 per year.

ANDEWANDTE CHEMIE (GESELLSCHAFT DEUTSCHER CHEMIKER, GW). V C H Verlagsgesellschaft mbH, Pappelallee 3, Postfach 1260, 6940 Weinheim, West Germany. Monthly. $300.00 per year.

CHEMICAL REVIEWS. American Chemical Society, 1155 Sixteenth Street, N.W., Washington, DC 20036. (800) 424-6747 or (202) 872-4700. Bimonthly. $83.00 per year.

CHEMICAL WEEK. McGraw-Hill Book, Inc., 1221 Avenue of the Americas, New York, NY 10020. (212) 512-2000. Weekly. $30.00 per year.

CHEMTECH. American Chemical Society, 1155 Sixteenth Street, N.W., Washington, DC 20036. (800) 424-6747 or (202) 872-4700. Monthly. $40.00 per year to individuals.

FLUORIDE. International Society for Fluoride Research, P.O. Box 692, Warren, MI 48090. Quarterly. $30.00.

INORGANIC CHEMISTRY. American Chemical Society, 1155 Sixteenth Street, N.W., Washington, DC 20036. (800) 424-6747 or (202) 872-4700. Monthly. $400.00 per year.

JOURNAL OF FLUORINE CHEMISTRY. Elsevier Sequoia, S.A., Box 851, CH-1001, Lausanne 1, Switzerland. Sixteen times per year. $460.00 per year.

JOURNAL OF THE AMERICAN CHEMICAL SOCIETY. American Chemical Society, 1155 Sixteenth Street, N.W., Washington, DC 20036. (800) 424-6747 or (202) 872-4700. Biweekly. $330.00 per year.

POLYHEDRON. Pergamon Journals, Inc., Maxwell House, Fairview Park, Elmsford, NY 10523. (914) 592-7700. Monthly. $595.00 per year.

RESEARCH CENTERS AND INSTITUTES

HARVARD UNIVERSITY. Chemical Laboratories, Oxford Street, Cambridge, MA 02138. (617) 495-4283.

RENSSELAER POLYTECHNICAL INSTITUTE. Chemistry Laboratories, Cogswell Laboratory, Troy, NY 13181. (518) 266-8462.

SOUTHERN ILLINOIS UNIVERSITY AT CORBONDALE. Research Programm in Chemistry and Biochemistry, Carbondale, IL 62901. (618) 453-5721.

UNIVERSITY OF WISCONSIN, MADISON. Theoretical Chemistry Institute, 1101 University Avenue, Madison, WI 53706. (608) 262-1511.

HAM RADIO

See: AMATEUR RADIO

HARBORS

See: COASTAL ENGINEERING

HARD DISKS

See: COMPUTER MEMORY AND STORAGE

HEAT EXCHANGERS

See also: COOLING TOWERS, FLUID MECHANICS, HEAT TRANSFER, MECHANICAL ENGINEERING, PRESSURE VESSELS, THERMODYNAMICS

ABSTRACT SERVICES AND INDEXES

APPLIED MECHANICS REVIEW. American Society of Mechanical Engineers, 345 East 47th Street, New York, NY 10017. (212) 705-7703. 1948 to present. Monthly. $360.00 per year.

APPLIED SCIENCE AND TECHNOLOGY INDEX. H.W. Wilson and Company, 950 University Avenue, Bronx, NY 10452. (800) 367-6670 or (212) 588-8400. Monthly. Inquire as to cost and availability.

CHEMICAL ABSTRACTS. American Chemical Society, Chemical Abstracts Service, Box 3012, Columbus, OH 43210. (614) 421-3600. 1907 to present. Weekly. $9500.00 per year.

CURRENT CONTENTS: ENGINEERING, TECHNOLOGY AND APPLIED SCIENCES. Institute for Scientific Information, 3501 Market Street, Philadelphia, PA 19104. (800) 523-1850 or (215) 386-0100. Weekly. $275.00 per year.

ENGINEERING INDEX MONTHLY AND AUTHOR INDEX. Engineering Information Inc., 345 East 47th Street, New York, NY 10017. (212) 705-7600. Monthly. $1560.00 per year.

ISMEC BULLETIN (Information Service in Mechanical Engineering). Cambridge Scientific Abstracts, 5161 River Road, Bethesda, MD 20816. (301) 951-1400. 1973 to present. Monthly. $450.00 per year.

PHYSICS ABSTRACTS. Institution of Electrical Engineers. Available from: IEEE Service Center, 445 Hoes Lane, Piscataway, NJ 08854. 1898 to present. Bimonthly. $1700.00 per year.

PHYSICS BRIEFS. Physik Verlag GmbH, Postfach 1260/1280, D-6940 Weinheim, West Germany. (212) 661-9404. 1920 to present. Twenty-six times per year. $1250.00 per year.

SCIENCE CITATION INDEX. Institute for Scientific Information, 3501 Market Street, Philadelphia, PA 19104. (800) 523-1850 or (215) 386-0100. Six times per year. $6200. 00 per year.

ASSOCIATIONS AND PROFESSIONAL SOCIETIES

AMERICAN INSTITUTE OF CHEMICAL ENGINEERS. 345 East 47th Street, New York, NY 10017. (212) 705-7703.

AMERICAN INSTITUTE OF PHYSICS. 335 East 45th Street, New York, NY 10017. (212) 661-9494.

AMERICAN SOCIETY OF MECHANICAL ENGINEERS. 345 East 47th Street, New York, NY 10017. (212) 705-7703.

DIRECTORIES AND BIOGRAPHICAL SOURCES

AMERICAN MEN AND WOMEN OF SCIENCE. R.R. Bowker, Inc., Order Department, 245 West 17th Street, New York, NY 10011. (800) 521-8110. Eight volumes. 1986. $595.00 for set.

1987 DIRECTORY OF ENGINEERING SOCIETIES AND RELATED ORGANIZATIONS. Gordon Davis, editor. Hemisphere Publishing Corporation, 1010 Vermont Avenue, NW, Washington, DC 20005. (800) 526-0275. 12th edition. 1987. $100.00.

RESEARCH CENTERS DIRECTORY 1988. Gale Research Company, Book Tower, Detroit, MI 48226. (800) 521-0707. 12th edition. 1987. $365.00 for set.

WHO'S WHO IN ENGINEERING. Gordon Davis, editor. Hemisphere Publishing Corporation, 1010 Vermont Avenue, NW, Washington, DC 20005. (800) 526-0275. 6th edition. 1985. $200.00.

GENERAL WORKS

FUNDAMENTALS OF HEAT EXCHANGER AND PRESSURE VESSEL TECHNOLOGY. J.P. Gupta. Hemisphere Publishing Corporation, 79 Madison Avenue, New York, NY 10016-7892. (800) 821-8312. 1986. $49.95.

HEAT TRANSFER EQUIPMENT DESIGN. R.K. Shah and others. Hemisphere Publishing Corporation, 79 Madison Avenue, New York, NY 10016-7892. 1988. $135.00.

INTRODUCTION TO HEAT TRANSFER. F. Incropera and D. Dewitt. John Wiley and Sons, Inc., 605 Third Avenue, New York, NY 10158. (800) 526-5368. 1985. $39.95.

HANDBOOKS AND MANUALS

HANDBOOK OF HEAT TRANSFER APPLICATIONS. W.M. Rohsenow and others. McGraw-Hill Book Company, 1221 Avenue of the Americas, New York, NY 10020. (212) 512-2000. Second edition. 1986. $79.50.

HEAT EXCHANGER SOURCEBOOK. J.W. Palen, editor. Hemisphere Publishing Corporation, 79 Madison Avenue, New York, NY 10016-7892. (800) 821-8312. 1986. $59.95.

ONLINE DATA BASES

CA SEARCH. Chemical Abstracts Service, P.O. Box 3012, Columbus, OH 43120. (800) 848-6538 or (614) 421-3600. Comprehensive guide to chemical literature, 1972 to present. Inquire as to online cost and availability.

COMPENDEX. Engineering Information, Inc., 345 East 47th Street, New York, NY 10017. (800) 221-1044 or (212) 705-7615. Engineering and technical literature, 1975 to present. Inquire as to online cost and availability.

DISSERTATION ABSTRACTS ONLINE. University Microfilms International, 300 North Zeeb Road, Ann Arbor, MI 48106. (800) 521-0600 or (313) 761-4700. Scope includes virtually all doctoral dissertations accepted at accredited American institutions from 1861 to present in over 250 subject areas. Inquire as to online cost and availability.

INSPEC. INSPEC Marketing Department, Institution of Electrical Engineers. Available from IEEE Service Center, 445 Hoes Lane, Piscataway, NJ 08854. (201) 981-0060. Online version of Physics Abstracts. Inquire as to online cost and availability.

NTIS. National Technical Information Service, 5285 Port Royal Road, Springfield, VA 22161. (703) 487-4630. Broad coverage of government sponsored research reports, 1964 to present. Inquire as to online cost and availability.

SCISEARCH. Institute for Scientific Information, 3501 Market Street, Philadelphia, PA 19104. (800) 523-1850 or (215) 386-0100. Broad multidisciplinary title and author index to the international literature of science and technology, 1974 to present. Inquire as to online cost and availability.

WILSONLINE. H.W. Wilson and Company, 950 University Avenue, Bronx, NY 10452. (800) 367-6770 or (212) 588-8400. Makes available online versions of the H.W. Wilson indexes including Applied Science and Technology Index, Business Periodicals Index and Readers' Guide to Periodical Literature. Approximately 1980 to present. Inquire as to online cost and availability.

PERIODICALS

EXPERIMENTAL THERMAL AND FLUID SCIENCE. Elsevier Science Publishing Company, Inc., 52 Vanderbilt Avenue, New York, NY 10017. (212) 370-5520. 1988 to present. Quarterly. $125.00 per year.

HEAT TRANSFER ENGINEERING. Hemisphere Publishing Corporation, 79 Madison Avenue, New York, NY 10016-7892. (800) 821-8312. 1979 to present. Quarterly. $97.50.

INTERNATIONAL COMMUNICATIONS IN HEAT AND MASS TRANSFER. Pergamon Press, Inc., Maxwell House, Fairview Park, Elmsford, NY 10523. (914) 592-7700. 1974 to present. Bimonthly. $160.00 per year.

INTERNATIONAL JOURNAL OF HEAT AND MASS TRANSFER. Pergamon Press, Inc., Maxwell House, Fairview Park, Elmsford, NY 10523. (914) 592-7700. 1960 to present. Monthly. $500.00 per year.

JOURNAL OF HEAT TRANSFER. American Society of Mechanical Engineers, 345 East 47th Street, New York, NY 10017. (212) 705-7703. 1970 to present. Quarterly. $100.00 per year.

NUMERICAL HEAT TRANSFER: AN INTERNATIONAL JOURNAL OF COMPUTATION AND METHODLOGY. Hemisphere Publishing Corporation, 79 Madison Avenue, New York, NY 10016-7892. (800) 821-8312. 1978 to present. Monthly. $370.00 per year.

RESEARCH CENTERS AND INSTITUTES

HEAT TRANSFER LABORATORY. Massachusetts Institute of Technology, 77 Massachusetts Avenue, Cambridge, MA 02139. (716) 253-2248.

HEAT TRANSFER RESEARCH. University of Wisconsin, Milwaukee, P.O. Box 784, Milwaukee, WI 53201. (414) 963-5001.

HEAT TRANSFER RESEARCH FACILITY. Columbia University, 632 West 125th Street, New York, NY 10027. (212) 280-4163.

INSTITUTE OF THERMO-FLUID ENGINEERING AND SCIENCE. Lehigh University, Whitaker Laboratory, Bethleham, PA 18015. (215) 861-4091.

HEAT PIPES

See also: FLUID MECHANICS, HEAT EXCHANGERS, HEAT TRANSFER, MECHANICAL ENGINEERING, THERMODYNAMICS

ABSTRACT SERVICES AND INDEXES

APPLIED MECHANICS REVIEW. American Society of Mechanical Engineers, 345 East 47th Street, New York, NY 10017. (212) 705-7703. 1948 to present. Monthly. $360.00 per year.

APPLIED SCIENCE AND TECHNOLOGY INDEX. H.W. Wilson and Company, 950 University Avenue, Bronx, NY 10452. (800) 367-6670 or (212) 588-8400. Monthly. Inquire as to cost and availability.

CHEMICAL ABSTRACTS. American Chemical Society, Chemical Abstracts Service, Box 3012, Columbus, OH 43210. (614)

421-3600. 1907 to present. Weekly. $9500.00 per year.

CONFERENCE PAPERS INDEX. Cambridge Scientific Abstracts, 5161 River Road, Bethesda, MD 20816. 1972 to present. Monthly. Inquire as to cost and availability.

CURRENT CONTENTS: ENGINEERING, TECHNOLOGY AND APPLIED SCIENCES. Institute for Scientific Information, 3501 Market Street, Philadelphia, PA 19104. (800) 523-1850 or (215) 386-0100. Weekly. $275.00 per year.

ENGINEERING INDEX MONTHLY AND AUTHOR INDEX. Engineering Information Inc., 345 East 47th Street, New York, NY 10017. (212) 705-7600. Monthly. $1560.00 per year.

ISMEC BULLETIN (Information Service in Mechanical Engineering). Cambridge Scientific Abstracts, 5161 River Road, Bethesda, MD 20816. (301) 951-1400. 1973 to present. Monthly. $450.00 per year.

PHYSICS ABSTRACTS. Institution of Electrical Engineers. Available from: IEEE Service Center, 445 Hoes Lane, Piscataway, NJ 08854. 1898 to present. Bimonthly. $1700.00 per year.

PHYSICS BRIEFS. Physik Verlag GmbH, Postfach 1260/1280, D-6940 Weinheim, West Germany. (212) 661-9404. 1920 to present. Twenty-six times per year. $1250.00 per year.

SCIENCE CITATION INDEX. Institute for Scientific Information, 3501 Market Street, Philadelphia, PA 19104. (800) 523-1850 or (215) 386-0100. Six times per year. $6200.00 per year.

ASSOCIATIONS AND PROFESSIONAL SOCIETIES

AMERICAN INSTITUTE OF PHYSICS. 335 East 45th Street, New York, NY 10017. (212) 661-9494.

AMERICAN SOCIETY OF MECHANICAL ENGINEERS. 345 East 47th Street, New York, NY 10017. (212) 705-7703.

DIRECTORIES AND BIOGRAPHICAL SOURCES

AMERICAN MEN AND WOMEN OF SCIENCE. R.R. Bowker, Inc., Order Department, 245 West 17th Street, New York, NY 10011. (800) 521-8110. Eight volumes. 1986. $595.00 for set.

INTERNATIONAL RESEARCH CENTERS DIRECTORY 1988-89. Darren L. Smith, editor. Gale Research Company, Book Tower, Detroit, MI 48226. (800) 521-0707. 4th edition. 1987. $360.00.

1987 DIRECTORY OF ENGINEERING SOCIETIES AND RELATED ORGANIZATIONS. Gordon Davis, editor. Hemisphere Publishing Corporation, 1010 Vermont Avenue, NW, Washington, DC 20005. (800) 526-0275. 12th edition. 1987. $100.00.

RESEARCH CENTERS DIRECTORY 1988. Gale Research Company, Book Tower, Detroit, MI 48226. (800) 521-0707. 12th edition. 1987. $365.00 for set.

WHO'S WHO IN ENGINEERING. Gordon Davis, editor. Hemisphere Publishing Corporation, 1010 Vermont Avenue, NW, Washington, DC 20005. (800) 526-0275. 6th edition. 1985. $200.00.

GENERAL WORKS

ANALYSIS OF HEAT AND MASS TRANSFER. E.R.G. Eckert and R.M. Drake. Hemisphere Publishing Corporation, 79 Madison Avenue, New York, NY 10016-7892. (800) 821-8312. 1987. $75.00.

HEAT CONDUCTION. S. Kakac. Hemisphere Publishing Corporation, 79 Madison Avenue, New York, NY 10016-7892. (800) 821-8312. Second edition. 1985.

HEAT PIPES. P.D. Dunn and D.A. Reay. Pergamon Press, Inc., Maxwell House, Fairview Park, Elmsford, NY 10523. (914) 592-7700. Third edition. 1982. $22.00.

INTRODUCTION TO HEAT TRANSFER. F. Incropera and D. Dewitt. John Wiley and Sons, Inc., 605 Third Avenue, New York, NY 10158. (800) 526-5368. 1985. $39.95.

PHYSICAL PRINCIPLES OF HEAT PIPES. M.N. Ivanovski and others. Oxford University Press, 200 Madison Avenue, New York, NY 10016. (800) 458-5833. 1982. $69.00.

HANDBOOKS AND MANUALS

HANDBOOK OF HEAT TRANSFER FUNDAMENTALS. W.M. Rohsenow and others. McGraw-Hill Book Company, 1221 Avenue of the Americas, New York, NY 10020. (212) 512-2000. Second edition. 1985. $95.00.

HEAT EXCHANGER SOURCEBOOK. J.W. Palen, editor. Hemisphere Publishing Corporation, 79 Madison Avenue, New York, NY 10016-7892. (800) 821-8312. 1986. $59.95.

ONLINE DATA BASES

CA SEARCH. Chemical Abstracts Service, P.O. Box 3012, Columbus, OH 43120. (800) 848-6538 or (614) 421-3600. Comprehensive guide to chemical literature, 1972 to present. Inquire as to online cost and availability.

COMPENDEX. Engineering Information, Inc., 345 East 47th Street, New York, NY 10017. (800) 221-1044 or (212) 705-7615. Engineering and technical literature, 1975 to present. Inquire as to online cost and availability.

DISSERTATION ABSTRACTS ONLINE. University Microfilms International, 300 North Zeeb Road, Ann Arbor, MI 48106. (800) 521-0600 or (313) 761-4700. Scope includes virtually all doctoral dissertations accepted at accredited American institutions from 1861 to present in over 250 subject areas. Inquire as to online cost and availability.

INSPEC. INSPEC Marketing Department, Institution of Electrical Engineers. Available from IEEE Service Center, 445 Hoes Lane, Piscataway, NJ 08854. (201) 981-0060. Online version of Physics Abstracts. Inquire as to online cost and availability.

NTIS. National Technical Information Service, 5285 Port Royal Road, Springfield, VA 22161. (703) 487-4630. Broad coverage of government sponsored research reports, 1964 to present. Inquire as to online cost and availability.

SCISEARCH. Institute for Scientific Information, 3501 Market Street, Philadelphia, PA 19104. (800) 523-1850 or (215) 386-0100. Broad multidisciplinary title and author index to the international literature of science and technology, 1974 to present. Inquire as to online cost and availability.

WILSONLINE. H.W. Wilson and Company, 950 University Avenue, Bronx, NY 10452. (800) 367-6770 or (212) 588-8400. Makes available online versions of the H.W. Wilson indexes including Applied Science and Technology Index, Business Periodicals Index and Readers' Guide to Periodical Literature. Approximately 1980 to present. Inquire as to online cost and availability.

PERIODICALS

EXPERIMENTAL THERMAL AND FLUID SCIENCE. Elsevier Science Publishing Company, Inc., 52 Vanderbilt Avenue, New York, NY 10017. (212) 370-5520. 1988 to present. Quarterly. $125.00 per year.

HEAT TRANSFER ENGINEERING. Hemisphere Publishing Corporation, 79 Madison Avenue, New York, NY 10016-7892. (800) 821-8312. 1979 to present. Quarterly. $97.50.

INTERNATIONAL COMMUNICATIONS IN HEAT AND MASS TRANSFER. Pergamon Press, Inc., Maxwell House, Fairview Park, Elmsford, NY 10523. (914) 592-7700. 1974 to present. Bimonthly. $160.00 per year.

INTERNATIONAL JOURNAL OF HEAT AND MASS TRANSFER. Pergamon Press, Inc., Maxwell House, Fairview Park, Elmsford, NY 10523. (914) 592-7700. 1960 to present. Monthly. $500.00 per year.

JOURNAL OF HEAT TRANSFER. American Society of Mechanical Engineers, 345 East 47th Street, New York, NY 10017. (212) 705-7703. 1970 to present. Quarterly. $100.00 per year.

NUMERICAL HEAT TRANSFER: AN INTERNATIONAL JOURNAL OF COMPUTATION AND METHODLOGY. Hemisphere Publishing Corporation, 79 Madison Avenue, New York, NY 10016-7892. (800) 821-8312. 1978 to present. Monthly. $370.00 per year.

RESEARCH CENTERS AND INSTITUTES

HEAT TRANSFER LABORATORY. Massachusetts Institute of Technology, 77 Massachusetts Avenue, Cambridge, MA 02139. (716) 253-2248.

HEAT TRANSFER RESEARCH. University of Wisconsin, Milwaukee, P.O. Box 784, Milwaukee, WI 53201. (414) 963-5001.

HEAT TRANSFER RESEARCH FACILITY. Columbia University, 632 West 125th Street, New York, NY 10027. (212) 280-4163.

INSTITUTE OF THERMO-FLUID ENGINEERING AND SCIENCE. Lehigh University, Whitaker Laboratory, Bethlehem, PA 18015. (215) 861-4091.

HEAT PUMPS

See also: AIR CONDITIONING, HEAT EXCHANGERS, HEAT PIPES, HEAT TRANSFER, HEATING AND VENTILATION, MECHANICAL ENGINEERING, THERMODYNAMICS

ABSTRACT SERVICES AND INDEXES

APPLIED MECHANICS REVIEW. American Society of Mechanical Engineers, 345 East 47th Street, New York, NY 10017. (212) 705-7703. 1948 to present. Monthly. $360.00 per year.

APPLIED SCIENCE AND TECHNOLOGY INDEX. H.W. Wilson and Company, 950 University Avenue, Bronx, NY 10452. (800) 367-6670 or (212) 588-8400. Monthly. Inquire as to cost and availability.

CURRENT CONTENTS: ENGINEERING, TECHNOLOGY AND APPLIED SCIENCES. Institute for Scientific Information, 3501 Market Street, Philadelphia, PA 19104. (800) 523-1850 or (215) 386-0100. Weekly. $275.00 per year.

ENGINEERING INDEX MONTHLY AND AUTHOR INDEX. Engineering Information Inc., 345 East 47th Street, New York, NY 10017. (212) 705-7600. Monthly. $1560.00 per year.

ISMEC BULLETIN (Information Service in Mechanical Engineering). Cambridge Scientific Abstracts, 5161 River Road, Bethesda, MD 20816. (301) 951-1400. 1973 to present. Monthly. $450.00 per year.

ASSOCIATIONS AND PROFESSIONAL SOCIETIES

AIR CONDITIONING AND REFRIGERATION INSTITUTE. 1501 Wilson Boulevard, Arlington, VA 22209. (703) 524-8800.

AMERICAN SOCIETY OF HEATING, REFRIGERATING AND AIR CONDITIONING ENGINEERS. 1791 Tullie Circle, N.E., Atlanta, GA 30329. (404) 636-8400.

AMERICAN SOCIETY OF MECHANICAL ENGINEERS. 345 East 47th Street, New York, NY 10017. (212) 705-7703.

EDISON ELECTRIC INSTITUTE. 1111 19th Street, N.W., Washington, DC 20036. (202) 828-7400.

DIRECTORIES AND BIOGRAPHICAL SOURCES

AMERICAN SOCIETY OF HEATING, REFRIGERATING AND AIR CONDITIONING ENGINEERS MEMBERSHIP ROSTER. 1791 Tullie Circle, N.E., Atlanta, GA 30329. (404) 636-8400. Biennial.

1987 DIRECTORY OF ENGINEERING SOCIETIES AND RELATED ORGANIZATIONS. Gordon Davis, editor. Hemisphere Publishing Corporation, 1010 Vermont Avenue, NW, Washington, DC 20005. (800) 526-0275. 12th edition. 1987. $100.00.

RESEARCH CENTERS DIRECTORY 1988. Gale Research Company, Book Tower, Detroit, MI 48226. (800) 521-0707. 12th edition. 1987. $365.00 for set.

WHO'S WHO IN ENGINEERING. Gordon Davis, editor. Hemisphere Publishing Corporation, 1010 Vermont Avenue, NW, Washington, DC 20005. (800) 526-0275. 6th edition. 1985. $200.00.

GENERAL WORKS

HEAT CONDUCTION. S. Kakac. Hemisphere Publishing Corporation, 79 Madison Avenue, New York, NY 10016-7892. (800) 821-8312. Second edition. 1985.

HEAT PUMP SYSTEMS. Harry J. Sauer and R.H. Howell. John Wiley and Sons, Inc., 605 Third Avenue, New York, NY 10158. (800) 526-5368. 1983. $52.95.

HEAT PUMP TECHNOLOGY. Billy. C. Langley. Reston Press, c/o Prentice-Hall Publishing, Inc., Englewood Cliffs, NJ 07632. (800) 562-0245. 1983. $29.95.

INTRODUCTION TO HEAT TRANSFER. F. Incropera and D. Dewitt. John Wiley and Sons, Inc., 605 Third Avenue, New York, NY 10158. (800) 526-5368. 1985. $39.95.

HANDBOOKS AND MANUALS

ASHRAE HANDBOOK. American Society of Heating, Refrigerating and Air Conditioning Engineers, 1791 Tullie Circle, N.E., Atlanta, GA 30329. (404) 636-8400. Annual. $80.00.

HANDBOOK OF HEAT TRANSFER FUNDAMENTALS. W.M. Rohsenow and others. McGraw-Hill Book Company, 1221 Avenue of the Americas, New York, NY 10020. (212) 512-2000. Second edition. 1985. $95.00.

HEAT EXCHANGER SOURCEBOOK. J.W. Palen, editor. Hemisphere Publishing Corporation, 79 Madison Avenue, New York, NY 10016-7892. (800) 821-8312. 1986. $59.95.

THERMODYNAMIC DESIGN DATA FOR HEAT PUMP SYSTEMS: A COMPREHENSIVE DATA BASE AND DESIGN MANUAL. F.A. Holland and others. Pergamon Press, Inc., Maxwell House, Fairview Park, Elmsford, NY 10523. (914) 592-7700. 1982. $165.00.

ONLINE DATA BASES

CA SEARCH. Chemical Abstracts Service, P.O. Box 3012, Columbus, OH 43120. (800) 848-6538 or (614) 421-3600. Comprehensive guide to chemical literature, 1972 to present. Inquire as to online cost and availability.

COMPENDEX. Engineering Information, Inc., 345 East 47th Street, New York, NY 10017. (800) 221-1044 or (212) 705-7615. Engineering and technical literature, 1975 to present. Inquire as to online cost and availability.

NTIS. National Technical Information Service, 5285 Port Royal Road, Springfield, VA 22161. (703) 487-4630. Broad coverage of government sponsored research reports, 1964 to present. Inquire as to online cost and availability.

WILSONLINE. H.W. Wilson and Company, 950 University Avenue, Bronx, NY 10452. (800) 367-6770 or (212) 588-8400. Makes available online versions of the H.W. Wilson indexes including Applied Science and Technology Index, Business Periodicals Index and Readers' Guide to Periodical Literature. Approximately 1980 to present. Inquire as to online cost and availability.

PERIODICALS

ASHRAE JOURNAL. American Society of Heating, Refrigerating and Air Conditioning Engineers, 1791 Tullie Circle, N.E., Atlanta, GA 30329. (404) 636-8400. 1914 to present. Monthly. $35.00 per year.

HEAT TRANSFER ENGINEERING. Hemisphere Publishing Corporation, 79 Madison Avenue, New York, NY 10016-7892. (800) 821-8312. 1979 to present. Quarterly. $97.50.

HEATING/PIPING/AIR CONDITIONING. Penton-IPC, 1100 Superior Avenue, Cleveland, OH 44114. (216) 696-7000. 1929 to present. Monthly. $35.00 per year.

JOURNAL OF HEAT TRANSFER. American Society of Mechanical Engineers, 345 East 47th Street, New York, NY 10017. (212) 705-7703. 1970 to present. Quarterly. $100.00 per year.

HEAT RESISTANT MATERIALS

See: MATERIALS SCIENCE

HEAT TRANSFER

See also: COOLING TOWERS, FLUID MECHANICS, HEAT EXCHANGERS, MECHANICAL ENGINEERING, THERMODYNAMICS

ABSTRACT SERVICES AND INDEXES

APPLIED MECHANICS REVIEW. American Society of Mechanical Engineers, 345 East 47th Street, New York, NY 10017. (212) 705-7703. 1948 to present. Monthly. $360.00 per year.

APPLIED SCIENCE AND TECHNOLOGY INDEX. H.W. Wilson and Company, 950 University Avenue, Bronx, NY 10452. (800) 367-6670 or (212) 588-8400. Monthly. Inquire as to cost and availability.

CHEMICAL ABSTRACTS. American Chemical Society, Chemical Abstracts Service, Box 3012, Columbus, OH 43210. (614) 421-3600. 1907 to present. Weekly. $9500.00 per year.

CONFERENCE PAPERS INDEX. Cambridge Scientific Abstracts, 5161 River Road, Bethesda, MD 20816. 1972 to present. Monthly. Inquire as to cost and availability.

CURRENT CONTENTS: ENGINEERING, TECHNOLOGY AND APPLIED SCIENCES. Institute for Scientific Information, 3501 Market Street, Philadelphia, PA 19104. (800) 523-1850 or (215) 386-0100. Weekly. $275.00 per year.

ENGINEERING INDEX MONTHLY AND AUTHOR INDEX. Engineering Information Inc., 345 East 47th Street, New York, NY 10017. (212) 705-7600. Monthly. $1560.00 per year.

ISMEC BULLETIN (Information Service in Mechanical Engineering). Cambridge Scientific Abstracts, 5161 River Road, Bethesda, MD 20816. (301) 951-1400. 1973 to present. Monthly. $450.00 per year.

INDEX TO SCIENTIFIC AND TECHNICAL PROCEEDINGS. Institute for Scientific Information, 3501 Market Street, Philadelphia, PA 19104. (800) 523-1850 or (215) 386-0100. 1978 to present. Monthly. $775.00 per year.

INDEX TO SCIENTIFIC REVIEWS. Institute for Scientific Information, 3501 Market Street, Philadelphia, PA 19104. (800) 523-1850 or (215) 386-0100. 1974 to present. Semi-annual. $550.00 per year.

PHYSICS ABSTRACTS. Institution of Electrical Engineers. Available from: IEEE Service Center, 445 Hoes Lane, Piscataway, NJ 08854. 1898 to present. Bimonthly. $1700.00 per year.

PHYSICS BRIEFS. Physik Verlag GmbH, Postfach 1260/1280, D-6940 Weinheim, West Germany. (212) 661-9404. 1920 to present. Twenty-six times per year. $1250.00 per year.

SCIENCE CITATION INDEX. Institute for Scientific Information, 3501 Market Street, Philadelphia, PA 19104. (800) 523-1850 or (215) 386-0100. Six times per year. $6200.00 per year.

ASSOCIATIONS AND PROFESSIONAL SOCIETIES

AMERICAN INSTITUTE OF PHYSICS. 335 East 45th Street, New York, NY 10017. (212) 661-9494.

AMERICAN SOCIETY OF MECHANICAL ENGINEERS. 345 East 47th Street, New York, NY 10017. (212) 705-7703.

DIRECTORIES AND BIOGRAPHICAL SOURCES

AMERICAN MEN AND WOMEN OF SCIENCE. R.R. Bowker, Inc., Order Department, 245 West 17th Street, New York, NY 10011. (800) 521-8110. Eight volumes. 1986. $595.00 for set.

INTERNATIONAL RESEARCH CENTERS DIRECTORY 1988-89. Darren L. Smith, editor. Gale Research Company, Book Tower, Detroit, MI 48226. (800) 521-0707. 4th edition. 1987. $360.00.

1987 DIRECTORY OF ENGINEERING SOCIETIES AND RELATED ORGANIZATIONS. Gordon Davis, editor. Hemisphere Publishing Corporation, 1010 Vermont Avenue, NW, Washington, DC 20005. (800) 526-0275. 12th edition. 1987. $100.00.

RESEARCH CENTERS DIRECTORY 1988. Gale Research Company, Book Tower, Detroit, MI 48226. (800) 521-0707. 12th edition. 1987. $365.00 for set.

WHO'S WHO IN ENGINEERING. Gordon Davis, editor. Hemisphere Publishing Corporation, 1010 Vermont Avenue, NW, Washington, DC 20005. (800) 526-0275. 6th edition. 1985. $200.00.

ENCYCLOPEDIAS AND DICTIONARIES

CONCISE SCIENCE DICTIONARY. Oxford University Press, 200 Madison Avenue, New York, NY 10016. (800) 458-5833. 1987. $9.95 in paper.

DICTIONARY OF THE PHYSICAL SCIENCES: TERMS, FORMULAS, DATA. Cesare Emiliani. Oxford University Press, 200 Madison Avenue, New York, NY 10016. (800) 458-5833. 1987. $19.95 in paper.

GENERAL WORKS

ANALYSIS OF HEAT AND MASS TRANSFER. E.R.G. Eckert and R.M. Drake. Hemisphere Publishing Corporation, 79 Madison Avenue, New York, NY 10016-7892. (800) 821-8312. 1987. $75.00.

HEAT CONDUCTION. S. Kakac. Hemisphere Publishing Corporation, 79 Madison Avenue, New York, NY 10016-7892. (800) 821-8312. Second edition. 1985.

INTRODUCTION TO HEAT TRANSFER. F. Incropera and D. Dewitt. John Wiley and Sons, Inc., 605 Third Avenue, New York, NY 10158. (800) 526-5368. 1985. $39.95.

HANDBOOKS AND MANUALS

HANDBOOK OF HEAT TRANSFER FUNDAMENTALS. W.M. Rohsenow and others. McGraw-Hill Book Company, 1221 Avenue of the Americas, New York, NY 10020. (212) 512-2000. Second edition. 1985. $95.00.

HEAT EXCHANGER SOURCEBOOK. J.W. Palen, editor. Hemisphere Publishing Corporation, 79 Madison Avenue, New York, NY 10016-7892. (800) 821-8312. 1986. $59.95.

ONLINE DATA BASES

CA SEARCH. Chemical Abstracts Service, P.O. Box 3012, Columbus, OH 43120. (800) 848-6538 or (614) 421-3600. Comprehensive guide to chemical literature, 1972 to present. Inquire as to online cost and availability.

COMPENDEX. Engineering Information, Inc., 345 East 47th Street, New York, NY 10017. (800) 221-1044 or (212) 705-7615. Engineering and technical literature, 1975 to present. Inquire as to online cost and availability.

DISSERTATION ABSTRACTS ONLINE. University Microfilms International, 300 North Zeeb Road, Ann Arbor, MI 48106. (800) 521-0600 or (313) 761-4700. Scope includes virtually all doctoral dissertations accepted at accredited American institutions from 1861 to present in over 250 subject areas. Inquire as to online cost and availability.

INSPEC. INSPEC Marketing Department, Institution of Electrical Engineers. Available from IEEE Service Center, 445 Hoes Lane, Piscataway, NJ 08854. (201) 981-0060. Online version of Physics Abstracts. Inquire as to online cost and availability.

NTIS. National Technical Information Service, 5285 Port Royal Road, Springfield, VA 22161. (703) 487-4630. Broad coverage of government sponsored research reports, 1964 to present. Inquire as to online cost and availability.

SCISEARCH. Institute for Scientific Information, 3501 Market Street, Philadelphia, PA 19104. (800) 523-1850 or (215) 386-0100. Broad multidisciplinary title and author index to the international literature of science and technology, 1974 to present. Inquire as to online cost and availability.

WILSONLINE. H.W. Wilson and Company, 950 University Avenue, Bronx, NY 10452. (800) 367-6770 or (212) 588-8400. Makes available online versions of the H.W. Wilson indexes including Applied Science and Technology Index, Business Periodicals Index and Readers' Guide to Periodical Literature. Approximately 1980 to present. Inquire as to online cost and availability.

PERIODICALS

EXPERIMENTAL THERMAL AND FLUID SCIENCE. Elsevier Science Publishing Company, Inc., 52 Vanderbilt Avenue, New York, NY 10017. (212) 370-5520. 1988 to present. Quarterly. $125.00 per year.

HEAT TRANSFER ENGINEERING. Hemisphere Publishing Corporation, 79 Madison Avenue, New York, NY 10016-7892. (800) 821-8312. 1979 to present. Quarterly. $97.50.

INTERNATIONAL COMMUNICATIONS IN HEAT AND MASS TRANSFER. Pergamon Press, Inc., Maxwell House, Fairview Park, Elmsford, NY 10523. (914) 592-7700. 1974 to present. Bimonthly. $160.00 per year.

INTERNATIONAL JOURNAL OF HEAT AND MASS TRANSFER. Pergamon Press, Inc., Maxwell House, Fairview Park, Elmsford, NY 10523. (914) 592-7700. 1960 to present. Monthly. $500.00 per year.

JOURNAL OF HEAT TRANSFER. American Society of Mechanical Engineers, 345 East 47th Street, New York, NY 10017. (212) 705-7703. 1970 to present. Quarterly. $100.00 per year.

NUMERICAL HEAT TRANSFER: AN INTERNATIONAL JOURNAL OF COMPUTATION AND METHODLOGY. Hemisphere Publishing Corporation, 79 Madison Avenue, New York, NY 10016-7892. (800) 821-8312. 1978 to present. Monthly. $370.00 per year.

RESEARCH CENTERS AND INSTITUTES

HEAT TRANSFER LABORATORY. Massachusetts Institute of Technology, 77 Massachusetts Avenue, Cambridge, MA 02139. (716) 253-2248.

HEAT TRANSFER RESEARCH. University of Wisconsin, Milwaukee, P.O. Box 784, Milwaukee, WI 53201. (414) 963-5001.

HEAT TRANSFER RESEARCH FACILITY. Columbia University, 632 West 125th Street, New York, NY 10027. (212) 280-4163.

INSTITUTE OF THERMO-FLUID ENGINEERING AND SCIENCE. Lehigh University, Whitaker Laboratory, Bethlehem, PA 18015. (215) 861-4091.

HEATING AND VENTILATION

See: AIR CONDITIONING

HEAVY METAL ALLOYS

See: ALLOYS

HEAVY WATER

See: HYDROGEN

HELICOPTERS

See also: AERODYNAMICS, AERONAUTICAL ENGINEERING, AERONAUTICS, AEROSPACE ENGINEERING

ABSTRACT SERVICES AND INDEXES

ALLOYS INDEX. American Society for Metals, 9639 Kinsman Road, Metals Park, OH 44073. (216) 338-5151. 1974 to present. Monthly. $225.00 per year.

APPLIED MECHANICS REVIEWS. American Society of Mechanical Engineers, 345 East 47th Street, New York, NY 10017. (212) 705-7703. 1948 to present. Monthly. $360.00 per year.

APPLIED SCIENCE AND TECHNOLOGY INDEX. H.W. Wilson Company, 950 University Avenue, Bronx, NY 10452. (800) 367-6670 or (212) 588-8400. 1958 to present. Monthly. Inquire as to cost and availability.

CHEMICAL ABSTRACTS. Chemical Abstracts Service, Box 3012, Columbus, OH 43210. (614) 421-3600. 1907 to present. Weekly. $9200.00 per year.

CURRENT CONTENTS: ENGINEERING, TECHNOLOGY AND APPLIED SCIENCES. Institute for Scientific Information, 3501 Market Street, Philadelphia, PA 19104. (800) 523-1850 or (215) 386-0100. 1970 to present. Weekly. $272.00 per year.

ENGINEERING INDEX MONTHLY AND AUTHOR INDEX. Engineering Information, Inc., 345 East 47th Street, New York, NY 10017. (800) 221-1044 or (212) 705-7600. Monthly, with annual cumulation. $1560.00 per year.

INTERNATIONAL AEROSPACE ABSTRACTS. AIAA/TIS, 1633 Broadway, New York, NY 10019. (212) 581-4300. Semimonthly. $700.00 per year.

METALS ABSTRACTS. American Society for Metals, 9639 Kinsman Road, Metals Park. OH 44073. (216) 338-5151. Monthly. $890.00.

METALS ABSTRACTS INDEX. American Society for Metals, 9639 Kinsman Road, Metals Park, OH 44073. (216) 338-5151. Monthly. (sold only to subscribers of Metals Abstracts).

SCIENCE ABSTRACTS. Section A: Physics; Section B: Electrical and Electronics Abstracts; Section C: Computer and Control Abstracts. Institute of Electrical Engineers, London, United Kingdom. Available from: Institute of Electrical and Electronic Engineers (IEEE), 445 Hoes Lane, Piscataway, NJ 08854. Inquire as to cost and availability.

SCIENCE CITATION INDEX. Institute for Scientific Information, 3501 Market Street, Philadelphia, PA 19104. (800) 523-1850 or (215) 386-0100.

STAR. (Scientific and Technical Aerospacce Reports). U.S. National Aeronautics and Space Administration, Scientific and Technical Information Facility, Box 8757, Baltimore-Washington International Airport, MD 21240. (202) 755-2210. Semimonthly, with semiannual and annual indexes. $85.00 per year.

WORLD ALUMINUM ABSTRACTS. Aluminum Association, Inc., 818 Connecticut Avenue, NW, Washington, DC 20006. (202) 862-5156. 1968 to present. Monthly. $240.00.

ASSOCIATIONS AND PROFESSIONAL SOCIETIES

AIRCRAFT ELECTRONICS ASSOCIATION, Box 1981, Independence, MO 64055. (816) 373-6565.

AMERICAN HELICOPTER SOCIETY, INC., 217 North Washington Street, Alexandria, Va 22314. (703) 684-6777.

AMERICAN INSTITUTE OF AERONAUTICS AND ASTRONAUTICS, 1633 Broadway, New York, New York 10019. (212) 581-4300.

FLIGHT SAFETY FOUNDATION, INC., 5510 Columbia Pike, Arlington, VA 22204-3194. (703) 820-2777.

SOCIETY OF AUTOMOTIVE ENGINEERS. SAE, Inc., 400 Commonwealth Drive, Warrendale, PA 15096. (412) 776-4841.

BIBLIOGRAPHIES

AERONAUTICAL ENGINEERING: A SPECIAL BIBLIOGRAPHY WITH INDEXES. U.S. National Aeronautics and Space Administration, Washington, DC 20546. (202) 755-2320. Available from National Technical Information Service, Springfield, VA 22161. 1970 to present. Monthly. Inquire as to cost and availability.

AIRCRAFT, ENGINES, AND AIRMEN: A SELECTIVE REVIEW OF THE PERIODICAL LITERATURE, 1930-1969. A. Hanniball. Scarecrow Press, Inc., 52 Liberty Street, Methuchen, NJ 08840. (201) 548-8600. 1972. $39.50.

NATIONAL AIR AND SPACE MUSEUM AND SMITHSONIAN INSTITUTION. Aerospace Periodical Index, 1973-1982. 1983. G.K. Hall and Company, 70 Lincoln Street, Boston, MA 02111. (800) 343-2806. $100.00.

SCIENTIFIC AND TECHNICAL BOOKS AND SERIALS IN PRINT, 1988; An Index to Literature in Science and Technology. R.R. Bowker Co., E. 42nd Street, New York, NY 10017. (800) 521-8110 or (212) 916-1600. $175.00.

DIRECTORIES AND BIOGRAPHICAL SOURCES

AAS DIRECTORY. American Astronautical Society, 6212-B Old Keene Mill Court, Springfield, VA 22152. (703) 866-0020. Annual. $35.00 per year.

RESEARCH CENTERS DIRECTORY. Gale Research Company, Detroit, Mi 48226. (800) 521-0707. Eleventh edition. 1987.

WHO'S WHO IN ENGINEERING. Engineers Joint Council, 345 East 47th Street, New York, NY 10017. (212) 705-7010. 1985. $200.00.

WORLD AVIATION DIRECTORY. Murdoch Magazines, 1156 15th Street, NW, Washington, DC 20005. (202) 822-4600. Semiannual. $75.00 per year.

ENCYCLOPEDIAS AND DICTIONARIES

DICTIONARY OF AEROSPACE ENGINEERING. M.G. Kotik. Elsevier Science Publishing Company, Inc., 52 Vanderbilt Avenue, New York, NY 10017. (212) 370-5520. 1986. $170.00.

ENCYCLOPEDIA OF PHYSICAL SCIENCE AND TECHNOLOGY. Academic Press, Inc., Orlando, FL 32887. (800) 321-5068 or (305) 345-2734. Fifteen volumes, 1986.

JANE'S AEROSPACE DICTIONARY. Jane's Publishing Inc., 135 West 50th Street, New York, NY 10020. (212) 586-7745. Second edition. 1986. $39.95.

MCGRAW-HILL DICTIONARY OF ENGINEERING. Sybil P. Parker, editor. McGraw-Hill Book, Inc., 1221 Avenue of the Americas, New York, NY 10020. 1984. $39.95.

MCGRAW-HILL ENCYCLOPEDIA OF SCIENCE AND TECHNOLOGY. McGraw-Hill Book, Inc., 1221 Avenue of the Americas, New York, NY 10020. (212) 512-2000.

GENERAL WORKS

FOUNDATIONS OF AERODYNAMICS: BASES OF AERODYNAMIC DESIGN. A.M. Kuethe and C. Chow. John Wiley & Sons, 605 Third Avenue, New York, NY 10158. (800) 526-5368 or (212) 850-6000. Fourth edition. 1986. $43.95.

HELICOPTER AERODYNAMICS. Society of Automotive Engineers. SAE, Inc., 400 Commonwealth Drive, Warrendale, PA 15096. (412) 776-4841. 1985. $19.95.

HELICOPTER THEORY. Wayne Johnson. Princeton University Press, Princeton, NJ 08540. (609) 452-4900. 1980. $115.00.

ILLUSTRATED GUIDE TO AERODYNAMICS. Hubert Smith. Tab Books, Inc., Monterey Lane, Blue Ridge Summit, PA 17214. (71&) 794-2191. 1986. $14.95.

INTRODUCTION TO THEORETICAL AND COMPUTATIONAL AERODYNAMICS. Jack Moran. John Wiley & Sons, 605 Third

Avenue, New York, NY 10158. (800) 526-5368 or (212) 850-6000. 1984. $40.95.

ROTARY-WING AERODYNAMICS. W.Z. Stepniewski and C.N. Keys. Dover Publications, Inc., 31 East Second Street, Mineola, NY 11501. (516) 294-7000. 1984. $14.50 in paper.

VERTICAL FLIGHT: THE AGE OF THE HELICOPTER. Walter J. Boyne and Donald S. Lopez, editors. Smithsonian Institution Press, 955 L'Enfant Plaza, Suite 210, Washington, DC 20560. 1984. $17.50 in paper.

HANDBOOKS AND MANUALS

JANE'S ALL THE WORLD'S AIRCRAFT 1987-88. John W.R. Taylor. Janes's Publishing, Inc., 115 Fifth Avenue, New York, NY 10003. (214) 254-9097. 1987.

COMPENDEX. Engineering Information, Inc., 345 East 47th Street, New York, NY 10017. (800) 221-1044 or (212) 705-7615. Engineering and technical literature, 1975 to present. Inquire as to cost and availability.

NASA. National Aeronautics and Space Administration, Scientific and Technical Information Branch, 300 7th Street, S.W., Washington, DC 20546.

NTIS. National Technical Information Service, 5285 Port Royal Road, Springfield, VA 22161. (703) 487-4630. Broad coverage of government sponsored research reports, 1964 to present. Inquire as to cost and availability.

PERIODICALS

AERONAUTICAL JOURNAL. Aeronautical Society, 4 Hamilton Place, London W1V OBQ, England. Ten times per year. $175.00 per year.

AEROSPACE AMERICA. American Institute of Aeronautics and Astronautics, 1633 Broadway, New York, NY 10019. (212) 581-4300. Monthly. $51.00 per year.

AEROSPACE ENGINEERING MAGAZINE (SOCIETY OF AUTOMOTIVE ENGINEERS). SAE, Inc., 400 Commonwealth Drive, Warrendale, PA 15096. (412) 776-4841. Monthly. $30.00 per year.

AIAA JOURNAL. AIAA/TIS, 1633 Broadway, New York, NY 10019. (212) 581-4300. Monthly. $205.00 per year.

AIRCRAFT ENGINEERING. Bunhill Publications Limited, 127 Stanstead Road, Forest Hill, London, SE23 JE1, England. 1929 to present. Monthly. $51.00 per year.

AMERICAN HELICOPTER SOCIETY JOURNAL. American Helicopter Society, Inc., 217 North Washington Street, Alexandria, Va 22314. (703) 684-6777. Quarterly. $25.00 per year.

AVIATION MECHANICS BULLETIN. Flight Safety Foundation, Inc., 5510 Columbia Pike, Arlington, Va 22204-3194. (703) 820-2777. Bimonthly. $6.50 per year.

AVIATION WEEK AND SPACE TECHNOLOGY. McGraw-Hill Book, Inc., 1221 Avenue of the Americas, New York, NY 10020. (212) 512-2000. Weekly. $55.00 per year.

CANADIAN AERONAUTICS AND SPACE JOURNAL. Canadian Aeronautics and Space Institute, 222 Somerset Street, West, Suite 601, Ottawa, ON K2P OJ1, Canada. (613) 234-0191. Quarterly. $35.00 per year.

FLYING. CBS Magazines, 1515 Broadway, New York, NY 10036. (212) 503-4200. Monthly. $14.00 per year.

IEEE TRANSACTIONS ON AEROSPACE. Institute of Electrical and Electronics Engineers, 345 East 47th Street, New York, NY 10017.

INTERNATIONAL JOURNAL OF TURBO AND JET ENGINES. Martinus Nijhoff Publishers. Available from: Kluwer Academic Publishers Group, Distribution Centre, Postbus 322, 3300 AH Dordrecht, The Netherlands. Quarterly. $125.00 per year.

JOURNAL OF AIRCRAFT. American Institute of Aeronautics and Astronautics, AIAA/TIS, 1633 Broadway, New York, NY 10019. (212) 581-4300. Monthly. $185.00 per year.

JOURNAL OF GUIDANCE AND CONTROL. American Institute of Aeronautics and Astronautics, AIAA/TIS, 1633 Broadway, New York, NY 10019. (212) 581-4300. Bimonthly. $95.00 per year.

JOURNAL OF PROPULSION AND POWER. American Institute of Aeronautics and Astronautics, AIAA/TIS, 1633 Broadway, New York, NY 10019. (212) 581-4300. Bimonthly. $170.00 per year.

ROTOR AND WING INTERNATIONAL. PJS Publications, Inc., New Plaza, Box 1790, Peoria, IL 61656. (309) 682-6626. Monthly. $24.00 per year.

VERTICA: THE INTERNATIONAL JOURNAL OF ROTORCRAFT AND POWERLIFT AIRCRAFT. Pergamon Press, Inc., Journals Division, Maxwell House, Fairview Park, Elmsford, NY 10523. (914) 592-7700. Quarterly. $135.00 per year.

VERTIFLITE. American Helicopter Society, Inc., 217 North Washington Street, Alexandria, VA 22314. (703) 684-6777. Bimonthly. $25.00 per year.

RESEARCH CENTERS AND INSTITUTES

FLIGHT RESEARCH LABORATORY. University of Kansas, Raymond Nichols Hall, Lawrence, KS 66045. (913) 864-3043.

OHIO STATE UNIVERSITY. Aeronautical and Astronautical Research Laboratory, 2300 West Case Road, Columbus, OH 43220. (614) 422-1241.

PRINCETON UNIVERSITY. Flight Mechanics Laboratory, Department of Mechanical and Aerospace, James Forrestal Campus, Princeton, NJ 08544. (609) 452-5149.

STANFORD UNIVERSITY. Aero Structures Laboratory, Department of Aeronautics and Astronomy, Stanford, CA 94305. (415) 497-3317.

UNIVERSITY OF ALABAMA. Flight Dynamics Laboratory, Box 2901, Tuscaloosa, Al 35487. (205) 348-7300.

UNIVERSITY OF TEXAS AT AUSTIN. Aeronautical Research Center, Aerospace Engineering and Engineering Mechanics, WRW 217, Austin, TX 78712. (512) 471-5962.

UNIVERSITY OF WASHINGTON. Aeronautical Laboratory, FS-10, Seattle, WA 98105. (206) 543-0439.

HELIUM

See also: AIRSHIPS, CRYOGENICS, ELEMENTS, INORGANIC CHEMISTRY

ABSTRACT SERVICES AND INDEXES

APPLIED SCIENCE AND TECHNOLOGY INDEX. H.W. Wilson and Company, 950 University Avenue, Bronx, NY 10452. (800) 367-6670 or (212) 588-8400. Monthly. Inquire as to cost and availability.

CHEMICAL ABSTRACTS. American Chemical Society, Chemical Abstracts Service, Box 3012, Columbus, OH 43210. (614) 421-3600. 1907 to present. Weekly. $9500.00 per year.

CONFERENCE PAPERS INDEX. Cambridge Scientific Abstracts, 5161 River Road, Bethesda, MD 20816. (301) 951-1400. 1972 to present. Monthly. Inquire as to cost and availability.

ENGINEERING INDEX MONTHLY AND AUTHOR INDEX. Engineering Information Inc., 345 East 47th Street, New York, NY 10017. (212) 705-7600. Monthly. $1560.00 per year.

GENERAL SCIENCE INDEX. H.W. Wilson and Company, 950 University Avenue, Bronx, NY 10452. (800) 367-6670 or (212) 588-8400. 1978 to present. Monthly. Inquire as to cost and availability.

INDEX TO SCIENTIFIC AND TECHNICAL PROCEEDINGS. Institute for Scientific Information, 3501 Market Street, Philadelphia, PA 19104. (800) 523-1850 or (215) 386-0100. 1978 to present. Monthly. $775.00 per year.

INDEX TO SCIENTIFIC REVIEWS. Institute for Scientific Information, 3501 Market Street, Philadelphia, PA 19104. (800) 523-1850 or (215) 386-0100. 1974 to present. Semi-annual. $550.00 per year.

PHYSICS ABSTRACTS. Institution of Electrical Engineers. Available from: IEEE Service Center, 445 Hoes Lane, Piscataway, NJ 08854. 1898 to present. Bimonthly. $1700.00 per year.

PHYSICS BRIEFS. Physik Verlag GmbH, Postfach 1260/1280, D-6940 Weinheim, West Germany. (212) 661-9404. 1920 to present. Twenty-six times per year. $1250.00 per year.

SCIENCE CITATION INDEX. Institute for Scientific Information, 3501 Market Street, Philadelphia, PA 19104. (800) 523-1850 or (215) 386-0100. Six times per year. $6200.00 per year.

ASSOCIATIONS AND PROFESSIONAL SOCIETIES

AMERICAN CHEMICAL SOCIETY. 1155 16th Street, N.W., Washington, DC 20036. (800) 424-6747.

AMERICAN INSTITUTE OF PHYSICS. 335 East 45th Street, New York, NY 10017. (212) 661-9494.

AMERICAN SOCIETY FOR METALS. Metals Park, OH 44073. (216) 338-5151.

DIRECTORIES AND BIOGRAPHICAL SOURCES

AMERICAN MEN AND WOMEN OF SCIENCE. R.R. Bowker, Inc., Order Department, 245 West 17th Street, New York, NY 10011. (800) 521-8110. Eight volumes. 1986. $595.00 for set.

INTERNATIONAL RESEARCH CENTERS DIRECTORY 1988-89. Darren L. Smith, editor. Gale Research Company, Book Tower, Detroit, MI 48226. (800) 521-0707. 4th edition. 1987. $360.00.

1987 DIRECTORY OF ENGINEERING SOCIETIES AND RELATED ORGANIZATIONS. Gordon Davis, editor. Hemisphere Publishing Corporation, 79 Madison Avenue, New York, NY 10016-7892. (800) 821-8312. 12th edition. 1987. $100.00.

RESEARCH CENTERS DIRECTORY 1988. Gale Research Company, Book Tower, Detroit, MI 48226. (800) 521-0707. 12th edition. 1987. $365.00 for set.

SCIENTIFIC AND TECHNICAL ORGANIZATIONS AND AGENCIES DIRECTORY. Margaret Labash Young, editor. Gale Research Company, Book Tower, Detroit, MI 48226. (800) 521-0707. 2nd edition. 1987. $185.00.

WHO'S WHO IN ENGINEERING. Gordon Davis, editor. Hemisphere Publishing Corporation, 79 Madison Avenue, New York, NY 10016-7892. (800) 821-8312. 6th edition. 1985. $200.00.

ENCYCLOPEDIAS AND DICTIONARIES

MCGRAW-HILL ENCYCLOPEDIA OF SCIENCE AND TECHNOLOGY. McGraw-Hill Book Company, 1221 Avenue of the Americas, New York, NY 10020. (212) 512-2000. 6th edition. 1987. $1600.00.

THESAURUS OF SCIENTIFIC, TECHNICAL, AND ENGINEERING TERMS. Hemisphere Publishing Corporation, 1010 Vermont Avenue, NW, Washington, DC 20005. (800) 526-0275. 1988. $125.00.

GENERAL WORKS

CHEMISTRY OF THE ELEMENTS. N.N. Greenwood and A. Earnshaw. Pergamon Press, Inc., Maxwell House, Fairview Park, Elmsford, NY 10523. (914) 592-7700. 1984. $145.00.

HELIUM ISOTOPES IN NATURE. B.A. Mamyrin and I.N. Tolstikhin. Elsevier Science Publishing Company, Inc., 52 Vanderbilt Avenue, New York, NY 10017. (212) 370-5520. 1984. $54.00.

INORGANIC CHEMISTRY. William W. Porterfield. Addison-Wesley Publishing Company, Inc., 1 Jacob Way, Reading, MA 01867. (617) 944-3700. 1983. $36.95.

INORGANIC CHEMISTRY: AN INTRODUCTION. T. Moeller. John Wiley and Sons, Inc., 605 Third Avenue, New York, NY 10158. (800) 526-5368. 1982. $44.95.

THE PHYSICS OF SOLID AND LIQUID HELIUM. K.H. Benneman and J.B. Ketterson, editors. Robert E. Krieger Publishing Company, Inc., P.O. Box 9542, Melbourne, FL 32902-9542. (305) 724-9542. 1978. Part 1, $46.50. Part 2, $80.95.

HANDBOOKS AND MANUALS

CRC HANDBOOK OF CHEMISTRY AND PHYSICS. Robert C. Weast, editor. CRC Press, 2000 Corporate Boulevard, Boca Raton, FL 33431. (800) 272-7737. 68th edition. 1987. $69.95.

ONLINE DATA BASES

CA SEARCH. Chemical Abstracts Service, P.O. Box 3012, Columbus, OH 43120. (800) 848-6538 or (614) 421-3600. Comprehensive guide to chemical literature, 1972 to present. Inquire as to online cost and availability.

COMPENDEX. Engineering Information, Inc., 345 East 47th Street, New York, NY 10017. (800) 221-1044 or (212) 705-7615. Engineering and technical literature, 1975 to present. Inquire as to online cost and availability.

DISSERTATION ABSTRACTS ONLINE. University Microfilms International, 300 North Zeeb Road, Ann Arbor, MI 48106. (800) 521-0600 or (313) 761-4700. Scope includes virtually all doctoral dissertations accepted at accredited American institutions from 1861 to present in over 250 subject areas. Inquire as to online cost and availability.

ENERGYLINE. EIC/Intelligence, Inc., 48 West 38th Street, New York, NY 10018. (212) 944-8500. A database of resources on the scientific, engineering, political, and socioeconomic aspects of energy resources. 1976 to present. Inquire as to online cost and availability.

INSPEC. INSPEC Marketing Department, Institution of Electrical Engineers. Available from IEEE Service Center, 445 Hoes Lane, Piscataway, NJ 08854. (201) 981-0060. Online version of Physics Abstracts. Inquire as to online cost and availability.

NTIS. National Technical Information Service, 5285 Port Royal Road, Springfield, VA 22161. (703) 487-4630. Broad coverage of government sponsored research reports, 1964 to present. Inquire as to online cost and availability.

SCISEARCH. Institute for Scientific Information, 3501 Market Street, Philadelphia, PA 19104. (800) 523-1850 or (215) 386-0100. Broad multidisciplinary title and author index to the international literature of science and technology, 1974 to present. Inquire as to online cost and availability.

WILSONLINE. H.W. Wilson and Company, 950 University Avenue, Bronx, NY 10452. (800) 367-6770 or (212) 588-8400. Makes available online versions of the H.W. Wilson indexes including Applied Science and Technology Index, Business Periodicals Index and Readers' Guide to Periodical Literature. Approximately 1980 to present. Inquire as to online cost and availability.

PERIODICALS

INORGANIC CHEMISTRY. American Chemical Society, 1155 16th Street, N.W., Washington, DC 20036. (800) 424-6747. 1962 to present. Biweekly. $300.00 per year.

INORGANICA CHIMICA ACTA. Elsevier Science Publishing Company, Inc., 52 Vanderbilt Avenue, New York, NY 10017. (212) 370-5520. 1967 to present. Biweekly. $1500.00 per year.

JOURNAL OF ENERGY RESOURCES TECHNOLOGY. American Society of Mechanical Engineers, 345 East 47th Street, New York, NY 10017. (212) 705-7703. 1979 to present. $24.00 per year.

RESEARCH CENTERS AND INSTITUTES

MICROKELVIN LABORATORY. University of Florida, Department of Physics, 118 Williamson, Gainesville, FL 32611. (904) 392-0485.

APPLIED SUPERCONDUCTIVITY RESEARCH CENTER. University of Wisconsin - Madison, Engineering Research Building, 1500 Johnson Drive, Madison, WI 537-6. (608) 263-5026.

HERSCHEL CASSEGRAIN TELESCOPES

See: TELESCOPES

HETEROCYCLIC CHEMISTRY

See also: BENZENE, CHEMISTRY, CHEMICAL ENGINEERING, HYDROCARBONS, ORGANIC CHEMISTRY, PLASTICS, PETROLEUM CHEMISTRY

ABSTRACT SERVICES AND INDEXES

CHEMICAL ABSTRACTS. American Chemical Society, Chemical Abstracts Service, Box 3012, Columbus, OH 43210. (614) 421-3600. 1907 to present. Weekly. $9500.00 per year.

CONFERENCE PAPERS INDEX. Cambridge Scientific Abstracts, 5161 River Road, Bethesda, MD 20816. 1972 to present. Monthly. Inquire as to cost and availability.

CURRENT CONTENTS: PHYSICAL, CHEMICAL AND EARTH SCIENCES. Institute for Scientific Information, 3501 Market Street, Philadelphia, PA 19104. (800) 523-1850 or (215) 386-0100. Weekly. $275.00 per year.

INDEX TO SCIENTIFIC AND TECHNICAL PROCEEDINGS. Institute for Scientific Information, 3501 Market Street, Philadelphia, PA 19104. (800) 523-1850 or (215) 386-0100. 1978 to present. Monthly. $775.00 per year.

INDEX TO SCIENTIFIC REVIEWS. Institute for Scientific Information, 3501 Market Street, Philadelphia, PA 19104. (800) 523-1850 or (215) 386-0100. 1974 to present. Semi-annual. $550.00 per year.

SCIENCE CITATION INDEX. Institute for Scientific Information, 3501 Market Street, Philadelphia, PA 19104. (800) 523-1850 or (215) 386-0100. Six times per year. $6200.00 per year.

ANNUAL REVIEWS AND YEARBOOKS

ADVANCES IN PHYSICAL ORGANIC CHEMISTRY. Academic Press, Inc., 6277 Sea Harbor Drive, Orlando, FL 32821. (800) 321-5068. 1963 to present. Irregular. Price varies, inquire.

STUDIES IN ORGANIC CHEMISTRY. Marcel Dekker Inc., 270 Madison Avenue, New York, NY 10016. (800) 228-1160. 1973 to present. Irregular. Price varies, inquire.

ASSOCIATIONS AND PROFESSIONAL SOCIETIES

AMERICAN CARBON SOCIETY. The Stackpole Corporation, St. Marys, PA 15857. (814) 781-8410.

AMERICAN CHEMICAL SOCIETY. 1155 16th Street, N.W., Washington, DC 20036. (202) 872-4600.

AMERICAN INSTITUTE OF CHEMICAL ENGINEERS. 345 East 47th Street, New York, NY 10017. (212) 705-7338.

AMERICAN OIL CHEMISTS' SOCIETY. 508 South Sixth Street, Champaign, IL 61820. (217) 359-2344.

ASSOCIATION OF OFFICIAL ANALYTICAL CHEMISTS. 1111 North 19th Street, Suite 210, Arlington, VA 22209. (703) 522-3032.

DIRECTORIES AND BIOGRAPHICAL SOURCES

AMERICAN MEN AND WOMEN OF SCIENCE. R.R. Bowker, Inc., Order Department, 245 West 17th Street, New York, NY 10011. (800) 521-8110. Eight volumes. 1986. $595.00 for set.

INTERNATIONAL RESEARCH CENTERS DIRECTORY 1988-89. Darren L. Smith, editor. Gale Research Company, Book Tower, Detroit, MI 48226. (800) 521-0707. 4th edition. 1987. $360.00.

1987 DIRECTORY OF ENGINEERING SOCIETIES AND RELATED ORGANIZATIONS. Gordon Davis, editor. Hemisphere Publishing Corporation, 1010 Vermont Avenue, NW, Washington, DC 20005. (800) 526-0275. 12th edition. 1987. $100.00.

RESEARCH CENTERS DIRECTORY 1988. Gale Research Company, Book Tower, Detroit, MI 48226. (800) 521-0707. 12th edition. 1987. $365.00 for set.

SCIENTIFIC AND TECHNICAL ORGANIZATIONS AND AGENCIES DIRECTORY. Margaret Labash Young, editor. Gale Research Company, Book Tower, Detroit, MI 48226. (800) 521-0707. 2nd edition. 1987. $185.00.

WHO'S WHO IN ENGINEERING. Gordon Davis, editor. Hemisphere Publishing Corporation, 1010 Vermont Avenue, NW, Washington, DC 20005. (800) 526-0275. 6th edition. 1985. $200.00.

ENCYCLOPEDIAS AND DICTIONARIES

DICTIONARY OF ORGANIC COMPOUNDS. John Buckingham, editor. Methuen, Inc., 29 West 35th Street, New York, NY 10001. (212) 244-3336. Fifth edition. 1988. $675.00.

DICTIONARY OF ORGANOMETALLIC COMPOUNDS. J.E. MacIntyre, editor. Methuen, Inc., 29 West 35th Street, New York, NY 10001. (212) 244-3336. 1988. $325.00.

ENCYCLOPEDIA OF CHEMISTRY. Douglas M. Considine, editor. Van Nostrand Reinhold Company, Inc., 135 West 50th Street, New York, NY 10020. (800) 543-2681. 4th edition. 1984.

$98.95.

THESAURUS OF SCIENTIFIC, TECHNICAL, AND ENGINEERING TERMS. Hemisphere Publishing Corporation, 1010 Vermont Avenue, NW, Washington, DC 20005. (800) 526-0275. 1988. $125.00.

GENERAL WORKS

BASIC ORGANIC CHEMISTRY. F.L. Wiseman. McGraw-Hill Book Company, 1221 Avenue of the Americas, New York, NY 10020. (212) 512-2000. 1988. $37.95.

BENZENE: BASIC AND HAZARDOUS PROPERTIES. M. Cherimisinoff. Marcel Dekker Inc., 270 Madison Avenue, New York, NY 10016. (800) 228-1160. 1979. $49.75.

BENZENE: SCIENTIFIC UPDATE. Myron A. Mehlman. Alan R. Liss, Inc., 150 Fifth Avenue, New York, NY 10011. (212) 741-2515. 1985. $24.00.

BENZOPYRENES. M.R.Osborne and N.T. Crosby. Cambridge University Press, 32 East 57th Street, New York, NY 10022. (800) 872-7423. 1987. $79.50.

CONTEMPORARY HETEROCYCLIC CHEMISTRY: SYNTHESIS, REACTIONS AND APPLICATIONS. G.R. Newkome and W.W. Paudler. John Wiley and Sons, Inc., 605 Third Avenue, New York, NY 10158. (800) 526-5368. 1982. $39.50.

FUNDAMENTALS OF ORGANIC CHEMISTRY. T.W.G. Solomon. John Wiley and Sons, Inc., 605 Third Avenue, New York, NY 10158. (800) 526-5368. 2nd edition. 1986. $39.95.

ORGANIC SYNTHESIS REACTIONS AND MECHANISMS. B. Christoph and others. Springer-Verlag New York, Inc., 175 Fifth Avenue, New York, NY 10010. (800) 526-7254. 1986. $60.00.

REDUCTIONS IN ORGANIC CHEMISTRY. M. Hudlicky. John Wiley and Sons, Inc., 605 Third Avenue, New York, NY 10158. (800) 526-5368. 1984. $45.00.

HANDBOOKS AND MANUALS

CRC HANDBOOK OF CHEMISTRY AND PHYSICS. Robert C. Weast, editor. CRC Press, 2000 Corporate Boulevard, Boca Raton, FL 33431. (800) 272-7737. 69th edition. 1988. $94.00.

CRC HANDBOOK OF DATA ON ORGANIC COMPOUNDS. R.C. Weast and M.J. Astle, editors. CRC Press, 2000 Corporate Boulevard, Boca Raton, FL 33431. (800) 272-7737. 1985. Two volumes. $270.00 for set.

A GUIDEBOOK TO MECHANISMS IN ORGANIC CHEMISTRY. P. Sykes. John Wiley and Sons, Inc., 605 Third Avenue, New York, NY 10158. (800) 526-5368. 6th edition. 1986. $21.95.

HANDBOOK OF APPLIED CHEMISTRY. Vollrath Hopp and Ingo Hennig. Hemisphere Publishing Corporation, 79 Madison Avenue, New York, NY 10016-7892. Order from: Taylor and Francis/Hemisphere Distribution Center, 242 Cherry Street, Philadelphia, PA 19106-1906. (800) 821-8312. 1983. $49.95.

HANDBOOK OF ORGANIC CHEMISTRY. John A. Dean. McGraw-Hill Book Company, 1221 Avenue of the Americas, New York, NY 10020. (212) 512-2000. 1987. $64.50.

HAZARDOUS CHEMICALS DESK REFERENCE. N.I. Sax and R.J. Lewis, editors. Van Nostrand Reinhold Company, Inc., 135 West 50th Street, New York, NY 10020. (800) 543-2681. 1987. $69.95.

LANGE'S HANDBOOK OF CHEMISTRY. John A. Dean, editor. McGraw-Hill Book Company, 1221 Avenue of the Americas, New York, NY 10020. (212) 512-2000. 1985. $59.50.

ONLINE DATA BASES

CA SEARCH. Chemical Abstracts Service, P.O. Box 3012, Columbus, OH 43120. (800) 848-6538 or (614) 421-3600. Comprehensive guide to chemical literature, 1972 to present. Inquire as to online cost and availability.

COMPENDEX. Engineering Information, Inc., 345 East 47th Street, New York, NY 10017. (800) 221-1044 or (212) 705-7615. Engineering and technical literature, 1975 to present. Inquire as to online cost and availability.

DISSERTATION ABSTRACTS ONLINE. University Microfilms International, 300 North Zeeb Road, Ann Arbor, MI 48106. (800) 521-0600 or (313) 761-4700. Scope includes virtually all doctoral dissertations accepted at accredited American institutions from 1861 to present in over 250 subject areas. Inquire as to online cost and availability.

NTIS. National Technical Information Service, 5285 Port Royal Road, Springfield, VA 22161. (703) 487-4630. Broad coverage of government sponsored research reports, 1964 to present. Inquire as to online cost and availability.

SCISEARCH. Institute for Scientific Information, 3501 Market Street, Philadelphia, PA 19104. (800) 523-1850 or (215) 386-0100. Broad multidisciplinary title and author index to the international literature of science and technology, 1974 to present. Inquire as to online cost and availability.

PERIODICALS

AMERICAN OIL CHEMISTS' SOCIETY. JOURNAL. American Oil Chemists' Society, 508 South Sixth Street, Champaign, IL 61820. (217) 359-2344. 1917 to present. Monthly. $60.00 per year.

CARBON. Pergamon Press, Inc., Maxwell House, Fairview Park, Elmsford, NY 10523. (914) 592-7700. 1963 to present. Bimonthly. $235.00 per year.

HYDROCARBON PROCESSING. Gulf Publishing Company, P.O. Box 2608, Houston, TX 77001. (713) 520-4444. Monthly. $10.00 per year.

JOURNAL OF HETEROCYCLIC CHEMISTRY. HeteroCorporation, Box 16000 MH, Tampa, FL 33687. 1964 to present. 6 times per year. $160.00 per year.

JOURNAL OF ORGANIC CHEMISTRY. American Chemical Society, 1155 16th Street, N.W., Washington, DC 20036. (800) 424-6747. 1936 to present. Semi-monthly. $218.00 per year.

JOURNAL OF POLYMER SCIENCE. POLYMER CHEMISTRY EDITION. John Wiley and Sons, Inc., 605 Third Avenue, New York, NY 10158. (800) 526-5368. 1962 to present. Monthly. $895.00 per year, includes all editions.

ORGANOMETALLICS. American Chemical Society, 1155 16th Street, N.W., Washington, DC 20036. (800) 424-6747. 1982 to present. Monthly. $195.00 per year.

TETRAHEDRON. Pergamon Press, Inc., Maxwell House, Fairview Park, Elmsford, NY 10523. (914) 592-7700. 1957 to present. 24 times per year. $1400.00 per year.

TETRAHEDRON LETTERS. Pergamon Press, Inc., Maxwell House, Fairview Park, Elmsford, NY 10523. (914) 592-7700. 1959 to present. Weekly. $1500.00 per year.

RESEARCH CENTERS AND INSTITUTES

CHEMICAL LABORATORIES. Harvard University, Oxford Street, Cambridge, MA 02138. (617) 495-4283.

CHEMICAL RESEARCH LABORATORY. Brown University, Providence, RI 02912. (401) 863-2256.

HIGH CARBON STEEL

See: STEEL AND STEEL MAKING

HIGH ENERGY PHYSICS

See: PARTICLE PHYSICS

HIGH EXPLOSIVES

See: EXPLOSIVES

HIGH FIDELITY

See: AUDIO ENGINEERING

HIGH TEMPERATURE MATERIALS

See: MATERIALS SCIENCE

HIGH VOLTAGE

See also: ELECTRIC POWER, ELECTRICAL ENGINEERING, HYDROELECTRIC POWER

ABSTRACT SERVICES AND INDEXES

APPLIED SCIENCE AND TECHNOLOGY INDEX. H.W. Wilson and Company, 950 University Avenue, Bronx, NY 10452. (800) 367-6670 or (212) 588-8400. Monthly. Inquire as to cost and availability.

CURRENT CONTENTS: ENGINEERING, TECHNOLOGY AND APPLIED SCIENCES. Institute for Scientific Information, 3501 Market Street, Philadelphia, PA 19104. (800) 523-1850 or (215) 386-0100. Weekly. $275.00 per year.

ELECTRIC POWER INDUSTRY ABSTRACTS. Edison Electric Institute, c/o Utility Date Institute, 2011 I Street, N.W., Suite 700, Washington, DC 20006. 1975 to present. Bimonthly. Inquire as to cost and availability.

ELECTRICAL AND ELECTRONICS ABSTRACTS. Institution of Electrical Engineers. Available from: Institute of Electrical and Electronics Engineers. IEEE Service Center, 445 Hoes Lane, Piscataway, NJ 08854. Monthly. $1250.00 per year.

ENGINEERING INDEX MONTHLY AND AUTHOR INDEX. Engineering Information Inc., 345 East 47th Street, New York, NY 10017. (212) 705-7600. Monthly. $1560.00 per year.

IEEE PUBLICATIONS BULLETIN. Institute of Electrical and Electronics Engineers. Institute of Electrical and Electronics Engineers. IEEE Service Center, 445 Hoes Lane, Piscataway, NJ 08854. Quarterly. Free.

PHYSICS ABSTRACTS. Institution of Electrical Engineers. Available from: IEEE Service Center, 445 Hoes Lane, Piscataway, NJ 08854. 1898 to present. Bimonthly. $1700.00 per year.

PHYSICS BRIEFS. Physik Verlag GmbH, Postfach 1260/1280, D-6940 Weinheim, West Germany. (212) 661-9404. 1920 to present. Twenty-six times per year. $1250.00 per year.

SCIENCE CITATION INDEX. Institute for Scientific Information, 3501 Market Street, Philadelphia, PA 19104. (800) 523-1850 or (215) 386-0100. Six times per year. $6200.00 per year.

ASSOCIATIONS AND PROFESSIONAL SOCIETIES

AMERICAN ELECTRONICS ASSOCIATION. P.O. Box 10045, 2670 Hanover Street, Palo Alto, CA 94303. (415) 857-9300.

EDISON ELECTRIC INSTITUTE. 1111 19th Street, N.W., Washington, DC 20036. (202) 828-7400.

INSTITUTE OF ELECTRICAL AND ELECTRONICS ENGINEERS. 345 East 47th Street, New York, NY 10017. (212) 705-7900.

DIRECTORIES AND BIOGRAPHICAL SOURCES

ELECTRICAL WORLD DIRECTORY OF ELECTRICAL UTILITIES. McGraw-Hill Book Company, 1221 Avenue of the Americas, New York, NY 10020. (212) 512-2000. Annual. $275.00.

INTERNATIONAL RESEARCH CENTERS DIRECTORY 1988-89. Darren L. Smith, editor. Gale Research Company, Book Tower, Detroit, MI 48226. (800) 521-0707. 4th edition. 1987. $360.00.

1987 DIRECTORY OF ENGINEERING SOCIETIES AND RELATED ORGANIZATIONS. Gordon Davis, editor. Hemisphere Publishing Corporation, 1010 Vermont Avenue, NW, Washington, DC 20005. (800) 526-0275. 12th edition. 1987. $100.00.

RESEARCH CENTERS DIRECTORY 1988. Gale Research Company, Book Tower, Detroit, MI 48226. (800) 521-0707. 12th edition. 1987. $365.00 for set.

SCIENTIFIC AND TECHNICAL ORGANIZATIONS AND AGENCIES DIRECTORY. Margaret Labash Young, editor. Gale Research Company, Book Tower, Detroit, MI 48226. (800) 521-0707. 2nd edition. 1987. $185.00.

WHO'S WHO IN ELECTRONICS. Harris Publishing Company, 2057-2 Aurora Road, Twinsburg, OH 44087. (216) 425-9143. Annual. $90.00.

WHO'S WHO IN ENGINEERING. Gordon Davis, editor. Hemisphere Publishing Corporation, 1010 Vermont Avenue, NW, Washington, DC 20005. (800) 526-0275. 6th edition. 1985. $200.00.

GENERAL WORKS

ELECTRIC POWER TRANSMISSION SYSTEM ENGINEERING: ANALYSIS AND DESIGN. Turan Gonen. John Wiley and Sons, Inc., 605 Third Avenue, New York, NY 10158. (800) 526-5368. 1987. $75.00.

ELECTRIC UTILITY SYSTEMS AND PRACTICES. H.M. Rustebakke. John Wiley and Sons, Inc., 605 Third Avenue, New York, NY 10158. (800) 526-5368. 1983. $45.95.

ELECTRICAL ENGINEERING. W.H. Roadstrum and Dan H. Wolaver. Harper and Row Publishers, Inc., 10 East 53rd Street, New York, NY 10022. (800) 242-7737. 1986. $37.50.

HIGH VOLTAGE ENGINEERING. E. Kuffel and W.S. Zaengl. Pergamon Press, Inc., Maxwell House, Fairview Park, Elmsford, NY 10523. (914) 592-7700. 1984. $19.50 in paper.

HIGH VOLTAGE: MEASUREMENT, TESTING AND DESIGN. T.J. Gallagher and A.J. Pearmain. John Wiley and Sons, Inc., 605 Third Avenue, New York, NY 10158. (800) 526-5368. 1983. $54.95.

IEEE STANDARD TECHNIQUES FOR HIGH VOLTAGE TESTING. Institute of Electrical and Electronics Engineers. IEEE Service Center, 445 Hoes Lane, Piscataway, NJ 08854. Sixth edition. 1978. $19.95.

HIGH VOLTAGE

HANDBOOKS AND MANUALS

HANDBOOK OF MODERN ELECTRONICS AND ELECTRICAL ENGINEERING. Charles Belove, editor. John Wiley and Sons, Inc., 605 Third Avenue, New York, NY 10158. (800) 526-5368. 1986. $88.95.

STANDARD HANDBOOK FOR ELECTRICAL ENGINEERS. D.G. Fink and H.W. Beaty. McGraw-Hill Book Company, 1221 Avenue of the Americas, New York, NY 10020. (212) 512-2000. 12th edition. 1987. $86.50.

ONLINE DATA BASES

COMPENDEX. Engineering Information, Inc., 345 East 47th Street, New York, NY 10017. (800) 221-1044 or (212) 705-7615. Engineering and technical literature, 1975 to present. Inquire as to online cost and availability.

INSPEC. INSPEC Marketing Department, Institution of Electrical Engineers. Available from IEEE Service Center, 445 Hoes Lane, Piscataway, NJ 08854. (201) 981-0060. Online version of Physics Abstracts. Inquire as to online cost and availability.

NTIS. National Technical Information Service, 5285 Port Royal Road, Springfield, VA 22161. (703) 487-4630. Broad coverage of government sponsored research reports, 1964 to present. Inquire as to online cost and availability.

SCISEARCH. Institute for Scientific Information, 3501 Market Street, Philadelphia, PA 19104. (800) 523-1850 or (215) 386-0100. Broad multidisciplinary title and author index to the international literature of science and technology, 1974 to present. Inquire as to online cost and availability.

WILSONLINE. H.W. Wilson and Company, 950 University Avenue, Bronx, NY 10452. (800) 367-6770 or (212) 588-8400. Makes available online versions of the H.W. Wilson indexes including Applied Science and Technology Index, Business Periodicals Index and Readers' Guide to Periodical Literature. Approximately 1980 to present. Inquire as to online cost and availability.

OTHER SOURCES

A GUIDE TO THE LITERATURE OF ELECTRICAL AND ELECTRONICS ENGINEERING. Susan B. Ardis. Libraries Unlimited Inc., P.O. Box 263, Littleton, CO 80160. (303) 770-1220. 1987. $37.50.

PERIODICALS

ELECTRIC LIGHT AND POWER. Technical Publishing Company, 875 Third Avenue, New York, NY 10022. (212) 605-9400. 1922 to present. Monthly. $38.00 per year.

ELECTRIC MACHINES AND POWER SYSTEMS. Hemisphere Publishing Corporation, 1010 Vermont Avenue, NW, Washington, DC 20005. (800) 526-0275. 1976 to present. Bimonthly. $134.50 per year.

ELECTRICAL WORLD. McGraw-Hill Book Company, 1221 Avenue of the Americas, New York, NY 10020. (212) 512-2000. 1874 to present. Monthly. $11.00 per year.

ELECTRONIC DESIGN. Hayden Publishing Company, 10 Mulholland Drive, Hasbrouck Heights, NJ 07604. (201) 288-7520. 1952 to present. Biweekly. $40.00 per year.

ELECTRONICS. McGraw-Hill Book Company, 1221 Avenue of the Americas, New York, NY 10020. (212) 512-2000. 1930 to present. Weekly. $32.00 per year.

IEEE POWER ENGINEERING REVIEW. Institute of Electrical and Electronics Engineers. IEEE Service Center, 445 Hoes Lane, Piscataway, NJ 08854. 1981 to present. Monthly. $75.00 per year.

IEEE TRANSACTIONS ON POWER DELIVERY. Institute of Electrical and Electronics Engineers. IEEE Service Center, 445 Hoes Lane, Piscataway, NJ 08854. 1986 to present. Quarterly. $100.00 per year.

INSTITUTE OF ELECTRICAL AND ELECTRONICS ENGINEERS PROCEEDINGS. Institute of Electrical and Electronics Engineers. IEEE Service Center, 445 Hoes Lane, Piscataway, NJ 08854. 1913 to present. Monthly. $140.00 per year.

RESEARCH CENTERS AND INSTITUTES

EDISON ELECTRIC INSTITUTE. 1111 19th Street, N.W., Washington, DC 20036. (202) 778-6778.

ELECTRIC POWER INSTITUTE. Texas A&M University, Department of Electrical Engineering, College Station, TX 77843.

ELECTRICAL ENGINEERING RESEARCH LABORATORIES. Purdue University, Electrical Engineering Building, West Lafayette, IN 47907. (317) 494-3536.

LABORATORY FOR ELECTROMAGNETIC AND ELECTRONIC SYSTEMS. 77 Massachusetts Avenue, Cambridge, MA 02139. (617) 253-4631.

HIGHWAY ENGINEERING

See also: ASPHALT, BRIDGES, CIVIL ENGINEERING, CONCRETE, CONSTRUCTION ENGINEERING, TRANSPORTATION ENGINEERING

ABSTRACT SERVICES AND INDEXES

APPLIED SCIENCE AND TECHNOLOGY INDEX. H.W. Wilson and Company, 950 University Avenue, Bronx, NY 10452. (800) 367-6670 or (212) 588-8400. Monthly. Inquire as to cost and availability.

CURRENT CONTENTS: ENGINEERING, TECHNOLOGY AND APPLIED SCIENCES. Institute for Scientific Information, 3501 Market Street, Philadelphia, PA 19104. (800) 523-1850 or (215) 386-0100. Weekly. $275.00 per year.

ENGINEERING INDEX MONTHLY AND AUTHOR INDEX. Engineering Information Inc., 345 East 47th Street, New York, NY 10017. (212) 705-7600. Monthly. $1560.00 per year.

INDEX TO SCIENTIFIC AND TECHNICAL PROCEEDINGS. Institute for Scientific Information, 3501 Market Street, Philadelphia, PA 19104. (800) 523-1850 or (215) 386-0100. 1978 to present. Monthly. $775.00 per year.

SCIENCE CITATION INDEX. Institute for Scientific Information, 3501 Market Street, Philadelphia, PA 19104. (800) 523-1850 or (215) 386-0100. Six times per year. $6200.00 per year.

ASSOCIATIONS AND PROFESSIONAL SOCIETIES

AMERICAN ASSOCIATION OF STATE HIGHWAY AND TRANSPORTATION OFFICIALS. 444 North Capitol Street, Washington, DC 20001. (202) 624-5800.

AMERICAN ROAD AND TRANSPORTATION BUILDERS ASSOCIATION. 525 School Street, S.W., Washington, DC 20024. (202) 488-2722.

AMERICAN SOCIETY FOR TESTING AND MATERIALS. 1916 Race Street, Philadelphia, PA 19103. (215) 299-5400.

AMERICAN SOCIETY OF CIVIL ENGINEERS. 345 East 47th Street, New York, NY 10017. (212) 705-7420.

INSTITUTE OF TRANSPORTATION ENGINEERS. 525 School Street, N.W., Suite 410, Washington, DC 20024. (202) 554-8050.

BIBLIOGRAPHIES

SCIENTIFIC AND TECHNICAL BOOKS AND SERIALS IN PRINT 1988; AN INDEX TO LITERATURE IN SCIENCE AND TECHNOLOGY. R.R. Bowker Company, 205 East 42nd Street, New York, NY 10017. (800) 521-8110. $175.00.

DIRECTORIES AND BIOGRAPHICAL SOURCES

AMERICAN ROAD AND TRANSPORTATION BUILDERS ASSOCIATION - OFFICIALS AND ENGINEERS DIRECTORY. American Road and Transportation Builders Association, 525 School Street, S.W., Washington, DC 20024. (202) 488-2722. Annual. $15.00.

INTERNATIONAL RESEARCH CENTERS DIRECTORY 1988-89. Darren L. Smith, editor. Gale Research Company, Book Tower, Detroit, MI 48226. (800) 521-0707. 4th edition. 1987. $360.00.

1987 DIRECTORY OF ENGINEERING SOCIETIES AND RELATED ORGANIZATIONS. Gordon Davis, editor. Hemisphere Publishing Corporation, 1010 Vermont Avenue, NW, Washington, DC 20005. (800) 526-0275. 12th edition. 1987. $100.00.

PUBLIC WORKS MANUAL. Public Works Journal Corporation, 200 South Broad Street, Ridgewood, NJ 07451. (201) 445-5800. Annual. $20.00.

RESEARCH CENTERS DIRECTORY 1988. Gale Research Company, Book Tower, Detroit, MI 48226. (800) 521-0707. 12th edition. 1987. $365.00 for set.

SCIENTIFIC AND TECHNICAL ORGANIZATIONS AND AGENCIES DIRECTORY. Margaret Labash Young, editor. Gale Research Company, Book Tower, Detroit, MI 48226. (800) 521-0707. 2nd edition. 1987. $185.00.

WHO'S WHO IN ENGINEERING. Gordon Davis, editor. Hemisphere Publishing Corporation, 1010 Vermont Avenue, NW, Washington, DC 20005. (800) 526-0275. 6th edition. 1985. $200.00.

ENCYCLOPEDIAS AND DICTIONARIES

CONSTRUCTION GLOSSARY: AN ENCYCLOPEDIC REFERENCE AND MANUAL. J.S. Stein. John Wiley and Sons, Inc., 605 Third Avenue, New York, NY 10158. (800) 526-5368. 1986. $35.95.

DICTIONARY OF CIVIL ENGINEERING. John S. Scott. John Wiley and Sons, Inc., 605 Third Avenue, New York, NY 10158. (800) 526-5368. Third edition. 1981. $26.95.

MCGRAW-HILL ENCYCLOPEDIA OF SCIENCE AND TECHNOLOGY. McGraw-Hill Book Company, 1221 Avenue of the Americas, New York, NY 10020. (212) 512-2000. 6th edition. 1987. $1600.00.

THESAURUS OF SCIENTIFIC, TECHNICAL, AND ENGINEERING TERMS. Hemisphere Publishing Corporation, 1010 Vermont Avenue, NW, Washington, DC 20005. (800) 526-0275. 1988. $125.00.

GENERAL WORKS

DESIGN OF REINFORCED CONCRETE. Samuel E. French. Prentice-Hall Publishing, Inc., Englewood Cliffs, NJ 07632. (800) 562-0245. 1987. $25.00.

ELEMENTARY THEORY OF STRUCTURES. Yu H. Yuan. Prentice-Hall Publishing, Inc., Englewood Cliffs, NJ 07632. (800) 562-0245. Third edition. 1988. $42.95.

FOUNDATION DESIGN AND CONSTRUCTION. M.J. Tomlinson. John Wiley and Sons, Inc., 605 Third Avenue, New York, NY 10158. (800) 526-5368. Fifth edition. 1986. $45.95.

HIGHWAY ENGINEERING. Paul H. Paquette and Randor J. Paquette. John Wiley and Sons, Inc., 605 Third Avenue, New York, NY 10158. (800) 526-5368. Fifth edition. 1986. $45.00.

HIGHWAYS: HIGHWAY ENGINEERING. C.A. O'Flaherty. Edward Arnold Publishers, Limited, 300 North Charles Street, Baltimore, MD 21201. (301) 539-1529. Volume 2. Third edition. 1987. $29.95 in paper.

MATERIALS FOR CIVIL AND HIGHWAY ENGINEERS. K.N. Derucher and G.P. Korgiatis. Prentice-Hall Publishing, Inc., Englewood Cliffs, NJ 07632. (800) 562-0245. Second edition. 1988. $42.50.

HANDBOOKS AND MANUALS

CIVIL ENGINEERING CALCULATIONS REFERENCE GUIDE. Tyler G. Hicks, editor. McGraw-Hill Book Company, 1221 Avenue of the Americas, New York, NY 10020. (212) 512-2000. 1987. $29.50.

CIVIL ENGINEERING PRACTICE. Paul N. Cheremisinoff and others, editors. Technomic Publishing Company, Inc., 851 Holland Avenue, Box 3535, Lancaster, PA 17604. (800) 233-9936. Five volumes. 1987-1988. $750.00 for set.

HANDBOOK OF CONCRETE ENGINEERING. Mark Fintel. Van Nostrand Reinhold Company, Inc., 135 West 50th Street, New York, NY 10020. (800) 543-2681. 1985. $89.95.

STANDARD HANDBOOK FOR CIVIL ENGINEERS. F. S. Merritt, editor. McGraw-Hill Book Company, 1221 Avenue of the Americas, New York, NY 10020. (212) 512-2000. Third edition. 1983. $89.50.

SURVEYING READY-REFERENCE MANUAL. Guy O. Stenstrom. McGraw-Hill Book Company, 1221 Avenue of the Americas, New York, NY 10020. (212) 512-2000. 1987. $26.50.

ONLINE DATA BASES

COMPENDEX. Engineering Information, Inc., 345 East 47th Street, New York, NY 10017. (800) 221-1044 or (212) 705-7615. Engineering and technical literature, 1975 to present. Inquire as to online cost and availability.

DISSERTATION ABSTRACTS ONLINE. University Microfilms International, 300 North Zeeb Road, Ann Arbor, MI 48106. (800) 521-0600 or (313) 761-4700. Scope includes virtually all doctoral dissertations accepted at accredited American institutions from 1861 to present in over 250 subject areas. Inquire as to online cost and availability.

GEOREF. Online version of the BIBLIOGRAPHY AND INDEX OF GEOLOGY. American Geological Institute, 4220 King Street, Alexandria, VA 22302. (703) 379-2480. 1969 to present. Inquire as to online cost and availability.

INSPEC. INSPEC Marketing Department, Institution of Electrical Engineers. Available from IEEE Service Center, 445 Hoes Lane, Piscataway, NJ 08854. (201) 981-0060. Online version of Physics Abstracts. Inquire as to online cost and availability.

NTIS. National Technical Information Service, 5285 Port Royal Road, Springfield, VA 22161. (703) 487-4630. Broad coverage of government sponsored research reports, 1964 to present. Inquire as to online cost and availability.

SCISEARCH. Institute for Scientific Information, 3501 Market Street, Philadelphia, PA 19104. (800) 523-1850 or (215) 386-0100. Broad multidisciplinary title and author index to the international literature of science and technology, 1974 to present. Inquire as to online cost and availability.

TRIS. National Academy of Sciences. Transportation Research, 2101 Constitution Avenue, N.W., Washington, DC 20418. (202) 334-2000. Covers highway and transportation research. 1968 to present. Inquire as to cost and availability.

HIGHWAY ENGINEERING

WILSONLINE. H.W. Wilson and Company, 950 University Avenue, Bronx, NY 10452. (800) 367-6770 or (212) 588-8400. Makes available online versions of the H.W. Wilson indexes including Applied Science and Technology Index, Business Periodicals Index and Readers' Guide to Periodical Literature. Approximately 1980 to present. Inquire as to online cost and availability.

OTHER SOURCES

WHAT EVERY ENGINEER SHOULD KNOW ABOUT ENGINEERING SOURCES. Marcel Dekker Inc., 270 Madison Avenue, New York, NY 10016. (800) 228-1160. 1984. $24.95.

PERIODICALS

AMERICAN CONCRETE INSTITUTE JOURNAL. P.O. Box 19150, Redford Station, Detroit, MI 48219. (513) 532-2600. 1929 to present. Bimonthly. $69.00 per year.

CIVIL ENGINEERING. American Society of Civil Engineers, 345 East 47th Street, New York, NY 10017. (212) 705-7420. 1930 to present. Monthly. $48.00 per year.

CONCRETE. Harcourt, Brace Jovanovich, Inc., 7500 Old Oak Boulevard, Cleveland, OH 44130. 1937 to present. Monthly. $25.00 per year.

HIGHWAY AND HEAVY CONSTRUCTION. Technical Publishing Company, 875 Third Avenue, New York, NY 10022. (212) 605-9400. 1892 to present. Monthly. $35.00 per year.

JOURNAL OF CONSTRUCTION ENGINEERING AND MANAGEMENT. American Society of Civil Engineers, 345 East 47th Street, New York, NY 10017. (212) 705-7420. 1956 to present. Quarterly. $48.00 per year.

JOURNAL OF STRUCTURAL MECHANICS. Marcel Dekker Inc., 270 Madison Avenue, New York, NY 10016. (800) 228-1160. 1972 to present. Quarterly. $75.00 per year.

JOURNAL OF SURVEYING ENGINEERING. American Society of Civil Engineers, 345 East 47th Street, New York, NY 10017. (212) 705-7420. 1956 to present. Three times per year. $35.00 per year.

RESEARCH CENTERS AND INSTITUTES

HIGHWAY RESEARCH CENTER. Auburn University, Auburn, AL 36849. (205) 826-5250.

JOINT HIGHWAY RESEARCH PROJECT. Purdue University, Civil Engineering Building, West Lafayette, IN 47907. (317) 494-2159.

TRANSPORTATION RESEARCH BOARD. 2101 Constitution Avenue, N.W., Washington, DC 20418. (202) 334-2934.

TURNER-FAIRBANK HIGHWAY RESEARCH CENTER. U.S. Department of Transportation, Federal Highway Administration, 6300 Georgetown Pike, McLean, VA 22101.

HOLOGRAPHY

See also: LASERS, OPTICAL COMMUNICATIONS, OPTICS, OPTOELECTRONICS, PHYSICS

ABSTRACT SERVICES AND INDEXES

APPLIED SCIENCE AND TECHNOLOGY INDEX. H.W. Wilson and Company, 950 University Avenue, Bronx, NY 10452. (800) 367-6670 or (212) 588-8400. Monthly. Inquire as to cost and availability.

CHEMICAL ABSTRACTS. American Chemical Society, Chemical Abstracts Service, Box 3012, Columbus, OH 43210. (614) 421-3600. 1907 to present. Weekly. $9500.00 per year.

CURRENT CONTENTS: ENGINEERING, TECHNOLOGY AND APPLIED SCIENCES. Institute for Scientific Information, 3501 Market Street, Philadelphia, PA 19104. (800) 523-1850 or (215) 386-0100. Weekly. $275.00 per year.

ELECTRICAL AND ELECTRONICS ABSTRACTS. Institution of Electrical Engineers. Available from: Institute of Electrical and Electronics Engineers. IEEE Service Center, 445 Hoes Lane, Piscataway, NJ 08854. Monthly. $1250.00 per year.

ENGINEERING INDEX MONTHLY AND AUTHOR INDEX. Engineering Information Inc., 345 East 47th Street, New York, NY 10017. (212) 705-7600. Monthly. $1560.00 per year.

GENERAL SCIENCE INDEX. H.W. Wilson and Company, 950 University Avenue, Bronx, NY 10452. (800) 367-6670 or (212) 588-8400. 1978 to present. Monthly. Inquire as to cost and availability.

GOVERNMENT REPORTS ANNOUNCEMENT AND INDEX. National Technical Information Service, 5285 Port Royal Road, Springfield, VA 22161. (703) 487-4929. Summaries of United States government sponsored research reports. 1946 to present. Biweekly. Inquire as to cost and availability.

PHYSICS ABSTRACTS. Institution of Electrical Engineers. Available from: IEEE Service Center, 445 Hoes Lane, Piscataway, NJ 08854. 1898 to present. Bimonthly. $1700.00 per year.

PHYSICS BRIEFS. Physik Verlag GmbH, Postfach 1260/1280, D-6940 Weinheim, West Germany. (212) 661-9404. 1920 to present. Twenty-six times per year. $1250.00 per year.

SCIENCE CITATION INDEX. Institute for Scientific Information, 3501 Market Street, Philadelphia, PA 19104. (800) 523-1850 or (215) 386-0100. Six times per year. $6200.00 per year.

ANNUAL REVIEWS AND YEARBOOKS

LASER APPLICATIONS. Academic Press, Inc., 6277 Sea Harbor Drive, Orlando, FL 32821. (800) 321-5068. 1971 to present. Irregular. Price varies, inquire.

ASSOCIATIONS AND PROFESSIONAL SOCIETIES

AMERICAN INSTITUTE OF PHYSICS. 335 East 45th Street, New York, NY 10017. (212) 661-9404.

INSTITUTION OF ELECTRICAL AND ELECTRONICS ENGINEERS (IEEE). 345 East 47th Street, New York, NY 10017. (212) 705-7900.

LASER INSTITUTE OF AMERICA. 5151 Monroe Street, Suite 103, Toledo, OH 43623. (419) 882-8706.

OPTICAL SOCIETY OF AMERICA. 1816 Jefferson Place, N.W., Washington, DC 20036. (202) 223-8130.

SOCIETY OF PHOTO-OPTICAL INSTRUMENTATION ENGINEERS - THE INTERNATIONAL SOCIETY OF OPTICAL ENGINEERING. P.O. Box 10, 1022 19th Street, Bellingham, WA 98227. (206) 676-3290.

DIRECTORIES AND BIOGRAPHICAL SOURCES

AMERICAN MEN AND WOMEN OF SCIENCE. R.R. Bowker, Inc., Order Department, 245 West 17th Street, New York, NY 10011. (800) 521-8110. Eight volumes. 1986. $595.00 for set.

DIRECTORY OF CONSULTANTS IN LASERS AND PHYSICS. J. Dick Publishing, Research Publications, 801 Green Bay Road, Lake Bluff, IL 60044. (312) 234-1220. Biennial. $85.00.

INTERNATIONAL RESEARCH CENTERS DIRECTORY 1988-89. Darren L. Smith, editor. Gale Research Company, Book Tower, Detroit, MI 48226. (800) 521-0707. 4th edition. 1987. $360.00.

LASER FOCUS/ELECTRO-OPTICS BUYERS' GUIDE. Advanced Technology Publications, Inc., PennWell Publishing, 119 Russell Street, Littleton, MA 01460. (617) 486-9501. Annual, January. $60.00.

1987 DIRECTORY OF ENGINEERING SOCIETIES AND RELATED ORGANIZATIONS. Gordon Davis, editor. Hemisphere Publishing Corporation, 1010 Vermont Avenue, NW, Washington, DC 20005. (800) 526-0275. 12th edition. 1987. $100.00.

RESEARCH CENTERS DIRECTORY 1988. Gale Research Company, Book Tower, Detroit, MI 48226. (800) 521-0707. 12th edition. 1987. $365.00 for set.

SCIENTIFIC AND TECHNICAL ORGANIZATIONS AND AGENCIES DIRECTORY. Margaret Labash Young, editor. Gale Research Company, Book Tower, Detroit, MI 48226. (800) 521-0707. 2nd edition. 1987. $185.00.

WHO'S WHO IN ENGINEERING. Gordon Davis, editor. Hemisphere Publishing Corporation, 1010 Vermont Avenue, NW, Washington, DC 20005. (800) 526-0275. 6th edition. 1985. $200.00.

GENERAL WORKS

ANALYTICAL APPLICATIONS OF LASERS. Edward H. Piepmeir, editor. John Wiley and Sons, Inc., 605 Third Avenue, New York, NY 10158. (800) 526-5368. 1986. $89.95.

LASERS AND HOLOGRAPHY: AN INTRODUCTION TO COHERENT OPTICS. Winston E. Kock. Dover Publications, Inc., 31 East Second Street, Mineola, NY 11501. (516) 294-7000. 1981. $3.50 in paper.

LASERS: INVENTIONS TO APPLICATION. Jesse H. Ausubel and H.D. Langford, editors. National Academy Press, 2101 Constitution Avenue, Washington, DC 20418. (202) 334-3313. 1987. $14.95 in paper.

LASERS: THEORY AND APPLICATIONS. K. Thyagarajan and A.K. Ghatak. Plenum Publishing Corporation, 233 Spring Street, New York, NY 10013. (800) 221-9369. 1981. $59.50.

HANDBOOKS AND MANUALS

HANDBOOK OF OPTICAL HOLOGRAPHY. H.J. Caulfield, editor. Academic Press, Inc., 6277 Sea Harbor Drive, Orlando, FL 32821. (800) 321-5068. 1979. $72.50.

SAFETY WITH LASERS AND OTHER OPTICAL SOURCES: A COMPREHENSIVE HANDBOOK. D. Sliney and M. Wolbarsht. Plenum Publishing Corporation, 233 Spring Street, New York, NY 10013. (800) 221-9369. 1980. $65.00.

ONLINE DATA BASES

CA SEARCH. Chemical Abstracts Service, P.O. Box 3012, Columbus, OH 43120. (800) 848-6538 or (614) 421-3600. Comprehensive guide to chemical literature, 1972 to present. Inquire as to online cost and availability.

COMPENDEX. Engineering Information, Inc., 345 East 47th Street, New York, NY 10017. (800) 221-1044 or (212) 705-7615. Engineering and technical literature, 1975 to present. Inquire as to online cost and availability.

DISSERTATION ABSTRACTS ONLINE. University Microfilms International, 300 North Zeeb Road, Ann Arbor, MI 48106. (800) 521-0600 or (313) 761-4700. Scope includes virtually all doctoral dissertations accepted at accredited American institutions from 1861 to present in over 250 subject areas. Inquire as to online cost and availability.

INSPEC. INSPEC Marketing Department, Institution of Electrical Engineers. Available from IEEE Service Center, 445 Hoes Lane, Piscataway, NJ 08854. (201) 981-0060. Online version of Physics Abstracts. Inquire as to online cost and availability.

NTIS. National Technical Information Service, 5285 Port Royal Road, Springfield, VA 22161. (703) 487-4630. Broad coverage of government sponsored research reports, 1964 to present. Inquire as to online cost and availability.

SCISEARCH. Institute for Scientific Information, 3501 Market Street, Philadelphia, PA 19104. (800) 523-1850 or (215) 386-0100. Broad multidisciplinary title and author index to the international literature of science and technology, 1974 to present. Inquire as to online cost and availability.

WILSONLINE. H.W. Wilson and Company, 950 University Avenue, Bronx, NY 10452. (800) 367-6770 or (212) 588-8400. Makes available online versions of the H.W. Wilson indexes including Applied Science and Technology Index, Business Periodicals Index and Readers' Guide to Periodical Literature. Approximately 1980 to present. Inquire as to online cost and availability.

PERIODICALS

APPLIED OPTICS. Optical Society of America. 1816 Jefferson Place, N.W., Washington, DC 20036. (202) 223-8130. Order from: American Institute of Physics, 335 East 45th Street, New York, NY 10017. (212) 661-9404. 1962 to present. Semi-monthly. $330.00 per year.

JOURNAL OF APPLIED PHYSICS. American Institute of Physics, 335 East 45th Street, New York, NY 10017. (212) 661-9494. 1931 to present. Semi-monthly. $495.00 per year.

JOURNAL OF LIGHTWAVE TECHNOLOGY. Institute of Electrical and Electronics Engineers. IEEE Service Center, 445 Hoes Lane, Piscataway, NJ 08854. 1983 to present. Monthly. $145.00 per year.

LASER FOCUS. PennWell Publishing Company, 119 Russell Street, Littleton, MA 01460. (617) 486-9501. 1965 to present. Monthly. $55.00 per year.

OPTICAL ENGINEERING. Society of Photo-Optical Instrumentation Engineers (SPIE), P.O. Box 10, 1022 19th Street, Bellingham, WA 98227. (206) 676-3290. 1962 to present. Monthly. $95.00 per year.

OPTICAL SOCIETY OF AMERICA, JOURNAL, PART B: OPTICAL PHYSICS. Optical Society of America. 1816 Jefferson Place, N.W., Washington, DC 20036. (202) 223-8130. Order from: American Institute of Physics, 335 East 45th Street, New York, NY 10017. (212) 661-9404. 1917 to present. Monthly. $180.00 per year.

OPTICS AND LASERS IN ENGINEERING. Elsevier Science Publishing Company, Inc., 52 Vanderbilt Avenue, New York, NY 10017. (212) 370-5520. 1980 to present. Quarterly. $95.00 per year.

OPTICS COMMUNICATIONS. Elsevier Science Publishing Company, Inc., 52 Vanderbilt Avenue, New York, NY 10017. (212) 370-5520. 1969 to present. 24 times per year. $425.00 per year.

OPTICS LETTERS. Optical Society of America. Order from: American Institute of Physics, 335 East 45th Street, New York, NY 10017. (212) 661-9404. 1977 to present. Monthly. $150.00 per year.

SOCIETY OF PHOTO-OPTICAL INSTRUMENTATION ENGINEERS (SPIE) PROCEEDINGS. Society of Photo-Optical Instrumentation Engineers (SPIE), P.O. Box 10, 1022 19th Street, Bellingham, WA 98227. (206) 676-3290. 1963 to present. Approximately 50 numbers per year. Approximately $40.00 per number.

RESEARCH CENTERS AND INSTITUTES

CENTER FOR OPTICS, LASERS AND HOLOGRAPHY. 100 Glen Cove Avenue, Glen Cove, NY 11542. (516) 686-7863.

CENTER FOR RESEARCH IN ELECTRO-OPTICS AND LASERS. University of Central Florida, Department of Physics, P.O. Box 25000 CEBA 419, Orlando, FL 32816-0001. (305) 275-2325.

LABORATORY FOR LASER ENERGETICS. University of Rochester, 250 East River Road, Rochester, NY 14623-1299. (716) 275-5101.

LASER RESEARCH CENTER. Massachusetts Institute of Technology, 77 Massachusetts Avenue, Cambridge, MA 02139. (617) 253-7700.

HOMING

See: NAVIGATION

HOVERCRAFT

See: AERONAUTICS

HULLS

See: NAVAL ARCHITECTURE

HUMAN ENGINEERING

See: ERGONOMICS

HURRICANES

See also: METEOROLOGY

ABSTRACT SERVICES AND INDEXES

GENERAL SCIENCE INDEX. H.W. Wilson Company, 950 University Avenue, Bronx, NY 10452. (800) 367-6770 or (212) 588-8400. Inquire as to cost and availability.

METEOROLOGICAL AND GEOASTROPHYSICAL ABSTRACTS. American Meteorological Society, 45 Beacon Street, Boston, MA 02108. (617) 227-2425. 1950 to present. Monthly. $450.00 per year.

SCIENCE CITATION INDEX. Institute for Scientific Information, 3501 Market Street, Philadelphia, PA 19104. (800) 523-1850 or (215) 386-0100. Inquire as to cost and availability.

ASSOCIATIONS AND PROFESSIONAL SOCIETIES

AMERICAN ASSOCIATION STATE CLIMATOLOGISTS. c/o Professor John Griffiths, Meteorology Department, O and M Building, Texas A and M University, College Station, TX 77843. (409) 845-7320.

AMERICAN METEOROLOGICAL SOCIETY. 45 Beacon Street, Boston, MA 02108. (617) 227-2425.

INTERNATIONAL ASSOCIATION OF METEOROLOGY AND ATMOSPHERIC PHYSICS. UCAR, Post Office Box 3000, Boulder, CO 80307.

NATIONAL ENVIRONMENTAL SATELLITE DATA, AND INFORMATION SERVICE. 3300 Whitehaven Street, NW, Washington, DC 20235. (202) 634-7318.

NATIONAL WEATHER ASSOCIATION. 4400 Stamp Road, Room 404, Temple Hills, MD 20748. (301) 899-3784.

UNIVERSITY CORPORATION FOR ATMOSPHERIC RESEARCH. Box 3000, 1850 Table Mesa Drive, Boulder, CO 80307. (303) 497-1000.

WEATHER MODIFICATION ASSOCIATION. Post Office Box 8116, Fresno, CA 93747. (209) 291-8466.

DIRECTORIES AND BIOGRAPHICAL SOURCES

AMERICAN MEN AND WOMEN OF SCIENCE. Physical and Biological Sciences. Sixteenth edition. R.R. Bowker Company, 205 East 42nd Street, New York, NY 10017. (800) 521-8110 or (212) 916-1600. 1986. $595.00.

GOVERNMENT RESEARCH DIRECTORY. Gale Research Company, Book Tower, Detroit, MI 48226. (800) 521-0707. Fourth edition. 1987. $350.00.

INTERDOC: DIRECTORY OF PUBLISHED PROCEEDINGS, SERIES. SEMT-Science/Engineering/Medicine/Technology. Interdoc Corporation, 173 Halstead Avenue, Box 326, Harrison, NY 10528. (014) 835-3506. Ten times per year. $325.00 per year.

METEOROLOGICAL SERVICES OF THE WORLD. World Meteorological Organization. Available from: American Meteorological Society, 45 Beacon Street, Boston, MA 02108. (617) 227-2425. Annual. $35.00.

NATIONAL WEATHER SERVICE OFFICES AND STATIONS. National Oceanic and Atmospheric Administration, Department of Commerce, Silver Spring, MD 20910. (301) 427-7698. Annual. Free.

RESEARCH CENTERS DIRECTORY. Gale Research Company, Book Tower, Detroit, MI 48226. (800) 521-0707. Twelfth edition. 1987. $365.00 for set.

ENCYCLOPEDIAS AND DICTIONARIES

ENCYCLOPEDIA OF CLIMATOLOGY. John E. Oliver and Rhodes W. Fairbridge, editors. Van Nostrand Reinhold, Incorporated, 115 Fifth Avenue, New York, NY 10003. (800) 543-2681. 1987. $89.95.

ENCYCLOPEDIA OF PHYSICAL SCIENCE AND TECHNOLOGY. Academic Press, Incorporated, Orlando, FL 32887. (800) 321-5068 or (305) 345-2734. Fifteen volumes, 1986.

GENERAL WORKS

HURRICANES, STORMS AND TORNADOS: GEOGRAPHIC CHARACTERISTICS AND GEOLOGICAL ACTIVITY. B.B. Battacharya. International Publishing Service, Post Office Box 230, Accord, MA 02018. (617) 749-3628. 1983. $26.50.

ONLINE DATA BASES

DISSERTATION ABSTRACTS ONLINE. University Microfilms International, 300 North Zeeb Road, Ann Arbor, MI 48106. (800) 521-0600 or (313) 761-4700. Scope includes virtually all doctoral dissertations accepted at accredited American institutions from 1861 to present in 252 subject areas. Inquire as to cost and availability.

METEOROLOGICAL AND GEOASTROPHYSICAL ABSTRACTS. American Meteorological Society, 45 Beacon Street, Boston, MA 02108. (617) 227-2425. 1950 to present. Monthly. $450.00 per year.

NTIS. National Technical Information Service, 5285 Port Royal Road, Springfield, VA 22161. (703) 487-4630. Broad coverage of government-sponsored research reports, 1964 to present. Inquire as to cost and availability.

SCISEARCH. Institute for Scientific Information, 3501 Market Street, Philadelphia, PA 19104. (800) 523-1850 or (215) 386-0100. Broad multidisciplinary title and author index to the international literature of science and technology, 1974 to present. Inquire as to cost and availability.

PERIODICALS

AGRICULTURAL AND FOREST METEOROLOGY. Elsevier Science Publishing Company, Incorporated, 52 Vanderbilt Avenue, New York, NY 10017. (212) 370-5520. 1964 to present. Monthly. $260.00 per year.

AMERICAN METEOROLOGICAL SOCIETY BULLETIN. American Meteorological Society, 45 Beacon Street, Boston, MA 02108. (617) 227-2425. 1920 to present. Monthly. $60.00 per year.

CLIMATIC CHANGE: AN INTERDISCIPLINARY, INTERNATIONAL JOURNAL DEVOTED TO THE DESCRIPTION, CAUSES AND IMPLICATIONS OF CLIMATIC CHANGE. D. Reidel Publishing Company, 190 Old Derby Street, Hingham, MA 02043. (617) 871-6600. 1977 to present. Six times per year. $125.00 per year.

JOURNAL OF THE ATMOSPHERIC SCIENCES. American Meteorological Society, 45 Beacon Street, Boston, MA 02108. (617) 227-2425. 1944 to present. Semimonthly. $220.00 per year.

MONTHLY WEATHER REVIEW. American Meteorological Society, 45 Beacon Street, Boston, MA 02108. (617) 227-2425. 1872 to present. Monthly. $120.00 per year.

NATIONAL WEATHER DIGEST. National Weather Association, 4400 Stamp Road, Room 404, Temple Hills, MD 20748. (301) 899-3784. 1976 to present. Quarterly. $20.00 per year.

WEATHER. Royal Meteorological Society, James Glaisher House, Grenville Place, Bracknell Berkshire, RG12 1BX, England. 1946 to present. $30.00 per year.

WEATHERWISE. Heldref Publications, 4000 Albemarle Street, NW, Washington, DC 20016. (202) 362-6445. 1948 to present. Bimonthly. $20.00 per year.

RESEARCH CENTERS AND INSTITUTES

COOPERATIVE INSTITUTE FOR MESOCALE METEOROLOGICAL STUDIES. University of Oklahoma, 401 East Boyd, Norman, OK 73019. (405) 325-3041.

NATIONAL CENTER FOR ATMOSPHERIC RESEARCH. Box 3000, Boulder, CO 80307. (303) 497-1000.

NATIONAL SEVERE STORMS LABORATORY. 1313 Halley Circle, Norman, OK 73069. (405) 360-3620.

UNIVERSITY OF FLORIDA. Coastal and Oceanographic Engineering Laboratory, 336 Weil Hall, Gainesville, FL 32607. (904) 392-1436.

UNIVERSITY OF MIAMI. Remote Sensing Laboratory, Post Office Box 248003, Coral Gables, FL 33124. (305) 284-2881.

HYDRAULIC ENGINEERING

See also: FLUID DYNAMICS, HYDRAULICS, HYDROLOGY, HYDRODYNAMICS

ABSTRACT SERVICES AND INDEXES

APPLIED MECHANICS REVIEWS. American Society of Mechanical Engineers, 345 East 47th Street, New York, NY 10017. (212) 705-7703. 1948 to present. Monthly. $360.00 per year.

APPLIED SCIENCE AND TECHNOLOGY INDEX. H.W. Wilson Company, 950 University Avenue, Bronx, NY 10452. (800) 367-6670 or (212) 588-8400. Inquire as to cost and availability.

CHEMICAL ABSTRACTS. Chemical Abstracts Service, 2540 Olentangy Road, Post Office Box 3012, Columbus, OH 43210. (800) 848-6538 or (614) 421-3600. Weekly. $9200.00 per year.

CIVIL ENGINEERING HYDRAULICS ABSTRACTS. BHRA Fluid Engineering, Cranfield, Bedford MK43 0AJ, England. Distributed by Learned Information Incorporated, 143 Old Marlton Pike, Medford, NJ 08055. 1968 to present. Monthly. $225.00 per year.

CURRENT CONTENTS: ENGINEERING, TECHNOLOGY AND APPLIED SCIENCES. Institute for Scientific Information, 3501 Market Street, Philadelphia, PA 19104. (800) 523-1850 or (215) 386-0100. 1970 to present. Weekly. $272.00 per year.

ENGINEERING INDEX MONTHLY AND AUTHOR INDEX. Engineering Information, Incorporated, 345 East 47th Street, New York, NY 10017. (800) 221-1044 or (212) 705-7600. Monthly, with annual cumulation. $1560.00 per year.

FLUID POWER ABSTRACTS. BHRA Fluid Engineering, Cranfield, Bedford MK43 0AJ, England. Distributed by Learned Information Incorporated, 143 Old Marlton Pike, Meford, NJ 08055. 1970 to present. Bimonthly. $196.00 per year.

INDUSTRIAL AERODYNAMICS ABSTRACTS. BHRA Fluid Engineering, Cranfield, Bedford MK43 0AJ, England. Distributed by Learned Information Incorporated, 143 Old Marlton Pike, Medford, NJ 08055. 1970 to present. Monthly. $196.00 per year.

INTERNATIONAL AEROSPACE ABSTRACTS. AIAA/TIS, 1633 Broadway, New York, NY 10019. (212) 581-4300. Semimonthly. $700.00 per year.

ISMEC BULLETIN. (Information Service in Mechanical Engineering). Cambridge Scientific Abstracts, 5161 River Road, Bethesda, MD 20816. (301) 951-1400. 1973 to present. Monthly. $475.00 per year.

SCIENCE CITATION INDEX. Institute for Scientific Information, 3501 Market Street, Philadelphia, PA 19104. (800) 523-1850 or (215) 386-0100. Inquire as to cost and availability.

STAR. (Scientific and Technical Aerospace Reports. United States National Aeronautics and Space Administration, Scientific and Technical Information Facility, Box 8757, Baltimore-Washington International Airport, MD 21240. (202) 755-2210. Semimonthly, with semiannual and annual indexes. $85.00 per year.

ANNUAL REVIEWS AND YEARBOOKS

ADVANCE IN HYDROSCIENCE. Ven Te Chow, editor. Academic Press, Incorporated, Orlando, FL 32887. (800) 321-5068 or (305) 345-2734. 1964-1981. Inquire, price varies.

DEVELOPMENTS IN HYDRAULIC ENGINEERING. P. Novak, editor. Elsevier Science Publishing Company, Incorporated, 52 Vanderbilt Avenue, New York, NY 10017. (212) 370-5520. 1983-85. $66.00 per volume.

ASSOCIATIONS AND PROFESSIONAL SOCIETIES

AMERICAN INSTITUTE OF AERONAUTICS AND ASTRONAUTICS. 1633 Broadway, New York, NY 10019. (212) 581-4300.

AMERICAN SOCIETY OF CIVIL ENGINEERS. 345 East 47th Street, New York, NY 10017-2398. (212) 705-7520.

AMERICAN SOCIETY OF MECHANICAL ENGINEERS. 345 East 47th Street, New York, NY 10017-2398. (212) 705-7722.

FLUID CONTROLS INSTITUTE. Post Office Box 9036, Morristown, NJ 07960. (201) 829-0990.

FLUID POWER SOCIETY. 3333 North Mayfair Road, Milwaukee, WI 53222. (414) 778-3377.

HYDRAULIC INSTITUTE. 712 Lakewood Center North, 14600 Detroit Avenue, Cleveland, OH 44107. (216) 226-7700.

INTERNATIONAL ASSOCIATION FOR HYDRAULIC RESEARCH. 185 Rotterdamseweg, Box 177, Delft, The Netherlands.

NATIONAL CONFERENCE ON FLUID POWER. IIT Research Institute, Ten West 35th Street, Chicago, IL 60616. (312) 567-4414.

NATIONAL FLUID POWER ASSOCIATION. 3333 North Mayfair Road, Suite 311, Milwaukee, WI 53222. (414) 778-3344.

SOCIETY OF AUTOMOTIVE ENGINEERS. SAE4, Incorporated, 400 Commonwealth Drive, Warrendale, PA 15096. (412) 776-4841.

DIRECTORIES AND BIOGRAPHICAL SOURCES

AAS DIRECTORY. American Astronautical Society, 6212-B Old Keene Mill Court, Springfield, VA 22152. (703) 866-0020. Annual. $35.00 per year.

DIRECTORY OF HYDRAULIC RESEARCH INSTITUTES AND LABORATORIES. International Association for Hydraulic Research, 185 Rotterdamseweg, Box 177, Delft, The Netherlands. 1986. 60 Dutch florins.

FLUID POWER HANDBOOK AND DIRECTORY. Penton/IPC, Incorporated, 1111 Chester Avenue, Cleveland, OH 44114. (216) 606-7000. Biennial. $45.00.

FLUID POWER SOCIETY MEMBERSHIP DIRECTORY. Fluid Power Society, 3333 North Mayfair Road, Milwaukee, WI 53222. (414) 778-3377. Annual. $50.00.

NATIONAL FLUID POWER ASSOCIATION MEMBERSHIP DIRECTORY. 3333 North Mayfair Road, Suite 311, Milwaukee, WI 53222. (414) 778-3344. Annual. $150.00.

RESEARCH CENTERS DIRECTORY. Gale Research Company, Book Tower, Detroit, MI 48226. (800) 521-0707. Twelfth edition. 1987. $365.00 for set.

WHO'S WHO IN ENGINEERING. Engineers Joint Council, 345 East 47th Street, New York, NY 10017. (212) 705-7010. 1985. $200.00.

ENCYCLOPEDIAS AND DICTIONARIES

ENCYCLOPEDIA OF FLUID MECHANICS. Nicholas P. Cheremisinnoff, editor. Gulf Publishing Company, Book Division, Post Office Box 2608, Houston, TX 77001. (713) 520-4444. Six volumes. 1987. $1000.00 for set.

ENCYCLOPEDIA OF PHYSICAL SCIENCE AND TECHNOLOGY. Academic Press, Incorporated, Orlando, FL 32887. (800) 321-5068 or (305) 345-2734. Fifteen volumes, 1986.

MCGRAW-HILL DICTIONARY OF ENGINEERING. Sybil P. Parker, editor. McGraw-Hill Book Company, 1221 Avenue of the Americas, New York, NY 10020. (212) 512-2000. 1984. $39.95.

MCGRAW-HILL ENCYCLOPEDIA OF SCIENCE AND TECHNOLOGY. McGraw-Hill Book, Incorporated, 1221 Avenue of the Americas, New York, NY 10020. (212) 997-3675. Fifth edition, 15 volumes. $1100.00.

GENERAL WORKS

APPLIED HYDRAULICS FOR TECHNOLOGY. John D. Kanen. Holt, Rinehart and Winston, Incorporated, 383 Madison Avenue, New York, NY 10017. (212) 872-2000. 1986. $28.95.

CIVIL ENGINEERING HYDRAULICS: ESSENTIAL THEORY WITH WORKED EXAMPLES. R.E. Featherstone and C. Nalluri. Sheridan House, Incorporated, 145 Palisade Street, Dobbs Ferry, NY 10522. (914) 693-2410. 1982. $25.00.

CLASSIC PAPERS IN HYDRAULICS. American Society of Civil Engineers, 345 East 47th Street, New York, NY 10017-2398. (212) 705-7520. 1982. $49.00.

ENGINEERING FLUID MECHANICS. John J. Bertin. Prentice-Hall, Incorporated, Englewood Cliffs, NJ 07632. (201) 592-2000. Second edition. 1987. $34.95.

FLOW VISUALISATION. Japan Society of Mechanical Engineers. Pergamon Press, Incorporated, Maxwell House, Fairview Park, Elmsford, NY 10523. (914) 592-7700. 1986. $65.00.

FUNDAMENTALS OF FLUID MECHANICS AND HYDRAULICS. J. Evett and C. Liu. McGraw-Hill Book Company, 1221 Avenue of the Americas, New York, NY 10020. (212) 512-2000. 1987. $49.95.

FUNDAMENTALS OF HYDRAULIC ENGINEERING SYSTEMS. Ned H. Hwang and Carlos E. Hita. Prentice-Hall, Incorporated, Englewood Cliffs, NJ 07632. (800) 562-0245. 1987. $41.95.

HYDRAULICS IN CIVIL ENGINEERING. Andrew Chadwick and John Morfett. Allan and Unwin, Incorporated, 8 Winchester Place, Winchester, MA 01890. (800) 547-8889. 1986. $50.00.

ILLUSTRATED GUIDE TO AERODYNAMICS. Hubert Smith. Tab Books, Incorporated, Monterey Lane, Blue Ridge Summit, PA 17214. (717) 794-2191. 1986. $14.95.

INDUSTRIAL HYDRAULICS. T.C. Frankenfield. Penton/IPC, Incorporated, 1111 Chester Avenue, Cleveland, OH 44114. (216) 696-7000. 1985. $40.00.

HANDBOOKS AND MANUALS

APPLIED FLUID DYNAMICS HANDBOOK. Robert D. Blevins. McGraw-Hill Book Company, 1221 Avenue of the Americas, New York, NY 10020. (212) 512-2000. 1984. $52.95.

HANDBOOK OF APPLIED HYDRAULICS. Calvin Davis and K.E. Sorensen. McGraw-Hill Book Company, 1221 Avenue of the Americas, New York, NY 10020. (212) 512-2000. Third edition. 1968. $95.00.

HANDBOOK OF HYDRAULIC RESISTANCE. I.E. Idelchik and others. Hemisphere Publishing Corporation, 79 Madison Avenue, New York, NY 10016. (800) 526-0275. Second edition. 1986. $89.95.

THE WILEY ENGINEER'S DESK REFERENCE. Sanford I. Heisler. John Wiley and Sons, Incorporated, 605 Third Avenue, New York, NY 10158. (800) 526-5368 or (212) 850-6418. 1984. $36.00.

ONLINE DATA BASES

COMPENDEX. Engineering Information, Incorporated, 345 East 47th Street, New York, NY 10017. (800) 221-1044 or (212) 705-7615. Engineering and technical literature, 1975 to present. Inquire as to cost and availability.

GEOREF. American Geological Institute, 4220 King Street, Alexandria, VA 22302. (703) 379-2480. 1967 to present. Inquire as to online cost and availability.

ISMEC (INFORMATION SERVICE IN MECHANICAL ENGINEERING). Cambridge Scientific Abstracts, 5161 River Road, Bethesda, MD 20816. (301) 951-1400. 1973 to present. Inquire as to online cost and availability.

NASA. National Aeronautics and Space Administration, Scientific and Technical Information Branch, 300 7th Street, SW, Washington, DC 20546. Citations and abstracts of aerospace literature, 1962 to present. Inquire as to cost and availability.

NTIS. National Technical Information Service, 5285 Port Royal Road, Springfield, VA 22161. (703) 487-4630. Broad coverage of government-sponsored research reports, 1964 to present. Inquire as to cost and availability.

PERIODICALS

EXPERIMENTS IN FLUIDS. Springer-Verlag New York, 175 Fifth Avenue, New York, NY 10010. (212) 460-1500. Bimonthly. $115.00 per year.

FLUID MECHANICS. Scripts Publishing Company, 7961 Eastern Avenue, Silver Spring, MD 20910. (301) 588-0484. Translation of Soviet research papers in the fields of applied and theoretical fluid mechanics. Bimonthly. $300.00 per year.

HYDRAULIC AND AIR ENGINEERING. Applied Technology Publications Limited, 15 Coombe Road, New Malden, Surrey KT3 4PX, England. Bimonthly. $15.00 per year.

HYDRAULICS AND PNEUMATICS. Penton/IPC, Incorporated, 1100 Superior Avenue, Cleveland, OH 44114. (216) 696-7000. monthly. $35.00 per year.

JOURNAL OF FLUID CONTROL (FLUIDICS QUARTERLY). Delbridge Publishing Company, Box 2989, Stanford, CA 94305. (408) 446-3131. Quarterly. $125.00 per year.

JOURNAL OF FLUID MECHANICS. Cambridge University Press, 32 East 57th Street, New York, NY 10022. (800) 872-7423. Twelve time per year. $380.00 per year.

JOURNAL OF FLUID ENGINEERING. American Society of Mechanical Engineers, 345 East 47th Street, New York, NY 10017-2398. (212) 705-7722. Quarterly. $100.00 per year.

JOURNAL OF HYDRAULIC ENGINEERING. American Society of Civil Engineers, 345 East 47th Street, New York, NY 10017-2398. (212) 705-7520. Monthly. $115.00 per year.

RESEARCH CENTERS AND INSTITUTES

CALIFORNIA INSTITUTE OF TECHNOLOGY. W.M. Keck Engineering Laboratory of Hydraulics and Water Resources, 1201 East California Boulevard, Pasadena, CA 91125. (818) 356-4404.

COLORADO STATE UNIVERSITY. Hydraulics and Hydromachinery Research Laboratories, Engineering Research Center, Fort Collins, CO 80523. (303) 491-8655.

UNIVERSITY OF ILLINOIS. Hydrosystems Laboratory, Department of Civil Engineering, Urbana, IL 61801. (217) 333-0107.

UNIVERSITY OF IOWA. Institute of Hydraulic Research, Iowa City, IA 52242. (319) 353-4679.

UNIVERSITY OF MINNESOTA. St. Anthony Falls Hydraulic Laboratory, Mississippi River at Third Avenue, SE, Minneapolis, MN 55414. (612) 627-4010.

SPECIFICATIONS AND STANDARDS

HYDRAULIC INSTITUTE STANDARDS. Hydraulic Institute, 712 Lakewood Center North, 14600 Detroit Avenue, Cleveland, OH 44107. (216) 226-7700. 1983.

HYDRAULICS

See also: FLUID DYNAMICS, HYDRAULIC ENGINEERING, HYDROLOGY, HYDRODYNAMICS

ABSTRACT SERVICES AND INDEXES

APPLIED MECHANICS REVIEWS. American Society of Mechanical Engineers, 345 East 47th Street, New York, NY 10017. (212) 705-7703. 1948 to present. Monthly. $360.00 per year.

APPLIED SCIENCE AND TECHNOLOGY INDEX. H.W. Wilson Company, 950 University Avenue, Bronx, NY 10452. (800) 367-6670 or (212) 588-8400. Inquire as to cost and availability.

CHEMICAL ABSTRACTS. Chemical Abstracts Service, 2540 Olentangy Road, Post Office Box 3012, Columbus, OH 43210. (800) 848-6538 or (614) 421-3600. Weekly. $9200.00 per year.

CIVIL ENGINEERING HYDRAULICS ABSTRACTS. BHRA Fluid Engineering, Cranfield, Bedford MK43 0AJ, England. Distributed by Learned Information Incorporated, 143 Old Marlton Pike, Medford, NJ 08055. 1968 to present. Monthly. $225.00 per year.

CURRENT CONTENTS: ENGINEERING, TECHNOLOGY AND APPLIED SCIENCES. Institute for Scientific Information, 3501 Market Street, Philadelphia, PA 19104. (800) 523-1850 or (215) 386-0100. 1970 to present. Weekly. $272.00 per year.

DELFT HYDROSCIENCE ABSTRACTS. Delft Hydraulics Laboratory, Rotterdamseweg 185, Postbus 177, NL-2600 MH Delft, The Netherlands. Monthly. $177.00 per year.

ENGINEERING INDEX MONTHLY AND AUTHOR INDEX. Engineering Information, Incorporated, 345 East 47th Street, New York, NY 10017. (800) 221-1044 or (212) 705-7600. Monthly, with annual cumulation. $1560.00 per year.

FLUID POWER ABSTRACTS. BHRA Fluid Engineering, Cranfield, Bedford MK43 0AJ, England. Distributed by Learned Information Incorporated, 143 Old Marlton Pike, Meford, NJ 08055. 1970 to present. Bimonthly. $196.00 per year.

INDUSTRIAL AERODYNAMICS ABSTRACTS. BHRA Fluid Engineering, Cranfield, Bedford MK43 0AJ, England. Distributed by Learned Information Incorporated, 143 Old Marlton Pike, Medford, NJ 08055. 1970 to present. Monthly. $196.00 per year.

INTERNATIONAL AEROSPACE ABSTRACTS. AIAA/TIS, 1633 Broadway, New York, NY 10019. (212) 581-4300. Semimonthly. $700.00 per year.

ISMEC BULLETIN. (Information Service in Mechanical Engineering). Cambridge Scientific Abstracts, 5161 River Road, Bethesda, MD 20816. (301) 951-1400. 1973 to present. Monthly. $475.00 per year.

SCIENCE CITATION INDEX. Institute for Scientific Information, 3501 Market Street, Philadelphia, PA 19104. (800) 523-1850 or (215) 386-0100. Inquire as to cost and availability.

STAR. (Scientific and Technical Aerospace Reports. United States National Aeronautics and Space Administration, Scientific and Technical Information Facility, Box 8757, Baltimore-Washington International Airport, MD 21240. (202) 755-2210. Semimonthly, with semiannual and annual indexes. $85.00 per year.

ANNUAL REVIEWS AND YEARBOOKS

ADVANCE IN HYDROSCIENCE. Ven Te Chow, editor. Academic Press, Incorporated, Orlando, FL 32887. (800) 321-5068 or (305) 345-2734. 1964-1981. Inquire, price varies.

DEVELOPMENTS IN HYDRAULIC ENGINEERING. P. Novak, editor. Elsevier Science Publishing Company, Incorporated, 52 Vanderbilt Avenue, New York, NY 10017. (212) 370-5520. 1983-85. $66.00 per volume.

ASSOCIATIONS AND PROFESSIONAL SOCIETIES

AMERICAN INSTITUTE OF AERONAUTICS AND ASTRONAUTICS. 1633 Broadway, New York, NY 10019. (212) 581-4300.

AMERICAN SOCIETY OF CIVIL ENGINEERS. 345 East 47th Street, New York, NY 10017-2398. (212) 705-7520.

AMERICAN SOCIETY OF MECHANICAL ENGINEERS. 345 East 47th Street, New York, NY 10017-2398. (212) 705-7722.

FLUID CONTROLS INSTITUTE. Post Office Box 9036, Morristown, NJ 07960. (201) 829-0990.

FLUID POWER SOCIETY. 3333 North Mayfair Road, Milwaukee, WI 53222. (414) 778-3377.

HYDRAULIC INSTITUTE. 712 Lakewood Center North, 14600 Detroit Avenue, Cleveland, OH 44107. (216) 226-7700.

INTERNATIONAL ASSOCIATION FOR HYDRAULIC RESEARCH. 185 Rotterdamseweg, Box 177, Delft, The Netherlands.

NATIONAL CONFERENCE ON FLUID POWER. IIT Research Institute, Ten West 35th Street, Chicago, IL 60616. (312) 567-4414.

NATIONAL FLUID POWER ASSOCIATION. 3333 North Mayfair Road, Suite 311, Milwaukee, WI 53222. (414) 778-3344.

SOCIETY OF AUTOMOTIVE ENGINEERS. SAE4, Incorporated, 400 Commonwealth Drive, Warrendale, PA 15096. (412) 776-4841.

DIRECTORIES AND BIOGRAPHICAL SOURCES

AAS DIRECTORY. American Astronautical Society, 6212-B Old Keene Mill Court, Springfield, VA 22152. (703) 866-0020. Annual. $35.00 per year.

DIRECTORY OF HYDRAULIC RESEARCH INSTITUTES AND LABORATORIES. International Association for Hydraulic Research, 185 Rotterdamseweg, Box 177, Delft, The Netherlands. 1986. 60 Dutch florins.

FLUID POWER HANDBOOK AND DIRECTORY. Penton/IPC, Incorporated, 1111 Chester Avenue, Cleveland, OH 44114. (216) 606-7000. Biennial. $45.00.

FLUID POWER SOCIETY MEMBERSHIP DIRECTORY. Fluid Power Society, 3333 North Mayfair Road, Milwaukee, WI 53222. (414) 778-3377. Annual. $50.00.

NATIONAL FLUID POWER ASSOCIATION MEMBERSHIP DIRECTORY. 3333 North Mayfair Road, Suite 311, Milwaukee, WI 53222. (414) 778-3344. Annual. $150.00.

RESEARCH CENTERS DIRECTORY. Gale Research Company, Book Tower, Detroit, MI 48226. (800) 521-0707. Twelfth edition. 1987. $365.00 for set.

WHO'S WHO IN ENGINEERING. Engineers Joint Council, 345 East 47th Street, New York, NY 10017. (212) 705-7010. 1985. $200.00.

ENCYCLOPEDIAS AND DICTIONARIES

ENCYCLOPEDIA OF FLUID MECHANICS. Nicholas P. Cheremisinnoff, editor. Gulf Publishing Company, Book Division, Post Office Box 2608, Houston, TX 77001. (713) 520-4444. Six volumes. 1987. $1000.00 for set.

ENCYCLOPEDIA OF PHYSICAL SCIENCE AND TECHNOLOGY. Academic Press, Incorporated, Orlando, FL 32887. (800) 321-5068 or (305) 345-2734. Fifteen volumes, 1986.

MCGRAW-HILL DICTIONARY OF ENGINEERING. Sybil P. Parker, editor. McGraw-Hill Book Company, 1221 Avenue of the Americas, New York, NY 10020. (212) 512-2000. 1984. $39.95.

MCGRAW-HILL ENCYCLOPEDIA OF SCIENCE AND TECHNOLOGY. McGraw-Hill Book, Incorporated, 1221 Avenue of the Americas, New York, NY 10020. (212) 997-3675. Fifth edition, 15 volumes. $1100.00.

GENERAL WORKS

APPLIED HYDRAULICS FOR TECHNOLOGY. John D. Kanen. Holt, Rinehart and Winston, Incorporated, 383 Madison Avenue, New York, NY 10017. (212) 872-2000. 1986. $28.95.

CIVIL ENGINEERING HYDRAULICS: ESSENTIAL THEORY WITH WORKED EXAMPLES. R.E. Featherstone and C. Nalluri. Sheridan House, Incorporated, 145 Palisade Street, Dobbs Ferry, NY 10522. (914) 693-2410. 1982. $25.00.

CLASSIC PAPERS IN HYDRAULICS. American Society of Civil Engineers, 345 East 47th Street, New York, NY 10017-2398. (212) 705-7520. 1982. $49.00.

COMPUTATIONAL HYDRAULICS. C.A. Brebbia. Butterworth, Incorporated, 80 Montvale Avenue, Stoneham, MA 02108. (617) 438-8464. 1983. $59.95.

ENGINEERING FLUID MECHANICS. John J. Bertin. Prentice-Hall, Incorporated, Englewood Cliffs, NJ 07632. (201) 592-2000. Second edition. 1987. $34.95.

FLOW VISUALIZATION. Japan Society of Mechanical Engineers. Pergamon Press, Incorporated, Maxwell House, Fairview Park, Elmsford, NY 10523. (914) 592-7700. 1986. $65.00.

FLUVIAL HYDROLOGY. S. Lawrence Dingman. W.H. Freeman, Incorporated, 41 Madison Avenue, New York, NY 10010. (212) 532-7660. 1983. $31.95.

FUNDAMENTALS OF FLUID MECHANICS AND HYDRAULICS. J. Evett and C. Liu. McGraw-Hill Book Company, 1221 Avenue of the Americas, New York, NY 10020. (212) 512-2000. 1987. $49.95.

FUNDAMENTALS OF HYDRAULIC ENGINEERING SYSTEMS. Ned H. Hwang and CArlos E. Hita. Prentice-Hall, Incorporated, Englewood Cliffs, NJ 07632. (800) 562-0245. 1987. $41.95.

HYDRAULICS IN CIVIL ENGINEERING. Andrew Chadwick and John Morfett. Allan and Unwin, Incorporated, 8 Winchester Place, Winchester, MA 01890. (800) 547-8889. 1986. $50.00.

ILLUSTRATED GUIDE TO AERODYNAMICS. Hubert Smith. Tab Books, Incorporated, Monterey Lane, Blue Ridge Summit, PA 17214. (717) 794-2191. 1986. $14.95.

INDUSTRIAL HYDRAULICS. T.C. Frankenfield. Penton/IPC, Incorporated, 1111 Chester Avenue, Cleveland, OH 44114. (216) 696-7000. 1985. $40.00.

HANDBOOKS AND MANUALS

APPLIED FLUID DYNAMICS HANDBOOK. Robert D. Blevins. McGraw-Hill Book Company, 1221 Avenue of the Americas, New York, NY 10020. (212) 512-2000. 1984. $52.95.

HANDBOOK OF APPLIED HYDRAULICS. CAlvin Davis and K.E. Sorensen. McGraw-Hill Book Company, 1221 Avenue of the Americas, New York, NY 10020. (212) 512-2000. Third edition. 1968. $95.00.

HANDBOOK OF HYDRAULIC RESISTANCE. I.E. Idelchik and others. Hemisphere Publishing Corporation, 79 Madison Avenue, New York, NY 10016. (800) 526-0275. Second edition. 1986. $89.95.

HANDBOOK OF HYDRAULICS. E.F. Brater. McGraw-Hill Book Company, 1221 Avenue of the Americas, New York, NY 10020. (212) 512-2000. Sixth edition. 1976. $57.50.

THE WILEY ENGINEER'S DESK REFERENCE. Sanford I. Heisler. John Wiley and Sons, Incorporated, 605 Third Avenue, New York, NY 10158. (800) 526-5368 or (212) 850-6418. 1984. $36.00.

ONLINE DATA BASES

COMPENDEX. Engineering Information, Incorporated, 345 East 47th Street, New York, NY 10017. (800) 221-1044 or (212) 705-7615. Engineering and technical literature, 1975 to present. Inquire as to cost and availability.

GEOREF. American Geological Institute, 4220 King Street, Alexandria, VA 22302. (703) 379-2480. 1967 to present. Inquire as to online cost and availability.

ISMEC (INFORMATION SERVICE IN MECHANICAL ENGINEERING). CAmbridge Scientific Abstracts, 5161 River Road, Bethesda, MD 20816. (301) 951-1400. 1973 to present. Inquire as to online cost and availability.

NASA. National Aeronautics and Space Administration, Scientific and Technical Information Branch, 300 7th Street, SW, Washington, DC 20546. Citations and abstracts of aerospace literature, 1962 to present. Inquire as to cost and availability.

NTIS. National Technical Information Service, 5285 Port Royal Road, Springfield, VA 22161. (703) 487-4630. Broad coverage of government-sponsored research reports, 1964 to present. Inquire as to cost and availability.

PERIODICALS

EXPERIMENTS IN FLUIDS. Springer-Verlag New York, 175 Fifth Avenue, New York, NY 10010. (212) 460-1500. Bimonthly. $115.00 per year.

FLUID MECHANICS. Scripts Publishing Company, 7961 Eastern Avenue, Silver Spring, MD 20910. (301) 588-0484. Translation of Soviet research papers in the fields of applied and theoretical fluid mechanics. Bimonthly. $300.00 per year.

HYDRAULIC AND AIR ENGINEERING. Applied Technology Publications Limited, 15 Coombe Road, New Malden, Surrey KT3 4PX, England. Bimonthly. $15.00 per year.

HYDRAULICS AND PNEUMATICS. Penton/IPC, Incorporated, 1100 Superior Avenue, Cleveland, OH 44114. (216) 696-7000. monthly. $35.00 per year.

JOURNAL OF FLUID CONTROL (FLUIDICS QUARTERLY). Delbridge Publishing Company, Box 2989, Stanford, CA 94305. (408) 446-3131. Quarterly. $125.00 per year.

JOURNAL OF FLUID MECHANICS. CAmbridge University Press, 32 East 57th Street, New York, NY 10022. (800) 872-7423. Twelve time per year. $380.00 per year.

JOURNAL OF FLUID ENGINEERING. American Society of Mechanical Engineers, 345 East 47th Street, New York, NY 10017-2398. (212) 705-7722. Quarterly. $100.00 per year.

JOURNAL OF HYDROLOGY. Elsevier Science Publishers B.V., Post Office Box 211, 1000 AE Amsterdam, The Netherlands. Thirty-two times per year. $675.00 per year.

JOURNAL OF HYDRAULIC ENGINEERING. American Society of Civil Engineers, 345 East 47th Street, New York, NY 10017-2398. (212) 705-7520. Monthly. $115.00 per year.

RESEARCH CENTERS AND INSTITUTES

CALIFORNIA INSTITUTE OF TECHNOLOGY. W.M. Keck Engineering Laboratory of Hydraulics and Water Resources, 1201 East CAlifornia Boulevard, Pasadena, CA 91125. (818) 356-4404.

COLORADO STATE UNIVERSITY. Hydraulics and Hydromachinery Research Laboratories, Engineering Research Center, Fort Collins, CO 80523. (303) 491-8655.

UNIVERSITY OF ILLINOIS. Hydrosystems Laboratory, Department of Civil Engineering, Urbana, IL 61801. (217) 333-0107.

UNIVERSITY OF IOWA. Institute of Hydraulic Research, Iowa City, IA 52242. (319) 353-4679.

UNIVERSITY OF MINNESOTA. St. Anthony Falls Hydraulic Laboratory, Mississippi River at Third Avenue, SE, Minneapolis, MN 55414. (612) 627-4010.

SPECIFICATIONS AND STANDARDS

HYDRAULIC INSTITUTE STANDARDS. Hydraulic Institute, 712 Lakewood Center North, 14600 Detroit Avenue, Cleveland, OH 44107. (216) 226-7700. 1983.

HYDROCARBONS

See: PETROLEUM CHEMISTRY

HYDRODYNAMICS

See also: FLUID DYNAMICS, HYDRAULIC ENGINEERING, HYDRAULICS, HYDROLOGY

ABSTRACT SERVICES AND INDEXES

APPLIED MECHANICS REVIEWS. American Society of Mechanical Engineers, 345 East 47th Street, New York, NY 10017. (212) 705-7703. 1948 to present. Monthly. $360.00 per year.

APPLIED SCIENCE AND TECHNOLOGY INDEX. H.W. Wilson Company, 950 University Avenue, Bronx, NY 10452. (800) 367-6670 or (212) 588-8400. Inquire as to cost and availability.

CHEMICAL ABSTRACTS. Chemical Abstracts Service, 2540 Olentangy Road, Post Office Box 3012, Columbus, OH 43210. (800) 848-6538 or (614) 421-3600. Weekly. $9200.00 per year.

CIVIL ENGINEERING HYDRAULICS ABSTRACTS. BHRA Fluid Engineering, Cranfield, Bedford MK43 0AJ, England. Distributed by Learned Information Incorporated, 143 Old Marlton Pike, Medford, NJ 08055. 1968 to present. Monthly. $225.00 per

HYDRODYNAMICS

year.

CURRENT CONTENTS: ENGINEERING, TECHNOLOGY AND APPLIED SCIENCES. Institute for Scientific Information, 3501 Market Street, Philadelphia, PA 19104. (800) 523-1850 or (215) 386-0100. 1970 to present. Weekly. $272.00 per year.

ENGINEERING INDEX MONTHLY AND AUTHOR INDEX. Engineering Information, Incorporated, 345 East 47th Street, New York, NY 10017. (800) 221-1044 or (212) 705-7600. Monthly, with annual cumulation. $1560.00 per year.

INTERNATIONAL AEROSPACE ABSTRACTS. AIAA/TIS, 1633 Broadway, New York, NY 10019. (212) 581-4300. Semimonthly. $700.00 per year.

ISMEC BULLETIN. (Information Service in Mechanical Engineering). Cambridge Scientific Abstracts, 5161 River Road, Bethesda, MD 20816. (301) 951-1400. 1973 to present. Monthly. $475.00 per year.

METEOROLOGICAL AND GEOASTROPHYSICAL ABSTRACTS. American Meteorological Society, 45 Beacon Street, Boston, MA 02108. (617) 227-2425. 1950 to present. Monthly. $450.00 per year.

SCIENCE CITATION INDEX. Institute for Scientific Information, 3501 Market Street, Philadelphia, PA 19104. (800) 523-1850 or (215) 386-0100. Inquire as to cost and availability.

STAR. (Scientific and Technical Aerospace Reports. United States National Aeronautics and Space Administration, Scientific and Technical Information Facility, Box 8757, Baltimore-Washington International Airport, MD 21240. (202) 755-2210. Semimonthly, with semiannual and annual indexes. $85.00 per year.

ANNUAL REVIEWS AND YEARBOOKS

ADVANCE IN HYDROSCIENCE. Ven Te Chow, editor. Academic Press, Incorporated, Orlando, FL 32887. (800) 321-5068 or (305) 345-2734. 1964-1981. Inquire, price varies.

DEVELOPMENTS IN HYDRAULIC ENGINEERING. P. Novak, editor. Elsevier Science Publishing Company, Incorporated, 52 Vanderbilt Avenue, New York, NY 10017. (212) 370-5520. 1983-85. $66.00 per volume.

ASSOCIATIONS AND PROFESSIONAL SOCIETIES

AMERICAN INSTITUTE OF AERONAUTICS AND ASTRONAUTICS. 1633 Broadway, New York, NY 10019. (212) 581-4300.

AMERICAN SOCIETY OF CIVIL ENGINEERS. 345 East 47th Street, New York, NY 10017-2398. (212) 705-7520.

AMERICAN SOCIETY OF MECHANICAL ENGINEERS. 345 East 47th Street, New York, NY 10017-2398. (212) 705-7722.

FLUID CONTROLS INSTITUTE. Post Office Box 9036, Morristown, NJ 07960. (201) 829-0990.

FLUID POWER SOCIETY. 3333 North Mayfair Road, Milwaukee, WI 53222. (414) 778-3377.

HYDRAULIC INSTITUTE. 712 Lakewood Center North, 14600 Detroit Avenue, Cleveland, OH 44107. (216) 226-7700.

INTERNATIONAL ASSOCIATION FOR HYDRAULIC RESEARCH. 185 Rotterdamseweg, Box 177, Delft, The Netherlands.

NATIONAL CONFERENCE ON FLUID POWER. IIT Research Institute, Ten West 35th Street, Chicago, IL 60616. (312) 567-4414.

NATIONAL FLUID POWER ASSOCIATION. 3333 North Mayfair Road, Suite 311, Milwaukee, WI 53222. (414) 778-3344.

DIRECTORIES AND BIOGRAPHICAL SOURCES

AAS DIRECTORY. American Astronautical Society, 6212-B Old Keene Mill Court, Springfield, VA 22152. (703) 866-0020. Annual. $35.00 per year.

DIRECTORY OF HYDRAULIC RESEARCH INSTITUTES AND LABORATORIES. International Association for Hydraulic Research, 185 Rotterdamseweg, Box 177, Delft, The Netherlands. 1986. 60 Dutch florins.

FLUID POWER HANDBOOK AND DIRECTORY. Penton/IPC, Incorporated, 1111 Chester Avenue, Cleveland, OH 44114. (216) 606-7000. Biennial. $45.00.

FLUID POWER SOCIETY MEMBERSHIP DIRECTORY. Fluid Power Society, 3333 North Mayfair Road, Milwaukee, WI 53222. (414) 778-3377. Annual. $50.00.

NATIONAL FLUID POWER ASSOCIATION MEMBERSHIP DIRECTORY. 3333 North Mayfair Road, Suite 311, Milwaukee, WI 53222. (414) 778-3344. Annual. $150.00.

RESEARCH CENTERS DIRECTORY. Gale Research Company, Book Tower, Detroit, MI 48226. (800) 521-0707. Twelfth edition. 1987. $365.00 for set.

WHO'S WHO IN ENGINEERING. Engineers Joint Council, 345 East 47th Street, New York, NY 10017. (212) 705-7010. 1985. $200.00.

ENCYCLOPEDIAS AND DICTIONARIES

ENCYCLOPEDIA OF FLUID MECHANICS. Nicholas P. Cheremisinnoff, editor. Gulf Publishing Company, Book Division, Post Office Box 2608, Houston, TX 77001. (713) 520-4444. Six volumes. 1987. $1000.00 for set.

ENCYCLOPEDIA OF PHYSICAL SCIENCE AND TECHNOLOGY. Academic Press, Incorporated, Orlando, FL 32887. (800) 321-5068 or (305) 345-2734. Fifteen volumes, 1986.

MCGRAW-HILL DICTIONARY OF ENGINEERING. Sybil P. Parker, editor. McGraw-Hill Book Company, 1221 Avenue of the Americas, New York, NY 10020. (212) 512-2000. 1984. $39.95.

MCGRAW-HILL ENCYCLOPEDIA OF SCIENCE AND TECHNOLOGY. McGraw-Hill Book, Incorporated, 1221 Avenue of the Americas, New York, NY 10020. (212) 997-3675. Fifth edition, 15 volumes. $1100.00.

GENERAL WORKS

CIVIL ENGINEERING HYDRAULICS: ESSENTIAL THEORY WITH WORKED EXAMPLES. R.E. Featherstone and C. Nalluri. Sheridan House, Incorporated, 145 Palisade Street, Dobbs Ferry, NY 10522. (914) 693-2410. 1982. $25.00.

ENGINEERING FLUID MECHANICS. John J. Bertin. Prentice-Hall, Incorporated, Englewood Cliffs, NJ 07632. (201) 592-2000. Second edition. 1987. $34.95.

FUNDAMENTALS OF FLUID MECHANICS AND HYDRAULICS. J. Evett and C. Liu. McGraw-Hill Book Company, 1221 Avenue of the Americas, New York, NY 10020. (212) 512-2000. 1987. $49.95.

FUNDAMENTALS OF HYDRAULIC ENGINEERING SYSTEMS. Ned H. Hwang and Carlos E. Hita. Prentice-Hall, Incorporated, Englewood Cliffs, NJ 07632. (800) 562-0245. 1987. $41.95.

HYDRAULICS IN CIVIL ENGINEERING. Andrew Chadwick and John Morfett. Allan and Unwin, Incorporated, 8 Winchester

Place, Winchester, MA 01890. (800) 547-8889. 1986. $50.00.

HYDRODYNAMIC INSTABILITIES AND THE TRANSITION TO TURBULENCE. H.L. Swinney and J.P. Gollub, editors. Springer Verlag New York, Incorporated, 175 Fifth Avenue, New York, NY 10010. (800) 526-7254. 1985. $19.50 in paper.

HYDRODYNAMIC STABILITY THEORY. Adelina Georgescu. Kluwer Academic Publishers, 101 Philip Drive, Assinippi Park, Norwell, MA 02061. (617) 871-6600. 1985. $58.00.

HYDRODYNAMICS. Horace Lamb. Dover Publications, Incorporated, 31 East Second Street, Mineola, NY 11501. (516) 294-7000. Sixth edition. 1932. $12.00 in paper.

LOW REYNOLDS NUMBER HYDRODYNAMICS. John Happel and Howard Brenner. Kluwer Academic Publishers, 101 Philip Drive, Assinippi Park, Norwell, MA 02061. (617) 871-6600. Second edition. 1986. $32.50.

HANDBOOKS AND MANUALS

APPLIED FLUID DYNAMICS HANDBOOK. Robert D. Blevins. McGraw-Hill Book Company, 1221 Avenue of the Americas, New York, NY 10020. (212) 512-2000. 1984. $52.95.

THE WILEY ENGINEER'S DESK REFERENCE. Sanford I. Heisler. John Wiley and Sons, Incorporated, 605 Third Avenue, New York, NY 10158. (800) 526-5368 or (212) 850-6418. 1984. $36.00.

ONLINE DATA BASES

COMPENDEX. Engineering Information, Incorporated, 345 East 47th Street, New York, NY 10017. (800) 221-1044 or (212) 705-7615. Engineering and technical literature, 1975 to present. Inquire as to cost and availability.

GEOREF. American Geological Institute, 4220 King Street, Alexandria, VA 22302. (703) 379-2480. 1967 to present. Inquire as to online cost and availability.

ISMEC (INFORMATION SERVICE IN MECHANICAL ENGINEERING). Cambridge Scientific Abstracts, 5161 River Road, Bethesda, MD 20816. (301) 951-1400. 1973 to present. Inquire as to online cost and availability.

NASA. National Aeronautics and Space Administration, Scientific and Technical Information Branch, 300 7th Street, SW, Washington, DC 20546. Citations and abstracts of aerospace literature, 1962 to present. Inquire as to cost and availability.

NTIS. National Technical Information Service, 5285 Port Royal Road, Springfield, VA 22161. (703) 487-4630. Broad coverage of government-sponsored research reports, 1964 to present. Inquire as to cost and availability.

PERIODICALS

EXPERIMENTS IN FLUIDS. Springer-Verlag New York, 175 Fifth Avenue, New York, NY 10010. (212) 460-1500. Bimonthly. $115.00 per year.

FLUID MECHANICS. Scripts Publishing Company, 7961 Eastern Avenue, Silver Spring, MD 20910. (301) 588-0484. Translation of Soviet research papers in the fields of applied and theoretical fluid mechanics. Bimonthly. $300.00 per year.

HYDRAULIC AND AIR ENGINEERING. Applied Technology Publications Limited, 15 Coombe Road, New Malden, Surrey KT3 4PX, England. Bimonthly. $15.00 per year.

HYDRAULICS AND PNEUMATICS. Penton/IPC, Incorporated, 1100 Superior Avenue, Cleveland, OH 44114. (216) 696-7000. monthly. $35.00 per year.

JOURNAL OF FLUID CONTROL (FLUIDICS QUARTERLY). Delbridge Publishing Company, Box 2989, Stanford, CA 94305. (408) 446-3131. Quarterly. $125.00 per year.

JOURNAL OF FLUID MECHANICS. Cambridge University Press, 32 East 57th Street, New York, NY 10022. (800) 872-7423. Twelve time per year. $380.00 per year.

JOURNAL OF FLUID ENGINEERING. American Society of Mechanical Engineers, 345 East 47th Street, New York, NY 10017-2398. (212) 705-7722. Quarterly. $100.00 per year.

JOURNAL OF HYDRAULIC ENGINEERING. American Society of Civil Engineers, 345 East 47th Street, New York, NY 10017-2398. (212) 705-7520. Monthly. $115.00 per year.

RESEARCH CENTERS AND INSTITUTES

MASSACHUSETTS INSTITUTE OF TECHNOLOGY. Ralph M. Parsons Laboratory, 15 Vassar Street, Cambridge, MA 02139. (617) 253-2726.

STEVENS INSTITUTE OF TECHNOLOGY. Davidson Laboratory, Castle Point Station, Hoboken, NJ 07030. (201) 420-5300.

UNIVERSITY OF MICHIGAN. Ship Hydrodynamics Laboratory, 126 West Engineering Building, Ann Arbor, MI 48109. (313) 764-9432.

HYDROELECTRIC POWER

See also: DAMS, ELECTRIC POWER ENGINEERING, ELECTRICAL ENGINEERING, HIGH VOLTAGE

ABSTRACT SERVICES AND INDEXES

APPLIED SCIENCE AND TECHNOLOGY INDEX. H.W. Wilson and Company, 950 University Avenue, Bronx, NY 10452. (800) 367-6670 or (212) 588-8400. Monthly. Inquire as to cost and availability.

CURRENT CONTENTS: ENGINEERING, TECHNOLOGY AND APPLIED SCIENCES. Institute for Scientific Information, 3501 Market Street, Philadelphia, PA 19104. (800) 523-1850 or (215) 386-0100. Weekly. $275.00 per year.

ELECTRIC POWER INDUSTRY ABSTRACTS. Edison Electric Institute, c/o Utility Date Institute, 2011 I Street, N.W., Suite 700, Washington, DC 20006. 1975 to present. Bimonthly. Inquire as to cost and availability.

ELECTRICAL AND ELECTRONICS ABSTRACTS. Institution of Electrical Engineers. Available from: Institute of Electrical and Electronics Engineers. IEEE Service Center, 445 Hoes Lane, Piscataway, NJ 08854. Monthly. $1250.00 per year.

ENGINEERING INDEX MONTHLY AND AUTHOR INDEX. Engineering Information Inc., 345 East 47th Street, New York, NY 10017. (212) 705-7600. Monthly. $1560.00 per year.

IEEE PUBLICATIONS BULLETIN. Institute of Electrical and Electronics Engineers. Institute of Electrical and Electronics Engineers. IEEE Service Center, 445 Hoes Lane, Piscataway, NJ 08854. Quarterly. Free.

SCIENCE CITATION INDEX. Institute for Scientific Information, 3501 Market Street, Philadelphia, PA 19104. (800) 523-1850 or (215) 386-0100. Six times per year. $6200. 00 per year.

ASSOCIATIONS AND PROFESSIONAL SOCIETIES

AMERICAN ELECTRONICS ASSOCIATION. P.O. Box 10045, 2670 Hanover Street, Palo Alto, CA 94303. (415) 857-9300.

HYDROELECTRIC POWER

AMERICAN SOCIETY OF CIVIL ENGINEERS. 345 East 47th Street, New York, NY 10017. (212) 705-7420.

EDISON ELECTRIC INSTITUTE. 1111 19th Street, N.W., Washington, DC 20036. (202) 828-7400.

INSTITUTE OF ELECTRICAL AND ELECTRONICS ENGINEERS. 345 East 47th Street, New York, NY 10017. (212) 705-7900.

DIRECTORIES AND BIOGRAPHICAL SOURCES

ELECTRICAL WORLD DIRECTORY OF ELECTRICAL UTILITIES. McGraw-Hill Book Company, 1221 Avenue of the Americas, New York, NY 10020. (212) 512-2000. Annual. $275.00.

INTERNATIONAL RESEARCH CENTERS DIRECTORY 1988-89. Darren L. Smith, editor. Gale Research Company, Book Tower, Detroit, MI 48226. (800) 521-0707. 4th edition. 1987. $360.00.

MAJOR DAMS, RESERVOIRS, AND HYDROELECTRIC PLANTS WORLDWIDE. U.S. Bureau of Reclamation, 18th and C Streets, N.W., Washington, DC 20240. (202) 343-1100. 1983. Free.

1987 DIRECTORY OF ENGINEERING SOCIETIES AND RELATED ORGANIZATIONS. Gordon Davis, editor. Hemisphere Publishing Corporation, 1010 Vermont Avenue, NW, Washington, DC 20005. (800) 526-0275. 12th edition. 1987. $100.00.

RESEARCH CENTERS DIRECTORY 1988. Gale Research Company, Book Tower, Detroit, MI 48226. (800) 521-0707. 12th edition. 1987. $365.00 for set.

SCIENTIFIC AND TECHNICAL ORGANIZATIONS AND AGENCIES DIRECTORY. Margaret Labash Young, editor. Gale Research Company, Book Tower, Detroit, MI 48226. (800) 521-0707. 2nd edition. 1987. $185.00.

WHO'S WHO IN ENGINEERING. Gordon Davis, editor. Hemisphere Publishing Corporation, 1010 Vermont Avenue, NW, Washington, DC 20005. (800) 526-0275. 6th edition. 1985. $200.00.

GENERAL WORKS

ELECTRIC POWER TRANSMISSION SYSTEM ENGINEERING: ANALYSIS AND DESIGN. Turan Gonen. John Wiley and Sons, Inc., 605 Third Avenue, New York, NY 10158. (800) 526-5368. 1987. $75.00.

ELECTRIC UTILITY SYSTEMS AND PRACTICES. H.M. Rustebakke. John Wiley and Sons, Inc., 605 Third Avenue, New York, NY 10158. (800) 526-5368. 1983. $45.95.

ELECTRICAL ENGINEERING. W.H. Roadstrum and Dan H. Wolaver. Harper and Row Publishers, Inc., 10 East 53rd Street, New York, NY 10022. (800) 242-7737. 1986. $37.50.

HIGH VOLTAGE ENGINEERING. E. Kuffel and W.S. Zaengl. Pergamon Press, Inc., Maxwell House, Fairview Park, Elmsford, NY 10523. (914) 592-7700. 1984. $19.50 in paper.

HIGH VOLTAGE: MEASUREMENT, TESTING AND DESIGN. T.J. Gallagher and A.J. Pearmain. John Wiley and Sons, Inc., 605 Third Avenue, New York, NY 10158. (800) 526-5368. 1983. $54.95.

IEEE STANDARD TECHNIQUES FOR HIGH VOLTAGE TESTING. Institute of Electrical and Electronics Engineers. IEEE Service Center, 445 Hoes Lane, Piscataway, NJ 08854. Sixth edition. 1978. $19.95.

HANDBOOKS AND MANUALS

HANDBOOK OF MODERN ELECTRONICS AND ELECTRICAL ENGINEERING. Charles Belove, editor. John Wiley and Sons, Inc., 605 Third Avenue, New York, NY 10158. (800) 526-5368. 1986. $88.95.

STANDARD HANDBOOK FOR ELECTRICAL ENGINEERS. D.G. Fink and H.W. Beaty. McGraw-Hill Book Company, 1221 Avenue of the Americas, New York, NY 10020. (212) 512-2000. 12th edition. 1987. $86.50.

ONLINE DATA BASES

COMPENDEX. Engineering Information, Inc., 345 East 47th Street, New York, NY 10017. (800) 221-1044 or (212) 705-7615. Engineering and technical literature, 1975 to present. Inquire as to online cost and availability.

INSPEC. INSPEC Marketing Department, Institution of Electrical Engineers. Available from IEEE Service Center, 445 Hoes Lane, Piscataway, NJ 08854. (201) 981-0060. Online version of Physics Abstracts. Inquire as to online cost and availability.

NTIS. National Technical Information Service, 5285 Port Royal Road, Springfield, VA 22161. (703) 487-4630. Broad coverage of government sponsored research reports, 1964 to present. Inquire as to online cost and availability.

SCISEARCH. Institute for Scientific Information, 3501 Market Street, Philadelphia, PA 19104. (800) 523-1850 or (215) 386-0100. Broad multidisciplinary title and author index to the international literature of science and technology, 1974 to present. Inquire as to online cost and availability.

WILSONLINE. H.W. Wilson and Company, 950 University Avenue, Bronx, NY 10452. (800) 367-6770 or (212) 588-8400. Makes available online versions of the H.W. Wilson indexes including Applied Science and Technology Index, Business Periodicals Index and Readers' Guide to Periodical Literature. Approximately 1980 to present. Inquire as to online cost and availability.

OTHER SOURCES

A GUIDE TO THE LITERATURE OF ELECTRICAL AND ELECTRONICS ENGINEERING. Susan B. Ardis. Libraries Unlimited Inc., P.O. Box 263, Littleton, CO 80160. (303) 770-1220. 1987. $37.50.

PERIODICALS

CIVIL ENGINEERING. American Society of Civil Engineers, 345 East 47th Street, New York, NY 10017. (212) 705-7420. 1930 to present. Monthly. $48.00 per year.

CONCRETE. Harcourt, Brace Jovanovich, Inc., 7500 Old Oak Boulevard, Cleveland, OH 44130. 1937 to present. Monthly. $25.00 per year.

ELECTRIC LIGHT AND POWER. Technical Publishing Company, 875 Third Avenue, New York, NY 10022. (212) 605-9400. 1922 to present. Monthly. $38.00 per year.

ELECTRIC MACHINES AND POWER SYSTEMS. Hemisphere Publishing Corporation, 1010 Vermont Avenue, NW, Washington, DC 20005. (800) 526-0275. 1976 to present. Bimonthly. $134.50 per year.

ELECTRICAL WORLD. McGraw-Hill Book Company, 1221 Avenue of the Americas, New York, NY 10020. (212) 512-2000. 1874 to present. Monthly. $11.00 per year.

IEEE POWER ENGINEERING REVIEW. Institute of Electrical and Electronics Engineers. IEEE Service Center, 445 Hoes Lane, Piscataway, NJ 08854. 1981 to present. Monthly. $75.00 per year.

IEEE TRANSACTIONS ON POWER DELIVERY. Institute of Electrical and Electronics Engineers. IEEE Service Center, 445 Hoes Lane, Piscataway, NJ 08854. 1986 to present. Quarterly. $100.00 per year.

IEEE TRANSACTIONS ON POWER SYSTEMS. Institute of Electrical and Electronics Engineers. IEEE Service Center, 445 Hoes Lane, Piscataway, NJ 08854. 1986 to present. Quarterly. $100.00 per year.

INSTITUTE OF ELECTRICAL AND ELECTRONICS ENGINEERS PROCEEDINGS. Institute of Electrical and Electronics Engineers. IEEE Service Center, 445 Hoes Lane, Piscataway, NJ 08854. 1913 to present. Monthly. $140.00 per year.

RESEARCH CENTERS AND INSTITUTES

EDISON ELECTRIC INSTITUTE. 1111 19th Street, N.W., Washington, DC 20036. (202) 778-6778.

ELECTRIC POWER INSTITUTE. Texas A&M University, Department of Electrical Engineering, College Station, TX 77843.

ELECTRICAL ENGINEERING RESEARCH LABORATORIES. Purdue University, Electrical Engineering Building, West Lafayette, IN 47907. (317) 494-3536.

GEOTHERMAL AND HYDROELECTRIC TECHNOLOGIES DIVISION. U.S. DEPARTMENT OF ENERGY. 1000 Independence Avenue, S.W., Washington, DC 20585. (202) 252-5340.

STATISTICS SOURCES

INTERNATIONAL ENERGY ANNUAL. U.S. Department of Energy, Energy Information Administration, 1000 Independence Avenue, S.W., Washington, DC 20585. Annual.

HYDROFOIL CRAFT

See: NAVAL ARCHITECTURE

HYDROGEN

See also: ELEMENTS, ENERGY, INORGANIC CHEMISTRY

ABSTRACT SERVICES AND INDEXES

ABSTRACT NEWSLETTER: ENERGY. National Technical Information Service, 5285 Port Royal Road, Springfield, VA 22161. (703) 487-4929. Weekly. $95.00 per year.

APPLIED SCIENCE AND TECHNOLOGY INDEX. H.W. Wilson and Company, 950 University Avenue, Bronx, NY 10452. (800) 367-6670 or (212) 588-8400. Monthly. Inquire as to cost and availability.

CHEMICAL ABSTRACTS. American Chemical Society, Chemical Abstracts Service, Box 3012, Columbus, OH 43210. (614) 421-3600. 1907 to present. Weekly. $9500.00 per year.

CONFERENCE PAPERS INDEX. Cambridge Scientific Abstracts, 5161 River Road, Bethesda, MD 20816. (301) 951-1400. 1972 to present. Monthly. Inquire as to cost and availability.

ENERGY RESEARCH ABSTRACTS. U.S. Department of Energy, Office of Scientific and Technical Information, Box 62, Oak Ridge, TN 37831. (615) 576-1155. Subscribe to: U.S. Superintendent of Documents, Washington, DC 20402. 1976 to present. Semimonthly. $146.00 per year.

ENGINEERING INDEX MONTHLY AND AUTHOR INDEX. Engineering Information Inc., 345 East 47th Street, New York, NY 10017. (212) 705-7600. Monthly. $1560.00 per year.

GENERAL SCIENCE INDEX. H.W. Wilson and Company, 950 University Avenue, Bronx, NY 10452. (800) 367-6670 or (212) 588-8400. 1978 to present. Monthly. Inquire as to cost and availability.

INDEX TO SCIENTIFIC AND TECHNICAL PROCEEDINGS. Institute for Scientific Information, 3501 Market Street, Philadelphia, PA 19104. (800) 523-1850 or (215) 386-0100. 1978 to present. Monthly. $775.00 per year.

INDEX TO SCIENTIFIC REVIEWS. Institute for Scientific Information, 3501 Market Street, Philadelphia, PA 19104. (800) 523-1850 or (215) 386-0100. 1974 to present. Semi-annual. $550.00 per year.

PHYSICS ABSTRACTS. Institution of Electrical Engineers. Available from: IEEE Service Center, 445 Hoes Lane, Piscataway, NJ 08854. 1898 to present. Bimonthly. $1700.00 per year.

PHYSICS BRIEFS. Physik Verlag GmbH, Postfach 1260/1280, D-6940 Weinheim, West Germany. (212) 661-9404. 1920 to present. Twenty-six times per year. $1250.00 per year.

SCIENCE CITATION INDEX. Institute for Scientific Information, 3501 Market Street, Philadelphia, PA 19104. (800) 523-1850 or (215) 386-0100. Six times per year. $6200.00 per year.

ASSOCIATIONS AND PROFESSIONAL SOCIETIES

AMERICAN CHEMICAL SOCIETY. 1155 16th Street, N.W., Washington, DC 20036. (800) 424-6747.

AMERICAN INSTITUTE OF PHYSICS. 335 East 45th Street, New York, NY 10017. (212) 661-9494.

AMERICAN SOCIETY FOR METALS. Metals Park, OH 44073. (216) 338-5151.

ASSOCIATION OF ENERGY ENGINEERS. 4025 Pleasantdale Road, Suite 340, Atlanta, GA 30340. (404) 447-5083.

INTERNATIONAL ASSOCIATION FOR HYDROGEN ENERGY. P.O. Box 248266, Coral Gables, FL 33124. (305) 284-4666.

DIRECTORIES AND BIOGRAPHICAL SOURCES

AMERICAN MEN AND WOMEN OF SCIENCE. R.R. Bowker, Inc., Order Department, 245 West 17th Street, New York, NY 10011. (800) 521-8110. Eight volumes. 1986. $595.00 for set.

ENERGY INFORMATION CENTERS DIRECTORY. Public Affairs and Information Program, Atomic Industrial Forum, 7101 Wisconsin Avenue, Bethesda, MD 20814. (301) 654-9260. 1985. Free.

INTERNATIONAL RESEARCH CENTERS DIRECTORY 1988-89. Darren L. Smith, editor. Gale Research Company, Book Tower, Detroit, MI 48226. (800) 521-0707. 4th edition. 1987. $360.00.

1987 DIRECTORY OF ENGINEERING SOCIETIES AND RELATED ORGANIZATIONS. Gordon Davis, editor. Hemisphere Publishing Corporation, 79 Madison Avenue, New York, NY 10016-7892. (800) 821-8312. 12th edition. 1987. $100.00.

RESEARCH CENTERS DIRECTORY 1988. Gale Research Company, Book Tower, Detroit, MI 48226. (800) 521-0707. 12th edition. 1987. $365.00 for set.

SCIENTIFIC AND TECHNICAL ORGANIZATIONS AND AGENCIES DIRECTORY. Margaret Labash Young, editor. Gale Research Company, Book Tower, Detroit, MI 48226. (800) 521-0707. 2nd edition. 1987. $185.00.

WHO'S WHO IN ENGINEERING. Gordon Davis, editor. Hemisphere Publishing Corporation, 79 Madison Avenue, New York, NY 10016-7892. (800) 821-8312. 6th edition. 1985. $200.00.

HYDROGEN

ENCYCLOPEDIAS AND DICTIONARIES

MCGRAW-HILL ENCYCLOPEDIA OF SCIENCE AND TECHNOLOGY. McGraw-Hill Book Company, 1221 Avenue of the Americas, New York, NY 10020. (212) 512-2000. 6th edition. 1987. $1600.00.

THESAURUS OF SCIENTIFIC, TECHNICAL, AND ENGINEERING TERMS. Hemisphere Publishing Corporation, 1010 Vermont Avenue, NW, Washington, DC 20005. (800) 526-0275. 1988. $125.00.

GENERAL WORKS

CHEMISTRY OF THE ELEMENTS. N.N. Greenwood and A. Earnshaw. Pergamon Press, Inc., Maxwell House, Fairview Park, Elmsford, NY 10523. (914) 592-7700. 1984. $145.00.

ENERGY OPTIONS. David Merrick and Richard Marshall. John Wiley and Sons, Inc., 605 Third Avenue, New York, NY 10158. (800) 526-5368. 1984. $75.95.

HYDROGEN POWER: AN INTRODUCTION TO HYDROGEN ENERGY AND ITS APPLICATIONS. L.O. Williams. Pergamon Press, Inc., Maxwell House, Fairview Park, Elmsford, NY 10523. (914) 592-7700. 1980. $35.20.

INORGANIC CHEMISTRY. William W. Porterfield. Addison-Wesley Publishing Company, Inc., 1 Jacob Way, Reading, MA 01867. (617) 944-3700. 1983. $36.95.

INORGANIC CHEMISTRY: AN INTRODUCTION. T. Moeller. John Wiley and Sons, Inc., 605 Third Avenue, New York, NY 10158. (800) 526-5368. 1982. $44.95.

HANDBOOKS AND MANUALS

HANDBOOK OF ENERGY SYSTEMS ENGINEERING. Leslie C. Wilbur, editor. John Wiley and Sons, Inc., 605 Third Avenue, New York, NY 10158. (800) 526-5368. 1985. $74.95.

HANDBOOK OF ENERGY TECHNOLOGY AND ECONOMICS. R.A. Meyers, editor. John Wiley and Sons, Inc., 605 Third Avenue, New York, NY 10158. (800) 526-5368. 1983. $79.95.

ONLINE DATA BASES

CA SEARCH. Chemical Abstracts Service, P.O. Box 3012, Columbus, OH 43120. (800) 848-6538 or (614) 421-3600. Comprehensive guide to chemical literature, 1972 to present. Inquire as to online cost and availability.

COMPENDEX. Engineering Information, Inc., 345 East 47th Street, New York, NY 10017. (800) 221-1044 or (212) 705-7615. Engineering and technical literature, 1975 to present. Inquire as to online cost and availability.

DISSERTATION ABSTRACTS ONLINE. University Microfilms International, 300 North Zeeb Road, Ann Arbor, MI 48106. (800) 521-0600 or (313) 761-4700. Scope includes virtually all doctoral dissertations accepted at accredited American institutions from 1861 to present in over 250 subject areas. Inquire as to online cost and availability.

DOE ENERGY DATA BASE. U.S. Department of Energy, Office of Scientific and Technical Information, P.O. Box 62, Oak Ridge, TN 37831. (615) 576-6837. A database that covers all aspects of energy including the science and technology of energy. 1948 to present. Available through the DIALOG search service or DOE/RECON. Inquire as to online cost and availability.

ENERGYLINE. EIC/Intelligence, Inc., 48 West 38th Street, New York, NY 10018. (212) 944-8500. A database of resources on the scientific, engineering, political, and socioeconomic aspects of energy resources. 1976 to present. Inquire as to online cost and availability.

INSPEC. INSPEC Marketing Department, Institution of Electrical Engineers. Available from IEEE Service Center, 445 Hoes Lane, Piscataway, NJ 08854. (201) 981-0060. Online version of Physics Abstracts. Inquire as to online cost and availability.

NTIS. National Technical Information Service, 5285 Port Royal Road, Springfield, VA 22161. (703) 487-4630. Broad coverage of government sponsored research reports, 1964 to present. Inquire as to online cost and availability.

SCISEARCH. Institute for Scientific Information, 3501 Market Street, Philadelphia, PA 19104. (800) 523-1850 or (215) 386-0100. Broad multidisciplinary title and author index to the international literature of science and technology, 1974 to present. Inquire as to online cost and availability.

WILSONLINE. H.W. Wilson and Company, 950 University Avenue, Bronx, NY 10452. (800) 367-6770 or (212) 588-8400. Makes available online versions of the H.W. Wilson indexes including Applied Science and Technology Index, Business Periodicals Index and Readers' Guide to Periodical Literature. Approximately 1980 to present. Inquire as to online cost and availability.

PERIODICALS

ALTERNATIVE SOURCES OF ENERGY. Alternative Sources of Energy, Inc., 107 South Central Avenue, Milaca, MN 56353. (612) 983-6892. 1971 to present. Ten times per year. $48.00 per year.

ENERGY AND FUELS. American Chemical Society, 1155 16th Street, N.W., Washington, DC 20036. (800) 424-6747. 1987 to present. Bimonthly. $269.00 per year.

HYDROGEN PROGRESS. Billings Corporation, 6030 Connecticut Avenue, Kansas City, MO 64120-1333. 1976 to present. Quarterly. $12.50 per year.

HYDROGEN TIMES. Clean Fuel Institute, 3600 Lime Street, Riverside, CA 92506. (714) 682-5864. 1982 to present. Quarterly. $25.00 per year.

INORGANIC CHEMISTRY. American Chemical Society, 1155 16th Street, N.W., Washington, DC 20036. (800) 424-6747. 1962 to present. Biweekly. $300.00 per year.

INORGANICA CHIMICA ACTA. Elsevier Science Publishing Company, Inc., 52 Vanderbilt Avenue, New York, NY 10017. (212) 370-5520. 1967 to present. Biweekly. $1500.00 per year.

JOURNAL OF ENERGY RESOURCES TECHNOLOGY. American Society of Mechanical Engineers, 345 East 47th Street, New York, NY 10017. (212) 705-7703. 1979 to present. $24.00 per year.

RESEARCH CENTERS AND INSTITUTES

CENTER FOR RESEARCH ON ENERGY ALTERNATIVES. Florida Institute of Technology, 150 West University Boulevard, Melbourne, FL 32901. (305) 768-8092.

CLEAN ENERGY RESEARCH INSTITUTE. University of Miami, 218 McArthur Building, Coral Gables, FL 33124. (305) 284-4666.

ENERGY RESEARCH/DEVELOPMENT CENTER. University of Kansas, 345 Nichols Hall, 2291 Irving Hill Drive, Lawrence, KS 66054. (913) 864-4079.

HYDROGEN RESEARCH CENTER. Texas A&M University, Department of Chemistry, College Station, TX 77843. (409) 845-5335.

HYDROGENATION

See also: CATALYSIS, COAL GASIFICATION AND LIQUEFACTION, HYDROGEN

ABSTRACT SERVICES AND INDEXES

APPLIED SCIENCE AND TECHNOLOGY INDEX. H.W. Wilson and Company, 950 University Avenue, Bronx, NY 10452. (800) 367-6670 or (212) 588-8400. Monthly. Inquire as to cost and availability.

CHEMICAL ABSTRACTS. American Chemical Society, Chemical Abstracts Service, Box 3012, Columbus, OH 43210. (614) 421-3600. 1907 to present. Weekly. $9500.00 per year.

CONFERENCE PAPERS INDEX. Cambridge Scientific Abstracts, 5161 River Road, Bethesda, MD 20816. (301) 951-1400. 1972 to present. Monthly. Inquire as to cost and availability.

ENGINEERING INDEX MONTHLY AND AUTHOR INDEX. Engineering Information Inc., 345 East 47th Street, New York, NY 10017. (212) 705-7600. Monthly. $1560.00 per year.

INDEX TO SCIENTIFIC AND TECHNICAL PROCEEDINGS. Institute for Scientific Information, 3501 Market Street, Philadelphia, PA 19104. (800) 523-1850 or (215) 386-0100. 1978 to present. Monthly. $775.00 per year.

INDEX TO SCIENTIFIC REVIEWS. Institute for Scientific Information, 3501 Market Street, Philadelphia, PA 19104. (800) 523-1850 or (215) 386-0100. 1974 to present. Semi-annual. $550.00 per year.

SCIENCE CITATION INDEX. Institute for Scientific Information, 3501 Market Street, Philadelphia, PA 19104. (800) 523-1850 or (215) 386-0100. Six times per year. $6200.00 per year.

ASSOCIATIONS AND PROFESSIONAL SOCIETIES

AMERICAN CHEMICAL SOCIETY. 1155 16th Street, N.W., Washington, DC 20036. (800) 424-6747.

AMERICAN INSTITUTE OF CHEMICAL ENGINEERS. 345 East 47th Street, New York, NY 10017. (212) 705-7338.

AMERICAN OIL CHEMISTS SOCIETY. 508 South Sixth Street, Champaign, IL 61820. (217) 359-2344.

DIRECTORIES AND BIOGRAPHICAL SOURCES

AMERICAN MEN AND WOMEN OF SCIENCE. R.R. Bowker, Inc., Order Department, 245 West 17th Street, New York, NY 10011. (800) 521-8110. Eight volumes. 1986. $595.00 for set.

INTERNATIONAL RESEARCH CENTERS DIRECTORY 1988-89. Darren L. Smith, editor. Gale Research Company, Book Tower, Detroit, MI 48226. (800) 521-0707. 4th edition. 1987. $360.00.

1987 DIRECTORY OF ENGINEERING SOCIETIES AND RELATED ORGANIZATIONS. Gordon Davis, editor. Hemisphere Publishing Corporation, 79 Madison Avenue, New York, NY 10016-7892. (800) 821-8312. 12th edition. 1987. $100.00.

RESEARCH CENTERS DIRECTORY 1988. Gale Research Company, Book Tower, Detroit, MI 48226. (800) 521-0707. 12th edition. 1987. $365.00 for set.

SCIENTIFIC AND TECHNICAL ORGANIZATIONS AND AGENCIES DIRECTORY. Margaret Labash Young, editor. Gale Research Company, Book Tower, Detroit, MI 48226. (800) 521-0707. 2nd edition. 1987. $185.00.

WHO'S WHO IN ENGINEERING. Gordon Davis, editor. Hemisphere Publishing Corporation, 79 Madison Avenue, New York, NY 10016-7892. (800) 821-8312. 6th edition. 1985. $200.00.

GENERAL WORKS

HYDROGENATION METHODS. Paul N. Rylander. Academic Press, Inc., 6277 Sea Harbor Drive, Orlando, FL 32821. (800) 321-5068. 1985. $50.00.

PHOTOGENERATION OF HYDROGEN. A. Harriman and M.A. West, editors. Academic Press, Inc., 6277 Sea Harbor Drive, Orlando, FL 32821. (800) 321-5068. 1983. $29.00.

PRACTICAL CATALYTIC HYDROGENATION: TECHNIQUES AND APPLICATIONS. Morris Freifelder. John Wiley and Sons, Inc., 605 Third Avenue, New York, NY 10158. (800) 526-5368. 1971. $42.50.

ONLINE DATA BASES

CA SEARCH. Chemical Abstracts Service, P.O. Box 3012, Columbus, OH 43120. (800) 848-6538 or (614) 421-3600. Comprehensive guide to chemical literature, 1972 to present. Inquire as to online cost and availability.

COMPENDEX. Engineering Information, Inc., 345 East 47th Street, New York, NY 10017. (800) 221-1044 or (212) 705-7615. Engineering and technical literature, 1975 to present. Inquire as to online cost and availability.

NTIS. National Technical Information Service, 5285 Port Royal Road, Springfield, VA 22161. (703) 487-4630. Broad coverage of government sponsored research reports, 1964 to present. Inquire as to online cost and availability.

SCISEARCH. Institute for Scientific Information, 3501 Market Street, Philadelphia, PA 19104. (800) 523-1850 or (215) 386-0100. Broad multidisciplinary title and author index to the international literature of science and technology, 1974 to present. Inquire as to online cost and availability.

WILSONLINE. H.W. Wilson and Company, 950 University Avenue, Bronx, NY 10452. (800) 367-6770 or (212) 588-8400. Makes available online versions of the H.W. Wilson indexes including Applied Science and Technology Index, Business Periodicals Index and Readers' Guide to Periodical Literature. Approximately 1980 to present. Inquire as to online cost and availability.

HYDROGEOLOGY

See also: GROUND WATER, HYDROLOGY, HYDRODYNAMICS, WATER RESOURCES

ABSTRACT SERVICES AND INDEXES

BIBLIOGRAPHY AND INDEX OF GEOLOGY. American Geological Institute, 4220 King Street, Alexandria, VA 22302. (703) 379-2480. 1969 to present. Monthly. $1100.00 per year.

CHEMICAL ABSTRACTS. Chemical Abstracts Service, 2540 Olentangy Road, Post Office Box 3012, Columbus, OH 43210. (800) 848-6538 or (614) 421-3600. Weekly. $9200.00 per year.

CURRENT CONTENTS: PHYSICAL, CHEMICAL AND EARTH SCIENCES. Institute for Scientific Information, 3501 Market Street, Philadelphia, PA 19104. (800) 523-1850 or (215) 386-0100. 1970 to present. Weekly. $272.00 per year.

DELFT HYDROSCIENCE ABSTRACTS. Delft Hydraulics Laboratory, Rotterdamseweg 185, Postbus 177, NL-2600 MH Delft, The Netherlands. Monthly. $177.00 per year.

ENGINEERING INDEX MONTHLY AND AUTHOR INDEX. Engineering Information, Incorporated, 345 East 47th Street, New York, NY 10017. (800) 221-1044 or (212) 705-7600. Monthly, with annual cumulation. $1560.00 per year.

METEOROLOGICAL AND GEOASTROPHYSICAL ABSTRACTS. American Meteorological Society, 45 Beacon Street, Boston, MA 02108. (617) 227-2425. 1950 to present. Monthly. $450.00 per year.

SCIENCE CITATION INDEX. Institute for Scientific Information, 3501 Market Street, Philadelphia, PA 19104. (800) 523-1850 or (215) 386-0100. Inquire as to cost and availability.

SELECTED WATER RESOURCES ABSTRACTS. United States Geological Survey, Water Resources Scientific Information Center. Available from: National Technical Information Service, Springfield, VA 22161. (703) 860-7455. 1968 to present. Monthly. $115.00 per year.

ANNUAL REVIEWS AND YEARBOOKS

ADVANCES IN HYDROSCIENCE. Ven Te Chow, editor. Academic Press, Incorporated, Orlando, FL 32887. (800) 321-5068 or (305) 345-2734. 1964-1981. Inquire, price varies.

ASSOCIATIONS AND PROFESSIONAL SOCIETIES

AMERICAN INSTITUTE OF HYDROLOGY. Post Office Box 14251, St. Paul, MN 55114. (612) 379-1030.

AMERICAN SOCIETY OF CIVIL ENGINEERS. 345 East 47th Street, New York, NY 10017-2398. (212) 705-7520.

INTERNATIONAL ASSOCIATION OF HYDROGEOLOGISTS. Postbus 9090, NL-6800 GX Arnhem, The Netherlands.

DIRECTORIES AND BIOGRAPHICAL SOURCES

RESEARCH CENTERS DIRECTORY. Gale Research Company, Book Tower, Detroit, MI 48226. (800) 521-0707. Twelfth edition. 1987. $365.00 for set.

WHO'S WHO IN ENGINEERING. Engineers Joint Council, 345 East 47th Street, New York, NY 10017. (212) 705-7010. 1985. $200.00.

ENCYCLOPEDIAS AND DICTIONARIES

ENCYCLOPEDIA OF PHYSICAL SCIENCE AND TECHNOLOGY. Academic Press, Incorporated, Orlando, FL 32887. (800) 321-5068 or (305) 345-2734. Fifteen volumes, 1986. Inquire as to price.

MCGRAW-HILL ENCYCLOPEDIA OF SCIENCE AND TECHNOLOGY. McGraw-Hill Book, 1221 Avenue of the Americas, New York, NY 10020. (212) 512-2000. Inquire as to price.

GENERAL WORKS

APPLIED HYDROGEOLOGY. Charles W. Fetter. Charles E. Merrill Publishing Company, 1300 Alum Creek Drive, Columbus, OH 43216. (614) 890-1111. 1980. $38.95.

BASIC HYDROLOGY. James J. Sharp and P.G. Sawden. Butterworth's, Incorporated, 80 Montvale Avenue, Stoneham, MA 02180. (617) 438-8464. 1984. $15.95 in paper.

CHEMICAL HYDROGEOLOGY. William R. Back and Allan Freeze, editors. Van Nostrand Reinhold Company, Incorporated, 135 West 50th Street, New York, NY 10020. (212) 265-8700. 1983. $55.00.

FLUVIAL HYDROLOGY. S. Lawrence Dingman. W.H. Freeman, Incorporated, 41 Madison Avenue, New York, NY 10010. (212) 532-7660. 1983. $31.95.

GROUND WATER HYDROLOGY. David K. Todd. John Wiley and Sons, Incorporated, 605 Third Avenue, New York, NY 10158. (800) 526-5368 or (212) 850-6000. Second edition. 1980. $48.00.

HYDROLOGY AND QUALITY OF WATER RESOURCES. Mark J. Hammer. John Wiley and Sons, Incorporated, 605 Third Avenue, New York, NY 10158. (800) 526-5368 or (212) 850-6000. Second edition. 1986. $42.95.

HYDROLOGY FOR ENGINEERS. Ray K. Linsley and others. McGraw-Hill Book Company, 1221 Avenue of the Americas, New York, NY 10020. (212) 512-2000. Third edition. 1982. $46.95.

HYDROLOGY IN PRACTICE. Elizabeth M. Shaw. Van Nostrand Reinhold Company, Incorporated, 135 West 50th Street, New York, NY 10020. (212) 265-8700. 1983. $45.95.

HYDROLOGY: PRINCIPLES, ANALYSIS AND DESIGN. H.M. Raghunath. John Wiley and Sons, Incorporated, 605 Third Avenue, New York, NY 10158. (800) 526-5368 or (212) 850-6000. 1985. $34.95.

INTRODUCTION TO HYDROMETEOROLOGY. J.P. Bruce and R.H. Clark. Pergamon Press, Incorporated, Maxwell House, Fairview Park, Elmsford, NY 10523. (914) 592-7700. 1987. $50.00.

PHYSICAL HYDROGEOLOGY. R.A. Freeze and W. Back, editors. Van Nostrand Reinhold Company, Incorporated, 135 West 50th Street, New York, NY 10020. (212) 265-8700. 1983. $52.00.

ONLINE DATA BASES

COMPENDEX. Engineering Information, Incorporated, 345 East 47th Street, New York, NY 10017. (800) 221-1044 or (212) 705-7615. Engineering and technical literature, 1975 to present. Inquire as to cost and availability.

GEOREF. American Geological Institute, 4220 King Street, Alexandria, VA 22302. (703) 379-2480. 1967 to present. Inquire as to online cost and availability.

NTIS. National Technical Information Service, 5285 Port Royal Road, Springfield, VA 22161. (703) 487-4630. Broad coverage of government-sponsored research reports, 1964 to present. Inquire as to cost and availability.

WATER DATA BANK. United States Department of Agriculture, Agricultural Research Service, Hydrology Laboratory, Water Dat Center, Room 139, Building 007, BARC-West, Beltsville, MD 20705. (301) 344-4411. Inquire as to online cost and availability.

PERIODICALS

GROUND WATER. National Water Well Association. Water Well Journal Publishing Company, 500 West Wilson Bridge Road, Worthington, OH 43085. (614) 846-4967. Bimonthly. $53.00 per year.

JOURNAL OF GROUND WATER. National Water Well Association. Water Well Journal Publishing Company, 500 West Wilson Bridge Road, Worthington, OH 43085. (614) 846-4967. Quarterly. $16.00 per year.

JOURNAL OF HYDROLOGY. Elsevier Science Publishers B.V., Post Office Box 211, 1000 AE Amsterdam, The Netherlands. Thirty-two times per year. $675.00 per year.

JOURNAL OF IRRIGATION AND DRAINAGE. American Society of Civil Engineers, 345 East 47th Street, New York, NY 10017-2398. (212) 705-7520. Quarterly. $44.00 per year.

JOURNAL OF WATER RESOURCES PLANNING AND MANAGEMENT. American Society of Civil Engineers, 345 East 47th Street, New York, NY 10017-2398. (212) 705-7520. Quarterly. $56.00 per year.

LIMNOLOGY AND OCEANOGRAPHY. American Society of Limnology and Oceanography, 1530 12th Avenue, Grafton, WI 53024. (414) 377-4871. Bimonthly. $60.00 per year.

WATER RESOURCES BULLETIN. American Water Resources Association, 5410 Grosvenor Lane, Suite 220, Bethesda, MD 20814. (301) 493-8600. Bimonthly. $65.00 per year.

WATER RESOURCES RESEARCH. American Geophysical Union, 2000 Florida Avenue, NW, Washington, DC 20009. (202) 462-6903. Monthly. $295.00 per year.

RESEARCH CENTERS AND INSTITUTES

CALIFORNIA INSTITUTE OF TECHNOLOGY. W.M. Keck Engineering Laboratory of Hydraulics and Water Resources, 1201 East California Boulevard, Pasadena, CA 91125. (818) 356-4404.

ILLINOIS STATE GEOLOGICAL SURVEY. Natural Resources Building, 615 East Peabody Drive, Champaign, IL 61820. (217) 344-1481.

TEXAS A AND M UNIVERSITY. Center for Engineering Geosciences, College Station, TX 77843. (409) 845-3224.

UNITED STATES ARMY HYDROLOGIC ENGINEERING CENTER. 609 Center Street, Davis, CA 95616. (916) 440-3285.

UNIVERSITY OF MINNESOTA. Minnesota Geological Survey, 2642 University Avenue, St. Paul, MN 55114. (612) 373-3372.

HYDROMECHANICS

See: HYDRODYNAMICS

HYDROSTATICS

See: HYDRODYNAMICS

HYPERBOLIC NAVIGATION SYSTEM

See: NAVIGATION

HYPERSONIC FLIGHT

See: AERODYNAMICS

HYPERSONICS

See: ACOUSTICS

I

ICE

See also: METEOROLOGY

ABSTRACT SERVICES AND INDEXES

APPLIED SCIENCE AND TECHNOLOGY INDEX. H.W. Wilson Company, 950 University Avenue, Bronx, NY 10452. (800) 367-6670 or (212) 588-8400. Inquire as to cost and availability.

CHEMICAL ABSTRACTS. Chemical Abstracts Service, 2540 Olentangy Road, Post Office Box 3012, Columbus, OH 43210. (800) 848-6538 or (614) 421-3600. Weekly. $9200.00 per year.

CURRENT CONTENTS: PHYSICAL AND CHEMICAL SCIENCES. Institute for Scientific Information, 3501 Market Street, Philadelphia, PA 19104. (800) 523-1850 or (215) 386-0100. $272.00 per year.

GENERAL SCIENCE INDEX. H.W. Wilson Company, 950 University Avenue, Bronx, NY 10452. (800) 367-6770 or (212) 588-8400. Inquire as to cost and availability.

METEOROLOGICAL AND GEOASTROPHYSICAL ABSTRACTS. American Meteorological Society, 45 Beacon Street, Boston, MA 02108. (617) 227-2425. 1950 to present. Monthly. $450.00 per year.

SCIENCE CITATION INDEX. Institute for Scientific Information, 3501 Market Street, Philadelphia, PA 19104. (800) 523-1850 or (215) 386-0100. Inquire as to cost and availability.

ASSOCIATIONS AND PROFESSIONAL SOCIETIES

AMERICAN METEOROLOGICAL SOCIETY. 45 Beacon Street, Boston, MA 02108. (617) 227-2425.

INTERNATIONAL ASSOCIATION OF METEOROLOGY AND ATMOSPHERIC PHYSICS. UCAR, Post Office Box 3000, Boulder, CO 80307.

NATIONAL WEATHER ASSOCIATION. 4400 Stamp Road, Room 404, Temple Hills, MD 20748. (301) 899-3784.

UNIVERSITY CORPORATION FOR ATMOSPHERIC RESEARCH. Box 3000, 1850 Table Mesa Drive, Boulder, CO 80307. (303) 497-1000.

WEATHER MODIFICATION ASSOCIATION. Post Office Box 8116, Fresno, CA 93747. (209) 291-8466.

BIBLIOGRAPHIES

SCIENCE BOOKS AND FILMS. American Association for the Advancement of Science, 1333 H Street, NW, Washington, DC 20005. Five times per year. $20.00 per year.

SCIENTIFIC AND TECHNICAL BOOKS AND SERIALS IN PRINT 1988; AN INDEX TO LITERATURE IN SCIENCE AND TECHNOLOGY. R.R. Bowker Company, 205 East 42nd Street, New York, NY 10017. (800) 521-8110 or (212) 916-1600. $175.00.

AMERICAN MEN AND WOMEN OF SCIENCE. Physical and Biological Sciences. Sixteenth edition. R.R. Bowker Company, 205 East 42nd Street, New York, NY 10017. (800) 521-8110 or (212) 916-1600. 1986. $595.00.

INTERDOC: DIRECTORY OF PUBLISHED PROCEEDINGS, SERIES. SEMT-Science/Engineering/Medicine/Technology. Interdoc Corporation, 173 Halstead Avenue, Box 326, Harrison, NY 10528. (014) 835-3506. Ten times per year. $325.00 per year.

INTERNATIONAL RESEARCH CENTERS DIRECTORY 1988-1989. Gale Research Company, Book Tower, Detroit, MI 48226. (800) 521-0707. Fourth edition. 1987. $360.00.

METEOROLOGICAL SERVICES OF THE WORLD. World Meteorological Organization. Available from: American Meteorological Society, 45 Beacon Street, Boston, MA 02108. (617) 227-2425. Annual. $35.00.

NATIONAL WEATHER SERVICE OFFICES AND STATIONS. National Oceanic and Atmospheric Administration, Department of Commerce, Silver Spring, MD 20910. (301) 427-7698. Annual. Free.

RESEARCH CENTERS DIRECTORY. Gale Research Company, Book Tower, Detroit, MI 48226. (800) 521-0707. Twelfth edition. 1987. $365.00 for set.

ENCYCLOPEDIAS AND DICTIONARIES

ENCYCLOPEDIA OF CLIMATOLOGY. John E. Oliver and Rhodes W. Fairbridge, editors. Van Nostrand Reinhold, Incorporated, 115 Fifth Avenue, New York, NY 10003. (800) 543-2681. 1987. $89.95.

GENERAL WORKS

THE CLIMACTIC RECORD IN POLAR ICE SHEETS. Gordon De Q. Robin. Cambridge University Press, 32 East 57th Street, New York, NY 10022. (212) 688-8888. 1983. $65.00.

DYNAMICS OF SNOW AND ICE MASSES. Samuel C. Colbeck. Academic Press, Incorporated, Orlando, FL 32887. (305) 345-4100. 1980. $60.50.

ICE TECHNOLOGY. T.K. Murthy, et al. Springer-Verlag New York, Incorporated, 175 Fifth Avenue, New York, NY 10010. (212) 460-1500. 1986. $125.50.

PHYSICS AND MECHANICS OF ICE: PROCEEDINGS. P. Tryde. Springer-Verlag New York, Incorporated, 175 Fifth Avenue, New York, NY 10010. (212) 460-1500. 1980. $43.70.

ICE

ONLINE DATA BASES

CA SEARCH. Chemical Abstracts Service, Post Office Box 3012, Columbus, OH 43210. Guide to chemical literature, 1972 to present. Inquire as to cost and availability.

DISSERTATION ABSTRACTS ONLINE. University Microfilms International, 300 North Zeeb Road, Ann Arbor, MI 48106. (800) 521-0600 or (313) 761-4700. Scope includes virtually all doctoral dissertations accepted at accredited American institutions from 1861 to present in 252 subject areas. Inquire as to cost and availability.

METEOROLOGICAL AND GEOASTROPHYSICAL ABSTRACTS. American Meteorological Society, 45 Beacon Street, Boston, MA 02108. (617) 227-2425. 1950 to present. Monthly. $450.00 per year.

SCISEARCH. Institute for Scientific Information, 3501 Market Street, Philadelphia, PA 19104. (800) 523-1850 or (215) 386-0100. Broad multidisciplinary title and author index to the international literature of science and technology, 1974 to present. Inquire as to cost and availability.

PERIODICALS

AMERICAN METEOROLOGICAL SOCIETY BULLETIN. American Meteorological Society, 45 Beacon Street, Boston, MA 02108. (617) 227-2425. 1920 to present. Monthly. $60.00 per year.

JOURNAL OF CLIMATE AND APPLIED METEOROLOGY. American Meteorological Society, 45 Beacon Street, Boston, MA 02108. (617) 227-2425. 1962 to present. Monthly. $120.00 per year.

JOURNAL OF THE ATMOSPHERIC SCIENCES. American Meteorological Society, 45 Beacon Street, Boston, MA 02108. (617) 227-2425. 1944 to present. Semimonthly. $220.00 per year.

MONTHLY WEATHER REVIEW. American Meteorological Society, 45 Beacon Street, Boston, MA 02108. (617) 227-2425. 1872 to present. Monthly. $120.00 per year.

NATIONAL WEATHER DIGEST. National Weather Association, 4400 Stamp Road, Room 404, Temple Hills, MD 20748. (301) 899-3784. 1976 to present. Quarterly. $20.00 per year.

WEATHER. Royal Meteorological Society, James Glaisher House, Grenville Place, Bracknell Berkshire, RG12 1BX, England. 1946 to present. $30.00 per year.

RESEARCH CENTERS AND INSTITUTES

DARTMOUTH COLLEGE. Ice Research Laboratory. Thayer School of Engineering, Hanover, NH 03755. (603) 646-2888.

STATE UNIVERSITY OF NEW YORK AT BUFFALO. Ice Core Laboratory, Department of Geological Sciences, 4240 Ridge Lea Road, Amherst, NY 14226. (716) 831-3054.

UNIVERSITY OF NEVADA. Atmospheric Ice Laboratory, Post Office Box 60220, Reno, NV 89506. (702) 972-1676.

UNIVERSITY OF NEW HAMPSHIRE. Glacier Research Group, Durham, NH 03824. (603) 862-1718.

ICEBERGES

See: ICE

IGNEOUS ROCKS

See also: CRUST, EARTH SCIENCES, GEOLOGY, GEOPHYSICS, GEOCHEMISTRY

ABSTRACT SERVICES AND INDEXES

BIBLIOGRAPHY AND INDEX OF GEOLOGY. American Geological Institue, 4220 King Street, Alexandria, Va 22302. (703) 379-2480. 1969 to present. Monthly. $1100.00 per year.

CHEMICAL ABSTRACTS. Chemical Abstracts Service, 2540 Olentangy Road, P.O. Box 3012, Columbus, OH 43210. (800) 848-6538 or (614) 421-3600. Weekly. $9200.00 per year.

DEEP-SEA RESEARCH WITH OCEANOGRAPHIC LITERATURE REVIEW. Pergamon Press, Inc., Maxwell House, Fairview Park, Elmsford, NY 10523. (914) 592-7700. 1953 to present. Twenty-four times per year. $600.00 per year.

MINERALOGICAL ABSTRACTS. Mineralogical Society and the Mineralogical Society of America, 41 Queen's Gate, London, SW7 5HR, England. 1959 to present. Quarterly. $190.00 per year.

OCEANIC ABSTRACTS. Cambridge Scientific Abstracts, 5161 River Road, Bethesda, MD 20816. (301) 951-1400. 1964 to present. Bimonthly. $652.00 per year.

SCIENCE CITATION INDEX. Institute for Scientific Information, 3501 Market Street, Philadelphia, PA 19104. (800) 523-1850 or (215) 386-0100. Inquire as to price and availability.

GENERAL SCIENCE INDEX. H.W. Wilson Company, 950 University Avenue, Bronx, NY 10452. (800) 367-6670 or (212) 588-8400. Inquire as to price and availability.

ANNUAL REVIEWS AND YEARBOOKS

ANNUAL REVIEW AND EARTH AND PLANETARY SCIENCES. Annual Reviews, Inc., 4139 El Camino Way, Palo Alto, CA 94306. (415) 493-4400.

MINERALS YEARBOOK. Bureau of Mines, U.S. Department of the Interior. Available from U.S. Government Printing Office, Washington, DC 20402. (202) 783-3238. Annual. Three volumes. $45.00.

ASSOCIATIONS AND PROFESSIONAL SOCIETIES

AMERICAN ASSOCIATION OF PETROLEUM GEOLOGISTS, P.O. Box 979, Tulsa, OK 74101. (918) 584-2555.

AMERICAN GEOLOGICAL INSTITUTE. 4220 King Street, Alexandria, Va 22302. (703) 379-2480.

AMERICAN INSTITUTE OF PROFESSIONAL GEOLOGISTS. 7828 Vance Drive, Suite 103, Arvada, Co 80003. (303) 431-0831.

ASSOCIATION OF ENGINEERING GEOLOGISTS. P.O. Box 1068, Brentwood, TN 37027. (615) 377-3578.

GEOLOGICAL SOCIETY OF AMERICA. Box 9140, Boulder, Co 80301. (303) 447-2020.

BIBLIOGRAPHIES

SCIENTIFIC AND TECHNICAL BOOKS AND SERIALS IN PRINT 1988; An Index to Literature in Science and TEchnology. R.R. Bowker Co., E. 42nd Street, New York, NY 10017. (800) 521-8110 or (212) 916-1600. $175.00.

DIRECTORIES AND BIOGRAPHICAL SOURCES

AMERICAN MEN AND WOMEN OF SCIENCE: PHYSICAL AND BIOLOGICAL SCIENCES. Sixteenth edition. R.R. Bowker Co., E. 42nd Street, New York, NY 10017. (800) 521-8110 or (212) 916-1600. $595.00.

AMERICAN INSTITUTE OF PROFESSIONAL GEOLOGISTS. Membership Directory. American Institute of Professional Geologists, 7828 Vance Drive, Suite 103, Arvada, CO 80003. (303) 431-0831. Annual, $15.00.

ASSOCIATION OF ENGINEERING GEOLOGISTS DIRECTORY. Association of Engineering Geologists, Dr. G. Lee Christensen, Civil Engineering Department, Villanova University, Villanova, PA 19085. (215) 645-4960. Annual. $15.00.

GEOLOGICAL SOCIETY OF AMERICA. Membership Directory. Geological Society of America, 3300 Penrose Place, Boulder, CO 80301. (303) 447-2020. Annual. Available to members only.

RESEARCH CENTERS DIRECTORY. Gale Research Company, Book Tower, Detroit, MI 48226. (800) 521-0707. 12th edition, 1987. $365.00 for set.

WHO'S WHO IN FRONTIER SCIENCE AND TECHNOLOGY. Marquis Who's Who, Inc., 200 East Ohio Street, Chicago, IL 60611. (800) 428-3898 or (312) 787-2008.

ENCYCLOPEDIAS AND DICTIONAIRES

DICTIONARY OF GEOLOGICAL TERMS. American Geological Institute. Doubleday and Company, Inc., 245 Park Avenue, New York, NY 10017. (800) 645-6156 or (212) 953-4561. Third edition. 1984. $19.95.

DICTIONARY OF PETROLOGY. S.I. Tomkeieff. John Wiley & Sons, 605 Third Avenue, New York, NY 10158. (800) 526-5368 or (212) 850-6000. 1983. $140.00.

ENCYCLOPEDIA OF PHYSICAL SCIENCE AND TECHNOLOGY. Academic Press, Inc., Orlando, FL 32887. (800) 321-5068.

GLOSSARY OF GEOLOGY. Robert L. Bates and Julia A. Jackson. American Geological Institute, 4220 King Street, Alexandria, VA 22032. (703) 379-2480. 1980. $60.00.

GENERAL WORKS

ATLAS OF IGNEOUS ROCKS AND THEIR TEXTURES. W.S. MacKenzie, C.H. Donaldson, C. Guilford. John Wiley & Sons, 605 Third Avenue, New York, NY 10158. (800) 526-5368 or (212) 850-6000. 1982. $29.95.

INTRODUCTION TO ROCK FORMING MINERALS. W.A. Deer, J. Zussman and R.A. Howie. John Wiley & Sons, 605 Third Avenue, New York, NY 10158. (800) 526-5368 or (212) 850-6000. 1966. $29.95.

PHYSICAL GEOLOGY. R.F. Flint and B.J. Skinner. John Wiley & Sons, 605 Third Avenue, New York, NY 10158. (800) 526-5368 or (212) 850-6000. Second edition. 1977. $40.95.

GEOLOGY IN THE FIELD. R.R. Bowker Co., E. 42nd Street, New York, NY 10017. (800) 521-8110 or (212) 916-1600. 1985. $23.95.

ONLINE DATA BASES

GEOREF. American Geological Institute, 4220 King Street, Alexandria, VA 22302. (800) 336-4764 or (703) 379-2480. Geology and geosciences literature, 1961 to present. Inquire as to online cost and availability.

GEOARCHIVE. Geosystems, P.O. Box 1024, Westminster, London, England, SW1 P 2JL. Citations to literature on geoscience, 1969 to present. Inquire as to online cost and availability.

NTIS. National Technical Information Service, 5285 Port Royal Road, Springfield, VA 22161. (703) 487-4630. Broad coverage of government sponsored research reports, 1964 to present. Inquire as to online cost and availability.

SCISEARCH. Institute for Scientific Information, 3501 Market Street, Philadelphia, PA 19104. (800) 523-1850 or (215) 386-0100. Broad interdisciplinary index to the literature of science and technology, 1965 to present. Inquire as to online cost and availability.

PERIODICALS

AMERICAN JOURNAL OF SCIENCE. Kline Geology laboratory, Yale University, New Haven, CT 06520. Ten times per year. $80.00 per year.

GEOCHIMICA ET COSMOCHIMICA ACTA. Pergamon Journals, Inc., Maxwell House, Fairview Park, Elmsford, NY 10523. (914) 592-7700. Monthly. $340.00 per year.

GEOLOGICAL MAGAZINE. Cambridge University Press, 32 East 57th Street, New York, NY 10022. (800) 872-7423. Bimonthly. $165.00 per year.

GEOLOGICAL SOCIETY JOURNAL. Geological Society of London. Blackwell Scientific Publications, Inc., 667 Lytton Avenue, Palo Alto, CA 94301. (415) 324-1688. 6 times per year. $290.00.

GEOLOGICAL SOCIETY OF AMERICA BULLETIN. P.O. Box 9140, 3300 Penrose Place, Boulder, CO 80301. (303) 447-2020. Monthly. $80.00.

GEOLOGY. Geological Society of America. P.O. Box 9140, 3300 Penrose Place, Boulder, CO 80301. (303) 447-2020. Monthly. $55.00 per year.

JOURNAL OF GEOLOGY. University of Chicago Press, 5801 South Ellis Street, Chicago, IL 60637. (312) 962-7600. Bimonthly. $30.00 per year.

JOURNAL OF PETROLOGY. Oxford University Press, Walton Street, Oxford, OX2 6DP, England. Quarterly. $135.00 per year.

JOURNAL OF STRUCTURAL GEOLOGY. Pergamon Journals, Inc., Maxwell House, Fairview Park, Elmsford, NY 10523. (914) 592-7700. Eight times per year. $160.00 per year.

LITHOS; An International Journal of Mineralogy, Petrology, and Geochemistry. Elsevier Science Publishers, P.O. Box 330, Irving-on-Hudson, NY 10533. Quarterly. $73.00 per year.

MOUNTAIN GEOLOGIST. Rocky Mountain Association of Geologists, 4201 West 51st Avenue, Denver, CO 80212-2902. Quarterly. $15.00 per year.

ROCK MECHANICS AND ROCK ENGINEERING. Springer-Verlag New York, Inc., 175 Fifth Avenue, New York, NY 10010. (800) 526-7254 or (212) 460-1500. Quarterly. $61.00 per year.

TECTONICS. American Geophysical Union, 2000 Florida Avenue, N.W., Washington, DC 20009. (202) 462-6903. Bimonthly. $30.00 per year to individuals.

TECTONOPHYSICS; an international journal of geotectonics and the geology and physics of the interior of the Earth. Elsevier Science Publishers, P.O. Box 330, Irving-on-Hudson, NY 10533. Forty-four times per year. $870.00 per year.

TEXTURES AND MICROSTRUCTURES. Gordon and Breach Science Publishers, Limited, P.O. Box 197, London WC2E 9PX, England. Eight times per year. $288.00 per year.

VOLCANOLOGY AND SEISMOLOGY. Gordon and Breach Science Publishers, 50 West 23rd Street, New York NY 10010.

(212) 206-8900. Monthly. $498.00 per year.

RESEARCH CENTERS AND INSTITUTES

U.S. GEOLOGICAL SURVEY, National Center, 12201 Sunrise Valley Drive, Reston, VA 22092. The major geological research agency of the federal government conducting research in most areas of pure and applied research in the geosciences.

GEOLOGICAL SOCIETY OF AMERICA. P.O. Box 9140, 3300 Penrose Place Boulder, CO 80301. (303) 447-2020.

SOCIETY OF ECONOMIC GEOLOGISTS. P.O. Box 571, Golden, CO 80402. (303) 279-1899.

SOCIETY OF ECONOMIC PALEONTOLOGISTS AND MINERALOGISTS. P.O. Box 4756, Tulsa, OK 74159. (918) 743-9765.

ILLUMINATION

See also: ELECTRICAL ENGINEERING, POWER ENGINEERING

ABSTRACT SERVICES AND INDEXES

APPLIED SCIENCE AND TECHNOLOGY INDEX. H.W. Wilson and Company, 950 University Avenue, Bronx, NY 10452. (800) 367-6670 or (212) 588-8400. Monthly. Inquire as to cost and availability.

CHEMICAL ABSTRACTS. American Chemical Society, Chemical Abstracts Service, Box 3012, Columbus, OH 43210. (614) 421-3600. 1907 to present. Weekly. $9500.00 per year.

CURRENT CONTENTS: ENGINEERING, TECHNOLOGY AND APPLIED SCIENCES. Institute for Scientific Information, 3501 Market Street, Philadelphia, PA 19104. (800) 523-1850 or (215) 386-0100. Weekly. $275.00 per year.

ELECTRICAL AND ELECTRONICS ABSTRACTS. Institution of Electrical Engineers. Available from: Institute of Electrical and Electronics Engineers. IEEE Service Center, 445 Hoes Lane, Piscataway, NJ 08854. Monthly. $1250.00 per year.

ENGINEERING INDEX MONTHLY AND AUTHOR INDEX. Engineering Information Inc., 345 East 47th Street, New York, NY 10017. (212) 705-7600. Monthly. $1560.00 per year.

IEEE PUBLICATIONS BULLETIN. Institute of Electrical and Electronics Engineers. Institute of Electrical and Electronics Engineers. IEEE Service Center, 445 Hoes Lane, Piscataway, NJ 08854. Quarterly. Free.

ASSOCIATIONS AND PROFESSIONAL SOCIETIES

AMERICAN ELECTRONICS ASSOCIATION. P.O. Box 10045, 2670 Hanover Street, Palo Alto, CA 94303. (415) 857-9300.

AMERICAN INSTITUTE OF PHYSICS. 335 East 45th Street, New York, NY 10017. (212) 661-9494.

EDISON ELECTRIC INSTITUTE. 1111 19th Street, N.W., Washington, DC 20036. (202) 828-7400.

ELECTRONICS INDUSTRIES ASSOCIATION. 2001 Eye Street, N.W., Washington, DC 20006. (202) 457-4900.

ILLUMINATING ENGINEERING SOCIETY OF NORTH AMERICA. 345 East 47th Street, New York, NY 10017. (212) 705-7925.

INSTITUTE OF ELECTRICAL AND ELECTRONICS ENGINEERS. 345 East 47th Street, New York, NY 10017. (212) 705-7900.

DIRECTORIES AND BIOGRAPHICAL SOURCES

AMERICAN MEN AND WOMEN OF SCIENCE. R.R. Bowker, Inc., Order Department, 245 West 17th Street, New York, NY 10011. (800) 521-8110. Eight volumes. 1986. $595.00 for set.

IEEE MEMBERSHIP DIRECTORY. Institute of Electrical and Electronics Engineers. IEEE Service Center, 445 Hoes Lane, Piscataway, NJ 08854. Annual. $7.00.

INTERNATIONAL RESEARCH CENTERS DIRECTORY 1988-89. Darren L. Smith, editor. Gale Research Company, Book Tower, Detroit, MI 48226. (800) 521-0707. 4th edition. 1987. $360.00.

1987 DIRECTORY OF ENGINEERING SOCIETIES AND RELATED ORGANIZATIONS. Gordon Davis, editor. Hemisphere Publishing Corporation, 1010 Vermont Avenue, NW, Washington, DC 20005. (800) 526-0275. 12th edition. 1987. $100.00.

RESEARCH CENTERS DIRECTORY 1988. Gale Research Company, Book Tower, Detroit, MI 48226. (800) 521-0707. 12th edition. 1987. $365.00 for set.

SCIENTIFIC AND TECHNICAL ORGANIZATIONS AND AGENCIES DIRECTORY. Margaret Labash Young, editor. Gale Research Company, Book Tower, Detroit, MI 48226. (800) 521-0707. 2nd edition. 1987. $185.00.

WHO'S WHO IN ENGINEERING. Gordon Davis, editor. Hemisphere Publishing Corporation, 1010 Vermont Avenue, NW, Washington, DC 20005. (800) 526-0275. 6th edition. 1985. $200.00.

GENERAL WORKS

HUMAN FACTORS IN LIGHTING. P.R. Boyce. Macmillan Publishing Company, Inc., 866 Third Avenue, New York, NY 10022. (800) 257-5755. 1981. $49.00.

ILLUMINATION ENGINEERING. Joseph B. Murdoch. Macmillan Publishing Company, Inc., 866 Third Avenue, New York, NY 10022. (800) 257-5755. 1985. $55.00.

ILLUMINATION ENGINEERING FOR ENERGY EFFICIENT LUMINOUS ENGINEERING. Ronald N. Helms. Prentice-Hall Publishing, Inc., Englewood Cliffs, NJ 07632. (800) 562-0245. 1980. $58.95.

INTRODUCTORY LIGHTING EDUCATION. Illuminating Engineering Society of North America, 345 East 47th Street, New York, NY 10017. (212) 705-7925. 1985. $125.00 for set.

LAMPS AND LIGHTING. M.S. Cayles and A.M. Marsden. Edward Arnold Publications, Limited, 300 North Carles Street, Baltimore, MD 21201. (301) 539-1529. Third edition. 1983. $67.95.

HANDBOOKS AND MANUALS

IES LIGHTING READY REFERENCE. Illuminating Engineering Society of North America, 345 East 47th Street, New York, NY 10017. (212) 705-7925. 1985. $50.00.

ONLINE DATA BASES

CA SEARCH. Chemical Abstracts Service, P.O. Box 3012, Columbus, OH 43120. (800) 848-6538 or (614) 421-3600. Comprehensive guide to chemical literature, 1972 to present. Inquire as to online cost and availability.

COMPENDEX. Engineering Information, Inc., 345 East 47th Street, New York, NY 10017. (800) 221-1044 or (212) 705-7615. Engineering and technical literature, 1975 to present. Inquire as to online cost and availability.

NTIS. National Technical Information Service, 5285 Port Royal Road, Springfield, VA 22161. (703) 487-4630. Broad coverage of government sponsored research reports, 1964 to present. Inquire

as to online cost and availability.

WILSONLINE. H.W. Wilson and Company, 950 University Avenue, Bronx, NY 10452. (800) 367-6770 or (212) 588-8400. Makes available online versions of the H.W. Wilson indexes including Applied Science and Technology Index, Business Periodicals Index and Readers' Guide to Periodical Literature. Approximately 1980 to present. Inquire as to online cost and availability.

PERIODICALS

IEEE TRANSACTIONS ON POWER SYSTEMS. Power Engineering Society. Institute of Electrical and Electronics Engineers. IEEE Service Center, 445 Hoes Lane, Piscataway, NJ 08854. 1959 to present. Monthly. $175.00 per year.

INSTITUTE OF ELECTRICAL AND ELECTRONICS ENGINEERS PROCEEDINGS. Institute of Electrical and Electronics Engineers. IEEE Service Center, 445 Hoes Lane, Piscataway, NJ 08854. 1913 to present. Monthly. $140.00 per year.

JOURNAL OF THE ILLUMINATING ENGINEERING SOCIETY. Illuminating Engineering Society of North America, 345 East 47th Street, New York, NY 10017. (212) 705-7925. 1971 to present. Semi-annual. $195.00 per year.

L D AND A: LIGHTING DESIGN AND APPLICATION. Illuminating Engineering Society of North America, 345 East 47th Street, New York, NY 10017. (212) 705-7925. 1906 to present. Monthly. $35.00 per year.

RESEARCH CENTERS AND INSTITUTES

LAWRENCE BERKELEY LABORATORY, APPLIED SCIENCES DIVISION. 1 Cyclotron Road, Berkeley, CA 94720. (415) 486-5001.

LIGHTING RESEARCH INSTITUTE. 1170 Broadway, Room 1114, New York, NY 10001. (212) 685-8680.

OFFICE OF BUILDING RESEARCH. Louisiana State University, School of Architecture, Baton Rouge, LA 70803. (504) 388-6885.

IMAGE ANALYSIS

See: CAD (COMPUTER-AIDED DESIGN)

IMAGE PROCESSING

See: CAD (COMPUTER-AIDED DESIGN)

IMPACT TESTING

See also: DESIGN ENGINEERING, ERGONOMICS, EXPLOSIVES, MATERIALS SCIENCE, METALLURGICAL ENGINEERING, NONDESTRUCTIVE TESTING, STRUCTURAL ENGINEERING

ABSTRACT SERVICES AND INDEXES

APPLIED MECHANICS REVIEW. American Society of Mechanical Engineers, 345 East 47th Street, New York, NY 10017. (212) 705-7703. 1948 to present. Monthly. $360.00 per year.

APPLIED SCIENCE AND TECHNOLOGY INDEX. H.W. Wilson and Company, 950 University Avenue, Bronx, NY 10452. (800) 367-6670 or (212) 588-8400. Monthly. Inquire as to cost and availability.

CERAMIC ABSTRACTS. American Ceramic Society, Inc., 65 Ceramic Drive, Columbus, OH 43214. (614) 268-8645. 1922 to present. Bimonthly. $185 per year.

CHEMICAL ABSTRACTS. American Chemical Society, Chemical Abstracts Service, Box 3012, Columbus, OH 43210. (614) 421-3600. 1907 to present. Weekly. $9500.00 per year.

CURRENT CONTENTS: ENGINEERING, TECHNOLOGY AND APPLIED SCIENCES. Institute for Scientific Information, 3501 Market Street, Philadelphia, PA 19104. (800) 523-1850 or (215) 386-0100. Weekly. $275.00 per year.

ENGINEERING INDEX MONTHLY AND AUTHOR INDEX. Engineering Information Inc., 345 East 47th Street, New York, NY 10017. (212) 705-7600. Monthly. $1560.00 per year.

INTERNATIONAL AEROSPACE ABSTRACTS. American Institute of Aeronautics and Astronautics, Technical Information Service, 555 West 57th Street, New York, NY 10019. (212) 247-6500. 1961 to present. Semi-monthly. $700.00 per year.

ISMEC BULLETIN (Information Service in Mechanical Engineering). Cambridge Scientific Abstracts, 5161 River Road, Bethesda, MD 20816. (301) 951-1400. 1973 to present. Monthly. $450.00 per year.

METALS ABSTRACTS AND METALS ABSTRACTS INDEX. American Society for Metals, Metals Park, OH 44073. (216) 338-5151. 1968 to present. Monthly. Abstracts are $1100.00 per year and Index is $460.00 per year.

PHYSICS ABSTRACTS. Institution of Electrical Engineers. Available from: IEEE Service Center, 445 Hoes Lane, Piscataway, NJ 08854. 1898 to present. Bimonthly. $1700.00 per year.

SCIENCE CITATION INDEX. Institute for Scientific Information, 3501 Market Street, Philadelphia, PA 19104. (800) 523-1850 or (215) 386-0100. Six times per year. $6200. 00 per year.

WORLD ALUMINUM ABSTRACTS. Aluminum Association, 818 Connecticut Avenue, NW, Washington, DC 20006. (202) 862-5156. 1968 to present. Monthly. $240.00 per year.

ASSOCIATIONS AND PROFESSIONAL SOCIETIES

AMERICAN CERAMIC SOCIETY, Inc., 65 Ceramic Drive, Columbus, OH 43214. (614) 268-8645. 1922 to present. Bimonthly. $185 per year.

AMERICAN POWDER METALLURGY INSTITUTE. 105 College Road, East, Princeton, NJ 08540. (609) 452-7700.

AMERICAN SOCIETY FOR METALS. Metals Park, OH 44073. (216) 338-5151.

AMERICAN SOCIETY FOR NONDESTRUCTIVE TESTING. 4153 Arlingate Plaza, Caller #28518, Columbus, OH 43228. (614) 274-6003.

AMERICAN SOCIETY FOR TESTING AND MATERIALS. 1916 Race Street, Philadelphia, PA 19103. (215) 299-5400.

AMERICAN SOCIETY OF MECHANICAL ENGINEERS. 345 47th Street, New York, NY 10017. (212) 705-7722.

MATERIALS PROPERTIES COUNCIL. 345 East 47th Street, New York, NY 10017. (212) 705-7693.

MATERIALS RESEARCH SOCIETY. 9800 McKnight Road, Suite 327, Pittsburgh, PA 15237. (412) 367-3003.

THE METALLURGICAL SOCIETY. 420 Commonwealth Drive, Warrendale, PA 15086. (412) 776-9000.

NATIONAL ASSOCIATION OF CORROSION ENGINEERS. Box 218340, Houston, TX 77218. (713) 492-0535.

NATIONAL MATERIALS ADVISORY BOARD. 2101 Constitution Avenue, NW, Washington, DC 20418. (202) 334-3505.

SOCIETY FOR THE ADVANCEMENT OF MATERIAL AND PROCESS ENGINEERING (SAMPE). P.O. Box 2459, 843 West Glentana Street, Covina, CA 91722. (818) 331-0616.

DIRECTORIES AND BIOGRAPHICAL SOURCES

MATERIALS DIRECTORY. Cahners Publishing Company, 275 Washington Street, Newton, MA 02158. Annual. $15.00.

1987 DIRECTORY OF ENGINEERING SOCIETIES AND RELATED ORGANIZATIONS. Gordon Davis, editor. Hemisphere Publishing Corporation, 79 Madison Avenue, New York, NY 10016-7892. (800) 821-8312. 12th edition. 1987. $100.00.

RESEARCH CENTERS DIRECTORY 1988. Gale Research Company, Book Tower, Detroit, MI 48226. (800) 521-0707. 12th edition. 1987. $365.00 for set.

SCIENTIFIC AND TECHNICAL ORGANIZATIONS AND AGENCIES DIRECTORY. Margaret Labash Young, editor. Gale Research Company, Book Tower, Detroit, MI 48226. (800) 521-0707. 2nd edition. 1987. $185.00.

WHO'S WHO IN ENGINEERING. Gordon Davis, editor. Hemisphere Publishing Corporation, 79 Madison Avenue, New York, NY 10016-7892. (800) 821-8312. 6th edition. 1985. $200.00.

GENERAL WORKS

ENGINEERING MATERIALS: AN INTRODUCTION TO THEIR PROPERTIES AND APPLICATIONS. Michael F. Ashby and David R.H. Jones. Pergamon Press, Inc., Maxwell House, Fairview Park, Elmsford, NY 10523. (914) 592-7700. 1980. $91.50.

IMPACT DYNAMICS. Jonas A. Zukas and others. John Wiley and Sons, Inc., 605 Third Avenue, New York, NY 10158. (800) 526-5368. 1982. $61.95.

INSTRUMENTED IMPACT TESTING. American Society for Testing and Materials, 1916 Race Street, Philadelphia, PA 19103. (215) 299-5400. STP 563. 1974. $21.75.

INTRODUCTION TO MATERIALS SCIENCE FOR ENGINEERS. James F. Shackelford. Macmillan Publishing Company, Inc., 866 Third Avenue, New York, NY 10022. (800) 257-5755. 2nd edition. 1988. Inquire.

MATERIALS AND PROCESSES FOR NDT TECHNOLOGY. Harry D. Moore, editor. American Society for Nondestructive Testing. 4153 Arlingate Plaza, Caller #28518, Columbus, OH 43228. (614) 274-6003. 1984. $30.75.

PRINCIPLES OF MATERIALS SCIENCE AND ENGINEERING. William F. Smith. McGraw-Hill Book Company, 1221 Avenue of the Americas, New York, NY 10020. (212) 512-2000. 1986. $45.95.

STATISTICS AND MECHANICS OF MATERIALS. Braja M. Das and Paul C. Hassler. Prentice-Hall Publishing, Inc., Englewood Cliffs, NJ 07632. (800) 562-0245. 1988. $42.50.

HANDBOOKS AND MANUALS

MATERIALS HANDBOOK. George S. Brady and Henry R. Clauser. McGraw-Hill Book Company, 1221 Avenue of the Americas, New York, NY 10020. (212) 512-2000. 12th edition. 1986. $59.50.

ONLINE DATA BASES

CA SEARCH. Chemical Abstracts Service, P.O. Box 3012, Columbus, OH 43120. (800) 848-6538 or (614) 421-3600. Comprehensive guide to chemical literature, 1972 to present. Inquire as to online cost and availability.

COMPENDEX. Engineering Information, Inc., 345 East 47th Street, New York, NY 10017. (800) 221-1044 or (212) 705-7615. Engineering and technical literature, 1975 to present. Inquire as to online cost and availability.

INSPEC. INSPEC Marketing Department, Institution of Electrical Engineers. Available from IEEE Service Center, 445 Hoes Lane, Piscataway, NJ 08854. (201) 981-0060. Online version of Physics Abstracts. Inquire as to online cost and availability.

ISMEC. Cambridge Scientific Abstracts, 5161 River Road, Besthda, MD 20816. (800) 638-8076 or (301) 951-1400. Literature of mechanical and production engineering, 1973 to present. Inquire as to online cost and availability.

NTIS. National Technical Information Service, 5285 Port Royal Road, Springfield, VA 22161. (703) 487-4630. Broad coverage of government sponsored research reports, 1964 to present. Inquire as to online cost and availability.

SCISEARCH. Institute for Scientific Information, 3501 Market Street, Philadelphia, PA 19104. (800) 523-1850 or (215) 386-0100. Broad multidisciplinary title and author index to the international literature of science and technology, 1974 to present. Inquire as to online cost and availability.

WILSONLINE. H.W. Wilson and Company, 950 University Avenue, Bronx, NY 10452. (800) 367-6770 or (212) 588-8400. Makes available online versions of the H.W. Wilson indexes including Applied Science and Technology Index, Business Periodicals Index and Readers' Guide to Periodical Literature. Approximately 1980 to present. Inquire as to online cost and availability.

PERIODICALS

JOURNAL OF ENGINEERING FOR INDUSTRY. American Society of Mechanical Engineers, 345 East 47th Street, New York, NY 10017. (212) 705-7703. 1970 to present. Quarterly. $100.00 per year.

JOURNAL OF MATERIALS RESEARCH. Materials Research Society. 9800 McKnight Road, Suite 327, Pittsburgh, PA 15237. (412) 367-3003.

JOURNAL OF METALS. American Institute of Mining, Metallurgical, and Petroleum Engineers, Inc., Metallurgical Society, 420 Commonwealth Drive, Warrendale, PA 15086. (412) 776-9086. 1949 to present. Monthly. $40.00 per year.

JOURNAL OF NONDESTRUCTIVE EVALUATION. Plenum Publishing Corporation, 233 Spring Street, New York, NY 10013. (800) 221-9369. 1980 to present. Quarterly. $85.00 per year.

JOURNAL OF TESTING AND EVALUATION. American Society for Testing and Materials. 1916 Race Street, Philadelphia, PA 19103. (215) 299-5400. 1966 to present. Bimonthly. $40.00 per year.

MATERIALS ENGINEERING. Penton-IPC, 1100 Superior Avenue, Cleveland, OH 44114. (216) 696-7000. 1929 to present. Monthly. $40.00.

MATERIALS EVALUATION. American Society for Nondestructive Testing, 4153 Arlingate Plaza, Caller #28518, Columbus, OH 43228. (614) 274-6003. 1942 to present. Monthly. $50.00 per year.

MATERIALS PERFORMANCE. National Association of Corrosion Engineers, P.O. Box 218340, Houston, TX 77218. (713) 492-0535. 1962 to present. Monthly. $50.00 per year.

SAMPE JOURNAL. Society for the Advancement of Material and Process Engineering (SAMPE), P.O. Box 2459, 843 West Glentana Street, Covina, CA 91722. (818) 331-0616. 1965 to present. Bimonthly. $31.00 per year.

RESEARCH CENTERS AND INSTITUTES

CENTER FOR MATERIALS SCIENCE AND ENGINEERING. Massachusetts Institute of Technology, Building 13, Room 2090, 77 Massachusetts Avenue, Cambridge, MA 02139. (617) 253-6801.

CENTER FOR RESEARCH IN MATERIALS SCIENCE AND ENGINEERING. University of Texas at Austin, ETC 9.104, Austin, TX 78712. (512) 471-1504.

MATERIALS RESEARCH LABORATORY. Ohio State University, 3063 Alpheus Smith Laboratory, 174 West 18th Avenue, Columbus, OH 43210. (614) 422-5190.

MATERIALS RESEARCH LABORATORY. University of Illinois, Urbana, IL 61801. (217) 333-1371.

MATERIALS SCIENCE CENTER. Cornell University, Clark Hall of Science, Ithaca, NY 14853. (607) 255-4272.

INDUSTRIAL ENGINEERING

See also: CHEMICAL ENGINEERING, CONTROL SYSTEMS, DESIGN ENGINEERING, MATERIALS HANDLING, MECHANICAL ENGINEERING, QUALITY CONTROL ENGINEERING, ROBOTICS

ABSTRACT SERVICES AND INDEXES

APPLIED MECHANICS REVIEW. American Society of Mechanical Engineers, 345 East 47th Street, New York, NY 10017. (212) 705-7703. 1948 to present. Monthly. $360.00 per year.

APPLIED SCIENCE AND TECHNOLOGY INDEX. H.W. Wilson and Company, 950 University Avenue, Bronx, NY 10452. (800) 367-6670 or (212) 588-8400. Monthly. Inquire as to cost and availability.

CHEMICAL ABSTRACTS. American Chemical Society, Chemical Abstracts Service, Box 3012, Columbus, OH 43210. (614) 421-3600. 1907 to present. Weekly. $9500.00 per year.

CONFERENCE PAPERS INDEX. Cambridge Scientific Abstracts, 5161 River Road, Bethesda, MD 20816. 1972 to present. Monthly. Inquire as to cost and availability.

CURRENT CONTENTS: ENGINEERING, TECHNOLOGY AND APPLIED SCIENCES. Institute for Scientific Information, 3501 Market Street, Philadelphia, PA 19104. (800) 523-1850 or (215) 386-0100. Weekly. $275.00 per year.

ENGINEERING INDEX MONTHLY AND AUTHOR INDEX. Engineering Information Inc., 345 East 47th Street, New York, NY 10017. (212) 705-7600. Monthly. $1560.00 per year.

ISMEC BULLETIN (Information Service in Mechanical Engineering). Cambridge Scientific Abstracts, 5161 River Road, Bethesda, MD 20816. (301) 951-1400. 1973 to present. Monthly. $450.00 per year.

INDEX TO SCIENTIFIC AND TECHNICAL PROCEEDINGS. Institute for Scientific Information, 3501 Market Street, Philadelphia, PA 19104. (800) 523-1850 or (215) 386-0100. 1978 to present. Monthly. $775.00 per year.

SCIENCE CITATION INDEX. Institute for Scientific Information, 3501 Market Street, Philadelphia, PA 19104. (800) 523-1850 or (215) 386-0100. Six times per year. $6200.00 per year.

ANNUAL REVIEWS AND YEARBOOKS

MATERIALS HANDLING ENGINEERING HANDBOOK. Penton-IPC, 1100 Superior Avenue, Cleveland, OH 44114. (216) 696-7000. Annual. Inquire.

ASSOCIATIONS AND PROFESSIONAL SOCIETIES

AMERICAN INSTITUTE OF PLANT ENGINEERS. 3975 Erie Avenue, Cincinnati, OH 45208. (513) 561-6000.

AMERICAN PRODUCTION AND INVENTORY CONTROL SOCIETY. 500 West Annandale Road, Falls Church, VA 22046. (703) 237-8344.

AMERICAN SOCIETY FOR QUALITY CONTROL. 230 West Wells Street, Suite 700, Milwaukee, WI 53203. (414) 272-8575.

AMERICAN SOCIETY OF MECHANICAL ENGINEERS. 345 East 47th Street, New York, NY 10017. (212) 705-7703.

INDUSTRIAL DESIGNERS SOCIETY OF AMERICA. 1360 Beverly Road, McLean, VA 22101. (703) 556-0919.

INSTITUTE OF INDUSTRIAL ENGINEERS. 25 Technology Park/Atlanta, Norcross, GA 30092. (404) 449-0460.

INTERNATIONAL ASSOCIATION OF QUALITY CIRCLES. 801-B West Eighth Street, Suite 301, Cincinnati, OH 45203. (513) 381-1959.

SOCIETY OF AMERICAN VALUE ENGINEERS. 221 North LaSalle Street, Suite 2026, Chicago, IL 60601. (312) 346-3265.

SOCIETY OF LOGISTICS ENGINEERS. 303 Williams Avenue, Suite 922, Huntsville, AL 35801. (205) 539-3800.

SOCIETY OF MANUFACTURING ENGINEERS. One SME Drive, Box 930, Dearborn, MI 48121. (313) 271-1500.

SOCIETY OF PACKING AND HANDLING ENGINEERS. Reston International Center, Reston, VA 22091. (703) 620-9380.

DIRECTORIES AND BIOGRAPHICAL SOURCES

AMERICAN MEN AND WOMEN OF SCIENCE. R.R. Bowker, Inc., Order Department, 245 West 17th Street, New York, NY 10011. (800) 521-8110. Eight volumes. 1986. $595.00 for set.

INTERNATIONAL RESEARCH CENTERS DIRECTORY 1988-89. Darren L. Smith, editor. Gale Research Company, Book Tower, Detroit, MI 48226. (800) 521-0707. 4th edition. 1987. $360.00.

1987 DIRECTORY OF ENGINEERING SOCIETIES AND RELATED ORGANIZATIONS. Gordon Davis, editor. Hemisphere Publishing Corporation, 1010 Vermont Avenue, NW, Washington, DC 20005. (800) 526-0275. 12th edition. 1987. $100.00.

PLANT ENGINEERING DIRECTORY. Technical Publishing Company, 1301 South Grove Avenue, Chicago, IL 60010. (312) 381-1840. Annual. $35.00.

RESEARCH CENTERS DIRECTORY 1988. Gale Research Company, Book Tower, Detroit, MI 48226. (800) 521-0707. 12th edition. 1987. $365.00 for set.

SCIENTIFIC AND TECHNICAL ORGANIZATIONS AND AGENCIES DIRECTORY. Margaret Labash Young, editor. Gale Research Company, Book Tower, Detroit, MI 48226. (800) 521-0707. 2nd edition. 1987. $185.00.

SOCIETY OF LOGISTICS ENGINEERS MEMBERSHIP DIRECTORY. 303 Williams Avenue, Suite 922, Huntsville, AL 35801. (205) 539-3800. Annual. Inquire.

SOCIETY OF PACKING AND HANDLING ENGINEERS DIRECTORY. Reston International Center, Reston, VA 22091. (703) 620-9380. Annual. Inquire.

WHO'S WHO IN ENGINEERING. Gordon Davis, editor. Hemisphere Publishing Corporation, 79 Madison Avenue, New York, NY 10016-7892. (800) 821-8312. Sixth edition. 1985. $200.00.

INDUSTRIAL ENGINEERING

ENCYCLOPEDIAS AND DICTIONARIES

THESAURUS OF SCIENTIFIC, TECHNICAL, AND ENGINEERING TERMS. Hemisphere Publishing Corporation, 1010 Vermont Avenue, NW, Washington, DC 20005. (800) 526-0275. 1988. $125.00.

GENERAL WORKS

APPLIED RELIABILITY. Paul A. Tobias and David C. Trindade. Van Nostrand Reinhold Company, Inc., 135 West 50th Street, New York, NY 10020. (800) 543-2681. 1986. $36.95.

INDUSTRIAL ENERGY MANAGEMENT AND UTILIZATION. Larry C. White and others. Hemisphere Publishing Corporation, 79 Madison Avenue, New York, NY 10016-7892. (800) 821-8312. 1988. $49.00.

INTEGRATED MANUFACTURING: STRATEGY, PLANNING, AND IMPLEMENTATION. E.R.G. Gerelle and J. Stark. McGraw-Hill Book Company, 1221 Avenue of the Americas, New York, NY 10020. (212) 512-2000. 1988. $34.50.

MANUFACTURING ENGINEERING: PRINCIPLES OF OPTIMIZATION. Daniel T. Koenig. Hemisphere Publishing Corporation, 79 Madison Avenue, New York, NY 10016-7892. (800) 821-8312. 1987. $45.75.

OPERATIONS RESEARCH: PRINCIPLES AND PRACTICE. A. Ravindran and others. John Wiley and Sons, Inc., 605 Third Avenue, New York, NY 10158. (800) 526-5368. Second edition. 1987. $49.95.

ROBOTICS: AN INTRODUCTION. D. McCloy and D. Harris. John Wiley and Sons, Inc., 605 Third Avenue, New York, NY 10158. (800) 526-5368. 1986. $39.95.

HANDBOOKS AND MANUALS

COMPUTER-INTEGRATED MANUFACTURING HANDBOOK. E. Telcholz and J.N. Orr. McGraw-Hill Book Company, 1221 Avenue of the Americas, New York, NY 10020. (212) 512-2000. 1987. $59.95.

HANDBOOK OF INDUSTRIAL ENGINEERING. Gavriel Salvendy. John Wiley and Sons, Inc., 605 Third Avenue, New York, NY 10158. (800) 526-5368. 1982. $99.50.

MAINTENANCE ENGINEERING HANDBOOK. Lindley R. Higgins. McGraw-Hill Book Company, 1221 Avenue of the Americas, New York, NY 10020. (212) 512-2000. 1987. $79.50.

PRODUCTION HANDBOOK. J.A. White. John Wiley and Sons, Inc., 605 Third Avenue, New York, NY 10158. (800) 526-5368. 4th edition. 1986. $78.50.

ONLINE DATA BASES

CA SEARCH. Chemical Abstracts Service, P.O. Box 3012, Columbus, OH 43120. (800) 848-6538 or (614) 421-3600. Comprehensive guide to chemical literature, 1972 to present. Inquire as to online cost and availability.

COMPENDEX. Engineering Information, Inc., 345 East 47th Street, New York, NY 10017. (800) 221-1044 or (212) 705-7615. Engineering and technical literature, 1975 to present. Inquire as to online cost and availability.

DISSERTATION ABSTRACTS ONLINE. University Microfilms International, 300 North Zeeb Road, Ann Arbor, MI 48106. (800) 521-0600 or (313) 761-4700. Scope includes virtually all doctoral dissertations accepted at accredited American institutions from 1861 to present in over 250 subject areas. Inquire as to online cost and availability.

INSPEC. INSPEC Marketing Department, Institution of Electrical Engineers. Available from IEEE Service Center, 445 Hoes Lane, Piscataway, NJ 08854. (201) 981-0060. Online version of Physics Abstracts. Inquire as to online cost and availability.

NTIS. National Technical Information Service, 5285 Port Royal Road, Springfield, VA 22161. (703) 487-4630. Broad coverage of government sponsored research reports, 1964 to present. Inquire as to online cost and availability.

SCISEARCH. Institute for Scientific Information, 3501 Market Street, Philadelphia, PA 19104. (800) 523-1850 or (215) 386-0100. Broad multidisciplinary title and author index to the international literature of science and technology, 1974 to present. Inquire as to online cost and availability.

WILSONLINE. H.W. Wilson and Company, 950 University Avenue, Bronx, NY 10452. (800) 367-6770 or (212) 588-8400. Makes available online versions of the H.W. Wilson indexes including Applied Science and Technology Index, Business Periodicals Index and Readers' Guide to Periodical Literature. Approximately 1980 to present. Inquire as to online cost and availability.

OTHER SOURCES

WHAT EVERY ENGINEER SHOULD KNOW ABOUT ENGINEERING SOURCES. Marcel Dekker Inc., 270 Madison Avenue, New York, NY 10016. (800) 228-1160. 1984. $24.95.

PERIODICALS

ADVANCED MANUFACTURING PROCESSES. Marcel Dekker Inc., 270 Madison Avenue, New York, NY 10016. (800) 228-1160. 1986 to present. Quarterly. $85.00 per year.

ADVANCED MANUFACTURING TECHNOLOGY. Technical Insights, Inc., Box 1304, Fort Lee, NJ 07024. (201) 568-4744. 1977 to present. Biweekly. $320.00 per year.

AMERICAN INDUSTRY. Publications for Industry, 21 Russell Woods Road, Great Neck, NY 11021. (516) 487-0990. 1946 to present. Monthly. $20.00 per year.

ASSEMBLY ENGINEERING. Hitchcock Publishing Company, 25 West 550 Geneva Road, Wheaton, IL 60188. (312) 665-1000. 1958 to present. Monthly. $65.00 per year.

IEEE TRANSACTIONS ON INDUSTRY APPLICATIONS. Institute of Electrical and Electronics Engineers. IEEE Service Center, 445 Hoes Lane, Piscataway, NJ 08854. 1965 to present. Bimonthly. $105.00 per year.

IIE TRANSACTIONS: INDUSTRIAL ENGINEERING RESEARCH AND DEVELOPMENT. Instuture of Industrial Engineers, 25 Technology Park/Atlanta, Norcross, GA 30092. (404) 449-0460. 1969 to present. Quarterly. $55.00 per year.

JOURNAL OF ENGINEERING FOR INDUSTRY. American Society of Mechanical Engineers, 345 East 47th Street, New York, NY 10017. (212) 705-7703. 1928 to present. Quarterly. $80.00 per year.

JOURNAL OF QUALITY TECHNOLOGY. American Society for Quality Control, 230 West Wells Street, Suite 700, Milwaukee, WI 53203. (414) 272-8575. 1969 to present. Quarterly. $10.00 per year.

JOURNAL OF OPERATIONS MANAGEMENT. American Production and Inventory Control Society, 500 West Annandale Road, Falls Church, VA 22046. (703) 237-8344. 1980 to present. Quarterly. $15.00 per year.

MANUFACTURING ENGINEERING. Society of Manufactruing Engineers, One SME Drive, Box 930, Dearborn, MI 48121. (313) 271-1500. 1932 to present. Monthly. $60.00 per year.

MATERIAL HANDLING ENGINEERING. Penton-IPC, 1100 Superior Avenue, Cleveland, OH 44114. (216) 696-7000. 1945 to present. Monthly. $35.00 per year.

MATERIALS ENGINEERING. Penton-IPC, 1100 Superior Avenue, Cleveland, OH 44114. (216)696-7000. 1929 to present. Monthly. $40.00 per year.

PLANT ENGINEERING MAGAZINE. Technical Publishing Company, 1301 South Grove Avenue, Chicago, IL 60010. (312) 381-1840. 1947 to present. Semimonthly. $50.00 per year.

PRODUCTION. Production Publishing, 44 East Long Lake Road, Bloomfield Hills, MI 48303. (313) 647-8400. 1934 to present. Monthly. $35.00 per year.

PRODUCTION AND INVENTORY MANAGEMENT JOURNAL. American Production and Inventory Control Society, 500 West Annandale Road, Falls Church, VA 22046. (703) 237-8344. 1960 to present. Quarterly. $30.00 per year.

QUALITY ASSURANCE. Hitchcock Publishing Company, 25 West 550 Geneva Road, Wheaton, IL 60188. (312) 665-1000. 1962 to present. Monthly. $45.00 per year.

RESEARCH CENTERS AND INSTITUTES

ADVANCED MANUFACTURING CENTER. Cleveland State University, Euclid Avenue at East 24th Street, Cleveland, OH 44115. (216) 687-4643.

CENTER FOR APPLIED ENGINEERING. University of Missouri - Rolla, 206 Harris Hall, Rolla, MO 65401. (314) 4559.

CENTER OF DESIGN AND MANUFACTURING INNOVATION. Lehigh University, H.S. Mohler Laboratory, Room 200, Bethleham, PA 18015. (215) 758-4114.

CENTER FOR QUALITY AND PRODUCTIVITY IMPROVEMENT. University of Wisconsin - Madison, 610 North Walnut Street, Madison, WI 53704. (608) 263-2520.

ENGINEERING AND INDUSTRIAL EXPERIMENT STATION. University of Florida, 300 Weil Hall, Gainesvill, FL 32611. (904) 392-0941.

SPECIFICATIONS AND STANDARDS

AUTOMATED MANUFACTURING. L.B. Garner, editor. American Society for Testing and Materials, 1916 Race Street, Philadelphia, PA 19103. (215) 299-5400. 1985. $38.00.

INDUSTRIAL ROBOTS

See: ROBOTICS

INERTIAL GUIDANCE SYSTEMS

See: GUIDANCE SYSTEMS

INERTIAL NAVIGATION

See: NAVIGATION

INERTIAL WELDING

See: WELDING

INFRARED ASTRONOMY

See also: ASTRONOMY, ASTROPHYSICS

ABSTRACT SERVICES AND INDEXES

ASTRONOMY AND ASTROPHYSICS ABSTRACTS. Springer-Verlag New York, Incorporated, 175 Fifth Avenue, New York, NY 10010. (212) 460-1500. $70.00 per year.

ASTRONOMY AND ASTROPHYSICS MONTHLY INDEX. Olivetree Associates, Post Office Box 236, Sierra Madre, CA 91024. $212.00 per year. Complimentary copies available on request.

PHYSICS ABSTRACTS. Institute of Electrical Engineers, London, United Kingdom. Available from: Institute of Electrical and Electronic Engineers (IEEE), 345 East 47th Street, New York, NY 10017. (212) 705-7900. 1898 to present. Monthly. $1670.00 per year.

SCIENCE CITATION INDEX. Institute for Scientific Information, 3501 Market Street, Philadelphia, PA 19104. (800) 523-1850 or (215) 386-0100. Inquire as to cost and availability.

STAR. (Scientific and Technical Aerospace Reports. United States National Aeronautics and Space Administration, Scientific and Technical Information Facility, Box 8757, Baltimore-Washington International Airport, MD 21240. (202) 755-2210. Semimonthly, with semiannual and annual indexes. $85.00 per year.

ANNUAL REVIEWS AND YEARBOOKS

ANNUAL REVIEW OF ASTRONOMY AND ASTROPHYSICS. Annual Reviews, Incorporated, 4139 El Camino Way, Palo Alto, CA 94306. (415) 493-4400.

ASSOCIATIONS AND PROFESSIONAL SOCIETIES

AMERICAN ASTRONOMICAL SOCIETY. 2000 Florida Avenue, NW, Suite 300, Washington, DC 20009. (202) 659-0134.

ASTRONOMICAL LEAGUE. Post Office Box 12821, Tucson, AZ 85732. (602) 790-8471.

ASTRONOMICAL SOCIETY OF THE PACIFIC. 1290 24th Avenue, San Francisco, CA 94122. (415) 661-8660.

BIBLIOGRAPHIES

A BIBLIOGRAPHY OF ASTRONOMY, 1970-1979. R.A. Seal and S.S. Martin. Libraries Unlimited, Incorporated, Littleton, CO 80160. 1982. $37.50.

SCIENTIFIC AND TECHNICAL BOOKS AND SERIALS IN PRINT 1988; AN INDEX TO LITERATURE IN SCIENCE AND TECHNOLOGY. R.R. Bowker Company, 205 East 42nd Street, New York, NY 10017. (800) 521-8110 or (212) 916-1600. $175.00.

DIRECTORIES AND BIOGRAPHICAL SOURCES

AMERICAN MEN AND WOMEN OF SCIENCE. Physical and Biological Sciences. Sixteenth edition. R.R. Bowker Company, 205 East 42nd Street, New York, NY 10017. (800) 521-8110 or (212) 916-1600. 1986. $595.00.

THE BIOGRAPHICAL DICTIONARY OF SCIENTISTS: ASTRONOMERS. D. Abbott, editor. Peter Bedrick Books, 125 East 23rd Street, New York, NY 10010. 1984.

DIRECTORY OF PHYSICS AND ASTRONOMY STAFF MEMBERS. American Institute of Physics, 335 East 45th Street, New York, NY 10017. Annual.

RESEARCH CENTERS DIRECTORY. Gale Research Company, Book Tower, Detroit, MI 48226. (800) 521-0707. Twelfth edition. 1987. $365.00 for set.

WHO'S WHO IN FRONTIER SCIENCE AND TECHNOLOGY. Marquis Who's Who, Incorporated, 200 East Ohio Street, Chicago, IL 60611. (800) 428-3898 or (312) 787-2008.

GENERAL WORKS

THE CAMBRIDGE ASTRONOMY GUIDE: A PRACTICAL INTRODUCTION TO ASTRONOMY. Cambridge University Press, 32 East 57th Street, New York, NY 10022. (212) 688-8888. 1985. $24.95.

INFRARED ASTRONOMY. Gaincarlo Setti, editor. Kluwer Academic Publishers, 190 Old Derby Street, Hingham, MA 02043. (617) 749-5262. 1978. $42.00.

OPTICAL AND INFRARED DETECTORS. R.J. Keyes, editor. Springer-Verlag, 175 Fifth Avenue, New York, NY 10010. (800) 526-7254. 1980. $30.00 in paper.

ONLINE DATA BASES

CA SEARCH. Chemical Abstracts Service, Post Office Box 3012, Columbus, OH 43210. Guide to chemical literature, 1972 to present. Inquire as to cost and availability.

DISSERTATION ABSTRACTS ONLINE. University Microfilms International, 300 North Zeeb Road, Ann Arbor, MI 48106. (800) 521-0600 or (313) 761-4700. Scope includes virtually all doctoral dissertations accepted at accredited American institutions from 1861 to present in 252 subject areas. Inquire as to cost and availability.

INSPEC. INSPEC Marketing Department, Institute of Electrical and Electronics Engineers, Incorporated, IEEE Service Department, 445 Hoes Lane, Piscataway, NJ 08854. (201) 981-0060. Inquire as to on-line cost and availability.

NASA. National Aeronautics and Space Administration, Scientific and Technical Information Branch, 300 7th Street, SW, Washington, DC 20546. Citations and abstracts of aerospace literature, 1962 to present. Inquire as to cost and availability.

NTIS. National Technical Information Service, 5285 Port Royal Road, Springfield, VA 22161. (703) 487-4630. Broad coverage of government-sponsored research reports, 1964 to present. Inquire as to cost and availability.

SCISEARCH. Institute for Scientific Information, 3501 Market Street, Philadelphia, PA 19104. (800) 523-1850 or (215) 386-0100. Broad multidisciplinary title and author index to the international literature of science and technology, 1974 to present. Inquire as to cost and availability.

PERIODICALS

ASTRONOMICAL JOURNAL. American Astronomical Society. Available from: American Institute of Physics, 335 East 45th Street, New York, NY 10017. (212) 661-9404. $125.00 per year.

ASTRONOMICAL SOCIETY OF THE PACIFIC. Publications. Astronomical Society of the Pacific, 1290 24th Avenue, San Francisco, CA 94122. (415) 661-8660. Monthly. $38.00.

ASTRONOMY. Astro Media Corporation, 625 East Paul Avenue, Milwaukee, WI 53202. Monthly. $18.00 per year.

ASTRONOMY AND ASTROPHYSICS ABSTRACTS. Springer-Verlag New York, Incorporated, 175 Fifth Avenue, New York, NY 10010. (212) 460-1500. $680.00 per year.

ASTROPHYSICAL JOURNAL. American Astronomical Society, University of Chicago Press, 5801 Ellis Avenue, Chicago, IL 60637. Biweekly. $305.00 per year.

ASTROPHYSICS AND SPACE SCIENCE. D. Reidel Publishing Company, 190 Older Derby Street, Hingham, MA 02043. Monthly. $101.00 per year.

EARTH, MOON AND PLANETS: AN INTERNATIONAL JOURNAL OF COMPARATIVE PLANETOLOGY. D. Reidel Publishing Company, 190 Old Derby Street, Hingham, MA 02043. Nine times per year. $275.00 per year.

ICARUS: INTERNATIONAL JOURNAL OF THE SOLAR SYSTEM STUDIES. Academic Press, Incorporated, Orlando, FL 32887. (305) 345-4100. Monthly. $484.00 per year.

MERCURY. Astronomical Society of the Pacific, 1290 245h Avenue, San Francisco, CA 94122. (415) 661-8660. Bimonthly. $21.00 per year.

MONTHLY NOTICES OF THE ROYAL ASTRONOMICAL SOCIETY. Blackwell Science Publications, Incorporated, 667 Lytton Avenue, Palo Alto, CA 94301. (415) 324-1688. Monthly. $134.00 per year to individuals.

PLANETARY AND SPACE SCIENCE. Pergamon Press, Incorporated, Maxwell House, Fairview Park, Elmsford, NY 10523. (914) 592-7700. Monthly. $430.00 per year.

SKY AND TELESCOPE. Sky Publishing Corporation, 49 Bay State Road, Cambridge, MA 02238. (617) 864-7360. Monthly. $18.00 per year.

SOLAR PHYSICS. D. Reidel Publishing Company, 190 Old Derby Street, Hingham, MA 02043. Monthly. $620.00 per year.

SOVIET ASTRONOMY (TRANSLATION OF ASTRO-NOMICHESKII ZHURNAL). American Institute of Physics, 335 East 45th Street, New York, NY 10017. (212) 661-9404. Bimonthly. $425.00 per year.

SPACE SCIENCE REVIEWS. D. Reidel Publishing Company, 190 Old Derby Street, Hingham, MA 02043. Monthly. $305.00 per year.

VISTAS IN ASTRONOMY. Pergamon Press, Incorporated, Maxwell House, Fairview Park, Elmsford, NY 10523. (914) 592-7700. Quarterly. $145.00 per year.

RESEARCH CENTERS AND INSTITUTES

HARVARD-SMITHSONIAN CENTER FOR ASTROPHYSICS. Fred L. Whipple Observatory, Post Office Box 97, Amado, AZ 85645. (602) 629-6741.

NATIONAL ASTRONOMY AND IONOSPHERE CENTER. Cornell University, Space Sciences Building, Ithaca, NY 14853. (607) 256-3734.

NATIONAL OPTICAL ASTRONOMY OBSERVATORIES. 1002 North Warren Avenue, Tucson, AZ 85719. (602) 325-9230.

UNIVERSITY OF WYOMING. Wyoming Infrared Observatory, University Station, Box 3905, Laramie, WY 82071. (307) 742-8666.

INFRARED PHOTOGRAPHY

See also: AERIAL PHOTOGRAPHY, PHOTOCHEMISTRY, PHOTOGRAMMETRY, PHOTOGRAPHIC FILM, PHOTOGRAPHY

ABSTRACT SERVICES AND INDEXES

APPLIED SCIENCE AND TECHNOLOGY INDEX. H.W. Wilson and Company, 950 University Avenue, Bronx, NY 10452. (800) 367-6670 or (212) 588-8400. Monthly. Inquire as to cost and availability.

CHEMICAL ABSTRACTS. American Chemical Society, Chemical Abstracts Service, Box 3012, Columbus, OH 43210. (614) 421-3600. 1907 to present. Weekly. $9500.00 per year.

ENGINEERING INDEX MONTHLY AND AUTHOR INDEX. Engineering Information Inc., 345 East 47th Street, New York, NY 10017. (212) 705-7600. Monthly. $1560.00 per year.

PHOTOGRAPHIC ABSTRACTS. Royal Photographic Society of Great Britain, Scientific and Technical Group, 62 Chelmsford Road, Shenfield, Brentwood, Essex, England. 1921 to present. Six times per year. $140.00 per year.

PHYSICS ABSTRACTS. Institution of Electrical Engineers. Available from: IEEE Service Center, 445 Hoes Lane, Piscataway, NJ 08854. 1898 to present. Bimonthly. $1700.00 per year.

SCIENCE CITATION INDEX. Institute for Scientific Information, 3501 Market Street, Philadelphia, PA 19104. (800) 523-1850 or (215) 386-0100. Six times per year. $6200.00 per year.

ASSOCIATIONS AND PROFESSIONAL SOCIETIES

AMERICAN SOCIETY FOR PHOTOGRAMMETRY AND REMOTE SENSING. 210 Little Falls Street, Falls Church, VA 22046-4398. (703) 534-6617.

OPTICAL SOCIETY OF AMERICA. 1816 Jefferson Place, N.W., Washington, DC 20036. (202) 223-8130.

SOCIETY OF PHOTOGRAPHIC SCIENTISTS AND ENGINEERS. 7003 Kilworth Lane, Springfield, VA 22151. (703) 642-9090.

SPIE - THE INTERNATIONAL SOCIETY FOR OPTICAL ENGINEERING. P.O. Box 10, 1022 19th Street, Bellingham, WA 98227. (206) 676-3290.

DIRECTORIES AND BIOGRAPHICAL SOURCES

INTERNATIONAL RESEARCH CENTERS DIRECTORY 1988-89. Darren L. Smith, editor. Gale Research Company, Book Tower, Detroit, MI 48226. (800) 521-0707. 4th edition. 1987. $360.00.

1987 DIRECTORY OF ENGINEERING SOCIETIES AND RELATED ORGANIZATIONS. Gordon Davis, editor. Hemisphere Publishing Corporation, 1010 Vermont Avenue, NW, Washington, DC 20005. (800) 526-0275. 12th edition. 1987. $100.00.

RESEARCH CENTERS DIRECTORY 1988. Gale Research Company, Book Tower, Detroit, MI 48226. (800) 521-0707. 12th edition. 1987. $365.00 for set.

SCIENTIFIC AND TECHNICAL ORGANIZATIONS AND AGENCIES DIRECTORY. Margaret Labash Young, editor. Gale Research Company, Book Tower, Detroit, MI 48226. (800) 521-0707. 2nd edition. 1987. $185.00.

WHO'S WHO IN ENGINEERING. Gordon Davis, editor. Hemisphere Publishing Corporation, 79 Madison Avenue, New York, NY 10016-7892. (800) 821-8312. 6th edition. 1985. $200.00.

ENCYCLOPEDIAS AND DICTIONARIES

THESAURUS OF PHOTOGRAPHIC SCIENCE AND ENGINEERING. Society of Photographic Scientists and Engineers. Books on Demand, 300 North Zeeb Road, Ann Arbor, MI 48106. (313) 761-4700. $34.50 in paper.

GENERAL WORKS

APPLIED INFRARED PHOTOGRAPHY. Eastman Kodak Company. Eastman Kodak Company, 343 State Street, Rochester, NY 14650. (716) 724-4000. 1981. $6.95 in paper.

THE ART OF INFRARED PHOTOGRAPHY: A COMPREHENSIVE GUIDE TO THE USE OF BLACK AND WHITE INFRARED FILM. Joseph Paduano. Morgan and Morgan, Inc., 145 Palisades Street, Dobbs Ferry, NY 10522. (914) 693-0023. 1984. $12.95 in paper.

PHOTOCHEMISTRY: AN INTRODUCTION. D.R. Arnold and others. Academic Press, Inc., 6277 Sea Harbor Drive, Orlando, FL 32821. (800) 321-5068. 1974. $45.00.

PHOTOGRAPHIC SCIENCE. Earl N. Mitchell. John Wiley and Sons, Inc., 605 Third Avenue, New York, NY 10158. (800) 526-5368. 1984. $37.50.

PHOTOGRAPHY FOR THE SCIENTIST. Richard A. Morton. Academic Press, Inc., 6277 Sea Harbor Drive, Orlando, FL 32821. (800) 321-5068. 2nd edition. 1984. $102.50.

HANDBOOKS AND MANUALS

HANDBOOK OF PHOTOGRAPHIC SCIENCE AND ENGINEERING. Society of Photographic Scientists and Engineers. 7003 Kilworth Lane, Springfield, VA 22151. (703) 642-9090. Inquire.

THE INFRARED HANDBOOK. William J. Wolfe and George J. Zissis, editors. Office of Naval Research. National Technical Information Service, 5285 Port Royal Road, Springfield, VA 22161. (703) 487-4929. 1986.

KODAK PROFESSIONAL PHOTOGUIDE. Carolyn Grimes, editor. Eastman Kodak Company, 343 State Street, Rochester, NY 14650. (716) 724-4000. 1986. $19.95 in paper.

ONLINE DATA BASES

CA SEARCH. Chemical Abstracts Service, P.O. Box 3012, Columbus, OH 43120. (800) 848-6538 or (614) 421-3600. Comprehensive guide to chemical literature, 1972 to present. Inquire as to online cost and availability.

COMPENDEX. Engineering Information, Inc., 345 East 47th Street, New York, NY 10017. (800) 221-1044 or (212) 705-7615. Engineering and technical literature, 1975 to present. Inquire as to online cost and availability.

DISSERTATION ABSTRACTS ONLINE. University Microfilms International, 300 North Zeeb Road, Ann Arbor, MI 48106. (800) 521-0600 or (313) 761-4700. Scope includes virtually all doctoral dissertations accepted at accredited American institutions from 1861 to present in over 250 subject areas. Inquire as to online cost and availability.

INSPEC. INSPEC Marketing Department, Institution of Electrical Engineers. Available from IEEE Service Center, 445 Hoes Lane, Piscataway, NJ 08854. (201) 981-0060. Online version of Physics Abstracts. Inquire as to online cost and availability.

NTIS. National Technical Information Service, 5285 Port Royal Road, Springfield, VA 22161. (703) 487-4630. Broad coverage of government sponsored research reports, 1964 to present. Inquire as to online cost and availability.

SCISEARCH. Institute for Scientific Information, 3501 Market Street, Philadelphia, PA 19104. (800) 523-1850 or (215) 386-0100. Broad multidisciplinary title and author index to the international literature of science and technology, 1974 to present. Inquire as to online cost and availability.

WILSONLINE. H.W. Wilson and Company, 950 University Avenue, Bronx, NY 10452. (800) 367-6770 or (212) 588-8400. Makes available online versions of the H.W. Wilson indexes including Applied Science and Technology Index, Business Periodicals Index and Readers' Guide to Periodical Literature. Approximately 1980 to present. Inquire as to online cost and availability.

INFRARED PHOTOGRAPHY

PERIODICALS

BRITISH JOURNAL OF PHOTOGRAPHY. Henry Greenwood and Company, Limited, 28 Great James Street, London WC1N 3HL, England. 1854 to present. Weekly. $70.00 per year.

FUNCTIONAL PHOTOGRAPHY. PTN Publishing Corporation, 210 Crossways Park Drive, Woodbury, NY 11797. (516) 496-8000. 1967 to present. Bimonthly. $7.50 per year.

INDUSTRIAL PHOTOGRAPHY. United Business Publications, Inc., 475 Park Avenue South, New York, NY 10016. (212) 725-2300. 1952 to present. Monthly. $15.00 per year.

JOURNAL OF IMAGING SCIENCE. Society of Photographic Scientists and Engineers. 7003 Kilworth Lane, Springfield, VA 22151. (703) 642-9090. 1956 to present. Bimonthly. $70.00 per year.

JOURNAL OF IMAGING TECHNOLOGY. Society of Photographic Scientists and Engineers. 7003 Kilworth Lane, Springfield, VA 22151. (703) 642-9090. 1975 to present. Bimonthly. $70.00 per year.

JOURNAL OF PHOTOGRAPHIC SCIENCE. Royal Photographic Society of Great Britain, 7 Ladbroke Walk, London W11, England. 1953 to present. Bimonthly. $40.00 per year.

KODAK TECH BITS. Eastman Kodak Company, 343 State Street, Rochester, NY 14650. (716) 724-4000. 1963 to present. Quarterly. Free to qualified personnel.

PHOTOGRAMMETRIC ENGINEERING AND REMOTE SENSING. American Society for Photogrammetry and Remote Sensing. 210 Little Falls Street, Falls Church, VA 22046-4398. (703) 534-6617. Order from: Allen Press, Inc., 1041 New Hampshire Street, Box 368, Lawrence, KS 66044. 1934 to present. Monthly. $80.00 per year.

SCIENTIFIC AND APPLLIED PHOTOGRAPHY AND CINEMATOGRAPHY. Gordon and Breach Science Publishers, Inc., 50 West 23rd Street, New York, NY 10010. (212) 206-8900. 12 times per year. $496.00 per year.

RESEARCH CENTERS AND INSTITUTES

IMAGE PERMANENCE INSTITUTE. Rochester Institute of Technology, RIT City Center, 50 West Main Street, Rochester, NY 14614. (716) 475-5199.

NATIONAL RESEARCH COUNCIL OF CANADA, DIVISION OF PHYSICS. Ottawa, ON, Canada K1A OR6. (613) 993-1053.

INFRARED SPECTROSCOPY

See: SPECTROSCOPY

INFRARED STARS

See: STARS

INORGANIC CHEMISTRY

See also: ANALYTICAL CHEMISTRY, CHEMISTRY, ORGANIC CHEMISTRY, PHYSICAL CHEMISTRY

ABSTRACT SERVICES AND INDEXES

APPLIED SCIENCE AND TECHNOLOGY INDEX. H.W. Wilson and Company, 950 University Avenue, Bronx, NY 10452. (800) 367-6670 or (212) 588-8400. Monthly. Inquire as to cost and availability.

CHEMICAL ABSTRACTS. American Chemical Society, Chemical Abstracts Service, Box 3012, Columbus, OH 43210. (614) 421-3600. 1907 to present. Weekly. $9500.00 per year.

CONFERENCE PAPERS INDEX. Cambridge Scientific Abstracts, 5161 River Road, Bethesda, MD 20816. 1972 to present. Monthly. Inquire as to cost and availability.

CURRENT CONTENTS: PHYSICAL, CHEMICAL AND EARTH SCIENCES. Institute for Scientific Information, 3501 Market Street, Philadelphia, PA 19104. (800) 523-1850 or (215) 386-0100. Weekly. $275.00 per year.

GENERAL SCIENCE INDEX. H.W. Wilson and Company, 950 University Avenue, Bronx, NY 10452. (800) 367-6670 or (212) 588-8400. 1978 to present. Monthly. Inquire as to cost and availability.

INDEX TO SCIENTIFIC AND TECHNICAL PROCEEDINGS. Institute for Scientific Information, 3501 Market Street, Philadelphia, PA 19104. (800) 523-1850 or (215) 386-0100. 1978 to present. Monthly. $775.00 per year.

INDEX TO SCIENTIFIC REVIEWS. Institute for Scientific Information, 3501 Market Street, Philadelphia, PA 19104. (800) 523-1850 or (215) 386-0100. 1974 to present. Semi-annual. $550.00 per year.

SCIENCE CITATION INDEX. Institute for Scientific Information, 3501 Market Street, Philadelphia, PA 19104. (800) 523-1850 or (215) 386-0100. Six times per year. $6200. 00 per year.

ANNUAL REVIEWS AND YEARBOOKS

ADVANCES IN INORGANIC AND RADIOCHEMISTRY. Academic Press, Inc., 6277 Sea Harbor Drive, Orlando, FL 32821. (800) 321-5068. 1959 to present. Irregular. Price varies, inquire.

ASSOCIATIONS AND PROFESSIONAL SOCIETIES

AMERICAN CHEMICAL SOCIETY. 1155 16th Street, N.W., Washington, DC 20036. (202) 872-4600.

AMERICAN INSTITUTE OF CHEMICAL ENGINEERS. 345 East 47th Street, New York, NY 10017. (212) 705-7338.

ASSOCIATION OF OFFICIAL ANALYTICAL CHEMISTS. 1111 North 19th Street, Suite 210, Arlington, VA 22209. (703) 522-3032.

BIBLIOGRAPHIES

NEW TECHNICAL BOOKS: A SELECTIVE LIST WITH DESCRIPTIVE ANNOTATIONS. New York Public Library, Science and Technology Research Center, Fifth Avenue and 42nd Street, New York, NY 10018. (212) 930-0800. 1915 to present. Monthly. $15.00 per year.

SCIENCE BOOKS AND FILMS. American Association for the Advancement of Science, 1333 H Street, NW, Washington, DC 20005. (202) 326-6454. Five times per year. $20.00 per year.

SCIENTIFIC AND TECHNICAL BOOKS AND SERIALS IN PRINT 1988; AN INDEX TO LITERATURE IN SCIENCE AND TECHNOLOGY. R.R. Bowker Company, 205 East 42nd Street, New York, NY 10017. (800) 521-8110. $175.00.

DIRECTORIES AND BIOGRAPHICAL SOURCES

AMERICAN MEN AND WOMEN OF SCIENCE. R.R. Bowker, Inc., Order Department, 245 West 17th Street, New York, NY 10011. (800) 521-8110. Eight volumes. 1986. $595.00 for set.

INTERNATIONAL RESEARCH CENTERS DIRECTORY 1988-89. Darren L. Smith, editor. Gale Research Company, Book Tower, Detroit, MI 48226. (800) 521-0707. 4th edition. 1987. $360.00.

1987 DIRECTORY OF ENGINEERING SOCIETIES AND RELATED ORGANIZATIONS. Gordon Davis, editor. Hemisphere Publishing Corporation, 1010 Vermont Avenue, NW, Washington, DC 20005. (800) 526-0275. 12th edition. 1987. $100.00.

RESEARCH CENTERS DIRECTORY 1988. Gale Research Company, Book Tower, Detroit, MI 48226. (800) 521-0707. 12th edition. 1987. $365.00 for set.

SCIENTIFIC AND TECHNICAL ORGANIZATIONS AND AGENCIES DIRECTORY. Margaret Labash Young, editor. Gale Research Company, Book Tower, Detroit, MI 48226. (800) 521-0707. 2nd edition. 1987. $185.00.

WHO'S WHO IN ENGINEERING. Gordon Davis, editor. Hemisphere Publishing Corporation, 1010 Vermont Avenue, NW, Washington, DC 20005. (800) 526-0275. 6th edition. 1985. $200.00.

ENCYCLOPEDIAS AND DICTIONARIES

ENCYCLOPEDIA OF CHEMISTRY. Douglas M. Considine, editor. Van Nostrand Reinhold Company, Inc., 135 West 50th Street, New York, NY 10020. (800) 543-2681. 4th edition. 1984. $98.95.

MCGRAW-HILL ENCYCLOPEDIA OF SCIENCE AND TECHNOLOGY. McGraw-Hill Book Company, 1221 Avenue of the Americas, New York, NY 10020. (212) 512-2000. 6th edition. 1987. $1600.00.

THESAURUS OF SCIENTIFIC, TECHNICAL, AND ENGINEERING TERMS. Hemisphere Publishing Corporation, 1010 Vermont Avenue, NW, Washington, DC 20005. (800) 526-0275. 1988. $125.00.

GENERAL WORKS

CHEMISTRY OF THE ELEMENTS. N.N. Greenwood and A. Earnshaw. Pergamon Press, Inc., Maxwell House, Fairview Park, Elmsford, NY 10523. (914) 592-7700. 1984. $145.00.

CONCEPTS AND MODELS OF INORGANIC CHEMISTRY. B.E. Douglas and others. John Wiley and Sons, Inc., 605 Third Avenue, New York, NY 10158. (800) 526-5368. Second edition. 1982. $42.50.

INORGANIC CHEMISTRY. William W. Porterfield. Addison-Wesley Publishing Company, Inc., 1 Jacob Way, Reading, MA 01867. (617) 944-3700. 1983. $36.95.

INORGANIC CHEMISTRY: AN INTRODUCTION. T. Moeller. John Wiley and Sons, Inc., 605 Third Avenue, New York, NY 10158. (800) 526-5368. 1982. $44.95.

HANDBOOKS AND MANUALS

CRC HANDBOOK OF CHEMISTRY AND PHYSICS. Robert C. Weast, editor. CRC Press, 2000 Corporate Boulevard, Boca Raton, FL 33431. (800) 272-7737. 68th edition. 1987. $69.95.

HANDBOOK OF APPLIED CHEMISTRY. Vollrath Hopp and Ingo Hennig. Hemisphere Publishing Corporation, 79 Madison Avenue, New York, NY 10016-7892. Order from: Taylor and Francis/Hemisphere Distribution Center, 242 Cherry Street, Philadelphia, PA 19106-1906. (800) 821-8312. 1983. $49.95.

LANGE'S HANDBOOK OF CHEMISTRY. John A. Dean, editor. McGraw-Hill Book Company, 1221 Avenue of the Americas, New York, NY 10020. (212) 512-2000. 1985. $59.50.

ONLINE DATA BASES

CA SEARCH. Chemical Abstracts Service, P.O. Box 3012, Columbus, OH 43120. (800) 848-6538 or (614) 421-3600. Comprehensive guide to chemical literature, 1972 to present. Inquire as to online cost and availability.

COMPENDEX. Engineering Information, Inc., 345 East 47th Street, New York, NY 10017. (800) 221-1044 or (212) 705-7615. Engineering and technical literature, 1975 to present. Inquire as to online cost and availability.

DISSERTATION ABSTRACTS ONLINE. University Microfilms International, 300 North Zeeb Road, Ann Arbor, MI 48106. (800) 521-0600 or (313) 761-4700. Scope includes virtually all doctoral dissertations accepted at accredited American institutions from 1861 to present in over 250 subject areas. Inquire as to online cost and availability.

NTIS. National Technical Information Service, 5285 Port Royal Road, Springfield, VA 22161. (703) 487-4630. Broad coverage of government sponsored research reports, 1964 to present. Inquire as to online cost and availability.

SCISEARCH. Institute for Scientific Information, 3501 Market Street, Philadelphia, PA 19104. (800) 523-1850 or (215) 386-0100. Broad multidisciplinary title and author index to the international literature of science and technology, 1974 to present. Inquire as to online cost and availability.

PERIODICALS

AMERICAN CHEMICAL SOCIETY JOURNAL. American Chemical Society, 1155 16th Street, N.W., Washington, DC 20036. (800) 424-6747. 1879 to present. Biweekly. $275.00 per year.

INORGANIC CHEMISTRY. American Chemical Society, 1155 16th Street, N.W., Washington, DC 20036. (800) 424-6747. 1962 to present. Biweekly. $300.00 per year.

INORGANICA CHIMICA ACTA. Elsevier Science Publishing Company, Inc., 52 Vanderbilt Avenue, New York, NY 10017. (212) 370-5520. 1967 to present. Biweekly. $1500.00 per year.

JOURNAL OF THE CHEMICAL SOCIETY. CHEMICAL COMMUNICATIONS. Royal Society of Chemistry, Burlington House, London, W1V OBN, England. 1965 to present. Semi-monthly. $385.00 per year.

POLYHEDRON. Pergamon Press, Inc., Maxwell House, Fairview Park, Elmsford, NY 10523. (914) 592-7700. 1982 to present. Monthly. $595.00.

ROYAL SOCIETY OF CHEMISTRY. JOURNAL. DALTON TRANSACTIONS. Royal Society of Chemistry, Burlington House, London, W1V OBN, England. 1972 to present. Monthly. $585.00 per year.

RESEARCH CENTERS AND INSTITUTES

CHEMICAL LABORATORIES. Harvard University, Oxford Street, Cambridge, MA 02138. (617) 495-4283.

CHEMICAL RESEARCH LABORATORY. Brown University, Providence, RI 02912. (401) 863-2256.

UNIVERSITY/INDUSTRY CHEMICAL RESEARCH CENTER. Mississippi State University, Department of Chemistry, P.O. Drawer CH, Mississippi State, MS 39762. (601) 325-3584.

INSTRUMENT FLIGHT

See: AERONAUTICS

INSTRUMENTATION

See also: METROLOGY, STANDARDS

ABSTRACT SERVICES AND INDEXES

APPLIED MECHANICS REVIEW. American Society of Mechanical Engineers, 345 East 47th Street, New York, NY 10017. (212) 705-7703. 1948 to present. Monthly. $360.00 per year.

APPLIED SCIENCE AND TECHNOLOGY INDEX. H.W. Wilson and Company, 950 University Avenue, Bronx, NY 10452. (800) 367-6670 or (212) 588-8400. Monthly. Inquire as to cost and availability.

CHEMICAL ABSTRACTS. American Chemical Society, Chemical Abstracts Service, Box 3012, Columbus, OH 43210. (614) 421-3600. 1907 to present. Weekly. $9500.00 per year.

ENGINEERING INDEX MONTHLY AND AUTHOR INDEX. Engineering Information Inc., 345 East 47th Street, New York, NY 10017. (212) 705-7600. Monthly. $1560.00 per year.

ISMEC BULLETIN (Information Service in Mechanical Engineering). Cambridge Scientific Abstracts, 5161 River Road, Bethesda, MD 20816. (301) 951-1400. 1973 to present. Monthly. $450.00 per year.

INDEX TO SCIENTIFIC AND TECHNICAL PROCEEDINGS. Institute for Scientific Information, 3501 Market Street, Philadelphia, PA 19104. (800) 523-1850 or (215) 386-0100. 1978 to present. Monthly. $775.00 per year.

INDEX TO SCIENTIFIC REVIEWS. Institute for Scientific Information, 3501 Market Street, Philadelphia, PA 19104. (800) 523-1850 or (215) 386-0100. 1974 to present. Semi-annual. $550.00 per year.

KEY ABSTRACTS - ELECTRICAL MEASUREMENTS AND INSTRUMENTATION. Institution of Electrical Engineers. Available from: IEEE Service Center, 445 Hoes Lane, Piscataway, NJ 08854. 1976 to present. Monthly. $105.00 per year.

KEY ABSTRACTS - PHYSICAL MEASUREMENTS AND INSTRUMENTATION. Institution of Electrical Engineers. Available from: IEEE Service Center, 445 Hoes Lane, Piscataway, NJ 08854. 1976 to present. Monthly. $105.00 per year.

PHYSICS ABSTRACTS. Institution of Electrical Engineers. Available from: IEEE Service Center, 445 Hoes Lane, Piscataway, NJ 08854. 1898 to present. Bimonthly. $1700.00 per year.

PHYSICS BRIEFS. Physik Verlag GmbH, Postfach 1260/1280, D-6940 Weinheim, West Germany. (212) 661-9404. 1920 to present. Twenty-six times per year. $1250.00 per year.

SCIENCE CITATION INDEX. Institute for Scientific Information, 3501 Market Street, Philadelphia, PA 19104. (800) 523-1850 or (215) 386-0100. Six times per year. $6200.00 per year.

STANDARDS ACTIONS. American National Standards Institute, 1430 Broadway, New York, NY 10018. (212) 354-3300. 1970 to present. Twice per month. $25.00 per year.

ASSOCIATIONS AND PROFESSIONAL SOCIETIES

AMERICAN INSTITUTE OF PHYSICS. 335 East 45th Street, New York, NY 10017. (212) 661-9494.

AMERICAN SOCIETY FOR TESTING AND MATERIALS. 1916 Race Street, Philadelphia, PA 19103. (215) 299-5400.

EDISON ELECTRIC INSTITUTE. 1111 19th Street, N.W., Washington, DC 20036. (202) 828-7400.

INSTITUTE OF ELECTRICAL AND ELECTRONICS ENGINEERS. 345 East 47th Street, New York, NY 10017. (212) 705-7900.

INSTRUMENT SOCIETY OF AMERICA. P.O. Box 12277, 67 Alexander Drive, Research Triangle Park, NC 27709. (919) 549-8411.

DIRECTORIES AND BIOGRAPHICAL SOURCES

AMERICAN MEN AND WOMEN OF SCIENCE. R.R. Bowker, Inc., Order Department, 245 West 17th Street, New York, NY 10011. (800) 521-8110. Eight volumes. 1986. $595.00 for set.

INTERNATIONAL RESEARCH CENTERS DIRECTORY 1988-89. Darren L. Smith, editor. Gale Research Company, Book Tower, Detroit, MI 48226. (800) 521-0707. 4th edition. 1987. $360.00.

1987 DIRECTORY OF ENGINEERING SOCIETIES AND RELATED ORGANIZATIONS. Gordon Davis, editor. Hemisphere Publishing Corporation, 79 Madison Avenue, New York, NY 10016-7892. (800) 821-8312. 12th edition. 1987. $100.00.

RESEARCH CENTERS DIRECTORY 1988. Gale Research Company, Book Tower, Detroit, MI 48226. (800) 521-0707. 12th edition. 1987. $365.00 for set.

SCIENTIFIC AND TECHNICAL ORGANIZATIONS AND AGENCIES DIRECTORY. Margaret Labash Young, editor. Gale Research Company, Book Tower, Detroit, MI 48226. (800) 521-0707. 2nd edition. 1987. $185.00.

WHO'S WHO IN ENGINEERING. Gordon Davis, editor. Hemisphere Publishing Corporation, 79 Madison Avenue, New York, NY 10016-7892. (800) 821-8312. Sixth edition. 1985. $200.00.

GENERAL WORKS

INSTRUMENTATION FOR ENGINEERING MEASUREMENTS. J.W. Dally and others. John Wiley and Sons, Inc., 605 Third Avenue, New York, NY 10158. (800) 526-5368. 1984. $44.50.

INSTRUMENTATION FUNDAMENTALS AND APPLICATIONS. R. Morrison. John Wiley and Sons, Inc., 605 Third Avenue, New York, NY 10158. (800) 526-5368. 1984. $21.95.

PRACTICAL PROCESS INSTRUMENTATION AND CONTROL. Jay Matley and others, editor. McGraw-Hill Book Company, 1221 Avenue of the Americas, New York, NY 10020. (212) 512-2000. 1986. $42.50.

HANDBOOKS AND MANUALS

HANDBOOK OF MEASUREMENT SCIENCE. P.H. Sydeham. John Wiley and Sons, Inc., 605 Third Avenue, New York, NY 10158. (800) 526-5368. Two volumes. Volume 1, 1982, $100.00. Volume 2, 1983, $135.00.

PROCESS INSTRUMENTS AND CONTROLS HANDBOOK. Douglas M. Considine, editor. McGraw-Hill Book Company, 1221 Avenue of the Americas, New York, NY 10020. (212) 512-2000. Third edition. 1985. $99.50.

ONLINE DATA BASES

CA SEARCH. Chemical Abstracts Service, P.O. Box 3012, Columbus, OH 43120. (800) 848-6538 or (614) 421-3600. Comprehensive guide to chemical literature, 1972 to present. Inquire as to online cost and availability.

COMPENDEX. Engineering Information, Inc., 345 East 47th Street, New York, NY 10017. (800) 221-1044 or (212) 705-7615. Engineering and technical literature, 1975 to present. Inquire as to online cost and availability.

DISSERTATION ABSTRACTS ONLINE. University Microfilms International, 300 North Zeeb Road, Ann Arbor, MI 48106. (800) 521-0600 or (313) 761-4700. Scope includes virtually all doctoral dissertations accepted at accredited American institutions from 1861 to present in over 250 subject areas. Inquire as to online cost and availability.

INSPEC. INSPEC Marketing Department, Institution of Electrical Engineers. Available from IEEE Service Center, 445 Hoes Lane, Piscataway, NJ 08854. (201) 981-0060. Online version of Physics Abstracts. Inquire as to online cost and availability.

NTIS. National Technical Information Service, 5285 Port Royal Road, Springfield, VA 22161. (703) 487-4630. Broad coverage of government sponsored research reports, 1964 to present. Inquire as to online cost and availability.

SCISEARCH. Institute for Scientific Information, 3501 Market Street, Philadelphia, PA 19104. (800) 523-1850 or (215) 386-0100. Broad multidisciplinary title and author index to the international literature of science and technology, 1974 to present. Inquire as to online cost and availability.

WILSONLINE. H.W. Wilson and Company, 950 University Avenue, Bronx, NY 10452. (800) 367-6770 or (212) 588-8400. Makes available online versions of the H.W. Wilson indexes including Applied Science and Technology Index, Business Periodicals Index and Readers' Guide to Periodical Literature. Approximately 1980 to present. Inquire as to online cost and availability.

OTHER SOURCES

WHAT EVERY ENGINEER SHOULD KNOW ABOUT ENGINEERING SOURCES. Marcel Dekker Inc., 270 Madison Avenue, New York, NY 10016. (800) 228-1160. 1984. $24.95.

PERIODICALS

ASTM STANDARDIZATION NEWS. American Society for Testing and Materials, 1916 Race Street, Philadelphia, PA 19103. (215) 299-5400. 1973 to present. Monthly. $20.00 per year.

CONTROL ENGINEERING: INSTRUMENTATION AND AUTOMATIC CONTROL SYSTEMS. Technical Publishing Company (New York), 875 Third Avenue, New York, NY 10022. (212) 605-9400. 1954 to present. Monthly. $50.00 per year.

IEEE TRANSACTIONS ON INSTRUMENTATION AND MEASUREMENT. Institute of Electrical and Electronics Engineers. IEEE Service Center, 445 Hoes Lane, Piscataway, NJ 08854. 1952 to present. Quarterly. $65.00 per year.

INDUSTRIAL AND SCIENTIFIC INSTRUMENTS. Hanover Press, Limited, 80 Highgate Road, London, NW5 1PB England. 1960 to present. Monthly. Inquire.

INTECH; THE INTERNATIONAL JOURNAL OF INSTRUMENTATION AND CONTROL. Instrument Society of America, P.O. Box 12277, 67 Alexander Drive, Research Triangle Park, NC 27709. (919) 549-8411. 1954 to present. Monthly. Price varies, inquire.

ISA TRANSACTIONS. Instrument Society of America, P.O. Box 12277, 67 Alexander Drive, Research Triangle Park, NC 27709. (919) 549-8411. 1961 to present. Quarterly. $75.00 per year.

JOURNAL OF PHYSICS E: SCIENTIFIC INSTRUMENTS. Institute of Physics, England. Available from: American Institute of Physics, 335 East 45th Street, New York, NY 10017. (212) 661-9494. 1968 to present. Monthly. $225.00 per year.

MEASUREMENTS AND CONTROL. Measurements and Date Corporation, 2994 West Liberty Avenue, Pittsburgh, PA 15216. (412) 343-9666. 1967 to present. Bimonthly. $20.00 per year.

MEASUREMENTS AND CONTROL NEWS. Measurements and Date Corporation, 2994 West Liberty Avenue, Pittsburgh, PA 15216. (412) 343-9666. 1967 to present. 6 times per year. Inquire.

METROLOGIA; INTERNATIONAL JOURNAL OF SCIENTIFIC METROLOGY. Springer-Verlag New York, Inc., 175 Fifth Avenue, New York, NY 10010. (800) 526-7254. 1965 to present. Quarterly. $95.00 per year.

REVIEW OF SCIENTIFIC INSTRUMENTS. American Institute of Physics, 335 East 45th Street, New York, NY 10017. (212) 661-9494. 1930 to present. Monthly. $300.00 per year.

STANDARDS ENGINEERING. Standards Engineering Society, 6700 Pennsylvania Avenue South, Minneapolis, MN 55423. Bimothly. $20.00 per year.

U.S. NATIONAL BUREAR OF STANDARDS. JOURNAL OF RESEARCH. Order from: Superintendent of Documents, Washington, DC 20402. 1959 to present. Bimonthly. $17.00 per year.

RESEARCH CENTERS AND INSTITUTES

INSTRUMENTATION AND CONTROL LABORATORY. Princeton University, Department of MAE, English Quadrangle, Princeton, NJ 08544. (609) 452-5154.

INSTRUMENTATION SYSTEMS CENTER. University of Wisconsin - Madison, 1500 Johnson Drive, Madison, WI 53706. (608) 263-1552.

MAJOR ANALYTICAL INSTRUMENTATION CENTER. University of Florida, 217 MAE, Gainesville, FL 32611. (904) 392-6985.

SPECIFICATIONS AND STANDARDS

STANDARDS AND PRACTICES FOR INSTRUMENTATION. Instrument Society of America, P.O. Box 12277, 67 Alexander Drive, Research Triangle Park, NC 27709. (919) 549-8411. 7th edition. 1983. $135.00.

INTEGRAL CALCULUS

See: CALCULUS

INTEGRATED CIRCUITS

See also: ELECTRONIC CIRCUITS AND COMPONENTS, ELECTRONICS, ELECTRONICS ENGINEERING, MICROELECTRONICS

ABSTRACT SERVICES AND INDEXES

APPLIED SCIENCE AND TECHNOLOGY INDEX. H.W. Wilson and Company, 950 University Avenue, Bronx, NY 10452. (800) 367-6670 or (212) 588-8400. Monthly. Inquire as to cost and availability.

CHEMICAL ABSTRACTS. American Chemical Society, Chemical Abstracts Service, Box 3012, Columbus, OH 43210. (614) 421-3600. 1907 to present. Weekly. $9500.00 per year.

CONFERENCE PAPERS INDEX. Cambridge Scientific Abstracts, 5161 River Road, Bethesda, MD 20816. (301) 951-1400. 1972 to present. Monthly. Inquire as to cost and availability.

CURRENT CONTENTS: ENGINEERING, TECHNOLOGY AND APPLIED SCIENCES. Institute for Scientific Information, 3501 Market Street, Philadelphia, PA 19104. (800) 523-1850 or (215) 386-0100. Weekly. $275.00 per year.

ELECTRICAL AND ELECTRONICS ABSTRACTS. Institution of Electrical Engineers. Available from: Institute of Electrical and Electronics Engineers. IEEE Service Center, 445 Hoes Lane, Piscataway, NJ 08854. Monthly. $1250.00 per year.

ELECTRONICS AND COMMUNICATIONS ABSTRACTS. Cambridge Scientific Abstracts, 5161 River Road, Bethesda, MD 20816. (301) 951-1400. Bimonthly. Inquire as to cost and availability.

ENGINEERING INDEX MONTHLY AND AUTHOR INDEX. Engineering Information Inc., 345 East 47th Street, New York, NY 10017. (212) 705-7600. Monthly. $1560.00 per year.

INDEX TO SCIENTIFIC AND TECHNICAL PROCEEDINGS. Institute for Scientific Information, 3501 Market Street, Philadelphia, PA 19104. (800) 523-1850 or (215) 386-0100. 1978 to present. Monthly. $775.00 per year.

INDEX TO SCIENTIFIC REVIEWS. Institute for Scientific Information, 3501 Market Street, Philadelphia, PA 19104. (800) 523-1850 or (215) 386-0100. 1974 to present. Semi-annual. $550.00 per year.

IEEE PUBLICATIONS BULLETIN. Institute of Electrical and Electronics Engineers. Institute of Electrical and Electronics Engineers. IEEE Service Center, 445 Hoes Lane, Piscataway, NJ 08854. Quarterly. Free.

PHYSICS ABSTRACTS. Institution of Electrical Engineers. Available from: IEEE Service Center, 445 Hoes Lane, Piscataway, NJ 08854. 1898 to present. Bimonthly. $1700.00 per year.

PHYSICS BRIEFS. Physik Verlag GmbH, Postfach 1260/1280, D-6940 Weinheim, West Germany. (212) 661-9404. 1920 to present. Twenty-six times per year. $1250.00 per year.

SCIENCE CITATION INDEX. Institute for Scientific Information, 3501 Market Street, Philadelphia, PA 19104. (800) 523-1850 or (215) 386-0100. Six times per year. $6200.00 per year.

SOLID STATE ABSTRACTS: AN ABSTRACT JOURNAL INVOLVING THE PHYSICS, METALLURGY, CRYSTALLOGRAPHY, CHEMISTRY, AND DEVICE TECHNOLOGY OF SOLIDS. Cambridge Scientific Abstracts, 5161 River Road, Bethesda, MD 20816. (301) 951-1400. 1957 to present. Bimonthly. $550.00 per year.

ANNUAL REVIEWS AND YEARBOOKS

ADVANCES IN ELECTRONICS AND ELECTRON PHYSICS. Academic Press, Inc., 6277 Sea Harbor Drive, Orlando, FL 32821. (800) 321-5068. Irregular. Approximately $80.00 per volume.

CRC CRITICAL REVIEWS IN SOLID STATE AND MATERIALS SCIENCES. CRC Press, 2000 Corporate Boulevard, Boca Raton, FL 33431. (800) 272-7737. Irregular. Inquire.

ASSOCIATIONS AND PROFESSIONAL SOCIETIES

AMERICAN ELECTRONICS ASSOCIATION. P.O. Box 10045, 2670 Hanover Street, Palo Alto, CA 94303. (415) 857-9300.

AMERICAN INSTITUTE OF PHYSICS. 335 East 45th Street, New York, NY 10017. (212) 661-9494.

EDISON ELECTRIC INSTITUTE. 1111 19th Street, N.W., Washington, DC 20036. (202) 828-7400.

ELECTRONICS INDUSTRIES ASSOCIATION. 2001 Eye Street, N.W., Washington, DC 20006. (202) 457-4900.

INSTITUTE OF ELECTRICAL AND ELECTRONICS ENGINEERS. 345 East 47th Street, New York, NY 10017. (212) 705-7900.

INTERNATIONAL SOCIETY FOR HYBRID MICRO-ELECTRONICS. P.O. Box 2698, 1861 Wiehle Avenue, Suite 340, Reston, VA 22090. (703) 471-0066.

DIRECTORIES AND BIOGRAPHICAL SOURCES

AMERICAN MEN AND WOMEN OF SCIENCE. R.R. Bowker, Inc., Order Department, 245 West 17th Street, New York, NY 10011. (800) 521-8110. Eight volumes. 1986. $595.00 for set.

IEEE MEMBERSHIP DIRECTORY. Institute of Electrical and Electronics Engineers. IEEE Service Center, 445 Hoes Lane, Piscataway, NJ 08854. Annual. $7.00.

INTERNATIONAL RESEARCH CENTERS DIRECTORY 1988-89. Darren L. Smith, editor. Gale Research Company, Book Tower, Detroit, MI 48226. (800) 521-0707. 4th edition. 1987. $360.00.

1987 DIRECTORY OF ENGINEERING SOCIETIES AND RELATED ORGANIZATIONS. Gordon Davis, editor. Hemisphere Publishing Corporation, 79 Madison Avenue, New York, NY 10016-7892. (800) 821-8312. 12th edition. 1987. $100.00.

RESEARCH CENTERS DIRECTORY 1988. Gale Research Company, Book Tower, Detroit, MI 48226. (800) 521-0707. 12th edition. 1987. $365.00 for set.

SCIENTIFIC AND TECHNICAL ORGANIZATIONS AND AGENCIES DIRECTORY. Margaret Labash Young, editor. Gale Research Company, Book Tower, Detroit, MI 48226. (800) 521-0707. 2nd edition. 1987. $185.00.

WHO'S WHO IN ELECTRONICS. Harris Publishing Company, 2057-2 Aurora Road, Twinsburg, OH 44087. (216) 425-9143. Annual. $90.00.

WHO'S WHO IN ENGINEERING. Gordon Davis, editor. Hemisphere Publishing Corporation, 79 Madison Avenue, New York, NY 10016-7892. (800) 821-8312. 6th edition. 1985. $200.00.

ENCYCLOPEDIAS AND DICTIONARIES

IEEE STANDARD DICTIONARY OF ELECTRICAL AND ELECTRONICS TERMS. Frank Jay, editor. John Wiley and Sons, Inc., 605 Third Avenue, New York, NY 10158. (800) 526-5368. 3rd edition. 1984. $49.95.

MCGRAW-HILL ENCYCLOPEDIA OF SCIENCE AND TECHNOLOGY. McGraw-Hill Book Company, 1221 Avenue of the Americas, New York, NY 10020. (212) 512-2000. 6th edition. 1987. $1600.00.

THESAURUS OF SCIENTIFIC, TECHNICAL, AND ENGINEERING TERMS. Hemisphere Publishing Corporation, 79 Madison Avenue, New York, NY 10016-7892. (800) 821-8312. 1988. $125.00.

GENERAL WORKS

DEVICE ELECTRONICS FOR INTEGRATED CIRCUITS. S. Muller and T.I. Kamins. John Wiley and Sons, Inc., 605 Third Avenue, New York, NY 10158. (800) 526-5368. Second edition. 1986. $41.95.

INTEGRATED ELECTRONIC CIRCUITS AND SYSTEMS. Robert King. Van Nostrand Reinhold Company, Inc., 135 West 50th Street, New York, NY 10020. (800) 543-2681. 1983. $37.95.

MICROPROCESSOR TECHNOLOGY AND MICROCOMPUTERS. E.J. Pasahow. McGraw-Hill Book Company, 1221 Avenue of the Americas, New York, NY 10020. (212) 512-2000. 1987. $31.95.

POWER INTEGRATED CIRCUITS: PHYSICS, DESIGN AND APPLICTIONS. Paolo Antognetti. McGraw-Hill Book Company, 1221 Avenue of the Americas, New York, NY 10020. (212) 512-2000. 1986. $42.50.

SEMICONDUCTOR CIRCUIT APPROXIMATIONS; AN INTRODUCTION TO TRANSISTORS AND INTEGRATED CIRCUITS. Albert P. Malvino. McGraw-Hill Book Company, 1221 Avenue of the Americas, New York, NY 10020. (212) 512-2000. Fourth edition. 1985. $34.95.

HANDBOOKS AND MANUALS

CONTEMPORARY ELECTRONICS CIRCUITS DESKBOOK. Harry Helms. McGraw-Hill Book Company, 1221 Avenue of the Americas, New York, NY 10020. (212) 512-2000. 1986. $29.95.

ELECTRONIC ENGINEERS HANDBOOK. Donald G. Fink, editor. McGraw-Hill Book Company, 1221 Avenue of the Americas, New York, NY 10020. (212) 512-2000. 2nd edition. 1982. $89.00.

HANDBOOK OF MODERN ELECTRONICS AND ELECTRICAL ENGINEERING. Charles Belove, editor. John Wiley and Sons, Inc., 605 Third Avenue, New York, NY 10158. (800) 526-5368. 1986. $88.95.

VLSI HANDBOOK. Norman G. Einspruch, editor. Academic Press, Inc., 6277 Sea Harbor Drive, Orlando, FL 32821. (800) 321-5068. 1985. $125.00.

ONLINE DATA BASES

CA SEARCH. Chemical Abstracts Service, P.O. Box 3012, Columbus, OH 43120. (800) 848-6538 or (614) 421-3600. Comprehensive guide to chemical literature, 1972 to present. Inquire as to online cost and availability.

COMPENDEX. Engineering Information, Inc., 345 East 47th Street, New York, NY 10017. (800) 221-1044 or (212) 705-7615. Engineering and technical literature, 1975 to present. Inquire as to online cost and availability.

DISSERTATION ABSTRACTS ONLINE. University Microfilms International, 300 North Zeeb Road, Ann Arbor, MI 48106. (800) 521-0600 or (313) 761-4700. Scope includes virtually all doctoral dissertations accepted at accredited American institutions from 1861 to present in over 250 subject areas. Inquire as to online cost and availability.

INSPEC. INSPEC Marketing Department, Institution of Electrical Engineers. Available from IEEE Service Center, 445 Hoes Lane, Piscataway, NJ 08854. (201) 981-0060. Online version of Physics Abstracts. Inquire as to online cost and availability.

NTIS. National Technical Information Service, 5285 Port Royal Road, Springfield, VA 22161. (703) 487-4630. Broad coverage of government sponsored research reports, 1964 to present. Inquire as to online cost and availability.

SCISEARCH. Institute for Scientific Information, 3501 Market Street, Philadelphia, PA 19104. (800) 523-1850 or (215) 386-0100. Broad multidisciplinary title and author index to the international literature of science and technology, 1974 to present. Inquire as to online cost and availability.

WILSONLINE. H.W. Wilson and Company, 950 University Avenue, Bronx, NY 10452. (800) 367-6770 or (212) 588-8400. Makes available online versions of the H.W. Wilson indexes including Applied Science and Technology Index, Business Periodicals Index and Readers' Guide to Periodical Literature. Approximately 1980 to present. Inquire as to online cost and availability.

OTHER SOURCES

A GUIDE TO THE LITERATURE OF ELECTRICAL AND ELECTRONICS ENGINEERING. Susan B. Ardis. Libraries Unlimited Inc., P.O. Box 263, Littleton, CO 80160. (303) 770-1220. 1987. $37.50.

PERIODICALS

ELECTRONIC DESIGN. Hayden Publishing Company, 10 Mulholland Drive, Hasbrouck Heights, NJ 07604. (201) 288-7520. 1952 to present. Biweekly. $40.00 per year.

ELECTRONICS. McGraw-Hill Book Company, 1221 Avenue of the Americas, New York, NY 10020. (212) 512-2000. 1930 to present. Weekly. $32.00 per year.

IEEE CIRCUITS AND DEVICES MAGAZINE. Institute of Electrical and Electronics Engineers. IEEE Service Center, 445 Hoes Lane, Piscataway, NJ 08854. Bimonthly. $70.00 per year.

IEEE JOURNAL OF SOLID STATE CIRCUITS. Institute of Electrical and Electronics Engineers. IEEE Service Center, 445 Hoes Lane, Piscataway, NJ 08854. 1966 to present. Bimonthly. $113.00 per year.

IEEE TRANSACTIONS ON ELECTRON DEVICES. Institute of Electrical and Electronics Engineers. IEEE Service Center, 445 Hoes Lane, Piscataway, NJ 08854. 1959 to present. Monthly. $175.00 per year.

INSTITUTE OF ELECTRICAL AND ELECTRONICS ENGINEERS PROCEEDINGS. Institute of Electrical and Electronics Engineers. IEEE Service Center, 445 Hoes Lane, Piscataway, NJ 08854. 1913 to present. Monthly. $140.00 per year.

SEMICONDUCTOR INTERNATIONAL. Cahners Publishing Company, Inc., Cahners Plaza, 1350 East Touhy Avenue, Des Plaines, IL 60018. (312) 635-8800. 1978 to present. Monthly. $55.00 per year.

SOLID STATE ELECTRONICS. Pergamon Press, Inc., Maxwell House, Fairview Park, Elmsford, NY 10523. (914) 592-7700. 1960 to present. Monthly. $330.00 per year.

RESEARCH CENTERS AND INSTITUTES

ELECTRICAL ENGINEERING RESEARCH LABORATORIES. Purdue University, Electrical Engineering Building, West Lafayette, IN 47907. (317) 494-3536.

ELECTRONICS RESEARCH CENTER. University of Texas at Austin, 132 Engineering Science Building, Austin, TX 78712. (512) 471-3954.

ELECTRONICS RESEARCH LABORATORY. University of California, Berkeley, 253 Cory Hall, Berkeley, CA 94720. (415) 642-2301.

LABORATORY FOR ELECTROMAGNETIC AND ELECTRONIC SYSTEMS. Massachusetts Institute of Technology, 77 Massachusetts Avenue, Cambridge, MA 02139. (617) 253-4631.

INTERNAL COMBUSTION ENGINE

See: ENGINES

INTERSTELLAR MATTER

See also: ASTRONOMY, ASTROPHYSICS, DARK MATTER, COSMOLOGY

ABSTRACT SERVICES AND INDEXES

ASTRONOMY AND ASTROPHYSICS ABSTRACTS. Springer-Verlag New York, Incorporated, 175 Fifth Avenue, New York, NY 10010. (212) 460-1500. $70.00 per year.

ASTRONOMY AND ASTROPHYSICS MONTHLY INDEX. Olivetree Associates, Post Office Box 236, Sierra Madre, CA 91024. $212.00 per year. Complimentary copies available on request.

PHYSICS ABSTRACTS. Institute of Electrical Engineers, London, United Kingdom. Available from: Institute of Electrical and Electronic Engineers (IEEE), 345 East 47th Street,

New York, NY 10017. (212) 705-7900. 1898 to present. Monthly. $1670.00 per year.

SCIENCE CITATION INDEX. Institute for Scientific Information, 3501 Market Street, Philadelphia, PA 19104. (800) 523-1850 or (215) 386-0100. Inquire as to cost and availability.

STAR. (Scientific and Technical Aerospace Reports. United States National Aeronautics and Space Administration, Scientific and Technical Information Facility, Box 8757, Baltimore-Washington International Airport, MD 21240. (202) 755-2210. Semimonthly, with semiannual and annual indexes. $85.00 per year.

ANNUAL REVIEWS AND YEARBOOKS

THE ASTRONOMICAL ALMANAC. Superintendent of Documents, United States Government Printing Office, Washington, DC 20402. (202) 783-3238. Yearly.

ANNUAL REVIEW OF ASTRONOMY AND ASTROPHYSICS. Annual Reviews, Incorporated, 4139 El Camino Way, Palo Alto, CA 94306. (415) 493-4400.

ASSOCIATIONS AND PROFESSIONAL SOCIETIES

AMERICAN ASTRONOMICAL SOCIETY. 2000 Florida Avenue, NW, Suite 300, Washington, DC 20009. (202) 659-0134.

AMERICAN ASSOCIATION OF VARIABLE STAR OBSERVERS. 187 Concord Avenue, Cambridge, MA 02138. (617) 354-0484.

ASTRONOMICAL LEAGUE. Post Office Box 12821, Tucson, AZ 85732. (602) 790-8471.

ASTRONOMICAL SOCIETY OF THE PACIFIC. 1290 24th Avenue, San Francisco, CA 94122. (415) 661-8660.

BIBLIOGRAPHIES

A BIBLIOGRAPHY OF ASTRONOMY, 1970-1979. R.A. Seal and S.S. Martin. Libraries Unlimited, Incorporated, Littleton, CO 80160. 1982. $37.50.

SCIENCE BOOKS AND FILMS. American Association for the Advancement of Science, 1333 H Street, NW, Washington, DC 20005. Five times per year. $20.00 per year.

SCIENTIFIC AND TECHNICAL BOOKS AND SERIALS IN PRINT 1988; AN INDEX TO LITERATURE IN SCIENCE AND TECHNOLOGY. R.R. Bowker Company, 205 East 42nd Street, New York, NY 10017. (800) 521-8110 or (212) 916-1600. $175.00.

DIRECTORIES AND BIOGRAPHICAL SOURCES

AMERICAN MEN AND WOMEN OF SCIENCE. Physical and Biological Sciences. Sixteenth edition. R.R. Bowker Company, 205 East 42nd Street, New York, NY 10017. (800) 521-8110 or (212) 916-1600. 1986. $595.00.

THE BIOGRAPHICAL DICTIONARY OF SCIENTISTS: ASTRONOMERS. D. Abbott, editor. Peter Bedrick Books, 125 East 23rd Street, New York, NY 10010. 1984.

DIRECTORY OF PHYSICS AND ASTRONOMY STAFF MEMBERS. American Institute of Physics, 335 East 45th Street, New York, NY 10017. Annual.

RESEARCH CENTERS DIRECTORY. Gale Research Company, Book Tower, Detroit, MI 48226. (800) 521-0707. Twelfth edition. 1987. $365.00 for set.

WHO'S WHO IN FRONTIER SCIENCE AND TECHNOLOGY. Marquis Who's Who, Incorporated, 200 East Ohio Street, Chicago, IL 60611. (800) 428-3898 or (312) 787-2008.

ENCYCLOPEDIAS AND DICTIONARIES

ENCYCLOPEDIA OF PHYSICAL SCIENCE AND TECHNOLOGY. Academic Press, Incorporated, Orlando, FL 32887. (800) 321-5068 or (305) 345-2734. Inquire as to cost and availability.

MCGRAW-HILL ENCYCLOPEDIA OF SCIENCE AND TECHNOLOGY. McGraw-Hill Book, Incorporated, 1221 Avenue of the Americas, New York, NY 10020. (212) 997-3675. Fifth edition, 15 volumes. $1100.00.

GENERAL WORKS

THE CAMBRIDGE ASTRONOMY GUIDE: A PRACTICAL INTRODUCTION TO ASTRONOMY. Cambridge University Press, 32 East 57th Street, New York, NY 10022. (212) 688-8888. 1985. $24.95.

DARK MATTER IN THE UNIVERSE. J. Kormendy and G.R. Knapp, editors. D. Reidel, 101 Philip Drive, Assinippi Park, Norwell, MA 02061. (617) 871-6600. 1987. $104.00.

THE INTERACTION OF SUPERNOVA REMNANTS WITH THE INTERSTELLAR MEDIUM. R.S. Roger and T.L. Landecker, editors. Cambridge University Press, 32 East 57th Street, New York, NY 10022. (212) 688-8888. 1988. $59.95.

INTERSTELLAR CHEMISTRY. Walter Duley and David A. Williams. Academic Press, Incorporated, 6277 Sea Harbor Drive, Orlando, FL 32821. (800) 321-5068. 1984. $49.50.

STARS, NEBULAE, AND THE INTERSTELLAR MEDIUM. C.R. Kitchen. Bosont: Adam Hilger. 1987. $33.00.

ONLINE DATA BASES

CA SEARCH. Chemical Abstracts Service, Post Office Box 3012, Columbus, OH 43210. Guide to chemical literature, 1972 to present. Inquire as to cost and availability.

DISSERTATION ABSTRACTS ONLINE. University Microfilms International, 300 North Zeeb Road, Ann Arbor, MI 48106. (800) 521-0600 or (313) 761-4700. Scope includes virtually all doctoral dissertations accepted at accredited American institutions from 1861 to present in 252 subject areas. Inquire as to cost and availability.

INSPEC. INSPEC Marketing Department, Institute of Electrical and Electronics Engineers, Incorporated, IEEE Service Department, 445 Hoes Lane, Piscataway, NJ 08854. (201) 981-0060. Inquire as to on-line cost and availability.

NASA. National Aeronautics and Space Administration, Scientific and Technical Information Branch, 300 7th Street, SW, Washington, DC 20546. Citations and abstracts of aerospace literature, 1962 to present. Inquire as to cost and availability.

NTIS. National Technical Information Service, 5285 Port Royal Road, Springfield, VA 22161. (703) 487-4630. Broad coverage of government-sponsored research reports, 1964 to present. Inquire as to cost and availability.

SCISEARCH. Institute for Scientific Information, 3501 Market Street, Philadelphia, PA 19104. (800) 523-1850 or (215) 386-0100. Broad multidisciplinary title and author index to the international literature of science and technology, 1974 to present. Inquire as to cost and availability.

OTHER SOURCES

ATLAS OF DEEP-SKY SPLENDORS. H. Vehrenberg. Cambridge University Press, 32 East 57th Street, New York, NY 10022. (800) 431-1580 or (212) 688-8888. Fourth edition, 1984. $44.50.

SKY CATALOGUE 2000.0. A. Hirshfeld and R. Sinnott. Cambridge University Press, 32 East 57th Street, New York, NY 10022. (800) 431-1580 or (212) 688-8888.

PERIODICALS

ASTRONOMICAL JOURNAL. American Astronomical Society. Available from: American Institute of Physics, 335 East 45th Street, New York, NY 10017. (212) 661-9404. $125.00 per year.

ASTRONOMICAL SOCIETY OF THE PACIFIC. Publications. Astronomical Society of the Pacific, 1290 24th Avenue, San Francisco, CA 94122. (415) 661-8660. Monthly. $38.00.

ASTRONOMY. Astro Media Corporation, 625 East Paul Avenue, Milwaukee, WI 53202. Monthly. $18.00 per year.

ASTRONOMY AND ASTROPHYSICS ABSTRACTS. Springer-Verlag New York, Incorporated, 175 Fifth Avenue, New York, NY 10010. (212) 460-1500. $680.00 per year.

ASTROPHYSICAL JOURNAL. American Astronomical Society, University of Chicago Press, 5801 Ellis Avenue, Chicago, IL 60637. Biweekly. $305.00 per year.

ASTROPHYSICS AND SPACE SCIENCE. D. Reidel Publishing Company, 190 Older Derby Street, Hingham, MA 02043. Monthly. $101.00 per year.

CELESTIAL MECHANICS: AN INTERNATIONAL JOURNAL OF SPACE DYNAMICS. Academic Press, Incorporated, Orlando, FL 32887. (305) 345-4100. Monthly. $484.00 per year.

ICARUS: INTERNATIONAL JOURNAL OF THE SOLAR SYSTEM STUDIES. Academic Press, Incorporated, Orlando, FL 32887. (305) 345-4100. Monthly. $484.00 per year.

MERCURY. Astronomical Society of the Pacific, 1290 245h Avenue, San Francisco, CA 94122. (415) 661-8660. Bimonthly. $21.00 per year.

MONTHLY NOTICES OF THE ROYAL ASTRONOMICAL SOCIETY. Blackwell Science Publications, Incorporated, 667 Lytton Avenue, Palo Alto, CA 94301. (415) 324-1688. Monthly. $134.00 per year to individuals.

PLANETARY AND SPACE SCIENCE. Pergamon Press, Incorporated, Maxwell House, Fairview Park, Elmsford, NY 10523. (914) 592-7700. Monthly. $430.00 per year.

SKY AND TELESCOPE. Sky Publishing Corporation, 49 Bay State Road, Cambridge, MA 02238. (617) 864-7360. Monthly. $18.00 per year.

SOVIET ASTRONOMY (TRANSLATION OF ASTRO-NOMICHESKII ZHURNAL). American Institute of Physics, 335 East 45th Street, New York, NY 10017. (212) 661-9404. Bimonthly. $425.00 per year.

SPACE SCIENCE REVIEWS. D. Reidel Publishing Company, 190 Old Derby Street, Hingham, MA 02043. Monthly. $305.00 per year.

VISTAS IN ASTRONOMY. Pergamon Press, Incorporated, Maxwell House, Fairview Park, Elmsford, NY 10523. (914) 592-7700. Quarterly. $145.00 per year.

RESEARCH CENTERS AND INSTITUTES

NATIONAL ASTRONOMY AND IONOSPHERE CENTER. Cornell University, Space Sciences Building, Ithaca, NY 14853. (607) 256-3734.

HARVARD-SMITHSONIAN CENTER FOR ASTROPHYSICS. Fred L. Whipple Observatory, Post Office Box 97, Amado, AZ 85645. (602) 629-6741.

NATIONAL OPTICAL ASTRONOMY OBSERVATORIES. 1002 North Warren Avenue, Tucson, AZ 85719. (602) 325-9230.

NATIONAL RADIO ASTRONOMY OBSERVATORY. Edgemont Road, Charlottesville, VA 22903. (804) 296-0211.

SPACE TELESCOPE SCIENCE INSTITUTE. 3700 San Martin Drive, Baltimore, MD 21218. (301) 338-4700.

IODINE

See also: BROMINE, CHLORINE, FLUORINE, HALIDES

ABSTRACT SERVICES AND INDEXES

APPLIED SCIENCE AND TECHNOLOGY INDEX. H.W. Wilson Company, 950 University Avenue, Bronx, NY 10452. (800) 367-6670 or (212) 588-8400. Inquire as to cost and availability.

CHEMICAL ABSTRACTS. Chemical Abstracts Service, 2540 Olentangy Road, P.O. Box 3012, Columbus, OH 43210. (800) 848-6538 or (614) 421-3600. Weekly. $9200.00 per year.

GENERAL SCIENCE INDEX. H.W. Wilson Company, 950 University Avenue, Bronx, NY 10452. (800) 367-6670 or (212) 588-8400. Inquire as to cost and availability.

PHYSICS ABSTRACTS. Institute of Electrical Engineers, London, United Kingdom. Available from: Institute of Electrical and Electronic Engineers (IEEE), 345 East 47th Street, New York, NY 10017. (212) 705-7900.

SCIENCE CITATION INDEX. Institute for Scientific Information, 3501 Market Street, Philadelphia, PA 19104. (800) 523-1850 or (215) 386-0100.

ASSOCIATIONS AND PROFESSIONAL SOCIETIES

AMERICAN CHEMICAL SOCIETY. 1155 16th Street, NW, Washington, DC 20036. (202) 872-4600.

ASSOCIATION OF CONSULTING CHEMISTS AND CHEMICAL ENGINEERS. 50 East 41st Street, Suite 92, New York, NY 10017. (212) 684-6255.

BIBLIOGRAPHIES

SCIENTIFIC AND TECHNICAL BOOKS IN PRINT; An Index to Literature in Science and TEchnology. R.R. Bowker Co., E. 42nd Street, New York, NY 10017. (800) 521-8110 or (212) 916-1600.

DIRECTORIES AND BIOGRAPHICAL SOURCES

AMERICAN INSTITUTE OF CHEMISTS. American Institute of Chemists, 7315 Wisconsin Avenue, Bethesda, MD 20814. (301) 652-2447. 1986. $35.00.

AMERICAN MEN AND WOMEN OF SCIENCE. Physical and Biological Sciences. Fifteenth edition. R.R. Bowker Co., E. 42nd Street, New York, NY 10017. (800) 521-8110 or (212) 916-1600.

BIOGRAPHICAL DICTIONARY OF SCIENTISTS: CHEMISTS. David Abbott, editor. P. Bedrick Books, 125 East 23rd Street, New York, NY 10010. (212) 777-1187. 1984. $18.95.

CHEMICAL WEEK - BUYERS GUIDE ISSUE. McGraw-Hill Book, Inc., 1221 Avenue of the Americas, New York, NY 10020. (800) 628-0004. Annual, October. $50.00.

CONSULTING SERVICES: CHEMISTS AND CHEMICAL ENGINEERS. Association of Consulting Chemists and Chemical Engineers, 50 East 41st Street, New York, NY 10017. (212)

684-6255. Annual. 1986. $45.00.

GOVERNMENT RESEARCH DIRECTORY. Gale Research Company, Book Tower, Detroit, MI 48226. (800) 521-0707. Fourth edition. 1987. $350.00.

INTERNATIONAL RESEARCH CENTERS DIRECTORY 1986-1987. Gale Research Company, Book Tower, Detroit, MI 48226. (800) 521-0707. Third edition. 1986. $330.00.

RESEARCH CENTERS DIRECTORY. Gale Research Company, Book Tower, Detroit, MI 48226. (800) 521-0707. Eleventh edition. 1987. $355.00.

SCIENTIFIC AND TECHNICAL ORGANIZATIONS AND AGENCIES DIRECTORY. Gale Research Company, Book Tower, Detroit, MI 48226. (800) 521-0707. 1085. $150.00.

WORLD GUIDE TO SCIENTIFIC ASSOCIATIONS AND LEARNED SOCIETIES. K.G. Saur Inc., 175 Fifth Avenue, New York, NY 10010. (800) 521-0707 or (212) 982-1302. Fourth edition. 1984. $112.00.

WHO'S WHO IN FRONTIER SCIENCE AND TECHNOLOGY. Marquis Who's Who, Inc., 200 East Ohio Street, Chicago, Illinois 60611. (800) 428-3898 or (312) 787-2008.

WHO'S WHO IN TECHNOLOGY TODAY. Reston Publishing Company, Inc., c/o Prentice-Hall, Inc., Englewood Cliffs, NJ 07632. (800) 262-6868. Biennial. Five volumes. $425.00. Covers the fields of electronics, computer science, physics, optics, chemistry, biotechnology, mechanics, energy, and earth science.

ENCYCLOPEDIAS AND DICTIONARIES

CONCISE ENCYCLOPEDIA OF CHEMICAL TECHNOLOGY. Kirt-Othmer. John Wiley & Sons, 605 Third Avenue, New York, NY 10158. (800) 526-5368 or (212) 850-6000. Third edition. 1985. $129.95.

CONDENSED CHEMICAL DICTIONARY. Gessner Hawley. Van Nostrand Reinhold, 115 Fifth Avenue, New York, NY 10003. Tenth edition. 1981. $49.95.

ENCYCLOPEDIA OF PHYSICAL SCIENCE AND TECHNOLOGY. Academic Press, Inc., Orlando, FL 32887. (800) 321-5068 or (305) 345-2734. Fifteen volumes, 1986. Inquire as to cost and availability.

GLOSSARY OF CHEMICAL TERMS. Clifford A. Hampel and Gessner G. Hawley. Van Nostrand Reinhold Company, 115 Fifth Avenue, New York, NY 10003. (800) 543-2681 or (212) 254-3232.

MCGRAW-HILL ENCYCLOPEDIA OF SCIENCE AND TECHNOLOGY. McGraw-Hill Book, Inc., 1221 Avenue of the Americas, New York, NY 10020. (212) 997-3675. Inquire as to cost and availability.

VAN NOSTRAND REINHOLD ENCYCLOPEDIA OF CHEMISTRY. Douglas M. Considine and Glenn D. Considine. Van Nostrand Reinhold Company, 115 Fifth Avenue, New York, NY 10003. (800) 543-2681 or (212) 254-3232 or (212) 254-3232. 1984. $97.95.

GENERAL WORKS

CHEMISTRY OF HALIDES, PSEUDOHALIDES AND AZIDES: SUPPLEMENT D. Chemistry of Functional Groups Series, Part 1 and 2. Saul Patai. John Wiley & Sons, 605 Third Avenue, New York, NY 10158. (800) 526-5368 or (212) 850-6000. 1983. $745.00 set.

CHEMISTRY OF THE ELEMENTS. N.N. Greenwood and A. Earnshaw. Pergamon Publishing, Inc., Maxwell House, Fairview Park, Elmsford, NY 10523. (914) 592-7700. 1984. $143.00.

HALOGEN CHEMISTRY. V. Gutman, editor. Academic Press, Inc., 6277 Sea Harbor Drive, Orlando, FL 32821. (800) 321-5068. Three volumes. 1967. $90.00 each.

HAZARDOUS AND TOXIC CHEMICAL: SAFE HANDLING AND DISPOSAL. Howard Fawcett. John Wiley & Sons, 605 Third Avenue, New York, NY 10158. (800) 526-5368 or (212) 850-6000. 1984. $37.00.

SAFE STORAGE OF LABORTORY CHEMICALS. David A. Pipitone. John Wiley & Sons, 605 Third Avenue, New York, NY 10158. (800) 526-5368 or (212) 850-6000. 1984. $60.00.

TOXICOLOGY OF HALOGENATED HYDROCARBONS: HEALTH AND ECOLOGICAL EFFECTS. M.A. Khan and R.H. Stanton, editors. Pergamon Press, Inc., Maxwell House, Fairview Park, Elmsford, NY 10523. (914) 592-7700. 1981. $72.50.

HANDBOOKS AND MANUALS

THE CHEMIST'S COMPANION: A HANDBOOK OF PRACTICAL DATA, TECHNIQUES, AND REFERENCES. Arnold J. Gordon and Richard A. Ford. John Wiley & Sons, 605 Third Avenue, New York, NY 10158. (800) 526-5368. 1973. $49.95.

CRC HANDBOOK OF CHEMISTRY AND PHYSICS. CRC Press, Inc., 2000 Corporate Blvd., N.W., Boca RAton, FL 33431. Sixty-seventh edition. 1986. $69.95.

HANDBOOK OF APPLIED CHEMISTRY: FACTS FOR ENGINEERS, SCIENTISTS, TECHNICIANS, AND TECHNICAL MANAGERS. Vollrath Hopp and Ingp Hennig. McGraw-Hill Book, Inc., 1221 Avenue of the Americas, New York, NY 10020. (800) 628-0004. 1983. $54.00.

HANDBOOK OF COMPUTATIONAL CHEMISTRY: A PRACTICAL GUIDE TO CHEMICAL STRUCTURE AND ENERGY CALCULATIONS. Tim Clark. John Wiley & Sons, 605 Third Avenue, New York, NY 10158. (800) 526-5368 or (212) 850-6000. 1985. $35.00.

LANGE'S HANDBOOK OF CHEMISTRY. John A. Dean, editor. McGraw-Hill Book, Inc., 1221 Avenue of the Americas, New York, NY 10020. (800) 628)-0004. 1985. $59.50.

ONLINE DATA BASES

CA SEARCH. Chemical Abstracts Service, P.O. Box 3012, Columbus, OH 43210. Guide to chemical literature, 1972 to present. Inquire as to cost and availability.

DISSERTATION ABSTRACTS ONLINE. University Microfilms International, 300 North Zeeb Road, Ann Arbor, MI 48106. (800) 521-0600 or (313) 761-4700. Scope includes virtually all doctoral dissertations accepted at accredited American institutions from 1861 to present in 252 subject areas. Inquire as to online cost and availability.

INSPEC. INSPEC Marketing Department, Institute of Electrical and Electronics Engineers, Inc., IEEE Service Department, 445 Hoes Lane, Piscataway, NJ 08854. (201) 981-0060. Inquire as to online cost and availability.

NTIS. National Technical Information Service, 5285 Port Royal Road, Springfield, VA 22161. (703) 487-4630. Broad coverage of government sponsored research reports, 1964 to present. Inquire as to online cost and availability.

SCISEARCH. Institute for Scientific Information, 3501 Market Street, Philadelphia, PA 19104. (800) 523-1850 or (215) 386-0100. Broad multidisciplinary title and author index to the international literature of science and technology, 1974 to present. Inquire as to online cost and availability.

OTHER SOURCES

ANNUAL ENERGY REVIEW. U.S. Department of Energy, Energy Information Administration, Washington, DC 20585. Annual.

CHEMICAL NOMENCLATURE USEAGE. Ronald Lees and Arthur F. Smith. John Wiley & Sons, 605 Third Avenue, New York, NY 10158. (800) 526-5368 or (212) 850-6000. 1983. $52.95.

GUIDE TO BASIC INFORMATION SOURCES IN CHEMISTRY. Arthur Antony. John Wiley & Sons, 605 Third Avenue, New York, NY 10158. (800) 526-5368 or (212) 850-6000. 1979. $26.95.

HOW TO FIND CHEMICAL INFORMATION: A GUIDE FOR PRACTICING CHEMISTS, TEACHERS, AND STUDENTS. John Wiley & Sons, 605 Third Avenue, New York, NY 10158. (800) 526-5368 or (212) 850-6000. 1986. $35.00.

MINERALS YEARBOOK. U.S. Department of the Interior, Bureau of Mines, C Street between Eighteenth and Nineteenth Streets, N.W., Washington, DC 20240. Annual.

PERIODICALS

ANALYTICAL CHEMISTRY. American Chemical Society, 1155 Sixteenth Street, N.W., Washington, DC 20036. (800) 424-6747 or (202) 872-4700. Monthly. $33.00 per year.

ANDEWANDTE CHEMIE (GESELLSCHAFT DEUTSCHER CHEMIKER, GW). VCH Verlagsgesellschaft mbH, Pappelallee 3, Postfach 1260, 6940 Weinheim, West Germany. Monthly. $300.00 per year.

CHEMICAL REVIEWS. American Chemical Society, 1155 Sixteenth Street, N.W., Washington, DC 20036. (800) 424-6747 or (202) 872-4700. Bimonthly. $83.00 per year.

CHEMICAL WEEK. McGraw-Hill Book, Inc., 1221 Avenue of the Americas, New York, NY 10020. (212) 512-2000. Weekly. $30.00 per year.

CHEMTECH. American Chemical Society, 1155 Sixteenth Street, N.W., Washington, DC 20036. (800) 424-6747 or (202) 872-4700. Monthly. $40.00 per year to individuals.

INORGANIC CHEMISTRY. American Chemical Society, 1155 Sixteenth Street, N.W., Washington, DC 20036. (800) 424-6747 or (202) 872-4700. Monthly. $400.00 per year.

JOURNAL OR THE AMERICAN CHEMICAL SOCIETY. American Chemical Society, 1155 Sixteenth Street, N.W., Washington, DC 20036. (800) 424-6747 or (202) 872-4700. Biweekly. $330.00 per year.

POLYHEDRON. Pergamon Journals, Inc., Maxwell House, Fairview Park, Elmsford, NY 10523. (914) 592-7700. Monthly. $595.00 per year.

RESEARCH CENTERS AND INSTITUTES

HARVARD UNIVERSITY. Chemical Laboratories, Oxford Street, Cambridge, MA 02138. (617) 495-4283.

RENSSELAER POLYTECHNICAL INSTITUTE. Chemistry Laboratories, Cogswell Laboratory, Troy, NY 13181. (518) 266-8462.

SOUTHERN ILLINOIS UNIVERSITY AT CARBONDALE. Research Program in Chemistry and Biochemistry, Carbondale, IL 62901. (618) 453-5721.

UNIVERSITY OF WISCONSIN, MADISON. Theoretical Chemistry Institute, 1101 University Avenue, Madison, WI 53706. (608) 262-1511.

ION ACCELERATOR

See: PARTICLE ACCELERATORS

ION LASER

See: LASERS

IONIZATION

See also: CHEMISTRY, IONS, PHYSICAL CHEMISTRY

ABSTRACT SERVICES AND INDEXES

APPLIED SCIENCE AND TECHNOLOGY INDEX. H.W. Wilson Company, 950 University Avenue, Bronx, NY 10452. (800) 367-6670 or (212) 588-8400. Inquire as to cost and availability.

CHEMICAL ABSTRACTS. Chemical Abstracts Service, 2540 Olentangy Road, Post Office Box 3012, Columbus, OH 43210. (800) 848-6538 or (614) 421-3600. Weekly. $9200.00 per year.

GENERAL SCIENCE INDEX. H.W. Wilson Company, 950 University Avenue, Bronx, NY 10452. (800) 367-6770 or (212) 588-8400. Inquire as to cost and availability.

MASS SPECTROMETRY BULLETIN. Royal Society of Chemistry. Order from: Distribution Centre, Blackhorse Road, Letchworth, Herts SG6 1HN, England. 1966 to present. Monthly. $350.00 per year.

PHYSICS ABSTRACTS. Institute of Electrical Engineers, London, United Kingdom. Available from: Institute of Electrical and Electronic Engineers (IEEE), 345 East 47th Street, New York, NY 10017. (212) 705-7900.

SCIENCE CITATION INDEX. Institute for Scientific Information, 3501 Market Street, Philadelphia, PA 19104. (800) 523-1850 or (215) 386-0100.

ANNUAL REVIEWS AND YEARBOOKS

ANNUAL REVIEW OF PHYSICAL CHEMISTRY. Annual Reviews Incorporated, 4139 El Camino Way, Palo Alto, CA 94306. (415) 493-4400. Annual.

ASSOCIATIONS AND PROFESSIONAL SOCIETIES

AMERICAN CHEMICAL SOCIETY. 1155 16th Street, NW, Washington, DC 20036. (202) 872-4600.

ASSOCIATION OF CONSULTING CHEMISTS AND CHEMICAL ENGINEERS. 50 East 41st Street, Suite 92, New York, NY 10017. (212) 684-6255.

BIBLIOGRAPHIES

SCIENCE BOOKS AND FILMS. American Association for the Advancement of Science, 1333 H Street, NW, Washington, DC 20005.

SCIENTIFIC AND TECHNICAL BOOKS AND SERIALS IN PRINT 1988; AN INDEX TO LITERATURE IN SCIENCE AND TECHNOLOGY. R.R. Bowker Company, 205 East 42nd Street, New York, NY 10017. (800) 521-8110 or (212) 916-1600. $175.00.

IONIZATION

DIRECTORIES AND BIOGRAPHICAL SOURCES

AMERICAN INSTITUTE OF CHEMISTS. American Institute of Chemists, 7315 Wisconsin Avenue, Bethesda, MD 20814. (301) 652-2447. 1986. $35.00.

AMERICAN MEN AND WOMEN OF SCIENCE. Physical and Biological Sciences. R.R. Bowker Company, 205 East 42nd Street, New York, NY 10017. (800) 521-8110 or (212) 916-1600. Fifteenth edition. $565.00.

BIOGRAPHICAL DICTIONARY OF SCIENTISTS: CHEMISTS. David Abbott, editor. P. Bedrick Books, 125 East 23rd Street, New York, NY 10010. (212) 777-1187. 1984. $18.95.

GOVERNMENT RESEARCH DIRECTORY. Gale Research Company, Book Tower, Detroit, MI 48226. (800) 521-0707. Fourth edition. 1987. $350.00.

INTERNATIONAL RESEARCH CENTERS DIRECTORY 1986-1987. Gale Research Company, Book Tower, Detroit, MI 48226. (800) 521-0707. Third edition. 1986. $330.00.

RESEARCH CENTERS DIRECTORY. Gale Research Company, Book Tower, Detroit, MI 48226. (800) 521-0707. Eleventh edition. 1987. $355.00.

RESEARCH SERVICES DIRECTORY. Robert J. Huffman and Mary M. Watkins, editors. Gale Research Company, Book Tower, Detroit, MI 48226. (800) 521-0707. Third edition. $290.00.

SCIENTIFIC AND TECHNICAL ORGANIZATIONS AND AGENCIES DIRECTORY. Gale Research Company, Book Tower, Detroit, MI 48226. (800) 521-0707. 1985. $150.00.

WHO'S WHO IN FRONTIER SCIENCE AND TECHNOLOGY. Marquis Who's Who, Incorporated, 200 East Ohio Street, Chicago, IL 60611. (800) 428-3898 or (312) 787-2008.

WHO'S WHO IN TECHNOLOGY TODAY. Reston Publishing Company, Incorporated, c/o Prentice-Hall, Incorporated, Englewood Cliffs, NJ 07632. (800) 262-6868. Biennial. Five volumes. $425.00. Covers the fields of electronics, computer science, physics, optics, chemistry, biotechnology, mechanics, energy, and earth science.

WORLD GUIDE TO SCIENTIFIC ASSOCIATIONS AND LEARNED SOCIETIES. K.G. Saur, Incorporated, 175 Fifth Avenue, New York, NY 10010. (800) 521-0707 or (212) 982-1302. Fourth edition. 1984. $112.00.

ENCYCLOPEDIAS AND DICTIONARIES

CONCISE ENCYCLOPEDIA OF CHEMICAL TECHNOLOGY. Kirk-Othmer. John Wiley and Sons, Incorporated, 605 Third Avenue, New York, NY 10158. (800) 526-5368 or (212) 850-6000. Third edition. 1985. $129.95.

CONDENSED CHEMICAL DICTIONARY. Gessner Hawley. Van Nostrand Reinhold, 115 Fifth Avenue, New York, NY 10003. Tenth edition. 1981. $49.95.

ENCYCLOPEDIA OF PHYSICAL SCIENCE AND TECHNOLOGY. Academic Press, Incorporated, Orlando, FL 32887. (800) 321-5068 or (305) 345-2734. Fifteen volumes, 1986.

GLOSSARY OF CHEMICAL TERMS. Clifford A. Hampel and Gessner G. Hawley. Van Nostrand Reinhold Company, 115 Fifth Avenue, New York, NY 10003. (800) 543-2681 or (212) 254-3232. Second edition. 1982. $21.95.

HAWLEY'S CONDENSED CHEMICAL DICTIONARY. N. Irving Sax and Richard J. Lewis, Sr., editors. Van Nostrand Reinhold, Incorporated, 115 Fifth Avenue, New York, NY 10003. (800) 543-2681. Eleventh edition. 1987. $52.95.

MCGRAW-HILL ENCYCLOPEDIA OF SCIENCE AND TECHNOLOGY. McGraw-Hill Book Company, Incorporated, 1221 Avenue of the Americas, New York, NY 10020. (212) 997-3675.

VAN NOSTRAND REINHOLD ENCYCLOPEDIA OF CHEMISTRY. Douglas M. Considine and Glenn D. Considine. Van Nostrand Reinhold Publishing Company, Incorporated, 115 Fifth Avenue, New York, NY 10003. (800) 543-2681 or (212) 254-3232. 1984. $97.95.

GENERAL WORKS

THE CHEMICAL BOND. John N. Murrell and others. John Wiley and Sons, Incorporated, 605 Third Avenue, New York, NY 10158. (800) 526-5368 or (212) 850-6000. 1985. $31.95.

CHEMICAL BONDING MODELS, VOLUME 1. J.F. Liebman and A. Greenberg, editors. VCH Publishing, Incorporated, 303 NW, 12th Avenue, Deerfield Beach, FL 33442-1705. (305) 428-5566. 1986. $77.50.

CHEMISTRY. Charles E. Mortimer. Wadsworth Publishing Company, 10 Davis Drive, Belmont, CA 94002. (415) 595-2350. Sixth edition. 1986. $49.95.

CHEMISTRY OF THE ELEMENTS. N.N. Greenwood and A. Earnshaw. Pergamon Publishing, Incorporated, Maxwell House, Fairview Park, Elmsford, NY 10523. (914) 592-7700. 1984. $143.00.

INORGANIC CHEMISTRY: PRINCIPLES OF STRUCTURE AND REACTIVITY. James E. Huheey. Harper and Row Publishers, Incorporated, 10 East 53rd Street, New York, NY 10022. (212) 207-7655. Third edition. $36.50.

INTRODUCTION TO PHYSICAL CHEMISTRY. M.F.C. Ladde and W.H. Lee. Cambridge University Press, 32 East 57th Street, New York, NY 10022. (212) 688-8885. 1986. $19.95 in paper.

INTRODUCTION TO SYMMETRY IN BONDING AND SPECTRA. Bodie Douglas and Charles A. Hollingsworth. Academic Press, Incorporated, 6277 Sea Harbor Drive, Orlando, FL 32821. (800) 321-5068. 1985. $40.00.

INVESTIGATION OF RATES AND MECHANISMS OF REACTIONS, PART 1. Claude F. Bernasconi, editor. John Wiley and Sons, Incorporated, 605 Third Avenue, New York, NY 10158. (800) 526-5368 or (212) 850-6000. Fourth edition. 1986. $175.00.

ION-MOLECULE REACTIONS. J.L. Franklin, editor. Van Nostrand Reinhold, Incorporated, 115 Fifth Avenue, New York, NY 10003. (800) 543-2681. Two volumes. 1983. $120.00 for set.

IONIZATION POTENTIALS: SOME VARIATIONS, IMPLICATIONS AND APPLICATIONS. L.H. Ahrens. Pergamon Press, Incorporated, Maxwell House, Fairview Park, Elmsford, NY 10523. (914) 592-7700. 1983. $32.00.

QUANTUM CHEMISTRY OF ATOMS AND MOLECULES. Philip S.C. Matthews. Cambridge University Press, 32 East 57th Street, New York, NY 10022. (212) 688-8885. 1986. $44.50.

PHYSICS OF HIGHLY CHARGED IONS. R.K. Janev and others. Springer-Verlag, 175 Fifth Avenue, New York, NY 10010. (800) 526-7254. 1985. $52.00.

SYMMETRY AND STRUCTURE. S.A. Kettle. John Wiley and Sons, Incorporated, 605 Third Avenue, New York, NY 10158. (800) 526-5368 or (212) 850-6000. 1985. $37.95.

HANDBOOKS AND MANUALS

THE CHEMIST'S COMPANION: A HANDBOOK OF PRACTICAL DATA, TECHNIQUES, AND REFERENCES. Arnold J. Gordon and Richard A. Ford. John Wiley and Sons, Incorporated, 605 Third Avenue, New York, NY 10158. (800) 526-5368. 1973. $49.95.

CRC HANDBOOK OF CHEMISTRY AND PHYSICS. CRC Press, Incorporated, 2000 Corporate Boulevard, NW, Boca Raton, FL 33431. Sixty-seventh edition. 1986. $69.95.

THE DETERMINATION OF IONIZATION CONSTANTS: A LABORATORY MANUAL. Adrien Albert and E.P. Serjeant. Chapman and Hall, Incorporated, 29 West 35th Street, New York, NY 10001. (212) 688-8885. Third edition. 1984. $39.95.

HANDBOOK OF APPLIED CHEMISTRY: FACTS FOR ENGINEERS, SCIENTISTS, TECHNICIANS, AND TECHNICAL MANAGERS. Vollrath Hopp and Ingp Hennig. McGraw-Hill Book Company, Incorporated, 1221 Avenue of the Americas, New York, NY 10020. (800) 628-0004. 1983. $54.00.

HANDBOOK OF COMPUTATIONAL CHEMISTRY: A PRACTICAL GUIDE TO CHEMICAL STRUCTURE AND ENERGY CALCULATIONS. Tim Clark. John Wiley and Sons, Incorporated, 605 Third Avenue, New York, NY 10158. (800) 526-5368 or (212) 850-6000. 1985. $35.00.

LANGE'S HANDBOOK OF CHEMISTRY. John A. Dean, editor. McGraw-Hill Book Company, Incorporated, 1221 Avenue of the Americas, New York, NY 10020. (800) 628-0004. 1985. $59.50.

TABLES OF PHYSICAL AND CHEMICAL CONSTANTS; AND SOME MATHEMATICAL FUNCTIONS. G.W.C. Kaye and T.H. Laby, editors. Longman, Incorporated 95 Church Street, White Plains, NY 10601. (914) 993-5000. 1986. $39.95.

ONLINE DATA BASES

CA SEARCH. Chemical Abstracts Service, Post Office Box 3012, Columbus, OH 43210. Guide to chemical literature, 1972 to present. Inquire as to cost and availability.

DISSERTATION ABSTRACTS ONLINE. University Microfilms International, 300 North Zeeb Road, Ann Arbor, MI 48106. (800) 521-0600 or (313) 761-4700. Scope includes virtually all doctoral dissertations accepted at accredited American institutions from 1861 to present in 252 subject areas. Inquire as to online cost and availability.

INSPEC. INSPEC Marketing Department, Institute of Electrical and Electronics Engineers, Incorporated, IEEE Service Department, 445 Hoes Lane, Piscataway, NJ 08854. (201) 981-0060. Inquire as to online cost and availability.

NTIS. National Technical Information Service, 5285 Port Royal Road, Springfield, VA 22161. (703) 487-4630. Broad coverage of government sponsored research reports, 1964 to present. Inquire as to online cost and availability.

SCISEARCH. Institute for Scientific Information, 3501 Market Street, Philadelphia, PA 19104. (800) 523-1850 or (215) 386-0100. Broad multidisciplinary title and author index to the international literature of science and technology, 1974 to present. Inquire as to online cost and availability.

OTHER SOURCES

GUIDE TO BASIC INFORMATION SOURCES IN CHEMISTRY. Arthur Antony. John Wiley and Sons, Incorporated, 605 Third Avenue, New York, NY 10158. (800) 526-5368 or (212) 850-6000. 1979. $26.95.

HOW TO FIND CHEMICAL INFORMATION: A GUIDE FOR PRACTICING CHEMISTS, TEACHERS AND STUDENTS. John Wiley and Sons, Incorporated, 605 Third Avenue, New York, NY 10158. (800) 526-5368 or (212) 850-6000. 1986. $35.00.

PERIODICALS

ANALYTICAL CHEMISTRY. American Chemical Society, 1155 Sixteenth Street, NW, Washington, DC 20036. (800) 424-6747 or (202) 872-4700. Monthly. $33.00 per year.

ANDEWANDTE CHEMIE (GESELLSCHAFT DEUTSCHER CHEMIKER, GW). V C H Verlagsgesellschaft mbH, Pappelallee 3, Postfach 1260, 6940 Weinheim, West Germany. Monthly. $300.00 per year.

BIOCHEMISTRY. American Chemical Society, 1155 Sixteenth Street, NW, Washington, DC 20036. (800) 424-6747 or (202) 872-4700. Biweekly. $303.00 per year.

CHEMICAL PHYSICS. Elsevier Science Publishers B.V., Box 211, 1000 AE Amsterdam, The Netherlands. Thirty times per year. $1200.00 per year.

CHEMICAL PHYSICS LETTERS. Elsevier Science Publishers B.V., Box 211, 1000 AE Amsterdam, The Netherlands. Sixty-six times per year. $1500.00 per year.

CHEMICAL REVIEWS. American Chemical Society, 1155 Sixteenth Street, NW, Washington, DC 20036. (800) 424-6747 or (202) 872-4700. Bimonthly. $83.00 per year.

CHEMICAL WEEK. McGraw-Hill Book Company, Incorporated, 1221 Avenue of the Americas, New York, NY 10020. (212) 512-2000. Weekly. $30.00 per year.

CHEMTECH. American Chemical Society, 1155 Sixteenth Street, NW, Washington, DC 20036. (800) 424-6747 or (202) 872-4700. Monthly. $40.00 per year to individuals.

INTERNATIONAL JOURNAL OF MASS SPECTROMETRY AND ION PROCESSES. Elsevier Science Publishing Company, Incorporated, 52 Vanderbilt Avenue, New York, NY 10017. (212) 370-5520. Twenty times per year. $590.00 per year.

INTERNATIONAL REVIEWS IN PHYSICAL CHEMISTRY. Taylor and Francis, Limited, 252 Cherry Street, Philadelphia, PA 19106-1906. Three times per year. $143.00 per year.

INORGANIC CHEMISTRY. American Chemical Society, 1155 Sixteenth Street, NW, Washington, DC 20036. (800) 424-6747 or (202) 872-4700. Monthly. $400.00 per year.

JOURNAL OF ORGANIC CHEMISTRY. American Chemical Society, 1155 Sixteenth Street, NW, Washington, DC 20036. (800) 424-6747 or (202) 872-4700. Biweekly. $265.00 per year.

JOURNAL OF PHYSICAL CHEMISTRY. American Chemical Society, 1155 Sixteenth Street, NW, Washington, DC 20036. (800) 424-6747 or (202) 872-4700. Biweekly. $369.00 per year.

JOURNAL OF THE AMERICAN CHEMICAL SOCIETY. American Chemical Society, 1155 Sixteenth Street, NW, Washington, DC 20036. (800) 424-6747 or (202) 872-4700. Biweekly. $330.00 per year.

RESEARCH CENTERS AND INSTITUTES

BROWN UNIVERSITY. Chemical Research Laboratory, Providence, RI 02912. (401) 863-2256.

CORNELL UNIVERSITY. Materials Science Center, Clark Hall of Science, Ithaca, NY 14853. (607) 255-4272.

HARVARD UNIVERSITY. Chemical Laboratories, Oxford Street, Cambridge, MA 02138. (617) 495-5283.

UNIVERSITY OF PENNSYLVANIA. Laboratory for Research on the Structure of Matter, 3231 Walnut Street, Philadelphia, PA 19104. (215) 898-8571.

IONSPHERE

See: METEOROLOGY, GEOPHYSICS

IONS

See: IONIZATION

IRON

See: CAST IRON, STEEL AND
AND STEEL MAKING

ISOMERS

See: ORGANIC CHEMISTRY

ISOTOPES

See: IONS

J

J PARTICLE

See: PARTICLE PHYSICS

JAHN-TELLER EFFECT

See also: PHYSICAL CHEMISTRY

ABSTRACT SERVICES AND INDEXES

CHEMICAL ABSTRACTS. American Chemical Society, Chemical Abstracts Service, Box 3012, Columbus, OH 43210. (614) 421-3600. 1907 to present. Weekly. $9500.00 per year.

CONFERENCE PAPERS INDEX. Cambridge Scientific Abstracts, 5161 River Road, Bethesda, MD 20816. 1972 to present. Monthly. Inquire as to cost and availability.

CURRENT CONTENTS: PHYSICAL, CHEMICAL AND EARTH SCIENCES. Institute for Scientific Information, 3501 Market Street, Philadelphia, PA 19104. (800) 523-1850 or (215) 386-0100. Weekly. $275.00 per year.

SCIENCE CITATION INDEX. Institute for Scientific Information, 3501 Market Street, Philadelphia, PA 19104. (800) 523-1850 or (215) 386-0100. Six times per year. $6200.00 per year.

ANNUAL REVIEWS AND YEARBOOKS

ADVANCES IN PHYSICAL ORGANIC CHEMISTRY. Academic Press, Inc., 6277 Sea Harbor Drive, Orlando, FL 32821. (800) 321-5068. 1963 to present. Irregular. Price varies, inquire.

ANNUAL REVIEW OF PHYSICAL CHEMISTRY. Annual Reviews, Inc., 4139 El Camino Way, Palo Alto, CA 94306. (415) 493-4400. Annual. Inquire.

ASSOCIATIONS AND PROFESSIONAL SOCIETIES

AMERICAN CHEMICAL SOCIETY. 1155 16th Street, N.W., Washington, DC 20036. (202) 872-4600.

AMERICAN INSTITUTE OF PHYSICS. 335 East 45th Street, New York, NY 10017. (212) 661-9494.

BIBLIOGRAPHIES

THE JAHN-TELLER EFFECT: A BIBLIOGRAPHIC REVIEW. I.B. Bersuker. Plenum Publishing Corporation, 233 Spring Street, New York, NY 10013. (800) 221-9369. 1984. $85.00.

DIRECTORIES AND BIOGRAPHICAL SOURCES

AMERICAN INSTITUTE OF CHEMISTS DIRECTORY. American Institute of Chemists, 7315 Wisconsin Avenue, Bethesda, MD 20814. (301) 652-2447. 1986. $35.00.

AMERICAN MEN AND WOMEN OF SCIENCE. R.R. Bowker, Inc., Order Department, 245 West 17th Street, New York, NY 10011. (800) 521-8110. Eight volumes. 1986. $595.00 for set.

INTERNATIONAL RESEARCH CENTERS DIRECTORY 1988-89. Darren L. Smith, editor. Gale Research Company, Book Tower, Detroit, MI 48226. (800) 521-0707. 4th edition. 1987. $360.00.

RESEARCH CENTERS DIRECTORY 1988. Gale Research Company, Book Tower, Detroit, MI 48226. (800) 521-0707. 12th edition. 1987. $365.00 for set.

SCIENTIFIC AND TECHNICAL ORGANIZATIONS AND AGENCIES DIRECTORY. Margaret Labash Young, editor. Gale Research Company, Book Tower, Detroit, MI 48226. (800) 521-0707. 2nd edition. 1987. $185.00.

ENCYCLOPEDIAS AND DICTIONARIES

ENCYCLOPEDIA OF CHEMISTRY. Douglas M. Considine, editor. Van Nostrand Reinhold Company, Inc., 135 West 50th Street, New York, NY 10020. (800) 543-2681. 4th edition. 1984. $98.95.

THESAURUS OF SCIENTIFIC, TECHNICAL, AND ENGINEERING TERMS. Hemisphere Publishing Corporation, 1010 Vermont Avenue, NW, Washington, DC 20005. (800) 526-0275. 1988. $125.00.

GENERAL WORKS

THE JAHN-TELLER EFFECT AND VIBRONIC INTERACTIONS IN MODERN CHEMISTRY. I.B. Bersuker. Plenum Publishing Corporation, 233 Spring Street, New York, NY 10013. (800) 221-9369. 1984. $45.00.

PHYSICAL CHEMISTRY. Robert Alberty. John Wiley and Sons, Inc., 605 Third Avenue, New York, NY 10158. (800) 526-5368. Seventh edition. 1987. $49.95.

HANDBOOKS AND MANUALS

CRC HANDBOOK OF CHEMISTRY AND PHYSICS. Robert C. Weast, editor. CRC Press, 2000 Corporate Boulevard, Boca Raton, FL 33431. (800) 272-7737. 68th edition. 1987. $69.95.

A GUIDEBOOK TO MECHANISMS IN ORGANIC CHEMISTRY. P. Sykes. John Wiley and Sons, Inc., 605 Third Avenue, New York, NY 10158. (800) 526-5368. 6th edition. 1986. $21.95.

LANGE'S HANDBOOK OF CHEMISTRY. John A. Dean, editor. McGraw-Hill Book Company, 1221 Avenue of the Americas,

New York, NY 10020. (212) 512-2000. 1985. $59.50.

TABLES OF PHYSICAL AND CHEMICAL CONSTANTS; AND SOME MATHEMATICAL FUNCTIONS. G.W.C. Kaye and T.H. Laby, editors. Longman, Inc., 95 Church Street, White Plains, NY 10601. (914) 993-5000. 1986. $39.95.

ONLINE DATA BASES

CA SEARCH. Chemical Abstracts Service, P.O. Box 3012, Columbus, OH 43120. (800) 848-6538 or (614) 421-3600. Comprehensive guide to chemical literature, 1972 to present. Inquire as to online cost and availability.

COMPENDEX. Engineering Information, Inc., 345 East 47th Street, New York, NY 10017. (800) 221-1044 or (212) 705-7615. Engineering and technical literature, 1975 to present. Inquire as to online cost and availability.

DISSERTATION ABSTRACTS ONLINE. University Microfilms International, 300 North Zeeb Road, Ann Arbor, MI 48106. (800) 521-0600 or (313) 761-4700. Scope includes virtually all doctoral dissertations accepted at accredited American institutions from 1861 to present in over 250 subject areas. Inquire as to online cost and availability.

NTIS. National Technical Information Service, 5285 Port Royal Road, Springfield, VA 22161. (703) 487-4630. Broad coverage of government sponsored research reports, 1964 to present. Inquire as to online cost and availability.

SCISEARCH. Institute for Scientific Information, 3501 Market Street, Philadelphia, PA 19104. (800) 523-1850 or (215) 386-0100. Broad multidisciplinary title and author index to the international literature of science and technology, 1974 to present. Inquire as to online cost and availability.

PERIODICALS

CHEMICAL PHYSICS. Elsevier Science Publishing Company, Inc., 52 Vanderbilt Avenue, New York, NY 10017. (212) 370-5520. Thirty times per year. $1200.00 per year.

CHEMICAL PHYSICS LETTERS. Elsevier Science Publishing Company, Inc., 52 Vanderbilt Avenue, New York, NY 10017. (212) 370-5520. Sixty-six times per year. $1500.00 per year.

INTERNATIONAL REVIEWS IN PHYSICAL CHEMISTRY. Taylor and Francis/Hemisphere Distribution Center, 242 Cherry Street, Philadelphia, PA 19106-1906. (800) 821-8312. Three times per year. $143.00 per year.

JOURNAL OF PHYSICAL CHEMISTRY. American Chemical Society, 1155 16th Street, N.W., Washington, DC 20036. (800) 424-6747. Biweekly. $375.00 per year.

RESEARCH CENTERS AND INSTITUTES

CHEMICAL LABORATORIES. Harvard University, Oxford Street, Cambridge, MA 02138. (617) 495-4283.

CHEMICAL RESEARCH LABORATORY. Brown University, Providence, RI 02912. (401) 863-2256.

JET PROPULSION

See also: AERODYNAMICS, AERONAUTICS, AERONAUTICAL ENGINEERING, AEROSPACE ENGINEERING

ABSTRACT SERVICES AND INDEXES

ALLOYS INDEX. American Society for Metals, 9639 Kinsman Road, Metals Park, OH 44073. (216) 338-5151. 1974 to present. Monthly. $225.00 per year.

APPLIED MECHANICS REVIEWS. American Society of Mechanical Engineers, 345 East 47th Street, New York, NY 10017. (212) 705-7703. 1948 to present. Monthly. $360.00 per year.

APPLIED SCIENCE AND TECHNOLOGY INDEX. H.W. Wilson Company, 950 University Avenue, Bronx, NY 10452. (800) 367-6670 or (212) 588-8400. 1958 to present. Monthly. Inquire as to cost and availability.

CURRENT CONTENTS: ENGINEERING, TECHNOLOGY AND APPLIED SCIENCES. Institute for Scientific Information, 3501 Market Street, Philadelphia, PA 19104. (800) 523-1850 or (215) 386-0100. 1970 to present. Weekly. $272.00 per year.

ENGINEERING INDEX MONTHLY AND AUTHOR INDEX. Engineering Information, Inc., 345 East 47th Street, New York, NY 10017. (800) 221-1044 or (212) 705-7600. Monthly, with annual cumulation. $1560.00 per year.

INTERNATIONAL AEROSPACE ABSTRACTS. AIAA/TIS, 1633 Broadway, New York, NY 10019. (212) 581-4300. Semimonthly. $700.00 per year.

METALS ABSTRACTS. American Society for Metals, 9639 Kinsman Road, Metals Park, OH 44073. (216) 338-5151. Monthly. $890.00.

METALS ABSTRACTS INDEX. American Society for Metals, 9639 Kinsman Road, Metals Park, OH 44073. (216) 338-5151. Monthly. (sold only to subscribers of Metals Abstracts).

SCIENCE ABSTRACTS. Section A: Physics; Section B: Electrical and Electronics Abstracts; Section D: Computer and Control Abstracts. Institute of Electrical Engineers, London, United Kingdom. Available from: Institute of Electrical and Electronic Engineers (IEEE), 445 Hoes Lane, Piscataway, NJ 08854. Inquire as to cost and availability.

SCIENCE CITATION INDEX. Institute for Scientific Information, 3501 Market Street, Philadelphia, PA 19104. (800) 523-1850 or (215) 386-0100.

STAR. (Scientific and Technical Aerospace Reports). U.S. National Aeronautics and Space Administration, Scientific and Technical Information Branch, Box 8757, Baltimore-Washington International Airport, MD 21240. (202) 755-2210. Semimonthly, with semiannual and annual indexes. $85.00 per year.

ASSOCIATIONS AND PROFESSIONAL SOCIETIES

AMERICAN HELICOPTER SOCIETY, INC., 217 North Washington Street, Alexandria, VA 22314. (703) 684-6777/

AMERICAN INSTITUTE OF AERONAUTICS AND ASTRONAUTICS, 1633 Broadway, New York, NY 10019. (212) 581-4300.

FLIGHT SAFETY FOUNDATION, INC., 5510 Columbia Pike, Arlington, VA 22204-3194. (703) 820-2777.

SOCIETY OF AUTOMOTIVE ENGINEERS. SAE, Inc., 400 Commonwealth Drive, Warrendale, PA 15096. (412) 776-4841.

BIBLIOGRAPHIES

AERONAUTICAL ENGINEERING: A SPECIAL BIBLIOGRAPHY WITH INDEXES. U.S. National Aeronautics and Space Administration, Washington, DC 20546. (202) 755-2320. Available from National Technical Information Service, Springfield, VA 22161. 1970 to present. Monthly. Inquire as to cost and availability.

AIRCRAFT, ENGINES, AND AIRMEN: A SELECTIVE REVIEW OF THE PERIODICAL LITERATURE, 1930-1969. A. Hanniball. Scarecrow Press, Inc., 52 Liberty Street, Methuchen, NJ 08840.

(201) 548-8600. 1972. $39.50.

NATIONAL AIR AND SPACE MUSEUM AND SMITHSONIAN INSTITUTION. Aerospacce Periodical Index, 1973-1982. 1983. G.K. Hall and Company, 70 Lincoln Street, Boston, MA 02111. (800) 343-2806. $100.00.

SCIENTIFIC AND TECHNICAL BOOKS AND SERIALS IN PRINT, 1988; An Index to Literature in Science and Technology. R.R. Bowker Co., E. 42nd Street, New York, NY 10017. (800) 521-8110 or (212) 916-1600. $175.00.

DIRECTORIES AND BIOGRAPHICAL SOURCES

AAS DIRECTORY. American Astronautical Society, 6212-B Old Keene Mill Court, Springfield, VA 22152. (703) 866-0020. Annual. $35.00 per year.

RESEARCH CENTERS DIRECTORY. Gale Research Company, Book Tower, Detroit, MI 48226. (800) 521-0707. Eleventh edition. 1987.

WHO'S WHO IN ENGINEERING. Engineers Joint Council, 345 East 47th Street, New York, NY 10017. (212) 705-7010. 1985. $200.00.

WORLD AVIATION DIRECTORY. Murdoch Magazine, 1156 15th Street, NW, Washington, DC 20005. (202) 822-4600. Semiannual. $75.00 per year.

ENCYCLOPEDIAS AND DICTIONAIRES

DICTIONARY OF AEROSPACE ENGINEERING. M.G. Kotik. Elsevier Science Publishing Company, Inc., 52 Vanderbilt Avenue, New York, NY 10017. (212) 370-5520. 1986. $170.00.

ENCYCLOPEDIA OF PHYSICAL SCIENCE AND TECHNOLOGY. Academic Press, Inc., Orlando, FL 32887. (800) 321-5068 or (305) 345-2734. Fifteen volumes, 1986.

JANE'S AEROSPACE DICTIONARY. Jane's Publishing Inc., 135 West 50th Street, New York, NY 10020. (212) 586-7745. Second edition. 1986. $39.95.

MCGRAW-HILL DICTIONARY OF ENGINEERING. Sybil P. Parker, editor. McGraw-Hill Book, Inc., 1221 Avenue of the Americas, New York, NY 10020. (212) 512-2000. 1984. $39.95.

MCGRAW-HILL ENCLYCLOPEDIA OF SCIENCE AND TECHNOLOGY. McGraw-Hill Book, Inc., 1221 Avenue of the Americas, New York, NY 10020. (212) 512-2000.

GENERAL WORKS

AIRCRAFT BASIC SCIENCE. R.D. Bent and J.L. McKinley. McGraw-Hill Book Company, 1221 Avenue of the Americas, New York, NY 10020. (212) 512-2000. 1980. $32.95.

FOUNDATIONS OF AERODYNAMICS: BASES OF AERODYNAMIC DESIGN. A.M. Kuethe and C. Chow. John Wiley & Sons, 605 Third Avenue, New York, NY 10158. (800) 526-5368 or (212) 850-6000. Fourth edition. 1986. $43.95.

ILLUSTRATED GUIDE TO AERODYNAMICS. Hubert Smith. TAb Books, Inc., Monterey Lane, Blue Ridge Summit, PA 17214. (717) 794-2191. 1986. $14.95.

INTRODUCTION AEROSPACE STRUCTURAL ANALYSIS. David H. Allen and Walter E. Haisler. John Wiley and Sons, Inc., 605 Third Avenue, New York, NY 10158. (800) 526-5368 or (212) 850-6000. 1985. $42.95.

INTRODUCTION TO AIRCRAFT PERFORMANCE, SELECTION AND DESIGN. F.J. Hale. John Wiley & Sons, 605 Third Avenue, New York, NY 10158. (800) 526-5368 or (212) 850-6000. 1984. $39.95.

INTRODUCTION TO THEORETICAL AND COMPUTATIONAL AERODYNAMICS. Jack Moran. John Wiley & Sons, 605 Third Avenue, New York, NY 10158. (800) 526-5368 or (212) 850-6000. 1984. $40.95.

MECHANICS AND THERMODYNAMICS OF PROPULSION. Philip G. Hill and C.R. Pererson. Addison-Wesley Publishing Company, Inc., 1 Jacob Way, Reading, MA 01867. (617) 944-3700. 1965. $43.95.

HANDBOOKS AND MANUALS

JANE'S ALL THE WORLD'S AIRCRAFT 1987-88. John W.R. Taylor. Jane's Publishing Inc., 115 Fifth Avenue, New York, NY 10003. (214) 254-9097. 1987.

ONLINE DATA BASES

COMPENDEX. Engineering Information, Inc., 345 East 47th Street, New York, NY 10017. (800) 221-1044 or (212) 705-7615. Engineering and technical literature, 1975 to present. Inquire as to cost and availability.

NASA. National Aeronautics and Space Administration, Scientific and Technical Information Branch, 300 7th Street, S.W., Washington, DC 20546. Citations and abstracts of aerospace literature, 1962 to present. Inquire as to cost and availability.

NTIS. National Technical Information Service, 5285 Port Royal Road, Springfield, VA 22161. (703) 487-4630. Broad coverage of government-sponsored research reports, 1964 to present. Inquire as to cost and availability.

PERIODICALS

AERONAUTICAL JOURNAL. Aeronautical Society, 4 Hamilton Place, London W1V OBQ, England. Ten times per year. $175.00 per year.

AEROSPACE AMERICA. American Institute of Aeronautics and Astronautics, 1633 Broadway, New York, NY 10019. (212) 581-4300. Monthly. $51.00 per year.

AEROSPACE ENGINEERING MAGAZINE (SOCIETY OF AUTOMOTIVE ENGINEERS). SAE, INC., 400 Commonwealth Drive, Warrendale, PA 15096. (412) 776-4841. Monthly. $30.00 per year.

AIAA JOURNAL. AIAA/TIS, 1633 Broadway, New York NY 10019. (212) 581-4300. Monthly. $205.00 per year.

AIRCRAFT ENGINEERING. Bunhill Publications Limited, 127 Stanstead Road, Forest Hill, London, SE23 JE1, England. 1929 to present. Monthly. $51.00 per year.

AMERICAN HELICOPTER SOCIETY JOURNAL. American Helicopter Society, Inc., 217 North Washington Street, 217 North Washington Street, Alexandria, Va 22314. (703) 684-6777. Quarterly. $25.00 per year.

AVIATION MECHANICS BULLETIN. Flight Safety Foundation, Inc., 5510 Columbia Pike, Arlington, Va 22204-3194. (703) 820-2777. Bimonthly. $6.50 per year.

AVIATION WEEK AND SPACE TECHNOLOGY. McGraw-Hill Book, Inc., 1221 Avenue of the Americas, New York, NY 10020. Weekly. $55.00 per year.

CANADIAN AERONAUTICS AND SPACE JOURNAL. Canadian Aeronautics and Space Institute, 222 Somerset Street, West, Suite 601, Ottawa, ON K2P OJ1, Canada. (613) 234-0191. Quarterly. $35.00 per year.

FLYING. CBS Magazines, 1515 Broadway, New York, NY 10036. (212) 503-4200. Monthly. $14.00 per year.

IEEE TRANSACTIONS ON AEROSPACE. Institute of Electrical and Electronics Engineers, 345 East 47th Street, New York, NY 10017. Bimonthly. $108.00 per year.

INTERNATIONAL JOURNAL OF TURBO AND JET ENGINES. Martinus Nijhoff Publishers. Available from: Kluwer Academic Publishers Group, Distribution Centre, Postbus 322, 3300 AH Dordrecht, The Netherlands. Quarterly. $125.00 per year.

JOURNAL OF AIRCRAFT. American Institute of Aeronautics and Astronautics, AIAA/TIS, 1633 Broadway, New York, NY 10019. (212) 581-4300. Monthly. $185.00 per year.

JOURNAL OF GUIDANCE AND CONTROL. American Institute of Aeronautics and Astronautics, AIAA/TIS, 1633 Broadway, New York, NY 10019. (212) 581-4300. Bimonthly. $95.00 per year.

JOURNAL OF PROPULSION AND POWER. American Institute of Aeronautics and Astronautics, AIAA/TIS, 1633 Broadway, New York, NY 10019. (212) 581-4300. Bimonthly. $170.00 per year.

ROTOR AND WING INTERNATIONAL. PJS Publications, Inc., New Plaza, Box 1790, Peoria, IL 61656. (309) 682-6626. Monthly. $24.00 per year.

VERTICA: THE INTERNATIONAL JOURNAL OF ROTORCCRAFT AND POWERLIFT AIRCRAFT. Pergamon Press, Inc., Journals Division, Maxwell House, Fairview Park, Elmsford, NY 10523. (914) 592-7700. Quarterly. $135.00 per year.

VERTIFLITE. American Helicopter Society, Inc., 217 North Washington Street, Alexandria, VA 22314. (703) 684-6777. Bimonthly. $25.00 per year.

RESEARCH CENTERS AND INSTITUTES

FLIGHT RESEARCH LABORATORY. University of Kansas, Raymond Nichols Hall, Lawrence, KS 66045. (913) 864-3043.

JOINT INSTITUTE FOR ADVANCEMENT OF LIGHT SCIENCES. Langley Research Center, Mail Stop 269, Hampton, VA 23665. (804) 865-3124.

OHIO STATE UNIVERSITY. Aeronautical and Astronautical Research Laboratory, 2300 West Case Road, Columbus, OH 43220. (614) 422-1241.

UNIVERSITY OF ALABAMA. Flight Dynamics Laboratory, Box 2901, Tuscaloosa, AL 35487. (205) 348-7300.

JET STREAM

See also: METEOROLOGY

ABSTRACT SERVICES AND INDEXES

GENERAL SCIENCE INDEX. H.W. Wilson Company, 950 University Avenue, Bronx, NY 10452. (800) 367-6770 or (212) 588-8400.

METEOROLOGICAL AND GEOASTROPHYSICAL ABSTRACTS. American Meteorological Society, 45 Beacon Street, Boston, MA 02108. (617) 227-2425. 1950 to present. Monthly. $450.00 per year.

OCEANIC ABSTRACTS. Cambridge Scientif Abstracts, Incorporated, 5161 River Road, Bethesda, MD 20816. (301) 951-1400. 1964 to present. Bimonthly. $660.00 per year.

OCEANOGRAPHIC LITERATURE REVIEW. Pergamon Press, Incorporated, Maxwell House, Fairview Park, Elmsford, NY 10523. (914) 592-7700. 1979 to present. Inquire as to cost and availability.

PHYSICS ABSTRACTS. Institution of Electrical Engineers (London). Available from: IEEE Service Center, 445 Hoes Lane, Piscataway, NJ 08854. 1898 to present. Bimonthly. $1700.00 per year.

SCIENCE CITATION INDEX. Institute for Scientific Information, 3501 Market Street, Philadelphia, PA 19104. (800) 523-1850 or (215) 386-0100.

ASSOCIATIONS AND PROFESSIONAL SOCIETIES

AMERICAN GEOPHYSICAL UNION. 2000 Florida Avenue, NW, Washington, DC 20009. (202) 462-6903.

AMERICAN METEOROLOGICAL SOCIETY. 45 Beacon Street, Boston, MA 02108. (617) 227-2425.

INTERNATIONAL ASSOCIATION OF METEOROLOGY AND ATMOSPHERIC PHYSICS. UCAR, Post Office Box 3000, Boulder, CO 80307.

NATIONAL WEATHER ASSOCIATION. 4400 Stamp Road, Room 404, Temple Hills, MD 10748. (301) 899-3784.

UNIVERSITY CORPORATION FOR ATMOSPHERIC RESEARCH. Box 3000, 1850 Table Mesa Drive, Boulder, CO 80307. (303) 497-1000.

DIRECTORIES AND BIOGRAPHICAL SOURCES

BIOGRAPHICAL DICTIONARY OF SCIENTISTS. T.I. Williams. Halsted Press, 605 Third Avenue, New York, NY 10158. (800) 526-5368 or (212) 850-6418. Third edition. 1982. $29.95.

GOVERNMENT RESEARCH DIRECTORY. Gale Research Company, Book Tower, Detroit, MI 48226. (800) 521-0707. Fourth edition. 1987. $350.00.

INTERDOC: DIRECTORY OF PUBLISHED PROCEEDINGS, SERIES. SEMT-Science/Engineering/Medicine/Technology. Interdoc Corporation, 173 Halstead Avenue, Box 326, Harrison, NY 10528. (914) 835-3506. Ten times per year. $325.00 per year.

METEOROLOGICAL SERVICES OF THE WORLD. World Meteorological Organization. Available from: American Meteorological Society, 45 Beacon Street, Boston, MA 02108. (617) 227-2425. Annual. $35.00.

NATIONAL WEATHER SERVICE OFFICES AND STATIONS. National Oceanic and Atmospheric Administration, Department of Commerce, Silver Spring, MD 20910. (301) 427-7698. Annual. Free.

ENCYCLOPEDIA AND DICTIONARIES

ENCYCLOPEDIA OF CLIMATOLOGY. John E. Oliver and Rhodes W. Fairbridge, editors. Van Nostrand Reinhold, Incorporated, 115 Fifth Avenue, New York, NY 10003. (800) 543-2681. 1987. $89.95.

GENERAL WORKS

JET-STREAM METEOROLOGY. Elmar R. Reiter. Books on Demand, 300 North Zeeb Street, Ann Arbor, MI 48106. (313) 761-4700. $132.30.

ONLINE DATA BASES

DISSERTATION ABSTRACTS ONLINE. University Microfilms International, 300 North Zeeb Road, Ann Arbor, MI 48106. (800) 521-0600 or (313) 761-4700. Scope includes virtually all doctoral dissertations accepted at accredited American institutions from 1861 to present in 252 subject areas. Inquire as to cost and availability..

METEOROLOGICAL AND GEOASTROPHYSICAL ABSTRACTS. American Meteorological Society, 45 Beacon Street, Boston, MA 02108. (617) 227-2425. 1972 to present. Inquire as to online cost and availability.

NTIS. National Technical Information Service, 5285 Port Royal Road, Springfield, VA 22161. (703) 487-4630. Broad coverage of government sponsored research reports, 1964 to present. Inquire as to cost and availability.

PERIODICALS

AMERICAN METEOROLOGICAL SOCIETY BULLETIN. American Meteorological Society, 45 Beacon Street, Boston, MA 02108. (617) 227-2425. 1920 to present. Monthly. $60.00 per year.

CLIMATIC CHANGE: AN INTERDISCIPLINARY, INTERNATIONAL JOURNAL DEVOTED TO THE DESCRIPTION, CAUSES AND IMPLICATIONS OF CLIMATIC CHANGE. D. Reidel Publishing Company, 190 Old Derby Street, Hingham, MA 02043. (617) 871-6600. 1977 to present. Six times per year. $125.00 per year.

DYNAMICS OF ATMOSPHERES AND OCEANS. Elsevier Science Publishing Company, Incorporated, 52 Vanderbilt Avenue, New York, NY 10017. (212) 370-5520. 1977 to present. Quarterly. $90.00 per year.

JOURNAL OF THE ATMOSPHERIC SCIENCES. American Meteorological Society, 45 Beacon Street, Boston, MA 02108. (617) 227-2425. 1944 to present. Semimonthly. $220.00 per year.

MONTHLY WEATHER REVIEW. American Meteorological Society, 45 Beacon Street, Boston, MA 02108. (617) 227-2425. 1872 to present. Monthly. $120.00 per year.

NATIONAL WEATHER DIGEST. National Weather Association, 4400 Stamp Road, Room 404, Temple Hills, MD 20748. (301) 899-3784. 1976 to present. Quarterly. $20.00 per year.

WEATHER. Royal Meteorological Society, James Glaisher House, Grenville Place, Bracknell Berkshire, RG12 1BX, England. 1946 to present. $30.00 per year.

WEATHERWISE. Heldref Publications, 4000 Albemarle Street, NW, Washington, DC 20016. (202) 362-6445. 1948 to present, Bimonthly. $20.00 per year.

RESEARCH CENTERS AND INSTITUTES

NATIONAL CENTER FOR ATMOSPHERIC RESEARCH. Box 3000, Boulder, CO 80307. (303) 497-1000.

JEWEL BEARINGS

See: BEARINGS AND BALL BEARINGS

JOINTS

See also: BUILDING MATERIALS, CIVIL ENGINEERING, CONSTRUCTION ENGINEERING, STRESS AND STRAIN, STRUCTURAL ENGINEERING

ABSTRACT SERVICES AND INDEXES

APPLIED MECHANICS REVIEW. American Society of Mechanical Engineers, 345 East 47th Street, New York, NY 10017. (212) 705-7703. 1948 to present. Monthly. $360.00 per year.

APPLIED SCIENCE AND TECHNOLOGY INDEX. H.W. Wilson and Company, 950 University Avenue, Bronx, NY 10452. (800) 367-6670 or (212) 588-8400. Monthly. Inquire as to cost and availability.

CURRENT CONTENTS: ENGINEERING, TECHNOLOGY AND APPLIED SCIENCES. Institute for Scientific Information, 3501 Market Street, Philadelphia, PA 19104. (800) 523-1850 or (215) 386-0100. Weekly. $275.00 per year.

ENGINEERING INDEX MONTHLY AND AUTHOR INDEX. Engineering Information Inc., 345 East 47th Street, New York, NY 10017. (212) 705-7600. Monthly. $1560.00 per year.

ISMEC BULLETIN (Information Service in Mechanical Engineering). Cambridge Scientific Abstracts, 5161 River Road, Bethesda, MD 20816. (301) 951-1400. 1973 to present. Monthly. $450.00 per year.

ASSOCIATIONS AND PROFESSIONAL SOCIETIES

AMERICAN SOCIETY OF CIVIL ENGINEERS. 345 East 47th Street, New York, NY 10017. (212) 705-7420.

AMERICAN SOCIETY OF MECHANICAL ENGINEERS. 345 East 47th Street, New York, NY 10017. (212) 705-7703.

RESEARCH COUNCIL ON STRUCTURAL CONNECTIONS. c/o Professor Karl H. Frank, Department of Civil Engineering, University of Texas, Austin, TX 78712. (512) 471-7259.

SOCIETY FOR EXPERIMENTAL MECHANICS. Seven School Street, Bethel, CT 06801. (203) 790-6373.

DIRECTORIES AND BIOGRAPHICAL SOURCES

RESEARCH CENTERS DIRECTORY 1988. Gale Research Company, Book Tower, Detroit, MI 48226. (800) 521-0707. 12th edition. 1987. $365.00 for set.

SCIENTIFIC AND TECHNICAL ORGANIZATIONS AND AGENCIES DIRECTORY. Margaret Labash Young, editor. Gale Research Company, Book Tower, Detroit, MI 48226. (800) 521-0707. 2nd edition. 1987. $185.00.

WHO'S WHO IN ENGINEERING. Gordon Davis, editor. Hemisphere Publishing Corporation, 1010 Vermont Avenue, NW, Washington, DC 20005. (800) 526-0275. 6th edition. 1985. $200.00.

GENERAL WORKS

ADHESIVE JOINTS. K.L. Mittal, editor. Plenum Publishing Corporation, 233 Spring Street, New York, NY 10013. (800) 221-9369. 1984. $125.00.

COUPLINGS AND JOINTS: DESIGN, SELECTION AND APPLICATIONS. J. Mancuso. Marcel Dekker Inc., 270 Madison Avenue, New York, NY 10016. (800) 228-1160. 1986. $49.95.

INTRODUCTION TO THE DESIGN AND BEHAVIOR OF BOLTED JOINTS. C. Bickford. Marcel Dekker Inc., 270 Madison Avenue, New York, NY 10016. (800) 228-1160. 1981. $67.00.

MECHANICS OF MATERIALS. A. Higdon, and others. John Wiley and Sons, Inc., 605 Third Avenue, New York, NY 10158. (800) 526-5368. Fourth edition. 1985. $45.95.

ONLINE DATA BASES

COMPENDEX. Engineering Information, Inc., 345 East 47th Street, New York, NY 10017. (800) 221-1044 or (212) 705-7615. Engineering and technical literature, 1975 to present. Inquire as to online cost and availability.

NTIS. National Technical Information Service, 5285 Port Royal Road, Springfield, VA 22161. (703) 487-4630. Broad coverage of

government sponsored research reports, 1964 to present. Inquire as to online cost and availability.

SCISEARCH. Institute for Scientific Information, 3501 Market Street, Philadelphia, PA 19104. (800) 523-1850 or (215) 386-0100. Broad multidisciplinary title and author index to the international literature of science and technology, 1974 to present. Inquire as to online cost and availability.

WILSONLINE. H.W. Wilson and Company, 950 University Avenue, Bronx, NY 10452. (800) 367-6770 or (212) 588-8400. Makes available online versions of the H.W. Wilson indexes including Applied Science and Technology Index, Business Periodicals Index and Readers' Guide to Periodical Literature. Approximately 1980 to present. Inquire as to online cost and availability.

PERIODICALS

JOURNAL OF ENGINEERING MECHANICS. American Society of Civil Engineers, 345 East 47th Street, New York, NY 10017. (212) 705-7420. 1956 to present. Monthly. $120.00 per year.

JOURNAL OF STRUCTURAL ENGINEERING. American Society of Civil Engineers, 345 East 47th Street, New York, NY 10017. (212) 705-7420. 1956 to present. Monthly. $140.00 per year.

JOSEPHSON EFFECT

See: CRYOGENICS

JUPITER

See also: ASTRONOMY, PLANETARY SCIENCE, SOLAR SYSTEM

ABSTRACT SERVICES AND INDEXES

ASTRONOMY AND ASTROPHYSICS ABSTRACTS. Springer-Verlag New York, Incorporated, 175 Fifth Avenue, New York, NY 10010. (212) 460-1500. Approximately $75.00 per year.

ASTRONOMY AND ASTROPHYSICS MONTHLY INDEX. Olivetree Associates, Post Office Box 236, Sierre Madre, Ca 91024. $212.00 per year. Complimentary copies available on request.

METEOROLOGICAL AND GEOASTROPHYSICAL. Abstracts. American Meteorological Society, 45 Beacon Street, Boston, MA 02108. (617) 227-2425. 1950 to present. Monthly. $450.00 per year.

STAR. (Scientific and Technical Aerospace Reports.) United States National Aeronautics and Space Administration, Scientific and Technical Information Facility, Box 8757, Baltimore-Washington International Airport, MD 21240. (202) 755-2210. Semimonthly, with semiannual and annual indexes. $85.00 per year.

ANNUAL REVIEWS AND YEARBOOKS

THE ASTRONOMICAL ALMANAC. Superintendent of Documents, U.S. Government Printing Office, Washington, DC 20402. (202) 783-3238. Yearly.

ANNUAL REVIEW OF ASTRONOMY AND ASTROPHYSICS. Annual Reviews, Incorporated, 4139 El Camino Way, Palo Alto, CA 94306. (415) 493-4400.

ANNUAL REVIEW OF EARTH AND PLANETARY SCIENCES. Annual Reviews, Incorporated, 4139 El Camino Way, Palo Alto, CA 94306. (415) 493-4400.

ASSOCIATIONS AND PROFESSIONAL SOCIETIES

AMERICAN ASTRONOMICAL SOCIETY. 2000 Florida Avenue, NW, Suite 300, Washington, DC 20009. (202) 659-0134.

AMERICAN GEOPHYSICAL UNION. 2000 Florida Avenue, NW, Washington, DC 20009. (202) 462-6903.

ASTRONOMICAL LEAGUE. P.O. Box 12821, Tucson, Az 85732. (602) 790-8471.

ASTRONOMICAL SOCIETY OF THE PACIFIC. 1290 24th Avenue, San Francisco, CA 94122. (415) 661-8660.

PLANETARY SOCIETY. 65 North Catalina Avenue, Pasadena, CA 91106-2301. (818) 793-5100.

BIBLIOGRAPHIES

A BIBLIOGRAPHY OF ASTRONOMY, 1970-1979. R.A. Seal and S.S. Martin. Libraries Unlimited, Incorporated, Littleton, CO 80160. 1982. $37.50.

DIRECTORIES AND BIOGRAPHICAL SOURCES

AMERICAN ASTRONOMICAL SOCIETY MEMBERS. 2000 Florida Avenue, NW, Suite 300, Washington, DC 20009. (202) 659-0134. Annual. Available to members only.

THE BIOGRAPHICAL DICTIONARY OF SCIENTISTS: ASTRONOMERS. D. Abbott, editor. Peter Bedrick Books, 125 East 23rd Street, New York, NY 10010. 1984. $24.95.

DIRECTORY OF PHYSICS AND ASTRONOMY STAFF MEMBERS. American Institute of Physics, 335 East 45th Street, New York, NY 10017. Annual.

GENERAL WORKS

DISTANT ENCOUNTERS: EXPLORATION OF JUPITER AND SATURN. Mark Washburn. Harcourt, Brace and Jovanovich, Incorporated, 1250 Sixth Avenue, San Diego, Ca 92101. (619) 231-616 . 1983. $12.95.

GALILEO: EXPLORATION OF JUPITER'S SYSTEM. C. Yeates, et al. NASA Special Publications. U.S. Government Printing Office, Washington, DC 20402. (202) 783-3238. 1985.

PLANETARY RINGS. Richard Greenberg, editor. University of Arizona Press, 1615 East Speedway, Tucson, AZ 85719. (602) 621-1441. 1984. $49.50.

SATELLITES OF JUPITER. David Morrison. University of Arizona Press, 1615 East Speedway, Tucson, AZ 85719. (602) 621-1441. 1982. $49.50.

ONLINE DATA BASES

CA SEARCH. Chemical Abstracts Service, P.O. Box 3012, Columbus, Oh 43210. Guide to chemical literature, 1972 to present. Inquire as to cost and availability.

DISSERTATION ABSTRACTS ONLINE. University Microfilms International, 300 North Zeeb Road, Ann Arbor, MI 48106. (800) 521-0600 or (313) 761-4700. Scope includes virtually all doctoral dissertations accepted at accredited American institutions from 1861 to present in 252 subject areas. Inquire as to cost and availability.

NASA. National Aeronautics and Space Administration, Scientific and Technical Information Branch, 300 7th Street, SW, Washington, DC 20546. Citations and abstracts of aerospace literature, 1962 to present. Inquire as to cost and availability.

NTIS. National Technical Information Service, 5285 Port Royal Road, Springfield, Va 22161. (703) 487-4630. Broad coverage of government sponsored research reports, 1964 to present. Inquire as to cost and availability.

SCISEARCH. Institute for Scientific Information, 3501 Market Street, Philadelphia, PA 19104. (800) 523-1850 or (215) 386-0100. Broad multidisciplinary title and author index to the international literature of science and technology, 1974 to present. Inquire as to cost and availability.

PERIODICALS

ASSOCIATION OF LUNAR AND PLANETARY OBSERVERS. Journal. Association of Lunar and Planetary Observers, Box 16131, San Francisco, CA 94116. (415) 566-5786. 1947 to present. Quarterly. $15.00 per year.

ASTRONOMICAL JOURNAL. American Astronomical Society. Available from: American Institute of Physics, 335 East 45th Street, New York, NY 10017. (212) 661-9404. $125.00 per year.

EARTH, MOON AND PLANETS; AN INTERNATIONAL JOURNAL OF COMPARATIVE PLANETOLOGY. D. Reidel Publishing Company, 190 Old Derby Street, Hingham, MA 02043. Nine times per year. $275.00 per year.

ICARUS: INTERNATIONAL JOURNAL OF THE SOLAR SYSTEM STUDIES. Academic Press, Incorporated, Orlando, FL 32887. (305) 345-4100. Monthly. $484.00 per year.

MERCURY. Astronomical Society of the Pacific, 1290 24th Avenue, San Francisco, CA 94122. (15) 661-8660. Bimonthly. $21.00 per year.

PLANETARY AN SPACE SCIENCE. Pergamon Press, Incorporated, Maxwell House, Fairview Park, Elmsford, NY 10523. (914) 592-7700. Monthly. $430.00 per year.

PLANETARY REPORT. Planetary Society, 65 North Catalina Avenue, Pasadena, CA 91106-2301. (818) 793-5100. 1980 to present. Bimonthly. $20.00 per year.

SOLAR SYSTEM RESEARCH (ENGLISH TRANSLATION OF ASTRONOMICHESKII VESTNIK). Consultants Bureau, 233 Spring Street, New York, NY 10013. (212) 620-8000. 1967 to present. Quarterly. $425.00 per year.

RESEARCH CENTERS AND INSTITUTES

LABORATORY FOR PLANETARY ATMOSPHERES RESEARCH. State University of New York at Stony Brook, Stony Brook, NY 11794-2300. (516) 632-8321.

LABORATORY FOR PLANETARY STUDIES. Cornell University, 302 Space Sciences Building, Ithaca, NY 14853. (607) 256-4971.

LUNAR AND PLANETARY INSTITUTE. 3303 NASA Road One, Houston, TX 77058. (713) 486-2139.

LUNAR AND PLANETARY LABORATORY. University of Arizona, Tucson, AZ 85721. (602) 621-6962.

UNIVERSITY OF ALABAMA IN HUNTSVILLE. Center for Space Plasma and Aeronomic Research. Engineering Building, Huntsville, AL 35899. (205) 895-6268.

UNIVERSITY OF FLORIDA. Radio Observatory. 211 Space Sciences Research Building, Gainesville, FL 36211. (904) 392-2052.

K

KARST

See also: CARBONATES, CAVES, GEOLOGY, GEOMORPHOLOGY, SEDIMENTARY ROCKS, SPELEOLOGY

ABSTRACT SERVICES AND INDEXES

BIBLIOGRAPHY AND INDEX OF GEOLOGY. American Geological Institute, 4220 King Street, Alexandria, VA 22302. (703) 379-2480. 1969 to present. Monthly. $1100.00 per year.

CHEMICAL ABSTRACTS. American Chemical Society, Chemical Abstracts Service, Box 3012, Columbus, OH 43210. (614) 421-3600. 1907 to present. Weekly. $9500.00 per year.

GENERAL SCIENCE INDEX. H.W. Wilson and Company, 950 University Avenue, Bronx, NY 10452. (800) 367-6670 or (212) 588-8400. 1978 to present. Monthly. Inquire as to cost and availability.

SCIENCE CITATION INDEX. Institute for Scientific Information, 3501 Market Street, Philadelphia, PA 19104. (800) 523-1850 or (215) 386-0100. Six times per year. $6200.00 per year.

ASSOCIATIONS AND PROFESSIONAL SOCIETIES

AMERICAN GEOLOGICAL INSTITUTE. 4220 King Street, Alexandria, VA 22302. (703) 379-2480.

ASSOCIATION OF AMERICAN GEOGRAPHERS. 1710 16th Street, N.W., Washington, DC 20009. (202) 234-1450.

GEOLOGICAL SOCIETY OF AMERICA. 3300 Penrose Place, Boulder, CO 80301. (303) 447-2020.

DIRECTORIES AND BIOGRAPHICAL SOURCES

AMERICAN MEN AND WOMEN OF SCIENCE. R.R. Bowker, Inc., Order Department, 245 West 17th Street, New York, NY 10011. (800) 521-8110. Eight volumes. 1986. $595.00 for set.

ASSOCIATION OF AMERICAN GEOGRAPHERS DIRECTORY. 1710 16th Street, N.W., Washington, DC 20009. (202) 234-1450. Annual.

INTERNATIONAL RESEARCH CENTERS DIRECTORY 1988-89. Darren L. Smith, editor. Gale Research Company, Book Tower, Detroit, MI 48226. (800) 521-0707. 4th edition. 1987. $360.00.

RESEARCH CENTERS DIRECTORY 1988. Gale Research Company, Book Tower, Detroit, MI 48226. (800) 521-0707. 12th edition. 1987. $365.00 for set.

SCIENTIFIC AND TECHNICAL ORGANIZATIONS AND AGENCIES DIRECTORY. Margaret Labash Young, editor. Gale Research Company, Book Tower, Detroit, MI 48226. (800) 521-0707. 2nd edition. 1987. $185.00.

GENERAL WORKS

GEOMORPHOLOGY; PURE AND APPLIED. M.G. Hart. Allen and Unwin, Inc., 8 Winchester Place, Winchester, MA 01890. (800) 547-8889. 1986. $35.95.

INTRODUCTION TO CARBONATE SEDIMENTS AND ROCKS. T.P. Scoffin. Methuen, Inc., 29 West 35th Street, New York, NY 10001. (212) 244-3336. 1987. $17.95 in paper.

KARST GEOMORPHOLOGY. J.N. Jennings. Blackwell Scientific Publications, Inc., 52 Beacon Street, Boston, MA 02108. (800) 325-4177. 1985. $14.95 in paper.

KARST HYDROGEOLOGY: ENGINEERING AND ENVIRONMENTAL APPLICATIONS. B.F. Beck and W.L. Wilson, editors. A.A. Balkema, P.O. Box 230, Accord, MA 02018. 1987. $105.00.

KARST LANDFORMS. M.M. Sweeting. Columbia University Press, 562 West 113th Street, New York, NY 10025. (212) 316-7100. 1973. $60.00.

SEDIMENTARY ENVIRONMENTS AND FACIES. H.G. Reading, editors. Blackwell Scientific Publications, Inc., 52 Beacon Street, Boston, MA 02108. (800) 325-4177. 2nd edition. 1986. $40.00 in paper.

HANDBOOKS AND MANUALS

ENCYCLOPEDIA OF FIELD GEOLOGY. Charles W. Finkl, editor. Van Nostrand Reinhold Company, Inc., 135 West 50th Street, New York, NY 10020. (800) 543-2681. 1988. $89.95.

ONLINE DATA BASES

CA SEARCH. Chemical Abstracts Service, P.O. Box 3012, Columbus, OH 43120. (800) 848-6538 or (614) 421-3600. Comprehensive guide to chemical literature, 1972 to present. Inquire as to online cost and availability.

GEOREF. Online version of the BIBLIOGRAPHY AND INDEX OF GEOLOGY. American Geological Institute, 4220 King Street, Alexandria, VA 22302. (703) 379-2480. 1969 to present. Inquire as to online cost and availability.

NTIS. National Technical Information Service, 5285 Port Royal Road, Springfield, VA 22161. (703) 487-4630. Broad coverage of government sponsored research reports, 1964 to present. Inquire as to online cost and availability.

SCISEARCH. Institute for Scientific Information, 3501 Market Street, Philadelphia, PA 19104. (800) 523-1850 or (215) 386-0100. Broad multidisciplinary title and author index to the international literature of science and technology, 1974 to present. Inquire as to online cost and availability.

WILSONLINE. H.W. Wilson and Company, 950 University Avenue, Bronx, NY 10452. (800) 367-6770 or (212) 588-8400.

Makes available online versions of the H.W. Wilson indexes including Applied Science and Technology Index, Business Periodicals Index and Readers' Guide to Periodical Literature. Approximately 1980 to present. Inquire as to online cost and availability.

PERIODICALS

CAVES AND CAVING. British Cave Research Association, 30 Main Road, Westonzoyland, Bridgewater, Somerset TA7 OEB, England. 1973 to present. Quarterly. $10.00 per year.

EARTH SURFACE PROCESSES AND LANDFORMS. British Geomorphological Research Group. Distributed by: John Wiley and Sons, Inc., 605 Third Avenue, New York, NY 10158. (800) 526-5368. 1976 to present. Bimonthly. $280.00 per year.

GEOLOGICAL SOCIETY OF AMERICA. BULLETIN. Geological Society of America. 3300 Penrose Place, Boulder, CO 80301. (303) 447-2020. 1888 to present. Monthly. $80.00 per year.

JOURNAL OF GEOLOGY. University of Chicago Press, 5801 Ellis Avenue, Chicago, IL 60637. (800) 621-2736. 1893 to present. Bimonthly. $30.00 per year.

PROGRESS IN PHYSICAL GEOGRAPHY. Cambridge University Press, 32 East 57th Street, New York, NY 10022. (800) 872-7423. 1976 to present. $50.00 per year.

SEDIMENTARY GEOLOGY. Elsevier Science Publishing Company, Inc., 52 Vanderbilt Avenue, New York, NY 10017. (212) 370-5520. 1967 to present. Twenty times per year. $420.00 per year.

SEDIMENTOLOGY. Blackwell Scientific Publications, Inc., 52 Beacon Street, Boston, MA 02108. (800) 325-4177. 1952 to present. 6 times per year. $184.00 per year.

KINEMATICS

See also: KINETICS, MECHANICS, STATICS

ABSTRACT SERVICES AND INDEXES

APPLIED MECHANICS REVIEW. American Society of Mechanical Engineers, 345 East 47th Street, New York, NY 10017. (212) 705-7703. 1948 to present. Monthly. $360.00 per year.

APPLIED SCIENCE AND TECHNOLOGY INDEX. H.W. Wilson and Company, 950 University Avenue, Bronx, NY 10452. (800) 367-6670 or (212) 588-8400. Monthly. Inquire as to cost and availability.

CHEMICAL ABSTRACTS. American Chemical Society, Chemical Abstracts Service, Box 3012, Columbus, OH 43210. (614) 421-3600. 1907 to present. Weekly. $9500.00 per year.

CONFERENCE PAPERS INDEX. Cambridge Scientific Abstracts, 5161 River Road, Bethesda, MD 20816. 1972 to present. Monthly. Inquire as to cost and availability.

CURRENT CONTENTS: PHYSICAL, CHEMICAL AND EARTH SCIENCES. Institute for Scientific Information, 3501 Market Street, Philadelphia, PA 19104. (800) 523-1850 or (215) 386-0100. Weekly. $275.00 per year.

ENGINEERING INDEX MONTHLY AND AUTHOR INDEX. Engineering Information Inc., 345 East 47th Street, New York, NY 10017. (212) 705-7600. Monthly. $1560.00 per year.

ISMEC BULLETIN (Information Service in Mechanical Engineering). Cambridge Scientific Abstracts, 5161 River Road, Bethesda, MD 20816. (301) 951-1400. 1973 to present. Monthly. $450.00 per year.

PHYSICS ABSTRACTS. Institution of Electrical Engineers. Available from: IEEE Service Center, 445 Hoes Lane, Piscataway, NJ 08854. 1898 to present. Bimonthly. $1700.00 per year.

PHYSICS BRIEFS. Physik Verlag GmbH, Postfach 1260/1280, D-6940 Weinheim, West Germany. (212) 661-9404. 1920 to present. Twenty-six times per year. $1250.00 per year.

SCIENCE CITATION INDEX. Institute for Scientific Information, 3501 Market Street, Philadelphia, PA 19104. (800) 523-1850 or (215) 386-0100. Six times per year. $6200.00 per year.

ASSOCIATIONS AND PROFESSIONAL SOCIETIES

AMERICAN CHEMICAL SOCIETY. 1155 16th Street, N.W., Washington, DC 20036. (800) 424-6747.

AMERICAN INSTITUTE OF PHYSICS. 335 East 45th Street, New York, NY 10017. (212) 661-9404.

AMERICAN PHYSICAL SOCIETY. 335 East 45th Street, New York, NY 10017. (212) 682-7341.

AMERICAN SOCIETY OF CIVIL ENGINEERS. 345 East 47th Street, New York, NY 10017. (212) 705-7420.

AMERICAN SOCIETY OF MECHANICAL ENGINEERS. 345 East 47th Street, New York, NY 10017. (212) 705-7703.

DIRECTORIES AND BIOGRAPHICAL SOURCES

AMERICAN MEN AND WOMEN OF SCIENCE. R.R. Bowker, Inc., Order Department, 245 West 17th Street, New York, NY 10011. (800) 521-8110. Eight volumes. 1986. $595.00 for set.

INTERNATIONAL RESEARCH CENTERS DIRECTORY 1988-89. Darren L. Smith, editor. Gale Research Company, Book Tower, Detroit, MI 48226. (800) 521-0707. 4th edition. 1987. $360.00.

1987 DIRECTORY OF ENGINEERING SOCIETIES AND RELATED ORGANIZATIONS. Gordon Davis, editor. Hemisphere Publishing Corporation, 1010 Vermont Avenue, NW, Washington, DC 20005. (800) 526-0275. 12th edition. 1987. $100.00.

RESEARCH CENTERS DIRECTORY 1988. Gale Research Company, Book Tower, Detroit, MI 48226. (800) 521-0707. 12th edition. 1987. $365.00 for set.

SCIENTIFIC AND TECHNICAL ORGANIZATIONS AND AGENCIES DIRECTORY. Margaret Labash Young, editor. Gale Research Company, Book Tower, Detroit, MI 48226. (800) 521-0707. 2nd edition. 1987. $185.00.

WHO'S WHO IN ENGINEERING. Gordon Davis, editor. Hemisphere Publishing Corporation, 1010 Vermont Avenue, NW, Washington, DC 20005. (800) 526-0275. 6th edition. 1985. $200.00.

ENCYCLOPEDIAS AND DICTIONARIES

CONCISE SCIENCE DICTIONARY. Oxford University Press, 200 Madison Avenue, New York, NY 10016. (800) 458-5833. 1987. $9.95 in paper.

DICTIONARY OF THE PHYSICAL SCIENCES: TERMS, FORMULAS, DATA. Cesare Emiliani. Oxford University Press, 200 Madison Avenue, New York, NY 10016. (800) 458-5833. 1987. $19.95 in paper.

MCGRAW-HILL ENCYCLOPEDIA OF SCIENCE AND TECHNOLOGY. McGraw-Hill Book Company, 1221 Avenue of the Americas, New York, NY 10020. (212) 512-2000. 6th edition. 1987. $1600.00.

THESAURUS OF SCIENTIFIC, TECHNICAL, AND ENGINEERING TERMS. Hemisphere Publishing Corporation, 1010 Vermont Avenue, NW, Washington, DC 20005. (800) 526-0275. 1988. $125.00.

GENERAL WORKS

CLASSICAL MECHANICS. E.A. Desloge. John Wiley and Sons, Inc., 605 Third Avenue, New York, NY 10158. (800) 526-5368. 1982. Two volumes. $110.00 for set.

DYNAMICS. S.N. Rasband. John Wiley and Sons, Inc., 605 Third Avenue, New York, NY 10158. (800) 526-5368. 1983. $37.50.

ELEMENTARY ENGINEERING MECHANICS. G.E. Drabble. Macmillan Publishing Company, Inc., 866 Third Avenue, New York, NY 10022. (800) 257-5755. 1986. $22.50.

KINEMATICS. Joseph S. Beggs. Hemisphere Publishing Corporation, 79 Madison Avenue, New York, NY 10016-7892. Order from: Taylor and Francis/Hemisphere Distribution Center, 242 Cherry Street, Philadelphia, PA 19106-1906. (800) 821-8312. 1983. $28.50.

SPACECRAFT DYNAMICS. Thomas R. Kane and others. McGraw-Hill Book Company, 1221 Avenue of the Americas, New York, NY 10020. (212) 512-2000. 1983. $51.95.

ONLINE DATA BASES

CA SEARCH. Chemical Abstracts Service, P.O. Box 3012, Columbus, OH 43120. (800) 848-6538 or (614) 421-3600. Comprehensive guide to chemical literature, 1972 to present. Inquire as to online cost and availability.

COMPENDEX. Engineering Information, Inc., 345 East 47th Street, New York, NY 10017. (800) 221-1044 or (212) 705-7615. Engineering and technical literature, 1975 to present. Inquire as to online cost and availability.

DISSERTATION ABSTRACTS ONLINE. University Microfilms International, 300 North Zeeb Road, Ann Arbor, MI 48106. (800) 521-0600 or (313) 761-4700. Scope includes virtually all doctoral dissertations accepted at accredited American institutions from 1861 to present in over 250 subject areas. Inquire as to online cost and availability.

INSPEC. INSPEC Marketing Department, Institution of Electrical Engineers. Available from IEEE Service Center, 445 Hoes Lane, Piscataway, NJ 08854. (201) 981-0060. Online version of Physics Abstracts. Inquire as to online cost and availability.

NTIS. National Technical Information Service, 5285 Port Royal Road, Springfield, VA 22161. (703) 487-4630. Broad coverage of government sponsored research reports, 1964 to present. Inquire as to online cost and availability.

SCISEARCH. Institute for Scientific Information, 3501 Market Street, Philadelphia, PA 19104. (800) 523-1850 or (215) 386-0100. Broad multidisciplinary title and author index to the international literature of science and technology, 1974 to present. Inquire as to online cost and availability.

WILSONLINE. H.W. Wilson and Company, 950 University Avenue, Bronx, NY 10452. (800) 367-6770 or (212) 588-8400. Makes available online versions of the H.W. Wilson indexes including Applied Science and Technology Index, Business Periodicals Index and Readers' Guide to Periodical Literature. Approximately 1980 to present. Inquire as to online cost and availability.

PERIODICALS

ARCHIVE FOR RATIONAL MECHANICS AND ANALYSIS. Springer-Verlag New York, Inc., 175 Fifth Avenue, New York, NY 10010. (800) 526-7254. 1957 to present. Quarterly. $450.00 per year.

INTERNATIONAL JOURNAL OF SOLIDS AND STRUCTURES. Pergamon Press, Inc., Maxwell House, Fairview Park, Elmsford, NY 10523. (914) 592-7700. 1965 to present. Monthly. $375.00 per year.

JOURNAL OF APPLIED MECHANICS. American Society of Mechanical Engineers, 345 East 47th Street, New York, NY 10017. (212) 705-7703. 1935 to present. Quarterly. $100.00 per year.

JOURNAL OF ENGINEERING MECHANICS. American Society of Civil Engineers, 345 East 47th Street, New York, NY 10017. (212) 705-7420. 1956 to present. Monthly. $150.00 per year.

JOURNAL OF MECHANICS AND PHYSICS OF SOLIDS. Pergamon Press, Inc., Maxwell House, Fairview Park, Elmsford, NY 10523. (914) 592-7700. 1952 to present. Bimonthly. $200.00 per year.

QUARTERLY JOURNAL OF MECHANICS AND APPLIED MATHEMATICS. Oxford University Press, 200 Madison Avenue, New York, NY 10016. (800) 458-5833. 1948 to present. Quaterly. $150.00 per year.

KINETICS

See also: CHAOS, DYNAMICS, KINEMATICS, MECHANICS, PHYSICAL CHEMISTRY

ABSTRACT SERVICES AND INDEXES

APPLIED MECHANICS REVIEW. American Society of Mechanical Engineers, 345 East 47th Street, New York, NY 10017. (212) 705-7703. 1948 to present. Monthly. $360.00 per year.

APPLIED SCIENCE AND TECHNOLOGY INDEX. H.W. Wilson and Company, 950 University Avenue, Bronx, NY 10452. (800) 367-6670 or (212) 588-8400. Monthly. Inquire as to cost and availability.

CHEMICAL ABSTRACTS. American Chemical Society, Chemical Abstracts Service, Box 3012, Columbus, OH 43210. (614) 421-3600. 1907 to present. Weekly. $9500.00 per year.

CONFERENCE PAPERS INDEX. Cambridge Scientific Abstracts, 5161 River Road, Bethesda, MD 20816. 1972 to present. Monthly. Inquire as to cost and availability.

CURRENT CONTENTS: PHYSICAL, CHEMICAL AND EARTH SCIENCES. Institute for Scientific Information, 3501 Market Street, Philadelphia, PA 19104. (800) 523-1850 or (215) 386-0100. Weekly. $275.00 per year.

ENGINEERING INDEX MONTHLY AND AUTHOR INDEX. Engineering Information Inc., 345 East 47th Street, New York, NY 10017. (212) 705-7600. Monthly. $1560.00 per year.

ISMEC BULLETIN (Information Service in Mechanical Engineering). Cambridge Scientific Abstracts, 5161 River Road, Bethesda, MD 20816. (301) 951-1400. 1973 to present. Monthly. $450.00 per year.

PHYSICS ABSTRACTS. Institution of Electrical Engineers. Available from: IEEE Service Center, 445 Hoes Lane, Piscataway, NJ 08854. 1898 to present. Bimonthly. $1700.00 per year.

PHYSICS BRIEFS. Physik Verlag GmbH, Postfach 1260/1280, D-6940 Weinheim, West Germany. (212) 661-9404. 1920 to present. Twenty-six times per year. $1250.00 per year.

SCIENCE CITATION INDEX. Institute for Scientific Information, 3501 Market Street, Philadelphia, PA 19104. (800) 523-1850 or (215) 386-0100. Six times per year. $6200.00 per year.

ASSOCIATIONS AND PROFESSIONAL SOCIETIES

AMERICAN CHEMICAL SOCIETY. 1155 16th Street, N.W., Washington, DC 20036. (800) 424-6747.

AMERICAN INSTITUTE OF PHYSICS. 335 East 45th Street, New York, NY 10017. (212) 661-9404.

AMERICAN PHYSICAL SOCIETY. 335 East 45th Street, New York, NY 10017. (212) 682-7341.

AMERICAN SOCIETY OF CIVIL ENGINEERS. 345 East 47th Street, New York, NY 10017. (212) 705-7420.

AMERICAN SOCIETY OF MECHANICAL ENGINEERS. 345 East 47th Street, New York, NY 10017. (212) 705-7703.

DIRECTORIES AND BIOGRAPHICAL SOURCES

AMERICAN MEN AND WOMEN OF SCIENCE. R.R. Bowker, Inc., Order Department, 245 West 17th Street, New York, NY 10011. (800) 521-8110. Eight volumes. 1986. $595.00 for set.

INTERNATIONAL RESEARCH CENTERS DIRECTORY 1988-89. Darren L. Smith, editor. Gale Research Company, Book Tower, Detroit, MI 48226. (800) 521-0707. 4th edition. 1987. $360.00.

1987 DIRECTORY OF ENGINEERING SOCIETIES AND RELATED ORGANIZATIONS. Gordon Davis, editor. Hemisphere Publishing Corporation, 1010 Vermont Avenue, NW, Washington, DC 20005. (800) 526-0275. 12th edition. 1987. $100.00.

RESEARCH CENTERS DIRECTORY 1988. Gale Research Company, Book Tower, Detroit, MI 48226. (800) 521-0707. 12th edition. 1987. $365.00 for set.

SCIENTIFIC AND TECHNICAL ORGANIZATIONS AND AGENCIES DIRECTORY. Margaret Labash Young, editor. Gale Research Company, Book Tower, Detroit, MI 48226. (800) 521-0707. 2nd edition. 1987. $185.00.

WHO'S WHO IN ENGINEERING. Gordon Davis, editor. Hemisphere Publishing Corporation, 1010 Vermont Avenue, NW, Washington, DC 20005. (800) 526-0275. 6th edition. 1985. $200.00.

ENCYCLOPEDIAS AND DICTIONARIES

CONCISE SCIENCE DICTIONARY. Oxford University Press, 200 Madison Avenue, New York, NY 10016. (800) 458-5833. 1987. $9.95 in paper.

DICTIONARY OF THE PHYSICAL SCIENCES: TERMS, FORMULAS, DATA. Cesare Emiliani. Oxford University Press, 200 Madison Avenue, New York, NY 10016. (800) 458-5833. 1987. $19.95 in paper.

MCGRAW-HILL ENCYCLOPEDIA OF SCIENCE AND TECHNOLOGY. McGraw-Hill Book Company, 1221 Avenue of the Americas, New York, NY 10020. (212) 512-2000. 6th edition. 1987. $1600.00.

THESAURUS OF SCIENTIFIC, TECHNICAL, AND ENGINEERING TERMS. Hemisphere Publishing Corporation, 1010 Vermont Avenue, NW, Washington, DC 20005. (800) 526-0275. 1988. $125.00.

GENERAL WORKS

CHAOTIC DYNAMICS AND FRACTALS. Michael F. Barnsley and Stephen G. Demko. Academic Press, Inc., 6277 Sea Harbor Drive, Orlando, FL 32821. (800) 321-5068. 1986. $29.95 in paper.

CLASSICAL MECHANICS. E.A. Desloge. John Wiley and Sons, Inc., 605 Third Avenue, New York, NY 10158. (800) 526-5368. 1982. Two volumes. $110.00 for set.

DYNAMICS. S.N. Rasband. John Wiley and Sons, Inc., 605 Third Avenue, New York, NY 10158. (800) 526-5368. 1983. $37.50.

ELEMENTARY ENGINEERING MECHANICS. G.E. Drabble. Macmillan Publishing Company, Inc., 866 Third Avenue, New York, NY 10022. (800) 257-5755. 1986. $22.50.

KINEMATICS. Joseph S. Beggs. Hemisphere Publishing Corporation, 79 Madison Avenue, New York, NY 10016-7892. Order from: Taylor and Francis/Hemisphere Distribution Center, 242 Cherry Street, Philadelphia, PA 19106-1906. (800) 821-8312. 1983. $28.50.

KINETICS AND CHEMICAL TECHNOLOGY. C.H. Bamford and others. Elsevier Science Publishing Company, Inc., 52 Vanderbilt Avenue, New York, NY 10017. (212) 370-5520. 1985. $100.00.

STATISTICAL MECHANICS AND DYNAMICS. Henry Eyring and others. John Wiley and Sons, Inc., 605 Third Avenue, New York, NY 10158. (800) 526-5368. 1982. $44.95.

ONLINE DATA BASES

CA SEARCH. Chemical Abstracts Service, P.O. Box 3012, Columbus, OH 43120. (800) 848-6538 or (614) 421-3600. Comprehensive guide to chemical literature, 1972 to present. Inquire as to online cost and availability.

COMPENDEX. Engineering Information, Inc., 345 East 47th Street, New York, NY 10017. (800) 221-1044 or (212) 705-7615. Engineering and technical literature, 1975 to present. Inquire as to online cost and availability.

DISSERTATION ABSTRACTS ONLINE. University Microfilms International, 300 North Zeeb Road, Ann Arbor, MI 48106. (800) 521-0600 or (313) 761-4700. Scope includes virtually all doctoral dissertations accepted at accredited American institutions from 1861 to present in over 250 subject areas. Inquire as to online cost and availability.

INSPEC. INSPEC Marketing Department, Institution of Electrical Engineers. Available from IEEE Service Center, 445 Hoes Lane, Piscataway, NJ 08854. (201) 981-0060. Online version of Physics Abstracts. Inquire as to online cost and availability.

NTIS. National Technical Information Service, 5285 Port Royal Road, Springfield, VA 22161. (703) 487-4630. Broad coverage of government sponsored research reports, 1964 to present. Inquire as to online cost and availability.

SCISEARCH. Institute for Scientific Information, 3501 Market Street, Philadelphia, PA 19104. (800) 523-1850 or (215) 386-0100. Broad multidisciplinary title and author index to the international literature of science and technology, 1974 to present. Inquire as to online cost and availability.

WILSONLINE. H.W. Wilson and Company, 950 University Avenue, Bronx, NY 10452. (800) 367-6770 or (212) 588-8400. Makes available online versions of the H.W. Wilson indexes including Applied Science and Technology Index, Business Periodicals Index and Readers' Guide to Periodical Literature. Approximately 1980 to present. Inquire as to online cost and availability.

PERIODICALS

ARCHIVE FOR RATIONAL MECHANICS AND ANALYSIS. Springer-Verlag New York, Inc., 175 Fifth Avenue, New York, NY 10010. (800) 526-7254. 1957 to present. Quarterly. $450.00 per year.

INTERNATIONAL JOURNAL OF CHEMICAL KINETICS. John Wiley and Sons, Inc., 605 Third Avenue, New York, NY 10158. (800) 526-5368. 1968 to present. Monthly. $300.00 per year.

INTERNATIONAL JOURNAL OF SOLIDS AND STRUCTURES. Pergamon Press, Inc., Maxwell House, Fairview Park, Elmsford, NY 10523. (914) 592-7700. 1965 to present. Monthly. $375.00 per year.

JOURNAL OF APPLIED MECHANICS. American Society of Mechanical Engineers, 345 East 47th Street, New York, NY 10017. (212) 705-7703. 1935 to present. Quarterly. $100.00 per year.

JOURNAL OF ENGINEERING MECHANICS. American Society of Civil Engineers, 345 East 47th Street, New York, NY 10017. (212) 705-7420. 1956 to present. Monthly. $150.00 per year.

JOURNAL OF MECHANICS AND PHYSICS OF SOLIDS. Pergamon Press, Inc., Maxwell House, Fairview Park, Elmsford, NY 10523. (914) 592-7700. 1952 to present. Bimonthly. $200.00 per year.

QUARTERLY JOURNAL OF MECHANICS AND APPLIED MATHEMATICS. Oxford University Press, 200 Madison Avenue, New York, NY 10016. (800) 458-5833. 1948 to present. Quaterly. $150.00 per year.

L

LANDSAT

See: ARTIFICIAL SATELLITES

LASER DISK

See: OPTICAL DISKS

LASERS

See also: FIBER OPTICS, HOLOGRAPHY, OPTICAL COMMUNICATIONS, OPTICAL DISKS, OPTICS, OPTOELECTRONICS, PHYSICS, SPECTROSCOPY

ABSTRACT SERVICES AND INDEXES

APPLIED SCIENCE AND TECHNOLOGY INDEX. H.W. Wilson and Company, 950 University Avenue, Bronx, NY 10452. (800) 367-6670 or (212) 588-8400. Monthly. Inquire as to cost and availability.

CHEMICAL ABSTRACTS. American Chemical Society, Chemical Abstracts Service, Box 3012, Columbus, OH 43210. (614) 421-3600. 1907 to present. Weekly. $9500.00 per year.

CONFERENCE PAPERS INDEX. Cambridge Scientific Abstracts, 5161 River Road, Bethesda, MD 20816. 1972 to present. Monthly. Inquire as to cost and availability.

CURRENT CONTENTS: ENGINEERING, TECHNOLOGY AND APPLIED SCIENCES. Institute for Scientific Information, 3501 Market Street, Philadelphia, PA 19104. (800) 523-1850 or (215) 386-0100. Weekly. $275.00 per year.

ELECTRICAL AND ELECTRONICS ABSTRACTS. Institution of Electrical Engineers. Available from: Institute of Electrical and Electronics Engineers. IEEE Service Center, 445 Hoes Lane, Piscataway, NJ 08854. Monthly. $1250.00 per year.

ENGINEERING INDEX MONTHLY AND AUTHOR INDEX. Engineering Information Inc., 345 East 47th Street, New York, NY 10017. (212) 705-7600. Monthly. $1560.00 per year.

GENERAL SCIENCE INDEX. H.W. Wilson and Company, 950 University Avenue, Bronx, NY 10452. (800) 367-6670 or (212) 588-8400. 1978 to present. Monthly. Inquire as to cost and availability.

GOVERNMENT REPORTS ANNOUNCEMENT AND INDEX. National Technical Information Service, 5285 Port Royal Road, Springfield, VA 22161. (703) 487-4929. Summaries of United States government sponsored research reports. 1946 to present. Biweekly. Inquire as to cost and availability.

INDEX TO SCIENTIFIC AND TECHNICAL PROCEEDINGS. Institute for Scientific Information, 3501 Market Street, Philadelphia, PA 19104. (800) 523-1850 or (215) 386-0100. 1978 to present. Monthly. $775.00 per year.

INDEX TO SCIENTIFIC REVIEWS. Institute for Scientific Information, 3501 Market Street, Philadelphia, PA 19104. (800) 523-1850 or (215) 386-0100. 1974 to present. Semi-annual. $550.00 per year.

PHYSICS ABSTRACTS. Institution of Electrical Engineers. Available from: IEEE Service Center, 445 Hoes Lane, Piscataway, NJ 08854. 1898 to present. Bimonthly. $1700.00 per year.

PHYSICS BRIEFS. Physik Verlag GmbH, Postfach 1260/1280, D-6940 Weinheim, West Germany. (212) 661-9404. 1920 to present. Twenty-six times per year. $1250.00 per year.

SCIENCE CITATION INDEX. Institute for Scientific Information, 3501 Market Street, Philadelphia, PA 19104. (800) 523-1850 or (215) 386-0100. Six times per year. $6200.00 per year.

ANNUAL REVIEWS AND YEARBOOKS

LASER APPLICATIONS. Academic Press, Inc., 6277 Sea Harbor Drive, Orlando, FL 32821. (800) 321-5068. 1971 to present. Irregular. Price varies, inquire.

SPRINGER SERIES IN OPTICAL SCIENCES. Springer-Verlag New York, Inc., 175 Fifth Avenue, New York, NY 10010. (800) 526-7254. 1976 to present. Irregular. Price varies, inquire.

ASSOCIATIONS AND PROFESSIONAL SOCIETIES

AMERICAN INSTITUTE OF PHYSICS. 335 East 45th Street, New York, NY 10017. (212) 661-9404.

INSTITUTION OF ELECTRICAL AND ELECTRONICS ENGINEERS (IEEE). 345 East 47th Street, New York, NY 10017. (212) 705-7900.

LASER INSTITUTE OF AMERICA. 5151 Monroe Street, Suite 103, Toledo, OH 43623. (419) 882-8706.

OPTICAL SOCIETY OF AMERICA. 1816 Jefferson Place, N.W., Washington, DC 20036. (202) 223-8130.

SOCIETY OF PHOTO-OPTICAL INSTRUMENTATION ENGINEERS - THE INTERNATIONAL SOCIETY OF OPTICAL ENGINEERING. P.O. Box 10, 1022 19th Street, Bellingham, WA 98227. (206) 676-3290.

DIRECTORIES AND BIOGRAPHICAL SOURCES

AMERICAN MEN AND WOMEN OF SCIENCE. R.R. Bowker, Inc., Order Department, 245 West 17th Street, New York, NY 10011. (800) 521-8110. Eight volumes. 1986. $595.00 for set.

DIRECTORY OF CONSULTANTS IN LASERS AND PHYSICS. J. Dick Publishing, Research Publications, 801 Green Bay Road, Lake Bluff, IL 60044. (312) 234-1220. Biennial. $85.00.

INTERNATIONAL RESEARCH CENTERS DIRECTORY 1988-89. Darren L. Smith, editor. Gale Research Company, Book Tower, Detroit, MI 48226. (800) 521-0707. 4th edition. 1987. $360.00.

LASER FOCUS/ELECTRO-OPTICS BUYERS' GUIDE. Advanced Technology Publications, Inc., PennWell Publishing, 119 Russell Street, Littleton, MA 01460. (617) 486-9501. Annual, January. $60.00.

1987 DIRECTORY OF ENGINEERING SOCIETIES AND RELATED ORGANIZATIONS. Gordon Davis, editor. Hemisphere Publishing Corporation, 1010 Vermont Avenue, NW, Washington, DC 20005. (800) 526-0275. 12th edition. 1987. $100.00.

RESEARCH CENTERS DIRECTORY 1988. Gale Research Company, Book Tower, Detroit, MI 48226. (800) 521-0707. 12th edition. 1987. $365.00 for set.

SCIENTIFIC AND TECHNICAL ORGANIZATIONS AND AGENCIES DIRECTORY. Margaret Labash Young, editor. Gale Research Company, Book Tower, Detroit, MI 48226. (800) 521-0707. 2nd edition. 1987. $185.00.

WHO'S WHO IN ENGINEERING. Gordon Davis, editor. Hemisphere Publishing Corporation, 1010 Vermont Avenue, NW, Washington, DC 20005. (800) 526-0275. 6th edition. 1985. $200.00.

GENERAL WORKS

ANALYTICAL APPLICATIONS OF LASERS. Edward H. Piepmeir, editor. John Wiley and Sons, Inc., 605 Third Avenue, New York, NY 10158. (800) 526-5368. 1986. $89.95.

ELECTROMAGNETIC PRINCIPLES OF INTEGRATED OPTICS. D.L. Lee. John Wiley and Sons, Inc., 605 Third Avenue, New York, NY 10158. (800) 526-5368. 1986. $90.00.

LASER-INDUCED CHEMICAL PROCESSES. J.L. Steinfeld, editor. Plenum Publishing Corporation, 233 Spring Street, New York, NY 10013. (800) 221-9369. 1981. $42.50.

LASER PROCESSING AND ANALYSIS OF MATERIALS. W.W. Duley. Plenum Publishing Corporation, 233 Spring Street, New York, NY 10013. (800) 221-9369. 1983. $65.00.

LASERS: INVENTIONS TO APPLICATION. Jesse H. Ausubel and H.D. Langford, editors. National Academy Press, 2101 Constitution Avenue, Washington, DC 20418. (202) 334-3313. 1987. $14.95 in paper.

LASERS: THEORY AND APPLICATIONS. K. Thyagarajan and A.K. Ghatak. Plenum Publishing Corporation, 233 Spring Street, New York, NY 10013. (800) 221-9369. 1981. $59.50.

HANDBOOKS AND MANUALS

HANDBOOK OF OPTICS. Optical Society of America. McGraw-Hill Book Company, 1221 Avenue of the Americas, New York, NY 10020. (212) 512-2000. 1978. $100.00.

SAFETY WITH LASERS AND OTHER OPTICAL SOURCES: A COMPREHENSIVE HANDBOOK. D. Sliney and M. Wolbarsht. Plenum Publishing Corporation, 233 Spring Street, New York, NY 10013. (800) 221-9369. 1980. $65.00.

ONLINE DATA BASES

CA SEARCH. Chemical Abstracts Service, P.O. Box 3012, Columbus, OH 43120. (800) 848-6538 or (614) 421-3600. Comprehensive guide to chemical literature, 1972 to present. Inquire as to online cost and availability.

COMPENDEX. Engineering Information, Inc., 345 East 47th Street, New York, NY 10017. (800) 221-1044 or (212) 705-7615. Engineering and technical literature, 1975 to present. Inquire as to online cost and availability.

DISSERTATION ABSTRACTS ONLINE. University Microfilms International, 300 North Zeeb Road, Ann Arbor, MI 48106. (800) 521-0600 or (313) 761-4700. Scope includes virtually all doctoral dissertations accepted at accredited American institutions from 1861 to present in over 250 subject areas. Inquire as to online cost and availability.

INSPEC. INSPEC Marketing Department, Institution of Electrical Engineers. Available from IEEE Service Center, 445 Hoes Lane, Piscataway, NJ 08854. (201) 981-0060. Online version of Physics Abstracts. Inquire as to online cost and availability.

NTIS. National Technical Information Service, 5285 Port Royal Road, Springfield, VA 22161. (703) 487-4630. Broad coverage of government sponsored research reports, 1964 to present. Inquire as to online cost and availability.

SCISEARCH. Institute for Scientific Information, 3501 Market Street, Philadelphia, PA 19104. (800) 523-1850 or (215) 386-0100. Broad multidisciplinary title and author index to the international literature of science and technology, 1974 to present. Inquire as to online cost and availability.

WILSONLINE. H.W. Wilson and Company, 950 University Avenue, Bronx, NY 10452. (800) 367-6770 or (212) 588-8400. Makes available online versions of the H.W. Wilson indexes including Applied Science and Technology Index, Business Periodicals Index and Readers' Guide to Periodical Literature. Approximately 1980 to present. Inquire as to online cost and availability.

PERIODICALS

APPLIED OPTICS. Optical Society of America. 1816 Jefferson Place, N.W., Washington, DC 20036. (202) 223-8130. Order from: American Institute of Physics, 335 East 45th Street, New York, NY 10017. (212) 661-9404. 1962 to present. Semi-monthly. $330.00 per year.

APPLIED PHYSICS B: PHOTOPHYSICS AND LASER CHEMISTRY. Springer-Verlag New York, Inc., 175 Fifth Avenue, New York, NY 10010. (800) 526-7254. 1962 to present. Monthly. $325.00 per year.

FIBER AND INTEGRATED OPTICS. Crane Russak and Company, Inc., 3 East 44th Street, New York, NY 10017. (212) 867-1490. 1977 to present. Quarterly. $86.00 per year.

IEEE JOURNAL OF QUANTUM ELECTRONICS. Institute of Electrical and Electronics Engineers. IEEE Service Center, 445 Hoes Lane, Piscataway, NJ 08854. 1962 to present. Monthly. $180.00 per year.

JOURNAL OF APPLIED PHYSICS. American Institute of Physics, 335 East 45th Street, New York, NY 10017. (212) 661-9494. 1931 to present. Semi-monthly. $495.00 per year.

JOURNAL OF LIGHTWAVE TECHNOLOGY. Institute of Electrical and Electronics Engineers. IEEE Service Center, 445 Hoes Lane, Piscataway, NJ 08854. 1983 to present. Monthly. $145.00 per year.

LASER FOCUS. PennWell Publishing Company, 119 Russell Street, Littleton, MA 01460. (617) 486-9501. 1965 to present. Monthly. $55.00 per year.

OPTICAL ENGINEERING. Society of Photo-Optical Instrumentation Engineers (SPIE), P.O. Box 10, 1022 19th Street, Bellingham, WA 98227. (206) 676-3290. 1962 to present. Monthly. $95.00 per year.

OPTICAL SOCIETY OF AMERICA, JOURNAL, PART B: OPTICAL PHYSICS. Optical Society of America. 1816 Jefferson Place, N.W., Washington, DC 20036. (202) 223-8130. Order from:

American Institute of Physics, 335 East 45th Street, New York, NY 10017. (212) 661-9404. 1917 to present. Monthly. $180.00 per year.

OPTICS AND LASERS IN ENGINEERING. Elsevier Science Publishing Company, Inc., 52 Vanderbilt Avenue, New York, NY 10017. (212) 370-5520. 1980 to present. Quarterly. $95.00 per year.

OPTICS COMMUNICATIONS. Elsevier Science Publishing Company, Inc., 52 Vanderbilt Avenue, New York, NY 10017. (212) 370-5520. 1969 to present. 24 times per year. $425.00 per year.

OPTICS LETTERS. Optical Society of America. Order from: American Institute of Physics, 335 East 45th Street, New York, NY 10017. (212) 661-9404. 1977 to present. Monthly. $150.00 per year.

SOCIETY OF PHOTO-OPTICAL INSTRUMENTATION ENGINEERS (SPIE) PROCEEDINGS. Society of Photo-Optical Instrumentation Engineers (SPIE), P.O. Box 10, 1022 19th Street, Bellingham, WA 98227. (206) 676-3290. 1963 to present. Approximately 50 numbers per year. Approximately $40.00 per number.

RESEARCH CENTERS AND INSTITUTES

CENTER FOR OPTICS, LASERS AND HOLOGRAPHY. 100 Glen Cove Avenue, Glen Cove, NY 11542. (516) 686-7863.

CENTER FOR RESEARCH IN ELECTRO-OPTICS AND LASERS. University of Central Florida, Department of Physics, P.O. Box 25000 CEBA 419, Orlando, FL 32816-0001. (305) 275-2325.

LABORATORY FOR LASER ENERGETICS. University of Rochester, 250 East River Road, Rochester, NY 14623-1299. (716) 275-5101.

LASER RESEARCH CENTER. Massachusetts Institute of Technology, 77 Massachusetts Avenue, Cambridge, MA 02139. (617) 253-7700.

LENSES

See: OPTICS

LEPTONS

See: PARTICLE PHYSICS

LIGHTING

See: ILLUMINATION

LIGHTNING

See also: METEOROLOGY

ABSTRACT SERVICES AND INDEXES

CHEMICAL ABSTRACTS. American Chemical Society, Chemical Abstracts Service, Box 3012, Columbus, OH 43210. (614) 421-3600. 1907 to present. Weekly. $9500.00 per year.

CURRENT CONTENTS: PHYSICAL, CHEMICAL, EARTH SCIENCES. Institute for Scientific Information, 3501 Market Street, Philadelphia, PA 19104. (800) 523-1850 or (215) 386-0100. Weekly. $275.00 per year.

ENGINEERING INDEX MONTHLY AND AUTHOR INDEX. Engineering Information Inc., 345 East 47th Street, New York, NY 10017. (212) 705-7600. Monthly. $1560.00 per year.

GENERAL SCIENCE INDEX. H.W. Wilson and Company, 950 University Avenue, Bronx, NY 10452. (800) 367-6670 or (212) 588-8400. 1978 to present. Monthly. Inquire as to cost and availability.

METEOROLOGICAL AND GEOASTROPHYSICAL ABSTRACTS. American Meteorological Society, 45 Beacon Street, Boston, MA 02108. (617) 227-2425. 1950 to present. Monthly. $450.00 per year.

OCEAN ABSTRACTS. Cambridge Scientific Abstracts, 5161 River Road, Bethesda, MD 20816. (301) 951-1400. 1963 to present. Bimonthly. $450.00 per year.

PHYSICS ABSTRACTS. Institution of Electrical Engineers. Available from: IEEE Service Center, 445 Hoes Lane, Piscataway, NJ 08854. 1898 to present. Bimonthly. $1700.00 per year.

PHYSICS BRIEFS. Physik Verlag GmbH, Postfach 1260/1280, D-6940 Weinheim, West Germany. (212) 661-9404. 1920 to present. Twenty-six times per year. $1250.00 per year.

SCIENCE CITATION INDEX. Institute for Scientific Information, 3501 Market Street, Philadelphia, PA 19104. (800) 523-1850 or (215) 386-0100. Six times per year. $6200.00 per year.

ASSOCIATIONS AND PROFESSIONAL SOCIETIES

AMERICAN ASSOCIATION OF STATE CLIMATOLOGISTS. C/O Professor John Griffiths, Meteorology Department, O and M Building, Texas A&M University, College Station, TX 77843. (409) 845-7320.

AMERICAN METEOROLOGY SOCIETY. 45 Beacon Street, Boston, MA 02108. (617) 227-2425.

INTERNATIONAL ASSOCIATION FO METEOROLOGY AND ATMOSPHERIC PHYSICS. University Corporation for Atmospheric Research, P.O. Box 3000, Boulder, CO 80307.

NATIONAL ENVIRONMENTAL SATELLITE DATA AND INFORMATION SERVICE. 3000 Whitehaven Street, N.W., Washington, DC 20255. (202) 634-7318.

NATIONAL WEATHER ASSOCIATION. 4400 Stamp Road, Room 404, Temple Hills, MD 20748. (301) 899-3784.

WEATHER MODIFICATION ASSOCIATION. P.O. Box 8116, Fresno, CA 93747. (209) 291-8466.

DIRECTORIES AND BIOGRAPHICAL SOURCES

AMERICAN MEN AND WOMEN OF SCIENCE. R.R. Bowker, Inc., Order Department, 245 West 17th Street, New York, NY 10011. (800) 521-8110. Eight volumes. 1986. $595.00 for set.

GOVERNMENT RESEARCH DIRECTORY. Gale Research Company, Book Tower, Detroit, MI 48226. (800) 521-0707. 4th edition. 1987. $350.00.

INTERNATIONAL RESEARCH CENTERS DIRECTORY 1988-89. Darren L. Smith, editor. Gale Research Company, Book Tower, Detroit, MI 48226. (800) 521-0707. 4th edition. 1987. $360.00.

METEOROLOGICAL SERVICES OF THE WORLD. World Meteorological Organization. Available from: American Meteorology Society, 45 Beacon Street, Boston, MA 02108. (617) 227-2425.

RESEARCH CENTERS DIRECTORY 1988. Gale Research Company, Book Tower, Detroit, MI 48226. (800) 521-0707. 12th edition. 1987. $365.00 for set.

LIGHTNING

SCIENTIFIC AND TECHNICAL ORGANIZATIONS AND AGENCIES DIRECTORY. Margaret Labash Young, editor. Gale Research Company, Book Tower, Detroit, MI 48226. (800) 521-0707. 2nd edition. 1987. $185.00.

GENERAL WORKS

BALL LIGHTNING AND BEAD LIGTHNING: EXTREME FORMS OF ATMOSPHERIC ELECTRICITY. James D. Barry. Plenum Publishing Corporation, 233 Spring Street, New York, NY 10013. (800) 221-9369. 1980. $35.00.

LIGHTNING AND ITS SPECTRUM: AN ATLAS OF PHOTOGRAPHS. Leon E. Salanave. University of Arizona Press, 1615 East Speedway, Tucson, AZ 86719. (602) 621-1441. 1980. $30.00

LIGHTNING, AURORAS, NOCTUAL LIGHTS AND RELATED LUMINOUS PHENOMENA. William R. Corliss. The Sourcebook Project, P.O. Box 107, Glen Arm, MD 21057. (301) 668-6047. 1982. $11.95.

ONLINE DATA BASES

CA SEARCH. Chemical Abstracts Service, P.O. Box 3012, Columbus, OH 43120. (800) 848-6538 or (614) 421-3600. Comprehensive guide to chemical literature, 1972 to present. Inquire as to online cost and availability.

COMPENDEX. Engineering Information, Inc., 345 East 47th Street, New York, NY 10017. (800) 221-1044 or (212) 705-7615. Engineering and technical literature, 1975 to present. Inquire as to online cost and availability.

DISSERTATION ABSTRACTS ONLINE. University Microfilms International, 300 North Zeeb Road, Ann Arbor, MI 48106. (800) 521-0600 or (313) 761-4700. Scope includes virtually all doctoral dissertations accepted at accredited American institutions from 1861 to present in over 250 subject areas. Inquire as to online cost and availability.

GEOREF. Online version of the BIBLIOGRAPHY AND INDEX OF GEOLOGY. American Geological Institute, 4220 King Street, Alexandria, VA 22302. (703) 379-2480. 1969 to present. Inquire as to online cost and availability.

INSPEC. INSPEC Marketing Department, Institution of Electrical Engineers. Available from IEEE Service Center, 445 Hoes Lane, Piscataway, NJ 08854. (201) 981-0060. Online version of Physics Abstracts. Inquire as to online cost and availability.

METEOROLOGICAL AND GEOASTROPHYSICAL ABSTRACTS. American Meteorological Society, 45 Beacon Street, Boston, MA 02108. (617) 227-2425. 1950 to present. Inquire as to cost and availability.

NTIS. National Technical Information Service, 5285 Port Royal Road, Springfield, VA 22161. (703) 487-4630. Broad coverage of government sponsored research reports, 1964 to present. Inquire as to online cost and availability.

SCISEARCH. Institute for Scientific Information, 3501 Market Street, Philadelphia, PA 19104. (800) 523-1850 or (215) 386-0100. Broad multidisciplinary title and author index to the international literature of science and technology, 1974 to present. Inquire as to online cost and availability.

PERIODICALS

AGRICULTURAL AND FOREST METEOROLOGY. Elsevier Science Publishing Company, Inc., 52 Vanderbilt Avenue, New York, NY 10017. (212) 370-5520. 1964 to present. $260.00 per year.

AMERICAN METEOROLOGICAL SOCIETY BULLETIN. American Meteorological Society, 45 Beacon Street, Boston, MA 02108. (617) 227-2425. 1920 to present. Monthly. $60.00 per year.

DYNAMICS OF ATMOSPHERE AND OCEANS. Elsevier Science Publishing Company, Inc., 52 Vanderbilt Avenue, New York, NY 10017. (212) 370-5520. 1977 to present. Quarterly. $90.00 per year.

JOURNAL OF ATMOSPHERIC SCIENCES. American Meteorological Society, 45 Beacon Street, Boston, MA 02108. (617) 227-2425. 1944 to present. Semimonthly. $220.00 per year.

MONTHLY WEARTHER REVIEW. American Meteorological Society, 45 Beacon Street, Boston, MA 02108. (617) 227-2425. 1872 to present. Monthly. $120.00 per year.

WEATHERWISE. Heldref Publications, 4000 Albemerle Street, N.W., Washington, DC 20016. (202) 362-6445. 1948 to present. Bimonthly. $20.00 per year.

RESEARCH CENTERS AND INSTITUTES

NATIONAL CENTER FOR ATMOSPHERIC RESEARCH. Box 3000, Boulder, CO 80307. (303) 497-1000.

NATIONAL SEVERE STORMS LABORATORY. 1313 Halley Circle, Norman, OK 73069. (405) 360-3620.

NEW MEXICO INSTITUTE OF MINING AND TECHNOLOGY. Irving Langmuir Laboratory, Socorro, NM 87801. (505) 835-5423.

LINEAR ACCELERATORS

See: PARTICLE ACCELERATORS

LINEAR ALGEBRA

See: ALGEBRA

LINEAR PROGRAMMING

See: MATHEMATICS

LOCAL AREA NETWORKS

See also: COMPUTERS, COMPUTER COMMUNICATION

ABSTRACT SERVICES AND INDEXES

APPLIED SCIENCE AND TECHNOLOGY INDEX. H.W. Wilson and Company, 950 University Avenue, Bronx, NY 10452. (800) 367-6670 or (212) 588-8400. Monthly. Inquire as to cost and availability.

COMPUMATH CITATION INDEX. Institute for Scientific Information, 3501 Market Street, Philadelphia, PA 19104. (800) 523-1850 or (215) 386-0100. Three times per year. $875.00 per year.

COMPUTER AND CONTROL ABSTRACTS. Institute of Electrical Engineers. Institute of Electrical and Electronics Engineers. IEEE Service Center, 445 Hoes Lane, Piscataway, NJ 08854. Semimonthly. $775.00 per year.

COMPUTER CONTENTS: THE BIWEEKLY COMPILATION OF TABLES OF CONTENTS FROM COMPUTER, ELECTRONIC AND TELECOMMUNICATIONS MAGAZINES, JOURNALS AND TRANSACTIONS. Find/SVP, 500 Fifth Avenue, New York, NY 10110. (800) 346-3787 or (212) 354-2424. Biweekly. $115.00 per year.

COMPUTING REVIEWS. Association of Computing Machinery, 11 West 42nd Street, New York, NY 10036. (212) 869-7440. Monthly. $60.00 per year.

ENGINEERING INDEX MONTHLY AND AUTHOR INDEX. Engineering Information Inc., 345 East 47th Street, New York, NY 10017. (212) 705-7600. Monthly. $1560.00 per year.

GOVERNMENT REPORTS ANNOUNCEMENT AND INDEX. National Technical Information Service, 5285 Port Royal Road, Springfield, VA 22161. (703) 487-4929. Summaries of United States government sponsored research reports. 1946 to present. Biweekly. Inquire as to cost and availability.

SCIENCE CITATION INDEX. Institute for Scientific Information, 3501 Market Street, Philadelphia, PA 19104. (800) 523-1850 or (215) 386-0100. Six times per year. $6200.00 per year.

ANNUAL REVIEWS AND YEARBOOKS

ADVANCES IN COMPUTERS. Academic Press, Inc., 6277 Sea Harbor Drive, Orlando, FL 32821. (800) 321-5068. Annual. Approximately $50.00 per volume.

ASSOCIATIONS AND PROFESSIONAL SOCIETIES

AMERICAN FEDERATION OF INFORMATION PROCESSING SOCIETIES. 1899 Preston White Drive, Reston, VA 22091. (703) 620-8900.

ASSOCIATION OF COMPUTING MACHINERY (ACM). 11 West 42nd Street, New York, NY 10036. (212) 869-7440.

IEEE COMPUTER SOCIETY. 1730 Massachusetts Avenue, N.W., Washington, DC 20036. (202) 371-0101.

INSTITUTE OF ELECTRICAL AND ELECTRONIC ENGINEERS (IEEE). 345 East 47th Street, New York, NY 10017. (212) 705-7900.

DIRECTORIES AND BIOGRAPHICAL SOURCES

AMERICAN MEN AND WOMEN OF SCIENCE. R.R. Bowker, Inc., Order Department, 245 West 17th Street, New York, NY 10011. (800) 521-8110. Eight volumes. 1986. $595.00 for set.

COMPUTER PERIPHERALS REVIEW. GML Information Services, 594 Marrett Road, Lexington, MA 02173. (617) 861-0515. Directory of computer peripheral equipment manufacturers. Two issues per year. $215.00 per year.

DATAPRO REPORTS ON DATA COMMUNICATIONS. DataPro Research Corporation, 1805 Underwood Boulevard, Delran, NJ 08075. Three base volumes, with monthly updates. $780.00.

1987 DIRECTORY OF ENGINEERING SOCIETIES AND RELATED ORGANIZATIONS. Gordon Davis, editor. Hemisphere Publishing Corporation, 1010 Vermont Avenue, NW, Washington, DC 20005. (800) 526-0275. 12th edition. 1987. $100.00.

RESEARCH CENTERS DIRECTORY 1988. Gale Research Company, Book Tower, Detroit, MI 48226. (800) 521-0707. 12th edition. 1987. $365.00 for set.

SCIENTIFIC AND TECHNICAL ORGANIZATIONS AND AGENCIES DIRECTORY. Margaret Labash Young, editor. Gale Research Company, Book Tower, Detroit, MI 48226. (800) 521-0707. 2nd edition. 1987. $185.00.

WHO'S WHO IN ENGINEERING. Gordon Davis, editor. Hemisphere Publishing Corporation, 1010 Vermont Avenue, NW, Washington, DC 20005. (800) 526-0275. 6th edition. 1985. $200.00.

ENCYCLOPEDIAS AND DICTIONARIES

COMPUTER AND TELECOMMUNICATIONS ACRONYMS. Julie E. Towell and Helen E. Sheppard, editors. Gale Research Company, Book Tower, Detroit, MI 48226. (800) 521-0707. 1986. $60.00.

ENCYCLOPEDIA OF INFORMATION SYSTEMS AND SERVICES 1988. Amy Lucas and Annette Novallo, editors. Gale Research Company, Book Tower, Detroit, MI 48226. (800) 521-0707. 8th edition. 1987. $400.00 for set.

PRENTICE-HALL ENCYCLOPEDIA OF INFORMATION TECHNOLOGY. Robert A. Edmunds. Prentice-Hall Publishing, Inc., Englewood Cliffs, NJ 07632. (800) 562-0245. 1987. $49.95.

GENERAL WORKS

DATA COMMUNICATIONS AND TELEPROCESSING SYSTEMS. Trevor Housley. Prentice-Hall Publishing, Inc., Englewood Cliffs, NJ 07632. (800) 562-0245. 1987. $38.95.

DATA COMMUNICATIONS NETWORKING DEVICES: CHARACTERISTICS, OPERATIONS, APPLICATIONS. Gilbert Held. John Wiley and Sons, Inc., 605 Third Avenue, New York, NY 10158. (800) 526-5368. 1986. $29.95

INTRODUCTION TO DATA COMMUNICATIONS AND COMPUTER NETWORKS. Fred Halsall. Addison-Wesley Publishing Company, Inc., 1 Jacob Way, Reading, MA 01867. (617) 944-3700. 1986. $31.95.

LOCAL AREA NETWORKS: ISSUES, PRODUCTS, AND DEVELOPMENTS. V.E. Cheong and R.A. Hirschheim. John Wiley and Sons, Inc., 605 Third Avenue, New York, NY 10158. (800) 526-5368. 1983. $34.95.

HANDBOOKS AND MANUALS

HANDBOOK OF DATA COMMUNICATION AND COMPUTER NETWORKS. Dimitris N. Chorafas. Petrocelli Books, 251 Wall Street, Princeton, NJ 08540. (609) 924-5851. 1985. $59.95.

ONLINE DATA BASES

COMPENDEX. Engineering Information, Inc., 345 East 47th Street, New York, NY 10017. (800) 221-1044 or (212) 705-7615. Engineering and technical literature, 1975 to present. Inquire as to online cost and availability.

DISSERTATION ABSTRACTS ONLINE. University Microfilms International, 300 North Zeeb Road, Ann Arbor, MI 48106. (800) 521-0600 or (313) 761-4700. Scope includes virtually all doctoral dissertations accepted at accredited American institutions from 1861 to present in over 250 subject areas. Inquire as to online cost and availability.

NTIS. National Technical Information Service, 5285 Port Royal Road, Springfield, VA 22161. (703) 487-4630. Broad coverage of government sponsored research reports, 1964 to present. Inquire as to online cost and availability.

SCISEARCH. Institute for Scientific Information, 3501 Market Street, Philadelphia, PA 19104. (800) 523-1850 or (215) 386-0100. Broad multidisciplinary title and author index to the international literature of science and technology, 1974 to present. Inquire as to online cost and availability.

WILSONLINE. H.W. Wilson and Company, 950 University Avenue, Bronx, NY 10452. (800) 367-6770 or (212) 588-8400. Makes available online versions of the H.W. Wilson indexes including Applied Science and Technology Index, Business Periodicals Index and Readers' Guide to Periodical Literature. Approximately 1980 to present. Inquire as to online cost and availability.

LOCAL AREA NETWORKS

PERIODICALS

BYTE; THE SMALL SYSTEMS JOURNAL. McGraw-Hill, Inc., 70 Main Street, Peterborough, NH 03458. (603) 924-9281. Subscription to: Box 590, Martinsville, NJ 08836. 1975 to present. Monthly. $21.00 per year.

COMPUTER COMMUNICATIONS. Butterworth's Publishing, 80 Montvale Avenue, Stoneham, MA 02180. (800) 325-4177. Bimonthly. $169.00 per year.

DATA COMMUNICATIONS. McGraw-Hill Book Company, 1221 Avenue of the Americas, New York, NY 10020. (212) 512-2000. Monthly. $30.00 per year.

DATAMATION. Technical Publishing Company, 875 Third Avenue, New York, NY 10022. (212) 605-9400. Semi-monthly. $50.00 per year.

JOURNAL OF TELECOMMUNICATION NETWORKS. Computer Science Press, Inc., 1803 Research Boulevard, Suite 500, Rockville, MD 20850-3155. (301) 251-9050. Quarterly. $100.00 per year.

LOCAL AREA NETWORKS (L.A.N. NEWSLETTER). Information Gatekeepers, Inc., 214 Harvard Avenue, Boston, MA 02134. (617) 232-3111. Monthly. $250.00 per year.

MICRO COMMUNICATIONS. Miller Freeman Publications, Inc., 500 Howard Street, San Francisco, CA 94105. (415) 397-1881. Bimonthly. $28.00 per year.

RESEARCH CENTERS AND INSTITUTES

COMPUTER COMMUNICATION NETWORK GROUP. University of Waterloo, CPH 2369A, Waterloo, ON, Canada N2L 3G1. (519) 885-1211.

DATABASE SYSTEMS RESEARCH CENTER. University of Maryland, College of Business and Management, Tydings Hall, College Park, MD 20742. (301) 454-6258.

SRI INTERNATIONAL CENTER FOR INTELLIGENT COMPUTER SYSTEMS. 333 Ravenswood Avenue, Menlo Park, CA 94025. (415) 859-4771.

LORAN

See: NAVIGATION

LOW TEMPERATURE PHYSICS

See: CRYOGENICS

LUBRICATION

See also: AUTOMOTIVE ENGINEERING, MACHINERY

ABSTRACT SERVICES AND INDEXES

APPLIED MECHANICS REVIEW. American Society of Mechanical Engineers, 345 East 47th Street, New York, NY 10017. (212) 705-7703. 1948 to present. Monthly. $360.00 per year.

APPLIED SCIENCE AND TECHNOLOGY INDEX. H.W. Wilson and Company, 950 University Avenue, Bronx, NY 10452. (800) 367-6670 or (212) 588-8400. Monthly. Inquire as to cost and availability.

CHEMICAL ABSTRACTS. American Chemical Society, Chemical Abstracts Service, Box 3012, Columbus, OH 43210. (614) 421-3600. 1907 to present. Weekly. $9500.00 per year.

CORROSION ABSTRACTS: ABSTRACTS OF THE WORLD'S LITERATURE ON CORROSION AND CORROSION MITIGATION. National Association of Corrosion Engineers, Box 218340, Houston, TX 77218. (713) 492-0535. 1962 to present. Bimonthly. $250.00 per year.

CURRENT CONTENTS: ENGINEERING, TECHNOLOGY AND APPLIED SCIENCES. Institute for Scientific Information, 3501 Market Street, Philadelphia, PA 19104. (800) 523-1850 or (215) 386-0100. Weekly. $275.00 per year.

ENGINEERING INDEX MONTHLY AND AUTHOR INDEX. Engineering Information Inc., 345 East 47th Street, New York, NY 10017. (212) 705-7600. Monthly. $1560.00 per year.

ISMEC BULLETIN (Information Service in Mechanical Engineering). Cambridge Scientific Abstracts, 5161 River Road, Bethesda, MD 20816. (301) 951-1400. 1973 to present. Monthly. $450.00 per year.

PETROLEUM ABSTRACTS. University of Tulsa, Information Services Division, 600 South College Street, Tulsa, OK 74104. (918) 592-6000. 1961 to present. Weekly. Inquire as to cost and availability.

SCIENCE CITATION INDEX. Institute for Scientific Information, 3501 Market Street, Philadelphia, PA 19104. (800) 523-1850 or (215) 386-0100. Six times per year. $6200.00 per year.

ASSOCIATIONS AND PROFESSIONAL SOCIETIES

AMERICAN SOCIETY FOR METALS. Metals Park, OH 44073. (216) 338-5151.

AMERICAN SOCIETY OF LUBRICATION ENGINEERS. 838 Busse Highway, Park Ridge, IL 60068. (312) 825-5536.

AMERICAN SOCIETY OF MECHANICAL ENGINEERS. 345 East 47th Street, New York, NY 10017. (212) 705-7703.

METALLURGICAL SOCIETY OF THE AIME (AMERICAN INSTITUTE OF MINING, METALLURGICAL AND PETROLEUM ENGINEERS). 420 Commonwealth Drive, Warrendale, PA 15086. (412) 776-9080.

NATIONAL ASSOCIATION OF CORROSION ENGINEERS. Box 218340, Houston, TX 77218. (713) 492-0535.

NATIONAL LUBRICATING GREASE INSTITUTE. 4635 Wyandotte Street, Kansas City, MO 64112. (816) 931-9480.

SOCIETY OF AUTOMOTIVE ENGINEERS (SAE). 400 Commonwealth Drive, Warrendale, PA 15096. (412) 776-4841.

DIRECTORIES AND BIOGRAPHICAL SOURCES

AMERICAN SOCIETY OF LUBRICATION ENGINEERS MEMBERSHIP ROSTER. American Society of Lubrication Engineers, 838 Busse Highway, Park Ridge, IL 60068. (312) 825-5536. Annual. $20.00.

1987 DIRECTORY OF ENGINEERING SOCIETIES AND RELATED ORGANIZATIONS. Gordon Davis, editor. Hemisphere Publishing Corporation, 1010 Vermont Avenue, NW, Washington, DC 20005. (800) 526-0275. 12th edition. 1987. $100.00.

RESEARCH CENTERS DIRECTORY 1988. Gale Research Company, Book Tower, Detroit, MI 48226. (800) 521-0707. 12th edition. 1987. $365.00 for set.

SCIENTIFIC AND TECHNICAL ORGANIZATIONS AND AGENCIES DIRECTORY. Margaret Labash Young, editor. Gale Research Company, Book Tower, Detroit, MI 48226. (800) 521-0707. 2nd edition. 1987. $185.00.

WHO'S WHO IN ENGINEERING. Gordon Davis, editor. Hemisphere Publishing Corporation, 1010 Vermont Avenue, NW, Washington, DC 20005. (800) 526-0275. 6th edition. 1985. $200.00.

ENCYCLOPEDIAS AND DICTIONARIES

THESAURUS OF SCIENTIFIC, TECHNICAL, AND ENGINEERING TERMS. Hemisphere Publishing Corporation, 1010 Vermont Avenue, NW, Washington, DC 20005. (800) 526-0275. 1988. $125.00.

GENERAL WORKS

BASIC LUBRICATION THEORY. A. Cameron and C.M. Ettles. Halsted Press, a division of John Wiley and Sons, Inc., 605 Third Avenue, New York, NY 10158. (800) 526-5368. Third edition. 1982. $37.95.

LUBRICATION: A PRACTICAL GUIDE TO LUBRICANT SELECTION. A.R. Lansdown. Pergamon Press, Inc., Maxwell House, Fairview Park, Elmsford, NY 10523. (914) 592-7700. 1982. $19.95 in paper.

NEW DIRECTIONS IN LUBRICATION, MATERIALS, WEAR AND SURFACE INTERACTION: TRIBOLOGY IN THE 80'S. William R. Loomis, editor. Noyes Data Corporation, Mill Road at Grand Avenue, Park Ridge, NJ 07656. (201) 391-8484. 1985. $69.00.

THEORY AND PRACTICE OF LUBRICATION FOR ENGINEERS. D.D. Fuller. John Wiley and Sons, Inc., 605 Third Avenue, New York, NY 10158. (800) 526-5368. 2nd edition. 1984. $64.95.

HANDBOOKS AND MANUALS

FRICTION, WEAR, AND LUBRICATION: A COMPLETE HANDBOOK OF TRIBOLOGY. I.V. Kregelsky and V.V. Alisin, editors. Pergamon Press, Inc., Maxwell House, Fairview Park, Elmsford, NY 10523. (914) 592-7700. Three volumes. 1982. $130.00 for set.

HANDBOOK OF LUBRICATION. American Society of Lubrication Engineers, 838 Busse Highway, Park Ridge, IL 60068. (312) 825-5536. Three volumes. 1983. $315.00 for set.

ONLINE DATA BASES

APILIT. American Petroleum Institute, Central Abstracting and Indexing Service, 2101 L Street, N.W., Washington, DC 20037. (202) 682-8000. Worldwide petroleum literature, 1964 to present. Inquire as to online cost and availability.

CA SEARCH. Chemical Abstracts Service, P.O. Box 3012, Columbus, OH 43120. (800) 848-6538 or (614) 421-3600. Comprehensive guide to chemical literature, 1972 to present. Inquire as to online cost and availability.

COMPENDEX. Engineering Information, Inc., 345 East 47th Street, New York, NY 10017. (800) 221-1044 or (212) 705-7615. Engineering and technical literature, 1975 to present. Inquire as to online cost and availability.

NTIS. National Technical Information Service, 5285 Port Royal Road, Springfield, VA 22161. (703) 487-4630. Broad coverage of government sponsored research reports, 1964 to present. Inquire as to online cost and availability.

SCISEARCH. Institute for Scientific Information, 3501 Market Street, Philadelphia, PA 19104. (800) 523-1850 or (215) 386-0100. Broad multidisciplinary title and author index to the international literature of science and technology, 1974 to present. Inquire as to online cost and availability.

WILSONLINE. H.W. Wilson and Company, 950 University Avenue, Bronx, NY 10452. (800) 367-6770 or (212) 588-8400. Makes available online versions of the H.W. Wilson indexes including Applied Science and Technology Index, Business Periodicals Index and Readers' Guide to Periodical Literature. Approximately 1980 to present. Inquire as to online cost and availability.

OTHER SOURCES

WHAT EVERY ENGINEER SHOULD KNOW ABOUT ENGINEERING SOURCES. Marcel Dekker Inc., 270 Madison Avenue, New York, NY 10016. (800) 228-1160. 1984. $24.95.

PERIODICALS

JOURNAL OF TRIBOLOGY. American Society of Mechanical Engineers, 345 East 47th Street, New York, NY 10017. (212) 705-7703. 1967 to present. Quarterly. $80.00 per year.

LUBRICATION ENGINEERING. American Society of Lubrication Engineers, 838 Busse Highway, Park Ridge, IL 60068. (312) 825-5536. Monthly. $39.00 per year.

N.L.G.I. SPOKESMAN. National Lubrication Grease Institute. 4635 Wyandotte Street, Kansas City, MO 64112. (816) 931-9480. 1937 to present. Monthly. $10.00 per year.

RESEARCH CENTERS AND INSTITUTES

COORDINATING RESEARCH COUNCIL, INC. 219 Perimeter Center Parkway, Atlanta, GA 30346. (404) 396-3400.

FLUID POWER RESEARCH CENTER. Oklahoma State University, Stillwater, OK 74074. (405) 624-7375.

INSTITUTE FOR WEAR CONTROL AND TRIBOLOGY. Rensselaer Polytechnic Institute, Mechanical Engineering Department, Jonsson Engineering Center, Troy, NY 12181. (518) 266-6000.

LUNAR GEOLOGY

See: ASTROLOGY, PLANETARY SCIENCES

M

MACHINE DESIGN

See also: CAD (COMPUTER-AIDED DESIGN), MACHINERY, MECHANICAL ENGINEERING

ABSTRACT SERVICES AND INDEXES

APPLIED MECHANICS REVIEW. American Society of Mechanical Engineers, 345 East 47th Street, New York, NY 10017. (212) 705-7703. 1948 to present. Monthly. $360.00 per year.

APPLIED SCIENCE AND TECHNOLOGY INDEX. H.W. Wilson and Company, 950 University Avenue, Bronx, NY 10452. (800) 367-6670 or (212) 588-8400. Monthly. Inquire as to cost and availability.

CAD/CAM ABSTRACTS. EIC/Intelligence, Inc., 48 West 38th Street, New York, NY 10018. (800) 223-6275 or (212) 944-8500. Monthly. $365.00 per year.

CHEMICAL ABSTRACTS. American Chemical Society, Chemical Abstracts Service, Box 3012, Columbus, OH 43210. (614) 421-3600. 1907 to present. Weekly. $9500.00 per year.

CURRENT CONTENTS: ENGINEERING, TECHNOLOGY AND APPLIED SCIENCES. Institute for Scientific Information, 3501 Market Street, Philadelphia, PA 19104. (800) 523-1850 or (215) 386-0100. Weekly. $275.00 per year.

ENGINEERING INDEX MONTHLY AND AUTHOR INDEX. Engineering Information Inc., 345 East 47th Street, New York, NY 10017. (212) 705-7600. Monthly. $1560.00 per year.

ISMEC BULLETIN (Information Service in Mechanical Engineering). Cambridge Scientific Abstracts, 5161 River Road, Bethesda, MD 20816. (301) 951-1400. 1973 to present. Monthly. $450.00 per year.

PHYSICS ABSTRACTS. Institution of Electrical Engineers. Available from: IEEE Service Center, 445 Hoes Lane, Piscataway, NJ 08854. 1898 to present. Bimonthly. $1700.00 per year.

SCIENCE CITATION INDEX. Institute for Scientific Information, 3501 Market Street, Philadelphia, PA 19104. (800) 523-1850 or (215) 386-0100. Six times per year. $6200. 00 per year.

ASSOCIATIONS AND PROFESSIONAL SOCIETIES

AMERICAN SOCIETY OF MECHANICAL ENGINEERS. 345 47th Street, New York, NY 10017. (212) 705-7722.

ASSOCIATION OF COMPUTING MACHINERY (ACM). 11 West 42nd Street, New York, NY 10036. (212) 896-7440.

INSTITUTE OF ELECTRICAL AND ELECTRONIC ENGINEERS (IEEE). 345 East 47th Street, New York, NY 10017. (212) 705-7900.

NATIONAL CONFERENCE ON FLUID POWER. IIT Research Institute, Ten West 35th Street, Chicago, IL 60616. (312) 567-4414.

VIBRATION INSTITUTE. 101 West 55th Street, Suite 206, Clarendon Hills, IL 60514. (312) 654-2254.

BIBLIOGRAPHIES

SCIENTIFIC AND TECHNICAL BOOKS AND SERIALS IN PRINT 1988; AN INDEX TO LITERATURE IN SCIENCE AND TECHNOLOGY. R.R. Bowker Company, 205 East 42nd Street, New York, NY 10017. (800) 521-8110. $175.00.

DIRECTORIES AND BIOGRAPHICAL SOURCES

COMPUTER-AIDED DESIGN (CAD) DIRECTORY. Technical Database Corporation, 1300 South Frazier, Conroe, TX 77305. (409) 539-9688.

INDUSTRIAL EQUIPMENT AND SUPPLIES DIRECTORY. American Business Directories, Inc., 5707 South 86th Circle, Omaha, NE 68127. (402) 331-7169.

MACHINERY BUYERS GUIDE. Findlay Publications Limited, Maitland House, Warrior Square, Southend-on-Sea, Essex SS5 5AR England. Annual. $45.00.

1987 DIRECTORY OF ENGINEERING SOCIETIES AND RELATED ORGANIZATIONS. Gordon Davis, editor. Hemisphere Publishing Corporation, 1010 Vermont Avenue, NW, Washington, DC 20005. (800) 526-0275. 12th edition. 1987. $100.00.

RESEARCH CENTERS DIRECTORY 1988. Gale Research Company, Book Tower, Detroit, MI 48226. (800) 521-0707. 12th edition. 1987. $365.00 for set.

SCIENTIFIC AND TECHNICAL ORGANIZATIONS AND AGENCIES DIRECTORY. Margaret Labash Young, editor. Gale Research Company, Book Tower, Detroit, MI 48226. (800) 521-0707. 2nd edition. 1987. $185.00.

WHO'S WHO IN ENGINEERING. Gordon Davis, editor. Hemisphere Publishing Corporation, 1010 Vermont Avenue, NW, Washington, DC 20005. (800) 526-0275. 6th edition. 1985. $200.00.

ENCYCLOPEDIAS AND DICTIONARIES

THESAURUS OF SCIENTIFIC, TECHNICAL, AND ENGINEERING TERMS. Hemisphere Publishing Corporation, 1010 Vermont Avenue, NW, Washington, DC 20005. (800) 526-0275. 1988. $125.00.

GENERAL WORKS

CAD: PRINCIPLES AND APPLICATIONS. Paul Barr and others. Prentice-Hall Publishing, Inc., Englewood Cliffs, NJ 07632. (800) 562-0245. 1985. $22.95.

COMPUTER-AIDED DESIGN AND MANUFACTURE. C.B. Besant and C.W.K. Lui. John Wiley and Sons, Inc., 605 Third Avenue, New York, NY 10158. (800) 526-5368. 3rd edition. 1986. $35.95 in paper.

DESIGN OF AUTOMATIC MACHINERY. Kendrick W. Lentz. Van Nostrand Reinhold Company, Inc., 135 West 50th Street, New York, NY 10020. (800) 543-2681. 1984. $42.95.

ENGINEERING MECHANICS. J.L. Meriam and L.G. Kraige. John Wiley and Sons, Inc., 605 Third Avenue, New York, NY 10158. (800) 526-5368. 2nd edition. 1986. $52.95.

INTRODUCTION TO DYNAMICS AND CONTROL. Leonard Meirovitch. John Wiley and Sons, Inc., 605 Third Avenue, New York, NY 10158. (800) 526-5368. 1985. $42.95.

HANDBOOKS AND MANUALS

APPLIED FLUID DYNAMICS HANDBOOK. Robert D. Blevins. Van Nostrand Reinhold Company, Inc., 135 West 50th Street, New York, NY 10020. (800) 543-2681. 1984. $52.95.

HANDBOOK OF MECHANICS, MATERIALS, AND STRUCTURES. A. Blake. John Wiley and Sons, Inc., 605 Third Avenue, New York, NY 10158. (800) 526-5368. 1985. $65.95.

MARK'S STANDARD HANDBOOK FOR MECHANICAL ENGINEERS. T. Baumeister, editor. McGraw-Hill Book Company, 1221 Avenue of the Americas, New York, NY 10020. (212) 512-2000. 8th edition. 1978. $96.00.

MECHANICAL DESIGN AND SYSTEMS HANDBOOK. Harold A. Rothbart, editor. McGraw-Hill Book Company, 1221 Avenue of the Americas, New York, NY 10020. (212) 512-2000. 2nd edition. 1985. $96.50.

MECHANICAL ENGINEERS' HANDBOOK. Myer Kutz, editor. John Wiley and Sons, Inc., 605 Third Avenue, New York, NY 10158. (800) 526-5368. 1986. $79.95.

PRESSURE VESSEL DESIGN HANDBOOK. Henry H. Bednar. Van Nostrand Reinhold Company, Inc., 135 West 50th Street, New York, NY 10020. (800) 543-2681. 2nd edition. 1986. $49.95.

STANDARD HANDBOOK OF MACHINE DESIGN. Joseph E. Shigley and Charles R. Mischke. McGraw-Hill Book Company, 1221 Avenue of the Americas, New York, NY 10020. (212) 512-2000. 1986. $89.00.

ONLINE DATA BASES

CA SEARCH. Chemical Abstracts Service, P.O. Box 3012, Columbus, OH 43120. (800) 848-6538 or (614) 421-3600. Comprehensive guide to chemical literature, 1972 to present. Inquire as to online cost and availability.

COMPENDEX. Engineering Information, Inc., 345 East 47th Street, New York, NY 10017. (800) 221-1044 or (212) 705-7615. Engineering and technical literature, 1975 to present. Inquire as to online cost and availability.

DISSERTATION ABSTRACTS ONLINE. University Microfilms International, 300 North Zeeb Road, Ann Arbor, MI 48106. (800) 521-0600 or (313) 761-4700. Scope includes virtually all doctoral dissertations accepted at accredited American institutions from 1861 to present in over 250 subject areas. Inquire as to online cost and availability.

INSPEC. INSPEC Marketing Department, Institution of Electrical Engineers. Available from IEEE Service Center, 445 Hoes Lane, Piscataway, NJ 08854. (201) 981-0060. Online version of Physics Abstracts. Inquire as to online cost and availability.

ISMEC. Cambridge Scientific Abstracts, 5161 River Road, Besthda, MD 20816. (800) 638-8076 or (301) 951-1400. Literature of mechanical and production engineering, 1973 to present. Inquire as to online cost and availability.

NTIS. National Technical Information Service, 5285 Port Royal Road, Springfield, VA 22161. (703) 487-4630. Broad coverage of government sponsored research reports, 1964 to present. Inquire as to online cost and availability.

SCISEARCH. Institute for Scientific Information, 3501 Market Street, Philadelphia, PA 19104. (800) 523-1850 or (215) 386-0100. Broad multidisciplinary title and author index to the international literature of science and technology, 1974 to present. Inquire as to online cost and availability.

WILSONLINE. H.W. Wilson and Company, 950 University Avenue, Bronx, NY 10452. (800) 367-6770 or (212) 588-8400. Makes available online versions of the H.W. Wilson indexes including Applied Science and Technology Index, Business Periodicals Index and Readers' Guide to Periodical Literature. Approximately 1980 to present. Inquire as to online cost and availability.

PERIODICALS

CIME (COMPUTERS IN MECHANICAL ENGINEERING). American Society of Mechanical Engineers, 345 East 47th Street, New York, NY 10017. (212) 705-7703.

INTERNATIONAL COMMUNICATIONS IN HEAT AND MASS TRANSFER. Pergamon Press, Inc., Maxwell House, Fairview Park, Elmsford, NY 10523. (914) 592-7700. 1974 to present. Bimonthly. $160.00 per year.

INTERNATIONAL JOURNAL FOR NUMERICAL METHODS IN ENGINEERING. John Wiley and Sons, Inc., 605 Third Avenue, New York, NY 10158. (800) 526-5368.

INTERNATIONAL JOURNAL OF HEAT AND MASS TRANSFER. Pergamon Press, Inc., Maxwell House, Fairview Park, Elmsford, NY 10523. (914) 592-7700. 1960 to present. Monthly. $500.00 per year.

INTERNATIONAL JOURNAL OF MULTIPHASE FLOW. Pergamon Press, Inc., Maxwell House, Fairview Park, Elmsford, NY 10523. (914) 592-7700. (212) 705-7703. 1974 to present. Bimonthly. $250.00 per year.

JOURNAL OF APPLIED MECHANICS. American Society of Mechanical Engineers, 345 East 47th Street, New York, NY 10017. (212) 705-7703. 1935 to present. Quarterly. $100.00 per year.

JOURNAL OF ENGINEERING FOR INDUSTRY. American Society of Mechanical Engineers, 345 East 47th Street, New York, NY 10017. (212) 705-7703. 1970 to present. Quarterly. $100.00 per year.

JOURNAL OF FLUID CONTROL. Delbridge Publishing Company, Box 2989, Stanford, CA 94305. (408) 446-3131. 1967 to present. Quarterly. Inquire as to cost and availability.

JOURNAL OF HEAT TRANSFER. American Society of Mechanical Engineers, 345 East 47th Street, New York, NY 10017. (212) 705-7703. 1970 to present. Quarterly. $100.00 per year.

JOURNAL OF MECHANISMS, TRANSMISSIONS AND AUTOMATION IN DESIGN. American Society of Mechanical Engineers, 345 East 47th Street, New York, NY 10017. (212) 705-7703. 1983 to present. Quarterly. $80.00 per year.

JOURNAL OF PRESSURE VESSEL TECHNOLOGY. American Society of Mechanical Engineers, 345 East 47th Street, New York, NY 10017. (212) 705-7703. 1974 to present. Quarterly. $80.00 per year.

LUBRICATION ENGINEERING. American Society of Lubrication Engineers, 838 Busse Highway, Park Ridge, IL 60068. 1945 to present. Monthly. $39.00 per year.

MACHINE DESIGN. Penton-IPC, 1100 Superior Avenue, Cleveland, OH 44114. (216) 696-7000. 1929 to present.

Twenty-eight times per year. $60.00 per year.

MECHANICAL ENGINEERING. American Society of Mechanical Engineers, 345 East 47th Street, New York, NY 10017. (212) 705-7722.

RESEARCH CENTERS AND INSTITUTES

LABORATORY FOR EXIPERMENTAL MECHANICS RESEARCH. State University of New York at Stony Brook, Stony Brook, NY 11794-2300. (516) 632-8311.

MACHINE SYSTEMS GROUP. Stevens Institute of Technology, Department of Mechanical Engineering, Castle Point Station, Hoboken, NJ 07030. (201) 420-5591.

MECHANICAL ENGINEERING DESIGN LABORATORY. University of Florida, 237 Mechanical Engineering Building, Gainesville, FL 32611. (904) 392-0827

MECHANICAL ENGINEERING RESEARCH LABORATORIES. Kansas State University, Durland Hall, Manhattan, KS 66506. (913) 532-5610.

MACHINERY

See also: CAD (COMPUTER-AIDED DESIGN), MACHINE DESIGN, MECHANICAL ENGINEERING

ABSTRACT SERVICES AND INDEXES

APPLIED MECHANICS REVIEW. American Society of Mechanical Engineers, 345 East 47th Street, New York, NY 10017. (212) 705-7703. 1948 to present. Monthly. $360.00 per year.

APPLIED SCIENCE AND TECHNOLOGY INDEX. H.W. Wilson and Company, 950 University Avenue, Bronx, NY 10452. (800) 367-6670 or (212) 588-8400. Monthly. Inquire as to cost and availability.

CAD/CAM ABSTRACTS. EIC/Intelligence, Inc., 48 West 38th Street, New York, NY 10018. (800) 223-6275 or (212) 944-8500. Monthly. $365.00 per year.

CHEMICAL ABSTRACTS. American Chemical Society, Chemical Abstracts Service, Box 3012, Columbus, OH 43210. (614) 421-3600. 1907 to present. Weekly. $9500.00 per year.

CURRENT CONTENTS: ENGINEERING, TECHNOLOGY AND APPLIED SCIENCES. Institute for Scientific Information, 3501 Market Street, Philadelphia, PA 19104. (800) 523-1850 or (215) 386-0100. Weekly. $275.00 per year.

ENGINEERING INDEX MONTHLY AND AUTHOR INDEX. Engineering Information Inc., 345 East 47th Street, New York, NY 10017. (212) 705-7600. Monthly. $1560.00 per year.

ISMEC BULLETIN (Information Service in Mechanical Engineering). Cambridge Scientific Abstracts, 5161 River Road, Bethesda, MD 20816. (301) 951-1400. 1973 to present. Monthly. $450.00 per year.

SCIENCE CITATION INDEX. Institute for Scientific Information, 3501 Market Street, Philadelphia, PA 19104. (800) 523-1850 or (215) 386-0100. Six times per year. $6200.00 per year.

ASSOCIATIONS AND PROFESSIONAL SOCIETIES

AMERICAN SOCIETY OF MECHANICAL ENGINEERS. 345 47th Street, New York, NY 10017. (212) 705-7722.

ASSOCIATION OF COMPUTING MACHINERY (ACM). 11 West 42nd Street, New York, NY 10036. (212) 896-7440.

INSTITUTE OF ELECTRICAL AND ELECTRONIC ENGINEERS (IEEE). 345 East 47th Street, New York, NY 10017. (212) 705-7900.

NATIONAL CONFERENCE ON FLUID POWER. IIT Research Institute, Ten West 35th Street, Chicago, IL 60616. (312) 567-4414.

VIBRATION INSTITUTE. 101 West 55th Street, Suite 206, Clarendon Hills, IL 60514. (312) 654-2254.

BIBLIOGRAPHIES

SCIENTIFIC AND TECHNICAL BOOKS AND SERIALS IN PRINT 1988; AN INDEX TO LITERATURE IN SCIENCE AND TECHNOLOGY. R.R. Bowker Company, 205 East 42nd Street, New York, NY 10017. (800) 521-8110. $175.00.

DIRECTORIES AND BIOGRAPHICAL SOURCES

COMPUTER-AIDED DESIGN (CAD) DIRECTORY. Technical Database Corporation, 1300 South Frazier, Conroe, TX 77305. (409) 539-9688.

INDUSTRIAL EQUIPMENT AND SUPPLIES DIRECTORY. American Business Directories, Inc., 5707 South 86th Circle, Omaha, NE 68127. (402) 331-7169.

MACHINERY BUYERS GUIDE. Findlay Publications Limited, Maitland House, Warrior Square, Southend-on-Sea, Essex SS5 5AR England. Annual. $45.00.

1987 DIRECTORY OF ENGINEERING SOCIETIES AND RELATED ORGANIZATIONS. Gordon Davis, editor. Hemisphere Publishing Corporation, 1010 Vermont Avenue, NW, Washington, DC 20005. (800) 526-0275. 12th edition. 1987. $100.00.

RESEARCH CENTERS DIRECTORY 1988. Gale Research Company, Book Tower, Detroit, MI 48226. (800) 521-0707. 12th edition. 1987. $365.00 for set.

SCIENTIFIC AND TECHNICAL ORGANIZATIONS AND AGENCIES DIRECTORY. Margaret Labash Young, editor. Gale Research Company, Book Tower, Detroit, MI 48226. (800) 521-0707. 2nd edition. 1987. $185.00.

WHO'S WHO IN ENGINEERING. Gordon Davis, editor. Hemisphere Publishing Corporation, 1010 Vermont Avenue, NW, Washington, DC 20005. (800) 526-0275. 6th edition. 1985. $200.00.

ENCYCLOPEDIAS AND DICTIONARIES

THESAURUS OF SCIENTIFIC, TECHNICAL, AND ENGINEERING TERMS. Hemisphere Publishing Corporation, 1010 Vermont Avenue, NW, Washington, DC 20005. (800) 526-0275. 1988. $125.00.

GENERAL WORKS

COMPUTER-AIDED DESIGN AND MANUFACTURE. C.B. Besant and C.W.K. Lui. John Wiley and Sons, Inc., 605 Third Avenue, New York, NY 10158. (800) 526-5368. 3rd edition. 1986. $35.95 in paper.

DESIGN OF AUTOMATIC MACHINERY. Kendrick W. Lentz. Van Nostrand Reinhold Company, Inc., 135 West 50th Street, New York, NY 10020. (800) 543-2681. 1984. $42.95.

ELECTRIC MACHINERY. Peter F. Ryff. Prentice-Hall Publishing, Inc., Englewood Cliffs, NJ 07632. (800) 562-0245. 1988. $34.00.

ENGINEERING MECHANICS. J.L. Meriam and L.G. Kraige. John Wiley and Sons, Inc., 605 Third Avenue, New York, NY 10158. (800) 526-5368. 2nd edition. 1986. $52.95.

INTRODUCTION TO DYNAMICS AND CONTROL. Leonard Meirovitch. John Wiley and Sons, Inc., 605 Third Avenue, New

York, NY 10158. (800) 526-5368. 1985. $42.95.

MECHANISMS AND DYNAMICS OF MACHINERY. Hamilton H. Mabie and Charles F. Reinholtz. John Wiley and Sons, Inc., 605 Third Avenue, New York, NY 10158. (800) 526-5368. 4th edition. 1987. $39.95.

HANDBOOKS AND MANUALS

APPLIED FLUID DYNAMICS HANDBOOK. Robert D. Blevins. Van Nostrand Reinhold Company, Inc., 135 West 50th Street, New York, NY 10020. (800) 543-2681. 1984. $52.95.

HANDBOOK OF MECHANICS, MATERIALS, AND STRUCTURES. A. Blake. John Wiley and Sons, Inc., 605 Third Avenue, New York, NY 10158. (800) 526-5368. 1985. $65.95.

MARK'S STANDARD HANDBOOK FOR MECHANICAL ENGINEERS. T. Baumeister, editor. McGraw-Hill Book Company, 1221 Avenue of the Americas, New York, NY 10020. (212) 512-2000. 8th edition. 1978. $96.00.

MECHANICAL DESIGN AND SYSTEMS HANDBOOK. Harold A. Rothbart, editor. McGraw-Hill Book Company, 1221 Avenue of the Americas, New York, NY 10020. (212) 512-2000. 2nd edition.1985. $96.50.

MECHANICAL ENGINEERS' HANDBOOK. Myer Kutz, editor. John Wiley and Sons, Inc., 605 Third Avenue, New York, NY 10158. (800) 526-5368. 1986. $79.95.

PRESSURE VESSEL DESIGN HANDBOOK. Henry H. Bednar. Van Nostrand Reinhold Company, Inc., 135 West 50th Street, New York, NY 10020. (800) 543-2681. 2nd edition. 1986. $49.95.

STANDARD HANDBOOK OF MACHINE DESIGN. Joseph E. Shigley and Charles R. Mischke. McGraw-Hill Book Company, 1221 Avenue of the Americas, New York, NY 10020. (212) 512-2000. 1986. $89.00.

ONLINE DATA BASES

CA SEARCH. Chemical Abstracts Service, P.O. Box 3012, Columbus, OH 43120. (800) 848-6538 or (614) 421-3600. Comprehensive guide to chemical literature, 1972 to present. Inquire as to online cost and availability.

COMPENDEX. Engineering Information, Inc., 345 East 47th Street, New York, NY 10017. (800) 221-1044 or (212) 705-7615. Engineering and technical literature, 1975 to present. Inquire as to online cost and availability.

DISSERTATION ABSTRACTS ONLINE. University Microfilms International, 300 North Zeeb Road, Ann Arbor, MI 48106. (800) 521-0600 or (313) 761-4700. Scope includes virtually all doctoral dissertations accepted at accredited American institutions from 1861 to present in over 250 subject areas. Inquire as to online cost and availability.

INSPEC. INSPEC Marketing Department, Institution of Electrical Engineers. Available from IEEE Service Center, 445 Hoes Lane, Piscataway, NJ 08854. (201) 981-0060. Online version of Physics Abstracts. Inquire as to online cost and availability.

ISMEC. Cambridge Scientific Abstracts, 5161 River Road, Besthda, MD 20816. (800) 638-8076 or (301) 951-1400. Literature of mechanical and production engineering, 1973 to present. Inquire as to online cost and availability.

NTIS. National Technical Information Service, 5285 Port Royal Road, Springfield, VA 22161. (703) 487-4630. Broad coverage of government sponsored research reports, 1964 to present. Inquire as to online cost and availability.

SCISEARCH. Institute for Scientific Information, 3501 Market Street, Philadelphia, PA 19104. (800) 523-1850 or (215) 386-0100. Broad multidisciplinary title and author index to the international literature of science and technology, 1974 to present. Inquire as to online cost and availability.

WILSONLINE. H.W. Wilson and Company, 950 University Avenue, Bronx, NY 10452. (800) 367-6770 or (212) 588-8400. Makes available online versions of the H.W. Wilson indexes including Applied Science and Technology Index, Business Periodicals Index and Readers' Guide to Periodical Literature. Approximately 1980 to present. Inquire as to online cost and availability.

PERIODICALS

CIME (COMPUTERS IN MECHANICAL ENGINEERING). American Society of Mechanical Engineers, 345 East 47th Street, New York, NY 10017. (212) 705-7703.

INTERNATIONAL COMMUNICATIONS IN HEAT AND MASS TRANSFER. Pergamon Press, Inc., Maxwell House, Fairview Park, Elmsford, NY 10523. (914) 592-7700. 1974 to present. Bimonthly. $160.00 per year.

INTERNATIONAL JOURNAL FOR NUMERICAL METHODS IN ENGINEERING. John Wiley and Sons, Inc., 605 Third Avenue, New York, NY 10158. (800) 526-5368.

INTERNATIONAL JOURNAL OF HEAT AND MASS TRANSFER. Pergamon Press, Inc., Maxwell House, Fairview Park, Elmsford, NY 10523. (914) 592-7700. 1960 to present. Monthly. $500.00 per year.

INTERNATIONAL JOURNAL OF MULTIPHASE FLOW. Pergamon Press, Inc., Maxwell House, Fairview Park, Elmsford, NY 10523. (914) 592-7700. (212) 705-7703. 1974 to present. Bimonthly. $250.00 per year.

JOURNAL OF APPLIED MECHANICS. American Society of Mechanical Engineers, 345 East 47th Street, New York, NY 10017. (212) 705-7703. 1935 to present. Quarterly. $100.00 per year.

JOURNAL OF ENGINEERING FOR INDUSTRY. American Society of Mechanical Engineers, 345 East 47th Street, New York, NY 10017. (212) 705-7703. 1970 to present. Quarterly. $100.00 per year.

JOURNAL OF FLUID CONTROL. Delbridge Publishing Company, Box 2989, Stanford, CA 94305. (408) 446-3131. 1967 to present. Quarterly. Inquire as to cost and availability.

JOURNAL OF HEAT TRANSFER. American Society of Mechanical Engineers, 345 East 47th Street, New York, NY 10017. (212) 705-7703. 1970 to present. Quarterly. $100.00 per year.

JOURNAL OF MECHANISMS, TRANSMISSIONS AND AUTOMATION IN DESIGN. American Society of Mechanical Engineers, 345 East 47th Street, New York, NY 10017. (212) 705-7703. 1983 to present. Quarterly. $80.00 per year.

JOURNAL OF PRESSURE VESSEL TECHNOLOGY. American Society of Mechanical Engineers, 345 East 47th Street, New York, NY 10017. (212) 705-7703. 1974 to present. Quarterly. $80.00 per year.

LUBRICATION ENGINEERING. American Society of Lubrication Engineers, 838 Busse Highway, Park Ridge, IL 60068. 1945 to present. Monthly. $39.00 per year.

MACHINE DESIGN. Penton-IPC, 1100 Superior Avenue, Cleveland, OH 44114. (216) 696-7000. 1929 to present. Twenty-eight times per year. $60.00 per year.

MECHANICAL ENGINEERING. American Society of Mechanical Engineers, 345 East 47th Street, New York, NY 10017. (212) 705-7722.

RESEARCH CENTERS AND INSTITUTES

LABORATORY FOR EXIPERMENTAL MECHANICS RESEARCH. State University of New York at Stony Brook, Stony Brook, NY 11794-2300. (516) 632-8311.

MACHINE SYSTEMS GROUP. Stevens Institute of Technology, Department of Mechanical Engineering, Castle Point Station, Hoboken, NJ 07030. (201) 420-5591.

MECHANICAL ENGINEERING DESIGN LABORATORY. University of Florida, 237 Mechanical Engineering Building, Gainesville, FL 32611. (904) 392-0827

MECHANICAL ENGINEERING RESEARCH LABORATORIES. Kansas State University, Durland Hall, Manhattan, KS 66506. (913) 532-5610.

MACHINING

See also: CAD (COMPUTER-AIDED DESIGN), MACHINE DESIGN, MACHINERY, MECHANICAL ENGINEERING

ABSTRACT SERVICES AND INDEXES

APPLIED MECHANICS REVIEW. American Society of Mechanical Engineers, 345 East 47th Street, New York, NY 10017. (212) 705-7703. 1948 to present. Monthly. $360.00 per year.

APPLIED SCIENCE AND TECHNOLOGY INDEX. H.W. Wilson and Company, 950 University Avenue, Bronx, NY 10452. (800) 367-6670 or (212) 588-8400. Monthly. Inquire as to cost and availability.

CAD/CAM ABSTRACTS. EIC/Intelligence, Inc., 48 West 38th Street, New York, NY 10018. (800) 223-6275 or (212) 944-8500. Monthly. $365.00 per year.

CHEMICAL ABSTRACTS. American Chemical Society, Chemical Abstracts Service, Box 3012, Columbus, OH 43210. (614) 421-3600. 1907 to present. Weekly. $9500.00 per year.

CURRENT CONTENTS: ENGINEERING, TECHNOLOGY AND APPLIED SCIENCES. Institute for Scientific Information, 3501 Market Street, Philadelphia, PA 19104. (800) 523-1850 or (215) 386-0100. Weekly. $275.00 per year.

ENGINEERING INDEX MONTHLY AND AUTHOR INDEX. Engineering Information Inc., 345 East 47th Street, New York, NY 10017. (212) 705-7600. Monthly. $1560.00 per year.

ISMEC BULLETIN (Information Service in Mechanical Engineering). Cambridge Scientific Abstracts, 5161 River Road, Bethesda, MD 20816. (301) 951-1400. 1973 to present. Monthly. $450.00 per year.

ASSOCIATIONS AND PROFESSIONAL SOCIETIES

AMERICAN SOCIETY OF MECHANICAL ENGINEERS. 345 47th Street, New York, NY 10017. (212) 705-7722.

AMERICAN SOCIETY FOR METALS. Metals Park, OH 44073. (216) 338-5151.

BIBLIOGRAPHIES

SCIENTIFIC AND TECHNICAL BOOKS AND SERIALS IN PRINT 1988; AN INDEX TO LITERATURE IN SCIENCE AND TECHNOLOGY. R.R. Bowker Company, 205 East 42nd Street, New York, NY 10017. (800) 521-8110. $175.00.

DIRECTORIES AND BIOGRAPHICAL SOURCES

DIRECTORY OF METALWORKING MACHINERY. U.S. Government Printing Office, Division of Public Documents, Washington, DC 20402. (202) 275-2035. $7.50.

INDUSTRIAL EQUIPMENT AND SUPPLIES DIRECTORY. American Business Directories, Inc., 5707 South 86th Circle, Omaha, NE 68127. (402) 331-7169.

MACHINERY BUYERS GUIDE. Findlay Publications Limited, Maitland House, Warrior Square, Southend-on-Sea, Essex SS5 5AR England. Annual. $45.00.

WHO'S WHO IN ENGINEERING. Gordon Davis, editor. Hemisphere Publishing Corporation, 1010 Vermont Avenue, NW, Washington, DC 20005. (800) 526-0275. 6th edition. 1985. $200.00.

GENERAL WORKS

MACHINING SOURCE BOOK. Mel M. Schwartz, editor. American Society for Metals, Metals Park, OH 44073. (216) 338-5151. 1988. $55.00.

METALS HANDBOOK. Volume 3: Machining. American Society for Metals, Metals Park, OH 44073. (216) 338-5151. 8th edition. $95.00.

HANDBOOKS AND MANUALS

HANDBOOK OF MECHANICS, MATERIALS, AND STRUCTURES. A. Blake. John Wiley and Sons, Inc., 605 Third Avenue, New York, NY 10158. (800) 526-5368. 1985. $65.95.

MACHINERY'S HANDBOOK. Erik Oberg. Industrial Press, Inc., P.O. Box C-772, Brooklyn, NY 15206. 22nd edition. 1984. $48.00.

MACHINING DATA HANDBOOK. Distributed by: American Society for Metals, Metals Park, OH 44073. (216) 338-5151. 1980. $160.00.

MARK'S STANDARD HANDBOOK FOR MECHANICAL ENGINEERS. T. Baumeister, editor. McGraw-Hill Book Company, 1221 Avenue of the Americas, New York, NY 10020. (212) 512-2000. 8th edition. 1978. $96.00.

MECHANICAL DESIGN AND SYSTEMS HANDBOOK. Harold A. Rothbart, editor. McGraw-Hill Book Company, 1221 Avenue of the Americas, New York, NY 10020. (212) 512-2000. 2nd edition.1985. $96.50.

MECHANICAL ENGINEERS' HANDBOOK. Myer Kutz, editor. John Wiley and Sons, Inc., 605 Third Avenue, New York, NY 10158. (800) 526-5368. 1986. $79.95.

STANDARD HANDBOOK OF MACHINE DESIGN. Joseph E. Shigley and Charles R. Mischke. McGraw-Hill Book Company, 1221 Avenue of the Americas, New York, NY 10020. (212) 512-2000. 1986. $89.00.

ONLINE DATA BASES

CA SEARCH. Chemical Abstracts Service, P.O. Box 3012, Columbus, OH 43120. (800) 848-6538 or (614) 421-3600. Comprehensive guide to chemical literature, 1972 to present. Inquire as to online cost and availability.

COMPENDEX. Engineering Information, Inc., 345 East 47th Street, New York, NY 10017. (800) 221-1044 or (212) 705-7615. Engineering and technical literature, 1975 to present. Inquire as to online cost and availability.

ISMEC. Cambridge Scientific Abstracts, 5161 River Road, Besthda, MD 20816. (800) 638-8076 or (301) 951-1400. Literature of mechanical and production engineering, 1973 to present. Inquire as to online cost and availability.

NTIS. National Technical Information Service, 5285 Port Royal Road, Springfield, VA 22161. (703) 487-4630. Broad coverage of government sponsored research reports, 1964 to present. Inquire as to online cost and availability.

WILSONLINE. H.W. Wilson and Company, 950 University Avenue, Bronx, NY 10452. (800) 367-6770 or (212) 588-8400. Makes available online versions of the H.W. Wilson indexes including Applied Science and Technology Index, Business Periodicals Index and Readers' Guide to Periodical Literature. Approximately 1980 to present. Inquire as to online cost and availability.

PERIODICALS

AUTOMATIC MACHINING. Screw Machine Publishing Company, Inc., 100 Seneca Avenue, Rochester, NY 14621. (716) 338-1522. 1939 to present. Monthly. $15.00.

CIME (COMPUTERS IN MECHANICAL ENGINEERING). American Society of Mechanical Engineers, 345 East 47th Street, New York, NY 10017. (212) 705-7703.

JOURNAL OF APPLIED MECHANICS. American Society of Mechanical Engineers, 345 East 47th Street, New York, NY 10017. (212) 705-7703. 1935 to present. Quarterly. $100.00 per year.

JOURNAL OF APPLIED METALWORKING. American Society for Metals, Metals Park, OH 44073. (216) 338-5151. 1979 to present. Semiannual. $87.00.

JOURNAL OF ENGINEERING FOR INDUSTRY. American Society of Mechanical Engineers, 345 East 47th Street, New York, NY 10017. (212) 705-7703. 1970 to present. Quarterly. $100.00 per year.

MACHINE DESIGN. Penton-IPC, 1100 Superior Avenue, Cleveland, OH 44114. (216) 696-7000. 1929 to present. Twenty-eight times per year. $60.00 per year.

MECHANICAL ENGINEERING. American Society of Mechanical Engineers, 345 East 47th Street, New York, NY 10017. (212) 705-7722.

METAL WORKING PRODUCTION. Action Communications, Inc., 7507 Kennedy Road, Box 406, Milliken, ON LOH 1K0, Canada. (416) 297-3222. 1974 to present. Bimonthly. $20.00 per year.

METIFAX MAGAZINE. Huebner Publications, Inc., 6521 Davis Industrial Parkway, Solon, OH 44139. (216) 248-1125. 1956 to present. Monthly. $40.00.

RESEARCH CENTERS AND INSTITUTES

MACHINE SYSTEMS GROUP. Stevens Institute of Technology, Department of Mechanical Engineering, Castle Point Station, Hoboken, NJ 07030. (201) 420-5591.

MECHANICAL ENGINEERING DESIGN LABORATORY. University of Florida, 237 Mechanical Engineering Building, Gainesville, FL 32611. (904) 392-0827

MECHANICAL ENGINEERING RESEARCH LABORATORIES. Kansas State University, Durland Hall, Manhattan, KS 66506. (913) 532-5610.

MAGNESIUM

See also: ALLOYS, MATERIALS SCIENCE

ABSTRACT SERVICES AND INDEXES

ALLOYS INDEX. American Society for Metals, 9639 Kinsman Road, Metals Park, OH 44073. (216) 338-5151. $130.00 per year.

APPLIED SCIENCE AND TECHNOLOGY INDEX. H.W. Wilson Company, 950 University Avenue, Bronx, NY 10452. (800) 367-6670 or (212) 588-8400. Inquire as to cost and availability.

CHEMICAL ABSTRACTS. Chemical Abstracts Service, 2540 Olentangy Road, Post Office Box 3012, Columbus, OH 43210. (800) 848-6538 or (614) 421-3600. Weekly. $9200.00 per year.

CURRENT CONTENTS: ENGINEERING, TECHNOLOGY, AND APPLIED SCIENCE. Institute for Scientific Information, 3501 Market Street, Philadelphia, PA 19104. (800) 523-1850 or (215) 386-0100. $272.00 per year.

ENGINEERING INDEX MONTHLY. Engineering Information, Incorporated, 345 East 47th Street, New York, NY 10017. (800) 221-1044 or (212) 705-7600. Monthly, with annual cumulation. $1560.00 per year.

METALS ABSTRACTS. American Society for Metals, 9639 Kinsman Road, Metals Park, OH 44073. (216) 338-5151. Monthly. $890.00.

METALS ABSTRACTS INDEX. American Society for Metals, 9639 Kinsman Road, Metals Park, OH 44073. (216) 338-5151. Monthly. (Sold only to subscribers of Metals Abstracts).

WORLD ALUMINUM ABSTRACTS. Aluminum Association, 818 Connecticut Avenue, NW, Washington, DC 20006. (202) 862-5100. 1968 to present. Monthly. $165.00 per year.

ANNUAL REVIEWS AND YEARBOOKS

ANNUAL REVIEW OF MATERIALS SCIENCE. Annual Reviews, Incorporated, 4139 El Camino Way, Palo Alto, CA 94306. (415) 493-4400.

ASSOCIATIONS AND PROFESSIONAL SOCIETIES

AMERICAN SOCIETY FOR METALS. Metals Park, OH 44073. (216) 338-5151.

INTERNATIONAL MAGNESIUM ASSOCIATION. Lancaster Building, Suite 400, 7927 Jones Branch Drive, McLean, VA 22102. (703) 442-8888.

METALLURGICAL SOCIETY OF THE AIME (AMERICAN INSTITUTE OF MINING, METALLURGICAL AND PETROLEUM ENGINEERS). 420 Commonwealth Drive, Warrendale, PA 15086. (412) 776-9080.

BIBLIOGRAPHIES

SUPERALLOYS: SOURCE BOOK. M.J. Donachie, Jr., editor. American Society for Metals, Metals Park, OH 44073. (216) 5151. 1983. $60.00.

DIRECTORIES AND BIOGRAPHICAL SOURCES

DUN'S INDUSTRIAL GUIDE - THE METAL WORKING DIRECTORY. Dun and Bradstreet, Incorporated, Three Century Drive, Parsippany, NJ 07054. (201) 455-0900. Annual. $550.00.

INTERNATIONAL MAGNESIUM ASSOCIATION - BUYER'S GUIDE. International Magnesium Association, Lancaster Building, Suite 400, 7927 Jones Branch Drive, McLean, VA

22102. (703) 442-8888. 1984. Free.

INTERNATIONAL RESEARCH CENTERS DIRECTORY 1986-1987. Gale Research Company, Book Tower, Detroit, MI 48226. (800) 521-0707. Third edition. 1986.

METAL PRODUCTS DIRECTORY. American Business Directories, Incorporated, Division of American Business Lists, Incorporated, 5707 South 86th Circle, Omaha, NE 68127. (402) 331-7169. 1986. $80.00.

RESEARCH CENTERS DIRECTORY. Gale Research Company, Book Tower, Detroit, MI 48226. (800) 521-0707. Eleventh edition. 1987.

SCIENTIFIC AND TECHNICAL ORGANIZATIONS AND AGENCIES DIRECTORY. Gale Research Company, Book Tower, Detroit, MI 48226. (800) 521-0707. 1985.

WHO'S WHO IN ENGINEERING. Engineers Joint Council, 345 East 47th Street, New York, NY 10017. (212) 705-7010. 1985. $200.00.

WHO'S WHO IN FRONTIER SCIENCE AND TECHNOLOGY. Marquis Who's Who, Incorporated, 200 East Ohio Street, Chicago, IL 60611. (800) 428-3898 or (312) 787-2008.

WHO'S WHO IN TECHNOLOGY TODAY. Reston Publishing Company, Incorporated, c/o Prentice-Hall, Incorporated, Englewood Cliffs, NJ 07632. (800) 262-6868. Biennial. Five volumes. $425.00. Covers the fields of electronics, computer science, physics, optics, chemistry, biotechnology, mechanics, energy, and earth science.

ENCYCLOPEDIAS AND DICTIONARIES

ENCYCLOPEDIA OF PHYSICAL SCIENCE AND TECHNOLOGY. Academic Press, Incorporated, Orlando, FL 32887. (800) 321-5068 or (305) 345-2734. Fifteen volumes, 1986.

MCGRAW-HILL DICTIONARY OF ENGINEERING. Sybil P. Parker, editor. McGraw-Hill Book Company, 1221 Avenue of the Americas, New York, NY 10020. (212) 512-2000. 1984. $39.95.

MCGRAW-HILL ENCYCLOPEDIA OF SCIENCE AND TECHNOLOGY. McGraw-Hill Book Company, Incorporated, 1221 Avenue of the Americas, New York, NY 10020. (212) 512-2000.

GENERAL WORKS

ESSENTIAL METALLURGY FOR ENGINEERS. W. Alexander, G. Davies, and K. Reynolds. Van Nostrand Reinhold, 115 Fifth Avenue, New York, NY 10003. (800) 543-2681. 1985. $17.95.

LIGHT ALLOYS METALLURGY OF THE LIGHT METALS. I.J. Polmear. State Mutual Books and Periodical Service, Limited, 521 Fifth Avenue, New York, NY 10017. (212) 682-5844. 1981. $45.00.

MAGNESIUM CASTING TECHNOLOGY. Arthur W. Brace and F.A. Allen. Books on Demand, 300 North Zeeb Road, Ann Arbor, MI 48106. (313) 761-4700. Reprint of 1958 edition. $43.50 in paper.

METALLURGY BASICS. D.V. Brown. Van Nostrand Reinhold, 115 Fifth Avenue, New York, NY 10003. (800) 543-2681. 1985. $17.95.

PHYSICAL METALLURGY. P. Haasen. Cambridge University Press, 32 East 57th Street, New York, NY 10022. (212) 688-8885. 1986. $24.95 in paper.

STRUCTURE AND PROPERTIES OF ENGINEERING ALLOYS. W.F. Smith. McGraw-Hill Book Company, Incorporated, 1221 Avenue of the Americas, New York, NY 10020. (212) 512-2000. 1980. $46.95.

SUPERALLOYS. C.T. Sims and W.C. Hagel. John Wiley and Sons, Incorporated, 605 Third Avenue, New York, NY 10158. (212) 850-6000. 1972. $74.95.

THEORY OF STRUCTURAL TRANSFORMATIONS IN SOLIDS. A.G. Khachaturyan. John Wiley and Sons, Incorporated, 605 Third Avenue, New York, NY 10158. (800) 526-5368 or (212) 850-6000. 1983. $69.95.

HANDBOOKS AND MANUALS

CRC HANDBOOK OF CHEMISTRY AND PHYSICS. CRC Press, Incorporated, 2000 Corporate Boulevard, Boca Raton, Florida 33341. (305) 994-0555. Sixth-seventh edition. 1986. $69.95.

SMITHELL'S METALS REFERENCE BOOK. Eric A. Brandes, editor. Butterworth Publishers, 80 Montvale Avenue, Stoneham, MA 02180. (800) 325-4177. Sixth edition. 1983. $210.00.

ONLINE DATA BASES

COMPENDEX. Engineering Information, Incorporated, 345 East 47th Street, New York, NY 10017. (800) 221-1044 or (212) 705-7615. Engineering and technical literature, 1975 to present. Inquire as to cost and availability.

INSPEC. INSPEC Marketing Department, Institute of Electrical and Electronics Engineers, Incorporated, IEEE Service Department, 445 Hoes Lane, Piscataway, NJ 08854. (201) 981-0060. Inquire as to on-line cost and availability.

METADEX. Metals Information, American Society for Metals, Metals Park, OH 44073. (216) 338-5151. (Metals Abstracts/Alloys Index). A worldwide literature on the science and practice of metallurgy, 1966 to present. Inquire as to online cost and availability.

NASA. National Aeronautics and Space Administration, Scientific and Technical Information Branch, 300 7th Street, SW, Washington, DC 20546. Citations and abstracts of aerospace literature, 1962 to present. Inquire as to cost and availability.

NON-FERROUS METALS ABSTRACTS. British Non-Ferrous Metals Technology Centre, Grove Laboratories, Denchworth Road, Wantage, Oxfordshire, England OX12 9 BJ. Citations and abstracts on non-ferrous metallurgy and technology, 1961 to present. Inquire as to online cost and availability.

NTIS. National Technical Information Service, 5285 Port Royal Road, Springfield, VA 22161. (703) 487-4630. Broad coverage of government sponsored research reports, 1964 to present. Inquire as to cost and availability.

WILSONLINE. H.W. Wilson Company, 950 University Avenue, Bronx, NY 10452. (800) 367-6770 or (212) 588-8400. Makes available online versions of the printed H.W. Wilson Indexes including Applied Science and Technology Index, Business Periodicals Index, and Reader's Guide to Periodical Literature. Period covered is generally 1983 to present. Inquire as to cost and availability.

OTHER SOURCES

INTERDOC: DIRECTORY OF PUBLISHED PROCEEDINGS, SERIES. SEMT-Science/Engineering/Medicine/Technology. Interdoc Corporation, 173 Halstead Avenue, Box 326, Harrison, NY 10528. (014) 835-3506. Ten times per year. $325.00 per year.

MATERIALS SCIENCE AND METALLURGY. National Technical Information Service (NTIS), 5285 Port Royal Road, Springfield, VA 22161. (703) 487-4630. Translations and abstracts of foreign language technical media. Irregular. $40.00 per year.

MAGNESIUM

PERIODICALS

ALLOY DIGEST. Engineering Publications Incorporated, Box 823, Upper Montclair, NJ 07043. (201) 746-7930. Monthly. $50.00 per year.

BULLETIN OF ALLOY PHASE DIAGRAMS. American Society for Metals, Metals Park, OH 44073. (216) 338-5151. Bimonthly. $90.00 per year.

JOURNAL OF METALS. Metallurgical Society of the AIME (American Institute of Mining, Metallurgical and Petroleum Engineers), 420 Commonwealth Drive, Warrendale, PA 15086. (412) 776-9080. Monthly. $40.00 per year.

LIGHT METAL AGE. Fellom Publishing Company, 693 Mission Street, San Francisco, CA 94105. (415) 781-1431. Bimonthly. $20.00 per year.

MAGNESIUM. International Magnesium Association, Lancaster Building, Suite 400, 7927 Jones Branch Drive, McLean, VA 22102. (703) 442-8888. Monthly. Inquire as to cost and availability.

MAGNESIUM MONTHLY REVIEW. 106 Spring Forest Road, Greenville, SC 29615-2241. (803) 244-5718. Monthly. $45.00 per year.

METALLURGICAL TRANSACTIONS. Metallurgical Society of the AIME (American Institute of Mining, Metallurgical and Petroleum Engineers), 420 Commonwealth Drive, Warrendale, PA 15086. (412) 776-9080. Monthly. $95.00 per year.

METALS WEEK. McGraw-Hill Book Company, Incorporated, 1221 Avenue of the Americas, New York, NY 10020. (212) 997-2823. Weekly. $527.00 per year.

RESEARCH CENTERS AND INSTITUTES

GANNON UNIVERSITY. Engineering Research Institute, University Square, Erie, PA 16541. (814) 871-7619.

STATISTICS SOURCES

MAGNESIUM MILL PRODUCTS. United States Bureau of the Census, United States Department of Commerce, Washington, DC 20233. (301) 763-7800. Statistical data on production, consumption, inventories and orders. Annual. $1.50.

MAGNETIC RESONANCE

See: NUCLEAR MAGNET RESONANCE

MAGNETISM

See: ELECTROMAGNETISM

MAGNETOHYDRODYNAMIC GENERATOR

See: MAGNETOHYDRODYNAMICS

MAGNETOHYDRODYNAMICS

See also: ELECTROMAGNETISM, PHYSICS, PLASMA PHYSICS

ABSTRACT SERVICES AND INDEXES

APPLIED MECHANICS REVIEW. American Society of Mechanical Engineers, 345 East 47th Street, New York, NY 10017. (212) 705-7703. 1948 to present. Monthly. $360.00 per year.

APPLIED SCIENCE AND TECHNOLOGY INDEX. H.W. Wilson and Company, 950 University Avenue, Bronx, NY 10452. (800) 367-6670 or (212) 588-8400. Monthly. Inquire as to cost and availability.

CHEMICAL ABSTRACTS. American Chemical Society, Chemical Abstracts Service, Box 3012, Columbus, OH 43210. (614) 421-3600. 1907 to present. Weekly. $9500.00 per year.

CONFERENCE PAPERS INDEX. Cambridge Scientific Abstracts, 5161 River Road, Bethesda, MD 20816. 1972 to present. Monthly. Inquire as to cost and availability.

CURRENT CONTENTS: PHYSICAL, CHEMICAL AND EARTH SCIENCES. Institute for Scientific Information, 3501 Market Street, Philadelphia, PA 19104. (800) 523-1850 or (215) 386-0100. Weekly. $275.00 per year.

ENGINEERING INDEX MONTHLY AND AUTHOR INDEX. Engineering Information Inc., 345 East 47th Street, New York, NY 10017. (212) 705-7600. Monthly. $1560.00 per year.

INDEX TO SCIENTIFIC AND TECHNICAL PROCEEDINGS. Institute for Scientific Information, 3501 Market Street, Philadelphia, PA 19104. (800) 523-1850 or (215) 386-0100. 1978 to present. Monthly. $775.00 per year.

ISMEC BULLETIN (Information Service in Mechanical Engineering). Cambridge Scientific Abstracts, 5161 River Road, Bethesda, MD 20816. (301) 951-1400. 1973 to present. Monthly. $450.00 per year.

MATHEMATICAL REVIEWS: A REVIEWING JOURNAL COVERING THE WORLD LITERATURE OF MATHEMATICAL RESEARCH. American Mathematical Society, P.O. Box 6248, Providence, RI 02940. (800) 7774 or (401) 272-9500. 1940 to present. Monthly. $2800.00 per year.

PHYSICS ABSTRACTS. Institution of Electrical Engineers. Available from: IEEE Service Center, 445 Hoes Lane, Piscataway, NJ 08854. 1898 to present. Bimonthly. $1700.00 per year.

SCIENCE CITATION INDEX. Institute for Scientific Information, 3501 Market Street, Philadelphia, PA 19104. (800) 523-1850 or (215) 386-0100. Six times per year. $6200.00 per year.

ASSOCIATIONS AND PROFESSIONAL SOCIETIES

AMERICAN INSTITUTE OF PHYSICS. 335 East 45th Street, New York, NY 10017. (212) 661-9404.

AMERICAN PHYSICAL SOCIETY. 335 East 45th Street, New York, NY 10017. (212) 682-7341.

IEEE (INSTITUTE OF ELECTRICAL AND ELECTRONICS ENGINEERS). 345 East 47th Street, New York, NY 10017. (212) 705-7900.

DIRECTORIES AND BIOGRAPHICAL SOURCES

INTERNATIONAL RESEARCH CENTERS DIRECTORY 1988-89. Darren L. Smith, editor. Gale Research Company, Book Tower, Detroit, MI 48226. (800) 521-0707. 4th edition. 1987. $360.00.

1987 DIRECTORY OF ENGINEERING SOCIETIES AND RELATED ORGANIZATIONS. Gordon Davis, editor. Hemisphere

Publishing Corporation, 1010 Vermont Avenue, NW, Washington, DC 20005. (800) 526-0275. 12th edition. 1987. $100.00.

RESEARCH CENTERS DIRECTORY 1988. Gale Research Company, Book Tower, Detroit, MI 48226. (800) 521-0707. 12th edition. 1987. $365.00 for set.

SCIENTIFIC AND TECHNICAL ORGANIZATIONS AND AGENCIES DIRECTORY. Margaret Labash Young, editor. Gale Research Company, Book Tower, Detroit, MI 48226. (800) 521-0707. 2nd edition. 1987. $185.00.

WHO'S WHO IN ENGINEERING. Gordon Davis, editor. Hemisphere Publishing Corporation, 1010 Vermont Avenue, NW, Washington, DC 20005. (800) 526-0275. 6th edition. 1985. $200.00.

ENCYCLOPEDIAS AND DICTIONARIES

THESAURUS OF SCIENTIFIC, TECHNICAL, AND ENGINEERING TERMS. Hemisphere Publishing Corporation, 1010 Vermont Avenue, NW, Washington, DC 20005. (800) 526-0275. 1988. $125.00.

GENERAL WORKS

AN INTRODUCTION TO MAGNETOHYDRODYNAMICS. Paul H. Roberts. Books on Demand, 300 North Zeeb Road, Ann Arbor, MI 48106. (800) 521-0600. $68.50 in paper.

MAGNETOHYDRODYNAMIC ENERGY CONVERSION. Hemisphere Publishing Corporation, 1010 Vermont Avenue, NW, Washington, DC 20005. (800) 526-0275. 1987. $50.00.

MHD AND FUSION MAGNETS: FIELD AND FORCE DESIGN CONCEPTS. R.J. Thome and J.M. Tarrh. John Wiley and Sons, Inc., 605 Third Avenue, New York, NY 10158. (800) 526-5368. 1982. $48.50.

ONLINE DATA BASES

CA SEARCH. Chemical Abstracts Service, P.O. Box 3012, Columbus, OH 43120. (800) 848-6538 or (614) 421-3600. Comprehensive guide to chemical literature, 1972 to present. Inquire as to online cost and availability.

COMPENDEX. Engineering Information, Inc., 345 East 47th Street, New York, NY 10017. (800) 221-1044 or (212) 705-7615. Engineering and technical literature, 1975 to persent. Inquire as to online cost and availability.

DISSERTATION ABSTRACTS ONLINE. University Microfilms International, 300 North Zeeb Road, Ann Arbor, MI 48106. (800) 521-0600 or (313) 761-4700. Scope includes virtually all doctoral dissertations accepted at accredited American institutions from 1861 to present in over 250 subject areas. Inquire as to online cost and availability.

INSPEC. INSPEC Marketing Department, Institution of Electrical Engineers. Available from IEEE Service Center, 445 Hoes Lane, Piscataway, NJ 08854. (201) 981-0060. Online version of Physics Abstracts. Inquire as to online cost and availability.

MATHFILE. American Mathematical Society, P.O. Box 6248, Providence, RI 02940. (800) 556-7774 or (401) 272-9500. An online version of Mathematical Reviews. 1973 to present. Inquire as to online cost and availability.

NTIS. National Technical Information Service, 5285 Port Royal Road, Springfield, VA 22161. (703) 487-4630. Broad coverage of government sponsored research reports, 1964 to present. Inquire as to online cost and availability.

SCISEARCH. Institute for Scientific Information, 3501 Market Street, Philadelphia, PA 19104. (800) 523-1850 or (215) 386-0100. Broad multidisciplinary title and author index to the international literature of science and technology, 1974 to present. Inquire as to online cost and availability.

PERIODICALS

IEEE TRANSACTIONS ON ELECTROMAGNETIC COMPATIBILITY. Institute of Electrical and Electronic Engineers. Available from: IEEE Service Center, 445 Hoes Lane, Piscataway, NJ 08854. 1959 to present. Quarterly. $75.00 per year.

IEEE TRANSACTIONS ON MAGNETICS. Institute of Electrical and Electronics Engineers. IEEE Service Center, 445 Hoes Lane, Piscataway, NJ 08854. 1965 to present. Bimonthly. $136.00 per year.

JOURNAL OF MAGNETISM AND MAGNETIC MATERIALS. Elsevier Science Publishing Company, Inc., 52 Vanderbilt Avenue, New York, NY 10017. (212) 370-5520. 1976 to present. Thirty-times per year. $1000.00 per year.

MAGNETOHYDRODYNAMICS. English translation of Magnitnaya Gidrodinamika. Consultants Burear, 233 Spring Street, New York, NY 10013. (212) 620-8000. 1965 to present. Quarterly. $495.00 per year.

MAGNETOHYDRODYNAMICS; AN INTERNATIONAL QUARTERLY. Hemisphere Publishing Corporation, 1010 Vermont Avenue, NW, Washington, DC 20005. (800) 526-0275. 1987 to present. Quarterly. $175.00 per year.

RESEARCH CENTERS AND INSTITUTES

AEROSPACE SCIENCES LABORATORY. Purdue University, Hanger #3, Purdue Airport, West Lafayette, IN 47906. (317) 494-3340.

HIGH TEMPERATURE GASDYNAMICS LABORATORY. Stanford University, Mechanical Engineering Department, Stanford, CA 94305. (415) 723-1745.

MHD ENERGY CENTER. Mississippi State University, College of Engineering, P.O. Drawer DE, Mississippi State, MS 39762. (601) 325-2105.

MAINFRAME COMPUTERS

See: COMPUTERS

MANNED SPACE FLIGHT

See also: ROCKETS, SPACECRAFT

ABSTRACT SERVICES AND INDEXES

APPLIED SCIENCE AND TECHNOLOGY INDEX. H.W. Wilson and Company, 950 University Avenue, Bronx, NY 10452. (800) 367-6670 or (212) 588-8400. Monthly. Inquire as to cost and availability.

ENGINEERING INDEX MONTHLY AND AUTHOR INDEX. Engineering Information Inc., 345 East 47th Street, New York, NY 10017. (212) 705-7600. Monthly. $1560.00 per year.

GENERAL SCIENCE INDEX. H.W. Wilson and Company, 950 University Avenue, Bronx, NY 10452. (800) 367-6670 or (212) 588-8400. 1978 to present. Monthly. Inquire as to cost and availability.

INTERNATIONAL AEROSPACE ABSTRACTS. American Institute of Aeronautics and Astronautics, Technical Information Service, 370 L'Enfant Promenade, S.W., Washington, DC 20024. (202) 646-7400. 1961 to present. Semi-monthly. $700.00 per year.

SCIENCE CITATION INDEX. Institute for Scientific Information, 3501 Market Street, Philadelphia, PA 19104. (800) 523-1850 or (215) 386-0100. Six times per year. $6200.00 per year.

STAR (SCIENTIFIC AND TECHNICAL AEROSPACE REPORTS). National Aeronautics and Space Administration, Scientific and Technical Information Facility, Box 8757, Baltimore-Washington International Airport, MD 21240. (202) 453-8545. 1963 to present. Semi-monthly. $85.00 per year with indexes.

ASSOCIATIONS AND PROFESSIONAL SOCIETIES

AMERICAN ASTRONAUTICAL SOCIETY. 6212-B Old Keene Mill Court, Springfield, VA 22152. (703) 866-0020.

AMERICAN INSTITUTE OF AERONAUTICS AND ASTRONAUTICS. 370 L'Enfant Promenade, S.W., Washington, DC 20024. (202) 646-7400.

AMERICAN SPACE FOUNDATION. 111 Massachusetts Avenue, N.W., Suite 200, Washington, DC 20002. (202) 289-2293.

NATIONAL SPACE INSTITUTE. West Wing, Suite 203, 600 Maryland Avenue, S.W., Washington, DC 20024. (202) 484-1111.

PLANETARY SOCIETY. 65 North Catalina Avenue, Pasadena, CA 91106. (818) 793-5100.

SOCIETY OF AUTOMOTIVE ENGINEERS. 400 Commonwealth Drive, Warrendale, PA 15096. (412) 776-4841.

SPACE STUDIES INSTITUTE. P.O. Box 82, 285 Rosedale, Road, Princeton, NJ 08540. (609) 921-0377.

DIRECTORIES AND BIOGRAPHICAL SOURCES

AMERICAN ASTRONAUTICAL SOCIETY DIRECTORY. American Astronautical Society, 6212-B Old Keene Mill Court, Springfield, VA 22152. (703) 866-0020. 1986. $30.00.

AMERICAN INSTITUTE OF AERONAUTICS AND ASTRONAUTICS ROSTER. American Institute of Aeronatics and Astronautics, 370 L'Enfant Promenade, S.W., Washington, DC 20024. (202) 646-7400. 1984. $75.00.

INTERNATIONAL RESEARCH CENTERS DIRECTORY 1988-89. Darren L. Smith, editor. Gale Research Company, Book Tower, Detroit, MI 48226. (800) 521-0707. 4th edition. 1987. $360.00.

RESEARCH CENTERS DIRECTORY 1988. Gale Research Company, Book Tower, Detroit, MI 48226. (800) 521-0707. 12th edition. 1987. $365.00 for set.

SCIENTIFIC AND TECHNICAL ORGANIZATIONS AND AGENCIES DIRECTORY. Margaret Labash Young, editor. Gale Research Company, Book Tower, Detroit, MI 48226. (800) 521-0707. 2nd edition. 1987. $185.00.

WHO'S WHO IN ENGINEERING. Gordon Davis, editor. Hemisphere Publishing Corporation, 1010 Vermont Avenue, NW, Washington, DC 20005. (800) 526-0275. 6th edition. 1985. $200.00.

GENERAL WORKS

ENVOYS OF MANKIND. George S. Robinson and H.M. White. Smithsonian Institution Press, 955 L'Enfant Plaza, Suite 2100, Washingtin, DC 20560. (202) 287-3388. 1986. $19.95.

INTERSTELLAR MIGRATION AND THE HUMAN EXPERIENCE. Ben R. Finney and Eric M. Jones. University of California Press, 2120 Berkeley Way, Berkeley, CA 94720. (415) 642-6683. 1985. $19.95.

LIVING ALOFT: HUMAN REQUIREMENTS FOR EXTENDED SPACEFLIGHT. NASA Special Publication SP-483. National Aeronautics and Space Administration, Washington, DC 20546. Available from: Superintendent of Documents, U.S. Government Printing Office, Washington, DC 20402. 1985.

SPACE: THE NEXT TWENTY-FIVE YEARS. Thomas R. McDonough. John Wiley and Sons, Inc., 605 Third Avenue, New York, NY 10158. (800) 526-5368. 1987. $17.95.

SPACEFARERS OF THE 80'S AND 90'S. Alcestis R. Oberg. Columbia University Press, 562 West 113th Street, New York, NY 10022. (212) 688-8885. 1985. $24.95.

ONLINE DATA BASES

CA SEARCH. Chemical Abstracts Service, P.O. Box 3012, Columbus, OH 43120. (800) 848-6538 or (614) 421-3600. Comprehensive guide to chemical and related literature, 1972 to present. Inquire as to online cost and availability.

COMPENDEX. Engineering Information, Inc., 345 East 47th Street, New York, NY 10017. (800) 221-1044 or (212) 705-7615. Engineering and technical literature, 1975 to present. Inquire as to online cost and availability.

INSPEC. INSPEC Marketing Department, Institution of Electrical Engineers. Available from IEEE Service Center, 445 Hoes Lane, Piscataway, NJ 08854. (201) 981-0060. Online version of Physics Abstracts. Inquire as to online cost and availability.

NTIS. National Technical Information Service, 5285 Port Royal Road, Springfield, VA 22161. (703) 487-4630. Broad coverage of government sponsored research reports, 1964 to present. Inquire as to online cost and availability.

SCISEARCH. Institute for Scientific Information, 3501 Market Street, Philadelphia, PA 19104. (800) 523-1850 or (215) 386-0100. Broad multidisciplinary title and author index to the international literature of science and technology, 1974 to present. Inquire as to online cost and availability.

WILSONLINE. H.W. Wilson and Company, 950 University Avenue, Bronx, NY 10452. (800) 367-6770 or (212) 588-8400. Makes available online versions of the H.W. Wilson indexes including Applied Science and Technology Index, Business Periodicals Index and Readers' Guide to Periodical Literature. Approximately 1980 to present. Inquire as to online cost and availability.

PERIODICALS

AEROSPACE AMERICA. American Institute of Aeronautics and Astronautics, Technical Information Service, 370 L'Enfant Promenade, S.W., Washington, DC 20024. (202) 646-7400. 1932 to present. Monthly. $56.00 per year.

AEROSPACE ENGINEERING MAGAZINE. Society of Automotive Engineers, 400 Commonwealth Drive, Warrendale, PA 15096. (412) 776-4841. 1981 to present. Monthly. $24.00 per year.

AIAA JOURNAL. American Institute of Aeronautics and Astronautics, Technical Information Service, 370 L'Enfant Promenade, S.W., Washington, DC 20024. (202) 646-7400. 1963 to present. Monthly. $205.00 per year.

AVIATION WEEK AND SPACE TECHNOLOGY. McGraw-Hill Book Company, 1221 Avenue of the Americas, New York, NY 10020. (212) 512-2000. 1916 to present. Weekly. $56.00 per year.

JOURNAL OF SPACECRAFT AND ROCKETS. American Institute of Aeronautics and Astronautics, Technical Information Service, 370 L'Enfant Promenade, S.W., Washington, DC 20024. (202) 646-7400. 1964 to present. Bimonthly. $95.00 per year.

PLANETARY REPORT. Planetary Society, 65 North Catalina Avenue, Pasadena, CA 91106. (818) 793-5100. 1980 to present. Bimonthly. $20.00 per year.

SPACEFLIGHT. British Interplanetary Society, 27-29 South Lambeth Road, London, SW8 1SZ, England. 1956 to present.

$50.00 per year.

RESEARCH CENTERS AND INSTITUTES

CENTER FOR SPACE RESEARCH AND APPLICATIONS. University of Texas at Austin, WRW 402, Austin, TX 78712. (512) 471-1356.

GODDARD SPACE FLIGHT CENTER. National Aeronautics and Space Administration, Greenbelt Road, Greenbelt, MD 20771. (301) 344-5121.

MARSHALL SPACE FLIGHT CENTER. National Aeronautics and Space Administration, Huntsville, AL 35812. (205) 453-2121.

SPACE SCIENCE AND ENGINEERING CENTER. University of Wisconsin, Madison, 1225 West Dayton Street, Madison, WI 53706. (608) 262-0544.

MAPPING

See: CARTOGRAPHY

MARINE ENGINEERING

See also: NAVAL ARCHITECTURE, SHIPBUILDING

ABSTRACT SERVICES AND INDEXES

APPLIED MECHANICS REVIEW. American Society of MechanicalEngineers, 345 East 47th Street, New York, NY 10017. (212) 705-7703. 1948 to present. Monthly. $360.00 per year.

APPLIED SCIENCE AND TECHNOLOGY INDEX. H.W. Wilson and Company, 950 University Avenue, Bronx, NY 10452. (800) 367-6670 or (212) 588-8400. Monthly. Inquire as to cost and availability.

CAD/CAM ABSTRACTS. EIC/Intelligence, Inc., 48 West 38th Street, New York, NY 10018. (800) 223-6275 or (212) 944-8500. Monthly. $365.00 per year.

CHEMICAL ABSTRACTS. American Chemical Society, Chemical Abstracts Service, Box 3012, Columbus, OH 43210. (614) 421-3600. 1907 to present. Weekly. $9500.00 per year.

CURRENT CONTENTS: ENGINEERING, TECHNOLOGY AND APPLIED SCIENCES. Institute for Scientific Information, 3501 Market Street, Philadelphia, PA 19104. (800) 523-1850 or (215) 386-0100. Weekly. $275.00 per year.

ENGINEERING INDEX MONTHLY AND AUTHOR INDEX. Engineering Information Inc., 345 East 47th Street, New York, NY 10017. (212) 705-7600. Monthly. $1560.00 per year.

ISMEC BULLETIN (Information Service in Mechanical Engineering). Cambridge Scientific Abstracts, 5161 River Road, Bethesda, MD 20816. (301) 951-1400. 1973 to present. Monthly. $450.00 per year.

PHYSICS ABSTRACTS. Institution of Electrical Engineers. Available from: IEEE Service Center, 445 Hoes Lane, Piscataway, NJ 08854. 1898 to present. Bimonthly. $1700.00 per year.

SCIENCE CITATION INDEX. Institute for Scientific Information, 3501 Market Street, Philadelphia, PA 19104. (800) 523-1850 or (215) 386-0100. Six times per year. $6200. 00 per year.

ASSOCIATIONS AND PROFESSIONAL SOCIETIES

AMERICAN SOCIETY OF MECHANICAL ENGINEERS. 345 47th Street, New York, NY 10017. (212) 705-7722.

AMERICAN SOCIETY OF NAVAL ENGINEERS. 1452 Duke Street, Alexandria, VA 22314. (703) 836-6727.

INSTITUTE OF MARINE ENGINEERS. 76 Mark Lane, London, EC3R 7JN, England.

SOCIETY OF NAVAL ARCHITECTS AND MARINE ENGINEERS. One World Trade Center, Suite 1369, New York, NY 10048. (212) 432-0310.

VIBRATION INSTITUTE. 101 West 55th Street, Suite 206, Clarendon Hills, IL 60514. (312) 654-2254.

DIRECTORIES AND BIOGRAPHICAL SOURCES

DIRECTORY OF SHIPOWNERS, SHIPBUILDERS, AND MARINE ENGINEERS. Industrial Press, Division of Business Press International, Limited. Available from: Robert E. Raynor, Business Press International, 205 East 42nd Street, New York, NY 10017. Annual. $75.00.

INDUSTRIAL EQUIPMENT AND SUPPLIES DIRECTORY. American Business Directories, Inc., 5707 South 86th Circle, Omaha, NE 68127. (402) 331-7169.

MARINE EQUIPMENT CATALOG. Maritime Reporter and Engineering News, 118 East 25th Street, New York, NY 10010. (212) 477-6700. Annual. $65.00.

1987 DIRECTORY OF ENGINEERING SOCIETIES AND RELATED ORGANIZATIONS. Gordon Davis, editor. Hemisphere Publishing Corporation, 1010 Vermont Avenue, NW, Washington, DC 20005. (800) 526-0275. 12th edition. 1987. $100.00.

RESEARCH CENTERS DIRECTORY 1988. Gale Research Company, Book Tower, Detroit, MI 48226. (800) 521-0707. 12th edition. 1987. $365.00 for set.

SCIENTIFIC AND TECHNICAL ORGANIZATIONS AND AGENCIES DIRECTORY. Margaret Labash Young, editor. Gale Research Company, Book Tower, Detroit, MI 48226. (800) 521-0707. 2nd edition. 1987. $185.00.

WHO'S WHO IN ENGINEERING. Gordon Davis, editor. Hemisphere Publishing Corporation, 1010 Vermont Avenue, NW, Washington, DC 20005. (800) 526-0275. 6th edition. 1985. $200.00.

GENERAL WORKS

ENGINEERING MECHANICS. J.L. Meriam and L.G. Kraige. John Wiley and Sons, Inc., 605 Third Avenue, New York, NY 10158. (800) 526-5368. 2nd edition. 1986. $52.95.

LOW SPEED MARINE DIESEL ENGINES. John B. Woodward. John Wiley and Sons, Inc., 605 Third Avenue, New York, NY 10158. (800) 526-5368. 1981. $55.50.

MARINE AUXILLARY MACHINERY. David W. Smith. Butterworth Press, Inc., 80 Montvale Avenue, Stoneham, MA 02180. (800) 325-4177. 6th edition. 1983. $59.95.

MARINE STEAM ENGINES AND TURBINES. S.C. McBirnie and W.J. Fox. Butterworth Press, Inc., 80 Montvale Avenue, Stoneham, MA 02180. (800) 325-4177. 4th edition. 1980. $64.95.

HANDBOOKS AND MANUALS

MARINE ENGINE ROOM BLUE BOOK. W.B. Paterson. Cornell Maritime Press, Inc., P.O. Box 456, Centreville, MD 21617. (301) 758-1075. 3rd edition. 1984. $15.00 in paper.

HANDBOOK OF MECHANICS, MATERIALS, AND STRUCTURES. A. Blake. John Wiley and Sons, Inc., 605 Third Avenue, New York, NY 10158. (800) 526-5368. 1985. $65.95.

MECHANICAL DESIGN AND SYSTEMS HANDBOOK. Harold A. Rothbart, editor. McGraw-Hill Book Company, 1221 Avenue of the Americas, New York, NY 10020. (212) 512-2000. 2nd edition.1985. $96.50.

MECHANICAL ENGINEERS' HANDBOOK. Myer Kutz, editor. John Wiley and Sons, Inc., 605 Third Avenue, New York, NY 10158. (800) 526-5368. 1986. $79.95.

PRESSURE VESSEL DESIGN HANDBOOK. Henry H. Bednar. Van Nostrand Reinhold Company, Inc., 135 West 50th Street, New York, NY 10020. (800) 543-2681. 2nd edition. 1986. $49.95.

STANDARD HANDBOOK OF MACHINE DESIGN. Joseph E. Shigley and Charles R. Mischke. McGraw-Hill Book Company, 1221 Avenue of the Americas, New York, NY 10020. (212) 512-2000. 1986. $89.00.

ONLINE DATA BASES

CA SEARCH. Chemical Abstracts Service, P.O. Box 3012, Columbus, OH 43120. (800) 848-6538 or (614) 421-3600. Comprehensive guide to chemical literature, 1972 to present. Inquire as to online cost and availability.

COMPENDEX. Engineering Information, Inc., 345 East 47th Street, New York, NY 10017. (800) 221-1044 or (212) 705-7615. Engineering and technical literature, 1975 to present. Inquire as to online cost and availability.

INSPEC. INSPEC Marketing Department, Institution of Electrical Engineers. Available from IEEE Service Center, 445 Hoes Lane, Piscataway, NJ 08854. (201) 981-0060. Online version of Physics Abstracts. Inquire as to online cost and availability.

ISMEC. Cambridge Scientific Abstracts, 5161 River Road, Besthda, MD 20816. (800) 638-8076 or (301) 951-1400. Literature of mechanical and production engineering, 1973 to present. Inquire as to online cost and availability.

NTIS. National Technical Information Service, 5285 Port Royal Road, Springfield, VA 22161. (703) 487-4630. Broad coverage of government sponsored research reports, 1964 to present. Inquire as to online cost and availability.

SCISEARCH. Institute for Scientific Information, 3501 Market Street, Philadelphia, PA 19104. (800) 523-1850 or (215) 386-0100. Broad multidisciplinary title and author index to the international literature of science and technology, 1974 to present. Inquire as to online cost and availability.

WILSONLINE. H.W. Wilson and Company, 950 University Avenue, Bronx, NY 10452. (800) 367-6770 or (212) 588-8400. Makes available online versions of the H.W. Wilson indexes including Applied Science and Technology Index, Business Periodicals Index and Readers' Guide to Periodical Literature. Approximately 1980 to present. Inquire as to online cost and availability.

PERIODICALS

JOURNAL OF APPLIED MECHANICS. American Society of Mechanical Engineers, 345 East 47th Street, New York, NY 10017. (212) 705-7703. 1935 to present. Quarterly. $100.00 per year.

JOURNAL OF ENGINEERING FOR INDUSTRY. American Society of Mechanical Engineers, 345 East 47th Street, New York, NY 10017. (212) 705-7703. 1970 to present. Quarterly. $100.00 per year.

JOURNAL OF MECHANISMS, TRANSMISSIONS AND AUTOMATION IN DESIGN. American Society of Mechanical Engineers, 345 East 47th Street, New York, NY 10017. (212) 705-7703. 1983 to present. Quarterly. $80.00 per year.

JOURNAL OF PRESSURE VESSEL TECHNOLOGY. American Society of Mechanical Engineers, 345 East 47th Street, New York, NY 10017. (212) 705-7703. 1974 to present. Quarterly. $80.00 per year.

JOURNAL OF SHIP RESEARCH. Society of Naval Architects and Marine Engineers. One World Trade Center, Suite 1369, New York, NY 10048. (212) 432-0310. 1957 to present. Quarterly. $40.00 per year.

MACHINE DESIGN. Penton-IPC, 1100 Superior Avenue, Cleveland, OH 44114. (216) 696-7000. 1929 to present. Twenty-eight times per year. $60.00 per year.

MARINE ENGINEERING/LOG. Simmons-Boardman Publishing Corporation, 345 Hudson Street, New York, NY 10014. (212) 620-7200. 1878 to present. Monthly. $30.00 per year.

MARINE ENGINEERS REVIEW. Marine Management Holdings, Limited, 76 Mark Lane, London EC3R 7JN, England. 1970 to present. Monthly. $40.00 per year.

MARINE TECHNOLOGY. Society of Naval Architects and Marine Engineers. One World Trade Center, Suite 1369, New York, NY 10048. (212) 432-0310. 1964 to present. Quarterly. $40.00 per year.

MECHANICAL ENGINEERING. American Society of Mechanical Engineers, 345 East 47th Street, New York, NY 10017. (212) 705-7722.

MOTOR SHIP. Industrial Press, Division of Business Press International, Limited. Available from: Robert E. Raynor, Business Press International, 205 East 42nd Street, New York, NY 10017. 1920 to present. Monthly. $115.00 per two years.

NAVAL ARCHITECT. Royal Institution of Naval Architects, 10 Upper Belgrave Street, London, SW1X 8BQ, England. 1971 to present. Bimonthly. $75.00 per year.

SOCIETY OF NAVAL ARCHITECTS AND MARINE ENGINEERS TRANSACTIONS. Society of Naval Architects and Marine Engineers. One World Trade Center, Suite 1369, New York, NY 10048. (212) 432-0310. 1893 to present. Annual. $45.00 per year.

RESEARCH CENTERS AND INSTITUTES

INSTITUTE FOR MARINE DYNAMICS. National Research Council of Canada, P.O. Box 12093, Postal Station A, St. John's, NF, Canada A1B 3T5. (709) 772-2469.

MARITIME RESEARCH DEPARTMENT. Webb Institute of Naval Architecture, Crescent Beach Road, Glen Cove, NY 11542. (516) 671-2356.

SHIP HYDRODYNAMICS LABORATORY. University of Michigan, 126 West Engineering Building, Ann Arbor, MI 48109. (313) 764-9432.

MARINE NAVIGATION

See: NAVIGATION

MARKOV PROCESSES

See also: STATISTICS, STOCHASTIC PROCESSES

ABSTRACT SERVICES AND INDEXES

COMPUMATH CITATION INDEX. Institute for Scientific Information, 3501 Market Street, Philadelphia, PA 19104. (800) 523-1850 or (215) 386-0100. Three times per year. $875.00 per year.

CURRENT MATHEMATICAL PUBLICATIONS. American Mathematical Society, P.O. Box 6248, Providence, RI 02940. (800) 556-7774 or (401) 272-9500. 1969 to present. Seventeen times per year. $230.00 per year.

ENGINEERING INDEX MONTHLY AND AUTHOR INDEX. Engineering Information Inc., 345 East 47th Street, New York, NY 10017. (212) 705-7600. Monthly. $1560.00 per year.

MATHEMATICAL REVIEWS: A REVIEWING JOURNAL COVERING THE WORLD LITERATURE OF MATHEMATICAL RESEARCH. American Mathematical Society, P.O. Box 6248, Providence, RI 02940. (800) 7774 or (401) 272-9500. 1940 to present. Monthly. $2800.00 per year.

SCIENCE CITATION INDEX. Institute for Scientific Information, 3501 Market Street, Philadelphia, PA 19104. (800) 523-1850 or (215) 386-0100. Six times per year. $6200.00 per year.

ASSOCIATIONS AND PROFESSIONAL SOCIETIES

AMERICAN MATHEMATICAL SOCIETY. P.O. Box 6248, Providence, RI 02940. (401) 272-9500.

MATHEMATICAL ASSOCIATION OF AMERICA. 1529 18th Street, N.W., Washington, DC 20036. (202) 387-5200.

SOCIETY FOR INDUSTRIAL AND APPLIED MATHEMATICS. 1400 Architects Building, 117 South 17th Street, Philadelphia, PA 19103. (215) 564-2929.

DIRECTORIES AND BIOGRAPHICAL SOURCES

INTERNATIONAL RESEARCH CENTERS DIRECTORY 1988-89. Darren L. Smith, editor. Gale Research Company, Book Tower, Detroit, MI 48226. (800) 521-0707. 4th edition. 1987. $360.00.

RESEARCH CENTERS DIRECTORY 1988. Gale Research Company, Book Tower, Detroit, MI 48226. (800) 521-0707. 12th edition. 1987. $365.00 for set.

ENCYCLOPEDIAS AND DICTIONARIES

ENCYCLOPEDIC DICTIONARY OF MATHEMATICS. Kiyosi Ito, editor. MIT Press, 55 Howard Street, Cambridge, MA 02142. (617) 253-2884. 2nd edition. 1987. Four volumes. $350.00 for set.

GENERAL WORKS

FUNDAMENTALS OF QUEQUEING THEORY. Donald Gross and Carl M. Harris. John Wiley and Sons, Inc., 605 Third Avenue, New York, NY 10158. (800) 526-5368. 2nd edition. 1985. $45.95.

MARKOV PROCESSES AND RELATED PROBLEMS OF ANALYSIS. E.B. Dynkin. Cambridge University Press, 32 East 57th Street, New York, NY 10022. (800) 872-7423. 1982. $29.95 in paper.

MARKOV RANDOM FIELDS. Y.A. Rozanov. Springer-Verlag New York, Inc., 175 Fifth Avenue, New York, NY 10010. (800) 526-7254. 1982. $46.50.

NUMERICAL METHODS IN ENGINEERING AND APPLIED SCIENCE. Bruce Irons. John Wiley and Sons, Inc., 605 Third Avenue, New York, NY 10158. (800) 526-5368. 1987. $32.95.

STATISTICS FOR RESEARCH. S. Dowdy and S. Wearden. John Wiley and Sons, Inc., 605 Third Avenue, New York, NY 10158. (800) 526-5368. 1983. $38.95.

ONLINE DATA BASES

DISSERTATION ABSTRACTS ONLINE. University Microfilms International, 300 North Zeeb Road, Ann Arbor, MI 48106. (800) 521-0600 or (313) 761-4700. Scope includes virtually all doctoral dissertations accepted at accredited American institutions from 1861 to present in over 250 subject areas. Inquire as to online cost and availability.

MATHFILE. American Mathematical Society, P.O. Box 6248, Providence, RI 02940. (800) 556-7774 or (401) 272-9500. An online version of Mathematical Reviews. 1973 to present. Inquire as to online cost and availability.

NTIS. National Technical Information Service, 5285 Port Royal Road, Springfield, VA 22161. (703) 487-4630. Broad coverage of government sponsored research reports, 1964 to present. Inquire as to online cost and availability.

SCISEARCH. Institute for Scientific Information, 3501 Market Street, Philadelphia, PA 19104. (800) 523-1850 or (215) 386-0100. Broad multidisciplinary title and author index to the international literature of science and technology, 1974 to present. Inquire as to online cost and availability.

PERIODICALS

AMERICAN JOURNAL OF MATHEMATICS. Johns Hopkis University Press, Journals Publishing Division, 701 West 40th Street, Suite 275, Baltimore, MD 21211. (301) 338-7864. Bimonthly. 1878 to present. $115.00 per year.

AMERICAN MATHEMATICAL MONTHLY. Mathematical Association of America, 1529 18th Street, N.W., Washington, DC 20036. (202) 387-5200. 1894 to present. Ten times per year. $70.00 per year.

APPLIED MATHEMATICS AND COMPUTATION. Elsevier Science Publishing Company, Inc., 52 Vanderbilt Avenue, New York, NY 10017. (212) 370-5520. 1975 to present. Monthly. $355.00 per year.

SIAM JOURNAL OF APPLIED MATHEMATICS. Society for Industrial and Applied Mathematics, 1405 Architects Building, 117 South 17th Street. Philadelphia, PA 19103. (215) 564-2929. 1953 to present. Bimonthly. $130.00 per year.

SIAM JOURNAL OF MATHEMATICAL ANALYSIS. Society for Industrial and Applied Mathematics, 1405 Architects Building, 117 South 17th Street. Philadelphia, PA 19103. (215) 564-2929. 1964 to present. Bimonthly. $120.00 per year.

SIAM REVIEW. Society for Industrial and Applied Mathematics, 1405 Architects Building, 117 South 17th Street. Philadelphia, PA 19103. (215) 564-2929. 1959 to present. Quarterly. $82.00 per year.

STOCHASTIC ANALYSIS AND APPLICATIONS. Marcel Dekker Inc., Journals Division, 270 Madison Avenue, New York, NY 10016. (800) 228-1160. 1983 to present. Quarterly. $125.00 per year.

STOCHASTIC PROCESSES AND THEIR APPLICATIONS. Elsevier Science Publishing Company, Inc., 52 Vanderbilt Avenue, New York, NY 10017. (212) 370-5520. 1973 to present. Six times per year. $250.00 per year.

STUDIES IN APPLIED MATHEMATICS. Elsevier Science Publishing Company, Inc., 52 Vanderbilt Avenue, New York, NY 10017. (212) 370-5520. 1921 to present. Six times per year. $126.00 per year.

RESEARCH CENTERS AND INSTITUTES

CENTER FOR APPLIED MATHEMATICS. University of Georgia, Tucker Hall, Athens, GA 30602. (404) 542-3491.

CENTER FOR MATHEMATICAL SCIENCE RESEARCH. Rutgers University, New Brunswick, NJ 08903. (201) 932-3117.

ENGINEERING SYSTEMS RESEARCH CENTER. University of California, Berkeley, 3115 Etcheverry Hall, Berkeley, CA 94720. (415) 642-4993.

INSTITUTE FOR MATHEMATICS AND ITS APPLICATIONS. University of Minnesota, 514 Vincent, 206 Church Street, S.E., Minneapolis, MN 55455. (612) 624-6066.

INSTITUTE OF APPLIED MATHEMATICS. University of Missouri - Rolla, Rolla, MO 65401. (314) 341-4151.

STATISTICAL LABORATORY. Kansas State University, Dickens Hall, Manhattan, KS 66506. (913) 532-6883.

MARS

See also: PLANETARY SCIENCE, SOLAR SYSTEM

ABSTRACT SERVICES AND INDEXES

ASTRONOMY AND ASTROPHYSICS ABSTRACTS. Springer-Verlag New York, Incorporated, 175 Fifth Avenue, New York, NY 10010. (212) 460-1500. $70.00 per year.

ASTRONOMY AND ASTROPHYSICS MONTHLY INDEX. Olivetree Associates, Post Office Box 236, Sierra Madre, CA 91024. $212.00 per year. Complimentary copies available on request.

METEOROLOGICAL AND GEOASTROPHYSICAL ABSTRACTS. American Meteorological Society, 45 Beacon Street, Boston, MA 02108. (617) 227-2425. 1950 to present. Monthly. $450.00 per year.

STAR. (Scientific and Technical Aerospace Reports. United States National Aeronautics and Space Administration, Scientific and Technical Information Facility, Box 8757, Baltimore-Washington International Airport, MD 21240. (202) 755-2210. Semimonthly, with semiannual and annual indexes. $85.00 per year.

ANNUAL REVIEWS AND YEARBOOKS

THE ASTRONOMICAL ALMANAC. Superintendent of Documents, United States Government Printing Office, Washington, DC 20402. (202) 783-3238. Yearly.

ANNUAL REVIEW OF ASTRONOMY AND ASTROPHYSICS. Annual Reviews, Incorporated, 4139 El Camino Way, Palo Alto, CA 94306. (415) 493-4400. Annual. Inquire.

ANNUAL REVIEW OF EARTH AND PLANETARY SCIENCES. Annual Reviews, Incorporated, 4139 El Camino Way, Palo Alto, CA 94306. (415) 493-4400. Annual. Inquire.

ASSOCIATIONS AND PROFESSIONAL SOCIETIES

AMERICAN ASTRONOMICAL SOCIETY. 2000 Florida Avenue, NW, Suite 300, Washington, DC 20009. (202) 659-0134.

AMERICAN GEOPHYSICAL UNION. 2000 Florida Avenue, NW, Washington, DC 20009. (202) 462-6903.

ASTRONOMICAL LEAGUE. Post Office Box 12821, Tucson, AZ 85732. (602) 790-8471.

ASTRONOMICAL SOCIETY OF THE PACIFIC. 1290 24th Avenue, San Francisco, CA 94122. (415) 661-8660.

PLANETARY SOCIETY. 65 North Catalina Avenue, Pasadena, CA 91106-2301. (818) 793-5100.

BIBLIOGRAPHIES

A BIBLIOGRAPHY OF ASTRONOMY, 1970-1979. R.A. Seal and S.S. Martin. Libraries Unlimited, Incorporated, Littleton, CO 80160. 1982. $37.50.

SCIENTIFIC AND TECHNICAL BOOKS AND SERIALS IN PRINT 1988: AN INDEX TO LITERATURE IN SCIENCE AND TECHNOLOGY. R.R. Bowker Company, 205 East 42nd Street, New York, NY 10017. (800) 521-8110 or (212) 916-1600. $175.00.

DIRECTORIES AND BIOGRAPHICAL SOURCES

AMERICAN ASTRONOMICAL SOCIETY MEMBERS. 2000 Florida Avenue, NW, Suite 300, Washington, DC 20009. (202) 659-0134. Annual. Available to members only.

THE BIOGRAPHICAL DICTIONARY OF SCIENTISTS: ASTRONOMERS. D. Abbott, editor. Peter Bedrick Books, 125 East 23rd Street, New York, NY 10010. 1984. $24.95.

DIRECTORY OF PHYSICS AND ASTRONOMY STAFF MEMBERS. American Institute of Physics, 335 East 45th Street, New York, NY 10017. Annual.

GENERAL WORKS

THE CASE FOR MARS II. C. McKay. American Astronautical Society, available from Univelt, Post Office Box 28130, San Diego, CA 92128. (619) 746-4005. 1985. $40.00.

ON MARS: EXPLORATION OF THE RED PLANET, 1958-1978. E. and L Ezell. NASA Special Publication SP-4212. 1984.

PHOTOCHEMISTRY OF THE ATMOSPHERE OF MARS AND VENUS. V.A. Krasnopolsky. Springer-Verlag New York, Incorporated, 175 Fifth Avenue, New York, NY 10010. (212) 460-1500. 1986. $90.00.

ONLINE DATA BASES

CA SEARCH. Chemical Abstracts Service, Post Office Box 3012, Columbus, OH 43210. Guide to chemical literature, 1972 to present. Inquire as to cost and availability.

DISSERTATION ABSTRACTS ONLINE. University Microfilms International, 300 North Zeeb Road, Ann Arbor, MI 48106. (800) 521-0600 or (313) 761-4700. Scope includes virtually all doctoral dissertations accepted at accredited American institutions from 1861 to present in 252 subject areas. Inquire as to cost and availability.

NASA. National Aeronautics and Space Administration, Scientific and Technical Information Branch, 300 7th Street, SW, Washington, DC 20546. Citations and abstracts of aerospace literature, 1962 to present. Inquire as to cost and availability.

NTIS. National Technical Information Service, 5285 Port Royal Road, Springfield, VA 22161. (703) 487-4630. Broad coverage of government sponsored research reports, 1964 to present. Inquire as to cost and availability.

SCISEARCH. Institute for Scientific Information, 3501 Market Street, Philadelphia, PA 19104. (800) 523-1850 or (215) 386-0100. Broad multidisciplinary title and author index to the international literature of science and technology, 1974 to present. Inquire as to cost and availability.

PERIODICALS

ASSOCIATION OF LUNAR AND PLANETARY OBSERVERS. Journal Association of Lunar and Planetary Observers, Box 16131, San Francisco, CA 94116. (415) 566-5786. 1947 to present. Quarterly. $15.00 per year.

ASTRONOMICAL JOURNAL. American Astronomical Society. Available from: American Institute of Physics, 335 East 45th Street, New York, NY 10017. 9212) 661-9404. $125.00 per year.

EARTH, MOON AND PLANETS: AN INTERNATIONAL JOURNAL OF COMPARATIVE PLANETOLOGY. D. Reidel

Publishing Company, 190 Old Derby Street, Hingham, MA 02043. Nine times per year. $275.00 per year.

ICARUS: INTERNATIONAL JOURNAL OF THE SOLAR SYSTEM STUDIES. Academic Press, Incorporated, Orlando, FL 32887. (305) 345-4100. Monthly. $484.00 per year.

MERCURY. Astronomical Society of the Pacific, 1290 24th Avenue, San Francisco, CA 94122. (415) 661-8660. Bimonthly. $21.00 per year.

PLANETARY AND SPACE SCIENCE. Pergamon Press, Incorporated, Maxwell House, Fairview Park, Elmsford, NY 10523. (914) 592-7700. Monthly. $430.00 per year.

PLANETARY REPORT. Planetary Society, 65 North Catalina Avenue, Pasadena, CA 91106-2301. (818) 793-5100. 1980 to present. Bimonthly. $20.00 per year.

SOLAR SYSTEM RESEARCH (ENGLISH TRANSLATION OF ASTRONOMICHESKII VESTNIK). Consultants Bureau, 233 Spring Street, New York, NY 10013. (212) 620-8000. 1967 to present. Quarterly. $425.00 per year.

RESEARCH CENTERS AND INSTITUTES

LABORATORY FOR PLANETARY ATMOSPHERES RESEARCH. State University of New York at Stony Brook, Stony Brook, NY 11794-2300. (516) 632-8321.

LABORATORY FOR PLANETARY STUDIES. Cornell University, 302 Space Sciences Building, Ithaca, NY 14853. (607) 256-4971.

LUNAR AND PLANETARY INSTITUTE. 3303 NASA Road One, Houston, TX 77058. (713) 486-2139.

LUNAR AND PLANETARY LABORATORY. University of Arizona, Tucson, AZ 85721. (602) 621-6962.

MIDSOUTH ASTRONOMICAL RESEARCH SOCIETY, INCORPORATED. Arkansas Sky Observatory, North Little Rock, AR 72116. (501) 835-8476.

WASHINGTON UNIVERSITY. Earth and Planetary Remote Sensing Laboratory. Department of Earth and Planetary Sciences, Campus Box 1169, St. Louis, MO 63130. (314) 889-5679.

MASER

See: LASERS

MASKING

See: SEMICONDUCTORS

MASS SPECTROMETRY

See also: SPECTROSCOPY

ABSTRACT SERVICES AND INDEXES

APPLIED SCIENCE AND TECHNOLOGY INDEX. H.W. Wilson and Company, 950 University Avenue, Bronx, NY 10452. (800) 367-6670 or (212) 588-8400. Monthly. Inquire as to cost and availability.

CHEMICAL ABSTRACTS. American Chemical Society, Chemical Abstracts Service, Box 3012, Columbus, OH 43210. (614) 421-3600. 1907 to present. Weekly. $9500.00 per year.

CURRENT CONTENTS: ENGINEERING, TECHNOLOGY AND APPLIED SCIENCES. Institute for Scientific Information, 3501 Market Street, Philadelphia, PA 19104. (800) 523-1850 or (215) 386-0100. Weekly. $275.00 per year.

ENGINEERING INDEX MONTHLY AND AUTHOR INDEX. Engineering Information Inc., 345 East 47th Street, New York, NY 10017. (212) 705-7600. Monthly. $1560.00 per year.

INDEX TO SCIENTIFIC AND TECHNICAL PROCEEDINGS. Institute for Scientific Information, 3501 Market Street, Philadelphia, PA 19104. (800) 523-1850 or (215) 386-0100. 1978 to present. Monthly. $775.00 per year.

MASS SPECTROMETRY BULLETIN. Royal Society of Chemistry. Available from: Distribution Centre, Blackhorse Road, Letchworth, Herts SG6 1HN, England. 1966 to present. Monthly. $350.00 per year.

PHYSICS ABSTRACTS. Institution of Electrical Engineers. Available from: IEEE Service Center, 445 Hoes Lane, Piscataway, NJ 08854. 1898 to present. Bimonthly. $1700.00 per year.

SCIENCE CITATION INDEX. Institute for Scientific Information, 3501 Market Street, Philadelphia, PA 19104. (800) 523-1850 or (215) 386-0100. Six times per year. $6200.00 per year.

ASSOCIATIONS AND PROFESSIONAL SOCIETIES

AMERICAN SOCIETY FOR MASS SPECTROSCOPY. P.O. Box 1508, East Lansing, MI 48823. (517) 337-2548.

SOCIETY FOR APPLIED SPECTROSCOPY. P.O. Box 1438, Frederick, MD 21701. (301) 694-8122.

DIRECTORIES AND BIOGRAPHICAL SOURCES

INTERNATIONAL RESEARCH CENTERS DIRECTORY 1988-89. Darren L. Smith, editor. Gale Research Company, Book Tower, Detroit, MI 48226. (800) 521-0707. 4th edition. 1987. $360.00.

RESEARCH CENTERS DIRECTORY 1988. Gale Research Company, Book Tower, Detroit, MI 48226. (800) 521-0707. 12th edition. 1987. $365.00 for set.

SCIENTIFIC AND TECHNICAL ORGANIZATIONS AND AGENCIES DIRECTORY. Margaret Labash Young, editor. Gale Research Company, Book Tower, Detroit, MI 48226. (800) 521-0707. 2nd edition. 1987. $185.00.

SOCIETY FOR APPLIED SPECTROSCOPY MEMBERSHIP DIRECTORY. Society for Applied Spectroscopy, P.O. Box 1438, Frederick, MD 21701. (301) 694-8122. 1988. $75.00.

ENCYCLOPEDIAS AND DICTIONARIES

DICTIONARY OF THE PHYSICAL SCIENCES: TERMS, FORMULAS, DATA. Cesare Emiliani. Oxford University Press, 200 Madison Avenue, New York, NY 10016. (800) 458-5833. 1987. $19.95 in paper.

THESAURUS OF SCIENTIFIC, TECHNICAL, AND ENGINEERING TERMS. Hemisphere Publishing Corporation, 1010 Vermont Avenue, NW, Washington, DC 20005. (800) 526-0275. 1988. $125.00.

GENERAL WORKS

INTRODUCTION TO MASS SPECTROMETRY. H.C. Hill. John Wiley and Sons, Inc., 605 Third Avenue, New York, NY 10158. (800) 526-5368. $39.95 in paper.

MASS SPECTROMETRY. H.E. Duckworth and others. Cambridge University Press, 32 East 57th Street, New York, NY 10022. (800) 872-7423. 2nd edition. 1986. $69.50.

MASS SPECTROMETRY: APPLICATIONS IN SCIENCE AND ENGINEERING. F.A. White and G.M. Wood. John Wiley and Sons, Inc., 605 Third Avenue, New York, NY 10158. (800) 526-5368. 1986. $72.50.

MASS SPECTROMETRY FOR CHEMISTS AND BIOCHEMISTS. Cambridge University Press, 32 East 57th Street, New York, NY 10022. (800) 872-7423. 1982. $25.95 in paper.

PRACTICAL ORGANIC MASS SPECTROMETRY. J.R. Chapman. John Wiley and Sons, Inc., 605 Third Avenue, New York, NY 10158. (800) 526-5368. 1985. $32.95.

SECONDARY ION MASS SPECTROMETRY: BASIC CONCEPTS, INSTRUMENTAL ASPECTS, APPLICATIONS AND TRENDS. A. Benninghoven. John Wiley and Sons, Inc., 605 Third Avenue, New York, NY 10158. (800) 526-5368. 1987. $150.00.

ONLINE DATA BASES

CA SEARCH. Chemical Abstracts Service, P.O. Box 3012, Columbus, OH 43120. (800) 848-6538 or (614) 421-3600. Comprehensive guide to chemical literature, 1972 to present. Inquire as to online cost and availability.

COMPENDEX. Engineering Information, Inc., 345 East 47th Street, New York, NY 10017. (800) 221-1044 or (212) 705-7615. Engineering and technical literature, 1975 to present. Inquire as to online cost and availability.

DISSERTATION ABSTRACTS ONLINE. University Microfilms International, 300 North Zeeb Road, Ann Arbor, MI 48106. (800) 521-0600 or (313) 761-4700. Scope includes virtually all doctoral dissertations accepted at accredited American institutions from 1861 to present in over 250 subject areas. Inquire as to online cost and availability.

INSPEC. INSPEC Marketing Department, Institution of Electrical Engineers. Available from IEEE Service Center, 445 Hoes Lane, Piscataway, NJ 08854. (201) 981-0060. Online version of Physics Abstracts. Inquire as to online cost and availability.

NTIS. National Technical Information Service, 5285 Port Royal Road, Springfield, VA 22161. (703) 487-4630. Broad coverage of government sponsored research reports, 1964 to present. Inquire as to online cost and availability.

SCISEARCH. Institute for Scientific Information, 3501 Market Street, Philadelphia, PA 19104. (800) 523-1850 or (215) 386-0100. Broad multidisciplinary title and author index to the international literature of science and technology, 1974 to present. Inquire as to online cost and availability.

WILSONLINE. H.W. Wilson and Company, 950 University Avenue, Bronx, NY 10452. (800) 367-6770 or (212) 588-8400. Makes available online versions of the H.W. Wilson indexes including Applied Science and Technology Index, Business Periodicals Index and Readers' Guide to Periodical Literature. Approximately 1980 to present. Inquire as to online cost and availability.

PERIODICALS

APPLIED SPECTROSCOPY. Society for Applied Spectroscopy, P.O. Box 1438, Frederick, MD 21701. (301) 694-8122. Monthly. $145.00 per year.

APPLIED SPECTROSCOPY REVIEWS. Marcel Dekker Inc., 270 Madison Avenue, New York, NY 10016. (800) 228-1160. 1964 to present. Quarterly. $185.00 per year.

INTERNATIONAL JOURNAL OF MASS SPECTROMETRY AND ION PROCESSES. Elsevier Science Publishing Company, Inc., 52 Vanderbilt Avenue, New York, NY 10017. (212) 370-5520. 1968 to present. Twenty-one times per year. $380.00 per year.

JOURNAL OF QUANTITATIVE SPECTROSCOPY AND RADIATIVE TRANSFER. Pergamon Press, Inc., Maxwell House, Fairview Park, Elmsford, NY 10523. (914) 592-7700. 1961 to present. Monthly. $485.00 per year.

MASS SPECTROMETRY REVIEWS. John Wiley and Sons, Inc., 605 Third Avenue, New York, NY 10158. (800) 526-5368. 1982 to present. Quarterly. $135.00 per year.

SPECTROSCOPY LETTERS. Marcel Dekker Inc., 270 Madison Avenue, New York, NY 10016. (800) 228-1160. 1968 to present. Ten times per year. $280.00 per year.

RESEARCH CENTERS AND INSTITUTES

CHROMATOGRAPHY-MASS SPECTROMETRY FACILITY. University of Missouri - Columbia, Room 4, Agricultural Building, Columbia, MO 65211. (314) 882-2608.

MASS SPECTROMETRY LABORATORY. University of Minnesota, Department of Chemistry, 207 Pleasant Street, Minneapolis, MN 55455. (612) 625-3053.

MIDWEST CENTER FOR MASS SPECTROMETRY. University of Nebraska - Lincoln, Room 18, Hamilton Hall, Department of Chemistry, Lincoln, NE 68588. (402) 472-3507.

MATERIALS HANDLING

See also: INDUSTRIAL ENGINEERING

ABSTRACT SERVICES AND INDEXES

APPLIED MECHANICS REVIEW. American Society of Mechanical Engineers, 345 East 47th Street, New York, NY 10017. (212) 705-7703. 1948 to present. Monthly. $360.00 per year.

APPLIED SCIENCE AND TECHNOLOGY INDEX. H.W. Wilson and Company, 950 University Avenue, Bronx, NY 10452. (800) 367-6670 or (212) 588-8400. Monthly. Inquire as to cost and availability.

CHEMICAL ABSTRACTS. American Chemical Society, Chemical Abstracts Service, Box 3012, Columbus, OH 43210. (614) 421-3600. 1907 to present. Weekly. $9500.00 per year.

CURRENT CONTENTS: ENGINEERING, TECHNOLOGY AND APPLIED SCIENCES. Institute for Scientific Information, 3501 Market Street, Philadelphia, PA 19104. (800) 523-1850 or (215) 386-0100. Weekly. $275.00 per year.

ENGINEERING INDEX MONTHLY AND AUTHOR INDEX. Engineering Information Inc., 345 East 47th Street, New York, NY 10017. (212) 705-7600. Monthly. $1560.00 per year.

ISMEC BULLETIN (Information Service in Mechanical Engineering). Cambridge Scientific Abstracts, 5161 River Road, Bethesda, MD 20816. (301) 951-1400. 1973 to present. Monthly. $450.00 per year.

SCIENCE CITATION INDEX. Institute for Scientific Information, 3501 Market Street, Philadelphia, PA 19104. (800) 523-1850 or (215) 386-0100. Six times per year. $6200.00 per year.

ASSOCIATIONS AND PROFESSIONAL SOCIETIES

AMERICAN SOCIETY OF MECHANICAL ENGINEERS. 345 47th Street, New York, NY 10017. (212) 705-7722.

INDUSTRIAL TRUCK ASSOCIATION. 1750 K Street, NW, Suite 210, Washington, DC 20006. (202) 296-9880.

INTERNATIONAL MATERIAL MANAGEMENT SOCIETY. 650 East Higgins Road, Schaumburg, IL 60195. (312) 310-9570.

MATERIAL HANDLING EQUIPMENT DISTRIBUTORS ASSOCIATION. 201 Route 45, Vernon Hills, IL 60061. (312)

680-3500.

MATERIAL HANDLING INSTITUTE. 1326 Freeport Road, Pittsburgh, PA 15238. (412) 782-1624.

DIRECTORIES AND BIOGRAPHICAL SOURCES

MATERIAL HANDLING ENGINEERING HANDBOOK AND DIRECTORY. Penton-IPC, 1100 Superior Avenue, Cleveland, OH 44114. (216)696-7000. Biennial. $35.00.

MODERN MATERIALS HANDLING CASEBOOK AND DIRECTORY ISSUE. Cahners Publishing Company, Inc., 275 Washington Street, Newton, MA 02158. (617) 964-3030. Annual. $10.00.

1987 DIRECTORY OF ENGINEERING SOCIETIES AND RELATED ORGANIZATIONS. Gordon Davis, editor. Hemisphere Publishing Corporation, 1010 Vermont Avenue, NW, Washington, DC 20005. (800) 526-0275. 12th edition. 1987. $100.00.

RESEARCH CENTERS DIRECTORY 1988. Gale Research Company, Book Tower, Detroit, MI 48226. (800) 521-0707. 12th edition. 1987. $365.00 for set.

SCIENTIFIC AND TECHNICAL ORGANIZATIONS AND AGENCIES DIRECTORY. Margaret Labash Young, editor. Gale Research Company, Book Tower, Detroit, MI 48226. (800) 521-0707. 2nd edition. 1987. $185.00.

WHO'S WHO IN ENGINEERING. Gordon Davis, editor. Hemisphere Publishing Corporation, 1010 Vermont Avenue, NW, Washington, DC 20005. (800) 526-0275. 6th edition. 1985. $200.00.

GENERAL WORKS

MARINE CARGO OPERATIONS. C.L. Sauerbier and R.J. Meurn. John Wiley and Sons, Inc., 605 Third Avenue, New York, NY 10158. (800) 526-5368. 1985. $49.95.

MATERIAL HANDLING SYSTEMS DESIGN. J.M. Apple. John Wiley and Sons, Inc., 605 Third Avenue, New York, NY 10158. (800) 526-5368. 1972. $51.50.

MATERIALS HANDLING: PRINCIPLE AND APPLICATIONS. Theodore H. Allegri. Van Nostrand Reinhold Company, Inc., 135 West 50th Street, New York, NY 10020. (800) 543-2681. 1984. $44.95.

HANDBOOKS AND MANUALS

BULK MATERIALS HANDLING HANDBOOK. Van Nostrand Reinhold Company, Inc., 135 West 50th Street, New York, NY 10020. (800) 543-2681. 1987. $59.95.

HANDBOOK OF MATERIALS HANADLING. R.G.T. Lindkvist. John Wiley and Sons, Inc., 605 Third Avenue, New York, NY 10158. (800) 526-5368. 1985. $69.95.

ONLINE DATA BASES

CA SEARCH. Chemical Abstracts Service, P.O. Box 3012, Columbus, OH 43120. (800) 848-6538 or (614) 421-3600. Comprehensive guide to chemical literature, 1972 to present. Inquire as to online cost and availability.

COMPENDEX. Engineering Information, Inc., 345 East 47th Street, New York, NY 10017. (800) 221-1044 or (212) 705-7615. Engineering and technical literature, 1975 to present. Inquire as to online cost and availability.

NTIS. National Technical Information Service, 5285 Port Royal Road, Springfield, VA 22161. (703) 487-4630. Broad coverage of government sponsored research reports, 1964 to present. Inquire as to online cost and availability.

SCISEARCH. Institute for Scientific Information, 3501 Market Street, Philadelphia, PA 19104. (800) 523-1850 or (215) 386-0100. Broad multidisciplinary title and author index to the international literature of science and technology, 1974 to present. Inquire as to online cost and availability.

WILSONLINE. H.W. Wilson and Company, 950 University Avenue, Bronx, NY 10452. (800) 367-6770 or (212) 588-8400. Makes available online versions of the H.W. Wilson indexes including Applied Science and Technology Index, Business Periodicals Index and Readers' Guide to Periodical Literature. Approximately 1980 to present. Inquire as to online cost and availability.

PERIODICALS

MATERIAL HANDLING ENGINEERING. Penton-IPC, 1100 Superior Avenue, Cleveland, OH 44114. (216) 696-7000. 1945 to present. Monthly.

MODERN MATERIALS HANDLING. Cahners Publishing Company, Inc., 275 Washington Street, Newton, MA 02158. (617) 964-3030. Monthly. $45.00 per year.

RESEARCH CENTERS AND INSTITUTES

CENTER FOR DESIGN AND MANUFACTURING INNOVATION. Lehigh University, H.S. Mohler Laboratory, 200, Bethleham, PA 18015. (215) 758-4114.

COAL AND MINERAL PROCESSING LABORATORY. University of British Columbia, 2332 West Mall, Vancouver, BC, Canada V6T 1W5. (604) 228-2540.

WEB HANDLING RESEARCH CENTER. Oklahoma State University, 111 Engineering North, Stillwater, OK 74078-0535. (405) 624-5140.

MATERIALS SCIENCE

See also: ALLOYS, FERROALLOYS, METALLURGICAL ENGINEERING, METALLURGY, METALS AND METALWORKING, NONDESTRUCTIVE TESTING

ABSTRACT SERVICES AND INDEXES

APPLIED MECHANICS REVIEW. American Society of Mechanical Engineers, 345 East 47th Street, New York, NY 10017. (212) 705-7703. 1948 to present. Monthly. $360.00 per year.

APPLIED SCIENCE AND TECHNOLOGY INDEX. H.W. Wilson and Company, 950 University Avenue, Bronx, NY 10452. (800) 367-6670 or (212) 588-8400. Monthly. Inquire as to cost and availability.

CERAMIC ABSTRACTS. American Ceramic Society, Inc., 65 Ceramic Drive, Columbus, OH 43214. (614) 268-8645. 1922 to present. Bimonthly. $185 per year.

CHEMICAL ABSTRACTS. American Chemical Society, Chemical Abstracts Service, Box 3012, Columbus, OH 43210. (614) 421-3600. 1907 to present. Weekly. $9500.00 per year.

CURRENT CONTENTS: ENGINEERING, TECHNOLOGY AND APPLIED SCIENCES. Institute for Scientific Information, 3501 Market Street, Philadelphia, PA 19104. (800) 523-1850 or (215) 386-0100. Weekly. $275.00 per year.

ENGINEERING INDEX MONTHLY AND AUTHOR INDEX. Engineering Information Inc., 345 East 47th Street, New York, NY 10017. (212) 705-7600. Monthly. $1560.00 per year.

INTERNATIONAL AEROSPACE ABSTRACTS. American Institute of Aeronautics and Astronautics, Technical Information

Service, 555 West 57th Street, New York, NY 10019. (212) 247-6500. 1961 to present. Semi-monthly. $700.00 per year.

ISMEC BULLETIN (Information Service in Mechanical Engineering). Cambridge Scientific Abstracts, 5161 River Road, Bethesda, MD 20816. (301) 951-1400. 1973 to present. Monthly. $450.00 per year.

METALS ABSTRACTS AND METALS ABSTRACTS INDEX. American Society for Metals, Metals Park, OH 44073. (216) 338-5151. 1968 to present. Monthly. Abstracts are $1100.00 per year and Index is $460.00 per year.

PHYSICS ABSTRACTS. Institution of Electrical Engineers. Available from: IEEE Service Center, 445 Hoes Lane, Piscataway, NJ 08854. 1898 to present. Bimonthly. $1700.00 per year.

SCIENCE CITATION INDEX. Institute for Scientific Information, 3501 Market Street, Philadelphia, PA 19104. (800) 523-1850 or (215) 386-0100. Six times per year. $6200.00 per year.

WORLD ALUMINUM ABSTRACTS. Aluminum Association, 818 Connecticut Avenue, NW, Washington, DC 20006. (202) 862-5156. 1968 to present. Monthly. $240.00 per year.

ASSOCIATIONS AND PROFESSIONAL SOCIETIES

AMERICAN CERAMIC SOCIETY, Inc., 65 Ceramic Drive, Columbus, OH 43214. (614) 268-8645.

AMERICAN POWDER METALLURGY INSTITUTE. 105 College Road, East, Princeton, NJ 08540. (609) 452-7700.

AMERICAN SOCIETY FOR METALS. Metals Park, OH 44073. (216) 338-5151.

AMERICAN SOCIETY FOR NONDESTRUCTIVE TESTING. 4153 Arlingate Plaza, Caller #28518, Columbus, OH 43228. (614) 274-6003.

AMERICAN SOCIETY FOR TESTING AND MATERIALS. 1916 Race Street, Philadelphia, PA 19103. (215) 299-5400.

AMERICAN SOCIETY OF MECHANICAL ENGINEERS. 345 47th Street, New York, NY 10017. (212) 705-7722.

MATERIALS PROPERTIES COUNCIL. 345 East 47th Street, New York, NY 10017. (212) 705-7693.

MATERIALS RESEARCH SOCIETY. 9800 McKnight Road, Suite 327, Pittsburgh, PA 15237. (412) 367-3003.

THE METALLURGICAL SOCIETY. 420 Commonwealth Drive, Warrendale, PA 15086. (412) 776-9000.

NATIONAL ASSOCIATION OF CORROSION ENGINEERS. Box 218340, Houston, TX 77218. (713) 492-0535.

NATIONAL MATERIALS ADVISORY BOARD. 2101 Constitution Avenue, NW, Washington, DC 20418. (202) 334-3505.

SOCIETY FOR THE ADVANCEMENT OF MATERIAL AND PROCESS ENGINEERING (SAMPE). P.O. Box 2459, 843 West Glentana Street, Covina, CA 91722. (818) 331-0616.

DIRECTORIES AND BIOGRAPHICAL SOURCES

MATERIALS DIRECTORY. Cahners Publishing Company, 275 Washington Street, Newton, MA 02158. Annual. $15.00.

1987 DIRECTORY OF ENGINEERING SOCIETIES AND RELATED ORGANIZATIONS. Gordon Davis, editor. Hemisphere Publishing Corporation, 1010 Vermont Avenue, NW, Washington, DC 20005. (800) 526-0275. 12th edition. 1987. $100.00.

RESEARCH CENTERS DIRECTORY 1988. Gale Research Company, Book Tower, Detroit, MI 48226. (800) 521-0707. 12th edition. 1987. $365.00 for set.

SCIENTIFIC AND TECHNICAL ORGANIZATIONS AND AGENCIES DIRECTORY. Margaret Labash Young, editor. Gale Research Company, Book Tower, Detroit, MI 48226. (800) 521-0707. 2nd edition. 1987. $185.00.

WHO'S WHO IN ENGINEERING. Gordon Davis, editor. Hemisphere Publishing Corporation, 1010 Vermont Avenue, NW, Washington, DC 20005. (800) 526-0275. 6th edition. 1985. $200.00.

ENCYCLOPEDIAS AND DICTIONARIES

THESAURUS OF SCIENTIFIC, TECHNICAL, AND ENGINEERING TERMS. Hemisphere Publishing Corporation, 1010 Vermont Avenue, NW, Washington, DC 20005. (800) 526-0275. 1988. $125.00.

GENERAL WORKS

ENGINEERING MATERIALS: AN INTRODUCTION TO THEIR PROPERTIES AND APPLICATIONS. Michael F. Ashby and David R.H. Jones. Pergamon Press, Inc., Maxwell House, Fairview Park, Elmsford, NY 10523. (914) 592-7700. 1980. $91.50.

INTRODUCTION TO MATERIALS SCIENCE FOR ENGINEERS. James F. Shackelford. Macmillan Publishing Company, Inc., 866 Third Avenue, New York, NY 10022. (800) 257-5755. 2nd edition. 1988. Inquire.

MATERIALS TO RESIST WEAR: A GUIDE TO THEIR SELECTION. A.R. Lansdown and A.L. Price. Pergamon Press, Inc., Maxwell House, Fairview Park, Elmsford, NY 10523. (914) 592-7700. 1987. $11.50 in paper.

PRINCIPLES OF MATERIALS SCIENCE AND ENGINEERING. William F. Smith. McGraw-Hill Book Company, 1221 Avenue of the Americas, New York, NY 10020. (212) 512-2000. 1986. $45.95.

STATISTICS AND MECHANICS OF MATERIALS. Braja M. Das and Paul C. Hassler. Prentice-Hall Publishing, Inc., Englewood Cliffs, NJ 07632. (800) 562-0245. 1988. $42.50.

STATISTICS AND STRENGTH OF MATERIALS. Karl K. Stevens. Prentice-Hall Publishing, Inc., Englewood Cliffs, NJ 07632. (800) 562-0245. 1987. $39.95.

HANDBOOKS AND MANUALS

MATERIALS HANDBOOK. George S. Brady and Henry R. Clauser. McGraw-Hill Book Company, 1221 Avenue of the Americas, New York, NY 10020. (212) 512-2000. 12th edition. 1986. $59.50.

PHYSICAL PROPERTIES OF MATERIALS FOR ENGINEERS. CRC Press, 2000 Corporate Boulevard, Boca Raton, FL 33431. (800) 272-7737. Three volumes. 1982. $275.00 for set.

ONLINE DATA BASES

CA SEARCH. Chemical Abstracts Service, P.O. Box 3012, Columbus, OH 43120. (800) 848-6538 or (614) 421-3600. Comprehensive guide to chemical literature, 1972 to present. Inquire as to online cost and availability.

COMPENDEX. Engineering Information, Inc., 345 East 47th Street, New York, NY 10017. (800) 221-1044 or (212) 705-7615. Engineering and technical literature, 1975 to present. Inquire as to online cost and availability.

INSPEC. INSPEC Marketing Department, Institution of Electrical Engineers. Available from IEEE Service Center, 445 Hoes Lane, Piscataway, NJ 08854. (201) 981-0060. Online version of Physics Abstracts. Inquire as to online cost and availability.

ISMEC. Cambridge Scientific Abstracts, 5161 River Road, Besthda, MD 20816. (800) 638-8076 or (301) 951-1400. Literature of mechanical and production engineering, 1973 to present. Inquire as to online cost and availability.

NTIS. National Technical Information Service, 5285 Port Royal Road, Springfield, VA 22161. (703) 487-4630. Broad coverage of government sponsored research reports, 1964 to present. Inquire as to online cost and availability.

SCISEARCH. Institute for Scientific Information, 3501 Market Street, Philadelphia, PA 19104. (800) 523-1850 or (215) 386-0100. Broad multidisciplinary title and author index to the international literature of science and technology, 1974 to present. Inquire as to online cost and availability.

WILSONLINE. H.W. Wilson and Company, 950 University Avenue, Bronx, NY 10452. (800) 367-6770 or (212) 588-8400. Makes available online versions of the H.W. Wilson indexes including Applied Science and Technology Index, Business Periodicals Index and Readers' Guide to Periodical Literature. Approximately 1980 to present. Inquire as to online cost and availability.

PERIODICALS

JOURNAL OF ENGINEERING FOR INDUSTRY. American Society of Mechanical Engineers, 345 East 47th Street, New York, NY 10017. (212) 705-7703. 1970 to present. Quarterly. $100.00 per year.

JOURNAL OF MATERIALS RESEARCH. Materials Research Society. 9800 McKnight Road, Suite 327, Pittsburgh, PA 15237. (412) 367-3003.

JOURNAL OF METALS. American Institute of Mining, Metallurgical, and Petroleum Engineers, Inc., Metallurgical Society, 420 Commonwealth Drive, Warrendale, PA 15086. (412) 776-9086. 1949 to present. Monthly. $40.00 per year.

JOURNAL OF NONDESTRUCTIVE EVALUATION. Plenum Publishing Corporation, 233 Spring Street, New York, NY 10013. (800) 221-9369. 1980 to present. Quarterly. $85.00 per year.

JOURNAL OF TESTING AND EVALUATION. American Society for Testing and Materials. 1916 Race Street, Philadelphia, PA 19103. (215) 299-5400. 1966 to present. Bimonthly. $40.00 per year.

MATERIALS ENGINEERING. Penton-IPC, 1100 Superior Avenue, Cleveland, OH 44114. (216)696-7000. 1929 to present. Monthly. $40.00.

MATERIALS EVALUATION. American Society for Nondestructive Testing, 4153 Arlingate Plaza, Caller #28518, Columbus, OH 43228. (614) 274-6003. 1942 to present. Monthly. $50.00 per year.

MATERIALS PERFORMANCE. National Association of Corrosion Engineers, P.O. Box 218340, Houston, TX 77218. (713) 492-0535. 1962 to present. Monthly. $50.00 per year.

SAMPE JOURNAL. Society for the Advancement of Material and Process Engineering (SAMPE), P.O. Box 2459, 843 West Glentana Street, Covina, CA 91722. (818) 331-0616. 1965 to present. Bimonthly. $31.00 per year.

RESEARCH CENTERS AND INSTITUTES

CENTER FOR MATERIALS SCIENCE AND ENGINEERING. Massachusetts Institute of Technology, Building 13, Room 2090, 77 Massachusetts Avenue, Cambridge, MA 02139. (617) 253-6801.

CENTER FOR RESEARCH IN MATERIALS SCIENCE AND ENGINEERING. University of Texas at Austin, ETC 9.104, Austin, TX 78712. (512) 471-1504.

MATERIALS RESEARCH LABORATORY. Ohio State University, 3063 Alpheus Smith Laboratory, 174 West 18th Avenue, Columbus, OH 43210. (614) 422-5190.

MATERIALS RESEARCH LABORATORY. University of Illinois, Urbana, IL 61801. (217) 333-1371.

MATERIALS SCIENCE CENTER. Cornell University, Clark Hall of Science, Ithaca, NY 14853. (607) 255-4272.

MATHEMATICAL PHYSICS

See also: PHYSICS

ABSTRACT SERVICES AND INDEXES

CHEMICAL ABSTRACTS. American Chemical Society, Chemical Abstracts Service, Box 3012, Columbus, OH 43210. (614) 421-3600. 1907 to present. Weekly. $9500.00 per year.

CONFERENCE PAPERS INDEX. Cambridge Scientific Abstracts, 5161 River Road, Bethesda, MD 20816. 1972 to present. Monthly. Inquire as to cost and availability.

CURRENT CONTENTS: PHYSICAL, CHEMICAL AND EARTH SCIENCES. Institute for Scientific Information, 3501 Market Street, Philadelphia, PA 19104. (800) 523-1850 or (215) 386-0100. Weekly. $275.00 per year.

CURRENT MATHEMATICAL PUBLICATIONS. American Mathematical Society, P.O. Box 6248, Providence, RI 02940. (800) 556-7774 or (401) 272-9500. 1969 to present. Seventeen times per year. $230.00 per year.

INDEX TO SCIENTIFIC AND TECHNICAL PROCEEDINGS. Institute for Scientific Information, 3501 Market Street, Philadelphia, PA 19104. (800) 523-1850 or (215) 386-0100. 1978 to present. Monthly. $775.00 per year.

INDEX TO SCIENTIFIC REVIEWS. Institute for Scientific Information, 3501 Market Street, Philadelphia, PA 19104. (800) 523-1850 or (215) 386-0100. 1974 to present. Semi-annual. $550.00 per year.

MATHEMATICAL REVIEWS: A REVIEWING JOURNAL COVERING THE WORLD LITERATURE OF MATHEMATICAL RESEARCH. American Mathematical Society, P.O. Box 6248, Providence, RI 02940. (800) 7774 or (401) 272-9500. 1940 to present. Monthly. $2800.00 per year.

PHYSICS ABSTRACTS. Institution of Electrical Engineers. Available from: IEEE Service Center, 445 Hoes Lane, Piscataway, NJ 08854. 1898 to present. Bimonthly. $1700.00 per year.

PHYSICS BRIEFS. Physik Verlag GmBH, Postfach 1260/1280, D-6940 Weinheim, West Germany. 1920 to present. Twenty-six times per year. $1200.00 per year.

SCIENCE CITATION INDEX. Institute for Scientific Information, 3501 Market Street, Philadelphia, PA 19104. (800) 523-1850 or (215) 386-0100. Six times per year. $6200. 00 per year.

ASSOCIATIONS AND PROFESSIONAL SOCIETIES

AMERICAN INSTITUTE OF PHYSICS. 335 East 45th Street, New York, NY 10017. (212) 661-9404.

AMERICAN MATHEMATICAL SOCIETY. P.O. Box 6248, Providence, RI 02940. (800) 7774 or (401) 272-9500.

AMERICAN PHYSICAL SOCIETY. 335 East 45th Street, New York, NY 10017. (212) 682-7341.

DIRECTORIES AND BIOGRAPHICAL SOURCES

INTERNATIONAL RESEARCH CENTERS DIRECTORY 1988-89. Darren L. Smith, editor. Gale Research Company, Book Tower, Detroit, MI 48226. (800) 521-0707. 4th edition. 1987. $360.00.

RESEARCH CENTERS DIRECTORY 1988. Gale Research Company, Book Tower, Detroit, MI 48226. (800) 521-0707. 12th edition. 1987. $365.00 for set.

SCIENTIFIC AND TECHNICAL ORGANIZATIONS AND AGENCIES DIRECTORY. Margaret Labash Young, editor. Gale Research Company, Book Tower, Detroit, MI 48226. (800) 521-0707. 2nd edition. 1987. $185.00.

GENERAL WORKS

INTRODUCTION TO PHYSICAL MATHEMATICS. P.G. Harper and Denis Weaire. Cambridge University Press, 32 East 57th Street, New York, NY 10022. (800) 872-7423. 1985. $14.95 in paper.

MATHEMATICAL METHODS IN THE PHYSICAL SCIENCES. M.L. Boas. John Wiley and Sons, Inc., 605 Third Avenue, New York, NY 10158. (800) 526-5368. 1983. $42.95.

MATHEMATICAL PHYSICS. Robert Geroch. University of Chicago Press, 5801 Ellis Avenue, Chicago, IL 60637. (800) 621-2736. 1985. $15.00 in paper.

SPECIAL FUNCTIONS OF MATHEMATICAL PHYSICS. Arnold F. Nikiforov and V.B. Uvarov. Birkhauser Boston, Inc., 380 Green Street, Cambridge, MA 02139. (617) 876-2333. 1987. $55.00.

ONLINE DATA BASES

CA SEARCH. Chemical Abstracts Service, P.O. Box 3012, Columbus, OH 43120. (800) 848-6538 or (614) 421-3600. Comprehensive guide to chemical literature, 1972 to present. Inquire as to online cost and availability.

DISSERTATION ABSTRACTS ONLINE. University Microfilms International, 300 North Zeeb Road, Ann Arbor, MI 48106. (800) 521-0600 or (313) 761-4700. Scope includes virtually all doctoral dissertations accepted at accredited American institutions from 1861 to present in over 250 subject areas. Inquire as to online cost and availability.

INSPEC. INSPEC Marketing Department, Institution of Electrical Engineers. Available from IEEE Service Center, 445 Hoes Lane, Piscataway, NJ 08854. (201) 981-0060. Online version of Physics Abstracts. Inquire as to online cost and availability.

MATHFILE. American Mathematical Society, P.O. Box 6248, Providence, RI 02940. (800) 556-7774 or (401) 272-9500. An online version of Mathematical Reviews. 1973 to present. Inquire as to online cost and availability.

NTIS. National Technical Information Service, 5285 Port Royal Road, Springfield, VA 22161. (703) 487-4630. Broad coverage of government sponsored research reports, 1964 to present. Inquire as to online cost and availability.

SCISEARCH. Institute for Scientific Information, 3501 Market Street, Philadelphia, PA 19104. (800) 523-1850 or (215) 386-0100. Broad multidisciplinary title and author index to the international literature of science and technology, 1974 to present. Inquire as to online cost and availability.

PERIODICALS

COMPUTERS IN PHYSICS. American Institute of Physics. 335 East 45th Street, New York, NY 10017. (212) 661-9404. 1987 to present. Bimonthly. $250.00 per year.

JOURNAL OF COMPUTATIONAL PHYSICS. Academic Press, Inc., 6277 Sea Harbor Drive, Orlando, FL 32821. (800) 321-5068. 1966 to present. Monthly. $685.00 per year.

JOURNAL OF MATHEMATICAL PHYSICS. American Institute of Physics. 335 East 45th Street, New York, NY 10017. (212) 661-9404. 1960 to present. Monthly. $550.00 per year.

JOURNAL OF PHYSICS A: MATHEMATICAL AND GENERAL. Institute of Physics, Britol, England. Available from: American Institute of Physics. 335 East 45th Street, New York, NY 10017. (212) 661-9404. 1968 to present. 18 times per year. $510.00 per year.

JOURNAL OF STATISTICAL PHYSICS. Plenum Publishing Corporation, 233 Spring Street, New York, NY 10013. (800) 221-9369. 1969 to present. Twelve times per year. $900.00 per year.

LETTERS IN MATEMATICAL PHYSICS. D. Reidel Publishing Company, 190 Old Derby Street, Hingham, MA 02043. (617) 871-6600. 1975 to present. Eight times per year. $165.00 per year.

PHYSICAL REVIEW LETTERS. American Institute of Physics. 335 East 45th Street, New York, NY 10017. (212) 661-9404. 1958 to present. Weekly. $430.00 per year.

RESEARCH CENTERS AND INSTITUTES

CENTER FOR THEORETICAL PHYSICS. Massachusetts Institution of Technology, Massachusetts Avenue, Cambridge, MA 02139. (617) 253-4852.

COURANT INSTITUTE OF MATHEMATICAL SCIENCES. New York Univsersity, 251 Mercer Street, New York, NY 10012. (212) 460-7100.

INSTITUTE FOR THEORETICAL PHYSICS. State Univsersity of New York at Stony Brook. Stony Brook, NY 11794. (516) 689-6000.

INSTITUTE OF THEORETICAL SCIENCE. University of Oregon, Eugene, OR 97403. (503) 686-5204.

MATHEMATICAL SCIENCES INSTITUTE. Cornell University, 294 Caldwell Hall, Ithaca, NY 14853. (607) 255-8005.

MATHEMATICS (APPLIED)

See also: ALGEBRA, CALCULUS, STATISTICS

ABSTRACT SERVICES AND INDEXES

APPLIED MECHANICS REVIEW. American Society of Mechanical Engineers, 345 East 47th Street, New York, NY 10017. (212) 705-7703. 1948 to present. Monthly. $360.00 per year.

APPLIED SCIENCE AND TECHNOLOGY INDEX. H.W. Wilson and Company, 950 University Avenue, Bronx, NY 10452. (800) 367-6670 or (212) 588-8400. Monthly. Inquire as to cost and availability.

CHEMICAL ABSTRACTS. American Chemical Society, Chemical Abstracts Service, Box 3012, Columbus, OH 43210. (614) 421-3600. 1907 to present. Weekly. $9500.00 per year.

COMPUMATH CITATION INDEX. Institute for Scientific Information, 3501 Market Street, Philadelphia, PA 19104. (800) 523-1850 or (215) 386-0100. Three times per year. $875.00 per year.

CURRENT MATHEMATICAL PUBLICATIONS. American Mathematical Society, P.O. Box 6248, Providence, RI 02940. (800) 556-7774 or (401) 272-9500. 1969 to present. Seventeen times per year. $230.00 per year.

ENGINEERING INDEX MONTHLY AND AUTHOR INDEX. Engineering Information Inc., 345 East 47th Street, New York, NY 10017. (212) 705-7600. Monthly. $1560.00 per year.

GENERAL SCIENCE INDEX. H.W. Wilson and Company, 950 University Avenue, Bronx, NY 10452. (800) 367-6670 or (212) 588-8400. 1978 to present. Monthly. Inquire as to cost and availability.

INDEX TO SCIENTIFIC REVIEWS. Institute for Scientific Information, 3501 Market Street, Philadelphia, PA 19104. (800) 523-1850 or (215) 386-0100. 1974 to present. Semi-annual. $550.00 per year.

MATHEMATICAL REVIEWS: A REVIEWING JOURNAL COVERING THE WORLD LITERATURE OF MATHEMATICAL RESEARCH. American Mathematical Society, P.O. Box 6248, Providence, RI 02940. (800) 7774 or (401) 272-9500. 1940 to present. Monthly. $2800.00 per year.

PHYSICS ABSTRACTS. Institution of Electrical Engineers. Available from: IEEE Service Center, 445 Hoes Lane, Piscataway, NJ 08854. 1898 to present. Bimonthly. $1700.00 per year.

SCIENCE CITATION INDEX. Institute for Scientific Information, 3501 Market Street, Philadelphia, PA 19104. (800) 523-1850 or (215) 386-0100. Six times per year. $6200.00 per year.

ANNUAL REVIEWS AND YEARBOOKS

ADVANCES IN APPLIED MATHEMATICS. Academic Press, Inc., 6277 Sea Harbor Drive, Orlando, FL 32821. (800) 321-5068. Irregular. Price varies, inquire.

ASSOCIATIONS AND PROFESSIONAL SOCIETIES

AMERICAN MATHEMATICAL SOCIETY. P.O. Box 6248, Providence, RI 02940. (401) 272-9500.

MATHEMATICAL ASSOCIATION OF AMERICA. 1529 18th Street, N.W., Washington, DC 20036. (202) 387-5200.

SOCIETY FOR INDUSTRIAL AND APPLIED MATHEMATICS. 1400 Architects Building, 117 South 17th Street, Philadelphia, PA 19103. (215) 564-2929.

BIBLIOGRAPHIES

SCIENTIFIC AND TECHNICAL BOOKS AND SERIALS IN PRINT 1988; AN INDEX TO LITERATURE IN SCIENCE AND TECHNOLOGY. R.R. Bowker Company, 205 East 42nd Street, New York, NY 10017. (800) 521-8110. $175.00.

DIRECTORIES AND BIOGRAPHICAL SOURCES

INTERNATIONAL RESEARCH CENTERS DIRECTORY 1988-89. Darren L. Smith, editor. Gale Research Company, Book Tower, Detroit, MI 48226. (800) 521-0707. 4th edition. 1987. $360.00.

RESEARCH CENTERS DIRECTORY 1988. Gale Research Company, Book Tower, Detroit, MI 48226. (800) 521-0707. 12th edition. 1987. $365.00 for set.

SCIENTIFIC AND TECHNICAL ORGANIZATIONS AND AGENCIES DIRECTORY. Margaret Labash Young, editor. Gale Research Company, Book Tower, Detroit, MI 48226. (800) 521-0707. 2nd edition. 1987. $185.00.

ENCYCLOPEDIAS AND DICTIONARIES

DICTIONARY OF THE PHYSICAL SCIENCES: TERMS, FORMULAS, DATA. Cesare Emiliani. Oxford University Press, 200 Madison Avenue, New York, NY 10016. (800) 458-5833. 1987. $19.95 in paper.

ENCYCLOPEDIA OF INFORMATION SYSTEMS AND SERVICES 1988. Amy Lucas and Annette Novallo, editors. Gale Research Company, Book Tower, Detroit, MI 48226. (800) 521-0707. 8th edition. 1987. $400.00 for set.

ENCYCLOPEDIC DICTIONARY OF MATHEMATICS. Kiyosi Ito, editor. MIT Press, 55 Howard Street, Cambridge, MA 02142. (617) 253-2884. 2nd edition. 1987. Four volumes. $350.00 for set.

THESAURUS OF SCIENTIFIC, TECHNICAL, AND ENGINEERING TERMS. Hemisphere Publishing Corporation, 1010 Vermont Avenue, NW, Washington, DC 20005. (800) 526-0275. 1988. $125.00.

GENERAL WORKS

FROM ONE TO ZERO: A UNIVERSAL HISTORY OF NUMBERS. Georges Irfah. Viking-Penguin, Inc., 40 West 23rd Street, New York, NY 10010. (212) 807-7300. 1986. $35.00.

FUNCTIONAL ANALYSIS. L.V. Kantorovich. Pergamon Press, Inc., Maxwell House, Fairview Park, Elmsford, NY 10523. (914) 592-7700. 2nd edition. 1982. $24.00.

FUNDAMENTALS OF MATHEMATICS. H. Behnke, F. Bachman, K. Fladt, W. Suss, and H. Kunle, editors. MIT Press, 55 Howard Street, Cambridge, MA 02142. (617) 253-2884. Volume 1, Foundations of Mathematics: Real Number System and Algebra; Volume 2, Geometry; Volume 3, Analysis. Inquire.

GRAPHS AND ALGORITHMS. M. Gondran and M. Minoux. John Wiley and Sons, Inc., 605 Third Avenue, New York, NY 10158. (800) 526-5368. 1984. $97.95.

MATHEMATICAl MODELS IN APPLIED MECHANICS. A.B. Taylor. Oxford University Press, 200 Madison Avenue, New York, NY 10016. (800) 458-5833. 1986. $29.95.

MATHEMATICS AND THE SEARCH FOR KNOWLEDGE. Morris Kline. Oxford University Press, 200 Madison Avenue, New York, NY 10016. (800) 458-5833. 1986. $19.95.

MIND TOOLS: THE FIVE LEVELS OF MATHEMATICAL REALITY. Rudy Rucker. Houghton Mifflin Company, 2 Park Street, Boston, MA 02108. (617) 725-5000. 1987. $17.95.

NUMERICAL METHODS IN ENGINEERING AND APPLIED SCIENCE. Bruce Irons. John Wiley and Sons, Inc., 605 Third Avenue, New York, NY 10158. (800) 526-5368. 1987. $32.95.

ON NEW METHODS OF ANALYSIS AND ITS APPLICATIONS. P. Turan. John Wiley and Sons, Inc., 605 Third Avenue, New York, NY 10158. (800) 526-5368. 1984. $52.50.

STATISTICS FOR RESEARCH. S. Dowdy and S. Wearden. John Wiley and Sons, Inc., 605 Third Avenue, New York, NY 10158. (800) 526-5368. 1983. $38.95.

HANDBOOKS AND MANUALS

CRC HANDBOOK OF MATHEMATICAL SCIENCES. William H. Beyer, editor. CRC Press, 2000 Corporate Boulevard, Boca Raton, FL 33431. (800) 272-7737. 6th edition. 1987. $64.95.

CRC HANDBOOK OF TABLES FOR PROBABILITY AND STATISTICS. CRC Press, 2000 Corporate Boulevard, Boca Raton, FL 33431. (800) 272-7737. 2nd edition. 1968. $49.95.

HANDBOOK OF MATHEMATICAL FUNCTIONS WITH FORMULAS, GRAPHS, AND MATHEMATICAL TABLES. John Wiley and Sons, Inc., 605 Third Avenue, New York, NY 10158. (800) 526-5368. 1964. $47.50.

INTEGRALS AND SERIES. A.P. Prudnikov and others. Gordon and Breach Science Publishers, Inc., 50 West 23rd Street, New York, NY 10010. (212) 206-8900. Inquire.

STANDARD MATHEMATICAL TABLES. William Beyer, editor. CRC Press, 2000 Corporate Boulevard, Boca Raton, FL 33431. (800) 272-7737. 28th edition. 1987. $24.95.

TABLES OF INTEGRALS, SERIES AND PRODUCTS: CORRECTED AND ENLARGED EDITION. I.S. Gradshteyn and

MATHEMATICS (APPLIED)

I.M. Ryzhik. Academic Press, Inc., 6277 Sea Harbor Drive, Orlando, FL 32821. (800) 321-5068. 1979. $25.00.

ONLINE DATA BASES

CA SEARCH. Chemical Abstracts Service, P.O. Box 3012, Columbus, OH 43120. (800) 848-6538 or (614) 421-3600. Comprehensive guide to chemical literature, 1972 to present. Inquire as to online cost and availability.

COMPENDEX. Engineering Information, Inc., 345 East 47th Street, New York, NY 10017. (800) 221-1044 or (212) 705-7615. Engineering and technical literature, 1975 to present. Inquire as to online cost and availability.

DISSERTATION ABSTRACTS ONLINE. University Microfilms International, 300 North Zeeb Road, Ann Arbor, MI 48106. (800) 521-0600 or (313) 761-4700. Scope includes virtually all doctoral dissertations accepted at accredited American institutions from 1861 to present in over 250 subject areas. Inquire as to online cost and availability.

INSPEC. INSPEC Marketing Department, Institution of Electrical Engineers. Available from IEEE Service Center, 445 Hoes Lane, Piscataway, NJ 08854. (201) 981-0060. Online version of Physics Abstracts. Inquire as to online cost and availability.

MATHFILE. American Mathematical Society, P.O. Box 6248, Providence, RI 02940. (800) 556-7774 or (401) 272-9500. An online version of Mathematical Reviews. 1973 to present. Inquire as to online cost and availability.

NTIS. National Technical Information Service, 5285 Port Royal Road, Springfield, VA 22161. (703) 487-4630. Broad coverage of government sponsored research reports, 1964 to present. Inquire as to online cost and availability.

SCISEARCH. Institute for Scientific Information, 3501 Market Street, Philadelphia, PA 19104. (800) 523-1850 or (215) 386-0100. Broad multidisciplinary title and author index to the international literature of science and technology, 1974 to present. Inquire as to online cost and availability.

WILSONLINE. H.W. Wilson and Company, 950 University Avenue, Bronx, NY 10452. (800) 367-6770 or (212) 588-8400. Makes available online versions of the H.W. Wilson indexes including Applied Science and Technology Index, Business Periodicals Index and Readers' Guide to Periodical Literature. Approximately 1980 to present. Inquire as to online cost and availability.

PERIODICALS

AMERICAN JOURNAL OF MATHEMATICS. Johns Hopkins University Press, Journals Publishing Division, 701 West 40th Street, Suite 275, Baltimore, MD 21211. (301) 338-7864. Bimonthly. 1878 to present. $115.00 per year.

AMERICAN MATHEMATICAL MONTHLY. Mathematical Association of America, 1529 18th Street, N.W., Washington, DC 20036. (202) 387-5200. 1894 to present. Ten times per year. $70.00 per year.

APPLIED MATHEMATICS AND COMPUTATION. Elsevier Science Publishing Company, Inc., 52 Vanderbilt Avenue, New York, NY 10017. (212) 370-5520. 1975 to present. Monthly. $355.00 per year.

COLLEGE MATHEMATICS JOURNAL. Mathematical Association of America, 1529 18th Street, N.W., Washington, DC 20036. Five times per year. $35.00 per year.

JOURNAL OF APPLIED MATHEMATICS AND SIMULATION (J.A.M.S.). University of Pittsburgh at Bradford, Bradford, PA 16701. 1987 to present. Quarterly. $75.00 per year.

MATHEMATICAL INTELLIGENCER. Springer-Verlag New York, Inc., 175 Fifth Avenue, New York, NY 10010. (800) 526-7254. 1978 to present. Quarterly. $25.00 per year.

PROBABILITY IN THE ENGINEERING AND INFORMATIONAL SCIENCES. Cambridge University Press, 32 East 57th Street, New York, NY 10022. (800) 872-7423. 1987 to present. Quarterly. $110.00 per year.

SIAM JOURNAL OF APPLIED MATHEMATICS. Society for Industrial and Applied Mathematics, 1405 Architects Building, 117 South 17th Street. Philadelphia, PA 19103. (215) 564-2929. 1953 to present. Bimonthly. $130.00 per year.

SIAM JOURNAL OF MATHEMATICAL ANALYSIS. Society for Industrial and Applied Mathematics, 1405 Architects Building, 117 South 17th Street. Philadelphia, PA 19103. (215) 564-2929. 1964 to present. Bimonthly. $120.00 per year.

SIAM REVIEW. Society for Industrial and Applied Mathematics, 1405 Architects Building, 117 South 17th Street. Philadelphia, PA 19103. (215) 564-2929. 1959 to present. Quarterly. $82.00 per year.

RESEARCH CENTERS AND INSTITUTES

CENTER FOR APPLIED MATHEMATICS. University of Georgia, Tucker Hall, Athens, GA 30602. (404) 542-3491.

CENTER FOR MATHEMATICAL SCIENCE RESEARCH. Rutgers University, New Brunswick, NJ 08903. (201) 932-3117.

CENTER FOR PURE AND APPLIED MATHEMATICS. University of California, Berkeley, 977 Evans Hall, Berkeley, CA 94720. (415) 642-0116.

INSTITUTE FOR MATHEMATICS AND ITS APPLICATIONS. University of Minnesota, 514 Vincent, 206 Church Street, S.E., Minneapolis, MN 55455. (612) 624-6066.

INSTITUTE OF APPLIED MATHEMATICS. University of Missouri - Rolla, Rolla, MO 65401. (314) 341-4151.

MECHANICAL ENGINEERING

ABSTRACT SERVICES AND INDEXES

APPLIED MECHANICS REVIEW. American Society of Mechanical Engineers, 345 East 47th Street, New York, NY 10017. (212) 705-7703. 1948 to present. Monthly. $360.00 per year.

APPLIED SCIENCE AND TECHNOLOGY INDEX. H.W. Wilson and Company, 950 University Avenue, Bronx, NY 10452. (800) 367-6670 or (212) 588-8400. Monthly. Inquire as to cost and availability.

CHEMICAL ABSTRACTS. American Chemical Society, Chemical Abstracts Service, Box 3012, Columbus, OH 43210. (614) 421-3600. 1907 to present. Weekly. $9500.00 per year.

CURRENT CONTENTS: ENGINEERING, TECHNOLOGY AND APPLIED SCIENCES. Institute for Scientific Information, 3501 Market Street, Philadelphia, PA 19104. (800) 523-1850 or (215) 386-0100. Weekly. $275.00 per year.

ENGINEERING INDEX MONTHLY AND AUTHOR INDEX. Engineering Information Inc., 345 East 47th Street, New York, NY 10017. (212) 705-7600. Monthly. $1560.00 per year.

ISMEC BULLETIN (Information Service in Mechanical Engineering). Cambridge Scientific Abstracts, 5161 River Road, Bethesda, MD 20816. (301) 951-1400. 1973 to present. Monthly. $450.00 per year.

INDEX TO SCIENTIFIC AND TECHNICAL PROCEEDINGS. Institute for Scientific Information, 3501 Market Street, Philadelphia, PA 19104. (800) 523-1850 or (215) 386-0100. 1978 to present. Monthly. $775.00 per year.

PHYSICS ABSTRACTS. Institution of Electrical Engineers. Available from: IEEE Service Center, 445 Hoes Lane, Piscataway, NJ 08854. 1898 to present. Bimonthly. $1700.00 per year.

SCIENCE CITATION INDEX. Institute for Scientific Information, 3501 Market Street, Philadelphia, PA 19104. (800) 523-1850 or (215) 386-0100. Six times per year. $6200.00 per year.

ASSOCIATIONS AND PROFESSIONAL SOCIETIES

AMERICAN SOCIETY OF MECHANICAL ENGINEERS. 345 47th Street, New York, NY 10017. (212) 705-7722.

VIBRATION INSTITUTE. 101 West 55th Street, Suite 206, Clarendon Hills, IL 60514. (312) 654-2254.

BIBLIOGRAPHIES

NEW TECHNICAL BOOKS: A SELECTIVE LIST WITH DESCRIPTIVE ANNOTATIONS. New York Public Library, Science and Technology Research Center, Fifth Avenue and 42nd Street, New York, NY 10018. (212) 930-0800. 1915 to present. Monthly. $15.00 per year.

SCIENCE BOOKS AND FILMS. American Association for the Advancement of Science, 1333 H Street, NW, Washington, DC 20005. (202) 326-6454. Five times per year. $20.00 per year.

SCIENTIFIC AND TECHNICAL BOOKS AND SERIALS IN PRINT 1988; AN INDEX TO LITERATURE IN SCIENCE AND TECHNOLOGY. R.R. Bowker Company, 205 East 42nd Street, New York, NY 10017. (800) 521-8110. $175.00.

DIRECTORIES AND BIOGRAPHICAL SOURCES

1987 DIRECTORY OF ENGINEERING SOCIETIES AND RELATED ORGANIZATIONS. Gordon Davis, editor. Hemisphere Publishing Corporation, 1010 Vermont Avenue, NW, Washington, DC 20005. (800) 526-0275. 12th edition. 1987. $100.00.

RESEARCH CENTERS DIRECTORY 1988. Gale Research Company, Book Tower, Detroit, MI 48226. (800) 521-0707. 12th edition. 1987. $365.00 for set.

SCIENTIFIC AND TECHNICAL ORGANIZATIONS AND AGENCIES DIRECTORY. Margaret Labash Young, editor. Gale Research Company, Book Tower, Detroit, MI 48226. (800) 521-0707. 2nd edition. 1987. $185.00.

WHO'S WHO IN ENGINEERING. Gordon Davis, editor. Hemisphere Publishing Corporation, 1010 Vermont Avenue, NW, Washington, DC 20005. (800) 526-0275. 6th edition. 1985. $200.00.

ENCYCLOPEDIAS AND DICTIONARIES

ENCYCLOPEDIA OF FLUID MECHANICS. Gulf Publishing Company, P.O. Box 2608, Houston, TX 77001. (713) 520-4444. 6 volumes. 1986. $1000.00 for set.

THESAURUS OF SCIENTIFIC, TECHNICAL, AND ENGINEERING TERMS. Hemisphere Publishing Corporation, 1010 Vermont Avenue, NW, Washington, DC 20005. (800) 526-0275. 1988. $125.00.

GENERAL WORKS

ANALYSIS OF HEAT AND MASS TRANSFER. E.R.G. Eckart and Robert M. Drake. Hemisphere Publishing Corporation, 1010 Vermont Avenue, NW, Washington, DC 20005. (800) 526-0275. 1987. $75.00.

ENGINEERING MECHANICS. J.L. Meriam and L.G. Kraige. John Wiley and Sons, Inc., 605 Third Avenue, New York, NY 10158. (800) 526-5368. 2nd edition. 1986. $52.95.

FUNDAMENTALS OF HEAT EXCHANGE AND PRESSURE VESSEL TECHNOLOGY. J.P. Gupta. Hemisphere Publishing Corporation, 1010 Vermont Avenue, NW, Washington, DC 20005. (800) 526-0275. 1986. $49.95.

HEAT CONDUCTION. Sadik Kakac and Yaman Yener. Hemisphere Publishing Corporation, 1010 Vermont Avenue, NW, Washington, DC 20005. (800) 526-0275. 1985. $44.00.

HANDBOOKS AND MANUALS

APPLIED FLUID DYNAMICS HANDBOOK. Robert D. Blevins. Van Nostrand Reinhold Company, Inc., 135 West 50th Street, New York, NY 10020. (800) 543-2681. 1984. $52.95.

HANDBOOK OF MECHANICS, MATERIALS, AND STRUCTURES. A. Blake. John Wiley and Sons, Inc., 605 Third Avenue, New York, NY 10158. (800) 526-5368. 1985. $65.95.

MARK'S STANDARD HANDBOOK FOR MECHANICAL ENGINEERS. T. Baumeister, editor. McGraw-Hill Book Company, 1221 Avenue of the Americas, New York, NY 10020. (212) 512-2000. 8th edition. 1978. $96.00.

MECHANICAL DESIGN AND SYSTEMS HANDBOOK. Harold A. Rothbart, editor. McGraw-Hill Book Company, 1221 Avenue of the Americas, New York, NY 10020. (212) 512-2000. 2nd edition.1985. $96.50.

MECHANICAL ENGINEERS' HANDBOOK. Myer Kutz, editor. John Wiley and Sons, Inc., 605 Third Avenue, New York, NY 10158. (800) 526-5368. 1986. $79.95.

PRESSURE VESSEL DESIGN HANDBOOK. Henry H. Bednar. Van Nostrand Reinhold Company, Inc., 135 West 50th Street, New York, NY 10020. (800) 543-2681. 2nd edition. 1986. $49.95.

STANDARD HANDBOOK OF MACHINE DESIGN. Joseph E. Shigley and Charles R. Mischke. McGraw-Hill Book Company, 1221 Avenue of the Americas, New York, NY 10020. (212) 512-2000. 1986. $89.00.

ONLINE DATA BASES

CA SEARCH. Chemical Abstracts Service, P.O. Box 3012, Columbus, OH 43120. (800) 848-6538 or (614) 421-3600. Comprehensive guide to chemical literature, 1972 to present. Inquire as to online cost and availability.

COMPENDEX. Engineering Information, Inc., 345 East 47th Street, New York, NY 10017. (800) 221-1044 or (212) 705-7615. Engineering and technical literature, 1975 to present. Inquire as to online cost and availability.

DISSERTATION ABSTRACTS ONLINE. University Microfilms International, 300 North Zeeb Road, Ann Arbor, MI 48106. (800) 521-0600 or (313) 761-4700. Scope includes virtually all doctoral dissertations accepted at accredited American institutions from 1861 to present in over 250 subject areas. Inquire as to online cost and availability.

INSPEC. INSPEC Marketing Department, Institution of Electrical Engineers. Available from IEEE Service Center, 445 Hoes Lane, Piscataway, NJ 08854. (201) 981-0060. Online version of Physics Abstracts. Inquire as to online cost and availability.

ISMEC. Cambridge Scientific Abstracts, 5161 River Road, Besthda, MD 20816. (800) 638-8076 or (301) 951-1400. Literature of mechanical and production engineering, 1973 to present. Inquire as to online cost and availability.

NTIS. National Technical Information Service, 5285 Port Royal Road, Springfield, VA 22161. (703) 487-4630. Broad coverage of government sponsored research reports, 1964 to present. Inquire as to online cost and availability.

SCISEARCH. Institute for Scientific Information, 3501 Market Street, Philadelphia, PA 19104. (800) 523-1850 or (215) 386-0100.

Broad multidisciplinary title and author index to the international literature of science and technology, 1974 to present. Inquire as to online cost and availability.

WILSONLINE. H.W. Wilson and Company, 950 University Avenue, Bronx, NY 10452. (800) 367-6770 or (212) 588-8400. Makes available online versions of the H.W. Wilson indexes including Applied Science and Technology Index, Business Periodicals Index and Readers' Guide to Periodical Literature. Approximately 1980 to present. Inquire as to online cost and availability.

PERIODICALS

CIME (COMPUTERS IN MECHANICAL ENGINEERING). American Society of Mechanical Engineers, 345 East 47th Street, New York, NY 10017. (212) 705-7703.

INTERNATIONAL COMMUNICATIONS IN HEAT AND MASS TRANSFER. Pergamon Press, Inc., Maxwell House, Fairview Park, Elmsford, NY 10523. (914) 592-7700. 1974 to present. Bimonthly. $160.00 per year.

INTERNATIONAL JOURNAL FOR NUMERICAL METHODS IN ENGINEERING. John Wiley and Sons, Inc., 605 Third Avenue, New York, NY 10158. (800) 526-5368.

INTERNATIONAL JOURNAL OF HEAT AND MASS TRANSFER. Pergamon Press, Inc., Maxwell House, Fairview Park, Elmsford, NY 10523. (914) 592-7700. 1960 to present. Monthly. $500.00 per year.

INTERNATIONAL JOURNAL OF MULTIPHASE FLOW. Pergamon Press, Inc., Maxwell House, Fairview Park, Elmsford, NY 10523. (914) 592-7700. (212) 705-7703. 1974 to present. Bimonthly. $250.00 per year.

JOURNAL OF APPLIED MECHANICS. American Society of Mechanical Engineers, 345 East 47th Street, New York, NY 10017. (212) 705-7703. 1935 to present. Quarterly. $100.00 per year.

JOURNAL OF ENGINEERING FOR INDUSTRY. American Society of Mechanical Engineers, 345 East 47th Street, New York, NY 10017. (212) 705-7703. 1970 to present. Quarterly. $100.00 per year.

JOURNAL OF FLUID CONTROL. Delbridge Publishing Company, Box 2989, Stanford, CA 94305. (408) 446-3131. 1967 to present. Quarterly. Inquire as to cost and availability.

JOURNAL OF HEAT TRANSFER. American Society of Mechanical Engineers, 345 East 47th Street, New York, NY 10017. (212) 705-7703. 1970 to present. Quarterly. $100.00 per year.

JOURNAL OF MECHANISMS, TRANSMISSIONS AND AUTOMATION IN DESIGN. American Society of Mechanical Engineers, 345 East 47th Street, New York, NY 10017. (212) 705-7703. 1983 to present. Quarterly. $80.00 per year.

JOURNAL OF PRESSURE VESSEL TECHNOLOGY. American Society of Mechanical Engineers, 345 East 47th Street, New York, NY 10017. (212) 705-7703. 1974 to present. Quarterly. $80.00 per year.

LUBRICATION ENGINEERING. American Society of Lubrication Engineers, 838 Busse Highway, Park Ridge, IL 60068. 1945 to present. Monthly. $39.00 per year.

MECHANICAL ENGINEERING. American Society of Mechanical Engineers, 345 East 47th Street, New York, NY 10017. (212) 705-7722.

RESEARCH CENTERS AND INSTITUTES

LABORATORY FOR EXIPERMENTAL MECHANICS RESEARCH. State University of New York at Stony Brook, Stony Brook, NY 11794-2300. (516) 632-8311.

MECHANICAL ENGINEERING DESIGN LABORATORY. University of Florida, 237 Mechanical Engineering Building, Gainesville, FL 32611. (904) 392-0827

MECHANICAL ENGINEERING LABORATORIES. Stevens Institute of Technology, Hoboken, NJ 07030. (201) 420-5591.

MECHANICAL ENGINEERING RESEARCH LABORATORIES. Kansas State University, Durland Hall, Manhattan, KS 66506. (913) 532-5610.

MECHANICS

See also: FLUID MECHANICS, KINEMATICS, MECHANICAL ENGINEERING, PHYSICS, QUANTUM MECHANICS, STATISTICAL MECHANICS

ABSTRACT SERVICES AND INDEXES

APPLIED MECHANICS REVIEW. American Society of Mechanical Engineers, 345 East 47th Street, New York, NY 10017. (212) 705-7703. 1948 to present. Monthly. $360.00 per year.

APPLIED SCIENCE AND TECHNOLOGY INDEX. H.W. Wilson and Company, 950 University Avenue, Bronx, NY 10452. (800) 367-6670 or (212) 588-8400. Monthly. Inquire as to cost and availability.

CHEMICAL ABSTRACTS. American Chemical Society, Chemical Abstracts Service, Box 3012, Columbus, OH 43210. (614) 421-3600. 1907 to present. Weekly. $9500.00 per year.

CONFERENCE PAPERS INDEX. Cambridge Scientific Abstracts, 5161 River Road, Bethesda, MD 20816. 1972 to present. Monthly. Inquire as to cost and availability.

CURRENT CONTENTS: PHYSICAL, CHEMICAL, AND EARTH SCIENCES. Institute for Scientific Information, 3501 Market Street, Philadelphia, PA 19104. (800) 523-1850 or (215) 386-0100. Weekly. $275.00 per year.

ENGINEERING INDEX MONTHLY AND AUTHOR INDEX. Engineering Information Inc., 345 East 47th Street, New York, NY 10017. (212) 705-7600. Monthly. $1560.00 per year.

GENERAL SCIENCE INDEX. H.W. Wilson and Company, 950 University Avenue, Bronx, NY 10452. (800) 367-6670 or (212) 588-8400. 1978 to present. Monthly. Inquire as to cost and availability.

ISMEC BULLETIN (Information Service in Mechanical Engineering). Cambridge Scientific Abstracts, 5161 River Road, Bethesda, MD 20816. (301) 951-1400. 1973 to present. Monthly. $450.00 per year.

INDEX TO SCIENTIFIC AND TECHNICAL PROCEEDINGS. Institute for Scientific Information, 3501 Market Street, Philadelphia, PA 19104. (800) 523-1850 or (215) 386-0100. 1978 to present. Monthly. $775.00 per year.

INDEX TO SCIENTIFIC REVIEWS. Institute for Scientific Information, 3501 Market Street, Philadelphia, PA 19104. (800) 523-1850 or (215) 386-0100. 1974 to present. Semi-annual. $550.00 per year.

PHYSICS ABSTRACTS. Institution of Electrical Engineers. Available from: IEEE Service Center, 445 Hoes Lane, Piscataway, NJ 08854. 1898 to present. Bimonthly. $1700.00 per year.

PHYSICS BRIEFS. Physik Verlag GmbH, Postfach 1260/1280, D-6940 Weinheim, West Germany. (212) 661-9404. 1920 to present. Twenty-six times per year. $1250.00 per year.

SCIENCE CITATION INDEX. Institute for Scientific Information, 3501 Market Street, Philadelphia, PA 19104. (800) 523-1850 or

(215) 386-0100. Six times per year. $6200.00 per year.

ASSOCIATIONS AND PROFESSIONAL SOCIETIES

AMERICAN CHEMICAL SOCIETY. 1155 16th Street, N.W., Washington, DC 20036. (800) 424-6747.

AMERICAN INSTITUTE OF PHYSICS. 335 East 45th Street, New York, NY 10017. (212) 661-9494.

AMERICAN SOCIETY OF MECHANICAL ENGINEERS. 345 East 47th Street, New York, NY 10017. (212) 705-7703.

DIRECTORIES AND BIOGRAPHICAL SOURCES

AMERICAN MEN AND WOMEN OF SCIENCE. R.R. Bowker, Inc., Order Department, 245 West 17th Street, New York, NY 10011. (800) 521-8110. Eight volumes. 1986. $595.00 for set.

INTERNATIONAL RESEARCH CENTERS DIRECTORY 1988-89. Darren L. Smith, editor. Gale Research Company, Book Tower, Detroit, MI 48226. (800) 521-0707. 4th edition. 1987. $360.00.

1987 DIRECTORY OF ENGINEERING SOCIETIES AND RELATED ORGANIZATIONS. Gordon Davis, editor. Hemisphere Publishing Corporation, 79 Madison Avenue, New York, NY 10016-7892. (800) 821-8312. 12th edition. 1987. $100.00.

RESEARCH CENTERS DIRECTORY 1988. Gale Research Company, Book Tower, Detroit, MI 48226. (800) 521-0707. 12th edition. 1987. $365.00 for set.

SCIENTIFIC AND TECHNICAL ORGANIZATIONS AND AGENCIES DIRECTORY. Margaret Labash Young, editor. Gale Research Company, Book Tower, Detroit, MI 48226. (800) 521-0707. 2nd edition. 1987. $185.00.

WHO'S WHO IN ENGINEERING. Gordon Davis, editor. Hemisphere Publishing Corporation, 79 Madison Avenue, New York, NY 10016-7892. (800) 821-8312. Sixth edition. 1985. $200.00.

ENCYCLOPEDIAS AND DICTIONARIES

MCGRAW-HILL DICTIONARY OF PHYSICS. McGraw-Hill Book Company, 1221 Avenue of the Americas, New York, NY 10020. (212) 512-2000. 1986. $17.95.

THESAURUS OF SCIENTIFIC, TECHNICAL, AND ENGINEERING TERMS. Hemisphere Publishing Corporation, 1010 Vermont Avenue, NW, Washington, DC 20005. (800) 526-0275. 1988. $125.00.

GENERAL WORKS

APPLIED ENGINEERING MECHANICS. A.C. Jensen and H.H. Chenoweth. McGraw-Hill Book Company, 1221 Avenue of the Americas, New York, NY 10020. (212) 512-2000. Fourth edition. 1983. $37.95.

CLASSICAL MECHANICS. E.A. Desloge. John Wiley and Sons, Inc., 605 Third Avenue, New York, NY 10158. (800) 526-5368. Two volumes. 1982. $59.95 for each volume.

QUANTUM MECHANICS. V.K. Thankappan. John Wiley and Sons, Inc., 605 Third Avenue, New York, NY 10158. (800) 526-5368. 1985. $45.95.

ONLINE DATA BASES

CA SEARCH. Chemical Abstracts Service, P.O. Box 3012, Columbus, OH 43120. (800) 848-6538 or (614) 421-3600. Comprehensive guide to chemical literature, 1972 to present. Inquire as to online cost and availability.

COMPENDEX. Engineering Information, Inc., 345 East 47th Street, New York, NY 10017. (800) 221-1044 or (212) 705-7615. Engineering and technical literature, 1975 to present. Inquire as to online cost and availability.

DISSERTATION ABSTRACTS ONLINE. University Microfilms International, 300 North Zeeb Road, Ann Arbor, MI 48106. (800) 521-0600 or (313) 761-4700. Scope includes virtually all doctoral dissertations accepted at accredited American institutions from 1861 to present in over 250 subject areas. Inquire as to online cost and availability.

INSPEC. INSPEC Marketing Department, Institution of Electrical Engineers. Available from IEEE Service Center, 445 Hoes Lane, Piscataway, NJ 08854. (201) 981-0060. Online version of Physics Abstracts. Inquire as to online cost and availability.

NTIS. National Technical Information Service, 5285 Port Royal Road, Springfield, VA 22161. (703) 487-4630. Broad coverage of government sponsored research reports, 1964 to present. Inquire as to online cost and availability.

SCISEARCH. Institute for Scientific Information, 3501 Market Street, Philadelphia, PA 19104. (800) 523-1850 or (215) 386-0100. Broad multidisciplinary title and author index to the international literature of science and technology, 1974 to present. Inquire as to online cost and availability.

WILSONLINE. H.W. Wilson and Company, 950 University Avenue, Bronx, NY 10452. (800) 367-6770 or (212) 588-8400. Makes available online versions of the H.W. Wilson indexes including Applied Science and Technology Index, Business Periodicals Index and Readers' Guide to Periodical Literature. Approximately 1980 to present. Inquire as to online cost and availability.

PERIODICALS

INTERNATIONAL JOURNAL OF SOLIDS AND STRUCTURES. Pergamon Press, Inc., Maxwell House, Fairview Park, Elmsford, NY 10523. (914) 592-7700. 1965 to present. Monthly. $375.00 per year.

JOURNAL OF RHEOLOGY. John Wiley and Sons, Inc., 605 Third Avenue, New York, NY 10158. (800) 526-5368. 1957 to present. Bimonthly. $152.00 per year.

JOURNAL OF THE MECHANICS AND PHYSICS OF SOLIDS. Pergamon Press, Inc., Maxwell House, Fairview Park, Elmsford, NY 10523. (914) 592-7700. 1956 to present. Bimonthly. $200.00 per year.

PHYSICAL REVIEW LETTERS. American Institute of Physics, 335 East 45th Street, New York, NY 10017. (212) 661-9494. 1958 to present. Weekly. $430.00 per year.

RESEARCH CENTERS AND INSTITUTES

CENTER FOR THE ADVANCEMENT OF COMPUTATIONAL MECHANICS. Georgia Institute of Technology, Atlanta, GA 30332. (404) 894-2758.

LABORATORY FOR EXPERIMENTAL MECHANICS RESEARCH. State University of New York at Stony Brook, Stony Brook, NY 11794-2300.

SOLID MECHANICS LABORATORY. University of Michigan, Department of Mechanical Engineering and Applied Mechanics, 2246A G.G. Brown Building, Ann Arbor, MI 48109. (313) 763-0684.

MERCURY

See also: PLANETARY SCIENCE, SOLAR SYSTEM

ABSTRACT SERVICES AND INDEXES

ASTRONOMY AND ASTROPHYSICS ABSTRACTS. Springer-Verlag New York, Incorporated, 175 Fifth Avenue, New York, NY 10010. (212) 460-1500. $70.00 per year.

ASTRONOMY AND ASTROPHYSICS MONTHLY INDEX. Olivetree Associates, Post Office Box 236, Sierra Madre, CA 91024. $212.00 per year. Complimentary copies available on request.

METEOROLOGICAL AND GEOASTROPHYSICAL ABSTRACTS. American Meteorological Society, 45 Beacon Street, Boston, MA 02108. (617) 227-2425. 1950 to present. Monthly. $450.00 per year.

STAR. (Scientific and Technical Aerospace Reports. United States National Aeronautics and Space Administration, Scientific and Technical Information Facility, Box 8757, Baltimore-Washington International Airport, MD 21240. (202) 755-2210. Semimonthly, with semiannual and annual indexes. $85.00 per year.

ANNUAL REVIEWS AND YEARBOOKS

THE ASTRONOMICAL ALMANAC. Superintendent of Documents, United States Government Printing Office, Washington, DC 20402. (202) 783-3238. Yearly.

ANNUAL REVIEW OF ASTRONOMY AND ASTROPHYSICS. Annual Reviews, Incorporated, 4139 El Camino Way, Palo Alto, CA 94306. (415) 493-4400. Annual. Inquire.

ANNUAL REVIEW OF EARTH AND PLANETARY SCIENCES. Annual Reviews, Incorporated, 4139 El Camino Way, Palo Alto, CA 94306. (415) 493-4400. Annual. Inquire.

ASSOCIATIONS AND PROFESSIONAL SOCIETIES

AMERICAN ASTRONOMICAL SOCIETY. 2000 Florida Avenue, NW, Suite 300, Washington, DC 20009. (202) 659-0134.

AMERICAN GEOPHYSICAL UNION. 2000 Florida Avenue, NW, Washington, DC 20009. (202) 462-6903.

ASTRONOMICAL LEAGUE. Post Office Box 12821, Tucson, AZ 85732. (602) 790-8471.

ASTRONOMICAL SOCIETY OF THE PACIFIC. 1290 24th Avenue, San Francisco, CA 94122. (415) 661-8660.

PLANETARY SOCIETY. 65 North Catalina Avenue, Pasadena, CA 91106-2301. (818) 793-5100.

BIBLIOGRAPHIES

A BIBLIOGRAPHY OF ASTRONOMY, 1970-1979. R.A. Seal and S.S. Martin. Libraries Unlimited, Incorporated, Littleton, CO 80160. 1982. $37.50.

SCIENTIFIC AND TECHNICAL BOOKS AND SERIALS IN PRINT 1988: AN INDEX TO LITERATURE IN SCIENCE AND TECHNOLOGY. R.R. Bowker Company, 205 East 42nd Street, New York, NY 10017. (800) 521-8110 or (212) 916-1600. $175.00.

DIRECTORIES AND BIOGRAPHICAL SOURCES

AMERICAN ASTRONOMICAL SOCIETY MEMBERS. 2000 Florida Avenue, NW, Suite 300, Washington, DC 20009. (202) 659-0134. Annual. Available to members only.

THE BIOGRAPHICAL DICTIONARY OF SCIENTISTS: ASTRONOMERS. D. Abbott, editor. Peter Bedrick Books, 125 East 23rd Street, New York, NY 10010. 1984. $24.95.

DIRECTORY OF PHYSICS AND ASTRONOMY STAFF MEMBERS. American Institute of Physics, 335 East 45th Street, New York, NY 10017. Annual.

GENERAL WORKS

MERCURY'S PERIHELION FROM LE VERRIER TO EINSTEIN. N.T. Roseveare. Oxford University Press, 200 Madison Avenue, New York, NY 10016. (212) 564-6680. 1982. $47.50.

ONLINE DATA BASES

CA SEARCH. Chemical Abstracts Service, Post Office Box 3012, Columbus, OH 43210. Guide to chemical literature, 1972 to present. Inquire as to cost and availability.

DISSERTATION ABSTRACTS ONLINE. University Microfilms International, 300 North Zeeb Road, Ann Arbor, MI 48106. (800) 521-0600 or (313) 761-4700. Scope includes virtually all doctoral dissertations accepted at accredited American institutions from 1861 to present in 252 subject areas. Inquire as to cost and availability.

NASA. National Aeronautics and Space Administration, Scientific and Technical Information Branch, 300 7th Street, SW, Washington, DC 20546. Citations and abstracts of aerospace literature, 1962 to present. Inquire as to cost and availability.

NTIS. National Technical Information Service, 5285 Port Royal Road, Springfield, VA 22161. (703) 487-4630. Broad coverage of government sponsored research reports, 1964 to present. Inquire as to cost and availability.

SCISEARCH. Institute for Scientific Information, 3501 Market Street, Philadelphia, PA 19104. (800) 523-1850 or (215) 386-0100. Broad multidisciplinary title and author index to the international literature of science and technology, 1974 to present. Inquire as to cost and availability.

PERIODICALS

ASSOCIATION OF LUNAR AND PLANETARY OBSERVERS. Journal Association of Lunar and Planetary Observers, Box 16131, San Francisco, CA 94116. (415) 566-5786. 1947 to present. Quarterly. $15.00 per year.

ASTRONOMICAL JOURNAL. American Astronomical Society. Available from: American Institute of Physics, 335 East 45th Street, New York, NY 10017. 9212) 661-9404. $125.00 per year.

EARTH, MOON AND PLANETS: AN INTERNATIONAL JOURNAL OF COMPARATIVE PLANETOLOGY. D. Reidel Publishing Company, 190 Old Derby Street, Hingham, MA 02043. Nine times per year. $275.00 per year.

ICARUS: INTERNATIONAL JOURNAL OF THE SOLAR SYSTEM STUDIES. Academic Press, Incorporated, Orlando, FL 32887. (305) 345-4100. Monthly. $484.00 per year.

MERCURY. Astronomical Society of the Pacific, 1290 24th Avenue, San Francisco, CA 94122. (415) 661-8660. Bimonthly. $21.00 per year.

PLANETARY AND SPACE SCIENCE. Pergamon Press, Incorporated, Maxwell House, Fairview Park, Elmsford, NY 10523. (914) 592-7700. Monthly. $430.00 per year.

PLANETARY REPORT. Planetary Society, 65 North Catalina Avenue, Pasadena, CA 91106-2301. (818) 793-5100. 1980 to present. Bimonthly. $20.00 per year.

SOLAR SYSTEM RESEARCH (ENGLISH TRANSLATION OF ASTRONOMICHESKII VESTNIK). Consultants Bureau, 233 Spring Street, New York, NY 10013. (212) 620-8000. 1967 to present. Quarterly. $425.00 per year.

RESEARCH CENTERS AND INSTITUTES

LABORATORY FOR PLANETARY ATMOSPHERES RESEARCH. State University of New York at Stony Brook, Stony Brook, NY 11794-2300. (516) 632-8321.

LABORATORY FOR PLANETARY STUDIES. Cornell University, 302 Space Sciences Building, Ithaca, NY 14853. (607) 256-4971.

LUNAR AND PLANETARY INSTITUTE. 3303 NASA Road One, Houston, TX 77058. (713) 486-2139.

LUNAR AND PLANETARY LABORATORY. University of Arizona, Tucson, AZ 85721. (602) 621-6962.

MESON

See: PARTICLE PHYSICS

METALLURGICAL ENGINEERING

See also: ALLOYS, FERROALLOYS, MACHINING, METALS AND METAL WORKING, METALLURGY

ABSTRACT SERVICES AND INDEXES

APPLIED MECHANICS REVIEW. American Society of Mechanical Engineers, 345 East 47th Street, New York, NY 10017. (212) 705-7703. 1948 to present. Monthly. $360.00 per year.

APPLIED SCIENCE AND TECHNOLOGY INDEX. H.W. Wilson and Company, 950 University Avenue, Bronx, NY 10452. (800) 367-6670 or (212) 588-8400. Monthly. Inquire as to cost and availability.

CHEMICAL ABSTRACTS. American Chemical Society, Chemical Abstracts Service, Box 3012, Columbus, OH 43210. (614) 421-3600. 1907 to present. Weekly. $9500.00 per year.

CORROSION ABSTRACTS. National Association of Corrosion Engineers, Box 218340, Houston, TX 77218. (713) 492-0535. 1962 to present. Bimonthly. $200.00 per year.

CURRENT CONTENTS: ENGINEERING, TECHNOLOGY AND APPLIED SCIENCES. Institute for Scientific Information, 3501 Market Street, Philadelphia, PA 19104. (800) 523-1850 or (215) 386-0100. Weekly. $275.00 per year.

ENGINEERING INDEX MONTHLY AND AUTHOR INDEX. Engineering Information Inc., 345 East 47th Street, New York, NY 10017. (212) 705-7600. Monthly. $1560.00 per year.

ISMEC BULLETIN (Information Service in Mechanical Engineering). Cambridge Scientific Abstracts, 5161 River Road, Bethesda, MD 20816. (301) 951-1400. 1973 to present. Monthly. $450.00 per year.

LEAD ABSTRACTS. Lead Development Association, 34 Berkeley Square, London, W1X 6AJ, England. 1958 to present. Quarterly. $70.00 per year.

METALS ABSTRACTS AND METALS ABSTRACTS INDEX. American Society for Metals, Metals Park, OH 44073. (216) 338-5151. 1968 to present. Monthly. Abstracts are $1100.00 per year and Index is $460.00 per year.

PHYSICS ABSTRACTS. Institution of Electrical Engineers. Available from: IEEE Service Center, 445 Hoes Lane, Piscataway, NJ 08854. 1898 to present. Bimonthly. $1700.00 per year.

SCIENCE CITATION INDEX. Institute for Scientific Information, 3501 Market Street, Philadelphia, PA 19104. (800) 523-1850 or (215) 386-0100. Six times per year. $6200.00 per year.

WORLD ALUMINUM ABSTRACTS. Aluminum Association, 818 Connecticut Avenue, NW, Washington, DC 20006. (202) 862-5156. 1968 to present. Monthly. $240.00 per year.

ASSOCIATIONS AND PROFESSIONAL SOCIETIES

ALUMINUM ASSOCIATION. 818 Connecticut Avenue, NW, Washington, DC 20006. (202) 862-5156.

AMERICAN INSTITUTE OF MINING, METALLURGICAL AND PETROLEUM ENGINEERS (AIME). 420 Commonwealth Drive, Warrendale, PA 15086. (412) 776-9086.

AMERICAN POWDER METALLURGY INSTITUTE. 105 College Road, East, Princeton, NJ 08540. (609) 452-7700.

AMERICAN SOCIETY FOR METALS. Metals Park, OH 44073. (216) 338-5151.

AMERICAN SOCIETY FOR TESTING AND MATERIALS. 1916 Race Street, Philadelphia, PA 19103. (215) 299-5400.

AMERICAN SOCIETY OF MECHANICAL ENGINEERS. 345 47th Street, New York, NY 10017. (212) 705-7722.

ASSOCIATION OF IRON AND STEEL ENGINEERS. Three Gateway Center, Suite 2350, Pittsburgh, PA 15222. (412) 281-6323.

THE METALLURGICAL SOCIETY. 420 Commonwealth Drive, Warrendale, PA 15086. (412) 776-9000.

NATIONAL ASSOCIATION OF CORROSION ENGINEERS. Box 218340, Houston, TX 77218. (713) 492-0535.

DIRECTORIES AND BIOGRAPHICAL SOURCES

INDUSTRIAL EQUIPMENT AND SUPPLIES DIRECTORY. American Business Directories, Inc., 5707 South 86th Circle, Omaha, NE 68127. (402) 331-7169.

MACHINERY BUYERS GUIDE. Findlay Publications Limited, Maitland House, Warrior Square, Southend-on-Sea, Essex SS5 5AR England. Annual. $45.00.

METALLURGICAL SOCIETY OF AIME - MEMBERSHIP LIST. American Institute of Mining, Metallurgical and Petroleum Engineers (AIME). 345 East 47th Street, New York, NY 10017. (212) 705-7695. 1984.

1987 DIRECTORY OF ENGINEERING SOCIETIES AND RELATED ORGANIZATIONS. Gordon Davis, editor. Hemisphere Publishing Corporation, 1010 Vermont Avenue, NW, Washington, DC 20005. (800) 526-0275. 12th edition. 1987. $100.00.

RESEARCH CENTERS DIRECTORY 1988. Gale Research Company, Book Tower, Detroit, MI 48226. (800) 521-0707. 12th edition. 1987. $365.00 for set.

SCIENTIFIC AND TECHNICAL ORGANIZATIONS AND AGENCIES DIRECTORY. Margaret Labash Young, editor. Gale Research Company, Book Tower, Detroit, MI 48226. (800) 521-0707. 2nd edition. 1987. $185.00.

WHO'S WHO IN ENGINEERING. Gordon Davis, editor. Hemisphere Publishing Corporation, 1010 Vermont Avenue, NW, Washington, DC 20005. (800) 526-0275. 6th edition. 1985. $200.00.

METALLURGICAL ENGINEERING

ENCYCLOPEDIAS AND DICTIONARIES

THESAURUS OF SCIENTIFIC, TECHNICAL, AND ENGINEERING TERMS. Hemisphere Publishing Corporation, 1010 Vermont Avenue, NW, Washington, DC 20005. (800) 526-0275. 1988. $125.00.

GENERAL WORKS

ESSENTIAL METALLURGY FOR ENGINEERS. William Alexander. Van Nostrand Reinhold Company, Inc., 135 West 50th Street, New York, NY 10020. (800) 543-2681. 1985. $17.95.

INTRODUCTION TO METALLURGICAL THERMODYNAMICS. David R. Gaskell. Hemisphere Publishing Corporation, 1010 Vermont Avenue, NW, Washington, DC 20005. (800) 526-0275. 1981. $51.00.

METALLURGY BASICS. Donald V. Brown. Van Nostrand Reinhold Company, Inc., 135 West 50th Street, New York, NY 10020. (800) 543-2681. 1983. $19.95 in paper.

HANDBOOKS AND MANUALS

HANDBOOK OF METAL FORMING. Kurt Lange, editor. John Wiley and Sons, Inc., 605 Third Avenue, New York, NY 10158. (800) 526-5368. 1985. $89.50.

HANDBOOK OF METAL FORMING PROCESSES. B. Avitzur. John Wiley and Sons, Inc., 605 Third Avenue, New York, NY 10158. (800) 526-5368. 1983. $105.00.

ONLINE DATA BASES

CA SEARCH. Chemical Abstracts Service, P.O. Box 3012, Columbus, OH 43120. (800) 848-6538 or (614) 421-3600. Comprehensive guide to chemical literature, 1972 to present. Inquire as to online cost and availability.

COMPENDEX. Engineering Information, Inc., 345 East 47th Street, New York, NY 10017. (800) 221-1044 or (212) 705-7615. Engineering and technical literature, 1975 to present. Inquire as to online cost and availability.

INSPEC. INSPEC Marketing Department, Institution of Electrical Engineers. Available from IEEE Service Center, 445 Hoes Lane, Piscataway, NJ 08854. (201) 981-0060. Online version of Physics Abstracts. Inquire as to online cost and availability.

ISMEC. Cambridge Scientific Abstracts, 5161 River Road, Besthda, MD 20816. (800) 638-8076 or (301) 951-1400. Literature of mechanical and production engineering, 1973 to present. Inquire as to online cost and availability.

NTIS. National Technical Information Service, 5285 Port Royal Road, Springfield, VA 22161. (703) 487-4630. Broad coverage of government sponsored research reports, 1964 to present. Inquire as to online cost and availability.

SCISEARCH. Institute for Scientific Information, 3501 Market Street, Philadelphia, PA 19104. (800) 523-1850 or (215) 386-0100. Broad multidisciplinary title and author index to the international literature of science and technology, 1974 to present. Inquire as to online cost and availability.

WILSONLINE. H.W. Wilson and Company, 950 University Avenue, Bronx, NY 10452. (800) 367-6770 or (212) 588-8400. Makes available online versions of the H.W. Wilson indexes including Applied Science and Technology Index, Business Periodicals Index and Readers' Guide to Periodical Literature. Approximately 1980 to present. Inquire as to online cost and availability.

PERIODICALS

IRON AND STEEL ENGINEER. Association of Iron and Steel Engineers, Suite 2350, Three Gateway Center, Pittsburgh, PA 15222. (412) 281-6323. 1924 to present. Monthly. $34.00 per year.

JOURNAL OF APPLIED MECHANICS. American Society of Mechanical Engineers, 345 East 47th Street, New York, NY 10017. (212) 705-7703. 1935 to present. Quarterly. $100.00 per year.

JOURNAL OF ENGINEERING FOR INDUSTRY. American Society of Mechanical Engineers, 345 East 47th Street, New York, NY 10017. (212) 705-7703. 1970 to present. Quarterly. $100.00 per year.

JOURNAL OF METALS. American Institute of Mining, Metallurgical, and Petroleum Engineers, Inc., Metallurgical Society, 420 Commonwealth Drive, Warrendale, PA 15086. (412) 776-9086. 1949 to present. Monthly. $40.00 per year.

METAL PROGRESS. American Society for Metals, Metals Park, OH 44073. (216) 338-5151. 1930 to present. Monthly. $40.00.

METIFAX MAGAZINE. Huebner Publications Inc., 6521 Davis Industrial Parkway, Solon, OH 44139. (216) 248-1125. 1956 to present. Monthly. $40.00 per year.

RESEARCH CENTERS AND INSTITUTES

CANADIAN INSTITUTE OF METALWORKING. 1276 Sandhill Drive, P.O. Box 7317, Ancaster, ON, Canada L9G 3N6.

COOPERATIVE PROGRAM IN METALLURGY. Pennsylvania State University, 208A Steidle Building, University Park, PA 16802. (814) 865-5446.

DEPARTMENT OF MATERIALS SCIENCES AND ENGINEERING. University of Florida, Gainesville, FL 32601. (904) 392-1454.

INSTITUTE OF MATERIALS SCIENCE. University of Connecticut, Storrs, CT 06268. (203) 486-4623.

METALLURGY

See also: ALLOYS, FERROALLOYS, MACHINING, METALLURGICAL ENGINEERING, METALS AND METALWORKING, WELDING

ABSTRACT SERVICES AND INDEXES

ALLOYS INDEX. American Society for Metals, Metals Park, OH 44073. (216) 338-5151. 1974 to present. Monthly. $225.00.

APPLIED MECHANICS REVIEW. American Society of Mechanical Engineers, 345 East 47th Street, New York, NY 10017. (212) 705-7703. 1948 to present. Monthly. $360.00 per year.

APPLIED SCIENCE AND TECHNOLOGY INDEX. H.W. Wilson and Company, 950 University Avenue, Bronx, NY 10452. (800) 367-6670 or (212) 588-8400. Monthly. Inquire as to cost and availability.

CHEMICAL ABSTRACTS. American Chemical Society, Chemical Abstracts Service, Box 3012, Columbus, OH 43210. (614) 421-3600. 1907 to present. Weekly. $9500.00 per year.

CORROSION ABSTRACTS. National Association of Corrosion Engineers, Box 218340, Houston, TX 77218. (713) 492-0535. 1962 to present. Bimonthly. $200.00 per year.

CURRENT CONTENTS: ENGINEERING, TECHNOLOGY AND APPLIED SCIENCES. Institute for Scientific Information, 3501 Market Street, Philadelphia, PA 19104. (800) 523-1850 or (215) 386-0100. Weekly. $275.00 per year.

ENGINEERING INDEX MONTHLY AND AUTHOR INDEX. Engineering Information Inc., 345 East 47th Street, New York, NY 10017. (212) 705-7600. Monthly. $1560.00 per year.

ISMEC BULLETIN (Information Service in Mechanical Engineering). Cambridge Scientific Abstracts, 5161 River Road, Bethesda, MD 20816. (301) 951-1400. 1973 to present. Monthly. $450.00 per year.

LEAD ABSTRACTS. Lead Development Association, 34 Berkeley Square, London, W1X 6AJ, England. 1958 to present. Quarterly. $70.00 per year.

METALS ABSTRACTS AND METALS ABSTRACTS INDEX. American Society for Metals, Metals Park, OH 44073. (216) 338-5151. 1968 to present. Monthly. Abstracts are $1100.00 per year and Index is $460.00 per year.

PHYSICS ABSTRACTS. Institution of Electrical Engineers. Available from: IEEE Service Center, 445 Hoes Lane, Piscataway, NJ 08854. 1898 to present. Bimonthly. $1700.00 per year.

SCIENCE CITATION INDEX. Institute for Scientific Information, 3501 Market Street, Philadelphia, PA 19104. (800) 523-1850 or (215) 386-0100. Six times per year. $6200.00 per year.

WORLD ALUMINUM ABSTRACTS. Aluminum Association, 818 Connecticut Avenue, NW, Washington, DC 20006. (202) 862-5156. 1968 to present. Monthly. $240.00 per year.

ASSOCIATIONS AND PROFESSIONAL SOCIETIES

ALUMINUM ASSOCIATION. 818 Connecticut Avenue, NW, Washington, DC 20006. (202) 862-5156.

AMERICAN INSTITUTE OF MINING, METALLURGICAL AND PETROLEUM ENGINEERS (AIME). 420 Commonwealth Drive, Warrendale, PA 15086. (412) 776-9086.

AMERICAN POWDER METALLURGY INSTITUTE. 105 College Road, East, Princeton, NJ 08540. (609) 452-7700.

AMERICAN SOCIETY FOR METALS. Metals Park, OH 44073. (216) 338-5151.

AMERICAN SOCIETY FOR TESTING AND MATERIALS. 1916 Race Street, Philadelphia, PA 19103. (215) 299-5400.

AMERICAN SOCIETY OF MECHANICAL ENGINEERS. 345 47th Street, New York, NY 10017. (212) 705-7722.

ASSOCIATION OF IRON AND STEEL ENGINEERS. Three Gateway Center, Suite 2350, Pittsburgh, PA 15222. (412) 281-6323.

THE METALLURGICAL SOCIETY. 420 Commonwealth Drive, Warrendale, PA 15086. (412) 776-9000.

NATIONAL ASSOCIATION OF CORROSION ENGINEERS. Box 218340, Houston, TX 77218. (713) 492-0535.

DIRECTORIES AND BIOGRAPHICAL SOURCES

DUN'S INDUSTRIAL GUIDE: THE METALWORKING DIRECTORY. Dun and Bradstreet Corporation, 49 Old Bloomfield Road, Mountain Lakes, NJ 07046. (201) 953-0300. Annual. $610.00.

INDUSTRIAL EQUIPMENT AND SUPPLIES DIRECTORY. American Business Directories, Inc., 5707 South 86th Circle, Omaha, NE 68127. (402) 331-7169.

MACHINERY BUYERS GUIDE. Findlay Publications Limited, Maitland House, Warrior Square, Southend-on-Sea, Essex SS5 5AR England. Annual. $45.00.

METALLURGICAL SOCIETY OF AIME - MEMBERSHIP LIST. American Institute of Mining, Metallurgical and Petroleum Engineers (AIME). 345 East 47th Street, New York, NY 10017. (212) 705-7695. 1984.

1987 DIRECTORY OF ENGINEERING SOCIETIES AND RELATED ORGANIZATIONS. Gordon Davis, editor. Hemisphere Publishing Corporation, 1010 Vermont Avenue, NW, Washington, DC 20005. (800) 526-0275. 12th edition. 1987. $100.00.

RESEARCH CENTERS DIRECTORY 1988. Gale Research Company, Book Tower, Detroit, MI 48226. (800) 521-0707. 12th edition. 1987. $365.00 for set.

SCIENTIFIC AND TECHNICAL ORGANIZATIONS AND AGENCIES DIRECTORY. Margaret Labash Young, editor. Gale Research Company, Book Tower, Detroit, MI 48226. (800) 521-0707. 2nd edition. 1987. $185.00.

WHO'S WHO IN ENGINEERING. Gordon Davis, editor. Hemisphere Publishing Corporation, 1010 Vermont Avenue, NW, Washington, DC 20005. (800) 526-0275. 6th edition. 1985. $200.00.

ENCYCLOPEDIAS AND DICTIONARIES

THESAURUS OF SCIENTIFIC, TECHNICAL, AND ENGINEERING TERMS. Hemisphere Publishing Corporation, 1010 Vermont Avenue, NW, Washington, DC 20005. (800) 526-0275. 1988. $125.00.

GENERAL WORKS

ESSENTIAL METALLURGY FOR ENGINEERS. William Alexander. Van Nostrand Reinhold Company, Inc., 135 West 50th Street, New York, NY 10020. (800) 543-2681. 1985. $17.95.

METALLURGY BASICS. Donald V. Brown. Van Nostrand Reinhold Company, Inc., 135 West 50th Street, New York, NY 10020. (800) 543-2681. 1983. $19.95 in paper.

METALS ENGINEERING: A TECHNICAL GUIDE. L.E. Samuals. American Society for Metals, Metals Park, OH 44073. (216) 338-5151. 1988. $68.00.

HANDBOOKS AND MANUALS

HANDBOOK OF METAL FORMING. Kurt Lange, editor. John Wiley and Sons, Inc., 605 Third Avenue, New York, NY 10158. (800) 526-5368. 1985. $89.50.

HANDBOOK OF METAL FORMING PROCESSES. B. Avitzur. John Wiley and Sons, Inc., 605 Third Avenue, New York, NY 10158. (800) 526-5368. 1983. $105.00.

METALS HANDBOOK. American Society for Metals, Metals Park, OH 44073. (216) 338-5151. 9th edition. 14 volumes. 1988. $1310.00 for set.

WOLDMAN'S ENGINEERING ALLOYS. Robert C. Gibbons, editor. American Society for Metals, Metals Park, OH 44073. (216) 338-5151. 6th edition. 1979. $112.00.

ONLINE DATA BASES

CA SEARCH. Chemical Abstracts Service, P.O. Box 3012, Columbus, OH 43120. (800) 848-6538 or (614) 421-3600. Comprehensive guide to chemical literature, 1972 to present. Inquire as to online cost and availability.

COMPENDEX. Engineering Information, Inc., 345 East 47th Street, New York, NY 10017. (800) 221-1044 or (212) 705-7615. Engineering and technical literature, 1975 to present. Inquire as to online cost and availability.

INSPEC. INSPEC Marketing Department, Institution of Electrical Engineers. Available from IEEE Service Center, 445 Hoes Lane,

METALLURGY

Piscataway, NJ 08854. (201) 981-0060. Online version of Physics Abstracts. Inquire as to online cost and availability.

ISMEC. Cambridge Scientific Abstracts, 5161 River Road, Besthda, MD 20816. (800) 638-8076 or (301) 951-1400. Literature of mechanical and production engineering, 1973 to present. Inquire as to online cost and availability.

NTIS. National Technical Information Service, 5285 Port Royal Road, Springfield, VA 22161. (703) 487-4630. Broad coverage of government sponsored research reports, 1964 to present. Inquire as to online cost and availability.

SCISEARCH. Institute for Scientific Information, 3501 Market Street, Philadelphia, PA 19104. (800) 523-1850 or (215) 386-0100. Broad multidisciplinary title and author index to the international literature of science and technology, 1974 to present. Inquire as to online cost and availability.

WILSONLINE. H.W. Wilson and Company, 950 University Avenue, Bronx, NY 10452. (800) 367-6770 or (212) 588-8400. Makes available online versions of the H.W. Wilson indexes including Applied Science and Technology Index, Business Periodicals Index and Readers' Guide to Periodical Literature. Approximately 1980 to present. Inquire as to online cost and availability.

PERIODICALS

ALLOY DIGEST. Engineering Alloys Digest, Inc., Box 823, Upper Montclair, NJ 07043 . (201) 746-7930. 1952 to present. Monthly. $50.00 per year.

IRON AND STEEL ENGINEER. Association of Iron and Steel Engineers, Suite 2350, Three Gateway Center, Pittsburgh, PA 15222. (412) 281-6323. 1924 to present. Monthly. $34.00 per year.

JOURNAL OF APPLIED MECHANICS. American Society of Mechanical Engineers, 345 East 47th Street, New York, NY 10017. (212) 705-7703. 1935 to present. Quarterly. $100.00 per year.

JOURNAL OF ENGINEERING FOR INDUSTRY. American Society of Mechanical Engineers, 345 East 47th Street, New York, NY 10017. (212) 705-7703. 1970 to present. Quarterly. $100.00 per year.

JOURNAL OF METALS. American Institute of Mining, Metallurgical, and Petroleum Engineers, Inc., Metallurgical Society, 420 Commonwealth Drive, Warrendale, PA 15086. (412) 776-9086. 1949 to present. Monthly. $40.00 per year.

LIGHT METAL AGE. Fellom Publishing Company, 693 Mission Street, San Francisco, CA 94105. (415) 781-1431. 1942 to present. Bimonthly. $20.00.

METAL PROGRESS. American Society for Metals, Metals Park, OH 44073. (216) 338-5151. 1930 to present. Monthly. $40.00.

METALLURGICAL TRANSACTIONS. Metallurgical Society of AIME, 420 Commonwealth Drive, Warrendale, PA 15086. (412) 776-9080. 1970 to present. Monthly. $95.00 per year.

METALS WEEK. McGraw-Hill Book Company, 1221 Avenue of the Americas, New York, NY 10020. (212) 997-2823. 1930 to present. Weekly. $597.00 per year.

METALWORKING DIGEST. Philos Publications, Inc., 1 East Chase Street, Baltimore, MD 21202. (301) 361-9060. 1969 to present. Nine times per year. $9.00 per year.

METIFAX MAGAZINE. Huebner Publications Inc., 6521 Davis Industrial Parkway, Solon, OH 44139. (216) 248-1125. 1956 to present. Monthly. $40.00 per year.

RESEARCH CENTERS AND INSTITUTES

CANADIAN INSTITUTE OF METALWORKING. 1276 Sandhill Drive, P.O. Box 7317, Ancaster, ON, Canada L9G 3N6.

COOPERATIVE PROGRAM IN METALLURGY. Pennsylvania State University, 208A Steidle Building, University Park, PA 16802. (814) 865-5446.

DEPARTMENT OF MATERIALS SCIENCES AND ENGINEERING. University of Florida, Gainesville, FL 32601. (904) 392-1454.

INSTITUTE OF MATERIALS SCIENCE. University of Connecticut, Storrs, CT 06268. (203) 486-4623.

METALS AND METALWORKING

See also: ALLOYS, FERROALLOYS, MACHINING, METALLURGICAL ENGINEERING, METALLURGY, WELDING

ABSTRACT SERVICES AND INDEXES

APPLIED MECHANICS REVIEW. American Society of Mechanical Engineers, 345 East 47th Street, New York, NY 10017. (212) 705-7703. 1948 to present. Monthly. $360.00 per year.

APPLIED SCIENCE AND TECHNOLOGY INDEX. H.W. Wilson and Company, 950 University Avenue, Bronx, NY 10452. (800) 367-6670 or (212) 588-8400. Monthly. Inquire as to cost and availability.

CHEMICAL ABSTRACTS. American Chemical Society, Chemical Abstracts Service, Box 3012, Columbus, OH 43210. (614) 421-3600. 1907 to present. Weekly. $9500.00 per year.

CORROSION ABSTRACTS. National Association of Corrosion Engineers, Box 218340, Houston, TX 77218. (713) 492-0535. 1962 to present. Bimonthly. $200.00 per year.

CURRENT CONTENTS: ENGINEERING, TECHNOLOGY AND APPLIED SCIENCES. Institute for Scientific Information, 3501 Market Street, Philadelphia, PA 19104. (800) 523-1850 or (215) 386-0100. Weekly. $275.00 per year.

ENGINEERING INDEX MONTHLY AND AUTHOR INDEX. Engineering Information Inc., 345 East 47th Street, New York, NY 10017. (212) 705-7600. Monthly. $1560.00 per year.

ISMEC BULLETIN (Information Service in Mechanical Engineering). Cambridge Scientific Abstracts, 5161 River Road, Bethesda, MD 20816. (301) 951-1400. 1973 to present. Monthly. $450.00 per year.

LEAD ABSTRACTS. Lead Development Association, 34 Berkeley Square, London, W1X 6AJ, England. 1958 to present. Quarterly. $70.00 per year.

METALS ABSTRACTS AND METALS ABSTRACTS INDEX. American Society for Metals, Metals Park, OH 44073. (216) 338-5151. 1968 to present. Monthly. Abstracts are $1100.00 per year and Index is $460.00 per year.

PHYSICS ABSTRACTS. Institution of Electrical Engineers. Available from: IEEE Service Center, 445 Hoes Lane, Piscataway, NJ 08854. 1898 to present. Bimonthly. $1700.00 per year.

SCIENCE CITATION INDEX. Institute for Scientific Information, 3501 Market Street, Philadelphia, PA 19104. (800) 523-1850 or (215) 386-0100. Six times per year. $6200. 00 per year.

WORLD ALUMINUM ABSTRACTS. Aluminum Association, 818 Connecticut Avenue, NW, Washington, DC 20006. (202) 862-5156. 1968 to present. Monthly. $240.00 per year.

ASSOCIATIONS AND PROFESSIONAL SOCIETIES

ALUMINUM ASSOCIATION. 818 Connecticut Avenue, NW, Washington, DC 20006. (202) 862-5156.

AMERICAN INSTITUTE OF MINING, METALLURGICAL AND PETROLEUM ENGINEERS (AIME). 420 Commonwealth Drive, Warrendale, PA 15086. (412) 776-9086.

AMERICAN POWDER METALLURGY INSTITUTE. 105 College Road, East, Princeton, NJ 08540. (609) 452-7700.

AMERICAN SOCIETY FOR METALS. Metals Park, OH 44073. (216) 338-5151.

AMERICAN SOCIETY FOR TESTING AND MATERIALS. 1916 Race Street, Philadelphia, PA 19103. (215) 299-5400.

AMERICAN SOCIETY OF MECHANICAL ENGINEERS. 345 47th Street, New York, NY 10017. (212) 705-7722.

ASSOCIATION OF IRON AND STEEL ENGINEERS. Three Gateway Center, Suite 2350, Pittsburgh, PA 15222. (412) 281-6323.

THE METALLURGICAL SOCIETY. 420 Commonwealth Drive, Warrendale, PA 15086. (412) 776-9000.

NATIONAL ASSOCIATION OF CORROSION ENGINEERS. Box 218340, Houston, TX 77218. (713) 492-0535.

DIRECTORIES AND BIOGRAPHICAL SOURCES

DUN'S INDUSTRIAL GUIDE: THE METALWORKING DIRECTORY. Dun and Bradstreet Corporation, 49 Old Bloomfield Road, Mountain Lakes, NJ 07046. (201) 953-0300. Annual. $610.00.

INDUSTRIAL EQUIPMENT AND SUPPLIES DIRECTORY. American Business Directories, Inc., 5707 South 86th Circle, Omaha, NE 68127. (402) 331-7169.

MACHINERY BUYERS GUIDE. Findlay Publications Limited, Maitland House, Warrior Square, Southend-on-Sea, Essex SS5 5AR England. Annual. $45.00.

METALLURGICAL SOCIETY OF AIME - MEMBERSHIP LIST. American Institute of Mining, Metallurgical and Petroleum Engineers (AIME). 345 East 47th Street, New York, NY 10017. (212) 705-7695. 1984.

1987 DIRECTORY OF ENGINEERING SOCIETIES AND RELATED ORGANIZATIONS. Gordon Davis, editor. Hemisphere Publishing Corporation, 1010 Vermont Avenue, NW, Washington, DC 20005. (800) 526-0275. 12th edition. 1987. $100.00.

RESEARCH CENTERS DIRECTORY 1988. Gale Research Company, Book Tower, Detroit, MI 48226. (800) 521-0707. 12th edition. 1987. $365.00 for set.

SCIENTIFIC AND TECHNICAL ORGANIZATIONS AND AGENCIES DIRECTORY. Margaret Labash Young, editor. Gale Research Company, Book Tower, Detroit, MI 48226. (800) 521-0707. 2nd edition. 1987. $185.00.

WHO'S WHO IN ENGINEERING. Gordon Davis, editor. Hemisphere Publishing Corporation, 1010 Vermont Avenue, NW, Washington, DC 20005. (800) 526-0275. 6th edition. 1985. $200.00.

ENCYCLOPEDIAS AND DICTIONARIES

THESAURUS OF SCIENTIFIC, TECHNICAL, AND ENGINEERING TERMS. Hemisphere Publishing Corporation, 1010 Vermont Avenue, NW, Washington, DC 20005. (800) 526-0275. 1988. $125.00.

GENERAL WORKS

ESSENTIAL METALLURGY FOR ENGINEERS. William Alexander. Van Nostrand Reinhold Company, Inc., 135 West 50th Street, New York, NY 10020. (800) 543-2681. 1985. $17.95.

METALLURGY BASICS. Donald V. Brown. Van Nostrand Reinhold Company, Inc., 135 West 50th Street, New York, NY 10020. (800) 543-2681. 1983. $19.95 in paper.

METALS ENGINEERING: A TECHNICAL GUIDE. L.E. Samuals. American Society for Metals, Metals Park, OH 44073. (216) 338-5151. 1988. $68.00.

HANDBOOKS AND MANUALS

HANDBOOK OF METAL FORMING. Kurt Lange, editor. John Wiley and Sons, Inc., 605 Third Avenue, New York, NY 10158. (800) 526-5368. 1985. $89.50.

HANDBOOK OF METAL FORMING PROCESSES. B. Avitzur. John Wiley and Sons, Inc., 605 Third Avenue, New York, NY 10158. (800) 526-5368. 1983. $105.00.

METALS HANDBOOK. American Society for Metals, Metals Park, OH 44073. (216) 338-5151. 9th edition. 14 volumes. 1988. $1310.00 for set.

WOLDMAN'S ENGINEERING ALLOYS. Robert C. Gibbons, editor. American Society for Metals, Metals Park, OH 44073. (216) 338-5151. 6th edition. 1979. $112.00.

ONLINE DATA BASES

CA SEARCH. Chemical Abstracts Service, P.O. Box 3012, Columbus, OH 43120. (800) 848-6538 or (614) 421-3600. Comprehensive guide to chemical literature, 1972 to present. Inquire as to online cost and availability.

COMPENDEX. Engineering Information, Inc., 345 East 47th Street, New York, NY 10017. (800) 221-1044 or (212) 705-7615. Engineering and technical literature, 1975 to present. Inquire as to online cost and availability.

INSPEC. INSPEC Marketing Department, Institution of Electrical Engineers. Available from IEEE Service Center, 445 Hoes Lane, Piscataway, NJ 08854. (201) 981-0060. Online version of Physics Abstracts. Inquire as to online cost and availability.

ISMEC. Cambridge Scientific Abstracts, 5161 River Road, Besthda, MD 20816. (800) 638-8076 or (301) 951-1400. Literature of mechanical and production engineering, 1973 to present. Inquire as to online cost and availability.

NTIS. National Technical Information Service, 5285 Port Royal Road, Springfield, VA 22161. (703) 487-4630. Broad coverage of government sponsored research reports, 1964 to present. Inquire as to online cost and availability.

SCISEARCH. Institute for Scientific Information, 3501 Market Street, Philadelphia, PA 19104. (800) 523-1850 or (215) 386-0100. Broad multidisciplinary title and author index to the international literature of science and technology, 1974 to present. Inquire as to online cost and availability.

WILSONLINE. H.W. Wilson and Company, 950 University Avenue, Bronx, NY 10452. (800) 367-6770 or (212) 588-8400. Makes available online versions of the H.W. Wilson indexes including Applied Science and Technology Index, Business Periodicals Index and Readers' Guide to Periodical Literature. Approximately 1980 to present. Inquire as to online cost and availability.

PERIODICALS

IRON AND STEEL ENGINEER. Association of Iron and Steel Engineers, Suite 2350, Three Gateway Center, Pittsburgh, PA

15222. (412) 281-6323. 1924 to present. Monthly. $34.00 per year.

JOURNAL OF APPLIED MECHANICS. American Society of Mechanical Engineers, 345 East 47th Street, New York, NY 10017. (212) 705-7703. 1935 to present. Quarterly. $100.00 per year.

JOURNAL OF ENGINEERING FOR INDUSTRY. American Society of Mechanical Engineers, 345 East 47th Street, New York, NY 10017. (212) 705-7703. 1970 to present. Quarterly. $100.00 per year.

JOURNAL OF METALS. American Institute of Mining, Metallurgical, and Petroleum Engineers, Inc., Metallurgical Society, 420 Commonwealth Drive, Warrendale, PA 15086. (412) 776-9086. 1949 to present. Monthly. $40.00 per year.

METAL PROGRESS. American Society for Metals, Metals Park, OH 44073. (216) 338-5151. 1930 to present. Monthly. $40.00.

METALS WEEK. McGraw-Hill Book Company, 1221 Avenue of the Americas, New York, NY 10020. (212) 997-2823. 1930 to present. Weekly. $597.00 per year.

METALWORKING DIGEST. Philos Publications, Inc., 1 East Chase Street, Baltimore, MD 21202. (301) 361-9060. 1969 to present. Nine times per year. $9.00 per year.

METIFAX MAGAZINE. Huebner Publications Inc., 6521 Davis Industrial Parkway, Solon, OH 44139. (216) 248-1125. 1956 to present. Monthly. $40.00 per year.

RESEARCH CENTERS AND INSTITUTES

CANADIAN INSTITUTE OF METALWORKING. 1276 Sandhill Drive, P.O. Box 7317, Ancaster, ON, Canada L9G 3N6.

COOPERATIVE PROGRAM IN METALLURGY. Pennsylvania State University, 208A Steidle Building, University Park, PA 16802. (814) 865-5446.

DEPARTMENT OF MATERIALS SCIENCES AND ENGINEERING. University of Florida, Gainesville, FL 32601. (904) 392-1454.

INSTITUTE OF MATERIALS SCIENCE. University of Connecticut, Storrs, CT 06268. (203) 486-4623.

METAMORPHIC ROCKS

See also: GEOLOGY, GEOPHYSICS, GEOCHEMISTRY, IGNEOUS ROCKS, SEDIMENTARY ROCKS

ABSTRACT SERVICES AND INDEXES

BIBLIOGRAPHY AND INDEX OF GEOLOGY. American Geological Institute, 4220 King Street, Alexandria, VA 22302. (703) 379-2480. 1969 to present. Monthly. $1100.00 per year.

CHEMICAL ABSTRACTS. Chemical Abstracts Service, 2540 Olentangy Road, Post Office Box 3012, Columbus, OH 43210. (800) 848-6538 or (614) 421-3600. Weekly. $9200.00 per year.

DEEP-SEA RESEARCH WITH OCEANOGRAPHIC LITERATURE REVIEW. Pergamon Press, Incorporated, Maxwell House, Fairview Park, Elmsford, NY 10523. (914) 592-7700. 1953 to present. Twenty-four times per year. $600.00 per year.

GENERAL SCIENCE INDEX. H.W. Wilson Company, 950 University Avenue, Bronx, NY 10452. (800) 367-6770 or (212) 588-8400. Inquire as to price and availability.

MINERALOGICAL ABSTRACTS. Mineralogical Society and the Mineralogical Society of America, 41 Queen's Gate, London, SW7 5HR, England. 1959 to present. Quarterly. $190.00 per year.

OCEANIC ABSTRACTS. Cambridge Scientific Abstracts, 5161 River Road, Bethesda, MD 20816. (301) 951-1400. 1964 to present. Bimonthly. $652.00 per year.

SCIENCE CITATION INDEX. Institute for Scientific Information, 3501 Market Street, Philadelphia, PA 19104. (800) 523-1850 or (215) 386-0100. Inquire as to price and availability.

ANNUAL REVIEWS AND YEARBOOKS

ANNUAL REVIEW AND EARTH AND PLANETARY SCIENCES. Annual Reviews, Incorporated, 4139 El Camino Way, Palo Alto, CA 94306. (415) 493-4400.

MINERALS YEARBOOK. Bureau of Mines, United States Department of the Interior. Available from United States Government Printing Office, Washington, DC 20402. (202) 783-3238. Annual. Three volumes. $45.00.

ASSOCIATIONS AND PROFESSIONAL SOCIETIES

AMERICAN GEOLOGICAL INSTITUTE. 4220 King Street, Alexandria, VA 22302. (703) 379-2480.

AMERICAN GEOPHYSICAL UNION. 2000 Florida Avenue, NW, Washington, DC 20009. (202) 462-6903.

AMERICAN INSTITUTE OF PROFESSIONAL GEOLOGISTS. 7828 Vance Drive, Suite 103, Arvada, CO 80003. (303) 431-0831.

ASSOCIATION OF ENGINEERING GEOLOGISTS. Post Office Box 1068, Brentwood, TN 37027. (615) 377-3578.

GEOLOGICAL SOCIETY OF AMERICA. Box 9140, Boulder, CO 80301. (303) 447-2020.

DIRECTORIES AND BIOGRAPHICAL SOURCES

AMERICAN MEN AND WOMEN OF SCIENCE: PHYSICAL AND BIOLOGICAL SCIENCES. Sixteenth edition. R.R. Bowker Company, 245 West 17th Street, New York, NY 10011. (800) 521-8810 or (212) 916-1600. $595.00.

AMERICAN INSTITUTE OF PROFESSIONAL GEOLOGISTS. Membership Directory. American Institute of Professional Geologists, 7828 Vance Drive, Suite 103, Arvada, CO 80003. (303) 431-0831. Annual, $15.00.

ASSOCIATION OF ENGINEERING GEOLOGISTS DIRECTORY. Association of Engineering Geologists, Dr. G. Lee Christensen, Civil Engineering Department, Villanova University, Villanova, PA 19085. (215) 645-4960. Annual. $15.00.

GEOLOGICAL SOCIETY OF AMERICA. Membership Directory. Geological Society of America, 3300 Penrose Place, Boulder, CO 80301. (303) 447-2020. Annual. Available to members only.

RESEARCH CENTERS DIRECTORY. Gale Research Company, Book Tower, Detroit, MI 48226. (800) 521-0707. Twelfth edition. $365.00 for set.

WHO'S WHO IN FRONTIER SCIENCE AND TECHNOLOGY. Marquis Who's Who, Incorporated, 200 East Ohio Street, Chicago, IL 60611. (800) 428-3898 or (312) 787-2008.

ENCYCLOPEDIAS AND DICTIONARIES

DICTIONARY OF GEOLOGICAL TERMS. American Geological Institute. Doubleday and Company, Incorporated, 245 Park Avenue, New York, NY 10017. (800) 645-6156 or (212) 953-4561. Third edition. 1984. $19.95.

DICTIONARY OF PETROLOGY. S.I. Tomkeieff. John Wiley and Sons, Incorporated, 605 Third Avenue, New York, NY 10158. (800) 526-5368 or (212) 850-6000. 1983. $140.00.

ENCYCLOPEDIA OF PHYSICAL SCIENCE AND TECHNOLOGY. Academic Press, Incorporated, Orlando, FL 32887. (800) 321-5068.

GLOSSARY OF GEOLOGY. Robert L. Bates and Julia A. Jackson. American Geological Institute, 4220 King Street, Alexandria, VA 22032. (703) 379-2480. 1980. $60.00.

GENERAL WORKS

FLUID-ROCK INTERACTIONS DURING METAMORPHISM. J.V. Walther, B.J. Wood, editors. Springer-Verlag Science Publishers, 175 Fifth Avenue, New York, NY 10010. (800) 526-7254. 1986. $44.00.

INTRODUCTION TO ROCK FORMING MINERALS. W.A. Deer, J. Zussman and R.A. Howie. John Wiley and Sons, Incorporated, 605 Third Avenue, New York, NY 10158. (800) 526-5368 or (212) 850-6000. 1966. $29.95.

METAMORPHIC REACTIONS: KINETICS, TEXTURES AND DEFORMATION. A.B. Thompson and D.C. Rubie, editors. Springer-Verlag Science Publishers, 175 Fifth Avenue, NEw York, NY 10010. (800) 526-7254. 1985. $48.00.

PHYSICAL GEOLOGY. R.F. Flint and B.J. Skinner. John Wiley and Sons, Incorporated, 605 Third Avenue, New York, NY 10158. (800) 526-5368 or (212) 850-6000. Second edition. 1977. $40.95.

HANDBOOKS AND MANUALS

GEOLOGY IN THE FIELD. R.R. Compton. John Wiley and Sons, Incorporated, 605 Third Avenue, New York, NY 10158. (800) 526-5368 or (212) 850-6418. 1985. $23.95.

ONLINE DATA BASES

GEOREF. American Geological Institute, 4220 King Street, Alexandria, VA 22302. (800) 336-4764 or (703) 379-2480. Geology and geosciences literature, 1961 to present. Inquire as to online cost and availability.

GEOARCHIVE. Geosystems, Post Office Box 1024, Westminster, London, England, SW1 P 2JL. Citations to literature on geoscience, 1969 to present. Inquire as to online cost and availability.

NTIS. National Technical Information Service, 5285 Port Royal Road, Springfield, VA 22161. (703) 487-4630. Broad coverage of government sponsored research reports, 1964 to present. Inquire as to online cost and availability.

SCISEARCH. Institute for Scientific Information, 3501 Market Street, Philadelphia, PA 19104. (800) 523-1850 or (215) 386-0100. Broad interdisciplinary index to the literature of science and technology, 1965 to present. Inquire as to online cost and availability.

PERIODICALS

AMERICAN JOURNAL OF SCIENCE. Kline Geology Laboratory, Yale University, New Haven, CT 06520. Ten times per year. $80.00 per year.

GEOCHIMICA ET COSMOCHIMICA ACTA. Pergamon Journals, Incorporated, Maxwell House, Fairview Park, Elmsford, NY 10523. (014) 592-7700. Monthly. $340.00 per year.

GEOLOGICAL MAGAZINE. Cambridge University Press, 32 East 57th Street, New York, NY 10022. (800) 872-7423. Bimonthly. $165.00 per year.

GEOLOGICAL SOCIETY JOURNAL. Geological Society of London. Blackwell Scientific Publications, Incorporated, 667 Lytton Avenue, Palo Alto, CA 94301. (415) 324-1688. Six times per year. $290.00.

GEOLOGICAL SOCIETY OF AMERICA BULLETIN. Post Office Box 9140, 3300 Penrose Place, Boulder, CO 80301. (303) 447-2020. Monthly. $80.00.

GEOLOGY. Geological Society of America. Post Office Box 9140, 3300 Penrose Place, Boulder, CO 80301. (303) 447-2020. Monthly. $55.00 per year.

JOURNAL OF GEOLOGY. University of Chicago Press, 5801 South Ellis Street, Chicago, IL 60637. (312) 962-7600. Bimonthly. $30.00 per year.

JOURNAL OF METAMORPHIC GEOLOGY. Blackwell Scientific Publications, Limited, Osney Mead, Oxford, OX2 0EL, England. Quarterly. $100.00 per year.

JOURNAL OF STRUCTURAL GEOLOGY. Pergamon Journals, Incorporated, Maxwell House, Fairview Park, Elmsford, NY 10523. (014) 592-7700. Eight times per year. $160.00 per year.

LITHOS: AN INTERNATIONAL JOURNAL OF MINERALOGY, PETROLOGY, AND GEOCHEMISTRY. Elsevier Science Publishers, Post Office Box 330, Irving-on-Hudson, NY 10533. Quarterly. $73.00 per year.

MOUNTAIN GEOLOGIST. Rocky Mountain Association of Geologists, 4201 West 51st Avenue, Denver, CO 80212-2902. Quarterly. $15.00 per year.

PROFESSIONAL GEOLOGIST. American Institute of Professional Geologists, 7828 Vance Drive, Suite 103, Arvada, CO 80003. Monthly.

ROCK MECHANICS AND ROCK ENGINEERING. Springer-Verlag New York, Incorporated, 175 Fifth Avenue, New York, NY 10010. (800) 526-7254 or (212) 460-1500. Quarterly. $61.00 per year.

TECTONICS. American Geophysical Union, 2000 Florida Avenue, NW, Washington, DC 20009. (202) 462-6903. Bimonthly. $30.00 per year to individuals.

TECTONOPHYSICS: AN INTERNATIONAL JOURNAL OF GEOTECTONICS AND THE GEOLOGY AND PHYSICS OF THE INTERIOR OF THE EARTH. Elsevier Science Publishers, Post Office Box 330, Irving-on-Hudson, NY 10533. Forty-four times per year. $870.00 per year.

TEXTURES AND MICROSTRUCTURES. Gordon and Breach Science Publishers, Limited, Post Office Box 197, London, WC2E 9PX, England. Eight times per year. $290.00 per year.

VOLCANOLOGY AND SEISMOLOGY. Gordon and Breach Science Publishers, 50 West 23rd Street, New York, NY 10010. (212) 206-8900. Monthly. $498.00 per year.

RESEARCH CENTERS AND INSTITUTES

UNITED STATES GEOLOGICAL SURVEY. National Center, 12201 Sunrise Valley Drive, Reston, VA 22092. The major geological research agency of the federal government conducting research in most areas of pure and applied research in the geosciences.

GEOLOGICAL SOCIETY OF AMERICA. Post Office Box 9140, 3300 Penrose Place, Boulder, CO 80301. (303) 447-2020.

SOCIETY OF ECONOMIC GEOLOGISTS. Post Office Box 571, Golden, CO 80402. (303) 279-1899.

SOCIETY OF ECONOMIC PALEONTOLOGISTS AND MINERALOGISTS. Post Office Box 4756, Tulsa, OK 74159. (918) 743-9765.

METEORITES

See: METEORS

METEOROLOGICAL SATELLITES

See: ARTIFICIAL SATELLITES

METEOROLOGY

See also: CLIMATE, CLOUDS, DROUGHT, FOG, GREENHOUSE EFFECT, HAIL, HURRICANES, JET STREAM, RAIN, SNOW, THUNDERSTORMS, TORNADOS, WEATHER MODIFICATION

ABSTRACT SERVICES AND INDEXES

APPLIED SCIENCE AND TECHNOLOGY INDEX. H.W. Wilson Company, 950 University Avenue, Bronx, NY 10452. (800) 367-6670 or (212) 588-8400. Inquire as to cost and availability.

BIBLIOGRAPHY AND INDEX OF GEOLOGY. American Geological Institute, 4220 King Street, Alexandria, VA 22302. (703) 379-2480. 1969 to present. Monthly. $1100.00 per year.

CHEMICAL ABSTRACTS. Chemical Abstracts Service, 2540 Olentangy Road, Post Office Box 3012, Columbus, OH 43210. (800) 848-6538 or (614) 421-3600. Weekly. $9200.00 per year.

CURRENT CONTENTS: PHYSICAL AND CHEMICAL SCIENCES. Institute for Scientific Information, 3501 Market Street, Philadelphia, PA 19104. (800) 523-1850 or (215) 386-0100. $272.00 per year.

GENERAL SCIENCE INDEX. H.W. Wilson Company, 950 University Avenue, Bronx, NY 10452. (800) 367-6770 or (212) 588-8400. Inquire as to cost and availability.

INDEX TO SCIENTIFIC REVIEWS. Institute for Scientific Information, 3501 Market Street, Philadelphia, PA 19104. (800) 523-1850 or (215) 386-0100. Semiannual. $550.00 per year.

METEOROLOGICAL AND GEOASTROPHYSICAL ABSTRACTS. American Meteorological Society, 45 Beacon Street, Boston, MA 02108. (617) 227-2425. 1950 to present. Monthly. $450.00 per year.

OCEANIC ABSTRACTS. Cambridge Scientific Abstracts, Incorporated, 5161 River Road, Bethesda, MD 20816. (301) 951-1400. 1964 to present. Inquire as to online cost and availability.

OCEANOGRAPHIC LITERATURE REVIEW. Pergamon Press Incorporated, Maxwell House, Fairview Park, Elmsford, NY 10523. (914) 592-7700. 1979 to present. Inquire as to cost and availability.

PHYSICS ABSTRACTS. Institute of Electrical Engineers, London, United Kingdom. Available from: Institute of Electrical and Electronic Engineers (IEEE), 345 East 47th Street, New York, NY 10017. (212) 705-7900. 1898 to present. Monthly. $1670.00 per year.

SCIENCE CITATION INDEX. Institute for Scientific Information, 3501 Market Street, Philadelphia, PA 19104. (800) 523-1850 or (215) 386-0100. Inquire as to cost and availability.

ANNUAL REVIEWS AND YEARBOOKS

OCEAN YEARBOOK. Elizabeth M. Borgese and Norton Ginsberg, editors. University of Chicago Press, 5801 Ellis Avenue, Chicago, IL 60637. (312) 962-7906. 1979 to present. $49.00 per volume.

ASSOCIATIONS AND PROFESSIONAL SOCIETIES

AMERICAN METEOROLOGICAL SOCIETY. 45 Beacon Street, Boston, MA 02108. (617) 227-2425.

INTERNATIONAL ASSOCIATION OF METEOROLOGY AND ATMOSPHERIC PHYSICS. UCAR, Post Office Box 3000, Boulder, CO 80307.

NATIONAL WEATHER SERVICE OFFICES AND STATIONS. National Oceanic and Atmospheric Administration, Department of Commerce, Silver Spring, MD 20910. (301) 427-7698. Annual. Free.

UNIVERSITY CORPORATION FOR ATMOSPHERIC RESEARCH. Box 3000, 1850 Table Mesa Drive, Boulder, CO 80307. (303) 497-1000.

WEATHER MODIFICATION ASSOCIATION. Post Office Box 8116, Fresno, CA 93747. (209) 291-8466.

BIBLIOGRAPHIES

SCIENCE BOOKS AND FILMS. American Association for the Advancement of Science, 1333 H Street, NW, Washington, DC 20005. Five times per year. $20.00 per year.

SCIENTIFIC AND TECHNICAL BOOKS AND SERIALS IN PRINT 1988: AN INDEX TO LITERATURE IN SCIENCE AND TECHNOLOGY. R.R. Bowker Company, 205 East 42nd Street, New York, NY 10017. (800) 521-8110 or (212) 916-1600. $175.00.

DIRECTORIES AND BIOGRAPHICAL SOURCES

AMERICAN MEN AND WOMEN OF SCIENCE. Physical and Biological Sciences. Sixteenth edition. R.R. Bowker Company, 205 East 42nd Street, New York, NY 10017. (800) 521-8110 or (212) 916-1600. 1986. $595.00.

BIOGRAPHICAL DICTIONARY OF SCIENTISTS. T.I. Williams. Halsted Press, 605 Third Avenue, New York, NY 10158. (800) 526-5368 or (212) 850-6418. Third edition. 1982. $29.95.

GOVERNMENT RESEARCH DIRECTORY. Gale Research Company, Book Tower, Detroit, MI 48226. (800) 521-0707. Fourth edition. 1987. $350.00.

INTERDOC: DIRECTORY OF PUBLISHED PROCEEDINGS, SERIES. SEMT-Science/Engineering/Medicine/Technology. Interdoc Corporation, 173 Halstead Avenue, Box 326, Harrison, NY 10528. (014) 835-3506. Ten times per year. $325.00 per year.

INTERNATIONAL RESEARCH CENTERS DIRECTORY 1988-1989. Gale Research Company, Book Tower, Detroit, MI 48226. (800) 521-0707. Fourth edition. 1987. $360.00.

METEOROLOGICAL SERVICES OF THE WORLD. World Meteorological Organization. Available from: American Meteorological Society, 45 Beacon Street, Boston, MA 02108. (617) 227-2425. Annual. $35.00.

NATIONAL WEATHER SERVICE OFFICES AND STATIONS. National Oceanic and Atmospheric Administration, Department of Commerce, Silver Spring, MD 20910. (301) 427-7698. Annual. Free.

RESEARCH CENTERS DIRECTORY. Gale Research Company, Book Tower, Detroit, MI 48226. (800) 521-0707. Twelfth edition. 1987. $365.00 for set.

SCIENTIFIC AND TECHNICAL ORGANIZATIONS AND AGENCIES DIRECTORY. Gale Research Company, Book Tower, Detroit, MI 48226. (800) 521-0707. Second edition. 1987. $185.00.

WHO'S WHO IN FRONTIER SCIENCE AND TECHNOLOGY. Marquis Who's Who, Incorporated, 200 East Ohio Street, Chicago, IL 60611. (800) 428-3898 or (312) 787-2008.

WHO'S WHO IN TECHNOLOGY TODAY. Reston Publishing Company, Incorporated, c/o Prentice-Hall, Incorporated, Englewood Cliffs, NJ 07632. (800) 262-6868. Biennial. Five volumes. $425.00. Covers the fields of electronics, computer science, physics, optics, chemistry, biotechnology, mechanics, energy, and earth science.

WORLD GUIDE TO SCIENTIFIC ASSOCIATIONS AND LEARNED SOCIETIES. K.G. Saur Incorporated, 175 Fifth Avenue, New York, NY 10010. (800) 521-0707 or (212) 982-1302. Fourth edition. 1984. $112.00.

ENCYCLOPEDIAS AND DICTIONARIES

ENCYCLOPEDIA OF CLIMATOLOGY. John E. Oliver and Rhodes W. Fairbridge, editors. Van Nostrand Reinhold, Incorporated, 115 Fifth Avenue, New York, NY 10003. (800) 543-2681. 1987. $89.95.

ENCYCLOPEDIA OF PHYSICAL SCIENCE AND TECHNOLOGY. Academic Press, Incorporated, Orlando, FL 32887. (800) 321-5068 or (305) 345-2734. Inquire as to cost and availability.

MCGRAW-HILL ENCYCLOPEDIA OF SCIENCE AND TECHNOLOGY. McGraw-Hill Book, Incorporated, 1221 Avenue of the Americas, New York, NY 10020. (212) 997-3675. Fifth edition, 15 volumes. $1100.00.

GENERAL WORKS

THE ATMOSPHERE: AN INTRODUCTION TO METEOROLOGY. Frederick K. Lutgens and Edward J. Tarbuck. Prentice-Hall Publishing, Incorporated, Englewood Cliffs, NJ 07632. (800) 526-0245. Third edition. 1986. $34.95.

ATMOSPHERIC CHEMISTRY: FUNDAMENTALS AND EXPERIMENTAL TECHNIQUES. B.J. Finlayson-Pitts and J.N. Pitts. John Wiley and Sons, Incorporated, 605 Third Avenue, New York, NY 10158. (800) 526-5368 or (212) 850-6000. 1986. $59.95.

GENERAL CIRCULATION OF THE OCEAN. Henry d.I. Abarbanel and W.R. Young, editors. Springer-Verlag, 175 Fifth Avenue, New York, NY 10010. (800) 526-7254. 1987. $69.00.

INTRODUCTION TO HYDROMETEOROLOGY. J.P. Bruce and R.H. Clark. Pergamon Press Incorporated, Maxwell House, Fairview Park, Elmsford, NY 10523. (914) 592-7700. 1984. $60.00.

PHYSICAL METEOROLOGY. Henry G. Houghton. MIT Press, 28 Carlton Street, Cambridge, MA 02142. (617) 253-5646. 1985. $37.50.

HANDBOOKS AND MANUALS

APPLIED FLUID DYNAMICS HANDBOOK. Robert D. Blevins. Van Nostrand Reinhold, Incorporated, 115 Fifth Avenue, New York, NY. (800) 543-2681. 1984. $51.95.

ONLINE DATA BASES

CA SEARCH. Chemical Abstracts Service, Post Office Box 3012, Columbus, OH 43210. Guide to chemical literature, 1972 to present. Inquire as to cost and availability.

DISSERTATION ABSTRACTS ONLINE. University Microfilms International, 300 North Zeeb Road, Ann Arbor, MI 48106. (800) 521-0600 or (313) 761-4700. Scope includes virtually all doctoral dissertations accepted at accredited American institutions from 1861 to present in 252 subject areas. Inquire as to cost and availability.

INSPEC. INSPEC Marketing Department, Institute of Electrical and Electronics Engineers, Incorporated, IEEE Service Department, 445 Hoes Lane, Piscataway, NJ 08854. (201) 981-0060. Inquire as to on-line cost and availability.

METEOROLOGICAL AND GEOASTROPHYSICAL ABSTRACTS. American Meteorological Society, 45 Beacon Street, Boston, MA 02108. (617) 227-2425. 1950 to present. Monthly. $450.00 per year.

NTIS. National Technical Information Service, 5285 Port Royal Road, Springfield, VA 22161. (703) 487-4630. Broad coverage of government-sponsored research reports, 1964 to present. Inquire as to cost and availability.

OCEANIC ABSTRACTS. Cambridge Scientific Abstracts, Incorporated, 5161 River Road, Bethesda, MD 20816. (301) 951-1400. 1964 to present. Inquire as to online cost and availability.

SCISEARCH. Institute for Scientific Information, 3501 Market Street, Philadelphia, PA 19104. (800) 523-1850 or (215) 386-0100. Broad multidisciplinary title and author index to the international literature of science and technology, 1974 to present. Inquire as to cost and availability.

WILSONLINE. H.W. Wilson Company, 950 University Avenue, Bronx, NY 10452. (800) 367-6770 or (212) 588-8400. Makes available online versions of the printed H.W. Wilson Indexes including Applied Science and Technology Index, Business Periodicals Index, and Readers' Guide to Periodical Literature. Period covered is generally 1983 to present. Inquire as to cost and availability.

PERIODICALS

AGRICULTURAL AND FOREST METEOROLOGY. Elsevier Science Publishing Company, Incorporated, 52 Vanderbilt Avenue, New York, NY 10017. (212) 370-5520. 1964 to present. Monthly. $260.00 per year.

AMERICAN METEOROLOGICAL SOCIETY BULLETIN. American Meteorological Society, 45 Beacon Street, Boston, MA 02108. (617) 227-2425.

BOUNDARY-LAYER METEOROLOGY: AN INTERNATIONAL JOURNAL OF PHYSICAL AND BIOLOGICAL PROCESSES IN THE ATMOSPHERIC BOUNDARY LAYER. D. Reidel Publishing Company, 190 Old Derby Street, Hingham, MA 02043. (617) 871-6600. 1970 to present. Sixteen times per year. $425.00 per year.

CLIMATIC CHANGE: AN INTERDISCIPLINARY, INTERNATIONAL JOURNAL DEVOTED TO THE DESCRIPTION, CAUSES AND IMPLICATIONS OF CLIMATIC CHANGE. D. Reidel Publishing Company, 190 Old Derby Street, Hingham, MA 02043. (617) 871-6600. 1977 to present. Six times per year. $125.00 per year.

DYNAMICS OF ATMOSPHERES AND OCEANS. Elsevier Science Publishing Company, Incorporated, 52 Vanderbilt Avenue, New York, NY 10017. (212) 370-5520. 1977 to present. Quarterly. $90.00 per year.

JOURNAL OF ATMOSPHERIC AND OCEANIC TECHNOLOGY. American Meteorological Society, 45 Beacon Street, Boston, MA 02108. (617) 227-2425. 1984 to present. Quarterly. $80.00 per year.

JOURNAL OF CLIMATE AND APPLIED METEOROLOGY. American Meteorological Society, 45 Beacon Street, Boston, MA 02108. (617) 227-2425. 1962 to present. Monthly. $120.00 per year.

JOURNAL OF PHYSICAL OCEANOGRAPHY. American Meteorological Society, 45 Beacon Street, Boston, MA 02108. (617) 227-2425. 1971 to present. Monthly. $120.00 per year.

JOURNAL OF THE ATMOSPHERIC SCIENCES. American Meteorological Society, 45 Beacon Street, Boston, MA 02108. (617) 227-2425. 1944 to present. Semimonthly. $220.00 per year.

MONTHLY WEATHER REVIEW. American Meteorological Society, 45 Beacon Street, Boston, MA 02108. (617) 227-2425. 1872 to present. Monthly. $120.00 per year.

NATIONAL WEATHER DIGEST. National Weather Association, 4400 Stamp Road, Room 404, Temple Hills, MD 20748. (301) 899-3784. 1976 to present. Quarterly. $20.00 per year.

WEATHER. Royal Meteorological Society, James Glaisher House, Grenville Place, Bracknell Berkshire, RG12 1BX, England. 1946 to present. $30.00 per year.

WEATHERWISE. Heldref Publications, 4000 Albemarle Street, NW, Washington, DC 20016. (202) 362-6445. 1948 to present. Bimonthly. $20.00 per year.

RESEARCH CENTERS AND INSTITUTES

COOPERATIVE INSTITUTE FOR MESOSCALE METEOROLOGICAL STUDIES. University of Oklahoma, 401 East Boyd, Norman, OK 73019. (405) 325-3041.

MOUNT WASHINGTON OBSERVATORY. 1 Washington Street, Gorham, NH 80307. (303) 497-1000.

NATIONAL CENTER FOR ATMOSPHERIC RESEARCH. Box 3000, Boulder, CO 80307. (303) 497-1000.

UNIVERSITY OF ARIZONA. Institute of Atmospheric Physics, Tucson, AZ 85721. (602) 626-6831.

UNIVERSITY OF MIAMI. Rosenstiel School of Marine and Atmospheric Science, 4600 Rickenbacker Causeway, Miami, FL 33149. (305) 361-4000.

WOODS HOLE OCEANOGRAPHIC INSTITUTION. Woods Hole, MA 02543. (617) 548-1400.

METEORS

See also: ASTEROIDS, PLANETARY SCIENCE, SOLAR SYSTEM

ABSTRACT SERVICES AND INDEXES

BIBLIOGRAPHY AND INDEX OF GEOLOGY. American Geological Institute, 4220 King Street, Alexandria, VA 22302. (703) 379-2480. 1969 to present. Monthly. $1100.00 per year.

CHEMICAL ABSTRACTS. American Chemical Society, Chemical Abstracts Service, Box 3012, Columbus, OH 43210. (614) 421-3600. 1907 to present. Weekly. $9500.00 per year.

CURRENT CONTENTS: PHYSICAL, CHEMICAL AND EARTH SCIENCES. Institute for Scientific Information, 3501 Market Street, Philadelphia, PA 19104. (800) 523-1850 or (215) 386-0100. Weekly. $275.00 per year.

METEOROLOGICAL AND GEOASTROPHYSICAL ABSTRACTS. American Meteorological Society, 45 Beacon Street, Boston, MA 02108. (617) 227-2425. 1950 to present. Monthly. $450.00 per year.

SCIENCE CITATION INDEX. Institute for Scientific Information, 3501 Market Street, Philadelphia, PA 19104. (800) 523-1850 or (215) 386-0100. Six times per year. $6200.00 per year.

ASSOCIATIONS AND PROFESSIONAL SOCIETIES

AMERICAN ASTRONOMICAL SOCIETY. 1816 Jefferson Place, N.W., Washington, DC 20036. (202) 659-0134.

AMERICAN GEOLOGICAL INSTITUTE. 4220 King Street, Alexandria, VA 22302. (703) 379-2480.

AMERICAN GEOPHYSICAL UNION. 2000 Florida Avenue, N.W., Washington, DC 20009. (202) 462-6903.

AMERICAN METEOR SOCIETY. Department of Physics and Astronomy, State University College, Geneseo, NY 14454. (716) 245-5284.

ASTRONOMICAL SOCIETY OF THE PACIFIC. 1290 24th Avenue, San Francisco, CA 94122. (415) 661-8660.

GEOLOGICAL SOCIETY OF AMERICA. 3300 Penrose Place, Boulder, CO 80301. (303) 447-2020.

METEORITICAL SOCIETY. c/o Donald D. Bogard, SN 4, Geochemistry, NASA Johnson Space Center, Houston, TX 77058. (713) 483-2296.

DIRECTORIES AND BIOGRAPHICAL SOURCES

AMERICAN MEN AND WOMEN OF SCIENCE. R.R. Bowker, Inc., Order Department, 245 West 17th Street, New York, NY 10011. (800) 521-8110. Eight volumes. 1986. $595.00 for set.

INTERNATIONAL RESEARCH CENTERS DIRECTORY 1988-89. Darren L. Smith, editor. Gale Research Company, Book Tower, Detroit, MI 48226. (800) 521-0707. 4th edition. 1987. $360.00.

RESEARCH CENTERS DIRECTORY 1988. Gale Research Company, Book Tower, Detroit, MI 48226. (800) 521-0707. 12th edition. 1987. $365.00 for set.

SCIENTIFIC AND TECHNICAL ORGANIZATIONS AND AGENCIES DIRECTORY. Margaret Labash Young, editor. Gale Research Company, Book Tower, Detroit, MI 48226. (800) 521-0707. 2nd edition. 1987. $185.00.

GENERAL WORKS

CATALOG OF METEOR SHOWERS. Gary W. Kronk. Enslow Publications, Inc., Bloy Street and Ramsey Avenue, Box 777, Hillside, NJ 07205. (201) 964-4116. 1987. $29.95.

CATALOGUE OF METEORITES. A.L. Graham and others, editors. University of Arizona Press, 1615 East Speedway, Tucson, AZ 85719. (602) 621-1441. 1985. $50.00.

COMETS, METEORS AND ASTEROIDS: HOW THEY AFFECT THE EARTH. Stan Gibilisco. TAB Books, Monterey Lane, Blue Ridge Summit, PA 17214. (717) 794-2191. 1985. $12.95 in paper.

METEORITE CRATERS. Kathleen Mark. University of Arizona Press, 1615 East Speedway, Tucson, AZ 85719. (602) 621-1441. 1987. $29.95.

METEORITES: THEIR RECORD ON EARLY SOLAR SYSTEM HISTORY. John T. Wasson. W.H. Freeman and Company, 41 Madison Avenue, New York, NY 10010. (212) 532-7660. 1985. $29.95.

HANDBOOKS AND MANUALS

HANDBOOK OF ELEMENTAL ABUNDANCES IN METEORITES: REVIEWS IN COSMOCHEMISTRY AND ALLIED SUBJECTS. Brian Mason. Gordon and Breach Science Publishers, Inc., 50 West 23rd Street, New York, NY 10010. (212) 206-8900. 1971. $149.50.

ONLINE DATA BASES

CA SEARCH. Chemical Abstracts Service, P.O. Box 3012, Columbus, OH 43120. (800) 848-6538 or (614) 421-3600. Comprehensive guide to chemical literature, 1972 to present. Inquire as to online cost and availability.

GEOREF. Online version of the BIBLIOGRAPHY AND INDEX OF GEOLOGY. American Geological Institute, 4220 King Street, Alexandria, VA 22302. (703) 379-2480. 1969 to present. Inquire as to online cost and availability.

NTIS. National Technical Information Service, 5285 Port Royal Road, Springfield, VA 22161. (703) 487-4630. Broad coverage of government sponsored research reports, 1964 to present. Inquire as to online cost and availability.

SCISEARCH. Institute for Scientific Information, 3501 Market Street, Philadelphia, PA 19104. (800) 523-1850 or (215) 386-0100. Broad multidisciplinary title and author index to the international literature of science and technology, 1974 to present. Inquire as to online cost and availability.

PERIODICALS

ASTRONOMY. AstroMedia Corporation, 625 E Street, Box 92788, Milwaukee, WI 53202. (414) 276-2689. 1973 to present. Monthly. $21.00 per year.

EARTH, MOON AND PLANETS. D. Reidel Publishing Company, 190 Old Derby Street, Hingham, MA 02043. 1969 to present. Nine times per year. $275.00 per year.

GEOCHEMICA ET COSMOCHIMICA ACTA. Pergamon Press, Inc., Maxwell House, Fairview Park, Elmsford, NY 10523. (914) 592-7700. 1950 to present. Monthly. $340.00 per year.

MERCURY. Astronomical Society of the Pacific, 1290 24th Avenue, San Francisco, CA 94122. (415) 661-8660. 1972 to present. Bimonthly. $21.00 per year.

METEORITICS. Center for Meteorite Studies, Arizona State University, Tempe, AZ 85287. (602) 965-3576. 1955 to present. Quarterly. $40.00 per year.

RESEARCH CENTERS AND INSTITUTES

CENTER FOR METEORITE STUDIES. Arizona State University, Tempe, AZ 85287. (602) 965-3576.

METROLOGY

See also: GEODESY, INSTRUMENTATION, STANDARDS, SURVEYING

ABSTRACT SERVICES AND INDEXES

APPLIED MECHANICS REVIEW. American Society of Mechanical Engineers, 345 East 47th Street, New York, NY 10017. (212) 705-7703. 1948 to present. Monthly. $360.00 per year.

APPLIED SCIENCE AND TECHNOLOGY INDEX. H.W. Wilson and Company, 950 University Avenue, Bronx, NY 10452. (800) 367-6670 or (212) 588-8400. Monthly. Inquire as to cost and availability.

CHEMICAL ABSTRACTS. American Chemical Society, Chemical Abstracts Service, Box 3012, Columbus, OH 43210. (614) 421-3600. 1907 to present. Weekly. $9500.00 per year.

ENGINEERING INDEX MONTHLY AND AUTHOR INDEX. Engineering Information Inc., 345 East 47th Street, New York, NY 10017. (212) 705-7600. Monthly. $1560.00 per year.

ISMEC BULLETIN (Information Service in Mechanical Engineering). Cambridge Scientific Abstracts, 5161 River Road, Bethesda, MD 20816. (301) 951-1400. 1973 to present. Monthly. $450.00 per year.

INDEX TO SCIENTIFIC AND TECHNICAL PROCEEDINGS. Institute for Scientific Information, 3501 Market Street, Philadelphia, PA 19104. (800) 523-1850 or (215) 386-0100. 1978 to present. Monthly. $775.00 per year.

INDEX TO SCIENTIFIC REVIEWS. Institute for Scientific Information, 3501 Market Street, Philadelphia, PA 19104. (800) 523-1850 or (215) 386-0100. 1974 to present. Semi-annual. $550.00 per year.

KEY ABSTRACTS - ELECTRICAL MEASUREMENTS AND INSTRUMENTATION. Institution of Electrical Engineers. Available from: IEEE Service Center, 445 Hoes Lane, Piscataway, NJ 08854. 1976 to present. Monthly. $105.00 per year.

KEY ABSTRACTS - PHYSICAL MEASUREMENTS AND INSTRUMENTATION. Institution of Electrical Engineers. Available from: IEEE Service Center, 445 Hoes Lane, Piscataway, NJ 08854. 1976 to present. Monthly. $105.00 per year.

PHYSICS ABSTRACTS. Institution of Electrical Engineers. Available from: IEEE Service Center, 445 Hoes Lane, Piscataway, NJ 08854. 1898 to present. Bimonthly. $1700.00 per year.

PHYSICS BRIEFS. Physik Verlag GmbH, Postfach 1260/1280, D-6940 Weinheim, West Germany. (212) 661-9404. 1920 to present. Twenty-six times per year. $1250.00 per year.

SCIENCE CITATION INDEX. Institute for Scientific Information, 3501 Market Street, Philadelphia, PA 19104. (800) 523-1850 or (215) 386-0100. Six times per year. $6200.00 per year.

STANDARDS ACTIONS. American National Standards Institute, 1430 Broadway, New York, NY 10018. (212) 354-3300. 1970 to present. Twice per month. $25.00 per year.

ASSOCIATIONS AND PROFESSIONAL SOCIETIES

AMERICAN INSTITUTE OF PHYSICS. 335 East 45th Street, New York, NY 10017. (212) 661-9494.

AMERICAN SOCIETY FOR TESTING AND MATERIALS. 1916 Race Street, Philadelphia, PA 19103. (215) 299-5400.

EDISON ELECTRIC INSTITUTE. 1111 19th Street, N.W., Washington, DC 20036. (202) 828-7400.

INSTITUTE OF ELECTRICAL AND ELECTRONICS ENGINEERS. 345 East 47th Street, New York, NY 10017. (212) 705-7900.

INSTRUMENT SOCIETY OF AMERICA. P.O. Box 12277, 67 Alexander Drive, Research Triangle Park, NC 27709. (919) 549-8411.

DIRECTORIES AND BIOGRAPHICAL SOURCES

AMERICAN MEN AND WOMEN OF SCIENCE. R.R. Bowker, Inc., Order Department, 245 West 17th Street, New York, NY 10011. (800) 521-8110. Eight volumes. 1986. $595.00 for set.

INTERNATIONAL RESEARCH CENTERS DIRECTORY 1988-89. Darren L. Smith, editor. Gale Research Company, Book Tower, Detroit, MI 48226. (800) 521-0707. 4th edition. 1987. $360.00.

1987 DIRECTORY OF ENGINEERING SOCIETIES AND RELATED ORGANIZATIONS. Gordon Davis, editor. Hemisphere Publishing Corporation, 79 Madison Avenue, New York, NY 10016-7892. (800) 821-8312. 12th edition. 1987. $100.00.

RESEARCH CENTERS DIRECTORY 1988. Gale Research Company, Book Tower, Detroit, MI 48226. (800) 521-0707. 12th edition. 1987. $365.00 for set.

SCIENTIFIC AND TECHNICAL ORGANIZATIONS AND AGENCIES DIRECTORY. Margaret Labash Young, editor. Gale Research Company, Book Tower, Detroit, MI 48226. (800) 521-0707. 2nd edition. 1987. $185.00.

WHO'S WHO IN ENGINEERING. Gordon Davis, editor. Hemisphere Publishing Corporation, 79 Madison Avenue, New

York, NY 10016-7892. (800) 821-8312. Sixth edition. 1985. $200.00.

GENERAL WORKS

ENGINEERING METROLOGY. D.M. Anthony. Pergamon Press, Inc., Maxwell House, Fairview Park, Elmsford, NY 10523. (914) 592-7700. 1986. $30.00.

EXPERIMENTAL MEASUREMENTS: PRECISION, ERROR AND TRUTH. N.C. Bradford. John Wiley and Sons, Inc., 605 Third Avenue, New York, NY 10158. (800) 526-5368. Second edition. 1985. $29.95.

LANDMARKS IN METROLOGY. Instrument Society of America, P.O. Box 12277, 67 Alexander Drive, Research Triangle Park, NC 27709. (919) 549-8411. 1983. $30.00 in paper.

MODERN TECHNIQUES IN METROLOGY. Paul L. Hewitt. Taylor and Francis, Inc., 242 Cherry Street, Philadelphia, PA 19106-1906. (215) 238-0939. 1984. $40.00.

HANDBOOKS AND MANUALS

HANDBOOK OF MEASUREMENT SCIENCE. P.H. Sydeham. John Wiley and Sons, Inc., 605 Third Avenue, New York, NY 10158. (800) 526-5368. Two volumes. Volume 1, 1982, $100.00. Volume 2, 1983, $135.00.

ONLINE DATA BASES

CA SEARCH. Chemical Abstracts Service, P.O. Box 3012, Columbus, OH 43120. (800) 848-6538 or (614) 421-3600. Comprehensive guide to chemical literature, 1972 to present. Inquire as to online cost and availability.

COMPENDEX. Engineering Information, Inc., 345 East 47th Street, New York, NY 10017. (800) 221-1044 or (212) 705-7615. Engineering and technical literature, 1975 to present. Inquire as to online cost and availability.

DISSERTATION ABSTRACTS ONLINE. University Microfilms International, 300 North Zeeb Road, Ann Arbor, MI 48106. (800) 521-0600 or (313) 761-4700. Scope includes virtually all doctoral dissertations accepted at accredited American institutions from 1861 to present in over 250 subject areas. Inquire as to online cost and availability.

INSPEC. INSPEC Marketing Department, Institution of Electrical Engineers. Available from IEEE Service Center, 445 Hoes Lane, Piscataway, NJ 08854. (201) 981-0060. Online version of Physics Abstracts. Inquire as to online cost and availability.

NTIS. National Technical Information Service, 5285 Port Royal Road, Springfield, VA 22161. (703) 487-4630. Broad coverage of government-sponsored research reports, 1964 to present. Inquire as to online cost and availability.

SCISEARCH. Institute for Scientific Information, 3501 Market Street, Philadelphia, PA 19104. (800) 523-1850 or (215) 386-0100. Broad multidisciplinary title and author index to the international literature of science and technology, 1974 to present. Inquire as to online cost and availability.

WILSONLINE. H.W. Wilson and Company, 950 University Avenue, Bronx, NY 10452. (800) 367-6770 or (212) 588-8400. Makes available online versions of the H.W. Wilson indexes including Applied Science and Technology Index, Business Periodicals Index and Readers' Guide to Periodical Literature. Approximately 1980 to present. Inquire as to online cost and availability.

OTHER SOURCES

WHAT EVERY ENGINEER SHOULD KNOW ABOUT ENGINEERING SOURCES. Marcel Dekker Inc., 270 Madison Avenue, New York, NY 10016. (800) 228-1160. 1984. $24.95.

PERIODICALS

ASTM STANDARDIZATION NEWS. American Society for Testing and Materials, 1916 Race Street, Philadelphia, PA 19103. (215) 299-5400. 1973 to present. Monthly. $20.00 per year.

IEEE TRANSACTIONS ON INSTRUMENTATION AND MEASUREMENT. Institute of Electrical and Electronics Engineers. IEEE Service Center, 445 Hoes Lane, Piscataway, NJ 08854. 1952 to present. Quarterly. $65.00 per year.

METROLOGIA; INTERNATIONAL JOURNAL OF SCIENTIFIC METROLOGY. Springer-Verlag New York, Inc., 175 Fifth Avenue, New York, NY 10010. (800) 526-7254. 1965 to present. Quarterly. $95.00 per year.

STANDARDS ENGINEERING. Standards Engineering Society, 6700 Pennsylvania Avenue South, Minneapolis, MN 55423. Bimothly. $20.00 per year.

U.S. NATIONAL BUREAU OF STANDARDS. JOURNAL OF RESEARCH. Order from: Superintendent of Documents, Washington, DC 20402. 1959 to present. Bimonthly. $17.00 per year.

RESEARCH CENTERS AND INSTITUTES

INSTRUMENTATION AND CONTROL LABORATORY. Princeton University, Department of MAE, English Quadrangle, Princeton, NJ 08544. (609) 452-5154.

INSTRUMENTATION SYSTEMS CENTER. University of Wisconsin - Madison, 1500 Johnson Drive, Madison, WI 53706. (608) 263-1552.

MAJOR ANALYTICAL INSTRUMENTATION CENTER. University of Florida, 217 MAE, Gainesville, FL 32611. (904) 392-6985.

SPECIFICATIONS AND STANDARDS

STANDARDS AND PRACTICES FOR INSTRUMENTATION. Instrument Society of America, P.O. Box 12277, 67 Alexander Drive, Research Triangle Park, NC 27709. (919) 549-8411. 7th edition. 1983. $135.00.

MHD

See: MAGNETOHYDRODYNAMICS

MICROCIRCUITRY

See: MICROELECTRONICS

MICROCOMPUTERS

See: COMPUTERS

MICROELECTRONICS

See also: ELECTRICAL ENGINEERING, ELECTRICITY, ELECTRONIC CIRCUITS AND COMPONENTS, ELECTRONIC ENGINEERING, MICROPROCESSORS

ABSTRACT SERVICES AND INDEXES

APPLIED SCIENCE AND TECHNOLOGY INDEX. H.W. Wilson Company, 950 University Avenue, Bronx, NY 10452. (800) 367-6670 or (212) 588-8400. Inquire as to cost and availability.

ELECTRONICS AND COMMUNICATIONS ABSTRACTS JOURNAL. Cambridge Scientific Abstracts, 5161 River Road, Bethesda, MD 20816. (301) 951-1400. Bimonthly. Inquire as to cost and availability.

ENGINEERING INDEX MONTHLY. Engineering Information, Incorporated, 345 East 47th Street, New York, NY 10017. (800) 221-1044 or (212) 705-7600. Monthly, with annual cumulation. $1425.00 per year.

PHYSICS ABSTRACTS. Institute of Electrical Engineers, London, United Kingdom. Available from: Institute of Electrical and Electronic Engineers (Ieee), 345 East 47th Street, New York, NY 10017. (212) 705-7900.

SCIENCE CITATION INDEX. Institute for Scientific Information, 3501 Market Street, Philadelphia, PA 19104. (800) 523-1850 or (215) 386-0100. Inquire as to cost and availability.

SOLID STATE ABSTRACTS: AN ABSTRACT JOURNAL INVOLVING THE PHYSICS, METALLURGY, CRYSTALLOGRAPHY, CHEMISTRY, AND DEVICE TECHNOLOGY OF SOLIDS. Cambridge Scientific Abstracts, 5161 River Road, Bethesda, MD 20816. (301) 951-1400. 1957 to present. Bimonthly. $550.00 per year.

ASSOCIATIONS AND PROFESSIONAL SOCIETIES

AMERICAN ELECTRONICS ASSOCIATION. Post Office Box 10045, 2670 Hanover Street, Palo Alto, CA 94303. (415) 857-9300.

ELECTRONIC INDUSTRIES ASSOCIATION. 2001 Eye Street, NW, Washington, DC 20006. (202) 457-4900.

IEEE (INSTITUTE OF ELECTRICAL AND ELECTRONICS ENGINEERS). 345 East 47th Street, New York, NY 10017. (212) 705-7900.

INTERNATIONAL SOCIETY FOR HYBRID MICROELECTRONICS. Post Office Box 2698, 1861 Wiehle Avenue, Suite 340, Reston, VA 22090. (703) 471-0066.

BIBLIOGRAPHIES

HANDBOOKS AND TABLES IN SCIENCE AND TECHNOLOGY. Russell H. Powell, editor. Oryx Press, 2214 North Central Avenue, Phoenix, AZ 85004-1483. (602) 254-6156. Second edition. 1983. $55.00.

SCIENTIFIC AND TECHNICAL BOOKS IN PRINT: AN INDEX TO LITERATURE IN SCIENCE AND TECHNOLOGY. R.R. Bowker Company, 205 East 42nd Street, New York, NY 10017. (800) 521-8110 or (212) 916-1600.

DIRECTORIES AND BIOGRAPHICAL SOURCES

AMERICAN ELECTRONICS ASSOCIATION DIRECTORY. American Electronics Association, Post Office Box 10045, 2670 Hanover Street, Palo Alto, CA 94303. (415) 857-9300. Annual.

EEM - ELECTRONIC ENGINEERS MASTER. Hearst Business Communications/UTP Division, 645 Stewart Avenue, Garden City, NY 11530. (516) 227-1300. Annual. $75.00 per copy.

IEEE MEMBERSHIP DIRECTORY. Institute of Electrical and Electronics Engineers, IEEE Service Center, 445 Hoes Lane, Piscataway, NJ 08854. (212) 705-7900. Annual. $7.00.

RESEARCH CENTERS DIRECTORY. Gale Research Company, Book Tower, Detroit, MI 48226. (800) 521-0707. Twelfth edition. 1987. $365.00 for set.

SCIENTIFIC AND TECHNICAL ORGANIZATIONS AND AGENCIES DIRECTORY. Gale Research Company, Book Tower, Detroit, MI 48226. (800) 521-0707. 1985.

WHO'S WHO IN ELECTRONICS. Harris Publishing Company, 2057-2 Aurora Road, Twinsburg, OH 44087. (216) 425-9000. Annual. $89.00.

WHO'S WHO IN ENGINEERING. Engineers Joint Council, 345 East 47th Street, New York, NY 10017. (212) 705-7010. 1985. $200.00.

WHO'S WHO IN FRONTIER SCIENCE AND TECHNOLOGY. Marquis Who's Who, Incorporated, 200 East Ohio Street, Chicago, IL 60611. (800) 428-3898 or (312) 787-2008.

WHO'S WHO IN TECHNOLOGY TODAY. Reston Publishing Company, Incorporated, c/o Prentice-Hall, Incorporated, Englewood Cliffs, NJ 07632. (800) 262-6868. Biennial. Five volumes. $425.00. Covers the fields of electronics, computer science, physics, optics, chemistry, biotechnology, mechanics, energy, and earth science.

ENCYCLOPEDIAS AND DICTIONARIES

ENCYCLOPEDIA OF PHYSICAL SCIENCE AND TECHNOLOGY. Academic Press, Incorporated, Orlando, FL 32887. (800) 321-5068 or (305) 345-2734. Fifteen volumes, 1986.

IEEE STANDARD DICTIONARY OF ELECTRICAL AND ELECTRONICS TERMS. Frank Jay, editor. John Wiley and Sons, Incorporated, 605 Third Avenue, New York, NY 10158. (800) 526-5368 or (212) 850-6000. Third edition. 1984. $49.95.

MCGRAW-HILL ENCYCLOPEDIA OF SCIENCE AND TECHNOLOGY. McGraw-Hill Book, Incorporated, 1221 Avenue of the Americas, New York, NY 10020. (212) 997-3675. Fifth edition, 15 volumes. $1100.00.

GENERAL WORKS

CIRCUITS AND SOFTWARE FOR ELECTRONICS ENGINEERS. Howard Bierman. McGraw-Hill Book Company, 1221 Avenue of the Americas, New York, NY 10020. (800) 628-0004. 1984. $39.50.

CIRCUITS, DEVICES AND SYSTEMS: A FIRST COURSE IN ELECTRICAL ENGINEERING. R.J. Smith. John Wiley and Sons, Incorporated, 605 Third Avenue, New York, NY 10158. (800) 526-5368 or (212) 850-6000. Fourth edition. 1984. $41.45.

CIRCUITS, SIGNALS AND SYSTEMS. William M. Siebert. McGraw-Hill Book Company, 1221 Avenue of the Americas, New York, NY 10020. (800) 628-0004. 1986. $37.95.

ELECTRONIC CIRCUIT ANALYSIS: BASIC PRINCIPLES. Roy A. Colclaser, Donald A. Neaman, Charles F. Hawkins. John Wiley and Sons, Incorporated, 605 Third Avenue, New York, NY 10158. (800) 526-5368 or (212) 850-6000. 1984. $42.50.

ELECTRONIC CIRCUITS AND APPLICATIONS. Stephen D. Senturia and B.D. Wedlock. John Wiley and Sons, Incorporated, 605 Third Avenue, New York, NY 10158. (800) 526-5368 or (212) 850-6000. 1975. $44.00.

ELECTRONIC DEVICES AND COMPONENTS. J. Seymour. John Wiley and Sons, Incorporated, 605 Third Avenue, New York, NY 10158. (800) 526-5368 or (212) 850-6000. 1981.

$37.00 in paper.

ELECTRONIC INVENTIONS AND DISCOVERIES: ELECTRONICS FROM ITS EARLIEST BEGINNINGS TO THE PRESENT DAY. G.W.A. Dummer. Pergamon Press, Incorporated, Maxwell House, Fairview Park, Elmsford, NY 10523. (914) 592-7700. 1983. $48.50.

MICROELECTRONICS. J. Millman. McGraw-Hill Book, 1221 Avenue of the Americas, New York, NY 10020. (212) 512-2000. (212) 512-2000. Second edition. 1987. $39.95.

MICROPROCESSOR ENGINEERING. Brian Holdsworth. Butterworth Publishing, Incorporated, 80 Montvale Avenue, Stoneham, MA 02180. (617) 438-8464. 1986. $39.95.

SEMICONDUCTOR CIRCUIT APPROXIMATIONS: AN INTRODUCTION TO TRANSISTORS AND INTEGRATED CIRCUITS. Albert Paul Malvino. McGraw-Hill Book Company, 1221 Avenue of the Americas, New York, NY 10020. (800) 628-0004. Fourth edition. 1985. $34.95.

HANDBOOKS AND MANUALS

CONTEMPORARY ELECTRONICS CIRCUITS DESKBOOK. Harry Helms. McGraw-Hill Book Company, 1221 Avenue of the Americas, New York, NY 10020. (800) 628-0004. 1986. $29.95.

CRC HANDBOOK OF TABLES FOR APPLIED ENGINEERING SCIENCE. R.E. Bolz and G.L. Tuve, editors. CRC Press, Incorporated, 2000 Corporate Boulevard, NW, Boca Raton, FL 33431. Second edition. 1973. $69.00.

ELECTRONIC ENGINEERS HANDBOOK. Donald G. Fink, editor. McGraw-Hill Book Company, 1221 Avenue of the Americas, New York, NY 10020. (800) 628-0004. Second edition. 1982. $89.00.

HANDBOOK OF ELECTRONICS INDUSTRY COST ESTIMATING DATA. Theodore Taylor. John Wiley and Sons, Incorporated, 605 Third Avenue, New York, NY 10158. (800) 526-5368 or (212) 850-6000. 1985. $54.95.

HANDBOOK OF ELECTRONICS MANUFACTURING ENGINEERING. B.S. Matisoff. Van Nostrand Reinhold Company, Incorporated, 115 Fifth Avenue, New York, NY 10003. (800) 543-2681. Second edition. 1986. $52.95.

HANDBOOK OF MODERN ELECTRONICS AND ELECTRICAL ENGINEERING. Charles Belove. John Wiley and Sons, Incorporated, 605 Third Avenue, New York, NY 10158. (800) 526-5368 or (212) 850-6000. 1986. $85.00.

REFERENCE MANUAL FOR TELECOMMUNICATIONS. R.L. Freeman. John Wiley and Sons, Incorporated, 605 Third Avenue, New York, NY 10158. (800) 526-5368 or (212) 850-6000. 1985. $79.95.

STANDARD HANDBOOK FOR ELECTRICAL ENGINEERS. Donald G. Fink and H. Wayne Beaty. McGraw-Hill Book Company, 1221 Avenue of the Americas, New York, NY 10020. (212) 512-2000. Eleventh edition. 1978. $85.00.

VLSI HANDBOOK. Norman G. Einspruch, editor. Academic Press, Incorporated, 6277 Sea Harbor Drive, Orlando, FL 32821. (800) 321-5068. 1985. $125.00.

THE WILEY ENGINEER'S DESK REFERENCE. Sanford I. Heisler. John Wiley and Sons, Incorporated, 605 Third Avenue, New York, NY 10158. (800) 526-5368 or (212) 850-6418. 1984. $36.00.

ONLINE DATA BASES

COMPENDEX. Engineering Information, Incorporated, 345 East 47th Street, New York, NY 10017. (800) 221-1044 or (212) 705-7615. Engineering and technical literature, 1975 to present. Inquire as to cost and availability.

DISSERTATION ABSTRACTS ONLINE. University Microfilms International, 300 North Zeeb Road, Ann Arbor, MI 48106. (800) 521-0600 or (313) 761-4700. Scope includes virtually all doctoral dissertations accepted at accredited American institutions from 1861 to present in 252 subject areas. Inquire as to cost and availability.

INSPEC. INSPEC Marketing Department, Institute of Electrical and Electronics Engineers, Incorporated, IEEE Service Department, 445 Hoes Lane, Piscataway, NJ 08854. (201) 981-0060. Inquire as to on-line cost and availability.

NTIS. National Technical Information Service, 5285 Port Royal Road, Springfield, VA 22161. (703) 487-4630. Broad coverage of government sponsored research reports, 1964 to present. Inquire as to cost and availability.

WILSONLINE. H.W. Wilson Company, 950 University Avenue, Bronx, NY 10452. (800) 367-6770 or (212) 588-8400. Makes available online versions of the printed H.W. Wilson Indexes including Applied Science and Technology Index, Business Periodicals Index, and Readers' Guide to Periodical Literature. Period covered is generally 1983 to present. Inquire as to cost and availability.

OTHER SOURCES

A GUIDE TO THE LITERATURE OF ELECTRICAL AND ELECTRONIC ENGINEERING. Susan B. Ardis. Libraries Unlimited, Incorporated, Post Office Box 263, Littleton, CO 80160. (303) 770-1220. 1986. $37.50.

PERIODICALS

CANADAIAN ELECTRONICS ENGINEERING. Maclean Hunter Research Bureau, 777 Bay Street, Toronto, ON M5W 1A7 Canada. (416) 596-5729.

CIRCUITS, SYSTEMS AND SIGNAL PROCESSING. Birkhauser Boston, Incorporated, 380 Green Street, Post Office Box 2007, Cambridge, MA 02139. (617) 876-2333. Quarterly. $175.00 per year.

ELECTROCOMPONENT SCIENCE. Gordon Breach Science Publishers, Post Office Box 786, Cooper Station, New York, NY 10276. (212) 206-8900. Quarterly. $224.00 per year.

ELECTRONIC DESIGN. Heyden Publishing Company, Incorporated, 10 Mulholland Drive, Hasbrouck Heights, NJ 07604. (201) 393-6000. Biweekly. $65.00.

ELECTRONIC ENGINEERING TIMES. CMP Publications, Incorporated, 600 Community Drive, Manhasset, NY 11030. (516) 365-4600. Weekly. $65.00 per year.

ELECTRONICS WEEK. McGraw-Hill Book Company, 1221 Avenue of the Americas, New York, NY 10020. (800) 628-0004. Weekly. $18.00.

IEEE CIRCUITS AND DEVICES MAGAZINE. Institute of Electrical and Electronics Engineers, IEEE Service Center, 445 Hoes Lane, Piscataway, NJ 08854. (212) 705-7900. Bimonthly. $70.00 per year.

IEEE ELECTRON DEVICE LETTERS. Institute of Electrical and Electronics Engineers, IEEE Service Center, 445 Hoes Lane, Piscataway, NJ 08854. (212) 705-7900. Monthly. $70.00 per year.

IEEE JOURNAL OF SOLID STATE CIRCUITS. Institute of Electrical and Electronics Engineers, IEEE Service Center, 445 Hoes Lane, Piscataway, NJ 08854. (212) 705-7900. Bimonthly. $113.00 per year.

IEEE TRANSACTIONS ON CIRCUITS AND SYSTEMS. Institute of Electrical and Electronics Engineers, IEEE Service Center, 445 Hoes Lane, Piscataway, NJ 08854. (212) 705-7900. Monthly. $108.00 per year.

IEEE TRANSACTIONS ON ELECTRON DEVICES. Institute of Electrical and Electronics Engineers, IEEE Service Center, 445 Hoes Lane, Piscataway, NJ 08854. (212) 705-7900. Monthly. $159.00 per year.

MICROELECTRONIC ENGINEERING: AN INTERDISCIPLINARY JOURNAL OF SEMICONDUCTOR MANUFACTURING TECHNOLOGY. Elsevier Science Publishing Company, Incorporated, 52 Vanderbilt Avenue, New York, NY 10017. (212) 370-5520. Eight times per year. $90.00 per year.

MICROELECTRONIC MANUFACTURING AND TESTING. Lake Publishing Corporation, 17730 West Peterson Road, Libertyville, IL 60048. (312) 362-8711. Monthly. $60.00 per year.

MICROELECTRONICS AND RELIABILITY. Pergamon Press, Incorporated, Journals Division, Maxwell House, Fairview Park, Elmsford, NY 10523. (914) 592-7700. Monthly. $250.00 per year.

MICROELECTRONICS JOURNAL. Benn Electronics Publications, Limited, Box 28, Luton, Beds, LU2 0ED, England. Bimonthly. $225.00 per year.

RADIO ELECTRONICS. Gernsback Publications, Incorporated, 500-B BiCounty Boulevard, Farmingdale, NY 11735. (516) 293-3000. Monthly. $16.00 per year.

SEMICONDUCTOR INTERNATIONAL. Cahners Publishing Company, Incorporated, Division of Reed Holdings, Incorporated, Cahners Plaza, 1350 East Touhy Avenue, Des Plaines, IL 60018. (312) 635-8800. Monthly. $55.00 per year.

VLSI SYSTEMS DESIGN. CMP Publications, Incorporated, 600 Community Drive, Manhasset, NY 11030. (516) 365-4600. Monthly. $20.00 per year.

RESEARCH CENTERS AND INSTITUTES

MASSACHUSETTS INSTITUTE OF TECHNOLOGY. Laboratory for Electromagnetic and Electronic Systems, 77 Massachusetts Avenue, Cambridge, MA 02139. (617) 253-4631.

NORTH CAROLINA STATE UNIVERSITY. Solid State Electronics Laboratory, 432 Daniels Hall, Raleigh, NC 27695. (919) 737-2336.

OHIO STATE UNIVERSITY. Electroscience Laboratory, 1320 Kinnear Road, Columbus, OH 43212. (614) 422-7981.

PENNSYLVANIA STATE UNIVERSITY. Solid State Device Laboratory, 210 Electrical Engineering, West Building, University Park, PA 16802. (814) 865-1666.

MICROPROCESSORS

See also: ELECTRICAL ENGINEERING, ELECTRICITY, ELECTRONIC CIRCUITS AND COMPONENTS, ELECTRONIC ENGINEERING, MICROELECTRONICS

ABSTRACT SERVICES AND INDEXES

APPLIED SCIENCE AND TECHNOLOGY INDEX. H.W. Wilson Company, 950 University Avenue, Bronx, NY 10452. (800) 367-6670 or (212) 588-8400. Inquire as to cost and availability.

ELECTRONICS AND COMMUNICATIONS ABSTRACTS JOURNAL. Cambridge Scientific Abstracts, 5161 River Road, Bethesda, MD 20816. (301) 951-1400. Bimonthly. Inquire as to cost and availability.

ENGINEERING INDEX MONTHLY. Engineering Information, Incorporated, 345 East 47th Street, New York, NY 10017. (800) 221-1044 or (212) 705-7600. Monthly, with annual cumulation. $1425.00 per year.

PHYSICS ABSTRACTS. Institute of Electrical Engineers, London, United Kingdom. Available from: Institute of Electrical and Electronic Engineers (Ieee), 345 East 47th Street, New York, NY 10017. (212) 705-7900.

SCIENCE CITATION INDEX. Institute for Scientific Information, 3501 Market Street, Philadelphia, PA 19104. (800) 523-1850 or (215) 386-0100. Inquire as to cost and availability.

SOLID STATE ABSTRACTS: AN ABSTRACT JOURNAL INVOLVING THE PHYSICS, METALLURGY, CRYSTALLOGRAPHY, CHEMISTRY, AND DEVICE TECHNOLOGY OF SOLIDS. Cambridge Scientific Abstracts, 5161 River Road, Bethesda, MD 20816. (301) 951-1400. 1957 to present. Bimonthly. $550.00 per year.

ASSOCIATIONS AND PROFESSIONAL SOCIETIES

AMERICAN ELECTRONICS ASSOCIATION. Post Office Box 10045, 2670 Hanover Street, Palo Alto, CA 94303. (415) 857-9300.

ELECTRONIC INDUSTRIES ASSOCIATION. 2001 Eye Street, NW, Washington, DC 20006. (202) 457-4900.

IEEE (INSTITUTE OF ELECTRICAL AND ELECTRONICS ENGINEERS). 345 East 47th Street, New York, NY 10017. (212) 705-7900.

INTERNATIONAL SOCIETY FOR HYBRID MICRO-ELECTRONICS. Post Office Box 2698, 1861 Wiehle Avenue, Suite 340, Reston, VA 22090. (703) 471-0066.

BIBLIOGRAPHIES

HANDBOOKS AND TABLES IN SCIENCE AND TECHNOLOGY. Russell H. Powell, editor. Oryx Press, 2214 North Central Avenue, Phoenix, AZ 85004-1483. (602) 254-6156. Second edition. 1983. $55.00.

SCIENTIFIC AND TECHNICAL BOOKS IN PRINT: AN INDEX TO LITERATURE IN SCIENCE AND TECHNOLOGY. R.R. Bowker Company, 205 East 42nd Street, New York, NY 10017. (800) 521-8110 or (212) 916-1600.

DIRECTORIES AND BIOGRAPHICAL SOURCES

AMERICAN ELECTRONICS ASSOCIATION DIRECTORY. American Electronics Association, Post Office Box 10045, 2670 Hanover Street, Palo Alto, CA 94303. (415) 857-9300. Annual.

EEM - ELECTRONIC ENGINEERS MASTER. Hearst Business Communications/UTP Division, 645 Stewart Avenue, Garden City, NY 11530. (516) 227-1300. Annual. $75.00 per copy.

IEEE MEMBERSHIP DIRECTORY. Institute of Electrical and Electronics Engineers, IEEE Service Center, 445 Hoes Lane, Piscataway, NJ 08854. (212) 705-7900. Annual. $7.00.

RESEARCH CENTERS DIRECTORY. Gale Research Company, Book Tower, Detroit, MI 48226. (800) 521-0707. Twelfth edition. 1987. $365.00 for set.

SCIENTIFIC AND TECHNICAL ORGANIZATIONS AND AGENCIES DIRECTORY. Gale Research Company, Book Tower, Detroit, MI 48226. (800) 521-0707. 1985.

WHO'S WHO IN ELECTRONICS. Harris Publishing Company, 2057-2 Aurora Road, Twinsburg, OH 44087. (216) 425-9000. Annual. $89.00.

WHO'S WHO IN ENGINEERING. Engineers Joint Council, 345 East 47th Street, New York, NY 10017. (212) 705-7010. 1985. $200.00.

WHO'S WHO IN FRONTIER SCIENCE AND TECHNOLOGY. Marquis Who's Who, Incorporated, 200 East Ohio Street,

Chicago, IL 60611. (800) 428-3898 or (312) 787-2008.

WHO'S WHO IN TECHNOLOGY TODAY. Reston Publishing Company, Incorporated, c/o Prentice-Hall, Incorporated, Englewood Cliffs, NJ 07632. (800) 262-6868. Biennial. Five volumes. $425.00. Covers the fields of electronics, computer science, physics, optics, chemistry, biotechnology, mechanics, energy, and earth science.

ENCYCLOPEDIAS AND DICTIONARIES

ENCYCLOPEDIA OF PHYSICAL SCIENCE AND TECHNOLOGY. Academic Press, Incorporated, Orlando, FL 32887. (800) 321-5068 or (305) 345-2734. Fifteen volumes, 1986.

IEEE STANDARD DICTIONARY OF ELECTRICAL AND ELECTRONICS TERMS. Frank Jay, editor. John Wiley and Sons, Incorporated, 605 Third Avenue, New York, NY 10158. (800) 526-5368 or (212) 850-6000. Third edition. 1984. $49.95.

MCGRAW-HILL ENCYCLOPEDIA OF SCIENCE AND TECHNOLOGY. McGraw-Hill Book, Incorporated, 1221 Avenue of the Americas, New York, NY 10020. (212) 997-3675. Fifth edition, 15 volumes. $1100.00.

GENERAL WORKS

BEGINNER'S GUIDE TO MICROPROCESSORS. Charles M. Gilmore, TAB Books, Incorporated, Monterey Lane, Blue Ridge Summit, PA 17214. (717) 794-2191. Second edition. 1984. $14.95.

CIRCUITS AND SOFTWARE FOR ELECTRONICS ENGINEERS. Howard Bierman. McGraw-Hill Book Company, 1221 Avenue of the Americas, New York, NY 10020. (800) 628-0004. 1984. $39.50.

CIRCUITS, DEVICES AND SYSTEMS: A FIRST COURSE IN ELECTRICAL ENGINEERING. R.J. Smith. John Wiley and Sons, Incorporated, 605 Third Avenue, New York, NY 10158. (800) 526-5368 or (212) 850-6000. Fourth edition. 1984. $41.45.

CIRCUITS, SIGNALS AND SYSTEMS. William M. Siebert. McGraw-Hill Book Company, 1221 Avenue of the Americas, New York, NY 10020. (800) 628-0004. 1986. $37.95.

ELECTRONIC CIRCUIT ANALYSIS: BASIC PRINCIPLES. Roy A. Colclaser, Donald A. Neaman, Charles F. Hawkins. John Wiley and Sons, Incorporated, 605 Third Avenue, New York, NY 10158. (800) 526-5368 or (212) 850-6000. 1984. $42.50.

ELECTRONIC CIRCUITS AND APPLICATIONS. Stephen D. Senturia and B.D. Wedlock. John Wiley and Sons, Incorporated, 605 Third Avenue, New York, NY 10158. (800) 526-5368 or (212) 850-6000. 1975. $44.00.

ELECTRONIC DEVICES AND COMPONENTS. J. Seymour. John Wiley and Sons, Incorporated, 605 Third Avenue, New York, NY 10158. (800) 526-5368 or (212) 850-6000. 1981. $37.00 in paper.

ELECTRONIC INVENTIONS AND DISCOVERIES: ELECTRONICS FROM ITS EARLIEST BEGINNINGS TO THE PRESENT DAY. G.W.A. Dummer. Pergamon Press, Incorporated, Maxwell House, Fairview Park, Elmsford, NY 10523. (914) 592-7700. 1983. $48.50.

MICROELECTRONICS. J. Millman. McGraw-Hill Book, 1221 Avenue of the Americas, New York, NY 10020. (212) 512-2000. (212) 512-2000. Second edition. 1987. $39.95.

MICROPROCESSOR ENGINEERING. Brian Holdsworth. Butterworth Publishing, Incorporated, 80 Montvale Avenue, Stoneham, MA 02180. (617) 438-8464. 1986. $39.95.

MICROPROCESSOR PROGRAMMING AND APPLICATIONS FOR SCIENTISTS AND ENGINEERS. R.R. Smardzewski. Elsevier Science Publishing Company, Incorporated, 52 Vanderbilt Avenue, New York, NY 10017. (212) 370-5520. 1984. $36.50.

SEMICONDUCTOR CIRCUIT APPROXIMATIONS: AN INTRODUCTION TO TRANSISTORS AND INTEGRATED CIRCUITS. Albert Paul Malvino. McGraw-Hill Book Company, 1221 Avenue of the Americas, New York, NY 10020. (800) 628-0004. Fourth edition. 1985. $34.95.

HANDBOOKS AND MANUALS

CONTEMPORARY ELECTRONICS CIRCUITS DESKBOOK. Harry Helms. McGraw-Hill Book Company, 1221 Avenue of the Americas, New York, NY 10020. (800) 628-0004. 1986. $29.95.

ELECTRONIC ENGINEERS HANDBOOK. Donald G. Fink, editor. McGraw-Hill Book Company, 1221 Avenue of the Americas, New York, NY 10020. (800) 628-0004. Second edition. 1982. $89.00.

HANDBOOK OF ELECTRONICS INDUSTRY COST ESTIMATING DATA. Theodore Taylor. John Wiley and Sons, Incorporated, 605 Third Avenue, New York, NY 10158. (800) 526-5368 or (212) 850-6000. 1985. $54.95.

HANDBOOK OF ELECTRONICS MANUFACTURING ENGINEERING. B.S. Matisoff. Van Nostrand Reinhold Company, Incorporated, 115 Fifth Avenue, New York, NY 10003. (800) 543-2681. Second edition. 1986. $52.95.

HANDBOOK OF MODERN ELECTRONICS AND ELECTRICAL ENGINEERING. Charles Belove. John Wiley and Sons, Incorporated, 605 Third Avenue, New York, NY 10158. (800) 526-5368 or (212) 850-6000. 1986. $85.00.

MICROPROCESSOR HANDBOOK. Joseph D. Greenfield, editor. John Wiley and Sons, Incorporated, 605 Third Avenue, New York, NY 10158. (800) 526-5368 or (212) 850-6000. 1985. $44.95.

REFERENCE MANUAL FOR TELECOMMUNICATIONS. R.L. Freeman. John Wiley and Sons, Incorporated, 605 Third Avenue, New York, NY 10158. (800) 526-5368 or (212) 850-6000. 1985. $79.95.

VLSI HANDBOOK. Norman G. Einspruch, editor. Academic Press, Incorporated, 6277 Sea Harbor Drive, Orlando, FL 32821. (800) 321-5068. 1985. $125.00.

ONLINE DATA BASES

COMPENDEX. Engineering Information, Incorporated, 345 East 47th Street, New York, NY 10017. (800) 221-1044 or (212) 705-7615. Engineering and technical literature, 1975 to present. Inquire as to cost and availability.

DISSERTATION ABSTRACTS ONLINE. University Microfilms International, 300 North Zeeb Road, Ann Arbor, MI 48106. (800) 521-0600 or (313) 761-4700. Scope includes virtually all doctoral dissertations accepted at accredited American institutions from 1861 to present in 252 subject areas. Inquire as to cost and availability.

INSPEC. INSPEC Marketing Department, Institute of Electrical and Electronics Engineers, Incorporated, IEEE Service Department, 445 Hoes Lane, Piscataway, NJ 08854. (201) 981-0060. Inquire as to on-line cost and availability.

NTIS. National Technical Information Service, 5285 Port Royal Road, Springfield, VA 22161. (703) 487-4630. Broad coverage of government sponsored research reports, 1964 to present. Inquire as to cost and availability.

WILSONLINE. H.W. Wilson Company, 950 University Avenue, Bronx, NY 10452. (800) 367-6770 or (212) 588-8400. Makes available online versions of the printed H.W. Wilson Indexes including Applied Science and Technology Index, Business Periodicals Index, and Readers' Guide to Periodical Literature.

Period covered is generally 1983 to present. Inquire as to cost and availability.

OTHER SOURCES

A GUIDE TO THE LITERATURE OF ELECTRICAL AND ELECTRONIC ENGINEERING. Susan B. Ardis. Libraries Unlimited, Incorporated, Post Office Box 263, Littleton, CO 80160. (303) 770-1220. 1986. $37.50.

PERIODICALS

CANADAIAN ELECTRONICS ENGINEERING. Maclean Hunter Research Bureau, 777 Bay Street, Toronto, ON M5W 1A7 Canada. (416) 596-5729.

CIRCUITS, SYSTEMS AND SIGNAL PROCESSING. Birkhauser Boston, Incorporated, 380 Green Street, Post Office Box 2007, Cambridge, MA 02139. (617) 876-2333. Quarterly. $175.00 per year.

ELECTROCOMPONENT SCIENCE. Gordon Breach Science Publishers, Post Office Box 786, Cooper Station, New York, NY 10276. (212) 206-8900. Quarterly. $224.00 per year.

ELECTRONIC DESIGN. Heyden Publishing Company, Incorporated, 10 Mulholland Drive, Hasbrouck Heights, NJ 07604. (201) 393-6000. Biweekly. $65.00.

ELECTRONIC ENGINEERING TIMES. CMP Publications, Incorporated, 600 Community Drive, Manhasset, NY 11030. (516) 365-4600. Weekly. $65.00 per year.

ELECTRONICS WEEK. McGraw-Hill Book Company, 1221 Avenue of the Americas, New York, NY 10020. (800) 628-0004. Weekly. $18.00.

IEEE CIRCUITS AND DEVICES MAGAZINE. Institute of Electrical and Electronics Engineers, IEEE Service Center, 445 Hoes Lane, Piscataway, NJ 08854. (212) 705-7900. Bimonthly. $70.00 per year.

IEEE ELECTRON DEVICE LETTERS. Institute of Electrical and Electronics Engineers, IEEE Service Center, 445 Hoes Lane, Piscataway, NJ 08854. (212) 705-7900. Monthly. $70.00 per year.

IEEE JOURNAL OF SOLID STATE CIRCUITS. Institute of Electrical and Electronics Engineers, IEEE Service Center, 445 Hoes Lane, Piscataway, NJ 08854. (212) 705-7900. Bimonthly. $113.00 per year.

IEEE TRANSACTIONS ON CIRCUITS AND SYSTEMS. Institute of Electrical and Electronics Engineers, IEEE Service Center, 445 Hoes Lane, Piscataway, NJ 08854. (212) 705-7900. Monthly. $108.00 per year.

IEEE TRANSACTIONS ON ELECTRON DEVICES. Institute of Electrical and Electronics Engineers, IEEE Service Center, 445 Hoes Lane, Piscataway, NJ 08854. (212) 705-7900. Monthly. $159.00 per year.

MICROELECTRONIC ENGINEERING: AN INTERDISCIPLINARY JOURNAL OF SEMICONDUCTOR MANUFACTURING TECHNOLOGY. Elsevier Science Publishing Company, Incorporated, 52 Vanderbilt Avenue, New York, NY 10017. (212) 370-5520. Eight times per year. $90.00 per year.

MICROELECTRONIC MANUFACTURING AND TESTING. Lake Publishing Corporation, 17730 West Peterson Road, Libertyville, IL 60048. (312) 362-8711. Monthly. $60.00 per year.

MICROELECTRONICS AND RELIABILITY. Pergamon Press, Incorporated, Journals Division, Maxwell House, Fairview Park, Elmsford, NY 10523. (914) 592-7700. Monthly. $250.00 per year.

MICROELECTRONICS JOURNAL. Benn Electronics Publications, Limited, Box 28, Luton, Beds, LU2 0ED, England. Bimonthly. $225.00 per year.

RADIO ELECTRONICS. Gernsback Publications, Incorporated, 500-B BiCounty Boulevard, Farmingdale, NY 11735. (516) 293-3000. Monthly. $16.00 per year.

SEMICONDUCTOR INTERNATIONAL. Cahners Publishing Company, Incorporated, Division of Reed Holdings, Incorporated, Cahners Plaza, 1350 East Touhy Avenue, Des Plaines, IL 60018. (312) 635-8800. Monthly. $55.00 per year.

VLSI SYSTEMS DESIGN. CMP Publications, Incorporated, 600 Community Drive, Manhasset, NY 11030. (516) 365-4600. Monthly. $20.00 per year.

RESEARCH CENTERS AND INSTITUTES

MASSACHUSETTS INSTITUTE OF TECHNOLOGY. Laboratory for Electromagnetic and Electronic Systems, 77 Massachusetts Avenue, Cambridge, MA 02139. (617) 253-4631.

NORTH CAROLINA STATE UNIVERSITY. Solid State Electronics Laboratory, 432 Daniels Hall, Raleigh, NC 27695. (919) 737-2336.

OHIO STATE UNIVERSITY. Electroscience Laboratory, 1320 Kinnear Road, Columbus, OH 43212. (614) 422-7981.

PENNSYLVANIA STATE UNIVERSITY. Solid State Device Laboratory, 210 Electrical Engineering, West Building, University Park, PA 16802. (814) 865-1666.

STANFORD UNIVERSITY. Stanford Electronics Laboratories, Stanford, CA 94305. (415) 497-1013.

MICROSCOPY

See also: ELECTRON MICROSCOPY, OPTICS

ABSTRACT SERVICES AND INDEXES

APPLIED SCIENCE AND TECHNOLOGY INDEX. H.W. Wilson and Company, 950 University Avenue, Bronx, NY 10452. (800) 367-6670 or (212) 588-8400. Monthly. Inquire as to cost and availability.

BIBLIOGRAPHY AND INDEX OF GEOLOGY. American Geological Institute, 4220 King Street, Alexandria, VA 22302. (703) 379-2480. 1969 to present. Monthly. $1100.00 per year.

CHEMICAL ABSTRACTS. American Chemical Society, Chemical Abstracts Service, Box 3012, Columbus, OH 43210. (614) 421-3600. 1907 to present. Weekly. $9500.00 per year.

CURRENT CONTENTS: ENGINEERING, TECHNOLOGY AND APPLIED SCIENCES. Institute for Scientific Information, 3501 Market Street, Philadelphia, PA 19104. (800) 523-1850 or (215) 386-0100. Weekly. $275.00 per year.

ENGINEERING INDEX MONTHLY AND AUTHOR INDEX. Engineering Information Inc., 345 East 47th Street, New York, NY 10017. (212) 705-7600. Monthly. $1560.00 per year.

GENERAL SCIENCE INDEX. H.W. Wilson and Company, 950 University Avenue, Bronx, NY 10452. (800) 367-6670 or (212) 588-8400. 1978 to present. Monthly. Inquire as to cost and availability.

PHYSICS ABSTRACTS. Institution of Electrical Engineers. Available from: IEEE Service Center, 445 Hoes Lane, Piscataway, NJ 08854. 1898 to present. Bimonthly. $1700.00 per year.

PHYSICS BRIEFS. Physik Verlag GmbH, Postfach 1260/1280, D-6940 Weinheim, West Germany. (212) 661-9404. 1920 to present.

Twenty-six times per year. $1250.00 per year.

SCIENCE CITATION INDEX. Institute for Scientific Information, 3501 Market Street, Philadelphia, PA 19104. (800) 523-1850 or (215) 386-0100. Six times per year. $6200.00 per year.

ANNUAL REVIEWS AND YEARBOOKS

ADVANCES IN OPTICAL AND ELECTRON MICROSCOPY. V.E. Cosslett and R. Barer, editors. Academic Press, Inc., 6277 Sea Harbor Drive, Orlando, FL 32821. (800) 321-5068. 1966 to present. Irregular. Inquire.

ASSOCIATIONS AND PROFESSIONAL SOCIETIES

ELECTRON MICROSCOPY SOCIETY OF AMERICA. c/o Linda L. Horton, 1497 Chain Bridge Road, Suite 104, McLean, VA 22101. (703) 827-0498.

MICROBEAM ANALYSIS SOCIETY. c/o John Armstrong, Department of Geology, 170-25, California Institute of Technology, Pasadena, CA 91125. (818) 356-6253.

OPTICAL SOCIETY OF AMERICA. 1816 Jefferson Place, N.W., Washington, DC 20036. (202) 223-8130.

SPIE - THE INTERNATIONAL SOCIETY FOR OPTICAL ENGINEERING. P.O. Box 10, Bellington, WA 98227. (206) 676-3290.

DIRECTORIES AND BIOGRAPHICAL SOURCES

AMERICAN MEN AND WOMEN OF SCIENCE. R.R. Bowker, Inc., Order Department, 245 West 17th Street, New York, NY 10011. (800) 521-8110. Eight volumes. 1986. $595.00 for set.

INTERNATIONAL RESEARCH CENTERS DIRECTORY 1988-89. Darren L. Smith, editor. Gale Research Company, Book Tower, Detroit, MI 48226. (800) 521-0707. 4th edition. 1987. $360.00.

1987 DIRECTORY OF ENGINEERING SOCIETIES AND RELATED ORGANIZATIONS. Gordon Davis, editor. Hemisphere Publishing Corporation, 1010 Vermont Avenue, NW, Washington, DC 20005. (800) 526-0275. 12th edition. 1987. $100.00.

RESEARCH CENTERS DIRECTORY 1988. Gale Research Company, Book Tower, Detroit, MI 48226. (800) 521-0707. 12th edition. 1987. $365.00 for set.

SCIENTIFIC AND TECHNICAL ORGANIZATIONS AND AGENCIES DIRECTORY. Margaret Labash Young, editor. Gale Research Company, Book Tower, Detroit, MI 48226. (800) 521-0707. 2nd edition. 1987. $185.00.

SPIE - THE INTERNATIONAL SOCIETY FOR OPTICAL ENGINEERING ANNUAL MEMBERSHIP DIRECTORY. P.O. Box 10, Bellington, WA 98227. (206) 676-3290. Annual. Inquire.

WHO'S WHO IN ENGINEERING. Gordon Davis, editor. Hemisphere Publishing Corporation, 1010 Vermont Avenue, NW, Washington, DC 20005. (800) 526-0275. 6th edition. 1985. $200.00.

GENERAL WORKS

INTRODUCTION TO ELECTRON MICROSCOPY. Saul Wischnitzer. Pergamon Press, Inc., Maxwell House, Fairview Park, Elmsford, NY 10523. (914) 592-7700. 3rd edition. 1981. $43.00.

INTRODUCTION TO MICROSCOPY BY MEANS OF LIGHT, ELECTRONS, X-RAYS, OR ULTRASOUND. T.G. Rochow and E.G. Rochow. Plenum Publishing Corporation, 233 Spring Street, New York, NY 10013. (800) 221-9369. 1979. $32.50.

INTRODUCTION TO THE OPTICAL MICROSCOPE. Savile Bradbury. Oxford University Press, 200 Madison Avenue, New York, NY 10016. (800) 458-5833. 1984. $9.95 in paper.

MICROSCOPES AND THEIR USES. Claude Marmasse. Gordon and Breach Science Publishers, Inc., 50 West 23rd Street, New York, NY 10010. (212) 206-8900. 1980. $29.00.

THEORY AND PRACTICE OF SCANNING OPTICAL MICROSCOPY. Tony C. Wilson and J.R. Sheppard. Academic Press, Inc., 6277 Sea Harbor Drive, Orlando, FL 32821. (800) 321-5068. 1984. $41.00.

ONLINE DATA BASES

CA SEARCH. Chemical Abstracts Service, P.O. Box 3012, Columbus, OH 43120. (800) 848-6538 or (614) 421-3600. Comprehensive guide to chemical literature, 1972 to present. Inquire as to online cost and availability.

COMPENDEX. Engineering Information, Inc., 345 East 47th Street, New York, NY 10017. (800) 221-1044 or (212) 705-7615. Engineering and technical literature, 1975 to present. Inquire as to online cost and availability.

DISSERTATION ABSTRACTS ONLINE. University Microfilms International, 300 North Zeeb Road, Ann Arbor, MI 48106. (800) 521-0600 or (313) 761-4700. Scope includes virtually all doctoral dissertations accepted at accredited American institutions from 1861 to present in over 250 subject areas. Inquire as to online cost and availability.

GEOREF. Online version of the BIBLIOGRAPHY AND INDEX OF GEOLOGY. American Geological Institute, 4220 King Street, Alexandria, VA 22302. (703) 379-2480. 1969 to present. Inquire as to online cost and availability.

INSPEC. INSPEC Marketing Department, Institution of Electrical Engineers. Available from IEEE Service Center, 445 Hoes Lane, Piscataway, NJ 08854. (201) 981-0060. Online version of Physics Abstracts. Inquire as to online cost and availability.

NTIS. National Technical Information Service, 5285 Port Royal Road, Springfield, VA 22161. (703) 487-4630. Broad coverage of government sponsored research reports, 1964 to present. Inquire as to online cost and availability.

SCISEARCH. Institute for Scientific Information, 3501 Market Street, Philadelphia, PA 19104. (800) 523-1850 or (215) 386-0100. Broad multidisciplinary title and author index to the international literature of science and technology, 1974 to present. Inquire as to online cost and availability.

WILSONLINE. H.W. Wilson and Company, 950 University Avenue, Bronx, NY 10452. (800) 367-6770 or (212) 588-8400. Makes available online versions of the H.W. Wilson indexes including Applied Science and Technology Index, Business Periodicals Index and Readers' Guide to Periodical Literature. Approximately 1980 to present. Inquire as to online cost and availability.

PERIODICALS

JOURNAL OF MICROSCOPY. Blackwell Scientific Publications, Inc., 52 Beacon Street, Boston, MA 02108. (800) 325-4177. 1878 to present. Monthly. $315.00 per year.

SCANNING ELECTRON MICROSCOPY. Scanning Microscopy Inc., Box 66507, AMF O'Hare, IL 60666. (312) 529-6677. 1968 to present. Quarterly. $109.00 per year.

RESEARCH CENTERS AND INSTITUTES

CENTER FOR ELECTRON MICROSCOPY. University of Illinois, 74 Bevier Hall, 905 South Goodwin Avenue, Urbana, IL 61801. (217) 333-2108.

ELECTRON MICROBEAM ANALYSIS LABORATORY. 413 Space Research Building, Ann Arbor, MI 48109. (313) 764-3360.

MICROWAVE SPECTROSCOPY

See: SPECTROSCOPY

MICROWAVES

See also: ELECTRICAL ENGINEERING, ELECTRICITY AND MAGNETISM

ABSTRACT SERVICES AND INDEXES

APPLIED SCIENCE AND TECHNOLOGY INDEX. H.W. Wilson and Company, 950 University Avenue, Bronx, NY 10452. (800) 367-6670 or (212) 588-8400. Monthly. Inquire as to cost and availability.

CHEMICAL ABSTRACTS. American Chemical Society, Chemical Abstracts Service, Box 3012, Columbus, OH 43210. (614) 421-3600. 1907 to present. Weekly. $9500.00 per year.

CURRENT CONTENTS: ENGINEERING, TECHNOLOGY AND APPLIED SCIENCES. Institute for Scientific Information, 3501 Market Street, Philadelphia, PA 19104. (800) 523-1850 or (215) 386-0100. Weekly. $275.00 per year.

ELECTRICAL AND ELECTRONIC ABSTRACTS. Institution of Electrical Engineers. Available from: IEEE Service Center, 445 Hoes Lane, Piscataway, NJ 08854. 1898 to present. Monthly. $1250.00 per year.

ENGINEERING INDEX MONTHLY AND AUTHOR INDEX. Engineering Information Inc., 345 East 47th Street, New York, NY 10017. (212) 705-7600. Monthly. $1560.00 per year.

PHYSICS ABSTRACTS. Institution of Electrical Engineers. Available from: IEEE Service Center, 445 Hoes Lane, Piscataway, NJ 08854. 1898 to present. Bimonthly. $1700.00 per year.

PHYSICS BRIEFS. Physik Verlag GmbH, Postfach 1260/1280, D-6940 Weinheim, West Germany. (212) 661-9404. 1920 to present. Twenty-six times per year. $1250.00 per year.

SCIENCE CITATION INDEX. Institute for Scientific Information, 3501 Market Street, Philadelphia, PA 19104. (800) 523-1850 or (215) 386-0100. Six times per year. $6200. 00 per year.

ASSOCIATIONS AND PROFESSIONAL SOCIETIES

IEEE MICROWAVE THEORY AND THECHNIQUES SOCIETY. Institute of Electrical and Electronics Engineers. IEEE Service Center, 445 Hoes Lane, Piscataway, NJ 08854.

INSTITUTE OF ELECTRICAL AND ELECTRONICS ENGINEERS. 345 East 47th Street, New York, NY 10017. (212) 705-7900.

INTERNATIONAL MICROWAVE POWER INSTITUTE. 13542 Union Village Circle, Clifton, VA 22024. (703) 830-5588.

BIBLIOGRAPHIES

NEW TECHNICAL BOOKS: A SELECTIVE LIST WITH DESCRIPTIVE ANNOTATIONS. New York Public Library, Science and Technology Research Center, Fifth Avenue and 42nd Street, New York, NY 10018. (212) 930-0800. 1915 to present. Monthly. $15.00 per year.

SCIENTIFIC AND TECHNICAL BOOKS AND SERIALS IN PRINT 1988; AN INDEX TO LITERATURE IN SCIENCE AND TECHNOLOGY. R.R. Bowker Company, 205 East 42nd Street, New York, NY 10017. (800) 521-8110. $175.00.

DIRECTORIES AND BIOGRAPHICAL SOURCES

AMERICAN MEN AND WOMEN OF SCIENCE. R.R. Bowker, Inc., Order Department, 245 West 17th Street, New York, NY 10011. (800) 521-8110. Eight volumes. 1986. $595.00 for set.

INTERNATIONAL DIRECTORY OF ELECTROMAGNETIC HEATING AND INSTRUMENTATION. International Microwave Power Institute, 13542 Union Village Circle, Clifton, VA 22024. (703) 830-5588. Annual. $12.00.

MICROWAVES AND RF PRODUCT DATA DIRECTORY. Hayden Publishing Company, Inc., 10 Mulholland Drive, Hasbrouck Heights, NJ 07604. Annual. $25.00.

1987 DIRECTORY OF ENGINEERING SOCIETIES AND RELATED ORGANIZATIONS. Gordon Davis, editor. Hemisphere Publishing Corporation, 1010 Vermont Avenue, NW, Washington, DC 20005. (800) 526-0275. 12th edition. 1987. $100.00.

RESEARCH CENTERS DIRECTORY 1988. Gale Research Company, Book Tower, Detroit, MI 48226. (800) 521-0707. 12th edition. 1987. $365.00 for set.

SCIENTIFIC AND TECHNICAL ORGANIZATIONS AND AGENCIES DIRECTORY. Margaret Labash Young, editor. Gale Research Company, Book Tower, Detroit, MI 48226. (800) 521-0707. 2nd edition. 1987. $185.00.

WHO'S WHO IN ENGINEERING. Gordon Davis, editor. Hemisphere Publishing Corporation, 1010 Vermont Avenue, NW, Washington, DC 20005. (800) 526-0275. 6th edition. 1985. $200.00.

ENCYCLOPEDIAS AND DICTIONARIES

THESAURUS OF SCIENTIFIC, TECHNICAL, AND ENGINEERING TERMS. Hemisphere Publishing Corporation, 1010 Vermont Avenue, NW, Washington, DC 20005. (800) 526-0275. 1988. $125.00.

GENERAL WORKS

MICROWAVE COMMUNICATIONS DEVICES AND CIRCUITS. Edgar Hund. McGraw-Hill Book Company, 1221 Avenue of the Americas, New York, NY 10020. (212) 512-2000. 1987. $28.95.

MICROWAVE ENGINEERING AND APPLIACATIONS. Om. P. Ganhi. Pergamon Press, Inc., Maxwell House, Fairview Park, Elmsford, NY 10523. (914) 592-7700. 1981. $32.50 in paper.

MICROWAVE PASSIVE DIRECTION FINDING. S. Lipsky. John Wiley and Sons, Inc., 605 Third Avenue, New York, NY 10158. (800) 526-5368. 1987. $44.95.

MICROWAVE THEORY, COMPONENTS AND DEVICES. John A. Seeger. Prentice-Hall Publishing, Inc., Englewood Cliffs, NJ 07632. (800) 562-0245. 1986. $90.00.

MICROWAVES: AN INTRODUCTION TO MICROWAVE THEORY AND TECHNIQUES. A.J. Baden Fuller. Pergamon Press, Inc., Maxwell House, Fairview Park, Elmsford, NY 10523. (914) 592-7700. 2nd edition. 1979. $19.85 in paper.

HANDBOOKS AND MANUALS

MICROWAVE TECHNIQUES AND LABORATORY MANUAL. M.L. Sisodia and G.S. Raghuvanshi. John Wiley and Sons, Inc., 605 Third Avenue, New York, NY 10158. (800) 526-5368. 1987. $21.95.

MICROWAVES

ONLINE DATA BASES

CA SEARCH. Chemical Abstracts Service, P.O. Box 3012, Columbus, OH 43120. (800) 848-6538 or (614) 421-3600. Comprehensive guide to chemical literature, 1972 to present. Inquire as to online cost and availability.

COMPENDEX. Engineering Information, Inc., 345 East 47th Street, New York, NY 10017. (800) 221-1044 or (212) 705-7615. Engineering and technical literature, 1975 to present. Inquire as to online cost and availability.

DISSERTATION ABSTRACTS ONLINE. University Microfilms International, 300 North Zeeb Road, Ann Arbor, MI 48106. (800) 521-0600 or (313) 761-4700. Scope includes virtually all doctoral dissertations accepted at accredited American institutions from 1861 to present in over 250 subject areas. Inquire as to online cost and availability.

INSPEC. INSPEC Marketing Department, Institution of Electrical Engineers. Available from IEEE Service Center, 445 Hoes Lane, Piscataway, NJ 08854. (201) 981-0060. Online version of Physics Abstracts. Inquire as to online cost and availability.

NTIS. National Technical Information Service, 5285 Port Royal Road, Springfield, VA 22161. (703) 487-4630. Broad coverage of government sponsored research reports, 1964 to present. Inquire as to online cost and availability.

SCISEARCH. Institute for Scientific Information, 3501 Market Street, Philadelphia, PA 19104. (800) 523-1850 or (215) 386-0100. Broad multidisciplinary title and author index to the international literature of science and technology, 1974 to present. Inquire as to online cost and availability.

WILSONLINE. H.W. Wilson and Company, 950 University Avenue, Bronx, NY 10452. (800) 367-6770 or (212) 588-8400. Makes available online versions of the H.W. Wilson indexes including Applied Science and Technology Index, Business Periodicals Index and Readers' Guide to Periodical Literature. Approximately 1980 to present. Inquire as to online cost and availability.

OTHER SOURCES

WHAT EVERY ENGINEER SHOULD KNOW ABOUT ENGINEERING SOURCES. Marcel Dekker Inc., 270 Madison Avenue, New York, NY 10016. (800) 228-1160. 1984. $24.95.

PERIODICALS

IEEE TRANSACTIONS ON MAGNETICS. Institute of Electrical and Electronics Engineers. IEEE Service Center, 445 Hoes Lane, Piscataway, NJ 08854. 1965 to present. Bimonthly. $136.00 per year.

IEEE TRANSACTIONS ON MICROWAVE THEORY AND TECHNIQUES. Institute of Electrical and Electronics Engineers. IEEE Service Center, 445 Hoes Lane, Piscataway, NJ 08854. 1953 to present. Monthly. $126.00 per year.

IEE PROCEEDINGS PART H: MICROWAVES, ANTENNAS AND PROPAGATION. Institution of Electrical Engineers, Savoy Place, London, WC2R OBL, England. 1980 to present. Bimonthly. $140.00 per year.

JOURNAL OF MICROWAVE POWER. International Microwave Power Institute, 13542 Union Village Circle, Clifton, VA 22024-2305. (703) 830-5588. 1966 to present. Quarterly. $75.00.

MICROWAVE JOURNAL. Horizon House, 610 Washington Street, Dedham, MA 02026. (617) 326-8220. 1958 to present. Monthly. $40.00 per year. Free to qualified personnel.

MICROWAVE SYSTEMS NEWS AND TECHNOLOGY. E.W. Communications, Inc., 1170 East Meadow Drive, Palo Alto, CA 94303. (415) 494-2800. 1971 to present. Monthly. Inquire.

MICROWAVES AND RF. Hayden Publishing Company, Inc., 10 Mulholland Drive, Hasbrouck Heights, NJ 07604. 1962 to present. Monthly. $30.00 per year.

RESEARCH CENTERS AND INSTITUTES

EDWARD L. GINZTON LABORATORY. Stanford University, Via Palou, Stanford, CA 94305. (415) 723-0200.

MICROWAVE DEVICE AND PHYSICAL ELECTRONICS LABORATORY. University of Utah, Electrical Engineering Department, Salt Lake City, UT 84112. (801) 581-7634.

MICROWAVE MICROELECTRONICS LABORATORY. Texas A&M University, Electrical Engineering Department, College Station, TX 77843. (409) 845-5285.

WEBER RESEARCH INSTITUTE. Polytechnic University, Route 110, Farmingdale, NY 11735. (516) 752-9701.

MILKY WAY

See also: ASTRONOMY, ASTROPHYSICS, GALAXIES, STARS

ABSTRACT SERVICES AND INDEXES

ASTRONOMY AND ASTROPHYSICS ABSTRACTS. Springer-Verlag New York, Inc., 175 Fifth Avenue, New York, NY 10010. (800) 526-7254. 1969 to present. Approximately $70.00 per year.

ASTRONOMY AND ASTROPHYSICS MONTHLY INDEX. Olivetree Associates, P.O. Box 236, Sierre Madre, CA 91024. $220.00 per year. Complementary copies available on request.

CHEMICAL ABSTRACTS. American Chemical Society, Chemical Abstracts Service, Box 3012, Columbus, OH 43210. (614) 421-3600. 1907 to present. Weekly. $9500.00 per year.

CONFERENCE PAPERS INDEX. Cambridge Scientific Abstracts, 5161 River Road, Bethesda, MD 20816. 1972 to present. Monthly. Inquire as to cost and availability.

CURRENT CONTENTS: PHYSICAL, CHEMICAL, AND EARTH SCIENCES. Institute for Scientific Information, 3501 Market Street, Philadelphia, PA 19104. (800) 523-1850 or (215) 386-0100. Weekly. $275.00 per year.

INDEX TO SCIENTIFIC AND TECHNICAL PROCEEDINGS. Institute for Scientific Information, 3501 Market Street, Philadelphia, PA 19104. (800) 523-1850 or (215) 386-0100. 1978 to present. Monthly. $775.00 per year.

INDEX TO SCIENTIFIC REVIEWS. Institute for Scientific Information, 3501 Market Street, Philadelphia, PA 19104. (800) 523-1850 or (215) 386-0100. 1974 to present. Semi-annual. $550.00 per year.

METEOROLOGICAL AND GEOASTROPHYSICAL ABSTRACTS. American Meteorological Society, 45 Beacon Street, Boston, MA 02108. (617) 227-2425. 1950 to present. Monthly. $450.00 per year.

PHYSICS ABSTRACTS. Institution of Electrical Engineers. Available from: IEEE Service Center, 445 Hoes Lane, Piscataway, NJ 08854. 1898 to present. Bimonthly. $1700.00 per year.

PHYSICS BRIEFS. Physik Verlag GmbH, Postfach 1260/1280, D-6940 Weinheim, West Germany. (212) 661-9404. 1920 to present. Twenty-six times per year. $1250.00 per year.

SCIENCE CITATION INDEX. Institute for Scientific Information, 3501 Market Street, Philadelphia, PA 19104. (800) 523-1850 or (215) 386-0100. Six times per year. $6200.00 per year.

STAR. (SCIENTIFIC AND TECHNICAL AEROSPACE REPORTS. U.S. National Aeronautics and Space Administration, Scientific and Technical Information Facility, Box 8757, Baltimore-Washington International Airport, MD 21240. (202) 755-2210. Semimonthly, with semiannual and annual indexes. $85.00 per year.

ANNUAL REVIEWS AND YEARBOOKS

ANNUAL REVIEW OF ASTRONOMY AND ASTROPHYSICS. Annual Reviews, Inc., 4139 El Camino Way, Palo Alto, CA 94306. (415) 493-4400. Annual. Inquire as to cost and availability.

ASSOCIATIONS AND PROFESSIONAL SOCIETIES

AMERICAN ASTRONOMICAL SOCIETY. 1816 Jefferson Place, N.W., Washington, DC 20036. (202) 659-0134.

ASTRONOMICAL SOCIETY OF THE PACIFIC. 1290 24th Avenue, San Francisco, CA 94122. (415) 661-8660.

BIBLIOGRAPHIES

A BIBLIOGRAPHY OF ASTRONOMY, 1970-1979. R.A. Seal and S.S. Martin. Libraries Unlimited Inc., P.O. Box 263, Littleton, CO 80160. (303) 770-1220. 1982. $37.50.

SCIENCE BOOKS AND FILMS. American Association for the Advancement of Science, 1333 H Street, NW, Washington, DC 20005. (202) 326-6454. Five times per year. $20.00 per year.

SCIENTIFIC AND TECHNICAL BOOKS AND SERIALS IN PRINT 1988; AN INDEX TO LITERATURE IN SCIENCE AND TECHNOLOGY. R.R. Bowker Company, 205 East 42nd Street, New York, NY 10017. (800) 521-8110. $175.00.

DIRECTORIES AND BIOGRAPHICAL SOURCES

AMERICAN MEN AND WOMEN OF SCIENCE. R.R. Bowker, Inc., Order Department, 245 West 17th Street, New York, NY 10011. (800) 521-8110. Eight volumes. 1986. $595.00 for set.

THE BIOGRAPHICAL DICTIONARY OF SCIENTISTS: ASTRONOMERS. D. Abbott, editor. Peter Bedrick Books, 125 East 23rd Street, New York, NY 10010. 1984.

DIRECTORY OF PHYSICS AND ASTRONOMY STAFF MEMBERS. American Institute of Physics, 335 East 45th Street, New York, NY 10017. (212) 661-9494. Annual.

INTERNATIONAL RESEARCH CENTERS DIRECTORY 1988-89. Darren L. Smith, editor. Gale Research Company, Book Tower, Detroit, MI 48226. (800) 521-0707. 4th edition. 1987. $360.00.

RESEARCH CENTERS DIRECTORY 1988. Gale Research Company, Book Tower, Detroit, MI 48226. (800) 521-0707. 12th edition. 1987. $365.00 for set.

SCIENTIFIC AND TECHNICAL ORGANIZATIONS AND AGENCIES DIRECTORY. Margaret Labash Young, editor. Gale Research Company, Book Tower, Detroit, MI 48226. (800) 521-0707. 2nd edition. 1987. $185.00.

ENCYCLOPEDIAS AND DICTIONARIES

ILLUSTRATED ENCYCLOPEDIA OF THE UNIVERSE: UNDERSTANDING AND EXPLORING THE COSMOS. Richard S. Lewis. Crown Publishers Inc., 1 Park Avenue, New York, NY 10016. (800) 526-4264. 1986. $24.95.

GENERAL WORKS

CONSTRUCTING THE UNIVERSE. David Layzer. W.H. Freeman and Company, 41 Madison Avenue, New York, NY 10010. (212) 532-7660. 1984. $29.95.

THE FORMATION AND EVOLUTION OF GALAXIES AND LARGE STRUCTURES IN THE UNIVERSE. Jean Audouze and Thanh Van Tran. Kluwer Academic Publishers, 190 Old Derby Street, Hingham, MA 02043. (617) 749-5262. 1984. $58.00.

THE GALAXY AND THE SOLAR SYSTEM. Roman Smoluchowski and others. University of Arizona Press, 1615 East Speedway, Tucson, AZ 85719. (602) 621-1441. 1986. $29.95.

GALAXIES. Paul W. Hodge. Harvard University Press, 79 Garden Street, Cambridge, MA 02138. (617) 495-2600. 1986. $22.50.

LARGE SCALE STRUCTURE OF THE UNIVERSE. P.J. Peebles. Princeton University Press, 41 William Street, Princeton, NJ 08540. (609) 452-4122. 1980. $15.95 in paper.

THE MILKY WAY. B.J. Bok and P.F. Bok. Harvard University Press, 79 Garden Street, Cambridge, MA 02138. (617) 495-2600. 1981. $29.95.

NEARBY GALAXIES ATLAS. R. Tully and others. Cambridge University Press, 32 East 57th Street, New York, NY 10022. (800) 872-7423. 1987. $59.50.

NEARBY GALAXIES CATALOG. R. Tully and others. Cambridge University Press, 32 East 57th Street, New York, NY 10022. (800) 872-7423. 1987. $49.50.

ONLINE DATA BASES

CA SEARCH. Chemical Abstracts Service, P.O. Box 3012, Columbus, OH 43120. (800) 848-6538 or (614) 421-3600. Comprehensive guide to chemical literature, 1972 to present. Inquire as to online cost and availability.

DISSERTATION ABSTRACTS ONLINE. University Microfilms International, 300 North Zeeb Road, Ann Arbor, MI 48106. (800) 521-0600 or (313) 761-4700. Scope includes virtually all doctoral dissertations accepted at accredited American institutions from 1861 to present in over 250 subject areas. Inquire as to online cost and availability.

INSPEC. INSPEC Marketing Department, Institution of Electrical Engineers. Available from IEEE Service Center, 445 Hoes Lane, Piscataway, NJ 08854. (201) 981-0060. Online version of Physics Abstracts. Inquire as to online cost and availability.

NTIS. National Technical Information Service, 5285 Port Royal Road, Springfield, VA 22161. (703) 487-4630. Broad coverage of government-sponsored research reports, 1964 to present. Inquire as to online cost and availability.

SCISEARCH. Institute for Scientific Information, 3501 Market Street, Philadelphia, PA 19104. (800) 523-1850 or (215) 386-0100. Broad multidisciplinary title and author index to the international literature of science and technology, 1974 to present. Inquire as to online cost and availability.

OTHER SOURCES

ATLAS OF DEEP-SKY SPLENDORS. H. Vehrenberh. Cambridge University Press, 32 East 57th Street, New York, NY 10022. (800) 872-7423. Fourth edition. 1984. $44.50.

PERIODICALS

ASTRONOMICAL JOURNAL. American Astronomical Society. Available from: American Institute of Physics, 335 East 45th Street, New York, NY 10017. (212) 661-9494. Monthly. $125.00 per year.

ASTRONOMICAL SOCIETY OF THE PACIFIC PUBLICATIONS. Astronomical Society of the Pacific, 1290 24th Avenue, San Francisco, CA 94122. (415) 661-8660. Monthly. $40.00 per year.

ASTRONOMY. AstroMedia Corporation, 625 E Street, Box 92788, Milwaukee, WI 53202. (414) 276-2689. 1973 to present. Monthly. $21.00 per year.

ASTRONOMY AND ASTROPHYSICS. Springer-Verlag New York, Inc., 175 Fifth Avenue, New York, NY 10010. (800) 526-7254. Monthly. $680.00 per year.

ASTROPHYSICAL JOURNAL. University of Chicago Press, 5801 Ellis Avenue, Chicago, IL 60637. (800) 621-2736. Biweekly. $315.00 per year.

ASTROPHYSICS AND SPACE SCIENCE. D. Reidel Publishing Company, 190 Old Derby Street, Hingham, MA 02043. Monthly. $101.00 per year.

GEOCHEMICA ET COSMOCHIMICA ACTA. Pergamon Press, Inc., Maxwell House, Fairview Park, Elmsford, NY 10523. (914) 592-7700. 1950 to present. Monthly. $340.00 per year.

SKY AND TELESCOPE. Sky Publishing Corporation, 49 Bay State Road, Cambridge, MA 02238. (617) 864-7360. Monthly. $18.00 per year.

MERCURY. Astronomical Society of the Pacific, 1290 24th Avenue, San Francisco, CA 94122. (415) 661-8660. 1972 to present. Bimonthly. $21.00 per year.

RESEARCH CENTERS AND INSTITUTES

CENTER FOR SPACE SCIENCE AND ASTROPHYSICS. Stanford University, 325 Durand Building, Stanford, CA 94305. (415) 497-3582.

HARVARD-SMITHSONIAN CENTER FOR ASTROPHYSICS. 60 Garden Street, Cambridge, MA 02138. (617) 495-7461.

MOUNT WILSON AND LAS CAMPANAS OBSERVATORIES. 813 Santa Barbara Street, Pasadena, CA 91101. (818) 577-1122.

NATIONAL OPTICAL ASTRONOMY OBSERVATORIES. 1002 North Warren Avenue, Tucson, AZ 85719. (602) 325-9230.

MINERAL EXPLORATION

See also: GEOLOGY, GEOPHYSICS, GEOCHEMISTRY, MINERALOGY

ABSTRACT SERVICES AND INDEXES

BIBLIOGRAPHY AND INDEX OF GEOLOGY. American Geological Institute, 4220 King Street, Alexandria, VA 22302. (703) 379-2480. 1969 to present. Monthly. $1100.00 per year.

CHEMICAL ABSTRACTS. Chemical Abstracts Service, 2540 Olentangy Road, Post Office Box 3012, Columbus, OH 43210. (800) 848-6538 or (614) 421-3600. Weekly. $9200.00 per year.

GENERAL SCIENCE INDEX. H.W. Wilson Company, 950 University Avenue, Bronx, NY 10452. (800) 367-6770 or (212) 588-8400. Inquire as to price and availability.

MINERAL ECONOMICS ABSTRACTS. Post Office Box 678, Green Farms, CT 06436. Monthly. $250.00 per year.

MINERALOGICAL ABSTRACTS. Mineralogical Society and the Mineralogical Society of America, 41 Queen's Gate, London, SW7 5HR, England. 1959 to present. Quarterly. $190.00 per year.

SCIENCE CITATION INDEX. Institute for Scientific Information, 3501 Market Street, Philadelphia, PA 19104. (800) 523-1850 or (215) 386-0100. Inquire as to price and availability.

ANNUAL REVIEWS AND YEARBOOKS

ANNUAL REVIEW AND EARTH AND PLANETARY SCIENCES. Annual Reviews, Incorporated, 4139 El Camino Way, Palo Alto, CA 94306. (415) 493-4400.

MINERALS YEARBOOK. Bureau of Mines, United States Department of the Interior. Available from United States Government Printing Office, Washington, DC 20402. (202) 783-3238. Annual. Three volumes. $45.00.

ASSOCIATIONS AND PROFESSIONAL SOCIETIES

AMERICAN ASSOCIATION OF PETROLEUM GEOLOGISTS. Box 979, Tulsa, OK 74101. (018) 584-2555.

AMERICAN FEDERATION OF MINERALOGICAL SOCIETIES. 1203 East Hillsboro, Tampa, FL 33604. (813) 238-0427.

AMERICAN GEOLOGICAL INSTITUTE. 4220 King Street, Alexandria, VA 22302. (703) 379-2480.

AMERICAN INSTITUTE OF PROFESSIONAL GEOLOGISTS. 7828 Vance Drive, Suite 103, Arvada, CO 80003. (303) 431-0831.

CLAY MINERALS SOCIETY. c/o William F. Moll, 22149 North Pet Lane, Prairie View, IL 60069.

GEOLOGICAL SOCIETY OF AMERICA. Box 9140, Boulder, CO 80301. (303) 447-2020.

MINERALOGICAL SOCIETY OF AMERICA. 2000 Florida Avenue, NW, Washington, DC 20009. (202) 462-6913.

SOCIETY OF ECONOMIC PALEONTOLOGISTS AND MINERALOGISTS. Post Office Box 4756, Tulsa, OK 74159. (918) 743-9765.

DIRECTORIES AND BIOGRAPHICAL SOURCES

AMERICAN MEN AND WOMEN OF SCIENCE: PHYSICAL AND BIOLOGICAL SCIENCES. Sixteenth edition. R.R. Bowker Company, 245 West 17th Street, New York, NY 10011. (800) 521-8810 or (212) 916-1600. $595.00.

AMERICAN INSTITUTE OF PROFESSIONAL GEOLOGISTS. Membership Directory. American Institute of Professional Geologists, 7828 Vance Drive, Suite 103, Arvada, CO 80003. (303) 431-0831. Annual, $15.00.

GEOLOGICAL SOCIETY OF AMERICA. Membership Directory. Geological Society of America, 3300 Penrose Place, Boulder, CO 80301. (303) 447-2020. Annual. Available to members only.

RESEARCH CENTERS DIRECTORY. Gale Research Company, Book Tower, Detroit, MI 48226. (800) 521-0707. Twelfth edition. 1987. $365.00.

SOCIETY OF ECONOMIC PALEONTOLOGISTS AND MINERALOGISTS - MEMBERSHIP LIST. Society of Economic Paleontologists and Mineralogists, 3530 East 31st Street, Tulsa, OK 74159. Every two years. $10.00.

WHO'S WHO IN FRONTIER SCIENCE AND TECHNOLOGY. Marquis Who's Who, Incorporated, 200 East Ohio Street, Chicago, IL 60611. (800) 428-3898 or (312) 787-2008.

ENCYCLOPEDIAS AND DICTIONARIES

DICTIONARY OF GEOLOGICAL TERMS. American Geological Institute. Doubleday and Company, Incorporated, 245 Park Avenue, New York, NY 10017. (800) 645-6156 or (212) 953-4561. Third edition. 1984. $19.95.

DICTIONARY OF PETROLOGY. S.I. Tomkeieff. John Wiley and Sons, Incorporated, 605 Third Avenue, New York, NY 10158. (800) 526-5368 or (212) 850-6000. 1983. $140.00.

GLOSSARY OF GEOLOGY. Robert L. Bates and Julia A. Jackson. American Geological Institute, 4220 King Street, Alexandria, VA 22032. (703) 379-2480. 1980. $60.00.

GENERAL WORKS

BIOGEOCHEMICAL METHODS OF PROSPECTING. Dmitril P. Malyuga. Plenum Publishing Corporation, 233 Spring Street, New York, NY 10013. (800) 221-9369. 1964. $32.50.

DESIGNING OPTIMAL STRATEGIES FOR MINERAL EXPLORATION. J.G. De Geoffroy and T.K. Wignall. Plenum Publishing Corporation, 233 Spring Street, New York, NY 10013. (800) 221-9369. 1985. $55.00.

GEOPHYSICAL AND GEOCHEMICAL TECHNIQUES FOR EXPLORATION OF HYDROCARBONS AND MINERALS. Marshall Sittig, editor. Noyes Data Corporation, Mill Road at Grand Avenue, Park Ridge, NJ 07656. (201) 391-8484. 1980. $40.00.

INTRODUCTION TO ROCK FORMING MINERALS. W.A. Deer, J. Zussman and R.A. Howie. John Wiley and Sons, Incorporated, 605 Third Avenue, New York, NY 10158. (800) 526-5368 or (212) 850-6000. 1966. $29.95.

MINERALOGY: CONCEPTS, DESCRIPTIONS, DETERMINATIONS. L.G. Berry and Brian Mason. W.H. Freeman and Company, 41 Madison Avenue, New York, NY 10010. (202) 532-7660. Second edition. 1983. $41.95.

PROSPECTING AND EXPLORATION OF MINERAL DEPOSITS. M. Kuzvart and M. Bohmer. elsevier Science Publishing Company, Incorporated, 52 Vanderbilt Avenue, New York, NY 10017. (212) 370-5520. Second edition. 1986. $106.00.

HANDBOOKS AND MANUALS

MANUAL OF MINERALOGY (AFTER JAMES D. DANA). Cornelis Klein and Cornelius S. Hurlbut, Jr. John Wiley and Sons, Incorporated, 605 Third Avenue, New York, NY 10158. (800) 526-5368 or (212) 850-6000. Twentieth edition. 1985. $39.95.

GEOLOGY IN THE FIELD. R.R. Compton. John Wiley and Sons, Incorporated, 605 Third Avenue, New York, NY 10158. (800) 526-5368 or (212) 850-6418. 1985. $23.95.

MINERAL PROSPECTING MANUAL. J.B. Chaussier and J. Morer. Elsevier Science Publishing Company, Incorporated, 52 Vanderbilt Avenue, New York, NY 10017. (212) 370-5520. 1986. $40.00.

ONLINE DATA BASES

GEOREF. American Geological Institute, 4220 King Street, Alexandria, VA 22302. (800) 336-4764 or (703) 379-2480. Geology and geosciences literature, 1961 to present. Inquire as to online cost and availability.

GEOARCHIVE. Geosystems, Post Office Box 1024, Westminister, London, England, SW1 P 2JL. Citations to literature on geoscience, 1969 to present. Inquire as to online cost and availability.

NTIS. National Technical Information Service, 5285 Port Royal Road, Springfield, VA 22161. (703) 487-4630. Broad coverage of government sponsored research reports, 1964 to present. Inquire as to online cost and availability.

SCISEARCH. Institute for Scientific Information, 3501 Market Street, Philadelphia, PA 19104. (800) 523-1850 or (215) 386-0100. Broad interdisciplinary index to the literature of science and technology, 1965 to present. Inquire as to online cost and availability.

PERIODICALS

AAPG BULLETIN. American Association of Petroleum Geologists, Box 979, Tulsa, OK 74101. (918) 584-2555. Monthly. $70.00 per year.

AMERICAN JOURNAL OF SCIENCE. Kline Geology Laboratory, Yale University, New Haven, CT 06520. Ten times per year. $80.00 per year.

AMERICAN MINERALOGIST. Mineralogical Society of America, 2000 Florida Avenue, NW, Washington, DC 20009. (202) 462-6913. Bimonthly. $115.00 per year.

COLORADO SCHOOL OF MINES QUARTERLY. Colorado School of Mines, Golden, CO 80401. Quaterly. $50.00 per year.

ECONOMIC GEOLOGY. Society of Economic Geologists, Post Office Box 571, Golden, CO 80402. (303) 279-1899. Eight times per year. $25.00. Monthly. $340.00 per year.

GEOCHIMICA ET COSMOCHIMICA ACTA. Pergamon Journals, Incorporated, Maxwell House, Fairview Park, Elmsford, NY 10523. (014) 592-7700. Monthly. $340.00 per year.

GEOLOGICAL MAGAZINE. Cambridge University Press, 32 East 57th Street, New York, NY 10022. (800) 872-7423. Bimonthly. $165.00 per year.

GEOLOGICAL SOCIETY JOURNAL. Geological Society of London. Blackwell Scientific Publications, Incorporated, 667 Lytton Avenue, Palo Alto, CA 94301. (415) 324-1688. Six times per year. $290.00.

GEOLOGICAL SOCIETY OF AMERICA BULLETIN. Post Office Box 9140, 3300 Penrose Place, Boulder, CO 80301. (303) 447-2020. Monthly. $80.00.

GEOLOGY. Geological Society of America. Post Office Box 9140, 3300 Penrose Place, Boulder, CO 80301. (303) 447-2020. Monthly. $55.00 per year.

GEOTIMES. American Geological Institute, 4220 King Street, Alexandria, VA 22032. Monthly. $18.00 per year.

JOURNAL OF GEOLOGY. University of Chicago Press, 5801 South Ellis Street, Chicago, IL 60637. (312) 962-7600. Bimonthly. $30.00 per year.

LITHOS: AN INTERNATIONAL JOURNAL OF MINERALOGY, PETROLOGY, AND GEOCHEMISTRY. Elsevier Science Publishers, Post Office Box 330, Irving-on-Hudson, NY 10533. Quarterly. $73.00 per year.

MINERALIUM DEPOSITA. Springer-Verlag New York, Incorporated, 175 Fifth Avenue, New York, NY 10010. (212) 460-1500. Quarterly. $118.00 per year.

ROCKS AND MINERALS. Heldref Publications, 4000 Albemerle Street, NW, Washington, DC 20016. (202) 362-6445. Bimonthly. $23.00 per year.

VOLCANOLOGY AND SEISMOLOGY. Gordon and Breach Science Publishers, 50 West 23rd Street, New York, NY 10010. (212) 206-8900. Monthly. $498.00 per year.

RESEARCH CENTERS AND INSTITUTES

GEOLOGICAL SOCIETY OF AMERICA. Post Office Box 9140, 3300 Penrose Place, Boulder, CO 80301. (303) 447-2020.

INSTITUT DE RECHERCHE EN EXPLORATION MINERALE. C.P. 6079, Station A, Montreal, PQ, Canada H3C 3A7. (514) 340-4991.

MINERAL INFORMATION INSTITUTE. 425 South Cherry Street, Suite 200, Denver, CO 80222. (303) 393-7211.

NEW MEXICO BUREAU OF MINES AND MINERAL RESOURCES. Institute of Mining and Technology, Campus Station, Socorro, NM 87801. (505) 835-5420.

SOCIETY OF ECONOMIC GEOLOGISTS. Post Office Box 571, Golden, CO 80402. (303) 279-1899.

SOCIETY OF ECONOMIC PALEONTOLOGISTS AND MINERALOGISTS. Post Office Box 4756, Tulsa, OK 74159. (918) 743-9765.

UNITED STATES GEOLOGICAL SURVEY. National Center, 12201 Sunrise Valley Drive, Reston, VA 22092. The major geological research agency of the federal government conducting research in most areas of pure and applied research in the geosciences.

MINERALOGY

See also: GEOLOGY, GEOPHYSICS, GEOCHEMISTRY

ABSTRACT SERVICES AND INDEXES

BIBLIOGRAPHY AND INDEX OF GEOLOGY. American Geological Institute, 4220 King Street, Alexandria, VA 22302. (703) 379-2480. 1969 to present. Monthly. $1100.00 per year.

CHEMICAL ABSTRACTS. Chemical Abstracts Service, 2540 Olentangy Road, Post Office Box 3012, Columbus, OH 43210. (800) 848-6538 or (614) 421-3600. Weekly. $9200.00 per year.

GENERAL SCIENCE INDEX. H.W. Wilson Company, 950 University Avenue, Bronx, NY 10452. (800) 367-6770 or (212) 588-8400. Inquire as to price and availability.

MINERALOGICAL ABSTRACTS. Mineralogical Society and the Mineralogical Society of America, 41 Queen's Gate, London, SW7 5HR, England. 1959 to present. Quarterly. $190.00 per year.

SCIENCE CITATION INDEX. Institute for Scientific Information, 3501 Market Street, Philadelphia, PA 19104. (800) 523-1850 or (215) 386-0100. Inquire as to price and availability.

ANNUAL REVIEWS AND YEARBOOKS

ANNUAL REVIEW AND EARTH AND PLANETARY SCIENCES. Annual Reviews, Incorporated, 4139 El Camino Way, Palo Alto, CA 94306. (415) 493-4400.

MINERALS YEARBOOK. Bureau of Mines, United States Department of the Interior. Available from United States Government Printing Office, Washington, DC 20402. (202) 783-3238. Annual. Three volumes. $45.00.

ASSOCIATIONS AND PROFESSIONAL SOCIETIES

AMERICAN FEDERATION OF MINERALOGICAL SOCIETIES. 1203 East Hillsboro, Tampa, FL 33604. (813) 238-0427.

AMERICAN GEOLOGICAL INSTITUTE. 4220 King Street, Alexandria, VA 22302. (703) 379-2480.

AMERICAN INSTITUTE OF PROFESSIONAL GEOLOGISTS. 7828 Vance Drive, Suite 103, Arvada, CO 80003. (303) 431-0831.

CLAY MINERALS SOCIETY. c/o William F. Moll, 22149 North Pet Lane, Prairie View, IL 60069.

GEOLOGICAL SOCIETY OF AMERICA. Box 9140, Boulder, CO 80301. (303) 447-2020.

MINERALOGICAL SOCIETY OF AMERICA. 2000 Florida Avenue, NW, Washington, DC 20009. (202) 462-6913.

SOCIETY OF ECONOMIC PALEONTOLOGISTS AND MINERALOGISTS. Post Office Box 4756, Tulsa, OK 74159. (918) 743-9765.

DIRECTORIES AND BIOGRAPHICAL SOURCES

AMERICAN MEN AND WOMEN OF SCIENCE: PHYSICAL AND BIOLOGICAL SCIENCES. Sixteenth edition. R.R. Bowker Company, 245 West 17th Street, New York, NY 10011. (800) 521-8810 or (212) 916-1600. $595.00.

AMERICAN INSTITUTE OF PROFESSIONAL GEOLOGISTS. Membership Directory. American Institute of Professional Geologists, 7828 Vance Drive, Suite 103, Arvada, CO 80003. (303) 431-0831. Annual, $15.00.

GEOLOGICAL SOCIETY OF AMERICA. Membership Directory. Geological Society of America, 3300 Penrose Place, Boulder, CO 80301. (303) 447-2020. Annual. Available to members only.

RESEARCH CENTERS DIRECTORY. Gale Research Company, Book Tower, Detroit, MI 48226. (800) 521-0707. Twelfth edition. 1987. $365.00.

WHO'S WHO IN FRONTIER SCIENCE AND TECHNOLOGY. Marquis Who's Who, Incorporated, 200 East Ohio Street, Chicago, IL 60611. (800) 428-3898 or (312) 787-2008.

ENCYCLOPEDIAS AND DICTIONARIES

DICTIONARY OF GEOLOGICAL TERMS. American Geological Institute. Doubleday and Company, Incorporated, 245 Park Avenue, New York, NY 10017. (800) 645-6156 or (212) 953-4561. Third edition. 1984. $19.95.

DICTIONARY OF PETROLOGY. S.I. Tomkeieff. John Wiley and Sons, Incorporated, 605 Third Avenue, New York, NY 10158. (800) 526-5368 or (212) 850-6000. 1983. $140.00.

ENCYCLOPEDIA OF PHYSICAL SCIENCE AND TECHNOLOGY. Academic press, Incorporated, Orlando, FL 32887. (800) 321-5068.

GLOSSARY OF GEOLOGY. Robert L. Bates and Julia A. Jackson. American Geological Institute, 4220 King Street, Alexandria, VA 22032. (703) 379-2480. 1980. $60.00.

GENERAL WORKS

THE EARTH. Peter J. Smith. Macmillan Publishing Company, 866 Third Avenue, New York, NY 10022. (800) 257-5755 or (212) 702-2000. 1986. $40.00.

INTRODUCTION TO ROCK FORMING MINERALS. W.A. Deer, J. Zussman and R.A. Howie. John Wiley and Sons, Incorporated, 605 Third Avenue, New York, NY 10158. (800) 526-5368 or (212) 850-6000. 1966. $29.95.

MINERALOGY: CONCEPTS, DESCRIPTIONS, DETERMINATIONS. L.G. Berry and Brian Mason. W.H. Freeman and Company, 41 Madison Avenue, New York, NY 10010. (202) 532-7660. Second edition. 1983. $41.95.

HANDBOOKS AND MANUALS

MANUAL OF MINERALOGY (AFTER JAMES D. DANA). Cornelis Klein and Cornelius S. Hurlbut, Jr. John Wiley and Sons, Incorporated, 605 Third Avenue, New York, NY 10158. (800) 526-5368 or (212) 850-6000. Twentieth edition. 1985. $39.95.

GEOLOGY IN THE FIELD. R.R. Compton. John Wiley and Sons, Incorporated, 605 Third Avenue, New York, NY 10158. (800) 526-5368 or (212) 850-6418. 1985. $23.95.

ONLINE DATA BASES

GEOREF. American Geological Institute, 4220 King Street, Alexandria, VA 22302. (800) 336-4764 or (703) 379-2480. Geology and geosciences literature, 1961 to present. Inquire as to online cost and availability.

GEOARCHIVE. Geosystems, Post Office Box 1024, Westminister, London, England, SW1 P 2JL. Citations to literature on geoscience, 1969 to present. Inquire as to online cost and availability.

NTIS. National Technical Information Service, 5285 Port Royal Road, Springfield, VA 22161. (703) 487-4630. Broad coverage of government sponsored research reports, 1964 to present. Inquire as to online cost and availability.

SCISEARCH. Institute for Scientific Information, 3501 Market Street, Philadelphia, PA 19104. (800) 523-1850 or (215) 386-0100. Broad interdisciplinary index to the literature of science and technology, 1965 to present. Inquire as to online cost and availability.

PERIODICALS

AMERICAN JOURNAL OF SCIENCE. Kline Geology Laboratory, Yale University, New Haven, CT 06520. Ten times per year. $80.00 per year.

AMERICAN MINERALOGIST. Mineralogical Society of America, 2000 Florida Avenue, NW, Washington, DC 20009. (202) 462-6913. Bimonthly. $115.00 per year.

CLAYS AND CLAY MINERALS (CLAY MINERALS SOCIETY). Allen Press, Incorporated, 1041 New Hampshire Street, Box 368, Lawrence, KS 66044. (812) 332-9600. Bimonthly. $96.00 per year.

ECONOMIC GEOLOGY. Society of Economic Geologists, Post Office Box 571, Golden, CO 80402. (303) 279-1899. Eight times per year. $25.00. Monthly. $340.00 per year.

GEOCHIMICA ET COSMOCHIMICA ACTA. Pergamon Journals, Incorporated, Maxwell House, Fairview Park, Elmsford, NY 10523. (014) 592-7700. Monthly. $340.00 per year.

GEOLOGICAL MAGAZINE. Cambridge University Press, 32 East 57th Street, New York, NY 10022. (800) 872-7423. Bimonthly. $165.00 per year.

GEOLOGICAL SOCIETY JOURNAL. Geological Society of London. Blackwell Scientific Publications, Incorporated, 667 Lytton Avenue, Palo Alto, CA 94301. (415) 324-1688. Six times per year. $290.00.

GEOLOGICAL SOCIETY OF AMERICA BULLETIN. Post Office Box 9140, 3300 Penrose Place, Boulder, CO 80301. (303) 447-2020. Monthly. $80.00.

GEOLOGY. Geological Society of America. Post Office Box 9140, 3300 Penrose Place, Boulder, CO 80301. (303) 447-2020. Monthly. $55.00 per year.

GEOTIMES. American Geological Institute, 4220 King Street, Alexandria, VA 22032. Monthly. $18.00 per year.

JOURNAL OF GEOLOGY. University of Chicago Press, 5801 South Ellis Street, Chicago, IL 60637. (312) 962-7600. Bimonthly. $30.00 per year.

JOURNAL OF SEDIMENTARY PETROLOGY. Society of Economic Paleontologists and Mineralogists, Post Office Box 4756, Tulsa, OK 47159. (918) 743-9765. Bimonthly.

LITHOS: AN INTERNATIONAL JOURNAL OF MINERALOGY, PETROLOGY, AND GEOCHEMISTRY. Elsevier Science Publishers, Post Office Box 330, Irving-on-Hudson, NY 10533. Quarterly. $73.00 per year.

MINERALIUM DEPOSITA. Springer-Verlag New York, Incorporated, 175 Fifth Avenue, New York, NY 10010. (212) 460-1500. Quarterly. $118.00 per year.

ROCKS AND MINERALS. Heldref Publications, 4000 Albemerle Street, NW, Washington, DC 20016. (202) 362-6445. Bimonthly. $23.00 per year.

SEDIMENTARY GEOLOGY. Elsevier Science Publishers, Post Office Box 330, Irving-on-Hudson, NY 10533. Twenty times per year. $421.00 per year.

VOLCANOLOGY AND SEISMOLOGY. Gordon and Breach Science Publishers, 50 West 23rd Street, New York, NY 10010. (212) 206-8900. Monthly. $498.00 per year.

RESEARCH CENTERS AND INSTITUTES

GEOLOGICAL SOCIETY OF AMERICA. Post Office Box 9140, 3300 Penrose Place, Boulder, CO 80301. (303) 447-2020.

MINERAL INFORMATION INSTITUTE. 425 South Cherry Street, Suite 200, Denver, CO 80222. (303) 393-7211.

NEW MEXICO BUREAU OF MINES AND MINERAL RESOURCES. Institute of Mining and Technology, Campus Station, Socorro, NM 87801. (505) 835-5420.

SOCIETY OF ECONOMIC GEOLOGISTS. Post Office Box 571, Golden, CO 80402. (303) 279-1899.

SOCIETY OF ECONOMIC PALEONTOLOGISTS AND MINERALOGISTS. Post Office Box 4756, Tulsa, OK 74159. (918) 743-9765.

UNITED STATES GEOLOGICAL SURVEY. National Center, 12201 Sunrise Valley Drive, Reston, VA 22092. The major geological research agency of the federal government conducting research in most areas of pure and applied research in the geosciences.

MINICOMPUTERS

See: COMPUTERS

MINING ENGINEERING

See also: GEOTECHNICAL ENGINEERING, PETROLEUM ENGINEERING, QUARRYING, ROCK MECHANICS

ABSTRACT SERVICES AND INDEXES

APPLIED MECHANICS REVIEW. American Society of Mechanical Engineers, 345 East 47th Street, New York, NY 10017. (212) 705-7703. 1948 to present. Monthly. $360.00 per year.

APPLIED SCIENCE AND TECHNOLOGY INDEX. H.W. Wilson and Company, 950 University Avenue, Bronx, NY 10452. (800) 367-6670 or (212) 588-8400. Monthly. Inquire as to cost and availability.

BIBLIOGRAPHY AND INDEX OF GEOLOGY. American Geological Institute, 4220 King Street, Alexandria, VA 22302. (703) 379-2480. 1969 to present. Monthly. $1100.00 per year.

CHEMICAL ABSTRACTS. American Chemical Society, Chemical Abstracts Service, Box 3012, Columbus, OH 43210. (614) 421-3600. 1907 to present. Weekly. $9500.00 per year.

CURRENT CONTENTS: ENGINEERING, TECHNOLOGY AND APPLIED SCIENCES. Institute for Scientific Information, 3501 Market Street, Philadelphia, PA 19104. (800) 523-1850 or (215) 386-0100. Weekly. $275.00 per year.

ENGINEERING INDEX MONTHLY AND AUTHOR INDEX. Engineering Information Inc., 345 East 47th Street, New York, NY 10017. (212) 705-7600. Monthly. $1560.00 per year.

INTERNATIONAL JOURNAL OF ROCK MECHANICS AND MINING SCIENCES AND GEOMECHANICS ABSTRACTS. Pergamon Press, Inc., Maxwell House, Fairview Park, Elmsford, NY 10523. (914) 592-7700. 1964 to present. Bimonthly. $385.00 per year.

ISMEC BULLETIN (Information Service in Mechanical Engineering). Cambridge Scientific Abstracts, 5161 River Road, Bethesda, MD 20816. (301) 951-1400. 1973 to present. Monthly. $450.00 per year.

OCEAN ABSTRACTS. Cambridge Scientific Abstracts, 5161 River Road, Bethesda, MD 20816. (301) 951-1400. 1963 to present. Bimonthly. $450.00 per year.

ASSOCIATIONS AND PROFESSIONAL SOCIETIES

AMERICAN INSTITUTE OF MINING, METALLURGICAL AND PETROLEUM ENGINEERS. 345 East 47th Street, New York, NY 10017. (212) 705-7695.

AMERICAN SOCIETY OF CIVIL ENGINEERS. 345 East 47th Street, New York, NY 10017. (212) 705-7496.

MINERALOGICAL SOCIETY OF AMERICA. 1625 I Street, N.W., Suite 414, Washington, DC 20006. (202) 775-4344.

MINING AND METALLURGICAL SOCIETY OF AMERICA. 275 Madison Avenue, Room 2301, New York, NY 10016. (212) 684-4150.

SOCIETY OF MINING ENGINEERS OF AIME. Caller Number D, Littleton, CO 80127. (303) 973-9550.

BIBLIOGRAPHIES

NEW TECHNICAL BOOKS: A SELECTIVE LIST WITH DESCRIPTIVE ANNOTATIONS. New York Public Library, Science and Technology Research Center, Fifth Avenue and 42nd Street, New York, NY 10018. (212) 930-0800. 1915 to present. Monthly. $15.00 per year.

SCIENTIFIC AND TECHNICAL BOOKS AND SERIALS IN PRINT 1988; AN INDEX TO LITERATURE IN SCIENCE AND TECHNOLOGY. R.R. Bowker Company, 205 East 42nd Street, New York, NY 10017. (800) 521-8110. $175.00.

DIRECTORIES AND BIOGRAPHICAL SOURCES

MINING ENGINEERING - SOCIETY OF MINING ENGINEERS DIRECTORY ISSUE. Society of Mining Engineers, Caller Number D, Littleton, CO 80127. Annual. $100.00.

1987 DIRECTORY OF ENGINEERING SOCIETIES AND RELATED ORGANIZATIONS. Gordon Davis, editor. Hemisphere Publishing Corporation, 1010 Vermont Avenue, NW, Washington, DC 20005. (800) 526-0275. 12th edition. 1987. $100.00.

QUARRIES DIRECTORY. American Business Directories, Inc., 5707 South 86th Street, Omaha, NE 68127. (402) 331-7169. 1986. $95.00.

RESEARCH CENTERS DIRECTORY 1988. Gale Research Company, Book Tower, Detroit, MI 48226. (800) 521-0707. 12th edition. 1987. $365.00 for set.

SCIENTIFIC AND TECHNICAL ORGANIZATIONS AND AGENCIES DIRECTORY. Margaret Labash Young, editor. Gale Research Company, Book Tower, Detroit, MI 48226. (800) 521-0707. 2nd edition. 1987. $185.00.

WESTERN MINING DIRECTORY. Howell Publishing Company, 311 Steele Street, Suite 208, Denver, CO 80206. Anuual. $45.00.

WHO'S WHO IN ENGINEERING. Gordon Davis, editor. Hemisphere Publishing Corporation, 1010 Vermont Avenue, NW, Washington, DC 20005. (800) 526-0275. 6th edition. 1985. $200.00.

WHO'S WHO IN MINERAL ENGINEERING. Society of Mining Engineers, Caller Number D, Littleton, CO 80127. (303) 973-9550. Annual. $100.00.

ENCYCLOPEDIAS AND DICTIONARIES

MCGRAW-HILL ENCYCLOPEDIA OF SCIENCE AND TECHNOLOGY. McGraw-Hill Book Company, 1221 Avenue of the Americas, New York, NY 10020. (212) 512-2000. 6th edition. 1987. $1600.00.

THESAURUS OF SCIENTIFIC, TECHNICAL, AND ENGINEERING TERMS. Hemisphere Publishing Corporation, 1010 Vermont Avenue, NW, Washington, DC 20005. (800) 526-0275. 1988. $125.00.

GENERAL WORKS

COAL MINE GROUND CONTROL. Syd S. Peng. John Wiley and Sons, Inc., 605 Third Avenue, New York, NY 10158. (800) 526-5368. 2nd edition. 1986. $79.95.

COAL MINING TECHNOLOGY: THEORY AND PRACTICE. Robert Stefanko. Society of Mining Engineers, Caller Number D, Littleton, CO 80127. (303) 973-9550. 1983. $45.00.

ELEMENTS OF MINING. Robert S Lewis. Books on Demand, 300 North Zeeb Road, Ann Arbor, MI 48106. (313) 761-4700. 3rd edition. $160.00 in paper.

LONGWALL MINING. Syd S. Peng. John Wiley and Sons, Inc., 605 Third Avenue, New York, NY 10158. (800) 526-5368. 1984. $79.95.

MINING GEOLOGY. Willard C. Lacy, editor. Van Nostrand Reinhold Company, Inc., 135 West 50th Street, New York, NY 10020. (800) 543-2681. 1983. $59.95.

UNDERGROUND MINING METHODS AND TECHNOLOGY. A.B. Szwilski and M.J. Richards, editors. Elsevier Science Publishing Company, Inc., 52 Vanderbilt Avenue, New York, NY 10017. (212) 370-5520. 1986. $112.25.

HANDBOOKS AND MANUALS

MINING ENGINEERS' HANDBOOK. Robert Peele. John Wiley and Sons, Inc., 605 Third Avenue, New York, NY 10158. (800) 526-5368. 3rd edition. Two volumes. 1941. $112.50 for set.

UNDERGROUND MINING METHODS HANDBOOK. William H. Hustrulid, editor. Society of Mining Engineers, Caller Number D, Littleton, CO 80127. (303) 973-9550. 1982. $120.00.

ONLINE DATA BASES

CA SEARCH. Chemical Abstracts Service, P.O. Box 3012, Columbus, OH 43120. (800) 848-6538 or (614) 421-3600. Comprehensive guide to chemical literature, 1972 to present. Inquire as to online cost and availability.

COMPENDEX. Engineering Information, Inc., 345 East 47th Street, New York, NY 10017. (800) 221-1044 or (212) 705-7615. Engineering and technical literature, 1975 to present. Inquire as to online cost and availability.

DISSERTATION ABSTRACTS ONLINE. University Microfilms International, 300 North Zeeb Road, Ann Arbor, MI 48106. (800) 521-0600 or (313) 761-4700. Scope includes virtually all doctoral dissertations accepted at accredited American institutions from 1861 to present in over 250 subject areas. Inquire as to online cost and availability.

GEOREF. Online version of the BIBLIOGRAPHY AND INDEX OF GEOLOGY. American Geological Institute, 4220 King Street, Alexandria, VA 22302. (703) 379-2480. 1969 to present. Inquire as to online cost and availability.

NTIS. National Technical Information Service, 5285 Port Royal Road, Springfield, VA 22161. (703) 487-4630. Broad coverage of government sponsored research reports, 1964 to present. Inquire as to online cost and availability.

SCISEARCH. Institute for Scientific Information, 3501 Market Street, Philadelphia, PA 19104. (800) 523-1850 or (215) 386-0100. Broad multidisciplinary title and author index to the international literature of science and technology, 1974 to present. Inquire as to online cost and availability.

WILSONLINE. H.W. Wilson and Company, 950 University Avenue, Bronx, NY 10452. (800) 367-6770 or (212) 588-8400. Makes available online versions of the H.W. Wilson indexes including Applied Science and Technology Index, Business Periodicals Index and Readers' Guide to Periodical Literature. Approximately 1980 to present. Inquire as to online cost and availability.

OTHER SOURCES

WHAT EVERY ENGINEER SHOULD KNOW ABOUT ENGINEERING SOURCES. Marcel Dekker Inc., 270 Madison Avenue, New York, NY 10016. (800) 228-1160. 1984. $24.95.

PERIODICALS

AMERICAN MINING CONGRESS JOURNAL. American Mining Congress, 1920 N Street, N.W., Suite 300, Washington, DC 20036. (202) 861-2800. 1915 to present. Semimonthly. $50.00 per year.

CIM BULLETIN. Canadian Institute of Mining and Metallurgy, 400-1130 Sherbrooke West, Montreal, Quebec H3A 2M8, Canada. 1898 to present. Monthly. $60.00 per year.

CANADIAN MINING JOURNAL. Southam Communications Limited, 1450 Don Mills Road, Don Mills, Ontario, M3B 2X7, Canada. (416) 445-6641. 1879 to present. Monthly. $47.50 per year.

COAL AGE. McGraw-Hill Book Company, 1221 Avenue of the Americas, New York, NY 10020. (212) 512-2000. 1911 to present. Monthly. $18.00 per year.

COLORADO SCHOOL OF MINES QUARTERLY. Colorado School of Mines Press, Golden, CO 80401. 1905 to present. Quarterly. $50.00 per year.

ENGINEERING AND MINING JOURNAL (E & M J). McGraw-Hill Book Company, 1221 Avenue of the Americas, New York, NY 10020. (212) 512-2000. 1866 to present. Monthly. $20.00 per year.

MARINE MINING. Crane Russak and Company, Inc., 3 East 44th Street, New York, NY 10017. (212) 867-1490. 1977 to present. Quarterly. $72.00 per year.

MINING ENGINEER. Institution of Mining Engineers, Danum House, South Parade, Doncaster DN1 2DY England. 1960 to present. Monthly. $60.00 per year.

MINING ENGINEERING. Society of Mining Engineers, Caller Number D, Littleton, CO 80127. (303) 973-9550. 1949 to present. Monthly. $40.00 per year.

NORTHERN MINER. Northern Miner Press Limited, 7 Labatt Avenue, Toronto, M5A 3P2 Canada. 1915 to present. Weekly. $40.00 per year.

PIT AND QUARRY. Harcourt Brace Jovanovich, Inc., 7500 Old Oak Boulevard, Cleveland, OH 44130. 1916 to present. Monthly. $15.00 per year.

RESEARCH CENTERS AND INSTITUTES

BUREAU OF MINES. U.S. Department of the Interior, 2401 E Street, N.W., Washington, DC 20241. (202) 634-1004.

EXCAVATION ENGINEERING AND EARTH MECHANICS INSTITUTE. Colorado School of Mines, Golden, CO 80401. (303) 273-3400.

MINING AND MINERAL RESOURCES INSTITUTE. New Mexico Institute of Mining and Technology, Campus Station, Socorro, NM 87801. (505) 835-5142.

MINING AND MINERAL RESOURCES RESEARCH INSTITUTE. Ohio State University, 233 Koffolt Laboratory, 140 West 19th Avenue, Columbus, OH 43210-1110. (614) 292-0102.

MIRROR OPTICS

See: OPTICS

MISSILES

See: GUIDED MISSILES

MODEMS

See: COMPUTER COMMUNICATIONS

MOLECULAR PHYSICS

See also: CHEMICAL BONDING, KINETICS, PHYSICAL CHEMISTRY, PHYSICS

ABSTRACT SERVICES AND INDEXES

CHEMICAL ABSTRACTS. American Chemical Society, Chemical Abstracts Service, Box 3012, Columbus, OH 43210. (614) 421-3600. 1907 to present. Weekly. $9500.00 per year.

CONFERENCE PAPERS INDEX. Cambridge Scientific Abstracts, 5161 River Road, Bethesda, MD 20816. 1972 to present. Monthly. Inquire as to cost and availability.

CURRENT CONTENTS: PHYSICAL, CHEMICAL AND EARTH SCIENCES. Institute for Scientific Information, 3501 Market Street, Philadelphia, PA 19104. (800) 523-1850 or (215) 386-0100. Weekly. $275.00 per year.

INDEX TO SCIENTIFIC AND TECHNICAL PROCEEDINGS. Institute for Scientific Information, 3501 Market Street, Philadelphia, PA 19104. (800) 523-1850 or (215) 386-0100. 1978 to present. Monthly. $775.00 per year.

INDEX TO SCIENTIFIC REVIEWS. Institute for Scientific Information, 3501 Market Street, Philadelphia, PA 19104. (800) 523-1850 or (215) 386-0100. 1974 to present. Semi-annual. $550.00 per year.

PHYSICS ABSTRACTS. Institution of Electrical Engineers. Available from: IEEE Service Center, 445 Hoes Lane, Piscataway, NJ 08854. 1898 to present. Bimonthly. $1700.00 per year.

PHYSICS BRIEFS. Physik Verlag GmbH, Postfach 1260/1280, D-6940 Weinheim, West Germany. (212) 661-9404. 1920 to present. Twenty-six times per year. $1250.00 per year.

SCIENCE CITATION INDEX. Institute for Scientific Information, 3501 Market Street, Philadelphia, PA 19104. (800) 523-1850 or (215) 386-0100. Six times per year. $6200.00 per year.

ASSOCIATIONS AND PROFESSIONAL SOCIETIES

AMERICAN CHEMICAL SOCIETY. 1155 16th Street, N.W., Washington, DC 20036. (202) 872-4600.

AMERICAN INSTITUTE OF PHYSICS. 335 East 45th Street, New York, NY 10017. (212) 661-9494.

BIBLIOGRAPHIES

SCIENTIFIC AND TECHNICAL BOOKS AND SERIALS IN PRINT 1988; AN INDEX TO LITERATURE IN SCIENCE AND TECHNOLOGY. R.R. Bowker Company, 205 East 42nd Street, New York, NY 10017. (800) 521-8110. $175.00.

DIRECTORIES AND BIOGRAPHICAL SOURCES

AMERICAN MEN AND WOMEN OF SCIENCE. R.R. Bowker, Inc., Order Department, 245 West 17th Street, New York, NY 10011. (800) 521-8110. Eight volumes. 1986. $595.00 for set.

INTERNATIONAL RESEARCH CENTERS DIRECTORY 1988-89. Darren L. Smith, editor. Gale Research Company, Book Tower, Detroit, MI 48226. (800) 521-0707. 4th edition. 1987. $360.00.

RESEARCH CENTERS DIRECTORY 1988. Gale Research Company, Book Tower, Detroit, MI 48226. (800) 521-0707. 12th edition. 1987. $365.00 for set.

SCIENTIFIC AND TECHNICAL ORGANIZATIONS AND AGENCIES DIRECTORY. Margaret Labash Young, editor. Gale Research Company, Book Tower, Detroit, MI 48226. (800) 521-0707. 2nd edition. 1987. $185.00.

ENCYCLOPEDIAS AND DICTIONARIES

CONCISE SCIENCE DICTIONARY. Oxford University Press, 200 Madison Avenue, New York, NY 10016. (800) 458-5833. 1987. $9.95 in paper.

DICTIONARY OF THE PHYSICAL SCIENCES: TERMS, FORMULAS, DATA. Cesare Emiliani. Oxford University Press, 200 Madison Avenue, New York, NY 10016. (800) 458-5833. 1987. $19.95 in paper.

MCGRAW-HILL ENCYCLOPEDIA OF SCIENCE AND TECHNOLOGY. McGraw-Hill Book Company, 1221 Avenue of the Americas, New York, NY 10020. (212) 512-2000. 6th edition. 1987. $1600.00.

THESAURUS OF SCIENTIFIC, TECHNICAL, AND ENGINEERING TERMS. Hemisphere Publishing Corporation, 1010 Vermont Avenue, NW, Washington, DC 20005. (800) 526-0275. 1988. $125.00.

GENERAL WORKS

INTRODUCTION TO ATOMIC AND MOLECULAR COLLISIONS. R.E. Johnson. Plenum Publishing Corporation, 233 Spring Street, New York, NY 10013. (800) 221-9369. 1982. $32.50.

INTRODUCTION TO MOLECULAR ENERGY TRANSFER. James T. Yardley. Academic Press, Inc., 6277 Sea Harbor Drive, Orlando, FL 32821. (800) 321-5068. 1980. $39.50.

MOLECULAR PHYSICS. A.N. Matveev. Imported Publications, 320 West Ohio Street, Chicago, IL 60610. (312) 787-9017. 1985. $10.95.

HANDBOOKS AND MANUALS

CRC HANDBOOK OF CHEMISTRY AND PHYSICS. Robert C. Weast, editor. CRC Press, 2000 Corporate Boulevard, Boca Raton, FL 33431. (800) 272-7737. 68th edition. 1987. $69.95.

ONLINE DATA BASES

CA SEARCH. Chemical Abstracts Service, P.O. Box 3012, Columbus, OH 43120. (800) 848-6538 or (614) 421-3600. Comprehensive guide to chemical literature, 1972 to present. Inquire as to online cost and availability.

DISSERTATION ABSTRACTS ONLINE. University Microfilms International, 300 North Zeeb Road, Ann Arbor, MI 48106. (800) 521-0600 or (313) 761-4700. Scope includes virtually all doctoral dissertations accepted at accredited American institutions from 1861 to present in over 250 subject areas. Inquire as to online cost and availability.

INSPEC. INSPEC Marketing Department, Institution of Electrical Engineers. Available from IEEE Service Center, 445 Hoes Lane, Piscataway, NJ 08854. (201) 981-0060. Online version of Physics Abstracts. Inquire as to online cost and availability.

NTIS. National Technical Information Service, 5285 Port Royal Road, Springfield, VA 22161. (703) 487-4630. Broad coverage of government sponsored research reports, 1964 to present. Inquire as to online cost and availability.

SCISEARCH. Institute for Scientific Information, 3501 Market Street, Philadelphia, PA 19104. (800) 523-1850 or (215) 386-0100. Broad multidisciplinary title and author index to the international literature of science and technology, 1974 to present. Inquire as to online cost and availability.

PERIODICALS

JOURNAL OF PHYSICS B: ATOMIC AND MOLECULAR PHYSICS. Institute of Physics. Available from: American Institute of Physics, 335 East 45th Street, New York, NY 10017. (212) 661-9494. 1968 to present. Twenty-four times year. $650.00 per year.

MOLECULAR CRYSTALS AND LIQUID CRYSTALS. Gordon and Breach Science Publishers, Inc., 50 West 23rd Street, New York, NY 10010. (212) 206-8900. 1966 to present. Biweekly. $300.00 per year.

MOLECULAR PHYSICS. Taylor and Francis Limited, Rankine Road, Basingstoke, Hants RG24 OPR, England. 1958 to present. Eighteen times per year. $860.00 per year.

RESEARCH CENTERS AND INSTITUTES

CENTER FOR CHEMICAL PHYSICS. University of Florida, Williamson Hall, Gainesville, FL 32611. (904) 392-7545.

CHEMICAL PHYSICS INSTITUTE. University of Oregon, Eugene, OR 97403. (503) 686-4773.

MONOPOLE

See: ELECTROMAGNETISM

MONORAIL TECHNOLOGY

See: RAILROAD ENGINEERING

MONTE CARLO METHOD

See: STATISTICAL METHODS

MOON

See: SATELLITES (NATURAL)

MOSSBAUER SPECTROSCOPY

See: SPECTROSCOPY

MOTORS

See also: DIESEL ENGINES, ELECTRIC MOTORS, ENGINES, INTERNAL COMBUSTION ENGINES, MACHINERY, STEAM TURBINES, TURBINES

ABSTRACT SERVICES AND INDEXES

APPLIED MECHANICS REVIEW. American Society of Mechanical Engineers, 345 East 47th Street, New York, NY 10017. (212) 705-7703. 1948 to present. Monthly. $360.00 per year.

APPLIED SCIENCE AND TECHNOLOGY INDEX. H.W. Wilson and Company, 950 University Avenue, Bronx, NY 10452. (800) 367-6670 or (212) 588-8400. Monthly. Inquire as to cost and availability.

CAD/CAM ABSTRACTS. EIC/Intelligence, Inc., 48 West 38th Street, New York, NY 10018. (800) 223-6275 or (212) 944-8500. Monthly. $365.00 per year.

CHEMICAL ABSTRACTS. American Chemical Society, Chemical Abstracts Service, Box 3012, Columbus, OH 43210. (614) 421-3600. 1907 to present. Weekly. $9500.00 per year.

CURRENT CONTENTS: ENGINEERING, TECHNOLOGY AND APPLIED SCIENCES. Institute for Scientific Information, 3501 Market Street, Philadelphia, PA 19104. (800) 523-1850 or (215) 386-0100. Weekly. $275.00 per year.

ENGINEERING INDEX MONTHLY AND AUTHOR INDEX. Engineering Information Inc., 345 East 47th Street, New York, NY 10017. (212) 705-7600. Monthly. $1560.00 per year.

ISMEC BULLETIN (Information Service in Mechanical Engineering). Cambridge Scientific Abstracts, 5161 River Road, Bethesda, MD 20816. (301) 951-1400. 1973 to present. Monthly. $450.00 per year.

PHYSICS ABSTRACTS. Institution of Electrical Engineers. Available from: IEEE Service Center, 445 Hoes Lane, Piscataway, NJ 08854. 1898 to present. Bimonthly. $1700.00 per year.

SAE TECHNICAL LITERATURE ABSTRACTS. Society of Automotive Engineers (SAE), 400 Commonwealth Drive, Warrendale, PA 15096. (412) 776-4841. 1970 to present. Quarterly. $55.00 per year.

SCIENCE CITATION INDEX. Institute for Scientific Information, 3501 Market Street, Philadelphia, PA 19104. (800) 523-1850 or (215) 386-0100. Six times per year. $6200.00 per year.

ASSOCIATIONS AND PROFESSIONAL SOCIETIES

AMERICAN SOCIETY OF MECHANICAL ENGINEERS. 345 47th Street, New York, NY 10017. (212) 705-7722.

INSTITUTE OF ELECTRICAL AND ELECTRONIC ENGINEERS (IEEE). 345 East 47th Street, New York, NY 10017. (212) 705-7900.

NATIONAL CONFERENCE ON FLUID POWER. IIT Research Institute, Ten West 35th Street, Chicago, IL 60616. (312) 567-4414.

SOCIETY OF AUTOMOTIVE ENGINEERS (SAE). 400 Commonwealth Drive, Warrendale, PA 15096. (412) 776-4841.

VIBRATION INSTITUTE. 101 West 55th Street, Suite 206, Clarendon Hills, IL 60514. (312) 654-2254.

BIBLIOGRAPHIES

SCIENTIFIC AND TECHNICAL BOOKS AND SERIALS IN PRINT 1988; AN INDEX TO LITERATURE IN SCIENCE AND TECHNOLOGY. R.R. Bowker Company, 205 East 42nd Street, New York, NY 10017. (800) 521-8110. $175.00.

DIRECTORIES AND BIOGRAPHICAL SOURCES

INDUSTRIAL EQUIPMENT AND SUPPLIES DIRECTORY. American Business Directories, Inc., 5707 South 86th Circle, Omaha, NE 68127. (402) 331-7169.

MACHINERY BUYERS GUIDE. Findlay Publications Limited, Maitland House, Warrior Square, Southend-on-Sea, Essex SS5 5AR England. Annual. $45.00.

1987 DIRECTORY OF ENGINEERING SOCIETIES AND RELATED ORGANIZATIONS. Gordon Davis, editor. Hemisphere Publishing Corporation, 1010 Vermont Avenue, NW, Washington, DC 20005. (800) 526-0275. 12th edition. 1987. $100.00.

RESEARCH CENTERS DIRECTORY 1988. Gale Research Company, Book Tower, Detroit, MI 48226. (800) 521-0707. 12th edition. 1987. $365.00 for set.

SCIENTIFIC AND TECHNICAL ORGANIZATIONS AND AGENCIES DIRECTORY. Margaret Labash Young, editor. Gale Research Company, Book Tower, Detroit, MI 48226. (800) 521-0707. 2nd edition. 1987. $185.00.

WHO'S WHO IN ENGINEERING. Gordon Davis, editor. Hemisphere Publishing Corporation, 1010 Vermont Avenue, NW, Washington, DC 20005. (800) 526-0275. 6th edition. 1985. $200.00.

ENCYCLOPEDIAS AND DICTIONARIES

THESAURUS OF SCIENTIFIC, TECHNICAL, AND ENGINEERING TERMS. Hemisphere Publishing Corporation, 1010 Vermont Avenue, NW, Washington, DC 20005. (800) 526-0275. 1988. $125.00.

GENERAL WORKS

ENGINEERING MECHANICS. J.L. Meriam and L.G. Kraige. John Wiley and Sons, Inc., 605 Third Avenue, New York, NY 10158. (800) 526-5368. 2nd edition. 1986. $52.95.

INTRODUCTION TO DYNAMICS AND CONTROL. Leonard Meirovitch. John Wiley and Sons, Inc., 605 Third Avenue, New York, NY 10158. (800) 526-5368. 1985. $42.95.

POLYPHASE MOTORS: A DIRECT APPROACH TO THEIR DESIGN. Enrico Levi. John Wiley and Sons, Inc., 605 Third Avenue, New York, NY 10158. (800) 526-5368. 1984. $52.95.

STEPPING MOTORS AND THEIR MICROPROCESSOR CONTROL. Takashi Kenjo. Oxford University Press, 200 Madison Avenue, New York, NY 10016. (800) 458-5833. 1984. $49.95.

HANDBOOKS AND MANUALS

HANDBOOK OF MECHANICS, MATERIALS, AND STRUCTURES. A. Blake. John Wiley and Sons, Inc., 605 Third Avenue, New York, NY 10158. (800) 526-5368. 1985. $65.95.

MARK'S STANDARD HANDBOOK FOR MECHANICAL ENGINEERS. T. Baumeister, editor. McGraw-Hill Book Company, 1221 Avenue of the Americas, New York, NY 10020. (212) 512-2000. 8th edition. 1978. $96.00.

MECHANICAL DESIGN AND SYSTEMS HANDBOOK. Harold A. Rothbart, editor. McGraw-Hill Book Company, 1221 Avenue of the Americas, New York, NY 10020. (212) 512-2000. 2nd edition.1985. $96.50.

MECHANICAL ENGINEERS' HANDBOOK. Myer Kutz, editor. John Wiley and Sons, Inc., 605 Third Avenue, New York, NY 10158. (800) 526-5368. 1986. $79.95.

STANDARD HANDBOOK OF MACHINE DESIGN. Joseph E. Shigley and Charles R. Mischke. McGraw-Hill Book Company, 1221 Avenue of the Americas, New York, NY 10020. (212) 512-2000. 1986. $89.00.

ONLINE DATA BASES

CA SEARCH. Chemical Abstracts Service, P.O. Box 3012, Columbus, OH 43120. (800) 848-6538 or (614) 421-3600. Comprehensive guide to chemical literature, 1972 to present. Inquire as to online cost and availability.

COMPENDEX. Engineering Information, Inc., 345 East 47th Street, New York, NY 10017. (800) 221-1044 or (212) 705-7615. Engineering and technical literature, 1975 to present. Inquire as to online cost and availability.

DISSERTATION ABSTRACTS ONLINE. University Microfilms International, 300 North Zeeb Road, Ann Arbor, MI 48106. (800) 521-0600 or (313) 761-4700. Scope includes virtually all doctoral dissertations accepted at accredited American institutions from 1861 to present in over 250 subject areas. Inquire as to online cost and availability.

INSPEC. INSPEC Marketing Department, Institution of Electrical Engineers. Available from IEEE Service Center, 445 Hoes Lane, Piscataway, NJ 08854. (201) 981-0060. Online version of Physics Abstracts. Inquire as to online cost and availability.

ISMEC. Cambridge Scientific Abstracts, 5161 River Road, Besthda, MD 20816. (800) 638-8076 or (301) 951-1400. Literature of mechanical and production engineering, 1973 to present. Inquire as to online cost and availability.

NTIS. National Technical Information Service, 5285 Port Royal Road, Springfield, VA 22161. (703) 487-4630. Broad coverage of government sponsored research reports, 1964 to present. Inquire as to online cost and availability.

SCISEARCH. Institute for Scientific Information, 3501 Market Street, Philadelphia, PA 19104. (800) 523-1850 or (215) 386-0100. Broad multidisciplinary title and author index to the international literature of science and technology, 1974 to present. Inquire as to online cost and availability.

WILSONLINE. H.W. Wilson and Company, 950 University Avenue, Bronx, NY 10452. (800) 367-6770 or (212) 588-8400. Makes available online versions of the H.W. Wilson indexes including Applied Science and Technology Index, Business Periodicals Index and Readers' Guide to Periodical Literature. Approximately 1980 to present. Inquire as to online cost and availability.

PERIODICALS

AUTOMOTIVE ENGINEERING. Society of Automotive Engineers (SAE), 400 Commonwealth Drive, Warrendale, PA 15096. (412) 776-4841. 1917 to present. Monthly. $33.00 per year.

CIME (COMPUTERS IN MECHANICAL ENGINEERING). American Society of Mechanical Engineers, 345 East 47th Street, New York, NY 10017. (212) 705-7703.

INTERNATIONAL COMMUNICATIONS IN HEAT AND MASS TRANSFER. Pergamon Press, Inc., Maxwell House, Fairview Park, Elmsford, NY 10523. (914) 592-7700. 1974 to present. Bimonthly. $160.00 per year.

INTERNATIONAL JOURNAL FOR NUMERICAL METHODS IN ENGINEERING. John Wiley and Sons, Inc., 605 Third Avenue, New York, NY 10158. (800) 526-5368.

INTERNATIONAL JOURNAL OF HEAT AND MASS TRANSFER. Pergamon Press, Inc., Maxwell House, Fairview Park, Elmsford, NY 10523. (914) 592-7700. 1960 to present. Monthly. $500.00 per year.

INTERNATIONAL JOURNAL OF MULTIPHASE FLOW. Pergamon Press, Inc., Maxwell House, Fairview Park, Elmsford, NY 10523. (914) 592-7700. (212) 705-7703. 1974 to present. Bimonthly. $250.00 per year.

JOURNAL OF APPLIED MECHANICS. American Society of Mechanical Engineers, 345 East 47th Street, New York, NY 10017. (212) 705-7703. 1935 to present. Quarterly. $100.00 per year.

JOURNAL OF ENGINEERING FOR INDUSTRY. American Society of Mechanical Engineers, 345 East 47th Street, New York, NY 10017. (212) 705-7703. 1970 to present. Quarterly. $100.00 per year.

JOURNAL OF FLUID CONTROL. Delbridge Publishing Company, Box 2989, Stanford, CA 94305. (408) 446-3131. 1967 to present. Quarterly. Inquire as to cost and availability.

JOURNAL OF HEAT TRANSFER. American Society of Mechanical Engineers, 345 East 47th Street, New York, NY 10017. (212) 705-7703. 1970 to present. Quarterly. $100.00 per year.

JOURNAL OF MECHANISMS, TRANSMISSIONS AND AUTOMATION IN DESIGN. American Society of Mechanical Engineers, 345 East 47th Street, New York, NY 10017. (212) 705-7703. 1983 to present. Quarterly. $80.00 per year.

JOURNAL OF PRESSURE VESSEL TECHNOLOGY. American Society of Mechanical Engineers, 345 East 47th Street, New York, NY 10017. (212) 705-7703. 1974 to present. Quarterly. $80.00 per year.

LUBRICATION ENGINEERING. American Society of Lubrication Engineers, 838 Busse Highway, Park Ridge, IL 60068. 1945 to present. Monthly. $39.00 per year.

MACHINE DESIGN. Penton-IPC, 1100 Superior Avenue, Cleveland, OH 44114. (216) 696-7000. 1929 to present. Twenty-eight times per year. $60.00 per year.

MECHANICAL ENGINEERING. American Society of Mechanical Engineers, 345 East 47th Street, New York, NY 10017. (212) 705-7722.

RESEARCH CENTERS AND INSTITUTES

LABORATORY FOR EXPERIMENTAL MECHANICS RESEARCH. State University of New York at Stony Brook, Stony Brook, NY 11794-2300. (516) 632-8311.

MACHINE SYSTEMS GROUP. Stevens Institute of Technology, Department of Mechanical Engineering, Castle Point Station, Hoboken, NJ 07030. (201) 420-5591.

MECHANICAL ENGINEERING DESIGN LABORATORY. University of Florida, 237 Mechanical Engineering Building, Gainesville, FL 32611. (904) 392-0827

MECHANICAL ENGINEERING RESEARCH LABORATORIES. Kansas State University, Durland Hall, Manhattan, KS 66506. (913) 532-5610.

MULTIVARIATE ANALYSIS

See: MATHEMATICS

N

NATURAL GAS

See also: PETROLEUM GEOLOGY

ABSTRACT SERVICES AND INDEXES

APPLIED SCIENCE AND TECHNOLOGY INDEX. H.W. Wilson and Company, 950 University Avenue, Bronx, NY 10452. (800) 367-6670 or (212) 588-8400. Monthly. Inquire as to cost and availability.

BIBLIOGRAPHY AND INDEX OF GEOLOGY. American Geological Institute, 4220 King Street, Alexandria, VA 22302. (703) 379-2480. 1969 to present. Monthly. $1100.00 per year.

CHEMICAL ABSTRACTS. American Chemical Society, Chemical Abstracts Service, Box 3012, Columbus, OH 43210. (614) 421-3600. 1907 to present. Weekly. $9500.00 per year.

CURRENT CONTENTS: ENGINEERING, TECHNOLOGY AND APPLIED SCIENCES. Institute for Scientific Information, 3501 Market Street, Philadelphia, PA 19104. (800) 523-1850 or (215) 386-0100. Weekly. $275.00 per year.

ENGINEERING INDEX MONTHLY AND AUTHOR INDEX. Engineering Information Inc., 345 East 47th Street, New York, NY 10017. (212) 705-7600. Monthly. $1560.00 per year.

GENERAL SCIENCE INDEX. H.W. Wilson and Company, 950 University Avenue, Bronx, NY 10452. (800) 367-6670 or (212) 588-8400. 1978 to present. Monthly. Inquire as to cost and availability.

INDEX TO SCIENTIFIC AND TECHNICAL PROCEEDINGS. Institute for Scientific Information, 3501 Market Street, Philadelphia, PA 19104. (800) 523-1850 or (215) 386-0100. 1978 to present. Monthly. $775.00 per year.

OCEAN ABSTRACTS. Cambridge Scientific Abstracts, 5161 River Road, Bethesda, MD 20816. (301) 951-1400. 1963 to present. Bimonthly. $450.00 per year.

SCIENCE CITATION INDEX. Institute for Scientific Information, 3501 Market Street, Philadelphia, PA 19104. (800) 523-1850 or (215) 386-0100. Six times per year. $6200. 00 per year.

ASSOCIATIONS AND PROFESSIONAL SOCIETIES

AMERICAN ASSOCIATION OF PETROLEUM GEOLOGISTS. Box 979, Tulsa, OK 74101. (918) 584-2555.

AMERICAN GAS ASSOCIATION. 1515 Wilson Boulevard, Arlington, VA 22209. (703) 841-8400.

AMERICAN GEOPHYSICAL UNION. 2000 Florida Avenue, N.W., Washington, DC 20009. (202) 462-6903.

AMERICAN INSTITUTE OF GEOLOGY. 4220 King Street, Alexandria, VA 22302. (703) 379-2480.

CANADIAN SOCIETY OF PETROLEUM GEOLOGISTS. 206 7th Avenue, S.W., Suite 505, Calgary, AB T2P OW7, Canada. (403) 264-5610.

GAS PROCESSORS ASSOCIATION. 1812 First National Bank Building, Tulsa, OK 74103. (918) 582-5112.

SOCIETY OF EXPLORATION GEOPHYSICISTS. Box 702740, Tulsa, OK 74170-2740.

DIRECTORIES AND BIOGRAPHICAL SOURCES

AAPG BULLETIN MEMBERSHIP DIRECTORY. American Association of Petroleum Geologists, P.O. Box 979, Tulsa, OK 74101. Included with AAPG Bulletin subscription.

AMERICAN MEN AND WOMEN OF SCIENCE. R.R. Bowker, Inc., Order Department, 245 West 17th Street, New York, NY 10011. (800) 521-8110. Eight volumes. 1986. $595.00 for set.

INTERNATIONAL RESEARCH CENTERS DIRECTORY 1988-89. Darren L. Smith, editor. Gale Research Company, Book Tower, Detroit, MI 48226. (800) 521-0707. 4th edition. 1987. $360.00.

1987 DIRECTORY OF ENGINEERING SOCIETIES AND RELATED ORGANIZATIONS. Gordon Davis, editor. Hemisphere Publishing Corporation, 1010 Vermont Avenue, NW, Washington, DC 20005. (800) 526-0275. 12th edition. 1987. $100.00.

RESEARCH CENTERS DIRECTORY 1988. Gale Research Company, Book Tower, Detroit, MI 48226. (800) 521-0707. 12th edition. 1987. $365.00 for set.

SCIENTIFIC AND TECHNICAL ORGANIZATIONS AND AGENCIES DIRECTORY. Margaret Labash Young, editor. Gale Research Company, Book Tower, Detroit, MI 48226. (800) 521-0707. 2nd edition. 1987. $185.00.

WHO'S WHO IN ENGINEERING. Gordon Davis, editor. Hemisphere Publishing Corporation, 1010 Vermont Avenue, NW, Washington, DC 20005. (800) 526-0275. 6th edition. 1985. $200.00.

ENCYCLOPEDIAS AND DICTIONARIES

DICTIONARY OF GEOLOGICAL TERMS. American Geological Institute. Doubleday and Company, Inc., 245 Park Avenue, New York, NY 10017. (800) 645-6156. Third edition. 1984. $19.95.

GENERAL WORKS

NATURAL GAS. E.N. Tiratsoo. Gulf Publishing Company, P.O. Box 2608, Houston, TX 77001. (713) 520-4444. Third edition. 1980. $55.00.

NATURAL GAS PRODUCTION ENGINEERING. Chi U. Ikoku. John Wiley and Sons, Inc., 605 Third Avenue, New York, NY 10158. (800) 526-5368. 1984. $45.00.

PETROLEUM DEVELOPMENT GEOLOGY. Parke A. Dickey. PennWell Books, Box 1260, Tulsa, OK 74101. (918) 663-4225.

PRACTICAL NATURAL GAS ENGINEERING. R.V. Smith. PennWell Publishing Company, P.O. Box 1260, Tulsa, OK 74101. (918) 663-4225. 1983. $49.95.

UNDERGROUND STORAGE OF NATURAL GAS. M.R. Tek. Gulf Publishing Company, P.O. Box 2608, Houston, TX 77001. (713) 520-4444. 1987. $65.00.

HANDBOOKS AND MANUALS

HANDBOOK OF NATURAL GAS ENGINEERING. D.L. Katz. McGraw-Hill Book Company, 1221 Avenue of the Americas, New York, NY 10020. (212) 512-2000. 1959. $100.00.

ONLINE DATA BASES

CA SEARCH. Chemical Abstracts Service, P.O. Box 3012, Columbus, OH 43120. (800) 848-6538 or (614) 421-3600. Comprehensive guide to chemical literature, 1972 to present. Inquire as to online cost and availability.

COMPENDEX. Engineering Information, Inc., 345 East 47th Street, New York, NY 10017. (800) 221-1044 or (212) 705-7615. Engineering and technical literature, 1975 to present. Inquire as to online cost and availability.

GEOARCHIVE. Geosystems, P.O. Box 1024, Westminister, London SW1 P 2JL, England. Citations to literature on geosciences, 1969 to present. Inquire as to online costs and availability.

GEOREF. Online version of the BIBLIOGRAPHY AND INDEX OF GEOLOGY. American Geological Institute, 4220 King Street, Alexandria, VA 22302. (703) 379-2480. 1969 to present. Inquire as to online cost and availability.

NTIS. National Technical Information Service, 5285 Port Royal Road, Springfield, VA 22161. (703) 487-4630. Broad coverage of government sponsored research reports, 1964 to present. Inquire as to online cost and availability.

SCISEARCH. Institute for Scientific Information, 3501 Market Street, Philadelphia, PA 19104. (800) 523-1850 or (215) 386-0100. Broad multidisciplinary title and author index to the international literature of science and technology, 1974 to present. Inquire as to online cost and availability.

WILSONLINE. H.W. Wilson and Company, 950 University Avenue, Bronx, NY 10452. (800) 367-6770 or (212) 588-8400. Makes available online versions of the H.W. Wilson indexes including Applied Science and Technology Index, Business Periodicals Index and Readers' Guide to Periodical Literature. Approximately 1980 to present. Inquire as to online cost and availability.

PERIODICALS

AAPG BULLETIN. American Association of Petroleum Geologists, P.O. Box 979, Tulsa, OK 74101. (918) 584-2555. Monthly. $40.00 per year.

AMERICAN GAS ASSOCIATION MONTHLY. American Gas Association. 1515 Wilson Boulevard, Arlington, VA 22209. (703) 841-8400. 1919 to present. Eleven times per year. $30.00.

BULLETIN OF CANADIAN PETROLEUM GEOLOGY. Canadian Society of Petroleum Geologists, 206 7th Avenue, S.W., Suite 505, Calgary, AB T2P OW7, Canada. (403) 264-5610. Quarterly. $50.00 per year.

GAS ENGINEERING AND MANAGEMENT. Institution of Gas Engineers, 17 Grosvenor Crescent, London SW1X 7ES, England. 1961 to present. Ten times per year. $60.00 per year.

GAS INDUSTRIES. Gas Industries Equipment and Appliance News, Inc., Box 558 Park Ridge, IL 60068. 1956 to present. Monthly. $15.00 per year.

NATURAL GAS MONTHLY. U.S. Energy Information Administration, Forrestal Building, Washington, DC 20585. (202) 252-8800. Monthly. $55.00 per year.

OIL AND GAS JOURNAL. PennWell Publishing Company, P.O. Box 1260, Tulsa, OK 74101. (918) 835-3161. 1902 to present. Weekly. $34.00 per year.

RESEARCH CENTERS AND INSTITUTES

CENTER FOR ENERGY AND MINERAL RESOURCES. Texas A&M University, Bizzell Hall, College Station, TX 77843. (409) 845-8025.

GAS RESEARCH INSTITUTE. 8600 West Bryn Mawr Avenue, Chicago, IL 60631. (312) 399-8100.

INSTITUTE FOR ENERGY RESOURCE STUDIES - POTENTIAL GAS AGENCY. Colorado School of Mines, Golden, CO 80401. (303) 279-4320.

NAVAL ARCHITECTURE

See also: MARINE ENGINEERING, SHIPBUILDING

ABSTRACT SERVICES AND INDEXES

APPLIED MECHANICS REVIEW. American Society of Mechanical Engineers, 345 East 47th Street, New York, NY 10017. (212) 705-7703. 1948 to present. Monthly. $360.00 per year.

APPLIED SCIENCE AND TECHNOLOGY INDEX. H.W. Wilson and Company, 950 University Avenue, Bronx, NY 10452. (800) 367-6670 or (212) 588-8400. Monthly. Inquire as to cost and availability.

CAD/CAM ABSTRACTS. EIC/Intelligence, Inc., 48 West 38th Street, New York, NY 10018. (800) 223-6275 or (212) 944-8500. Monthly. $365.00 per year.

CHEMICAL ABSTRACTS. American Chemical Society, Chemical Abstracts Service, Box 3012, Columbus, OH 43210. (614) 421-3600. 1907 to present. Weekly. $9500.00 per year.

CURRENT CONTENTS: ENGINEERING, TECHNOLOGY AND APPLIED SCIENCES. Institute for Scientific Information, 3501 Market Street, Philadelphia, PA 19104. (800) 523-1850 or (215) 386-0100. Weekly. $275.00 per year.

ENGINEERING INDEX MONTHLY AND AUTHOR INDEX. Engineering Information Inc., 345 East 47th Street, New York, NY 10017. (212) 705-7600. Monthly. $1560.00 per year.

ISMEC BULLETIN (Information Service in Mechanical Engineering). Cambridge Scientific Abstracts, 5161 River Road, Bethesda, MD 20816. (301) 951-1400. 1973 to present. Monthly. $450.00 per year.

OCEANIC ABSTRACTS. Cambridge Scientific Abstracts, 5161 River Road, Bethesda, MD 20816. (301) 951-1400. 1964 to present. Bimonthly. $655.00 per year.

SCIENCE CITATION INDEX. Institute for Scientific Information, 3501 Market Street, Philadelphia, PA 19104. (800) 523-1850 or (215) 386-0100. Six times per year. $6200.00 per year.

ASSOCIATIONS AND PROFESSIONAL SOCIETIES

AMERICAN SOCIETY OF MECHANICAL ENGINEERS. 345 47th Street, New York, NY 10017. (212) 705-7722.

AMERICAN SOCIETY OF NAVAL ENGINEERS. 1452 Duke Street, Alexandria, VA 22314. (703) 836-6727.

INSTITUTE OF MARINE ENGINEERS. 76 Mark Lane, London, EC3R 7JN, England.

SOCIETY OF NAVAL ARCHITECTS AND MARINE ENGINEERS. One World Trade Center, Suite 1369, New York, NY 10048. (212) 432-0310.

VIBRATION INSTITUTE. 101 West 55th Street, Suite 206, Clarendon Hills, IL 60514. (312) 654-2254.

DIRECTORIES AND BIOGRAPHICAL SOURCES

DIRECTORY OF SHIPOWNERS, SHIPBUILDERS, AND MARINE ENGINEERS. Industrial Press, Division of Business Press International, Limited. Available from: Robert E. Raynor, Business Press International, 205 East 42nd Street, New York, NY 10017. Annual. $75.00.

INDUSTRIAL EQUIPMENT AND SUPPLIES DIRECTORY. American Business Directories, Inc., 5707 South 86th Circle, Omaha, NE 68127. (402) 331-7169.

MARINE EQUIPMENT CATALOG. Maritime Reporter and Engineering News, 118 East 25th Street, New York, NY 10010. (212) 477-6700. Annual. $65.00.

1987 DIRECTORY OF ENGINEERING SOCIETIES AND RELATED ORGANIZATIONS. Gordon Davis, editor. Hemisphere Publishing Corporation, 1010 Vermont Avenue, NW, Washington, DC 20005. (800) 526-0275. 12th edition. 1987. $100.00.

RESEARCH CENTERS DIRECTORY 1988. Gale Research Company, Book Tower, Detroit, MI 48226. (800) 521-0707. 12th edition. 1987. $365.00 for set.

SCIENTIFIC AND TECHNICAL ORGANIZATIONS AND AGENCIES DIRECTORY. Margaret Labash Young, editor. Gale Research Company, Book Tower, Detroit, MI 48226. (800) 521-0707. 2nd edition. 1987. $185.00.

WHO'S WHO IN ENGINEERING. Gordon Davis, editor. Hemisphere Publishing Corporation, 1010 Vermont Avenue, NW, Washington, DC 20005. (800) 526-0275. 6th edition. 1985. $200.00.

GENERAL WORKS

HYDRONAUTICS. Herman E. Sheets and Victor T. Boatwright, editors. Academic Press, Inc., 6277 Sea Harbor Drive, Orlando, FL 32821. (800) 321-5068. 1970. $75.00.

MODERN SHIP DESIGN. Thomas C. Gillmer. Naval Institute Press, U.S. Naval Institute, Annapolis, MD 21402. (301) 268-6110. 2nd edition. 1975. $18.95.

HANDBOOKS AND MANUALS

MARINE ENGINE ROOM BLUE BOOK. W.B. Paterson. Cornell Maritime Press, Inc., P.O. Box 456, Centreville, MD 21617. (301) 758-1075. 3rd edition. 1984. $15.00 in paper.

ONLINE DATA BASES

CA SEARCH. Chemical Abstracts Service, P.O. Box 3012, Columbus, OH 43120. (800) 848-6538 or (614) 421-3600. Comprehensive guide to chemical literature, 1972 to present. Inquire as to online cost and availability.

COMPENDEX. Engineering Information, Inc., 345 East 47th Street, New York, NY 10017. (800) 221-1044 or (212) 705-7615. Engineering and technical literature, 1975 to present. Inquire as to online cost and availability.

INSPEC. INSPEC Marketing Department, Institution of Electrical Engineers. Available from IEEE Service Center, 445 Hoes Lane, Piscataway, NJ 08854. (201) 981-0060. Online version of Physics Abstracts. Inquire as to online cost and availability.

ISMEC. Cambridge Scientific Abstracts, 5161 River Road, Besthda, MD 20816. (800) 638-8076 or (301) 951-1400. Literature of mechanical and production engineering, 1973 to present. Inquire as to online cost and availability.

NTIS. National Technical Information Service, 5285 Port Royal Road, Springfield, VA 22161. (703) 487-4630. Broad coverage of government sponsored research reports, 1964 to present. Inquire as to online cost and availability.

SCISEARCH. Institute for Scientific Information, 3501 Market Street, Philadelphia, PA 19104. (800) 523-1850 or (215) 386-0100. Broad multidisciplinary title and author index to the international literature of science and technology, 1974 to present. Inquire as to online cost and availability.

WILSONLINE. H.W. Wilson and Company, 950 University Avenue, Bronx, NY 10452. (800) 367-6770 or (212) 588-8400. Makes available online versions of the H.W. Wilson indexes including Applied Science and Technology Index, Business Periodicals Index and Readers' Guide to Periodical Literature. Approximately 1980 to present. Inquire as to online cost and availability.

PERIODICALS

HIGH-SPEED SURFACE CRAFT. Capstan Publishing Company, Limited, Box 8, Tadworth, Surrey, KT20 5QR, England. 1961 to present. Bimonthly. $50.00 per year.

JOURNAL OF SHIP RESEARCH. Society of Naval Architects and Marine Engineers. One World Trade Center, Suite 1369, New York, NY 10048. (212) 432-0310. 1957 to present. Quarterly. $40.00 per year.

MARINE ENGINEERING/LOG. Simmons-Boardman Publishing Corporation, 345 Hudson Street, New York, NY 10014. (212) 620-7200. 1878 to present. Monthly. $30.00 per year.

MARINE ENGINEERS REVIEW. Marine Management Holdings, Limited, 76 Mark Lane, London EC3R 7JN, England. 1970 to present. Monthly. $40.00 per year.

MARINE TECHNOLOGY. Society of Naval Architects and Marine Engineers. One World Trade Center, Suite 1369, New York, NY 10048. (212) 432-0310. 1964 to present. Quarterly. $40.00 per year.

MOTOR SHIP. Industrial Press, Division of Business Press International, Limited. Available from: Robert E. Raynor, Business Press International, 205 East 42nd Street, New York, NY 10017. 1920 to present. Monthly. $115.00 per two years.

NAVAL ARCHITECT. Royal Institution of Naval Architects, 10 Upper Belgrave Street, London, SW1X 8BQ, England. 1971 to present. Bimonthly. $75.00 per year.

SOCIETY OF NAVAL ARCHITECTS AND MARINE ENGINEERS TRANSACTIONS. Society of Naval Architects and Marine Engineers. One World Trade Center, Suite 1369, New York, NY 10048. (212) 432-0310. 1893 to present. Annual. $45.00 per year.

RESEARCH CENTERS AND INSTITUTES

INSTITUTE FOR MARINE DYNAMICS. National Research Council of Canada, P.O. Box 12093, Postal Station A, St. John's, NF, Canada A1B 3T5. (709) 772-2469.

MARITIME RESEARCH DEPARTMENT. Webb Institute of Naval Architecture, Crescent Beach Road, Glen Cove, NY 11542. (516) 671-2356.

SHIP HYDRODYNAMICS LABORATORY. University of Michigan, 126 West Engineering Building, Ann Arbor, MI 48109. (313) 764-9432.

NAVIGATION

See also: AVIONICS, BUOYS, COMPASSES, GUIDANCE SYSTEMS, GYROSCOPES, RADAR

ABSTRACT SERVICES AND INDEXES

APPLIED SCIENCE AND TECHNOLOGY INDEX. H.W. Wilson and Company, 950 University Avenue, Bronx, NY 10452. (800) 367-6670 or (212) 588-8400. Monthly. Inquire as to cost and availability.

ENGINEERING INDEX MONTHLY AND AUTHOR INDEX. Engineering Information Inc., 345 East 47th Street, New York, NY 10017. (212) 705-7600. Monthly. $1560.00 per year.

GENERAL SCIENCE INDEX. H.W. Wilson and Company, 950 University Avenue, Bronx, NY 10452. (800) 367-6670 or (212) 588-8400. 1978 to present. Monthly. Inquire as to cost and availability.

INTERNATIONAL AEROSPACE ABSTRACTS. American Institute of Aeronautics and Astronautics, Technical Information Service, 370 L'Enfant Promenade, S.W., Washington, DC 20024. (202) 646-7400. 1961 to present. Semi-monthly. $700.00 per year.

ASSOCIATIONS AND PROFESSIONAL SOCIETIES

INSTITUTE OF NAVIGATION. 815 15th Street, N.W., Suite 832, Washington, DC 20005. (202) 783-4121.

INTERNATIONAL OMEGA ASSOCIATION. P.O. Box 2324, 1720 South Eads Street, Arlington, VA 22202. (301) 593-4144.

PERMANENT INTERNATIONAL ASSOCIATION OF NAVIGATION CONGRESSES, UNITED STATES SECTION. c/o U.S. Army Corps of Engineers, Water Resources Support Center, Casey Building, Fort Belvoir, VA 22060. (202) 355-2096.

WILD GOOSE ASSOCIATION. P.O. Box 556, Bedford, MA 01730.

DIRECTORIES AND BIOGRAPHICAL SOURCES

INTERNATIONAL RESEARCH CENTERS DIRECTORY 1988-89. Darren L. Smith, editor. Gale Research Company, Book Tower, Detroit, MI 48226. (800) 521-0707. 4th edition. 1987. $360.00.

1987 DIRECTORY OF ENGINEERING SOCIETIES AND RELATED ORGANIZATIONS. Gordon Davis, editor. Hemisphere Publishing Corporation, 1010 Vermont Avenue, NW, Washington, DC 20005. (800) 526-0275. 12th edition. 1987. $100.00.

RESEARCH CENTERS DIRECTORY 1988. Gale Research Company, Book Tower, Detroit, MI 48226. (800) 521-0707. 12th edition. 1987. $365.00 for set.

SCIENTIFIC AND TECHNICAL ORGANIZATIONS AND AGENCIES DIRECTORY. Margaret Labash Young, editor. Gale Research Company, Book Tower, Detroit, MI 48226. (800) 521-0707. 2nd edition. 1987. $185.00.

WHO'S WHO IN ENGINEERING. Gordon Davis, editor. Hemisphere Publishing Corporation, 79 Madison Avenue, New York, NY 10016-7892. (800) 821-8312. Sixth edition. 1985. $200.00.

GENERAL WORKS

AUTOMATIC CONTROL OF AIRCRAFT AND MISSILES. John H. Blakelock. John Wiley and Sons, Inc., 605 Third Avenue, New York, NY 10158. (800) 526-5368. 1965. $64.50.

AUTOMATIC FLIGHT CONTROL. E.H. Pallett. Sheridan Publishers, Inc., 145 Palisade Street, Dobbs Ferry, NY 10522. (914) 693-2410. Second edition. 1983. $32.50.

COCKPIT COMPUTERS AND NAVIGATION AVIONICS. P. Garrison. McGraw-Hill Book Company, 1221 Avenue of the Americas, New York, NY 10020. (212) 512-2000. 1982. $29.95.

DUTTON'S NAVIGATION AND PILOTING. Elbert S. Maloney. Naval Institute Press, U.S. Naval Institute, Annapolis, MD 21402. (301) 268-6110. Fourth edition. 1985. $32.95.

GYROSCOPIC THEORY. George Greehill. Chelsea Publishing Company, 15 East 26th Street, New York, NY 10010. (212) 889-8095. $22.50.

MARINE ELECTRONIC NAVIGATION. S.F. Appleyard. Methuen, Inc., 29 West 35th Street, New York, NY 10001. (212) 244-3336. 1980. $34.95.

SEAMANSHIP: FUNDAMENTALS FOR THE DECK OFFICER. David O. Dodge and S.E. Kyriss. Naval Institute Press, U.S. Naval Institute, Annapolis, MD 21402. (301) 268-6110. 1981. $16.95.

SHIPBOARD ANTENNAS. Preston E. Law. Artech House, 685 Canton Street, Norwood, MA 02062. (800) 225-9977. 1986. $61.00.

ONLINE DATA BASES

COMPENDEX. Engineering Information, Inc., 345 East 47th Street, New York, NY 10017. (800) 221-1044 or (212) 705-7615. Engineering and technical literature, 1975 to present. Inquire as to online cost and availability.

NTIS. National Technical Information Service, 5285 Port Royal Road, Springfield, VA 22161. (703) 487-4630. Broad coverage of government sponsored research reports, 1964 to present. Inquire as to online cost and availability.

WILSONLINE. H.W. Wilson and Company, 950 University Avenue, Bronx, NY 10452. (800) 367-6770 or (212) 588-8400. Makes available online versions of the H.W. Wilson indexes including Applied Science and Technology Index, Business Periodicals Index and Readers' Guide to Periodical Literature. Approximately 1980 to present. Inquire as to online cost and availability.

PERIODICALS

AEROSPACE AMERICA. American Institute of Aeronautics and Astronautics, Technical Information Service, 370 L'Enfant Promenade, S.W., Washington, DC 20024. (202) 646-7400. 1932 to present. Monthly. $56.00 per year.

AEROSPACE ENGINEERING MAGAZINE. Society of Automotive Engineers, 400 Commonwealth Drive, Warrendale, PA 15096. (412) 776-4841. 1981 to present. Monthly. $24.00 per year.

AIAA JOURNAL. American Institute of Aeronautics and Astronautics, Technical Information Service, 370 L'Enfant Promenade, S.W., Washington, DC 20024. (202) 646-7400. 1963 to present. Monthly. $205.00 per year.

AVIATION WEEK AND SPACE TECHNOLOGY. McGraw-Hill Book Company, 1221 Avenue of the Americas, New York, NY 10020. (212) 512-2000. 1916 to present. Weekly. $56.00 per year.

JOURNAL OF NAVIGATION. Royal Institute of Navigation. Cambridge University Press, 32 East 57th Street, New York, NY 10022. (800) 872-7423. 1947 to present. Three per year. $100.00 per year.

JOURNAL OF SHIP RESEARCH. Society of Naval Architects and Marine Engineers, One World Trade Center, Suite 1369,

New York, NY 10048. (212) 432-0310. 1957 to present. Quarterly. $40.00 per year.

JOURNAL OF SPACECRAFT AND ROCKETS. American Institute of Aeronautics and Astronautics, Technical Information Service, 370 L'Enfant Promenade, S.W., Washington, DC 20024. (202) 646-7400. 1964 to present. Bimonthly. $95.00 per year.

RESEARCH CENTERS AND INSTITUTES

AVIATION SAFETY INSTITUTE. 33 East North Street, P.O. Box 304, Worthington, OH 43085. (614) 885-4242.

AVIONICS ENGINEERING CENTER. Ohio University, 239 Stocker Center, Athens, OH 45701. (614) 593-1533.

CHARLES STARK DRAPER LABORATORY, INC. 555 Technology Square, Cambridge, MA 02139. (617) 258-1000.

NDT

See: NONDESTRUCTIVE TESTING

NEPTUNE

See also: PLANETARY SCIENCE, SOLAR SYSTEM

ABSTRACT SERVICES AND INDEXES

ASTRONOMY AND ASTROPHYSICS ABSTRACTS. Springer-Verlag New York, Incorporated, 175 Fifth Avenue, New York, NY 10010. (212) 460-1500. $70.00 per year.

ASTRONOMY AND ASTROPHYSICS MONTHLY INDEX. Olivetree Associates, Post Office Box 236, Sierra Madre, CA 91024. $212.00 per year. Complimentary copies available on request.

METEOROLOGICAL AND GEOASTROPHYSICAL ABSTRACTS. American Meteorological Society, 45 Beacon Street, Boston, MA 02108. (617) 227-2425. 1950 to present. Monthly. $450.00 per year.

STAR. (Scientific and Technical Aerospace Reports. United States National Aeronautics and Space Administration, Scientific and Technical Information Facility, Box 8757, Baltimore-Washington International Airport, MD 21240. (202) 755-2210. Semimonthly, with semiannual and annual indexes. $85.00 per year.

ANNUAL REVIEWS AND YEARBOOKS

THE ASTRONOMICAL ALMANAC. Superintendent of Documents, United States Government Printing Office, Washington, DC 20402. (202) 783-3238. Yearly.

ANNUAL REVIEW OF ASTRONOMY AND ASTROPHYSICS. Annual Reviews, Incorporated, 4139 El Camino Way, Palo Alto, CA 94306. (415) 493-4400. Annual. Inquire.

ANNUAL REVIEW OF EARTH AND PLANETARY SCIENCES. Annual Reviews, Incorporated, 4139 El Camino Way, Palo Alto, CA 94306. (415) 493-4400. Annual. Inquire.

ASSOCIATIONS AND PROFESSIONAL SOCIETIES

AMERICAN ASTRONOMICAL SOCIETY. 2000 Florida Avenue, NW, Suite 300, Washington, DC 20009. (202) 659-0134.

AMERICAN GEOPHYSICAL UNION. 2000 Florida Avenue, NW, Washington, DC 20009. (202) 462-6903.

ASTRONOMICAL LEAGUE. Post Office Box 12821, Tucson, AZ 85732. (602) 790-8471.

ASTRONOMICAL SOCIETY OF THE PACIFIC. 1290 24th Avenue, San Francisco, CA 94122. (415) 661-8660.

PLANETARY SOCIETY. 65 North Catalina Avenue, Pasadena, CA 91106-2301. (818) 793-5100.

BIBLIOGRAPHIES

A BIBLIOGRAPHY OF ASTRONOMY, 1970-1979. R.A. Seal and S.S. Martin. Libraries Unlimited, Incorporated, Littleton, CO 80160. 1982. $37.50.

SCIENTIFIC AND TECHNICAL BOOKS AND SERIALS IN PRINT 1988: AN INDEX TO LITERATURE IN SCIENCE AND TECHNOLOGY. R.R. Bowker Company, 205 East 42nd Street, New York, NY 10017. (800) 521-8110 or (212) 916-1600. $175.00.

DIRECTORIES AND BIOGRAPHICAL SOURCES

AMERICAN ASTRONOMICAL SOCIETY MEMBERS. 2000 Florida Avenue, NW, Suite 300, Washington, DC 20009. (202) 659-0134. Annual. Available to members only.

THE BIOGRAPHICAL DICTIONARY OF SCIENTISTS: ASTRONOMERS. D. Abbott, editor. Peter Bedrick Books, 125 East 23rd Street, New York, NY 10010. 1984. $24.95.

DIRECTORY OF PHYSICS AND ASTRONOMY STAFF MEMBERS. American Institute of Physics, 335 East 45th Street, New York, NY 10017. Annual.

GENERAL WORKS

URANUS AND NEPTUNE. J. Bergstrahl. NASA Conference Publication, 2330. Available from National Technical Information Service, 5285 Port Royla Road, Springfield, VA 22161. (703) 487-4838. 1984.

ONLINE DATA BASES

CA SEARCH. Chemical Abstracts Service, Post Office Box 3012, Columbus, OH 43210. Guide to chemical literature, 1972 to present. Inquire as to cost and availability.

DISSERTATION ABSTRACTS ONLINE. University Microfilms International, 300 North Zeeb Road, Ann Arbor, MI 48106. (800) 521-0600 or (313) 761-4700. Scope includes virtually all doctoral dissertations accepted at accredited American institutions from 1861 to present in 252 subject areas. Inquire as to cost and availability.

NASA. National Aeronautics and Space Administration, Scientific and Technical Information Branch, 300 7th Street, SW, Washington, DC 20546. Citations and abstracts of aerospace literature, 1962 to present. Inquire as to cost and availability.

NTIS. National Technical Information Service, 5285 Port Royal Road, Springfield, VA 22161. (703) 487-4630. Broad coverage of government sponsored research reports, 1964 to present. Inquire as to cost and availability.

SCISEARCH. Institute for Scientific Information, 3501 Market Street, Philadelphia, PA 19104. (800) 523-1850 or (215) 386-0100. Broad multidisciplinary title and author index to the international literature of science and technology, 1974 to present. Inquire as to cost and availability.

PERIODICALS

ASSOCIATION OF LUNAR AND PLANETARY OBSERVERS. Journal. Association of Lunar and Planetary Observers, Box

16131, San Francisco, CA 94116. (415) 566-5786. 1947 to present. Quarterly. $15.00 per year.

ASTRONOMICAL JOURNAL. American Astronomical Society. Available from: American Institute of Physics, 335 East 45th Street, New York, NY 10017. (212) 661-9404. $125.00 per year.

EARTH, MOON AND PLANETS: AN INTERNATIONAL JOURNAL OF COMPARATIVE PLANETOLOGY. D. Reidel Publishing Company, 190 Old Derby Street, Hingham, MA 02043. Nine times per year. $275.00 per year.

ICARUS: INTERNATIONAL JOURNAL OF THE SOLAR SYSTEM STUDIES. Academic Press, Incorporated, Orlando, FL 32887. (305) 345-4100. Monthly. $484.00 per year.

MERCURY. Astronomical Society of the Pacific, 1290 24th Avenue, San Francisco, CA 94122. (415) 661-8660. Bimonthly. $21.00 per year.

PLANETARY AND SPACE SCIENCE. Pergamon Press, Incorporated, Maxwell House, Fairview Park, Elmsford, NY 10523. (914) 592-7700. Monthly. $430.00 per year.

PLANETARY REPORT. Planetary Society, 65 North Catalina Avenue, Pasadena, CA 91106-2301. (818) 793-5100. 1980 to present. Bimonthly. $20.00 per year.

SOLAR SYSTEM RESEARCH (ENGLISH TRANSLATION OF ASTRONOMICHESKII VESTNIK). Consultants Bureau, 233 Spring Street, New York, NY 10013. (212) 620-8000. 1967 to present. Quarterly. $425.00 per year.

RESEARCH CENTERS AND INSTITUTES

LABORATORY FOR PLANETARY ATMOSPHERES RESEARCH. State University of New York at Stony Brook, Stony Brook, NY 11794-2300. (516) 632-8321.

LABORATORY FOR PLANETARY STUDIES. Cornell University, 302 Space Sciences Building, Ithaca, NY 14853. (607) 256-4971.

LUNAR AND PLANETARY INSTITUTE. 3303 NASA Road One, Houston, TX 77058. (713) 486-2139.

LUNAR AND PLANETARY LABORATORY. University of Arizona, Tucson, AZ 85721. (602) 621-6962.

NETWORKS (COMPUTER)

See also: COMPUTERS, COMPUTER COMMUNICATION, LOCAL AREA NETWORKS

ABSTRACT SERVICES AND INDEXES

APPLIED SCIENCE AND TECHNOLOGY INDEX. H.W. Wilson and Company, 950 University Avenue, Bronx, NY 10452. (800) 367-6670 or (212) 588-8400. Monthly. Inquire as to cost and availability.

COMPUMATH CITATION INDEX. Institute for Scientific Information, 3501 Market Street, Philadelphia, PA 19104. (800) 523-1850 or (215) 386-0100. Three times per year. $875.00 per year.

COMPUTER AND CONTROL ABSTRACTS. Institute of Electrical Engineers. Institute of Electrical and Electronics Engineers. IEEE Service Center, 445 Hoes Lane, Piscataway, NJ 08854. Semi-monthly. $775.00 per year.

COMPUTER CONTENTS: THE BIWEEKLY COMPILATION OF TABLES OF CONTENTS FROM COMPUTER, ELECTRONIC AND TELECOMMUNICATIONS MAGAZINES, JOURNALS AND TRANSACTIONS. Find/SVP, 500 Fifth Avenue, New York, NY 10110. (800) 346-3787 or (212) 354-2424. Biweekly. $115.00 per year.

COMPUTING REVIEWS. Association of Computing Machinery, 11 West 42nd Street, New York, NY 10036. (212) 869-7440. Monthly. $60.00 per year.

ENGINEERING INDEX MONTHLY AND AUTHOR INDEX. Engineering Information Inc., 345 East 47th Street, New York, NY 10017. (212) 705-7600. Monthly. $1560.00 per year.

GOVERNMENT REPORTS ANNOUNCEMENT AND INDEX. National Technical Information Service, 5285 Port Royal Road, Springfield, VA 22161. (703) 487-4929. Summaries of United States government sponsored research reports. 1946 to present. Biweekly. Inquire as to cost and availability.

SCIENCE CITATION INDEX. Institute for Scientific Information, 3501 Market Street, Philadelphia, PA 19104. (800) 523-1850 or (215) 386-0100. Six times per year. $6200.00 per year.

ANNUAL REVIEWS AND YEARBOOKS

ADVANCES IN COMPUTERS. Academic Press, Inc., 6277 Sea Harbor Drive, Orlando, FL 32821. (800) 321-5068. Annual. Approximately $50.00 per volume.

ASSOCIATIONS AND PROFESSIONAL SOCIETIES

AMERICAN FEDERATION OF INFORMATION PROCESSING SOCIETIES. 1899 Preston White Drive, Reston, VA 22091. (703) 620-8900.

ASSOCIATION OF COMPUTING MACHINERY (ACM). 11 West 42nd Street, New York, NY 10036. (212) 869-7440.

IEEE COMPUTER SOCIETY. 1730 Massachusetts Avenue, N.W., Washington, DC 20036. (202) 371-0101.

INSTITUTE OF ELECTRICAL AND ELECTRONIC ENGINEERS (IEEE). 345 East 47th Street, New York, NY 10017. (212) 705-7900.

DIRECTORIES AND BIOGRAPHICAL SOURCES

AMERICAN MEN AND WOMEN OF SCIENCE. R.R. Bowker, Inc., Order Department, 245 West 17th Street, New York, NY 10011. (800) 521-8110. Eight volumes. 1986. $595.00 for set.

COMPUTER PERIPHERALS REVIEW. GML Information Services, 594 Marrett Road, Lexington, MA 02173. (617) 861-0515. Directory of computer peripheral equipment manufacturers. Two issues per year. $215.00 per year.

DATAPRO REPORTS ON DATA COMMUNICATIONS. DataPro Research Corporation, 1805 Underwood Boulevard, Delran, NJ 08075. Three base volumes, with monthly updates. $780.00.

1987 DIRECTORY OF ENGINEERING SOCIETIES AND RELATED ORGANIZATIONS. Gordon Davis, editor. Hemisphere Publishing Corporation, 1010 Vermont Avenue, NW, Washington, DC 20005. (800) 526-0275. 12th edition. 1987. $100.00.

RESEARCH CENTERS DIRECTORY 1988. Gale Research Company, Book Tower, Detroit, MI 48226. (800) 521-0707. 12th edition. 1987. $365.00 for set.

SCIENTIFIC AND TECHNICAL ORGANIZATIONS AND AGENCIES DIRECTORY. Margaret Labash Young, editor. Gale Research Company, Book Tower, Detroit, MI 48226. (800) 521-0707. 2nd edition. 1987. $185.00.

WHO'S WHO IN ENGINEERING. Gordon Davis, editor. Hemisphere Publishing Corporation, 1010 Vermont Avenue, NW, Washington, DC 20005. (800) 526-0275. 6th edition. 1985. $200.00.

ENCYCLOPEDIAS AND DICTIONARIES

COMPUTER AND TELECOMMUNICATIONS ACROYMS. Julie E. Towell and Helen E. Sheppard, editors. Gale Research Company, Book Tower, Detroit, MI 48226. (800) 521-0707. 1986. $60.00.

ENCYCLOPEDIA OF INFORMATION SYSTEMS AND SERVICES 1988. Amy Lucas and Annette Novallo, editors. Gale Research Company, Book Tower, Detroit, MI 48226. (800) 521-0707. 8th edition. 1987. $400.00 for set.

PRENTICE-HALL ENCYCLOPEDIA OF INFORMATION TECHNOLOGY. Robert A. Edmunds. Prentice-Hall Publishing, Inc., Englewood Cliffs, NJ 07632. (800) 562-0245. 1987. $49.95.

GENERAL WORKS

DATA COMMUNICATIONS AND TELEPROCESSING SYSTEMS. Trevor Housley. Prentice-Hall Publishing, Inc., Englewood Cliffs, NJ 07632. (800) 562-0245. 1987. $38.95.

DATA COMMUNICATIONS NETWORKING DEVICES: CHARACTERISTICS, OPERATIONS, APPLICATIONS. Gilbert Held. John Wiley and Sons, Inc., 605 Third Avenue, New York, NY 10158. (800) 526-5368. 1986. $29.95

INTRODUCTION TO DATA COMMUNICATIONS AND COMPUTER NETWORKS. Fred Halsall. Addison-Wesley Publishing Company, Inc., 1 Jacob Way, Reading, MA 01867. (617) 944-3700. 1986. $31.95.

LOCAL AREA NETWORKS: ISSUES, PRODUCTS, AND DEVELOPMENTS. V.E. Cheong and R.A. Hirschheim. John Wiley and Sons, Inc., 605 Third Avenue, New York, NY 10158. (800) 526-5368. 1983. $34.95.

A PRACTICAL GUIDE TO COMPUTER COMMUNICATIONS AND NETWORKING. R.J. Deasington. John Wiley and Sons, Inc., 605 Third Avenue, New York, NY 10158. (800) 526-5368. 1984. $26.95 in paper.

PRINCIPLES OF COMPUTER COMMUNICATION NETWORK DESIGN. J. Seider. John Wiley and Sons, Inc., 605 Third Avenue, New York, NY 10158. (800) 526-5368. 1983. $101.95.

HANDBOOKS AND MANUALS

HANDBOOK OF DATA COMMUNICATION AND COMPUTER NETWORKS. Dimitris N. Chorafas. Petrocelli Books, 251 Wall Street, Princeton, NJ 08540. (609) 924-5851. 1985. $59.95.

REFERENCE MANUAL FOR TELECOMMUNICATIONS ENGINEERING. Roger L. Freeman. John Wiley and Sons, Inc., 605 Third Avenue, New York, NY 10158. (800) 526-5368. 1985. $88.95.

ONLINE DATA BASES

COMPENDEX. Engineering Information, Inc., 345 East 47th Street, New York, NY 10017. (800) 221-1044 or (212) 705-7615. Engineering and technical literature, 1975 to present. Inquire as to online cost and availability.

DISSERTATION ABSTRACTS ONLINE. University Microfilms International, 300 North Zeeb Road, Ann Arbor, MI 48106. (800) 521-0600 or (313) 761-4700. Scope includes virtually all doctoral dissertations accepted at accredited American institutions from 1861 to present in over 250 subject areas. Inquire as to online cost and availability.

NTIS. National Technical Information Service, 5285 Port Royal Road, Springfield, VA 22161. (703) 487-4630. Broad coverage of government sponsored research reports, 1964 to present. Inquire as to online cost and availability.

SCISEARCH. Institute for Scientific Information, 3501 Market Street, Philadelphia, PA 19104. (800) 523-1850 or (215) 386-0100. Broad multidisciplinary title and author index to the international literature of science and technology, 1974 to present. Inquire as to online cost and availability.

WILSONLINE. H.W. Wilson and Company, 950 University Avenue, Bronx, NY 10452. (800) 367-6770 or (212) 588-8400. Makes available online versions of the H.W. Wilson indexes including Applied Science and Technology Index, Business Periodicals Index and Readers' Guide to Periodical Literature. Approximately 1980 to present. Inquire as to online cost and availability.

PERIODICALS

BYTE; THE SMALL SYSTEMS JOURNAL. McGraw-Hill, Inc., 70 Main Street, Peterborough, NH 03458. (603) 924-9281. Subscription to: Box 590, Martinsville, NJ 08836. 1975 to present. Monthly. $21.00 per year.

COMPUTER COMMUNICATIONS. Butterworth's Publishing, 80 Montvale Avenue, Stoneham, MA 02180. (800) 325-4177. Bimonthly. $169.00 per year.

DATA COMMUNICATIONS. McGraw-Hill Book Company, 1221 Avenue of the Americas, New York, NY 10020. (212) 512-2000. Monthly. $30.00 per year.

DATAMATION. Technical Publishing Company, 875 Third Avenue, New York, NY 10022. (212) 605-9400. Semi-monthly. $50.00 per year.

JOURNAL OF TELECOMMUNICATION NETWORKS. Computer Science Press, Inc., 1803 Research Boulevard, Suite 500, Rockville, MD 20850-3155. (301) 251-9050. Quarterly. $100.00 per year.

LOCAL AREA NETWORKS (L.A.N. NEWSLETTER). Information Gatekeepers, Inc., 214 Harvard Avenue, Boston, MA 02134. (617) 232-3111. Monthly. $250.00 per year.

MICRO COMMUNICATIONS. Miller Freeman Publications, Inc., 500 Howard Street, San Francisco, CA 94105. (415) 397-1881. Bimonthly. $28.00 per year.

RESEARCH CENTERS AND INSTITUTES

COMPUTER COMMUNICATION NETWORK GROUP. University of Waterloo, CPH 2369A, Waterloo, ON, Canada N2L 3G1. (519) 885-1211.

DATABASE SYSTEMS RESEARCH CENTER. University of Maryland, College of Business and Management, Tydings Hall, College Park, MD 20742. (301) 454-6258.

SRI INTERNATIONAL CENTER FOR INTELLIGENT COMPUTER SYSTEMS. 333 Ravenswood Avenue, Menlo Park, CA 94025. (415) 859-4771.

SPECIFICATIONS AND STANDARDS

INSTITUTE OF ELECTRICAL AND ELECTRONICS ENGINEERS. LOCAL AREA NETWORKS: TOKEN RING ACCESS METHOD AND PHYSICAL LAYER SPECIFICATION - STANDARD 802.5. John Wiley and Sons, Inc., 605 Third Avenue, New York, NY 10158. (800) 526-5368. 1985. $19.95.

INSTITUTE OF ELECTRICAL AND ELECTRONICS ENGINEERS. LOCAL AREA NETWORKS: STANDARD LOGICAL LINK CONTROL: LOCAL AREA NETWORKS - STANDARD 802.2. John Wiley and Sons, Inc., 605 Third Avenue, New York, NY 10158. (800) 526-5368. 1985. $19.95.

NEUTRINOS

See: PARTICLE PHYSICS

NEUTRON OPTICS

See: OPTICS

NEUTRON SPECTROSCOPY

See: SPECTROSCOPY

NEUTRON STAR

See: STARS

NMR

See: NUCLEAR MAGNETIC RESONANCE

NOISE CONTROL

See also: ACOUSTICS, SOUND ENGINEERING, VIBRATION

ABSTRACT SERVICES AND INDEXES

APPLIED SCIENCE AND TECHNOLOGY INDEX. H.W. Wilson and Company, 950 University Avenue, Bronx, NY 10452. (800) 367-6670 or (212) 588-8400. Monthly. Inquire as to cost and availability.

CURRENT CONTENTS: ENGINEERING, TECHNOLOGY AND APPLIED SCIENCES. Institute for Scientific Information, 3501 Market Street, Philadelphia, PA 19104. (800) 523-1850 or (215) 386-0100. Weekly. $275.00 per year.

ENGINEERING INDEX MONTHLY AND AUTHOR INDEX. Engineering Information Inc., 345 East 47th Street, New York, NY 10017. (212) 705-7600. Monthly. $1560.00 per year.

POLLUTION ABSTRACTS. Cambridge Scientific Abstracts, Inc., 5161 River Road, Bethesda, MD 20816. (301) 951-1400. Six times per year. $550.00 per year.

ASSOCIATIONS AND PROFESSIONAL SOCIETIES

ACOUSTICAL SOCIETY OF AMERICA. 335 East 45th Street, New York, NY 10017. (212) 661-9404.

INSTITUTE OF NOISE CONTROL ENGINEERING. P.O. Box 3206, Arlington Branch, Poughkeepsie, NY 12603.

NOISE CONTROL PRODUCTS AND MATERIALS ASSOCIATION. 2506 Gross Point Road, Evanston, IL 60201. (312) 475-7300.

NOISE (NATIONAL ORGANIZATION TO INSURE A SOUND-CONTROLLED ENVIRONMENT). 1620 Eye Street, N.W., Suite 300, Washington, DC 20006. (202) 429-0166.

DIRECTORIES AND BIOGRAPHICAL SOURCES

1987 DIRECTORY OF ENGINEERING SOCIETIES AND RELATED ORGANIZATIONS. Gordon Davis, editor. Hemisphere Publishing Corporation, 1010 Vermont Avenue, NW, Washington, DC 20005. (800) 526-0275. 12th edition. 1987. $100.00.

RESEARCH CENTERS DIRECTORY 1988. Gale Research Company, Book Tower, Detroit, MI 48226. (800) 521-0707. 12th edition. 1987. $365.00 for set.

SCIENTIFIC AND TECHNICAL ORGANIZATIONS AND AGENCIES DIRECTORY. Margaret Labash Young, editor. Gale Research Company, Book Tower, Detroit, MI 48226. (800) 521-0707. 2nd edition. 1987. $185.00.

WHO'S WHO IN ENGINEERING. Gordon Davis, editor. Hemisphere Publishing Corporation, 79 Madison Avenue, New York, NY 10016-7892. (800) 821-8312. Sixth edition. 1985. $200.00.

GENERAL WORKS

NOISE AND VIBRATION IN BUILDINGS. Robert S. Jones. McGraw-Hill Book Company, 1221 Avenue of the Americas, New York, NY 10020. (212) 512-2000. 1984. $47.50.

NOISE CONTROL. M.J. Crocker, editor. Van Nostrand Reinhold Company, Inc., 135 West 50th Street, New York, NY 10020. (800) 543-2681. 1984. $66.95.

HANDBOOKS AND MANUALS

NOISE CONTROL: HANDBOOK OF PRINCIPLES AND PRACTICES. David M. Libscomb and Arthur C. Taylor, editors. Van Nostrand Reinhold Company, Inc., 135 West 50th Street, New York, NY 10020. (800) 543-2681. 1978. $37.95.

ONLINE DATA BASES

COMPENDEX. Engineering Information, Inc., 345 East 47th Street, New York, NY 10017. (800) 221-1044 or (212) 705-7615. Engineering and technical literature, 1975 to present. Inquire as to online cost and availability.

ENVIROLINE. EIC Intelligence, Inc., 48 West 38th Street, New York, NY 10018. (212) 944-8500. Worldwide environmental literature. 1970 to present. Inquire as to online cost and availability.

NTIS. National Technical Information Service, 5285 Port Royal Road, Springfield, VA 22161. (703) 487-4630. Broad coverage of government sponsored research reports, 1964 to present. Inquire as to online cost and availability.

WILSONLINE. H.W. Wilson and Company, 950 University Avenue, Bronx, NY 10452. (800) 367-6770 or (212) 588-8400. Makes available online versions of the H.W. Wilson indexes including Applied Science and Technology Index, Business Periodicals Index and Readers' Guide to Periodical Literature. Approximately 1980 to present. Inquire as to online cost and availability.

PERIODICALS

NOISE CONTROL ENGINEERING JOURNAL. Institute of Noise Control Engineering, Department of Mechanical Engineering, Auburn University, Auburn, AL 36849-3501. (205) 826-4820. Subscribe to: Box 3206, Arlington Branch, Poughkeepsie, NY 12603. 1973 to present. Bimonthly. $40.00 per year.

NOISE CONTROL REPORT. Business Publishers, Inc., 951 Pershing Drive, Silver Spring, MD 20910. (301) 587-6300. 1971 to present. Every two weeks. $150.00 per year.

NOISE/NEWS. Noise Control Foundation, Box 3206, Arlington Branch, Poughkeepsie, NY 12603. 1972 to present. Bimonthly.

RESEARCH CENTERS AND INSTITUTES

ACOUSTICS AND VIBRATION LABORATORY. University of Hartford, College of Engineering, West Hartford, CT 06117. (203) 243-4614.

CENTER FOR CONTROL AND VIBRATION. North Carolina State University, Campus Box 7910, Raleigh, NC 27695. (919) 737-3024.

NOISE CONTROL LABORATORY. Pennsylvannia State University, 30 Hammond Building, University Park, PA 16802. (814) 865-2761.

NONDESTRUCTIVE TESTING

See also: ALLOYS, FERROALLOYS, MATERIALS SCIENCE, METALLURGICAL ENGINEERING, METALLURGY, METALS AND METALWORKING

ABSTRACT SERVICES AND INDEXES

APPLIED MECHANICS REVIEW. American Society of Mechanical Engineers, 345 East 47th Street, New York, NY 10017. (212) 705-7703. 1948 to present. Monthly. $360.00 per year.

APPLIED SCIENCE AND TECHNOLOGY INDEX. H.W. Wilson and Company, 950 University Avenue, Bronx, NY 10452. (800) 367-6670 or (212) 588-8400. Monthly. Inquire as to cost and availability.

CERAMIC ABSTRACTS. American Ceramic Society, Inc., 65 Ceramic Drive, Columbus, OH 43214. (614) 268-8645. 1922 to present. Bimonthly. $185 per year.

CHEMICAL ABSTRACTS. American Chemical Society, Chemical Abstracts Service, Box 3012, Columbus, OH 43210. (614) 421-3600. 1907 to present. Weekly. $9500.00 per year.

CURRENT CONTENTS: ENGINEERING, TECHNOLOGY AND APPLIED SCIENCES. Institute for Scientific Information, 3501 Market Street, Philadelphia, PA 19104. (800) 523-1850 or (215) 386-0100. Weekly. $275.00 per year.

ENGINEERING INDEX MONTHLY AND AUTHOR INDEX. Engineering Information Inc., 345 East 47th Street, New York, NY 10017. (212) 705-7600. Monthly. $1560.00 per year.

INTERNATIONAL AEROSPACE ABSTRACTS. American Institute of Aeronautics and Astronautics, Technical Information Service, 555 West 57th Street, New York, NY 10019. (212) 247-6500. 1961 to present. Semi-monthly. $700.00 per year.

ISMEC BULLETIN (Information Service in Mechanical Engineering). Cambridge Scientific Abstracts, 5161 River Road, Bethesda, MD 20816. (301) 951-1400. 1973 to present. Monthly. $450.00 per year.

METALS ABSTRACTS AND METALS ABSTRACTS INDEX. American Society for Metals, Metals Park, OH 44073. (216) 338-5151. 1968 to present. Monthly. Abstracts are $1100.00 per year and Index is $460.00 per year.

PHYSICS ABSTRACTS. Institution of Electrical Engineers. Available from: IEEE Service Center, 445 Hoes Lane, Piscataway, NJ 08854. 1898 to prEsent. Bimonthly. $1700.00 per year.

SCIENCE CITATION INDEX. Institute for Scientific Information, 3501 Market Street, Philadelphia, PA 19104. (800) 523-1850 or (215) 386-0100. Six times per year. $6200. 00 per year.

WORLD ALUMINUM ABSTRACTS. Aluminum Association, 818 Connecticut Avenue, NW, Washington, DC 20006. (202) 862-5156. 1968 to present. Monthly. $240.00 per year.

ASSOCIATIONS AND PROFESSIONAL SOCIETIES

AMERICAN CERAMIC SOCIETY, Inc., 65 Ceramic Drive, Columbus, OH 43214. (614) 268-8645. 1922 to present. Bimonthly. $185 per year.

AMERICAN POWDER METALLURGY INSTITUTE. 105 College Road, East, Princeton, NJ 08540. (609) 452-7700.

AMERICAN SOCIETY FOR METALS. Metals Park, OH 44073. (216) 338-5151.

AMERICAN SOCIETY FOR NONDESTRUCTIVE TESTING. 4153 Arlingate Plaza, Caller #28518, Columbus, OH 43228. (614) 274-6003.

AMERICAN SOCIETY FOR TESTING AND MATERIALS. 1916 Race Street, Philadelphia, PA 19103. (215) 299-5400.

AMERICAN SOCIETY OF MECHANICAL ENGINEERS. 345 47th Street, New York, NY 10017. (212) 705-7722.

MATERIALS PROPERTIES COUNCIL. 345 East 47th Street, New York, NY 10017. (212) 705-7693.

MATERIALS RESEARCH SOCIETY. 9800 McKnight Road, Suite 327, Pittsburgh, PA 15237. (412) 367-3003.

THE METALLURGICAL SOCIETY. 420 Commonwealth Drive, Warrendale, PA 15086. (412) 776-9000.

NATIONAL ASSOCIATION OF CORROSION ENGINEERS. Box 218340, Houston, TX 77218. (713) 492-0535.

NATIONAL MATERIALS ADVISORY BOARD. 2101 Constitution Avenue, NW, Washington, DC 20418. (202) 334-3505.

SOCIETY FOR THE ADVANCEMENT OF MATERIAL AND PROCESS ENGINEERING (SAMPE). P.O. Box 2459, 843 West Glentana Street, Covina, CA 91722. (818) 331-0616.

DIRECTORIES AND BIOGRAPHICAL SOURCES

MATERIALS DIRECTORY. Cahners Publishing Company, 275 Washington Street, Newton, MA 02158. Annual. $15.00.

1987 DIRECTORY OF ENGINEERING SOCIETIES AND RELATED ORGANIZATIONS. Gordon Davis, editor. Hemisphere Publishing Corporation, 1010 Vermont Avenue, NW, Washington, DC 20005. (800) 526-0275. 12th edition. 1987. $100.00.

RESEARCH CENTERS DIRECTORY 1988. Gale Research Company, Book Tower, Detroit, MI 48226. (800) 521-0707. 12th edition. 1987. $365.00 for set.

SCIENTIFIC AND TECHNICAL ORGANIZATIONS AND AGENCIES DIRECTORY. Margaret Labash Young, editor. Gale Research Company, Book Tower, Detroit, MI 48226. (800) 521-0707. 2nd edition. 1987. $185.00.

WHO'S WHO IN ENGINEERING. Gordon Davis, editor. Hemisphere Publishing Corporation, 1010 Vermont Avenue, NW, Washington, DC 20005. (800) 526-0275. 6th edition. 1985. $200.00.

ENCYCLOPEDIAS AND DICTIONARIES

THESAURUS OF SCIENTIFIC, TECHNICAL, AND ENGINEERING TERMS. Hemisphere Publishing Corporation, 1010 Vermont Avenue, NW, Washington, DC 20005. (800) 526-0275. 1988. $125.00.

GENERAL WORKS

ENGINEERING MATERIALS: AN INTRODUCTION TO THEIR PROPERTIES AND APPLICATIONS. Michael F. Ashby and David R.H. Jones. Pergamon Press, Inc., Maxwell House,

Fairview Park, Elmsford, NY 10523. (914) 592-7700. 1980. $91.50.

INTRODUCTION TO MATERIALS SCIENCE FOR ENGINEERS. James F. Shackelford. Macmillan Publishing Company, Inc., 866 Third Avenue, New York, NY 10022. (800) 257-5755. 2nd edition. 1988. Inquire.

MATERIALS AND PROCESSES FOR NDT TECHNOLOGY. Harry D. Moore, editor. American Society for Nondestructive Testing. 4153 Arlingate Plaza, Caller #28518, Columbus, OH 43228. (614) 274-6003. 1984. $30.75.

PRINCIPLES OF MATERIALS SCIENCE AND ENGINEERING. William F. Smith. McGraw-Hill Book Company, 1221 Avenue of the Americas, New York, NY 10020. (212) 512-2000. 1986. $45.95.

STATISTICS AND MECHANICS OF MATERIALS. Braja M. Das and Paul C. Hassler. Prentice-Hall Publishing, Inc., Englewood Cliffs, NJ 07632. (800) 562-0245. 1988. $42.50.

STATISTICS AND STRENGTH OF MATERIALS. Karl K. Stevens. Prentice-Hall Publishing, Inc., Englewood Cliffs, NJ 07632. (800) 562-0245. 1987. $39.95.

ULTRASONIC SPECTRAL ANALYSIS FOR NONDESTRUCTIVE EVALUATION. Dale Fitting and Laszlo Adler. Plenum Publishing Corporation, 233 Spring Street, New York, NY 10013. (800) 221-9369. 1981. $59.50.

HANDBOOKS AND MANUALS

MATERIALS HANDBOOK. George S. Brady and Henry R. Clauser. McGraw-Hill Book Company, 1221 Avenue of the Americas, New York, NY 10020. (212) 512-2000. 12th edition. 1986. $59.50.

ONLINE DATA BASES

CA SEARCH. Chemical Abstracts Service, P.O. Box 3012, Columbus, OH 43120. (800) 848-6538 or (614) 421-3600. Comprehensive guide to chemical literature, 1972 to present. Inquire as to online cost and availability.

COMPENDEX. Engineering Information, Inc., 345 East 47th Street, New York, NY 10017. (800) 221-1044 or (212) 705-7615. Engineering and technical literature, 1975 to present. Inquire as to online cost and availability.

INSPEC. INSPEC Marketing Department, Institution of Electrical Engineers. Available from IEEE Service Center, 445 Hoes Lane, Piscataway, NJ 08854. (201) 981-0060. Online version of Physics Abstracts. Inquire as to online cost and availability.

ISMEC. Cambridge Scientific Abstracts, 5161 River Road, Besthda, MD 20816. (800) 638-8076 or (301) 951-1400. Literature of mechanical and production engineering, 1973 to present. Inquire as to online cost and availability.

NTIS. National Technical Information Service, 5285 Port Royal Road, Springfield, VA 22161. (703) 487-4630. Broad coverage of government sponsored research reports, 1964 to present. Inquire as to online cost and availability.

SCISEARCH. Institute for Scientific Information, 3501 Market Street, Philadelphia, PA 19104. (800) 523-1850 or (215) 386-0100. Broad multidisciplinary title and author index to the international literature of science and technology, 1974 to present. Inquire as to online cost and availability.

WILSONLINE. H.W. Wilson and Company, 950 University Avenue, Bronx, NY 10452. (800) 367-6770 or (212) 588-8400. Makes available online versions of the H.W. Wilson indexes including Applied Science and Technology Index, Business Periodicals Index and Readers' Guide to Periodical Literature. Approximately 1980 to present. Inquire as to online cost and availability.

PERIODICALS

JOURNAL OF ENGINEERING FOR INDUSTRY. American Society of Mechanical Engineers, 345 East 47th Street, New York, NY 10017. (212) 705-7703. 1970 to present. Quarterly. $100.00 per year.

JOURNAL OF MATERIALS RESEARCH. Materials Research Society. 9800 McKnight Road, Suite 327, Pittsburgh, PA 15237. (412) 367-3003.

JOURNAL OF METALS. American Institute of Mining, Metallurgical, and Petroleum Engineers, Inc., Metallurgical Society, 420 Commonwealth Drive, Warrendale, PA 15086. (412) 776-9086. 1949 to present. Monthly. $40.00 per year.

JOURNAL OF NONDESTRUCTIVE EVALUATION. Plenum Publishing Corporation, 233 Spring Street, New York, NY 10013. (800) 221-9369. 1980 to present. Quarterly. $85.00 per year.

JOURNAL OF TESTING AND EVALUATION. American Society for Testing and Materials. 1916 Race Street, Philadelphia, PA 19103. (215) 299-5400. 1966 to present. Bimonthly. $40.00 per year.

MATERIALS ENGINEERING. Penton-IPC, 1100 Superior Avenue, Cleveland, OH 44114. (216)696-7000. 1929 to present. Monthly. $40.00.

MATERIALS EVALUATION. American Society for Nondestructive Testing, 4153 Arlingate Plaza, Caller #28518, Columbus, OH 43228. (614) 274-6003. 1942 to present. Monthly. $50.00 per year.

MATERIALS PERFORMANCE. National Association of Corrosion Engineers, P.O. Box 218340, Houston, TX 77218. (713) 492-0535. 1962 to present. Monthly. $50.00 per year.

SAMPE JOURNAL. Society for the Advancement of Material and Process Engineering (SAMPE), P.O. Box 2459, 843 West Glentana Street, Covina, CA 91722. (818) 331-0616. 1965 to present. Bimonthly. $31.00 per year.

RESEARCH CENTERS AND INSTITUTES

CENTER FOR MATERIALS SCIENCE AND ENGINEERING. Massachusetts Institute of Technology, Building 13, Room 2090, 77 Massachusetts Avenue, Cambridge, MA 02139. (617) 253-6801.

CENTER FOR RESEARCH IN MATERIALS SCIENCE AND ENGINEERING. University of Texas at Austin, ETC 9.104, Austin, TX 78712. (512) 471-1504.

MATERIALS RESEARCH LABORATORY. Ohio State University, 3063 Alpheus Smith Laboratory, 174 West 18th Avenue, Columbus, OH 43210. (614) 422-5190.

MATERIALS RESEARCH LABORATORY. University of Illinois, Urbana, IL 61801. (217) 333-1371.

MATERIALS SCIENCE CENTER. Cornell University, Clark Hall of Science, Ithaca, NY 14853. (607) 255-4272.

NUCLEAR CHEMISTRY

See also: NUCLEAR ENERGY, NUCLEAR ENGINEERING

ABSTRACT SERVICES AND INDEXES

APPLIED SCIENCE AND TECHNOLOGY INDEX. H.W. Wilson and Company, 950 University Avenue, Bronx, NY 10452. (800) 367-6670 or (212) 588-8400. Monthly. Inquire as to cost and availability.

CHEMICAL ABSTRACTS. American Chemical Society, Chemical Abstracts Service, Box 3012, Columbus, OH 43210. (614) 421-3600. 1907 to present. Weekly. $9500.00 per year.

CURRENT CONTENTS: ENGINEERING, TECHNOLOGY AND APPLIED SCIENCES. Institute for Scientific Information, 3501 Market Street, Philadelphia, PA 19104. (800) 523-1850 or (215) 386-0100. Weekly. $275.00 per year.

ENGINEERING INDEX MONTHLY AND AUTHOR INDEX. Engineering Information Inc., 345 East 47th Street, New York, NY 10017. (212) 705-7600. Monthly. $1560.00 per year.

INDEX TO SCIENTIFIC AND TECHNICAL PROCEEDINGS. Institute for Scientific Information, 3501 Market Street, Philadelphia, PA 19104. (800) 523-1850 or (215) 386-0100. 1978 to present. Monthly. $775.00 per year.

PHYSICS ABSTRACTS. Institution of Electrical Engineers. Available from: IEEE Service Center, 445 Hoes Lane, Piscataway, NJ 08854. 1898 to present. Bimonthly. $1700.00 per year.

SCIENCE CITATION INDEX. Institute for Scientific Information, 3501 Market Street, Philadelphia, PA 19104. (800) 523-1850 or (215) 386-0100. Six times per year. $6200.00 per year.

ASSOCIATIONS AND PROFESSIONAL SOCIETIES

AMERICAN CHEMICAL SOCIETY. 1155 16th Street, N.W., Washington, DC 20036. (202) 872-4600.

AMERICAN INSTITUTE OF CHEMICAL ENGINEERS. 345 East 47th Street, New York, NY 10017. (212) 705-7338.

AMERICAN INSTITUTE OF PHYSICS. 335 East 45th Street, New York, NY 10017. (212) 661-9404.

AMERICAN NUCLEAR SOCIETY. 555 North Kensington Avenue, La Grange, IL 60525. (312) 352-6611.

FUSION ENERGY FOUNDATION. P.O. Box 17149, Washington, DC 20041. (703) 689-2490.

BIBLIOGRAPHIES

NEW TECHNICAL BOOKS: A SELECTIVE LIST WITH DESCRIPTIVE ANNOTATIONS. New York Public Library, Science and Technology Research Center, Fifth Avenue and 42nd Street, New York, NY 10018. (212) 930-0800. 1915 to present. Monthly. $15.00 per year.

SCIENCE BOOKS AND FILMS. American Association for the Advancement of Science, 1333 H Street, NW, Washington, DC 20005. (202) 326-6454. Five times per year. $20.00 per year.

SCIENTIFIC AND TECHNICAL BOOKS AND SERIALS IN PRINT 1988; AN INDEX TO LITERATURE IN SCIENCE AND TECHNOLOGY. R.R. Bowker Company, 205 East 42nd Street, New York, NY 10017. (800) 521-8110. $175.00.

DIRECTORIES AND BIOGRAPHICAL SOURCES

ENERGY INFORMATION CENTERS DIRECTORY. Public Affairs and Information Program, Atomic Industrial Forum, 7101 Wisconsin Avenue, Bethesda, MD 20814. (301) 654-9260. 1985. Free.

INTERNATIONAL DIRECTORY OF NUCLEAR UTILITIES. Lotte, Limited, Box 237, Contract Station 27, Lakewood, CO 80215. (303) 232-3026. Annual. $160.00.

INTERNATIONAL RESEARCH CENTERS DIRECTORY 1988-89. Darren L. Smith, editor. Gale Research Company, Book Tower, Detroit, MI 48226. (800) 521-0707. 4th edition. 1987. $360.00.

1987 DIRECTORY OF ENGINEERING SOCIETIES AND RELATED ORGANIZATIONS. Gordon Davis, editor. Hemisphere Publishing Corporation, 1010 Vermont Avenue, NW, Washington, DC 20005. (800) 526-0275. 12th edition. 1987. $100.00.

NUCLEAR REACTORS BUILT, BEING BUILT, OR PLANNED IN THE UNITED STATES. Office of Scientific and Technical Information, Department of Energy, Box 62, Oak Ridge, TN 37831. (615) 576-5637. Annual. $11.00. Send orders to: National Technical Information Service, Springfield, VA 22161.

RESEARCH CENTERS DIRECTORY 1988. Gale Research Company, Book Tower, Detroit, MI 48226. (800) 521-0707. 12th edition. 1987. $365.00 for set.

SCIENTIFIC AND TECHNICAL ORGANIZATIONS AND AGENCIES DIRECTORY. Margaret Labash Young, editor. Gale Research Company, Book Tower, Detroit, MI 48226. (800) 521-0707. 2nd edition. 1987. $185.00.

WHO'S WHO IN ENGINEERING. Gordon Davis, editor. Hemisphere Publishing Corporation, 1010 Vermont Avenue, NW, Washington, DC 20005. (800) 526-0275. 6th edition. 1985. $200.00.

ENCYCLOPEDIAS AND DICTIONARIES

THESAURUS OF SCIENTIFIC, TECHNICAL, AND ENGINEERING TERMS. Hemisphere Publishing Corporation, 1010 Vermont Avenue, NW, Washington, DC 20005. (800) 526-0275. 1988. $125.00.

GENERAL WORKS

ESSENTIALS OF NUCLEAR CHEMISTRY. H.J. Arnikar. John Wiley and Sons, Inc., 605 Third Avenue, New York, NY 10158. (800) 526-5368. 1982. $21.95.

NUCLEAR AND RADIOCHEMISTRY. G. Friedlander and others. John Wiley and Sons, Inc., 605 Third Avenue, New York, NY 10158. (800) 526-5368. 3rd edition. 1981. $34.50 in paper.

NUCLEAR CHEMICAL ENGINEERING. Manson Benedict and others. McGraw-Hill Book Company, 1221 Avenue of the Americas, New York, NY 10020. (212) 512-2000. 2nd edition. 1981. $52.95.

NUCLEAR MATERIALS AND APPLICATIONS. Benjamin M. Ma. Van Nostrand Reinhold Company, Inc., 135 West 50th Street, New York, NY 10020. (800) 543-2681. 1982. $45.95.

HANDBOOKS AND MANUALS

CRC HANDBOOK OF CHEMISTRY AND PHYSICS. Robert C. Weast, editor. CRC Press, 2000 Corporate Boulevard, Boca Raton, FL 33431. (800) 272-7737. 68th edition. 1987. $69.95.

CRC HANDBOOK OF RADIATION MEASUREMENT AND PROTECTION. Volume 1: Physical Science and Engineering Data. Allen Brodsky, editor. CRC Press, 2000 Corporate Boulevard, Boca Raton, FL 33431. (800) 272-7737. 1979. $112.50.

NUCLEAR ENGINEERING DATA BASES, STANDARDS AND NUMERICAL ANALYSIS. Jack Jedruch. Van Nostrand Reinhold Company, Inc., 135 West 50th Street, New York, NY 10020. (800) 543-2681. 1985. $59.95.

ONLINE DATA BASES

CA SEARCH. Chemical Abstracts Service, P.O. Box 3012, Columbus, OH 43120. (800) 848-6538 or (614) 421-3600. Comprehensive guide to chemical literature, 1972 to present. Inquire as to online cost and availability.

COMPENDEX. Engineering Information, Inc., 345 East 47th Street, New York, NY 10017. (800) 221-1044 or (212) 705-7615. Engineering and technical literature, 1975 to present. Inquire as to online cost and availability.

DISSERTATION ABSTRACTS ONLINE. University Microfilms International, 300 North Zeeb Road, Ann Arbor, MI 48106. (800) 521-0600 or (313) 761-4700. Scope includes virtually all doctoral dissertations accepted at accredited American institutions from 1861 to present in over 250 subject areas. Inquire as to online cost and availability.

DOE ENERGY DATA BASE. U.S. Department of Energy, Office of Scientific and Technical Information, P.O. Box 62, Oak Ridge, TN 37831. (615) 576-6837. A database that covers all aspects of energy including the science and technology of energy. 1948 to present. Available through the DIALOG search service or DOE/RECON. Inquire as to online cost and availability.

INSPEC. INSPEC Marketing Department, Institution of Electrical Engineers. Available from IEEE Service Center, 445 Hoes Lane, Piscataway, NJ 08854. (201) 981-0060. Online version of Physics Abstracts. Inquire as to online cost and availability.

NTIS. National Technical Information Service, 5285 Port Royal Road, Springfield, VA 22161. (703) 487-4630. Broad coverage of government sponsored research reports, 1964 to present. Inquire as to online cost and availability.

SCISEARCH. Institute for Scientific Information, 3501 Market Street, Philadelphia, PA 19104. (800) 523-1850 or (215) 386-0100. Broad multidisciplinary title and author index to the international literature of science and technology, 1974 to present. Inquire as to online cost and availability.

WILSONLINE. H.W. Wilson and Company, 950 University Avenue, Bronx, NY 10452. (800) 367-6770 or (212) 588-8400. Makes available online versions of the H.W. Wilson indexes including Applied Science and Technology Index, Business Periodicals Index and Readers' Guide to Periodical Literature. Approximately 1980 to present. Inquire as to online cost and availability.

PERIODICALS

AMERICAN NUCLEAR SOCIETY TRANSACTIONS. American Nuclear Society, 555 North Kensington Avenue, La Grange, IL 60525. (312) 352-6611. 1958 to present. Semiannual. $255.00 per year.

ANNALS OF NUCLEAR ENERGY. Pergamon Press, Inc., Maxwell House, Fairview Park, Elmsford, NY 10523. (914) 592-7700. 1954 to present. Monthly. $280.00 per year.

PROGRESS IN NUCLEAR ENERGY. Pergamon Press, Inc., Maxwell House, Fairview Park, Elmsford, NY 10523. (914) 592-7700. 1977 to present. Bimonthly. $250.00 per year.

RESEARCH CENTERS AND INSTITUTES

CHEMISTRY LABORATORIES. Rensselaer Polytechnic Institute, Cogswell Laboratory, Troy, NY 13181. (518) 266-8462.

LABORATORY OF BASIC AND APPLIED NUCLEAR RESEARCH. University of Cincinnati, Department of Chemistry, Cincinnati, OH 45221. (513) 475-3652.

NUCLEAR RESEARCH LABORATORY. Florida State University, Tallahassee, FL 32306. (904) 644-6584.

NUCLEAR ENERGY

See also: NUCLEAR ENGINEERING

ABSTRACT SERVICES AND INDEXES

APPLIED MECHANICS REVIEW. American Society of Mechanical Engineers, 345 East 47th Street, New York, NY 10017. (212) 705-7703. 1948 to present. Monthly. $360.00 per year.

APPLIED SCIENCE AND TECHNOLOGY INDEX. H.W. Wilson and Company, 950 University Avenue, Bronx, NY 10452. (800) 367-6670 or (212) 588-8400. Monthly. Inquire as to cost and availability.

CHEMICAL ABSTRACTS. American Chemical Society, Chemical Abstracts Service, Box 3012, Columbus, OH 43210. (614) 421-3600. 1907 to present. Weekly. $9500.00 per year.

CURRENT CONTENTS: ENGINEERING, TECHNOLOGY AND APPLIED SCIENCES. Institute for Scientific Information, 3501 Market Street, Philadelphia, PA 19104. (800) 523-1850 or (215) 386-0100. Weekly. $275.00 per year.

ENGINEERING INDEX MONTHLY AND AUTHOR INDEX. Engineering Information Inc., 345 East 47th Street, New York, NY 10017. (212) 705-7600. Monthly. $1560.00 per year.

ISMEC BULLETIN (Information Service in Mechanical Engineering). Cambridge Scientific Abstracts, 5161 River Road, Bethesda, MD 20816. (301) 951-1400. 1973 to present. Monthly. $450.00 per year.

INDEX TO SCIENTIFIC AND TECHNICAL PROCEEDINGS. Institute for Scientific Information, 3501 Market Street, Philadelphia, PA 19104. (800) 523-1850 or (215) 386-0100. 1978 to present. Monthly. $775.00 per year.

PHYSICS ABSTRACTS. Institution of Electrical Engineers. Available from: IEEE Service Center, 445 Hoes Lane, Piscataway, NJ 08854. 1898 to present. Bimonthly. $1700.00 per year.

SCIENCE CITATION INDEX. Institute for Scientific Information, 3501 Market Street, Philadelphia, PA 19104. (800) 523-1850 or (215) 386-0100. Six times per year. $6200.00 per year.

ANNUAL REVIEWS AND YEARBOOKS

ADVANCES IN NUCLEAR SCIENCE AND TECHNOLOGY. J.Lewins, editor. Plenum Publishing Corporation, 233 Spring Street, New York, NY 10013. (800) 221-9369. 1977 to present. Irregular. Inquire as to cost and availability.

ASSOCIATIONS AND PROFESSIONAL SOCIETIES

AMERICAN NUCLEAR SOCIETY. 555 North Kensington Avenue, La Grange, IL 60525. (312) 352-6611.

FUSION ENERGY FOUNDATION. P.O. Box 17149, Washington, DC 20041. (703) 689-2490.

INSTITUTE OF NUCLEAR MATERIALS MANAGEMENT. 60 Revere Drive, Northbrook, IL 60062. (312) 480-9080.

BIBLIOGRAPHIES

NEW TECHNICAL BOOKS: A SELECTIVE LIST WITH DESCRIPTIVE ANNOTATIONS. New York Public Library, Science and Technology Research Center, Fifth Avenue and 42nd Street, New York, NY 10018. (212) 930-0800. 1915 to present. Monthly. $15.00 per year.

SCIENCE BOOKS AND FILMS. American Association for the Advancement of Science, 1333 H Street, NW, Washington, DC 20005. (202) 326-6454. Five times per year. $20.00 per year.

SCIENTIFIC AND TECHNICAL BOOKS AND SERIALS IN PRINT 1988; AN INDEX TO LITERATURE IN SCIENCE AND TECHNOLOGY. R.R. Bowker Company, 205 East 42nd Street, New York, NY 10017. (800) 521-8110. $175.00.

DIRECTORIES AND BIOGRAPHICAL SOURCES

ENERGY INFORMATION CENTERS DIRECTORY. Public Affairs and Information Program, Atomic Industrial Forum, 7101 Wisconsin Avenue, Bethesda, MD 20814. (301) 654-9260. 1985. Free.

INTERNATIONAL DIRECTORY OF NUCLEAR UTILITIES. Lotte, Limited, Box 237, Contract Station 27, Lakewood, CO 80215. (303) 232-3026. Annual. $160.00.

INTERNATIONAL RESEARCH CENTERS DIRECTORY 1988-89. Darren L. Smith, editor. Gale Research Company, Book Tower, Detroit, MI 48226. (800) 521-0707. 4th edition. 1987. $360.00.

1987 DIRECTORY OF ENGINEERING SOCIETIES AND RELATED ORGANIZATIONS. Gordon Davis, editor. Hemisphere Publishing Corporation, 1010 Vermont Avenue, NW, Washington, DC 20005. (800) 526-0275. 12th edition. 1987. $100.00.

NUCLEAR REACTORS BUILT, BEING BUILT, OR PLANNED IN THE UNITED STATES. Office of Scientific and Technical Information, Department of Energy, Box 62, Oak Ridge, TN 37831. (615) 576-5637. Annual. $11.00. Send orders to: National Technical Information Service, Springfield, VA 22161.

RESEARCH CENTERS DIRECTORY 1988. Gale Research Company, Book Tower, Detroit, MI 48226. (800) 521-0707. 12th edition. 1987. $365.00 for set.

SCIENTIFIC AND TECHNICAL ORGANIZATIONS AND AGENCIES DIRECTORY. Margaret Labash Young, editor. Gale Research Company, Book Tower, Detroit, MI 48226. (800) 521-0707. 2nd edition. 1987. $185.00.

WHO'S WHO IN ENGINEERING. Gordon Davis, editor. Hemisphere Publishing Corporation, 1010 Vermont Avenue, NW, Washington, DC 20005. (800) 526-0275. 6th edition. 1985. $200.00.

ENCYCLOPEDIAS AND DICTIONARIES

THESAURUS OF SCIENTIFIC, TECHNICAL, AND ENGINEERING TERMS. Hemisphere Publishing Corporation, 1010 Vermont Avenue, NW, Washington, DC 20005. (800) 526-0275. 1988. $125.00.

GENERAL WORKS

INTRODUCTION TO NUCLEAR POWER. John G. Collier. Hemisphere Publishing Corporation, 1010 Vermont Avenue, NW, Washington, DC 20005. (800) 526-0275. 1987. $49.95.

NUCLEAR ENERGY TECHNOLOGY. Ronald Allen Knief. Hemisphere Publishing Corporation, 1010 Vermont Avenue, NW, Washington, DC 20005. (800) 526-0275. 1983. $48.00.

NUCLEAR FISSION REACTORS. I.R. Cameron. Plenum Publishing Corporation, 233 Spring Street, New York, NY 10013. (800) 221-9369. 1982. $49.50.

NUCLEAR PHYSICS FOR ENGINEERS AND SCIENTISTS: LOW ENERGY THEORY WITH APPLICATIONS INCLUDING REACTORS AND THEIR ENVIRONMENTAL IMPACT. S.E. Hunt. John Wiley and Sons, Inc., 605 Third Avenue, New York, NY 10158. (800) 526-5368. 1987. $129.95.

NUCLEAR REACTOR ENGINEERING. Samuel Glasstone. Van Nostrand Reinhold Company, Inc., 135 West 50th Street, New York, NY 10020. (800) 543-2681. 1980. $49.95.

NUCLEAR MATERIALS AND APPLICATIONS. Benjamin M. Ma. Van Nostrand Reinhold Company, Inc., 135 West 50th Street, New York, NY 10020. (800) 543-2681. 1982. $45.95.

HANDBOOKS AND MANUALS

CRC HANDBOOK OF CHEMISTRY AND PHYSICS. Robert C. Weast, editor. CRC Press, 2000 Corporate Boulevard, Boca Raton, FL 33431. (800) 272-7737. 68th edition. 1987. $69.95.

A GUIDE TO NUCLEAR POWER TECHNOLOGY: A RESOURCE FOR DECISION MAKING. F.J. Rahn and others. John Wiley and Sons, Inc., 605 Third Avenue, New York, NY 10158. (800) 526-5368. 1984. $85.95.

NUCLEAR ENGINEERING DATA BASES, STANDARDS AND NUMERICAL ANALYSIS. Jack Jedruch. Van Nostrand Reinhold Company, Inc., 135 West 50th Street, New York, NY 10020. (800) 543-2681. 1985. $59.95.

ONLINE DATA BASES

CA SEARCH. Chemical Abstracts Service, P.O. Box 3012, Columbus, OH 43120. (800) 848-6538 or (614) 421-3600. Comprehensive guide to chemical literature, 1972 to present. Inquire as to online cost and availability.

COMPENDEX. Engineering Information, Inc., 345 East 47th Street, New York, NY 10017. (800) 221-1044 or (212) 705-7615. Engineering and technical literature, 1975 to present. Inquire as to online cost and availability.

DISSERTATION ABSTRACTS ONLINE. University Microfilms International, 300 North Zeeb Road, Ann Arbor, MI 48106. (800) 521-0600 or (313) 761-4700. Scope includes virtually all doctoral dissertations accepted at accredited American institutions from 1861 to present in over 250 subject areas. Inquire as to online cost and availability.

DOE ENERGY DATA BASE. U.S. Department of Energy, Office of Scientific and Technical Information, P.O. Box 62, Oak Ridge, TN 37831. (615) 576-6837. A database that covers all aspects of energy including the science and technology of energy. 1948 to present. Available through the DIALOG search service or DOE/RECON. Inquire as to online cost and availability.

ENERGYLINE. EIC/Intelligence, Inc., 48 West 38th Street, New York, NY 10018. (212) 944-8500. A database of resources on the scientific, engineering, political, and socioeconomic aspects of energy resources. 1976 to present. Inquire as to online cost and availability.

INSPEC. INSPEC Marketing Department, Institution of Electrical Engineers. Available from IEEE Service Center, 445 Hoes Lane, Piscataway, NJ 08854. (201) 981-0060. Online version of Physics Abstracts. Inquire as to online cost and availability.

NTIS. National Technical Information Service, 5285 Port Royal Road, Springfield, VA 22161. (703) 487-4630. Broad coverage of government sponsored research reports, 1964 to present. Inquire as to online cost and availability.

SCISEARCH. Institute for Scientific Information, 3501 Market Street, Philadelphia, PA 19104. (800) 523-1850 or (215) 386-0100. Broad multidisciplinary title and author index to the international literature of science and technology, 1974 to present. Inquire as to online cost and availability.

WILSONLINE. H.W. Wilson and Company, 950 University Avenue, Bronx, NY 10452. (800) 367-6770 or (212) 588-8400. Makes available online versions of the H.W. Wilson indexes including Applied Science and Technology Index, Business Periodicals Index and Readers' Guide to Periodical Literature. Approximately 1980 to present. Inquire as to online cost and availability.

PERIODICALS

AMERICAN NUCLEAR SOCIETY TRANSACTIONS. American Nuclear Society, 555 North Kensington Avenue, La Grange, IL 60525. (312) 352-6611. 1958 to present. Semiannual. $255.00 per year.

ANNALS OF NUCLEAR ENERGY. Pergamon Press, Inc., Maxwell House, Fairview Park, Elmsford, NY 10523. (914) 592-7700. 1954 to present. Monthly. $280.00 per year.

BULLETIN OF THE ATOMIC SCIENTISTS. Educational Foundation for Nuclear Science, 5801 South Kenwood Avenue, Chicago, IL 60637. (312) 363-5225. 1945 to present. Ten times per year. $22.50 per year.

FUSION TECHNOLOGY. American Nuclear Society, 555 North Kensington Avenue, La Grange, IL 60525. (312) 352-6611. 1981 to present. Bimonthly. $250.00 per year.

JOURNAL OF FUSION ENERGY. Plenum Publishing Corporation, 233 Spring Street, New York, NY 10013. (800) 221-9369. 1981 to present. Bimonthly. $105.00 per year.

NUCLEAR ENGINEER. Institution of Nuclear Engineers, 1 Penerley Road, Nondon SE6 2LQ, England. 1959 to present. Bimonthly. $110.00 per year.

NUCLEAR ENGINEERING AND DESIGN. Elsevier Science Publishing Company, Inc., 52 Vanderbilt Avenue, New York, NY 10017. (212) 370-5520. 1965 to present. $160.00 per year.

NUCLEAR ENGINEERING INTERNATIONAL. Electrical-Electronic Press, Quadrant House, The Quadrant, Sutton, Surrey, SM2 5AS, England. 1956 to present. Monthly. $210.00 per year.

NUCLEAR SCIENCE AND ENGINEERING. American Nuclear Society, 555 North Kensington Avenue, La Grange, IL 60525. (312) 352-6611. 1956 to present. Monthly. $220.00 per year.

NUCLEAR TECHNOLOGY. American Nuclear Society, 555 North Kensington Avenue, La Grange, IL 60525. (312) 352-6611. 1965 to present. Monthly. $345.00 per year.

RESEARCH CENTERS AND INSTITUTES

ASSISTANT SECRETARY FOR NUCLEAR ENERGY. U.S. Department of Energy, 1000 Independence Avenue, SW, Washington, DC 20585. (202) 252-6450.

DEPARTMENT OF NUCLEAR ENGINEERING AND ENGINEERING PHYSICS. University of Wisconsin - Madison, 1500 Johnson Drive, Madison, WI 53706. (608) 263-1648.

INSTITUTE OF NUCLEAR SCIENCE AND ENGINEERING. Oregon State University, Radiation Center, 35th and Jefferson Streets, Corvallis, OR 97331. (503) 754-2341.

LABORATORY OF BASIC AND APPLIED NUCLEAR RESEARCH. University of Cincinnati, Department of Chemistry, Cincinnati, OH 45221. (513) 475-3652.

WHITESHELL NUCLEAR RESEARCH ESTABLISHMENT. Research Company, Atomic Energy of Canada Limited, Pinawa, MB, Canada ROE 1LO. (204) 753-2311.

SPECIFICATIONS AND STANDARDS

INFORMATION CENTER ON NUCLEAR STANDARDS. American Nuclear Society, 555 North Kensington Avenue, La Grange, IL 60525. (312) 352-6611. Standards for all aspects of the design, construction, operation, and maintenance of nuclear power plants.

NUCLEAR ENGINEERING

See also: NUCLEAR CHEMISTRY, NUCLEAR ENERGY, NUCLEAR REACTORS

ABSTRACT SERVICES AND INDEXES

APPLIED MECHANICS REVIEW. American Society of Mechanical Engineers, 345 East 47th Street, New York, NY 10017. (212) 705-7703. 1948 to present. Monthly. $360.00 per year.

APPLIED SCIENCE AND TECHNOLOGY INDEX. H.W. Wilson and Company, 950 University Avenue, Bronx, NY 10452. (800) 367-6670 or (212) 588-8400. Monthly. Inquire as to cost and availability.

CHEMICAL ABSTRACTS. American Chemical Society, Chemical Abstracts Service, Box 3012, Columbus, OH 43210. (614) 421-3600. 1907 to present. Weekly. $9500.00 per year.

CURRENT CONTENTS: ENGINEERING, TECHNOLOGY AND APPLIED SCIENCES. Institute for Scientific Information, 3501 Market Street, Philadelphia, PA 19104. (800) 523-1850 or (215) 386-0100. Weekly. $275.00 per year.

ENGINEERING INDEX MONTHLY AND AUTHOR INDEX. Engineering Information Inc., 345 East 47th Street, New York, NY 10017. (212) 705-7600. Monthly. $1560.00 per year.

ISMEC BULLETIN (Information Service in Mechanical Engineering). Cambridge Scientific Abstracts, 5161 River Road, Bethesda, MD 20816. (301) 951-1400. 1973 to present. Monthly. $450.00 per year.

INDEX TO SCIENTIFIC AND TECHNICAL PROCEEDINGS. Institute for Scientific Information, 3501 Market Street, Philadelphia, PA 19104. (800) 523-1850 or (215) 386-0100. 1978 to present. Monthly. $775.00 per year.

PHYSICS ABSTRACTS. Institution of Electrical Engineers. Available from: IEEE Service Center, 445 Hoes Lane, Piscataway, NJ 08854. 1898 to present. Bimonthly. $1700.00 per year.

SCIENCE CITATION INDEX. Institute for Scientific Information, 3501 Market Street, Philadelphia, PA 19104. (800) 523-1850 or (215) 386-0100. Six times per year. $6200.00 per year.

ANNUAL REVIEWS AND YEARBOOKS

ADVANCES IN NUCLEAR SCIENCE AND TECHNOLOGY. J. Lewins, editor. Plenum Publishing Corporation, 233 Spring Street, New York, NY 10013. (800) 221-9369. 1977 to present. Irregular. Inquire as to cost and availability.

ASSOCIATIONS AND PROFESSIONAL SOCIETIES

AMERICAN NUCLEAR SOCIETY. 555 North Kensington Avenue, La Grange, IL 60525. (312) 352-6611.

FUSION ENERGY FOUNDATION. P.O. Box 17149, Washington, DC 20041. (703) 689-2490.

INSTITUTE OF NUCLEAR MATERIALS MANAGEMENT. 60 Revere Drive, Northbrook, IL 60062. (312) 480-9080.

BIBLIOGRAPHIES

SCIENTIFIC AND TECHNICAL BOOKS AND SERIALS IN PRINT 1988; AN INDEX TO LITERATURE IN SCIENCE AND TECHNOLOGY. R.R. Bowker Company, 205 East 42nd Street, New York, NY 10017. (800) 521-8110. $175.00.

DIRECTORIES AND BIOGRAPHICAL SOURCES

ENERGY INFORMATION CENTERS DIRECTORY. Public Affairs and Information Program, Atomic Industrial Forum, 7101 Wisconsin Avenue, Bethesda, MD 20814. (301) 654-9260. 1985. Free.

INTERNATIONAL DIRECTORY OF NUCLEAR UTILITIES. Lotte, Limited, Box 237, Contract Station 27, Lakewood, CO 80215. (303) 232-3026. Annual. $160.00.

INTERNATIONAL RESEARCH CENTERS DIRECTORY 1988-89. Darren L. Smith, editor. Gale Research Company, Book Tower, Detroit, MI 48226. (800) 521-0707. 4th edition. 1987. $360.00.

1987 DIRECTORY OF ENGINEERING SOCIETIES AND RELATED ORGANIZATIONS. Gordon Davis, editor. Hemisphere Publishing Corporation, 1010 Vermont Avenue, NW, Washington, DC 20005. (800) 526-0275. 12th edition. 1987. $100.00.

NUCLEAR REACTORS BUILT, BEING BUILT, OR PLANNED IN THE UNITED STATES. Office of Scientific and Technical Information, Department of Energy, Box 62, Oak Ridge, TN 37831. (615) 576-5637. Annual. $11.00. Send orders to: National Technical Information Service, Springfield, VA 22161.

RESEARCH CENTERS DIRECTORY 1988. Gale Research Company, Book Tower, Detroit, MI 48226. (800) 521-0707. 12th edition. 1987. $365.00 for set.

SCIENTIFIC AND TECHNICAL ORGANIZATIONS AND AGENCIES DIRECTORY. Margaret Labash Young, editor. Gale Research Company, Book Tower, Detroit, MI 48226. (800) 521-0707. 2nd edition. 1987. $185.00.

WHO'S WHO IN ENGINEERING. Gordon Davis, editor. Hemisphere Publishing Corporation, 1010 Vermont Avenue, NW, Washington, DC 20005. (800) 526-0275. 6th edition. 1985. $200.00.

GENERAL WORKS

INTRODUCTION TO NUCLEAR POWER. John G. Collier. Hemisphere Publishing Corporation, 1010 Vermont Avenue, NW, Washington, DC 20005. (800) 526-0275. 1987. $49.95.

NUCLEAR ENERGY TECHNOLOGY. Ronald Allen Knief. Hemisphere Publishing Corporation, 1010 Vermont Avenue, NW, Washington, DC 20005. (800) 526-0275. 1983. $48.00.

NUCLEAR FISSION REACTORS. I.R. Cameron. Plenum Publishing Corporation, 233 Spring Street, New York, NY 10013. (800) 221-9369. 1982. $49.50.

NUCLEAR PHYSICS FOR ENGINEERS AND SCIENTISTS: LOW ENERGY THEORY WITH APPLICATIONS INCLUDING REACTORS AND THEIR ENVIRONMENTAL IMPACT. S.E. Hunt. John Wiley and Sons, Inc., 605 Third Avenue, New York, NY 10158. (800) 526-5368. 1987. $129.95.

NUCLEAR REACTOR ENGINEERING. Samuel Glasstone. Van Nostrand Reinhold Company, Inc., 135 West 50th Street, New York, NY 10020. (800) 543-2681. 1980. $49.95.

NUCLEAR MATERIALS AND APPLICATIONS. Benjamin M. Ma. Van Nostrand Reinhold Company, Inc., 135 West 50th Street, New York, NY 10020. (800) 543-2681. 1982. $45.95.

HANDBOOKS AND MANUALS

CRC HANDBOOK OF NUCLEAR REACTORS CALCULATIONS. CRC Press, 2000 Corporate Boulevard, Boca Raton, FL 33431. (800) 272-7737. Three volumes. 1986. $750.00 for set.

A GUIDE TO NUCLEAR POWER TECHNOLOGY: A RESOURCE FOR DECISION MAKING. F.J. Rahn and others. John Wiley and Sons, 605 Third Avenue, New York, NY 10158. (800) 526-5368. 1984. $85.95.

NUCLEAR ENGINEERING DATA BASES, STANDARDS AND NUMERICAL ANALYSIS. Jack Jedruch. Van Nostrand Reinhold Company, Inc., 135 West 50th Street, New York, NY 10020. (800) 543-2681. 1985. $59.95.

ONLINE DATA BASES

CA SEARCH. Chemical Abstracts Service, P.O. Box 3012, Columbus, OH 43120. (800) 848-6538 or (614) 421-3600. Comprehensive guide to chemical literature, 1972 to present. Inquire as to online cost and availability.

COMPENDEX. Engineering Information, Inc., 345 East 47th Street, New York, NY 10017. (800) 221-1044 or (212) 705-7615. Engineering and technical literature, 1975 to present. Inquire as to online cost and availability.

DISSERTATION ABSTRACTS ONLINE. University Microfilms International, 300 North Zeeb Road, Ann Arbor, MI 48106. (800) 521-0600 or (313) 761-4700. Scope includes virtually all doctoral dissertations accepted at accredited American institutions from 1861 to present in over 250 subject areas. Inquire as to online cost and availability.

DOE ENERGY DATA BASE. U.S. Department of Energy, Office of Scientific and Technical Information, P.O. Box 62, Oak Ridge, TN 37831. (615) 576-6837. A database that covers all aspects of energy including the science and technology of energy. 1948 to present. Available through the DIALOG search service or DOE/RECON. Inquire as to online cost and availability.

ENERGYLINE. EIC/Intelligence, Inc., 48 West 38th Street, New York, NY 10018. (212) 944-8500. A database of resources on the scientific, engineering, political, and socioeconomic aspects of energy resources. 1976 to present. Inquire as to online cost and availability.

INSPEC. INSPEC Marketing Department, Institution of Electrical Engineers. Available from IEEE Service Center, 445 Hoes Lane, Piscataway, NJ 08854. (201) 981-0060. Online version of Physics Abstracts. Inquire as to online cost and availability.

NTIS. National Technical Information Service, 5285 Port Royal Road, Springfield, VA 22161. (703) 487-4630. Broad coverage of government sponsored research reports, 1964 to present. Inquire as to online cost and availability.

SCISEARCH. Institute for Scientific Information, 3501 Market Street, Philadelphia, PA 19104. (800) 523-1850 or (215) 386-0100. Broad multidisciplinary title and author index to the international literature of science and technology, 1974 to present. Inquire as to online cost and availability.

WILSONLINE. H.W. Wilson and Company, 950 University Avenue, Bronx, NY 10452. (800) 367-6770 or (212) 588-8400. Makes available online versions of the H.W. Wilson indexes including Applied Science and Technology Index, Business Periodicals Index and Readers' Guide to Periodical Literature. Approximately 1980 to present. Inquire as to online cost and availability.

PERIODICALS

AMERICAN NUCLEAR SOCIETY TRANSACTIONS. American Nuclear Society, 555 North Kensington Avenue, La Grange, IL 60525. (312) 352-6611. 1958 to present. Semiannual. $255.00 per year.

ANNALS OF NUCLEAR ENERGY. Pergamon Press, Inc., Maxwell House, Fairview Park, Elmsford, NY 10523. (914) 592-7700. 1954 to present. Monthly. $280.00 per year.

BULLETIN OF THE ATOMIC SCIENTISTS. Educational Foundation for Nuclear Science, 5801 South Kenwood Avenue, Chicago, IL 60637. (312) 363-5225. 1945 to present. Ten times per year. $22.50 per year.

FUSION TECHNOLOGY. American Nuclear Society, 555 North Kensington Avenue, La Grange, IL 60525. (312) 352-6611. 1981 to present. Bimonthly. $250.00 per year.

JOURNAL OF FUSION ENERGY. Plenum Publishing Corporation, 233 Spring Street, New York, NY 10013. (800) 221-9369. 1981 to present. Bimonthly. $105.00 per year.

NUCLEAR ENGINEER. Institution of Nuclear Engineers, 1 Penerley Road, Nondon SE6 2LQ, England. 1959 to present. Bimonthly. $110.00 per year.

NUCLEAR ENGINEERING AND DESIGN. Elsevier Science Publishing Company, Inc., 52 Vanderbilt Avenue, New York, NY 10017. (212) 370-5520. 1965 to present. $160.00 per year.

NUCLEAR ENGINEERING INTERNATIONAL. Electrical-Electronic Press, Quadrant House, The Quadrant, Sutton, Surrey, SM2 5AS, England. 1956 to present. Monthly. $210.00 per year.

NUCLEAR SCIENCE AND ENGINEERING. American Nuclear Society, 555 North Kensington Avenue, La Grange, IL 60525. (312) 352-6611. 1956 to present. Monthly. $220.00 per year.

NUCLEAR TECHNOLOGY. American Nuclear Society, 555 North Kensington Avenue, La Grange, IL 60525. (312) 352-6611. 1965 to present. Monthly. $345.00 per year.

RESEARCH CENTERS AND INSTITUTES

ASSISTANT SECRETARY FOR NUCLEAR ENERGY. U.S. Department of Energy, 1000 Independence Avenue, SW, Washington, DC 20585. (202) 252-6450.

DEPARTMENT OF NUCLEAR ENGINEERING AND ENGINEERING PHYSICS. University of Wisconsin - Madison, 1500 Johnson Drive, Madison, WI 53706. (608) 263-1648.

INSTITUTE OF NUCLEAR SCIENCE AND ENGINEERING. Oregon State University, Radiation Center, 35th and Jefferson Streets, Corvallis, OR 97331. (503) 754-2341.

LABORATORY OF BASIC AND APPLIED NUCLEAR RESEARCH. University of Cincinnati, Department of Chemistry, Cincinnati, OH 45221. (513) 475-3652.

WHITESHELL NUCLEAR RESEARCH ESTABLISHMENT. Research Company, Atomic Energy of Canada Limited, Pinawa, MB, Canada ROE 1LO. (204) 753-2311.

SPECIFICATIONS AND STANDARDS

INFORMATION CENTER ON NUCLEAR STANDARDS. American Nuclear Society, 555 North Kensington Avenue, La Grange, IL 60525. (312) 352-6611. Standards for all aspects of the design, construction, operation, and maintenance of nuclear power plants.

NUCLEAR FUELS

See: NUCLEAR ENERGY

NUCLEAR MAGNETIC RESONANCE

See also: NUCLEAR ENERGY, SPECTROSCOPY

ABSTRACT SERVICES AND INDEXES

APPLIED MECHANICS REVIEW. American Society of Mechanical Engineers, 345 East 47th Street, New York, NY 10017. (212) 705-7703. 1948 to present. Monthly. $360.00 per year.

APPLIED SCIENCE AND TECHNOLOGY INDEX. H.W. Wilson and Company, 950 University Avenue, Bronx, NY 10452. (800) 367-6670 or (212) 588-8400. Monthly. Inquire as to cost and availability.

CHEMICAL ABSTRACTS. American Chemical Society, Chemical Abstracts Service, Box 3012, Columbus, OH 43210. (614) 421-3600. 1907 to present. Weekly. $9500.00 per year.

CURRENT CONTENTS: ENGINEERING, TECHNOLOGY AND APPLIED SCIENCES. Institute for Scientific Information, 3501 Market Street, Philadelphia, PA 19104. (800) 523-1850 or (215) 386-0100. Weekly. $275.00 per year.

ENGINEERING INDEX MONTHLY AND AUTHOR INDEX. Engineering Information Inc., 345 East 47th Street, New York, NY 10017. (212) 705-7600. Monthly. $1560.00 per year.

INDEX TO SCIENTIFIC AND TECHNICAL PROCEEDINGS. Institute for Scientific Information, 3501 Market Street, Philadelphia, PA 19104. (800) 523-1850 or (215) 386-0100. 1978 to present. Monthly. $775.00 per year.

NUCLEAR MAGNETIC RESONANCE SPECTROMETRY ABSTRACTS. PRM Science and Technology Agency Limited, 261A Finchley Road, Hampstead, London, NW3 6LU, England. 1971 to present. Bimonthly. $125.00 per year.

PHYSICS ABSTRACTS. Institution of Electrical Engineers. Available from: IEEE Service Center, 445 Hoes Lane, Piscataway, NJ 08854. 1898 to present. Bimonthly. $1700.00 per year.

SCIENCE CITATION INDEX. Institute for Scientific Information, 3501 Market Street, Philadelphia, PA 19104. (800) 523-1850 or (215) 386-0100. Six times per year. $6200.00 per year.

ANNUAL REVIEWS AND YEARBOOKS

PROGRESS IN NUCLEAR MAGNETIC RESONANCE. J.W. Emsley and others, editors. Pergamon Press, Inc., Maxwell House, Fairview Park, Elmsford, NY 10523. (914) 592-7700. 1962-1985. Price varies, inquire.

ASSOCIATIONS AND PROFESSIONAL SOCIETIES

AMERICAN SOCIETY FOR MASS SPECTROSCOPY. P.O. Box 1508, East Lansing, MI 48823. (517) 337-2548.

SOCIETY FOR APPLIED SPECTROSCOPY. P.O. Box 1438, Frederick, MD 21701. (301) 694-8122.

BIBLIOGRAPHIES

NMR IMAGING: A COMPREHENSIVE BIBLIOGRAPHY. Addison-Wesley Publishing Company, Inc., 1 Jacob Way, Reading, MA 01867. (617) 944-3700. 1983. $42.95.

DIRECTORIES AND BIOGRAPHICAL SOURCES

INTERNATIONAL RESEARCH CENTERS DIRECTORY 1988-89. Darren L. Smith, editor. Gale Research Company, Book Tower, Detroit, MI 48226. (800) 521-0707. 4th edition. 1987. $360.00.

1987 DIRECTORY OF ENGINEERING SOCIETIES AND RELATED ORGANIZATIONS. Gordon Davis, editor. Hemisphere Publishing Corporation, 1010 Vermont Avenue, NW, Washington, DC 20005. (800) 526-0275. 12th edition. 1987. $100.00.

RESEARCH CENTERS DIRECTORY 1988. Gale Research Company, Book Tower, Detroit, MI 48226. (800) 521-0707. 12th edition. 1987. $365.00 for set.

SCIENTIFIC AND TECHNICAL ORGANIZATIONS AND AGENCIES DIRECTORY. Margaret Labash Young, editor. Gale Research Company, Book Tower, Detroit, MI 48226. (800) 521-0707. 2nd edition. 1987. $185.00.

WHO'S WHO IN ENGINEERING. Gordon Davis, editor. Hemisphere Publishing Corporation, 1010 Vermont Avenue, NW, Washington, DC 20005. (800) 526-0275. 6th edition. 1985. $200.00.

GENERAL WORKS

MODERN NMR SPECTROSCOPY. J.K.M. Sanders and B.K. Hunter. Oxford University Press, 200 Madison Avenue, New York, NY 10016. (800) 458-5833. 1987. $70.00.

NMR SPECTROSCOPY: AN INTRODUCTION. Harold Guenther. John Wiley and Sons, Inc., 605 Third Avenue, New York, NY 10158. (800) 526-5368. 1980. $32.95 in paper.

NUCLEAR MAGNETIC RESONANCE. Atta-ur-Rahman. Springer-Verlag New York, Inc., 175 Fifth Avenue, New York, NY 10010. (800) 526-7254. 1986. $49.00.

PRINCIPLES OF NUCLEAR MAGNETIC RESONANCE IN ONE AND TWO DIMENSIONS. Richard R. Ernst and others. Oxford University Press, 200 Madison Avenue, New York, NY 10016. (800) 458-5833. 1986. $120.00.

THEORY OF MAGNETIC RESONANCE. C.A. Poole and H.A. Farach. John Wiley and Sons, Inc., 605 Third Avenue, New York, NY 10158. (800) 526-5368. 2nd edition. 1987. $120.00.

HANDBOOKS AND MANUALS

CRC HANDBOOK OF CHEMISTRY AND PHYSICS. Robert C. Weast, editor. CRC Press, 2000 Corporate Boulevard, Boca Raton, FL 33431. (800) 272-7737. 68th edition. 1987. $69.95.

ONLINE DATA BASES

CA SEARCH. Chemical Abstracts Service, P.O. Box 3012, Columbus, OH 43120. (800) 848-6538 or (614) 421-3600. Comprehensive guide to chemical literature, 1972 to present. Inquire as to online cost and availability.

COMPENDEX. Engineering Information, Inc., 345 East 47th Street, New York, NY 10017. (800) 221-1044 or (212) 705-7615. Engineering and technical literature, 1975 to present. Inquire as to online cost and availability.

DISSERTATION ABSTRACTS ONLINE. University Microfilms International, 300 North Zeeb Road, Ann Arbor, MI 48106. (800) 521-0600 or (313) 761-4700. Scope includes virtually all doctoral dissertations accepted at accredited American institutions from 1861 to present in over 250 subject areas. Inquire as to online cost and availability.

INSPEC. INSPEC Marketing Department, Institution of Electrical Engineers. Available from IEEE Service Center, 445 Hoes Lane, Piscataway, NJ 08854. (201) 981-0060. Online version of Physics Abstracts. Inquire as to online cost and availability.

NTIS. National Technical Information Service, 5285 Port Royal Road, Springfield, VA 22161. (703) 487-4630. Broad coverage of government sponsored research reports, 1964 to present. Inquire as to online cost and availability.

SCISEARCH. Institute for Scientific Information, 3501 Market Street, Philadelphia, PA 19104. (800) 523-1850 or (215) 386-0100. Broad multidisciplinary title and author index to the international literature of science and technology, 1974 to present. Inquire as to online cost and availability.

PERIODICALS

APPLIED SPECTROSCOPY. Marcel Dekker Journals, 270 Madison Avenue, New York, NY 10016. (212) 696-9000. 1964 to present. Four times per year. $185.00 per year.

SPECTROSCOPY LETTERS. Marcel Dekker Journals, 270 Madison Avenue, New York, NY 10016. (212) 696-9000. 1968 to present. Ten times per year. $280.00 per year.

RESEARCH CENTERS AND INSTITUTES

NUCLEAR MAGNETIC RESONANCE FACILITY. University of Missouri - Columbia, Chemical Building, Columbia, MO 65211. (314) 882-7725.

SOUTH CAROLINA NUCLEAR MAGNETIC RESONANCE LABORATORY. University of South Carolina, Department of Chemistry, Columbia, SC 29208. (803) 777-7341.

SOUTHERN CALIFORNIA REGIONAL CENTER FOR NUCLEAR MAGNETIC RESONANCE SPECTROSCOPY. California Institute of Technology, Mail Code 164-30, Pasadena, CA 91125. (818) 356-6241.

NUCLEAR PHYSICS

See: NUCLEAR ENERGY

NUCLEAR POWER PLANTS

See: NUCLEAR ENERGY

NUCLEAR REACTORS

See also: NUCLEAR ENERGY, NUCLEAR ENGINEERING

ABSTRACT SERVICES AND INDEXES

APPLIED MECHANICS REVIEW. American Society of Mechanical Engineers, 345 East 47th Street, New York, NY 10017. (212) 705-7703. 1948 to present. Monthly. $360.00 per year.

APPLIED SCIENCE AND TECHNOLOGY INDEX. H.W. Wilson and Company, 950 University Avenue, Bronx, NY 10452. (800) 367-6670 or (212) 588-8400. Monthly. Inquire as to cost and availability.

CHEMICAL ABSTRACTS. American Chemical Society, Chemical Abstracts Service, Box 3012, Columbus, OH 43210. (614) 421-3600. 1907 to present. Weekly. $9500.00 per year.

CURRENT CONTENTS: ENGINEERING, TECHNOLOGY AND APPLIED SCIENCES. Institute for Scientific Information, 3501 Market Street, Philadelphia, PA 19104. (800) 523-1850 or (215) 386-0100. Weekly. $275.00 per year.

ENGINEERING INDEX MONTHLY AND AUTHOR INDEX. Engineering Information Inc., 345 East 47th Street, New York, NY 10017. (212) 705-7600. Monthly. $1560.00 per year.

ISMEC BULLETIN (Information Service in Mechanical Engineering). Cambridge Scientific Abstracts, 5161 River Road, Bethesda, MD 20816. (301) 951-1400. 1973 to present. Monthly. $450.00 per year.

PHYSICS ABSTRACTS. Institution of Electrical Engineers. Available from: IEEE Service Center, 445 Hoes Lane, Piscataway, NJ 08854. 1898 to present. Bimonthly. $1700.00 per year.

SCIENCE CITATION INDEX. Institute for Scientific Information, 3501 Market Street, Philadelphia, PA 19104. (800) 523-1850 or (215) 386-0100. Six times per year. $6200.00 per year.

ANNUAL REVIEWS AND YEARBOOKS

ADVANCES IN NUCLEAR SCIENCE AND TECHNOLOGY. J.Lewins, editor. Plenum Publishing Corporation, 233 Spring Street, New York, NY 10013. (800) 221-9369. 1977 to present. Irregular. Inquire as to cost and availability.

ASSOCIATIONS AND PROFESSIONAL SOCIETIES

AMERICAN NUCLEAR SOCIETY. 555 North Kensington Avenue, La Grange, IL 60525. (312) 352-6611.

FUSION ENERGY FOUNDATION. P.O. Box 17149, Washington, DC 20041. (703) 689-2490.

INSTITUTE OF NUCLEAR MATERIALS MANAGEMENT. 60 Revere Drive, Northbrook, IL 60062. (312) 480-9080.

BIBLIOGRAPHIES

SCIENTIFIC AND TECHNICAL BOOKS AND SERIALS IN PRINT 1988; AN INDEX TO LITERATURE IN SCIENCE AND TECHNOLOGY. R.R. Bowker Company, 205 East 42nd Street, New York, NY 10017. (800) 521-8110. $175.00.

DIRECTORIES AND BIOGRAPHICAL SOURCES

ENERGY INFORMATION CENTERS DIRECTORY. Public Affairs and Information Program, Atomic Industrial Forum, 7101 Wisconsin Avenue, Bethesda, MD 20814. (301) 654-9260. 1985. Free.

INTERNATIONAL DIRECTORY OF NUCLEAR UTILITIES. Lotte, Limited, Box 237, Contract Station 27, Lakewood, CO 80215. (303) 232-3026. Annual. $160.00.

INTERNATIONAL RESEARCH CENTERS DIRECTORY 1988-89. Darren L. Smith, editor. Gale Research Company, Book Tower, Detroit, MI 48226. (800) 521-0707. 4th edition. 1987. $360.00.

1987 DIRECTORY OF ENGINEERING SOCIETIES AND RELATED ORGANIZATIONS. Gordon Davis, editor. Hemisphere Publishing Corporation, 1010 Vermont Avenue, NW, Washington, DC 20005. (800) 526-0275. 12th edition. 1987. $100.00.

NUCLEAR REACTORS BUILT, BEING BUILT, OR PLANNED IN THE UNITED STATES. Office of Scientific and Technical Information, Department of Energy, Box 62, Oak Ridge, TN 37831. (615) 576-5637. Annual. $11.00. Send orders to: National Technical Information Service, Springfield, VA 22161.

RESEARCH CENTERS DIRECTORY 1988. Gale Research Company, Book Tower, Detroit, MI 48226. (800) 521-0707. 12th edition. 1987. $365.00 for set.

SCIENTIFIC AND TECHNICAL ORGANIZATIONS AND AGENCIES DIRECTORY. Margaret Labash Young, editor. Gale Research Company, Book Tower, Detroit, MI 48226. (800) 521-0707. 2nd edition. 1987. $185.00.

WHO'S WHO IN ENGINEERING. Gordon Davis, editor. Hemisphere Publishing Corporation, 1010 Vermont Avenue, NW, Washington, DC 20005. (800) 526-0275. 6th edition. 1985. $200.00.

ENCYCLOPEDIAS AND DICTIONARIES

THESAURUS OF SCIENTIFIC, TECHNICAL, AND ENGINEERING TERMS. Hemisphere Publishing Corporation, 1010 Vermont Avenue, NW, Washington, DC 20005. (800) 526-0275. 1988. $125.00.

GENERAL WORKS

INTRODUCTION TO NUCLEAR POWER. John G. Collier. Hemisphere Publishing Corporation, 1010 Vermont Avenue, NW, Washington, DC 20005. (800) 526-0275. 1987. $49.95.

NUCLEAR ENERGY TECHNOLOGY. Ronald Allen Knief. Hemisphere Publishing Corporation, 1010 Vermont Avenue, NW, Washington, DC 20005. (800) 526-0275. 1983. $48.00.

NUCLEAR FISSION REACTORS. I.R. Cameron. Plenum Publishing Corporation, 233 Spring Street, New York, NY 10013. (800) 221-9369. 1982. $49.50.

NUCLEAR PHYSICS FOR ENGINEERS AND SCIENTISTS: LOW ENERGY THEORY WITH APPLICATIONS INCLUDING REACTORS AND THEIR ENVIRONMENTAL IMPACT. S.E. Hunt. John Wiley and Sons, Inc., 605 Third Avenue, New York, NY 10158. (800) 526-5368. 1987. $129.95.

NUCLEAR REACTOR ENGINEERING. Samuel Glasstone. Van Nostrand Reinhold Company, Inc., 135 West 50th Street, New York, NY 10020. (800) 543-2681. 1980. $49.95.

NUCLEAR MATERIALS AND APPLICATIONS. Benjamin M. Ma. Van Nostrand Reinhold Company, Inc., 135 West 50th Street, New York, NY 10020. (800) 543-2681. 1982. $45.95.

HANDBOOKS AND MANUALS

CRC HANDBOOK OF NUCLEAR REACTORS CALCULATIONS. CRC Press, 2000 Corporate Boulevard, Boca Raton, FL 33431. (800) 272-7737. Three volumes. 1986. $750.00 for set.

A GUIDE TO NUCLEAR POWER TECHNOLOGY: A RESOURCE FOR DECISION MAKING. F.J. Rahn and others. John Wiley and Sons, Inc., 605 Third Avenue, New York, NY 10158. (800) 526-5368. 1984. $85.95.

NUCLEAR ENGINEERING DATA BASES, STANDARDS AND NUMERICAL ANALYSIS. Jack Jedruch. Van Nostrand Reinhold Company, Inc., 135 West 50th Street, New York, NY 10020. (800) 543-2681. 1985. $59.95.

ONLINE DATA BASES

CA SEARCH. Chemical Abstracts Service, P.O. Box 3012, Columbus, OH 43120. (800) 848-6538 or (614) 421-3600. Comprehensive guide to chemical literature, 1972 to present. Inquire as to online cost and availability.

COMPENDEX. Engineering Information, Inc., 345 East 47th Street, New York, NY 10017. (800) 221-1044 or (212) 705-7615. Engineering and technical literature, 1975 to present. Inquire as to online cost and availability.

DOE ENERGY DATA BASE. U.S. Department of Energy, Office of Scientific and Technical Information, P.O. Box 62, Oak Ridge, TN 37831. (615) 576-6837. A database that covers all aspects of energy including the science and technology of energy. 1948 to present. Available through the DIALOG search service or DOE/RECON. Inquire as to online cost and availability.

ENERGYLINE. EIC/Intelligence, Inc., 48 West 38th Street, New York, NY 10018. (212) 944-8500. A database of resources on the scientific, engineering, political, and socioeconomic aspects of energy resources. 1976 to present. Inquire as to online cost and availability.

INSPEC. INSPEC Marketing Department, Institution of Electrical Engineers. Available from IEEE Service Center, 445 Hoes Lane, Piscataway, NJ 08854. (201) 981-0060. Online version of Physics Abstracts. Inquire as to online cost and availability.

NTIS. National Technical Information Service, 5285 Port Royal Road, Springfield, VA 22161. (703) 487-4630. Broad coverage of government sponsored research reports, 1964 to present. Inquire as to online cost and availability.

SCISEARCH. Institute for Scientific Information, 3501 Market Street, Philadelphia, PA 19104. (800) 523-1850 or (215) 386-0100. Broad multidisciplinary title and author index to the international literature of science and technology, 1974 to present. Inquire as to online cost and availability.

WILSONLINE. H.W. Wilson and Company, 950 University Avenue, Bronx, NY 10452. (800) 367-6770 or (212) 588-8400. Makes available online versions of the H.W. Wilson indexes including Applied Science and Technology Index, Business Periodicals Index and Readers' Guide to Periodical Literature. Approximately 1980 to present. Inquire as to online cost and availability.

PERIODICALS

AMERICAN NUCLEAR SOCIETY TRANSACTIONS. American Nuclear Society, 555 North Kensington Avenue, La Grange, IL 60525. (312) 352-6611. 1958 to present. Semiannual. $255.00 per year.

ANNALS OF NUCLEAR ENERGY. Pergamon Press, Inc., Maxwell House, Fairview Park, Elmsford, NY 10523. (914) 592-7700. 1954 to present. Monthly. $280.00 per year.

BULLETIN OF THE ATOMIC SCIENTISTS. Educational Foundation for Nuclear Science, 5801 South Kenwood Avenue, Chicago, IL 60637. (312) 363-5225. 1945 to present. Ten times per year. $22.50 per year.

FUSION TECHNOLOGY. American Nuclear Society, 555 North Kensington Avenue, La Grange, IL 60525. (312) 352-6611. 1981 to present. Bimonthly. $250.00 per year.

JOURNAL OF FUSION ENERGY. Plenum Publishing Corporation, 233 Spring Street, New York, NY 10013. (800) 221-9369. 1981 to present. Bimonthly. $105.00 per year.

NUCLEAR ENGINEER. Institution of Nuclear Engineers, 1 Penerley Road, Nondon SE6 2LQ, England. 1959 to present. Bimonthly. $110.00 per year.

NUCLEAR ENGINEERING AND DESIGN. Elsevier Science Publishing Company, Inc., 52 Vanderbilt Avenue, New York, NY 10017. (212) 370-5520. 1965 to present. $160.00 per year.

NUCLEAR ENGINEERING INTERNATIONAL. Electrical-Electronic Press, Quadrant House, The Quadrant, Sutton, Surrey, SM2 5AS, England. 1956 to present. Monthly. $210.00 per year.

NUCLEAR SCIENCE AND ENGINEERING. American Nuclear Society, 555 North Kensington Avenue, La Grange, IL 60525. (312) 352-6611. 1956 to present. Monthly. $220.00 per year.

NUCLEAR TECHNOLOGY. American Nuclear Society, 555 North Kensington Avenue, La Grange, IL 60525. (312) 352-6611. 1965 to present. Monthly. $345.00 per year.

RESEARCH CENTERS AND INSTITUTES

ASSISTANT SECRETARY FOR NUCLEAR ENERGY. U.S. Department of Energy, 1000 Independence Avenue, SW, Washington, DC 20585. (202) 252-6450.

DEPARTMENT OF NUCLEAR ENGINEERING AND ENGINEERING PHYSICS. University of Wisconsin - Madison, 1500 Johnson Drive, Madison, WI 53706. (608) 263-1648.

INSTITUTE OF NUCLEAR SCIENCE AND ENGINEERING. Oregon State University, Radiation Center, 35th and Jefferson Streets, Corvallis, OR 97331. (503) 754-2341.

LABORATORY OF BASIC AND APPLIED NUCLEAR RESEARCH. University of Cincinnati, Department of Chemistry, Cincinnati, OH 45221. (513) 475-3652.

WHITESHELL NUCLEAR RESEARCH ESTABLISHMENT. Research Company, Atomic Energy of Canada Limited, Pinawa, MB, Canada ROE 1LO. (204) 753-2311.

SPECIFICATIONS AND STANDARDS

INFORMATION CENTER ON NUCLEAR STANDARDS. American Nuclear Society, 555 North Kensington Avenue, La Grange, IL 60525. (312) 352-6611. Standards for all aspects of the design, construction, operation, and maintenance of nuclear power plants.

NUCLEAR SPECTROSCOPY

See: SPECTROSCOPY

NUCLEONICS

See: NUCLEAR PHYSICS

NUMERICAL ANALYSIS

See: MATHEMATICS

O

OBSERVATORIES

See also: ASTRONOMY, ASTROPHYSICS, TELESCOPES

ABSTRACT SERVICES AND INDEXES

ASTRONOMY AND ASTROPHYSICS ABSTRACTS. Springer-Verlag New York, Inc., 175 Fifth Avenue, New York, NY 10010. (800) 526-7254. 1969 to present. Approximately $70.00 per year.

ASTRONOMY AND ASTROPHYSICS MONTHLY INDEX. Olivetree Associates, P.O. Box 236, Sierre Madre, CA 91024. $220.00 per year. Complementary copies available on request.

CURRENT CONTENTS: PHYSICAL, CHEMICAL, AND EARTH SCIENCES. Institute for Scientific Information, 3501 Market Street, Philadelphia, PA 19104. (800) 523-1850 or (215) 386-0100. Weekly. $275.00 per year.

METEOROLOGICAL AND GEOASTROPHYSICAL ABSTRACTS. American Meteorological Society, 45 Beacon Street, Boston, MA 02108. (617) 227-2425. 1950 to present. Monthly. $450.00 per year.

PHYSICS ABSTRACTS. Institution of Electrical Engineers. Available from: IEEE Service Center, 445 Hoes Lane, Piscataway, NJ 08854. 1898 to present. Bimonthly. $1700.00 per year.

PHYSICS BRIEFS. Physik Verlag GmbH, Postfach 1260/1280, D-6940 Weinheim, West Germany. (212) 661-9404. 1920 to present. Twenty-six times per year. $1250.00 per year.

SCIENCE CITATION INDEX. Institute for Scientific Information, 3501 Market Street, Philadelphia, PA 19104. (800) 523-1850 or (215) 386-0100. Six times per year. $6200.00 per year.

STAR. (SCIENTIFIC AND TECHNICAL AEROSPACE REPORTS. U.S. National Aeronautics and Space Administration, Scientific and Technical Information Facility, Box 8757, Baltimore-Washington International Airport, MD 21240. (202) 755-2210. Semimonthly, with semiannual and annual indexes. $85.00 per year.

ANNUAL REVIEWS AND YEARBOOKS

ANNUAL REVIEW OF ASTRONOMY AND ASTROPHYSICS. Annual Reviews, Inc., 4139 El Camino Way, Palo Alto, CA 94306. (415) 493-4400. Annual. Inquire as to cost and availability.

ASSOCIATIONS AND PROFESSIONAL SOCIETIES

AMERICAN ASTRONOMICAL SOCIETY. 1816 Jefferson Place, N.W., Washington, DC 20036. (202) 659-0134.

ASTRONOMICAL SOCIETY OF THE PACIFIC. 1290 24th Avenue, San Francisco, CA 94122. (415) 661-8660.

INSTITUTION OF ELECTRICAL AND ELECTRONICS ENGINEERS (IEEE). 345 East 47th Street, New York, NY 10017. (212) 705-7900.

OPTICAL SOCIETY OF AMERICA. 1816 Jefferson Place, N.W., Washington, DC 20036. (202) 223-8130.

SOCIETY OF PHOTO-OPTICAL INSTRUMENTATION ENGINEERS - THE INTERNATIONAL SOCIETY OF OPTICAL ENGINEERING. P.O. Box 10, 1022 19th Street, Bellingham, WA 98227. (206) 676-3290.

BIBLIOGRAPHIES

A BIBLIOGRAPHY OF ASTRONOMY, 1970-1979. R.A. Seal and S.S. Martin. Libraries Unlimited Inc., P.O. Box 263, Littleton, CO 80160. (303) 770-1220. 1982. $37.50.

DIRECTORIES AND BIOGRAPHICAL SOURCES

AMERICAN MEN AND WOMEN OF SCIENCE. R.R. Bowker, Inc., Order Department, 245 West 17th Street, New York, NY 10011. (800) 521-8110. Eight volumes. 1986. $595.00 for set.

THE BIOGRAPHICAL DICTIONARY OF SCIENTISTS: ASTRONOMERS. D. Abbott, editor. Peter Bedrick Books, 125 East 23rd Street, New York, NY 10010. 1984. $28.00.

DIRECTORY OF PHYSICS AND ASTRONOMY STAFF MEMBERS. American Institute of Physics, 335 East 45th Street, New York, NY 10017. (212) 661-9494. Annual.

INTERNATIONAL RESEARCH CENTERS DIRECTORY 1988-89. Darren L. Smith, editor. Gale Research Company, Book Tower, Detroit, MI 48226. (800) 521-0707. 4th edition. 1987. $360.00.

RESEARCH CENTERS DIRECTORY 1988. Gale Research Company, Book Tower, Detroit, MI 48226. (800) 521-0707. 12th edition. 1987. $365.00 for set.

SCIENTIFIC AND TECHNICAL ORGANIZATIONS AND AGENCIES DIRECTORY. Margaret Labash Young, editor. Gale Research Company, Book Tower, Detroit, MI 48226. (800) 521-0707. 2nd edition. 1987. $185.00.

ENCYCLOPEDIAS AND DICTIONARIES

ILLUSTRATED ENCYCLOPEDIA OF THE UNIVERSE: UNDERSTANDING AND EXPLORING THE COSMOS. Richard S. Lewis. Crown Publishers Inc., 1 Park Avenue, New York, NY 10016. (800) 526-4264. 1986. $24.95.

GENERAL WORKS

ASTRONOMICAL OBSERVATORIES: AN OPTICAL PERSPECTIVE. Gordon Walker. Cambridge University Press, 32 East 57th Street, New York, NY 10022. (800) 872-7423. 1987. $80.00.

INFINITE VISTAS: NEW TOOLS FOR ASTRONOMY. J. Cornell and J. Carr, editors. Charles Scribner's and Sons, 115 Fifth Avenue, New York, NY 10003. (800) 257-5755. 1985. $18.95.

THE INVISIBLE UNIVERSE. F. Field and E. Chaisson. Birkhauser Boston, Inc., 380 Green Street, Cambridge, MA 02139. (617) 876-2333. 1984. $19.95.

ONLINE DATA BASES

DISSERTATION ABSTRACTS ONLINE. University Microfilms International, 300 North Zeeb Road, Ann Arbor, MI 48106. (800) 521-0600 or (313) 761-4700. Scope includes virtually all doctoral dissertations accepted at accredited American institutions from 1861 to present in over 250 subject areas. Inquire as to online cost and availability.

INSPEC. INSPEC Marketing Department, Institution of Electrical Engineers. Available from IEEE Service Center, 445 Hoes Lane, Piscataway, NJ 08854. (201) 981-0060. Online version of Physics Abstracts. Inquire as to online cost and availability.

NTIS. National Technical Information Service, 5285 Port Royal Road, Springfield, VA 22161. (703) 487-4630. Broad coverage of government sponsored research reports, 1964 to present. Inquire as to online cost and availability.

SCISEARCH. Institute for Scientific Information, 3501 Market Street, Philadelphia, PA 19104. (800) 523-1850 or (215) 386-0100. Broad multidisciplinary title and author index to the international literature of science and technology, 1974 to present. Inquire as to online cost and availability.

OTHER SOURCES

ASTRONOMICAL CENTERS OF THE WORLD. K. Krisciunas. Cambridge University Press, 32 East 57th Street, New York, NY 10022. (800) 872-7423. 1987. $45.00.

ATLAS OF DEEP-SKY SPLENDORS. H. Vehrenberh. Cambridge University Press, 32 East 57th Street, New York, NY 10022. (800) 872-7423. Fourth edition. 1984. $44.50.

PERIODICALS

ASTRONOMICAL JOURNAL. American Astronomical Society. Available from: American Institute of Physics, 335 East 45th Street, New York, NY 10017. (212) 661-9494. Monthly. $125.00 per year.

ASTRONOMICAL SOCIETY OF THE PACIFIC PUBLICATIONS. Astronomical Society of the Pacific, 1290 24th Avenue, San Francisco, CA 94122. (415) 661-8660. Monthly. $40.00 per year.

ASTRONOMY. AstroMedia Corporation, 625 E Street, Box 92788, Milwaukee, WI 53202. (414) 276-2689. 1973 to present. Monthly. $21.00 per year.

ASTRONOMY AND ASTROPHYSICS. Springer-Verlag New York, Inc., 175 Fifth Avenue, New York, NY 10010. (800) 526-7254. Monthly. $680.00 per year.

ASTROPHYSICAL JOURNAL. University of Chicago Press, 5801 Ellis Avenue, Chicago, IL 60637. (800) 621-2736. Biweekly. $315.00 per year.

ASTROPHYSICS AND SPACE SCIENCE. D. Reidel Publishing Company, 190 Old Derby Street, Hingham, MA 02043. Monthly. $101.00 per year.

SKY AND TELESCOPE. Sky Publishing Corporation, 49 Bay State Road, Cambridge, MA 02238. (617) 864-7360. Monthly. $18.00 per year.

MERCURY. Astronomical Society of the Pacific, 1290 24th Avenue, San Francisco, CA 94122. (415) 661-8660. 1972 to present. Bimonthly. $21.00 per year.

RESEARCH CENTERS AND INSTITUTES

MOUNT WILSON AND LAS CAMPANAS OBSERVATORIES. 813 Santa Barbara Street, Pasadena, CA 91101. (818) 577-1122.

MULTIPLE MIRROR TELESCOPE OBSERVATORY. University of Arizona, Tucson, AZ 85721. (602) 621-1558.

NATIONAL OPTICAL ASTRONOMY OBSERVATORIES. 950 North Cherry Avenue, Tucson, AZ 85719. (602) 325-9230.

NATIONAL RADIO ASTRONOMY OBSERVATORY. Edgemont Road, Charlottesville, VA 22903. (804) 296-0211.

OCEAN-ATMOSPHERE BOUNDARY

See: METEOROLOGY

OCEAN CURRENTS

See: OCEANOGRAPHY

OCEAN ENGINEERING

See also: CIVIL ENGINEERING, COASTAL ENGINEERING, COASTS, CONTINENTAL MARGINS

ABSTRACT SERVICES AND INDEXES

APPLIED SCIENCE AND TECHNOLOGY INDEX. H.W. Wilson and Company, 950 University Avenue, Bronx, NY 10452. (800) 367-6670 or (212) 588-8400. Monthly. Inquire as to cost and availability.

BIBLIOGRAPHY AND INDEX OF GEOLOGY. American Geological Institute, 4220 King Street, Alexandria, VA 22302. (703) 379-2480. 1969 to present. Monthly. $1100.00 per year.

CHEMICAL ABSTRACTS. American Chemical Society, Chemical Abstracts Service, Box 3012, Columbus, OH 43210. (614) 421-3600. 1907 to present. Weekly. $9500.00 per year.

CURRENT CONTENTS: ENGINEERING, TECHNOLOGY AND APPLIED SCIENCES. Institute for Scientific Information, 3501 Market Street, Philadelphia, PA 19104. (800) 523-1850 or (215) 386-0100. Weekly. $275.00 per year.

ENGINEERING INDEX MONTHLY AND AUTHOR INDEX. Engineering Information Inc., 345 East 47th Street, New York, NY 10017. (212) 705-7600. Monthly. $1560.00 per year.

OCEAN ABSTRACTS. Cambridge Scientific Abstracts, 5161 River Road, Bethesda, MD 20816. (301) 951-1400. 1963 to present. Bimonthly. $450.00 per year.

SCIENCE CITATION INDEX. Institute for Scientific Information, 3501 Market Street, Philadelphia, PA 19104. (800) 523-1850 or (215) 386-0100. Six times per year. $6200.00 per year.

ANNUAL REVIEWS AND YEARBOOKS

COASTAL ENGINEERING. American Society of Civil Engineers, 345 East 47th Street, New York, NY 10017. (212) 705-7420. 1974 to 1980. Multiple volume sets for each year. Price varies, inquire.

OFFSHORE STRUCTURES ENGINEERING. F.L. Carneiro and others, editors. Gulf Publishing Company, P.O. Box 2608, Houston, TX 77001. (713) 520-4444. Volumes 1-5. 1979-1984. Approximately $49.50 per volume.

ASSOCIATIONS AND PROFESSIONAL SOCIETIES

AMERICAN SOCIETY OF CIVIL ENGINEERS. 345 East 47th Street, New York, NY 10017. (212) 705-7420.

AMERICAN SOCIETY OF MECHANICAL ENGINEERS. 345 East 47th Street, New York, NY 10017. (212) 705-7703.

COASTAL ENGINEERING RESEARCH COUNCIL. 207 East Bay Street, Suite 311, Charleston, SC 29401. (803) 723-4864.

THE COASTAL SOCIETY. 5410 Grosvenor Lane, Suite 110, Bethesda, MD 20814. (301) 897-8616.

GEOLOGICAL SOCIETY OF AMERICA. 3300 Penrose Place, Boulder, CO 80301. (303) 447-2020.

DIRECTORIES AND BIOGRAPHICAL SOURCES

1987 DIRECTORY OF ENGINEERING SOCIETIES AND RELATED ORGANIZATIONS. Gordon Davis, editor. Hemisphere Publishing Corporation, 1010 Vermont Avenue, NW, Washington, DC 20005. (800) 526-0275. 12th edition. 1987. $100.00.

RESEARCH CENTERS DIRECTORY 1988. Gale Research Company, Book Tower, Detroit, MI 48226. (800) 521-0707. 12th edition. 1987. $365.00 for set.

SCIENTIFIC AND TECHNICAL ORGANIZATIONS AND AGENCIES DIRECTORY. Margaret Labash Young, editor. Gale Research Company, Book Tower, Detroit, MI 48226. (800) 521-0707. 2nd edition. 1987. $185.00.

WHO'S WHO IN ENGINEERING. Gordon Davis, editor. Hemisphere Publishing Corporation, 1010 Vermont Avenue, NW, Washington, DC 20005. (800) 526-0275. 6th edition. 1985. $200.00.

GENERAL WORKS

ADVANCED DYNAMICS OF MARINE STRUCTURES. Jan P. Hooft. John Wiley and Sons, Inc., 605 Third Avenue, New York, NY 10158. (800) 526-5368. 1982. $58.95.

BASIC COASTAL ENGINEERING. R.M. Sorensen. John Wiley and Sons, Inc., 605 Third Avenue, New York, NY 10158. (800) 526-5368. 1978. $39.95.

DYNAMICS OF OFFSHORE STRUCTURES. James F. Wilson, editor. John Wiley and Sons, Inc., 605 Third Avenue, New York, NY 10158. (800) 526-5368. 1984. $59.95.

THE MARINE ENVIRONMENT AND STRUCTURAL DESIGN. John W. Gaythwaite. Van Nostrand Reinhold Company, Inc., 135 West 50th Street, New York, NY 10020. (800) 543-2681. 1981. $28.95.

OFFSHORE STRUCTURAL ENGINEERING. Thomas H. Dawson. Prentice-Hall Publishing, Inc., Englewood Cliffs, NJ 07632. (800) 562-0245. 1983. $45.95.

SEA BED MECHANICS. J.F. Sleath. John Wiley and Sons, Inc., 605 Third Avenue, New York, NY 10158. (800) 526-5368. 1984. $47.50.

ONLINE DATA BASES

CA SEARCH. Chemical Abstracts Service, P.O. Box 3012, Columbus, OH 43120. (800) 848-6538 or (614) 421-3600. Comprehensive guide to chemical literature, 1972 to present. Inquire as to online cost and availability.

COMPENDEX. Engineering Information, Inc., 345 East 47th Street, New York, NY 10017. (800) 221-1044 or (212) 705-7615. Engineering and technical literature, 1975 to present. Inquire as to online cost and availability.

DISSERTATION ABSTRACTS ONLINE. University Microfilms International, 300 North Zeeb Road, Ann Arbor, MI 48106. (800) 521-0600 or (313) 761-4700. Scope includes virtually all doctoral dissertations accepted at accredited American institutions from 1861 to present in over 250 subject areas. Inquire as to online cost and availability.

GEOREF. Online version of the BIBLIOGRAPHY AND INDEX OF GEOLOGY. American Geological Institute, 4220 King Street, Alexandria, VA 22302. (703) 379-2480. 1969 to present. Inquire as to online cost and availability.

INSPEC. INSPEC Marketing Department, Institution of Electrical Engineers. Available from IEEE Service Center, 445 Hoes Lane, Piscataway, NJ 08854. (201) 981-0060. Online version of Physics Abstracts. Inquire as to online cost and availability.

NTIS. National Technical Information Service, 5285 Port Royal Road, Springfield, VA 22161. (703) 487-4630. Broad coverage of government sponsored research reports, 1964 to present. Inquire as to online cost and availability.

SCISEARCH. Institute for Scientific Information, 3501 Market Street, Philadelphia, PA 19104. (800) 523-1850 or (215) 386-0100. Broad multidisciplinary title and author index to the international literature of science and technology, 1974 to present. Inquire as to online cost and availability.

WILSONLINE. H.W. Wilson and Company, 950 University Avenue, Bronx, NY 10452. (800) 367-6770 or (212) 588-8400. Makes available online versions of the H.W. Wilson indexes including Applied Science and Technology Index, Business Periodicals Index and Readers' Guide to Periodical Literature. Approximately 1980 to present. Inquire as to online cost and availability.

PERIODICALS

COASTAL ENGINEERING: AN INTERNATIONAL JOURNAL FOR COASTAL, HARBOUR, AND OFFSHORE ENGINEERS. Elsevier Science Publishing Company, Inc., 52 Vanderbilt Avenue, New York, NY 10017. (212) 370-5520. 1977 to present. Quarterly. $95.00 per year.

COASTAL RESEARCH. Florida State University, Geology Department, Tallahassee, FL 32306. (904) 644-5860. 1962 to present. Three times per year. $5.00 per year.

COASTAL ZONE MANAGEMENT JOURNAL. Crane Russak and Company, Inc., 3 East 44th Street, New York, NY 10017. (212) 867-1490. 1973 to present. Quarterly. $72.00 per year.

COASTWATCH. University of North Carolina, Sea Grant College Program, Box 8605, North Carolina State University, Raleigh, NC 27695-8605. (919) 737-2454. 1970 to present. Monthly. Free.

ESTUARINE, COASTAL AND SHELF SCIENCE. Academic Press, Inc., 6277 Sea Harbor Drive, Orlando, FL 32821. (800) 321-5068. 1973 to present. Monthly. $315.00 per year.

JOURNAL OF WATERWAY, PORT, COASTAL AND OCEAN ENGINEERING. American Society of Civil Engineers, 345 East 47th Street, New York, NY 10017. (212) 705-7420. 1956 to present. Bimonthly. $72.00 per year.

OCEAN ENGINEERING. Pergamon Press, Inc., Maxwell House, Fairview Park, Elmsford, NY 10523. (914) 592-7700. 1968 to present. Bimonthly. $200.00 per year.

OCEAN SCIENCE AND ENGINEERING. Marcel Dekker Inc., 270 Madison Avenue, New York, NY 10016. (800) 228-1160. 1974 to present. Quarterly. $165.00 per year.

U.S. COASTAL ENGINEERING RESEARCH CENTER. QUARTERLY CIRCULAR INFORMATION BULLETIN. U.S. Coastal Engineering Research Center, Kingman Building, Fort Belvoir, VA 22060. (703) 325-7429. Quarterly. Inquire.

RESEARCH CENTERS AND INSTITUTES

COASTAL AND OCEANOGRAPHIC ENGINEERING LABORATORY. University of Florida, 336 Weil Hall, Gainesville, FL 32607. (904) 392-1436.

COASTAL ENGINEERING RESEARCH COUNCIL. 207 East Bay Street, Suite 311, Charleston, SC 29401. (803) 723-4864.

COASTAL LABORATORY. Queen's University at Kingston, Ellis Hall, Kingston, ON, Canada K7L 3N6. (613) 545-6265.

OCEAN ENGINEERING LABORATORY. University of Washington, Mail Stop FU-10, Mechanical Engineering Building, Seattle, WA 98195. (206) 543-7446.

OCEAN TRENCHES

See: OCEANOGRAPHY

OCEANOGRAPHY

See also: COASTS, CONTINENTAL MARGIN, GEOLOGY, METEOROLOGY, OCEAN ENGINEERING, TSUNAMIS

ABSTRACT SERVICES AND INDEXES

APPLIED SCIENCE AND TECHNOLOGY INDEX. H.W. Wilson Company, 950 University Avenue, Bronx, NY 10452. (800) 367-6670 or (212) 588-8400. Inquire as to cost and availability.

BIBLIOGRAPHY AND INDEX OF GEOLOGY. American Geological Institute, 4220 King Street, Alexandria, VA 22302. (703) 379-2480. 1969 to present. Monthly. $1100.00 per year.

CHEMICAL ABSTRACTS. Chemical Abstracts Service, 2540 Olentangy Road, Post Office Box 3012, Columbus, OH 43210. (800) 848-6538 or (614) 421-3600. Weekly. $9200.00 per year.

CURRENT CONTENTS: PHYSICAL AND CHEMICAL SCIENCES. Institute for Scientific Information, 3501 Market Street, Philadelphia, PA 19104. (800) 523-1850 or (215) 386-0100. $272.00 per year.

DEEP SEA RESEARCH WITH OCEANOGRAPHIC LITERATURE REVIEW. Pergamon Press Incorporated, Journals Division, Maxwell House, Fairview Park, Elmsford, NY 10523. (914) 592-7700. 1953 to present. Twenty-four times per year. $600.00 per year.

GENERAL SCIENCE INDEX. H.W. Wilson Company, 950 University Avenue, Bronx, NY 10452. (800) 367-6770 or (212) 588-8400. Inquire as to cost and availability.

INDEX TO SCIENTIFIC REVIEWS. Institute for Scientific Information, 3501 Market Street, Philadelphia, PA 19104. (800) 523-1850 or (215) 386-0100. Semiannual. $550.00 per year.

METEOROLOGICAL AND GEOASTROPHYSICAL ABSTRACTS. American Meteorological Society, 45 Beacon Street, Boston, MA 02108. (617) 227-2425. 1950 to present. Monthly. $450.00 per year.

OCEANOGRAPHIC LITERATURE REVIEW. Pergamon Press Incorporated, Maxwell House, Fairveiw Park, Elmsford, NY 10523. (914) 592-7700. 1979 to present. Inquire as to cost and availability.

PHYSICS ABSTRACTS. Institute of Electrical Engineers, London, United Kingdom. Available from: Institute of Electrical and Electronic Engineers (IEEE), 345 East 47th Street, New York, NY 10017. (212) 705-7900. 1898 to present. Monthly. $1670.00 per year.

SCIENCE CITATION INDEX. Institute for Scientific Information, 3501 Market Street, Philadelphia, PA 19104. (800) 523-1850 or (215) 386-0100. Inquire as to cost and availability.

ANNUAL REVIEWS AND YEARBOOKS

OCEAN YEARBOOK. Elizabeth M. Borgese and Norton Ginsberg, editors. University of Chicago Press, 5801 Ellis Avenue, Chicago, IL 60637. (312) 962-7906. 1979 to present. $49.00 per volume.

ASSOCIATIONS AND PROFESSIONAL SOCIETIES

AMERICAN LITTORAL SOCIETY. Sandy Hook, Highlands, NJ 07732. (201) 291-0055.

AMERICAN METEOROLOGICAL SOCIETY. 45 Beacon Street, Boston, MA 02108. (617) 227-2425.

AMERICAN SOCIETY OF LIMNOLOGY AND OCEANOGRAPHY. 1530 12th Avenue, Grafton, WI 53024. (414) 377-4871.

INTERNATIONAL OCEANOGRAPHIC FOUNDATION, 3979 Rickenbacker Causeway, Virginia Key, Miami, FL 33149. (305) 361-5786.

MARINE TECHNOLOGY SOCIETY. 2000 Florida Avenue, NW, Suite 500, Washington, DC 20009. (202) 462-7557.

NATIONAL OCEAN INDUSTRIES. 1050 17th Street, NW, Washington, DC 20036. (202) 785-5116.

OCEANIC SOCIETY. 1536 16th Street, NW, Washington, DC 20036. (202) 328-0098.

BIBLIOGRAPHIES

MARINE SCIENCE JOURNALS AND SERIALS: AN ANALYTICAL GUIDE. Greenwood Press, 88 Post Road West, Post Office Box 5007, Westport, CT 06881. (203) 226-3571. 1986. $36.95.

SCIENCE BOOKS AND FILMS. American Association for the Advancement of Science, 1333 H Street, NW, Washington, DC 20005. Five times per year. $20.00 per year.

SCIENTIFIC AND TECHNICAL BOOKS AND SERIALS IN PRINT 1988: AN INDEX TO LITERATURE IN SCIENCE AND TECHNOLOGY R.R. Bowker Company, 205 East 42nd Street, New York, NY 10017. (800) 521-8110 or (212) 916-1600. $175.00.

DIRECTORIES AND BIOGRAPHICAL SOURCES

AMERICAN MEN AND WOMEN OF SCIENCE. Physical and Biological Sciences. Sixteenth edition. R.R. Bowker Company, 205 East 42nd Street, New York, NY 10017. (800) 521-8110 or (212) 916-1600. 1986. $595.00.

AMERICAN SOCIETY OF LIMNOLOGY AND OCEANOGRAPHY MEMBERSHIP DIRECTORY. American Society of Limnology and Oceanography, 1530 12th Avenue, Grafton, WI 53024. (414) 377-4871. 1988. $10.00.

BIOGRAPHICAL DICTIONARY OF SCIENTISTS. T.I. Williams. Halsted Press, 605 Third Avenue, New York, NY 10158. (800) 526-5368 or (212) 850-6418. Third edition. 1982. $29.95.

GOVERNMENT RESEARCH DIRECTORY. Gale Research Company, Book Tower, Detroit, MI 48226. (800) 521-0707. Fourth edition. 1987. $350.00.

INTERDOC: DIRECTORY OF PUBLISHED PROCEEDINGS, SERIES. SEMT-Science/Engineering/Medicine/Technology. Interdoc Corporation, 173 Halstead Avenue, Box 326, Harrison,

NY 10528. (014) 835-3506. Ten times per year. $325.00 per year.

INTERNATIONAL RESEARCH CENTERS DIRECTORY 1988-1989. Gale Research Company, Book Tower, Detroit, MI 48226. (800) 521-0707. Fourth edition. 1987. $360.00.

NATIONAL OCEAN INDUSTRIES ASSOCIATION DIRECTORY OF MEMBERSHIP. National Ocean Industries, 1050 17th Street, NW, Washington, DC 20036. (202) 785-5116. Annual. $20.00.

RESEARCH CENTERS DIRECTORY. Gale Research Company, Book Tower, Detroit, MI 48226. (800) 521-0707. Twelfth edition. 1987. $365.00 for set.

SCIENTIFIC AND TECHNICAL ORGANIZATIONS AND AGENCIES DIRECTORY. Gale Research Company, Book Tower, Detroit, MI 48226. (800) 521-0707. Second edition. 1987. $185.00.

WHO'S WHO IN FRONTIER SCIENCE AND TECHNOLOGY. Marquis Who's Who, Incorporated, 200 East Ohio Street, Chicago, IL 60611. (800) 428-3898 or (312) 787-2008.

WHO'S WHO IN TECHNOLOGY TODAY. Reston Publishing Company, Incorporated, c/o Prentice-Hall, Incorporated, Englewood Cliffs, NJ 07632. (800) 262-6868. Biennial. Five volumes. $425.00. Covers the fields of electronics, computer science, physics, optics, chemistry, biotechnology, mechanics, energy, and earth science.

WORLD GUIDE TO SCIENTIFIC ASSOCIATIONS AND LEARNED SOCIETIES. K.G. Saur Incorporated, 175 Fifth Avenue, New York, NY 10010. (800) 521-0707 or (212) 982-1302. Fourth edition. 1984. $112.00.

ENCYCLOPEDIAS AND DICTIONARIES

ENCYCLOPEDIA OF PHYSICAL SCIENCE AND TECHNOLOGY. Academic Press, Incorporated, Orlando, FL 32887. (800) 321-5068 or (305) 345-2734. Fifteen volumes, 1986.

MCGRAW-HILL ENCYCLOPEDIA OF SCIENCE AND TECHNOLOGY. McGraw-Hill Book, Incorporated, 1221 Avenue of the Americas, New York, NY 10020. (212) 997-3675. Fifth edition, 15 volumes. $1100.00.

GENERAL WORKS

COASTAL AND ESTUARINE SEDIMENT DYNAMICS. K.R. Dyer. John Wiley and Sons, Incorporated, 605 Third Avenue, New York, NY 10158. (800) 526-5368 or (212) 850-6000. 1986. $69.95.

ENVIRONMENTAL OCEANOGRAPHY: AN INTRODUCTION TO THE BEHAVIOR OF COASTAL WATERS. T. Beer. Pergamon Press Incorporated, Maxwell House, Fairview Park, Elmsford, NY 10523. (914) 592-7700. 1983. $35.00.

GENERAL CIRCULATION OF THE OCEAN. Henry D.I. Abarbanel and W.R. Young, editors. Springer-Verlag, 175 Fifth Avenue, New York, NY 10010. (800) 526-7254. 1987. $69.00.

INTRODUCTION TO OCEANOGRAPHY. David A. Ross. Prentice-Hall Publishing, Incorporated, Englewood Cliffs, NJ 07632. (800) 562-0245. 1988. $39.95.

INTRODUCTORY DYNAMIC OCEANOGRAPHY. G.L. Pickard and S. Pond. Pergamon Press Incorporated, Maxwell House, Fairview Park, Elmsford, NY 10523. (914) 592-7700. Second edition. 1983. $55.00.

OCEAN SCIENCE. Keith Stowe. John Wiley and Sons, Incorporated, 605 Third Avenue, New York, NY 10158. (800) 526-5368 or (212) 850-6000. Second edition. 1983. $37.95 in paper.

OCEAN TIDES: MATHEMATICAL MODELS AND NUMERICAL EXPERIMENTS. G.I. Marchuk and B.A. Kagan. Pergamon Press Incorporated, Maxwell House, Fairview Park, Elmsford, NY 10523. (914) 592-7700. 1984. $80.00.

SATELLITE OCEANOGRAPHY: AN INTRODUCTION FOR OCEANOGRAPHERS AND REMOTE SENSING SCIENTISTS. John Wiley and Sons, Incorporated, 605 Third Avenue, New York, NY 10158. (800) 526-5368 or (212) 850-6000. 1985. $59.95.

SEA BED MECHANICS. J.F. Sleath. John Wiley and Sons, Incorporated, 605 Third Avenue, New York, NY 10158. (800) 526-5368 or (212) 850-6000. 1984. $49.95.

WANDERING CONTINENTS AND SPREADING SEA FLOORS ON AN EXPANDING EARTH. Lester King. John Wiley and Sons, Incorporated, 605 Third Avenue, New York, NY 10158. (800) 526-5368 or (212) 850-6000. 1983. $49.95.

HANDBOOKS AND MANUALS

APPLIED FLUID DYNAMICS HANDBOOK. Robert D. Blevins. Van Nostrand Reinhold, Incorporated, 115 Fifth Avenue, New York, NY. (800) 543-2681. 1984. $51.95.

COMMERCIAL DIVING REFERENCE AND OPERATIONS HANDBOOK. M. Freitag and A. Woods. John Wiley and Sons, Incorporated, 605 Third Avenue, New York, NY 10158. (800) 526-5368 or (212) 850-6000. 1983. $69.95.

CRC HANDBOOK OF CHEMISTRY AND PHYSICS. CRC Press, Incorporated, 2000 Corporation Boulevard, Boca Raton, FL 33341. (305) 994-0555. Sixth-seventh edition. 1986. $69.95.

MARITIME AFFAIRS: A WORLD HANDBOOK. Henry W. Degenhardt. Longman Press. Distributed by Gale Research Company, Book Tower, Detroit, MI 48226. (800) 521-0707. 1985. $90.00.

ONLINE DATA BASES

CA SEARCH. Chemical Abstracts Service, Post Office Box 3012, Columbus, OH 43210. Guide to chemical literature, 1972 to present. Inquire as to cost and availability.

DISSERTATION ABSTRACTS ONLINE. University Microfilms International, 300 North Zeeb Road, Ann Arbor, MI 48106. (800) 521-0600 or (313) 761-4700. Scope includes virtually all doctoral dissertations accepted at accredited American institutions from 1861 to present in 252 subject areas. Inquire as to cost and availability.

INSPEC. INSPEC Marketing Department, Institute of Electrical and Electronics Engineers, Incorporated, IEEE Service Department, 445 Hoes Lane, Piscataway, NJ 08854. (201) 981-0060. Inquire as to on-line cost and availability.

METEOROLOGICAL AND GEOASTROPHYSICAL ABSTRACTS. American Meteorological Society, 45 Beacon Street, Boston, MA 02108. (617) 227-2425. 1950 to present. Monthly. $450.00 per year.

NTIS. National Technical Information Service, 5285 Port Royal Road, Springfield, VA 22161. (703) 487-4630. Broad coverage of government sponsored research reports, 1964 to present. Inquire as to cost and availability.

OCEANIC ABSTRACTS. Cambridge Scientific Abstracts, Incorporated, 5161 River Road, Bethesda, MD 20816. (301) 951-1400. 1964 to present. Inquire as to online cost and availability.

SCISEARCH. Institute for Scientific Information, 3501 Market Street, Philadelphia, PA 19104. (800) 523-1850 or (215) 386-0100. Broad multidisciplinary title and author index to the international literature of science and technology, 1974 to present. Inquire as to cost and availability.

WILSONLINE. H.W. Wilson Company, 950 University Avenue, Bronx, NY 10452. (800) 367-6770 or (212) 588-8400. Makes available online versions of the printed H.W. Wilson Indexes including Applied Science and Technology Index, Business Periodicals Index, and Readers' Guide to Periodical Literature. Period covered is generally 1983 to present. Inquire as to cost and availability.

PERIODICALS

BULLETIN OF MARINE SCIENCE. Rosenstiel School of Marine and Atmospheric Science, 4600 Richenbacker Causeway, Miami, FL 33149. (305) 361-4190. 1950 to present. Bimonthly. $65.00 per year.

COASTAL ZONE MANAGEMENT JOURNAL. Crane, Russak and Company, Incorporated, 3 East 44th Street, New York, NY 10017. (212) 867-1490. 1973 to present. Quarterly. $75.00 per year.

CONTINENTAL SHELF RESEARCH. Pergamon Press Incorporated, Maxwell House, Fairview Park, Elmsford, NY 10523. (914) 592-7700. 1982 to present. Bimonthly. $1200.00 per year. Includes Oceanographic Literature Review.

DYNAMICS OF ATMOSPHERES AND OCEANS. Elsevier Science Publishing Company, Incorporated, 52 Vanderbilt Avenue, New York, NY 10017. (212) 370-5520. 1977 to present. Quarterly. $90.00 per year.

ESTUARINE, COASTAL AND SHELF SCIENCE. Academic Press, Incorporated, 111 Fifth Avenue, New York, NY 10003. 1973 to present. Monthly. $325.00 per year.

JOURNAL OF ATMOSPHERIC AND OCEANIC TECHNOLOGY. American Meteorological Society, 45 Beacon Street, Boston, MA 02108. (617) 227-2425. 1984 to present. Quarterly. $80.00 per year.

JOURNAL OF MARINE RESEARCH. Sears Foundation for Marine Research, Kline Geology Laboratory, Box 6666, New Haven, CT 06511. (203) 436-3715. 1937 to present. Quarterly. $30.00 per year.

JOURNAL OF PHYSICAL OCEANOGRAPHY. American Meteorological Society, 45 Beacon Street, Boston, MA 02108. (617) 227-2425. 1971 to present. Monthly. $120.00 per year.

LIMNOLOGY AND OCEANOGRAPHY. America Society of Limnology and Oceanography, 1530 12th Avenue, Grafton, WI 53024. (414) 377-4871. 1956 to present. Bimonthly. $60.00 per year.

MARINE GEOLOGY: INTERNATIONAL JOURNAL OF MARINE GEOLOGY, GEOCHEMISTRY AND GEOPHYSICS. Elsevier Science Publishing Company, Incorporated, 52 Vanderbilt Avenue, New York, NY 10017. (212) 370-5520. 1964 to present. Twenty-four times per year. $525.00 per year.

MARINE TECHNOLOGY SOCIETY JOURNAL. Marine Technology Society, Incorporated, 2000 Florida Avenue, NW, Suite 500, Washington, DC 20009. (202) 462-7557. 1966 to presenQuarterly. $50.00 per year.

OCEAN ENGINEERING. Pergamon Press Incorporated, Maxwell House, Fairview Park, Elmsford, NY 10523. (914) 592-7700. 1968 to present. Bimonthly. $200.00 per year.

OCEAN SCIENCE AND ENGINEERING. Marcel Dekker Journals, Incorporated, 270 Madison Avenue, New York, NY 10016. (800) 228-1160. 1974 to present. Quarterly. $165.00 per year.

OCEANUS. Woods Hole Oceanographic Institution. Order from: Allen Press, Incorporated, 1041 New Hampshire Street, Lawrence, KS 66044. 1952 to present. Quarterly. $20.00 per year.

SEA FRONTIERS. International Oceanographic Foundation, 3979 Rickenbacker Causeway, Virginia Key, Miami, FL 33149. (305) 361-5786. $1956 to present. Bimonthly. $18.00 per year.

RESEARCH CENTERS AND INSTITUTES

SCRIPPS INSTITUTION OF OCEANOGRAPHY. La Jolla, CA 92093. (619) 534-1744.

UNITED STATES COASTAL ENGINEERING RESEARCH CENTER. Kingman Building, Fort Belvoir, VA 22060. (703) 325-7429.

UNIVERSITY OF HAWAII. Hawaii Undersea Research Laboratory, Marine Sciences Building, 1000 Pope Road, Honolulu, HI 96822. (808) 948-6335.

UNIVERSITY OF MIAMI. Rosenstiel School of Marine and Atmospheric Science, 4600 Rickenbacker Causeway, Miami, FL 33149. (305) 361-4000.

WOODS HOLE OCEANOGRAPHIC INSTITUTION. Woods Hole, MA 02543. (617) 548-1400.

OFFSHORE ENGINEERING

See: OCEAN ENGINEERING

OPTICAL COMMUNICATIONS

See also: FIBER OPTICS, HOLOGRAPHY, LASERS, OPTICAL DISKS, OPTICS, OPTOELECTRONICS, PHYSICS, SPECTROSCOPY

ABSTRACT SERVICES AND INDEXES

APPLIED SCIENCE AND TECHNOLOGY INDEX. H.W. Wilson and Company, 950 University Avenue, Bronx, NY 10452. (800) 367-6670 or (212) 588-8400. Monthly. Inquire as to cost and availability.

CHEMICAL ABSTRACTS. American Chemical Society, Chemical Abstracts Service, Box 3012, Columbus, OH 43210. (614) 421-3600. 1907 to present. Weekly. $9500.00 per year.

CONFERENCE PAPERS INDEX. Cambridge Scientific Abstracts, 5161 River Road, Bethesda, MD 20816. 1972 to present. Monthly. Inquire as to cost and availability.

CURRENT CONTENTS: ENGINEERING, TECHNOLOGY AND APPLIED SCIENCES. Institute for Scientific Information, 3501 Market Street, Philadelphia, PA 19104. (800) 523-1850 or (215) 386-0100. Weekly. $275.00 per year.

ELECTRICAL AND ELECTRONICS ABSTRACTS. Institution of Electrical Engineers. Available from: Institute of Electrical and Electronics Engineers. IEEE Service Center, 445 Hoes Lane, Piscataway, NJ 08854. Monthly. $1250.00 per year.

ENGINEERING INDEX MONTHLY AND AUTHOR INDEX. Engineering Information Inc., 345 East 47th Street, New York, NY 10017. (212) 705-7600. Monthly. $1560.00 per year.

GENERAL SCIENCE INDEX. H.W. Wilson and Company, 950 University Avenue, Bronx, NY 10452. (800) 367-6670 or (212) 588-8400. 1978 to present. Monthly. Inquire as to cost and availability.

GOVERNMENT REPORTS ANNOUNCEMENT AND INDEX. National Technical Information Service, 5285 Port Royal Road, Springfield, VA 22161. (703) 487-4929. Summaries of United States government sponsored research reports. 1946 to present. Biweekly. Inquire as to cost and availability.

INDEX TO SCIENTIFIC AND TECHNICAL PROCEEDINGS. Institute for Scientific Information, 3501 Market Street, Philadelphia, PA 19104. (800) 523-1850 or (215) 386-0100. 1978 to present. Monthly. $775.00 per year.

INDEX TO SCIENTIFIC REVIEWS. Institute for Scientific Information, 3501 Market Street, Philadelphia, PA 19104. (800) 523-1850 or (215) 386-0100. 1974 to present. Semi-annual. $550.00 per year.

PHYSICS ABSTRACTS. Institution of Electrical Engineers. Available from: IEEE Service Center, 445 Hoes Lane, Piscataway, NJ 08854. 1898 to present. Bimonthly. $1700.00 per year.

PHYSICS BRIEFS. Physik Verlag GmbH, Postfach 1260/1280, D-6940 Weinheim, West Germany. (212) 661-9404. 1920 to present. Twenty-six times per year. $1250.00 per year.

SCIENCE CITATION INDEX. Institute for Scientific Information, 3501 Market Street, Philadelphia, PA 19104. (800) 523-1850 or (215) 386-0100. Six times per year. $6200.00 per year.

ASSOCIATIONS AND PROFESSIONAL SOCIETIES

AMERICAN INSTITUTE OF PHYSICS. 335 East 45th Street, New York, NY 10017. (212) 661-9404.

INSTITUTION OF ELECTRICAL AND ELECTRONICS ENGINEERS (IEEE). 345 East 47th Street, New York, NY 10017. (212) 705-7900.

LASER INSTITUTE OF AMERICA. 5151 Monroe Street, Suite 103, Toledo, OH 43623. (419) 882-8706.

OPTICAL SOCIETY OF AMERICA. 1816 Jefferson Place, N.W., Washington, DC 20036. (202) 223-8130.

SOCIETY OF PHOTO-OPTICAL INSTRUMENTATION ENGINEERS - THE INTERNATIONAL SOCIETY OF OPTICAL ENGINEERING. P.O. Box 10, 1022 19th Street, Bellingham, WA 98227. (206) 676-3290.

DIRECTORIES AND BIOGRAPHICAL SOURCES

AMERICAN MEN AND WOMEN OF SCIENCE. R.R. Bowker, Inc., Order Department, 245 West 17th Street, New York, NY 10011. (800) 521-8110. Eight volumes. 1986. $595.00 for set.

INTERNATIONAL RESEARCH CENTERS DIRECTORY 1988-89. Darren L. Smith, editor. Gale Research Company, Book Tower, Detroit, MI 48226. (800) 521-0707. 4th edition. 1987. $360.00.

1987 DIRECTORY OF ENGINEERING SOCIETIES AND RELATED ORGANIZATIONS. Gordon Davis, editor. Hemisphere Publishing Corporation, 1010 Vermont Avenue, NW, Washington, DC 20005. (800) 526-0275. 12th edition. 1987. $100.00.

RESEARCH CENTERS DIRECTORY 1988. Gale Research Company, Book Tower, Detroit, MI 48226. (800) 521-0707. 12th edition. 1987. $365.00 for set.

SCIENTIFIC AND TECHNICAL ORGANIZATIONS AND AGENCIES DIRECTORY. Margaret Labash Young, editor. Gale Research Company, Book Tower, Detroit, MI 48226. (800) 521-0707. 2nd edition. 1987. $185.00.

WHO'S WHO IN ENGINEERING. Gordon Davis, editor. Hemisphere Publishing Corporation, 1010 Vermont Avenue, NW, Washington, DC 20005. (800) 526-0275. 6th edition. 1985. $200.00.

GENERAL WORKS

ELECTROMAGNETIC PRINCIPLES OF INTEGRATED OPTICS. D.L. Lee. John Wiley and Sons, Inc., 605 Third Avenue, New York, NY 10158. (800) 526-5368. 1986. $90.00.

INTRODUCTION TO OPTICAL ELECTRONICS. Kenneth A. Jones. Harper and Row Publishers, Inc., 10 East 53rd Street, New York, NY 10022. (800) 242-7737. 1987. $34.95.

INTRODUCTION TO OPTICS. Frank L. Pedrotti and Leno S. Pedrotti. Prentice-Hall Publishing, Inc., Englewood Cliffs, NJ 07632. (800) 562-0245. 1987. $49.95.

LASERS: INVENTIONS TO APPLICATION. Jesse H. Ausubel and H.D. Langford, editors. National Academy Press, 2101 Constitution Avenue, Washington, DC 20418. (202) 334-3313. 1987. $14.95 in paper.

OPTICAL FIBRES. J. Geisler and others. Pergamon Press, Inc., Maxwell House, Fairview Park, Elmsford, NY 10523. (914) 592-7700. 1986. $165.00.

OPTICAL INFORMATION PROCESSING. Francis Yu. John Wiley and Sons, Inc., 605 Third Avenue, New York, NY 10158. (800) 526-5368. 1982. $65.95.

OPTICAL SYSTEM DESIGN. Rudolf Kingslake. Academic Press, Inc., 6277 Sea Harbor Drive, Orlando, FL 32821. (800) 321-5068. 1983. $32.00.

OPTICS. M.V. Klein and T.E. Furtak. John Wiley and Sons, Inc., 605 Third Avenue, New York, NY 10158. (800) 526-5368. 2nd edition. 1983. $44.95.

PRINCIPLES OF OPTICS. M. Born and E. Wolf. Pergamon Press, Inc., Maxwell House, Fairview Park, Elmsford, NY 10523. (914) 592-7700. 6th edition. 1980. $35.00 in paper.

HANDBOOKS AND MANUALS

HANDBOOK OF OPTICS. Optical Society of America. McGraw-Hill Book Company, 1221 Avenue of the Americas, New York, NY 10020. (212) 512-2000. 1978. $100.00.

ONLINE DATA BASES

CA SEARCH. Chemical Abstracts Service, P.O. Box 3012, Columbus, OH 43120. (800) 848-6538 or (614) 421-3600. Comprehensive guide to chemical literature, 1972 to present. Inquire as to online cost and availability.

COMPENDEX. Engineering Information, Inc., 345 East 47th Street, New York, NY 10017. (800) 221-1044 or (212) 705-7615. Engineering and technical literature, 1975 to present. Inquire as to online cost and availability.

DISSERTATION ABSTRACTS ONLINE. University Microfilms International, 300 North Zeeb Road, Ann Arbor, MI 48106. (800) 521-0600 or (313) 761-4700. Scope includes virtually all doctoral dissertations accepted at accredited American institutions from 1861 to present in over 250 subject areas. Inquire as to online cost and availability.

INSPEC. INSPEC Marketing Department, Institution of Electrical Engineers. Available from IEEE Service Center, 445 Hoes Lane, Piscataway, NJ 08854. (201) 981-0060. Online version of Physics Abstracts. Inquire as to online cost and availability.

NTIS. National Technical Information Service, 5285 Port Royal Road, Springfield, VA 22161. (703) 487-4630. Broad coverage of government sponsored research reports, 1964 to present. Inquire as to online cost and availability.

SCISEARCH. Institute for Scientific Information, 3501 Market Street, Philadelphia, PA 19104. (800) 523-1850 or (215) 386-0100. Broad multidisciplinary title and author index to the international literature of science and technology, 1974 to present. Inquire as to online cost and availability.

WILSONLINE. H.W. Wilson and Company, 950 University Avenue, Bronx, NY 10452. (800) 367-6770 or (212) 588-8400. Makes available online versions of the H.W. Wilson indexes including Applied Science and Technology Index, Business

Periodicals Index and Readers' Guide to Periodical Literature. Approximately 1980 to present. Inquire as to online cost and availability.

PERIODICALS

APPLIED OPTICS. Optical Society of America. 1816 Jefferson Place, N.W., Washington, DC 20036. (202) 223-8130. Order from: American Institute of Physics, 335 East 45th Street, New York, NY 10017. (212) 661-9404. 1962 to present. Semi-monthly. $330.00 per year.

FIBER AND INTEGRATED OPTICS. Crane Russak and Company, Inc., 3 East 44th Street, New York, NY 10017. (212) 867-1490. 1977 to present. Quarterly. $86.00 per year.

IEEE JOURNAL OF QUANTUM ELECTRONICS. Institute of Electrical and Electronics Engineers. IEEE Service Center, 445 Hoes Lane, Piscataway, NJ 08854. 1962 to present. Monthly. $180.00 per year.

JOURNAL OF LIGHTWAVE TECHNOLOGY. Institute of Electrical and Electronics Engineers. IEEE Service Center, 445 Hoes Lane, Piscataway, NJ 08854. 1983 to present. Monthly. $145.00 per year.

LASER FOCUS. PennWell Publishing Company, 119 Russell Street, Littleton, MA 01460. (617) 486-9501. 1965 to present. Monthly. $55.00 per year.

OPTICAL ENGINEERING. Society of Photo-Optical Instrumentation Engineers (SPIE), P.O. Box 10, 1022 19th Street, Bellingham, WA 98227. (206) 676-3290. 1962 to present. Monthly. $95.00 per year.

OPTICAL SOCIETY OF AMERICA, JOURNAL, PARTS A AND B. Optical Society of America. 1816 Jefferson Place, N.W., Washington, DC 20036. (202) 223-8130. Order from: American Institute of Physics, 335 East 45th Street, New York, NY 10017. (212) 661-9404. 1917 to present. Monthly. $180.00 each part per year.

OPTICS COMMUNICATIONS. Elsevier Science Publishing Company, Inc., 52 Vanderbilt Avenue, New York, NY 10017. (212) 370-5520. 1969 to present. 24 times per year. $425.00 per year.

OPTICS LETTERS. Optical Society of America. Order from: American Institute of Physics, 335 East 45th Street, New York, NY 10017. (212) 661-9404. 1977 to present. Monthly. $150.00 per year.

SOCIETY OF PHOTO-OPTICAL INSTRUMENTATION ENGINEERS (SPIE) PROCEEDINGS. Society of Photo-Optical Instrumentation Engineers (SPIE), P.O. Box 10, 1022 19th Street, Bellingham, WA 98227. (206) 676-3290. 1963 to present. Approximately 50 numbers per year. Approximately $40.00 per number.

RESEARCH CENTERS AND INSTITUTES

CENTER FOR APPLIED OPTICS. University of Alabama in Huntsville, Research Institute Building, Huntsville, AL 35899. (205) 895-6102.

CENTER FOR APPLIED OPTICS. University of Texas at Dallas, P.O. Box 830688, Richardson, TX 75083. (214) 690-2868.

INSTITUTE OF OPTICS. University of Rochester, Rochester, NY 14627. (716) 275-2314.

OPTICAL PHYSICS LABORATORY. University of Miami, Coral Gables, FL 33124. (305) 284-2324.

OPTICAL DISKS

See also: LASERS, OPTICS, TELEVISION, VIDEO TECHNOLOGY

ABSTRACT SERVICES AND INDEXES

APPLIED SCIENCE AND TECHNOLOGY INDEX. H.W. Wilson and Company, 950 University Avenue, Bronx, NY 10452. (800) 367-6670 or (212) 588-8400. Monthly. Inquire as to cost and availability.

ELECTRICAL AND ELECTRONICS ABSTRACTS. Institution of Electrical Engineers. Available from: Institute of Electrical and Electronics Engineers. IEEE Service Center, 445 Hoes Lane, Piscataway, NJ 08854. Monthly. $1250.00 per year.

ELECTRONICS AND COMMUNICATIONS ABSTRACTS. Cambridge Scientific Abstracts, 5161 River Road, Bethesda, MD 20816. (301) 951-1400. Bimonthly. Inquire as to cost and availability.

ENGINEERING INDEX MONTHLY AND AUTHOR INDEX. Engineering Information Inc., 345 East 47th Street, New York, NY 10017. (212) 705-7600. Monthly. $1560.00 per year.

IEEE PUBLICATIONS BULLETIN. Institute of Electrical and Electronics Engineers. Institute of Electrical and Electronics Engineers. IEEE Service Center, 445 Hoes Lane, Piscataway, NJ 08854. Quarterly. Free.

PHYSICS ABSTRACTS. Institution of Electrical Engineers. Available from: IEEE Service Center, 445 Hoes Lane, Piscataway, NJ 08854. 1898 to present. Bimonthly. $1700.00 per year.

PHYSICS BRIEFS. Physik Verlag GmbH, Postfach 1260/1280, D-6940 Weinheim, West Germany. (212) 661-9404. 1920 to present. Twenty-six times per year. $1250.00 per year.

SCIENCE CITATION INDEX. Institute for Scientific Information, 3501 Market Street, Philadelphia, PA 19104. (800) 523-1850 or (215) 386-0100. Six times per year. $6200.00 per year.

ANNUAL REVIEWS AND YEARBOOKS

LASER APPLICATIONS. Academic Press, Inc., 6277 Sea Harbor Drive, Orlando, FL 32821. (800) 321-5068. 1971 to present. Irregular. Price varies, inquire.

SPRINGER SERIES IN OPTICAL SCIENCES. Springer-Verlag New York, Inc., 175 Fifth Avenue, New York, NY 10010. (800) 526-7254. 1976 to present. Irregular. Price varies, inquire.

ASSOCIATIONS AND PROFESSIONAL SOCIETIES

AMERICAN ELECTRONICS ASSOCIATION. P.O. Box 10045, 2670 Hanover Street, Palo Alto, CA 94303. (415) 857-9300.

AMERICAN INSTITUTE OF PHYSICS. 335 East 45th Street, New York, NY 10017. (212) 661-9494.

ELECTRONICS INDUSTRIES ASSOCIATION. 2001 Eye Street, N.W., Washington, DC 20006. (202) 457-4900.

INSTITUTE OF ELECTRICAL AND ELECTRONICS ENGINEERS. 345 East 47th Street, New York, NY 10017. (212) 705-7900.

LASER INSTITUTE OF AMERICA. 5151 Monroe Street, Suite 103, Toledo, OH 43623. (419) 882-8706.

NATIONAL ASSOCIATION OF RADIO AND TELECOMMUNICATIONS ENGINEERS. P.O. Box 15029, Salem, OR 97309. (503) 581-7653.

OPTICAL SOCIETY OF AMERICA. 1816 Jefferson Place, N.W., Washington, DC 20036. (202) 223-8130.

SOCIETY OF PHOTO-OPTICAL INSTRUMENTATION ENGINEERS - THE INTERNATIONAL SOCIETY OF OPTICAL ENGINEERING. P.O. Box 10, 1022 19th Street, Bellingham, WA 98227. (206) 676-3290.

DIRECTORIES AND BIOGRAPHICAL SOURCES

AMERICAN MEN AND WOMEN OF SCIENCE. R.R. Bowker, Inc., Order Department, 245 West 17th Street, New York, NY 10011. (800) 521-8110. Eight volumes. 1986. $595.00 for set.

IEEE MEMBERSHIP DIRECTORY. Institute of Electrical and Electronics Engineers. IEEE Service Center, 445 Hoes Lane, Piscataway, NJ 08854. Annual. $7.00.

INTERNATIONAL RESEARCH CENTERS DIRECTORY 1988-89. Darren L. Smith, editor. Gale Research Company, Book Tower, Detroit, MI 48226. (800) 521-0707. 4th edition. 1987. $360.00.

1987 DIRECTORY OF ENGINEERING SOCIETIES AND RELATED ORGANIZATIONS. Gordon Davis, editor. Hemisphere Publishing Corporation, 1010 Vermont Avenue, NW, Washington, DC 20005. (800) 526-0275. 12th edition. 1987. $100.00.

OPTICAL INFORMATION SYSTEMS - VIDEODISC PROJECTS DIRECTORY SECTION. Meckler Publishing, 11 Ferry Lane West, Westport, CT 06880. (203) 226-6967. Every two months. $75.00 per year.

RESEARCH CENTERS DIRECTORY 1988. Gale Research Company, Book Tower, Detroit, MI 48226. (800) 521-0707. 12th edition. 1987. $365.00 for set.

SCIENTIFIC AND TECHNICAL ORGANIZATIONS AND AGENCIES DIRECTORY. Margaret Labash Young, editor. Gale Research Company, Book Tower, Detroit, MI 48226. (800) 521-0707. 2nd edition. 1987. $185.00.

WHO'S WHO IN ELECTRONICS. Harris Publishing Company, 2057-2 Aurora Road, Twinsburg, OH 44087. (216) 425-9143. Annual. $90.00.

WHO'S WHO IN ENGINEERING. Gordon Davis, editor. Hemisphere Publishing Corporation, 1010 Vermont Avenue, NW, Washington, DC 20005. (800) 526-0275. 6th edition. 1985. $200.00.

ENCYCLOPEDIAS AND DICTIONARIES

IEEE STANDARD DICTIONARY OF ELECTRICAL AND ELECTRONICS TERMS. Frank Jay, editor. John Wiley and Sons, Inc., 605 Third Avenue, New York, NY 10158. (800) 526-5368. 3rd edition. 1984. $49.95.

GENERAL WORKS

DIGITAL VIDEO I. Frank Davidoff and John Rossi, editors. Society of Motion Picture and Television Engineers, 595 West Hartsdale Avenue, White Plains, NY 10607. (914) 472-6606. 1982. $25.00.

VIDEODISC SYSTEMS: THEORY AND APPLICATIONS. Jordan Isailovic. Prentice-Hall Publishing, Inc., Englewood Cliffs, NJ 07632. (800) 562-0245. 1987. $42.95.

HANDBOOKS AND MANUALS

HANDBOOK OF MODERN ELECTRONICS AND ELECTRICAL ENGINEERING. Charles Belove, editor. John Wiley and Sons, Inc., 605 Third Avenue, New York, NY 10158. (800) 526-5368. 1986. $88.95.

ONLINE DATA BASES

CA SEARCH. Chemical Abstracts Service, P.O. Box 3012, Columbus, OH 43120. (800) 848-6538 or (614) 421-3600. Comprehensive guide to chemical literature, 1972 to present. Inquire as to online cost and availability.

COMPENDEX. Engineering Information, Inc., 345 East 47th Street, New York, NY 10017. (800) 221-1044 or (212) 705-7615. Engineering and technical literature, 1975 to present. Inquire as to online cost and availability.

INSPEC. INSPEC Marketing Department, Institution of Electrical Engineers. Available from IEEE Service Center, 445 Hoes Lane, Piscataway, NJ 08854. (201) 981-0060. Online version of Physics Abstracts. Inquire as to online cost and availability.

NTIS. National Technical Information Service, 5285 Port Royal Road, Springfield, VA 22161. (703) 487-4630. Broad coverage of government sponsored research reports, 1964 to present. Inquire as to online cost and availability.

SCISEARCH. Institute for Scientific Information, 3501 Market Street, Philadelphia, PA 19104. (800) 523-1850 or (215) 386-0100. Broad multidisciplinary title and author index to the international literature of science and technology, 1974 to present. Inquire as to online cost and availability.

WILSONLINE. H.W. Wilson and Company, 950 University Avenue, Bronx, NY 10452. (800) 367-6770 or (212) 588-8400. Makes available online versions of the H.W. Wilson indexes including Applied Science and Technology Index, Business Periodicals Index and Readers' Guide to Periodical Literature. Approximately 1980 to present. Inquire as to online cost and availability.

OTHER SOURCES

A GUIDE TO THE LITERATURE OF ELECTRICAL AND ELECTRONICS ENGINEERING. Susan B. Ardis. Libraries Unlimited Inc., P.O. Box 263, Littleton, CO 80160. (303) 770-1220. 1987. $37.50.

PERIODICALS

ELECTRONIC DESIGN. Hayden Publishing Company, 10 Mulholland Drive, Hasbrouck Heights, NJ 07604. (201) 288-7520. 1952 to present. Biweekly. $40.00 per year.

ELECTRONICS. McGraw-Hill Book Company, 1221 Avenue of the Americas, New York, NY 10020. (212) 512-2000. 1930 to present. Weekly. $32.00 per year.

OPTICAL INFORMATION SYSTEMS MAGAZINE. Meckler Publishing, 11 Ferry Lane West, Westport, CT 06880. (203) 226-6967. 1981 to present. Bimonthly. $75.00 per year.

RESEARCH CENTERS AND INSTITUTES

AMERICAN VIDEO INSTITUTE. Rochester Institute of Technology, One Lomb Memorial Drive, Building 7, P.O. Box 9887, Rochester, NY 14623. (716) 475-6625.

COMMUNICATIONS AND INFORMATION PROCESSING GROUP. Rensselaer Polytechnic Institute, Electrical, Computer and Systems Engineering Department, Troy, NY 12180. (518) 266-6486.

OPTICAL FIBERS

See: FIBER OPTICS

OPTICS

See also: FIBER OPTICS, HOLOGRAPHY, OPTICAL COMMUNICATIONS, OPTICAL DISKS, OPTOELECTRONICS, PHYSICS, SPECTROSCOPY

ABSTRACT SERVICES AND INDEXES

APPLIED SCIENCE AND TECHNOLOGY INDEX. H.W. Wilson and Company, 950 University Avenue, Bronx, NY 10452. (800) 367-6670 or (212) 588-8400. Monthly. Inquire as to cost and availability.

CHEMICAL ABSTRACTS. American Chemical Society, Chemical Abstracts Service, Box 3012, Columbus, OH 43210. (614) 421-3600. 1907 to present. Weekly. $9500.00 per year.

CONFERENCE PAPERS INDEX. Cambridge Scientific Abstracts, 5161 River Road, Bethesda, MD 20816. 1972 to present. Monthly. Inquire as to cost and availability.

CURRENT CONTENTS: ENGINEERING, TECHNOLOGY AND APPLIED SCIENCES. Institute for Scientific Information, 3501 Market Street, Philadelphia, PA 19104. (800) 523-1850 or (215) 386-0100. Weekly. $275.00 per year.

ENGINEERING INDEX MONTHLY AND AUTHOR INDEX. Engineering Information Inc., 345 East 47th Street, New York, NY 10017. (212) 705-7600. Monthly. $1560.00 per year.

GENERAL SCIENCE INDEX. H.W. Wilson and Company, 950 University Avenue, Bronx, NY 10452. (800) 367-6670 or (212) 588-8400. 1978 to present. Monthly. Inquire as to cost and availability.

INDEX TO SCIENTIFIC AND TECHNICAL PROCEEDINGS. Institute for Scientific Information, 3501 Market Street, Philadelphia, PA 19104. (800) 523-1850 or (215) 386-0100. 1978 to present. Monthly. $775.00 per year.

INDEX TO SCIENTIFIC REVIEWS. Institute for Scientific Information, 3501 Market Street, Philadelphia, PA 19104. (800) 523-1850 or (215) 386-0100. 1974 to present. Semi-annual. $550.00 per year.

PHYSICS ABSTRACTS. Institution of Electrical Engineers. Available from: IEEE Service Center, 445 Hoes Lane, Piscataway, NJ 08854. 1898 to present. Bimonthly. $1700.00 per year.

PHYSICS BRIEFS. Physik Verlag GmbH, Postfach 1260/1280, D-6940 Weinheim, West Germany. (212) 661-9404. 1920 to present. Twenty-six times per year. $1250.00 per year.

SCIENCE CITATION INDEX. Institute for Scientific Information, 3501 Market Street, Philadelphia, PA 19104. (800) 523-1850 or (215) 386-0100. Six times per year. $6200.00 per year.

ASSOCIATIONS AND PROFESSIONAL SOCIETIES

AMERICAN INSTITUTE OF PHYSICS. 335 East 45th Street, New York, NY 10017. (212) 661-9404.

INSTITUTION OF ELECTRICAL AND ELECTRONICS ENGINEERS (IEEE). 345 East 47th Street, New York, NY 10017. (212) 705-7900.

OPTICAL SOCIETY OF AMERICA. 1816 Jefferson Place, N.W., Washington, DC 20036. (202) 223-8130.

SOCIETY OF PHOTOGRAPHIC SCIENTISTS AND ENGINEERS. 7003 Kilworth Lane, Springfield, VA 22151. (703) 642-9090.

SOCIETY OF PHOTO-OPTICAL INSTRUMENTATION ENGINEERS - THE INTERNATIONAL SOCIETY OF OPTICAL ENGINEERING. P.O. Box 10, 1022 19th Street, Bellingham, WA 98227. (206) 676-3290.

BIBLIOGRAPHIES

NEW TECHNICAL BOOKS: A SELECTIVE LIST WITH DESCRIPTIVE ANNOTATIONS. New York Public Library, Science and Technology Research Center, Fifth Avenue and 42nd Street, New York, NY 10018. (212) 930-0800. 1915 to present. Monthly. $15.00 per year.

SCIENCE BOOKS AND FILMS. American Association for the Advancement of Science, 1333 H Street, NW, Washington, DC 20005. (202) 326-6454. Five times per year. $20.00 per year.

SCIENTIFIC AND TECHNICAL BOOKS AND SERIALS IN PRINT 1988; AN INDEX TO LITERATURE IN SCIENCE AND TECHNOLOGY. R.R. Bowker Company, 205 East 42nd Street, New York, NY 10017. (800) 521-8110. $175.00.

DIRECTORIES AND BIOGRAPHICAL SOURCES

AMERICAN MEN AND WOMEN OF SCIENCE. R.R. Bowker, Inc., Order Department, 245 West 17th Street, New York, NY 10011. (800) 521-8110. Eight volumes. 1986. $595.00 for set.

INTERNATIONAL RESEARCH CENTERS DIRECTORY 1988-89. Darren L. Smith, editor. Gale Research Company, Book Tower, Detroit, MI 48226. (800) 521-0707. 4th edition. 1987. $360.00.

1987 DIRECTORY OF ENGINEERING SOCIETIES AND RELATED ORGANIZATIONS. Gordon Davis, editor. Hemisphere Publishing Corporation, 1010 Vermont Avenue, NW, Washington, DC 20005. (800) 526-0275. 12th edition. 1987. $100.00.

RESEARCH CENTERS DIRECTORY 1988. Gale Research Company, Book Tower, Detroit, MI 48226. (800) 521-0707. 12th edition. 1987. $365.00 for set.

SCIENTIFIC AND TECHNICAL ORGANIZATIONS AND AGENCIES DIRECTORY. Margaret Labash Young, editor. Gale Research Company, Book Tower, Detroit, MI 48226. (800) 521-0707. 2nd edition. 1987. $185.00.

WHO'S WHO IN ENGINEERING. Gordon Davis, editor. Hemisphere Publishing Corporation, 1010 Vermont Avenue, NW, Washington, DC 20005. (800) 526-0275. 6th edition. 1985. $200.00.

ENCYCLOPEDIAS AND DICTIONARIES

CONCISE SCIENCE DICTIONARY. Oxford University Press, 200 Madison Avenue, New York, NY 10016. (800) 458-5833. 1987. $9.95 in paper.

DICTIONARY OF THE PHYSICAL SCIENCES: TERMS, FORMULAS, DATA. Cesare Emiliani. Oxford University Press, 200 Madison Avenue, New York, NY 10016. (800) 458-5833. 1987. $19.95 in paper.

MCGRAW-HILL ENCYCLOPEDIA OF SCIENCE AND TECHNOLOGY. McGraw-Hill Book Company, 1221 Avenue of the Americas, New York, NY 10020. (212) 512-2000. 6th edition. 1987. $1600.00.

THESAURUS OF SCIENTIFIC, TECHNICAL, AND ENGINEERING TERMS. Hemisphere Publishing Corporation, 1010 Vermont Avenue, NW, Washington, DC 20005. (800) 526-0275. 1988. $125.00.

GENERAL WORKS

ELECTROMAGNETIC PRINCIPLES OF INTEGRATED OPTICS. D.L. Lee. John Wiley and Sons, Inc., 605 Third Avenue, New York, NY 10158. (800) 526-5368. 1986. $90.00.

FOURIER OPTICS: AN INTRODUCTION. E.G. Stewart. John Wiley and Sons, Inc., 605 Third Avenue, New York, NY 10158. (800) 526-5368. 1983. $27.95.

INTRODUCTION TO OPTICAL ELECTRONICS. Kenneth A. Jones. Harper and Row Publishers, Inc., 10 East 53rd Street, New York, NY 10022. (800) 242-7737. 1987. $34.95.

INTRODUCTION TO OPTICS. Frank L. Pedrotti and Leno S. Pedrotti. Prentice-Hall Publishing, Inc., Englewood Cliffs, NJ 07632. (800) 562-0245. 1987. $49.95.

LASERS: INVENTIONS TO APPLICATION. Jesse H. Ausubel and H.D. Langford, editors. National Academy Press, 2101 Constitution Avenue, Washington, DC 20418. (202) 334-3313. 1987. $14.95 in paper.

OPTICAL FIBRES. J. Geisler and others. Pergamon Press, Inc., Maxwell House, Fairview Park, Elmsford, NY 10523. (914) 592-7700. 1986. $165.00.

OPTICS. M.V. Klein and T.E. Furtak. John Wiley and Sons, Inc., 605 Third Avenue, New York, NY 10158. (800) 526-5368. 2nd edition. 1983. $44.95.

PRINCIPLES OF OPTICS. M. Born and E. Wolf. Pergamon Press, Inc., Maxwell House, Fairview Park, Elmsford, NY 10523. (914) 592-7700. 6th edition. 1980. $35.00 in paper.

HANDBOOKS AND MANUALS

HANDBOOK OF OPTICS. Optical Society of America. McGraw-Hill Book Company, 1221 Avenue of the Americas, New York, NY 10020. (212) 512-2000. 1978. $100.00.

OPTICS SOURCE BOOK. Sybil P. Parker, editor-in-chief. McGraw-Hill Book Company, 1221 Avenue of the Americas, New York, NY 10020. (212) 512-2000. 1988. $45.00.

ONLINE DATA BASES

CA SEARCH. Chemical Abstracts Service, P.O. Box 3012, Columbus, OH 43120. (800) 848-6538 or (614) 421-3600. Comprehensive guide to chemical literature, 1972 to present. Inquire as to online cost and availability.

COMPENDEX. Engineering Information, Inc., 345 East 47th Street, New York, NY 10017. (800) 221-1044 or (212) 705-7615. Engineering and technical literature, 1975 to present. Inquire as to online cost and availability.

DISSERTATION ABSTRACTS ONLINE. University Microfilms International, 300 North Zeeb Road, Ann Arbor, MI 48106. (800) 521-0600 or (313) 761-4700. Scope includes virtually all doctoral dissertations accepted at accredited American institutions from 1861 to present in over 250 subject areas. Inquire as to online cost and availability.

INSPEC. INSPEC Marketing Department, Institution of Electrical Engineers. Available from IEEE Service Center, 445 Hoes Lane, Piscataway, NJ 08854. (201) 981-0060. Online version of Physics Abstracts. Inquire as to online cost and availability.

NTIS. National Technical Information Service, 5285 Port Royal Road, Springfield, VA 22161. (703) 487-4630. Broad coverage of government sponsored research reports, 1964 to present. Inquire as to online cost and availability.

SCISEARCH. Institute for Scientific Information, 3501 Market Street, Philadelphia, PA 19104. (800) 523-1850 or (215) 386-0100. Broad multidisciplinary title and author index to the international literature of science and technology, 1974 to present. Inquire as to online cost and availability.

WILSONLINE. H.W. Wilson and Company, 950 University Avenue, Bronx, NY 10452. (800) 367-6770 or (212) 588-8400. Makes available online versions of the H.W. Wilson indexes including Applied Science and Technology Index, Business Periodicals Index and Readers' Guide to Periodical Literature. Approximately 1980 to present. Inquire as to online cost and availability.

OTHER SOURCES

WHAT EVERY ENGINEER SHOULD KNOW ABOUT ENGINEERING SOURCES. Marcel Dekker Inc., 270 Madison Avenue, New York, NY 10016. (800) 228-1160. 1984. $24.95.

PERIODICALS

APPLIED OPTICS. Optical Society of America. 1816 Jefferson Place, N.W., Washington, DC 20036. (202) 223-8130. Order from: American Institute of Physics, 335 East 45th Street, New York, NY 10017. (212) 661-9404. 1962 to present. Semi-monthly. $330.00 per year.

APPLIED SPECTROSCOPY REVIEWS. Marcel Dekker Inc., 270 Madison Avenue, New York, NY 10016. (800) 228-1160. 1964 tp present. Quarterly. $185.00 per year.

FIBER AND INTEGRATED OPTICS. Crane Russak and Company, Inc., 3 East 44th Street, New York, NY 10017. (212) 867-1490. 1977 to present. Quarterly. $86.00 per year.

JOURNAL OF LIGHTWAVE TECHNOLOGY. Institute of Electrical and Electronics Engineers. IEEE Service Center, 445 Hoes Lane, Piscataway, NJ 08854. 1983 to present. Monthly. $145.00 per year.

LASER FOCUS. PennWell Publishing Company, 119 Russell Street, Littleton, MA 01460. (617) 486-9501. 1965 to present. Monthly. $55.00 per year.

OPTICAL ENGINEERING. Society of Photo-Optical Instrumentation Engineers (SPIE), P.O. Box 10, 1022 19th Street, Bellingham, WA 98227. (206) 676-3290. 1962 to present. Monthly. $95.00 per year.

OPTICAL SOCIETY OF AMERICA, JOURNAL, PARTS A AND B. Optical Society of America. 1816 Jefferson Place, N.W., Washington, DC 20036. (202) 223-8130. Order from: American Institute of Physics, 335 East 45th Street, New York, NY 10017. (212) 661-9404. 1917 to present. Monthly. $180.00 each part per year.

OPTICS COMMUNICATIONS. Elsevier Science Publishing Company, Inc., 52 Vanderbilt Avenue, New York, NY 10017. (212) 370-5520. 1969 to present. 24 times per year. $425.00 per year.

OPTICS LETTERS. Optical Society of America. Order from: American Institute of Physics, 335 East 45th Street, New York, NY 10017. (212) 661-9404. 1977 to present. Monthly. $150.00 per year.

SOCIETY OF PHOTO-OPTICAL INSTRUMENTATION ENGINEERS (SPIE) PROCEEDINGS. Society of Photo-Optical Instrumentation Engineers (SPIE), P.O. Box 10, 1022 19th Street, Bellingham, WA 98227. (206) 676-3290. 1963 to present. Approximately 50 numbers per year. Approximately $40.00 per number.

RESEARCH CENTERS AND INSTITUTES

CENTER FOR APPLIED OPTICS. University of Alabama in Huntsville, Research Institute Building, Huntsville, AL 35899. (205) 895-6102.

CENTER FOR APPLIED OPTICS. University of Texas at Dallas, P.O. Box 830688, Richardson, TX 75083. (214) 690-2868.

INSTITUTE OF OPTICS. University of Rochester, Rochester, NY 14627. (716) 275-2314.

OPTICAL PHYSICS LABORATORY. University of Miami, Coral Gables, FL 33124. (305) 284-2324.

OPTOELECTRONICS

See also: ELECTRONICS, ELECTRONICS ENGINEERING, FIBER OPTICS, HOLOGRAPHY, LASERS, OPTICAL COMMUNICATION, OPTICAL DISKS, OPTICS

ABSTRACT SERVICES AND INDEXES

APPLIED SCIENCE AND TECHNOLOGY INDEX. H.W. Wilson and Company, 950 University Avenue, Bronx, NY 10452. (800) 367-6670 or (212) 588-8400. Monthly. Inquire as to cost and availability.

CHEMICAL ABSTRACTS. American Chemical Society, Chemical Abstracts Service, Box 3012, Columbus, OH 43210. (614) 421-3600. 1907 to present. Weekly. $9500.00 per year.

CONFERENCE PAPERS INDEX. Cambridge Scientific Abstracts, 5161 River Road, Bethesda, MD 20816. 1972 to present. Monthly. Inquire as to cost and availability.

CURRENT CONTENTS: ENGINEERING, TECHNOLOGY AND APPLIED SCIENCES. Institute for Scientific Information, 3501 Market Street, Philadelphia, PA 19104. (800) 523-1850 or (215) 386-0100. Weekly. $275.00 per year.

ELECTRICAL AND ELECTRONICS ABSTRACTS. Institution of Electrical Engineers. Available from: Institute of Electrical and Electronics Engineers. IEEE Service Center, 445 Hoes Lane, Piscataway, NJ 08854. Monthly. $1250.00 per year.

ENGINEERING INDEX MONTHLY AND AUTHOR INDEX. Engineering Information Inc., 345 East 47th Street, New York, NY 10017. (212) 705-7600. Monthly. $1560.00 per year.

GOVERNMENT REPORTS ANNOUNCEMENT AND INDEX. National Technical Information Service, 5285 Port Royal Road, Springfield, VA 22161. (703) 487-4929. Summaries of United States government sponsored research reports. 1946 to present. Biweekly. Inquire as to cost and availability.

INDEX TO SCIENTIFIC AND TECHNICAL PROCEEDINGS. Institute for Scientific Information, 3501 Market Street, Philadelphia, PA 19104. (800) 523-1850 or (215) 386-0100. 1978 to present. Monthly. $775.00 per year.

INDEX TO SCIENTIFIC REVIEWS. Institute for Scientific Information, 3501 Market Street, Philadelphia, PA 19104. (800) 523-1850 or (215) 386-0100. 1974 to present. Semi-annual. $550.00 per year.

PHYSICS ABSTRACTS. Institution of Electrical Engineers. Available from: IEEE Service Center, 445 Hoes Lane, Piscataway, NJ 08854. 1898 to present. Bimonthly. $1700.00 per year.

PHYSICS BRIEFS. Physik Verlag GmbH, Postfach 1260/1280, D-6940 Weinheim, West Germany. (212) 661-9404. 1920 to present. Twenty-six times per year. $1250.00 per year.

SCIENCE CITATION INDEX. Institute for Scientific Information, 3501 Market Street, Philadelphia, PA 19104. (800) 523-1850 or (215) 386-0100. Six times per year. $6200.00 per year.

ASSOCIATIONS AND PROFESSIONAL SOCIETIES

AMERICAN INSTITUTE OF PHYSICS. 335 East 45th Street, New York, NY 10017. (212) 661-9404.

INSTITUTION OF ELECTRICAL AND ELECTRONICS ENGINEERS (IEEE). 345 East 47th Street, New York, NY 10017. (212) 705-7900.

LASER INSTITUTE OF AMERICA. 5151 Monroe Street, Suite 103, Toledo, OH 43623. (419) 882-8706.

OPTICAL SOCIETY OF AMERICA. 1816 Jefferson Place, N.W., Washington, DC 20036. (202) 223-8130.

SOCIETY OF PHOTO-OPTICAL INSTRUMENTATION ENGINEERS - THE INTERNATIONAL SOCIETY OF OPTICAL ENGINEERING. P.O. Box 10, 1022 19th Street, Bellingham, WA 98227. (206) 676-3290.

DIRECTORIES AND BIOGRAPHICAL SOURCES

AMERICAN MEN AND WOMEN OF SCIENCE. R.R. Bowker, Inc., Order Department, 245 West 17th Street, New York, NY 10011. (800) 521-8110. Eight volumes. 1986. $595.00 for set.

INTERNATIONAL RESEARCH CENTERS DIRECTORY 1988-89. Darren L. Smith, editor. Gale Research Company, Book Tower, Detroit, MI 48226. (800) 521-0707. 4th edition. 1987. $360.00.

1987 DIRECTORY OF ENGINEERING SOCIETIES AND RELATED ORGANIZATIONS. Gordon Davis, editor. Hemisphere Publishing Corporation, 1010 Vermont Avenue, NW, Washington, DC 20005. (800) 526-0275. 12th edition. 1987. $100.00.

RESEARCH CENTERS DIRECTORY 1988. Gale Research Company, Book Tower, Detroit, MI 48226. (800) 521-0707. 12th edition. 1987. $365.00 for set.

SCIENTIFIC AND TECHNICAL ORGANIZATIONS AND AGENCIES DIRECTORY. Margaret Labash Young, editor. Gale Research Company, Book Tower, Detroit, MI 48226. (800) 521-0707. 2nd edition. 1987. $185.00.

WHO'S WHO IN ENGINEERING. Gordon Davis, editor. Hemisphere Publishing Corporation, 1010 Vermont Avenue, NW, Washington, DC 20005. (800) 526-0275. 6th edition. 1985. $200.00.

GENERAL WORKS

ELECTROMAGNETIC PRINCIPLES OF INTEGRATED OPTICS. D.L. Lee. John Wiley and Sons, Inc., 605 Third Avenue, New York, NY 10158. (800) 526-5368. 1986. $90.00.

INTRODUCTION TO OPTICAL ELECTRONICS. Kenneth A. Jones. Harper and Row Publishers, Inc., 10 East 53rd Street, New York, NY 10022. (800) 242-7737. 1987. $34.95.

INTRODUCTION TO OPTICS. Frank L. Pedrotti and Leno S. Pedrotti. Prentice-Hall Publishing, Inc., Englewood Cliffs, NJ 07632. (800) 562-0245. 1987. $49.95.

LASERS: INVENTIONS TO APPLICATION. Jesse H. Ausubel and H.D. Langford, editors. National Academy Press, 2101 Constitution Avenue, Washington, DC 20418. (202) 334-3313. 1987. $14.95 in paper.

OPTICAL FIBRES. J. Geisler and others. Pergamon Press, Inc., Maxwell House, Fairview Park, Elmsford, NY 10523. (914) 592-7700. 1986. $165.00.

OPTICAL INFORMATION PROCESSING. Francis Yu. John Wiley and Sons, Inc., 605 Third Avenue, New York, NY 10158. (800) 526-5368. 1982. $65.95.

OPTICAL SYSTEM DESIGN. Rudolf Kingslake. Academic Press, Inc., 6277 Sea Harbor Drive, Orlando, FL 32821. (800) 321-5068. 1983. $32.00.

OPTICS. M.V. Klein and T.E. Furtak. John Wiley and Sons, Inc., 605 Third Avenue, New York, NY 10158. (800) 526-5368. 2nd edition. 1983. $44.95.

PRINCIPLES OF OPTICS. M. Born and E. Wolf. Pergamon Press, Inc., Maxwell House, Fairview Park, Elmsford, NY 10523. (914) 592-7700. 6th edition. 1980. $35.00 in paper.

HANDBOOKS AND MANUALS

HANDBOOK OF OPTICS. Optical Society of America. McGraw-Hill Book Company, 1221 Avenue of the Americas, New York,

NY 10020. (212) 512-2000. 1978. $100.00.

ONLINE DATA BASES

CA SEARCH. Chemical Abstracts Service, P.O. Box 3012, Columbus, OH 43120. (800) 848-6538 or (614) 421-3600. Comprehensive guide to chemical literature, 1972 to present. Inquire as to online cost and availability.

COMPENDEX. Engineering Information, Inc., 345 East 47th Street, New York, NY 10017. (800) 221-1044 or (212) 705-7615. Engineering and technical literature, 1975 to present. Inquire as to online cost and availability.

DISSERTATION ABSTRACTS ONLINE. University Microfilms International, 300 North Zeeb Road, Ann Arbor, MI 48106. (800) 521-0600 or (313) 761-4700. Scope includes virtually all doctoral dissertations accepted at accredited American institutions from 1861 to present in over 250 subject areas. Inquire as to online cost and availability.

INSPEC. INSPEC Marketing Department, Institution of Electrical Engineers. Available from IEEE Service Center, 445 Hoes Lane, Piscataway, NJ 08854. (201) 981-0060. Online version of Physics Abstracts. Inquire as to online cost and availability.

NTIS. National Technical Information Service, 5285 Port Royal Road, Springfield, VA 22161. (703) 487-4630. Broad coverage of government sponsored research reports, 1964 to present. Inquire as to online cost and availability.

SCISEARCH. Institute for Scientific Information, 3501 Market Street, Philadelphia, PA 19104. (800) 523-1850 or (215) 386-0100. Broad multidisciplinary title and author index to the international literature of science and technology, 1974 to present. Inquire as to online cost and availability.

WILSONLINE. H.W. Wilson and Company, 950 University Avenue, Bronx, NY 10452. (800) 367-6770 or (212) 588-8400. Makes available online versions of the H.W. Wilson indexes including Applied Science and Technology Index, Business Periodicals Index and Readers' Guide to Periodical Literature. Approximately 1980 to present. Inquire as to online cost and availability.

PERIODICALS

APPLIED OPTICS. Optical Society of America. 1816 Jefferson Place, N.W., Washington, DC 20036. (202) 223-8130. Order from: American Institute of Physics, 335 East 45th Street, New York, NY 10017. (212) 661-9404. 1962 to present. Semi-monthly. $330.00 per year.

FIBER AND INTEGRATED OPTICS. Crane Russak and Company, Inc., 3 East 44th Street, New York, NY 10017. (212) 867-1490. 1977 to present. Quarterly. $86.00 per year.

IEEE JOURNAL OF QUANTUM ELECTRONICS. Institute of Electrical and Electronics Engineers. IEEE Service Center, 445 Hoes Lane, Piscataway, NJ 08854. 1962 to present. Monthly. $180.00 per year.

JOURNAL OF LIGHTWAVE TECHNOLOGY. Institute of Electrical and Electronics Engineers. IEEE Service Center, 445 Hoes Lane, Piscataway, NJ 08854. 1983 to present. Monthly. $145.00 per year.

LASER FOCUS. PennWell Publishing Company, 119 Russell Street, Littleton, MA 01460. (617) 486-9501. 1965 to present. Monthly. $55.00 per year.

OPTICAL ENGINEERING. Society of Photo-Optical Instrumentation Engineers (SPIE), P.O. Box 10, 1022 19th Street, Bellingham, WA 98227. (206) 676-3290. 1962 to present. Monthly. $95.00 per year.

OPTICAL SOCIETY OF AMERICA, JOURNAL, PARTS A AND B. Optical Society of America. 1816 Jefferson Place, N.W., Washington, DC 20036. (202) 223-8130. Order from: American Institute of Physics, 335 East 45th Street, New York, NY 10017. (212) 661-9404. 1917 to present. Monthly. $180.00 each part per year.

OPTICS COMMUNICATIONS. Elsevier Science Publishing Company, Inc., 52 Vanderbilt Avenue, New York, NY 10017. (212) 370-5520. 1969 to present. 24 times per year. $425.00 per year.

OPTICS LETTERS. Optical Society of America. Order from: American Institute of Physics, 335 East 45th Street, New York, NY 10017. (212) 661-9404. 1977 to present. Monthly. $150.00 per year.

SOCIETY OF PHOTO-OPTICAL INSTRUMENTATION ENGINEERS (SPIE) PROCEEDINGS. Society of Photo-Optical Instrumentation Engineers (SPIE), P.O. Box 10, 1022 19th Street, Bellingham, WA 98227. (206) 676-3290. 1963 to present. Approximately 50 numbers per year. Approximately $40.00 per number.

RESEARCH CENTERS AND INSTITUTES

CENTER FOR APPLIED OPTICS. University of Alabama in Huntsville, Research Institute Building, Huntsville, AL 35899. (205) 895-6102.

CENTER FOR APPLIED OPTICS. University of Texas at Dallas, P.O. Box 830688, Richardson, TX 75083. (214) 690-2868.

INSTITUTE OF OPTICS. University of Rochester, Rochester, NY 14627. (716) 275-2314.

OPTICAL PHYSICS LABORATORY. University of Miami, Coral Gables, FL 33124. (305) 284-2324.

ORGANIC CHEMISTRY

See also: CARBON, CHEMISTRY

ABSTRACT SERVICES AND INDEXES

CHEMICAL ABSTRACTS. American Chemical Society, Chemical Abstracts Service, Box 3012, Columbus, OH 43210. (614) 421-3600. 1907 to present. Weekly. $9500.00 per year.

CONFERENCE PAPERS INDEX. Cambridge Scientific Abstracts, 5161 River Road, Bethesda, MD 20816. 1972 to present. Monthly. Inquire as to cost and availability.

CURRENT CONTENTS: PHYSICAL, CHEMICAL AND EARTH SCIENCES. Institute for Scientific Information, 3501 Market Street, Philadelphia, PA 19104. (800) 523-1850 or (215) 386-0100. Weekly. $275.00 per year.

GENERAL SCIENCE INDEX. H.W. Wilson and Company, 950 University Avenue, Bronx, NY 10452. (800) 367-6770 or (212) 588-8400. 1978 to present. Monthly. Inquire as to cost and availability.

INDEX TO SCIENTIFIC AND TECHNICAL PROCEEDINGS. Institute for Scientific Information, 3501 Market Street, Philadelphia, PA 19104. (800) 523-1850 or (215) 386-0100. 1978 to present. Monthly. $775.00 per year.

INDEX TO SCIENTIFIC REVIEWS. Institute for Scientific Information, 3501 Market Street, Philadelphia, PA 19104. (800) 523-1850 or (215) 386-0100. 1974 to present. Semi-annual. $550.00 per year.

SCIENCE CITATION INDEX. Institute for Scientific Information, 3501 Market Street, Philadelphia, PA 19104. (800) 523-1850 or (215) 386-0100. Six times per year. $6200.00 per year.

ANNUAL REVIEWS AND YEARBOOKS

ADVANCES IN PHYSICAL ORGANIC CHEMISTRY. Academic Press, Inc., 6277 Sea Harbor Drive, Orlando, FL 32821. (800) 321-5068. 1963 to present. Irregular. Price varies, inquire.

STUDIES IN ORGANIC CHEMISTRY. Marcel Dekker Inc., 270 Madison Avenue, New York, NY 10016. (800) 228-1160. 1973 to present. Irregular. Price varies, inquire.

ASSOCIATIONS AND PROFESSIONAL SOCIETIES

AMERICAN CARBON SOCIETY. The Stackpole Corporation, St. Marys, PA 15857. (814) 781-8410.

AMERICAN CHEMICAL SOCIETY. 1155 16th Street, N.W., Washington, DC 20036. (202) 872-4600.

AMERICAN INSTITUTE OF CHEMICAL ENGINEERS. 345 East 47th Street, New York, NY 10017. (212) 705-7338.

AMERICAN OIL CHEMISTS' SOCIETY. 508 South Sixth Street, Champaign, IL 61820. (217) 359-2344.

ASSOCIATION OF OFFICIAL ANALYTICAL CHEMISTS. 1111 North 19th Street, Suite 210, Arlington, VA 22209. (703) 522-3032.

BIBLIOGRAPHIES

NEW TECHNICAL BOOKS: A SELECTIVE LIST WITH DESCRIPTIVE ANNOTATIONS. New York Public Library, Science and Technology Research Center, Fifth Avenue and 42nd Street, New York, NY 10018. (212) 930-0800. 1915 to present. Monthly. $15.00 per year.

SCIENCE BOOKS AND FILMS. American Association for the Advancement of Science, 1333 H Street, NW, Washington, DC 20005. (202) 326-6454. Five times per year. $20.00 per year.

SCIENTIFIC AND TECHNICAL BOOKS AND SERIALS IN PRINT 1988; AN INDEX TO LITERATURE IN SCIENCE AND TECHNOLOGY. R.R. Bowker Company, 205 East 42nd Street, New York, NY 10017. (800) 521-8110. $175.00.

DIRECTORIES AND BIOGRAPHICAL SOURCES

AMERICAN MEN AND WOMEN OF SCIENCE. R.R. Bowker, Inc., Order Department, 245 West 17th Street, New York, NY 10011. (800) 521-8110. Eight volumes. 1986. $595.00 for set.

INTERNATIONAL RESEARCH CENTERS DIRECTORY 1988-89. Darren L. Smith, editor. Gale Research Company, Book Tower, Detroit, MI 48226. (800) 521-0707. 4th edition. 1987. $360.00.

1987 DIRECTORY OF ENGINEERING SOCIETIES AND RELATED ORGANIZATIONS. Gordon Davis, editor. Hemisphere Publishing Corporation, 1010 Vermont Avenue, NW, Washington, DC 20005. (800) 526-0275. 12th edition. 1987. $100.00.

RESEARCH CENTERS DIRECTORY 1988. Gale Research Company, Book Tower, Detroit, MI 48226. (800) 521-0707. 12th edition. 1987. $365.00 for set.

SCIENTIFIC AND TECHNICAL ORGANIZATIONS AND AGENCIES DIRECTORY. Margaret Labash Young, editor. Gale Research Company, Book Tower, Detroit, MI 48226. (800) 521-0707. 2nd edition. 1987. $185.00.

WHO'S WHO IN ENGINEERING. Gordon Davis, editor. Hemisphere Publishing Corporation, 1010 Vermont Avenue, NW, Washington, DC 20005. (800) 526-0275. 6th edition. 1985. $200.00.

ENCYCLOPEDIAS AND DICTIONARIES

CONCISE SCIENCE DICTIONARY. Oxford University Press, 200 Madison Avenue, New York, NY 10016. (800) 458-5833. 1987. $9.95 in paper.

DICTIONARY OF ORGANIC COMPOUNDS. John Buckingham, editor. Methuen, Inc., 29 West 35th Street, New York, NY 10001. (212) 244-3336. Fifth edition. 1988. $675.00.

DICTIONARY OF ORGANOMETALLIC COMPOUNDS. J.E. MacIntyre, editor. Methuen, Inc., 29 West 35th Street, New York, NY 10001. (212) 244-3336. 1988. $325.00.

ENCYCLOPEDIA OF CHEMISTRY. Douglas M. Considine, editor. Van Nostrand Reinhold Company, Inc., 135 West 50th Street, New York, NY 10020. (800) 543-2681. 4th edition. 1984. $98.95.

MCGRAW-HILL ENCYCLOPEDIA OF SCIENCE AND TECHNOLOGY. McGraw-Hill Book Company, 1221 Avenue of the Americas, New York, NY 10020. (212) 512-2000. 6th edition. 1987. $1600.00.

THESAURUS OF SCIENTIFIC, TECHNICAL, AND ENGINEERING TERMS. Hemisphere Publishing Corporation, 1010 Vermont Avenue, NW, Washington, DC 20005. (800) 526-0275. 1988. $125.00.

GENERAL WORKS

THE ART OF PROBLEM SOLVING IN ORGANIC CHEMISTRY. M.E. Alonso. John Wiley and Sons, Inc., 605 Third Avenue, New York, NY 10158. (800) 526-5368. 1987. $29.95.

BASIC ORGANIC CHEMISTRY. F.L. Wiseman. McGraw-Hill Book Company, 1221 Avenue of the Americas, New York, NY 10020. (212) 512-2000. 1988. $37.95.

FUNDAMENTALS OF ORGANIC CHEMISTRY. T.W.G. Solomon. John Wiley and Sons, Inc., 605 Third Avenue, New York, NY 10158. (800) 526-5368. 2nd edition. 1986. $39.95.

ORGANIC SYNTHESIS REACTIONS AND MECHANISMS. B. Christoph and others. Springer-Verlag New York, Inc., 175 Fifth Avenue, New York, NY 10010. (800) 526-7254. 1986. $60.00.

REDUCTIONS IN ORGANIC CHEMISTRY. M. Hudlicky. John Wiley and Sons, Inc., 605 Third Avenue, New York, NY 10158. (800) 526-5368. 1984. $45.00.

HANDBOOKS AND MANUALS

CRC HANDBOOK OF CHEMISTRY AND PHYSICS. Robert C. Weast, editor. CRC Press, 2000 Corporate Boulevard, Boca Raton, FL 33431. (800) 272-7737. 68th edition. 1987. $69.95.

CRC HANDBOOK OF DATA ON ORGANIC COMPOUNDS. R.C. Weast and M.J. Astle, editors. CRC Press, 2000 Corporate Boulevard, Boca Raton, FL 33431. (800) 272-7737. 1985. Two volumes. $270.00 for set.

A GUIDEBOOK TO MECHANISMS IN ORGANIC CHEMISTRY. P. Sykes. John Wiley and Sons, Inc., 605 Third Avenue, New York, NY 10158. (800) 526-5368. 6th edition. 1986. $21.95.

HANDBOOK OF APPLIED CHEMISTRY. Vollrath Hopp and Ingo Hennig. Hemisphere Publishing Corporation, 79 Madison Avenue, New York, NY 10016-7892. Order from: Taylor and Francis/Hemisphere Distribution Center, 242 Cherry Street, Philadelphia, PA 19106-1906. (800) 821-8312. 1983. $49.95.

HANDBOOK OF ORGANIC CHEMISTRY. John A. Dean. McGraw-Hill Book Company, 1221 Avenue of the Americas, New York, NY 10020. (212) 512-2000. 1987. $64.50.

LANGE'S HANDBOOK OF CHEMISTRY. John A. Dean, editor. McGraw-Hill Book Company, 1221 Avenue of the Americas,

New York, NY 10020. (212) 512-2000. 1985. $59.50.

ONLINE DATA BASES

CA SEARCH. Chemical Abstracts Service, P.O. Box 3012, Columbus, OH 43120. (800) 848-6538 or (614) 421-3600. Comprehensive guide to chemical literature, 1972 to present. Inquire as to online cost and availability.

COMPENDEX. Engineering Information, Inc., 345 East 47th Street, New York, NY 10017. (800) 221-1044 or (212) 705-7615. Engineering and technical literature, 1975 to present. Inquire as to online cost and availability.

DISSERTATION ABSTRACTS ONLINE. University Microfilms International, 300 North Zeeb Road, Ann Arbor, MI 48106. (800) 521-0600 or (313) 761-4700. Scope includes virtually all doctoral dissertations accepted at accredited American institutions from 1861 to present in over 250 subject areas. Inquire as to online cost and availability.

NTIS. National Technical Information Service, 5285 Port Royal Road, Springfield, VA 22161. (703) 487-4630. Broad coverage of government sponsored research reports, 1964 to present. Inquire as to online cost and availability.

SCISEARCH. Institute for Scientific Information, 3501 Market Street, Philadelphia, PA 19104. (800) 523-1850 or (215) 386-0100. Broad multidisciplinary title and author index to the international literature of science and technology, 1974 to present. Inquire as to online cost and availability.

PERIODICALS

AMERICAN OIL CHEMISTS' SOCIETY. JOURNAL. American Oil Chemists' Society, 508 South Sixth Street, Champaign, IL 61820. (217) 359-2344. 1917 to present. Monthly. $60.00 per year.

CARBON. Pergamon Press, Inc., Maxwell House, Fairview Park, Elmsford, NY 10523. (914) 592-7700. 1963 to present. Bimonthly. $235.00 per year.

HYDROCARBON PROCESSING. Gulf Publishing Company, P.O. Box 2608, Houston, TX 77001. (713) 520-4444. Monthly. $10.00 per year.

JOURNAL OF HETEROCYCLIC CHEMISTRY. Hetero-Corporation, Box 16000 MH, Tampa, FL 33687. 1964 to present. 6 times per year. $160.00 per year.

JOURNAL OF ORGANIC CHEMISTRY. American Chemical Society, 1155 16th Street, N.W., Washington, DC 20036. (800) 424-6747. 1936 to present. Semi-monthly. $218.00 per year.

JOURNAL OF POLYMER SCIENCE. POLYMER CHEMISTRY EDITION. John Wiley and Sons, Inc., 605 Third Avenue, New York, NY 10158. (800) 526-5368. 1962 to present. Monthly. $895.00 per year, includes all editions.

ORGANOMETALLICS. American Chemical Society, 1155 16th Street, N.W., Washington, DC 20036. (800) 424-6747. 1982 to present. Monthly. $195.00 per year.

TETRAHEDRON. Pergamon Press, Inc., Maxwell House, Fairview Park, Elmsford, NY 10523. (914) 592-7700. 1957 to present. 24 times per year. $1400.00 per year.

TETRAHEDRON LETTERS. Pergamon Press, Inc., Maxwell House, Fairview Park, Elmsford, NY 10523. (914) 592-7700. 1959 to present. Weekly. $1500.00 per year.

RESEARCH CENTERS AND INSTITUTES

CHEMICAL LABORATORIES. Harvard University, Oxford Street, Cambridge, MA 02138. (617) 495-4283.

CHEMICAL RESEARCH LABORATORY. Brown University, Providence, RI 02912. (401) 863-2256.

UNIVERSITY/INDUSTRY CHEMICAL RESEARCH CENTER. Mississippi State University, Department of Chemistry, P.O. Drawer CH, Mississippi State, MS 39762. (601) 325-3584.

ORGANOMETALLIC COMPOUNDS

See: ORGANIC CHEMISTRY

OSCILLOSCOPE

See: INSTRUMENTATION

OZONE

See also: METEOROLOGY

ABSTRACT SERVICES AND INDEXES

APPLIED SCIENCE AND TECHNOLOGY INDEX. H.W. Wilson Company, 950 University Avenue, Bronx, NY 10452. (800) 367-6670 or (212) 588-8400. Inquire as to cost and availability.

CHEMICAL ABSTRACTS. Chemical Abstracts Service, 2540 Olentangy Road, Post Office Box 3012, Columbus, OH 43210. (800) 848-6538 or (614) 421-3600. Weekly. $9200.00 per year.

GENERAL SCIENCE INDEX. H.W. Wilson Company, 950 University Avenue, Bronx, NY 10452. (800) 367-6770 or (212) 588-8400. Inquire as to cost and availability.

INDEX TO SCIENTIFIC REVIEWS. Institute for Scientific Information, 3501 Market Street, Philadelphia, PA 19104. (800) 523-1850 or (215) 386-0100. Semiannual. $550.00 per year.

METEOROLOGICAL AND GEOASTROPHYSICAL ABSTRACTS. American Meteorological Society, 45 Beacon Street, Boston, MA 02108. (617) 227-2425. 1950 to present. Monthly. $450.00 per year.

OCEANIC ABSTRACTS. Cambridge Scientific Abstracts, Incorporated, 5161 River Road, Bethesda, MD 20816. (301) 951-1400. 1964 to present. Inquire as to online cost and availability.

OCEANOGRAPHIC LITERATURE REVIEW. Pergamon Press Incorporated, Maxwell House, Fairview Park, Elmsford, NY 10523. (914) 592-7700. 1979 to present. Inquire as to cost and availability.

SCIENCE CITATION INDEX. Institute for Scientific Information, 3501 Market Street, Philadelphia, PA 19104. (800) 523-1850 or (215) 386-0100. Inquire as to cost and availability.

ANNUAL REVIEWS AND YEARBOOKS

OCEAN YEARBOOK. Elizabeth M. Borgese and Norton Ginsberg, editors. University of Chicago Press, 5801 Ellis Avenue, Chicago, IL 60637. (312) 962-7906. 1979 to present. $49.00 per volume.

ASSOCIATIONS AND PROFESSIONAL SOCIETIES

AMERICAN METEOROLOGICAL SOCIETY. 45 Beacon Street, Boston, MA 02108. (617) 227-2425.

INTERNATIONAL ASSOCIATION OF METEOROLOGY AND ATMOSPHERIC PHYSICS. UCAR, Post Office Box 3000,

Boulder, CO 80307.

NATIONAL WEATHER ASSOCIATION. 4400 Stamp Road, Room 404, Temple Hills, MD 20748. (301) 899-3784.

UNIVERSITY CORPORATION FOR ATMOSPHERIC RESEARCH. Box 3000, 1850 Table Mesa Drive, Boulder, CO 80307. (303) 497-1000.

BIBLIOGRAPHIES

SCIENCE BOOKS AND FILMS. American Association for the Advancement of Science, 1333 H Street, NW, Washington, DC 20005. Five times per year. $20.00 per year.

SCIENTIFIC AND TECHNICAL BOOKS AND SERIALS IN PRINT 1988: AN INDEX TO LITERATURE IN SCIENCE AND TECHNOLOGY. R.R. Bowker Company, 205 East 42nd Street, New York, NY 10017. (800) 521-8110 or (212) 916-1600. $175.00.

DIRECTORIES AND BIOGRAPHICAL SOURCES

GOVERNMENT RESEARCH DIRECTORY. Gale Research Company, Book Tower, Detroit, MI 48226. (800) 521-0707. Fourth edition. 1987. $350.00.

INTERDOC: DIRECTORY OF PUBLISHED PROCEEDINGS, SERIES. SEMT-Science/Engineering/Medicine/Technology. Interdoc Corporation, 173 Halstead Avenue, Box 326, Harrison, NY 10528. (014) 835-3506. Ten times per year. $325.00 per year.

INTERNATIONAL RESEARCH CENTERS DIRECTORY 1988-1989. Gale Research Company, Book Tower, Detroit, MI 48226. (800) 521-0707. Fourth edition. 1987. $360.00.

METEOROLOGICAL SERVICES OF THE WORLD. World Meteorological Organization. Available from: American Meteorological Society, 45 Beacon Street, Boston, MA 02108. (617) 227-2425. Annual. $35.00.

RESEARCH CENTERS DIRECTORY. Gale Research Company, Book Tower, Detroit, MI 48226. (800) 521-0707. Twelfth edition. 1987. $365.00 for set.

ENCYCLOPEDIAS AND DICTIONARIES

ENCYCLOPEDIA OF CLIMATOLOGY. John E. Oliver and Rhodes W. Fairbridge, editors. Van Nostrand Reinhold, Incorporated, 115 Fifth Avenue, New York, NY 10003. (800) 543-2681. 1987. $89.95.

ENCYCLOPEDIA OF PHYSICAL SCIENCE AND TECHNOLOGY. Academic Press, Incorporated, Orlando, FL 32887. (800) 321-5068 or (305) 345-2734. Inquire as to cost and availability.

MCGRAW-HILL ENCYCLOPEDIA OF SCIENCE AND TECHNOLOGY. McGraw-Hill Book, Incorporated, 1221 Avenue of the Americas, New York, NY 10020. (212) 997-3675. Fifth edition, 15 volumes. $1100.00.

GENERAL WORKS

ATMOSPHERIC OZONE. C.S. Zerefos and A. Ghazi. Kluwer Academic, 190 Old Derby Street, Hingham, MA 02043. (617) 749-5262. 1985. $99.00.

HANDBOOK OF OZONE TECHNOLOGY AND APPLICATIONS, VOL. 2. Rip G. Rice and Aharon Netzer. Butterworth's, 80 Montvale Avenue, Stoneham, MA 02180. (617) 438-8464. 1984. $59.95.

OZONE. M. Horvarth, et al. Elsevier Science Publishing Company, Incorporated, 52 Vanderbilt Avenue, New York, NY 10017. (212) 370-5520. 1985. $72.25.

STRATOSPHERIC OZONE REDUCTION, SOLAR ULTRAVIOLET RADIATION AND PLANT LIFE. R.C. Worrest and M.M. Caldwell. Springer-Verlag, New York, Incorporated, 175 Fifth Avenue, New York, NY 10010. (212) 460-1500. 1986. $72.00.

ONLINE DATA BASES

CA SEARCH. Chemical Abstracts Service, Post Office Box 3012, Columbus, OH 43210. Guide to chemical literature, 1972 to present. Inquire as to cost and availability.

DISSERTATION ABSTRACTS ONLINE. University Microfilms International, 300 North Zeeb Road, Ann Arbor, MI 48106. (800) 521-0600 or (313) 761-4700. Scope includes virtually all doctoral dissertations accepted at accredited American institutions from 1861 to present in 252 subject areas. Inquire as to cost and availability.

INSPEC. INSPEC Marketing Department, Institute of Electrical and Electronics Engineers, Incorporated, IEEE Service Department, 445 Hoes Lane, Piscataway, NJ 08854. (201) 981-0060. Inquire as to on-line cost and availability.

METEOROLOGICAL AND GEOASTROPHYSICAL ABSTRACTS. American Meteorological Society, 45 Beacon Street, Boston, MA 02108. (617) 227-2425. 1950 to present. Monthly. $450.00 per year.

NTIS. National Technical Information Service, 5285 Port Royal Road, Springfield, VA 22161. (703) 487-4630. Broad coverage of government sponsored research reports, 1964 to present. Inquire as to cost and availability.

OCEANIC ABSTRACTS. Cambridge Scientific Abstracts, Incorporated, 5161 River Road, Bethesda, MD 20816. (301) 951-1400. 1964 to present. Inquire as to online cost and availability.

SCISEARCH. Institute for Scientific Information, 3501 Market Street, Philadelphia, PA 19104. (800) 523-1850 or (215) 386-0100. Broad multidisciplinary title and author index to the international literature of science and technology, 1974 to present. Inquire as to cost and availability.

WILSONLINE. H.W. Wilson Company, 950 University Avenue, Bronx, NY 10452. (800) 367-6770 or (212) 588-8400. Makes available online versions of the printed H.W. Wilson Indexes including Applied Science and Technology Index, Business Periodicals Index, and Readers' Guide to Periodical Literature. Period covered is generally 1983 to present. Inquire as to cost and availability.

PERIODICALS

AGRICULTURAL AND FOREST METEOROLOGY. Elsevier Science Publishing Company, Incorporated, 52 Vanderbilt Avenue, New York, NY 10017. (212) 370-5520. 1964 to present. Monthly. $260.00 per year.

AMERICAN METEOROLOGICAL SOCIETY BULLETIN. American Meteorological Society, 45 Beacon Street, Boston, MA 02108. (617) 227-2425. 1920 to present. Monthly. $60.00 per year.

BOUNDARY-LAYER METEOROLOGY: AN INTERNATIONAL JOURNAL OF PHYSICAL AND BIOLOGICAL PROCESSES IN THE ATMOSPHERIC BOUNDARY LAYER. D. Reidel Publishing Company, 190 Old Derby Street, Hingham, MA 02043. (617) 871-6600. 1970 to present. Sixteen times per year. $425.00 per year.

CLIMATIC CHANGE: AN INTERDISCIPLINARY, INTERNATIONAL JOURNAL DEVOTED TO THE DESCRIPTION, CAUSES AND IMPLICATIONS OF CLIMATIC CHANGE. D. Reidel Publishing Company, 190 Old Derby Street, Hingham, MA 02043. (617) 871-6600. 1977 to present.

Six times per year. $125.00 per year.

DYNAMICS OF ATMOSPHERES AND OCEANS. Elsevier Science Publishing Company, Incorporated, 52 Vanderbilt Avenue, New York, NY 10017. (212) 370-5520. 1977 to present. Quarterly. $90.00 per year.

JOURNAL OF ATMOSPHERIC AND OCEANIC TECHNOLOGY. American Meteorological Society, 45 Beacon Street, Boston, MA 02108. (617) 227-2425. 1984 to present. Quarterly. $80.00 per year.

JOURNAL OF CLIMATE AND APPLIED METEOROLOGY. American Meteorological Society, 45 Beacon Street, Boston, MA 02108. (617) 227-2425. 1962 to present. Monthly. $120.00 per year.

JOURNAL OF THE ATMOSPHERIC SCIENCES. American Meteorological Society, 45 Beacon Street, Boston, MA 02108. (617) 227-2425. 1944 to present. Semimonthly. $220.00 per year.

RESEARCH CENTERS AND INSTITUTES

NATIONAL CENTER FOR ATMOSPHERIC RESEARCH. Box 3000, Boulder, CO 80307. (303) 497-1000.

UNIVERSITY OF FLORIDA. Interdisciplinary Center for Aeronomy and other Atmospheric Sciences, 311 Space Sciences Research Building, Gainesville, FL 32611. (904) 392-2001.

P

PAPER CHEMISTRY

ABSTRACT SERVICES AND INDEXES

ABSTRACT BULLETIN. Institute of Paper Chemistry, 1043 East South River Street, Appleton, WI 54911. Monthly. $600.00 per year.

APPLIED SCIENCE AND TECHNOLOGY INDEX. H.W. Wilson Company, 950 University Avenue, Bronx, NY 10452. (800) 367-6670 or (212) 588-8400. Inquire as to cost and availability.

CHEMICAL ABSTRACTS. Chemical Abstracts Service, 2540 Olentangy Road, Post Office Box 3012, Columbus, OH 43210. (800) 848-6538 or (614) 421-3600. Weekly. $9200.00 per year.

PAPER AND BOARD ABSTRACTS. Pergamon Journals, Incorporated, Maxwell House, Fairview Park, Elmsford, NY 10523. (914) 592-7700. Monthly. $385.00 per year.

SCIENCE CITATION INDEX. Institute for Scientific Information, 3501 Market Street, Philadelphia, PA 19104. (800) 523-1850 or (215) 386-0100.

ANNUAL REVIEWS AND YEARBOOKS

PAPER YEARBOOK. Harcourt, Brace and Jovanovich, Incorporated, 757 Third Avenue, New York, NY 10164. (212) 614-3000. Annual. $55.00.

ASSOCIATIONS AND PROFESSIONAL SOCIETIES

AMERICAN CHEMICAL SOCIETY. 1155 16th Street, NW, Washington, DC 20036. (202) 872-4600.

AMERICAN PAPER INSTITUTE. 260 Madison Avenue, New York, NY 10016. (212) 340-0600.

INSTITUTE OF PAPER CHEMISTRY. 1043 East South River Street, Appleton, WI 54911. (414) 734-9251.

TECHNICAL ASSOCIATION OF THE PULP AND PAPER INDUSTRY. Technology Park/Atlanta, Post Office Box 105113, Atlanta, GA 60005. (312) 956-0250.

DIRECTORIES AND BIOGRAPHICAL SOURCES

CONSULTING SERVICES: CHEMISTS AND CHEMICAL ENGINEERS. Association of Consulting Chemists and Chemical Engineers, 50 East 41st Street, New York, NY 10017. (212) 684-6255. Annual. 1986. $45.00.

INTERNATIONAL RESEARCH CENTERS DIRECTORY 1986-1987. Gale Research Company, Book Tower, Detroit, MI 48226. (800) 521-0707. Fourth edition. 1987. $360.00 for set.

RESEARCH CENTERS DIRECTORY. Gale Research Company, Book Tower, Detroit, MI 48226. (800) 521-0707. Twelfth edition. 1987.

TAPPI DIRECTORY. Technical Association of the Pulp and Paper Industry. Technology Park/Atlanta, Post Office Box 105113, Atlanta, Ga 60005. (312) 956-0250. Annual. Inquire as to cost and availability.

WORLD GUIDE TO SCIENTIFIC ASSOCIATIONS AND LEARNED SOCIETIES. K.G. Saur Incorporated, 175 Fifth Avenue, New York, NY 10010. (800) 521-0707 or (212) 982-1302. Fourth edition. 1984. $112.00.

WHO'S WHO IN FRONTIER SCIENCE AND TECHNOLOGY. Marquis Who's Who, Incorporated, 200 East Ohio Street, Chicago, IL 60611. (800) 428-3898 or (312) 787-2008.

WHO'S WHO IN TECHNOLOGY TODAY. Reston Publishing Company, Incorporated, c/o Prentice-Hall, Incorporated, Englewood Cliffs, NJ 07632. (800) 262-6868. Biennial. Five volumes.. $425.00. Covers the fields of electronics, computer science, physics, optics, chemistry, biotechnology, mechanics, energy, and earth science.

ENCYCLOPEDIAS AND DICTIONARIES

CONCISE ENCYCLOPEDIA OF CHEMICAL TECHNOLOGY. Kirk-Othmer. John Wiley and Sons, Incorporated, 605 Third Avenue, New York, NY 10158. (800) 526-5368 or (212) 850-6000. Third edition. 1985. $129.95.

CONDENSED CHEMICAL DICTIONARY. Gessner Hawley. Van Nostrand Reinhold, 115 Fifth Avenue, New York, NY 10003. Tenth edition. 1981. $49.95.

ENCYCLOPEDIA OF PHYSICAL SCIENCE AND TECHNOLOGY. Academic Press, Incorporated, Orlando, FL 32887. (800) 321-5068 or (305) 345-2734. Fifteen volumes, 1986.

GLOSSARY OF CHEMICAL TERMS. Clifford A. Hampel and Gessner G. Hawley. Van Nostrand Reinhold Company, 115 Fifth Avenue, New York, NY 10003. (800) 543-2681 or (212) 254-3232. Second edition. 1982. $21.95.

MCGRAW-HILL ENCYCLOPEDIA OF SCIENCE AND TECHNOLOGY. McGraw-Hill Book Company, Incorporated, 1221 Avenue of the Americas, New York, NY 10020. (212) 997-3675.

VAN NOSTRAND REINHOLD ENCYCLOPEDIA OF CHEMISTRY. Douglas M. Considine and Glenn D. Considine. Van Nostrand Reinhold Publishing Company, Incorporated, 115 Fifth Avenue, New York, NY 10003. (800) 543-2681 or (212) 254-3232. 1984. $97.95.

GENERAL WORKS

PULP AND PAPER: CHEMISTRY AND CHEMICAL TECHNOLOGY. James P. Casey, editor. John Wiley and Sons, Incorporated, 605 Third Avenue, New York, NY 10158. (800) 526-5368 or (212) 850-6000. Third edition. 1983. Four volumes, $305.00 for set.

STRUCTURE AND PHYSICAL PROPERTIES OF PAPER. H.F. Rance. Elsevier Science Publishing Company, Incorporated, 52 Vanderbilt Avenue, New York, NY 10017. (212) 370-5520. 1982. $85.00.

HANDBOOKS AND MANUALS

LANGE'S HANDBOOK OF CHEMISTRY. John A. Dean, editor. McGraw-Hill Book Company, Incorporated, 1221 Avenue of the Americas, New York, NY 10020. (800) 628-0004. 1985. $59.50.

ONLINE DATA BASES

CA SEARCH. Chemical Abstracts Service, Post Office Box 3012, Columbus, OH 43210. Guide to chemical literature, 1972 to present. Inquire as to cost and availability.

DISSERTATION ABSTRACTS ONLINE. University Microfilms International, 300 North Zeeb Road, Ann Arbor, MI 48106. (800) 521-0600 or (313) 761-4700. Scope includes virtually all doctoral dissertations accepted at accredited American institutions from 1861 to present in 252 subject areas. Inquire as to online cost and availability.

NTIS. National Technical Information Service, 5285 Port Royal Road, Springfield, VA 22161. (703) 487-4630. Broad coverage of government sponsored research reports, 1964 to present. Inquire as to online cost and availability.

PAPERCHEM. Institute of Paper Chemistry, 1043 East South River Street, Appleton, WI 54911. (414) 734-9251. Worldwide coverage of the scientific and technical paper industry chemical literature. 1968 to present. Inquire as to online cost and availability.

PIRA. Research Association for the Paper and board, Printing and Packaging Industries, Randalls Road, Leatherhead, Surrey, England KT22 7RU. Citations and abstracts pertaining to bookbinding nd other pulp, paper, and packaging industries. 1975 to present. Inquire as to online cost and availability.

SCISEARCH. Institute for Scientific Information, 3501 Market Street, Philadelphia, PA 19104. (800) 523-1850 or (215) 386-0100. Broad multidisciplinary title and author index to the international literature of science and technology, 1974 to present. Inquire as to online cost and availability.

OTHER SOURCES

GUIDE TO BASIC INFORMATION SOURCES IN CHEMISTRY. Arthur Antony. John Wiley and Sons, Incorporated, 605 Third Avenue, New York, NY 10158. (800) 526-5368 or (212) 850-6000. 1979. $26.95.

HOW TO FIND CHEMICAL INFORMATION: A GUIDE FOR PRACTICING CHEMISTS, TEACHERS, AND STUDENTS. John Wiley and Sons, Incorporated, 605 Third Avenue, New York, NY 10158. (800) 526-5368 or (212) 850-6000. 1986. $35.00.

PERIODICALS

JOURNAL OF PULP AND PAPER SCIENCE. Canadian Pulp, Sun Life Building, 1155 Metcalfe Street, Montreal, PQ H3B 2X9, Canada. (514) 866-6621. Bimonthly. $55.00 per year.

PULP AND PAPER. Miller Freeman Publications, 500 Howard Street, San Francisco, CA 94105. (415) 397-1881. Monthly. $50.00 per year.

TAPPI JOURNAL. Technical Association of the Pulp and Paper Industry. Technology Park/Atlanta, Post Office Box 105113, Atlanta, GA 60005. (312) 956-0250. Monthly.

RESEARCH CENTERS AND INSTITUTES

NORTH CAROLINA STATE UNIVERSITY. Pulp and Paper Laboratory, Box 8005, Raleigh, NC 27965. (919) 737-3181.

PULP AND PAPER RESEARCH INSTITUTE OF CANADA. 570 St. John Boulevard, Pointe Claire, PQ, Canada, H9R 3J9. (514) 630-4100.

TECHNICAL ASSOCIATION OF THE PULP AND PAPER INDUSTRY. Technology Park/Atlanta, Post Office Box 105113, Atlanta, GA 60005. (312) 956-0250.

UNIVERSITY OF WISCONSIN AT MADISON. Wisconsin Center for Structural and Materials Testing, 1415 Johnson Drive, Madison, WI 53706. (608) 262-3205.

PARALLEL COMPUTERS

See also: ARTIFICIAL INTELLIGENCE, COMPUTER OPERATING SYSTEMS, COMPUTER PROGRAMMING, COMPUTERS, SOFTWARE, SOFTWARE ENGINEERING, SYSTEMS ANALYSIS, SYSTEMS ENGINEERING

ABSTRACT SERVICES AND INDEXES

APPLIED SCIENCE AND TECHNOLOGY INDEX. H.W. Wilson and Company, 950 University Avenue, Bronx, NY 10452. (800) 367-6670 or (212) 588-8400. Monthly. Inquire as to cost and availability.

COMPUTER AND CONTROL ABSTRACTS. Institute of Electrical Engineers. Available from: Institute of Electrical and Electronics Engineers. IEEE Service Center, 445 Hoes Lane, Piscataway, NJ 08854. Semimonthly. $775.00 per year.

COMPUTER AND INFORMATION SYSTEMS: AN ABSTRACT JOURNAL PERTAINING TO THE THEORY, DESIGN, FABRICATION AND APPLICATION OF COMPUTER AND INFORMATION SYSTEMS. Cambridge Scientific Abstracts, 5161 River Road, Bethesda, MD 20816. 1972 to present. Semi-monthly. Inquire as to cost and availability.

COMPUTER CONTENTS: THE BIWEEKLY COMPILATION OF TABLES OF CONTENTS FROM COMPUTER, ELECTRONIC AND TELECOMMUNICATIONS MAGAZINES, JOURNALS AND TRANACTIONS. Find/SVP, 500 Fifth Avenue, New York, NY 10111. (800) 346-3787 or (212) 354-2424. Biweekly. $115.00 per year.

COMPUTER LITERATURE INDEX. Applied Computer Research, Inc., P.O. Box 9280, Phoenix, AZ 85068. (602) 995-5929. Quarterly. $125.00 per year.

COMPUTER PROGRAMS ABSTRACTS. U.S. National Aeronautics and Space Administration. Available from: U.S. Government Printing Office, Washington, DC 20402. Quarterly. $10.00 per year.

COMPUTING REVIEWS. Association of Computing Machinery, 11 West 42nd Street, New York, NY 10036. (212) 869-7440. Monthly. $60.00 per year.

CONFERENCE PAPERS INDEX. Cambridge Scientific Abstracts, 5161 River Road, Bethesda, MD 20816. 1972 to present. Monthly. Inquire as to cost and availability.

CURRENT CONTENTS: ENGINEERING, TECHNOLOGY AND APPLIED SCIENCES. Institute for Scientific Information, 3501

Market Street, Philadelphia, PA 19104. (800) 523-1850 or (215) 386-0100. Weekly. $275.00 per year.

ENGINEERING INDEX MONTHLY AND AUTHOR INDEX. Engineering Information Inc., 345 East 47th Street, New York, NY 10017. (212) 705-7600. Monthly. $1560.00 per year.

SCIENCE CITATION INDEX. Institute for Scientific Information, 3501 Market Street, Philadelphia, PA 19104. (800) 523-1850 or (215) 386-0100. Six times per year. $6200.00 per year.

ANNUAL REVIEWS AND YEARBOOKS

ADVANCES IN COMPUTERS. Academic Press, Inc., 6277 Sea Harbor Drive, Orlando, FL 32821. (800) 321-5068. Yearly. Approximately $50.00 per volume.

ASSOCIATIONS AND PROFESSIONAL SOCIETIES

AMERICAN FEDERATION OF INFORMATION PROCESSING SOCIETIES. 1899 Preston White Drive, Reston, VA 22091. (703) 620-8900.

ASSOCIATION OF COMPUTING MACHINERY (ACM). 11 West 42nd Street, New York, NY 10036. (212) 869-7440.

IEEE COMPUTER SOCIETY. 1730 Massachusetts Avenue, N.W., Washington, DC 20036. (202) 371-0101.

INSTITUTE OF ELECTRICAL AND ELECTRONICS ENGINEERS. IEEE Service Center, 445 Hoes Lane, Piscataway, NJ 08854.

SOCIETY FOR COMPUTER SIMULATION. P.O. Box 17900, San Diego, CA 92117. (619) 277-3888.

DIRECTORIES AND BIOGRAPHICAL SOURCES

AMERICAN MEN AND WOMEN OF SCIENCE. R.R. Bowker, Inc., Order Department, 245 West 17th Street, New York, NY 10011. (800) 521-8110. Eight volumes. 1986. $595.00 for set.

AMERICAN SOCIETY FOR INFORMATION SCIENCE HANDBOOK AND DIRECTORY. American Society for Information Science, 1424 16th Street, N.W., Suite 404, Washington, DC 20036. (202) 462-1000. $50.00.

COMPUTERS AND COMPUTING INFORMATION RESOURCES DIRECTORY. Martin Connors, editor. Gale Research Company, Book Tower, Detroit, MI 48226. (800) 521-0707. 1987. $165.00. Supplement available at $85.00.

INTERNATIONAL RESEARCH CENTERS DIRECTORY 1988-89. Darren L. Smith, editor. Gale Research Company, Book Tower, Detroit, MI 48226. (800) 521-0707. 4th edition. 1987. $360.00.

1987 DIRECTORY OF ENGINEERING SOCIETIES AND RELATED ORGANIZATIONS. Gordon Davis, editor. Hemisphere Publishing Corporation, 1010 Vermont Avenue, NW, Washington, DC 20005. (800) 526-0275. 12th edition. 1987. $100.00.

RESEARCH CENTERS DIRECTORY 1988. Gale Research Company, Book Tower, Detroit, MI 48226. (800) 521-0707. 12th edition. 1987. $365.00 for set.

SCIENTIFIC AND TECHNICAL ORGANIZATIONS AND AGENCIES DIRECTORY. Margaret Labash Young, editor. Gale Research Company, Book Tower, Detroit, MI 48226. (800) 521-0707. 2nd edition. 1987. $185.00.

WHO'S WHO IN ENGINEERING. Gordon Davis, editor. Hemisphere Publishing Corporation, 1010 Vermont Avenue, NW, Washington, DC 20005. (800) 526-0275. 6th edition. 1985. $200.00.

ENCYCLOPEDIAS AND DICTIONARIES

COMPUTER AND TELECOMMUNICATIONS ACRONYMS. Julie E. Towell and Helen E. Sheppard, editors. Gale Research Company, Book Tower, Detroit, MI 48226. (800) 521-0707. 1986. $60.00.

DICTIONARY OF COMPUTING. Oxford University Press, 200 Madison Avenue, New York, NY 10016. (800) 458-5833. Second edition. 1986. $29.95.

GENERAL WORKS

COMPUTER ARCHITECTURE AND ORGANIZATION. J.P. Hayes. McGraw-Hill Book Company, 1221 Avenue of the Americas, New York, NY 10020. (212) 512-2000. Second edition. 1988. $49.95.

THE CONNECTION MACHINE. W. Daniel Hills. MIT Press, 28 Carleton Street, Cambridge, MA 02142. (617) 253-2884. 1985. $39.95.

FIFTH GENERATION COMPUTERS. Richard K. Miller, editor. The Fiarmont Press, Inc., 4025 Pleasantdale Road, N.W., Suite 420, Atlanta, GA 30340. (404) 447-5314. 1987. $44.95.

PRINCIPLES OF PARALLEL AND MULTI PROCESSING. G.R. Desrochers. McGraw-Hill Book Company, 1221 Avenue of the Americas, New York, NY 10020. (212) 512-2000. 1987. $49.50.

ONLINE DATA BASES

COMPENDEX. Engineering Information, Inc., 345 East 47th Street, New York, NY 10017. (800) 221-1044 or (212) 705-7615. Engineering and technical literature, 1975 to present. Inquire as to online cost and availability.

DISSERTATION ABSTRACTS ONLINE. University Microfilms International, 300 North Zeeb Road, Ann Arbor, MI 48106. (800) 521-0600 or (313) 761-4700. Scope includes virtually all doctoral dissertations accepted at accredited American institutions from 1861 to present in over 250 subject areas. Inquire as to online cost and availability.

INSPEC. INSPEC Marketing Department, Institution of Electrical Engineers. Available from IEEE Service Center, 445 Hoes Lane, Piscataway, NJ 08854. (201) 981-0060. Online version of Physics Abstracts. Inquire as to online cost and availability.

NTIS. National Technical Information Service, 5285 Port Royal Road, Springfield, VA 22161. (703) 487-4630. Broad coverage of government sponsored research reports, 1964 to present. Inquire as to online cost and availability.

SCISEARCH. Institute for Scientific Information, 3501 Market Street, Philadelphia, PA 19104. (800) 523-1850 or (215) 386-0100. Broad multidisciplinary title and author index to the international literature of science and technology, 1974 to present. Inquire as to online cost and availability.

WILSONLINE. H.W. Wilson and Company, 950 University Avenue, Bronx, NY 10452. (800) 367-6770 or (212) 588-8400. Makes available online versions of the H.W. Wilson indexes including Applied Science and Technology Index, Business Periodicals Index and Readers' Guide to Periodical Literature. Approximately 1980 to present. Inquire as to online cost and availability.

PERIODICALS

COMMUNICATIONS OF THE ACM. Association of Computing Machinery, 11 West 42nd Street, New York, NY 10036. (212) 869-7440. Monthly. $80.00 per year.

COMPUTER. IEEE Computer Society. Institute of Electrical and Electronics Engineers. IEEE Service Center, 445 Hoes Lane, Piscataway, NJ 08854. 1966 to present. Monthly. $90.00 per year.

IEEE TRANSACTIONS ON COMPUTERS. Institute of Electrical and Electronics Engineers. IEEE Service Center, 445 Hoes Lane, Piscataway, NJ 08854. 1952 to present. Monthly. $160.00.

JOURNAL OF PARALLEL AND DISTRIBUTED COMPUTING. Academic Press, Inc., 6277 Sea Harbor Drive, Orlando, FL 32821. (800) 321-5068. 1984 to present. Quarterly. $110.00 per year.

JOURNAL OF SYSTEMS AND SOFTWARE. Elsevier Science Publishing Company, Inc., 52 Vanderbilt Avenue, New York, NY 10017. (212) 370-5520. 1979 to present. Quarterly. $95.00 per year.

SIGSOFT SOFTWARE ENGINEERING NOTICES. Association of Computing Machinery Special Interest Group on Software Engineering. 11 West 42nd Street, New York, NY 10036. (212) 869-7440. Quarterly. $12.00 per year.

SOFTWARE ENGINEERING JOURNAL. Institute of Electrical Engineers, Savoy Place, London, WC2R OBL, England. 1981 to present. Bimonthly. $85.00 per year.

RESEARCH CENTERS AND INSTITUTES

CENTER FOR ADVANCED COMPUTATIONAL SCIENCE. Temple University, North Bend Street, Philadelphia, PA 19122. (215) 787-8631.

COMPUTER AND INFORMATION TECHNOLOGY INSTITUTE. Rice University, Department of Computer Science, P.O. Box 1892, Houston, TX 77251. (713) 527-4834.

INSTITUTE FOR INFORMATION SCIENCE AND TECHNOLOGY. George Washington University, 801 22nd Street, N.W., Washington, DC 20052. (202) 676-4921.

PURDUE CENTER FOR PARALLEL AND VECTOR COMPUTING. Purdue University, Computer Science Department, West Lafayette, IN 47907. (317) 494-6003.

RESEARCH INSTITUTE FOR COMPUTING AND INFORMATION SYSTEMS. University of Houston at Clear Lake, 2700 Bay Area Boulevard, Houston, TX 77058. (713) 488-9392.

PARTICLE ACCELERATORS

See also: FUSION, PARTICLE PHYSICS, PHYSICS

ABSTRACT SERVICES AND INDEXES

CHEMICAL ABSTRACTS. American Chemical Society, Chemical Abstracts Service, Box 3012, Columbus, OH 43210. (614) 421-3600. 1907 to present. Weekly. $9500.00 per year.

CONFERENCE PAPERS INDEX. Cambridge Scientific Abstracts, 5161 River Road, Bethesda, MD 20816. 1972 to present. Monthly. Inquire as to cost and availability.

CURRENT CONTENTS: PHYSICAL, CHEMICAL AND EARTH SCIENCES. Institute for Scientific Information, 3501 Market Street, Philadelphia, PA 19104. (800) 523-1850 or (215) 386-0100. Weekly. $275.00 per year.

ENGINEERING INDEX MONTHLY AND AUTHOR INDEX. Engineering Information Inc., 345 East 47th Street, New York, NY 10017. (212) 705-7600. Monthly. $1560.00 per year.

INDEX TO SCIENTIFIC AND TECHNICAL PROCEEDINGS. Institute for Scientific Information, 3501 Market Street, Philadelphia, PA 19104. (800) 523-1850 or (215) 386-0100. 1978 to present. Monthly. $775.00 per year.

INDEX TO SCIENTIFIC REVIEWS. Institute for Scientific Information, 3501 Market Street, Philadelphia, PA 19104. (800) 523-1850 or (215) 386-0100. 1974 to present. Semi-annual. $550.00 per year.

MATHEMATICAL REVIEWS: A REVIEWING JOURNAL COVERING THE WORLD LITERATURE OF MATHEMATICAL RESEARCH. American Mathematical Society, P.O. Box 6248, Providence, RI 02940. (800) 7774 or (401) 272-9500. 1940 to present. Monthly. $2800.00 per year.

PHYSICS ABSTRACTS. Institution of Electrical Engineers. Available from: IEEE Service Center, 445 Hoes Lane, Piscataway, NJ 08854. 1898 to present. Bimonthly. $1700.00 per year.

PHYSICS BRIEFS. Physik Verlag GmbH, Postfach 1260/1280, D-6940 Weinheim, West Germany. Twenty-six times per year. $1200.00 per year.

SCIENCE CITATION INDEX. Institute for Scientific Information, 3501 Market Street, Philadelphia, PA 19104. (800) 523-1850 or (215) 386-0100. Six times per year. $6200.00 per year.

ANNUAL REVIEWS AND YEARBOOKS

ANNUAL REVIEW OF NUCLEAR AND PARTICLE SCIENCE. J.D. Jackson and others, editors. Annual Reviews, Inc., 4139 El Camino Way, Palo Alto, CA 94306. (415) 493-4400. Annual. Price varies. Volume 36, $34.00.

ASSOCIATIONS AND PROFESSIONAL SOCIETIES

AMERICAN INSTITUTE OF PHYSICS. 335 East 45th Street, New York, NY 10017. (212) 661-9404.

AMERICAN PHYSICAL SOCIETY. 335 East 45th Street, New York, NY 10017. (212) 682-7341.

DIRECTORIES AND BIOGRAPHICAL SOURCES

INTERNATIONAL RESEARCH CENTERS DIRECTORY 1988-89. Darren L. Smith, editor. Gale Research Company, Book Tower, Detroit, MI 48226. (800) 521-0707. 4th edition. 1987. $360.00.

RESEARCH CENTERS DIRECTORY 1988. Gale Research Company, Book Tower, Detroit, MI 48226. (800) 521-0707. 12th edition. 1987. $365.00 for set.

SCIENTIFIC AND TECHNICAL ORGANIZATIONS AND AGENCIES DIRECTORY. Margaret Labash Young, editor. Gale Research Company, Book Tower, Detroit, MI 48226. (800) 521-0707. 2nd edition. 1987. $185.00.

ENCYCLOPEDIAS AND DICTIONARIES

CONCISE SCIENCE DICTIONARY. Oxford University Press, 200 Madison Avenue, New York, NY 10016. (800) 458-5833. 1987. $9.95 in paper.

DICTIONARY OF THE PHYSICAL SCIENCES: TERMS, FORMULAS, DATA. Cesare Emiliani. Oxford University Press, 200 Madison Avenue, New York, NY 10016. (800) 458-5833. 1987. $19.95 in paper.

MCGRAW-HILL ENCYCLOPEDIA OF SCIENCE AND TECHNOLOGY. McGraw-Hill Book Company, 1221 Avenue of the Americas, New York, NY 10020. (212) 512-2000. 6th edition. 1987. $1600.00.

THESAURUS OF SCIENTIFIC, TECHNICAL, AND ENGINEERING TERMS. Hemisphere Publishing Corporation, 1010 Vermont Avenue, NW, Washington, DC 20005. (800) 526-0275. 1988. $125.00.

GENERAL WORKS

CONCEPTS OF PARTICLE PHYSICS. Kurt Gottfried and Victor F. Weisskopf. Oxford University Press, 200 Madison Avenue,

New York, NY 10016. (800) 458-5833. Two volumes. 1986-87. Volume 1, $20.00 in paper; volume 2, $70.00.

THE EXPERIMENTAL FOUNDATIONS OF PARTICLE PHYSICS. Robert N. Cahn and Gerson Goldhaber. Cambridge University Press, 32 East 57th Street, New York, NY 10022. (800) 872-7423. 1987. $50.00.

PARTICLE AND NUCLEAR PHYSICS. N. Hu and C.S. Wu, editors. Taylor and Francis, Inc., 242 Cherry Street, Philadelphia, PA 19106-1906. 1987. $65.00.

THE PARTICLE EXPLOSION. Frank Close and Christine Sutton. Oxford University Press, 200 Madison Avenue, New York, NY 10016. (800) 458-5833. 1987. $35.00.

PARTICLE PHYSICS. Necia Grant Cooper and Geoffrey B. West, editors. Cambridge University Press, 32 East 57th Street, New York, NY 10022. (800) 872-7423. 1987. $39.50.

PRINCIPLES OF CHARGED PARTICLES ACCELERATION. Stanley Humphries. John Wiley and Sons, Inc., 605 Third Avenue, New York, NY 10158. (800) 526-5368. 1986. $65.95.

ONLINE DATA BASES

CA SEARCH. Chemical Abstracts Service, P.O. Box 3012, Columbus, OH 43120. (800) 848-6538 or (614) 421-3600. Comprehensive guide to chemical literature, 1972 to present. Inquire as to online cost and availability.

COMPENDEX. Engineering Information, Inc., 345 East 47th Street, New York, NY 10017. (800) 221-1044 or (212) 705-7615. Engineering and technical literature, 1975 to present. Inquire as to online cost and availability.

DISSERTATION ABSTRACTS ONLINE. University Microfilms International, 300 North Zeeb Road, Ann Arbor, MI 48106. (800) 521-0600 or (313) 761-4700. Scope includes virtually all doctoral dissertations accepted at accredited American institutions from 1861 to present in over 250 subject areas. Inquire as to online cost and availability.

INSPEC. INSPEC Marketing Department, Institution of Electrical Engineers. Available from IEEE Service Center, 445 Hoes Lane, Piscataway, NJ 08854. (201) 981-0060. Online version of Physics Abstracts. Inquire as to online cost and availability.

MATHFILE. American Mathematical Society, P.O. Box 6248, Providence, RI 02940. (800) 556-7774 or (401) 272-9500. An online version of Mathematical Reviews. 1973 to present. Inquire as to online cost and availability.

NTIS. National Technical Information Service, 5285 Port Royal Road, Springfield, VA 22161. (703) 487-4630. Broad coverage of government sponsored research reports, 1964 to present. Inquire as to online cost and availability.

SCISEARCH. Institute for Scientific Information, 3501 Market Street, Philadelphia, PA 19104. (800) 523-1850 or (215) 386-0100. Broad multidisciplinary title and author index to the international literature of science and technology, 1974 to present. Inquire as to online cost and availability.

PERIODICALS

CANADIAN JOURNAL OF PHYSICS. National Research Council of Canada, Research Journals, Ottawa K1A OR6, Canada. Order from: Allen Press Inc., 1041 New Hampshire Street, Box 368, Lawrence, KS 66044. 1929 to present. Monthly. $200.00 per year.

COMMENTS ON NUCLEAR AND PARTICLE PHYSICS. Gordon and Breach Science Publishers, Inc., 50 West 23rd Street, New York, NY 10010. (212) 206-8900. 1967 to present. Six times per year. $175.00 per year.

PARTICLE ACCELERATORS. Gordon and Breach Science Publishers, Inc., 50 West 23rd Street, New York, NY 10010. (212) 206-8900. 1969 to present. Eight times per year. $290.00 per year.

PHYSICAL REVIEW LETTERS. American Institute of Physics, 335 East 45th Street, New York, NY 10017. (212) 661-9404. 1958 to present. Weekly. $450.00 per year.

PHYSICS LETTERS, SECTION B: NUCLEAR, ELEMENTARY PARTICLE AND HIGH-ENERGY PHYSICS. Elsevier Science Publishing Company, Inc., 52 Vanderbilt Avenue, New York, NY 10017. (212) 370-5520. Seventy-two times per year. $1700.00 per year.

PROGRESS IN PARTICLE AND NUCLEAR PHYSICS. Pergamon Press, Inc., Maxwell House, Fairview Park, Elmsford, NY 10523. (914) 592-7700. Two per year. $210.00 per year.

RESEARCH CENTERS AND INSTITUTES

ARGONNE NATIONAL LABORATORY. 9700 South Cass Avenue, Argonne, IL 60439. (312) 972-2000.

FERMI NATIONAL ACCELERATOR LABORATORY. P.O. Box 500, Batavia, IL 60510. (312) 840-3000.

HIGH ENERGY PARTICLE PHYSICS GROUP. Florida State University, Keen Building, Tallahassee, FL 32306. (904) 644-1492.

HIGH ENERGY PHYSICS AND ELEMENTARY PARTICLE RESEARCH PROGRAM. University of Pennsylvania, 33rd and Walnut Streets, Philadelphia, PA 19104. (215) 898-5960.

LAWRENCE BERKELEY LABORATORY, PHYSICS DIVISION. Building 50, Room 256, 1 Cyclotron Road, Berkeley, CA 94720. (415) 486-5421.

PARTICLE PHYSICS

See also: PARTICLE ACCELERATORS, PHYSICS

ABSTRACT SERVICES AND INDEXES

CHEMICAL ABSTRACTS. American Chemical Society, Chemical Abstracts Service, Box 3012, Columbus, OH 43210. (614) 421-3600. 1907 to present. Weekly. $9500.00 per year.

CONFERENCE PAPERS INDEX. Cambridge Scientific Abstracts, 5161 River Road, Bethesda, MD 20816. 1972 to present. Monthly. Inquire as to cost and availability.

CURRENT CONTENTS: PHYSICAL, CHEMICAL AND EARTH SCIENCES. Institute for Scientific Information, 3501 Market Street, Philadelphia, PA 19104. (800) 523-1850 or (215) 386-0100. Weekly. $275.00 per year.

INDEX TO SCIENTIFIC AND TECHNICAL PROCEEDINGS. Institute for Scientific Information, 3501 Market Street, Philadelphia, PA 19104. (800) 523-1850 or (215) 386-0100. 1978 to present. Monthly. $775.00 per year.

INDEX TO SCIENTIFIC REVIEWS. Institute for Scientific Information, 3501 Market Street, Philadelphia, PA 19104. (800) 523-1850 or (215) 386-0100. 1974 to present. Semi-annual. $550.00 per year.

MATHEMATICAL REVIEWS: A REVIEWING JOURNAL COVERING THE WORLD LITERATURE OF MATHEMATICAL RESEARCH. American Mathematical Society, P.O. Box 6248, Providence, RI 02940. (800) 7774 or (401) 272-9500. 1940 to present. Monthly. $2800.00 per year.

PHYSICS ABSTRACTS. Institution of Electrical Engineers. Available from: IEEE Service Center, 445 Hoes Lane, Piscataway, NJ 08854. 1898 to present. Bimonthly. $1700.00 per year.

PHYSICS BRIEFS. Physik Verlag GmbH, Postfach 1260/1280, D-6940 Weinheim, West Germany. Twenty-six times per year.

$1200.00 per year.

SCIENCE CITATION INDEX. Institute for Scientific Information, 3501 Market Street, Philadelphia, PA 19104. (800) 523-1850 or (215) 386-0100. Six times per year. $6200.00 per year.

ANNUAL REVIEWS AND YEARBOOKS

ANNUAL REVIEW OF NUCLEAR AND PARTICLE SCIENCE. J.D. Jackson and others, editors. Annual Reviews, Inc., 4139 El Camino Way, Palo Alto, CA 94306. (415) 493-4400. Annual. Price varies. Volume 36, $34.00.

ASSOCIATIONS AND PROFESSIONAL SOCIETIES

AMERICAN INSTITUTE OF PHYSICS. 335 East 45th Street, New York, NY 10017. (212) 661-9404.

AMERICAN PHYSICAL SOCIETY. 335 East 45th Street, New York, NY 10017. (212) 682-7341.

BIBLIOGRAPHIES

SCIENTIFIC AND TECHNICAL BOOKS AND SERIALS IN PRINT 1988; AN INDEX TO LITERATURE IN SCIENCE AND TECHNOLOGY. R.R. Bowker Company, 205 East 42nd Street, New York, NY 10017. (800) 521-8110. $175.00.

DIRECTORIES AND BIOGRAPHICAL SOURCES

INTERNATIONAL RESEARCH CENTERS DIRECTORY 1988-89. Darren L. Smith, editor. Gale Research Company, Book Tower, Detroit, MI 48226. (800) 521-0707. 4th edition. 1987. $360.00.

RESEARCH CENTERS DIRECTORY 1988. Gale Research Company, Book Tower, Detroit, MI 48226. (800) 521-0707. 12th edition. 1987. $365.00 for set.

SCIENTIFIC AND TECHNICAL ORGANIZATIONS AND AGENCIES DIRECTORY. Margaret Labash Young, editor. Gale Research Company, Book Tower, Detroit, MI 48226. (800) 521-0707. 2nd edition. 1987. $185.00.

ENCYCLOPEDIAS AND DICTIONARIES

CONCISE SCIENCE DICTIONARY. Oxford University Press, 200 Madison Avenue, New York, NY 10016. (800) 458-5833. 1987. $9.95 in paper.

DICTIONARY OF THE PHYSICAL SCIENCES: TERMS, FORMULAS, DATA. Cesare Emiliani. Oxford University Press, 200 Madison Avenue, New York, NY 10016. (800) 458-5833. 1987. $19.95 in paper.

MCGRAW-HILL ENCYCLOPEDIA OF SCIENCE AND TECHNOLOGY. McGraw-Hill Book Company, 1221 Avenue of the Americas, New York, NY 10020. (212) 512-2000. 6th edition. 1987. $1600.00.

THESAURUS OF SCIENTIFIC, TECHNICAL, AND ENGINEERING TERMS. Hemisphere Publishing Corporation, 1010 Vermont Avenue, NW, Washington, DC 20005. (800) 526-0275. 1988. $125.00.

GENERAL WORKS

CONCEPTS OF PARTICLE PHYSICS. Kurt Gottfried and Victor F. Weisskopf. Oxford University Press, 200 Madison Avenue, New York, NY 10016. (800) 458-5833. Two volumes. 1986-87. Volume 1, $20.00 in paper; volume 2, $70.00.

THE EXPERIMENTAL FOUNDATIONS OF PARTICLE PHYSICS. Robert N. Cahn and Gerson Goldhaber. Cambridge University Press, 32 East 57th Street, New York, NY 10022. (800) 872-7423. 1987. $50.00.

FUNDAMENTAL PARTICLES: AN INTRODUCTION TO QUARKS AND LEPTONS. Brian G. Duff. Taylor and Francis, Inc., 242 Cherry Street, Philadelphia, PA 19106-1906. 1986. $22.00.

THE GREAT DESIGN: PARTICLES, FIELDS AND CREATION. Robert K. Adair. Oxford University Press, 200 Madison Avenue, New York, NY 10016. (800) 458-5833. 1987. $24.95.

PARTICLE AND NUCLEAR PHYSICS. N. Hu and C.S. Wu, editors. Taylor and Francis, Inc., 242 Cherry Street, Philadelphia, PA 19106-1906. 1987. $65.00.

THE PARTICLE EXPLOSION. Frank Close and Christine Sutton. Oxford University Press, 200 Madison Avenue, New York, NY 10016. (800) 458-5833. 1987. $35.00.

THE PARTICLE HUNTERS. Yuval Ne'eman and Yoram Kirsh. Cambridge University Press, 32 East 57th Street, New York, NY 10022. (800) 872-7423. 1986. $13.95 in paper.

PARTICLE PHYSICS. Necia Grant Cooper and Geoffrey B. West, editors. Cambridge University Press, 32 East 57th Street, New York, NY 10022. (800) 872-7423. 1987. $39.50.

QUARKS AND LEPTONS: AN INTRODUCTION TO MODERN PARTICLE PHYSICS. Francis Halzen and Alan D. Martin. John Wiley and Sons, Inc., 605 Third Avenue, New York, NY 10158. (800) 526-5368. 1984. $36.95.

ONLINE DATA BASES

CA SEARCH. Chemical Abstracts Service, P.O. Box 3012, Columbus, OH 43120. (800) 848-6538 or (614) 421-3600. Comprehensive guide to chemical literature, 1972 to present. Inquire as to online cost and availability.

DISSERTATION ABSTRACTS ONLINE. University Microfilms International, 300 North Zeeb Road, Ann Arbor, MI 48106. (800) 521-0600 or (313) 761-4700. Scope includes virtually all doctoral dissertations accepted at accredited American institutions from 1861 to present in over 250 subject areas. Inquire as to online cost and availability.

INSPEC. INSPEC Marketing Department, Institution of Electrical Engineers. Available from IEEE Service Center, 445 Hoes Lane, Piscataway, NJ 08854. (201) 981-0060. Online version of Physics Abstracts. Inquire as to online cost and availability.

MATHFILE. American Mathematical Society, P.O. Box 6248, Providence, RI 02940. (800) 556-7774 or (401) 272-9500. An online version of Mathematical Reviews. 1973 to present. Inquire as to online cost and availability.

NTIS. National Technical Information Service, 5285 Port Royal Road, Springfield, VA 22161. (703) 487-4630. Broad coverage of government sponsored research reports, 1964 to present. Inquire as to online cost and availability.

SCISEARCH. Institute for Scientific Information, 3501 Market Street, Philadelphia, PA 19104. (800) 523-1850 or (215) 386-0100. Broad multidisciplinary title and author index to the international literature of science and technology, 1974 to present. Inquire as to online cost and availability.

PERIODICALS

CANADIAN JOURNAL OF PHYSICS. National Research Council of Canada, Research Journals, Ottawa K1A OR6, Canada. Order from: Allen Press Inc., 1041 New Hampshire Street, Box 368, Lawrence, KS 66044. 1929 to present. Monthly. $200.00 per year.

COMMENTS ON NUCLEAR AND PARTICLE PHYSICS. Gordon and Breach Science Publishers, Inc., 50 West 23rd Street, New York, NY 10010. (212) 206-8900. 1967 to present. Six times per year. $175.00 per year.

PARTICLE ACCELERATORS. Gordon and Breach Science Publishers, Inc., 50 West 23rd Street, New York, NY 10010. (212) 206-8900. 1969 to present. Eight times per year. $290.00 per year.

PHYSICAL REVIEW LETTERS. American Institute of Physics, 335 East 45th Street, New York, NY 10017. (212) 661-9404. 1958 to present. Weekly. $450.00 per year.

PHYSICS LETTERS, SECTION B: NUCLEAR, ELEMENTARY PARTICLE AND HIGH-ENERGY PHYSICS. Elsevier Science Publishing Company, Inc., 52 Vanderbilt Avenue, New York, NY 10017. (212) 370-5520. Seventy-two times per year. $1700.00 per year.

PROGRESS IN PARTICLE AND NUCLEAR PHYSICS. Pergamon Press, Inc., Maxwell House, Fairview Park, Elmsford, NY 10523. (914) 592-7700. Two per year. $210.00 per year.

RESEARCH CENTERS AND INSTITUTES

ARGONNE NATIONAL LABORATORY. 9700 South Cass Avenue, Argonne, IL 60439. (312) 972-2000.

FERMI NATIONAL ACCELERATOR LABORATORY. P.O. Box 500, Batavia, IL 60510. (312) 840-3000.

HIGH ENERGY PARTICLE PHYSICS GROUP. Florida State University, Keen Building, Tallahassee, FL 32306. (904) 644-1492.

HIGH ENERGY PHYSICS AND ELEMENTARY PARTICLE RESEARCH PROGRAM. University of Pennsylvania, 33rd and Walnut Streets, Philadelphia, PA 19104. (215) 898-5960.

LAWRENCE BERKELEY LABORATORY, PHYSICS DIVISION. Building 50, Room 256, 1 Cyclotron Road, Berkeley, CA 94720. (415) 486-5421.

PATTERN RECOGNITION

See: COMPUTER VISION

PASCAL

See: PROGRAMMING LANGUAGES

PETROLEUM ENGINEERING

See also: PETROLEUM GEOLOGY

ABSTRACT SERVICES AND INDEXES

BIBLIOGRAPHY AND INDEX OF GEOLOGY. American Geological Institute, 4220 King Street, Alexandria, VA 22302. (703) 379-2480. 1969 to present. Monthly. $1100.00 per year.

CHEMICAL ABSTRACTS. Chemical Abstracts Service, 2540 Olentangy Road, Post Office Box 3012, Columbus, OH 43210. (800) 848-6538 or (614) 421-3600. Weekly. $9200.00 per year.

DEEP-SEA RESEARCH WITH OCEANOGRAPHIC LITERATURE REVIEW. Pergamon Press, Incorporated, Maxwell House, Fairview Park, Elmsford, NY 10523. (914) 592-7700. 1953 to present. Twenty-four times per year. $600.00 per year.

OCEANIC ABSTRACTS. Cambridge Scientific Abstracts, 5161 River Road, Bethesda, MD 20816. (301) 951-1400. 1964 to present. Bimonthly. $652.00 per year.

SCIENCE CITATION INDEX. Institute for Scientific Information, 3501 Market Street, Philadelphia, PA 19104. (800) 523-1850 or (215) 386-0100. Inquire as to price and availability.

ASSOCIATIONS AND PROFESSIONAL SOCIETIES

AMERICAN ASSOCIATION OF PETROLEUM GEOLOGISTS. Post Office Box 979, Tulsa, OK 74101. (918) 584-2555.

AMERICAN GEOLOGICAL INSTITUTE. 4220 King Street, Alexandria, VA 22302. (703) 379-2480.

AMERICAN GEOPHYSICAL UNION. 2000 Florida Avenue, NW, Washington, DC 20009. (202) 462-6903.

AMERICAN INSTITUTE OF PROFESSIONAL GEOLOGISTS. 7828 Vance Drive, Suite 103, Arvada, CO 80003. (303) 431-0831.

CANADIAN SOCIETY OF PETROLEUM GEOLOGISTS. 206 7th Avenue, SW, Suite 505, Calgary, AB T2P 0W7, Canada. (403) 264-5610.

GEOLOGICAL SOCIETY OF AMERICA. Box 9140, Boulder, CO 80301. (303) 447-2020.

SOCIETY OF EXPLORATION GEOPHYSICISTS. Post Office Box 702740, Tulsa, OK 74170. (918) 493-3516.

DIRECTORIES AND BIOGRAPHICAL SOURCES

AAPG BULLETIN MEMBERSHIP DIRECTORY. American Association of Petroleum Geologists, Post Office Box 979, Tulsa, OK 74101. (918) 584-2555. Included with AAPG Bulletin.

AMERICAN INSTITUTE OF PROFESSIONAL GEOLOGISTS. Membership Directory. American Institute of Professional Geologists, 7828 Vance Drive, Suite 103, Arvada, CO 80003. (303) 431-0831. Annual. $15.00.

GEOLOGICAL SOCIETY OF AMERICA. Membership Directory. Geological Society of America, 3300 Penrose Place, Boulder, CO 80301. (303) 447-2020. Annual. Available to members only.

RESEARCH CENTERS DIRECTORY. Gale Research Company, Detroit, MI 48226. (800) 521-0707. Twelfth edition. 1987. $365.00.

WHO WHO'S IN ENGINEERING. Gordon Davis, editor. Hemisphere Publishing Corporation, 79 Madison Avenue, New York, NY 10016-7892. 6th edition. 1985. $200.00.

WHO'S WHO IN FRONTIER SCIENCE AND TECHNOLOGY. Marquis Who's Who, Incorporated, 200 East Ohio Street, Chicago, IL 60611. (800) 428-3898 or (312) 787-2008.

ENCYCLOPEDIAS AND DICTIONARIES

DICTIONARY OF GEOLOGICAL TERMS. American Geological Institute. Doubleday and Company, Incorporated, 245 Park Avenue, New York, NY 10017. (800) 645-6156 or (212) 953-4561. Third edition. 1984. $19.95.

DICTIONARY OF PETROLOGY. S.I. Tomkeieff. John Wiley and Sons, Incorporated, 605 Third Avenue, New York, NY 10158. (800) 526-5368 or (212) 850-6000. 1983. $140.00.

ENCYCLOPEDIA OF PHYSICAL SCIENCE AND TECHNOLOGY. Academic Press, Incorporated, Orlando, FL 32887. (800) 321-5068. Fifteen volumes. $2500.00.

GLOSSARY OF GEOLOGY. Robert L. Bates and Julia A. Jackson. American Geological Institute, 4220 King Street, Alexandria, VA 22032. (703) 379-2480. 1980. $60.00.

GENERAL WORKS

ELEMENTS OF PETROLEUM GEOLOGY. Richard C. Shelley. W.H. Freeman and Company, 41 Madison Avenue, New York, NY 10010. (212) 532-7660. 1985. $44.95.

GEOLOGIC WELL LOG ANALYSIS. Sylvain J. Pirson. Gulf Publishing Company, Post Office Box 2608, Houston, TX 77001. (713) 529-4301. Third edition. 1983. $35.00.

INTRODUCTION TO PETROLEUM GEOLOGY. G.D. Hobson and E.N. Tiratsoo. Gulf Publishing Company, Post Office Box 2608, Houston, TX 77001. (713) 529-4301. Second edition. 1985. $55.00.

PETROLEUM GEOLOGY. R.E. Chapman. Elsevier Science Publishing Company, Incorporated, 52 Vanderbilt Avenue, New York, NY 10017. (212) 370-5520. 1983. $44.25.

HANDBOOKS AND MANUALS

GEOLOGY IN THE FIELD. R.R. Compton. John Wiley and Sons, Incorporated, 605 Third Avenue, New York, NY 10158. (800) 526-5368 or (212) 850-6418. 1985. $23.95.

ONLINE DATA BASES

GEOREF. American Geological Institute, 4220 King Street, Alexandria, VA 22302. (800) 336-4764 or (703) 379-2480. Geology and geosciences literature, 1961 to present. Inquire as to online cost and availability.

GEOARCHIVE. Geosystems, Post Office Box 1024, Westminister, London, England, SW1 P 2JL. Citations to literature on geoscience, 1969 to present. Inquire as to online cost and availability.

NTIS. National Technical Information Service, 5285 Port Royal Road, Springfield, VA 22161. (703) 487-4630. Broad coverage of government sponsored research reports, 1964 to present. Inquire as to online cost and availability.

SCISEARCH. Institute for Scientific Information, 3501 Market Street, Philadelphia, PA 19104. (800) 523-1850 or (215) 386-0100. Broad interdisciplinary index to the literature of science and technology, 1965 to present. Inquire as to online cost and availability.

PERIODICALS

AAPG BULLETIN. American Association of Petroleum Geologists, Post Office Box 979, Tulsa, OK 74101. (918) 584-2555. Monthly. $36.00 per year.

AAPG EXPLORER. American Association of Petroleum Geologists, Post Office Box 979, Tulsa, OK 74101. (918) 584-2555. Monthly. $15.00 per year.

BULLETIN OF CANADIAN PETROLEUM GEOLOGY. Canadian Society of Petroleum Geologists, 206 7th Avenue, SW, Suite 505, Calgary, AB T2P OW7, Canada. (403) 264-5610. Quarterly. $45.00 per year.

ECONOMIC GEOLOGY. Society of Economic Geologists, Post Office Box 571, Golden, CO 80402. (303) 279-1899. 8 times per year. $25.00.

GEOPHYSICS. Society of Exploration Geophysicists, Post Office Box 702740, Tulsa, OK 74170. (918) 493-3516. Monthly. $45.00 per year.

GEOTIMES. American Geological Institute, 4220 King Street, Alexandria, VA 22032. Monthly. $18.00 per year.

INTERNATIONAL OIL SCOUTS ASSOCIATION. Official Publication. International Oil Scouts Association, 5000 East Ben White Boulevard, Austin, TX 78741. (512) 448-4088. Quarterly. Inquire as to cost and availability.

JOURNAL OF GEOPHYSICS. Springer-Verlag, Incorporated, 175 Fifth Avenue, New York, NY 10010. (800) 526-7254 or (212) 460-1500. Bimonthly. $175.00 per year.

JOURNAL OF PETROLEUM GEOLOGY. Scientific Press, Limited, Box 21, Beaconsfield, Bucks, HP9 1HW, England. Quarterly. $130.00 per year.

JOURNAL OF SEDIMENTARY PETROLOGY. Society of Economic Paleontologists and Mineralogists, Post Office Box 4756, Tulsa, OK 74159. (918) 743-9765. Bimonthly.

JOURNAL OF STRUCTURAL GEOLOGY. Pergamon Journals, Incorporated, Maxwell House, Fairview Park, Elmsford, NY 10523. (914) 592-7700. Eight times per year. $160.00 per year.

MOUNTAIN GEOLOGIST. Rocky Mountain Association of Geologists, 4201 West 51st Avenue, Denver, CO 80212-2902. Quarterly. $15.00 per year.

OIL AND GAS JOURNAL. PennWell Publishing Company, Box 1260, Tulsa, OK 74101. Weekly. $34.00 per year.

PROFESSIONAL GEOLOGIST. American Institute of Professional Geologists, 7828 Vance Drive, Suite 103, Arvada, CO 80003. Monthly.

SEDIMENTARY GEOLOGY. Elsevier Science Publishers, Post Office Box 330, Irving-on-Hudson, NY 10533. Twenty times per year. $421.00 per year.

SEDIMENTOLOGY. Blackwell Scientific Publications, Incorporated, 52 Beacon Street, Boston, MA 02108. (617) 720-0761. Six times per year. $185.00 per year.

RESEARCH CENTERS AND INSTITUTES

LOUISIANA GEOLOGICAL SURVEY. Box G, University Station, Baton Rouge, LA 70893.

SOCIETY OF ECONOMIC GEOLOGISTS. Post Office Box 571, Golden, CO 80402. (303) 279-1899.

SOCIETY OF ECONOMIC PALEONTOLOGISTS AND MINERALOGISTS. Post Office Box 4756, Tulsa, OK 74159. (918) 743-9765.

UNIVERSITY OF TEXAS AT AUSTIN. Bureau of Economic Geology, Box X, University Station, Austin, Texas 78713. (512) 471-7721.

PETROLEUM EXPLORATION

See: MINERAL EXPLORATION

PETROLEUM GEOLOGY

See also: GEOLOGY, GEOPHYSICS, GEOCHEMISTRY, PETROLEUM ENGINEERING

ABSTRACT SERVICES AND INDEXES

BIBLIOGRAPHY AND INDEX OF GEOLOGY. American Geological Institute, 4220 King Street, Alexandria, VA 22302. (703) 379-2480. 1969 to present. Monthly. $1100.00 per year.

CHEMICAL ABSTRACTS. Chemical Abstracts Service, 2540 Olentangy Road, Post Office Box 3012, Columbus, OH 43210. (800) 848-6538 or (614) 421-3600. Weekly. $9200.00 per year.

DEEP-SEA RESEARCH WITH OCEANOGRAPHIC LITERATURE REVIEW. Pergamon Press, Incorporated, Maxwell House, Fairview Park, Elmsford, NY 10523. (914) 592-7700. 1953 to present. Twenty-four times per year. $600.00 per year.

GENERAL SCIENCE INDEX. H.W. Wilson Company, 950 University Avenue, Bronx, NY 10452. (800) 367-6770 or (212)

588-8400. Inquire as to price and availability.

MINERALOGICAL ABSTRACTS. Mineralogical Society and the Mineralogical Society of America, 41 Queen's Gate, London, SW7 5HR, England. 1959 to present. Quarterly. $190.00 per year.

OCEANIC ABSTRACTS. Cambridge Scientific Abstracts, 5161 River Road, Bethesda, MD 20816. (301) 951-1400. 1964 to present. Bimonthly. $652.00 per year.

SCIENCE CITATION INDEX. Institute for Scientific Information, 3501 Market Street, Philadelphia, PA 19104. (800) 523-1850 or (215) 386-0100. Inquire as to price and availability.

ANNUAL REVIEWS AND YEARBOOKS

ADVANCES IN GEOPHYSICS. Academic Press, Incorporated, 6277 Sea Harbor Drive, Orlando, FL 32821. (800) 321-5068. Irregular. $62.00 per volume.

ANNUAL REVIEW AND EARTH AND PLANETARY SCIENCES. Annual Reviews, Incorporated, 4139 El Camino Way, Palo Alto, CA 94306. (415) 493-4400.

MINERALS YEARBOOK. Bureau of Mines, United States Department of the Interior. Available from United States Government Printing Office, Washington, DC 20402. (202) 783-3238. Annual. Three volumes. $45.00.

ASSOCIATIONS AND PROFESSIONAL SOCIETIES

AMERICAN ASSOCIATION OF PETROLEUM GEOLOGISTS. Post Office Box 979, Tulsa, OK 74101. (918) 584-2555.

AMERICAN GEOLOGICAL INSTITUTE. 4220 King Street, Alexandria, VA 22302. (703) 379-2480.

AMERICAN GEOPHYSICAL UNION. 2000 Florida Avenue, NW, Washington, DC 20009. (202) 462-6903.

AMERICAN INSTITUTE OF PROFESSIONAL GEOLOGISTS. 7828 Vance Drive, Suite 103, Arvada, CO 80003. (303) 431-0831.

CANADIAN SOCIETY OF PETROLEUM GEOLOGISTS. 206 7th Avenue, SW, Suite 505, Calgary, AB T2P 0W7, Canada. (403) 264-5610.

GEOLOGICAL SOCIETY OF AMERICA. Box 9140, Boulder, CO 80301. (303) 447-2020.

SOCIETY OF EXPLORATION GEOPHYSICISTS. Post Office Box 702740, Tulsa, OK 74170. (918) 493-3516.

DIRECTORIES AND BIOGRAPHICAL SOURCES

AAPG BULLETIN MEMBERSHIP DIRECTORY. American Association of Petroleum Geologists, Post Office Box 979, Tulsa, OK 74101. (918) 584-2555. Included with AAPG Bulletin.

AMERICAN INSTITUTE OF PROFESSIONAL GEOLOGISTS. Membership Directory. American Institute of Professional Geologists, 7828 Vance Drive, Suite 103, Arvada, CO 80003. (303) 431-0831. Annual. $15.00.

GEOLOGICAL SOCIETY OF AMERICA. Membership Directory. Geological Society of America, 3300 Penrose Place, Boulder, CO 80301. (303) 447-2020. Annual. Available to members only.

RESEARCH CENTERS DIRECTORY. Gale Research Company, Detroit, MI 48226. (800) 521-0707. Twelfth edition. 1987. $365.00.

WHO'S WHO IN FRONTIER SCIENCE AND TECHNOLOGY. Marquis Who's Who, Incorporated, 200 East Ohio Street, Chicago, IL 60611. (800) 428-3898 or (312) 787-2008.

ENCYCLOPEDIAS AND DICTIONARIES

DICTIONARY OF GEOLOGICAL TERMS. American Geological Institute. Doubleday and Company, Incorporated, 245 Park Avenue, New York, NY 10017. (800) 645-6156 or (212) 953-4561. Third edition. 1984. $19.95.

DICTIONARY OF PETROLOGY. S.I. Tomkeieff. John Wiley and Sons, Incorporated, 605 Third Avenue, New York, NY 10158. (800) 526-5368 or (212) 850-6000. 1983. $140.00.

ENCYCLOPEDIA OF PHYSICAL SCIENCE AND TECHNOLOGY. Academic Press, Incorporated, Orlando, FL 32887. (800) 321-5068. Fifteen volumes. $2500.00.

GLOSSARY OF GEOLOGY. Robert L. Bates and Julia A. Jackson. American Geological Institute, 4220 King Street, Alexandria, VA 22032. (703) 379-2480. 1980. $60.00.

GENERAL WORKS

ELEMENTS OF PETROLEUM GEOLOGY. Richard C. Shelley. W.H. Freeman and Company, 41 Madison Avenue, New York, NY 10010. (212) 532-7660. 1985. $44.95.

GEOLOGIC WELL LOG ANALYSIS. Sylvain J. Pirson. Gulf Publishing Company, Post Office Box 2608, Houston, TX 77001. (713) 529-4301. Third edition. 1983. $35.00.

INTRODUCTION TO PETROLEUM GEOLOGY. G.D. Hobson and E.N. Tiratsoo. Gulf Publishing Company, Post Office Box 2608, Houston, TX 77001. (713) 529-4301. Second edition. 1985. $55.00.

PETROLEUM GEOLOGY. R.E. Chapman. Elsevier Science Publishing Company, Incorporated, 52 Vanderbilt Avenue, New York, NY 10017. (212) 370-5520. 1983. $44.25.

HANDBOOKS AND MANUALS

GEOLOGY IN THE FIELD. R.R. Compton. John Wiley and Sons, Incorporated, 605 Third Avenue, New York, NY 10158. (800) 526-5368 or (212) 850-6418. 1985. $23.95.

ONLINE DATA BASES

GEOREF. American Geological Institute, 4220 King Street, Alexandria, VA 22302. (800) 336-4764 or (703) 379-2480. Geology and geosciences literature, 1961 to present. Inquire as to online cost and availability.

GEOARCHIVE. Geosystems, Post Office Box 1024, Westminster, London, England, SW1 P 2JL. Citations to literature on geoscience, 1969 to present. Inquire as to online cost and availability.

NTIS. National Technical Information Service, 5285 Port Royal Road, Springfield, VA 22161. (703) 487-4630. Broad coverage of government sponsored research reports, 1964 to present. Inquire as to online cost and availability.

SCISEARCH. Institute for Scientific Information, 3501 Market Street, Philadelphia, PA 19104. (800) 523-1850 or (215) 386-0100. Broad interdisciplinary index to the literature of science and technology, 1965 to present. Inquire as to online cost and availability.

PERIODICALS

AAPG BULLETIN. American Association of Petroleum Geologists, Post Office Box 979, Tulsa, OK 74101. (918) 584-2555. Monthly. $36.00 per year.

AAPG EXPLORER. American Association of Petroleum Geologists, Post Office Box 979, Tulsa, OK 74101. (918) 584-2555. Monthly. $15.00 per year.

BULLETIN OF CANADIAN PETROLEUM GEOLOGY. Canadian Society of Petroleum Geologists, 206 7th Avenue, SW, Suite 505, Calgary, AB T2P OW7, Canada. (403) 264-5610. Quarterly. $45.00 per year.

ECONOMIC GEOLOGY. Society of Economic Geologists, Post Office Box 571, Golden, CO 80402. (303) 279-1899. 8 times per year. $25.00.

GEOCHIMICA ET COSMOCHIMICA ACTA. Pergamon Journals, Incorporated, Maxwell House, Fairview Park, Elmsford, NY 10523. (014) 592-7700. Monthly. $340.00 per year.

GEOLOGICAL SOCIETY OF AMERICA BULLETIN. Post Office Box 9140, 3300 Penrose Place, Boulder, CO 80301. (303) 447-2020. Monthly. $80.00.

GEOLOGY. Geological Society of America, Post Office Box 9140, 3300 Penrose Place, Boulder, CO 80301. (303) 447-2020. Monthly. $55.00 per year.

GEOPHYSICS. Society of Exploration Geophysicists, Post Office Box 702740, Tulsa, OK 74170. (918) 493-3516. Monthly. $45.00 per year.

GEOTIMES. American Geological Institute, 4220 King Street, Alexandria, VA 22032. Monthly. $18.00 per year.

INTERNATIONAL OIL SCOUTS ASSOCIATION. Official Publication. International Oil Scouts Association, 5000 East Ben White Boulevard, Austin, TX 78741. (512) 448-4088. Quarterly. Inquire as to cost and availability.

JOURNAL OF GEOPHYSICS. Springer-Verlag, Incorporated, 175 Fifth Avenue, New York, NY 10010. (800) 526-7254 or (212) 460-1500. Bimonthly. $175.00 per year.

JOURNAL OF PETROLEUM GEOLOGY. Scientific Press, Limited, Box 21, Beaconsfield, Bucks, HP9 1HW, England. Quarterly. $130.00 per year.

JOURNAL OF SEDIMENTARY PETROLOGY. Society of Economic Paleontologists and Mineralogists, Post Office Box 4756, Tulsa, OK 74159. (918) 743-9765. Bimonthly.

JOURNAL OF STRUCTURAL GEOLOGY. Pergamon Journals, Incorporated, Maxwell House, Fairview Park, Elmsford, NY 10523. (914) 592-7700. Eight times per year. $160.00 per year.

MOUNTAIN GEOLOGIST. Rocky Mountain Association of Geologists, 4201 West 51st Avenue, Denver, CO 80212-2902. Quarterly. $15.00 per year.

OIL AND GAS JOURNAL. PennWell Publishing Company, Box 1260, Tulsa, OK 74101. Weekly. $34.00 per year.

PROFESSIONAL GEOLOGIST. American Institute of Professional Geologists, 7828 Vance Drive, Suite 103, Arvada, CO 80003. Monthly.

SEDIMENTARY GEOLOGY. Elsevier Science Publishers, Post Office Box 330, Irving-on-Hudson, NY 10533. Twenty times per year. $421.00 per year.

SEDIMENTOLOGY. Blackwell Scientific Publications, Incorporated, 52 Beacon Street, Boston, MA 02108. (617) 720-0761. Six times per year. $185.00 per year.

RESEARCH CENTERS AND INSTITUTES

LOUISIANA GEOLOGICAL SURVEY. Box G, University Station, Baton Rouge, LA 70893.

SOCIETY OF ECONOMIC GEOLOGISTS. Post Office Box 571, Golden, CO 80402. (303) 279-1899.

SOCIETY OF ECONOMIC PALEONTOLOGISTS AND MINERALOGISTS. Post Office Box 4756, Tulsa, OK 74159. (918) 743-9765.

UNIVERSITY OF TEXAS AT AUSTIN. Bureau of Economic Geology, Box X, University Station, Austin, Texas 78713. (512) 471-7721.

PETROLOGY

See: GEOLOGY, IGNEOUS ROCKS, METAMORPHIC ROCKS, SEDIMENTARY ROCKS

PHONOGRAPHS

See: AUDIO ENGINEERING

PHOTOCHEMISTRY

See also: COLOR, LASERS, PHOTOGRAPHIC FILM, PHOTOGRAPHY, PHYSICAL CHEMISTRY

ABSTRACT SERVICES AND INDEXES

APPLIED SCIENCE AND TECHNOLOGY INDEX. H.W. Wilson and Company, 950 University Avenue, Bronx, NY 10452. (800) 367-6670 or (212) 588-8400. Monthly. Inquire as to cost and availability.

CHEMICAL ABSTRACTS. American Chemical Society, Chemical Abstracts Service, Box 3012, Columbus, OH 43210. (614) 421-3600. 1907 to present. Weekly. $9500.00 per year.

CONFERENCE PAPERS INDEX. Cambridge Scientific Abstracts, 5161 River Road, Bethesda, MD 20816. 1972 to present. Monthly. Inquire as to cost and availability.

CURRENT CONTENTS: ENGINEERING, TECHNOLOGY AND APPLIED SCIENCES. Institute for Scientific Information, 3501 Market Street, Philadelphia, PA 19104. (800) 523-1850 or (215) 386-0100. Weekly. $275.00 per year.

ENGINEERING INDEX MONTHLY AND AUTHOR INDEX. Engineering Information Inc., 345 East 47th Street, New York, NY 10017. (212) 705-7600. Monthly. $1560.00 per year.

GENERAL SCIENCE INDEX. H.W. Wilson and Company, 950 University Avenue, Bronx, NY 10452. (800) 367-6670 or (212) 588-8400. 1978 to present. Monthly. Inquire as to cost and availability.

INDEX TO SCIENTIFIC AND TECHNICAL PROCEEDINGS. Institute for Scientific Information, 3501 Market Street, Philadelphia, PA 19104. (800) 523-1850 or (215) 386-0100. 1978 to present. Monthly. $775.00 per year.

INDEX TO SCIENTIFIC REVIEWS. Institute for Scientific Information, 3501 Market Street, Philadelphia, PA 19104. (800) 523-1850 or (215) 386-0100. 1974 to present. Semi-annual. $550.00 per year.

PHOTOGRAPHIC ABSTRACTS. Royal Photographic Society of Great Britain, Scientific and Technical Group, 62 Chelmsford Road, Shenfield, Brentwood, Essex, England. 1921 to present. Six times per year. $140.00 per year.

PHYSICS ABSTRACTS. Institution of Electrical Engineers. Available from: IEEE Service Center, 445 Hoes Lane, Piscataway, NJ 08854. 1898 to present. Bimonthly. $1700.00 per year.

SCIENCE CITATION INDEX. Institute for Scientific Information, 3501 Market Street, Philadelphia, PA 19104. (800) 523-1850 or (215) 386-0100. Six times per year. $6200.00 per year.

ANNUAL REVIEWS AND YEARBOOKS

ADVANCES IN PHOTOCHEMISTRY. David Volman and others, editors. John Wiley and Sons, Inc., 605 Third Avenue, New York, NY 10158. (800) 526-5368. Irregular. Volume 13, 1986. $79.95.

ASSOCIATIONS AND PROFESSIONAL SOCIETIES

AMERICAN CHEMICAL SOCIETY. 1155 16th Street, N.W., Washington, DC 20036. (202) 872-4600.

AMERICAN INSTITUTE OF CHEMICAL ENGINEERS. 345 East 47th Street, New York, NY 10017. (212) 705-7338.

OPTICAL SOCIETY OF AMERICA. 1816 Jefferson Place, N.W., Washington, DC 20036. (202) 223-8130.

SOCIETY OF PHOTOGRAPHIC SCIENTISTS AND ENGINEERS. 7003 Kilworth Lane, Springfield, VA 22151. (703) 642-9090.

SPIE - THE INTERNATIONAL SOCIETY FOR OPTICAL ENGINEERING. P.O. Box 10, 1022 19th Street, Bellingham, WA 98227. (206) 676-3290.

DIRECTORIES AND BIOGRAPHICAL SOURCES

INTERNATIONAL RESEARCH CENTERS DIRECTORY 1988-89. Darren L. Smith, editor. Gale Research Company, Book Tower, Detroit, MI 48226. (800) 521-0707. 4th edition. 1987. $360.00.

1987 DIRECTORY OF ENGINEERING SOCIETIES AND RELATED ORGANIZATIONS. Gordon Davis, editor. Hemisphere Publishing Corporation, 1010 Vermont Avenue, NW, Washington, DC 20005. (800) 526-0275. 12th edition. 1987. $100.00.

RESEARCH CENTERS DIRECTORY 1988. Gale Research Company, Book Tower, Detroit, MI 48226. (800) 521-0707. 12th edition. 1987. $365.00 for set.

SCIENTIFIC AND TECHNICAL ORGANIZATIONS AND AGENCIES DIRECTORY. Margaret Labash Young, editor. Gale Research Company, Book Tower, Detroit, MI 48226. (800) 521-0707. 2nd edition. 1987. $185.00.

WHO'S WHO IN ENGINEERING. Gordon Davis, editor. Hemisphere Publishing Corporation, 1010 Vermont Avenue, NW, Washington, DC 20005. (800) 526-0275. 6th edition. 1985. $200.00.

ENCYCLOPEDIAS AND DICTIONARIES

THESAURUS OF PHOTOGRAPHIC SCIENCE AND ENGINEERING. Society of Photographic Scientists and Engineers. Books on Demand, 300 North Zeeb Road, Ann Arbor, MI 48106. (313) 761-4700. $34.50 in paper.

THESAURUS OF SCIENTIFIC, TECHNICAL, AND ENGINEERING TERMS. Hemisphere Publishing Corporation, 1010 Vermont Avenue, NW, Washington, DC 20005. (800) 526-0275. 1988. $125.00.

GENERAL WORKS

APPLICATIONS OF PHOTOCHEMISTRY. Ralph Roberts. Technomic Publishing Company, 851 New Holland Avenue, Box 3535, Lancaster, PA 17604. (717) 291-5609. 1984. $25.00 in paper.

ELEMENTS OF INORGANIC PHOTOCHEMISTRY. D. Ferraudi. John Wiley and Sons, Inc., 605 Third Avenue, New York, NY 10158. (800) 526-5368. 1987. $45.95.

INTRODUCTION TO ORGANIC PHOTOCHEMISTRY. J.D. Coyle. John Wiley and Sons, Inc., 605 Third Avenue, New York, NY 10158. (800) 526-5368. 1986. $31.95.

LIGHT AND COLOR. R.D. Overheim and D.L. Wagner. John Wiley and Sons, Inc., 605 Third Avenue, New York, NY 10158. (800) 526-5368. 1982. $28.50.

PHOTOCHEMISTRY: AN INTRODUCTION. D.R. Arnold and others. Academic Press, Inc., 6277 Sea Harbor Drive, Orlando, FL 32821. (800) 321-5068. 1974. $45.00.

PHOTOGRAPHIC SCIENCE. Earl. N. Mitchell. John Wiley and Sons, Inc., 605 Third Avenue, New York, NY 10158. (800) 526-5368. 1984. $37.50.

PHOTOGRAPHY FOR THE SCIENTIST. Richard A. Morton. Academic Press, Inc., 6277 Sea Harbor Drive, Orlando, FL 32821. (800) 321-5068. 2nd edition. 1984. $102.50.

HANDBOOKS AND MANUALS

HANDBOOK OF PHOTOGRAPHIC SCIENCE AND ENGINEERING. Society of Photographic Scientists and Engineers. 7003 Kilworth Lane, Springfield, VA 22151. (703) 642-9090. Inquire.

KODAK PROFESSIONAL PHOTOGUIDE. Carolyn Grimes, editor. Eastman Kodak Company, 343 State Street, Rochester, NY 14650. (716) 724-4000. 1986. $19.95 in paper.

ONLINE DATA BASES

CA SEARCH. Chemical Abstracts Service, P.O. Box 3012, Columbus, OH 43120. (800) 848-6538 or (614) 421-3600. Comprehensive guide to chemical literature, 1972 to present. Inquire as to online cost and availability.

COMPENDEX. Engineering Information, Inc., 345 East 47th Street, New York, NY 10017. (800) 221-1044 or (212) 705-7615. Engineering and technical literature, 1975 to present. Inquire as to online cost and availability.

INSPEC. INSPEC Marketing Department, Institution of Electrical Engineers. Available from IEEE Service Center, 445 Hoes Lane, Piscataway, NJ 08854. (201) 981-0060. Online version of Physics Abstracts. Inquire as to online cost and availability.

NTIS. National Technical Information Service, 5285 Port Royal Road, Springfield, VA 22161. (703) 487-4630. Broad coverage of government sponsored research reports, 1964 to present. Inquire as to online cost and availability.

SCISEARCH. Institute for Scientific Information, 3501 Market Street, Philadelphia, PA 19104. (800) 523-1850 or (215) 386-0100. Broad multidisciplinary title and author index to the international literature of science and technology, 1974 to present. Inquire as to online cost and availability.

WILSONLINE. H.W. Wilson and Company, 950 University Avenue, Bronx, NY 10452. (800) 367-6770 or (212) 588-8400. Makes available online versions of the H.W. Wilson indexes including Applied Science and Technology Index, Business Periodicals Index and Readers' Guide to Periodical Literature. Approximately 1980 to present. Inquire as to online cost and availability.

PERIODICALS

BRITISH JOURNAL OF PHOTOGRAPHY. Henry Greenwood and Company, Limited, 28 Great James Street, London WC1N 3HL, England. 1854 to present. Weekly. $70.00 per year.

JOURNAL OF IMAGING SCIENCE. Society of Photographic Scientists and Engineers. 7003 Kilworth Lane, Springfield, VA 22151. (703) 642-9090. 1956 to present. Bimonthly. $70.00 per year.

JOURNAL OF IMAGING TECHNOLOGY. Society of Photographic Scientists and Engineers. 7003 Kilworth Lane, Springfield, VA 22151. (703) 642-9090. 1975 to present.

Bimonthly. $70.00 per year.

JOURNAL OF PHOTOGRAPHIC SCIENCE. Royal Photographic Society of Great Britain, 7 Ladbroke Walk, London W11, England. 1953 to present. Bimonthly. $40.00 per year.

KODAK TECH BITS. Eastman Kodak Company, 343 State Street, Rochester, NY 14650. (716) 724-4000. 1963 to present. Quarterly. Free to qualified personnel.

SCIENTIFIC AND APPLIED PHOTOGRAPHY AND CINEMATOGRAPHY. Gordon and Breach Science Publishers, Inc., 50 West 23rd Street, New York, NY 10010. (212) 206-8900. 12 times per year. $496.00 per year.

RESEARCH CENTERS AND INSTITUTES

IMAGE PERMANENCE INSTITUTE. Rochester Institute of Technology, RIT City Center, 50 West Main Street, Rochester, NY 14614. (716) 475-5199.

LAWRENCE BERKELEY LABORATORY, MATERIALS AND CHEMICALS SCIENCES DIVISION. Building 62, 1 Cyclotron Road, Berkeley, CA 94720. (415) 486-6062.

PHOTOCHEMISTRY UNIT. University of Western Ontario, London, ON, Canada N6A 5B7. (519) 679-2111.

RADIATION CHEMISTRY DATA CENTER. University of Norte Dame, Notre Dame, IN 46556. (219) 239-6528.

PHOTOELECTRICITY

See also: ELECTRICITY,
ENERGY, PHYSICAL CHEMISTRY, PHYSICS

ABSTRACT SERVICES AND INDEXES

APPLIED SCIENCE AND TECHNOLOGY INDEX. H.W. Wilson and Company, 950 University Avenue, Bronx, NY 10452. (800) 367-6670 or (212) 588-8400. Monthly. Inquire as to cost and availability.

CHEMICAL ABSTRACTS. American Chemical Society, Chemical Abstracts Service, Box 3012, Columbus, OH 43210. (614) 421-3600. 1907 to present. Weekly. $9500.00 per year.

CURRENT CONTENTS: ENGINEERING, TECHNOLOGY AND APPLIED SCIENCES. Institute for Scientific Information, 3501 Market Street, Philadelphia, PA 19104. (800) 523-1850 or (215) 386-0100. Weekly. $275.00 per year.

ELECTRICAL AND ELECTRONICS ABSTRACTS. Published by the Institute of Electrical Engineers, London. Available from IEEE Service Center, 445 Hoes Lane, Piscataway, NJ 08854. Monthly. $1250.00 per year.

ENGINEERING INDEX MONTHLY AND AUTHOR INDEX. Engineering Information Inc., 345 East 47th Street, New York, NY 10017. (212) 705-7600. Monthly. $1560.00 per year.

GENERAL SCIENCE INDEX. H.W. Wilson and Company, 950 University Avenue, Bronx, NY 10452. (800) 367-6670 or (212) 588-8400. 1978 to present. Monthly. Inquire as to cost and availability.

INDEX TO SCIENTIFIC AND TECHNICAL PROCEEDINGS. Institute for Scientific Information, 3501 Market Street, Philadelphia, PA 19104. (800) 523-1850 or (215) 386-0100. 1978 to present. Monthly. $775.00 per year.

INDEX TO SCIENTIFIC REVIEWS. Institute for Scientific Information, 3501 Market Street, Philadelphia, PA 19104. (800) 523-1850 or (215) 386-0100. 1974 to present. Semi-annual. $550.00 per year.

PHYSICS ABSTRACTS. Institution of Electrical Engineers. Available from: IEEE Service Center, 445 Hoes Lane, Piscataway, NJ 08854. 1898 to present. Bimonthly. $1700.00 per year.

SCIENCE CITATION INDEX. Institute for Scientific Information, 3501 Market Street, Philadelphia, PA 19104. (800) 523-1850 or (215) 386-0100. Six times per year. $6200.00 per year.

ANNUAL REVIEWS AND YEARBOOKS

CURRENT TOPICS IN PHOTOVOLTAICS. T.J. Coutts and J.O. Meakin, editors. Academic Press, Inc., 6277 Sea Harbor Drive, Orlando, FL 32821. (800) 321-5068. Irregular. 1985. $68.50.

ASSOCIATIONS AND PROFESSIONAL SOCIETIES

AMERICAN CHEMICAL SOCIETY. 1155 16th Street, N.W., Washington, DC 20036. (202) 872-4600.

AMERICAN INSTITUTE OF CHEMICAL ENGINEERS. 345 East 47th Street, New York, NY 10017. (212) 705-7338.

ASSOCIATION OF ENERGY ENGINEERS. 4025 Pleasantdale Road, Suite 340, Atlanta, GA 30340. (404) 447-5083.

IEEE (Institute of Electrical and Electronics Engineers). 345 East 47th Street, New York, NY 10017. (212) 705-7900.

OPTICAL SOCIETY OF AMERICA. 1816 Jefferson Place, N.W., Washington, DC 20036. (202) 223-8130.

SPIE - THE INTERNATIONAL SOCIETY FOR OPTICAL ENGINEERING. P.O. Box 10, 1022 19th Street, Bellingham, WA 98227. (206) 676-3290.

DIRECTORIES AND BIOGRAPHICAL SOURCES

INTERNATIONAL RESEARCH CENTERS DIRECTORY 1988-89. Darren L. Smith, editor. Gale Research Company, Book Tower, Detroit, MI 48226. (800) 521-0707. 4th edition. 1987. $360.00.

1987 DIRECTORY OF ENGINEERING SOCIETIES AND RELATED ORGANIZATIONS. Gordon Davis, editor. Hemisphere Publishing Corporation, 1010 Vermont Avenue, NW, Washington, DC 20005. (800) 526-0275. 12th edition. 1987. $100.00.

RESEARCH CENTERS DIRECTORY 1988. Gale Research Company, Book Tower, Detroit, MI 48226. (800) 521-0707. 12th edition. 1987. $365.00 for set.

SCIENTIFIC AND TECHNICAL ORGANIZATIONS AND AGENCIES DIRECTORY. Margaret Labash Young, editor. Gale Research Company, Book Tower, Detroit, MI 48226. (800) 521-0707. 2nd edition. 1987. $185.00.

WHO'S WHO IN ENGINEERING. Gordon Davis, editor. Hemisphere Publishing Corporation, 1010 Vermont Avenue, NW, Washington, DC 20005. (800) 526-0275. 6th edition. 1985. $200.00.

GENERAL WORKS

ALTERNATIVE SOURCES OF ENERGY. Larry Stoiaken and Donald Marier, editors. Alternative Sources of Energy, Inc., 107 South Central Avenue, Milaca, MN 56353. (612) 983-6892. 1987. $4.95 in paper.

PHOTOVOLTAIC ENERGY SYSTEMS: DESIGN AND INSTALLATION. M. Buresch. McGraw-Hill Book Company, 1221 Avenue of the Americas, New York, NY 10020. (212) 512-2000. 1983. $32.50.

SUNLIGHT TO ELECTRICITY: PHOTOVOLTAIC TECHNOLOGY AND PROSPECTS. Joseph A. Merrigan. MIT Press, 28 Carleton Street, Cambridge, MA 02142. (617) 253-2884. 2nd edition. 1982. $30.00.

ONLINE DATA BASES

CA SEARCH. Chemical Abstracts Service, P.O. Box 3012, Columbus, OH 43120. (800) 848-6538 or (614) 421-3600. Comprehensive guide to chemical literature, 1972 to present. Inquire as to online cost and availability.

COMPENDEX. Engineering Information, Inc., 345 East 47th Street, New York, NY 10017. (800) 221-1044 or (212) 705-7615. Engineering and technical literature, 1975 to present. Inquire as to online cost and availability.

INSPEC. INSPEC Marketing Department, Institution of Electrical Engineers. Available from IEEE Service Center, 445 Hoes Lane, Piscataway, NJ 08854. (201) 981-0060. Online version of Physics Abstracts. Inquire as to online cost and availability.

NTIS. National Technical Information Service, 5285 Port Royal Road, Springfield, VA 22161. (703) 487-4630. Broad coverage of government sponsored research reports, 1964 to present. Inquire as to online cost and availability.

SCISEARCH. Institute for Scientific Information, 3501 Market Street, Philadelphia, PA 19104. (800) 523-1850 or (215) 386-0100. Broad multidisciplinary title and author index to the international literature of science and technology, 1974 to present. Inquire as to online cost and availability.

WILSONLINE. H.W. Wilson and Company, 950 University Avenue, Bronx, NY 10452. (800) 367-6770 or (212) 588-8400. Makes available online versions of the H.W. Wilson indexes including Applied Science and Technology Index, Business Periodicals Index and Readers' Guide to Periodical Literature. Approximately 1980 to present. Inquire as to online cost and availability.

OTHER SOURCES

A GUIDE TO THE LITERATURE OF ELECTRICAL AND ELECTRONICS ENGINEERING. Susan B. Ardis. Libraries Unlimited, Inc., P.O. Box 263, Littleton, CO 80160. (303) 770-1220. 1986. $37.50.

PERIODICALS

NEW DEVICES AND MATERIALS IN PHOTOELECTRICS. Multi-Science Publishing Company, Limited, 42-45 New Broad Street, London EC2M 1QY, England. 1986 to present. Quarterly. $100.00 per year.

PHOTOVOLTAICS: THE SOLAR ELECTRIC MAGAZINE. Fore Publishers, Inc., Box 3269, Scottsdale, AZ 85257. (602) 994-1923. 1982 to present. Monthly. $24.00 per year.

SOLAR CELLS. Elsevier Science Publishing Company, Inc., 52 Vanderbilt Avenue, New York, NY 10017. (212) 370-5520. 1979 to present. Monthly. $325.00 per year.

RESEARCH CENTERS AND INSTITUTES

LABORATORY OF ELECTROCHEMICAL RESEARCH. University of Texas at Austin, Austin, TX 78712. (512) 471-3761.

VISIBILITY LABORATORY. University of California, San Diego, Scripps Institution of Oceanography, La Jolla, CA 92093. (619) 534-1744.

PHOTOGRAMMETRY

See also: AERIAL PHOTOGRAPHY, CARTOGRAPHY, REMOTE SENSING, SURVEYING

ABSTRACT SERVICES AND INDEXES

APPLIED SCIENCE AND TECHNOLOGY INDEX. H.W. Wilson and Company, 950 University Avenue, Bronx, NY 10452. (800) 367-6670 or (212) 588-8400. Monthly. Inquire as to cost and availability.

BIBLIOGRAPHY AND INDEX OF GEOLOGY. American Geological Institute, 4220 King Street, Alexandria, VA 22302. (703) 379-2480. 1969 to present. Monthly. $1100.00 per year.

GEOREF. Online version of the BIBLIOGRAPHY AND INDEX OF GEOLOGY. American Geological Institute, 4220 King Street, Alexandria, VA 22302. (703) 379-2480. 1969 to present. Inquire as to online cost and availability.

CHEMICAL ABSTRACTS. American Chemical Society, Chemical Abstracts Service, Box 3012, Columbus, OH 43210. (614) 421-3600. 1907 to present. Weekly. $9500.00 per year.

CURRENT CONTENTS: ENGINEERING, TECHNOLOGY AND APPLIED SCIENCES. Institute for Scientific Information, 3501 Market Street, Philadelphia, PA 19104. (800) 523-1850 or (215) 386-0100. Weekly. $275.00 per year.

ENGINEERING INDEX MONTHLY AND AUTHOR INDEX. Engineering Information Inc., 345 East 47th Street, New York, NY 10017. (212) 705-7600. Monthly. $1560.00 per year.

METEOROLOGICAL AND GEOASTROPHYSICAL ABSTRACTS. American Meteorological Society, 45 Beacon Street, Boston, MA 02108. (617) 227-2425. 1950 to present. Monthly. $450.00 per year.

PHOTOGRAPHIC ABSTRACTS. Royal Photographic Society of Great Britain, Scientific and Technical Group, 62 Chelmsford Road, Shenfield, Brentwood, Essex, England. 1921 to present. Six times per year. $140.00 per year.

PHYSICS ABSTRACTS. Institution of Electrical Engineers. Available from: IEEE Service Center, 445 Hoes Lane, Piscataway, NJ 08854. 1898 to present. Bimonthly. $1700.00 per year.

SCIENCE CITATION INDEX. Institute for Scientific Information, 3501 Market Street, Philadelphia, PA 19104. (800) 523-1850 or (215) 386-0100. Six times per year. $6200.00 per year.

ASSOCIATIONS AND PROFESSIONAL SOCIETIES

AMERICAN SOCIETY FOR PHOTOGRAMMETRY AND REMOTE SENSING. 210 Little Falls Street, Falls Church, VA 22046-4398. (703) 534-6617.

OPTICAL SOCIETY OF AMERICA. 1816 Jefferson Place, N.W., Washington, DC 20036. (202) 223-8130.

SOCIETY OF PHOTOGRAPHIC SCIENTISTS AND ENGINEERS. 7003 Kilworth Lane, Springfield, VA 22151. (703) 642-9090.

SPIE - THE INTERNATIONAL SOCIETY FOR OPTICAL ENGINEERING. P.O. Box 10, 1022 19th Street, Bellingham, WA 98227. (206) 676-3290.

BIBLIOGRAPHIES

SCIENTIFIC AND TECHNICAL BOOKS AND SERIALS IN PRINT 1988; AN INDEX TO LITERATURE IN SCIENCE AND TECHNOLOGY. R.R. Bowker Company, 205 East 42nd Street, New York, NY 10017. (800) 521-8110. $175.00.

DIRECTORIES AND BIOGRAPHICAL SOURCES

INTERNATIONAL RESEARCH CENTERS DIRECTORY 1988-89. Darren L. Smith, editor. Gale Research Company, Book Tower, Detroit, MI 48226. (800) 521-0707. 4th edition. 1987. $360.00.

1987 DIRECTORY OF ENGINEERING SOCIETIES AND RELATED ORGANIZATIONS. Gordon Davis, editor. Hemisphere Publishing Corporation, 1010 Vermont Avenue, NW, Washington, DC 20005. (800) 526-0275. 12th edition. 1987. $100.00.

RESEARCH CENTERS DIRECTORY 1988. Gale Research Company, Book Tower, Detroit, MI 48226. (800) 521-0707. 12th edition. 1987. $365.00 for set.

SCIENTIFIC AND TECHNICAL ORGANIZATIONS AND AGENCIES DIRECTORY. Margaret Labash Young, editor. Gale Research Company, Book Tower, Detroit, MI 48226. (800) 521-0707. 2nd edition. 1987. $185.00.

WHO'S WHO IN ENGINEERING. Gordon Davis, editor. Hemisphere Publishing Corporation, 1010 Vermont Avenue, NW, Washington, DC 20005. (800) 526-0275. 6th edition. 1985. $200.00.

ENCYCLOPEDIAS AND DICTIONARIES

MCGRAW-HILL ENCYCLOPEDIA OF SCIENCE AND TECHNOLOGY. McGraw-Hill Book Company, 1221 Avenue of the Americas, New York, NY 10020. (212) 512-2000. 6th edition. 1987. $1600.00.

THESAURUS OF PHOTOGRAPHIC SCIENCE AND ENGINEERING. Society of Photographic Scientists and Engineers. Books on Demand, 300 North Zeeb Road, Ann Arbor, MI 48106. (313) 761-4700. $34.50 in paper.

THESAURUS OF SCIENTIFIC, TECHNICAL, AND ENGINEERING TERMS. Hemisphere Publishing Corporation, 1010 Vermont Avenue, NW, Washington, DC 20005. (800) 526-0275. 1988. $125.00.

GENERAL WORKS

CLOSE-RANGE PHOTOGRAMMETRY AND SURVEYING: STATE OF THE ART. American Society for Photogrammetry and Remote Sensing. 210 Little Falls Street, Falls Church, VA 22046-4398. (703) 534-6617. 1985. $65.00 in paper.

ELEMENTS OF PHOTOGRAMMETRY. P.R. Wolf. McGraw-Hill Book Company, 1221 Avenue of the Americas, New York, NY 10020. (212) 512-2000. 2nd edition. 1983. $49.95.

PHOTOGRAMMETRY. Francis H. Moffitt and Edward M. Mikhail. Harper and Row Publishers, Inc., 10 East 53rd Street, New York, NY 10022. (212) 207-7655. 3rd edition. 1980. $41.95.

PHOTOGRAPHIC SCIENCE. Earl. N. Mitchell. John Wiley and Sons, Inc., 605 Third Avenue, New York, NY 10158. (800) 526-5368. 1984. $37.50.

SURVEY OF THE PROFESSION: PHOTOGRAMMETRY, SURVEYING, MAPPING, REMOTE SENSING. American Society for Photogrammetry and Remote Sensing. 210 Little Falls Street, Falls Church, VA 22046-4398. (703) 534-6617. 1982. $35.00 in paper.

HANDBOOKS AND MANUALS

HANDBOOK OF PHOTOGRAPHIC SCIENCE AND ENGINEERING. Society of Photographic Scientists and Engineers. 7003 Kilworth Lane, Springfield, VA 22151. (703) 642-9090. Inquire.

MANUAL OF PHOTOGRAMMETRY. Chester C. Slama, editor. American Society for Photogrammetry and Remote Sensing. 210 Little Falls Street, Falls Church, VA 22046-4398. (703) 534-6617. 4th edition. 1980. $59.00.

ONLINE DATA BASES

CA SEARCH. Chemical Abstracts Service, P.O. Box 3012, Columbus, OH 43120. (800) 848-6538 or (614) 421-3600. Comprehensive guide to chemical literature, 1972 to present. Inquire as to online cost and availability.

COMPENDEX. Engineering Information, Inc., 345 East 47th Street, New York, NY 10017. (800) 221-1044 or (212) 705-7615. Engineering and technical literature, 1975 to present. Inquire as to online cost and availability.

GEOREF. Online version of the BIBLIOGRAPHY AND INDEX OF GEOLOGY. American Geological Institute, 4220 King Street, Alexandria, VA 22302. (703) 379-2480. 1969 to present. Inquire as to online cost and availability.

NTIS. National Technical Information Service, 5285 Port Royal Road, Springfield, VA 22161. (703) 487-4630. Broad coverage of government sponsored research reports, 1964 to present. Inquire as to online cost and availability.

SCISEARCH. Institute for Scientific Information, 3501 Market Street, Philadelphia, PA 19104. (800) 523-1850 or (215) 386-0100. Broad multidisciplinary title and author index to the international literature of science and technology, 1974 to present. Inquire as to online cost and availability.

WILSONLINE. H.W. Wilson and Company, 950 University Avenue, Bronx, NY 10452. (800) 367-6770 or (212) 588-8400. Makes available online versions of the H.W. Wilson indexes including Applied Science and Technology Index, Business Periodicals Index and Readers' Guide to Periodical Literature. Approximately 1980 to present. Inquire as to online cost and availability.

PERIODICALS

INDUSTRIAL PHOTOGRAPHY. United Business Publications, Inc., 475 Park Avenue South, New York, NY 10016. (212) 725-2300. 1952 to present. Monthly. $15.00 per year.

JOURNAL OF IMAGING SCIENCE. Society of Photographic Scientists and Engineers. 7003 Kilworth Lane, Springfield, VA 22151. (703) 642-9090. 1956 to present. Bimonthly. $70.00 per year.

JOURNAL OF IMAGING TECHNOLOGY. Society of Photographic Scientists and Engineers. 7003 Kilworth Lane, Springfield, VA 22151. (703) 642-9090. 1975 to present. Bimonthly. $70.00 per year.

JOURNAL OF PHOTOGRAPHIC SCIENCE. Royal Photographic Society of Great Britain, 7 Ladbroke Walk, London W11, England. 1953 to present. Bimonthly. $40.00 per year.

PHOTOGRAMMETRIA. Elsevier Science Publishing Company, Inc., 52 Vanderbilt Avenue, New York, NY 10017. (212) 370-5520. 1949 to present. Quarterly. $65.00 per year.

PHOTOGRAMMETRIC ENGINEERING AND REMOTE SENSING. American Society for Photogrammetry and Remote Sensing. 210 Little Falls Street, Falls Church, VA 22046-4398. (703) 534-6617. Order from: Allen Press, Inc., 1041 New Hampshire Street, Box 368, Lawrence, KS 66044. 1934 to present. Monthly. $80.00 per year.

PHOTOGRAMMETRIC RECORD. Photogrammetry Society, Department of Photogrammetry and Surveying, University College London, Gower Street, London WC1E 6BT, England. Semiannual. $37.50 per year.

SCIENTIFIC AND APPLIED PHOTOGRAPHY AND CINEMATOGRAPHY. Gordon and Breach Science Publishers, Inc., 50 West 23rd Street, New York, NY 10010. (212) 206-8900. 12 times per year. $496.00 per year.

RESEARCH CENTERS AND INSTITUTES

CENTER FOR REMOTE SENSING AND CARTOGRAPHY. 420 Chipta Way, Salt Lake City, UT 84112. (801) 581-8218.

GEOPHOTOGRAPHY AND REMOTE SENSING CENTER. University of Idaho, Geology Department, Moscow, ID 83843. (208) 885-7977.

NATIONAL RESEARCH COUNCIL OF CANADA, DIVISION OF PHYSICS. Ottawa, ON, Canada K1A OR6. (613) 993-1053.

PHOTOGRAMMETRY AND REMOTE SENSING SECTION. Tennessee Valley Authority, Office of Natural Resources and Economic Development, Haney Building, Chattanooga, TN 37401. (615) 755-2148.

PHOTOGRAPHIC FILM

See also: AERIAL PHOTOGRAPHY, CAMERAS, COLOR, PHOTOCHEMISTRY, PHOTOGRAMMETRY, PHOTOGRAPHY

ABSTRACT SERVICES AND INDEXES

APPLIED SCIENCE AND TECHNOLOGY INDEX. H.W. Wilson and Company, 950 University Avenue, Bronx, NY 10452. (800) 367-6670 or (212) 588-8400. Monthly. Inquire as to cost and availability.

CHEMICAL ABSTRACTS. American Chemical Society, Chemical Abstracts Service, Box 3012, Columbus, OH 43210. (614) 421-3600. 1907 to present. Weekly. $9500.00 per year.

CURRENT CONTENTS: ENGINEERING, TECHNOLOGY AND APPLIED SCIENCES. Institute for Scientific Information, 3501 Market Street, Philadelphia, PA 19104. (800) 523-1850 or (215) 386-0100. Weekly. $275.00 per year.

ENGINEERING INDEX MONTHLY AND AUTHOR INDEX. Engineering Information Inc., 345 East 47th Street, New York, NY 10017. (212) 705-7600. Monthly. $1560.00 per year.

PHOTOGRAPHIC ABSTRACTS. Royal Photographic Society of Great Britain, Scientific and Technical Group, 62 Chelmsford Road, Shenfield, Brentwood, Essex, England. 1921 to present. Six times per year. $140.00 per year.

PHYSICS ABSTRACTS. Institution of Electrical Engineers. Available from: IEEE Service Center, 445 Hoes Lane, Piscataway, NJ 08854. 1898 to present. Bimonthly. $1700.00 per year.

SCIENCE CITATION INDEX. Institute for Scientific Information, 3501 Market Street, Philadelphia, PA 19104. (800) 523-1850 or (215) 386-0100. Six times per year. $6200.00 per year.

ASSOCIATIONS AND PROFESSIONAL SOCIETIES

OPTICAL SOCIETY OF AMERICA. 1816 Jefferson Place, N.W., Washington, DC 20036. (202) 223-8130.

SOCIETY OF PHOTOGRAPHIC SCIENTISTS AND ENGINEERS. 7003 Kilworth Lane, Springfield, VA 22151. (703) 642-9090.

SPIE - THE INTERNATIONAL SOCIETY FOR OPTICAL ENGINEERING. P.O. Box 10, 1022 19th Street, Bellingham, WA 98227. (206) 676-3290.

DIRECTORIES AND BIOGRAPHICAL SOURCES

INTERNATIONAL RESEARCH CENTERS DIRECTORY 1988-89. Darren L. Smith, editor. Gale Research Company, Book Tower, Detroit, MI 48226. (800) 521-0707. 4th edition. 1987. $360.00.

1987 DIRECTORY OF ENGINEERING SOCIETIES AND RELATED ORGANIZATIONS. Gordon Davis, editor. Hemisphere Publishing Corporation, 1010 Vermont Avenue, NW, Washington, DC 20005. (800) 526-0275. 12th edition. 1987. $100.00.

RESEARCH CENTERS DIRECTORY 1988. Gale Research Company, Book Tower, Detroit, MI 48226. (800) 521-0707. 12th edition. 1987. $365.00 for set.

SCIENTIFIC AND TECHNICAL ORGANIZATIONS AND AGENCIES DIRECTORY. Margaret Labash Young, editor. Gale Research Company, Book Tower, Detroit, MI 48226. (800) 521-0707. 2nd edition. 1987. $185.00.

WHO'S WHO IN ENGINEERING. Gordon Davis, editor. Hemisphere Publishing Corporation, 1010 Vermont Avenue, NW, Washington, DC 20005. (800) 526-0275. 6th edition. 1985. $200.00.

ENCYCLOPEDIAS AND DICTIONARIES

THESAURUS OF PHOTOGRAPHIC SCIENCE AND ENGINEERING. Society of Photographic Scientists and Engineers. Books on Demand, 300 North Zeeb Road, Ann Arbor, MI 48106. (313) 761-4700. $34.50 in paper.

THESAURUS OF SCIENTIFIC, TECHNICAL, AND ENGINEERING TERMS. Hemisphere Publishing Corporation, 1010 Vermont Avenue, NW, Washington, DC 20005. (800) 526-0275. 1988. $125.00.

GENERAL WORKS

KODAK FILMS - COLOR AND BLACK AND WHITE. Eastman Kodak Company, 343 State Street, Rochester, NY 14650. (716) 724-4000. 1985. $8.95 in paper.

LIGHT AND COLOR. R.D. Overheim and D.L. Wagner. John Wiley and Sons, Inc., 605 Third Avenue, New York, NY 10158. (800) 526-5368. 1982. $28.50.

PHOTOCHEMISTRY: AN INTRODUCTION. D.R. Arnold and others. Academic Press, Inc., 6277 Sea Harbor Drive, Orlando, FL 32821. (800) 321-5068. 1974. $45.00.

PHOTOGRAPHIC SCIENCE. Earl. N. Mitchell. John Wiley and Sons, Inc., 605 Third Avenue, New York, NY 10158. (800) 526-5368. 1984. $37.50.

PHOTOGRAPHY FOR THE SCIENTIST. Richard A. Morton. Academic Press, Inc., 6277 Sea Harbor Drive, Orlando, FL 32821. (800) 321-5068. 2nd edition. 1984. $102.50.

USER'S GUIDE TO PHOTOGRAPHIC FILMS. Dan O'Neill. Watson-Guptill Publishers, Inc., 1 Astro Plaza, 1515 Broadway, New York, NY 10036. (212) 764-7518. 1984. $24.95.

HANDBOOKS AND MANUALS

HANDBOOK OF PHOTOGRAPHIC SCIENCE AND ENGINEERING. Society of Photographic Scientists and Engineers. 7003 Kilworth Lane, Springfield, VA 22151. (703) 642-9090. Inquire.

KODAK PROFESSIONAL PHOTOGUIDE. Carolyn Grimes, editor. Eastman Kodak Company, 343 State Street, Rochester, NY 14650. (716) 724-4000. 1986. $19.95 in paper.

ONLINE DATA BASES

CA SEARCH. Chemical Abstracts Service, P.O. Box 3012, Columbus, OH 43120. (800) 848-6538 or (614) 421-3600. Comprehensive guide to chemical literature, 1972 to present. Inquire as to online cost and availability.

COMPENDEX. Engineering Information, Inc., 345 East 47th Street, New York, NY 10017. (800) 221-1044 or (212) 705-7615.

Engineering and technical literature, 1975 to present. Inquire as to online cost and availability.

INSPEC. INSPEC Marketing Department, Institution of Electrical Engineers. Available from IEEE Service Center, 445 Hoes Lane, Piscataway, NJ 08854. (201) 981-0060. Online version of Physics Abstracts. Inquire as to online cost and availability.

NTIS. National Technical Information Service, 5285 Port Royal Road, Springfield, VA 22161. (703) 487-4630. Broad coverage of government sponsored research reports, 1964 to present. Inquire as to online cost and availability.

SCISEARCH. Institute for Scientific Information, 3501 Market Street, Philadelphia, PA 19104. (800) 523-1850 or (215) 386-0100. Broad multidisciplinary title and author index to the international literature of science and technology, 1974 to present. Inquire as to online cost and availability.

WILSONLINE. H.W. Wilson and Company, 950 University Avenue, Bronx, NY 10452. (800) 367-6770 or (212) 588-8400. Makes available online versions of the H.W. Wilson indexes including Applied Science and Technology Index, Business Periodicals Index and Readers' Guide to Periodical Literature. Approximately 1980 to present. Inquire as to online cost and availability.

PERIODICALS

BRITISH JOURNAL OF PHOTOGRAPHY. Henry Greenwood and Company, Limited, 28 Great James Street, London WC1N 3HL, England. 1854 to present. Weekly. $70.00 per year.

FUNCTIONAL PHOTOGRAPHY. PTN Publishing Corporation, 210 Crossways Park Drive, Woodbury, NY 11797. (516) 496-8000. 1967 to present. Bimonthly. $7.50 per year.

INDUSTRIAL PHOTOGRAPHY. United Business Publications, Inc., 475 Park Avenue South, New York, NY 10016. (212) 725-2300. 1952 to present. Monthly. $15.00 per year.

JOURNAL OF IMAGING SCIENCE. Society of Photographic Scientists and Engineers. 7003 Kilworth Lane, Springfield, VA 22151. (703) 642-9090. 1956 to present. Bimonthly. $70.00 per year.

JOURNAL OF IMAGING TECHNOLOGY. Society of Photographic Scientists and Engineers. 7003 Kilworth Lane, Springfield, VA 22151. (703) 642-9090. 1975 to present. Bimonthly. $70.00 per year.

JOURNAL OF PHOTOGRAPHIC SCIENCE. Royal Photographic Society of Great Britain, 7 Ladbroke Walk, London W11, England. 1953 to present. Bimonthly. $40.00 per year.

KODAK TECH BITS. Eastman Kodak Company, 343 State Street, Rochester, NY 14650. (716) 724-4000. 1963 to present. Quarterly. Free to qualified personnel.

SCIENTIFIC AND APPLIED PHOTOGRAPHY AND CINEMATOGRAPHY. Gordon and Breach Science Publishers, Inc., 50 West 23rd Street, New York, NY 10010. (212) 206-8900. 12 times per year. $496.00 per year.

RESEARCH CENTERS AND INSTITUTES

GRAPHIC ARTS TECHNICAL FOUNDATION, RESEARCH DEPARTMENT. 4615 Forbes Avenue, Pittsburgh, PA 15213. (412) 621-6941.

IMAGE PERMANENCE INSTITUTE. Rochester Institute of Technology, RIT City Center, 50 West Main Street, Rochester, NY 14614. (716) 475-5199.

PHOTOGRAPHY

See also: AERIAL PHOTOGRAPHY, CAMERAS, COLOR, PHOTOCHEMISTRY, PHOTOGRAMMETRY, PHOTOGRAPHIC FILM

ABSTRACT SERVICES AND INDEXES

APPLIED SCIENCE AND TECHNOLOGY INDEX. H.W. Wilson and Company, 950 University Avenue, Bronx, NY 10452. (800) 367-6670 or (212) 588-8400. Monthly. Inquire as to cost and availability.

CHEMICAL ABSTRACTS. American Chemical Society, Chemical Abstracts Service, Box 3012, Columbus, OH 43210. (614) 421-3600. 1907 to present. Weekly. $9500.00 per year.

CONFERENCE PAPERS INDEX. Cambridge Scientific Abstracts, 5161 River Road, Bethesda, MD 20816. 1972 to present. Monthly. Inquire as to cost and availability.

CURRENT CONTENTS: ENGINEERING, TECHNOLOGY AND APPLIED SCIENCES. Institute for Scientific Information, 3501 Market Street, Philadelphia, PA 19104. (800) 523-1850 or (215) 386-0100. Weekly. $275.00 per year.

ENGINEERING INDEX MONTHLY AND AUTHOR INDEX. Engineering Information Inc., 345 East 47th Street, New York, NY 10017. (212) 705-7600. Monthly. $1560.00 per year.

GENERAL SCIENCE INDEX. H.W. Wilson and Company, 950 University Avenue, Bronx, NY 10452. (800) 367-6670 or (212) 588-8400. 1978 to present. Monthly. Inquire as to cost and availability.

INDEX TO SCIENTIFIC AND TECHNICAL PROCEEDINGS. Institute for Scientific Information, 3501 Market Street, Philadelphia, PA 19104. (800) 523-1850 or (215) 386-0100. 1978 to present. Monthly. $775.00 per year.

INDEX TO SCIENTIFIC REVIEWS. Institute for Scientific Information, 3501 Market Street, Philadelphia, PA 19104. (800) 523-1850 or (215) 386-0100. 1974 to present. Semi-annual. $550.00 per year.

PHOTOGRAPHIC ABSTRACTS. Royal Photographic Society of Great Britain, Scientific and Technical Group, 62 Chelmsford Road, Shenfield, Brentwood, Essex, England. 1921 to present. Six times per year. $140.00 per year.

PHYSICS ABSTRACTS. Institution of Electrical Engineers. Available from: IEEE Service Center, 445 Hoes Lane, Piscataway, NJ 08854. 1898 to present. Bimonthly. $1700.00 per year.

SCIENCE CITATION INDEX. Institute for Scientific Information, 3501 Market Street, Philadelphia, PA 19104. (800) 523-1850 or (215) 386-0100. Six times per year. $6200.00 per year.

ASSOCIATIONS AND PROFESSIONAL SOCIETIES

AMERICAN SOCIETY FOR PHOTOGRAMMETRY AND REMOTE SENSING. 210 Little Falls Street, Falls Church, VA 22046-4398. (703) 534-6617.

OPTICAL SOCIETY OF AMERICA. 1816 Jefferson Place, N.W., Washington, DC 20036. (202) 223-8130.

SOCIETY OF PHOTOGRAPHIC SCIENTISTS AND ENGINEERS. 7003 Kilworth Lane, Springfield, VA 22151. (703) 642-9090.

SPIE - THE INTERNATIONAL SOCIETY FOR OPTICAL ENGINEERING. P.O. Box 10, 1022 19th Street, Bellingham, WA 98227. (206) 676-3290.

BIBLIOGRAPHIES

NEW TECHNICAL BOOKS: A SELECTIVE LIST WITH DESCRIPTIVE ANNOTATIONS. New York Public Library, Science and Technology Research Center, Fifth Avenue and 42nd Street, New York, NY 10018. (212) 930-0800. 1915 to present. Monthly. $15.00 per year.

SCIENTIFIC AND TECHNICAL BOOKS AND SERIALS IN PRINT 1988; AN INDEX TO LITERATURE IN SCIENCE AND TECHNOLOGY. R.R. Bowker Company, 205 East 42nd Street, New York, NY 10017. (800) 521-8110. $175.00.

DIRECTORIES AND BIOGRAPHICAL SOURCES

INTERNATIONAL RESEARCH CENTERS DIRECTORY 1988-89. Darren L. Smith, editor. Gale Research Company, Book Tower, Detroit, MI 48226. (800) 521-0707. 4th edition. 1987. $360.00.

1987 DIRECTORY OF ENGINEERING SOCIETIES AND RELATED ORGANIZATIONS. Gordon Davis, editor. Hemisphere Publishing Corporation, 1010 Vermont Avenue, NW, Washington, DC 20005. (800) 526-0275. 12th edition. 1987. $100.00.

RESEARCH CENTERS DIRECTORY 1988. Gale Research Company, Book Tower, Detroit, MI 48226. (800) 521-0707. 12th edition. 1987. $365.00 for set.

SCIENTIFIC AND TECHNICAL ORGANIZATIONS AND AGENCIES DIRECTORY. Margaret Labash Young, editor. Gale Research Company, Book Tower, Detroit, MI 48226. (800) 521-0707. 2nd edition. 1987. $185.00.

WHO'S WHO IN ENGINEERING. Gordon Davis, editor. Hemisphere Publishing Corporation, 1010 Vermont Avenue, NW, Washington, DC 20005. (800) 526-0275. 6th edition. 1985. $200.00.

ENCYCLOPEDIAS AND DICTIONARIES

CONCISE SCIENCE DICTIONARY. Oxford University Press, 200 Madison Avenue, New York, NY 10016. (800) 458-5833. 1987. $9.95 in paper.

DICTIONARY OF THE PHYSICAL SCIENCES: TERMS, FORMULAS, DATA. Cesare Emiliani. Oxford University Press, 200 Madison Avenue, New York, NY 10016. (800) 458-5833. 1987. $19.95 in paper.

MCGRAW-HILL ENCYCLOPEDIA OF SCIENCE AND TECHNOLOGY. McGraw-Hill Book Company, 1221 Avenue of the Americas, New York, NY 10020. (212) 512-2000. 6th edition. 1987. $1600.00.

THESAURUS OF PHOTOGRAPHIC SCIENCE AND ENGINEERING. Society of Photographic Scientists and Engineers. Books on Demand, 300 North Zeeb Road, Ann Arbor, MI 48106. (313) 761-4700. $34.50 in paper.

THESAURUS OF SCIENTIFIC, TECHNICAL, AND ENGINEERING TERMS. Hemisphere Publishing Corporation, 1010 Vermont Avenue, NW, Washington, DC 20005. (800) 526-0275. 1988. $125.00.

GENERAL WORKS

LIGHT AND COLOR. R.D. Overheim and D.L. Wagner. John Wiley and Sons, Inc., 605 Third Avenue, New York, NY 10158. (800) 526-5368. 1982. $28.50.

PHOTOCHEMISTRY: AN INTRODUCTION. D.R. Arnold and others. Academic Press, Inc., 6277 Sea Harbor Drive, Orlando, FL 32821. (800) 321-5068. 1974. $45.00.

PHOTOGRAPHIC SCIENCE. Earl. N. Mitchell. John Wiley and Sons, Inc., 605 Third Avenue, New York, NY 10158. (800) 526-5368. 1984. $37.50.

PHOTOGRAPHY FOR THE SCIENTIST. Richard A. Morton. Academic Press, Inc., 6277 Sea Harbor Drive, Orlando, FL 32821. (800) 321-5068. 2nd edition. 1984. $102.50.

HANDBOOKS AND MANUALS

HANDBOOK OF PHOTOGRAPHIC SCIENCE AND ENGINEERING. Society of Photographic Scientists and Engineers. 7003 Kilworth Lane, Springfield, VA 22151. (703) 642-9090. Inquire.

KODAK PROFESSIONAL PHOTOGUIDE. Carolyn Grimes, editor. Eastman Kodak Company, 343 State Street, Rochester, NY 14650. (716) 724-4000. 1986. $19.95 in paper.

ONLINE DATA BASES

CA SEARCH. Chemical Abstracts Service, P.O. Box 3012, Columbus, OH 43120. (800) 848-6538 or (614) 421-3600. Comprehensive guide to chemical literature, 1972 to present. Inquire as to online cost and availability.

COMPENDEX. Engineering Information, Inc., 345 East 47th Street, New York, NY 10017. (800) 221-1044 or (212) 705-7615. Engineering and technical literature, 1975 to present. Inquire as to online cost and availability.

DISSERTATION ABSTRACTS ONLINE. University Microfilms International, 300 North Zeeb Road, Ann Arbor, MI 48106. (800) 521-0600 or (313) 761-4700. Scope includes virtually all doctoral dissertations accepted at accredited American institutions from 1861 to present in over 250 subject areas. Inquire as to online cost and availability.

INSPEC. INSPEC Marketing Department, Institution of Electrical Engineers. Available from IEEE Service Center, 445 Hoes Lane, Piscataway, NJ 08854. (201) 981-0060. Online version of Physics Abstracts. Inquire as to online cost and availability.

NTIS. National Technical Information Service, 5285 Port Royal Road, Springfield, VA 22161. (703) 487-4630. Broad coverage of government sponsored research reports, 1964 to present. Inquire as to online cost and availability.

SCISEARCH. Institute for Scientific Information, 3501 Market Street, Philadelphia, PA 19104. (800) 523-1850 or (215) 386-0100. Broad multidisciplinary title and author index to the international literature of science and technology, 1974 to present. Inquire as to online cost and availability.

WILSONLINE. H.W. Wilson and Company, 950 University Avenue, Bronx, NY 10452. (800) 367-6770 or (212) 588-8400. Makes available online versions of the H.W. Wilson indexes including Applied Science and Technology Index, Business Periodicals Index and Readers' Guide to Periodical Literature. Approximately 1980 to present. Inquire as to online cost and availability.

PERIODICALS

BRITISH JOURNAL OF PHOTOGRAPHY. Henry Greenwood and Company, Limited, 28 Great James Street, London WC1N 3HL, England. 1854 to present. Weekly. $70.00 per year.

FUNCTIONAL PHOTOGRAPHY. PTN Publishing Corporation, 210 Crossways Park Drive, Woodbury, NY 11797. (516) 496-8000. 1967 to present. Bimonthly. $7.50 per year.

INDUSTRIAL PHOTOGRAPHY. United Business Publications, Inc., 475 Park Avenue South, New York, NY 10016. (212) 725-2300. 1952 to present. Monthly. $15.00 per year.

JOURNAL OF IMAGING SCIENCE. Society of Photographic Scientists and Engineers. 7003 Kilworth Lane, Springfield, VA 22151. (703) 642-9090. 1956 to present. Bimonthly. $70.00 per year.

JOURNAL OF IMAGING TECHNOLOGY. Society of Photographic Scientists and Engineers. 7003 Kilworth Lane, Springfield, VA 22151. (703) 642-9090. 1975 to present. Bimonthly. $70.00 per year.

JOURNAL OF PHOTOGRAPHIC SCIENCE. Royal Photographic Society of Great Britain, 7 Ladbroke Walk, London W11, England. 1953 to present. Bimonthly. $40.00 per year.

KODAK TECH BITS. Eastman Kodak Company, 343 State Street, Rochester, NY 14650. (716) 724-4000. 1963 to present. Quarterly. Free to qualified personnel.

PHOTOGRAMMETRIC ENGINEERING AND REMOTE SENSING. American Society for Photogrammetry and Remote Sensing. 210 Little Falls Street, Falls Church, VA 22046-4398. (703) 534-6617. Order from: Allen Press, Inc., 1041 New Hampshire Street, Box 368, Lawrence, KS 66044. 1934 to present. Monthly. $80.00 per year.

SCIENTIFIC AND APPLIED PHOTOGRAPHY AND CINEMATOGRAPHY. Gordon and Breach Science Publishers, Inc., 50 West 23rd Street, New York, NY 10010. (212) 206-8900. 12 times per year. $496.00 per year.

RESEARCH CENTERS AND INSTITUTES

GEOPHOTOGRAPHY AND REMOTE SENSING CENTER. University of Idaho, Geology Department, Moscow, ID 83843. (208) 885-7977.

GRAPHIC ARTS TECHNICAL FOUNDATION, RESEARCH DEPARTMENT. 4615 Forbes Avenue, Pittsburgh, PA 15213. (412) 621-6941.

IMAGE PERMANENCE INSTITUTE. Rochester Institute of Technology, RIT City Center, 50 West Main Street, Rochester, NY 14614. (716) 475-5199.

NATIONAL RESEARCH COUNCIL OF CANADA, DIVISION OF PHYSICS. Ottawa, ON, Canada K1A OR6. (613) 993-1053.

PHOTOMETRY

See: ASTRONOMY

PHYSICAL CHEMISTRY

See also: CATALYSIS, CHEMICAL BONDING, CHEMISTRY, COLLOIDS, CRYSTALLOGRAPHY, ELECTROCHEMISTRY, PHOTOCHEMISTRY, POLYMERS, QUANTUM CHEMISTRY

ABSTRACT SERVICES AND INDEXES

APPLIED SCIENCE AND TECHNOLOGY INDEX. H.W. Wilson Company, 950 University Avenue, Bronx, NY 10452. (800) 367-6670 or (212) 588-8400. Inquire as to cost and availability.

CHEMICAL ABSTRACTS. Chemical Abstracts Service, 2540 Olentangy Road, Post Office Box 3012, Columbus, OH 43210. (800) 848-6538 or (614) 421-3600. Weekly. $9200.00 per year.

GENERAL SCIENCE INDEX. H.W. Wilson Company, 950 University Avenue, Bronx, NY 10452. (800) 367-6770 or (212) 588-8400. Inquire as to cost and availability.

PHYSICS ABSTRACTS. Institute of Electrical Engineers, London, United Kingdom. Available from: Institute of Electrical and Electronic Engineers (IEEE), 345 East 47th Street, New York, NY 10017. (212) 705-7900.

SCIENCE CITATION INDEX. Institute for Scientific Information, 3501 Market Street, Philadelphia, PA 19104. (800) 523-1850 or (215) 386-0100.

ANNUAL REVIEWS AND YEARBOOKS

ANNUAL REVIEW OF PHYSICAL CHEMISTRY. Annual Reviews Incorporated, 4139 El Camino Way, Palo Alto, CA 94306. (415) 493-4400. Annual.

ASSOCIATIONS AND PROFESSIONAL SOCIETIES

AMERICAN CHEMICAL SOCIETY. 1155 16th Street, NW, Washington, DC 20036. (202) 872-4600.

ASSOCIATION OF CONSULTING CHEMISTS AND CHEMICAL ENGINEERS. 50 East 41st Street, Suite 92, New York, NY 10017. (212) 684-6255.

DIVISION OF PHYSICAL CHEMISTRY (A DIVISION OF THE AMERICAN CHEMICAL SOCIETY). c/o Dr. James Kinsey, Chemistry Department, Room 6-215, Massachsuetts Institute of Technology, Cambridge, MA 02139.

BIBLIOGRAPHIES

SCIENCE BOOKS AND FILMS. American Association for the Advancement of Science, 1333 H Street, NW, Washington, DC 20005.

SCIENTIFIC AND TECHNICAL BOOKS AND SERIALS IN PRINT 1988: AN INDEX TO LITERATURE IN SCIENCE AND TECHNOLOGY. R.R. Bowker Company, 205 East 42nd Street, New York, NY 10017. (800) 521-8110 or (212) 916-1600. $175.00.

DIRECTORIES AND BIOGRAPHICAL SOURCES

AMERICAN INSTITUTE OF CHEMISTS. American Institute of Chemists, 7315 Wisconsin Avenue, Bethesda, MD 20814. (301) 652-2447. 1986. $35.00.

AMERICAN MEN AND WOMEN OF SCIENCE. Physical and Biological Sciences. R.R. Bowker Company, 205 East 42nd Street, New York, NY 10017. (800) 521-8110 or 9212) 916-1600. Fifteenth edition. $565.00.

BIOGRAPHICAL DICTIONARY OF SCIENTISTS: CHEMISTS. David Abbott, editor. P. Bedrick Books, 125 East 23rd Street, New York, NY 10010. (212) 777-1187. 1984. $18.95.

CHEMICAL WEEK - BUYERS GUIDE ISSUE. McGraw-Hill Book Company, Incorporated, 1221 Avenue of the Americas, New York, NY 10020. (800) 628-0004. Annual, October. $50.00.

CONSULTING SERVICES: CHEMISTS AND CHEMICAL ENGINEERS. Association of Consulting Chemists and Chemical Engineers, 50 East 41st Street, New York, NY 10017. (212) 684-6255. Annual. 1986. $45.00.

GOVERNMENT RESEARCH DIRECTORY. Gale Research Company, Book Tower, Detroit, MI 48226. (800) 521-0707. Fourth edition. 1987. $350.00.

INTERNATIONAL RESEARCH CENTERS DIRECTORY 1986-1987. Gale Research Company, Book Tower, Detroit, MI 48226. (800) 521-0707. Third edition. 1986. $330.00.

RESEARCH CENTERS DIRECTORY. Gale Research Company, Book Tower, Detroit, MI 48226. (800) 521-0707. Eleventh edition. 1987. $355.00.

RESEARCH SERVICES DIRECTORY. Robert J. Huffman and Mary M. Watkins, editors. Gale Research Company, Book Tower, Detroit, MI 48226. (800) 521-0707. Third edition.

$290.00.

SCIENTIFIC AND TECHNICAL ORGANIZATIONS AND AGENCIES DIRECTORY. Gale Research Company, Book Tower, Detroit, MI 48226. (800) 521-0707. 1985. $150.00.

WHO'S WHO IN FRONTIER SCIENCE AND TECHNOLOGY. Marquis Who's Who, Incorporated, 200 East Ohio Street, Chicago, IL 60611. (800) 428-3898 or (312) 787-2008.

WHO'S WHO IN TECHNOLOGY TODAY. Reston Publishing Company, Incorporated, c/o Prentice-Hall, Incorporated, Englewood Cliffs, NJ 07632. (800) 262-6868. Biennial. Five volumes. $425.00. Covers the fields of biotechnology, mechanics, energy, and earth science.

WORLD GUIDE TO SCIENTIFIC ASSOCIATIONS AND LEARNED SOCIETIES. K.G. Saur Incorporated, 175 Fifth Avenue, New York, NY 10010. (800) 521-0707 or (212) 982-1302. Fourth edition. 1984. $112.00.

ENCYCLOPEDIAS AND DICTIONARIES

CONCISE ENCYCLOPEDIA OF CHEMICAL TECHNOLOGY. Kirk-Othmer. John Wiley and Sons, Incorporated, 605 Third Avenue, New York, NY 10158. (800) 526-5368 or (212) 850-6000. Third edition. 1985. $129.95.

CONDENSED CHEMICAL DICTIONARY. Gessner Hawley. Van Nostrand Reinhold, 115 Fifth Avenue, New York, NY 10003. Tenth edition. 1981. $49.95.

ENCYCLOPEDIA OF PHYSICAL SCIENCE AND TECHNOLOGY. Academic Press, Incorporated, Orlando, FL 32887. (800) 321-5068 or (305) 345-2734. Fifteen volumes. 1986.

GLOSSARY OF CHEMICAL TERMS. Clifford A. Hampel and Gessner G. Hawley. Van Nostrand Reinhold Company, 115 Fifth Avenue, New York, NY 10003. (800) 543-2681 or (212) 254-3232. Second edition. 1982. $21.95.

HAWLEY'S CONDENSED CHEMICAL DICTIONARY. N. Irving Sax and Richard J. Lewis, Sr., editors. Van Nostrand Reinhold, Incorporated, 115 Fifth Avenue, New York, NY 10003. (800) 543-2681. Eleventh edition. 1987. $52.95.

MCGRAW-HILL ENCYCLOPEDIA OF SCIENCE AND TECHNOLOGY. McGraw-Hill Book Company, Incorporated, 1221 Avenue of the Americas, New York, NY 10020. (212) 997-3675.

VAN NOSTRAND REINHOLD ENCYCLOPEDIA OF CHEMISTRY. Douglas M. Considine and Glenn D. Considine. Van Nostrand Reinhold Publishing Company, Incorporated, 115 Fifth Avenue, New York, NY 10003. (800) 543-2681 or (212) 254-3232. 1984. $97.95.

GENERAL WORKS

THE CHEMICAL BOND. John N. Murrell and others. John Wiley and Sons, Incorporated, 605 Third Avenue, New York, NY 10158. (800) 526-5368 or (212) 850-6000. 1985. $31.95.

CHEMICAL BONDING MODELS, VOLUME 1. J.F. Liebman and A. Greenberg, editors. VCH Publishing, Incorporated, 303 NW 12th Avenue, Deerfield Beach, FL 33442-1705. (305) 428-5566. 1986. $77.50.

CHEMISTRY. Charles E. Mortimer. Wadsworth Publishing Company, 10 Davis Drive, Belmont, CA 94002. (415) 595-2350. Sixth edition. 1986. $49.95.

CHEMISTRY OF THE ELEMENTS. N.N. Greenwood and A. Earnshaw. Pergamon Publishing, Incorporated, Maxwell House, Fairview Park, Elmsford, NY 10523. (914) 592-7700. 1984. $143.00.

INORGANIC CHEMISTRY: PRINCIPLES OF STRUCTURE AND REACTIVITY. James E. Huheey. Harper and Row Publishers Incorporated, 10 East 53rd Street, New York, NY 10022. (212) 207-7655. Third edition. $36.50.

INTRODUCTION TO PHYSICAL CHEMISTRY. M.F.C. Ladde and W.H. Lee. Cambridge University Press, 32 East 57th Street, New York, NY 10022. (212) 688-8885. 1986. $19.95 in paper.

INTRODUCTION TO SYMMETRY IN BONDING AND SPECTRA. Bodie Douglas and Charles A. Hollingsworth. Academic Press, Incorporated, 6277 Sea Harbor Drive, Orlando, FL 32821. (800) 321-5068. 1985. $40.00.

INVESTIGATION OF RATES AND MECHANISMS OF REACTIONS, PART 1. Claude F. Bernasconi, editor. John Wiley and Sons, Incorporated, 605 Third Avenue, New York, NY 10158. (800) 526-5368 or (212) 850-6000. Fourth edition. 1986. $175.00.

PHYSICAL CHEMISTRY. Robert A. Alberty. John Wiley and Sons, Incorporated, 605 Third Avenue, New York, NY 10158. (800) 526-5368 or (212) 850-6000. 1987. Inquire as to cost and availability.

QUANTUM CHEMISTRY OF ATOMS AND MOLECULES. Philip S. C. Matthews. Cambridge University Press, 32 East 57th Street, New York, NY 10022. (212) 688-8885. 1986. $44.50.

SYMMETRY AND STRUCTURE. S.A. Kettle. John Wiley and Sons, Incorporated, 605 Third Avenue, New York, NY 10158. (800) 526-5368 or (212) 850-6000. 1985. $37.95.

HANDBOOKS AND MANUALS

THE CHEMIST'S COMPANION: A HANDBOOK OF PRACTICAL DATA, TECHNIQUES, AND REFERENCES. Arnold J. Gordon and Richard A. Ford. John Wiley and Sons, Incorporated, 605 Third Avenue, New York, NY 10158. (800) 526-5368. 1973. $49.95.

CRC HANDBOOK OF CHEMISTRY AND PHYSICS. CRC Press, Incorporated, 2000 Corporate Boulevard, NW, Boca Raton, FL 33431. Sixty-seventh edition. 1986. $69.95.

HANDBOOK OF APPLIED CHEMISTRY: FACTS FOR ENGINEERS, SCIENTISTS, TECHNICIANS, AND TECHNICAL MANAGERS. Vollrath Hopp and Ingp Hennig. McGraw-Hill Book Company, 1221 Avenue of the Americas, New York, NY 10020. (800) 628-0004. 1983. $54.00.

HANDBOOK OF COMPUTATIONAL CHEMISTRY: A PRACTICAL GUIDE TO CHEMICAL STRUCTURE AND ENERGY CALCULATIONS. Tim Clark. John Wiley and Sons, Incorporated, 605 Third Avenue, New York, NY 10158. (800) 526-5368 or (212) 850-6000. 1985. $35.00.

LANGE'S HANDBOOK OF CHEMISTRY. John A. Dean, editor. McGraw-Hill Book Company, 1221 Avenue of the Americas, New York, NY 10020. (800) 628-0004. 1985. $59.50.

TABLES OF PHYSICAL AND CHEMICAL CONSTANTS; AND SOME MATHEMATICAL FUNCTIONS. G.W.C. Kaye and T.H. Laby, editors. Longman, Incorporated, 95 Church Street, White Plains, NY 10601. (914) 993-5000. 1986. $39.95.

ONLINE DATA BASES

CA SEARCH. Chemical Abstracts Service, Post Office Box 3012, Columbus, OH 43210. Guide to chemical literature, 1972 to present. Inquire as to cost and availability.

DISSERTATION ABSTRACTS ONLINE. University Microfilms International, 300 North Zeeb Road, Ann Arbor, MI 48106. (800) 521-0600 or (313) 761-4700. Scope includes virtually all doctoral dissertations accepted at accredited American institutions from 1861 to present in 252 subject areas. Inquire

as to online cost and availability.

INSPEC. INSPEC Marketing Department, Institute of Electrical and Electronics Engineers, Incorporated, IEEE Service Department, 445 Hoes Lane, Piscataway, NJ 08854. (201) 981-0060. Inquire as to online cost and availability.

NTIS. National Technical Information Service, 5285 Port Royal Road, Springfield, VA 22161. (703) 487-4630. Broad coverage of government sponsored research reports, 1964 to present. Inquire as to online cost and availability.

SCISEARCH. Institute for Scientific Information, 3501 Market Street, Philadelphia, PA 19104. (800) 523-1850 or (215) 386-0100. Broad multidisciplinary title and author index to the international literature of science and technology, 1974 to present. Inquire as to online cost and availability.

OTHER SOURCES

GUIDE TO BASIC INFORMATION SOURCES IN CHEMISTRY. Arthur Antony. John Wiley and Sons, Incorporated, 605 Third Avenue, New York, NY 10158. (800) 526-5368 or (212) 850-6000. 1979. $26.95.

HOW TO FIND CHEMICAL INFORMATION: A GUIDE FOR PRACTICING CHEMISTS, TEACHERS, AND STUDENTS. John Wiley and Sons, Incorporated, 605 Third Avenue, New York, NY 10158. (800) 526-5368 or (212) 850-6000. 1986. $35.00.

PERIODICALS

ANDEWANDTE CHEMIE (GESELLSCHAFT DEUTSCHER CHEMIKER, GW). V C H Verlagsgesellschaft mbH, Pappelallee 3, Postfach 1260, 6940 Weinheim, West Germany. Monthly. $300.00 per year.

CHEMICAL PHYSICS. Elsevier Science Publishers B.V, Box 211, 1000 AE Amsterdam, The Netherlands. Thirty times per year. $1200.00 per year.

CHEMICAL PHYSICS LETTERS. Elsevier Science Publishers B.V., Box 211, 1000 AE Amsterdam, The Netherlands. Sixty-six times per year. $1500.00 per year.

CHEMICAL REVIEWS. American Chemical Society, 1155 Sixteenth Street, NW, Washington, DC 20036. (800) 424-6747 or (202) 872-4700. Bimonthly. $83.00 per year.

CHEMTECH. American Chemical Society, 1155 Sixteenth Street, NW, Washington, DC 20036. (800) 424-6747 or (202) 872-4700. Monthly. $40.00 per year to individuals.

INTERNATIONAL REVIEWS IN PHYSICAL CHEMISTRY. Taylor and Francis, Limited, 242 Cherry Street, Philadelphia, PA 19106-1906. Three times per year. $143.00 per year.

INORGANIC CHEMISTRY. American Chemical Society, 1155 Sixteenth Street, NW, Washington, DC 20036. (800) 424-6747 or (202) 872-4700. Monthly. $400.00 per year.

JOURNAL OF CATALYSIS. Academic Press, Incorporated, Journals Division, 1250 Sixth Avenue, San Diego, Ca 92101. (619) 230-1840. Monthly. $654.00 per year.

JOURNAL OF PHYSICAL CHEMISTRY. American Chemical Society, 1155 Sixteenth Street, NW, Washington, DC 20036. (800) 424-6747 or (202) 872-4700. Biweekly. $369.00 per year.

JOURNAL OF SOLID STATE CHEMISTRY. Academic Press, Incorporated, Journals ?Division, 1250 Sixth Avenue, San Diego, CA 92101. (619) 230-1840. Fifteen times per year. $530.00 per year.

JOURNAL OF THE AMERICAN CHEMICAL SOCIETY. American Chemical Society, 1155 Sixteenth Street, NW, Washington, DC 20036. (800) 424-6747 or (202) 872-4700. Biweekly. $330.00 per year.

RESEARCH CENTERS AND INSTITUTES

BROWN UNIVERSITY. Chemical Research Laboratory, Providence, RI 02912. (401) 863-2256.

CORNELL UNIVERSITY. Materials Science Center, Clark Hall of Science, Ithaca, NY 14853. (607) 255-4272.

HARVARD UNIVERSITY. Chemical Laboratories, Oxford Street, Cambridge, MA 02138. (617) 495-4283.

UNIVERSITY OF PENNSYLVANIA. Laboratory for Research on the Structure of Matter, 3231 Walnut Street, Philadelphia, PA 19104. (215) 898-8571.

UNIVERSITY OF TEXAS AT AUSTIN. Institute of Theoretical Chemistry, Department of Chemistry, WEL 2.204, Austin, TX 78712. (512) 471-3114.

PHYSICAL GEOLOGY

See also: GEOLOGY, GEOPHYSICS, GEOCHEMISTRY

ABSTRACT SERVICES AND INDEXES

BIBLIOGRAPHY AND INDEX OF GEOLOGY. American Geological Institute, 4220 King Street, Alexandria, VA 22302. (703) 379-2480. 1969 to present. Monthly. $1100.00 per year.

CHEMICAL ABSTRACTS. Chemical Abstracts Service, 2540 Olentangy Road, Post Office Box 3012, Columbus, OH 43210. (800) 848-6538 or (614) 421-3600. Weekly. $9200.00 per year.

DEEP-SEA RESEARCH WITH OCEANOGRAPHIC LITERATURE REVIEW. Pergamon Press, Incorporated, Maxwell House, Fairview Park, Elmsford, NY 10523. (914) 592-7700. 1953 to present. Twenty-four times per year. $600.00 per year.

GENERAL SCIENCE INDEX. H.W. Wilson Company, 950 University Avenue, Bronx, NY 10452. (800) 367-6770 or (212) 588-8400. Inquire as to price and availability.

MINERALOGICAL ABSTRACTS. Mineralogical Society and the Mineralogical Society of America, 41 Queen's Gate, London, SW7 5HR, England. 1959 to present. Quarterly. $190.00 per year.

OCEANIC ABSTRACTS. Cambridge Scientific Abstracts, 5161 River Road, Bethesda, MD 20816. (301) 951-1400. 1964 to present. Bimonthly. $652.00 per year.

SCIENCE CITATION INDEX. Institute for Scientific Information, 3501 Market Street, Philadelphia, PA 19104. (800) 523-1850 or (215) 386-0100. Inquire as to price and availability.

ANNUAL REVIEWS AND YEARBOOKS

ADVANCES IN GEOPHYSICS. Academic Press, Incorporated, 6277 Sea Harbor Drive, Orlando, FL 32821. (800) 321-5068. Irregular. $62.00 per volume.

ANNUAL REVIEW AND EARTH AND PLANETARY SCIENCES. Annual Reviews, Incorporated, 4139 El Camino Way, Palo Alto, CA 94306. (415) 493-4400.

MINERALS YEARBOOK. Bureau of Mines, United States Department of the Interior. Available from United States Government Printing Office, Washington, DC 20402. (202) 783-3238. Annual. Three volumes. $45.00.

ASSOCIATIONS AND PROFESSIONAL SOCIETIES

AMERICAN ASSOCIATION OF PETROLEUM GEOLOGISTS. Post Office Box 979, Tulsa, OK 74101. (918) 584-2555.

AMERICAN GEOLOGICAL INSTITUTE. 4220 King Street, Alexandria, VA 22302. (703) 379-2480.

AMERICAN GEOPHYSICAL UNION. 2000 Florida Avenue, NW, Washington, DC 20009. (202) 462-6903.

AMERICAN INSTITUTE OF PROFESSIONAL GEOLOGISTS. 7828 Vance Drive, Suite 103, Arvada, CO 80003. (303) 431-0831.

ASSOCIATION OF ENGINEERING GEOLOGISTS. Post Office Box 1068, Brentwood, TN 37027. (615) 377-3578.

GEOLOGICAL SOCIETY OF AMERICA. Box 9140, Boulder, CO 80301. (303) 447-2020.

DIRECTORIES AND BIOGRAPHICAL SOURCES

AMERICAN MEN AND WOMEN OF SCIENCE: PHYSICAL AND BIOLOGICAL SCIENCES. Sixteenth edition. R.R. Bowker Company, 245 West 17th Street, New York, NY 10011. (800) 521-8810 or (212) 916-1600. $595.00.

AMERICAN INSTITUTE OF PROFESSIONAL GEOLOGISTS. Membership Directory. American Institute of Professional Geologists, 7828 Vance Drive, Suite 103, Arvada, CO 80003. (303) 431-0831. Annual. $15.00.

ASSOCIATION OF ENGINEERING GEOLOGISTS DIRECTORY. Association of Engineering Geologists, Dr. G. Lee Christensen, Civil Engineering Department, Villanova University, Villanova, PA 19085. (215) 645-4960. Annual. $15.00.

GEOLOGICAL SOCIETY OF AMERICA. Membership Directory. Geological Society of America, 3300 Penrose Place, Boulder, CO 80301. (303) 447-2020. Annual. Available to members only.

RESEARCH CENTERS DIRECTORY. Gale Research Company, Detroit, MI 48226. (800) 521-0707. Twelfth edition. 1987. $365.00.

WHO'S WHO IN FRONTIER SCIENCE AND TECHNOLOGY. Marquis Who's Who, Incorporated, 200 East Ohio Street, Chicago, IL 60611. (800) 428-3898 or (312) 787-2008.

ENCYCLOPEDIAS AND DICTIONARIES

DICTIONARY OF GEOLOGICAL TERMS. American Geological Institute. Doubleday and Company, Incorporated, 245 Park Avenue, New York, NY 10017. (800) 645-6156 or (212) 953-4561. Third edition. 1984. $19.95.

DICTIONARY OF PETROLOGY. S.I. Tomkeieff. John Wiley and Sons, Incorporated, 605 Third Avenue, New York, NY 10158. (800) 526-5368 or (212) 850-6000. 1983. $140.00.

ENCYCLOPEDIA OF PHYSICAL SCIENCE AND TECHNOLOGY. Academic Press, Incorporated, Orlando, FL 32887. (800) 321-5068. Fifteen volumes. $2500.00.

GLOSSARY OF GEOLOGY. Robert L. Bates and Julia A. Jackson. American Geological Institute, 4220 King Street, Alexandria, VA 22032. (703) 379-2480. 1980. $60.00.

GENERAL WORKS

THE EARTH. Peter J. Smith. Macmillan Publishing Company, 866 Third Avenue, New York, NY 10022. (800) 257-5755 or (212) 702-2000. 1986. $40.00.

PHYSICAL GEOLOGY. R.F. Flint and B.J. Skinner. John Wiley and Sons, Incorporated, 605 Third Avenue, New York, NY 10158. (800) 526-5368 or (212) 850-6000. Third edition. 1987. $45.95.

PHYSICAL GEOLOGY. Sheldon Judson and others. Prentice-Hall, Incorporated, Englewood Cliffs, NJ 07632. (800) 562-0245. Seventh edition. 1987. $36.95.

HANDBOOKS AND MANUALS

GEOLOGY IN THE FIELD. R.R. Compton. John Wiley and Sons, Incorporated, 605 Third Avenue, New York, NY 10158. (800) 526-5368 or (212) 850-6418. 1985. $23.95.

ONLINE DATA BASES

GEOREF. American Geological Institute, 4220 King Street, Alexandria, VA 22302. (800) 336-4764 or (703) 379-2480. Geology and geosciences literature, 1961 to present. Inquire as to online cost and availability.

GEOARCHIVE. Geosystems, Post Office Box 1024, Westminister, London, England, SW1 P 2JL. Citations to literature on geoscience, 1969 to present. Inquire as to online cost and availability.

NTIS. National Technical Information Service, 5285 Port Royal Road, Springfield, VA 22161. (703) 487-4630. Broad coverage of government sponsored research reports, 1964 to present. Inquire as to online cost and availability.

SCISEARCH. Institute for Scientific Information, 3501 Market Street, Philadelphia, PA 19104. (800) 523-1850 or (215) 386-0100. Broad interdisciplinary index to the literature of science and technology, 1965 to present. Inquire as to online cost and availability.

PERIODICALS

AAPG BULLETIN. American Association of Petroleum Geologists, Post Office Box 979, Tulsa, OK 74101. (918) 584-2555. Monthly. $36.00 per year.

AMERICAN JOURNAL OF SCIENCE. Kline Geology Laboratory, Yale University, New Haven, CT 06520. Ten times per year. $80.00 per year.

ECONOMIC GEOLOGY. Society of Economic Geologists, Post Office Box 571, Golden, CO 80402. (303) 279-1899. 8 times per year. $25.00.

GEOCHIMICA ET COSMOCHIMICA ACTA. Pergamon Journals, Incorporated, Maxwell House, Fairview Park, Elmsford, NY 10523. (014) 592-7700. Monthly. $340.00 per year.

GEOLOGICAL MAGAZINE. Cambridge University Press, 32 East 57th Street, New York, NY 10022. (800) 872-7423. Bimonthly. $165.00 per year.

GEOLOGICAL SOCIETY JOURNAL. Geological Society of London. Blackwell Scientific Publications, Incorporated, 667 Lytton Avenue, Palo Alto, CA 94301. (415) 324-1688. Six times per year. $290.00.

GEOLOGICAL SOCIETY OF AMERICA BULLETIN. Post Office Box 9140, 3300 Penrose Place, Boulder, CO 80301. (303) 447-2020. Monthly. $80.00.

GEOLOGY. Geological Society of America, Post Office Box 9140, 3300 Penrose Place, Boulder, CO 80301. (303) 447-2020. Monthly. $55.00 per year.

GEOPHYSICS. Society of Exploration Geophysicists, Post Office Box 702740, Tulsa, OK 74170. (918) 493-3516. Monthly. $45.00 per year.

GEOTIMES. American Geological Institute, 4220 King Street, Alexandria, VA 22032. Monthly. $18.00 per year.

INTERNATIONAL ASSOCIATION FOR MATHEMATICAL GEOLOGY. Journal. Plenum Publishing Corporation, 233 Spring Street, New York, NY 10013. (800) 221-9369. Six times per year. $225.00 per year.

JOURNAL OF GEOLOGY. University of Chicago Press, 5801 South Ellis Street, Chicago, IL 60637. (312) 962-7600.

PHYSICAL GEOLOGY

Bimonthly. $30.00 per year.

JOURNAL OF GEOPHYSICAL RESEARCH. American Geophysical Union, 2000 Florida Avenue, NW, Washington, DC 20009. (202) 462-6903. Weekly. $680.00 per year to individuals.

JOURNAL OF GEOPHYSICS. Springer-Verlag, Incorporated, 175 Fifth Avenue, New York, NY 10010. (800) 526-7254 or (212) 460-1500. Bimonthly. $175.00 per year.

JOURNAL OF SEDIMENTARY PETROLOGY. Society of Economic Paleontologists and Mineralogists, Post Office Box 4756, Tulsa, OK 74159. (918) 743-9765. Bimonthly.

JOURNAL OF STRUCTURAL GEOLOGY. Pergamon Journals, Incorporated, Maxwell House, Fairview Park, Elmsford, NY 10523. (914) 592-7700. Eight times per year. $160.00 per year.

LITHOS: AN INTERNATIONAL JOURNAL OF MINERALOGY, PETROLOGY, AND GEOCHEMISTRY. Elsevier Science Publishers, Post Office Box 330, Irving-on-Hudson, NY 10533. Quarterly. $73.00 per year.

MOUNTAIN GEOLOGIST. Rocky Mountain Association of Geologists, 4201 West 51st Avenue, Denver, CO 80212-2902. Quarterly. $15.00 per year.

PROFESSIONAL GEOLOGIST. American Institute of Professional Geologists, 7828 Vance Drive, Suite 103, Arvada, CO 80003. Monthly.

REVIEWS OF GEOPHYSICS. American Geophysical Union, 2000 Florida Avenue, NW, Washington, DC 20009. (202) 462-6903. Quarterly. $240.00 per year.

ROCK MECHANICS AND ROCK ENGINEERING. Springer-Verlag New York, Incorporated, 175 Fifth Avenue, New York, NY 10010. (800) 526-7254 or (212) 460-1500. Quarterly. $61.00 per year.

SEDIMENTARY GEOLOGY. Elsevier Science Publishers, Post Office Box 330, Irving-on-Hudson, NY 10533. Twenty times per year. $421.00 per year.

SEDIMENTOLOGY. Blackwell Scientific Publications, Incorporated, 52 Beacon Street, Boston, MA 02108. (617) 720-0761. Six times per year. $185.00 per year.

TECTONICS. American Geophysical Union, 2000 Florida Avenue, NW, Washington, DC 20009. (202) 462-6903. Bimonthly. $30.00 per year to individuals.

TECTONOPHYSICS: AN INTERNATIONAL JOURNAL OF GEOTECTONICS AND THE GEOLOGY AND PHYSICS OF THE INTERIOR OF THE EARTH. Elsevier Science Publishers, Post Office Box 330, Irving-on-Hudson, NY 10533. Forty-four times per year. $870.00 per year.

VOLCANOLOGY AND SEISMOLOGY. Gordon and Breach Science Publishers, 50 West 23rd Street, New York, NY 10010. (212) 206-8900. Monthly. $498.00 per year.

RESEARCH CENTERS AND INSTITUTES

GEOLOGICAL SOCIETY OF AMERICA. Post Office Box 9140, 3300 Penrose Place, Boulder, CO 80301. (303) 447-2020.

SOCIETY OF ECONOMIC GEOLOGISTS. Post Office Box 571, Golden, CO 80402. (303) 279-1899.

SOCIETY OF ECONOMIC PALEONTOLOGISTS AND MINERALOGISTS. Post Office Box 4756, Tulsa, OK 74159. (918) 743-9765.

UNITED STATES GEOLOGICAL SURVEY. National Center, 12201 Sunrise Valley Drive, Reston, VA 22092. The major geological research agency of the federal government conducting research in most areas of pure and applied research in the geosciences.

PHYSICS

ABSTRACT SERVICES AND INDEXES

ACOUSTICS ABSTRACTS. Multi-Science Publishing Company, Limited, 42-45 New Broad Street, London EC2M 1QY, England. Monthly. $295.00 per year.

APPLIED SCIENCE AND TECHNOLOGY INDEX. H.W. Wilson Company, 950 University Avenue, Bronx, NY 10452. (800) 367-6670 or (212) 588-8400. Inquire as to cost and availability.

CHEMICAL ABSTRACTS. Chemical Abstracts Service, 2540 Olentangy Road, Post Office Box 3012, Columbus, OH 43210. (800) 848-6538 or (614) 421-3600. Weekly. $9200.00 per year.

CURRENT CONTENTS: PHYSICAL, CHEMICAL, AND EARTH SCIENCE SCIENCES. Institute for Scientific Information, 3501 Market Street, Philadelphia, PA 19104. (800) 523-1850 or (215) 386-0100. $272.00 per year.

GENERAL PHYSICS ADVANCE ABSTRACTS. American Institute of Physics, 335 East 45th Street, New York, NY 10017. (212) 661-9404. 1985 to present. Semimonthly. $150.00 per year.

GENERAL SCIENCE INDEX. H.W. Wilson Company, 950 University Avenue, Bronx, NY 10452. (800) 367-6770 or (212) 588-8400. Inquire as to cost and availability.

JOURNAL OF CURRENT LASER ABSTRACTS. Institute for Laser Documentation, Box 2070, Rolling Hills, CA 90274. 1964 to present. Monthly. $350.00 per year.

MATHEMATICAL REVIEWS. American Mathematical Society, Post Office Box 6248, Providence, RI 02940. (800) 556-7774 or (401) 272-9500.

METEOROLOGICAL AND GEOASTROPHYSICAL ABSTRACTS. American Meteorological Society, 45 Beacon Street, Boston, MA 02108. (617) 227-2425. 1950 to present. Monthly. $450.00 per year.

PHYSICAL REVIEW ABSTRACTS. American Institute of Physics, 335 East 45th Street, New York, NY 10017. (212) 661-9404. 1970 to present. Semimonthly. $140.00 per year.

PHYSICS ABSTRACTS. Institute of Electrical Engineers, London, United Kingdom. Available from: Institute of Electrical and Electronic Engineers (IEEE), 345 East 47th Street, New York, NY 10017. (212) 705-7900. 1898 to present. Monthly. $1670.00 per year.

PHYSICS BRIEFS. Physics Verlag GmbH, Postfach 1260/1280, D-6940, Weinheim, West Germany. 1920 to present. Twenty-six times per year. $1200.00 per year.

SCIENCE CITATION INDEX. Institute for Scientific Information, 3501 Market Street, Philadelphia, PA 19104. (800) 523-1850 or (215) 386-0100. Inquire as to cost and availability.

SOLID STATE ABSTRACTS: AN ABSTRACT JOURNAL INVOLVING THE PHYSICS, METALLURGY, CRYSTALLOGRAPHY, CHEMISTRY, AND DEVICE TECHNOLOGY OF SOLIDS. Cambridge Scientific Abstracts, 5161 River Road, Bethesda, MD 20816. (301) 951-1400. 1957 to present. Bimonthly. $550.00 per year.

STAR. (Scientific and Technical Aerospace Reports. United States National Aeronautics and Space Administration, Scientific and Technical Information Facility, Box 8757, Baltimore-Washington International Airport, MD 21240. (202) 755-2210. Semimonthly, with semiannual and annual indexes. $85.00 per year.

ANNUAL REVIEWS AND YEARBOOKS

ADVANCES IN APPLIED MECHANICS. Academic Press, Incorporated, 6277 Sea Harbor Drive, Orlando, FL 32821. (800) 321-5068.

ADVANCES IN ATOMIC AND MOLECULAR PHYSICS. Academic Press, Incorporated, 6277 Sea Harbor Drive, Orlando, FL 32821. (800) 321-5068.

ADVANCES IN ELECTRONICS AND ELECTRON PHYSICS. Academic Press, Incorporated, 6277 Sea Harbor Drive, Orlando, FL 32821. (800) 321-5068.

ADVANCES IN NUCLEAR PHYSICS. Plenum Publishing Corporation, 233 Spring Street, New York, NY 10013. (800) 221-9369.

ADVANCES IN PHYSICS. Taylor and Francis, Incorporated, 242 Cherry Street, Philadelphia, PA 19106-1906. (215) 238-0939.

ANNUAL REVIEW OF MATERIALS SCIENCE. Annual Reviews, Incorporated, 4139 El Camino Way, Palo Alto, CA 94306. (415) 493-4400.

ANNUAL REVIEW OF NUCLEAR AND PARTICLE SCIENCE. Annual Reviews, Incorporated, 4139 El Camino Way, Palo Alto, CA 94306. (415) 493-4400.

COMPUTER PHYSICS REPORTS. Elsevier Science Publishing Company, Incorporated, 52 Vanderbilt Avenue, New York, NY 10017. (212) 370-5520.

CRC CRITICAL REVIEWS IN SOLID STATE AND MATERIALS SCIENCES. CRC Press, Incorporated, 2000 Corporate Boulevard, NW, Boca Raton, FL 33431.

PHYSICS REPORTS - REVIEW SECTION OF PHYSICS LETTERS. Elsevier Science Publishing Company, Incorporated, 52 Vanderbilt Avenue, New York, NY 10017. (212) 370-5520.

PHYSICS AND CHEMISTRY OF THE EARTH. Pergamon Press, Incorporated, Maxwell House, Fairview Park, Elmsford, NY 10523. (914) 592-7700.

PROGRESS IN CRYSTAL GROWTH AND CHARACTERIZATION. Pergamon Press, Incorporated, Maxwell House, Fairview Park, Elmsford, NY 10523. (914) 592-7700.

PROGRESS IN OPTICS. Elsevier Science Publishing Company, Incorporated, 52 Vanderbilt Avenue, New York, NY 10017. (212) 370-5520.

PROGRESS IN QUANTUM ELECTRONICS. Pergamon Press, Incorporated, Maxwell House, Fairview Park, Elmsford, NY 10523. (914) 592-7700.

SOLID STATE PHYSICS - ADVANCED IN RESEARCH AND APPLICATIONS. Academic Press, Incorporated, 6277 Sea Harbor Drive, Orlando, FL 32821. (800) 321-5068.

TOPICS IN APPLIED PHYSICS. Spring-Verlag New York, Incorporated, 175 Fifth Avenue, New York, NY 10010. (800) 526-7254.

ASSOCIATIONS AND PROFESSIONAL SOCIETIES

AMERICAN INSTITUTE OF PHYSICS. 335 East 45th Street, New York, NY 10017. (212) 661-9404.

AMERICAN PHYSICAL SOCIETY. 335 East 45th Street, New York, NY 10017. (212) 682-7341.

BIBLIOGRAPHIES

SCIENCE BOOKS AND FILMS. American Association for the Advancement of Science, 1333 H Street, NW, Washington, DC 20005. Five times per year. $20.00 per year.

SCIENTIFIC AND TECHNICAL BOOKS AND SERIALS IN PRINT 1988: AN INDEX TO LITERATURE IN SCIENCE AND TECHNOLOGY R.R. Bowker Company, 205 East 42nd Street, New York, NY 10017. (800) 521-8110 or (212) 916-1600. $175.00.

DIRECTORIES AND BIOGRAPHICAL SOURCES

AMERICAN ASSOCIATION OF PHYSICS TEACHERS - Directory of Members. American Association Physics Teachers, 5110 Roanoke Place, College Park, MD 20740. Biennial. $2.00.

AMERICAN MEN AND WOMEN OF SCIENCE. Physical and Biological Sciences. Sixteenth edition. R.R. Bowker Company, 205 East 42nd Street, New York, NY 10017. (800) 521-8110 or (212) 916-1600. 1986. $595.00.

AMERICAN PHYSICAL SOCIETY MEMBERSHIP DIRECTORY BULLETIN ISSUE. American Physical Society, 335 East 45th Street, New York, NY 10017. (212) 682-7341. Biennial. $30.00.

BIOGRAPHICAL DICTIONARY OF SCIENTISTS. T.I. Williams. Halsted Press, 605 Third Avenue, New York, NY 10158. (800) 526-5368 or (212) 850-6418. Third edition. 1982. $29.95.

DIRECTORY OF PHYSICS AND ASTRONOMY STAFF MEMBERS. American Institute of Physics, 335 East 45th Street, New York, NY 10017. (212) 661-9404. Biennial. $30.00.

GOVERNMENT RESEARCH DIRECTORY. Gale Research Company, Book Tower, Detroit, MI 48226. (800) 521-0707. Fourth edition. 1987. $350.00.

INTERNATIONAL RESEARCH CENTERS DIRECTORY 1988-1989. Gale Research Company, Book Tower, Detroit, MI 48226. (800) 521-0707. Fourth edition. 1987. $360.00.

RESEARCH CENTERS DIRECTORY. Gale Research Company, Book Tower, Detroit, MI 48226. (800) 521-0707. Twelfth edition. 1987. $365.00 for set.

SCIENTIFIC AND TECHNICAL ORGANIZATIONS AND AGENCIES DIRECTORY. Gale Research Company, Book Tower, Detroit, MI 48226. (800) 521-0707. Second edition. 1987. $185.00.

WHO'S WHO IN FRONTIER SCIENCE AND TECHNOLOGY. Marquis Who's Who, Incorporated, 200 East Ohio Street, Chicago, IL 60611. (800) 428-3898 or (312) 787-2008.

WHO'S WHO IN TECHNOLOGY TODAY. Reston Publishing Company, Incorporated, c/o Prentice-Hall, Incorporated, Englewood Cliffs, NJ 07632. (800) 262-6868. Biennial. Five volumes. $425.00. Covers the fields of electronics, computer science, physics, optics, chemistry, biotechnology, mechanics, energy, and earth science.

WORLD GUIDE TO SCIENTIFIC ASSOCIATIONS AND LEARNED SOCIETIES. K.G. Saur Incorporated, 175 Fifth Avenue, New York, NY 10010. (800) 521-0707 or (212) 982-1302. Fourth edition. 1984. $112.00.

ENCYCLOPEDIAS AND DICTIONARIES

ENCYCLOPEDIA OF PHYSICAL SCIENCE AND TECHNOLOGY. Academic Press, Incorporated, Orlando, FL 32887. (800) 321-5068 or (305) 345-2734. Fifteen volumes, 1986.

ENCYCLOPEDIA OF PHYSICS. Robert M. Besancon, editor. Van Nostrand Reinhold, Incorporated, 115 Fifth Avenue, New York, NY 10003. (800) 543-2681. Third edition. 1985. $99.95.

MCGRAW-HILL ENCYCLOPEDIA OF SCIENCE AND TECHNOLOGY. McGraw-Hill Book, Incorporated, 1221 Avenue of the Americas, New York, NY 10020. (212) 997-3675. Fifth edition, 15 volumes. $1100.00.

PHYSICS

GENERAL WORKS

CHAOS: MAKING A NEW SCIENCE. James Gleick. Viking-Penguin, Incorporated, 40 West 23rd Street, New York, NY 10010. (800) 631-3577. 1987. $19.95.

FEARFUL SYMMETRY: THE SEARCH FOR BEAUTY IN MODERN PHYSICS. A. Zee. Macmillan Publishing Company, Incorporated, 866 Third Avenue, New York, NY 10022. (800) 257-5755. 1987. $22.50.

FUNDAMENTALS OF PHYSICS. D. Halliday and R. Resnik. John Wiley and Sons, Incorporated, 605 Third Avenue, New York, NY 10158. (800) 526-5368 or (212) 850-6000. Second edition. 1986. $42.95.

LONGING FOR HARMONIES: THEMES AND VARIATIONS FROM MODERN PHYSICS. Frank Wilczek and Betsy Devine. W.W. Norton and Company, Incorporated, 500 Fifth Avenue, New York, NY 10110. (800) 223-2584. 1988. $19.95.

MAKING OF THE ATOMIC BOMB. Richard Rhodes. Simon and Schuster, Incorporated, 1230 Avenue of the Americas, New York, NY 10020. (800) 223-2336. 1987. $22.95.

HANDBOOKS AND MANUALS

CRC HANDBOOK OF CHEMISTRY AND PHYSICS. CRC Press, Incorporated, 2000 Corporation Boulevard, Boca Raton, FL 33341. (305) 994-0555. Sixth-seventh edition. 1986. $69.95.

HANDBOOKS AND TABLES IN SCIENCE AND TECHNOLOGY. Russell H. Powell, editor. Oryx Press, 2214 North Central Avenue, Phoenix, AZ 85004-1483. (602) 254-6156. Second edition. 1983. $55.00.

MATHEMATICAL METHODS FOR PHYSICISTS. George B. Arfken. Academic Press, Incorporated, 6277 Sea Harbor Drive, Orlando, FL 32821. (800) 321-5068. Third edition. 1985. $44.50

ONLINE DATA BASES

CA SEARCH. Chemical Abstracts Service, Post Office Box 3012, Columbus, OH 43210. Guide to chemical literature, 1972 to present. Inquire as to cost and availability.

DISSERTATION ABSTRACTS ONLINE. University Microfilms International, 300 North Zeeb Road, Ann Arbor, MI 48106. (800) 521-0600 or (313) 761-4700. Scope includes virtually all doctoral dissertations accepted at accredited American institutions from 1861 to present in 252 subject areas. Inquire as to cost and availability.

INSPEC. INSPEC Marketing Department, Institute of Electrical and Electronics Engineers, Incorporated, IEEE Service Department, 445 Hoes Lane, Piscataway, NJ 08854. (201) 981-0060. Inquire as to on-line cost and availability.

MATHFILE. American Mathematical Society, Post Office Box 6248, Providence, RI 02940. (800) 556-7774 or (401) 272-9500. Scope includes pure and applied mathematics and related areas of physics, statistics, engineering, computer science, and operations research literature since 1973. Inquire as to cost and availability

NASA. National Aeronautics and Space Administration, Scientific and Technical Information Branch, 300 7th Street, SW, Washington, DC 20546. Citations and abstracts of aerospace literature, 1962 to present. Inquire as to cost and availability.

NTIS. National Technical Information Service, 5285 Port Royal Road, Springfield, VA 22161. (703) 487-4630. Broad coverage of government-sponsored research reports, 1964 to present. Inquire as to cost and availability.

SCISEARCH. Institute for Scientific Information, 3501 Market Street, Philadelphia, PA 19104. (800) 523-1850 or (215) 386-0100. Broad multidisciplinary title and author index to the international literature of science and technology, 1974 to present. Inquire as to cost and availability.

WILSONLINE. H.W. Wilson Company, 950 University Avenue, Bronx, NY 10452. (800) 367-6770 or (212) 588-8400. Makes available online versions of the printed H.W. Wilson Indexes including Applied Science and Technology Index, Business Periodicals Index, and Readers' Guide to Periodical Literature. Period covered is generally 1983 to present. Inquire as to cost and availability.

PERIODICALS

AMERICAN JOURNAL OF PHYSICS. American Association of Physics Teachers, 5100 Roanoke Place, College Park, MD 20740. Monthly. $104.00 per year.

CANADIAN JOURNAL OF PHYSICS. National Research Council of Canada, Research Journals, Ottawa, ON K1A 0R6, Canada. Monthly. $78.00 per year.

CONTEMPORARY PHYSICS. Taylor and Francis, Incorporated, 242 Cherry Street, Philadelphia, PA 19106-1906. (215) 238-0939. Bimonthly. $184.00 per year.

JOURNAL OF APPLIED PHYSICS. American Institute of Physics, 335 East 45th Street, New York, NY 10017. (212) 661-9404. Semimonthly. $495.00 per year.

JOURNAL OF PHYSICS. A. Mathematical and General Physics. American Institute of Physics, 335 East 45th Street, New York, NY 10017. (212) 661-9404. Eighteen times per year. $510.00 per year.

JOURNAL OF PHYSICS. B. Atomic and Molecular Physics. American Institute of Physics, 335 East 45th Street, New York, NY 10017. (212) 661-9404. Semimonthly. $620.00 per year.

JOURNAL OF PHYSICS. C. Solid State Physics. American Institute of Physics, 335 East 45th Street, New York, NY 10017. (212) 661-9404. Thirty-six times per year. $840.00 per year.

JOURNAL OF PHYSICS. D. Applied Physics. American Institute of Physics, 335 East 45th Street, New York, NY 10017. (212) 661-9404. Monthly. $310.00 per year.

JOURNAL OF PHYSICS. E. Scientific Instruments. American Institute of Physics, 335 East 45th Street, New York, NY 10017. (212) 661-9404. Monthly. $225.00 per year.

JOURNAL OF PHYSICS. F. Metal Physics. American Institute of Physics, 335 East 45th Street, New York, NY 10017. (212) 661-9404. Monthly. $435.00 per year.

JOURNAL OF PHYSICS. G. Nuclear Physics. American Institute of Physics, 335 East 45th Street, New York, NY 10017. (212) 661-9404. Monthly. $380.00 per year.

OPTICAL SOCIETY OF AMERICA. Journal Part A: Optics and Image Science. American Institute of Physics, 335 East 45th Street, New York, NY 10017. 9212) 661-9404. Monthly. $180.00 per year.

OPTICAL SOCIETY OF AMERICA. Journal Part B: Optical Physics. American Institute of Physics, 335 East 45th Street, New York, NY 10017. (212) 661-9404. Monthly. $180.00 per year.

PHYSICAL REVIEW A (GENERAL PHYSICS). American Institute of Physics, 335 East 45th Street, New York, NY 10017. (212) 661-9404. Monthly. $445.00 per year.

PHYSICAL REVIEW B (CONDENSED MATTER). American Institute of Physics, 335 East 45th Street, New York, NY 10017. (212) 661-9404. Semimonthly. $850.00 per year.

PHYSICAL REVIEW C (NUCLEAR PHYSICS). American Institute of Physics, 335 East 45th Street, New York, NY 10017.

(212) 661-9404. Monthly. $345.00 per year.

PHYSICAL REVIEW D (PARTICLES AND FIELDS). American Institute of Physics, 335 East 45th Street, New York, NY 10017. (212) 661-9404. Semimonthly. $510.00 per year.

PHYSICAL REVIEW LETTERS. American Institute of Physics, 335 East 45th Street, New York, NY 10017. (212) 661-9404. Weekly. $430.00 per year.

PHYSICS TODAY. American Institute of Physics, 335 East 45th Street, New York, NY 10017. (212) 661-9404. Monthly. $50.00 per year.

REVIEWS OF MODERN PHYSICS. American Institute of Physics, 335 East 45th Street, New York, NY 10017. (212) 661-9404. Quarterly. $145.00 per year.

SOVIET PHYSICS (TRANSLATION OF DOKLADY AKADEMIL NAUK SSSR). American Institute of Physics, 335 East 45th Street, New York, NY 10017. 9212) 661-9404. Monthly. $4485.00 per year.

PIEZOELECTRICITY

See also: CRYSTALLOGRAPHY, ELECTRICITY, AND SOLID STATE PHYSICS

ABSTRACT SERVICES AND INDEXES

APPLIED SCIENCE AND TECHNOLOGY INDEX. H.W. Wilson and Company, 950 University Avenue, Bronx, NY 10452. (800) 367-6670 or (212) 588-8400. Monthly. Inquire as to cost and availability.

CHEMICAL ABSTRACTS. American Chemical Society, Chemical Abstracts Service, Box 3012, Columbus, OH 43210. (614) 421-3600. 1907 to present. Weekly. $9500.00 per year.

CURRENT CONTENTS: ENGINEERING, TECHNOLOGY AND APPLIED SCIENCES. Institute for Scientific Information, 3501 Market Street, Philadelphia, PA 19104. (800) 523-1850 or (215) 386-0100. Weekly. $275.00 per year.

ENGINEERING INDEX MONTHLY AND AUTHOR INDEX. Engineering Information Inc., 345 East 47th Street, New York, NY 10017. (212) 705-7600. Monthly. $1560.00 per year.

PHYSICS ABSTRACTS. Institution of Electrical Engineers. Available from: IEEE Service Center, 445 Hoes Lane, Piscataway, NJ 08854. 1898 to present. Bimonthly. $1700.00 per year.

PHYSICS BRIEFS. Physik Verlag GmbH, Postfach 1260/1280, D-6940 Weinheim, West Germany. (212) 661-9404. 1920 to present. Twenty-six times per year. $1250.00 per year.

SCIENCE CITATION INDEX. Institute for Scientific Information, 3501 Market Street, Philadelphia, PA 19104. (800) 523-1850 or (215) 386-0100. Six times per year. $6200.00 per year.

SOLID STATE ABSTRACTS JOURNAL. Cambridge Scientific Abstracts, 5161 River Road, Bethesda, MD 201816. (301) 951-1400. 1957 to present. Bimonthly. $550.00 per year.

ASSOCIATIONS AND PROFESSIONAL SOCIETIES

INSTITUTE OF ELECTRICAL AND ELECTRONICS ENGINEERS (IEEE). 345 East 47th Street, New York, NY 10017. (212) 705-7900.

BIBLIOGRAPHIES

SOURCEBOOK OF PYROELECTRICITY. Sidney B. Lang. Gordon and Breach Science Publishers, Inc., 50 West 23rd Street, New York, NY 10010. (212) 206-8900. 1974. $125.00.

DIRECTORIES AND BIOGRAPHICAL SOURCES

1987 DIRECTORY OF ENGINEERING SOCIETIES AND RELATED ORGANIZATIONS. Gordon Davis, editor. Hemisphere Publishing Corporation, 1010 Vermont Avenue, NW, Washington, DC 20005. (800) 526-0275. 12th edition. 1987. $100.00.

RESEARCH CENTERS DIRECTORY 1988. Gale Research Company, Book Tower, Detroit, MI 48226. (800) 521-0707. 12th edition. 1987. $365.00 for set.

SCIENTIFIC AND TECHNICAL ORGANIZATIONS AND AGENCIES DIRECTORY. Margaret Labash Young, editor. Gale Research Company, Book Tower, Detroit, MI 48226. (800) 521-0707. 2nd edition. 1987. $185.00.

WHO'S WHO IN ENGINEERING. Gordon Davis, editor. Hemisphere Publishing Corporation, 1010 Vermont Avenue, NW, Washington, DC 20005. (800) 526-0275. 6th edition. 1985. $200.00.

GENERAL WORKS

PIEZOELECTRICITY. J.J. Gagnepain and T.R. Meeker, editors. Gordon and Breach Science Publishers, Inc., 50 West 23rd Street, New York, NY 10010. (212) 206-8900. 1982. $125.00.

PIEZOELECTRICITY. G.W. Taylor and others, editors. Gordon and Breach Science Publishers, Inc., 50 West 23rd Street, New York, NY 10010. (212) 206-8900. 1985. $84.00.

ONLINE DATA BASES

CA SEARCH. Chemical Abstracts Service, P.O. Box 3012, Columbus, OH 43120. (800) 848-6538 or (614) 421-3600. Comprehensive guide to chemical literature, 1972 to present. Inquire as to online cost and availability.

COMPENDEX. Engineering Information, Inc., 345 East 47th Street, New York, NY 10017. (800) 221-1044 or (212) 705-7615. Engineering and technical literature, 1975 to present. Inquire as to online cost and availability.

INSPEC. INSPEC Marketing Department, Institution of Electrical Engineers. Available from IEEE Service Center, 445 Hoes Lane, Piscataway, NJ 08854. (201) 981-0060. Online version of Physics Abstracts. Inquire as to online cost and availability.

NTIS. National Technical Information Service, 5285 Port Royal Road, Springfield, VA 22161. (703) 487-4630. Broad coverage of government sponsored research reports, 1964 to present. Inquire as to online cost and availability.

SCISEARCH. Institute for Scientific Information, 3501 Market Street, Philadelphia, PA 19104. (800) 523-1850 or (215) 386-0100. Broad multidisciplinary title and author index to the international literature of science and technology, 1974 to present. Inquire as to online cost and availability.

WILSONLINE. H.W. Wilson and Company, 950 University Avenue, Bronx, NY 10452. (800) 367-6770 or (212) 588-8400. Makes available online versions of the H.W. Wilson indexes including Applied Science and Technology Index, Business Periodicals Index and Readers' Guide to Periodical Literature. Approximately 1980 to present. Inquire as to online cost and availability.

PERIODICALS

FERROELECTRICS LETTERS. Gordon and Breach Science Publishers, Inc., 50 West 23rd Street, New York, NY 10010. (212) 206-8900. Monthly. $145.00 per year.

JOURNAL OF PHYSICS AND CHEMISTRY OF SOLIDS. Pergamon Press, Inc., Maxwell House, Fairview Park, Elmsford, NY 10523. (914) 592-7700. 1956 to present. Monthly. $550.00 per year.

JOURNAL OF PHYSICS C: SOLID STATE PHYSICS. Institute of Physics, London. Available from: American Institute of Physics, 335 East 45th Street, New York, NY 10017. 1968 to present. Thirty-six times per year. $850.00 per year.

RESEARCH CENTERS AND INSTITUTES

GROUPE DE RECHERCHE SUR LES DIELECTRIQUES. University of Quebec at Trois-Rivieres, Case Postale 500, Boulevard des Forges, Trois-Rivieres, PQ, Canada G9A 5H7. (819) 376-5107.

MICROELECTRONICS RESEARCH CENTER. Iowa State University, 1925 Scholl Road, Ames, IA 50011. (515) 294-7732.

PIEZOELECTRICITY RESEARCH LABORATORY. York University, 4700 Keele Street, North York, ON, Canada M3J 1P3. (416) 736-5250.

PIPELINE TECHNOLOGY

See also: PETROLEUM ENGINEERING

ABSTRACT SERVICES AND INDEXES

APPLIED SCIENCE AND TECHNOLOGY INDEX. H.W. Wilson and Company, 950 University Avenue, Bronx, NY 10452. (800) 367-6670 or (212) 588-8400. Monthly. Inquire as to cost and availability.

CHEMICAL ABSTRACTS. American Chemical Society, Chemical Abstracts Service, Box 3012, Columbus, OH 43210. (614) 421-3600. 1907 to present. Weekly. $9500.00 per year.

CURRENT CONTENTS: ENGINEERING, TECHNOLOGY AND APPLIED SCIENCES. Institute for Scientific Information, 3501 Market Street, Philadelphia, PA 19104. (800) 523-1850 or (215) 386-0100. Weekly. $275.00 per year.

ENGINEERING INDEX MONTHLY AND AUTHOR INDEX. Engineering Information Inc., 345 East 47th Street, New York, NY 10017. (212) 705-7600. Monthly. $1560.00 per year.

ASSOCIATIONS AND PROFESSIONAL SOCIETIES

AMERICAN INSTITUTE OF CHEMICAL ENGINEERS. 345 East 47th Street, New York, NY 10017. (212) 705-7338.

AMERICAN INSTITUTE OF MINING, METALLURGICAL AND PETROLEUM ENGINEERS, 345 East 47th Street, New York, NY 10017. (212) 705-7695.

AMERICAN SOCIETY OF CIVIL ENGINEERS. 345 East 47th Street, New York, NY 10017. (212) 705-7496.

SOCIETY OF PETROLEUM ENGINEERS. P.O. Box 833836, Richardson, TX 75083. (214) 669-3377.

DIRECTORIES AND BIOGRAPHICAL SOURCES

INTERNATIONAL RESEARCH CENTERS DIRECTORY 1988-89. Darren L. Smith, editor. Gale Research Company, Book Tower, Detroit, MI 48226. (800) 521-0707. 4th edition. 1987. $360.00.

1987 DIRECTORY OF ENGINEERING SOCIETIES AND RELATED ORGANIZATIONS. Gordon Davis, editor. Hemisphere Publishing Corporation, 1010 Vermont Avenue, NW, Washington, DC 20005. (800) 526-0275. 12th edition. 1987. $100.00.

RESEARCH CENTERS DIRECTORY 1988. Gale Research Company, Book Tower, Detroit, MI 48226. (800) 521-0707. 12th edition. 1987. $365.00 for set.

SCIENTIFIC AND TECHNICAL ORGANIZATIONS AND AGENCIES DIRECTORY. Margaret Labash Young, editor. Gale Research Company, Book Tower, Detroit, MI 48226. (800) 521-0707. 2nd edition. 1987. $185.00.

WHO'S WHO IN ENGINEERING. Gordon Davis, editor. Hemisphere Publishing Corporation, 1010 Vermont Avenue, NW, Washington, DC 20005. (800) 526-0275. 6th edition. 1985. $200.00.

GENERAL WORKS

FUNDAMENTALS OF PIPELINE ENGINEERING. J. Vincent-Genod. Gulf Publishing Company, P.O. Box 2608, Houston, TX 77001. (713) 520-4444. 1984. $39.00.

PIPELINES IN ADVERSE ENVIRONMENTS II. Mark B. Pickell. American Society of Civil Engineers, 345 East 47th Street, New York, NY 10017. (212) 705-7496. 1983. $60.00.

PIPING AND PIPE SUPPORT SYSTEMS: DESIGN AND ENGINEERING. R.P. Smith and T. Van Laan. McGraw-Hill Book Company, 1221 Avenue of the Americas, New York, NY 10020. (212) 512-2000. 1987. $49.95.

WORLD PIPELINES. J.N. Tiratsoo, editor. Gulf Publishing Company, P.O. Box 2608, Houston, TX 77001. (713) 520-4444. 1983. $59.00.

HANDBOOKS AND MANUALS

BASIC PIPELINE ENGINEERING MANUAL. John L. Cranmer. PennWell Books, P.O. Box 1260, Tulsa, OK 74101. (918) 835-3161. 1983. $65.95.

PIPING STRESS HANDBOOK. Victor Helguero. Gulf Publishing Company, P.O. Box 2608, Houston, TX 77001. (713) 520-4444. 2nd edition. 1985. $67.00.

ONLINE DATA BASES

CA SEARCH. Chemical Abstracts Service, P.O. Box 3012, Columbus, OH 43120. (800) 848-6538 or (614) 421-3600. Comprehensive guide to chemical literature, 1972 to present. Inquire as to online cost and availability.

COMPENDEX. Engineering Information, Inc., 345 East 47th Street, New York, NY 10017. (800) 221-1044 or (212) 705-7615. Engineering and technical literature, 1975 to present. Inquire as to online cost and availability.

NTIS. National Technical Information Service, 5285 Port Royal Road, Springfield, VA 22161. (703) 487-4630. Broad coverage of government sponsored research reports, 1964 to present. Inquire as to online cost and availability.

SCISEARCH. Institute for Scientific Information, 3501 Market Street, Philadelphia, PA 19104. (800) 523-1850 or (215) 386-0100. Broad multidisciplinary title and author index to the international literature of science and technology, 1974 to present. Inquire as to online cost and availability.

WILSONLINE. H.W. Wilson and Company, 950 University Avenue, Bronx, NY 10452. (800) 367-6770 or (212) 588-8400. Makes available online versions of the H.W. Wilson indexes including Applied Science and Technology Index, Business Periodicals Index and Readers' Guide to Periodical Literature. Approximately 1980 to present. Inquire as to online cost and availability.

PERIODICALS

PIPE LINE INDUSTRY. Gulf Publishing Company, P.O. Box 2608, Houston, TX 77001. (713) 520-4444. 1954 to present. Monthly. Free.

PIPELINE AND GAS JOURNAL. Energy Publications, Box 1589, Dallas, TX 75221. (214) 691-3911. 14 times per year. $15.00 per year.

PIPELINE AND UNDERGROUND UTILITIES CONSTRUCTION. Oildom Publishing Company, 3314 Mercer Street, Houston, TX 77027. (713) 622-0676. 13 times per year. $28.00.

PIPELINE DIGEST. Universal News, Inc., Box 55225, Houston, TX 77255-5225. (713) 468-2626. 1963 to present. Semi-monthly. $35.00 per year.

PIPES AND PIPELINES INTERNATIONAL. Scientific Surveys Limited, 4 Burkes Parade, Beaconsfield, Bucks, England. 1956 to present. Bimonthly. $67.00 per year.

RESEARCH CENTERS AND INSTITUTES

HYDRAULIC TRANSIENTS UNIT. University of Michigan, Ann Arbor, MI 48109. (313) 764-7148.

PLANETS

See: SOLAR SYSTEM

PLANETARY SCIENCES

See also: ASTROGEOLOGY, ASTRONOMY, ASTROPHYSICS, JUPITER, MARS, MERCURY, METEORS, NEPTUNE, PLUTO, SATELLITES, SATURN, SEISMOLOGY, SOLAR SYSTEM, URANUS, VENUS, VOLCANOLOGY

ABSTRACT SERVICES AND INDEXES

ASTRONOMY AND ASTROPHYSICS ABSTRACTS. Springer-Verlag New York, Inc., 175 Fifth Avenue, New York, NY 10010. (800) 526-7254. 1969 to present. Approximately $70.00 per year.

ASTRONOMY AND ASTROPHYSICS MONTHLY INDEX. Olivetree Associates, P.O. Box 236, Sierre Madre, CA 91024. $220.00 per year. Complementary copies available on request.

CHEMICAL ABSTRACTS. American Chemical Society, Chemical Abstracts Service, Box 3012, Columbus, OH 43210. (614) 421-3600. 1907 to present. Weekly. $10,000.00 per year.

CONFERENCE PAPERS INDEX. Cambridge Scientific Abstracts, 5161 River Road, Bethesda, MD 20816. 1972 to present. Monthly. Inquire as to cost and availability.

CURRENT CONTENTS: PHYSICAL, CHEMICAL, AND EARTH SCIENCES. Institute for Scientific Information, 3501 Market Street, Philadelphia, PA 19104. (800) 523-1850 or (215) 386-0100. Weekly. $275.00 per year.

INDEX TO SCIENTIFIC AND TECHNICAL PROCEEDINGS. Institute for Scientific Information, 3501 Market Street, Philadelphia, PA 19104. (800) 523-1850 or (215) 386-0100. 1978 to present. Monthly. $775.00 per year.

INDEX TO SCIENTIFIC REVIEWS. Institute for Scientific Information, 3501 Market Street, Philadelphia, PA 19104. (800) 523-1850 or (215) 386-0100. 1974 to present. Semi-annual. $550.00 per year.

METEOROLOGICAL AND GEOASTROPHYSICAL ABSTRACTS. American Meteorological Society, 45 Beacon Street, Boston, MA 02108. (617) 227-2425. 1950 to present. Monthly. $450.00 per year.

PHYSICS ABSTRACTS. Institution of Electrical Engineers. Available from: IEEE Service Center, 445 Hoes Lane, Piscataway, NJ 08854. 1898 to present. Bimonthly. $1700.00 per year.

PHYSICS BRIEFS. Physik Verlag GmbH, Postfach 1260/1280, D-6940 Weinheim, West Germany. (212) 661-9404. 1920 to present. Twenty-six times per year. $1250.00 per year.

SCIENCE CITATION INDEX. Institute for Scientific Information, 3501 Market Street, Philadelphia, PA 19104. (800) 523-1850 or (215) 386-0100. Six times per year. $6200. 00 per year.

STAR. (SCIENTIFIC AND TECHNICAL AEROSPACE REPORTS). U.S. National Aeronautics and Space Administration, Scientific and Technical Information Facility, Box 8757, Baltimore-Washington International Airport, MD 21240. (202) 755-2210. Semimonthly, with semiannual and annual indexes. $85.00 per year.

ANNUAL REVIEWS AND YEARBOOKS

ANNUAL REVIEW OF ASTRONOMY AND ASTROPHYSICS. Annual Reviews, Inc., 4139 El Camino Way, Palo Alto, CA 94306. (415) 493-4400. Annual. Inquire as to cost and availability.

ANNUAL REVIEW OF EARTH AND PLANETARY SCIENCES. Annual Reviews, Inc., 4139 El Camino Way, Palo Alto, CA 94306. (415) 493-4400. Annual. Inquire as to cost and availability.

ASSOCIATIONS AND PROFESSIONAL SOCIETIES

AMERICAN ASTRONOMICAL SOCIETY. 1816 Jefferson Place, N.W., Washington, DC 20036. (202) 659-0134.

AMERICAN GEOPHYSICAL UNION. 2000 Florida Avenue, N.W., Washington, DC 20009. (202) 462-6903.

AMERICAN INSTITUTE OF PHYSICS. 335 East 45th Street, New York, NY 10017. (212) 661-9494.

ASTRONOMICAL SOCIETY OF THE PACIFIC. 1290 24th Avenue, San Francisco, CA 94122. (415) 661-8660.

GEOLOGICAL SOCIETY OF AMERICA. 3300 Penrose Place, Boulder, CO 80301. (303) 447-2020.

PLANETARY SOCIETY. 65 North Catalina Avenue, Pasadena, CA 91106. (818) 793-5100.

BIBLIOGRAPHIES

A BIBLIOGRAPHY OF ASTRONOMY, 1970-1979. R.A. Seal and S.S. Martin. Libraries Unlimited Inc., P.O. Box 263, Littleton, CO 80160. (303) 770-1220. 1982. $37.50.

SCIENCE BOOKS AND FILMS. American Association for the Advancement of Science, 1333 H Street, NW, Washington, DC 20005. (202) 326-6454. Five times per year. $20.00 per year.

SCIENTIFIC AND TECHNICAL BOOKS AND SERIALS IN PRINT 1988; AN INDEX TO LITERATURE IN SCIENCE AND TECHNOLOGY. R.R. Bowker Company, 205 East 42nd Street, New York, NY 10017. (800) 521-8110. $175.00.

DIRECTORIES AND BIOGRAPHICAL SOURCES

AMERICAN ASTRONOMICAL SOCIETY MEMBERSHIP DIRECTORY. 1816 Jefferson Place, N.W., Washington, DC 20036. (202) 659-0134. Annual. Inquire.

AMERICAN MEN AND WOMEN OF SCIENCE. R.R. Bowker, Inc., Order Department, 245 West 17th Street, New York, NY 10011. (800) 521-8110. Eight volumes. 1986. $595.00 for set.

THE BIOGRAPHICAL DICTIONARY OF SCIENTISTS: ASTRONOMERS. D. Abbott, editor. Peter Bedrick Books, 125 East 23rd Street, New York, NY 10010. 1984.

DIRECTORY OF PHYSICS AND ASTRONOMY STAFF MEMBERS. American Institute of Physics, 335 East 45th Street, New York, NY 10017. (212) 661-9494. Annual.

INTERNATIONAL RESEARCH CENTERS DIRECTORY 1988-89. Darren L. Smith, editor. Gale Research Company, Book Tower, Detroit, MI 48226. (800) 521-0707. 4th edition. 1987. $360.00.

RESEARCH CENTERS DIRECTORY 1988. Gale Research Company, Book Tower, Detroit, MI 48226. (800) 521-0707. 12th edition. 1987. $365.00 for set.

SCIENTIFIC AND TECHNICAL ORGANIZATIONS AND AGENCIES DIRECTORY. Margaret Labash Young, editor. Gale Research Company, Book Tower, Detroit, MI 48226. (800) 521-0707. 2nd edition. 1987. $185.00.

GENERAL WORKS

ATMOSPHERES AND IONSPHERES OF THE OUTER PLANETS AND THEIR SATELLITES. S.K. Atreya. Springer-Verlag New York, Inc., 175 Fifth Avenue, New York, NY 10010. (800) 526-7254. 1987. $69.50.

THE BIRTH OF THE EARTH. David E. Fisher. Columbia University Press, 562 West 113th Street, New York, NY 10025. (212) 316-7100. 1987. $24.95.

CHEMISTRY AND PHYSICS OF THE TERRESTRIAL PLANETS. Surendra K. Saxena. Springer-Verlag New York, Inc., 175 Fifth Avenue, New York, NY 10010. (800) 526-7254. 1986. $59.00.

THE GALAXY AND THE SOLAR SYSTEM. Roman Smoluchowski and others. University of Arizona Press, 1615 East Speedway, Tucson, AZ 85719. (602) 621-1441. 1986. $29.95.

GEOLOGY OF THE TERRESTRIAL PLANETS. M. Carr, editor. National Aeronautics and Space Administration Special Publication 469. National Technical Information Service, 5285 Port Royal Road, Springfield, VA 22161. (703) 487-4929. 1984. $16.00.

MERCURY: THE ELUSIVE PLANET. Robert Strom. Cambridge University Press, 32 East 57th Street, New York, NY 10022. (800) 872-7423. 1986. $30.00.

METEORITES: THEIR RECORD OF EARLY SOLAR-SYSTEM HISTORY. J. Wasson. W.H. Freeman and Company, 41 Madison Avenue, New York, NY 10010. (212) 532-7660. 1985. $34.95.

PLANETS AND THEIR ATMOSPHERES: ORIGIN AND EVOLUTION. J. Lewis and R. Prinn. Academic Press, Inc., 6277 Sea Harbor Drive, Orlando, FL 32821. (800) 321-5068. 1983. $59.00.

RECENT ADVANCES IN PLANETARY METEOROLOGY. G. Hunt, editor. Cambridge University Press, 32 East 57th Street, New York, NY 10022. (800) 872-7423. 1985. $39.50.

THE PLANETARY SYSTEM. Tobias Owen and David Morrison. Addison-Wesley Publishing Company, Inc., 1 Jacob Way, Reading, MA 01867. (617) 944-3700. 1987. $39.95.

RINGS: DISCOVERIES FROM GALILEO TO VOYAGER. MIT Press, 28 Carleton Street, Cambridge, MA 02142. (617) 253-2884. 1985. $17.50.

SATELLITES. J.A. Burns and M.S. Matthews, editors. University of Arizona Press, 1615 East Speedway, Tucson, AZ 85719. (602) 621-1441. 1986. $55.00.

THE STORY OF THE EARTH. P. Cattermole and P. Moore. Cambridge University Press, 32 East 57th Street, New York, NY 10022. (800) 872-7423. 1985. $29.95.

THEORY OF PLANETARY ATMOSPHERES: AN INTRODUCTION TO THEIR PHYSICS AND CHEMISTRY. J.W. Chamberlain and D.M. Hunten. Academic Press, Inc., 6277 Sea Harbor Drive, Orlando, FL 32821. (800) 321-5068. 1987. $49.50.

URANUS AND NEPTUNE. J. Bergstrahl, editor. National Aeronautics and Space Administration Conference Paper 2330. National Technical Information Service, 5285 Port Royal Road, Springfield, VA 22161. (703) 487-4929. 1984. $25.00.

VENUS: AN ARRANT TWIN. E. Burgess. Columbia University Press, 562 West 113th Street, New York, NY 10025. (212) 316-7100. 1985. $29.95.

ONLINE DATA BASES

CA SEARCH. Chemical Abstracts Service, P.O. Box 3012, Columbus, OH 43120. (800) 848-6538 or (614) 421-3600. Comprehensive guide to chemical literature, 1972 to present. Inquire as to online cost and availability.

DISSERTATION ABSTRACTS ONLINE. University Microfilms International, 300 North Zeeb Road, Ann Arbor, MI 48106. (800) 521-0600 or (313) 761-4700. Scope includes virtually all doctoral dissertations accepted at accredited American institutions from 1861 to present in over 250 subject areas. Inquire as to online cost and availability.

INSPEC. INSPEC Marketing Department, Institution of Electrical Engineers. Available from IEEE Service Center, 445 Hoes Lane, Piscataway, NJ 08854. (201) 981-0060. Online version of Physics Abstracts. Inquire as to online cost and availability.

NTIS. National Technical Information Service, 5285 Port Royal Road, Springfield, VA 22161. (703) 487-4630. Broad coverage of government sponsored research reports, 1964 to present. Inquire as to online cost and availability.

SCISEARCH. Institute for Scientific Information, 3501 Market Street, Philadelphia, PA 19104. (800) 523-1850 or (215) 386-0100. Broad multidisciplinary title and author index to the international literature of science and technology, 1974 to present. Inquire as to online cost and availability.

OTHER SOURCES

ATLAS OF THE SOLAR SYSTEM. P. Moore and G. Hunt. Rand McNally and Company, P.O. Box 7600, Chicago, IL 60680. (800) 323-4070. 1983. $40.00.

CAMBRIDGE PHOTOGRAPHIC ATLAS OF THE PLANETS. Cambridge University Press, 32 East 57th Street, New York, NY 10022. (800) 872-7423. 1986.

PERIODICALS

ASTRONOMICAL JOURNAL. American Astronomical Society. Available from: American Institute of Physics, 335 East 45th Street, New York, NY 10017. (212) 661-9494. Monthly. $125.00 per year.

ASTRONOMICAL SOCIETY OF THE PACIFIC PUBLICATIONS. Astronomical Society of the Pacific, 1290 24th Avenue, San Francisco, CA 94122. (415) 661-8660. Monthly. $40.00 per year.

ASTRONOMY. AstroMedia Corporation, 625 E Street, Box 92788, Milwaukee, WI 53202. (414) 276-2689. 1973 to present. Monthly. $21.00 per year.

ASTRONOMY AND ASTROPHYSICS. Springer-Verlag New York, Inc., 175 Fifth Avenue, New York, NY 10010. (800) 526-7254. Monthly. $680.00 per year.

EARTH, MOON AND PLANETS. D. Reidel Publishing Company, 190 Old Derby Street, Hingham, MA 02043. 1969 to present. Nine times per year. $275.00 per year.

GEOCHEMICA ET COSMOCHIMICA ACTA. Pergamon Press, Inc., Maxwell House, Fairview Park, Elmsford, NY 10523. (914) 592-7700. 1950 to present. Monthly. $340.00 per year.

LUNAR AND PLANETARY INFORMATION BULLETIN. Lunar and Planetary Institute, 3303 NASA Road One, Houston, TX 77058-4399. (713) 486-2135. 1970 to present. Three times per year. Free.

MERCURY. Astronomical Society of the Pacific, 1290 24th Avenue, San Francisco, CA 94122. (415) 661-8660. 1972 to present. Bimonthly. $21.00 per year.

METEORITICS. Center for Meteorite Studies, Arizona State University, Tempe, AZ 85287. (602) 965-3576. 1955 to present. Quarterly. $40.00 per year.

PLANETARY AND SPACE SCIENCE. Pergamon Press, Inc., Maxwell House, Fairview Park, Elmsford, NY 10523. (914) 592-7700. 1959 to present. Monthly. $430.00 per year.

PLANETARY REPORT. Planetary Society, 65 North Catalina, Pasadena, CA 91106-2301. 1980 to present. Bimonthly. $20.00 per year.

SKY AND TELESCOPE. Sky Publishing Corporation, 49 Bay State Road, Cambridge, MA 02238. (617) 864-7360. Monthly. $18.00 per year.

RESEARCH CENTERS AND INSTITUTES

CENTER FOR EARTH AND PLANETARY STUDIES. National Air and Space Museum, Smithsonian Institution, Washington, DC 20560. 357-1424.

LABORATORY FOR PLANETARY ATMOSPHERES RESEARCH. State University of New York at Stony Brook, Stony Brook, NY 11794-2300. (516) 632-8321.

LABORATORY FOR PLANETARY GEOLOGY. Arizona State University, Department of Geology, Tempe, AZ 85281. (602) 965-7029.

LABORATORY FOR PLANETARY STUDIES. Cornell University, 302 Space Sciences Building, Ithaca, NY 14853. (607) 256-4971.

LUNAR AND PLANETARY INSTITUTE. 3303 NASA Road One, Houston, TX 77058. (713) 486-2139.

PLASMA PHYSICS

See also: ASTROPHYSICS, FUSION, PARTICLE ACCELERATORS, PHYSICS

ABSTRACT SERVICES AND INDEXES

CHEMICAL ABSTRACTS. American Chemical Society, Chemical Abstracts Service, Box 3012, Columbus, OH 43210. (614) 421-3600. 1907 to present. Weekly. $9500.00 per year.

CONFERENCE PAPERS INDEX. Cambridge Scientific Abstracts, 5161 River Road, Bethesda, MD 20816. 1972 to present. Monthly. Inquire as to cost and availability.

CURRENT CONTENTS: PHYSICAL, CHEMICAL AND EARTH SCIENCES. Institute for Scientific Information, 3501 Market Street, Philadelphia, PA 19104. (800) 523-1850 or (215) 386-0100. Weekly. $275.00 per year.

INDEX TO SCIENTIFIC AND TECHNICAL PROCEEDINGS. Institute for Scientific Information, 3501 Market Street, Philadelphia, PA 19104. (800) 523-1850 or (215) 386-0100. 1978 to present. Monthly. $775.00 per year.

INDEX TO SCIENTIFIC REVIEWS. Institute for Scientific Information, 3501 Market Street, Philadelphia, PA 19104. (800) 523-1850 or (215) 386-0100. 1974 to present. Semi-annual. $550.00 per year.

MATHEMATICAL REVIEWS: A REVIEWING JOURNAL COVERING THE WORLD LITERATURE OF MATHEMATICAL RESEARCH. American Mathematical Society, P.O. Box 6248, Providence, RI 02940. (800) 7774 or (401) 272-9500. 1940 to present. Monthly. $2800.00 per year.

PHYSICS ABSTRACTS. Institution of Electrical Engineers. Available from: IEEE Service Center, 445 Hoes Lane, Piscataway, NJ 08854. 1898 to present. Bimonthly. $1700.00 per year.

PHYSICS BRIEFS. Physik Verlag GmbH, Postfach 1260/1280, D-6940 Weinheim, West Germany. Twenty-six times per year. $1200.00 per year.

SCIENCE CITATION INDEX. Institute for Scientific Information, 3501 Market Street, Philadelphia, PA 19104. (800) 523-1850 or (215) 386-0100. Six times per year. $6200.00 per year.

ANNUAL REVIEWS AND YEARBOOKS

REVIEWS OF PLASMA PHYSICS. M.A. Leontovich, editor. Plenum Publishing Corporation, 233 Spring Street, New York, NY 10013. (800) 221-9369. Volumes 1-8. 1965-1980. Price varies. Inquire.

ASSOCIATIONS AND PROFESSIONAL SOCIETIES

AMERICAN INSTITUTE OF PHYSICS. 335 East 45th Street, New York, NY 10017. (212) 661-9404.

AMERICAN NUCLEAR SOCIETY. 555 North Kensington Avenue, La Grange Park, IL 60525. (312) 352-6611.

AMERICAN PHYSICAL SOCIETY. 335 East 45th Street, New York, NY 10017. (212) 682-7341.

FUSION ENERGY FOUNDATION. Box 17149, Washington, DC 20041-0149. (703) 689-2490.

IEEE NUCLEAR AND PLASMA SCIENCES SOCIETY. Institute of Electrical and Electronics Engineers, 345 East 47th Street, New York, NY 10017. (212) 705-7867.

DIRECTORIES AND BIOGRAPHICAL SOURCES

AMERICAN MEN AND WOMEN OF SCIENCE. R.R. Bowker, Inc., Order Department, 245 West 17th Street, New York, NY 10011. (800) 521-8110. Eight volumes. 1986. $595.00 for set.

DIRECTORY OF PHYSICS AND ASTRONOMY STAFF MEMBERS. American Institute of Physics, 335 East 45th Street, New York, NY 10017. (516) 346-7800. Every other year. $30.00.

INTERNATIONAL RESEARCH CENTERS DIRECTORY 1988-89. Darren L. Smith, editor. Gale Research Company, Book Tower, Detroit, MI 48226. (800) 521-0707. 4th edition. 1987. $360.00.

RESEARCH CENTERS DIRECTORY 1988. Gale Research Company, Book Tower, Detroit, MI 48226. (800) 521-0707. 12th edition. 1987. $365.00 for set.

SCIENTIFIC AND TECHNICAL ORGANIZATIONS AND AGENCIES DIRECTORY. Margaret Labash Young, editor. Gale Research Company, Book Tower, Detroit, MI 48226. (800) 521-0707. 2nd edition. 1987. $185.00.

ENCYCLOPEDIAS AND DICTIONARIES

CONCISE SCIENCE DICTIONARY. Oxford University Press, 200 Madison Avenue, New York, NY 10016. (800) 458-5833. 1987. $9.95 in paper.

DICTIONARY OF THE PHYSICAL SCIENCES: TERMS, FORMULAS, DATA. Cesare Emiliani. Oxford University Press, 200 Madison Avenue, New York, NY 10016. (800) 458-5833. 1987. $19.95 in paper.

ENCYCLOPEDIA OF PHYSICS. A.M. Prokhorov and S. Chomet, editors. Hemisphere Publishing Corporation, 1010 Vermont Avenue, NW, Washington, DC 20005. (800) 526-0275. 1988. $155.00.

MCGRAW-HILL ENCYCLOPEDIA OF SCIENCE AND TECHNOLOGY. McGraw-Hill Book Company, 1221 Avenue of the Americas, New York, NY 10020. (212) 512-2000. 6th edition. 1987. $1600.00.

THESAURUS OF SCIENTIFIC, TECHNICAL, AND ENGINEERING TERMS. Hemisphere Publishing Corporation, 1010 Vermont Avenue, NW, Washington, DC 20005. (800) 526-0275. 1988. $125.00.

GENERAL WORKS

APPLIED ATOMIC COLLISION PHYSICS, VOLUME 2, PLASMAS. C.F. Barnet and others, editors. Academic Press, Inc., 6277 Sea Harbor Drive, Orlando, FL 32821. (800) 321-5068. 1984. $88.00.

FUSION AND TECHNOLOGY: AN INTRODUCTION TO THE PHYSICS AND TECHNOLOGY OF MAGNETIC CONFINEMENT FUSION. W.M. Stacey. John Wiley and Sons, Inc., 605 Third Avenue, New York, NY 10158. (800) 526-5368. 1984. $45.95.

INTRODUCTION TO PLASMA PHYSICS AND CONTROLLED FUSION. Volume 1: Plasma Physics. Francis F. Chen. Plenum Publishing Corporation, 233 Spring Street, New York, NY 10013. (800) 221-9369. 2nd edition. 1984. $24.50.

INTRODUCTION TO PLASMA THEORY. D.R. Nicholson. John Wiley and Sons, Inc., 605 Third Avenue, New York, NY 10158. (800) 526-5368. 1983. $41.50.

TOPICS ON NONLINEAR WAVE-PLASMA INTERACTION. Klaus Baumgartel and Sauer Konrad. Birkhauser Boston, Inc., 380 Green Street, Cambridge, MA 02139. (617) 876-2333. 1987. $36.00.

ONLINE DATA BASES

CA SEARCH. Chemical Abstracts Service, P.O. Box 3012, Columbus, OH 43120. (800) 848-6538 or (614) 421-3600. Comprehensive guide to chemical literature, 1972 to present. Inquire as to online cost and availability.

DISSERTATION ABSTRACTS ONLINE. University Microfilms International, 300 North Zeeb Road, Ann Arbor, MI 48106. (800) 521-0600 or (313) 761-4700. Scope includes virtually all doctoral dissertations accepted at accredited American institutions from 1861 to present in over 250 subject areas. Inquire as to online cost and availability.

INSPEC. INSPEC Marketing Department, Institution of Electrical Engineers. Available from IEEE Service Center, 445 Hoes Lane, Piscataway, NJ 08854. (201) 981-0060. Online version of Physics Abstracts. Inquire as to online cost and availability.

MATHFILE. American Mathematical Society, P.O. Box 6248, Providence, RI 02940. (800) 556-7774 or (401) 272-9500. An online version of Mathematical Reviews. 1973 to present. Inquire as to online cost and availability.

NTIS. National Technical Information Service, 5285 Port Royal Road, Springfield, VA 22161. (703) 487-4630. Broad coverage of government sponsored research reports, 1964 to present. Inquire as to online cost and availability.

SCISEARCH. Institute for Scientific Information, 3501 Market Street, Philadelphia, PA 19104. (800) 523-1850 or (215) 386-0100. Broad multidisciplinary title and author index to the international literature of science and technology, 1974 to present. Inquire as to online cost and availability.

PERIODICALS

CANADIAN JOURNAL OF PHYSICS. National Research Council of Canada, Research Journals, Ottawa K1A OR6, Canada. Order from: Allen Press Inc., 1041 New Hampshire Street, Box 368, Lawrence, KS 66044. 1929 to present. Monthly. $200.00 per year.

FUSION (NEW YORK). Fusion Energy Foundation, Box 17149, Washington, DC 20041-0149. (703) 689-2490. 1975 to present. Bimonthly. $40.00 per year.

FUSION TECHNOLOGY. American Nuclear Society, 555 North Kensington Avenue, La Grange Park, IL 60525. (312) 352-6611. 1981 to present. Bimonthly. $250.00 per year.

INTERNATIONAL JOURNAL OF FUSION ENERGY. Fusion Energy Foundation, Box 17149, Washington, DC 20041-0149. (703) 689-2490. 1977 to present. Quarterly. $80.00 per year.

INTERNATIONAL JOURNAL THERMOPHYSICS. Plenum Publishing Corporation, 233 Spring Street, New York, NY 10013. (800) 221-9369. 1980 to present. Six times per year. $150.00 per year.

JOURNAL OF FUSION ENERGY. Plenum Publishing Corporation, 233 Spring Street, New York, NY 10013. (800) 221-9369. 1981 to present. Bimonthly. $105.00 per year.

PHYSICAL REVIEW LETTERS. American Institute of Physics, 335 East 45th Street, New York, NY 10017. (212) 661-9404. 1958 to present. Weekly. $450.00 per year.

PHYSICS LETTERS, SECTION B: NUCLEAR, ELEMENTARY PARTICLE AND HIGH-ENERGY PHYSICS. Elsevier Science Publishing Company, Inc., 52 Vanderbilt Avenue, New York, NY 10017. (212) 370-5520. Seventy-two times per year. $1700.00 per year.

PLASMA CHEMISTRY AND PLASMA PROCESSING. Plenum Publishing Corporation, 233 Spring Street, New York, NY 10013. (800) 221-9369. 1981 to present. Quarterly. $115.00 per year.

PLASMA PHYSICS AND CONTROLLED FUSION. Pergamon Press, Inc., Maxwell House, Fairview Park, Elmsford, NY 10523. (914) 592-7700. 1959 to present. Monthly. $325.00 per year.

RESEARCH CENTERS AND INSTITUTES

NATIONAL SCIENCE FOUNDATION. Directorate of Engineering, Plasmas Section, 1800 G Street, N.W., Washington, DC 20550. (202) 357-9618.

PLASMA PHYSICS GROUP. University of Wisconsin - Madison, Department of Physics, 1150 University Avenue, Madison, WI 53706. (608) 262-3595.

PLASMA PHYSICS LABORATORY. Columbia University, Applied Physics and Nuclear Engineering Department, 520 West 120th Street, New York, NY 10027. (212) 280-4457.

PLASMA PHYSICS LABORATORY. Princeton University, James Forrestal Research Campus, P.O. Box 451, Princeton, NJ 08544. (609) 683-2000.

PLASTICS

See also: ADHESIVES, CERAMICS, CHEMISTRY, COMPOSITES, MATERIALS SCIENCE, ORGANIC CHEMISTRY, POLYMERS, THERMOPLASTICS

ABSTRACT SERVICES AND INDEXES

APPLIED SCIENCE AND TECHNOLOGY INDEX. H.W. Wilson and Company, 950 University Avenue, Bronx, NY 10452. (800) 367-6670 or (212) 588-8400. Monthly. Inquire as to cost and availability.

CHEMICAL ABSTRACTS. American Chemical Society, Chemical Abstracts Service, Box 3012, Columbus, OH 43210. (614) 421-3600. 1907 to present. Weekly. $9500.00 per year.

CONFERENCE PAPERS INDEX. Cambridge Scientific Abstracts, 5161 River Road, Bethesda, MD 20816. 1972 to present. Monthly. Inquire as to cost and availability.

CURRENT CONTENTS: ENGINEERING, TECHNOLOGY AND APPLIED SCIENCES. Institute for Scientific Information, 3501 Market Street, Philadelphia, PA 19104. (800) 523-1850 or (215) 386-0100. Weekly. $275.00 per year.

ENGINEERING INDEX MONTHLY AND AUTHOR INDEX. Engineering Information Inc., 345 East 47th Street, New York, NY 10017. (212) 705-7600. Monthly. $1560.00 per year.

GENERAL SCIENCE INDEX. H.W. Wilson and Company, 950 University Avenue, Bronx, NY 10452. (800) 367-6670 or (212) 588-8400. 1978 to present. Monthly. Inquire as to cost and availability.

INDEX TO SCIENTIFIC AND TECHNICAL PROCEEDINGS. Institute for Scientific Information, 3501 Market Street, Philadelphia, PA 19104. (800) 523-1850 or (215) 386-0100. 1978 to present. Monthly. $775.00 per year.

INDEX TO SCIENTIFIC REVIEWS. Institute for Scientific Information, 3501 Market Street, Philadelphia, PA 19104. (800) 523-1850 or (215) 386-0100. 1974 to present. Semi-annual. $550.00 per year.

SCIENCE CITATION INDEX. Institute for Scientific Information, 3501 Market Street, Philadelphia, PA 19104. (800) 523-1850 or (215) 386-0100. Six times per year. $6200.00 per year.

ASSOCIATIONS AND PROFESSIONAL SOCIETIES

AMERICAN CHEMICAL SOCIETY. 1155 16th Street, N.W., Washington, DC 20036. (800) 424-6747.

PLASTICS INSTITUTE OF AMERICA. Stevens Institute of Technology, Castle Point Station, Hoboken, NJ 07030. (201) 420-5553.

SOCIETY OF PLASTICS ENGINEERS, INC. 14 Fairfield Drive, Brookfield Center, CT 06805. (203) 775-0471.

SOCIETY OF THE PLASTICS INDUSTRY. 355 Lexington Avenue, New York, NY 10017. (212) 503-0600.

DIRECTORIES AND BIOGRAPHICAL SOURCES

AMERICAN MEN AND WOMEN OF SCIENCE. R.R. Bowker, Inc., Order Department, 245 West 17th Street, New York, NY 10011. (800) 521-8110. Eight volumes. 1986. $595.00 for set.

INTERNATIONAL RESEARCH CENTERS DIRECTORY 1988-89. Darren L. Smith, editor. Gale Research Company, Book Tower, Detroit, MI 48226. (800) 521-0707. 4th edition. 1987. $360.00.

1987 DIRECTORY OF ENGINEERING SOCIETIES AND RELATED ORGANIZATIONS. Gordon Davis, editor. Hemisphere Publishing Corporation, 1010 Vermont Avenue, NW, Washington, DC 20005. (800) 526-0275. 12th edition. 1987. $100.00.

RESEARCH CENTERS DIRECTORY 1988. Gale Research Company, Book Tower, Detroit, MI 48226. (800) 521-0707. 12th edition. 1987. $365.00 for set.

SCIENTIFIC AND TECHNICAL ORGANIZATIONS AND AGENCIES DIRECTORY. Margaret Labash Young, editor. Gale Research Company, Book Tower, Detroit, MI 48226. (800) 521-0707. 2nd edition. 1987. $185.00.

SOCIETY OF THE PLASTICS INDUSTRY - MEMBERSHIP DIRECTORY AND BUYERS GUIDE. 355 Lexington Avenue, New York, NY 10017. (212) 503-0600. Annual. $90.00.

WHO'S WHO IN ENGINEERING. Gordon Davis, editor. Hemisphere Publishing Corporation, 79 Madison Avenue, New York, NY 10016-7892. (800) 821-8312. Sixth edition. 1985. $200.00.

WHO'S WHO IN TECHNOLOGY. Research Publications, 12 Lunar Drive, Woodbridge, CT 06525. (203) 397-2600. 1986. Seven volume set, $545.00.

ENCYCLOPEDIAS AND DICTIONARIES

ENCYCLOPEDIA OF POLYMER SCIENCE AND ENGINEERING. John Wiley and Sons, Inc., 605 Third Avenue, New York, NY 10158. (800) 526-5368. Second edition. 1985 to present. $3500.00 for set.

THESAURUS OF SCIENTIFIC, TECHNICAL, AND ENGINEERING TERMS. Hemisphere Publishing Corporation, 1010 Vermont Avenue, NW, Washington, DC 20005. (800) 526-0275. 1988. $125.00.

GENERAL WORKS

FUTURE TRENDS IN POLYMER SCIENCE AND TECHNOLOGY. Ezio Martuscelli and others. Technomic Publishing Company, Inc., 851 New Holland Avenue, Box 3535, Lancaster, PA 17604. (717) 291-5609. 1987. $49.00.

PLASTICS MATERIALS AND PROCESSING. S. Schwartz and S. Goodman. Van Nostrand Reinhold Company, Inc., 135 West 50th Street, New York, NY 10020. (800) 543-2681. 1982. $100.00.

HANDBOOKS AND MANUALS

PLASTICS ENGINEERING HANDBOOK OF THE SOCIETY OF THE PLASTICS INDUSTRY, INC. Joel Frados, editor. Van Nostrand Reinhold Company, Inc., 135 West 50th Street, New York, NY 10020. (800) 543-2681. 1985. $54.95.

PLASTICS MOLD ENGINEERING HANDBOOK. J. Harry DuBois and W. I. Pribble, editors. Van Nostrand Reinhold Company, Inc., 135 West 50th Street, New York, NY 10020. (800) 543-2681. Second edition. 1987. $59.95.

PLASTICS TECHNOLOGY HANDBOOK. M. Chanda and S.K. Roy. Marcel Dekker Inc., 270 Madison Avenue, New York, NY 10016. (800) 228-1160. 1987. $99.75.

ONLINE DATA BASES

CA SEARCH. Chemical Abstracts Service, P.O. Box 3012, Columbus, OH 43120. (800) 848-6538 or (614) 421-3600. Comprehensive guide to chemical literature, 1972 to present. Inquire as to online cost and availability.

COMPENDEX. Engineering Information, Inc., 345 East 47th Street, New York, NY 10017. (800) 221-1044 or (212) 705-7615. Engineering and technical literature, 1975 to present. Inquire as to online cost and availability.

DISSERTATION ABSTRACTS ONLINE. University Microfilms International, 300 North Zeeb Road, Ann Arbor, MI 48106. (800) 521-0600 or (313) 761-4700. Scope includes virtually all doctoral dissertations accepted at accredited American institutions from 1861 to present in over 250 subject areas. Inquire as to online cost and availability.

INSPEC. INSPEC Marketing Department, Institution of Electrical Engineers. Available from IEEE Service Center, 445 Hoes Lane, Piscataway, NJ 08854. (201) 981-0060. Online version of Physics Abstracts. Inquire as to online cost and availability.

NTIS. National Technical Information Service, 5285 Port Royal Road, Springfield, VA 22161. (703) 487-4630. Broad coverage of government sponsored research reports, 1964 to present. Inquire

as to online cost and availability.

SCISEARCH. Institute for Scientific Information, 3501 Market Street, Philadelphia, PA 19104. (800) 523-1850 or (215) 386-0100. Broad multidisciplinary title and author index to the international literature of science and technology, 1974 to present. Inquire as to online cost and availability.

WILSONLINE. H.W. Wilson and Company, 950 University Avenue, Bronx, NY 10452. (800) 367-6770 or (212) 588-8400. Makes available online versions of the H.W. Wilson indexes including Applied Science and Technology Index, Business Periodicals Index and Readers' Guide to Periodical Literature. Approximately 1980 to present. Inquire as to online cost and availability.

OTHER SOURCES

WHAT EVERY ENGINEER SHOULD KNOW ABOUT PLASTICS. Marcel Dekker Inc., 270 Madison Avenue, New York, NY 10016. (800) 228-1160. 1986. $24.95.

PERIODICALS

ADVANCES IN POLYMER TECHNOLOGY. John Wiley and Sons, Inc., 605 Third Avenue, New York, NY 10158. (800) 526-5368. 1977 to present. Quarterly. $100.00 per year.

CANADIAN PLASTICS. Southam Communications, Limited, 1450 Don Mills Road, Don Mills, ON, M3B 2X7, Canada. 1943 to present. Nine times per year. $45.00 per year.

HIGH PERFORMANCE PLASTICS. Elsevier Science Publishing Company, Inc., 52 Vanderbilt Avenue, New York, NY 10017. (212) 370-5520. 1983 to present. Monthly. $135.00 per year.

INTERNATIONAL JOURNAL OF ADHESION AND ADHESIVES. Butterworth's Publishing, 80 Montvale Avenue, Stoneham, MA 02180. (800) 325-4177. 1980 to present. Quarterly. $155.00 per year.

JOURNAL OF CELLULAR PLASTICS. Technomic Publishing Company, Inc., 851 New Holland Avenue, Box 3535, Lancaster, PA 17604. (717) 291-5609. 1965 to present. Bimonthly. $70.00 per year.

JOURNAL OF ELASTOMERS AND PLASTICS. Technomic Publishing Company, Inc., 851 New Holland Avenue, Box 3535, Lancaster, PA 17604. (717) 291-5609. 1969 to present. Quarterly. $115.00 per year.

JOURNAL OF PLASTIC FILM AND SHEETING. Technomic Publishing Company, Inc., 851 New Holland Avenue, Box 3535, Lancaster, PA 17604. (717) 291-5609. 1985 to present. Quarterly. $125.00 per year.

MODERN PLASTICS. McGraw-Hill Book Company, 1221 Avenue of the Americas, New York, NY 10020. (212) 512-2000. 1925 to present. Monthly. $30.00 per year.

PLASTICS COMPOUNDING: FOR RESIN PRODUCERS, FORMULATORS AND COMPOUNDERS. Harcourt, Brace and Jovanovich Publications, 262 Main Street, Chatham, NJ 07928. (201) 635-1671. 1978 to present. Bimonthly. $21.00 per year.

PLASTICS DESIGN FORUM. Harcourt, Brace and Jovanovich Publications, 262 Main Street, Chatham, NJ 07928. (201) 635-1671. 1976 to present. Bimonthly. $20.00 per year.

PLASTICS ENGINEERING. Society of Plastics Engineers, Inc., 14 Fairfield Drive, Brookfield Center, CT 06805. (203) 775-0471. 1945 to present. Monthly. $30.00 per year.

PLASTICS TECHNOLOGY. Bill Communications, Inc., 633 Third Avenue, New York, NY 10017. (212) 986-4800. 1955 to present. 13 times per year. $36.00 per year.

PLASTICS WORLD. Cahners Publishing Company, Inc., 221 Columbus Avenue, Boston, MA 02116. (617) 536-7780. 1942 to present. Monthly. $45.00 per year.

POLYMER COMPOSITES. Society of Plastics Engineers, Inc., 14 Fairfield Drive, Brookfield Center, CT 06805. (203) 775-0471. 1980 to present. Six times per year. $100.00 per year.

RESEARCH CENTERS AND INSTITUTES

PLASTICS INSTITUTE OF AMERICA. Stevens Institute of Technology, Castle Point Station, Hoboken, NJ 07030. (201) 420-5553.

POLYMER RESEARCH CENTER. University of Cincinnati, Mail Location 172, Cincinnati, OH 45221. (513) 475-2453.

POLYMER RESEARCH LABORATORY. University of Michigan, Dow Building, Ann Arbor, MI 48109. (313) 763-2240.

PLATE TECTONICS

See also: CONTINTENTAL DRIFT, CONTINTENTAL MARGINS, GEOLOGY, GEOPHYSICS, MARINE GEOLOGY

ABSTRACT SERVICES AND INDEXES

BIBLIOGRAPHY AND INDEX OF GEOLOGY. American Geological Institute, 4220 King Street, Alexandria, VA 22302. (703) 379-2480. 1969 to present. Monthly. $1100.00 per year.

CHEMICAL ABSTRACTS. American Chemical Society, Chemical Abstracts Service, Box 3012, Columbus, OH 43210. (614) 421-3600. 1907 to present. Weekly. $9500.00 per year.

CONFERENCE PAPERS INDEX. Cambridge Scientific Abstracts, 5161 River Road, Bethesda, MD 20816. 1972 to present. Monthly. Inquire as to cost and availability.

CURRENT CONTENTS: PHYSICAL, CHEMICAL AND EARTH SCIENCES. Institute for Scientific Information, 3501 Market Street, Philadelphia, PA 19104. (800) 523-1850 or (215) 386-0100. Weekly. $275.00 per year.

GENERAL SCIENCE INDEX. H.W. Wilson and Company, 950 University Avenue, Bronx, NY 10452. (800) 367-6670 or (212) 588-8400. 1978 to present. Monthly. Inquire as to cost and availability.

INDEX TO SCIENTIFIC AND TECHNICAL PROCEEDINGS. Institute for Scientific Information, 3501 Market Street, Philadelphia, PA 19104. (800) 523-1850 or (215) 386-0100. 1978 to present. Monthly. $775.00 per year.

INDEX TO SCIENTIFIC REVIEWS. Institute for Scientific Information, 3501 Market Street, Philadelphia, PA 19104. (800) 523-1850 or (215) 386-0100. 1974 to present. Semi-annual. $550.00 per year.

METEOROLOGICAL AND GEOASTROPHYSICAL ABSTRACTS. American Meteorological Society, 45 Beacon Street, Boston, MA 02108. (617) 227-2425. 1950 to present. Monthly. $450.00 per year.

OCEAN ABSTRACTS. Cambridge Scientific Abstracts, 5161 River Road, Bethesda, MD 20816. (301) 951-1400. 1963 to present. Bimonthly. $450.00 per year.

SCIENCE CITATION INDEX. Institute for Scientific Information, 3501 Market Street, Philadelphia, PA 19104. (800) 523-1850 or (215) 386-0100. Six times per year. $6200.00 per year.

ANNUAL REVIEWS AND YEARBOOKS

ANNUAL REVIEW OF EARTH AND PLANETARY SCIENCES. Annual Reviews, Inc., 4139 El Camino Way, Palo Alto, CA 94306. (415) 493-4400. Annual. Inquire as to cost and availability.

ASSOCIATIONS AND PROFESSIONAL SOCIETIES

AMERICAN GEOLOGICAL INSTITUTE. 4220 King Street, Alexandria, VA 22302. (703) 379-2480.

AMERICAN GEOPHYSICAL UNION. 2000 Florida Avenue, N.W., Washington, DC 20009. (202) 462-6903.

GEOLOGICAL SOCIETY OF AMERICA. 3300 Penrose Place, Boulder, CO 80301. (303) 447-2020.

DIRECTORIES AND BIOGRAPHICAL SOURCES

AMERICAN MEN AND WOMEN OF SCIENCE. R.R. Bowker, Inc., Order Department, 245 West 17th Street, New York, NY 10011. (800) 521-8110. Eight volumes. 1986. $595.00 for set.

INTERNATIONAL RESEARCH CENTERS DIRECTORY 1988-89. Darren L. Smith, editor. Gale Research Company, Book Tower, Detroit, MI 48226. (800) 521-0707. 4th edition. 1987. $360.00.

RESEARCH CENTERS DIRECTORY 1988. Gale Research Company, Book Tower, Detroit, MI 48226. (800) 521-0707. 12th edition. 1987. $365.00 for set.

SCIENTIFIC AND TECHNICAL ORGANIZATIONS AND AGENCIES DIRECTORY. Margaret Labash Young, editor. Gale Research Company, Book Tower, Detroit, MI 48226. (800) 521-0707. 2nd edition. 1987. $185.00.

ENCYCLOPEDIAS AND DICTIONARIES

MCGRAW-HILL DICTIONARY OF EARTH SCIENCES. McGraw-Hill Book Company, 1221 Avenue of the Americas, New York, NY 10020. (212) 512-2000. 1984. $36.00.

GENERAL WORKS

CONTINENTAL DRIFT. James H. Shea, editor. Van Nostrand Reinhold Company, Inc., 135 West 50th Street, New York, NY 10020. (800) 543-2681. 1985. $49.50.

THE EVOLVING CONTINENTS. B.F. Windley. John Wiley and Sons, Inc., 605 Third Avenue, New York, NY 10158. (800) 526-5368. Second edition. 1984. $31.95 in paper.

MARINE GEOLOGY: AN ADVENTURE INTO THE UNKNOWN. R.N. Anderson. John Wiley and Sons, Inc., 605 Third Avenue, New York, NY 10158. (800) 526-5368. 1984. $25.95.

MECHANISMS OF CONTINENTAL DRIFT AND PLATE TECTONICS. P.A. Davies and S.K. Runcorn, editors. Academic Press, Inc., 6277 Sea Harbor Drive, Orlando, FL 32821. (800) 321-5068. 1981. $75.50.

NEW VIEWS ON AN OLD PLANET: CONTINENTAL DRIFT AND THE HISTORY OF THE EARTH. T.H. Van Andel. Cambridge University Press, 32 East 57th Street, New York, NY 10022. (800) 872-7423. 1985. $19.95.

WANDERING CONTINENTS AND SPREADING SEA FLOORS ON AN EXPANDING EARTH. Lester C. King. John Wiley and Sons, Inc., 605 Third Avenue, New York, NY 10158. (800) 526-5368. 1984. $47.95.

ONLINE DATA BASES

DISSERTATION ABSTRACTS ONLINE. University Microfilms International, 300 North Zeeb Road, Ann Arbor, MI 48106. (800) 521-0600 or (313) 761-4700. Scope includes virtually all doctoral dissertations accepted at accredited American institutions from 1861 to present in over 250 subject areas. Inquire as to online cost and availability.

GEOREF. Online version of the BIBLIOGRAPHY AND INDEX OF GEOLOGY. American Geological Institute, 4220 King Street, Alexandria, VA 22302. (703) 379-2480. 1969 to present. Inquire as to online cost and availability.

NTIS. National Technical Information Service, 5285 Port Royal Road, Springfield, VA 22161. (703) 487-4630. Broad coverage of government sponsored research reports, 1964 to present. Inquire as to online cost and availability.

SCISEARCH. Institute for Scientific Information, 3501 Market Street, Philadelphia, PA 19104. (800) 523-1850 or (215) 386-0100. Broad multidisciplinary title and author index to the international literature of science and technology, 1974 to present. Inquire as to online cost and availability.

WILSONLINE. H.W. Wilson and Company, 950 University Avenue, Bronx, NY 10452. (800) 367-6770 or (212) 588-8400. Makes available online versions of the H.W. Wilson indexes including Applied Science and Technology Index, Business Periodicals Index and Readers' Guide to Periodical Literature. Approximately 1980 to present. Inquire as to online cost and availability.

PERIODICALS

CONTINENTAL SHELF RESEARCH. Pergamon Press, Inc., Maxwell House, Fairview Park, Elmsford, NY 10523. (914) 592-7700. 1982 to present. Bimonthly. $1200.00 per year. Includes Oceanographic Literature Review.

GEOLOGICAL SOCIETY OF AMERICA BULLETIN. Geological Society of America, 3300 Penrose Place, Boulder, CO 80301. (303) 447-2020. 1888 to present. Monthly. $80.00 per year.

GEO-MARINE LETTERS. Springer-Verlag New York, Inc., 175 Fifth Avenue, New York, NY 10010. (800) 526-7254. 1981 to present. Bimonthly. $100.00 per year.

JOURNAL OF GEOLOGY. University of Chicago Press, 5801 South Ellis Avenue, Chicago, IL 60637. (800) 621-2736. 1893 to present. Bimonthly. $30.00 per year.

MARINE GEOLOGY; INTERNATIONAL JOURNAL OF MARINE GEOLOGY, GEOCHEMISTRY AND GEOPHYSICS. Elsevier Science Publishing Company, Inc., 52 Vanderbilt Avenue, New York, NY 10017. (212) 370-5520. 1964 to present. Twenty-four times per year. $510.00 per year.

TECTONOPHYSICS; INTERNATIONAL JOURNAL OF GEOTECTONICS AND THE GEOLOGY AND PHYSICS OF THE INTERIOR OF THE EARTH. Elsevier Science Publishing Company, Inc., 52 Vanderbilt Avenue, New York, NY 10017. (212) 370-5520. Forty-four times per year. $900.00 per year.

RESEARCH CENTERS AND INSTITUTES

GEOLOGICAL RESEARCH DIVISION. University of California, San Diego, A-020, Scripps Institution of Oceanography, La Jolla, CA 92093. (619) 534-1830.

INSTITUTE FOR THE STUDY OF THE CONTINENTS. Cornell University, 3120 Snee Hall, Ithaca, NY 14853-1504. (607) 255-3474.

LAMONT-DOHERTY GEOLOGICAL OBSERVATORY. Columbia University, Palisades, NY 10964. (914) 359-2900.

PLUTO

See also: PLANETARY SCIENCES, SOLAR SYSTEM

ABSTRACT SERVICES AND INDEXES

ASTRONOMY AND ASTROPHYSICS ABSTRACTS. Springer-Verlag New York, Incorporated, 175 Fifth Avenue, New York, NY 10010. (212) 460-1500. $70.00 per year.

ASTRONOMY AND ASTROPHYSICS MONTHLY INDEX. Olivetree Associates, Post Office Box 236, Sierra Madre, CA 91024. $212.00 per year. Complimentary copies available on request.

METEOROLOGICAL AND GEOASTROPHYSICAL ABSTRACTS. American Meteorological Society, 45 Beacon Street, Boston, MA 02108. (617) 227-2425. 1950 to present. Monthly. $450.00 per year.

STAR. (Scientific and Technical Aerospace Reports. United States National Aeronautics and Space Administration, Scientific and Technical Information Facility, Box 8757, Baltimore-Washington International Airport, MD 21240. (202) 755-2210. Semimonthly, with semiannual and annual indexes. $85.00 per year.

ANNUAL REVIEWS AND YEARBOOKS

THE ASTRONOMICAL ALMANAC. Superintendent of Documents, United States Government Printing Office, Washington, DC 20402. (202) 783-3238. Yearly.

ANNUAL REVIEW OF ASTRONOMY AND ASTROPHYSICS. Annual Reviews, Incorporated, 4139 El Camino Way, Palo Alto, CA 94306. (415) 493-4400. Annual. Inquire.

ANNUAL REVIEW OF EARTH AND PLANETARY SCIENCES. Annual Reviews, Incorporated, 4139 El Camino Way, Palo Alto, CA 94306. (415) 493-4400. Annual. Inquire.

ASSOCIATIONS AND PROFESSIONAL SOCIETIES

AMERICAN ASTRONOMICAL SOCIETY. 2000 Florida Avenue, NW, Suite 300, Washington, DC 20009. (202) 659-0134.

AMERICAN GEOPHYSICAL UNION. 2000 Florida Avenue, NW, Washington, DC 20009. (202) 462-6903.

ASTRONOMICAL LEAGUE. Post Office Box 12821, Tucson, AZ 85732. (602) 790-8471.

ASTRONOMICAL SOCIETY OF THE PACIFIC. 1290 24th Avenue, San Francisco, CA 94122. (415) 661-8660.

PLANETARY SOCIETY. 65 North Catalina Avenue, Pasadena, CA 91106-2301. (818) 793-5100.

BIBLIOGRAPHIES

A BIBLIOGRAPHY OF ASTRONOMY, 1970-1979. R.A. Seal and S.S. Martin. Libraries Unlimited, Incorporated, Littleton, CO 80160. 1982. $37.50.

SCIENTIFIC AND TECHNICAL BOOKS AND SERIALS IN PRINT 1988: AN INDEX TO LITERATURE IN SCIENCE AND TECHNOLOGY. R.R. Bowker Company, 205 East 42nd Street, New York, NY 10017. (800) 521-8110 or (212) 916-1600. $175.00.

DIRECTORIES AND BIOGRAPHICAL SOURCES

AMERICAN ASTRONOMICAL SOCIETY MEMBERS. 2000 Florida Avenue, NW, Suite 300, Washington, DC 20009. (202) 659-0134. Annual. Available to members only.

THE BIOGRAPHICAL DICTIONARY OF SCIENTISTS: ASTRONOMERS. D. Abbott, editor. Peter Bedrick Books, 125 East 23rd Street, New York, NY 10010. 1984. $24.95.

DIRECTORY OF PHYSICS AND ASTRONOMY STAFF MEMBERS. American Institute of Physics, 335 East 45th Street, New York, NY 10017. Annual.

GENERAL WORKS

THE PLANET PLUTO. Anthony J. Whyte and Herbert A. Wise. Pergamon Press, Incorporated, Maxwell House, Fairview Park, Elmsford, NY 10523. (914) 592-7700. 1980. $25.50.

ONLINE DATA BASES

CA SEARCH. Chemical Abstracts Service, Post Office Box 3012, Columbus, OH 43210. Guide to chemical literature, 1972 to present. Inquire as to cost and availability.

DISSERTATION ABSTRACTS ONLINE. University Microfilms International, 300 North Zeeb Road, Ann Arbor, MI 48106. (800) 521-0600 or (313) 761-4700. Scope includes virtually all doctoral dissertations accepted at accredited American institutions from 1861 to present in 252 subject areas. Inquire as to cost and availability.

NASA. National Aeronautics and Space Administration, Scientific and Technical Information Branch, 300 7th Street, SW, Washington, DC 20546. Citations and abstracts of aerospace literature, 1962 to present. Inquire as to cost and availability.

NTIS. National Technical Information Service, 5285 Port Royal Road, Springfield, VA 22161. (703) 487-4630. Broad coverage of government sponsored research reports, 1964 to present. Inquire as to cost and availability.

SCISEARCH. Institute for Scientific Information, 3501 Market Street, Philadelphia, PA 19104. (800) 523-1850 or (215) 386-0100. Broad multidisciplinary title and author index to the international literature of science and technology, 1974 to present. Inquire as to cost and availability.

PERIODICALS

ASSOCIATION OF LUNAR AND PLANETARY OBSERVERS. Journal Association of Lunar and Planetary Observers, Box 16131, San Francisco, CA 94116. (415) 566-5786. 1947 to present. Quarterly. $15.00 per year.

ASTRONOMICAL JOURNAL. American Astronomical Society. Available from: American Institute of Physics, 335 East 45th Street, New York, NY 10017. 9212) 661-9404. $125.00 per year.

EARTH, MOON AND PLANETS: AN INTERNATIONAL JOURNAL OF COMPARATIVE PLANETOLOGY. D. Reidel Publishing Company, 190 Old Derby Street, Hingham, MA 02043. Nine times per year. $275.00 per year.

ICARUS: INTERNATIONAL JOURNAL OF THE SOLAR SYSTEM STUDIES. Academic Press, Incorporated, Orlando, FL 32887. (305) 345-4100. Monthly. $484.00 per year.

MERCURY. Astronomical Society of the Pacific, 1290 24th Avenue, San Francisco, CA 94122. (415) 661-8660. Bimonthly. $21.00 per year.

PLANETARY AND SPACE SCIENCE. Pergamon Press, Incorporated, Maxwell House, Fairview Park, Elmsford, NY 10523. (914) 592-7700. Monthly. $430.00 per year.

PLANETARY REPORT. Planetary Society, 65 North Catalina Avenue, Pasadena, CA 91106-2301. (818) 793-5100. 1980 to present. Bimonthly. $20.00 per year.

SOLAR SYSTEM RESEARCH (ENGLISH TRANSLATION OF ASTRONOMICHESKII VESTNIK). Consultants Bureau, 233

Spring Street, New York, NY 10013. (212) 620-8000. 1967 to present. Quarterly. $425.00 per year.

RESEARCH CENTERS AND INSTITUTES

LABORATORY FOR PLANETARY STUDIES. Cornell University, 302 Space Sciences Building, Ithaca, NY 14853. (607) 256-4971.

LUNAR AND PLANETARY INSTITUTE. 3303 NASA Road One, Houston, TX 77058. (713) 486-2139.

LUNAR AND PLANETARY LABORATORY. University of Arizona, Tucson, AZ 85721. (602) 621-6962.

PLUTONIUM

See also: ELEMENTS (CHEMICAL), NUCLEAR CHEMISTRY, NUCLEAR ENGINEERING, NUCLEAR REACTORS

ABSTRACT SERVICES AND INDEXES

APPLIED MECHANICS REVIEW. American Society of Mechanical Engineers, 345 East 47th Street, New York, NY 10017. (212) 705-7703. 1948 to present. Monthly. $360.00 per year.

APPLIED SCIENCE AND TECHNOLOGY INDEX. H.W. Wilson and Company, 950 University Avenue, Bronx, NY 10452. (800) 367-6670 or (212) 588-8400. Monthly. Inquire as to cost and availability.

CHEMICAL ABSTRACTS. American Chemical Society, Chemical Abstracts Service, Box 3012, Columbus, OH 43210. (614) 421-3600. 1907 to present. Weekly. $9500.00 per year.

CURRENT CONTENTS: ENGINEERING, TECHNOLOGY AND APPLIED SCIENCES. Institute for Scientific Information, 3501 Market Street, Philadelphia, PA 19104. (800) 523-1850 or (215) 386-0100. Weekly. $275.00 per year.

ENGINEERING INDEX MONTHLY AND AUTHOR INDEX. Engineering Information Inc., 345 East 47th Street, New York, NY 10017. (212) 705-7600. Monthly. $1560.00 per year.

INDEX TO SCIENTIFIC AND TECHNICAL PROCEEDINGS. Institute for Scientific Information, 3501 Market Street, Philadelphia, PA 19104. (800) 523-1850 or (215) 386-0100. 1978 to present. Monthly. $775.00 per year.

PHYSICS ABSTRACTS. Institution of Electrical Engineers. Available from: IEEE Service Center, 445 Hoes Lane, Piscataway, NJ 08854. 1898 to present. Bimonthly. $1700.00 per year.

SCIENCE CITATION INDEX. Institute for Scientific Information, 3501 Market Street, Philadelphia, PA 19104. (800) 523-1850 or (215) 386-0100. Six times per year. $6200.00 per year.

ANNUAL REVIEWS AND YEARBOOKS

ADVANCES IN NUCLEAR SCIENCE AND TECHNOLOGY. J.Lewins, editor. Plenum Publishing Corporation, 233 Spring Street, New York, NY 10013. (800) 221-9369. 1977 to presnet. Irregular. Inquire as to cost and availability.

ASSOCIATIONS AND PROFESSIONAL SOCIETIES

AMERICAN CHEMICAL SOCIETY. 1155 16th Street, N.W., Washington, DC 20036. (800) 424-6747.

AMERICAN INSTITUTE OF PHYSICS. 335 East 45th Street, New York, NY 10017. (212) 661-9494.

AMERICAN NUCLEAR SOCIETY. 555 North Kensington Avenue, La Grange, IL 60525. (312) 352-6611.

INSTITUTE OF NUCLEAR MATERIALS MANAGEMENT. 60 Revere Drive, Northbrook, IL 60062. (312) 480-9080.

DIRECTORIES AND BIOGRAPHICAL SOURCES

ENERGY INFORMATION CENTERS DIRECTORY. Public Affairs and Information Program, Atomic Industrial Forum, 7101 Wisconsin Avenue, Bethesda, MD 20814. (301) 654-9260. 1985. Free.

INTERNATIONAL RESEARCH CENTERS DIRECTORY 1988-89. Darren L. Smith, editor. Gale Research Company, Book Tower, Detroit, MI 48226. (800) 521-0707. 4th edition. 1987. $360.00.

1987 DIRECTORY OF ENGINEERING SOCIETIES AND RELATED ORGANIZATIONS. Gordon Davis, editor. Hemisphere Publishing Corporation, 1010 Vermont Avenue, NW, Washington, DC 20005. (800) 526-0275. 12th edition. 1987. $100.00.

RESEARCH CENTERS DIRECTORY 1988. Gale Research Company, Book Tower, Detroit, MI 48226. (800) 521-0707. 12th edition. 1987. $365.00 for set.

SCIENTIFIC AND TECHNICAL ORGANIZATIONS AND AGENCIES DIRECTORY. Margaret Labash Young, editor. Gale Research Company, Book Tower, Detroit, MI 48226. (800) 521-0707. 2nd edition. 1987. $185.00.

WHO'S WHO IN ENGINEERING. Gordon Davis, editor. Hemisphere Publishing Corporation, 1010 Vermont Avenue, NW, Washington, DC 20005. (800) 526-0275. 6th edition. 1985. $200.00.

GENERAL WORKS

INTRODUCTION TO NUCLEAR POWER. John G. Collier. Hemisphere Publishing Corporation, 1010 Vermont Avenue, NW, Washington, DC 20005. (800) 526-0275. 1987. $49.95.

NUCLEAR ENERGY TECHNOLOGY. Ronald Allen Knief. Hemisphere Publishing Corporation, 1010 Vermont Avenue, NW, Washington, DC 20005. (800) 526-0275. 1983. $48.00.

NUCLEAR FISSION REACTORS. I.R. Cameron. Plenum Publishing Corporation, 233 Spring Street, New York, NY 10013. (800) 221-9369. 1982. $49.50.

NUCLEAR REACTOR ENGINEERING. Samuel Glasstone. Van Nostrand Reinhold Company, Inc., 135 West 50th Street, New York, NY 10020. (800) 543-2681. 1980. $49.95.

NUCLEAR MATERIALS AND APPLICATIONS. Benjamin M. Ma. Van Nostrand Reinhold Company, Inc., 135 West 50th Street, New York, NY 10020. (800) 543-2681. 1982. $45.95.

PLUTONIUM CHEMISTRY. William T. Carnall and G.R. Choppin, editors. American Chemical Society, 1155 16th Street, N.W., Washington, DC 20036. (800) 424-6747. 1983. $54.95.

HANDBOOKS AND MANUALS

A GUIDE TO NUCLEAR POWER TECHNOLOGY: A RESOURCE FOR DECISION MAKING. F.J. Rahn and others. John Wiley and Sons, Inc., 605 Third Avenue, New York, NY 10158. (800) 526-5368. 1984. $85.95.

NUCLEAR ENGINEERING DATA BASES, STANDARDS AND NUMERICAL ANALYSIS. Jack Jedruch. Van Nostrand Reinhold Company, Inc., 135 West 50th Street, New York, NY 10020. (800) 543-2681. 1985. $59.95.

PLUTONIUM HANDBOOK. O.J. Wick. American Nuclear Society, 555 North Kensington Avenue, La Grange, IL 60525. (312) 352-6611. Two volume set. 1980. $98.00.

PLUTONIUM

ONLINE DATA BASES

CA SEARCH. Chemical Abstracts Service, P.O. Box 3012, Columbus, OH 43120. (800) 848-6538 or (614) 421-3600. Comprehensive guide to chemical literature, 1972 to present. Inquire as to online cost and availability.

COMPENDEX. Engineering Information, Inc., 345 East 47th Street, New York, NY 10017. (800) 221-1044 or (212) 705-7615. Engineering and technical literature, 1975 to present. Inquire as to online cost and availability.

DISSERTATION ABSTRACTS ONLINE. University Microfilms International, 300 North Zeeb Road, Ann Arbor, MI 48106. (800) 521-0600 or (313) 761-4700. Scope includes virtually all doctoral dissertations accepted at accredited American institutions from 1861 to present in over 250 subject areas. Inquire as to online cost and availability.

DOE ENERGY DATA BASE. U.S. Department of Energy, Office of Scientific and Technical Information, P.O. Box 62, Oak Ridge, TN 37831. (615) 576-6837. A database that covers all aspects of energy including the science and technology of energy. 1948 to present. Available through the DIALOG search service or DOE/RECON. Inquire as to online cost and availability.

ENERGYLINE. EIC/Intelligence, Inc., 48 West 38th Street, New York, NY 10018. (212) 944-8500. A database of resources on the scientific, engineering, political, and socioeconomic aspects of energy resources. 1976 to present. Inquire as to online cost and availability.

INSPEC. INSPEC Marketing Department, Institution of Electrical Engineers. Available from IEEE Service Center, 445 Hoes Lane, Piscataway, NJ 08854. (201) 981-0060. Online version of Physics Abstracts. Inquire as to online cost and availability.

NTIS. National Technical Information Service, 5285 Port Royal Road, Springfield, VA 22161. (703) 487-4630. Broad coverage of government sponsored research reports, 1964 to present. Inquire as to online cost and availability.

SCISEARCH. Institute for Scientific Information, 3501 Market Street, Philadelphia, PA 19104. (800) 523-1850 or (215) 386-0100. Broad multidisciplinary title and author index to the international literature of science and technology, 1974 to present. Inquire as to online cost and availability.

WILSONLINE. H.W. Wilson and Company, 950 University Avenue, Bronx, NY 10452. (800) 367-6770 or (212) 588-8400. Makes available online versions of the H.W. Wilson indexes including Applied Science and Technology Index, Business Periodicals Index and Readers' Guide to Periodical Literature. Approximately 1980 to present. Inquire as to online cost and availability.

PERIODICALS

AMERICAN NUCLEAR SOCIETY TRANSACTIONS. American Nuclear Society, 555 North Kensington Avenue, La Grange, IL 60525. (312) 352-6611. 1958 to present. Semiannual. $255.00 per year.

ANNALS OF NUCLEAR ENERGY. Pergamon Press, Inc., Maxwell House, Fairview Park, Elmsford, NY 10523. (914) 592-7700. 1954 to present. Monthly. $280.00 per year.

NUCLEAR ENGINEER. Institution of Nuclear Engineers, 1 Penerley Road, Nondon SE6 2LQ, England. 1959 to present. Bimonthly. $110.00 per year.

NUCLEAR ENGINEERING AND DESIGN. Elsevier Science Publishing Company, Inc., 52 Vanderbilt Avenue, New York, NY 10017. (212) 370-5520. 1965 to present. $160.00 per year.

NUCLEAR ENGINEERING INTERNATIONAL. Electrical-Electronic Press, Quadrant House, The Quadrant, Sutton, Surrey, SM2 5AS, England. 1956 to present. Monthly. $210.00 per year.

NUCLEAR SCIENCE AND ENGINEERING. American Nuclear Society, 555 North Kensington Avenue, La Grange, IL 60525. (312) 352-6611. 1956 to present. Monthly. $220.00 per year.

NUCLEAR TECHNOLOGY. American Nuclear Society, 555 North Kensington Avenue, La Grange, IL 60525. (312) 352-6611. 1965 to present. Monthly. $345.00 per year.

RESEARCH CENTERS AND INSTITUTES

ASSISTANT SECRETARY FOR NUCLEAR ENERGY. U.S. Department of Energy, 1000 Independence Avenue, SW, Washington, DC 20585. (202) 252-6450.

DEPARTMENT OF NUCLEAR ENGINEERING AND ENGINEERING PHYSICS. University of Wisconsin - Madison, 1500 Johnson Drive, Madison, WI 53706. (608) 263-1648.

INSTITUTE OF NUCLEAR SCIENCE AND ENGINEERING. Oregon State University, Radiation Center, 35th and Jefferson Streets, Corvallis, OR 97331. (503) 754-2341.

LABORATORY OF BASIC AND APPLIED NUCLEAR RESEARCH. University of Cincinnati, Department of Chemistry, Cincinnati, OH 45221. (513) 475-3652.

WHITESHELL NUCLEAR RESEARCH ESTABLISHMENT. Research Company, Atomic Energy of Canada Limited, Pinawa, MB, Canada ROE 1LO. (204) 753-2311.

SPECIFICATIONS AND STANDARDS

INFORMATION CENTER ON NUCLEAR STANDARDS. American Nuclear Society, 555 North Kensington Avenue, La Grange, IL 60525. (312) 352-6611. Standards for all aspects of the design, construction, operation, and maintenance of nuclear power plants.

POLYMERS

See also: ADHESIVES, CERAMICS, CHEMISTRY, COMPOSITES, MATERIALS SCIENCE, ORGANIC CHEMISTRY, PLASTICS, THERMOPLASTICS

ABSTRACT SERVICES AND INDEXES

APPLIED SCIENCE AND TECHNOLOGY INDEX. H.W. Wilson and Company, 950 University Avenue, Bronx, NY 10452. (800) 367-6670 or (212) 588-8400. Monthly. Inquire as to cost and availability.

CHEMICAL ABSTRACTS. American Chemical Society, Chemical Abstracts Service, Box 3012, Columbus, OH 43210. (614) 421-3600. 1907 to present. Weekly. $9500.00 per year.

CONFERENCE PAPERS INDEX. Cambridge Scientific Abstracts, 5161 River Road, Bethesda, MD 20816. 1972 to present. Monthly. Inquire as to cost and availability.

CURRENT CONTENTS: ENGINEERING, TECHNOLOGY AND APPLIED SCIENCES. Institute for Scientific Information, 3501 Market Street, Philadelphia, PA 19104. (800) 523-1850 or (215) 386-0100. Weekly. $275.00 per year.

ENGINEERING INDEX MONTHLY AND AUTHOR INDEX. Engineering Information Inc., 345 East 47th Street, New York, NY 10017. (212) 705-7600. Monthly. $1560.00 per year.

GENERAL SCIENCE INDEX. H.W. Wilson and Company, 950 University Avenue, Bronx, NY 10452. (800) 367-6670 or (212) 588-8400. 1978 to present. Monthly. Inquire as to cost and availability.

INDEX TO SCIENTIFIC AND TECHNICAL PROCEEDINGS. Institute for Scientific Information, 3501 Market Street,

Philadelphia, PA 19104. (800) 523-1850 or (215) 386-0100. 1978 to present. Monthly. $775.00 per year.

INDEX TO SCIENTIFIC REVIEWS. Institute for Scientific Information, 3501 Market Street, Philadelphia, PA 19104. (800) 523-1850 or (215) 386-0100. 1974 to present. Semi-annual. $550.00 per year.

SCIENCE CITATION INDEX. Institute for Scientific Information, 3501 Market Street, Philadelphia, PA 19104. (800) 523-1850 or (215) 386-0100. Six times per year. $6200.00 per year.

ASSOCIATIONS AND PROFESSIONAL SOCIETIES

AMERICAN CHEMICAL SOCIETY. 1155 16th Street, N.W., Washington, DC 20036. (800) 424-6747.

PLASTICS INSTITUTE OF AMERICA. Stevens Institute of Technology, Castle Point Station, Hoboken, NJ 07030. (201) 420-5553.

SOCIETY OF PLASTICS ENGINEERS, INC. 14 Fairfield Drive, Brookfield Center, CT 06805. (203) 775-0471.

SOCIETY OF THE PLASTICS INDUSTRY. 355 Lexington Avenue, New York, NY 10017. (212) 503-0600.

DIRECTORIES AND BIOGRAPHICAL SOURCES

AMERICAN MEN AND WOMEN OF SCIENCE. R.R. Bowker, Inc., Order Department, 245 West 17th Street, New York, NY 10011. (800) 521-8110. Eight volumes. 1986. $595.00 for set.

INTERNATIONAL RESEARCH CENTERS DIRECTORY 1988-89. Darren L. Smith, editor. Gale Research Company, Book Tower, Detroit, MI 48226. (800) 521-0707. 4th edition. 1987. $360.00.

1987 DIRECTORY OF ENGINEERING SOCIETIES AND RELATED ORGANIZATIONS. Gordon Davis, editor. Hemisphere Publishing Corporation, 1010 Vermont Avenue, NW, Washington, DC 20005. (800) 526-0275. 12th edition. 1987. $100.00.

RESEARCH CENTERS DIRECTORY 1988. Gale Research Company, Book Tower, Detroit, MI 48226. (800) 521-0707. 12th edition. 1987. $365.00 for set.

SCIENTIFIC AND TECHNICAL ORGANIZATIONS AND AGENCIES DIRECTORY. Margaret Labash Young, editor. Gale Research Company, Book Tower, Detroit, MI 48226. (800) 521-0707. 2nd edition. 1987. $185.00.

SOCIETY OF THE PLASTICS INDUSTRY - MEMBERSHIP DIRECTORY AND BUYERS GUIDE. 355 Lexington Avenue, New York, NY 10017. (212) 503-0600. Annual. $90.00.

WHO'S WHO IN ENGINEERING. Gordon Davis, editor. Hemisphere Publishing Corporation, 79 Madison Avenue, New York, NY 10016-7892. (800) 821-8312. Sixth edition. 1985. $200.00.

WHO'S WHO IN TECHNOLOGY. Research Publications, 12 Lunar Drive, Woodbridge, CT 06525. (203) 397-2600. 1986. Seven volume set, $545.00.

ENCYCLOPEDIAS AND DICTIONARIES

ENCYCLOPEDIA OF POLYMER SCIENCE AND ENGINEERING. John Wiley and Sons, Inc., 605 Third Avenue, New York, NY 10158. (800) 526-5368. Second edition. 1985 to present. $3500.00 for set.

THESAURUS OF SCIENTIFIC, TECHNICAL, AND ENGINEERING TERMS. Hemisphere Publishing Corporation, 1010 Vermont Avenue, NW, Washington, DC 20005. (800) 526-0275. 1988. $125.00.

GENERAL WORKS

FUNDAMENTAL PRINCIPLES OF POLYMERIC MATERIALS. Stephen L. Rosen. John Wiley and Sons, Inc., 605 Third Avenue, New York, NY 10158. (800) 526-5368. 1982. $35.95.

FUTURE TRENDS IN POLYMER SCIENCE AND TECHNOLOGY. Ezio Martuscelli and others. Technomic Publishing Company, Inc., 851 New Holland Avenue, Box 3535, Lancaster, PA 17604. (717) 291-5609. 1987. $49.00.

PLASTICS MATERIALS AND PROCESSING. S. Schwartz and S. Goodman. Van Nostrand Reinhold Company, Inc., 135 West 50th Street, New York, NY 10020. (800) 543-2681. 1982. $100.00.

POLYMER PROCESSING AND PROPERTIES. G. Astarita. Plenum Publishing Corporation, 233 Spring Street, New York, NY 10013. (800) 221-9369. 1984. $69.50.

TEXTBOOK OF POLYMER SCIENCE. Fred W. Billmeyer. John Wiley and Sons, Inc., 605 Third Avenue, New York, NY 10158. (800) 526-5368. 1984. $35.95.

HANDBOOKS AND MANUALS

PLASTICS ENGINEERING HANDBOOK OF THE SOCIETY OF THE PLASTICS INDUSTRY, INC. Joel Frados, editor. Van Nostrand Reinhold Company, Inc., 135 West 50th Street, New York, NY 10020. (800) 543-2681. 1985. $54.95.

PLASTICS TECHNOLOGY HANDBOOK. M. Chanda and S.K. Roy. Marcel Dekker Inc., 270 Madison Avenue, New York, NY 10016. (800) 228-1160. 1987. $99.75.

POLYMER HANDBOOK. Johannes Brandrup and E.H. Immergut. John Wiley and Sons, Inc., 605 Third Avenue, New York, NY 10158. (800) 526-5368. Second edition. 1975. $89.50.

ONLINE DATA BASES

CA SEARCH. Chemical Abstracts Service, P.O. Box 3012, Columbus, OH 43120. (800) 848-6538 or (614) 421-3600. Comprehensive guide to chemical literature, 1972 to present. Inquire as to online cost and availability.

COMPENDEX. Engineering Information, Inc., 345 East 47th Street, New York, NY 10017. (800) 221-1044 or (212) 705-7615. Engineering and technical literature, 1975 to present. Inquire as to online cost and availability.

DISSERTATION ABSTRACTS ONLINE. University Microfilms International, 300 North Zeeb Road, Ann Arbor, MI 48106. (800) 521-0600 or (313) 761-4700. Scope includes virtually all doctoral dissertations accepted at accredited American institutions from 1861 to present in over 250 subject areas. Inquire as to online cost and availability.

INSPEC. INSPEC Marketing Department, Institution of Electrical Engineers. Available from IEEE Service Center, 445 Hoes Lane, Piscataway, NJ 08854. (201) 981-0060. Online version of Physics Abstracts. Inquire as to online cost and availability.

NTIS. National Technical Information Service, 5285 Port Royal Road, Springfield, VA 22161. (703) 487-4630. Broad coverage of government sponsored research reports, 1964 to present. Inquire as to online cost and availability.

SCISEARCH. Institute for Scientific Information, 3501 Market Street, Philadelphia, PA 19104. (800) 523-1850 or (215) 386-0100. Broad multidisciplinary title and author index to the international literature of science and technology, 1974 to present. Inquire as to online cost and availability.

WILSONLINE. H.W. Wilson and Company, 950 University Avenue, Bronx, NY 10452. (800) 367-6770 or (212) 588-8400. Makes available online versions of the H.W. Wilson indexes including Applied Science and Technology Index, Business Periodicals Index and Readers' Guide to Periodical Literature.

Approximately 1980 to present. Inquire as to online cost and availability.

OTHER SOURCES

WHAT EVERY ENGINEER SHOULD KNOW ABOUT PLASTICS. Marcel Dekker Inc., 270 Madison Avenue, New York, NY 10016. (800) 228-1160. 1986. $24.95.

PERIODICALS

ADVANCES IN POLYMER TECHNOLOGY. John Wiley and Sons, Inc., 605 Third Avenue, New York, NY 10158. (800) 526-5368. 1977 to present. Quarterly. $100.00 per year.

CANADIAN PLASTICS. Southam Communications, Limited, 1450 Don Mills Road, Don Mills, ON, M3B 2X7, Canada. 1943 to present. Nine times per year. $45.00 per year.

HIGH PERFORMANCE PLASTICS. Elsevier Science Publishing Company, Inc., 52 Vanderbilt Avenue, New York, NY 10017. (212) 370-5520. 1983 to present. Monthly. $135.00 per year.

INTERNATIONAL JOURNAL OF ADHESION AND ADHESIVES. Butterworth's Publishing, 80 Montvale Avenue, Stoneham, MA 02180. (800) 325-4177. 1980 to present. Quarterly. $155.00 per year.

JOURNAL OF CELLULAR PLASTICS. Technomic Publishing Company, Inc., 851 New Holland Avenue, Box 3535, Lancaster, PA 17604. (717) 291-5609. 1965 to present. Bimonthly. $70.00 per year.

JOURNAL OF ELASTOMERS AND PLASTICS. Technomic Publishing Company, Inc., 851 New Holland Avenue, Box 3535, Lancaster, PA 17604. (717) 291-5609. 1969 to present. Quarterly. $115.00 per year.

JOURNAL OF PLASTIC FILM AND SHEETING. Technomic Publishing Company, Inc., 851 New Holland Avenue, Box 3535, Lancaster, PA 17604. (717) 291-5609. 1985 to present. Quarterly. $125.00 per year.

JOURNAL OF POLYMER SCIENCE: POLYMER CHEMISTRY EDITION; POLYMER PHYSICS EDITION; POLYMER LETTERS EDITION. John Wiley and Sons, Inc., 605 Third Avenue, New York, NY 10158. (800) 526-5368. 1962 to present. Monthly. $895.00 per year for all editions.

MODERN PLASTICS. McGraw-Hill Book Company, 1221 Avenue of the Americas, New York, NY 10020. (212) 512-2000. 1925 to present. Monthly. $30.00 per year.

PLASTICS COMPOUNDING: FOR RESIN PRODUCERS, FORMULATORS AND COMPOUNDERS. Harcourt, Brace and Jovanovich Publications, 262 Main Street, Chatham, NJ 07928. (201) 635-1671. 1978 to present. Bimonthly. $21.00 per year.

PLASTICS DESIGN FORUM. Harcourt, Brace and Jovanovich Publications, 262 Main Street, Chatham, NJ 07928. (201) 635-1671. 1976 to present. Bimonthly. $20.00 per year.

PLASTICS ENGINEERING. Society of Plastics Engineers, Inc., 14 Fairfield Drive, Brookfield Center, CT 06805. (203) 775-0471. 1945 to present. Monthly. $30.00 per year.

PLASTICS TECHNOLOGY. Bill Communications, Inc., 633 Third Avenue, New York, NY 10017. (212) 986-4800. 1955 to present. 13 times per year. $36.00 per year.

PLASTICS WORLD. Cahners Publishing Company, Inc., 221 Columbus Avenue, Boston, MA 02116. (617) 536-7780. 1942 to present. Monthly. $45.00 per year.

POLYMER COMPOSITES. Society of Plastics Engineers, Inc., 14 Fairfield Drive, Brookfield Center, CT 06805. (203) 775-0471. 1980 to present. Six times per year. $100.00 per year.

POLYMER PLASTICS TECHNOLOGY AND ENGINEERING. Marcel Dekker Inc., 270 Madison Avenue, New York, NY 10016. (800) 228-1160. 1970 to present. Quarterly. $220.00 per year.

POLYMERIC MATERIALS SCIENCE AND ENGINEERING. American Chemical Society, 1155 16th Street, N.W., Washington, DC 20036. (800) 424-6747. Semi-annual. $30.00 per year.

RESEARCH CENTERS AND INSTITUTES

PLASTICS INSTITUTE OF AMERICA. Stevens Institute of Technology, Castle Point Station, Hoboken, NJ 07030. (201) 420-5553.

POLYMER RESEARCH CENTER. University of Cincinnati, Mail Location 172, Cincinnati, OH 45221. (513) 475-2453.

POLYMER RESEARCH LABORATORY. University of Michigan, Dow Building, Ann Arbor, MI 48109. (313) 763-2240.

PORTLAND CEMENT

See: CEMENT

POWDER METALLURGY

See also: METALLURGY

ABSTRACT SERVICES AND INDEXES

APPLIED MECHANICS REVIEW. American Society of Mechanical Engineers, 345 East 47th Street, New York, NY 10017. (212) 705-7703. 1948 to present. Monthly. $360.00 per year.

APPLIED SCIENCE AND TECHNOLOGY INDEX. H.W. Wilson and Company, 950 University Avenue, Bronx, NY 10452. (800) 367-6670 or (212) 588-8400. Monthly. Inquire as to cost and availability.

CHEMICAL ABSTRACTS. American Chemical Society, Chemical Abstracts Service, Box 3012, Columbus, OH 43210. (614) 421-3600. 1907 to present. Weekly. $9500.00 per year.

CURRENT CONTENTS: ENGINEERING, TECHNOLOGY AND APPLIED SCIENCES. Institute for Scientific Information, 3501 Market Street, Philadelphia, PA 19104. (800) 523-1850 or (215) 386-0100. Weekly. $275.00 per year.

ENGINEERING INDEX MONTHLY AND AUTHOR INDEX. Engineering Information Inc., 345 East 47th Street, New York, NY 10017. (212) 705-7600. Monthly. $1560.00 per year.

ISMEC BULLETIN (Information Service in Mechanical Engineering). Cambridge Scientific Abstracts, 5161 River Road, Bethesda, MD 20816. (301) 951-1400. 1973 to present. Monthly. $450.00 per year.

INDEX TO SCIENTIFIC AND TECHNICAL PROCEEDINGS. Institute for Scientific Information, 3501 Market Street, Philadelphia, PA 19104. (800) 523-1850 or (215) 386-0100. 1978 to present. Monthly. $775.00 per year.

METALS ABSTRACTS AND METALS ABSTRACTS INDEX. American Society for Metals, Metals Park, OH 44073. (216) 338-5151. 1968 to present. Monthly. $1250.00 per year.

SCIENCE CITATION INDEX. Institute for Scientific Information, 3501 Market Street, Philadelphia, PA 19104. (800) 523-1850 or (215) 386-0100. Six times per year. $6200.00 per year.

WORLD ALUMINUM ABSTRACTS. Aluminum Association, Inc., 818 Connecticut Avenue, N.W., Washington, DC 20006. (202) 862-5156. 1968 to present. Monthly. $250.00 per year.

ASSOCIATIONS AND PROFESSIONAL SOCIETIES

AMERICAN POWDER METALLURGY INSTITUTE. 105 College Road, East, Princeton, NJ 08540. (609) 452-7700.

AMERICAN SOCIETY FOR METALS. Metals Park, OH 44073. (216) 338-5151.

METALLURICAL SOCIETY OF THE AMERICAN INSTITUTE OF MINING, METALLURGICAL AND PETROLEUM ENGINEERS, P.O. Box 411, 410 Commonwealth Drive, Warrendale, PA 15086. (412) 776-1535.

DIRECTORIES AND BIOGRAPHICAL SOURCES

INTERNATIONAL RESEARCH CENTERS DIRECTORY 1988-89. Darren L. Smith, editor. Gale Research Company, Book Tower, Detroit, MI 48226. (800) 521-0707. 4th edition. 1987. $360.00.

1987 DIRECTORY OF ENGINEERING SOCIETIES AND RELATED ORGANIZATIONS. Gordon Davis, editor. Hemisphere Publishing Corporation, 1010 Vermont Avenue, NW, Washington, DC 20005. (800) 526-0275. 12th edition. 1987. $100.00.

RESEARCH CENTERS DIRECTORY 1988. Gale Research Company, Book Tower, Detroit, MI 48226. (800) 521-0707. 12th edition. 1987. $365.00 for set.

SCIENTIFIC AND TECHNICAL ORGANIZATIONS AND AGENCIES DIRECTORY. Margaret Labash Young, editor. Gale Research Company, Book Tower, Detroit, MI 48226. (800) 521-0707. 2nd edition. 1987. $185.00.

WHO'S WHO IN ENGINEERING. Gordon Davis, editor. Hemisphere Publishing Corporation, 1010 Vermont Avenue, NW, Washington, DC 20005. (800) 526-0275. 6th edition. 1985. $200.00.

WHO'S WHO IN POWDER METALLURGY. American Powder Metallurgy Institute. 105 College Road, East, Princeton, NJ 08540. (609) 452-7700. Annual. $50.00.

ENCYCLOPEDIAS AND DICTIONARIES

THESAURUS OF SCIENTIFIC, TECHNICAL, AND ENGINEERING TERMS. Hemisphere Publishing Corporation, 1010 Vermont Avenue, NW, Washington, DC 20005. (800) 526-0275. 1988. $125.00.

GENERAL WORKS

MODERN DEVELOPMENTS IN POWDER METALLURGY. E.N. Aqua and C.I. Whitman, editors. American Powder Metallurgy Institute, 105 College Road, East, Princeton, NJ 08540. 1985. Three volume set. $200.00 for set.

POWDER METALLURGY: APPLICATIONS, ADVANTAGES, AND LIMITATIONS. American Society for Metals, Metals Park, OH 44073. (800) 368-9800. 1983. $60.00.

POWDER METALLURGY PROCESSING: NEW TECHNIQUES AND ANALYSES. Howard A. Kuhn. Academic Press, Inc., 6277 Sea Harbor Drive, Orlando, FL 32821. (800) 321-5068. 1978. $56.50.

POWDER METALLURGY SCIENCE. Metal Powder Industries Federation, 105 College Road, East, Princeton, NJ 08540. (609) 452-7700. 1984. $35.00.

HANDBOOKS AND MANUALS

HANDBOOK OF POWDER METALLURGY. H.H. Hausner and M.K. Mal. Chemical Publishing Company, 80 8th Avenue, New York, NY 10011. (212) 255-1950. 2nd edition. 1982. $85.00.

METALS HANDBOOK. American Society for Metals, Metals Park, OH 44073. (800) 368-9800 or (216) 338-4634. Nineth edition. 1988. Fourteen volumes. $1310.00 for set.

ONLINE DATA BASES

CA SEARCH. Chemical Abstracts Service, P.O. Box 3012, Columbus, OH 43120. (800) 848-6538 or (614) 421-3600. Comprehensive guide to chemical literature, 1972 to present. Inquire as to online cost and availability.

COMPENDEX. Engineering Information, Inc., 345 East 47th Street, New York, NY 10017. (800) 221-1044 or (212) 705-7615. Engineering and technical literature, 1975 to present. Inquire as to online cost and availability.

DISSERTATION ABSTRACTS ONLINE. University Microfilms International, 300 North Zeeb Road, Ann Arbor, MI 48106. (800) 521-0600 or (313) 761-4700. Scope includes virtually all doctoral dissertations accepted at accredited American institutions from 1861 to present in over 250 subject areas. Inquire as to online cost and availability.

INSPEC. INSPEC Marketing Department, Institution of Electrical Engineers. Available from IEEE Service Center, 445 Hoes Lane, Piscataway, NJ 08854. (201) 981-0060. Online version of Physics Abstracts. Inquire as to online cost and availability.

NTIS. National Technical Information Service, 5285 Port Royal Road, Springfield, VA 22161. (703) 487-4630. Broad coverage of government sponsored research reports, 1964 to present. Inquire as to online cost and availability.

SCISEARCH. Institute for Scientific Information, 3501 Market Street, Philadelphia, PA 19104. (800) 523-1850 or (215) 386-0100. Broad multidisciplinary title and author index to the international literature of science and technology, 1974 to present. Inquire as to online cost and availability.

WILSONLINE. H.W. Wilson and Company, 950 University Avenue, Bronx, NY 10452. (800) 367-6770 or (212) 588-8400. Makes available online versions of the H.W. Wilson indexes including Applied Science and Technology Index, Business Periodicals Index and Readers' Guide to Periodical Literature. Approximately 1980 to present. Inquire as to online cost and availability.

PERIODICALS

JOURNAL OF METALS. American Institute of Mining, Metallurgical, and Petroleum Engineers, Inc., Metallurgical Society, 420 Commonwealth Drive, Warrendale, PA 15086. (412) 776-9086. 1949 to present. Monthly. $40.00 per year.

LIGHT METAL AGE. Fellom Publishing Company, 693 Mission Street, San Francisco, CA 94105. (415) 781-1431. 1942 to present. Bimonthly. $20.00.

METAL PROGRESS. American Society for Metals, Metals Park, OH 44073. (216) 338-5151. 1930 to present. Monthly. $40.00.

METALLURGICAL TRANSACTIONS. Metallurgical Society of AIME, 420 Commonwealth Drive, Warrendale, PA 15086. (412) 776-9080. 1970 to present. Monthly. $95.00 per year.

METALS WEEK. McGraw-Hill Book Company, 1221 Avenue of the Americas, New York, NY 10020. (212) 997-2823. 1930 to present. Weekly. $597.00 per year.

METALWORKING DIGEST. Philos Publications, Inc., 1 East Chase Street, Baltimore, MD 21202. (301) 361-9060. 1969 to present. Nine times per year. $9.00 per year.

METIFAX MAGAZINE. Huebner Publications Inc., 6521 Davis Industrial Parkway, Solon, OH 44139. (216) 248-1125. 1956 to present. Monthly. $40.00 per year.

POWDER METALLURGY. Institute of Metals, 1 Carlton House Terrace, London, SW1Y 5DB, England. 1958 to present. Quarterly. $65.00 per year.

RESEARCH CENTERS AND INSTITUTES

CANADIAN INSTITUTE OF METALWORKING. 1276 Sandhill Drive, P.O. Box 7317, Ancaster, ON, Canada L9G 3N6.

COOPERATIVE PROGRAM IN METALLURGY. Pennsylvania State University, 208A Steidle Building, University Park, PA 16802. (814) 865-5446.

DEPARTMENT OF MATERIALS SCIENCES AND ENGINEERING. University of Florida, Gainesville, FL 32601. (904) 392-1454.

INSTITUTE OF MATERIALS SCIENCE. University of Connecticut, Storrs, CT 06268. (203) 486-4623.

POWER SYSTEMS

See: ELECTRIC POWER ENGINEERING

PRECIPITATION

See: HAIL, RAIN, SNOW

PRESSURE VESSELS

See also: BOILERS, ELECTRIC POWER ENGINEERING, HEAT EXCHANGERS, HEAT TRANSFER, MECHANICAL ENGINEERING, NUCLEAR ENGINEERING, THERMODYNAMICS

ABSTRACT SERVICES AND INDEXES

APPLIED MECHANICS REVIEW. American Society of Mechanical Engineers, 345 East 47th Street, New York, NY 10017. (212) 705-7703. 1948 to present. Monthly. $360.00 per year.

APPLIED SCIENCE AND TECHNOLOGY INDEX. H.W. Wilson and Company, 950 University Avenue, Bronx, NY 10452. (800) 367-6670 or (212) 588-8400. Monthly. Inquire as to cost and availability.

CHEMICAL ABSTRACTS. American Chemical Society, Chemical Abstracts Service, Box 3012, Columbus, OH 43210. (614) 421-3600. 1907 to present. Weekly. $9500.00 per year.

CURRENT CONTENTS: ENGINEERING, TECHNOLOGY AND APPLIED SCIENCES. Institute for Scientific Information, 3501 Market Street, Philadelphia, PA 19104. (800) 523-1850 or (215) 386-0100. Weekly. $275.00 per year.

ENGINEERING INDEX MONTHLY AND AUTHOR INDEX. Engineering Information Inc., 345 East 47th Street, New York, NY 10017. (212) 705-7600. Monthly. $1560.00 per year.

ISMEC BULLETIN (Information Service in Mechanical Engineering). Cambridge Scientific Abstracts, 5161 River Road, Bethesda, MD 20816. (301) 951-1400. 1973 to present. Monthly. $450.00 per year.

PHYSICS ABSTRACTS. Institution of Electrical Engineers. Available from: IEEE Service Center, 445 Hoes Lane, Piscataway, NJ 08854. 1898 to present. Bimonthly. $1700.00 per year.

PHYSICS BRIEFS. Physik Verlag GmbH, Postfach 1260/1280, D-6940 Weinheim, West Germany. (212) 661-9404. 1920 to present. Twenty-six times per year. $1250.00 per year.

SCIENCE CITATION INDEX. Institute for Scientific Information, 3501 Market Street, Philadelphia, PA 19104. (800) 523-1850 or (215) 386-0100. Six times per year. $6200.00 per year.

ASSOCIATIONS AND PROFESSIONAL SOCIETIES

AMERICAN INSTITUTE OF CHEMICAL ENGINEERS. 345 East 47th Street, New York, NY 10017. (212) 705-7703.

AMERICAN SOCIETY OF HEATING, REFRIGERATING AND AIR CONDITIONING ENGINEERS, INC., 1791 Tullie Circle, Atlanta, GA 30329. (404) 636-8400.

AMERICAN SOCIETY OF MECHANICAL ENGINEERS. 345 East 47th Street, New York, NY 10017. (212) 705-7703.

DIRECTORIES AND BIOGRAPHICAL SOURCES

AMERICAN MEN AND WOMEN OF SCIENCE. R.R. Bowker, Inc., Order Department, 245 West 17th Street, New York, NY 10011. (800) 521-8110. Eight volumes. 1986. $595.00 for set.

1987 DIRECTORY OF ENGINEERING SOCIETIES AND RELATED ORGANIZATIONS. Gordon Davis, editor. Hemisphere Publishing Corporation, 1010 Vermont Avenue, NW, Washington, DC 20005. (800) 526-0275. 12th edition. 1987. $100.00.

RESEARCH CENTERS DIRECTORY 1988. Gale Research Company, Book Tower, Detroit, MI 48226. (800) 521-0707. 12th edition. 1987. $365.00 for set.

WHO'S WHO IN ENGINEERING. Gordon Davis, editor. Hemisphere Publishing Corporation, 1010 Vermont Avenue, NW, Washington, DC 20005. (800) 526-0275. 6th edition. 1985. $200.00.

GENERAL WORKS

FUNDAMENTALS OF HEAT EXCHANGER AND PRESSURE VESSEL TECHNOLOGY. J.P. Gupta. Hemisphere Publishing Corporation, 79 Madison Avenue, New York, NY 10016-7892. (800) 821-8312. 1986. $49.95.

HEAT TRANSFER EQUIPMENT DESIGN. R.K. Shah and others. Hemisphere Publishing Corporation, 79 Madison Avenue, New York, NY 10016-7892. 1988. $135.00.

INTRODUCTION TO HEAT TRANSFER. F. Incropera and D. Dewitt. John Wiley and Sons, Inc., 605 Third Avenue, New York, NY 10158. (800) 526-5368. 1985. $39.95.

MANAGING STEAM: AN ENGINEERING GUIDE TO INDUSTRIAL, COMMERCIAL AND UTILITY SYSTEMS. Jason Makansi, editor. Hemisphere Publishing Corporation, 79 Madison Avenue, New York, NY 10016-7892. Order from: Taylor and Francis/Hemisphere Distribution Center, 242 Cherry Street, Philadelphia, PA 19106-1906. (800) 821-8312. 1986. $35.00.

POWER PLANT SYSTEM DESIGN. K.W. Li and A.P. Priddy. John Wiley and Sons, Inc., 605 Third Avenue, New York, NY 10158. (800) 526-5368. 1985. $44.50.

PRESSURE VESSEL SYSTEMS; A USER'S GUIDE TO SAFE OPERATIONS AND MAINTENANCE. Anthony L. Kohan. McGraw-Hill Book Company, 1221 Avenue of the Americas, New York, NY 10020. (212) 512-2000. 1987. $42.50.

THEORY AND DESIGN OF PRESSURE VESSELS. John F. Harvey. Van Nostrand Reinhold Company, Inc., 135 West 50th Street, New York, NY 10020. (800) 543-2681. 1985. $51.95.

HANDBOOKS AND MANUALS

HANDBOOK OF HEAT TRANSFER APPLICATIONS. W.M. Rohsenow and others. McGraw-Hill Book Company, 1221 Avenue of the Americas, New York, NY 10020. (212) 512-2000. Second edition. 1986. $79.50.

HEAT EXCHANGER SOURCEBOOK. J.W. Palen, editor. Hemisphere Publishing Corporation, 79 Madison Avenue, New York, NY 10016-7892. (800) 821-8312. 1986. $59.95.

PRESSURE DESIGN HANDBOOK. Henry Bedner. McGraw-Hill Book Company, 1221 Avenue of the Americas, New York, NY 10020. (212) 512-2000. Second edition. 1985. $49.95.

STEAM TABLES: THERMODYNAMICS PROPERTIES OF WATER INCLUDING VAPOR, LIQUID, AND SOLID PHASES. J.H. Keenan and others. John Wiley and Sons, Inc., 605 Third Avenue, New York, NY 10158. (800) 526-5368. 1969. $44.95.

ONLINE DATA BASES

CA SEARCH. Chemical Abstracts Service, P.O. Box 3012, Columbus, OH 43120. (800) 848-6538 or (614) 421-3600. Comprehensive guide to chemical literature, 1972 to present. Inquire as to online cost and availability.

COMPENDEX. Engineering Information, Inc., 345 East 47th Street, New York, NY 10017. (800) 221-1044 or (212) 705-7615. Engineering and technical literature, 1975 to present. Inquire as to online cost and availability.

INSPEC. INSPEC Marketing Department, Institution of Electrical Engineers. Available from IEEE Service Center, 445 Hoes Lane, Piscataway, NJ 08854. (201) 981-0060. Online version of Physics Abstracts. Inquire as to online cost and availability.

NTIS. National Technical Information Service, 5285 Port Royal Road, Springfield, VA 22161. (703) 487-4630. Broad coverage of government sponsored research reports, 1964 to present. Inquire as to online cost and availability.

SCISEARCH. Institute for Scientific Information, 3501 Market Street, Philadelphia, PA 19104. (800) 523-1850 or (215) 386-0100. Broad multidisciplinary title and author index to the international literature of science and technology, 1974 to present. Inquire as to online cost and availability.

WILSONLINE. H.W. Wilson and Company, 950 University Avenue, Bronx, NY 10452. (800) 367-6770 or (212) 588-8400. Makes available online versions of the H.W. Wilson indexes including Applied Science and Technology Index, Business Periodicals Index and Readers' Guide to Periodical Literature. Approximately 1980 to present. Inquire as to online cost and availability.

PERIODICALS

ASHRAE JOURNAL. American Society of Heating, Refrigerating and Air Conditioning Engineers, Inc., 1791 Tullie Circle, Atlanta, GA 30329. (404) 636-8400. 1914 to present. Monthly. $35.00 per year.

HEAT TRANSFER ENGINEERING. Hemisphere Publishing Corporation, 79 Madison Avenue, New York, NY 10016-7892. (800) 821-8312. 1979 to present. Quarterly. $97.50.

HEATING/PIPING/AIR CONDITIONING. Penton-IPC, Reinhold Publishing Division, 600 Summer Street, Box 1361, Stamford, CT 06904. (203) 348-7531. 1929 to present. Monthly. $35.00 per year.

JOURNAL OF HEAT TRANSFER. American Society of Mechanical Engineers, 345 East 47th Street, New York, NY 10017. (212) 705-7703. 1970 to present. Quarterly. $100.00 per year.

JOURNAL OF PRESSURE VESSEL TECHNOLOGY. American Society of Mechanical Engineers, 345 East 47th Street, New York, NY 10017. (212) 705-7703. 1974 to present. Quarterly. $80.00 per year.

PLANT ENGINEERING. Technical Publishing, Box 1030, 1301 South Grove Avenue, Barrington, IL 60010. (312) 381-1840. 1947 to present. Semi-monthly. $50.00 per year.

POWER. McGraw-Hill Book Company, 1221 Avenue of the Americas, New York, NY 10020. (212) 512-2000. 1882 to present. Monthly. $12.00 per year.

STEAM POWER. Kirk Enterprises, Limited, Midlands Steam Centre, 106a Derby Road, Loughborough LE11 OAG, England. 1949 to present. Quarterly. $20.00 to present.

RESEARCH CENTERS AND INSTITUTES

HEAT TRANSFER LABORATORY. Massachusetts Institute of Technology, 77 Massachusetts Avenue, Cambridge, MA 02139. (716) 253-2248.

HEAT TRANSFER RESEARCH. University of Wisconsin, Milwaukee, P.O. Box 784, Milwaukee, WI 53201. (414) 963-5001.

HEAT TRANSFER RESEARCH FACILITY. Columbia University, 632 West 125th Street, New York, NY 10027. (212) 280-4163.

INSTITUTE OF THERMO-FLUID ENGINEERING AND SCIENCE. Lehigh University, Whitaker Laboratory, Bethlehem, PA 18015. (215) 861-4091.

SPECIFICATIONS AND STANDARDS

PRESSURE VESSELS: THE ASME CODE SIMPLIFIED. Robert Chuse. McGraw-Hill Book Company, 1221 Avenue of the Americas, New York, NY 10020. (212) 512-2000. Sixth edition. 1984. $35.00.

PRESTRESSED CONCRETE

See: CONCRETE

PROBABILITY

See also: MARKOV PROCESSES, MATHEMATICS, STATISTICS, STOCHASTIC PROCESSES

ABSTRACT SERVICES AND INDEXES

APPLIED MECHANICS REVIEW. American Society of Mechanical Engineers, 345 East 47th Street, New York, NY 10017. (212) 705-7703. 1948 to present. Monthly. $360.00 per year.

APPLIED SCIENCE AND TECHNOLOGY INDEX. H.W. Wilson and Company, 950 University Avenue, Bronx, NY 10452. (800) 367-6670 or (212) 588-8400. Monthly. Inquire as to cost and availability.

COMPUMATH CITATION INDEX. Institute for Scientific Information, 3501 Market Street, Philadelphia, PA 19104. (800) 523-1850 or (215) 386-0100. Three times per year. $875.00 per year.

CURRENT MATHEMATICAL PUBLICATIONS. American Mathematical Society, P.O. Box 6248, Providence, RI 02940. (800) 556-7774 or (401) 272-9500. 1969 to present. Seventeen times per year. $230.00 per year.

ENGINEERING INDEX MONTHLY AND AUTHOR INDEX. Engineering Information Inc., 345 East 47th Street, New York, NY 10017. (212) 705-7600. Monthly. $1560.00 per year.

MATHEMATICAL REVIEWS: A REVIEWING JOURNAL COVERING THE WORLD LITERATURE OF MATHEMATICAL RESEARCH. American Mathematical Society, P.O. Box 6248, Providence, RI 02940. (800) 7774 or (401) 272-9500. 1940 to present. Monthly. $2800.00 per year.

PHYSICS ABSTRACTS. Institution of Electrical Engineers. Available from: IEEE Service Center, 445 Hoes Lane, Piscataway, NJ 08854. 1898 to present. Bimonthly. $1700.00 per year.

SCIENCE CITATION INDEX. Institute for Scientific Information, 3501 Market Street, Philadelphia, PA 19104. (800) 523-1850 or (215) 386-0100. Six times per year. $6200.00 per year.

ANNUAL REVIEWS AND YEARBOOKS

ADVANCES IN APPLIED MATHEMATICS. Academic Press, Inc., 6277 Sea Harbor Drive, Orlando, FL 32821. (800) 321-5068. Irregular. Price varies, inquire.

ASSOCIATIONS AND PROFESSIONAL SOCIETIES

AMERICAN MATHEMATICAL SOCIETY. P.O. Box 6248, Providence, RI 02940. (401) 272-9500.

AMERICAN STATISTICAL ASSOCIATION. 806 15th Street, N.W., Suite 640, Washington, DC 20005. (202) 393-3253.

INSTITUTE OF MATHEMATICAL STATISTICS, 3401 Investment Boulevard, Suite 7, Hayward, CA 94545. (415) 783-8141. MATHEMATICAL ASSOCIATION OF AMERICA. 1529 18th Street, N.W., Washington, DC 20036. (202) 387-5200.

SOCIETY FOR INDUSTRIAL AND APPLIED MATHEMATICS. 1400 Architects Building, 117 South 17th Street, Philadelphia, PA 19103. (215) 564-2929.

DIRECTORIES AND BIOGRAPHICAL SOURCES

INTERNATIONAL RESEARCH CENTERS DIRECTORY 1988-89. Darren L. Smith, editor. Gale Research Company, Book Tower, Detroit, MI 48226. (800) 521-0707. 4th edition. 1987. $360.00.

RESEARCH CENTERS DIRECTORY 1988. Gale Research Company, Book Tower, Detroit, MI 48226. (800) 521-0707. 12th edition. 1987. $365.00 for set.

SCIENTIFIC AND TECHNICAL ORGANIZATIONS AND AGENCIES DIRECTORY. Margaret Labash Young, editor. Gale Research Company, Book Tower, Detroit, MI 48226. (800) 521-0707. 2nd edition. 1987. $185.00.

ENCYCLOPEDIAS AND DICTIONARIES

ENCYCLOPEDIA OF STATISTICAL SCIENCES. Samuel Kotz and others, editors. John Wiley and Sons, Inc., 605 Third Avenue, New York, NY 10158. (800) 526-5368. 1982. Two volumes. $175.00 for set.

ENCYCLOPEDIC DICTIONARY OF MATHEMATICS. Kiyosi Ito, editor. MIT Press, 55 Howard Street, Cambridge, MA 02142. (617) 253-2884. 2nd edition. 1987. Four volumes. $350.00 for set.

GENERAL WORKS

NUMERICAL METHODS IN ENGINEERING AND APPLIED SCIENCE. Bruce Irons. John Wiley and Sons, Inc., 605 Third Avenue, New York, NY 10158. (800) 526-5368. 1987. $32.95.

PROBABILITY AND STATISTICS FOR ENGINEERING AND THE SCIENCES. Jay L. Devore. Wadsworth Publishing Company, 10 Davis Drive, Belmont, CA 94002. (415) 595-2350. 1986. $32.50.

PROBABILITY AND STOCHASTIC PROCESSES FOR ENGINEERS. Carl W. Helstrom. Macmillan Publishing Company, Inc., 866 Third Avenue, New York, NY 10022. (800) 257-5755. 1984. $39.95.

PROBABILITY THEORY AND MATHEMATICAL STATISTICS WITH APPLICATIONS. W. Grossman and others. Kluwer Academic Publishers, 190 Old Derby Street, Hingham, MA 02043. (617) 749-5262. 1987. $89.00.

STOCHASTIC PROCESSES. Sheldon M. Ross. John Wiley and Sons, Inc., 605 Third Avenue, New York, NY 10158. (800) 526-5368. 1982. $42.95.

HANDBOOKS AND MANUALS

CRC HANDBOOK OF MATHEMATICAL SCIENCES. William H. Beyer, editor. CRC Press, 2000 Corporate Boulevard, Boca Raton, FL 33431. (800) 272-7737. 6th edition. 1987. $64.95.

CRC HANDBOOK OF TABLES FOR PROBABILITY AND STATISTICS. CRC Press, 2000 Corporate Boulevard, Boca Raton, FL 33431. (800) 272-7737. 2nd edition. 1968. $49.95.

STANDARD MATHEMATICAL TABLES. William Beyer, editor. CRC Press, 2000 Corporate Boulevard, Boca Raton, FL 33431. (800) 272-7737. 28th edition. 1987. $24.95.

ONLINE DATA BASES

COMPENDEX. Engineering Information, Inc., 345 East 47th Street, New York, NY 10017. (800) 221-1044 or (212) 705-7615. Engineering and technical literature, 1975 to present. Inquire as to online cost and availability.

DISSERTATION ABSTRACTS ONLINE. University Microfilms International, 300 North Zeeb Road, Ann Arbor, MI 48106. (800) 521-0600 or (313) 761-4700. Scope includes virtually all doctoral dissertations accepted at accredited American institutions from 1861 to present in over 250 subject areas. Inquire as to online cost and availability.

INSPEC. INSPEC Marketing Department, Institution of Electrical Engineers. Available from IEEE Service Center, 445 Hoes Lane, Piscataway, NJ 08854. (201) 981-0060. Online version of Physics Abstracts. Inquire as to online cost and availability.

MATHFILE. American Mathematical Society, P.O. Box 6248, Providence, RI 02940. (800) 556-7774 or (401) 272-9500. An online version of Mathematical Reviews. 1973 to present. Inquire as to online cost and availability.

NTIS. National Technical Information Service, 5285 Port Royal Road, Springfield, VA 22161. (703) 487-4630. Broad coverage of government sponsored research reports, 1964 to present. Inquire as to online cost and availability.

SCISEARCH. Institute for Scientific Information, 3501 Market Street, Philadelphia, PA 19104. (800) 523-1850 or (215) 386-0100. Broad multidisciplinary title and author index to the international literature of science and technology, 1974 to present. Inquire as to online cost and availability.

PERIODICALS

ANNALS OF STATISTICS. Institute of Mathematical Statistics, 3401 Investment Boulevard, Suite 7, Hayward, CA 94545. (415) 783-8141. 1973 to present. Quarterly. $85.00 per year.

APPLIED MATHEMATICS AND COMPUTATION. Elsevier Science Publishing Company, Inc., 52 Vanderbilt Avenue, New York, NY 10017. (212) 370-5520. 1975 to present. Monthly. $355.00 per year.

COMMUNICATIONS IN STATISTICS, PART A: THEORY AND METHODS. Marcel Dekker Inc., Journals Division, 270 Madison Avenue, New York, NY 10016. (800) 228-1160. 1973 to present. Monthly. $550.00 per year.

JASA. JOURNAL OF THE AMERICAN STATISTICAL ASSOCIATION. American Statistical Association. 806 15th Street, N.W., Suite 640, Washington, DC 20005. (202) 393-3253. 1888 to present. Quarterly. $70.00 per year.

PROBABILITY AND RELATED FIELDS. Springer-Verlag New York, Inc., 175 Fifth Avenue, New York, NY 10010. (800) 526-7254. 1962 to present. Monthly. $660.00 per year.

PROBABILITY IN THE ENGINEERING AND INFORMATIONAL SCIENCES. Cambridge University Press, 32 East 57th Street, New York, NY 10022. (800) 872-7423. 1987 to present. Quarterly. $110.00 per year.

SIAM JOURNAL OF APPLIED MATHEMATICS. Society for Industrial and Applied Mathematics, 1405 Architects Building, 117 South 17th Street. Philadelphia, PA 19103. (215) 564-2929. 1953 to present. Bimonthly. $130.00 per year.

SIAM JOURNAL OF MATHEMATICAL ANALYSIS. Society for Industrial and Applied Mathematics, 1405 Architects Building, 117 South 17th Street. Philadelphia, PA 19103. (215) 564-2929. 1964 to present. Bimonthly. $120.00 per year.

SIAM JOURNAL OF SCIENTIFIC AND STATISTICAL COMPUTING. Society for Industrial and Applied Mathematics, 1405 Architects Building, 117 South 17th Street. Philadelphia, PA 19103. (215) 564-2929. 1980 to present. Quarterly. $78.00 per year.

SIAM REVIEW. Society for Industrial and Applied Mathematics, 1405 Architects Building, 117 South 17th Street. Philadelphia, PA 19103. (215) 564-2929. 1959 to present. Quarterly. $82.00 per year.

THEORY OF PROBABILITY AND ITS APPLICATIONS. Society for Industrial and Applied Mathematics, 1405 Architects Building, 117 South 17th Street. Philadelphia, PA 19103. (215) 564-2929. 1956 to present. Quarterly. $216.00 per year.

THEORY OF PROBABILITY AND MATHEMATICAL STATISTICS. American Mathematical Society, P.O. Box 6248, Providence, RI 02940. (401) 272-9500. 1974 to present. Biannual. $225.00 per year.

RESEARCH CENTERS AND INSTITUTES

CENTER FOR APPLIED MATHEMATICS. University of Georgia, Tucker Hall, Athens, GA 30602. (404) 542-3491.

CENTER FOR MATHEMATICAL SCIENCE RESEARCH. Rutgers University, New Brunswick, NJ 08903. (201) 932-3117.

CENTER FOR PURE AND APPLIED MATHEMATICS. University of Californai, Berkeley, 977 Evans Hall, Berkeley, CA 94720. (415) 642-0116.

INSTITUTE FOR MATHEMATICS AND ITS APPLICATIONS. University of Minnesota, 514 Vincent, 206 Church Street, S.E., Minneapolis, MN 55455. (612) 624-6066.

INSTITUTE OF APPLIED MATHEMATICS. University of Missouri - Rolla, Rolla, MO 65401. (314) 341-4151.

LABORATORY FOR RESEARCH IN STATISTICS AND PROBABILITY. Carleton University, Room 611, Art Tower, Ottawa, ON, Canada K1S 5B6. (613) 564-6752.

PRODUCTION ENGINEERING

See: INDUSTRIAL ENGINEERING

PROGRAMMING

See: COMPUTER PROGRAMMING, PROGRAMMING LANGUAGES

PROGRAMMING LANGUAGES

See also: ARTIFICIAL INTELLIGENCE, COMPUTER OPERATING SYSTEMS, COMPUTER PROGRAMMING, COMPUTERS, PARELLEL COMPUTERS, SOFTWARE, SOFTWARE ENGINEERING, SYSTEMS ANALYSIS

ABSTRACT SERVICES AND INDEXES

APPLIED SCIENCE AND TECHNOLOGY INDEX. H.W. Wilson and Company, 950 University Avenue, Bronx, NY 10452. (800) 367-6670 or (212) 588-8400. Monthly. Inquire as to cost and availability.

COMPUTER AND CONTROL ABSTRACTS. Institute of Electrical Engineers. Available from: Institute of Electrical and Electronics Engineers. IEEE Service Center, 445 Hoes Lane, Piscataway, NJ 08854. Semimonthly. $775.00 per year.

COMPUTER AND INFORMATION SYSTEMS: AN ABSTRACT JOURNAL PERTAINING TO THE THEORY, DESIGN, FABRICATION AND APPLICATION OF COMPUTER AND INFORMATION SYSTEMS. Cambridge Scientific Abstracts, 5161 River Road, Bethesda, MD 20816. 1972 to present. Semi-monthly. Inquire as to cost and availability.

COMPUTER CONTENTS: THE BIWEEKLY COMPILATION OF TABLES OF CONTENTS FROM COMPUTER, ELECTRONIC AND TELECOMMUNICATIONS MAGAZINES, JOURNALS AND TRANACTIONS. Find/SVP, 500 Fifth Avenue, New York, NY 101110. (800) 346-3787 or (212) 354-2424. Biweekly. $115.00 per year.

COMPUTER LITERATURE INDEX. Applied Computer Research, Inc., P.O. Box 9280, Phoenix, AZ 85068. (602) 995-5929. Quarterly. $125.00 per year.

COMPUTER PROGRAMS ABSTRACTS. U.S. National Aeronautics and Space Administration. Available from: U.S. Government Printing Office, Washington, DC 20402. Quarterly. $10.00 per year.

COMPUTING REVIEWS. Association of Computing Machinery, 11 West 42nd Street, New York, NY 10036. (212) 869-7440. Monthly. $60.00 per year.

CONFERENCE PAPERS INDEX. Cambridge Scientific Abstracts, 5161 River Road, Bethesda, MD 20816. 1972 to present. Monthly. Inquire as to cost and availability.

CURRENT MATHEMATICAL PUBLICATIONS. American Mathematical Society, P.O. Box 6248, Providence, RI 02940. (800) 556-7774 or (401) 272-9500. 1969 to present. Seventeen times per year. $230.00 per year.

ENGINEERING INDEX MONTHLY AND AUTHOR INDEX. Engineering Information Inc., 345 East 47th Street, New York, NY 10017. (212) 705-7600. Monthly. $1560.00 per year.

INDEX TO SCIENTIFIC AND TECHNICAL PROCEEDINGS. Institute for Scientific Information, 3501 Market Street, Philadelphia, PA 19104. (800) 523-1850 or (215) 386-0100. 1978 to present. Monthly. $775.00 per year.

INDEX TO SCIENTIFIC REVIEWS. Institute for Scientific Information, 3501 Market Street, Philadelphia, PA 19104. (800) 523-1850 or (215) 386-0100. 1974 to present. Semi-annual. $550.00 per year.

MATHEMATICAL REVIEWS: A REVIEWING JOURNAL COVERING THE WORLD LITERATURE OF MATHEMATICAL RESEARCH. American Mathematical Society, P.O. Box 6248, Providence, RI 02940. (800) 7774 or (401) 272-9500. 1940 to present. Monthly. $2800.00 per year.

PCR-2: PERSONAL COMPUTER REVIEW - SQUARED. Toolbox Publications, Inc., P.O. Box 5451, 2514 Birch Creek Lane, Orchard Lake, MI 48033. 1987 to present. Bimonthly. $60.00 per year.

PHYSICS ABSTRACTS. Institution of Electrical Engineers. Available from: IEEE Service Center, 445 Hoes Lane, Piscataway, NJ 08854. 1898 to present. Bimonthly. $1700.00 per year.

PHYSICS BRIEFS. Physik Verlag GmbH, Postfach 1260/1280, D-6940 Weinheim, West Germany. (212) 661-9404. 1920 to present. Twenty-six times per year. $1250.00 per year.

SCIENCE CITATION INDEX. Institute for Scientific Information, 3501 Market Street, Philadelphia, PA 19104. (800) 523-1850 or (215) 386-0100. Six times per year. $6200.00 per year.

ANNUAL REVIEWS AND YEARBOOKS

ADVANCES IN COMPUTERS. Academic Press, Inc., 6277 Sea Harbor Drive, Orlando, FL 32821. (800) 321-5068. Yearly. Approximately $50.00 per volume.

COMPUTER PUBLISHERS AND PUBLICATIONS 1988-89: AN INTERNATIONAL DIRECTORY AND YEARBOOK. Efrem Sigel and Frederica Evan, editors. Gale Research Company, Book Tower, Detroit, MI 48226. (800) 521-0707. Third edition. $140.00.

ASSOCIATIONS AND PROFESSIONAL SOCIETIES

AMERICAN FEDERATION OF INFORMATION PROCESSING SOCIETIES. 1899 Preston White Drive, Reston, VA 22091. (703) 620-8900.

ASSOCIATION OF COMPUTER PROGRAMMERS AND ANALYSTS. 2108-C Gallows Road, Vienna, VA 22180. (703) 790-0490.

ASSOCIATION OF COMPUTING MACHINERY (ACM). 11 West 42nd Street, New York, NY 10036. (212) 869-7440.

IEEE COMPUTER SOCIETY. 1730 Massachusetts Avenue, N.W., Washington, DC 20036. (202) 371-0101.

INSTITUTE OF ELECTRICAL AND ELECTRONICS ENGINEERS. IEEE Service Center, 445 Hoes Lane, Piscataway, NJ 08854.

MACHINE VISION ASSOCIATION. P.O. Box 930, One SME Drive, Dearborn, MI 48121. (313) 271-1500.

SOCIETY FOR COMPUTER SIMULATION. P.O. Box 17900, San Diego, CA 92117. (619) 277-3888.

SOCIETY FOR INFORMATION DISPLAY. 8055 Manchester Avenue, Suite 615, Playa Del Rey, CA 90293. (213) 305-1502.

BIBLIOGRAPHIES

NEW TECHNICAL BOOKS: A SELECTIVE LIST WITH DESCRIPTIVE ANNOTATIONS. New York Public Library, Science and Technology Research Center, Fifth Avenue and 42nd Street, New York, NY 10018. (212) 930-0800. 1915 to present. Monthly. $15.00 per year.

SCIENCE BOOKS AND FILMS. American Association for the Advancement of Science, 1333 H Street, NW, Washington, DC 20005. (202) 326-6454. Five times per year. $20.00 per year.

SCIENTIFIC AND TECHNICAL BOOKS AND SERIALS IN PRINT 1988; AN INDEX TO LITERATURE IN SCIENCE AND TECHNOLOGY. R.R. Bowker Company, 205 East 42nd Street, New York, NY 10017. (800) 521-8110. $175.00.

DIRECTORIES AND BIOGRAPHICAL SOURCES

AMERICAN MEN AND WOMEN OF SCIENCE. R.R. Bowker, Inc., Order Department, 245 West 17th Street, New York, NY 10011. (800) 521-8110. Eight volumes. 1986. $595.00 for set.

AMERICAN SOCIETY FOR INFORMATION SCIENCE HANDBOOK AND DIRECTORY. American Society for Information Science, 1424 16th Street, N.W., Suite 404, Washington, DC 20036. (202) 462-1000. $50.00.

COMPUTERS AND COMPUTING INFORMATION RESOURCES DIRECTORY. Martin Connors, editor. Gale Research Company, Book Tower, Detroit, MI 48226. (800) 521-0707. 1987. $165.00. Supplement available at $85.00.

INTERNATIONAL RESEARCH CENTERS DIRECTORY 1988-89. Darren L. Smith, editor. Gale Research Company, Book Tower, Detroit, MI 48226. (800) 521-0707. 4th edition. 1987. $360.00.

1987 DIRECTORY OF ENGINEERING SOCIETIES AND RELATED ORGANIZATIONS. Gordon Davis, editor. Hemisphere Publishing Corporation, 1010 Vermont Avenue, NW, Washington, DC 20005. (800) 526-0275. 12th edition. 1987. $100.00.

RESEARCH CENTERS DIRECTORY 1988. Gale Research Company, Book Tower, Detroit, MI 48226. (800) 521-0707. 12th edition. 1987. $365.00 for set.

SCIENTIFIC AND TECHNICAL ORGANIZATIONS AND AGENCIES DIRECTORY. Margaret Labash Young, editor. Gale Research Company, Book Tower, Detroit, MI 48226. (800) 521-0707. 2nd edition. 1987. $185.00.

WHO'S WHO IN ENGINEERING. Gordon Davis, editor. Hemisphere Publishing Corporation, 1010 Vermont Avenue, NW, Washington, DC 20005. (800) 526-0275. 6th edition. 1985. $200.00.

ENCYCLOPEDIAS AND DICTIONARIES

COMPUTER AND TELECOMMUNICATIONS ACRONYMS. Julie E. Towell and Helen E. Sheppard, editors. Gale Research Company, Book Tower, Detroit, MI 48226. (800) 521-0707. 1986. $60.00.

DICTIONARY OF COMPUTING. Oxford University Press, 200 Madison Avenue, New York, NY 10016. (800) 458-5833. Second edition. 1986. $29.95.

ENCYCLOPEDIA OF INFORMATION SYSTEMS AND SERVICES 1988. Amy Lucas and Annette Novallo, editors. Gale Research Company, Book Tower, Detroit, MI 48226. (800) 521-0707. 8th edition. 1987. $400.00 for set.

PRENTICE-HALL ENCYCLOPEDIA OF INFORMATION TECHNOLOGY. Robert A. Edmunds. Prentice-Hall Publishing, Inc., Englewood Cliffs, NJ 07632. (800) 562-0245. 1987. $49.95.

SOFTWARE ENCYCLOPEDIA. R.R. Bowker Company, 205 East 42nd Street, New York, NY 10017. (800) 521-8110. Tow volumes. 1987. $125.00 for set.

GENERAL WORKS

COMPUTERS AND COMPUTER LANGUAGES. G. Silverman and D.B. Turklew. McGraw-Hill Book Company, 1221 Avenue of the Americas, New York, NY 10020. (212) 512-2000. 1988. $33.95.

FORTRAN 77: A STRUCTURED, DISCIPLINED STYLE. G. Davis and T. Hoffman. McGraw-Hill Book Company, 1221 Avenue of the Americas, New York, NY 10020. (212) 512-2000. 1988. $32.95.

GENERAL PURPOSE PROGRAMMING LANGUAGES. John V. Cugini. Petrocelli Books, available from: Van Nostrand Reinhold Company, Inc., 135 West 50th Street, New York, NY 10020. (800) 543-2681. 1986. $24.95.

MICROPROCESSORS AND MICROCOMPUTERS: HARDWARE AND SOFTWARE. R.J. Tocci and L.P. Laskowski. Prentice-Hall Publishing, Inc., Englewood Cliffs, NJ 07632. (800) 562-0245. Third edition. 1987. $37.95.

PRINCIPLES OF PROGRAMMING LANGUAGES: DESIGN, EVALUATION AND IMPLEMENTATION. Bruce MacLennan.

Holt, Rinehart and Winston, Inc., 383 Madison Avenue, New York, NY 10017. (212) 872-2000. Second edition. 1986. $38.75.

PROGRAMMING LANGUAGES. A.B. Tucker. McGraw-Hill Book Company, 1221 Avenue of the Americas, New York, NY 10020. (212) 512-2000. Second edition. 1986. $43.95.

PROGRAMMING LANGUAGES: A GRAND TOUR. Ellis Horowitz, editor. Computer Science Press, 11 Taft Court, Rockville, MD 20850. (800) 242-7737. Third Edition. 1987. $39.95.

HANDBOOKS AND MANUALS

HIGH LEVEL LANGUAGE AND SOFTWARE APPLICATIONS REFERENCE. W.J. Birnes. McGraw-Hill Book Company, 1221 Avenue of the Americas, New York, NY 10020. (212) 512-2000. 1988. $29.95.

ONLINE DATA BASES

COMPENDEX. Engineering Information, Inc., 345 East 47th Street, New York, NY 10017. (800) 221-1044 or (212) 705-7615. Engineering and technical literature, 1975 to present. Inquire as to online cost and availability.

DISSERTATION ABSTRACTS ONLINE. University Microfilms International, 300 North Zeeb Road, Ann Arbor, MI 48106. (800) 521-0600 or (313) 761-4700. Scope includes virtually all doctoral dissertations accepted at accredited American institutions from 1861 to present in over 250 subject areas. Inquire as to online cost and availability.

INSPEC. INSPEC Marketing Department, Institution of Electrical Engineers. Available from IEEE Service Center, 445 Hoes Lane, Piscataway, NJ 08854. (201) 981-0060. Online version of Physics Abstracts. Inquire as to online cost and availability.

MATHFILE. American Mathematical Society, P.O. Box 6248, Providence, RI 02940. (800) 556-7774 or (401) 272-9500. An online version of Mathematical Reviews. 1973 to present. Inquire as to online cost and availability.

NTIS. National Technical Information Service, 5285 Port Royal Road, Springfield, VA 22161. (703) 487-4630. Broad coverage of government sponsored research reports, 1964 to present. Inquire as to online cost and availability.

SCISEARCH. Institute for Scientific Information, 3501 Market Street, Philadelphia, PA 19104. (800) 523-1850 or (215) 386-0100. Broad multidisciplinary title and author index to the international literature of science and technology, 1974 to present. Inquire as to online cost and availability.

WILSONLINE. H.W. Wilson and Company, 950 University Avenue, Bronx, NY 10452. (800) 367-6770 or (212) 588-8400. Makes available online versions of the H.W. Wilson indexes including Applied Science and Technology Index, Business Periodicals Index and Readers' Guide to Periodical Literature. Approximately 1980 to present. Inquire as to online cost and availability.

PERIODICALS

ACM TRANSACTIONS ON MATHEMATICAL SOFTWARE. Association of Computing Machinery, 11 West 42nd Street, New York, NY 10036. (212) 869-7440. 1975 to present. Quarterly. $55.00 per year.

ACM TRANSACTIONS ON PROGRAMMING LANGUAGES AND SYSTEMS. Association of Computing Machinery, 11 West 42nd Street, New York, NY 10036. (212) 869-7440. 1979 to present. Quarterly. $55.00 per year.

ADVANCES IN ENGINEERING SOFTWARE. CML Publications, 400 West Cummings Park, Suite 6200, Woburn, MA 01801. (617) 933-7374. 1979 to present. Quarterly. $130.00 per year.

BYTE. Byte Publications, Inc., 70 Main Street, Petersborough, NH 03458. (603) 924-9281. Monthly. $21.00 per year.

C JOURNAL. InfoPro Systems, 3108 Route 10, Denville, NJ 07834. (201) 989-0570. 1985 to present. Quarterly. $28.00 per year.

COMMUNICATIONS OF THE ACM. Association of Computing Machinery, 11 West 42nd Street, New York, NY 10036. (212) 869-7440. Monthly. $80.00 per year.

COMPUTER LANGUAGES. Pergamon Press, Inc., Maxwell House, Fairview Park, Elmsford, NY 10523. (914) 592-7700. 1976 to present. Quarterly. $195.00 per year.

DR. DOBB'S JOURNAL OF SOFTWARE TOOLS. M & T Publishing, Inc., 2464 Embarcadero Way, Palo Alto, CA 94303. (415) 424-0600. 1976 to present. Monthly. $25.00 per year.

IEEE SOFTWARE. Institution of Electrical and Electronics Engineers. IEEE Service Center, 445 Hoes Lane, Piscataway, NJ 08854. 1984 to present. Quarterly. $15.00 per issue.

IEEE TRANSACTIONS ON SOFTWARE ENGINEERING. Institute of Electrical and Electronics Engineers. IEEE Service Center, 445 Hoes Lane, Piscataway, NJ 08854. 1975 to present. Monthly. $160.00 per year.

INTERFACE. International Computer Programs, Inc., 9000 Keystone Crossing, Indianapolis, IN 46240. (317) 844-7461. 1975 to present. Quarterly. $10.00 per year.

JOURNAL OF LOGIC PROGRAMMING. Elsevier Science Publishing Company, Inc., 52 Vanderbilt Avenue, New York, NY 10017. (212) 370-5520. 1984 to present. Quarterly. $95.00 per year.

JOURNAL OF PASCAL, ADA, AND MODULA-2. John Wiley and Sons, Inc., 605 Third Avenue, New York, NY 10158. (800) 526-5368. 1982 to present. Bimonthly. $20.00 per year.

JOURNAL OF SYSTEMS AND SOFTWARE. Elsevier Science Publishing Company, Inc., 52 Vanderbilt Avenue, New York, NY 10017. (212) 370-5520. 1979 to present. Quarterly. $95.00 per year.

PASCAL AND MODULA-2. Pascal Users Group, Box 538, Chesterland, OH 44026. (216) 729-3227. Quarterly. $25.00 per year.

SIGPLAN NOTICES. Association of Computing Machinery Special Interest Group on Programming Languages, 11 West 42nd Street, New York, NY 10036. (212) 869-7440. 1965 to present. Monthly. $25.00 per year.

SIGSOFT SOFTWARE ENGINEERING NOTICES. Association of Computing Machinery Special Interest Group on Software Engineering. 11 West 42nd Street, New York, NY 10036. (212) 869-7440. Quarterly. $12.00 per year.

SOFTWARE DEVELOPER'S MONTHLY. SourceView Press, 835 Castro Street, Martinez, CA 94553. (415) 228-6220. 1985 to present. Monthly. $144.00 per year.

SOFTWARE ENGINEERING JOURNAL. Institute of Electrical Engineers, Savoy Place, London, WC2R OBL, England. 1981 to present. Bimonthly. $85.00 per year.

SOFTWARE PRACTICE AND EXPERIENCE. John Wiley and Sons, Inc., 605 Third Avenue, New York, NY 10158. (800) 526-5368. 1971 to present. Monthly. $260.00 per year.

UNIX REVIEW (SAN FRANCISCO). Miller Freeman Publications, Inc., 500 Howard Street, San Francisco, CA 94105. (415) 397-1881. 1983 to present. Monthly. $35.00 per year.

RESEARCH CENTERS AND INSTITUTES

INSTITUTE FOR INFORMATION SCIENCE AND TECHNOLOGY. George Washington University, 801 22nd Street,

N.W., Washington, DC 20052. (202) 676-4921.

RESEARCH INSTITUTE FOR COMPUTING AND INFORMATION SYSTEMS. University of Houston at Clear Lake, 2700 Bay Area Boulevard, Houston, TX 77058. (713) 488-9392.

SOFTWARE ENGINEERING INSTITUTE. Carnegie-Mellon University, Pittsburgh, PA 15213. (412) 268-6700.

SOFTWARE ENGINEERING RESEARCH INSTITUTE. Georgia Institute of Technology, 258 Fourth Street, N.W., Atlanta, GA 30332. (404) 894-3180.

PROGRAMS

See: SOFTWARE

PROJECTIVE GEOMETRY

See: GEOMETRY

PROPELLANTS

See also: AERONAUTICAL ENGINEERING, AEROSPACE ENGINEERING, CHEMICAL ENGINEERING, EXPLOSIVES, FUELS, ORDNANCE, PROPULSION SYSTEMS, ROCKETS

ABSTRACT SERVICES AND INDEXES

APPLIED SCIENCE AND TECHNOLOGY INDEX. H.W. Wilson and Company, 950 University Avenue, Bronx, NY 10452. (800) 367-6670 or (212) 588-8400. Monthly. Inquire as to cost and availability.

CHEMICAL ABSTRACTS. American Chemical Society, Chemical Abstracts Service, Box 3012, Columbus, OH 43210. (614) 421-3600. 1907 to present. Weekly. $9500.00 per year.

ENGINEERING INDEX MONTHLY AND AUTHOR INDEX. Engineering Information Inc., 345 East 47th Street, New York, NY 10017. (212) 705-7600. Monthly. $1560.00 per year.

INTERNATIONAL AEROSPACE ABSTRACTS. American Institute of Aeronautics and Astronautics, Technical Information Service, 370 L'Enfant Promenade, S.W., Washington, DC 20024. (202) 646-7400. 1961 to present. Semi-monthly. $700.00 per year.

SCIENCE CITATION INDEX. Institute for Scientific Information, 3501 Market Street, Philadelphia, PA 19104. (800) 523-1850 or (215) 386-0100. Six times per year. $6200.00 per year.

STAR (SCIENTIFIC AND TECHNICAL AEROSPACE REPORTS). National Aeronautics and Space Administration, Scientific and Technical Information Facility, Box 8757, Baltimore-Washington International Airport, MD 21240. (202) 453-8545. 1963 to present. Semi-monthly. $85.00 per year with indexes.

ASSOCIATIONS AND PROFESSIONAL SOCIETIES

AMERICAN ASTRONAUTICAL SOCIETY. 6212-B Old Keene Mill Court, Springfield, VA 22152. (703) 866-0020.

AMERICAN CHEMICAL SOCIETY. 1155 16th Street, N.W., Washington, DC 20036. (800) 424-6747.

AMERICAN INSTITUTE OF AERONAUTICS AND ASTRONAUTICS. 370 L'Enfant Promenade, S.W., Washington, DC 20024. (202) 646-7400.

AMERICAN SPACE FOUNDATION. 111 Massachusetts Avenue, N.W., Suite 200, Washington, DC 20002. (202) 289-2293.

NATIONAL SPACE INSTITUTE. West Wing, Suite 203, 600 Maryland Avenue, S.W., Washington, DC 20024. (202) 484-1111.

SOCIETY OF AUTOMOTIVE ENGINEERS. 400 Commonwealth Drive, Warrendale, PA 15096. (412) 776-4841.

DIRECTORIES AND BIOGRAPHICAL SOURCES

AMERICAN ASTRONAUTICAL SOCIETY DIRECTORY. American Astronautical Society, 6212-B Old Keene Mill Court, Springfield, VA 22152. (703) 866-0020. 1986. $30.00.

AMERICAN INSTITUTE OF AERONAUTICS AND ASTRONAUTICS ROSTER. American Institute of Aeronautics and Astronautics, 370 L'Enfant Promenade, S.W., Washington, DC 20024. (202) 646-7400. 1984. $75.00.

INTERNATIONAL RESEARCH CENTERS DIRECTORY 1988-89. Darren L. Smith, editor. Gale Research Company, Book Tower, Detroit, MI 48226. (800) 521-0707. 4th edition. 1987. $360.00.

RESEARCH CENTERS DIRECTORY 1988. Gale Research Company, Book Tower, Detroit, MI 48226. (800) 521-0707. 12th edition. 1987. $365.00 for set.

SCIENTIFIC AND TECHNICAL ORGANIZATIONS AND AGENCIES DIRECTORY. Margaret Labash Young, editor. Gale Research Company, Book Tower, Detroit, MI 48226. (800) 521-0707. 2nd edition. 1987. $185.00.

WHO'S WHO IN ENGINEERING. Gordon Davis, editor. Hemisphere Publishing Corporation, 1010 Vermont Avenue, NW, Washington, DC 20005. (800) 526-0275. 6th edition. 1985. $200.00.

GENERAL WORKS

CHEMICAL ROCKET PROPULSION AND COMBUSTION RESEARCH. S.S. Penner. Gordon and Breach Science Publishers, Inc., 50 West 23rd Street, New York, NY 10010. (212) 206-8900. 1962. $45.25.

EXPLOSIVES AND PROPELLANTS FROM COMMONLY AVAILABLE MATERIALS. Gordon Press Publishers, P.O. Box 459, Bowling Green Station, New York, NY 10004. (212) 668-8819. 1986. $79.95.

PROPELLANTS: MANUFACTURE, HAZARDS AND TESTING. Carl Boyars and Karl Klager, editors. American Chemical Society, 1155 16th Street, N.W., Washington, DC 20036. (800) 424-6747. 1969. $34.95.

ROCKET PROPULSION ELEMENTS: AN INTRODUCTION TO THE ENGINEERING OF ROCKETS. George A. Sutton and Donald M. Ross. John Wiley and Sons, Inc., 605 Third Avenue, New York, NY 10158. (800) 526-5368. Fourth edition. 1976. $52.95.

ROCKET PROPULSION TECHNOLOGY. D.S. Carton and others, editors. Plenum Publishing Corporation, 233 Spring Street, New York, NY 10013. (800) 221-9369. 1961. $32.50.

ONLINE DATA BASES

CA SEARCH. Chemical Abstracts Service, P.O. Box 3012, Columbus, OH 43120. (800) 848-6538 or (614) 421-3600. Comprehensive guide to chemical and related literature, 1972 to present. Inquire as to online cost and availability.

COMPENDEX. Engineering Information, Inc., 345 East 47th Street, New York, NY 10017. (800) 221-1044 or (212) 705-7615. Engineering and technical literature, 1975 to present. Inquire as to online cost and availability.

INSPEC. INSPEC Marketing Department, Institution of Electrical Engineers. Available from IEEE Service Center, 445 Hoes Lane, Piscataway, NJ 08854. (201) 981-0060. Online version of Physics Abstracts. Inquire as to online cost and availability.

NTIS. National Technical Information Service, 5285 Port Royal Road, Springfield, VA 22161. (703) 487-4630. Broad coverage of government sponsored research reports, 1964 to present. Inquire as to online cost and availability.

SCISEARCH. Institute for Scientific Information, 3501 Market Street, Philadelphia, PA 19104. (800) 523-1850 or (215) 386-0100. Broad multidisciplinary title and author index to the international literature of science and technology, 1974 to present. Inquire as to online cost and availability.

WILSONLINE. H.W. Wilson and Company, 950 University Avenue, Bronx, NY 10452. (800) 367-6770 or (212) 588-8400. Makes available online versions of the H.W. Wilson indexes including Applied Science and Technology Index, Business Periodicals Index and Readers' Guide to Periodical Literature. Approximately 1980 to present. Inquire as to online cost and availability.

PERIODICALS

AEROSPACE ENGINEERING MAGAZINE. Society of Automotive Engineers, 400 Commonwealth Drive, Warrendale, PA 15096. (412) 776-4841. 1981 to present. Monthly. $24.00 per year.

AIAA JOURNAL. American Institute of Aeronautics and Astronautics, Technical Information Service, 370 L'Enfant Promenade, S.W., Washington, DC 20024. (202) 646-7400. 1963 to present. Monthly. $205.00 per year.

AVIATION WEEK AND SPACE TECHNOLOGY. McGraw-Hill Book Company, 1221 Avenue of the Americas, New York, NY 10020. (212) 512-2000. 1916 to present. Weekly. $56.00 per year.

JOURNAL OF SPACECRAFT AND ROCKETS. American Institute of Aeronautics and Astronautics, Technical Information Service, 370 L'Enfant Promenade, S.W., Washington, DC 20024. (202) 646-7400. 1964 to present. Bimonthly. $95.00 per year.

PROPELLANTS, EXPLOSIVES, PYROTECHNICS. International Pyrotechnics Society. Available from: VCH Publishers, Inc., 303 N.W. 12th Avenue, Deerfield Beach, FL 33441. 1976 to present. Bimonthly. $175.00 per year.

RESEARCH CENTERS AND INSTITUTES

GODDARD SPACE FLIGHT CENTER. National Aeronautics and Space Administration, Greenbelt Road, Greenbelt, MD 20771. (301) 344-5121.

MARSHALL SPACE FLIGHT CENTER. National Aeronautics and Space Administration, Huntsville, AL 35812. (205) 453-2121.

THERMAL SCIENCES AND PROPULSION CENTER. Purdue University, Lafayette, IN 47907. (317) 494-1500.

PROPJET

See: JET PROPULSION

PROPULSION SYSTEMS

See also: AERONAUTICAL ENGINEERING, AEROSPACE ENGINEERING, GUIDED MISSILES, JET PROPULSION, PROPELLANTS, ROCKETS

ABSTRACT SERVICES AND INDEXES

APPLIED SCIENCE AND TECHNOLOGY INDEX. H.W. Wilson and Company, 950 University Avenue, Bronx, NY 10452. (800) 367-6670 or (212) 588-8400. Monthly. Inquire as to cost and availability.

ENGINEERING INDEX MONTHLY AND AUTHOR INDEX. Engineering Information Inc., 345 East 47th Street, New York, NY 10017. (212) 705-7600. Monthly. $1560.00 per year.

INTERNATIONAL AEROSPACE ABSTRACTS. American Institute of Aeronautics and Astronautics, Technical Information Service, 370 L'Enfant Promenade, S.W., Washington, DC 20024. (202) 646-7400. 1961 to present. Semi-monthly. $700.00 per year.

SCIENCE CITATION INDEX. Institute for Scientific Information, 3501 Market Street, Philadelphia, PA 19104. (800) 523-1850 or (215) 386-0100. Six times per year. $6200.00 per year.

STAR (SCIENTIFIC AND TECHNICAL AEROSPACE REPORTS). National Aeronautics and Space Administration, Scientific and Technical Information Facility, Box 8757, Baltimore-Washington International Airport, MD 21240. (202) 453-8545. 1963 to present. Semi-monthly. $85.00 per year with indexes.

ASSOCIATIONS AND PROFESSIONAL SOCIETIES

AMERICAN ASTRONAUTICAL SOCIETY. 6212-B Old Keene Mill Court, Springfield, VA 22152. (703) 866-0020.

AMERICAN INSTITUTE OF AERONAUTICS AND ASTRONAUTICS. 370 L'Enfant Promenade, S.W., Washington, DC 20024. (202) 646-7400.

AMERICAN SPACE FOUNDATION. 111 Massachusetts Avenue, N.W., Suite 200, Washington, DC 20002. (202) 289-2293.

NATIONAL SPACE INSTITUTE. West Wing, Suite 203, 600 Maryland Avenue, S.W., Washington, DC 20024. (202) 484-1111.

SOCIETY OF AUTOMOTIVE ENGINEERS. 400 Commonwealth Drive, Warrendale, PA 15096. (412) 776-4841.

SPACE STUDIES INSTITUTE. P.O. Box 82, 285 Rosedale, Road, Princeton, NJ 08540. (609) 921-0377.

DIRECTORIES AND BIOGRAPHICAL SOURCES

AMERICAN ASTRONAUTICAL SOCIETY DIRECTORY. American Astronautical Society, 6212-B Old Keene Mill Court, Springfield, VA 22152. (703) 866-0020. 1986. $30.00.

AMERICAN INSTITUTE OF AERONAUTICS AND ASTRONAUTICS ROSTER. American Institute of Aeronautics and Astronautics, 370 L'Enfant Promenade, S.W., Washington, DC 20024. (202) 646-7400. 1984. $75.00.

INTERNATIONAL RESEARCH CENTERS DIRECTORY 1988-89. Darren L. Smith, editor. Gale Research Company, Book Tower, Detroit, MI 48226. (800) 521-0707. 4th edition. 1987. $360.00.

RESEARCH CENTERS DIRECTORY 1988. Gale Research Company, Book Tower, Detroit, MI 48226. (800) 521-0707. 12th edition. 1987. $365.00 for set.

SCIENTIFIC AND TECHNICAL ORGANIZATIONS AND AGENCIES DIRECTORY. Margaret Labash Young, editor. Gale Research Company, Book Tower, Detroit, MI 48226. (800) 521-0707. 2nd edition. 1987. $185.00.

WHO'S WHO IN ENGINEERING. Gordon Davis, editor. Hemisphere Publishing Corporation, 1010 Vermont Avenue, NW, Washington, DC 20005. (800) 526-0275. 6th edition. 1985. $200.00.

GENERAL WORKS

CHEMICAL ROCKET PROPULSION AND COMBUSTION RESEARCH. S.S. Penner. Gordon and Breach Science Publishers, Inc., 50 West 23rd Street, New York, NY 10010. (212) 206-8900. 1962. $45.25.

GUIDED WEAPONS. J.E. Lee. Pergamon Press, Inc., Maxwell House, Fairview Park, Elmsford, NY 10523. (914) 592-7700. 1983. $13.00 in paper.

RADAR HOMING GUIDANCE FOR TACTICAL MISSILES. D.A. James. Halsted Press, a division of John Wiley and Sons, Inc., 605 Third Avenue, New York, NY 10158. (800) 526-5368. 1986. $34.95.

ROCKET PROPULSION ELEMENTS: AN INTRODUCTION TO THE ENGINEERING OF ROCKETS. George A. Sutton and Donald M. Ross. John Wiley and Sons, Inc., 605 Third Avenue, New York, NY 10158. (800) 526-5368. Fourth edition. 1976. $52.95.

ROCKET PROPULSION TECHNOLOGY. D.S. Carton and others, editors. Plenum Publishing Corporation, 233 Spring Street, New York, NY 10013. (800) 221-9369. 1961. $32.50.

ONLINE DATA BASES

CA SEARCH. Chemical Abstracts Service, P.O. Box 3012, Columbus, OH 43120. (800) 848-6538 or (614) 421-3600. Comprehensive guide to chemical and related literature, 1972 to present. Inquire as to online cost and availability.

COMPENDEX. Engineering Information, Inc., 345 East 47th Street, New York, NY 10017. (800) 221-1044 or (212) 705-7615. Engineering and technical literature, 1975 to present. Inquire as to online cost and availability.

INSPEC. INSPEC Marketing Department, Institution of Electrical Engineers. Available from IEEE Service Center, 445 Hoes Lane, Piscataway, NJ 08854. (201) 981-0060. Online version of Physics Abstracts. Inquire as to online cost and availability.

NTIS. National Technical Information Service, 5285 Port Royal Road, Springfield, VA 22161. (703) 487-4630. Broad coverage of government sponsored research reports, 1964 to present. Inquire as to online cost and availability.

SCISEARCH. Institute for Scientific Information, 3501 Market Street, Philadelphia, PA 19104. (800) 523-1850 or (215) 386-0100. Broad multidisciplinary title and author index to the international literature of science and technology, 1974 to present. Inquire as to online cost and availability.

WILSONLINE. H.W. Wilson and Company, 950 University Avenue, Bronx, NY 10452. (800) 367-6770 or (212) 588-8400. Makes available online versions of the H.W. Wilson indexes including Applied Science and Technology Index, Business Periodicals Index and Readers' Guide to Periodical Literature. Approximately 1980 to present. Inquire as to online cost and availability.

PERIODICALS

AEROSPACE AMERICA. American Institute of Aeronautics and Astronautics, Technical Information Service, 370 L'Enfant Promenade, S.W., Washington, DC 20024. (202) 646-7400. 1932 to present. Monthly. $56.00 per year.

AEROSPACE ENGINEERING MAGAZINE. Society of Automotive Engineers, 400 Commonwealth Drive, Warrendale, PA 15096. (412) 776-4841. 1981 to present. Monthly. $24.00 per year.

AIAA JOURNAL. American Institute of Aeronautics and Astronautics, Technical Information Service, 370 L'Enfant Promenade, S.W., Washington, DC 20024. (202) 646-7400. 1963 to present. Monthly. $205.00 per year.

AVIATION WEEK AND SPACE TECHNOLOGY. McGraw-Hill Book Company, 1221 Avenue of the Americas, New York, NY 10020. (212) 512-2000. 1916 to present. Weekly. $56.00 per year.

JOURNAL OF SPACECRAFT AND ROCKETS. American Institute of Aeronautics and Astronautics, Technical Information Service, 370 L'Enfant Promenade, S.W., Washington, DC 20024. (202) 646-7400. 1964 to present. Bimonthly. $95.00 per year.

RESEARCH CENTERS AND INSTITUTES

CENTER FOR SPACE RESEARCH AND APPLICATIONS. University of Texas at Austin, WRW 402, Austin, TX 78712. (512) 471-1356.

GODDARD SPACE FLIGHT CENTER. National Aeronautics and Space Administration, Greenbelt Road, Greenbelt, MD 20771. (301) 344-5121.

GRADUATE AERONAUTICAL LABORATORIES. California Institute of Technology, 105-50, Pasadena, CA 91125. (818) 356-4551.

MARSHALL SPACE FLIGHT CENTER. National Aeronautics and Space Administration, Huntsville, AL 35812. (205) 453-2121.

SPACE SCIENCE AND ENGINEERING CENTER. University of Wisconsin, Madison, 1225 West Dayton Street, Madison, WI 53706. (608) 262-0544.

THERMAL SCIENCES AND PROPULSION CENTER. Purdue University, Lafayette, IN 47907. (317) 494-1500.

PROSPECTING

See: MINERAL EXPLORATION

PULSARS

See also: ASTRONOMY, ASTROPHYSICS, GALAXIES, RADIO ASTRONOMY, STARS

ABSTRACT SERVICES AND INDEXES

ASTRONOMY AND ASTROPHYSICS ABSTRACTS. Springer-Verlag New York, Inc., 175 Fifth Avenue, New York, NY 10010. (800) 526-7254. 1969 to present. Approximately $70.00 per year.

ASTRONOMY AND ASTROPHYSICS MONTHLY INDEX. Olivetree Associates, P.O. Box 236, Sierre Madre, CA 91024. $220.00 per year. Complementary copies available on request.

CHEMICAL ABSTRACTS. American Chemical Society, Chemical Abstracts Service, Box 3012, Columbus, OH 43210. (614) 421-3600. 1907 to present. Weekly. $9500.00 per year.

CONFERENCE PAPERS INDEX. Cambridge Scientific Abstracts, 5161 River Road, Bethesda, MD 20816. 1972 to present. Monthly. Inquire as to cost and availability.

CURRENT CONTENTS: PHYSICAL, CHEMICAL, AND EARTH SCIENCES. Institute for Scientific Information, 3501 Market Street, Philadelphia, PA 19104. (800) 523-1850 or (215) 386-0100. Weekly. $275.00 per year.

INDEX TO SCIENTIFIC AND TECHNICAL PROCEEDINGS. Institute for Scientific Information, 3501 Market Street, Philadelphia, PA 19104. (800) 523-1850 or (215) 386-0100. 1978 to present. Monthly. $775.00 per year.

INDEX TO SCIENTIFIC REVIEWS. Institute for Scientific Information, 3501 Market Street, Philadelphia, PA 19104. (800) 523-1850 or (215) 386-0100. 1974 to present. Semi-annual. $550.00 per year.

METEOROLOGICAL AND GEOASTROPHYSICAL ABSTRACTS. American Meteorological Society, 45 Beacon Street, Boston, MA 02108. (617) 227-2425. 1950 to present. Monthly. $450.00 per year.

PHYSICS ABSTRACTS. Institution of Electrical Engineers. Available from: IEEE Service Center, 445 Hoes Lane, Piscataway, NJ 08854. 1898 to present. Bimonthly. $1700.00 per year.

PHYSICS BRIEFS. Physik Verlag GmbH, Postfach 1260/1280, D-6940 Weinheim, West Germany. (212) 661-9404. 1920 to present. Twenty-six times per year. $1250.00 per year.

SCIENCE CITATION INDEX. Institute for Scientific Information, 3501 Market Street, Philadelphia, PA 19104. (800) 523-1850 or (215) 386-0100. Six times per year. $6200.00 per year.

STAR. (SCIENTIFIC AND TECHNICAL AEROSPACE REPORTS. U.S. National Aeronautics and Space Administration, Scientific and Technical Information Facility, Box 8757, Baltimore-Washington International Airport, MD 21240. (202) 755-2210. Semimonthly, with semiannual and annual indexes. $85.00 per year.

ANNUAL REVIEWS AND YEARBOOKS

ANNUAL REVIEW OF ASTRONOMY AND ASTROPHYSICS. Annual Reviews, Inc., 4139 El Camino Way, Palo Alto, CA 94306. (415) 493-4400. Annual. Inquire as to cost and availability.

ASSOCIATIONS AND PROFESSIONAL SOCIETIES

AMERICAN ASTRONOMICAL SOCIETY. 1816 Jefferson Place, N.W., Washington, DC 20036. (202) 659-0134.

AMERICAN INSTITUTE OF PHYSICS. 335 East 45th Street, New York, NY 10017. (212) 661-9494.

ASTRONOMICAL SOCIETY OF THE PACIFIC. 1290 24th Avenue, San Francisco, CA 94122. (415) 661-8660.

BIBLIOGRAPHIES

A BIBLIOGRAPHY OF ASTRONOMY, 1970-1979. R.A. Seal and S.S. Martin. Libraries Unlimited Inc., P.O. Box 263, Littleton, CO 80160. (303) 770-1220. 1982. $37.50.

DIRECTORIES AND BIOGRAPHICAL SOURCES

AMERICAN ASTRONOMICAL SOCIETY MEMBERSHIP DIRECTORY. 1816 Jefferson Place, N.W., Washington, DC 20036. (202) 659-0134. Annual. Inquire.

AMERICAN MEN AND WOMEN OF SCIENCE. R.R. Bowker, Inc., Order Department, 245 West 17th Street, New York, NY 10011. (800) 521-8110. Eight volumes. 1986. $595.00 for set.

THE BIOGRAPHICAL DICTIONARY OF SCIENTISTS: ASTRONOMERS. D. Abbott, editor. Peter Bedrick Books, 125 East 23rd Street, New York, NY 10010. 1984.

DIRECTORY OF PHYSICS AND ASTRONOMY STAFF MEMBERS. American Institute of Physics, 335 East 45th Street, New York, NY 10017. (212) 661-9494. Annual.

INTERNATIONAL RESEARCH CENTERS DIRECTORY 1988-89. Darren L. Smith, editor. Gale Research Company, Book Tower, Detroit, MI 48226. (800) 521-0707. 4th edition. 1987. $360.00.

RESEARCH CENTERS DIRECTORY 1988. Gale Research Company, Book Tower, Detroit, MI 48226. (800) 521-0707. 12th edition. 1987. $365.00 for set.

SCIENTIFIC AND TECHNICAL ORGANIZATIONS AND AGENCIES DIRECTORY. Margaret Labash Young, editor. Gale Research Company, Book Tower, Detroit, MI 48226. (800) 521-0707. 2nd edition. 1987. $185.00.

ENCYCLOPEDIAS AND DICTIONARIES

ILLUSTRATED ENCYCLOPEDIA OF THE UNIVERSE: UNDERSTANDING AND EXPLORING THE COSMOS. Richard S. Lewis. Crown Publishers Inc., 1 Park Avenue, New York, NY 10016. (800) 526-4264. 1986. $24.95.

GENERAL WORKS

BLACK HOLES, WHITE DWARFS, AND NEUTRON STARS: THE PHYSICS OF COMPACT OBJECTS. Stuart L. Shapiro and Saul A. Teukolsky. John Wiley and Sons, Inc., 605 Third Avenue, New York, NY 10158. (800) 526-5368. 1983. $49.95.

CONSTRUCTING THE UNIVERSE. David Layzer. W.H. Freeman and Company, 41 Madison Avenue, New York, NY 10010. (212) 532-7660. 1984. $29.95.

THE MILKY WAY. B.J. Bok and P.F. Bok. Harvard University Press, 79 Garden Street, Cambridge, MA 02138. (617) 495-2600. 1981. $29.95.

NEUTRON STARS. J.M. Irvine. Oxford University Press, 200 Madison Avenue, New York, NY 10016. (800) 458-5833. $39.50.

THE PHYSICS OF STARS. S.A. Kaplan. John Wiley and Sons, Inc., 605 Third Avenue, New York, NY 10158. (800) 526-5368. 1982. $39.95.

PULSARS. R.N. Manchester and J.H. Taylor. W.H. Freeman and Company, 41 Madison Avenue, New York, NY 10010. (212) 532-7660. 1977. $39.95.

PULSARS. W. Sieber and W.R. Wielebinski, editors. Kluwer Academic Publishers, 190 Old Derby Street, Hingham, MA 02043. (617) 749-5262. 1981. $60.50.

ONLINE DATA BASES

CA SEARCH. Chemical Abstracts Service, P.O. Box 3012, Columbus, OH 43120. (800) 848-6538 or (614) 421-3600. Comprehensive guide to chemical literature, 1972 to present. Inquire as to online cost and availability.

DISSERTATION ABSTRACTS ONLINE. University Microfilms International, 300 North Zeeb Road, Ann Arbor, MI 48106. (800) 521-0600 or (313) 761-4700. Scope includes virtually all doctoral dissertations accepted at accredited American institutions from 1861 to present in over 250 subject areas. Inquire as to online cost and availability.

INSPEC. INSPEC Marketing Department, Institution of Electrical Engineers. Available from IEEE Service Center, 445 Hoes Lane, Piscataway, NJ 08854. (201) 981-0060. Online version of Physics Abstracts. Inquire as to online cost and availability.

NTIS. National Technical Information Service, 5285 Port Royal Road, Springfield, VA 22161. (703) 487-4630. Broad coverage of government sponsored research reports, 1964 to present. Inquire as to online cost and availability.

SCISEARCH. Institute for Scientific Information, 3501 Market Street, Philadelphia, PA 19104. (800) 523-1850 or (215) 386-0100. Broad multidisciplinary title and author index to the international literature of science and technology, 1974 to present. Inquire as to online cost and availability.

OTHER SOURCES

ATLAS OF DEEP-SKY SPLENDORS. H. Vehrenberh. Cambridge University Press, 32 East 57th Street, New York, NY 10022. (800) 872-7423. Fourth edition. 1984. $44.50.

PERIODICALS

ASTRONOMICAL JOURNAL. American Astronomical Society. Available from: American Institute of Physics, 335 East 45th Street, New York, NY 10017. (212) 661-9494. Monthly. $125.00 per year.

ASTRONOMICAL SOCIETY OF THE PACIFIC PUBLICATIONS. Astronomical Society of the Pacific, 1290 24th Avenue, San Francisco, CA 94122. (415) 661-8660. Monthly. $40.00 per year.

ASTRONOMY. AstroMedia Corporation, 625 E Street, Box 92788, Milwaukee, WI 53202. (414) 276-2689. 1973 to present. Monthly. $21.00 per year.

ASTRONOMY AND ASTROPHYSICS. Springer-Verlag New York, Inc., 175 Fifth Avenue, New York, NY 10010. (800) 526-7254. Monthly. $680.00 per year.

ASTROPHYSICAL JOURNAL. University of Chicago Press, 5801 Ellis Avenue, Chicago, IL 60637. (800) 621-2736. Biweekly. $315.00 per year.

ASTROPHYSICS AND SPACE SCIENCE. D. Reidel Publishing Company, 190 Old Derby Street, Hingham, MA 02043. Monthly. $101.00 per year.

SKY AND TELESCOPE. Sky Publishing Corporation, 49 Bay State Road, Cambridge, MA 02238. (617) 864-7360. Monthly. $18.00 per year.

MERCURY. Astronomical Society of the Pacific, 1290 24th Avenue, San Francisco, CA 94122. (415) 661-8660. 1972 to present. Bimonthly. $21.00 per year.

RESEARCH CENTERS AND INSTITUTES

CENTER FOR SPACE SCIENCE AND ASTROPHYSICS. Stanford University, 325 Durand Building, Stanford, CA 94305. (415) 497-3582.

HARVARD-SMITHSONIAN CENTER FOR ASTROPHYSICS. 60 Garden Street, Cambridge, MA 02138. (617) 495-7461.

MOUNT WILSON AND LAS CAMPANAS OBSERVATORIES. 813 Santa Barbara Street, Pasadena, CA 91101. (818) 577-1122.

NATIONAL OPTICAL ASTRONOMY OBSERVATORIES. 1002 North Warren Avenue, Tucson, AZ 85719. (602) 325-9230.

PUMPING MACHINERY

See also: PIPELINE TECHNOLOGY

ABSTRACT SERVICES AND INDEXES

APPLIED MECHANICS REVIEW. American Society of Mechanical Engineers, 345 East 47th Street, New York, NY 10017. (212) 705-7703. 1948 to present. Monthly. $360.00 per year.

APPLIED SCIENCE AND TECHNOLOGY INDEX. H.W. Wilson and Company, 950 University Avenue, Bronx, NY 10452. (800) 367-6670 or (212) 588-8400. Monthly. Inquire as to cost and availability.

CHEMICAL ABSTRACTS. American Chemical Society, Chemical Abstracts Service, Box 3012, Columbus, OH 43210. (614) 421-3600. 1907 to present. Weekly. $9500.00 per year.

CURRENT CONTENTS: ENGINEERING, TECHNOLOGY AND APPLIED SCIENCES. Institute for Scientific Information, 3501 Market Street, Philadelphia, PA 19104. (800) 523-1850 or (215) 386-0100. Weekly. $275.00 per year.

ENGINEERING INDEX MONTHLY AND AUTHOR INDEX. Engineering Information Inc., 345 East 47th Street, New York, NY 10017. (212) 705-7600. Monthly. $1560.00 per year.

ISMEC BULLETIN (Information Service in Mechanical Engineering). Cambridge Scientific Abstracts, 5161 River Road, Bethesda, MD 20816. (301) 951-1400. 1973 to present. Monthly. $450.00 per year.

OCEAN ABSTRACTS. Cambridge Scientific Abstracts, 5161 River Road, Bethesda, MD 20816. (301) 951-1400. 1963 to present. Bimonthly. $450.00 per year.

ASSOCIATIONS AND PROFESSIONAL SOCIETIES

AMERICAN SOCIETY OF HEATING, REFRIGRATION AND AIR CONDITIONING ENGINEERS INC. 1791 Tullie Circle, Atlanta, GA 30329. (404) 636-8400.

AMERICAN SOCIETY OF MECHANICAL ENGINEERS. 345 East 47th Street, New York, NY 10017. (212) 705-7703.

AMERICAN WATER WORKS ASSOCIATION. 6666 West Qunicey Avenue, Denver, CO 80235. (303) 794-7711.

NATIONAL WATER WELL ASSOCIATION. 6375 Riverside Drive, Dublin, OH 43017. (614) 761-1711.

DIRECTORIES AND BIOGRAPHICAL SOURCES

1987 DIRECTORY OF ENGINEERING SOCIETIES AND RELATED ORGANIZATIONS. Gordon Davis, editor. Hemisphere Publishing Corporation, 1010 Vermont Avenue, NW, Washington, DC 20005. (800) 526-0275. 12th edition. 1987. $100.00.

PLANT ENGINEERING. Technical Publishing Company, Box 1030, 1301 South Grove Avenue, Barrington, IL 60010. (312) 381-1840. 1947 to present. Annual. $50.00. Listings of industrial equipment suppliers.

RESEARCH CENTERS DIRECTORY 1988. Gale Research Company, Book Tower, Detroit, MI 48226. (800) 521-0707. 12th edition. 1987. $365.00 for set.

WHO'S WHO IN ENGINEERING. Gordon Davis, editor. Hemisphere Publishing Corporation, 1010 Vermont Avenue, NW, Washington, DC 20005. (800) 526-0275. 6th edition. 1985. $200.00.

GENERAL WORKS

PIPING AND PIPE SUPPORT SYSTEMS: DESIGN AND ENGINEERING. R.P. Smith and T. Van Laan. McGraw-Hill Book Company, 1221 Avenue of the Americas, New York, NY 10020. (212) 512-2000. 1987. $49.95.

PUMP APPLICATION ENGINEERING. E. Hicks. McGraw-Hill Book Company, 1221 Avenue of the Americas, New York, NY 10020. (212) 512-2000. 1970. $49.50.

VERTICAL TURBINE, MIXED FLOW, AND PROPELLER PUMPS. John L. Dicmas. McGraw-Hill Book Company, 1221 Avenue of the Americas, New York, NY 10020. (212) 512-2000. 1987. $45.00.

HANDBOOKS AND MANUALS

PUMP HANDBOOK. Igor J. Karassik and others. McGraw-Hill Book Company, 1221 Avenue of the Americas, New York, NY 10020. (212) 512-2000. 2nd edition. 1986. $92.50.

ONLINE DATA BASES

CA SEARCH. Chemical Abstracts Service, P.O. Box 3012, Columbus, OH 43120. (800) 848-6538 or (614) 421-3600. Comprehensive guide to chemical literature, 1972 to present. Inquire as to online cost and availability.

COMPENDEX. Engineering Information, Inc., 345 East 47th Street, New York, NY 10017. (800) 221-1044 or (212) 705-7615. Engineering and technical literature, 1975 to present. Inquire as to online cost and availability.

NTIS. National Technical Information Service, 5285 Port Royal Road, Springfield, VA 22161. (703) 487-4630. Broad coverage of government sponsored research reports, 1964 to present. Inquire as to online cost and availability.

SCISEARCH. Institute for Scientific Information, 3501 Market Street, Philadelphia, PA 19104. (800) 523-1850 or (215) 386-0100. Broad multidisciplinary title and author index to the international literature of science and technology, 1974 to present. Inquire as to online cost and availability.

WILSONLINE. H.W. Wilson and Company, 950 University Avenue, Bronx, NY 10452. (800) 367-6770 or (212) 588-8400. Makes available online versions of the H.W. Wilson indexes including Applied Science and Technology Index, Business Periodicals Index and Readers' Guide to Periodical Literature. Approximately 1980 to present. Inquire as to online cost and availability.

PERIODICALS

INDUSTRIAL PUMP AND VALVE. Industrial Pump and Valve Company, 31505 Grand River, Suite 1, Farmington, MI 48024. (313) 474-7778. 1983 to present. Bimonthly. Inquire.

PUMP NEWS. Impact Publications, Box 1972, Eates Park, CO 80517. (303) 586-5636. 1974 to present. Monthly. $15.00 per year.

PUMPS

See: PIPELINE TECHNOLOGY, PUMPING MACHINERY

PYROMETALLURGY

See: METALLURGY

Q

QUALITY CONTROL ENGINEERING

See also: CHEMICAL ENGINEERING, CONTROL SYSTEMS, DESIGN ENGINEERING, INDUSTRIAL ENGINEERING, MATERIALS HANDLING, MECHANICAL ENGINEERING, OPERATIONS RESEARCH, ROBOTICS

ABSTRACT SERVICES AND INDEXES

APPLIED MECHANICS REVIEW. American Society of Mechanical Engineers, 345 East 47th Street, New York, NY 10017. (212) 705-7703. 1948 to present. Monthly. $360.00 per year.

APPLIED SCIENCE AND TECHNOLOGY INDEX. H.W. Wilson and Company, 950 University Avenue, Bronx, NY 10452. (800) 367-6670 or (212) 588-8400. Monthly. Inquire as to cost and availability.

CHEMICAL ABSTRACTS. American Chemical Society, Chemical Abstracts Service, Box 3012, Columbus, OH 43210. (614) 421-3600. 1907 to present. Weekly. $9500.00 per year.

CONFERENCE PAPERS INDEX. Cambridge Scientific Abstracts, 5161 River Road, Bethesda, MD 20816. 1972 to present. Monthly. Inquire as to cost and availability.

CURRENT CONTENTS: ENGINEERING, TECHNOLOGY AND APPLIED SCIENCES. Institute for Scientific Information, 3501 Market Street, Philadelphia, PA 19104. (800) 523-1850 or (215) 386-0100. Weekly. $275.00 per year.

ENGINEERING INDEX MONTHLY AND AUTHOR INDEX. Engineering Information Inc., 345 East 47th Street, New York, NY 10017. (212) 705-7600. Monthly. $1560.00 per year.

ISMEC BULLETIN (Information Service in Mechanical Engineering). Cambridge Scientific Abstracts, 5161 River Road, Bethesda, MD 20816. (301) 951-1400. 1973 to present. Monthly. $450.00 per year.

INDEX TO SCIENTIFIC AND TECHNICAL PROCEEDINGS. Institute for Scientific Information, 3501 Market Street, Philadelphia, PA 19104. (800) 523-1850 or (215) 386-0100. 1978 to present. Monthly. $775.00 per year.

SCIENCE CITATION INDEX. Institute for Scientific Information, 3501 Market Street, Philadelphia, PA 19104. (800) 523-1850 or (215) 386-0100. Six times per year. $6200.00 per year.

ASSOCIATIONS AND PROFESSIONAL SOCIETIES

AMERICAN PRODUCTION AND INVENTORY CONTROL SOCIETY. 500 West Annandale Road, Falls Church, VA 22046. (703) 237-8344.

AMERICAN SOCIETY FOR QUALITY CONTROL. 230 West Wells Street, Suite 700, Milwaukee, WI 53203. (414) 272-8575.

AMERICAN SOCIETY OF MECHANICAL ENGINEERS. 345 East 47th Street, New York, NY 10017. (212) 705-7703.

INDUSTRIAL DESIGNERS SOCIETY OF AMERICA. 1360 Beverly Road, McLean, VA 22101. (703) 556-0919.

INSTITUTE OF INDUSTRIAL ENGINEERS. 25 Technology Park/Atlanta, Norcross, GA 30092. (404) 449-0460.

INTERNATIONAL ASSOCIATION OF QUALITY CIRCLES. 801-B West Eighth Street, Suite 301, Cincinnati, OH 45203. (513) 381-1959.

SOCIETY OF AMERICAN VALUE ENGINEERS. 221 North LaSalle Street, Suite 2026, Chicago, IL 60601. (312) 346-3265.

SOCIETY OF LOGISTICS ENGINEERS. 303 Williams Avenue, Suite 922, Huntsville, AL 35801. (205) 539-3800.

SOCIETY OF MANUFACTURING ENGINEERS. One SME Drive, Box 930, Dearborn, MI 48121. (313) 271-1500.

SOCIETY OF PACKING AND HANDLING ENGINEERS. Reston International Center, Reston, VA 22091. (703) 620-9380.

DIRECTORIES AND BIOGRAPHICAL SOURCES

AMERICAN MEN AND WOMEN OF SCIENCE. R.R. Bowker, Inc., Order Department, 245 West 17th Street, New York, NY 10011. (800) 521-8110. Eight volumes. 1986. $595.00 for set.

INTERNATIONAL RESEARCH CENTERS DIRECTORY 1988-89. Darren L. Smith, editor. Gale Research Company, Book Tower, Detroit, MI 48226. (800) 521-0707. 4th edition. 1987. $360.00.

1987 DIRECTORY OF ENGINEERING SOCIETIES AND RELATED ORGANIZATIONS. Gordon Davis, editor. Hemisphere Publishing Corporation, 1010 Vermont Avenue, NW, Washington, DC 20005. (800) 526-0275. 12th edition. 1987. $100.00.

PLANT ENGINEERING DIRECTORY. Technical Publishing Company, 1301 South Grove Avenue, Chicago, IL 60010. (312) 381-1840. Annual. $35.00.

RESEARCH CENTERS DIRECTORY 1988. Gale Research Company, Book Tower, Detroit, MI 48226. (800) 521-0707. 12th edition. 1987. $365.00 for set.

SCIENTIFIC AND TECHNICAL ORGANIZATIONS AND AGENCIES DIRECTORY. Margaret Labash Young, editor. Gale Research Company, Book Tower, Detroit, MI 48226. (800) 521-0707. 2nd edition. 1987. $185.00.

QUALITY BUYERS GUIDE FOR TEST AND MEASUREMENT EQUIPMENT ISSURE. Hitchcock Publishing Company, 25 West 550 Geneva Road, Wheaton, IL 60188. (312) 665-1000. Annual. $25.00.

SOCIETY OF LOGISTICS ENGINEERS MEMBERSHIP DIRECTORY. 303 Williams Avenue, Suite 922, Huntsville, AL

35801. (205) 539-3800. Annual. Inquire.

SOCIETY OF PACKING AND HANDLING ENGINEERS DIRECTORY. Reston International Center, Reston, VA 22091. (703) 620-9380. Annual. Inquire.

WHO'S WHO IN ENGINEERING. Gordon Davis, editor. Hemisphere Publishing Corporation, 79 Madison Avenue, New York, NY 10016-7892. (800) 821-8312. Sixth edition. 1985. $200.00.

GENERAL WORKS

APPLIED RELIABILITY. Paul A. Tobias and David C. Trindade. Van Nostrand Reinhold Company, Inc., 135 West 50th Street, New York, NY 10020. (800) 543-2681. 1986. $36.95.

MANUFACTURING ENGINEERING: PRINCIPLES OF OPTIMIZATION. Daniel T. Koenig. Hemisphere Publishing Corporation, 79 Madison Avenue, New York, NY 10016-7892. (800) 821-8312. 1987. $45.75.

OPERATIONS RESEARCH: PRINCIPLES AND PRACTICE. A. Ravindran and others. John Wiley and Sons, Inc., 605 Third Avenue, New York, NY 10158. (800) 526-5368. Second edition. 1987. $49.95.

QUALITY CONTROL SYSTEMS: PROCEDURES FOR PLANNING QUALITY PROGRAMS. J.R. Taylor. McGraw-Hill Book Company, 1221 Avenue of the Americas, New York, NY 10020. (212) 512-2000. 1988. $49.50.

ROBOTICS: AN INTRODUCTION. D. McCloy and D. Harris. John Wiley and Sons, Inc., 605 Third Avenue, New York, NY 10158. (800) 526-5368. 1986. $39.95.

STATISTICAL QUALITY CONTROL. E.L. Grant and R.S. Leavenworth. McGraw-Hill Book Company, 1221 Avenue of the Americas, New York, NY 10020. (212) 512-2000. Sixth edition. 1988. $47.95.

HANDBOOKS AND MANUALS

COMPUTER-INTEGRATED MANUFACTURING HANDBOOK. E. Telcholz and J.N. Orr. McGraw-Hill Book Company, 1221 Avenue of the Americas, New York, NY 10020. (212) 512-2000. 1987. $59.95.

HANDBOOK OF INDUSTRIAL ENGINEERING. Gavriel Salvendy. John Wiley and Sons, Inc., 605 Third Avenue, New York, NY 10158. (800) 526-5368. 1982. $99.50.

JURAN'S QUALITY CONTROL HANDBOOK. J.M. Juran and F.M. Gryna. McGraw-Hill Book Company, 1221 Avenue of the Americas, New York, NY 10020. (212) 512-2000. Fourth edition. $79.50.

PRODUCTION HANDBOOK. J.A. White. John Wiley and Sons, Inc., 605 Third Avenue, New York, NY 10158. (800) 526-5368. 4th edition. 1986. $78.50.

ONLINE DATA BASES

CA SEARCH. Chemical Abstracts Service, P.O. Box 3012, Columbus, OH 43120. (800) 848-6538 or (614) 421-3600. Comprehensive guide to chemical literature, 1972 to present. Inquire as to online cost and availability.

COMPENDEX. Engineering Information, Inc., 345 East 47th Street, New York, NY 10017. (800) 221-1044 or (212) 705-7615. Engineering and technical literature, 1975 to present. Inquire as to online cost and availability.

DISSERTATION ABSTRACTS ONLINE. University Microfilms International, 300 North Zeeb Road, Ann Arbor, MI 48106. (800) 521-0600 or (313) 761-4700. Scope includes virtually all doctoral dissertations accepted at accredited American institutions from 1861 to present in over 250 subject areas. Inquire as to online cost and availability.

INSPEC. INSPEC Marketing Department, Institution of Electrical Engineers. Available from IEEE Service Center, 445 Hoes Lane, Piscataway, NJ 08854. (201) 981-0060. Online version of Physics Abstracts. Inquire as to online cost and availability.

NTIS. National Technical Information Service, 5285 Port Royal Road, Springfield, VA 22161. (703) 487-4630. Broad coverage of government sponsored research reports, 1964 to present. Inquire as to online cost and availability.

SCISEARCH. Institute for Scientific Information, 3501 Market Street, Philadelphia, PA 19104. (800) 523-1850 or (215) 386-0100. Broad multidisciplinary title and author index to the international literature of science and technology, 1974 to present. Inquire as to online cost and availability.

WILSONLINE. H.W. Wilson and Company, 950 University Avenue, Bronx, NY 10452. (800) 367-6770 or (212) 588-8400. Makes available online versions of the H.W. Wilson indexes including Applied Science and Technology Index, Business Periodicals Index and Readers' Guide to Periodical Literature. Approximately 1980 to present. Inquire as to online cost and availability.

PERIODICALS

ADVANCED MANUFACTURING PROCESSES. Marcel Dekker Inc., 270 Madison Avenue, New York, NY 10016. (800) 228-1160. 1986 to present. Quarterly. $85.00 per year.

ADVANCED MANUFACTURING TECHNOLOGY. Technical Insights, Inc., Box 1304, Fort Lee, NJ 07024. (201) 568-4744. 1977 to present. Biweekly. $320.00 per year.

AMERICAN INDUSTRY. Publications for Industry, 21 Russell Woods Road, Great Neck, NY 11021. (516) 487-0990. 1946 to present. Monthly. $20.00 per year.

ASSEMBLY ENGINEERING. Hitchcock Publishing Company, 25 West 550 Geneva Road, Wheaton, IL 60188. (312) 665-1000. 1958 to present. Monthly. $65.00 per year.

IEEE TRANSACTIONS ON INDUSTRY APPLICATIONS. Institute of Electrical and Electronics Engineers. IEEE Service Center, 445 Hoes Lane, Piscataway, NJ 08854. 1965 to present. Bimonthly. $105.00 per year.

IIE TRANSACTIONS: INDUSTRIAL ENGINEERING RESEARCH AND DEVELOPMENT. Instuture of Industrial Engineers, 25 Technology Park/Atlanta, Norcross, GA 30092. (404) 449-0460. 1969 to present. Quarterly. $55.00 per year.

JOURNAL OF ENGINEERING FOR INDUSTRY. American Society of Mechanical Engineers, 345 East 47th Street, New York, NY 10017. (212) 705-7703. 1928 to present. Quarterly. $80.00 per year.

JOURNAL OF QUALITY TECHNOLOGY. American Society for Quality Control, 230 West Wells Street, Suite 700, Milwaukee, WI 53203. (414) 272-8575. 1969 to present. Quarterly. $14.00 per year.

JOURNAL OF OPERATIONS MANAGEMENT. American Production and Inventory Control Society, 500 West Annandale Road, Falls Church, VA 22046. (703) 237-8344. 1980 to present. Quarterly. $15.00 per year.

MANUFACTURING ENGINEERING. Society of Manufactruing Engineers, One SME Drive, Box 930, Dearborn, MI 48121. (313) 271-1500. 1932 to present. Monthly. $60.00 per year.

MATERIAL HANDLING ENGINEERING. Penton-IPC, 1100 Superior Avenue, Cleveland, OH 44114. (216) 696-7000. 1945 to present. Monthly. $35.00 per year.

MATERIALS ENGINEERING. Penton-IPC, 1100 Superior Avenue, Cleveland, OH 44114. (216)696-7000. 1929 to present. Monthly.

$40.00 per year.

PRODUCTION. Production Publishing, 44 East Long Lake Road, Bloomfield Hills, MI 48303. (313) 647-8400. 1934 to present. Monthly. $35.00 per year.

PRODUCTION AND INVENTORY MANAGEMENT JOURNAL. American Production and Inventory Control Society, 500 West Annandale Road, Falls Church, VA 22046. (703) 237-8344. 1960 to present. Quarterly. $30.00 per year.

QUALITY ASSURANCE. Hitchcock Publishing Company, 25 West 550 Geneva Road, Wheaton, IL 60188. (312) 665-1000. 1962 to present. Monthly. $45.00 per year.

QUALITY PROGRESS. American Society for Quality Control, 230 West Wells Street, Suite 700, Milwaukee, WI 53203. (414) 272-8575. 1944 to present. Monthly. Inquire.

RESEARCH CENTERS AND INSTITUTES

ADVANCED MANUFACTURING CENTER. Cleveland State University, Euclid Avenue at East 24th Street, Cleveland, OH 44115. (216) 687-4643.

CENTER FOR APPLIED ENGINEERING. University of Missouri - Rolla, 206 Harris Hall, Rolla, MO 65401. (314) 4559.

CENTER OF DESIGN AND MANUFACTURING INNOVATION. Lehigh University, H.S. Mohler Laboratory, Room 200, Bethleham, PA 18015. (215) 758-4114.

CENTER FOR QUALITY AND PRODUCTIVITY IMPROVEMENT. University of Wisconsin - Madison, 610 North Walnut Street, Madison, WI 53704. (608) 263-2520.

ENGINEERING AND INDUSTRIAL EXPERIMENT STATION. University of Florida, 300 Weil Hall, Gainesvill, FL 32611. (904) 392-0941.

QUANTUM CHEMISTRY

See also: CHEMICAL BONDING, CHEMISTRY, PHYSICAL CHEMISTRY

ABSTRACT SERVICES AND INDEXES

APPLIED SCIENCE AND TECHNOLOGY INDEX. H.W. Wilson Company, 950 University Avenue, Bronx, NY 10452. (800) 367-6670 or (212) 588-8400. Inquire as to cost and availability.

CHEMICAL ABSTRACTS. Chemical Abstracts Service, 2540 Olentangy Road, P.O. Box 3012, Columbus, OH 43210. (800) 848-6538 or (614) 421-3600. Weekly. $9200.00 per year.

GENERAL SCIENCE INDEX. H.W. Wilson Company, 950 University Avenue, Bronx, NY 10452. (800) 367-6770 or (212) 588-8400. Inquire as to cost and availability.

PHYSICS ABSTRACTS. Institute of Electrical Engineers, London, United Kingdom. Available from: Institute of Electrical and Electronic Engineers (IEEE), 345 East 47th Street, New York, NY 10017. (212) 705-7900.

SCIENCE CITATION INDEX. Institute for Scientific Information, 3501 Market Street, Philadelphia, PA 19104. (800) 523-1850 or (215) 386-0100.

ANNUAL REVIEWS AND YEARBOOKS

ADVANCES IN QUANTUM CHEMISTRY. Per-Olov Lowdin, editor. Academic Press, Incorporated, Orlando, FL 32887. (800) 321-5068 or (305) 345-2734. Irregular. Price varies.

ANNUAL REVIEW OF PHYSICAL CHEMISTRY. Annual Reviews Incorporated, 4139 El Camino Way, Palo Alto, CA 94306. (415) 493-4400. Annual.

ASSOCIATIONS AND PROFESSIONAL SOCIETIES

AMERICAN CHEMICAL SOCIETY. 1155 Sixteenth Street, NW, Washington, DC 20036. (202) 872-4600.

ASSOCIATION OF CONSULTING CHEMISTS AND CHEMICAL ENGINEERS. 50 East 41st Street, Suite 92, New York, NY 10017. (212) 684-6255.

DIVISION OF PHYSICAL CHEMISTRY (A DIVISION OF THE AMERICAN CHEMICAL SOCIETY). c/o Dr. James Kinsey, Chemistry Department, Room 6-215, Massachusetts Institute of Technology, Cambridge, MA 02139.

BIBLIOGRAPHIES

SCIENCE BOOKS AND FILMS. American Association for the Advancement of Science, 1333 H Street, NW, Washington, DC 20005.

SCIENTIFIC AND TECHNICAL BOOKS AND SERIALS IN PRINT 1988: AN INDEX TO LITERATURE IN SCIENCE AND TECHNOLOGY. R.R. Bowker Company 205 East 42nd Street, New York, NY 10017. (800) 521-8110 or (212) 916-1600. $175.00.

DIRECTORIES AND BIOGRAPHICAL SOURCES

AMERICAN INSTITUTE OF CHEMISTS. American Institute of Chemists, 7315 Wisconsin Avenue, Bethesda, MD 20814. (301) 652-2447. 1986. $35.00.

AMERICAN MEN AND WOMEN OF SCIENCE. Physical and Bilogical Sciences. R.R. Bowker Company, 205 East 42nd Street, New York, NY 10017. (800) 521-8110 or (212) 916-1600. Fifteenth edition. $565.00.

BIOGRAPHICAL DICTIONARY OF SCIENTISTS: CHEMISTS. David Abbott, editor. P. Bedrick Books, 125 East 23rd Street, New York, NY 10010. (212) 777-1187. 1984. $18.95.

CONSULTING SERVICES: CHEMISTS AND CHEMICAL ENGINEERS. Association of Consulting Chemists and Chemical Engineers, 50 East 41st Street, New York, NY 10017. (212) 684-6255. Annual. 1986. $45.00.

GOVERNMENT RESEARCH DIRECTORY. Gale Research Company, Book Tower, Detroit, MI 48226. (800) 521-0707. Fourth edition. 1987. $350.00.

INTERNATIONAL RESEARCH CENTERS DIRECTORY 1986-1987. Gale Research Company, Book Tower, Detroit, MI 48226. (800) 521-0707. Third edition. 1986. $330.00.

RESEARCH CENTERS DIRECTORY. Gale Research Company, Book Tower, Detroit, MI 48226. (800) 521-0707. Eleventh edition. 1987. $355.00.

RESEARCH SERVICES DIRECTORY. Robert J. Huffman and Mary M. Watkins, editors. Gale Research Company, Book Towers, Detroit, MI 48226. (800) 521-0707. Third edition. $290.00.

SCIENTIFIC AND TECHNICAL ORGANIZATIONS AND AGENCIES DIRECTORY. Gale Research Company, Book Tower, Detroit, MI 48226. (800) 521-0707. 1985. $150.00.

WHO'S WHO IN FRONTIER SCIENCE AND TECHNOLOGY. Marquis Who's Who, Incorporated, 200 East Ohio Street, Chicago, IL 60611. (800) 428-3898 or (312) 787-2008.

WHO'S WHO IN TECHNOLOGY TODAY. Reston Publishing Company, Incorporated, c/o Prentice-Hall, Incorporated,

Englewood Cliffs, NJ 07632. (800) 262-6868. Biennial. Five volumes. $425.00. Covers the fields of electronics, computer science, physics, optics, chemistry, biotechnology, mechanics, energy, and earth science.

WORLD GUIDE TO SCIENTIFIC ASSOCIATIONS AND LEARNED SOCIETIES. K.G. Saur Incorporated, 175 Fifth Avenue, New York, NY 10010. (800) 521-0707 or (212) 982-1302. Fourth edition. 1984. $112.00.

ENCYCLOPEDIAS AND DICTIONARIES

CONCISE ENCYCLOPEDIA OF CHEMICAL TECHNOLOGY. Kirk-Othmer. John Wiley and Sons, Incorporated, 605 Third Avenue, New York, NY 10158. (800) 526-5368 or (212) 850-6000. Third edition. 1985. $129.95.

CONDENSED CHEMICAL DICTIONARY. Gessner Hawley. Van Nostrand Reinhold, 115 Fifth Avenue, New York, NY 10003. Tenth edition. 1981. $49.95.

ENCYCLOPEDIA OF PHYSICAL SCIENCE AND TECHNOLOGY. Academic Press, Incorporated, Orlando, FL 32887. (800) 321-5068 or (305) 345-2734. Fifteen volumes, 1986.

GLOSSARY OF CHEMICAL TERMS. Clifford A. Hampel and Gessner G. Hawley. Van Nostrand Reinhold Company, 115 Fifth Avenue, New York, NY 10003. (800) 543-2681 or (212) 254-3232. Second edition. 1987. $52.95.

MCGRAW-HILL ENCYCLOPEDIA OF SCIENCE AND TECHNOLOGY. McGraw-Hill Book, Incorporated, 1221 Avenue of the Americas, New York, NY 10020. (212) 997-3675.

VAN NOSTRAND REINHOLD ENCYCLOPEDIA OF CHEMISTRY. Douglas M. Considine and Glenn D. Considine. Van Nostrand Reinhold Publishing Company, Incorporated, 115 Fifth Avenue, New York, NY 10003. (800) 543-2681 or (212) 254-3232. 1984. $97.95.

GENERAL WORKS

APPLIED QUANTUM CHEMISTRY. Vedene H. Smith. Kluwer Academic Publishers, 190 Old Derby Street, Hingham, MA 02043. (617) 749-5262. 1986. $88.00.

THE CHEMICAL BOND. John N. Murrell and others. John Wiley and Sons, Incorporated, 605 Third Avenue, New York, NY 10158. (800) 526-5368 or (212) 850-6000. 1985. $31.95.

INTRODUCTION TO PHYSICAL CHEMISTRY. M.F.C. Ladde and W.H. Lee. Cambridge University Press, 32 East 57th Street, New York, NY 10022. (212) 688-8885. 1986. $19.95 in paper.

INTRODUCTION TO SYMMETRY IN BONDING AND SPECTRA. Bodie Douglas and Charles A. Hollingsworth. Academic Press, Incorporated, 6277 Sea Harbor Drive, Orlando, FL 32821. (800) 321-5068. 1985. $40.00.

INVESTIGATION OF RATES AND MECHANISMS OF REACTIONS, PART 1. Claude F. Bernasconi, editor. John Wiley and Sons, Incorporated, 605 Third Avenue, New York, NY 10158. (800) 526-5368 or (212) 850-6000. Fourth edition. 1986. $175.00.

PHYSICAL CHEMISTRY. Robert A. Alberty. John Wiley and Sons, Incorporated, 605 Third Avenue, New York, NY 10158. (800) 526-5368 or (212) 850-6000. 1987. Inquire as to cost and availability.

QUANTUM CHEMISTRY. Raymond Daudel and others. John Wiley and Sons, Incorporated, 605 Third Avenue, New York, NY 10158. (800) 526-5368 or (212) 850-6000. 1984. $139.95.

QUANTUM CHEMISTRY OF ATOMS AND MOLECULES. Philip S. C. Matthews. Cambridge University Press, 32 East 57th Street, New York, NY 10022. (212) 688-8885. 1986. $44.50.

SYMMETRY AND STRUCTURE. S.A. Kettle. John Wiley and Sons, Incorporated, 605 Third Avenue, New York, NY 10158. (800) 526-5368 or (212) 850-6000. 1985. $37.95.

HANDBOOKS AND MANUALS

THE CHEMIST'S COMPANION: A HANDBOOK OF PRACTICAL DATA, TECHNIQUES, AND REFERENCES. Arnold J. Gordon and Richard A. Ford. John Wiley and Sons, Incorporated, 605 Third Avenue, New York, NY 10158. (800) 526-5368. 1973. $49.95.

CRC HANDBOOK OF CHEMISTRY AND PHYSICS. CRC Press, Incorporated, 2000 Corporate Boulevard, NW, Boca Raton, FL 33431. Sixty-seventh edition. 1986. $69.95.

HANDBOOK OF APPLIED CHEMISTRY: FACTS FOR ENGINEERS, SCIENTISTS. TECHNICIANS, AND TECHNICAL MANAGERS. Vollrath Hopp and Ingp Hennig. McGraw-Hill Book Company, 1221 Avenue of the Americas, New York, NY 10020. (800) 628-0004. 1983. $54.00.

HANDBOOK OF COMPUTATIONAL CHEMISTRY: A PRACTICAL GUIDE TO CHEMICAL STRUCTURE AND ENERGY CALCULATIONS. Tim Clark. John Wiley and Sons, Incorporated, 605 Third Avenue, New York, NY 10158. (800) 526-5368 or (212) 850-6000. 1985. $35.00.

LANGE'S HANDBOOK OF CHEMISTRY. John A. Dean, editor. McGraw-Hill Book Company, 1221 Avenue of the Americas, New York, NY 10020. (800) 628-0004. 1985. $59.50.

TABLES OF PHYSICAL AND CHEMICAL CONSTANTS; AND SOME MATHEMATICAL FUNCTIONS. G.W.C. Kaye and T.H. Laby, editors. Longman, Incorporated, 95 Church Street, White Plains, NY 10601. (914) 993-5000. 1986. $39.95.

ONLINE DATA BASES

CA SEARCH. Chemical Abstracts Service, P.O. Box 3012, Columbus, OH 43210. Guide to chemical literature, 1972 to present. Inquire as to cost and availability.

DISSERTATION ABSTRACTS ONLINE. University Microfilms International, 300 North Zeeb Road, Ann Arbor, MI 48106. (800) 521-0600 or (313) 761-4700. Scope includes virtually all doctoral dissertations accepted at accredited American Institutions from 1861 to present in 252 subject areas. Inquire as to online cost and availability.

INSPEC. INSPEC Marketing Department, Institute of Electrical and Electronics Engineers, Incorporated, IEEE Service Department, 445 Hoes Lane, Piscataway, NJ 08854. (201) 981-0060. Inquire as to online cost and availability.

NTIS. National Technical Information Service, 5285 Port Royal Road, Springfield, VA 22161. (703) 487-4630. Broad coverage of government sponsored research reports, 1964 to present. Inquire as to cost and availability.

SCISEARCH. Institute for Scientific Information, 3501 Market Street, Philadelphia, PA 19104. (800) 523-1850 or (215) 386-0100. Broad multidisciplinary title and author index to the international literature of science and technology, 1974 to present. Inquire as to online cost and availability.

OTHER SOURCES

GUIDE TO BASIC INFORMATION SOURCES IN CHEMISTRY. Arthur Antony. John Wiley and Sons, Incorporated, 605 Third Avenue, New York, NY 10158. (800) 526-5368 or (212) 850-6000. 1979. $26.95.

HOW TO FIND CHEMICAL INFORMATION: A GUIDE FOR PRACTICING CHEMISTS, TEACHERS, AND STUDENTS. John Wiley and Sons, Incorporated, 605 Third Avenue, New York, NY 10158. (800) 526-5368 or (212) 850-6000. 1986. $35.00.

PERIODICALS

ANDEWANDTE CHEMI (GESELLSCHAFT DEUTSCHER CHEMIKER, GW). V C H Verlagsgesellschaft mbH, Pappelallee 3, Postfach 1260, 6940 Weinheim, West Germany. Monthly. $300.00 per year.

CHEMICAL PHYSICS. Elsevier Science Publishers B.V., Box 211, 1000 AE Amsterdam, The Netherlands. Thirty times per year. $1200.00 per year.

CHEMICAL PHYSICS LETTERS. Elsevier Science Publishers B.V., Box 211, 1000 AE Amsterdam, The Netherlands. Sixty-six times per year. $1500.00 per year.

CHEMICAL REVIEWS. American Chemical Society, 1155 Sixteenth Street, NW, Washington, DC 20036. (800) 424-6747 or (202) 872-4700. Bimonthly. $83.00 per year.

CHEMTECH. American Chemical Society, 1155 Sixteenth Street, NW, Washington, DC 20036. (800) 424-6747 or (202) 872-4700. Monthly. $40.00 per year to individuals.

INTERNATIONAL REVIEWS IN PHYSICAL CHEMISTRY. Taylor and Francis, Limited, 242 Cherry Street, Philadelphia, PA 19106-1906. Three times per year. $143.00 per year.

INORGANIC CHEMISTRY. American Chemical Society, 1155 Sixteenth Street, NW, Washington, DC 20036. (800) 424-6747 or (202) 872-4700. Monthly. $400.00 per year.

JOURNAL OF CATALYSIS. Academic Press, Incorporated, Journals Division, 1250 Sixth Avenue, San Diego, CA 92101. (619) 20-1840. Monthly. $654.00 per year.

JOURNAL OF PHYSICAL CHEMISTRY. American Chemical Society, 1155 Sixteenth Street, NW, Washington, DC 20036. (800) 424-6747 or (202) 872-4700. Biweekly. $369.00 per year.

JOURNAL OF SOLID STATE CHEMISTRY. Academic Press, Incorporated, Journals Division, 1250 Sixth Avenue, San Diego, CA 92101. (619) 230-1840. Fifteen times per year. $530.00 per year.

JOURNAL OF THE AMERICAN CHEMICAL SOCIETY. American Chemical Society, 1155 Sixteenth Street, NW, Washington, DC 20036. (800) 424-6747 or (202) 872-4700. Biweekly. $330.00 per year.

RESEARCH CENTERS AND INSTITUTES

BROWN UNIVERSITY. Chemical Research Laboratory, Providence, RI 02912. (401) 863-2256.

CORNELL UNIVERSITY. Materials Science Center, Clark Hall of Science, Ithaca, NY 14853. (607) 255-4272.

HARVARD UNIVERSITY. Chemical Laboratories, Oxford Street, Cambridge, MA 02138. (617) 495-4283.

INSTITUTE FOR SCIENTIFIC RESEARCH, INCORPORATED. 271 Main Street, Suite 302, Stoneham, MA 02180. (617) 438-7894.

UNIVERSITY OF PENNSYLVANIA. Laboratory for Research on the Structure of Matter, 3231 Walnut Street, Philadelphia, PA 19104. (215) 898-8571.

UNIVERSITY OF TEXAS AT AUSTIN. Institute of Theoretical Chemistry, Department of Chemistry, WEL 2.204, Austin, TX 78712. (512) 471-3114.

QUANTUM ELECTRODYNAMICS

See: QUANTUM MECHANICS

QUANTUM MECHANICS

See also: MECHANICS, PHYSICS, STATISTICAL MECHANICS

ABSTRACT SERVICES AND INDEXES

APPLIED MECHANICS REVIEW. American Society of Mechanical Engineers, 345 East 47th Street, New York, NY 10017. (212) 705-7703. 1948 to present. Monthly. $360.00 per year.

CHEMICAL ABSTRACTS. American Chemical Society, Chemical Abstracts Service, Box 3012, Columbus, OH 43210. (614) 421-3600. 1907 to present. Weekly. $9500.00 per year.

CONFERENCE PAPERS INDEX. Cambridge Scientific Abstracts, 5161 River Road, Bethesda, MD 20816. 1972 to present. Monthly. Inquire as to cost and availability.

CURRENT CONTENTS: PHYSICAL, CHEMICAL, AND EARTH SCIENCES. Institute for Scientific Information, 3501 Market Street, Philadelphia, PA 19104. (800) 523-1850 or (215) 386-0100. Weekly. $275.00 per year.

ENGINEERING INDEX MONTHLY AND AUTHOR INDEX. Engineering Information Inc., 345 East 47th Street, New York, NY 10017. (212) 705-7600. Monthly. $1560.00 per year.

GENERAL SCIENCE INDEX. H.W. Wilson and Company, 950 University Avenue, Bronx, NY 10452. (800) 367-6670 or (212) 588-8400. 1978 to present. Monthly. Inquire as to cost and availability.

ISMEC BULLETIN (Information Service in Mechanical Engineering). Cambridge Scientific Abstracts, 5161 River Road, Bethesda, MD 20816. (301) 951-1400. 1973 to present. Monthly. $450.00 per year.

INDEX TO SCIENTIFIC AND TECHNICAL PROCEEDINGS. Institute for Scientific Information, 3501 Market Street, Philadelphia, PA 19104. (800) 523-1850 or (215) 386-0100. 1978 to present. Monthly. $775.00 per year.

INDEX TO SCIENTIFIC REVIEWS. Institute for Scientific Information, 3501 Market Street, Philadelphia, PA 19104. (800) 523-1850 or (215) 386-0100. 1974 to present. Semi-annual. $550.00 per year.

PHYSICS ABSTRACTS. Institution of Electrical Engineers. Available from: IEEE Service Center, 445 Hoes Lane, Piscataway, NJ 08854. 1898 to present. Bimonthly. $1700.00 per year.

PHYSICS BRIEFS. Physik Verlag GmbH, Postfach 1260/1280, D-6940 Weinheim, West Germany. (212) 661-9404. 1920 to present. Twenty-six times per year. $1250.00 per year.

SCIENCE CITATION INDEX. Institute for Scientific Information, 3501 Market Street, Philadelphia, PA 19104. (800) 523-1850 or (215) 386-0100. Six times per year. $6200. 00 per year.

ASSOCIATIONS AND PROFESSIONAL SOCIETIES

AMERICAN CHEMICAL SOCIETY. 1155 16th Street, N.W., Washington, DC 20036. (800) 424-6747.

AMERICAN INSTITUTE OF PHYSICS. 335 East 45th Street, New York, NY 10017. (212) 661-9494.

AMERICAN SOCIETY OF MECHANICAL ENGINEERS. 345 East 47th Street, New York, NY 10017. (212) 705-7703.

DIRECTORIES AND BIOGRAPHICAL SOURCES

AMERICAN MEN AND WOMEN OF SCIENCE. R.R. Bowker, Inc., Order Department, 245 West 17th Street, New York, NY 10011. (800) 521-8110. Eight volumes. 1986. $595.00 for set.

INTERNATIONAL RESEARCH CENTERS DIRECTORY 1988-89. Darren L. Smith, editor. Gale Research Company, Book Tower, Detroit, MI 48226. (800) 521-0707. 4th edition. 1987. $360.00.

1987 DIRECTORY OF ENGINEERING SOCIETIES AND RELATED ORGANIZATIONS. Gordon Davis, editor. Hemisphere Publishing Corporation, 79 Madison Avenue, New York, NY 10016-7892. (800) 821-8312. 12th edition. 1987. $100.00.

RESEARCH CENTERS DIRECTORY 1988. Gale Research Company, Book Tower, Detroit, MI 48226. (800) 521-0707. 12th edition. 1987. $365.00 for set.

SCIENTIFIC AND TECHNICAL ORGANIZATIONS AND AGENCIES DIRECTORY. Margaret Labash Young, editor. Gale Research Company, Book Tower, Detroit, MI 48226. (800) 521-0707. 2nd edition. 1987. $185.00.

WHO'S WHO IN ENGINEERING. Gordon Davis, editor. Hemisphere Publishing Corporation, 79 Madison Avenue, New York, NY 10016-7892. (800) 821-8312. Sixth edition. 1985. $200.00.

ENCYCLOPEDIAS AND DICTIONARIES

MCGRAW-HILL DICTIONARY OF PHYSICS. McGraw-Hill Book Company, 1221 Avenue of the Americas, New York, NY 10020. (212) 512-2000. 1986. $17.95.

THESAURUS OF SCIENTIFIC, TECHNICAL, AND ENGINEERING TERMS. Hemisphere Publishing Corporation, 1010 Vermont Avenue, NW, Washington, DC 20005. (800) 526-0275. 1988. $125.00.

GENERAL WORKS

GHOST IN THE ATOM: A DISCUSSION OF THE MYSTERIES OF QUANTUM PHYSICS. P.C.W. Davies and J.R. Brown. Cambridge University Press, 32 East 57th Street, New York, NY 10022. (800) 872-7423. 1986. $29.95.

QUANTUM MECHANICS. V.K. Thankappan. John Wiley and Sons, Inc., 605 Third Avenue, New York, NY 10158. (800) 526-5368. 1985. $45.95.

QUANTUM MECHANICS FROM GENERAL RELATIVITY. Mendel Sachs. Kluwer Academic Publishers, 190 Old Derby Street, Hingham, MA 02043. (617) 749-5262. 1986. $72.95.

QUANTUM REALITY: BEYOND THE NEW PHYSICS. Nick Herbert. Doubleday and Company, Inc., 245 Park Avenue, New York, NY 10017. (212) 953-4561. 1986. $15.95.

THE QUANTUM UNIVERSE. A.J.G. Hey and others. Cambridge University Press, 32 East 57th Street, New York, NY 10022. (800) 872-7423. 1987. $47.50.

ONLINE DATA BASES

CA SEARCH. Chemical Abstracts Service, P.O. Box 3012, Columbus, OH 43120. (800) 848-6538 or (614) 421-3600. Comprehensive guide to chemical literature, 1972 to present. Inquire as to online cost and availability.

COMPENDEX. Engineering Information, Inc., 345 East 47th Street, New York, NY 10017. (800) 221-1044 or (212) 705-7615. Engineering and technical literature, 1975 to present. Inquire as to online cost and availability.

DISSERTATION ABSTRACTS ONLINE. University Microfilms International, 300 North Zeeb Road, Ann Arbor, MI 48106. (800) 521-0600 or (313) 761-4700. Scope includes virtually all doctoral dissertations accepted at accredited American institutions from 1861 to present in over 250 subject areas. Inquire as to online cost and availability.

INSPEC. INSPEC Marketing Department, Institution of Electrical Engineers. Available from IEEE Service Center, 445 Hoes Lane, Piscataway, NJ 08854. (201) 981-0060. Online version of Physics Abstracts. Inquire as to online cost and availability.

NTIS. National Technical Information Service, 5285 Port Royal Road, Springfield, VA 22161. (703) 487-4630. Broad coverage of government sponsored research reports, 1964 to present. Inquire as to online cost and availability.

SCISEARCH. Institute for Scientific Information, 3501 Market Street, Philadelphia, PA 19104. (800) 523-1850 or (215) 386-0100. Broad multidisciplinary title and author index to the international literature of science and technology, 1974 to present. Inquire as to online cost and availability.

WILSONLINE. H.W. Wilson and Company, 950 University Avenue, Bronx, NY 10452. (800) 367-6770 or (212) 588-8400. Makes available online versions of the H.W. Wilson indexes including Applied Science and Technology Index, Business Periodicals Index and Readers' Guide to Periodical Literature. Approximately 1980 to present. Inquire as to online cost and availability.

PERIODICALS

INTERNATIONAL JOURNAL OF SOLIDS AND STRUCTURES. Pergamon Press, Inc., Maxwell House, Fairview Park, Elmsford, NY 10523. (914) 592-7700. 1965 to present. Monthly. $375.00 per year.

JOURNAL OF NON-NEWTONIAN FLUID MECHANICS. Elsevier Science Publishing Company, Inc., 52 Vanderbilt Avenue, New York, NY 10017. (212) 370-5520. 1976 to present. Nine times per year. $180.00 per year.

JOURNAL OF RHEOLOGY. John Wiley and Sons, Inc., 605 Third Avenue, New York, NY 10158. (800) 526-5368. 1957 to present. Bimonthly. $152.00 per year.

JOURNAL OF THE MECHANICS AND PHYSICS OF SOLIDS. Pergamon Press, Inc., Maxwell House, Fairview Park, Elmsford, NY 10523. (914) 592-7700. 1956 to present. Bimonthly. $200.00 per year.

PHYSICAL REVIEW LETTERS. American Institute of Physics, 335 East 45th Street, New York, NY 10017. (212) 661-9494. 1958 to present. Weekly. $430.00 per year.

RESEARCH CENTERS AND INSTITUTES

CENTER FOR THE ADVANCEMENT OF COMPUTATIONAL MECHANICS. Georgia Institute of Technology, Atlanta, GA 30332. (404) 894-2758.

ENRICO FERMI INSTITUTE. University of Chicago, 5630 South Ellis Avenue, Chicago, IL 60637. (312) 962-7823.

LABORATORY FOR EXPERIMENTAL MECHANICS RESEARCH. State University of New York at Stony Brook, Stony Brook, NY 11794-2300.

SOLID MECHANICS LABORATORY. University of Michigan, Department of Mechanical Engineering and Applied Mechanics, 2246A G.G. Brown Building, Ann Arbor, MI 48109. (313) 763-0684.

QUANTUM THEORY

See: QUANTUM MECHANICS

QUARKS

See: PARTICLE PHYSICS

QUARRYING

See also: DRILLING, MINING ENGINEERING, ROCK MECHANICS

ABSTRACT SERVICES AND INDEXES

APPLIED SCIENCE AND TECHNOLOGY INDEX. H.W. Wilson and Company, 950 University Avenue, Bronx, NY 10452. (800) 367-6670 or (212) 588-8400. Monthly. Inquire as to cost and availability.

BIBLIOGRAPHY AND INDEX OF GEOLOGY. American Geological Institute, 4220 King Street, Alexandria, VA 22302. (703) 379-2480. 1969 to present. Monthly. $1100.00 per year.

CHEMICAL ABSTRACTS. American Chemical Society, Chemical Abstracts Service, Box 3012, Columbus, OH 43210. (614) 421-3600. 1907 to present. Weekly. $9500.00 per year.

CURRENT CONTENTS: ENGINEERING, TECHNOLOGY AND APPLIED SCIENCES. Institute for Scientific Information, 3501 Market Street, Philadelphia, PA 19104. (800) 523-1850 or (215) 386-0100. Weekly. $275.00 per year.

ENGINEERING INDEX MONTHLY AND AUTHOR INDEX. Engineering Information Inc., 345 East 47th Street, New York, NY 10017. (212) 705-7600. Monthly. $1560.00 per year.

INTERNATIONAL JOURNAL OF ROCK MECHANICS AND MINING SCIENCES AND GEOMECHANICS ABSTRACTS. Pergamon Press, Inc., Maxwell House, Fairview Park, Elmsford, NY 10523. (914) 592-7700. 1964 to present. Bimonthly. $385.00 per year.

ASSOCIATIONS AND PROFESSIONAL SOCIETIES

AMERICAN INSTITUTE OF MINING, METALLURGICAL AND PETROLEUM ENGINEERS. 345 East 47th Street, New York, NY 10017. (212) 705-7695.

AMERICAN SOCIETY OF CIVIL ENGINEERS. 345 East 47th Street, New York, NY 10017. (212) 705-7496.

MINERALOGICAL SOCIETY OF AMERICA. 1625 I Street, N.W., Suite 414, Washington, DC 20006. (202) 775-4344.

MINING AND METALLURGICAL SOCIETY OF AMERICA. 275 Madison Avenue, Room 2301, New York, NY 10016. (212) 684-4150.

SOCIETY OF MINING ENGINEERS OF AIME. Caller Number D, Littleton, CO 80127. (303) 973-9550.

DIRECTORIES AND BIOGRAPHICAL SOURCES

MINING ENGINEERING - SOCIETY OF MINING ENGINEERS DIRECTORY ISSUE. Society of Mining Engineers, Caller Number D, Littleton, CO 80127. Annual. $100.00.

QUARRIES DIRECTORY. American Business Directories, Inc., 5707 South 86th Street, Omaha, NE 68127. (402) 331-7169. 1986. $95.00.

RESEARCH CENTERS DIRECTORY 1988. Gale Research Company, Book Tower, Detroit, MI 48226. (800) 521-0707. 12th edition. 1987. $365.00 for set.

SCIENTIFIC AND TECHNICAL ORGANIZATIONS AND AGENCIES DIRECTORY. Margaret Labash Young, editor. Gale Research Company, Book Tower, Detroit, MI 48226. (800) 521-0707. 2nd edition. 1987. $185.00.

WESTERN MINING DIRECTORY. Howell Publishing Company, 311 Steele Street, Suite 208, Denver, CO 80206. Anuual. $45.00.

WHO'S WHO IN ENGINEERING. Gordon Davis, editor. Hemisphere Publishing Corporation, 1010 Vermont Avenue, NW, Washington, DC 20005. (800) 526-0275. 6th edition. 1985. $200.00.

WHO'S WHO IN MINERAL ENGINEERING. Society of Mining Engineers, Caller Number D, Littleton, CO 80127. (303) 973-9550. Annual. $100.00.

GENERAL WORKS

ELEMENTS OF MINING. Robert S Lewis. Books on Demand, 300 North Zeeb Road, Ann Arbor, MI 48106. (313) 761-4700. 3rd edition. $160.00 in paper.

MINING GEOLOGY. Willard C. Lacy, editor. Van Nostrand Reinhold Company, Inc., 135 West 50th Street, New York, NY 10020. (800) 543-2681. 1983. $59.95.

QUARRYING, OPENCAST AND ALLUVIAL MINING. John Sinclair. Elsevier Science Publishing Company, Inc., 52 Vanderbilt Avenue, New York, NY 10017. (212) 370-5520. 1969. $50.00.

STONE COUNTRY. Scott R. Sanders. Indiana University Press, 10th and Morton Streets, Bloomington, IN 74403. (812) 335-8287. 1985. $24.95.

ONLINE DATA BASES

CA SEARCH. Chemical Abstracts Service, P.O. Box 3012, Columbus, OH 43120. (800) 848-6538 or (614) 421-3600. Comprehensive guide to chemical literature, 1972 to present. Inquire as to online cost and availability.

COMPENDEX. Engineering Information, Inc., 345 East 47th Street, New York, NY 10017. (800) 221-1044 or (212) 705-7615. Engineering and technical literature, 1975 to present. Inquire as to online cost and availability.

DISSERTATION ABSTRACTS ONLINE. University Microfilms International, 300 North Zeeb Road, Ann Arbor, MI 48106. (800) 521-0600 or (313) 761-4700. Scope includes virtually all doctoral dissertations accepted at accredited American institutions from 1861 to present in over 250 subject areas. Inquire as to online cost and availability.

GEOREF. Online version of the BIBLIOGRAPHY AND INDEX OF GEOLOGY. American Geological Institute, 4220 King Street, Alexandria, VA 22302. (703) 379-2480. 1969 to present. Inquire as to online cost and availability.

NTIS. National Technical Information Service, 5285 Port Royal Road, Springfield, VA 22161. (703) 487-4630. Broad coverage of government sponsored research reports, 1964 to present. Inquire as to online cost and availability.

SCISEARCH. Institute for Scientific Information, 3501 Market Street, Philadelphia, PA 19104. (800) 523-1850 or (215) 386-0100. Broad multidisciplinary title and author index to the international literature of science and technology, 1974 to present. Inquire as to online cost and availability.

WILSONLINE. H.W. Wilson and Company, 950 University Avenue, Bronx, NY 10452. (800) 367-6770 or (212) 588-8400. Makes available online versions of the H.W. Wilson indexes including Applied Science and Technology Index, Business Periodicals Index and Readers' Guide to Periodical Literature. Approximately 1980 to present. Inquire as to online cost and availability.

PERIODICALS

AMERICAN MINING CONGRESS JOURNAL. American Mining Congress, 1920 N Street, N.W., Suite 300, Washington, DC 20036. (202) 861-2800. 1915 to present. Semimonthly. $50.00 per year.

QUARRYING

CIM BULLETIN. Canadian Institute of Mining and Metallurgy, 400-1130 Sherbrooke West, Montreal, Quebec H3A 2M8, Canada. 1898 to present. Monthly. $60.00 per year.

CANADIAN MINING JOURNAL. Southam Communications Limited, 1450 Don Mills Road, Don Mills, Ontario, M3B 2X7, Canada. (416) 445-6641. 1879 to present. Monthly. $47.50 per year.

COAL AGE. McGraw-Hill Book Company, 1221 Avenue of the Americas, New York, NY 10020. (212) 512-2000. 1911 to present. Monthly. $18.00 per year.

COLORADO SCHOOL OF MINES QUARTERLY. Colorado School of Mines Press, Golden, CO 80401. 1905 to present. Quarterly. $50.00 per year.

ENGINEERING AND MINING JOURNAL (E & M J). McGraw-Hill Book Company, 1221 Avenue of the Americas, New York, NY 10020. (212) 512-2000. 1866 to present. Monthly. $20.00 per year.

MINING ENGINEER. Institution of Mining Engineers, Danum House, South Parade, Doncaster DN1 2DY England. 1960 to present. Monthly. $60.00 per year.

MINING ENGINEERING. Society of Mining Engineers, Caller Number D, Littleton, CO 80127. (303) 973-9550. 1949 to present. Monthly. $40.00 per year.

NORTHERN MINER. Northern Miner Press Limited, 7 Labatt Avenue, Toronto, M5A 3P2 Canada. 1915 to present. Weekly. $40.00 per year.

PIT AND QUARRY. Harcourt Brace Jovanovich, Inc., 7500 Old Oak Boulevard, Cleveland, OH 44130. 1916 to present. Monthly. $15.00 per year.

RESEARCH CENTERS AND INSTITUTES

BUREAU OF MINES. U.S. Department of the Interior, 2401 E Street, N.W., Washington, DC 20241. (202) 634-1004.

EXCAVATION ENGINEERING AND EARTH MECHANICS INSTITUTE. Colorado School of Mines, Golden, CO 80401. (303) 273-3400.

MINING AND MINERAL RESOURCES INSTITUTE. New Mexico Institute of Mining and Technology, Campus Station, Socorro, NM 87801. (505) 835-5142.

MINING AND MINERAL RESOURCES RESEARCH INSTITUTE. Ohio State University, 233 Koffolt Laboratory, 140 West 19th Avenue, Columbus, OH 43210-1110. (614) 292-0102.

QUASARS

See also: ASTRONOMY, ASTROPHYSICS, BIG BANG, COSMOCHEMISTRY, COSMOLOGY, RADIO ASTRONOMY, STARS

ABSTRACT SERVICES AND INDEXES

ASTRONOMY AND ASTROPHYSICS ABSTRACTS. Springer-Verlag New York, Incorporated, 175 Fifth Avenue, New York, NY 10010. (212) 460-1500. $70.00 per year.

ASTRONOMY AND ASTROPHYSICS ABSTRACTS. Springer-Verlag New York, Incorporated, 175 Fifth Avenue, New York, NY 10010. (212) 460-1500. $70.00 per year.

PHYSICS ABSTRACTS. Institute of Electrical Engineers, London, United Kingdom. Available from: Institute of Electrical and Electronic Engineers (IEEE), 345 East 47th Street, New York, NY 10017. (212) 705-7900. 1898 to present. Monthly. $1670.00 per year.

SCIENCE CITATION INDEX. Institute for Scientific Information, 3501 Market Street, Philadelphia, PA 19104. (800) 523-1850 or (215) 386-0100. Inquire as to cost and availability.

ANNUAL REVIEWS AND YEARBOOKS

ANNUAL REVIEW OF ASTRONOMY AND ASTROPHYSICS. Annual Reviews, Incorporated, 4139 El Camino Way, Palo Alto, CA 94306. (415) 493-4400.

ASSOCIATIONS AND PROFESSIONAL SOCIETIES

AMERICAN ASTRONOMICAL SOCIETY. 2000 Florida Avenue, NW, Suite 300, Washington, DC 20009. (202) 659-0134.

AMERICAN ASSOCIATION OF VARIABLE STAR OBSERVERS. 187 Concord Avenue, Cambridge, MA 02138. (617) 354-0484.

ASTRONOMICAL LEAGUE. Post Office Box 12821, Tucson, AZ 85732. (602) 790-8471.

ASTRONOMICAL SOCIETY OF THE PACIFIC. 1290 24th Avenue, San Francisco, CA 94122. (415) 661-8660.

BIBLIOGRAPHIES

A BIBLIOGRAPHY OF ASTRONOMY, 1970-1979. R.A. Seal and S.S. Martin. Libraries Unlimited, Incorporated, Littleton, CO 80160. 1982. $37.50.

DIRECTORIES AND BIOGRAPHICAL SOURCES

AMERICAN MEN AND WOMEN OF SCIENCE. Physical and Biological Sciences. Sixteenth edition. R.R. Bowker Company, 205 East 42nd Street, New York, NY 10017. (800) 521-8110 or (212) 916-1600. 1986. $595.00.

THE BIOGRAPHICAL DICTIONARY OF SCIENTISTS: ASTRONOMERS. D. Abbott, editor. Peter Bedrick Books, 125 East 23rd Street, New York, NY 10010. 1984.

DIRECTORY OF PHYSICS AND ASTRONOMY STAFF MEMBERS. American Institute of Physics, 335 East 45th Street, New York, NY 10017. Annual.

RESEARCH CENTERS DIRECTORY. Gale Research Company, Book Tower, Detroit, MI 48226. (800) 521-0707. Twelfth edition. 1987. $365.00 for set.

WHO'S WHO IN FRONTIER SCIENCE AND TECHNOLOGY. Marquis Who's Who, Incorporated, 200 East Ohio Street, Chicago, IL 60611. (800) 428-3898 or (312) 787-2008.

ENCYCLOPEDIAS AND DICTIONARIES

ASTEROIDS TO QUASARS. Phyllis Luggar. Cambridge University Press, 32 East 57th Street, New York, NY 10022. (800) 872-7423. 1988. $37.50.

ILLUSTRATED ENCYCLOPEDIA OF THE UNIVERSE: UNDERSTANDING AND EXPLORING THE COSMOS. Richard S. Lewis. Crown Publishers Incorporated, 1 Park Avenue, New York, NY 10016. (800) 526-4264. 1986. $24.95.

GENERAL WORKS

CONSTRUCTING THE UNIVERSE. David Layzer. W.H. Freeman and Company, 41 Madison Avenue, New York, NY 10010. (212) 532-7660. 1984. $29.95.

FIRST LIGHT: THE SEARCH FOR THE EDGE OF THE UNIVERSE. Richard Preston. Atlantic Monthly Press, Distributed by Little Brown and Company, 34 Beacon Street, Boston, MA 02108. (617) 536-9500. 1987. $18.95.

FIRST THREE MINUTES: A MODERN VIEW OF THE ORIGIN OF THE UNIVERSE. Steven Weinberg. Basic Books, Incorporated, 10 East 53rd Street, New York, NY 10022. (800) 242-7737. 1976. $14.95.

QUARK TO QUASAR. Peter H. Cadogan. Cambridge University Press, 32 East 57th Street, New York, NY 10022. (800) 872-7423. 1985. $24.95.

QUASAR ASTRONOMY. Daniel Weedman. Cambridge University Press, 32 East 57th Street, New York, NY 10022. (800) 872-7423. 1988. $14.95.

QUASARS. G. Swarup and V.K. Kapahi, editors. Kluwer Academic Publishers, 190 Old Derby Street, Hingham, MA 02043. (617) 749-5262. 1986. $110.00.

ONLINE DATA BASES

DISSERTATION ABSTRACTS ONLINE. University Microfilms International, 300 North Zeeb Road, Ann Arbor, MI 48106. (800) 521-0600 or (313) 761-4700. Scope includes virtually all doctoral dissertations accepted at accredited American institutions from 1861 to present in 252 subject areas. Inquire as to cost and availability.

INSPEC. INSPEC Marketing Department, Institute of Electrical and Electronics Engineers, Incorporated, IEEE Service Department, 445 Hoes Lane, Piscataway, NJ 08854. (201) 981-0060. Inquire as to on-line cost and availability.

NASA. National Aeronautics and Space Administration, Scientific and Technical Information Branch, 300 7th Street, SW, Washington, DC 20546. Citations and abstracts of aerospace literature, 1962 to present. Inquire as to cost and availability.

NTIS. National Technical Information Service, 5285 Port Royal Road, Springfield, VA 22161. (703) 487-4630. Broad coverage of government-sponsored research reports, 1964 to present. Inquire as to cost and availability.

SCISEARCH. Institute for Scientific Information, 3501 Market Street, Philadelphia, PA 19104. (800) 523-1850 or (215) 386-0100. Broad multidisciplinary title and author index to the international literature of science and technology, 1974 to present. Inquire as to cost and availability.

OTHER SOURCES

ATLAS OF DEEP-SKY SPLENDORS. H. Vehrenberg. Cambridge University Press, 32 East 57th Street, New York, NY 10022. (800) 431-1580 or (212) 688-8888. Fourth edition, 1984. $44.50.

SKY CATALOGUE 2000.0. A. Hirshfeld and R. Sinnott. Cambridge University Press, 32 East 57th Street, New York, NY 10022. (800) 431-1580 or (212) 688-8888.

PERIODICALS

ASTRONOMICAL JOURNAL. American Astronomical Society. Available from: American Institute of Physics, 335 East 45th Street, New York, NY 10017. (212) 661-9404. $125.00 per year.

ASTRONOMICAL SOCIETY OF THE PACIFIC. Publications. Astronomical Society of the Pacific, 1290 24th Avenue, San Francisco, CA 94122. (415) 661-8660. Monthly. $38.00.

ASTRONOMY. Astro Media Corporation, 625 East Paul Avenue, Milwaukee, WI 53202. Monthly. $18.00 per year.

ASTRONOMY AND ASTROPHYSICS. Springer-Verlag New York, Incorporated, 175 Fifth Avenue, New York, NY 10010. (800) 526-7254 or (212) 460-1500. $680.00 per year.

ASTROPHYSICAL JOURNAL. American Astronomical Society, University of Chicago Press, 5801 Ellis Avenue, Chicago, IL 60637. Biweekly. $305.00 per year.

ASTROPHYSICS AND SPACE SCIENCE. D. Reidel Publishing Company, 190 Old Derby Street, Hingham, MA 02043. Monthly. $101.00 per year.

CELESTIAL MECHANICS: AN INTERNATIONAL JOURNAL OF SPACE DYNAMICS. D. Reidel Publishing Company, 190 Old Derby Street, Hingham, MA 02043. Monthly. $310.00 per year.

FUNDAMENTALS OF COSMIC PHYSICS. Gordon and Breach Science Publishers, Limited, Post Office Box 197, London, WC2E 9PX. Monthly. $335.00 per year.

MERCURY. Astronomical Society of the Pacific, 1290 24th Avenue, San Francisco, CA 94122. (415) 661-8660. Bimonthly. $21.00 per year.

MONTHLY NOTICES OF THE ROYAL ASTRONOMICAL SOCIETY. Blackwell Science Publications, Incorporated, 667 Lytton Avenue, Palo Alto, CA 94301. (415) 324-1688. Monthly. $134.00 per year to individuals.

SKY AND TELESCOPE. Sky Publishing Corporation, 49 Bay State Road, Cambridge, MA 02238. (617) 864-7360. Monthly. $18.00 per year.

SOVIET ASTRONOMY (TRANSLATION OF ASTRO-NOMICHESKII ZHURNAL). American Institute of Physics, 335 East 45th Street, New York, NY 10017. (212) 661-9404. Bimonthly. $425.00 per year.

SPACE SCIENCE REVIEWS. D. Reidel Publishing Company, 190 Old Derby Street, Hingham, MA 02043. Monthly. $305.00 per year.

VISTAS IN ASTRONOMY. Pergamon Press, Incorporated, Maxwell House, Fairview Park, Elmsford, NY 10523. (914) 592-7700. Quarterly. $145.00 per year.

RESEARCH CENTERS AND INSTITUTES

CALIFORNIA INSTITUTE OF TECHNOLOGY. Palomar Observatory, 105-24, Pasadena, CA 91125. (818) 356-4033.

FRED L. WHIPPLE OBSERVATORY. Smithsonian Institution, Post Office Box 97, Amado, AZ 85645. (602) 629-6741.

JOINT INSTITUTE FOR LABORATORY ASTROPHYSICS (JILA). University of Colorado - Boulder, Boulder, CO 80309. (303) 492-7789.

UNIVERSITY OF MARYLAND. Astronomy Program, College Park, MD 20742. (301) 454-3005.

UNIVERSITY OF TEXAS AT AUSTIN. McDonald Observatory, Post Office Box 1337, Fort Davis, TX 79734. (915) 426-3263.

QUATERNIONS

See: PARTICLE PHYSICS

QUENCHING

See: MATALLURGY

R

RADAR

See also: ANTENNAS, ELECTRICAL ENGINEERING, ELECTRONIC CIRCUITS AND COMPONENTS, ELECTRONICS, ELECTRONICS ENGINEERING, MICROWAVES, RADIO

ABSTRACT SERVICES AND INDEXES

APPLIED SCIENCE AND TECHNOLOGY INDEX. H.W. Wilson and Company, 950 University Avenue, Bronx, NY 10452. (800) 367-6670 or (212) 588-8400. Monthly. Inquire as to cost and availability.

CURRENT CONTENTS: ENGINEERING, TECHNOLOGY AND APPLIED SCIENCES. Institute for Scientific Information, 3501 Market Street, Philadelphia, PA 19104. (800) 523-1850 or (215) 386-0100. Weekly. $275.00 per year.

ELECTRICAL AND ELECTRONICS ABSTRACTS. Institution of Electrical Engineers. Available from: Institute of Electrical and Electronics Engineers. IEEE Service Center, 445 Hoes Lane, Piscataway, NJ 08854. Monthly. $1250.00 per year.

ELECTRONICS AND COMMUNICATIONS ABSTRACTS. Cambridge Scientific Abstracts, 5161 River Road, Bethesda, MD 20816. (301) 951-1400. Bimonthly. Inquire as to cost and availability.

ENGINEERING INDEX MONTHLY AND AUTHOR INDEX. Engineering Information Inc., 345 East 47th Street, New York, NY 10017. (212) 705-7600. Monthly. $1560.00 per year.

GENERAL SCIENCE INDEX. H.W. Wilson and Company, 950 University Avenue, Bronx, NY 10452. (800) 367-6670 or (212) 588-8400. 1978 to present. Monthly. Inquire as to cost and availability.

IEEE PUBLICATIONS BULLETIN. Institute of Electrical and Electronics Engineers. Institute of Electrical and Electronics Engineers. IEEE Service Center, 445 Hoes Lane, Piscataway, NJ 08854. Quarterly. Free.

PHYSICS ABSTRACTS. Institution of Electrical Engineers. Available from: IEEE Service Center, 445 Hoes Lane, Piscataway, NJ 08854. 1898 to present. Bimonthly. $1700.00 per year.

PHYSICS BRIEFS. Physik Verlag GmbH, Postfach 1260/1280, D-6940 Weinheim, West Germany. (212) 661-9404. 1920 to present. Twenty-six times per year. $1250.00 per year.

SCIENCE CITATION INDEX. Institute for Scientific Information, 3501 Market Street, Philadelphia, PA 19104. (800) 523-1850 or (215) 386-0100. Six times per year. $6200.00 per year.

SOLID STATE ABSTRACTS: AN ABSTRACT JOURNAL INVOLVING THE PHYSICS, METALLURGY, CRYSTALLOGRAPHY, CHEMISTRY, AND DEVICE TECHNOLOGY OF SOLIDS. Cambridge Scientific Abstracts, 5161 River Road, Bethesda, MD 20816. (301) 951-1400. 1957 to present. Bimonthly. $550.00 per year.

ANNUAL REVIEWS AND YEARBOOKS

ADVANCES IN ELECTRONICS AND ELECTRON PHYSICS. Academic Press, Inc., 6277 Sea Harbor Drive, Orlando, FL 32821. (800) 321-5068. Irregular. Approximately $80.00 per volume.

ASSOCIATIONS AND PROFESSIONAL SOCIETIES

AMERICAN ELECTRONICS ASSOCIATION. P.O. Box 10045, 2670 Hanover Street, Palo Alto, CA 94303. (415) 857-9300.

AMERICAN INSTITUTE OF PHYSICS. 335 East 45th Street, New York, NY 10017. (212) 661-9494.

ELECTRONICS INDUSTRIES ASSOCIATION. 2001 Eye Street, N.W., Washington, DC 20006. (202) 457-4900.

INSTITUTE OF ELECTRICAL AND ELECTRONICS ENGINEERS. 345 East 47th Street, New York, NY 10017. (212) 705-7900.

INSTITUTE OF RADIO ENGINEERS. Institute of Electrical and Electronics Engineers, 345 East 47th Street, New York, NY 10017. (212) 705-7900.

NATIONAL ASSOCIATION OF RADIO AND TELECOMMUNICATIONS ENGINEERS. P.O. Box 15029, Salem, OR 97309. (503) 581-7653.

UNITED STATES NATIONAL COMMITTEE FOR THE INTERNATIONAL UNION OF RADIO SCIENCE. Board on Physics and Astronomy, National Research Council, 2101 Constitution Avenue, N.W., Washington, DC 20418. (202) 334-3559.

DIRECTORIES AND BIOGRAPHICAL SOURCES

AMERICAN MEN AND WOMEN OF SCIENCE. R.R. Bowker, Inc., Order Department, 245 West 17th Street, New York, NY 10011. (800) 521-8110. Eight volumes. 1986. $595.00 for set.

IEEE MEMBERSHIP DIRECTORY. Institute of Electrical and Electronics Engineers. IEEE Service Center, 445 Hoes Lane, Piscataway, NJ 08854. Annual. $7.00.

INTERNATIONAL RESEARCH CENTERS DIRECTORY 1988-89. Darren L. Smith, editor. Gale Research Company, Book Tower, Detroit, MI 48226. (800) 521-0707. 4th edition. 1987. $360.00.

1987 DIRECTORY OF ENGINEERING SOCIETIES AND RELATED ORGANIZATIONS. Gordon Davis, editor. Hemisphere Publishing Corporation, 1010 Vermont Avenue, NW, Washington, DC 20005. (800) 526-0275. 12th edition. 1987. $100.00.

RESEARCH CENTERS DIRECTORY 1988. Gale Research Company, Book Tower, Detroit, MI 48226. (800) 521-0707. 12th edition. 1987. $365.00 for set.

SCIENTIFIC AND TECHNICAL ORGANIZATIONS AND AGENCIES DIRECTORY. Margaret Labash Young, editor. Gale Research Company, Book Tower, Detroit, MI 48226. (800) 521-0707. 2nd edition. 1987. $185.00.

WHO'S WHO IN ELECTRONICS. Harris Publishing Company, 2057-2 Aurora Road, Twinsburg, OH 44087. (216) 425-9143. Annual. $90.00.

WHO'S WHO IN ENGINEERING. Gordon Davis, editor. Hemisphere Publishing Corporation, 1010 Vermont Avenue, NW, Washington, DC 20005. (800) 526-0275. 6th edition. 1985. $200.00.

ENCYCLOPEDIAS AND DICTIONARIES

IEEE STANDARD DICTIONARY OF ELECTRICAL AND ELECTRONICS TERMS. Frank Jay, editor. John Wiley and Sons, Inc., 605 Third Avenue, New York, NY 10158. (800) 526-5368. 3rd edition. 1984. $49.95.

GENERAL WORKS

ANTENNAS AND RADIO WAVE PROPAGATION. R.E. Collin. McGraw-Hill Book Company, 1221 Avenue of the Americas, New York, NY 10020. (212) 512-2000. 1985. $49.95.

ELECTROMAGNETIC CONCEPTS AND PRINCIPLES. Stanley V. Marshall and Gabriel G. Skitek. Prentice-Hall Publishing, Inc., Englewood Cliffs, NJ 07632. (800) 562-0245. 2nd edition. 1987. $42.95.

PRINCIPLES OF ANTENNA THEORY. K.F. Lee. John Wiley and Sons, Inc., 605 Third Avenue, New York, NY 10158. (800) 526-5368. 1984. $32.95.

RADAR PRINCIPLES FOR THE NON-SPECIALIST. John C. Toomay. Van Nostrand Reinhold Company, Inc., 135 West 50th Street, New York, NY 10020. (800) 543-2681. 1982. $24.95.

UNDERSTANDING ANTENNAS FOR RADAR, COMMUNICATIONS AND AVIONICS. B. Rulf and G.A. Robertshaw. Van Nostrand Reinhold Company, Inc., 135 West 50th Street, New York, NY 10020. (800) 543-2681. 1987. $46.95.

HANDBOOKS AND MANUALS

ANTENNA HANDBOOK: THEORY, APPLICATIONS AND DESIGN. Y.T. Lo and S.W. Lee. Van Nostrand Reinhold Company, Inc., 135 West 50th Street, New York, NY 10020. (800) 543-2681. 1988. $129.95.

ELECTRONIC ENGINEERS HANDBOOK. Donald G. Fink, editor. McGraw-Hill Book Company, 1221 Avenue of the Americas, New York, NY 10020. (212) 512-2000. 2nd edition. 1982. $89.00.

HANDBOOK OF MODERN ELECTRONICS AND ELECTRICAL ENGINEERING. Charles Belove, editor. John Wiley and Sons, Inc., 605 Third Avenue, New York, NY 10158. (800) 526-5368. 1986. $88.95.

RADAR HANDBOOK. K. Skolnik. McGraw-Hill Book Company, 1221 Avenue of the Americas, New York, NY 10020. (212) 512-2000. 1970. $100.00.

RADAR: HANDBOOK OF THEORY AND PRACTICE. Daniel P. Meyer and Herbert A. Mayer. Academic Press, Inc., 6277 Sea Harbor Drive, Orlando, FL 32821. (800) 321-5068. 1973. $84.50.

ONLINE DATA BASES

COMPENDEX. Engineering Information, Inc., 345 East 47th Street, New York, NY 10017. (800) 221-1044 or (212) 705-7615. Engineering and technical literature, 1975 to present. Inquire as to online cost and availability.

INSPEC. INSPEC Marketing Department, Institution of Electrical Engineers. Available from IEEE Service Center, 445 Hoes Lane, Piscataway, NJ 08854. (201) 981-0060. Online version of Physics Abstracts. Inquire as to online cost and availability.

NTIS. National Technical Information Service, 5285 Port Royal Road, Springfield, VA 22161. (703) 487-4630. Broad coverage of government sponsored research reports, 1964 to present. Inquire as to online cost and availability.

SCISEARCH. Institute for Scientific Information, 3501 Market Street, Philadelphia, PA 19104. (800) 523-1850 or (215) 386-0100. Broad multidisciplinary title and author index to the international literature of science and technology, 1974 to present. Inquire as to online cost and availability.

WILSONLINE. H.W. Wilson and Company, 950 University Avenue, Bronx, NY 10452. (800) 367-6770 or (212) 588-8400. Makes available online versions of the H.W. Wilson indexes including Applied Science and Technology Index, Business Periodicals Index and Readers' Guide to Periodical Literature. Approximately 1980 to present. Inquire as to online cost and availability.

OTHER SOURCES

A GUIDE TO THE LITERATURE OF ELECTRICAL AND ELECTRONICS ENGINEERING. Susan B. Ardis. Libraries Unlimited Inc., P.O. Box 263, Littleton, CO 80160. (303) 770-1220. 1987. $37.50.

PERIODICALS

ELECTRONIC DESIGN. Hayden Publishing Company, 10 Mulholland Drive, Hasbrouck Heights, NJ 07604. (201) 288-7520. 1952 to present. Biweekly. $40.00 per year.

ELECTRONICS. McGraw-Hill Book Company, 1221 Avenue of the Americas, New York, NY 10020. (212) 512-2000. 1930 to present. Weekly. $32.00 per year.

ELECTRONICS AND WIRELESS WORLD. I.P.C. Electrical-Electronic Press, Ltd., Quadrant House, The Quadrant, Sutton, Surrey, SM2 5AS England. 1911 to present. Monthly. $105.00 per year.

IEE PROCEEDINGS PART H: MICROWAVES, ANTENNAS AND PROPAGATION. Institution of Electrical Engineers (London). Available from: Institute of Electrical and Electronics Engineers. IEEE Service Center, 445 Hoes Lane, Piscataway, NJ 08854. 1980 to present. Bimonthly. Inquire.

IEEE CIRCUITS AND DEVICES MAGAZINE. Institute of Electrical and Electronics Engineers. IEEE Service Center, 445 Hoes Lane, Piscataway, NJ 08854. Bimonthly. $70.00 per year.

IEEE JOURNAL OF SOLID STATE CIRCUITS. Institute of Electrical and Electronics Engineers. IEEE Service Center, 445 Hoes Lane, Piscataway, NJ 08854. 1966 to present. Bimonthly. $113.00 per year.

IEEE TRANSACTIONS ON ANTENNAS AND PROPAGATION. Institute of Electrical and Electronics Engineers. IEEE Service Center, 445 Hoes Lane, Piscataway, NJ 08854. 1952 to present. Monthly. $140.00 per year.

RESEARCH CENTERS AND INSTITUTES

RADAR AND INSTRUMENTATION LABORATORY. Georgia Institute of Technology, Georgia Tech Research Institute, Atlanta, GA 30332. (404) 424-9621.

RADAR SYSTEMS AND REMOTE SENSING LABORATORY. 2291 Irving Hill Drive, Campus West, Lawrence, KS 66045. (913) 864-4832.

RADIATION

See also: GAMMA-RAYS, NUCLEAR ENERGY, NUCLEAR ENGINEERING, RADIATION SHIELDING

ABSTRACT SERVICES AND INDEXES

APPLIED SCIENCE AND TECHNOLOGY INDEX. H.W. Wilson Company, 950 University Avenue, Bronx, NY 10452. (800) 367-6670 or (212) 588-8400. Inquire as to cost and availability.

CHEMICAL ABSTRACTS. Chemical Abstracts Service, 2540 Olentangy Road, Post Office Box 3012, Columbus, OH 43210. (800) 848-6538 or (614) 421-3600. Weekly. $9200.00 per year.

CURRENT CONTENTS: PHYSICAL AND CHEMICAL SCIENCES. Institute for Scientific Information, 3501 Market Street, Philadelphia, PA 19104. (800) 523-1850 or (215) 386-0100. $272.00 per year.

GENERAL SCIENCE INDEX. H.W. Wilson Company, 950 University Avenue, Bronx, NY 10452. (800) 367-6770 or (212) 588-8400. Inquire as to cost and availability.

INDEX TO SCIENTIFIC REVIEWS. Institute for Scientific Information, 3501 Market Street, Philadelphia, PA 19104. (800) 523-1850 or (215) 386-0100. Semiannual. $550.00 per year.

PHYSICS ABSTRACTS. Institute of Electrical Engineers, London, United Kingdom. Available from: Institute of Electrical and Electronic Engineers (IEEE), 345 East 47th Street, New York, NY 10017. (212) 705-7900.

SCIENCE CITATION INDEX. Institute for Scientific Information, 3501 Market Street, Philadelphia, PA 19104. (800) 523-1850 or (215) 386-0100. Inquire as to cost and availability.

ASSOCIATIONS AND PROFESSIONAL SOCIETIES

AMERICAN NUCLEAR SOCIETY. 555 North Kensington Avenue, La Grange Park, IL 60525. (312) 352-6611.

AMERICAN SOCIETY OF NONDESTRUCTIVE TESTING. 4153 Arlingate Plaza, Caller #28518, Columbus, OH 43228. (614) 274-6003.

NATIONAL COUNCIL ON RADIATION PROTECTION AND MEASUREMENTS. 7910 Woodmont Avenue, Suite 1016, Bethesda, MD 20814. (301) 657-2652.

RADIATION RESEARCH SOCIETY. 925 Chestnut Street, Philadelphia, PA 19107. (215) 574-3153.

BIBLIOGRAPHIES

SCIENTIFIC AND TECHNICAL BOOKS AND SERIALS IN PRINT 1988: AN INDEX TO LITERATURE IN SCIENCE AND TECHNOLOGY. R.R. Bowker Company, 205 East 42nd Street, New York, NY 10017. (800) 521-8110 or (212) 916-1600. $175.00.

DIRECTORIES AND BIOGRAPHICAL SOURCES

AMERICAN MEN AND WOMEN OF SCIENCE. Physical and Biological Sciences. Sixteenth edition. R.R. Bowker Company, 205 East 42nd Street, New York, NY 10017. (800) 521-8110 or (212) 916-1600. 1986. $595.00.

DIRECTORY OF PHYSICS AND ASTRONOMY STAFF MEMBERS. American Institute of Physics, 335 East 45th Street, New York, NY 10017. Annual.

RADIATION RESEARCH SOCIETY - MEMBERSHIP DIRECTORY. Radiation Research Society, 925 Chestnut Street, Philadelphia, PA 19107. (215) 574-3153. 1987. Available to members only.

RESEARCH CENTERS DIRECTORY. Gale Research Company, Book Tower, Detroit, MI 48226. (800) 521-0707. Twelfth edition. 1987. $365.00 for set.

WHO'S WHO IN FRONTIER SCIENCE AND TECHNOLOGY. Marquis Who's Who, Incorporated, 200 East Ohio Street, Chicago, IL 60611. (800) 428-3898 or (312) 787-2008.

ENCYCLOPEDIA AND DICTIONARIES

ENCYCLOPEDIA OF PHYSICAL SCIENCE AND TECHNOLOGY. Academic Press, Incorporated, Orlando, FL 32887. (800) 321-5068 or (305) 345-2734. Inquire as to cost and availability.

MCGRAW-HILL ENCYCLOPEDIA OF SCIENCE AND TECHNOLOGY. McGraw-Hill Book, Incorporated, 1221 Avenue of the Americas, New York, NY 10020. (212) 997-3675. Fifth edition, 15 volumes. $1100.00.

GENERAL WORKS

INTRODUCTION TO ATMOSPHERIC RADIATION. Kuo-Nan Liou. Academic Press, Incorporated, 6277 Sea Harbor Drive, Orlando, FL 32821. (800) 321-5068. 1980. $37.50.

OUR RADIANT WORLD. David Little. Iowa State University Press, 2121 South State Avenue, Ames, IA 50010. (515) 294-5280. 1986. $19.95.

RADIATION DETECTION AND MEASUREMENT. G.F. Knoll. John Wiley and Sons, Incorporated, 605 Third Avenue, New York, NY 10158. (800) 526-5368 or (212) 850-6000. 1979. $55.00.

RADIOACTIVITY AND ITS MEASUREMENT. W.B. Mann, B.L. Ayers and S.B. Garfinkel. Pergamon Press, Incorporated, Maxwell House, Fairview Park, Elmsford, NY 10523. (914) 592-7700. Second edition. 1980. $18.00.

TECHNIQUES OF RADIATION DOSIMETRY. K. Mahesk and D.R. Vij. John Wiley and Sons, Incorporated, 605 Third Avenue, New York, NY 10158. (800) 526-5368 or (212) 850-6000. 1985. $29.95.

HANDBOOKS AND MANUALS

CRC HANDBOOK OF RADIATION MEASUREMENT AND PROTECTION. Allen Broadsky, editor. CRC Press, 2000 Corporate Boulevard, Boca Raton, FL 33431. (305) 994-0555. Two volumes. 1979-82. $200.00.

ONLINE DATA BASES

CA SEARCH. Chemical Abstracts Service, Post Office Box 3012, Columbus, OH 43210. Guide to chemical literature, 1972 to present. Inquire as to cost and availability.

DISSERTATION ABSTRACTS ONLINE. University Microfilms International, 300 North Zeeb Road, Ann Arbor, MI 48106. (800) 521-0600 or (313) 761-4700. Scope includes virtually all doctoral dissertations accepted at accredited American institutions from 1861 to present in 252 subject areas. Inquire as to cost and availability.

INSPEC. INSPEC Marketing Department, Institute of Electrical and Electronics Engineers, Incorporated, IEEE Service Department, 445 Hoes Lane, Piscataway, NJ 08854. (201) 981-0060. Inquire as to on-line cost and availability.

NASA. National Aeronautics and Space Administration, Scientific and Technical Information Branch, 300 7th Street, SW, Washington, DC 20546. Citations and abstracts of aerospace literature, 1962 to present. Inquire as to cost and availability.

NTIS. National Technical Information Service, 5285 Port Royal Road, Springfield, VA 22161. (703) 487-4630. Broad coverage of government sponsored research reports, 1964 to present. Inquire as to cost and availability.

SCISEARCH. Institute for Scientific Information, 3501 Market Street, Philadelphia, PA 19104. (800) 523-1850 or (215) 386-0100. Broad multidisciplinary title and author index to the international literature of science and technology, 1974 to present. Inquire as to cost and availability.

PERIODICALS

BULLETIN OF THE ATOMIC SCIENTISTS. Education Foundation for Nuclear Science, 5801 South Kenwood Avenue, Chicago, IL 60637. (312) 363-5225. Ten times per year. $22.50 per year.

JOURNAL OF ENVIRONMENTAL RADIOACTIVITY. Elsevier Science Publishing Company, Incorporated, 52 Vanderbilt Avenue, New York, NY 10017. (212) 370-5520. Semimonthly. $81.00 per year.

NUCLEAR SCIENCE AND ENGINEERING. American Nuclear Society, 555 North Kensington Avenue, La Grange Park, IL 60525. (312) 352-6611. Monthly. $340.00 per year.

RADIATION RESEARCH. Academic Press, Incorporated, 111 Fifth Avenue, New York, NY 10003. (212) 741-6800. Monthly. $300.00 per year.

RESEARCH CENTERS AND INSTITUTES

INDIANA STATE UNIVERSITY. Radiation Laboratory, 6th and Chestnut Streets, Science Building, Room 112, Physics Department, Terre Haute, IN 47809. (812) 237-2045.

NORTH CAROLINA STATE UNIVERSITY. Center for Engineering Applications of Radioisotopes, Box 7909, Department of Nuclear Engineering, Raleigh, NC 27695-7909. (919) 737-3378.

STANFORD SYNCHROTRON RADIATION LABORATORY. SLAC BIN #69, Post Office Box 4349, Stanford, CA 94305. (415) 854-3300.

UNIVERSITY OF MICHIGAN. Ford Nuclear Reactor - Phoenix Memorial Laboratory, 2301 Bonisteel Boulevard, Ann Arbor, MI 48109-2100. (313) 764-6223.

UNIVERSITY OF WASHINGTON. Nuclear Physics Laboratory, Seattle, WA 98195. (206) 543-4080.

RADIATION SHIELDING

See also: GAMMA-RAYS, NUCLEAR ENERGY, NUCLEAR ENGINEERING, RADIATION

ABSTRACT SERVICES AND INDEXES

APPLIED SCIENCE AND TECHNOLOGY INDEX. H.W. Wilson Company, 950 University Avenue, Bronx, NY 10452. (800) 367-6670 or (212) 588-8400. Inquire as to cost and availability.

CHEMICAL ABSTRACTS. Chemical Abstracts Service, 2540 Olentangy Road, Post Office Box 3012, Columbus, OH 43210. (800) 848-6538 or (614) 421-3600. Weekly. $9200.00 per year.

CURRENT CONTENTS: PHYSICAL AND CHEMICAL SCIENCES. Institute for Scientific Information, 3501 Market Street, Philadelphia, PA 19104. (800) 523-1850 or (215) 386-0100. $272.00 per year.

GENERAL SCIENCE INDEX. H.W. Wilson Company, 950 University Avenue, Bronx, NY 10452. (800) 367-6770 or (212) 588-8400. Inquire as to cost and availability.

INDEX TO SCIENTIFIC REVIEWS. Institute for Scientific Information, 3501 Market Street, Philadelphia, PA 19104. (800) 523-1850 or (215) 386-0100. Semiannual. $550.00 per year.

PHYSICS ABSTRACTS. Institute of Electrical Engineers, London, United Kingdom. Available from: Institute of Electrical and Electronic Engineers (IEEE), 345 East 47th Street, New York, NY 10017. (212) 705-7900.

SCIENCE CITATION INDEX. Institute for Scientific Information, 3501 Market Street, Philadelphia, PA 19104. (800) 523-1850 or (215) 386-0100. Inquire as to cost and availability.

ASSOCIATIONS AND PROFESSIONAL SOCIETIES

AMERICAN NUCLEAR SOCIETY. 555 North Kensington Avenue, La Grange Park, IL 60525. (312) 352-6611.

AMERICAN SOCIETY OF NONDESTRUCTIVE TESTING. 4153 Arlingate Plaza, Caller #28518, Columbus, OH 43228. (614) 274-6003.

NATIONAL COUNCIL ON RADIATION PROTECTION AND MEASUREMENTS. 7910 Woodmont Avenue, Suite 1016, Bethesda, MD 20814. (301) 657-2652.

RADIATION RESEARCH SOCIETY. 925 Chestnut Street, Philadelphia, PA 19107. (215) 574-3153.

DIRECTORIES AND BIOGRAPHICAL SOURCES

AMERICAN MEN AND WOMEN OF SCIENCE. Physical and Biological Sciences. Sixteenth edition. R.R. Bowker Company, 205 East 42nd Street, New York, NY 10017. (800) 521-8110 or (212) 916-1600. 1986. $595.00.

DIRECTORY OF PHYSICS AND ASTRONOMY STAFF MEMBERS. American Institute of Physics, 335 East 45th Street, New York, NY 10017. Annual.

RADIATION RESEARCH SOCIETY - MEMBERSHIP DIRECTORY. Radiation Research Society, 925 Chestnut Street, Philadelphia, PA 19107. (215) 574-3153. 1987. Available to members only.

RESEARCH CENTERS DIRECTORY. Gale Research Company, Book Tower, Detroit, MI 48226. (800) 521-0707. Twelfth edition. 1987. $365.00 for set.

WHO'S WHO IN FRONTIER SCIENCE AND TECHNOLOGY. Marquis Who's Who, Incorporated, 200 East Ohio Street, Chicago, IL 60611. (800) 428-3898 or (312) 787-2008.

ENCYCLOPEDIA AND DICTIONARIES

ENCYCLOPEDIA OF PHYSICAL SCIENCE AND TECHNOLOGY. Academic Press, Incorporated, Orlando, FL 32887. (800) 321-5068 or (305) 345-2734. Inquire as to cost and availability.

MCGRAW-HILL ENCYCLOPEDIA OF SCIENCE AND TECHNOLOGY. McGraw-Hill Book, Incorporated, 1221 Avenue of the Americas, New York, NY 10020. (212) 997-3675. Fifth edition, 15 volumes. $1100.00.

GENERAL WORKS

INTRODUCTION TO ATMOSPHERIC RADIATION. Kuo-Nan Liou. Academic Press, Incorporated, 6277 Sea Harbor Drive, Orlando, FL 32821. (800) 321-5068. 1980. $37.50.

AN INTRODUCTION TO RADIATION PROTECTION. A. Martin and S.A. Harbinson. Metheun, Incorporated, 29 West 35th Street, New York, NY 10001. (212) 244-3336. 1986.

$29.95.

OUR RADIANT WORLD. David Little. Iowa State University Press, 2121 South State Avenue, Ames, IA 50010. (515) 294-5280. 1986. $19.95.

RADIATION DETECTION AND MEASUREMENT. G.F. Knoll. John Wiley and Sons, Incorporated, 605 Third Avenue, New York, NY 10158. (800) 526-5368 or (212) 850-6000. 1979. $55.00.

RADIOACTIVITY AND ITS MEASUREMENT. W.B. Mann, B.L. Ayers and S.B. Garfinkel. Pergamon Press, Incorporated, Maxwell House, Fairview Park, Elmsford, NY 10523. (914) 592-7700. Second edition. 1980. $18.00.

HANDBOOKS AND MANUALS

CRC HANDBOOK OF MANAGEMENT OF RADIATION PROTECTION PROGRAMS. Kenneth L. Miller and William A. Weidner, editors. CRC Press, 2000 Corporate Boulevard, Boca Raton, FL 33431. (305) 994-0555. 1986. $99.00.

CRC HANDBOOK OF RADIATION MEASUREMENT AND PROTECTION. Allen Broadsky, editor. CRC Press, 2000 Corporate Boulevard, Boca Raton, FL 33431. (305) 994-0555. Two volumes. 1979-82. $200.00.

ONLINE DATA BASES

CA SEARCH. Chemical Abstracts Service, Post Office Box 3012, Columbus, OH 43210. Guide to chemical literature, 1972 to present. Inquire as to cost and availability.

DISSERTATION ABSTRACTS ONLINE. University Microfilms International, 300 North Zeeb Road, Ann Arbor, MI 48106. (800) 521-0600 or (313) 761-4700. Scope includes virtually all doctoral dissertations accepted at accredited American institutions from 1861 to present in 252 subject areas. Inquire as to cost and availability.

INSPEC. INSPEC Marketing Department, Institute of Electrical and Electronics Engineers, Incorporated, IEEE Service Department, 445 Hoes Lane, Piscataway, NJ 08854. (201) 981-0060. Inquire as to on-line cost and availability.

NASA. National Aeronautics and Space Administration, Scientific and Technical Information Branch, 300 7th Street, SW, Washington, DC 20546. Citations and abstracts of aerospace literature, 1962 to present. Inquire as to cost and availability.

NTIS. National Technical Information Service, 5285 Port Royal Road, Springfield, VA 22161. (703) 487-4630. Broad coverage of government sponsored research reports, 1964 to present. Inquire as to cost and availability.

SCISEARCH. Institute for Scientific Information, 3501 Market Street, Philadelphia, PA 19104. (800) 523-1850 or (215) 386-0100. Broad multidisciplinary title and author index to the international literature of science and technology, 1974 to present. Inquire as to cost and availability.

PERIODICALS

BULLETIN OF THE ATOMIC SCIENTISTS. Education Foundation for Nuclear Science, 5801 South Kenwood Avenue, Chicago, IL 60637. (312) 363-5225. Ten times per year. $22.50 per year.

JOURNAL OF ENVIRONMENTAL RADIOACTIVITY. Elsevier Science Publishing Company, Incorporated, 52 Vanderbilt Avenue, New York, NY 10017. (212) 370-5520. Semimonthly. $81.00 per year.

NUCLEAR SCIENCE AND ENGINEERING. American Nuclear Society, 555 North Kensington Avenue, La Grange Park, IL 60525. (312) 352-6611. Monthly. $340.00 per year.

RADIATION RESEARCH. Academic Press, Incorporated, 111 Fifth Avenue, New York, NY 10003. (212) 741-6800. Monthly. $300.00 per year.

RESEARCH CENTERS AND INSTITUTES

GEORGIA INSTITUTE OF TECHNOLOGY. Frank H. Neely Nuclear Research Center, 900 Atlantic Drive, NW, Atlanta, GA 30332. (404) 894-3620.

INDIANA STATE UNIVERSITY. Radiation Laboratory, 6th and Chestnut Streets, Science Building, Room 112, Physics Department, Terre Haute, IN 47809. (812) 237-2045.

LOUISIANA STATE UNIVERSITY. Nuclear Science Center, Baton Rouge, LA 70803-5820. (504) 388-2163.

STANFORD SYNCHROTRON RADIATION LABORATORY. SLAC BIN #69, Post Office Box 4349, Stanford, CA 94305. (415) 854-3300.

UNITED STATES NUCLEAR REGULATORY COMMISSION. Occupational Radiation Protection Branch, Mail Stop 1130-SS, Washington, DC 20555.

UNIVERSITY OF MICHIGAN. Ford Nuclear Reactor - Phoenix Memorial Laboratory, 2301 Bonisteel Boulevard, Ann Arbor, MI 48109-2100. (313) 764-6223.

UNIVERSITY OF WASHINGTON. Nuclear Physics Laboratory, Seattle, WA 98195. (206) 543-4080.

RADIO

See also: AMATEUR RADIO, ANTENNAS, ELECTRICAL ENGINEERING, ELECTRONIC CIRCUITS AND COMPONENTS, ELECTRONICS, ELECTRONICS ENGINEERING, RADAR, TELEVISION

ABSTRACT SERVICES AND INDEXES

APPLIED SCIENCE AND TECHNOLOGY INDEX. H.W. Wilson and Company, 950 University Avenue, Bronx, NY 10452. (800) 367-6670 or (212) 588-8400. Monthly. Inquire as to cost and availability.

CHEMICAL ABSTRACTS. American Chemical Society, Chemical Abstracts Service, Box 3012, Columbus, OH 43210. (614) 421-3600. 1907 to present. Weekly. $9500.00 per year.

CURRENT CONTENTS: ENGINEERING, TECHNOLOGY AND APPLIED SCIENCES. Institute for Scientific Information, 3501 Market Street, Philadelphia, PA 19104. (800) 523-1850 or (215) 386-0100. Weekly. $275.00 per year.

ELECTRICAL AND ELECTRONICS ABSTRACTS. Institution of Electrical Engineers. Available from: Institute of Electrical and Electronics Engineers. IEEE Service Center, 445 Hoes Lane, Piscataway, NJ 08854. Monthly. $1250.00 per year.

ELECTRONICS AND COMMUNICATIONS ABSTRACTS. Cambridge Scientific Abstracts, 5161 River Road, Bethesda, MD 20816. (301) 951-1400. Bimonthly. Inquire as to cost and availability.

ENGINEERING INDEX MONTHLY AND AUTHOR INDEX. Engineering Information Inc., 345 East 47th Street, New York, NY 10017. (212) 705-7600. Monthly. $1560.00 per year.

GENERAL SCIENCE INDEX. H.W. Wilson and Company, 950 University Avenue, Bronx, NY 10452. (800) 367-6670 or (212) 588-8400. 1978 to present. Monthly. Inquire as to cost and availability.

IEEE PUBLICATIONS BULLETIN. Institute of Electrical and Electronics Engineers. Institute of Electrical and Electronics

Engineers. IEEE Service Center, 445 Hoes Lane, Piscataway, NJ 08854. Quarterly. Free.

PHYSICS ABSTRACTS. Institution of Electrical Engineers. Available from: IEEE Service Center, 445 Hoes Lane, Piscataway, NJ 08854. 1898 to present. Bimonthly. $1700.00 per year.

PHYSICS BRIEFS. Physik Verlag GmbH, Postfach 1260/1280, D-6940 Weinheim, West Germany. (212) 661-9404. 1920 to present. Twenty-six times per year. $1250.00 per year.

SCIENCE CITATION INDEX. Institute for Scientific Information, 3501 Market Street, Philadelphia, PA 19104. (800) 523-1850 or (215) 386-0100. Six times per year. $6200.00 per year.

SOLID STATE ABSTRACTS: AN ABSTRACT JOURNAL INVOLVING THE PHYSICS, METALLURGY, CRYSTALLOGRAPHY, CHEMISTRY, AND DEVICE TECHNOLOGY OF SOLIDS. Cambridge Scientific Abstracts, 5161 River Road, Bethesda, MD 20816. (301) 951-1400. 1957 to present. Bimonthly. $550.00 per year.

ANNUAL REVIEWS AND YEARBOOKS

ADVANCES IN ELECTRONICS AND ELECTRON PHYSICS. Academic Press, Inc., 6277 Sea Harbor Drive, Orlando, FL 32821. (800) 321-5068. Irregular. Approximately $80.00 per volume.

ASSOCIATIONS AND PROFESSIONAL SOCIETIES

AMERICAN ELECTRONICS ASSOCIATION. P.O. Box 10045, 2670 Hanover Street, Palo Alto, CA 94303. (415) 857-9300.

AMERICAN INSTITUTE OF PHYSICS. 335 East 45th Street, New York, NY 10017. (212) 661-9494.

ELECTRONICS INDUSTRIES ASSOCIATION. 2001 Eye Street, N.W., Washington, DC 20006. (202) 457-4900.

INSTITUTE OF ELECTRICAL AND ELECTRONICS ENGINEERS. 345 East 47th Street, New York, NY 10017. (212) 705-7900.

INSTITUTE OF RADIO ENGINEERS. Institute of Electrical and Electronics Engineers, 345 East 47th Street, New York, NY 10017. (212) 705-7900.

NATIONAL ASSOCIATION OF RADIO AND TELECOMMUNICATIONS ENGINEERS. P.O. Box 15029, Salem, OR 97309. (503) 581-7653.

RADIO AMATEUR SATELLITE CORPORATION. P.O. Box 27, Washington, DC 20044. (301) 589-6062.

UNITED STATES NATIONAL COMMITTEE FOR THE INTERNATIONAL UNION OF RADIO SCIENCE. Board on Physics and Astronomy, National Research Council, 2101 Constitution Avenue, N.W., Washington, DC 20418. (202) 334-3559.

DIRECTORIES AND BIOGRAPHICAL SOURCES

AMERICAN MEN AND WOMEN OF SCIENCE. R.R. Bowker, Inc., Order Department, 245 West 17th Street, New York, NY 10011. (800) 521-8110. Eight volumes. 1986. $595.00 for set.

BROADCAST ENGINEERING BUYERS GUIDE/SPEC BOOK ISSUE. Intertec Publishing Corporation, Box 12901, Overland Park, KS 66212. (913) 888-4664. Annual. $20.00 per year.

IEEE MEMBERSHIP DIRECTORY. Institute of Electrical and Electronics Engineers. IEEE Service Center, 445 Hoes Lane, Piscataway, NJ 08854. Annual. $7.00.

INTERNATIONAL RESEARCH CENTERS DIRECTORY 1988-89. Darren L. Smith, editor. Gale Research Company, Book Tower, Detroit, MI 48226. (800) 521-0707. 4th edition. 1987. $360.00.

1987 DIRECTORY OF ENGINEERING SOCIETIES AND RELATED ORGANIZATIONS. Gordon Davis, editor. Hemisphere Publishing Corporation, 1010 Vermont Avenue, NW, Washington, DC 20005. (800) 526-0275. 12th edition. 1987. $100.00.

RADIO AMATEUR CALLBOOK. Radio Amateur Callbook, Inc., 925 Sherwood Drive, Lake Bluff, IL 60044. (312) 234-6600. Annual. North American Edition, $22.00. International Edition, $21.00, plus shipping and handling.

RESEARCH CENTERS DIRECTORY 1988. Gale Research Company, Book Tower, Detroit, MI 48226. (800) 521-0707. 12th edition. 1987. $365.00 for set.

SCIENTIFIC AND TECHNICAL ORGANIZATIONS AND AGENCIES DIRECTORY. Margaret Labash Young, editor. Gale Research Company, Book Tower, Detroit, MI 48226. (800) 521-0707. 2nd edition. 1987. $185.00.

WHO'S WHO IN ELECTRONICS. Harris Publishing Company, 2057-2 Aurora Road, Twinsburg, OH 44087. (216) 425-9143. Annual. $90.00.

WHO'S WHO IN ENGINEERING. Gordon Davis, editor. Hemisphere Publishing Corporation, 1010 Vermont Avenue, NW, Washington, DC 20005. (800) 526-0275. 6th edition. 1985. $200.00.

ENCYCLOPEDIAS AND DICTIONARIES

DICTIONARY OF AUDIO, RADIO AND VIDEO. R.S. Roberts. Butterworth's Publishing, 80 Montvale Avenue, Stoneham, MA 02180. (800) 325-4177. 1981. $45.00.

IEEE STANDARD DICTIONARY OF ELECTRICAL AND ELECTRONICS TERMS. Frank Jay, editor. John Wiley and Sons, Inc., 605 Third Avenue, New York, NY 10158. (800) 526-5368. 3rd edition. 1984. $49.95.

GENERAL WORKS

AMATEUR RADIO EQUIPMENT FUNDAMENTALS. Albert D. Helfrick. Prentice-Hall Publishing, Inc., Englewood Cliffs, NJ 07632. (800) 562-0245. 1982. $31.95.

AMATEUR RADIO, SUPER HOBBY: WHAT IS IT, WHO WE ARE, HOW TO JOIN. Vince Luciani. McGraw-Hill Book Company, 1221 Avenue of the Americas, New York, NY 10020. (212) 512-2000. 1984. $9.95 in paper.

ELECTROMAGNETIC CONCEPTS AND PRINCIPLES. Stanley V. Marshall and Gabriel G. Skitek. Prentice-Hall Publishing, Inc., Englewood Cliffs, NJ 07632. (800) 562-0245. 2nd edition. 1987. $42.95.

ELECTRONIC INVENTIONS AND DISCOVERIES: ELECTRONICS FROM ITS EARLIEST BEGINNINGS TO PRESENT DAY. G.W.A. Dummer. Pergamon Press, Inc., Maxwell House, Fairview Park, Elmsford, NY 10523. (914) 592-7700. 1983. $49.50.

SOLID STATE RADIO ENGINEERING. Herbert L. Krauss and Charles W. Bostian. John Wiley and Sons, Inc., 605 Third Avenue, New York, NY 10158. (800) 526-5368. 1980. $45.00.

TROUBLESHOOTING, SERVICING AND THEORY OF AM, FM AND FM STEREO RECEIVERS. Clarence Green and Robert Bourque. Prentice-Hall Publishing, Inc., Englewood Cliffs, NJ 07632. (800) 562-0245. 1987. $29.95.

UNDERSTANDING RADIO ELECTRONICS. M. Kaufmann and others. McGraw-Hill Book Company, 1221 Avenue of the Americas, New York, NY 10020. (212) 512-2000. Fourth edition. 1972. $33.50.

HANDBOOKS AND MANUALS

ELECTRONIC ENGINEERS HANDBOOK. Donald G. Fink, editor. McGraw-Hill Book Company, 1221 Avenue of the Americas, New York, NY 10020. (212) 512-2000. 2nd edition. 1982. $89.00.

HANDBOOK FOR RADIO ENGINEERING. John F. Ross. Butterworth's Publishing, 80 Montvale Avenue, Stoneham, MA 02180. (800) 325-4177. 1980. $125.00.

HANDBOOK OF MODERN ELECTRONICS AND ELECTRICAL ENGINEERING. Charles Belove, editor. John Wiley and Sons, Inc., 605 Third Avenue, New York, NY 10158. (800) 526-5368. 1986. $88.95.

ONLINE DATA BASES

CA SEARCH. Chemical Abstracts Service, P.O. Box 3012, Columbus, OH 43120. (800) 848-6538 or (614) 421-3600. Comprehensive guide to chemical literature, 1972 to present. Inquire as to online cost and availability.

COMPENDEX. Engineering Information, Inc., 345 East 47th Street, New York, NY 10017. (800) 221-1044 or (212) 705-7615. Engineering and technical literature, 1975 to present. Inquire as to online cost and availability.

INSPEC. INSPEC Marketing Department, Institution of Electrical Engineers. Available from IEEE Service Center, 445 Hoes Lane, Piscataway, NJ 08854. (201) 981-0060. Online version of Physics Abstracts. Inquire as to online cost and availability.

NTIS. National Technical Information Service, 5285 Port Royal Road, Springfield, VA 22161. (703) 487-4630. Broad coverage of government sponsored research reports, 1964 to present. Inquire as to online cost and availability.

SCISEARCH. Institute for Scientific Information, 3501 Market Street, Philadelphia, PA 19104. (800) 523-1850 or (215) 386-0100. Broad multidisciplinary title and author index to the international literature of science and technology, 1974 to present. Inquire as to online cost and availability.

WILSONLINE. H.W. Wilson and Company, 950 University Avenue, Bronx, NY 10452. (800) 367-6770 or (212) 588-8400. Makes available online versions of the H.W. Wilson indexes including Applied Science and Technology Index, Business Periodicals Index and Readers' Guide to Periodical Literature. Approximately 1980 to present. Inquire as to online cost and availability.

OTHER SOURCES

A GUIDE TO THE LITERATURE OF ELECTRICAL AND ELECTRONICS ENGINEERING. Susan B. Ardis. Libraries Unlimited Inc., P.O. Box 263, Littleton, CO 80160. (303) 770-1220. 1987. $37.50.

PERIODICALS

BROADCASTER ENGINEERING. Intertec Publishing Corporation, Box 12901, Overland Park, KS 66212. (913) 888-4664. 1959 to present. Monthly. $25.00 per year.

COMMUNICATIONS: FOR THE PROFESSIONAL IN LAND MOBILE RADIO. Cardiff Publishing Company, 6530 South Yosemite, Enlewood, CO 80111. (303) 694-1522. 1963 to present. Monthly. $20.00 per year.

CQ; THE RADIO AMATEUR'S JOURNAL. CQ Publishing, Inc., 76 North Broadway, Hicksville, NY 11801. (516) 681-2922. 1945 to present. Monthly. $16.00 per year.

ELECTRONIC DESIGN. Hayden Publishing Company, 10 Mulholland Drive, Hasbrouck Heights, NJ 07604. (201) 288-7520. 1952 to present. Biweekly. $40.00 per year.

ELECTRONICS. McGraw-Hill Book Company, 1221 Avenue of the Americas, New York, NY 10020. (212) 512-2000. 1930 to present. Weekly. $32.00 per year.

ELECTRONICS AND WIRELESS WORLD. I.P.C. Electrical-Electronic Press, Ltd., Quadrant House, The Quadrant, Sutton, Surrey, SM2 5AS England. 1911 to present. Monthly. $105.00 per year.

IEEE CIRCUITS AND DEVICES MAGAZINE. Institute of Electrical and Electronics Engineers. IEEE Service Center, 445 Hoes Lane, Piscataway, NJ 08854. Bimonthly. $70.00 per year.

IEEE JOURNAL OF SOLID STATE CIRCUITS. Institute of Electrical and Electronics Engineers. IEEE Service Center, 445 Hoes Lane, Piscataway, NJ 08854. 1966 to present. Bimonthly. $113.00 per year.

IEEE TRANSACTIONS ON BROADCASTING. Institute of Electrical and Electronics Engineers. IEEE Service Center, 445 Hoes Lane, Piscataway, NJ 08854. 1955 to present. Quarterly. $37.00 per year.

RADIO COMMUNICATIONS REPORT. Titsch Publishing Company, 2500 Curtis Street, Suite 200, Denver, CO 80217. (303) 573-1433. 1981 to present. Monthly. $15.00 per year.

RESEARCH CENTERS AND INSTITUTES

CENTER FOR ADVANCED TECHNOLOGY IN TELECOMMUNICATIONS. Polytechnic University, 333 Jay Street, Brooklyn, NY 11201. (718) 643-5160.

COMMUNICATIONS RESEARCH LABORATORY. McMaster University, 1280 Main Street West, Hamilton, ON, Canada L8S 4K1. (416) 525-9140.

ELECTRONICS RESEARCH LABORATORY. Montana State University, Bozeman, MT 59717. (406) 994-2505.

RADIO ASTRONOMY

See also: ASTRONOMY, ASTROPHYSICS, RADIO TELESCOPES

ABSTRACT SERVICES AND INDEXES

ASTRONOMY AND ASTROPHYSICS ABSTRACTS. Springer-Verlag New York, Incorporated, 175 Fifth Avenue, New York, NY 10010. (212) 460-1500. $70.00 per year.

ASTRONOMY AND ASTROPHYSICS MONTHLY INDEX. Olivetree Associates, P.O. Box 236, Sierre Madre, CA 91024. $212.00 per year. Complimentary copies available on request.

PHYSICS ABSTRACTS. Institute of Electrical Engineers, London, United Kingdom. Available from: Institute of Electrical and Electronic Engineers (IEEE), 345 East 47th Street, New York, NY 10017. (212) 705-7900.

SCIENCE CITATION INDEX. Institute for Scientific Information, 3501 Market Street, Philadelphia, PA 19104. (800) 523-1850 or (215) 386-0100.

STAR. (SCIENTIFIC AND TECHNICAL AEROSPACE REPORTS). U.S. National Aeronautics and Space Administration, Scientific and Technical Information Facility, Box 8757, Baltimore-Washington International Airport, MD 21240. (202) 755-2210. Semimonthly, with semiannual and annual indexes. $85.00 per year.

ANNUAL REVIEWS AND YEARBOOKS

THE ASTRONOMICAL ALMANAC. Superintendent of Documents, U.S. Government Printing Office, Washington, DC

20402. (202) 783-3238. Yearly.

ANNUAL REVIEW OF ASTRONOMY AND ASTROPHYSICS. Annual Reviews, Incorporated, 4139 El Camino Way, Palo Alto, CA 94306. (415) 493-4400.

ASSOCIATIONS AND PROFESSIONAL SOCIETIES

AMERICAN ASTRONOMICAL SOCIETY. 2000 Florida Avenue, NW, Suite 300, Washington, DC 20009. (202) 659-0134.

AMERICAN ASSOCIATION OF VARIABLE STAR OBSERVERS. 187 Concord Avenue, Cambridge, MA 02138. (617) 354-0484.

ASTRONOMICAL LEAGUE. P.O. Box 12821, Tucson, AZ 85732. (602) 790-8471.

ASTRONOMICAL SOCIETY OF THE PACIFIC. 1290 24th Avenue, San Francisco, CA 94122. (415) 661-8660.

BIBLIOGRAPHIES

A BIBLIOGRAPHY OF ASTRONOMY, 1970-1979. R.A. Seal and S.S. Martin. Libraries Unlimited, Incorporated, Littleton, CO 80160. 1982. $37.50.

SCIENTIFIC AND TECHNICAL BOOKS IN PRINT 1988: AN INDEX TO LITERATURE IN SCIENCE AND TECHNOLOGY. R.R. Bowker Company, 205 East 42nd Street, New York, NY 10017. (800) 521-8110 or (212) 916-1600. $175.00.

DIRECTORIES AND BIOGRAPHICAL SOURCES

AMERICAN MEN AND WOMEN OF SCIENCE. Physical and Biological Sciences. Fifteenth edition. R.R. Bowker Company, 205 East 42nd Street, New York, NY 10017. (800) 521-8110 or (212) 916-1600.

THE BIOGRAPHICAL DICTIONARY OF SCIENTISTS: ASTRONOMERS. D. Abbott, editor. Peter Bedrick Books, 125 East 23rd Street, New York, NY 10010. 1984.

DIRECTORY OF PHYSICS AND ASTRONOMY STAFF MEMBERS. American Institute of Physics, 335 East 45th Street, New York, NY 10017. Annual.

RESEARCH CENTERS DIRECTORY. Gale Research Company, Detroit, MI 48226. Eleventh edition, 1987. (800) 521-0707. $340.00.

WHO'S WHO IN FRONTIER SCIENCE AND TECHNOLOGY. Marquis Who's Who, Incorporated, 200 East Ohio Street, Chicago, IL 60611. (800) 428-3898 or (312) 787-2008.

GENERAL WORKS

THE CAMBRIDGE ASTRONOMY GUIDE: A PRACTICAL INTRODUCTION TO ASTRONOMY. Cambridge University Press, 32 East 57th Street, New York, NY 10022. (212) 688-8888. 1985. $24.95.

CLASSICS IN RADIO ASTRONOMY. W.T. Sullivan. Kluwer Academic Publishers, 190 Old Derby Street, Hingham, MA 02043. (617) 749-5262. 1982. $59.50.

THE INVISIBLE UNIVERSE REVEALED: THE STORY OF RADIO ASTRONOMY. G.L. Vershuur. Springer-Verlag, 175 Fifth Avenue, New York, NY 10010. (800) 526-7254. 1987. $55.00.

HANDBOOKS AND MANUALS

THE AMATEUR RADIO ASTRONOMY OBSERVER'S HANDBOOK. John P. Shields. Crown Publishers, Incorporated, 1 Park Avenue, New York, NY 10016. (212) 532-9200. 1986.

$17.95.

ONLINE DATA BASES

CA SEARCH. Chemical Abstracts Service, P.O. Box 3012, Columbus, OH 43210. Guide to chemical literature, 1972 to present. Inquire as to cost and availability.

DISSERTATION ABSTRACTS ONLINE. University Microfilms International, 300 North Zeeb Road, Ann Arbor, MI 48106. (800) 521-0600 or (313) 761-4700. Scope includes virtually all doctoral dissertations accepted at accredited American institutions from 1861 to present in 252 subject areas. Inquire as to cost and availability.

INSPEC. INSPEC Marketing Department, Institute of Electrical and Electronics Engineers, Incorporated, IEEE Service Department, 445 Hoes Lane, Piscataway, NJ 08854. (201) 981-0060. Inquire as to on-line cost and availability.

NASA. National Aeronautics and Space Administration, Scientific and Technical Information Branch, 300 7th Street, SW, Washington, DC 20546. Citations and abstracts of aerospace literature, 1962 to present. Inquire as to cost and availability.

NTIS. National Technical Information Service, 5285 Port Royal Road, Springfield, VA 22161. (703) 487-4630. Broad coverage of government sponsored research reports, 1964 to present. Inquire as to cost and availability.

SCISEARCH. Institute for Scientific Information, 3501 Market Street, Philadelphia, PA 19104. (800) 523-1850 or (215) 386-0100. Broad multidisciplinary title and author index to the international literature of science and technology, 1974 to present. Inquire as to cost and availability.

PERIODICALS

ASTRONOMICAL JOURNAL. American Astronomical Society. Available from: American Institute of Physics, 335 East 45th Street, New York, NY 10017. (212) 661-9404. $125.00 per year.

ASTRONOMICAL SOCIETY OF THE PACIFIC. Publications. Astronomical Society of the Pacific, 1290 24th Avenue, San Francisco, CA 94122. (415) 661-8660. Monthly. $38.00.

ASTRONOMY. Astro Media Corporation, 625 East Paul Avenue, Milwaukee, WI 53202. Monthly. 418.00 per year.

ASTRONOMY AND ASTROPHYSICS. Springer-Verlag New York, Incorporated, 175 Fifth Avenue, New York, NY 10010. (800) 526-7254 or (212) 460-1500. $680.00 per year.

ASTROPHYSICAL JOURNAL. American Astronomical Society, University of Chicago Press, 5801 Ellis Avenue, Chicago, IL 60637. Biweekly. $305.00 per year.

ASTROPHYSICS AND SPACE SCIENCE. D. Reidel Publishing Company, 190 Old Derby Street, Hingham, MA 02043. Monthly. $101.00 per year.

CELESTIAL MECHANICS; AN INTERNATIONAL JOURNAL OF SPACE DYNAMICS. D. Reidel Publishing Company, 190 Old Derby Street, Hingham, MA 02043. Monthly. $310.00 per year.

ICARUS: INTERNATIONAL JOURNAL OF THE SOLAR SYSTEM STUDIES. Academic Press, Incorporated, Orlando, FL 32887. (305) 345-4100. Monthly. $484.00 per year.

MERCURY. Astronomical Society of the Pacific, 1290 24th Avenue, San Francisco, CA 94122. (15) 661-8660. Bimonthly. $21.00 per year.

MONTHLY NOTICES OF THE ROYAL ASTRONOMICAL SOCIETY. Blackwell Science Publications, Incorporated, 667 Lytton Avenue, Palo Alto, CA 94301. (415) 324-1688. Monthly. $850.00 per year.

PLANETARY AN SPACE SCIENCE. Pergamon Press, Incorporated, Maxwell House, Fairview Park, Elmsford, NY 10523. (914) 592-7700. Monthly. $430.00 per year.

SKY AND TELESCOPE. Sky Publishing Corporation, 49 Bay State Road, Cambridge, MA 02238. (617) 864-7360. Monthly. $18.00 per year.

SOLAR PHYSICS. D. Reidel Publishing Company, 190 Old Derby Street, Hingham, MA 02043. Monthly. $620.00 per year.

SOVIET ASTRONOMY (TRANSLATION OF ASTRONOMICHESKII ZHURNALL). American Institute of Physics, 335 East 45th Street, New York, NY 10017. (212) 661-9404. Bimonthly. $425.00 per year.

SPACE SCIENCE REVIEWS. D. Reidel Publishing Company, 190 Old Derby Street, Hingham, MA 02043. Monthly. $305.00 per year.

VISTAS IN ASTRONOMY. Pergamon Press, Incorporated, Maxwell House, Fairview Park, Elmsford, NY 10523. (914) 592-7700. Quarterly. $145.00 per year.

RESEARCH CENTERS AND INSTITUTES

CORNELL UNIVERSITY. Arechibo Observatory, Box 995, Arecibo, PR 00613. (809) 878-2612.

NATIONAL ASTRONOMY AND IONOSPHERE CENTER. Cornell University, Space Sciences Building, Ithaca, NY 14853. (607) 256-3734.

NATIONAL RADIO ASTRONOMY OBSERVATORY. Edgemont Road, Charlottesville, VA 22903. (804) 296-0211.

UNIVERSITY OF MICHIGAN. Radio Astronomy Observatory, 937 Dennison Building, Ann Arbor, MI 48109. (313) 426-8441.

UNIVERSITY OF TEXAS AT AUSTIN. Radio Astronomy Observatory, Austin, TX 78712. (512) 471-6600.

RADIO TELESCOPES

See also: ASTRONOMY, ASTROPHYSICS, RADIO ASTRONOMY

ABSTRACT SERVICES AND INDEXES

ASTRONOMY AND ASTROPHYSICS ABSTRACTS. Springer-Verlag New York, Incorporated, 175 Fifth Avenue, New York, NY 10010. (212) 460-1500. $70.00 per year.

ASTRONOMY AND ASTROPHYSICS MONTHLY INDEX. Olivetree Associates, P.O. Box 236, Sierre Madre, CA 91024. $212.00 per year. Complimentary copies available on request.

PHYSICS ABSTRACTS. Institute of Electrical Engineers, London, United Kingdom. Available from: Institute of Electrical and Electronic Engineers (IEEE), 345 East 47th Street, New York, NY 10017. (212) 705-7900.

SCIENCE CITATION INDEX. Institute for Scientific Information, 3501 Market Street, Philadelphia, PA 19104. (800) 523-1850 or (215) 386-0100.

STAR. (SCIENTIFIC AND TECHNICAL AEROSPACE REPORTS). U.S. National Aeronautics and Space Administration, Scientific and Technical Information Facility, Box 8757, Baltimore-Washington International Airport, MD 21240. (202) 755-2210. Semimonthly, with semiannual and annual indexes. $85.00 per year.

ANNUAL REVIEWS AND YEARBOOKS

ANNUAL REVIEW OF ASTRONOMY AND ASTROPHYSICS. Annual Reviews, Incorporated, 4139 El Camino Way, Palo Alto, CA 94306. (415) 493-4400.

ASSOCIATIONS AND PROFESSIONAL SOCIETIES

AMERICAN ASTRONOMICAL SOCIETY. 2000 Florida Avenue, NW, Suite 300, Washington, DC 20009. (202) 659-0134.

AMERICAN ASSOCIATION OF VARIABLE STAR OBSERVERS. 187 Concord Avenue, Cambridge, MA 02138. (617) 354-0484.

ASTRONOMICAL LEAGUE. P.O. Box 12821, Tucson, AZ 85732. (602) 790-8471.

ASTRONOMICAL SOCIETY OF THE PACIFIC. 1290 24th Avenue, San Francisco, CA 94122. (415) 661-8660.

BIBLIOGRAPHIES

A BIBLIOGRAPHY OF ASTRONOMY, 1970-1979. R.A. Seal and S.S. Martin. Libraries Unlimited, Incorporated, Littleton, CO 80160. 1982. $37.50.

DIRECTORIES AND BIOGRAPHICAL SOURCES

AMERICAN MEN AND WOMEN OF SCIENCE. Physical and Biological Sciences. Fifteenth edition. R.R. Bowker Company, 205 East 42nd Street, New York, NY 10017. (800) 521-8110 or (212) 916-1600.

THE BIOGRAPHICAL DICTIONARY OF SCIENTISTS: ASTRONOMERS. D. Abbott, editor. Peter Bedrick Books, 125 East 23rd Street, New York, NY 10010. 1984.

DIRECTORY OF PHYSICS AND ASTRONOMY STAFF MEMBERS. American Institute of Physics, 335 East 45th Street, New York, NY 10017. Annual.

RESEARCH CENTERS DIRECTORY. Gale Research Company, Detroit, MI 48226. Eleventh edition, 1987. (800) 521-0707. $340.00.

WHO'S WHO IN FRONTIER SCIENCE AND TECHNOLOGY. Marquis Who's Who, Incorporated, 200 East Ohio Street, Chicago, IL 60611. (800) 428-3898 or (312) 787-2008.

GENERAL WORKS

THE CAMBRIDGE ASTRONOMY GUIDE: A PRACTICAL INTRODUCTION TO ASTRONOMY. Cambridge University Press, 32 East 57th Street, New York, NY 10022. (212) 688-8888. 1985. $24.95.

CLASSICS IN RADIO ASTRONOMY. W.T. Sullivan. Kluwer Academic Publishers, 190 Old Derby Street, Hingham, MA 02043. (617) 749-5262. 1982. $59.50.

THE INVISIBLE UNIVERSE REVEALED: THE STORY OF RADIO ASTRONOMY. G.L. Vershuur. Springer-Verlag, 175 Fifth Avenue, New York, NY 10010. (800) 526-7254. 1987. $55.00.

RADIO TELESCOPES. D.V. Skobel'tsyn, editor. Plenum Publishers, Incorporated, 233 Spring Street, New York, NY 10013. (800) 221-9369. 1966. $35.00.

HANDBOOKS AND MANUALS

THE AMATEUR RADIO ASTRONOMY OBSERVER'S HANDBOOK. John P. Shields. Crown Publishers, Incorporated, 1 Park Avenue, New York, NY 10016. (212) 532-9200. 1986. $17.95.

ONLINE DATA BASES

DISSERTATION ABSTRACTS ONLINE. University Microfilms International, 300 North Zeeb Road, Ann Arbor, MI 48106. (800) 521-0600 or (313) 761-4700. Scope includes virtually all doctoral dissertations accepted at accredited American institutions from 1861 to present in 252 subject areas. Inquire as to cost and availability.

INSPEC. INSPEC Marketing Department, Institute of Electrical and Electronics Engineers, Incorporated, IEEE Service Department, 445 Hoes Lane, Piscataway, NJ 08854. (201) 981-0060. Inquire as to on-line cost and availability.

NASA. National Aeronautics and Space Administration, Scientific and Technical Information Branch, 300 7th Street, SW, Washington, DC 20546. Citations and abstracts of aerospace literature, 1962 to present. Inquire as to cost and availability.

NTIS. National Technical Information Service, 5285 Port Royal Road, Springfield, VA 22161. (703) 487-4630. Broad coverage of government sponsored research reports, 1964 to present. Inquire as to cost and availability.

SCISEARCH. Institute for Scientific Information, 3501 Market Street, Philadelphia, PA 19104. (800) 523-1850 or (215) 386-0100. Broad multidisciplinary title and author index to the international literature of science and technology, 1974 to present. Inquire as to cost and availability.

PERIODICALS

ASTRONOMICAL JOURNAL. American Astronomical Society. Available from: American Institute of Physics, 335 East 45th Street, New York, NY 10017. (212) 661-9404. $125.00 per year.

ASTRONOMICAL SOCIETY OF THE PACIFIC. Publications. Astronomical Society of the Pacific, 1290 24th Avenue, San Francisco, CA 94122. (415) 661-8660. Monthly. $38.00.

ASTRONOMY. Astro Media Corporation, 625 East Paul Avenue, Milwaukee, WI 53202. Monthly. 418.00 per year.

ASTRONOMY AND ASTROPHYSICS. Springer-Verlag New York, Incorporated, 175 Fifth Avenue, New York, NY 10010. (800) 526-7254 or (212) 460-1500. $680.00 per year.

ASTROPHYSICAL JOURNAL. American Astronomical Society, University of Chicago Press, 5801 Ellis Avenue, Chicago, IL 60637. Biweekly. $305.00 per year.

ASTROPHYSICS AND SPACE SCIENCE. D. Reidel Publishing Company, 190 Old Derby Street, Hingham, MA 02043. Monthly. $101.00 per year.

CELESTIAL MECHANICS; AN INTERNATIONAL JOURNAL OF SPACE DYNAMICS. D. Reidel Publishing Company, 190 Old Derby Street, Hingham, MA 02043. Monthly. $310.00 per year.

ICARUS: INTERNATIONAL JOURNAL OF THE SOLAR SYSTEM STUDIES. Academic Press, Incorporated, Orlando, FL 32887. (305) 345-4100. Monthly. $484.00 per year.

MERCURY. Astronomical Society of the Pacific, 1290 24th Avenue, San Francisco, CA 94122. (15) 661-8660. Bimonthly. $21.00 per year.

MONTHLY NOTICES OF THE ROYAL ASTRONOMICAL SOCIETY. Blackwell Science Publications, Incorporated, 667 Lytton Avenue, Palo Alto, CA 94301. (415) 324-1688. Monthly. $850.00 per year.

PLANETARY AN SPACE SCIENCE. Pergamon Press, Incorporated, Maxwell House, Fairview Park, Elmsford, NY 10523. (914) 592-7700. Monthly. $430.00 per year.

SKY AND TELESCOPE. Sky Publishing Corporation, 49 Bay State Road, Cambridge, MA 02238. (617) 864-7360. Monthly. $18.00 per year.

SOLAR PHYSICS. D. Reidel Publishing Company, 190 Old Derby Street, Hingham, MA 02043. Monthly. $620.00 per year.

SOVIET ASTRONOMY (TRANSLATION OF ASTRO-NOMICHESKII ZHURNAL). American Institute of Physics, 335 East 45th Street, New York, NY 10017. (212) 661-9404. Bimonthly. $425.00 per year.

SPACE SCIENCE REVIEWS. D. Reidel Publishing Company, 190 Old Derby Street, Hingham, MA 02043. Monthly. $305.00 per year.

VISTAS IN ASTRONOMY. Pergamon Press, Incorporated, Maxwell House, Fairview Park, Elmsford, NY 10523. (914) 592-7700. Quarterly. $145.00 per year.

RESEARCH CENTERS AND INSTITUTES

CORNELL UNIVERSITY. Arechibo Observatory, Box 995, Arecibo, PR 00613. (809) 878-2612.

NATIONAL ASTRONOMY AND IONOSPHERE CENTER. Cornell University, Space Sciences Building, Ithaca, NY 14853. (607) 256-3734.

NATIONAL RADIO ASTRONOMY OBSERVATORY. Edgemont Road, Charlottesville, VA 22903. (804) 296-0211.

UNIVERSITY OF MICHIGAN. Radio Astronomy Observatory, 937 Dennison Building, Ann Arbor, MI 48109. (313) 426-8441.

UNIVERSITY OF TEXAS AT AUSTIN. Radio Astronomy Observatory, Austin, TX 78712. (512) 471-6600.

RADIOACTIVITY

See: RADIATION

RADIOCARBON DATING

See also: ANALYTICAL CHEMISTRY, CHEMISTRY, CARBON

ABSTRACT SERVICES AND INDEXES

CHEMICAL ABSTRACTS. American Chemical Society, Chemical Abstracts Service, Box 3012, Columbus, OH 43210. (614) 421-3600. 1907 to present. Weekly. $9500.00 per year.

CONFERENCE PAPERS INDEX. Cambridge Scientific Abstracts, 5161 River Road, Bethesda, MD 20816. 1972 to present. Monthly. Inquire as to cost and availability.

INDEX TO SCIENTIFIC AND TECHNICAL PROCEEDINGS. Institute for Scientific Information, 3501 Market Street, Philadelphia, PA 19104. (800) 523-1850 or (215) 386-0100. 1978 to present. Monthly. $775.00 per year.

INDEX TO SCIENTIFIC REVIEWS. Institute for Scientific Information, 3501 Market Street, Philadelphia, PA 19104. (800) 523-1850 or (215) 386-0100. 1974 to present. Semi-annual. $550.00 per year.

SCIENCE CITATION INDEX. Institute for Scientific Information, 3501 Market Street, Philadelphia, PA 19104. (800) 523-1850 or (215) 386-0100. Six times per year. $6200.00 per year.

ANNUAL REVIEWS AND YEARBOOKS

CHEMISTRY AND PHYSICS OF CARBON. Philip J. Walker and Peter A. Thrower, editors. Marcel Dekker Inc., 270 Madison Avenue, New York, NY 10016. (800) 228-1160. 1966 to 1984. Irregular. Price varies. Inquire.

ASSOCIATIONS AND PROFESSIONAL SOCIETIES

AMERICAN CARBON SOCIETY. The Stackpole Corporation, St. Marys, PA 15857. (814) 781-8410.

AMERICAN CHEMICAL SOCIETY. 1155 16th Street, N.W., Washington, DC 20036. (202) 872-4600.

AMERICAN INSTITUTE OF CHEMICAL ENGINEERS. 345 East 47th Street, New York, NY 10017. (212) 705-7338.

ASSOCIATION OF OFFICIAL ANALYTICAL CHEMISTS. 1111 North 19th Street, Suite 210, Arlington, VA 22209. (703) 522-3032.

DIRECTORIES AND BIOGRAPHICAL SOURCES

AMERICAN MEN AND WOMEN OF SCIENCE. R.R. Bowker, Inc., Order Department, 245 West 17th Street, New York, NY 10011. (800) 521-8110. Eight volumes. 1986. $595.00 for set.

INTERNATIONAL RESEARCH CENTERS DIRECTORY 1988-89. Darren L. Smith, editor. Gale Research Company, Book Tower, Detroit, MI 48226. (800) 521-0707. 4th edition. 1987. $360.00.

RESEARCH CENTERS DIRECTORY 1988. Gale Research Company, Book Tower, Detroit, MI 48226. (800) 521-0707. 12th edition. 1987. $365.00 for set.

SCIENTIFIC AND TECHNICAL ORGANIZATIONS AND AGENCIES DIRECTORY. Margaret Labash Young, editor. Gale Research Company, Book Tower, Detroit, MI 48226. (800) 521-0707. 2nd edition. 1987. $185.00.

GENERAL WORKS

CARBON-FOURTEEN DATING. R. Burleigh. Academic Press, Inc., 6277 Sea Harbor Drive, Orlando, FL 32821. (800) 321-5068. 1987. Inquire.

RADIOCARBON DATING. Willard F. Libby. Books on Demand, 300 North Zeeb Road, Ann Arbor, MI 48106. (313) 761-4700. Second edition. $46.80.

ONLINE DATA BASES

CA SEARCH. Chemical Abstracts Service, P.O. Box 3012, Columbus, OH 43120. (800) 848-6538 or (614) 421-3600. Comprehensive guide to chemical literature, 1972 to present. Inquire as to online cost and availability.

DISSERTATION ABSTRACTS ONLINE. University Microfilms International, 300 North Zeeb Road, Ann Arbor, MI 48106. (800) 521-0600 or (313) 761-4700. Scope includes virtually all doctoral dissertations accepted at accredited American institutions from 1861 to present in over 250 subject areas. Inquire as to online cost and availability.

NTIS. National Technical Information Service, 5285 Port Royal Road, Springfield, VA 22161. (703) 487-4630. Broad coverage of government sponsored research reports, 1964 to present. Inquire as to online cost and availability.

SCISEARCH. Institute for Scientific Information, 3501 Market Street, Philadelphia, PA 19104. (800) 523-1850 or (215) 386-0100. Broad multidisciplinary title and author index to the international literature of science and technology, 1974 to present. Inquire as to online cost and availability.

PERIODICALS

RADIOCARBON. (Yale University, Kline Geology Laboratory) American Journal of Science, Box 6666, New Haven, CT 06511. (203) 436-3827. 1959 to present. $50.00 per year.

RESEARCH CENTERS AND INSTITUTES

LABORATORY OF ISOTOPE GEOCHEMISTRY. University of Arizona, 136 Gould-Simpson Building, Tucson, AZ 85721. (602) 621-6014.

RADIOCHEMISTRY

See: NUCLEAR CHEMISTRY

RAILROAD ENGINEERING

ABSTRACT SERVICES AND INDEXES

APPLIED SCIENCE AND TECHNOLOGY INDEX. H.W. Wilson Company, 950 University Avenue, Bronx, NY 10452. (800) 367-6670 or (212) 588-8400. Inquire as to cost and availability.

CHEMICAL ABSTRACTS. Chemical Abstracts Service, 2540 Olentangy Road, Post Office Box 3012, Columbus, OH 43210. (800) 848-6538 or (614) 421-3600. Weekly. $9200.00 per year.

ENGINEERING INDEX MONTHLY. Engineering Information, Incorporated 345 East 47th Street, New York, NY 10017. (800) 221-1044 or (212) 705-7600. Monthly, with annual cumulation. $1425.00 per year.

GENERAL SCIENCE INDEX. H.W. Wilson Company, 950 University Avenue, Bronx, NY 10452. (800) 367-6770 or (212) 588-8400. Inquire as to cost and availability.

METALS ABSTRACTS. American Society for Metals, 9639 Kinsman Road, Metals Park, OH 44073. (216) 338-5151. Monthly. $890.00.

METALS ABSTRACTS INDEX. American Society for Metals, 9639 Kinsman Road, Metals Park, OH 44073. (216) 338-5151. Monthly. (Sold only to subscribers of Metals Abstracts).

DIRECTORIES AND BIOGRAPHICAL SOURCES

JANE'S WORLD RAILWAYS. Jane's Publishing Incorporated, 115 5th Avenue, New York, NY 10003. (214) 254-9097. Annual. $150.00.

RAILROADS DIRECTORY. American Business Directories, Incorporated, 5707 South 86th Circle, Omaha, NE 68127. (402) 331-7169. 1986. $75.00.

RAILWAY TRACK AND STRUCTURES - RAILROAD TRACK CONTRACTORS DIRECTORY ISSUE. Simmons-Boardman Publishing Corporation, 29 East Madison Avenue, Chicago, IL 60602. (312) 641-5815. Annual. $2.00.

RESEARCH CENTERS DIRECTORY. Gale Research Company, Book Tower, Detroit, MI 48226. (800) 521-0707. Twelfth edition. 1987. $365.00 for set.

WHO'S WHO IN RAILROADING AND RAPID TRANSIT. National Railway Publication Company, 424 West 33rd Street, New York, NY 10001. (212) 714-3100. Irregular. $70.00.

RAILROAD ENGINEERING

ENCYCLOPEDIAS AND DICTIONARIES

CAR AND LOCOMOTIVE CYCLOPEDIA. Simmons-Boardman Publishing Corporation, 345 Hudson Street, New York, NY 10014. (212) 620-7200. 1984. $70.00.

RAILWAY AGE'S COMPREHENSIVE RAILROAD DICTIONARY. Simmons-Boardman Publishing Corporation, 345 Hudson Street, New York, NY 10014. (212) 620-7200. 1984. $18.00.

GENERAL WORKS

RAILROAD ENGINEERING. William W. Hay. John Wiley and Sons, Incorporated, 605 Third Avenue, New York, NY 10158. (800) 526-5368 or (212) 850-6000. Second edition. 1982. $59.95.

RAILROAD TRACK THEORY AND PRACTICE: MATERIAL PROPERTIES, CROSS SECTIONS, WELDING AND TREATMENT. Fritz Fastenrath, editor. Frederick Unger Publishing Company, Incorporated, 39 Cooper Square, New York, NY 10003. (212) 473-7885. 1980. $85.00.

ONLINE DATA BASES

CA SEARCH. Chemical Abstracts Service, Post Office Box 3012, Columbus, OH 43210. Guide to chemical literature, 1972 to present. Inquire as to cost and availability.

COMPENDEX. Engineering Information, Incorporated, 345 East 47th Street, New York, NY 10017. (800) 221-1044 or (212) 705-7615. Engineering and technical literature, 1975 to present. Inquire as to cost and availability.

NTIS. National Technical Information Service, 5285 Port Royal Road, Springfield, VA 22161. (703) 487-4630. Broad coverage of government sponsored research reports, 1964 to present. Inquire as to cost and availability.

WILSONLINE. H.W. Wilson Company, 950 University Avenue, Bronx, NY 10452. (800) 367-6770 or (212) 588-8400. Makes available online versions of the printed H.W. Wilson Indexes including Applied Science and Technology Index, Business Periodicals Index, and Readers' Guide to Periodical Literature. Period covered is generally 1983 to present. Inquire as to cost and availability.

PERIODICALS

CROSS TIES. Railway Tie Association, 314 North Broadway, St. Louis, MO 63102. (314) 231-8099. Monthly. $24.00 per year.

INTERNATIONAL RAIL JOURNAL AND RAPID TRANSIT. Simmons-Boardman Publishing Corporation, 345 Hudson Street, NEw York, NY 10014. (212) 620-7200. Monthly. $25.00 per year.

MODERN RAILROADS. Enright-Reilly Publishing Company, 2020 West Oakton Street, Park Ridge, IL 60068. (312) 399-0202. Monthly. $35.00 per year.

PROGRESSIVE RAILROADING. Murphy-Richter Publishing Company, 2 North Riverside Plaza, Chicago, IL 60606. (312) 454-9155. Monthly. $25.00 per year.

RAILWAY AGE. Simmons-Boardman Publishing Corporation, 345 Hudson Street, New York, NY 10014. (212) 620-7200. Monthly. $30.00 per year.

RAILWAY GAZETTE INTERNATIONAL: A JOURNAL OF MANAGEMENT, ENGINEERING AND OPERATION. Business Press International, Limited, Quadrant House, The Quadrant, Sutton, Surrey SM2 5As, England. Monthly. $35.00 per year.

RAILWAY TRACK AND STRUCTURES. Simmons-Boardman Publishing Corporation, 345 Hudson Street, New York, NY 10014. (212) 620-7200. Monthly. $12.00 per year.

RESEARCH CENTERS AND INSTITUTES

ASSOCIATION OF AMERICAN RAILROADS RESEARCH AND TEST DEPARTMENT. 50 F Street, NW, Washington, DC 20001. (202) 639-2250.

CARNEGIE-MELLON UNIVERSITY. Rail Systems Center, Institute, 4400 Fifth Avenue, Pittsburgh, PA 15213. (412) 268-2960.

UNIVERSITY OF ILLINOIS. Railway Wheel Research Program, 224 Talbot Laboratory, 104 South Wright Street, Urbana, IL 61801. (217) 333-2313.

RAIN

See also: METEOROLOGY

ABSTRACT SERVICES AND INDEXES

APPLIED SCIENCE AND TECHNOLOGY INDEX. H.W. Wilson Company, 950 University Avenue, Bronx, NY 10452. (800) 367-6670 or (212) 588-8400. Inquire as to cost and availability.

CHEMICAL ABSTRACTS. Chemical Abstracts Service, 2540 Olentangy Road, Post Office Box 3012, Columbus, OH 43210. (800) 848-6538 or (614) 421-3600. Weekly. $9200.00 per year.

CURRENT CONTENTS: PHYSICAL AND CHEMICAL SCIENCES. Institute for Scientific Information, 3501 Market Street, Philadelphia, PA 19104. (800) 523-1850 or (215) 386-0100. $272.00 per year.

GENERAL SCIENCE INDES. H.W. Wilson Company, 950 University Avenue, Bronx, NY 10452. (800) 367-6770 or (212) 588-8400. Inquire as to cost and availability.

METEOROLOGICAL AND GEOASTROPHYSICAL ABSTRACTS. American Meteorological Society, 45 Beacon Street, Boston, MA 02108. (617) 227-2425. 1950 to present. Monthly. $450.00 per year.

SCIENCE CITATION INDEX. Institute for Scientific Information, 3501 Market Street, Philadelphia, PA 19104. (800) 523-1850 or (215) 386-0100. Inquire as to cost and availability.

ASSOCIATION AND PROFESSIONAL SOCIETIES

AMERICAN METEOROLOGICAL SOCIETY. 45 Beacon Street, Boston, MA 02108. (617) 227-2425.

INTERNATIONAL ASSOCIATION OF METEOROLOGY AND ATMOSPHERIC PHYSICS. UCAR, Post Office Box 3000, Boulder, CO 80307.

NATIONAL WEATHER ASSOCIATION. 4400 Stamp Road, Room 404, Temple Hills, MD 20748. (301) 899-3784.

UNIVERSITY CORPORATION FOR ATMOSPHERIC RESEARCH. Box 3000, 1850 Table Mesa Drive, Boulder, CO 80307. (303) 497-1000.

WEATHER MODIFICATION ASSOCIATION. Post Office Box 8116, Fresno, CA 93747. (209) 291-8466.

BIBLIOGRAPHIES

SCIENCE BOOKS AND FILMS. American Association for the Advancement of Science, 1333 H Street, NW, Washington, DC 20005. Five times per year. $20.00 per year.

SCIENTIFIC AND TECHNICAL BOOKS AND SERIALS IN PRINT 1988: AN INDEX TO LITERATURE IN SCIENCE AND TECHNOLOGY. R.R. Bowker Company, 205 East 42nd Street, New York, NY 10017. (800) 521-8110 or (212) 916-1600.

$175.00.

DIRECTORIES AND BIOGRAPHICAL SOURCES

AMERICAN MEN AND WOMEN OF SCIENCE. Physical and Biological Sciences. Sixteenth edition. R.R. Bowker Company, 205 East 42nd Street, New York, NY 10017. (800) 521-8110 or (212) 916-1600. 1986. $595.00.

INTERDOC: DIRECTORY OF PUBLISHED PROCEEDINGS, SERIES. SEMT-Science/Engineering/Medicine/Technology. Interdoc Corporation, 173 Halstead Avenue, Box 326, Harrison, NY 10528. (014) 835-3506. Ten times per year. $325.00 per year.

INTERNATIONAL RESEARCH CENTERS DIRECTORY 1988-1989. Gale Research Company, Book Tower, Detroit, MI 48226. (800) 521-0707. Fourth edition. 1987. $360.00.

METEOROLOGICAL SERVICES OF THE WORLD. World Meteorological Organization. Available from: American Meteorological Society, 45 Beacon Street, Boston, MA 02108. (617) 227-2425. Annual. $35.00.

NATIONAL WEATHER SERVICE OFFICES AND STATIONS. National Oceanic and Atmospheric Administration, Department of Commerce, Silver Spring, MD 20910. (301) 427-7698. Annual. Free.

RESEARCH CENTERS DIRECTORY. Gale Research Company, Book Tower, Detroit, MI 48226. (800) 521-0707. Twelfth edition. 1987. $365.00 for set.

ENCYCLOPEDIAS AND DICTIONARIES

ENCYCLOPEDIA OF CLIMATOLOGY. John E. Oliver and Rhodes W. Fairbridge, editors. Van Nostrand Reinhold, Incorporated, 115 Fifth Avenue, New York, NY 10003. (800) 543-2681. 1987. $89.95.

GENERAL WORKS

RAINFALL AS THE BASIS FOR URBAN RUN-OFF DESIGN AND ANALYSIS: PROCEEDINGS OF A SPECIALIZED SEMINAR HELD IN COPENHAGEN, DENMARK, 24-26 AUGUST, 1983. P. Harremoes, editor. Pergamon Press, Incorporated, Maxwell House, Fairview Park, Elmsford, NY 10523. (914) 592-7700. 1984. $80.00.

SAMPLING AND ANALYSIS OF RAIN-STP 823. S. Campbell, editor. American Society for Testing and Materials, 1916 Race Street, Philadelphia, PA 19103. (215) 299-5400. 1984. $18.00.

ONLINE DATA BASES

CA SEARCH. Chemical Abstracts Service, Post Office Box 3012, Columbus, OH 43210. Guide to chemical literature, 1972 to present. Inquire as to cost and availability.

DISSERTATION ABSTRACTS ONLINE. University Microfilms International, 300 North Zeeb Road, Ann Arbor, MI 48106. (800) 521-0600 or (313) 761-4700. Scope includes virtually all doctoral dissertations accepted at accredited American institutions from 1861 to present in 252 subject areas. Inquire as to cost and availability.

METEOROLOGICAL AND GEOASTROPHYSICAL ABSTRACTS. American Meteorological Society, 45 Beacon Street, Boston, MA 02108. (617) 227-2425. 1950 to present. Monthly. $450.00 per year.

SCISEARCH. Institute for Scientific Information, 3501 Market Street, Philadelphia, PA 19104. (800) 523-1850 or (215) 386-0100. Broad multidisciplinary title and author index to the international literature of science and technology, 1974 to present. Inquire as to cost and availability.

PERIODICALS

AGRICULTURAL AND FOREST METEOROLOGY. Elsevier Science Publishing Company, Incorporated, 52 Vanderbilt Avenue, New York, NY 10017. (212) 370-5520. 1964 to present. Monthly. $260.00 per year.

AMERICAN METEOROLOGICAL SOCIETY BULLETIN. American Meteorological Society, 45 Beacon Street, Boston, MA 02108. (617) 227-2425. 1920 to present. Monthly. $60.00 per year.

JOURNAL OF CLIMATE AND APPLIED METEOROLOGY. American Meteorological Society, 45 Beacon Street, Boston, MA 02108. (617) 227-2425. 1962 to present. Monthly. $120.00 per year.

JOURNAL OF THE ATMOSPHERIC SCIENCES. American Meteorological Society, 45 Beacon Street, Boston, MA 02108. (617) 227-2425. 1944 to present. Semimonthly. $220.00 per year.

MONTHLY WEATHER REVIEW. American Meteorological Society, 45 Beacon Street, Boston, MA 02108. (617) 227-2425. 1872 to present. Monthly. $120.00 per year.

NATIONAL WEATHER DIGEST. National Weather Association, 4400 Stamp Road, Room 404, Temple Hills, MD 20748. (301) 899-3784. 1976 to present. Quarterly. $20.00 per year.

WEATHER. Royal Meteorological Society, James Glaisher House, Grenville Place, Bracknell Berkshire, RG12 1BX, England. 1946 to present. $30.00 per year.

WEATHERWISE. Heldref Publications, 4000 Albemarle Street, NW, Washington, DC 20016. (202) 362-6445. 1948 to present. Bimonthly. $20.00 per year.

RESEARCH CENTERS AND INSTITUTES

COOPERATIVE INSTITUTE FOR MESOSCALE METEOROLOGICAL STUDIES. University of Oklahoma, 401 East Boyd, Norman, OK 73019. (405) 325-3041.

NATIONAL CENTER FOR ATMOSPHERIC RESEARCH. Box 3000, Boulder, CO 80307. (303) 497-1000.

UNIVERSITY OF NEVADA. Atmospheric Water Resources Laboratory. Desert Research Laboratory, Post Office Box 60220, Reno, NV 89505. (702) 972-1676.

RAMAN SPECTROSCOPY

See: SPECTROSCOPY

RAMJETS

See: JET PROPULSION

REACTORS

See: NUCLEAR REACTORS

RECURSIVE PROGRAMMING

See: COMPUTER PROGRAMMING

RED DWARFS

See: STARS

RED GIANTS

See: STARS

RED SHIFT

See: ASTROPHYSICS

REFRIGERATION

See: AIR CONDITIONING

RELATIVITY

See: CELESTRIAL MECHANICS, PHYSICS

RELAYS

See also: CIRCUIT BREAKERS, ELECTRICAL ENGINEERING, ELECTRICITY, ELECTROMAGNETISM, ELECTRONIC CIRCUITS AND COMPONENTS, ELECTRONICS, ELECTRONICS ENGINEERING

ABSTRACT SERVICES AND INDEXES

APPLIED SCIENCE AND TECHNOLOGY INDEX. H.W. Wilson and Company, 950 University Avenue, Bronx, NY 10452. (800) 367-6670 or (212) 588-8400. Monthly. Inquire as to cost and availability.

CURRENT CONTENTS: ENGINEERING, TECHNOLOGY AND APPLIED SCIENCES. Institute for Scientific Information, 3501 Market Street, Philadelphia, PA 19104. (800) 523-1850 or (215) 386-0100. Weekly. $275.00 per year.

ELECTRIC POWER INDUSTRY ABSTRACTS. Edison Electric Institute, c/o Utility Date Institute, 2011 I Street, N.W., Suite 700, Washington, DC 20006. 1975 to present. Bimonthly. Inquire as to cost and availability.

ELECTRICAL AND ELECTRONICS ABSTRACTS. Institution of Electrical Engineers. Available from: Institute of Electrical and Electronics Engineers. IEEE Service Center, 445 Hoes Lane, Piscataway, NJ 08854. Monthly. $1250.00 per year.

ENGINEERING INDEX MONTHLY AND AUTHOR INDEX. Engineering Information Inc., 345 East 47th Street, New York, NY 10017. (212) 705-7600. Monthly. $1560.00 per year.

IEEE PUBLICATIONS BULLETIN. Institute of Electrical and Electronics Engineers. Institute of Electrical and Electronics Engineers. IEEE Service Center, 445 Hoes Lane, Piscataway, NJ 08854. Quarterly. Free.

ASSOCIATIONS AND PROFESSIONAL SOCIETIES

AMERICAN ELECTRONICS ASSOCIATION. P.O. Box 10045, 2670 Hanover Street, Palo Alto, CA 94303. (415) 857-9300.

AMERICAN INSTITUTE OF PHYSICS. 335 East 45th Street, New York, NY 10017. (212) 661-9494.

EDISON ELECTRIC INSTITUTE. 1111 19th Street, N.W., Washington, DC 20036. (202) 828-7400.

INSTITUTE OF ELECTRICAL AND ELECTRONICS ENGINEERS. 345 East 47th Street, New York, NY 10017. (212) 705-7900.

DIRECTORIES AND BIOGRAPHICAL SOURCES

1987 DIRECTORY OF ENGINEERING SOCIETIES AND RELATED ORGANIZATIONS. Gordon Davis, editor. Hemisphere Publishing Corporation, 79 Madison Avenue, New York, NY 10016-7892. (800) 821-8312. 12th edition. 1987. $100.00.

RESEARCH CENTERS DIRECTORY 1988. Gale Research Company, Book Tower, Detroit, MI 48226. (800) 521-0707. 12th edition. 1987. $365.00 for set.

SCIENTIFIC AND TECHNICAL ORGANIZATIONS AND AGENCIES DIRECTORY. Margaret Labash Young, editor. Gale Research Company, Book Tower, Detroit, MI 48226. (800) 521-0707. 2nd edition. 1987. $185.00.

WHO'S WHO IN ELECTRONICS. Harris Publishing Company, 2057-2 Aurora Road, Twinsburg, OH 44087. (216) 425-9143. Annual. $90.00.

WHO'S WHO IN ENGINEERING. Gordon Davis, editor. Hemisphere Publishing Corporation, 79 Madison Avenue, New York, NY 10016-7892. (800) 821-8312. 6th edition. 1985. $200.00.

GENERAL WORKS

BASIC ELECTRIC CIRCUITS. Donald P. Leach. John Wiley and Sons, Inc., 605 Third Avenue, New York, NY 10158. (800) 526-5368. 1984. $34.95.

ELECTRICAL ENGINEERING. W.H. Roadstrum and Dan H. Wolaver. Harper and Row Publishers, Inc., 10 East 53rd Street, New York, NY 10022. (800) 242-7737. 1986. $37.50.

ELECTRICITY AND MAGNETISM. M.H. Nayfeh and M.K. Brussel. John Wiley and Sons, Inc., 605 Third Avenue, New York, NY 10158. (800) 526-5368. 1985. $32.95.

FUNDAMENTALS OF ELECTRIC CIRCUITS. David A. Bell. Prentice-Hall Publishing, Inc., Englewood Cliffs, NJ 07632. (800) 562-0245. 4th edition. 1988. $30.25.

INTRODUCTION TO ELECTRICITY AND ELECTRONICS FOR THE COMPUTER AGE. R. Rosen. John Wiley and Sons, Inc., 605 Third Avenue, New York, NY 10158. (800) 526-5368. 1987. $35.95.

PROTECTIVE RELAYS: THEIR THEORY AND PRACTICE. A.R. Warrington. Methuen, Inc., 29 West 35th Street, New York, NY 10001. (212) 244-3336. Two volumes. Volume one, second edition, 1968, $44.95. Volume two, third edition, 1978, $44.95.

HANDBOOKS AND MANUALS

HANDBOOK OF MODERN ELECTRONICS AND ELECTRICAL ENGINEERING. Charles Belove, editor. John Wiley and Sons, Inc., 605 Third Avenue, New York, NY 10158. (800) 526-5368. 1986. $88.95.

ONLINE DATA BASES

COMPENDEX. Engineering Information, Inc., 345 East 47th Street, New York, NY 10017. (800) 221-1044 or (212) 705-7615. Engineering and technical literature, 1975 to present. Inquire as to online cost and availability.

INSPEC. INSPEC Marketing Department, Institution of Electrical Engineers. Available from IEEE Service Center, 445 Hoes Lane, Piscataway, NJ 08854. (201) 981-0060. Online version of Physics Abstracts. Inquire as to online cost and availability.

NTIS. National Technical Information Service, 5285 Port Royal Road, Springfield, VA 22161. (703) 487-4630. Broad coverage of government sponsored research reports, 1964 to present. Inquire as to online cost and availability.

SCISEARCH. Institute for Scientific Information, 3501 Market Street, Philadelphia, PA 19104. (800) 523-1850 or (215) 386-0100. Broad multidisciplinary title and author index to the international literature of science and technology, 1974 to present. Inquire as to online cost and availability.

WILSONLINE. H.W. Wilson and Company, 950 University Avenue, Bronx, NY 10452. (800) 367-6770 or (212) 588-8400. Makes available online versions of the H.W. Wilson indexes including Applied Science and Technology Index, Business Periodicals Index and Readers' Guide to Periodical Literature. Approximately 1980 to present. Inquire as to online cost and availability.

OTHER SOURCES

A GUIDE TO THE LITERATURE OF ELECTRICAL AND ELECTRONICS ENGINEERING. Susan B. Ardis. Libraries Unlimited Inc., P.O. Box 263, Littleton, CO 80160. (303) 770-1220. 1987. $37.50.

PERIODICALS

ELECTRIC LIGHT AND POWER. Technical Publishing Company, 875 Third Avenue, New York, NY 10022. (212) 605-9400. 1922 to present. Monthly. $38.00 per year.

ELECTRIC MACHINES AND POWER SYSTEMS. Hemisphere Publishing Corporation, 1010 Vermont Avenue, NW, Washington, DC 20005. (800) 526-0275. 1976 to present. Bimonthly. $134.50 per year.

ELECTRICAL WORLD. McGraw-Hill Book Company, 1221 Avenue of the Americas, New York, NY 10020. (212) 512-2000. 1874 to present. Monthly. $11.00 per year.

ELECTRONIC DESIGN. Hayden Publishing Company, 10 Mulholland Drive, Hasbrouck Heights, NJ 07604. (201) 288-7520. 1952 to present. Biweekly. $40.00 per year.

ELECTRONICS. McGraw-Hill Book Company, 1221 Avenue of the Americas, New York, NY 10020. (212) 512-2000. 1930 to present. Weekly. $32.00 per year.

IEEE POWER ENGINEERING REVIEW. Institute of Electrical and Electronics Engineers. IEEE Service Center, 445 Hoes Lane, Piscataway, NJ 08854. 1981 to present. Monthly. $75.00 per year.

IEEE TRANSACTIONS ON POWER DELIVERY. Institute of Electrical and Electronics Engineers. IEEE Service Center, 445 Hoes Lane, Piscataway, NJ 08854. 1986 to present. Quarterly. $100.00 per year.

INSTITUTE OF ELECTRICAL AND ELECTRONICS ENGINEERS PROCEEDINGS. Institute of Electrical and Electronics Engineers. IEEE Service Center, 445 Hoes Lane, Piscataway, NJ 08854. 1913 to present. Monthly. $140.00 per year.

RESEARCH CENTERS AND INSTITUTES

EDISON ELECTRIC INSTITUTE. 1111 19th Street, N.W., Washington, DC 20036. (202) 778-6778.

ELECTRIC POWER INSTITUTE. Texas A&M University, Department of Electrical Engineering, College Station, TX 77843.

ELECTRICAL ENGINEERING RESEARCH LABORATORIES. Purdue University, Electrical Engineering Building, West Lafayette, IN 47907. (317) 494-3536.

LABORATORY FOR ELECTROMAGNETIC AND ELECTRONIC SYSTEMS. 77 Massachusetts Avenue, Cambridge, MA 02139. (617) 253-4631.

RELIABILITY ENGINEERING

See: INDUSTRIAL ENGINEERING, QUALITY CONTROL ENGINEERING

REMOTE CONTROL SYSTEMS

See: CONTROL SYSTEMS

REMOTE SENSING

See also: AERIAL PHOTOGRAPHY, CARTOGRAPHY, PHOTOGRAMMETRY, SURVEYING

ABSTRACT SERVICES AND INDEXES

APPLIED SCIENCE AND TECHNOLOGY INDEX. H.W. Wilson and Company, 950 University Avenue, Bronx, NY 10452. (800) 367-6670 or (212) 588-8400. Monthly. Inquire as to cost and availability.

BIBLIOGRAPHY AND INDEX OF GEOLOGY. American Geological Institute, 4220 King Street, Alexandria, VA 22302. (703) 379-2480. 1969 to present. Monthly. $1100.00 per year.

CHEMICAL ABSTRACTS. American Chemical Society, Chemical Abstracts Service, Box 3012, Columbus, OH 43210. (614) 421-3600. 1907 to present. Weekly. $9500.00 per year.

CURRENT CONTENTS: ENGINEERING, TECHNOLOGY AND APPLIED SCIENCES. Institute for Scientific Information, 3501 Market Street, Philadelphia, PA 19104. (800) 523-1850 or (215) 386-0100. Weekly. $275.00 per year.

ENGINEERING INDEX MONTHLY AND AUTHOR INDEX. Engineering Information Inc., 345 East 47th Street, New York, NY 10017. (212) 705-7600. Monthly. $1560.00 per year.

METEOROLOGICAL AND GEOASTROPHYSICAL ABSTRACTS. American Meteorological Society, 45 Beacon Street, Boston, MA 02108. (617) 227-2425. 1950 to present. Monthly. $450.00 per year.

PHYSICS ABSTRACTS. Institution of Electrical Engineers. Available from: IEEE Service Center, 445 Hoes Lane, Piscataway, NJ 08854. 1898 to present. Bimonthly. $1700.00 per year.

REMOTE SENSING OF NATURAL RESOURCES: A QUARTERLY LITERATURE REVIEW. University of New Mexico, Technology Application Center, Albuquerque, NM 87131. (505) 277-3622. 1974 to present. Quarterly. $150.00. Available to qualified agencies only.

SCIENCE CITATION INDEX. Institute for Scientific Information, 3501 Market Street, Philadelphia, PA 19104. (800) 523-1850 or (215) 386-0100. Six times per year. $6200.00 per year.

ASSOCIATIONS AND PROFESSIONAL SOCIETIES

AMERICAN SOCIETY FOR PHOTOGRAMMETRY AND REMOTE SENSING. 210 Little Falls Street, Falls Church, VA 22046 4398. (703) 534-6617.

OPTICAL SOCIETY OF AMERICA. 1816 Jefferson Place, N.W., Washington, DC 20036. (202) 223-8130.

SOCIETY OF PHOTOGRAPHIC SCIENTISTS AND ENGINEERS. 7003 Kilworth Lane, Springfield, VA 22151. (703) 642-9090.

SPIE - THE INTERNATIONAL SOCIETY FOR OPTICAL ENGINEERING. P.O. Box 10, 1022 19th Street, Bellingham, WA 98227. (206) 676-3290.

BIBLIOGRAPHIES

SCIENTIFIC AND TECHNICAL BOOKS AND SERIALS IN PRINT 1988; AN INDEX TO LITERATURE IN SCIENCE AND TECHNOLOGY. R.R. Bowker Company, 205 East 42nd Street, New York, NY 10017. (800) 521-8110. $175.00.

DIRECTORIES AND BIOGRAPHICAL SOURCES

INTERNATIONAL RESEARCH CENTERS DIRECTORY 1988-89. Darren L. Smith, editor. Gale Research Company, Book Tower, Detroit, MI 48226. (800) 521-0707. 4th edition. 1987. $360.00.

1987 DIRECTORY OF ENGINEERING SOCIETIES AND RELATED ORGANIZATIONS. Gordon Davis, editor. Hemisphere Publishing Corporation, 1010 Vermont Avenue, NW, Washington, DC 20005. (800) 526-0275. 12th edition. 1987. $100.00.

RESEARCH CENTERS DIRECTORY 1988. Gale Research Company, Book Tower, Detroit, MI 48226. (800) 521-0707. 12th edition. 1987. $365.00 for set.

SCIENTIFIC AND TECHNICAL ORGANIZATIONS AND AGENCIES DIRECTORY. Margaret Labash Young, editor. Gale Research Company, Book Tower, Detroit, MI 48226. (800) 521-0707. 2nd edition. 1987. $185.00.

WHO'S WHO IN ENGINEERING. Gordon Davis, editor. Hemisphere Publishing Corporation, 1010 Vermont Avenue, NW, Washington, DC 20005. (800) 526-0275. 6th edition. 1985. $200.00.

ENCYCLOPEDIAS AND DICTIONARIES

MCGRAW-HILL ENCYCLOPEDIA OF SCIENCE AND TECHNOLOGY. McGraw-Hill Book Company, 1221 Avenue of the Americas, New York, NY 10020. (212) 512-2000. 6th edition. 1987. $1600.00.

THESAURUS OF PHOTOGRAPHIC SCIENCE AND ENGINEERING. Society of Photographic Scientists and Engineers. Books on Demand, 300 North Zeeb Road, Ann Arbor, MI 48106. (313) 761-4700. $34.50 in paper.

THESAURUS OF SCIENTIFIC, TECHNICAL, AND ENGINEERING TERMS. Hemisphere Publishing Corporation, 1010 Vermont Avenue, NW, Washington, DC 20005. (800) 526-0275. 1988. $125.00.

GENERAL WORKS

CLOSE-RANGE PHOTOGRAMMETRY AND SURVEYING: STATE OF THE ART. American Society for Photogrammetry and Remote Sensing. 210 Little Falls Street, Falls Church, VA 22046-4398. (703) 534-6617. 1985. $65.00 in paper.

ELEMENTS OF PHOTOGRAMMETRY. P.R. Wolf. McGraw-Hill Book Company, 1221 Avenue of the Americas, New York, NY 10020. (212) 512-2000. 2nd edition. 1983. $49.95.

PHOTOGRAMMETRY. Francis H. Moffitt and Edward M. Mikhail. Harper and Row Publishers, Inc., 10 East 53rd Street, New York, NY 10022. (212) 207-7655. 3rd edition. 1980. $41.95.

PRINCIPLES OF REMOTE SENSING. Paul Curran. Halstead Press, division of John Wiley and Sons, Inc., 605 Third Avenue, New York, NY 10158. (800) 526-5368. 1986. $35.95.

REMOTE SENSING. Floyd F. Sabins. W.H. Freeman and Company, 41 Madison Avenue, New York, NY 10010. (212) 532-7660. 2nd edition. 1986. $39.95.

REMOTE SENSING METHODS AND APPLICATIONS. R. Hord. John Wiley and Sons, Inc., 605 Third Avenue, New York, NY 10158. (800) 526-5368. 1986. $39.95.

SURVEY OF THE PROFESSION: PHOTOGRAMMETRY, SURVEYING, MAPPING, REMOTE SENSING. American Society for Photogrammetry and Remote Sensing. 210 Little Falls Street, Falls Church, VA 22046-4398. (703) 534-6617. 1982. $35.00 in paper.

HANDBOOKS AND MANUALS

MANUAL OF PHOTOGRAMMETRY. Chester C. Slama, editor. American Society for Photogrammetry and Remote Sensing. 210 Little Falls Street, Falls Church, VA 22046-4398. (703) 534-6617. 4th edition. 1980. $59.00.

MANUAL OF REMOTE SENSING. Robert N. Colwell, editor. American Society for Photogrammetry and Remote Sensing. 210 Little Falls Street, Falls Church, VA 22046-4398. (703) 534-6617. 2nd edition. 1983. $106.00 for set.

ONLINE DATA BASES

CA SEARCH. Chemical Abstracts Service, P.O. Box 3012, Columbus, OH 43120. (800) 848-6538 or (614) 421-3600. Comprehensive guide to chemical literature, 1972 to present. Inquire as to online cost and availability.

COMPENDEX. Engineering Information, Inc., 345 East 47th Street, New York, NY 10017. (800) 221-1044 or (212) 705-7615. Engineering and technical literature, 1975 to present. Inquire as to online cost and availability.

GEOREF. Online version of the BIBLIOGRAPHY AND INDEX OF GEOLOGY. American Geological Institute, 4220 King Street, Alexandria, VA 22302. (703) 379-2480. 1969 to present. Inquire as to online cost and availability.

NTIS. National Technical Information Service, 5285 Port Royal Road, Springfield, VA 22161. (703) 487-4630. Broad coverage of government sponsored research reports, 1964 to present. Inquire as to online cost and availability.

SCISEARCH. Institute for Scientific Information, 3501 Market Street, Philadelphia, PA 19104. (800) 523-1850 or (215) 386-0100. Broad multidisciplinary title and author index to the international literature of science and technology, 1974 to present. Inquire as to online cost and availability.

WILSONLINE. H.W. Wilson and Company, 950 University Avenue, Bronx, NY 10452. (800) 367-6770 or (212) 588-8400. Makes available online versions of the H.W. Wilson indexes including Applied Science and Technology Index, Business Periodicals Index and Readers' Guide to Periodical Literature. Approximately 1980 to present. Inquire as to online cost and availability.

PERIODICALS

IEEE TRANSACTIONS ON GEOSCIENCE AND REMOTE SENSING. IEEE Geoscience and Remote Sensing Society. Institute of Electrical and Electronics Engineers, 345 East 47th Street, New York, NY 10017. (212) 705-7900. Order from: IEEE Service Center, 445 Hoes Lane, Piscataway, NJ 08854. 1963 to present. Bimonthly. $110.00 per year.

JOURNAL OF IMAGING SCIENCE. Society of Photographic Scientists and Engineers. 7003 Kilworth Lane, Springfield, VA 22151. (703) 642-9090. 1956 to present. Bimonthly. $70.00 per year.

JOURNAL OF IMAGING TECHNOLOGY. Society of Photographic Scientists and Engineers. 7003 Kilworth Lane, Springfield, VA 22151. (703) 642-9090. 1975 to present. Bimonthly. $70.00 per year.

PHOTOGRAMMETRIA. Elsevier Science Publishing Company, Inc., 52 Vanderbilt Avenue, New York, NY 10017. (212) 370-5520. 1949 to present. Quarterly. $65.00 per year.

PHOTOGRAMMETRIC ENGINEERING AND REMOTE SENSING. American Society for Photogrammetry and Remote Sensing. 210 Little Falls Street, Falls Church, VA 22046-4398. (703) 534-6617. Order from: Allen Press, Inc., 1041 New Hampshire Street, Box 368, Lawrence, KS 66044. 1934 to present. Monthly. $80.00 per year.

PHOTOGRAMMETRIC RECORD. Photogrammetry Society, Department of Photogrammetry and Surveying, University College London, Gower Street, London WC1E 6BT, England. Semiannual. $37.50 per year.

SCIENTIFIC AND APPLIED PHOTOGRAPHY AND CINEMATOGRAPHY. Gordon and Breach Science Publishers, Inc., 50 West 23rd Street, New York, NY 10010. (212) 206-8900. 12 times per year. $496.00 per year.

REMOTE SENSING OF ENVIRONMENT. Elsevier Science Publishing Company, Inc., 52 Vanderbilt Avenue, New York, NY 10017. (212) 370-5520. 1968 to present. Six times per year. $210.00 per year.

REMOTE SENSING REVIEWS. Harwood Academic Publishers, 50 West 23rd Street, New York, NY 10010. (212) 206-8900. Quarterly. $160.00 per year.

RESEARCH CENTERS AND INSTITUTES

CENTER FOR REMOTE SENSING AND CARTOGRAPHY. 420 Chipta Way, Salt Lake City, UT 84112. (801) 581-8218.

GEOPHOTOGRAPHY AND REMOTE SENSING CENTER. University of Idaho, Geology Department, Moscow, ID 83843. (208) 885-7977.

NATIONAL RESEARCH COUNCIL OF CANADA, DIVISION OF PHYSICS. Ottawa, ON, Canada K1A OR6. (613) 993-1053.

PHOTOGRAMMETRY AND REMOTE SENSING SECTION. Tennessee Valley Authority, Office of Natural Resources and Economic Development, Haney Building, Chattanooga, TN 37401. (615) 755-2148.

RESERVOIRS

See: DAMS

RESINS

See: PLASTICS

RESISTORS

See also: ELECTRICAL ENGINEERING, ELECTRICITY, ELECTRONIC CIRCUITS AND COMPONENTS, ELECTRONICS, ELECTRONICS ENGINEERING, MICROELECTRONICS

ABSTRACT SERVICES AND INDEXES

APPLIED SCIENCE AND TECHNOLOGY INDEX. H.W. Wilson and Company, 950 University Avenue, Bronx, NY 10452. (800) 367-6670 or (212) 588-8400. Monthly. Inquire as to cost and availability.

CHEMICAL ABSTRACTS. American Chemical Society, Chemical Abstracts Service, Box 3012, Columbus, OH 43210. (614) 421-3600. 1907 to present. Weekly. $9500.00 per year.

CURRENT CONTENTS: ENGINEERING, TECHNOLOGY AND APPLIED SCIENCES. Institute for Scientific Information, 3501 Market Street, Philadelphia, PA 19104. (800) 523-1850 or (215) 386-0100. Weekly. $275.00 per year.

ELECTRICAL AND ELECTRONICS ABSTRACTS. Institution of Electrical Engineers. Available from: Institute of Electrical and Electronics Engineers. IEEE Service Center, 445 Hoes Lane, Piscataway, NJ 08854. Monthly. $1250.00 per year.

ELECTRONICS AND COMMUNICATIONS ABSTRACTS. Cambridge Scientific Abstracts, 5161 River Road, Bethesda, MD 20816. (301) 951-1400. Bimonthly. Inquire as to cost and availability.

ENGINEERING INDEX MONTHLY AND AUTHOR INDEX. Engineering Information Inc., 345 East 47th Street, New York, NY 10017. (212) 705-7600. Monthly. $1560.00 per year.

IEEE PUBLICATIONS BULLETIN. Institute of Electrical and Electronics Engineers. Institute of Electrical and Electronics Engineers. IEEE Service Center, 445 Hoes Lane, Piscataway, NJ 08854. Quarterly. Free.

PHYSICS ABSTRACTS. Institution of Electrical Engineers. Available from: IEEE Service Center, 445 Hoes Lane, Piscataway, NJ 08854. 1898 to present. Bimonthly. $1700.00 per year.

PHYSICS BRIEFS. Physik Verlag GmbH, Postfach 1260/1280, D-6940 Weinheim, West Germany. (212) 661-9404. 1920 to present. Twenty-six times per year. $1250.00 per year.

SCIENCE CITATION INDEX. Institute for Scientific Information, 3501 Market Street, Philadelphia, PA 19104. (800) 523-1850 or (215) 386-0100. Six times per year. $6200. 00 per year.

SOLID STATE ABSTRACTS: AN ABSTRACT JOURNAL INVOLVING THE PHYSICS, METALLURGY, CRYSTALLOGRAPHY, CHEMISTRY, AND DEVICE TECHNOLOGY OF SOLIDS. Cambridge Scientific Abstracts, 5161 River Road, Bethesda, MD 20816. (301) 951-1400. 1957 to present. Bimonthly. $550.00 per year.

ANNUAL REVIEWS AND YEARBOOKS

ADVANCES IN ELECTRONICS AND ELECTRON PHYSICS. Academic Press, Inc., 6277 Sea Harbor Drive, Orlando, FL 32821. (800) 321-5068. Irregular. Approximately $80.00 per volume.

CRC CRITICAL REVIEWS IN SOLID STATE AND MATERIALS SCIENCES. CRC Press, 2000 Corporate Boulevard, Boca Raton, FL 33431. (800) 272-7737. Irregular. Inquire.

ASSOCIATIONS AND PROFESSIONAL SOCIETIES

AMERICAN ELECTRONICS ASSOCIATION. P.O. Box 10045, 2670 Hanover Street, Palo Alto, CA 94303. (415) 857-9300.

AMERICAN INSTITUTE OF PHYSICS. 335 East 45th Street, New York, NY 10017. (212) 661-9494.

EDISON ELECTRIC INSTITUTE. 1111 19th Street, N.W., Washington, DC 20036. (202) 828-7400.

ELECTRONICS INDUSTRIES ASSOCIATION. 2001 Eye Street, N.W., Washington, DC 20006. (202) 457-4900.

INSTITUTE OF ELECTRICAL AND ELECTRONICS ENGINEERS. 345 East 47th Street, New York, NY 10017. (212) 705-7900.

INTERNATIONAL SOCIETY FOR HYBRID MICROELECTRONICS. P.O. Box 2698, 1861 Wiehle Avenue, Suite 340, Reston, VA 22090. (703) 471-0066.

DIRECTORIES AND BIOGRAPHICAL SOURCES

IEEE MEMBERSHIP DIRECTORY. Institute of Electrical and Electronics Engineers. IEEE Service Center, 445 Hoes Lane, Piscataway, NJ 08854. Annual. $7.00.

1987 DIRECTORY OF ENGINEERING SOCIETIES AND RELATED ORGANIZATIONS. Gordon Davis, editor. Hemisphere Publishing Corporation, 79 Madison Avenue, New York, NY 10016-7892. (800) 821-8312. 12th edition. 1987. $100.00.

RESEARCH CENTERS DIRECTORY 1988. Gale Research Company, Book Tower, Detroit, MI 48226. (800) 521-0707. 12th edition. 1987. $365.00 for set.

SCIENTIFIC AND TECHNICAL ORGANIZATIONS AND AGENCIES DIRECTORY. Margaret Labash Young, editor. Gale Research Company, Book Tower, Detroit, MI 48226. (800) 521-0707. 2nd edition. 1987. $185.00.

WHO'S WHO IN ELECTRONICS. Harris Publishing Company, 2057-2 Aurora Road, Twinsburg, OH 44087. (216) 425-9143. Annual. $90.00.

ENCYCLOPEDIAS AND DICTIONARIES

IEEE STANDARD DICTIONARY OF ELECTRICAL AND ELECTRONICS TERMS. Frank Jay, editor. John Wiley and Sons, Inc., 605 Third Avenue, New York, NY 10158. (800) 526-5368. 3rd edition. 1984. $49.95.

GENERAL WORKS

ELECTRONIC INVENTIONS AND DISCOVERIES: ELECTRONICS FROM ITS EARLIEST BEGINNINGS TO PRESENT DAY. G.W.A. Dummer. Pergamon Press, Inc., Maxwell House, Fairview Park, Elmsford, NY 10523. (914) 592-7700. 1983. $49.50.

ELECTRONICS. H. Kybett. John Wiley and Sons, Inc., 605 Third Avenue, New York, NY 10158. (800) 526-5368. Second edition. 1986. $12.95 in paper.

INTRODUCTION TO ELECTRICITY AND ELECTRONICS FOR THE COMPUTER AGE. R. Rosen. John Wiley and Sons, Inc., 605 Third Avenue, New York, NY 10158. (800) 526-5368. 1987. $35.95.

POWER ELECTRONICS. B.M. Bird and K.G. King. John Wiley and Sons, Inc., 605 Third Avenue, New York, NY 10158. (800) 526-5368. 1983. $29.95 in paper.

HANDBOOKS AND MANUALS

ELECTRONIC ENGINEERS HANDBOOK. Donald G. Fink, editor. McGraw-Hill Book Company, 1221 Avenue of the Americas, New York, NY 10020. (212) 512-2000. 2nd edition. 1982. $89.00.

HANDBOOK OF MODERN ELECTRONICS AND ELECTRICAL ENGINEERING. Charles Belove, editor. John Wiley and Sons, Inc., 605 Third Avenue, New York, NY 10158. (800) 526-5368. 1986. $88.95.

ONLINE DATA BASES

CA SEARCH. Chemical Abstracts Service, P.O. Box 3012, Columbus, OH 43120. (800) 848-6538 or (614) 421-3600. Comprehensive guide to chemical literature, 1972 to present. Inquire as to online cost and availability.

COMPENDEX. Engineering Information, Inc., 345 East 47th Street, New York, NY 10017. (800) 221-1044 or (212) 705-7615. Engineering and technical literature, 1975 to present. Inquire as to online cost and availability.

INSPEC. INSPEC Marketing Department, Institution of Electrical Engineers. Available from IEEE Service Center, 445 Hoes Lane, Piscataway, NJ 08854. (201) 981-0060. Online version of Physics Abstracts. Inquire as to online cost and availability.

NTIS. National Technical Information Service, 5285 Port Royal Road, Springfield, VA 22161. (703) 487-4630. Broad coverage of government sponsored research reports, 1964 to present. Inquire as to online cost and availability.

SCISEARCH. Institute for Scientific Information, 3501 Market Street, Philadelphia, PA 19104. (800) 523-1850 or (215) 386-0100. Broad multidisciplinary title and author index to the international literature of science and technology, 1974 to present. Inquire as to online cost and availability.

WILSONLINE. H.W. Wilson and Company, 950 University Avenue, Bronx, NY 10452. (800) 367-6770 or (212) 588-8400. Makes available online versions of the H.W. Wilson indexes including Applied Science and Technology Index, Business Periodicals Index and Readers' Guide to Periodical Literature. Approximately 1980 to present. Inquire as to online cost and availability.

OTHER SOURCES

A GUIDE TO THE LITERATURE OF ELECTRICAL AND ELECTRONICS ENGINEERING. Susan B. Ardis. Libraries Unlimited Inc., P.O. Box 263, Littleton, CO 80160. (303) 770-1220. 1987. $37.50.

PERIODICALS

ELECTRONIC DESIGN. Hayden Publishing Company, 10 Mulholland Drive, Hasbrouck Heights, NJ 07604. (201) 288-7520. 1952 to present. Biweekly. $40.00 per year.

ELECTRONICS. McGraw-Hill Book Company, 1221 Avenue of the Americas, New York, NY 10020. (212) 512-2000. 1930 to present. Weekly. $32.00 per year.

IEEE CIRCUITS AND DEVICES MAGAZINE. Institute of Electrical and Electronics Engineers. IEEE Service Center, 445 Hoes Lane, Piscataway, NJ 08854. Bimonthly. $70.00 per year.

IEEE JOURNAL OF SOLID STATE CIRCUITS. Institute of Electrical and Electronics Engineers. IEEE Service Center, 445 Hoes Lane, Piscataway, NJ 08854. 1966 to present. Bimonthly. $113.00 per year.

IEEE TRANSACTIONS ON ELECTRON DEVICES. Institute of Electrical and Electronics Engineers. IEEE Service Center, 445 Hoes Lane, Piscataway, NJ 08854. 1959 to present. Monthly. $175.00 per year.

INSTITUTE OF ELECTRICAL AND ELECTRONICS ENGINEERS PROCEEDINGS. Institute of Electrical and Electronics Engineers. IEEE Service Center, 445 Hoes Lane, Piscataway, NJ 08854. 1913 to present. Monthly. $140.00 per year.

SEMICONDUCTOR INTERNATIONAL. Cahners Publishing Company, Inc., Cahners Plaza, 1350 East Touhy Avenue, Des Plaines, IL 60018. (312) 635-8800. 1978 to present. Monthly. $55.00 per year.

SOLID STATE ELECTRONICS. Pergamon Press, Inc., Maxwell House, Fairview Park, Elmsford, NY 10523. (914) 592-7700. 1960 to present. Monthly. $330.00 per year.

RESEARCH CENTERS AND INSTITUTES

ELECTRICAL ENGINEERING RESEARCH LABORATORIES. Purdue University, Electrical Engineering Building, West Lafayette, IN 47907. (317) 494-3536.

ELECTRONICS RESEARCH CENTER. University of Texas at Austin, 132 Engineering Science Building, Austin, TX 78712.

(512) 471-3954.

ELECTRONICS RESEARCH LABORATORY. University of California, Berkeley, 253 Cory Hall, Berkeley, CA 94720. (415) 642-2301.

LABORATORY FOR ELECTROMAGNETIC AND ELECTRONIC SYSTEMS. Massachusetts Institute of Technology, 77 Massachusetts Avenue, Cambridge, MA 02139. (617) 253-4631.

RHEOLOGY

See: FLUID MECHANICS

RINGS (PLANETARY)

See: PLANETARY SCIENCES

ROAD BUILDING

See: HIGHWAY ENGINEERING

ROBOTICS

See also: ARTIFICIAL INTELLIGENCE, CAM, COMPUTERS, CONTROL SYSTEMS, DESIGN ENGINEERING, INDUSTRIAL ENGINEERING, MATERIALS HANDLING, MECHANICAL ENGINEERING

ABSTRACT SERVICES AND INDEXES

APPLIED MECHANICS REVIEW. American Society of Mechanical Engineers, 345 East 47th Street, New York, NY 10017. (212) 705-7703. 1948 to present. Monthly. $360.00 per year.

APPLIED SCIENCE AND TECHNOLOGY INDEX. H.W. Wilson and Company, 950 University Avenue, Bronx, NY 10452. (800) 367-6670 or (212) 588-8400. Monthly. Inquire as to cost and availability.

CHEMICAL ABSTRACTS. American Chemical Society, Chemical Abstracts Service, Box 3012, Columbus, OH 43210. (614) 421-3600. 1907 to present. Weekly. $9500.00 per year.

CONFERENCE PAPERS INDEX. Cambridge Scientific Abstracts, 5161 River Road, Bethesda, MD 20816. 1972 to present. Monthly. Inquire as to cost and availability.

CURRENT CONTENTS: ENGINEERING, TECHNOLOGY AND APPLIED SCIENCES. Institute for Scientific Information, 3501 Market Street, Philadelphia, PA 19104. (800) 523-1850 or (215) 386-0100. Weekly. $275.00 per year.

ENGINEERING INDEX MONTHLY AND AUTHOR INDEX. Engineering Information Inc., 345 East 47th Street, New York, NY 10017. (212) 705-7600. Monthly. $1560.00 per year.

ISMEC BULLETIN (Information Service in Mechanical Engineering). Cambridge Scientific Abstracts, 5161 River Road, Bethesda, MD 20816. (301) 951-1400. 1973 to present. Monthly. $450.00 per year.

INDEX TO SCIENTIFIC AND TECHNICAL PROCEEDINGS. Institute for Scientific Information, 3501 Market Street, Philadelphia, PA 19104. (800) 523-1850 or (215) 386-0100. 1978 to present. Monthly. $775.00 per year.

SCIENCE CITATION INDEX. Institute for Scientific Information, 3501 Market Street, Philadelphia, PA 19104. (800) 523-1850 or (215) 386-0100. Six times per year. $6200.00 per year.

ASSOCIATIONS AND PROFESSIONAL SOCIETIES

AMERICAN SOCIETY OF MECHANICAL ENGINEERS. 345 East 47th Street, New York, NY 10017. (212) 705-7703.

ASSOCIATION FOR UNMANNED VEHICLE SYSTEMS. 1133 15th Street, N.W. Washington, DC 20005. (202) 429-9440.

INDUSTRIAL DESIGNERS SOCIETY OF AMERICA. 1360 Beverly Road, McLean, VA 22101. (703) 556-0919.

INSTITUTE OF INDUSTRIAL ENGINEERS. 25 Technology Park/Atlanta, Norcross, GA 30092. (404) 449-0460.

INSTRUMENT SOCIETY OF AMERICA. 67 Alexander Drive, Research Triangle Park, NC 27709. (919) 549-8411.

ROBOTIC INDUSTIRES ASSOCIATION, P.O. Box 3724, 900 Victors Way, Ann Arbor, MI 48106. (313) 994-6088.

ROBOTICS INTERNATIONAL OF SME (Society of Manufacturing Engineers). P.O. Box 930, One SME Drive, Dearborn, MI 48121. (313) 271-1500.

SOCIETY OF AUTOMOTIVE ENGINEERS (SAE). 400 Commonwealth Drive, Warrendale, PA 15096. (412) 776-4841.

SOCIETY OF LOGISTICS ENGINEERS. 303 Williams Avenue, Suite 922, Huntsville, AL 35801. (205) 539-3800.

SOCIETY OF MANUFACTURING ENGINEERS. One SME Drive, Box 930, Dearborn, MI 48121. (313) 271-1500.

DIRECTORIES AND BIOGRAPHICAL SOURCES

AMERICAN MEN AND WOMEN OF SCIENCE. R.R. Bowker, Inc., Order Department, 245 West 17th Street, New York, NY 10011. (800) 521-8110. Eight volumes. 1986. $595.00 for set.

INTERNATIONAL RESEARCH CENTERS DIRECTORY 1988-89. Darren L. Smith, editor. Gale Research Company, Book Tower, Detroit, MI 48226. (800) 521-0707. 4th edition. 1987. $360.00.

INDUSTRIAL ROBOTICS HANDBOOK. Industrial Press, Inc., Box C-772, Brooklyn, NY 11205. (212) 505-2600. Inquire.

1987 DIRECTORY OF ENGINEERING SOCIETIES AND RELATED ORGANIZATIONS. Gordon Davis, editor. Hemisphere Publishing Corporation, 1010 Vermont Avenue, NW, Washington, DC 20005. (800) 526-0275. 12th edition. 1987. $100.00.

PLANT ENGINEERING DIRECTORY. Technical Publishing Company, 1301 South Grove Avenue, Chicago, IL 60010. (312) 381-1840.
Annual. $35.00.

RESEARCH CENTERS DIRECTORY 1988. Gale Research Company, Book Tower, Detroit, MI 48226. (800) 521-0707. 12th edition. 1987. $365.00 for set.

ROBOTICS TECHNICAL DIRECTORY. Instrument Society of America, 67 Alexander Drive, Research Triangle Park, NC 27709. (919) 549-8411. Annual. $49.95.

ROBOTICS WORLD DIRECTORY. Communications Channels, Inc., 6255 Barfield Road, Atlanta, GA 30328. (404) 256-9800. Annual. $35.00.

SCIENTIFIC AND TECHNICAL ORGANIZATIONS AND AGENCIES DIRECTORY. Margaret Labash Young, editor. Gale Research Company, Book Tower, Detroit, MI 48226. (800) 521-0707. 2nd edition. 1987. $185.00.

SOCIETY OF LOGISTICS ENGINEERS MEMBERSHIP DIRECTORY. 303 Williams Avenue, Suite 922, Huntsville, AL

35801. (205) 539-3800. Annual. Inquire.

WHO'S WHO IN ENGINEERING. Gordon Davis, editor. Hemisphere Publishing Corporation, 79 Madison Avenue, New York, NY 10016-7892. (800) 821-8312. Sixth edition. 1985. $200.00.

GENERAL WORKS

ENGINEERING FOUNDATIONS OF ROBOTICS. F.N. Nagy and A. Siegler. Prentice-Hall Publishing, Inc., Englewood Cliffs, NJ 07632. (800) 562-0245. 1987. $28.95.

MANUFACTURING ENGINEERING: PRINCIPLES OF OPTIMIZATION. Daniel T. Koenig. Hemisphere Publishing Corporation, 79 Madison Avenue, New York, NY 10016-7892. (800) 821-8312. 1987. $45.75.

MECHANICAL DESIGN OF ROBOTS. Eugene I. Rivin. McGraw-Hill Book Company, 1221 Avenue of the Americas, New York, NY 10020. (212) 512-2000. 1987. $49.50.

ROBOTICS. Marvin Minsky, editor. Doubleday and Company, Inc., 245 Park Avenue, New York, NY 10017. (800) 645-6156. 1986. $19.95.

ROBOTICS: AN INTRODUCTION. D. McCloy and D. Harris. John Wiley and Sons, Inc., 605 Third Avenue, New York, NY 10158. (800) 526-5368. 1986. $39.95.

ROBOTICS FOR ENGINEERS. Y. Koren. McGraw-Hill Book Company, 1221 Avenue of the Americas, New York, NY 10020. (212) 512-2000. 1985. $46.00.

HANDBOOKS AND MANUALS

COMPUTER-INTEGRATED MANUFACTURING HANDBOOK. E. Telcholz and J.N. Orr. McGraw-Hill Book Company, 1221 Avenue of the Americas, New York, NY 10020. (212) 512-2000. 1987. $59.95.

HANDBOOK OF INDUSTRIAL ENGINEERING. Gavriel Salvendy. John Wiley and Sons, Inc., 605 Third Avenue, New York, NY 10158. (800) 526-5368. 1982. $99.50.

ROBOT DESIGN HANDBOOK. G.B. Andeen and SRI International. McGraw-Hill Book Company, 1221 Avenue of the Americas, New York, NY 10020. (212) 512-2000. 1988. $42.50.

ONLINE DATA BASES

AEROSPACE DATABASE. American Institute of Aeronautics and Astronautics, 370 L'Enfant Promenade, S.W., Washington, DC 20024. (202) 646-7400. 1963 to present. Inquire as to online cost and availability.

CA SEARCH. Chemical Abstracts Service, P.O. Box 3012, Columbus, OH 43120. (800) 848-6538 or (614) 421-3600. Comprehensive guide to chemical literature, 1972 to present. Inquire as to online cost and availability.

COMPENDEX. Engineering Information, Inc., 345 East 47th Street, New York, NY 10017. (800) 221-1044 or (212) 705-7615. Engineering and technical literature, 1975 to present. Inquire as to online cost and availability.

DISSERTATION ABSTRACTS ONLINE. University Microfilms International, 300 North Zeeb Road, Ann Arbor, MI 48106. (800) 521-0600 or (313) 761-4700. Scope includes virtually all doctoral dissertations accepted at accredited American institutions from 1861 to present in over 250 subject areas. Inquire as to online cost and availability.

INSPEC. INSPEC Marketing Department, Institution of Electrical Engineers. Available from IEEE Service Center, 445 Hoes Lane, Piscataway, NJ 08854. (201) 981-0060. Online version of Physics Abstracts. Inquire as to online cost and availability.

NTIS. National Technical Information Service, 5285 Port Royal Road, Springfield, VA 22161. (703) 487-4630. Broad coverage of government sponsored research reports, 1964 to present. Inquire as to online cost and availability.

ROBOMATIX INFORMATION SYSTEM. EIC/Intelligence, Inc., 48 West 38th Street, New York, NY 10018. (212) 944-8500. Provides access to worldwide information on robotics and automation. 1973 to present. Inquire as to online cost and availability.

SCISEARCH. Institute for Scientific Information, 3501 Market Street, Philadelphia, PA 19104. (800) 523-1850 or (215) 386-0100. Broad multidisciplinary title and author index to the international literature of science and technology, 1974 to present. Inquire as to online cost and availability.

WILSONLINE. H.W. Wilson and Company, 950 University Avenue, Bronx, NY 10452. (800) 367-6770 or (212) 588-8400. Makes available online versions of the H.W. Wilson indexes including Applied Science and Technology Index, Business Periodicals Index and Readers' Guide to Periodical Literature. Approximately 1980 to present. Inquire as to online cost and availability.

PERIODICALS

ADVANCED MANUFACTURING PROCESSES. Marcel Dekker Inc., 270 Madison Avenue, New York, NY 10016. (800) 228-1160. 1986 to present. Quarterly. $85.00 per year.

ADVANCED MANUFACTURING TECHNOLOGY. Technical Insights, Inc., Box 1304, Fort Lee, NJ 07024. (201) 568-4744. 1977 to present. Biweekly. $320.00 per year.

AMERICAN INDUSTRY. Publications for Industry, 21 Russell Woods Road, Great Neck, NY 11021. (516) 487-0990. 1946 to present. Monthly. $20.00 per year.

ASSEMBLY ENGINEERING. Hitchcock Publishing Company, 25 West 550 Geneva Road, Wheaton, IL 60188. (312) 665-1000. 1958 to present. Monthly. $65.00 per year.

IEEE TRANSACTIONS ON INDUSTRY APPLICATIONS. Institute of Electrical and Electronics Engineers. IEEE Service Center, 445 Hoes Lane, Piscataway, NJ 08854. 1965 to present. Bimonthly. $105.00 per year.

INDUSTRIAL ROBOT: AN INTERNATIONAL QUARTERLY JOURNAL ON INDUSTRIAL ROBOT TECHNOLOGY. International Fluidics Services Limited, 39 High Street, Kempston, Bedford MK42 7BT, England. 1973 to present. Quarterly. $100.00 per year.

IIE TRANSACTIONS: INDUSTRIAL ENGINEERING RESEARCH AND DEVELOPMENT. Insuture of Industrial Engineers, 25 Technology Park/Atlanta, Norcross, GA 30092. (404) 449-0460. 1969 to present. Quarterly. $55.00 per year.

INTERNATIONAL JOURNAL OF ROBOTICS RESEARCH. MIT Press, 28 Carleton Street, Cambridge, MA 02142. (617) 253-2884. 1982 to present. Quarterly. $50.00 per year.

JOURNAL OF ENGINEERING FOR INDUSTRY. American Society of Mechanical Engineers, 345 East 47th Street, New York, NY 10017. (212) 705-7703. 1928 to present. Quarterly. $80.00 per year.

JOURNAL OF ROBOTIC SYSTEMS. John Wiley and Sons, Inc., 605 Third Avenue, New York, NY 10158. (800) 526-5368. 1984 to present. Quarterly. $95.00 per year.

MANUFACTURING ENGINEERING. Society of Manufactruing Engineers, One SME Drive, Box 930, Dearborn, MI 48121. (313) 271-1500. 1932 to present. Monthly. $60.00 per year.

ROBOTICA: THE INTERNATIONAL JOURNAL OF INFORMATION, EDUCATION AND RESEARCH IN ROBOTICS AND ARTIFICIAL INTELLIGENCE. Cambridge University Press,

32 East 57th Street, New York, NY 10022. (800) 872-7423. 1983 to present. Quarterly. $50.00 per year.

ROBOTICS. Elsevier Science Publishing Company, Inc., 52 Vanderbilt Avenue, New York, NY 10017. (212) 370-5520. 1985 to present. Quarterly. $101.00 per year.

ROBOTICS AND COMPUTER-INTEGRATED MANUFACTURING. Pergamon Press, Inc., Maxwell House, Fairview Park, Elmsford, NY 10523. (914) 592-7700. 1984 to present. Quarterly. $125.00 per year.

ROBOTICS ENGINEERING; THE JOURNAL OF INTELLIGENT MACHINES. North American Technology, Inc., 174 Concord Street, Peterborough, NH 03458. (603) 924-7136. 1979 to present. Monthly. $25.00 per year.

RESEARCH CENTERS AND INSTITUTES

GENERAL ROBOTICS AND ACTIVE SENSORY PROCESSING LABORATORY. University of Pennsylvania, Room 100, Moore School of Electrial Engineering/D2, Philadelphia, PA 19104. (215) 898-6222.

INTELLIGENT COMPUTER SYSTEMS RESEARCH INSTITUTE. University of Miami, P.O. Box 248235, Coral Gables, FL 33124. (305) 284-5195.

ROBOTIC RESEARCH LABORATORY. New York Institute of Technology, Old Westbury, NY 11568. (516) 686-7669.

ROBOTICS AND AUTOMATION LABORATORY. Rensselaer Polytechnic Institute, School of Engineering, Troy, NY 12180. (518) 266-6076.

ROBOTICS INSTITUTE. Carnegie-Mellon University, Schenley Park, 4630, Pittsburgh, PA 15213. (412) 268-3818.

ROCK MECHANICS

See also: CIVIL ENGINEERING, EARTHQUAKE ENGINEERING, GEOTECHNICAL ENGINEERING, MINING ENGINEERING, QUARRYING, SOIL MECHANICS

ABSTRACT SERVICES AND INDEXES

APPLIED SCIENCE AND TECHNOLOGY INDEX. H.W. Wilson and Company, 950 University Avenue, Bronx, NY 10452. (800) 367-6670 or (212) 588-8400. Monthly. Inquire as to cost and availability.

BIBLIOGRAPHY AND INDEX OF GEOLOGY. American Geological Institute, 4220 King Street, Alexandria, VA 22302. (703) 379-2480. 1969 to present. Monthly. $1100.00 per year.

CURRENT CONTENTS: ENGINEERING, TECHNOLOGY AND APPLIED SCIENCES. Institute for Scientific Information, 3501 Market Street, Philadelphia, PA 19104. (800) 523-1850 or (215) 386-0100. Weekly. $275.00 per year.

ENGINEERING INDEX MONTHLY AND AUTHOR INDEX. Engineering Information Inc., 345 East 47th Street, New York, NY 10017. (212) 705-7600. Monthly. $1560.00 per year.

INTERNATIONAL JOURNAL OF ROCK MECHANICS AND MINING SCIENCES AND GEOMECHANICS ABSTRACTS. Pergamon Press, Inc., Maxwell House, Fairview Park, Elmsford, NY 10523. (914) 592-7700. 1964 to present. Bimonthly. $385.00 per year.

SCIENCE CITATION INDEX. Institute for Scientific Information, 3501 Market Street, Philadelphia, PA 19104. (800) 523-1850 or (215) 386-0100. Six times per year. $6200. 00 per year.

ASSOCIATIONS AND PROFESSIONAL SOCIETIES

AMERICAN GEOLOGICAL INSTITUTE. 4220 King Street, Alexandria, VA 22302. (703) 379-2480.

AMERICAN SOCIETY OF CIVIL ENGINEERS. 345 East 47th Street, New York, NY 10017. (212) 705-7420.

ASSOCIATION OF ENGINEERING GEOLOGISTS. P.O. Box 1068, Brentwood, TN 37027. (615) 377-3578.

GEOLOGICAL SOCIETY OF AMERICA. 3300 Penrose Place, Boulder, CO 80301. (303) 447-2020.

DIRECTORIES AND BIOGRAPHICAL SOURCES

AMERICAN MEN AND WOMEN OF SCIENCE. R.R. Bowker, Inc., Order Department, 245 West 17th Street, New York, NY 10011. (800) 521-8110. Eight volumes. 1986. $595.00 for set.

INTERNATIONAL RESEARCH CENTERS DIRECTORY 1988-89. Darren L. Smith, editor. Gale Research Company, Book Tower, Detroit, MI 48226. (800) 521-0707. 4th edition. 1987. $360.00.

1987 DIRECTORY OF ENGINEERING SOCIETIES AND RELATED ORGANIZATIONS. Gordon Davis, editor. Hemisphere Publishing Corporation, 1010 Vermont Avenue, NW, Washington, DC 20005. (800) 526-0275. 12th edition. 1987. $100.00.

RESEARCH CENTERS DIRECTORY 1988. Gale Research Company, Book Tower, Detroit, MI 48226. (800) 521-0707. 12th edition. 1987. $365.00 for set.

SCIENTIFIC AND TECHNICAL ORGANIZATIONS AND AGENCIES DIRECTORY. Margaret Labash Young, editor. Gale Research Company, Book Tower, Detroit, MI 48226. (800) 521-0707. 2nd edition. 1987. $185.00.

WHO'S WHO IN ENGINEERING. Gordon Davis, editor. Hemisphere Publishing Corporation, 79 Madison Avenue, New York, NY 10016-7892. (800) 821-8312. Sixth edition. 1985. $200.00.

GENERAL WORKS

INTRODUCTION TO ROCK MECHANICS. R.E. Goodman. John Wiley and Sons, Inc., 605 Third Avenue, New York, NY 10158. (800) 526-5368. 1980. $49.95.

OUTLINE OF SOIL AND ROCK MECHANICS. Pierre Habib. Cambridge University Press, 32 East 57th Street, New York, NY 10022. (800) 872-7423. 1983. $11.95 in paper.

ROCK MECHANICS. L. Mueller, editor. Springer-Verlag New York, Inc., 175 Fifth Avenue, New York, NY 10010. (800) 526-7254. 1982. $34.90.

ROCK MECHANICS IN ENGINEERING PRACTICE. K.G. Stagg and O.C. Zienkiewicz. John Wiley and Sons, Inc., 605 Third Avenue, New York, NY 10158. (800) 526-5368. 1968. $75.00.

HANDBOOKS AND MANUALS

GEOTECHNICAL ENGINEERING PRACTICES. R.E. Hunt. McGraw-Hill Book Company, 1221 Avenue of the Americas, New York, NY 10020. (212) 512-2000. 1986. $65.00.

HANDBOOK OF GEOLOGY IN CIVIL ENGINEERING. R.F. Legget and P. Karrow. McGraw-Hill Book Company, 1221 Avenue of the Americas, New York, NY 10020. (212) 512-2000. 1983. $82.50.

ONLINE DATA BASES

CA SEARCH. Chemical Abstracts Service, P.O. Box 3012, Columbus, OH 43120. (800) 848-6538 or (614) 421-3600. Comprehensive guide to chemical literature, 1972 to present.

Inquire as to online cost and availability.

COMPENDEX. Engineering Information, Inc., 345 East 47th Street, New York, NY 10017. (800) 221-1044 or (212) 705-7615. Engineering and technical literature, 1975 to present. Inquire as to online cost and availability.

DISSERTATION ABSTRACTS ONLINE. University Microfilms International, 300 North Zeeb Road, Ann Arbor, MI 48106. (800) 521-0600 or (313) 761-4700. Scope includes virtually all doctoral dissertations accepted at accredited American institutions from 1861 to present in over 250 subject areas. Inquire as to online cost and availability.

GEOREF. Online version of the BIBLIOGRAPHY AND INDEX OF GEOLOGY. American Geological Institute, 4220 King Street, Alexandria, VA 22302. (703) 379-2480. 1969 to present. Inquire as to online cost and availability.

INSPEC. INSPEC Marketing Department, Institution of Electrical Engineers. Available from IEEE Service Center, 445 Hoes Lane, Piscataway, NJ 08854. (201) 981-0060. Online version of Physics Abstracts. Inquire as to online cost and availability.

NTIS. National Technical Information Service, 5285 Port Royal Road, Springfield, VA 22161. (703) 487-4630. Broad coverage of government sponsored research reports, 1964 to present. Inquire as to online cost and availability.

SCISEARCH. Institute for Scientific Information, 3501 Market Street, Philadelphia, PA 19104. (800) 523-1850 or (215) 386-0100. Broad multidisciplinary title and author index to the international literature of science and technology, 1974 to present. Inquire as to online cost and availability.

WILSONLINE. H.W. Wilson and Company, 950 University Avenue, Bronx, NY 10452. (800) 367-6770 or (212) 588-8400. Makes available online versions of the H.W. Wilson indexes including Applied Science and Technology Index, Business Periodicals Index and Readers' Guide to Periodical Literature. Approximately 1980 to present. Inquire as to online cost and availability.

OTHER SOURCES

WHAT EVERY ENGINEER SHOULD KNOW ABOUT ENGINEERING SOURCES. Marcel Dekker Inc., 270 Madison Avenue, New York, NY 10016. (800) 228-1160. 1984. $24.95.

PERIODICALS

ENGINEERING GEOLOGY. Elsevier Science Publishing Company, Inc., 52 Vanderbilt Avenue, New York, NY 10017. (212) 370-5520. 1965 to present. Eight times per year. $250.00 per year.

HIGHWAY AND HEAVY CONSTRUCTION. Technical Publishing Company, 875 Third Avenue, New York, NY 10022. (212) 605-9400. 1892 to present. Monthly. $35.00 per year.

JOURNAL OF GEOTECHNICAL ENGINEERING. American Society of Civil Engineers, 345 East 47th Street, New York, NY 10017. (212) 705-7420. 1956 to present. Monthly. $100.00 per year.

JOURNAL OF IRRIGATION AND DRAINAGE. American Society of Civil Engineers, 345 East 47th Street, New York, NY 10017. (212) 705-7420. 1956 to present. Monthly. $115.00 per year.

ROCK MECHANICS AND ROCK ENGINEERING. Springer-Verlag New York, Inc., 175 Fifth Avenue, New York, NY 10010. (800) 526-7254. 1963 to present. Quarterly. $65.00 per year.

RESEARCH CENTERS AND INSTITUTES

CENTER FOR CONCRETE AND GEOMATERIALS. Northwestern University, Technological Institute, Room 2474, 2145 Sheridan Road, Evanston, IL 60201. (312) 491-3858.

GEOTECHNICAL ENGINEERING CENTER. University of Texas at Austin, Department of Civil Engineering, ECJ 6.200, Austin, TX. (512) 471-1555.

ROCK MECHANICS AND GROUND CONTROL LABORATORY. West Virginia University, Department of Mining Engineering, P.O. Box 6070, Morgantown, WV 26506. (304) 293-5695.

UNDERGROUND SPACE CENTER. University of Minnesota, 790 Civil and Mineral Engineering Building, 500 Pillsbury Drive, S.E., Minneapolis, MN 55455. (612) 624-0066.

ROCKET ASTRONOMY

See: ASTRONOMY

ROCKET ENGINES

See: ROCKETS

ROCKETS

See also: AEROSPACE ENGINEERING, GUIDED MISSILES, ORDNANCE

ABSTRACT SERVICES AND INDEXES

APPLIED SCIENCE AND TECHNOLOGY INDEX. H.W. Wilson and Company, 950 University Avenue, Bronx, NY 10452. (800) 367-6670 or (212) 588-8400. Monthly. Inquire as to cost and availability.

ENGINEERING INDEX MONTHLY AND AUTHOR INDEX. Engineering Information Inc., 345 East 47th Street, New York, NY 10017. (212) 705-7600. Monthly. $1560.00 per year.

INTERNATIONAL AEROSPACE ABSTRACTS. American Institute of Aeronautics and Astronautics, Technical Information Service, 370 L'Enfant Promenade, S.W., Washington, DC 20024. (202) 646-7400. 1961 to present. Semi-monthly. $700.00 per year.

SCIENCE CITATION INDEX. Institute for Scientific Information, 3501 Market Street, Philadelphia, PA 19104. (800) 523-1850 or (215) 386-0100. Six times per year. $6200.00 per year.

STAR (SCIENTIFIC AND TECHNICAL AEROSPACE REPORTS). National Aeronautics and Space Administration, Scientific and Technical Information Facility, Box 8757, Baltimore-Washington International Airport, MD 21240. (202) 453-8545. 1963 to present. Semi-monthly. $85.00 per year with indexes.

ASSOCIATIONS AND PROFESSIONAL SOCIETIES

AMERICAN ASTRONAUTICAL SOCIETY. 6212-B Old Keene Mill Court, Springfield, VA 22152. (703) 866-0020.

AMERICAN INSTITUTE OF AERONAUTICS AND ASTRONAUTICS. 370 L'Enfant Promenade, S.W., Washington, DC 20024. (202) 646-7400.

AMERICAN SPACE FOUNDATION. 111 Massachusetts Avenue, N.W., Suite 200, Washington, DC 20002. (202) 289-2293.

NATIONAL SPACE INSTITUTE. West Wing, Suite 203, 600 Maryland Avenue, S.W., Washington, DC 20024. (202) 484-1111.

SOCIETY OF AUTOMOTIVE ENGINEERS. 400 Commonwealth Drive, Warrendale, PA 15096. (412) 776-4841.

SPACE STUDIES INSTITUTE. P.O. Box 82, 285 Rosedale, Road, Princeton, NJ 08540. (609) 921-0377.

DIRECTORIES AND BIOGRAPHICAL SOURCES

AMERICAN ASTRONAUTICAL SOCIETY DIRECTORY. American Astronautical Society, 6212-B Old Keene Mill Court, Springfield, VA 22152. (703) 866-0020. 1986. $30.00.

AMERICAN INSTITUTE OF AERONAUTICS AND ASTRONAUTICS ROSTER. American Institute of Aeronatics and Astronautics, 370 L'Enfant Promenade, S.W., Washington, DC 20024. (202) 646-7400. 1984. $75.00.

INTERNATIONAL RESEARCH CENTERS DIRECTORY 1988-89. Darren L. Smith, editor. Gale Research Company, Book Tower, Detroit, MI 48226. (800) 521-0707. 4th edition. 1987. $360.00.

RESEARCH CENTERS DIRECTORY 1988. Gale Research Company, Book Tower, Detroit, MI 48226. (800) 521-0707. 12th edition. 1987. $365.00 for set.

SCIENTIFIC AND TECHNICAL ORGANIZATIONS AND AGENCIES DIRECTORY. Margaret Labash Young, editor. Gale Research Company, Book Tower, Detroit, MI 48226. (800) 521-0707. 2nd edition. 1987. $185.00.

WHO'S WHO IN ENGINEERING. Gordon Davis, editor. Hemisphere Publishing Corporation, 1010 Vermont Avenue, NW, Washington, DC 20005. (800) 526-0275. 6th edition. 1985. $200.00.

GENERAL WORKS

CHEMICAL ROCKET PROPULSION AND COMBUSTION RESEARCH. S.S. Penner. Gordon and Breach Science Publishers, Inc., 50 West 23rd Street, New York, NY 10010. (212) 206-8900. 1962. $45.25.

ROCKET PROPULSION ELEMENTS: AN INTRODUCTION TO THE ENGINEERING OF ROCKETS. George A. Sutton and Donald M. Ross. John Wiley and Sons, Inc., 605 Third Avenue, New York, NY 10158. (800) 526-5368. Fourth edition. 1976. $52.95.

ROCKET PROPULSION TECHNOLOGY. D.S. Carton and others, editors. Plenum Publishing Corporation, 233 Spring Street, New York, NY 10013. (800) 221-9369. 1961. $32.50.

THE ROCKET TEAM. Frederick I. Ordway and Mitchell R. Sharpe. MIT Press, 28 Carleton Street, Cambridge, MA 02142. (617) 253-2884. 1982. $9.95 in paper.

ONLINE DATA BASES

CA SEARCH. Chemical Abstracts Service, P.O. Box 3012, Columbus, OH 43120. (800) 848-6538 or (614) 421-3600. Comprehensive guide to chemical and related literature, 1972 to present. Inquire as to online cost and availability.

COMPENDEX. Engineering Information, Inc., 345 East 47th Street, New York, NY 10017. (800) 221-1044 or (212) 705-7615. Engineering and technical literature, 1975 to present. Inquire as to online cost and availability.

INSPEC. INSPEC Marketing Department, Institution of Electrical Engineers. Available from IEEE Service Center, 445 Hoes Lane, Piscataway, NJ 08854. (201) 981-0060. Online version of Physics Abstracts. Inquire as to online cost and availability.

NTIS. National Technical Information Service, 5285 Port Royal Road, Springfield, VA 22161. (703) 487-4630. Broad coverage of government sponsored research reports, 1964 to present. Inquire as to online cost and availability.

SCISEARCH. Institute for Scientific Information, 3501 Market Street, Philadelphia, PA 19104. (800) 523-1850 or (215) 386-0100. Broad multidisciplinary title and author index to the international literature of science and technology, 1974 to present. Inquire as to online cost and availability.

WILSONLINE. H.W. Wilson and Company, 950 University Avenue, Bronx, NY 10452. (800) 367-6770 or (212) 588-8400. Makes available online versions of the H.W. Wilson indexes including Applied Science and Technology Index, Business Periodicals Index and Readers' Guide to Periodical Literature. Approximately 1980 to present. Inquire as to online cost and availability.

PERIODICALS

AEROSPACE AMERICA. American Institute of Aeronautics and Astronautics, Technical Information Service, 370 L'Enfant Promenade, S.W., Washington, DC 20024. (202) 646-7400. 1932 to present. Monthly. $56.00 per year.

AEROSPACE ENGINEERING MAGAZINE. Society of Automotive Engineers, 400 Commonwealth Drive, Warrendale, PA 15096. (412) 776-4841. 1981 to present. Monthly. $24.00 per year.

AIAA JOURNAL. American Institute of Aeronautics and Astronautics, Technical Information Service, 370 L'Enfant Promenade, S.W., Washington, DC 20024. (202) 646-7400. 1963 to present. Monthly. $205.00 per year.

AVIATION WEEK AND SPACE TECHNOLOGY. McGraw-Hill Book Company, 1221 Avenue of the Americas, New York, NY 10020. (212) 512-2000. 1916 to present. Weekly. $56.00 per year.

JOURNAL OF SPACECRAFT AND ROCKETS. American Institute of Aeronautics and Astronautics, Technical Information Service, 370 L'Enfant Promenade, S.W., Washington, DC 20024. (202) 646-7400. 1964 to present. Bimonthly. $95.00 per year.

PLANETARY REPORT. Planetary Society, 65 North Catalina Avenue, Pasadena, CA 91106. (818) 793-5100. 1980 to present. Bimonthly. $20.00 per year.

SPACEFLIGHT. British Interplanetary Society, 27-29 South Lambeth Road, London, SW8 1SZ, England. 1956 to present. $50.00 per year.

RESEARCH CENTERS AND INSTITUTES

CENTER FOR SPACE RESEARCH AND APPLICATIONS. University of Texas at Austin, WRW 402, Austin, TX 78712. (512) 471-1356.

GODDARD SPACE FLIGHT CENTER. National Aeronautics and Space Administration, Greenbelt Road, Greenbelt, MD 20771. (301) 344-5121.

MARSHALL SPACE FLIGHT CENTER. National Aeronautics and Space Administration, Huntsville, AL 35812. (205) 453-2121.

SPACE SCIENCE AND ENGINEERING CENTER. University of Wisconsin, Madison, 1225 West Dayton Street, Madison, WI 53706. (608) 262-0544.

ROTARY ENGINES

See: AUTOMOTIVE ENGINEERING

S

SAFETY ENGINEERING

See also: ERGONOMICS, FIRE PROTECTION, INDUSTRIAL ENGINEERING, NOISE CONTROL

ABSTRACT SERVICES AND INDEXES

APPLIED MECHANICS REVIEW. American Society of Mechanical Engineers, 345 East 47th Street, New York, NY 10017. (212) 705-7703. 1948 to present. Monthly. $360.00 per year.

APPLIED SCIENCE AND TECHNOLOGY INDEX. H.W. Wilson and Company, 950 University Avenue, Bronx, NY 10452. (800) 367-6670 or (212) 588-8400. Monthly. Inquire as to cost and availability.

CHEMICAL ABSTRACTS. American Chemical Society, Chemical Abstracts Service, Box 3012, Columbus, OH 43210. (614) 421-3600. 1907 to present. Weekly. $9500.00 per year.

CURRENT CONTENTS: ENGINEERING, TECHNOLOGY AND APPLIED SCIENCES. Institute for Scientific Information, 3501 Market Street, Philadelphia, PA 19104. (800) 523-1850 or (215) 386-0100. Weekly. $275.00 per year.

ENGINEERING INDEX MONTHLY AND AUTHOR INDEX. Engineering Information Inc., 345 East 47th Street, New York, NY 10017. (212) 705-7600. Monthly. $1560.00 per year.

SAFETY SCIENCE ABSTRACTS JOURNAL. Cambridge Scientific Abstracts, 5161 River Road, Bethesda, MD 20816. (301) 951-1400. 1973 to present. Quarterly. $365.00 per year.

SCIENCE CITATION INDEX. Institute for Scientific Information, 3501 Market Street, Philadelphia, PA 19104. (800) 523-1850 or (215) 386-0100. Six times per year. $6200.00 per year.

ASSOCIATIONS AND PROFESSIONAL SOCIETIES

AMERICAN SOCIETY OF SAFETY ENGINEERS. 1800 East Oakton Street, Des Plaines, IL 60018-2187. (312) 692-4121.

NATIONAL SAFETY COUNCIL. 444 North Michigan Avenue, Chicago, IL 60611. (312) 527-4800.

DIRECTORIES AND BIOGRAPHICAL SOURCES

AMERICAN MEN AND WOMEN OF SCIENCE. R.R. Bowker, Inc., Order Department, 245 West 17th Street, New York, NY 10011. (800) 521-8110. Eight volumes. 1986. $595.00 for set.

INTERNATIONAL RESEARCH CENTERS DIRECTORY 1988-89. Darren L. Smith, editor. Gale Research Company, Book Tower, Detroit, MI 48226. (800) 521-0707. 4th edition. 1987. $360.00.

1987 DIRECTORY OF ENGINEERING SOCIETIES AND RELATED ORGANIZATIONS. Gordon Davis, editor. Hemisphere Publishing Corporation, 1010 Vermont Avenue, NW, Washington, DC 20005. (800) 526-0275. 12th edition. 1987. $100.00.

RESEARCH CENTERS DIRECTORY 1988. Gale Research Company, Book Tower, Detroit, MI 48226. (800) 521-0707. 12th edition. 1987. $365.00 for set.

SCIENTIFIC AND TECHNICAL ORGANIZATIONS AND AGENCIES DIRECTORY. Margaret Labash Young, editor. Gale Research Company, Book Tower, Detroit, MI 48226. (800) 521-0707. 2nd edition. 1987. $185.00.

WHO'S WHO IN ENGINEERING. Gordon Davis, editor. Hemisphere Publishing Corporation, 79 Madison Avenue, New York, NY 10016-7892. (800) 821-8312. Sixth edition. 1985. $200.00.

ENCYCLOPEDIAS AND DICTIONARIES

THESAURUS OF SCIENTIFIC, TECHNICAL, AND ENGINEERING TERMS. Hemisphere Publishing Corporation, 1010 Vermont Avenue, NW, Washington, DC 20005. (800) 526-0275. 1988. $125.00.

GENERAL WORKS

ENGINEERING DESIGN FOR THE CONTROL OF WORKPLACE HAZARDS. R.A. Wadden and P.A. Scheff. McGraw-Hill Book Company, 1221 Avenue of the Americas, New York, NY 10020. (212) 512-2000. 1987. $69.50.

SAFETY MANAGEMENT: IMPROVING PERFORMANCE. D.K. Denton. McGraw-Hill Book Company, 1221 Avenue of the Americas, New York, NY 10020. (212) 512-2000. 1982. $46.95.

SYSTEM SAFETY ENGINEERING AND MANAGEMENT. H.E. Roland and B. Moriarty. John Wiley and Sons, Inc., 605 Third Avenue, New York, NY 10158. (800) 526-5368. 1983. $44.95.

HANDBOOKS AND MANUALS

HANDBOOK OF NOISE CONTROL. C.M. Harris. McGraw-Hill Book Company, 1221 Avenue of the Americas, New York, NY 10020. (212) 512-2000. 2nd edition. 1979. $73.50.

ONLINE DATA BASES

CA SEARCH. Chemical Abstracts Service, P.O. Box 3012, Columbus, OH 43120. (800) 848-6538 or (614) 421-3600. Comprehensive guide to chemical literature, 1972 to present. Inquire as to online cost and availability.

COMPENDEX. Engineering Information, Inc., 345 East 47th Street, New York, NY 10017. (800) 221-1044 or (212) 705-7615. Engineering and technical literature, 1975 to present. Inquire as to online cost and availability.

DISSERTATION ABSTRACTS ONLINE. University Microfilms International, 300 North Zeeb Road, Ann Arbor, MI 48106. (800) 521-0600 or (313) 761-4700. Scope includes virtually all doctoral dissertations accepted at accredited American institutions from 1861 to present in over 250 subject areas. Inquire as to online cost and availability.

NTIS. National Technical Information Service, 5285 Port Royal Road, Springfield, VA 22161. (703) 487-4630. Broad coverage of government sponsored research reports, 1964 to present. Inquire as to online cost and availability.

SCISEARCH. Institute for Scientific Information, 3501 Market Street, Philadelphia, PA 19104. (800) 523-1850 or (215) 386-0100. Broad multidisciplinary title and author index to the international literature of science and technology, 1974 to present. Inquire as to online cost and availability.

WILSONLINE. H.W. Wilson and Company, 950 University Avenue, Bronx, NY 10452. (800) 367-6770 or (212) 588-8400. Makes available online versions of the H.W. Wilson indexes including Applied Science and Technology Index, Business Periodicals Index and Readers' Guide to Periodical Literature. Approximately 1980 to present. Inquire as to online cost and availability.

OTHER SOURCES

WHAT EVERY ENGINEER SHOULD KNOW ABOUT ENGINEERING SOURCES. Marcel Dekker Inc., 270 Madison Avenue, New York, NY 10016. (800) 228-1160. 1984. $24.95.

PERIODICALS

ACCIDENT ANALYSIS AND PREVENTION. Pergamon Press, Inc., Maxwell House, Fairview Park, Elmsford, NY 10523. (914) 592-7700. 1969 to present. Bimonthly. $185.00.

HAZARD PREVENTION. System Safety Society, Inc. Gallant Charger Publishing Company, 14252 Culva Drive, Suite A-261, Irvine, CA 92714. (714) 474-0330. 1965 to present. Bimonthly. $30.00 per year.

INDUSTRIAL HEALTH. Merton Allen Associates, 2307 Dean Street, Schenectady, NY 12309. (518) 393-1933. 1984 to present. Monthly. $149.00 per year.

INDUSTRIAL SAFETY. (Industrial Safety Product News). Ames Publishing Company, 201 King of Prussia Road, Radnor, PA 19087. (215) 964-4000. 1967 to present. Monthly. $24.00 per year.

JOURNAL OF SAFETY RESEARCH. Pergamon Press, Inc., Maxwell House, Fairview Park, Elmsford, NY 10523. (914) 592-7700. 1982 to present. Quarterly. $75.00 per year.

NATIONAL SAFETY AND HEALTH NEWS. National Safety Council, 444 North Michigan Avenue, Chicago, IL 60611. (312) 527-4800. 1919 to present. Monthly. $26.00 per year.

PROFESSIONAL SAFETY. American Society of Safety Engineers, 1800 East Oakton Street, Des Plaines, IL 60018-2187. (312) 692-4121. 1956 to present. Monthly. $30.00 per year.

SOCIETY UPDATE. AMERICAN SOCIETY OF SAFETY ENGINEERS. 1800 East Oakton Street, Des Plaines, IL 60018-2187. (312) 692-4121. 1911 to present. Quarterly. Membership only, inquire.

RESEARCH CENTERS AND INSTITUTES

CENTER FOR ERGONOMICS. University of Michigan, 1205 Beal Street, I.O.E. Building, Ann Arbor, MI 48109-2117.

FACTORY MUTUAL RESEARCH CORPORATION. 1151 Boston-Providence Turnpike, Norwood, MA 02062. (617) 762-4300.

INSTITUTE FOR ADVANCED SAFETY STUDIES. 5950 West Touhy Avenue, Niles, IL 60648. (312) 647-1101.

SALT DOMES

See: PETROLEUM GEOLOGY

SANITARY ENGINEERING

See: ENVIRONMENTAL ENGINEERING

SATELLITES (NATURAL)

See also: ARTIFICIAL SATALLITES, ASTROGEOLOGY, ASTRONOMY, ASTROPHYSICS, JUPITER, MARS, METEORS, NEPTUNE, PLANETARY SCIENCES, SATURN, URANUS

ABSTRACT SERVICES AND INDEXES

ASTRONOMY AND ASTROPHYSICS ABSTRACTS. Springer-Verlag New York, Inc., 175 Fifth Avenue, New York, NY 10010. (800) 526-7254. 1969 to present. Approximately $70.00 per year.

ASTRONOMY AND ASTROPHYSICS MONTHLY INDEX. Olivetree Associates, P.O. Box 236, Sierre Madre, CA 91024. $220.00 per year. Complementary copies available on request.

CHEMICAL ABSTRACTS. American Chemical Society, Chemical Abstracts Service, Box 3012, Columbus, OH 43210. (614) 421-3600. 1907 to present. Weekly. $10,000.00 per year.

CONFERENCE PAPERS INDEX. Cambridge Scientific Abstracts, 5161 River Road, Bethesda, MD 20816. 1972 to present. Monthly. Inquire as to cost and availability.

CURRENT CONTENTS: PHYSICAL, CHEMICAL, AND EARTH SCIENCES. Institute for Scientific Information, 3501 Market Street, Philadelphia, PA 19104. (800) 523-1850 or (215) 386-0100. Weekly. $275.00 per year.

INDEX TO SCIENTIFIC AND TECHNICAL PROCEEDINGS. Institute for Scientific Information, 3501 Market Street, Philadelphia, PA 19104. (800) 523-1850 or (215) 386-0100. 1978 to present. Monthly. $775.00 per year.

INDEX TO SCIENTIFIC REVIEWS. Institute for Scientific Information, 3501 Market Street, Philadelphia, PA 19104. (800) 523-1850 or (215) 386-0100. 1974 to present. Semi-annual. $550.00 per year.

METEOROLOGICAL AND GEOASTROPHYSICAL ABSTRACTS. American Meteorological Society, 45 Beacon Street, Boston, MA 02108. (617) 227-2425. 1950 to present. Monthly. $450.00 per year.

PHYSICS ABSTRACTS. Institution of Electrical Engineers. Available from: IEEE Service Center, 445 Hoes Lane, Piscataway, NJ 08854. 1898 to present. Bimonthly. $1700.00 per year.

PHYSICS BRIEFS. Physik Verlag GmbH, Postfach 1260/1280, D-6940 Weinheim, West Germany. (212) 661-9404. 1920 to present. Twenty-six times per year. $1250.00 per year.

SCIENCE CITATION INDEX. Institute for Scientific Information, 3501 Market Street, Philadelphia, PA 19104. (800) 523-1850 or (215) 386-0100. Six times per year. $6200.00 per year.

STAR. (SCIENTIFIC AND TECHNICAL AEROSPACE REPORTS). U.S. National Aeronautics and Space Administration, Scientific and Technical Information Facility, Box 8757, Baltimore-Washington International Airport, MD 21240. (202) 755-2210.

Semimonthly, with semiannual and annual indexes. $85.00 per year.

ANNUAL REVIEWS AND YEARBOOKS

ANNUAL REVIEW OF ASTRONOMY AND ASTROPHYSICS. Annual Reviews, Inc., 4139 El Camino Way, Palo Alto, CA 94306. (415) 493-4400. Annual. Inquire as to cost and availability.

ANNUAL REVIEW OF EARTH AND PLANETARY SCIENCES. Annual Reviews, Inc., 4139 El Camino Way, Palo Alto, CA 94306. (415) 493-4400. Annual. Inquire as to cost and availability.

ASSOCIATIONS AND PROFESSIONAL SOCIETIES

AMERICAN ASTRONOMICAL SOCIETY. 1816 Jefferson Place, N.W., Washington, DC 20036. (202) 659-0134.

AMERICAN GEOPHYSICAL UNION. 2000 Florida Avenue, N.W., Washington, DC 20009. (202) 462-6903.

AMERICAN INSTITUTE OF PHYSICS. 335 East 45th Street, New York, NY 10017. (212) 661-9494.

ASTRONOMICAL SOCIETY OF THE PACIFIC. 1290 24th Avenue, San Francisco, CA 94122. (415) 661-8660.

GEOLOGICAL SOCIETY OF AMERICA. 3300 Penrose Place, Boulder, CO 80301. (303) 447-2020.

PLANETARY SOCIETY. 65 North Catalina Avenue, Pasadena, CA 91106. (818) 793-5100.

BIBLIOGRAPHIES

A BIBLIOGRAPHY OF ASTRONOMY, 1970-1979. R.A. Seal and S.S. Martin. Libraries Unlimited Inc., P.O. Box 263, Littleton, CO 80160. (303) 770-1220. 1982. $37.50.

SCIENCE BOOKS AND FILMS. American Association for the Advancement of Science, 1333 H Street, NW, Washington, DC 20005. (202) 326-6454. Five times per year. $20.00 per year.

SCIENTIFIC AND TECHNICAL BOOKS AND SERIALS IN PRINT 1988; AN INDEX TO LITERATURE IN SCIENCE AND TECHNOLOGY. R.R. Bowker Company, 205 East 42nd Street, New York, NY 10017. (800) 521-8110. $175.00.

DIRECTORIES AND BIOGRAPHICAL SOURCES

AMERICAN ASTRONOMICAL SOCIETY MEMBERSHIP DIRECTORY. 1816 Jefferson Place, N.W., Washington, DC 20036. (202) 659-0134. Annual. Inquire.

AMERICAN MEN AND WOMEN OF SCIENCE. R.R. Bowker, Inc., Order Department, 245 West 17th Street, New York, NY 10011. (800) 521-8110. Eight volumes. 1986. $595.00 for set.

THE BIOGRAPHICAL DICTIONARY OF SCIENTISTS: ASTRONOMERS. D. Abbott, editor. Peter Bedrick Books, 125 East 23rd Street, New York, NY 10010. 1984.

DIRECTORY OF PHYSICS AND ASTRONOMY STAFF MEMBERS. American Institute of Physics, 335 East 45th Street, New York, NY 10017. (212) 661-9494. Annual.

INTERNATIONAL RESEARCH CENTERS DIRECTORY 1988-89. Darren L. Smith, editor. Gale Research Company, Book Tower, Detroit, MI 48226. (800) 521-0707. 4th edition. 1987. $360.00.

RESEARCH CENTERS DIRECTORY 1988. Gale Research Company, Book Tower, Detroit, MI 48226. (800) 521-0707. 12th edition. 1987. $365.00 for set.

SCIENTIFIC AND TECHNICAL ORGANIZATIONS AND AGENCIES DIRECTORY. Margaret Labash Young, editor. Gale Research Company, Book Tower, Detroit, MI 48226. (800) 521-0707. 2nd edition. 1987. $185.00.

GENERAL WORKS

THE BIRTH OF THE EARTH. David E. Fisher. Columbia University Press, 562 West 113th Street, New York, NY 10025. (212) 316-7100. 1987. $24.95.

CHEMISTRY AND PHYSICS OF THE TERRESTRIAL PLANETS. Surendra K. Saxena. Springer-Verlag New York, Inc., 175 Fifth Avenue, New York, NY 10010. (800) 526-7254. 1986. $59.00.

THE GALAXY AND THE SOLAR SYSTEM. Roman Smoluchowski and others. University of Arizona Press, 1615 East Speedway, Tucson, AZ 85719. (602) 621-1441. 1986. $29.95.

GEOLOGY OF THE TERRESTRIAL PLANETS. M. Carr, editor. National Aeronautics and Space Administration Special Publication 469. National Technical Information Service, 5285 Port Royal Road, Springfield, VA 22161. (703) 487-4929. 1984. $16.00.

METEORITES: THEIR RECORD OF EARLY SOLAR-SYSTEM HISTORY. J. Wasson. W.H. Freeman and Company, 41 Madison Avenue, New York, NY 10010. (212) 532-7660. 1985. $34.95.

PLANETS AND THEIR ATMOSPHERES: ORIGIN AND EVOLUTION. J. Lewis and R. Prinn. Academic Press, Inc., 6277 Sea Harbor Drive, Orlando, FL 32821. (800) 321-5068. 1983. $59.00.

RECENT ADVANCES IN PLANETARY METEOROLOGY. G. Hunt, editor. Cambridge University Press, 32 East 57th Street, New York, NY 10022. (800) 872-7423. 1985. $39.50.

THE PLANETARY SYSTEM. Tobias Owen and David Morrison. Addison-Wesley Publishing Company, Inc., 1 Jacob Way, Reading, MA 01867. (617) 944-3700. 1987. $39.95.

RINGS: DISCOVERIES FROM GALILEO TO VOYAGER. MIT Press, 28 Carleton Street, Cambridge, MA 02142. (617) 253-2884. 1985. $17.50.

SATELLITES. J.A. Burns and M.S. Matthews, editors. University of Arizona Press, 1615 East Speedway, Tucson, AZ 85719. (602) 621-1441. 1986. $55.00.

THE STORY OF THE EARTH. P. Cattermole and P. Moore. Cambridge University Press, 32 East 57th Street, New York, NY 10022. (800) 872-7423. 1985. $29.95.

THEORY OF PLANETARY ATMOSPHERES: AN INTRODUCTION TO THEIR PHYSICS AND CHEMISTRY. J.W. Chamberlain and D.M. Hunten. Academic Press, Inc., 6277 Sea Harbor Drive, Orlando, FL 32821. (800) 321-5068. 1987. $49.50.

URANUS AND NEPTUNE. J. Bergstrahl, editor. National Aeronautics and Space Administration Conference Paper 2330. National Technical Information Service, 5285 Port Royal Road, Springfield, VA 22161. (703) 487-4929. 1984. $25.00.

ONLINE DATA BASES

CA SEARCH. Chemical Abstracts Service, P.O. Box 3012, Columbus, OH 43120. (800) 848-6538 or (614) 421-3600. Comprehensive guide to chemical literature, 1972 to present. Inquire as to online cost and availability.

DISSERTATION ABSTRACTS ONLINE. University Microfilms International, 300 North Zeeb Road, Ann Arbor, MI 48106. (800) 521-0600 or (313) 761-4700. Scope includes virtually all doctoral dissertations accepted at accredited American institutions from 1861 to present in over 250 subject areas. Inquire as to online cost and availability.

INSPEC. INSPEC Marketing Department, Institution of Electrical Engineers. Available from IEEE Service Center, 445 Hoes Lane,

SATELLITES (NATURAL)

Piscataway, NJ 08854. (201) 981-0060. Online version of Physics Abstracts. Inquire as to online cost and availability.

NTIS. National Technical Information Service, 5285 Port Royal Road, Springfield, VA 22161. (703) 487-4630. Broad coverage of government sponsored research reports, 1964 to present. Inquire as to online cost and availability.

SCISEARCH. Institute for Scientific Information, 3501 Market Street, Philadelphia, PA 19104. (800) 523-1850 or (215) 386-0100. Broad multidisciplinary title and author index to the international literature of science and technology, 1974 to present. Inquire as to online cost and availability.

OTHER SOURCES

ATLAS OF THE SOLAR SYSTEM. P. Moore and G. Hunt. Rand McNally and Company, P.O. Box 7600, Chicago, IL 60680. (800) 323-4070. 1983. $40.00.

CAMBRIDGE PHOTOGRAPHIC ATLAS OF THE PLANETS. Cambridge University Press, 32 East 57th Street, New York, NY 10022. (800) 872-7423. 1986.

PERIODICALS

ASTRONOMICAL JOURNAL. American Astronomical Society. Available from: American Institute of Physics, 335 East 45th Street, New York, NY 10017. (212) 661-9494. Monthly. $125.00 per year.

ASTRONOMICAL SOCIETY OF THE PACIFIC PUBLICATIONS. Astronomical Society of the Pacific, 1290 24th Avenue, San Francisco, CA 94122. (415) 661-8660. Monthly. $40.00 per year.

ASTRONOMY. AstroMedia Corporation, 625 E Street, Box 92788, Milwaukee, WI 53202. (414) 276-2689. 1973 to present. Monthly. $21.00 per year.

ASTRONOMY AND ASTROPHYSICS. Springer-Verlag New York, Inc., 175 Fifth Avenue, New York, NY 10010. (800) 526-7254. Monthly. $680.00 per year.

EARTH, MOON AND PLANETS. D. Reidel Publishing Company, 190 Old Derby Street, Hingham, MA 02043. 1969 to present. Nine times per year. $275.00 per year.

GEOCHEMICA ET COSMOCHIMICA ACTA. Pergamon Press, Inc., Maxwell House, Fairview Park, Elmsford, NY 10523. (914) 592-7700. 1950 to present. Monthly. $340.00 per year.

LUNAR AND PLANETARY INFORMATION BULLETIN. Lunar and Planetary Institute, 3303 NASA Road One, Houston, TX 77058-4399. (713) 486-2135. 1970 to present. Three times per year. Free.

MERCURY. Astronomical Society of the Pacific, 1290 24th Avenue, San Francisco, CA 94122. (415) 661-8660. 1972 to present. Bimonthly. $21.00 per year.

METEORITICS. Center for Meteorite Studies, Arizona State University, Tempe, AZ 85287. (602) 965-3576. 1955 to present. Quarterly. $40.00 per year.

PLANETARY AND SPACE SCIENCE. Pergamon Press, Inc., Maxwell House, Fairview Park, Elmsford, NY 10523. (914) 592-7700. 1959 to present. Monthly. $430.00 per year.

PLANETARY REPORT. Planetary Society, 65 North Catalina, Pasadena, CA 91106-2301. 1980 to present. Bimonthly. $20.00 per year.

SKY AND TELESCOPE. Sky Publishing Corporation, 49 Bay State Road, Cambridge, MA 02238. (617) 864-7360. Monthly. $18.00 per year.

RESEARCH CENTERS AND INSTITUTES

LABORATORY FOR PLANETARY ATMOSPHERES RESEARCH. State University of New York at Stony Brook, Stony Brook, NY 11794-2300. (516) 632-8321.

LABORATORY FOR PLANETARY GEOLOGY. Arizona State University, Department of Geology, Tempe, AZ 85281. (602) 965-7029.

LABORATORY FOR PLANETARY STUDIES. Cornell University, 302 Space Sciences Building, Ithaca, NY 14853. (607) 256-4971.

LUNAR AND PLANETARY INSTITUTE. 3303 NASA Road One, Houston, TX 77058. (713) 486-2139.

SATURN

See also: PLANETARY SCIENCE, SOLAR SYSTEM

ABSTRACT SERVICES AND INDEXES

ASTRONOMY AND ASTROPHYSICS ABSTRACTS. Springer-Verlag New York, Incorporated, 175 Fifth Avenue, New York, NY 10010. (212) 460-1500. $70.00 per year.

ASTRONOMY AND ASTROPHYSICS MONTHLY INDEX. Olivetree Associates, Post Office Box 236, Sierra Madre, CA 91024. $212.00 per year. Complimentary copies available on request.

METEOROLOGICAL AND GEOASTROPHYSICAL ABSTRACTS. American Meteorological Society, 45 Beacon Street, Boston, MA 02108. (617) 227-2425. 1950 to present. Monthly. $450.00 per year.

STAR. (Scientific and Technical Aerospace Reports. United States National Aeronautics and Space Administration, Scientific and Technical Information Facility, Box 8757, Baltimore-Washington International Airport, MD 21240. (202) 755-2210. Semimonthly, with semiannual and annual indexes. $85.00 per year.

ANNUAL REVIEWS AND YEARBOOKS

THE ASTRONOMICAL ALMANAC. Superintendent of Documents, United States Government Printing Office, Washington, DC 20402. (202) 783-3238. Yearly.

ANNUAL REVIEW OF ASTRONOMY AND ASTROPHYSICS. Annual Reviews, Incorporated, 4139 El Camino Way, Palo Alto, CA 94306. (415) 493-4400. Annual. Inquire.

ANNUAL REVIEW OF EARTH AND PLANETARY SCIENCES. Annual Reviews, Incorporated, 4139 El Camino Way, Palo Alto, CA 94306. (415) 493-4400. Annual. Inquire.

ASSOCIATIONS AND PROFESSIONAL SOCIETIES

AMERICAN ASTRONOMICAL SOCIETY. 2000 Florida Avenue, NW, Suite 300, Washington, DC 20009. (202) 659-0134.

AMERICAN GEOPHYSICAL UNION. 2000 Florida Avenue, NW, Washington, DC 20009. (202) 462-6903.

ASTRONOMICAL LEAGUE. Post Office Box 12821, Tucson, AZ 85732. (602) 790-8471.

ASTRONOMICAL SOCIETY OF THE PACIFIC. 1290 24th Avenue, San Francisco, CA 94122. (415) 661-8660.

PLANETARY SOCIETY. 65 North Catalina Avenue, Pasadena, CA 91106-2301. (818) 793-5100.

BIBLIOGRAPHIES

A BIBLIOGRAPHY OF ASTRONOMY, 1970-1979. R.A. Seal and S.S. Martin. Libraries Unlimited, Incorporated, Littleton, CO 80160. 1982. $37.50.

SCIENTIFIC AND TECHNICAL BOOKS AND SERIALS IN PRINT 1988: AN INDEX TO LITERATURE IN SCIENCE AND TECHNOLOGY. R.R. Bowker Company, 205 East 42nd Street, New York, NY 10017. (800) 521-8110 or (212) 916-1600. $175.00.

DIRECTORIES AND BIOGRAPHICAL SOURCES

AMERICAN ASTRONOMICAL SOCIETY MEMBERS. 2000 Florida Avenue, NW, Suite 300, Washington, DC 20009. (202) 659-0134. Annual. Available to members only.

THE BIOGRAPHICAL DICTIONARY OF SCIENTISTS: ASTRONOMERS. D. Abbott, editor. Peter Bedrick Books, 125 East 23rd Street, New York, NY 10010. 1984. $24.95.

DIRECTORY OF PHYSICS AND ASTRONOMY STAFF MEMBERS. American Institute of Physics, 335 East 45th Street, New York, NY 10017. Annual.

GENERAL WORKS

SATURN. Tom Gehrels and Mildred S. Matthews. University of Arizona Press, 1615 East Speedway, Tucson, AZ 85719. (602) 621-1441. 1984. $37.50.

ONLINE DATA BASES

CA SEARCH. Chemical Abstracts Service, Post Office Box 3012, Columbus, OH 43210. Guide to chemical literature, 1972 to present. Inquire as to cost and availability.

DISSERTATION ABSTRACTS ONLINE. University Microfilms International, 300 North Zeeb Road, Ann Arbor, MI 48106. (800) 521-0600 or (313) 761-4700. Scope includes virtually all doctoral dissertations accepted at accredited American institutions from 1861 to present in 252 subject areas. Inquire as to cost and availability.

NASA. National Aeronautics and Space Administration, Scientific and Technical Information Branch, 300 7th Street, SW, Washington, DC 20546. Citations and abstracts of aerospace literature, 1962 to present. Inquire as to cost and availability.

NTIS. National Technical Information Service, 5285 Port Royal Road, Springfield, VA 22161. (703) 487-4630. Broad coverage of government-sponsored research reports, 1964 to present. Inquire as to cost and availability.

SCISEARCH. Institute for Scientific Information, 3501 Market Street, Philadelphia, PA 19104. (800) 523-1850 or (215) 386-0100. Broad multidisciplinary title and author index to the international literature of science and technology, 1974 to present. Inquire as to cost and availability.

PERIODICALS

ASSOCIATION OF LUNAR AND PLANETARY OBSERVERS. Journal Association of Lunar and Planetary Observers, Box 16131, San Francisco, CA 94116. (415) 566-5786. 1947 to present. Quarterly. $15.00 per year.

ASTRONOMICAL JOURNAL. American Astronomical Society. Available from: American Institute of Physics, 335 East 45th Street, New York, NY 10017. 9212) 661-9404. $125.00 per year.

EARTH, MOON AND PLANETS: AN INTERNATIONAL JOURNAL OF COMPARATIVE PLANETOLOGY. D. Reidel Publishing Company, 190 Old Derby Street, Hingham, MA 02043. Nine times per year. $275.00 per year.

ICARUS: INTERNATIONAL JOURNAL OF THE SOLAR SYSTEM STUDIES. Academic Press, Incorporated, Orlando, FL 32887. (305) 345-4100. Monthly. $484.00 per year.

MERCURY. Astronomical Society of the Pacific, 1290 24th Avenue, San Francisco, CA 94122. (415) 661-8660. Bimonthly. $21.00 per year.

PLANETARY AND SPACE SCIENCE. Pergamon Press, Incorporated, Maxwell House, Fairview Park, Elmsford, NY 10523. (914) 592-7700. Monthly. $430.00 per year.

PLANETARY REPORT. Planetary Society, 65 North Catalina Avenue, Pasadena, CA 91106-2301. (818) 793-5100. 1980 to present. Bimonthly. $20.00 per year.

SOLAR SYSTEM RESEARCH (ENGLISH TRANSLATION OF ASTRONOMICHESKII VESTNIK). Consultants Bureau, 233 Spring Street, New York, NY 10013. (212) 620-8000. 1967 to present. Quarterly. $425.00 per year.

RESEARCH CENTERS AND INSTITUTES

LABORATORY FOR PLANETARY ATMOSPHERES RESEARCH. State University of New York at Stony Brook, Stony Brook, NY 11794-2300. (516) 632-8321.

LABORATORY FOR PLANETARY STUDIES. Cornell University, 302 Space Sciences Building, Ithaca, NY 14853. (607) 256-4971.

LUNAR AND PLANETARY INSTITUTE. 3303 NASA Road One, Houston, TX 77058. (713) 486-2139.

LUNAR AND PLANETARY LABORATORY. University of Arizona, Tucson, AZ 85721. (602) 621-6962.

UNIVERSITY OF ALABAMA IN HUNTSVILLE. Center for Space Plasma and Aeronomic Research. Engineering Building, Huntsville, AL 35899. (205) 895-6268.

SCIENTIFIC INSTRUMENTATION

See: INSTRUMENTATION, METROLOGY

SCINTILLATION

See: RADIATION

SCRAMJETS

See: JET PROPULSION

SCRUBBERS

See: AIR POLLUTION

SEA ICE

See: ICE

SEA WATER

See: OCEANOGRAPHY

SEDIMENTARY ROCKS

See also: CARBONATES,
GEOLOGY, GEOPHYSICS, GEOCHEMISTRY,
IGNEOUS ROCKS, METAMORPHIC ROCKS

ABSTRACT SERVICES AND INDEXES

BIBLIOGRAPHY AND INDEX OF GEOLOGY. American Geological Institute, 4220 King Street, Alexandria, VA 22302. (703) 379-2480. 1969 to present. Monthly. $1100.00 per year.

CHEMICAL ABSTRACTS. Chemical Abstracts Service, 25440 Olentangy Road, Post Office Box 3012, Columbus, OH 43210. (800) 848-6538 or (614) 421-3600. Weekly. $9200.00 per year.

DEEP-SEA RESEARCH WITH OCEANOGRAPHIC LITERATURE REVIEW. Pergamon Press, Incorporated, Maxwell House, Fairview Park, Elmsford, NY 10523. (914) 592-7700. 1953 to present. Twenty-four times per year. $600.00 per year.

GENERAL SCIENCE INDEX. H.W. Wilson Company, 950 University Avenue, Bronx, NY 10452. (800) 367-6770 or (212) 588-8400. Inquire as to price and availability.

MINERALOGICAL ABSTRACTS. Mineralogical society and the Mineralogical society of America, 41 Queen's Gate, London, SW7 5HR, England. 1959 to present. Quarterly $190.00 per year.

OCEANIC ABSTRACTS. Cambridge Scientific Abstracts, 5161 River Road, Bethesda, MD 20816. (301) 951-1400. 1964 to present. Bimonthly. $652.00 per year.

SCIENCE CITATION INDEX. Institute for Scientific Information, 3501 Market Street, Philadelphia, PA 19104. (800) 523-1850 or (215) 386-0100. Inquire as to price and availability.

ANNUAL REVIEWS AND YEARBOOKS

ANNUAL REVIEW AND EARTH AND PLANETARY SCIENCES. Annual Reviews, Incorporated, 4139 El Camino Way, Palo Alto, CA 94306. (415) 493-4400.

MINERALS YEARBOOK. Bureau of Mines, United States Department of the Interior. Available from United States Government Printing Office, Washington, DC 20402. (202) 783-3238. Annual. Three volumes. $45.00.

ASSOCIATIONS AND PROFESSIONAL SOCIETIES

AMERICAN ASSOCIATION OF PETROLEUM GEOLOGISTS. Post Office Box 979, Tulsa, OK 74101. (918) 584-2555.

AMERICAN GEOLOGICAL INSTITUTE. 4220 King Street, Alexandria, VA 22302. (703) 379-2480.

AMERICAN GEOPHYSICAL UNION. 2000 Florida Avenue, NW, Washington, DC 20009. (202) 462-6903.

AMERICAN INSTITUTE OF PROFESSIONAL GEOLOGISTS. 7828 Vance Drive, Suite 103, Arvada, CO 80003. (303) 431-0831.

ASSOCIATION OF ENGINEERING GEOLOGISTS. Post Office Box 1068, Brentwood, TN 37027. (615) 377-3578.

GEOLOGICAL SOCIETY OF AMERICA. Box 9140, Boulder, CO 80301. (303) 447-2020.

SOCIETY OF ECONOMIC PALEONTOLOGISTS AND MINERALOGISTS. Box 4756, Tulsa, OK 74159. (018) 743-9765.

DIRECTORIES AND BIOGRAPHICAL SOURCES

AMERICAN MEN AND WOMEN OF SCIENCE: PHYSICAL AND BIOLOGICAL SCIENCES. Sixteenth edition. R.R. Bowker Company, 245 West 17th Street, New York, NY 10011. (800) 521-8810 ro (212) 916-1600. $595.00.

AMERICAN INSTITUTE OF PROFESSIONAL GEOLOGISTS. Membership Directory. American Institute of Professional Geologists, 7828 Vance Drive, Suite 103, Arvada, CO 80003. (303) 431-0831. Annual, $15.00.

ASSOCIATION OF ENGINEERING GEOLOGISTS DIRECTORY. Association of Engineering Geologists, Dr. G. Lee Christensen, Civil Engineering Department, Villanova University, Villanova, PA 19085. (215) 645-4960. Annual. $15.00.

GEOLOGICAL SOCIETY OF AMERICA. Membership Directory. Geological Society of America, 3300 Penrose Place, Boulder, CO 80301. (303) 447-2020. Annual. Available to members only.

RESEARCH CENTERS DIRECTORY. Gale Research Company, Book Tower, Detroit, Mi 48226. (800) 521-0707. Twelfth edition. 1987. $365.00.

WHO'S WHO IN FRONTIER SCIENCE AND TECHNOLOGY. Marquis Who's Who, Incorporated, 200 East Ohio Street, Chicago, IL 60611. (800) 428-3898 or (312) 787-2008.

ENCYCLOPEDIAS AND DICTIONARIES

DICTIONARY OF GEOLOGICAL TERMS. American Geological Institute. Doubleday and Company, Incorporated, 245 Park Avenue, New York, NY 10017. (800) 645-6156 or (212) 953-4561. Third edition. 1984. $19.95.

DICTIONARY OF PETROLOGY. S.I. Tomkeieff. John Wiley and Sons, Incorporated, 605 Third Avenue, New York, NY 10158. (800) 526-5368 or (212) 850-6000. 1983. $140.00.

GLOSSARY OF GEOLOGY. Robert L. Bates and Julia A. Jackson. American Geological Institute, 4220 King Street, Alexandria, VA 22032. (703) 379-2480. 1980. $60.00.

GENERAL WORKS

ATLAS OF SEDIMENTARY ROCKS UNDER THE MICROSCOPE. A.E. Adams, W.S. MacKenzie, C. Guilford. John Wiley and Sons, Incorporated, 605 Third Avenue, New York, NY 10158. (800) 526-5368 or (212) 850-6000. 1984. $29.95.

A FIELD DESCRIPTION OF SEDIMENTARY ROCKS. Maurice E. Tucker. John Wiley and Sons, Incorporated, 605 Third Avenue, New York, NY 10158. (800) 526-5368 or (212) 850-6000. 1982. $14.95 in paper.

INTRODUCTION TO ROCK FORMING MINERALS. W.A. Deer, J. Zussman and R.A. Howie. John Wiley and Sons, Incorporated, 605 Third Avenue, New York, NY 10158. (800) 526-5368 or (212) 850-6000. 1966. $29.95.

INTRODUCTION TO SEDIMENTOLOGY. R.C. Selley. Academic Press, Incorporated, 6277 Sea Harbor Drive, Orlando, FL 32821. (800) 321-5068. Second edition. 1982. $47.50.

SEDIMENTARY PETROLOGY: AN INTRODUCTION. M.E. Tucker. John Wiley and Sons, Incorporated, 605 Third Avenue, New York, NY 10158. (800) 526-5368 or (212) 850-6000. 1981. $29.95.

HANDBOOKS AND MANUALS

GEOLOGY IN THE FIELD. R.R. Compton. John Wiley and Sons, Incorporated, 605 Third Avenue, New York, NY 10158. (800) 526-5368 or (212) 850-6418. 1985. $23.95.

ONLINE DATA BASES

GEOREF. American Geological Institute, 4220 King Street, Alexandria, VA 22302. (800) 336-4764 or (703) 379-2480.

Geology and geosciences literature, 1961 to present. Inquire as to online cost and availability.

GEOARCHIVE. Geosystems, Post Office Box 1024, Westminister, London, England, SW1 P 2JL. Citations to literature on geoscience, 1969 to present. Inquire as to online cost and availability.

NTIS. National Technical Information Service, 5285 Port Royal Road, Springfield, VA 22161. (703) 487-4630. Broad coverage of government sponsored research reports, 1964 to present. Inquire as to online cost and availability.

SCISEARCH. Institute for Scientific Information, 3501 Market Street, Philadelphia, PA 19104. (800) 523-1850 or (215) 386-0100. Broad interdisciplinary index to the literature of science and technology, 1956 to present. Inquire as to online cost and availability.

PERIODICALS

AAPG BULLETIN. American Association of Petroleum Geologists, Post Office Box 979, Tulsa, OK 74101. (918) 584-2555.

AMERICAN JOURNAL OF SCIENCE. Kline Geology Laboratory, Yale University, New Haven, CT 06520. Ten times per year. $80.00 per year.

ECONOMIC GEOLOGY. Society of Economic Geologists, Post Office Box 571, Golden, CO 80402. (303) 279-1899. Eight times per year. $25.00.

GEOCHIMICA ET COSMOCHIMICA ACTA. Pergamon Journals, Incorporated, Maxwell House, Fairview Park, Elmsford, NY 10523. (914) 592-7700. Monthly. $340.00 per year.

GEOLOGICAL MAGAZINE. Cambridge University Press, 32 East 57th Street, New York, NY 10022. (800) 872-7423. Bimonthly. $165.00 per year.

GEOLOGICAL SOCIETY JOURNAL. Geological Society of London. Blackwell Scientific Publications, Incorporated, 667 Lytton Avenue, Palo Alto, CA 94301. (415) 324-1688. Six times per year. $290.00.

GEOLOGICAL SOCIETY OF AMERICA BULLETIN. Post Office Box 9140, 3300 Penrose Place, Boulder, CO 80301. (303) 447-2020. Monthly. $80.00.

GEOLOGY. Geological Society of America, Post Office Box 9140, 3300 Penrose Place, Boulder, CO 80301. (303) 447-2020. Monthly. $55.00 per year.

JOURNAL OF GEOLOGY. University of Chicago Press, 5801 South Ellis Street, Chicago, IL 60637. (312) 962-7600. Bimonthly. $30.00 per year.

JOURNAL OF SEDIMENTARY PETROLOGY. Society of Economic Paleontologists and Mineralogists, Post Office Box 4756, Tulsa, OK 74159. (918) 743-9765. Bimonthly. $95.00 per year.

LITHOS: AN INTERNATIONAL JOURNAL OF MINERALOGY, PETROLOGY, AND GEOCHEMISTRY. Elsevier Science Publishers, Post Office Box 330, Irving-on-Hudson, NY 10533. Quarterly. $73.00 per year.

MOUNTAIN GEOLOGISTS. Rocky Mountain Association of Geologists, 4201 West 51st Avenue, Denver, CO 80212-2902. Quarterly. $15.00 per year.

PROFESSIONAL GEOLOGISTS. American Institute of Professional Geologists, 7828 Vance Drive, Suite 103, Arvada, CO 80003. Monthly.

SEDIMENTARY GEOLOGY. Elsevier Science Publishers, Post Office Box 330, Irving-on-Hudson, NY 10533. Twenty times per year. $421.00 per year.

SEDIMENTOLOGY. Blackwell Scientific Publications, Incorporated, 52 Beacon Street, Boston, MA 02108. (617) 720-0761. Six times per year. $185.00 per year.

RESEARCH CENTERS AND INSTITUTES

UNITED STATES GEOLOGICAL SURVEY. National Center, 12201 Sunrise Valley Drive, Reston, VA 22092. The major geological research agency of the federal government conducting research in most areas of pure and applied research in the geosciences.

GEOLOGICAL SOCIETY OF AMERICA. Post Office Box 9140, 3300 Penrose Place, Boulder, CO 80301. (303) 447-2020.

SOCIETY OF ECONOMIC GEOLOGISTS. Post Office Box 571, Golden, CO 80402. (303) 279-1899.

SOCIETY OF ECONOMIC PALEONTOLOGISTS AND MINERALOGISTS. Post Office Box 4756, Tulsa, OK 74159. (918) 743-9765.

SEISMOLOGY

See also: EARTHQUAKES, EARTHQUAKE ENGINEERING, GEOLOGY, GEOPHYSICS, MINERAL EXPLORATION, OCEANOGRAPHY, PLANETARY SCIENCES, PLATE TECTONICS, VOLCANOLOGY

ABSTRACT SERVICES AND INDEXES

APPLIED SCIENCE AND TECHNOLOGY INDEX. H.W. Wilson and Company, 950 University Avenue, Bronx, NY 10452. (800) 367-6670 or (212) 588-8400. Monthly. Inquire as to cost and availability.

BIBLIOGRAPHY AND INDEX OF GEOLOGY. American Geological Institute, 4220 King Street, Alexandria, VA 22302. (703) 379-2480. 1969 to present. Monthly. $1100.00 per year.

CHEMICAL ABSTRACTS. American Chemical Society, Chemical Abstracts Service, Box 3012, Columbus, OH 43210. (614) 421-3600. 1907 to present. Weekly. $9500.00 per year.

CONFERENCE PAPERS INDEX. Cambridge Scientific Abstracts, 5161 River Road, Bethesda, MD 20816. 1972 to present. Monthly. Inquire as to cost and availability.

CURRENT CONTENTS: ENGINEERING, TECHNOLOGY AND APPLIED SCIENCES. Institute for Scientific Information, 3501 Market Street, Philadelphia, PA 19104. (800) 523-1850 or (215) 386-0100. Weekly. $275.00 per year.

ENGINEERING INDEX MONTHLY AND AUTHOR INDEX. Engineering Information Inc., 345 East 47th Street, New York, NY 10017. (212) 705-7600. Monthly. $1560.00 per year.

INDEX TO SCIENTIFIC AND TECHNICAL PROCEEDINGS. Institute for Scientific Information, 3501 Market Street, Philadelphia, PA 19104. (800) 523-1850 or (215) 386-0100. 1978 to present. Monthly. $775.00 per year.

INDEX TO SCIENTIFIC REVIEWS. Institute for Scientific Information, 3501 Market Street, Philadelphia, PA 19104. (800) 523-1850 or (215) 386-0100. 1974 to present. Semi-annual. $550.00 per year.

METEOROLOGICAL AND GEOASTROPHYSICAL ABSTRACTS. American Meteorological Society, 45 Beacon Street, Boston, MA 02108. (617) 227-2425. 1950 to present. Monthly. $450.00 per year.

OCEAN ABSTRACTS. Cambridge Scientific Abstracts, 5161 River Road, Bethesda, MD 20816. (301) 951-1400. 1963 to present. Bimonthly. $450.00 per year.

PHYSICS ABSTRACTS. Institution of Electrical Engineers. Available from: IEEE Service Center, 445 Hoes Lane, Piscataway, NJ 08854. 1898 to present. Bimonthly. $1700.00 per year.

PHYSICS BRIEFS. Physik Verlag GmbH, Postfach 1260/1280, D-6940 Weinheim, West Germany. (212) 661-9404. 1920 to present. Twenty-six times per year. $1250.00 per year.

SCIENCE CITATION INDEX. Institute for Scientific Information, 3501 Market Street, Philadelphia, PA 19104. (800) 523-1850 or (215) 386-0100. Six times per year. $6200. 00 per year.

ASSOCIATIONS AND PROFESSIONAL SOCIETIES

AMERICAN ASTRONOMICAL SOCIETY. 1816 Jefferson Place, N.W., Washington, DC 20036. (202) 659-0134.

AMERICAN GEOPHYSICAL UNION. 2000 Florida Avenue, N.W., Washington, DC 20009. (202) 462-6903.

AMERICAN INSTITUTE OF PHYSICS. 335 East 45th Street, New York, NY 10017. (212) 661-9494.

EARTHQUAKE ENGINEERING RESEARCH INSTITUTE. 2620 Telegraph Avenue, Berkeley, CA 94704. (415) 848-0972.

GEOLOGICAL SOCIETY OF AMERICA. 3300 Penrose Place, Boulder, CO 80301. (303) 447-2020.

SEISMOLOGICAL SOCIETY OF AMERICA. 6431 Fairmont Avenue, No. 7, El Cerrito, CA 94530. (415) 525-5474.

SOCIETY OF EXPLORATION GEOPHYSICISTS. P.O. Box 702740, Tulsa, OK 74170. (918) 493-3516.

DIRECTORIES AND BIOGRAPHICAL SOURCES

AMERICAN MEN AND WOMEN OF SCIENCE. R.R. Bowker, Inc., Order Department, 245 West 17th Street, New York, NY 10011. (800) 521-8110. Eight volumes. 1986. $595.00 for set.

INTERNATIONAL RESEARCH CENTERS DIRECTORY 1988-89. Darren L. Smith, editor. Gale Research Company, Book Tower, Detroit, MI 48226. (800) 521-0707. 4th edition. 1987. $360.00.

1987 DIRECTORY OF ENGINEERING SOCIETIES AND RELATED ORGANIZATIONS. Gordon Davis, editor. Hemisphere Publishing Corporation, 1010 Vermont Avenue, NW, Washington, DC 20005. (800) 526-0275. 12th edition. 1987. $100.00.

RESEARCH CENTERS DIRECTORY 1988. Gale Research Company, Book Tower, Detroit, MI 48226. (800) 521-0707. 12th edition. 1987. $365.00 for set.

SCIENTIFIC AND TECHNICAL ORGANIZATIONS AND AGENCIES DIRECTORY. Margaret Labash Young, editor. Gale Research Company, Book Tower, Detroit, MI 48226. (800) 521-0707. 2nd edition. 1987. $185.00.

SOCIETY OF EXPLORATION GEOPHYSICISTS ROSTER. P.O. Box 702740, Tulsa, OK 74170. (918) 493-3516. Annual. Inquire.

WHO'S WHO IN ENGINEERING. Gordon Davis, editor. Hemisphere Publishing Corporation, 79 Madison Avenue, New York, NY 10016-7892. (800) 821-8312. Sixth edition. 1985. $200.00.

ENCYCLOPEDIAS AND DICTIONARIES

MCGRAW-HILL ENCYCLOPEDIA OF THE GEOLOGICAL SCIENCES. McGraw-Hill Book Company, 1221 Avenue of the Americas, New York, NY 10020. (212) 512-2000. Second edition. 1988. $85.00.

GENERAL WORKS

THE GREAT WAVES. Douglas Myles. McGraw-Hill Book Company, 1221 Avenue of the Americas, New York, NY 10020. (212) 512-2000. 1985. $16.95.

INTRODUCTION TO GEOPHYSICAL PROSPECTING. M.B. Dobrin and C.H. Savit. McGraw-Hill Book Company, 1221 Avenue of the Americas, New York, NY 10020. (212) 512-2000. 4th edition. 1988. $39.95.

AN INTRODUCTION TO SEISMIC INTERPRETATION. R. McQuillin and others. Gulf Publishing Company, P.O. Box 2608, Houston, TX 77001. (713) 520-4444. Second edition. 1985. $48.00.

AN INTRODUCTION TO THE THEORY OF SEISMOLOGY. K.E. Bulletin and A.B. Bolt. Cambridge University Press, 32 East 57th Street, New York, NY 10022. (800) 872-7423. 1985. $69.50.

PRACTICAL APPROACHES TO EARTHQUAKE PREDICTION AND WARNING. C. Kisslinger and T. Rikitake, editors. Kluwer Academic Publishers, 190 Old Derby Street, Hingham, MA 02043. (617) 749-5262. 1986. $79.00.

SEISMIC WAVES AND SOURCES. A. Ben-Menahem and S. Singh. Springer-Verlag New York, Inc., 175 Fifth Avenue, New York, NY 10010. (800) 526-7254. 1981. $99.00.

ONLINE DATA BASES

CA SEARCH. Chemical Abstracts Service, P.O. Box 3012, Columbus, OH 43120. (800) 848-6538 or (614) 421-3600. Comprehensive guide to chemical literature, 1972 to present. Inquire as to online cost and availability.

COMPENDEX. Engineering Information, Inc., 345 East 47th Street, New York, NY 10017. (800) 221-1044 or (212) 705-7615. Engineering and technical literature, 1975 to present. Inquire as to online cost and availability.

DISSERTATION ABSTRACTS ONLINE. University Microfilms International, 300 North Zeeb Road, Ann Arbor, MI 48106. (800) 521-0600 or (313) 761-4700. Scope includes virtually all doctoral dissertations accepted at accredited American institutions from 1861 to present in over 250 subject areas. Inquire as to online cost and availability.

GEOREF. Online version of the BIBLIOGRAPHY AND INDEX OF GEOLOGY. American Geological Institute, 4220 King Street, Alexandria, VA 22302. (703) 379-2480. 1969 to present. Inquire as to online cost and availability.

INSPEC. INSPEC Marketing Department, Institution of Electrical Engineers. Available from IEEE Service Center, 445 Hoes Lane, Piscataway, NJ 08854. (201) 981-0060. Online version of Physics Abstracts. Inquire as to online cost and availability.

NTIS. National Technical Information Service, 5285 Port Royal Road, Springfield, VA 22161. (703) 487-4630. Broad coverage of government sponsored research reports, 1964 to present. Inquire as to online cost and availability.

SCISEARCH. Institute for Scientific Information, 3501 Market Street, Philadelphia, PA 19104. (800) 523-1850 or (215) 386-0100. Broad multidisciplinary title and author index to the international literature of science and technology, 1974 to present. Inquire as to online cost and availability.

WILSONLINE. H.W. Wilson and Company, 950 University Avenue, Bronx, NY 10452. (800) 367-6770 or (212) 588-8400. Makes available online versions of the H.W. Wilson indexes including Applied Science and Technology Index, Business Periodicals Index and Readers' Guide to Periodical Literature. Approximately 1980 to present. Inquire as to online cost and availability.

PERIODICALS

EARTHQUAKE INFORMATION BULLETIN. U.S. Geological Survey, 12201 Sunrise Valley Drive, Reston, VA 22092. Order from: Superintendent of Documents, U.S. Government Printing Office, 20402. Bimonthly. $15.00 per year.

EARTHQUAKE PREDICTION RESEARCH. D. Reidel Publishing Company, 190 Old Derby Street, Hingham, MA 02043. 1982 to present. Quarterly. $85.00 per year.

GEOPHYSICAL AND ASTROPHYSICAL FLUID DYNAMICS. Gordon and Breach Science Publishers, Inc., 50 West 23rd Street, New York, NY 10010. (212) 206-8900. 1970 to present. 16 times per year. $260.00 per year.

GEOPHYSICAL RESEARCH LETTERS. American Geophysical Union, 2000 Florida Avenue, N.W., Washington, DC 20009. (202) 462-6903. 1974 to present. Monthly. $185.00 per year.

GEOPHYSICS. Society of Exploration Geophysicists, P.O. Box 702740, Tulsa, OK 74170. (918) 493-3516. 1936 to present. Monthly. $45.00 per year.

JGR: JOURNAL OF GEOPHYSICAL RESEARCH: SOLID EARTH AND PLANETS. American Geophysical Union, 2000 Florida Avenue, N.W., Washington, DC 20009. (202) 462-6903. Monthly. $760.00 per year.

JOURNAL OF VOLCANOLOGY AND GEOTHERMAL RESEARCH. Elsevier Science Publishing Company, Inc., 52 Vanderbilt Avenue, New York, NY 10017. (212) 370-5520. 1976 to present. 16 times per year. $360.00 per year.

SEISMOLOGICAL SOCIETY OF AMERICA BULLETIN. Seismologycal Society of America, 6431 Fairmont Avenue, No. 7, El Cerrito, CA 94530. (415) 525-5474. 1911 to present. Bimonthly. $90.00 per year.

TECTONOPHYSICS: AN INTERNATIONAL JOURNAL OF GEOTECTONICS AND THE GEOLOGY AND PHYSICS OF THE INTERIOR OF THE EARTH. Elsevier Science Publishing Company, Inc., 52 Vanderbilt Avenue, New York, NY 10017. (212) 370-5520. 1964 to present. $1200.00 per year.

VOLCANOLOGY AND SEISMOLOGY. Gordon and Breach Science Publishers, Inc., 50 West 23rd Street, New York, NY 10010. (212) 206-8900. Monthly. $500.00 per year.

RESEARCH CENTERS AND INSTITUTES

EARTHQUAKE ENGINEERING RESEARCH INSTITUTE. 2620 Telegraph Avenue, Berkeley, CA 94704. (415) 848-0972.

LAMONT-DOHERTY GEOLOGICAL OBSERVATORY. Columbia University, Palisades, NY 10964. (914) 359-2900.

SEISMOGRAPHIC STATIONS. University of California at Berkeley, Berkeley, CA 94720. (415) 642-3977.

SEISMOLOGICAL LABORATORY. California Institute of Technology, Pasadena, CA 91125. (818) 356-6912.

SEMICONDUCTORS

See also: ELECTRICAL ENGINEERING, ELECTRICITY, ELECTRONIC CIRCUITS AND COMPONENTS, ELECTRONIC ENGINEERING

ABSTRACT SERVICES AND INDEXES

APPLIED SCIENCE AND TECHNOLOGY INDEX. H.W. Wilson Company, 950 University Avenue, Bronx, NY 10452. (800) 367-6670 or (212) 588-8400. Inquire as to cost and availability.

ELECTRONICS AND COMMUNICATIONS ABSTRACTS JOURNAL. Cambridge Scientific Abstracts, 5161 River Road, Bethesda, MD 20816. (301) 951-1400. Bimonthly. Inquire as to cost and availability.

ENGINEERING INDEX MONTHLY. Engineering Information, Incorporated, 345 East 47th Street, New York, NY 10017. (800) 221-1044 or (212) 705-7600. Monthly, with annual cumulation. $1425.00 per year.

PHYSICS ABSTRACTS. Institute of Electrical Engineers, London, United Kingdom. Available from: Institute of Electrical and Electronic Engineers (IEEE), 345 East 47th Street, New York, NY 10017. (212) 705-7900.

SCIENCE CITATION INDEX. Institute for Scientific Information, 3501 Market Street, Philadelphia, PA 19104. (800) 523-1850 or (215) 386-0100. Inquire as to cost and availability.

SOLID STATE ABSTRACTS: AN ABSTRACT JOURNAL INVOLVING THE PHYSICS, METALLURGY, CRYSTALLOGRAPHY, CHEMISTRY, AND DEVICE TECHNOLOGY OF SOLIDS. Cambridge Scientific Abstracts, 5161 River Road, Bethesda, MD 20816. (301) 951-1400. 1957 to present. Bimonthly. $550.00 per year.

ANNUAL REVIEWS AND YEARBOOKS

CRC CRITICAL REVIEWS IN SOLID STATE AND MATERIALS SCIENCES. CRC Press, Incorporated, 2000 Corporate Boulevard, NW, Boca Raton, FL 33431. Irregular. Inquire as to cost and availability.

PROGRESS IN CRYSTAL GROWTH AND CHARACTERIZATION. Pergamon Press, Incorporated, Maxwell House, Fairview Park, Elmsford, NY 10523. (914) 592-7700. Irregular. Inquire as to cost and availability.

ASSOCIATIONS AND PROFESSIONAL SOCIETIES

AMERICAN ELECTRONICS ASSOCIATION. Post Office Box 10045, 2670 Hanover Street, Palo Alto, CA 94303. (415) 857-9300.

ELECTRONIC INDUSTRIES ASSOCIATION. 2001 Eye Street, NW, Washington, DC 20006. (202) 457-4900.

IEEE (INSTITUTE OF ELECTRICAL AND ELECTRONICS ENGINEERS). 345 East 47th Street, New York, NY 10017. (212) 705-7900.

INTERNATIONAL SOCIETY FOR HYBRID MICROELECTRONICS. Post Office Box 2698, 1861 Wiehle Avenue, Suite 340, Reston, VA 22090. (703) 471-0066.

BIBLIOGRAPHIES

HANDBOOKS AND TABLES IN SCIENCE AND TECHNOLOGY. Russell H. Powell, editor. Oryx Press, 2214 North Central Avenue, Phoenix, AZ 85004-1483. (602) 254-6156. Second edition. 1983. $55.00.

SCIENTIFIC AND TECHNICAL BOOKS IN PRINT: AN INDEX TO LITERATURE IN SCIENCE AND TECHNOLOGY. R.R. Bowker Company, 205 East 42nd Street, New York, NY 10017. (800) 521-8110 or (212) 916-1600.

DIRECTORIES AND BIOGRAPHICAL SOURCES

AMERICAN ELECTRONICS ASSOCIATION DIRECTORY. American Electronics Association, Post Office Box 10045, 2670 Hanover Street, Palo Alto, CA 94303. (415) 857-9300. Annual.

EEM - ELECTRONIC ENGINEERS MASTER. Hearst Business Communications/UTP Division, 645 Stewart Avenue, Garden City, NY 11530. (516) 227-1300. Annual. $75.00 per copy.

IEEE MEMBERSHIP DIRECTORY. Institute of Electrical and Electronics Engineers, IEEE Service Center, 445 Hoes Lane, Piscataway, NJ 08854. (212) 705-7900. Annual. $7.00.

RESEARCH CENTERS DIRECTORY. Gale Research Company, Book Tower, Detroit, MI 48226. (800) 521-0707. Twelfth edition. 1987. $365.00 for set.

SCIENTIFIC AND TECHNICAL ORGANIZATIONS AND AGENCIES DIRECTORY. Gale Research Company, Book Tower, Detroit, MI 48226. (800) 521-0707. 1985.

WHO'S WHO IN ELECTRONICS. Harris Publishing Company, 2057-2 Aurora Road, Twinsburg, OH 44087. (216) 425-9000. Annual. $89.00.

WHO'S WHO IN ENGINEERING. Engineers Joint Council, 345 East 47th Street, New York, NY 10017. (212) 705-7010. 1985. $200.00.

WHO'S WHO IN FRONTIER SCIENCE AND TECHNOLOGY. Marquis Who's Who, Incorporated, 200 East Ohio Street, Chicago, IL 60611. (800) 428-3898 or (312) 787-2008.

WHO'S WHO IN TECHNOLOGY TODAY. Reston Publishing Company, Incorporated, c/o Prentice-Hall, Incorporated, Englewood Cliffs, NJ 07632. (800) 262-6868. Biennial. Five volumes. $425.00. Covers the fields of electronics, computer science, physics, optics, chemistry, biotechnology, mechanics, energy, and earth science.

ENCYCLOPEDIAS AND DICTIONARIES

ENCYCLOPEDIA OF PHYSICAL SCIENCE AND TECHNOLOGY. Academic Press, Incorporated, Orlando, FL 32887. (800) 321-5068 or (305) 345-2734. Fifteen volumes, 1986.

IEEE STANDARD DICTIONARY OF ELECTRICAL AND ELECTRONICS TERMS. Frank Jay, editor. John Wiley and Sons, Incorporated, 605 Third Avenue, New York, NY 10158. (800) 526-5368 or (212) 850-6000. Third edition. 1984. $49.95.

MCGRAW-HILL ENCYCLOPEDIA OF SCIENCE AND TECHNOLOGY. McGraw-Hill Book, Incorporated, 1221 Avenue of the Americas, New York, NY 10020. (212) 997-3675. Fifth edition, 15 volumes. $1100.00.

GENERAL WORKS

CIRCUITS, DEVICES AND SYSTEMS: A FIRST COURSE IN ELECTRICAL ENGINEERING. R.J. Smith. John Wiley and Sons, Incorporated, 605 Third Avenue, New York, NY 10158. (800) 526-5368 or (212) 850-6000. Fourth edition. 1984. $41.45.

ELECTRONIC CIRCUIT ANALYSIS: BASIC PRINCIPLES. Roy A. Colclaser, Donald A. Neaman, Charles F. Hawkins. John Wiley and Sons, Incorporated, 605 Third Avenue, New York, NY 10158. (800) 526-5368 or (212) 850-6000. 1984. $42.50.

ELECTRONIC CIRCUITS AND APPLICATIONS. Stephen D. Senturia and B.D. Wedlock. John Wiley and Sons, Incorporated, 605 Third Avenue, New York, NY 10158. (800) 526-5368 or (212) 850-6000. 1975. $44.00.

ELECTRONIC DEVICES AND COMPONENTS. J. Seymour. John Wiley and Sons, Incorporated, 605 Third Avenue, New York, NY 10158. (800) 526-5368 or (212) 850-6000. 1981. $37.00 in paper.

ELECTRONIC INVENTIONS AND DISCOVERIES: ELECTRONICS FROM ITS EARLIEST BEGINNINGS TO THE PRESENT DAY. G.W.A. Dummer. Pergamon Press, Incorporated, Maxwell House, Fairview Park, Elmsford, NY 10523. (914) 592-7700. 1983. $48.50.

INTRODUCTION TO SEMICONDUCTOR DEVICE MODELLING. C.M. Snowden. Taylor and Francis, Incorporated, 242 Cherry Street, Philadelphia, PA 19106-1906. (215) 238-0939. 1986. $37.00.

THE PHYSICS OF SEMICONDUCTOR DEVICES. D.A. Fraser. Oxford University Press, 200 Madison Avenue, New York, NY 10016. (800) 458-5833. 1983. $14.95.

SEMICONDUCTOR CIRCUIT APPROXIMATIONS: AN INTRODUCTION TO TRANSISTORS AND INTEGRATED CIRCUITS. Albert Paul Malvino. McGraw-Hill Book Company, 1221 Avenue of the Americas, New York, NY 10020. (800) 628-0004. Fourth edition. 1985. $34.95.

SEMICONDUCTOR DEVICES. Simon M. Sze. John Wiley and Sons, Incorporated, 605 Third Avenue, New York, NY 10158. (800) 526-5368 or (212) 850-6000. 1985. $37.95.

SOLID STATE DEVICES AND INTEGRATED CIRCUITS. William Cooper and Henry Wiesbecker. Reston Publishing Company, Incorporated, c/o Prentice-Hall, Incorporated, Englewood Cliffs, NJ 07632. (800) 262-6868. 1982. $32.95.

HANDBOOKS AND MANUALS

CONTEMPORARY ELECTRONICS CIRCUITS DESKBOOK. Harry Helms. McGraw-Hill Book Company, 1221 Avenue of the Americas, New York, NY 10020. (800) 628-0004. 1986. $29.95.

CRC HANDBOOK OF TABLES FOR APPLIED ENGINEERING SCIENCE. R.E. Bolz and G.L. Tuve, editors. CRC Press, Incorporated, 2000 Corporated Boulevard, NW, Boca Raton, FL 33431. Second edition. 1973. $69.00.

ELECTRONIC ENGINEERS HANDBOOK. Donald G. Fink, editor. McGraw-Hill Book Company, 1221 Avenue of the Americas, New York, NY 10020. (800) 628-0004. Second edition. 1982. $89.00.

HANDBOOK OF ELECTRONICS MANUFACTURING ENGINEERING. B.S. Matisoff. Van Nostrand Reinhold Company, Incorporated, 115 Fifth Avenue, New York, NY 10003. (800) 543-2681. Second edition. 1986. $52.95.

HANDBOOK OF MODERN ELECTRONICS AND ELECTRICAL ENGINEERING. Charles Belove. John Wiley and Sons, Incorporated, 605 Third Avenue, New York, NY 10158. (800) 526-5368 or (212) 850-6000. 1986. $85.00.

REFERENCE MANUAL FOR TELECOMMUNICATIONS. R.L. Freeman. John Wiley and Sons, Incorporated, 605 Third Avenue, New York, NY 10158. (800) 526-5368 or (212) 850-6000. 1985. $79.95.

STANDARD HANDBOOK FOR ELECTRICAL ENGINEERS. Donald G. Fink and H. Wayne Beaty. McGraw-Hill Book Company, 1221 Avenue of the Americas, New York, NY 10020. (212) 512-2000. Eleventh edition. 1978. $85.00.

VLSI HANDBOOK. Norman G. Einspruch, editor. Academic Press, Incorporated, 6277 Sea Harbor Drive, Orlando, FL 32821. (800) 321-5068. 1985. $125.00.

THE WILEY ENGINEER'S DESK REFERENCE. Sanford I. Heisler. John Wilet and Sons, Incorporated, 605 Third Avenue, New York, NY 10158. (800) 526-5368 or (212) 850-6418. $36.00.

ONLINE DATA BASES

COMPENDEX. Engineering Information, Incorporated, 345 East 47th Street, New York, NY 10017. (800) 221-1044 or (212) 705-7615. Engineering and technical literature, 1975 to present. Inquire as to cost and availability.

DISSERTATION ABSTRACTS ONLINE. University Microfilms International, 300 North Zeeb Road, Ann Arbor, MI 48106. (800) 521-0600 or (313) 761-4700. Scope includes virtually all doctoral dissertations accepted at accredited American institutions from 1861 to present in 252 subject areas. Inquire as to cost and availability.

INSPEC. INSPEC Marketing Department, Institute of Electrical and Electronics Engineers, Incorporated, IEEE Service Department, 445 Hoes Lane, Piscataway, NJ 08854. (201) 981-0060. Inquire as to on-line cost and availability.

NTIS. National Technical Information Service, 5285 Port Royal Road, Springfield, VA 22161. (703) 487-4630. Broad coverage of government-sponsored research reports, 1964 to present. Inquire as to cost and availability.

WILSONLINE. H.W. Wilson Company, 950 University Avenue, Bronx, NY 10452. (800) 367-6770 or (212) 588-8400. Makes available online versions of the printed H.W. Wilson Indexes including Applied Science and Technology Index, Business Periodicals Index, and Readers' Guide to Periodical Literature. Period covered is generally 1983 to present. Inquire as to cost and availability.

OTHER SOURCES

A GUIDE TO THE LITERATURE OF ELECTRICAL AND ELECTRONIC ENGINEERING. Susan B. Ardis. Libraries Unlimited, Incorporated, Post Office Box 263, Littleton, CO 80160. (303) 770-1220. 1986. $37.50.

PERIODICALS

CANADAIAN ELECTRONICS ENGINEERING. Maclean Hunter Research Bureau, 777 Bay Street, Toronto, ON M5W 1A7 Canada. (416) 596-5729.

ELECTROCOMPONENT SCIENCE. Gordon Breach Science Publishers, Post Office Box 786, Cooper Station, New York, NY 10276. (212) 206-8900. Quarterly. $224.00 per year.

ELECTRONIC DESIGN. Heyden Publishing Company, Incorporated, 10 Mulholland Drive, Hasbrouck Heights, NJ 07604. (201) 393-6000. Biweekly. $65.00.

ELECTRONIC ENGINEERING TIMES. CMP Publications, Incorporated, 600 Community Drive, Manhasset, NY 11030. (516) 365-4600. Weekly. $65.00 per year.

ELECTRONICS WEEK. McGraw-Hill Book Company, 1221 Avenue of the Americas, New York, NY 10020. (800) 628-0004. Weekly. $18.00.

IEEE CIRCUITS AND DEVICES MAGAZINE. Institute of Electrical and Electronics Engineers, IEEE Service Center, 445 Hoes Lane, Piscataway, NJ 08854. (212) 705-7900. Bimonthly. $70.00 per year.

IEEE ELECTRON DEVICE LETTERS. Institute of Electrical and Electronics Engineers, IEEE Service Center, 445 Hoes Lane, Piscataway, NJ 08854. (212) 705-7900. Monthly. $70.00 per year.

IEEE JOURNAL OF SOLID STATE CIRCUITS. Institute of Electrical and Electronics Engineers, IEEE Service Center, 445 Hoes Lane, Piscataway, NJ 08854. (212) 705-7900. Bimonthly. $113.00 per year.

IEEE TRANSACTIONS ON CIRCUITS AND SYSTEMS. Institute of Electrical and Electronics Engineers, IEEE Service Center, 445 Hoes Lane, Piscataway, NJ 08854. (212) 705-7900. Monthly. $108.00 per year.

IEEE TRANSACTIONS ON ELECTRON DEVICES. Institute of Electrical and Electronics Engineers, IEEE Service Center, 445 Hoes Lane, Piscataway, NJ 08854. (212) 705-7900. Monthly. $159.00 per year.

JOURNAL OF PHYSICS. C. Solid State Physics. American Institute of Physics, 335 East 45th Street, New York, NY 10017. (212) 661-9404. Thirty-six times per year. $840.00 per year.

JOURNAL OF PHYSICS AND CHEMISTRY OF SOLIDS. Pergamon Press, Incorporated, Journals Division, Maxwell House, Fairview Park, Elmsford, NY 10523. (914) 592-7700. Monthly. $530.00 per year.

PHYSICAL REVIEW B (CONDENSED MATTER). American Institute of Physics, 335 East 45th Street, New York, NY 10017. (212) 661-9404. Semimonthly. $850.00 per year.

RADIO ELECTRONICS. Gernsback Publications, Incorporated, 500-B BiCounty Boulevard, Farmingdale, NY 11735. (516) 293-3000. Monthly. $16.00 per year.

SEMICONDUCTOR INDUSTRY AND BUSINESS SURVEY. HTE Management, Incorporated, 4575 Scotts Valley Drive, Suite 105, Scotts Valley, CA 95066. Monthly. Eighteen times per year. $400.00 per year.

SEMICONDUCTOR INTERNATIONAL. Cahners Publishing Company, Incorporated, Division of Reed Holdings, Incorporated, Cahners Plaza, 1350 East Touhy Avenue, Des Plaines, IL 60018. (312) 635-8800. Monthly. $55.00 per year.

SEMICONDUCTOR SCIENCE AND TECHNOLOGY. Institute of Physics, Techno House, Ratcliffe Way, Bristol BS1 6NX, England. Monthly. $100.00 per year.

SEMICONDUCTORS AND INSULATORS. Gordon and Breach Science Publishers, Limited, Post Office Box 197, London, WC2E 9PX, England. Eight times per year. $230.00 per year.

SOLID-STATE ELECTRONICS. Pergamon Press, Incorporated, Journals Division, Maxwell House, Fairview Park, Elmsford, NY 10523. (914) 592-7700. Monthly. $330.00 per year.

SOLID STATE TECHNOLOGY. Cowan Publishing Corporation, 14 Vanderventer Avenue, Port Washington, NY 11050. Monthly. $18.00.

VLSI SYSTEMS DESIGN. CMP Publications, Incorporated, 600 Community Drive, Manhasset, NY 11030. (516) 365-4600. Monthly. $20.00 per year.

RESEARCH CENTERS AND INSTITUTES

MASSACHUSETTS INSTITUTE OF TECHNOLOGY. Laboratory for Electromagnetic and Electronic Systems, 77 Massachusetts Avenue, Cambridge, MA 02139. (617) 253-4631.

NORTH CAROLINA STATE UNIVERSITY. Solid State Electronics Laboratory, 432 Daniels Hall, Raleigh, NC 27695. (919) 737-2336.

OHIO STATE UNIVERSITY. Electroscience Laboratory, 1320 Kinnear Road, Columbus, OH 43212. (614) 422-7981.

PENNSYLVANIA STATE UNIVERSITY. Solid State Device Laboratory, 210 Electrical Engineering, West Building, University Park, PA 16802. (814) 865-1666.

STANFORD UNIVERSITY. Stanford Electronics Laboratories, Stanford, CA 94305. (415) 497-1013.

SERVOMECHANISMS

See: ROBOTICS

SEWAGE TREATMENT

See also: ENVIRONMENTAL ENGINEERING, GROUND WATER POLLUTION, SLUDGE, WATER POLLUTION, WATER TREATMENT

ABSTRACT SERVICES AND INDEXES

CHEMICAL ABSTRACTS. Chemical Abstracts Service, 2540 Olentangy Road, Post Office Box 3012, Columbus, OH 43210. (800) 848-6538 or (614) 421-3600. Weekly. $9200.00 per year.

CURRENT CONTENTS: PHYSICAL, CHEMICAL AND EARTH SCIENCES. Institute for Scientific Information, 3501 Market Street, Philadelphia, PA 19104. (800) 523-1850 or (215) 386-0100. 1970 to present. Weekly. $272.00 per year.

ENGINEERING INDEX MONTHLY AND AUTHOR INDEX. Engineering Information, Incorporated, 345 East 47th Street, New York, NY 10017. (800) 221-1044 or (212) 705-7600. Monthly, with annual cumulation. $1560.00 per year.

POLLUTION ABSTRACTS. Cambridge Scientific Abstracts, 5161 River Road, Bethesda, MD 20816. (301) 951-1400. 1970 to present. Bimonthly. $465.00 per year.

SCIENCE CITATION INDEX. Institute for Scientific Information, 3501 Market Street, Philadelphia, PA 19104. (800) 523-1850 or (215) 386-0100. Inquire as to cost and availability.

SELECTED WATER RESOURCES ABSTRACTS. United States Geological Survey, Water Resources Scientific Information Center. Available from: National Technical Information Service, Springfield, VA 22161. (703) 860-7455. 1968 to present. Monthly. $115.00 per year.

WATER QUALITY CONTROL DIGEST. University Digest Services, Post Office Box 343, Troy, MI 48099. (313) 651-2528. 1969 to present. Bimonthly. $87.00 per year.

ASSOCIATIONS AND PROFESSIONAL SOCIETIES

AMERICAN PUBLIC WORKS ASSOCIATION. 1313 East 60th Street, Chicago, IL 60637. (312) 667-2200.

AMERICAN SOCIETY OF CIVIL ENGINEERS. 345 East 47th Street, New York, NY 10017-2398. (212) 705-7520.

AMERICAN WATER RESOURCES ASSOCIATION. 5410 Grosvenor Lane, Suite, Bethesda, MD 20814. (301) 493-8600.

AMERICAN WATER WORKS ASSOCIATION. 666 West Quincy Avenue, Denver, CO 80235. (303) 794-7711.

ASSOCIATION OF GROUND WATER SCIENTISTS AND ENGINEERS. 6375 Riverside Drive, Dublin, OH 43017. (614) 761-1711.

ASSOCIATION OF METROPOLITAN SEWERAGE AGENCIES. 1015 18th Street, NW, Suite 1002, Washington, DC 20036. (202) 659-9161.

GROUND WATER INSTITUTE. Post Office Box 981, Minneapolis, MN 55440. (612) 698-4395.

NATIONAL WATER RESOURCES ASSOCIATION. 955 L'Enfant Plaza, SW, Suite 1202N, Washington, DC 20024. (202) 488-0610.

NATIONAL WATER WELL ASSOCIATION. 6375 Riverside Drive, Dublin, OH 43017. (614) 761-1711.

WATER POLLUTION CONTROL FEDERATION. 2626 Pennsylvania Avenue, Washington, DC 20037. (337-2500.

WATER QUALITY ASSOCIATION. 4151 Naperville Road, Lisle, IL 60532. (312) 369-1600.

BIBLIOGRAPHIES

STABILISATION, DISINFECTION, AND ODOUR CONTROL IN SEWAGE SLUDGE TREATMENT: AN ANNOTATED BIBLIOGRAPHY COVERING THE PERIOD 1950-1983. E.S. Connor and A.M. Bruce. John Wiley and Sons, Incorporated, 605 Third Avenue, New York, NY 10158. (800) 526-5368 or (212) 850-6000. 1984. $63.95.

WATER POLLUTION: A GUIDE TO INFORMATION SOURCES. Allen W. Knight and Mary Ann Simmons, editors. Gale Research Company, Book Tower, Detroit, MI 48226. (800) 521-0707. 1980. $62.00.

DIRECTORIES AND BIOGRAPHICAL SOURCES

AMERICAN WATER WORKS ASSOCIATION - OFFICERS AND COMMITTEE DIRECTORY. American Water Works Association, 6666 West Quincy Avenue, Denver, CO 80235. (303) 794-7711. Annual. $15.00.

JOURNAL OF THE AMERICAN WATER WORKS ASSOCIATION BUYERS GUIDE ISSUE. American Water Works Association, 6666 West Quincy Avenue, Denver, CO 80235. (303) 794-7711. Annual, with subscription.

PUBLIC WORKS MANUAL. Public Works Journal Corporation, 200 South Broad Street, Ridgewood, NJ 07451. (201) 445-5800. Annual. $20.00.

SCIENTIFIC AND TECHNICAL ORGANIZATIONS AND AGENCIES DIRECTORY. Gale Research Company, Book Tower, Detroit, MI 48226. (800) 521-0707. Second edition. Two volumes. 1987. $185.00 set.

WATER QUALITY ASSOCIATION DIRECTORY. Water Quality Association, 4151 Naperville Road, Lisle, IL 60532. (312) 369-1600. Annual. Available to members only.

GENERAL WORKS

APPLIED STREAM SANITATION. C.J. Velz. John Wiley and Sons, Incorporated, 605 Third Avenue, New York, NY 10158. (800) 526-5368 or (212) 850-6000. Second edition. 1984. $79.95.

CHEMICAL PROCESSES IN WASTE WATER TREATMENT. W.J. Eilbeck and G. Mattock. E. Horwood. Distributed by: john Wiley and Sons, Incorporated, 605 Third Avenue, New York, Ny 10158. (800) 526-5368 or (212) 850-6000. 1987. $85.00.

ENVIRONMENTAL ENGINEERING AND SANITATION. J.A. Salvato. John Wiley and Sons, Incorporated, 605 Third Avenue, New York, NY 10158. (800) 526-5368 or (212) 850-6000. Third edition. 1982. $70.00.

GROUND WATER CONTAMINATION. J.H. Guswa and others. Noyes Data Corporation, Mill Road at Grand Avenue, Park Ridge, NJ 07656. (201) 391-8484. 1985. $48.00.

GROUND WATER CONTAMINATION. National Research Council. National Academy Press, 2101 Constitution Avenue, NW, Washington, DC 20418. (202) 334-3313. 1984. $17.95.

GROUND WATER POLLUTION CONTROL. Larry W. Canter and R.C. Knox. Lewis Publishing, Incorporated, 121 South Main Street, Chelsea, MI 48118. (313) 475-8619. 1985. $49.95.

GROUND WATER QUALITY. C.H. Ward and others. John Wiley and Sons, Incorporated, 605 Third Avenue, New York, NY 10158. (800) 526-5368 or (212) 850-6000. 1985. $45.00.

GROUND WATER QUALITY. C.H. Ward and others. John Wiley and Sons, Incorporated, 605 Third Avenue, New York, NY 10158. (800) 526-5368 or (212) 850-6000. 1985. $45.00.

GROUND WATER QUALITY PROTECTION. L.W. Canter and others. Lewis Publishing, Incorporated, 121 South Main Street, Chelsea, MI 48118. (313) 475-8619. 1986. $49.95.

PROCESSING AND USE OF SEWAGE SLUDGE. P. L'Hermite and H. Ott, editors. Kluwer Academic, 190 Old Derby Street, Hingham, MA 02043. (617) 749-5262. 1984. $79.50.

URBAN WATER INFRASTRUCTURE: PLANNING, MANAGEMENT, AND OPERATIONS. N.S. Grigg. John Wiley and Sons, Incorporated, 605 Third Avenue, New York, NY 10158. (800) 526-5368 or (212) 850-6000. 1986. $39.95.

WATER AND WASTE-WATER TECHNOLOGY. M.J. Hammer. John Wiley and Sons, Incorporated, 605 Third Avenue, New York, NY 10158. (800) 526-5368 or (212) 850-6000. Second edition. 1986. $35.95.

WATER SUPPLY AND POLLUTION CONTROL. Warren Viessman and Mark J. Hammer. Harper and Row Publishers, Incorporated, 10 East 53rd Street, New York, NY 10022. (800) 242-7737. Fourth edition. 1985. $45.00.

WATER TREATMENT PRINCIPLES AND DESIGN. J.M. Montgomery. John Wiley and Sons, Incorporated, 605 Third Avenue, New York, NY 10158. (800) 526-5368 or (212) 850-6000. 1985. $49.95.

HANDBOOKS AND MANUALS

HANDBOOK OF URBAN DRAINAGE AND SEWAGE DISPOSAL. Klaus Imhoff and others. John Wiley and Sons, Incorporated, 605 Third Avenue, New York, NY 10158. (800) 526-5368 or (212) 850-6000. Fourth edition. 1987. $49.95.

WATER TREATMENT HANDBOOK. Degremont Company. John Wiley and Sons, Incorporated, 605 Third Avenue, New York, NY 10158. (800) 526-5368 or (212) 850-6000. Fifth edition. 1979. $95.00.

ONLINE DATA BASES

COMPENDEX. Engineering Information, Incorporated, 345 East 47th Street, New York, NY 10017. (800) 221-1044 or (212) 705-7615. Engineering and technical literature, 1975 to present. Inquire as to cost and availability.

ENVIROLINE. EIC Intelligence, Incorporated, 48 West 38th Street, New York, NY 10018. (212) 944-8500. Worldwide environmental literature, 1979 to present. Inquire as to online cost and availability.

NTIS. National Technical Information Service, 5285 Port Royal Road, Springfield, VA 22161. (703) 487-4630. Broad coverage of government-sponsored research reports, 1964 to present. Inquire as to cost and availability.

POLLUTION ABSTRACTS. Cambridge Scientific Abstracts, 5161 River Road, Bethesda, MD 20816. (301) 951-1400. 1970 to present. Available for online searching through DIALOG Information Services and BRS Information Technologies. Inquire as to online cost and availability.

WATER DATA BANK. United States Department of Agriculture, Agricultural Research Service, Hydrology Laboratory, Water Data Center, Room 139, Building 007, BARC-West, Beltsville, MD 20705. (301) 344-4411. Inquire as to online cost and availability.

WATERNET. American Water Works Association, 6666 West Quincy Avenue, Denver, CO 80235. (303) 794-7711. A data base providing abstracts and indexing of information published in the American Water Works Association Journal. 1971 to present. Inquire as to online cost and availability.

PERIODICALS

AMERICAN WATER WORKS ASSOCIATION. Journal. American Water Works Association, 6666 West Quincy Avenue, Denver, CO 80235. (303) 794-7711. Monthly. $50.00 per year.

ASSOCIATION OF METROPOLITAN SEWERAGE AGENCIES BULLETIN. Association of Metropolitan Sewerage Agencies, 1015 18th Street, NW, Suite 1002, Washington, DC 20036. (202) 659-9161. Monthly.

GROUND WATER. National Water Well Association. Water Well Journal Publishing Company, 500 West Wilson Bridge Road, Worthington, OH 43085. (614) 846-4967. Bimonthly. $53.00 per year.

JOURNAL OF ENVIRONMENTAL ENGINEERING. American Society of Civil Engineers, 345 East 47th Street, New York, Ny 10017. (212) 705-7275. 1956 to present. Bimonthly. $80.00 per year.

JOURNAL OF ENVIRONMENTAL SCIENCES. Institute of Environmental Sciences, 940 East Northwest Highway, Moutn Prospect, IL 60056. (312) 255-1561. Bimonthly. $25.00 per year.

JOURNAL OF GROUND WATER. National Water Well Association. Water Well Journal Publishing Company, 500 West Wilson Bridge Road, Worthington, OH 43085. (614) 846-4967. Quarterly. $16.00 per year.

JOURNAL OF WATER RESOURCES PLANNING AND MANAGEMENT. American Society of Civil Engineers, 345 East 47th Street, New York, NY 10017-2398. (212) 705-7520. Quarterly. $56.00 per year.

THE MANAGEMENT OF WORLD WASTES. Communication Channels, Incorporated, 6255 Barfield Road, Atlanta, Ga 30328. (404) 256-9800. Monthly. $27.00 per year.

PUBLIC WORKS MAGAZINE. Public Works Journal Corporation, 200 South Broad Street, Ridgewood, NJ 07451. (202) 445-5800. Monthly. 430.00 per year.

SANITARY MAINTENANCE. Trade Press Publishing Company, 2100 West Florist Avenue, Milwaukee, WI 53209. (414) 228-7701. Monthly. $30.00 per year.

SLUDGE NEWSLETTER. Business Publishers, 951 Pershing Drive, Silver Spring, MD 20910. (301) 587-6300. Biweekly. $172.72 per year.

WATER, AIR AND SOIL POLLUTION. D. Reidel Publishing Company, 190 Old Derby Street, Hingham, MA 02043. Twenty times per year. $550.00 per year.

WATER AND WASTES DIGEST. Scranton Gillette Communications, Incorporated, 380 Northwest Highway, Des Plaines, IL 60016. (312) 298-6622. Bimonthly. $10.00 per year.

WATER: ENGINEERING AND MANAGEMENT. Scranton Gillette Communications, Incorporated, 380 Northwest Highway, Des Plaines, IL 60016. (312) 298-6622. Monthly. $20.00 per year.

WATER POLLUTION CONTROL. The Bureau of National Affairs, Incorporated, 1231 25th Street, NW, Washington, DC 20037. (202) 452-4200. Biweekly. $360.00 per year.

WATER POLLUTION CONTROL FEDERATION. Journal. Water Pollution Control Federation, 2626 Pennsylvania Avenue, NW, Washington, DC 20037. (202) 337-2500. Monthly. $120.00 per year.

WATER RESEARCH. International Association on Water Pollution Research. Pergamon Press Incorporated, Journals Division, Maxwell House, Fairview Park, Elmsford, NY 10523. (914) 592-7700. Monthly. $550.00 per year.

RESEARCH CENTERS AND INSTITUTES

CALIFORNIA INSTITUTE OF TECHNOLOGY. W.M. Keck Engineering Laboratory of Hydraulics and Water Resources, 1201 East California Boulevard, Pasadena, CA 91125. (818) 356-4404.

LENOX INSTITUTE FOR RESEARCH, INCORPORATED. 101 Yokum Avenue, Lenox, MA 01240. (413) 637-3025.

PENNSYLVANIA STATE UNIVERSITY. Environmental Engineering Laboratory, 212 Sackett Building, University Park, PA 16802. (814) 863-4385.

UNIVERSITY OF ILLINOIS. Advanced Environmental Control Technology Research Center, 3230 Newmark C.E. Laboratory, 208 North Romine Street, Urbana, IL 61801. (217) 333-3822.

UNIVERSITY OF TEXAS. Institute of Environmental Health, Post Office Box 20186, Houston, TX 77225-0186. (713) 792-4425.

WASHINGTON STATE UNIVERSITY. Environmental Engineering Research Laboratory, 141 Sloan, Pullman, WA 99164. (509) 335-3175.

SHIP DESIGN

See: NAVAL ARCHITECTURE

SHIPBUILDING

See also: MARINE ENGINEERING, NAVAL ARCHITECTURE

ABSTRACT SERVICES AND INDEXES

APPLIED MECHANICS REVIEW. American Society of MechanicalEngineers, 345 East 47th Street, New York, NY 10017. (212) 705-7703. 1948 to present. Monthly. $360.00 per year.

APPLIED SCIENCE AND TECHNOLOGY INDEX. H.W. Wilson and Company, 950 University Avenue, Bronx, NY 10452. (800) 367-6670 or (212) 588-8400. Monthly. Inquire as to cost and availability.

CAD/CAM ABSTRACTS. EIC/Intelligence, Inc., 48 West 38th Street, New York, NY 10018. (800) 223-6275 or (212) 944-8500. Monthly. $365.00 per year.

CHEMICAL ABSTRACTS. American Chemical Society, Chemical Abstracts Service, Box 3012, Columbus, OH 43210. (614) 421-3600. 1907 to present. Weekly. $9500.00 per year.

CURRENT CONTENTS: ENGINEERING, TECHNOLOGY AND APPLIED SCIENCES. Institute for Scientific Information, 3501 Market Street, Philadelphia, PA 19104. (800) 523-1850 or (215) 386-0100. Weekly. $275.00 per year.

ENGINEERING INDEX MONTHLY AND AUTHOR INDEX. Engineering Information Inc., 345 East 47th Street, New York, NY 10017. (212) 705-7600. Monthly. $1560.00 per year.

ISMEC BULLETIN (Information Service in Mechanical Engineering). Cambridge Scientific Abstracts, 5161 River Road, Bethesda, MD 20816. (301) 951-1400. 1973 to present. Monthly. $450.00 per year.

OCEANIC ABSTRACTS. Cambridge Scientific Abstracts, 5161 River Road, Bethesda, MD 20816. (301) 951-1400. 1964 to present. Bimonthly. $655.00 per year.

SCIENCE CITATION INDEX. Institute for Scientific Information, 3501 Market Street, Philadelphia, PA 19104. (800) 523-1850 or (215) 386-0100. Six times per year. $6200.00 per year.

ASSOCIATIONS AND PROFESSIONAL SOCIETIES

AMERICAN SOCIETY OF NAVAL ENGINEERS. 1452 Duke Street, Alexandria, VA 22314. (703) 836-6727.

INSTITUTE OF MARINE ENGINEERS. 76 Mark Lane, London, EC3R 7JN, England.

SOCIETY OF NAVAL ARCHITECTS AND MARINE ENGINEERS. One World Trade Center, Suite 1369, New York, NY 10048. (212) 432-0310.

DIRECTORIES AND BIOGRAPHICAL SOURCES

DIRECTORY OF SHIPOWNERS, SHIPBUILDERS, AND MARINE ENGINEERS. Industrial Press, Division of Business Press International, Limited. Available from: Robert E. Raynor, Business Press International, 205 East 42nd Street, New York, NY 10017. Annual. $75.00.

MARINE EQUIPMENT CATALOG. Maritime Reporter and Engineering News, 118 East 25th Street, New York, NY 10010. (212) 477-6700. Annual. $65.00.

1987 DIRECTORY OF ENGINEERING SOCIETIES AND RELATED ORGANIZATIONS. Gordon Davis, editor. Hemisphere Publishing Corporation, 1010 Vermont Avenue, NW, Washington, DC 20005. (800) 526-0275. 12th edition. 1987. $100.00.

RESEARCH CENTERS DIRECTORY 1988. Gale Research Company, Book Tower, Detroit, MI 48226. (800) 521-0707. 12th edition. 1987. $365.00 for set.

SCIENTIFIC AND TECHNICAL ORGANIZATIONS AND AGENCIES DIRECTORY. Margaret Labash Young, editor. Gale Research Company, Book Tower, Detroit, MI 48226. (800) 521-0707. 2nd edition. 1987. $185.00.

WHO'S WHO IN ENGINEERING. Gordon Davis, editor. Hemisphere Publishing Corporation, 1010 Vermont Avenue, NW, Washington, DC 20005. (800) 526-0275. 6th edition. 1985. $200.00.

GENERAL WORKS

MERCHANT SHIP CONSTRUCTION. H.J. Pursey. Sheridan House, Inc., 145 Palisade Street, Dobbs Ferry, NY 10522. (914) 693-2410. 7th edition. 1983. $26.50.

SHIP DESIGN AND CONSTRUCTION. Robert Taggert, editor. Society of Naval Architects and Marine Engineers. One World Trade Center, Suite 1369, New York, NY 10048. (212) 432-0310. 3rd edition. 1980. $75.00.

SHIP STRUCTURAL DESIGN: A RATIONALLY-BASED, COMPUTER AIDED, OPTIMIZATION APPROACH. Owen F. Hughes. John Wiley and Sons, Inc., 605 Third Avenue, New York, NY 10158. (800) 526-5368. 1983. $77.95.

SHIP STRUCTURAL DESIGN CONCEPTS. J. Harvey Evans, editor. Cornell Maritime Press Inc., P.O. Box 456, Centreville, MD 21617. (301) 758-1075. 1975. $45.00.

HANDBOOKS AND MANUALS

MARINE ENGINE ROOM BLUE BOOK. W.B. Paterson. Cornell Maritime Press, Inc., P.O. Box 456, Centreville, MD 21617. (301) 758-1075. 3rd edition. 1984. $15.00 in paper.

ONLINE DATA BASES

CA SEARCH. Chemical Abstracts Service, P.O. Box 3012, Columbus, OH 43120. (800) 848-6538 or (614) 421-3600. Comprehensive guide to chemical literature, 1972 to present. Inquire as to online cost and availability.

COMPENDEX. Engineering Information, Inc., 345 East 47th Street, New York, NY 10017. (800) 221-1044 or (212) 705-7615. Engineering and technical literature, 1975 to present. Inquire as to online cost and availability.

NTIS. National Technical Information Service, 5285 Port Royal Road, Springfield, VA 22161. (703) 487-4630. Broad coverage of government sponsored research reports, 1964 to present. Inquire as to online cost and availability.

SCISEARCH. Institute for Scientific Information, 3501 Market Street, Philadelphia, PA 19104. (800) 523-1850 or (215) 386-0100. Broad multidisciplinary title and author index to the international literature of science and technology, 1974 to present. Inquire as to online cost and availability.

WILSONLINE. H.W. Wilson and Company, 950 University Avenue, Bronx, NY 10452. (800) 367-6770 or (212) 588-8400. Makes available online versions of the H.W. Wilson indexes including Applied Science and Technology Index, Business Periodicals Index and Readers' Guide to Periodical Literature. Approximately 1980 to present. Inquire as to online cost and availability.

PERIODICALS

JOURNAL OF SHIP RESEARCH. Society of Naval Architects and Marine Engineers. One World Trade Center, Suite 1369, New York, NY 10048. (212) 432-0310. 1957 to present. Quarterly. $40.00 per year.

MARINE ENGINEERING/LOG. Simmons-Boardman Publishing Corporation, 345 Hudson Street, New York, NY 10014. (212) 620-7200. 1878 to present. Monthly. $30.00 per year.

MARINE ENGINEERS REVIEW. Marine Management Holdings, Limited, 76 Mark Lane, London EC3R 7JN, England. 1970 to present. Monthly. $40.00 per year.

MARINE TECHNOLOGY. Society of Naval Architects and Marine Engineers. One World Trade Center, Suite 1369, New York, NY 10048. (212) 432-0310. 1964 to present. Quarterly. $40.00 per year.

MOTOR SHIP. Industrial Press, Division of Business Press International, Limited. Available from: Robert E. Raynor, Business Press International, 205 East 42nd Street, New York, NY 10017. 1920 to present. Monthly. $115.00 per two years.

NAVAL ARCHITECT. Royal Institution of Naval Architects, 10 Upper Belgrave Street, London, SW1X 8BQ, England. 1971 to present. Bimonthly. $75.00 per year.

SHIPBUILDING AND MARINE ENGINEERING INTERNATIONAL. Lloyds of London Press, Limited, Sheepen Place, Colchester, Essex, CO3 3LP, England. 1879 to present. Ten times per year. $65.00 per year.

SOCIETY OF NAVAL ARCHITECTS AND MARINE ENGINEERS TRANSACTIONS. Society of Naval Architects and Marine Engineers. One World Trade Center, Suite 1369, New York, NY 10048. (212) 432-0310. 1893 to present. Annual. $45.00 per year.

RESEARCH CENTERS AND INSTITUTES

INSTITUTE FOR MARINE DYNAMICS. National Research Council of Canada, P.O. Box 12093, Postal Station A, St. John's, NF, Canada A1B 3T5. (709) 772-2469.

MARITIME RESEARCH DEPARTMENT. Webb Institute of Naval Architecture, Crescent Beach Road, Glen Cove, NY 11542. (516) 671-2356.

SHIP HYDRODYNAMICS LABORATORY. University of Michigan, 126 West Engineering Building, Ann Arbor, MI 48109. (313) 764-9432.

SIDE LOOKING RADAR

See: RADAR

SIGNAL PROCESSING

See also: ELECTRICAL ENGINEERING, ELECTRONIC CIRCUITS AND COMPONENTS, ELECTRONIC ENGINEERING

ABSTRACT SERVICES AND INDEXES

APPLIED SCIENCE AND TECHNOLOGY INDEX. H.W. Wilson Company, 950 University Avenue, Bronx, NY 10452. (800) 367-6670 or (212) 588-8400. Inquire as to cost and availability.

ELECTRONICS AND COMMUNICATIONS ABSTRACTS JOURNAL. Cambridge Scientific Abstracts, 5161 River Road, Bethesda, MD 20816. (301) 951-1400. Bimonthly. Inquire as to cost and availability.

ENGINEERING INDEX MONTHLY. Engineering Information, Incorporated, 345 East 47th Street, New York, NY 10017. (800) 221-1044 or (212) 705-7600. Monthly, with annual cumulation. $1425.00 per year.

PHYSICS ABSTRACTS. Institute of Electrical Engineers, London, United Kingdom. Available from: Institute of Electrical and Electronic Engineers (Ieee), 345 East 47th Street, New York, NY 10017. (212) 705-7900.

SCIENCE CITATION INDEX. Institute for Scientific Information, 3501 Market Street, Philadelphia, PA 19104. (800) 523-1850 or (215) 386-0100. Inquire as to cost and availability.

SOLID STATE ABSTRACTS: AN ABSTRACT JOURNAL INVOLVING THE PHYSICS, METALLURGY, CRYSTALLOGRAPHY, CHEMISTRY, AND DEVICE TECHNOLOGY OF SOLIDS. Cambridge Scientific Abstracts, 5161 River Road, Bethesda, MD 20816. (301) 951-1400. 1957 to present. Bimonthly. $550.00 per year.

ANNUAL REVIEWS AND YEARBOOKS

ADVANCES IN ELECTRONICS AND ELECTRON PHYSICS. Academic Press, Incorporated, 6277 Sea Harbor Drive, Orlando, FL 32821. (800) 321-5068. Irregular. Approximately $80.00 per volume.

CRC CRITICAL REVIEWS IN SOLID STATE AND MATERIALS SCIENCES. CRC Press, Incorporated, 2000 Corporate Boulevard, NW, Boca Raton, FL 33431. Irregular. Inquire as to cost and availability.

ASSOCIATIONS AND PROFESSIONAL SOCIETIES

AMERICAN ELECTRONICS ASSOCIATION. Post Office Box 10045, 2670 Hanover Street, Palo Alto, CA 94303. (415) 857-9300.

ASSOCIATION OF OLD CROWS (ELECTRONIC WARFARE). 2300 Ninth Street, South, Suite 300, Arlington, VA 22204. (703) 920-1600.

ELECTRONIC INDUSTRIES ASSOCIATION. 2001 Eye Street, NW, Washington, DC 20006. (202) 457-4900.

IEEE (INSTITUTE OF ELECTRICAL AND ELECTRONICS ENGINEERS). 345 East 47th Street, New York, NY 10017. (212) 705-7900.

INTERNATIONAL SOCIETY FOR HYBRID MICRO-ELECTRONICS. Post Office Box 2698, 1861 Wiehle Avenue, Suite 340, Reston, VA 22090. (703) 471-0066.

BIBLIOGRAPHIES

HANDBOOKS AND TABLES IN SCIENCE AND TECHNOLOGY. Russell H. Powell, editor. Oryx Press, 2214 North Central Avenue, Phoenix, AZ 85004-1483. (602) 254-6156.

Second edition. 1983. $55.00.

SCIENTIFIC AND TECHNICAL BOOKS IN PRINT: AN INDEX TO LITERATURE IN SCIENCE AND TECHNOLOGY. R.R. Bowker Company, 205 East 42nd Street, New York, NY 10017. (800) 521-8110 or (212) 916-1600.

DIRECTORIES AND BIOGRAPHICAL SOURCES

IEEE MEMBERSHIP DIRECTORY. Institute of Electrical and Electronics Engineers, IEEE Service Center, 445 Hoes Lane, Piscataway, NJ 08854. (212) 705-7900. Annual. $7.00.

RESEARCH CENTERS DIRECTORY. Gale Research Company, Book Tower, Detroit, MI 48226. (800) 521-0707. Twelfth edition. 1987. $365.00 for set.

SCIENTIFIC AND TECHNICAL ORGANIZATIONS AND AGENCIES DIRECTORY. Gale Research Company, Book Tower, Detroit, MI 48226. (800) 521-0707. 1985.

WHO'S WHO IN ELECTRONICS. Harris Publishing Company, 2057-2 Aurora Road, Twinsburg, OH 44087. (216) 425-9000. Annual. $89.00.

WHO'S WHO IN ENGINEERING. Engineers Joint Council, 345 East 47th Street, New York, NY 10017. (212) 705-7010. 1985. $200.00.

WHO'S WHO IN FRONTIER SCIENCE AND TECHNOLOGY. Marquis Who's Who, Incorporated, 200 East Ohio Street, Chicago, IL 60611. (800) 428-3898 or (312) 787-2008.

WHO'S WHO IN TECHNOLOGY TODAY. Reston Publishing Company, Incorporated, c/o Prentice-Hall, Incorporated, Englewood Cliffs, NJ 07632. (800) 262-6868. Biennial. Five volumes. $425.00. Covers the fields of electronics, computer science, physics, optics, chemistry, biotechnology, mechanics, energy, and earth science.

ENCYCLOPEDIAS AND DICTIONARIES

ENCYCLOPEDIA OF PHYSICAL SCIENCE AND TECHNOLOGY. Academic Press, Incorporated, Orlando, FL 32887. (800) 321-5068 or (305) 345-2734. Fifteen volumes, 1986.

IEEE STANDARD DICTIONARY OF ELECTRICAL AND ELECTRONICS TERMS. Frank Jay, editor. John Wiley and Sons, Incorporated, 605 Third Avenue, New York, NY 10158. (800) 526-5368 or (212) 850-6000. Third edition. 1984. $49.95.

GENERAL WORKS

ANALOG SIGNAL PROCESSING AND INSTRUMENTATION. Arie Arbel. Cambridge University Press, 32 East 57th Street, New York, NY 10022. (800) 872-7423. 1984. $24.95.

CIRCUITS, DEVICES AND SYSTEMS: A FIRST COURSE IN ELECTRICAL ENGINEERING. R.J. Smith. John Wiley and Sons, Incorporated, 605 Third Avenue, New York, NY 10158. (800) 526-5368 or (212) 850-6000. Fourth edition. 1984. $41.45.

CIRCUITS, SIGNALS AND SYSTEMS. William M. Siebert. McGraw-Hill Book Company, 1221 Avenue of the Americas, New York, NY 10020. (800) 628-0004. 1986. $37.95.

DIGITAL SIGNAL PROCESSING. Richard A. Roberts and Clifford T. Mullis. Addison-Wesley Publishing Company, Incorporated, 1 Jacob Way Reading, MA 10867. (617) 944-3700. 1987. $44.95.

ELECTROMAGNETIC WAVE THEORY. J.A. Kong. John Wiley and Sons, Incorporated, 605 Third Avenue, New York, NY 10158. (800) 526-5368 or (212) 850-6000. 1985. $34.95.

FAST ALGORITHMS FOR DIGITAL SIGNAL PROCESSING. Richard E. Balhut. Addison-Wesley Publishing Company, Incorporated, 1 Jacob Way Reading, MA 01867. (617) 944-3700. 1985. $44.95.

FOUNDATIONS OF DIGITAL SIGNAL PROCESSING AND TIME SERIES ANALYSIS. James A. Cadzow. Macmillan Publishing Company Incorporated, 866 Third Avenue, New York, NY 10022. (800) 257-5755 or (212) 935-2000. 1987. $36.95.

FUNDAMENTALS OF SPEECH SIGNAL PROCESSING: MONOGRAPH. Shuzo Saito and Kazuo Nakata, editors. Academic Press, Incorporated, 6277 Sea Harbor Drive, Orlando, FL 32821. (800) 321-5068. 1985. $59.00.

INTRODUCTION TO RANDOM PROCESSES: WITH APPLICATION TO SIGNALS AND SYSTEMS. William A. Gardner. MacMillan Publishing Company Incorporated, 866 Third Avenue, New York, NY 10022. (800) 257-5755 or (212) 935-2000. 1986. $34.95.

SIGNAL PROCESSING: MODEL BASED APPROACH. J.V. Candy. McGraw-Hill Book, Incorporated, 1221 Avenue of the Americas, New York, NY 10020. (212) 512-2000. 1986. $42.95.

SIGNAL THEORY AND PROCESSING. Frederic De Coulon. Artech House, Incorporated, 610 Washington Street, Dedham, MA 02026. (800) 225-9977. 1986. $60.00.

TELECOMMUNICATION ENGINEERING: ANALOG AND DIGITAL NETWORK DESIGN. R.L. Freeman. John Wiley and Sons, Incorporated, 605 Third Avenue, New York, NY 10158. (800) 526-5368 or (212) 526-5368 or (212) 850-6000. 1980. $48.95.

HANDBOOKS AND MANUALS

CONTEMPORARY ELECTRONICS CIRCUITS DESKBOOK. Harry Helms. McGraw-Hill Book Company, 1221 Avenue of the Americas, New York, NY 10020. (800) 628-0004. 1986. $29.95.

CRC HANDBOOK OF TABLES FOR APPLIED ENGINEERING SCIENCE. R.E. Bolz and G.L. Tuve, editors. CRC Press, Incorporated, 2000 Corporated Boulevard, NW, Boca Raton, FL 33431. Second edition. 1973. $69.00.

ELECTRONIC ENGINEERS HANDBOOK. Donald G. Fink, editor. McGraw-Hill Book Company, 1221 Avenue of the Americas, New York, NY 10020. (800) 628-0004. Second edition. 1982. $89.00.

HANDBOOK OF MODERN ELECTRONICS AND ELECTRICAL ENGINEERING. Charles Belove. John Wiley and Sons, Incorporated, 605 Third Avenue, New York, NY 10158. (800) 526-5368 or (212) 850-6000. 1986. $85.00.

REFERENCE MANUAL FOR TELECOMMUNICATIONS. R.L. Freeman. John Wiley and Sons, Incorporated, 605 Third Avenue, New York, NY 10158. (800) 526-5368 or (212) 850-6000. 1985. $79.95.

STANDARD HANDBOOK FOR ELECTRICAL ENGINEERS. Donald G. Fink and H. Wayne Beaty. McGraw-Hill Book Company, 1221 Avenue of the Americas, New York, NY 10020. (212) 512-2000. Eleventh edition. 1978. $85.00.

VLSI HANDBOOK. Norman G. Einspruch, editor. Academic Press, Incorporated, 6277 Sea Harbor Drive, Orlando, FL 32821. (800) 321-5068. 1985. $125.00.

THE WILEY ENGINEER'S DESK REFERENCE. Sanford I. Heisler. John Wilet and Sons, Incorporated, 605 Third Avenue, New York, NY 10158. (800) 526-5368 or (212) 850-6418. $36.00.

ONLINE DATA BASES

COMPENDEX. Engineering Information, Incorporated, 345 East 47th Street, New York, NY 10017. (800) 221-1044 or (212) 705-7615. Engineering and technical literature, 1975 to present. Inquire as to cost and availability.

DISSERTATION ABSTRACTS ONLINE. University Microfilms International, 300 North Zeeb Road, Ann Arbor, MI 48106. (800) 521-0600 or (313) 761-4700. Scope includes virtually all doctoral dissertations accepted at accredited American institutions from 1861 to present in 252 subject areas. Inquire as to cost and availability.

INSPEC. INSPEC Marketing Department, Institute of Electrical and Electronics Engineers, Incorporated, IEEE Service Department, 445 Hoes Lane, Piscataway, NJ 08854. (201) 981-0060. Inquire as to on-line cost and availability.

NTIS. National Technical Information Service, 5285 Port Royal Road, Springfield, VA 22161. (703) 487-4630. Broad coverage of government-sponsored research reports, 1964 to present. Inquire as to cost and availability.

WILSONLINE. H.W. Wilson Company, 950 University Avenue, Bronx, NY 10452. (800) 367-6770 or (212) 588-8400. Makes available online versions of the printed H.W. Wilson Indexes including Applied Science and Technology Index, Business Periodicals Index, and Readers' Guide to Periodical Literature. Period covered is generally 1983 to present. Inquire as to cost and availability.

OTHER SOURCES

A GUIDE TO THE LITERATURE OF ELECTRICAL AND ELECTRONIC ENGINEERING. Susan B. Ardis. Libraries Unlimited, Incorporated, Post Office Box 263, Littleton, CO 80160. (303) 770-1220. 1986. $37.50.

PERIODICALS

CANADAIAN ELECTRONICS ENGINEERING. Maclean Hunter Research Bureau, 777 Bay Street, Toronto, ON M5W 1A7 Canada. (416) 596-5729.

CIRCUITS, SYSTEMS AND SIGNAL PROCESSING. Birkhauser Boston, Incorporated, 380 Green Street, Post Office Box 2007, Cambridge, MA 02139. (617) 876-2333. Quarterly. $175.00 per year.

ELECTRONIC DESIGN. Heyden Publishing Company, Incorporated, 10 Mulholland Drive, Hasbrouck Heights, NJ 07604. (201) 393-6000. Biweekly. $65.00.

ELECTRONICS WEEK. McGraw-Hill Book Company, 1221 Avenue of the Americas, New York, NY 10020. (800) 628-0004. Weekly. $18.00.

IEE PROCEEDINGS PART F: COMMUNICATIONS, RADAR AND SIGNAL PROCESSING. Institution of Electrical Engineers, Savoy Place, London WC2R 0BL, England. Bimonthly. Inquire as to cost and availability.

IEEE CIRCUITS AND DEVICES MAGAZINE. Institute of Electrical and Electronics Engineers, IEEE Service Center, 445 Hoes Lane, Piscataway, NJ 08854. (212) 705-7900. Bimonthly. $70.00 per year.

IEEE TRANSACTIONS ON ACOUSTICS, SPEECH AND SIGNAL PROCESSING. IEEE Service Center, 445 Hoes Lane, Piscataway, NJ 08854. (212) 705-7900. Bimonthly. $130.00 per year.

IEEE TRANSACTIONS ON CIRCUITS AND SYSTEMS. Institute of Electrical and Electronics Engineers, IEEE Service Center, 445 Hoes Lane, Piscataway, NJ 08854. (212) 705-7900. Monthly. $108.00 per year.

SIGNAL PROCESSING: A EUROPEAN JOURNAL DEVOTED TO THE METHODS AND APPLICATIONS OF SIGNAL PROCESSING. Elsevier Science Publishers B.V., Box 211, 1000 AE Amsterdam, The Netherlands. Eight times per year. $165.00 per year.

RESEARCH CENTERS AND INSTITUTES

ARIZONA STATE UNIVERSITY. Center for Research In Engineering and Applied Sciences, Tempe, AZ 85287. (602) 965-1725.

MASSACHUSETTS INSTITUTE OF TECHNOLOGY. Laboratory for Electromagnetic and Electronic Systems, 77 Massachusetts Avenue, Cambridge, MA 02139. (617) 253-4631.

OHIO STATE UNIVERSITY. Electroscience Laboratory, 1320 Kinnear Road, Columbus, OH 43212. (614) 422-7981.

PENNSYLVANIA STATE UNIVERSITY. Solid State Device Laboratory, 210 Electrical Engineering, West Building, University Park, PA 16802. (814) 865-1666.

PURDUE UNIVERSITY. Applied Ultrasonics and Electromagnetic Signal Processing Laboratory, School of Electrical Engineering, West Lafayette, IN 47907. (317) 494-3563

STANFORD UNIVERSITY. Stanford Electronics Laboratories, Stanford, CA 94305. (415) 497-1013.

UNIVERSITY OF CALIFORNIA AT BERKELEY. Electronics Research Laboratory, 253 Cory Hall, Berkeley, CA 94720. (415) 642-2301.

UNIVERSITY OF KANSAS. Telecommunications and Information Systems Laboratory, Nichols Hall, Lawrence, KS 66045. (913) 864-4832.

SILICON

See also: BORON, CHEMISTRY, ELECTRONIC CIRCUITS AND COMPONENTS, GLASS, INORGANIC CHEMISTRY, MATERIALS SCIENCE, MICROELECTRONICS, ORGANIC CHEMISTRY

ABSTRACT SERVICES AND INDEXES

APPLIED SCIENCE AND TECHNOLOGY INDEX. H.W. Wilson and Company, 950 University Avenue, Bronx, NY 10452. (800) 367-6670 or (212) 588-8400. Monthly. Inquire as to cost and availability.

CHEMICAL ABSTRACTS. American Chemical Society, Chemical Abstracts Service, Box 3012, Columbus, OH 43210. (614) 421-3600. 1907 to present. Weekly. $9500.00 per year.

CONFERENCE PAPERS INDEX. Cambridge Scientific Abstracts, 5161 River Road, Bethesda, MD 20816. 1972 to present. Monthly. Inquire as to cost and availability.

CURRENT CONTENTS: ENGINEERING, TECHNOLOGY AND APPLIED SCIENCES. Institute for Scientific Information, 3501 Market Street, Philadelphia, PA 19104. (800) 523-1850 or (215) 386-0100. Weekly. $275.00 per year.

ELECTRICAL AND ELECTRONICS ABSTRACTS. Institution of Electrical Engineers. Available from: Institute of Electrical and Electronics Engineers. IEEE Service Center, 445 Hoes Lane, Piscataway, NJ 08854. Monthly. $1250.00 per year.

ELECTRONICS AND COMMUNICATIONS ABSTRACTS. Cambridge Scientific Abstracts, 5161 River Road, Bethesda, MD 20816. (301) 951-1400. Bimonthly. Inquire as to cost and availability.

ENGINEERING INDEX MONTHLY AND AUTHOR INDEX. Engineering Information Inc., 345 East 47th Street, New York, NY 10017. (212) 705-7600. Monthly. $1560.00 per year.

SCIENCE CITATION INDEX. Institute for Scientific Information, 3501 Market Street, Philadelphia, PA 19104. (800) 523-1850 or (215) 386-0100. Six times per year. $6200. 00 per year.

SILICON

SOLID STATE ABSTRACTS: AN ABSTRACT JOURNAL INVOLVING THE PHYSICS, METALLURGY, CRYSTALLOGRAPHY, CHEMISTRY, AND DEVICE TECHNOLOGY OF SOLIDS. Cambridge Scientific Abstracts, 5161 River Road, Bethesda, MD 20816. (301) 951-1400. 1957 to present. Bimonthly. $550.00 per year.

ASSOCIATIONS AND PROFESSIONAL SOCIETIES

AMERICAN CHEMICAL SOCIETY. 1155 16th Street, N.W., Washington, DC 20036. (800) 424-6747.

AMERICAN INSTITUTE OF PHYSICS. 335 East 45th Street, New York, NY 10017. (212) 661-9494.

AMERICAN SOCIETY FOR TESTING AND MATERIALS. 1916 Race Street, Philadelphia, PA 19103. (215) 299-5400.

DIRECTORIES AND BIOGRAPHICAL SOURCES

AMERICAN MEN AND WOMEN OF SCIENCE. R.R. Bowker, Inc., Order Department, 245 West 17th Street, New York, NY 10011. (800) 521-8110. Eight volumes. 1986. $595.00 for set.

INTERNATIONAL RESEARCH CENTERS DIRECTORY 1988-89. Darren L. Smith, editor. Gale Research Company, Book Tower, Detroit, MI 48226. (800) 521-0707. 4th edition. 1987. $360.00.

1987 DIRECTORY OF ENGINEERING SOCIETIES AND RELATED ORGANIZATIONS. Gordon Davis, editor. Hemisphere Publishing Corporation, 1010 Vermont Avenue, NW, Washington, DC 20005. (800) 526-0275. 12th edition. 1987. $100.00.

RESEARCH CENTERS DIRECTORY 1988. Gale Research Company, Book Tower, Detroit, MI 48226. (800) 521-0707. 12th edition. 1987. $365.00 for set.

SCIENTIFIC AND TECHNICAL ORGANIZATIONS AND AGENCIES DIRECTORY. Margaret Labash Young, editor. Gale Research Company, Book Tower, Detroit, MI 48226. (800) 521-0707. 2nd edition. 1987. $185.00.

WHO'S WHO IN ENGINEERING. Gordon Davis, editor. Hemisphere Publishing Corporation, 1010 Vermont Avenue, NW, Washington, DC 20005. (800) 526-0275. 6th edition. 1985. $200.00.

GENERAL WORKS

STRUCTURAL CHEMISTY OF BORON AND SILICON. Springer-Verlag New York, Inc., 175 Fifth Avenue, New York, NY 10010. (800) 526-7254. 1986. $56.00.

ONLINE DATA BASES

CA SEARCH. Chemical Abstracts Service, P.O. Box 3012, Columbus, OH 43120. (800) 848-6538 or (614) 421-3600. Comprehensive guide to chemical literature, 1972 to present. Inquire as to online cost and availability.

COMPENDEX. Engineering Information, Inc., 345 East 47th Street, New York, NY 10017. (800) 221-1044 or (212) 705-7615. Engineering and technical literature, 1975 to present. Inquire as to online cost and availability.

DISSERTATION ABSTRACTS ONLINE. University Microfilms International, 300 North Zeeb Road, Ann Arbor, MI 48106. (800) 521-0600 or (313) 761-4700. Scope includes virtually all doctoral dissertations accepted at accredited American institutions from 1861 to present in over 250 subject areas. Inquire as to online cost and availability.

NTIS. National Technical Information Service, 5285 Port Royal Road, Springfield, VA 22161. (703) 487-4630. Broad coverage of government sponsored research reports, 1964 to present. Inquire as to online cost and availability.

SCISEARCH. Institute for Scientific Information, 3501 Market Street, Philadelphia, PA 19104. (800) 523-1850 or (215) 386-0100. Broad multidisciplinary title and author index to the international literature of science and technology, 1974 to present. Inquire as to online cost and availability.

WILSONLINE. H.W. Wilson and Company, 950 University Avenue, Bronx, NY 10452. (800) 367-6770 or (212) 588-8400. Makes available online versions of the H.W. Wilson indexes including Applied Science and Technology Index, Business Periodicals Index and Readers' Guide to Periodical Literature. Approximately 1980 to present. Inquire as to online cost and availability.

PERIODICALS

AMERICAN CHEMICAL SOCIETY JOURNAL. American Chemical Society, 1155 16th Street, N.W., Washington, DC 20036. (800) 424-6747. 1879 to present. Semi-monthly. $275.00 per year.

INDUSTRIAL AND ENGINEERING CHEMISTRY FUNDAMENTALS. American Chemical Society, 1155 16th Street, N.W., Washington, DC 20036. (800) 424-6747. 1962 to present. Quarterly. $60.00 per year.

INORGANIC CHEMISTRY. American Chemical Society, 1155 16th Street, N.W., Washington, DC 20036. (800) 424-6747. 1962 to present. Biweekly. $300.00 per year.

SEMICONDUCTOR INTERNATIONAL. Cahners Publishing Company, Inc., Cahners Plaza, 1350 East Touhy Avenue, Des Plaines, IL 60018. (312) 635-8800. 1978 to present. Monthly. $55.00 per year.

SOLID STATE ELECTRONICS. Pergamon Press, Inc., Maxwell House, Fairview Park, Elmsford, NY 10523. (914) 592-7700. 1960 to present. Monthly. $330.00 per year.

RESEARCH CENTERS AND INSTITUTES

MICROFABRICATION LABORATORY (MICROLAB). University of California, Berkeley, Department of Electrical Engineering and Computer Science, Berkeley, CA 94720. (415) 642-2716.

SOLID STATE ELECTRONICS LABORATORY. North Carolina State University, 432 Daniels Hall, Raleigh, NC 27695. (919) 737-2336.

SILICON CHIPS

See: INTEGRATED CIRCUITS

SINTERING

See: POWDER METALLURGY

SLUDGE

See also: ENVIRONMENTAL ENGINEERING, GROUND WATER POLLUTION, SEWAGE TREATMENT POLLUTION, WATER TREATMENT

ABSTRACT SERVICES AND INDEXES

CHEMICAL ABSTRACTS. Chemical Abstracts Service, 2540 Olentangy Road, Post Office Box 3012, Columbus, OH 43210. (800) 848-6538 or (614) 421-3600. Weekly. $9200.00 per year.

CURRENT CONTENTS: PHYSICAL, CHEMICAL AND EARTH SCIENCES. Institute for Scientific Information, 3501 Market Street, Philadelphia, PA 19104. (800) 523-1850 or (215) 386-0100. 1970 to present. Weekly. $272.00 per year.

ENGINEERING INDEX MONTHLY AND AUTHOR INDEX. Engineering Information, Incorporated, 345 East 47th Street, New York, NY 10017. (800) 221-1044 or (212) 705-7600. Monthly, with annual cumulation. $1560.00 per year.

POLLUTION ABSTRACTS. Cambridge Scientific Abstracts, 5161 River Road, Bethesda, MD 10816. (301) 951-1400. 1970 to present. Bimonthly. $465.00 per year.

SCIENCE CITATION INDEX. Institute for Scientific Information, 3501 Market Street, Philadelphia, PA 19104. (800) 523-1850 or (215) 386-0100. Inquire as to cost and availability.

SELECTED WATER RESOURCES ABSTRACTS. United States Geological Survey, Water Resources Scientific Information Center. Available from: National Technical Information Service, Springfield, VA 22161. (703) 860-7455. 1968 to present. Monthly. $115.00 per year.

WATER QUALITY CONTROL DIGEST. University Digest Services, Post Office Box 343, Troy, MI 48099. (313) 651-2528. 1969 to present. Bimonthly. $87.00 per year.

ASSOCIATIONS AND PROFESSIONAL SOCIETIES

AMERICAN PUBLIC WORKS ASSOCIATION. 1313 East 60th Street, Chicago, IL 60637. (312) 667-2200.

AMERICAN SOCIETY OF CIVIL ENGINEERS. 345 East 47th Street, New York, NY 10017-2398. (212) 705-7520.

AMERICAN WATER RESOURCES ASSOCIATION. 5410 Grosvenor Lane, Suite, Bethesda, MD 20814. (301) 493-8600.

AMERICAN WATER WORKS ASSOCIATION. 666 West Quincy Avenue, Denver, CO 80235. (303) 794-7711.

ASSOCIATION OF GROUND WATER SCIENTISTS AND ENGINEERS. 6375 Riverside Drive, Dublin, OH 43017. (614) 761-1711.

ASSOCIATION OF METROPOLITAN SEWERAGE AGENCIES. 1015 18th Street, NW, Suite 1002, Washington, DC 20036. (202) 659-9161.

NATIONAL WATER RESOURCES ASSOCIATION. 955 L'Enfant Plaza, SW, Suite 1202N, Washington, DC 20024. (202) 488-0610.

NATIONAL WATER WELL ASSOCIATION. 6375 Riverside Drive, Dublin, OH 43017. (614) 761-1711.

WATER AND WASTEWATER EQUIPMENT MANUFACTURERS ASSOCIATION. Post Office Box 17402, Dulles International Airport, Washington, DC 20041. (703) 661-8011.

WATER POLLUTION CONTROL FEDERATION. 2626 Pennsylvania Avenue, Washington, DC 20037. (337-2500.

WATER QUALITY ASSOCIATION. 4151 Naperville Road, Lisle, IL 60532. (312) 369-1600.

BIBLIOGRAPHIES

STABILISATION, DISINFECTION, AND ODOUR CONTROL IN SEWAGE SLUDGE TREATMENT: AN ANNOTATED BIBLIOGRAPHY COVERING THE PERIOD 1950-1983. E.S. Connor and A.M. Bruce. John Wiley and Sons, Incorporated, 605 Third Avenue, New York, NY 10158. (800) 526-5368 or (212) 850-6000. 1984. $63.95.

WATER POLLUTION: A GUIDE TO INFORMATION SOURCES. Allen W. Knight and Mary Ann Simmons, editors. Gale Research Company, Book Tower, Detroit, MI 48226. (800) 521-0707. 1980. $62.00.

DIRECTORIES AND BIOGRAPHICAL SOURCES

AMERICAN WATER WORKS ASSOCIATION - OFFICERS AND COMMITTEE DIRECTORY. American Water Works Association, 6666 West Quincy Avenue, Denver, CO 80235. (303) 794-7711. Annual. $15.00.

JOURNAL OF THE AMERICAN WATER WORKS ASSOCIATION BUYERS GUIDE ISSUE. American Water Works Association, 6666 West Quincy Avenue, Denver, CO 80235. (303) 794-7711. Annual, with subscription.

PUBLIC WORKS MANUAL. Public Works Journal Corporation, 200 South Broad Street, Ridgewood, NJ 07451. (201) 445-5800. Annual. $20.00.

SCIENTIFIC AND TECHNICAL ORGANIZATIONS AND AGENCIES DIRECTORY. Gale Research Company, Book Tower, Detroit, MI 48226. (800) 521-0707. Second edition. Two volumes. 1987. $185.00 set.

WATER QUALITY ASSOCIATION DIRECTORY. Water Quality Association, 4151 Naperville Road, Lisle, IL 60532. (312) 369-1600. Annual. Available to members only.

GENERAL WORKS

CHEMICAL PROCESSES IN WASTE WATER TREATMENT. W.J. Eilbeck and G. Mattock. E. Horwood. Distributed by: john Wiley and Sons, Incorporated, 605 Third Avenue, New York, Ny 10158. (800) 526-5368 or (212) 850-6000. 1987. $85.00.

ENVIRONMENTAL ENGINEERING AND SANITATION. J.A. Salvato. John Wiley and Sons, Incorporated, 605 Third Avenue, New York, NY 10158. (800) 526-5368 or (212) 850-6000. Third edition. 1982. $70.00.

PROCESSING AND USE OF SEWAGE SLUDGE. P. L'Hermite and H. Ott, editors. Kluwer Academic, 190 Old Derby Street, Hingham, MA 02043. (617) 749-5262. 1984. $79.50.

SEWAGE SLUDGE STABILIZATION AND DISINFECTION. A.M. Bruce. John Wiley and Sons, Incorporated, 605 Third Avenue, New York, NY 10158. (800) 526-5368 or (212) 850-6000. 1984. $75.00.

UTILIZATION OF SEWAGE SLUDGE ON LAND: Rates of Application and Long-term Effects on Metal. S. Berglund and others, editors. Kluwer Academic, 190 Old Derby Street, Hingham, MA 02043. (617) 749-5262. 1984. $39.00.

WATER AND WASTE-WATER TECHNOLOGY. M.J. Hammer. John Wiley and Sons, Incorporated, 605 Third Avenue, New York, NY 10158. (800) 526-5368 or (212) 850-6000. Second edition. 1986. $35.95.

WATER SUPPLY AND POLLUTION CONTROL. Warren Viessman and Mark J. Hammer. Harper and Row Publishers, Incorporated, 10 East 53rd Street, New York, NY 10022. (800) 242-7737. Fourth edition. 1985. $45.00.

WATER TREATMENT PRINCIPLES AND DESIGN. J.M. Montgomery. John Wiley and Sons, Incorporated, 605 Third Avenue, New York, NY 10158. (800) 526-5368 or (212) 850-6000. 1985. $49.95.

HANDBOOKS AND MANUALS

WATER TREATMENT HANDBOOK. Degremont Company. John Wiley and Sons, Incorporated, 605 Third Avenue, New York, NY 10158. (800) 526-5368 or (212) 850-6000. Fifth edition. 1979. $95.00.

ONLINE DATA BASES

COMPENDEX. Engineering Information, Incorporated, 345 East 47th Street, New York, NY 10017. (800) 221-1044 or (212) 705-7615. Engineering and technical literature, 1975 to present. Inquire as to cost and availability.

ENVIROLINE. EIC Intelligence, Incorporated, 48 West 38th Street, New York, NY 10018. (212) 944-8500. Worldwide environmental literature, 1979 to present. Inquire as to online cost and availability.

NTIS. National Technical Information Service, 5285 Port Royal Road, Springfield, VA 22161. (703) 487-4630. Broad coverage of government-sponsored research reports, 1964 to present. Inquire as to cost and availability.

POLLUTION ABSTRACTS. Cambridge Scientific Abstracts, 5161 River Road, Bethesda, MD 20816. (301) 951-1400. 1970 to present. Available for online searching through DIALOG Information Services and BRS Information Technologies. Inquire as to online cost and availability.

WATER DATA BANK. United States Department of Agriculture, Agricultural Research Service, Hydrology Laboratory, Water Data Center, Room 139, Building 007, BARC-West, Beltsville, MD 20705. (301) 344-4411. Inquire as to online cost and availability.

WATERNET. American Water Works Association, 6666 West Quincy Avenue, Denver, CO 80235. (303) 794-7711. A data base providing abstracts and indexing of information published in the American Water Works Association Journal. 1971 to present. Inquire as to online cost and availability.

PERIODICALS

AMERICAN WATER WORKS ASSOCIATION. Journal. American Water Works Association, 6666 West Quincy Avenue, Denver, CO 80235. (303) 794-7711. Monthly. $50.00 per year.

ASSOCIATION OF METROPOLITAN SEWERAGE AGENCIES BULLETIN. Association of Metropolitan Sewerage Agencies, 1015 18th Street, NW, Suite 1002, Washington, DC 20036. (202) 659-9161. Monthly.

JOURNAL OF ENVIRONMENTAL ENGINEERING. American Society of Civil Engineers, 345 East 47th Street, New York, NY 10017. (212) 705-7275. 1956 to present. Bimonthly. $80.00 per year.

JOURNAL OF ENVIRONMENTAL SCIENCES. Institute of Environmental Sciences, 940 East Northwest Highway, Mount Prospect, IL 60056. (312) 255-1561. Bimonthly. $25.00 per year.

JOURNAL OF WATER RESOURCES PLANNING AND MANAGEMENT. American Society of Civil Engineers, 345 East 47th Street, New York, NY 10017-2398. (212) 705-7520. Quarterly. $56.00 per year.

THE MANAGEMENT OF WORLD WASTES. Communication Channels, Incorporated, 6255 Barfield Road, Atlanta, GA 30328. (404) 256-9800. Monthly. $27.00 per year.

PUBLIC WORKS MAGAZINE. Public Works Journal Corporation, 200 South Broad Street, Ridgewood, NJ 07451. (202) 445-5800. Monthly. 430.00 per year.

SANITARY MAINTENANCE. Trade Press Publishing Company, 2100 West Florist Avenue, Milwaukee, WI 53209. (414) 228-7701. Monthly. $30.00 per year.

SLUDGE NEWSLETTER. Business Publishers, 951 Pershing Drive, Silver Spring, MD 20910. (301) 587-6300. Biweekly. $172.72 per year.

WATER, AIR AND SOIL POLLUTION. D. Reidel Publishing Company, 190 Old Derby Street, Hingham, MA 02043. Twenty times per year. $550.00 per year.

WATER AND WASTES DIGEST. Scranton Gillette Communications, Incorporated, 380 Northwest Highway, Des Plaines, IL 60016. (312) 298-6622. Bimonthly. $10.00 per year.

WATER: ENGINEERING AND MANAGEMENT. Scranton Gillette Communications, Incorporated, 380 Northwest Highway, Des Plaines, IL 60016. (312) 298-6622. Monthly. $20.00 per year.

WATER POLLUTION CONTROL. The Bureau of National Affairs, Incorporated, 1231 25th Street, NW, Washington, DC 20037. (202) 452-4200. Biweekly. $360.00 per year.

WATER POLLUTION CONTROL FEDERATION. Journal. Water Pollution Control Federation, 2626 Pennsylvania Avenue, NW, Washington, DC 20037. (202) 337-2500. Monthly. $120.00 per year.

WATER RESEARCH. International Association on Water Pollution Research. Pergamon Press Incorporated, Journals Division, Maxwell House, Fairview Park, Elmsford, NY 10523. (914) 592-7700. Monthly. $550.00 per year.

RESEARCH CENTERS AND INSTITUTES

CALIFORNIA INSTITUTE OF TECHNOLOGY. W.M. Keck Engineering Laboratory of Hydraulics and Water Resources, 1201 East California Boulevard, Pasadena, CA 91125. (818) 356-4404.

LENOX INSTITUTE FOR RESEARCH, INCORPORATED. 101 Yokum Avenue, Lenox, MA 01240. (413) 637-3025.

PENNSYLVANIA STATE UNIVERSITY. Environmental Engineering Laboratory, 212 Sackett Building, University Park, PA 16802. (814) 863-4385.

UNIVERSITY OF ILLINOIS. Advanced Environmental Control Technology Research Center, 3230 Newmark C.E. Laboratory, 208 North Romine Street, Urbana, IL 61801. (217) 333-3822.

UNIVERSITY OF TEXAS. Institute of Environmental Health, Post Office Box 20186, Houston, TX 77225-0186. (713) 792-4425.

WASHINGTON STATE UNIVERSITY. Environmental Engineering Research Laboratory, 141 Sloan, Pullman, WA 99164. (509) 335-3175.

SMELTING

See: METALLURGY

SNOW

See also: METEOROLOGY

ABSTRACT SERVICES AND INDEXES

APPLIED SCIENCE AND TECHNOLOGY INDEX. H.W. Wilson Company, 950 University Avenue, Bronx, NY 10452. (800) 367-6670 or (212) 588-8400. Inquire as to cost and availability.

CHEMICAL ABSTRACTS. Chemical Abstracts Service, 2540 Olentangy Road, Post Office Box 3012, Columbus, OH 43210. (800) 848-6538 or (614) 421-3600. Weekly. $9200.00 per year.

CURRENT CONTENTS: PHYSICAL AND CHEMICAL SCIENCES. Institute for Scientific Information, 3501 Market Street, Philadelphia, PA 19104. (800) 523-1850 or (215) 386-0100. $272.00 per year.

GENERAL SCIENCE INDEX. H.W. Wilson Company, 950 University Avenue, Bronx, NY 10452. (800) 367-6770 or (212) 588-8400. Inquire as to cost and availability.

METEOROLOGICAL AND GEOASTROPHYSICAL ABSTRACTS. American Meteorological Society, 45 Beacon Street, Boston, MA 02108. (617) 227-2425. 1950 to present. Monthly. $450.00 per year.

SCIENCE CITATION INDEX. Institute for Scientific Information, 3501 Market Street, Philadelphia, PA 19104. (800) 523-1850 or (215) 386-0100. Inquire as to cost and availability.

ASSOCIATION AND PROFESSIONAL SOCIETIES

AMERICAN METEOROLOGICAL SOCIETY. 45 Beacon Street, Boston, MA 02108. (617) 227-2425.

INTERNATIONAL ASSOCIATION OF METEOROLOGY AND ATMOSPHERIC PHYSICS. UCAR, Post Office Box 3000, Boulder, CO 80307.

NATIONAL WEATHER ASSOCIATION. 4400 Stamp Road, Room 404, Temple Hills, MD 10748. (301) 899-3784.

UNIVERSITY CORPORATION FOR ATMOSPHERIC RESEARCH. Box 3000, 1850 Table Mesa Drive, Boulder, CO 80307. (303) 497-1000.

WEATHER MODIFICATION ASSOCIATION. Post Office Box 8116, Fresno, CA 93747. (209) 291-8466.

BIBLIOGRAPHIES

SCIENCE BOOKS AND FILMS. American Association for the Advancement of Science, 1333 H Street, NW, Washington, DC 20005. Five times per year. $20.00 per year.

SCIENTIFIC AND TECHNICAL BOOKS AND SERIALS IN PRINT 1988; AN INDEX TO LITERATURE IN SCIENCE AND TECHNOLOGY. R.R. Bowker Company, 205 East 42nd Street, New York, NY 10017. (800) 521-8110 or (212) 916-1600. $175.00.

DIRECTORIES AND BIOGRAPHICAL SOURCES

AMERICAN MEN AND WOMEN OF SCIENCE. Physical and Biological Sciences. Sixteenth edition. R.R. Bowker Company, 205 East 42nd Street, New York, NY 10017. (800) 521-8110 or (212) 916-1600. 1986. $595.00.

INTERDOC: DIRECTORY OF PUBLISHED PROCEEDINGS, SERIES. SEMT-Science/Engineering/Medicine/Technology. Interdoc Corporation, 173 Halstead Avenue, Box 326, Harrison, NY 10528. (014) 835-3506. Ten times per year. $325.00 per year.

INTERNATIONAL RESEARCH CENTERS DIRECTORY 1988-1989. Gale Research Company, Book Tower, Detroit, MI 48226. (800) 521-0707. Fourth edition. 1987. $360.00.

METEOROLOGICAL SERVICES OF THE WORLD. World Meteorological Organization. Available from: American Meteorological Society, 45 Organization. Available from: American Meteorological Society, 45 Beacon Street, Boston, MA 02108. (617) 227-2425. Annual. $35.00.

NATIONAL WEATHER SERVICE OFFICES AND STATIONS. National Oceanic and Atmospheric Administration, Department of Commerce, Silver Spring, MD 20910. (301) 427-7698. Annual. Free.

RESEARCH CENTERS DIRECTORY. Gale Research Company, Book Tower, Detroit, MI 48226. (800) 521-0707. Twelfth edition. 1987. $365.00 for set.

ENCYCLOPEDIAS AND DICTIONARIES

ENCYCLOPEDIA OF CLIMATOLOGY. John E. Oliver and Rhodes W. Fairbridge, editors. Van Nostrand Reinhold, Incorporated, 115 Fifth Avenue, New York, NY 10003. (800) 543-2681. 1987. $89.95.

GENERAL WORKS

HANDBOOK OF SNOW: PRINCIPLES, PROCESSES, MANAGEMENT AND USE. D.M. Gray and D.H. Male, editors. Pergamon Press, Incorporated, Maxwell House, Fairview Park, Elmsford, NY 10523. (914) 592-7700. 1981. $72.50.

DYNAMICS OF SNOW AND ICE MASSES. Samuel C. Colbeck. Academic Press, Incorporated, Orlando, FL 32887. (305) 345-4100. $60.50.

ONLINE DATA BASES

CA SEARCH. Chemical Abstracts Service, Post Office Box 3012, Columbus, OH 43210. Guide to chemical literature, 1972 to present. Inquire as to cost and availability.

DISSERTATION ABSTRACTS ONLINE. University Microfilms International, 300 North Zeeb Road, Ann Arbor, MI 48106. (800) 521-0600 or (313) 761-4700. Scope includes virtually all doctoral dissertations accepted at accredited American institutions from 1861 to present in 252 subject areas. Inquire as to cost and availability.

METEOROLOGICAL AND GEOASTROPHYSICAL ABSTRACTS. American Meteorological Society, 45 Beacon Street, Boston, MA 02108. (617) 227-2425. 1950 to present. Monthly. $450.00 per year.

SCISEARCH. Institute for Scientific Information, 3501 Market Street, Philadelphia, PA 19104. (800) 523-1850 or (215) 386-0100. Broad multidisciplinary title and author index to the international literature of science and technology, 1974 to present. Inquire as to cost and availability.

PERIODICALS

AGRICULTURAL AND FOREST METEOROLOGY. Elsevier Science Publishing Company, Incorporated, 52 Vanderbilt Avenue, New York, NY 10017. (212) 370-5520. 1964 to present. Monthly. $260.00 per year.

AMERICAN METEOROLOGICAL SOCIETY BULLETIN. American Meteorological Society, 45 Beacon Street, Boston, Ma 02108. (617) 227-2425. 1920 to present. Monthly. $60.00 per year.

JOURNAL OF CLIMATE AND APPLIED METEOROLOGY. American Meteorological Society, 45 Beacon Street, Boston, MA 02108. (617) 227-2425. 1962 to present. Monthly. $120.00 per year.

JOURNAL OF THE ATMOSPHERIC SCIENCES. American Meteorological Society, 45 Beacon Street, Boston, MA 02108. (617) 227-2425. 1944 to present. Semimonthly. $220.00 per year.

MONTHLY WEATHER REVIEW. American Meteorological Society, 45 Beacon Street, Boston, MA 02108. (617) 227-2425. 1872 to present. Monthly. $120.00 per year.

NATIONAL WEATHER DIGEST. National Weather Association, 4400 Stamp Road, Room 404, Temple Hills, MD 20748. (301) 899-3784. 1976 to present. Quarterly. $20.00 per year.

WEATHER. Royal Meteorological Society, James Glaisher House, Grenville Place, Bracknell Berkshire, RG12 1BX, England. 1946 to present. $30.00 per year.

WEATHERWISE. Heldref Publications, 4000 Albemarle Street, NW, Washington, DC 20016. (202) 362-6445. 1948 to present. Bimonthly. $20.00 per year.

RESEARCH CENTERS AND INSTITUTES

MICHIGAN TECHNOLOGICAL UNIVERSITY. Institute of Snow Research, Houghton, MI 49931. (906) 487-2750.

NATIONAL CENTER FOR ATMOSPHERIC RESEARCH. Box 3000, Boulder, CO 80307. (303) 497-1000.

UNIVERSITY OF NEVADA. Atmospheric Water Resources Laboratory. Desert Research Institute, Post Office Box 60220, Reno, NV 89506. (702) 972-1676.

SOFTWARE

See also: ARTIFICIAL INTELLIGENCE, COMPUTER OPERATING SYSTEMS, COMPUTER PROGRAMMING, COMPUTERS, PARALLEL COMPUTERS, PROGRAMMING LANGUAGES, SOFTWARE ENGINEERING, SYSTEMS ANALYSIS

ABSTRACT SERVICES AND INDEXES

APPLIED SCIENCE AND TECHNOLOGY INDEX. H.W. Wilson and Company, 950 University Avenue, Bronx, NY 10452. (800) 367-6670 or (212) 588-8400. Monthly. Inquire as to cost and availability.

COMPUTER AND CONTROL ABSTRACTS. Institute of Electrical Engineers. Available from: Institute of Electrical and Electronics Engineers. IEEE Service Center, 445 Hoes Lane, Piscataway, NJ 08854. Semimonthly. $775.00 per year.

COMPUTER AND INFORMATION SYSTEMS: AN ABSTRACT JOURNAL PERTAINING TO THE THEORY, DESIGN, FABRICATION AND APPLICATION OF COMPUTER AND INFORMATION SYSTEMS. Cambridge Scientific Abstracts, 5161 River Road, Bethesda, MD 20816. 1972 to present. Semi-monthly. Inquire as to cost and availability.

COMPUTER CONTENTS: THE BIWEEKLY COMPILATION OF TABLES OF CONTENTS FROM COMPUTER, ELECTRONIC AND TELECOMMUNICATIONS MAGAZINES, JOURNALS AND TRANACTIONS. Find/SVP, 500 Fifth Avenue, New York, NY 101110. (800) 346-3787 or (212) 354-2424. Biweekly. $115.00 per year.

COMPUTER LITERATURE INDEX. Applied Computer Research, Inc., P.O. Box 9280, Phoenix, AZ 85068. (602) 995-5929. Quarterly. $125.00 per year.

COMPUTER PROGRAMS ABSTRACTS. U.S. National Aeronautics and Space Administration. Available from: U.S. Government Printing Office, Washington, DC 20402. Quarterly. $10.00 per year.

COMPUTING REVIEWS. Association of Computing Machinery, 11 West 42nd Street, New York, NY 10036. (212) 869-7440. Monthly. $60.00 per year.

CONFERENCE PAPERS INDEX. Cambridge Scientific Abstracts, 5161 River Road, Bethesda, MD 20816. 1972 to present. Monthly. Inquire as to cost and availability.

CURRENT MATHEMATICAL PUBLICATIONS. American Mathematical Society, P.O. Box 6248, Providence, RI 02940. (800) 556-7774 or (401) 272-9500. 1969 to present. Seventeen times per year. $230.00 per year.

ENGINEERING INDEX MONTHLY AND AUTHOR INDEX. Engineering Information Inc., 345 East 47th Street, New York, NY 10017. (212) 705-7600. Monthly. $1560.00 per year.

INDEX TO SCIENTIFIC AND TECHNICAL PROCEEDINGS. Institute for Scientific Information, 3501 Market Street, Philadelphia, PA 19104. (800) 523-1850 or (215) 386-0100. 1978 to present. Monthly. $775.00 per year.

INDEX TO SCIENTIFIC REVIEWS. Institute for Scientific Information, 3501 Market Street, Philadelphia, PA 19104. (800) 523-1850 or (215) 386-0100. 1974 to present. Semi-annual. $550.00 per year.

MATHEMATICAL REVIEWS: A REVIEWING JOURNAL COVERING THE WORLD LITERATURE OF MATHEMATICAL RESEARCH. American Mathematical Society, P.O. Box 6248, Providence, RI 02940. (800) 7774 or (401) 272-9500. 1940 to present. Monthly. $2800.00 per year.

PCR-2: PERSONAL COMPUTER REVIEW - SQUARED. Toolbox Publications, Inc., P.O. Box 5451, 2514 Birch Creek Lane, Orchard Lake, MI 48033. 1987 to present. Bimonthly. $60.00 per year.

PHYSICS ABSTRACTS. Institution of Electrical Engineers. Available from: IEEE Service Center, 445 Hoes Lane, Piscataway, NJ 08854. 1898 to present. Bimonthly. $1700.00 per year.

PHYSICS BRIEFS. Physik Verlag GmbH, Postfach 1260/1280, D-6940 Weinheim, West Germany. (212) 661-9404. 1920 to present. Twenty-six times per year. $1250.00 per year.

SCIENCE CITATION INDEX. Institute for Scientific Information, 3501 Market Street, Philadelphia, PA 19104. (800) 523-1850 or (215) 386-0100. Six times per year. $6200. 00 per year.

ANNUAL REVIEWS AND YEARBOOKS

ADVANCES IN COMPUTERS. Academic Press, Inc., 6277 Sea Harbor Drive, Orlando, FL 32821. (800) 321-5068. Yearly. Approximately $50.00 per volume.

COMPUTER PUBLISHERS AND PUBLICATIONS 1988-89: AN INTERNATIONAL DIRECTORY AND YEARBOOK. Efrem Sigel and Frederica Evan, editors. Gale Research Company, Book Tower, Detroit, MI 48226. (800) 521-0707. Third edition. $140.00.

ASSOCIATIONS AND PROFESSIONAL SOCIETIES

AMERICAN FEDERATION OF INFORMATION PROCESSING SOCIETIES. 1899 Preston White Drive, Reston, VA 22091. (703) 620-8900.

ASSOCIATION OF COMPUTER PROGRAMMERS AND ANALYSTS. 2108-C Gallows Road, Vienna, VA 22180. (703) 790-0490.

ASSOCIATION OF COMPUTING MACHINERY (ACM). 11 West 42nd Street, New York, NY 10036. (212) 869-7440.

IEEE COMPUTER SOCIETY. 1730 Massachusetts Avenue, N.W., Washington, DC 20036. (202) 371-0101.

INSTITUTE OF ELECTRICAL AND ELECTRONICS ENGINEERS. IEEE Service Center, 445 Hoes Lane, Piscataway, NJ 08854.

MACHINE VISION ASSOCIATION. P.O. Box 930, One SME Drive, Dearborn, MI 48121. (313) 271-1500.

SOCIETY FOR COMPUTER SIMULATION. P.O. Box 17900, San Diego, CA 92117. (619) 277-3888.

SOCIETY FOR INFORMATION DISPLAY. 8055 Manchester Avenue, Suite 615, Playa Del Rey, CA 90293. (213) 305-1502.

BIBLIOGRAPHIES

NEW TECHNICAL BOOKS: A SELECTIVE LIST WITH DESCRIPTIVE ANNOTATIONS. New York Public Library, Science and Technology Research Center, Fifth Avenue and 42nd Street, New York, NY 10018. (212) 930-0800. 1915 to present.

Monthly. $15.00 per year.

SCIENCE BOOKS AND FILMS. American Association for the Advancement of Science, 1333 H Street, NW, Washington, DC 20005. (202) 326-6454. Five times per year. $20.00 per year.

SCIENTIFIC AND TECHNICAL BOOKS AND SERIALS IN PRINT 1988; AN INDEX TO LITERATURE IN SCIENCE AND TECHNOLOGY. R.R. Bowker Company, 205 East 42nd Street, New York, NY 10017. (800) 521-8110. $175.00.

DIRECTORIES AND BIOGRAPHICAL SOURCES

AMERICAN MEN AND WOMEN OF SCIENCE. R.R. Bowker, Inc., Order Department, 245 West 17th Street, New York, NY 10011. (800) 521-8110. Eight volumes. 1986. $595.00 for set.

AMERICAN SOCIETY FOR INFORMATION SCIENCE HANDBOOK AND DIRECTORY. American Society for Information Science, 1424 16th Street, N.W., Suite 404, Washington, DC 20036. (202) 462-1000. $50.00.

COMPUTERS AND COMPUTING INFORMATION RESOURCES DIRECTORY. Martin Connors, editor. Gale Research Company, Book Tower, Detroit, MI 48226. (800) 521-0707. 1987. $165.00. Supplement available at $85.00.

INTERNATIONAL RESEARCH CENTERS DIRECTORY 1988-89. Darren L. Smith, editor. Gale Research Company, Book Tower, Detroit, MI 48226. (800) 521-0707. 4th edition. 1987. $360.00.

1987 DIRECTORY OF ENGINEERING SOCIETIES AND RELATED ORGANIZATIONS. Gordon Davis, editor. Hemisphere Publishing Corporation, 1010 Vermont Avenue, NW, Washington, DC 20005. (800) 526-0275. 12th edition. 1987. $100.00.

RESEARCH CENTERS DIRECTORY 1988. Gale Research Company, Book Tower, Detroit, MI 48226. (800) 521-0707. 12th edition. 1987. $365.00 for set.

SCIENTIFIC AND TECHNICAL ORGANIZATIONS AND AGENCIES DIRECTORY. Margaret Labash Young, editor. Gale Research Company, Book Tower, Detroit, MI 48226. (800) 521-0707. 2nd edition. 1987. $185.00.

WHO'S WHO IN ENGINEERING. Gordon Davis, editor. Hemisphere Publishing Corporation, 1010 Vermont Avenue, NW, Washington, DC 20005. (800) 526-0275. 6th edition. 1985. $200.00.

ENCYCLOPEDIAS AND DICTIONARIES

COMPUTER AND TELECOMMUNICATIONS ACRONYMS. Julie E. Towell and Helen E. Sheppard, editors. Gale Research Company, Book Tower, Detroit, MI 48226. (800) 521-0707. 1986. $60.00.

DICTIONARY OF COMPUTING. Oxford University Press, 200 Madison Avenue, New York, NY 10016. (800) 458-5833. Second edition. 1986. $29.95.

ENCYCLOPEDIA OF INFORMATION SYSTEMS AND SERVICES 1988. Amy Lucas and Annette Novallo, editors. Gale Research Company, Book Tower, Detroit, MI 48226. (800) 521-0707. 8th edition. 1987. $400.00 for set.

PRENTICE-HALL ENCYCLOPEDIA OF INFORMATION TECHNOLOGY. Robert A. Edmunds. Prentice-Hall Publishing, Inc., Englewood Cliffs, NJ 07632. (800) 562-0245. 1987. $49.95.

SOFTWARE ENCYCLOPEDIA. R.R. Bowker Company, 205 East 42nd Street, New York, NY 10017. (800) 521-8110. Two volumes. 1987. $125.00 for set.

GENERAL WORKS

CONCURRENT PROGRAMMING FOR SOFTWARE ENGINEERS. Dick Whiddett. Halsted Press, available from: John Wiley and Sons, Inc., 605 Third Avenue, New York, NY 10158. (800) 526-5368. 1987. $39.95.

GENERAL PURPOSE PROGRAMMING LANGUAGES. John V. Cugini. Petrocelli Books, available from: Van Nostrand Reinhold Company, Inc., 135 West 50th Street, New York, NY 10020. (800) 543-2681. 1986. $24.95.

HIGH LEVEL LANGUAGE AND SOFTWARE APPLICATIONS REFERENCE. W.J. Birnes. McGraw-Hill Book Company, 1221 Avenue of the Americas, New York, NY 10020. (212) 512-2000. 1988. $29.95.

MICROPROCESSORS AND MICROCOMPUTERS: HARDWARE AND SOFTWARE. R.J. Tocci and L.P. Laskowski. Prentice-Hall Publishing, Inc., Englewood Cliffs, NJ 07632. (800) 562-0245. Third edition. 1987. $37.95.

PRINCIPLES OF PROGRAMMING LANGUAGES: DESIGN, EVALUATION AND IMPLEMENTATION. Bruce MacLennan. Holt, Rinehart and Winston, Inc., 383 Madison Avenue, New York, NY 10017. (212) 872-2000. Second edition. 1986. $38.75.

PROGRAMMING LANGUAGES. A.B. Tucker. McGraw-Hill Book Company, 1221 Avenue of the Americas, New York, NY 10020. (212) 512-2000. Second edition. 1986. $43.95.

PROGRAMMING LANGUAGES: A GRAND TOUR. Ellis Horowitz, editor. Computer Science Press, 11 Taft Court, Rockville, MD 20850. (800) 242-7737. Third Edition. 1987. $39.95.

SOFTWARE ENGINEERING IN C. Philip E. Margolis and Peter A. Darnell. Springer-Verlag New York, Inc., 175 Fifth Avenue, New York, NY 10010. (800) 526-7254. 1988. $49.50.

HANDBOOKS AND MANUALS

HANDBOOK OF SOFTWARE ENGINEERING. Charles R. Vick, editor. Van Nostrand Reinhold Company, Inc., 135 West 50th Street, New York, NY 10020. (800) 543-2681. 1984. $66.95.

ONLINE DATA BASES

COMPENDEX. Engineering Information, Inc., 345 East 47th Street, New York, NY 10017. (800) 221-1044 or (212) 705-7615. Engineering and technical literature, 1975 to present. Inquire as to online cost and availability.

DISSERTATION ABSTRACTS ONLINE. University Microfilms International, 300 North Zeeb Road, Ann Arbor, MI 48106. (800) 521-0600 or (313) 761-4700. Scope includes virtually all doctoral dissertations accepted at accredited American institutions from 1861 to present in over 250 subject areas. Inquire as to online cost and availability.

INSPEC. INSPEC Marketing Department, Institution of Electrical Engineers. Available from IEEE Service Center, 445 Hoes Lane, Piscataway, NJ 08854. (201) 981-0060. Online version of Physics Abstracts. Inquire as to online cost and availability.

MATHFILE. American Mathematical Society, P.O. Box 6248, Providence, RI 02940. (800) 556-7774 or (401) 272-9500. An online version of Mathematical Reviews. 1973 to present. Inquire as to online cost and availability.

NTIS. National Technical Information Service, 5285 Port Royal Road, Springfield, VA 22161. (703) 487-4630. Broad coverage of government sponsored research reports, 1964 to present. Inquire as to online cost and availability.

SCISEARCH. Institute for Scientific Information, 3501 Market Street, Philadelphia, PA 19104. (800) 523-1850 or (215) 386-0100. Broad multidisciplinary title and author index to the international literature of science and technology, 1974 to present. Inquire as to online cost and availability.

WILSONLINE. H.W. Wilson and Company, 950 University Avenue, Bronx, NY 10452. (800) 367-6770 or (212) 588-8400. Makes available online versions of the H.W. Wilson indexes including Applied Science and Technology Index, Business Periodicals Index and Readers' Guide to Periodical Literature. Approximately 1980 to present. Inquire as to online cost and availability.

PERIODICALS

ACM TRANSACTIONS ON MATHEMATICAL SOFTWARE. Association of Computing Machinery, 11 West 42nd Street, New York, NY 10036. (212) 869-7440. 1975 to present. Quarterly. $55.00 per year.

ACM TRANSACTIONS ON PROGRAMMING LANGUAGES AND SYSTEMS. Association of Computing Machinery, 11 West 42nd Street, New York, NY 10036. (212) 869-7440. 1979 to present. Quarterly. $55.00 per year.

ADVANCES IN ENGINEERING SOFTWARE. CML Publications, 400 West Cummings Park, Suite 6200, Woburn, MA 01801. (617) 933-7374. 1979 to present. Quarterly. $130.00 per year.

BYTE. Byte Publications, Inc., 70 Main Street, Petersborough, NH 03458. (603) 924-9281. Monthly. $21.00 per year.

C JOURNAL. InfoPro Systems, 3108 Route 10, Denville, NJ 07834. (201) 989-0570. 1985 to present. Quarterly. $28.00 per year.

COMMUNICATIONS OF THE ACM. Association of Computing Machinery, 11 West 42nd Street, New York, NY 10036. (212) 869-7440. Monthly. $80.00 per year.

COMPUTER LANGUAGES. Pergamon Press, Inc., Maxwell House, Fairview Park, Elmsford, NY 10523. (914) 592-7700. 1976 to present. Quarterly. $195.00 per year.

DR. DOBB'S JOURNAL OF SOFTWARE TOOLS. M & T Publishing, Inc., 2464 Embarcadero Way, Palo Alto, CA 94303. (415) 424-0600. 1976 to present. Monthly. $25.00 per year.

IEEE SOFTWARE. Institution of Electrical and Electronics Engineers. IEEE Service Center, 445 Hoes Lane, Piscataway, NJ 08854. 1984 to present. Quarterly. $15.00 per issue.

IEEE TRANSACTIONS ON SOFTWARE ENGINEERING. Institute of Electrical and Electronics Engineers. IEEE Service Center, 445 Hoes Lane, Piscataway, NJ 08854. 1975 to present. Monthly. $160.00 per year.

INTERFACE. International Computer Programs, Inc., 9000 Keystone Crossing, Indianapolis, IN 46240. (317) 844-7461. 1975 to present. Quarterly. $10.00 per year.

JOURNAL OF LOGIC PROGRAMMING. Elsevier Science Publishing Company, Inc., 52 Vanderbilt Avenue, New York, NY 10017. (212) 370-5520. 1984 to present. Quarterly. $95.00 per year.

JOURNAL OF PASCAL, ADA, AND MODULA-2. John Wiley and Sons, Inc., 605 Third Avenue, New York, NY 10158. (800) 526-5368. 1982 to present. Bimonthly. $20.00 per year.

JOURNAL OF SYSTEMS AND SOFTWARE. Elsevier Science Publishing Company, Inc., 52 Vanderbilt Avenue, New York, NY 10017. (212) 370-5520. 1979 to present. Quarterly. $95.00 per year.

PASCAL AND MODULA-2. Pascal Users Group, Box 538, Chesterland, OH 44026. (216) 729-3227. Quarterly. $25.00 per year.

SIGPLAN NOTICES. Association of Computing Machinery Special Interest Group on Programming Languages, 11 West 42nd Street, New York, NY 10036. (212) 869-7440. 1965 to present. Monthly. $25.00 per year.

SIGSOFT SOFTWARE ENGINEERING NOTICES. Association of Computing Machinery Special Interest Group on Software Engineering. 11 West 42nd Street, New York, NY 10036. (212) 869-7440. Quarterly. $12.00 per year.

SOFTWARE DEVELOPER'S MONTHLY. SourceView Press, 835 Castro Street, Martinez, CA 94553. (415) 228-6220. 1985 to present. Monthly. $144.00 per year.

SOFTWARE ENGINEERING JOURNAL. Institute of Electrical Engineers, Savoy Place, London, WC2R OBL, England. 1981 to present. Bimonthly. $85.00 per year.

SOFTWARE PRACTICE AND EXPERIENCE. John Wiley and Sons, Inc., 605 Third Avenue, New York, NY 10158. (800) 526-5368. 1971 to present. Monthly. $260.00 per year.

UNIX REVIEW (SAN FRANCISCO). Miller Freeman Publications, Inc., 500 Howard Street, San Francisco, CA 94105. (415) 397-1881. 1983 to present. Monthly. $35.00 per year.

RESEARCH CENTERS AND INSTITUTES

INSTITUTE FOR INFORMATION SCIENCE AND TECHNOLOGY. George Washington University, 801 22nd Street, N.W., Washington, DC 20052. (202) 676-4921.

RESEARCH INSTITUTE FOR COMPUTING AND INFORMATION SYSTEMS. University of Houston at Clear Lake, 2700 Bay Area Boulevard, Houston, TX 77058. (713) 488-9392.

SOFTWARE ENGINEERING INSTITUTE. Carnegie-Mellon University, Pittsburgh, PA 15213. (412) 268-6700.

SOFTWARE ENGINEERING RESEARCH INSTITUTE. Georgia Institute of Technology, 258 Fourth Street, N.W., Atlanta, GA 30332. (404) 894-3180.

SOFTWARE ENGINEERING

See also: ARTIFICIAL INTELLIGENCE, COMPUTER PROGRAMMING, COMPUTER OPERATING SYSTEMS, COMPUTERS, PARALLEL COMPUTERS, PROGRAMMING LANGUAGES, SOFTWARE, SYSTEMS ANALYSIS, SYSTEMS ENGINEERING

ABSTRACT SERVICES AND INDEXES

APPLIED SCIENCE AND TECHNOLOGY INDEX. H.W. Wilson and Company, 950 University Avenue, Bronx, NY 10452. (800) 367-6670 or (212) 588-8400. Monthly. Inquire as to cost and availability.

COMPUTER AND CONTROL ABSTRACTS. Institute of Electrical Engineers. Available from: Institute of Electrical and Electronics Engineers. IEEE Service Center, 445 Hoes Lane, Piscataway, NJ 08854. Semimonthly. $775.00 per year.

COMPUTER AND INFORMATION SYSTEMS: AN ABSTRACT JOURNAL PERTAINING TO THE THEORY, DESIGN, FABRICATION AND APPLICATION OF COMPUTER AND INFORMATION SYSTEMS. Cambridge Scientific Abstracts, 5161 River Road, Bethesda, MD 20816. 1972 to present. Semi-monthly. Inquire as to cost and availability.

COMPUTER CONTENTS: THE BIWEEKLY COMPILATION OF TABLES OF CONTENTS FROM COMPUTER, ELECTRONIC AND TELECOMMUNICATIONS MAGAZINES, JOURNALS AND TRANACTIONS. Find/SVP, 500 Fifth Avenue, New York, NY 101110. (800) 346-3787 or (212) 354-2424. Biweekly. $115.00 per year.

COMPUTER LITERATURE INDEX. Applied Computer Research, Inc., P.O. Box 9280, Phoenix, AZ 85068. (602) 995-5929. Quarterly. $125.00 per year.

COMPUTER PROGRAMS ABSTRACTS. U.S. National Aeronautics and Space Administration. Available from: U.S. Government Printing Office, Washington, DC 20402. Quarterly. $10.00 per year.

COMPUTING REVIEWS. Association of Computing Machinery, 11 West 42nd Street, New York, NY 10036. (212) 869-7440. Monthly. $60.00 per year.

CONFERENCE PAPERS INDEX. Cambridge Scientific Abstracts, 5161 River Road, Bethesda, MD 20816. 1972 to present. Monthly. Inquire as to cost and availability.

CURRENT MATHEMATICAL PUBLICATIONS. American Mathematical Society, P.O. Box 6248, Providence, RI 02940. (800) 556-7774 or (401) 272-9500. 1969 to present. Seventeen times per year. $230.00 per year.

ENGINEERING INDEX MONTHLY AND AUTHOR INDEX. Engineering Information Inc., 345 East 47th Street, New York, NY 10017. (212) 705-7600. Monthly. $1560.00 per year.

INDEX TO SCIENTIFIC AND TECHNICAL PROCEEDINGS. Institute for Scientific Information, 3501 Market Street, Philadelphia, PA 19104. (800) 523-1850 or (215) 386-0100. 1978 to present. Monthly. $775.00 per year.

INDEX TO SCIENTIFIC REVIEWS. Institute for Scientific Information, 3501 Market Street, Philadelphia, PA 19104. (800) 523-1850 or (215) 386-0100. 1974 to present. Semi-annual. $550.00 per year.

MATHEMATICAL REVIEWS: A REVIEWING JOURNAL COVERING THE WORLD LITERATURE OF MATHEMATICAL RESEARCH. American Mathematical Society, P.O. Box 6248, Providence, RI 02940. (800) 7774 or (401) 272-9500. 1940 to present. Monthly. $2800.00 per year.

PCR-2: PERSONAL COMPUTER REVIEW - SQUARED. Toolbox Publications, Inc., P.O. Box 5451, 2514 Birch Creek Lane, Orchard Lake, MI 48033. 1987 to present. Bimonthly. $60.00 per year.

PHYSICS ABSTRACTS. Institution of Electrical Engineers. Available from: IEEE Service Center, 445 Hoes Lane, Piscataway, NJ 08854. 1898 to present. Bimonthly. $1700.00 per year.

PHYSICS BRIEFS. Physik Verlag GmbH, Postfach 1260/1280, D-6940 Weinheim, West Germany. (212) 661-9404. 1920 to present. Twenty-six times per year. $1250.00 per year.

SCIENCE CITATION INDEX. Institute for Scientific Information, 3501 Market Street, Philadelphia, PA 19104. (800) 523-1850 or (215) 386-0100. Six times per year. $6200. 00 per year.

ANNUAL REVIEWS AND YEARBOOKS

ADVANCES IN COMPUTERS. Academic Press, Inc., 6277 Sea Harbor Drive, Orlando, FL 32821. (800) 321-5068. Yearly. Approximately $50.00 per volume.

COMPUTER PUBLISHERS AND PUBLICATIONS 1988-89: AN INTERNATIONAL DIRECTORY AND YEARBOOK. Efrem Sigel and Frederica Evan, editors. Gale Research Company, Book Tower, Detroit, MI 48226. (800) 521-0707. Third edition. $140.00.

ASSOCIATIONS AND PROFESSIONAL SOCIETIES

AMERICAN FEDERATION OF INFORMATION PROCESSING SOCIETIES. 1899 Preston White Drive, Reston, VA 22091. (703) 620-8900.

ASSOCIATION OF COMPUTER PROGRAMMERS AND ANALYSTS. 2108-C Gallows Road, Vienna, VA 22180. (703) 790-0490.

ASSOCIATION OF COMPUTING MACHINERY (ACM). 11 West 42nd Street, New York, NY 10036. (212) 869-7440.

IEEE COMPUTER SOCIETY. 1730 Massachusetts Avenue, N.W., Washington, DC 20036. (202) 371-0101.

INSTITUTE OF ELECTRICAL AND ELECTRONICS ENGINEERS. IEEE Service Center, 445 Hoes Lane, Piscataway, NJ 08854.

SOCIETY FOR COMPUTER SIMULATION. P.O. Box 17900, San Diego, CA 92117. (619) 277-3888.

DIRECTORIES AND BIOGRAPHICAL SOURCES

AMERICAN MEN AND WOMEN OF SCIENCE. R.R. Bowker, Inc., Order Department, 245 West 17th Street, New York, NY 10011. (800) 521-8110. Eight volumes. 1986. $595.00 for set.

AMERICAN SOCIETY FOR INFORMATION SCIENCE HANDBOOK AND DIRECTORY. American Society for Information Science, 1424 16th Street, N.W., Suite 404, Washington, DC 20036. (202) 462-1000. $50.00.

COMPUTERS AND COMPUTING INFORMATION RESOURCES DIRECTORY. Martin Connors, editor. Gale Research Company, Book Tower, Detroit, MI 48226. (800) 521-0707. 1987. $165.00. Supplement available at $85.00.

INTERNATIONAL RESEARCH CENTERS DIRECTORY 1988-89. Darren L. Smith, editor. Gale Research Company, Book Tower, Detroit, MI 48226. (800) 521-0707. 4th edition. 1987. $360.00.

1987 DIRECTORY OF ENGINEERING SOCIETIES AND RELATED ORGANIZATIONS. Gordon Davis, editor. Hemisphere Publishing Corporation, 1010 Vermont Avenue, NW, Washington, DC 20005. (800) 526-0275. 12th edition. 1987. $100.00.

RESEARCH CENTERS DIRECTORY 1988. Gale Research Company, Book Tower, Detroit, MI 48226. (800) 521-0707. 12th edition. 1987. $365.00 for set.

SCIENTIFIC AND TECHNICAL ORGANIZATIONS AND AGENCIES DIRECTORY. Margaret Labash Young, editor. Gale Research Company, Book Tower, Detroit, MI 48226. (800) 521-0707. 2nd edition. 1987. $185.00.

WHO'S WHO IN ENGINEERING. Gordon Davis, editor. Hemisphere Publishing Corporation, 1010 Vermont Avenue, NW, Washington, DC 20005. (800) 526-0275. 6th edition. 1985. $200.00.

ENCYCLOPEDIAS AND DICTIONARIES

COMPUTER AND TELECOMMUNICATIONS ACRONYMS. Julie E. Towell and Helen E. Sheppard, editors. Gale Research Company, Book Tower, Detroit, MI 48226. (800) 521-0707. 1986. $60.00.

DICTIONARY OF COMPUTING. Oxford University Press, 200 Madison Avenue, New York, NY 10016. (800) 458-5833. Second edition. 1986. $29.95.

ENCYCLOPEDIA OF INFORMATION SYSTEMS AND SERVICES 1988. Amy Lucas and Annette Novallo, editors. Gale Research Company, Book Tower, Detroit, MI 48226. (800) 521-0707. 8th edition. 1987. $400.00 for set.

PRENTICE-HALL ENCYCLOPEDIA OF INFORMATION TECHNOLOGY. Robert A. Edmunds. Prentice-Hall Publishing, Inc., Englewood Cliffs, NJ 07632. (800) 562-0245. 1987. $49.95.

SOFTWARE ENCYCLOPEDIA. R.R. Bowker Company, 205 East 42nd Street, New York, NY 10017. (800) 521-8110. Tow volumes. 1987. $125.00 for set.

GENERAL WORKS

CONCURRENT PROGRAMMING FOR SOFTWARE ENGINEERS. Dick Whiddett. Halsted Press, available from: John Wiley and Sons, Inc., 605 Third Avenue, New York, NY 10158. (800) 526-5368. 1987. $39.95.

GENERAL PURPOSE PROGRAMMING LANGUAGES. John V. Cugini. Petrocelli Books, available from: Van Nostrand Reinhold Company, Inc., 135 West 50th Street, New York, NY 10020. (800) 543-2681. 1986. $24.95.

HIGH LEVEL LANGUAGE AND SOFTWARE APPLICATIONS REFERENCE. W.J. Birnes. McGraw-Hill Book Company, 1221 Avenue of the Americas, New York, NY 10020. (212) 512-2000. 1988. $29.95.

MICROPROCESSORS AND MICROCOMPUTERS: HARDWARE AND SOFTWARE. R.J. Tocci and L.P. Laskowski. Prentice-Hall Publishing, Inc., Englewood Cliffs, NJ 07632. (800) 562-0245. Third edition. 1987. $37.95.

PRINCIPLES OF PROGRAMMING LANGUAGES: DESIGN, EVALUATION AND IMPLEMENTATION. Bruce MacLennan. Holt, Rinehart and Winston, Inc., 383 Madison Avenue, New York, NY 10017. (212) 872-2000. Second edition. 1986. $38.75.

SOFTWARE ENGINEERING IN C. Philip E. Margolis and Peter A. Darnell. Springer-Verlag New York, Inc., 175 Fifth Avenue, New York, NY 10010. (800) 526-7254. 1988. $49.50.

HANDBOOKS AND MANUALS

HANDBOOK OF SOFTWARE ENGINEERING. Charles R. Vick, editor. Van Nostrand Reinhold Company, Inc., 135 West 50th Street, New York, NY 10020. (800) 543-2681. 1984. $66.95.

ONLINE DATA BASES

COMPENDEX. Engineering Information, Inc., 345 East 47th Street, New York, NY 10017. (800) 221-1044 or (212) 705-7615. Engineering and technical literature, 1975 to present. Inquire as to online cost and availability.

DISSERTATION ABSTRACTS ONLINE. University Microfilms International, 300 North Zeeb Road, Ann Arbor, MI 48106. (800) 521-0600 or (313) 761-4700. Scope includes virtually all doctoral dissertations accepted at accredited American institutions from 1861 to present in over 250 subject areas. Inquire as to online cost and availability.

INSPEC. INSPEC Marketing Department, Institution of Electrical Engineers. Available from IEEE Service Center, 445 Hoes Lane, Piscataway, NJ 08854. (201) 981-0060. Online version of Physics Abstracts. Inquire as to online cost and availability.

MATHFILE. American Mathematical Society, P.O. Box 6248, Providence, RI 02940. (800) 556-7774 or (401) 272-9500. An online version of Mathematical Reviews. 1973 to present. Inquire as to online cost and availability.

NTIS. National Technical Information Service, 5285 Port Royal Road, Springfield, VA 22161. (703) 487-4630. Broad coverage of government sponsored research reports, 1964 to present. Inquire as to online cost and availability.

SCISEARCH. Institute for Scientific Information, 3501 Market Street, Philadelphia, PA 19104. (800) 523-1850 or (215) 386-0100. Broad multidisciplinary title and author index to the international literature of science and technology, 1974 to present. Inquire as to online cost and availability.

WILSONLINE. H.W. Wilson and Company, 950 University Avenue, Bronx, NY 10452. (800) 367-6770 or (212) 588-8400. Makes available online versions of the H.W. Wilson indexes including Applied Science and Technology Index, Business Periodicals Index and Readers' Guide to Periodical Literature. Approximately 1980 to present. Inquire as to online cost and availability.

PERIODICALS

ACM TRANSACTIONS ON MATHEMATICAL SOFTWARE. Association of Computing Machinery, 11 West 42nd Street, New York, NY 10036. (212) 869-7440. 1975 to present. Quarterly. $55.00 per year.

ACM TRANSACTIONS ON PROGRAMMING LANGUAGES AND SYSTEMS. Association of Computing Machinery, 11 West 42nd Street, New York, NY 10036. (212) 869-7440. 1979 to present. Quarterly. $55.00 per year.

ADVANCES IN ENGINEERING SOFTWARE. CML Publications, 400 West Cummings Park, Suite 6200, Woburn, MA 01801. (617) 933-7374. 1979 to present. Quarterly. $130.00 per year.

BYTE. Byte Publications, Inc., 70 Main Street, Petersborough, NH 03458. (603) 924-9281. Monthly. $21.00 per year.

C JOURNAL. InfoPro Systems, 3108 Route 10, Denville, NJ 07834. (201) 989-0570. 1985 to present. Quarterly. $28.00 per year.

COMMUNICATIONS OF THE ACM. Association of Computing Machinery, 11 West 42nd Street, New York, NY 10036. (212) 869-7440. Monthly. $80.00 per year.

COMPUTER LANGUAGES. Pergamon Press, Inc., Maxwell House, Fairview Park, Elmsford, NY 10523. (914) 592-7700. 1976 to present. Quarterly. $195.00 per year.

IEEE SOFTWARE. Institution of Electrical and Electronics Engineers. IEEE Service Center, 445 Hoes Lane, Piscataway, NJ 08854. 1984 to present. Quarterly. $15.00 per issue.

IEEE TRANSACTIONS ON SOFTWARE ENGINEERING. Institute of Electrical and Electronics Engineers. IEEE Service Center, 445 Hoes Lane, Piscataway, NJ 08854. 1975 to present. Monthly. $160.00 per year.

INTERFACE. International Computer Programs, Inc., 9000 Keystone Crossing, Indianapolis, IN 46240. (317) 844-7461. 1975 to present. Quarterly. $10.00 per year.

JOURNAL OF LOGIC PROGRAMMING. Elsevier Science Publishing Company, Inc., 52 Vanderbilt Avenue, New York, NY 10017. (212) 370-5520. 1984 to present. Quarterly. $95.00 per year.

JOURNAL OF PASCAL, ADA, AND MODULA-2. John Wiley and Sons, Inc., 605 Third Avenue, New York, NY 10158. (800) 526-5368. 1982 to present. Bimonthly. $20.00 per year.

JOURNAL OF SYSTEMS AND SOFTWARE. Elsevier Science Publishing Company, Inc., 52 Vanderbilt Avenue, New York, NY 10017. (212) 370-5520. 1979 to present. Quarterly. $95.00 per year.

PASCAL AND MODULA-2. Pascal Users Group, Box 538, Chesterland, OH 44026. (216) 729-3227. Quarterly. $25.00 per year.

SIGPLAN NOTICES. Association of Computing Machinery Special Interest Group on Programming Languages, 11 West 42nd Street, New York, NY 10036. (212) 869-7440. 1965 to present. Monthly. $25.00 per year.

SIGSOFT SOFTWARE ENGINEERING NOTICES. Association of Computing Machinery Special Interest Group on Software Engineering. 11 West 42nd Street, New York, NY 10036. (212) 869-7440. Quarterly. $12.00 per year.

SOFTWARE ENGINEERING JOURNAL. Institute of Electrical Engineers, Savoy Place, London, WC2R OBL, England. 1981 to present. Bimonthly. $85.00 per year.

SOFTWARE PRACTICE AND EXPERIENCE. John Wiley and Sons, Inc., 605 Third Avenue, New York, NY 10158. (800) 526-5368. 1971 to present. Monthly. $260.00 per year.

UNIX REVIEW (SAN FRANCISCO). Miller Freeman Publications, Inc., 500 Howard Street, San Francisco, CA 94105. (415) 397-1881. 1983 to present. Monthly. $35.00 per year.

RESEARCH CENTERS AND INSTITUTES

INSTITUTE FOR INFORMATION SCIENCE AND TECHNOLOGY. George Washington University, 801 22nd Street, N.W., Washington, DC 20052. (202) 676-4921.

RESEARCH INSTITUTE FOR COMPUTING AND INFORMATION SYSTEMS. University of Houston at Clear Lake, 2700 Bay Area Boulevard, Houston, TX 77058. (713) 488-9392.

SOFTWARE ENGINEERING INSTITUTE. Carnegie-Mellon University, Pittsburgh, PA 15213. (412) 268-6700.

SOFTWARE ENGINEERING RESEARCH INSTITUTE. Georgia Institute of Technology, 258 Fourth Street, N.W., Atlanta, GA 30332. (404) 894-3180.

SOIL CHEMISTRY

See also: AGRICULTURAL ENGINEERING, CIVIL ENGINEERING, EARTHQUAKE ENGINEERING, GEOTECHNICAL ENGINEERING, ROCK MECHANICS, SOIL MECHANICS, SOIL SCIENCE

ABSTRACT SERVICES AND INDEXES

AGRICULTURAL ENGINEERING ABSTRACTS. Commonwealth Agricultural Bureaux. Distributed by: Unipub, Box 433, Murray Hill Station, New York, NY 10010. 1976 to present. Monthly. $220.00 per year.

APPLIED SCIENCE AND TECHNOLOGY INDEX. H.W. Wilson and Company, 950 University Avenue, Bronx, NY 10452. (800) 367-6670 or (212) 588-8400. Monthly. Inquire as to cost and availability.

BIBLIOGRAPHY AND INDEX OF GEOLOGY. American Geological Institute, 4220 King Street, Alexandria, VA 22302. (703) 379-2480. 1969 to present. Monthly. $1100.00 per year.

CHEMICAL ABSTRACTS. American Chemical Society, Chemical Abstracts Service, Box 3012, Columbus, OH 43210. (614) 421-3600. 1907 to present. Weekly. Inquire as to cost and availability.

CURRENT CONTENTS: ENGINEERING, TECHNOLOGY AND APPLIED SCIENCES. Institute for Scientific Information, 3501 Market Street, Philadelphia, PA 19104. (800) 523-1850 or (215) 386-0100. Weekly. $275.00 per year.

ENGINEERING INDEX MONTHLY AND AUTHOR INDEX. Engineering Information Inc., 345 East 47th Street, New York, NY 10017. (212) 705-7600. Monthly. $1560.00 per year.

INTERNATIONAL JOURNAL OF ROCK MECHANICS AND MINING SCIENCES AND GEOMECHANICS ABSTRACTS. Pergamon Press, Inc., Maxwell House, Fairview Park, Elmsford, NY 10523. (914) 592-7700. 1964 to present. Bimonthly. $385.00 per year.

IRRIGATION AND DRAINAGE ABSTRACTS. Commonwealth Agricultural Bureaux. Distributed by: Unipub, Box 433, Murray Hill Station, New York, NY 10010. 1975 to present. Quarterly. $120.00 per year.

SCIENCE CITATION INDEX. Institute for Scientific Information, 3501 Market Street, Philadelphia, PA 19104. (800) 523-1850 or (215) 386-0100. Six times per year. $6200.00 per year.

SOILS AND FERTILIZERS. Commonwealth Agricultural Bureaux. Distributed by: Unipub, Box 433, Murray Hill Station, New York, NY 10010. 1937 to present. Monthly. $450.00 per year.

ASSOCIATIONS AND PROFESSIONAL SOCIETIES

AMERICAN GEOLOGICAL INSTITUTE. 4220 King Street, Alexandria, VA 22302. (703) 379-2480.

AMERICAN SOCIETY OF AGRICULTURAL ENGINEERS. 2950 Niles Road, St. Joseph, MI 49085. (616) 429-0300.

AMERICAN SOCIETY OF AGRONOMY, Inc. 677 South Segoe Road, Madison, WI 53711. (608) 274-1212.

AMERICAN SOCIETY OF CIVIL ENGINEERS. 345 East 47th Street, New York, NY 10017. (212) 705-7420.

ASSOCIATION OF ENGINEERING GEOLOGISTS. P.O. Box 1068, Brentwood, TN 37027. (615) 377-3578.

ASSOCIATION OF SOIL AND FOUNDATION ENGINEERS. 8811 Colesville Road, Suite G106, Silver Spring, MD 20910. (301) 565-2733.

COMMISSION ON SOIL MINERALOGY. Soil and Crop Sciences Department, Texas A&M University, College Station, TX 77843. (409) 845-8322.

GEOLOGICAL SOCIETY OF AMERICA. 3300 Penrose Place, Boulder, CO 80301. (303) 447-2020.

SOIL CONSERVATION SOICETY OF AMERICA. 7515 Northeast Ankeny Road, Ankeny, IA 50021. (515) 289-2331.

SOIL SCIENCE SOCIETY OF AMERICA. 677 South Segoe Road, Madison, WI 53711. (608) 273-8080.

DIRECTORIES AND BIOGRAPHICAL SOURCES

AMERICAN MEN AND WOMEN OF SCIENCE. R.R. Bowker, Inc., Order Department, 245 West 17th Street, New York, NY 10011. (800) 521-8110. Eight volumes. 1986. $595.00 for set.

INTERNATIONAL RESEARCH CENTERS DIRECTORY 1988-89. Darren L. Smith, editor. Gale Research Company, Book Tower, Detroit, MI 48226. (800) 521-0707. 4th edition. 1987. $360.00.

1987 DIRECTORY OF ENGINEERING SOCIETIES AND RELATED ORGANIZATIONS. Gordon Davis, editor. Hemisphere Publishing Corporation, 1010 Vermont Avenue, NW, Washington, DC 20005. (800) 526-0275. 12th edition. 1987. $100.00.

RESEARCH CENTERS DIRECTORY 1988. Gale Research Company, Book Tower, Detroit, MI 48226. (800) 521-0707. 12th edition. 1987. $365.00 for set.

SCIENTIFIC AND TECHNICAL ORGANIZATIONS AND AGENCIES DIRECTORY. Margaret Labash Young, editor. Gale Research Company, Book Tower, Detroit, MI 48226. (800) 521-0707. 2nd edition. 1987. $185.00.

GENERAL WORKS

SOIL CHEMISTRY. H. Bohn and others. John Wiley and Sons, Inc., 605 Third Avenue, New York, NY 10158. (800) 526-5368. Second edition. 1985. $29.95.

SOIL CLASSIFICATION. Charles Finkl, editor. Van Nostrand Reinhold Company, Inc., 135 West 50th Street, New York, NY 10020. (800) 543-2681. 1982. $46.50.

SOIL PHYSICAL CHEMISTRY. Donald L. Sparks, editor. CRC Press, 2000 Corporate Boulevard, Boca Raton, FL 33431. (800) 272-7737. 1986. $110.00.

SOIL SCIENCE: PRINCIPLES AND PRACTICE. R.L. Hausenbuiller. William C. Brown Publications, 2460 Kerper Boulevard, Dubuque, IA 52001. (319) 589-2822. Third edition. 1985. $35.95.

SURFACE CHEMISTRY OF SOILS. G. Sposito. Oxford University Press, 200 Madison Avenue, New York, NY 10016. (800) 458-5833. 1984. $42.50.

ONLINE DATA BASES

AGRICOLA (AGRICULTURAL ON-LINE ACCESS). National Agricultural Library, Information Systems Division, 10301 Baltimore Boulevard, Beltsville, MD 20705. (301) 344-3813. An on-line bibliographic data base covering all aspects of agricultural sciences, agricultural engineering and soil science. 1970 to present. Inquire as to online cost and availability.

CA SEARCH. Chemical Abstracts Service, P.O. Box 3012, Columbus, OH 43120. (800) 848-6538 or (614) 421-3600. Comprehensive guide to chemical literature, 1972 to present. Inquire as to online cost and availability.

COMPENDEX. Engineering Information, Inc., 345 East 47th Street, New York, NY 10017. (800) 221-1044 or (212) 705-7615. Engineering and technical literature, 1975 to present. Inquire as to online cost and availability.

DISSERTATION ABSTRACTS ONLINE. University Microfilms International, 300 North Zeeb Road, Ann Arbor, MI 48106. (800) 521-0600 or (313) 761-4700. Scope includes virtually all doctoral dissertations accepted at accredited American institutions from 1861 to present in over 250 subject areas. Inquire as to online cost and availability.

GEOREF. Online version of the BIBLIOGRAPHY AND INDEX OF GEOLOGY. American Geological Institute, 4220 King Street, Alexandria, VA 22302. (703) 379-2480. 1969 to present. Inquire as to online cost and availability.

NTIS. National Technical Information Service, 5285 Port Royal Road, Springfield, VA 22161. (703) 487-4630. Broad coverage of government sponsored research reports, 1964 to present. Inquire as to online cost and availability.

SCISEARCH. Institute for Scientific Information, 3501 Market Street, Philadelphia, PA 19104. (800) 523-1850 or (215) 386-0100. Broad multidisciplinary title and author index to the international literature of science and technology, 1974 to present. Inquire as to online cost and availability.

WILSONLINE. H.W. Wilson and Company, 950 University Avenue, Bronx, NY 10452. (800) 367-6770 or (212) 588-8400. Makes available online versions of the H.W. Wilson indexes including Applied Science and Technology Index, Business Periodicals Index and Readers' Guide to Periodical Literature. Approximately 1980 to present. Inquire as to online cost and availability.

PERIODICALS

CANADIAN JOURNAL OF SOIL SCIENCE. Agricultural Institute of Canada, 151 Slater Street, Suite 907, Ottawa, ON K1P 5H4, Canada. 1921 to present. Quarterly. $56.00 per year.

COMMUNICATIONS IN SOIL SCIENCE AND PLANT ANALYSIS. Marcel Dekker Inc., 270 Madison Avenue, New York, NY 10016. (800) 228-1160. 1970 to present. Monthly. $300.00 per year.

CROPS AND SOILS MAGAZINE. American Society of Agronomy, Inc., 677 South Segoe Road, Madison, WI 53711. (608) 274-1212. 1948 to present. Monthly. $10.00 per year.

GEODERMA: INTERNATIONAL JOURNAL OF SOIL SCIENCE. Elsevier Science Publishing Company, Inc., 52 Vanderbilt Avenue, New York, NY 10017. (212) 370-5520. 1967 to present. Eight times per year. $250.00 per year.

JOURNAL OF IRRIGATION AND DRAINAGE. American Society of Civil Engineers, 345 East 47th Street, New York, NY 10017. (212) 705-7420. 1956 to present. Monthly. $115.00 per year.

JOURNAL OF SOIL AND WATER CONSERVATION. Soil Conservation Society of America, 7515 Northeast Ankeny Road, Ankeny, IA 50021. (515) 289-2331. 1946 to present. Bimonthly. $25.00 per year.

SOIL SCIENCE. Williams and Wilkins, 428 East Preston Street, Baltimore, MD 21202. (301) 528-4000. 1916 to present. Monthly. $45.00 per year to individuals.

SOIL SCIENCE SOCIETY OF AMERICA JOURNAL. Soil Science Society of America, 677 South Segoe Road, Madison, WI 53711. (608) 273-8080. 1936 to present. Bimonthly. $65.00 per year.

RESEARCH CENTERS AND INSTITUTES

AGRICULTURAL ENGINEERING DEPARTMENT. Michigan State University, East Lansing, MI 48824. (517) 353-7268.

GEOTECHNICAL/CIVIL ENGINEERING MATERIALS RESEARCH LABORATORY. Iowa State University, Town Engineering Building and Spangler Research Laboratory, Ames, IA 50011. (515) 294-7627.

SOIL AND ENVIRONMENTAL CHEMISTRY LABORATORY. Pennsylvania State University, 104 Research Unit A, University Park, PA 16802. (814) 865-1221.

SOIL MECHANICS LABORATORY. University of Michigan, Department of Mechanical Engineering and Applied Mechanics, 2246A G.G. Brown Building, Ann Arbor, MI 48109. (313) 763-0684.

SOIL MECHANICS

See also: AGRICULTURAL ENGINEERING, CIVIL ENGINEERING, EARTHQUAKE ENGINEERING, GEOTECHNICAL ENGINEERING, MINING ENGINEERING, QUARRYING, ROCK MECHANICS, SOIL CHEMISTRY, SOIL SCIENCE

ABSTRACT SERVICES AND INDEXES

APPLIED SCIENCE AND TECHNOLOGY INDEX. H.W. Wilson and Company, 950 University Avenue, Bronx, NY 10452. (800) 367-6670 or (212) 588-8400. Monthly. Inquire as to cost and availability.

BIBLIOGRAPHY AND INDEX OF GEOLOGY. American Geological Institute, 4220 King Street, Alexandria, VA 22302. (703) 379-2480. 1969 to present. Monthly. $1100.00 per year.

CURRENT CONTENTS: ENGINEERING, TECHNOLOGY AND APPLIED SCIENCES. Institute for Scientific Information, 3501 Market Street, Philadelphia, PA 19104. (800) 523-1850 or (215) 386-0100. Weekly. $275.00 per year.

ENGINEERING INDEX MONTHLY AND AUTHOR INDEX. Engineering Information Inc., 345 East 47th Street, New York, NY 10017. (212) 705-7600. Monthly. $1560.00 per year.

INTERNATIONAL JOURNAL OF ROCK MECHANICS AND MINING SCIENCES AND GEOMECHANICS ABSTRACTS. Pergamon Press, Inc., Maxwell House, Fairview Park, Elmsford, NY 10523. (914) 592-7700. 1964 to present. Bimonthly. $385.00 per year.

SCIENCE CITATION INDEX. Institute for Scientific Information, 3501 Market Street, Philadelphia, PA 19104. (800) 523-1850 or (215) 386-0100. Six times per year. $6200.00 per year.

ASSOCIATIONS AND PROFESSIONAL SOCIETIES

AMERICAN GEOLOGICAL INSTITUTE. 4220 King Street, Alexandria, VA 22302. (703) 379-2480.

AMERICAN SOCIETY OF CIVIL ENGINEERS. 345 East 47th Street, New York, NY 10017. (212) 705-7420.

ASSOCIATION OF ENGINEERING GEOLOGISTS. P.O. Box 1068, Brentwood, TN 37027. (615) 377-3578.

ASSOCIATION OF SOIL AND FOUNDATION ENGINEERS. 8811 Colesville Road, Suite G106, Silver Spring, MD 20910. (301) 565-2733.

COMMISSION ON SOIL MINERALOGY. Soil and Crop Sciences Department, Texas A&M University, College Station, TX 77843. (409) 845-8322.

GEOLOGICAL SOCIETY OF AMERICA. 3300 Penrose Place, Boulder, CO 80301. (303) 447-2020.

SOIL SCIENCE SOCIETY OF AMERICA. 677 South Segoe Road, Madison, WI 53711. (608) 273-8080.

U.S. NATIONAL SOCIETY FOR THE INTERNATIONAL SOCIETY OF SOIL MECHANICS AND FOUNDATION ENGINEERING. c/o Professor Harvey E. Wahls, Civil Engineering Department, Box 7908, North Carolina State University, Raleigh, NC 27695. (919) 737-2331.

DIRECTORIES AND BIOGRAPHICAL SOURCES

AMERICAN MEN AND WOMEN OF SCIENCE. R.R. Bowker, Inc., Order Department, 245 West 17th Street, New York, NY 10011. (800) 521-8110. Eight volumes. 1986. $595.00 for set.

INTERNATIONAL RESEARCH CENTERS DIRECTORY 1988-89. Darren L. Smith, editor. Gale Research Company, Book Tower, Detroit, MI 48226. (800) 521-0707. 4th edition. 1987. $360.00.

1987 DIRECTORY OF ENGINEERING SOCIETIES AND RELATED ORGANIZATIONS. Gordon Davis, editor. Hemisphere Publishing Corporation, 1010 Vermont Avenue, NW, Washington, DC 20005. (800) 526-0275. 12th edition. 1987. $100.00.

RESEARCH CENTERS DIRECTORY 1988. Gale Research Company, Book Tower, Detroit, MI 48226. (800) 521-0707. 12th edition. 1987. $365.00 for set.

SCIENTIFIC AND TECHNICAL ORGANIZATIONS AND AGENCIES DIRECTORY. Margaret Labash Young, editor. Gale Research Company, Book Tower, Detroit, MI 48226. (800) 521-0707. 2nd edition. 1987. $185.00.

WHO'S WHO IN ENGINEERING. Gordon Davis, editor. Hemisphere Publishing Corporation, 79 Madison Avenue, New York, NY 10016-7892. (800) 821-8312. Sixth edition. 1985. $200.00.

GENERAL WORKS

ENGINEERING PROPERTIES OF SOILS AND ROCKS. F. Bell. Butterworth's Publishing, 80 Montvale Avenue, Stoneham, MA 02180. (800) 325-4177. Second edition. 1983. $19.95 in paper.

OUTLINE OF SOIL AND ROCK MECHANICS. Pierre Habib. Cambridge University Press, 32 East 57th Street, New York, NY 10022. (800) 872-7423. 1983. $11.95 in paper.

SOIL MECHANICS. R.F. Craig. Van Nostrand Reinhold Company, Inc., 135 West 50th Street, New York, NY 10020. (800) 543-2681. Third edition. 1983. $19.95 in paper.

HANDBOOKS AND MANUALS

GEOTECHNICAL ENGINEERING PRACTICES. R.E. Hunt. McGraw-Hill Book Company, 1221 Avenue of the Americas, New York, NY 10020. (212) 512-2000. 1986. $65.00.

A GUIDE TO SOIL MECHANICS. Malcolm Bolton. John Wiley and Sons, Inc., 605 Third Avenue, New York, NY 10158. (800) 526-5368. 1980. $59.95.

HANDBOOK OF GEOLOGY IN CIVIL ENGINEERING. R.F. Legget and P. Karrow. McGraw-Hill Book Company, 1221 Avenue of the Americas, New York, NY 10020. (212) 512-2000. 1983. $82.50.

ONLINE DATA BASES

AGRICOLA (AGRICULTURAL ON-LINE ACCESS). National Agricultural Library, Information Systems Division, 10301 Baltimore Boulevard, Beltsville, MD 20705. (301) 344-3813. An on-line bibliographic data base covering all aspects of agricultural sciences, agricultural engineering and soil science. 1970 to present. Inquire as to online cost and availability.

CA SEARCH. Chemical Abstracts Service, P.O. Box 3012, Columbus, OH 43120. (800) 848-6538 or (614) 421-3600. Comprehensive guide to chemical literature, 1972 to present. Inquire as to online cost and availability.

COMPENDEX. Engineering Information, Inc., 345 East 47th Street, New York, NY 10017. (800) 221-1044 or (212) 705-7615. Engineering and technical literature, 1975 to present. Inquire as to online cost and availability.

DISSERTATION ABSTRACTS ONLINE. University Microfilms International, 300 North Zeeb Road, Ann Arbor, MI 48106. (800) 521-0600 or (313) 761-4700. Scope includes virtually all doctoral dissertations accepted at accredited American institutions from 1861 to present in over 250 subject areas. Inquire as to online cost and availability.

GEOREF. Online version of the BIBLIOGRAPHY AND INDEX OF GEOLOGY. American Geological Institute, 4220 King Street, Alexandria, VA 22302. (703) 379-2480. 1969 to present. Inquire as to online cost and availability.

INSPEC. INSPEC Marketing Department, Institution of Electrical Engineers. Available from IEEE Service Center, 445 Hoes Lane, Piscataway, NJ 08854. (201) 981-0060. Online version of Physics Abstracts. Inquire as to online cost and availability.

NTIS. National Technical Information Service, 5285 Port Royal Road, Springfield, VA 22161. (703) 487-4630. Broad coverage of government sponsored research reports, 1964 to present. Inquire as to online cost and availability.

SCISEARCH. Institute for Scientific Information, 3501 Market Street, Philadelphia, PA 19104. (800) 523-1850 or (215) 386-0100. Broad multidisciplinary title and author index to the international literature of science and technology, 1974 to present. Inquire as to online cost and availability.

WILSONLINE. H.W. Wilson and Company, 950 University Avenue, Bronx, NY 10452. (800) 367-6770 or (212) 588-8400. Makes available online versions of the H.W. Wilson indexes including Applied Science and Technology Index, Business Periodicals Index and Readers' Guide to Periodical Literature. Approximately 1980 to present. Inquire as to online cost and availability.

OTHER SOURCES

WHAT EVERY ENGINEER SHOULD KNOW ABOUT ENGINEERING SOURCES. Marcel Dekker Inc., 270 Madison Avenue, New York, NY 10016. (800) 228-1160. 1984. $24.95.

PERIODICALS

ENGINEERING GEOLOGY. Elsevier Science Publishing Company, Inc., 52 Vanderbilt Avenue, New York, NY 10017. (212) 370-5520. 1965 to present. Eight times per year. $250.00 per year.

HIGHWAY AND HEAVY CONSTRUCTION. Technical Publishing Company, 875 Third Avenue, New York, NY 10022.

(212) 605-9400. 1892 to present. Monthly. $35.00 per year.

JOURNAL OF GEOTECHNICAL ENGINEERING. American Society of Civil Engineers, 345 East 47th Street, New York, NY 10017. (212) 705-7420. 1956 to present. Monthly. $100.00 per year.

JOURNAL OF IRRIGATION AND DRAINAGE. American Society of Civil Engineers, 345 East 47th Street, New York, NY 10017. (212) 705-7420. 1956 to present. Monthly. $115.00 per year.

ROCK MECHANICS AND ROCK ENGINEERING. Springer-Verlag New York, Inc., 175 Fifth Avenue, New York, NY 10010. (800) 526-7254. 1963 to present. Quarterly. $65.00 per year.

SOIL DYNAMICS AND EARTHQUAKE ENGINEERING. CML Publications, 400 West Cummings Park, Suite 6200, Woburn, MA 01801. (617) 933-7374. 1981 to present. Quarterly. $130.00 per year.

SOIL MECHANICS ARCHIVES. University of Waterloo. Distributed by: Oxford University Press, 200 Madison Avenue, New York, NY 10016. (800) 458-5833. 1976 to present. Quarterly. $100.00 per year.

SOILS AND FOUNDATIONS. Japanese Society of Soil Mechanics and Foundation Engineering, Doshitsu Kogakkai, Sugayama Building, 4F, Kanda Awaji-cho, 2-23, Chiyoda-ku, Tokyo 101, Japan. 1960 to present. Quarterly. Inquire.

RESEARCH CENTERS AND INSTITUTES

AGRICULTURAL ENGINEERING DEPARTMENT. Michigan State University, East Lansing, MI 48824. (517) 353-7268.

GEOTECHNICAL/CIVIL ENGINEERING MATERIALS RESEARCH LABORATORIES. Iowa State University, Town Engineering Building and Spangler Research Laboratory, Ames, IA 50011. (515) 294-7627.

ROCK MECHANICS AND GROUND CONTROL LABORATORY. West Virginia University, Department of Mining Engineering, P.O. Box 6070, Morgantown, WV 26506. (304) 293-5695.

SOIL MECHANICS LABORATORY. University of Michigan, Department of Mechanical Engineering and Applied Mechanics, 2246A G.G. Brown Building, Ann Arbor, MI 48109. (313) 763-0684.

SOIL SCIENCE

See also: AGRICULTURAL ENGINEERING, CIVIL ENGINEERING, EARTHQUAKE ENGINEERING, GEOTECHNICAL ENGINEERING, ROCK MECHANICS, SOIL CHEMISTRY, SOIL MECHANICS

ABSTRACT SERVICES AND INDEXES

AGRICULTURAL ENGINEERING ABSTRACTS. Commonwealth Agricultural Bureaux. Distributed by: Unipub, Box 433, Murray Hill Station, New York, NY 10010. 1976 to present. Monthly. $220.00 per year.

APPLIED SCIENCE AND TECHNOLOGY INDEX. H.W. Wilson and Company, 950 University Avenue, Bronx, NY 10452. (800) 367-6670 or (212) 588-8400. Monthly. Inquire as to cost and availability.

BIBLIOGRAPHY AND INDEX OF GEOLOGY. American Geological Institute, 4220 King Street, Alexandria, VA 22302. (703) 379-2480. 1969 to present. Monthly. $1100.00 per year.

CHEMICAL ABSTRACTS. American Chemical Society, Chemical Abstracts Service, Box 3012, Columbus, OH 43210. (614) 421-3600. 1907 to present. Weekly. Inquire as to cost and availability.

CURRENT CONTENTS: ENGINEERING, TECHNOLOGY AND APPLIED SCIENCES. Institute for Scientific Information, 3501 Market Street, Philadelphia, PA 19104. (800) 523-1850 or (215) 386-0100. Weekly. $275.00 per year.

ENGINEERING INDEX MONTHLY AND AUTHOR INDEX. Engineering Information Inc., 345 East 47th Street, New York, NY 10017. (212) 705-7600. Monthly. $1560.00 per year.

INTERNATIONAL JOURNAL OF ROCK MECHANICS AND MINING SCIENCES AND GEOMECHANICS ABSTRACTS. Pergamon Press, Inc., Maxwell House, Fairview Park, Elmsford, NY 10523. (914) 592-7700. 1964 to present. Bimonthly. $385.00 per year.

IRRIGATION AND DRAINAGE ABSTRACTS. Commonwealth Agricultural Bureaux. Distributed by: Unipub, Box 433, Murray Hill Station, New York, NY 10010. 1975 to present. Quarterly. $120.00 per year.

SCIENCE CITATION INDEX. Institute for Scientific Information, 3501 Market Street, Philadelphia, PA 19104. (800) 523-1850 or (215) 386-0100. Six times per year. $6200.00 per year.

SOILS AND FERTILIZERS. Commonwealth Agricultural Bureaux. Distributed by: Unipub, Box 433, Murray Hill Station, New York, NY 10010. 1937 to present. Monthly. $450.00 per year.

ASSOCIATIONS AND PROFESSIONAL SOCIETIES

AMERICAN GEOLOGICAL INSTITUTE. 4220 King Street, Alexandria, VA 22302. (703) 379-2480.

AMERICAN SOCIETY OF AGRICULTURAL ENGINEERS. 2950 Niles Road, St. Joseph, MI 49085. (616) 429-0300.

AMERICAN SOCIETY OF AGRONOMY, Inc. 677 South Segoe Road, Madison, WI 53711. (608) 274-1212.

AMERICAN SOCIETY OF CIVIL ENGINEERS. 345 East 47th Street, New York, NY 10017. (212) 705-7420.

ASSOCIATION OF ENGINEERING GEOLOGISTS. P.O. Box 1068, Brentwood, TN 37027. (615) 377-3578.

ASSOCIATION OF SOIL AND FOUNDATION ENGINEERS. 8811 Colesville Road, Suite G106, Silver Spring, MD 20910. (301) 565-2733.

COMMISSION ON SOIL MINERALOGY. Soil and Crop Sciences Department, Texas A&M University, College Station, TX 77843. (409) 845-8322.

GEOLOGICAL SOCIETY OF AMERICA. 3300 Penrose Place, Boulder, CO 80301. (303) 447-2020.

SOIL CONSERVATION SOICETY OF AMERICA. 7515 Northeast Ankeny Road, Ankeny, IA 50021. (515) 289-2331.

SOIL SCIENCE SOCIETY OF AMERICA. 677 South Segoe Road, Madison, WI 53711. (608) 273-8080.

U.S. NATIONAL SOCIETY FOR THE INTERNATIONAL SOCIETY OF SOIL MECHANICS AND FOUNDATION ENGINEERING. c/o Professor Harvey E. Wahls, Civil Engineering Department, Box 7908, North Carolina State University, Raleigh, NC 27695. (919) 737-2331.

DIRECTORIES AND BIOGRAPHICAL SOURCES

AMERICAN MEN AND WOMEN OF SCIENCE. R.R. Bowker, Inc., Order Department, 245 West 17th Street, New York, NY 10011. (800) 521-8110. Eight volumes. 1986. $595.00 for set.

INTERNATIONAL RESEARCH CENTERS DIRECTORY 1988-89. Darren L. Smith, editor. Gale Research Company, Book Tower, Detroit, MI 48226. (800) 521-0707. 4th edition. 1987. $360.00.

1987 DIRECTORY OF ENGINEERING SOCIETIES AND RELATED ORGANIZATIONS. Gordon Davis, editor. Hemisphere Publishing Corporation, 1010 Vermont Avenue, NW, Washington, DC 20005. (800) 526-0275. 12th edition. 1987. $100.00.

RESEARCH CENTERS DIRECTORY 1988. Gale Research Company, Book Tower, Detroit, MI 48226. (800) 521-0707. 12th edition. 1987. $365.00 for set.

SCIENTIFIC AND TECHNICAL ORGANIZATIONS AND AGENCIES DIRECTORY. Margaret Labash Young, editor. Gale Research Company, Book Tower, Detroit, MI 48226. (800) 521-0707. 2nd edition. 1987. $185.00.

GENERAL WORKS

ENGINEERING PROPERTIES OF SOILS AND ROCKS. F. Bell. Butterworth's Publishing, 80 Montvale Avenue, Stoneham, MA 02180. (800) 325-4177. Second edition. 1983. $19.95 in paper.

OUTLINE OF SOIL AND ROCK MECHANICS. Pierre Habib. Cambridge University Press, 32 East 57th Street, New York, NY 10022. (800) 872-7423. 1983. $11.95 in paper.

SOIL CLASSIFICATION. Charles Finkl, editor. Van Nostrand Reinhold Company, Inc., 135 West 50th Street, New York, NY 10020. (800) 543-2681. 1982. $46.50.

SOIL MECHANICS. R.F. Craig. Van Nostrand Reinhold Company, Inc., 135 West 50th Street, New York, NY 10020. (800) 543-2681. Third edition. 1983. $19.95 in paper.

SOIL SCIENCE: PRINCIPLES AND PRACTICE. R.L. Hausenbuiller. William C. Brown Publications, 2460 Kerper Boulevard, Dubuque, IA 52001. (319) 589-2822. Third edition. 1985. $35.95.

HANDBOOKS AND MANUALS

A GUIDE TO SOIL MECHANICS. Malcolm Bolton. John Wiley and Sons, Inc., 605 Third Avenue, New York, NY 10158. (800) 526-5368. 1980. $59.95.

ONLINE DATA BASES

AGRICOLA (AGRICULTURAL ON-LINE ACCESS). National Agricultural Library, Information Systems Division, 10301 Baltimore Boulevard, Beltsville, MD 20705. (301) 344-3813. An on-line bibliographic data base covering all aspects of agricultural sciences, agricultural engineering and soil science. 1970 to present. Inquire as to online cost and availability.

CA SEARCH. Chemical Abstracts Service, P.O. Box 3012, Columbus, OH 43120. (800) 848-6538 or (614) 421-3600. Comprehensive guide to chemical literature, 1972 to present. Inquire as to online cost and availability.

COMPENDEX. Engineering Information, Inc., 345 East 47th Street, New York, NY 10017. (800) 221-1044 or (212) 705-7615. Engineering and technical literature, 1975 to present. Inquire as to online cost and availability.

DISSERTATION ABSTRACTS ONLINE. University Microfilms International, 300 North Zeeb Road, Ann Arbor, MI 48106. (800) 521-0600 or (313) 761-4700. Scope includes virtually all doctoral dissertations accepted at accredited American institutions from 1861 to present in over 250 subject areas. Inquire as to online cost and availability.

GEOREF. Online version of the BIBLIOGRAPHY AND INDEX OF GEOLOGY. American Geological Institute, 4220 King Street, Alexandria, VA 22302. (703) 379-2480. 1969 to present. Inquire as to online cost and availability.

INSPEC. INSPEC Marketing Department, Institution of Electrical Engineers. Available from IEEE Service Center, 445 Hoes Lane, Piscataway, NJ 08854. (201) 981-0060. Online version of Physics Abstracts. Inquire as to online cost and availability.

NTIS. National Technical Information Service, 5285 Port Royal Road, Springfield, VA 22161. (703) 487-4630. Broad coverage of government sponsored research reports, 1964 to present. Inquire as to online cost and availability.

SCISEARCH. Institute for Scientific Information, 3501 Market Street, Philadelphia, PA 19104. (800) 523-1850 or (215) 386-0100. Broad multidisciplinary title and author index to the international literature of science and technology, 1974 to present. Inquire as to online cost and availability.

WILSONLINE. H.W. Wilson and Company, 950 University Avenue, Bronx, NY 10452. (800) 367-6770 or (212) 588-8400. Makes available online versions of the H.W. Wilson indexes including Applied Science and Technology Index, Business Periodicals Index and Readers' Guide to Periodical Literature. Approximately 1980 to present. Inquire as to online cost and availability.

PERIODICALS

AGRICULTURAL ENGINEERING. American Society of Agricultural Engineers, 2950 Niles Road, St. Joseph, MI 49085. (616) 429-0300. 1920 to present. Monthly. $25.00 per year.

CANADIAN JOURNAL OF SOIL SCIENCE. Agricultural Institute of Canada, 151 Slater Street, Suite 907, Ottawa, ON K1P 5H4, Canada. 1921 to present. Quarterly. $56.00 per year.

COMMUNICATIONS IN SOIL SCIENCE AND PLANT ANALYSIS. Marcel Dekker Inc., 270 Madison Avenue, New York, NY 10016. (800) 228-1160. 1970 to present. Monthly. $300.00 per year.

CROPS AND SOILS MAGAZINE. American Society of Agronomy, Inc., 677 South Segoe Road, Madison, WI 53711. (608) 274-1212. 1948 to present. Monthly. $10.00 per year.

GEODERMA: INTERNATIONAL JOURNAL OF SOIL SCIENCE. Elsevier Science Publishing Company, Inc., 52 Vanderbilt Avenue, New York, NY 10017. (212) 370-5520. 1967 to present. Eight times per year. $250.00 per year.

JOURNAL OF IRRIGATION AND DRAINAGE. American Society of Civil Engineers, 345 East 47th Street, New York, NY 10017. (212) 705-7420. 1956 to present. Monthly. $115.00 per year.

JOURNAL OF SOIL AND WATER CONSERVATION. Soil Conservation Society of America, 7515 Northeast Ankeny Road, Ankeny, IA 50021. (515) 289-2331. 1946 to present. Bimonthly. $25.00 per year.

ROCK MECHANICS AND ROCK ENGINEERING. Springer-Verlag New York, Inc., 175 Fifth Avenue, New York, NY 10010. (800) 526-7254. 1963 to present. Quarterly. $65.00 per year.

SOIL DYNAMICS AND EARTHQUAKE ENGINEERING. CML Publications, 400 West Cummings Park, Suite 6200, Woburn, MA 01801. (617) 933-7374. 1981 to present. Quarterly. $130.00 per year.

SOIL MECHANICS ARCHIVES. University of Waterloo. Distributed by: Oxford University Press, 200 Madison Avenue, New York, NY 10016. (800) 458-5833. 1976 to present. Quarterly. $100.00 per year.

SOIL SCIENCE. Williams and Wilkins, 428 East Preston Street, Baltimore, MD 21202. (301) 528-4000. 1916 to present. Monthly. $45.00 per year to individuals.

SOIL SCIENCE SOCIETY OF AMERICA JOURNAL. Soil Science Society of America, 677 South Segoe Road, Madison, WI 53711. (608) 273-8080. 1936 to present. Bimonthly. $65.00 per year.

SOILS AND FOUNDATIONS. Japanese Society of Soil Mechanics and Foundation Engineering, Doshitsu Kogakkai, Sugayama

SOIL SCIENCE

Building, 4F, Kanda Awaji-cho, 2-23, Chiyoda-ku, Tokyo 101, Japan. 1960 to present. Quarterly. Inquire.

RESEARCH CENTERS AND INSTITUTES

AGRICULTURAL ENGINEERING DEPARTMENT. Michigan State University, East Lansing, MI 48824. (517) 353-7268.

GEOTECHNICAL/CIVIL ENGINEERING MATERIALS RESEARCH LABORATORY. Iowa State University, Town Engineering Building and Spangler Research Laboratory, Ames, IA 50011. (515) 294-7627.

SOIL MECHANICS LABORATORY. University of Michigan, Department of Mechanical Engineering and Applied Mechanics, 2246A G.G. Brown Building, Ann Arbor, MI 48109. (313) 763-0684.

SOLAR ENERGY

See also: COAL, ENERGY, FUSION, GEOTHERMAL ENERGY, HYDROELECTRIC POWER, HYDROGEN, NATURAL GAS, NUCLEAR ENERGY, PETROLEUM

ABSTRACT SERVICES AND INDEXES

ABSTRACT NEWSLETTER: ENERGY. National Technical Information Service, 5285 Port Royal Road, Springfield, VA 22161. (703) 487-4929. Weekly. $95.00 per year.

APPLIED SCIENCE AND TECHNOLOGY INDEX. H.W. Wilson and Company, 950 University Avenue, Bronx, NY 10452. (800) 367-6670 or (212) 588-8400. Monthly. Inquire as to cost and availability.

CHEMICAL ABSTRACTS. American Chemical Society, Chemical Abstracts Service, Box 3012, Columbus, OH 43210. (614) 421-3600. 1907 to present. Weekly. $9500.00 per year.

CURRENT CONTENTS: ENGINEERING, TECHNOLOGY AND APPLIED SCIENCES. Institute for Scientific Information, 3501 Market Street, Philadelphia, PA 19104. (800) 523-1850 or (215) 386-0100. Weekly. $275.00 per year.

ENERGY RESEARCH ABSTRACTS. U.S. Department of Energy, Office of Scientific and Technical Information, Box 62, Oak Ridge, TN 37831. (615) 576-1155. Subscribe to: U.S. Superintendent of Documents, Washington, DC 20402. 1976 to present. Semimonthly. $146.00 per year.

ENGINEERING INDEX MONTHLY AND AUTHOR INDEX. Engineering Information Inc., 345 East 47th Street, New York, NY 10017. (212) 705-7600. Monthly. $1560.00 per year.

SCIENCE CITATION INDEX. Institute for Scientific Information, 3501 Market Street, Philadelphia, PA 19104. (800) 523-1850 or (215) 386-0100. Six times per year. $6200.00 per year.

ANNUAL REVIEWS AND YEARBOOKS

ADVANCES IN ENERGY SYSTEMS AND TECHNOLOGY. Academic Press, Inc., 6277 Sea Harbor Drive, Orlando, FL 32821. (800) 321-5068. 1979 to present. Irregular. Inquire.

ANNUAL REVIEW OF ENERGY. Annual Reviews, Inc., 4139 El Camino Way, Palo Alto, CA 94306. (415) 493-4400. Annual. Inquire.

ASSOCIATIONS AND PROFESSIONAL SOCIETIES

AMERICAN SOLAR ENERGY SOCIETY. 2030 17th Street, Boulder, CO 80302. (303) 443-3130.

ASSOCIATION OF ENERGY ENGINEERS. 4025 Pleasantdale Road, Suite 340, Atlanta, GA 30340. (404) 447-5083.

EDISON ELECTRIC INSTITUTE. 1111 19th Street, N.W., Washington, DC 20036. (202) 828-7400.

INSTITUTE OF ELECTRICAL AND ELECTRONICS ENGINEERS. 345 East 47th Street, New York, NY 10017. (212) 705-7900.

DIRECTORIES AND BIOGRAPHICAL SOURCES

AMERICAN MEN AND WOMEN OF SCIENCE. R.R. Bowker, Inc., Order Department, 245 West 17th Street, New York, NY 10011. (800) 521-8110. Eight volumes. 1986. $595.00 for set.

ENERGY INFORMATION CENTERS DIRECTORY. Public Affairs and Information Program, Atomic Industrial Forum, 7101 Wisconsin Avenue, Bethesda, MD 20814. (301) 654-9260. 1985. Free.

IEEE MEMBERSHIP DIRECTORY. Institute of Electrical and Electronics Engineers. IEEE Service Center, 445 Hoes Lane, Piscataway, NJ 08854. Annual. $7.00.

INTERNATIONAL RESEARCH CENTERS DIRECTORY 1988-89. Darren L. Smith, editor. Gale Research Company, Book Tower, Detroit, MI 48226. (800) 521-0707. 4th edition. 1987. $360.00.

1987 DIRECTORY OF ENGINEERING SOCIETIES AND RELATED ORGANIZATIONS. Gordon Davis, editor. Hemisphere Publishing Corporation, 1010 Vermont Avenue, NW, Washington, DC 20005. (800) 526-0275. 12th edition. 1987. $100.00.

RESEARCH CENTERS DIRECTORY 1988. Gale Research Company, Book Tower, Detroit, MI 48226. (800) 521-0707. 12th edition. 1987. $365.00 for set.

SCIENTIFIC AND TECHNICAL ORGANIZATIONS AND AGENCIES DIRECTORY. Margaret Labash Young, editor. Gale Research Company, Book Tower, Detroit, MI 48226. (800) 521-0707. 2nd edition. 1987. $185.00.

WHO'S WHO IN ENGINEERING. Gordon Davis, editor. Hemisphere Publishing Corporation, 1010 Vermont Avenue, NW, Washington, DC 20005. (800) 526-0275. 6th edition. 1985. $200.00.

ENCYCLOPEDIAS AND DICTIONARIES

MCGRAW-HILL ENCYCLOPEDIA OF SCIENCE AND TECHNOLOGY. McGraw-Hill Book Company, 1221 Avenue of the Americas, New York, NY 10020. (212) 512-2000. 6th edition. 1987. $1600.00.

THESAURUS OF SCIENTIFIC, TECHNICAL, AND ENGINEERING TERMS. Hemisphere Publishing Corporation, 1010 Vermont Avenue, NW, Washington, DC 20005. (800) 526-0275. 1988. $125.00.

GENERAL WORKS

ENERGY OPTIONS. David Merrick and Richard Marshall. John Wiley and Sons, Inc., 605 Third Avenue, New York, NY 10158. (800) 526-5368. 1984. $75.95.

ENERGY 2000: AN OVERVIEW OF THE WORLD'S ENERGY RESOURCES IN THE DECADES TO COME. Heinz Knoepfel. Gordon and Breach Science Publishers, Inc., 50 West 23rd Street, New York, NY 10010. (212) 206-8900. 1986. $42.00.

SUN POWER: AN INTRODUCTION TO THE APPLICATION OF SOLAR ENERGY. J.C. McVeigh. Pergamon Press, Inc., Maxwell House, Fairview Park, Elmsford, NY 10523. (914) 592-7700. 2nd edition. 1983. $48.40.

HANDBOOKS AND MANUALS

HANDBOOK OF ENERGY SYSTEMS ENGINEERING. Leslie C. Wilbur, editor. John Wiley and Sons, Inc., 605 Third Avenue, New York, NY 10158. (800) 526-5368. 1985. $74.95.

HANDBOOK OF ENERGY TECHNOLOGY AND ECONOMICS. R.A. Meyers, editor. John Wiley and Sons, Inc., 605 Third Avenue, New York, NY 10158. (800) 526-5368. 1983. $79.95.

ONLINE DATA BASES

CA SEARCH. Chemical Abstracts Service, P.O. Box 3012, Columbus, OH 43120. (800) 848-6538 or (614) 421-3600. Comprehensive guide to chemical literature, 1972 to present. Inquire as to online cost and availability.

COMPENDEX. Engineering Information, Inc., 345 East 47th Street, New York, NY 10017. (800) 221-1044 or (212) 705-7615. Engineering and technical literature, 1975 to present. Inquire as to online cost and availability.

DISSERTATION ABSTRACTS ONLINE. University Microfilms International, 300 North Zeeb Road, Ann Arbor, MI 48106. (800) 521-0600 or (313) 761-4700. Scope includes virtually all doctoral dissertations accepted at accredited American institutions from 1861 to present in over 250 subject areas. Inquire as to online cost and availability.

DOE ENERGY DATA BASE. U.S. Department of Energy, Office of Scientific and Technical Information, P.O. Box 62, Oak Ridge, TN 37831. (615) 576-6837. A database that covers all aspects of energy including the science and technology of energy. 1948 to present. Available through the DIALOG search service or DOE/RECON. Inquire as to online cost and availability.

ENERGYLINE. EIC/Intelligence, Inc., 48 West 38th Street, New York, NY 10018. (212) 944-8500. A database of resources on the scientific, engineering, political, and socioeconomic aspects of energy resources. 1976 to present. Inquire as to online cost and availability.

INSPEC. INSPEC Marketing Department, Institution of Electrical Engineers. Available from IEEE Service Center, 445 Hoes Lane, Piscataway, NJ 08854. (201) 981-0060. Online version of Physics Abstracts. Inquire as to online cost and availability.

NTIS. National Technical Information Service, 5285 Port Royal Road, Springfield, VA 22161. (703) 487-4630. Broad coverage of government sponsored research reports, 1964 to present. Inquire as to online cost and availability.

SCISEARCH. Institute for Scientific Information, 3501 Market Street, Philadelphia, PA 19104. (800) 523-1850 or (215) 386-0100. Broad multidisciplinary title and author index to the international literature of science and technology, 1974 to present. Inquire as to online cost and availability.

WILSONLINE. H.W. Wilson and Company, 950 University Avenue, Bronx, NY 10452. (800) 367-6770 or (212) 588-8400. Makes available online versions of the H.W. Wilson indexes including Applied Science and Technology Index, Business Periodicals Index and Readers' Guide to Periodical Literature. Approximately 1980 to present. Inquire as to online cost and availability.

PERIODICALS

ALTERNATIVE SOURCES OF ENERGY. Alternative Sources of Energy, Inc., 107 South Central Avenue, Milaca, MN 56353. (612) 983-6892. 1971 to present. Ten times per year. $48.00 per year.

ENERGY AND FUELS. American Chemical Society, 1155 16th Street, N.W., Washington, DC 20036. (800) 424-6747. 1987 to present. Bimonthly. $269.00 per year.

ENERGY ENGINEERING. Association of Energy Engineers. Available from: Fairmont Press, Box 14227, Atlanta, GA 30324. (404) 447-5314. 1904 to present. Bimonthly. $44.00 per year.

ENERGY SOURCES. Crane Russak and Company, Inc., 3 East 44th Street, New York, NY 10017. (212) 867-1490. 1973. Quarterly. $73.00 per year.

INTERNATIONAL JOURNAL OF SOLAR ENERGY. Harwood Academic Publishers, 50 West 23rd Street, New York, NY 10010. (212) 206-8900. 1982 to present. 6 times per year. $132.00 per year.

JOURNAL OF ENERGY RESOURCES TECHNOLOGY. American Society of Mechanical Engineers, 345 East 47th Street, New York, NY 10017. (212) 705-7703. 1979 to present. $24.00 per year. SOLAR ENERGY. Pergamon Press, Inc., Maxwell House, Fairview Park, Elmsford, NY 10523. (914) 592-7700. 1957 to present. Monthly. $298.00 per year.

RESEARCH CENTERS AND INSTITUTES

CENTER FOR ENERGY STUDIES. University of Texas at Austin, 1318 EME Building, 10100 Burnet Road, Austin, TX 78758. (512) 471-7792.

ENERGY CENTER. Stevens Institute of Technology, Department of Mechanical Energy, Castle Point Station, Hoboken, NJ 07030. (201) 420-5560.

ENERGY LABORATORY. Massachusetts Institute of Technology, 123 Amherst Street, Building E-40-455, Cambridge, MA 02139. (617) 253-3400.

SOLAR PHYSICS

See: SUN

SOLAR SYSTEM

See also: ASTROGEOLOGY, ASTRONOMY, ASTROPHYSICS, JUPITER, MARS, MERCURY, METEORITICS, NEPTUNE, PLANETARY SCIENCES, PLUTO, SATELLITES, SATURN, URANUS, VENUS

ABSTRACT SERVICES AND INDEXES

ASTRONOMY AND ASTROPHYSICS ABSTRACTS. Springer-Verlag New York, Inc., 175 Fifth Avenue, New York, NY 10010. (800) 526-7254. 1969 to present. Approximately $70.00 per year.

ASTRONOMY AND ASTROPHYSICS MONTHLY INDEX. Olivetree Associates, P.O. Box 236, Sierre Madre, CA 91024. $220.00 per year. Complementary copies available on request.

CHEMICAL ABSTRACTS. American Chemical Society, Chemical Abstracts Service, Box 3012, Columbus, OH 43210. (614) 421-3600. 1907 to present. Weekly. $10,000.00 per year.

CONFERENCE PAPERS INDEX. Cambridge Scientific Abstracts, 5161 River Road, Bethesda, MD 20816. 1972 to present. Monthly. Inquire as to cost and availability.

CURRENT CONTENTS: PHYSICAL, CHEMICAL, AND EARTH SCIENCES. Institute for Scientific Information, 3501 Market Street, Philadelphia, PA 19104. (800) 523-1850 or (215) 386-0100. Weekly. $275.00 per year.

INDEX TO SCIENTIFIC AND TECHNICAL PROCEEDINGS. Institute for Scientific Information, 3501 Market Street, Philadelphia, PA 19104. (800) 523-1850 or (215) 386-0100. 1978 to present. Monthly. $775.00 per year.

INDEX TO SCIENTIFIC REVIEWS. Institute for Scientific Information, 3501 Market Street, Philadelphia, PA 19104. (800)

SOLAR SYSTEM

523-1850 or (215) 386-0100. 1974 to present. Semi-annual. $550.00 per year.

METEOROLOGICAL AND GEOASTROPHYSICAL ABSTRACTS. American Meteorological Society, 45 Beacon Street, Boston, MA 02108. (617) 227-2425. 1950 to present. Monthly. $450.00 per year.

PHYSICS ABSTRACTS. Institution of Electrical Engineers. Available from: IEEE Service Center, 445 Hoes Lane, Piscataway, NJ 08854. 1898 to present. Bimonthly. $1700.00 per year.

PHYSICS BRIEFS. Physik Verlag GmbH, Postfach 1260/1280, D-6940 Weinheim, West Germany. (212) 661-9404. 1920 to present. Twenty-six times per year. $1250.00 per year.

SCIENCE CITATION INDEX. Institute for Scientific Information, 3501 Market Street, Philadelphia, PA 19104. (800) 523-1850 or (215) 386-0100. Six times per year. $6200.00 per year.

STAR. (SCIENTIFIC AND TECHNICAL AEROSPACE REPORTS). U.S. National Aeronautics and Space Administration, Scientific and Technical Information Facility, Box 8757, Baltimore-Washington International Airport, MD 21240. (202) 755-2210. Semimonthly, with semiannual and annual indexes. $85.00 per year.

ANNUAL REVIEWS AND YEARBOOKS

ANNUAL REVIEW OF ASTRONOMY AND ASTROPHYSICS. Annual Reviews, Inc., 4139 El Camino Way, Palo Alto, CA 94306. (415) 493-4400. Annual. Inquire as to cost and availability.

ANNUAL REVIEW OF EARTH AND PLANETARY SCIENCES. Annual Reviews, Inc., 4139 El Camino Way, Palo Alto, CA 94306. (415) 493-4400. Annual. Inquire as to cost and availability.

ASSOCIATIONS AND PROFESSIONAL SOCIETIES

AMERICAN ASTRONOMICAL SOCIETY. 1816 Jefferson Place, N.W., Washington, DC 20036. (202) 659-0134.

AMERICAN GEOPHYSICAL UNION. 2000 Florida Avenue, N.W., Washington, DC 20009. (202) 462-6903.

AMERICAN INSTITUTE OF PHYSICS. 335 East 45th Street, New York, NY 10017. (212) 661-9494.

ASTRONOMICAL SOCIETY OF THE PACIFIC. 1290 24th Avenue, San Francisco, CA 94122. (415) 661-8660.

GEOLOGICAL SOCIETY OF AMERICA. 3300 Penrose Place, Boulder, CO 80301. (303) 447-2020.

PLANETARY SOCIETY. 65 North Catalina Avenue, Pasadena, CA 91106. (818) 793-5100.

BIBLIOGRAPHIES

A BIBLIOGRAPHY OF ASTRONOMY, 1970-1979. R.A. Seal and S.S. Martin. Libraries Unlimited Inc., P.O. Box 263, Littleton, CO 80160. (303) 770-1220. 1982. $37.50.

SCIENCE BOOKS AND FILMS. American Association for the Advancement of Science, 1333 H Street, NW, Washington, DC 20005. (202) 326-6454. Five times per year. $20.00 per year.

SCIENTIFIC AND TECHNICAL BOOKS AND SERIALS IN PRINT 1988; AN INDEX TO LITERATURE IN SCIENCE AND TECHNOLOGY. R.R. Bowker Company, 205 East 42nd Street, New York, NY 10017. (800) 521-8110. $175.00.

DIRECTORIES AND BIOGRAPHICAL SOURCES

AMERICAN ASTRONOMICAL SOCIETY MEMBERSHIP DIRECTORY. 1816 Jefferson Place, N.W., Washington, DC 20036. (202) 659-0134. Annual. Inquire.

AMERICAN MEN AND WOMEN OF SCIENCE. R.R. Bowker, Inc., Order Department, 245 West 17th Street, New York, NY 10011. (800) 521-8110. Eight volumes. 1986. $595.00 for set.

THE BIOGRAPHICAL DICTIONARY OF SCIENTISTS: ASTRONOMERS. D. Abbott, editor. Peter Bedrick Books, 125 East 23rd Street, New York, NY 10010. 1984.

DIRECTORY OF PHYSICS AND ASTRONOMY STAFF MEMBERS. American Institute of Physics, 335 East 45th Street, New York, NY 10017. (212) 661-9494. Annual.

INTERNATIONAL RESEARCH CENTERS DIRECTORY 1988-89. Darren L. Smith, editor. Gale Research Company, Book Tower, Detroit, MI 48226. (800) 521-0707. 4th edition. 1987. $360.00.

RESEARCH CENTERS DIRECTORY 1988. Gale Research Company, Book Tower, Detroit, MI 48226. (800) 521-0707. 12th edition. 1987. $365.00 for set.

SCIENTIFIC AND TECHNICAL ORGANIZATIONS AND AGENCIES DIRECTORY. Margaret Labash Young, editor. Gale Research Company, Book Tower, Detroit, MI 48226. (800) 521-0707. 2nd edition. 1987. $185.00.

GENERAL WORKS

ATMOSPHERES AND IONSPHERES OF THE OUTER PLANETS AND THEIR SATELLITES. S.K. Atreya. Springer-Verlag New York, Inc., 175 Fifth Avenue, New York, NY 10010. (800) 526-7254. 1987. $69.50.

THE BIRTH OF THE EARTH. David E. Fisher. Columbia University Press, 562 West 113th Street, New York, NY 10025. (212) 316-7100. 1987. $24.95.

CHEMISTRY AND PHYSICS OF THE TERRESTRIAL PLANETS. Surendra K. Saxena. Springer-Verlag New York, Inc., 175 Fifth Avenue, New York, NY 10010. (800) 526-7254. 1986. $59.00.

THE *GALAXY AND THE SOLAR SYSTEM. Roman Smoluchowski and others. University of Arizona Press, 1615 East Speedway, Tucson, AZ 85719. (602) 621-1441. 1986. $29.95.

GEOLOGY OF THE TERRESTRIAL PLANETS. M. Carr, editor. National Aeronautics and Space Administration Special Publication 469. National Technical Information Service, 5285 Port Royal Road, Springfield, VA 22161. (703) 487-4929. 1984. $16.00.

MERCURY: THE ELUSIVE PLANET. Robert Strom. Cambridge University Press, 32 East 57th Street, New York, NY 10022. (800) 872-7423. 1986. $30.00.

METEORITES: THEIR RECORD OF EARLY SOLAR-SYSTEM HISTORY. J. Wasson. W.H. Freeman and Company, 41 Madison Avenue, New York, NY 10010. (212) 532-7660. 1985. $34.95.

PLANETS AND THEIR ATMOSPHERES: ORIGIN AND EVOLUTION. J. Lewis and R. Prinn. Academic Press, Inc., 6277 Sea Harbor Drive, Orlando, FL 32821. (800) 321-5068. 1983. $59.00.

RECENT ADVANCES IN PLANETARY METEOROLOGY. G. Hunt, editor. Cambridge University Press, 32 East 57th Street, New York, NY 10022. (800) 872-7423. 1985. $39.50.

THE PLANETARY SYSTEM. Tobias Owen and David Morrison. Addison-Wesley Publishing Company, Inc., 1 Jacob Way, Reading, MA 01867. (617) 944-3700. 1987. $39.95.

RINGS: DISCOVERIES FROM GALILEO TO VOYAGER. MIT Press, 28 Carleton Street, Cambridge, MA 02142. (617) 253-2884. 1985. $17.50.

SATELLITES. J.A. Burns and M.S. Matthews, editors. University of Arizona Press, 1615 East Speedway, Tucson, AZ 85719. (602) 621-1441. 1986. $55.00.

THE STORY OF THE EARTH. P. Cattermole and P. Moore. Cambridge University Press, 32 East 57th Street, New York, NY 10022. (800) 872-7423. 1985. $29.95.

THEORY OF PLANETARY ATMOSPHERES: AN INTRODUCTION TO THEIR PHYSICS AND CHEMISTRY. J.W. Chamberlain and D.M. Hunten. Academic Press, Inc., 6277 Sea Harbor Drive, Orlando, FL 32821. (800) 321-5068. 1987. $49.50.

URANUS AND NEPTUNE. J. Bergstrahl, editor. National Aeronautics and Space Administration Conference Paper 2330. National Technical Information Service, 5285 Port Royal Road, Springfield, VA 22161. (703) 487-4929. 1984. $25.00.

VENUS: AN ARRANT TWIN. E. Burgess. Columbia University Press, 562 West 113th Street, New York, NY 10025. (212) 316-7100. 1985. $29.95.

ONLINE DATA BASES

CA SEARCH. Chemical Abstracts Service, P.O. Box 3012, Columbus, OH 43120. (800) 848-6538 or (614) 421-3600. Comprehensive guide to chemical literature, 1972 to present. Inquire as to online cost and availability.

DISSERTATION ABSTRACTS ONLINE. University Microfilms International, 300 North Zeeb Road, Ann Arbor, MI 48106. (800) 521-0600 or (313) 761-4700. Scope includes virtually all doctoral dissertations accepted at accredited American institutions from 1861 to present in over 250 subject areas. Inquire as to online cost and availability.

INSPEC. INSPEC Marketing Department, Institution of Electrical Engineers. Available from IEEE Service Center, 445 Hoes Lane, Piscataway, NJ 08854. (201) 981-0060. Online version of Physics Abstracts. Inquire as to online cost and availability.

NTIS. National Technical Information Service, 5285 Port Royal Road, Springfield, VA 22161. (703) 487-4630. Broad coverage of government sponsored research reports, 1964 to present. Inquire as to online cost and availability.

SCISEARCH. Institute for Scientific Information, 3501 Market Street, Philadelphia, PA 19104. (800) 523-1850 or (215) 386-0100. Broad multidisciplinary title and author index to the international literature of science and technology, 1974 to present. Inquire as to online cost and availability.

OTHER SOURCES

ATLAS OF THE SOLAR SYSTEM. P. Moore and G. Hunt. Rand McNally and Company, P.O. Box 7600, Chicago, IL 60680. (800) 323-4070. 1983. $40.00.

CAMBRIDGE PHOTOGRAPHIC ATLAS OF THE PLANETS. Cambridge University Press, 32 East 57th Street, New York, NY 10022. (800) 872-7423. 1986.

PERIODICALS

ASTRONOMICAL JOURNAL. American Astronomical Society. Available from: American Institute of Physics, 335 East 45th Street, New York, NY 10017. (212) 661-9494. Monthly. $125.00 per year.

ASTRONOMICAL SOCIETY OF THE PACIFIC PUBLICATIONS. Astronomical Society of the Pacific, 1290 24th Avenue, San Francisco, CA 94122. (415) 661-8660. Monthly. $40.00 per year.

ASTRONOMY. AstroMedia Corporation, 625 E Street, Box 92788, Milwaukee, WI 53202. (414) 276-2689. 1973 to present. Monthly. $21.00 per year.

ASTRONOMY AND ASTROPHYSICS. Springer-Verlag New York, Inc., 175 Fifth Avenue, New York, NY 10010. (800) 526-7254. Monthly. $680.00 per year.

EARTH, MOON AND PLANETS. D. Reidel Publishing Company, 190 Old Derby Street, Hingham, MA 02043. 1969 to present. Nine times per year. $275.00 per year.

GEOCHEMICA ET COSMOCHIMICA ACTA. Pergamon Press, Inc., Maxwell House, Fairview Park, Elmsford, NY 10523. (914) 592-7700. 1950 to present. Monthly. $340.00 per year.

LUNAR AND PLANETARY INFORMATION BULLETIN. Lunar and Planetary Institute, 3303 NASA Road One, Houston, TX 77058-4399. (713) 486-2135. 1970 to present. Three times per year. Free.

MERCURY. Astronomical Society of the Pacific, 1290 24th Avenue, San Francisco, CA 94122. (415) 661-8660. 1972 to present. Bimonthly. $21.00 per year.

METEORITICS. Center for Meteorite Studies, Arizona State University, Tempe, AZ 85287. (602) 965-3576. 1955 to present. Quarterly. $40.00 per year.

PLANETARY AND SPACE SCIENCE. Pergamon Press, Inc., Maxwell House, Fairview Park, Elmsford, NY 10523. (914) 592-7700. 1959 to present. Monthly. $430.00 per year.

PLANETARY REPORT. Planetary Society, 65 North Catalina, Pasadena, CA 91106-2301. 1980 to present. Bimonthly. $20.00 per year.

SKY AND TELESCOPE. Sky Publishing Corporation, 49 Bay State Road, Cambridge, MA 02238. (617) 864-7360. Monthly. $18.00 per year.

RESEARCH CENTERS AND INSTITUTES

LABORATORY FOR PLANETARY ATMOSPHERES RESEARCH. State University of New York at Stony Brook, Stony Brook, NY 11794-2300. (516) 632-8321.

LABORATORY FOR PLANETARY GEOLOGY. Arizona State University, Department of Geology, Tempe, AZ 85281. (602) 965-7029.

LABORATORY FOR PLANETARY STUDIES. Cornell University, 302 Space Sciences Building, Ithaca, NY 14853. (607) 256-4971.

LUNAR AND PLANETARY INSTITUTE. 3303 NASA Road One, Houston, TX 77058. (713) 486-2139.

SOLAR WIND

See also: SUN

ABSTRACT SERVICES AND INDEXES

ASTRONOMY AND ASTROPHYSICS ABSTRACTS. Springer-Verlag New York, Incorporated, 175 Fifth Avenue, New York, NY 10010. (212) 460-1500. $70.00 per year.

ASTRONOMY AND ASTROPHYSICS MONTHLY INDEX. Olivetree Associates, Post Office Box 236, Sierra Madre, CA 91024. $212.00 per year. Complimentary copies available on request.

SCIENCE CITATION INDEX. Institute for Scientific Information, 3501 Market Street, Philadelphia, PA 19104. (800) 523-1850 or (215) 386-0100. Inquire as to cost and availability.

STAR. (Scientific and Technical Aerospace Reports. United States National Aeronautics and Space Administration, Scientific and Technical Information Facility, Box 8757, Baltimore-Washington International Airport, MD 21240. (202) 755-2210. Semimonthly, with semiannual and annual indexes. $85.00 per year.

ANNUAL REVIEWS AND YEARBOOKS

THE ASTRONOMICAL ALMANAC. Superintendent of Documents, United States Government Printing Office, Washington, DC 20402. (202) 783-3238. Yearly.

ANNUAL REVIEW OF EARTH AND PLANETARY SCIENCES. Annual Reviews, Incorporated, 4139 East Camino Way, Palo Alto, CA 94306. (415) 493-4400.

ASSOCIATIONS AND PROFESSIONAL SOCIETIES

AMERICAN ASTRONOMICAL SOCIETY. 2000 Florida Avenue, NW, Suite 300, Washington, DC 20009. (202) 659-0134.

AMERICAN ASSOCIATION OF VARIABLE STAR OBSERVERS. 187 Concord Avenue, Cambridge, MA 02138. (617) 354-0484.

ASTRONOMICAL LEAGUE. Post Office Box 12821, Tucson, AZ 85732. (602) 790-8471.

ASTRONOMICAL SOCIETY OF THE PACIFIC. 1290 24th Avenue, San Francisco, CA 94122. (415) 661-8660.

BIBLIOGRAPHIES

A BIBLIOGRAPHY OF ASTRONOMY, 1970-1979. R.A. Seal and S.S. Martin. Libraries Unlimited, Incorporated, Littleton, CO 80160. 1982. $37.50.

SCIENCE BOOKS AND FILMS. American Association for the Advancement of Science, 1333 H Street, NW, Washington, DC 20005. 1965 to present. Five times per year. $20.00 per year.

SCIENTIFIC AND TECHNICAL BOOKS AND SERIALS IN PRINT 1988: AN INDEX TO LITERATURE IN SCIENCE AND TECHNOLOGY. R.R. Bowker Company, 205 East 42nd Street, New York, NY 10017. (800) 521-8110 or (212) 916-1600. $175.00.

DIRECTORIES AND BIOGRAPHICAL SOURCES

THE BIOGRAPHICAL DICTIONARY OF SCIENTISTS: ASTRONOMERS. D. Abbott, editor. Peter Bedrick Books, 125 East 23rd Street, New York, NY 10010. 1984. $24.95.

DIRECTORY OF PHYSICS AND ASTRONOMY STAFF MEMBERS. American Institute of Physics, 335 East 45th Street, New York, NY 10017. Annual.

RESEARCH CENTERS DIRECTORY. Gale Research Company, Book Tower, Detroit, MI 48226. (800) 521-0707. Twelfth edition. 1987. $365.00 for set.

WHO'S WHO IN FRONTIER SCIENCE AND TECHNOLOGY. Marquis Who's Who, Incorporated, 200 East Ohio Street, Chicago, IL 60611. (800) 428-3898 or (312) 787-2008.

ENCYCLOPEDIAS AND DICTIONARIES

ENCYCLOPEDIA OF PHYSICAL SCIENCE AND TECHNOLOGY. Academic Press, Incorporated, Orlando, FL 32887. (800) 321-5068 or (305) 345-2734. Inquire as to cost and availability.

MCGRAW-HILL ENCYCLOPEDIA OF SCIENCE AND TECHNOLOGY. McGraw-Hill Book, Incorporated, 1221 Avenue of the Americas, New York, NY 10020. (212) 997-3675. Fifth edition, 15 volumes. $1100.00.

GENERAL WORKS

CORONAL EXPANSION AND SOLAR WIND. A.J. Hundhausen. Springer-Verlag New York, Incorporated, 175 Fifth Avenue, New York, NY 10010. (212) 460-1500. 1972. $32.00.

SOLAR WIND: MAGNETOSPHERE COUPLING ASSL. Y. Kamide and J.A. Slavin. Kluwer Academic, 190 Old Derby Road, Hingham, MA 02043. (617) 749-5262. 1986. $154.00.

ONLINE DATA BASES

DISSERTATION ABSTRACTS ONLINE. University Microfilms International, 300 North Zeeb Road, Ann Arbor, MI 48106. (800) 521-0600 or (313) 761-4700. Scope includes virtually all doctoral dissertations accepted at accredited American institutions from 1861 to present in 252 subject areas. Inquire as to cost and availability.

NASA. National Aeronautics and Space Administration, Scientific and Technical Information Branch, 300 7th Street, SW, Washington, DC 20546. Citations and abstracts of aerospace literature, 1962 to present. Inquire as to cost and availability.

NTIS. National Technical Information Service, 5285 Port Royal Road, Springfield, VA 22161. (703) 487-4630. Broad coverage of government sponsored research reports, 1964 to present. Inquire as to cost and availability.

SCISEARCH. Institute for Scientific Information, 3501 Market Street, Philadelphia, PA 19104. (800) 523-1850 or (215) 386-0100. Broad multidisciplinary title and author index to the international literature of science and technology, 1974 to present. Inquire as to cost and availability.

PERIODICALS

ASTRONOMICAL JOURNAL. American Astronomical Society. Available from: American Institute of Physics, 335 East 45th Street, New York, NY 10017. 9212) 661-9404. $125.00 per year.

ASTRONOMICAL SOCIETY OF THE PACIFIC. Publications. Astronomical Society of the Pacific, 1290 24th Avenue, San Francisco, CA 94122. (415) 661-8660. Monthly. $38.00.

ASTRONOMY. Astro Media Corporation, 625 East Paul Avenue, Milwaukee, WI 53202. Monthly. $18.00 per year.

ASTRONOMY AND ASTROPHYSICS. Springer-Verlag New York, Incorporated, 175 Fifth Avenue, New York, NY 10010. (800) 526-7254 or (212) 460-1500. $680.00 per year.

ASTROPHYSICAL JOURNAL. American Astronomical Society, University of Chicago Press, 5801 Ellis Avenuew, Chicago, IL 60637. Biweekly. $305.00 per year.

ASTROPHYSICS AND SPACE SCIENCE. D. Reidel Publishing Company, 190 Old Derby Street, Hingham, MA 02043. Monthly. $101.00 per year.

ICARUS: INTERNATIONAL JOURNAL OF THE SOLAR SYSTEM STUDIES. Academic Press, Incorporated, Orlando, FL 32887. (305) 345-4100. Monthly. $484.00 per year.

MERCURY. Astronomical Society of the Pacific, 1290 24th Avenue, San Francisco, CA 94122. (415) 661-8660. Bimonthly. $21.00 per year.

MONTHLY NOTICES OF THE ROYAL ASTRONOMICAL SOCIETY. Blackwell Science Publications, Incorporated, 667 Lytton Avenue, Palo Alto, CA 94301. (415) 324-1688. Monthly. $134.00 per year.

SOLAR PHYSICS. D. Reidel Publishing Company, 190 Old Derby Street, Hingham, MA 02043. Monthly. $620.00 per year.

SOVIET ASTRONOMY (TRANSLATION OF ASTRONOMICHESKII ZHURNAL). American Institute of Physics, 335 East 45th Street, New York, NY 10017. (212) 661-9404. Bimonthly. $425.00 per year.

SPACE SCIENCE REVIEWS. D. Reidel Publishing Company, 190 Old Derby Street, Hingham, MA 02043. Monthly. $305.00 per year.

VISTAS IN ASTRONOMY. Pergamon Press, Incorporated, Maxwell House, Fairview Park, Elmsford, NY 10523. (914) 592-7700. Quarterly.

RESEARCH CENTERS AND INSTITUTES

CALIFORNIA INSTITUTE OF TECHNOLOGY. Palomar Observatory, 105-24, Pasadena, CA 91125. (818) 356-4033.

NATIONAL SOLAR OBSERVATORY. 950 North Cherry Avenue, Tucson, AZ 85726. (602) 327-5511.

UNIVERSITY OF ALABAMA IN HUNTSVILLE. Center for Space Plasma and Aeronomic Research. Engineering Building, Huntsville, AL 35899. (205) 895-6268.

UNIVERSITY OF NEW HAMPSHIRE. Space Science Center. Science and Engineering Research Building, Durham, NH 03824. (603) 862-2751.

SOLID STATE CHEMISTRY

See: PHYSICAL CHEMISTRY,
SOLID STATE PHYSICS

SOLID STATE PHYSICS

See also: MATERIALS SCIENCE, PHYSICS,
SOLIDS

ABSTRACT SERVICES AND INDEXES

ACOUSTICS ABSTRACTS. Multi-Science Publishing Company, Limited, 42-45 New Broad Street, London EC2M 1QY, England. Monthly. $295.00 per year.

APPLIED SCIENCE AND TECHNOLOGY INDEX. H.W. Wilson Company, 950 University Avenue, Bronx, NY 10452. (800) 367-6670 or (212) 588-8400. Inquire as to cost and availability.

CHEMICAL ABSTRACTS. Chemical Abstracts Service, 2540 Olentangy Road, Post Office Box 3012, Columbus, OH 43210. (800) 848-6538 or (614) 421-3600. Weekly. $9200.00 per year.

CURRENT CONTENTS: PHYSICAL, CHEMICAL AND EARTH SCIENCES. Institute for Scientific Information, 3501 Market Street, Philadelphia, PA 19104. (800) 523-1850 or (215) 386-0100. Weekly. $272.00 per year.

CURRENT PAPERS IN PHYSICS. Available from: IEEE Service Center, 445 Hoes Lane, Piscataway, NJ 08854. Biweekly. $215.00 per year.

GENERAL PHYSICS ADVANCE ABSTRACTS. American Institute of Physics, 335 East 45th Street, New York, NY 10017. (212) 661-9404. 1985 to present. Semimonthly. $150.00 per year.

GENERAL SCIENCE INDEX. H.W. Wilson Company, 950 University Avenue, Bronx, NY 10452. (800) 367-6770 or (212) 588-8400. Inquire as to cost and availability.

JOURNAL OF CURRENT LASER ABSTRACTS. Institute for Laser Documentation, Box 2070, Rolling Hills, CA 90274. 1964 to present. Monthly. $350.00 per year.

MATHEMATICAL REVIEWS. American Mathematical Society, Post Office Box 6248, Providence, RI 02940. (800) 556-7774 or (401) 272-9500.

PHYSICAL REVIEW ABSTRACTS. American Institute of Physics, 335 East 45th Street, New York, NY 10017. (212) 661-9404. 1970 to present. Semimonthly. $140.00 per year.

PHYSICS ABSTRACTS. Institute of Electrical Engineers, London, United Kingdom. Available from: Institute of Electrical and Electronic Engineers (IEEE), 345 East 47th Street, New York, NY 10017. (212) 705-7900. 1898 to present. Monthly. $1670.00 per year.

PHYSICS BRIEFS. Physics Verlag BmbH, Postfach 1260/1280, D-6940, Weinheim, West Germany. 1920 to present. Twenty-six times per year. $1200.00 per year.

SCIENCE CITATION INDEX. Institute for Scientific Information, 3501 Market Street, Philadelphia, PA 19104. (800) 523-1850 or (215) 386-0100. Inquire as to cost and availability.

SOLID STATE ABSTRACTS: AN ABSTRACT JOURNAL INVOLVING THE PHYSICS, METALLURGY, CRYSTALLOGRAPHY, CHEMISTRY, AND DEVICE TECHNOLOGY OF SOLIDS. Cambridge Scientific Abstracts, 5161 River Road, Bethesda, MD 20816. (301) 951-1400. 1957 to present. Bimonthly. $550.00 per year.

ANNUAL REVIEWS AND YEARBOOKS

ADVANCES IN APPLIED MECHANICS. Academic Press, Incorporated, 6277 Sea Harbor Drive, Orlando, FL 32821. (800) 321-5068. Irregular. Inquire as to cost and availability.

ADVANCES IN ATOMIC AND MOLECULAR PHYSICS. Academic Press, Incorporated, 6277 Sea Harbor Drive, Orlando, FL 32821. (800) 321-5068. Irregular. Inquire as to cost and availability.

ADVANCES IN ELECTRONICS AND ELECTRON PHYSICS. Academic Press, Incorporated, 6277 Sea Harbor Drive, Orlando, FL 32821. (800) 321-5068. Irregular. Inquire as to cost and availability.

ADVANCES IN NUCLEAR PHYSICS. Plenum Publishing Corporation, 233 Spring Street, New York, NY 10013. (800) 221-9369. Irregular. Inquire as to cost and availability.

ADVANCES IN PHYSICS. Taylor and Francis, Incorporated, 242 Cherry Street, Philadelphia, PA 19106-1906. (215) 238-0939. Irregular. Inquire as to cost and availability.

ANNUAL REVIEW OF MATERIALS SCIENCE. Annual Reviews, Incorporated, 4139 El Camino Way, Palo Alto, CA 94306. (415) 493-4400. Annual. Inquire as to cost and availability.

ANNUAL REVIEW OF NUCLEAR AND PARTICLE SCIENCE. Annual Reviews, Incorporated, 4139 El Camino Way, Palo Alto, CA 94306. (415) 493-4400. Annual. Inquire as to cost and availability.

CRC CRITICAL REVIEWS IN SOLID STATE AND MATERIALS SCIENCES. CRC Press, Incorporated, 2000 Corporate Boulevard, NW, Boca Raton, FL 33431. Irregular. Inquire as to cost and availability.

PROGRESS IN CRYSTAL GROWTH AND CHARACTERIZATION. Pergamon Press, Incorporated, Maxwell House, Fairview Park, Elmsford, NY 10523. (914) 592-7700. Irregular. Inquire as to cost and availability.

SOLID STATE PHYSICS - ADVANCES IN RESEARCH AND APPLICATIONS. Academic Press, Incorporated, 6277 Sea Harbor Drive, Orlando, FL 32821. (800) 321-5068. Irregular. as to cost and availability.

TOPICS IN APPLIED PHYSICS. Spring-Verlag New York, Incorporated, 175 Fifth Avenue, New York, NY 10010. (800) 526-7254. Irregular. Inquire as to cost and availability.

ASSOCIATIONS AND PROFESSIONAL SOCIETIES

AMERICAN INSTITUTE OF PHYSICS. 335 East 45th Street, New York, NY 10017. (212) 661-9404.

AMERICAN PHYSICAL SOCIETY. 335 East 45th Street, New York, NY 10017. (212) 682-7341.

BIBLIOGRAPHIES

SCIENCE BOOKS AND FILMS. American Association for the Advancement of Science, 1333 H Street, NW, Washington, DC 20005.

SCIENTIFIC AND TECHNICAL BOOKS AND SERIALS IN PRINT 1988: AN INDEX TO LITERATURE IN SCIENCE AND TECHNOLOGY. R.R. Bowker Company, 205 East 42nd Street, New York, NY 10017. (800) 521-8110 or (212) 916-1600. $175.00.

DIRECTORIES AND BIOGRAPHICAL SOURCES

AMERICAN ASSOCIATION OF PHYSICS TEACHERS - DIRECTORY OF MEMBERS. American Association Physics Teachers, 5110 Roanoke Place, College Park, MD 20740. Biennial. $2.00.

AMERICAN MEN AND WOMEN OF SCIENCE. Physical and Biological Sciences. Sixteenth edition. R.R. Bowker Company, 205 East 42nd Street, New York, NY 10017. (800) 521-8110 or (212) 916-1600. 1986. $595.00.

AMERICAN PHYSICAL SOCIETY MEMBERSHIP DIRECTORY BULLETIN ISSUE. American Physical Society, 335 East 45th Street, New York, NY 10017. (212) 682-7341. Biennial. $30.00.

BIOGRAPHICAL DICTIONARY OF SCIENTISTS: CHEMISTS. David Abbott, editor. P. Bedrick Books, 125 East 23rd Street, New York, NY 10010. (212) 777-1187. 1984. $18.95.

DIRECTORY OF PHYSICS AND ASTRONOMY STAFF MEMBERS. American Institute of Physics, 335 East 45th Street, New York, NY 10017. (212) 661-9404. Biennial. $30.00.

GOVERNMENT RESEARCH DIRECTORY. Gale Research Company, Book Tower, Detroit, MI 48226. (800) 521-0707. Fourth edition. 1987. $350.00.

INTERNATIONAL RESEARCH CENTERS DIRECTORY 1986-1987. Gale Research Company, Book Tower, Detroit, MI 48226. (800) 521-0707. Third edition. 1986. $330.00.

RESEARCH CENTERS DIRECTORY. Gale Research Company, Book Tower, Detroit, MI 48226. (800) 521-0707. Twelfth edition. 1987. $365.00 for set.

SCIENTIFIC AND TECHNICAL ORGANIZATIONS AND AGENCIES DIRECTORY. Gale Research Company, Book Tower, Detroit, MI 48226. (800) 521-0707. 1985. $150.00.

WHO'S WHO IN FRONTIER SCIENCE AND TECHNOLOGY. Marquis Who's Who, Incorporated, 200 East Ohio Street, Chicago, IL 60611. (800) 428-3898 or (312) 787-2008.

WHO'S WHO IN TECHNOLOGY TODAY. Reston Publishing Company, Incorporated, c/o Prentice-Hall, Incorporated, Englewood Cliffs, NJ 07632. (800) 262-6868. Biennial. Five volumes. $425.00. Covers the fields of electronics, computer science, physics, optics, chemistry, biotechnology, mechanics, energy, and earth science.

WORLD GUIDE TO SCIENTIFIC ASSOCIATIONS AND LEARNED SOCIETIES. K.G. Saur Incorporated, 175 Fifth Avenue, New York, NY 10010. (800) 521-0707 or (212) 982-1302. Fourth edition. 1984. $112.00.

ENCYCLOPEDIAS AND DICTIONARIES

ENCYCLOPEDIA OF PHYSICAL SCIENCE AND TECHNOLOGY. Academic Press, Incorporated, Orlando, FL 32887. (800) 321-5068 or (305) 345-2734. Inquire as to cost and availability.

ENCYCLOPEDIA OF PHYSICS. Robert M. Besancon, editor. Van Nostrand Reinhold, Incorporated, 115 Fifth Avenue, New York, NY 10003. (800) 543-2681. Third edition. 1985. $99.95.

MCGRAW-HILL ENCYCLOPEDIA OF SCIENCE AND TECHNOLOGY. McGraw-Hill Book, Incorporated, 1221 Avenue of the Americas, New York, NY 10020. (212) 997-3675. Fifth edition, 15 volumes. $1100.00.

GENERAL WORKS

FUNDAMENTALS OF PHYSICS. D. Halliday and R. Resnik. John Wiley and Sons, Incorporated, 605 Third Avenue, New York, NY 10158. (800) 526-5368 or (212) 850-6000. Second edition. 1986. $42.95.

HANDBOOKS AND MANUALS

CRC HANDBOOK OF CHEMISTRY AND PHYSICS. CRC Press, Incorporated, 2000 Corporate Boulevard, NW, Boca Raton, FL 33431. Sixty-seventh edition. 1986. $69.95.

HANDBOOKS AND TABLES IN SCIENCE AND TECHNOLOGY. Russell H. Powell, editor. Oryx Press, 2214 North Central Avenue, Phoenix, AZ 85004-1483. (602) 254-6156. Second edition. 1983. $55.00.

MATHEMATICAL METHODS FOR PHYSICSTS. George B. Arfken. Academic Press, Incorporated, 6277 Sea Harbor Drive, Orlando, FL 32821. (800) 321-5068. Third edition. 1985. $44.50.

ONLINE DATA BASES

CA SEARCH. Chemical Abstracts Service, Post Office Box 3012, Columbus, OH 43210. Guide to chemical literature, 1972 to present. Inquire as to cost and availability.

DISSERTATION ABSTRACTS ONLINE. University Microfilms International, 300 North Zeeb Road, Ann Arbor, MI 48106. (800) 521-0600 or (313) 761-4700. Scope includes virtually all doctoral dissertations accepted at accredited American institutions from 1861 to present in 252 subject areas. Inquire as to cost and availability.

INSPEC. INSPEC Marketing Department, Institute of Electrical and Electronics Engineers, Incorporated, IEEE Service Department, 445 Hoes Lane, Piscataway, NJ 08854. (201) 981-0060. Inquire as to on-line cost and availability.

MATHFILE. American Mathematical Society, Post Office Box 6248, Providence, RI 02940. (800) 556-7774 or (401) 272-9500. Scope includes pure and applied mathematics and related areas of physics, statistics, engineering, computer science, and operations research literature since 1973. Inquire as to cost and availability.

NASA. National Aeronautics and Space Administration, Scientific and Technical Information Branch, 300 7th Street, SW, Washington, DC 20546. Citations and abstracts of aerospace literature, 1962 to present. Inquire as to cost and availability.

NTIS. National Technical Information Service, 5285 Port Royal Road, Springfield, VA 22161. (703) 487-4630. Broad coverage of government-sponsored research reports, 1964 to present. Inquire as to cost and availability.

SCISEARCH. Institute for Scientific Information, 3501 Market Street, Philadelphia, PA 19104. (800) 523-1850 or (215) 386-0100. Broad multidisciplinary title and author index to the

international literature of science and technology, 1974 to present. Inquire as to cost and availability.

WILSONLINE. H.W. Wilson Company, 950 University Avenue, Bronx, NY 10452. (800) 367-6770 or (212) 588-8400. Makes available online versions of the printed H.W. Wilson Indexes including Applied Science and Technology Index, Business Periodicals Index, and Readers' Guide to Periodical Literature. Period covered is generally 1983 to present. Inquire as to cost and availability.

PERIODICALS

AMERICAN JOURNAL OF PHYSICS. Ameriacan Association of Physics Teachers, 5100 Roanoke Place, College Park, MD 20740. Monthly. $104.00 per year.

APPLIED PHYSICS A: SOLIDS AND SURFACES. (Deutsche Physikalische Gesellschaft, G.W.). Springer-Verlag, 175 Fifth Avenue, New York, NY 10010. (212) 460-1500. Twelve times per year. $325.00 per year.

CANADIAN JOURNAL OF PHYSICS. National Research Council of Canada, Research Journals, Ottawa, ON K1A OR6, Canada, Monthly. $78.00 per year.

CONTEMPORARY PHYSICS. Taylor and Francis, Incorporated, 242 Cherry Street, Philadelphia, PA 19106-1906. (215) 238-0939. Bimonthly. $184.00 per year.

JOURNAL OF APPLIED PHYSICS. American Institute of Physics, 335 East 45th Street, New York, NY 10017. (212) 661-9404. Semimonthly. $495.00 per year.

JOURNAL OF NON-CRYSTALLINE SOLIDS: A JOURNAL DEVOTED TO GLASSES AND AMORPHOUS MATERIALS. Elsevier Science Publishers B.V., Box 211, 1000 AE Amsterdam, The Netherlands. Thirty times per year. $1100.00 per year.

JOURNAL OF PHYSICS. A. Mathematical and General Physics. American Institute of Physics, 335 East 45th Street, New York, NY 10017. (212) 661-9404. Eighteen times per year. $510.00 per year.

JOURNAL OF PHYSICS. B. Atomic and Molecular Physics. American Institute of Physics, 335 East 45th Street, New York, NY 10017. (212) 661-9404. Semimonthly. $620.00 per year.

JOURNAL OF PHYSICS. C. Solid State Physics. American Institute of Physics, 335 East 45th Street, New York, NY 10017. (212) 661-9404. Thirty-six times per year. $840.00 per year.

JOURNAL OF PHYSICS AND CHEMISTRY OF SOLIDS. Pergamon Press, Incorporated, Journals Division, Maxwell House, Fairview Park, Elmsford, NY 10523. (914) 592-7700. Monthly. $530.00 per year.

OPTICAL SOCIETY OF AMERICA. Journal Part B: Optical Physics. American Institute of Physics, 335 East 45th Street, New York, NY 10017. (212) 661-9404. Monthly. $180.00 per year.

PHYSICAL REVIEW B (CONDENSED MATTER). American Institute of Physics, 335 East 45th Street, New York, NY 10017. (212) 661-9404. Semimonthly. $850.00 per year.

PHYSICAL REVIEW B (CONDENSED MATTER). American Institute of Physics, 335 East 45th Street, New York, NY 10017. (212) 661-9404. Monthly. $345.00 per year.

PHYSICAL REVIEW LETTERS. American Institute of Physics, 335 East 45th Street, New York, NY 10017. (212) 661-9404. Weekly. $430.00 per year.

PHYSICS REPORTS - REVIEW SECTION OF PHYSICS LETTERS. Elsevier Science Publishing Company, Incorporated, 52 Vanderbitl Avenue, New York, NY 10017. (212) 370-5520. Seventy-two times per year. $920.00 per year.

PHYSICS TODAY. American Institute of Physics, 335 East 45th Street, New York, NY 10017. (212) 661-9404. Monthly. $50.00 per year.

REVIEWS OF MODERN PHYSICS. American Institute of Physics, 335 East 45th Street, New York, NY 10017. (212) 661-9404. Quarterly. $145.00 per year.

SOLID STATE COMMUNICATIONS. Pergamon Press, Incorporated, Journals Division, Maxwell House, Fairview Park, Elmsford, NY 10523. (914) 592-7700. Forty-eight times per year. $690.00 per year.

SOVIET PHYSICS - SEMICONDUCTORS (TRANSLATION OF: FIZIKA I TECKNIKA POLUPROVODNIKOV). American Institute of Physics, 335 East 45th Street, New York, NY 10017. (212) 661-9404. Monthly. $810.00 per year.

SOVIET PHYSICS - SOLID STATE (TRANSLATION OF: FIZIKA TVERDOGE TELA.). American Institute of Physics, 335 East 45th Street, New York, NY 10017. (212) 661-9404. Monthly. $850.00 per year.

RESEARCH CENTERS AND INSTITUTES

CLINTON P. ANDERSON MESON PHYSICS FACILITY. Los Alamos, NM 87545. (505) 667-5907.

OHIO STATE UNIVERSITY. Engineering Experiment Station, 2070 Neil Avenue, Columbus, OH 43210. (614) 422-2411.

PENNSYLVANIA STATE UNIVERSITY. Center for Electronic Materials and Devices, 113 Electrical Engineering, West Building, University Park, PA 16802. (814) 865-1666.

STANFORD UNIVERSITY. Center for Integrated Systems, Stanford, CA 94305. (415) 725-3620.

UNIVERSITY OF ARIZONA. Microelectronics Laboratory, College of Engineering and Mines, Tucson, AZ 85721. (602) 621-6095.

UNIVERSITY OF WISCONSIN - MADISON. Theoretical Chemistry Institute, 1101 University Avenue, Madison, WI 53706. (608) 262-1511.

SOLID WASTE DISPOSAL

See also: ENVIRONMENTAL ENGINEERING, GROUND WATER POLLUTION, WATER POLLUTION, WATER TREATMENT

ABSTRACT SERVICES AND INDEXES

APPLIED SCIENCE AND TECHNOLOGY INDEX. H.W. Wilson Company, 950 University Avenue, Bronx, NY 10452. (800) 367-6670 or (212) 588-8400. Inquire as to cost and availability.

CHEMICAL ABSTRACTS. Chemical Abstracts Service, 2540 Olentangy Road, Post Office Box 3012, Columbus, OH 43210. (800) 848-6538 or (614) 421-3600. Weekly. $9200.00 per year.

CURRENT CONTENTS: PHYSICAL, CHEMICAL AND EARTH SCIENCES. Institute for Scientific Information, 3501 Market Street, Philadelphia, PA 19104. (800) 523-1850 or (215) 386-0100. 1970 to present. Weekly. $272.00 per year.

ENGINEERING INDEX MONTHLY AND AUTHOR INDEX. Engineering Information, Incorporated, 345 East 47th Street, New York, NY 10017. (800) 221-1044 or (212) 705-7600. Monthly, with annual cumulation. $1560.00 per year.

POLLUTION ABSTRACTS. Cambridge Scientific Abstracts, 5161 River Road, Bethesda, MD 10816. (301) 951-1400. 1970 to present. Bimonthly. $465.00 per year.

SCIENCE CITATION INDEX. Institute for Scientific Information, 3501 Market Street, Philadelphia, PA 19104. (800) 523-1850 or (215) 386-0100. Inquire as to cost and availability.

ANNUAL REVIEWS AND YEARBOOKS

SANITATION INDUSTRY YEARBOOK. Communications Channels, Incorporated, 6255 Barfield Road, Atlanta, GA 30328. (404) 256-9800. Annual. $18.00.

ASSOCIATIONS AND PROFESSIONAL SOCIETIES

AMERICAN PUBLIC WORKS ASSOCIATION. 1313 East 60th Street, Chicago, IL 60637. (312) 667-2200.

AMERICAN SOCIETY OF CIVIL ENGINEERS. 345 East 47th Street, New York, NY 10017-2398. (212) 705-7520.

ASSOCIATION OF STATE AND TERRITORIAL SOLID WASTE MANAGEMENT OFFICIALS. 444 North Capitol Street, NW, Suite 345, Washington, DC 20001. (202) 624-5828.

NATIONAL SOLID WASTES MANAGEMENT ASSOCIATION. 1730 Rhode Island Avenue, NW, Suite 1000, Washington, DC 20036. (202) 659-4613.

WATER POLLUTION CONTROL FEDERATION. 2626 Pennsylvania Avenue, Washington, DC 20037. (202) 337-2500.

WATER QUALITY ASSOCIATION. 4151 Naperville Road, Lisle, IL 60532. (312) 369-1600.

BIBLIOGRAPHIES

HAZARDOUS WASTE SERVICES DIRECTORY. J.J. Keller and Associates, Incorporated, 145 West Wisconsin Avenue, Nenah, WI 54956. (414) 722-2848. Semi-annual updates. $75.00, plus $55.00 per year.

MANAGEMENT OF WORLD WASTES BUYERS GUIDE ISSUE. Communications Channels, Incorporated, 6255 Barfield Road, Atlanta, GA 30328. (404) 256-9800. Annual. $2.50.

PUBLIC WORKS MANUAL. Public Works Journal Corporation, 200 South Broad Street, Ridgewood, NJ 07451. (201)

DIRECTORIES AND BIOGRAPHICAL SOURCES

AMERICAN WATER WORKS ASSOCIATION - OFFICERS AND COMMITTEE DIRECTORY. American Water Works Association, 6666 West Quincy Avenue, Denver, CO 80235. (303) 794-7711. Annual. $15.00.

JOURNAL OF THE AMERICAN WATER WORKS ASSOCIATION BUYERS GUIDE ISSUE. American Water Works Association, 6666 West Quincy Avenue, Denver, CO 80235. (303) 794-7711. Annual, with subscription.

PUBLIC WORKS MANUAL. Public Works Journal Corporation, 200 South Broad Street, Ridgewood, NJ 07451. (201) 445-5800. Annual. $20.00.

SCIENTIFIC AND TECHNICAL ORGANIZATIONS AND AGENCIES DIRECTORY. Gale Research Company, Book Tower, Detroit, MI 48226. (800) 521-0707. Second edition. Two volumes. 1987. $185.00 set.

TRASH AND GARBAGE REMOVAL DIRECTORY. American Business Directories, Incorporated, Division of American Business Lists, Incorporated, 5707 South 86th Street, Omaha, NE 68127. (402) 331-7169.

WATER QUALITY ASSOCIATION DIRECTORY. Water Quality Association, 4151 Naperville Road, Lisle, IL 60532. (312) 369-1600. Annual. Available to members only.

GENERAL WORKS

APPLIED STREAM SANITATION. C.J. Velz. John Wiley and Sons, Incorporated, 605 Third Avenue, New York, NY 10158. (800) 526-5368 or (212) 850-6000. Second edition. 1984. $79.95.

ENVIRONMENTAL ENGINEERING AND SANITATION. J.A. Salvato. John Wiley and Sons, Incorporated, 605 Third Avenue, New York, NY 10158. (800) 526-5368 or (212) 850-6000. Third edition. 1982. $70.00.

PRACTICAL WASTE MANAGEMENT. John R. Holmes, editor. John Wiley and Sons, Incorporated, 605 Third Avenue, New York, NY 10158. (800) 526-5368 or (212) 850-6000. 1983. $108.00.

SOLID WASTES: ORIGINS, COLLECTION, PROCESSING AND DISPOSAL. C.L. Mantell. John Wiley and Sons, Incorporated, 605 Third Avenue, New York, NY 10158. (800) 526-5368 or (212) 850-6000. 1975. $110.00.

UNIT OPERATIONS MODELS FOR SOLID WASTE PROCESSING. G.M. Savage. Noyes Data Corporation, Mill Road at Grand Avenue, Park Ridge, NJ 07656. (201) 391-8484. 1986. $36.00.

HANDBOOKS AND MANUALS

SOLID WASTE HANDBOOK: A PRACTICAL GUIDE. W.D. Robinson. John Wiley and Sons, Incorporated, 605 Third Avenue, New York, NY 10158. (800) 526-5368 or (212) 850-6000. 1986. $75.95.

ONLINE DATA BASES

COMPENDEX. Engineering Information, Incorporated, 345 East 47th Street, New York, NY 10017. (800) 221-1044 or (212) 705-7615. Engineering and technical literature, 1975 to present. Inquire as to cost and availability.

ENVIROLINE. EIC Intelligence, Incorporated, 48 West 38th Street, New York, NY 10018. (212) 944-8500. Worldwide environmental literature, 1979 to present. Inquire as to online cost and availability.

GEOREF. American Geological Institute, 4220 King Street, Alexandria, VA 22302. (703) 379-2480. 1967 to present. Inquire as to online cost and availability.

NTIS. National Technical Information Service, 5285 Port Royal Road, Springfield, VA 22161. (703) 487-4630. Broad coverage of government-sponsored research reports, 1964 to present. Inquire as to cost and availability.

POLLUTION ABSTRACTS. Cambridge Scientific Abstracts, 5161 River Road, Bethesda, MD 20816. (301) 951-1400. 1970 to present. Available for online searching through DIALOG Information Services and BRS Information Technologies. Inquire as to online cost and availability.

PERIODICALS

JOURNAL OF ENVIRONMENTAL ENGINEERING. American Society of Civil Engineers, 345 East 47th Street, New York, NY 10017. (212) 705-7275. 1956 to present. Bimonthly. $80.00 per year.

JOURNAL OF ENVIRONMENTAL SCIENCES. Institute of Environmental Sciences, 940 East Northwest Highway, Moutn Prospect, IL 60056. (312) 255-1561. Bimonthly. $25.00 per year.

MANAGEMENT OF WORLD WASTES. Communications Channels, Incorporated, 6255 Barfield Road, Atlanta, GA 30328. (404) 256-9800. Monthly. $27.00 per year.

SOLID WASTE REPORT. Business Publishers, Incorporated, 951 Pershing Drive, Silver Spring, MD 20910. (301) 587-6300. Biweekly. $197.00 per year.

WASTE AGE. National Solid Wastes Management Association. 1730 Rhode Island Avenue, NW, Suite 1000, Washington, DC 20036. (202) 659-4613. Monthly. $32.00 per year.

WATER, AIR AND SOIL POLLUTION. D. Reidel Publishing Company, 190 Old Derby Street, Hingham, MA 02043. Twenty times per year. $550.00 per year.

RESEARCH CENTERS AND INSTITUTES

CENTER HILL SOLID AND HAZARDOUS WASTE RESEARCH LABORATORY. Central Hill Laboratory, 5995 Center Hill Road, Cincinnati, OH 45224. (513) 569-7885.

ILLINOIS INSTITUTE OF TECHNOLOGY. Industrial Waste Elimination Research Center, 3200 South State Street, Chicago, IL 60616. (312) 567-3535.

LENOX INSTITUTE FOR RESEARCH, INCORPORATED. 101 Yokum Avenue, Lenox, MA 01240. (413) 637-3025.

MARYLAND DEPARTMENT OF NATURAL RESOURCES. Maryland Environmental Service, Tawes State Office Building, Annapolis, MD 21401.

PENNSYLVANIA STATE UNIVERSITY. Environmental Engineering Laboratory, 212 Sackett Building, University Park, PA 16802. (814) 863-4385.

UNIVERSITY OF ILLINOIS. Advanced Environmental Control Technology Research Center, 3230 Newmark C.E. Laboratory, 208 North Romine Street, Urbana, IL 61801. (217) 333-3822.

WASHINGTON STATE UNIVERSITY. Environmental Engineering Research Laboratory, 141 Sloan, Pullman, WA 99164. (509) 335-3175.

SONAR

See also: ACOUSTICS, ANTENNAS, ELECTRICAL ENGINEERING, ELECTRONIC CIRCUITS AND COMPONENTS, ELECTRONICS, ELECTRONICS ENGINEERING, MICROWAVES, RADAR, RADIO

ABSTRACT SERVICES AND INDEXES

APPLIED SCIENCE AND TECHNOLOGY INDEX. H.W. Wilson and Company, 950 University Avenue, Bronx, NY 10452. (800) 367-6670 or (212) 588-8400. Monthly. Inquire as to cost and availability.

CURRENT CONTENTS: ENGINEERING, TECHNOLOGY AND APPLIED SCIENCES. Institute for Scientific Information, 3501 Market Street, Philadelphia, PA 19104. (800) 523-1850 or (215) 386-0100. Weekly. $275.00 per year.

ELECTRICAL AND ELECTRONICS ABSTRACTS. Institution of Electrical Engineers. Available from: Institute of Electrical and Electronics Engineers. IEEE Service Center, 445 Hoes Lane, Piscataway, NJ 08854. Monthly. $1250.00 per year.

ELECTRONICS AND COMMUNICATIONS ABSTRACTS. Cambridge Scientific Abstracts, 5161 River Road, Bethesda, MD 20816. (301) 951-1400. Bimonthly. Inquire as to cost and availability.

ENGINEERING INDEX MONTHLY AND AUTHOR INDEX. Engineering Information Inc., 345 East 47th Street, New York, NY 10017. (212) 705-7600. Monthly. $1560.00 per year.

IEEE PUBLICATIONS BULLETIN. Institute of Electrical and Electronics Engineers. Institute of Electrical and Electronics Engineers. IEEE Service Center, 445 Hoes Lane, Piscataway, NJ 08854. Quarterly. Free.

PHYSICS ABSTRACTS. Institution of Electrical Engineers. Available from: IEEE Service Center, 445 Hoes Lane, Piscataway, NJ 08854. 1898 to present. Bimonthly. $1700.00 per year.

PHYSICS BRIEFS. Physik Verlag GmbH, Postfach 1260/1280, D-6940 Weinheim, West Germany. (212) 661-9404. 1920 to present. Twenty-six times per year. $1250.00 per year.

SCIENCE CITATION INDEX. Institute for Scientific Information, 3501 Market Street, Philadelphia, PA 19104. (800) 523-1850 or (215) 386-0100. Six times per year. $6200.00 per year.

SOLID STATE ABSTRACTS: AN ABSTRACT JOURNAL INVOLVING THE PHYSICS, METALLURGY, CRYSTALLOGRAPHY, CHEMISTRY, AND DEVICE TECHNOLOGY OF SOLIDS. Cambridge Scientific Abstracts, 5161 River Road, Bethesda, MD 20816. (301) 951-1400. 1957 to present. Bimonthly. $550.00 per year.

ANNUAL REVIEWS AND YEARBOOKS

ADVANCES IN ELECTRONICS AND ELECTRON PHYSICS. Academic Press, Inc., 6277 Sea Harbor Drive, Orlando, FL 32821. (800) 321-5068. Irregular. Approximately $80.00 per volume.

ASSOCIATIONS AND PROFESSIONAL SOCIETIES

AMERICAN ELECTRONICS ASSOCIATION. P.O. Box 10045, 2670 Hanover Street, Palo Alto, CA 94303. (415) 857-9300.

AMERICAN INSTITUTE OF PHYSICS. 335 East 45th Street, New York, NY 10017. (212) 661-9494.

ELECTRONICS INDUSTRIES ASSOCIATION. 2001 Eye Street, N.W., Washington, DC 20006. (202) 457-4900.

INSTITUTE OF ELECTRICAL AND ELECTRONICS ENGINEERS. 345 East 47th Street, New York, NY 10017. (212) 705-7900.

INSTITUTE OF RADIO ENGINEERS. Institute of Electrical and Electronics Engineers, 345 East 47th Street, New York, NY 10017. (212) 705-7900.

NATIONAL ASSOCIATION OF RADIO AND TELECOMMUNICATIONS ENGINEERS. P.O. Box 15029, Salem, OR 97309. (503) 581-7653.

UNITED STATES NATIONAL COMMITTEE FOR THE INTERNATIONAL UNION OF RADIO SCIENCE. Board on Physics and Astronomy, National Research Council, 2101 Constitution Avenue, N.W., Washington, DC 20418. (202) 334-3559.

DIRECTORIES AND BIOGRAPHICAL SOURCES

AMERICAN MEN AND WOMEN OF SCIENCE. R.R. Bowker, Inc., Order Department, 245 West 17th Street, New York, NY 10011. (800) 521-8110. Eight volumes. 1986. $595.00 for set.

IEEE MEMBERSHIP DIRECTORY. Institute of Electrical and Electronics Engineers. IEEE Service Center, 445 Hoes Lane, Piscataway, NJ 08854. Annual. $7.00.

INTERNATIONAL RESEARCH CENTERS DIRECTORY 1988-89. Darren L. Smith, editor. Gale Research Company, Book Tower, Detroit, MI 48226. (800) 521-0707. 4th edition. 1987. $360.00.

1987 DIRECTORY OF ENGINEERING SOCIETIES AND RELATED ORGANIZATIONS. Gordon Davis, editor. Hemisphere Publishing Corporation, 1010 Vermont Avenue, NW, Washington, DC 20005. (800) 526-0275. 12th edition. 1987. $100.00.

RESEARCH CENTERS DIRECTORY 1988. Gale Research Company, Book Tower, Detroit, MI 48226. (800) 521-0707. 12th edition. 1987. $365.00 for set.

SCIENTIFIC AND TECHNICAL ORGANIZATIONS AND AGENCIES DIRECTORY. Margaret Labash Young, editor. Gale Research Company, Book Tower, Detroit, MI 48226. (800) 521-0707. 2nd edition. 1987. $185.00.

WHO'S WHO IN ELECTRONICS. Harris Publishing Company, 2057-2 Aurora Road, Twinsburg, OH 44087. (216) 425-9143. Annual. $90.00.

WHO'S WHO IN ENGINEERING. Gordon Davis, editor. Hemisphere Publishing Corporation, 1010 Vermont Avenue, NW, Washington, DC 20005. (800) 526-0275. 6th edition. 1985. $200.00.

ENCYCLOPEDIAS AND DICTIONARIES

IEEE STANDARD DICTIONARY OF ELECTRICAL AND ELECTRONICS TERMS. Frank Jay, editor. John Wiley and Sons, Inc., 605 Third Avenue, New York, NY 10158. (800) 526-5368. 3rd edition. 1984. $49.95.

GENERAL WORKS

ANTENNAS AND RADIO WAVE PROPAGATION. R.E. Collin. McGraw-Hill Book Company, 1221 Avenue of the Americas, New York, NY 10020. (212) 512-2000. 1985. $49.95.

ELECTROMAGNETIC CONCEPTS AND PRINCIPLES. Stanley V. Marshall and Gabriel G. Skitek. Prentice-Hall Publishing, Inc., Englewood Cliffs, NJ 07632. (800) 562-0245. 2nd edition. 1987. $42.95.

PRINCIPLES OF UNDERWATER SOUND. R.J. Urick. McGraw-Hill Book Company, 1221 Avenue of the Americas, New York, NY 10020. (212) 512-2000. Third edition. 1983. $53.00.

RADAR PRINCIPLES FOR THE NON-SPECIALIST. John C. Toomay. Van Nostrand Reinhold Company, Inc., 135 West 50th Street, New York, NY 10020. (800) 543-2681. 1982. $24.95.

SHIPBOARD ELECTROMAGNETICS. Preston E. Law. Artech House, 685 Canton Street, Norwood, MA 02062. (800) 225-9977. Second edition. 1986. $67.00.

SONAR AND UNDERWATER SOUND. Albert W. Cox. Lexington Books, 125 Spring Street, Lexington, MA 02173. (800) 235-3565. 1975. $24.00.

SONAR IMAGES. Harold E. Egerton. Prentice-Hall Publishing, Inc., Englewood Cliffs, NJ 07632. (800) 562-0245. 1986. $36.95.

HANDBOOKS AND MANUALS

ANTENNA HANDBOOK: THEORY, APPLICATIONS AND DESIGN. Y.T. Lo and S.W. Lee. Van Nostrand Reinhold Company, Inc., 135 West 50th Street, New York, NY 10020. (800) 543-2681. 1988. $129.95.

ELECTRONIC ENGINEERS HANDBOOK. Donald G. Fink, editor. McGraw-Hill Book Company, 1221 Avenue of the Americas, New York, NY 10020. (212) 512-2000. 2nd edition. 1982. $89.00.

HANDBOOK OF MODERN ELECTRONICS AND ELECTRICAL ENGINEERING. Charles Belove, editor. John Wiley and Sons, Inc., 605 Third Avenue, New York, NY 10158. (800) 526-5368. 1986. $88.95.

ONLINE DATA BASES

COMPENDEX. Engineering Information, Inc., 345 East 47th Street, New York, NY 10017. (800) 221-1044 or (212) 705-7615. Engineering and technical literature, 1975 to present. Inquire as to online cost and availability.

INSPEC. INSPEC Marketing Department, Institution of Electrical Engineers. Available from IEEE Service Center, 445 Hoes Lane, Piscataway, NJ 08854. (201) 981-0060. Online version of Physics Abstracts. Inquire as to online cost and availability.

NTIS. National Technical Information Service, 5285 Port Royal Road, Springfield, VA 22161. (703) 487-4630. Broad coverage of government sponsored research reports, 1964 to present. Inquire as to online cost and availability.

SCISEARCH. Institute for Scientific Information, 3501 Market Street, Philadelphia, PA 19104. (800) 523-1850 or (215) 386-0100. Broad multidisciplinary title and author index to the international literature of science and technology, 1974 to present. Inquire as to online cost and availability.

WILSONLINE. H.W. Wilson and Company, 950 University Avenue, Bronx, NY 10452. (800) 367-6770 or (212) 588-8400. Makes available online versions of the H.W. Wilson indexes including Applied Science and Technology Index, Business Periodicals Index and Readers' Guide to Periodical Literature. Approximately 1980 to present. Inquire as to online cost and availability.

OTHER SOURCES

A GUIDE TO THE LITERATURE OF ELECTRICAL AND ELECTRONICS ENGINEERING. Susan B. Ardis. Libraries Unlimited Inc., P.O. Box 263, Littleton, CO 80160. (303) 770-1220. 1987. $37.50.

PERIODICALS

ELECTRONIC DESIGN. Hayden Publishing Company, 10 Mulholland Drive, Hasbrouck Heights, NJ 07604. (201) 288-7520. 1952 to present. Biweekly. $40.00 per year.

ELECTRONICS. McGraw-Hill Book Company, 1221 Avenue of the Americas, New York, NY 10020. (212) 512-2000. 1930 to present. Weekly. $32.00 per year.

ELECTRONICS AND WIRELESS WORLD. I.P.C. Electrical-Electronic Press, Ltd., Quadrant House, The Quadrant, Sutton, Surrey, SM2 5AS England. 1911 to present. Monthly. $105.00 per year.

IEE PROCEEDINGS PART H: MICROWAVES, ANTENNAS AND PROPAGATION. Institution of Electrical Engineers (London). Available from: Institute of Electrical and Electronics Engineers. IEEE Service Center, 445 Hoes Lane, Piscataway, NJ 08854. 1980 to present. Bimonthly. Inquire.

IEEE CIRCUITS AND DEVICES MAGAZINE. Institute of Electrical and Electronics Engineers. IEEE Service Center, 445 Hoes Lane, Piscataway, NJ 08854. Bimonthly. $70.00 per year.

IEEE JOURNAL OF SOLID STATE CIRCUITS. Institute of Electrical and Electronics Engineers. IEEE Service Center, 445 Hoes Lane, Piscataway, NJ 08854. 1966 to present. Bimonthly. $113.00 per year.

IEEE TRANSACTIONS ON ANTENNAS AND PROPAGATION. Institute of Electrical and Electronics Engineers. IEEE Service Center, 445 Hoes Lane, Piscataway, NJ 08854. 1952 to present. Monthly. $140.00 per year.

RESEARCH CENTERS AND INSTITUTES

APPLIED MICROELECTRONICS INSTITUTE. 1127 Barrington Street, Halifax, NS, Canada B3H 2P8. (902) 423-8227.

APPLIED RESEARCH LABORATORIES. University of Texas at Austin, P.O. Box 8029, University Station, Austin, TX 78713. (512) 835-3200.

RADAR AND INSTRUMENTATION LABORATORY. Georgia Institute of Technology, Georgia Tech Research Institute, Atlanta, GA 30332. (404) 424-9621.

RADAR SYSTEMS AND REMOTE SENSING LABORATORY. 2291 Irving Hill Drive, Campus West, Lawrence, KS 66045. (913) 864-4832.

SOUND RECORDING

See: AUDIO ENGINEERING

SPACE SHUTTLE

See also: AEROSPACE ENGINEERING, MANNED SPACE FLIGHT, SPACE STATION, ROCKETS

ABSTRACT SERVICES AND INDEXES

APPLIED SCIENCE AND TECHNOLOGY INDEX. H.W. Wilson and Company, 950 University Avenue, Bronx, NY 10452. (800) 367-6670 or (212) 588-8400. Monthly. Inquire as to cost and availability.

ENGINEERING INDEX MONTHLY AND AUTHOR INDEX. Engineering Information Inc., 345 East 47th Street, New York, NY 10017. (212) 705-7600. Monthly. $1560.00 per year.

GENERAL SCIENCE INDEX. H.W. Wilson and Company, 950 University Avenue, Bronx, NY 10452. (800) 367-6670 or (212) 588-8400. 1978 to present. Monthly. Inquire as to cost and availability.

INTERNATIONAL AEROSPACE ABSTRACTS. American Institute of Aeronautics and Astronautics, Technical Information Service, 370 L'Enfant Promenade, S.W., Washington, DC 20024. (202) 646-7400. 1961 to present. Semi-monthly. $700.00 per year.

SCIENCE CITATION INDEX. Institute for Scientific Information, 3501 Market Street, Philadelphia, PA 19104. (800) 523-1850 or (215) 386-0100. Six times per year. $6200.00 per year.

STAR (SCIENTIFIC AND TECHNICAL AEROSPACE REPORTS). National Aeronautics and Space Administration, Scientific and Technical Information Facility, Box 8757, Baltimore-Washington International Airport, MD 21240. (202) 453-8545. 1963 to present. Semi-monthly. $85.00 per year with indexes.

ASSOCIATIONS AND PROFESSIONAL SOCIETIES

AMERICAN ASTRONAUTICAL SOCIETY. 6212-B Old Keene Mill Court, Springfield, VA 22152. (703) 866-0020.

AMERICAN INSTITUTE OF AERONAUTICS AND ASTRONAUTICS. 370 L'Enfant Promenade, S.W., Washington, DC 20024. (202) 646-7400.

AMERICAN SPACE FOUNDATION. 111 Massachusetts Avenue, N.W., Suite 200, Washington, DC 20002. (202) 289-2293.

NATIONAL SPACE INSTITUTE. West Wing, Suite 203, 600 Maryland Avenue, S.W., Washington, DC 20024. (202) 484-1111.

PLANETARY SOCIETY. 65 North Catalina Avenue, Pasadena, CA 91106. (818) 793-5100.

SOCIETY OF AUTOMOTIVE ENGINEERS. 400 Commonwealth Drive, Warrendale, PA 15096. (412) 776-4841.

SPACE STUDIES INSTITUTE. P.O. Box 82, 285 Rosedale, Road, Princeton, NJ 08540. (609) 921-0377.

DIRECTORIES AND BIOGRAPHICAL SOURCES

AMERICAN ASTRONAUTICAL SOCIETY DIRECTORY. American Astronautical Society, 6212-B Old Keene Mill Court, Springfield, VA 22152. (703) 866-0020. 1986. $30.00.

AMERICAN INSTITUTE OF AERONAUTICS AND ASTRONAUTICS ROSTER. American Institute of Aeronautics and Astronautics, 370 L'Enfant Promenade, S.W., Washington, DC 20024. (202) 646-7400. 1984. $75.00.

INTERNATIONAL RESEARCH CENTERS DIRECTORY 1988-89. Darren L. Smith, editor. Gale Research Company, Book Tower, Detroit, MI 48226. (800) 521-0707. 4th edition. 1987. $360.00.

RESEARCH CENTERS DIRECTORY 1988. Gale Research Company, Book Tower, Detroit, MI 48226. (800) 521-0707. 12th edition. 1987. $365.00 for set.

SCIENTIFIC AND TECHNICAL ORGANIZATIONS AND AGENCIES DIRECTORY. Margaret Labash Young, editor. Gale Research Company, Book Tower, Detroit, MI 48226. (800) 521-0707. 2nd edition. 1987. $185.00.

WHO'S WHO IN ENGINEERING. Gordon Davis, editor. Hemisphere Publishing Corporation, 1010 Vermont Avenue, NW, Washington, DC 20005. (800) 526-0275. 6th edition. 1985. $200.00.

ENCYCLOPEDIAS AND DICTIONARIES

DICTIONARY OF SPACE TECHNOLOGY. J.A. Angelo, Jr. Facts on File, Inc., 460 Park Avenue South, New York, NY 10016. (212) 683-2244. 1982. $24.95.

GENERAL WORKS

CHALLENGER: A MAJOR MALFUNCTION. Malcolm McConnell. Doubleday and Company, 501 Franklin Avenue, Garden City, NY 11530. (516) 294-4561. 1987. $17.95.

LIVING ALOFT: HUMAN REQUIREMENTS FOR EXTENDED SPACEFLIGHT. NASA Special Publication SP-483. National Aeronautics and Space Administration, Washington, DC 20546. Available from: Superintendent of Documents, U.S. Government Printing Office, Washington, DC 20402. 1985.

SPACE: THE NEXT TWENTY-FIVE YEARS. Thomas R. McDonough. John Wiley and Sons, Inc., 605 Third Avenue, New York, NY 10158. (800) 526-5368. 1987. $17.95.

SPACE SHUTTLE. Melvyn Smith. Haynes Publishing, Inc., 861 Lawrence Drive, Newbury Park, CA 91320. (805) 498-6703. 1985. $22.95.

SPACE SHUTTLE AND SPACELAB UTILIZATION: NEAR-TERM AND LONG-TERM BENEFITS FOR MANKIND. G.W. Morgenthaler and M. Hollstein, editors. Univelt, Inc., P.O. Box 28130, San Diego, CA 92128. (619) 746-4005. 1978. $40.00.

SPACECRAFT DYNAMICS. T.R. Kane and others. McGraw-Hill Book Company, 1221 Avenue of the Americas, New York, NY 10020. (212) 512-2000. 1983. $55.00.

SPACEFARERS OF THE 80'S AND 90'S. Alcestis R. Oberg. Columbia University Press, 562 West 113th Street, New York, NY 10022. (212) 688-8885. 1985. $24.95.

ONLINE DATA BASES

CA SEARCH. Chemical Abstracts Service, P.O. Box 3012, Columbus, OH 43120. (800) 848-6538 or (614) 421-3600. Comprehensive guide to chemical and related literature, 1972 to present. Inquire as to online cost and availability.

COMPENDEX. Engineering Information, Inc., 345 East 47th Street, New York, NY 10017. (800) 221-1044 or (212) 705-7615.

Engineering and technical literature, 1975 to present. Inquire as to online cost and availability.

INSPEC. INSPEC Marketing Department, Institution of Electrical Engineers. Available from IEEE Service Center, 445 Hoes Lane, Piscataway, NJ 08854. (201) 981-0060. Online version of Physics Abstracts. Inquire as to online cost and availability.

NTIS. National Technical Information Service, 5285 Port Royal Road, Springfield, VA 22161. (703) 487-4630. Broad coverage of government sponsored research reports, 1964 to present. Inquire as to online cost and availability.

SCISEARCH. Institute for Scientific Information, 3501 Market Street, Philadelphia, PA 19104. (800) 523-1850 or (215) 386-0100. Broad multidisciplinary title and author index to the international literature of science and technology, 1974 to present. Inquire as to online cost and availability.

WILSONLINE. H.W. Wilson and Company, 950 University Avenue, Bronx, NY 10452. (800) 367-6770 or (212) 588-8400. Makes available online versions of the H.W. Wilson indexes including Applied Science and Technology Index, Business Periodicals Index and Readers' Guide to Periodical Literature. Approximately 1980 to present. Inquire as to online cost and availability.

PERIODICALS

AEROSPACE AMERICA. American Institute of Aeronautics and Astronautics, Technical Information Service, 370 L'Enfant Promenade, S.W., Washington, DC 20024. (202) 646-7400. 1932 to present. Monthly. $56.00 per year.

AEROSPACE ENGINEERING MAGAZINE. Society of Automotive Engineers, 400 Commonwealth Drive, Warrendale, PA 15096. (412) 776-4841. 1981 to present. Monthly. $24.00 per year.

AIAA JOURNAL. American Institute of Aeronautics and Astronautics, Technical Information Service, 370 L'Enfant Promenade, S.W., Washington, DC 20024. (202) 646-7400. 1963 to present. Monthly. $205.00 per year.

AVIATION WEEK AND SPACE TECHNOLOGY. McGraw-Hill Book Company, 1221 Avenue of the Americas, New York, NY 10020. (212) 512-2000. 1916 to present. Weekly. $56.00 per year.

JOURNAL OF SPACECRAFT AND ROCKETS. American Institute of Aeronautics and Astronautics, Technical Information Service, 370 L'Enfant Promenade, S.W., Washington, DC 20024. (202) 646-7400. 1964 to present. Bimonthly. $95.00 per year.

PLANETARY REPORT. Planetary Society, 65 North Catalina Avenue, Pasadena, CA 91106. (818) 793-5100. 1980 to present. Bimonthly. $20.00 per year.

SPACEFLIGHT. British Interplanetary Society, 27-29 South Lambeth Road, London, SW8 1SZ, England. 1956 to present. $50.00 per year.

RESEARCH CENTERS AND INSTITUTES

CENTER FOR SPACE RESEARCH AND APPLICATIONS. University of Texas at Austin, WRW 402, Austin, TX 78712. (512) 471-1356.

GODDARD SPACE FLIGHT CENTER. National Aeronautics and Space Administration, Greenbelt Road, Greenbelt, MD 20771. (301) 344-5121.

MARSHALL SPACE FLIGHT CENTER. National Aeronautics and Space Administration, Huntsville, AL 35812. (205) 453-2121.

SPACE SCIENCE AND ENGINEERING CENTER. University of Wisconsin, Madison, 1225 West Dayton Street, Madison, WI 53706. (608) 262-0544.

SPACE STATION

See also: AEROSPACE ENGINEERING, MANNED SPACE FLIGHT, SPACE SHUTTLE

ABSTRACT SERVICES AND INDEXES

APPLIED SCIENCE AND TECHNOLOGY INDEX. H.W. Wilson and Company, 950 University Avenue, Bronx, NY 10452. (800) 367-6670 or (212) 588-8400. Monthly. Inquire as to cost and availability.

ENGINEERING INDEX MONTHLY AND AUTHOR INDEX. Engineering Information Inc., 345 East 47th Street, New York, NY 10017. (212) 705-7600. Monthly. $1560.00 per year.

GENERAL SCIENCE INDEX. H.W. Wilson and Company, 950 University Avenue, Bronx, NY 10452. (800) 367-6670 or (212) 588-8400. 1978 to present. Monthly. Inquire as to cost and availability.

INTERNATIONAL AEROSPACE ABSTRACTS. American Institute of Aeronautics and Astronautics, Technical Information Service, 370 L'Enfant Promenade, S.W., Washington, DC 20024. (202) 646-7400. 1961 to present. Semi-monthly. $700.00 per year.

SCIENCE CITATION INDEX. Institute for Scientific Information, 3501 Market Street, Philadelphia, PA 19104. (800) 523-1850 or (215) 386-0100. Six times per year. $6200.00 per year.

STAR (SCIENTIFIC AND TECHNICAL AEROSPACE REPORTS). National Aeronautics and Space Administration, Scientific and Technical Information Facility, Box 8757, Baltimore-Washington International Airport, MD 21240. (202) 453-8545. 1963 to present. Semi-monthly. $85.00 per year with indexes.

ASSOCIATIONS AND PROFESSIONAL SOCIETIES

AMERICAN ASTRONAUTICAL SOCIETY. 6212-B Old Keene Mill Court, Springfield, VA 22152. (703) 866-0020.

AMERICAN INSTITUTE OF AERONAUTICS AND ASTRONAUTICS. 370 L'Enfant Promenade, S.W., Washington, DC 20024. (202) 646-7400.

AMERICAN SPACE FOUNDATION. 111 Massachusetts Avenue, N.W., Suite 200, Washington, DC 20002. (202) 289-2293.

NATIONAL SPACE INSTITUTE. West Wing, Suite 203, 600 Maryland Avenue, S.W., Washington, DC 20024. (202) 484-1111.

PLANETARY SOCIETY. 65 North Catalina Avenue, Pasadena, CA 91106. (818) 793-5100.

SOCIETY OF AUTOMOTIVE ENGINEERS. 400 Commonwealth Drive, Warrendale, PA 15096. (412) 776-4841.

SPACE STUDIES INSTITUTE. P.O. Box 82, 285 Rosedale, Road, Princeton, NJ 08540. (609) 921-0377.

DIRECTORIES AND BIOGRAPHICAL SOURCES

AMERICAN ASTRONAUTICAL SOCIETY DIRECTORY. American Astronautical Society, 6212-B Old Keene Mill Court, Springfield, VA 22152. (703) 866-0020. 1986. $30.00.

AMERICAN INSTITUTE OF AERONAUTICS AND ASTRONAUTICS ROSTER. American Institute of Aeronautics and Astronautics, 370 L'Enfant Promenade, S.W., Washington, DC 20024. (202) 646-7400. 1984. $75.00.

INTERNATIONAL RESEARCH CENTERS DIRECTORY 1988-89. Darren L. Smith, editor. Gale Research Company, Book Tower, Detroit, MI 48226. (800) 521-0707. 4th edition. 1987. $360.00.

RESEARCH CENTERS DIRECTORY 1988. Gale Research Company, Book Tower, Detroit, MI 48226. (800) 521-0707. 12th edition. 1987. $365.00 for set.

SCIENTIFIC AND TECHNICAL ORGANIZATIONS AND AGENCIES DIRECTORY. Margaret Labash Young, editor. Gale Research Company, Book Tower, Detroit, MI 48226. (800) 521-0707. 2nd edition. 1987. $185.00.

WHO'S WHO IN ENGINEERING. Gordon Davis, editor. Hemisphere Publishing Corporation, 1010 Vermont Avenue, NW, Washington, DC 20005. (800) 526-0275. 6th edition. 1985. $200.00.

GENERAL WORKS

LIVING ALOFT: HUMAN REQUIREMENTS FOR EXTENDED SPACEFLIGHT. NASA Special Publication SP-483. National Aeronautics and Space Administration, Washington, DC 20546. Available from: Superintendent of Documents, U.S. Government Printing Office, Washington, DC 20402. 1985.

THE SPACE STATION. T. Simpson, editor. Institute of Electrical and Electronics Engineers. IEEE Service Center, 445 Hoes Lane, Piscataway, NJ 08854. 1985. $19.95.

SPACE STATION PROGRAM: DESCRIPTION, APPLICATIONS, AND OPPORTUNITIES. Space Station Task Force, National Aeronautics and Space Administration. Noyes Data Corporation, Mill Road at Grand Avenue, Park Ridge, NJ 07656. (201) 391-8484. 1985. $67.00.

SPACE: THE NEXT TWENTY-FIVE YEARS. Thomas R. McDonough. John Wiley and Sons, Inc., 605 Third Avenue, New York, NY 10158. (800) 526-5368. 1987. $17.95.

SPACEFARERS OF THE 80'S AND 90'S. Alcestis R. Oberg. Columbia University Press, 562 West 113th Street, New York, NY 10022. (212) 688-8885. 1985. $24.95.

ONLINE DATA BASES

CA SEARCH. Chemical Abstracts Service, P.O. Box 3012, Columbus, OH 43120. (800) 848-6538 or (614) 421-3600. Comprehensive guide to chemical and related literature, 1972 to present. Inquire as to online cost and availability.

COMPENDEX. Engineering Information, Inc., 345 East 47th Street, New York, NY 10017. (800) 221-1044 or (212) 705-7615. Engineering and technical literature, 1975 to present. Inquire as to online cost and availability.

INSPEC. INSPEC Marketing Department, Institution of Electrical Engineers. Available from IEEE Service Center, 445 Hoes Lane, Piscataway, NJ 08854. (201) 981-0060. Online version of Physics Abstracts. Inquire as to online cost and availability.

NTIS. National Technical Information Service, 5285 Port Royal Road, Springfield, VA 22161. (703) 487-4630. Broad coverage of government sponsored research reports, 1964 to present. Inquire as to online cost and availability.

SCISEARCH. Institute for Scientific Information, 3501 Market Street, Philadelphia, PA 19104. (800) 523-1850 or (215) 386-0100. Broad multidisciplinary title and author index to the international literature of science and technology, 1974 to present. Inquire as to online cost and availability.

WILSONLINE. H.W. Wilson and Company, 950 University Avenue, Bronx, NY 10452. (800) 367-6770 or (212) 588-8400. Makes available online versions of the H.W. Wilson indexes including Applied Science and Technology Index, Business Periodicals Index and Readers' Guide to Periodical Literature. Approximately 1980 to present. Inquire as to online cost and availability.

PERIODICALS

AEROSPACE AMERICA. American Institute of Aeronautics and Astronautics, Technical Information Service, 370 L'Enfant Promenade, S.W., Washington, DC 20024. (202) 646-7400. 1932 to present. Monthly. $56.00 per year.

AEROSPACE ENGINEERING MAGAZINE. Society of Automotive Engineers, 400 Commonwealth Drive, Warrendale, PA 15096. (412) 776-4841. 1981 to present. Monthly. $24.00 per year.

AIAA JOURNAL. American Institute of Aeronautics and Astronautics, Technical Information Service, 370 L'Enfant Promenade, S.W., Washington, DC 20024. (202) 646-7400. 1963 to present. Monthly. $205.00 per year.

AVIATION WEEK AND SPACE TECHNOLOGY. McGraw-Hill Book Company, 1221 Avenue of the Americas, New York, NY 10020. (212) 512-2000. 1916 to present. Weekly. $56.00 per year.

JOURNAL OF SPACECRAFT AND ROCKETS. American Institute of Aeronautics and Astronautics, Technical Information Service, 370 L'Enfant Promenade, S.W., Washington, DC 20024. (202) 646-7400. 1964 to present. Bimonthly. $95.00 per year.

PLANETARY REPORT. Planetary Society, 65 North Catalina Avenue, Pasadena, CA 91106. (818) 793-5100. 1980 to present. Bimonthly. $20.00 per year.

SPACEFLIGHT. British Interplanetary Society, 27-29 South Lambeth Road, London, SW8 1SZ, England. 1956 to present. $50.00 per year.

RESEARCH CENTERS AND INSTITUTES

CENTER FOR SPACE RESEARCH AND APPLICATIONS. University of Texas at Austin, WRW 402, Austin, TX 78712. (512) 471-1356.

GODDARD SPACE FLIGHT CENTER. National Aeronautics and Space Administration, Greenbelt Road, Greenbelt, MD 20771. (301) 344-5121.

MARSHALL SPACE FLIGHT CENTER. National Aeronautics and Space Administration, Huntsville, AL 35812. (205) 453-2121.

SPACE SCIENCE AND ENGINEERING CENTER. University of Wisconsin, Madison, 1225 West Dayton Street, Madison, WI 53706. (608) 262-0544.

SPACE STRUCTURES

See: SPACE STATION

SPACECRAFT

See also: AEROSPACE ENGINEERING, MANNED SPACE FLIGHT, ROCKETS, SPACE SHUTTLE, SPACE STATION

ABSTRACT SERVICES AND INDEXES

APPLIED SCIENCE AND TECHNOLOGY INDEX. H.W. Wilson and Company, 950 University Avenue, Bronx, NY 10452. (800) 367-6670 or (212) 588-8400. Monthly. Inquire as to cost and availability.

ENGINEERING INDEX MONTHLY AND AUTHOR INDEX. Engineering Information Inc., 345 East 47th Street, New York, NY 10017. (212) 705-7600. Monthly. $1560.00 per year.

INTERNATIONAL AEROSPACE ABSTRACTS. American Institute of Aeronautics and Astronautics, Technical Information Service, 370 L'Enfant Promenade, S.W., Washington, DC 20024. (202) 646-7400. 1961 to present. Semi-monthly. $700.00 per year.

SCIENCE CITATION INDEX. Institute for Scientific Information, 3501 Market Street, Philadelphia, PA 19104. (800) 523-1850 or (215) 386-0100. Six times per year. $6200. 00 per year.

STAR (SCIENTIFIC AND TECHNICAL AEROSPACE REPORTS). National Aeronautics and Space Administration, Scientific and Technical Information Facility, Box 8757, Baltimore-Washington International Airport, MD 21240. (202) 453-8545. 1963 to present. Semi-monthly. $85.00 per year with indexes.

ASSOCIATIONS AND PROFESSIONAL SOCIETIES

AMERICAN ASTRONAUTICAL SOCIETY. 6212-B Old Keene Mill Court, Springfield, VA 22152. (703) 866-0020.

AMERICAN INSTITUTE OF AERONAUTICS AND ASTRONAUTICS. 370 L'Enfant Promenade, S.W., Washington, DC 20024. (202) 646-7400.

AMERICAN SPACE FOUNDATION. 111 Massachusetts Avenue, N.W., Suite 200, Washington, DC 20002. (202) 289-2293.

NATIONAL SPACE INSTITUTE. West Wing, Suite 203, 600 Maryland Avenue, S.W., Washington, DC 20024. (202) 484-1111.

PLANETARY SOCIETY. 65 North Catalina Avenue, Pasadena, CA 91106. (818) 793-5100.

SOCIETY OF AUTOMOTIVE ENGINEERS. 400 Commonwealth Drive, Warrendale, PA 15096. (412) 776-4841.

SPACE STUDIES INSTITUTE. P.O. Box 82, 285 Rosedale, Road, Princeton, NJ 08540. (609) 921-0377.

DIRECTORIES AND BIOGRAPHICAL SOURCES

AMERICAN ASTRONAUTICAL SOCIETY DIRECTORY. American Astronautical Society, 6212-B Old Keene Mill Court, Springfield, VA 22152. (703) 866-0020. 1986. $30.00.

AMERICAN INSTITUTE OF AERONAUTICS AND ASTRONAUTICS ROSTER. American Institute of Aeronautics and Astronautics, 370 L'Enfant Promenade, S.W., Washington, DC 20024. (202) 646-7400. 1984. $75.00.

INTERNATIONAL RESEARCH CENTERS DIRECTORY 1988-89. Darren L. Smith, editor. Gale Research Company, Book Tower, Detroit, MI 48226. (800) 521-0707. 4th edition. 1987. $360.00.

RESEARCH CENTERS DIRECTORY 1988. Gale Research Company, Book Tower, Detroit, MI 48226. (800) 521-0707. 12th edition. 1987. $365.00 for set.

SCIENTIFIC AND TECHNICAL ORGANIZATIONS AND AGENCIES DIRECTORY. Margaret Labash Young, editor. Gale Research Company, Book Tower, Detroit, MI 48226. (800) 521-0707. 2nd edition. 1987. $185.00.

WHO'S WHO IN ENGINEERING. Gordon Davis, editor. Hemisphere Publishing Corporation, 1010 Vermont Avenue, NW, Washington, DC 20005. (800) 526-0275. 6th edition. 1985. $200.00.

GENERAL WORKS

ENVOYS OF MANKIND. George S. Robinson and H.M. White. Smithsonian Institution Press, 955 L'Enfant Plaza, Suite 2100, Washingtin, DC 20560. (202) 287-3388. 1986. $19.95.

INTERSTELLAR MIGRATION AND THE HUMAN EXPERIENCE. Ben R. Finney and Eric M. Jones. University of California Press, 2120 Berkeley Way, Berkeley, CA 94720. (415) 642-6683. 1985. $19.95.

SPACE SHUTTLE. Melvyn Smith. Haynes Publishing, Inc., 861 Lawrence Drive, Newbury Park, CA 91320. (805) 498-6703. 1985. $22.95.

SPACE SHUTTLE AND SPACELAB UTILIZATION: NEAR-TERM AND LONG-TERM BENEFITS FOR MANKIND. G.W. Morgenthaler and M. Hollstein, editors. Univelt, Inc., P.O. Box 28130, San Diego, CA 92128. (619) 746-4005. 1978. $40.00.

SPACE: THE NEXT TWENTY-FIVE YEARS. Thomas R. McDonough. John Wiley and Sons, Inc., 605 Third Avenue, New York, NY 10158. (800) 526-5368. 1987. $17.95.

SPACECRAFT ATTITUDE DYNAMICS. P.C. Hughes. John Wiley and Sons, Inc., 605 Third Avenue, New York, NY 10158. (800) 526-5368. 1986. $47.95.

SPACECRAFT DYNAMICS. T.R. Kane and P.W. Likins and others. McGraw-Hill Book Company, 1221 Avenue of the Americas, New York, NY 10020. (212) 512-2000. 1983. $57.95.

SPACEFARERS OF THE 80'S AND 90'S. Alcestis R. Oberg. Columbia University Press, 562 West 113th Street, New York, NY 10022. (212) 688-8885. 1985. $24.95.

ONLINE DATA BASES

CA SEARCH. Chemical Abstracts Service, P.O. Box 3012, Columbus, OH 43120. (800) 848-6538 or (614) 421-3600. Comprehensive guide to chemical and related literature, 1972 to present. Inquire as to online cost and availability.

COMPENDEX. Engineering Information, Inc., 345 East 47th Street, New York, NY 10017. (800) 221-1044 or (212) 705-7615. Engineering and technical literature, 1975 to present. Inquire as to online cost and availability.

INSPEC. INSPEC Marketing Department, Institution of Electrical Engineers. Available from IEEE Service Center, 445 Hoes Lane, Piscataway, NJ 08854. (201) 981-0060. Online version of Physics Abstracts. Inquire as to online cost and availability.

NTIS. National Technical Information Service, 5285 Port Royal Road, Springfield, VA 22161. (703) 487-4630. Broad coverage of government sponsored research reports, 1964 to present. Inquire as to online cost and availability.

SCISEARCH. Institute for Scientific Information, 3501 Market Street, Philadelphia, PA 19104. (800) 523-1850 or (215) 386-0100. Broad multidisciplinary title and author index to the international literature of science and technology, 1974 to present. Inquire as to online cost and availability.

WILSONLINE. H.W. Wilson and Company, 950 University Avenue, Bronx, NY 10452. (800) 367-6770 or (212) 588-8400. Makes available online versions of the H.W. Wilson indexes including Applied Science and Technology Index, Business Periodicals Index and Readers' Guide to Periodical Literature. Approximately 1980 to present. Inquire as to online cost and availability.

PERIODICALS

AEROSPACE AMERICA. American Institute of Aeronautics and Astronautics, Technical Information Service, 370 L'Enfant Promenade, S.W., Washington, DC 20024. (202) 646-7400. 1932 to present. Monthly. $56.00 per year.

AEROSPACE ENGINEERING MAGAZINE. Society of Automotive Engineers, 400 Commonwealth Drive, Warrendale, PA 15096. (412) 776-4841. 1981 to present. Monthly. $24.00 per year.

AIAA JOURNAL. American Institute of Aeronautics and Astronautics, Technical Information Service, 370 L'Enfant Promenade, S.W., Washington, DC 20024. (202) 646-7400. 1963 to present. Monthly. $205.00 per year.

AVIATION WEEK AND SPACE TECHNOLOGY. McGraw-Hill Book Company, 1221 Avenue of the Americas, New York, NY 10020. (212) 512-2000. 1916 to present. Weekly. $56.00 per year.

JOURNAL OF SPACECRAFT AND ROCKETS. American Institute of Aeronautics and Astronautics, Technical Information Service, 370 L'Enfant Promenade, S.W., Washington, DC 20024. (202) 646-7400. 1964 to present. Bimonthly. $95.00 per year.

PLANETARY REPORT. Planetary Society, 65 North Catalina Avenue, Pasadena, CA 91106. (818) 793-5100. 1980 to present. Bimonthly. $20.00 per year.

SPACEFLIGHT. British Interplanetary Society, 27-29 South Lambeth Road, London, SW8 1SZ, England. 1956 to present. $50.00 per year.

RESEARCH CENTERS AND INSTITUTES

CENTER FOR SPACE RESEARCH AND APPLICATIONS. University of Texas at Austin, WRW 402, Austin, TX 78712. (512) 471-1356.

GODDARD SPACE FLIGHT CENTER. National Aeronautics and Space Administration, Greenbelt Road, Greenbelt, MD 20771. (301) 344-5121.

MARSHALL SPACE FLIGHT CENTER. National Aeronautics and Space Administration, Huntsville, AL 35812. (205) 453-2121.

SPACE SCIENCE AND ENGINEERING CENTER. University of Wisconsin, Madison, 1225 West Dayton Street, Madison, WI 53706. (608) 262-0544.

SPACECRAFT PROPULSION SYSTEM

See: SPACECRAFT

SPECTROSCOPY

See also: ANALYTICAL CHEMISTRY, ELECTRON SPECTROSCOPY, MASS SPECTROMETRY, NUCLEAR MAGNETIC RESONANCE

ABSTRACT SERVICES AND INDEXES

APPLIED SCIENCE AND TECHNOLOGY INDEX. H.W. Wilson and Company, 950 University Avenue, Bronx, NY 10452. (800) 367-6670 or (212) 588-8400. Monthly. Inquire as to cost and availability.

CHEMICAL ABSTRACTS. American Chemical Society, Chemical Abstracts Service, Box 3012, Columbus, OH 43210. (614) 421-3600. 1907 to present. Weekly. $9500.00 per year.

CURRENT CONTENTS: ENGINEERING, TECHNOLOGY AND APPLIED SCIENCES. Institute for Scientific Information, 3501 Market Street, Philadelphia, PA 19104. (800) 523-1850 or (215) 386-0100. Weekly. $275.00 per year.

ENGINEERING INDEX MONTHLY AND AUTHOR INDEX. Engineering Information Inc., 345 East 47th Street, New York, NY 10017. (212) 705-7600. Monthly. $1560.00 per year.

INDEX TO SCIENTIFIC AND TECHNICAL PROCEEDINGS. Institute for Scientific Information, 3501 Market Street, Philadelphia, PA 19104. (800) 523-1850 or (215) 386-0100. 1978 to present. Monthly. $775.00 per year.

MASS SPECTROMETRY BULLETIN. Royal Society of Chemistry. Available from: Distribution Centre, Blackhorse Road, Letchworth, Herts SG6 1HN, England. 1966 to present. Monthly. $350.00 per year.

PHYSICS ABSTRACTS. Institution of Electrical Engineers. Available from: IEEE Service Center, 445 Hoes Lane, Piscataway, NJ 08854. 1898 to present. Bimonthly. $1700.00 per year.

SCIENCE CITATION INDEX. Institute for Scientific Information, 3501 Market Street, Philadelphia, PA 19104. (800) 523-1850 or (215) 386-0100. Six times per year. $6200. 00 per year.

ANNUAL REVIEWS AND YEARBOOKS

ADVANCES IN INFRARED AND RAMAN SPECTROSCOPY. R.J.H. Clark and R.E. Hester, editors. John Wiley and Sons, Inc., 605 Third Avenue, New York, NY 10158. (800) 526-5368. 1975-1985. Price varies, inquire.

APPLIED SPECTROSCOPY REVIEWS. Marcel Dekker Inc., 270 Madison Avenue, New York, NY 10016. (800) 228-1160. Price varies, inquire.

PROGRESS IN ANALYTICAL ATOMIC SPECTROSCOPY. Pergamon Press, Inc., Maxwell House, Fairview Park, Elmsford, NY 10523. (914) 592-7700. Irregular. Price varies, inquire.

ASSOCIATIONS AND PROFESSIONAL SOCIETIES

AMERICAN CHEMICAL SOCIETY. 1155 16th Street, N.W., Washington, DC 20036. (202) 872-4600.

AMERICAN SOCIETY FOR MASS SPECTROSCOPY. P.O. Box 1508, East Lansing, MI 48823. (517) 337-2548.

COBLENTZ SOCIETY. c/o Robert W. Hannah, Perkin-Elmer Corporation, Main Avenue, Norwalk, CT 06859. (203) 431-7797.

SOCIETY FOR APPLIED SPECTROSCOPY. P.O. Box 1438, Frederick, MD 21701. (301) 694-8122.

DIRECTORIES AND BIOGRAPHICAL SOURCES

INTERNATIONAL RESEARCH CENTERS DIRECTORY 1988-89. Darren L. Smith, editor. Gale Research Company, Book Tower, Detroit, MI 48226. (800) 521-0707. 4th edition. 1987. $360.00.

RESEARCH CENTERS DIRECTORY 1988. Gale Research Company, Book Tower, Detroit, MI 48226. (800) 521-0707. 12th edition. 1987. $365.00 for set.

SCIENTIFIC AND TECHNICAL ORGANIZATIONS AND AGENCIES DIRECTORY. Margaret Labash Young, editor. Gale Research Company, Book Tower, Detroit, MI 48226. (800) 521-0707. 2nd edition. 1987. $185.00.

SOCIETY FOR APPLIED SPECTROSCOPY MEMBERSHIP DIRECTORY. Society for Applied Spectroscopy, P.O. Box 1438, Frederick, MD 21701. (301) 694-8122. 1988. $75.00.

ENCYCLOPEDIAS AND DICTIONARIES

DICTIONARY OF SPECTROSCOPY. R.C. Denney. John Wiley and Sons, Inc., 605 Third Avenue, New York, NY 10158. (800) 526-5368. 1982. $42.00.

DICTIONARY OF THE PHYSICAL SCIENCES: TERMS, FORMULAS, DATA. Cesare Emiliani. Oxford University Press, 200 Madison Avenue, New York, NY 10016. (800) 458-5833. 1987. $19.95 in paper.

THESAURUS OF SCIENTIFIC, TECHNICAL, AND ENGINEERING TERMS. Hemisphere Publishing Corporation, 1010 Vermont Avenue, NW, Washington, DC 20005. (800) 526-0275. 1988. $125.00.

GENERAL WORKS

ATOMIC ABSORPTION SPECTROSCOPY. J.W. Robinson. Marcel Dekker Inc., 270 Madison Avenue, New York, NY 10016. (800) 228-1160. Second edition. 1975. $34.50.

CHEMICAL MOSSBAUER SPECTROSCOPY. R.H. Herber, editor. Plenum Publishing Corporation, 233 Spring Street, New York, NY 10013. (800) 221-9369. 1984. $59.50.

HIGH RESOLUTION SPECTROSCOPY. J. Michael Hollas. Butterworth's Publishing, 80 Montvale Avenue, Stoneham, MA 02180. (800) 325-4177. 1982. $130.00.

INTRODUCTION TO APPLICATIONS OF SYMMETRY TO BONDING AND SPECTRA. B.E. Douglas and C.A. Hollingsworth. Academic Press, Inc., 6277 Sea Harbor Drive, Orlando, FL 32821. (800) 321-5068. 1985. $39.00.

MOLECULES AND RADIATION: AN INTRODUCTION TO MODERN MOLECULAR SPECTROSCOPY. Jeffrey I. Steinfeld. MIT Press, 28 Carleton Street, Cambridge, MA 02142. (617) 253-2884. 1985. $19.95.

PRINCIPLES OF SPECTROSCOPY. Raymond Chang. Robert E. Krieger Publishing Company, Inc., P.O. Box 9542, Melbourne, FL 32902-9542. (305) 724-9542. 1978. $21.50.

HANDBOOKS AND MANUALS

HANDBOOK OF SPECTROSCOPY. VOLUME 3. J.W. Robinson. CRC Press, 2000 Corporate Boulevard, Boca Raton, FL 33431. (800) 272-7737. 1981. $97.00.

LANGE'S HANDBOOK OF CHEMISTRY. John A. Dean, editor. McGraw-Hill Book Company, 1221 Avenue of the Americas, New York, NY 10020. (212) 512-2000. 13th edition. 1985. $59.50.

SPECTROSCOPY SOURCE BOOK. Sybil P. Parker, editor-in-chief. McGraw-Hill Book Company, 1221 Avenue of the Americas, New York, NY 10020. (212) 512-2000. 1988. $40.00.

ONLINE DATA BASES

CA SEARCH. Chemical Abstracts Service, P.O. Box 3012, Columbus, OH 43120. (800) 848-6538 or (614) 421-3600. Comprehensive guide to chemical literature, 1972 to present. Inquire as to online cost and availability.

COMPENDEX. Engineering Information, Inc., 345 East 47th Street, New York, NY 10017. (800) 221-1044 or (212) 705-7615. Engineering and technical literature, 1975 to present. Inquire as to online cost and availability.

DISSERTATION ABSTRACTS ONLINE. University Microfilms International, 300 North Zeeb Road, Ann Arbor, MI 48106. (800) 521-0600 or (313) 761-4700. Scope includes virtually all doctoral dissertations accepted at accredited American institutions from 1861 to present in over 250 subject areas. Inquire as to online cost and availability.

INSPEC. INSPEC Marketing Department, Institution of Electrical Engineers. Available from IEEE Service Center, 445 Hoes Lane, Piscataway, NJ 08854. (201) 981-0060. Online version of Physics Abstracts. Inquire as to online cost and availability.

NTIS. National Technical Information Service, 5285 Port Royal Road, Springfield, VA 22161. (703) 487-4630. Broad coverage of government sponsored research reports, 1964 to present. Inquire as to online cost and availability.

SCISEARCH. Institute for Scientific Information, 3501 Market Street, Philadelphia, PA 19104. (800) 523-1850 or (215) 386-0100. Broad multidisciplinary title and author index to the international literature of science and technology, 1974 to present. Inquire as to online cost and availability.

WILSONLINE. H.W. Wilson and Company, 950 University Avenue, Bronx, NY 10452. (800) 367-6770 or (212) 588-8400. Makes available online versions of the H.W. Wilson indexes including Applied Science and Technology Index, Business Periodicals Index and Readers' Guide to Periodical Literature. Approximately 1980 to present. Inquire as to online cost and availability.

PERIODICALS

APPLIED SPECTROSCOPY. Society for Applied Spectroscopy, P.O. Box 1438, Frederick, MD 21701. (301) 694-8122. 1946 to present. Monthly. $145.00 per year.

APPLIED SPECTROSCOPY REVIEWS. Marcel Dekker Inc., 270 Madison Avenue, New York, NY 10016. (800) 228-1160. 1964 to present. Quarterly. $185.00 per year.

INTERNATIONAL JOURNAL OF MASS SPECTROMETRY AND ION PROCESSES. Elsevier Science Publishing Company, Inc., 52 Vanderbilt Avenue, New York, NY 10017. (212) 370-5520. 1968 to present. Twenty-one times per year. $380.00 per year.

JOURNAL OF ELECTRON SPECTROSCOPY AND RELATED PHENOMENA. Elsevier Science Publishing Company, Inc., 52 Vanderbilt Avenue, New York, NY 10017. (212) 370-5520. 1972 to present. Monthly. $255.00 per year.

JOURNAL OF MOLECULAR SPECTROSCOPY. Academic Press, Inc., 6277 Sea Harbor Drive, Orlando, FL 32821. (800) 321-5068. 1957 to present. Monthly. $630.00 per year.

JOURNAL OF QUANTITATIVE SPECTROSCOPY AND RADIATIVE TRANSFER. Pergamon Press, Inc., Maxwell House, Fairview Park, Elmsford, NY 10523. (914) 592-7700. 1961 to present. Monthly. $485.00 per year.

MASS SPECTROMETRY REVIEWS. John Wiley and Sons, Inc., 605 Third Avenue, New York, NY 10158. (800) 526-5368. 1982 to present. Quarterly. $135.00 per year.

SPECTROCHIMICA ACTA. A: MOLECULAR SPECTROSCOPY. Pergamon Press, Inc., Maxwell House, Fairview Park, Elmsford, NY 10523. (914) 592-7700. 1939 to present. Monthly. $400.00 per year.

SPECTROCHIMICA ACTA. B: ATOMIC SPECTROSCOPY. Pergamon Press, Inc., Maxwell House, Fairview Park, Elmsford, NY 10523. (914) 592-7700. 1939 to present. Monthly. $350.00 per year.

SPECTROSCOPY LETTERS. Marcel Dekker Inc., 270 Madison Avenue, New York, NY 10016. (800) 228-1160. 1968 to present. Ten times per year. $280.00 per year.

RESEARCH CENTERS AND INSTITUTES

ATOMIC, MOLECULAR AND CHEMICAL PHYSICS LABORATORY. University of Nevada - Reno, Reno, NV 89507. (702) 784-4920.

GEORGE R. HARRISON SPECTROSCOPY LABORATORY. Massachusetts Institute of Technology, 77 Massachusetts Avenue, Cambridge, MA 02139. (617) 253-7700.

INSTITUTE FOR PHYSICAL SCIENCE AND TECHNOLOGY. University of Maryland, College Park, MD 20742. (301) 454-2636.

INSTITUTE OF OPTICS. University of Rochester, Rochester, NY 14627. (716) 275-2314.

SOUTHERN CALIFORNIA REGIONAL CENTER FOR NUCLEAR MAGNETIC RESONANCE SPECTROSCOPY. California Institute of Technology, Mail Code 164-30, Pasadena, CA 91125. (818) 356-6241.

SPECTRUM ANALYSIS

See: SPECTROSCOPY

SPELEOLOGY

See also: CARBONATES, CAVES AND CAVING, GEOLOGY, GEOMORPHOLOGY, KARST, SEDIMENTARY ROCKS

ABSTRACT SERVICES AND INDEXES

BIBLIOGRAPHY AND INDEX OF GEOLOGY. American Geological Institute, 4220 King Street, Alexandria, VA 22302. (703) 379-2480. 1969 to present. Monthly. $1100.00 per year.

CHEMICAL ABSTRACTS. American Chemical Society, Chemical Abstracts Service, Box 3012, Columbus, OH 43210. (614) 421-3600. 1907 to present. Weekly. $9500.00 per year.

GENERAL SCIENCE INDEX. H.W. Wilson and Company, 950 University Avenue, Bronx, NY 10452. (800) 367-6670 or (212) 588-8400. 1978 to present. Monthly. Inquire as to cost and availability.

SCIENCE CITATION INDEX. Institute for Scientific Information, 3501 Market Street, Philadelphia, PA 19104. (800) 523-1850 or (215) 386-0100. Six times per year. $6200.00 per year.

ANNUAL REVIEWS AND YEARBOOKS

CAVE RESEARCH FOUNDATION ANNUAL REPORT. 1019 Maplewood Drive, #211, Cedar Falls, IA 50613. Available from: Cave Books MO, 756 Harvard Avenue, St. Louis, MO 63130. (314) 862-7646. Annual. Inquire as to price and availability.

ASSOCIATIONS AND PROFESSIONAL SOCIETIES

AMERICAN GEOLOGICAL INSTITUTE. 4220 King Street, Alexandria, VA 22302. (703) 379-2480.

ASSOCIATION FOR MEXICAN CAVE STUDIES. P.O. Box 7037, Austin, TX 78712. (512) 847-2709.

CAVE RESEARCH FOUNDATION. 1019 Maplewood Drive, #211, Cedar Falls, IA 50613.

GEOLOGICAL SOCIETY OF AMERICA. 3300 Penrose Place, Boulder, CO 80301. (303) 447-2020.

NATIONAL SPELEOLOGICAL SOCIETY. Cave Avenue, Huntsville, AL 35810. (205) 852-1300.

SOCIETY OF ECONOMIC PALEONTOLOGISTS AND MINERALOGISTS. Box 4756, Tulsa, OK 74159. (918) 743-9765.

DIRECTORIES AND BIOGRAPHICAL SOURCES

AMERICAN MEN AND WOMEN OF SCIENCE. R.R. Bowker, Inc., Order Department, 245 West 17th Street, New York, NY 10011. (800) 521-8110. Eight volumes. 1986. $595.00 for set.

INTERNATIONAL RESEARCH CENTERS DIRECTORY 1988-89. Darren L. Smith, editor. Gale Research Company, Book Tower, Detroit, MI 48226. (800) 521-0707. 4th edition. 1987. $360.00.

NATIONAL SPELEOLOGICAL SOCIETY MEMBERSHIP LIST. Cave Avenue, Huntsville, AL 35810. (205) 852-1300. Annual. Inquire.

RESEARCH CENTERS DIRECTORY 1988. Gale Research Company, Book Tower, Detroit, MI 48226. (800) 521-0707. 12th edition. 1987. $365.00 for set.

SCIENTIFIC AND TECHNICAL ORGANIZATIONS AND AGENCIES DIRECTORY. Margaret Labash Young, editor. Gale Research Company, Book Tower, Detroit, MI 48226. (800) 521-0707. 2nd edition. 1987. $185.00.

SOCIETY OF ECONOMIC PALEONTOLOGISTS AND MINERALOGISTS MEMBERSHIP LIST. Box 4756, Tulsa, OK 74159. (918) 743-9765. Biennial. Inquire.

GENERAL WORKS

CAVERS, CAVES AND CAVING. Bruce Sloane. Rutgers University Press, 30 College Avenue, New Brunswick, NJ 08903. (201) 932-7764. 1977. $25.00.

EXPLORING CAVES: A GUIDE TO THE UNDERGROUND WILDERNESS. David McClurg. Stackpole Books, Inc., P.O. Box 1831, Harrisburg, PA 17105. (800) 732-3669. 1980. $11.95.

INTRODUCTION TO CARBONATE SEDIMENTS AND ROCKS. T.P. Scoffin. Methuen, Inc., 29 West 35th Street, New York, NY 10001. (212) 244-3336. 1987. $17.95 in paper.

KARST GEOMORPHOLOGY. J.N. Jennings. Blackwell Scientific Publications, Inc., 52 Beacon Street, Boston, MA 02108. (800) 325-4177. 1985. $14.95 in paper.

KARST LANDFORMS. M.M. Sweeting. Columbia University Press, 562 West 113th Street, New York, NY 10025. (212) 316-7100. 1973. $60.00.

SCIENCE OF SPELEOLOGY. T.D. Ford, editor. Academic Press, Inc., 6277 Sea Harbor Drive, Orlando, FL 32821. (800) 321-5068. 1976. $60.50.

SEDIMENTARY ENVIRONMENTS AND FACIES. H.G. Reading, editors. Blackwell Scientific Publications, Inc., 52 Beacon Street, Boston, MA 02108. (800) 325-4177. 2nd edition. 1986. $40.00 in paper.

SPELEOLOGY: THE STUDY OF CAVES. George W. Moore and G.N. Sullivan. Cave Books, 756 Harvard Avenue, St. Louis, MO 63130. (314) 862-7646. 1981. $5.95.

HANDBOOKS AND MANUALS

CAVE RESEARCH FOUNDATION PERSONNEL MANUAL. Diana O. Daunt-Mergens, editor. Available from: Cave Books, 756 Harvard Avenue, St. Louis, MO 63130. (314) 862-7646. 1981. $5.00.

CAVES AND CAVING: A HANDBOOK AND GUIDE TO AMERICAN CAVES. Don Jacobson and Lee Stral. Harbor House Publications, 221 Water Street, Boyne City, MI 49712. (616) 582-2814. 1986. $10.95.

ENCYCLOPEDIA OF FIELD GEOLOGY. Charles W. Finkl, editor. Van Nostrand Reinhold Company, Inc., 135 West 50th Street, New York, NY 10020. (800) 543-2681. 1988. $89.95.

ONLINE DATA BASES

CA SEARCH. Chemical Abstracts Service, P.O. Box 3012, Columbus, OH 43120. (800) 848-6538 or (614) 421-3600. Comprehensive guide to chemical literature, 1972 to present. Inquire as to online cost and availability.

GEOREF. Online version of the BIBLIOGRAPHY AND INDEX OF GEOLOGY. American Geological Institute, 4220 King Street, Alexandria, VA 22302. (703) 379-2480. 1969 to present. Inquire as to online cost and availability.

NTIS. National Technical Information Service, 5285 Port Royal Road, Springfield, VA 22161. (703) 487-4630. Broad coverage of government sponsored research reports, 1964 to present. Inquire as to online cost and availability.

SPELEOLOGY

SCISEARCH. Institute for Scientific Information, 3501 Market Street, Philadelphia, PA 19104. (800) 523-1850 or (215) 386-0100. Broad multidisciplinary title and author index to the international literature of science and technology, 1974 to present. Inquire as to online cost and availability.

WILSONLINE. H.W. Wilson and Company, 950 University Avenue, Bronx, NY 10452. (800) 367-6770 or (212) 588-8400. Makes available online versions of the H.W. Wilson indexes including Applied Science and Technology Index, Business Periodicals Index and Readers' Guide to Periodical Literature. Approximately 1980 to present. Inquire as to online cost and availability.

PERIODICALS

CAVES AND CAVING. British Cave Research Association, 30 Main Road, Westonzoyland, Bridgewater, Somerset TA7 OEB, England. 1973 to present. Quarterly. $10.00 per year.

EARTH SURFACE PROCESSES AND LANDFORMS. British Geomorphological Research Group. Distributed by: John Wiley and Sons, Inc., 605 Third Avenue, New York, NY 10158. (800) 526-5368. 1976 to present. Bimonthly. $280.00 per year.

GEOLOGICAL SOCIETY OF AMERICA. BULLETIN. Geological Society of America. 3300 Penrose Place, Boulder, CO 80301. (303) 447-2020. 1888 to present. Monthly. $80.00 per year.

JOURNAL OF GEOLOGY. University of Chicago Press, 5801 Ellis Avenue, Chicago, IL 60637. (800) 621-2736. 1893 to present. Bimonthly. $30.00 per year.

JOURNAL OF SEDIMENTARY PETROLOGY. Society of Economic Paleontologists and Mineralogists, Box 4756, Tulsa, OK 74159. (918) 743-9765. 1931 to present. Bimonthly. $41.00 per year to individuals.

JOURNAL OF SPELEAN HISTORY. American Spelean History Association, 711 East Atlantic Avenue, Altoona, PA 16602. (814) 946-3155. 1968 to present. Quarterly. $5.00 per year.

NATIONAL SPELEOLOGICAL SOCIETY BULLETIN. National Speleological Society, Cave Avenue, Huntsville, AL 35810. (205) 852-1300. 1941 to present. Semi-annual. $15.00 per year.

NATIONAL SPELEOLOGICAL SOCIETY NEWS. National Speleological Society, Cave Avenue, Huntsville, AL 35810. (205) 852-1300. 1943 to present. Monthly. $15.00 per year.

PROGRESS IN PHYSICAL GEOGRAPHY. Cambridge University Press, 32 East 57th Street, New York, NY 10022. (800) 872-7423. 1976 to present. $50.00 per year.

SEDIMENTARY GEOLOGY. Elsevier Science Publishing Company, Inc., 52 Vanderbilt Avenue, New York, NY 10017. (212) 370-5520. 1967 to present. Twenty times per year. $420.00 per year.

SEDIMENTOLOGY. Blackwell Scientific Publications, Inc., 52 Beacon Street, Boston, MA 02108. (800) 325-4177. 1952 to present. 6 times per year. $184.00 per year.

RESEARCH CENTERS AND INSTITUTES

ASSOCIATION FOR MEXICAN CAVE STUDIES. P.O. Box 7037, Austin, TX 78712. (512) 847-2709.

CAVE RESEARCH FOUNDATION. 1019 Maplewood Drive, #211, Cedar Falls, IA 50613.

SPIN GLASSES

See: SOLID STATE PHYSICS

STANDARDS

See also: INSTRUMENTATION, METROLOGY, QUALITY CONTROL ENGINEERING

ABSTRACT SERVICES AND INDEXES

APPLIED MECHANICS REVIEW. American Society of Mechanical Engineers, 345 East 47th Street, New York, NY 10017. (212) 705-7703. 1948 to present. Monthly. $360.00 per year.

APPLIED SCIENCE AND TECHNOLOGY INDEX. H.W. Wilson and Company, 950 University Avenue, Bronx, NY 10452. (800) 367-6670 or (212) 588-8400. Monthly. Inquire as to cost and availability.

CHEMICAL ABSTRACTS. American Chemical Society, Chemical Abstracts Service, Box 3012, Columbus, OH 43210. (614) 421-3600. 1907 to present. Weekly. $9500.00 per year.

ENGINEERING INDEX MONTHLY AND AUTHOR INDEX. Engineering Information Inc., 345 East 47th Street, New York, NY 10017. (212) 705-7600. Monthly. $1560.00 per year.

ISMEC BULLETIN (Information Service in Mechanical Engineering). Cambridge Scientific Abstracts, 5161 River Road, Bethesda, MD 20816. (301) 951-1400. 1973 to present. Monthly. $450.00 per year.

INDEX TO SCIENTIFIC AND TECHNICAL PROCEEDINGS. Institute for Scientific Information, 3501 Market Street, Philadelphia, PA 19104. (800) 523-1850 or (215) 386-0100. 1978 to present. Monthly. $775.00 per year.

INDEX TO SCIENTIFIC REVIEWS. Institute for Scientific Information, 3501 Market Street, Philadelphia, PA 19104. (800) 523-1850 or (215) 386-0100. 1974 to present. Semi-annual. $550.00 per year.

KEY ABSTRACTS - ELECTRICAL MEASUREMENTS AND INSTRUMENTATION. Institution of Electrical Engineers. Available from: IEEE Service Center, 445 Hoes Lane, Piscataway, NJ 08854. 1976 to present. Monthly. $105.00 per year.

KEY ABSTRACTS - PHYSICAL MEASUREMENTS AND INSTRUMENTATION. Institution of Electrical Engineers. Available from: IEEE Service Center, 445 Hoes Lane, Piscataway, NJ 08854. 1976 to present. Monthly. $105.00 per year.

PHYSICS ABSTRACTS. Institution of Electrical Engineers. Available from: IEEE Service Center, 445 Hoes Lane, Piscataway, NJ 08854. 1898 to present. Bimonthly. $1700.00 per year.

PHYSICS BRIEFS. Physik Verlag GmbH, Postfach 1260/1280, D-6940 Weinheim, West Germany. (212) 661-9404. 1920 to present. Twenty-six times per year. $1250.00 per year.

SCIENCE CITATION INDEX. Institute for Scientific Information, 3501 Market Street, Philadelphia, PA 19104. (800) 523-1850 or (215) 386-0100. Six times per year. $6200.00 per year.

STANDARDS ACTIONS. American National Standards Institute, 1430 Broadway, New York, NY 10018. (212) 354-3300. 1970 to present. Twice per month. $25.00 per year.

ASSOCIATIONS AND PROFESSIONAL SOCIETIES

AMERICAN INSTITUTE OF PHYSICS. 335 East 45th Street, New York, NY 10017. (212) 661-9494.

AMERICAN NATIONAL STANDARDS INSTITUTE. 1430 Broadway, New York, NY 10018. (212) 354-3300.

AMERICAN SOCIETY FOR TESTING AND MATERIALS. 1916 Race Street, Philadelphia, PA 19103. (215) 299-5400.

EDISON ELECTRIC INSTITUTE. 1111 19th Street, N.W., Washington, DC 20036. (202) 828-7400.

INSTITUTE OF ELECTRICAL AND ELECTRONICS ENGINEERS. 345 East 47th Street, New York, NY 10017. (212) 705-7900.

INSTRUMENT SOCIETY OF AMERICA. P.O. Box 12277, 67 Alexander Drive, Research Triangle Park, NC 27709. (919) 549-8411.

NATIONAL STANDARDS ASSOCIATION. 5161 River Road, Bethesda, MD 20816. (301) 951-1310.

DIRECTORIES AND BIOGRAPHICAL SOURCES

AMERICAN MEN AND WOMEN OF SCIENCE. R.R. Bowker, Inc., Order Department, 245 West 17th Street, New York, NY 10011. (800) 521-8110. Eight volumes. 1986. $595.00 for set.

INTERNATIONAL RESEARCH CENTERS DIRECTORY 1988-89. Darren L. Smith, editor. Gale Research Company, Book Tower, Detroit, MI 48226. (800) 521-0707. 4th edition. 1987. $360.00.

1987 DIRECTORY OF ENGINEERING SOCIETIES AND RELATED ORGANIZATIONS. Gordon Davis, editor. Hemisphere Publishing Corporation, 79 Madison Avenue, New York, NY 10016-7892. (800) 821-8312. 12th edition. 1987. $100.00.

RESEARCH CENTERS DIRECTORY 1988. Gale Research Company, Book Tower, Detroit, MI 48226. (800) 521-0707. 12th edition. 1987. $365.00 for set.

SCIENTIFIC AND TECHNICAL ORGANIZATIONS AND AGENCIES DIRECTORY. Margaret Labash Young, editor. Gale Research Company, Book Tower, Detroit, MI 48226. (800) 521-0707. 2nd edition. 1987. $185.00.

WHO'S WHO IN ENGINEERING. Gordon Davis, editor. Hemisphere Publishing Corporation, 79 Madison Avenue, New York, NY 10016-7892. (800) 821-8312. Sixth edition. 1985. $200.00.

HANDBOOKS AND MANUALS

HANDBOOK OF MEASUREMENT SCIENCE. P.H. Sydeham. John Wiley and Sons, Inc., 605 Third Avenue, New York, NY 10158. (800) 526-5368. Two volumes. Volume 1, 1982, $100.00. Volume 2, 1983, $135.00.

ONLINE DATA BASES

CA SEARCH. Chemical Abstracts Service, P.O. Box 3012, Columbus, OH 43120. (800) 848-6538 or (614) 421-3600. Comprehensive guide to chemical literature, 1972 to present. Inquire as to online cost and availability.

COMPENDEX. Engineering Information, Inc., 345 East 47th Street, New York, NY 10017. (800) 221-1044 or (212) 705-7615. Engineering and technical literature, 1975 to present. Inquire as to online cost and availability.

DISSERTATION ABSTRACTS ONLINE. University Microfilms International, 300 North Zeeb Road, Ann Arbor, MI 48106. (800) 521-0600 or (313) 761-4700. Scope includes virtually all doctoral dissertations accepted at accredited American institutions from 1861 to present in over 250 subject areas. Inquire as to online cost and availability.

INSPEC. INSPEC Marketing Department, Institution of Electrical Engineers. Available from IEEE Service Center, 445 Hoes Lane, Piscataway, NJ 08854. (201) 981-0060. Online version of Physics Abstracts. Inquire as to online cost and availability.

NTIS. National Technical Information Service, 5285 Port Royal Road, Springfield, VA 22161. (703) 487-4630. Broad coverage of government sponsored research reports, 1964 to present. Inquire as to online cost and availability.

SCISEARCH. Institute for Scientific Information, 3501 Market Street, Philadelphia, PA 19104. (800) 523-1850 or (215) 386-0100. Broad multidisciplinary title and author index to the international literature of science and technology, 1974 to present. Inquire as to online cost and availability.

STANDARDS AND SPECIFICATIONS DATABASE. National Standards Association, 5161 River Road, Bethesda, MD 20816. (301) 951-1310. Provides access to information on government and industry standards, specifications, and documents relating to terminology, performance, testing, safety, materials, products and the like. Updated monthly. Inquire as to online availability and cost.

WILSONLINE. H.W. Wilson and Company, 950 University Avenue, Bronx, NY 10452. (800) 367-6770 or (212) 588-8400. Makes available online versions of the H.W. Wilson indexes including Applied Science and Technology Index, Business Periodicals Index and Readers' Guide to Periodical Literature. Approximately 1980 to present. Inquire as to online cost and availability.

OTHER SOURCES

WHAT EVERY ENGINEER SHOULD KNOW ABOUT ENGINEERING SOURCES. Marcel Dekker Inc., 270 Madison Avenue, New York, NY 10016. (800) 228-1160. 1984. $24.95.

PERIODICALS

ASTM STANDARDIZATION NEWS. American Society for Testing and Materials, 1916 Race Street, Philadelphia, PA 19103. (215) 299-5400. 1973 to present. Monthly. $20.00 per year.

IEEE TRANSACTIONS ON INSTRUMENTATION AND MEASUREMENT. Institute of Electrical and Electronics Engineers. IEEE Service Center, 445 Hoes Lane, Piscataway, NJ 08854. 1952 to present. Quarterly. $65.00 per year.

METROLOGIA; INTERNATIONAL JOURNAL OF SCIENTIFIC METROLOGY. Springer-Verlag New York, Inc., 175 Fifth Avenue, New York, NY 10010. (800) 526-7254. 1965 to present. Quarterly. $95.00 per year.

STANDARDS ENGINEERING. Standards Engineering Society, 6700 Pennsylvania Avenue South, Minneapolis, MN 55423. Bimothly. $20.00 per year.

U.S. NATIONAL BUREAR OF STANDARDS. JOURNAL OF RESEARCH. Order from: Superintendent of Documents, Washington, DC 20402. 1959 to present. Bimonthly. $17.00 per year.

RESEARCH CENTERS AND INSTITUTES

NATIONAL BUREAU OF STANDARDS. National Engineering Laboratory, Office of Engineering Standards, Gaithersberg, MD 20899. (301) 921-3434.

SPECIFICATIONS AND STANDARDS

STANDARDS AND PRACTICES FOR INSTRUMENTATION. Instrument Society of America, P.O. Box 12277, 67 Alexander Drive, Research Triangle Park, NC 27709. (919) 549-8411. 7th edition. 1983. $135.00.

STARS

See also: ASTRONOMY, ASTROPHYSICS, BINARY STARS, COSMOLOGY, GALAXIES

ABSTRACT SERVICES AND INDEXES

ASTRONOMY AND ASTROPHYSICS ABSTRACTS. Springer-Verlag New York, Incorporated, 175 Fifth Avenue, New York, NY 10010. (212) 460-1500. $70.00 per year.

ASTRONOMY AND ASTROPHYSICS MONTHLY INDEX. Olivetree Associates, Post Office Box 236, Sierra Madre, CA 91024. $212.00 per year. Complimentary copies available on request.

GENERAL SCIENCE INDES. H.W. Wilson Company, 950 University Avenue, Bronx, NY 10452. (800) 367-6770 or (212) 588-8400. Inquire as to cost and availability.

PHYSICS ABSTRACTS. Institute of Electrical Engineers, London, United Kingdom. Available from: Institute of Electrical and Electronic Engineers (IEEE), 345 East 47th Street, New York, NY 10017. (212) 705-7900. 1898 to present. Monthly. $1670.00 per year.

SCIENCE CITATION INDEX. Institute for Scientific Information, 3501 Market Street, Philadelphia, PA 19104. (800) 523-1850 or (215) 386-0100. Inquire as to cost and availability.

STAR. (Scientific and Technical Aerospace Reports. United States National Aeronautics and Space Administration, Scientific and Technical Information Facility, Box 8757, Baltimore-Washington International Airport, MD 21240. (202) 755-2210. Semimonthly, with semiannual and annual indexes. $85.00 per year.

ANNUAL REVIEWS AND YEARBOOKS

THE ASTRONOMICAL ALMANAC. Superintendent of Documents, United States Government Printing Office, Washington, DC 20402. (202) 783-3238. Yearly.

ANNUAL REVIEW OF ASTRONOMY AND ASTROPHYSICS. Annual Reviews, Incorporated, 4139 El Camino Way, Palo Alto, CA 94306. (415) 493-4400. Annual. Inquire.

ASSOCIATIONS AND PROFESSIONAL SOCIETIES

AMERICAN ASTRONOMICAL SOCIETY. 2000 Florida Avenue, NW, Suite 300, Washington, DC 20009. (202) 659-0134.

AMERICAN ASSOCIATION OF VARIABLE STAR OBSERVERS. 187 Concord Avenue, Cambridge, MA 02138. (617) 354-0484.

ASTRONOMICAL LEAGUE. Post Office Box 12821, Tucson, AZ 85732. (602) 790-8471.

ASTRONOMICAL SOCIETY OF THE PACIFIC. 1290 24th Avenue, San Francisco, CA 94122. (415) 661-8660.

BIBLIOGRAPHIES

A BIBLIOGRAPHY OF ASTRONOMY, 1970-1979. R.A. Seal and S.S. Martin. Libraries Unlimited, Incorporated, Littleton, CO 80160. 1982. $37.50.

SCIENCE BOOKS AND FILMS. American Association for the Advancement of Science, 1333 H Street, NW, Washington, DC 20005. 1965 to present. Five times per year. $20.00 per year.

SCIENTIFIC AND TECHNICAL BOOKS AND SERIALS IN PRINT 1988: AN INDEX TO LITERATURE IN SCIENCE AND TECHNOLOGY. R.R. Bowker Company, 205 East 42nd Street, New York, NY 10017. (800) 521-8110 or (212) 916-1600. $175.00.

DIRECTORIES AND BIOGRAPHICAL SOURCES

AMERICAN MEN AND WOMEN OF SCIENCE. Physical and Biological Sciences. Sixteenth edition. R.R. Bowker Company, 205 East 42nd Street, New York, NY 10017. (800) 521-8110 or (212) 916-1600. 1986. $595.00.

THE BIOGRAPHICAL DICTIONARY OF SCIENTISTS: ASTRONOMERS. D. Abbott, editor. Peter Bedrick Books, 125 East 23rd Street, New York, NY 10010. 1984. $24.95.

DIRECTORY OF PHYSICS AND ASTRONOMY STAFF MEMBERS. American Institute of Physics, 335 East 45th Street, New York, NY 10017. Annual.

RESEARCH CENTERS DIRECTORY. Gale Research Company, Book Tower, Detroit, MI 48226. (800) 521-0707. Twelfth edition. 1987. $365.00 for set.

WHO'S WHO IN FRONTIER SCIENCE AND TECHNOLOGY. Marquis Who's Who, Incorporated, 200 East Ohio Street, Chicago, IL 60611. (800) 428-3898 or (312) 787-2008.

ENCYCLOPEDIAS AND DICTIONARIES

ENCYCLOPEDIA OF PHYSICAL SCIENCE AND TECHNOLOGY. Academic Press, Incorporated, Orlando, FL 32887. (800) 321-5068 or (305) 345-2734. Inquire as to cost and availability.

MCGRAW-HILL ENCYCLOPEDIA OF SCIENCE AND TECHNOLOGY. McGraw-Hill Book, Incorporated, 1221 Avenue of the Americas, New York, NY 10020. (212) 997-3675. Fifth edition, 15 volumes. $1100.00.

GENERAL WORKS

THE CAMBRIDGE ASTRONOMY GUIDE: A PRACTICAL INTRODUCTION TO ASTRONOMY. Cambridge University Press, 32 East 57th Street, New York, NY 10022. (212) 688-8888. 1985. $24.95.

ONE HUNDRED BILLION SUNS: THE BIRTH, LIFE AND DEATH OF THE STARS. Rudolf Kippenhahn. Basic Books, Incorporated, 10 East 53rd Street, New York, NY 10022. (800) 242-7737.

PHYSICS OF THE STARS. S.A. Kaplan. John Wiley and Sons, Incorporated, 605 Third Avenue, New York, NY 10158. (800) 526-5368 or (212) 850-6000. 1982. $34.95.

PROTOSTARS AND PLANETS: STUDIES OF STAR FORMATION AND THE ORIGIN OF THE SOLAR SYSTEM. Tom Gehrels, editor. University of Arizona Press, 1615 East Speedway, Tucson, AZ 85719. (602) 621-1441. 1978. $160.00.

PROTOSTARS AND PLANETS II. David C. Black and Mildred S. Matthews, editors. University of Arizona Press, 1615 East Speedway, Tucson, AZ 85719. (602) 621-1441. 1985. $45.00.

THE QUEST FOR SS433. David H. Clark. Penguin Books, Incorporated, 40 West 23rd Street, New York, NY 10010. (800) 631-3577. 1986. $6.95 in paper.

ONLINE DATA BASES

CA SEARCH. Chemical Abstracts Service, Post Office Box 3012, Columbus, OH 43210. Guide to chemical literature, 1972 to present. Inquire as to cost and availability.

DISSERTATION ABSTRACTS ONLINE. University Microfilms International, 300 North Zeeb Road, Ann Arbor, MI 48106. (800) 521-0600 or (313) 761-4700. Scope includes virtually all doctoral dissertations accepted at accredited American institutions from 1861 to present in 252 subject areas. Inquire as to cost and availability.

INSPEC. INSPEC Marketing Department, Institute of Electrical and Electronics Engineers, Incorporated, IEEE Service Department, 445 Hoes Lane, Piscataway, NJ 08854. (201) 981-0060. Inquire as to on-line cost and availability.

NASA. National Aeronautics and Space Administration, Scientific and Technical Information Branch, 300 7th Street, SW, Washington, DC 20546. Citations and abstracts of aerospace literature, 1962 to present. Inquire as to cost and availability.

NTIS. National Technical Information Service, 5285 Port Royal Road, Springfield, VA 22161. (703) 487-4630. Broad coverage of government sponsored research reports, 1964 to present. Inquire as to cost and availability.

SCISEARCH. Institute for Scientific Information, 3501 Market Street, Philadelphia, PA 19104. (800) 523-1850 or (215) 386-0100. Broad multidisciplinary title and author index to the international literature of science and technology, 1974 to present. Inquire as to cost and availability.

WILSONLINE. H.W. Wilson Company, 950 University Avenue, Bronx, NY 10452. (800) 367-6770 or (212) 588-8400. Makes available online versions of the printed H.W. Wilson Indexes including Applied Science and Technology Index, Business Periodicals Index, and Readers' Guide to Periodical Literature. Period covered is generally 1983 to present. Inquire as to cost and availability.

OTHER SOURCES

ATLAS OF DEEP-SKY SPLENDORS. H. Vehrenberg. Cambridge University Press, 32 East 57th Street, New York, NY 10022. (800) 431-1580 or (212) 688-8888. Fourth edition, 1984. $47.50.

THE BRIGHT STAR CATALOGUE AND SUPPLEMENT. D. Hoffleit and Carlos Jaschek. yale University Observatory, Post Office Box 6666, New Haven, CT 06511. (203) 436-3460. Fourth edition, 1982. $35.00.

THE CAMBRIDGE ATLAS OF ASTRONOMY. Jean Audouze and Guy Israel, editors. Cambridge University Press, 32 East 57th Street, New York, NY 10022. (212) 688-8888. 1985. $75.00.

SKY CATALOGUE 2000.0. A. Hirshfeld and R. Sinnott. Cambridge University Press, 32 East 57th Street, New York, NY 10022. (800) 431-1580 or (212) 688-8888.

STAR MAPS FOR BEGINNERS. I.M. Levitt and Roy K. Marshall. Simon and Schuster, Incorporated, 1230 Avenue of the Americas, New York, NY 10020. (800) 223-2336. 1983. $7.95.

STARS, GALAXIES, COSMOS: A CATALOG OF ASTRONOMICAL ANOMALIES. W.R. Corliss, compiler. Sourcebook Project, Post Office Box 107, Glen Arm, MD 21057. (301) 668-6047. 1987. $17.95.

PERIODICALS

ASTRONOMICAL JOURNAL. American Astronomical Society. Available from: American Institute of Physics, 335 East 45th Street, New York, NY 10017. 9212) 661-9404. $125.00 per year.

ASTRONOMICAL SOCIETY OF THE PACIFIC. Publications. Astronomical Society of the Pacific, 1290 24th Avenue, San Francisco, CA 94122. (415) 661-8660. Monthly. $38.00.

ASTRONOMY. Astro Media Corporation, 625 East Paul Avenue, Milwaukee, WI 53202. Monthly. $18.00 per year.

ASTRONOMY AND ASTROPHYSICS ABSTRACTS. Springer-Verlag New York, Incorporated, 175 Fifth Avenue, New York, NY 10010. (212) 460-1500. $70.00 per year.

ASTROPHYSICAL JOURNAL. American Astronomical Society, University of Chicago Press, 5801 Ellis Avenuew, Chicago, IL 60637. Biweekly. $305.00 per year.

ASTROPHYSICS AND SPACE SCIENCE. D. Reidel Publishing Company, 190 Old Derby Street, Hingham, MA 02043. Monthly. $101.00 per year.

CELESTIAL MECHANICS: AN INTERNATIONAL JOURNAL OF SPACE DYNAMICS. D. Reidel Publishing Company, 190 Old Derby Street, Hingham, MA 02043. Monthly. $310.00 per year.

MERCURY. Astronomical Society of the Pacific, 1290 24th Avenue, San Francisco, CA 94122. (415) 661-8660. Bimonthly. $21.00 per year.

MONTHLY NOTICES OF THE ROYAL ASTRONOMICAL SOCIETY. Blackwell Science Publications, Incorporated, 667 Lytton Avenue, Palo Alto, CA 94301. (415) 324-1688. Monthly. $134.00 per year.

SKY AND TELESCOPE. Sky Publishing Corporation, 49 Bay State Road, Cambridge, MA 02238. (617) 864-7360. Monthly. $18.00 per year.

SOLAR PHYSICS. D. Reidel Publishing Company, 190 Old Derby Street, Hingham, MA 02043. Monthly. $620.00 per year.

SOVIET ASTRONOMY (TRANSLATION OF ASTRO-NOMICHESKII ZHURNAL). American Institute of Physics, 335 East 45th Street, New York, NY 10017. (212) 661-9404. Bimonthly. $425.00 per year.

SPACE SCIENCE REVIEWS. D. Reidel Publishing Company, 190 Old Derby Street, Hingham, MA 02043. Monthly. $305.00 per year.

VISTAS IN ASTRONOMY. Pergamon Press, Incorporated, Maxwell House, Fairview Park, Elmsford, NY 10523. (914) 592-7700. Quarterly.

RESEARCH CENTERS AND INSTITUTES

CALIFORNIA INSTITUTE OF TECHNOLOGY. Palomar Observatory, 105-24, Pasadena, CA 91125. (818) 356-4033.

HARVARD-SMITHSONIAN CENTER FOR ASTROPHYSICS. 60 Garden Street, Cambridge, MA 02138. (617) 495-7461.

NATIONAL OPTICAL ASTRONOMY OBSERVATORIES. 1002 North Warren Avenue, Tucson, AZ 85719. (602) 325-9230.

NATIONAL RADIO ASTRONOMY OBSERVATORY. Edgemont Road, Charlottesville, VA 22903. (804) 296-0211.

SPACE TELESCOPE SCIENCE INSTITUTE. 3700 San Martin Drive, Baltimore, MD 21218. (301) 338-4700.

STATICS

See: MECHANICS

STATISTICAL MECHANICS

See: MECHANICS

STATISTICAL THERMODYNAMICS

See: MECHANICS

STATISTICS

See also: MARKOV PROCESSES, MATHEMATICS, PROBABILITY, STOCHASTIC PROCESSES

ABSTRACT SERVICES AND INDEXES

APPLIED MECHANICS REVIEW. American Society of Mechanical Engineers, 345 East 47th Street, New York, NY 10017. (212) 705-7703. 1948 to present. Monthly. $360.00 per year.

APPLIED SCIENCE AND TECHNOLOGY INDEX. H.W. Wilson and Company, 950 University Avenue, Bronx, NY 10452. (800) 367-6670 or (212) 588-8400. Monthly. Inquire as to cost and availability.

CHEMICAL ABSTRACTS. American Chemical Society, Chemical Abstracts Service, Box 3012, Columbus, OH 43210. (614) 421-3600. 1907 to present. Weekly. $9500.00 per year.

COMPUMATH CITATION INDEX. Institute for Scientific Information, 3501 Market Street, Philadelphia, PA 19104. (800) 523-1850 or (215) 386-0100. Three times per year. $875.00 per year.

CURRENT MATHEMATICAL PUBLICATIONS. American Mathematical Society, P.O. Box 6248, Providence, RI 02940. (800) 556-7774 or (401) 272-9500. 1969 to present. Seventeen times per year. $230.00 per year.

ENGINEERING INDEX MONTHLY AND AUTHOR INDEX. Engineering Information Inc., 345 East 47th Street, New York, NY 10017. (212) 705-7600. Monthly. $1560.00 per year.

GENERAL SCIENCE INDEX. H.W. Wilson and Company, 950 University Avenue, Bronx, NY 10452. (800) 367-6670 or (212) 588-8400. 1978 to present. Monthly. Inquire as to cost and availability.

INDEX TO SCIENTIFIC REVIEWS. Institute for Scientific Information, 3501 Market Street, Philadelphia, PA 19104. (800) 523-1850 or (215) 386-0100. 1974 to present. Semi-annual. $550.00 per year.

MATHEMATICAL REVIEWS: A REVIEWING JOURNAL COVERING THE WORLD LITERATURE OF MATHEMATICAL RESEARCH. American Mathematical Society, P.O. Box 6248, Providence, RI 02940. (800) 7774 or (401) 272-9500. 1940 to present. Monthly. $2800.00 per year.

PHYSICS ABSTRACTS. Institution of Electrical Engineers. Available from: IEEE Service Center, 445 Hoes Lane, Piscataway, NJ 08854. 1898 to present. Bimonthly. $1700.00 per year.

SCIENCE CITATION INDEX. Institute for Scientific Information, 3501 Market Street, Philadelphia, PA 19104. (800) 523-1850 or (215) 386-0100. Six times per year. $6200. 00 per year.

ANNUAL REVIEWS AND YEARBOOKS

ADVANCES IN APPLIED MATHEMATICS. Academic Press, Inc., 6277 Sea Harbor Drive, Orlando, FL 32821. (800) 321-5068. Irregular. Price varies, inquire.

ASSOCIATIONS AND PROFESSIONAL SOCIETIES

AMERICAN MATHEMATICAL SOCIETY. P.O. Box 6248, Providence, RI 02940. (401) 272-9500.

AMERICAN STATISTICAL ASSOCIATION. 806 15th Street, N.W., Suite 640, Washington, DC 20005. (202) 393-3253.

INSTITUTE OF MATHEMATICAL STATISTICS, 3401 Investment Boulevard, Suite 7, Hayward, CA 94545. (415) 783-8141.

MATHEMATICAL ASSOCIATION OF AMERICA. 1529 18th Street, N.W., Washington, DC 20036. (202) 387-5200.

SOCIETY FOR INDUSTRIAL AND APPLIED MATHEMATICS. 1400 Architects Building, 117 South 17th Street, Philadelphia, PA 19103. (215) 564-2929.

BIBLIOGRAPHIES

SCIENTIFIC AND TECHNICAL BOOKS AND SERIALS IN PRINT 1988; AN INDEX TO LITERATURE IN SCIENCE AND TECHNOLOGY. R.R. Bowker Company, 205 East 42nd Street, New York, NY 10017. (800) 521-8110. $175.00.

DIRECTORIES AND BIOGRAPHICAL SOURCES

INTERNATIONAL RESEARCH CENTERS DIRECTORY 1988-89. Darren L. Smith, editor. Gale Research Company, Book Tower, Detroit, MI 48226. (800) 521-0707. 4th edition. 1987. $360.00.

RESEARCH CENTERS DIRECTORY 1988. Gale Research Company, Book Tower, Detroit, MI 48226. (800) 521-0707. 12th edition. 1987. $365.00 for set.

SCIENTIFIC AND TECHNICAL ORGANIZATIONS AND AGENCIES DIRECTORY. Margaret Labash Young, editor. Gale Research Company, Book Tower, Detroit, MI 48226. (800) 521-0707. 2nd edition. 1987. $185.00.

ENCYCLOPEDIAS AND DICTIONARIES

DICTIONARY OF THE PHYSICAL SCIENCES: TERMS, FORMULAS, DATA. Cesare Emiliani. Oxford University Press, 200 Madison Avenue, New York, NY 10016. (800) 458-5833. 1987. $19.95 in paper.

ENCYCLOPEDIA OF INFORMATION SYSTEMS AND SERVICES 1988. Amy Lucas and Annette Novallo, editors. Gale Research Company, Book Tower, Detroit, MI 48226. (800) 521-0707. 8th edition. 1987. $400.00 for set.

ENCYCLOPEDIA OF STATISTICAL SCIENCES. Samuel Kotz and others, editors. John Wiley and Sons, Inc., 605 Third Avenue, New York, NY 10158. (800) 526-5368. 1982. Two volumes. $175.00 for set.

ENCYCLOPEDIC DICTIONARY OF MATHEMATICS. Kiyosi Ito, editor. MIT Press, 55 Howard Street, Cambridge, MA 02142. (617) 253-2884. 2nd edition. 1987. Four volumes. $350.00 for set.

THESAURUS OF SCIENTIFIC, TECHNICAL, AND ENGINEERING TERMS. Hemisphere Publishing Corporation, 1010 Vermont Avenue, NW, Washington, DC 20005. (800) 526-0275. 1988. $125.00.

GENERAL WORKS

INTRODUCTORY STATISTICS. Ronald J. Wonnacott and Thomas H. Wonnacott. John Wiley and Sons, Inc., 605 Third Avenue, New York, NY 10158. (800) 526-5368. 4th edition. 1985. $35.50.

MODERN ELEMENTARY STATISTICS. John E. Freund. Prentice-Hall Publishing, Inc., Englewood Cliffs, NJ 07632. (800) 562-0245. 7th edition. 1988. $35.95.

NUMERICAL METHODS IN ENGINEERING AND APPLIED SCIENCE. Bruce Irons. John Wiley and Sons, Inc., 605 Third Avenue, New York, NY 10158. (800) 526-5368. 1987. $32.95.

PROBABILITY AND STATISTICS FOR ENGINEERING AND THE SCIENCES. Jay L. Devore. Wadsworth Publishing Company, 10 Davis Drive, Belmont, CA 94002. (415) 595-2350. 1986. $32.50.

PROBABILITY THEORY AND MATHEMATICAL STATISTICS WITH APPLICATIONS. W. Grossman and others. Kluwer Academic Publishers, 190 Old Derby Street, Hingham, MA 02043. (617) 749-5262. 1987. $89.00.

STATISTICAL METHODS FOR ENGINEERS AND SCIENTISTS. Robert M. Bethea. Marcel Dekker Inc., 270 Madison Avenue, New York, NY 10016. (800) 228-1160. 1985. $24.95

STATISTICS. Donald J. Koosis. John Wiley and Sons, Inc., 605 Third Avenue, New York, NY 10158. (800) 526-5368. 3rd edition. 1985. $10.95 in paper.

STATISTICS FOR RESEARCH. S. Dowdy and S. Wearden. John Wiley and Sons, Inc., 605 Third Avenue, New York, NY 10158. (800) 526-5368. 1983. $38.95.

STATISTICS WITHOUT TEARS: A PRIMER FOR NON-MATHEMATICIANS. Derek Rowntree. Charles Scribner's Sons, 115 Fifth Avenue, New York, NY 10003. (800) 257-5755. 1982. $7.95 in paper.

HANDBOOKS AND MANUALS

CRC HANDBOOK OF MATHEMATICAL SCIENCES. William H. Beyer, editor. CRC Press, 2000 Corporate Boulevard, Boca Raton, FL 33431. (800) 272-7737. 6th edition. 1987. $64.95.

CRC HANDBOOK OF TABLES FOR PROBABILITY AND STATISTICS. CRC Press, 2000 Corporate Boulevard, Boca Raton, FL 33431. (800) 272-7737. 2nd edition. 1968. $49.95.

STANDARD MATHEMATICAL TABLES. William Beyer, editor. CRC Press, 2000 Corporate Boulevard, Boca Raton, FL 33431. (800) 272-7737. 28th edition. 1987. $24.95.

ONLINE DATA BASES

CA SEARCH. Chemical Abstracts Service, P.O. Box 3012, Columbus, OH 43120. (800) 848-6538 or (614) 421-3600. Comprehensive guide to chemical literature, 1972 to present. Inquire as to online cost and availability.

COMPENDEX. Engineering Information, Inc., 345 East 47th Street, New York, NY 10017. (800) 221-1044 or (212) 705-7615. Engineering and technical literature, 1975 to present. Inquire as to online cost and availability.

DISSERTATION ABSTRACTS ONLINE. University Microfilms International, 300 North Zeeb Road, Ann Arbor, MI 48106. (800) 521-0600 or (313) 761-4700. Scope includes virtually all doctoral dissertations accepted at accredited American institutions from 1861 to present in over 250 subject areas. Inquire as to online cost and availability.

INSPEC. INSPEC Marketing Department, Institution of Electrical Engineers. Available from IEEE Service Center, 445 Hoes Lane, Piscataway, NJ 08854. (201) 981-0060. Online version of Physics Abstracts. Inquire as to online cost and availability.

MATHFILE. American Mathematical Society, P.O. Box 6248, Providence, RI 02940. (800) 556-7774 or (401) 272-9500. An online version of Mathematical Reviews. 1973 to present. Inquire as to online cost and availability.

NTIS. National Technical Information Service, 5285 Port Royal Road, Springfield, VA 22161. (703) 487-4630. Broad coverage of government sponsored research reports, 1964 to present. Inquire as to online cost and availability.

SCISEARCH. Institute for Scientific Information, 3501 Market Street, Philadelphia, PA 19104. (800) 523-1850 or (215) 386-0100. Broad multidisciplinary title and author index to the international literature of science and technology, 1974 to present. Inquire as to online cost and availability.

WILSONLINE. H.W. Wilson and Company, 950 University Avenue, Bronx, NY 10452. (800) 367-6770 or (212) 588-8400. Makes available online versions of the H.W. Wilson indexes including Applied Science and Technology Index, Business Periodicals Index and Readers' Guide to Periodical Literature. Approximately 1980 to present. Inquire as to online cost and availability.

PERIODICALS

ANNALS OF STATISTICS. Institute of Mathematical Statistics, 3401 Investment Boulevard, Suite 7, Hayward, CA 94545. (415) 783-8141. 1973 to present. Quarterly. $85.00 per year.

APPLIED MATHEMATICS AND COMPUTATION. Elsevier Science Publishing Company, Inc., 52 Vanderbilt Avenue, New York, NY 10017. (212) 370-5520. 1975 to present. Monthly. $355.00 per year.

COMMUNICATIONS IN STATISTICS, PART A: THEORY AND METHODS. Marcel Dekker Inc., Journals Division, 270 Madison Avenue, New York, NY 10016. (800) 228-1160. 1973 to present. Monthly. $550.00 per year.

JASA. JOURNAL OF THE AMERICAN STATISTICAL ASSOCIATION. American Statistical Association. 806 15th Street, N.W., Suite 640, Washington, DC 20005. (202) 393-3253. 1888 to present. Quarterly. $70.00 per year.

SIAM JOURNAL OF APPLIED MATHEMATICS. Society for Industrial and Applied Mathematics, 1405 Architects Building, 117 South 17th Street. Philadelphia, PA 19103. (215) 564-2929. 1953 to present. Bimonthly. $130.00 per year.

SIAM JOURNAL OF MATHEMATICAL ANALYSIS. Society for Industrial and Applied Mathematics, 1405 Architects Building, 117 South 17th Street. Philadelphia, PA 19103. (215) 564-2929. 1964 to present. Bimonthly. $120.00 per year.

SIAM JOURNAL OF SCIENTIFIC AND STATISTICAL COMPUTING. Society for Industrial and Applied Mathematics, 1405 Architects Building, 117 South 17th Street. Philadelphia, PA 19103. (215) 564-2929. 1980 to present. Quarterly. $78.00 per year.

SIAM REVIEW. Society for Industrial and Applied Mathematics, 1405 Architects Building, 117 South 17th Street. Philadelphia, PA 19103. (215) 564-2929. 1959 to present. Quarterly. $82.00 per year.

STOCHASTICS. Gordon and Breach Science Publishers, Inc., 50 West 23rd Street, New York, NY 10010. (212) 206-8900. Eight times per year. $270.00 per year.

RESEARCH CENTERS AND INSTITUTES

CENTER FOR APPLIED MATHEMATICS. University of Georgia, Tucker Hall, Athens, GA 30602. (404) 542-3491.

CENTER FOR MATHEMATICAL SCIENCE RESEARCH. Rutgers University, New Brunswick, NJ 08903. (201) 932-3117.

CENTER FOR PURE AND APPLIED MATHEMATICS. University of Californai, Berkeley, 977 Evans Hall, Berkeley, CA 94720. (415) 642-0116.

INSTITUTE FOR MATHEMATICS AND ITS APPLICATIONS. University of Minnesota, 514 Vincent, 206 Church Street, S.E., Minneapolis, MN 55455. (612) 624-6066.

INSTITUTE OF APPLIED MATHEMATICS. University of Missouri - Rolla, Rolla, MO 65401. (314) 341-4151.

STEAM

See also: BOILERS, GEOTHERMAL ENERGY, HEAT EXCHANGERS, HEAT TRANSFER, MECHANICAL ENGINEERING, NUCLEAR ENGINEERING, PRESSURE VESSELS, STEAM ENGINES, STEAM TURBINES, THERMODYNAMICS

ABSTRACT SERVICES AND INDEXES

APPLIED MECHANICS REVIEW. American Society of Mechanical Engineers, 345 East 47th Street, New York, NY 10017. (212) 705-7703. 1948 to present. Monthly. $360.00 per year.

APPLIED SCIENCE AND TECHNOLOGY INDEX. H.W. Wilson and Company, 950 University Avenue, Bronx, NY 10452. (800) 367-6670 or (212) 588-8400. Monthly. Inquire as to cost and availability.

CHEMICAL ABSTRACTS. American Chemical Society, Chemical Abstracts Service, Box 3012, Columbus, OH 43210. (614) 421-3600. 1907 to present. Weekly. $9500.00 per year.

CURRENT CONTENTS: ENGINEERING, TECHNOLOGY AND APPLIED SCIENCES. Institute for Scientific Information, 3501 Market Street, Philadelphia, PA 19104. (800) 523-1850 or (215) 386-0100. Weekly. $275.00 per year.

ENGINEERING INDEX MONTHLY AND AUTHOR INDEX. Engineering Information Inc., 345 East 47th Street, New York, NY 10017. (212) 705-7600. Monthly. $1560.00 per year.

ISMEC BULLETIN (Information Service in Mechanical Engineering). Cambridge Scientific Abstracts, 5161 River Road, Bethesda, MD 20816. (301) 951-1400. 1973 to present. Monthly. $450.00 per year.

PHYSICS ABSTRACTS. Institution of Electrical Engineers. Available from: IEEE Service Center, 445 Hoes Lane, Piscataway, NJ 08854. 1898 to present. Bimonthly. $1700.00 per year.

PHYSICS BRIEFS. Physik Verlag GmbH, Postfach 1260/1280, D-6940 Weinheim, West Germany. (212) 661-9404. 1920 to present. Twenty-six times per year. $1250.00 per year.

SCIENCE CITATION INDEX. Institute for Scientific Information, 3501 Market Street, Philadelphia, PA 19104. (800) 523-1850 or (215) 386-0100. Six times per year. $6200.00 per year.

ASSOCIATIONS AND PROFESSIONAL SOCIETIES

AMERICAN INSTITUTE OF CHEMICAL ENGINEERS. 345 East 47th Street, New York, NY 10017. (212) 705-7703.

AMERICAN SOCIETY OF HEATING, REFRIGERATING AND AIR CONDITIONING ENGINEERS, INC., 1791 Tullie Circle, Atlanta, GA 30329. (404) 636-8400.

AMERICAN SOCIETY OF MECHANICAL ENGINEERS. 345 East 47th Street, New York, NY 10017. (212) 705-7703.

DIRECTORIES AND BIOGRAPHICAL SOURCES

AMERICAN MEN AND WOMEN OF SCIENCE. R.R. Bowker, Inc., Order Department, 245 West 17th Street, New York, NY 10011. (800) 521-8110. Eight volumes. 1986. $595.00 for set.

1987 DIRECTORY OF ENGINEERING SOCIETIES AND RELATED ORGANIZATIONS. Gordon Davis, editor. Hemisphere Publishing Corporation, 1010 Vermont Avenue, NW, Washington, DC 20005. (800) 526-0275. 12th edition. 1987. $100.00.

RESEARCH CENTERS DIRECTORY 1988. Gale Research Company, Book Tower, Detroit, MI 48226. (800) 521-0707. 12th edition. 1987. $365.00 for set.

WHO'S WHO IN ENGINEERING. Gordon Davis, editor. Hemisphere Publishing Corporation, 1010 Vermont Avenue, NW, Washington, DC 20005. (800) 526-0275. 6th edition. 1985. $200.00.

GENERAL WORKS

FUNDAMENTALS OF HEAT EXCHANGER AND PRESSURE VESSEL TECHNOLOGY. J.P. Gupta. Hemisphere Publishing Corporation, 79 Madison Avenue, New York, NY 10016-7892. (800) 821-8312. 1986. $49.95.

HEAT TRANSFER EQUIPMENT DESIGN. R.K. Shah and others. Hemisphere Publishing Corporation, 79 Madison Avenue, New York, NY 10016-7892. 1988. $135.00.

INTRODUCTION TO HEAT TRANSFER. F. Incropera and D. Dewitt. John Wiley and Sons, Inc., 605 Third Avenue, New York, NY 10158. (800) 526-5368. 1985. $39.95.

MANAGING STEAM: AN ENGINEERING GUIDE TO INDUSTRIAL, COMMERCIAL AND UTILITY SYSTEMS. Jason Makansi, editor. Hemisphere Publishing Corporation, 79 Madison Avenue, New York, NY 10016-7892. Order from: Taylor and Francis/Hemisphere Distribution Center, 242 Cherry Street, Philadelphia, PA 19106-1906. (800) 821-8312. 1986. $35.00.

POWER PLANT SYSTEM DESIGN. K.W. Li and A.P. Priddy. John Wiley and Sons, Inc., 605 Third Avenue, New York, NY 10158. (800) 526-5368. 1985. $44.50.

STATIONARY STEAM ENGINES. G. Haynes. Seven Hills Books, 519 West Third Street, Cincinnati, OH 45202. (513) 381-3881. 1983. $3.50 in paper.

STEAM ENGINE DESIGN. Lindsay Publications, P.O. Box 12, Bradley, IL 60915. 1983. $9.95.

TWO-PHASE STEAM FLOW IN TURBINES AND SEPARATORS. M.J. Moore. Hemisphere Publishing Corporation, 79 Madison Avenue, New York, NY 10016-7892. Order from: Taylor and Francis/Hemisphere Distribution Center, 242 Cherry Street, Philadelphia, PA 19106-1906. (800) 821-8312. 1976. $44.50.

HANDBOOKS AND MANUALS

HANDBOOK OF HEAT TRANSFER APPLICATIONS. W.M. Rohsenow and others. McGraw-Hill Book Company, 1221 Avenue of the Americas, New York, NY 10020. (212) 512-2000. Second edition. 1986. $79.50.

HEAT EXCHANGER SOURCEBOOK. J.W. Palen, editor. Hemisphere Publishing Corporation, 79 Madison Avenue, New York, NY 10016-7892. (800) 821-8312. 1986. $59.95.

STEAM TABLES: THERMODYNAMICS PROPERTIES OF WATER INCLUDING VAPOR, LIQUID, AND SOLID PHASES. J.H. Keenan and others. John Wiley and Sons, Inc., 605 Third Avenue, New York, NY 10158. (800) 526-5368. 1969. $44.95.

ONLINE DATA BASES

CA SEARCH. Chemical Abstracts Service, P.O. Box 3012, Columbus, OH 43120. (800) 848-6538 or (614) 421-3600. Comprehensive guide to chemical literature, 1972 to present. Inquire as to online cost and availability.

COMPENDEX. Engineering Information, Inc., 345 East 47th Street, New York, NY 10017. (800) 221-1044 or (212) 705-7615. Engineering and technical literature, 1975 to present. Inquire as to online cost and availability.

INSPEC. INSPEC Marketing Department, Institution of Electrical Engineers. Available from IEEE Service Center, 445 Hoes Lane, Piscataway, NJ 08854. (201) 981-0060. Online version of Physics Abstracts. Inquire as to online cost and availability.

NTIS. National Technical Information Service, 5285 Port Royal Road, Springfield, VA 22161. (703) 487-4630. Broad coverage of government sponsored research reports, 1964 to present. Inquire as to online cost and availability.

SCISEARCH. Institute for Scientific Information, 3501 Market Street, Philadelphia, PA 19104. (800) 523-1850 or (215) 386-0100. Broad multidisciplinary title and author index to the international literature of science and technology, 1974 to present. Inquire as to online cost and availability.

WILSONLINE. H.W. Wilson and Company, 950 University Avenue, Bronx, NY 10452. (800) 367-6770 or (212) 588-8400. Makes available online versions of the H.W. Wilson indexes including Applied Science and Technology Index, Business Periodicals Index and Readers' Guide to Periodical Literature. Approximately 1980 to present. Inquire as to online cost and

availability.

PERIODICALS

ASHRAE JOURNAL. American Society of Heating, Refrigerating and Air Conditioning Engineers, Inc., 1791 Tullie Circle, Atlanta, GA 30329. (404) 636-8400. 1914 to present. Monthly. $35.00 per year.

EXPERIMENTAL THERMAL AND FLUID SCIENCE. Elsevier Science Publishing Company, Inc., 52 Vanderbilt Avenue, New York, NY 10017. (212) 370-5520. 1988 to present. Quarterly. $125.00 per year.

HEAT TRANSFER ENGINEERING. Hemisphere Publishing Corporation, 79 Madison Avenue, New York, NY 10016-7892. (800) 821-8312. 1979 to present. Quarterly. $97.50.

HEATING/PIPING/AIR CONDITIONING. Penton-IPC, Reinhold Publishing Division, 600 Summer Street, Box 1361, Stamford, CT 06904. (203) 348-7531. 1929 to present. Monthly. $35.00 per year.

INTERNATIONAL COMMUNICATIONS IN HEAT AND MASS TRANSFER. Pergamon Press, Inc., Maxwell House, Fairview Park, Elmsford, NY 10523. (914) 592-7700. 1974 to present. Bimonthly. $160.00 per year.

INTERNATIONAL JOURNAL OF HEAT AND MASS TRANSFER. Pergamon Press, Inc., Maxwell House, Fairview Park, Elmsford, NY 10523. (914) 592-7700. 1960 to present. Monthly. $500.00 per year.

JOURNAL OF HEAT TRANSFER. American Society of Mechanical Engineers, 345 East 47th Street, New York, NY 10017. (212) 705-7703. 1970 to present. Quarterly. $100.00 per year.

JOURNAL OF PRESSURE VESSEL TECHNOLOGY. American Society of Mechanical Engineers, 345 East 47th Street, New York, NY 10017. (212) 705-7703. 1974 to present. Quarterly. $80.00 per year.

NUMERICAL HEAT TRANSFER: AN INTERNATIONAL JOURNAL OF COMPUTATION AND METHOLOGY. Hemisphere Publishing Corporation, 79 Madison Avenue, New York, NY 10016-7892. (800) 821-8312. 1978 to present. Monthly. $370.00 per year.

PLANT ENGINEERING. Technical Publishing, Box 1030, 1301 South Grove Avenue, Barrington, IL 60010. (312) 381-1840. 1947 to present. Semi-monthly. $50.00 per year.

POWER. McGraw-Hill Book Company, 1221 Avenue of the Americas, New York, NY 10020. (212) 512-2000. 1882 to present. Monthly. $12.00 per year.

POWER TRANSMISSION DESIGN. Penton-IPC, 1100 Superior Avenue, Cleveland, OH 44114. (216)696-7000. 1959 to present. Monthly. $35.00 per year.

STEAM POWER. Kirk Enterprises, Limited, Midlands Steam Centre, 106a Derby Road, Loughborough LE11 OAG, England. 1949 to present. Quarterly. $20.00 to present.

RESEARCH CENTERS AND INSTITUTES

HEAT TRANSFER LABORATORY. Massachusetts Institute of Technology, 77 Massachusetts Avenue, Cambridge, MA 02139. (716) 253-2248.

HEAT TRANSFER RESEARCH. University of Wisconsin, Milwaukee, P.O. Box 784, Milwaukee, WI 53201. (414) 963-5001.

HEAT TRANSFER RESEARCH FACILITY. Columbia University, 632 West 125th Street, New York, NY 10027. (212) 280-4163.

INSTITUTE OF THERMO-FLUID ENGINEERING AND SCIENCE. Lehigh University, Whitaker Laboratory, Bethleham, PA 18015. (215) 861-4091.

STEAM ENGINES

See also: ENGINES, STEAM TURBINES

ABSTRACT SERVICES AND INDEXES

APPLIED MECHANICS REVIEW. American Society of Mechanical Engineers, 345 East 47th Street, New York, NY 10017. (212) 705-7703. 1948 to present. Monthly. $360.00 per year.

APPLIED SCIENCE AND TECHNOLOGY INDEX. H.W. Wilson and Company, 950 University Avenue, Bronx, NY 10452. (800) 367-6670 or (212) 588-8400. Monthly. Inquire as to cost and availability.

ENGINEERING INDEX MONTHLY AND AUTHOR INDEX. Engineering Information Inc., 345 East 47th Street, New York, NY 10017. (212) 705-7600. Monthly. $1560.00 per year.

ISMEC BULLETIN (Information Service in Mechanical Engineering). Cambridge Scientific Abstracts, 5161 River Road, Bethesda, MD 20816. (301) 951-1400. 1973 to present. Monthly. $450.00 per year.

ASSOCIATIONS AND PROFESSIONAL SOCIETIES

AMERICAN SOCIETY OF MECHANICAL ENGINEERS. 345 East 47th Street, New York, NY 10017. (212) 705-7703.

SOCIETY OF AUTOMOTIVE ENGINEERS (SAE). 400 Commonwealth Drive, Warrendale, PA 15096. (412) 776-4841.

NATIONAL ASSOCIATION OF POWER ENGINEERS. 2350 East Devon Avenue, Suite 115, Des Plaines, IL 60018. (312) 298-0600.

DIRECTORIES AND BIOGRAPHICAL SOURCES

1987 DIRECTORY OF ENGINEERING SOCIETIES AND RELATED ORGANIZATIONS. Gordon Davis, editor. Hemisphere Publishing Corporation, 1010 Vermont Avenue, NW, Washington, DC 20005. (800) 526-0275. 12th edition. 1987. $100.00.

RESEARCH CENTERS DIRECTORY 1988. Gale Research Company, Book Tower, Detroit, MI 48226. (800) 521-0707. 12th edition. 1987. $365.00 for set.

SCIENTIFIC AND TECHNICAL ORGANIZATIONS AND AGENCIES DIRECTORY. Margaret Labash Young, editor. Gale Research Company, Book Tower, Detroit, MI 48226. (800) 521-0707. 2nd edition. 1987. $185.00.

WHO'S WHO IN ENGINEERING. Gordon Davis, editor. Hemisphere Publishing Corporation, 1010 Vermont Avenue, NW, Washington, DC 20005. (800) 526-0275. 6th edition. 1985. $200.00.

ENCYCLOPEDIAS AND DICTIONARIES

THESAURUS OF SCIENTIFIC, TECHNICAL, AND ENGINEERING TERMS. Hemisphere Publishing Corporation, 1010 Vermont Avenue, NW, Washington, DC 20005. (800) 526-0275. 1988. $125.00.

GENERAL WORKS

MANAGING STEAM: AN ENGINEERING GUIDE TO INDUSTRIAL, COMMERCIAL AND UTILITY SYSTEMS. Jason Makansi, editor. Hemisphere Publishing Corporation, 79 Madison Avenue, New York, NY 10016-7892. Order from: Taylor and Francis/Hemisphere Distribution Center, 242 Cherry Street, Philadelphia, PA 19106-1906. (800) 821-8312. 1986. $35.00.

STEAM ENGINES

POWER PLANT SYSTEM DESIGN. K.W. Li and A.P. Priddy. John Wiley and Sons, Inc., 605 Third Avenue, New York, NY 10158. (800) 526-5368. 1985. $44.50.

STATIONARY STEAM ENGINES. G. Haynes. Seven Hills Books, 519 West Third Street, Cincinnati, OH 45202. (513) 381-3881. 1983. $3.50 in paper.

STEAM ENGINE DESIGN. Lindsay Publications, P.O. Box 12, Bradley, IL 60915. 1983. $9.95.

TWO-PHASE STEAM FLOW IN TURBINES AND SEPARATORS. M.J. Moore. Hemisphere Publishing Corporation, 79 Madison Avenue, New York, NY 10016-7892. Order from: Taylor and Francis/Hemisphere Distribution Center, 242 Cherry Street, Philadelphia, PA 19106-1906. (800) 821-8312. 1976. $44.50.

HANDBOOKS AND MANUALS

STEAM TABLES: THERMODYNAMICS PROPERTIES OF WATER INCLUDING VAPOR, LIQUID, AND SOLID PHASES. J.H. Keenan and others. John Wiley and Sons, Inc., 605 Third Avenue, New York, NY 10158. (800) 526-5368. 1969. $44.95.

ONLINE DATA BASES

COMPENDEX. Engineering Information, Inc., 345 East 47th Street, New York, NY 10017. (800) 221-1044 or (212) 705-7615. Engineering and technical literature, 1975 to present. Inquire as to online cost and availability.

NTIS. National Technical Information Service, 5285 Port Royal Road, Springfield, VA 22161. (703) 487-4630. Broad coverage of government sponsored research reports, 1964 to present. Inquire as to online cost and availability.

WILSONLINE. H.W. Wilson and Company, 950 University Avenue, Bronx, NY 10452. (800) 367-6770 or (212) 588-8400. Makes available online versions of the H.W. Wilson indexes including Applied Science and Technology Index, Business Periodicals Index and Readers' Guide to Periodical Literature. Approximately 1980 to present. Inquire as to online cost and availability.

OTHER SOURCES

WHAT EVERY ENGINEER SHOULD KNOW ABOUT ENGINEERING SOURCES. Marcel Dekker Inc., 270 Madison Avenue, New York, NY 10016. (800) 228-1160. 1984. $24.95.

PERIODICALS

JOURNAL OF PRESSURE VESSEL TECHNOLOGY. American Society of Mechanical Engineers, 345 East 47th Street, New York, NY 10017. (212) 705-7703. 1974 to present. Quarterly. $80.00 per year.

PLANT ENGINEERING. Technical Publishing, Box 1030, 1301 South Grove Avenue, Barrington, IL 60010. (312) 381-1840. 1947 to present. Semi-monthly. $50.00 per year.

POWER. McGraw-Hill Book Company, 1221 Avenue of the Americas, New York, NY 10020. (212) 512-2000. 1882 to present. Monthly. $12.00 per year.

POWER TRANSMISSION DESIGN. Penton-IPC, 1100 Superior Avenue, Cleveland, OH 44114. (216)696-7000. 1959 to present. Monthly. $35.00 per year.

STEAM POWER. Kirk Enterprises, Limited, Midlands Steam Centre, 106a Derby Road, Loughborough LE11 OAG, England. 1949 to present. Quarterly. $20.00 to present.

TURBOMACHINERY MAINTENANCE NEWSLETTER. Business Journals, Inc., 22 South Smith Street, Norwalk, CT 06855. (203) 853-6015. 1983 to present. $95.00 per year.

RESEARCH CENTERS AND INSTITUTES

TURBOMACHINERY LABORATORY. Texas A&M University, Mechanical Engineering Department, College Station, TX 77843. (409) 845-7417.

STEAM TURBINES

See also ENGINES, GAS TURBINES, STEAM, STEAM ENGINES, TURBINES

ABSTRACT SERVICES AND INDEXES

APPLIED MECHANICS REVIEW. American Society of Mechanical Engineers, 345 East 47th Street, New York, NY 10017. (212) 705-7703. 1948 to present. Monthly. $360.00 per year.

APPLIED SCIENCE AND TECHNOLOGY INDEX. H.W. Wilson and Company, 950 University Avenue, Bronx, NY 10452. (800) 367-6670 or (212) 588-8400. Monthly. Inquire as to cost and availability.

ENGINEERING INDEX MONTHLY AND AUTHOR INDEX. Engineering Information Inc., 345 East 47th Street, New York, NY 10017. (212) 705-7600. Monthly. $1560.00 per year.

ISMEC BULLETIN (Information Service in Mechanical Engineering). Cambridge Scientific Abstracts, 5161 River Road, Bethesda, MD 20816. (301) 951-1400. 1973 to present. Monthly. $450.00 per year.

ASSOCIATIONS AND PROFESSIONAL SOCIETIES

AMERICAN SOCIETY OF MECHANICAL ENGINEERS. 345 East 47th Street, New York, NY 10017. (212) 705-7703.

INTERNATIONAL GAS TURBINE INSTITUTE. 4250 Perimeter Park, South, Atlanta, GA 30341. (404) 451-1905.

SOCIETY OF AUTOMOTIVE ENGINEERS (SAE). 400 Commonwealth Drive, Warrendale, PA 15096. (412) 776-4841.

NATIONAL ASSOCIATION OF POWER ENGINEERS. 2350 East Devon Avenue, Suite 115, Des Plaines, IL 60018. (312) 298-0600.

DIRECTORIES AND BIOGRAPHICAL SOURCES

1987 DIRECTORY OF ENGINEERING SOCIETIES AND RELATED ORGANIZATIONS. Gordon Davis, editor. Hemisphere Publishing Corporation, 1010 Vermont Avenue, NW, Washington, DC 20005. (800) 526-0275. 12th edition. 1987. $100.00.

RESEARCH CENTERS DIRECTORY 1988. Gale Research Company, Book Tower, Detroit, MI 48226. (800) 521-0707. 12th edition. 1987. $365.00 for set.

SCIENTIFIC AND TECHNICAL ORGANIZATIONS AND AGENCIES DIRECTORY. Margaret Labash Young, editor. Gale Research Company, Book Tower, Detroit, MI 48226. (800) 521-0707. 2nd edition. 1987. $185.00.

WHO'S WHO IN ENGINEERING. Gordon Davis, editor. Hemisphere Publishing Corporation, 1010 Vermont Avenue, NW, Washington, DC 20005. (800) 526-0275. 6th edition. 1985. $200.00.

ENCYCLOPEDIAS AND DICTIONARIES

THESAURUS OF SCIENTIFIC, TECHNICAL, AND ENGINEERING TERMS. Hemisphere Publishing Corporation, 1010 Vermont Avenue, NW, Washington, DC 20005. (800) 526-0275. 1988. $125.00.

GENERAL WORKS

AEROTHERMODYNAMICS OF LOW PRESSURE STEAM TURBINES AND CONDESERS. M.J. Moore and C.H. Sieverding, editors. Hemisphere Publishing Corporation, 79 Madison Avenue, New York, NY 10016-7892. Order from: Taylor and Francis/Hemisphere Distribution Center, 242 Cherry Street, Philadelphia, PA 19106-1906. (800) 821-8312. 1986. $65.95.

AERO-THERMODYNAMICS OF STEAM TURBINES. W.E. Steltz and A.M. Donaldson, editors. American Society of Mechanical Engineers, 345 East 47th Street, New York, NY 10017. (212) 705-7703. 1981. $24.00.

POWER PLANT SYSTEM DESIGN. K.W. Li and A.P. Priddy. John Wiley and Sons, Inc., 605 Third Avenue, New York, NY 10158. (800) 526-5368. 1985. $44.50.

TWO-PHASE STEAM FLOW IN TURBINES AND SEPARATORS. M.J. Moore. Hemisphere Publishing Corporation, 79 Madison Avenue, New York, NY 10016-7892. Order from: Taylor and Francis/Hemisphere Distribution Center, 242 Cherry Street, Philadelphia, PA 19106-1906. (800) 821-8312. 1976. $44.50.

HANDBOOKS AND MANUALS

STEAM TABLES: THERMODYNAMICS PROPERTIES OF WATER INCLUDING VAPOR, LIQUID, AND SOLID PHASES. J.H. Keenan and others. John Wiley and Sons, Inc., 605 Third Avenue, New York, NY 10158. (800) 526-5368. 1969. $44.95.

TURBOMACHINERY INTERNATIONAL HANDBOOK. Business Journals, Inc., 22 South Smith Street, Norwalk, CT 06855. (203) 853-6015. Annual. $50.00 per copy.

ONLINE DATA BASES

COMPENDEX. Engineering Information, Inc., 345 East 47th Street, New York, NY 10017. (800) 221-1044 or (212) 705-7615. Engineering and technical literature, 1975 to present. Inquire as to online cost and availability.

NTIS. National Technical Information Service, 5285 Port Royal Road, Springfield, VA 22161. (703) 487-4630. Broad coverage of government sponsored research reports, 1964 to present. Inquire as to online cost and availability.

WILSONLINE. H.W. Wilson and Company, 950 University Avenue, Bronx, NY 10452. (800) 367-6770 or (212) 588-8400. Makes available online versions of the H.W. Wilson indexes including Applied Science and Technology Index, Business Periodicals Index and Readers' Guide to Periodical Literature. Approximately 1980 to present. Inquire as to online cost and availability.

OTHER SOURCES

WHAT EVERY ENGINEER SHOULD KNOW ABOUT ENGINEERING SOURCES. Marcel Dekker Inc., 270 Madison Avenue, New York, NY 10016. (800) 228-1160. 1984. $24.95.

PERIODICALS

JOURNAL OF PRESSURE VESSEL TECHNOLOGY. American Society of Mechanical Engineers, 345 East 47th Street, New York, NY 10017. (212) 705-7703. 1974 to present. Quarterly. $80.00 per year.

PLANT ENGINEERING. Technical Publishing, Box 1030, 1301 South Grove Avenue, Barrington, IL 60010. (312) 381-1840. 1947 to present. Semi-monthly. $50.00 per year.

POWER. McGraw-Hill Book Company, 1221 Avenue of the Americas, New York, NY 10020. (212) 512-2000. 1882 to present. Monthly. $12.00 per year.

POWER TRANSMISSION DESIGN. Penton-IPC, 1100 Superior Avenue, Cleveland, OH 44114. (216)696-7000. 1959 to present. Monthly. $35.00 per year.

STEAM POWER. Kirk Enterprises, Limited, Midlands Steam Centre, 106a Derby Road, Loughborough LE11 OAG, England. 1949 to present. Quarterly. $20.00 to present.

TURBOMACHINERY MAINTENANCE NEWSLETTER. Business Journals, Inc., 22 South Smith Street, Norwalk, CT 06855. (203) 853-6015. 1983 to present. $95.00 per year.

RESEARCH CENTERS AND INSTITUTES

INTERNATIONAL GAS TURBINE INSTITUTE. 4250 Perimeter Park, South, Atlanta, GA 30341. (404) 451-1905.

TURBOMACHINERY LABORATORY. Texas A&M University, Mechanical Engineering Department, College Station, TX 77843. (409) 845-7417.

STEEL AND STEEL MAKING

See also: ALLOYS,
CORROSION, FERROALLOYS, MACHINING,
METALLURGICAL ENGINEERING, METALLURGY,
METALS AND METALWORKING

ABSTRACT SERVICES AND INDEXES

ALLOYS INDEX. American Society for Metals, Metals Park, OH 44073. (216) 338-5151. 1974 to present. Monthly. $225.00.

APPLIED MECHANICS REVIEW. American Society of Mechanical Engineers, 345 East 47th Street, New York, NY 10017. (212) 705-7703. 1948 to present. Monthly. $360.00 per year.

APPLIED SCIENCE AND TECHNOLOGY INDEX. H.W. Wilson and Company, 950 University Avenue, Bronx, NY 10452. (800) 367-6670 or (212) 588-8400. Monthly. Inquire as to cost and availability.

CHEMICAL ABSTRACTS. American Chemical Society, Chemical Abstracts Service, Box 3012, Columbus, OH 43210. (614) 421-3600. 1907 to present. Weekly. $9500.00 per year.

CORROSION ABSTRACTS. National Association of Corrosion Engineers, Box 218340, Houston, TX 77218. (713) 492-0535. 1962 to present. Bimonthly. $200.00 per year.

CURRENT CONTENTS: ENGINEERING, TECHNOLOGY AND APPLIED SCIENCES. Institute for Scientific Information, 3501 Market Street, Philadelphia, PA 19104. (800) 523-1850 or (215) 386-0100. Weekly. $275.00 per year.

ENGINEERING INDEX MONTHLY AND AUTHOR INDEX. Engineering Information Inc., 345 East 47th Street, New York, NY 10017. (212) 705-7600. Monthly. $1560.00 per year.

ISMEC BULLETIN (Information Service in Mechanical Engineering). Cambridge Scientific Abstracts, 5161 River Road, Bethesda, MD 20816. (301) 951-1400. 1973 to present. Monthly. $450.00 per year.

METALS ABSTRACTS AND METALS ABSTRACTS INDEX. American Society for Metals, Metals Park, OH 44073. (216) 338-5151. 1968 to present. Monthly. Abstracts are $1100.00 per year and Index is $460.00 per year.

PHYSICS ABSTRACTS. Institution of Electrical Engineers. Available from: IEEE Service Center, 445 Hoes Lane, Piscataway, NJ 08854. 1898 to present. Bimonthly. $1700.00 per year.

SCIENCE CITATION INDEX. Institute for Scientific Information, 3501 Market Street, Philadelphia, PA 19104. (800) 523-1850 or (215) 386-0100. Six times per year. $6200. 00 per year.

STEEL AND STEEL MAKING

ASSOCIATIONS AND PROFESSIONAL SOCIETIES

AMERICAN INSTITUTE OF MINING, METALLURGICAL AND PETROLEUM ENGINEERS (AIME). 420 Commonwealth Drive, Warrendale, PA 15086. (412) 776-9086.

AMERICAN IRON AND STEEL INSTITUTE. 1000 Sixteenth Street, N.W., Washington, DC 20036. (202) 452-7100.

AMERICAN POWDER METALLURGY INSTITUTE. 105 College Road, East, Princeton, NJ 08540. (609) 452-7700.

AMERICAN SOCIETY FOR METALS. Metals Park, OH 44073. (216) 338-5151.

AMERICAN SOCIETY FOR TESTING AND MATERIALS. 1916 Race Street, Philadelphia, PA 19103. (215) 299-5400.

AMERICAN SOCIETY OF MECHANICAL ENGINEERS. 345 47th Street, New York, NY 10017. (212) 705-7722.

ASSOCIATION OF IRON AND STEEL ENGINEERS. Three Gateway Center, Suite 2350, Pittsburgh, PA 15222. (412) 281-6323.

THE METALLURGICAL SOCIETY. 420 Commonwealth Drive, Warrendale, PA 15086. (412) 776-9000.

NATIONAL ASSOCIATION OF CORROSION ENGINEERS. Box 218340, Houston, TX 77218. (713) 492-0535.

DIRECTORIES AND BIOGRAPHICAL SOURCES

DUN'S INDUSTRIAL GUIDE: THE METALWORKING DIRECTORY. Dun and Bradstreet Corporation, 49 Old Bloomfield Road, Mountain Lakes, NJ 07046. (201) 953-0300. Annual. $610.00.

INDUSTRIAL EQUIPMENT AND SUPPLIES DIRECTORY. American Business Directories, Inc., 5707 South 86th Circle, Omaha, NE 68127. (402) 331-7169.

MACHINERY BUYERS GUIDE. Findlay Publications Limited, Maitland House, Warrior Square, Southend-on-Sea, Essex SS5 5AR England. Annual. $45.00.

METALLURGICAL SOCIETY OF AIME - MEMBERSHIP LIST. American Institute of Mining, Metallurgical and Petroleum Engineers (AIME). 345 East 47th Street, New York, NY 10017. (212) 705-7695. 1984.

1987 DIRECTORY OF ENGINEERING SOCIETIES AND RELATED ORGANIZATIONS. Gordon Davis, editor. Hemisphere Publishing Corporation, 79 Madison Avenue, New York, NY 10016-7892. (800) 821-8312. 12th edition. 1987. $100.00.

RESEARCH CENTERS DIRECTORY 1988. Gale Research Company, Book Tower, Detroit, MI 48226. (800) 521-0707. 12th edition. 1987. $365.00 for set.

SCIENTIFIC AND TECHNICAL ORGANIZATIONS AND AGENCIES DIRECTORY. Margaret Labash Young, editor. Gale Research Company, Book Tower, Detroit, MI 48226. (800) 521-0707. 2nd edition. 1987. $185.00.

WHO'S WHO IN ENGINEERING. Gordon Davis, editor. Hemisphere Publishing Corporation, 1010 Vermont Avenue, NW, Washington, DC 20005. (800) 526-0275. 6th edition. 1985. $200.00.

GENERAL WORKS

ENGINEERING PROPERTIES OF STEELS. Phillip Harvey, editor. American Society for Metals, Metals Park, OH 44073. (216) 338-5151. 1982. $95.00.

STAINLESS STEELS. R.A. Lula. American Society for Metals, Metals Park, OH 44073. (216) 338-5151. 1985. $55.00.

HANDBOOKS AND MANUALS

HANDBOOK OF METAL FORMING. Kurt Lange, editor. John Wiley and Sons, Inc., 605 Third Avenue, New York, NY 10158. (800) 526-5368. 1985. $89.50.

HANDBOOK OF METAL FORMING PROCESSES. B. Avitzur. John Wiley and Sons, Inc., 605 Third Avenue, New York, NY 10158. (800) 526-5368. 1983. $105.00.

METALS HANDBOOK. American Society for Metals, Metals Park, OH 44073. (216) 338-5151. 9th edition. 14 volumes. 1988. $1310.00 for set.

WOLDMAN'S ENGINEERING ALLOYS. Robert C. Gibbons, editor. American Society for Metals, Metals Park, OH 44073. (216) 338-5151. 6th edition. 1979. $112.00.

ONLINE DATA BASES

CA SEARCH. Chemical Abstracts Service, P.O. Box 3012, Columbus, OH 43120. (800) 848-6538 or (614) 421-3600. Comprehensive guide to chemical literature, 1972 to present. Inquire as to online cost and availability.

COMPENDEX. Engineering Information, Inc., 345 East 47th Street, New York, NY 10017. (800) 221-1044 or (212) 705-7615. Engineering and technical literature, 1975 to present. Inquire as to online cost and availability.

INSPEC. INSPEC Marketing Department, Institution of Electrical Engineers. Available from IEEE Service Center, 445 Hoes Lane, Piscataway, NJ 08854. (201) 981-0060. Online version of Physics Abstracts. Inquire as to online cost and availability.

ISMEC. Cambridge Scientific Abstracts, 5161 River Road, Besthda, MD 20816. (800) 638-8076 or (301) 951-1400. Literature of mechanical and production engineering, 1973 to present. Inquire as to online cost and availability.

NTIS. National Technical Information Service, 5285 Port Royal Road, Springfield, VA 22161. (703) 487-4630. Broad coverage of government sponsored research reports, 1964 to present. Inquire as to online cost and availability.

SCISEARCH. Institute for Scientific Information, 3501 Market Street, Philadelphia, PA 19104. (800) 523-1850 or (215) 386-0100. Broad multidisciplinary title and author index to the international literature of science and technology, 1974 to present. Inquire as to online cost and availability.

WILSONLINE. H.W. Wilson and Company, 950 University Avenue, Bronx, NY 10452. (800) 367-6770 or (212) 588-8400. Makes available online versions of the H.W. Wilson indexes including Applied Science and Technology Index, Business Periodicals Index and Readers' Guide to Periodical Literature. Approximately 1980 to present. Inquire as to online cost and availability.

PERIODICALS

ALLOY DIGEST. Engineering Alloys Digest, Inc., Box 823, Upper Montclair, NJ 07043 . (201) 746-7930. 1952 to present. Monthly. $50.00 per year.

INDUSTRY WEEK. Penton-IPC, 1100 Superior Avenue, Cleveland, OH 44114. (216)696-7000. 1882 to present. 26 times per year. $50.00 per year.

IRON AND STEEL ENGINEER. Association of Iron and Steel Engineers, Suite 2350, Three Gateway Center, Pittsburgh, PA 15222. (412) 281-6323. 1924 to present. Monthly. $34.00 per year.

I. S. AND M. (IRON AND STEELMAKER). Iron and Steel Society, 410 Commonwealth Drive, Warrendale, PA 15096. (412) 776-1535. 1974 to present. Monthly. $35.00 per year.

JOURNAL OF ENGINEERING FOR INDUSTRY. American Society of Mechanical Engineers, 345 East 47th Street, New York, NY 10017. (212) 705-7703. 1970 to present. Quarterly. $100.00 per year.

JOURNAL OF METALS. American Institute of Mining, Metallurgical, and Petroleum Engineers, Inc., Metallurgical Society, 420 Commonwealth Drive, Warrendale, PA 15086. (412) 776-9086. 1949 to present. Monthly. $40.00 per year.

METAL PROGRESS. American Society for Metals, Metals Park, OH 44073. (216) 338-5151. 1930 to present. Monthly. $40.00.

METALLURGICAL TRANSACTIONS. Metallurgical Society of AIME, 420 Commonwealth Drive, Warrendale, PA 15086. (412) 776-9080. 1970 to present. Monthly. $95.00 per year.

METALS WEEK. McGraw-Hill Book Company, 1221 Avenue of the Americas, New York, NY 10020. (212) 997-2823. 1930 to present. Weekly. $597.00 per year.

METALWORKING DIGEST. Philos Publications, Inc., 1 East Chase Street, Baltimore, MD 21202. (301) 361-9060. 1969 to present. Nine times per year. $9.00 per year.

METIFAX MAGAZINE. Huebner Publications Inc., 6521 Davis Industrial Parkway, Solon, OH 44139. (216) 248-1125. 1956 to present. Monthly. $40.00 per year.

RESEARCH CENTERS AND INSTITUTES

ADVANCED STEEL PROCESSING AND PRODUCTS RESEARCH CENTER. Colorado School of Mines, Golden, CO 80401. (303) 273-3774.

AMERICAN IRON AND STEEL INSTITUTE. 1000 Sixteenth Street, N.W., Washington, DC 20036. (202) 452-7100.

CENTER FOR IRON AND STEEL MAKING RESEARCH. Carnegie-Mellon University, MEMS Department, Pittsburgh, PA 15213. (412) 268-2677.

STATISTICS SOURCES

STEEL MILL PRODUCTS. Current Business Reports, Series M33B, U.S. Department of Commerce, Washington, DC 20233. (202) 655-4000. Annual. $2.00.

WORLD METAL STATISTICS. World Bureau of Metal Statistics, 41 Doughty Street, London, WC1N 2LF, England. 1948 to present. Monthly. $1250.00 per year.

STEREOCHEMISTRY

See: PHYSICAL CHEMISTRY

STOCHASTIC PROCESSES

See: MARKOV PROCESSES, STATISTICS

STRENGTH OF MATERIALS

See: MATERIALS SCIENCE

STRESS AND STRAIN

See also: AERONAUTICAL ENGINEERING, CIVIL ENGINEERING, DYNAMICS, FAILURE ANALYSIS, MATERIALS SCIENCE, MECHANICAL ENGINEERING, MECHANICS, STRUCTURAL ENGINEERING

ABSTRACT SERVICES AND INDEXES

ABSTRACT JOURNAL IN EARTHQUAKE ENGINEERING. University of California at Berkeley, Earthquake Engineering Research Center, 1301 South 46th Street, Richmond, CA 94804. (415) 231-9413. Semiannual. $70.00 per copy.

APPLIED SCIENCE AND TECHNOLOGY INDEX. H.W. Wilson and Company, 950 University Avenue, Bronx, NY 10452. (800) 367-6670 or (212) 588-8400. Monthly. Inquire as to cost and availability.

CURRENT CONTENTS: ENGINEERING, TECHNOLOGY AND APPLIED SCIENCES. Institute for Scientific Information, 3501 Market Street, Philadelphia, PA 19104. (800) 523-1850 or (215) 386-0100. Weekly. $275.00 per year.

ENGINEERING INDEX MONTHLY AND AUTHOR INDEX. Engineering Information Inc., 345 East 47th Street, New York, NY 10017. (212) 705-7600. Monthly. $1560.00 per year.

INDEX TO SCIENTIFIC AND TECHNICAL PROCEEDINGS. Institute for Scientific Information, 3501 Market Street, Philadelphia, PA 19104. (800) 523-1850 or (215) 386-0100. 1978 to present. Monthly. $775.00 per year.

INTERNATIONAL CIVIL ENGINEERING ABSTRACTS. CITIS Limited, 2 Rosemount Terrace, Blackrock, Dublin, Ireland. 1974 to present. Monthly. $350.00 per year.

INTERNATIONAL STRUCTURAL ENGINEERING ABSTRACTS. CITIS Limited, 2 Rosemount Terrace, Blackrock, Dublin, Ireland. 1986 to present. Quarterly. $95.00 per year.

PUBLICATIONS INFORMATION. American Society of Civil Engineers, 345 East 47th Street, New York, NY 10017. (212) 705-7420. Abstracts, subject and author indexes to the publications of the American Society of Civil Engineers. Bimonthly. $80.00 per year.

SCIENCE CITATION INDEX. Institute for Scientific Information, 3501 Market Street, Philadelphia, PA 19104. (800) 523-1850 or (215) 386-0100. Six times per year. $6200.00 per year.

ASSOCIATIONS AND PROFESSIONAL SOCIETIES

AMERICAN SOCIETY FOR TESTING AND MATERIALS. 1916 Race Street, Philadelphia, PA 19103. (215) 299-5400.

AMERICAN SOCIETY OF CIVIL ENGINEERS. 345 East 47th Street, New York, NY 10017. (212) 705-7420.

AMERICAN SOCIETY OF MECHANICAL ENGINEERS. 345 East 47th Street, New York, NY 10017. (212) 705-7703.

DIRECTORIES AND BIOGRAPHICAL SOURCES

AMERICAN SOCIETY OF CIVIL ENGINEERS DIRECTORY: OFFICIAL REGISTER. 345 East 47th Street, New York, NY 10017. (212) 705-7420. Annual. Free.

ENGINEERS - CIVIL. American Business Directories, Inc., 5707 South 86th Circle, Omaha, NE 68127. Annual. $80.00.

INTERNATIONAL RESEARCH CENTERS DIRECTORY 1988-89. Darren L. Smith, editor. Gale Research Company, Book Tower, Detroit, MI 48226. (800) 521-0707. 4th edition. 1987. $360.00.

1987 DIRECTORY OF ENGINEERING SOCIETIES AND RELATED ORGANIZATIONS. Gordon Davis, editor. Hemisphere Publishing Corporation, 1010 Vermont Avenue, NW, Washington, DC 20005. (800) 526-0275. 12th edition. 1987. $100.00.

RESEARCH CENTERS DIRECTORY 1988. Gale Research Company, Book Tower, Detroit, MI 48226. (800) 521-0707. 12th edition. 1987. $365.00 for set.

SCIENTIFIC AND TECHNICAL ORGANIZATIONS AND AGENCIES DIRECTORY. Margaret Labash Young, editor. Gale Research Company, Book Tower, Detroit, MI 48226. (800) 521-0707. 2nd edition. 1987. $185.00.

WHO'S WHO IN ENGINEERING. Gordon Davis, editor. Hemisphere Publishing Corporation, 1010 Vermont Avenue, NW, Washington, DC 20005. (800) 526-0275. 6th edition. 1985. $200.00.

ENCYCLOPEDIAS AND DICTIONARIES

MCGRAW-HILL ENCYCLOPEDIA OF SCIENCE AND TECHNOLOGY. McGraw-Hill Book Company, 1221 Avenue of the Americas, New York, NY 10020. (212) 512-2000. 6th edition. 1987. $1600.00.

THESAURUS OF SCIENTIFIC, TECHNICAL, AND ENGINEERING TERMS. Hemisphere Publishing Corporation, 1010 Vermont Avenue, NW, Washington, DC 20005. (800) 526-0275. 1988. $125.00.

GENERAL WORKS

ADVANCED MECHANICS OF MATERIALS. A.P. Boresi and O.M. Sidebottom. John Wiley and Sons, Inc., 605 Third Avenue, New York, NY 10158. (800) 526-5368. Fourth edition. 1985. $44.00.

ELEMENTARY THEORY OF STRUCTURES. Yu H. Yuan. Prentice-Hall Publishing, Inc., Englewood Cliffs, NJ 07632. (800) 562-0245. Third edition. 1988. $42.95.

FATIGUE AND FRACTURE IN STEEL BRIDGES: CASE STUDIES. John W. Fisher. John Wiley and Sons, Inc., 605 Third Avenue, New York, NY 10158. (800) 526-5368. 1984. $45.95.

INTRODUCTION TO EARTHQUAKE ENGINEERING. S. Okamoto. Columbia University Press, 562 West 113th Street, New York, NY 10025. (212) 316-7100. 1985. $75.00.

MICROCOMPUTER-AIDED ENGINEERING: STRUCTURAL DYNAMICS. Mario Paz. Van Nostrand Reinhold Company, Inc., 135 West 50th Street, New York, NY 10020. (800) 543-2681. 1986. $49.95.

PROBABILISTIC FRACTURE MECHANICS AND RELIABILITY. James W. Provan, editor. Martinus Nijhoff. Distributed by Kluwer Academic Publishers, 190 Old Derby Street, Hingham, MA 02043. (617) 749-5262. 1987. $135.50.

STRUCTURAL ANALYSIS. R.C. Coates and others. Van Nostrand Reinhold Company, Inc., 135 West 50th Street, New York, NY 10020. (800) 543-2681. Third edition. 1987. $47.95 in paper.

STRUCTURAL ENGINEERING ANALYSIS ON PERSONAL COMPUTERS. John F. Fleming. McGraw-Hill Book Company, 1221 Avenue of the Americas, New York, NY 10020. (212) 512-2000. 1986. $19.95.

STRUCTURAL STABILITY. Wai-Fah Chen and E.M. Lui. Elsevier Science Publishing Company, Inc., 52 Vanderbilt Avenue, New York, NY 10017. (212) 370-5520. 1987. $49.50.

WIND EFFECTS ON STRUCTURES: AN INTRODUCTION TO WIND ENGINEERING. Emil Simiu and Robert H. Scanlan. John Wiley and Sons, Inc., 605 Third Avenue, New York, NY 10158. (800) 526-5368. Second edition. 1985. $49.95.

HANDBOOKS AND MANUALS

CIVIL ENGINEERING CALCULATIONS REFERENCE GUIDE. Tyler G. Hicks, editor. McGraw-Hill Book Company, 1221 Avenue of the Americas, New York, NY 10020. (212) 512-2000. 1987. $29.50.

CIVIL ENGINEERING PRACTICE. Paul N. Cheremisinoff and others, editors. Technomic Publishing Company, Inc., 851 Holland Avenue, Box 3535, Lancaster, PA 17604. (800) 233-9936. Five volumes. 1987-1988. $750.00 for set.

HANDBOOK OF MECHANICS, MATERIALS, AND STRUCTURES. A. Blake. John Wiley and Sons, Inc., 605 Third Avenue, New York, NY 10158. (800) 526-5368. 1985. $64.50.

MECHANICAL ENGINEER'S HANDBOOK. Myer Kutz, editor. John Wiley and Sons, Inc., 605 Third Avenue, New York, NY 10158. (800) 526-5368. 1986. $79.95.

STRUCTURAL ENGINEERING HANDBOOK. E.H. Gaylord and C.N. Gaylord, editors. McGraw-Hill Book Company, 1221 Avenue of the Americas, New York, NY 10020. (212) 512-2000. Second edition. 1979. $76.50.

ONLINE DATA BASES

COMPENDEX. Engineering Information, Inc., 345 East 47th Street, New York, NY 10017. (800) 221-1044 or (212) 705-7615. Engineering and technical literature, 1975 to present. Inquire as to online cost and availability.

DISSERTATION ABSTRACTS ONLINE. University Microfilms International, 300 North Zeeb Road, Ann Arbor, MI 48106. (800) 521-0600 or (313) 761-4700. Scope includes virtually all doctoral dissertations accepted at accredited American institutions from 1861 to present in over 250 subject areas. Inquire as to online cost and availability.

GEOREF. Online version of the BIBLIOGRAPHY AND INDEX OF GEOLOGY. American Geological Institute, 4220 King Street, Alexandria, VA 22302. (703) 379-2480. 1969 to present. Inquire as to online cost and availability.

INSPEC. INSPEC Marketing Department, Institution of Electrical Engineers. Available from IEEE Service Center, 445 Hoes Lane, Piscataway, NJ 08854. (201) 981-0060. Online version of Physics Abstracts. Inquire as to online cost and availability.

NTIS. National Technical Information Service, 5285 Port Royal Road, Springfield, VA 22161. (703) 487-4630. Broad coverage of government sponsored research reports, 1964 to present. Inquire as to online cost and availability.

SCISEARCH. Institute for Scientific Information, 3501 Market Street, Philadelphia, PA 19104. (800) 523-1850 or (215) 386-0100. Broad multidisciplinary title and author index to the international literature of science and technology, 1974 to present. Inquire as to online cost and availability.

WILSONLINE. H.W. Wilson and Company, 950 University Avenue, Bronx, NY 10452. (800) 367-6770 or (212) 588-8400. Makes available online versions of the H.W. Wilson indexes including Applied Science and Technology Index, Business Periodicals Index and Readers' Guide to Periodical Literature. Approximately 1980 to present. Inquire as to online cost and availability.

PERIODICALS

JOURNAL OF ENGINEERING MECHANICS. American Society of Civil Engineers, 345 East 47th Street, New York, NY 10017. (212) 705-7420. 1956 to present. Monthly. $120.00 per year.

JOURNAL OF STRUCTURAL ENGINEERING. American Society of Civil Engineers, 345 East 47th Street, New York, NY 10017. (212) 705-7420. 1956 to present. Monthly. $140.00 per year.

JOURNAL OF STRUCTURAL MECHANICS. Marcel Dekker Inc., 270 Madison Avenue, New York, NY 10016. (800) 228-1160. 1972 to present. Quarterly. $75.00 per year.

JOURNAL OF TESTING AND EVALUATION. American Society for Testing and Materials, 1916 Race Street, Philadelphia, PA 19103. (215) 299-5400. 1966 to present. Bimonthly. $40.00 per year.

MATERIALS ENGINEERING. Penton-IPC, 1100 Superior Avenue, Cleveland, OH 44114. (216)696-7000. 1929 to present. Monthly. $40.00 per year.

MATERIALS EVALUATION. American Society for Nondestructive Testing, 4153 Arlingate Plaza, Caller 28518, Columbus, OH 43228. 1942 to present. Monthly. $50.00 per year.

SAMPE JOURNAL. Society for the Advancement of Material and Process Engineering, 843 West Glentana, Box 2459, Covina, CA 91722. (818) 331-0610. 1965 to present. Bimonthly. $31.00 per year.

STRAIN. British Society for Strain Measurement, Suite 7, Exchange Building, Ouayside, Newcastle-upon-Tyne, NE1 3BJ, England. 1965 to present. Quarterly. $45.00 per year.

STRUCTURAL ENGINEERING PRACTICE: ANALYSIS, DESIGN, MANAGEMENT. Marcel Dekker Inc., 270 Madison Avenue, New York, NY 10016. (800) 228-1160. 1981 to present. Quarterly. $75.00 per year.

RESEARCH CENTERS AND INSTITUTES

PHIL M. FERGUSON STRUCTURAL ENGINEERING LABORATORY. University of Texas at Austin, Balcones Research Center, 10100 Burnet Road, Building 24, Austin, TX 78758. (512) 471-3062.

STRUCTURAL ENGINEERING LABORATORY. University of Michigan, 2340 G.G. Brown Building, Ann Arbor, MI 48109. (313) 763-3046.

STRUCTURAL ENGINEERING MATERIALS LABORATORY. University of California, Berkeley, Davis Hall, Berkeley, CA 94720. (415) 642-3434.

STRUCTURAL STABILITY RESEARCH COUNCIL. Fritz Engineering Laboratory No. 13, Lehigh University, Bethlehem, PA 18105. (215) 861-3519.

WISCONSIN CENTER FOR STRUCTURAL AND MATERIALS TESTING. University of Wisconsin at Madison, 1415 Johnson Drive, Madison, WI 53706. (608) 262-3205.

SPECIFICATIONS AND STANDARDS

STRESS RELAXATION TESTING. A. Fox, editor. American Society for Testing and Materials, 1916 Race Street, Philadelphia, PA 19103. (215) 299-5400. STP No. 676. 1979. $23.75.

STRIP MINING

See: MINING ENGINEERING

STRUCTURAL ENGINEERING

See also: BRIDGES, CIVIL ENGINEERING, CONCRETE, CONSTRUCTION ENGINNERING, FAILURE ANALYSIS, HIGHWAY ENGINEERING, STEEL AND STEEL MAKING

ABSTRACT SERVICES AND INDEXES

ABSTRACT JOURNAL IN EARTHQUAKE ENGINEERING. University of California at Berkeley, Earthquake Engineering Research Center, 1301 South 46th Street, Richmond, CA 94804. (415) 231-9413. Semiannual. $70.00 per copy.

APPLIED SCIENCE AND TECHNOLOGY INDEX. H.W. Wilson and Company, 950 University Avenue, Bronx, NY 10452. (800) 367-6670 or (212) 588-8400. Monthly. Inquire as to cost and availability.

CURRENT CONTENTS: ENGINEERING, TECHNOLOGY AND APPLIED SCIENCES. Institute for Scientific Information, 3501 Market Street, Philadelphia, PA 19104. (800) 523-1850 or (215) 386-0100. Weekly. $275.00 per year.

ENGINEERING INDEX MONTHLY AND AUTHOR INDEX. Engineering Information Inc., 345 East 47th Street, New York, NY 10017. (212) 705-7600. Monthly. $1560.00 per year.

INDEX TO SCIENTIFIC AND TECHNICAL PROCEEDINGS. Institute for Scientific Information, 3501 Market Street, Philadelphia, PA 19104. (800) 523-1850 or (215) 386-0100. 1978 to present. Monthly. $775.00 per year.

INTERNATIONAL CIVIL ENGINEERING ABSTRACTS. CITIS Limited, 2 Rosemount Terrace, Blackrock, Dublin, Ireland. 1974 to present. Monthly. $350.00 per year.

INTERNATIONAL STRUCTURAL ENGINEERING ABSTRACTS. CITIS Limited, 2 Rosemount Terrace, Blackrock, Dublin, Ireland. 1986 to present. Quarterly. $95.00 per year.

PUBLICATIONS INFORMATION. American Society of Civil Engineers, 345 East 47th Street, New York, NY 10017. (212) 705-7420. Abstracts, subject and author indexes to the publications of the American Society of Civil Engineers. Bimonthly. $80.00 per year.

SCIENCE CITATION INDEX. Institute for Scientific Information, 3501 Market Street, Philadelphia, PA 19104. (800) 523-1850 or (215) 386-0100. Six times per year. $6200.00 per year.

ASSOCIATIONS AND PROFESSIONAL SOCIETIES

AMERICAN SOCIETY FOR TESTING AND MATERIALS. 1916 Race Street, Philadelphia, PA 19103. (215) 299-5400.

AMERICAN SOCIETY OF CIVIL ENGINEERS. 345 East 47th Street, New York, NY 10017. (212) 705-7420.

BIBLIOGRAPHIES

NEW TECHNICAL BOOKS: A SELECTIVE LIST WITH DESCRIPTIVE ANNOTATIONS. New York Public Library, Science and Technology Research Center, Fifth Avenue and 42nd Street, New York, NY 10018. (212) 930-0800. 1915 to present. Monthly. $15.00 per year.

SCIENTIFIC AND TECHNICAL BOOKS AND SERIALS IN PRINT 1988; AN INDEX TO LITERATURE IN SCIENCE AND TECHNOLOGY. R.R. Bowker Company, 205 East 42nd Street, New York, NY 10017. (800) 521-8110. $175.00.

DIRECTORIES AND BIOGRAPHICAL SOURCES

AMERICAN SOCIETY OF CIVIL ENGINEERS DIRECTORY: OFFICIAL REGISTER. 345 East 47th Street, New York, NY

10017. (212) 705-7420. Annual. Free.

ENGINEERS - CIVIL. American Business Directories, Inc., 5707 South 86th Circle, Omaha, NE 68127. Annual. $80.00.

INTERNATIONAL RESEARCH CENTERS DIRECTORY 1988-89. Darren L. Smith, editor. Gale Research Company, Book Tower, Detroit, MI 48226. (800) 521-0707. 4th edition. 1987. $360.00.

1987 DIRECTORY OF ENGINEERING SOCIETIES AND RELATED ORGANIZATIONS. Gordon Davis, editor. Hemisphere Publishing Corporation, 1010 Vermont Avenue, NW, Washington, DC 20005. (800) 526-0275. 12th edition. 1987. $100.00.

RESEARCH CENTERS DIRECTORY 1988. Gale Research Company, Book Tower, Detroit, MI 48226. (800) 521-0707. 12th edition. 1987. $365.00 for set.

SCIENTIFIC AND TECHNICAL ORGANIZATIONS AND AGENCIES DIRECTORY. Margaret Labash Young, editor. Gale Research Company, Book Tower, Detroit, MI 48226. (800) 521-0707. 2nd edition. 1987. $185.00.

WHO'S WHO IN ENGINEERING. Gordon Davis, editor. Hemisphere Publishing Corporation, 1010 Vermont Avenue, NW, Washington, DC 20005. (800) 526-0275. 6th edition. 1985. $200.00.

ENCYCLOPEDIAS AND DICTIONARIES

CONSTRUCTION GLOSSARY: AN ENCYCLOPEDIC REFERENCE AND MANUAL. J.S. Stein. John Wiley and Sons, Inc., 605 Third Avenue, New York, NY 10158. (800) 526-5368. 1986. $35.95.

DICTIONARY OF CIVIL ENGINEERING. John S. Scott. John Wiley and Sons, Inc., 605 Third Avenue, New York, NY 10158. (800) 526-5368. Third edition. 1981. $26.95.

MCGRAW-HILL ENCYCLOPEDIA OF SCIENCE AND TECHNOLOGY. McGraw-Hill Book Company, 1221 Avenue of the Americas, New York, NY 10020. (212) 512-2000. 6th edition. 1987. $1600.00.

THESAURUS OF SCIENTIFIC, TECHNICAL, AND ENGINEERING TERMS. Hemisphere Publishing Corporation, 1010 Vermont Avenue, NW, Washington, DC 20005. (800) 526-0275. 1988. $125.00.

GENERAL WORKS

APPLIED STRUCTURAL STEEL DESIGN. L. Spiegel and G.F. Limbrunner. Prentice-Hall Publishing, Inc., Englewood Cliffs, NJ 07632. (800) 562-0245. 1986. $34.95.

DESIGN OF REINFORCED CONCRETE. Samuel E. French. Prentice-Hall Publishing, Inc., Englewood Cliffs, NJ 07632. (800) 562-0245. 1987. $25.00.

ELEMENTARY THEORY OF STRUCTURES. Yu H. Yuan. Prentice-Hall Publishing, Inc., Englewood Cliffs, NJ 07632. (800) 562-0245. Third edition. 1988. $42.95.

FATIGUE AND FRACTURE IN STEEL BRIDGES: CASE STUDIES. John W. Fisher. John Wiley and Sons, Inc., 605 Third Avenue, New York, NY 10158. (800) 526-5368. 1984. $45.95.

FOUNDATION DESIGN AND CONSTRUCTION. M.J. Tomlinson. John Wiley and Sons, Inc., 605 Third Avenue, New York, NY 10158. (800) 526-5368. Fifth edition. 1986. $45.95.

INTRODUCTION TO EARTHQUAKE ENGINEERING. S. Okamoto. Columbia University Press, 562 West 113th Street, New York, NY 10025. (212) 316-7100. 1985. $75.00.

MICROCOMPUTER-AIDED ENGINEERING: STRUCTURAL DYNAMICS. Mario Paz. Van Nostrand Reinhold Company, Inc., 135 West 50th Street, New York, NY 10020. (800) 543-2681. 1986.

$49.95.

PROBABILISTIC FRACTURE MECHANICS AND RELIABILITY. James W. Provan, editor. Martinus Nijhoff. Distributed by Kluwer Academic Publishers, 190 Old Derby Street, Hingham, MA 02043. (617) 749-5262. 1987. $135.50.

STRUCTURAL ANALYSIS. R.C. Coates and others. Van Nostrand Reinhold Company, Inc., 135 West 50th Street, New York, NY 10020. (800) 543-2681. Third edition. 1987. $47.95 in paper.

STRUCTURAL ENGINEERING ANALYSIS ON PERSONAL COMPUTERS. John F. Fleming. McGraw-Hill Book Company, 1221 Avenue of the Americas, New York, NY 10020. (212) 512-2000. 1986. $19.95.

STRUCTURAL STABILITY. Wai-Fah Chen and E.M. Lui. Elsevier Science Publishing Company, Inc., 52 Vanderbilt Avenue, New York, NY 10017. (212) 370-5520. 1987. $49.50.

WIND EFFECTS ON STRUCTURES: AN INTRODUCTION TO WIND ENGINEERING. Emil Simiu and Robert H. Scanlan. John Wiley and Sons, Inc., 605 Third Avenue, New York, NY 10158. (800) 526-5368. Second edition. 1985. $49.95.

HANDBOOKS AND MANUALS

BUILDING STRUCTURES HANDBOOK. R.N. White and C.G. Salmon. John Wiley and Sons, Inc., 605 Third Avenue, New York, NY 10158. (800) 526-5368. 1986. $79.95.

CIVIL ENGINEERING CALCULATIONS REFERENCE GUIDE. Tyler G. Hicks, editor. McGraw-Hill Book Company, 1221 Avenue of the Americas, New York, NY 10020. (212) 512-2000. 1987. $29.50.

CIVIL ENGINEERING PRACTICE. Paul N. Cheremisinoff and others, editors. Technomic Publishing Company, Inc., 851 Holland Avenue, Box 3535, Lancaster, PA 17604. (800) 233-9936. Five volumes. 1987-1988. $750.00 for set.

HANDBOOK OF CONCRETE ENGINEERING. Mark Fintel. Van Nostrand Reinhold Company, Inc., 135 West 50th Street, New York, NY 10020. (800) 543-2681. 1986. $89.95.

HANDBOOK OF MECHANICS, MATERIALS, AND STRUCTURES. A. Blake. John Wiley and Sons, Inc., 605 Third Avenue, New York, NY 10158. (800) 526-5368. 1985. $64.50.

STANDARD HANDBOOK FOR CIVIL ENGINEERS. F. S. Merritt, editor. McGraw-Hill Book Company, 1221 Avenue of the Americas, New York, NY 10020. (212) 512-2000. Third edition. 1983. $89.50.

STRUCTURAL ENGINEERING HANDBOOK. E.H. Gaylord and C.N. Gaylord, editors. McGraw-Hill Book Company, 1221 Avenue of the Americas, New York, NY 10020. (212) 512-2000. Second edition. 1979. $76.50.

ONLINE DATA BASES

COMPENDEX. Engineering Information, Inc., 345 East 47th Street, New York, NY 10017. (800) 221-1044 or (212) 705-7615. Engineering and technical literature, 1975 to present. Inquire as to online cost and availability.

DISSERTATION ABSTRACTS ONLINE. University Microfilms International, 300 North Zeeb Road, Ann Arbor, MI 48106. (800) 521-0600 or (313) 761-4700. Scope includes virtually all doctoral dissertations accepted at accredited American institutions from 1861 to present in over 250 subject areas. Inquire as to online cost and availability.

GEOREF. Online version of the BIBLIOGRAPHY AND INDEX OF GEOLOGY. American Geological Institute, 4220 King Street, Alexandria, VA 22302. (703) 379-2480. 1969 to present. Inquire as to online cost and availability.

INSPEC. INSPEC Marketing Department, Institution of Electrical Engineers. Available from IEEE Service Center, 445 Hoes Lane, Piscataway, NJ 08854. (201) 981-0060. Online version of Physics Abstracts. Inquire as to online cost and availability.

NTIS. National Technical Information Service, 5285 Port Royal Road, Springfield, VA 22161. (703) 487-4630. Broad coverage of government sponsored research reports, 1964 to present. Inquire as to online cost and availability.

SCISEARCH. Institute for Scientific Information, 3501 Market Street, Philadelphia, PA 19104. (800) 523-1850 or (215) 386-0100. Broad multidisciplinary title and author index to the international literature of science and technology, 1974 to present. Inquire as to online cost and availability.

TRIS. National Academy of Sciences. Transportation Research, 2101 Constitution Avenue, N.W., Washington, DC 20418. (202) 334-2000. Covers highway and transportation research. 1968 to present. Inquire as to cost and availability.

WILSONLINE. H.W. Wilson and Company, 950 University Avenue, Bronx, NY 10452. (800) 367-6770 or (212) 588-8400. Makes available online versions of the H.W. Wilson indexes including Applied Science and Technology Index, Business Periodicals Index and Readers' Guide to Periodical Literature. Approximately 1980 to present. Inquire as to online cost and availability.

PERIODICALS

AMERICAN CONCRETE INSTITUTE JOURNAL. P.O. Box 19150, Redford Station, Detroit, MI 48219. (513) 532-2600. 1929 to present. Bimonthly. $69.00 per year.

CANADIAN JOURNAL OF CIVIL ENGINEERING. National Research Council of Canada, Montreal Road, Ottawa, ON, Canada K1A OR6. Quarterly. $40.00 per year.

CIVIL ENGINEERING. American Society of Civil Engineers, 345 East 47th Street, New York, NY 10017. (212) 705-7420. 1930 to present. Monthly. $48.00 per year.

CONCRETE. Harcourt, Brace Jovanovich, Inc., 7500 Old Oak Boulevard, Cleveland, OH 44130. 1937 to present. Monthly. $25.00 per year.

JOURNAL OF ENGINEERING MECHANICS. American Society of Civil Engineers, 345 East 47th Street, New York, NY 10017. (212) 705-7420. 1956 to present. Monthly. $120.00 per year.

JOURNAL OF STRUCTURAL ENGINEERING. American Society of Civil Engineers, 345 East 47th Street, New York, NY 10017. (212) 705-7420. 1956 to present. Monthly. $140.00 per year.

JOURNAL OF STRUCTURAL MECHANICS. Marcel Dekker Inc., 270 Madison Avenue, New York, NY 10016. (800) 228-1160. 1972 to present. Quarterly. $75.00 per year.

JOURNAL OF SURVEYING ENGINEERING. American Society of Civil Engineers, 345 East 47th Street, New York, NY 10017. (212) 705-7420. 1956 to present. Three times per year. $35.00 per year.

STRUCTURAL ENGINEER. PART A AND B. Institution of Structural Engineers, 11 Upper Belgrave Street, London SW1X 8BH, England. 1908 to present. Monthly. $140.00 per year.

STRUCTURAL ENGINEERING PRACTICE: ANALYSIS, DESIGN, MANAGEMENT. Marcel Dekker Inc., 270 Madison Avenue, New York, NY 10016. (800) 228-1160. 1981 to present. Quarterly. $75.00 per year.

RESEARCH CENTERS AND INSTITUTES

PHIL M. FERGUSON STRUCTURAL ENGINEERING LABORATORY. University of Texas at Austin, Balcones Research Center, 10100 Burnet Road, Building 24, Austin, TX 78758. (512) 471-3062.

STRUCTURAL ENGINEERING LABORATORY. University of Michigan, 2340 G.G. Brown Building, Ann Arbor, MI 48109. (313) 763-3046.

STRUCTURAL ENGINEERING MATERIALS LABORATORY. University of California, Berkeley, Davis Hall, Berkeley, CA 94720. (415) 642-3434.

STRUCTURAL STABILITY RESEARCH COUNCIL. Fritz Engineering Laboratory No. 13, Lehigh University, Bethlehem, PA 18105. (215) 861-3519.

SPECIFICATIONS AND STANDARDS

STRUCTURAL DESIGN GUIDE TO THE ACI BUILDING CODE. Paul Rice. Van Nostrand Reinhold Company, Inc., 135 West 50th Street, New York, NY 10020. (800) 543-2681. Third edition. 1985. $47.95.

STRUCTURAL GEOLOGY

See: PHYSICAL GEOLOGY

STRUCTURAL MATERIALS

See: STRUCTURAL ENGINEERING

STRUCTURAL PETROLOGY

See: PETROLOGY

STRUCTURES

See also: BRIDGES, CIVIL ENGINEERING, CONCRETE, CONSTRUCTION ENGINEERING, FAILURE ANALYSIS, HIGHWAY ENGINEERING, STEEL AND STEEL MAKING, STRUCTURAL ENGINEERING

ABSTRACT SERVICES AND INDEXES

ABSTRACT JOURNAL IN EARTHQUAKE ENGINEERING. University of California at Berkeley, Earthquake Engineering Research Center, 1301 South 46th Street, Richmond, CA 94804. (415) 231-9413. Semiannual. $70.00 per copy.

APPLIED SCIENCE AND TECHNOLOGY INDEX. H.W. Wilson and Company, 950 University Avenue, Bronx, NY 10452. (800) 367-6670 or (212) 588-8400. Monthly. Inquire as to cost and availability.

CURRENT CONTENTS: ENGINEERING, TECHNOLOGY AND APPLIED SCIENCES. Institute for Scientific Information, 3501 Market Street, Philadelphia, PA 19104. (800) 523-1850 or (215) 386-0100. Weekly. $275.00 per year.

ENGINEERING INDEX MONTHLY AND AUTHOR INDEX. Engineering Information Inc., 345 East 47th Street, New York, NY 10017. (212) 705-7600. Monthly. $1560.00 per year.

INTERNATIONAL CIVIL ENGINEERING ABSTRACTS. CITIS Limited, 2 Rosemount Terrace, Blackrock, Dublin, Ireland. 1974 to present. Monthly. $350.00 per year.

INTERNATIONAL STRUCTURAL ENGINEERING ABSTRACTS. CITIS Limited, 2 Rosemount Terrace, Blackrock, Dublin, Ireland.

STRUCTURES

1986 to present. Quarterly. $95.00 per year.

PUBLICATIONS INFORMATION. American Society of Civil Engineers, 345 East 47th Street, New York, NY 10017. (212) 705-7420. Abstracts, subject and author indexes to the publications of the American Society of Civil Engineers. Bimonthly. $80.00 per year.

SCIENCE CITATION INDEX. Institute for Scientific Information, 3501 Market Street, Philadelphia, PA 19104. (800) 523-1850 or (215) 386-0100. Six times per year. $6200.00 per year.

ASSOCIATIONS AND PROFESSIONAL SOCIETIES

AMERICAN SOCIETY FOR TESTING AND MATERIALS. 1916 Race Street, Philadelphia, PA 19103. (215) 299-5400.

AMERICAN SOCIETY OF CIVIL ENGINEERS. 345 East 47th Street, New York, NY 10017. (212) 705-7420.

DIRECTORIES AND BIOGRAPHICAL SOURCES

AMERICAN SOCIETY OF CIVIL ENGINEERS DIRECTORY: OFFICIAL REGISTER. 345 East 47th Street, New York, NY 10017. (212) 705-7420. Annual. Free.

ENGINEERS - CIVIL. American Business Directories, Inc., 5707 South 86th Circle, Omaha, NE 68127. Annual. $80.00.

INTERNATIONAL RESEARCH CENTERS DIRECTORY 1988-89. Darren L. Smith, editor. Gale Research Company, Book Tower, Detroit, MI 48226. (800) 521-0707. 4th edition. 1987. $360.00.

1987 DIRECTORY OF ENGINEERING SOCIETIES AND RELATED ORGANIZATIONS. Gordon Davis, editor. Hemisphere Publishing Corporation, 1010 Vermont Avenue, NW, Washington, DC 20005. (800) 526-0275. 12th edition. 1987. $100.00.

RESEARCH CENTERS DIRECTORY 1988. Gale Research Company, Book Tower, Detroit, MI 48226. (800) 521-0707. 12th edition. 1987. $365.00 for set.

SCIENTIFIC AND TECHNICAL ORGANIZATIONS AND AGENCIES DIRECTORY. Margaret Labash Young, editor. Gale Research Company, Book Tower, Detroit, MI 48226. (800) 521-0707. 2nd edition. 1987. $185.00.

WHO'S WHO IN ENGINEERING. Gordon Davis, editor. Hemisphere Publishing Corporation, 1010 Vermont Avenue, NW, Washington, DC 20005. (800) 526-0275. 6th edition. 1985. $200.00.

ENCYCLOPEDIAS AND DICTIONARIES

CONSTRUCTION GLOSSARY: AN ENCYCLOPEDIC REFERENCE AND MANUAL. J.S. Stein. John Wiley and Sons, Inc., 605 Third Avenue, New York, NY 10158. (800) 526-5368. 1986. $35.95.

DICTIONARY OF CIVIL ENGINEERING. John S. Scott. John Wiley and Sons, Inc., 605 Third Avenue, New York, NY 10158. (800) 526-5368. Third edition. 1981. $26.95.

GENERAL WORKS

APPLIED STRUCTURAL STEEL DESIGN. L. Spiegel and G.F. Limbrunner. Prentice-Hall Publishing, Inc., Englewood Cliffs, NJ 07632. (800) 562-0245. 1986. $34.95.

DESIGN OF REINFORCED CONCRETE. Samuel E. French. Prentice-Hall Publishing, Inc., Englewood Cliffs, NJ 07632. (800) 562-0245. 1987. $25.00.

ELEMENTARY THEORY OF STRUCTURES. Yu H. Yuan. Prentice-Hall Publishing, Inc., Englewood Cliffs, NJ 07632. (800) 562-0245. Third edition. 1988. $42.95.

FATIGUE AND FRACTURE IN STEEL BRIDGES: CASE STUDIES. John W. Fisher. John Wiley and Sons, Inc., 605 Third Avenue, New York, NY 10158. (800) 526-5368. 1984. $45.95.

FOUNDATION DESIGN AND CONSTRUCTION. M.J. Tomlinson. John Wiley and Sons, Inc., 605 Third Avenue, New York, NY 10158. (800) 526-5368. Fifth edition. 1986. $45.95.

INTRODUCTION TO EARTHQUAKE ENGINEERING. S. Okamoto. Columbia University Press, 562 West 113th Street, New York, NY 10025. (212) 316-7100. 1985. $75.00.

MICROCOMPUTER-AIDED ENGINEERING: STRUCTURAL DYNAMICS. Mario Paz. Van Nostrand Reinhold Company, Inc., 135 West 50th Street, New York, NY 10020. (800) 543-2681. 1986. $49.95.

PROBABILISTIC FRACTURE MECHANICS AND RELIABILITY. James W. Provan, editor. Martinus Nijhoff. Distributed by Kluwer Academic Publishers, 190 Old Derby Street, Hingham, MA 02043. (617) 749-5262. 1987. $135.50.

STRUCTURAL ANALYSIS. R.C. Coates and others. Van Nostrand Reinhold Company, Inc., 135 West 50th Street, New York, NY 10020. (800) 543-2681. Third edition. 1987. $47.95 in paper.

STRUCTURAL ENGINEERING ANALYSIS ON PERSONAL COMPUTERS. John F. Fleming. McGraw-Hill Book Company, 1221 Avenue of the Americas, New York, NY 10020. (212) 512-2000. 1986. $19.95.

STRUCTURAL STABILITY. Wai-Fah Chen and E.M. Lui. Elsevier Science Publishing Company, Inc., 52 Vanderbilt Avenue, New York, NY 10017. (212) 370-5520. 1987. $49.50.

WIND EFFECTS ON STRUCTURES: AN INTRODUCTION TO WIND ENGINEERING. Emil Simiu and Robert H. Scanlan. John Wiley and Sons, Inc., 605 Third Avenue, New York, NY 10158. (800) 526-5368. Second edition. 1985. $49.95.

HANDBOOKS AND MANUALS

BUILDING STRUCTURES HANDBOOK. R.N. White and C.G. Salmon. John Wiley and Sons, Inc., 605 Third Avenue, New York, NY 10158. (800) 526-5368. 1986. $79.95.

CIVIL ENGINEERING CALCULATIONS REFERENCE GUIDE. Tyler G. Hicks, editor. McGraw-Hill Book Company, 1221 Avenue of the Americas, New York, NY 10020. (212) 512-2000. 1987. $29.50.

CIVIL ENGINEERING PRACTICE. Paul N. Cheremisinoff and others, editors. Technomic Publishing Company, Inc., 851 Holland Avenue, Box 3535, Lancaster, PA 17604. (800) 233-9936. Five volumes. 1987-1988. $750.00 for set.

HANDBOOK OF CONCRETE ENGINEERING. Mark Fintel. Van Nostrand Reinhold Company, Inc., 135 West 50th Street, New York, NY 10020. (800) 543-2681. 1986. $89.95.

HANDBOOK OF MECHANICS, MATERIALS, AND STRUCTURES. A. Blake. John Wiley and Sons, Inc., 605 Third Avenue, New York, NY 10158. (800) 526-5368. 1985. $64.50.

STANDARD HANDBOOK FOR CIVIL ENGINEERS. F.S. Merritt, editor. McGraw-Hill Book Company, 1221 Avenue of the Americas, New York, NY 10020. (212) 512-2000. Third edition. 1983. $89.50.

STRUCTURAL ENGINEERING HANDBOOK. E.H. Gaylord and C.N. Gaylord, editors. McGraw-Hill Book Company, 1221 Avenue of the Americas, New York, NY 10020. (212) 512-2000. Second edition. 1979. $76.50.

ONLINE DATA BASES

COMPENDEX. Engineering Information, Inc., 345 East 47th Street, New York, NY 10017. (800) 221-1044 or (212) 705-7615. Engineering and technical literature, 1975 to present. Inquire as to online cost and availability.

DISSERTATION ABSTRACTS ONLINE. University Microfilms International, 300 North Zeeb Road, Ann Arbor, MI 48106. (800) 521-0600 or (313) 761-4700. Scope includes virtually all doctoral dissertations accepted at accredited American institutions from 1861 to present in over 250 subject areas. Inquire as to online cost and availability.

GEOREF. Online version of the BIBLIOGRAPHY AND INDEX OF GEOLOGY. American Geological Institute, 4220 King Street, Alexandria, VA 22302. (703) 379-2480. 1969 to present. Inquire as to online cost and availability.

INSPEC. INSPEC Marketing Department, Institution of Electrical Engineers. Available from IEEE Service Center, 445 Hoes Lane, Piscataway, NJ 08854. (201) 981-0060. Online version of Physics Abstracts. Inquire as to online cost and availability.

NTIS. National Technical Information Service, 5285 Port Royal Road, Springfield, VA 22161. (703) 487-4630. Broad coverage of government sponsored research reports, 1964 to present. Inquire as to online cost and availability.

SCISEARCH. Institute for Scientific Information, 3501 Market Street, Philadelphia, PA 19104. (800) 523-1850 or (215) 386-0100. Broad multidisciplinary title and author index to the international literature of science and technology, 1974 to present. Inquire as to online cost and availability.

TRIS. National Academy of Sciences. Transportation Research, 2101 Constitution Avenue, N.W., Washington, DC 20418. (202) 334-2000. Covers highway and transportation research. 1968 to present. Inquire as to cost and availability.

WILSONLINE. H.W. Wilson and Company, 950 University Avenue, Bronx, NY 10452. (800) 367-6770 or (212) 588-8400. Makes available online versions of the H.W. Wilson indexes including Applied Science and Technology Index, Business Periodicals Index and Readers' Guide to Periodical Literature. Approximately 1980 to present. Inquire as to online cost and availability.

PERIODICALS

AMERICAN CONCRETE INSTITUTE JOURNAL. P.O. Box 19150, Redford Station, Detroit, MI 48219. (513) 532-2600. 1929 to present. Bimonthly. $69.00 per year.

CANADIAN JOURNAL OF CIVIL ENGINEERING. National Research Council of Canada, Montreal Road, Ottawa, ON, Canada K1A OR6. Quarterly. $40.00 per year.

CIVIL ENGINEERING. American Society of Civil Engineers, 345 East 47th Street, New York, NY 10017. (212) 705-7420. 1930 to present. Monthly. $48.00 per year.

CONCRETE. Harcourt, Brace Jovanovich, Inc., 7500 Old Oak Boulevard, Cleveland, OH 44130. 1937 to present. Monthly. $25.00 per year.

JOURNAL OF ENGINEERING MECHANICS. American Society of Civil Engineers, 345 East 47th Street, New York, NY 10017. (212) 705-7420. 1956 to present. Monthly. $120.00 per year.

JOURNAL OF STRUCTURAL ENGINEERING. American Society of Civil Engineers, 345 East 47th Street, New York, NY 10017. (212) 705-7420. 1956 to present. Monthly. $140.00 per year.

JOURNAL OF STRUCTURAL MECHANICS. Marcel Dekker Inc., 270 Madison Avenue, New York, NY 10016. (800) 228-1160. 1972 to present. Quarterly. $75.00 per year.

STRUCTURAL ENGINEER. PART A AND B. Institution of Structural Engineers, 11 Upper Belgrave Street, London SW1X 8BH, England. 1908 to present. Monthly. $140.00 per year.

STRUCTURAL ENGINEERING PRACTICE: ANALYSIS, DESIGN, MANAGEMENT. Marcel Dekker Inc., 270 Madison Avenue, New York, NY 10016. (800) 228-1160. 1981 to present. Quarterly. $75.00 per year.

RESEARCH CENTERS AND INSTITUTES

PHIL M. FERGUSON STRUCTURAL ENGINEERING LABORATORY. University of Texas at Austin, Balcones Research Center, 10100 Burnet Road, Building 24, Austin, TX 78758. (512) 471-3062.

STRUCTURAL ENGINEERING LABORATORY. University of Michigan, 2340 G.G. Brown Building, Ann Arbor, MI 48109. (313) 763-3046.

STRUCTURAL ENGINEERING MATERIALS LABORATORY. University of California, Berkeley, Davis Hall, Berkeley, CA 94720. (415) 642-3434.

STRUCTURAL STABILITY RESEARCH COUNCIL. Fritz Engineering Laboratory No. 13, Lehigh University, Bethlehem, PA 18105. (215) 861-3519.

SPECIFICATIONS AND STANDARDS

STRUCTURAL DESIGN GUIDE TO THE ACI BUILDING CODE. Paul Rice. Van Nostrand Reinhold Company, Inc., 135 West 50th Street, New York, NY 10020. (800) 543-2681. Third edition. 1985. $47.95.

SUN

See also: SOLAR SYSTEM; SOLAR WIND

ABSTRACT SERVICES AND INDEXES

ASTRONOMY AND ASTROPHYSICS ABSTRACTS. Springer-Verlag New York, Incorporated, 175 Fifth Avenue, New York, NY 10010. (212) 460-1500. $70.00 per year.

ASTRONOMY AND ASTROPHYSICS MONTHLY INDEX. Olivetree Associates, Post Office Box 236, Sierra Madre, CA 91024. $212.00 per year. Complimentary copies available on request.

GENERAL SCIENCE INDES. H.W. Wilson Company, 950 University Avenue, Bronx, NY 10452. (800) 367-6770 or (212) 588-8400. Inquire as to cost and availability.

SCIENCE CITATION INDEX. Institute for Scientific Information, 3501 Market Street, Philadelphia, PA 19104. (800) 523-1850 or (215) 386-0100. Inquire as to cost and availability.

STAR. (Scientific and Technical Aerospace Reports. United States National Aeronautics and Space Administration, Scientific and Technical Information Facility, Box 8757, Baltimore-Washington International Airport, MD 21240. (202) 755-2210. Semimonthly, with semiannual and annual indexes. $85.00 per year.

ANNUAL REVIEWS AND YEARBOOKS

THE ASTRONOMICAL ALMANAC. Superintendent of Documents, United States Government Printing Office, Washington, DC 20402. (202) 783-3238. Yearly.

ANNUAL REVIEW OF ASTRONOMY AND ASTROPHYSICS. Annual Reviews, Incorporated, 4139 El Camino Way, Palo Alto, CA 94306. (415) 493-4400. Annual. Inquire.

ASSOCIATIONS AND PROFESSIONAL SOCIETIES

AMERICAN ASTRONOMICAL SOCIETY. 2000 Florida Avenue, NW, Suite 300, Washington, DC 20009. (202) 659-0134.

AMERICAN ASSOCIATION OF VARIABLE STAR OBSERVERS. 187 Concord Avenue, Cambridge, MA 02138. (617) 354-0484.

ASTRONOMICAL LEAGUE. Post Office Box 12821, Tucson, AZ 85732. (602) 790-8471.

ASTRONOMICAL SOCIETY OF THE PACIFIC. 1290 24th Avenue, San Francisco, CA 94122. (415) 661-8660.

BIBLIOGRAPHIES

A BIBLIOGRAPHY OF ASTRONOMY, 1970-1979. R.A. Seal and S.S. Martin. Libraries Unlimited, Incorporated, Littleton, CO 80160. 1982. $37.50.

SCIENCE BOOKS AND FILMS. American Association for the Advancement of Science, 1333 H Street, NW, Washington, DC 20005. 1965 to present. Five times per year. $20.00 per year.

SCIENTIFIC AND TECHNICAL BOOKS AND SERIALS IN PRINT 1988: AN INDEX TO LITERATURE IN SCIENCE AND TECHNOLOGY. R.R. Bowker Company, 205 East 42nd Street, New York, NY 10017. (800) 521-8110 or (212) 916-1600. $175.00.

DIRECTORIES AND BIOGRAPHICAL SOURCES

AMERICAN MEN AND WOMEN OF SCIENCE. Physical and Biological Sciences. Sixteenth edition. R.R. Bowker Company, 205 East 42nd Street, New York, NY 10017. (800) 521-8110 or (212) 916-1600. 1986. $595.00.

THE BIOGRAPHICAL DICTIONARY OF SCIENTISTS: ASTRONOMERS. D. Abbott, editor. Peter Bedrick Books, 125 East 23rd Street, New York, NY 10010. 1984. $24.95.

DIRECTORY OF PHYSICS AND ASTRONOMY STAFF MEMBERS. American Institute of Physics, 335 East 45th Street, New York, NY 10017. Annual.

RESEARCH CENTERS DIRECTORY. Gale Research Company, Book Tower, Detroit, MI 48226. (800) 521-0707. Twelfth edition. 1987. $365.00 for set.

WHO'S WHO IN FRONTIER SCIENCE AND TECHNOLOGY. Marquis Who's Who, Incorporated, 200 East Ohio Street, Chicago, IL 60611. (800) 428-3898 or (312) 787-2008.

ENCYCLOPEDIAS AND DICTIONARIES

ENCYCLOPEDIA OF PHYSICAL SCIENCE AND TECHNOLOGY. Academic Press, Incorporated, Orlando, FL 32887. (800) 321-5068 or (305) 345-2734. Inquire as to cost and availability.

MCGRAW-HILL ENCYCLOPEDIA OF SCIENCE AND TECHNOLOGY. McGraw-Hill Book, Incorporated, 1221 Avenue of the Americas, New York, NY 10020. (212) 997-3675. Fifth edition, 15 volumes. $1100.00.

GENERAL WORKS

AN INTRODUCTION TO SOLAR RADIATION. Iqbal Mohammad. Academic Press, Incorporated, Orlando, FL 32887. (305) 345-4100. 1983. $35.50.

PROGRESS IN SOLAR PHYSICS. C. De Jager and Z. Svestka. Kluwer Academic, 190 Old Derby Street, Hingham, MA 02043. (617) 749-5262. 1986. $99.50.

THE SUN, OUR STAR. Robert W. Noyes. Harvard University Press, 79 Garden Street, Cambridge, MA 02138. (617) 495-2600. 1982. $20.00.

ONLINE DATA BASES

DISSERTATION ABSTRACTS ONLINE. University Microfilms International, 300 North Zeeb Road, Ann Arbor, MI 48106. (800) 521-0600 or (313) 761-4700. Scope includes virtually all doctoral dissertations accepted at accredited American institutions from 1861 to present in 252 subject areas. Inquire as to cost and availability.

INSPEC. INSPEC Marketing Department, Institute of Electrical and Electronics Engineers, Incorporated, IEEE Service Department, 445 Hoes Lane, Piscataway, NJ 08854. (201) 981-0060. Inquire as to on-line cost and availability.

NASA. National Aeronautics and Space Administration, Scientific and Technical Information Branch, 300 7th Street, SW, Washington, DC 20546. Citations and abstracts of aerospace literature, 1962 to present. Inquire as to cost and availability.

NTIS. National Technical Information Service, 5285 Port Royal Road, Springfield, VA 22161. (703) 487-4630. Broad coverage of government-sponsored research reports, 1964 to present. Inquire as to cost and availability.

SCISEARCH. Institute for Scientific Information, 3501 Market Street, Philadelphia, PA 19104. (800) 523-1850 or (215) 386-0100. Broad multidisciplinary title and author index to the international literature of science and technology, 1974 to present. Inquire as to cost and availability.

WILSONLINE. H.W. Wilson Company, 950 University Avenue, Bronx, NY 10452. (800) 367-6770 or (212) 588-8400. Makes available online versions of the printed H.W. Wilson Indexes including Applied Science and Technology Index, Business Periodicals Index, and Readers' Guide to Periodical Literature. Period covered is generally 1983 to present. Inquire as to cost and availability.

PERIODICALS

ASTRONOMICAL JOURNAL. American Astronomical Society. Available from: American Institute of Physics, 335 East 45th Street, New York, NY 10017. 9212) 661-9404. $125.00 per year.

ASTRONOMICAL SOCIETY OF THE PACIFIC. Publications. Astronomical Society of the Pacific, 1290 24th Avenue, San Francisco, CA 94122. (415) 661-8660. Monthly. $38.00.

ASTRONOMY. Astro Media Corporation, 625 East Paul Avenue, Milwaukee, WI 53202. Monthly. $18.00 per year.

ASTRONOMY AND ASTROPHYSICS. Springer-Verlag New York, Incorporated, 175 Fifth Avenue, New York, NY 10010. (800) 526-7254 or (212) 460-1500. $680.00 per year.

ASTROPHYSICAL JOURNAL. American Astronomical Society, University of Chicago Press, 5801 Ellis Aveneuw, Chicago, IL 60637. Biweekly. $305.00 per year.

ASTROPHYSICS AND SPACE SCIENCE. D. Reidel Publishing Company, 190 Old Derby Street, Hingham, MA 02043. Monthly. $101.00 per year.

ICARUS: INTERNATIONAL JOURNAL OF THE SOLAR SYSTEM STUDIES. Academic Press, Incorporated, Orlando, FL 32887. (305) 345-4100. Monthly. $484.00 per year.

MERCURY. Astronomical Society of the Pacific, 1290 24th Avenue, San Francisco, CA 94122. (415) 661-8660. Bimonthly. $21.00 per year.

MONTHLY NOTICES OF THE ROYAL ASTRONOMICAL SOCIETY. Blackwell Science Publications, Incorporated, 667 Lytton Avenue, Palo Alto, CA 94301. (415) 324-1688.

Monthly. $134.00 per year.

PLANETARY AND SPACE SCIENCE. Pergamon Press, Incorporated, Maxwell House, Fairview Park, Elmsford, NY 10523. (914) 592-7700. Monthly. $430.00 per year.

SOLAR-GEOPHYSICAL DATA, PARTS 1 AND 2. United States Geophysical Data Center, 325 Broadway, NOAA E/GC2, Boulder, CO 80303. (303) 497-6135. Monthly. $35.00 per year.

SOLAR PHYSICS. D. Reidel Publishing Company, 190 Old Derby Street, Hingham, MA 02043. Monthly. $620.00 per year.

SOVIET ASTRONOMY (TRANSLATION OF ASTRONOMICHESKII ZHURNAL). American Institute of Physics, 335 East 45th Street, New York, NY 10017. (212) 661-9404. Bimonthly. $425.00 per year.

SPACE SCIENCE REVIEWS. D. Reidel Publishing Company, 190 Old Derby Street, Hingham, MA 02043. Monthly. $305.00 per year.

VISTAS IN ASTRONOMY. Pergamon Press, Incorporated, Maxwell House, Fairview Park, Elmsford, NY 10523. (914) 592-7700. Quarterly.

RESEARCH CENTERS AND INSTITUTES

CALIFORNIA INSTITUTE OF TECHNOLOGY. Palomar Observatory, 105-24, Pasadena, CA 91125. (818) 356-4033.

CALIFORNIA STATE UNIVERSITY, NORTHRIDGE. San Fernando Observatory. Department of Physics and Astronomy, 18111 Nordhoff Street, Northridge, CA 91330. (818) 367-9333.

HARVARD-SMITHSONIAN CENTER FOR ASTROPHYSICS. 60 Garden Street, Cambridge, MA 02138. (617) 495-7461.

HIGH ALTITUDE OBSERVATORY. National Center for Atmospheric Research, Post Office Box 3000, Boulder, CO 80307. (303) 497-1500.

NATIONAL ASTRONOMY AND IONOSPHERE CENTER. Cornell University, Space Sciences Building, Ithaca, NY 14853. (607) 256-3734.

NATIONAL OPTICAL ASTRONOMY OBSERVATORIES. 1002 North Warren Avenue, Tucson, AZ 85719. (602) 325-9230.

NATIONAL SOLAR OBSERVATORY. 950 North Cherry Avenue, Tucson, AZ 85726. (602) 327-5511.

UNIVERSITY OF ARIZONA. Santa Catalina Laboratory for Experimental Relativity by Astometry. Building 81, Department of Physics, Tucson, AZ 85721. (602) 621-6782.

SUPERCONDUCTING SUPER COLLIDER

See: PARTICLE ACCELERATORS

SUPERCONDUCTIVITY

See also: CRYOGENICS, ELECTRICAL ENGINEERING, ELECTRICITY, PHYSICAL CHEMISTRY, PHYSICS, SOLID STATE PHYSICS

ABSTRACT SERVICES AND INDEXES

APPLIED MECHANICS REVIEW. American Society of Mechanical Engineers, 345 East 47th Street, New York, NY 10017. (212) 705-7703. 1948 to present. Monthly. $360.00 per year.

APPLIED SCIENCE AND TECHNOLOGY INDEX. H.W. Wilson and Company, 950 University Avenue, Bronx, NY 10452. (800) 367-6670 or (212) 588-8400. Monthly. Inquire as to cost and availability.

CHEMICAL ABSTRACTS. American Chemical Society, Chemical Abstracts Service, Box 3012, Columbus, OH 43210. (614) 421-3600. 1907 to present. Weekly. $9500.00 per year.

CONFERENCE PAPERS INDEX. Cambridge Scientific Abstracts, 5161 River Road, Bethesda, MD 20816. 1972 to present. Monthly. Inquire as to cost and availability.

CURRENT CONTENTS: ENGINEERING, TECHNOLOGY AND APPLIED SCIENCES. Institute for Scientific Information, 3501 Market Street, Philadelphia, PA 19104. (800) 523-1850 or (215) 386-0100. Weekly. $275.00 per year.

CURRENT CONTENTS: PHYSICAL, CHEMICAL AND EARTH SCIENCES. Institute for Scientific Information, 3501 Market Street, Philadelphia, PA 19104. (800) 523-1850 or (215) 386-0100. Weekly. $275.00 per year.

ENGINEERING INDEX MONTHLY AND AUTHOR INDEX. Engineering Information Inc., 345 East 47th Street, New York, NY 10017. (212) 705-7600. Monthly. $1560.00 per year.

INDEX TO SCIENTIFIC AND TECHNICAL PROCEEDINGS. Institute for Scientific Information, 3501 Market Street, Philadelphia, PA 19104. (800) 523-1850 or (215) 386-0100. 1978 to present. Monthly. $775.00 per year.

INDEX TO SCIENTIFIC REVIEWS. Institute for Scientific Information, 3501 Market Street, Philadelphia, PA 19104. (800) 523-1850 or (215) 386-0100. 1974 to present. Semi-annual. $550.00 per year.

PHYSICS ABSTRACTS. Institution of Electrical Engineers. Available from: IEEE Service Center, 445 Hoes Lane, Piscataway, NJ 08854. 1898 to present. Bimonthly. $1700.00 per year.

PHYSICS BRIEFS. Physik Verlag GmbH, Postfach 1260/1280, D-6940 Weinheim, West Germany. (212) 661-9404. 1920 to present. Twenty-six times per year. $1250.00 per year.

SCIENCE CITATION INDEX. Institute for Scientific Information, 3501 Market Street, Philadelphia, PA 19104. (800) 523-1850 or (215) 386-0100. Six times per year. $6200.00 per year.

ANNUAL REVIEWS AND YEARBOOKS

ADVANCES IN CRYOGENIC ENGINEERING. R.P. Reed and others, editors. Plenum Publishing Corporation, 233 Spring Street, New York, NY 10013. (800) 221-9369. 1960-1984. Irregular. Price varies, inquire.

ASSOCIATIONS AND PROFESSIONAL SOCIETIES

AMERICAN CHEMICAL SOCIETY. 1155 16th Street, N.W., Washington, DC 20036. (800) 424-6747.

AMERICAN INSTITUTE OF CHEMICAL ENGINEERS. 345 East 47th Street, New York, NY 10017. (212) 705-7338.

AMERICAN INSTITUTE OF PHYSICS. 335 East 45th Street, New York, NY 10017. (212) 661-9494.

CRYOGENIC ENGINEERING CONFERENCE. 73 Vassar Street, Room 41-204, Cambridge, MA 02139. (617) 253-2296.

CRYOGENIC SOCIETY OF AMERICA. c/o Huget Advertising, 1033 South Boulevard, Oak Park, IL 60302. (312) 383-7053.

INSTITUTE OF ELECTRICAL AND ELECTRONICS ENGINEERS. IEEE Service Center, 445 Hoes Lane, Piscataway, NJ 08854.

DIRECTORIES AND BIOGRAPHICAL SOURCES

AMERICAN MEN AND WOMEN OF SCIENCE. R.R. Bowker, Inc., Order Department, 245 West 17th Street, New York, NY 10011. (800) 521-8110. Eight volumes. 1986. $595.00 for set.

INTERNATIONAL RESEARCH CENTERS DIRECTORY 1988-89. Darren L. Smith, editor. Gale Research Company, Book Tower, Detroit, MI 48226. (800) 521-0707. 4th edition. 1987. $360.00.

1987 DIRECTORY OF ENGINEERING SOCIETIES AND RELATED ORGANIZATIONS. Gordon Davis, editor. Hemisphere Publishing Corporation, 79 Madison Avenue, New York, NY 10016-7892. (800) 821-8312. 12th edition. 1987. $100.00.

RESEARCH CENTERS DIRECTORY 1988. Gale Research Company, Book Tower, Detroit, MI 48226. (800) 521-0707. 12th edition. 1987. $365.00 for set.

WHO'S WHO IN ENGINEERING. Gordon Davis, editor. Hemisphere Publishing Corporation, 79 Madison Avenue, New York, NY 10016-7892. (800) 821-8312. 6th edition. 1985. $200.00.

GENERAL WORKS

APPLIED THERMODYNAMICS FOR ENGINEERING TECHNOLOGIES. T.D. Eastop and A. McConkey. John Wiley and Sons, Inc., 605 Third Avenue, New York, NY 10158. (800) 526-5368. 1986. $29.95.

CRYOGENIC SYSTEMS. Randall F. Barron. Oxford University Press, 200 Madison Avenue, New York, NY 10016. (800) 458-5833. 1985. $59.00.

FUNDAMENTALS OF ENGINEERING THERMODYNAMICS. V. Hubka and others. McGraw-Hill Book Company, 1221 Avenue of the Americas, New York, NY 10020. (212) 512-2000. 1986. $39.95.

SUPERCONDUCTIVITY: THE THRESHOLD OF A NEW TECHNOLOGY. TAB Books Inc., P.O. Box 40, Blue Ridge Summitt, PA 17214. (800) 233-1128. 1988. $18.95.

SUPERFLUIDITY AND SUPERCONDUCTIVITY. D.R. Tilley. Taylor and Francis International Publishers, 242 Cherry Street, Philadelphia, PA 19106. (800) 821-8312. Second edition. 1986. $95.00.

TECHNIQUES IN LOW TEMPERATURE PHYSICS. R.C. Richardson. Benjamin-Cummings Publishing Company, 2727 Sand Hill Road, Menlo Park, CA 94025. (415) 854-6020. 1987. $59.95.

ONLINE DATA BASES

CA SEARCH. Chemical Abstracts Service, P.O. Box 3012, Columbus, OH 43120. (800) 848-6538 or (614) 421-3600. Comprehensive guide to chemical literature, 1972 to present. Inquire as to online cost and availability.

COMPENDEX. Engineering Information, Inc., 345 East 47th Street, New York, NY 10017. (800) 221-1044 or (212) 705-7615. Engineering and technical literature, 1975 to present. Inquire as to online cost and availability.

DISSERTATION ABSTRACTS ONLINE. University Microfilms International, 300 North Zeeb Road, Ann Arbor, MI 48106. (800) 521-0600 or (313) 761-4700. Scope includes virtually all doctoral dissertations accepted at accredited American institutions from 1861 to present in over 250 subject areas. Inquire as to online cost and availability.

INSPEC. INSPEC Marketing Department, Institution of Electrical Engineers. Available from IEEE Service Center, 445 Hoes Lane, Piscataway, NJ 08854. (201) 981-0060. Online version of Physics Abstracts. Inquire as to online cost and availability.

NTIS. National Technical Information Service, 5285 Port Royal Road, Springfield, VA 22161. (703) 487-4630. Broad coverage of government sponsored research reports, 1964 to present. Inquire as to online cost and availability.

SCISEARCH. Institute for Scientific Information, 3501 Market Street, Philadelphia, PA 19104. (800) 523-1850 or (215) 386-0100. Broad multidisciplinary title and author index to the international literature of science and technology, 1974 to present. Inquire as to online cost and availability.

WILSONLINE. H.W. Wilson and Company, 950 University Avenue, Bronx, NY 10452. (800) 367-6770 or (212) 588-8400. Makes available online versions of the H.W. Wilson indexes including Applied Science and Technology Index, Business Periodicals Index and Readers' Guide to Periodical Literature. Approximately 1980 to present. Inquire as to online cost and availability.

PERIODICALS

CRYOGENIC INFORMATION REPORT. Technical Economic Associates, Box 1972, Estes Park, CO 80517. (303) 586-5636. 1963 to present. Ten times per year. $85.00 per year.

CRYOGENICS: THE INTERNATIONAL JOURNAL OF LOW TEMPERATURE ENGINEERING AND RESEARCH. Butterworth Publishing, 80 Montvale Avenue, Stoneham, MA 02180. (800) 325-4177. 1960 to present. Monthly. $335.00 per year.

EXPERIMENTAL THERMAL AND FLUID SCIENCE. Elsevier Science Publishing Company, Inc., 52 Vanderbilt Avenue, New York, NY 10017. (212) 370-5520. 1988 to present. Quarterly. $125.00 per year.

HEAT TRANSFER AND FLUID FLOW DIGEST. Hemisphere Publishing Corporation, 79 Madison Avenue, New York, NY 10016-7892. (800) 821-8312. 1968 to present. Monthly. $137.50 per year.

HEAT TRANSFER ENGINEERING. Hemisphere Publishing Corporation, 79 Madison Avenue, New York, NY 10016-7892. (800) 821-8312. 1979 to present. Quarterly. $97.50.

INTERNATIONAL COMMUNICATIONS IN HEAT AND MASS TRANSFER. Pergamon Press, Inc., Maxwell House, Fairview Park, Elmsford, NY 10523. (914) 592-7700. 1974 to present. Bimonthly. $160.00 per year.

INTERNATIONAL JOURNAL OF HEAT AND FLUID FLOW. Butterworth Publishing, 80 Montvale Avenue, Stoneham, MA 02180. (800) 325-4177. 1979 to present. Quarterly. $140.00 per year.

INTERNATIONAL JOURNAL OF HEAT AND MASS TRANSFER. Pergamon Press, Inc., Maxwell House, Fairview Park, Elmsford, NY 10523. (914) 592-7700. 1960 to present. Monthly. $500.00 per year.

INTERNATIONAL JOURNAL OF THERMOPHYSICS. Plenum Publishing Corporation, 233 Spring Street, New York, NY 10013. (800) 221-9369. 1980 to present. Six times per year. $150.00 per year.

JOURNAL OF HEAT TRANSFER. American Society of Mechanical Engineers, 345 East 47th Street, New York, NY 10017. (212) 705-7703. 1970 to present. Quarterly. $100.00 per year.

JOURNAL OF LOW TEMPERATURE PHYSICS. Plenum Publishing Corporation, 233 Spring Street, New York, NY 10013. (800) 221-9369. 1969 to present. 24 times per year. $750.00 per year.

JOURNAL OF SUPERCONDUCTIVITY. Plenum Publishing Corporation, 233 Spring Street, New York, NY 10013. (800) 221-9369. 1988 to present. Quarterly. $125.00 per year.

NUMERICAL HEAT TRANSFER: AN INTERNATIONAL JOURNAL OF COMPUTATION AND METHOLOGY. Hemisphere Publishing Corporation, 79 Madison Avenue, New York, NY 10016-7892. (800) 821-8312. 1978 to present. Monthly. $370.00 per year.

RESEARCH CENTERS AND INSTITUTES

APPLIED SUPERCONDUCTIVITY RESEARCH CENTER. University of Wisconsin at Madison, Engineering Research Building, 1500 Johnson Drive, Madison, WI 53706. (608) 263-5026.

CENTER FOR THERMODYNAMICS. Brigham Young University, 226 ESC, Provo, UT 84602. (801) 378-3668.

HEAT TRANSFER LABORATORY. Massachusetts Institute of Technology, 77 Massachusetts Avenue, Cambridge, MA 02139. (716) 253-2248.

THERMODYNAMICS RESEARCH CENTER. Texas A&M University, Texas Engineering Experiment Station, College Station, TX 77843-3111. (409) 845-4940.

SUPERNOVA

See: STARS

SURFACE CHEMISTRY

See also: CATALYSIS, CHEMICAL BONDING, CHEMISTRY, COLLOIDS, ELECTROCHEMISTRY, PHOTOCHEMISTRY, PHYSICAL CHEMISTRY, POLYMERS

ABSTRACT SERVICES AND INDEXES

APPLIED SCIENCE AND TECHNOLOGY INDEX. H.W. Wilson Company, 950 University Avenue, Bronx, NY 10452. (800) 367-6670 or (212) 588-8400. Inquire as to cost and availability.

CHEMICAL ABSTRACTS. Chemical Abstracts Service, 2540 Olentangy Road, Post Office Box 3012, Columbus, OH 43210. (800) 848-6538 or (614) 421-3600. Weekly. $9200.00 per year.

GENERAL SCIENCE INDEX. H.W. Wilson Company, 950 University Avenue, Bronx, NY 10452. (800) 367-6770 or (212) 588-8400. Inquire as to cost and availability.

PHYSICS ABSTRACTS. Institute of Electrical Engineers, London, United Kingdom. Available from: Institute of Electrical and Electronic Engineers (IEEE), 345 East 47th Street, New York, NY 10017. (212) 705-7900.

SCIENCE CITATION INDEX. Institute for Scientific Information, 3501 Market Street, Philadelphia, PA 19104. (800) 523-1850 or (215) 386-0100.

ANNUAL REVIEWS AND YEARBOOKS

ANNUAL REVIEW OF PHYSICAL CHEMISTRY. Annual Reviews Incorporated, 4139 El Camino Way, Palo Alto, CA 94306. (415) 493-4400. Annual.

ASSOCIATIONS AND PROFESSIONAL SOCIETIES

AMERICAN CHEMICAL SOCIETY. 1155 16th Street, NW, Washington, DC 20036. (202) 872-4600.

ASSOCIATION OF CONSULTING CHEMISTS AND CHEMICAL ENGINEERS. 50 East 41st Street, Suite 92, New York, NY 10017. (212) 684-6255.

DIVISION OF PHYSICAL CHEMISTRY (A DIVISION OF THE AMERICAN CHEMICAL SOCIETY). c/o Dr. James Kinsey, Chemistry Department, Room 6-215, Massachusetts Institute of Technology, Cambridge, MA 02139.

BIBLIOGRAPHIES

SCIENCE BOOKS AND FILMS. American Association for the Advancement of Science, 1333 H Street, NW, Washington, DC 20005.

SCIENTIFIC AND TECHNICAL BOOKS AND SERIALS IN PRINT 1988: AN INDEX TO LITERATURE IN SCIENCE AND TECHNOLOGY. R.R. Bowker Company, 205 East 42nd Street, New York, NY 10017. (800) 521-8110 or (212) 916-1600. $175.00.

DIRECTORIES AND BIOGRAPHICAL SOURCES

AMERICAN INSTITUTE OF CHEMISTS. American Institute of Chemists, 7315 Wisconsin Avenue, Bethesda, MD 20814. (301) 652-2447. 1986. $35.00.

AMERICAN MEN AND WOMEN OF SCIENCE. Physical and Biological Sciences. R.R. Bowker Company, 205 East 42nd Street, New York, NY 10017. (800) 521-8110 or 9212) 916-1600. Fifteenth edition. $565.00.

BIOGRAPHICAL DICTIONARY OF SCIENTISTS: CHEMISTS. David Abbott, editor. P. Bedrick Books, 125 East 23rd Street, New York, NY 10010. (212) 777-1187. 1984. $18.95.

CONSULTING SERVICES: CHEMISTS AND CHEMICAL ENGINEERS. Association of Consulting Chemists and Chemical Engineers, 50 East 41st Street, New York, NY 10017. (212) 684-6255. Annual. 1986. $45.00.

GOVERNMENT RESEARCH DIRECTORY. Gale Research Company, Book Tower, Detroit, MI 48226. (800) 521-0707. Fourth edition. 1987. $350.00.

INTERNATIONAL RESEARCH CENTERS DIRECTORY 1986-1987. Gale Research Company, Book Tower, Detroit, MI 48226. (800) 521-0707. Third edition. 1986. $330.00.

RESEARCH CENTERS DIRECTORY. Gale Research Company, Book Tower, Detroit, MI 48226. (800) 521-0707. Eleventh edition. 1987. $355.00.

RESEARCH SERVICES DIRECTORY. Robert J. Huffman and Mary M. Watkins, editors. Gale Research Company, Book Tower, Detroit, MI 48226. (800) 521-0707. Third edition. $290.00.

SCIENTIFIC AND TECHNICAL ORGANIZATIONS AND AGENCIES DIRECTORY. Gale Research Company, Book Tower, Detroit, MI 48226. (800) 521-0707. 1985. $150.00.

WHO'S WHO IN FRONTIER SCIENCE AND TECHNOLOGY. Marquis Who's Who, Incorporated, 200 East Ohio Street, Chicago, IL 60611. (800) 428-3898 or (312) 787-2008.

WHO'S WHO IN TECHNOLOGY TODAY. Reston Publishing Company, Incorporated, c/o Prentice-Hall, Incorporated, Englewood Cliffs, NJ 07632. (800) 262-6868. Biennial. Five volumes. $425.00. Covers the fields of biotechnology, mechanics, energy, and earth science.

WORLD GUIDE TO SCIENTIFIC ASSOCIATIONS AND LEARNED SOCIETIES. K.G. Saur Incorporated, 175 Fifth Avenue, New York, NY 10010. (800) 521-0707 or (212) 982-1302. Fourth edition. 1984. $112.00.

ENCYCLOPEDIAS AND DICTIONARIES

CONCISE ENCYCLOPEDIA OF CHEMICAL TECHNOLOGY. Kirk-Othmer. John Wiley and Sons, Incorporated, 605 Third

Avenue, New York, NY 10158. (800) 526-5368 or (212) 850-6000. Third edition. 1985. $129.95.

CONDENSED CHEMICAL DICTIONARY. Gessner Hawley. Van Nostrand Reinhold, 115 Fifth Avenue, New York, NY 10003. Tenth edition. 1981. $49.95.

ENCYCLOPEDIA OF PHYSICAL SCIENCE AND TECHNOLOGY. Academic Press, Incorporated, Orlando, FL 32887. (800) 321-5068 or (305) 345-2734. Fifteen volumes. 1986.

GLOSSARY OF CHEMICAL TERMS. Clifford A. Hampel and Gessner G. Hawley. Van Nostrand Reinhold Company, 115 Fifth Avenue, New York, NY 10003. (800) 543-2681 or (212) 254-3232. Second edition. 1982. $21.95.

HAWLEY'S CONDENSED CHEMICAL DICTIONARY. N. Irving Sax and Richard J. Lewis, Sr., editors. Van Nostrand Reinhold, Incorporated, 115 Fifth Avenue, New York, NY 10003. (800) 543-2681. Eleventh edition. 1987. $52.95.

MCGRAW-HILL ENCYCLOPEDIA OF SCIENCE AND TECHNOLOGY. McGraw-Hill Book Company, Incorporated, 1221 Avenue of the Americas, New York, NY 10020. (212) 997-3675.

VAN NOSTRAND REINHOLD ENCYCLOPEDIA OF CHEMISTRY. Douglas M. Considine and Glenn D. Considine. Van Nostrand Reinhold Publishing Company, Incorporated, 115 Fifth Avenue, New York, NY 10003. (800) 543-2681 or (212) 254-3232. 1984. $97.95.

GENERAL WORKS

CHEMISTRY AND PHYSICS OF SOLID SURFACES. R. Vanselow and R. Howe, editors. Springer-Verlag New York, Incorporated, 175 Fifth Avenue, New York, NY 10010. (800) 526-7254. 1982. $49.00.

CHEMISTRY OF INTERFACES. M.J. Jaycock and G.D. Parfitt. John Wiley and Sons, Incorporated, 605 Third Avenue, New York, NY 10158. (800) 526-5368 or (212) 850-6000. 1983. $31.95 in paper.

COLLOID PHENOMENA: ADVANCED TOPICS. C.S. Hirtzel and Raj Rajagopalan. Noyes Data Corporation, Mill Road at Grand Avenue, Park Ridge, NJ 07656. 1985. $36.00.

INORGANIC CHEMISTRY: PRINCIPLES OF STRUCTURE AND REACTIVITY. James E. Huheey. Harper and Row Publishers Incorporated, 10 East 53rd Street, New York, NY 10022. (212) 207-7655. Third edition. $36.50.

INTRODUCTION TO COLLOID AND SURFACE CHEMISTRY. D.J. Shaw. Butterworths Publishing, 80 Montvale Avenue, Stoneham, MA 02180. (617) 720-0761.

INTRODUCTION TO PHYSICAL CHEMISTRY. M.F.C. Ladde and W.H. Lee. Cambridge University Press, 32 East 57th Street, New York, NY 10022. (212) 688-8885. 1986. $19.95 in paper.

MODERN TRENDS OF COLLOID SCIENCE IN CHEMISTRY AND BIOLOGY. Hansfriedrich Eicke, editor. Birkhauser Boston Incorporated, 380 Green Street, Cambridge, MA 02139. (617) 876-2333. 1985. $34.95.

PHYSICAL CHEMISTRY. Robert A. Alberty. John Wiley and Sons, Incorporated, 605 Third Avenue, New York, NY 10158. (800) 526-5368 or (212) 850-6000. 1987. Inquire as to cost and availability.

PHYSICAL CHEMISTRY OF SURFACES. Arthur W. Adamson. John Wiley and Sons, Incorporated, 605 Third Avenue, New York, NY 10158. (800) 526-5368 or (212) 850-6000. Fourth edition. 1982. $44.95.

SCIENCE AND TECHNOLOGY OF POLYMER COLLOIDS. R. Buscall and others, editors. Elsevier Science Publishing Company, Incorporated, 52 Vanderbilt Avenue, New York, NY 10017. (212) 370-5520. 1985. $63.00.

THE STRUCTURE OF SURFACES. M.A. Van Hove and S.Y. Tong, editors. Springer-Verlag New York, Incorporated, 175 Fifth Avenue, New York, NY 10010. (800) 526-7254. 1985. $49.00.

HANDBOOKS AND MANUALS

THE CHEMIST'S COMPANION: A HANDBOOK OF PRACTICAL DATA, TECHNIQUES, AND REFERENCES. Arnold J. Gordon and Richard A. Ford. John Wiley and Sons, Incorporated, 605 Third Avenue, New York, NY 10158. (800) 526-5368. 1973. $49.95.

CRC HANDBOOK OF CHEMISTRY AND PHYSICS. CRC Press, Incorporated, 2000 Corporate Boulevard, NW, Boca Raton, FL 33431. Sixty-seventh edition. 1986. $69.95.

HANDBOOK OF APPLIED CHEMISTRY: FACTS FOR ENGINEERS, SCIENTISTS, TECHNICIANS, AND TECHNICAL MANAGERS. Vollrath Hopp and Ingp Hennig. McGraw-Hill Book Company, 1221 Avenue of the Americas, New York, NY 10020. (800) 628-0004. 1983. $54.00.

HANDBOOK OF COMPUTATIONAL CHEMISTRY: A PRACTICAL GUIDE TO CHEMICAL STRUCTURE AND ENERGY CALCULATIONS. Tim Clark. John Wiley and Sons, Incorporated, 605 Third Avenue, New York, NY 10158. (800) 526-5368 or (212) 850-6000. 1985. $35.00.

LANGE'S HANDBOOK OF CHEMISTRY. John A. Dean, editor. McGraw-Hill Book Company, 1221 Avenue of the Americas, New York, NY 10020. (800) 628-0004. 1985. $59.50.

TABLES OF PHYSICAL AND CHEMICAL CONSTANTS; AND SOME MATHEMATICAL FUNCTIONS. G.W.C. Kaye and T.H. Laby, editors. Longman, Incorporated, 95 Church Street, White Plains, NY 10601. (914) 993-5000. 1986. $39.95.

ONLINE DATA BASES

CA SEARCH. Chemical Abstracts Service, Post Office Box 3012, Columbus, OH 43210. Guide to chemical literature, 1972 to present. Inquire as to cost and availability.

DISSERTATION ABSTRACTS ONLINE. University Microfilms International, 300 North Zeeb Road, Ann Arbor, MI 48106. (800) 521-0600 or (313) 761-4700. Scope includes virtually all doctoral dissertations accepted at accredited American institutions from 1861 to present in 252 subject areas. Inquire as to online cost and availability.

INSPEC. INSPEC Marketing Department, Institute of Electrical and Electronics Engineers, Incorporated, IEEE Service Department, 445 Hoes Lane, Piscataway, NJ 08854. (201) 981-0060. Inquire as to online cost and availability.

NTIS. National Technical Information Service, 5285 Port Royal Road, Springfield, VA 22161. (703) 487-4630. Broad coverage of government sponsored research reports, 1964 to present. Inquire as to online cost and availability.

SCISEARCH. Institute for Scientific Information, 3501 Market Street, Philadelphia, PA 19104. (800) 523-1850 or (215) 386-0100. Broad multidisciplinary title and author index to the international literature of science and technology, 1974 to present. Inquire as to online cost and availability.

OTHER SOURCES

GUIDE TO BASIC INFORMATION SOURCES IN CHEMISTRY. Arthur Antony. John Wiley and Sons, Incorporated, 605 Third Avenue, New York, NY 10158. (800) 526-5368 or (212) 850-6000. 1979. $26.95.

HOW TO FIND CHEMICAL INFORMATION: A GUIDE FOR PRACTICING CHEMISTS, TEACHERS, AND STUDENTS. John Wiley and Sons, Incorporated, 605 Third Avenue, New York, NY 10158. (800) 526-5368 or (212) 850-6000. 1986. $35.00.

PERIODICALS

ADVANCES IN COLLOID AND INTERFACE SCIENCE. Elsevier Science Publishers B.V., Box 211, 1000 AE Amsterdam, The Netherlands. Eight times per year. $190.00 per year.

ANDEWANDTE CHEMIE (GESELLSCHAFT DEUTSCHER CHEMIKER, GW). V C H Verlagsgesellschaft mbH, Pappelallee 3, Postfach 1260, 6940 Weinheim, West Germany. Monthly. $300.00 per year.

CHEMICAL PHYSICS. Elsevier Science Publishers B.V, Box 211, 1000 AE Amsterdam, The Netherlands. Thirty times per year. $1200.00 per year.

CHEMICAL PHYSICS LETTERS. Elsevier Science Publishers B.V., Box 211, 1000 AE Amsterdam, The Netherlands. Sixty-six times per year. $1500.00 per year.

COLLOID AND POLYMER SCIENCE. Dr. Dietrich Steinkopff Verlag, Saalbaustrr 12, Postfach, Darmstadt 11, West Germany. Monthly. $355.00 per year.

COLLOIDS AND SURFACES. Elsevier Science Publishers B.V., Box 211, 1000 AE Amsterdam, The Netherlands. Twenty-four times per year. $540.00 per year.

INTERNATIONAL REVIEWS IN PHYSICAL CHEMISTRY. Taylor and Francis, Limited, 242 Cherry Street, Philadelphia, PA 19106-1906. Three times per year. $143.00 per year.

INORGANIC CHEMISTRY. American Chemical Society, 1155 Sixteenth Street, NW, Washington, DC 20036. (800) 424-6747 or (202) 872-4700. Monthly. $400.00 per year.

JOURNAL OF CATALYSIS. Academic Press, Incorporated, Journals Division, 1250 Sixth Avenue, San Diego, Ca 92101. (619) 230-1840. Monthly. $654.00 per year.

JOURNAL OF COLLOID AND INTERFACE SCIENCE. Academic Press, Incorporated, Journals Division, 1250 Sixth Avenue, San Diego, CA 92101. (619) 230-1840. Monthly. $667.00 per year.

JOURNAL OF PHYSICAL CHEMISTRY. American Chemical Society, 1155 Sixteenth Street, NW, Washington, DC 20036. (800) 424-6747 or (202) 872-4700. Biweekly. $369.00 per year.

JOURNAL OF SOLID STATE CHEMISTRY. Academic Press, Incorporated, Journals Division, 1250 Sixth Avenue, San Diego, CA 92101. (619) 230-1840. Fifteen times per year. $530.00 per year.

JOURNAL OF THE AMERICAN CHEMICAL SOCIETY. American Chemical Society, 1155 Sixteenth Street, NW, Washington, DC 20036. (800) 424-6747 or (202) 872-4700. Biweekly. $330.00 per year.

SIA - SURFACE AND INTERFACE ANALYSIS. John Wiley and Sons, Limited, Baffins Lane, Chichester, Sussex PO19 1UD, England. Monthly. $240.00 per year.

SURFACE TECHNOLOGY. Elsevier Sequoia S.A., Post Office Box 851, CH-1001 Lausanne 1, Switzerland. Monthly. $332.00 per year.

RESEARCH CENTERS AND INSTITUTES

BROWN UNIVERSITY. Chemical Research Laboratory, Providence, RI 02912. (401) 863-2256.

CLARKSON UNIVERSITY. Institute of Colloid and Surface Science, Potsdam, NY 13676. (315) 268-2353.

COATINGS INDUSTRY EDUCATION FUND. 1315 Walnut Street, Philadelphia, PA 19107. (215) 545-1507.

LEHIGH UNIVERSITY. Center for Surface and Coatings Research, Bethleham, PA 18015. (215) 861-3570.

RICE UNIVERSITY. Quantum Institute, Post Office Box 1892, Houston, TX 77251. (713) 527-6028.

UNIVERSITY OF MICHIGAN. Colloid Stability Laboratorym, 3095 Dow Building, Ann Arbor, MI 48109. (313) 764-4313.

UNIVERSITY OF TEXAS AT AUSTIN. Institute of Theoretical Chemistry, Department of Chemistry, WEL 2.204, Austin, TX 78712. (512) 471-3114.

SURVEYING

See also: AERIAL PHOTOGRAPHY, ARTIFICIAL SATELLITES, CARTOGRAPHY, GEODESY, PHOTOGRAMMETRY, REMOTE SENSING, TOPOGRAPHIC MAPPING

ABSTRACT SERVICES AND INDEXES

APPLIED SCIENCE AND TECHNOLOGY INDEX. H.W. Wilson and Company, 950 University Avenue, Bronx, NY 10452. (800) 367-6670 or (212) 588-8400. Monthly. Inquire as to cost and availability.

BIBLIOGRAPHY AND INDEX OF GEOLOGY. American Geological Institute, 4220 King Street, Alexandria, VA 22302. (703) 379-2480. 1969 to present. Monthly. $1100.00 per year.

CURRENT CONTENTS: ENGINEERING, TECHNOLOGY AND APPLIED SCIENCES. Institute for Scientific Information, 3501 Market Street, Philadelphia, PA 19104. (800) 523-1850 or (215) 386-0100. Weekly. $275.00 per year.

ENGINEERING INDEX MONTHLY AND AUTHOR INDEX. Engineering Information Inc., 345 East 47th Street, New York, NY 10017. (212) 705-7600. Monthly. $1560.00 per year.

METEOROLOGICAL AND GEOASTROPHYSICAL ABSTRACTS. American Meteorological Society, 45 Beacon Street, Boston, MA 02108. (617) 227-2425. 1950 to present. Monthly. $450.00 per year.

REMOTE SENSING OF NATURAL RESOURCES: A QUARTERLY LITERATURE REVIEW. University of New Mexico, Technology Application Center, Albuquerque, NM 87131. (505) 277-3622. 1974 to present. Quarterly. $150.00. Available to qualified agencies only.

ASSOCIATIONS AND PROFESSIONAL SOCIETIES

AMERICAN ASSOCIATION FOR GEODETIC SURVEYING. c/o American Congress on Surveying and Mapping, 210 Little Falls Street, Falls Church, VA 22046. (703) 241-2446.

AMERICAN CARTOGRAPHIC ASSOCIATION. c/o American Congress on Surveying and Mapping, 210 Little Falls Street, Falls Church, VA 22046. (703) 241-2446.

AMERICAN CONGRESS ON SURVEYING AND MAPPING. 210 Little Falls Street, Falls Church, VA 22046. (703) 241-2446.

AMERICAN SOCIETY FOR PHOTOGRAMMETRY AND REMOTE SENSING. 210 Little Falls Street, Falls Church, VA 22046-4398. (703) 534-6617.

NATIONAL ASSOCIATION OF PROFESSIONAL SURVEYORS. c/o American Congress on Surveying and Mapping, 210 Little Falls Street, Falls Church, VA 22046. (703) 241-2446.

NORTH AMERICAN CARTOGRAPHIC ASSOCIATION. 6010 Executive Boulevard, Suite 100, Rockville, MD 20852. (301)

443-8075.

OPTICAL SOCIETY OF AMERICA. 1816 Jefferson Place, N.W., Washington, DC 20036. (202) 223-8130.

SOCIETY OF PHOTOGRAPHIC SCIENTISTS AND ENGINEERS. 7003 Kilworth Lane, Springfield, VA 22151. (703) 642-9090.

SPIE - THE INTERNATIONAL SOCIETY FOR OPTICAL ENGINEERING. P.O. Box 10, 1022 19th Street, Bellingham, WA 98227. (206) 676-3290.

DIRECTORIES AND BIOGRAPHICAL SOURCES

INTERNATIONAL RESEARCH CENTERS DIRECTORY 1988-89. Darren L. Smith, editor. Gale Research Company, Book Tower, Detroit, MI 48226. (800) 521-0707. 4th edition. 1987. $360.00.

1987 DIRECTORY OF ENGINEERING SOCIETIES AND RELATED ORGANIZATIONS. Gordon Davis, editor. Hemisphere Publishing Corporation, 1010 Vermont Avenue, NW, Washington, DC 20005. (800) 526-0275. 12th edition. 1987. $100.00.

RESEARCH CENTERS DIRECTORY 1988. Gale Research Company, Book Tower, Detroit, MI 48226. (800) 521-0707. 12th edition. 1987. $365.00 for set.

SCIENTIFIC AND TECHNICAL ORGANIZATIONS AND AGENCIES DIRECTORY. Margaret Labash Young, editor. Gale Research Company, Book Tower, Detroit, MI 48226. (800) 521-0707. 2nd edition. 1987. $185.00.

WHO'S WHO IN ENGINEERING. Gordon Davis, editor. Hemisphere Publishing Corporation, 1010 Vermont Avenue, NW, Washington, DC 20005. (800) 526-0275. 6th edition. 1985. $200.00.

GENERAL WORKS

CLOSE-RANGE PHOTOGRAMMETRY AND SURVEYING: STATE OF THE ART. American Society for Photogrammetry and Remote Sensing. 210 Little Falls Street, Falls Church, VA 22046-4398. (703) 534-6617. 1985. $65.00 in paper.

ELEMENTS OF CARTOGRAPHY. Arthur H. Robinson and others. John Wiley and Sons, Inc., 605 Third Avenue, New York, NY 10158. (800) 526-5368. 5th edition. 1984. $35.50.

ELEMENTS OF PHOTOGRAMMETRY. P.R. Wolf. McGraw-Hill Book Company, 1221 Avenue of the Americas, New York, NY 10020. (212) 512-2000. 2nd edition. 1983. $49.95.

INTRODUCTORY CARTOGRAPHY. John Campbell. Prentice-Hall Publishing, Inc., Englewood Cliffs, NJ 07632. (800) 562-0245. 1984. $37.95.

PHOTOGRAMMETRY. Francis H. Moffitt and Edward M. Mikhail. Harper and Row Publishers, Inc., 10 East 53rd Street, New York, NY 10022. (212) 207-7655. 3rd edition. 1980. $41.95.

PRINCIPLES OF REMOTE SENSING. Paul Curran. Halstead Press, division of John Wiley and Sons, Inc., 605 Third Avenue, New York, NY 10158. (800) 526-5368. 1986. $35.95.

REMOTE SENSING. Floyd F. Sabins. W.H. Freeman and Company, 41 Madison Avenue, New York, NY 10010. (212) 532-7660. 2nd edition. 1986. $39.95.

REMOTE SENSING METHODS AND APPPLICATIONS. R. Hord. John Wiley and Sons, Inc., 605 Third Avenue, New York, NY 10158. (800) 526-5368. 1986. $39.95.

SURVEY OF THE PROFESSION: PHOTOGRAMMETRY, SURVEYING, MAPPING, REMOTE SENSING. American Society for Photogrammetry and Remote Sensing. 210 Little Falls Street, Falls Church, VA 22046-4398. (703) 534-6617. 1982. $35.00 in paper.

SURVEYING: THEORY AND PRACTICE. R.E. Davis and F.S. Foote and others. McGraw-Hill Book Company, 1221 Avenue of the Americas, New York, NY 10020. (212) 512-2000. 6th edition. 1981. $49.95.

HANDBOOKS AND MANUALS

MANUAL OF PHOTOGRAMMETRY. Chester C. Slama, editor. American Society for Photogrammetry and Remote Sensing. 210 Little Falls Street, Falls Church, VA 22046-4398. (703) 534-6617. 4th edition. 1980. $59.00.

MANUAL OF REMOTE SENSING. Robert N. Colwell, editor. American Society for Photogrammetry and Remote Sensing. 210 Little Falls Street, Falls Church, VA 22046-4398. (703) 534-6617. 2nd edition. 1983. $106.00 for set.

PRACTICAL MANUAL OF SITE DEVELOPMENT. B.C. Colley. McGraw-Hill Book Company, 1221 Avenue of the Americas, New York, NY 10020. (212) 512-2000. 1986. $42.00.

ONLINE DATA BASES

CA SEARCH. Chemical Abstracts Service, P.O. Box 3012, Columbus, OH 43120. (800) 848-6538 or (614) 421-3600. Comprehensive guide to chemical literature, 1972 to present. Inquire as to online cost and availability.

COMPENDEX. Engineering Information, Inc., 345 East 47th Street, New York, NY 10017. (800) 221-1044 or (212) 705-7615. Engineering and technical literature, 1975 to present. Inquire as to online cost and availability.

GEOREF. Online version of the BIBLIOGRAPHY AND INDEX OF GEOLOGY. American Geological Institute, 4220 King Street, Alexandria, VA 22302. (703) 379-2480. 1969 to present. Inquire as to online cost and availability.

NTIS. National Technical Information Service, 5285 Port Royal Road, Springfield, VA 22161. (703) 487-4630. Broad coverage of government sponsored research reports, 1964 to present. Inquire as to online cost and availability.

SCISEARCH. Institute for Scientific Information, 3501 Market Street, Philadelphia, PA 19104. (800) 523-1850 or (215) 386-0100. Broad multidisciplinary title and author index to the international literature of science and technology, 1974 to present. Inquire as to online cost and availability.

WILSONLINE. H.W. Wilson and Company, 950 University Avenue, Bronx, NY 10452. (800) 367-6770 or (212) 588-8400. Makes available online versions of the H.W. Wilson indexes including Applied Science and Technology Index, Business Periodicals Index and Readers' Guide to Periodical Literature. Approximately 1980 to present. Inquire as to online cost and availability.

PERIODICALS

AMERICAN CARTOGRAPHER. American Congress on Surveying and Mapping, 210 Little Falls Street, Falls Church, VA 22046. (703) 241-2446. 1974 to present. Quarterly. $60.00 per year.

AMERICAN CONGRESS OF SURVEYING AND MAPPING. BULLETIN. American Congress on Surveying and Mapping, 210 Little Falls Street, Falls Church, VA 22046. (703) 241-2446. 1950 to present. Bimonthly. $50.00 per year.

AMERICAN CONGRESS OF SURVEYING AND MAPPING. PROCEEDINGS. American Congress on Surveying and Mapping, 210 Little Falls Street, Falls Church, VA 22046. (703) 241-2446. 1942 to present. Semi-annual. $25.00 per year.

ASSOCIATION OF AMERICAN GEOGRAPHERS. ANNALS. Association of American Geographers, 1710 16th Street, N.W., Washington, DC 20009. (202) 234-1450. 1911 to present.

Quarterly. $45.00 per year.

IEEE TRANSACTIONS ON GEOSCIENCE AND REMOTE SENSING. IEEE Geoscience and Remote Sensing Society. Institute of Electrical and Electronics Engineers, 345 East 47th Street, New York, NY 10017. (212) 705-7900. Order from: IEEE Service Center, 445 Hoes Lane, Piscataway, NJ 08854. 1963 to present. Bimonthly. $110.00 per year.

JOURNAL OF IMAGING SCIENCE. Society of Photographic Scientists and Engineers. 7003 Kilworth Lane, Springfield, VA 22151. (703) 642-9090. 1956 to present. Bimonthly. $70.00 per year.

JOURNAL OF IMAGING TECHNOLOGY. Society of Photographic Scientists and Engineers. 7003 Kilworth Lane, Springfield, VA 22151. (703) 642-9090. 1975 to present. Bimonthly. $70.00 per year.

MARINE GEODESY: AN INTERNATIONAL JOURNAL OF OCEAN SURVEYS, MAPPING, AND SENSING. Crane Russak and Company, Inc., 3 East 44th Street, New York, NY 10017. (212) 867-1490. 1977 to present. Quarterly. $86.00 per year.

PHOTOGRAMMETRIA. Elsevier Science Publishing Company, Inc., 52 Vanderbilt Avenue, New York, NY 10017. (212) 370-5520. 1949 to present. Quarterly. $65.00 per year.

PHOTOGRAMMETRIC ENGINEERING AND REMOTE SENSING. American Society for Photogrammetry and Remote Sensing. 210 Little Falls Street, Falls Church, VA 22046-4398. (703) 534-6617. Order from: Allen Press, Inc., 1041 New Hampshire Street, Box 368, Lawrence, KS 66044. 1934 to present. Monthly. $80.00 per year.

PHOTOGRAMMETRIC RECORD. Photogrammetry Society, Department of Photogrammetry and Surveying, University College London, Gower Street, London WC1E 6BT, England. Semiannual. $37.50 per year.

SCIENTIFIC AND APPLIED PHOTOGRAPHY AND CINEMATOGRAPHY. Gordon and Breach Science Publishers, Inc., 50 West 23rd Street, New York, NY 10010. (212) 206-8900. 12 times per year. $496.00 per year.

REMOTE SENSING OF ENVIRONMENT. Elsevier Science Publishing Company, Inc., 52 Vanderbilt Avenue, New York, NY 10017. (212) 370-5520. 1968 to present. Six times per year. $210.00 per year.

REMOTE SENSING REVIEWS. Harwood Academic Publishers, 50 West 23rd Street, New York, NY 10010. (212) 206-8900. Quarterly. $160.00 per year.

RESEARCH CENTERS AND INSTITUTES

CENTER FOR REMOTE SENSING AND CARTOGRAPHY. 420 Chipta Way, Salt Lake City, UT 84112. (801) 581-8218.

GEOPHOTOGRAPHY AND REMOTE SENSING CENTER. University of Idaho, Geology Department, Moscow, ID 83843. (208) 885-7977.

LAND INFORMATION AND COMPUTER GRAPHICS FACILITY. University of Wisconsin - Madison, Department of Landscape Architecture, 25 Ag Hall, Madison, WI 53706. (608) 263-5534.

U.S. ARMY ENGINEER TOPOGRAPHIC LABORATORIES. U.S. Army Corps of Engineers, Fort Belvoir, VA 22060. (202) 355-2634.

SWITCHING THEORY

See: ELECTRICAL ENGINEERING

SYNTHETIC FUELS

See also: CHEMICAL ENGINEERING, COAL, ENERGY, FUELS, NATURAL GAS, NUCLEAR ENERGY, PETROLEUM

ABSTRACT SERVICES AND INDEXES

ABSTRACT NEWSLETTER: ENERGY. National Technical Information Service, 5285 Port Royal Road, Springfield, VA 22161. (703) 487-4929. Weekly. $95.00 per year.

APPLIED SCIENCE AND TECHNOLOGY INDEX. H.W. Wilson and Company, 950 University Avenue, Bronx, NY 10452. (800) 367-6670 or (212) 588-8400. Monthly. Inquire as to cost and availability.

CHEMICAL ABSTRACTS. American Chemical Society, Chemical Abstracts Service, Box 3012, Columbus, OH 43210. (614) 421-3600. 1907 to present. Weekly. $9500.00 per year.

CONFERENCE PAPERS INDEX. Cambridge Scientific Abstracts, 5161 River Road, Bethesda, MD 20816. (301) 951-1400. 1972 to present. Monthly. Inquire as to cost and availability.

CURRENT CONTENTS: ENGINEERING, TECHNOLOGY AND APPLIED SCIENCES. Institute for Scientific Information, 3501 Market Street, Philadelphia, PA 19104. (800) 523-1850 or (215) 386-0100. Weekly. $275.00 per year.

ENERGY RESEARCH ABSTRACTS. U.S. Department of Energy, Office of Scientific and Technical Information, Box 62, Oak Ridge, TN 37831. (615) 576-1155. Subscribe to: U.S. Superintendent of Documents, Washington, DC 20402. 1976 to present. Semimonthly. $146.00 per year.

ENGINEERING INDEX MONTHLY AND AUTHOR INDEX. Engineering Information Inc., 345 East 47th Street, New York, NY 10017. (212) 705-7600. Monthly. $1560.00 per year.

GENERAL SCIENCE INDEX. H.W. Wilson and Company, 950 University Avenue, Bronx, NY 10452. (800) 367-6670 or (212) 588-8400. 1978 to present. Monthly. Inquire as to cost and availability.

INDEX TO SCIENTIFIC AND TECHNICAL PROCEEDINGS. Institute for Scientific Information, 3501 Market Street, Philadelphia, PA 19104. (800) 523-1850 or (215) 386-0100. 1978 to present. Monthly. $775.00 per year.

INDEX TO SCIENTIFIC REVIEWS. Institute for Scientific Information, 3501 Market Street, Philadelphia, PA 19104. (800) 523-1850 or (215) 386-0100. 1974 to present. Semi-annual. $550.00 per year.

PHYSICS ABSTRACTS. Institution of Electrical Engineers. Available from: IEEE Service Center, 445 Hoes Lane, Piscataway, NJ 08854. 1898 to present. Bimonthly. $1700.00 per year.

PHYSICS BRIEFS. Physik Verlag GmbH, Postfach 1260/1280, D-6940 Weinheim, West Germany. (212) 661-9404. 1920 to present. Twenty-six times per year. $1250.00 per year.

SCIENCE CITATION INDEX. Institute for Scientific Information, 3501 Market Street, Philadelphia, PA 19104. (800) 523-1850 or (215) 386-0100. Six times per year. $6200. 00 per year.

ANNUAL REVIEWS AND YEARBOOKS

ADVANCES IN ENERGY SYSTEMS AND TECHNOLOGY. Academic Press, Inc., 6277 Sea Harbor Drive, Orlando, FL 32821. (800) 321-5068. 1979 to present. Irregular. Inquire.

ANNUAL REVIEW OF ENERGY. Annual Reviews, Inc., 4139 El Camino Way, Palo Alto, CA 94306. (415) 493-4400. Annual. Inquire.

SYNTHETIC FUELS

ASSOCIATIONS AND PROFESSIONAL SOCIETIES

AMERICAN CHEMICAL SOCIETY. 1155 16th Street, N.W., Washington, DC 20036. (800) 424-6747.

AMERICAN INSTITUTE OF PHYSICS. 335 East 45th Street, New York, NY 10017. (212) 661-9494.

ASSOCIATION OF ENERGY ENGINEERS. 4025 Pleasantdale Road, Suite 340, Atlanta, GA 30340. (404) 447-5083.

COUNCIL ON SYNTHETIC FUELS. 1301 Pennsylvania Avenue, N.W., Suite 325, Washington, DC 20004. (202) 347-7069.

INSTITUTE OF ELECTRICAL AND ELECTRONICS ENGINEERS. 345 East 47th Street, New York, NY 10017. (212) 705-7900.

DIRECTORIES AND BIOGRAPHICAL SOURCES

AMERICAN MEN AND WOMEN OF SCIENCE. R.R. Bowker, Inc., Order Department, 245 West 17th Street, New York, NY 10011. (800) 521-8110. Eight volumes. 1986. $595.00 for set.

ENERGY INFORMATION CENTERS DIRECTORY. Public Affairs and Information Program, Atomic Industrial Forum, 7101 Wisconsin Avenue, Bethesda, MD 20814. (301) 654-9260. 1985. Free.

INTERNATIONAL RESEARCH CENTERS DIRECTORY 1988-89. Darren L. Smith, editor. Gale Research Company, Book Tower, Detroit, MI 48226. (800) 521-0707. 4th edition. 1987. $360.00.

1987 DIRECTORY OF ENGINEERING SOCIETIES AND RELATED ORGANIZATIONS. Gordon Davis, editor. Hemisphere Publishing Corporation, 79 Madison Avenue, New York, NY 10016-7892. (800) 821-8312. 12th edition. 1987. $100.00.

RESEARCH CENTERS DIRECTORY 1988. Gale Research Company, Book Tower, Detroit, MI 48226. (800) 521-0707. 12th edition. 1987. $365.00 for set.

SCIENTIFIC AND TECHNICAL ORGANIZATIONS AND AGENCIES DIRECTORY. Margaret Labash Young, editor. Gale Research Company, Book Tower, Detroit, MI 48226. (800) 521-0707. 2nd edition. 1987. $185.00.

SYNFUELS PROJECT DIRECTORY. Pasha Publications, Inc., 1401 Wilson Boulevard, Suite 1000, Arlington, VA 22209. (703) 528-1244. 1984. $125.00.

WHO'S WHO IN ENGINEERING. Gordon Davis, editor. Hemisphere Publishing Corporation, 79 Madison Avenue, New York, NY 10016-7892. (800) 821-8312. 6th edition. 1985. $200.00.

GENERAL WORKS

ENERGY 2000: AN OVERVIEW OF THE WORLD'S ENERGY RESOURCES IN THE DECADES TO COME. Heinz Knoepfel. Gordon and Breach Science Publishers, Inc., 50 West 23rd Street, New York, NY 10010. (212) 206-8900. 1986. $42.00.

FUEL AND ENERGY. J.H. Harker and J. Backhurst. Academic Press, Inc., 6277 Sea Harbor Drive, Orlando, FL 32821. (800) 321-5068. 1981. $33.00 in paper.

FUELS AND FUEL TECHNOLOGY. W. Francis and M.C. Peters. Pergamon Press, Inc., Maxwell House, Fairview Park, Elmsford, NY 10523. (914) 592-7700. 1980. $42.00 in paper.

PETROLEUM RESOURCES AND DEVELOPMENT. Kameel I. Kheir, editor. Columbia University Press, 562 West 113th Street, New York, NY 10025. (212) 316-7100. 1988. $30.00.

THE SCIENCE AND TECHNOLOGY OF COAL AND COAL UTILIZATION. B.R. Cooper, editor. Plenum Publishing Corporation, 233 Spring Street, New York, NY 10013. (800) 221-9369. 1984. $85.00.

SYNTHETIC FUELS. R.F. Probstein and R.E. Hicks. McGraw-Hill Book Company, 1221 Avenue of the Americas, New York, NY 10020. (212) 512-2000. 1982. $43.50.

HANDBOOKS AND MANUALS

HANDBOOK OF ENERGY SYSTEMS ENGINEERING. Leslie C. Wilbur, editor. John Wiley and Sons, Inc., 605 Third Avenue, New York, NY 10158. (800) 526-5368. 1985. $74.95.

HANDBOOK OF ENERGY TECHNOLOGY AND ECONOMICS. R.A. Meyers, editor. John Wiley and Sons, Inc., 605 Third Avenue, New York, NY 10158. (800) 526-5368. 1983. $79.95.

HANDBOOK OF SYNFUELS TECHNOLOGY. R.A. Meyers. McGraw-Hill Book Company, 1221 Avenue of the Americas, New York, NY 10020. (212) 512-2000. 1984. $98.00.

ONLINE DATA BASES

CA SEARCH. Chemical Abstracts Service, P.O. Box 3012, Columbus, OH 43120. (800) 848-6538 or (614) 421-3600. Comprehensive guide to chemical literature, 1972 to present. Inquire as to online cost and availability.

COMPENDEX. Engineering Information, Inc., 345 East 47th Street, New York, NY 10017. (800) 221-1044 or (212) 705-7615. Engineering and technical literature, 1975 to present. Inquire as to online cost and availability.

DISSERTATION ABSTRACTS ONLINE. University Microfilms International, 300 North Zeeb Road, Ann Arbor, MI 48106. (800) 521-0600 or (313) 761-4700. Scope includes virtually all doctoral dissertations accepted at accredited American institutions from 1861 to present in over 250 subject areas. Inquire as to online cost and availability.

DOE ENERGY DATA BASE. U.S. Department of Energy, Office of Scientific and Technical Information, P.O. Box 62, Oak Ridge, TN 37831. (615) 576-6837. A database that covers all aspects of energy including the science and technology of energy. 1948 to present. Available through the DIALOG search service or DOE/RECON. Inquire as to online cost and availability.

ENERGYLINE. EIC/Intelligence, Inc., 48 West 38th Street, New York, NY 10018. (212) 944-8500. A database of resources on the scientific, engineering, political, and socioeconomic aspects of energy resources. 1976 to present. Inquire as to online cost and availability.

INSPEC. INSPEC Marketing Department, Institution of Electrical Engineers. Available from IEEE Service Center, 445 Hoes Lane, Piscataway, NJ 08854. (201) 981-0060. Online version of Physics Abstracts. Inquire as to online cost and availability.

NTIS. National Technical Information Service, 5285 Port Royal Road, Springfield, VA 22161. (703) 487-4630. Broad coverage of government sponsored research reports, 1964 to present. Inquire as to online cost and availability.

SCISEARCH. Institute for Scientific Information, 3501 Market Street, Philadelphia, PA 19104. (800) 523-1850 or (215) 386-0100. Broad multidisciplinary title and author index to the international literature of science and technology, 1974 to present. Inquire as to online cost and availability.

WILSONLINE. H.W. Wilson and Company, 950 University Avenue, Bronx, NY 10452. (800) 367-6770 or (212) 588-8400. Makes available online versions of the H.W. Wilson indexes including Applied Science and Technology Index, Business Periodicals Index and Readers' Guide to Periodical Literature. Approximately 1980 to present. Inquire as to online cost and availability.

PERIODICALS

ALTERNATIVE SOURCES OF ENERGY. Alternative Sources of Energy, Inc., 107 South Central Avenue, Milaca, MN 56353. (612) 983-6892. 1971 to present. Ten times per year. $48.00 per year.

ENERGY AND FUELS. American Chemical Society, 1155 16th Street, N.W., Washington, DC 20036. (800) 424-6747. 1987 to present. Bimonthly. $269.00 per year.

ENERGY ENGINEERING. Association of Energy Engineers. Available from: Fairmont Press, Box 14227, Atlanta, GA 30324. (404) 447-5314. 1904 to present. Bimonthly. $44.00 per year.

ENERGY SOURCES. Crane Russak and Company, Inc., 3 East 44th Street, New York, NY 10017. (212) 867-1490. 1973. Quarterly. $73.00 per year.

JOURNAL OF ENERGY RESOURCES TECHNOLOGY. American Society of Mechanical Engineers, 345 East 47th Street, New York, NY 10017. (212) 705-7703. 1979 to present. $24.00 per year.

POWER ENGINEERING (NEW YORK). Technical Publishing Company, 875 Third Avenue, New York, NY 10022. (212) 605-9400. 1896 to present. Monthly. $24.00 per year.

SYNFUELS. McGraw-Hill Book Company, 1221 Avenue of the Americas, New York, NY 10020. (212) 512-2000. 1979 to present. Weekly. $725.00 per year.

RESEARCH CENTERS AND INSTITUTES

CENTER FOR ENERGY STUDIES. University of Texas at Austin, 1318 EME Building, 10100 Burnet Road, Austin, TX 78758. (512) 471-7792.

ENERGY CENTER. Stevens Institute of Technology, Department of Mechanical Energy, Castle Point Station, Hoboken, NJ 07030. (201) 420-5560.

SYNTHETIC FUELS CENTER. Massachusetts Institute of Technology, 123 Amherst Street, Building E-40-455, Cambridge, MA 02139. (617) 253-3478.

ENERGY RESEARCH/DEVELOPMENT CENTER. University of Kansas, 345 Nichols Hall, 2291 Irving Hill Drive, Lawrence, KS 66045. (913) 864-4079.

INSTITUTE OF GAS TECHNOLOGY. 3424 South State Street, Chicago, IL 60616. (312) 567-3650.

OAK RIDGE NATIONAL LABORATORY. U.S. Department of Energy, P.O. Box X, Oak Ridge, TN 39831. (615) 576-2900.

SYSTEMS ANALYSIS

See also: COMPUTER OPERATING SYSTEMS, COMPUTERS, SOFTWARE, SOFTWARE ENGINEERING, SYSTEMS ENGINEERING

ABSTRACT SERVICES AND INDEXES

APPLIED SCIENCE AND TECHNOLOGY INDEX. H.W. Wilson and Company, 950 University Avenue, Bronx, NY 10452. (800) 367-6670 or (212) 588-8400. Monthly. Inquire as to cost and availability.

COMPUTER AND CONTROL ABSTRACTS. Institute of Electrical Engineers. Available from: Institute of Electrical and Electronics Engineers. IEEE Service Center, 445 Hoes Lane, Piscataway, NJ 08854. Semimonthly. $775.00 per year.

COMPUTER AND INFORMATION SYSTEMS: AN ABSTRACT JOURNAL PERTAINING TO THE THEORY, DESIGN, FABRICATION AND APPLICATION OF COMPUTER AND INFORMATION SYSTEMS. Cambridge Scientific Abstracts, 5161 River Road, Bethesda, MD 20816. 1972 to present. Semi-monthly. Inquire as to cost and availability.

COMPUTER CONTENTS: THE BIWEEKLY COMPILATION OF TABLES OF CONTENTS FROM COMPUTER, ELECTRONIC AND TELECOMMUNICATIONS MAGAZINES, JOURNALS AND TRANACTIONS. Find/SVP, 500 Fifth Avenue, New York, NY 101110. (800) 346-3787 or (212) 354-2424. Biweekly. $115.00 per year.

COMPUTER LITERATURE INDEX. Applied Computer Research, Inc., P.O. Box 9280, Phoenix, AZ 85068. (602) 995-5929. Quarterly. $125.00 per year.

COMPUTER PROGRAMS ABSTRACTS. U.S. National Aeronautics and Space Administration. Available from: U.S. Government Printing Office, Washington, DC 20402. Quarterly. $10.00 per year.

COMPUTING REVIEWS. Association of Computing Machinery, 11 West 42nd Street, New York, NY 10036. (212) 869-7440. Monthly. $60.00 per year.

ENGINEERING INDEX MONTHLY AND AUTHOR INDEX. Engineering Information Inc., 345 East 47th Street, New York, NY 10017. (212) 705-7600. Monthly. $1560.00 per year.

INDEX TO SCIENTIFIC AND TECHNICAL PROCEEDINGS. Institute for Scientific Information, 3501 Market Street, Philadelphia, PA 19104. (800) 523-1850 or (215) 386-0100. 1978 to present. Monthly. $775.00 per year.

INDEX TO SCIENTIFIC REVIEWS. Institute for Scientific Information, 3501 Market Street, Philadelphia, PA 19104. (800) 523-1850 or (215) 386-0100. 1974 to present. Semi-annual. $550.00 per year.

PCR-2: PERSONAL COMPUTER REVIEW - SQUARED. Toolbox Publications, Inc., P.O. Box 5451, 2514 Birch Creek Lane, Orchard Lake, MI 48033. 1987 to present. Bimonthly. $60.00 per year.

SCIENCE CITATION INDEX. Institute for Scientific Information, 3501 Market Street, Philadelphia, PA 19104. (800) 523-1850 or (215) 386-0100. Six times per year. $6200.00 per year.

ANNUAL REVIEWS AND YEARBOOKS

COMPUTER PUBLISHERS AND PUBLICATIONS 1988-89: AN INTERNATIONAL DIRECTORY AND YEARBOOK. Efrem Sigel and Frederica Evan, editors. Gale Research Company, Book Tower, Detroit, MI 48226. (800) 521-0707. Third edition. $140.00.

ASSOCIATIONS AND PROFESSIONAL SOCIETIES

AMERICAN FEDERATION OF INFORMATION PROCESSING SOCIETIES. 1899 Preston White Drive, Reston, VA 22091. (703) 620-8900.

ASSOCIATION OF COMPUTER PROGRAMMERS AND ANALYSTS. 2108-C Gallows Road, Vienna, VA 22180. (703) 790-0490.

ASSOCIATION OF COMPUTING MACHINERY (ACM). 11 West 42nd Street, New York, NY 10036. (212) 869-7440.

IEEE COMPUTER SOCIETY. 1730 Massachusetts Avenue, N.W., Washington, DC 20036. (202) 371-0101.

INSTITUTE OF ELECTRICAL AND ELECTRONICS ENGINEERS. IEEE Service Center, 445 Hoes Lane, Piscataway, NJ 08854.

MACHINE VISION ASSOCIATION. P.O. Box 930, One SME Drive, Dearborn, MI 48121. (313) 271-1500.

SOCIETY FOR COMPUTER SIMULATION. P.O. Box 17900, San Diego, CA 92117. (619) 277-3888.

SOCIETY FOR INFORMATION DISPLAY. 8055 Manchester Avenue, Suite 615, Playa Del Rey, CA 90293. (213) 305-1502.

DIRECTORIES AND BIOGRAPHICAL SOURCES

AMERICAN SOCIETY FOR INFORMATION SCIENCE HANDBOOK AND DIRECTORY. American Society for Information Science, 1424 16th Street, N.W., Suite 404, Washington, DC 20036. (202) 462-1000. $50.00.

COMPUTERS AND COMPUTING INFORMATION RESOURCES DIRECTORY. Martin Connors, editor. Gale Research Company, Book Tower, Detroit, MI 48226. (800) 521-0707. 1987. $165.00. Supplement available at $85.00.

INTERNATIONAL RESEARCH CENTERS DIRECTORY 1988-89. Darren L. Smith, editor. Gale Research Company, Book Tower, Detroit, MI 48226. (800) 521-0707. 4th edition. 1987. $360.00.

1987 DIRECTORY OF ENGINEERING SOCIETIES AND RELATED ORGANIZATIONS. Gordon Davis, editor. Hemisphere Publishing Corporation, 1010 Vermont Avenue, NW, Washington, DC 20005. (800) 526-0275. 12th edition. 1987. $100.00.

RESEARCH CENTERS DIRECTORY 1988. Gale Research Company, Book Tower, Detroit, MI 48226. (800) 521-0707. 12th edition. 1987. $365.00 for set.

SCIENTIFIC AND TECHNICAL ORGANIZATIONS AND AGENCIES DIRECTORY. Margaret Labash Young, editor. Gale Research Company, Book Tower, Detroit, MI 48226. (800) 521-0707. 2nd edition. 1987. $185.00.

WHO'S WHO IN ENGINEERING. Gordon Davis, editor. Hemisphere Publishing Corporation, 1010 Vermont Avenue, NW, Washington, DC 20005. (800) 526-0275. 6th edition. 1985. $200.00.

ENCYCLOPEDIAS AND DICTIONARIES

COMPUTER AND TELECOMMUNICATIONS ACRONYMS. Julie E. Towell and Helen E. Sheppard, editors. Gale Research Company, Book Tower, Detroit, MI 48226. (800) 521-0707. 1986. $60.00.

DICTIONARY OF COMPUTING. Oxford University Press, 200 Madison Avenue, New York, NY 10016. (800) 458-5833. Second edition. 1986. $29.95.

ENCYCLOPEDIA OF INFORMATION SYSTEMS AND SERVICES 1988. Amy Lucas and Annette Novallo, editors. Gale Research Company, Book Tower, Detroit, MI 48226. (800) 521-0707. 8th edition. 1987. $400.00 for set.

PRENTICE-HALL ENCYCLOPEDIA OF INFORMATION TECHNOLOGY. Robert A. Edmunds. Prentice-Hall Publishing, Inc., Englewood Cliffs, NJ 07632. (800) 562-0245. 1987. $49.95.

SOFTWARE ENCYCLOPEDIA. R.R. Bowker Company, 205 East 42nd Street, New York, NY 10017. (800) 521-8110. Tow volumes. 1987. $125.00 for set.

GENERAL WORKS

DESIGNING KNOWLEDGE-BASED SYSTEMS. T.R. Addis. Prentice-Hall Publishing, Inc., Englewood Cliffs, NJ 07632. (800) 562-0245. 1986. $34.95.

FUNDAMENTALS OF SYSTEMS ANALYSIS. Jerry Fitzgerald and others. John Wiley and Sons, Inc., 605 Third Avenue, New York, NY 10158. (800) 526-5368. Second edition. 1981. $35.95.

HIGH LEVEL LANGUAGE AND SOFTWARE APPLICATIONS REFERENCE. W.J. Birnes. McGraw-Hill Book Company, 1221 Avenue of the Americas, New York, NY 10020. (212) 512-2000. 1988. $29.95.

PRINCIPLES OF PROGRAMMING LANGUAGES: DESIGN, EVALUATION AND IMPLEMENTATION. Bruce MacLennan. Holt, Rinehart and Winston, Inc., 383 Madison Avenue, New York, NY 10017. (212) 872-2000. Second edition. 1986. $38.75.

SYSTEM DEVELOPMENT STANDARDS. C. Candullo. McGraw-Hill Book Company, 1221 Avenue of the Americas, New York, NY 10020. (212) 512-2000. 1985. $49.95.

SYSTEMATIC SYSTEMS APPROACH: AN INTEGRATED METHOD FOR SOLVING SYSTEMS PROBLEMS. Thomas H. Athey. Prentice-Hall Publishing, Inc., Englewood Cliffs, NJ 07632. (800) 562-0245. 1982. $47.50.

SYSTEMS ANALYSIS AND DESIGN FOR COMPUTER APPLICATIONS. D. Millington. John Wiley and Sons, Inc., 605 Third Avenue, New York, NY 10158. (800) 526-5368. 1981. $52.95.

SYSTEMS ENGINEERING AND ANALYSIS. B. Blanchard and W. Fabrycky. Prentice-Hall Publishing, Inc., Englewood Cliffs, NJ 07632. (800) 562-0245. 1981. $39.95.

HANDBOOKS AND MANUALS

A HANDBOOK OF SYSTEMS ANALYSIS. John E. Bingham and Garth W. Davies. Halsted Press, a division of John Wiley and Sons, Inc., 605 Third Avenue, New York, NY 10158. (800) 526-5368. Second edition. 1980. $29.95.

ONLINE DATA BASES

COMPENDEX. Engineering Information, Inc., 345 East 47th Street, New York, NY 10017. (800) 221-1044 or (212) 705-7615. Engineering and technical literature, 1975 to present. Inquire as to online cost and availability.

DISSERTATION ABSTRACTS ONLINE. University Microfilms International, 300 North Zeeb Road, Ann Arbor, MI 48106. (800) 521-0600 or (313) 761-4700. Scope includes virtually all doctoral dissertations accepted at accredited American institutions from 1861 to present in over 250 subject areas. Inquire as to online cost and availability.

INSPEC. INSPEC Marketing Department, Institution of Electrical Engineers. Available from IEEE Service Center, 445 Hoes Lane, Piscataway, NJ 08854. (201) 981-0060. Online version of Physics Abstracts. Inquire as to online cost and availability.

NTIS. National Technical Information Service, 5285 Port Royal Road, Springfield, VA 22161. (703) 487-4630. Broad coverage of government sponsored research reports, 1964 to present. Inquire as to online cost and availability.

SCISEARCH. Institute for Scientific Information, 3501 Market Street, Philadelphia, PA 19104. (800) 523-1850 or (215) 386-0100. Broad multidisciplinary title and author index to the international literature of science and technology, 1974 to present. Inquire as to online cost and availability.

WILSONLINE. H.W. Wilson and Company, 950 University Avenue, Bronx, NY 10452. (800) 367-6770 or (212) 588-8400. Makes available online versions of the H.W. Wilson indexes including Applied Science and Technology Index, Business Periodicals Index and Readers' Guide to Periodical Literature. Approximately 1980 to present. Inquire as to online cost and availability.

PERIODICALS

ACM TRANSACTIONS ON PROGRAMMING LANGUAGES AND SYSTEMS. Association of Computing Machinery, 11 West 42nd Street, New York, NY 10036. (212) 869-7440. 1979 to present. Quarterly. $55.00 per year.

ADVANCES IN ENGINEERING SOFTWARE. CML Publications, 400 West Cummings Park, Suite 6200, Woburn, MA 01801. (617) 933-7374. 1979 to present. Quarterly. $130.00 per year.

BYTE. Byte Publications, Inc., 70 Main Street, Petersborough, NH 03458. (603) 924-9281. Monthly. $21.00 per year.

COMMUNICATIONS OF THE ACM. Association of Computing Machinery, 11 West 42nd Street, New York, NY 10036. (212) 869-7440. Monthly. $80.00 per year.

COMPUTER LANGUAGES. Pergamon Press, Inc., Maxwell House, Fairview Park, Elmsford, NY 10523. (914) 592-7700. 1976 to present. Quarterly. $195.00 per year.

IEEE SOFTWARE. Institution of Electrical and Electronics Engineers. IEEE Service Center, 445 Hoes Lane, Piscataway, NJ 08854. 1984 to present. Quarterly. $15.00 per issue.

IEEE TRANSACTIONS ON SOFTWARE ENGINEERING. Institute of Electrical and Electronics Engineers. IEEE Service Center, 445 Hoes Lane, Piscataway, NJ 08854. 1975 to present. Monthly. $160.00 per year.

INTERFACE. International Computer Programs, Inc., 9000 Keystone Crossing, Indianapolis, IN 46240. (317) 844-7461. 1975 to present. Quarterly. $10.00 per year.

JOURNAL OF LOGIC PROGRAMMING. Elsevier Science Publishing Company, Inc., 52 Vanderbilt Avenue, New York, NY 10017. (212) 370-5520. 1984 to present. Quarterly. $95.00 per year.

JOURNAL OF SYSTEMS AND SOFTWARE. Elsevier Science Publishing Company, Inc., 52 Vanderbilt Avenue, New York, NY 10017. (212) 370-5520. 1979 to present. Quarterly. $95.00 per year.

MINI-MICRO SYSTEMS. Cahners Publishing Company, Inc., 275 Washington Street, Newton, MA 02158. (617) 964-3030. 1968 to present. Monthly. $65.00 per year.

SIGPLAN NOTICES. Association of Computing Machinery Special Interest Group on Programming Languages, 11 West 42nd Street, New York, NY 10036. (212) 869-7440. 1965 to present. Monthly. $25.00 per year.

SIGSOFT SOFTWARE ENGINEERING NOTICES. Association of Computing Machinery Special Interest Group on Software Engineering. 11 West 42nd Street, New York, NY 10036. (212) 869-7440. Quarterly. $12.00 per year.

SOFTWARE DEVELOPER'S MONTHLY. SourceView Press, 835 Castro Street, Martinez, CA 94553. (415) 228-6220. 1985 to present. Monthly. $144.00 per year.

SOFTWARE ENGINEERING JOURNAL. Institute of Electrical Engineers, Savoy Place, London, WC2R OBL, England. 1981 to present. Bimonthly. $85.00 per year.

SOFTWARE PRACTICE AND EXPERIENCE. John Wiley and Sons, Inc., 605 Third Avenue, New York, NY 10158. (800) 526-5368. 1971 to present. Monthly. $260.00 per year.

SYSTEMS AND CONTROL LETTERS. Elsevier Science Publishing Company, Inc., 52 Vanderbilt Avenue, New York, NY 10017. (212) 370-5520. 1981 to present. Ten times per year.

RESEARCH CENTERS AND INSTITUTES

DATABASE SYSTEMS RESEARCH AND DEVELOPMENT CENTER. University of Florida, 512 Weil Hall, Gainesville, FL 32611. (904) 392-2371.

DEPARTMENT OF ENGINEERING-ECONOMIC SYSTEMS. Stanford University, Terman Engineering Building, Room 306, Stanford, CA 94305-4025. (415) 723-4168.

PROGRAM FOR RESEARCH IN INFORMATION SYSTEMS ENGINEERING. University of Michigan, 1205 Beale, 10 East Building, Ann Arbor, MI 48109. (313) 763-2238.

SYSTEMS ENGINEERING

See also: COMPUTERS, COMPUTER OPERATING SYSTEMS, SOFTWARE, SOFTWARE ENGINEERING, SYSTEMS ANALYSIS

ABSTRACT SERVICES AND INDEXES

APPLIED SCIENCE AND TECHNOLOGY INDEX. H.W. Wilson and Company, 950 University Avenue, Bronx, NY 10452. (800) 367-6670 or (212) 588-8400. Monthly. Inquire as to cost and availability.

COMPUTER AND CONTROL ABSTRACTS. Institute of Electrical Engineers. Available from: Institute of Electrical and Electronics Engineers. IEEE Service Center, 445 Hoes Lane, Piscataway, NJ 08854. Semimonthly. $775.00 per year.

COMPUTER AND INFORMATION SYSTEMS: AN ABSTRACT JOURNAL PERTAINING TO THE THEORY, DESIGN, FABRICATION AND APPLICATION OF COMPUTER AND INFORMATION SYSTEMS. Cambridge Scientific Abstracts, 5161 River Road, Bethesda, MD 20816. 1972 to present. Semi-monthly. Inquire as to cost and availability.

COMPUTER CONTENTS: THE BIWEEKLY COMPILATION OF TABLES OF CONTENTS FROM COMPUTER, ELECTRONIC AND TELECOMMUNICATIONS MAGAZINES, JOURNALS AND TRANACTIONS. Find/SVP, 500 Fifth Avenue, New York, NY 101110. (800) 346-3787 or (212) 354-2424. Biweekly. $115.00 per year.

COMPUTER LITERATURE INDEX. Applied Computer Research, Inc., P.O. Box 9280, Phoenix, AZ 85068. (602) 995-5929. Quarterly. $125.00 per year.

COMPUTER PROGRAMS ABSTRACTS. U.S. National Aeronautics and Space Administration. Available from: U.S. Government Printing Office, Washington, DC 20402. Quarterly. $10.00 per year.

COMPUTING REVIEWS. Association of Computing Machinery, 11 West 42nd Street, New York, NY 10036. (212) 869-7440. Monthly. $60.00 per year.

ENGINEERING INDEX MONTHLY AND AUTHOR INDEX. Engineering Information Inc., 345 East 47th Street, New York, NY 10017. (212) 705-7600. Monthly. $1560.00 per year.

INDEX TO SCIENTIFIC AND TECHNICAL PROCEEDINGS. Institute for Scientific Information, 3501 Market Street, Philadelphia, PA 19104. (800) 523-1850 or (215) 386-0100. 1978 to present. Monthly. $775.00 per year.

INDEX TO SCIENTIFIC REVIEWS. Institute for Scientific Information, 3501 Market Street, Philadelphia, PA 19104. (800) 523-1850 or (215) 386-0100. 1974 to present. Semi-annual. $550.00 per year.

PCR-2: PERSONAL COMPUTER REVIEW - SQUARED. Toolbox Publications, Inc., P.O. Box 5451, 2514 Birch Creek Lane, Orchard Lake, MI 48033. 1987 to present. Bimonthly. $60.00 per year.

SCIENCE CITATION INDEX. Institute for Scientific Information, 3501 Market Street, Philadelphia, PA 19104. (800) 523-1850 or (215) 386-0100. Six times per year. $6200.00 per year.

ASSOCIATIONS AND PROFESSIONAL SOCIETIES

AMERICAN FEDERATION OF INFORMATION PROCESSING SOCIETIES. 1899 Preston White Drive, Reston, VA 22091. (703) 620-8900.

ASSOCIATION OF COMPUTER PROGRAMMERS AND ANALYSTS. 2108-C Gallows Road, Vienna, VA 22180. (703) 790-0490.

SYSTEMS ENGINEERING

ASSOCIATION OF COMPUTING MACHINERY (ACM). 11 West 42nd Street, New York, NY 10036. (212) 869-7440.

IEEE COMPUTER SOCIETY. 1730 Massachusetts Avenue, N.W., Washington, DC 20036. (202) 371-0101.

INSTITUTE OF ELECTRICAL AND ELECTRONICS ENGINEERS. IEEE Service Center, 445 Hoes Lane, Piscataway, NJ 08854.

MACHINE VISION ASSOCIATION. P.O. Box 930, One SME Drive, Dearborn, MI 48121. (313) 271-1500.

SOCIETY FOR COMPUTER SIMULATION. P.O. Box 17900, San Diego, CA 92117. (619) 277-3888.

SOCIETY FOR INFORMATION DISPLAY. 8055 Manchester Avenue, Suite 615, Playa Del Rey, CA 90293. (213) 305-1502.

DIRECTORIES AND BIOGRAPHICAL SOURCES

AMERICAN SOCIETY FOR INFORMATION SCIENCE HANDBOOK AND DIRECTORY. American Society for Information Science, 1424 16th Street, N.W., Suite 404, Washington, DC 20036. (202) 462-1000. $50.00.

COMPUTERS AND COMPUTING INFORMATION RESOURCES DIRECTORY. Martin Connors, editor. Gale Research Company, Book Tower, Detroit, MI 48226. (800) 521-0707. 1987. $165.00. Supplement available at $85.00.

INTERNATIONAL RESEARCH CENTERS DIRECTORY 1988-89. Darren L. Smith, editor. Gale Research Company, Book Tower, Detroit, MI 48226. (800) 521-0707. 4th edition. 1987. $360.00.

1987 DIRECTORY OF ENGINEERING SOCIETIES AND RELATED ORGANIZATIONS. Gordon Davis, editor. Hemisphere Publishing Corporation, 79 Madison Avenue, New York, NY 10016-7892. (800) 821-8312. 12th edition. 1987. $100.00.

RESEARCH CENTERS DIRECTORY 1988. Gale Research Company, Book Tower, Detroit, MI 48226. (800) 521-0707. 12th edition. 1987. $365.00 for set.

SCIENTIFIC AND TECHNICAL ORGANIZATIONS AND AGENCIES DIRECTORY. Margaret Labash Young, editor. Gale Research Company, Book Tower, Detroit, MI 48226. (800) 521-0707. 2nd edition. 1987. $185.00.

WHO'S WHO IN ENGINEERING. Gordon Davis, editor. Hemisphere Publishing Corporation, 79 Madison Avenue, New York, NY 10016-7892. (800) 821-8312. 6th edition. 1985. $200.00.

ENCYCLOPEDIAS AND DICTIONARIES

COMPUTER AND TELECOMMUNICATIONS ACRONYMS. Julie E. Towell and Helen E. Sheppard, editors. Gale Research Company, Book Tower, Detroit, MI 48226. (800) 521-0707. 1986. $60.00.

DICTIONARY OF COMPUTING. Oxford University Press, 200 Madison Avenue, New York, NY 10016. (800) 458-5833. Second edition. 1986. $29.95.

ENCYCLOPEDIA OF INFORMATION SYSTEMS AND SERVICES 1988. Amy Lucas and Annette Novallo, editors. Gale Research Company, Book Tower, Detroit, MI 48226. (800) 521-0707. 8th edition. 1987. $400.00 for set.

PRENTICE-HALL ENCYCLOPEDIA OF INFORMATION TECHNOLOGY. Robert A. Edmunds. Prentice-Hall Publishing, Inc., Englewood Cliffs, NJ 07632. (800) 562-0245. 1987. $49.95.

SOFTWARE ENCYCLOPEDIA. R.R. Bowker Company, 205 East 42nd Street, New York, NY 10017. (800) 521-8110. Tow volumes. 1987. $125.00 for set.

GENERAL WORKS

APPLIED SYSTEMS ENGINEERING. A. Gheorghe. John Wiley and Sons, Inc., 605 Third Avenue, New York, NY 10158. (800) 526-5368. 1982. $63.95.

CONTROL SYSTEM PRINCIPLES AND DESIGN. E.O. Doebelin. John Wiley and Sons, Inc., 605 Third Avenue, New York, NY 10158. (800) 526-5368. 1986. $44.50.

DESIGNING KNOWLEDGE-BASED SYSTEMS. T.R. Addis. Prentice-Hall Publishing, Inc., Englewood Cliffs, NJ 07632. (800) 562-0245. 1986. $34.95.

FUNDAMENTALS OF SYSTEMS ANALYSIS. Jerry Fitzgerald and others. John Wiley and Sons, Inc., 605 Third Avenue, New York, NY 10158. (800) 526-5368. Second edition. 1981. $35.95.

HIGH LEVEL LANGUAGE AND SOFTWARE APPLICATIONS REFERENCE. W.J. Birnes. McGraw-Hill Book Company, 1221 Avenue of the Americas, New York, NY 10020. (212) 512-2000. 1988. $29.95.

PRINCIPLES OF PROGRAMMING LANGUAGES: DESIGN, EVALUATION AND IMPLEMENTATION. Bruce MacLennan. Holt, Rinehart and Winston, Inc., 383 Madison Avenue, New York, NY 10017. (212) 872-2000. Second edition. 1986. $38.75.

SYSTEM DEVELOPMENT STANDARDS. C. Candullo. McGraw-Hill Book Company, 1221 Avenue of the Americas, New York, NY 10020. (212) 512-2000. 1985. $49.95.

SYSTEMATIC SYSTEMS APPROACH: AN INTEGRATED METHOD FOR SOLVING SYSTEMS PROBLEMS. Thomas H. Athey. Prentice-Hall Publishing, Inc., Englewood Cliffs, NJ 07632. (800) 562-0245. 1982. $47.50.

SYSTEMS ANALYSIS AND DESIGN FOR COMPUTER APPLICATIONS. D. Millington. John Wiley and Sons, Inc., 605 Third Avenue, New York, NY 10158. (800) 526-5368. 1981. $52.95.

SYSTEMS ENGINEERING AND ANALYSIS. B. Blanchard and W. Fabrycky. Prentice-Hall Publishing, Inc., Englewood Cliffs, NJ 07632. (800) 562-0245. 1981. $39.95.

HANDBOOKS AND MANUALS

A HANDBOOK OF SYSTEMS ANALYSIS. John E. Bingham and Garth W. Davies. Halsted Press, a division of John Wiley and Sons, Inc., 605 Third Avenue, New York, NY 10158. (800) 526-5368. Second edition. 1980. $29.95.

ONLINE DATA BASES

COMPENDEX. Engineering Information, Inc., 345 East 47th Street, New York, NY 10017. (800) 221-1044 or (212) 705-7615. Engineering and technical literature, 1975 to present. Inquire as to online cost and availability.

DISSERTATION ABSTRACTS ONLINE. University Microfilms International, 300 North Zeeb Road, Ann Arbor, MI 48106. (800) 521-0600 or (313) 761-4700. Scope includes virtually all doctoral dissertations accepted at accredited American institutions from 1861 to present in over 250 subject areas. Inquire as to online cost and availability.

INSPEC. INSPEC Marketing Department, Institution of Electrical Engineers. Available from IEEE Service Center, 445 Hoes Lane, Piscataway, NJ 08854. (201) 981-0060. Online version of Physics Abstracts. Inquire as to online cost and availability.

NTIS. National Technical Information Service, 5285 Port Royal Road, Springfield, VA 22161. (703) 487-4630. Broad coverage of government sponsored research reports, 1964 to present. Inquire as to online cost and availability.

SCISEARCH. Institute for Scientific Information, 3501 Market Street, Philadelphia, PA 19104. (800) 523-1850 or (215) 386-0100.

Broad multidisciplinary title and author index to the international literature of science and technology, 1974 to present. Inquire as to online cost and availability.

WILSONLINE. H.W. Wilson and Company, 950 University Avenue, Bronx, NY 10452. (800) 367-6770 or (212) 588-8400. Makes available online versions of the H.W. Wilson indexes including Applied Science and Technology Index, Business Periodicals Index and Readers' Guide to Periodical Literature. Approximately 1980 to present. Inquire as to online cost and availability.

PERIODICALS

ACM TRANSACTIONS ON PROGRAMMING LANGUAGES AND SYSTEMS. Association of Computing Machinery, 11 West 42nd Street, New York, NY 10036. (212) 869-7440. 1979 to present. Quarterly. $55.00 per year.

ADVANCES IN ENGINEERING SOFTWARE. CML Publications, 400 West Cummings Park, Suite 6200, Woburn, MA 01801. (617) 933-7374. 1979 to present. Quarterly. $130.00 per year.

BYTE. Byte Publications, Inc., 70 Main Street, Petersborough, NH 03458. (603) 924-9281. Monthly. $21.00 per year.

COMMUNICATIONS OF THE ACM. Association of Computing Machinery, 11 West 42nd Street, New York, NY 10036. (212) 869-7440. Monthly. $80.00 per year.

COMPUTER LANGUAGES. Pergamon Press, Inc., Maxwell House, Fairview Park, Elmsford, NY 10523. (914) 592-7700. 1976 to present. Quarterly. $195.00 per year.

IEEE SOFTWARE. Institution of Electrical and Electronics Engineers. IEEE Service Center, 445 Hoes Lane, Piscataway, NJ 08854. 1984 to present. Quarterly. $15.00 per issue.

IEEE TRANSACTIONS ON SOFTWARE ENGINEERING. Institute of Electrical and Electronics Engineers. IEEE Service Center, 445 Hoes Lane, Piscataway, NJ 08854. 1975 to present. Monthly. $160.00 per year.

INTERFACE. International Computer Programs, Inc., 9000 Keystone Crossing, Indianapolis, IN 46240. (317) 844-7461. 1975 to present. Quarterly. $10.00 per year.

JOURNAL OF LOGIC PROGRAMMING. Elsevier Science Publishing Company, Inc., 52 Vanderbilt Avenue, New York, NY 10017. (212) 370-5520. 1984 to present. Quarterly. $95.00 per year.

JOURNAL OF SYSTEMS AND SOFTWARE. Elsevier Science Publishing Company, Inc., 52 Vanderbilt Avenue, New York, NY 10017. (212) 370-5520. 1979 to present. Quarterly. $95.00 per year.

MINI-MICRO SYSTEMS. Cahners Publishing Company, Inc., 275 Washington Street, Newton, MA 02158. (617) 964-3030. 1968 to present. Monthly. $65.00 per year.

SIGPLAN NOTICES. Association of Computing Machinery Special Interest Group on Programming Languages, 11 West 42nd Street, New York, NY 10036. (212) 869-7440. 1965 to present. Monthly. $25.00 per year.

SIGSOFT SOFTWARE ENGINEERING NOTICES. Association of Computing Machinery Special Interest Group on Software Engineering. 11 West 42nd Street, New York, NY 10036. (212) 869-7440. Quarterly. $12.00 per year.

SOFTWARE DEVELOPER'S MONTHLY. SourceView Press, 835 Castro Street, Martinez, CA 94553. (415) 228-6220. 1985 to present. Monthly. $144.00 per year.

SOFTWARE ENGINEERING JOURNAL. Institute of Electrical Engineers, Savoy Place, London, WC2R OBL, England. 1981 to present. Bimonthly. $85.00 per year.

SOFTWARE PRACTICE AND EXPERIENCE. John Wiley and Sons, Inc., 605 Third Avenue, New York, NY 10158. (800) 526-5368. 1971 to present. Monthly. $260.00 per year.

SYSTEMS AND CONTROL LETTERS. Elsevier Science Publishing Company, Inc., 52 Vanderbilt Avenue, New York, NY 10017. (212) 370-5520. 1981 to present. Ten times per year.

RESEARCH CENTERS AND INSTITUTES

DATABASE SYSTEMS RESEARCH AND DEVELOPMENT CENTER. University of Florida, 512 Weil Hall, Gainesville, FL 32611. (904) 392-2371.

DEPARTMENT OF ENGINEERING-ECONOMIC SYSTEMS. Stanford University, Terman Engineering Building, Room 306, Stanford, CA 94305-4025. (415) 723-4168.

PROGRAM FOR RESEARCH IN INFORMATION SYSTEMS ENGINEERING. University of Michigan, 1205 Beale, 10 East Building, Ann Arbor, MI 48109. (313) 763-2238.

T

TECTONICS

See: PLATE TECTONICS

TEKTITES

See also: ASTEROIDS, ASTROGEOLOGY, METEORS, PLANETARY SCIENCE, SOLAR SYSTEM

ABSTRACT SERVICES AND INDEXES

BIBLIOGRAPHY AND INDEX OF GEOLOGY. American Geological Institute, 4220 King Street, Alexandria, VA 22302. (703) 379-2480. 1969 to present. Monthly. $1100.00 per year.

CHEMICAL ABSTRACTS. American Chemical Society, Chemical Abstracts Service, Box 3012, Columbus, OH 43210. (614) 421-3600. 1907 to present. Weekly. $9500.00 per year.

CURRENT CONTENTS: PHYSICAL, CHEMICAL AND EARTH SCIENCES. Institute for Scientific Information, 3501 Market Street, Philadelphia, PA 19104. (800) 523-1850 or (215) 386-0100. Weekly. $275.00 per year.

SCIENCE CITATION INDEX. Institute for Scientific Information, 3501 Market Street, Philadelphia, PA 19104. (800) 523-1850 or (215) 386-0100. Six times per year. $6200. 00 per year.

ASSOCIATIONS AND PROFESSIONAL SOCIETIES

AMERICAN ASTRONOMICAL SOCIETY. 1816 Jefferson Place, N.W., Washington, DC 20036. (202) 659-0134.

AMERICAN GEOLOGICAL INSTITUTE. 4220 King Street, Alexandria, VA 22302. (703) 379-2480.

AMERICAN METEOR SOCIETY. Department of Physics and Astronomy, State University College, Geneseo, NY 14454. (716) 245-5284.

ASTRONOMICAL SOCIETY OF THE PACIFIC. 1290 24th Avenue, San Francisco, CA 94122. (415) 661-8660.

GEOLOGICAL SOCIETY OF AMERICA. 3300 Penrose Place, Boulder, CO 80301. (303) 447-2020.

METEORITICAL SOCIETY. c/o Donald D. Bogard, SN 4, Geochemistry, NASA Johnson Space Center, Houston, TX 77058. (713) 483-2296.

DIRECTORIES AND BIOGRAPHICAL SOURCES

AMERICAN MEN AND WOMEN OF SCIENCE. R.R. Bowker, Inc., Order Department, 245 West 17th Street, New York, NY 10011. (800) 521-8110. Eight volumes. 1986. $595.00 for set.

INTERNATIONAL RESEARCH CENTERS DIRECTORY 1988-89. Darren L. Smith, editor. Gale Research Company, Book Tower, Detroit, MI 48226. (800) 521-0707. 4th edition. 1987. $360.00.

RESEARCH CENTERS DIRECTORY 1988. Gale Research Company, Book Tower, Detroit, MI 48226. (800) 521-0707. 12th edition. 1987. $365.00 for set.

SCIENTIFIC AND TECHNICAL ORGANIZATIONS AND AGENCIES DIRECTORY. Margaret Labash Young, editor. Gale Research Company, Book Tower, Detroit, MI 48226. (800) 521-0707. 2nd edition. 1987. $185.00.

GENERAL WORKS

CATALOGUE OF METEORITES. A.L. Graham and others, editors. University of Arizona Press, 1615 East Speedway, Tucson, AZ 85719. (602) 621-1441. 1985. $50.00.

COMETS, METEORS AND ASTEROIDS: HOW THEY AFFECT THE EARTH. Stan Gibilisco. TAB Books, Monterey Lane, Blue Ridge Summit, PA 17214. (717) 794-2191. 1985. $12.95 in paper.

METEORITES: THEIR RECORD ON EARLY SOLAR SYSTEM HISTORY. John T. Wasson. W.H. Freeman and Company, 41 Madison Avenue, New York, NY 10010. (212) 532-7660. 1985. $29.95.

TEKTITES. Virgil Barnes and Mildred Barnes, editors. Van Nostrand Reinhold Company, Inc., 135 West 50th Street, New York, NY 10020. (800) 543-2681. 1973. $57.95.

HANDBOOKS AND MANUALS

HANDBOOK OF ELEMENTAL ABUNDANCES IN METEORITES: REVIEWS IN COSMOCHEMISTRY AND ALLIED SUBJECTS. Brian Mason. Gordon and Breach Science Publishers, Inc., 50 West 23rd Street, New York, NY 10010. (212) 206-8900. 1971. $149.50.

ONLINE DATA BASES

CA SEARCH. Chemical Abstracts Service, P.O. Box 3012, Columbus, OH 43120. (800) 848-6538 or (614) 421-3600. Comprehensive guide to chemical literature, 1972 to present. Inquire as to online cost and availability.

GEOREF. Online version of the BIBLIOGRAPHY AND INDEX OF GEOLOGY. American Geological Institute, 4220 King Street, Alexandria, VA 22302. (703) 379-2480. 1969 to present. Inquire as to online cost and availability.

NTIS. National Technical Information Service, 5285 Port Royal Road, Springfield, VA 22161. (703) 487-4630. Broad coverage of government sponsored research reports, 1964 to present. Inquire as to online cost and availability.

SCISEARCH. Institute for Scientific Information, 3501 Market Street, Philadelphia, PA 19104. (800) 523-1850 or (215) 386-0100. Broad multidisciplinary title and author index to the international literature of science and technology, 1974 to present. Inquire as to online cost and availability.

PERIODICALS

ASTRONOMY. AstroMedia Corporation, 625 E Street, Box 92788, Milwaukee, WI 53202. (414) 276-2689. 1973 to present. Monthly. $21.00 per year.

GEOCHEMICA ET COSMOCHIMICA ACTA. Pergamon Press, Inc., Maxwell House, Fairview Park, Elmsford, NY 10523. (914) 592-7700. 1950 to present. Monthly. $340.00 per year.

MERCURY. Astronomical Society of the Pacific, 1290 24th Avenue, San Francisco, CA 94122. (415) 661-8660. 1972 to present. Bimonthly. $21.00 per year.

METEORITICS. Center for Meteorite Studies, Arizona State University, Tempe, AZ 85287. (602) 965-3576. 1955 to present. Quarterly. $40.00 per year.

RESEARCH CENTERS AND INSTITUTES

CENTER FOR METEORITE STUDIES. Arizona State University, Tempe, AZ 85287. (602) 965-3576.

TELECOMMUNICATIONS

See also: ARTIFICIAL SATELLITES, RADIO, REMOTE SENSING, TELEVISION

ABSTRACT SERVICES AND INDEXES

APPLIED SCIENCE AND TECHNOLOGY INDEX. H.W. Wilson and Company, 950 University Avenue, Bronx, NY 10452. (800) 367-6670 or (212) 588-8400. Monthly. Inquire as to cost and availability.

CURRENT CONTENTS: ENGINEERING, TECHNOLOGY AND APPLIED SCIENCES. Institute for Scientific Information, 3501 Market Street, Philadelphia, PA 19104. (800) 523-1850 or (215) 386-0100. Weekly. $275.00 per year.

ELECTRICAL AND ELECTRONICS ABSTRACTS. Institution of Electrical Engineers. Available from: Institute of Electrical and Electronics Engineers. IEEE Service Center, 445 Hoes Lane, Piscataway, NJ 08854. Monthly. $1250.00 per year.

ELECTRONICS AND COMMUNICATIONS ABSTRACTS. Cambridge Scientific Abstracts, 5161 River Road, Bethesda, MD 20816. (301) 951-1400. Bimonthly. Inquire as to cost and availability.

ENGINEERING INDEX MONTHLY AND AUTHOR INDEX. Engineering Information Inc., 345 East 47th Street, New York, NY 10017. (212) 705-7600. Monthly. $1560.00 per year.

IEEE PUBLICATIONS BULLETIN. Institute of Electrical and Electronics Engineers. Institute of Electrical and Electronics Engineers. IEEE Service Center, 445 Hoes Lane, Piscataway, NJ 08854. Quarterly. Free.

INDEX TO SCIENTIFIC AND TECHNICAL PROCEEDINGS. Institute for Scientific Information, 3501 Market Street, Philadelphia, PA 19104. (800) 523-1850 or (215) 386-0100. 1978 to present. Monthly. $775.00 per year.

INDEX TO SCIENTIFIC REVIEWS. Institute for Scientific Information, 3501 Market Street, Philadelphia, PA 19104. (800) 523-1850 or (215) 386-0100. 1974 to present. Semi-annual. $550.00 per year.

METEOROLOGICAL AND GEOASTROPHYSICAL ABSTRACTS. American Meteorological Society, 45 Beacon Street, Boston, MA 02108. (617) 227-2425. 1950 to present. Monthly. $450.00 per year.

OCEAN ABSTRACTS. Cambridge Scientific Abstracts, 5161 River Road, Bethesda, MD 20816. (301) 951-1400. 1963 to present. Bimonthly. $450.00 per year.

PHYSICS ABSTRACTS. Institution of Electrical Engineers. Available from: IEEE Service Center, 445 Hoes Lane, Piscataway, NJ 08854. 1898 to present. Bimonthly. $1700.00 per year.

PHYSICS BRIEFS. Physik Verlag GmbH, Postfach 1260/1280, D-6940 Weinheim, West Germany. (212) 661-9404. 1920 to present. Twenty-six times per year. $1250.00 per year.

SCIENCE CITATION INDEX. Institute for Scientific Information, 3501 Market Street, Philadelphia, PA 19104. (800) 523-1850 or (215) 386-0100. Six times per year. $6200.00 per year.

ASSOCIATIONS AND PROFESSIONAL SOCIETIES

AMERICAN ELECTRONICS ASSOCIATION. P.O. Box 10045, 2670 Hanover Street, Palo Alto, CA 94303. (415) 857-9300.

AMERICAN INSTITUTE OF PHYSICS. 335 East 45th Street, New York, NY 10017. (212) 661-9494.

ELECTRONICS INDUSTRIES ASSOCIATION. 2001 Eye Street, N.W., Washington, DC 20006. (202) 457-4900.

INSTITUTE OF ELECTRICAL AND ELECTRONICS ENGINEERS. 345 East 47th Street, New York, NY 10017. (212) 705-7900.

INTERNATIONAL TELECOMMUNICATIONS SATELLITE ORGANIZATION. 3400 Winternational Drive, N.W., Washington, DC 20008. (202) 944-6800.

NATIONAL ASSOCIATION OF RADIO AND TELECOMMUNICATIONS ENGINEERS. P.O. Box 15029, Salem, OR 97309. (503) 581-7653.

NATIONAL ENVIRONMENTAL SATELLITE, DATA, AND INFORMATION SERVICE. 3300 Whitehaven Street, N.W., Washington, DC 20235. (202) 634-7318.

RADIO AMATEUR SATELLITE CORPORATION. P.O. Box 27, Washington, DC 20044. (301) 589-6062.

SOCIETY OF CABLE TELEVISION ENGINEERS. P.O. Box 2389, West Chester, PA 19380. (215) 363-6888.

SOCIETY OF MOTION PICTURE AND TELEVISION ENGINEERS. 595 West Hartsdale Avenue, White Plains, NY 10607. (914) 472-6606.

DIRECTORIES AND BIOGRAPHICAL SOURCES

BROADCAST ENGINEERING BUYERS GUIDE/SPEC BOOK ISSUE. Intertec Publishing Corporation, Box 12901, Overland Park, KS 66212. (913) 888-4664. Annual. $20.00 per year.

IEEE MEMBERSHIP DIRECTORY. Institute of Electrical and Electronics Engineers. IEEE Service Center, 445 Hoes Lane, Piscataway, NJ 08854. Annual. $7.00.

INTERNATIONAL RESEARCH CENTERS DIRECTORY 1988-89. Darren L. Smith, editor. Gale Research Company, Book Tower, Detroit, MI 48226. (800) 521-0707. 4th edition. 1987. $360.00.

INTERNATIONAL SATELLITE DIRECTORY. S.F.P. Designs, Inc., 369 Redwood Avenue, Corte Madera, CA 96925. (415) 927-0379.

1987 DIRECTORY OF ENGINEERING SOCIETIES AND RELATED ORGANIZATIONS. Gordon Davis, editor. Hemisphere Publishing Corporation, 1010 Vermont Avenue, NW, Washington, DC 20005. (800) 526-0275. 12th edition. 1987. $100.00.

RESEARCH CENTERS DIRECTORY 1988. Gale Research Company, Book Tower, Detroit, MI 48226. (800) 521-0707. 12th edition. 1987. $365.00 for set.

SATELLITE COMMUNICATIONS, SATELLITE INDUSTRY DIRECTORY ISSUE. Cardiff Publishing Company, 6530 South Yosemite Street, Englewood, CO 80111. (303) 694-1522.

SATELLITE DIRECTORY. Phillips Publishing, Inc., 7811 Montrose Road, Potomac, MD 20854. (301) 340-2100.

SCIENTIFIC AND TECHNICAL ORGANIZATIONS AND AGENCIES DIRECTORY. Margaret Labash Young, editor. Gale Research Company, Book Tower, Detroit, MI 48226. (800) 521-0707. 2nd edition. 1987. $185.00.

WHO'S WHO IN ENGINEERING. Gordon Davis, editor. Hemisphere Publishing Corporation, 1010 Vermont Avenue, NW, Washington, DC 20005. (800) 526-0275. 6th edition. 1985. $200.00.

GENERAL WORKS

INTRODUCTION TO SATELLITE COMMUNICATIONS. G.B. Bleazard. John Wiley and Sons, Inc., 605 Third Avenue, New York, NY 10158. (800) 526-5368. 1985. $39.95.

TELECOMMUNICATIONS ENGINEERING. J. Dunlop and D.G. Smith. Van Nostrand Reinhold Company, Inc., 135 West 50th Street, New York, NY 10020. (800) 543-2681. 1984. $37.95.

HANDBOOKS AND MANUALS

REFERENCE MANUAL FOR TELECOMMUNICATIONS. R.L. Freeman. John Wiley and Sons, Inc., 605 Third Avenue, New York, NY 10158. (800) 526-5368. 1985. $85.00.

WORLDWIDE TELECOMMUNICATIONS GUIDE FOR THE BUSINESS MANAGER. W.L. Vignault. John Wiley and Sons, Inc., 605 Third Avenue, New York, NY 10158. (800) 526-5368. 1987. $52.00.

ONLINE DATA BASES

COMPENDEX. Engineering Information, Inc., 345 East 47th Street, New York, NY 10017. (800) 221-1044 or (212) 705-7615. Engineering and technical literature, 1975 to present. Inquire as to online cost and availability.

DISSERTATION ABSTRACTS ONLINE. University Microfilms International, 300 North Zeeb Road, Ann Arbor, MI 48106. (800) 521-0600 or (313) 761-4700. Scope includes virtually all doctoral dissertations accepted at accredited American institutions from 1861 to present in over 250 subject areas. Inquire as to online cost and availability.

INSPEC. INSPEC Marketing Department, Institution of Electrical Engineers. Available from IEEE Service Center, 445 Hoes Lane, Piscataway, NJ 08854. (201) 981-0060. Online version of Physics Abstracts. Inquire as to online cost and availability.

NTIS. National Technical Information Service, 5285 Port Royal Road, Springfield, VA 22161. (703) 487-4630. Broad coverage of government sponsored research reports, 1964 to present. Inquire as to online cost and availability.

SCISEARCH. Institute for Scientific Information, 3501 Market Street, Philadelphia, PA 19104. (800) 523-1850 or (215) 386-0100. Broad multidisciplinary title and author index to the international literature of science and technology, 1974 to present. Inquire as to online cost and availability.

WILSONLINE. H.W. Wilson and Company, 950 University Avenue, Bronx, NY 10452. (800) 367-6770 or (212) 588-8400. Makes available online versions of the H.W. Wilson indexes including Applied Science and Technology Index, Business Periodicals Index and Readers' Guide to Periodical Literature.

Approximately 1980 to present. Inquire as to online cost and availability.

PERIODICALS

IEEE COMMUNICATIONS MAGAZINE. Institute of Electrical and Electronics Engineers. IEEE Service Center, 445 Hoes Lane, Piscataway, NJ 08854. 1953 to present. Monthly. $10.00 per year.

IEEE TRANSACTIONS IN COMMUNICATIONS TECHNOLOGY. Institute of Electrical and Electronics Engineers. IEEE Service Center, 445 Hoes Lane, Piscataway, NJ 08854. 1953 to present. Monthly. $115.00 per year.

IEEE TRANSACTIONS ON GEOSCIENCE AND REMOTE SENSING. IEEE Geoscience and Remote Sensing Society. Institute of Electrical and Electronics Engineers, 345 East 47th Street, New York, NY 10017. (212) 705-7900. Order from: IEEE Service Center, 445 Hoes Lane, Piscataway, NJ 08854. 1963 to present. Bimonthly. $110.00 per year.

PHOTOGRAMMETRIC ENGINEERING AND REMOTE SENSING. American Society for Photogrammetry and Remote Sensing. 210 Little Falls Street, Falls Church, VA 22046-4398. (703) 534-6617. Order from: Allen Press, Inc., 1041 New Hampshire Street, Box 368, Lawrence, KS 66044. 1934 to present. Monthly. $80.00 per year.

REMOTE SENSING OF ENVIRONMENT. Elsevier Science Publishing Company, Inc., 52 Vanderbilt Avenue, New York, NY 10017. (212) 370-5520. 1968 to present. Six times per year. $210.00 per year.

REMOTE SENSING REVIEWS. Harwood Academic Publishers, 50 West 23rd Street, New York, NY 10010. (212) 206-8900. Quarterly. $160.00 per year.

SATELLITE COMMUNICATIONS. Cardiff Publishing Company, 6530 South Yosemite, Englewood, CO 80111. (303) 694-1522. 1977 to present. Monthly. $27.00 per year.

RESEARCH CENTERS AND INSTITUTES

CENTER FOR ADVANCED TECHNOLOGY IN TELECOMMUNICATIONS. Polytechnic University, 333 Jay Street, Brooklyn, NY 11201. (718) 643-5160.

CENTER FOR SPACE RESEARCH. Massachusetts Institute of Technology, 77 Massachusetts Avenue, Cambridge, MA 02139. (617) 253-7501.

TELECOMMUNICATIONS AND INFORMATION SYSTEMS LABORATORY. University of Kansas, Nichols Hall, Lawrence, KS 66045. (913) 864-4832.

TELEGRAPHY

See: TELECOMMUNICATIONS

TELEPHONES

Se: TELECOMMUNICATIONS

TELESCOPES

See also: ASTRONOMY, ASTROPHYSICS, OBSERVATORIES, OPTICS, RADIO ASTRONOMY

ABSTRACT SERVICES AND INDEXES

ASTRONOMY AND ASTROPHYSICS ABSTRACTS. Springer-Verlag New York, Inc., 175 Fifth Avenue, New York, NY 10010. (800) 526-7254. 1969 to present. Approximately $70.00 per year.

ASTRONOMY AND ASTROPHYSICS MONTHLY INDEX. Olivetree Associates, P.O. Box 236, Sierre Madre, CA 91024. $220.00 per year. Complementary copies available on request.

CURRENT CONTENTS: PHYSICAL, CHEMICAL, AND EARTH SCIENCES. Institute for Scientific Information, 3501 Market Street, Philadelphia, PA 19104. (800) 523-1850 or (215) 386-0100. Weekly. $275.00 per year.

PHYSICS ABSTRACTS. Institution of Electrical Engineers. Available from: IEEE Service Center, 445 Hoes Lane, Piscataway, NJ 08854. 1898 to present. Bimonthly. $1700.00 per year.

PHYSICS BRIEFS. Physik Verlag GmbH, Postfach 1260/1280, D-6940 Weinheim, West Germany. (212) 661-9404. 1920 to present. Twenty-six times per year. $1250.00 per year.

SCIENCE CITATION INDEX. Institute for Scientific Information, 3501 Market Street, Philadelphia, PA 19104. (800) 523-1850 or (215) 386-0100. Six times per year. $6200.00 per year.

STAR. (SCIENTIFIC AND TECHNICAL AEROSPACE REPORTS). U.S. National Aeronautics and Space Administration, Scientific and Technical Information Facility, Box 8757, Baltimore-Washington International Airport, MD 21240. (202) 755-2210. Semimonthly, with semiannual and annual indexes. $85.00 per year.

ANNUAL REVIEWS AND YEARBOOKS

ANNUAL REVIEW OF ASTRONOMY AND ASTROPHYSICS. Annual Reviews, Inc., 4139 El Camino Way, Palo Alto, CA 94306. (415) 493-4400. Annual. Inquire as to cost and availability.

ASSOCIATIONS AND PROFESSIONAL SOCIETIES

AMERICAN ASTRONOMICAL SOCIETY. 1816 Jefferson Place, N.W., Washington, DC 20036. (202) 659-0134.

ASTRONOMICAL SOCIETY OF THE PACIFIC. 1290 24th Avenue, San Francisco, CA 94122. (415) 661-8660.

INSTITUTION OF ELECTRICAL AND ELECTRONICS ENGINEERS (IEEE). 345 East 47th Street, New York, NY 10017. (212) 705-7900.

OPTICAL SOCIETY OF AMERICA. 1816 Jefferson Place, N.W., Washington, DC 20036. (202) 223-8130.

SOCIETY OF PHOTO-OPTICAL INSTRUMENTATION ENGINEERS - THE INTERNATIONAL SOCIETY OF OPTICAL ENGINEERING. P.O. Box 10, 1022 19th Street, Bellingham, WA 98227. (206) 676-3290.

BIBLIOGRAPHIES

A BIBLIOGRAPHY OF ASTRONOMY, 1970-1979. R.A. Seal and S.S. Martin. Libraries Unlimited Inc., P.O. Box 263, Littleton, CO 80160. (303) 770-1220. 1982. $37.50.

DIRECTORIES AND BIOGRAPHICAL SOURCES

AMERICAN MEN AND WOMEN OF SCIENCE. R.R. Bowker, Inc., Order Department, 245 West 17th Street, New York, NY 10011. (800) 521-8110. Eight volumes. 1986. $595.00 for set.

THE BIOGRAPHICAL DICTIONARY OF SCIENTISTS: ASTRONOMERS. D. Abbott, editor. Peter Bedrick Books, 125 East 23rd Street, New York, NY 10010. 1984. $28.00.

DIRECTORY OF PHYSICS AND ASTRONOMY STAFF MEMBERS. American Institute of Physics, 335 East 45th Street, New York, NY 10017. (212) 661-9494. Annual.

INTERNATIONAL RESEARCH CENTERS DIRECTORY 1988-89. Darren L. Smith, editor. Gale Research Company, Book Tower, Detroit, MI 48226. (800) 521-0707. 4th edition. 1987. $360.00.

RESEARCH CENTERS DIRECTORY 1988. Gale Research Company, Book Tower, Detroit, MI 48226. (800) 521-0707. 12th edition. 1987. $365.00 for set.

SCIENTIFIC AND TECHNICAL ORGANIZATIONS AND AGENCIES DIRECTORY. Margaret Labash Young, editor. Gale Research Company, Book Tower, Detroit, MI 48226. (800) 521-0707. 2nd edition. 1987. $185.00.

ENCYCLOPEDIAS AND DICTIONARIES

ILLUSTRATED ENCYCLOPEDIA OF THE UNIVERSE: UNDERSTANDING AND EXPLORING THE COSMOS. Richard S. Lewis. Crown Publishers Inc., 1 Park Avenue, New York, NY 10016. (800) 526-4264. 1986. $24.95.

GENERAL WORKS

ASTRONOMICAL OBSERVATORIES: AN OPTICAL PERSPECTIVE. Gordon Walker. Cambridge University Press, 32 East 57th Street, New York, NY 10022. (800) 872-7423. 1987. $80.00.

DEEP-SKY OBSERVING WITH SMALL TELESCOPES. David J. Eicher. Enslow Publications, Inc., Bloy Street and Ramsey Avenue, Box 777, Hillside, NJ 07205. (201) 964-4116. 1988. $18.95.

HOW TO MAKE A TELESCOPE. J. Texereau. Willmann-Bell, Inc., P.O. Box 3125, Richmond, VA 23235. (804) 320-7016. Second edition. 1984. $19.95.

INFINITE VISTAS: NEW TOOLS FOR ASTRONOMY. J. Cornell and J. Carr, editors. Charles Scribner's and Sons, 115 Fifth Avenue, New York, NY 10003. (800) 257-5755. 1985. $18.95.

THE INVISIBLE UNIVERSE. F. Field and E. Chaisson. Birkhauser Boston, Inc., 380 Green Street, Cambridge, MA 02139. (617) 876-2333. 1984. $19.95.

VERY LARGE TELESCOPES, THEIR INSTRUMENTATION AND PROGRAMS. M. Ulrich and K. Kjar, editors. International Astronomical Union Colloquium 79. Kluwer Academic Publishers, 190 Old Derby Street, Hingham, MA 02043. (617) 749-5262. 1984.

ONLINE DATA BASES

DISSERTATION ABSTRACTS ONLINE. University Microfilms International, 300 North Zeeb Road, Ann Arbor, MI 48106. (800) 521-0600 or (313) 761-4700. Scope includes virtually all doctoral dissertations accepted at accredited American institutions from 1861 to present in over 250 subject areas. Inquire as to online cost and availability.

INSPEC. INSPEC Marketing Department, Institution of Electrical Engineers. Available from IEEE Service Center, 445 Hoes Lane, Piscataway, NJ 08854. (201) 981-0060. Online version of Physics Abstracts. Inquire as to online cost and availability.

NTIS. National Technical Information Service, 5285 Port Royal Road, Springfield, VA 22161. (703) 487-4630. Broad coverage of government sponsored research reports, 1964 to present. Inquire as to online cost and availability.

SCISEARCH. Institute for Scientific Information, 3501 Market Street, Philadelphia, PA 19104. (800) 523-1850 or (215) 386-0100. Broad multidisciplinary title and author index to the international literature of science and technology, 1974 to present. Inquire as to online cost and availability.

OTHER SOURCES

ASTRONOMICAL CENTERS OF THE WORLD. K. Krisciunas. Cambridge University Press, 32 East 57th Street, New York, NY 10022. (800) 872-7423. 1987. $45.00.

PERIODICALS

ASTRONOMICAL JOURNAL. American Astronomical Society. Available from: American Institute of Physics, 335 East 45th Street, New York, NY 10017. (212) 661-9494. Monthly. $125.00 per year.

ASTRONOMICAL SOCIETY OF THE PACIFIC PUBLICATIONS. Astronomical Society of the Pacific, 1290 24th Avenue, San Francisco, CA 94122. (415) 661-8660. Monthly. $40.00 per year.

ASTRONOMY. AstroMedia Corporation, 625 E Street, Box 92788, Milwaukee, WI 53202. (414) 276-2689. 1973 to present. Monthly. $21.00 per year.

ASTRONOMY AND ASTROPHYSICS. Springer-Verlag New York, Inc., 175 Fifth Avenue, New York, NY 10010. (800) 526-7254. Monthly. $680.00 per year.

ASTROPHYSICAL JOURNAL. University of Chicago Press, 5801 Ellis Avenue, Chicago, IL 60637. (800) 621-2736. Biweekly. $315.00 per year.

ASTROPHYSICS AND SPACE SCIENCE. D. Reidel Publishing Company, 190 Old Derby Street, Hingham, MA 02043. Monthly. $101.00 per year.

SKY AND TELESCOPE. Sky Publishing Corporation, 49 Bay State Road, Cambridge, MA 02238. (617) 864-7360. Monthly. $18.00 per year.

MERCURY. Astronomical Society of the Pacific, 1290 24th Avenue, San Francisco, CA 94122. (415) 661-8660. 1972 to present. Bimonthly. $21.00 per year.

RESEARCH CENTERS AND INSTITUTES

MOUNT WILSON AND LAS CAMPANAS OBSERVATORIES. 813 Santa Barbara Street, Pasadena, CA 91101. (818) 577-1122.

MULTIPLE MIRROR TELESCOPE OBSERVATORY. University of Arizona, Tucson, AZ 85721. (602) 621-1558.

NATIONAL OPTICAL ASTRONOMY OBSERVATORIES. 950 North Cherry Avenue, Tucson, AZ 85719. (602) 325-9230.

NATIONAL RADIO ASTRONOMY OBSERVATORY. Edgemont Road, Charlottesville, VA 22903. (804) 296-0211.

TELEVISION

See also: ANTENNAS, ELECTRICAL ENGINEERING, ELECTRONIC CIRCUITS AND COMPONENTS, ELECTRONICS, ELECTRONICS ENGINEERING, RADAR

ABSTRACT SERVICES AND INDEXES

APPLIED SCIENCE AND TECHNOLOGY INDEX. H.W. Wilson and Company, 950 University Avenue, Bronx, NY 10452. (800) 367-6670 or (212) 588-8400. Monthly. Inquire as to cost and availability.

ELECTRICAL AND ELECTRONICS ABSTRACTS. Institution of Electrical Engineers. Available from: Institute of Electrical and Electronics Engineers. IEEE Service Center, 445 Hoes Lane, Piscataway, NJ 08854. Monthly. $1250.00 per year.

ELECTRONICS AND COMMUNICATIONS ABSTRACTS. Cambridge Scientific Abstracts, 5161 River Road, Bethesda, MD 20816. (301) 951-1400. Bimonthly. Inquire as to cost and availability.

ENGINEERING INDEX MONTHLY AND AUTHOR INDEX. Engineering Information Inc., 345 East 47th Street, New York, NY 10017. (212) 705-7600. Monthly. $1560.00 per year.

IEEE PUBLICATIONS BULLETIN. Institute of Electrical and Electronics Engineers. Institute of Electrical and Electronics Engineers. IEEE Service Center, 445 Hoes Lane, Piscataway, NJ 08854. Quarterly. Free.

PHYSICS ABSTRACTS. Institution of Electrical Engineers. Available from: IEEE Service Center, 445 Hoes Lane, Piscataway, NJ 08854. 1898 to present. Bimonthly. $1700.00 per year.

PHYSICS BRIEFS. Physik Verlag GmbH, Postfach 1260/1280, D-6940 Weinheim, West Germany. (212) 661-9404. 1920 to present. Twenty-six times per year. $1250.00 per year.

SCIENCE CITATION INDEX. Institute for Scientific Information, 3501 Market Street, Philadelphia, PA 19104. (800) 523-1850 or (215) 386-0100. Six times per year. $6200.00 per year.

ANNUAL REVIEWS AND YEARBOOKS

ADVANCES IN ELECTRONICS AND ELECTRON PHYSICS. Academic Press, Inc., 6277 Sea Harbor Drive, Orlando, FL 32821. (800) 321-5068. Irregular. Approximately $80.00 per volume.

ASSOCIATIONS AND PROFESSIONAL SOCIETIES

AMERICAN ELECTRONICS ASSOCIATION. P.O. Box 10045, 2670 Hanover Street, Palo Alto, CA 94303. (415) 857-9300.

AMERICAN INSTITUTE OF PHYSICS. 335 East 45th Street, New York, NY 10017. (212) 661-9494.

ELECTRONICS INDUSTRIES ASSOCIATION. 2001 Eye Street, N.W., Washington, DC 20006. (202) 457-4900.

INSTITUTE OF ELECTRICAL AND ELECTRONICS ENGINEERS. 345 East 47th Street, New York, NY 10017. (212) 705-7900.

NATIONAL ASSOCIATION OF RADIO AND TELECOMMUNICATIONS ENGINEERS. P.O. Box 15029, Salem, OR 97309. (503) 581-7653.

SOCIETY OF CABLE TELEVISION ENGINEERS. P.O. Box 2389, West Chester, PA 19380. (215) 363-6888.

SOCIETY OF MOTION PICTURE AND TELEVISION ENGINEERS. 595 West Hartsdale Avenue, White Plains, NY 10607. (914) 472-6606.

DIRECTORIES AND BIOGRAPHICAL SOURCES

AMERICAN MEN AND WOMEN OF SCIENCE. R.R. Bowker, Inc., Order Department, 245 West 17th Street, New York, NY 10011. (800) 521-8110. Eight volumes. 1986. $595.00 for set.

BROADCAST ENGINEERING BUYERS GUIDE/SPEC BOOK ISSUE. Intertec Publishing Corporation, Box 12901, Overland Park, KS 66212. (913) 888-4664. Annual. $20.00 per year.

IEEE MEMBERSHIP DIRECTORY. Institute of Electrical and Electronics Engineers. IEEE Service Center, 445 Hoes Lane, Piscataway, NJ 08854. Annual. $7.00.

TELEVISION

INTERNATIONAL RESEARCH CENTERS DIRECTORY 1988-89. Darren L. Smith, editor. Gale Research Company, Book Tower, Detroit, MI 48226. (800) 521-0707. 4th edition. 1987. $360.00.

INTERNATIONAL SATELLITE DIRECTORY. S.F.P. Designs, Inc., 369 Redwood Avenue, Corte Madera, CA 96925. (415) 927-0379.

1987 DIRECTORY OF ENGINEERING SOCIETIES AND RELATED ORGANIZATIONS. Gordon Davis, editor. Hemisphere Publishing Corporation, 1010 Vermont Avenue, NW, Washington, DC 20005. (800) 526-0275. 12th edition. 1987. $100.00.

RESEARCH CENTERS DIRECTORY 1988. Gale Research Company, Book Tower, Detroit, MI 48226. (800) 521-0707. 12th edition. 1987. $365.00 for set.

SATELLITE COMMUNICATIONS, SATELLITE INDUSTRY DIRECTORY ISSUE. Cardiff Publishing Company, 6530 South Yosemite Street, Englewood, CO 80111. (303) 694-1522.

SATELLITE DIRECTORY. Phillips Publishing, Inc., 7811 Montrose Road, Potomac, MD 20854. (301) 340-2100.

SCIENTIFIC AND TECHNICAL ORGANIZATIONS AND AGENCIES DIRECTORY. Margaret Labash Young, editor. Gale Research Company, Book Tower, Detroit, MI 48226. (800) 521-0707. 2nd edition. 1987. $185.00.

WHO'S WHO IN ELECTRONICS. Harris Publishing Company, 2057-2 Aurora Road, Twinsburg, OH 44087. (216) 425-9143. Annual. $90.00.

WHO'S WHO IN ENGINEERING. Gordon Davis, editor. Hemisphere Publishing Corporation, 1010 Vermont Avenue, NW, Washington, DC 20005. (800) 526-0275. 6th edition. 1985. $200.00.

ENCYCLOPEDIAS AND DICTIONARIES

DICTIONARY OF AUDIO, RADIO AND VIDEO. R.S. Roberts. Butterworth's Publishing, 80 Montvale Avenue, Stoneham, MA 02180. (800) 325-4177. 1981. $45.00.

IEEE STANDARD DICTIONARY OF ELECTRICAL AND ELECTRONICS TERMS. Frank Jay, editor. John Wiley and Sons, Inc., 605 Third Avenue, New York, NY 10158. (800) 526-5368. 3rd edition. 1984. $49.95.

GENERAL WORKS

BASIC TELEVISION. Bernard Grob. McGraw-Hill Book Company, 1221 Avenue of the Americas, New York, NY 10020. (212) 512-2000. Fourth edition. 1975. $36.95.

ELECTROMAGNETIC CONCEPTS AND PRINCIPLES. Stanley V. Marshall and Gabriel G. Skitek. Prentice-Hall Publishing, Inc., Englewood Cliffs, NJ 07632. (800) 562-0245. 2nd edition. 1987. $42.95.

ELECTRONIC INVENTIONS AND DISCOVERIES: ELECTRONICS FROM ITS EARLIEST BEGINNINGS TO PRESENT DAY. G.W.A. Dummer. Pergamon Press, Inc., Maxwell House, Fairview Park, Elmsford, NY 10523. (914) 592-7700. 1983. $49.50.

TELEVISION PRODUCTION. Alan Wurtzel. McGraw-Hill Book Company, 1221 Avenue of the Americas, New York, NY 10020. (212) 512-2000. Second edition. 1983. $38.95.

TELEVISION: THEORY AND SERVICING. C. Buscombe. Reston Publishing Company, Inc., c/o Prentice-Hall Publishing, Inc., Englewood Cliffs, NJ 07632. (800) 562-0245. 1984. $34.95.

HANDBOOKS AND MANUALS

ELECTRONIC ENGINEERS HANDBOOK. Donald G. Fink, editor. McGraw-Hill Book Company, 1221 Avenue of the Americas, New York, NY 10020. (212) 512-2000. 2nd edition. 1982. $89.00.

HANDBOOK OF MODERN ELECTRONICS AND ELECTRICAL ENGINEERING. Charles Belove, editor. John Wiley and Sons, Inc., 605 Third Avenue, New York, NY 10158. (800) 526-5368. 1986. $88.95.

ONLINE DATA BASES

CA SEARCH. Chemical Abstracts Service, P.O. Box 3012, Columbus, OH 43120. (800) 848-6538 or (614) 421-3600. Comprehensive guide to chemical literature, 1972 to present. Inquire as to online cost and availability.

COMPENDEX. Engineering Information, Inc., 345 East 47th Street, New York, NY 10017. (800) 221-1044 or (212) 705-7615. Engineering and technical literature, 1975 to present. Inquire as to online cost and availability.

INSPEC. INSPEC Marketing Department, Institution of Electrical Engineers. Available from IEEE Service Center, 445 Hoes Lane, Piscataway, NJ 08854. (201) 981-0060. Online version of Physics Abstracts. Inquire as to online cost and availability.

NTIS. National Technical Information Service, 5285 Port Royal Road, Springfield, VA 22161. (703) 487-4630. Broad coverage of government sponsored research reports, 1964 to present. Inquire as to online cost and availability.

SCISEARCH. Institute for Scientific Information, 3501 Market Street, Philadelphia, PA 19104. (800) 523-1850 or (215) 386-0100. Broad multidisciplinary title and author index to the international literature of science and technology, 1974 to present. Inquire as to online cost and availability.

WILSONLINE. H.W. Wilson and Company, 950 University Avenue, Bronx, NY 10452. (800) 367-6770 or (212) 588-8400. Makes available online versions of the H.W. Wilson indexes including Applied Science and Technology Index, Business Periodicals Index and Readers' Guide to Periodical Literature. Approximately 1980 to present. Inquire as to online cost and availability.

OTHER SOURCES

A GUIDE TO THE LITERATURE OF ELECTRICAL AND ELECTRONICS ENGINEERING. Susan B. Ardis. Libraries Unlimited Inc., P.O. Box 263, Littleton, CO 80160. (303) 770-1220. 1987. $37.50.

PERIODICALS

BROADCASTER ENGINEERING. Intertec Publishing Corporation, Box 12901, Overland Park, KS 66212. (913) 888-4664. 1959 to present. Monthly. $25.00 per year.

ELECTRONIC DESIGN. Hayden Publishing Company, 10 Mulholland Drive, Hasbrouck Heights, NJ 07604. (201) 288-7520. 1952 to present. Biweekly. $40.00 per year.

ELECTRONICS. McGraw-Hill Book Company, 1221 Avenue of the Americas, New York, NY 10020. (212) 512-2000. 1930 to present. Weekly. $32.00 per year.

ELECTRONICS AND WIRELESS WORLD. I.P.C. Electrical-Electronic Press, Ltd., Quadrant House, The Quadrant, Sutton, Surrey, SM2 5AS England. 1911 to present. Monthly. $105.00 per year.

IEEE CIRCUITS AND DEVICES MAGAZINE. Institute of Electrical and Electronics Engineers. IEEE Service Center, 445 Hoes Lane, Piscataway, NJ 08854. Bimonthly. $70.00 per year.

IEEE JOURNAL OF SOLID STATE CIRCUITS. Institute of Electrical and Electronics Engineers. IEEE Service Center, 445 Hoes Lane, Piscataway, NJ 08854. 1966 to present. Bimonthly. $113.00 per year.

IEEE TRANSACTIONS ON BROADCASTING. Institute of Electrical and Electronics Engineers. IEEE Service Center, 445 Hoes Lane, Piscataway, NJ 08854. 1955 to present. Quarterly. $37.00 per year.

TELEVISION BROADCAST. Globecom Publishing Limited, 4451 West 107th, No. 210, Overland Park, KS 66207-4024. 1978 to present. Monthly. $40.00 per year.

RESEARCH CENTERS AND INSTITUTES

CENTER FOR ADVANCED TECHNOLOGY IN TELE-COMMUNICATIONS. Polytechnic University, 333 Jay Street, Brooklyn, NY 11201. (718) 643-5160.

COMMUNICATIONS RESEARCH LABORATORY. McMaster University, 1280 Main Street West, Hamilton, ON, Canada L8S 4K1. (416) 525-9140.

ELECTRONICS RESEARCH LABORATORY. Montana State University, Bozeman, MT 59717. (406) 994-2505.

TEMPERATURE MEASUREMENT

See: THERMOMETERS

TERRAIN SENSING

See: REMOTE SENSING

TERRESTRIAL MAGNETISM

See: GEOPHYSICS

TEXTILES

See also: MATERIALS SCIENCE

ABSTRACT SERVICES AND INDEXES

APPLIED SCIENCE AND TECHNOLOGY INDEX. H.W. Wilson and Company, 950 University Avenue, Bronx, NY 10452. (800) 367-6670 or (212) 588-8400. Monthly. Inquire as to cost and availability.

CHEMICAL ABSTRACTS. American Chemical Society, Chemical Abstracts Service, Box 3012, Columbus, OH 43210. (614) 421-3600. 1907 to present. Weekly. $9500.00 per year.

CURRENT CONTENTS: ENGINEERING, TECHNOLOGY AND APPLIED SCIENCES. Institute for Scientific Information, 3501 Market Street, Philadelphia, PA 19104. (800) 523-1850 or (215) 386-0100. Weekly. $275.00 per year.

ENGINEERING INDEX MONTHLY AND AUTHOR INDEX. Engineering Information Inc., 345 East 47th Street, New York, NY 10017. (212) 705-7600. Monthly. $1560.00 per year.

SCIENCE CITATION INDEX. Institute for Scientific Information, 3501 Market Street, Philadelphia, PA 19104. (800) 523-1850 or (215) 386-0100. Six times per year. $6200. 00 per year.

TEXTILE TECHNOLOGY DIGEST. Institute of Textile Technology, Charlottesville, VA 22902. (804) 296-5511. 1944 to present. Monthly. $300.00 per year.

WORLD TEXTILE ABSTRACTS. Shirley Institute, Didsbury, Manchester, England M20 8RX. 1969 to present. Semi-monthly. $280.00 per year.

ANNUAL REVIEWS AND YEARBOOKS

INTERNATIONAL TEXTILE REVIEW. McGraw-Hill Book Company, 1221 Avenue of the Americas, New York, NY 10020. (212) 512-2000. Annual. $120.00.

ASSOCIATIONS AND PROFESSIONAL SOCIETIES

AMERICAN ASSOCIATION FOR TEXTILE TECHNOLOGY. 1500 Broadway, Suite 1904, New York, NY 10036. (212) 575-8987.

AMERICAN ASSOCIATION OF TEXTILE CHEMISTS AND COLORISTS. P.O. Box 12215, Research Triangle Park, NC 27709. (919) 549-8141.

TEXTILE RESEARCH INSTITUTE. P.O. Box 625, Princeton, NJ 08542. (609) 924-3150.

DIRECTORIES AND BIOGRAPHICAL SOURCES

AMERICAN ASSOCIATION OF TEXTILE CHEMISTS AND COLORISTS MEMBERSHIP DIRECTORY. P.O. Box 12215, Research Triangle Park, NC 27709. (919) 549-8141. Annual. $46.00.

DAVISON'S TEXTILE BLUE BOOK. Davison Publishing Company, Inc., Box 477, Ridgewood, NJ 07451. (201) 445-3135. 1986. $90.00.

INTERNATIONAL RESEARCH CENTERS DIRECTORY 1988-89. Darren L. Smith, editor. Gale Research Company, Book Tower, Detroit, MI 48226. (800) 521-0707. 4th edition. 1987. $360.00.

RESEARCH CENTERS DIRECTORY 1988. Gale Research Company, Book Tower, Detroit, MI 48226. (800) 521-0707. 12th edition. 1987. $365.00 for set.

ENCYCLOPEDIAS AND DICTIONARIES

ENCYCLOPEDIA OF TEXTILES, FIBERS AND NON-WOVEN FABRICS. Martin Grayson, editor. John Wiley and Sons, Inc., 605 Third Avenue, New York, NY 10158. (800) 526-5368. 1984. $64.95.

GENERAL WORKS

INTRODUCTORY TEXTILE SCIENCE. Marjory L. Joseph. Holt, Reinhart and Winston, 383 Madison Avenue, New York, NY 10017. (212) 872-2000. Fourth edition. 1981. $29.95.

TEXTILES: FIBER TO FABRIC. B. Corbman. McGraw-Hill Book Company, 1221 Avenue of the Americas, New York, NY 10020. (212) 512-2000. Sixth edition. 1982. $32.50.

HANDBOOKS AND MANUALS

HANDBOOK OF CHEMICAL SPECIALTIES: TEXTILE FIBER PROCESSING, PREPARATION AND BLEACHING. John E. Nettles. John Wiley and Sons, Inc., 605 Third Avenue, New York, NY 10158. (800) 526-5368. 1983. $70.00.

ONLINE DATA BASES

CA SEARCH. Chemical Abstracts Service, P.O. Box 3012, Columbus, OH 43120. (800) 848-6538 or (614) 421-3600. Comprehensive guide to chemical literature, 1972 to present.

Inquire as to online cost and availability.

COMPENDEX. Engineering Information, Inc., 345 East 47th Street, New York, NY 10017. (800) 221-1044 or (212) 705-7615. Engineering and technical literature, 1975 to present. Inquire as to online cost and availability.

DISSERTATION ABSTRACTS ONLINE. University Microfilms International, 300 North Zeeb Road, Ann Arbor, MI 48106. (800) 521-0600 or (313) 761-4700. Scope includes virtually all doctoral dissertations accepted at accredited American institutions from 1861 to present in over 250 subject areas. Inquire as to online cost and availability.

NTIS. National Technical Information Service, 5285 Port Royal Road, Springfield, VA 22161. (703) 487-4630. Broad coverage of government sponsored research reports, 1964 to present. Inquire as to online cost and availability.

SCISEARCH. Institute for Scientific Information, 3501 Market Street, Philadelphia, PA 19104. (800) 523-1850 or (215) 386-0100. Broad multidisciplinary title and author index to the international literature of science and technology, 1974 to present. Inquire as to online cost and availability.

WILSONLINE. H.W. Wilson and Company, 950 University Avenue, Bronx, NY 10452. (800) 367-6770 or (212) 588-8400. Makes available online versions of the H.W. Wilson indexes including Applied Science and Technology Index, Business Periodicals Index and Readers' Guide to Periodical Literature. Approximately 1980 to present. Inquire as to online cost and availability.

WORLD TEXTILES. Shirley Institute, Didsbury, Manchester, England M20 8RX. Includes United States and British patent information. Inquire as to cost and availability.

PERIODICALS

HIGH PERFORMANCE TEXTILES. Elsevier Science Publishing Company, Inc., 52 Vanderbilt Avenue, New York, NY 10017. (212) 370-5520. 1980 to present. Monthly. $245.00 per year.

JOURNAL OF INDUSTRIAL FABRICS. Industrial Fabrics Association International, 345 Cedar Building, Suite 450, St Paul, MN 55101. (612) 222-2508. 1982 to present. Quarterly. $40.00 per year.

TEXTILE CHEMIST AND COLORIST. P.O. Box 12215, Research Triangle Park, NC 27709. (919) 549-8141. Monthly. $30.00 per year.

TEXTILE INSTITUTE JOURNAL. Textile Institute, 10 Blackfriars Street, Manchester, England M3 5DR. 1910 to present. Bimonthly. $85.00 per year.

TEXTILE RESEARCH JOURNAL. Textile Research Institute, P.O. Box 625, Princeton, NJ 08542. (609) 924-3150. 1930 to present. Monthly. $115.00 per year.

TEXTILE WORLD. McGraw-Hill Book Company, 1221 Avenue of the Americas, New York, NY 10020. (212) 512-2000. 1868 to present. Monthly. $35.00 per year.

RESEARCH CENTERS AND INSTITUTES

INSTITUTE OF TEXTILE TECHNOLOGY. P.O. Box 391, Charlottesville, VA 22902. (804) 296-5511.

TEXTILE RESEARCH INSTITUTE. P.O. Box 625, Princeton, NJ 08542. (609) 924-3150.

TEXTILE RESEARCH PROGRAM. Clemson University, Sirrine Hall, Clemson, SC 29631. (803) 656-3177.

SPECIFICATIONS AND STANDARDS

ASTM PERFORMANCE STANDARDS FOR TEXTILE FABRICS. American Society for Testing and Materials, 1916 Race Street, Philadelphia, PA 19103. (215) 299-5400. 1983. $17.00.

THEORETICAL PHYSICS

See: PHYSICS

THERMOCHEMISTRY

See: PHYSICAL CHEMISTRY

THERMOCOUPLES

See also: ELECTRICAL ENGINEERING, METROLOGY, THERMODYNAMICS

ABSTRACT SERVICES AND INDEXES

APPLIED SCIENCE AND TECHNOLOGY INDEX. H.W. Wilson and Company, 950 University Avenue, Bronx, NY 10452. (800) 367-6670 or (212) 588-8400. Monthly. Inquire as to cost and availability.

CHEMICAL ABSTRACTS. American Chemical Society, Chemical Abstracts Service, Box 3012, Columbus, OH 43210. (614) 421-3600. 1907 to present. Weekly. $9500.00 per year.

ELECTRICAL AND ELECTRONICS ABSTRACTS. Institution of Electrical Engineers. Available from: Institute of Electrical and Electronics Engineers. IEEE Service Center, 445 Hoes Lane, Piscataway, NJ 08854. Monthly. $1250.00 per year.

ENGINEERING INDEX MONTHLY AND AUTHOR INDEX. Engineering Information Inc., 345 East 47th Street, New York, NY 10017. (212) 705-7600. Monthly. $1560.00 per year.

PHYSICS ABSTRACTS. Institution of Electrical Engineers. Available from: IEEE Service Center, 445 Hoes Lane, Piscataway, NJ 08854. 1898 to present. Bimonthly. $1700.00 per year.

PHYSICS BRIEFS. Physik Verlag GmbH, Postfach 1260/1280, D-6940 Weinheim, West Germany. (212) 661-9404. 1920 to present. Twenty-six times per year. $1250.00 per year.

SCIENCE CITATION INDEX. Institute for Scientific Information, 3501 Market Street, Philadelphia, PA 19104. (800) 523-1850 or (215) 386-0100. Six times per year. $6200.00 per year.

ASSOCIATIONS AND PROFESSIONAL SOCIETIES

AMERICAN INSTITUTE OF PHYSICS. 335 East 45th Street, New York, NY 10017. (212) 661-9494.

AMERICAN SOCIETY FOR TESTING AND MATERIALS. 1916 Race Street, Philadelphia, PA 19103. (215) 299-5400.

EDISON ELECTRIC INSTITUTE. 1111 19th Street, N.W., Washington, DC 20036. (202) 828-7400.

INSTITUTE OF ELECTRICAL AND ELECTRONICS ENGINEERS. 345 East 47th Street, New York, NY 10017. (212) 705-7900.

INSTRUMENT SOCIETY OF AMERICA. P.O. Box 12277, 67 Alexander Drive, Research Triangle Park, NC 27709. (919) 549-8411.

DIRECTORIES AND BIOGRAPHICAL SOURCES

1987 DIRECTORY OF ENGINEERING SOCIETIES AND RELATED ORGANIZATIONS. Gordon Davis, editor. Hemisphere Publishing Corporation, 79 Madison Avenue, New York, NY 10016-7892. (800) 821-8312. 12th edition. 1987. $100.00.

RESEARCH CENTERS DIRECTORY 1988. Gale Research Company, Book Tower, Detroit, MI 48226. (800) 521-0707. 12th edition. 1987. $365.00 for set.

SCIENTIFIC AND TECHNICAL ORGANIZATIONS AND AGENCIES DIRECTORY. Margaret Labash Young, editor. Gale Research Company, Book Tower, Detroit, MI 48226. (800) 521-0707. 2nd edition. 1987. $185.00.

WHO'S WHO IN ELECTRONICS. Harris Publishing Company, 2057-2 Aurora Road, Twinsburg, OH 44087. (216) 425-9143. Annual. $90.00.

WHO'S WHO IN ENGINEERING. Gordon Davis, editor. Hemisphere Publishing Corporation, 1010 Vermont Avenue, NW, Washington, DC 20005. (800) 526-0275. 6th edition. 1985. $200.00.

GENERAL WORKS

HIGH TEMPERATURE NON-METALLIC THERMOCOUPLES AND SHEATHS. G.V. Samsonov and P.S. Kislyi. Plenum Publishing Corporation, 233 Spring Street, New York, NY 10013. (800) 221-9369. 1967. $29.50.

HANDBOOKS AND MANUALS

HANDBOOK OF MODERN ELECTRONICS AND ELECTRICAL ENGINEERING. Charles Belove, editor. John Wiley and Sons, Inc., 605 Third Avenue, New York, NY 10158. (800) 526-5368. 1986. $88.95.

MANUAL ON THE USE OF THERMOCOUPLES IN TEMPERATURE MEASUREMENT. American Society for Testing and Materials, 1916 Race Street, Philadelphia, PA 19103. (215) 299-5400. STP series number 470. Third edition. 1981. $24.00.

ONLINE DATA BASES

CA SEARCH. Chemical Abstracts Service, P.O. Box 3012, Columbus, OH 43120. (800) 848-6538 or (614) 421-3600. Comprehensive guide to chemical literature, 1972 to present. Inquire as to online cost and availability.

COMPENDEX. Engineering Information, Inc., 345 East 47th Street, New York, NY 10017. (800) 221-1044 or (212) 705-7615. Engineering and technical literature, 1975 to present. Inquire as to online cost and availability.

NTIS. National Technical Information Service, 5285 Port Royal Road, Springfield, VA 22161. (703) 487-4630. Broad coverage of government sponsored research reports, 1964 to present. Inquire as to online cost and availability.

SCISEARCH. Institute for Scientific Information, 3501 Market Street, Philadelphia, PA 19104. (800) 523-1850 or (215) 386-0100. Broad multidisciplinary title and author index to the international literature of science and technology, 1974 to present. Inquire as to online cost and availability.

WILSONLINE. H.W. Wilson and Company, 950 University Avenue, Bronx, NY 10452. (800) 367-6770 or (212) 588-8400. Makes available online versions of the H.W. Wilson indexes including Applied Science and Technology Index, Business Periodicals Index and Readers' Guide to Periodical Literature. Approximately 1980 to present. Inquire as to online cost and availability.

PERIODICALS

ELECTRICAL WORLD. McGraw-Hill Book Company, 1221 Avenue of the Americas, New York, NY 10020. (212) 512-2000. 1874 to present. Monthly. $11.00 per year.

REVIEW OF SCIENTIFIC INSTRUMENTS. American Institute of Physics, 335 East 45th Street, New York, NY 10017. (212) 661-9494. 1930 to present. Monthly. $300.00 per year.

RESEARCH CENTERS AND INSTITUTES

EDISON ELECTRIC INSTITUTE. 1111 19th Street, N.W., Washington, DC 20036. (202) 778-6778.

ELECTRICAL ENGINEERING RESEARCH LABORATORIES. Purdue University, Electrical Engineering Building, West Lafayette, IN 47907. (317) 494-3536.

SPECIFICATIONS AND STANDARDS

ASTM STANDARDS ON THERMOCOUPLES. American Society for Testing and Materials, 1916 Race Street, Philadelphia, PA 19103. (215) 299-5400. 1980. $14.00.

STANDARDS AND PRACTICES FOR INSTRUMENTATION. Instrument Society of America, P.O. Box 12277, 67 Alexander Drive, Research Triangle Park, NC 27709. (919) 549-8411. 7th edition. 1983. $135.00.

THERMODYNAMICS

See also: CRYOGENICS, FLUID MECHANICS, HEAT TRANSFER, PHYSICS, THERMOCHEMISTRY

ABSTRACT SERVICES AND INDEXES

APPLIED MECHANICS REVIEW. American Society of Mechanical Engineers, 345 East 47th Street, New York, NY 10017. (212) 705-7703. 1948 to present. Monthly. $360.00 per year.

APPLIED SCIENCE AND TECHNOLOGY INDEX. H.W. Wilson and Company, 950 University Avenue, Bronx, NY 10452. (800) 367-6670 or (212) 588-8400. Monthly. Inquire as to cost and availability.

CHEMICAL ABSTRACTS. American Chemical Society, Chemical Abstracts Service, Box 3012, Columbus, OH 43210. (614) 421-3600. 1907 to present. Weekly. $9500.00 per year.

CONFERENCE PAPERS INDEX. Cambridge Scientific Abstracts, 5161 River Road, Bethesda, MD 20816. 1972 to present. Monthly. Inquire as to cost and availability.

CURRENT CONTENTS: ENGINEERING, TECHNOLOGY AND APPLIED SCIENCES. Institute for Scientific Information, 3501 Market Street, Philadelphia, PA 19104. (800) 523-1850 or (215) 386-0100. Weekly. $275.00 per year.

CURRENT CONTENTS: PHYSICAL, CHEMICAL AND EARTH SCIENCES. Institute for Scientific Information, 3501 Market Street, Philadelphia, PA 19104. (800) 523-1850 or (215) 386-0100. Weekly. $275.00 per year.

ENGINEERING INDEX MONTHLY AND AUTHOR INDEX. Engineering Information Inc., 345 East 47th Street, New York, NY 10017. (212) 705-7600. Monthly. $1560.00 per year.

ISMEC BULLETIN (Information Service in Mechanical Engineering). Cambridge Scientific Abstracts, 5161 River Road, Bethesda, MD 20816. (301) 951-1400. 1973 to present. Monthly. $450.00 per year.

INDEX TO SCIENTIFIC AND TECHNICAL PROCEEDINGS. Institute for Scientific Information, 3501 Market Street,

THERMODYNAMICS

Philadelphia, PA 19104. (800) 523-1850 or (215) 386-0100. 1978 to present. Monthly. $775.00 per year.

INDEX TO SCIENTIFIC REVIEWS. Institute for Scientific Information, 3501 Market Street, Philadelphia, PA 19104. (800) 523-1850 or (215) 386-0100. 1974 to present. Semi-annual. $550.00 per year.

PHYSICS ABSTRACTS. Institution of Electrical Engineers. Available from: IEEE Service Center, 445 Hoes Lane, Piscataway, NJ 08854. 1898 to present. Bimonthly. $1700.00 per year.

PHYSICS BRIEFS. Physik Verlag GmbH, Postfach 1260/1280, D-6940 Weinheim, West Germany. (212) 661-9404. 1920 to present. Twenty-six times per year. $1250.00 per year.

SCIENCE CITATION INDEX. Institute for Scientific Information, 3501 Market Street, Philadelphia, PA 19104. (800) 523-1850 or (215) 386-0100. Six times per year. $6200.00 per year.

ASSOCIATIONS AND PROFESSIONAL SOCIETIES

AMERICAN CHEMICAL SOCIETY. 1155 16th Street, N.W., Washington, DC 20036. (800) 424-6747.

AMERICAN INSTITUTE OF CHEMICAL ENGINEERS. 345 East 47th Street, New York, NY 10017. (212) 705-7338.

AMERICAN INSTITUTE OF PHYSICS. 335 East 45th Street, New York, NY 10017. (212) 661-9494.

AMERICAN SOCIETY OF MECHANICAL ENGINEERS. 345 East 47th Street, New York, NY 10017. (212) 705-7703.

DIRECTORIES AND BIOGRAPHICAL SOURCES

AMERICAN MEN AND WOMEN OF SCIENCE. R.R. Bowker, Inc., Order Department, 245 West 17th Street, New York, NY 10011. (800) 521-8110. Eight volumes. 1986. $595.00 for set.

INTERNATIONAL RESEARCH CENTERS DIRECTORY 1988-89. Darren L. Smith, editor. Gale Research Company, Book Tower, Detroit, MI 48226. (800) 521-0707. 4th edition. 1987. $360.00.

1987 DIRECTORY OF ENGINEERING SOCIETIES AND RELATED ORGANIZATIONS. Gordon Davis, editor. Hemisphere Publishing Corporation, 1010 Vermont Avenue, NW, Washington, DC 20005. (800) 526-0275. 12th edition. 1987. $100.00.

RESEARCH CENTERS DIRECTORY 1988. Gale Research Company, Book Tower, Detroit, MI 48226. (800) 521-0707. 12th edition. 1987. $365.00 for set.

WHO'S WHO IN ENGINEERING. Gordon Davis, editor. Hemisphere Publishing Corporation, 1010 Vermont Avenue, NW, Washington, DC 20005. (800) 526-0275. 6th edition. 1985. $200.00.

ENCYCLOPEDIAS AND DICTIONARIES

CONCISE SCIENCE DICTIONARY. Oxford University Press, 200 Madison Avenue, New York, NY 10016. (800) 458-5833. 1987. $9.95 in paper.

DICTIONARY OF THE PHYSICAL SCIENCES: TERMS, FORMULAS, DATA. Cesare Emiliani. Oxford University Press, 200 Madison Avenue, New York, NY 10016. (800) 458-5833. 1987. $19.95 in paper.

GENERAL WORKS

ANALYSIS OF HEAT AND MASS TRANSFER. E.R.G. Eckert and R.M. Drake. Hemisphere Publishing Corporation, 79 Madison Avenue, New York, NY 10016-7892. (800) 821-8312. 1987. $75.00.

APPLIED THERMODYNAMICS FOR ENGINEERING TECHNOLOGIES. T.D. Eastop and A. McConkey. John Wiley and Sons, Inc., 605 Third Avenue, New York, NY 10158. (800) 526-5368. 1986. $29.95.

A COURSE IN THERMODYNAMICS. J. Kestin. Hemisphere Publishing Corporation, 79 Madison Avenue, New York, NY 10016-7892. (800) 821-8312. Two volumes. 1979. $45.00 each.

FUNDAMENTALS OF ENGINEERING THERMODYNAMICS. V. Hubka and others. McGraw-Hill Book Company, 1221 Avenue of the Americas, New York, NY 10020. (212) 512-2000. 1986. $39.95.

HEAT CONDUCTION. S. Kakac. Hemisphere Publishing Corporation, 79 Madison Avenue, New York, NY 10016-7892. (800) 821-8312. Second edition. 1985.

INTRODUCTION TO HEAT TRANSFER. F. Incropera and D. Dewitt. John Wiley and Sons, Inc., 605 Third Avenue, New York, NY 10158. (800) 526-5368. 1985. $39.95.

THERMODYNAMIC PROPERTIES OF INDIVIDUAL SUBSTANCES. V.P. Glushko and L.V. Gurvich, editors. Hemisphere Publishing Corporation, 79 Madison Avenue, New York, NY 10016-7892. (800) 821-8312. Volume 1, parts 1 and 2. 1988. $245.00.

HANDBOOKS AND MANUALS

HANDBOOK OF HEAT TRANSFER FUNDAMENTALS. W.M. Rohsenow and others. McGraw-Hill Book Company, 1221 Avenue of the Americas, New York, NY 10020. (212) 512-2000. Second edition. 1985. $95.00.

ONLINE DATA BASES

CA SEARCH. Chemical Abstracts Service, P.O. Box 3012, Columbus, OH 43120. (800) 848-6538 or (614) 421-3600. Comprehensive guide to chemical literature, 1972 to present. Inquire as to online cost and availability.

COMPENDEX. Engineering Information, Inc., 345 East 47th Street, New York, NY 10017. (800) 221-1044 or (212) 705-7615. Engineering and technical literature, 1975 to present. Inquire as to online cost and availability.

DISSERTATION ABSTRACTS ONLINE. University Microfilms International, 300 North Zeeb Road, Ann Arbor, MI 48106. (800) 521-0600 or (313) 761-4700. Scope includes virtually all doctoral dissertations accepted at accredited American institutions from 1861 to present in over 250 subject areas. Inquire as to online cost and availability.

INSPEC. INSPEC Marketing Department, Institution of Electrical Engineers. Available from IEEE Service Center, 445 Hoes Lane, Piscataway, NJ 08854. (201) 981-0060. Online version of Physics Abstracts. Inquire as to online cost and availability.

NTIS. National Technical Information Service, 5285 Port Royal Road, Springfield, VA 22161. (703) 487-4630. Broad coverage of government sponsored research reports, 1964 to present. Inquire as to online cost and availability.

SCISEARCH. Institute for Scientific Information, 3501 Market Street, Philadelphia, PA 19104. (800) 523-1850 or (215) 386-0100. Broad multidisciplinary title and author index to the international literature of science and technology, 1974 to present. Inquire as to online cost and availability.

WILSONLINE. H.W. Wilson and Company, 950 University Avenue, Bronx, NY 10452. (800) 367-6770 or (212) 588-8400. Makes available online versions of the H.W. Wilson indexes including Applied Science and Technology Index, Business Periodicals Index and Readers' Guide to Periodical Literature. Approximately 1980 to present. Inquire as to online cost and availability.

PERIODICALS

CRYOGENIC INFORMATION REPORT. Technical Economic Associates, Box 1972, Estes Park, CO 80517. (303) 586-5636. 1963 to present. Ten times per year. $85.00 per year.

CRYOGENICS: THE INTERNATIONAL JOURNAL OF LOW TEMPERATURE ENGINEERING AND RESEARCH. Butterworth Publishing, 80 Montvale Avenue, Stoneham, MA 02180. (800) 325-4177. 1960 to present. Monthly. $335.00 per year.

EXPERIMENTAL THERMAL AND FLUID SCIENCE. Elsevier Science Publishing Company, Inc., 52 Vanderbilt Avenue, New York, NY 10017. (212) 370-5520. 1988 to present. Quarterly. $125.00 per year.

HEAT TRANSFER AND FLUID FLOW DIGEST. Hemisphere Publishing Corporation, 79 Madison Avenue, New York, NY 10016-7892. (800) 821-8312. 1968 to present. Monthly. $137.50 per year.

HEAT TRANSFER ENGINEERING. Hemisphere Publishing Corporation, 79 Madison Avenue, New York, NY 10016-7892. (800) 821-8312. 1979 to present. Quarterly. $97.50.

INTERNATIONAL COMMUNICATIONS IN HEAT AND MASS TRANSFER. Pergamon Press, Inc., Maxwell House, Fairview Park, Elmsford, NY 10523. (914) 592-7700. 1974 to present. Bimonthly. $160.00 per year.

INTERNATIONAL JOURNAL OF HEAT AND FLUID FLOW. Butterworth Publishing, 80 Montvale Avenue, Stoneham, MA 02180. (800) 325-4177. 1979 to present. Quarterly. $140.00 per year.

INTERNATIONAL JOURNAL OF HEAT AND MASS TRANSFER. Pergamon Press, Inc., Maxwell House, Fairview Park, Elmsford, NY 10523. (914) 592-7700. 1960 to present. Monthly. $500.00 per year.

INTERNATIONAL JOURNAL OF THERMOPHYSICS. Plenum Publishing Corporation, 233 Spring Street, New York, NY 10013. (800) 221-9369. 1980 to present. Six times per year. $150.00 per year.

JOURNAL OF HEAT TRANSFER. American Society of Mechanical Engineers, 345 East 47th Street, New York, NY 10017. (212) 705-7703. 1970 to present. Quarterly. $100.00 per year.

JOURNAL OF LOW TEMPERATURE PHYSICS. Plenum Publishing Corporation, 233 Spring Street, New York, NY 10013. (800) 221-9369. 1969 to present. 24 times per year. $750.00 per year.

NUMERICAL HEAT TRANSFER: AN INTERNATIONAL JOURNAL OF COMPUTATION AND METHOLOGY. Hemisphere Publishing Corporation, 79 Madison Avenue, New York, NY 10016-7892. (800) 821-8312. 1978 to present. Monthly. $370.00 per year.

RESEARCH CENTERS AND INSTITUTES

CENTER FOR THERMODYNAMICS. Brigham Young University, 226 ESC, Provo, UT 84602. (801) 378-3668.

HEAT TRANSFER LABORATORY. Massachusetts Institute of Technology, 77 Massachusetts Avenue, Cambridge, MA 02139. (716) 253-2248.

THERMODYNAMICS RESEARCH CENTER. Texas A&M University, Texas Engineering Experiment Station, College Station, TX 77843-3111. (409) 845-4940.

THERMOELECTRICITY

See: ELECTRICITY

THERMOMETERS

See: INSTRUMENTATION

THERMONUCLEAR FUSION

See: FUSION

THERMOPLASTICS

See also: MATERIALS SCIENCE, ORGANIC CHEMISTRY, PACKAGING, PLASTICS, POLYMERS, RESINS

ABSTRACT SERVICES AND INDEXES

APPLIED SCIENCE AND TECHNOLOGY INDEX. H.W. Wilson and Company, 950 University Avenue, Bronx, NY 10452. (800) 367-6670 or (212) 588-8400. Monthly. Inquire as to cost and availability.

CHEMICAL ABSTRACTS. American Chemical Society, Chemical Abstracts Service, Box 3012, Columbus, OH 43210. (614) 421-3600. 1907 to present. Weekly. $9500.00 per year.

CURRENT CONTENTS: ENGINEERING, TECHNOLOGY AND APPLIED SCIENCES. Institute for Scientific Information, 3501 Market Street, Philadelphia, PA 19104. (800) 523-1850 or (215) 386-0100. Weekly. $275.00 per year.

ENGINEERING INDEX MONTHLY AND AUTHOR INDEX. Engineering Information Inc., 345 East 47th Street, New York, NY 10017. (212) 705-7600. Monthly. $1560.00 per year.

SCIENCE CITATION INDEX. Institute for Scientific Information, 3501 Market Street, Philadelphia, PA 19104. (800) 523-1850 or (215) 386-0100. Six times per year. $6200.00 per year.

ASSOCIATIONS AND PROFESSIONAL SOCIETIES

AMERICAN CHEMICAL SOCIETY. 1155 16th Street, N.W., Washington, DC 20036. (800) 424-6747.

PACKAGING INSTITUTE. 20 East 46th Street, New York, NY 10017. (212) 687-8874.

PLASTICS INSTITUTE OF AMERICA. Stevens Institute of Technology, Castle Point Station, Hoboken, NJ 07030. (201) 420-5553.

SOCIETY OF PLASTICS ENGINEERS, INC. 14 Fairfield Drive, Brookfield Center, CT 06805. (203) 775-0471.

SOCIETY OF THE PLASTICS INDUSTRY. 355 Lexington Avenue, New York, NY 10017. (212) 503-0600.

DIRECTORIES AND BIOGRAPHICAL SOURCES

AMERICAN MEN AND WOMEN OF SCIENCE. R.R. Bowker, Inc., Order Department, 245 West 17th Street, New York, NY 10011. (800) 521-8110. Eight volumes. 1986. $595.00 for set.

INTERNATIONAL RESEARCH CENTERS DIRECTORY 1988-89. Darren L. Smith, editor. Gale Research Company, Book Tower, Detroit, MI 48226. (800) 521-0707. 4th edition. 1987. $360.00.

1987 DIRECTORY OF ENGINEERING SOCIETIES AND RELATED ORGANIZATIONS. Gordon Davis, editor. Hemisphere Publishing Corporation, 1010 Vermont Avenue, NW, Washington, DC 20005. (800) 526-0275. 12th edition. 1987. $100.00.

RESEARCH CENTERS DIRECTORY 1988. Gale Research Company, Book Tower, Detroit, MI 48226. (800) 521-0707. 12th edition. 1987. $365.00 for set.

SCIENTIFIC AND TECHNICAL ORGANIZATIONS AND AGENCIES DIRECTORY. Margaret Labash Young, editor. Gale Research Company, Book Tower, Detroit, MI 48226. (800) 521-0707. 2nd edition. 1987. $185.00.

SOCIETY OF THE PLASTICS INDUSTRY - MEMBERSHIP DIRECTORY AND BUYERS GUIDE. 355 Lexington Avenue, New York, NY 10017. (212) 503-0600. Annual. $90.00.

WHO'S WHO IN ENGINEERING. Gordon Davis, editor. Hemisphere Publishing Corporation, 79 Madison Avenue, New York, NY 10016-7892. (800) 821-8312. Sixth edition. 1985. $200.00.

WHO'S WHO IN TECHNOLOGY. Research Publications, 12 Lunar Drive, Woodbridge, CT 06525. (203) 397-2600. 1986. Seven volume set, $545.00.

ENCYCLOPEDIAS AND DICTIONARIES

ENCYCLOPEDIA OF POLYMER SCIENCE AND ENGINEERING. John Wiley and Sons, Inc., 605 Third Avenue, New York, NY 10158. (800) 526-5368. Second edition. 1985 to present. $3500.00 for set.

THESAURUS OF SCIENTIFIC, TECHNICAL, AND ENGINEERING TERMS. Hemisphere Publishing Corporation, 1010 Vermont Avenue, NW, Washington, DC 20005. (800) 526-0275. 1988. $125.00.

GENERAL WORKS

ENGINEERING THERMOPLASTICS. James M. Margolis. Marcel Dekker Inc., 270 Madison Avenue, New York, NY 10016. (800) 228-1160. 1985. $65.00.

FUTURE TRENDS IN POLYMER SCIENCE AND TECHNOLOGY. Ezio Martuscelli and others. Technomic Publishing Company, Inc., 851 New Holland Avenue, Box 3535, Lancaster, PA 17604. (717) 291-5609. 1987. $49.00.

MECHANICAL PROPERTIES OF REINFORCED PLASTICS. D.W. Clegg and A.A. Collyer, editors. Elsevier Science Publishing Company, Inc., 52 Vanderbilt Avenue, New York, NY 10017. (212) 370-5520. 1986. $72.00.

PLASTICS MATERIALS AND PROCESSING. S. Schwartz and S. Goodman. Van Nostrand Reinhold Company, Inc., 135 West 50th Street, New York, NY 10020. (800) 543-2681. 1982. $100.00.

THERMOPLASTICS: MATERIALS ENGINEERING. L. Mascia. Elsevier Science Publishing Company, Inc., 52 Vanderbilt Avenue, New York, NY 10017. (212) 370-5520. 1983. $85.00.

HANDBOOKS AND MANUALS

HANDBOOK OF PACKAGING ENGINEERING. J.F. Hanlon. John Wiley and Sons, Inc., 605 Third Avenue, New York, NY 10158. (800) 526-5368. Second edition. 1983. $65.00.

PLASTICS ENGINEERING HANDBOOK OF THE SOCIETY OF THE PLASTICS INDUSTRY, INC. Joel Frados, editor. Van Nostrand Reinhold Company, Inc., 135 West 50th Street, New York, NY 10020. (800) 543-2681. 1985. $54.95.

PLASTICS MOLD ENGINEERING HANDBOOK. J. Harry DuBois and W. I. Pribble, editors. Van Nostrand Reinhold Company, Inc., 135 West 50th Street, New York, NY 10020. (800) 543-2681. Second edition. 1987. $59.95.

PLASTICS TECHNOLOGY HANDBOOK. M. Chanda and S.K. Roy. Marcel Dekker Inc., 270 Madison Avenue, New York, NY 10016. (800) 228-1160. 1987. $99.75.

ONLINE DATA BASES

CA SEARCH. Chemical Abstracts Service, P.O. Box 3012, Columbus, OH 43120. (800) 848-6538 or (614) 421-3600. Comprehensive guide to chemical literature, 1972 to present. Inquire as to online cost and availability.

COMPENDEX. Engineering Information, Inc., 345 East 47th Street, New York, NY 10017. (800) 221-1044 or (212) 705-7615. Engineering and technical literature, 1975 to present. Inquire as to online cost and availability.

NTIS. National Technical Information Service, 5285 Port Royal Road, Springfield, VA 22161. (703) 487-4630. Broad coverage of government sponsored research reports, 1964 to present. Inquire as to online cost and availability.

SCISEARCH. Institute for Scientific Information, 3501 Market Street, Philadelphia, PA 19104. (800) 523-1850 or (215) 386-0100. Broad multidisciplinary title and author index to the international literature of science and technology, 1974 to present. Inquire as to online cost and availability.

WILSONLINE. H.W. Wilson and Company, 950 University Avenue, Bronx, NY 10452. (800) 367-6770 or (212) 588-8400. Makes available online versions of the H.W. Wilson indexes including Applied Science and Technology Index, Business Periodicals Index and Readers' Guide to Periodical Literature. Approximately 1980 to present. Inquire as to online cost and availability.

OTHER SOURCES

WHAT EVERY ENGINEER SHOULD KNOW ABOUT PLASTICS. Marcel Dekker Inc., 270 Madison Avenue, New York, NY 10016. (800) 228-1160. 1986. $24.95.

PERIODICALS

ADVANCES IN POLYMER TECHNOLOGY. John Wiley and Sons, Inc., 605 Third Avenue, New York, NY 10158. (800) 526-5368. 1977 to present. Quarterly. $100.00 per year.

CANADIAN PLASTICS. Southam Communications, Limited, 1450 Don Mills Road, Don Mills, ON, M3B 2X7, Canada. 1943 to present. Nine times per year. $45.00 per year.

HIGH PERFORMANCE PLASTICS. Elsevier Science Publishing Company, Inc., 52 Vanderbilt Avenue, New York, NY 10017. (212) 370-5520. 1983 to present. Monthly. $135.00 per year.

INTERNATIONAL JOURNAL OF ADHESION AND ADHESIVES. Butterworth's Publishing, 80 Montvale Avenue, Stoneham, MA 02180. (800) 325-4177. 1980 to present. Quarterly. $155.00 per year.

JOURNAL OF CELLULAR PLASTICS. Technomic Publishing Company, Inc., 851 New Holland Avenue, Box 3535, Lancaster, PA 17604. (717) 291-5609. 1965 to present. Bimonthly. $70.00 per year.

JOURNAL OF ELASTOMERS AND PLASTICS. Technomic Publishing Company, Inc., 851 New Holland Avenue, Box 3535, Lancaster, PA 17604. (717) 291-5609. 1969 to present. Quarterly. $115.00 per year.

JOURNAL OF PLASTIC FILM AND SHEETING. Technomic Publishing Company, Inc., 851 New Holland Avenue, Box 3535, Lancaster, PA 17604. (717) 291-5609. 1985 to present. Quarterly. $125.00 per year.

MODERN PLASTICS. McGraw-Hill Book Company, 1221 Avenue of the Americas, New York, NY 10020. (212) 512-2000. 1925 to present. Monthly. $30.00 per year.

PLASTICS COMPOUNDING: FOR RESIN PRODUCERS, FORMULATORS AND COMPOUNDERS. Harcourt, Brace and Jovanovich Publications, 262 Main Street, Chatham, NJ 07928.

(201) 635-1671. 1978 to present. Bimonthly. $21.00 per year.

PLASTICS DESIGN FORUM. Harcourt, Brace and Jovanovich Publications, 262 Main Street, Chatham, NJ 07928. (201) 635-1671. 1976 to present. Bimonthly. $20.00 per year.

PLASTICS ENGINEERING. Society of Plastics Engineers, Inc., 14 Fairfield Drive, Brookfield Center, CT 06805. (203) 775-0471. 1945 to present. Monthly. $30.00 per year.

PLASTICS TECHNOLOGY. Bill Communications, Inc., 633 Third Avenue, New York, NY 10017. (212) 986-4800. 1955 to present. 13 times per year. $36.00 per year.

PLASTICS WORLD. Cahners Publishing Company, Inc., 221 Columbus Avenue, Boston, MA 02116. (617) 536-7780. 1942 to present. Monthly. $45.00 per year.

POLYMER COMPOSITES. Society of Plastics Engineers, Inc., 14 Fairfield Drive, Brookfield Center, CT 06805. (203) 775-0471. 1980 to present. Six times per year. $100.00 per year.

RESEARCH CENTERS AND INSTITUTES

PLASTICS INSTITUTE OF AMERICA. Stevens Institute of Technology, Castle Point Station, Hoboken, NJ 07030. (201) 420-5553.

POLYMER RESEARCH CENTER. University of Cincinnati, Mail Location 172, Cincinnati, OH 45221. (513) 475-2453.

POLYMER RESEARCH LABORATORY. University of Michigan, Dow Building, Ann Arbor, MI 48109. (313) 763-2240.

THIN FILMS

See: MICROELECTRONICS

THUNDERSTORMS

See also: METEOROLOGY

ABSTRACT SERVICES AND INDEXES

GENERAL SCIENCE INDEX. H.W. Wilson Company, 950 University Avenue, Bronx, NY 10452. (800) 367-6770 or (212) 588-8400. Inquire as to cost and availability.

METEOROLOGICAL AND GEOASTROPHYSICAL ABSTRACTS. American Meteorological Society, 45 Beacon Street, Boston, MA 02108. (617) 227-2425. 1950 to present. Monthly. $450.00 per year.

SCIENCE CITATION INDEX. Institute for Scientific Information, 3501 Market Street, Philadelphia, PA 19104. (800) 523-1850 or (215) 386-0100. Inquire as to cost and availability.

ASSOCIATIONS AND PROFESSIONAL SOCIETIES

AMERICAN ASSOCIATION STATE CLIMATOLOGISTS. c/o Professor John Griffiths, Meteorology Department, O and M Building, Texas A and M University, College Station, TX 77843. (409) 845-7320.

AMERICAN METEOROLOGICAL SOCIETY. 45 Beacon Street, Boston, MA 02108. (617) 227-2425.

INTERNATIONAL ASSOCIATION OF METEOROLOGY AND ATMOSPHERIC PHYSICS. UCAR, Post Office Box 3000, Boulder, CO 80307.

NATIONAL ENVIRONMENTAL SATELLITE DATA, AND INFORMATION SERVICE. 3300 Whitehaven Street, NW, Washington, DC 20235. (202) 634-7318.

NATIONAL WEATHER ASSOCIATION. 4400 Stamp Road, Room 404, Temple Hills, MD 20748. (301) 899-3784.

UNIVERSITY CORPORATION FOR ATMOSPHERIC RESEARCH. Box 3000, 1850 Table Mesa Drive, Boulder, CO 80307. (303) 497-1000.

WEATHER MODIFICATION ASSOCIATION. Post Office Box 8116, Fresno, CA 93747. (209) 291-8466.

DIRECTORIES AND BIOGRAPHICAL SOURCES

AMERICAN MEN AND WOMEN OF SCIENCE. Physical and Biological Sciences. Sixteenth edition. R.R. Bowker Company, 205 East 42nd Street, New York, NY 10017. (800) 521-8110 or (212) 916-1600. 1986. $595.00.

GOVERNMENT RESEARCH DIRECTORY. Gale Research Company, Book Tower, Detroit, MI 48226. (800) 521-0707. Fourth edition. 1987. $350.00.

INTERDOC: DIRECTORY OF PUBLISHED PROCEEDINGS, SERIES. SEMT-Science/Engineering/Medicine/Technology. Interdoc Corporation, 173 Halstead Avenue, Box 326, Harrison, NY 10528. (014) 835-3506. Ten times per year. $325.00 per year.

METEOROLOGICAL SERVICES OF THE WORLD. World Meteorological Organization. Available from: American Meteorological Society, 45 Beacon Street, Boston, MA 02108. (617) 227-2425. Annual. $35.00.

NATIONAL WEATHER SERVICE OFFICES AND STATIONS. National Oceanic and Atmospheric Administration, Department of Commerce, Silver Spring, MD 20910. (301) 427-7698. Annual. Free.

RESEARCH CENTERS DIRECTORY. Gale Research Company, Book Tower, Detroit, MI 48226. (800) 521-0707. Twelfth edition. 1987. $365.00 for set.

SCIENTIFIC AND TECHNICAL ORGANIZATIONS AND AGENCIES DIRECTORY. Gale Research Company, Book Tower, Detroit, MI 48226. (800) 521-0707. Second edition. 1987. $185.00.

ENCYCLOPEDIAS AND DICTIONARIES

ENCYCLOPEDIA OF CLIMATOLOGY. John E. Oliver and Rhodes W. Fairbridge, editors. Van Nostrand Reinhold, Incorporated, 115 Fifth Avenue, New York, NY 10003. (800) 543-2681. 1987. $89.95.

ENCYCLOPEDIA OF PHYSICAL SCIENCE AND TECHNOLOGY. Academic Press, Incorporated, Orlando, FL 32887. (800) 321-5068 or (305) 345-2734. Inquire as to cost and availability.

GENERAL WORKS

THUNDERSTORM MORPHOLOGY AND DYNAMICS. Edwin Kessler. University of Oklahoma Press, Post Office Box 1657, Hagerstown, MD 21741. (800) 638-3030. 1986. $68.50.

THUNDERSTORMS. C. Magono. Elsevier Science Publishing Company, Incorporated, 52 Vanderbilt Avenue, New York, NY 10017. (212) 370-5520. 1980. $70.25.

ONLINE DATA BASES

DISSERTATION ABSTRACTS ONLINE. University Microfilms International, 300 North Zeeb Road, Ann Arbor, MI 48106.

(800) 521-0600 or (313) 761-4700. Scope includes virtually all doctoral dissertations accepted at accredited American institutions from 1861 to present in 252 subject areas. Inquire as to cost and availability.

METEOROLOGICAL AND GEOASTROPHYSICAL ABSTRACTS. American Meteorological Society, 45 Beacon Street, Boston, MA 02108. (617) 227-2425. 1950 to present. Monthly. $450.00 per year.

NTIS. National Technical Information Service, 5285 Port Royal Road, Springfield, VA 22161. (703) 487-4630. Broad coverage of government-sponsored research reports, 1964 to present. Inquire as to cost and availability.

SCISEARCH. Institute for Scientific Information, 3501 Market Street, Philadelphia, PA 19104. (800) 523-1850 or (215) 386-0100. Broad multidisciplinary title and author index to the international literature of science and technology, 1974 to present. Inquire as to cost and availability.

WILSONLINE. H.W. Wilson Company, 950 University Avenue, Bronx, NY 10452. (800) 367-6770 or (212) 588-8400. Makes available online versions of the printed H.W. Wilson Indexes including Applied Science and Technology Index, Business Periodicals Index, and Readers' Guide to Periodical Literature. Period covered is generally 1983 to present. Inquire as to cost and availability.

PERIODICALS

AGRICULTURAL AND FOREST METEOROLOGY. Elsevier Science Publishing Company, Incorporated, 52 Vanderbilt Avenue, New York, NY 10017. (212) 370-5520. 1964 to present. Monthly. $260.00 per year.

AMERICAN METEOROLOGICAL SOCIETY BULLETIN. American Meteorological Society, 45 Beacon Street, Boston, MA 02108. (617) 227-2425.

CLIMATIC CHANGE: AN INTERDISCIPLINARY, INTERNATIONAL JOURNAL DEVOTED TO THE DESCRIPTION, CAUSES AND IMPLICATIONS OF CLIMATIC CHANGE. D. Reidel Publishing Company, 190 Old Derby Street, Hingham, MA 02043. (617) 871-6600. 1977 to present. Six times per year. $125.00 per year.

DYNAMICS OF ATMOSPHERES AND OCEANS. Elsevier Science Publishing Company, Incorporated, 52 Vanderbilt Avenue, New York, NY 10017. (212) 370-5520. 1977 to present. Quarterly. $90.00 per year.

JOURNAL OF THE ATMOSPHERIC SCIENCES. American Meteorological Society, 45 Beacon Street, Boston, MA 02108. (617) 227-2425. 1944 to present. Semimonthly. $220.00 per year.

MONTHLY WEATHER REVIEW. American Meteorological Society, 45 Beacon Street, Boston, MA 02108. (617) 227-2425. 1872 to present. Monthly. $120.00 per year.

NATIONAL WEATHER DIGEST. National Weather Association, 4400 Stamp Road, Room 404, Temple Hills, MD 20748. (301) 899-3784. 1976 to present. Quarterly. $20.00 per year.

WEATHER. Royal Meteorological Society, James Glaisher House, Grenville Place, Bracknell Berkshire, RG12 1BX, England. 1946 to present. $30.00 per year.

WEATHERWISE. Heldref Publications, 4000 Albemarle Street, NW, Washington, DC 20016. (202) 362-6445. 1948 to present. Bimonthly. $20.00 per year.

RESEARCH CENTERS AND INSTITUTES

NATIONAL CENTER FOR ATMOSPHERIC RESEARCH. Box 3000, Boulder, CO 80307. (303) 497-1000.

NATIONAL SEVERE STORMS LABORATORY. 1313 Halley Circle, Norman, OK 73069. (405) 360-3620.

NEW MEXICO INSTITUTE OF MINING AND TECHNOLOGY. Irving Langmuir Laboratory, Socorro, NM 87801. (505) 835-5423.

SOUTH DAKOTA SCHOOL OF MINES AND TECHNOLOGY. Institute of Atmospheric Sciences, 501 East St. Joseph Street, Rapid City, SD 57701-3995. (605) 394-2291.

UNIVERSITY OF ARIZONA. Institute of Atmospheric Physics, Tucson, AZ 85721. (602) 626-6831.

UNIVERSITY OF MIAMI. Rosenstiel School of Marine and Atmospheric Science, 4600 Rickenbacker Causeway, Miami, FL 33149. (305) 361-4000.

TIDAL WAVES

See: TSUNAMIS

TIN

See also: ALLOYS, BRASS AND BRONZE, COPPER, MATERIALS SCIENCE, ZINC

ABSTRACT SERVICES AND INDEXES

ALLOYS INDEX. American Society for Metals, 9639 Kinsman Road, Metals Park, OH 44073. (216) 338-5151. $130.00 per year.

APPLIED MECHANICS REVIEWS. American Society of Mechanical Engineers, 345 East 47th Street, New York, NY 10017. (212) 705-7722. Monthly. $380.00 per year. Critical reviews of the world literature in applied mechanics and related engineering science.

APPLIED SCIENCE AND TECHNOLOGY INDEX. H.W. Wilson Company, 950 University Avenue, Bronx, NY 10452. (800) 367-6670 or (212) 588-8400. Inquire as to cost and availability.

CHEMICAL ABSTRACTS. Chemical Abstracts Service, 2540 Olentangy Road, Post Office Box 3012, Columbus, OH 43210. (800) 848-6538 or (614) 421-3600. Weekly. $9200.00 per year.

CURRENT CONTENTS: ENGINEERING, TECHNOLOGY. Institute for Scientific Information, 3501 Market Street, Philadelphia, PA 19104. (800) 523-1850 or (215) 386-0100. $272.00 per year.

ENGINEERING INDEX MONTHLY. Engineering Information, Incorporated, 345 East 47th Street, New York, NY 10017. (800) 221-1044 or (212) 705-7600. Monthly, with annual cumulation. $1560.00 per year.

INTERNATIONAL COPPER INFORMATION BULLETIN. Copper Development Association, Orchard House, Mutton Lane, Potters Bar, Herts, EN6 3AP, England. 1976 to present. Quarterly. $35.00 per year.

METALS ABSTRACTS. American Society for Metals, 9639 Kinsman Road, Metals Park, OH 44073. (216) 338-5151. Monthly. $890.00.

METALS ABSTRACTS INDEX. American Society for Metals, 9639 Kinsman Road, Metals Park, OH 44073. (216) 338-5151. Monthly. (Sold only to subscribers of Metals Abstracts.)

WORLD ALUMINUM ABSTRACTS. Aluminum Association, 818 Connecticut Avenue, NW, Washington, DC 20006. (202) 862-5100. 1968 to present. Monthly. $165.00 per year.

ZINC ABSTRACTS. Zinc Development Association, 34 Berkeley Square, London W1X 6AJ, England. 1943 to present. Quarterly. $100.00 per year.

ANNUAL REVIEWS AND YEARBOOKS

ANNUAL REVIEW OF MATERIALS SCIENCE. Annual Reviews, Incorporated, 4139 El Camino Way, Palo Alto, CA 94306. (415) 493-4400.

ASSOCIATIONS AND PROFESSIONAL SOCIETIES

AMERICAN SOCIETY FOR METALS. Metals Park, OH 44073. (216) 338-5151.

AMERICAN TIN TRADE ASSOCIATION. Post Office Box 1347, New York, NY 10150. (212) 599-8300.

BRASS AND BRONZE INGOT INSTITUTE. 33 North LaSalle Street, Room 3500, Chicago, IL 60602. (312) 236-2715.

COPPER DEVELOPMENT ASSOCIATION. Box 1840, Greenwich Office Park 2, Greenwich, CT 06836. (203) 625-8210.

INTERNATIONAL TIN RESEARCH INSTITUTE. Fraser Road, Perivale, Greenford, Middlesex, UB6 7AQ, England.

MALAYSIAN TIN BUREAU. 1625 Eye Street, NW, Suite 913, Washington, DC 20006. (202) 331-7550.

METALLURGICAL SOCIETY OF THE AIME (AMERICAN INSTITUTE OF MINING, METALLURGICAL AND PETROLEUM ENGINEERS). 420 Commonwealth Drive, Warrendale, PA 15086. (412) 776-9080.

TIN RESEARCH INSTITUTE. 1353 Perry Street, Columbus, OH 43201. (614) 424-6200.

DIRECTORIES AND BIOGRAPHICAL SOURCES

DUN'S INDUSTRIAL GUIDE - THE METAL WORKING DIRECTORY. Dun and Bradstreet, Incorporated, Three Century Drive, Parsippany, NJ 07054. (201) 455-0900. Annual. $550.00.

INTERNATIONAL RESEARCH CENTERS DIRECTORY 1986-1987. Gale Research Company, Book Tower, Detroit, MI 48226. (800) 521-0707. Third edition. 1986.

METAL PRODUCTS DIRECTORY. American Business Directories, Incorporated, Division of American Business Lists, Incorporated, 5707 South 86th Circle, Omaha, NE 68127. (402) 331-7169. 1986. $80.00.

RESEARCH CENTERS DIRECTORY. Gale Research Company, Book Tower, Detroit, MI 48226. (800) 521-0707. Eleventh edition. 1987.

SCIENTIFIC AND TECHNICAL ORGANIZATIONS AND AGENCIES DIRECTORY. Gale Research Company, Book Tower, Detroit, MI 48226. (800) 521-0707. 1985.

WHO'S WHO IN ENGINEERING. Engineers Joint Council, 345 East 47th Street, New York, NY 10017. (212) 705-7010. 1985. $200.00.

WHO'S WHO IN FRONTIER SCIENCE AND TECHNOLOGY. Marquis Who's Who, Incorporated, 200 East Ohio Street, Chicago, IL 60611. (800) 428-3898 or (312) 787-2008.

WHO'S WHO IN TECHNOLOGY TODAY. Reston Publishing Company, Incorporated, c/o Prentice-Hall, Incorporated, Englewood Cliffs, NJ 07632. (800) 262-6868. Biennial. Five volumes. $425.00. Covers the fields of electronics, computer science, physics, optics, chemistry, biotechnology, mechanics, energy, and earth science.

ENCYCLOPEDIAS AND DICTIONARIES

ENCYCLOPEDIA OF PHYSICAL SCIENCE AND TECHNOLOGY. Academic Press, Incorporated, Orlando, FL 32887. (800) 321-5068 or (305) 345-2734. Fifteen volumes, 1986.

MCGRAW-HILL DICTIONARY OF ENGINEERING. Sybil P. Parker, editor. McGraw-Hill Book Company, 1221 Avenue of the Americas, New York, NY 10020. (212) 512-2000. 1984. $39.95.

MCGRAW-HILL ENCYCLOPEDIA OF SCIENCE AND TECHNOLOGY. McGraw-Hill Book Company, Incorporated, 1221 Avenue of the Americas, New York, NY 10020. (212) 512-2000.

GENERAL WORKS

ESSENTIAL METALLURGY FOR ENGINEERS. W. Alexander, G. Davies, and K. Reynolds. Van Nostrand Reinhold, 115 Fifth Avenue, New York, NY 10003. (800) 543-2681. 1985. $17.95.

PHYSICAL METALLURGY. P. Haasen. Cambridge University Press, 32 East 57th Street, New York, NY 10022. (212) 688-8885. 1986. $24.95 in paper.

METALLURGY BASICS. D.V. Brown. Van Nostrand Reinhold, 115 Fifth Avenue, New York, NY 10003. (800) 543-2681. 1985. $17.95.

STRUCTURE AND PROPERTIES OF ENGINEERING ALLOYS. W.F. Smith. McGraw-Hill Book Company, Incorporated, 1221 Avenue of the Americas, New York, NY 10020. (212) 512-2000. 1980. $46.95.

TIN AND ITS ALLOYS AND COMPOUNDS. B.T. Berry and C.G. Thwaites. John Wiley and Sons, Incorporated, 605 Third Avenue, New York, NY 10158. (800) 526-5368 or (212) 850-6000. 1983. $74.95.

TIN AND TIN MINING. R.L. Atkinson. Seven Hills Books, Incorporated, 519 West Third Street, Cincinnati, OH 45202. (513) 381-3881. 1985. $3.50 in paper.

TINPLATE AND MODERN CANMAKING TECHNOLOGY. E. Morgan. Pergamon Press, Incorporated, Maxwell House, Fairview Park, Elmsford, NY 10523. (914) 592-7700.

HANDBOOKS AND MANUALS

CRC HANDBOOK OF CHEMISTRY AND PHYSICS. CRC Press, Incorporated, 2000 Corporate Boulevard, Boca Raton, Florida 33341. (305) 994-0555. Sixth-seventh edition. 1986. $69.95.

SMITHELL'S METALS REFERENCE BOOK. Eric A. Brandes, editor. Butterworth Publishers, 80 Montvale Avenue, Stoneham, MA 02180. (800) 325-4177. Sixth edition. 1983. $210.00.

ONLINE DATA BASES

COMPENDEX. Engineering Information, Incorporated, 345 East 47th Street, New York, NY 10017. (800) 221-1044 or (212) 705-7615. Engineering and technical literature, 1975 to present. Inquire as to cost and availability.

INSPEC. INSPEC Marketing Department, Institute of Electrical and Electronics Engineers, Incorporated, IEEE Service Department, 445 Hoes Lane, Piscataway, NJ 08854. (201) 981-0060. Inquire as to on-line cost and availability.

METADEX. Metals Information, American Society for Metals, Metals Park, OH 44073. (216) 338-5151. (Metals Abstracts/Alloys Index). A worldwide literature on the science and practice of metallurgy, 1966 to present. Inquire as to online cost and availability.

TIN

NASA. National Aeronautics and Space Administration, Scientific and Technical Information Branch, 300 7th Street, SW, Washington, DC 20546. Citations and abstracts of aerospace literature, 1962 to present. Inquire as to cost and availability.

NON-FERROUS METALS ABSTRACTS. British Non-Ferrous Metals Technology Centre, Grove Laboratories, Denchworth Road, Wantage, Oxfordshire, England OX12 9 BJ. Citations and abstracts on non-ferrous metallurgy and technology, 1961 to present. Inquire as to online cost and availability.

NTIS. National Technical Information Service, 5285 Port Royal Road, Springfield, VA 22161. (703) 487-4630. Broad coverage of government sponsored research reports, 1964 to present. Inquire as to cost and availability.

WILSONLINE. H.W. Wilson Company, 950 University Avenue, Bronx, NY 10452. (800) 367-6770 or (212) 588-8400. Makes available online versions of the printed H.W. Wilson Indexes including Applied Science and Technology Index, Business Periodicals Index, and Reader's Guide to Periodical Literature. Period covered is generally 1983 to present. Inquire as to cost and availability.

OTHER SOURCES

MATERIALS SCIENCE AND METALLURGY. National Technical Information Service (NTIS), 5285 Port Royal Road, Springfield, VA 22161. (703) 487-4630. Translations and abstracts of foreign language technical media. Irregular. $40.00 per year.

PERIODICALS

ALLOY DIGEST. Engineering Publications Incorporated, Box 823, Upper Montclair, NJ 07043. (201) 746-7930. Monthly. $50.00 per year.

JOURNAL OF METALS. Metallurgical Society of the AIME (American Institute of Mining, Metallurgical and Petroleum Engineers), 420 Commonwealth Drive, Warrendale, PA 15086. (412) 776-9080. Monthly. $40.00 per year.

METALLURGICAL TRANSACTIONS. Metallurgical Society of the AIME (American Institute of Mining, Metallurgical and Petroleum Engineers), 420 Commonwealth Drive, Warrendale, PA 15086. (412) 776-9080. Monthly. $95.00 per year.

METALS WEEK. McGraw-Hill Book Company, Incorporated, 1221 Avenue of the Americas, New York, NY 10020. (212) 997-2823. Weekly. $527.00 per year.

TIN AND ITS USES. International Tin Research Institute, Fraser Road, Perivale, Greenford, Middlesex, UB6 7AQ, England. Quarterly. Free.

TIN INTERNATIONAL. Tin Publications, Limited, 222 Strand, London, WC2R 1BA, England. Monthly. $85.00 per year.

TIN NEWS. Malaysian Tin Bureau, 1625 Eye Street, NW, Suite 913, Washington, DC 20006. (202) 331-7550. Monthly. Free.

RESEARCH CENTERS AND INSTITUTES

PHYSICAL METALLURGY RESEARCH LABORATORIES. Centre for Mineral and Energy Technology, 555 Booth Street, Ottawa, ON Canada K1A 0G1. (613) 995-4807.

TEXAS A & M UNIVERSITY. Mechanics and Materials Center, ERC Building, College Station, TX 77843. (409) 845-7512.

TIN RESEARCH INSTITUTE. 1353 Perry Street, Columbus, OH 43201. (614) 424-6200.

UNIVERSITY OF CONNECTICUT. Institute of Materials Science, Storrs, CT 06268. (203) 486-4623.

UNIVERSITY OF FLORIDA. Department of Materials Science and Engineering, Gainesville, FL 32601. (904) 392-1454.

STATISTICAL SOURCES

METALS STATISTICS. Fairchild Publications, 7 East 12th Street, New York, NY 10003. (212) 741-4426. Annual. $50.00.

WORLD METAL STATISTICS. Bureau of Metal Statistics, 41 Doughty Street, London, WC1N 2LF, England. 1948 to present. Monthly. $1250.00 per year.

TITANIUM

See also: ALLOYS, ALUMINUM, MATERIALS SCIENCE, METALLURGY

ABSTRACT SERVICES AND INDEXES

ALLOYS INDEX. American Society for Metals, 9639 Kinsman Road, Metals Park, OH 44073. (216) 338-5151. $130.00 per year.

APPLIED MECHANICS REVIEWS. American Society of Mechanical Engineers, 345 East 47th Street, New York, NY 10017. (212) 705-7722. Monthly. $380.00 per year. Critical reviews of the world literature in applied mechanics and related engineering science.

APPLIED SCIENCE AND TECHNOLOGY INDEX. H.W. Wilson Company, 950 University Avenue, Bronx, NY 10452. (800) 367-6670 or (212) 588-8400. Inquire as to cost and availability.

CURRENT CONTENTS: ENGINEERING, TECHNOLOGY. Institute for Scientific Information, 3501 Market Street, Philadelphia, PA 19104. (800) 523-1850 or (215) 386-0100. $272.00 per year.

ENGINEERING INDEX MONTHLY. Engineering Information, Incorporated, 345 East 47th Street, New York, NY 10017. (800) 221-1044 or (212) 705-7600. Monthly, with annual cumulation. $1560.00 per year.

METALS ABSTRACTS. American Society for Metals, 9639 Kinsman Road, Metals Park, OH 44073. (216) 338-5151. Monthly. $890.00.

METALS ABSTRACTS INDEX. American Society for Metals, 9639 Kinsman Road, Metals Park, OH 44073. (216) 338-5151. Monthly. (Sold only to subscribers of Metals Abstracts.)

WORLD ALUMINUM ABSTRACTS. Aluminum Association, 818 Connecticut Avenue, NW, Washington, DC 20006. (202) 862-5100. 1968 to present. Monthly. $165.00 per year.

ANNUAL REVIEWS AND YEARBOOKS

ANNUAL REVIEW OF MATERIALS SCIENCE. Annual Reviews, Incorporated, 4139 El Camino Way, Palo Alto, CA 94306. (415) 493-4400.

ASSOCIATIONS AND PROFESSIONAL SOCIETIES

ALUMINUM ASSOCIATION. 818 Connecticut Avenue, NW, Washington, DC 20006. (202) 862-5100.

ALUMINUM EXTRUDERS COUNCIL. 4300-L Lincoln Avenue, Rolling Meadows, IL 60008. (312) 359-8160.

ALUMINUM RECYCLING ASSOCIATION. 900 17th Street, NW, Suite 504, Washington, DC 20006. (202) 785-0550.

AMERICAN SOCIETY FOR METALS. Metals Park, OH 44073. (216) 338-5151.

DIRECTORIES AND BIOGRAPHICAL SOURCES

DUN'S INDUSTRIAL GUIDE - THE METAL WORKING DIRECTORY. Dun and Bradstreet, Incorporated, Three Century Drive, Parsippany, NJ 07054. (201) 455-0900. Annual. $550.00.

RESEARCH CENTERS DIRECTORY. Gale Research Company, Book Tower, Detroit, MI 48226. (800) 521-0707. Eleventh edition. 1987.

SCIENTIFIC AND TECHNICAL ORGANIZATIONS AND AGENCIES DIRECTORY. Gale Research Company, Book Tower, Detroit, MI 48226. (800) 521-0707. 1985.

WHO'S WHO IN TECHNOLOGY TODAY. Reston Publishing Company, Incorporated, c/o Prentice-Hall, Incorporated, Englewood Cliffs, NJ 07632. (800) 262-6868. Biennial. Five volumes. $425.00. Covers the fields of electronics, computer science, physics, optics, chemistry, biotechnology, mechanics, energy, and earth science.

ENCYCLOPEDIAS AND DICTIONARIES

ENCYCLOPEDIA OF MATERIALS SCIENCE AND ENGINEERING. Michael B. Bever, editor. MIT Press, 28 Carlton Street, Cambridge, MA 02142. (617) 253-5646. Eight volumes. 1986. $1950.00.

ENCYCLOPEDIA OF PHYSICAL SCIENCE AND TECHNOLOGY. Academic Press, Incorporated, Orlando, FL 32887. (800) 321-5068 or (305) 345-2734. Fifteen volumes, 1986.

MCGRAW-HILL ENCYCLOPEDIA OF SCIENCE AND TECHNOLOGY. McGraw-Hill Book Company, Incorporated, 1221 Avenue of the Americas, New York, NY 10020. (212) 512-2000.

GENERAL WORKS

APPLIED SUPERCONDUCTIVITY, METALLURGY, AND PHYSICS OF TITANIUM ALLOYS, VOLUME 1: FUNDAMENTALS. E.W. Collins. Plenum Publishers, Incorporated, 233 Spring Street, New York, NY 10013. (800) 221-9369. 1985. $97.50.

ESSENTIAL METALLURGY FOR ENGINEERS. W. Alexander, G. Davies, and K. Reynolds. Van Nostrand Reinhold, 115 Fifth Avenue, New York, NY 10003. (800) 543-2681. 1985. $17.95.

METALLURGY BASICS. D.V. Brown. Van Nostrand Reinhold, 115 Fifth Avenue, New York, NY 10003. (800) 543-2681. 1985. $17.95.

SUPERALLOYS. C.T. Sims and W.C. Hagel. John Wiley and Sons, Incorporated, 605 Third Avenue, New York, NY 10158. (212) 850-6000. 1972. $74.95.

HANDBOOKS AND MANUALS

CRC HANDBOOK OF CHEMISTRY AND PHYSICS. CRC Press, Incorporated, 2000 Corporate Boulevard, Boca Raton, Florida 33341. (305) 994-0555. Sixth-seventh edition. 1986. $69.95.

SMITHELL'S METALS REFERENCE BOOK. Eric A. Brandes, editor. Butterworth Publishers, 80 Montvale Avenue, Stoneham, MA 02180. (800) 325-4177. Sixth edition. 1983. $210.00.

TITANIUM AND TITANIUM ALLOYS: SOURCE BOOK. Matthew Donachie, Jr., editor. American Society for Metals, 9639 Kinsman Road, Metals Park, OH 44073. (216) 338-5151. 1982. $60.00.

ONLINE DATA BASES

COMPENDEX. Engineering Information, Incorporated, 345 East 47th Street, New York, NY 10017. (800) 221-1044 or (212) 705-7615. Engineering and technical literature, 1975 to present. Inquire as to cost and availability.

INSPEC. INSPEC Marketing Department, Institute of Electrical and Electronics Engineers, Incorporated, IEEE Service Department, 445 Hoes Lane, Piscataway, NJ 08854. (201) 981-0060. Inquire as to on-line cost and availability.

METADEX. Metals Information, American Society for Metals, Metals Park, OH 44073. (216) 338-5151. (Metals Abstracts/Alloys Index). A worldwide literature on the science and practice of metallurgy, 1966 to present. Inquire as to online cost and availability.

NASA. National Aeronautics and Space Administration, Scientific and Technical Information Branch, 300 7th Street, SW, Washington, DC 20546. Citations and abstracts of aerospace literature, 1962 to present. Inquire as to cost and availability.

NON-FERROUS METALS ABSTRACTS. British Non-Ferrous Metals Technology Centre, Grove Laboratories, Denchworth Road, Wantage, Oxfordshire, England OX12 9BJ. Citations and abstracts on non-ferrous metallurgy and technology, 1961 to present. Inquire as to online cost and availability.

NTIS. National Technical Information Service, 5285 Port Royal Road, Springfield, VA 22161. (703) 487-4630. Broad coverage of government sponsored research reports, 1964 to present. Inquire as to cost and availability.

OTHER SOURCES

ALUMINUM DEVELOPMENTS DIGEST. Aluminum Association, Incorporated, 818 Connecticut Avenue, NW, Washington, DC 20006. (202) 862-5100. Three times per year.

MATERIALS SCIENCE AND METALLURGY. National Technical Information Service (NTIS), 5285 Port Royal Road, Springfield, VA 22161. (703) 487-4630. Translations and abstracts of foreign language technical media. Irregular. $40.00 per year.

PERIODICALS

ALLOY DIGEST. Engineering Publications Incorporated, Box 823, Upper Montclair, NJ 07043. (201) 746-7930. Monthly. $50.00 per year.

E&MJ: ENGINEERING AND MINING JOURNAL. McGraw-Hill Book Company, 1221 Avenue of the Americas, New York, NY 10020. (212) 512-2000. Monthly. $20.00 per year.

JOURNAL OF METALS. Metallurgical Society of the AIME (American Institute of Mining, Metallurgical and Petroleum Engineers), 420 Commonwealth Drive, Warrendale, PA 15086. (412) 776-9080. Monthly. $40.00 per year.

LIGHT METAL AGE. Fellom Publishing Company, 693 Mission Street, San Francisco, CA 94105. (415) 781-1431. Bimonthly. $20.00 per year.

RESEARCH CENTERS AND INSTITUTES

GENERAL ELECTRIC COMPANY. Research and Development Center, Post Office Box 8, Schenectady, NY 12301. (518) 385-8415.

LAWRENCE BERKELEY LABORATORY. Center for Advanced Materials, 1 Cyclotron Road, Berkeley, CA 94720. (415) 486-4755.

PHYSICAL METALLURGY RESEARCH LABORATORIES. Centre for Mineral and Energy Technology, 555 Booth Street, Ottawa, ON Canada K1A 0G1. (613) 995-4807.

TEXAS A & M UNIVERSITY. Mechanics and Materials Center, ERC Building, College Station, TX 77843. (409) 845-7512.

UNIVERSITY OF CONNECTICUT. Institute of Materials Science, Storrs, CT 06268. (203) 486-4623.

UNIVERSITY OF FLORIDA. Department of Materials Science and Engineering, Gainesville, FL 32601. (904) 392-1454.

UNIVERSITY OF WISCONSIN AT MADISON. Cast Metals Laboratory, 1509 University Avenue, Madison, WI 53706. (608) 262-2562.

SPECIFICATIONS AND STANDARDS

ALUMINUM STANDARDS AND DATA. Aluminum Association, Incorporated, 818 Connecticut Avenue, NW, Washington, DC 20006. (202) 862-5100. Biennial. $12.00.

STATISTICAL SOURCES

ALUMINUM INGOT. United States of the Census, United States Department of Commerce, Washington, DC 20233. (301) 763-7800. Monthly. $20.50 per year.

ALUMINUM SITUATION. Aluminum Association, Incorporated, 818 Connecticut Avenue, NW, Washington, DC 20006. (202) 862-5100. Monthly.

ALUMINUM STATISTICAL REVIEW. Aluminum Association, Incorporated, 818 Connecticut Avenue, NW, Washington, DC 20006. (202) 862-5100. Annual. $25.00.

AMERICAN BUREAU OF METAL STATISTICS YEAR BOOK. American Bureau of Metal Statistics, Post Office Box 1405, 400 Plaza Drive, Secaucus, NJ 07094. Annual.

TOPOGRAPHIC MAPPING

See also: AERIAL PHOTOGRAPHY, ARTIFICIAL SATELLITES, CARTOGRAPHY, GEODESY, PHOTOGRAMMETRY, REMOTE SENSING, SURVEYING

ABSTRACT SERVICES AND INDEXES

APPLIED SCIENCE AND TECHNOLOGY INDEX. H.W. Wilson and Company, 950 University Avenue, Bronx, NY 10452. (800) 367-6670 or (212) 588-8400. Monthly. Inquire as to cost and availability.

BIBLIOGRAPHY AND INDEX OF GEOLOGY. American Geological Institute, 4220 King Street, Alexandria, VA 22302. (703) 379-2480. 1969 to present. Monthly. $1100.00 per year.

CURRENT CONTENTS: ENGINEERING, TECHNOLOGY AND APPLIED SCIENCES. Institute for Scientific Information, 3501 Market Street, Philadelphia, PA 19104. (800) 523-1850 or (215) 386-0100. Weekly. $275.00 per year.

ENGINEERING INDEX MONTHLY AND AUTHOR INDEX. Engineering Information Inc., 345 East 47th Street, New York, NY 10017. (212) 705-7600. Monthly. $1560.00 per year.

METEOROLOGICAL AND GEOASTROPHYSICAL ABSTRACTS. American Meteorological Society, 45 Beacon Street, Boston, MA 02108. (617) 227-2425. 1950 to present. Monthly. $450.00 per year.

REMOTE SENSING OF NATURAL RESOURCES: A QUARTERLY LITERATURE REVIEW. University of New Mexico, Technology Application Center, Albuquerque, NM 87131. (505) 277-3622. 1974 to present. Quarterly. $150.00. Available to qualified agencies only.

SCIENCE CITATION INDEX. Institute for Scientific Information, 3501 Market Street, Philadelphia, PA 19104. (800) 523-1850 or (215) 386-0100. Six times per year. $6200.00 per year.

ASSOCIATIONS AND PROFESSIONAL SOCIETIES

AMERICAN ASSOCIATION FOR GEODETIC SURVEYING. c/o American Congress on Surveying and Mapping, 210 Little Falls Street, Falls Church, VA 22046. (703) 241-2446.

AMERICAN CARTOGRAPHIC ASSOCIATION. c/o American Congress on Surveying and Mapping, 210 Little Falls Street, Falls Church, VA 22046. (703) 241-2446.

AMERICAN CONGRESS ON SURVEYING AND MAPPING. 210 Little Falls Street, Falls Church, VA 22046. (703) 241-2446.

AMERICAN SOCIETY FOR PHOTOGRAMMETRY AND REMOTE SENSING. 210 Little Falls Street, Falls Church, VA 22046-4398. (703) 534-6617.

NATIONAL ASSOCIATION OF PROFESSIONAL SURVEYORS. c/o American Congress on Surveying and Mapping, 210 Little Falls Street, Falls Church, VA 22046. (703) 241-2446.

NORTH AMERICAN CARTOGRAPHIC ASSOCIATION. 6010 Executive Boulevard, Suite 100, Rockville, MD 20852. (301) 443-8075.

OPTICAL SOCIETY OF AMERICA. 1816 Jefferson Place, N.W., Washington, DC 20036. (202) 223-8130.

SOCIETY OF PHOTOGRAPHIC SCIENTISTS AND ENGINEERS. 7003 Kilworth Lane, Springfield, VA 22151. (703) 642-9090.

SPIE - THE INTERNATIONAL SOCIETY FOR OPTICAL ENGINEERING. P.O. Box 10, 1022 19th Street, Bellingham, WA 98227. (206) 676-3290.

BIBLIOGRAPHIES

BIBLIOGRAPHY OF CARTOGRAPHY. Library of Congress Geography and Map Division. G.K. Hall and Company, 70 Lincoln Street, Boston, MA 02111. (800) 343-2806. Five volumes and supplements. 1973-1979. Inquire for set price.

DIRECTORIES AND BIOGRAPHICAL SOURCES

INTERNATIONAL RESEARCH CENTERS DIRECTORY 1988-89. Darren L. Smith, editor. Gale Research Company, Book Tower, Detroit, MI 48226. (800) 521-0707. 4th edition. 1987. $360.00.

1987 DIRECTORY OF ENGINEERING SOCIETIES AND RELATED ORGANIZATIONS. Gordon Davis, editor. Hemisphere Publishing Corporation, 1010 Vermont Avenue, NW, Washington, DC 20005. (800) 526-0275. 12th edition. 1987. $100.00.

RESEARCH CENTERS DIRECTORY 1988. Gale Research Company, Book Tower, Detroit, MI 48226. (800) 521-0707. 12th edition. 1987. $365.00 for set.

SCIENTIFIC AND TECHNICAL ORGANIZATIONS AND AGENCIES DIRECTORY. Margaret Labash Young, editor. Gale Research Company, Book Tower, Detroit, MI 48226. (800) 521-0707. 2nd edition. 1987. $185.00.

WHO'S WHO IN ENGINEERING. Gordon Davis, editor. Hemisphere Publishing Corporation, 1010 Vermont Avenue, NW, Washington, DC 20005. (800) 526-0275. 6th edition. 1985. $200.00.

GENERAL WORKS

CLOSE-RANGE PHOTOGRAMMETRY AND SURVEYING: STATE OF THE ART. American Society for Photogrammetry and Remote Sensing. 210 Little Falls Street, Falls Church, VA 22046-4398. (703) 534-6617. 1985. $65.00 in paper.

ELEMENTS OF CARTOGRAPHY. Arthur H. Robinson and others. John Wiley and Sons, Inc., 605 Third Avenue, New York, NY 10158. (800) 526-5368. 5th edition. 1984. $35.50.

ELEMENTS OF PHOTOGRAMMETRY. P.R. Wolf. McGraw-Hill Book Company, 1221 Avenue of the Americas, New York, NY 10020. (212) 512-2000. 2nd edition. 1983. $49.95.

INTRODUCTORY CARTOGRAPHY. John Campbell. Prentice-Hall Publishing, Inc., Englewood Cliffs, NJ 07632. (800) 562-0245. 1984. $37.95.

PHOTOGRAMMETRY. Francis H. Moffitt and Edward M. Mikhail. Harper and Row Publishers, Inc., 10 East 53rd Street, New York, NY 10022. (212) 207-7655. 3rd edition. 1980. $41.95.

PRINCIPLES OF REMOTE SENSING. Paul Curran. Halstead Press, division of John Wiley and Sons, Inc., 605 Third Avenue, New York, NY 10158. (800) 526-5368. 1986. $35.95.

REMOTE SENSING. Floyd F. Sabins. W.H. Freeman and Company, 41 Madison Avenue, New York, NY 10010. (212) 532-7660. 2nd edition. 1986. $39.95.

REMOTE SENSING METHODS AND APPPLICATIONS. R. Hord. John Wiley and Sons, Inc., 605 Third Avenue, New York, NY 10158. (800) 526-5368. 1986. $39.95.

SURVEY OF THE PROFESSION: PHOTOGRAMMETRY, SURVEYING, MAPPING, REMOTE SENSING. American Society for Photogrammetry and Remote Sensing. 210 Little Falls Street, Falls Church, VA 22046-4398. (703) 534-6617. 1982. $35.00 in paper.

HANDBOOKS AND MANUALS

MANUAL OF PHOTOGRAMMETRY. Chester C. Slama, editor. American Society for Photogrammetry and Remote Sensing. 210 Little Falls Street, Falls Church, VA 22046-4398. (703) 534-6617. 4th edition. 1980. $59.00.

MANUAL OF REMOTE SENSING. Robert N. Colwell, editor. American Society for Photogrammetry and Remote Sensing. 210 Little Falls Street, Falls Church, VA 22046-4398. (703) 534-6617. 2nd edition. 1983. $106.00 for set.

ONLINE DATA BASES

CA SEARCH. Chemical Abstracts Service, P.O. Box 3012, Columbus, OH 43120. (800) 848-6538 or (614) 421-3600. Comprehensive guide to chemical literature, 1972 to present. Inquire as to online cost and availability.

COMPENDEX. Engineering Information, Inc., 345 East 47th Street, New York, NY 10017. (800) 221-1044 or (212) 705-7615. Engineering and technical literature, 1975 to present. Inquire as to online cost and availability.

GEOREF. Online version of the BIBLIOGRAPHY AND INDEX OF GEOLOGY. American Geological Institute, 4220 King Street, Alexandria, VA 22302. (703) 379-2480. 1969 to present. Inquire as to online cost and availability.

NTIS. National Technical Information Service, 5285 Port Royal Road, Springfield, VA 22161. (703) 487-4630. Broad coverage of government sponsored research reports, 1964 to present. Inquire as to online cost and availability.

SCISEARCH. Institute for Scientific Information, 3501 Market Street, Philadelphia, PA 19104. (800) 523-1850 or (215) 386-0100. Broad multidisciplinary title and author index to the international literature of science and technology, 1974 to present. Inquire as to online cost and availability.

WILSONLINE. H.W. Wilson and Company, 950 University Avenue, Bronx, NY 10452. (800) 367-6770 or (212) 588-8400. Makes available online versions of the H.W. Wilson indexes including Applied Science and Technology Index, Business Periodicals Index and Readers' Guide to Periodical Literature. Approximately 1980 to present. Inquire as to online cost and availability.

PERIODICALS

AMERICAN CARTOGRAPHER. American Congress on Surveying and Mapping, 210 Little Falls Street, Falls Church, VA 22046. (703) 241-2446. 1974 to present. Quarterly. $60.00 per year.

AMERICAN CONGRESS OF SURVEYING AND MAPPING. BULLETIN. American Congress on Surveying and Mapping, 210 Little Falls Street, Falls Church, VA 22046. (703) 241-2446. 1950 to present. Bimonthly. $50.00 per year.

AMERICAN CONGRESS OF SURVEYING AND MAPPING. PROCEEDINGS. American Congress on Surveying and Mapping, 210 Little Falls Street, Falls Church, VA 22046. (703) 241-2446. 1942 to present. Semi-annual. $25.00 per year.

ASSOCIATION OF AMERICAN GEOGRAPHERS. ANNALS. Association of American Geographers, 1710 16th Street, N.W., Washington, DC 20009. (202) 234-1450. 1911 to present. Quarterly. $45.00 per year.

IEEE TRANSACTIONS ON GEOSCIENCE AND REMOTE SENSING. IEEE Geoscience and Remote Sensing Society. Institute of Electrical and Electronics Engineers, 345 East 47th Street, New York, NY 10017. (212) 705-7900. Order from: IEEE Service Center, 445 Hoes Lane, Piscataway, NJ 08854. 1963 to present. Bimonthly. $110.00 per year.

JOURNAL OF IMAGING SCIENCE. Society of Photographic Scientists and Engineers. 7003 Kilworth Lane, Springfield, VA 22151. (703) 642-9090. 1956 to present. Bimonthly. $70.00 per year.

JOURNAL OF IMAGING TECHNOLOGY. Society of Photographic Scientists and Engineers. 7003 Kilworth Lane, Springfield, VA 22151. (703) 642-9090. 1975 to present. Bimonthly. $70.00 per year.

MARINE GEODESY: AN INTERNATIONAL JOURNAL OF OCEAN SURVEYS, MAPPING, AND SENSING. Crane Russak and Company, Inc., 3 East 44th Street, New York, NY 10017. (212) 867-1490. 1977 to present. Quarterly. $86.00 per year.

PHOTOGRAMMETRIA. Elsevier Science Publishing Company, Inc., 52 Vanderbilt Avenue, New York, NY 10017. (212) 370-5520. 1949 to present. Quarterly. $65.00 per year.

PHOTOGRAMMETRIC ENGINEERING AND REMOTE SENSING. American Society for Photogrammetry and Remote Sensing. 210 Little Falls Street, Falls Church, VA 22046-4398. (703) 534-6617. Order from: Allen Press, Inc., 1041 New Hampshire Street, Box 368, Lawrence, KS 66044. 1934 to present. Monthly. $80.00 per year.

PHOTOGRAMMETRIC RECORD. Photogrammetry Society, Department of Photogrammetry and Surveying, University College London, Gower Street, London WC1E 6BT, England. Semiannual. $37.50 per year.

SCIENTIFIC AND APPLIED PHOTOGRAPHY AND CINEMATOGRAPHY. Gordon and Breach Science Publishers, Inc., 50 West 23rd Street, New York, NY 10010. (212) 206-8900. 12 times per year. $496.00 per year.

REMOTE SENSING OF ENVIRONMENT. Elsevier Science Publishing Company, Inc., 52 Vanderbilt Avenue, New York, NY 10017. (212) 370-5520. 1968 to present. Six times per year. $210.00 per year.

REMOTE SENSING REVIEWS. Harwood Academic Publishers, 50 West 23rd Street, New York, NY 10010. (212) 206-8900. Quarterly. $160.00 per year.

RESEARCH CENTERS AND INSTITUTES

CENTER FOR REMOTE SENSING AND CARTOGRAPHY. 420 Chipta Way, Salt Lake City, UT 84112. (801) 581-8218.

GEOPHOTOGRAPHY AND REMOTE SENSING CENTER. University of Idaho, Geology Department, Moscow, ID 83843. (208) 885-7977.

NATIONAL RESEARCH COUNCIL OF CANADA, DIVISION OF PHYSICS. Ottawa, ON, Canada K1A OR6. (613) 993-1053.

PHOTOGRAMMETRY AND REMOTE SENSING SECTION. Tennessee Valley Authority, Office of Natural Resources and Economic Development, Haney Building, Chattanooga, TN 37401. (615) 755-2148.

U.S. ARMY ENGINEER TOPOGRAPHIC LABORATORIES. U.S. Army Corps of Engineers, Fort Belvoir, VA 22060. (202) 355-2634.

TORNADOS

See also: METEOROLOGY, THUNDERSTORMS

ABSTRACT SERVICES AND INDEXES

GENERAL SCIENCE INDEX. H.W. Wilson Company, 950 University Avenue, Bronx, NY 10452. (800) 367-6770 or (212) 588-8400. Inquire as to cost and availability.

METEOROLOGICAL AND GEOASTROPHYSICAL ABSTRACTS. American Meteorological Society, 45 Beacon Street, Boston, MA 02108. (617) 227-2425. 1950 to present. Monthly. $450.00 per year.

SCIENCE CITATION INDEX. Institute for Scientific Information, 3501 Market Street, Philadelphia, PA 19104. (800) 523-1850 or (215) 386-0100. Inquire as to cost and availability.

ASSOCIATIONS AND PROFESSIONAL SOCIETIES

AMERICAN ASSOCIATION STATE CLIMATOLOGISTS. c/o Professor John Griffiths, Meteorology Department, O and M Building, Texas A and M University, College Station, TX 77843. (409) 845-7320.

AMERICAN METEOROLOGICAL SOCIETY. 45 Beacon Street, Boston, MA 02108. (617) 227-2425.

INTERNATIONAL ASSOCIATION OF METEOROLOGY AND ATMOSPHERIC PHYSICS. UCAR, Post Office Box 3000, Boulder, CO 80307.

NATIONAL ENVIRONMENTAL SATELLITE DATA, AND INFORMATION SERVICE. 3300 Whitehaven Street, NW, Washington, DC 20235. (202) 634-7318.

NATIONAL WEATHER ASSOCIATION. 4400 Stamp Road, Room 404, Temple Hills, MD 20748. (301) 899-3784.

UNIVERSITY CORPORATION FOR ATMOSPHERIC RESEARCH. Box 3000, 1850 Table Mesa Drive, Boulder, CO 80307. (303) 497-1000.

WEATHER MODIFICATION ASSOCIATION. Post Office Box 8116, Fresno, CA 93747. (209) 291-8466.

DIRECTORIES AND BIOGRAPHICAL SOURCES

AMERICAN MEN AND WOMEN OF SCIENCE. Physical and Biological Sciences. Sixteenth edition. R.R. Bowker Company, 205 East 42nd Street, New York, NY 10017. (800) 521-8110 or (212) 916-1600. 1986. $595.00.

GOVERNMENT RESEARCH DIRECTORY. Gale Research Company, Book Tower, Detroit, MI 48226. (800) 521-0707. Fourth edition. 1987. $350.00.

INTERDOC: DIRECTORY OF PUBLISHED PROCEEDINGS, SERIES. SEMT-Science/Engineering/Medicine/Technology. Interdoc Corporation, 173 Halstead Avenue, Box 326, Harrison, NY 10528. (014) 835-3506. Ten times per year. $325.00 per year.

METEOROLOGICAL SERVICES OF THE WORLD. World Meteorological Organization. Available from: American Meteorological Society, 45 Beacon Street, Boston, MA 02108. (617) 227-2425. Annual. $35.00.

NATIONAL WEATHER SERVICE OFFICES AND STATIONS. National Oceanic and Atmospheric Administration, Department of Commerce, Silver Spring, MD 20910. (301) 427-7698. Annual. Free.

RESEARCH CENTERS DIRECTORY. Gale Research Company, Book Tower, Detroit, MI 48226. (800) 521-0707. Twelfth edition. 1987. $365.00 for set.

ENCYCLOPEDIAS AND DICTIONARIES

ENCYCLOPEDIA OF CLIMATOLOGY. John E. Oliver and Rhodes W. Fairbridge, editors. Van Nostrand Reinhold, Incorporated, 115 Fifth Avenue, New York, NY 10003. (800) 543-2681. 1987. $89.95.

ENCYCLOPEDIA OF PHYSICAL SCIENCE AND TECHNOLOGY. Academic Press, Incorporated, Orlando, FL 32887. (800) 321-5068 or (305) 345-2734. Inquire as to cost and availability.

GENERAL WORKS

HURRICANES, STORMS AND TORNADOS: GEOGRAPHIC CHARACTERISTICS AND GEOLOGICAL ACTIVITY. B.B. Battacharya. International Publishing Service, Post Office Box 230, Accord, MA 02018. (617) 749-3628. 1983. $26.50.

TORNADOES, DARK DAYS, ANOMALOUS PRECIPITATION AND RELATED WEATHER PHENOMENA. William R. Corliss. The Sourcebook Project, Post Office Box 107, Glen Arm, MD 21057. (301) 668-6047. 1983. $11.95.

ONLINE DATA BASES

DISSERTATION ABSTRACTS ONLINE. University Microfilms International, 300 North Zeeb Road, Ann Arbor, MI 48106. (800) 521-0600 or (313) 761-4700. Scope includes virtually all doctoral dissertations accepted at accredited American institutions from 1861 to present in 252 subject areas. Inquire as to cost and availability.

METEOROLOGICAL AND GEOASTROPHYSICAL ABSTRACTS. American Meteorological Society, 45 Beacon Street, Boston, MA 02108. (617) 227-2425. 1950 to present. Monthly. $450.00 per year.

NTIS. National Technical Information Service, 5285 Port Royal Road, Springfield, VA 22161. (703) 487-4630. Broad coverage of government-sponsored research reports, 1964 to present. Inquire as to cost and availability.

SCISEARCH. Institute for Scientific Information, 3501 Market Street, Philadelphia, PA 19104. (800) 523-1850 or (215) 386-0100. Broad multidisciplinary title and author index to the international literature of science and technology, 1974 to present. Inquire as to cost and availability.

PERIODICALS

AGRICULTURAL AND FOREST METEOROLOGY. Elsevier Science Publishing Company, Incorporated, 52 Vanderbilt Avenue, New York, NY 10017. (212) 370-5520. 1964 to present. Monthly. $260.00 per year.

AMERICAN METEOROLOGICAL SOCIETY BULLETIN. American Meteorological Society, 45 Beacon Street, Boston, MA 02108. (617) 227-2425.

CLIMATIC CHANGE: AN INTERDISCIPLINARY, INTERNATIONAL JOURNAL DEVOTED TO THE DESCRIPTION, CAUSES AND IMPLICATIONS OF CLIMATIC CHANGE. D. Reidel Publishing Company, 190 Old Derby Street, Hingham, MA 02043. (617) 871-6600. 1977 to present. Six times per year. $125.00 per year.

JOURNAL OF THE ATMOSPHERIC SCIENCES. American Meteorological Society, 45 Beacon Street, Boston, MA 02108. (617) 227-2425. 1944 to present. Semimonthly. $220.00 per year.

MONTHLY WEATHER REVIEW. American Meteorological Society, 45 Beacon Street, Boston, MA 02108. (617) 227-2425. 1872 to present. Monthly. $120.00 per year.

NATIONAL WEATHER DIGEST. National Weather Association, 4400 Stamp Road, Room 404, Temple Hills, MD 20748. (301) 899-3784. 1976 to present. Quarterly. $20.00 per year.

WEATHER. Royal Meteorological Society, James Glaisher House, Grenville Place, Bracknell Berkshire, RG12 1BX, England. 1946 to present. $30.00 per year.

WEATHERWISE. Heldref Publications, 4000 Albemarle Street, NW, Washington, DC 20016. (202) 362-6445. 1948 to present. Bimonthly. $20.00 per year.

RESEARCH CENTERS AND INSTITUTES

COOPERATIVE INSTITUTE FOR MESCOCALE METEROLOGICAL STUDIES. University of Oklahoma, 401 East Boyd, Norman, OK 73019. (405) 325-3041.

NATIONAL CENTER FOR ATMOSPHERIC RESEARCH. Box 3000, Boulder, CO 80307. (303) 497-1000.

NATIONAL SEVERE STORMS LABORATORY. 1313 Halley Circle, Norman, OK 73069. (405) 360-3620.

TRACE ANALYSIS

See: ANALYTICAL CHEMISTRY

TRAFFIC ENGINEERING

See: HIGHWAY ENGINEERING

TRANSFORMERS

See: ELECTRIC POWER ENGINEERING

TRANSISTORS

See: SEMICONDUCTORS

TRANSMISSION LINES

See: ELECTRIC POWER ENGINEERING

TRANSMISSIONS

See: AUTOMOTIVE ENGINEERING

TRIBOLOGY

See: LUBRICATION

TRIGONOMETRY

See: MATHEMATICS

TSUNAMIS

See also: EARTHQUAKE ENGINEERING, GEOLOGY, GEOPHYSICS, GEOTECHNICAL ENGINEERING, OCEANOGRAPHY, PLANETARY SCIENCE, PLATE TECTONICS, SEISMOLOGY, VOLCANOLOGY

ABSTRACT SERVICES AND INDEXES

BIBLIOGRAPHY AND INDEX OF GEOLOGY. American Geological Institute, 4220 King Street, Alexandria, VA 22302. (703) 379-2480. 1969 to present. Monthly. $1100.00 per year.

CHEMICAL ABSTRACTS. American Chemical Society, Chemical Abstracts Service, Box 3012, Columbus, OH 43210. (614) 421-3600. 1907 to present. Weekly. $9500.00 per year.

CONFERENCE PAPERS INDEX. Cambridge Scientific Abstracts, 5161 River Road, Bethesda, MD 20816. 1972 to present. Monthly. Inquire as to cost and availability.

CURRENT CONTENTS: PHYSICAL, CHEMICAL AND EARTH SCIENCES. Institute for Scientific Information, 3501 Market Street, Philadelphia, PA 19104. (800) 523-1850 or (215) 386-0100. Weekly. $295.00 per year.

INDEX TO SCIENTIFIC AND TECHNICAL PROCEEDINGS. Institute for Scientific Information, 3501 Market Street, Philadelphia, PA 19104. (800) 523-1850 or (215) 386-0100. 1978 to present. Monthly. $775.00 per year.

INDEX TO SCIENTIFIC REVIEWS. Institute for Scientific Information, 3501 Market Street, Philadelphia, PA 19104. (800) 523-1850 or (215) 386-0100. 1974 to present. Semi-annual. $550.00 per year.

METEOROLOGICAL AND GEOASTROPHYSICAL ABSTRACTS. American Meteorological Society, 45 Beacon Street, Boston, MA 02108. (617) 227-2425. 1950 to present. Monthly. $450.00 per year.

OCEAN ABSTRACTS. Cambridge Scientific Abstracts, 5161 River Road, Bethesda, MD 20816. (301) 951-1400. 1963 to present. Bimonthly. $450.00 per year.

PHYSICS ABSTRACTS. Institution of Electrical Engineers. Available from: IEEE Service Center, 445 Hoes Lane, Piscataway, NJ 08854. 1898 to present. Bimonthly. $1700.00 per year.

PHYSICS BRIEFS. Physik Verlag GmbH, Postfach 1260/1280, D-6940 Weinheim, West Germany. (212) 661-9404. 1920 to present. Twenty-six times per year. $1250.00 per year.

SCIENCE CITATION INDEX. Institute for Scientific Information, 3501 Market Street, Philadelphia, PA 19104. (800) 523-1850 or (215) 386-0100. Six times per year. $6200. 00 per year.

ASSOCIATIONS AND PROFESSIONAL SOCIETIES

AMERICAN GEOPHYSICAL UNION. 2000 Florida Avenue, N.W., Washington, DC 20009. (202) 462-6903.

EARTHQUAKE ENGINEERING RESEARCH INSTITUTE. 2620 Telegraph Avenue, Berkeley, CA 94704. (415) 848-0972.

GEOLOGICAL SOCIETY OF AMERICA. 3300 Penrose Place, Boulder, CO 80301. (303) 447-2020.

SEISMOLOGICAL SOCIETY OF AMERICA. 6431 Fairmont Avenue, No. 7, El Cerrito, CA 94530. (415) 525-5474.

DIRECTORIES AND BIOGRAPHICAL SOURCES

AMERICAN MEN AND WOMEN OF SCIENCE. R.R. Bowker, Inc., Order Department, 245 West 17th Street, New York, NY 10011. (800) 521-8110. Eight volumes. 1986. $595.00 for set.

INTERNATIONAL RESEARCH CENTERS DIRECTORY 1988-89. Darren L. Smith, editor. Gale Research Company, Book Tower, Detroit, MI 48226. (800) 521-0707. 4th edition. 1987. $360.00.

RESEARCH CENTERS DIRECTORY 1988. Gale Research Company, Book Tower, Detroit, MI 48226. (800) 521-0707. 12th edition. 1987. $365.00 for set.

SCIENTIFIC AND TECHNICAL ORGANIZATIONS AND AGENCIES DIRECTORY. Margaret Labash Young, editor. Gale Research Company, Book Tower, Detroit, MI 48226. (800) 521-0707. 2nd edition. 1987. $185.00.

SOCIETY OF EXPLORATION GEOPHYSICISTS ROSTER. P.O. Box 702740, Tulsa, OK 74170. (918) 493-3516. Annual. Inquire.

WHO'S WHO IN ENGINEERING. Gordon Davis, editor. Hemisphere Publishing Corporation, 79 Madison Avenue, New York, NY 10016-7892. (800) 821-8312. Sixth edition. 1985. $200.00.

ENCYCLOPEDIAS AND DICTIONARIES

MCGRAW-HILL ENCYCLOPEDIA OF THE GEOLOGICAL SCIENCES. McGraw-Hill Book Company, 1221 Avenue of the Americas, New York, NY 10020. (212) 512-2000. Second edition. 1988. $85.00.

GENERAL WORKS

EARTHQUAKE PREDICTION. K. Shimazaki and W.D. Stuart, editors. Birkhauser Boston, Inc., 380 Green Street, Cambridge, MA 02139. (617) 876-2333. 1985. $24.95.

EARTHQUAKES, TIDES, UNIDENTIFIED SOUNDS AND RELATED PHENOMENA. William R. Corliss. The Sourcebook Project, P.O. Box 107, Glen Arm, MD 21057. (301) 668-6047. 1983. $12.95.

THE GREAT WAVES. Douglas Myles. McGraw-Hill Book Company, 1221 Avenue of the Americas, New York, NY 10020. (212) 512-2000. 1985. $16.95.

PRACTICAL APPROACHES TO EARTHQUAKE PREDICTION AND WARNING. C. Kisslinger and T. Rikitake, editors. Kluwer Academic Publishers, 190 Old Derby Street, Hingham, MA 02043. (617) 749-5262. 1986. $79.00.

SEISMIC WAVES AND SOURCES. A. Ben-Menahem and S. Singh. Springer-Verlag New York, Inc., 175 Fifth Avenue, New York, NY 10010. (800) 526-7254. 1981. $99.00.

ONLINE DATA BASES

CA SEARCH. Chemical Abstracts Service, P.O. Box 3012, Columbus, OH 43120. (800) 848-6538 or (614) 421-3600. Comprehensive guide to chemical literature, 1972 to present. Inquire as to online cost and availability.

COMPENDEX. Engineering Information, Inc., 345 East 47th Street, New York, NY 10017. (800) 221-1044 or (212) 705-7615. Engineering and technical literature, 1975 to present. Inquire as to online cost and availability.

DISSERTATION ABSTRACTS ONLINE. University Microfilms International, 300 North Zeeb Road, Ann Arbor, MI 48106. (800) 521-0600 or (313) 761-4700. Scope includes virtually all doctoral dissertations accepted at accredited American institutions from 1861 to present in over 250 subject areas. Inquire as to online cost and availability.

GEOREF. Online version of the BIBLIOGRAPHY AND INDEX OF GEOLOGY. American Geological Institute, 4220 King Street, Alexandria, VA 22302. (703) 379-2480. 1969 to present. Inquire as to online cost and availability.

INSPEC. INSPEC Marketing Department, Institution of Electrical Engineers. Available from IEEE Service Center, 445 Hoes Lane, Piscataway, NJ 08854. (201) 981-0060. Online version of Physics Abstracts. Inquire as to online cost and availability.

NTIS. National Technical Information Service, 5285 Port Royal Road, Springfield, VA 22161. (703) 487-4630. Broad coverage of government sponsored research reports, 1964 to present. Inquire as to online cost and availability.

SCISEARCH. Institute for Scientific Information, 3501 Market Street, Philadelphia, PA 19104. (800) 523-1850 or (215) 386-0100. Broad multidisciplinary title and author index to the international literature of science and technology, 1974 to present. Inquire as to online cost and availability.

PERIODICALS

DISASTERS: THE INTERNATIONAL JOURNAL OF DISASTER STUDIES AND PRACTICE. International Disaster Institute. Subscribe to: IDI, 85 Maryleborn Hign Street, London, W1M 3DE, England. 1977 to present. Quarterly. $60.00 per year.

EARTHQUAKE INFORMATION BULLETIN. U.S. Geological Survey, 12201 Sunrise Valley Drive, Reston, VA 22092. Order from: Superintendent of Documents, U.S. Government Printing Office, 20402. Bimonthly. $15.00 per year.

EARTHQUAKE NOTES. Seismological Society of America, c/o Wilbur Rinehart, 1320 Meadow Avenue, Boulder, CO 80302. 1929 to present. Quarterly. $10.00 per year.

EARTHQUAKE PREDICTION RESEARCH. D. Reidel Publishing Company, 190 Old Derby Street, Hingham, MA 02043. 1982 to present. Quarterly. $85.00 per year.

GEOPHYSICAL AND ASTROPHYSICAL FLUID DYNAMICS. Gordon and Breach Science Publishers, Inc., 50 West 23rd Street, New York, NY 10010. (212) 206-8900. 1970 to present. 16 times per year. $260.00 per year.

GEOPHYSICAL RESEARCH LETTERS. American Geophysical Union, 2000 Florida Avenue, N.W., Washington, DC 20009. (202) 462-6903. 1974 to present. Monthly. $185.00 per year.

JGR: JOURNAL OF GEOPHYSICAL RESEARCH: SOLID EARTH AND PLANETS. American Geophysical Union, 2000 Florida Avenue, N.W., Washington, DC 20009. (202) 462-6903. Monthly. $760.00 per year.

SEISMOLOGICAL SOCIETY OF AMERICA BULLETIN. Seismologycal Society of America, 6431 Fairmont Avenue, No. 7, El Cerrito, CA 94530. (415) 525-5474. 1911 to present. Bimonthly. $90.00 per year.

VOLCANOLOGY AND SEISMOLOGY. Gordon and Breach Science Publishers, Inc., 50 West 23rd Street, New York, NY 10010. (212) 206-8900. Monthly. $500.00 per year.

RESEARCH CENTERS AND INSTITUTES

EARTHQUAKE ENGINEERING RESEARCH INSTITUTE. 2620 Telegraph Avenue, Berkeley, CA 94704. (415) 848-0972.

LAMONT-DOHERTY GEOLOGICAL OBSERVATORY. Columbia University, Palisades, NY 10964. (914) 359-2900.

SEISMOGRAPHIC STATIONS. University of California at Berkeley, Berkeley, CA 94720. (415) 642-3977.

SEISMOLOGICAL LABORATORY. California Institute of Technology, Pasadena, CA 91125. (818) 356-6912.

TURBINES

See also: AUTOMOTIVE ENGINEERING, ENGINES, GAS TURBINES, JET PROPULSION, STEAM ENGINES, STEAM TURBINES

ABSTRACT SERVICES AND INDEXES

APPLIED MECHANICS REVIEW. American Society of Mechanical Engineers, 345 East 47th Street, New York, NY 10017. (212) 705-7703. 1948 to present. Monthly. $360.00 per year.

APPLIED SCIENCE AND TECHNOLOGY INDEX. H.W. Wilson and Company, 950 University Avenue, Bronx, NY 10452. (800) 367-6670 or (212) 588-8400. Monthly. Inquire as to cost and availability.

ENGINEERING INDEX MONTHLY AND AUTHOR INDEX. Engineering Information Inc., 345 East 47th Street, New York, NY 10017. (212) 705-7600. Monthly. $1560.00 per year.

INTERNATIONAL AEROSPACE ABSTRACTS. American Institute of Aeronautics and Astronautics, Technical Information Service, 370 L'Enfant Promenade, S.W., Washington, DC 20024. (202) 646-7400. 1961 to present. Semi-monthly. $700.00 per year.

ISMEC BULLETIN (Information Service in Mechanical Engineering). Cambridge Scientific Abstracts, 5161 River Road, Bethesda, MD 20816. (301) 951-1400. 1973 to present. Monthly. $450.00 per year.

ASSOCIATIONS AND PROFESSIONAL SOCIETIES

AMERICAN INSTITUTE OF AERONAUTICS AND ASTRONAUTICS. 370 L'Enfant Promenade, S.W., Washington, DC 20024. (202) 646-7400.

AMERICAN SOCIETY OF MECHANICAL ENGINEERS. 345 East 47th Street, New York, NY 10017. (212) 705-7703.

INTERNATIONAL GAS TURBINE INSTITUTE. 4250 Perimeter Park, South, Atlanta, GA 30341. (404) 451-1905.

SOCIETY OF AUTOMOTIVE ENGINEERS (SAE). 400 Commonwealth Drive, Warrendale, PA 15096. (412) 776-4841.

NATIONAL ASSOCIATION OF POWER ENGINEERS. 2350 East Devon Avenue, Suite 115, Des Plaines, IL 60018. (312) 298-0600.

DIRECTORIES AND BIOGRAPHICAL SOURCES

ANNUAL WHO'S WHO IN GAS TURBINE TECHNOLOGY. Internaitonal Gas Turbine Institute, 4250 Perimeter Park, South, Atlanta, GA 30341. (404) 451-1905. Annual. Inquire.

1987 DIRECTORY OF ENGINEERING SOCIETIES AND RELATED ORGANIZATIONS. Gordon Davis, editor. Hemisphere Publishing Corporation, 1010 Vermont Avenue, NW, Washington, DC 20005. (800) 526-0275. 12th edition. 1987. $100.00.

RESEARCH CENTERS DIRECTORY 1988. Gale Research Company, Book Tower, Detroit, MI 48226. (800) 521-0707. 12th edition. 1987. $365.00 for set.

SCIENTIFIC AND TECHNICAL ORGANIZATIONS AND AGENCIES DIRECTORY. Margaret Labash Young, editor. Gale Research Company, Book Tower, Detroit, MI 48226. (800) 521-0707. 2nd edition. 1987. $185.00.

WHO'S WHO IN ENGINEERING. Gordon Davis, editor. Hemisphere Publishing Corporation, 1010 Vermont Avenue, NW, Washington, DC 20005. (800) 526-0275. 6th edition. 1985. $200.00.

ENCYCLOPEDIAS AND DICTIONARIES

THESAURUS OF SCIENTIFIC, TECHNICAL, AND ENGINEERING TERMS. Hemisphere Publishing Corporation, 1010 Vermont Avenue, NW, Washington, DC 20005. (800) 526-0275. 1988. $125.00.

GENERAL WORKS

AERODYNAMICS OF TURBINES AND COMPRESSORS. W.R. Hawthorne, editor. Princeton University Press, 41 William Street, Princeton, NJ 08540. (609) 452-4122. 1964. $66.50.

THE DESIGN OF HIGH-EFFICIENCY TURBOMACHINERY AND GAS TURBINES. David G. Wilson. MIT Press, 28 Carleton Street, Cambridge, MA 02142. (617) 253-2884. 1983. $39.95.

MARINE STEAM ENGINES AND TURBINES. S.C. McBirnie and W.J. Fox. Butterworth's Publishing, 80 Montvale Avenue, Stoneham, MA 02180. (800) 325-4177. 4th edition. 1980. $64.95.

TURBOMECHANICS: A GUIDE TO DESIGN, SELECTION AND THEORY. O.E. Balje, editor. John Wiley and Sons, Inc., 605 Third Avenue, New York, NY 10158. (800) 526-5368. 1981. $64.95.

TURBOMACHINERY. W. Logan. Marcel Dekker Inc., 270 Madison Avenue, New York, NY 10016. (800) 228-1160. 1981. $29.75.

HANDBOOKS AND MANUALS

SAWYER'S TURBOMACHINERY MAINTENANCE HANDBOOKS. John W. Sawyer and Kurt Hallberg, editors. Turbomachinery International Publications, P.O. Box 5550, Norwalk, CT 06856. (203) 853-6015. Three volumes. 1981. $180.00 for set.

ONLINE DATA BASES

COMPENDEX. Engineering Information, Inc., 345 East 47th Street, New York, NY 10017. (800) 221-1044 or (212) 705-7615. Engineering and technical literature, 1975 to present. Inquire as to online cost and availability.

NTIS. National Technical Information Service, 5285 Port Royal Road, Springfield, VA 22161. (703) 487-4630. Broad coverage of government sponsored research reports, 1964 to present. Inquire as to online cost and availability.

WILSONLINE. H.W. Wilson and Company, 950 University Avenue, Bronx, NY 10452. (800) 367-6770 or (212) 588-8400. Makes available online versions of the H.W. Wilson indexes including Applied Science and Technology Index, Business Periodicals Index and Readers' Guide to Periodical Literature. Approximately 1980 to present. Inquire as to online cost and availability.

OTHER SOURCES

WHAT EVERY ENGINEER SHOULD KNOW ABOUT ENGINEERING SOURCES. Marcel Dekker Inc., 270 Madison

Avenue, New York, NY 10016. (800) 228-1160. 1984. $24.95.

PERIODICALS

INTERNATIONAL JOURNAL OF TURBO AND JET ENGINES. Kluwer Academic Publishers, 190 Old Derby Street, Hingham, MA 02043. (617) 749-5262. Quarterly. $125.00 per year.

JOURNAL OF ENGINEERING FOR GAS TURBINES AND POWER. American Society of Mechanical Engineers, 345 East 47th Street, New York, NY 10017. (212) 705-7703. 1970 to present. Quarterly. $100.00 per year.

JOURNAL OF PROPULSION AND POWER. American Institute of Aeronautics and Astronautics, 370 L'Enfant Promenade, S.W., Washington, DC 20024. (202) 646-7400. Bimonthly. $170.00 per year.

TURBOMACHINERY INTERNATIONAL. Turbomachinery International Publications, P.O. Box 5550, Norwalk, CT 06856. (203) 853-6015. 1959 To present. 9 times per year. $38.00 per year

RESEARCH CENTERS AND INSTITUTES

INTERNATIONAL GAS TURBINE INSTITUTE. 4250 Perimeter Park, South, Atlanta, GA 30341. (404) 451-1905.

TURBOMACHINERY LABORATORY. Texas A&M University, Mechanical Engineering Department, College Station, TX 77843. (409) 845-7417.

TURBOPROP

See: JET PROPULSION

U

ULTRA LARGE SCALE INTEGRATION

See: VERY LARGE SCALE INTEGRATION

ULTRALIGHT AIRCRAFT

See: AIRCRAFT

ULTRASONICS

See: ACOUSTICS

ULTRAVIOLET ASTRONOMY

See: ASTRONOMY

ULTRAVIOLET SPECTROSCOPY

See: SPECTROSCOPY

UNCERTAINTY PRINCIPLE

See: QUANTUM MECHANICS

UNDERWATER PHOTOGRAPHY

See also: CAMERAS, PHOTOCHEMISTRY, PHOTOGRAPHIC FILM, PHOTOGRAPHY

ABSTRACT SERVICES AND INDEXES

APPLIED SCIENCE AND TECHNOLOGY INDEX. H.W. Wilson and Company, 950 University Avenue, Bronx, NY 10452. (800) 367-6670 or (212) 588-8400. Monthly. Inquire as to cost and availability.

CHEMICAL ABSTRACTS. American Chemical Society, Chemical Abstracts Service, Box 3012, Columbus, OH 43210. (614) 421-3600. 1907 to present. Weekly. $9500.00 per year.

ENGINEERING INDEX MONTHLY AND AUTHOR INDEX. Engineering Information Inc., 345 East 47th Street, New York, NY 10017. (212) 705-7600. Monthly. $1560.00 per year.

PHOTOGRAPHIC ABSTRACTS. Royal Photographic Society of Great Britain, Scientific and Technical Group, 62 Chelmsford Road, Shenfield, Brentwood, Essex, England. 1921 to present. Six times per year. $140.00 per year.

ASSOCIATIONS AND PROFESSIONAL SOCIETIES

OPTICAL SOCIETY OF AMERICA. 1816 Jefferson Place, N.W., Washington, DC 20036. (202) 223-8130.

SOCIETY OF PHOTOGRAPHIC SCIENTISTS AND ENGINEERS. 7003 Kilworth Lane, Springfield, VA 22151. (703) 642-9090.

SPIE - THE INTERNATIONAL SOCIETY FOR OPTICAL ENGINEERING. P.O. Box 10, 1022 19th Street, Bellingham, WA 98227. (206) 676-3290.

DIRECTORIES AND BIOGRAPHICAL SOURCES

INTERNATIONAL RESEARCH CENTERS DIRECTORY 1988-89. Darren L. Smith, editor. Gale Research Company, Book Tower, Detroit, MI 48226. (800) 521-0707. 4th edition. 1987. $360.00.

1987 DIRECTORY OF ENGINEERING SOCIETIES AND RELATED ORGANIZATIONS. Gordon Davis, editor. Hemisphere Publishing Corporation, 1010 Vermont Avenue, NW, Washington, DC 20005. (800) 526-0275. 12th edition. 1987. $100.00.

RESEARCH CENTERS DIRECTORY 1988. Gale Research Company, Book Tower, Detroit, MI 48226. (800) 521-0707. 12th edition. 1987. $365.00 for set.

SCIENTIFIC AND TECHNICAL ORGANIZATIONS AND AGENCIES DIRECTORY. Margaret Labash Young, editor. Gale Research Company, Book Tower, Detroit, MI 48226. (800) 521-0707. 2nd edition. 1987. $185.00.

WHO'S WHO IN ENGINEERING. Gordon Davis, editor. Hemisphere Publishing Corporation, 1010 Vermont Avenue, NW, Washington, DC 20005. (800) 526-0275. 6th edition. 1985. $200.00.

ENCYCLOPEDIAS AND DICTIONARIES

THESAURUS OF PHOTOGRAPHIC SCIENCE AND ENGINEERING. Society of Photographic Scientists and Engineers. Books on Demand, 300 North Zeeb Road, Ann Arbor, MI 48106. (313) 761-4700. $34.50 in paper.

THESAURUS OF SCIENTIFIC, TECHNICAL, AND ENGINEERING TERMS. Hemisphere Publishing Corporation, 1010 Vermont Avenue, NW, Washington, DC 20005. (800) 526-0275. 1988. $125.00.

GENERAL WORKS

PHOTOGRAPHIC SCIENCE. Earl. N. Mitchell. John Wiley and Sons, Inc., 605 Third Avenue, New York, NY 10158. (800)

526-5368. 1984. $37.50.

PHOTOGRAPHY FOR THE SCIENTIST. Richard A. Morton. Academic Press, Inc., 6277 Sea Harbor Drive, Orlando, FL 32821. (800) 321-5068. 2nd edition. 1984. $102.50.

UNDERWATER PHOTOGRAPHY. John Turner. Focal Press, 80 Montvale Avenue, Stoneham, MA 02180. (617) 438-8464. 1982. $39.95.

UNDERWATER PHOTOGRAPHY AND TELEVISION FOR SCIENTISTS. Oxford University Press, 200 Madison Avenue, New York, NY 10016. (800) 458-5833. 1985. $57.50.

HANDBOOKS AND MANUALS

HANDBOOK OF PHOTOGRAPHIC SCIENCE AND ENGINEERING. Society of Photographic Scientists and Engineers. 7003 Kilworth Lane, Springfield, VA 22151. (703) 642-9090. Inquire.

KODAK PROFESSIONAL PHOTOGUIDE. Carolyn Grimes, editor. Eastman Kodak Company, 343 State Street, Rochester, NY 14650. (716) 724-4000. 1986. $19.95 in paper.

ONLINE DATA BASES

CA SEARCH. Chemical Abstracts Service, P.O. Box 3012, Columbus, OH 43120. (800) 848-6538 or (614) 421-3600. Comprehensive guide to chemical literature, 1972 to present. Inquire as to online cost and availability.

COMPENDEX. Engineering Information, Inc., 345 East 47th Street, New York, NY 10017. (800) 221-1044 or (212) 705-7615. Engineering and technical literature, 1975 to present. Inquire as to online cost and availability.

NTIS. National Technical Information Service, 5285 Port Royal Road, Springfield, VA 22161. (703) 487-4630. Broad coverage of government sponsored research reports, 1964 to present. Inquire as to online cost and availability.

SCISEARCH. Institute for Scientific Information, 3501 Market Street, Philadelphia, PA 19104. (800) 523-1850 or (215) 386-0100. Broad multidisciplinary title and author index to the international literature of science and technology, 1974 to present. Inquire as to online cost and availability.

WILSONLINE. H.W. Wilson and Company, 950 University Avenue, Bronx, NY 10452. (800) 367-6770 or (212) 588-8400. Makes available online versions of the H.W. Wilson indexes including Applied Science and Technology Index, Business Periodicals Index and Readers' Guide to Periodical Literature. Approximately 1980 to present. Inquire as to online cost and availability.

PERIODICALS

BRITISH JOURNAL OF PHOTOGRAPHY. Henry Greenwood and Company, Limited, 28 Great James Street, London WC1N 3HL, England. 1854 to present. Weekly. $70.00 per year.

FUNCTIONAL PHOTOGRAPHY. PTN Publishing Corporation, 210 Crossways Park Drive, Woodbury, NY 11797. (516) 496-8000. 1967 to present. Bimonthly. $7.50 per year.

INDUSTRIAL PHOTOGRAPHY. United Business Publications, Inc., 475 Park Avenue South, New York, NY 10016. (212) 725-2300. 1952 to present. Monthly. $15.00 per year.

JOURNAL OF IMAGING SCIENCE. Society of Photographic Scientists and Engineers. 7003 Kilworth Lane, Springfield, VA 22151. (703) 642-9090. 1956 to present. Bimonthly. $70.00 per year.

JOURNAL OF IMAGING TECHNOLOGY. Society of Photographic Scientists and Engineers. 7003 Kilworth Lane, Springfield, VA 22151. (703) 642-9090. 1975 to present. Bimonthly. $70.00 per year.

JOURNAL OF PHOTOGRAPHIC SCIENCE. Royal Photographic Society of Great Britain, 7 Ladbroke Walk, London W11, England. 1953 to present. Bimonthly. $40.00 per year.

KODAK TECH BITS. Eastman Kodak Company, 343 State Street, Rochester, NY 14650. (716) 724-4000. 1963 to present. Quarterly. Free to qualified personnel.

PHOTOGRAMMETRIC ENGINEERING AND REMOTE SENSING. American Society for Photogrammetry and Remote Sensing. 210 Little Falls Street, Falls Church, VA 22046-4398. (703) 534-6617. Order from: Allen Press, Inc., 1041 New Hampshire Street, Box 368, Lawrence, KS 66044. 1934 to present. Monthly. $80.00 per year.

SCIENTIFIC AND APPLLIED PHOTOGRAPHY AND CINEMATOGRAPHY. Gordon and Breach Science Publishers, Inc., 50 West 23rd Street, New York, NY 10010. (212) 206-8900. 12 times per year. $496.00 per year.

UNIX

See: COMPUTER OPERATING SYSTEMS

UPPER ATMOSPHERE

See: METEOROLOGY

URANUS

See also: PLANETARY SCIENCE, SOLAR SYSTEM

ABSTRACT SERVICES AND INDEXES

ASTRONOMY AND ASTROPHYSICS ABSTRACTS. Springer-Verlag New York, Incorporated, 175 Fifth Avenue, New York, NY 10010. (212) 460-1500. $70.00 per year.

ASTRONOMY AND ASTROPHYSICS MONTHLY INDEX. Olivetree Associates, Post Office Box 236, Sierra Madre, CA 91024. $212.00 per year. Complimentary copies available on request.

METEOROLOGICAL AND GEOASTROPHYSICAL ABSTRACTS. American Meteorological Society, 45 Beacon Street, Boston, MA 02108. (617) 227-2425. 1950 to present. Monthly. $450.00 per year.

STAR. (Scientific and Technical Aerospace Reports. United States National Aeronautics and Space Administration, Scientific and Technical Information Facility, Box 8757, Baltimore-Washington International Airport, MD 21240. (202) 755-2210. Semimonthly, with semiannual and annual indexes. $85.00 per year.

ANNUAL REVIEWS AND YEARBOOKS

THE ASTRONOMICAL ALMANAC. Superintendent of Documents, United States Government Printing Office, Washington, DC 20402. (202) 783-3238. Yearly.

ANNUAL REVIEW OF ASTRONOMY AND ASTROPHYSICS. Annual Reviews, Incorporated, 4139 El Camino Way, Palo Alto, CA 94306. (415) 493-4400. Annual. Inquire.

ANNUAL REVIEW OF EARTH AND PLANETARY SCIENCES. Annual Reviews, Incorporated, 4139 El Camino Way, Palo Alto, CA 94306. (415) 493-4400. Annual. Inquire.

ASSOCIATIONS AND PROFESSIONAL SOCIETIES

AMERICAN ASTRONOMICAL SOCIETY. 2000 Florida Avenue, NW, Suite 300, Washington, DC 20009. (202) 659-0134.

AMERICAN GEOPHYSICAL UNION. 2000 Florida Avenue, NW, Washington, DC 20009. (202) 462-6903.

ASTRONOMICAL LEAGUE. Post Office Box 12821, Tucson, AZ 85732. (602) 790-8471.

ASTRONOMICAL SOCIETY OF THE PACIFIC. 1290 24th Avenue, San Francisco, CA 94122. (415) 661-8660.

PLANETARY SOCIETY. 65 North Catalina Avenue, Pasadena, CA 91106-2301. (818) 793-5100.

BIBLIOGRAPHIES

A BIBLIOGRAPHY OF ASTRONOMY, 1970-1979. R.A. Seal and S.S. Martin. Libraries Unlimited, Incorporated, Littleton, CO 80160. 1982. $37.50.

SCIENTIFIC AND TECHNICAL BOOKS AND SERIALS IN PRINT 1988: AN INDEX TO LITERATURE IN SCIENCE AND TECHNOLOGY. R.R. Bowker Company, 205 East 42nd Street, New York, NY 10017. (800) 521-8110 or (212) 916-1600. $175.00.

DIRECTORIES AND BIOGRAPHICAL SOURCES

AMERICAN ASTRONOMICAL SOCIETY MEMBERS. 2000 Florida Avenue, NW, Suite 300, Washington, DC 20009. (202) 659-0134. Annual. Available to members only.

THE BIOGRAPHICAL DICTIONARY OF SCIENTISTS: ASTRONOMERS. D. Abbott, editor. Peter Bedrick Books, 125 East 23rd Street, New York, NY 10010. 1984. $24.95.

DIRECTORY OF PHYSICS AND ASTRONOMY STAFF MEMBERS. American Institute of Physics, 335 East 45th Street, New York, NY 10017. Annual.

GENERAL WORKS

URANUS AND NEPTUNE. J. Bergstrahl. NASA Conference Publication, 2330. Available from National Technical Information Service, 5285 Port Royla Road, Springfield, VA 22161. (703) 487-4838. 1984.

ONLINE DATA BASES

CA SEARCH. Chemical Abstracts Service, Post Office Box 3012, Columbus, OH 43210. Guide to chemical literature, 1972 to present. Inquire as to cost and availability.

DISSERTATION ABSTRACTS ONLINE. University Microfilms International, 300 North Zeeb Road, Ann Arbor, MI 48106. (800) 521-0600 or (313) 761-4700. Scope includes virtually all doctoral dissertations accepted at accredited American institutions from 1861 to present in 252 subject areas. Inquire as to cost and availability.

NASA. National Aeronautics and Space Administration, Scientific and Technical Information Branch, 300 7th Street, SW, Washington, DC 20546. Citations and abstracts of aerospace literature, 1962 to present. Inquire as to cost and availability.

NTIS. National Technical Information Service, 5285 Port Royal Road, Springfield, VA 22161. (703) 487-4630. Broad coverage of government sponsored research reports, 1964 to present. Inquire as to cost and availability.

SCISEARCH. Institute for Scientific Information, 3501 Market Street, Philadelphia, PA 19104. (800) 523-1850 or (215) 386-0100. Broad multidisciplinary title and author index to the international literature of science and technology, 1974 to present. Inquire as to cost and availability.

PERIODICALS

ASSOCIATION OF LUNAR AND PLANETARY OBSERVERS. Journal Association of Lunar and Planetary Observers, Box 16131, San Francisco, CA 94116. (415) 566-5786. 1947 to present. Quarterly. $15.00 per year.

ASTRONOMICAL JOURNAL. American Astronomical Society. Available from: American Institute of Physics, 335 East 45th Street, New York, NY 10017. 9212) 661-9404. $125.00 per year.

EARTH, MOON AND PLANETS: AN INTERNATIONAL JOURNAL OF COMPARATIVE PLANETOLOGY. D. Reidel Publishing Company, 190 Old Derby Street, Hingham, MA 02043. Nine times per year. $275.00 per year.

ICARUS: INTERNATIONAL JOURNAL OF THE SOLAR SYSTEM STUDIES. Academic Press, Incorporated, Orlando, FL 32887. (305) 345-4100. Monthly. $484.00 per year.

MERCURY. Astronomical Society of the Pacific, 1290 24th Avenue, San Francisco, Ca 94122. (415) 661-8660. Bimonthly. $21.00 per year.

PLANETARY AND SPACE SCIENCE. Pergamon Press, Incorporated, Maxwell House, Fairview Park, Elmsford, NY 10523. (914) 592-7700. Monthly. $430.00 per year.

PLANETARY REPORT. Planetary Society, 65 North Catalina Avenue, Pasadena, CA 91106-2301. (818) 793-5100. 1980 to present. Bimonthly. $20.00 per year.

SOLAR SYSTEM RESEARCH (ENGLISH TRANSLATION OF ASTRONOMICHESKII VESTNIK). Consultants Bureau, 233 Spring Street, New York, NY 10013. (212) 620-8000. 1967 to present. Quarterly. $425.00 per year.

RESEARCH CENTERS AND INSTITUTES

LABORATORY FOR PLANETARY ATMOSPHERES RESEARCH. State University of New York at Stony Brook, Stony Brook, NY 11794-2300. (516) 632-8321.

LABORATORY FOR PLANETARY STUDIES. Cornell University, 302 Space Sciences Building, Ithaca, NY 14853. (607) 256-4971.

LUNAR AND PLANETARY INSTITUTE. 3303 NASA Road One, Houston, TX 77058. (713) 486-2139.

LUNAR AND PLANETARY LABORATORY. University of Arizona, Tucson, AZ 85721. (602) 621-6962.

UNIVERSITY OF ALABAMA IN HUNTSVILLE. Center for Space Plasma and Aeronomic Research. Engineering Building, Huntsville, AL 35899. (205) 895-6268

V

VACUUM METALLURGY

See: METALLURGY

VENTILATION

See: AIR CONDITIONING

VENUS

See also: PLANETARY SCIENCE, SOLAR SYSTEM

ABSTRACT SERVICES AND INDEXES

ASTRONOMY AND ASTROPHYSICS ABSTRACTS. Springer-Verlag New York, Incorporated, 175 Fifth Avenue, New York, NY 10010. (212) 460-1500. $70.00 per year.

ASTRONOMY AND ASTROPHYSICS MONTHLY INDEX. Olivetree Associates, Post Office Box 236, Sierra Madre, CA 91024. $212.00 per year. Complimentary copies available on request.

METEOROLOGICAL AND GEOASTROPHYSICAL ABSTRACTS. American Meteorological Society, 45 Beacon Street, Boston, MA 02108. (617) 227-2425. 1950 to present. Monthly. $450.00 per year.

STAR. (Scientific and Technical Aerospace Reports. United States National Aeronautics and Space Administration, Scientific and Technical Information Facility, Box 8757, Baltimore-Washington International Airport, MD 21240. (202) 755-2210. Semimonthly, with semiannual and annual indexes. $85.00 per year.

ANNUAL REVIEWS AND YEARBOOKS

THE ASTRONOMICAL ALMANAC. Superintendent of Documents, United States Government Printing Office, Washington, DC 20402. (202) 783-3238. Yearly.

ANNUAL REVIEW OF ASTRONOMY AND ASTROPHYSICS. Annual Reviews, Incorporated, 4139 El Camino Way, Palo Alto, CA 94306. (415) 493-4400. Annual. Inquire.

ANNUAL REVIEW OF EARTH AND PLANETARY SCIENCES. Annual Reviews, Incorporated, 4139 El Camino Way, Palo Alto, CA 94306. (415) 493-4400. Annual. Inquire.

ASSOCIATIONS AND PROFESSIONAL SOCIETIES

AMERICAN ASTRONOMICAL SOCIETY. 2000 Florida Avenue, NW, Suite 300, Washington, DC 20009. (202) 659-0134.

AMERICAN GEOPHYSICAL UNION. 2000 Florida Avenue, NW, Washington, DC 20009. (202) 462-6903.

ASTRONOMICAL LEAGUE. Post Office Box 12821, Tucson, AZ 85732. (602) 790-8471.

ASTRONOMICAL SOCIETY OF THE PACIFIC. 1290 24th Avenue, San Francisco, CA 94122. (415) 661-8660.

PLANETARY SOCIETY. 65 North Catalina Avenue, Pasadena, CA 91106-2301. (818) 793-5100.

BIBLIOGRAPHIES

A BIBLIOGRAPHY OF ASTRONOMY, 1970-1979. R.A. Seal and S.S. Martin. Libraries Unlimited, Incorporated, Littleton, CO 80160. 1982. $37.50.

SCIENTIFIC AND TECHNICAL BOOKS AND SERIALS IN PRINT 1988: AN INDEX TO LITERATURE IN SCIENCE AND TECHNOLOGY. R.R. Bowker Company, 205 East 42nd Street, New York, NY 10017. (800) 521-8110 or (212) 916-1600. $175.00.

DIRECTORIES AND BIOGRAPHICAL SOURCES

AMERICAN ASTRONOMICAL SOCIETY MEMBERS. 2000 Florida Avenue, NW, Suite 300, Washington, DC 20009. (202) 659-0134. Annual. Available to members only.

THE BIOGRAPHICAL DICTIONARY OF SCIENTISTS: ASTRONOMERS. D. Abbott, editor. Peter Bedrick Books, 125 East 23rd Street, New York, NY 10010. 1984. $24.95.

DIRECTORY OF PHYSICS AND ASTRONOMY STAFF MEMBERS. American Institute of Physics, 335 East 45th Street, New York, NY 10017. Annual.

GENERAL WORKS

THE ATMOSPHERE OF VENUS: RECENT FINDINGS: PROCEEDINGS OF WORKSHOP III OF THE COSPAR 25TH OLENARY MEETING HELD IN GRAZ, AUSTRIA, 28 JUNE-2 JULY 1984. G.M. Keating, et al. Pergamon Press, Incorporated, Maxwell House, Fairview Park, Elmsford, NY 10523. (914) 592-7700. 1985. $49.50.

PHOTOCHEMISTRY OF THE ATMOSPHERE OF MARS AND VENUS. V.A. Krasnopolsky. Springer-Verlag New York, Incorporated, 175 Fifth Avenue, New York, NY 10010. (212) 460-1500. 1986. $90.00.

VENUS: AN ERRANT TWIN. Eric Burgess. Columbia University Press, 562 West 113th Street, New York, NY 10025. (212) 316-7100. 1985. $29.95.

VENUS

ONLINE DATA BASES

CA SEARCH. Chemical Abstracts Service, Post Office Box 3012, Columbus, OH 43210. Guide to chemical literature, 1972 to present. Inquire as to cost and availability.

DISSERTATION ABSTRACTS ONLINE. University Microfilms International, 300 North Zeeb Road, Ann Arbor, MI 48106. (800) 521-0600 or (313) 761-4700. Scope includes virtually all doctoral dissertations accepted at accredited American institutions from 1861 to present in 252 subject areas. Inquire as to cost and availability.

NASA. National Aeronautics and Space Administration, Scientific and Technical Information Branch, 300 7th Street, SW, Washington, DC 20546. Citations and abstracts of aerospace literature, 1962 to present. Inquire as to cost and availability.

NTIS. National Technical Information Service, 5285 Port Royal Road, Springfield, VA 22161. (703) 487-4630. Broad coverage of government-sponsored research reports, 1964 to present. Inquire as to cost and availability.

SCISEARCH. Institute for Scientific Information, 3501 Market Street, Philadelphia, PA 19104. (800) 523-1850 or (215) 386-0100. Broad multidisciplinary title and author index to the international literature of science and technology, 1974 to present. Inquire as to cost and availability.

PERIODICALS

ASSOCIATION OF LUNAR AND PLANETARY OBSERVERS. Journal Association of Lunar and Planetary Observers, Box 16131, San Francisco, CA 94116. (415) 566-5786. 1947 to present. Quarterly. $15.00 per year.

ASTRONOMICAL JOURNAL. American Astronomical Society. Available from: American Institute of Physics, 335 East 45th Street, New York, NY 10017. 9212) 661-9404. $125.00 per year.

EARTH, MOON AND PLANETS: AN INTERNATIONAL JOURNAL OF COMPARATIVE PLANETOLOGY. D. Reidel Publishing Company, 190 Old Derby Street, Hingham, MA 02043. Nine times per year. $275.00 per year.

ICARUS: INTERNATIONAL JOURNAL OF THE SOLAR SYSTEM STUDIES. Academic Press, Incorporated, Orlando, FL 32887. (305) 345-4100. Monthly. $484.00 per year.

MERCURY. Astronomical Society of the Pacific, 1290 24th Avenue, San Francisco, CA 94122. (415) 661-8660. Bimonthly. $21.00 per year.

PLANETARY AND SPACE SCIENCE. Pergamon Press, Incorporated, Maxwell House, Fairview Park, Elmsford, NY 10523. (914) 592-7700. Monthly. $430.00 per year.

PLANETARY REPORT. Planetary Society, 65 North Catalina Avenue, Pasadena, CA 91106-2301. (818) 793-5100. 1980 to present. Bimonthly. $20.00 per year.

SOLAR SYSTEM RESEARCH (ENGLISH TRANSLATION OF ASTRONOMICHESKII VESTNIK). Consultants Bureau, 233 Spring Street, New York, NY 10013. (212) 620-8000. 1967 to present. Quarterly. $425.00 per year.

RESEARCH CENTERS AND INSTITUTES

LABORATORY FOR PLANETARY ATMOSPHERES RESEARCH. State University of New York at Stony Brook, Stony Brook, NY 11794-2300. (516) 632-8321.

LABORATORY FOR PLANETARY STUDIES. Cornell University, 302 Space Sciences Building, Ithaca, NY 14853. (607) 256-4971.

LUNAR AND PLANETARY INSTITUTE. 3303 NASA Road One, Houston, TX 77058. (713) 486-2139.

LUNAR AND PLANETARY LABORATORY. University of Arizona, Tucson, AZ 85721. (602) 621-6962.

WASHINGTON UNIVERSITY. Earth and Planetary Remote Sensing Laboratory, Department of Earth and Planetary Sciences, Campus Box 1169, St. Louis, MO 63130. (314) 889-5679.

VERTICAL TAKEOFF AND LANDING (VTOL)

See: AIRCRAFT

VERY LARGE ARRAY

See: RADIO TELESCOPES

VERY LARGE SCALE INTEGRATION (VLSI)

See also: ELECTRICAL ENGINEERING, ELECTRONIC CIRCUITS AND COMPONENTS, ELECTRONIC ENGINEERING, MICROELECTRONICS, MICROPROCESSORS

ABSTRACT SERVICES AND INDEXES

APPLIED SCIENCE AND TECHNOLOGY INDEX. H.W. Wilson Company, 950 University Avenue, Bronx, NY 10452. (800) 367-6670 or (212) 588-8400. Inquire as to cost and availability.

ELECTRONICS AND COMMUNICATIONS ABSTRACTS JOURNAL. Cambridge Scientific Abstracts, 5161 River Road, Bethesda, MD 20816. (301) 951-1400. Bimonthly. Inquire as to cost and availability.

ENGINEERING INDEX MONTHLY. Engineering Information, Incorporated, 345 East 47th Street, New York, NY 10017. (800) 221-1044 or (212) 705-7600. Monthly, with annual cumulation. $1425.00 per year.

PHYSICS ABSTRACTS. Institute of Electrical Engineers, London, United Kingdom. Available from: Institute of Electrical and Electronic Engineers (Ieee), 345 East 47th Street, New York, NY 10017. (212) 705-7900.

SCIENCE CITATION INDEX. Institute for Scientific Information, 3501 Market Street, Philadelphia, PA 19104. (800) 523-1850 or (215) 386-0100. Inquire as to cost and availability.

SOLID STATE ABSTRACTS: AN ABSTRACT JOURNAL INVOLVING THE PHYSICS, METALLURGY, CRYSTALLOGRAPHY, CHEMISTRY, AND DEVICE TECHNOLOGY OF SOLIDS. Cambridge Scientific Abstracts, 5161 River Road, Bethesda, MD 20816. (301) 951-1400. 1957 to present. Bimonthly. $550.00 per year.

ASSOCIATIONS AND PROFESSIONAL SOCIETIES

AMERICAN ELECTRONICS ASSOCIATION. Post Office Box 10045, 2670 Hanover Street, Palo Alto, CA 94303. (415) 857-9300.

ELECTRONIC INDUSTRIES ASSOCIATION. 2001 Eye Street, NW, Washington, DC 20006. (202) 457-4900.

IEEE (INSTITUTE OF ELECTRICAL AND ELECTRONICS ENGINEERS). 345 East 47th Street, New York, NY 10017. (212) 705-7900.

INTERNATIONAL SOCIETY FOR HYBRID MICROELECTRONICS. Post Office Box 2698, 1861 Wiehle Avenue, Suite 340, Reston, VA 22090. (703) 471-0066.

BIBLIOGRAPHIES

HANDBOOKS AND TABLES IN SCIENCE AND TECHNOLOGY. Russell H. Powell, editor. Oryx Press, 2214 North Central Avenue, Phoenix, AZ 85004-1483. (602) 254-6156. Second edition. 1983. $55.00.

SCIENTIFIC AND TECHNICAL BOOKS IN PRINT: AN INDEX TO LITERATURE IN SCIENCE AND TECHNOLOGY. R.R. Bowker Company, 205 East 42nd Street, New York, NY 10017. (800) 521-8110 or (212) 916-1600.

DIRECTORIES AND BIOGRAPHICAL SOURCES

AMERICAN ELECTRONICS ASSOCIATION DIRECTORY. American Electronics Association, Post Office Box 10045, 2670 Hanover Street, Palo Alto, CA 94303. (415) 857-9300. Annual.

EEM - ELECTRONIC ENGINEERS MASTER. Hearst Business Communications/UTP Division, 645 Stewart Avenue, Garden City, NY 11530. (516) 227-1300. Annual. $75.00 per copy.

IEEE MEMBERSHIP DIRECTORY. Institute of Electrical and Electronics Engineers, IEEE Service Center, 445 Hoes Lane, Piscataway, NJ 08854. (212) 705-7900. Annual. $7.00.

RESEARCH CENTERS DIRECTORY. Gale Research Company, Book Tower, Detroit, MI 48226. (800) 521-0707. Twelfth edition. 1987. $365.00 for set.

WHO'S WHO IN ELECTRONICS. Harris Publishing Company, 2057-2 Aurora Road, Twinsburg, OH 44087. (216) 425-9000. Annual. $89.00.

WHO'S WHO IN ENGINEERING. Engineers Joint Council, 345 East 47th Street, New York, NY 10017. (212) 705-7010. 1985. $200.00.

WHO'S WHO IN FRONTIER SCIENCE AND TECHNOLOGY. Marquis Who's Who, Incorporated, 200 East Ohio Street, Chicago, IL 60611. (800) 428-3898 or (312) 787-2008.

WHO'S WHO IN TECHNOLOGY TODAY. Reston Publishing Company, Incorporated, c/o Prentice-Hall, Incorporated, Englewood Cliffs, NJ 07632. (800) 262-6868. Biennial. Five volumes. $425.00. Covers the fields of electronics, computer science, physics, optics, chemistry, biotechnology, mechanics, energy, and earth science.

ENCYCLOPEDIAS AND DICTIONARIES

ENCYCLOPEDIA OF PHYSICAL SCIENCE AND TECHNOLOGY. Academic Press, Incorporated, Orlando, FL 32887. (800) 321-5068 or (305) 345-2734. Fifteen volumes, 1986.

IEEE STANDARD DICTIONARY OF ELECTRICAL AND ELECTRONICS TERMS. Frank Jay, editor. John Wiley and Sons, Incorporated, 605 Third Avenue, New York, NY 10158. (800) 526-5368 or (212) 850-6000. Third edition. 1984. $49.95.

MCGRAW-HILL ENCYCLOPEDIA OF SCIENCE AND TECHNOLOGY. McGraw-Hill Book, Incorporated, 1221 Avenue of the Americas, New York, NY 10020. (212) 997-3675. Fifth edition, 15 volumes. $1100.00.

GENERAL WORKS

CIRCUITS, DEVICES AND SYSTEMS: A FIRST COURSE IN ELECTRICAL ENGINEERING. R.J. Smith. John Wiley and Sons, Incorporated, 605 Third Avenue, New York, NY 10158. (800) 526-5368 or (212) 850-6000. Fourth edition. 1984. $41.45.

ELECTRONIC CIRCUIT ANALYSIS: BASIC PRINCIPLES. Roy A. Colclaser, Donald A. Neaman, Charles F. Hawkins. John Wiley and Sons, Incorporated, 605 Third Avenue, New York, NY 10158. (800) 526-5368 or (212) 850-6000. 1984. $42.50.

ELECTRONIC CIRCUITS AND APPLICATIONS. Stephen D. Senturia and B.D. Wedlock. John Wiley and Sons, Incorporated, 605 Third Avenue, New York, NY 10158. (800) 526-5368 or (212) 850-6000. 1975. $44.00.

ELECTRONIC DEVICES AND COMPONENTS. J. Seymour. John Wiley and Sons, Incorporated, 605 Third Avenue, New York, NY 10158. (800) 526-5368 or (212) 850-6000. 1981. $37.00 in paper.

ELECTRONIC INVENTIONS AND DISCOVERIES: ELECTRONICS FROM ITS EARLIEST BEGINNINGS TO THE PRESENT DAY. G.W.A. Dummer. Pergamon Press, Incorporated, Maxwell House, Fairview Park, Elmsford, NY 10523. (914) 592-7700. 1983. $48.50.

INTRODUCTION TO VLSI TESTING. Robert J. Feugate and Steven M. McIntyre. Prentice-Hall, Incorporated, Englewood Cliffs, NJ 07632. (800) 562-0245. 1988. $27.75.

MICROELECTRONICS. J. Millman. McGraw-Hill Book, 1221 Avenue of the Americas, New York, NY 10020. (212) 512-2000. (212) 512-2000. Second edition. 1987. $39.95.

MICROPROCESSOR ENGINEERING. Brian Holdsworth. Butterworth Publishing, Incorporated, 80 Montvale Avenue, Stoneham, MA 02180. (617) 438-8464. 1986. $39.95.

SEMICONDUCTOR CIRCUIT APPROXIMATIONS: AN INTRODUCTION TO TRANSISTORS AND INTEGRATED CIRCUITS. Albert Paul Malvino. McGraw-Hill Book Company, 1221 Avenue of the Americas, New York, NY 10020. (800) 628-0004. Fourth edition. 1985. $34.95.

ULTRA LARGE SCALE INTEGRATED MICROELECTRONICS. David K. Ferry. Prentice-Hall, Incorporated, Englewood Cliffs, NJ 07632. (800) 562-0245. 1988. $49.95.

VLSI ENGINEERING. Thomas E. Dillinger. Prentice-Hall, Incorporated, Englewood Cliffs, NJ 07632. (800) 562-0245. 1988. $49.95.

HANDBOOKS AND MANUALS

CONTEMPORARY ELECTRONICS CIRCUITS DESKBOOK. Harry Helms. McGraw-Hill Book Company, 1221 Avenue of the Americas, New York, NY 10020. (800) 628-0004. 1986. $29.95.

ELECTRONIC ENGINEERS HANDBOOK. Donald G. Fink, editor. McGraw-Hill Book Company, 1221 Avenue of the Americas, New York, NY 10020. (800) 628-0004. Second edition. 1982. $89.00.

HANDBOOK OF ELECTRONICS MANUFACTURING ENGINEERING. B.S. Matisoff. Van Nostrand Reinhold Company, Incorporated, 115 Fifth Avenue, New York, NY 10003. (800) 543-2681. Second edition. 1986. $52.95.

HANDBOOK OF MODERN ELECTRONICS AND ELECTRICAL ENGINEERING. Charles Belove. John Wiley and Sons, Incorporated, 605 Third Avenue, New York, NY 10158. (800) 526-5368 or (212) 850-6000. 1986. $85.00.

VLSI HANDBOOK. Norman G. Einspruch, editor. Academic Press, Incorporated, 6277 Sea Harbor Drive, Orlando, FL 32821. (800) 321-5068. 1985. $125.00.

THE WILEY ENGINEER'S DESK REFERENCE. Sanford I. Heisler. John Wiley and Sons, Incorporated, 605 Third Avenue, New York, NY 10158. (800) 526-5368 or (212) 850-6418. 1984. $36.00.

ONLINE DATA BASES

COMPENDEX. Engineering Information, Incorporated, 345 East 47th Street, New York, NY 10017. (800) 221-1044 or (212) 705-7615. Engineering and technical literature, 1975 to present. Inquire as to cost and availability.

DISSERTATION ABSTRACTS ONLINE. University Microfilms International, 300 North Zeeb Road, Ann Arbor, MI 48106. (800) 521-0600 or (313) 761-4700. Scope includes virtually all doctoral dissertations accepted at accredited American institutions from 1861 to present in 252 subject areas. Inquire as to cost and availability.

INSPEC. INSPEC Marketing Department, Institute of Electrical and Electronics Engineers, Incorporated, IEEE Service Department, 445 Hoes Lane, Piscataway, NJ 08854. (201) 981-0060. Inquire as to on-line cost and availability.

NTIS. National Technical Information Service, 5285 Port Royal Road, Springfield, VA 22161. (703) 487-4630. Broad coverage of government sponsored research reports, 1964 to present. Inquire as to cost and availability.

WILSONLINE. H.W. Wilson Company, 950 University Avenue, Bronx, NY 10452. (800) 367-6770 or (212) 588-8400. Makes available online versions of the printed H.W. Wilson Indexes including Applied Science and Technology Index, Business Periodicals Index, and Readers' Guide to Periodical Literature. Period covered is generally 1983 to present. Inquire as to cost and availability.

OTHER SOURCES

A GUIDE TO THE LITERATURE OF ELECTRICAL AND ELECTRONIC ENGINEERING. Susan B. Ardis. Libraries Unlimited, Incorporated, Post Office Box 263, Littleton, CO 80160. (303) 770-1220. 1986. $37.50.

PERIODICALS

CANADAIAN ELECTRONICS ENGINEERING. Maclean Hunter Research Bureau, 777 Bay Street, Toronto, ON M5W 1A7 Canada. (416) 596-5729.

CIRCUITS, SYSTEMS AND SIGNAL PROCESSING. Birkhauser Boston, Incorporated, 380 Green Street, Post Office Box 2007, Cambridge, MA 02139. (617) 876-2333. Quarterly. $175.00 per year.

ELECTROCOMPONENT SCIENCE. Gordon Breach Science Publishers, Post Office Box 786, Cooper Station, New York, NY 10276. (212) 206-8900. Quarterly. $224.00 per year.

ELECTRONIC DESIGN. Heyden Publishing Company, Incorporated, 10 Mulholland Drive, Hasbrouck Heights, NJ 07604. (201) 393-6000. Biweekly. $65.00.

ELECTRONIC ENGINEERING TIMES. CMP Publications, Incorporated, 600 Community Drive, Manhasset, NY 11030. (516) 365-4600. Weekly. $65.00 per year.

ELECTRONICS WEEK. McGraw-Hill Book Company, 1221 Avenue of the Americas, New York, NY 10020. (800) 628-0004. Weekly. $18.00.

IEEE CIRCUITS AND DEVICES MAGAZINE. Institute of Electrical and Electronics Engineers, IEEE Service Center, 445 Hoes Lane, Piscataway, NJ 08854. (212) 705-7900. Bimonthly. $70.00 per year.

IEEE ELECTRON DEVICE LETTERS. Institute of Electrical and Electronics Engineers, IEEE Service Center, 445 Hoes Lane, Piscataway, NJ 08854. (212) 705-7900. Monthly. $70.00 per year.

IEEE JOURNAL OF SOLID STATE CIRCUITS. Institute of Electrical and Electronics Engineers, IEEE Service Center, 445 Hoes Lane, Piscataway, NJ 08854. (212) 705-7900. Bimonthly. $113.00 per year.

IEEE TRANSACTIONS ON CIRCUITS AND SYSTEMS. Institute of Electrical and Electronics Engineers, IEEE Service Center, 445 Hoes Lane, Piscataway, NJ 08854. (212) 705-7900. Monthly. $108.00 per year.

IEEE TRANSACTIONS ON ELECTRON DEVICES. Institute of Electrical and Electronics Engineers, IEEE Service Center, 445 Hoes Lane, Piscataway, NJ 08854. (212) 705-7900. Monthly. $159.00 per year.

MICROELECTRONIC ENGINEERING: AN INTERDISCIPLINARY JOURNAL OF SEMICONDUCTOR MANUFACTURING TECHNOLOGY. Elsevier Science Publishing Company, Incorporated, 52 Vanderbilt Avenue, New York, NY 10017. (212) 370-5520. Eight times per year. $90.00 per year.

MICROELECTRONIC MANUFACTURING AND TESTING. Lake Publishing Corporation, 17730 West Peterson Road, Libertyville, IL 60048. (312) 362-8711. Monthly. $60.00 per year.

MICROELECTRONICS AND RELIABILITY. Pergamon Press, Incorporated, Journals Division, Maxwell House, Fairview Park, Elmsford, NY 10523. (914) 592-7700. Monthly. $250.00 per year.

MICROELECTRONICS JOURNAL. Benn Electronics Publications, Limited, Box 28, Luton, Beds, LU2 0ED, England. Bimonthly. $225.00 per year.

RADIO ELECTRONICS. Gernsback Publications, Incorporated, 500-B BiCounty Boulevard, Farmingdale, NY 11735. (516) 293-3000. Monthly. $16.00 per year.

SEMICONDUCTOR INTERNATIONAL. Cahners Publishing Company, Incorporated, Division of Reed Holdings, Incorporated, Cahners Plaza, 1350 East Touhy Avenue, Des Plaines, IL 60018. (312) 635-8800. Monthly. $55.00 per year.

VLSI SYSTEMS DESIGN. CMP Publications, Incorporated, 600 Community Drive, Manhasset, NY 11030. (516) 365-4600. Monthly. $20.00 per year.

RESEARCH CENTERS AND INSTITUTES

MASSACHUSETTS INSTITUTE OF TECHNOLOGY. Laboratory for Electromagnetic and Electronic Systems, 77 Massachusetts Avenue, Cambridge, MA 02139. (617) 253-4631.

NORTH CAROLINA STATE UNIVERSITY. Solid State Electronics Laboratory, 432 Daniels Hall, Raleigh, NC 27695. (919) 737-2336.

OHIO STATE UNIVERSITY. Electroscience Laboratory, 1320 Kinnear Road, Columbus, OH 43212. (614) 422-7981.

PENNSYLVANIA STATE UNIVERSITY. Solid State Device Laboratory, 210 Electrical Engineering, West Building, University Park, PA 16802. (814) 865-1666.

VIBRATION

See also: MACHINE DESIGN, MECHANICAL ENGINEERING

ABSTRACT SERVICES AND INDEXES

APPLIED MECHANICS REVIEW. American Society of Mechanical Engineers, 345 East 47th Street, New York, NY 10017. (212) 705-7703. 1948 to present. Monthly. $360.00 per year.

VS;IAPPLIED SCIENCE AND TECHNOLOGY INDEX. H.W. Wilson and Company, 950 University Avenue, Bronx, NY 10452. (800) 367-6770 or (212) 588-8400. Monthly. Inquire as to cost and availability.

CHEMICAL ABSTRACTS. American Chemical Society, Chemical Abstracts Service, Box 3012, Columbus, OH 43210. (614) 421-3600. 1907 to present. Weekly. $9500.00 per year.

CURRENT CONTENTS: ENGINEERING, TECHNOLOGY AND APPLIED SCIENCES. Institute for Scientific Information, 3501

Market Street, Philadelphia, PA 19104. (800) 523-1850 or (215) 386-0100. Weekly. $275.00 per year.

ENGINEERING INDEX MONTHLY AND AUTHOR INDEX. Engineering Information Inc., 345 East 47th Street, New York, NY 10017. (212) 705-7600. Monthly. $1560.00 per year.

ISMEC BULLETIN (Information Service in Mechanical Engineering). Cambridge Scientific Abstracts, 5161 River Road, Bethesda, MD 20816. (301) 951-1400. 1973 to present. Monthly. $450.00 per year.

INDEX TO SCIENTIFIC AND TECHNICAL PROCEEDINGS. Institute for Scientific Information, 3501 Market Street, Philadelphia, PA 19104. (800) 523-1850 or (215) 386-0100. 1978 to present. Monthly. $775.00 per year.

PHYSICS ABSTRACTS. Institution of Electrical Engineers. Available from: IEEE Service Center, 445 Hoes Lane, Piscataway, NJ 08854. 1898 to present. Bimonthly. $1700.00 per year.

SCIENCE CITATION INDEX. Institute for Scientific Information, 3501 Market Street, Philadelphia, PA 19104. (800) 523-1850 or (215) 386-0100. Six times per year. $6200.00 per year.

ASSOCIATIONS AND PROFESSIONAL SOCIETIES

AMERICAN SOCIETY OF MECHANICAL ENGINEERS. 345 47th Street, New York, NY 10017. (212) 705-7722.

VIBRATION INSTITUTE. 101 West 55th Street, Suite 206, Clarendon Hills, IL 60514. (312) 654-2254.

DIRECTORIES AND BIOGRAPHICAL SOURCES

1987 DIRECTORY OF ENGINEERING SOCIETIES AND RELATED ORGANIZATIONS. Gordon Davis, editor. Hemisphere Publishing Corporation, 1010 Vermont Avenue, NW, Washington, DC 20005. (800) 526-0275. 12th edition. 1987. $100.00.

RESEARCH CENTERS DIRECTORY 1988. Gale Research Company, Book Tower, Detroit, MI 48226. (800) 521-0707. 12th edition. 1987. $365.00 for set.

SCIENTIFIC AND TECHNICAL ORGANIZATIONS AND AGENCIES DIRECTORY. Margaret Labash Young, editor. Gale Research Company, Book Tower, Detroit, MI 48226. (800) 521-0707. 2nd edition. 1987. $185.00.

WHO'S WHO IN ENGINEERING. Gordon Davis, editor. Hemisphere Publishing Corporation, 1010 Vermont Avenue, NW, Washington, DC 20005. (800) 526-0275. 6th edition. 1985. $200.00.

ENCYCLOPEDIAS AND DICTIONARIES

THESAURUS OF SCIENTIFIC, TECHNICAL, AND ENGINEERING TERMS. Hemisphere Publishing Corporation, 1010 Vermont Avenue, NW, Washington, DC 20005. (800) 526-0275. 1988. $125.00.

GENERAL WORKS

CHAOTIC VIBRATIONS: AN INTRODUCTION FOR APPLIED SCIENTISTS AND ENGINEERS. Francis C. Moon. John Wiley and Sons, Inc., 605 Third Avenue, New York, NY 10158. (800) 526-5368. 1987. $39.95.

ENGINEERING MECHANICS. J.L. Meriam and L.G. Kraige. John Wiley and Sons, Inc., 605 Third Avenue, New York, NY 10158. (800) 526-5368. 2nd edition. 1986. $52.95.

VIBRATION IN MECHANICAL SYSTEMS. Maurice Roseau. Springer-Verlag New York, Inc., 175 Fifth Avenue, New York, NY 10010. (800) 526-7254. 1987. $69.00.

HANDBOOKS AND MANUALS

MARK'S STANDARD HANDBOOK FOR MECHANICAL ENGINEERS. T. Baumeister, editor. McGraw-Hill Book Company, 1221 Avenue of the Americas, New York, NY 10020. (212) 512-2000. 8th edition. 1978. $96.00.

MECHANICAL DESIGN AND SYSTEMS HANDBOOK. Harold A. Rothbart, editor. McGraw-Hill Book Company, 1221 Avenue of the Americas, New York, NY 10020. (212) 512-2000. 2nd edition. 1985. $96.50.

MECHANICAL ENGINEERS' HANDBOOK. Myer Kutz, editor. John Wiley and Sons, Inc., 605 Third Avenue, New York, NY 10158. (800) 526-5368. 1986. $79.95.

STANDARD HANDBOOK OF MACHINE DESIGN. Joseph E. Shigley and Charles R. Mischke. McGraw-Hill Book Company, 1221 Avenue of the Americas, New York, NY 10020. (212) 512-2000. 1986. $89.00.

ONLINE DATA BASES

CA SEARCH. Chemical Abstracts Service, P.O. Box 3012, Columbus, OH 43120. (800) 848-6538 or (614) 421-3600. Comprehensive guide to chemical literature, 1972 to present. Inquire as to online cost and availability.

COMPENDEX. Engineering Information, Inc., 345 East 47th Street, New York, NY 10017. (800) 221-1044 or (212) 705-7615. Engineering and technical literature, 1975 to present. Inquire as to online cost and availability.

DISSERTATION ABSTRACTS ONLINE. University Microfilms International, 300 North Zeeb Road, Ann Arbor, MI 48106. (800) 521-0600 or (313) 761-4700. Scope includes virtually all doctoral dissertations accepted at accredited American institutions from 1861 to present in over 250 subject areas. Inquire as to online cost and availability.

INSPEC. INSPEC Marketing Department, Institution of Electrical Engineers. Available from IEEE Service Center, 445 Hoes Lane, Piscataway, NJ 08854. (201) 981-0060. Online version of Physics Abstracts. Inquire as to online cost and availability.

ISMEC. Cambridge Scientific Abstracts, 5161 River Road, Besthda, MD 20816. (800) 638-8076 or (301) 951-1400. Literature of mechanical and production engineering, 1973 to present. Inquire as to online cost and availability.

NTIS. National Technical Information Service, 5285 Port Royal Road, Springfield, VA 22161. (703) 487-4630. Broad coverage of government sponsored research reports, 1964 to present. Inquire as to online cost and availability.

SCISEARCH. Institute for Scientific Information, 3501 Market Street, Philadelphia, PA 19104. (800) 523-1850 or (215) 386-0100. Broad multidisciplinary title and author index to the international literature of science and technology, 1974 to present. Inquire as to online cost and availability.

WILSONLINE. H.W. Wilson and Company, 950 University Avenue, Bronx, NY 10452. (800) 367-6770 or (212) 588-8400. Makes available online versions of the H.W. Wilson indexes including Applied Science and Technology Index, Business Periodicals Index and Readers' Guide to Periodical Literature. Approximately 1980 to present. Inquire as to online cost and availability.

PERIODICALS

CIME (COMPUTERS IN MECHANICAL ENGINEERING). American Society of Mechanical Engineers, 345 East 47th Street, New York, NY 10017. (212) 705-7703.

INTERNATIONAL JOURNAL FOR NUMERICAL METHODS IN ENGINEERING. John Wiley and Sons, Inc., 605 Third Avenue, New York, NY 10158. (800) 526-5368.

VIBRATION

JOURNAL OF APPLIED MECHANICS. American Society of Mechanical Engineers, 345 East 47th Street, New York, NY 10017. (212) 705-7703. 1935 to present. Quarterly. $100.00 per year.

JOURNAL OF ENGINEERING FOR INDUSTRY. American Society of Mechanical Engineers, 345 East 47th Street, New York, NY 10017. (212) 705-7703. 1970 to present. Quarterly. $100.00 per year.

JOURNAL OF HEAT TRANSFER. American Society of Mechanical Engineers, 345 East 47th Street, New York, NY 10017. (212) 705-7703. 1970 to present. Quarterly. $100.00 per year.

JOURNAL OF MECHANISMS, TRANSMISSIONS AND AUTOMATION IN DESIGN. American Society of Mechanical Engineers, 345 East 47th Street, New York, NY 10017. (212) 705-7703. 1983 to present. Quarterly. $80.00 per year.

MACHINE DESIGN. Penton-IPC, 1100 Superior Avenue, Cleveland, OH 44114. (216)696-7000. 1929 present. Twenty-eight times per year. $60.00 per year.

MECHANISM AND MACHINE THEORY. Pergamon Press, Inc., Maxwell House, Fairview Park, Elmsford, NY 10523. (914) 592-7700. 1966 to present. Bimonthly. $215.00 per year.

MECHANICAL ENGINEERING. American Society of Mechanical Engineers, 345 East 47th Street, New York, NY 10017. (212) 705-7722. 1906 to present. Monthly. $35.00 per year.

RESEARCH CENTERS AND INSTITUTES

LABORATORY FOR EXPERIMENTAL MECHANICS RESEARCH. State University of New York at Stony Brook, Stony Brook, NY 11794-2300. (516) 632-8311.

MECHANICAL ENGINEERING DESIGN LABORATORY. University of Florida, 237 Mechanical Engineering Building, Gainesville, FL 32611. (904) 392-0827

MECHANICAL ENGINEERING LABORATORIES. Stevens Institute of Technology, Hoboken, NJ 07030. (201) 420-5591.

MECHANICAL ENGINEERING RESEARCH LABORATORIES. Kansas State University, Durland Hall, Manhattan, KS 66506. (913) 532-5610.

VIBRATION INSTITUTE. 101 West 55th Street, Suite 206, Clarendon Hills, IL 60514. (312) 654-2254.

VIDEO DISKS

See: VIDEO TECHNOLOGY

VIDEO TAPE

See: VIDEO TECHNOLOGY

VIDEO TECHNOLOGY

See also: ELECTRONIC CIRCUITS AND COMPONENTS, ELECTRONICS, ELECTRONICS ENGINEERING, OPTICS, TELECOMMUNICATIONS, TELEVISION

ABSTRACT SERVICES AND INDEXES

APPLIED SCIENCE AND TECHNOLOGY INDEX. H.W. Wilson and Company, 950 University Avenue, Bronx, NY 10452. (800) 367-6670 or (212) 588-8400. Monthly. Inquire as to cost and availability.

ELECTRICAL AND ELECTRONICS ABSTRACTS. Institution of Electrical Engineers. Available from: Institute of Electrical and Electronics Engineers. IEEE Service Center, 445 Hoes Lane, Piscataway, NJ 08854. Monthly. $1250.00 per year.

ELECTRONICS AND COMMUNICATIONS ABSTRACTS. Cambridge Scientific Abstracts, 5161 River Road, Bethesda, MD 20816. (301) 951-1400. Bimonthly. Inquire as to cost and availability.

ENGINEERING INDEX MONTHLY AND AUTHOR INDEX. Engineering Information Inc., 345 East 47th Street, New York, NY 10017. (212) 705-7600. Monthly. $1560.00 per year.

IEEE PUBLICATIONS BULLETIN. Institute of Electrical and Electronics Engineers. Institute of Electrical and Electronics Engineers. IEEE Service Center, 445 Hoes Lane, Piscataway, NJ 08854. Quarterly. Free.

PHYSICS ABSTRACTS. Institution of Electrical Engineers. Available from: IEEE Service Center, 445 Hoes Lane, Piscataway, NJ 08854. 1898 to present. Bimonthly. $1700.00 per year.

PHYSICS BRIEFS. Physik Verlag GmbH, Postfach 1260/1280, D-6940 Weinheim, West Germany. (212) 661-9404. 1920 to present. Twenty-six times per year. $1250.00 per year.

SCIENCE CITATION INDEX. Institute for Scientific Information, 3501 Market Street, Philadelphia, PA 19104. (800) 523-1850 or (215) 386-0100. Six times per year. $6200. 00 per year.

ASSOCIATIONS AND PROFESSIONAL SOCIETIES

AMERICAN ELECTRONICS ASSOCIATION. P.O. Box 10045, 2670 Hanover Street, Palo Alto, CA 94303. (415) 857-9300.

AMERICAN INSTITUTE OF PHYSICS. 335 East 45th Street, New York, NY 10017. (212) 661-9494.

ELECTRONICS INDUSTRIES ASSOCIATION. 2001 Eye Street, N.W., Washington, DC 20006. (202) 457-4900.

INSTITUTE OF ELECTRICAL AND ELECTRONICS ENGINEERS. 345 East 47th Street, New York, NY 10017. (212) 705-7900.

NATIONAL ASSOCIATION OF RADIO AND TELECOMMUNICATIONS ENGINEERS. P.O. Box 15029, Salem, OR 97309. (503) 581-7653.

SOCIETY OF CABLE TELEVISION ENGINEERS. P.O. Box 2389, West Chester, PA 19380. (215) 363-6888.

SOCIETY OF MOTION PICTURE AND TELEVISION ENGINEERS. 595 West Hartsdale Avenue, White Plains, NY 10607. (914) 472-6606.

DIRECTORIES AND BIOGRAPHICAL SOURCES

AMERICAN MEN AND WOMEN OF SCIENCE. R.R. Bowker, Inc., Order Department, 245 West 17th Street, New York, NY 10011. (800) 521-8110. Eight volumes. 1986. $595.00 for set.

BROADCAST ENGINEERING BUYERS GUIDE/SPEC BOOK ISSUE. Intertec Publishing Corporation, Box 12901, Overland Park, KS 66212. (913) 888-4664. Annual. $20.00 per year.

IEEE MEMBERSHIP DIRECTORY. Institute of Electrical and Electronics Engineers. IEEE Service Center, 445 Hoes Lane, Piscataway, NJ 08854. Annual. $7.00.

INTERNATIONAL RESEARCH CENTERS DIRECTORY 1988-89. Darren L. Smith, editor. Gale Research Company, Book Tower, Detroit, MI 48226. (800) 521-0707. 4th edition. 1987. $360.00.

INTERNATIONAL SATELLITE DIRECTORY. S.F.P. Designs, Inc., 369 Redwood Avenue, Corte Madera, CA 96925. (415) 927-0379.

1987 DIRECTORY OF ENGINEERING SOCIETIES AND RELATED ORGANIZATIONS. Gordon Davis, editor. Hemisphere Publishing Corporation, 1010 Vermont Avenue, NW, Washington, DC 20005. (800) 526-0275. 12th edition. 1987. $100.00.

RESEARCH CENTERS DIRECTORY 1988. Gale Research Company, Book Tower, Detroit, MI 48226. (800) 521-0707. 12th edition. 1987. $365.00 for set.

SATELLITE COMMUNICATIONS, SATELLITE INDUSTRY DIRECTORY ISSUE. Cardiff Publishing Company, 6530 South Yosemite Street, Englewood, CO 80111. (303) 694-1522.

SATELLITE DIRECTORY. Phillips Publishing, Inc., 7811 Montrose Road, Potomac, MD 20854. (301) 340-2100.

SCIENTIFIC AND TECHNICAL ORGANIZATIONS AND AGENCIES DIRECTORY. Margaret Labash Young, editor. Gale Research Company, Book Tower, Detroit, MI 48226. (800) 521-0707. 2nd edition. 1987. $185.00.

WHO'S WHO IN ELECTRONICS. Harris Publishing Company, 2057-2 Aurora Road, Twinsburg, OH 44087. (216) 425-9143. Annual. $90.00.

WHO'S WHO IN ENGINEERING. Gordon Davis, editor. Hemisphere Publishing Corporation, 1010 Vermont Avenue, NW, Washington, DC 20005. (800) 526-0275. 6th edition. 1985. $200.00.

ENCYCLOPEDIAS AND DICTIONARIES

DICTIONARY OF AUDIO, RADIO AND VIDEO. R.S. Roberts. Butterworth's Publishing, 80 Montvale Avenue, Stoneham, MA 02180. (800) 325-4177. 1981. $45.00.

IEEE STANDARD DICTIONARY OF ELECTRICAL AND ELECTRONICS TERMS. Frank Jay, editor. John Wiley and Sons, Inc., 605 Third Avenue, New York, NY 10158. (800) 526-5368. 3rd edition. 1984. $49.95.

GENERAL WORKS

BASIC TELEVISION. Bernard Grob. McGraw-Hill Book Company, 1221 Avenue of the Americas, New York, NY 10020. (212) 512-2000. Fourth edition. 1975. $36.95.

DIGITAL VIDEO I. Frank Davidoff and John Rossi, editors. Society of Motion Picture and Television Engineers, 595 West Hartsdale Avenue, White Plains, NY 10607. (914) 472-6606. 1982. $25.00.

ELECTRONIC INVENTIONS AND DISCOVERIES: ELECTRONICS FROM ITS EARLIEST BEGINNINGS TO PRESENT DAY. G.W.A. Dummer. Pergamon Press, Inc., Maxwell House, Fairview Park, Elmsford, NY 10523. (914) 592-7700. 1983. $49.50.

SOLID STATE VIDEO CAMERAS. A. Cristol. Pergamon Press, Inc., Maxwell House, Fairview Park, Elmsford, NY 10523. (914) 592-7700. 1986. $55.00.

TELEVISION PRODUCTION. Alan Wurtzel. McGraw-Hill Book Company, 1221 Avenue of the Americas, New York, NY 10020. (212) 512-2000. Second edition. 1983. $38.95.

TELEVISION: THEORY AND SERVICING. C. Buscombe. Reston Publishing Company, Inc., c/o Prentice-Hall Publishing, Inc., Englewood Cliffs, NJ 07632. (800) 562-0245. 1984. $34.95.

VIDEODISC SYSTEMS: THEORY AND APPLICATIONS. Jordan Isailovic. Prentice-Hall Publishing, Inc., Englewood Cliffs, NJ 07632. (800) 562-0245. 1987. $42.95.

HANDBOOKS AND MANUALS

ELECTRONIC ENGINEERS HANDBOOK. Donald G. Fink, editor. McGraw-Hill Book Company, 1221 Avenue of the Americas, New York, NY 10020. (212) 512-2000. 2nd edition. 1982. $89.00.

HANDBOOK OF MODERN ELECTRONICS AND ELECTRICAL ENGINEERING. Charles Belove, editor. John Wiley and Sons, Inc., 605 Third Avenue, New York, NY 10158. (800) 526-5368. 1986. $88.95.

ONLINE DATA BASES

CA SEARCH. Chemical Abstracts Service, P.O. Box 3012, Columbus, OH 43120. (800) 848-6538 or (614) 421-3600. Comprehensive guide to chemical literature, 1972 to present. Inquire as to online cost and availability.

COMPENDEX. Engineering Information, Inc., 345 East 47th Street, New York, NY 10017. (800) 221-1044 or (212) 705-7615. Engineering and technical literature, 1975 to present. Inquire as to online cost and availability.

INSPEC. INSPEC Marketing Department, Institution of Electrical Engineers. Available from IEEE Service Center, 445 Hoes Lane, Piscataway, NJ 08854. (201) 981-0060. Online version of Physics Abstracts. Inquire as to online cost and availability.

NTIS. National Technical Information Service, 5285 Port Royal Road, Springfield, VA 22161. (703) 487-4630. Broad coverage of government sponsored research reports, 1964 to present. Inquire as to online cost and availability.

SCISEARCH. Institute for Scientific Information, 3501 Market Street, Philadelphia, PA 19104. (800) 523-1850 or (215) 386-0100. Broad multidisciplinary title and author index to the international literature of science and technology, 1974 to present. Inquire as to online cost and availability.

WILSONLINE. H.W. Wilson and Company, 950 University Avenue, Bronx, NY 10452. (800) 367-6770 or (212) 588-8400. Makes available online versions of the H.W. Wilson indexes including Applied Science and Technology Index, Business Periodicals Index and Readers' Guide to Periodical Literature. Approximately 1980 to present. Inquire as to online cost and availability.

OTHER SOURCES

A GUIDE TO THE LITERATURE OF ELECTRICAL AND ELECTRONICS ENGINEERING. Susan B. Ardis. Libraries Unlimited Inc., P.O. Box 263, Littleton, CO 80160. (303) 770-1220. 1987. $37.50.

PERIODICALS

BROADCASTER ENGINEERING. Intertec Publishing Corporation, Box 12901, Overland Park, KS 66212. (913) 888-4664. 1959 to present. Monthly. $25.00 per year.

ELECTRONIC DESIGN. Hayden Publishing Company, 10 Mulholland Drive, Hasbrouck Heights, NJ 07604. (201) 288-7520. 1952 to present. Biweekly. $40.00 per year.

ELECTRONICS. McGraw-Hill Book Company, 1221 Avenue of the Americas, New York, NY 10020. (212) 512-2000. 1930 to present. Weekly. $32.00 per year.

ELECTRONICS AND WIRELESS WORLD. I.P.C. Electrical-Electronic Press, Ltd., Quadrant House, The Quadrant, Sutton, Surrey, SM2 5AS England. 1911 to present. Monthly. $105.00 per year.

IEEE CIRCUITS AND DEVICES MAGAZINE. Institute of Electrical and Electronics Engineers. IEEE Service Center, 445 Hoes Lane, Piscataway, NJ 08854. Bimonthly. $70.00 per year.

IEEE JOURNAL OF SOLID STATE CIRCUITS. Institute of Electrical and Electronics Engineers. IEEE Service Center, 445 Hoes Lane, Piscataway, NJ 08854. 1966 to present. Bimonthly. $113.00 per year.

IEEE TRANSACTIONS ON BROADCASTING. Institute of Electrical and Electronics Engineers. IEEE Service Center, 445 Hoes Lane, Piscataway, NJ 08854. 1955 to present. Quarterly. $37.00 per year.

OPTICAL INFORMATION SYSTEMS MAGAZINE. Meckler Publishing, 11 Ferry Lane West, Westport, CT 06880. (203) 226-6967. 1981 to present. Bimonthly. $75.00 per year.

RESEARCH CENTERS AND INSTITUTES

CENTER FOR ADVANCED TECHNOLOGY IN TELECOMMUNICATIONS. Polytechnic University, 333 Jay Street, Brooklyn, NY 11201. (718) 643-5160.

COMMUNICATIONS RESEARCH LABORATORY. McMaster University, 1280 Main Street West, Hamilton, ON, Canada L8S 4K1. (416) 525-9140.

ELECTRONICS RESEARCH LABORATORY. Montana State University, Bozeman, MT 59717. (406) 994-2505.

VOLCANOLOGY

See also: EARTHQUAKES, EARTHQUAKE ENGINEERING, GEOLOGY, GEOPHYSICS, GEOTHERMAL ENERGY, IGNEOUS ROCKS, OCEANOGRAPHY, PLANETARY SCIENCE, PLATE TECTONICS, SEISMOLOGY

ABSTRACT SERVICES AND INDEXES

BIBLIOGRAPHY AND INDEX OF GEOLOGY. American Geological Institute, 4220 King Street, Alexandria, VA 22302. (703) 379-2480. 1969 to present. Monthly. $1100.00 per year.

CHEMICAL ABSTRACTS. American Chemical Society, Chemical Abstracts Service, Box 3012, Columbus, OH 43210. (614) 421-3600. 1907 to present. Weekly. $9500.00 per year.

CONFERENCE PAPERS INDEX. Cambridge Scientific Abstracts, 5161 River Road, Bethesda, MD 20816. 1972 to present. Monthly. Inquire as to cost and availability.

CURRENT CONTENTS: PHYSICAL, CHEMICAL AND EARTH SCIENCES. Institute for Scientific Information, 3501 Market Street, Philadelphia, PA 19104. (800) 523-1850 or (215) 386-0100. Weekly. $275.00 per year.

ENGINEERING INDEX MONTHLY AND AUTHOR INDEX. Engineering Information Inc., 345 East 47th Street, New York, NY 10017. (212) 705-7600. Monthly. $1560.00 per year.

INDEX TO SCIENTIFIC AND TECHNICAL PROCEEDINGS. Institute for Scientific Information, 3501 Market Street, Philadelphia, PA 19104. (800) 523-1850 or (215) 386-0100. 1978 to present. Monthly. $775.00 per year.

INDEX TO SCIENTIFIC REVIEWS. Institute for Scientific Information, 3501 Market Street, Philadelphia, PA 19104. (800) 523-1850 or (215) 386-0100. 1974 to present. Semi-annual. $550.00 per year.

METEOROLOGICAL AND GEOASTROPHYSICAL ABSTRACTS. American Meteorological Society, 45 Beacon Street, Boston, MA 02108. (617) 227-2425. 1950 to present. Monthly. $450.00 per year.

OCEAN ABSTRACTS. Cambridge Scientific Abstracts, 5161 River Road, Bethesda, MD 20816. (301) 951-1400. 1963 to present. Bimonthly. $450.00 per year.

SCIENCE CITATION INDEX. Institute for Scientific Information, 3501 Market Street, Philadelphia, PA 19104. (800) 523-1850 or (215) 386-0100. Six times per year. $6200.00 per year.

ASSOCIATIONS AND PROFESSIONAL SOCIETIES

AMERICAN ASTRONOMICAL SOCIETY. 1816 Jefferson Place, N.W., Washington, DC 20036. (202) 659-0134.

AMERICAN GEOPHYSICAL UNION. 2000 Florida Avenue, N.W., Washington, DC 20009. (202) 462-6903.

EARTHQUAKE ENGINEERING RESEARCH INSTITUTE. 2620 Telegraph Avenue, Berkeley, CA 94704. (415) 848-0972.

GEOLOGICAL SOCIETY OF AMERICA. 3300 Penrose Place, Boulder, CO 80301. (303) 447-2020.

SEISMOLOGICAL SOCIETY OF AMERICA. 6431 Fairmont Avenue, No. 7, El Cerrito, CA 94530. (415) 525-5474.

DIRECTORIES AND BIOGRAPHICAL SOURCES

AMERICAN MEN AND WOMEN OF SCIENCE. R.R. Bowker, Inc., Order Department, 245 West 17th Street, New York, NY 10011. (800) 521-8110. Eight volumes. 1986. $595.00 for set.

INTERNATIONAL RESEARCH CENTERS DIRECTORY 1988-89. Darren L. Smith, editor. Gale Research Company, Book Tower, Detroit, MI 48226. (800) 521-0707. 4th edition. 1987. $360.00.

RESEARCH CENTERS DIRECTORY 1988. Gale Research Company, Book Tower, Detroit, MI 48226. (800) 521-0707. 12th edition. 1987. $365.00 for set.

SCIENTIFIC AND TECHNICAL ORGANIZATIONS AND AGENCIES DIRECTORY. Margaret Labash Young, editor. Gale Research Company, Book Tower, Detroit, MI 48226. (800) 521-0707. 2nd edition. 1987. $185.00.

ENCYCLOPEDIAS AND DICTIONARIES

MCGRAW-HILL ENCYCLOPEDIA OF THE GEOLOGICAL SCIENCES. McGraw-Hill Book Company, 1221 Avenue of the Americas, New York, NY 10020. (212) 512-2000. Second edition. 1988. $85.00.

GENERAL WORKS

EXPLOSIVE VOLCANISM. M.F. Sheridan and F. Barberi, editors. Elsevier Science Publishing Company, Inc., 52 Vanderbilt Avenue, New York, NY 10017. (212) 370-5520. 1983. $85.00.

AN INTRODUCTION TO THE THEORY OF SEISMOLOGY. K.E. Bulletin and A.B. Bolt. Cambridge University Press, 32 East 57th Street, New York, NY 10022. (800) 872-7423. 1985. $69.50.

VOLCANIC ACTIVITY AND HUMAN ECOLOGY. Payson D. Sheets, and Donald K. Grayson, editors. Academic Press, Inc., 6277 Sea Harbor Drive, Orlando, FL 32821. (800) 321-5068. 1979. $63.00.

VOLCANOES OF THE WORLD: A REGIONAL GAZETTEER AND CHRONOLOGY OF VOLCANISM DURING THE LAST 10,000 YEARS. Tom Simkin and others. Van Nostrand Reinhold Company, Inc., 135 West 50th Street, New York, NY 10020. (800) 543-2681. 1981. $31.95.

VOLCANOES OF THE EARTH. Fred M. Bullard. University of Texas Press, P.O. Box 7819, Austin, TX 78713-7819. (512) 471-7233. Second edition. 1984. $35.00.

ONLINE DATA BASES

CA SEARCH. Chemical Abstracts Service, P.O. Box 3012, Columbus, OH 43120. (800) 848-6538 or (614) 421-3600. Comprehensive guide to chemical literature, 1972 to present. Inquire as to online cost and availability.

COMPENDEX. Engineering Information, Inc., 345 East 47th Street, New York, NY 10017. (800) 221-1044 or (212) 705-7615. Engineering and technical literature, 1975 to present. Inquire as to online cost and availability.

DISSERTATION ABSTRACTS ONLINE. University Microfilms International, 300 North Zeeb Road, Ann Arbor, MI 48106. (800) 521-0600 or (313) 761-4700. Scope includes virtually all doctoral dissertations accepted at accredited American institutions from 1861 to present in over 250 subject areas. Inquire as to online cost and availability.

GEOREF. Online version of the BIBLIOGRAPHY AND INDEX OF GEOLOGY. American Geological Institute, 4220 King Street, Alexandria, VA 22302. (703) 379-2480. 1969 to present. Inquire as to online cost and availability.

NTIS. National Technical Information Service, 5285 Port Royal Road, Springfield, VA 22161. (703) 487-4630. Broad coverage of government sponsored research reports, 1964 to present. Inquire as to online cost and availability.

SCISEARCH. Institute for Scientific Information, 3501 Market Street, Philadelphia, PA 19104. (800) 523-1850 or (215) 386-0100. Broad multidisciplinary title and author index to the international literature of science and technology, 1974 to present. Inquire as to online cost and availability.

PERIODICALS

EARTHQUAKE INFORMATION BULLETIN. U.S. Geological Survey, 12201 Sunrise Valley Drive, Reston, VA 22092. Order from: Superintendent of Documents, U.S. Government Printing Office, 20402. Bimonthly. $15.00 per year.

EARTHQUAKE PREDICTION RESEARCH. D. Reidel Publishing Company, 190 Old Derby Street, Hingham, MA 02043. 1982 to present. Quarterly. $85.00 per year.

GEOPHYSICAL AND ASTROPHYSICAL FLUID DYNAMICS. Gordon and Breach Science Publishers, Inc., 50 West 23rd Street, New York, NY 10010. (212) 206-8900. 1970 to present. 16 times per year. $260.00 per year.

GEOPHYSICAL RESEARCH LETTERS. American Geophysical Union, 2000 Florida Avenue, N.W., Washington, DC 20009. (202) 462-6903. 1974 to present. Monthly. $185.00 per year.

JGR: JOURNAL OF GEOPHYSICAL RESEARCH: SOLID EARTH AND PLANETS. American Geophysical Union, 2000 Florida Avenue, N.W., Washington, DC 20009. (202) 462-6903. Monthly. $760.00 per year.

JOURNAL OF VOLCANOLOGY AND GEOTHERMAL RESEARCH. Elsevier Science Publishing Company, Inc., 52 Vanderbilt Avenue, New York, NY 10017. (212) 370-5520. 1976 to present. 16 times per year. $360.00 per year.

SEISMOLOGICAL SOCIETY OF AMERICA BULLETIN. Seismologycal Society of America, 6431 Fairmont Avenue, No. 7, El Cerrito, CA 94530. (415) 525-5474. 1911 to present. Bimonthly. $90.00 per year.

TECTONOPHYSICS: AN INTERNATIONAL JOURNAL OF GEOTECTONICS AND THE GEOLOGY AND PHYSICS OF THE INTERIOR OF THE EARTH. Elsevier Science Publishing Company, Inc., 52 Vanderbilt Avenue, New York, NY 10017. (212) 370-5520. 1964 to present. $1200.00 per year.

VOLCANOLOGY AND SEISMOLOGY. Gordon and Breach Science Publishers, Inc., 50 West 23rd Street, New York, NY 10010. (212) 206-8900. Monthly. $500.00 per year.

RESEARCH CENTERS AND INSTITUTES

HAWAII INSTITUTE OF GEOPHYSICS. University of Hawaii, 2525 Correa Road, Honolulu, HI 96822. (808) 948-8760.

INCORPORATED RESEARCH INSTITUTIONS FOR SEISMOLOGY. 1616 North Fort Meyer Drive, Suite 1440, Arlington, VA 22209. (703) 524-6222.

LAMONT-DOHERTY GEOLOGICAL OBSERVATORY. Columbia University, Palisades, NY 10964. (914) 359-2900.

W

WANKEL ENGINE

See: ENGINES

WATER ANALYSIS

See: ENVIRONMENTAL ENGINEERING

WATER POLLUTION

See also: GROUND WATER, GROUND WATER POLLUTION, HYDROGEOLOGY, HYDROLOGY, WATER RESOURCES

ABSTRACT SERVICES AND INDEXES

CHEMICAL ABSTRACTS. Chemical Abstracts Service, 2540 Olentangy Road, Post Office Box 3012, Columbus, OH 43210. (800) 848-6538 or (614) 421-3600. Weekly. $9200.00 per year.

CURRENT CONTENTS: ENGINEERING, TECHNOLOGY, AND APPLIED SCIENCES. Institute for Scientific Information, 3501 Market Street, Philadelphia, PA 19104. (800) 523-1850 or (215) 386-0100. 1970 to present. Weekly. $272.00 per year.

CURRENT CONTENTS: PHYSICAL, CHEMICAL AND EARTH SCIENCES. Institute for Scientific Information, 3501 Market Street, Philadelphia, PA 19104. (800) 523-1850 or (215) 386-0100. 1970 to present. Weekly. $272.00 per year.

ENGINEERING INDEX MONTHLY AND AUTHOR INDEX. Engineering Information, Incorporated, 345 East 47th Street, New York, NY 10017. (800) 221-1044 or (212) 705-7600. Monthly, with annual cumulation. $1560.00 per year.

HYDRO-ABSTRACTS. Environmental Hydrology Corporation, Box 14701, Minneapolis, MN 55414. (612) 379-0901. 1968 to present. Monthly. $120.00 per year.

POLLUTION ABSTRACTS. Cambridge Scientific Abstracts, 5161 River Road, Bethesda, MD 20816. (301) 951-1400. 1970 to present. Bimonthly. $465.00 per year.

SCIENCE CITATION INDEX. Institute for Scientific Information, 3501 Market Street, Philadelphia, PA 19104. (800) 523-1850 or (215) 386-0100. Inquire as to cost and availability.

SELECTED WATER RESOURCES ABSTRACTS. United States Geological Survey, Water Resources Scientific Information Center. Available from: National Technical Information Service, Springfield, VA 22161. (703) 860-7455. 1968 to present. Monthly. $115.00 per year.

WATER QUALITY CONTROL DIGEST. University Digest Services, Post Office Box 343, Troy, MI 48099. (313) 651-2528. 1969 to present. Bimonthly. $87.00 per year.

ASSOCIATION AND PROFESSIONAL SOCIETIES

AMERICAN PUBLIC WORKS ASSOCIATION. 1313 East 605h Street, Chicago, IL 60637. (312) 667-2200.

AMERICAN SOCIETY OF CIVIL ENGINEERS. 345 East 47th Street, New York, NY 10017-2398. (212) 705-7520.

ASSOCIATION OF GROUND WATER SCIENTISTS AND ENGINEERS. 6375 Riverside Drive, Dublin, OH 43017. (614) 761-1711.

GROUND WATER INSTITUTE. Post Office Box 981, Minneapolis, MN 55440. (612) 698-4395.

NATIONAL WATER RESOURCES ASSOCIATION. 955 L'Enfant Plaza, SW, Suite 1202N, Washington, DC 20024. (202) 488-0610.

NATIONAL WATER WELL ASSOCIATION. 6375 Riverside Drive, Dublin, OH 43017. (614) 761-1711.

WATER POLLUTION CONTROL FEDERATION. 2626 Pennsylvania Avenue, Washington, DC 20037. (202) 337-2500.

WATER QUALITY ASSOCIATION. 4151 Naperville Road, Lisle, IL 60532. (312) 369-1600.

BIBLIOGRAPHIES

WATER POLLUTION: A GUIDE TO INFORMATION SOURCES. Allen W. Knight and Mary Ann Simmons, editors. Gale Research Company, Book Tower, Detroit, MI 48226. (800) 521-0707. 1980. $62.00.

DIRECTORIES AND BIOGRAPHICAL SOURCES

AMERICAN WATER WORKS ASSOCIATION - OFFICERS AND COMMITTEE DIRECTORY. American Water Works Association, 6666 West Quincy Avenue, Denver, CO 80235. (303) 794-7711. Annual. $15.00.

JOURNAL OF THE AMERICAN WATER WORKS ASSOCIATION BUYERS GUIDE ISSUE. American Water Works Association, 6666 West Quincy Avenue, Denver, CO 80235. (303) 794-7711. Annual, with subscription.

PUBLIC WORKS MANUAL. Public Works Journal Corporation, 200 South Broad Street, Ridgewood, NJ 07451. (201) 445-5800. Annual. $20.00.

SCIENTIFIC AND TECHNICAL ORGANIZATIONS AND AGENCIES DIRECTORY. Gale Research Company, Book Tower, Detroit, MI 48226. (800) 521-0707. Second edition. Two volumes. 1987. $185.00 set.

WATER QUALITY ASSOCIATION DIRECTORY. Water Quality Association, 4151 Naperville Road, Lisle, IL 60532. (312) 369-1600. Annual. Available to members only.

GENERAL WORKS

GROUND WATER CHEMISTRY. Richard Rice. Lewis Publishing, Incorporated, 121 South Main Street, Chelsea, MI 48118. (313) 475-8619. 1986. $45.00.

GROUND WATER CONTAMINATION. J.H. Guswa and others. Noyes Data Corporation, Mill Road at Grand Avenue, Park Ridge, NJ 07656. (201) 391-8484. 1985. $48.00.

GROUND WATER CONTAMINATION. National Research Council. National Academy Press, 2101 Constitution Avenue, NW, Washington, DC 20418. (202) 334-3313. 1984. $17.95.

GROUND WATER POLLUTION CONTROL. Larry W. Canter and R.C. Knox. Lewis Publishing, Incorporated, 121 South Main Street, Chelsea, MI 48118. (313) 475-8619. 1985. $49.95.

GROUND WATER QUALITY. C.H. Ward and others. John Wiley and Sons, Incorporated, 605 Third Avenue, New York, NY 10158. (800) 526-5368 or (212) 850-6000. 1985. $45.00.

GROUND WATER QUALITY PROTECTION. L.W. Canter and others. Lewis Publishing, Incorporated, 121 South Main Street, Chelsea, MI 48118. (313) 475-8619. 1986. $49.95.

HYDROLOGY AND QUALITY OF WATER RESOURCES. Mark J. Hammer. John Wiley and Sons, Incorporated, 605 Third Avenue, New York, NY 10158. (800) 526-5368 or (212) 850-6000. Second edition. 1986. $42.95.

WATER SUPPLY AND POLLUTION CONTROL. Warren Viessman and Mark J. Hammer. Harper and Row Publishers, Incorporated, 10 East 53rd Street, New York, NY 10022. (800) 242-7737. Fourth edition. 1985. $45.00.

HANDBOOKS AND MANUALS

WATER TREATMENT HANDBOOK. Degremont Company. John Wiley and Sons, Incorporated, 605 Third Avenue, New York, NY 10158. (800) 526-5368 or (212) 850-6000. Fifth edition. 1979. $95.00.

ONLINE DATA BASES

COMPENDEX. Engineering Information, Incorporated, 345 East 47th Street, New York, NY 10017. (800) 221-1044 or (212) 705-7615. Engineering and technical literature, 1975 to present. Inquire as to cost and availability.

GEOREF. American Geological Institute, 4220 King Street, Alexandria, VA 22302. (703) 379-2480. 1967 to present. Inquire as to online cost and availability.

NTIS. National Technical Information Service, 5285 Port Royal Road, Springfield, VA 22161. (703) 487-4630. Broad coverage of government sponsored research reports, 1964 to present. Inquire as to cost and availability.

POLLUTION ABSTRACTS. Cambridge Scientific Abstracts, 5161 River Road, Bethesda, MD 20816. (301) 951-1400. 1970 to present. Available for online searching through DIALOG Information Services and BRS Information Technologies. Inquire as to online cost and availability.

WATER DATA BANK. United States Department of Agriculture, Agricultural Research Service, Hydrology Laboratory, Water Data Center, Room 139, Building 007, BARC-West, Beltsville, MD 20705. (301) 344-4411. Inquire as to online cost and availability.

WATERNET. American Water Works Association, 6666 West Quincy Avenue, Denver, CO 80235. (303) 794-7711. A data base providing abstracts and indexing of information published in the American Water Works Association Journal. 1971 to present. Inquire as to online cost and availability.

PERIODICALS

AMERICAN WATER WORKS ASSOCIATION. Journal. American Water Works Association, 6666 West Quincy Avenue, Denver, CO 80235. (303) 794-7711. Monthly. $50.00 per year.

GROUND WATER. National Water Well Association. Water Well Journal Publishing Company, 500 West Wilson Bridge Road, Worthington, OH 43085. (614) 846-4967. Bimonthly. $53.00 per year.

JOURNAL OF ENVIRONMENTAL ENGINEERING. American Society of Civil Engineers, 345 East 47th Street, New York, NY 10017. (212) 705-7275. 1956 to present. Bimonthly. $80.00 per year.

JOURNAL OF GROUND WATER. National Water Well Association. Water Well Journal Publishing Company, 500 West Wilson Bridge Road, Worthington, OH 43085. (614) 846-4967. Quarterly. $16.00 per year.

JOURNAL OF HYDROLOGY. Elsevier Science Publishers B.V., Post Office Box 211, 1000 AE Amsterdam, The Netherlands. Thirty-two times per year. $675.00 per year.

JOURNAL OF HYDRAULIC ENGINEERING. American Society of Civil Engineers, 345 East 47th Street, New York, NY 10017-2398. (212) 705-7520. Monthly. $115.00 per year.

JOURNAL OF IRRIGATION AND DRAINAGE. American Society of Civil Engineers, 345 East 47th Street, New York, NY 10017-2398. (212) 705-7520. Quarterly. $44.00 per year.

JOURNAL OF WATER RESOURCES PLANNING AND MANAGEMENT. American Society of Civil Engineers, 345 East 47th Street, New York, NY 10017-2398. (212) 705-7520. Quarterly. $56.00 per year.

LIMINOLOGY AND OCEANOGRAPHY. American Society of Limnology and Oceanography, 1530 12th Avenue, Grafton, WI 53024. (414) 377-4871. Bimonthly. $60.00 per year.

WATER RESOURCES BULLETIN. American Water Resources Association, 5410 Grosvenor Lane, Suite 220, Bethesda, MD 20814. (301) 493-8600. Bimonthly. $65.00 per year.

WATER RESOURCES RESEARCH. American Geophysical Union, 2000 Florida Avenue, NW, Washington, DC 20009. (202) 462-6903. Monthly. $295.00 per year.

RESEARCH CENTERS AND INSTITUTES

CALIFORNIA INSTITUTE OF TECHNOLOGY. W.M. Keck Engineering Laboratory of Hydraulics and Water Resources, 1201 East California Boulevard, Pasadena, CA 91125. (818) 356-4404.

CALIFORNIA INSTITUTE OF TECHNOLOGY. W.M. Keck Engineering Laboratory of Hydraulics and Water Resources, 1201 East California Boulevard, Pasadena, CA 91125. (818) 356-4404.

COLORADO STATE UNIVERSITY. Hydraulics and Hydromachinery Research Laboratories, Engineering Research Center, Fort Collins, CO 80523. (303) 491-8655.

UNITED STATES ARMY HYDROLOGIC ENGINEERING CENTER. 609 Center Street, Davis, CA 95616. (916) 440-3285.

UNIVERSITY OF ILLINOIS. Hydrosystems Laboratory, Department of Civil Engineering, Urbana, IL 61801. (217) 333-0107.

WATER RESOURCES

See also: GROUND WATER, HYDROGEOLOGY, HYDROLOGY, WATER POLLUTION

ABSTRACT SERVICES AND INDEXES

BIBLIOGRAPHY AND INDEX OF GEOLOGY. American Geological Institute, 4220 King Street, Alexandria, VA 22302. (703) 379-2480. 1969 to present. Monthly. $1100.00 per year.

CHEMICAL ABSTRACTS. Chemical Abstracts Service, 2540 Olentangy Road, Post Office Box 3012, Columbus, OH 43210. (800) 848-6538 or (614) 421-3600. Weekly. $9200.00 per year.

CURRENT CONTENTS: PHYSICAL, CHEMICAL AND EARTH SCIENCES. Institute for Scientific Information, 3501 Market Street, Philadelphia, PA 19104. (800) 523-1850 or (215) 386-0100. 1970 to present. Weekly. $272.00 per year.

DELFT HYDROSCIENCE ABSTRACTS. Delft Hydraulics Laboratory, Rotterdamseweg 185, Postbus 177, NL-2600 MH Delft, The Netherlands. Monthly. $177.00 per year.

ENGINEERING INDEX MONTHLY AND AUTHOR INDEX. Engineering Information, Incorporated, 345 East 47th Street, New York, NY 10017. (800) 221-1044 or (212) 705-7600. Monthly, with annual cumulation. $1560.00 per year.

HYDRO-ABSTRACTS. Environmental Hydrology Corporation, Box 14701, Minneapolis, MN 55414. (612) 379-0901. 1968 to present. Monthly. $120.00 per year.

METEOROLOGICAL AND GEOASTROPHYSICAL ABSTRACTS. American Meteorological Society, 45 Beacon Street, Boston, MA 02108. (617) 227-2425. 1950 to present. Monthly. $450.00 per year.

SCIENCE CITATION INDEX. Institute for Scientific Information, 3501 Market Street, Philadelphia, PA 19104. (800) 523-1850 or (215) 386-0100. Inquire as to cost and availability.

SELECTED WATER RESOURCES ABSTRACTS. United States Geological Survey, Water Resources Scientific Information Center. Available from: National Technical Information Service, Springfield, VA 22161. (703) 860-7455. 1968 to present. Monthly. $115.00 per year.

WATER QUALITY CONTROL DIGEST. University Digest Services, Post Office Box 343, Troy, MI 48099. (313) 651-2528. 1969 to present. Bimonthly. $87.00 per year.

ANNUAL REVIEWS AND YEARBOOKS

ADVANCES IN HYDROSCIENCE. Ven Te Chow, editor. Academic Press, Incorporated, Orlando, FL 32887. (800) 321-5068 or (305) 345-2734. 1964-1981. Inquire, price varies.

ASSOCIATIONS AND PROFESSIONAL SOCIETIES

AMERICAN INSTITUTE OF HYDROLOGY. Post Office Box 14251, St. Paul, MN 55114. (612) 379-1030.

AMERICAN PUBLIC WORKS ASSOCIATION. 1313 East 60th Street, Chicago, IL 60637. (312) 667-2200.

AMERICAN SOCIETY OF CIVIL ENGINEERS. 345 East 47th Street, New York, NY 10017-2398. (212) 705-7520.

AMERICAN WATER WORKS ASSOCIATION. 6666 West Quincy Avenue, Denver, CO 80235. (303) 794-7711.

ASSOCIATION OF GROUND WATER SCIENTISTS AND ENGINEERS. 6375 Riverside Drive, Dublin, OH 43017. (614) 761-1711.

GROUND WATER INSTITUTE. Post Office Box 981, Minneapolis, MN 55440. (612) 698-4395.

NATIONAL WATER RESOURCES ASSOCIATION. 955 L'Enfant Plaza, SW, Suite 1202N, Washington, DC 20024. (212) 988-0610.

NATIONAL WATER WELL ASSOCIATION. 6375 Riverside Drive, Dublin, OH 43017. (614) 761-1711.

WATER POLLUTION CONTROL FEDERATION. 2626 Pennsylvania Avenue, Washington, DC 20037. (202) 337-2500.

WATER QUALITY ASSOCIATION. 4151 Naperville Road, Lisle, IL 60532. (312) 369-1600.

BIBLIOGRAPHIES

GERAGHTY AND MILLER'S GROUNDWATER BIBLIOGRAPHY. Frits Van Der Leeden. Water Information Center, Incorporated, 6800 Jericho Turnpike, Syosset, NY 11791. (516) 921-7690. Third edition. 1983. $22.00.

DIRECTORIES AND BIOGRAPHICAL SOURCES

AMERICAN WATER WORKS ASSOCIATION - OFFICERS AND COMMITTEE DIRECTORY. American Water Works Association, 6666 West Quincy Avenue, Denver, CO 80235. (303) 794-7711. Annual. $15.00.

JOURNAL OF THE AMERICAN WATER WORKS ASSOCIATION BUYERS GUIDE ISSUE. American Water Works Association, 6666 West Quincy Avenue, Denver, CO 80235. (303) 794-7711. Annual, with subscription.

PUBLIC WORKS MANUAL. Public Works Journal Corporation, 200 South Broad Street, Ridgewood, NJ 07451. (201) 445-5800. Annual. $20.00.

SCIENTIFIC AND TECHNICAL ORGANIZATIONS AND AGENCIES DIRECTORY. Gale Research Company, Book Tower, Detroit, MI 48226. (800) 521-0707. Second edition. Two volumes. 1987. $185.00 set.

WATER QUALITY ASSOCIATION DIRECTORY. Water Quality Association, 4151 Naperville Road, Lisle, IL 60532. (312) 369-1600. Annual. Available to members only.

GENERAL WORKS

BASIC HYDROLOGY. James J. Sharp and P.G. Sawden. Butterworth's, Incorporated, 80 Montvale Avenue, Stoneham, MA 02180. (617) 438-8464. 1984. $15.95 in paper.

GROUND WATER. R. Bowen. Elsevier Science Publishing Company, Incorporated, 52 Vanderbilt Avenue, New York, NY 10017. (212) 370-5520. Second edition. 1986. $79.25.

GROUND WATER CHEMISTRY. Richard Rice. Lewis Publishing, Incorporated, 121 South Main Street, Chelsea, MI 48118. (313) 475-8619. 1986. $45.00.

GROUND WATER HYDROLOGY. David K. Todd. John Wiley and Sons, Incorporated, 605 Third Avenue, New York, NY 10158. (800) 526-5368 or (212) 850-6000. Second edition. 1980. $48.00.

GROUND WATER POLLUTION CONTROL. Larry W. Canter and R.C. Knox. Lewis Publishing, Incorporated, 121 South Main Street, Chelsea, MI 48118. (313) 475-8619. 1985. $49.95.

GROUND WATER QUALITY. C.H. Ward and others. John Wiley and Sons, Incorporated, 605 Third Avenue, New York, NY 10158. (800) 526-5368 or (212) 850-6000. 1985. $45.00.

GROUND WATER SYSTEMS PLANNING AND MANAGEMENT. Robert Willis. Prentice Hall, Incorporated, Englewood Cliffs, NJ 07632. (800) 562-0245. 1987. $49.00.

HYDROLOGY AND QUALITY OF WATER RESOURCES. Mark J. Hammer. John Wiley and Sons, Incorporated, 605 Third Avenue, New York, NY 10158. (800) 526-5368 or (212) 850-6000. Second edition. 1986. $42.95.

ONLINE DATA BASES

COMPENDEX. Engineering Information, Incorporated, 345 East 47th Street, New York, NY 10017. (800) 221-1044 or (212) 705-7615. Engineering and technical literature, 1975 to present. Inquire as to cost and availability.

GEOREF. American Geological Institute, 4220 King Street, Alexandria, VA 22302. (703) 379-2480. 1967 to present. Inquire as to online cost and availability.

NTIS. National Technical Information Service, 5285 Port Royal Road, Springfield, VA 22161. (703) 487-4630. Broad coverage of government sponsored research reports, 1964 to present. Inquire as to cost and availability.

WATER DATA BANK. United States Department of Agriculture, Agricultural Research Service, Hydrology Laboratory, Water Dat Center, Room 139, Building 007, BARC-West, Beltsville, MD 20705. (301) 344-4411. Inquire as to online cost and availability.

PERIODICALS

AMERICAN WATER WORKS ASSOCIATION. Journal. American Water Works Association, 6666 West Quincy Avenue, Denver, CO 80235. (303) 794-7711. Monthly. $50.00 per year.

GROUND WATER. National Water Well Association. Water Well Journal Publishing Company, 500 West Wilson Bridge Road, Worthington, OH 43085. (614) 846-4967. Bimonthly. $53.00 per year.

JOURNAL OF GROUND WATER. National Water Well Association. Water Well Journal Publishing Company, 500 West Wilson Bridge Road, Worthington, OH 43085. (614) 846-4967. Quarterly. $16.00 per year.

JOURNAL OF HYDROLOGY. Elsevier Science Publishers B.V., Post Office Box 211, 1000 AE Amsterdam, The Netherlands. Thirty-two times per year. $675.00 per year.

JOURNAL OF IRRIGATION AND DRAINAGE. American Society of Civil Engineers, 345 East 47th Street, New York, NY 10017-2398. (212) 705-7520. Quarterly. $44.00 per year.

JOURNAL OF WATER RESOURCES PLANNING AND MANAGEMENT. American Society of Civil Engineers, 345 East 47th Street, New York, NY 10017-2398. (212) 705-7520. Quarterly. $56.00 per year.

LIMINOLOGY AND OCEANOGRAPHY. American Society of Limnology and Oceanography, 1530 12th Avenue, Grafton, WI 53024. (414) 377-4871. Bimonthly. $60.00 per year.

WATER RESOURCES BULLETIN. American Water Resources Association, 5410 Grosvenor Lane, Suite 220, Bethesda, MD 20814. (301) 493-8600. Bimonthly. $65.00 per year.

WATER RESOURCES RESEARCH. American Geophysical Union, 2000 Florida Avenue, NW, Washington, DC 20009. (202) 462-6903. Monthly. $295.00 per year.

RESEARCH CENTERS AND INSTITUTES

CALIFORNIA INSTITUTE OF TECHNOLOGY. W.M. Keck Engineering Laboratory of Hydraulics and Water Resources, 1201 East California Boulevard, Pasadena, CA 91125. (818) 356-4404.

ILLINOIS STATE GEOLOGICAL SURVEY. Natural Resources Building, 615 East Peabody Drive, Champaign, IL 61820. (217) 344-1481.

NATIONAL SCIENCE FOUNDATION. Hydraulics, Hydrology, and Water Resources Engineering Program, 1800 G Street, NW, Washington, DC 20550.

UNITED STATES GEOLOGICAL SURVEY. Water Resources Division, 421 National Center, Reston, VA 22092. (703) 860-6031.

UNIVERSITY OF MINNESOTA. Minnesota Geological Survey, 2642 University Avenue, St. Paul, MN 55114. (612) 373-3372.

WATER TREATMENT

See also: GROUND WATER, GROUND WATER POLLUTION, HYDROGEOLOGY, HYDROLOGY, WATER POLLUTION, WATER RESOURCES

ABSTRACT SERVICES AND INDEXES

CHEMICAL ABSTRACTS. Chemical Abstracts Service, 2540 Olentangy Road, Post Office Box 3012, Columbus, OH 43210. (800) 848-6538 or (614) 421-3600. Weekly. $9200.00 per year.

CURRENT CONTENTS: PHYSICAL, CHEMICAL AND EARTH SCIENCES. Institute for Scientific Information, 3501 Market Street, Philadelphia, PA 19104. (800) 523-1850 or (215) 386-0100. 1970 to present. Weekly. $272.00 per year.

DELFT HYDROSCIENCE ABSTRACTS. Delft Hydraulics Laboratory, Rotterdamseweg 185, Postbus 177, NL-2600 MH Delft, The Netherlands. Monthly. $177.00 per year.

ENGINEERING INDEX MONTHLY AND AUTHOR INDEX. Engineering Information, Incorporated, 345 East 47th Street, New York, NY 10017. (800) 221-1044 or (212) 705-7600. Monthly, with annual cumulation. $1560.00 per year.

HYDRO-ABSTRACTS. Environmental Hydrology Corporation, Box 14701, Minneapolis, MN 55414. (612) 379-0901. 1968 to present. Monthly. $120.00 per year.

POLLUTION ABSTRACTS. Cambridge Scientific Abstracts, 5161 River Road, Bethesda, MD 20816. (301) 951-1400. 1970 to present. Bimonthly. $465.00 per year.

SCIENCE CITATION INDEX. Institute for Scientific Information, 3501 Market Street, Philadelphia, PA 19104. (800) 523-1850 or (215) 386-0100. Inquire as to cost and availability.

SELECTED WATER RESOURCES ABSTRACTS. United States Geological Survey, Water Resources Scientific Information Center. Available from: National Technical Information Service, Springfield, VA 22161. (703) 860-7455. 1968 to present. Monthly. $115.00 per year.

WATER QUALITY CONTROL DIGEST. University Digest Services, Post Office Box 343, Troy, MI 48099. (313) 651-2528. 1969 to present. Bimonthly. $87.00 per year.

ASSOCIATIONS AND PROFESSIONAL SOCIETIES

AMERICAN PUBLIC WORKS ASSOCIATION. 1313 East 60th Street, Chicago, IL 60637. (312) 667-2200.

AMERICAN SOCIETY OF CIVIL ENGINEERS. 345 East 47th Street, New York, NY 10017-2398. (212) 705-7520.

AMERICAN WATER RESOURCES ASSOCIATION. 5410 Grosvenor Lane, Suite, Bethesda, MD 20814. (301) 493-8600.

AMERICAN WATER WORKS ASSOCIATION. 6666 West Quincy Avenue, Denver, CO 80235. (303) 794-7711.

ASSOCIATION OF GROUND WATER SCIENTISTS AND ENGINEERS. 6375 Riverside Drive, Dublin, OH 43017. (614) 761-1711.

GROUND WATER INSTITUTE. Post Office Box 981, Minneapolis, MN 55440. (612) 698-4395.

NATIONAL WATER RESOURCES ASSOCIATION. 955 L'Enfant Plaza, SW, Suite 1202N, Washington, DC 20024. (212) 988-0610.

NATIONAL WATER WELL ASSOCIATION. 6375 Riverside Drive, Dublin, OH 43017. (614) 761-1711.

WATER POLLUTION CONTROL FEDERATION. 2626 Pennsylvania Avenue, Washington, DC 20037. (202) 337-2500.

WATER QUALITY ASSOCIATION. 4151 Naperville Road, Lisle, IL 60532. (312) 369-1600.

BIBLIOGRAPHIES

GERAGHTY AND MILLER'S GROUNDWATER BIBLIOGRAPHY. Frits Van Der Leeden. Water Information Center, Incorporated, 6800 Jericho Turnpike, Syosset, NY 11791. (516) 921-7690. Third edition. 1983. $22.00.

WATER POLLUTION: A GUIDE TO INFORMATION SOURCES. Allen W. Knight and Mary Ann Simmons, editors. Gale Research Company, Book Tower, Detroit, MI 48226. (800) 521-0707. 1980. $62.00.

DIRECTORIES AND BIOGRAPHICAL SOURCES

AMERICAN WATER WORKS ASSOCIATION - OFFICERS AND COMMITTEE DIRECTORY. American Water Works Association, 6666 West Quincy Avenue, Denver, CO 80235. (303) 794-7711. Annual. $15.00.

JOURNAL OF THE AMERICAN WATER WORKS ASSOCIATION BUYERS GUIDE ISSUE. American Water Works Association, 6666 West Quincy Avenue, Denver, CO 80235. (303) 794-7711. Annual, with subscription.

PUBLIC WORKS MANUAL. Public Works Journal Corporation, 200 South Broad Street, Ridgewood, NJ 07451. (201) 445-5800. Annual. $20.00.

SCIENTIFIC AND TECHNICAL ORGANIZATIONS AND AGENCIES DIRECTORY. Gale Research Company, Book Tower, Detroit, MI 48226. (800) 521-0707. Second edition. Two volumes. 1987. $185.00 set.

WATER QUALITY ASSOCIATION DIRECTORY. Water Quality Association, 4151 Naperville Road, Lisle, IL 60532. (312) 369-1600. Annual. Available to members only.

GENERAL WORKS

GROUND WATER CONTAMINATION. J.H. Guswa and others. Noyes Data Corporation, Mill Road at Grand Avenue, Park Ridge, NJ 07656. (201) 391-8484. 1985. $48.00.

GROUND WATER CONTAMINATION. National Research Council. National Academy Press, 2101 Constitution Avenue, NW, Washington, DC 20418. (202) 334-3313. 1984. $17.95.

GROUND WATER POLLUTION CONTROL. Larry W. Canter and R.C. Knox. Lewis Publishing, Incorporated, 121 South Main Street, Chelsea, MI 48118. (313) 475-8619. 1985. $49.95.

GROUND WATER QUALITY. C.H. Ward and others. John Wiley and Sons, Incorporated, 605 Third Avenue, New York, NY 10158. (800) 526-5368 or (212) 850-6000. 1985. $45.00.

GROUND WATER QUALITY PROTECTION. L.W. Canter and others. Lewis Publishing, Incorporated, 121 South Main Street, Chelsea, MI 48118. (313) 475-8619. 1986. $49.95.

HYDROLOGY AND QUALITY OF WATER RESOURCES. Mark J. Hammer. John Wiley and Sons, Incorporated, 605 Third Avenue, New York, NY 10158. (800) 526-5368 or (212) 850-6000. Second edition. 1986. $42.95.

URBAN WATER INFRASTRUCTURE: PLANNING, MANAGEMENT, AND OPERATIONS. N.S. Grigg. John Wiley and Sons, Incorporated, 605 Third Avenue, New York, NY 10158. (800) 526-5368 or (212) 850-6000. 1986. $39.95.

WATER SUPPLY AND POLLUTION CONTROL. Warren Viesman and Mark J. Hammer. Harper and Row Publishers, Incorporated, 10 East 53rd Street, New York, NY 10022. (800) 242-7737. Fourth edition. 1985. $45.00.

WATER TREATMENT PRINCIPLES AND DESIGN. J.M. Montgomery. John Wiley and Sons, Incorporated, 605 Third Avenue, New York, NY 10158. (800) 526-5368 or (212) 850-6000. 1985. $49.95.

HANDBOOKS AND MANUALS

WATER TREATMENT HANDBOOK. Degremont Company. John Wiley and Sons, Incorporated, 605 Third Avenue, New York, NY 10158. (800) 526-5368 or (212) 850-6000. Fifth edition. 1979. $95.00.

ONLINE DATA BASES

COMPENDEX. Engineering Information, Incorporated, 345 East 47th Street, New York, NY 10017. (800) 221-1044 or (212) 705-7615. Engineering and technical literature, 1975 to present. Inquire as to cost and availability.

GEOREF. American Geological Institute, 4220 King Street, Alexandria, VA 22302. (703) 379-2480. 1967 to present. Inquire as to online cost and availability.

NTIS. National Technical Information Service, 5285 Port Royal Road, Springfield, VA 22161. (703) 487-4630. Broad coverage of government sponsored research reports, 1964 to present. Inquire as to cost and availability.

POLLUTION ABSTRACTS. Cambridge Scientific Abstracts, 5161 River Road, Bethesda, MD 20816. (301) 951-1400. 1970 to present. Available for online searching through DIALOG Information Services and BRS Information Technologies. Inquire as to online cost and availability.

WATER DATA BANK. United States Department of Agriculture, Agricultural Research Service, Hydrology Laboratory, Water Dat Center, Room 139, Building 007, BARC-West, Beltsville, MD 20705. (301) 344-4411. Inquire as to online cost and availability.

WATERNET. American Water Works Association, 6666 West Quincy Avenue, Denver, CO 80235. (303) 794-7711. A data base providing abstracts and indexing of information published in the American Water Works Association Journal. 1971 to present. Inquire as to online cost and availability.

PERIODICALS

AMERICAN WATER WORKS ASSOCIATION. Journal. American Water Works Association, 6666 West Quincy Avenue, Denver, CO 80235. (303) 794-7711. Monthly. $50.00 per year.

GROUND WATER. National Water Well Association. Water Well Journal Publishing Company, 500 West Wilson Bridge Road, Worthington, OH 43085. (614) 846-4967. Bimonthly. $53.00 per year.

JOURNAL OF ENVIRONMENTAL ENGINEERING. American Society of Civil Engineers, 345 East 47th Street, New York, NY 10017. (212) 705-7275. 1956 to present. Bimonthly. $80.00 per year.

JOURNAL OF GROUND WATER. National Water Well Association. Water Well Journal Publishing Company, 500 West Wilson Bridge Road, Worthington, OH 43085. (614) 846-4967.

Quarterly. $16.00 per year.

JOURNAL OF WATER RESOURCES PLANNING AND MANAGEMENT. American Society of Civil Engineers, 345 East 47th Street, New York, NY 10017-2398. (212) 705-7520. Quarterly. $56.00 per year.

WATER, AIR AND SOIL POLLUTION. D. Reidel Publishing Company, 190 Old Derby Street, Hingham, MA 02043. Twenty times per year. $550.00 per year.

WATER: ENGINEERING AND MANAGEMENT. Scranton Gillette Communications, Incorporated, 380 Northwest Highway, Des Plaines, IL 60016. (312) 298-6622. Monthly. $20.00 per year.

WATER POLLUTION CONTROL. The Bureau of National Affairs, Incorporated, 1231 25th Street, NW, Washington, DC 20037. (202) 452-4200. Biweekly. $360.00 per year.

WATER POLLUTION CONTROL FEDERATION. Journal. Water Pollution Control Federation, 2626 Pennsylvania Avenue, NW, Washington, DC 20037. (202) 337-2500. Monthly. $120.00 per year.

WATER RESEARCH. International Association on Water Pollution Research. Pergamon Press Incorporated, Journals Division, Maxwell House, Fairview Park, Elmsford, NY 10523. (914) 592-7700. Monthly. $550.00 per year.

WATER RESOURCES BULLETIN. American Water Resources Association, 5410 Grosvenor Lane, Suite 220, Bethesda, MD 20814. (301) 493-8600. Bimonthly. $65.00 per year.

WATER RESOURCES RESEARCH. American Geophysical Union, 2000 Florida Avenue, NW, Washington, DC 20009. (202) 462-6903. Monthly. $295.00 per year.

RESEARCH CENTERS AND INSTITUTES

CALIFORNIA INSTITUTE OF TECHNOLOGY. W.M. Keck Engineering Laboratory of Hydraulics and Water Resources, 1201 East California Boulevard, Pasadena, CA 91125. (818) 356-4404.

LENOX INSTITUTE FOR RESEARCH, INCORPORATED. 101 Yokum Avenue, Lenox, MA 01240. (413) 637-3025.

PENNSYLVANIA STATE UNIVERSITY. Environmental Engineering Laboratory, 212 Sackett Building, University Park, PA 16802. (814) 863-4385.

UNIVERSITY OF ILLINOIS. Advanced Environmental Control Technology Research Center, 3230 Newmark C.E. Laboratory, 208 North Romine Street, Urbana, Il 61801. (217) 333-3822.

UNIVERSITY OF TEXAS. Institute of Environmental Health, Post Office Box 20186, Houston, TX 77225-0186. (713) 792-4425.

WASHINGTON STATE UNIVERSITY. Environmental Engineering Research Laboratory, 141 Sloan, Pullman, WA 99164. (509) 335-3175.

WAVE OPTICS

See: OPTICS

WAVES

See: OCEANOGRAPHY, TSUNAMIS

WEATHER

See: METEOROLOGY

WEATHER FORECASTING

See: METEOROLOGY

WEATHER MODIFICATION

See also: METEOROLOGY

ABSTRACT SERVICES AND INDEXES

APPLIED SCIENCE AND TECHNOLOGY INDEX. H.W. Wilson Company, 950 University Avenue, Bronx, NY 10452. (800) 367-6670 or (212) 588-8400. Inquire as to cost and availability.

CURRENT CONTENTS: PHYSICAL AND CHEMICAL SCIENCES. Institute for Scientific Information, 3501 Market Street, Philadelphia, PA 19104. (800) 523-1850 or (215) 386-0100. $272.00 per year.

GENERAL SCIENCE INDEX. H.W. Wilson Company, 950 University Avenue, Bronx, NY 10452. (800) 367-6770 or (212) 588-8400. Inquire as to cost and availability.

METEOROLOGICAL AND GEOASTROPHYSICAL ABSTRACTS. American Meteorological Society, 45 Beacon Street, Boston, MA 02108. (617) 227-2425. 1950 to present. Monthly. $450.00 per year.

OCEANIC ABSTRACTS. Cambridge Scientific Abstracts, Incorporated, 5161 River Road, Bethesda, MD 20816. (301) 951-1400. 1964 to present. Inquire as to online cost and availability.

SCIENCE CITATION INDEX. Institute for Scientific Information, 3501 Market Street, Philadelphia, PA 19104. (800) 523-1850 or (215) 386-0100. Inquire as to cost and availability.

ASSOCIATIONS AND PROFESSIONAL SOCIETIES

AMERICAN METEOROLOGICAL SOCIETY. 45 Beacon Street, Boston, MA 02108. (617) 227-2425.

INTERNATIONAL ASSOCIATION OF METEOROLOGY AND ATMOSPHERIC PHYSICS. UCAR, Post Office Box 3000, Boulder, CO 80307.

NATIONAL WEATHER ASSOCIATION. 4400 Stamp Road, Room 404, Temple Hills, MD 20748. (301) 899-3784.

UNIVERSITY CORPORATION FOR ATMOSPHERIC RESEARCH. Box 3000, 1850 Table Mesa Drive, Boulder, CO 80307. (303) 497-1000.

WEATHER MODIFICATION ASSOCIATION. Post Office Box 8116, Fresno, CA 93747. (209) 291-8466.

BIBLIOGRAPHIES

SCIENCE BOOKS AND FILMS. American Association for the Advancement of Science, 1333 H Street, NW, Washington, DC 20005. Five times per year. $20.00 per year.

SCIENTIFIC AND TECHNICAL BOOKS AND SERIALS IN PRINT 1988: AN INDEX TO LITERATURE IN SCIENCE AND TECHNOLOGY. R.R. Bowker Company, 205 East 42nd Street, New York, NY 10017. (800) 521-8110 or (212) 916-1600. $175.00.

DIRECTORIES AND BIOGRAPHICAL SOURCES

AMERICAN MEN AND WOMEN OF SCIENCE. Physical and Biological Sciences. Sixteenth edition. R.R. Bowker Company, 205 East 42nd Street, New York, NY 10017. (800) 521-8110 or (212) 916-1600. 1986. $595.00.

BIOGRAPHICAL DICTIONARY OF SCIENTISTS. T.I. Williams. Halsted Press, 605 Third Avenue, New York, NY 10158. (800) 526-5368 or (212) 850-6418. Third edition. 1982. $29.95.

GOVERNMENT RESEARCH DIRECTORY. Gale Research Company, Book Tower, Detroit, MI 48226. (800) 521-0707. Fourth edition. 1987. $350.00.

INTERDOC: DIRECTORY OF PUBLISHED PROCEEDINGS, SERIES. SEMT-Science/Engineering/Medicine/Technology. Interdoc Corporation, 173 Halstead Avenue, Box 326, Harrison, NY 10528. (014) 835-3506. Ten times per year. $325.00 per year.

INTERNATIONAL RESEARCH CENTERS DIRECTORY 1988-1989. Gale Research Company, Book Tower, Detroit, MI 48226. (800) 521-0707. Fourth edition. 1987. $360.00.

METEOROLOGICAL SERVICES OF THE WORLD. World Meteorological Organization. Available from: American Meteorological Society, 45 Beacon Street, Boston, MA 02108. (617) 227-2425. Annual. $35.00.

NATIONAL WEATHER SERVICE OFFICES AND STATIONS. National Oceanic and Atmospheric Administration, Department of Commerce, Silver Spring, MD 20910. (301) 427-7698. Annual. Free.

RESEARCH CENTERS DIRECTORY. Gale Research Company, Book Tower, Detroit, MI 48226. (800) 521-0707. Twelfth edition. 1987. $365.00 for set.

SCIENTIFIC AND TECHNICAL ORGANIZATIONS AND AGENCIES DIRECTORY. Gale Research Company, Book Tower, Detroit, MI 48226. (800) 521-0707. Second edition. 1987. $185.00.

ENCYCLOPEDIAS AND DICTIONARIES

ENCYCLOPEDIA OF CLIMATOLOGY. John E. Oliver and Rhodes W. Fairbridge, editors. Van Nostrand Reinhold, Incorporated, 115 Fifth Avenue, New York, NY 10003. (800) 543-2681. 1987. $89.95.

MCGRAW-HILL ENCYCLOPEDIA OF SCIENCE AND TECHNOLOGY. McGraw-Hill Book, Incorporated, 1221 Avenue of the Americas, New York, NY 10020. (212) 997-3675. Fifth edition, 15 volumes. $1100.00.

GENERAL WORKS

STATISTICAL ANALYSIS OF WEATHER MODIFICATION EXPERIMENTS. Wegman and DePrist. Dekker, Marcel, Incorporated, 270 Madison Avenue, New York, NY 10016. (212) 696-9000. 1980. $29.75.

WEATHER MODIFICATION, PROSPECT AND PROBLEMS. Georg Breuer. Cambridge University Press, 32 East 57th Street, New York, NY 10022. (212) 688-8888. 1980. $34.50.

ONLINE DATA BASES

DISSERTATION ABSTRACTS ONLINE. University Microfilms International, 300 North Zeeb Road, Ann Arbor, MI 48106. (800) 521-0600 or (313) 761-4700. Scope includes virtually all doctoral dissertations accepted at accredited American institutions from 1861 to present in 252 subject areas. Inquire as to cost and availability.

METEOROLOGICAL AND GEOASTROPHYSICAL ABSTRACTS. American Meteorological Society, 45 Beacon Street, Boston, MA 02108. (617) 227-2425. 1950 to present. Monthly. $450.00 per year.

NTIS. National Technical Information Service, 5285 Port Royal Road, Springfield, VA 22161. (703) 487-4630. Broad coverage of government sponsored research reports, 1964 to present. Inquire as to cost and availability.

OCEANIC ABSTRACTS. Cambridge Scientific Abstracts, Incorporated, 5161 River Road, Bethesda, MD 20816. (301) 951-1400. 1964 to present. Inquire as to online cost and availability.

SCISEARCH. Institute for Scientific Information, 3501 Market Street, Philadelphia, PA 19104. (800) 523-1850 or (215) 386-0100. Broad multidisciplinary title and author index to the international literature of science and technology, 1974 to present. Inquire as to cost and availability.

PERIODICALS

AGRICULTURAL AND FOREST METEOROLOGY. Elsevier Science Publishing Company, Incorporated, 52 Vanderbilt Avenue, New York, NY 10017. (212) 370-5520. 1964 to present. Monthly. $260.00 per year.

AMERICAN METEOROLOGICAL SOCIETY BULLETIN. American Meteorological Society, 45 Beacon Street, Boston, MA 02108. (617) 227-2425.

BOUNDARY-LAYER METEOROLOGY: AN INTERNATIONAL JOURNAL OF PHYSICAL AND BIOLOGICAL PROCESSES IN THE ATMOSPHERIC BOUNDARY LAYER. D. Reidel Publishing Company, 190 Old Derby Street, Hingham, MA 02043. (617) 871-6600. 1970 to present. Sixteen times per year. $425.00 per year.

CLIMATIC CHANGE: AN INTERDISCIPLINARY, INTERNATIONAL JOURNAL DEVOTED TO THE DESCRIPTION, CAUSES AND IMPLICATIONS OF CLIMATIC CHANGE. D. Reidel Publishing Company, 190 Old Derby Street, Hingham, MA 02043. (617) 871-6600. 1977 to present. Six times per year. $125.00 per year.

DYNAMICS OF ATMOSPHERES AND OCEANS. Elsevier Science Publishing Company, Incorporated, 52 Vanderbilt Avenue, New York, NY 10017. (212) 370-5520. 1977 to present. Quarterly. $90.00 per year.

JOURNAL OF ATMOSPHERIC AND OCEANIC TECHNOLOGY. American Meteorological Society, 45 Beacon Street, Boston, MA 02108. (617) 227-2425. 1984 to present. Quarterly. $80.00 per year.

JOURNAL OF CLIMATE AND APPLIED METEOROLOGY. American Meteorological Society, 45 Beacon Street, Boston, MA 02108. (617) 227-2425. 1962 to present. Monthly. $120.00 per year.

JOURNAL OF THE ATMOSPHERIC SCIENCES. American Meteorological Society, 45 Beacon Street, Boston, MA 02108. (617) 227-2425. 1944 to present. Semimonthly. $220.00 per year.

MONTHLY WEATHER REVIEW. American Meteorological Society, 45 Beacon Street, Boston, MA 02108. (617) 227-2425. 1872 to present. Monthly. $120.00 per year.

NATIONAL WEATHER DIGEST. National Weather Association, 4400 Stamp Road, Room 404, Temple Hills, MD 20748. (301) 899-3784. 1976 to present. Quarterly. $20.00 per year.

WEATHER. Royal Meteorological Society, James Glaisher House, Grenville Place, Bracknell Berkshire, RG12 1BX, England. 1946 to present. $30.00 per year.

WEATHERWISE. Heldref Publications, 4000 Albemarle Street, NW, Washington, DC 20016. (202) 362-6445. 1948 to present. Bimonthly. $20.00 per year.

RESEARCH CENTERS AND INSTITUTES

COOPERATIVE INSTITUTE FOR MESOSCALE METEOROLOGICAL STUDIES. University of Oklahoma, 401 East Boyd, Norman, OK 73019. (405) 325-3041.

NATIONAL CENTER FOR ATMOSPHERIC RESEARCH. Box 3000, Boulder, CO 80307. (303) 497-1000.

PENNSYLVANIA STATE UNIVERSITY. Earth and Mineral Sciences Experiment Station, University Park, PA 16802. (814) 865-7659.

SOUTH DAKOTA SCHOOL OF MINES AND TECHNOLOGY. Institute of Atmospheric Studies, 502 East St. Joseph Street, Rapid City, SD 57701-3995. (605) 394-2291.

UNIVERSITY OF NEVADA. Atmospheric Sciences Center, Post Office Box 60220, Reno, NV. (702) 972-1676.

WELDING

See also: MACHINING, METALLURGICAL ENGINEERING, METALLURGY, METALS AND METALWORKING

ABSTRACT SERVICES AND INDEXES

APPLIED MECHANICS REVIEW. American Society of Mechanical Engineers, 345 East 47th Street, New York, NY 10017. (212) 705-7703. 1948 to present. Monthly. $360.00 per year.

APPLIED SCIENCE AND TECHNOLOGY INDEX. H.W. Wilson and Company, 950 University Avenue, Bronx, NY 10452. (800) 367-6670 or (212) 588-8400. Monthly. Inquire as to cost and availability.

CHEMICAL ABSTRACTS. American Chemical Society, Chemical Abstracts Service, Box 3012, Columbus, OH 43210. (614) 421-3600. 1907 to present. Weekly. $9500.00 per year.

ENGINEERING INDEX MONTHLY AND AUTHOR INDEX. Engineering Information Inc., 345 East 47th Street, New York, NY 10017. (212) 705-7600. Monthly. $1560.00 per year.

ISMEC BULLETIN (Information Service in Mechanical Engineering). Cambridge Scientific Abstracts, 5161 River Road, Bethesda, MD 20816. (301) 951-1400. 1973 to present. Monthly. $450.00 per year.

LEAD ABSTRACTS. Lead Development Association, 34 Berkeley Square, London, W1X 6AJ, England. 1958 to present. Quarterly. $70.00 per year.

METALS ABSTRACTS AND METALS ABSTRACTS INDEX. American Society for Metals, Metals Park, OH 44073. (216) 338-5151. 1968 to present. Monthly. Abstracts are $1100.00 per year and Index is $460.00 per year.

PHYSICS ABSTRACTS. Institution of Electrical Engineers. Available from: IEEE Service Center, 445 Hoes Lane, Piscataway, NJ 08854. 1898 to present. Bimonthly. $1700.00 per year.

SCIENCE CITATION INDEX. Institute for Scientific Information, 3501 Market Street, Philadelphia, PA 19104. (800) 523-1850 or (215) 386-0100. Six times per year. $6200.00 per year.

WORLD ALUMINUM ABSTRACTS. Aluminum Association, 818 Connecticut Avenue, NW, Washington, DC 20006. (202) 862-5156. 1968 to present. Monthly. $240.00 per year.

ASSOCIATIONS AND PROFESSIONAL SOCIETIES

ALUMINUM ASSOCIATION. 818 Connecticut Avenue, NW, Washington, DC 20006. (202) 862-5156.

AMERICAN INSTITUTE OF MINING, METALLURGICAL AND PETROLEUM ENGINEERS (AIME). 420 Commonwealth Drive, Warrendale, PA 15086. (412) 776-9086.

AMERICAN SOCIETY FOR METALS. Metals Park, OH 44073. (216) 338-5151.

AMERICAN SOCIETY FOR TESTING AND MATERIALS. 1916 Race Street, Philadelphia, PA 19103. (215) 299-5400.

AMERICAN SOCIETY OF MECHANICAL ENGINEERS. 345 47th Street, New York, NY 10017. (212) 705-7722.

AMERICAN WELDING SOCIETY. Box 351040, Miami, FL 33125. (305) 443-9353.

THE METALLURGICAL SOCIETY. 420 Commonwealth Drive, Warrendale, PA 15086. (412) 776-9000.

DIRECTORIES AND BIOGRAPHICAL SOURCES

DUN'S INDUSTRIAL GUIDE: THE METALWORKING DIRECTORY. Dun and Bradstreet Corporation, 49 Old Bloomfield Road, Mountain Lakes, NJ 07046. (201) 953-0300. Annual. $610.00.

INDUSTRIAL EQUIPMENT AND SUPPLIES DIRECTORY. American Business Directories, Inc., 5707 South 86th Circle, Omaha, NE 68127. (402) 331-7169.

METALLURGICAL SOCIETY OF AIME - MEMBERSHIP LIST. American Institute of Mining, Metallurgical and Petroleum Engineers (AIME). 345 East 47th Street, New York, NY 10017. (212) 705-7695. 1984.

1987 DIRECTORY OF ENGINEERING SOCIETIES AND RELATED ORGANIZATIONS. Gordon Davis, editor. Hemisphere Publishing Corporation, 79 Madison Avenue, New York, NY 10016-7892. (800) 821-8312. 12th edition. 1987. $100.00.

RESEARCH CENTERS DIRECTORY 1988. Gale Research Company, Book Tower, Detroit, MI 48226. (800) 521-0707. 12th edition. 1987. $365.00 for set.

SCIENTIFIC AND TECHNICAL ORGANIZATIONS AND AGENCIES DIRECTORY. Margaret Labash Young, editor. Gale Research Company, Book Tower, Detroit, MI 48226. (800) 521-0707. 2nd edition. 1987. $185.00.

WHO'S WHO IN ENGINEERING. Gordon Davis, editor. Hemisphere Publishing Corporation, 79 Madison Avenue, New York, NY 10016-7892. (800) 821-8312. 6th edition. 1985. $200.00.

GENERAL WORKS

BRAZING. Mel Schwartz. American Society for Metals, Metals Park, OH 44073. (216) 338-5151. 1987. $92.00.

WELDING CRAFT PRACTICE. N. Parkin and C.R. Flood. Pergamon Press, Inc., Maxwell House, Fairview Park, Elmsford, NY 10523. (914) 592-7700. Volume 1: Oxy-Acetylene Gas Welding and Related Studies, second edition, 1979, $12.00 in paper. Volume 2: Electric Arc Welding and Related Studies, second edition, 1980, $11.00 in paper.

WELDING: FUNDAMENTALS AND PROCEDURES. Jerry Galyen and others. John Wiley and Sons, Inc., 605 Third Avenue, New York, NY 10158. (800) 526-5368. 1984. $22.95.

WELDING PRACTICE AND PROCEDURES. Richard Carr and Robert O'Con. Prentice-Hall Publishing, Inc., Englewood Cliffs, NJ 07632. (800) 562-0245. 1983. $27.95.

HANDBOOKS AND MANUALS

METALS HANDBOOK. American Society for Metals, Metals Park, OH 44073. (216) 338-5151. 9th edition. 14 volumes. 1988. $1310.00 for set.

WELDER'S GUIDE. James Brumbaugh. Macmillan Publishing Company, Inc., 866 Third Avenue, New York, NY 10022. (800) 257-5755. Third edition. 1983. $20.00.

WELDER'S HANDBOOK. Richard Finch. Prentice-Hall Publishing, Inc., Englewood Cliffs, NJ 07632. (800) 562-0245. 1985. $12.95.

ONLINE DATA BASES

CA SEARCH. Chemical Abstracts Service, P.O. Box 3012, Columbus, OH 43120. (800) 848-6538 or (614) 421-3600. Comprehensive guide to chemical literature, 1972 to present. Inquire as to online cost and availability.

COMPENDEX. Engineering Information, Inc., 345 East 47th Street, New York, NY 10017. (800) 221-1044 or (212) 705-7615. Engineering and technical literature, 1975 to present. Inquire as to online cost and availability.

ISMEC. Cambridge Scientific Abstracts, 5161 River Road, Besthda, MD 20816. (800) 638-8076 or (301) 951-1400. Literature of mechanical and production engineering, 1973 to present. Inquire as to online cost and availability.

INSPEC. INSPEC Marketing Department, Institution of Electrical Engineers. Available from IEEE Service Center, 445 Hoes Lane, Piscataway, NJ 08854. (201) 981-0060. Online version of Physics Abstracts. Inquire as to online cost and availability.

NTIS. National Technical Information Service, 5285 Port Royal Road, Springfield, VA 22161. (703) 487-4630. Broad coverage of government sponsored research reports, 1964 to present. Inquire as to online cost and availability.

SCISEARCH. Institute for Scientific Information, 3501 Market Street, Philadelphia, PA 19104. (800) 523-1850 or (215) 386-0100. Broad multidisciplinary title and author index to the international literature of science and technology, 1974 to present. Inquire as to online cost and availability.

WILSONLINE. H.W. Wilson and Company, 950 University Avenue, Bronx, NY 10452. (800) 367-6770 or (212) 588-8400. Makes available online versions of the H.W. Wilson indexes including Applied Science and Technology Index, Business Periodicals Index and Readers' Guide to Periodical Literature. Approximately 1980 to present. Inquire as to online cost and availability.

PERIODICALS

CANADIAN WELDER AND FABRICATOR. Sanford Evans Communications, Limited, 1077 Saint James Street, Box 6900, Winnipeg, Manitoba R3C 3B1, Canada. (204) 775-0201. 1909 to present. Monthly. $18.00 Canadian per year.

JOURNAL OF METALS. American Institute of Mining, Metallurgical, and Petroleum Engineers, Inc., Metallurgical Society, 420 Commonwealth Drive, Warrendale, PA 15086. (412) 776-9086. 1949 to present. Monthly. $40.00 per year.

METAL PROGRESS. American Society for Metals, Metals Park, OH 44073. (216) 338-5151. 1930 to present. Monthly. $40.00.

METALS WEEK. McGraw-Hill Book Company, 1221 Avenue of the Americas, New York, NY 10020. (212) 997-2823. 1930 to present. Weekly. $597.00 per year.

METALWORKING DIGEST. Philos Publications, Inc., 1 East Chase Street, Baltimore, MD 21202. (301) 361-9060. 1969 to present. Nine times per year. $9.00 per year.

WELDING AND JOINING DIGEST. American Society for Metals, Metals Park, OH 44073. (216) 338-5151. Monthly. $90.00 per year.

WELDING DIGEST AND FABRICATION. Penton-IPC, 1100 Superior Avenue, Cleveland, OH 44114. (216) 696-7000. 1930 to present. Monthly. $35.00 per year.

WELDING JOURNAL. American Welding Society, Box 351040, Miami, FL 33125. (305) 443-9353. 1922 to present. Monthly. $30.00 per year.

WRC PROGRESS REPORTS. Welding Research Council, United Engineering Center, 345 East 47th Street, New York, NY 10017. Six times per year. Inquire.

RESEARCH CENTERS AND INSTITUTES

CENTER FOR WELDING RESEARCH. Colorado School of Mines, Golden, CO 80401. (303) 273-3767.

WELDING RESEARCH COUNCIL. United Engineering Center, 345 East 47th Street, New York, NY 10017.

WELLS

See: HYDROLOGY, PETROLEUM ENGINEERING

WIND POWER

See: ENERGY

WIND TUNNELS

See: AERONAUTICAL ENGINEERING

WINDMILLS

See: ENERGY

WIRE AND CABLE

See also: WIRING, ELECTRICAL

ABSTRACT SERVICES AND INDEXES

APPLIED SCIENCE AND TECHNOLOGY INDEX. H.W. Wilson Company, 950 University Avenue, Bronx, NY 10452. (800) 367-6670 or (212) 588-8400. Inquire as to cost and availability.

CURRENT CONTENTS: ENGINEERING, TECHNOLOGY. Institute for Scientific Information, 3501 Market Street, Philadelphia, PA 19104. (800) 523-1850 or (215) 386-0100. $272.00 per year.

ENGINEERING INDEX MONTHLY. Engineering Information, Incorporated, 345 East 47th Street, New York, NY 10017. (800) 221-1044 or (212) 705-7600. Monthly, with annual cumulation. $1560.00 per year.

INTERNATIONAL COPPER INFORMATION BULLETIN. Copper Development Association, Orchard House, Mutton Lane, Potters Bar, Herts. EN6 3AP, England. 1976 to present. Quarterly. $35.00 per year.

METALS ABSTRACTS. American Society for Metals, 9639 Kinsman Road, Metals Park, OH 44073. (216) 338-5151. Monthly. $890.00.

METALS ABSTRACTS INDEX. American Society for Metals, 9639 Kinsman Road, Metals Park, OH 44073. (216) 338-5151. Monthly. (Sold only to subscribers of Metals Abstracts).

ASSOCIATIONS AND PROFESSIONAL SOCIETIES

AMERICAN SOCIETY FOR METALS. Metals Park, OH 44073. (216) 338-5151.

AMERICAN WIRE PRODUCERS ASSOCIATION. 1101 Connecticut Avenue, Suite 700, NW, Washington, DC 20036. (202) 857-1155.

INSULATED CABLE ENGINEERS ASSOCIATION. Post Office Box P, South Yarmouth, MA 02664. (617) 394-4424.

WIRE ASSOCIATION INTERNATIONAL. 1570 Boston Post Road, Guildford, CT 06437. (203) 453-2777.

WIRE MACHINERY BUILDERS ASSOCIATION. 7297 Lee Highway, Suite N, Falls Church, VA 22042. (703) 533-9530.

DIRECTORIES AND BIOGRAPHICAL SOURCES

DIRECTORY OF WIRE COMPANIES OF NORTH AMERICA. Business Information Services, 20 Pine Mountain Road, Ridgefield, CT 06877. (203) 748-6529. Annual. $45.00 per year.

DUN'S INDUSTRIAL GUIDE - THE METAL WORKING DIRECTORY. Dun and Bradstreet, Incorporated, Three Century Drive, Parsippany, NJ 07054. (201) 455-0900. Annual. $550.00.

METAL PRODUCTS DIRECTORY. American Business Directories, Incorporated, Division of American Business Lists, Incorporated, 5707 South 86th Circle, Omaha, NE 68127. (402) 331-7169. 1986. $80.00.

RESEARCH CENTERS DIRECTORY. Gale Research Company, Book Tower, Detroit, MI 48226. (800) 521-0707. Eleventh edition. 1987.

WIRE JOURNAL INTERNATIONAL DIRECTORY/CATALOG. 1570 Boston Post Road, Guilford, CT 06437. (203) 453-2777. Annual. $35.00.

WIRE TECH BUYERS GUIDE. Huebner Publications, Incorporated, 6521 Davis Industrial Parkway, Solon, OH 44139. (216) 248-1125. Annual. $30.00.

WHO'S WHO IN ENGINEERING. Engineers Joint Council, 345 East 47th Street, New York, NY 10017. (212) 705-7010. 1985. $200.00.

WHO'S WHO IN TECHNOLOGY TODAY. Reston Publishing Company, Incorporated, c/o Prentice-Hall, Incorporated, Englewood Cliffs, NJ 07632. (800) 262-6868. Biennial. Five volumes. $425.00. Covers the fields of electronics, computer science, physics, optics, chemistry, biotechnology, mechanics, energy, and earth science.

ENCYCLOPEDIAS AND DICTIONARIES

ENCYCLOPEDIA OF PHYSICAL SCIENCE AND TECHNOLOGY. Academic Press, Incorporated, Orlando, FL 32887. (800) 321-5068 or (305) 345-2734. Fifteen volumes, 1986.

MCGRAW-HILL DICTIONARY OF ENGINEERING. Sybil P. Parker, editor. McGraw-Hill Book Company, 1221 Avenue of the Americas, New York, NY 10020. (212) 512-2000. 1984. $39.95.

MCGRAW-HILL ENCYCLOPEDIA OF SCIENCE AND TECHNOLOGY. McGraw-Hill Book, 1221 Avenue of the Americas, New York, NY 10020. (212) 512-2000.

GENERAL WORKS

ELECTRICAL WIRING - INDUSTRIAL. Robert L. Smith. Van Nostrand Reinhold Book Company, Incorporated, 115 Fifth Avenue, New York, NY 10003. (800) 543-2681. 1982. $17.95.

METALLURGY BASICS. D.V. Brown. Van Nostrand Reinhold Book Company, Incorporated, 115 Fifth Avenue, New York, NY 10003. (800) 543-2681. 1985. $17.95.

HANDBOOKS AND MANUALS

AMERICAN ELECTRICIANS HANDBOOK. Terrell Croft and others, editors. McGraw-Hill Book Company, Incorporated, 1221 Avenue of the Americas, New York, NY 10020. (212) 512-2000. Tenth edition. 1981. $55.00.

ELECTRIC CABLES HANDBOOK. D. McAllister. Sheridan House, Incorporated, 145 Palisade Street, Dobbs Ferry, NY 10522. (914) 693-2410. 1982. $120.00.

ONLINE DATA BASES

COMPENDEX. Engineering Information, Incorporated, 345 East 47th Street, New York, NY 10017. (800) 221-1044 or (212) 705-7615. Engineering and technical literature, 1975 to present. Inquire as to cost and availability.

INSPEC. INSPEC Marketing Department, Institute of Electrical and Electronics Engineers, Incorporated, IEEE Service Department, 445 Hoes Lane, Piscataway, NJ 08854. (201) 981-0060. Inquire as to on-line cost and availability.

METADEX. Metals Information, American Society for Metals, Metals Park, OH 44073. (216) 338-5151. (Metals Abstracts/Alloys/Index). A worldwide literature on the science and practice of metallurgy, 1966 to present. Inquire as to online cost and availability.

NASA. National Aeronautics and Space Administration, Scientific and Technical Information Branch, 300 7th Street, SW, Washington, DC 20546. Citations and abstracts of aerospace literature, 1962 to present. Inquire as to cost and availability.

NON-FERROUS METALS ABSTRACTS. British Non-Ferrous Metals Technology Centre, Grove Laboratoryies, Denchworth Road, Wantage, Oxfordshire, England OX12 9 BJ. Citations and abstracts on non-ferrous metallurgy and technology, 1961 to present. Inquire as to online cost and availability.

NTIS. National Technical Information Service, 5285 Port Royal Road, Springfield, VA 22161. (703) 487-4630. Broad coverage of government sponsored research reports, 1964 to present. Inquire as to cost and availability.

WILSONLINE. H.W. Wilson Company, 950 University Avenue, Bronx, NY 10452. (800) 367-6770 or (212) 588-8400. Makes available online versions of the printed H.W. Wilson Indexes including Applied Science and Technology Index, Business Periodicals Index, and Readers' Guide to Periodical Literature. Period covered is generally 1983 to present. Inquire as to cost and availability.

PERIODICALS

JOURNAL OF METALS. Metallurgical Society of the AIME (American Institute of Mining, Metallurgical and Petroleum Engineers), 420 Commonwealth Drive, Warrendale, PA 15086. (412) 776-9080. Monthly. $40.00 per year.

METALS WEEK. McGraw-Hill Book Company, Incorporated, 1221 Avenue of the Americas, New York, NY 10020. (212) 997-2823. Weekly. $527.00 per year.

WIRE INDUSTRY. Magnum Publications, Limited, 110-112 Station Road, East, Oxford, Surrey, RH8 0QA, England. Monthly. $42.50 per year.

WIRE INDUSTRY NEWSLETTER. Business Information Services, 7 Hampden Road, Stafford Springs, CT 06076. (203) 684-5877. Biweekly. $175.00 per year.

WIRE JOURNAL INTERNATIONAL. Wire Association International, 1570 Boston Post Road, Guilford, CT 06437. (203) 453-2777. Monthly. $50.00 per year.

WIRE LINE. American Wire Producers Association, 1101 Connecticut Avenue, NW, Suite 700, Washington, DC 20036. (202) 857-1155. Monthly. $100.00 per year.

WIRE WORLD INTERNATIONAL (ENGLISH LANGUAGE EDITION OF DRAHTWELT). Voga Verlag, Max-Planck Strasse, 7-9, Postfach 6740, 8700 Wuerzburg 1, West Germany. Bimonthly. $54.00 per year.

WIRING, ELECTRICAL

See also: ELECTRONIC CIRCUITS AND COMPONENTS, WIRE AND CABLE

ABSTRACT SERVICES AND INDEXES

APPLIED SCIENCE AND TECHNOLOGY INDEX. H.W. Wilson Company, 950 University Avenue, Bronx, NY 10452. (800) 367-6670 or (212) 588-8400. Inquire as to cost and availability.

CURRENT CONTENTS: ENGINEERING, TECHNOLOGY. Institute for Scientific Information, 3501 Market Street, Philadelphia, PA 19104. (800) 523-1850 or (215) 386-0100. $272.00 per year.

ENGINEERING INDEX MONTHLY. Engineering Information, Incorporated, 345 East 47th Street, New York, NY 10017. (800) 221-1044 or (212) 705-7600. Monthly, with annual cumulation. $1560.00 per year.

INTERNATIONAL COPPER INFORMATION BULLETIN. Copper Development Association, Orchard House, Mutton Lane, Potters Bar, Herts. EN6 3AP, England. 1976 to present. Quarterly. $35.00 per year.

METALS ABSTRACTS. American Society for Metals, 9639 Kinsman Road, Metals Park, OH 44073. (216) 338-5151. Monthly. $890.00.

METALS ABSTRACTS INDEX. American Society for Metals, 9639 Kinsman Road, Metals Park, OH 44073. (216) 338-5151. Monthly. (Sold only to subscribers of Metals Abstracts).

ANNUAL REVIEWS AND YEARBOOKS

E C AND M: ELECTRICAL CONSTRUCTION AND MAINTENANCE ELECTRICAL PRODUCTS YEARBOOK. McGraw-Hill Book Company, Incorporated, 1221 Avenue of the Americas, New York, NY 10020. (212) 512-2000. Annual. Inquire as to cost and availability.

ASSOCIATIONS AND PROFESSIONAL SOCIETIES

AMERICAN SOCIETY FOR METALS. Metals Park, OH 44073. (216) 338-5151.

AMERICAN WIRE PRODUCERS ASSOCIATION. 1101 Connecticut Avenue, Suite 700, NW, Washington, DC 20036. (202) 857-1155.

INSULATED CABLE ENGINEERS ASSOCIATION. Post Office Box P, South Yarmouth, MA 02664. (617) 394-4424.

WIRE ASSOCIATION INTERNATIONAL. 1570 Boston Post Road, Guildford, CT 06437. (203) 453-2777.

WIRE MACHINERY BUILDERS ASSOCIATION. 7297 Lee Highway, Suite N, Falls Church, VA 22042. (703) 533-9530.

DIRECTORIES AND BIOGRAPHICAL SOURCES

DIRECTORY OF WIRE COMPANIES OF NORTH AMERICA. Business Information Services, 20 Pine Mountain Road, Ridgefield, CT 06877. (203) 748-6529. Annual. $45.00 per year.

DUN'S INDUSTRIAL GUIDE - THE METAL WORKING DIRECTORY. Dun and Bradstreet, Incorporated, Three Century Drive, Parsippany, NJ 07054. (201) 455-0900. Annual. $550.00.

METAL PRODUCTS DIRECTORY. American Business Directories, Incorporated, Division of American Business Lists, Incorporated, 5707 South 86th Circle, Omaha, NE 68127. (402) 331-7169. 1986. $80.00.

RESEARCH CENTERS DIRECTORY. Gale Research Company, Book Tower, Detroit, MI 48226. (800) 521-0707. Eleventh edition. 1987.

WIRE JOURNAL INTERNATIONAL DIRECTORY/CATALOG. 1570 Boston Post Road, Guilford, CT 06437. (203) 453-2777. Annual. $35.00.

WIRE TECH BUYERS GUIDE. Huebner Publications, Incorporated, 6521 Davis Industrial Parkway, Solon, OH 44139. (216) 248-1125. Annual. $30.00.

WHO'S WHO IN ENGINEERING. Engineers Joint Council, 345 East 47th Street, New York, NY 10017. (212) 705-7010. 1985. $200.00.

WHO'S WHO IN TECHNOLOGY TODAY. Reston Publishing Company, Incorporated, c/o Prentice-Hall, Incorporated, Englewood Cliffs, NJ 07632. (800) 262-6868. Biennial. Five volumes. $425.00. Covers the fields of electronics, computer science, physics, optics, chemistry, biotechnology, mechanics, energy, and earth science.

ENCYCLOPEDIAS AND DICTIONARIES

ENCYCLOPEDIA OF PHYSICAL SCIENCE AND TECHNOLOGY. Academic Press, Incorporated, Orlando, FL 32887. (800) 321-5068 or (305) 345-2734. Fifteen volumes, 1986.

MCGRAW-HILL DICTIONARY OF ENGINEERING. Sybil P. Parker, editor. McGraw-Hill Book Company, 1221 Avenue of the Americas, New York, NY 10020. (212) 512-2000. 1984. $39.95.

MCGRAW-HILL ENCYCLOPEDIA OF SCIENCE AND TECHNOLOGY. McGraw-HIll Book Company, Incorporated, 1221 Avenue of the Americas, New York, NY 10020. (212) 512-2000.

GENERAL WORKS

ELECTRICAL WIRING - INDUSTRIAL. Robert L. Smith. Van Nostrand Reinhold Book Company, Incorporated, 115 Fifth Avenue, New York, NY 10003. (800) 543-2681. 1982. $17.95.

INDUSTRIAL ELECTRICAL WIRING. John T. Earl. Prentice-Hall, Incorporated, Englewood Cliffs, NJ 07632. (201) 592-2000. 1987. $29.95.

PRACTICAL ELECTRICAL WIRING. Herbert P. Richter and W.C. Schwan. McGraw-Hill Book Company, Incorporated, 1221 Avenue of the Americas, New York, NY 10020. (212) 512-2000. Thirteenth edition. 1984. $32.95.

RAPID ELECTRICAL ESTIMATING AND PRICING. C.K. Kolstad and G.V. Kohnert. McGraw-Hill Book Company,

Incorporated, 1221 Avenue of the Americas, New York, NY 10020. (212) 512-2000. 1986. $49.75.

HANDBOOKS AND MANUALS

AMERICAN ELECTRICIANS HANDBOOK. Terrell Croft and others, editors. McGraw-Hill Book Company, Incorporated, 1221 Avenue of the Americas, New York, NY 10020. (212) 512-2000. Tenth edition. 1981. $55.00.

ELECTRIC CABLES HANDBOOK. D. McAllister. Sheridan House, Incorporated, 145 Palisade Street, Dobbs Ferry, NY 10522. (914) 693-2410. 1982. $120.00.

LINEMAN'S AND CABLEMAN'S HANDBOOK. Edwin B. Kurtz and Thomas M. Shoemaker. McGraw-Hill Book Company, Incorporated, 1221 Avenue of the Americas, New York, NY 10020. (212) 512-2000. 1986. $54.50.

ONLINE DATA BASES

COMPENDEX. Engineering Information, Incorporated, 345 East 47th Street, New York, NY 10017. (800) 221-1044 or (212) 705-7615. Engineering and technical literature, 1975 to present. Inquire as to cost and availability.

INSPEC. INSPEC Marketing Department, Institute of Electrical and Electronics Engineers, Incorporated, IEEE Service Department, 445 Hoes Lane, Piscataway, NJ 08854. (201) 981-0060. Inquire as to on-line cost and availability.

METADEX. Metals Information, American Society for Metals, Metals Park, OH 44073. (216) 338-5151. (Metals Abstracts/Alloys/Index). A worldwide literature on the science and practice of metallurgy, 1966 to present. Inquire as to online cost and availability.

NASA. National Aeronautics and Space Administration, Scientific and Technical Information Branch, 300 7th Street, SW, Washington, DC 20546. Citations and abstracts of aerospace literature, 1962 to present. Inquire as to cost and availability.

NON-FERROUS METALS ABSTRACTS. British Non-Ferrous Metals Technology Centre, Grove Laboratoryies, Denchworth Road, Wantage, Oxfordshire, England OX12 9 BJ. Citations and abstracts on non-ferrous metallurgy and technology, 1961 to present. Inquire as to online cost and availability.

NTIS. National Technical Information Service, 5285 Port Royal Road, Springfield, VA 22161. (703) 487-4630. Broad coverage of government sponsored research reports, 1964 to present. Inquire as to cost and availability.

WILSONLINE. H.W. Wilson Company, 950 University Avenue, Bronx, NY 10452. (800) 367-6770 or (212) 588-8400. Makes available online versions of the printed H.W. Wilson Indexes including Applied Science and Technology Index, Business Periodicals Index, and Readers' Guide to Periodical Literature. Period covered is generally 1983 to present. Inquire as to cost and availability.

OTHER SOURCES

WIRING DEVICES AND SUPPLIES. United States Bureau of the Census, United States Department of Commerce, Washington, DC 20233. (202) 655-4000. Annual. $1.25.

PERIODICALS

E C AND M: ELECTRICAL CONSTRUCTION AND MAINTENANCE. McGraw-Hill Book Company, Incorporated, 1221 Avenue of the Americas, New York, NY 10020. (212) 512-2000. Monthly. $18.00 per year.

JOURNAL OF METALS. Metallurgical Society of the AIME (American Institute of Mining, Metallurgical and Petroleum Engineers), 420 Commonwealth Drive, Warrendale, PA 15086. (412) 776-9080. Monthly. $40.00 per year.

METALS WEEK. McGraw-Hill Book Company, Incorporated, 1221 Avenue of the Americas, New York, NY 10020. (212) 997-2823. Weekly. $527.00 per year.

WIRE INDUSTRY. Magnum Publications, Limited, 110-112 Station Road, East, Oxford, Surrey, RH8 0QA, England. Monthly. $42.50 per year.

WIRE INDUSTRY NEWSLETTER. Business Information Services, 7 Hampden Road, Stafford Springs, CT 06076. (203) 684-5877. Biweekly. $175.00 per year.

WIRE JOURNAL INTERNATIONAL. Wire Association International, 1570 Boston Post Road, Guildford, CT 06437. (203) 453-2777. Monthly. $50.00 per year.

WIRE LINE. American Wire Producers Association, 1101 Connecticut Avenue, NW, Suite 700, Washington, DC 20036. (202) 857-1155. Monthly. $100.00 per year.

WIRE WORLD INTERNATIONAL (ENGLISH LANGUAGE EDITION OF DRAHTWELT). Voga Verlag, Max-Planck Strasse, 7-9, Postfach 6740, 8700 Wuerzburg 1, West Germany. Bimonthly. $54.00 per year.

X - Y - Z

X-RAY ASTRONOMY

See: ASTRONOMY

X-RAYS

See also: ASTRONOMY, CRYSTALLOGRAPHY, MATERIALS SCIENCE, NONDESTRUCTIVE TESTING (NDT), PARTICLE PHYSICS

ABSTRACT SERVICES AND INDEXES

APPLIED SCIENCE AND TECHNOLOGY INDEX. H.W. Wilson and Company, 950 University Avenue, Bronx, NY 10452. (800) 367-6670 or (212) 588-8400. Monthly. Inquire as to cost and availability.

ASTRONOMY AND ASTROPHYSICS ABSTRACTS. Springer-Verlag New York, Inc., 175 Fifth Avenue, New York, NY 10010. (800) 526-7254. 1969 to present. Approximately $70.00 per year.

ASTRONOMY AND ASTROPHYSICS MONTHLY INDEX. Olivetree Associates, P.O. Box 236, Sierre Madre, CA 91024. $220.00 per year. Complementary copies available on request.

CHEMICAL ABSTRACTS. American Chemical Society, Chemical Abstracts Service, Box 3012, Columbus, OH 43210. (614) 421-3600. 1907 to present. Weekly. $9500.00 per year.

CONFERENCE PAPERS INDEX. Cambridge Scientific Abstracts, 5161 River Road, Bethesda, MD 20816. 1972 to present. Monthly. Inquire as to cost and availability.

CURRENT CONTENTS: PHYSICAL, CHEMICAL AND EARTH SCIENCES. Institute for Scientific Information, 3501 Market Street, Philadelphia, PA 19104. (800) 523-1850 or (215) 386-0100. Weekly. $275.00 per year.

ENGINEERING INDEX MONTHLY AND AUTHOR INDEX. Engineering Information Inc., 345 East 47th Street, New York, NY 10017. (212) 705-7600. Monthly. $1560.00 per year.

GENERAL SCIENCE INDEX. H.W. Wilson and Company, 950 University Avenue, Bronx, NY 10452. (800) 367-6670 or (212) 588-8400. 1978 to present. Monthly. Inquire as to cost and availability.

INDEX TO SCIENTIFIC AND TECHNICAL PROCEEDINGS. Institute for Scientific Information, 3501 Market Street, Philadelphia, PA 19104. (800) 523-1850 or (215) 386-0100. 1978 to present. Monthly. $775.00 per year.

INDEX TO SCIENTIFIC REVIEWS. Institute for Scientific Information, 3501 Market Street, Philadelphia, PA 19104. (800) 523-1850 or (215) 386-0100. 1974 to present. Semi-annual. $550.00 per year.

PHYSICS ABSTRACTS. Institution of Electrical Engineers. Available from: IEEE Service Center, 445 Hoes Lane, Piscataway, NJ 08854. 1898 to present. Bimonthly. $1700.00 per year.

PHYSICS BRIEFS. Physik Verlag GmbH, Postfach 1260/1280, D-6940 Weinheim, West Germany. (212) 661-9404. 1920 to present. Twenty-six times per year. $1250.00 per year.

SCIENCE CITATION INDEX. Institute for Scientific Information, 3501 Market Street, Philadelphia, PA 19104. (800) 523-1850 or (215) 386-0100. Six times per year. $6200. 00 per year.

X-RAY DIFFRACTION ABSTRACTS. PRM Science and Technology Agency, Limited, 261A Finchley Road, Hampstead, London, NW3 6LU, England. 1973 to present. Quarterly. $90.00 per year.

ASSOCIATIONS AND PROFESSIONAL SOCIETIES

AMERICAN CHEMICAL SOCIETY. 1155 16th Street, N.W., Washington, DC 20036. (800) 424-6747.

AMERICAN INSTITUTE OF PHYSICS. 335 East 45th Street, New York, NY 10017. (212) 661-9494.

AMERICAN SOCIETY FOR NONDESTRUCTIVE TESTING. 4153 Arlingate Plaza, Caller #28518, Columbus, OH 43228.

AMERICAN SOCIETY FOR TESTING AND MATERIALS. 1916 Race Street, Philadelphia, PA 19103. (215) 299-5400.

NATIONAL COUNCIL ON RADIATION PROTECTION AND MEASUREMENTS. 7910 Woodmont Avenue, Suite 1016, Bethesda, MD 20814. (301) 657-2652.

BIBLIOGRAPHIES

NEW TECHNICAL BOOKS: A SELECTIVE LIST WITH DESCRIPTIVE ANNOTATIONS. New York Public Library, Science and Technology Research Center, Fifth Avenue and 42nd Street, New York, NY 10018. (212) 930-0800. 1915 to present. Monthly. $15.00 per year.

SCIENTIFIC AND TECHNICAL BOOKS AND SERIALS IN PRINT 1988; AN INDEX TO LITERATURE IN SCIENCE AND TECHNOLOGY. R.R. Bowker Company, 205 East 42nd Street, New York, NY 10017. (800) 521-8110. $175.00.

DIRECTORIES AND BIOGRAPHICAL SOURCES

AMERICAN MEN AND WOMEN OF SCIENCE. R.R. Bowker, Inc., Order Department, 245 West 17th Street, New York, NY 10011. (800) 521-8110. Eight volumes. 1986. $595.00 for set.

INTERNATIONAL RESEARCH CENTERS DIRECTORY 1988-89. Darren L. Smith, editor. Gale Research Company, Book Tower, Detroit, MI 48226. (800) 521-0707. 4th edition. 1987. $360.00.

1987 DIRECTORY OF ENGINEERING SOCIETIES AND RELATED ORGANIZATIONS. Gordon Davis, editor. Hemisphere Publishing Corporation, 1010 Vermont Avenue, NW, Washington, DC 20005. (800) 526-0275. 12th edition. 1987. $100.00.

RESEARCH CENTERS DIRECTORY 1988. Gale Research Company, Book Tower, Detroit, MI 48226. (800) 521-0707. 12th edition. 1987. $365.00 for set.

SCIENTIFIC AND TECHNICAL ORGANIZATIONS AND AGENCIES DIRECTORY. Margaret Labash Young, editor. Gale Research Company, Book Tower, Detroit, MI 48226. (800) 521-0707. 2nd edition. 1987. $185.00.

WHO'S WHO IN ENGINEERING. Gordon Davis, editor. Hemisphere Publishing Corporation, 1010 Vermont Avenue, NW, Washington, DC 20005. (800) 526-0275. 6th edition. 1985. $200.00.

ENCYCLOPEDIAS AND DICTIONARIES

CONCISE SCIENCE DICTIONARY. Oxford University Press, 200 Madison Avenue, New York, NY 10016. (800) 458-5833. 1987. $9.95 in paper.

DICTIONARY OF THE PHYSICAL SCIENCES: TERMS, FORMULAS, DATA. Cesare Emiliani. Oxford University Press, 200 Madison Avenue, New York, NY 10016. (800) 458-5833. 1987. $19.95 in paper.

MCGRAW-HILL ENCYCLOPEDIA OF SCIENCE AND TECHNOLOGY. McGraw-Hill Book Company, 1221 Avenue of the Americas, New York, NY 10020. (212) 512-2000. 6th edition. 1987. $1600.00.

THESAURUS OF SCIENTIFIC, TECHNICAL, AND ENGINEERING TERMS. Hemisphere Publishing Corporation, 1010 Vermont Avenue, NW, Washington, DC 20005. (800) 526-0275. 1988. $125.00.

GENERAL WORKS

FUNDAMENTALS OF X-RAY AND RADIUM PHYSICS. Joseph Selman. C.C. Thomas, 2600 South First Street, Springfield, IL 62794-9265. (217) 789-8980. Seventh edition. 1985. $29.75.

PRINCIPLES OF X-RAY METALLURGY. T. Kovacs. Plenum Publishing Corporation, 233 Spring Street, New York, NY 10013. (800) 221-9369. 1969. $24.50.

X-RAY METHODS. Clive Winston. John Wiley and Sons, Inc., 605 Third Avenue, New York, NY 10158. (800) 526-5368. 1987. $75.00.

X-RAYS AND THEIR APPLICATIONS. A.G. Brown. Plenum Publishing Corporation, 233 Spring Street, New York, NY 10013. (800) 221-9369. 1975. $8.95 in paper.

HANDBOOKS AND MANUALS

CRC HANDBOOK OF CHEMISTRY AND PHYSICS. Robert C. Weast, editor. CRC Press, 2000 Corporate Boulevard, Boca Raton, FL 33431. (800) 272-7737. 68th edition. 1987. $69.95.

ONLINE DATA BASES

CA SEARCH. Chemical Abstracts Service, P.O. Box 3012, Columbus, OH 43120. (800) 848-6538 or (614) 421-3600. Comprehensive guide to chemical literature, 1972 to present. Inquire as to online cost and availability.

COMPENDEX. Engineering Information, Inc., 345 East 47th Street, New York, NY 10017. (800) 221-1044 or (212) 705-7615. Engineering and technical literature, 1975 to present. Inquire as to online cost and availability.

DISSERTATION ABSTRACTS ONLINE. University Microfilms International, 300 North Zeeb Road, Ann Arbor, MI 48106. (800) 521-0600 or (313) 761-4700. Scope includes virtually all doctoral dissertations accepted at accredited American institutions from 1861 to present in over 250 subject areas. Inquire as to online cost and availability.

GEOREF. Online version of the BIBLIOGRAPHY AND INDEX OF GEOLOGY. American Geological Institute, 4220 King Street, Alexandria, VA 22302. (703) 379-2480. 1969 to present. Inquire as to online cost and availability.

INSPEC. INSPEC Marketing Department, Institution of Electrical Engineers. Available from IEEE Service Center, 445 Hoes Lane, Piscataway, NJ 08854. (201) 981-0060. Online version of Physics Abstracts. Inquire as to online cost and availability.

NTIS. National Technical Information Service, 5285 Port Royal Road, Springfield, VA 22161. (703) 487-4630. Broad coverage of government sponsored research reports, 1964 to present. Inquire as to online cost and availability.

SCISEARCH. Institute for Scientific Information, 3501 Market Street, Philadelphia, PA 19104. (800) 523-1850 or (215) 386-0100. Broad multidisciplinary title and author index to the international literature of science and technology, 1974 to present. Inquire as to online cost and availability.

WILSONLINE. H.W. Wilson and Company, 950 University Avenue, Bronx, NY 10452. (800) 367-6770 or (212) 588-8400. Makes available online versions of the H.W. Wilson indexes including Applied Science and Technology Index, Business Periodicals Index and Readers' Guide to Periodical Literature. Approximately 1980 to present. Inquire as to online cost and availability.

PERIODICALS

APPLIED OPTICS. Optical Society of America. 1816 Jefferson Place, N.W., Washington, DC 20036. (202) 223-8130. Order from: American Institute of Physics, 335 East 45th Street, New York, NY 10017. (212) 661-9404. 1962 to present. Semi-monthly. $330.00 per year.

APPLIED SPECTROSCOPY REVIEWS. Marcel Dekker Inc., 270 Madison Avenue, New York, NY 10016. (800) 228-1160. 1964 to present. Quarterly. $185.00 per year.

OPTICAL ENGINEERING. Society of Photo-Optical Instrumentation Engineers (SPIE), P.O. Box 10, 1022 19th Street, Bellingham, WA 98227. (206) 676-3290. 1962 to present. Monthly. $95.00 per year.

OPTICAL SOCIETY OF AMERICA, JOURNAL, PARTS A AND B. Optical Society of America. 1816 Jefferson Place, N.W., Washington, DC 20036. (202) 223-8130. Order from: American Institute of Physics, 335 East 45th Street, New York, NY 10017. (212) 661-9404. 1917 to present. Monthly. $180.00 each part per year.

PHYSICAL REVIEW LETTERS. American Institute of Physics, 335 East 45th Street, New York, NY 10017. (212) 661-9494. 1958 to present. Weekly. $450.00 per year.

XRS-X-RAY SPECTROMETRY. John Wiley and Sons, Inc., 605 Third Avenue, New York, NY 10158. (800) 526-5368. 1972 to present. Quarterly. $240.00 per year.

RESEARCH CENTERS AND INSTITUTES

CENTER FOR X-RAY OPTICS. Lawrence Berkeley Laboratory, 1 Cyclotron Road, Berkeley, CA 94720. (415) 486-4985.

MAJOR ANALYTICAL INSTRUMENTATION CENTER. University of Florida, 217 MAE, Gainesville, FL 32611. (904) 392-6985.

WEBER RESEARCH INSTITUTE. Polytechnic University, Route 110, Farmingdale, NY 11735. (516) 752-9701.

ZINC

See also: ALLOYS, BRASS AND BRONZE, HEAVY METAL ALLOYS, MATERIALS SCIENCE, TIN

ABSTRACT SERVICES AND INDEXES

ALLOYS INDEX. American Society for Metals, 9639 Kinsman Road, Metals Park, OH 44073. (216) 338-5151. $130.00 per year.

APPLIED MECHANICS REVIEWS. American Society of Mechanical Engineers, 345 East 47th Street, New York, NY 10017. (212) 705-7722. Monthly. $380.00 per year. Critical reviews of the world literature in applied mechanics and related engineering science.

APPLIED SCIENCE AND TECHNOLOGY INDEX. H.W. Wilson Company, 950 University Avenue, Bronx, NY 10452. (800) 367-6670 or (212) 588-8400. Inquire as to cost and availability.

CHEMICAL ABSTRACTS. Chemical Abstracts Service, 2540 Olentangy Road, Post Office Box 3012, Columbus, OH 43210. (800) 848-6538 or (614) 421-3600. Weekly. $9200.00 per year.

CURRENT CONTENTS: ENGINEERING, TECHNOLOGY. Institute for Scientific Information, 3501 Market Street, Philadelphia, PA 19104. (800) 523-1850 or (215) 386-0100. $272.00 per year.

ENGINEERING INDEX MONTHLY. Engineering Information, Incorporated, 345 East 47th Street, New York, NY 10017. (800) 221-1044 or (212) 705-7600. Monthly, with annual cumulation. $1560.00 per year.

INTERNATIONAL COPPER INFORMATION BULLETIN. Copper Development Association, Orchard House, Mutton Lane, Potters Bar, Herts, EN6 3AP, England. 1976 to present. Quarterly. $35.00 per year.

METALS ABSTRACTS. American Society for Metals, 9639 Kinsman Road, Metals Park, OH 44073. (216) 338-5151. Monthly. $890.00.

METALS ABSTRACTS INDEX. American Society for Metals, 9639 Kinsman Road, Metals Park, OH 44073. (216) 338-5151. Monthly. (Sold only to subscribers of Metals Abstracts.)

ZINC ABSTRACTS. Zinc Development Association, 34 Berkeley Square, London W1X 6AJ, England. 1943 to present. Quarterly. $100.00 per year.

ANNUAL REVIEWS AND YEARBOOKS

ANNUAL REVIEW OF MATERIALS SCIENCE. Annual Reviews, Incorporated, 4139 El Camino Way, Palo Alto, CA 94306. (415) 493-4400.

ASSOCIATIONS AND PROFESSIONAL SOCIETIES

AMERICAN SOCIETY FOR METALS. Metals Park, OH 44073. (216) 338-5151.

BRASS AND BRONZE INGOT INSTITUTE. 33 North LaSalle Street, Room 3500, Chicago, IL 60602. (312) 236-2715.

COPPER DEVELOPMENT ASSOCIATION. Box 1840, Greenwich Office Park 2, Greenwich, CT 06836. (203) 625-8210.

INDEPENDENT ZINC ALLOYERS ASSOCIATION. 900 17th Street, NW, Suite 504, Washington, DC 20036. (202) 785-0550.

METALLURGICAL SOCIETY OF THE AIME (AMERICAN INSTITUTE OF MINING, METALLURGICAL AND PETROLEUM ENGINEERS). 420 Commonwealth Drive, Warrendale, PA 15086. (412) 776-9080.

ZINC DEVELOPMENT ASSOCIATION. 34 Berkeley Square, London W1X 6AJ, England.

ZINC INSTITUTE. 292 Madison Avenue, New York, NY 10017. (212) 578-4750.

BIBLIOGRAPHIES

SCIENTIFIC AND TECHNICAL BOOKS AND SERIALS IN PRINT 1988; AN INDEX TO LITERATURE IN SCIENCE AND TECHNOLOGY. R.R. Bowker Company, 205 East 42nd Street, New York, NY 10017. (800) 521-8110 or (212) 916-1600. $175.00.

DIRECTORIES AND BIOGRAPHICAL SOURCES

DUN'S INDUSTRIAL GUIDE - THE METAL WORKING DIRECTORY. Dun and Bradstreet, Incorporated, Three Century Drive, Parsippany, NJ 07054. (201) 455-0900. Annual. $550.00.

INTERNATIONAL RESEARCH CENTERS DIRECTORY 1986-1987. Gale Research Company, Book Tower, Detroit, MI 48226. (800) 521-0707. Third edition. 1986.

RESEARCH CENTERS DIRECTORY. Gale Research Company, Book Tower, Detroit, MI 48226. (800) 521-0707. Eleventh edition. 1987.

SCIENTIFIC AND TECHNICAL ORGANIZATIONS AND AGENCIES DIRECTORY. Gale Research Company, Book Tower, Detroit, MI 48226. (800) 521-0707. 1985.

WHO'S WHO IN ENGINEERING. Engineers Joint Council, 345 East 47th Street, New York, NY 10017. (212) 705-7010. 1985. $200.00.

WHO'S WHO IN FRONTIER SCIENCE AND TECHNOLOGY. Marquis Who's Who, Incorporated, 200 East Ohio Street, Chicago, IL 60611. (800) 428-3898 or (312) 787-2008.

WHO'S WHO IN TECHNOLOGY TODAY. Reston Publishing Company, Incorporated, c/o Prentice-Hall, Incorporated, Englewood Cliffs, NJ 07632. (800) 262-6868. Biennial. Five volumes. $425.00. Covers the fields of electronics, computer science, physics, optics, chemistry, biotechnology, mechanics, energy, and earth science.

ENCYCLOPEDIAS AND DICTIONARIES

ENCYCLOPEDIA OF PHYSICAL SCIENCE AND TECHNOLOGY. Academic Press, Incorporated, Orlando, FL 32887. (800) 321-5068 or (305) 345-2734. Fifteen volumes, 1986.

MCGRAW-HILL DICTIONARY OF ENGINEERING. Sybil P. Parker, editor. McGraw-Hill Book Company, 1221 Avenue of the Americas, New York, NY 10020. (212) 512-2000. 1984. $39.95.

MCGRAW-HILL ENCYCLOPEDIA OF SCIENCE AND TECHNOLOGY. McGraw-Hill Book Company, Incorporated, 1221 Avenue of the Americas, New York, NY 10020. (212) 512-2000.

GENERAL WORKS

ESSENTIAL METALLURGY FOR ENGINEERS. W. Alexander, G. Davies, and K. Reynolds. Van Nostrand Reinhold, 115 Fifth Avenue, New York, NY 10003. (800) 543-2681. 1985. $17.95.

PHYSICAL METALLURGY. P. Haasen. Cambridge University Press, 32 East 57th Street, New York, NY 10022. (212) 688-8885. 1986. $24.95 in paper.

METALLURGY BASICS. D.V. Brown. Van Nostrand Reinhold, 115 Fifth Avenue, New York, NY 10003. (800) 543-2681. 1985. $17.95.

STRUCTURE AND PROPERTIES OF ENGINEERING ALLOYS. W.F. Smith. McGraw-Hill Book Company, Incorporated, 1221 Avenue of the Americas, New York, NY 10020. (212) 512-2000. 1980. $46.95.

THEORY OF STRUCTURAL TRANSFORMATIONS IN SOLIDS. A.G. Khachaturyan. John Wiley and Sons, Incorporated, 605 Third Avenue, New York, NY 10158. (800) 526-5368 or (212) 850-6000. 1983. $69.95.

THE WORLD ZINC INDUSTRY. Satyadev Gupta. Lexington Books, 125 Spring Street, Lexington, MA 02173. (617) 862-6650. 1981. $29.00.

ZINC AND ITS ALLOYS AND COMPOUNDS. S.W. Morgan. John Wiley and Sons, Incorporated, 605 Third Avenue, New York, NY 10158. (800) 526-5368 or (212) 850-6000. 1985. $64.95.

HANDBOOKS AND MANUALS

CRC HANDBOOK OF CHEMISTRY AND PHYSICS. CRC Press, Incorporated, 2000 Corporate Boulevard, Boca Raton, Florida 33341. (305) 994-0555. Sixth-seventh edition. 1986. $69.95.

SMITHELL'S METALS REFERENCE BOOK. Eric A. Brandes, editor. Butterworth Publishers, 80 Montvale Avenue, Stoneham, MA 02180. (800) 325-4177. Sixth edition. 1983. $210.00.

ONLINE DATA BASES

COMPENDEX. Engineering Information, Incorporated, 345 East 47th Street, New York, NY 10017. (800) 221-1044 or (212) 705-7615. Engineering and technical literature, 1975 to present. Inquire as to cost and availability.

INSPEC. INSPEC Marketing Department, Institute of Electrical and Electronics Engineers, Incorporated, IEEE Service Department, 445 Hoes Lane, Piscataway, NJ 08854. (201) 981-0060. Inquire as to on-line cost and availability.

METADEX. Metals Information, American Society for Metals, Metals Park, OH 44073. (216) 338-5151. (Metals Abstracts/Alloys Index). A worldwide literature on the science and practice of metallurgy, 1966 to present. Inquire as to online cost and availability.

NASA. National Aeronautics and Space Administration, Scientific and Technical Information Branch, 300 7th Street, SW, Washington, DC 20546. Citations and abstracts of aerospace literature, 1962 to present. Inquire as to cost and availability.

NON-FERROUS METALS ABSTRACTS. British Non-Ferrous Metals Technology Centre, Grove Laboratories, Denchworth Road, Wantage, Oxfordshire, England OX12 9 BJ. Citations and abstracts on non-ferrous metallurgy and technology, 1961 to present. Inquire as to online cost and availability.

NTIS. National Technical Information Service, 5285 Port Royal Road, Springfield, VA 22161. (703) 487-4630. Broad coverage of government sponsored research reports, 1964 to present. Inquire as to cost and availability.

WILSONLINE. H.W. Wilson Company, 950 University Avenue, Bronx, NY 10452. (800) 367-6770 or (212) 588-8400. Makes available online versions of the printed H.W. Wilson Indexes including Applied Science and Technology Index, Business Periodicals Index, and Reader's Guide to Periodical Literature. Period covered is generally 1983 to present. Inquire as to cost and availability.

OTHER SOURCES

MATERIALS SCIENCE AND METALLURGY. National Technical Information Service (NTIS), 5285 Port Royal Road, Springfield, VA 22161. (703) 487-4630. Translations and abstracts of foreign language technical media. Irregular. $40.00 per year.

PERIODICALS

ALLOY DIGEST. Engineering Publications Incorporated, Box 823, Upper Montclair, NJ 07043. (201) 746-7930. Monthly. $50.00 per year.

JOURNAL OF METALS. Metallurgical Society of the AIME (American Institute of Mining, Metallurgical and Petroleum Engineers), 420 Commonwealth Drive, Warrendale, PA 15086. (412) 776-9080. Monthly. $40.00 per year.

METALLURGICAL TRANSACTIONS. Metallurgical Society of the AIME (American Institute of Mining, Metallurgical and Petroleum Engineers), 420 Commonwealth Drive, Warrendale, PA 15086. (412) 776-9080. Monthly. $95.00 per year.

METALS WEEK. McGraw-Hill Book Company, Incorporated, 1221 Avenue of the Americas, New York, NY 10020. (212) 997-2823. Weekly. $527.00 per year.

TIN INTERNATIONAL. Tin Publications, Limited, 222 Strand, London, WC2R 1BA, England. Monthly. $85.00 per year.

RESEARCH CENTERS AND INSTITUTES

COPPER DEVELOPMENT ASSOCIATION. Box 1840, Greenwhich Office Park 2, Greenwich, CT 06836. (203) 625-8210.

INTERNATIONAL LEAD-ZINC RESEARCH ORGANIZATION. 292 Madison Avenue, New York, NY 10017. (212) 532-2372.

UNIVERSITY OF CONNECTICUT. Institute of Materials Science, Storrs, CT 06268. (203) 486-4623.

UNIVERSITY OF FLORIDA. Department of Materials Science and Engineering, Gainesville, FL 32601. (904) 392-1454.

STATISTICAL SOURCES

METALS STATISTICS. Fairchild Publications, 7 East 12th Street, New York, NY 10003. (212) 741-4426. Annual. $50.00.

ZIRCONIUM

See: ELEMENTS (CHEMICAL)

Sci Ref Z 7401 .E56 1989
Encyclopedia of physical sciences and engineering

FEB 27 1989